THE CONCISE
OXFORD DICTIONARY OF
ENGLISH
PLACE-NAMES

THE CONCISE
OXFORD DICTIONARY OF
ENGLISH
PLACE-NAMES

BY

EILERT EKWALL

FOURTH EDITION

OXFORD
AT THE CLARENDON PRESS

Oxford University Press, Walton Street, Oxford OX2 6DP

Oxford New York Toronto
Delhi Bombay Calcutta Madras Karachi
Kuala Lumpur Singapore Hong Kong Tokyo
Nairobi Dar es Salaam Cape Town
Melbourne Auckland

and associated companies in
Beirut Berlin Ibadan Mexico City Nicosia

Oxford is a trade mark of Oxford University Press

Published in the United States by
Oxford University Press, New York

First edition 1936
Second edition 1940
Third edition 1947
Reprinted 1950
Fourth edition 1960
Reprinted 1964, 1966, 1970, 1974, 1977, 1980,
1981, 1984, 1985

ISBN 0 19 869103 3

Printed in Great Britain
at the University Press, Oxford
by David Stanford
Printer to the University

PREFACE TO THE FOURTH EDITION

IN the twenty-three years that have elapsed since the dictionary was first published important work has been done in the field of English name-study, and a good deal of fresh material has become available, in the first place thanks to the English Place-name Society, which throws new light on many place-names. A definite solution of more problems and better or more probable etymologies for many names can now be suggested than before. For the new edition the author has gone through the whole book carefully article by article, all along consulting available literature, especially that of later date. The dictionary has been reset, all material in appendixes having been worked into the text together with a considerable body of fresh additions and corrections. Many articles have been rewritten. The book has been brought up to date, and it is hoped that the revised edition, in all probability the last to be produced by the author, will be considered an advance.

For the articles on Cornish names invaluable help has been derived from Mr. Gover's great manuscript survey, and the author's sincere thanks are offered to him for permission to use it. Thanks are also due to many correspondents for friendly criticism and valuable items of local information, some of which will be found to have been included (e.g. under Humberstone Li, Kenyon. Staple Fitzpaine). Dr. P. H. Reaney has supplied helpful material or Suffolk names.

E. E.

Lund
March 1959

FROM THE PREFACE
TO THE FIRST EDITION

ENOUGH is being said in the Introduction on the aim and scope of this book, so that there is little to say here. It need hardly be pointed out that a Concise Dictionary of English Place-names can contain only a limited number of names, and can deal only briefly with those included. For fuller information the reader is referred to the monographs that have been, or will be, published, especially the volumes of the English Place-name Society.

I offer my grateful thanks to the editors of the English Place-name Society for the kind interest shown in the work while it was in its preparatory stage. Special thanks are due to Dr. Allen Mawer and Mr. J. E. B. Gover for having placed at my disposal valuable material from the collections of the Place-name Society, not least the important newly discovered Old English forms from a manuscript in the Bodleian Library (in this book referred to as Bodley MS). Professor F. M. Stenton kindly undertook to read a proof and offered many important suggestions.

I owe a particular debt of gratitude to Miss E. G. Withycombe, who read a first proof of the whole book, and whose criticism and numerous suggestions or additions have been of very great value.

My wife assisted both in the collection of material and in the proof-reading.

E. E.

Lund
October 1935

CONTENTS

CONTENTS

INTRODUCTION

I. GENERAL REMARKS

1. THIS dictionary, for obvious reasons, cannot lay claim to completeness. It would be an impossible task to deal with all English place-names in one volume. The principle has been to include what may be called the chief English place-names. The dictionary embraces names of the country, of the counties, and other important divisions (as CRAVEN, KESTEVEN, LINDSEY), towns (except those of late origin), parishes, villages, some names of estates and hamlets, or even farms whose names are old and etymologically interesting, rivers, lakes—also names of capes, hills,[1] bays for which early material is available. Names of hundreds, as being no longer in real use, have been omitted. The material includes most of the names listed in Bartholomew's *Gazetteer*, except for those that belong to Wales, Scotland, Ireland, the Isle of Man and the Channel Islands. Some names given by Bartholomew have been omitted, either because of the insignificance of the places or because the names are self-explanatory, or because no early forms were available. On the other hand, the collection includes some names not given by Bartholomew.

It may perhaps seem unnecessary to give such full lists of commonly occurring names like NEWTON, THORPE, STOKE, NORTON, SUTTON, EASTON, WESTON. Of the first three only a selection is given, though rather a full one. But even a name such as Newton has a certain interest that may not be immediately apparent. NEWTON is identical in origin with NEWINGTON, NEWNTON, NAUNTON, NITON, and it is evidently of importance to give readers an idea of the distribution of the name-form Newton. Further, the Newtons, like many other common names, often have a surname attached to them, a distinctive addition such as (NEWTON) BURGOLAND, FLOTMAN, &c., which must be explained. The local distribution of the names STOKE, THORPE, also has a good deal of interest. Thorpe is partly English, partly Scandinavian.

EASTON is not always 'eastern TŪN', and it is identical in most cases with ASTON. These two names had to be fully dealt with. For NORTON, SUTTON, WESTON similar special reasons for a full treatment cannot be given, but those with sur-names had to be dealt with, and those remaining do not take up a very great deal of space. Also it may interest readers to find early examples of the various Nortons, &c.

2. An etymological dictionary of English place-names without some early documentary material would be useless, and for each name some early forms are adduced. It is the first principle of place-name etymology that there must be early name-forms on which to found the explanation. It is true, of course, that some names, such as Norton and Weston, can be explained without early forms. But in general it would be useless to try to explain place-names on the

---

[1] The names of the hills in the Lake District are mostly not found in earlier sources than eighteenth-century maps or guide-books, where they generally appear in their present forms. Such names are Brandreth, Old Man, Saddleback, Steeple, Wetherlam. Some of these names probably arose comparatively late. They have not been included in the dictionary. Some of the names are, at least apparently, self-explanatory, while others are not. Old Man probably contains *man* in the dialectal sense 'a cairn or pile of stones marking a summit'.

strength of the modern name-form alone. It is a common experience that names now identical in form are often shown by the early forms to be totally different in origin. A few illustrative examples may find place here.

BROUGHTON is mostly Old English *Brōc-tūn* 'TŪN on a brook', but several Broughtons are *Burh-tūn* or even *Beorg-tūn* 'TŪN by a hill or barrow'. BURTON is generally *Burh-tūn*, but sometimes *Byrh-tūn* (with *burg* in the genitive form) or something quite different. HAMPTON has three sources, Old English *Hām-tūn*, *Hamm-tūn*, and *Hēa(n)-tūn*. HAUGHTON is mostly *Halh-tūn*, but not always. HOUGHTON, usually *Hōh-tūn*, is sometimes *Halh-tūn*, sometimes neither.

Of the three ATHERSTONES, one is '*Æpelheard*'s TŪN', one '*Æpelrēd*'s TŪN', the third '*Ēadrīc*'s TŪN'. ALFORD in Lincolnshire may be 'alder ford', but ALFORD in Somerset is '*Ealdgȳp*'s ford'. The four ALSTONS given in the dictionary all have different etymologies. ASHFORD is mostly Old English *Æsc-ford*, but the Kent and Middlesex names have different origins. The four BARNSLEYS dealt with in the dictionary have as many different etymologies. The three KIMBERLEYS in Norfolk, Notts, and Warwickshire mean respectively '*Cyneburg*'s, *Cynemær*'s, and *Cynebald*'s LĒAH'. MILTON is sometimes 'middle TŪN', sometimes 'TŪN with a mill'. Even MIDDLETON is not always 'middle TŪN'. At least two Middletons have different etymologies. SLAUGHTERFORD in Gloucestershire has as first element an Old English *slōhtre* 'marsh', SLAUGHTERFORD in Wilts an Old English word (*slāhporn*) for 'blackthorn'. The three TRAFFORDS have different origins. WING in Bucks and Rutland are both *Wenge* from the thirteenth century, but earlier forms show that the two names are totally different in origin.

Examples of this kind, which could easily be multiplied, give an idea of the pitfalls that beset the way of the place-name student. Two more cases may be quoted here in order to show what curious coincidences are sometimes met with. BRANT BROUGHTON is on the river BRANT, whose name is old and well established. It may seem self-evident that the distinctive addition is the river-name. But early name-forms show that this is not the case. BRANT is a modification of earlier *Brend* and the like, and Brant Broughton means 'Burnt Broughton', 'the Broughton that was burnt down at some time'. WYEGATE is near the river WYE, and has naturally been supposed to mean 'the gate by the river Wye'. But the name happens to appear in a tenth-century charter in the form *Uuiggangeat*, which shows that the meaning is '*Wicga*'s gate'. It is obvious that without early material place-name etymology is mere guess-work.

If good name-forms from Domesday Book or texts from the twelfth or early thirteenth century are available, a fairly safe etymology can generally be attained. But even then the etymology of many names remains more or less doubtful. An illustrative case is ALVESTON in Warwickshire. It is *Alvestone* and the like from Domesday on. Were it not for the isolated form *Eanulfestun* in an Old English source, we should have had no means of finding out that the name means '*Ēanwulf*'s TŪN', not '*Ælf*'s TŪN' or the like. Similarly it is only due to the isolated form *Hygeredingtun* of 831 that we know that the first element of HARLINGTON in Middlesex is derived from the personal name *Hygerēd*. All post-Conquest sources have forms like *Herdinton*. HILMARTON is *Helmerintone* in Domesday, and has been held to have as first element a derivative of an Old English *Helmær*. But an Old English form *Helmerdingtun* shows that the base is an unrecorded Old English *Helmheard*. ISLEWORTH is *Gisteleworde* in Domesday, which is difficult to explain definitely. But it is almost certainly found in the form *Gislheresuuyrth* in a charter of 695, which tells us that the name contains the personal name *Gislhere*. It is obvious that if strong reduction has taken place so early as the time of Domesday in the cases that are capable of proof, a similar phenomenon may be suspected to have occurred in names not recorded in earlier sources. A name such as ALVINGHAM (*Aluingeham* in Domesday) may quite well represent an Old English *Ēanwulfingahām*. It is indeed highly probable that names such as KILLINGHOLME, KILVINGTON contain derivatives in *-ing* of *Cynewulf*, KILMINGTON, one of *Cynehelm*, though this cannot be proved by the help of Old English forms to be so. *Hard-* in HARDINGHAM, HARDINGTON may very well represent Old English *Heardrēd* or *Hēahrēd*. HARDISWORTHY is probably 'Hererēd's WORPIG'. BARDSEA in Lancashire is fairly certainly 'Beornrēd's island'. It is very likely that BARDSEY in West Yorkshire is identical in origin and that BARDSLEY contains the same personal name. TYLDESLEY is best explained from *Tilwaldes lēah*, TILSTOCK from *Tidhilde stoc*, though the earliest forms are *Tildesle, Tildestok* and the like.

It has been objected against derivation of place-name elements from personal names that the more common full-names (dithematic names such as *Cynewulf*) are apparently fewer than so-called short-names in place-names. In reality full-names were far more common in place-names than appears at first sight. Many elements that look like short-names are in reality full-names that have been reduced in form. The common Old English names *Cynehelm, Cynewulf* are in fact quite frequently found in place-names, as we should expect them to be.

Many place-names are proved by Old English forms to go back to *-ing(a)tūn*, though post-Conquest forms show no trace of the *-ing-*. Thus one TIBBERTON (*Tidbertun* in Domesday) is *Tidbrihtingctun* in a charter of 972. It is only thanks to this example that the exact base can be established. The later forms seem to point to a compound of Old English *Tidbeorht* and TŪN. There is every reason to suppose that many names found in post-Conquest sources which appear to consist of a personal name in its uninflected form and TŪN, or the like, are in reality old names in *-ing(a)tūn*, &c. ALWALTON Hunts is *Æpelwoldingtun* in 955, but *Alwoltune* in Domesday and later without a trace of the *-ing-*. It seems extremely probable that KIMBOLTON in Hunts, which is found first in Domesday in the form *Chenebaltone*, goes back to *Cynebaldingtūn*. But we cannot be absolutely sure that this is so, for Kimbolton may represent an Old English *Cynebaldestūn*, which has lost its *-s-*. In the dictionary the etymologies of names of this type have generally been given in the form '*Cynebald*'s TŪN' and so on.

3. The importance for the place-name student of finding out the topography of the places whose names he is trying to explain etymologically has been frequently stressed and may now be looked upon as generally admitted. Frequently a definite etymology cannot be attained without a study of the local conditions of the place. The names MIDFORD, MITFORD, to take an example, may mean 'middle ford' and have been so explained, but the fact that both places are situated at the junction of streams renders it obvious that both names have as first element Old English *gemȳpe* 'junction of streams'. HOWLE HILL in Herefordshire is situated by a marked hill. The name cannot be derived from any known English word, but the situation suggests that we have here to do with an unrecorded Old English word for 'hill', cognate with German *Hügel*. The river-name HAMPS may be suspected to be identical with Welsh HAFHESP, which means 'summer-dry', 'river dry in summer'. The writer had made this conjecture before seeing

the river, but the correctness of the theory could not be proved until it had
been found out by a visit to the river that it actually does go dry in summer.
It has naturally been impossible to visit all the places whose names are dealt
with in the book. But maps often give as good information as a visit to the
place itself. It is often enough to be able to say whether a place is on a river or
not, whether it is on low land or on a hill, and so on. It is not necessary to go to a
place called SEATON to make sure if the name means 'village on the sea' or 'village
on a lake'. Besides, a great many places were named without reference to local
conditions—as the numerous names containing personal names or adjectives (*new*,
old, &c.). For the rest, the author has had to rely on the experience gained in the
course of many summers' travelling in all the various parts of England, which,
supplemented by the study of maps, must supply the want of special journeys of
exploration.

4. The modern pronunciation of place-names is sometimes of value for the
etymology. The derivation of the first element of WROTHAM from an old word
with $\bar{o}$, for instance, is proved to be correct by the modern pronunciation with
($\overline{oo}$). The early forms might quite well represent an Old English form with δ.
Similarly the pronunciation ($\overline{oo}$) for RODING shows that we have to start from a
base with $\bar{o}$. Derivation from Old English *rop(u)* 'clearing' is thus ruled out. The
fact that names such as ASPATRIA, CUMREW, CUMWHINTON have the accent on the
second element is a strong point in favour of the etymologies suggested. But
the importance of the modern pronunciation should not be overrated. In a great
many cases the historically correct form has been superseded by a spelling-
pronunciation. And the genuine local form is by no means always easy to get
hold of. In the dictionary, pronunciations have been given sparingly, the principle having been to include chiefly those that may be of value for etymological
purposes. Unfortunately it has been impossible to make systematic collections
of local pronunciations. Those given have partly been taken from available
sources, such as monographs on place-names and A. Lloyd James's *Broadcast
English*, ii. Hope's *Glossary* has not been often quoted, because the notation is
not clear enough and the information given may be antiquated. A good deal of
information has been collected by the author in the course of many years of place-
name study.

5. In this place attention may be drawn to the importance in dealing with
place-name etymology of taking into account the characteristics of the dialects
of the various districts. A name such as WEALD can be derived from Old English
weald 'wold' only if it comes from the Saxon or Kentish parts of England.
WEALD MOORS in Shropshire, for instance, must be explained in some other way.
In fact, early forms prove that Weald is here *wilde* 'wild, uncultivated'. The name
WILTON may be explained as 'TŪN by a well or spring', if the place is situated in
a West Saxon district, where Old English *wella* appears as *willa*, but not if it is
found outside that area. WALL in Shropshire may be, and probably should be,
derived from Old English *wælla*, the typical Mercian form of *wella*, but the same
name, when found in other parts of England than the West Midlands, must be
explained in some other way. ROCK in Worcestershire is proved by early forms
to represent Old English (æt) *þǣre āce* '(at) the oak', but ROCK in Northumberland
cannot be so explained, for in Northern dialects Old English $\bar{a}$ did not become
later *o*. BRATTON in Devon and BRETTON in Yorkshire may quite well be identical
in origin, both containing Old English *brǣc* 'newly cultivated land', Bratton the
Saxon form *brǣc*, Bretton the Anglian form *brēc*.

II. VARIOUS TYPES OF PLACE-NAMES

In this survey only so-called habitation-names, i.e. names of inhabited places, as villages, homesteads, and the like, will be considered. Nature-names, that is, names of rivers, hills, and so on, do not call for particular treatment in this section.

1. Folk-names

A comparatively small, but highly interesting, group is formed by place-names that were originally names of the inhabitants of the places—what may be called folk-names.

It is a common phenomenon that tribal names come to denote the district inhabited by the tribe. Well-known cases are the German names FRANKEN, PREUSSEN, SACHSEN, THÜRINGEN, which originally denoted 'the Franks, Prussians', and so on, but came to mean 'Franconia, Prussia', &c. ELSASS 'Alsace', formerly Elsâzzun, meant 'those dwelling outside' (that is, west of the Rhine). SWEDEN literally means 'the Swedes'. Similar names are found in France. AMIENS, RHEIMS, SOISSONS, come from the Gaulish tribal names Ambiani, Remi, Sues-siones. Of the same type are the English names DEVON, ESSEX, SUSSEX, WESSEX, WALES (OE Wealas 'Welsh people'), CORNWALL. See further, on these names, the articles in the dictionary.

It is not so well known that names of villages have sometimes arisen in the same way. There are many German examples of this type of name, for instance, the numerous names in -ingen, as SICKINGEN, SIGMARINGEN, which mean 'Siggo's (Sigimar's) people'. MÜNCHEN 'Munich' means 'the monks'. In England the most important group of names belonging here is formed by names in -ing, representing Old English -ingas, as BARKING, HASTINGS, READING, SONNING. Hæstingas is actually recorded as the name of a tribe in Sussex. Names in -ingas are mostly derived from personal names and originally denoted the descendants or dependants of a certain man. But some are derived from a place-name, as BLYTHING, the name of a hundred in Suffolk, 'people on the river Blyth', and it is possible that some other types occur. The names in -ingas are doubtless very ancient and date from the time of the earliest Anglo-Saxon settlements. They also throw some light on early Anglo-Saxon social conditions.

Corresponding to German names such as Elsass, which contain an old word, meaning 'dwellers, inhabitants', there are Old English names in -sætan (-sæte), as DORSET, SOMERSET, further Estursete in Domesday, 'dwellers on the river Stour' (the name of an old hundred in Kent), Mersete in Domesday, 'borderers' (the name of an old hundred in Shropshire, derived from OE gemære 'boundary'). Here belong Tempsiter (olim Temsete), the name of a large manor on the Teme, literally 'dwellers on the Teme', and the village-name GRANTCHESTER (olim Grantesete) 'dwellers on the river Granta'. It has been suggested that the numer-ous names in -sett in Norfolk and Suffolk belong here, but LETHERINGSETT, WETHERINGSETT can hardly be names in -sætan, and the probability is that other names in -sett contain Old English geset or sæte 'abode' or the like.

Old English -waru (-ware) 'dwellers' occurs in burgwaru 'townspeople', Cant-waru 'Kentish-men' and the like. Some formations of this kind have become place-names. ROMNEY MARSH is called Merscuuare in 774. The literal meaning is 'marsh-dwellers'. CLEWER in Berks and Somerset represents Old English Clif-ware 'cliff-dwellers'. RIDWARE may belong here.

NORFOLK and SUFFOLK originally meant 'the Northern and the Southern people', but came at an early date to denote the two parts into which East Anglia was divided. *Haliwerfolc* or *Cuthbertfolk* was the old name of the soke of the Bishop of Durham. A district in Worcestershire is called *Kinefolka* in a document of *c.* 1115. The name seems to be an Old English *cyne-folc* 'royal people'. The village-names FOLKE, FREEFOLK belong here.

BONHUNT and CHESHUNT seem to go back to Old English names in -*huntan* 'huntsmen'. See HUNTA in the dictionary, where German parallels are adduced.

To this group belong some isolated names of high antiquity, relics which give a hint as to the important part played by the tribe in primitive Anglo-Saxon society. JARROW in Durham is identical with the name *Gyrwe*, which denoted a tribe in Lincolnshire, Northamptonshire, &c. The name means 'marsh-dwellers'. Jarrow is mentioned already by Bede. NORTHILL and SOUTHILL in Bedfordshire mean 'the Northern and the Southern *Gifle*'. The *Gifle* were a tribe, mentioned in the seventh-century Tribal Hidage. The name means 'dwellers on the river Ivel'. HITCHIN is identical with the old tribal name *Hicce*, likewise mentioned in the Tribal Hidage. OUNDLE is an old tribal name, as shown by *Undalana mægð* in the Old English translation of Bede's *Historia ecclesiastica*. Bede calls the place *prouincia Undalum*. RIPON is likewise an old tribal name, which is found also in REPTON. No doubt other etymologically obscure place-names represent the names of forgotten tribes. Skeat suggested that MIMMS is a case in point, and he may well have been right.

2. *Habitation-names Proper*

Here belong the place-names which from the beginning denoted inhabited places, homesteads and villages, names such as LANGHAM, WHITTINGTON.

(*a*) Two groups may be distinguished. Names such as STANTON, WALTHAM denoted the homestead or group of homesteads (village) proper. Such names are habitation-names in the strictest sense. *Hām* from the first meant 'home, homestead'. *Tūn* originally meant 'enclosure' or even 'fence', but it must at an early date have developed the meaning 'enclosure round a homestead, toft', which soon passed into that of 'homestead' and 'village' or even 'town'. In English place-names *tūn* as a rule has the meaning 'homestead' or 'village'. A similar change has taken place in *worþ*. To this group also belong names containing such elements as *þorp, wīc, hāmstede, hāmtūn*, also such as *burg, hūs, cot, ærn*.

To the other subgroup belong names such as BUCKINGHAM, STRENSALL, BRADLEY, which contain the elements *hamm, halh, lēah*. These elements originally denoted, or could denote, a defined area of ground, a piece of flat land in a river bend, an enclosure or an open place in a wood. Place-names containing these or similar elements (e.g. *ēg* 'island' or 'land on a river', *feld*) were originally nature-names, but it is very probable that the meaning of the elements underwent a change. As a *hamm* or a *lēah* often coincided with the (cultivated) area of a settlement, the original meaning of the words would easily pass into that of 'land belonging to a homestead or village'. Even if no such change in meaning took place, it is better to class names of this kind with habitation-names. A name such as AVELEY rather means 'open land belonging to *Ælfgӯþ*' than 'homestead at *Ælfgӯþelēah*'.

Some elements are difficult to place definitely, because the exact meanings are not known. Such are *stede* and *stoc*, which originally meant 'place', but no doubt at an early date developed more special meanings.

(*b*) The two most important of these elements are *hām* and *tūn*, and some notes on the relations between them may find a place here. Of the two, *hām* is the earlier word for 'homestead, village', and names in -*ham* are on the whole earlier than those in -*tūn*. This does not mean that names in -*tūn* are all comparatively late. The remarks made in the article TŪN on names in -*ingatūn* found in France should be noticed. But on the whole, names in -*hām* represent an earlier stratum. This is indicated already by the fact that names in -*hām* (especially -*ingahām*) are far more frequent in the east than in the west of England. Very likely most names in -*ham* found in western (or at least south-western) counties are rather names in -*hamm* than in -*hām*. This is corroborated by the following facts.

In early Old English, adjectives were often joined to nouns in their uninflected form, as in *hēah-flōd, eald-spell, wīd-sǣ*. In place-names this type of compound is rare. *Middel* 'middle' is an exception, being generally uninflected, as in *Middeltūn*, but here *middel* may be really the noun *middel*. Otherwise adjectives generally appear in the weak form in place-names. NEWTON is always *Niwatun*, dat. *Niwantune* in Old English sources. But -*hām* frequently has an uninflected adjective joined to it. This can be demonstrated only in some special cases, for in names not recorded until after the Norman Conquest it is generally impossible to decide what was the original form of the adjective. But HIGHAM is a valuable test-word. It is a common place-name, being found in Kent (at least six[1] times), Essex, Suffolk (twice), Northampton (twice), Bedfordshire, Derbyshire, Leicestershire, Lancashire, Yorkshire, Northumberland; and HAUGHAM in Lincs is identical in origin. The Old English base is *Hēah-hām*, which is actually recorded for one name. On the other hand, *hēah* is always in the weak form when combined with *tūn*. It is Old English *Hēa-tūn*, in the dative *Hēan-tūne* (or in the north *Hēa-tūne*), whence the names HEATON, HAMPTON, HEANTON, HEMPTON, HENTON, HINTON, and the like. HEIGHTON in Sussex probably has Old English *hēah* 'height' as first element. It is obvious that Higham represents the earlier type, but this does not mean that the type *Hēa-tūn* is late. An example of *Heantun* occurs as early as 780. No doubt Higham belongs to the very earliest stratum of English place-names.

HENHAM occurs twice (Essex, Suffolk). There is no reason to doubt that it goes back to Old English *Hēa-hām*, in the dative *Hēan-hām*. If it does, it shows that *hām* continued in use as a living place-name element till the time when it became usual to give adjectives the weak form in forming place-names. This is obviously the explanation of the name *Nīwa-hām*, in the dative *Nīwan-hām*, which is the source of NEWNHAM, NUNEHAM. The very meaning of this name ('new *hām*') indicates that it is comparatively late. A village founded at the time of the first Anglo-Saxon settlements would not be called 'new village'. It may be added that when *hēah* is combined with the elements *dūn, lēah*, the weak form is used. HENDON is Old English (æt) *Hēan-dūne*. HENLEY, HANDLEY from Old English (æt) *Hēan-lēage* are common names. HEALEY is only found in the north and in Lancashire, where the dative of the weak form of *hēah* was *hēa* (with loss of *n*). HIGHLEY in Shropshire does not contain the adjective *hēah*.

It has been suggested that other adjectives besides *hēah* sometimes appear in the uninflected form. Thus ROFFEY has been derived from Old English *Rūh-hege*.

---

[1] Some Highams are not mentioned in the dictionary, because they now denote very unimportant places.

But this name represents Old English *rāh-hege* 'enclosure for roe-deer'. On ROUGHAM, ROUGHTON see the articles in the dictionary.

Another curious fact is that *hām* is hardly ever combined with *brōc* 'brook', while it is sometimes combined with *burna*. On the other hand, *tūn* is frequently combined with *brōc*, never with *burna*, except in the north. The BURNHAMS in Bucks, Essex, Norfolk seem to represent *Burn-hām*. BROCKHAM occurs once in Surrey, but it may be *Brocc-hamm* or *Brōc-hamm* just as well as *Brōc-hām*. BRUNTON is found twice in Northumberland. BROUGHTON is very common, some twenty being from *Brōc-tūn*, and the same is the etymology of several BROCKTONS, BROCTONS, BRATTONS, and one BROTTON. Evidently the normal types are Old English *Burnhām* and *Brōctūn*.

The explanation of this is as follows. The old word for 'brook' was *burna*, but it began at an early date to be replaced by *brōc*, which originally meant 'marsh', the meaning of the corresponding word in continental languages (Dutch *broek*, &c.). *Burn* has gone out of use in the greater part of England, and is now a north country word. *Burnhām* evidently dates from an early period, when *burna* was still a common word for 'brook, stream', while *Brōctūn* arose at a somewhat later period. *Burna* is common in names of streams and place-names derived from such, and *Bourne* is a common river-name. Except in the north, these names must be very old.

There are several other elements which are frequently combined with *tūn*, never with *hām*. For one of them an explanation similar to that given for *brōc* may be suggested, viz. *halh* (*healh*). Old English *H(e)alh-tūn* is the source of HALTON, HAIGHTON, HALLATON, HALLOUGHTON, HAUGHTON, HOLTON, HOUGHTON (or some of them, together at least 28 cases). *Halh* has several meanings, but one of the most important is 'haugh'. It is quite possible that this special meaning had not yet developed at the time when *hām* was still a common place-name element, and that this is one of the reasons why *halh* is never combined with *hām*. But a similar explanation would not do for some other elements, as Old English *ēa, ēg, clif, hyll, mersc, mōr, ōfer*. It is very curious that names such as EATON, EYTON, CLIFTON, HILTON (HULTON), MARSTON, MOR(E)TON, OVERTON are very common, but that EAHAM, CLIFHAM, &c., are totally absent. On the other hand, *hām* is combined, for instance, with *burg, mere, sǣ, wald* (as BURGHAM, MARHAM, SEAHAM, WALTHAM). All these elements are, of course, combined also with *tūn*, and far more frequently. It seems we must conclude that the earliest settlements were not very frequently named from topographical features.

(c) Some of the words belonging to this group are frequently found alone as place-names, while others only occur as elements of compound names. *Hām* and *tūn* are never used alone as place-names, while *stoc, þorp, wīc, worþ* are often so used (STOKE, THORPE, WEEK, WICK, WORTH, WORTHY). In several instances places originally called simply Stoke, Wick, &c., came later to be known by a composite name, a distinctive addition being tacked on, as STOGURSEY, STOKESAY, BISHOP-STOKE, WICK EPISCOPI, PAINSWICK, originally Stoke, Wick, &c. Of a similar type are really also *beretūn, berewīc, bōpltūn, hāmstede, hāmtūn*, though they are formally compounds. BARTON, BERWICK, BOLTON, HAMPSTEAD, HAMPTON are common names.

It is easy to understand why such words as *burg, ceaster* are frequently found alone as place-names (BURGH, BURY, &c., CHESTER). There would generally be only one fort or Roman station in a district. Such an explanation would not hold good for the names STOKE, WICK, &c. The reason why words such as *stoc, wīc* could be used alone as place-names was that they denoted a dependency of a mother village or manor, an outlying part, a dairy-farm, a cattle-fold, or the like. That *stoc* had some such meaning is shown by the fact that several names containing the element have as a first element the name of a neighbouring village, as CHARDSTOCK near CHARD, &c. See STOC in the dictionary. The *stoc*, *wīc*, &c., would be little known outside the village or manor to which it belonged, and the simple name *Stoc*, *Wīc* would generally suffice. If a more distinctive designation was sometimes necessary, the natural expedient was to call the place 'the *stoc* of such-and-such a village'. That *wīc* frequently denoted a dairy-farm or cattle-farm is shown by names such as BUTTERWICK, CHISWICK, COWICK, SHAPWICK. The name WICK is particularly common in districts where cattle-farming has played an important role from ancient times, as in the south-western counties.

THORPE holds a place of its own in so far as it is to a great extent of Scandinavian origin, but may yet be dealt with here. Both the English and the Scandinavian *þorp* are frequently found alone as place-names. Typically English are THROOP, THRUP(P). In Denmark a *thorp* was a daughter settlement, originally dependent on the mother village and therefore naturally called simply *Thorp*. See further ÞORP in the dictionary. In all probability Old English *þorp* was used in a similar way, and this explains why the word is fairly often used alone as a place-name. The name Thrupp cannot well mean simply 'village'. A name such as 'the village' is not distinctive enough.

BARTON, BERWICK and the like are quite analogous to STOKE, WICK. A *beretūn* or *berewīc* was a dependent farm, an outlying farm on which corn was grown and stored. It belonged to a village, or more probably to a manor. It could be called simply *Beretūn* or *Berewīc*, if the manor had only one such or if there was only one in the neighbourhood.

Names such as BOLTON, HAMPSTEAD, HAMPTON (from *hāmtūn*), also BOLD, BOOTLE hold a place of their own. These names mean literally 'dwelling, homestead', and evidently from the first denoted independent settlements. The probability seems to be that names of this kind were given to a village or estate that was the centre of a comparatively large settlement, as the mother village in contradistinction to its daughter settlements, outlying farms, &c., or it might be the demesne farm of a manor. Very likely, compounds such as *hāmstede*, *hāmtūn*, *bōþltūn* arose because the words *hām*, *tūn* had got a wider application than the old one of 'homestead', &c.

(d) Near habitation-names stand names that originally denoted a pasture-ground or a shelter for the protection of animals, a cow-house, a cattle-fold, &c. A pasture or a shelter for animals would often develop into a homestead or a village. Here belong names with such elements as *denn*, *bǣr*, *dūn*, *bȳre*, *hlōse*, *(ge)set* (at least partly). *Denn* 'a pasture, especially a swine-pasture', is particularly common in Kent and Sussex. It is the second element of a great number of village-names in Kent, as HALDEN, TENTERDEN. Originally the *denns* belonged to the various *lathes* and were often situated far from the district to which they belonged. Later they were assigned to villages in the *lathe*, and eventually in many cases became independent villages. Old English *dūn* 'hill' must have acquired the special sense 'hill pasture', whence perhaps even 'pasture' generally. This is doubtless the meaning of *dūn* in many place-names, especially such as have for their first element a personal name or words such as *king*, *bishop* (as KINGSDOWN, BISHOPTON in Warwickshire).

3. Original Nature-names

(a) Many names of villages or homesteads originally denoted a natural object, or more rarely some product of human activity near which, or sometimes in which, the place was situated. Names of this kind, when used of inhabited places, are elliptical or metonymical. OXFORD at first designated a ford over the Thames. When a village grew up at the ford it was named from the ford, and Oxford, when used of the village, originally meant 'the village at Oxford'. ARROW, the name of a village on the river Arrow, at first denoted the river, later the village on the Arrow, and finally Arrow village.

Here belong the numerous habitation-names that were in the first instance names of rivers and streams, fords, lakes, springs, bays, hills or ridges, fens, woods, &c. Most of the names are English compounds or English common nouns, as SHERBOURNE, BEXHILL, BOURNE, MERE. But many rivers had pre-English names, which often came to denote settlements on the river, as DARENTH, ISLE, TARRANT. Also some other pre-English nature-names became names of settlements, as BARR, CANNOCK, MELLOR, PENN.

It is obvious that CAMBRIDGE is really of the same type as Oxford. Cambridge originally meant 'the bridge over the Granta', but came to mean the 'village at *Grantanbrycg*', and finally '*Grantanbrycg* village'. Of the same kind are names that contain words for road, wall, ditch, pole, memorial-stone, &c., as BARN-STAPLE, WALL.

(b) There are indications that names of this kind were originally to a great extent elliptical, the nature-name being preceded by a preposition, and an expression for 'homestead' or 'village' being understood. TWYNING was originally *Bituinæum* 'between the rivers', i.e. 'the place between the rivers'. Many names still preserve the preposition, though its meaning is not now understood except in some names of late origin, as UNDERMILLBECK, UNDERSKIDDAW, the latter of which means 'the village at the foot of Skiddaw'.

The most common Old English preposition before place-names was *æt* 'at, by'. It may be preserved in ATTERCLIFFE. The *T-* sometimes prefixed to a name beginning in a vowel in early sources, as in *Taneburne, Ticelle* in Domesday for ENBORNE, ITCHEL, is evidently a remnant of *æt*. ENBORNE, the village-name, would seem to have been originally *æt Enedburnan*. Old English *bī* is found in BYFLEET, BYGRAVE; *binnan* 'within' in BINDON, BENWELL; *beneopan* 'beneath' in Neadon in Devon [*Beneadona* DB, *Bynythedon* 1333 Subs] '(place) below the hill'; *bufan* 'above' in BOVENEY, BOVERIDGE; *betwēon, betweox* 'between' in TWIN(E)HAM, TWYNING; *under* in UNDERBARROW, &c.; *uppan* in UPHILL. *Up, down*, when combined with river-names, as in UPAVON, DOWN AMPNEY, mean 'higher up, lower down'. Upavon is '(place) up the Avon'. Here belong expressions such as *be ēastan (westan)*, &c. One WESTWOOD in Kent is *Beuuestanuudan* in a charter of 805 (BCS 323). The name means '(place) west of the wood'. BESTWALL is 'east of the wall'. Southwhimple in Devon [*Bysouthewympel* 1333 Subs] means 'south of the river *Wimpel*'. EASTNOR is '(place) east of the ridge', the Cheshire SIDDINGTON 'south of the hill'. Probably many EASTBROOKS, WESTBROOKS really mean 'east of the brook', &c.

Elliptical are also names like ASHINGTON in Somerset, EASTINGTON, SIDDINGTON in Gloucestershire, SODINGTON, UPTON in Essex, which have often been held to contain derivatives in *-ing* (OE *Sūpingatūn* '*tūn* of the dwellers in the south'). Sodington is *Supintuna gemæru c.* 957 Birch, Cart. Sax. 1007. *Supintun* is really *sūp in tūne* 'south in the village', elliptical for 'land south in the village'.

In Old English sources, combinations of the type here dealt with are often found, but they are not always used elliptically. In Birch, Cart. Sax. 313 (A.D. 804) we read of *xxx manentium under Ofre* '30 hides under *Ofer*'. In Kemble, Cod. dipl. 770 (A.D. 1044) a place called *Benyðan Elddin* is also called *þære anre gyrde beniðan elddin* 'the yardland below Elddin'. Cf. also SIDBURY in the dictionary (under sūþ). Presumably a person was said to live *be ēastan wealle* 'at Best-wall' or *betwēon ēam* 'at Twyning'. In Old Norse it is said that a man *bió undir Karlsfelli* 'lived at the foot of Karlsfell' or *undir Brekkum* 'at the foot of the slopes'. But at a very early date these prepositional phrases came to be used elliptically as place-names. The original meaning was gradually obscured, at least in some cases, and a preposition could be placed before the elliptical name—as in *In Sudaneie* 676 Birch, Cart. Sax. 42 (not in an original charter, it is true).

Elliptical names are known in many other languages. They are common in Norway and Iceland. Landnáma has *under Brekkum* (*þar heiter nu under Brekkum* 'that is now called *under Brekkum*') and the like. Norwegian examples are *unde Bergo* 'below the hill', AUSTVATN from *austan vatn* 'east of the lake'. From Sweden may be adduced cases like VÄSTANÅ, VÄSTANSJÖ 'west of the stream, lake'. German street-names such as UNTER DEN LINDEN are well known. Welsh examples are ISCOED 'below the wood', TRAWSFYNYDD 'across the mountain'. ARFON is '(district) opposite to Anglesey'. Anglesey is Welsh *Mon*; *ar* means 'opposite to' Carnarvon is really Welsh *Caer yn Arfon* 'town in Arfon'.

(c) In the cases so far dealt with the preposition or prepositional phrase has been preserved to our days. In Old English a seemingly redundant preposition is very common before place-names. King Alfred in Orosius calls Sleswig *port þe mon hæt æt Hæþum* (lit. 'at the heaths'). Birch, Cart. Sax. 76 (A.D. 692), mentions 'locus cujus nomen est *Æt-stretfordæ*' (i.e. Stratford). The 'nominative' was *æt Hæþum*, *Æt-stretfordæ*. In the dictionary, name-forms of the type *Æt-strætforda* are regularly quoted in this form, while the preposition is placed between brackets if it has its ordinary locative function. Thus (æt) *Niwantune*, if the name occurs in such a phrase as 'ten hides *æt Niwantune*'. Here the name was *Niwatun*, not *æt Niwantune*.

Bede generally places a preposition before names of villages that were originally nature-names or the like. He uses the Latin preposition *ad*, evidently a translation of Old English *æt*. BARROW in Lincolnshire is *Adbaruae* 'at the grove', a lost TWYFORD *Adtuifyrdi* 'at the double ford', STONE in Hants *Ad Lapidem* 'at the stone'. GATESHEAD is *Ad Capræ Caput*. Gateshead originally denoted the hill there. YEAVERING is *Adgefrin*. *Gefrin* is the old name of Yeavering Bell. WALL-BOTTLE is *Ad Murum* 'at the Roman wall'. WHITHORN in Galloway is *Ad Candidam Casam*, but the church itself is *Candida Casa*. The English forms would be *Hwīte ærn* 'the white house' or 'stone church' and *æt Hwītan ærne* 'Whithorn village'. The preposition *in* is used in cases like *Inderauuda* or *In silua Derorum* 'Beverley', literally '(place) in *Derawudu*',[1] *Ingetlingum* 'Gilling', *Inhrypum* 'Ripon', literally '(place) in the province of the *Getlingas* or *Hrype*'. No preposition is found before names of meeting-places, as *Augustinaes ac*, *Clofeshoch*, *Herutford*. Otherwise the preposition is put almost without exception before nature-names used as habitation-names. In 'in loco, qui dicitur *Stanford*' or 'in loco, qui uocatur *Hreutford*', the phrase 'in loco qui dicitur (uocatur)' may be said to be equivalent to a preposition.

[1] *In Dera Wuda* c 890 OEBede. The name *Derawudu* means 'wood of the Deirans'.

A preposition is absent before names in -*hām*, -*tūn*, -*burg*, -*ceaster*, as *Godmund-dinggaham*, *Penneltun*, *Tilaburg*, *Grantacaestir*, before pre-English names such as *Domnoc* 'Dunwich', *Dorcic* 'Dorchester' (in Oxfordshire), also before names in -*ei* (-*eu*) 'island', as *Cerotaesei* 'Chertsey', -*feld*, -*halch*, -*hamm*, as *Haethfeld* 'Hatfield', *Streonaeshalch*, *Bosanhamm*. On the last three elements see p. xiv. It is absent also before names in -*dūn*, as in *Briudun* 'Breedon', *Uetadun* 'Watton', *Uilfaræsdun*. On *dūn* see also p. xvii.

If Bede does not keep up the distinction between prepositional and non-preposi-tional names with absolute consistency, there are at any rate very few exceptions.

In early Old English charters there are strong traces of a similar rule. Examples are: *æt Dene c.* 725 Birch, Cart. Sax. 144 (Dean Sx), *ad Hlidum* 774 ib. 214 (Lydd), *æt Segcesbearuue* 777 ib. 223 (Sedgeberrow), *æt Breodune*, *æt Wærsetfelda* 780 ib. 234, *æt Clife* 769–85 ib. 246 (Bishops Cleeve Gl), *on Lingahæse* 793 ib. 265 (Hayes Mx), *aet Burnan* 805 ib. 319 (Bishopsbourne), *æt Folcanstanæ* 824 ib. 378 (Folkestone), *æt Sceldesforda* 824 ib. 380, *on Cert* 843 ib. 442 (Chart K).

Names of meeting-places have no preposition, as *Clobeshoas* 793–6 Birch, Cart. Sax. 274, *Hacleah* 805 ib. 322, *Oslafeshlau* 825 ib. 384.

Before names in -*hām*, -*tūn* and the like the preposition is absent in early charters, as *Pecganham* 680 Birch, Cart. Sax. 50, *Budinhaam*, *Ricingahaam*, &c. 692 ib. 81, *Fledanburg* 691–2 ib. 76, *Wieghelmestun* 697 ib. 97, *Tuicanhom* 704 ib. 111, *Hehham* 774 ib. 213.

Apparent exceptions are *aet Liminge* 798 ib. 289, *aet Wigorna ceastre* 778–9 ib. 231. In both cases monasteries are referred to. These were called '(monastery) at Lyminge and Worcester'.

In later Old English charters a pleonastic preposition is common also before original habitation-names. The types *æt Strǣtforda* and *Ricingahām* were kept well apart till about 850, but later the preposition is often added also before the latter type. Probably we have here to do with a peculiarity of the official language, a custom of scribes. The preposition had begun to be dropped in everyday speech in the old prepositional type; *æt Strǣtforda* had become *Strǣtford*. But the old type was retained in the official language, and scribes naturally began to introduce the preposition also where it was not used in earlier days. Formulas such as *æt Stantune*, *æt Mortune* are very common in tenth-century charters. The fact that these phrases are sometimes stated to be used locally ('in loco quem solicolæ *æt Derantune* vocitant' 934 BCS 702) or are even held up to ridicule (as 'cui ruricolæ . . . ludibundisque vocabulis nomen indiderunt *æt Meaphám*' 939 ib. 741) should not be taken seriously. These phrases are used also when no preposition is put (as BCS 756). It was not the prepositional phrase the scribes found ridiculous, but the name itself.

The material brought forward indicates that when original nature-names were used as names of inhabited places, it was very common in the early Anglo-Saxon period to place a preposition before them, an elliptical name being the result. But this usage was not retained very long. In many cases the prepositional phrase would have to be used with its original import. In combination with a word for 'dwell' *æt Strǣtforda* would mean 'at Stratford', not 'Stratford'. The preposi-tional names would often be difficult to construe, as in the objective or genitive case. The common use in charters of derivatives denoting the inhabitants instead of the name of the place may well to some extent be due to the difficulties of construction. The boundary of OARE Brk (originally no doubt *æt Oran*) is given in a charter as *Orhæma gemære*.

But it cannot be looked upon as proved that the change of nature-names to habitation-names always took place *via* prepositional names. Partly at least it may be due to metonymy.

III. THE VARIOUS ORIGINS OF ENGLISH PLACE-NAMES

The English place-nomenclature is composite, place-names being derived from various sources. This sometimes renders a definite etymology difficult. The majority of place-names are English, but there are not a few Celtic names, a great many Scandinavian, and some Latin and French names. A number of hybrid names occur.

1. *The Celtic Element in English Place-names*

The majority of Celtic place-names in England are British, but some Gaelic names occur.

(a) *The British element in English place-names.* The earliest English place-names are those taken over by the Anglo-Saxons from the Britons at the time of the Anglo-Saxon settlements in Britain. It is possible that the Britons themselves, who were not the first inhabitants of Britain, had adopted some place-names from their predecessors, but if any such names have been preserved to our days, they cannot be picked out at the present stage of research, and except for possible Latin names, pre-English names are here looked upon as British, whatever the ultimate etymology.

Some of the names belonging to this group are recorded in sources anterior to the Anglo-Saxon invasion, in the works of classical writers such as Cæsar, Tacitus, Ptolemy. For such, British origin is obvious. Unfortunately the number of names so recorded is small. Most are found only in English, the majority even in post-Conquest sources.

The place-names of Cornwall are preponderatingly British, though there are not a few English place-names even in that county. Cornwall partly retained its British language (Cornish) till the eighteenth century, and the majority of place-names arose before the time when English superseded the ancient language. Cornish place-names thus differ in character from those of the rest of England, and there might have been reason to omit them, like those of Wales and Scotland. But the English place-names in Cornwall could not be omitted, and some place-names of British origin are widely known and will interest many people. For these reasons a selection of Cornish names has been included. A full treatment would be impossible. The problems offered by these British names are in many ways different from those bound up with place-names of Germanic origin, and etymologies would have had to be made rather full in order to be understood. Only the most important Cornish names have been included as a rule, chiefly names of towns, parishes, rivers. The parish names to a great extent consist of, or contain, saints' names. They are really rather uninteresting, when the etymology is found, but it will doubtless be of some interest to know that such a name as PHILLACK really means '(the church of) St. Felicitas' or PHILLEIGH '(the church of) St. Filius', or that MEVAGISSEY consists of two saints' names. The history and origin of Cornish saints' names are very obscure, and a full treatment of these names could not be attempted. Often all that has been possible is to give the saint's name generally held to be the source of place-names. Sometimes it has been possible to suggest an etymology for the saint's name or to adduce a Breton or Welsh parallel.

Apart from Cornwall, the British place-names mostly belong to certain definite categories.

Names of rivers and streams. These are to a great extent British, particularly names of important rivers, as AIRE, AVON, DEE, DERWENT (DARENT, DART), DON, ESK (AXE, EXE), OUSE, SEVERN, STOUR, TEES, THAMES, TRENT, WYE. But many small streams also have British names. They are particularly common in some districts, as in Dorset (as CALE, CERNE, CHAR, DIVELISH, FROME, IWERNE, LIDDEN, LODDEN, LYME, and the original stream-names WINFRITH, WYNFORD), Somerset (as BRUE, CAM BROOK, CHEW, DOWLISH, FROME, KENN, PARRET, WELLOW), Gloucester (as CAM, CARANT, FROME, LEADON), Hereford (as ARROW, DULAS, GARREN, LUGG, OLCHON, WORM), Salop (as COUND, DOWLES, MEOLE, RODEN, TERN), Cumberland (as CALDER, CAM BECK, COCKER, DACRE, ELLEN, LYNE).

But there are not a few river- and stream-names of English origin. Among names of rivers may be mentioned DEARNE, IRWELL, MERSEY, REDE, SWALE, WANTSUM, WENSUM, WISKE. Names of small streams are to a very great extent English, as BLACKWATER, ENBORNE, LAMBOURN in Berks, HAMBLE, MEDINA in Hants, PIDDLE in Dorset, LOUD in Lancashire, MEASE, SENCE in Leicestershire, GREET, SMITE in Notts, BLYTH in Suffolk, SHEAF in Yorkshire. Many arose at a late period, owing to back-formation, as PANG in Berks, CHELMER in Essex, ROTHER in Kent and Sussex, ANT, BURE in Norfolk, ADUR, ARUN in Sussex.

In the Scandinavian parts of England not a few names of streams have Scandinavian names, as GRETA, LIZA in Cumberland, WINSTER in Lancashire, BAIN in Lincolnshire, WREAK in Leicestershire, ROTHAY in Westmorland, BAIN, SKELL in Yorkshire.

Names of hills and forests. Several of these are British. Of hill-names (or original hill-names) may be mentioned BARR, BRENT, CANNOCK, CHEVIN, CREECH, CRICH, CRICK, LYDEARD, MALVERN, MELLOR, PENN, PENNARD. Some have had an explanatory English word for 'hill' added to the British name, as BREDON, BREEDON, BRILL, originally a simple name identical with Welsh *bre* 'hill', PENDLE, originally identical with Welsh *pen* 'head, top'. Of forest-names may be mentioned CHUTE, KINVER, MORFE, PENGE, SAVERNAKE. CHEETWOOD, CHETWODE have had Old English *wudu* 'wood' added to a name derived from British *cēto-* (Welsh *coed*) 'wood'. This word is the first element of CHATHAM, CHEETHAM. LICHFIELD has had *feld* added to the British name *Lētocēto-* 'grey wood'.

Names of British towns or forts. These are often preserved, but in many cases only as the first element of a compound name, whose second element is an explanatory English word for 'fort' or the like. The British name often appears in a clipped form. Here belong such names as LONDON, CARLISLE, DOVER, DUNWICH, LYMPNE, PENKRIDGE, RECULVER, YORK, which have developed from the British names; further DORCHESTER, GLOUCESTER, MANCHESTER, ROCHESTER, WINCHESTER, COUNTISBURY, SALISBURY and many others, which consist of the British name with addition of Old English *ceaster*, or *burg*. Names of this kind are found all over England.

Territorial names of British origin are found here and there in various parts of England. Here belong KENT, THANET, WIGHT; CRAVEN, ELMET, LEEDS (an old name of a district). LINDSEY has had Old English *ēg* 'island' added to the old name (*Lindissa* in Bede).

British names of villages and hamlets. These are on the whole rare, except in some districts, where a British language lived on for a comparatively long time, as in Herefordshire, Cumberland, and to some extent in Dorset, Wilts, Salop, Lancashire. Herefordshire remained partly Welsh till a late period, and the place-nomenclature west of the Wye is to a great extent Welsh. Among anglicized Welsh names may be mentioned DINEDOR, MOCCAS, TRETIRE, TREVILLE. In Cumberland are found some typically British names such as BLENCARN, CUMDIVOCK, CUMREW, CUMWHINTON, also such interesting names as BIRDOSWALD, CUMWHITTON, on which see dictionary. Of Shropshire names may be mentioned ERCALL, HODNET, PREES, WENLOCK. In Cheshire British names are remarkably few, but some do occur, as CREWE, INCE, LANDICAN, LISCARD, WERNETH. Lancashire has an exceptionally large number, as CULCHETH, ECCLES, HAYDOCK, INCE (2), PENKETH, PREESE, TREALES. Devonshire might be expected to have many British names, but they are not very numerous. In Somerset there are still fewer. An interesting name is WATCHET. Of British names in other counties may be mentioned CAMS, LISS in Hants, PENKRIDGE in Staffordshire.

British words with a topographical meaning. Some words of this kind were adopted by the English and used in forming place-names. It is generally held that Old English *cumb* 'valley' is a British loan-word. It is common in place-names. Of British origin is *tor* 'hill', found in dialects and in some place-names, as TORRE, DUNSTER, VOBSTER. The element *funta* found in BEDFONT, HAVANT, &c., is ultimately from Latin *fontana*, but the immediate source is British. Other such elements are Old English *carr* 'rock' (in CARHAM, CARHAMPTON), *luh* 'lake' (e.g. LUTTON in Lincolnshire), also *brocc* 'badger'. It is not improbable that Old British *eclēs* 'church' (Welsh *eglwys*, from Latin *ecclesia*) was adopted by the English and used for some time before the word *church* (OE *cirice*) came in. That element is a little too common in English place-names for it to be probable that each name of this kind shows the adoption of a British place-name.

(*b*) *The Gaelic element in English place-names.* Place-names of Gaelic (Goidelic), i.e. Irish or Scotch-Gaelic, origin are few in England. BECKERY (in Somerset) is due to special circumstances; according to a tradition it is the result of a settlement of Irish monks. Otherwise such names are probably only found in the border counties. GREYSOUTHEN, RAVENGLASS in Cumberland seem to be cases in point.

But some words of Irish-Gaelic origin were introduced and used in place-names. *Crag* probably belongs here. Not a few Irish-Gaelic personal names are found as in DOVENBY, GLASSONBY in Cumberland, BECCONSALL in Lancashire. But most of these were probably introduced by Scandinavians, or rather were borne by Irishmen or Gaels who came over with Scandinavians from Ireland, the Isle of Man, &c., or by partly celticized Scandinavians. These Scandinavians also introduced the words *cross*, *erg* 'a shieling' (in MANSERGH, MOSSER, &c.). They are also responsible for the curious 'inversion-compounds' sometimes met with in the English place-nomenclature, in which the elements are Scandinavian, but the order between them is of the Celtic type, the defining element coming last. A typical example is BRIGSTEER '*Styr*'s bridge'. See further ASPATRIA, KIRKOSWALD, &c.

2. *Place-names of English Origin*

These cannot be dealt with on the same lines as Celtic or Scandinavian names. A reference must suffice to what has been said in Section ii and the notes on river-names of native English origin on p. xxii.

3. *The Scandinavian Element in English Place-names*

(a) *Distribution and provenance of Scandinavian place-names*

The Scandinavian element in English place-names is very considerable. It is an outcome of the extensive Scandinavian settlements made in England from the latter half of the ninth century onwards. The Scandinavians were mostly Danes, but in the north-western part of the country (Cheshire, Lancashire, Westmorland, Cumberland, West Yorkshire) the settlers were mostly Norwegians. The Norwegian element is characterized by certain peculiarities, such as the elements *búð* (as against Danish *bōð*), *gil*, *skáli*, *brekka*, and in these districts are also found obvious traces of Irish-Gaelic influence, as in the common use of *erg* 'shieling' (see p. xxiii) and a sprinkling of Irish personal names. Traces of Norwegian influence are seen also in East Yorkshire, however, as in SCORBROUGH, which contains Old Norse *búð*, and several names in *erg*. Also the names NORMANBY, NORMANTON testify to isolated Norwegian settlements in the Danelaw. Criteria of Danish colonization are the elements *bōð* and *þorp* (in so far as not English), *hulm* for *holm*. *Þorp* is rare in Norway and hardly occurs in Iceland.

The Scandinavian element is particularly strong in parts of the Danelaw (Yorkshire, Lincolnshire, Leicestershire, Nottinghamshire, Rutland), further in the Lake District (Cumberland, Westmorland, Lancashire). A considerable Scandinavian influence is noticeable in Norfolk, parts of Derbyshire, Northamptonshire, Durham, Cheshire. Outside these districts it is, on the whole, slight. In some districts, as in Lincolnshire and the Lake District, Scandinavian place-names are very numerous; in some areas they even outnumber old English names. In Lincolnshire names of wapentakes are mostly Scandinavian. The word *wapentake* itself, and probably the division into wapentakes (found in Lincolnshire, Leicestershire, Rutland, Derbyshire, Nottinghamshire, Northamptonshire, Yorkshire, Lancashire, Durham) are Scandinavian. In Norfolk names of hundreds are to a great extent Scandinavian.

(b) *The Scandinavian element in English place-names varies in character*

Scandinavian names in the strictest sense. Many Scandinavian place-names in England must have been formed by people who spoke a Scandinavian language. These may be called Scandinavian place-names in the strictest sense. Most are compounds, both of whose elements are Scandinavian words. Many are shown by typically Scandinavian inflexional forms to be genuine Scandinavian names. Many of them contain Scandinavian genitives in *-ar*, as AMOTHERBY, AMOUNDERNESS, BORROWDALE (ON *Borgardalr*), BECKERMET, DALDERBY, HOLDERNESS, LITHERLAND, SCORBROUGH (ON *skógar-búð*). In Old Danish the genitive ending *-ar* generally became *-a* in compounds. This explains names such as HAWERBY, OSGODBY, OWMBY, which in the early twelfth-century Lindsey Survey appear as *Hawardabi, Osgotabi, Ounabi* (from Old Danish *Hāvarðabȳr*, earlier *Hāvarðarbȳr*, &c.). Another common Old Scandinavian genitive ending is *-s*, as against Old English *-es*. Before *s* a *d* (*ð*) became *t*, and *ts* often gave *s*. Also other consonants were apt to get lost before *s*. Typically Scandinavian are therefore names like BRACEBY (*Breizbi* in Domesday from *Breiðs bȳr*), LACEBY (from *Leifs bȳr*), GRASBY (from *Griōts-bȳr*). WASDALE comes from Old Norse *Vatnsdalr* 'valley with a lake'. Some names show other typically Scandinavian inflexions, as SAWREY (*Sourer* in 1336, from Old Norse *saurar*, the plural of *saurr* 'wet place'),

Stather in BURTON ON STATHER, which comes from a plural identical with Old Norse *stǫðvar* 'landing-place' (from *stǫð*).

Also pre-Scandinavian place-names adopted by Scandinavians sometimes got Scandinavian inflexions, when combined with Scandinavian elements, as ALLER-DALE (*Alnerdall* 11th cent., where *Alner-* represents *Alnar*, an Old Norse genitive of the old name of the Ellen), ENNERDALE 'the valley of the Ehen', NIDDERDALE 'the valley of the Nidd'. The names ELLEN, EHEN, NIDD are British. Remarkable examples are OSMOTHERLEY in Lancashire and Yorkshire, which have as second element Old English *hlāw* and *lēah*, the Old Norse personal name *Ásmundr* in the genitive form as first element. The words *hlāw*, *lēah* would seem to have been adopted early by the Scandinavians in the district.

Scandinavian names in the strictest sense are of various kinds. There are numerous names of rivers and streams (cf. p. xxii), lakes (mostly anglicized at an early period, as ELTERWATER, ULLSWATER, WINDERMERE), headlands (as RAVENSER ODD, SKEGNESS), hills (as ROSEBERRY TOPPING, SCAFELL, SKIDDAW). But the majority are names of villages, homesteads, dairy-farms, pastures, and the like. The most important Scandinavian elements in names of villages and home-steads are *by*, *toft*, *þorp*. Of words for 'shieling', &c., may be mentioned *sætr* (as in HAWKSHEAD), *skáli*, *bōð*. Names in *-thwaite* are, of course, Scandinavian.

Numerous nature-names, apart from those already mentioned, are among Scandinavian names, though most are preserved as names of inhabited places. Among elements found in such names, may be mentioned *gil* 'valley', *lundr* 'grove', *skōgr* 'wood', *viðr* 'forest', *kelda* 'spring', *mȳrr* 'mire'. Only as first elements occur such words as *gās* 'goose' (GAISGILL), *īkorni* 'squirrel' (ICKORN-SHAW), *refr* 'fox' (REAGILL), *rá* 'landmark' (RABY, ROBY), *blár* 'dark' (BLEA TARN), *forn* 'old' (FOREMARK).

Scandinavianized names. Many English names adopted by Scandinavians were changed in form so as to conform better to Scandinavian habits of speech. The names were 'Scandinavianized'. Many such forms were adopted by the English, and the names are preserved in their Scandinavianized form.

Certain English sounds or combinations of sounds were unfamiliar to Scan-dinavians, who unconsciously substituted a sound or sound-combination used in their own language. Old English *sc* had already become an *sh*-sound or some-thing similar by the time of the Scandinavian immigration, whereas in Scandi-navian *sk* was always pronounced with hard *k*. Hence *sc* was often replaced by *sk*. SKELTON, SKIPTON are Scandinavianized forms of *Scelftūn*, *Scīptūn*, which normally gave SHELTON, SHIPTON. But it is possible that *s* was sometimes sub-stituted for *sc*. SILPHO in North Yorkshire is very likely a case in point.

Old English *c*, *g* (*ȝ*), before front vowels or *j*, were palatalized and appear later as *ch*, *y*, *dg*, as in *child*, *yard*, *bridge* from *cild*, *geard*, *brycg*. In Old Scandinavian, palatalized sounds were unknown. Hence sound-substitution often took place in names containing these sounds. KESWICK, KILDWICK are Scandinavianized forms of *Cēsewīc*, *Cildawīc*, which otherwise became CHESWICK, CHILDWICK. Many names in *-brigg*, *-rigg* (for *-bridge*, *-ridge*) are to be similarly explained. But here also a different sound-substitution was possible. See BRIGSLEY, SESSAY in the dictionary.

A short *d* never occurred in Scandinavian between vowels, *ð* (dh) generally corresponding to Old English *d*. In such positions *d* was replaced by Old Scandinavian *ð*. Hence MYTHOP from original *Midhop*, LOUTH from original *Hlūde*. This is a very common phenomenon.

Initial *w* before *u* was lost early in Scandinavian. Old Norse *ulfr* corresponds to Old English *wulf*. In place-names beginning in *Wu-* the *W* was often dropped when they were adopted by Scandinavians. ULVERSTON is pretty certainly from Old English *Wulfherestūn*.

Many Old English place-names contained elements which may have offered no difficulties to Scandinavian speakers, but occurred in a somewhat different form in Scandinavian. Old English *āc* 'oak', *brād* 'broad', *cyning* 'king', *stān* 'stone', correspond to Old Norse *eik*, *breiðr*, *konungr*, *steinn*, &c. In many place-names Scandinavians substituted the Scandinavian word for the Old English one. In some cases it can be proved that this kind of substitution has taken place. CONISCLIFFE in Durham is *Ciningesclif* 778 ASC. *Conis-* is clearly from Scandinavian *konungs-*. STAINLEY in Yorkshire is recorded as *Stanleh c.* 972. HOWDEN is *Heafuddæne* in a charter of 959. The later form must be due to substitution of the Old Scandinavian word (ON *hǫfuð*, ODan *hoved*) for Old English *hēafod* 'head'. MELTON in Norfolk is recorded as *Middilton* in an Old English charter. The old form *Middeltūn* was replaced by a Scandinavianized *Meðaltūn* (*meðal* means 'middle'). One RAWCLIFFE is *Readeclive* in Symeon of Durham. The later form is due to substitution of Old Norse *rauðr* for Old English *rēad* 'red'. BECKWITH is *Becwuda c.* 972. Old English *wudu* has been replaced by Old Norse *viðr* 'wood'. There can be little doubt that names such as BRAITHWELL, STAINLAND, STAINTON were originally *Brāda wella*, *Stānland*, *Stāntūn*. Sometimes this is indicated by Domesday forms such as *Bradewelle*, *Stantone*.

Sometimes even an etymologically unrelated synonym was substituted for an Old English word. Thus HOLBECK in Notts was Old English *hola brōc*. Old Scandinavian *bekkr* was substituted for *brōc*. Old English *burg* has several times been superseded by *by*, as in BADBY, NASEBY. It is likely that the common name WILLOUGHBY, which consists of Old English *welig* and Old Scandinavian *bȳr*, is partly due to substitution of *by* for the *tūn* of an original *Weligtūn*.

Many apparently hybrid names are to be explained in the ways here indicated. The probability is that hybrid names were by no means so common as has been generally assumed.

Other apparent hybrids are due to the opposite phenomenon, the substitution of English sounds or synonyms for Scandinavian ones. Sometimes a mere sound-change is responsible for an apparent hybrid. The common name ASHBY, in early sources *Askebi* and the like, is Old Scandinavian *Askabȳr*. *Askebi* became *Askby* and *Asby*, the medial *k* being lost. *Asby* is a common twelfth- and thirteenth-century spelling, which is still preserved in a few names. But as a rule *Asby* was later 'restored' to *Ashby*, because the meaning of the name was understood. Substitution of English for Scandinavian sounds would not often take place, because few Scandinavian sounds offered any difficulty to English speakers. But we find *sh* substituted for Old Norse *hj* in SHAP, for *sj* in SHUNNER HOWE. Substitution of an English for a Scandinavian synonym has sometimes taken place. English *east* has replaced Scandinavian *aust* in EAST RIDING (*Oustredinc*, *Estreding* in Domesday). *Riding* is Old Scandinavian *þriðiungr*, and the old name was *Aust-þriðiungr*. English *new* has doubtless often replaced Scandinavian *nȳ(r)* in names such as NEWBALL, NEWBY. Newball is evidently from *Nȳ-bōle*, the source of the common Swedish names NYBBLE, NIBBLE, &c. In ELTERWATER, WINDERMERE it is fairly certain that the original second element was Old Scandinavian *siōr* or *vatn*, for the first element shows a typically Scandinavian genitive in *-ar*.

Scandinavian loan-words in English place-names. Many Scandinavian words and personal names were introduced into English at an early date and were used by English people in forming place-names. Many names containing elements such as *beck* 'brook', *bōþe, holm, kerr* (ON *kiarr*), *lund, toft* were evidently formed by English people, as indicated by the fact that the elements are combined with English words, as BESCAR, BLACKTOFT, OLDHAM, &c. Scandinavian personal names are frequently combined with English second elements. Such names must generally be looked upon as formed by English people. Here belong numerous names in *-tūn* (GRIMSTON, THURGARTON, &c.), and many others, as COPPINGFORD, THURSTONFIELD.

Scandinavian personal names are occasionally met with as elements of place-names in districts where Scandinavian settlements did not take place. Examples are EAST GARSTON in Berks, SWAINSTON in Wight, THURLOXTON in Somerset, TOLPUDDLE in Dorset. It is probable that some places with names like these were named from Scandinavians who belonged to the bodyguard (*thingmannalid*) of King Cnut or Edward the Confessor and received manors in various parts of the country. East Garston seems to have been named from the *Esgar stallere* who is mentioned in Domesday. *Stallere* is Old Norse *stallari* 'constable'.

4. *The Latin Element in English Place-names*

Latin names in the stricter sense are few and of various kinds.

Some early Latin loan-words with a topographical or similar sense occur in place-names. Some of these may have been adopted already before the Anglo-Saxon emigration, and these are not in reality distinct from native words. But some may have been introduced after the migration to Britain, and it is possible that some came into English through British. Here belong elements such as *ceaster, foss* 'ditch', *port* 'port' (Latin *portus*) and 'gate' (Latin *porta*), *wīc*. Latin *fossa, portus, vicus* were adopted early by the Britons (Old Welsh *fos, gwic, porth*). *Castra* was a common Latin name of forts, and CHESTER may in some instances represent adoption of the Latin *Castra* (as certainly in the case of Chester in Cheshire). PORTLAND, originally *Port*, may well have been taken direct from a Latin name *Portus*. Portsmouth Harbour was no doubt called *Portus* by the Romans, and PORTSMOUTH contains this Latin name. Other names in *Port-* probably have a similar explanation. AUST is Latin *Augusta*. LINCOLN is Latin *Lindum Colonia*. But in all these cases the names may have been introduced through the medium of British. For CATTERICK see dictionary.

In later times Latin names were rarely adopted. But latinized forms occur. In early records, names, especially of French origin, are sometimes latinized, and in a few cases these latinized forms have prevailed. Thus PONTEFRACT is in early records sometimes *Pontfreit*, the Old French form, which is no doubt original, sometimes *Pons fractus*, the latinized form. The spelling represents the Latin form, but the pronunciation pŭmfrĭt, which goes back to Old French *Pontfreit*, is still used locally. MONTACUTE is from French *Montaigu*, but the present form is a latinization of the French name. *Sarum*, the latinized form of SALISBURY, is preserved in the name OLD SARUM.

Many additions to place-names occur in a Latin form, as ASHBY PUERORUM, BROMPTON REGIS, TOLLER FRATRUM and PORCORUM, LINSTEAD MAGNA and PARVA, WESTON SUPER MARE, BOOTLE CUM LINACRE, STOKE SUB HAMDON, and the like.

5. *The French Element in English Place-names*

French place-names are far more common in England than Latin names, but the total number is not nearly that of Scandinavian names. French influence shows itself on English place-names also in other ways than in the adoption of names of French origin.

(*a*) Norman names were given by their Norman owners to a good many castles or manors, and Norman monks often gave French names to monasteries or to manors or granges belonging to monasteries. We have here cases of deliberate naming of places. These names often superseded earlier English names. Many were probably transplanted from the Continent. The names are frequently descriptive of beautiful scenery. The majority contain the adjective *beau* (*bel*), as BEAMISH, BEAUCHIEF, BEAUMANOR, BEAUMONT, BELASIS, BELPER, BELVOIR. BEAUMONT in Essex replaced the Old English *Fulanpettæ* 'foul pit', where the reason for the change is obvious. BEACHY HEAD is an example of a cape getting a French name. Other French names are BLANCHLAND, GROSMONT, KIRMOND, MALPAS, MONTACUTE, MOUNTSORREL, PLESHEY, RICHMOND, most, if not all, transplanted. FOUNTAINS was named from its springs, and RIEVAULX is a translation of *Ryedale*. Interesting are DIEULACRES, HALTEMPRICE, names of monasteries. BATTLE was named after the battle of Hastings. The forest-names FREEMANTLE and SALCEY are probably transplanted. BOULGE, COWDRAY, DEVIZES, and KEARSNEY contain Old French common nouns and do not seem to be transplanted. MISERDEN, earlier *Musardere*, is a derivative from the French family name *Musard* ('Musard's place').

Many place-names have a distinctive addition consisting of a French personal or family name, as HOLME LACY, HOPE MANSEL, SHEPTON MALLET, STOKESAY. The French definite article or a French preposition occurs in names such as CHAPEL EN LE FRITH, HAUGHTON LE SKERNE, LEAFIELD.

(*b*) Norman influence shows itself most strongly in its effect on the form of English place-names. We have here a counterpart of the Scandinavianization dealt with in a previous paragraph. But the French and the English languages differed far more widely from each other than Scandinavian and English, and the English names offered far greater difficulties to a Norman than to a Scandinavian. In many cases Normanized forms have superseded an English one.

Obvious cases are names in *-cester*, *-ceter* (GLOUCESTER, EXETER, &c.), where *c* for *ch* and *t* for *st* are due to Norman influence. The English *ch*-sound from Old English palatalized *č* (in *ceaster*, later *chester*) was replaced by French *c* (originally pronounced *ts*), and loss of *s* before *t* is a well-known French sound-change (*être* from *estre*, &c.). The same change of palatalized *č* is found in CIRENCESTER, DISS, WHISSENDINE, loss of *s* before *t* in TRAFFORD in Lancashire (from *Strētford*), before *n* in NOTTINGHAM from Old English *Snotingahām*.

Substitution of *t* for English *th* has often taken place, as in TARBOCK, TARLETON. Initial *y* has sometimes been replaced by *j*, as in JARROW, JESMOND.

Dissimilation often takes place. If a name contained two *r*'s one was often changed to *l*, as in BULSTRODE, BULVERHYTHE (first element OE *burg*), SALISBURY (OE *Searoburg*). Also in other cases *r* was apt to be exchanged for *l*, as in BULPHAN (originally *Burgefen*) or an *n* for *r*, as in DURHAM (earlier *Dūnholm*). Dissimilatory loss of *r* is seen in CAMBRIDGE (OE *Grantanbrycg*).

Whether the loss of an intervocalic *d* or *th* in cases like CHILDREY (OE *Cillan rīþ*), STEPNEY (earlier *Stebbenhethe*) is due to Norman influence or to an English sound-change, it is difficult to say.

The change of *a* to *au* in STAUNTON and the like is due to French influence. Before *n* the *a* was nasalized and became English *au*. CAMBRIDGE comes from a Normanized form *Cauntebrigge*.

In early post-Conquest records, especially such as were written by Norman scribes, as Domesday Book or early Pipe Rolls, Norman spellings are very common, and Norman influence has a far wider range than that indicated above. A prosthetic *e* is common before *st-*, &c., as in *Estou* for STOW. An *s* or *es* is substituted for Old English *h*, as in *Bristelmestune* for Old English *Brihthelmestūn*, *Slapeford*, *Esledes* for Old English forms beginning in *Hl-*. A *d* is often written for *th* (OE *þ*). A vowel is often inserted between consonants, as in *Chenistetone* for KNIGHTON, *Salopesberia* for SHREWSBURY, &c. In many cases such spellings hardly represent definite Norman forms of English names; they are rather to be looked upon as attempts of scribes to render English forms that they heard pronounced. However that may be, in the majority of cases the Norman forms have not prevailed, and in somewhat later medieval records names generally appear in genuine English forms. Much of what has been held to be due to Norman influence is probably the result of native English sound-development

A French derivative suffix, generally a diminutive one, has sometimes been added to English names. The suffix *-el* is found in early forms of BYTHAM (*Bihamel*), the suffix *-et* in CLARET, CRICKET, HAMPNETT. The occasional *-ville* for earlier *-feld*, as in ENVILLE, is partly due to a native change of *f* to *v*, partly to popular etymology. In some late names, as LANGRIVILLE, French *ville* has been used in forming place-names.

6. *Place-names from Other Sources*

There are very few English place-names from other sources than those already accounted for. An interesting case is BALDOCK, which comes from an earlier form of Bagdad. It commemorates the Crusades, particularly the part taken in them by the Knights Templars. SYON HOUSE, a monastery founded in 1414–15, was clearly named from the Sion of the Bible.

In later times, of course, names from all sorts of sources have been given to manors or country houses, but these fall outside the scope of the present study. BLENHEIM PARK near Oxford is one of the few mentioned in the dictionary.

IV. THE VALUE OF PLACE-NAME STUDY

The study of place-names is of value in itself, inasmuch as the meaning and history of English place-names must offer a good deal of interest to English people. Place-names form a part of the vocabulary and deserve as much attention as other words. But incidentally they give valuable information of a particular kind.

1. Place-names embody important material for the history of England. The various elements in the English place-nomenclature (British, Scandinavian, &c.) testify to changes in the population of the British Isles, and place-names often supplement the meagre data of recorded history. They give hints as to the districts where a British population preserved its language for a comparatively long time. Names such as WALTON, BIRKBY (from *Bretabȳr* 'village of the Britons') indicate places where Britons dwelt in independent settlements, after the Anglo-Saxon immigration. A systematic study of various types of place-names will,

it may be hoped, give some information on the history of Anglo-Saxon settlements and the distribution of the population at various periods. It has been shown that place-names in *-ing* (READING and the like) are very old, and probably belong to the period of the early settlements of Anglo-Saxons in England. These names, where they occur frequently, indicate early Anglo-Saxon communities. These and other names consisting of or containing folk-names give a hint as to the importance of the tribe and family in early Anglo-Saxon times. Names such as DENGE, EASTRY, LYMINGE, STURRY, which contain an Old English *gē* corresponding to German *Gau*, tell us that Kent was, at an early period, divided into districts called *gē*, which would be analogous to the German *Gau*'s.

The Scandinavian place-names in England are really the chief source for our knowledge of the Scandinavian settlements. They tell us what parts of England were most thickly populated by Scandinavians. They tell us that the Scandinavian population in the east of England was on the whole Danish, that in the north-west chiefly Norwegian. But some names (as HULME, FLIXTON, URMSTON) prove that there was once a Danish colony in South-East Lancashire.

2. Place-names have something to tell us about Anglo-Saxon religion and belief before the conversion to Christianity. Though the Christian religion was introduced about 150 years after the immigration into England, there had been time for many places to get names referring to heathen worship. Names such as WOODNESBOROUGH, WEDNESBURY, WEDNESFIELD, WANSDYKE denoted places where *Wōden* was worshipped or that were associated with him. THUNDERFIELD, THUNDER(S)LEY, probably THURSLEY in Surrey, must have been places where *Þunor* had a temple, and TUESLEY, TYSOE record places dedicated to the worship of *Tīw* (*Tīg*). FROYLE, FREEFOLK and others may contain the name of the goddess *Frīg* (*Frēo*). GRIM'S DITCH contains *Grīm*, a byname of Woden. Many names such as HARROW, WEEDON, WILLEY contain a word which meant a heathen temple (OE *hearg*, *wēoh*).

Numerous names throw some light on the belief in supernatural beings of a lower order, elves, sprites, goblins. Such are GRIMLEY, SHINCLIFFE, SHOBROOKE, SHUCKBURGH, THURSFORD, probably HASCOMBE, HESCOMBE. Interesting is *Niker-poll* in an Assize Roll of 1263, which contains Old English *nicor* 'water monster'. The pool was at Pershore. DRAKELOW is probably a reminiscence of a myth according to which the mound in question was inhabited by a dragon. Names like HOLYWELL, HOLY OAKES tell us of holy wells or trees; FRITWELL, ELWELL of wishing-wells.

3. Some place-names indicate familiarity with old heroic sagas in various parts of England. *Grendles mere* 931 Birch, Cart. Sax. 677, *Grendeles pytt* 739 ib. 1331, found in boundaries of Ham in Wilts and Crediton, show that the Grendel episode in Beowulf was known in Devon and Wilts, perhaps even that it was localized to definite places in these counties. The name WAYLAND SMITH'S CAVE (OE *Welandes smiððe* 955 ib. 908) shows that the Weland saga was popular in Wilts in the Old English time. There is a good deal of probability that some place-names in WADE- contain the name of the mythical hero Wade. *Hnæf* in *Hnæfes scylf* 973–4 ib. 1307 (in bounds of Crondall) and in NASEBY may well refer, not to an Anglo-Saxon *Hnæf*, but to the hero of Finnsburg. WITHINGTON in Gloucestershire (*Wudiandun* 737) may possibly have got its name from the *Widia* of heroic saga.

But it is obvious that the greatest caution must be exercised in the interpretation of names like these. In most cases, places were no doubt named from actual persons, not from heroes of popular legends and sagas. But it is quite possible that people were named from legendary heroes sometimes, and in that case names testify to familiarity with the sagas in question.

4. Place-names give important information on antiquities. Names such as STRATFORD, STRETTON contain Old English *stræt* 'Roman road', and speak of a Roman road running past or through the place. Such names give help in determining the line of Roman roads and often even tell of such roads that are now unknown. Names containing words for 'fort', 'Roman station', sometimes prove that an old fort must have once existed at a place where there are now no traces of it. The name ECCLES and names in ECCLES- indicate that there was an old church in the place, probably one of British origin. On FAWLER see dictionary.

Some place-names give information on early architecture or building-material. BERECHURCH, BRADKIRK, FELKIRK mean 'church made of boards or planks'. On the other hand, WHITCHURCH, literally 'white church', in reality seems to mean 'stone church'. VOWCHURCH is 'multicoloured church', and must refer to some decoration. HORNCHURCH is 'church with horns or hornlike gables'. STONEHOUSE is self-explanatory; it is a rare name. SHINGLE HALL and THATCHAM seem to have been named from the material of their roofs. LOFTHOUSE, fairly common in Yorkshire, refers to a house with an upper story. On BELCHAMP see dictionary.

Names in -*bridge*, of course, indicate the existence of an early bridge, but they sometimes give a hint as to the material or nature of the bridge, as STAMBRIDGE, RISBRIDGE (the name of a hundred in Suffolk), STALBRIDGE. See also BOW. BRIDGFORD is doubtless generally 'ford with a footbridge'. Other names tell of stepping stones or other contrivances to facilitate the crossing of a stream, as CLAPPERSGATE, BAMFORD, STAPLEFORD, or for crossing marshy land, as RUNCTON, SOCKBRIDGE, WARPSGROVE.

Numerous names indicate early systems of defence, look-out places, and the like. Examples are WARDLOW, WARTHILL 'watch place', TOTTERNHOE 'spur of land with a look-out house'. Many other names contain the stem *tōt-* 'to look out', as TOTHILL, TOTHAM, &c. To this group belong other names, as HALTON in Northumberland, very likely GLANTLEES and the like. *Beacon* is found in BEACONSFIELD and BECKNEY.

Prehistoric stone circles may be referred to by names such as QUARLES, WHARLES, WHEELTON. STANION may well refer to a cromlech, as do certainly FEATHERSTONE, SHILSTONE. THURLESTONE really denoted a stone with a hole in it.

5. Place-names give information on early institutions, social conditions, and the like. Important information is obtained on early meeting-places. Here belong such names as MUTLOW, SPELLOW, SPETCHLEY, STOULTON, THINGWALL. A study of such names, many preserved as names of hundreds or wapentakes, gives clues to Anglo-Saxon and Scandinavian customs. It is obvious that gravemounds were often used for meeting-places. In other cases a conspicuous hill, a prominent tree, a ford, was chosen for this purpose. The word *hlōse* found in the hundred-names Loes in Suffolk, Clackclose in Norfolk, probably refers to some temporary shelter set up at a meeting-place, and hardly has its usual meaning 'pigsty'. The hundred-name Gartree in Leicestershire should very likely be explained in connexion with the Longobardic *gairethinx* 'the common thing', whose first element is *gair* 'spear' (= Old English *gār*).

Names such as KINGSTON, QUINTON, ATHELNEY, tell us that the places once belonged to the king, the queen, some princes. REEPHAM must have been under the supervision of a reeve. DAMERHAM is 'the judges' manor'. ABBOTSTON(E) is 'manor belonging to an abbot or abbess'. KNIGHTON was a village held by knights, DROINTON one held by *drengs*. CHARLTON (CHORLTON) is 'village of free peasants'. The name indicates that other villages had some other status and tells of early manorialism. BUCKLAND was an estate held by charter.

Many place-names tell of early industry and the like. Many names, of course, contain the word *mill*, as MELLIS, MELLS, MELFORD, MILLBROOK. In some cases windmills must be meant. CROFT in Leicestershire (from OE *cræft*), WILD seem to mean 'mechanical contrivance'. A wind- or watermill may be referred to. ORGRAVE, STANDHILL and others testify to mining industry, as do names like QUERNMORE, QUORNDON. KIRBY OVERBLOW contains an otherwise unknown Old English *ōrblāwere* 'smelter'. SALT and names in SALT-, SALTER- tell of the salt-making industry. Many names contain words for trades or tradesmen, as HUN-TINGTON, FULLERTON, FISHERTON, WOODMANCOTE, COLSTERDALE, BICKERTON; SMEATON, SUTTERTON; HOPPERTON, SAPPERTON, POTTERTON; MANGERTON; BEMER-TON, HARPERLEY, HORNBLOTTON.

Names like PLAISTOW, PLAYFORD, HESKETH, perhaps FOLLIFOOT, tell of ancient sports or horse-racing. Ancient deer-parks are commemorated by names like HARTLIP, HINDLIP, DARTON, DASSETT, ROFFEY.

Traps for catching animals are referred to by such names as BAWDRIP, GILDER-DALE, SNARGATE, TRAFFORD in Northants, WOOKEY; KEPIER, YARM.

Something may be gathered from place-names on early agriculture and cattle-farming.

Many names contain words for cereals, as BARLEY, RYTON, WHEATLEY. It is remarkable that *oats* is very rare in place-names; OTELEY is an exception. But pilloats is found early in PILLATON, and very likely *Haver-* in place-names is partly an Old English *hæfera* 'oats' or Old Scandinavian *hafri* the same, as in HAVERHILL, HAVERBRACK. Fruit-trees must have been extensively grown, as indicated by numerous names containing the words for apple, pear, plum. *Cherry* seems to occur several times as a distinctive addition, as in CHERRY WILLINGHAM.

Names of domestic animals are common, as in COWLEY, OXTON, SHEPTON, SHAPWICK. It is of interest to find that the custom of grazing cattle at some distance from the home village or homestead, which is still common in Scan-dinavia, must have been prevalent in England, not only in Scandinavian districts, but also in purely English areas. In the north, names in -*sett* (ON *sætr*), -*erg* testify to it. In the south, numerous names in -*stoc*, -*wīc* doubtless belong here. SOMERTON must have been a place to which people moved in summer for the sake of better pasture. There is even a case where a SOMERTON and a WINTERTON are found close together. In Kent and Sussex, villages in low-lying districts had outlying pastures in the Weald districts. They were called *denn* and seem to have been mostly used for swine-pastures. Likewise marshland was allotted to various villages. Place-names sometimes give a hint as to the mother village of a *denn* or piece of marsh. TENTERDEN is 'the *denn* of the Thanet people', BURMARSH, in South Kent, 'the marsh of the Canterbury people'. Cf. also DUNGE MARSH.

6. Place-names are of great value for linguistic study.

(*a*) They frequently contain personal names, and are therefore a source of first-rate importance for our knowledge of the Anglo-Saxon personal nomencla-ture. A systematic study of personal names found in place-names will give im-portant information on the personal names of the early tribes. It is interesting to find, for instance, that names in *Sax-* and -*mund* are particularly common in

East Anglia (SAXLINGHAM, SAXMUNDHAM, MUNDHAM, MENDHAM, MENDLESHAM, &c.). Many personal names are only recorded in place-names. Thus *Godhelm*, the base of GODALMING, has only been found with certainty in this name, though very likely it is also the first element of GOLDSBOROUGH in the West Riding. Old English *god* (*gōd*) and *helm* are common in personal names, and it would be remarkable if they should not sometimes have been combined into *Godhelm*. THIRSTON contains an Old English *Þræsfriþ*. This is the only known example of the element *Thras-* so common in Continental names.

Women's names are quite common. Particularly frequent is *Ēadburg* (ABBERTON, ABRAM, &c.). Perhaps some of the places are named from Eadburg, wife of Beorhtric, king of Wessex. But many other women's names occur, as *Beaduburg* in BABRAHAM, *Cēngifu* in KNAYTON, KNIVETON and others, *Ēadgifu* in EDDINGTON Berks. An interesting group is formed by AUDLEY, BALTERLEY, BETLEY, and BARTHOMLEY, containing the women's names *Aldgȳþ*, *Baldþrȳþ*, *Bettu*, and probably *Beorhtwynn*. The four places are situated close together on the border of Cheshire and Staffordshire. All have for their second element Old English LĒAH.

(*b*) Place-names contain many old words not otherwise recorded. They show that Old English preserved several words found in other Germanic languages, but not found in Old English literature. Thus the names BEESTON, BESSACAR, &c., have as first element an Old English *bēos* 'reed, rush', corresponding to Dutch *Bies*, Low German *bēse*. BLEAN contains a word identical with Old High German *blacha* 'coarse cloth'. DOILEY shows that Old English had an adjective *diger* corresponding to Old Norse *digr* 'thick'. HARDRES contains an old word for 'forest' found in German HARZ. REDLYNCH, LYSCOMBE in Dorset show that Old English had a word *lisc* identical with Old High German *lisca*, Dutch *lisch* 'reeds'. SOMPTING, SUNT testify to an Old English *sumpt* corresponding to Old High German *sunft* 'marsh'. THE SWIN is identical with HET ZWIN in Holland and shows that Old English had a word *swin* 'creek', identical with Dutch *zwin*.

Other place-names prove the existence of Old English words otherwise unknown in any Germanic language. A very interesting case is offered by the names ENHAM and YEN HALL, which contain an Old English *ēan* 'lamb', corresponding to Latin *agnus*, but hitherto found only in derivatives (OE *ēanian* 'to lamb', *geēan* 'with lamb'). PEAK and several other names presuppose a word *pēac* 'hill', cognate with Swedish *pjuk* 'a mound'. PRAWLE has as first element an Old English *prāw* 'peering', cognate with *prīwan* 'to wink'. Several names such as WYTHAM in Berks must contain an Old English *wiht* 'bend', which is unknown outside place-names.

Some place-names testify to otherwise unknown side-forms of Old English words, as an *ēstre* by the side of *eowestre* 'sheepfold' (EASTER), or *hagga* by the side of *haga* 'haw' (HAGLEY). Some tell us of unknown derivatives with suffixes, as FRANT, ETCHELLS, NECHELLS.

(*c*) Place-names often afford far earlier references for words than those found in literature. Only a few examples can be given here. *Dimple* is recorded in OED from *c*. 1400. In the topographical sense 'depression in the ground' it is recorded *c*. 1205 (*Kerlingedimpel* in the Furness Cartulary, apparently 'ducking-pool'). *Dod* 'a round hill' is first exemplified in OED from 1843. As a place-name the word is found in 1230. *Hunter* (first quotation in OED *c*. 1250) occurs in a place-name recorded in Domesday (HUNSTON in Suffolk). *Potter* is found in a place-name in 951 (*Potteresleag* BCS 890), otherwise not until 1284 (OED). *Sprod* 'a salmon in its second year' is recorded in OED from 1617. It is evidently found in the place-name *Sprodpulhey* 1418 (Chetham Soc. 95).

(*d*) Only a few remarks can be made here on the value of place-names for the history of English sounds. They often help in the dating and localization of sound-changes, the fixing of dialect boundaries and the like. The distribution of place-name forms such as STRATFORD, STRATTON, and STRETFORD, STRETTON gives information on the Old English dialects in which the word *street* appears as *strǣt* and as *strēt*, that is, it helps in drawing the line between Saxon and Anglian territory. The distribution of 'fractured' forms such as Saxon-Kentish *ceald*, *cealf* as against Anglian *cald*, *calf*, which is an important dialect criterion, is illustrated by place-names. One of the chief tests for the distinction between Midland and Northern English is the development of Old English *ā*, which becomes Middle English *ō* in the Midlands, but remains in the North. The *ā–ō* line can be drawn with a good deal of accuracy by the help of place-names.

Many sound-changes can be exemplified in place-names that hardly occur in other words, partly because certain combinations of sounds found in them are rarely met with in other words. Loss of consonants, especially *r* and *l*, owing to dissimilation is common. Examples are WENTWORTH from earlier *Wintreworth* and the like, PUNCKNOWLE, which is pretty certainly Old English *plūm-cnoll* 'plum-tree knoll'. Change of *t*, *k*, to *d*, *g* between vowels is common, rare in other words. Change of *nk* to *ng* or *rt* to *rd* is often found, as in many names containing Old English *hlinc*, *ceart*. See e.g. LYNG, CHARD, CHADACRE. A change of *d* to *th* is well known in words like *father*, *mother* (OE *fæder*, *mōdor*). In place-names a similar change of *b* to *v* or *g* to *gh* is sometimes met with. See e.g. PAVENHAM from *Pabenham*, MOGGERHANGER.

V. ARRANGEMENT OF THE DICTIONARY

1. Place-names are given in alphabetical order. Where, for some reason, a name is dealt with in some other place, a reference is inserted at its place in the list, except for cases where the name of an unimportant place is adduced by way of illustration. If several examples of a name occur, they are arranged by counties, these being given in alphabetical order. Names with a distinctive addition are listed under the chief component, thus LONG COMPTON, CAPHEATON under COMPTON, HEATON. But references are often given from the longer forms, especially when there can be some doubt as to which is the chief component.

Elements that frequently occur are dealt with systematically in the body of the dictionary, the Old (Middle) English or Old Scandinavian form serving as head word. Examples of the names containing the various elements are here collected with more or less detailed information on meaning, variation in form, later development, &c. For elements of very frequent occurrence, only some typical instances are adduced. Elements of special interest, though comparatively rarely used, have been exemplified more fully. Sometimes under an element only a reference is given to the place-names containing it. Elements that can be easily found by consulting the book have been omitted (e.g. *elm*, *fisc*).

At an early stage of the work an attempt was made to arrange a number of names under elements, instead of dealing with them in their alphabetical place. It was found, however, that this arrangement necessitated a great many references which would swallow up the space saved and cause unnecessary trouble to the user of the dictionary. In some cases the earlier form has been allowed to remain, as under *bæce*, *skáli*. Here the number of names collected under chief elements is small. The variation in the spelling between COATES and COTES, COTON and COTTON is the reason for collecting many names under COT. Similarly COMB, COMBE, COOMBE are to be found under COMB.

2. The material has been, with few exceptions, collected from the sources by the writer. This is true even of counties whose place-names have been dealt with in monographs such as the publications of the Place-name Society. The collections for the dictionary were begun many years ago and practically completed before the publication of some recent place-name books, such as those on Dorset, Northamptonshire, Surrey, and Kent. Some examples have been taken from existing books on place-names, and these forms have often been very important for the etymology; but so far as possible, examples so adopted have been checked from the sources. When this was not possible, forms are mostly quoted, not from the ultimate source, but from the book from which they were taken.

The material is derived chiefly from the earliest sources—Old English, as Bede, the Anglo-Saxon Chronicle, Old English charters—and early post-Conquest sources, as Domesday, Pipe Rolls, Feet of Fines, Curia Regis Rolls, the Book of Fees, the earlier Charter, Close and Patent Rolls, &c. Of course, many special sources for particular counties or districts have been made use of, such as the *Danelaw Charters* for Lincolnshire, the *Early Yorkshire Charters*, the Lindsey, Leicestershire, and Northamptonshire Surveys, and so on. Later sources have been adduced chiefly where the earlier failed, and for reasons of economy no attempt has generally been made to trace the history of names in detail from the earliest down to the present form.

Much effort has been taken to reach the greatest accuracy possible in the information given. The forms from early records are given in the form they have in the text. Normalization has generally been avoided. But sometimes a contraction has been replaced by the full form without remark, as *Sanctus* for *Scũs* or the like. For *v* in Domesday *u* is given when it denotes a vowel. Some obviously erroneous readings have been silently corrected. It is a well-known fact that early editors often misread *u* as *n* or vice versa, or read *lb*, *lk* for what was evidently meant to be *bb*, *kk*. To save space, name-forms are sometimes abbreviated, being replaced by the initial letter and the sign ∼. This has been done when the form agrees exactly with that immediately preceding. Thus under HUTTON SCOUGH the forms are given as *Hoton* 1212 Fees, *H∼ in foresta* 1248 Cl. *H∼* represents *Hoton*, not *Hutton*.

3. Considerations of space have necessitated strict economy of expression.[1] It has been impossible to give much more than the results of the etymological investigations that have been necessary. It has also been impossible generally to acknowledge the debt owed to previous scholars under each name. It is obvious that the writer has derived very great help from his predecessors, not only as regards material, but also in the identification of early forms, and especially in the etymological work. But in each case the etymology has been tested by the author, who has attempted to be as independent of earlier research as possible. The dictionary contains a considerable number of new etymologies. Some of these have been fully worked out in previous publications, especially in *Studies on English Place and Personal Names* (1931) and *Studies on English Place-names*

---

[1] For this reason also the date (1086) has been omitted before DB (Domesday).

(1936), which are to be looked upon as preliminary studies for this work.[1] A number of further etymologies have been dealt with in a third study of a similar kind, which has been recently published.[2] It should be added that the etymologies are based on a very large amount of material, of which only a small portion can be here given. The idea has been to include two or three important forms for each name, as a rule the earliest found, but sometimes slightly later forms have been preferred as of greater value for the etymology. It should also be remembered that many Old English charters are preserved in later copies, where name-forms are often really Middle English.

The meanings of many common place-name elements, as *halh*, *hām*, *hamm*, *lēah*, *tūn*, varied a good deal, and it is often impossible to decide definitely what is the exact meaning in each particular case. For this reason such elements are often left untranslated, the form of the etymology being e.g. '*Cynemǣr*'s TŪN'. The small capitals of TŪN imply a reference to the article TŪN. Similarly the small capitals often used in printing forms of English place-names imply a reference to the article on the name in question. It may be added here that in an etymology such as '*Cynemǣr*'s TŪN', *Cynemǣr* is a male Old English personal name, and a general reference is here made to Searle's *Onomasticon* and Redin's *Uncompounded Personal Names*, where references are generally to be found.

Definite etymologies of river-names are often difficult to give, and they would frequently have to be rather full in order to be properly understood. A reference is made to the author's *English River-Names*, where the names are fully dealt with. When a definite etymology can be given, it is, of course, duly set down.

A good deal of work has been given to the elucidation of 'surnames' added to place-names. Indeed, these additions are here treated rather more fully than is generally done in monographs on place-names. Thus an attempt has been made to find the source of family names so often occurring in this function (BERWICK BASSETT, SUTTON SCOTNEY, and the like). Information of this nature is generally given in smaller type at the end of an article. Occasionally, however, it seemed more practical to deal with the surname in immediate connexion with each name. This has been done in long articles such as STOKE, THORPE, where surnames are particularly numerous.

VI. A SELECT LIST OF WORKS CONSULTED

See also Section vii (Abbreviations).

ALEXANDER, H. *The Place-names of Oxfordshire.* Oxford, 1912 (PNO).
ANDERSON (ARNGART), O. S. *The English Hundred-names.* Lund, 1934–6.
—— *Old English Material in the Leningrad Manuscript of Bede's Ecclesiastical History.* Lund, 1941.
ARMSTRONG, A. M., MAWER, A., STENTON, F. M., and DICKINS, BRUCE. *The Place-names of Cumberland.* Cambridge, 1950–2 (PNCu(S)).
ATTENBOROUGH, F. L. *The Laws of the Earliest English Kings.* Cambridge, 1922.
BADDELEY, W. ST. C. *The Place-names of Gloucestershire.* Gloucester, 1913 (PNGl).

[1] Other contributions are: 'Loss of a Nasal before Labial Consonants' (in the *Klaeber Volume*), 'Etymological Notes' (*Englische Studien*, 64, and *Studia Neophilologica*, i, ii), 'Drayton, Draycot, Drax, &c.' (*Namn och Bygd* 20), 'Names of Trades in English Place-names' (in *Historical Essays in Honour of James Tait*), 'Grim's Ditch' (in *Studia Germanica, tillägnade E. A. Kock*, Lund, 1934), 'Some Notes on English Place-names containing Names of Heathen Deities' (*Englische Studien*, 70).

[2] *Etymological Notes on English Place-names.* Lund, 1959.

BANNISTER, A. T. *The Place-names of Herefordshire.* Cambridge, 1916 (PNHe).

BARTHOLOMEW, J. G. *The Survey Gazetteer of the British Isles.* London, 1904 ff.

The Bayeux Tapestry. A comprehensive Survey. (Sir Frank Stenton, General Editor.) London, Phaidon Press, 1957.

BESTMANN, F. *Die lautliche Gestaltung englischer Ortsnamen im Altfranzösischen und Anglonormannischen.* Zürich–Leipzig, 1938.

BINZ, C. *Zeugnisse zur germanischen Sage in England.* Beiträge XX. Halle, 1895.

BJÖRKMAN, E. *Zur englischen Namenkunde.* Halle, 1912.

—— *Nordische Personennamen in England.* Halle, 1910.

—— *Scandinavian Loan-words in Middle English.* Halle, 1900, 1902.

BLOMÉ, B. *The Place-names of North Devonshire.* Uppsala, 1929.

BOWCOCK, E. W. *Place-names of Shropshire.* Shrewsbury, 1923 (PNSa).

BRADLEY, H. *Collected Papers.* Oxford, 1928.

BRANDL, A. *Zur Geographie der altenglischen Dialekte.* Berlin, 1915.

BRIDGES, JOHN. *History of Northamptonshire.* Oxford, 1791.

Broadcast English, ii. By A. Lloyd James. London, 1930, second ed. 1936.

BUGGE, A. *Vikingerne.* Copenhagen, 1904, 1906.

CAMERON, K. *Place-names of Derbyshire.* Cambridge, 1959 (PNDb(S)).

Catalogue of English Coins. Anglo-Saxon Series. London, 1887 ff.

CHADWICK, H. M. *The Origin of the English Nation.* Cambridge, 1907.

—— *Studies in Old English.* Cambridge, 1899.

CHAMBERS, R. W. *Beowulf.* Cambridge, 1921.

CHARLES, B. G. *Non-Celtic Place-names in Wales.* London, 1938.

CODRINGTON, TH. *Roman Roads in Britain.* London, 1919.

COLLINGWOOD, W. G. *Lake District History.* Kendal, 1925.

Danmarks gamle Personnavne. I. *Fornavne.* II. *Tilnavne.* Copenhagen, 1936 ff.

Danmarks stednavne. Copenhagen, 1922 ff.

DEXTER, T. F. G. *Cornish Names.* London, 1926.

Dictionnaire topographique de la France. Paris, 1861 ff.

DUIGNAN, W. H. *Staffordshire Place Names.* London, 1902 (PNSt).

—— *Warwickshire Place Names.* London, 1912 (PNWa).

EKBLOM, E. *The Place-names of Wiltshire.* Uppsala, 1917 (PNW).

EKWALL, E. *Contributions to the History of Old English Dialects.* Lund, 1917.

—— *Early London Personal Names.* Lund, 1947 (ELPN).

—— *English Place-names in -ing.* Lund, 1923.

—— *English River-Names.* Oxford, 1928 (ERN).

—— *The Place-names of Lancashire.* Manchester, 1922 (PNLa).

—— *The Population of Medieval London.* Lund, 1956 (SPML).

—— *Scandinavians and Celts.* Lund, 1918.

—— *Street-names of the City of London.* Oxford, 1954 (Street-names).

—— *Studies on English Place and Personal Names.* Lund, 1931.

—— *Studies on English Place-names.* Stockholm, 1936 (SPN).

—— *Two Early London Subsidy Rolls.* Lund, 1951 (ELSR).

ELLIS, A. J. *On Early English Pronunciation*, vol. v., 1889.

ELLIS, H. *A General Introduction to Domesday Book.* London, 1833.

ERNAULT, E. *Glossaire moyen-breton.* Paris, 1895 f.

EVANS, D. SILVAN. *A Dictionary of the Welsh Language.* Carmarthen, 1893 ff.

FABRICIUS, A. *Danske Minder i Normandiet.* Copenhagen, 1897.

FÄGERSTEN, A. *The Place-names of Dorset.* Uppsala, 1933 (PNDo).

FARRER, W. *Honors and Knights' Fees.* London, 1923–5.

VON FEILITZEN, O. *The Pre-Conquest Personal Names of Domesday Book.* Uppsala, 1937.

FEIST, S. *Etymologisches Wörterbuch der gotischen Sprache.* Halle, 1923.

FORSBERG, R. *A Contribution to a Dictionary of Old English Place-names.* Uppsala, 1950 (Forsberg).

FORSSNER, T. *Continental-Germanic Personal Names in England.* Uppsala, 1916.
FÖRSTEMANN, E. *Altdeutsches Namenbuch.* Bonn, 1900–16.
FÖRSTER, M. *Der Flussname Themse und seine Sippe.* München, 1941 (Förster, Themse).
—— *Zur Geschichte des Reliquienkultus in Altengland.* München, 1943 (Förster, Reliquienkultus).
—— *Keltisches Wortgut im Englischen.* Halle, 1921.
FRITZNER, J. *Ordbog over det gamle norske sprog.* Christiania, 1886–96.
Geirfa Barddoniaeth Gynnar Gymraeg. Gan J. Lloyd-Jones. Caerdydd, 1931 ff.
GELLING, M. *The Place-names of Oxfordshire, based on material collected by Doris Mary Stenton.* Cambridge, 1953 ff. (PNO(S)).
GOODALL, A. *Place-names of South-West Yorkshire.* Cambridge, 1914.
GOVER, J. E. B. *The Place-names of Middlesex.* London, 1922 (PNMx).
GOVER, J. E. B., MAWER, A., and STENTON, F. M. *The Place-names of Devon.* Cambridge, 1931–2 (PND).
—— *The Place-names of Hertfordshire.* Cambridge, 1938 (PNHrt (S)).
—— *The Place-names of Middlesex.* Cambridge, 1942 (PNMx(S)).
—— *The Place-names of Northamptonshire.* Cambridge, 1933 (PNNp).
—— *The Place-names of Nottinghamshire.* Cambridge, 1940 (PNNt(S)).
—— *The Place-names of Surrey.* Cambridge, 1934 (PNSr).
—— *The Place-names of Warwickshire.* Cambridge, 1936 (PNWa(S)).
—— *The Place-names of Wiltshire.* Cambridge, 1939 (PNW(S)).
GRÖHLER, H. *Ursprung und Bedeutung der französischen Ortsnamen.* Heidelberg, 1913.
GROSS, C. *The Sources and Literature of English History.* London, 1915.
GRUNDY, G. B. *On the Meanings of Certain Terms in the Anglo-Saxon Charters.* Essays and Studies, vii (1922).
HALLQVIST, H. *Studies in Old English fractured ea.* Lund, 1948.
HARMER, F. E. *Anglo-Saxon Writs.* Manchester, 1952 (ASWrits).
—— *Select English Historical Documents.* Cambridge, 1914.
HARRISON, H. *Surnames of the United Kingdom.* London, 1912–18.
HELLQUIST, E. *Studier över de svenska sjönamnen.* Stockholm, 1903–6.
—— *Svensk etymologisk ordbok.* Lund, 1920–2.
—— *De svenska ortnamnen på -by.* Gothenburg, 1919.
HENDERSON, C. *Essays in Cornish History.* Oxford, 1935.
HEUSER, W. *Alt-London.* Strassburg, 1914.
HILDEBRAND, B. E. *Anglosaxiska mynt i svenska kongliga myntkabinettet.* Stockholm, 1881.
HILL, JAMES S. *The Place-names of Somerset.* Bristol, 1914.
HOGAN, E. *Onomasticon Goedelicum.* Dublin, 1910.
HOLTHAUSEN, F. *Altenglisches Etymologisches Wörterbuch.* Halle, 1932–4.
INDREBØ, G. *Norske Innsjønamn.* Kristiania, 1924 &c.
Introduction to the Survey of English Place-names. Cambridge, 1924 (IPN).
JACKSON, K. *Language and History in Early Britain.* Edinburgh, 1953.
JELLINGHAUS, H. *Die westfälischen Ortsnamen.* Osnabrück, 1923.
JOHNSTON, J. B. *The Place-names of England and Wales.* London, 1914.
—— *Place-names of Scotland.* Edinburgh, 1903.
JOLLIFFE, J. E. A. *Pre-Feudal England.* London, 1933.
JONES, D. *An English Pronouncing Dictionary.*
JOYCE, P. W. *Irish Names of Places.* Dublin, 1910–13.
KARLSTRÖM, S. *Old English Compound Place-names in -ing.* Uppsala, 1927.
KEMBLE, J. M. *Saxons in England.* London, 1849.
KLUGE, F. *Etymologisches Wörterbuch der deutschen Sprache.* 1883 &c.
—— *Nominale Stammbildungslehre.* Halle, 1899.
KNEEN, J. J. *The Place-names of the Isle of Man.* Douglas, 1925–9.

KÖKERITZ, H. *The Place-names of the Isle of Wight*. Uppsala, 1940 (PNWt).

LANGENFELT, G. *Toponymics*. Uppsala, 1920.

LEBEL, P. *Principes et méthodes d'hydronomie française*. Paris, 1956.

LIEBERMANN, F. *Die Gesetze der Angelsachsen*. Halle, 1903–16.

LIND, E. H. *Norsk-isländska dopnamn och fingerade namn från medeltiden*. Uppsala, 1905–15. *Supplementband*. Oslo, 1931.

—— *Norsk-isländska personbinamn från medeltiden*. Uppsala, 1920–1.

LINDKVIST, H. *Middle English Place-names of Scandinavian Origin*. Uppsala, 1912.

LLOYD-JONES, J. *Enwau Lleoedd Sir Gaernarfon*. Caerdydd, 1928.

LÖFVENBERG, M. T. *Studies on Middle English Local Surnames*. Lund, 1942.

LONGNON, A. *Les noms de lieu de la France*. Paris, 1920–3.

LOTH, J. *Chrestomathie bretonne*. Paris, 1890.

—— *Contributions à l'étude des Romans de la Table Ronde*. Paris, 1912.

—— *Les Mabinogion du Livre Rouge de Hergest*. Paris, 1913.

—— *Les noms des saints bretons*. Paris, 1910.

LUNDGREN, M. F., and BRATE, E. *Personnamn från medeltiden*. Stockholm, 1892–1933.

MACBAIN, A. *An Etymological Dictionary of the Gaelic Language*. Stirling, 1911.

—— *Place-names Highlands & Islands of Scotland*. Stirling, 1922.

MAITLAND, F. *Domesday Book and Beyond*. Cambridge, 1907.

MANSION, J. *Oud-gentsche naamkunde*. 's-Gravenhage, 1924.

—— *De Voornaamste Bestanddeelen der Vlaamsche Plaatsnamen*. Brussel, 1935.

MAWER, A. *The Place-names of Northumberland and Durham*. Cambridge, 1920 (PNNb).

—— *Problems of Place-name Study*. Cambridge, 1929.

MAWER, A., and STENTON, F. M. *The Place-names of Bedfordshire & Huntingdonshire*. Cambridge, 1926 (PNBd).

—— *The Place-names of Buckinghamshire*. Cambridge, 1925 (PNBk).

—— *The Place-names of Sussex*. Cambridge, 1929, 1930 (PNSx).

—— *The Place-names of Worcestershire*. Cambridge, 1927 (PNWo).

—— *See also* Gover, Mawer, and Stenton.

McCLURE, E. *British Place-names in their Historical Setting*. London, 1910.

MIDDENDORFF, H. *Altenglisches Flurnamenbuch*. Halle, 1902.

MILLER, TH. *Place-names in the English Bede*. Strassburg, 1896.

MOERMAN, H. J. *Nederlandse Plaatsnamen*. Leiden, 1956.

MOORMAN, F. W. *The Place-names of the West Riding of Yorkshire*. Leeds, 1910.

MORRIS JONES, SIR J. *A Welsh Grammar*. Oxford, 1913.

MÜLLER, R. *Die Namen des nordhumbrischen Liber Vitae*. Berlin, 1901.

MUTSCHMANN, H. *The Place-names of Nottinghamshire*. Cambridge, 1913 (PNNt).

Namn och Bygd. Tidskrift för nordisk ortnamnsforskning. Uppsala, 1913 ff.

NICOLSON and BURN. *History . . . of Westmorland and Cumberland*. London, 1777.

NIELSEN, O. *Olddanske Personnavne*. Copenhagen, 1883.

Nomina Geographica Neerlandica. Amsterdam–Leiden, 1881 ff.

Owen's Pembrokeshire. London, 1892 ff.

PEDERSEN, H. *Vergleichende Grammatik der keltischen Sprachen*. Göttingen, 1909, 1913.

PHILIPPSON, E. A. *Germanisches Heidentum bei den Angelsachsen*. Leipzig, 1929.

Reallexikon der germanischen Altertumskunde. Ed. J. Hoops. Strassburg, 1911–19.

REANEY, P. H. *The Place-names of Cambridgeshire and the Isle of Ely*. Cambridge, 1943 (PNCa(S)).

—— *The Place-names of Essex*. Cambridge, 1935 (PNEss).

REDIN, M. *Uncompounded Personal Names in Old English*. Uppsala, 1919.

RHYS, J. *Celtic Britain*. London, 1904.

xl INTRODUCTION

RITTER, O. *Vermischte Beiträge zur englischen Sprachgeschichte.* Halle, 1922.

ROUND, J. H. *Feudal England.* London, 1909.

RYGH, O. *Norske Elvenavne.* Christiania, 1904.

SCHÖNFELD, M. *Wörterbuch der altgermanischen Personen- und Völkernamen.* Heidelberg, 1911.

SEARLE, W. G. *Onomasticon Anglo-Saxonicum.* Cambridge, 1897.

SEDGEFIELD, W. J. *The Place-names of Cumberland and Westmorland.* Manchester, 1915 (PNCu).

SKEAT, W. W. *The Place-names of Bedfordshire.* Cambridge, 1906.

—— *The Place-names of Berkshire.* Oxford, 1911.

—— *The Place-names of Cambridgeshire.* Cambridge, 1904 (PNCa).

—— *The Place-names of Hertfordshire.* Hertford, 1904 (PNHrt).

—— *The Place-names of Huntingdonshire.* Cambridge, 1904 (PNHu).

—— *The Place-names of Suffolk.* Cambridge, 1913 (PNSf).

SMITH, A. H. *English Place-name Elements.* Cambridge, 1956 (EPN(S)).

—— *The Place-names of the East Riding of Yorkshire and the City of York.* Cambridge, 1937 (PNER).

—— *The Place-names of the North Riding of Yorkshire.* Cambridge, 1928 (PNNR).

SOLMSEN, F. *Indogermanische Eigennamen als Spiegel der Kulturgeschichte.* Heidelberg, 1922.

Spurrell's Welsh-English Dictionary. Ed. J. B. Anwyl. Carmarthen, 1915.

STANTON, R. *A Menology of England and Wales.* London, 1887.

STEENSTRUP, J. *Studier over de ældste danske stednavnes bygning.* Copenhagen, 1909.

STENTON, F. M. *Anglo-Saxon England.* Oxford, 1943.

—— *English Feudalism.* Oxford, 1932.

—— *The Free Peasantry of the Northern Danelaw.* Lund, 1926.

—— *The Latin Charters of the Anglo-Saxon Period.* Oxford, 1955.

—— *The Place-names of Berkshire.* Reading, 1911.

—— *See also* Gover and Mawer.

STOKES, W. *Urkeltischer Sprachschatz.* Göttingen, 1894.

STREATFEILD, G. S. *Lincolnshire and the Danes.* London, 1884.

Studia Neophilologica. Uppsala, 1928 ff.

TANNER, T. *Notitia Monastica.* Oxford, 1695.

TAYLOR, I. *Words and Places.* 1864.

TENGSTRAND, E. *Genitival Composition in Old English Place-names.* Uppsala, 1940.

TENGVIK, G. *Old English Bynames.* Uppsala, 1938.

TOBLER, A. *Altfranzösisches Wörterbuch,* herausgegeben von E. Lommatzch. Berlin, 1925 ff.

TORP, A. *Nynorsk etymologisk ordbok.* Christiania, 1919.

—— *Wortschatz der germanischen Spracheinheit.* Göttingen, 1909.

WALDE, A. *Vergleichendes Wörterbuch der indogermanischen Sprachen.* Ed. J. Pokorny. Berlin and Leipzig, 1926–32.

WALKER, B. *The Place-names of Derbyshire.* DbAS xxxvi f. Derby, 1914 f. (PNDb).

WALLENBERG, J. K. *Kentish Place-names.* Uppsala, 1931.

—— *The Place-names of Kent.* Uppsala, 1934 (PNK).

WATSON, W. J. *The History of Celtic Place-names of Scotland.* Edinburgh, 1926.

WELLS, J. E. *A Manual of the Writings in Middle English.* New Haven, 1916.

WILLIAMS, R. *Lexicon Cornu-Britannicum.* Llandovery, 1865.

WYLD, H. C., and HIRST, T. O. *The Place-names of Lancashire.* London, 1911.

ZACHRISSON, R. E. *Anglo-Norman Influence on English Place-names.* Lund, 1909.

—— *Romans, Kelts, and Saxons in Ancient Britain.* Uppsala, 1927.

—— *Some Instances of Latin Influence on English Place-nomenclature.* Lund, 1910.

—— *Two Instances of French Influence on English Place-names.* Stockholm, 1914.

Zeitschrift für Ortsnamenforschung. 1925 ff.
ZEUSS, I. C. *Grammatica celtica.* Editio altera cur. H. Ebel. Berlin, 1871.
ZIMMER, H. *Nennius vindicatus.* Berlin, 1893.

VII. ABBREVIATIONS

The list is at the same time a list of sources of early forms of place-names.

a	ante.
Abbr	*Placitorum abbreviatio.* Record Com. 1811.
Abingd	*Chronicon monasterii de Abingdon.* Chr. & Mem. 2 (1858).
AC	*Ancient Charters.* Pipe Roll Soc. 1888.
AD	*Catalogue of Ancient Deeds.* London, 1890–1906.
AJ	*Royal Archaeological Institute. Archaeological Journal.* London, 1845 ff.
Alfred CP	*King Alfred's West-Saxon Version of Gregory's Pastoral Care.* EETS 45, 50.
Alfred Or	*King Alfred's Orosius.* EETS 79.
AN	Anglo-Norman.
Ann Cambr	*Annales Cambriæ.* Chr. & Mem. 20.
Ann Mon	*Annales Monastici.* Chr. & Mem. 36.
Ant Glast	William of Malmesbury, *De antiquitate Glastoniensis ecclesiæ.* Oxford, 1727.
Arch	*Archaeologia.* London, 1770 &c.
Arch Cambr	*Archaeologia Cambrensis.*
Arch Cant	*Archaeologia Cantiana.* London, 1858 ff.
ASC	The Anglo-Saxon Chronicle.
ASCh	*Anglo-Saxon Charters.* Ed. A. J. Robertson. Cambridge, 1939.
Ass	Assize Rolls (*Rolls of the Justices in Eyre for Yorkshire, 1218–19.* Ed. Doris M. Stenton. Selden Soc. 56, 1937. *Rolls of the Justices in Eyre for Gloucestershire, Warwickshire, and Staffordshire, 1221, 1222.* Ed. Doris M. Stenton. Selden Soc. 59, 1940).
Asser	*Asser's Life of King Alfred.* Oxford, 1904.
ASWrits	*See* Harmer in Works Consulted.
Athelney	*See* Muchelney.
Bardsley	Bardsley, C. W. *A Dictionary of English and Welsh Surnames.* London, 1901.
Bath	*Two Chartularies of Bath Priory.* So Rec. Soc. 7.
BBH	*The Black Book of Hexham.* Surt. Soc. 46.
BCS	*Cartularium saxonicum,* ed. Birch. London, 1885–93.
Bd	Bedfordshire.
Bede	Bede's *Historia ecclesiastica.*
Bede HAbb	Bede's *Historia abbatum.*
Berk	*Catalogue of the Charters…at Berkeley Castle.* Ed. I. H. Jeayes. Bristol, 1892.
Bk	Buckinghamshire.
Blomefield	Blomefield, F. *History of the County of Norfolk.* London, 1805–10.
BM	*Index to the Charters and Rolls in the British Museum.* London, 1900, 1912.
BMFacs	*Facsimiles of Royal and other Charters in the British Museum.* London, 1903.
BoB	*Boldon Book. Domesday Book* iv. Surt. Soc. 25.

Bodl	*Calendar of Charters and Rolls in the Bodleian Library.* Oxford, 1878.
Bodley MS	Extracts made in the 17th cent. from an ancient roll of the church of St. Paul (Bodl. James 23). See now Camden Soc. 3rd Ser. 58.
Bract	*Bracton's Note-book.* Ed. F. W. Maitland. London, 1887.
Brecon	Brecon Charters. Arch Cambr IV. xiii f.
Bret	Breton.
Bridl	*Bridlington Chartulary.* Ed. W. T. Lancaster. Leeds, 1912.
Brinkburn	*Brinkburn Chartulary.* Surt. Soc. 90.
Brit	British.
Brk	Berkshire.
Bruton	*Bruton & Montacute Chartularies.* So Rec. Soc. 8.
B-T	Bosworth and Toller, *Anglo-Saxon Dictionary.* Oxford, 1881–1921.
Buckfast	Buckfast Cartulary (Exeter Ep. Reg.).
Buckland	*Cartulary of Buckland Priory.* So Rec. Soc. 25.
Burghal Hidage	*See* BCS 1335, Maitland, *Domesday Book,* pp. 502 ff., *London Mediæval Studies,* i. 63.
Burton	*Abstract of Burton Cartulary.* Salt Soc. OS v.
Bury	*Documents from the Abbey of Bury St. Edmunds.* Brit. Ac. Records VIII.
Byrhtferth	*Byrhtferth's Manual.* Ed. S. J. Crawford. EETS 177.
Ca	Cambridgeshire.
Cal. of Wills	*Calendar of Wills ... in the Court of Husting, London.* London, 1889.
Calverley	*Calverley Charters.* Thoresby Soc. 6.
Cambr Bor Ch	*Charters of the Borough of Cambridge.* Cambridge, 1901.
Camden	Camden, *Britannia.* London, 1586.
Carew	Carew, R. *The Survey of Cornwall.* London, 1602.
CC	*Cockersand Chartulary.* Chetham Soc. NS, 38 ff.
CCC	List of St. Paul's manors from MS. 383 in Corpus Christi Coll. Cambridge. Herrig's Archiv civ.
Ch	Charter Rolls.
Chamb	*Accounts of the Chamberlains &c. of the County of Chester.* La Rec. Soc. 59.
Chester	*Chartulary of the Abbey of St. Werburgh, Chester.* Chetham Soc. NS, 79 ff.
Chr. & Mem.	*Chronicles and Memorials of Great Britain and Ireland during the Middle Ages.* Rolls Series. London, 1858 ff.
ChronEve	*Chronicon Abbatiæ de Evesham.* Chr. & Mem. 29.
ChronRams	*Chronicon Abbatiæ Ramesiensis.* Chr. & Mem. 83.
Chs	Cheshire.
Cl	Close Rolls.
Clerkenwell	*Cartulary of St. Mary Clerkenwell.* Ed. W. O. Hassall. Camden Soc. 3rd Ser. 71. London, 1949.
Co	Cornwall, Cornish.
CodWint	Codex Wintoniensis. BM MS. Add. 15350.
Coins	Anglo-Saxon Coins. See *Catalogue of English Coins* and Hildebrand in Works Consulted.
Colchester	*Colchester Cartulary.* Roxburghe Club. London, 1897.
Coll[ectanea]	*Collectanea topographica et genealogica.* London, 1834–43.
Collinson	Collinson, J. *History of ... Somerset.* 1791.
ContFr	Continental French.
Copinger	Copinger, W. A. *County of Suffolk.* London, 1904 f.
Cor	Coroners' Rolls.
Court	*County Court, City Court and Eyre Rolls of Chester.* Chetham Soc. NS, 84.

Crawf	*The Crawford Collection of Early Charters.* Oxford, 1895.
Crispin	Robinson, J. A. *Abbot Gilbert Crispin.* London, 1911.
Crondal	*Crondal Records.* Ha Rec. Soc. 3.
Ct	Court Rolls.
Cu	Cumberland.
Cur	*Curia Regis Rolls.*
CWNS, CWOS	*Transactions of the Cumberland and Westmorland Antiquarian and Archaeological Society.* Old and New Series. 1866 ff.
d.	died.
D	Devonshire.
Dale	Dale Abbey Charters. DbAS xxiv.
Dan	Danish.
Darley	Darley Charters. DbAS xxvi.
Db	Derbyshire.
DB	*Domesday Book.* London, 1783–1816. This reference also includes the Exon Domesday. The date of both is 1086.
DbAS	*Journal of the Derbyshire Archaeological Society.*
DC	*Danelaw Charters.* Ed. F. M. Stenton. Brit. Ac. Records V.
De Banco	Placita de Banco.
Derby	*Derbyshire Charters.* Ed. I. H. Jeayes. London, 1906.
dial.	dialect(al).
Dieulacres	*Chartulary of Dieulacres Abbey.* Salt Soc. ix.
DL	*Ducatus Lancastriæ.* Record Com. 1823.
DM	Domesday Monachorum. VHK III.
Do	Dorset.
Drayton	Drayton, M. *Polyolbion.* London, 1612.
DST	*Historiae Dunelmensis Scriptores Tres.* Surt. Soc. 9.
Du	Durham. Also Dutch.
Dugdale	Dugdale, W. *The Antiquities of Warwickshire.* London, 1656.
Duk	*Dukery Records.* Ed. Robert White. Worksop, 1904.
Dunster	*Documents . . . illustrating the History of Dunster.* So Rec. Soc. 33.
Durh Acc Rolls	*Account Rolls of the Abbey of Durham.* Surt. Soc. 99 ff.
E	Earle, J. *A Hand-Book to the Land-Charters &c.* Oxford, 1888.
E 1 &c.	temp. Edward I &c.
Eadmer	*Eadmeri Historia novorum.* Chr. & Mem. 81.
EDD	Wright, J. *The English Dialect Dictionary.*
Eddi	Eddi, *Vita Wilfridi Episcopi.* Chr. & Mem. 71.
EETS	*The Early English Text Society.*
EFris	East Frisian.
EHR	*The English Historical Review.*
EMidl	East Midland.
Ep	Episcopal Registers.
EPN	*The Chief Elements used in English Place-names,* by A. Mawer. EPS i.
EPN(S)	*See* Smith in Works Consulted.
EPS	*Publications of the English Place-name Society.*
ERN	Ekwall, E. *English River-Names.* Oxford, 1928.
ESax	East Saxon.
Ess	Essex.
Ethelwerd	Ethelwerd's Chronicle. In Petrie, *Monumenta.*
Ewias	Bannister, A. T. *History of Ewias Harold.* Hereford, 1902.
Ex	*The Exeter Book.* London, 1933.
Exon	Exon Domesday. *Domesday Book iv.*
Eynsham	*Cartulary of the Abbey of Eynsham.* Oxford Hist. Soc. 1906–8.
Eyton	Eyton, R. W. *Antiquities of Shropshire.* London, 1854–60.

FA	*Feudal Aids*. Rolls Ser. 1899 ff.
FC	*The Coucher Book of Furness Abbey*. Chetham Soc. NS, 9 ff.
Fees	*The Book of Fees*. Rolls Ser. 1920–31.
Felix	Vita S. Guthlaci auctore Felice. In *Memorials of Saint Guthlac*. Ed. W. de Gray Birch. Wisbech, 1881.
FF	Feet of Fines (*Feet of Fines for the County of Norfolk 1198–1202*. Ed. Barbara Dodwell. London, 1952 (Pipe Roll Soc.). *Feet of Fines for the County of Lincoln 1199–1216*. Ed. Margaret Walker. London, 1954 (Pipe Roll Soc.)).
Finchale	*Finchale Charters*. Surt. Soc. 6.
Fine	*Calendar of the Fine Rolls*. Rolls Ser. 1911 ff.
FineR	*Excerpta e Rotulis Finium*. Record Com. 1835–6.
Flaxley	*Cartulary of . . . Flaxley*. Ed. A. W. Crawley-Boevey. Exeter, 1887.
Flem	Flemish.
Flor	Florence of Worcester's Chronicle. In Petrie, *Monumenta*.
For	Forest Charters.
Fount	*Chartulary of . . . Fountains*. Ed. W. T. Lancaster. Leeds, 1915.
FountM	*Memorials of the Abbey of . . . Fountains*. Surt. Soc. 42 ff.
FPD	*Feodarium Prioratus Dunelmensis*. Surt. Soc. 58.
Fr	*Documents preserved in France*. Rolls Ser. 1899. Also French.
Fridesw	*Cartulary of the Monastery of St. Frideswide*. Oxford Hist. Soc. 1894–6.
Fris	Frisian.
G	German.
Gael	Gaelic.
Gaimar	Gaimar, *Lestorie des Engles*. Chr. & Mem. 91.
Gale	Gale, Thomas. *Historiæ Britannicæ scriptores xv*. Oxford, 1691.
Gaul	Gaulish.
Gaunt Reg	*John of Gaunt's Register*. Camden Soc. 1911.
GeldR	Geld Roll (the Geld Roll of 1084 for the SW. counties is printed in DB iv; that of 1066–75 for Northamptonshire in Ellis, *Introduction*).
Germ	Germanic.
Gervase	Gervase of Canterbury, *Mappa mundi*. Chr. & Mem. 73.
Gesta	*Gesta Abbatum mon. S. Albani*. Chr. & Mem. 28.
Gilb	*Gilbertine Charters*. Lincoln Rec. Soc. 18.
Gildas	Gildas, *De excidio et conquestu Britanniae*. MGH, Auct. ant. xiii.
Gir	*Giraldi Cambrensis Opera*. Chr. & Mem. 21.
Gk	Greek.
Gl	Gloucestershire.
Glaston	*Rentalia et custumaria . . . abbatum . . . Glastoniæ*. So Rec. Soc. 1891.
Glouc	*Historia et Cartularium monasterii S. Petri Gloucestriæ*. Chr. & Mem. 33.
Godric	*Vita S. Godrici*. Surt. Soc. 20.
Godstow	*Register of Godstow Nunnery*. EETS 129 ff.
Gospatric's ch.	See CWNS v, StB, p. 526, VHCu II, pp. 231 ff.
Goth	Gothic.
Guisb	*Cartularium prioratus de Gyseburne*. Surt. Soc. 86, 89.
Guthlac	*The Anglo-Saxon Version of the Life of St. Guthlac*. Ed. C. W. Goodwin. London, 1848.
h.	hamlet.
Ha	Hampshire.
Ha Rec. Soc.	*Hampshire Record Society*. 1889 ff.

Hall, Sheffield	Hall, T. W. *Charters . . . relative to Sheffield*. Sheffield, 1916.
Harl Ch	*Harley Charter* (BM).
Harrison	Harrison, W. *The Description of Britaine*, 1577.
Hasted	Hasted, E. *History . . . of Kent*. Canterbury, 1778–99.
Hatfield	*Bishop Hatfield's Survey*. Surt. Soc. 32.
HB	*Historia Brittonum*. Ed. Mommsen. MGH, Auct. ant. xiii.
hd	hundred.
He	Herefordshire.
Heimskringla	Snorri Sturluson, *Heimskringla*.
HEl	*Historia Eliensis* or *Liber Eliensis*. Anglia Christiana Soc. 1848.
Heming	*Liber de terris* (etc.) *Ecclesiæ Wigorniensis*. Ed. Hearne. 1723.
Hengwrt MS. 150	In the National Library of Wales.
Hereford	*Charters . . . of Hereford Cathedral*. Cantilupe Soc. 4.
Hexh	*The Priory of Hexham*. Surt. Soc. 44, 46.
HHunt	Henry of Huntingdon, *Historia Anglorum*. Chr. & Mem. 74.
Higden	Higden's *Polychronicon*. Chr. & Mem. 41.
HMC	Historical MSS. Commission.
Holder	Holder, A. *Alt-celtischer Sprachschatz*. Leipzig, 1896 ff.
Holme	*St. Benet of Holme* 1020–1210. Norfolk Rec. Soc. 2.
Holme C	*Register . . . of Holm Cultram*. Kendal, 1929.
Hope	Hope, R. C. *A Glossary of Dialectal Place-nomenclature*. London, 1883.
Hrt	Hertfordshire.
HSC	Historia de S. Cuthberto. Surt. Soc. 51, Chr. & Mem. 75.
Hu	Huntingdonshire.
Hy 1 &c.	temp. Henry I &c.
Hyda	*Liber monasterii de Hyda*. Chr. & Mem. 45.
Hyde	*Liber Vitae. Register . . . of . . . Hyde Abbey*. Ha Rec. Soc. 1892.
IA	*Itinerarium Antonini* (&c.). Ed. Parthey & Pinder. Berlin, 1848.
ICC	*Inquisitio Comitatus Cantabrigiensis*. Ed. N. E. A. S. Hamilton. London, 1876.
Icel	Icelandic.
IE	Inquisitio Eliensis. In *Domesday Book* iv.
IG	Indo-Germanic.
Ipm	Inquisitiones post mortem.
IPN	*Introduction to the Survey of English Place-names*. Cambridge, 1924 (EPS).
Ir	Irish.
J	temp. John.
JAA	*Journal of the British Archaeological Association*.
K	Kent.
KCD	*Codex diplomaticus ævi Saxonici*. Ed. J. M. Kemble. London, 1839–48.
Kendale	*Records of Kendale*. Kendal, 1923 ff.
KInq	*An Eleventh-century Inquisition of St. Augustine's, Canterbury*. Brit. Ac. Records IV.
Kirkst	*Coucher Book of Kirkstall*. Thoresby Soc. 8.
KnFees	Knights' Fees (Kent). *See* Arch Cant xii.
Knytlinga saga	In *Sǫgur Danakonunga*. Copenhagen, 1919 ff.
La	Lancashire.
LaCh	*Lancashire Pipe Rolls. Early Lancashire Charters*. Ed. W. Farrer. Liverpool, 1902.
Lacy	*Two 'Compoti' of the . . . Manors of Henry de Lacy*. Chetham Soc. OS, 112.
LaInq	*Lancashire Inquests, &c.* La Record Soc. 48, 54.

Lancaster	*Chartulary of Lancaster Priory.* Chetham Soc. NS, 26 ff.
Landnáma	*Landnámabók.* Copenhagen, 1900.
Lanercost	The Register of Lanercost (MS in Carlisle Chapter Library).
Lat	Latin.
Launceston	*The Histories of Launceston and Dunheved.* By R. & O. B. Peter. Plymouth, 1885.
Laws	Anglo-Saxon Laws. *See* Attenborough *and* Liebermann in Works Consulted.
Lay(amon)	*Layamon's Brut.* Ed. F. Madden. London, 1847.
Le	Leicestershire.
LeS	The Leicestershire Survey. In VHLe and Round, *Feudal England.*
LG	Low German.
Lhuyd	Lhuyd, H. *Commentarioli Britannicae Descriptionis Fragmentum.* 1572.
Li	Lincolnshire.
Lipscomb	Lipscomb, G. *History . . . of the County of Buckingham.* London, 1847.
LiS	*The Lincolnshire Survey.* Ed. J. Greenstreet. London, 1884. Also in Lincoln Rec. Soc. 19.
LL	*Liber Landavensis.* Oxford, 1893.
LN	Liber niger mon. S. Petri de Burgo. In *Chronicon Petroburgense.* Camden Soc. 1849.
LoPleas	*Calendar of Plea and Memoranda Rolls . . . at the Guildhall.* Ed. A. H. Thomas. Cambridge, 1926–32.
LP	*Letters and Papers of the Reign of Henry VIII.* London, 1862 &c.
LVD	*Liber Vitae Ecclesiae Dunelmensis.* Surt. Soc. 136.
Lyttelton	*Charters of the Lyttelton Family.* Ed. I. H. Jeayes. London, 1893.
m.	mile(s).
Mab	The Mabinogion.
Madox	Madox, Thomas. *Formulare Anglicanum.* London, 1702.
Malm	*Registrum Malmesburiense.* Chr. & Mem. 72.
Mamecestre	*Mamecestre.* Ed. John Hartland. Chetham Soc. OS, 53 &c.
Map	*Facsimile of the Ancient Map of Great Britain in the Bodleian Library, Oxford, A.D. 1325–50.* Southampton, 1870 (reprinted 1938).
Marden	*Account of the Manor of Marden.*
Marrick	Charters of Marrigg Priory. *Collectanea* v.
MBret	Middle Breton.
MDu	Middle Dutch.
ME	Middle English.
Mededelingen	*Mededelingen van de Vereniging voor Naamkunde te Leuven en de Commissie voor Naamkunde te Amsterdam.* Leuven (in progress).
MGH	*Monumenta Germaniae Historica.*
MHG	Middle High German.
Middleton	*MSS of Lord Middleton at Wollaton Hall.* HMC 1911.
Midl	Midland.
MIr	Middle Irish.
Misc	*Inquisitions Miscellaneous.* Rolls Ser. 1916 &c.
MLG	Middle Low German.
mon.	moneyer.
Mon	Dugdale, W. *Monasticon Anglicanum.* London, 1846.
Monm	Geoffrey of Monmouth, *Historia Regum Britanniæ.* Ed. A. Griscombe. London, 1929. Also Monmouthshire.
Montacute	*See* Bruton.
Muchelney	*Cartularies of . . . Muchelney and Athelney.* So Rec. Soc. 14.

MW	Middle Welsh.
Mx	Middlesex.
n.	neuter.
Nb	Northumberland.
ND	*Notitia Dignitatum*. Ed. Böcking. Bonn, 1839–53.
n.d.	no date (undated).
N & Q	Notes and Queries.
Newcastle	*Newcastle Deeds*. Surt. Soc. 137.
Newminster	*Newminster Cartulary*. Surt. Soc. 66.
Nf	Norfolk.
NG	*Norske Gaardnavne*. Christiania, 1897 ff.
NI	*Nonarum inquisitiones*. Record Com. 1807.
Nichols	Nichols, J. *History . . . of the County of Leicester*. London, 1795–1815.
Norw	Norwegian.
Np	Northamptonshire.
NpCh	*Facsimiles of Early Charters from Northamptonshire Collections*. Ed. F. M. Stenton. Np Rec. Soc. iv.
NS	The Northamptonshire Survey. VHNp I.
Nt	Nottinghamshire.
O	Oxfordshire.
obl.	oblique form.
Obl	*Rotuli de Oblatis et Finibus*. Record Com. 1835.
OBret	Old Breton.
OBrit	Old British.
OCo	Old Cornish.
ODan	Old Danish.
ODu	Old Dutch.
OE	Old English.
OEBede	The Old English Version of Bede's *Historia ecclesiastica*.
OED	*A New English Dictionary*. Oxford, 1884 &c.
OEScand	Old East Scandinavian.
OET	*The Oldest English Texts*. Ed. Sweet. EETS 83.
OFr	Old French.
OFris	Old Frisian.
OG	Old German.
OHG	Old High German.
Ol	Oliver, G. *Monasticon Diocesis Exoniensis*. Exeter, 1846.
OLG	Old Low German (Old Saxon).
ON	Old Norse.
ONFr	Old North French.
Ordericus	Ordericus Vitalis, *Historia Ecclesiastica*.
Orig	*Rotulorum originalium . . . abbreviatio*. Record Com. 1805–10.
Ormerod	Ormerod, G. *History of the County . . . of Chester*. London, 1875–82.
Os	*Cartulary of Oseney Abbey*. Ed. H. E. Salter. Oxford Hist. Soc. 89 ff. Oxford, 1929–36.
OScand	Old Scandinavian.
OSFacs	*Facsimiles of Anglo-Saxon MSS*. Southampton, 1878–84.
Osney	*Register of Oseney Abbey*. EETS 133 ff.
OSw	Old Swedish.
OW	Old Welsh.
Oxf	*Facsimiles of Early Charters in Oxford*. Oxford, 1929.
(p)	personal name.
P	Pipe Rolls.

par.	parish.
Pat	Patent Rolls.
Penshurst	*The Penshurst MSS.* HMC 1925.
Percy	*The Percy Chartulary.* Surt. Soc. 117.
pers. n.	personal name.
Petrie	Petrie, H. *Monumenta Historica Britannica.* London, 1848.
pl. n. (ns.)	place-name(s).
PN	Place-names (PnBd, PnBk, PNSx, PNWo, *see* Mawer and Stenton; PND, PNHrt(S), PNMx(S), PNNp, PNNt(S), PNSr, PNW(S), PNWa(S), *see* Gover, Mawer, and Stenton; PNCa(S), PNEss, *see* Reaney; PNER, PNNR, *see* Smith; PNCu(S), *see* Armstrong, Mawer, Stenton, and Dickins; PNDb(S), *see* Cameron; PNO(S), *see* Gelling; PNCa, PNHrt, PNHu, PNSf, *see* Skeat; PNCu, *see* Sedgefield; PNDb, *see* Walker; PNDo, *see* Fägersten; PNGl, *see* Baddeley; PNHe, *see* Bannister; PNK, *see* Wallenberg; PNLa, *see* Ekwall; PNMx, *see* Gover; PNNb, *see* Mawer; PNNt, *see* Mutschmann; PNO, *see* Alexander; PNSt, PNWa, *see* Duignan; PNW, *see* Ekblom; PNWt, *see* Kökeritz, all in Works Consulted).
Pont	*Pontefract Chartulary.* YAS 25, 30.
Pp	*Entries in . . . Papal Registers.* Rolls Ser. 1893 ff.
PT	Poll-tax Rolls.
Ptol(emy)	*Claudii Ptolemaei Geographia.* Ed. C. Müllerus. Paris, 1883.
Pudsay	*The Pudsay Deeds.* YAS 56.
PWint	*The Pipe Roll of the Bishopric of Winchester for 1208–9.* Ed. Hubert Hall. London, 1903.
QW	*Placita de quo Warranto.* Record Com. 1818.
R	river-name.
R 1 &c.	temp. Richard I &c.
RA	*Registrum antiquissimum.* Lincoln Rec. Soc. 27 ff.
Rams	*Cartularium monasterii de Rameseia.* Chr. & Mem. 79.
Rav	*Ravennatis Anonymi Cosmographia.* Berlin, 1860.
RBE	*The Red Book of the Exchequer.* Rolls Ser. 1896.
Rec Carn	*The Record of Caernarvon.* Record Com. 1838.
Red Book	The Red Book of Hergest.
Rees	Rees, W. J. *Lives of the Cambro British Saints.* Llandovery, 1853.
Reg	*Regesta Regum Anglo-Normannorum.* Oxford, 1913, 1956.
Reg Dun	*Reginaldi Monachi Dunelmensis Libellus.* Surt. Soc. 1.
Reg Roff	*Registrum Roffense.* Ed. J. Thorpe. London, 1769.
Reg Wilt	*Registrum Wiltunense.* Ed. R. C. Hoare. London, 1827.
Rep	Welsh Records. *Deputy Keeper's Report* 26.
RH	*Rotuli hundredorum.* Record Com. 1812–18.
Riev	*Rievaulx Cartulary.* Surt. Soc. 83.
Rit Dun	*Rituale Ecclesiae Dunelmensis.* Surt. Soc. 140.
Rob Br	*The Story of England by Robert Manning of Brunne.* Chr. & Mem. 87.
Rob Gl	*The Chronicle of Robert of Gloucester.* Chr. & Mem. 86.
Ronton	*The Ronton Chartulary.* Salt Soc. OS, iv.
Rot Cur	*Rotuli curiæ regis.* Record Com. 1835.
Rot dom	*Rotuli de dominabus.* Pipe Roll Soc.
RPD	*Registrum Palatinum Dunelmense.* London, 1873 ff.
Ru	Rutland.
Rudder	Rudder, S. *History of Gloucestershire.* Cirencester, 1779.
Rushw MS	The Rushworth MS. In Skeat, W. W., *The Gospel according to St. Matthew,* &c. Cambridge, 1871 &c.

Rutland	*MSS. of the Duke of Rutland*. HMC 1905.
RWrits	*Facsimiles of English Royal Writs to A.D. 1100*. Ed. T. A. M. Bishop and P. Chaplais. Oxford, 1957.
Rydeware	*Rydeware Cartulary*. Salt Soc. OS, xvi.
s.	son.
s.a.	sub anno.
Sa	Shropshire.
SaDeeds	Old Shropshire Deeds. Sa Arch. Soc. 1886.
Saints	The Saints of England. In *Die Heiligen Englands*. Ed. Liebermann. Hannover, 1889. Also in Hyde.
Salisbury	*Salisbury Charters*. Chr. & Mem. 97.
Sarum	*The Register of St. Osmund*. Chr. & Mem. 78.
Saxton	Saxton's maps, 1574 ff.
Sc	*Calendar of Documents relating to Scotland*. Edinburgh, 1881–8.
Scand	Scandinavian.
SD	Symeon of Durham, *Historia ecclesiæ Dunelmensis* and *Historia regum*. Chr. & Mem. 75, Surt. Soc. 51.
Seine-Inf.	Seine-Inférieure (France).
Selborne	*Charters . . . relating to Selborne*. Ha Rec. Soc. 1891 ff.
Selby	*The Coucher Book of Selby*. YAS 10, 13.
Sele	*Chartulary of the Priory of Sele*. Ed. Salzman. 1913.
Sf	Suffolk.
Sh	Shakespeare.
Shaw	Shaw, S. *History . . . of Staffordshire*. London, 1798–1801.
So	Somerset.
Speed	Speed's Maps. 1607 &c.
SPN	*See* Ekwall in Works Consulted.
Sr	Surrey.
SrAS	*Surrey Archaeological Society Collections*.
St	Staffordshire.
Stafford	*The Staffordshire Cartulary*. Salt Soc.
StAug	*The Register of St. Augustine's Abbey, Canterbury*. Brit. Ac. Records, II, III.
StB	*The Register of the Priory of St. Bees*. Surt. Soc. 126.
StEdm	*Memorials of St. Edmund's Abbey*. Chr. & Mem. 96.
Steph	temp. Stephen.
StPaul	*The Domesday of St. Paul's*. Camden Soc. 1858.
St Pauls	*Early Charters of the Cathedral Church of St. Paul, London*. Ed. Marion Gibbs. Camden Soc. 3rd Ser. 58. London, 1939.
Subs	Subsidy Rolls.
Surt. Soc.	The Surtees Society.
Surv	Early Worcestershire Surveys. VHWo I.
Sw	Swedish.
Swithun	*S. Swithunus, Miracula metrica auctore Wulfstano*. Ed. P. Michael Huber.
Sx	Sussex.
t.	tempore (time of).
Taliesin	*The Book of Taliesin*. Ed. J. G. Evans. Llandovery, 1910.
Tax	*Taxatio ecclesiastica*. Record Com. 1802.
Test Karl	*Testamenta Karleolensia*. Carlisle, 1893.
Text Roff	*Textus Roffensis*. Ed. Th. Hearne. Oxford, 1720.
Th	*Diplomatarium anglicum*. Ed. B. Thorpe. London, 1865.
Thame	*The Thame Cartulary*. Ed. H. E. Salter. Oxford Rec. Soc. Oxford, 1947–8.
Thorney Fragm	Memoranda of gifts to Thorney Abbey. Cambr. Phil. Soc. lxi–lxiii.

Thoroton   Thoroton, R. *The Antiquities of Nottinghamshire*. London, 1677.
Torre   Torre Abbey Cartulary. MS. in Public Record Office.
Totnes   Watkins, H. R. *History of Totnes*. Torquay, 1914.
TP   Tabula Peuteringiana.
TpR   *Records of the Templars in England*. Ed. Beatrice A. Lees.
                     Brit. Ac. Records IX. London, 1935.
Trev   Higden's *Polychronicon*, translated by John Trevisa. Chr. &
                     Mem. 41.
Trib Hid   The Tribal Hidage. BCS 297.
Val   *The Valuation of Norwich*. Ed. W. E. Lunt. Oxford, 1926.
Vale Royal   *The Ledger-Book of Vale Royal Abbey*. La Rec. Soc. 68.
VE   *Valor Ecclesiasticus*. Record Com. 1810 ff.
VH   *Victoria History of the Counties of England* (VHEss = Victoria
                     History of Essex, &c.).
vil.   village.
Vita Oswini   In *Miscellanea biographica*. Surt. Soc. 8.
W   Wiltshire.
W 1 &c.   temp. William I &c.
Wa   Warwickshire.
Wakef   *Wakefield Court Rolls*. YAS 29 ff.
wap.   wapentake.
We   Westmorland.
Wells   *Wells MSS*. HMC 1907, 1914.
Wendover   Roger of Wendover, *Flores Historiarum*. Chr. & Mem. 84.
WhC   *The Coucher Book of Whalley Abbey*. Chetham Soc. OS, 10 ff.
Whellan   Whellan, W. *History of Cumberland and Westmorland*. 1860.
Whitaker   Whitaker, T. D. *The History of . . . Whalley*. London, 1872–6.
Whitby   *Cartularium Abbathiæ de Whiteby*. Surt. Soc. 69, 72.
Wills   *Anglo-Saxon Wills*. Ed. Dorothy Whitelock. Cambridge, 1930.
Winchc   *Registrum Monasterii de Winchelcumba*. Exeter, 1892 ff.
WMalm   William of Malmesbury, *Gesta pontificum*. Chr. & Mem. 52. Id.
                     Gesta regum. Chr. & Mem. 90.
WMidl   West Midland.
Wo   Worcestershire.
WoCh   *Original Charters relating to the City of Worcester*. Wo. Hist. Soc.
                     1909.
WoP   *Registrum prioratus B.M. Wigorniensis*. Camden Soc. 1865.
Works   *Public Works in Mediaeval Law*. Selden Soc. 32, 40.
WR   *The Register of the Priory of Wetherhal*. London, 1897.
Wright   Wright, Thomas. *History of Essex*. London, 1831–6.
WSax   West Saxon.
Wt   Wight.
WWorc   William of Worcester's *Itinerarium*. Ed. J. Nasmith. Cambridge,
                     1778.
Y   Yorkshire.
YAS   The Yorkshire Archaeological Society. Record Series.
YCh   *Early Yorkshire Charters*. Ed. Farrer. Edinburgh, 1914 &c.
YCh iv, v   *Early Yorkshire Charters*, vol. iv, v. The Honour of Richmond,
                     ed. Charles Travis Clay. YAS, Extra Series, 1935, 1936.
YD   *Yorkshire Deeds*. YAS 39 ff.
YE   The East Riding of Yorkshire.
YInq   *Yorkshire Inquisitions*. YAS 12 ff.
YN   The North Riding of Yorkshire.
YW   The West Riding of Yorkshire.
*   An asterisk indicates a hypothetical form.

KEY TO THE PRONUNCIATION

THE notation is mainly that used in *The Concise Oxford Dictionary*. The pronunciation is shown by the following test-words.

mãte, mēte, mīte, mōte, moot, mūte, caw, cow, bah, boil.

mãre, mēre, mīre, mōre, moor, demure, dowry, part, pert, port.

răck, rĕck, rĭck, rŏck, rŭck, rŏŏk.

Italicized vowels have an indistinct sound, as in *a*gain, moment, fing*er*. Borough is (bŭr*u*).

dh = *th* in *then*, **th** = *th* in *think*, **j** as in *jet*, **ng** as in *sing*, **ngg** as in *finger*, **n-g** as in *unguarded*, **s** as in *sister*, **z** as in *zeal*, **ch** as in *chip* (chĭp), **zh** as in *fusion* (fūzhn).

The place of the accent is marked by a turned period placed after the accented vowel or diphthong, as **Blenca·rn.**

In the etymological discussions č, ǧ are sometimes used to designate palatalized *c* (*k*), *g*, which later became *ch* (ch) and *dg* (j), as in OE *dĭč*, *senǧan*, whence *ditch*, *singe*.

A

ON á, ODan, OSw ā 'river, stream' = OE
ēa. See BEELA, BRATHAY, GRETA, LIZA,
RAWTHEY, ROTHAY, RAUGHTON; ABY, AMBLE-
SIDE, AYRESOME, AYTON.

Abberley Wo [*Edboldelege* DB, *Alboldelega*
c 1180 Fr, *Abbedeslegh* 1216 Cl]. '*Ēadbeald*'s LĒAH.' Forms with *lb* no doubt
represent *bb*.

Abberton Ess [*Edburgetuna* DB, *-ton* 1204
FF, *Adburgetun* 1247 Ipm]. '*Ēadburg*'s
manor.' *Ēadburg* is a common OE woman's
name.

Abberton Wo [*Eadbrihtincgtun* 972 BCS
1282, *Edbretintune* DB]. 'The TŪN of *Ēadbeorht*'s people.'

Abberwick Nb [*Alburwic* 1170 P, *Alburck-
wick* 1279 Ass]. 'The WĪC of *Alu(h)burg*'
(a woman).

Abbeystead. See STEDE.

Abbey Town Cu. The site of Holme Cultram Abbey.

Abbotsbury Do [(æt) *Abbodesbyrig* 1045 Th,
Abedesberie DB]. 'The manor of the abbot
(OE *abbod*).' A monastery was founded
here c 1026.

Abbotsham D [*Hama* DB, *Abbudesham*
1238 Ass]. Originally *Hamm*; see HAMM.
The abbot was of Tavistock.

Abbotsley Hu [*Adboldesl'* 12 PNHu, *Ab-
botesle* 1276, *Adboldesle* 1279 RH]. '*Ēadbeald*'s LĒAH.' Cf. ABBERLEY.

Abbotston W [*Abbedeston* 1256 FF, *Ab-
bassetone* 1338, *Abbessetone* 1346 Ipm].
'The manor of the abbess' (of Wilton).
Abbess was OE *abbodisse*.

Abbotstone Ha [*Abbodestun* 11 BCS 1161,
Abedestune DB]. 'The manor of the abbot.'
A~ belonged to Winchester Cathedral.

Abdon Sa [*Abetune* DB, *Abbeton* 1227, 1240
FF]. '*Abba*'s TŪN.'

Aberford YW [*Ædburford* 1176, *Ædburgforð*
1177 P, *Abberford* 1251 Ch]. '*Ēadburg*'s
ford'; cf. ABBERTON Ess.

Abergave·nny, Welsh **Abergefenni** or **Y
Fenni,** Monm [*Gobannio* (abl.) 4 IA, *Aber-
gavenni* 1175 Ann Cambr, *Abergevenni* 1191
Gir, *Bergaven'* 1291 Tax]. The old name
Gobannion is derived from the Celtic word
found as Welsh *gof*, OIr *goba* (gen. *gobann*),
'smith', perhaps used as a pers. n. This
name was transferred to the river on which
A~ stands and which is *Gebenni, Geuenni*
c 1150 LL, *Gevenni* 1191 Gir. Later a new
name *Aber Gefenni* 'mouth of the Gavenny
(Gefenni)' was formed. *Aber* is Welsh for
'confluence, mouth of a river' and refers to
the confluence of the Gavenny with the Usk.

Abingdon (ă-) Brk [(mons) *Æbbandun* c 730,
Æbbanduna 811 BCS 155, 850, *Abbandun*
931 BCS 680, 961 BM, *Abbendone* DB].
'*Æbba*'s or *Æbbe*'s DŪN.' *Æbbe* is a woman's
name. The original Abingdon was on an
upland ridge near Boar's Hill, referred to as
Abbendun 955 BCS 906. It was later removed to the present site, which was
originally *Seouechesham* (Abingd i, p. 6);
cf. SEACOURT.

Abinger (ăbïnjer) Sr [*Abinceborne* DB,
Abingewurd 1191, *Abbingewurda* 1192, *Ebb-
ingwurðe* 1198 P]. 'The WORÞ of *Eabba*'s
or *Æbba*'s people.'

Abinghall Gl [*Abbenhale* 1165 Flaxley, *-hal*
1220 Fees, *Abenhale* 1221 Ass]. '*Abba*'s
HALH.'

Abington, Great & Little, Ca [*Abintona*
c 1080 ICC, *-tone* DB, *Abingtton Magna,
Abiton Parva* 1254 Val], A~ **Pigotts** Ca
[*Abintona* c 1080 ICC, *-tone* DB, *Abingeton*
1202 FF], A~ Np [*Habintun* 1066–75 Geld
R, *Abintone* DB, *Abbingeton* 1203 Cur].
'The TŪN of *Abba*'s people.'

A~ **Pigotts** was held by Picot vicecomes c 1080
(ICC), by John Pykot 1434 Pat. *Picot* is a nickname derived from OFr *picot* 'point, pointed
object'.

Abkettleby. See KETTLEBY.

Ablington Gl [*Eadbaldingtun* 855, 899 BCS
487, 580, *Abelinton* 1207 Cur], A~ W
[*Alboldintone* DB, *Ablinton* 1242 Fees,
Eblinton 1252 Ch]. 'The TŪN of *Ēadbeald*'s
people.'

Abney Db [*Habenai* DB, *Abbeneia* 1200 P].
'*Abba*'s island.' See ĒG.

Abram (ă-) La [*Adburgham* a 1199 CC,
Edburgham 1212 Fees]. '*Ēadburg*'s HĀM.'
Cf. ABBERTON Ess.

Abridge (ā-) Ess [*Affebrigg* 1203 Cur].
'*Æffa*'s bridge.'

Abson Gl [*Abbedeston* 1167, *Abodeston* 1176
P]. 'The manor of the abbot' (of Bath and
Gloucester).

Abthorpe Np [*Torp* DB, *Abetrop* 1190,
Abbethorp 1230 P]. '*Abba*'s thorp.'

Aby (ā-) Li [*Abi* DB, *Aby* 1219 Ep]. 'The
BY on the stream' (Calceby Brook). See Ā,
and cf. Sw ÅBY, Dan AABY.

OE **āc** 'oak' is often used alone as a pl. n.,
esp. in the plural form. See ACOMB, AIKE,
OAKE, OAKEN. In NOKE, ROCK Wo the initial
consonant is a relic of the def. art. (OE *þām,
þǣre*). The OE dat. sing. form *ǣc* (with
palatal *c*) is the source of EACH and is the
second element of some names, as BRAD-
NINCH, CRESSAGE, RADNAGE. OE *āc* is the
second el. also of BRADDOCK, HARROCK, HEN-
NOCK, HODSOCK, HOLY OAKES, MATLOCK and

is common as first el.; cf. ACTON (AIGHTON, AUGHTON), ACOL, AGDEN, EAGLE, OAKFORD, OAKLEY, OKEFORD, OKEOVER, KNOCKHOLT, &c. For OE *ācen, ǣcen* 'of oaks' see AKELEY, EACHWICK, OAKEN-.

Acaster Malbis & Selby YW [*Acastre* DB, 1228 Ep, *Acaster Malebisse* 1252 Ch, *Acastre Seleby* 1285 FA]. 'The Roman fort (OE *ceaster*) of *Aca*.'
The manors were held by the Malbis family (William Malebisse was in possession in the 12th cent.) and Selby Abbey (from c 1110; cf. YCh 462). *Malebisse* is a Norman surname, meaning 'ill beast' (Lat *Mala bestia*).

Accrington La [*Akarinton* a 1194 Kirkst, *Akerynton* 1258 Ass]. An OE *Ǣcerntūn* 'TŪN where acorns grew'.

Achurch, Thorpe, Np [(æt) *Asencircan, Asecyrcan* 972–92 BCS 1130, *Asechirce* DB]. The first el. is a pers. n., perhaps OE **Asa* or OScand *Āsi*. The same el. may be found in *Asendike* Li, the name of an old ditch [*Esendic* 656 ASC E].

Acklam YE [*Aclun, Hacle* DB, *Aclum* 1130 P, *Acclum* 1154–70 YCh 32], A~ YN [*Achelum, Aclun* DB]. OE *āc-lēum*, dat. plur. of *āclēah* 'oak wood'. Cf. ACLE.

Ackleton Sa [*Aclinton* 1176 P, *Adelacton* 1292 QW]. 'The TŪN of *Ēadlāc*'s people.'

Acklington Nb [*Eclinton* 1177, *Aclinton* 1187, 1190 P, *Aclington* 1242 Fees]. Probably identical in origin with ACKLETON.

Ackton YW [*Aitone, Acitone* DB, *Aicton* 12 Kirkst]. OE *Āctūn*; see ACTON.

Ackworth YW [*Aceuurde* DB, *Akeworth* 1201 Cur]. '*Acca*'s WORP.'

Acle (ā-) Nf [*Acle* DB, *Achelai* 1159 P, *Acleda* 1186–1210 Holme, *Aclee* 1197 FF]. OE *āclēah* 'oak wood'; cf. OAKLEY.

Acol K [*Acholt* 1270 Ass]. OE *ācholt* 'oak wood'.

Acomb Nb nr Corbridge [*Akum* 1268 Ipm], **East A~** Nb [*Acum* 1242 Fees], A~ YN [*Akum* 1222 FF], A~ YW [*Acum* DB]. OE *ācum*, dat. plur. of *āc* 'oak'.

Aconbury He [*Akornebir'* 1213 Cl, *-bury* 1218 Pat, *-bire* 1244 Misc, *Okernebur'* 1241 Cl]. 'Old fort inhabited by squirrels' (OE *ācweorna*).

Acre, Castle, South & West, Nf [*Acra, Acre* DB, *Accara* 1121 AC, *Castelacr'* 1235 Cl, *Sutacra* 1242 Fees, *Westacre* 1203 Ass]. OE *æcer* 'field'. If the source is OE *æcer* sing., and not the dat. plur. *æcrum*, AKER in Norway, ÅKER in Sweden may be compared.

Acrise K [*Acres* DB, *Aqus* (for *Acris*), *Hacris* 11 DM, *Hacrise* 1166 RBE]. OE *āc-hrīs* 'oak copse'.

Acton, a common name, is generally **1.** OE *Āctūn* 'TŪN by the oak(s)', e.g. A~ Chs nr Nantwich [*Actune* DB], **A~ Grange** Chs nr Runcorn [*Acton* 1260 Court], **Iron A~** Gl [*Actune* DB, *Irnacton* 1287 QW], A~ **Beau-**

champ He [*Aactune* 727 BCS 146], **A~** Mx [*Acton* 1232 Cl, 1242 Fees], **A~ Burnell & Pigot** Sa [*Actune, Ǣctune* DB, *Akton Burnill* 1198 FF, *Acton Picot* 1255 RH], **A~ Reynold** Sa [*Achetone* DB, *Acton Reyner* 1255 RH], **A~ Round** Sa [*Achetune* DB, *Acton la Runde* 1284 Ipm], **A~ Scott** Sa [*Actune* DB, *Scottes Acton* 1289 Misc], **Stone A~** Sa [*Staniacton* 1242 Fees], **A~ Trussell** St [? *Actun* 1002 Wills, *Actone* DB, *Actona Willelmi* 1167 P].

2. OE *Ac(c)an tūn* '*Ac(c)a*'s TŪN' : A~ **Turville** Gl [*Achetone* DB, *Akentona* 1169 P, *Acton Torvile* 1284 Ipm], A~ Nb [*Aketon* 1242 Fees, 1256 Ass], **A~ Sf** [*Acantun* c 995 BCS 1289, *Achetuna* DB].

3. **Acton** Do [*Tacatone* DB, *Taketon* 1305 Cl]. See TACKLEY.
The distinguishing elements are mostly names of local families. **A~ Beauchamp** He was held by the Beauchamps from the 12th cent. Beauchamp (Lat *Bellus campus*) is a family name taken from one of the Beauchamps in France. —**A~ Burnell** Sa was held by Gerin Burnell in 1183 (Eyton). Burnell is a family name, originally a byname from OFr *brunel* 'brown'.— **Iron A~** Gl from iron mines.—**A~ Pigot** Sa came to William Picot in the 12th cent. Cf. ABINGTON PIGOTTS.—**A~ Reynold** Sa was held by Reyner de Acton in 1203 (Ass). Reyner is OFr *Rainer*, a pers. n. of OG origin.—**Round A~** Sa must mean 'round A~'.—**A~ Scott** Sa was held by Walter le Scot c 1240 (Eyton).— **A~ Trussell** St was named from a local family. Trussel is a nickname derived from OFr *troussel* 'packet'.—**A~ Turville** Gl was held by Robert Turevil in 1236 (Fees). Turville is a family name derived from one of the Tourvilles or Trouvilles in Normandy.

Adbaston St [*Edboldestone* DB, *Ǣdbaldeston* 1175 P]. '*Ēadbald*'s TŪN.'

Adber Do [*Eátan beares* (gen.) 956 BCS 931, *Eattebera, Ettebere* DB]. '*Eata*'s grove.' See BEARU.

Adbolton Nt [*Alboltune* DB, *Albotton* 1197 P, *Adbolton* 1200 P, 1265 Misc]. '*Ēadbald*'s or *Ǣþelbald*'s TŪN.'

Adcot Sa [*Addecote* c 1241 Eyton]. '*Adda*'s COT.'

Adderbury O [(æt) *Eadburggebyrig* c 950 Wills, *Edburgberie* DB]. '*Ēadburg*'s BURG.' Cf. ABBERTON Ess.

Adderley Sa [*Eldredelei* DB, *Aldrideleye* 1283 Ch, *Addredeleye* 1284 Cl], A~ St [*Aldredeslega* 1130 P]. '*Aldrēd*'s (*Ealdrēd*'s) LĒAH.'

Adderstone Nb [*Edredeston* 1233 P]. '*Ēadrēd*'s TŪN.'

Addingham Cu [*Addingham* 1292 QW], A~ YW [*Haddincham* c 972 BCS 1278, *Odingehem* DB, *Addingeham* c 1130 SD]. 'The HĀM of *Adda*'s people.'

Addington Bk [*Edintone* DB, *Adinton* 1176 P], A~ K [*Eddintune* DB, *Edintona* 1175 P], **Great & Little A~** Np [*Edintone* DB, *-tona* 1130 P, *Maior, Minor Adinton* 1220 Fees], A~ Sr [*Eddintone* DB, *Adingeton* 1203 Cur,

Adinton 1219 Fees]. '*Ead(d)a*'s TŪN' or 'the TŪN of *Ead(d)a*'s or *Æddi*'s people'.

Addiscombe Sr [*Edescamp* 1229 FF, *Adiscaumpe* 1352 AD]. '*Æddi*'s field.' See CAMP. Addiscombe is c 2 m. from ADDINGTON Sr, which was probably named from the same *Æddi*.

Addlestone Sr [*Attelesdene* 1241, *Atelesdon* 1271 PNSr]. '*Ættel*'s DENU or valley.' *Ættel is related to *Ætta, Ætla.*

Addlethorpe Li [*Arduluetorp* DB, *Ardeltorp* 1212 Fees, *Addeltorp* 1202 Ass], A~ YW [*Ardulfestorp* DB]. '*Eardwulf*'s thorp.'

Adel (ă-) YW [*Adele* DB, *Adela* 1100–8 Fr]. OE *adela* 'filth, filthy place'.

Adforton He [*Alfertintune* DB, *Atfreton* 1256 Eyton, *Atforton* 1292 QW]. 'The TŪN of *Ealdfrið*'s or *Eadfrið*'s people.'

Adgarley La [*Eadgarlith* 1180–90 FC, *Adgareslith* 1212 Fees]. '*Eadgār*'s slope.' See HLIÞ.

Adgeston Wt [*Avicestone* DB, *Auicheston* 1198 FF]. '*Æfic*'s TŪN.'

Adisham K [*Adesham* 616 BCS 12 (late copy), *Edesham* 1006 KCD 715, DB, *Eades-, Edesham* 11 DM]. '*Ead*'s or *Æddi*'s HĀM.' *Ead is a short form of names in *Ead-*.

Adlestrop Gl [*Tatlestrop* 11 KCD 1367, *-trop* 1251 Ch, *Tedestrop* DB]. '*Tātel*'s or *Tǣtel's thorp.' Cf. TALTON. For the loss of *T-* cf. ELSTREE.

Adlingfleet YW [*Adelingesfluet* DB, *Athelingflet* 1230 P]. 'The stream of the *æpeling* or prince.' See FLĒOT.

Adlington Chs [*Edulvintune* DB, *Adelvinton* 1248 Ipm], A~ La [*Adelventon* 1202 FF]. 'The TŪN of *Eadwulf*'s people.' Cf. *Eadulfingtun* c 1000 Th (unidentified).

Admaston St [*Ædmundeston* 1176, 1178, *Edmodeston* 1177, *Ædmodeston* 1180 P]. '*Eadmund*'s TŪN.'

Admergill YW [*Admergyll* 14 Kirkst]. '*Eadmǣr*'s gill or valley.'

Admington Gl [*Edelmintone* DB, 1221 Ass, *Ethelmintona* 1175 Winchc, *Adilmington* 1251 Ch]. 'The TŪN of *Æpelhelm*'s people.'

Adney Sa [*Eduney* 1212 Fees, *Edeweny* 1292 Ch, *Addeney* 1327 Subs]. '*Eadwynn*'s island.' *Eadwynn* is a woman's name.

Adsett Gl [*Eddeseta* 1220 Fees, *Addesete* 1221 Ass]. '*Æddi*'s or *Eadda*'s fold.' See (GE)SET.

Adstock Bk [*Edestocha* DB, *Addestoke* 1221 Ep]. '*Æddi*'s or *Eadda*'s STOC.'

Adstone Np [*Atene-, Etenestone* DB, *Etteneston* c 1200 BM]. '*Ættin*'s TŪN.' *Ættin is an earlier form of *Ætti.*

Adstone Sa [*Edestan* c 1150, *Eadeston* 12, *Addeston* 1203 Eyton, *Ad(d)estan* 1221 Ass]. '*Eadda*'s or *Æddi*'s stone.'

Adur R Sx. A late back-formation from

Portus Adurni ND, which was supposed to be at the mouth of the Adur.

Advent Co [(capella) *Sancti Adweny* 1302 Ass (Gover)]. Elliptical for 'the church of St. Advent or Adwen'. Adwen seems to be identical with Bret *St. Aouen.*

Adwalton YW [*Athelwaldon* 1202 FF, *Adwelton* 1208 Cur]. '*Æþelwald*'s TŪN.'

Adwell O [*Advelle* DB, *Adewell* 1176 P, 1204 Cur, *Edewelle* 1279 RH]. '*Ead(d)a*'s spring or stream.'

Adwick (ădĭk) le **Street** YW [*Adewic* DB, 1269 Ch], A~ upon **Dearne** YW [*Adeuuic* DB]. '*Adda*'s or *Eadda*'s WĪC.'

One is on a Roman road, the other on the DEARNE.

OE **æcer** 'field, ploughed land' (= ON *akr*) is used alone in ACRE Nf and is the second el. of some names, as BARN-, BENACRE, BESSACAR, BICKNACRE, GATACRE, HALNAKER, LINACRE, WEDDIKER, also in ALSAGER, CLIVIGER, which show a curious assibilated form of the word. OScand *akr* is found in MUKER, ROSEACRE, STAINSACRE, TARNACRE.

OE **æppel** 'apple'. See APLEY, APPLE- (passim), EPPLETON.

OE **ærn** 'house', esp. 'storehouse', also in a more original form *renn* in *rendegn* (for *ærnþegn* 'house-officer'), is sometimes found in pl. ns. It is used alone in ARNE, and as a second el. e.g. in BREWERNE, CHITTERNE, COLERNE, COWARNE, CREWKERNE, HARDHORN, NEWERNE, SEASALTER, STANION, WALDRON, also in CHARD, DINDER, FINDERN, MARK, which contain the form *renn.*

OE **æsc** 'ash-tree' is a very common pl. n. element. It is often used alone, in the sing. or plur., to form pl. ns.; see ASH (also ASHBOCKING, -BRITTLE, -REIGNY, -WATER, ROSE ASH), ASHE, ESH, ASHEN, ASHTON (2), ASKHAM, NASH (with *N-* from the dat. sg. or pl. of the preceding def. art.). It forms the second el. of AVENAGE, BORROWASH, DODNASH, FRANCHE, HAMNISH, MATLASK, MONYASH, and others. It is very common as a first el., as in ASHLEY, ASHTON, AISHOLT, ASTON, ESHER, &c. Sometimes it is difficult to distinguish *æsc* from the corresponding OScand word (ON *askr* &c.). The latter is the first el. of ASHBY, ASKRIGG, ASKWITH. Occasionally the Scand form seems to have replaced original *æsc*, as in ASKE, ASKERN. OE *æscen* adj. 'of ash' is the first el. of ASHDON, ASHENDON, ASHINGTON Nb and some others.

OE **æspe** 'aspen' is sometimes the first el. of pl. ns., as in APSLEY, ASPALL, ASPLEY and the like, ESPLEY. See also APPS.

Afflington Do [*Alfrunetone* DB, *Alfrington* 1263 Ipm]. '*Ælfrūn*'s TŪN.' A lady of this name (*Alueron*) held the manor in the time of Edward the Confessor (DB).

Affpuddle. See PIDDLE.

Afton Wt [*Affetune* DB]. '*Æffa*'s TŪN.'

Agardsley St [*Edgareslege* DB, *Adgeresley* 1280 Ass]. '*Éadgār*'s LĒAH.'

Agden Hu [*Accedena* c 1124, *Accadena* 1136–8 RA]. '*Acca*'s valley.' A~ YW [*Akeden* 1246 FF]. OE *āc-denu* 'oak valley'.

Agglethorpe YN [*Aculestorp* DB, *Akolvesthorp* 1246 FF]. '*Ācwulf*'s thorp.'

Aglionby Cu [*Agyllunby* c 1200 WR]. '*Agyllun*'s BY.' Lawrence s. of Agyllun held land in Aglionby in the 12th cent. *Agyllun* is a Norman name, originally no doubt a nickname from Fr *aiguillon* 'point, thorn'.

Aigburth La [*Aykeberh* c 1200 CC]. ON *Eikiberg* 'oak hill'.

Aighton La [*Actun* DB]. A variant of ACTON.

Aikber YN [*Aykebergh* 1290 Ch, 1293 QW]. Identical with AIGBURTH.

Aike YE [*Ach* DB, *Ake* 1203 FF]. OE *āc* 'oak'.

Aikton Cu [*Aictun* c 1200 StB, *Aykton* 1232 FF]. Probably OE *Āctūn* 'oak TŪN', with OE *āc* replaced by OScand *eik*.

Ailby Li [*Alebi* DB, *Alabi*, *Alebi* Hy 2 DC]. 'The BY of *Ali*' (ODan *Ali*, ON *Áli*).

Ailsworth Np [*Ægelesuurð* 948 BCS 871, *Egleswurðe* 972 ib. 1281, *Eglesworde* DB]. '*Ægel*'s WORP.' OE *Ægel* is not evidenced in independent use, but must be postulated for several names, as AYLESFORD &c. Cf. ON *Egill*, Goth *Agil*.

Ainderby Mires YN [*Endrebi* DB, *Aynderby in le Myre* 1499 AD], A~ **Quernhow** YN [*Aiendrebi*, *Andrebi* DB, *Aynderby juxta Querenhou* 1301 Subs], A~ **Steeple** YN [*Eindrebi* DB, *Aynderby wyth Stepil* 1316 FA]. '*Eindriði*'s BY.' ON *Eindriði* is from *Einrǽði* 'sole ruler'.

Quernhow 'mill hill' is a neighbouring hill. Steeple refers to the church spire.

Ainsdale La [*Einuluesdel* DB]. '*Ægenwulf*'s valley.'

Ainstable Cu [*Ainstapelid* 1178 P, *Ainstapelith* 1227 FF]. 'Slope overgrown with bracken (ON *einstapi*).' See HLIP.

Ainsworth La [*Haineswrthe* c 1200 CC, *Aynesworth* 1285 Ass]. See WORP. First el. as in EYNSFORD.

Aintree La [*Ayntre* a 1220 CC]. ON *eintré* 'lonely tree'.

Aire R YW [*Yr* 959 BCS 1052, *Air* c 1160 YD, *Eir* 1175–7 YCh 1626]. Probably OCelt *Isara* 'strong river'. The river name enters into **Airedale**, **Airmyn** [*Ermenie* DB, *Eyreminne* 1100–8 YCh 470], **Airton** [*Airtone* DB]. Airmyn contains OScand *mynni* 'mouth of a river'.

Airyholme YN nr Malton [*Erghum* 1138 Mon], A~ YN in Ayton [*Ergun* DB]. 'The shielings', the dat. plur. of ERG.

Aisby Li (2) [both *Asebi* DB]. '*Asi*'s BY.' *Asi* (DB) is ON *Ási*, ODan *Asi*.

Aisholt So [*Æscholt* 854 BCS 476, *Ascholt* 1197 P]. 'Ash copse.'

Aiskew YN [*Echescol* DB, *Aykescogh* 1235 FF]. OScand *eikiskógr* 'oak wood'.

Aislaby Du [*Aselacby* c 1215 FPD], A~ YN nr Middleton [*Aslachesbi* DB, *Aslakebi* 1167 P]. '*Aslac*'s BY.' *Aslac* (DB, &c.) is ON *Áslákr*, ODan *Aslakr*.

Aislaby YN nr Whitby [*Asulvesby* DB]. '*Asulf*'s BY.' 'First el. ON *Ásulfr*, ODan, OSw *Asulf*.

Aismunderby YW [*Asmundrebi* DB, *Asmundby* 1242 Fees]. '*Asmund*'s BY.' First el. ON *Asmundr*, OSw *Asmunder*. Aismunder- is the OScand gen. *Asmundar*.

Aisthorpe Li [*Æstorp*, *Estorp* DB, *Esttorp* c 1115 LiS]. OE *Ēast-þorp* 'eastern thorp'.

Akeld (ā-) Nb [*Achelda* 1169 P, *Akelde* c 1225 Fees, *Akild* 1242 Fees]. OE *āc-helde* 'oak slope'.

Akeley Bk [*Achelei* DB, *Akileia* c 1155 Oxf, *Akelay* 1175 P]. OE *ācen-lēah* 'oak wood'; first el. OE *ācen* adj. 'of oak'.

Akeman Street, a Roman road [*Accemannestrete* 1151–4 Thame, *Akemannestrete* 1315 BM]. The same first el. is found in an old name of Bath: *Acemannes ceaster* 973 ASC (A), *urbs Achumanensis* 965 BCS 1164, *civitas Aquamania* 972 ib. 1287, and Akeman Street probably means 'the road to Bath'.

Akenham Sf [*Acreham*, *Acheham* DB, *Akenham* 1286 QW]. '*Aca*'s HĀM.'

Alberbury Sa [*Alberberie* DB, *Alberbur'* 1242 Fees]. 'The BURG of *Aluburg* or *Ealhburg*.' Both are known OE woman's names.

Albion [*Albion* Pliny, c 730 Bede], an old name of Great Britain. Generally held to mean 'white island'. Cf. Lat *albus* 'white'.

Albourne Sx [*Aleburn* 1177 P]. OE *alrburna* 'alder brook'.

Albright Hussy Sa [*Etbretone*, *Abretone* DB, *Adbricton* 1242 Fees, *Adbryghton Heose* 1327 Subs], **Albrightlee** Sa nr Albright Hussy [*Etbretelie* DB, *Edbricteleg* 1195 Eyton], **Albrighton** Sa nr Shrewsbury [*Etbritone* DB, *Adbrichton Monachorum* 1255 RH]. '*Éadbeorht*'s TŪN and LĒAH.' Albright is shortened from *Albrighton*. Albrightlee is the LĒAH or wood of the Eadbeorht whose name enters into Albright Hussy.

Hussy (in early records often *Hosatus*) is a Norman surname, originally a byname (OFr *housé* 'booted'). Walter Hussey held the manor c 1165.

Albrighton Sa nr Shifnal [*Albricstone* DB, *Albrictona* 1167 P]. '*Alubeorht*'s (or *Aldbeorht*'s or *Æþelbeorht*'s) TŪN.'

Alburgh Nf [*Aldeberga* DB]. 'Old barrow' or '*Alda*'s barrow'; see BEORG.

Albury Hrt [*Eldeberie* DB, *Audebir* 1230 P], A~ O [*Aldeberie* DB], A~ Sr [*Aldeburi* 675

BCS 39, *Ealdeburi* 1062 KCD 812, *Eldeberie* DB]. 'Old BURG.' '*Ealda*'s BURG' is a possible alternative.

Alby Nf [*Alabei, Alebei* DB, *Alebi* 1195 ff. P]. '*Ali*'s BY.' Cf. AILBY.

Alcaston Sa [*Ælmundestune* DB, *Alghamston* 1327 Subs]. '*Ealhmund*'s TŪN.'

Alcester (awlster) Wa [*Alencestr'* 1165, 1177 P]. 'Roman fort on R ALNE.' A~ Do [*Alyncestre* 1518 BM] was named from A~ Wa.

Alchester O in Wendlebury [*Alavna* c 650 Rav, *Alencestr'* c 1160 PNO(S)]. Brit *Alauna* (cf. ALNE YN), here used of a Roman station, with later addition of OE *ceaster*.

Alciston (ahstn) Sx [*Alsistone* DB, *Alsiestun* 1212 Fees]. '*Ælfsige*'s or *Ealhsige*'s TŪN.'

Alconbury Hu [*Acumesberie* DB, *Alcmundeberi* 1163, *Alchmundesberi* 1168 P], **Alcumlow** Chs [*Alkmundelowe* 14 BM]. '*Ealhmund*'s BURG and hill.' See HLĀW.

Aldborough Nf [*Aldeburg* DB], A~ YW [*Burg* DB, *Aldeburgh* 1316 FA, *Vetus Burgus* 1204 FF]. 'Old fort.' A~ Y is on the site of the Roman *Isurium*. See URE. Cf. ALBURY.

Aldbourne W [(æt) *Ealdincburnan* c 970 Wills, *Aldeborne* DB, *Aldiburna* 1182 P]. 'The stream of *Ealda*'s people.'

Aldbrough YE [*Aldenburg* DB, *Aldaburga* c 1160 YCh 1307], A~ YN [*Aldeburne* DB, *-burg* 1247 Ch], **Aldbury** Hrt [*Aldeberie* DB, *-bur* 1242 Ep]. 'Old fort.' Cf. ALBURY.

Aldcliffe La [*Aldeclif* DB, *-clive* 1212 Fees]. '*Alda*'s cliff.'

Alde R Sf. A back-formation from **Aldeburgh** [*Aldeburc* DB, *-burga* 1198 (1253) Ch], which means 'old fort'.

Aldeby Nf [*Aldebury* DB, *Aldeby* c 1180 Holme, *Aldebi* 13 BM]. Apparently identical with ALDBROUGH, &c., but with OE *-burg* replaced by OScand BY.

Aldenham Hrt [*Ældenham* 785 BCS 245, *Ealdenham* 1066 KCD 824, *Aldenham* 969 Crawf, *Eldeham* DB]. '*Ealda*'s HĀM' or 'old HĀM'.

Alderbury W [(to) *Æþelware byrig* 972 BCS 1286, *Athelwarabyrig* 10 Swithun, *Alwar(es)berie* DB]. 'The BURG of *Æþelwaru*.' This woman's name is not evidenced, but cf. OHG *Adalwara*.

Alderford Nf [*Alraforda* 1163 BM], **Alderholt** Do [*Alreholt* 1314 Ipm], **Alderley** Gl [*Alrelie* DB, *-leg* 1220 Fees]. 'Alder ford, copse, wood or clearing.'

Alderley Chs [*Aldredelie* DB, *Alderdel[ege]* 1275 Cl, *Oure Aldredeleg* 1281 Court]. '*Aldrēd*'s LĒAH.'

Aldermaston Brk [*Ældremanestone* DB, *Aldermannestun* 1167 P], **Alderminster** Wo [*Aldermannestun* 1167 P, *Aldermaston*

1787 PNWo]. 'The TŪN of the *ealdormann*.' A~ Wo shows a remarkable later change. Its old name was (in) *Sture* 972 BM, *Sture* DB.

Aldersey Chs [*Aldrisey* 1284 Ipm, *Alderiseye* 1289 Court]. 'The river land (OE *ēg*) of *Aldhere* or *Æþelrīc*.'

Aldershot Ha [*Alreshete* 1248 Crondal, *-shute* 1316 FA]. 'Alder copse'; see SCĒAT.

Alderton Ess [*Ælwartone, Alwartune* 1062 Th, *Alewardtun* R 1 (1246) Ch, *Alwardeton* 1250 Cl]. '*Ælfweard*'s TŪN.'

Alderton Gl [*Aldritone* DB, *Aldrinton* 1186 P, 1205 Cur], A~ Np [*Aldritone* DB, *Aldrinton* 1186 P], A~ W [*Aldrintone* DB, *Aldrinton* 1195 Cur]. 'The TŪN of *Ealdhere*'s people.'

Alderton Sa [*Olreton* 1309 Ipm], A~ Sf [*Alretuna* DB, *Alretun* c 1150 Crawf]. OE *Alratūn* 'alder TŪN'.

Alderwasley (ăl-) Db [*Alrewaseleg* 1251 Ch, *Alrewasseleye* 1282 FF]. 'LĒAH by an alder swamp'; cf. ALLERWASH, ALREWAS.

Aldfield YW [*Aldefeld* DB]. 'Old field.'

Aldford Chs [*Aldeford* 1253 Ch, 1265-91 Chester]. 'Old ford.'

Aldham Ess [*Aldeham* DB, *Aldenham* 1167 P], A~ Sf [*Aldeham* DB, *Ealdham* DB, c 1095 Bury]. '*Ealda*'s HĀM' or 'old HĀM'. Cf. ALDENHAM.

Aldingbourne Sx [(æt) *Ealdingburnan* c 880 BCS 553, *Aldingeborne* DB]. 'The stream of *Ealda*'s people.'

Aldingham La [*Aldingham* DB, *Aldingeham* 1292 QW]. 'The HĀM of *Alda*'s people.'

Aldington K nr Lympne [*Aldintone* DB, *Ealditun, Ealdintune* 11 DM, *Aldinton* 1197 FF], A~ Wo [*Aldintona* 709 BCS 125, *-tone* DB]. 'The TŪN of *Ealda*'s people.'

Aldon Sa [*Alledone* DB, *Euledon* 1230 P, *Eweldon* 1318 Ipm, *Overe, Nethere Eweledon* 1318 Ch]. OE *æwell* 'source of a river' and *dūn* 'a hill'.

Aldoth Cu [*Aldelathe* c 1230 Holme C]. 'The old lathe or barn.'

Aldreth Ca [*Alreheða* 1170 f., *-huða* 1172 P]. 'Landing-place by the alders.' See HȲÞ.

Aldridge St [*Alrewic* DB, *Alrewyz* 1236 Fees]. 'WĪC among alders.'

Aldringham Sf [*Alrincham* DB, *Alringeham* 1199 P, *Aldringham* 1275 RH]. 'The HĀM of *Aldhere*'s people.'

Aldrington Sx [*Eldretune* DB, *Aldringeton* 1200 Cur]. 'The TŪN of *Ealdhere*'s people.'

Aldsworth Gl [*Aldeswrde* DB, *-wrth* 1291 Tax], A~ Sx [*Aldeswerde* 1271 Ch]. '*Eald*'s WORÞ.' *Eald* is a short form of names in *Eald-*. Cf. AWSWORTH.

Aldwark Db [*Aldwerke* 1140 Derby], A~

YN [*Aldeuuerc* DB], A~ YW [*Aldewerk* 1226 FF]. 'Old fort'; cf. (GE)WEORC.

Aldwick Sx [*Aldewyc* 1236 FF]. 'Old wīc.'

Aldwinkle Np [*Eldewincle, Aldevincle* DB, *Aldewyncl' Sancti Petri* 1254 Val]. '*Ealda*'s nook'. The place is by a nook in a chain of hills. See WINCEL.

Aldworth Brk [*Wurda, Aldewurda* 1167 P, *Aldewrth* 1220 Fees]. Originally *Worþ* (cf. WORþ); the addition is probably the adjective *old*.

Aldthorpe Nf [*Alatorp* DB, *Aletorp* 1180 P]. '*Ali*'s thorp'; cf. AILBY.

Alfington D [*Alfinton* 1244 Ass]. 'The TŪN of *Æþelwulf*'s or *Eanulf*'s or *Ælf*'s people.'

Alfold Sr [*Alfold* 1228 Cl, *Eldefolde* 1257 FF, *Aldefold* 1304 Ep]. 'Old fold.'

Alford Li [*Alforde* DB, *Auford* 1175 P, 1202 Ass]. OE *alr-ford* 'alder ford' or OE *alh-ford* 'ford by a heathen temple'. Cf. *Alkham*.

Alford So [*Aldedeford* DB, *Aldicheford* 1227 FF]. 'The ford of *Ealdgȳþ*' (a woman).

Alfoxton So [*Alfagestone* DB, *Alfexton* 1249 FF]. '*Ælfhēah*'s TŪN.'

Alfreton Db [*Ælfredingtun* 1002 Wills, *Alfreton* 1236 Fees]. 'The TŪN of *Ælfrēd*'s people.'

Alfrick Wo [*Alcredeswike* 13 AD, *Alfrewike* 1275 Ass]. '*Ealhrēd*'s wīc.'

Alfriston Sx [*Alvricestone* DB, *Alfrichestuna* c 1150 Fr]. '*Ælfrīc*'s TŪN.'

Algarkirk Li [*Alfgare* DB, *Algarescherche* 1194 P, *Algercherch* 1202 Ass]. 'The church of *Alger*' (OE *Ælfgār* or ON *Alfgeirr*, ODan *Alfger*). Possibly named from the *Algar comes* mentioned BCS 409.

Alham R So [*Alum* 842 BCS 438]. Identical with ALN. On the stream is **Alhampton** [*Alentona* DB, *Alemton* 1177 P].

Alice Holt Forest Ha [*Alfsiholt* 1169 P, *Alfsiesholt* 1242 Cl]. '*Ælfsige*'s wood.'

Alkborough Li [*Alchebarge* DB, *Alchebarua* c 1115 LiS, *-barue* 1125–8 LN]. First el. may be OE *Al(u)ca* pers. n. The second may be OE *beorg* 'hill' or 'barrow' or *bearu* 'grove'. Often spelt *Hautebarg* &c., owing to association with Fr *haut* 'high'.

Alkerton Gl [*Alcrintone* DB, *-ton* 1220 Fees], A~ O [*Alcrintone* DB, *-ton* 1163 P, *Alkerington* 1259 FF]. 'The TŪN of *Ealhhere*'s people.'

Alkham K [*Ealhham* 11 DM, *Aukeham* 1204 Pp, *Alkam* 1242 Fees]. OE *Ealh-hām* 'HĀM by a heathen temple' (OE *ealh*).

Alkincoats La [*Altenecote* 1201 P, *-cot* 1241 Cl]. Etymology obscure.

Alkington Gl [*Alchmundingtuun, Alhmundingtun* 889 BCS 559 f., *Almintune* DB, *Alkminton* 1194 P], A~ Sa [*Alchetune* DB, *Alkyntone* 1327 Subs]. 'The TŪN of *Ealhmund*'s people.'

Alkmonton Db [*Alchementune* DB, *Alcmunton* 1242 Fees]. '*Alhmund*'s TŪN.'

Alkrington La [*Alkinton* 1212 Fees, *Alkeryngton* 1313 FF]. 'The TŪN of *Alhhere*'s people.'

Allaston Gl [*Alvredestone* DB, *Aluredestona* 1167 P]. '*Ælfrēd*'s TŪN.'

Allen R Co [*Alan* 1199 FF, 1200 P, *Aleyn* 1285 QW]. Identical with ALN.

Allen R Do [*Alen* 1577 Harrison]. A back-formation; cf. *Aldwynesbrigg* 1281 QW, *Aleyn Bridge* Leland (a bridge over the Allen): '*Ealdwine*'s bridge.' The old name was WIMBORNE.

Allen R Nb [*Alwent* 1275 ERN]. See AL-WIN. From Allen are derived **Allendale** [*Alwentedal* 1226 Hexh], **Allendale Town** [*Alewenton* 1245 Ep], **Allenheads**.

Allensmore He [*Mora Alani* 1241 Cl, *Alainesmor* 1265 Ch]. The moor was reclaimed by Alan de Plokenet (13th cent.).

Aller D [*Alre* DB], A~ So [(æt) *Alre* 878 ASC, *Alre* DB]. OE *alor* 'alder'.

Allerby Cu [*Aylewardby* c 1275 StB]. '*Æþelweard*'s BY.' Earlier *Crossebyaylward* 1258, *Aylwardcrosseby* 1260 PNCu(S). '*Æþelweard*'s Crosby.'

Allerdale Cu [*Alnerdall* 11 Gospatric's ch, *Alredala* 1191 P]. 'The valley of the ELLEN' (formerly *Alen*). The name is Scandinavian in form (ON *Alnardalr*).

Allerford So nr Taunton [*Alrford* 882 BCS 550], A~ So nr Minehead [*Alresford* DB]. 'Alder ford.'

Allerston YN [*Alurestan, -stain* DB, *Alverstan* 1208 FF]. '*Ælfhere*'s or *Ælfrīc*'s stone.'

Allerthorpe YE [*Aluuarestorp* DB, *Alwardthorp* 1235 FF]. '*Ælfweard*'s thorp.'

Allerthorpe YN [*Erleuestorp* DB, *Arlathorp* 1270 Ipm, *Arlaugthorpe* 1301 Subs]. The first element may be ON *Arnlaugr*, ODan *Arløgh*.

Allerton La [*Alretune* DB], A~ YW nr Bradford [*Alretune* DB], A~ **Bywater** YW [*Alretune* DB, *Allerton Bywater* 1257 Ipm], **Chapel** A~ & A~ **Gledhow** YW [*Alretun* DB, *Allerton Gledhowe* 1285 FA]. OE *Alratūn* 'alder TŪN'.

A~ **Bywater** is on the Aire.—Gledhow 'kite hill' is a pl. n.

Allerton, Chapel, So [*Aluuarditona, Alwarditone* DB, *Alwareton* 1170 P]. '*Ælfweard*'s TŪN.'

Allerton Mauleverer YW [*Aluretone, Alvertone* DB, *Alverton* 1242 Fees, *Aluerton Mauleuerer* 1231 Ass]. '*Ælfweard*'s or *Ælfrēd*'s TŪN.'

Mauleverer is a Norman nickname and surname meaning 'poor harrier'. Richard Malus Leporarius held *Alvertonia* c 1110 (YCh 729).

Allerwash Nb [*Alrewes* 1202 FF, *-was* 1280 Cl]. 'Alder swamp.' Cf. WÆSSE.

Allesley Wa [*Alleslega* 1176 P, *-leg* 1236 Fees, *Alvesley* 1540 Mon (VE)]. Probably '*Ælle*'s LĒAH' in spite of the 1540 form. A~ is near **Alspath** in Meriden [*Ailespede* DB, *Allespade* 1201 Cur]. '*Ælle*'s path.'

Allestree Db [*Adelardestreu* DB, *Athelardestre* 1208 FF]. '*Æpelheard*'s tree.'

Allexton Le [*Adelachestone* DB, *Adlakestone* 1226 Ep]. '*Éadlāc*'s TŪN.'

Allington, East, D [*Alintone* DB, *Allyngton* 1242 Fees], A~ K nr Maidstone [*Elentune* 11 DM, *Elentun* DB], A~ W nr Chippenham [*Allentone* DB, *Alinton* 1178 P, *Alynton* 1316 FA]. 'The TŪN of *Ælla*'s or *Ælle*'s people.'

Allington Do [*Adelingtone* DB, *Athelington* 1227 FF], A~ Li [*Adelingetone* DB, *Adelington* 1228 Ep], A~ W nr Devizes [*Adelingtone* DB, *Alingeton* 1195 Cur]. 'The TŪN of the æthelings or princes', or 'the TŪN of *Æpelhēah*'s people'.

Allington Ha [*Ellatune* DB, *Aldinton* 1187 ff. P], A~ W nr Amesbury [*Aldintona* 1178 BM]. 'The TŪN of *Ealda*'s people' or 'old TŪN'.

Allington K nr Lenham [*Alnoitone* DB, *Alnodentune* 11 DM, *Eilnothinton* 1242 Fees]. 'The TŪN of *Æpelnōp*'s people.'

Allithwaite La [*Hailiuethait* c 1170 FC]. 'The thwaite of *Eilífr*', a Norseman (ON *Eilífr*).

Allonby Cu [*Aleynby* 1274 Cl]. '*Aleyn*'s BY.' *Aleyn* is a Norman name.

Allostock Chs [*Allostok* 1312 Misc]. Possibly 'Old Lostock'. A~ is c 4 m. from Lostock Gralam.

Allowenshay So [*Aylwynesheye* 1315 Ipm]. '*Æpelwine*'s enclosure.' See (GE)HÆG.

Allscot Sa nr Wellington [*Aldedriscotam* 1176 Eyton, *Alderescote* 1291 Tax]. '*Aldrēd*'s COT.' A~ Sa nr Bridgnorth [*Eluescota* 1177 P, *Alvescote* 1448 AD]. '*Ælfwulf*'s, *Ælf*'s, *Æpelwulf*'s or *Éanwulf*'s COT.'

Allt Bough. See ALT.

Almeley He [*Elmelie* DB, *Elmel*' 1231 Cl]. OE *elmlēah* 'elm wood'.

Almer Do [*Almere* 1212 Fees]. OE *ælmere* 'eel lake'. The lake is *Elmere* 943 BCS 781.

Almholme YW [*Almeholme* c 1235 Selby]. 'Elm island.' First el. OScand *almr* 'elm'.

Almington St [*Almentone* DB, *Alcminton* 1242 Fees]. '*Alhmund*'s TŪN' or 'the TŪN of *Alhmund*'s people'.

Almondbury (āmbrī) YW [*Almaneberie* DB, *Almannebire* 1230 Ep]. Perhaps 'the BURG of all the men'; cf. ON *almannaping* 'assembly of all the men'.

Almondsbury Gl [*Almodesberie* DB, *-bure* 1154, *Alemundebere* 1233 Berk, *Almodebiria* 1221 Ass, *Almundesbur* 1285 Ch]. '*Æpelmōd*'s or *Ealhmund*'s BURG.'

Aln (āl, ǎln) R Nb [*Alaunos* c 150 Ptol, *Alne* c 730 Bede]. A British river-name. On the Aln are **Alnham** (ǎln-) [*Alneham* 1228 FPD], **Alnmouth** (āl-) [*Alnemuth* 1201 Ch], and **Alnwick** (ǎnīk) [*Alauna* c 150 Ptol, *Alnewich* 1178, *-wic* 1181 P].

Alne (awn) R Wa [*Æluuinnæ* c 730 BCS 157]. A British river-name, meaning 'very white' (cf. Welsh *gwyn* 'white'). The river gave its name to **Great Alne** vil. [*Alne* DB] and to ALCESTER.

Alne (awn) YN [*Alna* c 1050 HSC, 1230 Ep, *Alne* DB]. Probably a British *Alauna*, related to *Alaunos* (see ALN), but here used of a forest.

Alnesbourn Sf [*Aluesbrunna* DB, *Alnesburn* 1250 Cl]. '*Ælfwine*'s stream.'

Alney Gl [*Olanig* 1016 ASC D, E]. Identical with OLNEY Bk.

Alnham, Alnmouth, Alnwick. See ALN.

OE *alor* 'alder', a common element. OE *alor* sg. is the source of ALLER D, So, ARLE Gl. For *alor* as second el. cf. BICKNOLLER, LONGNER, LONGNOR; as first el. e.g. ALFORD, ALDERFORD, -LEY, -SHOT, -TON, ALDRIDGE, ALLERFORD, -WASH, AUB(O)URN, OLLERTON, ORLETON.

Alperton Mx [*Alprinton* 1199 FF, *Alpertone* 1282 PNMx(S)]. First el. probably a derivative in *-ingas* of a pers. n. in *Ealh-*, e.g. *Ealhburg*. Cf. ALPRAHAM.

Alphamstone Ess [*Alfelmestuna* DB, *Alfhameston* 1238 Subs]. '*Ælfhelm*'s TŪN.'

Alpheton Sf [*Alflede(s)ton* 1204 Cur, *Alfleton* 1254 Val]. 'The TŪN of *Ælflēd*.' *Ælflēd* is a woman's name.

Alphington D [*Alfintun* c 1060 E, *Alfintone* DB, *-ton* 1232 Cl]. 'The TŪN of *Ælfwulf*'s or *Ælf*'s people.'

Alpington Nf [*Alcmuntona, Algamundestuna* DB]. '*Alhmund*'s TŪN.'

Alport Db [*Aldeport* 1276 RH]. 'Old town'; see PORT.

Alpraham Chs [*Alburgham* DB, *Alpram* 1259 Court]. 'The HĀM of *Alhburg*' (a woman).

Alresford (ahlz-) Ess [*Ælesford* c 995 BCS 1289, *Eiles-, Elesforda* DB]. '*Ægel*'s ford'; cf. AILSWORTH.

Alresford (awls-) Ha [*Alresford* 701 BCS 102, *-e* DB]. 'Alder ford.' The river-name **Alre** is a back-formation.

Alrewas (awlras) St [*Alrewasse* 942 BCS 771, *-was* DB]. 'Alder swamp'; see WÆSSE.

Alsager (awlsajer) Chs [*Eleacier* DB, *Alisacher* 1288 Court]. '*Ælle*'s field.' See ÆCER.

Alscot Gl [*Ælfsiescota* 1188, 1190 P]. '*Ælfsige*'s COT.'

Alsop en le Dale Db [*Elleshope* DB, *Aleshop* 1241 RA]. '*Ælle*'s valley.' See HOP.

Alspath. See ALLESLEY.

Alston Cu [*Aldeneby* 1164–71 CWNS xi, *Aldeneston* 1210 CWNS xi, 1254 Val, *Aldeniston* 1232 Ch]. Perhaps '*Aldhūn*'s BY and TŪN'.

Alston D in Malborough [*Alwinestone* DB, *-ton* 1242 Fees]. '*Ælfwine*'s TŪN.'

Alston La [*Halfiston*, *Alleston* 1246 Ass]. '*Ælf*'s or *Ælfsige*'s TŪN.'

Alston Sutton So [*Alnodestuna* DB]. '*Æþelnōþ*'s TŪN.'

Alstonby Cu [*Astaneby* 1277 Ipm]. '*Astin*'s BY.' *Astin*, a form of *Asketin* (ON *Asketill*), is found in Cu and We in the 13th cent.

Alstone Gl [*Ælfsigestun* 969 BCS 1233, *Alsiston* 1221 Ass]. '*Ælfsige*'s TŪN.'

Alstone St [*Aluerdestone* DB, *Aluredeston* 1194 ff. P]. '*Ælfrēd*'s TŪN.'

Alstonfield St [*Ænestanefelt* DB, *Alfstanesfeld* 1179 P]. '*Ælfstān*'s FELD.'

Alt R La [*Alt* c 1190 CC, 1292 Ass]. A British river-name, derived from the root of Lat *palus*: 'muddy river'. **Altcar** [*Acrer* DB, *Altekar* 1251 FF]. 'Marsh on R Alt.' See KERR.

Alt La nr Manchester [*Alt* c 1200 LaCh]. Welsh *allt* 'a hill'. **Allt Bough** He [*Altebogh* 1474 BM]. 'Small hill' (Welsh *allt bach* or *bychan*). **Alt Went** He [*Altegwynt* 1419 BM] may be 'windy hill' (Welsh *gwynt* 'wind').

Altarnun Co [*Altrenune* c 1100 Montacute, *Alternon* c 1235 Ep]. 'The altar of St. Nonn', a British female saint.

Altcar. See ALT.

Altham La [*Elvetham* c 1150 Whitaker, *Alvetham* 1242 Fees]. 'HAMM inhabited by swans' (OE *elfetu*, *ælfetu*).

Althorne Ess [*Aledhorn* 1197 FF, *Alethorn* 1203 FF, 1208 Cur]. If the first form is trustworthy, the first el. seems to be OE *æled* 'fire'. The name would mean 'thornbush by a beacon'.

Althorp Np [*Olletorp* DB, *Olethorp* 1208 Cur]. '*Olla*'s thorp.' Cf. OLNEY.

Althorpe Li [*Aletorp* DB, *-thorp* 1234 Ep]. '*Ali*'s thorp.' Cf. AILBY.

Altofts YW [*Altoftis* Hy 2 (1230) Ch]. 'Old tofts.'

Alton Db nr Wirksworth [*Alton* 1296 Ipm]. **Alton Grange** Le [*Heletone* DB, *Alton* c 1125 LeS] may be OE *Aldatūn* 'old TŪN.'

Alton Pancras Do [(æt) *Awultune* 1002–12 ASWrits No. 63, *Awoltona* c 1160 BM, *Aulton Pancras* 1412 FA], A~ Ha [*Aultone* DB, *Aweltona* 1175 P], **A~ Barnes & Priors** W [*Aweltun* 825 BCS 390, *Aulton prioris* 1199 P, *Aweltun Berner*' 1275 RH]. 'The TŪN at the source (OE *ǣ-*, *āwiell*)' of the rivers Piddle, Wey, and Avon respectively.

A~ Pancras from the dedication of the church. **—A~ Barnes** and **Priors** were held by the Berners family (from BERNIÈRES in Normandy) and by the priory of St. Swithun, Winchester.

Alton St [*Elvetone* DB, *Alveton* 1283 Misc]. '*Ælfa*'s TŪN.'

Alton W nr Amesbury [*Eltone* DB, *Alletona* 1281 BM]. '*Ælla*'s TŪN.'

Alton Wo [*Eanulfintun* 1023 KCD 738, *Alvintune* DB]. 'The TŪN of *Ēanwulf*'s people.'

Altrincham (-nj-) Chs [*Aldringeham* 1290, *Aldringham* 1318 Ch, *Altrincham* 1386 AD]. 'The HĀM of *Aldhere*'s people.'

Alvanley Chs [*Elveldelie* DB, *Alvaldeleh* c 1220 BM]. '*Ælfwald*'s LĒAH.'

Alvaston Chs [*Alwaldeston* 13 BM]. '*Ælfwald*'s TŪN.'

Alvaston Db [*Alewaldestun* 1002 E, *Ælwoldestun* DB]. '*Alhwald*'s TŪN.'

Alvechurch (awlcherch) Wo [*Ælfgyðe cyrcan* 10 BCS 1320, *Ælfiðe cyrce* 11 Heming]. 'The church of *Ælfgyþ*' (a woman).

Alvecote Wa, an old priory [*Avecote* c 1160 PNWa(S)]. Called *Aucot* by Tanner (1695). '*Afa*'s COT.'

Alvediston W [*Alfwieteston* 1167 P, *Avitheton* 1203 Cur, *Alvitheston* 1222 FF]. Perhaps '*Ælfgēat*'s TŪN.'

Alveley Sa [*Alvidelege* DB, *Aluitheleg* 1195, *Alfithelea* 1196 P]. 'The LĒAH of *Ælfgyþ*' (a woman).

Alverdiscott D [*Alveredescote* DB]. '*Ælfrēd*'s COT.'

Alverstoke Ha [*æt Stoce* 948 BCS 865, *Alwarestoch* DB, *Alvardstok* 1316 FA]. '*Ælf-* or *Æþelweard*'s STOC.'

Alverstone Wt [*Alvrestone* DB, *Alvredeston* 1287–90 Fees]. '*Ælfrēd*'s TŪN.'

Alverthorpe YW [*Alvelthorpe* 1199 (1232) Ch, *Alvirthorpe* 1274 Wakef]. '*Ælfhere*'s thorp.'

Alverton Co in Madron [*Alwaretone* DB, *Alwarton*' 1229 Fees, *Alverton* 1303 FA]. '*Ælfwaru*'s TŪN' probably, in spite of the fact that the manor was held by one *Aluuard* before the Conquest (DB).

Alverton Nt [*Aluriton* DB, *Alurinton* 1221–30 Fees]. 'The TŪN of *Ælfhere*'s or *Ælfrīc*'s people.'

Alvescot (awls-) O [*Elfegescote* DB, *Ælfegescota* 1187 P]. '*Ælfhēah*'s COT.'

Alveston Gl [*Alwestan* DB, 1130 P, *Aloestan* 1156 P]. '*Alwīh*'s stone.'

Alveston Wa [*Eanulfestun* 966 BCS 1182, *Alvestone* DB]. '*Ēanwulf*'s TŪN.'

Alvingham Li [*Aluingeham* DB, 1218 Ass, *Alwingham* 1200 Cur], **Alvington, West,**

D [*Alvintone* DB, *Alfington* 1242 Fees]. 'The HĀM and TŪN of *Ælfwine*'s or *Ælf(a)*'s or *Ēanwulf*'s people.'

Alvington Gl [*Alwintone* 1221 Ass, *Alfinton* 1228 Cl]. Perhaps '*Ælfwynn*'s TŪN'. *Ælf-wynn* is a woman's name.

Alwalton Hu [*Æþelwoldingtun* 955 BCS 909, *Alwoltune* DB, *Aðelwoltun* 1158 P]. 'The TŪN of *Æþelwald*'s people.'

Alwin R Nb [*Alewent* c 1200, *Alwent* c 1240 Newminster]. A British river-name. Hence Alwinton Nb [*Alwenton* 1242 Fees, 1254 Val].

Alwington D [*Alwinetona* DB, *Alwynton* 1242 Fees]. 'The TŪN of *Ælfwynn*' (a woman).

Alwoodley YW [*Aluuoldelei* DB, *Adelwaldesleia* 1166 P]. '*Æþelwald*'s LĒAH.'

Amaston Sa [*Enbaldestune* DB, *Emboldeston* 1208 FF], Ambaston or Embaston Db [*Emboldestune* DB, *Ambaldestone* 1219 FF]. '*Ēanbald*'s TŪN.'

Amber R Db [*Ambre* 1191–9 BM]. A British river-name meaning 'the river' and related to Sanskr *ambhas* 'water', Lat *imber* 'shower'. Ambergate is on the river.

Amberden Ess nr Debden [*Amerdene* c 1050 KCD 907, *Amberdana* DB, *-den* 1176 P], Amberley Gl [*Amberlegh* 1242 Ass], A~ Sx [*Amberle* 957 BCS 997 (late copy), *Ambrelie* DB]. First el. very likely OE *amore* the name of a bird; cf. G *Ammer* and *yellowhammer*. 'Valley and wood frequented by this bird.'

Amberley He [*Amburlege* DB, *Amberleye* 1248 Ipm, *Aumbresle* 1243 Ch]. Perhaps 'the LĒAH of *Ēanburg*' (a woman).

Ambersham Sx [*Æmbres-*, *Embresham* 963 BCS 1114, *Ambresham* 1166 P]. '*Embre*'s HĀM.' This unrecorded pers. n. may be derived from that found in AMESBURY, OMBERSLEY.

Amble Nb [*Ambell* 1204 Ch, *Ambbill* a 1216 Percy, *Anebell* 1256 Ass, *Alnebill* 1279 Ass]. '*Anna*'s promontory.' Second el. *bill*, as in PORTLAND BILL, &c.

Amblecote St [*Elmelecote* DB, *Emelecote* 1236, *Amelecot* 1242 Fees]. Perhaps '*Æmela*'s COT'. For *Æmela cf. OE *Æmele* pers. n.

Ambleside We [*Amelsate* 1275 Ch, Cl]. 'Shieling (ON *sætr*) on a sandbank by a river.' First el. ON *ámelr* from *á* 'river' and *melr* 'sandbank'.

Ambrosden O [*Ambresdone* DB, *-don* 1239 Ep]. '*Ambr*'s hill' (cf. AMESBURY).

Amcotts Li [*Amecotes* DB, *Ammecotes* 1230 P]. See COT. First el. a pers. n. *Amma* cognate with OHG *Amo* &c., and presupposed by several place-names.

Amersham Bk [*Agmodesham* 1066 KCD 824, *Elmodesham* DB, *Augmodesham* 1197 FF, *Aumodeshame* 1222 Ep, *Amundesham* 1227 Ass]. '*Ealgmund*'s HĀM.'

Amerton St [*Aunbriton, Ambrihiton* 1230 P, *Ambricton* 1251 Ch]. '*Ēanbeorht*'s TŪN.'

Amesbury (ā-) W [(æt) *Ambresbyrig* c 880 BCS 553, c 1000 Saints, *-burch* 932 BCS 691]. '*Ambr*'s BURG.' This pers. n. may be postulated on the strength of AMBROSDEN, AMESBURY, OMBERSLEY, and OHG *Ambricus, Ambrico* pers. n.

Amington Wa [*Aminton* 1197 FF, 1212 Cur]. 'The TŪN of *Amma*'s people'; cf. AMCOTTS. But the forms *Ermendone* DB, *Arminton* 1221 PNWa(S), if trustworthy, may indicate that the original first el. was OE *Earnmund* (cf. APPLETON Brk), which interchanged with a short-name *Eamma*.

Amotherby YN [*Aimundrebi* DB]. An ON *Eymundarbýr* 'the BY of *Eymundr*'.

Amou·nderness La [*Aghemundesnes* 930 YCh 1, *Agemundrenesse* DB]. An ON *Agmundar-* or *Ǫgmundarnes* 'the headland of one *Agmundr* or *Ǫgmundr*'.

Ampleforth YN [*Ampreforde* DB, *Ampleford* 1167 P]. 'Ford where sorrel (OE *ampre*) grew.' The change of *r* to *l* is due to dissimilation.

Ampney Crucis, St. Mary, St. Peter, Down A~ Gl [*Omenie* DB, *Amenell, Dunamenell* 1205 Cur, *Ameneye Sancte Crucis* 1287 QW, *Amenel Sancti Petri* c 1275 Glouc, *Ammeneye Beate Marie* 1291 Tax]. Really a name of Ampney Brook, an OE *Amman-ēa* '*Amma*'s stream'. See AMCOTTS.

A~ Crucis is said to be named from a cross in the churchyard.—A~ St. Peter belonged to St. Peter's, Gloucester.—A~ St. Mary owing to the dedication of the church.—Down A~ is lower down Ampney Brook.

Amport Ha. See ANN.

Ampthill Bd [*Ammetelle* DB, *Amethull* 1202 Ass]. OE *æmethyll* 'ant-heap' or 'ant-infested hill'.

Ampton Sf [*Hametuna* DB, *Ametune* c 1095 Bury, *Ameton* 1196 FF]. '*Amma*'s TŪN.' See AMCOTTS.

Amwell Hrt [*Emmeuuelle* DB, *Eme(s)welle* R 1 Cur]. Apparently '*Emma*'s stream or spring'. Great A~ is on the Lea. *Emma (*Æmma*) may be a cognate of *Amma*.

OE *ān* 'one', *āna* 'lonely'. See ANCOATS, ONE- (passim), OLNEY Np, ONLEY, ONSTON.

Ancaster Li [*Anecastre* Hy 2 DC, 1196 FF]. 'The Roman fort of *Anna*.'

Ancholme (ăngk-) R Li [*Oncel* c 1000 Saints, *Ancolna* c 1155 DC]. A compound of a British *an-* 'marsh' (cf. Gaul *anam* 'marsh', Ir *an* 'water', Engl *fen*) and the river-name COLNE (1).

Ancoats La [*Einecot* 1212 Fees, *Hanekotys* 1242 Fees]. 'Lonely huts.' First el. OE *āna* 'lonely'.

Ancroft Nb [*Anecroft* 1195 (1335) Ch, 1208–10 Fees, *Ane(s)croft* 1254 Val]. 'Lonely croft.'

Anderby Li [*Andreby* c 1135 RA, -*bi* 12 BM]. 'The BY of *Eindriði* (cf. AINDERBY) or *Arndor*.' ODan *Arndor* (*Andor*) corresponds to ON *Arnþórr*.

Anderson Do. See WINTERBOURNE ANDERSON.

Anderton Chs [*Andrelton* 1183, *Anderton* 1185, *Enderton* 1186 P], A~ La [*Anderton* 1212 Fees]. '*Ēanrēd*'s TŪN.'

Andover Ha [*Andeferas* 955 BCS 912, (in) *Andeferan* 10 Swithun, *Andovere* DB]. From an old name of the ANTON, containing the river-name *Ann* (v. ANN) and a derivative of Brit *dubro*- 'water, river' (Welsh *dwfr* &c.). **Andoversford** Gl [*Onnanford* 759 BCS 187, *Andovre* 13 PNGl] is '*Anna*'s ford', later associated with ANDOVER Ha.

Andred (Forest) K, Sx [*Andredesleage* 477, *Andred* 755, 893 ASC, *Andredes Weald* 1018 BM, *Andret* DB]. Originally the name of a place [*Anderitos* or -*ridos* c 425 ND, *Andredescester* 491 ASC], generally identified with Pevensey. The name consists of the Brit prefix *ande*- and *ritu*-, Welsh *rhyd* 'ford'. Jackson renders it 'the great fords'.

Andwell Ha [*Hænedewella* c 1150 Fr, *Enedwelle* 1154-72 Oxf]. 'Duck stream' (first el. OE *æned* 'duck').

Angersleigh (änj-) So [*Lega* DB, *Legh Militis* 1290 Ep, *Aungerlegh* 1354 Ep]. Originally OE *Lēah*, later Angersleigh from an early owner. John Aungier held the manor before 1290 (Ep). *Aunger* is a Norman Christian name from OHG *Ansger*.

Angerton La [*Angertona* 1293 FC], A~ Nb [*Angerton* 1187 P, 1242 Fees]. 'Grazing-farm.' First el. an OE *angr* corresponding to OHG *angar*, G *anger* 'grassland' and found also in ANGRAM and ONGAR. In Angerton La ON *angr* 'bay' is a possible alternative.

Anglesey Ca, an old monastery [*Angleseia* 1242 Fees, *Angleseya* 1254 Val]. Possibly 'the island of the Angle'.

Anglezark La [*Andelevesarewe* 1202 FF, *Anlauesargh* 1224 FF]. 'The ERG or shieling of *Anlaf*.' OE *Anlaf* is from an early form of ON *Óláfr*.

Angmering Sx [(æt) *Angemæringum* c 880 BCS 553, *Angemare* DB, *Hangemera* 1176 P]. OE *Angemæringas* '*Angenmær*'s people'. OE *Angenmær* is unrecorded, but cf. *Angenlāf*.

Angram YN, YW [*Angrum* 13]. OE *angrum*, dat. plur. of *angr* 'grazing-land'. See ANGERTON.

Anick Nb [*Æilnewic* c 1160 Hexh, *Einewik* 1225 Ep]. Perhaps the WĪC of *Egelwin*, bishop of Durham in the 11th cent.

Anker R Wa [*Oncer* c 1000 Saints]. A British river-name meaning 'winding river' and

related to Gaul *anco*- 'hook', Lat *ancræ* 'valley'.

Ankerwyke Bk [*Ankerwich* 1182 P, -*wic* 1194 BM]. 'The nunnery', an OE *ancorwīc* (cf. OE *ancorsetl* 'hermitage', ME *ancre* 'nun').

Anlaby YE [*Umlouebi* DB, *Anlaweby* 1234 Cl]. '*Anlaf*'s BY.' ON *Óláfr* appears as OE *Anlaf, Onlaf, Unlaf*.

Anmer Nf [*Anemere* DB, -*mera* 1177 P, *Anedemere* 1291 Tax]. OE *æned-mere* 'duck mere'.

Ann, Abbotts & Little, Amport, Sarson Ha [*Anne, Annæ* 901 BCS 597, *Anna, Anne* DB, *Anne Abbatis, Anna de Port* c 1270 Ep, *Anna Savage* 1242 Fees. Ann is an old name of the brook that joins the Anton near these places, and very likely of the Anton itself; cf. ONNY.

Abbotts Ann belonged to Newminster Abbey, Winchester.—**Amport** was held by Adam de Portu in 1199 (Cur), **Sarson** 'Savage's ANN' by Richard le Salvage in 1203 (Cur). Cf. also MONXTON and NEEN SAVAGE.

Annaside Cu [*Ainreseta* c 1140 StB, *Aynerset* 1242 FF]. '*Einar*'s SÆTR or shieling.' ON *Einarr* is a common name.

Annesley Nt [*Aneslei* DB, -*lea* 1175 P]. May be '*An*'s LĒAH.' **An* pers. n. would belong to *ān* 'one' and seems to be found in **Onesacre** YW [*Anesacre* DB].

Anningsley Sr [*Annyngelegh* 1324 Ipm], **Annington** Sx [*æt Anningadune* 956 BCS 961, *Annigedona* 1153 Oxf]. 'The LĒAH and the DŪN of *Anna*'s people.'

Ansford So [*Almundesford* DB, 1219 FF]. '*Ealhmund*'s ford.'

Ansley Wa [*Hanslei* DB, *Anesteleye* 1325 Misc, *Ansteley* 1416 BM], **Anslow** St [*Ansedl*[*ega*] c 1180 Fr, *Ansedelee* 1300 BM]. OE *ānsetllēah* 'LĒAH with a hermitage' (OE *ānsetl*). Cf. *Ansætleh* c 972 BCS 1278 (YW).

Anstey, East & West, D [*Anestige, Anestinge* DB, *Estanesty* 1263 Ep, *Westanostige* 1234 Bract], A~ Do [*Anstigan* 942 BCS 775], A~ Ha [*Hanstige* DB, *Anesti* 1200 Cur], A~ Hrt [*Anestige* DB], A~ Le [*Anstige* DB, *Anesti* 1209-35 Ep], A~ W [*Anestige* DB, *Anestig* 1242 Fees]. OE *ānstiga* 'narrow footpath', probably esp. one up a hill. The Ansteys are generally on hills or on hill slopes.

Anston, North & South, YW [*Anestan, Litelanstan* DB, *Anestan* 1199 (1232) Ch, *Annestan* c 1180 YCh 1412, *North-, Suthanstan* 1297 Subs]. May be 'single stone' (OE *āna stān*) or '*Anna*'s stone'.

Ansty Wa [*Anestie* DB]. Identical with ANSTEY.

Anthorn Cu [*Aynthorn* 1289 Cl, 1332 Subs]. 'Single thorn-bush' (ON *einþorn*).

Antingham Nf [*Antingham* 1044–7 KCD 785, DB, *Antigeham* DB, *Entingeham* 1264 Ch]. 'The HĀM of *Anta*'s people.' *Anta* is found in *Antan hlaw* BCS 246. The river-name **Ant** is a back-formation.

Anton R Ha. A modern name due to wrong identification of Tacitus's corrupt *Antona* for *Trisantona* with ANDOVER.

Antony Co [*Antone* DB, *Anton* 1289 FF]. Perhaps '*Anna*'s or *Anta*'s TŪN'.

Antrobus Chs [*Entrebus* DB, *Anterbus* 1282 Ch]. Unexplained. Hardly English.

Anwick (ănĭk) Li [*Amuinc*, *Haniwic* DB, *Amewic* 1218 Ass, *-wyk* 1250 FF]. '*Amma*'s wīc.' Cf. AMCOTTS.

Apedale Hall St [*Apedal* 1277 Misc, *-e* 1283 Ipm]. Possibly OE *æppeldæl* 'apple valley' with loss of *l* owing to dissimilation. Or the first el. may be as in next name.

Apethorpe Np [*Patorp* (sic) DB, *Apetorp* 1163 P, *Appetorp* 1167 f. P]. '*Api*'s thorp.' First el. ODan, OSw *Api* pers. n.

Apeton St [*Abetone* DB, *Abbeton* 1242 Fees, *Abbenton* 1253 Ass]. '*Abba*'s TŪN.'

Apley Li [*Apeleia* DB, c 1115 LiS], **A~** Sa [*Eppelle* c 1195 Eyton, *Appeleg* 1242 Fees], **A~** So [*Appelie* DB], **A~** Wt [*Appelea* 1190 P]. OE *æppel-lēah* 'apple wood'.

Apperknowle Db [*Apelknol* 1317, *Appur-knoll* 1467 Derby]. OE *apuldor-cnoll* 'apple-tree hill'.

Apperley Gl [*Apperleg* 1212 Fees, *-leye* 1327 Subs], **A~ Bridge** YW [*Apperley* 1279 Ep]. OE *apuldor-lēah* 'apple-tree wood'. **Apperley** Nb [*Appeltreleg* 1201 FF, *Apil-treley* 1242 Fees]. Identical in meaning, but the first el. is OE *æppeltrēo*.

Appleby, Magna & Parva, Le [*Æppelby* 1002 E, *Aplebi* DB, *Parua Appilby* 1327 Subs], **A~** Li [*Aplebi* DB, 1130 P, *Appelbi* 1167 P], **A~** We [*Aplebi* 1130, *Appelbi* 1190 ff., *Vetus Appelbi* 1198 P, *Eppelbi* R 1 (1308) Ch]. 'Apple BY.' *Apple* is ON *epli*, OSw *æple*. So Appleby contains OE *æppel*, which may have replaced earlier Scand *epli*. Or Appleby may be a Scandinavianized form of APPLETON.

Appledore D nr Bideford [*le Apildore* 1335 AD], **A~** K [(æt) *Apuldre* 893 f. ASC, *Apeldres* DB]. OE *apuldor* 'apple-tree'.

Appledram Sx [*Apeldreham* 1126–33 AC, *Appeltrieham* 1198 P]. 'HĀM or HAMM with apple-trees' (OE *apuldor*).

Appleford Brk [*Æppelford* c 895 BCS 581, 940 ib. 760, *Apleford* DB]. 'Ford by which apples grew.' **A~** Wt [*Apledeforde* DB, *Appeltreford* 1287–90 Fees]. 'Ford by the apple-tree(s).'

Applegarth YN [*Appelgard* c 1160 Riev, *-garth* 1228 FF]. ON *apaldgarðr* 'apple-orchard'.

Applesham Sx [*Aplesham* DB, 1349 Ipm, *Appulham* 1371 Ipm]. Probably from OE

Æppel-hām (or *-hamm*) with later addition of a medial *s*.

Appleshaw Ha [*Appelsag* 1200 Cur, *-shawe* 1284 Ipm]. Self-explanatory.

Applethwaite Cu [*Appelthweit* 1223 P], **A~** We [*Applethwayt* 1256 Kendale]. 'Thwaite or clearing where apples grew.'

Appleton Brk [*Ærmundeslea* or *Æppeltun* 942 BCS 777, *Apletune* DB], **A~** Chs [*Epletune* DB], **A~** Cu [*Appelton in Allerdale* 1229 Ch], **A~** K [*Apletone* DB, *Apelthone* 1242 Fees], **A~** La [*Apelton* 1182 P], **A~** Nf [*Appletuna* DB], **East & West A~** YN [*Apelton* DB], **A~** le **Moor** YN [*Apeltun* DB], **A~** le **Street** YN [*Aple-*, *Apeltun* DB], **A~ Wiske** YN [*Apeltona*, *Apletune* DB], **Nun A~** & **A~ Roebuck** YW [*Æppel-tune* c 972 BCS 1279, *Apleton* DB]. OE *Æppeltūn* 'TŪN where apples grew'. OE *æppeltūn* is recorded in the sense 'orchard'.

Nun A~ was the seat of a nunnery founded in the time of Stephen (cf. *dominæ de Apeltun* c 1180 DC).—A~ **Roebuck** was apparently named from an early owner; the name *Rabuk* occurs in the Poll Tax of 1379 under Appleton. —A~ le **Street** is on a supposed Roman road. —A~ Brk was originally *Earnmundeslēah* '*Earn-mund*'s LĒAH'.

Appletree Np [*Apeltreya* 12 NS, *Appeltre* 1176 P]. OE *æppeltrēo* 'apple-tree'.

Appletreewick YW [*Apletrewic* DB]. 'The wīc by the apple-tree(s).'

Appley Bridge La [*Appelleie* 13 CC]. See APLEY.

Apps Court Sr [*Epse* 675 BCS 39, *Æpse* 1062 KCD 812, *Ebsa* DB]. OE *æspe*, *æpse* 'aspen'.

Appuldurcomb Wt [*Apeldurecumbe* 13 BM, *Appeldrecumbe* 1330 Ch]. 'Apple-tree coomb.' See APULDOR.

Apsley End Bd [*Aspele* 1253 Ch]. OE *æsp-lēah* 'aspen wood'.

OE **apuldor** 'apple-tree'. See APPERKNOWLE &c., APPLEDORE, APPLEDRAM, APPLEFORD. The cognate ON **apaldr** occurs in APPLE-GARTH.

Aqualate Mere St [*water of Mere* 1227 Ass]. OE *mere* 'lake'. The mere was named from a place **Aqualate** [*Aguilade* 1227 Ass, *Akilot* 1275, *-e* 1282 Ipm, *Aquilot* 1327 Subs]. Apparently an OE *āc-gelād* 'oak str:am', really the name of a stream.

Arborfield Brk [*Edburgefeld* 1220, *Erburge-feld* 1222 Sarum, *Hereburgfeld* 1230 P]. Perhaps 'the FELD of *Hereburg*' (a woman). Or the first el. may be OE *eorþburg*; cf. ARBURY.

Arbury La [*Erthbury* 1246 FF], **A~** Wa [*Ordburi* Hy 2, *Erdburia* 13 Mon]. OE *eorþburg* 'earth fortification'. The same is the origin of **Arbury** Ca, Hrt, the name of Roman camps.

Archenfield (deanery) He [*Ircingafeld* 918 ASC, *Arcenefelde* DB, *Erchenefeld* 1138 AC]. The first el. is the Welsh name of the district (*Erchin, Ercincg* &c. c 1150 LL), which was named from *Ariconium* (IA), a place somewhere in Dean Forest. The etymology of the name is obscure.

Ardeley or **Yardley** Hrt [*Eardeleage* 939 BCS 737, *Erdelei* DB, *Erdelegh* 1241 Ep]. '*Earda*'s LĒAH.' OE **Earda* is a regular short form of *Eardwulf. Eardeleage* 939 is in a copy of the charter.

Arden Chs [*Arderne* 1260, *Ardren* 1288 Court], A~ Wa [*Eardene* W 1 Abingd, *Ardena* 1130 P, *Arderne* 1200 Cur, 1236 Fees]. Perhaps OE **eardærn* 'dwelling-house'; cf. *eardwīc* 'dwelling-place'. The loss of the second *r* is due to dissimilation.

Arden YN [*Ardene* DB, *Erdene* 12 Riev, *Arden* 1244 Cl]. Second el. OE *denu* 'valley'. First el. OE *ēar* 'gravel' (cf. ERITH) or *earn* 'eagle'.

Ardingly· Sx [*Erdingelega* 1107–18 AC, *-lege* 1205 FF, *Herdingheleia* 1121 AC]. 'The woodland (LĒAH) of *Eorēd*'s people.'

Ardington Brk [*Æþeredingetun* 961 BCS 1079, *Ardintone* DB, *Ærdinton* 1192 P]. 'The TŪN of *Æþelrēd*'s people.'

Ardleigh Ess [*Erleia, -lega* DB, *Erdlega* 1170, *Ardlega* 1194 P, *Erdelega* 1195 P]. Hardly identical with ARDELEY. First el. may be OE *eard* 'dwelling-place' or OE *erþ* 'ploughing'. For *d* from *þ* cf. HADLEIGH. See LĒAH.

Ardley O [*Eardulfes lea* 995 KCD 1289, *Ardulveslie* DB, *Ardulflee* 1236 Fees]. '*Eardwulf*'s LĒAH.'

Ardsley, East & West, YW [*Erdeslawe* DB, 1208 FF, *Ardislawe* 1202 FF], A~ YW nr Barnsley [*Erdeslegh* 1202 FF, *Erdeslaia* 12, *-leie* 13 BM]. '*Eorēd*'s hill or mound (HLĀW) and LĒAH.' Or the first el. may be an OE **Eard*, a short form of *Eardwulf*.

Ardwick La [*Atheriswyke* 1282 Ipm, *Atherdwic* Mamecestre]. '*Æþelrēd*'s WĪC.'

Areley Kings Wo [*Erneleia* c 1138 BM, *Ernleȝe* 1205 Lay]. OE *earn-lēah* 'eagle wood'.

Argam YE [*Ergone* DB, *Ergum* 1218 FF]. 'The shielings'; see ERG.

Arkendale YW [*Arche-, Arghendene* DB, *Erkeden* 1177 P, *Arkendenn, Arkeden* 1200 Cur]. First el. possibly a stream-name derived from the adj. *eorcon, earcon* (in pers. ns. as *Eorconweald*), corresponding to Goth *airkns* 'holy', OHG *erchan* 'genuine'. The original meaning of the adj. was 'pure, clear'. Or it may be a short form **Eorcna, *Earcna* of names in *Eorcon-*.

Arkengarthdale YN [*Arkillesgarth* 1199 VH, *-gardh* 1201 Ch]. '*Arnkell*'s enclosure' (ON *garðr*). *Arnkell* (*-ketill*) is a common OScand name.

Arkesden Ess [*Archesdana* DB, *Arkedenn'* 1203 Cur, *-den* 1236 Fees, *Arkesdene* 1254 Val]. Etymology obscure.

Arkholme (ărum) La [*Ergune* DB, *Argum* 1196 P]. See ARGAM.

Arkle Beck R YN [*Arkelbek* 1226 FF], **Arkle Town** YN [*Arkilton* 1473 VH], **Arkleby** Cu [*Arkelby* 1298 Ipm]. '*Arnkell*'s beck, TŪN, BY.' See ARKENGARTHDALE.

Arksey YW [*Archeseia* DB, *Arkesia* 1184, *Arkeseye* 1230 P, *Archexea* 1276 Ep]. Perhaps '*Arnketill*'s island'. Cf. ARKENGARTHDALE.

Arkston He [*Archelestune* c 1170 Hereford, *Arketeleston* 1211 Cur]. '*Arnketill*'s TŪN.' Cf. ARKENGARTHDALE.

Arle Gl [*Alre* 1185 TpR]. OE *alor* 'alder'.

Arlecdon Cu [*Arlachadena, Arlauch-, Arlokedene* 12 StB]. Perhaps an OE *earnlacedenu* 'valley of the eagle stream'.

Arlescote Wa [*Orlavescote* DB, *Ordlavescot* c 1140 Fr]. '*Ordlāf*'s COT.'

Arlesey Bd [*Alricheseia* 1062 Th, *Alriceseie* DB, *Aluricheseya* 1247 Ass]. '*Ælfrīc*'s river land' (OE *ēg, īeg*).

Arleston Db [*Erlestune* DB, *Erleston* 1294 FF]. 'The TŪN of the earl.'

Arleston Sa [*Erdelveston* 1180 For, *Erdulveston* 1209 Eyton]. '*Eardwulf*'s TŪN.'

Arley Chs [*Arleye* 1599 AD], A~ La nr Bolton le Moors [*Erelegh* 1283 VH], A~ La nr Blackburn [*Ereley* 13 VH], A~ Wa [(æt) *Ernlege* 963 BCS 1100, *Earnlege* 1001 BM, *Arlei* DB], **Upper & Lower** A~ Wo [(of) *Earnleie* 996 Mon, *Ernlege, alia Ernlege* DB, *Arnleg* (wood) 1232 Cl]. OE *earn-lēah* 'eagle wood'.

Arlingham Gl [*Erlingeham* DB, 1220 Fees]. 'The HĀM of the people of the earl.'

Arlington D [*Alferdintona* DB, *Alfrintone* 1258–62 Ep]. 'The TŪN of *Ælffriþ*'s people.'

Arlington Gl [*Alvredintune* DB, *Alurintone* 1221 Ass]. 'The TŪN of *Ælfrēd*'s people.'

Arlington Sx [*Erlington, Herlintone* DB, *Erlington* 1230 P]. 'The TŪN of the people of the earl.'

Armathwaite Cu [*Ermitethwayt* 1232 P]. 'The clearing (*þveit*) of the hermit.' There are three places with this name in Cu, one in Ainstable, the site of a nunnery founded by William Rufus, one in Bassenthwaite, one in Hesket in the Forest. A~ in Ainstable is first recorded in 1212 (PNCu(S)).

Arminghall Nf [*Hameringahala* DB, *Ambringehale* c 1105, c 1140 BM, *Ameringehale* 1212 Fees]. If the form in *H-* is reliable, the name may mean 'HALH of the people on the hill'. Cf. HAMER. If not, the first el. may mean '*Ēanmǣr*'s people'.

Armitage St [*Armytage* 1520 BM]. 'The hermitage.'

Armley YW [*Ermelai* DB, *-leia* 12 Kirkst, *Armeslie* c 1170 YCh 1634, *Armelay* 1241 FF]. The first el. may belong to OE *earm* 'wretched', perhaps in the sense 'outlawed'. If so, 'the wood of the outlaw(s)'.

Armscott Wo [(æt) *Eadmundescotan* 1042 E]. '*Ēadmund*'s COTS.'

Armston Np [*Mermeston* DB, *Armeston* 1202 Ass, 13 BM, *Ermeston* 1227 Ch, *Armistone* 1232 Ep]. Perhaps '*Earnmund*'s TŪN'.

Armthorpe YW [*Ernulfestorp* DB, *Ernetorp* 1201 Cur, *Armethorp* 1237 Ep]. '*Earnwulf*'s thorp.'

Arnaby Cu [*Arnolvebi* c 1230 FC]. 'The BY of *Arnulf*' (ON *Ǫrnulfr*, OSw *Arnulf*).

Arncliffe YW [*Arneclif* DB, *Erneclive* 1223 FF]. OE *earna-clif* 'eagles' cliff'.

Arncot, Upper & Lower, O [(æt) *Earnigcotan* 983 KCD 1279, *Ernicote* DB, *Nether-*, *Overarnecote* 1283 Misc]. 'The COTS of *Earn*'s people.'

Arne Do [*Harne* 1285, *Arne* 1316 FA]. OE *ærn* 'house'.

Arnesby Le [*Erendesbi*, *-berie* DB, *Erendesbi* 1169 P, *Erndesby* 1233 Ch]. 'The BY of *Iarund*' (ODan *Iarund*, ON *Iǫrundr*).

Arnford YW [*Erneforde* DB, *Arneford* 1197 P]. OE *ærneford* 'ford fit for riding'. Cf. OE *ærneweg* 'road fit to ride on'.

Arnold Nt [*Ernehale* DB, *Ærnehala* 1169 P], A~ YE [*Ærnhale* 1190 P, *Arnhale* 1240 FF]. 'HALH frequented by eagles.' See EARN.

Arnside We [*Arnuluesheued* 1208 FF, *Arnolvesheued* 1246 FF]. '*Earnwulf*'s hill or headland.' Cf. HĒAFOD.

Arnwood Ha [*Ernemude* DB, *Arnemuda* 1106 Bath, *Ernewod* 1242 Fees]. OE *earnwudu* 'eagle wood'.

Arram YE nr Hornsea [*Argum* DB, *Erghum* 1285 FA]. See ARGAM.

Arras YE [*Herges* 1156 YCh 1388, *Erghes* 1316 FA]. 'The shielings'; see ERG.

Arrathorne YN [*Ergthorn* 13 Marrick]. 'Thorn by the shieling.' See ERG.

Arreton Wt [*Eaderingtun* c 880 BCS 553, *Adrintone* DB]. 'The TŪN of *Ēadhere*'s people.'

Arrington Ca [*Earnnington* c 950 Wills, *Erningtune* DB, *Earningatone* 1087 Fr]. 'The TŪN of *Earn(a)*'s people.' The same first el. is found in **Armingford** (hd) Ca [*Earnigaford* 970 BCS 1267] and **Ermine Street** [*Earninga stræt* 955, 957 BCS 909, 1003, *Ermingestrete* c 1090 Laws].

Arrow R He [*Erge* 958 BCS 1040, *Arewe* 1256 BM]. A British river-name cognate with Welsh *ariant* 'silver' &c. The base *arg-* meant 'white, bright'.

Arrow R Wa [*Arwan stream* 11 Heming, *Aruwe* 1247 Ass]. A British river-name

identical with *Arva* in Gaul (now AUVE, AVRE in France). The meaning may be 'running water, stream'. On the Arrow is **Arrow** vil. [*Arosætna* [land] 7 Trib Hid, really 'the dwellers on the Arrow', *Arue* 710 BCS 127, *Arve* DB].

Arrowe Chs [*Arwe* c 1245 Chester, *Harche* 1312 Ipm]. 'The shieling'; see ERG.

Arscott Sa [*Ardescote* 1255 RH, *-cot* 1276 Ipm]. The first el. is a pers. n., e.g. *Æþelrēd* or *Ēadrēd* or *Ēanrēd*.

Arthington YW [*Ardinton*, *Hardinctone* DB, *Arthington* c 1200 YCh 60, 1226 FF, *Arthigton* 1204 FF]. The first el. is a derivative with the suffix *-ing* from a pers. n., e.g. *Eardnōþ* or *Ēanrēd*.

Arthingworth Np [*Arniworde*, *Arningvorde* DB, *Erningwrth* 1202 Ass, *Arningwrth* 1220 Fees, *-worde* 1233 Ep, *Arthingworth* 1274 RH, *-wrth* 1291 Tax]. Apparently a reduction of a longer name, e.g. *Eardnōþingaworþ*. *Eardnōþ* is found as the name of a moneyer (t. Æthelred II). See WORP.

Arthuret Cu [*Arturet* c 1190 WR, 1267 Ch, *Arturede* 1202 FF]. Has been identified with *Ard eryd* in the Chron. of the Picts and Scots, *Armterid* 573 Ann Cambr, whose etymology is obscure. The identification is doubtful.

Artington Sr [*Artedena* 1177 P, *Ertedun* 1212 Fees, *Erdintona* 1173, *Erdinton* 1194 ff. P]. OE *Eardantūn* or *-dūn*, the first el. being the OE *Earda* postulated under ARDELEY.

Arun R Sx [*Aron*, *Arunus* 1577 Harrison]. A back-formation from **Arundel** (ăr-) [*Harundel* DB, *Arundell*' 1087 Fr]. This has been explained as OE *Hārhūn-dell* 'hoarhound valley' (OE *hārhūne*).

Asby Cu [*Asbie* 1654, *Ashby* 1713 PNCu(S)], **Great & Little A~** We [*Aschaby* c 1160 YCh 175, *Magna, Parva Askeby* 1292 QW]. OScand *Askabȳ(r)* 'BY where ash-trees grew'; cf. OSw *Askaby*, Dan *Askeby*.

Ascot Brk [*Estcota* 1177 P, *Astcote* 1348 BM], A~ O nr Watlington [*Estcote* c 1130 Oxf, *Ascote* 1316 FA], **A~ under Wychwood & A~ d'Oilly** O [*Estcota* 1190 P, *Ascote Doilly* 1327 Misc]. OE *Ēast-cot* 'eastern COT'.

See WYCHWOOD.—Wido de Oileio held *Escota* c 1100 (RA). The d'Oiley family took its name from one of the OUILLYS in Normandy.

Ascote Np. See ASTCOTE.

Ascote, Chapel, Wa [*Astanescote* 12 Mon]. '*Ēadstān*'s COT.'

Ascott Bk [*Estcota* 1220 PNBk]. See ASCOT.

Asenby YN [*Estanesbi* DB, *Aystaneby* 1182 Percy, *Eistanebi* 1204 Cur]. 'The BY of *Eysteinn*' (ON *Eysteinn*, ODan, OSw *Østen*).

Asfordby Le [*Osferdebie* DB, *Asfordebia* c 1125 LeS, *Asfordebi* 1184 P, *Essefordebi*

1204 RA]. '*Asford*'s BY.' *Asford*, *Asfort* DB is from ODan *Asfrith*.

Asgarby Li nr Spilsby [*Asgerebi* DB, *Asgerbi* c 1105, *An(s)gerbi* c 1142 RA], A~ Li nr Sleaford [*Asegarby* 1201 Cur, *Asgarbi* 1202 Ass]. '*Asgar*'s BY.' *Asgar* (DB &c.) is ODan *Asgair*, *Esger*, OSw *Asger*, ON *Asgeirr*.

Ash Db [*æt Æsce* 987 Hengwrt MS 150, *Eisse* DB, *Eyss* 1242 Fees], A~ Do in Stourpaine [*Aisse* DB], A~ or **Ashe Ingen** He [*Ascis* DB], A~ next Ridley K [*Eisse* DB, *Esse* 1205 Cur], A~ next Sandwich K [*Æsce* 11 DM], A~ **Magna & Parva** Sa [*Magna, Parva Asche* 1285 Eyton], A~ So nr Martock [*Esse* 1225 Ass], A~ **Priors** So [*Æsce* 1065 Wells, *Esse Prior* 1263 Ass], A~ Sr [*Essa* 1171, *Eisse* 1198 P], **Ashe** Ha [*Esse* DB]. OE *æsc* 'ash', partly perhaps in the dat. plur. form *æscum*.

For **Ash Ingen** see ASTON INGHAM.—A~ **Priors** belonged to the prior of Taunton at least from 1201 (Ass).

Ashampstead Brk [*Esshampsted* 1212 Fees, *Asshehamstede* 1309 Ch]. 'Homestead by the ash-tree(s).'

Ashbocking Sf [*Ass(i)a* DB, *Esse* 1198 FF, *Assh* 1296 Ch, *Bokkynge Assh* 1411 BM]. The original name was ASH.

Bocking was the name of a local family (from Bocking Ess). Ralph de Bocking held the manor in 1338 (Ch).

Ashbourne Db [*Esseburne* DB, *-burn* 1188 P, *Ascheburn* 12 BM]. Really an old name of HENMORE BROOK [*Esseb[urne]* 1275 RH, *Askebek* c 1200 ERN]. 'Stream where ash-trees grew.'

Ashbrittle So [*Aisse* DB, *Esse Britel* 1212 Fees]. Originally ASH.

The manor was held by Bretel (de St. Clair) in 1086 (DB).

Ashbrook Gl [*Esbroc, Estbroce* DB, *Astbrok* 1303 FA]. 'Eastern brook.'

Ashburn R Sx [*Asheburne water* 1577 Harrison]. On it is **Ashburnham** [*Esseborne* DB, *Esburnham* 12 (1432) Pat]. Cf. ASHBOURNE.

Ashburton D [*Æscburnan land* 1008–12 Crawf, *Essebretone* DB]. A~ is on a stream now called YEO, but formerly **Ashbourne** [*Æscburne* 10 BCS 1323]. The original name was identical with the river-name, but later TŪN was added. Cf. ASHBOURNE.

Ashbury Brk [*Æscesbyrig* 856 BCS 491, *Æscesburh* 953 ib. 899, *Eissesberie* DB]. '*Æsc*'s BURG.' Cf. ASHDOWN Brk.

Ashbury D [*Esseberie* DB, *Aysshebyry* 1276 Cl]. 'Fort where ash-trees grew.'

Ashby, a common name, is identical with ASBY (q.v.) and means 'BY where ash-trees grew'. A possible alternative in some cases, however, is '*Aski*'s BY'; ON *Aski* pers. n. is evidenced. **Ashby Folville** Le [*Ascbi* DB, *Essebia Fulconis de Foleuilla* Hy 2 DC, *Esseby Foleville* 1233 Ep], A~ **Magna & Parva** Le [*Essebi* DB, *Magna Essebi* 12 DC, *Parva Esseby* c 1225 Ep], A~ **de la Zouch** (zōosh) Le [*Ascebi* DB, *Esseby la Zusche* 1241 Ep], A~ Li nr Bottesford [*Aschebi* DB, c 1115 LiS], A~ **cum Fenby** Li [*Aschebi* DB, c 1115 LiS, *Askeby* 1205 Cur], A~ **de la Launde** Li [*Aschebi* DB, *Askebi* 1202 Ass], A~ **by Partney** Li [*Aschebi* DB, c 1115 LiS, *Estaskebi* 1208 Cur], A~ **Puerorum** Li [*Aschebi* DB, *Ascbi* c 1115 LiS, *Askeby . . parvorum chori Lincoln'* 1291 Tax], **West** A~ Li [*Aschebi* DB, c 1115 LiS, *Askebi* 1202 Ass], A~ Nf in Flegg [*Askeby* 1044–7 KCD 785, *Aschebei* DB], A~ Nf nr Norwich [*Ascebei* DB, *Asheby* 1291 Tax], **Canons** A~ Np [*Ascebi* DB, *Esseby Canonicorum* 1254 Val], **Castle** A~ Np [*Asebi* DB, *Esseby* 1220 Fees], **Cold** A~ Np [*Essebi* DB, *Caldessebi* c 1150 BM], **Mears** A~ Np [*Asbi* DB, *Esseby Mares* 1282 Ipm], A~ **St. Ledgers** Np [*Ascebi* DB, *Esseby Sancti Leodegarii* 1248 Ep, *Assheby Leger* 1339 AD], A~ Sf [*Aschebi* 1198 FF, *Askeby* 1254 Val].

Canons A~ Np was the seat of a priory.—A~ **Folville** Le from the local family, named from one of the Follevilles in France (one is in Calvados, Normandy).—A~ **de la Launde** Li was held by William de la Launde in 1346 (FA).—**Mears** A~ Np was given to Robert de Mares in 1242 (Ch). Mares is from MARAIS in France (Fr *marais* 'marsh').—A~ **Puerorum** Li was assigned to the support of the choristers of Lincoln.—A~ **St. Ledgers** Np from the dedication of the church to St. Leger (*Leodegarius*).—A~ **de la Zouch** Le was held by Roger de la Zuche in 1200 (FF). Zouch is a family name (from Fr *souche* 'stump').

Ashcombe D [*Aissecome* DB, *Ascumbe* 1200 Cur]. 'Ash-tree valley.'

Ashcott So [*Aissecote* DB, *Ascote* 1198 P, *Asshcote* 1327 Subs]. 'COT by ash-trees.'

Ashdon Ess [*Ascenduna* DB, *Essendona* 1121 AC, *-duna* c 1150 Fr]. OE *æscen dūn* 'hill overgrown with ash-trees'.

Ashdown Brk [*Æscesdun* 648 ff. ASC, c 894 Asser, *Æsces Dune* (mons) 955 BCS 908, *Assedone* DB]. '*Æsc*'s hill.' Cf. ASHBURY. Asser's translation 'mons fraxini' is a case of popular etymology.

Ashdown Forest Sx [*Essendon* 1165 P, *Asshendon* 1234 Cl]. See ASHDON Ess.

Ashe. See ASH.

Asheldham Ess [*Assildeham* c 1130 Bodl, 1206 FF, *Eissildeham* 1181 P]. '*Æschild*'s HĀM.' This woman's name is unrecorded, but cf. OHG *Aschilt*.

Ashen Ess [*Asce* DB, *Esse* 1166 RBE, *Asshen* 1428 FA]. OE *æscum*, dat. plur. of OE *æsc* 'ash-tree'.

Ashendon Bk [*Assedune* DB, *Assendon* 1232 Ch, *Essendon* 1242 Fees]. See ASHDON.

Ashey Wt [*Æsces hege* 982 Hyda, *Asseheye* 1291 Tax, *Assheseye* 1316 FA]. OE *æsces-*

hege 'ash enclosure' or *Æsces hege* 'the enclosure of *Æsc*'.

Ashfield Sa [*Assefeld* 1167 P, *Asshefeld* 1327 Subs], **A~** Sf nr Debenham [*Assefelda* DB, *Esfeld* 1198 FF], **Great A~** Sf [*Eascefelda, Escefella* DB, *Aysefeld Magna* 1291 Tax]. 'FELD with ash-trees.'

Ashford D [*Aisseford* DB, *Asford* 1242 Fees], **West A~** D [*Westesford* 1242 Fees], **A~** Db [æt *Æscforda* 926 BCS 658, *Aisseford* DB], **A~ Bowdler & Carbonell** Sa [*Esseford* DB, 1138 AC, *Asford Budlers, Aysford Carbonel* 1255 RH]. 'Ash-tree ford.'
Bowdler is a Norman family name, Henry de Boulers was a tenant 1211–12 RBE.—William Carbunel held Ashford c 1180 (Eyton). Carbonell is a Norman nickname and family name, perhaps from OFr *carbonel*, the name of a certain fish.

Ashford K [*Essetesford* 1046 Wills, DB, *Æscedesford* 11 DM, *Essheteforde* 1262 RBE]. The first el. may be OE *æsc-scēat* 'ash copse'; see SCĒAT.

Ashford Mx [*Ecelesford* 969 Crawf, *Echelesford* 1291 Tax; *Exford* 1062 KCD 812, *Exeforde* DB; *Esseford* 1198 P]. The brook over which the ford was is *Eclesbroc* 962 BCS 1085. The form *Echelesford* points to an OE first el. with palatal *c*; hence it is hardly identical with ECCLES. Possibly an unrecorded OE pers. n. *Eccel*, derived from *Ecca*. But cf. ECCHINSWELL.

Ashill Nf [*Asscelea, Essalai* DB, *Assele* 1208 Cur]. OE *æsclēah*; cf. ASHLEY.

Ashill So [*Aisselle* DB, *Esselle* 1212 Fees, *Asshulle* 1327 Subs]. 'Ash hill.'

Ashingdon Ess [*Assandun* 1016 ASC (E), *Nesenduna* DB, *Essendon* 1203 Cur]. 'The hill of the ass (OE *assa*)' or rather '*Assa*'s hill', though *Assa* pers. n. is unrecorded. Cf. ASSINGTON Sf.

Ashington Nb [*Essenden* 1205 Cur, *Essingden* 1242 Fees]. OE *æscen-denu* 'valley overgrown with ash'.

Ashington So [*Essentone* DB, *Estinton* 1186 ff. P, *Estington* 1225 Ass, *Astynton* 1291 Tax]. Identical with EASTINGTON D, Do.

Ashington Sx [*Essingetona* 1073 Fr, *Essington* 1235 Sele, *Ashshintone* 1305 FF]. 'The TŪN of *Æsc*'s people.'

Ashleworth Gl [*Esceleuuorde* DB, *Æisseleworda* 1130 P, *Esseleswrd* 1154 Berk, *Asshelesworth* 1291 Tax]. Probably '*Æscla*'s WORÞ'. This name is not on record, but a regular formation from *Æsc*. Cf. OHG *Ascalo, Ascila*.

Ashley Green Bk [*Essleie* 1227 Ass], **A~** Ca [*Esselie* DB, *Asle* 1242 Fees], **A~** Chs [*Ascelie* DB], **A~** D [*Aissheley* 1483 Pat], **A~** Do [*Asseleghe* 1246 FF], **A~** Ha nr Lymington [*Esselie* DB, *Est Assely* 1285 Ep], **A~** Ha nr Winchester [*Asselegh* 1275 Ipm], **North A~** Ha nr Ringwood [*Essileg* 1236 Fees, *Asshele* 1316 FA, *Northascheley* 1390 BM], **A~** K [*Asshelegh* 1279 Cl], **A~** Np [*Ascelei* DB, *Eissele* 1203 Ass], **A~** St [*Esselie* DB, *-legh* 1230 Ass], **A~** W nr Box [*Asseleye* 13 BM], **A~** W nr Bradford [*Assheley* 1494 Ipm], **A~** W nr Tetbury [*Esselie* DB, *-lega* 1195 Cur]. OE *æsclēah*, as a rule no doubt 'ash wood'. Cf. LĒAH.

Ashleyhay Db [*Asslewehay* 1255 Ipm, *Asshelehay* 1309 Ipm]. OE *æsclēah* (see ASHLEY) and *hege* or *gehæg* 'enclosure'.

Ashling Sx [*Estlinges* 1185 ff. P, *Estesshlinge* 1288 FF, *Est-, Westaslinge* 1296 Subs]. Either an OE *Æsclingas* '*Æscla*'s people' (cf. ASHLEWORTH) or OE *æschlinc* 'ash hill'.

Ashmanhaugh Nf [*Ashmanhaga* 1183 Holme, *Assemaneshawe* 1254 Val]. 'The enclosure (OE *haga*) of the *æscman* or pirate.' Possibly a nickname *Æscmann* may be assumed.

Ashmansworth Ha [*Æscmeres wierð* 909, *Æscmeresweorð* 934 BCS 624, 706, *Esmeresworth* 1171 Ep]. 'WORÞ belonging to *Æscmere*', a lost place [*Æscmere* 863 BCS 508]. *Æscmere* means 'lake where ash-trees grew'.

Ashmore Do [*Aisemare* DB, *Essemera* 1107 (1300) Ch, 1182 P, *Asshemere* 1305 FF]. Perhaps OE *æscmere* 'lake where ash-trees grew'. But the place is on the Wilts border, and OE *gemǽre* 'boundary' is a possible second el. If so, the first el. may be OE *Æsca* pers. n.

Ashness Cu [*Esknese* 1211 FC]. ON *Eskines* 'ness with an ash copse' (ON *eski*).

Ashop Db [*Essop* 1215 Ch, *Asshope* 1229 Cl]. OE *æsc-hop* 'ash valley'.

Ashorne Wa [*Asshonere* 1236 Fees, *Asshorne* 1370 BM]. OE *æsc-horn(a)* 'headland overgrown with ash-trees'.

Ashover Db [*Essovre* DB, *Esshovere* 1252 FF]. 'Ridge overgrown with ash-trees'; see OFER.

Ashow Wa [*Asceshot* DB, *Axeho* 1210 Cur, *Aisho* 1236 Fees]. OE *æsc-hōh* 'ash-tree hill or ridge'.

Ashperton He [*Spertune* DB, *Aspretonia* 1144, *-tuna* 1146 Fr, *Aspertun* 1242 Fees]. The first el. may be OE *æscburna* (cf. ASHBOURNE) or *æscbeorg* 'ash hill' or *Æscbeorht* pers. n.

Ashprington D [*Aisbertone* DB, *Asprintona* 1088 Totnes, *-ton* 1188 P, *Ayspringtone* 1309 Ep]. The first el. seems to be a derivative with the suffix *-ing* of a pers. n. (*Æscbeorn* or *-beorht*) or of OE *æscburna* or *æscbeorg*; cf. ASHPERTON.

Ashreigny or **Ringsash** D [*Aissa* DB, *Esshereingni* 1238 Ass, *Ringgesashe* 1306 Ep]. Originally ASH.

The addition -reigny, Rings- from the local family. William de Regny held *Esse* before 1219 (FF). Reigny from one REGNY in France.

Ashtead Sr [*Estede* c 1150 BM, *Asshstede* 1385 BM]. 'Place where ash grew.' Cf. STEDE.

Ashton, a common name, is 1. usually OE *Æsctūn* 'TŪN where ash grew': A~ **on Mersey** Chs [*Asshton* 1408 AD], A~ **by Tarvin** Chs [*Estone* DB], A~ **under Hill** Gl [*Æschæma gemæru* 1042 KCD 764, *Essetone* DB], A~ **La nr Lancaster** [*Estun* DB, *Esshton* 1332 Subs], A~ **La nr Preston** [*Estun* DB, *Assheton* 1326 FF], A~ **under Lyne** La [*Eston* 1212 Fees, *Asshton under Lyme* 1305 FF], A~ **in Makerfield** La [*Ashton* 1255 FF], A~ **Np nr Oundle** [*Ascetone* DB], A~ **Np in Bainton** [*Æsctun* c 960 BCS 1128], **Long** A~ **So** [*Estune* DB, *Ayston* 1256 Ch], A~ **Gifford** W [*Æiston* 1194 P, *Ashtone Giffard* 1354 Ipm], A~ **Keynes** W [*Æsctun* c 880 BCS 553, *Essitone* DB, *Eston* 1242 Fees], **Steeple, West & Rood** A~ W [*Æyston* 964 BCS 1127, *Aistone* DB, *Stepell Asschton* 1341 NI, *Westaston* 1257 Fine R, *Rode Asseton* 1475 Ipm].

2. Other origins: **Higher Ashton** D [*Aiserstone* DB, *Assherestone* 1275 Ep]. '*Æschere*'s TŪN.' A~ **Np nr Northampton** [*Asce* DB, *Eysse* 1220 Fees, *Asshen* 1339 BM]. OE *æscum* 'the ash-trees' (dat. plur. of *æsc*). A~ **He** [*Alston* 1237 Cl, *Aleston* 1303 FA]. '*Ælli*'s TŪN.'

A~ **Gifford** was held by Elias Giffard in 1242 (Fees). Giffard is a Norman nickname and family name (from OFr *gifard* 'bloated').—A~ **under Hill** is at Bredon Hill.—A~ **Keynes** was held by Henry Kaignel in 1242 (Fees), by Robert de Kaynes before 1282 (Ipm). Keynes from CAHAGNES in Normandy.—A~ **under Lyne** and A~ **in Makerfield**; see LYME, MAKERFIELD.—**Rood** A~ no doubt from the *rodestan* mentioned in bounds of Ashton BCS 1127.—**Steeple** A~ from the church steeple.

Ashurst K [*Esherst* 1268 Ipm], A~ **Sx** [*Essehurst* 1164 P, *Aishurst* [boscus] 1199 Obl, *Aschurst* 1283 BM]. 'Ash wood'; see HYRST.

Ashwater D [*Aissa* DB, *Esse Valteri* 1271 Ep]. Originally ASH. Walter de Doneheved held *Essewater* in 1282 (QW).

Ashwell Ess [*Asseuuella* DB], A~ Hrt [*Æscewelle* 1065 BM, *Asceuuelle* DB], A~ Ru [*Exewelle* DB, *Essewell* 1202 Ass, *Asshewell* 1291 Tax]. 'Ash stream.'

Ashwellthorpe Nf [*Aescewelle, Thorp* c 1066 Wills, *Aissewellethorp* 1254 Val]. 'Thorp belonging to Ashwell.' By the charter of c 1066 8 acres in Thorp were to go to *Aescewelle kirke*.

Ashwick So [(æt) *Æscwican* 1060–6 KCD 821, *Escewiche* DB]. 'WĪC by ash-trees.'

Ashwicken Nf [*Wiche* DB, *Wyken* 1254 Val, *Askiwiken* 1275 RH, *Asse Wykin* 1302 FA].

OE *wīcum*, dat. plur. of wĪC. *Ash-* may be *ash* the tree.

Ashwood St [*Aswude* 1232 Cl, *Asshewode* 1292 Ch]. Self-explanatory.

Ashworth La [*Assewrthe* 1236, *Asheworth* 1347 FF]. 'WORÞ by ash-trees.'

Aske Hall YN [*Hasse* DB, *Aske* 1218 FF]. OE *æsc* 'ash-tree', replaced by ON *askr*. Cf. ÆSC.

Askern YW [*Askern* 1195 ff. P, *-e* 1197 P, 1218 FF]. Perhaps OE *æsc-hyrne* 'corner where ash-trees grew', with *æsc* replaced by Scand *ask*.

Askerswell Do [*Oscherwille* DB, *Oskerwell* 1194 P, *Askereswell* 1208 Cur]. '*Ōsgār*'s stream.' Later *Asker-* may be due to the corresponding Scand name (ON *Āsgeirr*, ODan *Asger*).

Askerton Cu [*Askerton* 1256 FF, *Askreton* 1295 Cl]. '*Asger*'s TŪN.' Cf. ASGARBY.

Askham Nt [*Ascam* DB, *Ascham* 1167 P, *Askham* 1329 QW], A~ **Bryan & Richard** YW [*Asc(h)am* DB, *Askeham* 1200 Cur, *Ascam Bryan* 1285 FA, *Askham Ric[ard]i* 1291 Tax]. OE *Æschām* 'HĀM where ash-trees grew'.

Brian son of Scolland held *Ascam* t. Hy 2 (Misc). Cf. BRAMPTON BRYAN.—A~ **Richard** perhaps from Richard earl of Cornwall (13th cent.).

Askham We [*Askum* 1232 Pat, 1292 QW]. OE *æscum* or OScand *askum* dat. plur. 'the ash-trees'.

Askrigg YN [*Ascric* DB, *Askeric* 1218 FF, *Askric* 1330 Ch, *Askerigg* 1287 FA]. The early forms tell against the second el. being the word for ridge, ON *hryggr*. A~ is probably OE *Æscric* (cf. RIC) 'ash stream' with *sk* for *sh* owing to Scand influence. A~ is on a stream.

Askwith YW [*Ascvid* DB, *Ascwith* 1201 FF]. ON *ask-viðr* 'ash wood'.

Aslackby Li [*Aslachebi* DB, *Aslakesbi* 1190 P]. '*Aslac*'s BY'; see AISLABY.

Aslacton Nf [*Oslactuna, Aslactuna* DB, *Aselaketon* 1101–7 Holme, *Aslacton* 1208 FF], **Aslockton** Nt [*Aslache(s)tone* DB, *Aselakeston* 1185 P]. '*Aslac*'s TŪN'; cf. AISLABY. A~ Nf may have been originally OE *Ōslāc(es)tūn*.

Aspall Sf [*Espala* DB, *Aspala* DB, *Asphal'* 1208 Cur, *Aspehale* 1254 Val]. 'HALH overgrown with aspens.'

Aspā·tria Cu [*Ascpatric* c 1230 BM, *Askpatrik* 1291 Ch]. 'Patric's ash.' First el. ON *askr*. The order of the elements is due to Celtic influence. Cf. Introd. p. xxiii.

Aspenden Hrt [*Apseden* 1212 Fees, *Aspedene* 1220 Fees]. 'Aspen valley.'

Aspley Guise Bd [*Æpslea* 969 BCS 1229, *Aspeleia* DB, *Aspeleye Gyse* 1363 Cl], A~

St [*Haspeleia* DB, *Aspelega* 1227 Ass], A~ Wa [*æt Æpsleage* 963 BCS 1111, *Aspelea* 1180 P, -*ley* 1273 Ipm]. OE *æsp-lēah* 'aspen wood'.

Anselm de Gyse got A~ t. Hy 3. Guise from GUISE in France.

Asps Wa [*Aspes* 1203 Ass]. 'The aspens.'

Aspul La [*Aspul* 1212 Fees, *Apshull* 1246 Ass]. 'Aspen hill.'

Asselby YE [*Aschilebi* DB, *Eskilby* 1198 FF]. 'The BY of *Askel*' (ON *Áskell*, ODan *Eskil*, OSw *Æskil*).

Assendon O [*Assundene* 774 BCS 216]. 'The valley of the ass' (OE *assa*).

Asserby Li [*Asforthebi* c 1200 DC, *Esfordebi* 1202 Ass, *Asfordebi* 1212 Fees]. Identical with ASFORDBY.

Assington Sf [*Asetona* DB, *Asinton* 1173 f., *Assintona* 1175 P, *Essinton* 1219 FF]. The first el. is a pers. n., which may be identical with that in ACHURCH or an OE **Assa*.

Astbury Chs [*Esteburi* c 1100, *Asteburi* c 1180 Chester]. 'Eastern BURG.'

Astcote Np [*Aviescote* DB, *Hauekescote* 1198 P, *Hauescote* 1316 FA]. '*Æfic*'s COT.'

Asterby (ā-) Li [*Estrebi* DB, 1130 P, *Eisterby* 1189 BM, *Aisterby* 1212 Fees]. OScand *eystri bȳ(r)* 'eastern BY'.

Asterleigh O [*Estrelega* 1209–19 Ep, *Esterleg* 1220 Fees], **Asterley** Sa [*Asterlegh* 1316 FA]. OE *ēasterra lēah* 'eastern LĒAH'.

Asterton Sa [*Esthampton* 1255 RH]. 'East HĀMTŪN.'

Asthall O [*Esthale* DB, -*hal* 1163 P, -*halle* 1190 P], A~ **Lea** [*Esthallingeleies, Esthalluncgeleia* c 1200 BM]. Asthall may be OE *Ēastheall* 'eastern hall'. See HALL. But if (æt) *East Heolon* 11 Ælfric's Pentateuch belongs here, it is OE *Easthealas*. See HALH. *Heolon* must be for *Healon*. A~ Lea is OE *Ēastheallinga lēah* 'the woodland of the Asthall people'.

Astle Chs [*Asthull* 1245–50, 1267 Chester]. 'The eastern hill.'

Astley La [*Asteleghe* c 1210 CC, *Asteleg* 1246 Ass], A~ **Sa** nr Shrewsbury [*Hesleie* DB, *Estleg* 1203 Ass], A~ **Abbotts** Sa [*Hestlee* c 1150 BM, *Astleye Ab*[*ba*]*tis* 1327 Subs], A~ Wa [*Estleia* DB, *Astleg* 1242 Fees], A~ Wo [*Æstlæh* 11 Heming, *Eslei* DB]. OE *east-lēah* 'eastern LĒAH'.

A~ Abbotts belonged to Shrewsbury Abbey.

Astley YW nr Leeds [*Askelhale* 1300 Misc]. Apparently '*Askel*'s HALH'. Cf. ASSELBY.

Aston, a common name, is **1.** mostly OE *Ēast-tūn* 'eastern TŪN': A~ **Abbots** Bk [*Estone* DB, *Eston abbatis Sancti Albani* 1237–40 Fees, *Aston Abbatis* 1262 Ass], A~ **Clinton** Bk [*Estone* DB, *Eston et Clynton* 1244, *Aston Clinton* 1237–40 Fees], A~ **Mullins** Bk [*Eston* 1212 Cur, *Astone* 1346 FA], A~ **Sandford** Bk [*Estone* DB, *Aston Sanford* 1242 Fees], **Ivinghoe** A~ Bk [*Estone* DB, *Eston apud Ivyngho* 1282 Ep, *Ivyngho Aston* 1490 Ipm], A~ **Tirrold** Brk [*Eastun* 964 BCS 1143, *Estone* DB, *Aston Turald* 1380 Bodl], A~ **Upthorpe** Brk [*Aston et Upthrop* 1316 FA], A~ Chs nr Nantwich [*Aston* 1287 Court], A~ **by Budworth** Chs [*Estone* DB], A~ **juxta Mondrum** Chs [*Estone* DB, *Aston by Mondrem* 1350 Chamb], A~ **by Sutton** Chs [*Estone* DB], A~ Db nr Hope [*Estune* DB], A~ Db nr Sudbury [*Estune* DB, *Easton* c 1200 Fr], **Coal** A~ Db [*Estune* DB, *Cold Aston* c 1260 Derby], A~ **upon Trent** Db [*Easton* 958 BCS 1041, *Æstun* DB, *Aston super Trentam* 1330 QW], A~ **Blank** Gl [*Eastun* 716–43 BCS 165, *Estona frigida* 1275 BM], A~ **on Carrant** Gl [*Estone* DB, *Aston sup. Carent* 1327 Subs], A~ **Somerville** Gl [*Eastun* 930 BCS 667, *Eston Somervill* 1285 FA], A~ **Subedge** Gl [*Estune* DB, *Aston sub Egge* 1327 Subs], A~ **Ingham** He [*Estune* DB, *Estun Ingan* 1242 Fees], A~ Hrt [*Estone* DB], A~ **Flamville** Le [*Eston* c 1225 Ep, *Aston Flamuill* 1327 Subs], A~ **le Walls** Np [*Estone* DB, *Aston in the Walles* 1483 AD], A~ O nr Bampton [*Esttun* c 1070 Ex, *Estone* DB], A~ **Rowant** O [*Estone* DB, *Aston Roaud* 1318 Ipm], **Steeple, Mid & North** A~ O [*Estone* DB, *Stepelestone* 1219 Ep, *Midelestun* 1220 Fees, *Nort Eston* 1200 Cur], A~ **Sa** nr Oswestry [*Estone* DB], A~ **Sa** nr Wellington [*Eastun* 975 BCS 1315], A~ **Sa** nr Wem [*Estune* DB], A~ **Botterell** Sa [*Estone* DB, *Eston Boterel* 1263 Ch], **Chetwynd & Church** A~ Sa [*Eastun* 963 BCS 1119], A~ **Eyre** Sa [*Estone* DB, *Aston Aer* 1291 Tax], **Munslow** A~ Sa [*Estune* DB], A~ **Pigot** Sa [*Eston Pigot* Hy 3 Misc, *Aston Pigod* 1267 Eyton], A~ **Rogers** Sa [*Aston* 1242 Fees, *Astone Roger* 1327 Subs], A~ **St** nr Seighford [*Estone* DB], A~ & **Little** A~ St nr Stone [*Eastun* 957 BCS 987, *Estone* DB, *Little Aston* 1266 Ass], **Wheaton** A~ **St** [*Estone* DB, *Wetenaston* 1248 Ipm], A~ **Wa** nr Birmingham [*Estone* DB, *Aston juxta Burningham* 1293 Ct], A~ **Cantlow** Wa [*Estone* DB, *Aston Cantelou* 1273 Ipm], A~ **Fields** Wo [*Eastun* 767 BCS 202], A~ **Magna** Wo [*Estona* 1209 Fees], **White Ladies** A~ Wo [*Eastun* 977 KCD 615, *Estun* DB, *Whitladyaston* 1481 Ipm], A~ YW [*Eastone, Estone* DB, *Aston* 1218 FF].

2. OE *Æsctūn* (see ASHTON): **Cold** A~ Gl [*Æsctun* 931 BCS 670, *Escetone* DB, *Coldaston* 1316 Ch], A~ **He** in Kingsland [*Esscetuna* 1123 PNHe], A~ **on Clun** Sa [*Assheston* 1291 Tax, *Asseton* 1316 FA].

A~ **Abbots** Bk was held by St. Albans abbey.— Blank in A~ **Blank** Gl may be the adj. *blank* 'bare'.—A~ **Botterell** Sa was held by William Boterell in 1203 (Cur). Botterell is a Fr nickname and family name, perhaps from OFr *boterel* 'toad'.—A~ **Cantlow** Wa came to William de Cantilupo in 1204 (Dugdale). The family name is from CANTELOUP in Normandy.— A~ **on Carrant** Gl. See CAPANT.—**Chetwynd** A~ Sa is near CHETWYND.—A~ **Clinton**

Bk was named from the Clinton family, which seems to have taken its name from a place CLINTON near A~ Clinton. William de Clinton had seizin of *Eston* in 1194 (P).—**Coal A~** Db and **Cold A~** Gl; see CALD.—**A~ Eyre** Sa was held by Robert grandson of Aer in 1212 (Fees). Robert fil. Aier also occurs in 1166 (RBE). Possibly *Aer* (*Eyre*) is derived from *Alcher*, the name of the Domesday tenant.—**A~ Flamville** Le came to Robert de Flamvile c 1100 (Nichols). Flamville is from FLAMANVILLE in Normandy.— **A~ Ingham** He was held by Richard Ingan in 1212 (Fees). Ingan may be identical with Engaine (cf. COLNE Ess).—**A~ Mullins** Bk was held by John de Molyns in 1346 (FA). Mullins is from one of the places in France called MOULINES or MOULINS ('the mills').—**Munslow A~** Sa is nr MUNSLOW.—**A~ Pigot** Sa was acquired by Ralph fitz Picot in the late 12th cent. (Eyton). Cf. ABINGTON PIGOTTS.—**A~ Rogers** Sa was held by Roger de Aston in 1242 (Fees).—**A~ Rowant** O was held by Rowald de Eston in 1236 (Fees). *Rowald* is a Norman name (from OHG *Hruodwald*).—**A~ Sandford** Bk was held by John de Sanford in 1199 (FF).—**A~ Somerville** Gl was named from the local family, whose name is derived from SOMMEVILLE in France.— **Steeple A~** O from the church steeple.—**A~ Subedge** Gl is at the foot of a ridge.—**A~ Tirrold** Brk was held by Nicholas son of Turold in the 12th cent. *Turold* is a Norman name of Norse origin.—**A~ le Walls** Np from some early entrenchments.—**Wheaton A~** St is presumably 'wheaten A~'.—**White Ladies A~** Wo was held by the Cistercian nuns of Whitstones.

Astrop Np [*Estorp* 1200 Cur, *Estrop* 1270 Ipm]. 'Eastern thorp.'

Astwell Np [*Estwelle* DB, *-wella* 1163 P]. 'Eastern stream.'

Astwick Bd [*Estuuiche* DB, *Estwic* 1233 Ep], **A~** Np [*Estwich* 1202 Ass]. 'Eastern wīc.'

Astwood Bk [*Estwode* c 1152 Fr, *Astwode* 1242 Fees], **A~** Wo [*Estwod* 1208 Cur]. 'Eastern wood.'

Aswarby Li [*Asuuardebi* DB, *Assewartheby* 1219 Ep], **Aswardby** Li [*Asuuardebi* 1196 FF, *Aswardebi* 1212 Fees]. '*Asuert*'s BY.' *Asuert* (DB) is from ODan *Aswarth*.

Atcham or **Attingham** Sa [*Atingeham* DB, *Ettingham* 1199 (1285) Ch, *Attingeham* 1292 QW, *Hatincham* 1198 Fees]. 'The HĀM of *Ætti*'s or *Eata*'s people.'

Atch Lench. See LENCH.

Athelhampton Do [*Pidele* DB, *Pidel'athelamston* 1288 Cl, *Athelhameston* 1303 FA]. '*Æþelhelm*'s TŪN.' Cf. PIDDLE.

Athelington Sf [*Alinggeton* 1219, *Athelinton* 1234 FF, *Athelington* 1254 Val]. See ALLINGTON Do.

Athelney So [(æt) *Æþelingaeigge* 878 ASC, *Æthelingaeg* c 894 Asser, *Adelingi* DB]. 'The island of the *æþelingas* or princes.'

Atherfield Wt [*Aderingefelda* 959 Ann Mon, *Arefeld* 1287–90 Fees, *Atherfelde* 1324 Misc]. See FELD. First el. as in ARRETON.

Atherington D [*Hadrintone* 1272, *Adring-*

tone 1309 Ep, *Atherington* 1322 Cl]. 'The TŪN of *Eadhere*'s or *Æpelhere*'s people.'

Atherstone So [*Alardeston* 1225 Ass, *Athelardeston* 1252 FF]. '*Æpelheard*'s TŪN.' **A~** Wa nr Tamworth [*Aderestone* DB, *Atheredestone* 1221 Ass]. '*Æpelrēd*'s TŪN.' **A~ upon Stour** Wa [*Eadrichestone* 710 BCS 127, *Edricestone* DB, *Adricheston* 1226–8 Fees]. '*Eadrīc*'s TŪN.'

Atherton La [*Aderton* 1212 Fees, *Atherton* 1322 Ipm]. '*Æpelhere*'s TŪN.'

Atlow Db [*Etelawe* DB, *Attelawe* 12, *-lowe* 1272 Derby]. '*Eatta*'s HLĀW.' *Hlāw* apparently means burial-mound here.

Attenborough Nt [*Adinburcha* Hy 2 (1316) Ch, *Aedingburc* 1205 Pp, *Adingburg* 1229 Ep]. 'The BURG of *Eada* or of *Eada*'s or *Æddi*'s people.'

Atterby Li [*Adredebi* 1185 TpR, *Atheradeby* 1202 Ass]. '*Eadrēd*'s or *Æpelrēd*'s BY.'

Attercliffe YW [*Ateclive* DB, *Atterclive* 1296 YInq]. Apparently OE *æt þǣm clife* 'at the cliff' with later insertion of *r*.

Atterton K nr Dover [*Eadredestun* 11 DM, *Etretone* DB]. '*Eadrēd*'s TŪN.'

Atterton Le [?*Æperedes dun* c 972 BCS 1283, *Attreton* 1209–35 Ep, 1227 Cl]. If the form of c 972 belongs here '*Æpelrēd*'s DŪN'. Otherwise '*Æpelrēd*'s or *Eadrēd*'s TŪN'.

Attington O [*Attendone* DB, *Ettendun* 1209–12 Fees, *Attendon* 1279 RH, *Attyndon* 1291 Tax]. Perhaps OE *æt þǣm dūnum* 'at the hills'. Or else '*Eatta*'s DŪN'. The situation favours the first alternative.

Attleborough Nf [*Atleburc* DB, 1194 P]. '*Ætla*'s BURG.'

Attleborough Wa [*Atteleberga* 1155–9 Fr, *Atleberga* 12 DC]. '*Ætla*'s hill or barrow.' See BEORG.

Attlebridge Nf [*Atlebruge* DB, *-brigge* 1175 P]. '*Ætla*'s bridge.'

Atwick YE [*Attingewic* c 1130 BM, c 1137 YCh 1144]. 'The wīc of *Atta*'s people.'

Atworth W [*Attenwrõe* 1001 KCD 706, *Atewurthe* 1242 Fees]. '*Atta*'s worp.'

Aubourn Li [*Aburne* DB, *Aburn* 1194 ff. P, *Auburn* 1254 Val, *Alburn* 1275 Cl.] OE *alr-burna* 'alder stream'.

Auburn YE [*Eleburne* DB, *Alburnia* c 1135 YCh 1152]. Possibly identical with AUBOURN Li. But perhaps rather OE *æl-burna* 'eel stream' with OScand *āl* substituted for OE *æl*.

Auckland, Bishop, A~ St. Andrew & St. Helen, West A~ Du [*Alclit* c 1050 HSC, 1104–8 SD, *Alklet* 12 FPD, *Auclent* 1202 FF, *Aucland* 1254 Val]. Possibly transferred from ALCLYDE, the old name of Dumbarton, which means 'the cliff on the Clyde'. The OBrit form of Alclyde was

Alclut (*Alcluith* c 730 Bede), whence OE *Alclȳt*. Cf. *Stræcled* 875 ASC for Strathclyde. The name was later associated with Scand *aukland* 'additional land'.

Bishop A~ belonged to the Bishop of Durham.

Auckley YW [*Alcheslei, Alchelie* DB, *Alkeleg* 1240 FF]. Perhaps '*Alca*'s LĒAH'.

Audenshaw La [*Aldenesawe, Aldwynshawe* c 1200 LaCh]. '*Aldwine*'s shaw or copse.'

Audleby Li [*Alduluebi* DB, *Aldolby* c 1115 LiS]. '*Aldwulf*'s BY.'

Audlem Chs [*Aldelime* DB, *-lima* 1259 BM]. See LYME. Audlem is 'old Lyme' or 'the part of Lyme belonging to *Alda*'; cf. BURSLEM.

Audley St [*Aldidelege* DB, *Aldithelega* 1182 P]. 'The LĒAH of *Aldgȳþ*' (a woman).

Aughton (awtn) La nr Ormskirk [*Achetun* DB, *Acton* 1235 Fees], **A~** (âftn) La nr Lancaster [*Aghton* c 1330 PNLa], **A~** YE [*Actun* DB, *Acton* 1230 P], **A~** YW [*Actone* DB, *Acton* 1202 FF]. OE *Āctūn* 'oak TŪN'. Cf. ACTON.

Aunby Li [*Ounebi* 1219 Ass, *-by* 1228 Ep, *Ounesby* 1288 Ipm], **Aunsby** Li [*Ounesbi* DB, *Ounebi* 1202 Ass, *Outhenby* 1281 Ch]. '*Ouðen*'s BY.' *Ouðen* LVD, *Oudon* DB is ON *Auðun*, ODan *Øthen*. *Auðun* was often shortened to ON *Aun*, and this seems to be the form of the name in Aun(s)by.

Aune, Aunemouth. See 2. AVON.

Aust Gl [*æt Austan* 794, 929 BCS 269, 665, *Augusta* c 1105 Fr, *Auste* 1209 Fees], **Aust Cliff** [*Austrecliue* DB, *Austeclive* 1233 Cl]. Aust is Lat *Augusta*. The reason for the name is not clear. Aust is on the Severn.

Austerfield YW [*Oustrefeld* DB, *Ousterfeud* 1279 Ipm]. 'FELD with a sheepfold' (OE *eowestre*).

Austerson Chs [*Alstaniston* 1260 Court, *Alstonestona* c 1266 Chester]. '*Ælfstān*'s or *Æþelstān*'s or *Alhstān*'s TŪN.'

Austhorpe YW [*Ossetorp* DB, *Austorp* 1190 P, *Oustorp* c 1180 YCh 1619]. ON *Austþorp* 'eastern thorp'.

Austhwaite Cu [*Auestwait* c 1220 StB, *-thwayt* 1256 P]. See THWAITE. The first el. is a pers. n., perhaps OSw *Afaster*.

Austonley YW [*Alstanesleie* DB, *Alstanley* 1274 Wakef]. '*Ælfstān*'s or *Alhstān*'s LĒAH.'

Austrey Wa [*æt Alduluestreow* 958 BCS 1021, *Aldulfestreo* 1002 E, *Aldulvestrev* DB]. '*Ealdwulf*'s tree.'

Austwick YW [*Oustewic* DB, *Austwich, Estwich* 1175 P]. 'Eastern WĪC.' Very likely a Scandinavianized form of OE *Ēastwīc*.

Autby Li [*Alwoldebi* DB, *Aluoldabi* c 1115 LiS]. '*Alfwald*'s BY.' *Alfwald* is an ODan pers. n. Or the first el. may be OE *Ælfwald*.

Authorpe Li nr Louth [*Agetorp* DB, *Haghe-*

torp c 1115 LiS, *Haghethorp* 12 DC], **A~** Li nr Mumby [*Aghetorp* c 1115 LiS]. '*Aghi*'s thorp.' *Aghi* is an ODan pers. n.

Avebury (ābrī) W [*Avreberie* DB, *Avesbiria* 1114 Fr, *Aveberia* 1189 Fr, *Aueberia* 1195 Cur]. '*Afa*'s BURG.'

Aveley Ess [*Aluithelea* DB, *Alvithele* 1206 Cur, *Auvilers* 1212 Fees]. 'The LĒAH of *Ælfgȳþ*' (a woman).

Avely Hall Sf [*Aluenelega* DB, *Alwyneleye* 1298 BM]. 'The LĒAH of *Ælfwynn*' (a woman).

Avenage Gl [*Abenesse* c 1170 Cotton ch. x. 7, *Abbenessa* 1169 P]. '*Abba*'s ash-tree.'

Avenbury He [*Aveneburi* DB, *Avenebiri* 1242 Fees]. A~ is on the Frome, which may have been called alternatively Avon. If so, 'BURG on the Avon'.

Avening Gl [(to) *Æfeningum* 896 BCS 574, *Aveninge* DB, *Aueninges* 1192 P]. OE *Æfeningas* 'dwellers on the Avon'. Avon must be an old name of the stream on which A~ is.

Averham (ār-) Nt [*Aigrun* DB, *Ægrum* 12 DC, *Egrum* c 1200, 1231 BM, 1227 Ep, *Averam* 1316 FA]. Possibly an OE *ēagārum* (dat. plur.) 'the gores or strips of land on the river (Trent)'. See GĀRA and cf. LANGAR Nt.

Aveton Gifford (awtn) D [*Avetone* DB, *Aueton* 1230 P, *Aveton Giffard* 1276 Cl]. 'TŪN on R Avon' (2). Walter Giffard held A~ in 1242 (Fees). Cf. ASHTON GIFFORD.

Avill (ă-) So [*Avena* DB, *Auena* 1188 P, *Avele* 1233 Cl]. Avill is also the name of the stream on which the hamlet stands. It is an old river-name, identical with AVON.

Avington Brk [*Avintone* DB, *Avintona* 1167 P], **A~** Ha [*Afintun* 961 BCS 1068, *Avintune* DB, *Avingtone* c 1270 Ep]. 'The TŪN of *Afa*'s people.'

Avisford Sx [*Auesford* (hd) 1166 P, *Avesforde* 1332 AD]. '*Æfic*'s ford.'

Avon, a common river-name, goes back to OBrit *Abonā*, the source of Welsh *Afon* &c. The name is identical with Welsh *afon*, Co *avon*, Ir *abhann* 'river'. The following Avons are recorded in early sources: 1. **(East) Avon** W, Ha [*Abon* 688 Malm, *Afen* a 672 BCS 27]. On this are **Avon** Ha [*Avere* DB], **Netheravon** and **Upavon** W [*Nigravre* DB, *Nederavena* c 1150 AC; *Oppavrene* DB, *Upavene* c 1150 BM]. 2. **Avon** or **Aune** D [(on) *Afene* 847 BCS 451, *Auene* 1238 Ass]. On this are **Aveton** (q.v.) and **Avon-** or **Aunemouth** [*Avenemue* 1287 Ipm]. 3. **(Lower) Avon** W, So, Gl [*Abon* 688 BCS 71, c 894 Asser, *Afene* (obl.) 808 BCS 327]. On this are **Avon** W [*Auene* 1065 KCD 817] and **Avonmouth** Gl [*Afene muþan* 918 ASC]. 4. **(Upper) Avon** Np, Le, Wa, Gl, Wo [*Afen* 704-9, *Afene* 780

BCS 123, 235]. **Little** or **Middle Avon** Gl [*Avon* c 1540 Leland] seems to have had a different name in early times. See also AVENBURY, AVENING, and (Avon) DASSETT.

Awbridge Ha [*Abedric* DB, *Abrigge* 1167 P, *Abbeterigges* c 1286, *Abboteregg* 1291 Ep]. 'The ridge of the abbot' (OE *abbod*).

Awliscombe D [*Aulescome* DB, -*cumb* 1201 Cur, *Houelescumb* 1286 FA]. See CUMB. The first el. may be OE *ǽwiell*, *āwiell* 'source of a river'.

Awre Gl [?*Alre* 872 BCS 535, *Avre* DB, *Aura* 1156 ff., ?*Alra* 1184 P, *Oure* 1200 Cur]. Perhaps OE *alor* 'alder'; cf. ALLER. If so, the change of *Alre* to *Aure* is due to Norman influence.

Awsworth Nt [*Ealdeswurð* 1002 E, -*wyrð* 1004 KCD 710, *Eldesvorde* DB]. '*Eald*'s WORÞ.' OE *Eald is a normal short form of names in *Eald-*.

Axe R So [*Aesce* 712 BCS 128, *Axam* 693 ib. 121], A~ R Do, D [(on) *Axan* 1005 KCD 1301, *Axe* 1244 Ass]. A variant form of EXE (q.v.). On Axe So is **Axbridge** [*Axanbrycg* c 910 ASCh (Burghal Hidage), *Axebruge* 1084 GeldR, 1168 P]. On the other Axe are **Axminster** D [*Ascanmynster* 755 ASC, *Axanmynster* ib. (E), *Axemunster* 901 BCS 588]: 'the minster on the Axe', and **Axmouth** D [*Axanmuða* c 880 BCS 553, *Axamuða* 1046 ASC (E)]

Axford W [*Axeford* 1185, 1195 P, 1200 Cur]. Perhaps OE *æsca-ford* 'ford by the ash-trees'. In *æsca sc* would normally be preserved as *sk* and metathesis to *ks*, *x* would follow.

Axholme Li [*Haxeholm* c 1115 LiS, *Haxiholma* c 1150 Fr, *Axiholm* 1179 P, *Haxi-*, *Axiholm* 1233 Cl]. 'The holm or island of Haxey.' HAXEY is a vil. in Axholme; it may be the old name of the Isle of Axholme. The addition -*holm* is Scandinavian. The loss of initial *h*- is due to dissimilation. See further HAXEY.

Axminster, Axmouth. See AXE.

Axwell Park Du [*Aksheles* 1344 PNNb]. 'Shielings by the oaks.' Cf. SCHĒLE.

Aycliffe, Great, School A~ Du [*Heaclif* c 1090 SD, *Aclea* c 1085 LVD, *Acleia* 1195 (1335) Ch, *Acle* 1254 Val, *Aclyff* 1381 Pat, *Sculacle* 12 FPD]. Great Aycliffe may go back to OE *āclēah* 'oak wood or clearing', School A~ (which is on a hill) to OE *ácclif* 'oak hill', and *Heaclif* may be really *Hēa Ácclif* 'high Aycliffe'.

School A~ from a Scandinavian called *Scula* (ON *Skúli*), who was given land in Durham c 920 (HSC).

Aydon Nb nr Corbridge [*Ayden* 1242 Fees, 1256 Ass, *Eyden* 1285 Cl, *Hayden* 1242 Fees, 1322 Ipm], A~ Nb nr Alnwick [*Aydun* 1279 Ass, *Haydene* 1325 Percy, *Haydon*, *Ayden*

1346 FA]. The second el. of no. 1 is OE *denu* 'valley' or rather *denn* 'pasture', that of no. 2 OE *dūn* 'hill'. The first el. might be OE *ēg* 'island', but *hēg* 'hay' would be more suitable.

Aylburton Gl [*Alberton* 1176, *Ailbrichton* 1186 P, *Ailbricton* 1227 Ch]. '*Æþelbeorht*'s TŪN.'

Ayle Burn R Nb [*Alne* 1347 ERN]. Identical with ALN. On Ayle Burn is **Ayle.**

Aylesbeare D [*Eilcsberge* DB, *Aylesbere* 1227 Ch, -*bare* 1231 FF, *Aillesbir'* 1235 Cl]. The earliest forms point to OE *beorg* 'barrow' as the second el., later replaced by OE *bearu* 'grove'. The first el. is *Ægel* pers. n.; cf. AILSWORTH, AYLESBURY, &c.

Aylesbury Bk [*Ægelesburg* 571, *Æglesbyrig* 921 ASC, (æt) *Ægelesbyrig* c 970 BCS 1174, *Eilesberia* DB]. '*Ægel*'s BURG'; cf. AILSWORTH.

Aylesby Li [*Alesbi* DB, c 1115 LiS, 1202 Ass]. '*Ali*'s BY'; cf. AILBY.

Aylesford K [*Ægelesford* 455 ASC (W), *Æglesford*, *Æilesford* 10 BCS 1321 f., *Ailesford* DB]. '*Ægel*'s ford'; cf. AILSWORTH. *Agæles* (*Ægeles*) *þrep* 455 ASC (A, E) is probably an alternative name of Aylesford, meaning '*Ægel*'s thorp'.

Aylestone Le [*Ailestone* DB, -*tona* 1209–19 Ep, *Eyleston* 1254 Val]. '*Ægel*'s TŪN'; cf. AILSWORTH. ON *Egill* is also a possible first el.

Aylmerton Nf [*Almartune* DB, *Adelmērton* 1199, 1208 FF]. '*Æþelmǽr*'s TŪN.'

Aylsham Nf [*Eilessam*, *Ailesham* DB, *Ailesham* 1159, *Eilesham* 1168 P]. '*Ægel*'s HĀM'; cf. AILSWORTH.

Aylton He [*Ailenetona* 1138 AC, *Eleuetun* 1242, *Ayleue(n)tun* 1250 Fees]. 'The TŪN of *Æþelgifu*' (a woman).

Aylworth Gl [*Ailewrde* DB, *Eyleworth* 1220 Fees, *Eileworth* 1230 Cl]. Apparently '*Ægla*'s WORÞ'. *Ægla* may be a side-form of *Ægel*; cf. AILSWORTH.

Aymestrey He [*Elmodestreu* DB, *Ailmondestre* 1291 Tax]. '*Æþelmund*'s tree.'

Aynho Np [*Aienho* DB, *Eynho* 1220 Fees, *Ainho* 1226 Ep]. '*Æga*'s HŌH or hill.' OE *Æga* seems to be the first el. of *Æganstan* BCS 226. Cf. OHG *Aigo*, *Eigio*.

Ayot St. Lawrence & St. Peter Hrt [*Ægate*, *Aiegete* c 1060 ASWrits, *Ægete* 1065 BM, *Aiete* DB, *ecclesia Sancti Laurentii de Aieta* 1249 Ep]. OE *Ægan geat* '*Æga*'s pass'; cf. AYNHO (PNHrt(S)). Ayot St. Peter belonged to St. Peter's, Westminster.

Ayresom YN [*Arusum* c 1180 YCh 659]. ON *ár-húsum* dat. plur. 'the river houses' (ON *á* 'river', gen. *ár*, and *hús*). A~ is on the Tees.

Aysgarth YN [*Echescard* DB, *Aikeskerth* 1202, *-scarth* 1223 FF]. ON *eikiskarð* 'gap or pass with an oak wood'. See SKARÐ. ON *eiki* means 'oaks, oak wood'. The *scarð* is the gap in the chain of hills south of the Ure formed by Bishop Dale.

Ayshford D [*Æsc-*, *Escford* 958 BCS 1027, *Aisseford* DB]. See ASHFORD D.

Ayston Ru [*Æðelstanestun* 1046 KCD 785, *Adelstaneston* 1203 Cur]. '*Æþelstān*'s TŪN.'

Ayton, Great & Little, YN [*Atun* DB, *Etona* c 1162 YCh 881, *Aton*, *Parva Aton* 1279–81 QW], **East A~** YN [*Atun* DB, *Aton* 1231 FF], **West A~** YN [*Atun* DB, *Aton* 1234 FF]. 'The TŪN on the river' (Leven and Derwent). The original name was very likely OE *Ēatūn* (cf. EATON), with OE *ēa* replaced by the synonymous ON *á*.

Azerley YW [*Aserla*, *Asserle*, *Haserlai* DB, *Azerlagh* R 1 Fount, *Azerlay* 1277, *Hatherley* 1283 Ipm, *Atherlay* 1281 Cl]. See LĒAH. The first el. may be the OScand. pers. n. *Atzer*, *Azer* (OE coins, DB &c.) from ON *Ǫzurr*, *Azurr*, ODan *Azur*.

B

Babbacombe D [*Babbecumbe* c 1200 Torre]. '*Babba*'s valley.'

Babcary. See CARY.

Babingley Nf [*Babinghelea* DB, *Babbingele* 1212 Fees], **Babington** So [*Babbingtona* DB, *Babingtone* DB, *-ton* 1201 Cur]. 'The LĒAH and TŪN of *Babba*'s people.'

Bablock Hythe O [*Babbelak* 1279 RH, (water of) *Babbelak* 1317 Pat]. Bablock is OE *Babban lacu* '*Babba*'s stream'. Hythe is OE *hȳþ* 'landing-place'.

Babraham Ca [*Bathburgeham* c 1080 ICC, *Badburgham* DB, *Badburgeham* 1166 P]. 'The HĀM of *Beaduburg*.' This woman's name is unrecorded.

Babthorpe YE [*Babetorp* DB, *Babbethorp* c 1200 YCh 997], **Babworth** Nt [*Baburde* DB, *Babbeuurde* 12 DC, *-wrth* 1294 Ch]. '*Babba*'s thorp and WORÞ.'

Bache. See BÆCE.

Backbarrow La [*Bakbarowe* 1537 PNLa]. 'Hill with a backlike top' (OE *bæc* 'back' and BEORG).

Backford Chs [*Bacfort* c 1150 Chester, *-ford* 1260 Court]. 'Ford by a ridge.' See BÆC.

Backwell So [*Bacoile* DB, *Bacwell* 1202 Cur, 1241 Ass]. 'Stream coming from a ridge.' See BÆC.

Backworth Nb [*Bachisurda*, *Bacwrth* 12 (1271) Ch, *Backewrth* 1268 Ipm]. '*Bacca*'s WORÞ.'

Baconsthorpe Nf [*Torp*, *Baconstorp* DB, *Bacunestorp* 1203 Ass]. Originally THORP. *Bacons-* from a Norman owner. *Bacon* is a surname, originally a nickname from OFr *bacun* 'bacon', 'carcase of a pig'.

Bacton He [*Bachetune* DB, *Baketon* 1188 P, *Bakyntune* 1249 Fees], **B~** Nf [*Baketuna* DB, *-tun* c 1150 Crawf, *-ton* 1185 ff. P], **B~** Sf [*Bachetuna* DB, *Baketon* 1198 Cur]. '*Bacca*'s TŪN.'

Bācup La [*Fulebachope* c 1200 WhC, *Bacop* 1324 Ipm]. OE *bæc-hop* 'valley by a ridge'. See BÆC, HOP.

Badbury Do [(æt) *Baddanbyrig* 901 ASC, *Badeberi* (hd) 1182 P], **B~** W [*Badeburi*, 955 BCS 904, *Badeberie* DB]. '*Badda*'s BURG.' Both are at prehistoric camps, B~ Do at Badbury Rings, B~ W at Liddington Castle. B~ Hill Brk is a hill with a 'Danish' camp. BADBY, originally *Baddanburg*, is near an ancient earthwork. Also BAUMBER Li was OE *Bad(d)anburg*. The fact that *Baddanburg* refers at least in three cases to prehistoric camps is remarkable and may suggest that *Badda* was a legendary hero, who was associated with ancient camps.

Badby Np [*æt Baddan byrig*, *Baddan by* 944 BCS 792, *Baddanbyr(i)g* KCD 1356, *Badebi* DB]. Originally OE *Baddanburg* (cf. BADBURY), later changed to *Baddan by* with Scand BY for OE BURG.

Baddeley St [*Baddilige* 1227 Ch, *Badeleye* 1270, *Badilegh* 1271 Ass]. 'The LĒAH of *Badda*'s people.'

Baddesley, North, Ha [*Bedeslei* DB, *Baddesly* c 1270 Ep, *Badesleye* 1291 Tax], **South B~** Ha [*Bedeslei* DB, *Badeslie* 1212 Fees, *-legh* 1236 Ch, *Baddeslegh* 1235 Cl], **B~ Clinton** Wa [*Badesleia* 1166 RBE, *-leye* 1298 Ipm, *Baddesley Clynton* 1466 AD], **B~ Ensor** Wa [*Bedeslei* DB, *Badeleia* 1198 f. P, *Baddesleye Endeshover* 1327 Subs]. OE *Bæddes-lēah* '*Bæddi*'s woodland'. OE *Bæddi* is a regular formation from *Badda*.

N. and S. B~ Ha are at opposite sides of the New Forest, and *Bæddes-lēah* was doubtless the old name of the forest district later called the New Forest. B~ Clinton and Ensor are over 15 miles apart, but at each end of a long ridge, which was no doubt wooded in the old days. *Bæddes-lēah* may have been the name of that forest also. B~ Clinton was held by John de Clinton in 1298 (Ipm). Cf. ASTON CLINTON.— B~ Ensor came to Thomas de Ednesouere in 1259 (Ipm). His son Thomas died without issue in 1285 (Ipm). The family came from EDENSOR Db.

Baddiley Chs [*Bedelei* DB, *Badile* 1287 Court]. 'The LĒAH of *Bēada*' (in *Beadan healan* BCS 936) or 'the LĒAH of *Beadda*' (in *Bæaddan syla* BCS 961).

Baddington Chs [*Batington* 1283 Ipm,

Batinton 1283 Cl, 1309 Ipm]. 'The TŪN of *Bata*'s people'; cf. BATCOMBE.

Baddow Ess [*Beadewan* c 1000 BCS 1306, *Baduuen* DB, *Badewe* 1212 Fees, *Badewe Magna, Parva* 1238 Subs]. An old name of the CHELMER [*Beadewan ea* 1062 KCD 813].

Badenhall St [*Badehale* DB, *Badenhale* 1242 Fees]. '*Bada*'s HALH.'

Badgemore O [*Begevrde* DB, *Bagerigge* 1181 P, *Bagemercroft, -grene* 1425, *Baiemore crofte* 1594, *Badgmoor* 1797 PNO(S)]. Originally '*Bæcga*'s HRYCG' and '*Bæcga*'s WORÞ' (cf. BADGEWORTH Gl), probably the names of the two Badgemores shown in the map. *Badgemore* is late for *Badgemer*, and the latter no doubt represents an OE *Bæcghǽme* for the inhabitants of the place, used in combination with *croft* and *grene* (as *supra*). *Bageme-* became *Bagemer-*, perhaps owing to anticipation of the *r* of *croft* and *grene*, and a new name *Badgemer*, whence later *Badgemore*, arose.

Badger Sa [*Beghesovre* DB, *Begesour* c 1154 Eyton, *Bageshour* 1212 Fees, *Beggesor* 1229 Cl]. '*Bægi*'s shore.' See ŌFER. OE *Bægi is a regular derivative of *Bacga*.

Badgeworth Gl [*Beganwurþan* 872 BCS 535, *Becgwirðe* 1022 KCD 1317, *Beiewrde* DB, *Beggewurda* 1190 P], **Badgworth** So [*Bagewerre* DB, *Bæggewurda* 1158, *Beggewurda* 1159 P, *Baggeworth* 1225 Cl]. '*Bæcga*'s WORÞ.' *Bæcga is a derivative of *Bacga*.

Badingham Sf [*Badincham* DB, *Bedingham* 1203 Cur, *Badingham* 1254 Val]. 'The HĀM of *Bēada*'s people' (cf. *Beadan healan* BCS 936).

Badlesmere K [*Badelesmere* DB, 1163–5 Fr, *-mar* 1200 Cur, *Bæthdesmere* 11 DM, *Bedelesmera* 1170 P]. Second el. probably OE *mere* 'lake'. The first may be a pers. n. *Bæddel, a pet form of *Badda*.

Badley Sf [*Badelea* DB, *-le* 1200 Cur, *Baddelea* 1200 P]. '*Ba(d)da*'s LĒAH.'

Badlingham Ca [*Belincgesham* c 1080 ICC, *Bellingeham* DB, *Bethlingeham* 1086 IE, *Badlingham* 1232 FF]. The first el. is a derivative of a pers. n., e.g. *Bæddel* (cf. BADLESMERE). If so, 'the HĀM of *Bæddel*'s people'.

Badminton Gl [*Badimyncgtun* 972 BCS 1282, *Madmintune* DB, *Badmintun* c 1200 BM]. 'The TŪN of *Beadumund*'s people.'

Badmondisfield Sf [*Bademundesfelda* DB, *Badmundesfeld* 1270 Ipm]. '*Beadumund*'s FELD.'

Badsaddle Np [*Bateshasel* 12 NS, *Baddeshasel* 1220 Fees]. '*Bæddi*'s hazel-bush' (cf. BADDESLEY).

Badsey Wo [*Baddeseia* 709, *-ege* 714 BCS 125, 130, *Badesei* DB]. '*Bæddi*'s island' (cf. BADDESLEY). The same man gave his name to a stream near Badsey (*Bæddeswellan* 972 BCS 1282).

Badsworth YW [*Badesworde* DB, *Baddeswurth* 1226 FF]. '*Bæddi*'s WORÞ' (cf. BADDESLEY).

Badwell Ash Sf [*Badewell* 1254 Val, *Badewelle Asfelde* 13 BM, *Asshfeld Badewelle* 1320 AD]. '*Bada*'s stream.'
Ash is a shortening of Ashfield. Badwell Ash is near Gt Ashfield.

OE **bæc** 'back', probably also, like OHG *bah*, 'ridge'. See BAKE, BACKBARROW, -FORD, -WELL, BACUP, BASHALL, BECKHAMPTON, also BURBAGE Le.

OE **bæce, bece** 'stream', ME *bache* 'valley of a stream', dial. *bache, batch* &c. 'stream, valley'; cf. G *bach*, ON *bekkr* &c. OE *bæce* is the source of **Bache** Chs [*Bache* 1119, *Bechia* c 1150 Chester], B~ He several [one is *Becce* DB], B~ Sa [*La Bache* 13 BM]. First el. of BATCHCOTT, BEACHAMPTON, BECCLES, BECKTON, BEIGHTON, BETCHTON, BETTON. Second el. of BURBAGE, COLEBATCH, COTTESBACH, DEBACH, SANDBACH, HASELBECH, HAZLEBADGE, HOL-, LAND-, WATERBEACH, WISBECH. Cf. PINCHBECK.

OE **bǽr** 'pasture', esp. 'swine-pasture', found alone and in *den-, weald-, wudubǽr*. Sometimes difficult to distinguish from BEARU, BȲRE, BEORG. See BEER, BERE, also STOCK-, ORLINGBURY.

OE **bæþ** 'bath', probably in such a sense as 'hot spring', is found in BATH, BATHLEY &c., BALE, MOREBATH.

Bagborough So [*Bacganbeorg* 904 BCS 610, *Baggabeorc* 1065 Wells, *Bageberge* DB, *Little, West Baggebergh* 1243 Ass]. Probably '*Bacga*'s hill'; cf. however BAGLEY. Another *Bacgan beorg* is recorded as *Baggenbeorg* KCD 1368 (Gl or Wa).

Bagby YN [*Bag(h)ebi* DB, *Baggaby* c 1160 YCh 175, *Baggebi* J BM]. '*Baggi*'s BY.' *Baggi* is found in ON, ODan and OSw.

Bagendon Gl [*Benwedene* DB, *Baggingeden* 1220 Fees, *Bachingeden* c 1250 Eynsham, *Bagingeden* 1291 Tax]. 'The valley of *Bæcga*'s people'; cf. BADGEWORTH.

Bageridge Do [*Bagerug* 1250 Fees], B~ St [*Bagerugg* 1322 Ipm], **Baggridge** So [*Bagerug* 1314 Ipm]. See HRYCG. The first el. may be *Bacga* pers. n. or the element discussed under BAGLEY.

Baggrave Le [*Badegraue* DB, *Badesgraua* 1169, *Babbegraua* 1177, *Babegraue* 1190 P, *Babbegrave* 1226–8 Fees]. Probably '*Babba*'s grove'. See GRĀF. If the two earliest forms are reliable, however, the first el. might be OE *Beaduburg*; cf. BABRAHAM.

Baggrow Cu [*Baggerawe* 1332 Subs]. Identical with **Bagraw** Nb [*Bagraw* 1385 PNNb]. Second el. OE *rāw* 'a row', later 'a row of houses, a street'. The first seems to be ME *bagger*, an early form of *badger* 'hawker'.

Baginton (-ăg) Wa [*Badechitone* DB, *Bathekintona* Hy 2 (1314) Ch, *Batkinton* 1242 Fees, *Bakynton* 1285 Ipm]. 'The TŪN of *Badeca*'s people'; cf. BAKEWELL.

Bagley Brk [*Bacgan leah* 955 f. BCS 906, 924, *Bacheleia* (nemus) c 1075 Abingd], B~ Sa [*Bageleia* c 1090, *Baggeleg* 1225 Eyton], B~ So [*Bagelie*, *Bagaleia* DB, *Baggelega* 1176 P], B~ YW [*Bagalaia* 1148 YCh 179, *Baggelega* 1188 P, *-le* 1199 Obl]. To these should be added BAGULEY and (to) *Baggan leage* 909 BCS 624 (Ha). Second el. OE *lēah*, in some cases certainly 'woodland'. The·first may sometimes be *Bacga* pers. n. But the name Bagley is very common and there are several other names that contain a first el. *Bag-* (BAGERIDGE, BAGSHAW, BAGSHOT, BAGSLATE &c.). It is curious that the second element is usually a word for a natural feature, very rarely one for an inhabited place. There are no examples of *Bag-* combined with HĀM, TŪN, WĪC. This suggests that in some cases at least *Bag-* represents some significant word. In Scand languages *bagge* means 'a wether, a ram'. MDu *bagghe* means 'a small pig'. There may have been an OE word *bacga* denoting some animal. If BAWDRIP contains this element, *bacga* must have denoted some wild animal, for *-drip* is OE *treppe* 'trap'. We can only guess at the meaning of the hypothetical word. It might have been 'fox' or 'badger'.

Bagnall St [*Badegenhall* 1273, *Baginholt* 1271, *Bagenholt* 1281 Ass]. The first el. is OE *Badeca* pers. n.; cf. BAGINTON. The second may be HOLT or HALH.

Bagnor Brk [*Bagenore* DB, *Baggenore* 1230 P]. Second el. OE ŌRA 'bank'. For the first see BAGLEY.

Bagraw Nb. See BAGGROW.

Bagshaw Db [*Baggeshage*, *-shawe* 1318 PNDb]. See BAGLEY and SCAGA.

Bagshot Sr [*Bache-*, *Bagsheta* 1165 P, *Bacsete* 1204 Cur, *Bacsiete* 1212, *Bagsete* 1219 Fees, *-schate* 1221 FF, *Baggeshete* 1253 Ch]. Second el. OE SCĒAT. For the first see BAGLEY.

Bagshot W [*Bechesgete* DB, *Bechesieta* 1130 P]. '*Beocc*'s gate.' See GEAT. *Beocc is found in *Beoccesheal* 968 BCS 1213 (in the neighbouring Bedwyn).

Bagslate La [*Bagslade* 13 WhC]. The same name seems to be *Bacga slæd* 816 BCS 356 (Wo). Second el. SLÆD 'valley'. For the first cf. BAGLEY.

Bagthorpe Nf [*Bachestorp* DB, *Bakethorp* 1254 Val, *Baggetorp* 1206 Cur, *-thorp* 1291 Tax]. '*Bakki*'s thorp.' *Bakki* is an ODan pers. n. Or the first el. may be OE *Bacca* pers. n.

Bagtor D [*Bagetore* DB, *Baggetorre* 1242 Fees]. '*Bacga*'s hill.' See TORR. But cf. BAGLEY.

Baguley Chs [*Bagelei* DB, *Baggeleg* 1260 Court]. See BAGLEY.

Bagworth Le [*Bageworde* DB, *Baggeworth* 1270 Ch]. '*Bacga*'s WORÞ.'

Baildon YW [*Bægeltun*, *Bældun* c 1030 YCh 7, *Beldune* DB, *Beildon* 1182 P, *Baildun* c 1190 YCh 1657]. The place is near Baildon Hill, to which the name no doubt referred originally. Second el. DŪN 'hill'. The first is obscure. It might be an OE *bĕg-hyll* 'berry hill, hill where berries grew'.

Bailey La [*Baillee* 1204 FF, *Beyley* 1246 Ass]. OE *bĕg-lēah* 'wood or clearing where berries grew'.

Bain R Li [*Beina* c 1145, *Baina* Hy 2 DC], B~ R YN [*Bain*, *Bein* 1218 FF]. From ON *beinn* 'straight', but perhaps in some transferred sense, e.g. 'helpful'. **Bainbridge** YN [*Bainebrigg* 1218 FF] is on the Bain. **Bainbridge** YW may have been named from a lost river-name Bain, but *Bain-* may here be *bain* 'handy, direct' (of a road) from ON *beinn*.

Bainton Np [*Badingtun* 972–92 BCS 1130, *Badinton* 1176 P, 1200 Cur], B~ O [*Baditone* DB, *Badyngton* c 1205 Fridesw]. 'The TŪN of *Bada*'s people.' *Ælfnoð Badan sune* is mentioned in BCS 1130.

Bainton YE [*Bagentone* DB, *Baingtun* c 1155 YCh 1097, *Baenton* 1191 P, *Bainton* 1222 FF]. 'The TŪN of *Baga*'s people.'

Baisbrowne We [*Basebrun* Hy 3 Mon, *Baysbrowne* c 1512 Kendale]. ON *báss* 'cowshed' and *brún* 'edge, brink'.

Bake Co [*West Baak* 1325 AD]. OE BÆC 'back, ridge'.

Bakewell Db [(to) *Badecan wiellon* 924 ASC, *Badecanwelle* 949 BCS 884, *Badeqvella* DB, *Bauquell* 1236 Fees]. '*Badeca*'s spring(s) or stream.' *Badeca* is a later form of *Baduca*, found also in *Badecan dæne* BCS 1307. Cf. BAGINTON, BAGNALL.

Balby YW [*Balle*(s)*bi* DB, *Ballebi* 1173 P, *-by* c 1190 YCh 1006]. '*Balle*'s BY.' *Balle* is an ODan pers. n.

Balcombe Sx [*Balecumba* 1121 AC, 1189 P, *Belecumbe* 1279 Ass]. The first el. might be *Bægloc* pers. n. (cf. BASHLEY) or OE *bealu* 'evil, calamity'. See CUMB.

Balder R YN [*Baudre* 1240 FF, *Balder* 13 YD]. Either a back-formation from **Baldersdale** [*Baldersdale* 1327 Ipm], whose first el. is then the pers. n. *Baldhere*, or more likely an old river-name consisting of Welsh *bal* 'peak' and *dwfr* (from Brit *dubro-*) 'stream'.

Baldersby YN [*Baldrebi* DB, J Ass, *Baldeby* 1228 FF]. '*Baldhere*'s BY', though the absence of the genitive ending *-s* is curious.

Balderston La [*Balderestone* a 1172 Whitaker, *Baldreston* 1256 FF], **Balderstone** La [*Baldreston* 1323 Ct]. '*Baldhere*'s TŪN.'

Balderton Chs [*Baldreton* 1330 Ch], B~ Nt [*Baldretone* DB, *Baldertun* c 1160 RA]. '*Baldhere*'s or *Baldþrȳþ*'s TŪN.' Cf. BALTERLEY.

Baldock Hrt [*Baldac* 1168 P, 1197 FF, 1200 Cur, 1227 Ch]. *Baldac* is the OFr name of Bagdad. *Baldak* is used by Mandeville, *Baldock* by Skelton, of Bagdad. This name was given to Baldock Hrt by the Knights Templars, who held the manor.

Baldon, Toot & Marsh, O [*Bealddunheama gemǽre* 1050 KCD 793, *Bealdanhema gemǽre* 1054 ib. 800, *Balde(n)done* DB, *Mersse Baldindon* 1241 Ep, *Totbaldyndone* 1312 BM]. OE *Bealdan dūn* '*Bealda*'s hill'; cf. *Baldan geat* BCS 1282, *Bealdan ǽrsc* BCS 624.

Toot is ME *tote* 'an isolated conspicuous hill, a look-out hill'. Baldon is near a hill. Marsh is the common word *marsh*.

Baldslow Sx [*Baldeslei* (hd) DB, *-lawe* (hd) Hy 3 Misc, *-lowe* 1296 Subs]. '*Beald*'s hill.' *Beald* is found in *Bealdessol* 944 BCS 797.

Baldwinholme Cu [*Baldewynholm* 1332 Subs]. '*Baldwine*'s holm.' Or the first el. may be the corresponding OFr name (*Baudoin*) from OHG *Baldewin*.

Bale Nf [*Bathele* DB, 1177 P, *Bale* 1208 Cur]. Identical with BATHLEY Nt. For the reduction of *-lēah* cf. SALL.

Bǎlham Sr [*Bælgenham* 957 BCS 994, *Belgeham* DB, c 1080 Reg, *Belga-*, *Balgaham* c 1115 (1330) Ch]. The OE form seems to have been *Bealgan-hamm*. The form of 957 is in a late transcript. *Bealga* appears to be a pers. n., derived from the adj. *bealg* that is the source of ME *balgh* 'rounded, smooth'; cf. BALSHAW.

Balk YN [*Balk* 12 YD]. OE *balc* is found in the sense 'a ridge in ploughing'. Probably it had a wider sense of 'ridge'. Balk is under a ridge.

Balkholme YE [*Balcholm* 1199 FF, *Balkeholm* 1240, *Balkholm* 1246 FF]. See HOLM. The first el. may be OE *balc* or *balca* 'ridge, bank', but the place is in a low situation, and ON *Balki* pers. n. is perhaps more likely.

Balking Brk [*Bedalacinge* 948 BCS 873, *Baða-*, *Badalacing* 963 ib. 1121, *Badeleching* 1187 P, *-leking* c 1200 Berk, *Bathelking* c 1286 BM]. Perhaps an old stream-name; cf. LOCKINGE Brk. If the second el. is *Lācing*, identical with Lockinge, the first might be a form of OE *bæþ* 'bath' (*baþa* gen. plur.).

Ballidon Db [*Belidene* DB, 12 Derby, *Balidene* 12 BM]. OE *bælgdenu* 'sack-like valley'; first el. OE *belg*, *bælg* 'bag'.

Ballingdon Sf [*Belindune* DB, *Balidon* 1236

Fees, 1332 Ch]. Second el. OE *dūn* 'hill'. The first might be the adj. *bealg* 'rounded', mentioned under BALHAM and found in *Balgandun* 704–9 BCS 123.

Ballingham He [*Badelingeham* 1215 Cl, *Baldingham* 1251 Ch, *Balyngeham* 13 BM]. OE *Badelingaham* 'the HĀM of **Badela*'s people'; cf. *Badelan broc* KCD 714. B~ is held to be identical with *Lann Budgualan* c 1150 LL, but if so, there is hardly any etymological connexion.

Balne YW [(buscus de) *Balna* 1160 YCh 484, *Balna* c 1185 ib. 495, *Balne* 1175 P]. Balne was also the name of a district, to judge by the name THORPE IN BALNE. Identical with Balne is **Balm Green** in Sheffield [*the Balne* 1332, *Balne* 1333 Hall, Sheffield]. The name is obscure. It might possibly be OE *Bēan-lēah* 'bean clearing', whence *Banlēa* and with metathesis *Balne*.

Balsall, Temple, Wa [*Belesale* 1185 TpR, *Belessale* 1226 AD, *Beleshale* 1327 Ch, *Balleshale* 1353 BM]. '*Bælli*'s HALH.' *Bælli* is evidenced in *Bælles weg* BCS 814. B~ belonged to the Knights Templars at least in 1185.

Balscott O [*Berescote* DB, *Belescot* 1204, 1208 Cur, *Balescot* 1242 Fees]. '*Bælli*'s COT'; cf. BALSALL.

Balsdean Sx [*Baldesdene* c 1100 PNSx]. '*Beald*'s valley'; cf. BALDSLOW.

Balsham Ca [*Bellesham gemære* 974 BCS 1305, *Belesham* c 1050 KCD 907, DB, *Bælesham* 1086 IE, *Balesham* 1245 Ch]. '*Bælli*'s HĀM'; cf. BALSALL.

Balshaw La [*Balghschae* 1296 Lacy, *Balschagh* 1311 Ipm]. See SCAGA. The first el. is ME *balgh* 'rounded, smooth', OE **b(e)alg*. See BALHAM, BALLINGDON.

Balterley St [*Baltryðeleag* 1002 Wills, *Baltredelege* DB, *-legh* 1289 Court]. '*Baldþrȳþ*'s LĒAH.' The woman's name *B(e)aldþrȳþ* is not otherwise recorded.

Baltonsborough So [*Balteresberghe* 744 BCS 168, *Baltunesberge* DB, *Baltenesberge* 1196 f. P]. '*Bealdhūn*'s hill or barrow.' See BEORG.

Bamber Bridge La [*Bymbrig* n.d. VH]. First el. the pers. n. *Bimme* found 1246 Ass (La). See BRYCG.

Bamborough or **-burgh** Nb [*Bebbanburh* 547 ff. ASC (E), *-burg* c 890 OEBede, *Baenburg* 1130 P, *Banburg* 1212 Fees]. '*Bebbe*'s BURG.' Bamborough was built by king Ida in 547 according to ASC (E). It was named, according to Bede, after a former queen *Bebba*, evidently Bebbe, queen of Æthelfrith (593–617).

Bamford Db [*Banford* DB, *Bamford* 1228 Ch], B~ La [*Baunford* 1282 FF, *Bamford* 1322 La Inq]. 'Ford with a beam', i.e. a foot-bridge.

Bamfurlong La [*Banforthlang* 1442 VH]. 'Furlong or strip where beans were grown.'

Bampton, Little, & Kirkbampton Cu [*Bamton* 1201 Cur, 1224 P, *Banton* 1201 Cur, *Little Bampton* 1317 Misc], B∼ O [*Bemtun* c 1070 Ex, *Bentone* DB, *Bamton* 1212 BM], B∼ We [*Banton* 1201 Cur, *Bampton* 1291 Tax]. OE *Bēam-tūn* 'TŪN by a beam' or 'TŪN made of beams'.

Bampton D [*Badentone* DB, *Bathentona* 1156 Wells, *Baentona* 1130 P]. An OE *Bæphǣmatūn* 'the TŪN of the dwellers at a bath or hot spring'.

Banbury O [*Banesberie* DB, *Banabereie* c 1100 EHR ix, *Banebiria* 1146 RA, *Banneberia* 1109 Eynsham, c 1160 RA, -*bir'* 1229 Cl]. Perhaps '*Bana*'s BURG', though the normal *nn* is remarkable. OHG has *Banno* by the side of *Bano*. OE *Ban(n)a* is not actually evidenced, but cf. BANNINGHAM.

Bandon Hill Sr [*Bandon* 1203, 1206 Cur, *Bamdon* 1208 Obl]. OE *Bēandūn* 'hill where beans were grown'. Cf. *Beandun* 614 ASC (a different place).

Banham Nf [*Benham* DB, *Banham* 1168, 1174 P, 1212 Fees]. OE *Bēan-hām* 'HĀM where beans were grown'.

ME **banke** 'bank, ridge', from ODan *banke*, ON *bakke*, occurs in some names as the second el., mostly in the sense 'hill' (PICKUP BANK &c. La, FIRBANK, NINEBANKS).

Banningham Nf [*Banincham* DB, *Banningeham* 1170 P, *Baningham* 1212 Fees]. 'The HĀM of *Bana*'s or *Banna*'s people.' Cf. BANBURY.

Bannisdale We [*Banendesdala* c 1180, *Banandesdale* 1251, *Bannandesdalabec* 1198 Kendale]. The first el. is very likely an ON byname **Bannandi* 'the man who curses'.

Banstead Sr [*Benstede* 675 BCS 39, *Benestede* DB, *Banstede* c 1269 BM]. 'Place where beans were grown'; cf. STEDE and *Beanstede* 947 BCS 820 (Merstham Sr).

Banwell So [*Banuwille* c 894 Asser, (æt) *Bananwylle* 904, 968 BCS 612, 1219, *Banwelle* DB]. This may be '*Bana*'s stream' (cf. BANBURY). But a meaning 'felon stream' is also possible (first el. OE *bana* 'murderer'); cf. WARNBOROUGH.

Bapchild K [*Baccancelde* 696 f. BCS 91, 94, *Bacchechild* 1197 P, *Babchilde* 1572 BM]. '*Bacca*'s spring.' See CELDE.

Bapton W [*Babinton* 1220 FF, 1242 Fees]. 'The TŪN of *Babba*'s people.' B∼ is near BAVERSTOCK (q.v.).

OE **bār** 'boar'. See BAR-, BOAR- (passim), BORESFORD, BORLEY, BOSWORTH.

Barbon We [*Berebrune* DB, -*brunna* 1195 P]. 'The stream of the bear.' Really the name of Barbon Beck.

Barbourne Wo [*Beferburna* 904 BCS 608]. 'Beaver stream.'

Barbury Hill W [(æt) *Beran byrg* 556 ASC, *Bereberia* 1180 Fr]. '*Bera*'s BURG'; cf. BARHAM K.

Barby Np [*Berchebi* DB, *Bergebi* c 1200 BM, *Bereweby* 1236 Ep]. OScand *Bergabȳ(r)* 'BY on the hill(s)'. Cf. BARROWBY.

Barcheston Wa [*Berricestone, Bercestone* DB, *Bercheston* 1193 P]. As a name of a hundred it is *Bedriceston, -e* DB. These forms may point to a first el. OE *Bedrīc* rather than *Beornrīc*. *Bedrīc* is unrecorded, but cf. *Bedfriþ, -hæp, -helm*.

Barcombe Sx [*Bercham* DB, *Bercamp* 1200 Cur, *Berecomp* 1233 Cl]. OE *bere-camp* 'barley field'.

Barden YN [*Bernedan* DB, *Berdene* 1184 PNNR]. 'The valley of the barn' (OE *berern*), or 'valley where barley grew' (first el. OE *beren* adj. 'of barley').

Barden YW [*Berdene* 12 Mon, *Berden* 1314 Ipm]. 'Barley valley' (first el. OE *bere* 'barley').

Bardfield, Great & Little, B∼ Saling Ess [*Birde-, Byrdefelda* DB, *Berdefeld* 1191 P, *Berdefeud Magna, Parva* 1238 Subs, *Berdefeld Salynge* 1314 Ipm]. 'FELD on the bank' (of the Pant). First el. OE *byrde*, on which SEE STIBBARD.

B∼ **Saling** from the neighbouring SALING.

Bardley Sa [*Berdesleia* 1123 AC, -*lega* 1194 f., *Bordeslea* 1196, -*leg* 1197 f. P]. '*Beornrēd*'s LĒAH.'

Bardney Li [*Bearddanig* 716 ASC, *Beardaneu* c 730 Bede, *Beardanea* c 890 OE Bede, *Bardanege* c 1000 Saints, *Bardenai* DB]. '*Bearda*'s island.' **Bearda* corresponds to OHG *Bardo*.

Bardon Le. See STANTON UNDER BARDON.

Bardsea La [*Berretseige* DB, *Berdeseia* 1155 LaCh], **Bardsey** YW [*Berdesei* DB, -*eia* 1158 P, 1209 FF, -*eye* 1230 P]. '*Beornrēd*'s island.'

Bardsley La [*Berdesley* 1422 PNLa]. '*Beornrēd*'s LĒAH.'

Bardwell Sf [*Beordewella, Berdeuuella* DB, *Berdewelle* 1190 P, 1197 FF]. Perhaps '*Bearda*'s spring'; cf. BARDNEY. Or the first el. may be OE *brerd, breord* 'brim, bank', with dissimilatory loss of the first *r*.

Bare La [*Bare* DB]. OE *bearu* 'grove'.

Barford has at least three sources: 1. OE *bereford* 'barley ford', perhaps ford used at the time of the corn harvest; cf. HEYFORD. A certain example is *Bereford* 852 BCS 466. Here probably belong most Barfords. 2. OE *Beranford* '*Bera*'s ford' (cf. BARHAM K) or 'ford of the bear' (OE *bera*). Here certainly belongs **Barford** Sr [*Bæranford* 909 BCS 627 (*æ* often for *e* in the MS), *Bereford* c 1200 Ep] and probably **Barford** Wa [*Æþberanforda* 792 BCS 264, *Bereford* DB]. 3. OE *beorcford* 'birch ford': **Little Barford**

Bd [*Berkeford* 1202 Ass, *Berkford* 1269 FF]. To 1. or 2. belong: **Great Barford** Bd [*Bereforde* DB, *-ford* 1197 FF], B~ Nf [*Bereforda* DB, *-ford* c 1184 Fr], B~ Np [*Bereford* DB], B~ St. John & St. Michael O [*Bereford* DB, c 1130 Oxf], B~ W nr Downton [*Bereford* DB, 1266 Misc], B~ St. Martin W [*Bereford* DB, *Berevord St. Martin* 1304 Ch], also BARFORTH.

Barforth YN [*Bereford* c 1050 HSC, 1166 P]. A late modification of BARFORD.

Barfreston K [*Berfrestone* DB, *Berfredestune* 11 DM]. '*Beornfriþ*'s or *Beorhtfriþ*'s TŪN.'

Barham Ca [*Bercheham* c 1080 ICC, DB], B~ Hu [*Bercheham* 1086 IE, *Bergham* 1260 Ass], B~ Sf [*Bercheham* c 1050 KCD 907, *Bercham* DB, *Bergham* 1252 Ch]. OE *Beorghām* 'HĀM on the hill'.

Bărham K [*Bioraham* 799 BCS 293, *Beoraham* 805 ib. 319, *Berham* DB]. '*Biora*'s HĀM.' *Biora, Beora* (from *Bera*) is a pers. n., not recorded, but corresponding to OHG *Bæro*, ON *Biari*. It is really OE *bera* 'a bear'. Cf. BENEFIELD.

Barholm Li [*Berc(a)ham* DB, *Berham* 1138 NpCh, *Bercham* 1202 Ass]. OE *Beorghām*; cf. BARHAM. *Beorg* may here mean 'barrow'.

Barkby Le [*Barchebi* DB, *Barkeby* 13 BM]. '*Bark*'s BY.' *Bark* (*Barch* DB) is from ON *Bǫrkr* (gen. *Barkar*).

Barkestone Le [*Barchestone* DB, *-tona* 1114–16 RA]. '*Bark*'s TŪN'; cf. BARKBY.

Barkham Brk [*Beorcham* 952 BCS 895, *Bercheham* DB], B~ Sx [*Bercheham* DB, *Berchamme* 1332 Subs]. OE *beorc-hamm* 'HAMM where birch grew'.

Barking Ess [*Berecingas* 695 BCS 87, *Monasterium Bericinense, In Berecingum* c 730 Bede, *in Bercingum* (*Beorcengum*) c 890 OEBede], B~ Sf [*Berchinges* c 1050 KCD 907, *Berchingas* DB]. '*Berica*'s people.' For OE *Berica* cf. OHG *Berico*, Goth *Beric*. The early forms rule out derivation of B~ Ess from OE *beorc* 'birch', but the name was later associated with the word.

Barkisland YW [*Barkesland* 1275 Wakef]. '*Bark*'s land'; cf. BARKBY.

Barkston Li [*Barchestune* DB, *Barkeston* 1202 Ass], B~ YW [*Barcestun* c 1030 YCh 7, *Barchestun* DB]. Identical with BARKESTONE.

Barkway Hrt [*Bercheuuei* DB, *-weie* c 1182 BM, *Berchwei* 1176 P, *Berkeweya* 1212 Fees]. Seems to be OE *beorcweg* 'road through birches'. Or more likely perhaps the first el. is an OE *beorce* 'birch-grown hill'. The place is on a ridge, over which a road runs.

Barkwith Li [*Barcuurde* DB, *-worda* c 1115 LiS, *Westbarkeworth* 1202 Ass, *Barkewurthe* 1252 Ch]. See WORÞ. The first el. is difficult. The Scand pers. n. *Bark* (see

BARKBY) is a possible source. More likely *Bark-* is a British hill-name, identical with the first el. of BERKSHIRE and referring to the Wolds, on whose lower slope B~ is.

Barlaston St [*Beorelfestun* 1002 Wills, *Bernulvestone* DB, *Berleston* 1167 P]. '*Beornwulf*'s TŪN.'

Barlavington Sx. See LAVINGTON.

Barlborough Db [*Barleburh* 1002 Wills, *-burg* DB]. The first el. may be OE *Bārlēah* 'boar wood or clearing', an earlier name of the place.

Barlby YE [*Bardulbi* DB, *-beia* c 1105 Fr, *Barthelby* c 1190 YCh 996]. '*Bardwulf*'s BY.'

Barle R So [*Bergel* 1219, *Burewelle* 1279 For]. OE *beorgwella* 'hill stream'.

Barlestone Le [*Berulvestone* DB, *Berlestone* 1166 RBE]. '*Berwulf*'s or *Beornwulf*'s TŪN.'

Barley Hrt [*Beranlea* 11 Th, *Beoronleam* c 1050 PNHrt(S), *Berlai* DB, *Berle* 1253 Ch]. Apparently 'the LĒAH of *Be(o)ra*'. Cf. BARHAM K.

Barley La [*Bayrlegh* 1324 Ipm, *Barelegh* 1325 Ct], B~ YW [*Barle* 1219 FF]. OE *bār-lēah* 'boar wood or clearing' or *bær-lēah* 'barley clearing'. Cf. BERE.

Barlinch So [*Berliȝ* c 1180, *-linȝ* n.d. Buckland, *Berlyz* 1243 Ass, *Berlinch* 1339 Ch]. OE *bere-hlinc* 'barley hill'.

Barling Ess [(æt) *Bærlingum* 998 Crawf, *Berlinga* DB, *Berlinge* 1197 P], **Barlings** Li [*Berlinge* DB, *Barlinga* c 1115 LiS, *Berlinges* 1126 Fr]. '*Bærla*'s people.' *Bærla* may be cognate with OE *bār* 'boar' and *Bære* pers. n. BCS 194.

Barlow Db [*Barleie* DB, *-lee* 1203 Cur, *-leia* 1207 BM, *Berlegh* 1230 Ch]. OE *bār-lēah* 'boar LĒAH' or *bærlēah* 'barley clearing'. Cf. BARLEY, BERE.

Barlow La [*Barlowe* 1254 Abbr, *Berlawe* 1260 Ass]. OE *berehlāw* 'barley hill'.

Barlow YW [*Bernlege* c 1030 YCh 7, *Berlai* DB, *Berlay* 1205 FF]. First el. either OE *berern* 'barn' or *beren* 'of barley'.

Barmbrough YW [*Berne-, Barneburg* DB, *Barnaburc* 1148 YCh 179]. Perhaps '*Biarni*'s BURG'; cf. BARMBY.

Barmby on the Marsh YE [*Bærnabi* c 1050 YCh 9, *Barnebi* DB], B~ on the Moor YE [*Bernebi, Barnebi* DB, *Barneby* 1235 FF]. '*Biarni*'s BY.' *Biarni* is a well-evidenced OScand pers. n.

Barmer Nf [*Benemara* DB, *Beremere* 1202 FF, 1254 Val]. Second el. no doubt OE *mere* 'lake'. The first may be OE *bera* 'bear'.

Barming, East & West, K [*Bermelinge, Bermelie* DB, *Bermeling* 11 DM, *Bearmlinges* a 1150 Text Roff, *Barmelinges* 1197 P, *Bremling* 1277 Ipm]. Difficult in view of the variation of forms. Possibly in reality *brōmhlincas* or *brēmel-hlincas* 'broom or bramble hills' or a derivative with -ingas of OE brōm-lēah or brēmel-lēah 'broom or bramble LĒAH'.

Barmoor Nb [*Beiremor* 1231, *Beigermore* 1232 Cl]. 'Cranberry moor' (OE *bēger* 'berries' and MŌR).

Barmpton Du [*Bermetun* c 1090 SD, *Bermentun* Hy 1 FPD, *-tona* c 1120 YCh 933]. 'Beornmund's TŪN.'

Barmston Du [*Berneston* 1361, *Bermeston* 1471 PNNb], **B~** YE [*Benestone* DB, *Bernestuna* c 1175 YCh 379]. 'Beorn's TŪN.' But B~ Du may be 'Beornmund's TŪN.'

Barnaby YN [*Bernodebi* DB, *Bernaldeby* 12 Guisb, 1231 FF]. 'Beornnōþ's or Beornwald's BY.'

Barnack Np [(on) *Beornican* 972–92 BCS 1130, *Bernech, Bernak* c 1050 KCD 904, *Bernac* DB, *Berneca* 1163, *Bernech* 1167 P, *Bernek* 1202 Ass]. Probably from OE *Beornwīc* 'Beorn's wicks' (cf. BARNHAM) with early loss of *w* as in OE *ǣnetre* from *ān-wintre*. Cf. WINCH. The early forms of the type *Bernac* are due to influence from the Fr pl. n. *Bernac*.

Barnacle Wa [*Bernhangre* DB, *Bernangre* 1352 AD]. First el. OE *berern* 'barn' or *beren* 'of barley'. Second OE HANGRA 'slope'.

Barnacre La [*Berneacre* 1517 DL]. First el. OE *berern* 'barn'.

Barnard Castle Du [*Castellum Bern'* c 1150 Crawf, *C~ Bernardi* 1200 P, *Bernardescastell* 1399 AD]. 'The castle of Bernard Balliol' (d. a 1167).

Barnardiston Sf [*Bernardeston* 1194 P, 1242 Fees]. 'Beornheard's TŪN.'

Barnbow YW [*Barnesburc* c 1190 YCh 1651, *Barnebu* 1191 ib. (1653), 1195 P]. 'Biarni's homestead'; cf. BARMBY and BÚ.

Barnby Moor Nt [*Barnebi* DB], **B~ in the Willows** Nt [*Barnebi* DB, 12 DC], **B~** Sf [*Barnebei, -by* DB, *-by* 1254 Val], **B~** YN [*Barnebi* DB], **B~ upon Don** YW [*Barnebi* DB, *-by* 1202 FF, *Barneby super Don* 1285 FA]. Identical with BARMBY.

Barnes Sr [*Berne* 939 BCS 737, (of) *Bærnun* c 1000 CCC, *Berne* DB, *Bernes* 1222 St Paul]. 'The barns' (OE *berern*).

Barnet, Chipping, East, High & New, Hrt, **Friern Barnet** Mx [*Bernet* 1196 P, *Barnet* 1196 FF, *La Barnette* 1249 Misc, *Chepyng Barnet* 1343 Misc, *Freron Barnet* 1460 PNMx]. OE *bærnet* 'burning, place cleared by burning'.

Chipping means 'market'.—Friern is the gen. plur. of ME *frere* 'friar'. Friern B~ belonged to the Hospital of St. John of Jerusalem.

Barnetby le Wold Li [*Bernodebi* DB, *Bernetebi* DB, c 1115 LiS]. 'Beornnōþ's BY.'

Barney Nf [*Berlei* DB, *Berneie* 1198 FF, *Berneia* 1214 FF]. Second el. OE *ēg* 'island,

river land'. The first may be OE *beren* 'of barley' or *berern* 'barn'.

Barnham Broom Nf [*Bernham* DB, 1202 FF, 1276 Misc], **B~** Sf [*Byornham* c 1000 BCS 1306, *Bernham* DB, 1197 FF]. 'Beorn's HĀM.' *Beorn* here appears without gen. *-s* already in an OE text.

Barnham Sx [*Berneham* DB, *Bernham* 1105 Fr, 1230 Ch]. 'Beorna's HĀM.'

Barnhill YE [*Beornhyll* 959 YCh 4, *Bernehelt, -held* DB, *Bernhill* 1199 FF]. 'Beorn's hill.' Cf. BARNHAM.

Barningham, Little, B~ Norwood & Winter or **Town B~** Nf [*Berningeham* DB, 1166 RBE, *Beringham* 1203 Cur, *Tunberningham, Magna Berningham, B~ Parva* 1254 Val], **B~** Sf [*Bernincham* DB, *Berningeham* c 1095 Bury], **B~** YN [*Berningham* DB, *Bernigeham* 1214 FF]. 'The HĀM of *Beorn's* people.' Little B~ Nf may be *Berneswrde* DB ('Beorn's WORÞ').

Norwood means 'north of the wood' (OE *be norþan wuda*).

Barnoldby le Beck Li [*Bernulfbi* DB, *Bernetebi* c 1115 LiS, *Bernolesbi* 1202 BM]. 'Bernulf's BY.' *Bernulf* is ON *Biǫrnulfr* or OE *Beornwulf*.

The addition *le Beck* 'by the brook' in contradistinction to BARNETBY LE WOLD.

Barnoldswick YW [*Bernulfeswic* DB, *Bernolveswic* c 1172 YCh 1461]. 'Beornwulf's WĪC.'

Barnsdale Ru [*Bernardeshull* 1202 Ass]. 'Beornheard's hill.'

Barnshaw Chs [*Bernulisah* c 1200, *Bernulfschawe* c 1300 Chester]. 'Beornwulf's wood.' See SCAGA.

Barnsley Do [*Bernardeslega* 1178 P, *-le* 1212 Fees]. 'Beornheard's LĒAH.'

Barnsley Gl [(æt) *Bearmodeslea* c 802 BCS 304, (æt) *Beorondeslea* 855 ib. 487, *Bernesleis* DB]. 'Beornmōd's LĒAH.'

Barnsley Wt [*Benverdeslei* DB, *Bermardeslegh* 1287–90 Fees]. 'Beornfriþ's LĒAH.'

Barnsley YW [*Berneslai* DB, *-laya* c 1150 YCh 1771]. 'Beorn's LĒAH.'

Barnstaple D [*Beardastapol* 979 ff. Coins, 1018 Crawf, *Bardanstapol* 10 Coins, *Berdestaple* 1166 RBE, *Barnestaple* DB, *-stapl'* 1158 f. P]. Apparently 'Bearda's staple or post'; cf. BARDNEY. The early change to *Barne-* is remarkable, as is the fact that the same name seems to occur also as that of a hundred in Ess (Barstable hd). Alternatively it may be suggested that Barnstaple has as first el. OE *barda* 'beaked ship', the name meaning 'post to which a warship was moored'. The form *Barne-* might then represent the gen. plur. *bardna* 'of the warships'.

Barnston Chs [*Bernestone* DB, *Berlestona* c 1100 Chester]. 'Beornwulf's TŪN.'

Barnston Ess [*Bernestuna* DB, *-ton* 1221 BM], B~ Nt [*Bernestune* DB, *-ton* 1169 P]. '*Beorn*'s TŪN.' But if *Bernolueston* 1182 P refers to B~ Nt, it is '*Beornwulf*'s TŪN'.

Barnton Chs [*Bertintune* DB, *Berthyngton* 1300 &c. Ormerod]. Doubtful. Possibly 'the TŪN of *Beornnōþ*'s people'.

Barnwell Ca [*Beornewelle* 1060 Th, *Bernewell* 1170 P, 1198 Cur, 1230 P]. '*Beorna*'s spring or stream' or 'the spring (stream) of the warriors' (OE *beorn*). OE *Beorna* pers. n. is not well evidenced.

Barnwell St. Andrew & All Saints Np [*Byrnewilla* 972–92 BCS 1130, *Bernewelle* DB, 12 BM, *Bernewell Omnium Sanctorum, Sancti Andree* 1254 Val]. 'The stream by the burial-mound', an OE *byrg(e)ne-wella* (first el. OE *byrgen* 'burial-mound').

Barnwood Gl [*Berneuude* DB, *Bernwude* 1221 Ass, *Bernewod* 1291 Tax]. The first el. may be the gen. plur. of OE *beorn* 'warrior' or the pers. n. *Beorna*.

Barr. Great, St [*æt Bearre* 957 BCS 987, *Barre* DB, *Little Barre* 1208 FF, *Great Barre* 1322 Ipm]. Welsh *bar* 'top, summit', Ir *barr*, from OCelt *barro-*. Barr Beacon is a hill of 700 ft.

Barras We [?*Le Berghes* 1309 Ipm]. Whether the ex. of 1309 belongs here or not, OE *beorgas* 'the hills'.

Barrasford Nb [*Barwisford* 1242 Fees, *Barewesford* 1256 Ass]. 'Ford by a grove' (OE *bearu*, gen. *bearwes*).

Barrington Ca [*Barentone* DB, *-tona* 1086 IE, *-ton* 1242 Fees]. Apparently '*Bāra*'s TŪN'. No such OE name is recorded, but OHG has *Baro*. OE *baran tūn* 'the bare TŪN' does not seem likely.

Barrington Gl [*Berni(n)tone* DB, *Bernintun* 1156 P, *Magna Berningtone* 1221 Ass, *parua Bernynton* 1291 Tax]. 'The TŪN of *Beorn*'s people.'

Barrington So [*Barintone* DB, *Barinton* 1185 P, 1201 Ass, *Barington* 1225 Ass]. Apparently an OE *Bāring(a)tūn*, 'the TŪN of *Bāra*'s people'; cf. BARRINGTON Ca.

Barrow is 1. mostly OE *bearu*, dat. *bearwe* 'grove, wood': Great & Little Barrow Chs [*Barue* 958 BCS 1041, *Bero* DB], B~ upon Trent Db [*Barwe*, *Bareuue* DB, *Barewe* 1197 FF], B~ Gl [*Berwe*, *Berewe* 12 Winchc, *Barewe* 1232 Cl], B~ upon Soar Le [*Barhou* DB, *Baru* 1158 Fr, *Barwa* Hy 2 BM, *Barewe on Sore* 1311 Ipm], B~ upon Humber Li [*Adbaruae* i.e. *ad nemus* c 730 Bede, *Æt Bearwe* c 890 OEBede, *Bearuwe* 716–43 BCS 165, *Barewe* DB], B~ Sa [*Barwe* 1267 Ch, *Barewe* 1291 Tax], B~ Sf [*Baro* DB, *Barue* 1201 Cur], North & South B~ So [*Berue*, *Berrowene* DB, *Barewe* 1225 Ass, *Sud Berwe*, *Northberwe* 1242 Fees], B~ Gurney & Minchin So [*Berue* DB, *Barewe* 1269 Ipm, *Barwe Gurnay* 1283 Misc

2. OE *beorg* 'hill' or 'mound': Barrow Ru [*Berc*, *Berghes* 1206 Cur], Barrow So nr Twerton [*la Bergh* 1232 Cl].

3. Barrow in Furness La [*Barrai* 1190, *Barray* 1292 FC]. An ON *Barrey*, which may have been transferred to Barrow from BARRA, one of the Hebrides. Barrow in Furness was formerly an island; the second el. of the name is ON *ey* 'island'.

B~ Gurney was held by Nigel [de Gurnai] in 1086 (DB). Gurney from GOURNAY in Normandy.—B~ Minchin was the seat of a nunnery; *minchin* is OE *mynecen* 'nun'.

Barrowby Li [*Bergebi* DB, 1202 Ass, *Beruby* 1242 Fees], B~ YW nr Kirkby Overblow [*Berghebi* DB, *Bergeby* 1236 Ch], B~ YW nr Leeds. OScand *Berga-bȳ(r)* 'BY on the hill(s)'; cf. BARBY.

Barrowden Ru [*Berchedone* DB, *Bergendona* 1130 P, *Bergedon* 1202 Ass]. OE *beorgadūn* 'DŪN with barrows or burial-mounds'.

Barrowford La [*Barouforde* 1296 Lacy]. 'Ford by a grove' (OE *bearu*).

Barsby Le [*Barnesbi* DB, *-bia* 1177 P]. '*Barn*'s BY.' ON *barn* 'child' occurs as a byname and as a fictitious Christian name.

Barsham Nf [*Barsa-*, *Barseham*, *Norbarsam* DB, *Barsham* 1185 P, 1200 Cur, *Est-*, *Westbaresham*, *Northbarsham* 1254 Val], B~ Sf [*Bars-*, *Bersham* DB, *Barsham* 1196 FF, 1254 Val]. '*Bār*'s HĀM.' *Bār* pers. n. occurs in DB. It may be OE *bār* 'boar' used as a nickname. Cf. (Husbands) BOSWORTH.

Barston Wa [*Bertanestone* DB, *Berestanestona* 1185 TpR]. '*Beornstān*'s or *Beorhtstān*'s TŪN.'

Bartestree He [*Bertoldestrev* DB, *Berkwoldestre* 1206 Cur]. '*Beorhtweald*'s tree.'

Barthomley Chs [*Bertemelev* DB, *Bertumleg* 1260 Court, *-lega* c 1295 BM, *Berthoneleg(h)* 1286 Court]. '*Beorhtwynn*'s LĒAH.' OE *-wynn* became *-wunn* and *-un*; the change of *n* to *m* may be due to the initial labial. *Beorhtwynn* is a woman's name. Barthomley is close to Audley, Balterley, and Betley, whose names contain a woman's name.

Barthorpe YE [*Barchetorp*, *Barchertorp* DB, *Barkerthorp* 1232 FF]. OScand *Barkarþorp* 'the thorp of *Bǫrkr*'. ON *Bǫrkr* had the gen. *Barkar*. Cf. BARKBY.

Bartington Chs [*Bertintune* DB, *Bertenton*, *Berthinton* 1282 Court]. Perhaps 'the TŪN of *Beorht*'s people'. Cf. BARNTON.

Bartley Regis Ha [*Berchelai* 1107 (1300) Ch, *Barkeley* 1316 FA], B~ Green Wo [*Berchelai* DB]. OE *beorc-lēah* 'birch wood'.

Bartlow Ca [*Berkelawe* 1232 FF, *Berklawe* 1254 Val]. OE *beorc-hlāw* 'birch-clad tumulus'. Named from one of the tumuli called Bartlow Hills. Bartlow End Ess [*Berkelowe* 1330 Ch] is close to Bartlow Ca.

Barton, a common name, nearly always goes back to OE *beretūn* or **bærtūn* (cf. *bærlic* 'barley') from BERE 'barley, corn' and TŪN. OE *beretūn* is recorded in the sense of 'threshing floor'. But a meaning 'corn farm' must also have existed in OE. In BCS 808 the place-name *Bertuna* is rendered by 'villa frumentaria'. From 'corn farm' developed the later recorded meanings 'demesne farm' and 'outlying grange'. The last is probably the meaning of Barton in most cases. Several Bartons were actually granges belonging to monasteries. The distinctive additions in several cases refer to situation in a fertile district; cf. B~ in the Beans, B~ in Fabis, B~ in the Clay, B~ Turf.

1. Barton in the Clay Bd [*Bertone* DB], **B~ Hartshorn** Bk [*Bertone* DB, *Barton Hartshorn* c 1450 Ep], **B~** Brk in Abingdon [*Bertune* DB, *La Berton* 1242 Fees], **B~ Ca** [*Barton* 1060 KCD 809, *Bertone* DB], **B~** Chs [*Barton* c 1275 Chester], **B~ Blount** Db [*Barctune* DB, *Bartona* c 1165 BM, *Bartonblonte* 1535 VE], **B~ Gl** nr Bristol [*Bertune apud Bristov* DB, *Berton Bristoll* 1220 Fees], **B~ Gl** nr Guiting [*Berton* 1287 QW], **King's B~** Gl [*Bertune* DB, *la Berton* 1234 Cl], **Abbot's B~** Ha [*Bertona* 1243 Ep, *Abbotesberton* 1329 Ch], **B~ Peverel** Ha [*Berton* 1227 FineR, 1242 Fees], **B~ Stacey** Ha [*Bertun* c 1000 ASCh, *Bertune* DB, *Berton Sacy* 1302 Misc, 1316 FA], **B~** He [*Bertune* DB], **B~ K** [*Bertun* 833 BCS 408], **B~ La** nr Preston [*Bartun* DB, *Berton* 1226 LaInq], **B~ La** in Halsall [*Bartune* DB], **B~ on Irwell** La [*Barton* 1196 P], **B~ in the Beans** Le [*Bartone* DB, *-ton* 1202 FF], **B~ upon Humber** Li [*Bertone* DB, *Bartuna* c 1115 LiS, *Barton super Humbre* 1281 QW], **B~ Bendish** Nf [*Bertuna* DB, *Berton Binnedich* 1249 Ipm], **B~ Turf** Nf [*Berton* 1043 Wills, *Bertuna* DB], **Earls B~** Np [*Bartone* DB, *B~ comitis David* 1187 P, *Earls Barton* 1290 Ipm], **B~ Seagrave** Np [*Bertone* DB, *Barton Segrave* 1412 FA], **B~ in Fabis** Nt [*Bartone* DB, *Barton in le Benes* 1388 Cor], **Steeple, Westcott & Middle B~** O [*Bærtun* c 1050 KCD 950, *Bertone* DB, *Magna, Westcote Berton* 1242 Fees], **Great B~** Sf [*Bertuna* 945 BCS 808, *Bertun* c 950 Wills, *Bertuna* DB, *Magna Bertone* 1254 Val], **B~ Mills** Sf [*Bertona* DB, *-ton* c 1235 Bodl, *Parva Bertone* 1254 Val], **B~ So** nr Bath [*la Berton de Bathonia* 1212 Fees], **B~ St. David** So [*Bertone* DB], **B~ under Needwood** St [*Barton* 942 BCS 771, *Bertone* DB, *Barton sub Nedwode* Hy 3 BM], **B~ Wa** nr Stratford [*Berton* 1315 Ipm], **B~ on the Heath** Wa [*Bertone* DB, *Barton* 1236 Fees], **B~ We** [*Bartunaheved* c 1160 YCh 175, *Barton Heved* 12 (1290) Ch], **B~ YN** [*Barton* DB, 1228 Ep], **B~ le Street** YN [*Bartun* DB, *-ton* 1220 FF], **B~ le Willows** YN [*Bartun* DB].

2. Barton Wt [*Burtone* a 1260 Ep, *-ton* 1287–90 Fees]. OE *Burhtūn*. See BURTON.

On surnames such as *in the Beans* (*Clay*) see *supra*.—**Abbot's B~** Ha belonged to Hyde Abbey.—**B~ Bendish** Nf means 'B~ inside the ditch'. The place is west of DEVIL'S DITCH.— **B~ Blount** Db is called *Barton Bakepuys* 1428 FA and was held by R. de Bakepuz c 1160 (Derby), by Thomas Blount in 1428. *Bakepuz* is a family name derived from BACQEPUIS in Normandy. The manor was held by the Blounts from 1381. Blount is OFr *blund* 'blond' used as a byname.—**Earls B~** Np belonged to the earl of Huntingdon.—**B~ Hartshorn** Bk may have been named from a place of this name. Cf. HARTSHORNE Db.—**B~ under Needwood** St. See NEEDWOOD.—**B~ Peverel** Ha was held by Andrew Peverel in 1227 (FineR). Peverel is a Norman nickname and family name, probably meaning 'peppercorn'.—**B~ St. David** So from the dedication of the church.—**B~ Seagrave** Np was held in part by Nicholas de Segrave in 1314 (Ipm). Cf. SEAGRAVE.—**B~ Stacey** Ha was held by Rogo de Saci in 1199 (P). Sacey from SACY or SACEY in France. *Stacey* is a late corruption.—**B~ le Street** YN is on a supposed Roman road.—**B~ Turf** Nf contains the word *turf*. See above.

Barugh (barf) YN [*Berg, Berch* DB], **B~** YW [*Berg* DB]. OE *beorg* 'hill'.

Barway Ca [*Bergeia* 1156 ff. P, *Berweya* 1254 Ep]. OE *beorg-ēg* 'barrow island, island with tumuli on it'.

Barwell Le [*Barwalle* 1043 Th, *Barewelle* DB, 12 DC]. 'Boar stream' (first el. OE *bār*).

Barwick (bărĭk) Nf [*Bereuuica* DB, *-wic* 1195 FF], **B~ So** [*Berewyk* 1219 FF, 1327 Subs], **B~** (bărĭk) **in Elmet** YW [*Bereuuith* DB, *-wic* 13 BM]. OE *berewīc* (q.v.). See ELMET.

Barwise We [*Berwis* c 1235, 1490 WR]. OE *beorgas* 'hills'.

Barwythe Bd in Studham [*Bereuuorde* DB, *Baresworth* 1200 Cur, *Barewurthe* 1204–12 Fees]. Possibly identical with (Husbands) BOSWORTH.

Baschurch Sa [*Basecherche* (hd), *Bascherche* DB, *Baschirche* Hy 1 (1267) Ch, *Eglwysseu bassa* Red Book of Hergest]. OE *Basses cirice* 'Bass's church'.

Bascote Wa [*Bachecota* 1175, *Baskecota* 1190 f., *Baschecota* 1192 P, *Bace-, Bascecot* 1206 Cur]. First el. apparently an unrecorded pers. n. *Basuca* or the like; cf. BASING &c.

Basford Chs [*Berchesford* DB, *Barkisford* 1260 Court], **B~ St** [*Bechesword* DB, *Barkeford* 1199 (1265) Ch, *Barkisford* 1281 Ipm]. The first el. might be *Beorcol* as in BASWICH or a related name. Or B~ might be OE *beorcford* 'birch ford' with an intrusive *s*.

Basford (bās-) Nt [*Baseford* DB, 1226 Ep, *Basingef[ord]* 1201 Cur, *Basseford* 1230 P, *Base-, Beseford* 1252 Cl]. 'Basa's ford.' **Basa* corresponds to OHG *Baso*, and a pers. n. stem *Bas-* must be assumed for several names, e.g. BASING.

Bashall Eaves YW [*Baschelf* DB, *Bacshelf* 1251 Ch, *Bacsholf* 1304 Ch]. OE *bæc* 'back', probably also 'ridge', and SCYLF. The name may mean 'the slope of the ridge'. Cf. EFES.

Bashley Ha [*at Bageslucesleia* 1053 KCD 798, *Bailocheslei* DB, *Bailokesle* 1272 (1313) Ch]. '*Bægloc*'s LĒAH.'

Basildon Brk [*Bastedene* DB, *Bastlesden* 1175, *Baselesden* 1199 P, *Bastlesden* 1212, *Bestlesdene* 1242 Fees]. The first el. is no doubt a pers. n., which is also found in the name of a ford near Basildon [*Bestlesforda* 688–90, 699 BCS 74, 100, *Bæstlæs-*, *Bestlesford* 891 ib. 565]. The name may well have been OE *Bessel* (*Bæssel*), a derivative of *Bass* or *Bassa*. The gen. of this would be *Bessles*, whence *Bestles*; cf. OE *mistlic* from *mislic* and the like. The second el. is OE *denu* 'valley'.

Basildon Ess [*Berlesduna* DB, *Bert(h)lesdon* 1194 P, *Bartlesden* 1201 Cur, *-don* 1238 Subs]. '*Beorhtel*'s or *Beorhtwulf*'s DŪN.'

Basing (-ās-) Ha [(æt) *Basengum* 871 ASC, *ad Basyngum* 945 BCS 803, *Basengas* c 894 Asser, *Basinges* DB]. '*Basa*'s people'; cf. BASFORD Nt. **Basingstoke** Ha [*Basingastoc* 990 KCD 673, *Basingestoches* DB]. 'The STOC belonging to Basing.' The KCD form is *on Embasingastocæ* for *onem* (= *on efn) B~* (Tengstrand).

Baslow Db [*Basselau* DB, *-lawa* 1179 P, *-lowe* 1242 Fees, *Bassalawa* 1157 P]. '*Bassa*'s hill or burial-mound.' See HLĀW.

Bassenthwaite Cu [*Bastunthuait* 12 StB, 1225 CWNS xxi], **Bassenthwaite Lake** [*Bastunwater* c 1220 CWNS xxi]. '*Bastun*'s thwaite or clearing.' *Bastun*, no doubt a Norman nickname from *bastun* 'stick', occurs as a byname in England, e.g. *Ernaldus Bastun* 1191 ff. P (Nf or Sf), *Richard Bastun* 1203 Ass (Np).

Bassingbourn Ca [*Basingeburna* c 1080 ICC, *Basingborne* DB, *Bassingeburna* 1158, *-burn* 1195 P], **Bassingfield** Nt [*Basingfelt* DB, *Basingefeld* 1265 Ipm, *Bassingfeld* 1285 FA], **Bassingham** Li [*Basingeham* DB, 1176 P, *Basingham* 1202 Ass]. 'The stream (FELD, HĀM) of *Basa*'s or *Bassa*'s people.' See BASING and BURNA, FELD, HĀM.

Bassingthorpe Li [*Torp* DB, *Basewinttorp* 1202 Ass, *Basewyntorp* 1252 FF]. Originally THORP. Robert Basewin held Bassingthorpe in 1202 (Ass). *Basewin* is the pers. n. *Batsuen, Basuin* found in DB, from ON *bátsveinn* 'boatswain'.

Baston Li [*Bacstune, Bastune* DB, *Baston* 1167 P]. ON *Bakstūn* 'the TŪN of *Bak(r)*'. ON *Bak* at least occurs as a byname.

Bastonford Wo [*Berstanesford* 1275 Subs]. '*Beorht-* or *Beornstān*'s ford.'

Bastwick Nf [*Bastuuic* DB, *-wich* 1181 ff. P], **Woodbastwick** Nf [*Bastwik* 1044–7 KCD 785, *Bastuuic* DB, *Wodbastwyk* 1253 Ch]. OE *bæst-wīc* 'farm where bast was got'. But OE *bæst* is recorded in the sense 'lime-tree', and LG *bast* is used in the sense of 'lime-grove'. Bastwick might thus mean 'WĪC in a lime-grove'.

Baswich (băsij) St [*Bercheswic* DB, *Bercleswich* 12 PNSt, *-wyk* 1291 Tax]. '*Beorcol*'s WĪC.'

Batchcott Sa [*Bechecot* 1212 Fees, *Bachecot* 1255 RH]. 'COT in a valley'; see BÆCE.

Batchworth Hrt [*Bæcces wyrð* 1007 Crawf, *Bacewrth* 1220 Fees]. '*Bæcci*'s *wyrþ*.' See WORÞ. OE *Bæcci* is a derivative of *Bacca*. Cf. also BATSFORD.

Batcombe Do [*Batecumbe* 1201 Cur, 1274 Ipm], B~ So nr Bruton [(æt) *Batancumbæ* 965–71 BCS 1174, *at Batecombe* 940 ib. 749, *Batecumbe* DB, 1225 Ass], B~ So nr Axbridge [*Batecumb* 1263 Wells]. This seems to be '*Bata*'s valley'. OE *Bata* is recorded as a byname (*Ælfric Bata*), and OHG has *Bazzo, Bezilo*, &c. The same name may enter into BATLEY, **Battenhall** Wo [*Batenhale* 969 BCS 1240] and **Batton** D [*Bateton* 1254 Ass]. On the other hand the occurrence of three Batcombes is remarkable, and *Bat-* might be a common noun *bata*, corresponding to ON *bati*, OFris *bata* 'profit, gain', and used in some transferred sense, such as 'fat pasture', or even 'good husbandman'.

Bath So [*Hat Bathu* 676 BCS 43, *æt Baðum* 796 BCS 277, 906 ASC, *Bade* DB]. 'The baths', referring to the Roman baths.

Bathampton So [*Hamtun* 956 BCS 973, *Hantone* DB]. 'The HĀMTŪN by Bath.'

Bathampton W [*Bathamton* 1195 Cur, 1229 Pat]. OE *Bæþhæmatūn* 'the TŪN of the people at a hot spring' or the like. Cf. BAMPTON D. The place is called *Bathamewily* 1242 Fees. It is near Wylye.

Bathealton (bătltn) So [*Badeheltone* DB, *Badialton* 1196 P, 1225 Ass, *Badeyalton* 1249 FF]. The first el. has always *d* in early forms; *Bath-* seems due to late association with BATH. The original name may have been *Badan Eald-tūn* or *Healh-tūn*. *Bada* is a known pers. n. *Eald-tūn* would be 'old TŪN', *Healh-tūn* 'TŪN in a *healh*' (see HALH). Cf. DORMSTON.

Batheaston So nr Bath [*Estone* DB, *Batheneston* 1258 Wells, 1263 FF]. Originally OE *Ēast-tūn* 'eastern TŪN'; later the name BATH was added.

Batherton Chs [*Berdeltune* DB, *Bertherton* 1260 Court, 1283 Cl]. Perhaps 'the TŪN of *Beornþrÿþ*' (a woman).

Bathford So nr Bath [*æt Forda* 957 BCS 1001, *Forde* DB]. 'The ford.' *Bath* is a late addition.

Bathley Nt [*Badeleie* DB, *Bad(e)lee* 12 DC, *Batheleg* 1242 Fees]. In B~ was a locality called *Bath(e)ker* (Dale). The first el. of Bathley must be OE *bæþ* 'bath' in some sense, possibly 'hot spring'. The OE form would have been *Baþalēah*. Cf. BALE Nf.

Bathurst Sx [*Badeherste* c 1123 BM]. '*Bada*'s HYRST.'

Bathwick So nr Bath [*Wiche* DB]. 'wīc belonging to Bath.'

Batley YW [*Bateleia, Bathelie* DB, *Bateleia* c 1175 YCh 1480, *-leya* 1226 FF]. See LĒAH. First el. as in BATCOMBE.

Batsford Gl [æt *Bæccesore* 716–43 BCS 163, *Beceshore* DB, *Bechesoure* 1220, *-ofere* 1236 Fees]. '*Bæcci*'s ridge'; cf. BATCHWORTH. See OFER.

Battenhall. See BATCOMBE.

Battersby YN [*Badresbi* DB, *Baderesby* 1203 Cur, *Batheresby* 1240 FF]. 'The BY of *Bǫðvarr*' (ON *Bǫðvarr*).

Battersea Sr [*Badrices-, Batriceseg* 693 BCS 82, *Badoricesheah* 695 ib. 87, *Patricesy* DB, *Batricheseie* 1198 FF, *Batricheseia* 1200 Cur]. '*Beaduric*'s island.' The first el. was associated with the name *Patric*; hence often *Patricheseie* and the like in early sources. Modern *t* instead of original *d* may be due to this influence.

Battisford Sf [*Beteforda, Betesfort* DB, *Batesford* 1191 ff. P, 1242 Fees]. '*Bætti*'s ford.' *Bætti* is a side-form of *Betti*.

Battle Sx [*La batailge* DB, *þæt mynster æt þære Bataille* 1094 ASC (E)]. The monastery was founded in memory of the battle of Hastings.

Battlefield Sa [*St. Mary Magdalene of Batelfeld* 1415 Bodl]. A college of secular canons was founded here by Henry 4 in memory of the Battle of Shrewsbury in 1403.

Battlesden Bd [*Badelesdone* DB, 1227 Ass, *-don* c 1155 Oxf, *-donia* 1179 BM]. '*Bæddel*'s DŪN'; cf. BADLESMERE.

Batton. See BATCOMBE.

Battrix YW [*Bathirarghes* 1342 Whitaker's Whalley, *Batharar*' c 1367 Kirkst]. 'The shielings of *Bǫðvarr*'; cf. ERG and BATTERSBY.

Baughurst (-g-) Ha [*Beaggan hyrst* 909 BCS 624, *Baggeherst* 1176 P]. '*Beagga*'s HYRST.' *Beagga* is a short form of names like *Beagmund, Beagnōþ*.

Baumber Li [*Badeburg* DB, *Baburc* c 1115 LiS, *Baenburch* c 1145 DC, *Baumbur*' 1212 Fees]. '*Bada*'s or *Badda*'s BURG'; cf. BADBURY. Baumber is on a ridge along which runs a Roman road.

Baunton Gl [*Baudintone* DB, *Baldinton* 1208 Cur, *-e* 1221 Ass]. 'The TŪN of *Beald(a)*'s people.' Cf. BALDON &c.

Baverstock W [*Babanstoc* 968 Reg Wilt, *Babestoche* DB]. '*Babba*'s STOC.'

Bavington Nb [*Parva Babington* 1242 Fees, *Babinton* 1242 Cl, *Babbinton* 1257 Ch]. 'The TŪN of *Babba*'s people.'

Bawburgh Nf [*Bauenburc* DB, *Bauburg* c 1130 BM, *Baauburg* 1235 Cl]. '*Beawa*'s BURG.' A pers. n. *Beaw* is probably the first el. of BAWSEY Nf and of BEAUSALE Wa,

BEAUXFIELD K. It is no doubt a nickname, identical with OE *beaw* 'gadfly'. *Beawa* may be an extension of this. *Beawa* became *Beawa* and *Bawa*.

Bawdeswell Nf [*Baldereswella* DB, *Baldrusella* 1163 BM, *Baldeswell* 1208 FF]. '*Baldhere*'s spring.'

Bawdrip So [*Bagetrepe* DB, *Bakatripe* 1166 RBE, *Baketreppe* 1201 Cur, *Baggetrippe* 1243 Ass, *Bagedryþ* 1294 Ipm]. Second el. OE *treppe* 'trap'. The first may be *Bacga* pers. n. or, if there was an OE animal's name *bacga* (cf. BAGLEY), more likely this word.

Bawdsey Sf [*Baldeseia, Baldereseia* DB, *Baldreseia* 1109–31 BM, *Baudeseye* 1254 Val]. '*Baldhere*'s island.'

Bawsey Nf [*Boweseia* DB, *Bauseia* 1196 P, *Bauseye* 1254 Val]. '*Beaw*'s island'; cf. BAWBURGH.

Bawtry YW [*Baltry* 1199 (1232) Ch, *Bautre* 1247 Ipm, 1268 Ep]. '*B(e)alda*'s tree.'

Baxby YN [*Bachesbi* DB, *Baxebi* c 1175 YCh 790]. See BY. First el. as in BASTON.

Baxenden La [*Bacstanden* 1324 Ipm]. 'Bakestone valley.' A bakestone is a flat stone on which cakes are baked.

Baxterley Wa [*Basterleia* c 1170, *-le* c 1180 Middleton, *Bakesterleye* 1282 Cl]. 'The LĒAH of the baker (OE *bæcestre*).' Or the word may here be used as a family name.

Baycliff La [*Belleclive* 1212 Fees, *Belecliue* 1269 Ass]. The first el. might be that discussed under BELAUGH. Second el. OE *clif* 'slope'.

Baycliff W [*Ballochelie* DB, *Bail(l)eclive* 1242 Fees]. Perhaps '*Bægloc*'s cliff'; cf. BASHLEY.

Baydon W [*Beidona* 1146 Salisbury, *Beydon* 1246 Misc]. OE *beg-dūn* 'DŪN where berries grew'.

Bayfield Nf [*Baiafelda* DB, *Baifeld* 1180 P, 1200 Cur, *Beinfeld* 1200 Cur], **Bayford** Hrt [*Begesford* DB, *Begeford* c 1090 Reg, *Beiford* 1190, *Bei(n)ford* 1196 P]. '*Bæga*'s FELD and ford'; cf. BAYWORTH.

Bayford So [*Boyford* 1243 Ass, 1274 RH]. '*Boia*'s ford.'

Bayham Sx [*Begeham* 1228 Cl, 1235 Ch, *Begehamme* 1315 Ass]. '*Bæga*'s HAMM.'

Baylham Sf [*Beleham* DB, *Beilham* 1191 P, *Beylham* 1228 FF, 1254 Val]. The place is at a bend of the Gipping. The first el. of the name may be an OE *begel 'a bend', cognate with Norw *bøygel*, MLG *bögel* 'loop' (< *baugila-). Second el. HAMM (or HĀM).

Baynton W [*Beienton* 1199, *Beinton* 1200 Cur]. '*Bæga*'s TŪN.'

Baysdale YN [*Basdale* c 1200 YCh 564]. 'Valley with a cow-shed' (ON *báss*).

Bayston Sa [*Begestan* DB, *Beyston* 1255 RH, *Beystan* 1280 Cl]. '*Bæga*'s stone.' Cf. BAYWORTH.

Bayswater Mx [aqua vocata *Bayards Watering Place* 1380, *Bayards Watering* 1652, *Bayards Watering Place* 1654 PNMx(S)]. *Bayard* may be a family name of French origin, but more likely it is *bayard* '(bay) horse', here the *Bayard* of many proverbial sayings. *Watering* is 'watering-place'.

Baythorn Ess [*Babingbirne* c 950 Wills, (æt) *Babbingþyrnan* c 995 BCS 1289]. 'The thorn-bush of *Babba*'s people.'

Bayton Wo [*Beitone* c 1080 Fr, *Betune* DB]. '*Bæga*'s TŪN.'

Bayworth Brk [*Bægan wyrð*, æt *Bægenweorðe* 956 BCS 924, 932, *Begeurde* DB]. '*Bæga*'s WORÞ.' *Bæga* is identical with *Bægia* 718 BCS 139.

Beachampton Bk [*Bec(h)entone* DB, *Becchamton* 1176 P]. 'The TŪN of the dwellers at the brook', OE *Bec-hæmatūn*; cf. BÆCE.

Beachley Gl nr Chepstow [*Beteslega* Hy 2, *Betesleia* 13 (1307) Ch]. '*Betti*'s LĒAH.'

Beachy Head Sx [*Beuchef* 1279 QW]. Fr *beau chef* 'beautiful headland'.

Beaconsfield (-ĕ-) Bk [*Bekenesfeld* 1185 P, 1254 Val, *Bekenefeld* 1198 FF]. 'FELD by the beacon.'

Beadlam YN [*Bodlum* DB, *Bothlum* 12 Riev, *Bodhlum* 1240 FF]. The dat. plur. of OE *bōþl*, *bōtl* 'house, dwelling'.

Beadnell Nb [*Bedehal* 1161 P, *Bedenhala* 1177 P]. '*Bēda*'s HALH.'

Beaford D [*Baverdone* DB, *Beuford* 1242 Fees, *Beauforde* 1278 Ep]. 'Gadfly ford.' OE *bēaw* means 'gadfly'.

OE **bēag** 'ring, circle'. See BEAL YW, BEWHOLME, WILBY Sf.

Beal Nb [*Behil* 1208–10 Fees, *Behulle* 1248 Sc]. OE *bēo-hyll* (cf. *Byohyll* BCS 1027) 'bee hill'.

Beal YW [*Begale* DB, *Begehal* Hy 2 (1230) Ch]. Either '*Bēaga*'s or *Bēage*'s HALH' or 'HALH by the bends' (OE *bēag(a)halh* with OE *bēag* 'ring' as first el.). The place is at bends of the Aire.

Bealings Sf [*Belinges* DB, 1228 FF, *parua Belinges* DB, *Parva Beling* 1254 Val]. Possibly a derivative with the suffix *-ingas* from the element discussed under BELAUGH.

OE **bēam** 'tree, beam' occasionally occurs in pl. ns., as BAMPTON, BEMPTON. A probable meaning in pl. ns. is 'foot-bridge, bridge formed by a single beam', found in *pons de Dakenhambeem* 1330 For (DAGENHAM Ess). This sense is pretty certain in BAMFORD (cf. BRIDGFORD) and BENFLEET. **Beambridge** Chs may be 'bridge built with beams'.

Beaminster (-ĕ-) Do [*Bebingmynster* 872 BCS 535, *Beiminstre* DB, *Beministre* 1212 Fees]. 'The minster or church of *Bebbe* or her people.'

Beamish Du [*Bellus Mansus* 1251 Cl, *Bewmys* 1288 PNNb]. 'Beautiful mansion.' The second el. is OFr *mes*, *metz*, &c. from *mansus*. Cf. BEAUMETZ in France.

Beamsley YW [*Bedmesleia*, *Bemeslai*, *Bomeslai* DB, *Bethmesleia* 1182–5 YCh 199, *-le* c 1210 ib. 514]. The first el. is probably a derivative of OE *boþm* 'valley'; cf. BITTESWELL, LĒAH. The regular *e* suggests an OE *beþme*, analogous to *exen* from *oxa* and the like.

OE **bēan** 'bean' is a fairly common first el. See e.g. BAN-, BEAN-, BEN- (passim), BEENHAM, BINCKNOLL, BINCOMBE, BINEGAR, BINSTEAD.

Beane R Hrt [*Beneficcan* 913 ASC, *Beneficche* 13 Wendover, *Beane* 1577 Harrison]. A pre-English name of difficult history. From it are derived BENGEO and BENNINGTON Hrt.

Beanley Nb [*Benelega* c 1150 Percy, *-leg* 1256 Ass]. OE *bēan-lēah* 'bean clearing'.

Beard Db [*Berde* 1253 Cl, *Berd* 1316 Ipm]. OE *brerd* 'brim, bank' with dissimilatory loss of the first *r*. Beard Hall is on a hillside.

Beardwood La [*Berdewrthe* 1258 LaInq, *-worthe* 1296 Lacy]. '*Bearda*'s WORÞ'; cf. BARDNEY.

Bearl Nb [*Berhill* 1242 Fees, *Berehill* 1250 Ipm]. OE *bere-hyll* 'barley hill'.

Bearley Wa [*Burlei* DB, *-lea* 1169 P, *-lega*, *-leia* 13 BM]. See BURLEY.

Bear Park Du [*Beaurepayre* 1267 FPD]. A Fr name meaning 'beautiful retreat'.

Bearsted K [*Berghamstyde* c 700 Laws, *Bergestede* 1285 Ch]. 'Homestead on a hill.'

Bearstone Sa [*Bardestune* DB, *Bardeston* 1285 FA]. '*Beadurēd*'s or *Beard*'s TŪN.' OE *Beard* is unrecorded.

OE **bearu** (dat. *bearwe*) 'grove, wood' is the source of BARE, BARROW (1). It is the first el. of BARRASFORD, BARROWFORD and occurs sometimes as second el., e.g. in SEDGEBERROW. In some SW. dialects, esp. in Devon, the word acquired *u*-stem inflexion (gen., dat. *beara*) and *ea* became ME *ea*, *e*. Hence BEER D, ADBER, ROCKBEARE, SHEBBEAR, &c. Cf. HASELBURY.

Bearwardcote Db [*Beruerdes-*, *Bereuuardescote* DB, *Berwardecote* 1281 FF]. This might be 'the COT of the bearward', though *bearward* is first evidenced from 1399 (OED). More likely the first el. is a pers. n. *Beornweard; cf. *Beorward*, abbot of Glastonbury according to William of Malmesbury.

Beauchief Abbey (bē-) Db [*Beuchef* 1208 FF, *Beauchef* 1230 Cl, *Bellum Caput* 13

Derby]. 'Beautiful headland.' The name refers to a spur of hill.

Beaudesert St [*Beaudesert* 1293 Ass], **B~** Wa [*Beldesert* 1181 P, *Beaudesert* 1227 Ch]. 'Beautiful wild.'

Beaufront Castle Nb [*Beaufroun* 1356 BM]. 'Beautiful brow.'

Beaulieu (būli) Ha [*Bellus Locus Regis* 1205 BM, 1236 Ch]. 'Beautiful place.'

Beauma·nor (bō-) Le [*Beumaner* 1265 Ch]. 'Beautiful seat.'

Beaumont (bē-) Cu [*Beaumund* 1292 QW], **B~** (bō-) Ess [(æt) *Fulanpettæ* c 995 BCS 1289, *Fulepet* DB, *Bealmont* 1175–80 BM], **B~** (bō-) La [*Bellus Mons* 1190 FC, *Belmunt* 1212 Fees]. 'Beautiful hill.' Partly no doubt transferred from France, where BEAUMONT is common. The change of name in the case of Beaumont Ess (from 'foul pit' to 'beautiful hill') is noteworthy.

Beaurepaire Ha [*Beaurepeir* 1346 FA]. Cf. BEAR PARK.

Beausale (bū-) Wa [*Beoshelle* DB, *Beausala* 12 BM, *Biaushall* 1252 Ch]. '*Bēaw*'s HALH'; cf. BAWBURGH.

Beauvale Nt [*Bella Vallis* 1414 BM]. 'Beautiful vale.'

Beauworth (bū-) Ha [*Beowyrð* 938 BCS 731, c 965 ib. 1153, *Bewurthe* 1284 Ch]. 'Bee farm.' See WORÞ.

Beauxfield K [*Bewes-*, *Beasfeld* c 772 BCS 207, *Bewesfel* DB, *Bawesfeld* 1086 KInq]. '*Bēaw*'s FELD'; cf. BAWBURGH. Beauxfield is in **Bewsborough** hd [*Bevsberge* DB, *Beauuesberga* 1086 KInq]. '*Bēaw*'s hill or burial-mound.'

Beaworthy D [*Begeurde* DB, *Beghworthy* 1242 Fees]. '*Bēaga*'s or *Bēage*'s WORÞ.' *Bēaga* is not with certainty evidenced in independent use, but is a regular short form of *Bēagmund* and the like.

Bebington Chs [*Bebinton* c 1100, *-a* 1119 Chester, *Bebington* 1288 Court]. 'The TŪN of *Bebbe*'s people.' *Bebbe* is evidenced as a woman's name and as a man's name in *Bebbesham* 680 BCS 50.

Bebside Nb [*Bibeshet* 1198 (1271), 1204 Ch]. First el. *Bibba* pers. n. (in *Bibban þorn* KCD 1303). The second may be SCĒAT or (GE)SET.

Becca Hall YW [*Bechawe* 1275 Cl, *Bekhagh* 1285 FA]. Perhaps '*Beocca*'s enclosure' (OE HAGA).

Beccles Sf [*Becles* DB, *Beacles* c 1095 Bury, *Beclis* 1157, *Becclis* 1158 P]. Probably OE *bec-lǣs* 'pasture on the stream'; cf. BÆCE. Beccles is on the Waveney.

Becconsall La [*Bekaneshou* 1208 FF]. 'The mound (ON *haugr*) of *Bekan*.' *Bekan* is an ON pers. n. of Ir origin (OIr *Beccán*).

OE **bece**. See BÆCE. OE **bēce** 'beech'. See BEECH, BECKWITH, BEECHBURN, BITCHFIELD.

ME beck 'brook', from ON *bekkr*, ODan *bæk*, etymologically identical with OE *bece*, *bæce*, is the source of the stream-name **Beck** Sf, Li. It is common as the second el. of the names of brooks, some of which have become names of places, as BIRBECK, CALDBECK. Names in *-beck* are chiefly found in the Northern counties and in Db, La, Li, Nt. For the word as a first el. see BECKERMET, -MONDS.

Beckbury Sa [*Becheberie* DB, *Beckebir'* 1229 Cl]. '*Beocca*'s BURG.'

Beckenham K [*Biohhahema mearc* 862, *Beohhahammes gemæru* 973 BCS 506, 1295, *Beohhæma mearc* 987 BM, *Bacheham* DB, *Bekenham* 13 BM]. '*Beohha*'s HĀM.' *Beohha occurs also in *Biohhandun* BCS 702 (K) and may be explained as a short form of names in *Beorht-*.

Beckering Li [*Bechelinge* DB, *Becheringa* c 1115 LiS, *Bekering* 1170, 1196 P]. A derivative in *-ingas*, but the base is obscure.

Beckermet Cu [*Becheremet* 12 StB, *Bekyrmet* 1188 P, *Bekermet* 1242 FF], **Beckermonds** YW [*Bekeresmotes* 1241 FF]. ON *bekkiar-mót* 'junction of streams'. In Beckermet the actual element may be ON *mœti* 'meeting'. The *-es-* of Beckermonds in the early form is curious. Presumably the first el. was misunderstood as a pers. n. For the intrusive *n* cf. EAMONT.

Beckery Island So [*Bekeria, que parva Ybernia dicitur* 971 BCS 1277, *Beckerie* Hy 2 (1227) Ch]. OIr *Bec-Ériu* 'little Ireland', found as the name of islands also in Ireland. The name seems to have been given by Irish monks.

Beckett Brk [*Becote* DB, 1180 P, *Buccot* 1220 Fees]. OE *bēo-cot* 'bee COT'.

Beckett D [*Bikkecoth* 1242 Fees]. '*Bicca*'s COT.'

Beckford Gl [*Beccanford* 803 BCS 309, *Beceford* DB]. '*Becca*'s ford.' The pers. n. *Becca* is recorded in Widsith and in BECKLEY Sx.

Beckham Nf [*Beccheham, Becham* DB, *Becheham* 1175 P, *Est Bekkam* 1379, *West Becham* 1300 BM]. '*Be(o)cca*'s HĀM.'

Beckhampton W [*Bachentune* DB, *Bachamton* 1199 Rot Cur, *Bechampton* 1266 Pat]. OE *Bæchæmatūn* 'the TŪN of the dwellers by the ridge'; cf. BÆC.

Beckingham Li [*Beghingham* a 1184 DC, *Bekingeham* 1177, 1193 P, *Bekingham* 1206 Cl], **B~** Nt [*Bechingeham* DB, *Bekingeham* 1187 P, *Bekingham* 1204 Cur]. 'The HĀM of *Be(o)cca*'s or *Beohha*'s people'; cf. BECKENHAM.

Beckington So [*Bechintone* DB, *-ton* 1161 P, *Bekinton* 1186 P]. 'The TŪN of *Be(o)cca*'s people.'

Beckley K [*Bioccan lea* 889 BCS 562, *Bichelei* DB, *Beckele* 1242 Fees]. '*Beocca*'s LĒAH.'

Beckley O [*Beccalege* 1005–12 PNO(S), *Bechelie* DB, *Becheleia* 1197 P, *Bekele* c 1130 Oxf, *Beckele* 1272 Ipm]. '*Be(o)cca*'s lēah.'

Beckley Sx [(æt) *Beccanlea* c 880 BCS 553, *Becheleya* 1253 BM]. '*Becca*'s lēah.' Cf. BECKFORD.

Beckling Sf [*Becclinga* DB, *Beclinges* 1183 P]. '*Be(o)ccel*'s people.'

Beckney Ess [*Bacheneia* DB, *Bekkoney* 1310 Ipm]. OE *bēacen-(ī)ēg* 'beacon island'.

Beckton Ha in Pennington [*Becton* 1168 f. P]. 'TŪN in a valley.' Cf. BÆCE.

Beckwith YW [*Becwudu* c 972 BCS 1278, *Becvi* DB, *Bekwithshagh* 1323 Ipm]. OE *bēc-wudu* 'beech wood' (cf. BĒCE), later Scandinavianized with ON *viðr* instead of OE *wudu*.

Bedale (bē-) YN [*Bedale* DB, *Bedhal* 1256 Abbr]. '*Bēda*'s HALH'; cf. BEADNELL.

Bedburn Du [*Bedburn* 1291 Tax]. '*Bēda*'s stream.'

Bedchester Do [*Beteswirþe sled* 932 BCS 691, *Bedeshurst* 1372 FF]. *Beteswirþe* BCS 691 was at any rate very near Bedchester. It may not be identical with it, but the first el. is very likely the same. Bedchester is a late modification. The original form was *Bēdes-hyrst* or *-wyrþ*, whose first el. is OE *B(ī)ēdi*, found also in BIDDESTONE and BISTERNE.

Beddingham Sx [*Beadyngham* 801 BCS 302, *Beadingahamm* c 880 ib. 553, *Bedingham* DB]. 'The HAMM of *Bēada*'s people.' *Bēada* is found in *Beadan healan* BCS 936. Cf. also BEEDING.

Beddington Sr [*Beaddinctun* 901–8, c 965 BCS 618, 1155, *Beaddingtun* 909 ib. 620, *Beddintone* DB]. 'The TŪN of *Beadda*'s people.' *Beadda* is a short form of names in *Beald-*. Cf. BADDILEY.

Bedfield Sf [*Bedefeud* 1254 Val, *-feld* 1291 Tax]. '*Bē(a)da*'s FELD'; cf. BEDINGFIELD, which is c 3 miles distant.

Bedfont Mx [*Bedefunt, -funde, Westbedefund* DB, *Bedefunte* 1198 FF, *Estbedefont* 1236 Fees]. '*Bēda*'s spring'; cf. FUNTA.

Bedford Bd [*Bedanford* 880 Laws, 918 ASC, *Bydanford* c 1000 Saints, *Bedeford* DB], **B~** La [*Bedeford* 1201 P, 1258 Ass]. '*B(ī)ēda*'s ford.' **Bedfordshire** is *Bedanfordscir* 1011 ASC (E).

Bedhampton Ha [*Betametona* DB, *Bethametona* 1167 P, *Bedhampton* 1249 Ass]. Apparently a name in *-hæmatūn* (cf. HĀMTŪN). The first el. is obscure. It seems to have ended in *-t* rather than *-d*. OE *bēte* 'beet-root' might be thought of.

Bedingfield Sf [*Bedinge-, Badingafelda* DB, *Bedingefeld* c 1095 Bury, 1193 P]. 'The FELD of *Bē(a)da*'s people'; cf. BEDFIELD.

Bedingham Nf [*Bedingaham* DB, *Bedingham* 1254 Val, 1271 Ipm]. 'The HĀM of *Bēda*'s people.'

Bedlington Nb [*Bedlingtun* c 1050 HSC, *Betlingtun* 1104–8, *Betlingetun* c 1130 SD, *Betlington* 1254 Val]. Seems to be 'the TŪN of *Bēdla*'s or *Bētla*'s people'. Neither name is evidenced, but they are normal derivatives of *Bēda* and *Bōta*.

Bedminster So [*Beiminstre, Betministra* DB, *Bedmenistr(a)* 1156 ff. P, *Bedministr* 1194 P]. '*Bēda*'s minster or church.'

Bednall St [*Bedehala* DB, *Bedenhale* 1327 Subs]. '*Bēda*'s HALH.'

Bedstone Sa [*Betietetune* DB, *Bedeston* 1176 P]. '*Bedgēat*'s TŪN.' OE *Bedgēat* is unrecorded, but cf. *Ælf-, Sige-, Wulfgēat* and *Bedfriþ, -helm*.

Bedwardine Wo [*Bedewordine* 1235 FF]. '*Bēda*'s worþign.' See worþ.

Bedwell Hrt [*Bedewell* 1240 Fees, *Bidewelle* c 1330 Gesta]. 'Stream in a valley'; cf. BYDEN.

Bedworth Wa [*Bedeword* DB, *Beddewrða, Bedewurda* Hy 2 DC]. '*Bēda*'s worþ.'

Bedwyn W [*Bedewinde* 778 BCS 225, (æt) *Bedewindan* c 880 BCS 553, *Bedvinde* DB, *Estbedewinde* 1156 P]. From an OE form of dial. *bedwind* 'Clematis, Convolvulus'.

Beeby Le [*Bebi* DB, *-bia* c 1125 LeS]. 'BY where bees were kept.' OSw *Biby* may mean 'bee village'.

Beech Hill Brk [*La Beche* 1335 Ch], **B~** St [*Le Bech* 1285 FA]. OE *bēce* 'beech'.

Beechamwell Nf [*Bichamdic* c 1050 Th, *Bicham* DB, 1208 FF, *Bichham Welles* 1212 Fees]. The original name was *Bicham*, probably from OE *Biccan hām*, whose first el. seems to be a pers. n. *Bicca*, an assibilated form of *Bicca*. The place is near Devils Ditch; hence *Bichamdic*. The addition *-well* seems to refer to some spring(s).

Beechburn Du [*Bycheburn* 1304 Cl]. 'Beech stream' or '*Bicca*'s stream'; cf. prec. name.

Beeching Stoke. See STOKE.

Beeding, Lower & Upper, Sx [(æt) *Beadingum* c 880 BCS 553, *Bedinges* 1073 Fr, DB, *Bidinges* c 1230 Sele; *Netherbetynges* 1279 Ass]. OE *Bēadingas* '*Bēada*'s people'. Upper Beeding is on the lower Adur and in a lower position than Lower Beeding; it was the original settlement. It looks as if *Upper* here means 'Great' or the like, *Lower*, 'Little'. Cf. BEDDINGHAM.

Beedon Brk [*Bydenhæma gemære* 951 BCS 892, *Bydene* 965 ib. 1171, *Bedene* DB]. OE *byden* 'shallow valley'. See BYDEN.

Beeford YE [*Biworde* DB, *Beford* c 1150, *Biford* c 1180 YCh 1306, 1310]. OE *bēoford* 'ford at which bees were found'.

Beela R. See BEETHAM.

Beeleigh Ess [*Belegh* 1251, *Byleigh* 1291 Ch]. OE *bēo-lēah* 'bee wood'.

Beeley Db [*Begelie* DB, *Beegeleg* 1205 Abbr, *Begalaia* 13 Derby]. '*Bēage*'s LĒAH.'

Beelsby Li [*Belesbi* DB, c 1115 LiS, 1130 P]. '*Beli*'s BY.' *Beli* is an ON pers. n.

Beenham Brk [*Beneham* 1242 Fees, *Benham* 1291 Tax]. OE *bēan-hamm* 'HAMM where beans were grown'.

Beer D nr Colyton [*Bera* DB, *Bere* 1275 RH, *Beare* 1303 FA], **B~ Charter** D [*Bera* DB, *Bere* 1242 Fees, *Beare* 1303 FA, *Bairecharteray* 1288 Ass], **B~ Hackett** Do [*Bera* 1176 ff. P], **B~ Crocombe** So [*Bere* DB, 1232 Cl, *Craucombesbere* 1325 Misc], **Beere** So nr Cannington [*Bera* DB]. The Devon Beers are OE BEARU 'grove'. The Do and Somerset ones seem to be rather OE BÆR 'pasture'.

B~ Charter was held by Simon de Chartray in 1242 (Fees).—**B~ Crocombe** was granted to Godfrey de Craucombe in 1227 (Ch). Cf. CROW-COMBE So.—**B~ Hackett** was held by Haket de Bera in 1176 ff. (P). Haket is a Norman pers. n.

Beesby in the Marsh Li [*Besebi, Bizebi* DB, *Beisebi* 1162 P, *Besebi* Hy 2 DC, *Beseby* 1254 Val], **B~ Li** nr Hawerby [*Besebi, Basebi* DB, *Besabi* c 1115 LiS, *Beseby* 1254 Val]. First el. the pers. n. *Besy* DB (Li), which probably represents ODan *Bøsi*.

Beesthorpe Nt [*Bestorp* DB, 1182 P]. '*Bøsi*'s thorp'; cf. BEESBY.

Beeston Bd [*Bistone* DB, *Beston* 1201 Cur, *Buston* 1246 FF], **B~ Regis** Nf [*Besentuna, Besetune* DB, *Beeston* 1207 FF], **B~ St. Andrew** Nf [*Bes(e)tuna, Beofetuna* DB, *Beston* 1254 Val], **B~ St. Lawrence** Nf [*Bestone* 1044–7 KCD 785, *Besetuna* DB, *Bestone* 1254 Val], **B~** Nf nr E. Dereham [*Bestone* 1254 Val], **B~** Nt [*Bestune* DB, *Beston* 1169, *Beeston* 1181 P], **B~** YW [*Bestone* DB, *Beeston* c 1180 YCh 1620]. OE *Bēos-tūn* 'TŪN where reed or rush grew'. OE **bēos* corresponds to MLG *bēse*, MDu *biese*, Du *Bies* (base **beusō-*).

Beeston Chs [*Buistane* DB, *Bustan, Bestan* 1282 ff. Court]. Second el. OE *stān* 'stone'. If Beeston has been correctly identified with *Bovio* (abl.) IA, *Bee-* is a later form of this British name. Otherwise it might be OE *byge* 'traffic, commerce'. The name would then mean 'stone where commerce took place'.

Beetham (-ēdh-) We [*Biedun* DB, *Bethome* c 1095 Kendale, *Bethum* 1157 YCh 354]. The dat. plur. of ON *biö̃r* 'table' in a transferred sense such as 'flat area'. BEELA R was formerly **Beetha** [*Betha* c 1195 FC]. It goes back to an ON *Biöõ-á*, perhaps 'the Beetham river'.

Beetley Nf [*Betellea* DB, *Betel*' 1204 Cur].

If the DB spelling is reliable, the first el. contained an *l*. If so, the name probably means 'wood in which wooden mallets were got' (OE *bīetel* 'beetle' and LĒAH).

Befcote St [*Beficote* DB, *Beffecote* 1227 FF]. '*Beffa*'s COT.'

OE **bĕg, bĕger** 'berry'. See BAILDON, BAILEY, BAYDON, BARMOOR.

Begbrooke O [*Bechebroc* DB, *Bekebroc* 1190 P]. '*Be(o)cca*'s brook.'

Beighton Db [*Bectun* 1002 Wills, -*e* DB, *Becton* 1236 FF]. 'TŪN by the brook.' See BÆCE.

Beighton Nf [*Begetuna* DB, *Begeton* 1186 P, *Beghetun* 1202 FF]. '*Bēaga*'s or *Bēage*'s TŪN'; cf. BEAWORTHY.

Bekesbourne K [*Burnes* DB, *Burna* 11 DM, *Burnes* 1198 Fees, *Bekesborne* 1280 BM]. Originally *Burna* 'the stream'. Later often *Livingesburn* and the like (1206 FF &c.), from an owner called *Lēofing*. The manor was held in 1198 and 1208 by Willelmus de Beche or Becco (Fees, Cur), whose name is derived from BEC in Normandy.

Belah or **Beela** R We [*Belewe* 1292 Ass, *Belew* 1314 Ipm]. From an OE *Belge* 'roaring river', related by OE *bylgian*, ME *belwe* 'to bellow, roar'. Cf. BELPH.

Belasis Du nr Billingham [*Belasis* 1305 RPD], **Bellasis** Du nr Durham [*Belasis* 14 FPD], **B~** Nb [*Beleassis* 1279 Ass], **Bellasize** YE [*Belasise* 1312 Ipm]. 'Beautiful site' (OFr *assise*).

Belaugh Nf [*Belhae, Belahe* 1044–7 Holme, *Belaga* DB, *Belhag* 1147–9 Holme, *Belhagwe* 1249 Cl, *Belhage* 1254 Val]. The name is identical with **Bylaugh** Nf [*Belega* DB, *Belag* 1203 Cur, 1208 FF, *Belhawe* 1254 Val, *Belhaye* 1275 RH], further with *Belaga*, *Belahaye* 12 Glouc (pasture in Gl), **Bilhagh** Nt [*Billehah* 1244, *Bilhaye* 1252, *Belawe* 1292 Cl], **Bealeys** YE in Lockington [(grangia de) *Belaga* 1151 YCh 1383, *Belag* 1221 FF]. The common base is OE *Bel-haga*. The first el. is a word not evidenced in English as a common noun, but probably found in *Beel* (placea) 1276 RH (in YW) and in several other pl. ns., e.g. BELTON (q.v.). The element is evidently identical with one found in numerous Continental pl. ns., e.g. BEELEN in Münster [*Bele* 1146], BEELEN in Gelderland [*Bele* 1188]. The meaning of this Continental element is likewise unknown. It might belong to ON *bil*, Dan *bil, bæl* 'point of time', lit. 'interval, interspace'. If so, the word might naturally have been used of a glade in a forest or a piece of dry land in fenny country.

Belbroughton Wo [*Bellebrocton* 1290, *Belne Brocton* 1298 Ep]. Really Bell and Broughton; the two were once distinct villages. **Bell** [*Beolne* 817 BCS 360, *Bellem* DB, *Belne* 1212 Fees] is an old river-name

[*Beolne* 1300 For], which may be derived from OE *beolone* 'henbane'. **Broughton** [*Broctun* 817 BCS 360] is 'TŪN on the brook' (Bell Brook).

Belby YE [*Belleby* 959 YCh 4, *Ballebi* DB, *Bellebi* 1202 FF]. The first el. would seem to be a pers. n., but neither ODan *Balli* nor ON *Beli* seems probable. ON *bialla*, OSw *biælla* 'bell' might have been used as a nickname. Or the first el. may be simply the word for 'bell'.

Belchalwell Do [*Bell, Caldewell* 1207 Cur, *Belle* and *Chaldwell* 1286 Ch]. A combination of two names of the same type as BEL-BROUGHTON. Bell is dial. *bell* 'a hill', probably OE *belle* 'bell' in transferred use. Chalwell means 'cold stream' (OE *cealde-welle*).

Belchamp Otton, St. Paul's & Walter Ess [*Bylcham* 939 BCS 737, *Belhcham* c 1040 Wills, *Belcham, Belcamp* DB, *Belcham Otes* 1256 FF, (in) *Bello Campo Sancti Pauli* 1248 FF, *Waterbelcham* 1297 Ch]. The original second el. was HĀM (or possibly HAMM); *-champ* is due to a later change. The first may be OE *bylc* or *belc*. *Bylcham* 939 may be an inverse spelling for *Belcham*. Probably the correct form is *belc-*, which is a derivative of OE *balca* in its ME sense of 'beam'. The word is probably found in the OE Exodus in the form *bælce* (dat.). OE *bælc* (or *bælce*) probably meant 'a roof made of beams' or the like, and Belchamp is 'a house with a roof of a certain kind'.

B~ **Otton** was held by William fil. Otonis in 1212; Otto is a well-known German name, also used in OFr.—B~ **St. Paul's** belonged to St. Paul's in London.—B~ **Walter** was held by Walter de Teye in 1297 (Ch).

Belchford Li [*Beltesford* DB, 1297 BM]. '*Belt*'s ford.' The same pers. n. is found in **Beltisloe** hd Li [*Belteslawe* DB] and **Belthesholm** (nr Kirkstead Li) c 1155 DC. The name may be derived from the word *belt*.

Belford Nb [*Beleford* 1242 Fees, 1250 Ipm, *Belleford* 1300 Pat]. May be identical with *Bellanford* 848 BCS 454 (situation unknown). The latter name seems most likely to contain a pers. n., but a name *Bella* is not evidenced elsewhere. Belford Nb may contain the element discussed under BELAUGH.

Belgrave Le [*Merdegrave* DB, *Merthegrava, nunc Belegrava* c 1135 Ordericus, *Belegraue,* 1185, *-graua* 1191 P, *-greve* 1252 Ch]. Second el. OE *græf* 'grove'. The first was OE *mearþ* 'marten', which was associated with Fr *merde* 'faeces, filth' and replaced by OFr *bel, -e* 'fine'.

Bellasis, Bellasize. See BELASIS.

Belleau Li [*Elgelo* DB, *Helegelo* 1194 P, *Helegloue* 1202 Ass]. The original name meant '*Helgi*'s meadow' or 'the holy meadow'. *Helgi* is a well-known OScand pers. n. It is the definite form of the adj. (ON) *heilagr* 'holy'. Second el. ON *ló*

'meadow, glade', identical with OE LĒAH. The form *Belleau*, which is French and means 'beautiful water', is not found in early records.

Bellerby YN [*Belgebi* DB, *Belegerebi* 1167 P, *Belgereby* 1203 FF, *Belgreby* 1231 FF]. First el. the ON byname *Belgr* (gen. *Belgiar*), really *belgr* 'a bag'.

Bellingham K [*Beringahammes gemæru* 973 BCS 1295, *Beringaham* 998 KCD 700, *Belingeham* 1198 (1285) Ch]. 'The HĀM of *Be(o)ra*'s people'; cf. BARHAM K.

Bellingham (-nj) Nb [*Bainlingham* c 1170 Reg Dun, *Bel(l)ingham* 1254 Val]. In 1326 Ipm is mentioned a place *le Belles* in Bellingham. This may well be *bell* in the sense 'a hill' (cf. BELCHALWELL). If so, Bellingham may mean 'the HĀM of the dwellers at the hill'.

Bellister Nb [*Belester* 1279 Ass, *Belestre* 1306 Ipm, *Belecestre* 1355 Orig]. A French name, as suggested by Mrs. Gelling, *Namn och bygd* 45, pp. 170 f., but no doubt OF *bel estre* 'fine place (site)' rather than *bel ester* 'pleasant sojourn'.

Belmesthorpe Ru [*Beolmesδorp* 1066–9 KCD 927, *Belmestorp* DB, *Burmestorp* 1230 Ch, *Bulmestorp* 1233 Cl]. '*Beornhelm*'s thorp.'

Belper Db [*Beurepeir* 1231 Cl, *Beurepeyr* 1251 Ch]. 'Beautiful retreat.'

Belph Db [*Bolh, Belgh, Belhismere* 12 (1316) Ch]. Really the name of a stream; cf. BELAH.

Belsay Nb [*Bilesho* 1163, *Belesho* 1171 P, *Billesho* 1204 Ch, *Belesho* 1254 Val]. '*Bill*'s ridge.' A pers. n. *Bill* is presupposed by many pl. ns. It is a short form of *Bilfriþ, Bilheard* &c.

Belsize Hrt, B~ Np [*Bellasyse* 1254 Val]. See BELASIS.

Belstead Ess [*Belesteda* DB, *Belsted* 1204 Cur], B~ Sf [*Belesteda* DB, *Belstede* 1198 FF]. See STEDE. First el. as in BELAUGH.

Belstone D [*Belestham* DB, *Belestan* 1168 P, 1255 FF]. 'The logan stone', lit. 'bell stone', 'stone that rocks like a bell'.

Belswardyne Sa [*Bellevrdine* DB, *Bedleswurthe* 1228 FF, *-worth* 1237 Cl, *-wrthin* 1269, *-worthyn* 1281 Ch]. '*Bēdel*'s WORÞIGN.' See WORÞ. *Bēdel* is a derivative of *Bēda*.

Belthorpe YE [*Belkertorp* 1205 Obl, *-tho[r]p* 1240 FF, *Belgerthorp* 1242 P, 1246 FF]. 'The thorp of *Belgr*', if the form in *-g-* is correct. Cf. BELLERBY.

Beltoft Li [*Beltot* DB, *Beltoft* 1202 Ass]. See TOFT. Beltoft is near Belton (3). The first el. is as in BELTON.

Belton Le [*Beltona* c 1125 LeS, *Belton* c 1225, *Beletone* 1222 Ep], B~ Li nr Grantham [*Beltone* DB, *Beltuna* 1146 RA], B~ Li in Axholme [*Beltone* DB, *Bealton* 1179

P, *Bealton, Beltona* 1224 Ep], **B~** Ru [*Bealton* 1167 P, *Belton* 1205 Obl, 1232 Pat], **B~** Sf [*Beletuna* DB, *Beleton* 1198 FF, 1212 Fees]. For a discussion of the first el. see BELAUGH.

Belvoir (bĕver) Le [*Belveder* 1130 P, (de) *Bello Videre* 1145 BM, *Belueeir* Hy 2 DC]. 'Beautiful view.' A Fr name, identical in meaning with *belvedere*.

Belwood Li [*Belwode* c 1184 TpR]. Situated nr Belton (3). First el. as in BELTON.

Bembridge Wt [*Bynnebrygg* 1316 FA, *-brigge* 1345 Misc]. OE *binnan brycge* '(place) inside the bridge'.

Bemersley St [*Bemeresleg* 1252 Ch]. 'The LĒAH of the trumpeter'; cf. BEMERTON.

Bĕmerton W [*Bimertone* DB, *Bimerton* 1242 Fees, *Beomertona* 1107 (1300) Ch]. OE *bȳmera tūn* 'the TŪN of the trumpeters'. OE *bȳmere* is found also in *bymera cumb* a 672 BCS 27, in boundaries of Downton, which is not far from Bemerton. Cf. HORN-BLOTTON.

Bempton YE [*Bentone* DB, *Bemtona* c 1130 BM, c 1150 YCh 1154]. OE *Bēam-tūn*; cf. BĒAM and BAMPTON.

Benacre Sf [*Benagra* DB, *Beanacer* c 1095 Bury, *Benaker* 1229 Cl]. 'Bean field'; cf. ÆCER.

Bendish Hrt [*Beanedisc* 11 Gesta, *Benedis* DB], **Bendysh** Ess [*Benedisc* 1068 EHR xi, DB]. 'Bean field'; see EDISC.

Benefield (bĕni-) Np [*Beringafeld* c 970, 972–92 BCS 1129 f., *Benefeld* DB, *Benifeld* 1130 P, *Berifeld* 1236 Fees]. 'The FELD of *Bera*'s people'; cf. BARHAM K.

Benenden K [*Bing-, Bynidene* 993 Hyda, *Benindene* DB, *Binnigdaenne* 11 DM, *Bynindenn* 1242 Fees]. 'The DENN or swine-pasture of *Bionna*'s people.'

Benfieldside Du [*Benfeldside* 1297 Pp]. 'Slope with a bean-field or bent-grass field.'

Benfleet Ess [*Beamfleot* 894 ASC, 1067 BM, *Benflet* DB, *Bemflet* 1165 P, *Northbemflet* 1292 Ch]. 'Stream with a beam, i.e. a footbridge'; cf. BĒAM.

Bengeo (-nj-) Hrt [*Belingehou* DB, *Beningho* 1202 Cur, 1212 Fees]. 'The ridge of the dwellers on R BEANE.' See HŌH.

Bengeworth (-nj-) Wo [*Benningcuuyrð* 907 BCS 616, *Bynnyncgwyrð* 980 KCD 625, *Beningeorde* DB]. 'The WORÞ of *Beonna*'s people.'

Benhall Sf [*Benehala, Benenhala* DB, *Benhale* 1254 Val]. 'HALH where beans were grown.' First el. OE *bēan* and *bēanen* 'of beans'.

Benham Park, Hoe & Marsh B~, B~ Valence Brk [*Bennaham,* (æt) *Bennanhamme* 956, *Bennanham* 960 BCS 942, 1055, *Benneham* DB, *Holebenham* 1220 Fees,

Benham Valence 1316 Ipm]. '*Benna*'s HAMM.' OE *Beonna* is hardly to be assumed, as the ex. of 960 is in an original charter. *Benna* is not recorded in independent use.

Hoe from *hol* 'hollow'.—**B~ Valence** was given to William de Valencia in 1251 (Ch). Valence is a family name (from one of the Valences in France).

Benhilton Sr [*Benhull* 1392 PNSr]. The original name means 'bean hill'.

Beningbrough YN [*Benniburg* DB, *Beningeburc* 1167 P, *-burg* 1223 FF], **Benington** Li [*Benigtun* Hy 2 DC, *Benington* 1212 Fees], **Benningholme** YE [*Benincol* DB, *Benigholm* 1199 P, *Beningeholm* 1223 FF]. 'The BURG, TŪN & HOLM of *Beonna*'s people.'

Bennington Hrt [*Belintone, Benintone* DB, *Benington* 1239 Ep]. 'The TŪN of the dwellers on R BEANE.'

Bennington, Long, Li [*Beningtun* DB, 1212 Fees, *Bennington* 1163 Fr, *Bunington* 1209–19 Ep, *Byniton* 1240 FF], **Benniworth** Li [*Beningurde* DB, *Beningewurða* 1171 P, *-wurda* 1194 P]. 'The TŪN and WORÞ of *Beonna*'s people.'

Benridge Nb in Mitford [*Benerig* 1242 Fees]. 'Ridge where beans were grown.'

Bensham Du [*Benchelm* c 1245 PNNb]. The elements are *bench* in some topographical sense, e.g. 'ledge', and *helm* 'a shed for cattle'.

Benson or **Bensington** O [*Banesinga villa* c 730 BCS 155, *Bænesingtun* 571 ASC, *Beonsincg-, Bynsincgtun* 880 BCS 547, *Bensentun* 1156 P, *Bensinton* 1200 Cur]. 'The TŪN of the *Benesingas*.' The base of this appears to be an OE pers. n. *Benesa*, derived from *Bana (Bona)*.

Bentfield Ess [*Benedfelda* DB], **Benthall** Sa [*Benehale* DB, *Benethala* 1167 P], **Bentham** Gl [*Benetham* c 1250 Glouc], **High & Lower B~** YW [*Benetain* DB, *duæ Bentham* 1204 FC]. 'FELD, HALH, HĀM overgrown with bent-grass.' OE *beonet* (=OLG *binut*, OHG *binuz*) means 'bent-grass'.

Bentley, Fenny, Db [*Benedlege* DB, *Fennibenetlegh* 1272 FF], **Hungry B~** Db [*Beneleie* DB, *Hungre Bentley* 1431 FA], **B~** Ess °[*Benetleye* c 1040 Wills, *Benetlea* DB, *Benteleye Magna, Parva* 1254 Val], **B~** Ha nr Alton [*Beonetleh* c 965, *Beonetlegæ gemære* 973–4 BCS 1154, 1307, *Benedlei* DB], **B~** Ha in Mottisfont [*Beneclege* DB, *Benetleg* 1227 Ch], **B~ Priory** Mx [*Benetleg'* 1243 Cl], **B~** Sa [*Benetleg* 1233 Cl], **B~** Sf [*Benetleia* DB], **B~** St [*Benetlea* 1165 P, *-leg* 1198 Fees], **B~** Wa [*Beonetlege* 1005 Eynsham], **B~** Wo in Holt [*Beonetlæage* 962 BCS 1087], **B~ Pauncefote** Wo [*Beneslei* DB, *Benetlega* 1185 P, *Benetleg Pancevot* 1212 Fees], **B~** YE [*Benedlage* DB, *Bentelay* 1229 Ep], **B~** YW nr Arksey [*Benedleia* DB], **B~** YW nr Emley [*Benetleia* J BM], **B~** YW nr Chapel Allerton [*Bentelay* E 1

BM]. OE *beonet-lēah* 'clearing overgrown with bent-grass'; cf. BEONET.

Hungry B~ perhaps because the soil needed much manure.—**B~ Pauncefote** was held by Richard Panzeuot in 1185. Pauncefote is a Norman nickname meaning 'round belly'.

Benton, Little, and **Longbenton** Nb [*Bentune* c 1190 Godric, *Magna Beneton* 1256 Ass, *Parva Bentona* 1236 Fees]. OE *Beonet-tūn* 'TŪN where bent-grass grew'; cf. BENT-FIELD. Or perhaps better OE *Bēantūn* 'bean farm'.

Bentworth Ha [*Binteworda* 1130, *-wurða* 1167 f. P, *Binthevorda* 1155 Fr]. The first el. is connected with **Binton** Sr in Seale [(in) *Bintungom* 688 BCS 72]. Derivation from OE *beonet* 'bent-grass' is improbable. Probably Bentworth has as first el. a pers. n. **Binta*, corresponding to OHG *Binizo* (now *Binz*). From this also *Bintungom* is derived.

Benwell Nb [*Bynnewalle* c 1050 HSC, *Benewell* 1242 Fees]. OE *bionnan walle* '(the place) inside the (Roman) Wall'. B~ is between the Wall and the Tyne.

Benwick Ca [*Beymwich* 1221, *Benewich* 1244, *Beyn-*, *Beinwyke*, *Bemwyk* 1251 &c. PNCa(S)]. The low situation on the Nene tells against OE *bēan* 'bean' as a first el. Possibly OE *Bēamwīc* 'wīc by a beam or plank bridge'. Cf. BENFLEET.

OE **bēo** 'bee' is a fairly common first el. See BEE-, BEO- (passim), BEAL Nb, BECKETT, BEAUWORTH. OE **bēocere** 'bee-keeper'. See BICKERSHAW &c.

Beobridge Sa [*Beebrugge* 1194 Eyton, *Bebrig'* 1200 Cur, *Bewbrugge* c 1275 PNSa]. Perhaps 'bridge near which bees were found'; cf. BEEFORD. Or '*Bēofa*'s bridge'. Cf. BEVINGTON.

OE **beofor** 'beaver'. See BARBOURNE, BEVER-COTES &c., BEWERLEY.

Beoley (-ē-) Wo [*Beoleah* 972 BCS 1282, *Beolege* DB]. OE *bēo-lēah* 'bee wood'.

OE **beonet** is only used in pl. ns. It corresponds to G *Binse* (OHG *binuz*), OLG *binut* 'juncus, scirpus'. LG *bent* is used of a kind of grass (*Molinia cœrulea*), and ME and Mod *bent* of grasses of a reedy or rush-like habit, or which have persistent stiff or rigid stems. The exact meaning of *beonet* in pl. ns. is thus doubtful, and the term *bent-grass* includes reeds and rushes and other coarse grasses. It is noteworthy that names in *beonet* are absent or rare in districts where names in *bēos* occur.

OE **beorc** 'birch' is a common first el. See BARK- BERK- (passim), BARFORD (3), BART-LEY, BARTLOW.

OE **beorg** means 'mountain, hill, hillock, mound, esp. grave-mound', and all these senses are represented in pl. ns. *Beorg* corresponds to OScand *berg, biarg*, which is found in some names, as BARBY, BARROWBY, BORROWBY, BERRIER. OE *beorg* is the source of BARROW (2), BARUGH, BERROW, BURGH (2). It is the first el. of some names, e.g. BAR-ROWDEN, BEARSTED, BERGHOLT, BERSTED, BURGHFIELD, (?) BURFORD O, BROUGHTON (3). As a second el. it should have appeared as *-barrow* or *-berrow*, but mostly the modern spelling has *-borough* or *-burgh*, as ALK-, FARN-, HAR-, LASBOROUGH, SMALLBURGH. Also *-berry* occurs, as in BLACKBERRY.

OE **beorht** 'bright'. See BIRTLEY, BREIGHT-MET, BRIGHT-, BRITWELL.

Bepton Sx [*Babintone* DB, *-ton* 1176 P, *Bebbinton* 1241 FF]. '*Bebba*'s or *Bebbe*'s TŪN.' For *Bebbe* see BEBINGTON. *Bebba* is unrecorded.

Berden Ess [*Berdane* DB, *Berden* 1198 FF, *Bereden* 1267 Ch, *Bierden* 1428 FA]. 'Valley with a cow-house.' Cf. BȲRE.

OE **bere** 'barley', perhaps also 'corn', is a common first el. There must also have been a shorter form *bær-* (found in OE *bærlic* 'barley'). See BAR- passim (especially BAR-FORD, BARTON), BEARL, BEREWĪC. The adj. *beren* 'of barley' may be found in some names, as BARNACLE, BARNEY. For OE *berern* 'barn' see BARNES, BARNACRE, BERNE and the like, WHITBURN.

Bere Ferrers D [*Birland* DB, *Ber* 1242 Fees, *Byr* 1281 Ass, 1318 Ep, *Byrfferers* 1239 Ass]. OE *byrh*, a dat. form of *burh*. The *-h* would be particularly apt to be lost in the compound *Byrhland*, whence *Birland*.

The manor was held by William de Ferers in 1242 (Fees). Ferrers is from one of the FERRIÈRES in Normandy (lit. 'the smithies'). **Bere Alston** was originally Alston [*Alphameston* 1339 Ass]. '*Ælfhelm*'s TŪN.'

Bere Forest Ha [*La Bera* 1168, 1185 P, *foresta de Bera* 1190 ff. P]. OE *bær* 'pasture'.

Bere Regis Do [*Bere* DB, 1244 Fees, *Bycre* 1259 Pat, *Kingesbere* 1280 Ch]. Perhaps OE BEARU 'grove'.

Berechurch Ess [*Berdechirche* c 1270 Colchester]. 'Church made of boards.' First el. OE *bred* 'board' or an adj. *byrden* 'of boards' (from OE *bord* 'board'; cf. ME *borden*).

OE **berern.** See BERE.

OE **beretūn.** See BARTON.

OE **berewīc**, lit. 'corn farm', was used in the same senses as *beretūn*, i.e. 'berewick, grange, an outlying part of an estate'. See BARWICK, BERRICK, BERWICK, BORWICK.

OScand **berg.** See BEORG.

Bergholt, West, Ess [*Bercolt* DB, *Bercholt* 1273 BM], **East B~** Sf [*Bercolt* DB, *Bercholt* 1130 P, *Bergholt*, *Berghout* 1228 FF &c.]. Both seem to be OE *beorg-holt* 'copse by a hill' to judge by the old pronunciation with (rf).

Bericote Wa [*Bericote* DB, *Biricot* 1252

Fees]. OE *byrg-cot* 'COT belonging to the BURG'.

Berkeley (-ark-) Gl [*Berclingas* 804 BCS 313, (æt) *Beorclea* 824 ib. 379, *Berchelai* DB, 1130 P]. OE *beorc-lēah* 'birch wood'. *Berclingas* means 'the monks of Berkeley'.

Berkesdon Hrt [*Berchedene* DB, *Berkeden* 1198 FF, 1212 Fees]. OE *beorc-denu* 'birch valley'.

Berkhamstead (berk-), **Great**, Hrt [*Beorhþanstædæ* (sic) 966–75 Wills, *Beorhhamstede* 1066 ASC (D), *Berchehamstede* DB, *Berchamstede* 1156, *Berkamsted* 1230 P], **Little B~** Hrt [*Berchehamstede* DB, *Parva Berkamested* 1238 Ep]. Gt. B~ is 'homestead on a hill' (OE *beorghāmstede*). Li. B~ is probably identical but might be 'homestead among birches' (OE *beorchāmstede*).

Berkley (-ark-) So [*Berchelei* DB]. OE *beorc-lēah* 'birch wood'.

Berkshire (-ark-) [*Bearrucscir* 860 ASC, *Be(a)rrocscire* c 894 Asser, *Berruc-*, *Bærrucscir* 931 BCS 687 f., *Bearrucscira* 11 Th]. Asser says *Berrocscire* was named after *Berroc silva*, 'ubi buxus abundantissime nascitur'. This may be right, but if so, the forest was no doubt named from a hill *B(e)arruc*. This name is derived from Brit **barro-*, Welsh *bar* 'top, summit'; cf. BARR.

Berkswell (-ark-) Wa [*Berchewelle* DB, *Berkeswelle* 1236 Fees, *Bercleswelle* 1221 ff. PNWa(S)]. '*Beorcol*'s spring.'

Bermondsey Sr [*Vermundesei* 708–15 BCS 133, *Bermundesye* DB, *-ia* 1130 P]. '*Beornmund*'s island.'

Berne Do [*La Berne* 1281 Ipm, *Le Bernes* 1431 FA]. OE *berern* 'barn'.

Bernwood Bk [*Byrnewuda* 921 ASC, *Bernewude* 1197 P]. Cf. BICESTER. The first el. is OE *byrgen* 'burial mound'.

Berrick Prior & Salome O [*Berewiche* DB, *-wic* 1228 Cl]. See BEREWIC and BRITWELL SALOME.

B~ Prior belonged to the prior of Canterbury.

Berrier Cu [*Berghgerge* 1166 P]. 'Shieling on a hill'; see BEORG, ERG.

Berrington Gl [*Berinton* 1205 Obl, *Biryton* 1274 Cl], **B~** He [*Beri(n)ton* 1236 Brecon], **B~** Sa [*Beritune* DB, *Biriton* 1242, *Byrinton* 1236 Fees], **B~ Green** Wo [*Beritune* DB, *Beri-*, *Biriton* 1275 Subs]. OE *Byr(i)gtūn* 'TŪN by or belonging to a BURG'. Cf. BURTON (2). A correspondent notes that B~ Gl was the main Saxon settlement at Chipping Campden. This may explain the meaning of *byrigtūn* here.

Berrington Nb [*Berigdon* 1208–10 Fees, *Beringdon* 1269 Ass]. OE *byrigdūn* 'hill with a BURG'. There is an old camp near by.

Berrow So [*at Burgh'* 973 BCS 1291, *Berges* 1196 P, *Bergh* 1249 FF]. OE *beorg* or *beorgas* 'the hill(s) or mound(s)'. *Burgh* 973 is in a late transcript.

Berrow Wo [*la berge, la Berwe* 1190 Hearne's Heming, *Berge* 1203 Cur]. OE *beorg* 'hill'.

Berry Pomeroy D [*Berie* DB, *Beri* 1242 Fees, *Bury Pomerey* 1281 Ipm], **Berrynarbor** D [*Biria* c 1150 Fr, *Bery Narberd* 1244 FF]. OE *byrig*, dat. of *burg* 'fort' &c.

Radulf de Pomerei held *Berie* in 1086 (DB). Pomeroy is OFr *pommeraie* 'apple orchard', also used as a pl. n., as LA POMERAYE in Calvados, Normandy.—**Berrynarbor** was held by Philip de Nerebert in 1210 (FF). The family name, *Nerbert* 1172 P, was no doubt taken from NARBERTH in Pembroke [*Nerberth* 1414 BM].

Bersted Sx [*Beorganstede, north Beorganstede* 680 BCS 50, *Beorganstedinga mearc* 988 KCD 663]. Very likely OE *Beorghāmstede* with assimilation of *m* to *n* before *s*. Cf. HĀMSTEDE. *Beorg* must refer to a tumulus.

Berwick Do [*Berewich* 1194 P], **B~** Ess [*Berwyk* 1412 FA], **B~** K [(æt) *Berwican* 1032 Th, *Berewic* DB], **B~ Hill** Nb [*Berewic* 1205 Cur], **B~** (bĕrĭk) **upon Tweed** Nb [*Berewich* 1167 BM, *Berewicum super Twedam* 1229 Cl], **Great B~** Sa [*Berewic* DB], **B~ Maviston** Sa [*Berewic* DB], **B~** (berwĭk) Sx [*Berewice* DB], **B~** (bărĭk) **Bassett** W [*Berewic* 1185 TpR, *Berwyk Basset* 1325 Ch], **B~ St. James** W [*Berewyk Sancti Jacobi* c 1191 Salisbury], **B~ St. John** W [*Berewic* 1196 P, *Berewyke S. Johannis* 1267 Salisbury], **B~ St. Leonard** W [*Berewyk Sancti Leonardi* 1291 Tax], **B~** YW [*Berewic* 1195 P]. See BEREWĪC.

B~ Bassett was held by Alan Basset in 1212 (RBE). Basset is a Norman nickname from OFr *basset* 'of low stature'.—**B~ Maviston** was held by Henry Malveisin in 1166 (Eyton). Malveisin is a Norman nickname meaning 'bad neighbour'.

Bescaby Le [*Bersaltebi* 1194, *Berscandebi* 1195, *Bergaldebi* 1196 P, *Berscaldeby* 13 Fees]. B~ is 1½ miles from SALTBY and higher up the hill. Its name is *Saltby* with *beorg* 'hill' prefixed. Thus 'Hill Saltby'.

Bescar La [*Birchecar* 1331 PNLa]. 'Birch carr'; see KERR. **B~ Lane.** See BESTWOOD.

Bescot St [*Bresmundescote* DB, *Bermundescote* 1273 Ass]. '*Beorhtmund*'s COT.'

Besford Sa [*Betford* DB, *Besseford* 1254 Ipm], **B~** Wo [*Bettesford* 972 BCS 1282, *Bezceford* 1176 P]. '*Betti*'s ford.'

Beslow Sa [*Beteslawe* DB, *Beszelawæ* 1176 P]. '*Betti*'s hill or burial-mound'; see HLĀW.

Bessacar YW [*Beseacra* 1182 P, *Besacra* c 1190 YCh 817, *-acre* 1202 FF]. 'Rush-grown field'; cf. BEESTON.

Bessels Leigh. See LEIGH.

Bessingby YE [*Basinghebi* DB, *Basingebi* c 1130 BM, *Besingebi* 1166 YCh 1139]. The forms suggest an original *Bæsingabȳr* or the like. The first el. is apparently a derivative of the pers. n. discussed under BASING. But B~ is probably a Scandinavianized form of an OE *Bæsingaburh* or *-tūn*. Cf. BADBY, SELBY.

Bessingham Nf [*Basingeham* DB, 1177 P, *Basingham* 1254 Val]. 'The HĀM of the *Basingas*'; cf. BASING.

Besthorpe Nf [*Besethorp* DB, *Bestorp* 1198 FF], **B~** Nt [*Besthorp* c 1163 RA, 1316, 1402 FA]. Identical with BEESTHORPE.

Bestwall Do [*Beastewelle* DB, *Byestewalle juxta Warham* 1310 FF]. OE *be ēastan wealle* '(the place) east of the wall' (of Wareham).

Bestwood Nt [*Beskewuda* 1178 P, *Bestwud* 1227 Cl, *Beskewode* 1249 Cl]. The first el. is no doubt a derivative of the OE **beos* discussed under BEESTON, i.e. an OE *bēosuc* 'a tuft of reeds', analogous to OE *rixuc* 'a clump of rushes' and the like. The same el. seems to enter into **Bescar Lane** Nt [*Beskhale* (boscus) c 1200 (1316) Ch].

Beswick La [*Bexwic* 1200–30 CC]. This might be '*Bēac*'s WĪC'; cf. *Beaces hlaw* BCS 917.

Beswick YE [*Basewic* DB, *Besewic* R 1 Cur, 1233 FF]. In view of the DB form the first el. can hardly be the el. discussed under BEESTON. It is possibly an OE pers. n. *Basa* (cf. BASING) or *Besi* as in BEESBY.

Betchton Chs [*Becheton* 1260 Court, *Becheton*, *Bechinton* 1308 Ipm]. 'TŪN by the brook.' See BÆCE. The place is close to Sandbach.

Betchworth Sr [*Becesworde* DB, *Becheswrde* 1198 FF, *-wurth* 1242 Fees]. The first el. is more probably a pers. n. **Becci*, derived from *Bacca*, than OE *bece* 'brook'. OE *Bæcci* pers. n. is actually evidenced in BATCHWORTH.

Bethersden K [*Baedericesdaenne* 11 DM, *Beatrichesden* 1316 FA]. '*Beadurīc*'s DENN or swine-pasture.'

Bethnal Green Mx [*Blithehale* 13, *Blithenhale* 1341 PNMx]. See HALH. The first el. may be a stream-name *Blīþe* (see BLYTHE) or a pers. n. **Blīþa*.

Betley St [*Betelege* DB, *Bettelega* 1175 P]. '*Bette*'s LĒAH.' *Bette* is a woman's name. The neighbouring AUDLEY, BALTERLEY, and BARTHOMLEY are all named from women.

Betterton Brk [*Bedretone* DB, *Beterintona* c 1130 Oxf, *Betreton* 1185 P, *Baterinton* 1236 Fees]. Perhaps identical with BATHERTON Chs (*Beornþryþetūn*). Or better the first el. may be an unrecorded *Beaduþryþ*, a woman's name.

Betteshanger K [*Betlesnangre* 1176 BM Facs, *Betlesengre* 1198 FF, *-angre* 1242 Fees]. Second el. HANGRA. The first might be OE *gebytle* 'house', with Kentish change of *y* to *e*.

Bettiscombe Do [*Bethescomme* 1190 (1332) Ch, *Bettescumbe* 1291 Tax]. '*Betti*'s coomb.'

Betton Abbots or **Strange** Sa [*Betune* DB, *Beiton* 1197 FF, *Betton Extranea* 1285 FA, *B~ Abbatis* 1301 For], **B~ in Hales** Sa [*Baitune* DB, *Beitona* c 1090 Eyton, *Betton* 1276 Cl]. The two Bettons seem to have influenced each other. **B~** Abbots is near BAYSTON, and the situation does not suit OE *bece* 'valley' as first el. It is probably '*Bæga*'s TŪN'. **B~** in Hales is near the Tern and its name probably means 'TŪN by the stream'; cf. BÆCE.

Strange is a family name, derived from OFr *estrange*, Lat *extraneus*. Hamo le Strange made a grant in Betton c 1160. **B~ Abbots** belonged to Shrewsbury Abbey.

Bettws-y-Crwyn Sa [*Betteus* 1256, (Chapel of) *Betteus* 1276 Ep]. Welsh *bettws* 'chapel' (from OE *bed-hūs* 'oratory'). *Crwyn* may be Welsh *crowyn* 'pigsty'.

Beult R K [*Beule* 1612 Drayton]. Etymology obscure.

Bevercotes Nt [*Beurecote* 1165 P, *Beuercote* 12 DC, *-cot* 1230 P]. 'Beavers' huts.'

Beverley YE [*Beferlic* c 1000 Saints, *Beoferlic* ASC (E) (sub anno 721), *Bevreli* DB, *Beuerlacum* c 1067 YCh 88]. Probably 'beaver stream'. The second el. may be a form of OE **lecc*, the source of ME *leche*, *lache* 'a stream'. Cf. Introd. p. xix.

Beversbrook W [*Bevresbroc* DB]. 'Beaver stream.'

Beverstone Gl [*Byferes stan* 1048 ASC (E), *Beurestane* DB]. Though the place is in a fairly high situation and no stream of any size is found near by, the name can hardly mean anything else than 'beaver's stone'. OE *beofor* 'beaver' has a side-form *byfor*.

Bevington Wa [*Bovynton* 1291 Tax, *Bynynton* 1316 FA]. '*Bēofa*'s TŪN.' OE *Bēofa* occurs written *Beoba* BCS 108, 145, 211 and in *Beófan stan* 901 BCS 596.

Bewaldeth Cu [*Bualdith* 1256 FF, *Boaldith* 1278 Ch, *Boualdith* 1319 Ipm]. 'The BŪ or homestead of *Aldgȳþ*.' For the order of the elements cf. ASPATRIA. *Aldgȳþ* or *Ealdgȳþ* is a woman's name.

Bewcastle Cu [*Buchecastre* c 1178 WR, 1250 Ipm, *Bothecaster* 1272 Ipm, *-castell* 1327 Cl]. The elements are ON *būð* 'booth' and OE *ceaster* 'Roman fort'. The meaning of the name is possibly 'old fort used as a shelter for sheep'.

Bewdley Wo [*Beuleu* 1275 Subs, *Buleye* 1315, *Beudle*, *Beaudele* 1335 Ipm]. Identical with BEAULIEU.

Bewerley YW [*Bevrelie* DB, *Beuerli*, *-lai* 12 Fount, *-leia* 1166 P, *-lay* 1297 Subs]. OE *beoforlēah* 'clearing inhabited by beavers'.

Bewholme YE [*Begun* DB, *Beghum* c 1150 YCh 1331, 1333]. OE *bēagum*, dat. plur. of *bēag* 'ring'. The name may refer to some stone circles.

Bewick Nb [*Bowich* 1167 P, *Bewic* 1201 Cur], B~ YE [*Biuuich* DB, *Bewic* 1196 FF]. OE *bēo-wīc* 'bee farm'.

Bewley Du [*Bellus Locus* c 1335 FPD), B~ Castle We [*Bellus locus* 1256 WR]. Identical with BEAULIEU.

Bexhill Sx [*Bex-*, *Bixlea* 772 BCS 208, *Bexelei* DB]. 'Box wood.' First el. an OE *byxe*, derived from *box* and found also in BEXLEY and BIX, perhaps BIXLEY. Second el. OE LĒAH.

Bexington Do [*Bessintone* DB, *Buxinton* 1212 Fees, *Bixinton* 1234 BM]. First el. OE *byxen* 'of box'.

Bexley K [*Bixle* 765–91 BCS 260, *Byxlea* 814 ib. 346]. See BEXHILL.

Bexton Chs [*Bexton* 1260 Court], **Bexwell** Nf [*Bekeswella* DB, *Bekeswell* 1177, 1196 P]. '*Bēac*'s TŪN and stream.' Cf. BESWICK La.

Beyton Sf [*Begatona* DB, *Beketon* 1208 FF]. See BEIGHTON Nf.

Bibury Gl [*Beaganbyrig* 721–43, 10 BCS 166, 1320, *Begeberie* DB]. '*Bēage*'s BURG.' By the charter in BCS 166 land was given in Bibury to earl Leppa and his daughter Beage.

Bicester (bister) O [*Bernecestre* DB, c 1130 Oxf, *Burnecestre* 1219 Fees]. Cf. BERNWOOD. First el. OE *byrgen* 'burial-mound'. Second el. OE *ceaster* 'Roman fort'. B~ is on Akeman Street.

Bickenhall So [*Bichehalle* DB, *Bikehilla* 1186 ff. P, *Bikehull* 1243 Ass, *Bikehal* 1201, *Bikenhal* 1243 Ass]. '*Bica*'s hill or hall.' The interchange of OE HYLL and HALL is remarkable.

Bickenhill Wa [*Bichehelle* DB, *Bikehilla* 1187 P, *Bykenhull* Hy 3 BM]. '*Bica*'s hill.'

Bicker Li [*Bichere* DB, *Bikere* 1194 P, *Bicre* 1206 Cur]. Probably OScand *bý-kiarr* 'the village marsh'. Cf. KERR. Another Bicker in Li is mentioned in Selby ii. 165: (turbary in) *Biker*. The place was in Stallingborough.

Bickershaw La [*Bikersah* c 1200 CC], **Bickerstaffe** La [*Bikerstad* a 1190 CC, *-stath* 1226 LaInq]. 'The copse and the landing-place of the bee-keepers'; cf. SCAGA and STÆP, and see BICKERTON.

Bickerston Nf [*Bicherstuna* DB, *Bikereston* 1197 P, 1203 Ass]. 'The TŪN of the bee-keeper'; see BICKERTON.

Bickerton Chs [*Bicretone* DB, *Bikirton* 1260 Court, *Bikerton* 1290 Ipm], B~ He [*Bicretune* DB, *Bichertunia* 12 Fr], B~ Nb [*Bikertun* 1236, *-ton* 1242 Fees], B~ YW [*Bicretone* DB, *Bykerton* 1244 Ch]. A lost Bickerton YW nr Otley is *Biceratun* c 1030 YCh 7, *Bichertun* DB. 'The TŪN of the bee-

keepers.' First el. OE *bēocere* 'bee-keeper', also in the form *bycere* in *Bycera fald* 972 BCS 1282.

Bickford St [*Bigeford* DB, *Bicford* 1251 Ass]. '*Bica*'s ford.'

Bickington D nr Newton Abbot [*Bechintona* 1107 (1300) Ch, *Buketon* 1219 Ass, 1228 FF, *Bukyngton* 1303 FA], **Abbots B~** D [*Bicatona* DB, *Bukington* 1291 Tax], **High B~** D [*Bichentona* DB, *Bukint[on]* 1212 Fees, *Bukynton* 1238 FF]. Apparently 'the TŪN of (the people of) *Beocca*'. The abbot was that of Hartland.

Bickleigh D nr Exeter [*Bicanleag* 904 BCS 613, *Bichelia* DB, *Bikelegh* 1238 Ass], B~ D nr Plymouth [*Bicheleia* DB, *Bikeleg* 1225 Pat], **Bickley** Chs [*Bichelei* DB, *Bikeleg* 1259 Court], B~ K [*Byckeleye* 1279 FF, *Bykeleye* 1292 Ass], B~ Wo [*Bykeleg* 1240 WoP]. Probably '*Bica*'s or *Bicca*'s LĒAH'. The number of Bickleighs and Bickleys is rather considerable, but *Bica* and *Bicca* are well evidenced. What may suggest that we have to reckon also with some other element is *Bicheneleia* DB, identified in VHD with BICKINGLEIGH. This looks as if the first el. were the gen. plur. of an OE *bic(c)a* or *bic(c)e*. No such word is evidenced, but it might conceivably have been a word for 'woodpecker' related to *bicker*, OHG *anabicchan*, MHG *bicken* 'to prod'; cf. PURBECK. The identification of *Bicanleag* BCS 613 is doubtful, but it was very likely one of the Devon Bickleighs. It is called 'villa venatoria'.

Bickmarsh Wa [*at Bicanmersce* 967 BCS 1201, *Bichemerse* DB], **Bicknacre** Ess [*Bikenacher* 1186, *-acra* 1190 P, *-acre* 1235 FF], **Bicknoller** So [*Bykenalre* 1291 FF, *Bikenalre* 1334 Ep]. '*Bica*'s marsh, field, alder.' See ÆCER, ALOR.

Bicknor, English, Gl [*Bicanofre* DB, *Bikenoura* 1190 ff. P, *Bykenore Anglican*' 1291 Tax], **Welsh B~** He [*Bykenore Walens*' 1291 Tax]. Both names mean '*Bica*'s ridge'. The places are not far apart, but on opposite sides of the Wye on prominent spurs of hills. See OFER. The places may have belonged to the same *Bica*.

Bicknor K [*Bikenora* 1186 P, *-ore* 1195 P, 1242 Fees]. OE *Bican ōra* 'slope belonging to *Bica*'. See ŌRA.

Bickton Ha [*Bichetone* DB, *Biketon* 1242 Fees]. '*Bica*'s TŪN.'

Bicton D [*Bechatona* DB, *Buketon* 1228 FF, 1259 Ipm]. '*Beocca*'s TŪN.'

Bicton Sa nr Shrewsbury [*Bichetone* DB, *Bykedon* 1248 FF, *-e* 1274 RH], B~ Sa nr Clun [*Bikedon* 1302 Ipm]. '*Bica*'s DŪN.'

Bidborough K [*Bitteberga* c 1100 Text Roff, *-bergh* c 1280 Bodl, *Bytberghe* 1346 FA]. '*Bitta*'s hill or mound.' OE *Bitta* pers. n. seems to be evidenced in *Bitan cnol*, *Bittan cnoll* BCS 594, 1080 (Hurstbourne

Ha), *Bitan dene* BCS 917 (W). Cf. also OHG *Pizzo*.

Biddenden K [*Bidingden* 993 Hyda, *Bidindaenne* 11 DM, -*denne* 1204 Ch]. 'The DENN or swine-pasture of *Bida* or *Bidda*.' *Bida, Bidda* pers. ns. are recorded in pl. ns., as *Bidan holt* BCS 834 (Sx), *Bidan ham* ib. 1067 (W), *Biddanstiorf* ib. 502 (K). Cf. also OHG *Bito*.

Biddenham Bd [*Bide(n)ham* DB, *Bedeham* 1198 FF, *Budeham* 1247 Ass, *Bydenham* 1254 Val]. '*Bīeda*'s HĀM.'

Biddestone W [*Bedestone* DB, -*ston* 1182 P, *Bedeneston* 1187 ff. P, *Buddeston* 1215 Cl, *Budeston* 1259 Ipm]. 'The TŪN of *Biedīn, *Biede*.' This is a derivative with a suffix -*īn* from *Bīeda* pers. n.

Biddick Du [*Bidich* c 1190 Godric]. OE *bī dīc* 'by the ditch'.

Biddisham So [*Biddesham* 1065 KCD 816, 1203 FF, *Bidesham* 1209 FF]. '*Biddi*'s HAMM.' *Biddi* is a derivative of *Bidda*; see BIDDENDEN.

Biddlesden Bk [*Betesdene* DB, *Bethlesdena* c 1160 NpCh, 12 DC, *Butlesden* 1231 FF, *Bitlesdena* c 1150 Harl MS 4714]. Second el. OE DENU 'valley'. The first seems to be an OE *byple*, a side-form of *gebytle* 'dwelling', a derivative of BOPL. Cf. BETTESHANGER.

Biddlestone Nb [*Bitnesden* 1181 Newminster, *Bidlisden* 1242 Fees, *Bitellisden* 1265 Misc]. Apparently identical with prec. name.

Biddulph (bĭdl) St [*Bidolf* DB, 1208 FF, 1227 Ch, *Bydulf* 1291 Tax]. Apparently OE *bī* 'by' and an unrecorded OE *dulf* or the like 'digging, mine' (from *delfan* 'to dig'); cf. DILHORNE. If this is right the name means 'the place by the mine'.

Bideford (bĭdĭ-) D [*Bedeford* DB, *Bediforda* DB, -*ford* 1202 FF, *Budiford* 1232 Pat]. Perhaps '*Bīeda*'s ford', but cf. BIDFORD.

Bidford Wa [*Budiford* 710 BCS 127, *Bedeford* DB, *Bediford* 1156, *Budiford* 1164, *Bideford* 1230 P]. The first el. seems to be identical with that of *Bydictun* 822 BCS 370 (held to be an original); the place is unidentified. *Bydic* seems to be some significant term. It might be related to OE *byden* 'vessel, tub', OHG *butîn*, G *Bottich* 'coop, tub'. The meaning might be 'trough', here used in a transferred sense 'valley, dip in the ground', or 'deep place in a river'. There are similar names in Germany. BÜTTGEN is *Budica* 9, *Budeche* 1197.

Bidston Chs [*Budeston, Bediston* 1260 Court, *Bideston* 13 Chester]. The first el. might be an OE *Byddi*, derived from *Budda*.

Bidwell Bd [*Budewelle* 1228 FF, *Bedewell* 1279 RH]. 'Stream in a valley.' Cf. BYDEN.

Bielby YE [*Belebi* DB, -*beia* c 1200 YCh 1130]. '*Beli*'s BY.' Cf. BEELSBY.

Bierley YW [*Birle* DB, 1195 YCh (1767), 1240 FF]. OE *byrh-lēah* 'LĒAH by the BURG'.

Bierton (-ēr̄-) Bk [*Bortone* DB, *Burton* 1237–40 Fees, *Birton* 1290 RA]. OE *Byrhtūn* 'TŪN by the BURG', i.e. Aylesbury.

Bigbury D [*Bicheberie* DB, *Bickeberi* 1201 Cur]. '*Bic(c)a*'s BURG.' Cf. (to) *Bicanbyrig* 1005 KCD 714 (a lost place in O).

Bigby Li [*Bechebi* DB, c 1115 LiS, *Bekebi* 1212 Fees]. See BY. First el. ODan *Bekki* pers. n.

Biggin, a common pl. n. B~ Hu is *Bigging* 1361 AD. From ME *bigging* 'building, house', a derivative of *big* 'to build' (OScand *byggia*).

Biggleswade Bd [*Pichelesuuade* DB, *Bicheleswada* 1132 RA, *Bikeleswada* 1181 P]. OE *Bicc(e)les gewæd* '*Biccel*'s ford'. *Biccel* is a normal derivative of *Bicca*.

Bighton Ha [*Bicin(c)gtun* 959 BCS 1045, *Bighetone* DB, *Bicentona* 1158 Fr]. 'The TŪN of *Bica*'s people.'

Biglands Cu [*bygland* 1490 PNCu(S)]. 'Barley land' (ON *bygg* 'barley').

Bignor Sx [*Bigenevre* DB, *Begenoura* 1167 P, *Bykeneuere, Bygeneuere* 1314 Ipm]. Perhaps 'hill slope belonging to *Bica*'. See YFER.

Bigrigg Cu [*Bigrig* 1235 StB]. 'Barley ridge' (ON *bygg* 'barley' and *hryggr* 'back, ridge').

Bilborough Nt [*Bileburch* DB, -*burc* 1166 P]. '*Bila*'s or *Billa*'s BURG.' *Bila* and *Billa* are not recorded in independent use, but are presupposed by many pl. ns., e.g. *Billan cumb* BCS 1286, *Billan dene* ib. 757, *Bilanleag* ib. 917.

Bilbrook So [*Bilrebroc* 1227 Pat, *Byllerebroke* 1526 AD], B~ St [*Bilrebroch* DB, -*broc* 1167 P, 1227 Ass, *Billebroc* 1167 P]. 'Brook in which billers or watercress grew.' Billers (*Byllerne* 1440) is held to be of Celtic origin (Gael *biolaire*, Ir *biolar*, Co, Bret *beler*).

Bilbrough YW [*Mileburg* DB, *Billeburc* 1167 P, -*burg* 1226 FF]. See BILBOROUGH.

Bilby Nt [*Billebi* DB, *Byleby* 1242 Fees]. '*Bille*'s BY.' First el. OSw *Bille*, ON *Bili* pers. n.

Bildeston Sf [*Bilestuna* DB, *Billestona* 1130 P, *Bildestone* 1166 RBE, *Bildeston* 1219, 1242 Fees]. '*Bild*'s TŪN.' First el. ON *Bildr*, ODan *Bild* pers. n. Cf. BILSDALE.

Bilhagh Nt. See BELAUGH.

Bilham YW [*Bileham* DB, *Bilham* 1230 P, *Bylleham* 1285 FA]. '*Billa*'s HĀM.' Cf. BILBOROUGH.

Billeri·cay (-ĭkĭ) Ess [*Billerika* 1313 AD, *Billerica* 1343 Ipm]. The same name is found in Kent, now apparently lost, *Byllerica, Beleryca* 1316 Ipm, *Billirica* 1313 QW, and Collinson mentions *Billerica, Bellerica*, a grange belonging to Witham Friary So. The curious name is unexplained.

Billesdon Le [*Billesdone* DB, *-don* 1190 P, 1203 Ass], **Billesley** Wa [*Billeslæh* (*-leah*) 704–9 BCS 123, *Billeslei* DB, *Bileslega* 1158 P]. 'The DŪN and the LĒAH belonging to *Bill*.' *Bil* is found in DB and a pers. n. *Bill* is presupposed by many other pl. ns., e.g. *Billesham* 955–9 BCS 936, *Bileshamm* 990 KCD 675. It is a regular short-form of *Bilheard* &c.

Billing Np [*Bel*(*l*)*inge* DB, *Billingge Magna, Parva* 12 NS, *Billinges* 1223 BM]. OE *Billingas* '*Bill*(*a*)'s people'. Or *Billingas* might be derived from OE *bill* 'sword'.

Billingbear Brk [*le Pyllingeber*' 1238, *Pillin*(*g*)*ber*' 1240 Cl]. Second el. BÆR 'pasture'. The first appears to be an OE *Pillingas*, which might mean 'the people at (a lost place) PILLEY'.

Billingborough Li [*Billingeburg, Bellingeburg* DB, *Billingburga* c 1160 DC]. 'The BURG of the *Billingas*'; cf. BILLING. Identical is (to) *Billingabyrig* c 725 BCS 144 (Sx).

Billinge La [*Billing* 1202, 1206 P, 1246 Ass, *Bulling* c 1200 CC, 1204 P]. Probably originally the name of Billinge Hill. If the *u*-forms can be disregarded, the name may be derived from *bill* 'sword'.

Billingford Nf nr E. Dereham [*Billingeforda* DB, *-ford* 1212 Fees]. 'The ford of the *Billingas*'; cf. BILLING. B~ Nf nr Diss is *Prelestuna* DB, *Prillestone* 1346, *Pirliston* 1428 FA.

Billingham Du [*Billingham* c 1050 HSC, *Billingaham* 1085 LVD, *Billingeham* c 1130 SD]. 'The HĀM of the *Billingas*'; cf. BILLING.

Billinghay Li [*Belingei* DB, *Billingeia* 1190 P, *Bilingéé* 1202 Ass]. 'The stream or island of the *Billingas*'; cf. BILLINGBOROUGH and BILLING.

Billingley YW [*Bilingeleia* DB, *Billinglea* 1167, *Billingelea* 1178, 1190 f. P]. 'The LĒAH of the *Billingas*.' Cf. BILLING.

Billingshurst Sx [*Bellingesherst* 1203 Cur, *Billingeshurst* 1249 FF]. '*Billing*'s HYRST.'

Billingsley Sa [*Bylgeslege* 1055 ASC (C), *Billingsleya* 1147 LaCh]. This must be compared with *Byli*(*g*)*esdyne* c 995 BCS 1289 (Sf). The gen. forms suggest as first el. an OE pers. n., which might be related to *Bylga* in *Byligan fen* 972 BCS 1282; cf. *Bylian pól* 949 ib. 883, *Bilian wyrth* 933 ib. 696. OE *Bylga* and *Bylgi*, if they are pers. ns., might be related to OE *bolgen* (in *bolgenmōd* 'angry').

Billington Bd [*Billendon* 1196 P, *Bilindon* 1196 FF]. '*Billa*'s DŪN.' Cf. BILBOROUGH.

Billington La [*Billingduna* 1196 YCh 1524, *-don* 1203 P]. Really the name of **Billing-**ton Moor, a long ridge, referred to c 1130 by SD as *Billingahoth* 'the HŌH or ridge of the *Billingas*'; cf. BILLING. Second el. DŪN, later supplanted by TŪN.

Billington St [*Belintone* DB, *Billenton* 1208 Cur]. '*Billa*'s TŪN.' Cf. BILBOROUGH.

Billockby Nf [*Bit*(*h*)*lakebei* DB, *Billokebi* 1204 Cur, *Billokesbi* 1202 FF]. The first el. is evidently a pers. n., but no similar name is known elsewhere. It might be OScand *Áki* with ON *biðill* 'wooer' prefixed (*Biðil-Áki*).

Bilney, East, Nf [*Billneye* 1254 Val, *Bylneye* 1316 FA], **West Bilney** Nf [*Benelai, Binelai, Bilenei* DB, *Bilneia* 1205 Cur, *Westbilneye* E 1 BM]. Probably OE *Billan ēa* (or *ēg*) '*Billa*'s stream or island'.

Bilsborough La [*Billesbure* 1187 ff. P, *-burgh* 1212 Fees]. '*Bill*'s BURG'; cf. BILLESDON.

Bilsby Li [*Billesbi* DB, 1212 Fees]. '*Bille*'s BY'; cf. BILBY.

Bilsdale YN [*Bildesdala* c 1155 Riev, *Bildisdal* 1208 FF]. '*Bild*'s valley'; cf. BILDESTON.

Bilsington K [*Bilsvitone* DB, *Bilswithetun* 11 DM]. '*Bilswiþ*'s TŪN.' *Bilswiþ* is the name of a queen and abbess.

Bilsthorpe Nt [*Bildestorp* DB]. '*Bild*'s thorp'; cf. BILDESTON.

Bilston St [*Bilsetnatun* 996 Mon, *Bilsatena gemæro* 985 KCD 650, *Billestune* DB]. 'The TŪN of the *Bilsætan*.' *Bilsætan* means 'the dwellers at *Bil*' or the like. *Bil* might be elliptical for a longer name, but no such name is known in the neighbourhood. Or *Bill* may have been the name of the hill near which Bilston is (from OE *bill* 'sword').

Bilstone Le [*Bildestone* DB, *Bildiston* 1242 Fees]. See BILDESTON.

Bilton Nb [*Bylton* 1242 Fees], **B~** YE [*Bil*(*l*)*etone* DB, *Biltona* 1297 Subs], **B~** YW nr York [*Biletone* DB, *Bilton* 1291 Tax], **B~** YW nr Harrogate [*Bil*(*l*)*etone* DB, *Biltona* c 1185 YCh 515]. '*Bil*(*l*)*a*'s TŪN'; cf. BILBOROUGH.

Bilton Wa [*Beltone, Bentone* DB, *Belton, Bulthon* 1225–30 BM, *Belton* Hy 2 (1235) Ch, *Bulton* 1276 RH]. Possibly 'TŪN where henbane (OE *beolone*) grew'. Or perhaps better the first el. is a stream-name *Beolne*, as suggested in PNWa(S). Cf. BEL-BROUGHTON.

Binbrook Li [*Binnibroc* DB, *Binnabroc* c 1115 LiS, *Binne broke* c 1090 YCh 350]. '*Bynna*'s brook.' OE *binnan brōce* 'inside the brook' does not seem to be suitable.

Binchester Du [*Bynceastre* c 1050 HSC, *Bincestre* 1104–8 SD, *Bincestr*' 1196 P]. First el. perhaps OE *binn* 'manger', later also 'stall'. The old fort may have been used as a shelter for cattle.

Bincknoll W [*Biencnoll* 1242 Fees, *Benecnoll* 1251 Ch]. OE *bēancnoll* 'knoll where beans grew'.

Bincombe Do [*Beuncumbe* 987 KCD 656, *Beincome* DB, *Biemecomma*, *Biencomme* 1157 Fr, *Bencumbe* 1252 Ch]. OE *bēancumb* 'coomb where beans grew'.

Binderton Sx [*Bertredtone* DB, *Bendriton* 1233 Cl, *Bendreton* 1237 Fees]. 'The TŪN of *Beornþrȳþ*' (a woman).

Bindon Do [*Binnedon* 1199, *Binnendon* 1200 P, *Binedon* 1212 Fees, *Binnedun* 1227 FF]. OE *binnan dūne* '(the place) inside the DŪN'. Bindon is south of Bindon Hill.

Binegar So [*Begenhangra* 1065 Wells, *Behenhanger* 1243 Ass, *Benhangre* 1176 Wells]. The later forms point to OE *bēan-hangra* 'slope where beans grew', but the earliest may point to a first el. *Bēage* pers. n.

Binfield Brk [*Benetfeld* a 1162 Oxf, 1176 P, 1230 P]. 'Bent-grass FELD'; cf. BENTFIELD.

Binfield Heath O [*Beonan feld* 963 BCS 1123, *Benetfeld* 1188 P]. If the form of 963 belongs here, '*Beona*'s FELD'. Otherwise 'bent-grass field'. *Beonan* may be a poor spelling for *Beonnan*, but it occurs twice.

Bingfield Nb [*Bingefeld* 1181 P, 1293 QW], **Bingham** Nt [*Bingheham* DB, *Bingeham* 1165 P], **Bingley** YW [*Bingheleia* DB, *Binggeleia* 1182–5 YCh 199, *Bingelea* 1196 P]. There are an ON *bingr*, Norw *binge* 'a stall', G *Binge*, Engl *bing* 'a heap'. The three names might contain this word, but the regular early *Binge-* may point rather to a contracted earlier *Bynninga-* or the like. If so, 'the FELD (HĀM, LĒAH) of *Bynna*'s people.'

Binham Nf [*Binneham*, *Benincham* DB, *Binham* 1156 Holme, 1200 Cur]. '*Bynna*'s HĀM.'

Binley Wa [*Bilnei* DB, *Bilneia* 1202 Ass, *Bilneya* 1236 Fees, *Binlea* R 1 Cur]. Identical with BILNEY.

Binnington YE [*Binne-*, *Bigneton* DB, *Benitona* c 1136 YCh 1144]. 'The TŪN of *Bynna*'s people.'

Binsey O [*Beneseye* 1122, *Buneseie* 1141 Fridesw, *Beneseya* 1291 Tax]. '*Byni*'s island.'

Binstead Wt [*Benestede* DB], **Binsted** Ha [*Ben(e)stede* DB], **B~** Sx [*Benestede* DB, *Biensted* 1332 Subs]. OE *bēan-stede* 'place where beans were grown'.

Binton Wa [*Bynningtun* c 1012 KCD 898, *Benintone* DB, *Buninton* 1236 Fees]. 'The TŪN of *Bynna*'s people.' **B~** Sr. See BENTWORTH.

Bintree Nf [*Binnetre* DB, *Binetre* 1180 P]. '*Bynna*'s tree.'

Binweston Sa. See WESTON.

Birbeck or **Birkbeck** We [*Birkbeke* 1496 CWNS xxviii]. ON *birki-bekkr* 'brook where birch grew'.

OE **birce** 'birch'. See BIRCH, BIRCH-, BIRTLE(S), BESCAR. OE **bircen** 'of birch' is found in BIRKENHEAD, BIRKENSHAW, BIRKENSIDE. Cf. BIRKIN. For OScand **birki** 'birch grove' see BIRBECK, BIRK-, BRISCOE.

Birch Ess [*Bricceia*, *Bricia* DB, *Bricha* 1177 P, *Breche* 1204 Cur, *Brich* 1238 Subs]. The regular early *Bricha* &c. (never *Birch*) shows that this is not OE *birce* 'birch', but OE *bryce* 'breaking', here in the sense 'newly cultivated land'. Cf. **Bruche** La [*Bruch* 1280, *Briche* 1314 PNLa]. **Birch Hall** Ess in Kirby is *Birichou* DB, *Bircho* Hy 2 (1286) Ch. It is 'birch hill'.

Birch, Much & Little, He [*Birches* 1252 Fees]. OE *birce* 'birch' (in the plur. form).

Bircham, Great, B~ Tofts, B~ Newton Nf [*Brecham* DB, 1179, 1190 ff. P, *Breccham* 1195 P, *Parva Brecham* 1254 Val]. First el. OE BRÆC 'newly cultivated land'. Hence 'newly cultivated HĀM'.
B~ Newton is *Niwetuna* DB. B~ Tofts is *Toftes* 1205 Cur, *Brechamtof[t]es* 1272 Ch. Cf. TOFT.

Birchanger Ess [*Bilichangra* DB, *Birecengre* c 1130 Oxf, *Birchangre* 1222 FF]. 'Birch slope'; cf. HANGRA.

Bircher He [*Burchoure* 1212 RBE]. 'Ridge overgrown with birch'; cf. OFER.

Birchill Db [*Berceles* DB, *Birchulles* 1347 BM], **Birchills** St [*Byrchehull* 1275 Cl]. 'Birch-covered hills.'

Birchington K [*Bircheton* 1275 RH]. 'TŪN among birches.'

Bircholt K [*Birichalt* 11 DM, *Bircheholt* (hd) 1219 Fees]. 'Birch grove.'

Birchover Db [*Barcouere* DB, *Birchoure* 1226 FF, *Birchover* 1265 Misc]. See BIRCHER.

Birdbrook Ess [*Bridebroc* DB, *Briddebrok* 1292 Ch]. 'Brook frequented by birds' (OE *bridd*).

Birdforth YN [*Bruthewrthescire* 1088 LVD, *Brudeford* 1166 P, 1199 Pp, 1219 FF, *Burdeford* 1226 Pat]. OE *Brȳda-ford* 'ford of the brides'; cf. BRITFORD. The name has been influenced by ON *brúðr* 'bride'.

Birdham Sx [*Bridham* 683 BCS 64, *Brideham* DB]. 'HAMM frequented by birds.'

Birdingbury Wa [*Byrtingabirig* 1043 KCD 916, *Berdingeberie* DB, *Birthingbir* 1197 f. P]. 'The BURG of *Beorht*'s people.'

Birdlip Gl [*Bridelepe* 1221 Ass, *Brudelep* 1262 PNGl, *Bredelepe* 1288 Cl]. Second el. OE *hlīep* 'leap'. The place is situated over a steep slope, so *hlīep* may here mean 'abyss'. The first el. may be OE *brȳd* 'bride', and the name may refer to some (?legendary) incident. Or it may be *bridd* 'bird'. If so, we may compare (on) *swealewan hlypan* 1023 KCD 739.

Birdoswald Cu [*Borddosewald* c 1200 WR, *Burthoswald* Lanercost]. '*Oswald*'s fold' (Welsh *buarth*). The place-name is Welsh, though *Oswald* is English.

Birdsall YE [*Briteshale* DB, *Brideshala* DB, c 1110 YCh 25, *Briddeshala* 1187 P, -*hal* 1219 Fees]. '*Bridd*'s HALH' rather than 'the HALH of the birds'. OE *Brid* is at least the name of a moneyer.

Birkby Cu nr Maryport [*Brettebi* 1163 P, 1297 Holme C, *Brictebi* 1190 P], **B~** Cu nr Ravenglass [*Bretteby* 13 StB, 1292 QW], **B~** YN [*Bretebi* DB, 1088 LVD], **B~** YW [*Bretebi* DB]. ON *Bretabýr* 'the BY of the Britons'.

Birkdale La [*Birkedale* c 1200 CC]. 'Birch valley.'

Birkenhead Chs [*Birkened* c 1150 Mon, *Bircheuet* 1260 Court]. 'Headland overgrown with birch'; first el. OE *bircen* 'birchen'.

Birkenshaw YW [*Birkenschawe* 1274 Wakef]. 'Birch grove.'

Birkenside Nb [*Byrkinside* 1262 Ipm]. 'Birch-covered slope'; cf. SĪDE.

Birker Cu [*Bircherhebec* c 1200, *Birker* 1292 StB]. 'Shieling in a birch grove'; cf. ERG. First el. ON *birki* 'birches, birch grove'.

Birkin YW [*Byrcene* c 1030 YCh 7, *Berchine*, *Berchinge* DB, *Birchin* c 1200 YCh 822]. A substantivized form of OE *bircen* 'birchen': 'birch grove'. Cf. BRACKEN.

Birklands Nt in Edwinstowe [*Birchwude* 1188 P, *Birkelund* 1251 Cl]. An interesting case of an OE name (*Bircwudu* 'birch wood') being exchanged for its OScand synonym (*Birkilundr*).

Birley Db nr Beighton [*Birlay* 1280 Beauchief Cart, -*leye* 1291 Tax], **B~** Db nr Brampton [*Bir-*, *Burleia* Hy 2 (1314) Ch], **B~** Db nr Hathersage [*Birlay* 1371 DbAS 48]. Either OE *bȳr-lēah* 'LĒAH with a BȲRE or cowshed' or *byrh-lēah* 'LĒAH by a BURG'.

Birley He [*Burlei* DB, *Burleg* 1230 P, -*lege* 1242 Fees]. Identical with BURLEY.

Birling K [*Boerlingas* 788 BCS 253, (of) *Bærlingan* c 980 ib. 1132, *Berlingæs* 10 ib. 1321, *Berlinge* DB], **B~** Nb [*Berlinga* 1187 P, *Birling* 1242 Fees]. Identical with BARLING.

Birlingham Wo [*Byrlingahamm* 972 BCS 1282, *Berlingeham* DB]. 'The HAMM of *Byrla*'s people.' No such name is recorded, but it may be a derivative of names in *Burg-*, originally *Byrgla.

Birmingham Wa [*Bermingeham* DB, *Bremingeham* 1166 RBE, 1167 P, *Burmingeham* 1167, *Brimingeham* 1169 P]. The early forms vary a good deal. Probably 'the HĀM of *Beornmund*'s people' (OE *Beornmundingahām*). Or the direct base may be a pet-form *Beorma from *Beornmund*. *Brumma-*

gem represents the old pronunciation of the name.

Birstal YW [*Birstale* 12 YCh 1636, *Byrstall* c 1200 ib. 1701, *Burstall* 1202 FF]. OE *byrg-steall*, identical in meaning with *burg-steall*. Cf. next name.

Birstall Le [*Burstelle* DB, -*stal* 1166, *Burcstal* 1167 P]. OE *burg-steall* is found in the sense 'a city'. Here it more likely means 'site of a BURG', i.e. either 'old disused fort' or simply 'fort'.

Birstwith YW [*Beristade* DB]. OE *byrg-stæp* 'landing-place belonging to the BURG'; cf. STÆP. Alternatively B~ may be ON *bȳiarstaðr* 'farmstead' (PNLa 151).

Birthorpe Li [*Berchetorp* DB, *Birketorp* 1195 Cur, 1202 Ass]. First el. OScand *birki* 'birches, birch grove'.

Birtle La [*Birkel* 1246 Ass, *Birkehill* 1347 FF], **Birtles** Chs [*Birchelis* 1260 Court, 1268 Chester]. 'Birch hill(s).'

Birtley Du [*Britleia* 1183 BoB, *Birteley* 1344 RPD], **B~** Nb [*Birtleye* 1229 Pat, *Birteley* 1242 Fees]. 'Bright LĒAH' (OE *beorhte lēah*).

Birtsmorton. See MORTON.

Bisbrooke Ru [*Bitlesbroch* DB, -*broc* 1167 P, -*broch* 1238 Ep]. '*Bitel*'s brook.' The pers. n. *Bitel is a regular derivative of *Bita*; cf. BIDBOROUGH. It may be the first el. of BITTISCOMBE So. The weak *Bitela* is found in *Bitelanwyrth* 956 BCS 952.

Biscathorpe Li [*Biscopetorp* DB, c 1115 LiS]. 'The thorp of the bishop.' For OE *biscop* see further the next names and BUSH- (passim).

Biscott Bd [*Bissopescote* DB]. 'The bishop's COT.'

Bisham Brk [*Bistesham* DB, *Bustlesham* 1196 FF, *Bistlesham* 1199 P, 1203 FF, *Bestlesham* 1230 P]. The first el. is an unrecorded pers. n., which may have been OE *Byssel, a derivative of *Byssa* (in *Byssan broc* 1014 KCD 1309; cf. also BISLEY Sr). *Byssles* (gen.) would become *Bystles*; cf. BASILDON Brk.

Bishampton Wo [*Bisantune* DB, *Bishamtone* 11 Heming, *Bisshopeshampton* 1381 Pat]. 'The bishop's HĀMTŪN.'

Bishopdale YN [*Biscopedal* 1202 FF]. 'The valley belonging to the bishop or to *Biscop*.' OE *Biscop* occurs as a pers. n.

Bishopsbourne K [*aet Burnan* 799, 805 BCS 293, 319, *Burnes* DB, *Biscopesburne* 11 DM]. Originally *Burna* 'the stream'. The manor belonged to the Archbishop of Canterbury.

Bishops Castle Sa [*Bissopes Castell* 1269 Eyton, *Castrum Episcopi* 13 BM]. Founded by one of the bishops of Hereford a 1154.

Bishopsteignton. See TEIGNTON.

Bishopstoke Ha [*?æt Stóce* 928 BCS 663, *Stoches* DB, *Stoke Episcopi* c 1270 Ep]. See STOC. Belonged to the Bishop of Winchester.

Bishopstone Bk [*Bissopeston* 1227 FF], B~ He [*Biscopestone* 1166 RBE], B~ Sx [*Biscopestone* DB], B~ W nr Swindon [*Bissopeston* 1223 Sarum], B~ W nr Wilton [*Bissopeston* 1167, 1190 P]. Cf. BISHTON, BUSHTON. 'The bishop's manor.' The bishop is generally that of the respective diocese. B~ W nr Wilton belonged to the Bishop of Winchester.

Bishopstrow W [*Biscopestrev* DB, *Bissopestreu* 1121 AC]. 'The bishop's tree.' It has been suggested that the tree was a wooden cross erected in memory of Bishop Aldhelm, to whom the church is dedicated.

Bishopsworth So [*Biscopewrde* DB, *Bisscopewurth* 1243 Ass]. 'The bishop's WORÞ.'

Bishopthorpe YW [*Torp* DB, *Thorp super Usam* 1194 YCh 1859, *Biscupthorp* 1275 Ipm]. 'The THORP held by the bishop' (the Archbishop of York).

Bishopton Du [*Biscoptun* 1104–8 SD, *Bisopeston* 1196 FF], B~ YW [*Biscoptun* c 1030 YCh 7]. 'The bishop's manor' (the Bishop of Durham and the Archbishop of York respectively).

Bishopton Wa [*Biscopesdun* 1016 KCD 724, *Bissopesdon* 1197 P]. 'The DŪN belonging to the bishop' (of Worcester).

Bishton Gl [*Bispestun* 11 BCS 928], B~ Monm [*Bysshopston* 1440 Pat], B~ Sa [*Bispetone* DB, *Bispeston* 1221 Eyton], B~ St [*Bispestone* DB]. 'The bishop's manor.' B~ Gl seems to have been named from Archbishop Stigand (cf. BCS 929). B~ Monm belonged to the bishop of Llandaff.

Bisley Gl [(to) *Bislege* 896 BCS 574, *Biselege* DB, *Biseleia* 1130 P, c 1177 BM, *Bisselega* 1157 P]. Perhaps '*Bise*'s LĒAH.' The place is not on a stream; hence the first el. cannot be a river-name.

Bisley Sr [*Busseleghe* 933, 967 BCS 697, 1195, *Bussely* c 1270 Ep]. '*Byssa*'s LĒAH'; cf. BISHAM.

Bispham La in Croston [*Bispam* 1219 Ass], **Great & Little** B~ La [*Biscopham* DB]. 'The bishop's manor.' OE *bisceophām* is recorded in the sense 'episcopal estate' (Crawf 23).

Biss Brook R W [*Bis, Bys* 964 BCS 1127]. A British river-name, identical with Welsh *bys* 'finger' &c., from *bissi-, originally 'a twig' and used to denote a tributary stream or a river fork.

Bisterne Ha [*Betestre* DB, *Budestorn* 1187 ff., *Bedenestorn* 1190 P]. Second el. OE ÞORN 'thorn bush'. First el. as in BIDDESTONE.

Bitchfield Li [*Billesfelt* DB, *-feld* c 1172

BM, 1202 Ass]. '*Bill*'s FELD'; cf. BILLESLEY, BILSBOROUGH.

Bitchfield Nb [*Bechefeud* 1242 Cl, *-feld* 1242 Fees]. 'Beech FELD.'

Bittadon D [*Bedendone* DB, *Bettenden* 1205 FF, *Bittedene* 1242 Fees]. '*Beotta*'s valley.' *Beotting* 'son of *Beotta*' occurs BCS 417. *Beotta* is a short form of names in *Beorht-*.

Bittering Nf [*Britringa* DB, *Bit(t)ringe* 1202 FF, *Bitering* 1252 Ch]. OE *Brihtheringas* 'the people of *Brihthere* (*Beorhthere*)'. The first *r* was lost owing to dissimilation.

Bitterley Sa [*Buterlie* DB, *Buterle* 1242 Fees]. 'Place where butter was made'; cf. LĒAH.

Bitterne Ha [*Byterne* 1172 Ep, *Byterne* 1242 Cl, *Bitterne* 1284 Ep]. Second el. apparently OE *ærn* 'house'. The first might be OE *byht* 'bend', referring to the horseshoe-shaped ridge close by.

Bittesby Le [*Bichesbie* DB, *Bittlesby* 1258 Ch, *Butlisby* 1270, *Butlesby* 1274 Ipm]. '*Byttel*'s BY.' *Byttel* is unrecorded, but *Byttic* occurs.

Bitteswell Le [*Betmeswelle* DB, *Buthmeswelle* 12 DC, *Bidmeswell* 1242 Fees]. See WELLA. The first el. is OE *byþme* 'head of a valley'. The word may also have denoted 'a valley' or the like. OE *byþme* (*bytme*) is fem. in the instances on record, but there may have been a side-form *byþme* neut. Cf. also BYTHAM and BEAMSLEY.

Bittiscombe So [*Bitelescumba* 1180 P]. '*Bitel*'s CUMB or valley'; cf. BISBROOKE.

Bitton Gl [*Betune* DB, *Bettun* 1159 P, *Button* 1211–13 Fees, *Boyton* 1275 Ipm]. 'TŪN on R BOYD.'

Bix O [*Bixa* DB, *Bixe* 1235 Cl]. OE *byxe* 'box wood', derived from *box*. Cf. BEXHILL.

Bixley Nf [*Biskele, Bichesle* DB, *Biskele* 1196 FF, *Bixle* 1228 Cl], B~ **Heath** Sf [*Bischelea* DB]. Probably identical with BEXLEY (OE *byxe-lēah*).

OScand **blā(r)** 'blue, dark', ME *bla, blo*, dial. *blae, bloe* occurs in some pl. ns., as **Blea Tarn** Cu, We, **Blea Water** We. Blea Tarn means 'dark tarn'. It goes back to earlier *Blatern(e)*, common in early records, e.g. *Blaterne* 1227 Ch (a grange nr Shap We), c 1240 WR (Cu). The corresponding Scand name is found in Sweden and Norway (*Blåtjärn* in Sweden, *Blaatjernet* in Norway). Cf. BLACKER, BLAWITH, BLOWICK, and BLOFIELD.

Blaby Le [*Bladi* DB, *Blabi* 1175 P]. The first el. is probably a byname derived from BLĀ(R) 'dark'. Cf. OSw *Joar blaa*, ONorw *Iusse bla, Randulf Bla* 1202 Ass (Li).

Blackawton D [*Auetone* DB, *Aveton* 1259 FF, *Blakaueton* 1286 FA]. '*Afa*'s TŪN.' *Afa* pers. n. is evidenced in Devon. *Black-* for distinction from AVETON GIFFORD.

Blackberry Hill Le [*Blakebergwe* 1252 Misc], **Blackborough** D [*Blacaberga, Blacheberge* DB, *Blakeberga* 1194 P], B~ Nf [*Blakeberge* c 1150 Crawf, *-berg* 1205 Cur]. 'Black hill.'

Blackburn La [*Blacheburne* DB, *Blakeburn* 1187 P]. 'Dark-coloured stream.' The stream is now called the BLACKWATER.

Blackden Chs [*Blak(e)den* 1304 Chamb]. 'Dark valley.'

Blackdown Do [*Blakedon* 1275 RH], B~ Wa [*Blakedon* 1317 Ipm]. 'Black hill.'

Blacker YW [*Blaker* 1266 Misc]. An OScand *blā-kiarr* 'dark marsh'; cf. BLĀ(R), KERR.

Blackford So nr Sherborne [*Blacheford* DB, *Blakeford* 1276 RH], B~ So in Wedmore [*Blacford* 1227, *Blakeford* 1257 Ch]. 'Black ford.'

Blackfordby Le [*Blakefordebi* c 1125 LeS, *Blacfordebi* 1200 Cur]. 'The BY at the black ford.'

Blackheath K [*Blachehedfeld* 1166 P, *Blakeheth* 1275 RH]. 'Black heath.'

Blackland W [*Blakeland* 1194 P]. 'Black land', referring no doubt to dark soil.

Blackley (-āk-) La [*Blakeley* 1282 Ipm]. 'Black wood or clearing.' See LĒAH.

Blackmanstone K [*Blachemenestone* DB, *Blakemannestun* 1265 Misc]. '*Blacman*'s (*Blæcman*'s) TŪN.'

Blackmoor Do [*Blakemor* 1205 Cl, 1212 Fees, *Blakemore* (forest) 1270 Ipm]. 'Black moor', probably in the sense 'forest'.

Blackmoor Ha [?*Blackemere* (stagnum) 931–40 BCS 640, *Blachemere* 1168 P, *Blakemera* 1170 P], **Blackmore** Hrt [*Blachemene* DB, *Blakemere* 1198 FF, 1204–12 Fees]. 'Black mere.'

Blackmore Ess [*Blakemore* 1232 Ch, 1254 Val], B~ **Forest** W [*Blakemor* 1243 ff. Cl], B~ Wo [*Blakemor* 1314 Ipm]. 'Black moor.'

Blacko La [*Blacho* 12, *Blakhow* 1329 Kirkst]. 'Black hill.' Second el. ON HAUGR 'hill'.

Blackpool La [*Blacke Pull* 1661 La Wills]. Named from a peaty-coloured pool, called *Pul* c 1260 CC.

Blackrod La [*Blakerode* 1201 ff. P]. 'Black clearing'; see ROD.

Blackshaw YW [*Blakschey* 1539 Halifax Wills]. 'Black grove.'

Blacksnape La [*Blakesnape* 1614 PNLa]. 'Black pasture.' See SNAPE.

Blackstone Edge La [*Blakeston edge hill* 1577 Saxton]. 'The edge or ridge with the black stone' (a boundary stone between Lancashire and Yorkshire). See ECG.

Blackthorn O [*Blaketorn* 1190 f., *Blachetorn* 1192 f. P]. OE *blæcþorn*, ME *blakthorn* 'blackthorn'.

Blacktoft YE [*Blaketofte* 1199 FF]. 'The black TOFT' or '*Blaca*'s TOFT'.

Blackwater R Ess [*Blackwater* 1576 Saxton], B~ R Brk, Ha [*la Blakewatere* 1279 For], B~ Do, Ha [*Blackwater* 1577 Harrison]. Cf. BLACKBURN. 'Dark-coloured stream.' There are several places called Blackwater. One in Ha is on the Blackwater [*la Blakewat*' 1256 Ass]. Another is on Blackwater Do.

Blackwell Db nr Bakewell [*Blacheuuelle* DB], B~ Db nr Alfreton [*Blacwelle* 13 BM], B~ Du [*Blakewell* 1183 BoB, B~ Wo [*Blacwælle* 964 BCS 1135]. 'Black stream.'

Blackwood YE [*Blakwod* 1280 Misc], B~ YW [*le Blacwode* 1308 Wakef]. 'Black wood.'

Blacon Chs [*Blachehol* DB, *Blachenot* c 1100 Chester, *Blakene* 1260 Court]. OE *blacanhol* 'dark hollow or pool'. The loss of *-l* is due to dissimilation.

Bladon O [*Bibladene* 872 BCS 535, *Blade* DB, *Bladen* 1231 Ch]. B~ is on the EVEN-LODE, whose old name was **Bladon** [*Bladon* 675 BCS 37, *Bladaen* 718 ib. 139]. *Bibladene* means '(the village) on the Bladon'. The etymology of the river-name is obscure.

OE **blæc** 'black' is a common first el. in pl. ns., especially names of streams and hills. The meaning is no doubt generally 'dark-coloured', referring to the colour of the water, the surface of hills or the soil. In names of forests the reference is rather to darkness produced by denseness of the forest. The exact meaning of *Black* in cases like BLACKAWTON, BLACK BOURTON is not clear. **Black Brook, Blackbrook, Black Burn, Blackburn** are common names. **Black Brook** Db is *Blakebroc* 1230 P. See further BLACK-, BLAKE- (passim), BLACON, BLAGDON, BLAGRAVE.

OE **blæcþorn** 'blackthorn'. See BLACK-THORN, BLATHERWYCKE.

Blagdon D [*Blakedone* 1242 Fees], B~ Do [*Blakedon* 1234 Cl], B~ So nr Axbridge [*Blachedone* DB, *Blakedune* R 1 Berk], B~ So nr Pitminster [*Blakedona* 1155–8 (1334) Ch, *-done* 1225 FF]. 'Black hill'; see DŪN.

Blagdon Nb [*Blakedenn* 1203 Cur, *-den* 1242 Fees, 1256 Ass]. 'Black valley.' See DENU.

Blagrave Brk [*Blacgraua, -graue* 1194 f. P, *Blacgrave* 1255 Ch]. 'Black grove.' See GRÆF.

Blaisdon Gl [*Blechedon* 1186 P, *Blechesdon* 1200, 1202 Cur, 1220 Fees]. 'The DŪN of *Blæcci* (*Blecci*).' This unrecorded pers. n. is a derivative of *Blæcca*. Cf. BLETCHING-DON.

Blakehow or **Blakey** YN [*Blakehou* 1223, 1234 FF]. 'Black hill'; cf. BLACKO.

Blakemere He [*Blakemere* 1249 AD]. 'Black mere.'

Blakeney Gl [*Blachen'* 1185, *Blakeneia* 1196 P], **B~** Nf [*Blakenye* 1242, *Blakene* 1248 Cl]. 'Black island' or '*Blaca*'s island'. B~ Nf was formerly **Snitterley** [*Esnuterle, Snuterlea* DB, *Sniterle* 1242 Fees]. '*Snytra*'s LĒAH'; cf. SNETTERTON, SNITTERTON.

Blakenhall Chs [*Blachenhale* DB, *the Blake Halch* 1260 Court]. 'Black HALH.'

Blakenham Sf [*Blacham, Blacheham* DB, *Blakeham* 1190 FF, *Blakenham Parva* 1254 Val, *Blakenham Magna* 1291 Tax]. '*Blaca*'s HĀM.'

Blakesley Np [*Blaculveslea* DB, *Blaculfeslea* 1190 P, *Blacolvesle* 1203 Cur]. A pers. n. *Blæcwulf* is not known, and *blæc* is rare in pers. ns. Still such a name may have existed, perhaps as an occasional formation to distinguish one *Wulf* from others. But the name may quite well mean 'the wood of the black wolf'.

Blakeston Du [*Bleikestuna* c 1100 DST, *Bleichestona* Hy 1 FPD]. '*Bleik*'s TŪN.' ON *bleikr* 'pale' is used as a byname.

Blanchland Nb [*Blanchelande* 1165 PNNb, *Alba Landa* 1203 Cur]. 'The white laund', but transferred from BLANCHE-LANDE in Normandy. Second el. Fr *lande* 'glade' (whence Engl *laund, lawn*).

Bland YW [*Bland* 1226 FF, 1257 Ch]. Perhaps originally a stream-name, from *Blanda*, which may occur as the name of a stream in Norway. But the regular monosyllabic form may point to OE *gebland* 'commotion, storm' (in compounds). B~ is in a high and exposed situation. For a similar name cf. SNITTER. The meaning would be 'windy place'.

Blandford St. Mary, B~ Forum, Langton Long Blandford Do [*Blane-, Bleneford* DB, *Blæneford* DB (Exon), *Bleinefort* 1201 BM, *Longa Bladeneford* 1180 P, *Langeton Blaneford* 1310 Cl, *Blaneford St. Mary* 1262 FF, *Cheping Blaneford* 1288 FF, *Blaneford forum* 1291 Tax]. May be OE *blægna ford* 'ford where gudgeons were seen' (first el. OE *blæge* 'gudgeon').

Forum refers to a market in the town.

Blankney Li [*Blachene* DB, *Blancaneia* 1157 YCh 354, *Blankenei* 1202 Ass]. OE *Blancan ēg*, more likely '*Blanca*'s island' than 'the island of the horse'. OE *blanca* 'horse' is evidenced, but not *Blanca* pers. n. Cf., however, OHG *Blanko*.

Blaston Le [*Bladestone, Blauestone* DB, *Blathiston* c 1225 Ep, *Blatheston* 1254 Val, *Blaeston* 1163 P]. '*Blēaþ*'s TŪN.' *Blēaþ* may be a nickname formed from OE *blēaþ* 'timid, sluggish'. Early shortened *ēa* would give ME *a*.

Blatchington, East, Sx [*Blechinton* 1169 f. P, *Blachington* 1226 FF], **West** B~ Sx [*Blacinctona* 1121 AC, *Blechinton* 1242 Fees]. 'The TŪN of *Blæcca*'s people.'

Blatchinworth La [*Blackenworthe* 1276 Ass]. '*Blæcca*'s WORÞ.'

Blatherwycke Np [*Blarewiche* DB, *Blatherwic* 1203 Ass, 1227 Ch, *Bladrewyc* 1230 P]. *Blather-* is probably a worn-down form of OE *blæcþorn*; cf. BLACKTHORN and WĪC.

Blawith La [*Blawit* 1276, *-with* 1341 FC]. ON *blávið̄r* 'black forest'. Cf. BLĀ(R).

Blaxhall Sf [*Blachessala, Blaccheshala* DB, *Blakeshal* 1270 Ipm]. '*Blæc*'s HALH.' *Blac(h)* pers. n. is found in DB.

Blaxton YW [*Blacstan* 1293 YInq]. 'Black stone.'

Blaydon Du [*Bladon* 1340 RPD]. Perhaps 'black DŪN' with ON *blár* or even OE *blæc* as first el. Cf. however BLOFIELD.

Bleadon So [*Bleodun* 956 BCS 959, *Bledone* 975 ib. 1313, DB, *Bleodon* 1236 BM]. First el. OE *blēo* 'colour', possibly here in adjectival use 'coloured'. The name means 'coloured hill' and refers to the variegated appearance of the hill-side, green parts interchanging with white, where the limestone comes to the surface.

Blean (wood) K [(in) *Blean* 786 BCS 248, (on) *Blean* 814 ib. 348, (to) *Blean ðem wiada* 858 ib. 496, *Bleen* 1230 P, *Le Blee* 1314 Ipm, *Boghton under Blee* Chaucer, Cant Tales G, 556]. The OE form of the name was *Blēa*, dat. *Blēan*. This goes back to earlier *Blahwōn. Like BLEE in Germany [*Blahe* 10], it is identical with OHG *blacha*, MHG *blahe* 'coarse cloth, rug', which is here used in a transferred sense of rough ground, very likely of broken ground overgrown with shrub, or else in that of 'patch of tilled ground'. In the latter case the name originally denoted a farm or the like and was transferred to the forest.

Blean YN [*Bleyng* 13 (1307) Ch, 1301 Subs], **Blean Beck** [*Blainbec* 1218 FF]. ON *Blæingr*, a derivative of *blár* 'blue, dark' and denoting either Blean Beck or Semer Water, near which Blean is. *Blæingr* is only recorded as a pers. n. in ON.

Bleasby Li [*Blesebi, Blasebi* DB, *Blesebi* c 1115 LiS, *Blisebi* 1203 Cur]. '*Blesi*'s BY.' *Blesi* is an ON byname, derived from Norw *bles*, Dan *blis*, Sw *bläs* 'a white spot'.

Bleasby Nt [*Blisetune* 958 YCh 2, *Blisemere* 958 BCS 1029, *Blesby* 1275 RH, *Bleseby* 1303 FA]. Perhaps the first el. is ON *Blesi* as in BLEASBY Li. There were two forms of the name, an OE *Blisatūn* (*Blesatūn*) and an OScand *Blisabȳr* or *Blesabȳr*. *Blisemere* is probably a copyist's error for *Blisetun*, due to the preceding *Gypesmere*.

Bleasdale La [*Blesedale* 1228 Cl]. *Bleas-* may be the pers. n. *Blesi*; cf. BLEASBY. But more likely it is ON *blesi* 'a light spot, blaze', Norw *blesa* 'a bare spot on a hillside'.

Blea Tarn, Blea Water. See BLĀ(R).

Blechingley Sr [*Blachingelei* DB, *Blecchingeleg* 1195 P]. 'The LĒAH of *Blecca*'s (*Blæcca*'s) people.'

Bledington Gl [*Bladintun* DB, *-tona* 1169 P, *-ton* 1291 Tax]. 'TŪN on R BLADON.'

Bledisloe Gl [*Bliteslav* DB, *Blicheslawe* 1220 Fees, *Blithelauwe* 1221 Ass]. '*Bliþ*'s or *Blīþe*'s burial-mound or hill.' See HLĀW. OE *Bliþ and *Blīþe are regular short forms of names like *Blīþhere, -weald*.

Bledlow Bk [*Bleddanhlæw* 966–75 Wills, *Bleddehlæwe* c 1033 Th, *Bledelawa* 1174 P]. '*Bledda*'s burial-mound'; cf. HLĀW. *Bledda is found also in **Bledney** So nr Wedmore [*Bledenithe* 712 BCS 128, *Bleddanhid'* 725 ib. 142]. Cf. HYTHE.

Blenca·rn Cu [*Blenecarn* 1210 FF, *Blencarn* 1211 FF]. A British name meaning 'hill with a cairn' (Welsh *blaen* 'top' and *carn* 'cairn').

Blencogo Cu [*Plenecogo* 1195 P, *Blenecoghow* c 1275 StB, *Blencoghow* 1305 Ipm]. A British pl. n. *Blencog*, to which was added ON *haugr* 'hill'. *Blencog* contains Welsh *blaen* 'top' and possibly Welsh *cog* 'cuckoo': 'cuckoo hill'.

Blencow Cu [*Blenco* 1232 P, *Blenkhaw* 1254 Ipm]. Obscure. The name seems to contain ON *haugr* 'hill', and the first el. may be a contracted form of a British name containing Welsh *blaen* 'top'.

Blendworth Ha [*Blednewrthie* c 1170 Add Ch, *Bledenewrth* 1256 Ass (Gover), *Blaneworth* c 1270 Ep, *Blendeworth* 1291 Tax]. See WORÞ. First el. apparently a pers. n. *Blǣdna*, related to *Blǣdla*.

Bleng R Cu [*Bleng* 1576 Saxton]. Identical with BLEAN YW.

Blenheim Park O. Named from the victory of Blenheim in 1704.

Blenkinsopp Nb [*Blencheneshopa* 1178 P, *Blenkeneshop* 1256 Ass, *Blencanhishop* 1236 Fees]. Second el. HOP 'valley'. The first is obscure. It may be a pers. n. But it might also be a pl. n. containing Welsh *blaen* (cf. BLENCARN &c.).

Blennerhasset Cu [*Blendherseta, Blennerheiseta* 1188, *Blenhersete* 1190, *Blenherseta* 1194 P, *Blenerheyset* 1235 FF]. Apparently a hybrid name, the second part being an ON *heysætr* 'hay shieling'. The first seems to contain Welsh *blaen* (cf. BLENCARN). It might be a Welsh *Blaen-dre* 'hill farm'. Cf. TREF.

Bletchingdon O [*Blecesdone* DB, *Blechesdune* c 1130 Oxf, *Blechesdon* 1197 FF, *Blechindon* 1279 Abbr]. Originally *Blecces dūn*, identical with BLAISDON Gl, later changed to *Blechindon* &c., as if from *Blecca*. A ford near Whitchurch O is called *Blecces ford* and *Bleccan ford* 1012 KCD 1307.

Bletchley Bk [*Blechelai* 1106–9 Fr, c 1155 Oxf, *Blechelee* 1212 FF]. '*Blecca*'s LĒAH.'

Bletchley Sa [*Blecheslee* 1222, *Bleccheleg* 1254 Eyton, *Blecheleg* 1255 RH]. 'The LĒAH of *Blecca* or *Blecci*'; cf. BLETCHINGDON.

Bletsoe Bd [*Blechesho, Blacheshou* DB, *Blechesho* 1199 FF, *Blechenesho* 1247 Ass]. '*Bleecci*'s hill.' Cf. HŌH, BLAISDON.

Blewbury Brk [*Bleoburg* 944 BCS 801, *Bleobyrig dun* 964 ib. 1143, *Blidberia* DB, *Bleberia* c 1145 BM]. See BURG. First el. as in BLEADON.

Blickling Nf [*Blikelinges* DB, *Bliccling* 1166 RBE]. '*Blicla*'s people', OE *Bliclingas*. *Blicla is a short form of once existing names in *Blic*-; cf. OG *Blicger* &c. *Blic* is OHG *blich* 'splendour', OE *blice* 'gleam'.

Blidworth Nt [*Blideworde* DB, *Blieswurde* 1158 f. P, *Bleðewurda* 1164, *Bledewurda* 1180, 1194 P]. See WORÞ. The first el. is a pers. n. belonging to OE *blīþe* or *blēaþ*. Cf. BLEDISLOE, BLASTON.

Blindcrake Cu [*Blenecreyc* 12 Lanercost, *Blenckrayk* 1246 FF]. A British name consisting of Welsh *blaen* 'top' and *craig* (earlier *creic*) 'crag, rock'.

Blisland Co [*Blislond* 1291 Tax, *-e* 1300 FF]. An earlier name is *Glustone* (for *Blustone*) DB, *Bloiston* 1177 P, 1198 Cur, *Blieston, Bleston* 1195, *Bluston* 1196 P. The first el. is no doubt the old Cornish name of the place. Its etymology is obscure.

Blisworth Np [*Blidesworde* DB, *Bliðeswurde* 1198 P]. See WORÞ and BLEDISLOE.

Blithbury St [*Blidebire* 1200 P, *Blythebury* n.d. Rydeware], **Blithfield** St [*Blidevelt* DB], **Blithford** St [*Blitheford* 1322 Ipm]. 'BURG, FELD, FORD on R BLYTHE.'

Blockley Wo [*Bloccanleeh* 855 BCS 488, *-lea* 10 BCS (1320), 978 KCD 620, *Blochelei* DB]. *Blocca* must be a pers. n., though not otherwise evidenced. Cf. *Blocc* in BLOXHAM &c.

Blofield Nf [*Blafelda, Blawefelle, Blauuefelda* DB, *Blafeld* 1156 P, *Blofeld* 1294 Bodl]. First el. OE *blǣw*, cognate with ON *blár* 'blue'. But OE *blǣw* is only found in the sense 'pigment'. Possibly there was an OE *blǣw* sb. in the sense of 'woad', the name of the plant chiefly used for making dyes. Cf. BLAUFELDEN in Germany [*Blauelden* 12].

Bloomsbury Mx [*Blemondisberi* 1281, *Blemundisbury* 1335 AD]. Named from some member of the Blemund family resident in London. William de Blemunt in Tottenham is mentioned 1201 Cur, and William de Blemund 1230 P. BLÉMONT is a place in Vienne (France).

Blore St nr Ashbourne [*Blora* DB, *Blore* 1227 Ass, 1240 FF], **B~** St nr Market Drayton [*Blore* 1293 Ass, QW], **Blurton** St [*Blorton* 1195 f. P, 1248 FF]. *Blore* represents an unrecorded OE *blōr*, identical with ME *blure, bloure* 'blister, swelling' (here used in the sense 'hill') or else a word meaning 'bare spot', related to MHG *blas*, MDu *blaer* 'bare', Du *blaar* 'cow with a white spot', ON *blesi* (cf. BLEASDALE), OE *blerig, blere* 'bald'. The Blores and Blurton are in high exposed places.

Blowick La. Probably 'dark bay' (ON *blár* 'dark' and *vík* 'bay').

Bloxham O [*Blockesham* 1067 BM, *Blochesham* DB], **Bloxholm** Li [*Blochesham* DB, *Bloxeham* 1130 P, *Bloksham* 1229 Ep], **Bloxwich** St [*Blocheswic* DB, *Blockswich* 1271 For], **Bloxworth** Do [*Blocheshorde* DB, *Blokeswurthe* 1201 FF]. The first el. is evidently an unrecorded OE pers. n. *Blocc*. Cf. *Blocca* in BLOCKLEY.

Blubberhouses YW [*Bluberhusum* 1172 YCh 511, *-huse* 1227 FF]. The first el. is related to ME *blubber* 'the foaming or boiling of the sea; bubble, foam', but the exact meaning is obscure. It may be a name of the lake on which the place stands.

Blundeston Sf [*Blundeston* 1203 Cur, c 1220 Bodl, *Blunteston* 1205 Cur, *Bluntiston* 1231 FF]. '*Blunt*'s TŪN.' OE *Blunt* is found in *Bluntesig* KCD 666.

Blunham Bd [*Blunham* DB, c 1190 Reg, *Bluneham* DB, c 1095 Bury]. First el. possibly a pers. n. **Blūwa*, corresponding to OHG *Pluwo*.

Blunsdon St. Andrew, Broad B~ W [*Bluntesdone* DB, *Brodebluntesdon* 1263 Ipm, *Bluntesdone Sancti Andree* 1316 FA]. '*Blunt*'s DŪN'; cf. BLUNDESTON, and see BRĀD.

Bluntisham Hu [*Bluntesham* c 1050 KCD 907, DB, 1158 P]. '*Blunt*'s HĀM'; cf. BLUNDESTON.

Blunt's Hall Ess [*Blundeshala* DB, *Blunteshala* 1167 P, *-hal* 1230 Ch]. '*Blunt*'s HALH'; cf. BLUNDESTON.

Blurton St. See BLORE.

Blyborough Li [*Bliburg* DB, c 1200 DC, *-burc* c 1115 LiS, *Blieburc* 1203 Cur]. Blyborough is c 5 m. from **Blyton** Li [*Blitone*, *Blittone* DB, *Blituna* c 1115 LiS, *Bliton* 1223 Ep]. The two names presumably have the same first el. They are not on the same stream, so a stream-name *Blīþe* is excluded. Possibly the first el. is a pers. n. **Blīþa*, derived from *blīþe* adj. Another possibility would be the pers. n. *Blīh* c 1050 YCh 9, held to be from ON *Blígr*. Neither is quite satisfactory, and one might think of the same first el. as for BLEADON, BLEWBURY.

Blyford Sf [*Blitleford* c 1060 Th, *Blideforda* DB]. 'The ford over the BLYTH.'

Blymhill St [*Brumhelle* DB, *Blumehil* 1167 P, *Blimenhul* 1236 Fees]. The first el. is probably OE *plȳme* 'plum-tree'. The change of *p-* to *b-* is analogous to *t-* > *d-* in DUNSTALL.

Blyth (-īdh) R Nb [*Blitha* 1204 Ch]. See BLYTHE. Blyth, the place, is *Blida* 1130 P, but *Blithmuth* 1236, *Blithemuth* 1250 Newminster ('the mouth of the Blyth').

Blyth Nt [*Blide* DB, *Blida* 1130 P]. Named from the river RYTON, which was formerly Blythe [*Blide* 958 BCS 1044].

Blyth R Sf [*Blith* 1586 Camden]; see BLYTHE. On the Blyth is **Blythburgh** Sf [*Blideburh* DB, *Blieburc* 1157 P]. The district round the river is **Blything** hd [*Blidinga* DB]. The name means 'the dwellers on the Blyth'.

Blythe R St [*Blithe* 12 Mon], **B~** or **Blithe** Wa [*Blitha* Hy 2 Mon, *Blithe* 1276 Ipm]. Blythe and Blyth are derived from OE *blīþe* adj. 'gentle' or 'merry'.

Blyton. See BLYBOROUGH.

Boarhunt Ha [(Æt) *Byrhfunt* 10 BCS 1161, *Bor(e)hunte* DB, *Burhunt* 1170 P]. 'The spring by the BURG'; cf. FUNTA. The first el. appears first in the gen. form *byrh*, later in the form *bur(h)*.

Boar's Hill Brk [*Boreshulla* Hy 2 Abingd]. Lit. 'boar's hill'.

Boarstall Bk [*Burchestala* 1158 f., *Burcstala* 1161 P]. See BIRSTALL, BURSTALL.

Boarzell Sx [*Baresselle* a 1123 PNSx, *Borsel* 1253 Ch, *Borshull* 1279 Ass]. 'Shelter for the boar'; second el. OE (GE)SELL.

Bobbing K [*Bobinge* 11 DM, *-s* 1179 P, *Bobbing* 1227, *-e* 1241 Ass]. '*Bobba*'s people.'

Bobbington St [*Bubintone* DB, *Bubington* 1236 Fees], **Bobbingworth** Ess [*Bubingeorda* DB, *Bubingewrthe* Hy 3 BM]. 'The TŪN and WORÞ of *Bubba*'s people.'

OE **bōc** 'beech'. See next names and BOOKHAM, BOUGHTON, BUCKHOLT, -HURST.

Bockenfield Nb [*Bokenfeld* 1242 Fees]. 'Beech FELD.' First el. OE **bōcen* 'of beech', derived from *bōc* 'beech'.

Bockhampton Brk [*Bochentone* DB, *Bochamton* 1179 P, *Bochampton* 1220 Fees], **B~** Do [?*Buchæmatun* 1002–14 KCD 708, *Bochehamtone* DB, *Bocameton* 1212 Fees], **B~** Ha [*Northbocamtone* 13, *Bochamton* 1295 BM]. OE *Bōchǣmatūn*, 'the TŪN of the *Bōchǣme*'. *Bōchǣme* may be 'the people of Buckland' (an earlier name of the places) or 'dwellers by the beeches'. Elliptical formations like *Bōc-hǣme* from *Bōcland* are common in OE.

Bocking Ess [(on) *Boccinge* a 995, (æt) *Boccinge* 997 Wills, *Boccing* c 1050 KCD 896, *Bochinges* DB]. Apparently a singular name in *-ing*, not OE *Boccingas*. *Boccing* may be derived from a pers. n. *Bocca*, a side-form of *Bucca*. Or *Boccing* may be for earlier *Beoccing* (from *Beocca*). But a possible instance of a plural *Boccinges* t. Æthelred II is pointed out PNEss lxi.

Bockingham Hall Ess [*Botingham* DB, 1205 Cur, *Botingeham* 1206 FF]. 'The HĀM of *Bōta*'s people.'

Bockleton Sa [*Boclinton* 1291 Ch, *-e* 1292 Ep, *Boklyntone* 1280 Ep], B~ Wo [*Boclintun* DB, *-ton* 1176 P]. Possibly an OE *Bōchlinc-tūn* 'TŪN by the beech hill'. Or the first el. may be a tribal name *Boccelingas*, derived from a pers. n. *B(e)occel*; cf. BOCK-ING.

OE bōcland. See BUCKLAND.

Boco·nnoc Co [*Botchonod* DB, *Bokonnecke*, *Boccunneke* 1310 ff. Ep]. '*Conoc*'s dwelling.' First el. Co *bod* (later *bos*) 'house, dwelling', second a Cornish pers. n. corresponding to Welsh *Cynog*.

Boddington Gl [*Botintone*, *Botingtune* DB, *Botinton* 1212 Fees]. 'The TŪN of *Bōta*'s people.' B~ Np [*Botendone* DB, *-don* 1199 P]. '*Bōta*'s DŪN.'

Bodenham W [*Boteham*, *Buteham* 1249 PNW(S), *Botteham* 1255 RH, *Botenham* 1260 FF]. Probably '*Bōta*'s HĀM or HAMM'.

Bŏdenham He [*Bodeham* DB, 1180 P, *Bodenham* 1249 Fees], **Bodham** Nf [*Bodham*, *Bodenham* DB, *Bodeham* 1175 P]. '*Boda*'s HĀM.' *Boda* is the name of a moneyer and occurs in DB. Cf. OHG *Bodo*, *Boto*. OE *boda* 'messenger' is a less likely first el.

Bodiam (-ŏj-) Sx [*Bodeham* DB, *Bodiham* 1170, 1230 P, *-hamme*, *Bodingham* 1259 FF]. '*Boda*'s HAMM.' Cf. BODENHAM.

Bodicote O [*Bodicote* DB, *Bodicota* 1194 P]. OE *Bodan cot*; cf. BODENHAM. Or 'the COT of *Boda*'s people' (OE *Bodingacot*).

Bodmin Co [(on) *Bodmine* 11 E, *Bodmine* DB, *Botmenei* 12 Life of St. Cadoc]. Perhaps 'the house of the monks', Co *bod* (earlier *bot*) 'house, dwelling' and *meneich* 'monks' (plur. of *manach*). There was a monastery at Bodmin, said to have been founded by King Æthelstan in 926. For the loss of final *-ch*, cf. LOOE. Henderson's derivation of B~ from an OCo *bot menehi* may be formally preferable, but his rendering 'dwelling by the sanctuary' is improbable. *Menehi* is a derivative of *manach* 'monk' and meant 'monastery'. The derived meaning 'sanctuary' found in MBret can hardly be assumed for OCo. *Bot menehi* would mean 'house of the monastery' or about the same as *bot meneich*.

Bodney Nf [*Budeneia*, *Bodeneia* DB, *Bodeneie* 1199 P, *Bedeneye*, *Bodeneye* 1254 Val]. If the forms are reliable, the first el. seems to be OE *Beoda* pers. n., not recorded, but cf. *Beoduca*. For the change of *eo* to *o*, cf. MODENEY.

OE boga 'bow' is found in pl. ns. in various senses, as 'bow' in BOWDEN Db, BOWDON, BOWOOD, 'arched bridge' in BOW, BOWES, 'bend of a river' in BOWLAND.

Bognor Sx [*Bucganora* 680 BCS 50, *Bugenor* 1275 RH]. OE *Bucge*, a woman's name, and *ōra* 'shore', here in the sense 'a (gravelly) landing-place'. **Little Bognor** Sx [*Boganora* 953 BCS 898] was probably a dependency of Bognor and named from it.

Bolam Du [*Bolom* 1317 Ipm], B~ Nb [*Bolum* 1155 BM, 1212 Fees, 1254 Val], **Bole** Nt [*Bolun* DB, *Bolum* 1240 FF], **Bolham** or **Bollom** Nt [*Bolum* 1335 BM]. All go back to OE *bolum*, which must be the plural of a word such as *bol*. It is probably identical with ME *bole*, ON *bolr*, 'trunk of a tree', MHG *bole* 'a plank'. ME *bole* is held to be a Scand loan-word, but it may well be native English. The names may mean 'the tree-trunks' or (in the case of Bole and Bolham) 'the planks' (referring to a plank bridge).

Bolas Sa [*Belewas* 1198 Fees, *Bowlewas* 1255 RH, *Boulewas* 1265 Ch, *Bowelewas* 1293 QW, *Magna Boulwas* 1327 Subs]. Second el. OE WÆSSE. The first might be an OE *Bog(e)lēah* 'wood where bows were got', or 'wood by a bend'. Cf. BOWDON, BOWOOD D.

Bold La [*Bolde* 1204 P, *Bold* 1257 Ass], **The Bold** Sa [*Bolde* 13 Misc, 1327 Subs]. OE *bold* 'house, palace'. See BŌPL.

Boldon Du [*Boldun* c 1170 FPD, *Boldon* 1291 Tax]. OE *bōþl-dūn* 'hill with a homestead'.

Boldre Ha [*Bovre*, *Bovreford* DB, *Bolre* 1236 BM, 1253 Cl]. Apparently an old name of Lymington River [*Bolre* J Christchurch Cart (Gover)]. Possibly derived from an old form of dial. *boulder* 'bulrush'.

Boldron YN [*Bolrum* c 1180 PNNR, *Bolerum* 1204 FF, *Bulerun* 1280 YInq]. 'Bull clearing'; second el. OScand *rūm* 'room, cleared space'.

Bole, Bolham. See BOLAM.

Bolingbroke Li [*Bolinbroc* DB, *Bulincbroc* c 1156 RA, *Bulingbroc* 1202 Ass]. OE *Bul(l)-ingabrōc* 'the brook of the people of *Bul(l)a*'. OE *Bula*, *Bulla* are recorded in OE pl. ns., as *Bulan ham* 774 BCS 213, *Bulan setl* 955 ib. 903.

Bollin R Chs [*Bolyn* c 1275 Chester]. From the river are named: **Bollin Fee** Chs [*Bolyn* Hy 3 Ipm, *Bolynne* 1286 Court], **Bollington** Chs nr Macclesfield [*Bolinton*, *Bolington* 1285 Court], **Bollington** Chs nr Altrincham [*Bolington* 1287 Court]. The etymology of the river-name Bollin is obscure.

Bolney O [*Bollehede* DB, *Buleheðe* 1176 P, *-hith* 1236 Ep]. Either '*Bula*'s HȲÞ or landing-place' or, more likely, 'landing-place for bullocks'; cf. ROTHERHITHE. The OE form in the latter case would be *bulena hȳþ*. See BULA.

Bolney (-ōn-) Sx [*Bolneye* 1263 FF, *Boleney* 1271 Ass]. '*Bol(l)a*'s island.'

Bolnhurst (-ōn-) Bd [*Bulehestre* DB,

Boleherst 1179 P, *Bolneherst* 1240 FF].
OE *bulena hyrst* 'bullocks' HYRST'. See BULA.

Bolsover Db [*Belesovre* DB, *Bolesoura* 1167
f. P, *Bulesoūres* 1197 f. P, *Bolesor* 1230 P].
Second el. OE *ofer* 'edge, slope'. The first
is obscure. An OE *bulan-læs* 'bullock pasture' might be suggested. Cf. BOULSDON.

Bolsterstone YW [*Bolstyrston* 1398 YD].
Possibly 'stone resembling a bolster', but
bolsterstone may have had some technical
meaning.

Bolstone or **Boulstone** He [*Boleston* 1193
ff., *Bolestan* 1194, 1200 P]. '*Bula*'s stone.'

Bolt Head D [*Bult Poynt* 1577 Saxton].
Near Bolt Head is **Boltbury** or **Bolberry**
D [*Boltesberie* DB, *Boltebir* 1228 FF, 1230
Cl, 1242 Fees]. The headland or the long
ridge at Bolberry had a name *Bolt*, from OE
bolt 'an arrow', or else a name such as *Bolte*,
derived from the same word. See BURG.

Boltby YN [*Boltebi* DB, *-by* 1209 FF].
'*Bolt*'s BY.' *Boltr* is a common ON byname. ON *Bolti* would suit better formally,
but it is not evidenced.

Bolton Cu nr Wigton [*Bothilton* c 1227
Holme C], **B~ in Copeland** Cu [*Bothelton*
1251 FF, *Boutona* c 1200 StB], **B~** La nr
Urswick [*Bodeltun* DB, *Botheltun* c 1185
FC], **Little B~** La [*Bothelton* 1212 Fees],
B~ le Moors La [*Boelton* 1185 P, *Bolton
on the Mores* 1331 FF], **B~ le Sands** La
[*Bodeltone* DB, *Bothelton* 1201 P], **B~**
Nb [*Bolton* 1200 P, *Boulton* 1227 Ch], **B~**
We [*Botelton* c 1180 WR, *Bolton* 1328 FF],
B~ YE [*Bodelton* DB, *Boelton* 1252 Ch],
Castle B~ YN [*Bodelton* DB, *Bodeltona*
1197 (1252) Ch], **B~ upon Swale** YN
[*Boletone* DB, *Boelton* 1208 FF], **B~** YW nr
Bradford [*Bodeltone* DB], **B~ Abbey** YW
[*Bodeltone* DB, (prior de) *Botheltona* 12
StB], **B~ by Bowland** YW [*Bodeltone*
DB, *Boelton in Boland* 12 Pudsay], **B~
upon Dearne** YW [*Bodeltone* DB, *Boulton
on Dyrn* 1339 FF], **B~ Percy** YW [*Bodel-
tune* DB, *Boulton Perci* 1318 Ipm]. OE
Bōþltūn, consisting of OE *bōþl* (*bōtl*, *bold*)
'dwelling, house' and TŪN. There was
probably an OE common noun *bōþltūn*,
with about the same meaning as *hāmtūn*,
perhaps, like OSw *bolbyr*, the village proper
in contradistinction to the outlying parts.
Cf. Introd. p. xvi f.
B~ Percy was held by William de Percy in 1086
(DB). Percy from PERCY in Normandy.

Bonby Li [*Bundebi* DB, *Bondebi* c 1115 LiS].
Most likely '*Bondi*'s BY'. *Bondi* occurs in
DB; it is ODan *Bundi*.

Bonchurch Wt [*Bonecerce* DB, *-church* c
1270 Ep, *Bunchurche* 1382 Cl]. '*Buna*'s
church.' *Buna* is a known OE name. *Bon-*
is pronounced *Bŭn-* (PNWt).

Bondleigh D [*Bolenei* DB, *Bonlege* 1205 FF,
Bonelegh 1242 Fees]. The forms suggest
the development *Bolan-īeg* ('*Bola*'s island')

> *Boleney* > *Boneley*. But the situation
points rather to *Bolan-lēah*, as suggested in
PND.

Bonehill St [*Bolenhull* 1230 P, 1271 For,
Bulenhull 1230 P]. '*Bula*'s hill' or 'hill of
the bullock'. Cf. BULA.

Bongate We at Appleby [*Bondegate* 1292
WR]. Cf. *Bondegate* at Ripon c 1300 Fount.
The name means 'the street of the bondmen
or villeins'. In a grant of King John to
Robert de Veteripont is mentioned 'vetus
Appulbi ubi villani manent' (CWNS ix,
325).

Bonhunt Ess (now in Wicken Bonhunt), an
old vil. [*Banhunta* DB, 1141 BM, *Bonhunte*
1262 FF]. Cf. HUNTA. Either '*Bana*'s
huntsmen' or OE (*ge*)*bann-huntan* 'huntsmen liable to be summoned'. For *o* from *a*
before *n*, cf. *Pont*, a form of PANT.

Boningale Sa [*Bolynghale* 12 Eyton, *Boling-
hale* 1285 Ch]. 'The HALH of *Bol(l)a*'s
people.'

Bonnington K [*Bonintone* DB, *Bunnigtun*
11 DM, *Boninton* 1206 Cur]. 'The TŪN of
Buna's people.' See SUTTON BONNINGTON
Nt.

Bonsall Db [*Bunteshale* DB, *Bontishale* 13
Derby, *Bontesale* 1291 Tax]. '*Bunt*'s HALH.'
Bunt is not evidenced, but the cognate
Buntel is found in *Bunteles pyt* BCS 674.

Bonthorpe Li [*Brunetorþ* DB, *Burnetorþ*
c 1115 LiS, *Bruntorþ* 1212, *-thorp* 1242
Fees]. It is difficult to decide if this is thorp
on the stream (OE *burna*) or the thorp of
one *Brūna* (OE) or *Brūni* (OScand).

Bonwick YE [*Bounewic* 1219 Ass]. '*Buna*'s
WĪC.' Cf. the neighbouring ATWICK, CAT-
WICK, which are named from persons.

Bookham Sr [*Bocham* 675 BCS 39, 933 ib.
697, *Bocheham* DB, *Bokham Magna* 1273
BM, *Littille Bokham* 1263 FF]. OE *Bōc-
hām* 'village by beeches'.

Booley Sa [*Boleley* c 1100 Eyton]. 'Pasture
where bullocks grazed'; see LĒAH, BULA.

Bootham (-ōōdh-) York [*Buthum* c 1155,
c 1175 YCh 260 f., 1251 Ass]. ON *bùðum*
dat. plur. 'the (market) booths'.

Boothby Li at Welton le Marsh [*Bodebi* DB,
Botheby 13 BM], **B~ Graffoe** Li [*Bodebi*
DB, *Bobi* 1202 Ass], **B~ Pagnell** Li [*Bodebi*
DB, *Boebi* 1138 NpCh, *Bodebi* a 1183 DC].
Perhaps OScand *Bōþabȳr* 'BY with booths'.
B~ Graffoe from Graffoe wapentake or the
ridge that was its meeting-place. **B~ Pagnell**
was held by Johannes Paynel in 1346 (FA).
Paynel (Lat *Paganellus*) is a diminutive of *Pain*
(Lat *Paganus*), an OFr pers. n., lit. 'the heathen'
(OFr *paien*).

Boothorpe Le [*Bocthorp* c 1125 LeS, *Bothorp* 13 Fees]. First el. ODan, OSw *Bo*
pers. n. (= ON *Búi*).

Bootle Cu [*Bodele* DB, *Botle* 12 StB, *Botele*

1251 FF], **B~** La [*Boltelai* DB, *Botle* 1212 Fees]. OE *bōtl* 'dwelling'; cf. BŌÞL.

Booton Nf [*Botuna* DB, *Boton, Boiton* 1203 Cur]. '*Bōta*'s TŪN' or possibly '*Bō*'s TŪN'. Cf. BOOTHORPE.

Boraston Sa [*Bureston* 1188 P, *Burston* 1208 FF]. Possibly 'the TŪN of the *gebūr* or peasant'.

Borden K [*Bordena* 1177 P, *-den* 1190 P, 1206 FF, *-denne* 1260 FF]. Second el. OE DENN 'swine-pasture'. The first is doubtful. Hardly OE *bār* 'boar'. It may be OE *bord* 'board' or the *bor* discussed under BOREHAM.

Bordesley Wa [*Bordeslegh* 1226–8 Fees, *-leye* 1285 QW], **B~** Wo [*Bordeslega* 1138 (1266) Ch, *-lea* 1156, 1159, *Bordelea* 1159 P]. The two Bordesleys are only some 7 or 8 m. apart. If LĒAH means 'wood', they may have been named from the same wood. The name may have been OE *Brordes lēah* 'wood belonging to *Brord*'. One of the *r*'s would be apt to be lost owing to dissimilation. If the first el. is OE *bord* 'board', the name means 'wood where boards were got'.

Bordley YW [*Borelaie* DB, *Bordeleia* c 1140 FC, *Bordelay* 1162 YCh 81]. 'Wood or clearing belonging to *Brorda*' or 'wood from which boards were got'. The form is rather in favour of the first alternative. OE *Brorda* is not uncommon and may be expected in some pl. ns. Owing to dissimilation it would tend to become *Borda*, a form found in DB. OE *Borda* is not recorded.

Boreatton Park Sa [*Acton* c 1245 Eyton]. See ACTON. The prefix may be the word *boar*.

Boreham Ess [*Borham* c 1040 Wills, DB, 1191 P, 1203 Cur, *Borreham* 1198 FF, *Boorham* 1254 Val], **B~ Wood** Hrt [*Borham* 1278 QW, *-wode* 1329 Misc], **B~ Street** Sx [*Borham* 12, 1321 AD]. The first el. must be an OE *bor*, which may be identical with OHG *bor* 'upper room', also in *empor* 'up'. The word may be the base of OE *borlice* 'excellently'. OE *bor* would seem to have meant 'elevation, hill'. The corresponding Scand word meant 'a portage', but this hardly suits the English names.

Boresford He [*Boresforde* 1379 BM]. 'Boar's ford.'

Borley Ess [*Barlea* DB, *Barlee* 1253 Ch, *Borle* 1238 Ass, *Borlee* 1254 Val], **Bor(e)ley** Wo [*Barlege* c 1190 PNWo]. OE *bār-lēah* 'boars' wood'.

Boroughbridge YW [*pons de Burgo* 1155 YCh 75, *Ponteburc* 1169, *Puntdeburc* 1194 P, *Burbrigg* 1293 Cl]. 'The bridge of the BURG' (Aldborough).

Borrowash (bŭrŏwŏsh) Db [*Burysasch* 1272 FF, *Burwishasshe* 1275 RH]. 'The ash by Burgh.' The earlier name seems to have been BURGH [*Burgh upon Derwent* 1269 Dale].

Borrowby YN nr Leake [*Bergebi* DB], **B~** YN nr Lythe [*Berge(s)bi* DB]. See BARBY.

Borrowdale Cu [*Borgordale* 12 StB, *Borcheredale* c 1209 FC], **B~** We [*Borgheredale* Hy 2 Kendale, *Borgherdal* Hy 2 (1247) Ch]. ON *Borgardalr* or *Borgarárdalr* 'valley with a fort' or 'the valley of *Borgará*', i.e. stream by a fort (ON *borg*, gen. *borgar*, see Ā). B~ Cu is on the upper Derwent, called *Borghra* 1211 FC. B~ We is on **Borrow Beck** [*Burgra* 1198 Kendale, *Borghra* 1235 FF].

Borstal K [*Borhsteall* 10 BCS 339, *Borcsteall* 10 ib. 1322, *Borchetelle* DB]. OE *borgsteall* meant 'a place of refuge', but may have already developed the sense 'pathway up a steep hill' found in modern dialectal *borstal*. The elements are OE *borg* 'security' and *steall* 'place'.

Borwick (bŏrĭk) La [*Bereuuic* DB, *Berwik* 1228 Cl]. See BEREWĪC.

Bosbury He [*Bosanbirig* a 1118 Flor, *Boseberge* DB, *-bir* 1230 Cl]. '*Bōsa*'s BURG.'

Boscastle Co [*Boterelescastel* 1302 QW, *Chastelboterel* 1312 BM]. '*Boterel*'s castle.' William de *Botereus* held the manor in 1302 (QW). The family presumably took its name from LES BOTTEREAUX in Normandy.

Boscawen Co [*Boscauan* 1332 AD, *Boscawen* 1356 FF]. Co *bos* 'house, dwelling' (earlier *bod, bot*) and a Cornish equivalent of Welsh *ysgawen*, Bret *scaven* 'elder-tree'.

Boscobel Sa. B~ House was originally a hunting-lodge in Brewood Forest built c 1580 by the Giffards of Chillington (DNB, under Richard Penderel). The name is apparently a form of Lat *boscus bellus* 'beautiful wood' (*bosco bello* abl. or an Italianized *Boscobello*).

Boscombe W [*Boscumbe* DB, *-cumba* 1178 BM, *Boxcumbe* 1200 Cur, *Borrescumb* 1256 Cl, *Borscumbe* 1275 RH]. First el. apparently as in BOWSLEY rather than OE *box* 'box-tree'.

Bosden Chs [*Bosedon* 1248 Ipm, *Bosedun* 1286 Court]. '*Bōsa*'s DŪN.'

Bosham (-ŏz-) Sx [*Bosanham(m)* c 730 Bede, *-ham* c 890 OEBede, *Boseham* DB]. '*Bōsa*'s HĀM or HAMM.'

Bosley Chs [*Boselega* DB, *Boseleg* 1278 Ipm]. '*Bōsa*'s LĒAH.'

Bosmere Sf [*Bosemera* DB, *Bosemere* c 1230 Bodl]. '*Bōsa*'s lake.' Bosmere is a lake and a place on it, also the name of a hundred.

Bossall YN [*Boscele, Bosciale* DB, *Botshal* c 1170, *Bozhale* c 1180 Middleton, *Bozhal* 1225 Ep]. Apparently OE *Bōtsiges halh*, though *Bōtsige* is unrecorded.

Bossi·ney Co [*Botcinnii* DB, *Boccyny, Bocciny* 1236 Ch]. Co *bod* (earlier *bot*) 'dwelling, house' and the old Cornish pers. n. *Cini*, corresponding to Welsh *Cini* c 1150 LL.

Bossington Ha [*Bosintone* DB, *-ton* 1167 P], **B~** K [*ad Bosingtune* 873 BCS 536, *-tun* 941 ib. 766], **B~** So [*Bosintone* DB, *Bosinton* 1256 Ass]. 'The TŪN of *Bōsa*'s people.'

Bostock Chs [*Botestoch* DB, *Bostoc* 1260 Court]. '*Bōta*'s STOC.'

Boston Li [*Botuluestan* 1130 P, *Botolfston* 13 Dame Siriz]. 'St. Botulf's stone.' The earlier name was *Icanho* 654 ASC, '*Ica*'s HŌH or hill'. Boston may have got its name from a stone or a stone cross at which St. Botulf preached Christianity to the Middle Anglians. Or OE *stān*, like ON *steinn*, ODan *sten*, may have meant 'stone house'. If so, 'Botulf's stone church'. Cf. HOLYSTONE. St. Botulf was not buried in Boston.

Bosworth, Husbands, Le [*Baresworde* DB, *Bareswurþe* 12 DC, *Boreswurth* 1230 P]. 'The WORÞ of the boar' (OE *bār*) or, more likely, '*Bār*'s WORÞ'. Cf. BARSHAM.

Husbands in contradistinction to MARKET BOSWORTH. Perhaps 'husbandmen's B~', i.e. 'Rural B~'.

Bosworth, Market, Le [*Boseworde* DB, *-wurda* 1192 P]. '*Bōsa*'s WORÞ.'

Botcherby Cu [*Bocherby* c 1175 WR, *Boschardebi* 1191 P], **Botchergate** Cu in Carlisle [*Vicus Bochardi* 1245 WR, *Bochergate* 1363 Cl], **Botcheston** Le [*Buchardeston* 1285, *Bocherdston* 1428 FA, *Borchardeston* 1327 Subs]. '*Bochard*'s BY, street, TŪN.' *Bochard* is a Fr form of OG *Burchard*. In Botchergate *-gate* is ON *gata* 'road, street'.

Botesdale Sf [*Botholuesdal* 1275 RH, *Botulfesdale* 1313 FF]. '*Bōtwulf*'s valley.'

ME **bōþ(e)**, from OEScand *bōþ*, also ME *bouþe*, from ON *búð*, is common in pl. ns. in senses such as 'cow-house, herdsman's hut'. See BEWCASTLE, BOOTHAM, BOOTHBY, BOUTH, BOWDERDALE, SCORBROUGH. Common as a second element.

Bothal (bŏtl) Nb [*Bothala* Hy 2 (1271) Ch, *Bothale* 1212 Fees, *Botehale* 1256 Ass]. '*Bōta*'s HALH.'

Bothamsall (-ŏdh-) Nt [*Bodmescel* DB, *Bodmeshil, Botmeshil* 12 DC, *Bothmeshill* 1211–13 Fees, *Bothemessel* 1247 P]. OE BOÞM 'valley' and SCYLF 'slope, bank'.

Bothel (booĕl) Cu [*Bothle* c 1125 StB, *Bothele* 1301 Ch]. OE *bōþl* 'dwelling, house'. **Bothel** Nb is probably identical.

Bōthenhampton (-th-) Do [*Bolem'tona* 1107 (1330) Ch, *Bothenamtone* 1285 FA]. OE *Boþmhǣmatūn* 'the TŪN of the dwellers in the valley'. See BOÞM.

OE ***bōþl, bōtl, bold** 'dwelling, house, palace' is common in pl. ns. The form of the el. varies a good deal. See BOLD, BOOTLE, BOTHEL, BEADLAM. It is the first el. of *bōþltūn*, whence BOLTON (q.v.), BOTTESFORD. As a second el. it appears as *-bottle* in HAR-, NEW-, SHILBOTTLE &c., as *-bald* in NEWBALD, as *-bold* in NEW-, PAR-, WYCHBOLD.

OE **boþm, botm** 'bottom' is also used in the sense 'valley, dell'. See BOTHAMSALL, BOTHENHAMPTON, BOTTOMSTEAD, BROADBOTTOM. On the derivative *byþme, bytme, beþme*, see BYTHAM, BEAMSLEY, BITTESWELL.

Botley Bk [*Bottlea* 1167 P, *Bottelag* 1195 Cur], **B~** Brk [*Botelea* W 2, *-leia* Hy 1 Abingd, *Boteley* 1242 Fees], **B~** Ha [*Botelie* DB, *Botlai* 1185 P, *Boteleg* 1236 Fees], **B~** Wa [*Butte-, Botteleia* 1199 FF, *Botleye* 1335 Ch]. Looks like '*Bōta*'s LĒAH'. But Botley is a little too common for it to be quite likely that the first el. is in all cases *Bōta* pers. n. *Bōtlēah* may have been a wood in which tenants had a right to take timber for *bōt*, i.e. repair, or firewood. Cf. LEEBOTWOOD. Botley Bk and Wa may contain *Botta* or *Beotta* pers. n.

ON **botn** 'bottom, the innermost part of a valley, a bay' is found in **Botton** La [*Bottun* c 1230 CC] and as a second el. in STARBOTTON, WYTHBURN.

Botolphs Sx [*Sanctus Botulphus* 1288 Ass]. 'St. Botulph's church.'

Bottesford Le [*Botesford* DB, *Botlesford* c 1125 LeS, 1203 Ass, *Bodlesford* 1236 Fees], **B~** Li [*Budlesforde* DB, *Botlesforda* c 1115 LiS, *-ford* 1202 Ass, *Botnesford* 1272 FF]. 'Ford belonging to the *bōtl* or house.'

Bottisham Ca [*Bodekesham* 1060 KCD 809, *Bodichessham* DB, *Bodkesham* c 1210 Bodl, *Botekesham* 1254 Val]. '*Boduc*'s HĀM.' *Boduc* is not recorded, but *Boda* must have been in use in OE times; cf. BODENHAM. But the original form of the name may well have been *Beoduc; cf. *Beoduca*, which is recorded.

Bottlesford W [*Botewelleford* 1348 Cor (PNW(S))]. The first el. is a river-name, found as *Botanwælle* 892, *-wylle* 934 BCS 567, 699: '*Bōta*'s stream'.

Bottomstead Brk [*Botelhamested* 1237 Cl, *Bodenhampstede* 1317 Ch]. Cf. HĀMSTEDE. First el. very likely OE *boþm* 'valley'.

Botton. See BOTN.

Botus Fleming Co [*Boflumiet* 1261 Ep, *Botfleming* 1291 Tax, *Bodflumiet* 1318 Ep]. Co *bot* (*bod, bos*) 'house, dwelling' and an obscure element, later replaced by the family name *Fleming*.

Botwell Mx [*Botewælle, -uuelle* 831 BCS 400 f.]. The charter BCS 400 is held to be an original. If so, *Bote-* cannot be a worndown form of *Bōtan*, gen. of *Bōta* pers. n. A fem. name *Bōt* is possible. More likely *Bot-* represents OE *bōt* 'mending', 'remedy'. The name would then mean 'healing spring or brook'.

Boughton, Great, Chs [*Bocstone* DB, *Bocthona* c 1100 Chester], **B~** (-aw-) (or **Bockton**) **Aluph** K [*Boctune* 11 DM, *Boltune* DB, *Botun Alou* 1237 Cl], **B~** under **Blean** K [*Boltune* DB, *Bocton* 1247 Ch, *Bocton juxta la Blen* 1288 Ep], **B~ Malherbe** K

[*Boctun* 11 DM, *Boltone* DB, *Boctun Malerbe* 1275 RH], **B~ Monchelsea** K [*Boctune* 11 DM, *Bocton* 1242 Fees, *Bocton Munchanesy* 1280 Cl]. OE *Bōc-tūn* 'TŪN where beeches grew'.

B~ Aluph was held by one Alulf 1211–12 (RBE). The name represents OG *Adalulf*.—**B~ under Blean.** See BLEAN.—**B~ Malherbe** from the Malherbe family. The name is an uncomplimentary one, which means 'evil herb'.—**B~ Monchelsea** belonged to Warin de Montecan(isio) in 1242 (Fees). The name is French and taken from MONT-CANISI in Calvados, Normandy.

Boughton Hu [*Buchetone* DB, *Bugheton* 1225 FF, *Buweton* 1220 Fees], **B~** Li [*Buchetun* Hy 2 DC, *Buketon* 1195 FF], **B~** Nf [*Buchetuna* DB, *Buche-, Bugeton* 1180, *Buketon* 1197 P], **B~** (-ow-) Np nr Northampton [*Buchetone* DB, *Buketon* 1203 Cur, 1220 Fees], **B~** Np nr Kettering [*Boctone* DB, *Buketon* 1201 Cur, 1236 Fees], **B~** (-ōō-) Nt [*Buchetone* DB, *Buketon* 12 DC]. OE *Buccan tūn* '*Bucca*'s TŪN'.

Boughton Park Wo [*Bolton* 1275 Subs]. Perhaps a variant of BOLTON.

Bouldon Sa [*Bolledone* DB, -*don* 1199 FF, 13 Misc; *Bullardone* 1166 RBE, *Bullardune* 1205 FF, *Bollardedon* 1242 Fees]. The two types of forms seem undoubtedly to refer to Bouldon. Perhaps the original name was *Bulan* or *Bullan dūn*, the first el. being OE *bula* 'bullock' or *Bula, Bulla* pers. n. (cf. *Bulan ham* BCS 213, *Bullanholt* ib. 565). *Bullardone* might then represent an alternative name *Bulwara dūn* 'the DŪN of the people at Bouldon'. Cf. -WARU.

Boulge Sf [*Bulges* DB, *Bulge(s)* 1254 Val]. OFr *bouge* 'uncultivated ground covered with heather and the like'.

Boulmer (-ōōm-) Nb [*Bulemer* 1161 P]. 'Bullock mere.'

Boulsdon Gl [*Buleston* 1220 Fees, *Bolesdon* 1301 Misc]. Second el. DŪN 'hill'. The first might be an OE *bulan-lǣs* 'pasture for bullocks'. Cf. BOLSOVER.

Boulsworth Hill La [*Bulswyre* 14 WhC]. 'Bull's neck.' Cf. SWĪRA.

Boultham Li [*Buletham* DB, *Bulteham* 1202 Ass, *Bultham* 1254 Val]. OE *bulothamm* 'HAMM overgrown with ragged robin'. OE *bulot, bulut* is a plant-name; it is rendered by 'ragged robin' or 'cuckoo-flower'.

Boulton Db [*Boletune* DB, -*ton* 1176 P, *Bolton* c 1250 BM]. '*Bola*'s TŪN.'

Bourn Ca [*Brune* DB, *Brunna* 1194 P, *Brunne* 1196 P], **St. Mary Bourne** Ha [*Burne* 1316 FA], **Bourne** Li [*Brunne* c 960 BCS 1060, *Brune* DB, *Brunna* 1138 NpCh]. OE BURNA 'stream'. *Burna* is from earlier *brunna*, and this form was no doubt in use in some dialects in the OE period. **Bourne** R. See BURNA.

Bournemouth Ha [*la Bournemowþe* 1407

Christchurch Cart]. 'The mouth of the stream' (OE *burna*). There is a small stream here.

Bourton (-oor-) Bk [*Burtone* DB, -*ton* c 1155 Oxf, *Borton* 1248 Cl], **B~** Do [*Bureton* 1212 Fees, *Burton, Buriton* 1243 Ass], **B~ on the Hill** Gl [*Bortune* DB, *Burton* 1206 Cur, *Bourton in Henemerssh* 1415 AD], **B~** (-ōr-) **on the Water** Gl [*Burchtun* 714 BCS 130, *Burgtun, Burhtun* 949 BCS 882, *Burgton* 1251 Ch, *Bortune* DB], **Black B~** O [?*Burtun* 1005 KCD 714, *Burtone* DB, *Burtona* c 1130 Oxf, *Burton* 1196 P], **Great & Little B~** O [*Burton* 1209–12 Fees, *Mucheleburton* 1323 Ipm], **B~** Sa [*Burtune* DB, *Burnton* 1251 Cl], **B~** So nr Wick St. Lawrence [*Burton* 1274 Ipm], **B~ on Dunsmore** Wa [*Bortone* DB, *Burton super Dunnesmor* 1236 Fees]. OE *Burh-tūn*, identical with BURTON (1); cf. BURG.

The exact meaning of the name cannot be determined in each case. **B~ on the Water**, however, is close to an ancient earthwork, **Salmonsbury Camp** [*Sulmonnesburg* 779 BCS 230]. The name may mean 'the BURG of the ploughman', from the enclosure having been used for keeping oxen. OE **sulhman* 'ploughman' may have preceded *ploughman*. **B~ on the Water** is on the Windrush, while **B~ on the Hill** is in higher country.—**B~** Do is near Gillingham, an old borough.—For **Black B~** see BLÆC.—For **B~ on Dunsmore**, see DUNSMORE.

Bourton, Flax, So [?*Bricton* 1276 RH, *Bratton* 1316 FA, *Boryton* 1327 Subs]. If the form of 1276 belongs here, probably OE *Beorg-tūn* 'TŪN by the hill'. In 1276 RH is mentioned *mons de Bricton*. *Flax* presumably refers to flax being grown at the place.

Bouth (bowdh) La [*Bouthe* 1336 FC]. ON *būð* 'booth'.

Bouthwaite YW [*Burtheit* 1184, -*thweit* 1198 Fount]. 'Clearing with a *būr* or storehouse.'

Boveney (bŭvnĭ) Bk [*Bouenie* DB, *Boueneie* 1156 P]. OE *bufan īege* '(the place) above the island'. **B~** is above a small island in the Thames.

Bōveridge Do [*Bovehric* DB, *Boverig* 1256 FF]. OE *bufan hrycge* '(the place) on the ridge'. The place is on a narrow ridge.

Bovey (-ŭ-, -ŏ-) **Tracy, North B~** D [*Bovi* DB, *Buui* a 1093 E, *Northebovy* 1199, *Sutbovi* 1219 FF, *Bovy Tracy* 1276 RH]. The places are on the river **Bovey** [*Boui* 1238 Ass], whose name is British and very likely identical with BOBBIO in Italy (*Bobium* in ancient sources), but of difficult etymology.

B~ Tracy was held by Eva de Tracy in 1219 (Fees); Tracy from TRACY-BOCAGE near Caen or TRACY-SUR-MER near Bayeux.

Bovingdon Hrt [*Bovenden* 1248 Ep, *Buuendon* Hy 3 BM, *Bovindon* 1291 Ch]. OE *bufan dūne* '(the place) on the DŪN or hill'.

Bovington Do [*Bovintone* DB, -*ton* 1236

Fees, 1280 Ch]. 'The TŪN of *Bōfa* or his people.'

Bow D [*Limet* DB, *Nymetboghe* 1270 Ass]. 'Arched bridge.' Bow is on the Yeo, formerly *Nymet*. *Bow* 'arch of a bridge, an arched bridge' is from OE *boga*. Bow Mx was formerly **Stratford le Bow** [*Strafford* 1203 Cur, *Stratford atte Bowe* 1279 FF, Chaucer's Prol.]. *Bow* refers to a bridge built in the time of Henry I. **Bow Brook** Do, Ha, Wo may we contain *bow* in the sense 'bridge'.

Bow Fell Cu [*Bowesfel* 1242 FC]. Cf. also *Bowesscard, Bouescarth* ib. (near Bow Fell). The old name was no doubt OE *Boga* 'the bow'. The hill forms roughly a bow. Cf. FELL, SKARÐ.

Bowcombe Wt [*Bovecome* DB, *Bouecumba* 1186 P]. '*Bōfa*'s valley'; see CUMB.

Bowden Edge Db [*Boudone* 1275 RH, *Boudon* 1339 ff. Derby]. OE *Bog-dūn* or *Boge-dūn* 'arched hill', i.e. 'rounded hill'. Cf. BOWDON. First el. OE *boga* 'bow'.

Bowden, Great & Little, Le [*Bugedone* DB, *-don* 1208 BM, *Buedon* 1169 P, *Buwedon* 1230 P, *Buggedon* 1173 P, 1226–8 Fees, *Magna Bugedon* 1204 Cur, *Parva Bughedon* 1220 Fees]. Either OE *Bucgan dūn* 'the DŪN belonging to *Bucge*' or *Būgan dūn* 'the DŪN belonging to *Būga*'. *Bucge* is a woman's name. The numerous forms with *-gg-* tell in favour of the first alternative.

Bowden W [*Bovedone* c 1240 PNW(S), *Bouedon* 1371 AD]. Identical with BOVINGDON.

Bowderdale Cu [*Boutherdalbeck* 1322 Cl]. ON *Búðar-dalr* 'valley with a booth'; cf. BŌÞE. *Búðar* is the gen. of *búð*.

Bowdon (-ō-) Chs [*Bogedone* DB, *Bowdon* c 1150 Mon]. See BOWDEN Db.

Bower, East & West, So [*Bur, Bure* DB, *Little Bures* 1208 FF, *Estboure* 1391 BM]. OE *būr* 'cottage'.

Bowers Gifford Ess [*Bura* DB, *Bures* 1203 Cur, 1238 Subs]. OE *būr* 'cottage'.

B~ was held by William Giffard in 1243 (FF). Cf. ASHTON GIFFORD.

Bowes YN [*Bogas* 1148 YCh 179, *Bogis* 1171 P, *Bouis* 1241 Ch]. The plural of OE *boga* 'bow', probably in the sense 'arched bridge'.

Bowland (bŏl-) **Forest** La, YW [*Boelanda* 1102 LaCh, *Bouland* c 1140 ib., *Bochlande* a 1194 Kirkst]. 'The land by the bow', probably in the sense 'bend of the river Ribble'. From the forest was named **Bowland** La [*Boelanda* 1102 LaCh].

Bowley He [*Bolelei* DB]. Identical with BOOLEY.

Bowling YW [*Bollinc* DB, *Bolling* c 1150 Crawf, *-es* 1166 P]. It is doubtful if this is a name in *-ing*. It may be a compound

of OE *bolla* and *hlinc* 'hill'. *Bolla* 'bowl' might be used in the sense 'dip in the ground' or the like. There are ponds at Bowling.

Bowmont (bō-) R Nb [*Bolbenda* c 1050 HSC, *-bent* 1293 Ass, *Boubent* c 1540 Leland]. The second el. is probably connected with the subst. *bend*. The first is obscure.

Bowness (-ō-) Cu on Solway Firth [*Bounes* c 1225 Holme C, 1287 Cl]. 'Rounded headland'; first el. OE *boga* or ON *bogi* 'bow'.

Bowness (-ōō-) We [*Bulnysperke* 1390–4, *Bulnes* 1391 Kendale]. 'Bulls' headland.'

Bowood Do [*Bovewode* DB, *Buuewod* 1230 P, *Buwod* 1207 Cur]. '(The place) above the wood', OE *bufan wuda*. B~ W [*Bouewode* 1319 Pat] may be the same name. B~ or **Bowd** D [*Boghewode* 1281 Ep] seems to be OE *Boge-wudu* 'bow wood, wood where bows were got'.

Bowscale Cu [*Bowscalez* 1486 Ipm]. Second el. ON *skáli* 'hut'. The first may be an old name of Bowscale Fell; cf. BOW FELL.

Bowsden Nb [*Bolesdon* 1195 (1335) Ch, *-den* 1196 P, *Bollisdon* 1242 Fees]. Second el. apparently DŪN. The first is possibly OE *bōþl* 'dwelling'.

Bowsley or **Boasley** D [(æt) *Borslea* c 970 BCS 1247 f., *Bosleia* DB]. The first el., as suggested in PND, may be a plant-name *bors*, cognate with *bristle, burr* &c.

Bowston We [*Bolteston* c 1190, 1243 Kendale]. '*Bolt*'s TŪN.' *Boltr* is at least well evidenced as a byname in Scandinavia.

Bowthorpe Nf [*Bowe-, Boethorp* DB, *Boytorp* 1183 Holme, *Bugetorp* 1230 Ch]. '*Bō*'s or *Bōi*'s thorp.' *Bō(i)* is ODan, OSw *Bō*, ON *Búi*.

Bowthorpe YE [*Boletorp* 1200 FF, *Buletorp* 1200 Cur, *Bolethorp* 1241 FF]. '*Bula*'s thorp.' Cf. BOLINGBROKE.

Box Gl [*La Boxe* 1261 Ipm], B~ **Hall** Hrt [*Boxe* DB, 1197 P], B~ W [*Bocza* 1144 AC, *La Boxe* 1182 P]. OE *box* 'box-tree'.

Boxford Brk [*Boxora* 821 BCS 366, *æt Boxoran* 958 ib. 1022, *Bousore* DB]. 'Slope or shore overgrown with box.' Cf. ŌRA.

Boxford Sf [*Boxford* 12 BM, 1254 Val]. 'Ford where box grew.'

Boxgrove Sx [*Bosgrave* DB, *Boxgrava* 1225 BM, *-grove* 1337 Ipm]. Self-explanatory. See GRÁF.

Boxley K [*Boxlea* 11 DM, *Boxeleia* 1130, *-lega* 1157 P]. OE *box-lēah* 'box wood'.

Boxted Ess [*Bocchesteda, Bucchesteda* DB, *Bochesteda* c 1130 Bodl, *Bokestede* 1180 P]. Looks like OE *Bōc-hāmstede* 'homestead among beeches'. See BŌC, HĀMSTEDE.

Boxted Sf [*Boesteda* DB, *Boxsted* 1196 FF, 1242 Fees]. 'Place where box grew.' See STEDE.

Boxwell Gl [*Boxewelle* DB, *Boxwell* 1291 Tax]. 'Stream where box grew.'

Boxworth Ca [*Bochesuuorde* DB, *-wrth* 1199, 1203 Cur, *Buckeswrth* 1200 Cur, *Bukeswrth* 1228 FF]. '*Bucc*'s worp.' Cf. BUXHALL.

Boyatt Ha [*Boviete* DB, *Bouieta* 1167 P, *Boveiat* 1246 Cl]. Probably '*Bōfa*'s gate'; cf. GEAT.

Boycott Bk [*Boicote* DB, *-cota* 1167 P], B~ Sa [*Boicot* 1203 Ass]. '*Boia*'s COT.'

Boyd R Gl [*Byd* 950, 972 BCS 887, 1282]. A British river-name.

Boyland Nf [*Boielund* DB, *Boylund* 1228 FF, 1242 Fees]. '*Boia*'s grove.' Cf. LUND.

Boylestone Db [*Boilestun* DB, *Boilestunia* 12 BM, *Bodeleston* 1203 Cur, *Boyleston* 1256 Ipm]. The first el. looks like a pers. n., but perhaps it is an OE *Bog-hyll* (cf. BOWDEN Db), a name of the ridge on which the place is.

Boynton YE [*Bouintone* DB, *Bovingtona* c 1150 YCh 1306, *Buvington* 1206 FF, *Boyngtona* 1297 Subs]. 'The TŪN of *Bōfa*'s people.'

Boythorpe Db [*Buitorp* DB, *Boythorp* 1234 Fees, 1251 ff. Derby], B~ YE [*Buitorp* DB, *Buithorp* 1166, *Boitorp* 1194 P, *Bothorp* 1284 Cl]. Possibly '*Boia*'s thorp'.

Boyton Co [*Boietone* DB, *Boyton* 1291 Tax], B~ Ess [*Boituna* DB, *Boitton* 1240 FF], B~ Sf [*Boituna* DB, *Boiton* 1196 FF], B~ W [*Boientone* DB, *Boiton* 1167 P, *-a* 1242 Fees]. '*Boia*'s TŪN.'

Bozeat (bōzhet) Np [*Bosiete* DB, *Bosegete* c 1155, *-ʒate* 13 BM]. '*Bōsa*'s gate.' See GEAT.

Brābourne K [*Bradanburna* c 860 BCS 501, *Bradeburne* DB]. 'Broad stream.'

Braceborough Li [*Braseborg*, *Breseburc* DB, *Bresseburc* 1180, *Bressenburc* 1194 P, *Bresseburg* 1206 Ass, *Bressingburg* 1212 Fees, *Bresingburg* 1254 Val]. The first el. is doubtless the OE poetic word *bresne*, *bræsne* 'bold, mighty, strong', used as an epithet of a king and of a city (*bresne Babiloniʒe*). OE *bræsne burg* 'strong fort' is possible, but more likely *bræsne* is here substantivized, meaning 'the bold one(s)', 'the warrior(s)', or used as a byname or Christian name. An OE *Bræsnan burg* or *Bræsnena burg* would explain all later forms.

Bracebridge Li [*Brachebrige*, *Bragebruge* DB, *Bracebrig* c 1160 RA, 1212 Fees, *-brigg* 1198 P, *Brascebrigg* 1230 Ch]. The name seems to be identical in meaning with Ris-bridge, the first el. being dial. *brash* 'small branches, twigs', a word cognate with Sw *bräske* 'dried twigs', Norw *brisk* 'juniper'. The forms show Norman influence.

Braceby Li [*Breiʒbi*, *Breʒbi* DB, *Breicebi* 1212 Fees]. OScand *Breiðsbȳr*. First el.

ON *Breiðr*, ODan *Breth*, OSw *Bredh* pers. n., from *breiðr* 'broad'.

Bracewell YW [*Braisuelle* DB, *Braycewell* 1147–50 YCh 1471]. Either '*Brægd*'s spring or stream' (cf. BREADSALL) or '*Breið*'s spring or stream' (cf. BRACEBY).

Bracken YE [(æt) *Bracenan* c 972 BCS 1279, *Brachen* DB, *Brakene* 1195 Cur]. 'The brackens', the dat. plur. of OE **bræcen* or OScand *brakni* 'bracken'. The word *bracken* is generally held to be a Scandinavian loan-word.

Brackenborough Li [*Brachenberg* DB, *Brakenberga* 1150–60 DC], **Brackenfield** Db [*Brakinweyt* 1273, *Brakenthweyt* 1329 Ipm], **Brackenholm** YE [*Bracheneholm* DB]. 'Hill, thwaite or clearing, holm or island overgrown with bracken.'

Brackenthwaite Cu [*Bracanethuait* 12 StB, *Brakanthuayt* 1332 Subs]. 'Bracken clearing.' There are other Brackenthwaites in Cu. One nr Wigton is *Brankanthweyt* 1255 P.

Brackley Np [*Brachelai* DB, *Brackelea* 1173, *Bracchelea* 1182, *Brackeley* 1230 P], **Bracknell** Brk [*Braccan heal*, (of) *Brachan heale* 942 BCS 778, *Brakenhale* 15 AD]. See LĒAH, HALH. First el. a pers. n. *Bracca* or *Breahha*. The latter is the source of BRAUGHING.

Bracon Ash Nf [*Brachene* 1175 P, *Brakene* 1230 Ch]. Identical with BRACKEN. Ash is no doubt *ash* the tree.

OE **brād** 'broad, wide' is common as a first el. in pl. ns. The meaning is sometimes 'broad' as in BRADWELL, BROADWAY, sometimes 'wide, spacious', as in BRADFIELD, BRADLEY. See BRABOURNE, names in BRAD- and BROAD-, BRAYBROOKE, BREDGAR. When used as a distinguishing epithet, as in BROAD BLUNSDON, *Broad* seems to have much the same meaning as *Great, Much*. In BRAITH-WELL, BRAYSTONES, BRAYTON OE *brād* seems to have been replaced by OScand *breiðr*, which is the first el. of BRAITHWAITE, BRATHAY, BRATOFT, BRAWITH.

Bradbourne Db [*Bradeburne* DB, Hy 2 DC]. 'Broad stream.' Cf. *aqua de Brade-burn* 1281 Ass.

Bradbury Du [*Brydbyrig* c 1050 HSC, 1104–8 SD, *Bradbery* 1183 BoB]. OE *Bred-burg* 'BURG built of boards'. The early *y* for *e* has an analogy in *Cytel* for ON *Ketill* in HSC.

Bradden Np [*Bradene* DB, *Braddene* 12 NS, 1230 Ep, *Bradenden* 1186 P, *Bradden* 1220 Fees]. OE *brāde denu* 'broad valley'.

Braddock or **Broadoak** Co [*Brodehoc* DB, *Bretohk* 1224, *Brothok* 1291 FF, *Brothek* 1316 Ipm]. Possibly an English name meaning 'broad oak'. If so, the early *o* for OE *ā* may be due to substitution of Co *ō* for OE *ā*.

Brădenham Bk [*Bradeham* DB, 1195 Cur, *Bradenham* 1242 Fees], **B~** (-ăd-) Nf [*Brade(n)ham* DB, *Westbradeham* 1197 P, *Estbradeham* 1242 Fees]. 'Wide or extensive HĂM or HAMM.'

Brădenstoke W [*Bradenestoche* DB, -*stoke* 1195 Cur]. 'The STOC belonging to BRAY-DON.' B~ was the site of a monastery, and *stoc* may here mean 'holy place'. See STOC.

Bradeston or **Braydeston** Nf [*Bregestuna* DB, *Breydeston* 1252 Ch, 1322 AD]. '*Brægd*'s TŪN'; cf. BREADSALL.

Bradfield Brk [*in Bradanfelda* 688–90 BCS 74, *Bradefelt* DB], **B~** Ess [*Bradefelda* DB, -*feld* 1196 P], **B~** Nf [*Bradefeld* 1212 Fees], **B~ Combust, St. Clare & St. George** Sf [*Bradefelda* DB, -*felde* c 1095 Bury, -*feld* 1197 FF, *Bradefeud Sencler, Monachorum, Parva* 1254 Val], **B~** YW [*Bradefeld* 1275 Cl]. 'The wide FELD.'

B~ St. Clair was held by John le Seyncler 1253 Ipm.—**B~ St. George** presumably from the dedication of the church.—**Combust** means 'burnt'. The place must have been burnt down at some early period.

Bradford Chs [*Bradeford* 1119 Chester], **B~** D [*Bradeford* DB], **B~ Abbas** Do [*æt Bradanforda* 933 BCS 695, *Bradeford* DB], **B~ Bryan** Do [*Bradeford* 1236 Fees, *B~ Brian* 1289 FF], **B~ Peverell** Do [*Bradeford* DB, 1196 f. P, *Bradeford Peverel* 1275 FF], **B~ La** [*Bradeford* 1196 FF], **B~** Nb nr Bamburgh [*Bradeford* 1212 Fees, 1266 Ipm], **B~** Nb nr Bolam [*Bradeford* 1242 Fees, 1272 Ipm], **B~** So [(fram) *Bradan forda* 882 BCS 550, *Bradeford* DB], **B~ on Avon** W [(æt) *Bradanforda be Afne* 652 ASC, *Bradeford* DB], **B~** YW [*Bradeford* DB], **West B~** YW [*Bradeforde* DB, *Braford in Bouland* 1251 Cl]. 'Broad ford.'

B~ Abbas was held by Sherborne Abbey.—**B~ Bryan** from Brian de Insula. Cf. BRYANS-TON.—**B~ Peverell** was given to Robert Peverell in 1200 (Ch). Cf. BARTON PEVEREL.

Bradgate Le [*Bradegate* c 1275 AD]. 'Broad gate'; cf. GEAT.

Bradiford D [*Bradeford* 1222 FF]. Identical with BRADFORD.

Brăding Wt [*Berardinz* DB, *Brerdinges* 1254 Ep, *Brerding* 1287–90 Fees]. Probably an OE *Brerding*, a derivative of *brerd* 'brim, border'. It may be a name of Brading Down, a ridge near Brading.

Bradkirk La [*Bredekirke* 1235 FF]. 'Plank church.' First el. OE *bred* 'board'. The second el., OScand *kirkia*, has probably replaced OE *cirice*.

Brădle Do [*Bradelege* DB, -*leye* 1285 FA], **Bradley** Brk [*Bradeleia* W 1 Abingd, *Bradelea* 1167 P], **B~** Chs nr Frodsham [*Bradeleg* 1278 Ipm], **B~** Chs nr Whitchurch [*Bradeleg* 1259 Court, 1290 Ipm], **B~** Db [*Braidelei* DB, *Bradleye, Bradele* c 1200 Derby], **B~** Du [*Bradeley* 1322 Ipm], **B~** Gl [*Bradele* 1271 Ch], **B~** Ha [(æt) *Bradan-*

leáge 909 BCS 625, *Bradelie* DB], **B~** Le [*Bradele* 1254 Val], **B~** Li [*Bredelou, Bradelai* DB, *Bredelai* c 1115 LiS, *Bradelai* 1170, 1197 P], **Great & Little B~** Sf [*Bradeleia* DB, *Parva Bradel'* 1199 Cur, *Bradeleya Magna, Parva* 1254 Val], **West B~** So [*Bradelega* 1196 P], **B~** St nr Stafford [*Bradelea* DB], **B~** St nr Burslem [*Bradelie* DB], **B~** St nr Dudley [*Bradeleg* DB, -*leye* 1290 Misc], **Maiden B~** W [*Bradelie* DB, *Meydenebradele* 1267 Ch], **North B~** W [*Bradeleye* 1291 Tax], **B~ Field** or **Brad-leyfield** We [*Brathelaf* 1272 Ipm, *Braithlagh* 1324 Ipm, *Bradelay* 1292 Ch], **B~** Wo [*Bradanlæh* c 730 BCS 153, (æt) *Bradan-leage* 803 BCS 308], **B~** YW nr Hudders-field [*Bradeleia* DB, c 1180 YCh 1692], **B~** YW nr Skipton [*Bradelei* DB]. 'Wide LĒAH.'

The early forms of **B~** We show much Scandi-navian influence. **Maiden B~** means 'maidens' Bradley'. There was a monastery in the place with a hospital. The infirm women of *Bradele* are mentioned 1227 Ch.

Bradley in the Moors St [*Bretlei* DB, *Bredleye* 1274 Ipm, 1327 Subs]. OE *bred-lēah* 'wood where boards were got'.

Bradmore Nt [*Brademere* DB, -*mar* 1226–8, 1242 Fees]. 'Wide lake.' There is no longer a lake here.

Bradney So in Bawdrip [*Bredeneia* DB, *Bradenye* 1243 Ass, *Bradeneye* 1324 Ipm]. 'Broad island.'

Bradninch D [*Bradenese* DB, -*nes* 1227 Ch, *Bradenech* 1238 Cl, *Bradenesshe* 1349 Ep]. Probably 'broad oak' (OE *brādan āc*, dat. of *āc* 'oak').

Bradnop St [*Bradenhop* 1219 FF, 1233 BM]. 'Broad valley.' See HOP.

Bradnor He [*Bradenore* 1337 Ipm]. 'Broad slope or hill.' Cf. ŌRA, OFER.

Bradon, South, So [*Bredde, Brede, Bredene* DB, *Bradene* 1266 Ep, *Sut Braden* 1266 Ep, *North-, Suthbradene* 1320 Ipm], **Goose Bradon** So [*Brade* DB, *Gosebraden* 1292 FF]. Bradon is apparently identical in origin with BRAYDON.

Goose is a family name, spelt (de) *Gouiz* or *Guuiz* in early sources. Cf. GOUVIX in Calvados (Normandy).

Bradpole Do [*Bratepolle* DB, *Bradepol* 12 Fr, 1212 Fees]. 'Broad pool.'

Bradshaw Db [*Bradschag* 1345 Derby], **B~ La** [*Bradeshaghe* 1246 Ass], **B~** YW. 'Extensive grove.' See SCAGA.

Bradsole K, old abbey [*Bradesole* 1204, 1227 Ch, 1234 Cl]. 'Broad mire or wallow-ing-place.' Cf. SOL.

Bradstone D [(æt) *Bradan stane* c 970 BCS 1247, *Bradestana* DB]. 'Broad stone.'

Bradwall Chs [*Bradewell* 1281, -*wall* 1289 Court]. 'Broad stream.' Cf. WELLA.

Bradway Db [*Bradeweye* 1300 Ipm]. 'Broad road.'

Bradwell Bk [*Bradeuuelle* DB, *-welle* 1151–4 Fr], **B~** Db [*Bradewelle* DB, *-well* 1230 P], **B~** Ess nr Braintree [*Bradewell* 1238 Subs, *-e* 1254 Val], **B~ juxta Mare** or **B~ near the Sea** Ess [*Bradewella* 1194 P, *-well* 1212 Fees], **B~** Sf [*Bradewell* 1211 FF, *Bradwell* c 1210 Bodl], **B~** St [*Bradewull* 1227 Ch]. 'Broad stream.'

The two Bradwells Ess are miles apart, but both on the Blackwater. Evidently the Blackwater was once known as Bradwell. Bradwell juxta Mare is *Effecestra* DB. This is the old name of the place, *Ythancaestir* c 730 Bede, *Yŏpanceaster* c 890 OEBede. This name has been identified with *Othona* in ND, but more likely *Yppan* is the gen. of a pers. n. *Yppa* or *Yppe* (fem.). Such a name would be a short form of OE names in *Yp-*. Only one is known, the fem. name *Ypswip*, but in OHG the corresponding name-stem *Und-* is common.—**B~** Sf is near Breydon Water, a widening of the Yare.

Bradworthy D [*Brawordine* DB, *Bradewurŏa* 1175 P, *Brawrthi* 1198 FF]. 'Broad WORP(IGN).'

OE **bræc** occurs occasionally in charters, but the exact meaning of the word is not clear. Very likely two words are represented here, OE *bræc* (with short vowel), the source of modern *brake* 'a copse, thicket', and OE *bræc* (with long vowel), identical with OHG *brāhha*, MLG *brāke* 'ground broken up for cultivation' and the source of ME *breche* in the Owl and the Nightingale. The latter word is probably found in BIRCHAM, BRAXTED, BRECKLES and in some BRATTONS and BRETTONS.

OE **brǣdu.** See BREDE, BREDFIELD, SHULBREDE, WEYBREAD.

OE **brǣr, brēr** 'briar' is not rare in pl. ns. as the first el. Cf. BREARTON, BREARY, BRERETON, BRIER- (passim).

Brafferton Du [*Bradfortuna* 1091 FPD], **B~** YN [*Bratfortune, Bradfortune* DB]. 'TŪN by the broad ford.'

Brāfield on the Green Np [*Bragefelde, Brache(s)feld* DB, *Bragefeld* 1163 P, *Brawefeld* 1220 Fees, *Braunfeld* 12 NS], **Cold Brayfield** Bk [*æt Bragenfelda* 967 BCS 1209, *Bragefeld* 1185 P]. The two places are near each other and were named from the same FELD. The name means 'FELD by the hill'. The first el. is OE *bragen* 'brain', probably also 'crown of the head' and in transferred use 'hill'.

Braham YW nr Spofforth [*Michel-, Litelbram* DB, *Bram* 1198 P, *Braham* 1242 Fees]. Perhaps OE *bram-hamm* from *brōm-hamm* 'HAMM covered with broom'. Cf. BRŌM.

Brailes Wa [*Brailes* DB, 1130 P, *Bragels* Hy 1 Abingd, *Brailis* 1175 P], **Brail(e)s-ford** Db nr Derby [*Brailesford* DB, *Braylesford* 1242 Fees], **Brailesford** Db nr N. Wingfield [*Braylesford* 1330 QW, 1356 BM], **Brailsham** Sx [*Breilesham* 1230 P]. Brailes appears to represent an OE **brægels*, which may be from **bærgels*, a side-form of

byrgels 'burial-place'. The meaning would be 'tumulus'. For the metathesis cf. BRAINSHAUGH, BROXTON. Brailsford, Brailsham contain the same element.

Brainshaugh Nb [*Bregesne* 1104–8 SD, *Brainesleie* n.d. FPD]. From a side-form of *borrans* 'burial-mound'. *Borrans* goes back to OE **burgæsn* or the like. *Bregesne* 1104–8 seems to presuppose an OE **bergesn* from **bargisnō*, a derivative of *beorgan* 'to protect'.

Braintree (-ahn-) Ess [*Branchetreu* DB, *Branketre* 1274 RH]. Second el. OE *trēo* 'tree'. The first is probably a pers. n., possibly related to *Branuc* in BRANSCOMBE D.

Braiseworth Sf [*Briseworde* DB, *-wurd* 1196 P, *-wrth* 1201 Cur, *Briseworth, Bruswrthe* 1254 Val]. The first el. is OE *briosa* 'gadfly', but probably here used as a nickname. See WORP.

Braishfield All Saints Ha [*Braifeld* 1201 Cur, *Braysfeld* 1282 Ep, *Braisfelde* 1346 FA, *Breysfelde* c 1286 Ep]. First el. possibly identical with BRAINSHAUGH.

Braithwaite, High & Low, Cu [*Braythawayt* 1317 Pat], **B~** Cu nr Portinscale [*Braithait* 1210–16 CWNS xxi, *Braythwayt* 1292 Cl], **B~** Cu nr Brigham [*Braythweit* 1315 Ipm], **B~** YN [*Braytwayt* 1301 Subs]. 'Broad clearing'; see THWAITE. The name is Scandinavian. First el. ON *breiŏr* 'broad'. **Braithwaite** YW (2 different) are no doubt identical in origin.

Braithwell YW [*Bradewelle* DB, *Braythewelle* 1289 Ch]. A Scandinavianized form of BRADWELL.

Bramber Sx [*Bremre* 956 BCS 961, *Brembre* DB, *Brenbria* 1144 Oxf]. OE *brēmer* 'bramble thicket'.

Brambridge Ha [(æt) *Brombrigce* 909, *Brombrygc* 10 BCS 620, 1160]. 'Broom bridge.'

Bramcote Nt [*Broncote, Brunecote* DB, *Bramcote* 1197 P], **B~** Wa [*Brancote* DB, *Bromcote* 1285 QW]. 'COT among the broom.'

Bramdean Ha [(to) *Brómdæne* 932 BCS 689, 1045 KCD 780, *Brondene* DB]. 'Valley where broom grew.'

Bramerton Nf [*Brambretuna* DB, *Bramerton* 1254 Val]. 'TŪN among brambles.' Cf. BRAMBER.

Bramfield Hrt [*Brandefelle* DB, *Brantefeld* 1209–19 Ep, *Brantesfeld* 1254 Val]. 'Steep FELD' (first el. OE *brant* 'steep') or '**Brant's* FELD'.

Bramfield Sf [*Brunfelda, Brumfella* DB, *Bramfeld* 1166 P, *Bramfeud* 1254 Val]. 'Broom FELD.'

Bramford Sf [*Bromford* 1040 Wills, *Branfort* DB, *Bramford* 1198 FF]. 'Broom ford.'

Bramhall Chs [*Bramale* DB]. 'Broom HALH.'

Bramham YW [*Brame-, Bramham, Braham* DB, *Brumham* 1081 YCh 1002]. 'HĀM or HAMM where broom grew.'

Bramhope YW [*Bram-, Brahop* DB, *Bramhope* 1240 FF]. 'Broom HOP.'

Bramingham Bd [*Bramblehangre* 1240 Ass]. 'Bramble slope.' See HANGRA.

Bramley Db nr Baslow [*Bromleye* 1239 Derby], B~ Ha [*Brumelai* DB, *Bromelege* c 1160 Oxf], B~ Sr [*Brun-, Bronlei* DB, *Bromlega* 1170 P], B~ YW nr Leeds [*Brameleia* DB, *Bramley* 1218 Kirkst], B~ **Grange** YW nr Kirkby Malzeard [*Brameleia* DB, *Bramleia* 12 Fount], B~ YW nr Rotherham [*Bramelei* DB, *Bramele* 1218 FF]. OE *brōm-lēah* 'clearing overgrown with broom'. For the Yks Bramleys, derivation from OE *brǣmel-lēah* (cf. *brēmel, brǣmel* 'bramble') is also possible.

Brampford Speke D [*Branford* DB, *Bramford* 1194 P, *Bramford Spec* 1275 RH]. Apparently OE *brōmford* 'ford where broom grew'. But if B~ is identical with *Brenteforlond* 944 BCS 799 (see PNNt(S) xxxvii), the old name was *Brenteford*, and *Brente* may have been the name of a stream here. Cf. BRENT R. The name was then later refashioned to *Brōmford*.

The manor was held by Richard de Espec c 1170. The usual form of the name is *Lespec* or *Le Espek*, from OFr *espech* 'wood-pecker'.

Brampton Cu [*Braunton* 1252 Ch, *Brampton* 1291 Tax], B~ Db [*Brantune* DB, *Bramton* 1179 P], B~ **Abbotts** He [*Bruntune* DB, *Bromtun* 1242 Fees], B~ **Bryan** He [*Brantune* DB, *Bramptone Brian* 1275 Ep], **Great** B~ He [*Bruntune* DB, *Bramtona* 1132 PNHe], **Little** B~ He [*Bruntune* DB, *Brompton* 1287 Ipm], B~ Hu [*Brantone* DB, *-tona* 1130, *Bramtona* 1168 P], B~ Li [*Brantune* DB, *Brantuna* c 1115 LiS], B~ Nf [*Brantuna* DB, *Bramptone* 1254 Val], B~ **Ash** Np [*Brantone* DB, *Bramton* 1220 Fees], **Chapel & Church** B~ Np [*Brantone* DB, *Bramton et alia Bramton* 1220 Fees], B~ Sf [*Brantuna* DB, *Branton* 1198 Cur], B~ We [*Branton* 1208 FF, *Brampton* 1283 Cl], B~ **Bierlow** YW [*Brantone* DB, *Bramton* 1234 FF, *Bramptonbyryalgh* 1364 BM], B~ **en le Morthen** YW [*Brantone* DB, *Bramton in Moring* 1297 Subs]. OE *Brōm-tūn* 'TŪN where broom grew'.

B~ **Abbotts** belonged to St. Peter's, Gloucester.—*Bierlow* in B~ **Bierlow** is an ON *býiarlǫg*, corresponding to OSw *byalag* 'village community'.—B~ **Bryan** was held by Brian de *Brantone* in 1185 (TpR). *Brian* is a Norman Christian name of Breton origin.—B~ **en le Morthen**. See MORTHING.

Bramshall St [*Branselle* DB, *Brumeshel* 1195 P, *Bromschulf* 1327 Subs]. OE *brōm-scylf* 'broom-covered slope'.

Bramshaw Ha [*Bramessage* DB, *Bremblessath* 1212 Fees, *Brembelshawe* 1341 Misc]. 'Bramble wood.'

Bramshill Ha [*Bromeselle* DB, *Bromeshell* 1167 P, *-hill, Bremmeshill* 1205 Cur]. Perhaps identical with BRAMSHALL, though the second el. may also be OE *gesell* 'shelter'.

Bramshott Ha [*Brenbresete* DB, *Brembelsite* 1242 Fees, *Brembelshete* 1272 Ass]. OE *brǣmel-scēat* 'SCĒAT where brambles grew'.

Bramwith YW [*Branuuode, Branuuithe* DB, *Bramvith* 1201 FF]. OE *brōmvuudu* 'broom wood', later with OScand *viðr* 'wood' for OE *wudu*.

Brancaster Nf [*Bramcestria* c 960 BCS 1059, *Broncestra* DB, *Bramcestre* c 1110 BM, *Brancestr'* 1170 P]. See CEASTER. The first el. is generally taken to be the OBrit *Branodunon*, which is held to have been at Brancaster. At any rate, if this is right, early association with OE *brōm* took place.

Brancepeth Du [*Brentespethe* 1085 DST, *Brandespethe* 1155 FPD]. B~ is situated at the hill by BRANDON. The probability is that Brancepeth means 'the path to BRANDON', OE *Brōmdunes-pæþ*.

Brandesburton. See BURTON.

Brandeston Sf [*Brantestona* DB, *Branteston* 1195 FF, 1235 Fees], **Brandiston** Nf [*Brantestuna* DB, *-ton* 1203 Cur]. '*Brant*'s TŪN.' OE *Brant* is found in many pl. ns. Cf. *Branteswyrð* 937 BCS 712, BRANSTON Le &c.

Brandon Du [*Bromdune* c 1190 Godric, *Brandun* 1217 Pp], B~ **Parva** Nf [*Brandun* DB, *Brandon* 1199 FF], B~ Sf [*Brandune* c 1050 KCD 907, *Bromdun* 10 Thorney Fragm, *Brandona* DB], B~ Wa [*Brandune* DB, *Brandon* 1236 Fees]. OE *brōm-d:in* 'broom hill'.

Brandon Li [*Branðon* 1060–6 KCD 818, *Brandune* DB]. This may be OE *brōm-dūn* like most Brandons. But this Brandon is on the river BRANT, and the first el. may be the river-name Brant.

Brandon Nb [*Bremdona* c 1150 Percy, *-don* 1293 QW, *Bromdun* 1236, 1242 Fees]. Partly OE *brōm-dūn* 'broom hill', partly an OE *brēmen-dūn* with an adj. *brēmen* 'broomy' as first el.

Brandsby YN [*Branzbi* DB, *Branndesby* 1227 Ep]. '*Brand*'s BY.' *Brand* (in DB &c.) is ON *Brandr*, ODan, OSw *Brand*.

Brandwood La [*Brendewod* c 1200 WhC]. 'Burnt wood.'

Branksea. See BROWNSEA.

Bransbury Ha [*Brandesburi* 1046 KCD 1335, *Brandesberee* DB]. '*Brand*'s BURG.' The name *Brand* is possibly Scandinavian.

Bransby Li [*Branzbi* DB, c 1115 LiS, *Brandesby* 13 BM]. See BRANDSBY.

Branscombe D [*Branecescumb* c 880 BCS 553, *Brancescumb* c 1070 Ex, *Branchescome* DB]. '*Branoc*'s valley.' *Branoc* is a British name (OWelsh *Branoc*) derived from *bran* 'raven'.

Bransdale YN [*Brandesdal* 1219 Ass]. '*Brand*'s valley'; cf. BRANDSBY.

Bransford Wo [*Bregnesford* 963 BCS 1106, *Bradnesforde* DB]. First el. possibly identical with BRAINSHAUGH. But *Bragenmonna broc* BCS 1107 may be connected with the name Bransford. If so, it has as first el. OE *brægen* 'brain' in the sense 'crown of the head, hill'. Cf. BRAFIELD.

Branston Le [*Brantestone* DB, *Branteston*, *Brandeston* 1190 P, *Brantiston* 1254 Val], B~ St [*Brontiston* 942 BCS 771, *Brantestone* DB, *Branteston* 1230 P]. '*Brant*'s TŪN'; cf. BRANDESTON.

Branston Li [*Branztune*, *Branztone* DB, *Branteston* 1200 Cur]. Possibly identical with the other Branstons, but more likely OScand *Brands-tūn* '*Brand*'s TŪN'; cf. BRANDSBY.

Brant R Li [*Brant* 1316 Ipm]. From OE *brant* 'steep'.

Brantham Sf [*Brantham* DB, *Braham* 1198 Cur, 1254 Val]. Apparently '*Brant*'s or **Branta*'s HĀM'. Cf. BRANDESTON.

Branthwaite Cu nr Workington [*Bromthweit* 1210 Cur, *-twetth* c 1230 StB], B~ Cu nr Caldbeck [*Braunthwait* 1332 Subs]. 'Broom-covered clearing'; cf. THWAITE.

Brantingham YE [*Brenting(e)ham* DB, *Brantingeham* 1202 FF]. 'The HĀM of *Brant*'s people.' Cf. BRANDESTON.

Branton Nb [*Bremetona* c 1150 Percy, *Bremtun* 1236, *-ton* 1242 Fees]. 'TŪN where broom grew.' First el. OE **brēmen* 'broomy'; cf. BRANDON Nb.

Branton YW nr Doncaster [*Brantune* DB, *Bramton* 1240 FF], B~ Green YW [*Brantona* 1157 YCh 354, *Brampton* 1285 FA]. OE *Brōm-tūn* 'TŪN where broom grew'.

Brantwood La [?*Brentwode* 1356 FC]. 'Burnt wood.'

Branxton Nb [*Brankeston* 1195 (1335) Ch, 1242 Fees, *-a* 1202 FF]. '*Branoc*'s TŪN'; cf. BRANSCOMBE.

Brascote Le [*Brocardescote* DB, *Brokardescote* 1280 Misc]. '*Brocheard*'s COT.' OE *Brocheard* is found in *Brochardes ford* BCS 1282, *Brocheardes hamm* ib. 1343.

Brassington Db [*Branzinctun* DB, *Brancinton* 1195 Cur, *Bracynton* 1251 Derby]. This may be OE *Brantstīg-tūn* 'TŪN by the steep path'. The place is on a steep slope. Cf. **Bransty** Cu [*Bransty* 1200–13 StB, *Branci*, *Brancy* 1279 Ass, *Brantsty* 1322 Ipm]. 'Steep path.'

Brāsted K [*Bradestede* 11 DM, *-steda* 1184 P, *-sted* 1235 Ch]. 'Broad place.'

Brathay (-ādh-) R La, We [*Braitha* c 1160 LaCh, 1196 FF]. ON *Breið-á* 'broad river' (ON *breiðr* 'broad' and *á* 'river').

Brātoft Li [*Breietoft* DB, *Breitoft* c 1115 LiS]. 'Broad toft.'

Brattleby Li [*Brotulbi* DB, *Brotolbi*, *Brotulebi* c 1115 LiS]. The first el. looks like an OScand *Brotulfr* pers. n. Such a name is not on record, but it may be *Ulfr* with a prefix *Brot*; cf. the ON byname *Brotamaðr*, which is explained either as 'the epileptic' or 'the boisterous fellow'. ON *brot* means lit. 'breaking'.

Bratton Clovelly D [*Bratona* DB, *Bratton* 1229 FF, *Bracton* 1330 Ipm, *Bratton Clavyle* 1280 Ipm], B~ Fleming D [*Bratona* DB, *Bratton* 1242 Fees, *Bretton* 1285 Pat], B~ So nr Minehead [*Bratone* DB, *Bracton* 1194, *Bratton* 1195 P], B~ W [*Bratton* 1178 P, 1242 Fees, *Bracton* 1195 Cur, *-e* 1212 RBE, *Bretton* 1250 Fees]. OE *Bræc-tūn* 'newly cultivated TŪN'; cf. BRÆC.

B~ **Clovelly** was held by Roger de Clavill in 1254 (Cl). Claville is a Fr family name (from CLAVILLE Eure or Seine-Inf.).—B~ **Fleming** was held by Baldwin le Flemeng in 1242 (Fees).

Bratton Sa [*Brochetone* DB], B~ Seymour So [*Broctune* DB, *Brocton* 1195 P, 1242 Fees]. 'TŪN on the brook.'
Seymour is the Fr family name Saint Maur (from one ST.-MAUR in France). Roger de St. Maur is said to have acquired the manor c 1400.

Braughing (-ăf-) Hrt [*Breahingas* 825–8 Bodley MS, (ad) *Brahcingum* 944–6 BCS 812, *Brachinges* DB, *Brahing* 1200, *Brakinges* 1208 Cur]. '**Breahha*'s people'; cf. BRACKLEY.

Brauncewell Li [*Branzewelle* DB, *Brancewella* c 1158 Add Ch 6038]. '*Brand*'s stream.' Cf. BRANDSBY.

Braunston Np [*Brantestun* 956 BCS 978, *Brandestone* DB, *Branteston* 1220 Fees], B~ Ru [*Branteston* 1167 P, 1206 Cur], **Braunstone** Le [*Brantestone* DB, *Branteston* 1258 Ipm]. Identical with BRANSTON (1) and BRANDESTON.

Braunton D [*Brantona* DB, 1158 P, *Bramtona* 1168 P]. OE *Brōm-tūn*; see BRAMPTON.

Brawby YN [*Bragebi* DB, c 1150 YCh 782]. '*Bragi*'s BY.' *Bragi* is an ON pers. n.

Brawith (-ĕ-) YN [*Braithewat* 1208 Cur]. 'Broad ford' (ON *breiðr* 'broad' and *vað* 'ford').

Braxted Ess [*Brachesteda* DB, *Magna Bracsted* 1206 Cur, *Parva Bracstede* 1254 Val]. See STEDE. First el. either OE *bræc* 'thicket' or *bræc* 'newly cultivated land'. See BRÆC.

Bray Brk [*Brai* DB, 1156 P, *Bray* 1190 P]. OE *brēg* 'brow', here 'brow of a hill'.

Bray, High, D [*Bræg* c 970 BCS 1253, *Braia* DB, *Brai* 1121 AC, *Hautebray* 1242 Fees]. Either identical with BRAY Brk, or from Welsh, Co *bre* (< **breg*) 'hill'. The river-name **Bray** [*Bray* 1249 Ass] is a back-formation.

Braybrooke Np [*Bradebroc* DB, *Braibroc*

1163 P, *Brabroc* 1197 FF]. OE *Brāda-brōc* 'the broad brook'. The later development *Brade- > Brae- > Braie-* is abnormal.

Braydeston Nf. See BRADESTON.

Braydon W [*Bradon* (silva) 688 BCS 70, *Bradene* 956 ib. 922, (on) *Bradene* 905 ASC, *Bradenebroc* 1228 Cl, (foresta de) *Bradene* 1250 Fees]. Apparently identical with BRADON. Probably a pre-English name. Etymology obscure.

Brayfield Bk. See BRAFIELD.

Braystones Cu [*Bradestanes* 1247 StB, 1279 Ass, *Braythestanes* 1294 Cl]. OE *brādan stānas* 'broad stones', Scandinavianized.

Brayton Cu [*Breyton* 1255 P, *Brayton* c 1275 StB], B~ YW [*Breiðe-tun* c 1030, *Braiþatun* c 1050 YCh 7, 9, *Bretone, Brettan* DB, *Brattuna* c 1080 YCh 468]. 'Broad TŪN.' First el. OScand *breiðr* 'broad', which may have replaced OE *brād*.

Breadsall Db [(æt) *Bregdeshale* 1002 Wills, *Brægdesheale* 1004 KCD 710, *Braideshale* DB, *Breideshale* Hy 2 DC]. '*Brægd*'s HALH.' OE *Brægd* pers. n. is not recorded, but is presupposed also by BRADESTON Nf and BREASTON Db. The name is related to OE *brægd* 'trick, deceit', *brægden* 'crafty'.

Breadstone Gl [*Bradeneston* 1273 Misc]. Probably 'broad stone'.

Breage (-ēg) Co [*Vicaria Sancte Breace* 1264 Ep, *St. Breaca* 1326 Ch]. A Cornish saint's name. Breage is elliptical for St. Breage's church.

Breamish R Nb [*Bromic* c 1050 HSC, *Bremyz* 1293 Ass]. A British river-name related to Welsh *brefu* 'to roar', Lat *fremo*; cf. Welsh *Afon Brefi*.

Breamore Ha [*Brumore* DB, *Brummora* 1167 P, 1219 Fees, *Brommore* 13 BM]. OE *brōm-mōr* 'moor covered with broom'.

Brean So [*Brien* DB, *Breene* 1243 Ass, *Broen* 1254 Ass, *Bren* 1212 RBE, *Breon* 1334 Ep]. Seems to presuppose an OE **Brēon*, perhaps **Brēo*, gen. *Brēon*. This looks like the el. *Brēo-* in early forms of BRE(E)DON, which comes from OBrit **brigā*, Welsh *bre* 'hill'. Brean may go back to a derivative of Brit *brigā*. The name probably referred originally to BREAN DOWN, a high headland.

Brearton YW [*Braretone* DB, *Brierton* 1187 P, *Brertona* 1202 FF]. OE *Brēr-tūn* 'TŪN where briars grew'. Cf. BRÆR.

Breary YW [*Brerehagh* 1285 Ch]. 'Briar enclosure.' Cf. BRÆR, HAGA.

Breaston Db [*Braidestune* DB, *Breydiston* 1242 Fees, -*e* 1282 BM]. '*Brægd*'s TŪN'; cf. BREADSALL.

Breckles Nf [*Brecc(h)les* DB, *Brecles* 1254 Val]. OE *bræc-læs* 'meadow by newly-cleared land'; cf. BRÆC.

OE **bred** 'board, plank' is the first el. of BRAD-, BREDBURY, BRADKIRK, BRADLEY (in the Moors), BREDHURST, BURDALE, perhaps BERECHURCH, BRETFORD. The derivative **briden** 'of boards' is found in BREDENBURY.

Bredbury Chs [*Bretberie* DB, *Bredbury* 1248 Ipm]. See BRADBURY.

Brede Sx [*Brade* 1161 P, *Brede* 1202 Fr]. OE *brædu* 'breadth', here in a concrete sense 'plain' or 'flat expanse'.

Bredenbury He [*Brideneberie* DB, -*burch* 1180 P, *Brudenebury* c 1275 Ep]. OE *bridene burg* 'BURG made of boards' (OE *briden* 'of boards').

Bredfield Sf [*Bredefelda, Berdefelda, Bradefelda* DB, *Brade-, Bredefeud* 1254 Val]. First el. OE *brædu* in the later recorded sense 'a space in a field' or some such sense as 'broad strip of cultivated land'.

Bredgar K [*Bradegare* 11 DM, 1219 Fees]. 'Broad strip'; cf. GĀRA.

Bredhurst K [*Bredehurst* 1240, *Bredherst* 1270 Ass]. 'HYRST where boards were got.'

Bredicot Wo [(æt) *Bradigcotan* 840 BCS 428, (to) *Bradingccotan* c 980 KCD 683, *Bradecote* DB]. Perhaps 'the COTS of *Brāda*'s people'.

Bredon Wo [*Breodun in Huic'* 772, *Breodun* 841 BCS 209, 434, *Breoduninga gemære* 984 E, *Breodun* DB], **Breedon on the Hill** Le [*Briudun* c 730 Bede, *Breodun* c 890 OEBede, *Bredona* 12 DC]. The name consists of a British hill-name, identical with Welsh *bre* 'hill' (from OBrit **brigā*), and OE DŪN. Bredon is close to Bredon Hill, Breedon close to Breedon Hill. The addition *in Huic'* 772 stands for *in Huiccum* 'in the territory of the *Hwicce*'. See WHICHFORD.

Bredwardine He [*Brerdewordin* 12 BM, *Bredworthin* 1227 Ch]. See WORÞ(IGN). First el. OE *brerd* 'brim, bank'. The place is on the slope of a steep ridge.

Bredy, Long & Little, Do [*Brydiga* 11 Coins, (to) *Brydian* c 910 Burghal Hidage, *Langebridia, Litelbride* DB]. Named from the river **Bride** [*Brydie* 1288 Ass]. Bride is a British river-name related to Welsh *brydio* 'to boil, throb'.

Breedon Le. See BREDON.

OScand **breiðr.** See BRĀD.

Breightmet (-āt-) La [*Brihtmede* 1257 FF]. 'Bright meadow.'

Breighton YE [*Bristun, Bricstune* DB, *Britton* 1242 Fees, *Brighton* 1313 BM]. OE *Brycg-tūn* 'TŪN by the bridge'.

Breinton He [*Bruntune, Breuntuna* c 1210, *Brahintone* 1252 Hereford, *Broyntun* 1242 Fees]. OE *Brȳning(a)tūn* 'the TŪN of *Brȳni*'s people'.

ON **brekka**, Norw *brekka, brekk* 'slope, hill' is found as the second el. of some pl.

ns., as NORBRECK, SCARISBRICK, WARBRECK La, HAVERBRACK We.

OE brēmel, brǣmel, 'bramble, blackberry-bush' occurs as the first el. of some names, as BRAMINGHAM, BRAMLEY, BRAMSHAW, BRAMSHOTT, BREMHILL &c. Brēmel is a derivative of brōm. OE brēmer 'bramble, bramble thicket' is found in BRAMBER, BRAMERTON.

Bremeridge W [Bremelrigge 1276 Ipm]. 'Bramble ridge.'

Bremhill W [Bre(o)mel, Broemel 937 BCS 716–19, Bremela 1065 KCD 817, Breme DB, Bremel 1190 P]. OE brēmel 'bramble' &c., in collective use.

Bremilham W [Bremelham 1065 KCD 817, Brumilham 1178 BM]. 'HĀM or HAMM where brambles grew.'

Brenchley K [Braencesle c 1100 Text Roff, Brencheslega 1185 P, -le 1242 Fees, Branches-legh 1230 Cl]. See LĒAH. The first el. appears to be a pers. n. Brænci, which is of obscure history.

Brendon D [Brandone DB, Bremdon, Bram-don, Brundon 12 &c. Buckland]. OE brōm-dūn 'broom hill'.

Brendon Hills So [Brunedun 1204 (1313) Ch, Brundon 1227 FF]. The original name was Brūna or Brūne, preserved in the pl. n. Brown [(in) Brunan 854 BCS 476, Brune DB]. This is a derivative of OE brūn and means 'the brown one'.

Brenkley Nb [Brinchelawa 1178 P, Brinke-lawe 1242 Fees]. 'Brynca's mound or hill.'

Brent R Mx [Brægente 959 BCS 1351, (of) Brægentan 972 ib. 1290, Brainte 1202 FF]. A British river-name *Brigantiā, the source also of BRAINT in Anglesea and meaning 'the high or holy river'. On the Brent is Brent-ford [Breguntford 705, (æt) Bregentforda 781 BCS 115, 241].

Brent, South, D [Brenta DB, 1240 For, Brente 1275 RH], East Brent So [Brente 663, 725 BCS 121, 142, Brentemerse DB, Est-, Sudbrente 1196 P, Est, Suth Brunte 1305 FF], Brentor D [Brenta 12 Ol, Brienta 1228 FF, Brentetor 1232 Ch]. Brent is an old hill-name. All three places are at prominent hills. At East Brent D is Brent Knoll [Brenteknol 1289 FF]. Brente in the earliest instance really refers to Brent Knoll. Brente might be a derivative of OE brant 'steep'. But forms such as Brienta, Brunte suggest that it is rather a British name, identical with OBrit Brigantiā 'high place'.

Brentford Bk [Braynford 1336 Ipm]. First el. as in BRAFIELD, though with the OE form brægen. Brentford Mx. See BRENT R.

Brentingby Le [Brantingbia c 1125 LeS, Brentingebi Hy 2 DC, Brentengebi 1170 P]. Brenting occurs as a surety in BCS 1130

(Peterborough). Perhaps this is the first el. of Brentingby. Some forms rather suggest original Brentingabȳr 'the BY of Brant's people'.

Brentor D. See BRENT and TORR.

Brentwood Ess [Boscus Arsus 1176 P, Bois Ars 1227 Ch, Brendewode 1274 Cl]. 'Burnt wood.'

Brenzett K [?Bennede circe 11 DM, Bren-sete DB, 1271 BM, 1295 Ep, Bretseta R 1 BM]. Bennede seems to be corrupt for Bernede 'burnt'. Brenzett may be 'burnt house' (OE sǣte 'house').

OE brerd 'rim, border; bank'. See BEARD, BRADING, BREDWARDINE.

Brereton Chs [Bretone DB, Brereton 1289 Court]. 'TŪN where briars grew' or 'briar enclosure'.

Brereton St [Breredon 1279 Cl, 1317 Ch]. 'Briar hill.' See DŪN.

Bressingham Nf [Bresing(a)ham, Brasinc-ham DB, Brisingeham c 1095 Bury, Brising-ham 1202 FF, 1254 Val]. Apparently OE Brīosingahām. Cf. BRAISEWORTH.

Bret R. See BRETTENHAM.

Bretby Db [Bretebi DB, c 1150 Chester, Brettebi 1166 P, -by 1202 FF]. 'The BY of the Britons'; cf. BIRKBY.

Bretford Wa [Bretford c 1180 Fr, 1237 Cl, 1273 Ipm], Bretforton Wo [Bretfertona 709, Brotfortun 714 BCS 125, 130, Brad-ferdtuna c 860 KCD 289, Bratfortune DB, Bretforton 1275 Subs]. OE bred-ford 'ford provided with planks'. There may have been a plank bridge for foot-passengers or boards may have been laid down on the bottom of the river. Bretforton is 'TŪN at Bretford'.

Bretherdale We [Britherdal Hy 2 (1247) Ch, Bretheredale 1291 FF], Bretherton La [Bretherton a 1190 CC]. 'The valley and the TŪN belonging to the brothers or brother.' In the latter case the first el. is the ON gen. sing. brœðr.

Brettenham Nf [Bretham DB, 1201 Cur, Breteham 1170 P, Bretenham 1257 Ipm], B~ Sf [Bretham, Bretenhama DB, Brethen-ham c 1095 Bury, 1191 FF, Bretenham c 1095 Bury]. The first el. is no doubt a pers. n. Bretta. This may mean 'the Briton'; cf. OE Brettas 'Britons'. Or rather it may be a hypocoristic form of names in Beorht-, Breht- (> Briht-). The river-name Bret is a back-formation.

Bretton Db [Brecton 1301 Ipm, Brettone 1240 FF], Monk B~ YW [Brettone, Bretone DB, Brettona c 1157 YCh 1665, Monk-bretton 1291 Tax], West B~ YW [Bre-tone DB, Brecton, Breton 1256 Pp, West Bretton c 1200 YCh 1525]. The occasional spellings with ct suggest that this is not OE Bretta-tūn 'the Britons' TŪN', but OE

Brēc-tūn, identical with BRATTON D. Cf. BRÆC. Monk B~ was a monastery.

Brewerne Gl [*Brewere* DB, *Bruurne* DB, *Bruerne* 1227 Flaxley]. OE *brēowærn* 'brewery'.

Brewham So [*Briweham* DB, 1251 Ch]. 'HĀM on R BRUE.'

Brewood (broōd) St [*Brewde* DB, *-wuda* 1188 P]. 'Wood on the hill.' First el. as in BREDON.

Breydon Water Nf [*fluvius Braydyng* 1325–50 Map, *Breything* 15 ERN]. ODan *breithing*, Dan dial. *bredding* 'place where a narrow piece of water widens itself'. B~ Water is a tidal mere near the mouth of the Yare.

Brīcett Sf [*Brieseta* DB, *Brisete* 1198 FF, *Breset* 1203 Ass, 1229 Ch, *Brissete* 1212 RBE, *Parva Briset* 1212 Fees, *Magna Brisete* 1235 FF]. Second el. OE (*ge*)*set* 'fold'. The first may be OE *briosa* 'gadfly'.

Brickendon Hrt [*Brycandun* c 1000 BCS 1306, *Brikandun* 959 ib. 1050, *Brichendone*, *Brichedone* DB]. It is possible that the first el. is an OE hill-name *Brice*, *dūn* being then an explanatory addition. *Brice* might be related to the first element of BRICKHILL.

Brickhampton Gl [*Breithelminton* 1220 Fees]. 'The TŪN of *Beorhthelm*'s people.'

Brickhill, Great, Little & Bow, Bk [*Brichelle* DB, *Bolle Brichulle*, *Magna Brikehille* 1197, *Parua Brichull* 1198 FF]. First el. Welsh *brig* 'top, summit' (from **brīk-*); second OE HYLL. Bow is OE *Bolla* pers. n.

Bricklehampton Wo [*Brihtulfingtun* 972 BCS 1282, *Bricstelmestune* DB]. If the first example belongs here, the place was first named after one *Beorhtwulf*. The present name means '*Beorhthelm*'s TŪN'.

OE **bridd** 'bird'. See BIRDHAM, BRIDGE-MERE.

Briddlesford Wt [*Breilesforde* DB, *Bridlesford* 1168 P]. The first el. is OE *brigdels* 'bridle', though the meaning of 'bridle ford' is not obvious. Cf., however, *bridle-path*.

Bride R. See BREDY.

Bridekirk Cu [*Bricekirk* 1291 Tax, *Brethekirke* 1292 QW]. 'The church of St. Bride.' *Bride* or *Bridget* is OIr *Brigit*.

Bridestowe (brīdi-) D [*Bridestov* DB, *Brightesstowe* 1242 Fees, *Brigide-*, *Briʒtestowe* 1259 ff. Ep], **Bridstow** He [*Lann San Frʒit*, *Lann San Bregit* c 1130 LL, *Bridestowe* 1277 Ep]. 'The holy place dedicated to St. Bride.'

Bridford D [*Bridaford* c 1080 E, *Brideforda* DB, *Brideford* 1206 Cur, *Brudeford* 1253 Ch]. OE *brȳda ford* 'ford of the brides'; cf. BRITFORD W and *brydelades ford* 909 BCS 627 (Sr). A bride-ford would be one used

when going to the wedding. Such a ford would be a safe one.

Bridge Sollers He [*Bricge* DB, *Bruges Solers* 1291 Tax], B~ K [*Brygge* 11 DM, *Bregge* 13 BM]. 'The bridge.'
Sollers is a Norman family name, very likely from SOLIERS nr Caen. Henry de Solers is mentioned in connexion with *Brugge* c 1160 (Hereford).

Bridg(e)hampton So [*Brugamton* 1281 Ipm]. 'The HĀMTŪN by the bridge.'

Bridgemere Chs [*Briddismere* 1260 Court, (fishery in) *Briddesmere* 1308 Ipm]. 'Mere frequented by birds', *bridd* being used in a collective sense.

Bridgerule D [*Brige* DB, *Bruge Ruardi* 1242 Fees]. Originally *Brycg* 'the bridge'. B~ belonged to Ruald Adobat in 1086 (DB). *Ruald* is a Fr Christian name of OG origin.

Bridgford, East, Nt [*Brugeford* DB, *Estbrigeford* 1291 Tax], **West** B~ Nt [*Brigeforde* DB, *Brigeford ad pontem* 1291 Tax], B~ St [*Brigeford* DB, *Brugeford* 1246 Ch]. 'Ford by a bridge' or 'ford with a foot-bridge'. In the former case the reference would be to an ancient disused bridge.

Bridgham Nf [*Brugeham* c 1050 KCD 907, *Briggeham* 1230 P]. 'HĀM by a bridge.'

Bridgnorth Sa [*Brug* 1156 P, *Brugg' Norht* 1282 Cl]. Originally *Brycg* 'the bridge'; *-north* for distinction from QUATBRIDGE (lost).

Bridgwater So [*Brugie* DB, *Brigewaltier* 1194 P]. Originally *Brycg* 'the bridge'. B~ belonged to the fee of Walter de Dowai.

Bridlington YE [*Bretlinton* DB, *Bridelington* c 1135 YCh 102, *Bretlington* 1196, *Bridlington* 1197 P]. The first el. is a derivative in *-ingas* from some pers. n., perhaps OE *Beorhtel*.

Bridmore W [*Brudemere* 1185 P, 1242 Fees, *Bridemere* 1203 Cur]. The first el. is connected with *Brydinga dic* 955 BCS 917, which must denote a locality nr Bridmore, and BURCOMBE W. Its origin is obscure. The second el. is OE *mere* 'lake'. There is a pool at the place.

Bridport Do [*Brideport* DB, *Bridiport* c 1150 Fr]. 'The port or borough belonging to BREDY.' Bredy itself was once a borough.

Bridstow He. See BRIDESTOWE.

Briercliffe La [*Brerecleve* a 1193 Whitaker, *-clive* 1258 LaInq], **Brierley** St [*Brereley* 14 PNSt], **Brierly** YW [*Breselai* DB, *Brerelay* 1194 f. P], **Brierton** Du [*Brereton* 1317 Misc]. 'Cliff, LĒAH, TŪN where briars grew.'

Brigg Li. See GLANDFORD.

Brigham Cu [*Bricgaham* 12 StB, *Briggeham* 1210 P], B~ YE [*Bringeham* DB, *Brichem*

1190 ff. P, *Brigham* 1226–8 Fees]. OE *Brycg-hām* 'HĀM by the bridge'. The spellings with *ch* denote palatal *g*.

Brighouse YW [*Brighuses* 1240 FF]. 'Houses by the bridge.' The place is on the Calder where it is crossed by an ancient road. *Brig-* is the Scandinavian form (ON *bryggia*).

Brighstone or **Brixton** Wt [*Brihtwiston* 1212 P, *Brightestone* 1284 Ep, *Briceston* 1284 Ch]. '*Beorhtwīg*'s TŪN.'

Brighthampton O [*Byrhtelmingtun* 984 Hengwrt MS 150, *Bristelmestone* DB, *Brihtelmeston* 1161 P]. '*Beorhthelm*'s TŪN.'

Brightling Sx [(æt) *Byrhtlingan* 1016–20 KCD 732, *Brislinga* DB]. '*Beorhthelm*'s or *Beorhtel*'s people.'

Brightlingsea (briklzī) Ess [*Brictriceseia* DB, *Bric(h)tricheseye, -eie* 1212 RBE, *Brihtlenggesseya* 1230 P, *Brychtlingeseye* 1253 Ch]. Originally '*Brihtrīc*'s (*Beorhtrīc*'s) island'. Later *Brightling* may be a hypocoristic form of *Brihtrīc*, or it may mean '*Brihtrīc*'s descendant'.

Brighton Sx [*Bristelmestune* DB, *-tuna* 1107–18 AC]. '*Beorhthelm*'s TŪN.'

Brightside Bierlow YW [*Brikeserd* 1297 Subs, *Brekesherth* 1345 FF, *Brikserth* 1383 Goodall, *Brightside* 1577 FF]. Possibly Brightside is a different name from *Brikeserd* &c., and means 'bright slope'. *Brikeserd* has as second el. OE *erþ* 'ploughed land'. The first looks like a pers. n., e.g. *Brihtrīc*. Cf. BRAMPTON BIERLOW.

Brightwalton Brk [*æt Beorhtwaldingtune* 939 BCS 743, *Bristwoldintona* 1087 BM]. 'The TŪN of *Beorhtweald*'s people.'

Brightwell Brk [*Beorhtawille*, (æt) *Brihtanwylle* 854 BCS 474, *æt Beorhttanwille* 945 ib. 810, *Bricsteuuelle* DB], **B~ Baldwin** O [*æt Berhtan-, Byrhtanwellan* 880 BCS 547, *Bretewelle* DB], **B~** Sf [*Briðwelle* c 1050 KCD 907, *Brihtewella* DB]. 'Bright spring.' In BCS 830 *Beorhtanwille* (B~ Brk) is rendered 'declaratam fontem'.

B~ Baldwin came to Sir Baldwin de Bereford in 1373 (PNO(S)).

Brigmerston W [*Brismartone* DB, *Bricimariston* 1199 Ch]. '*Beorhtmǽr*'s TŪN.'

Brignall YN [*Bring(en)hale* DB, *Briganhala* 1150–4 YCh 185, *Bricgehala* 1176 P, *Briggenhale* 1218 FF, *Bringenhale* 1280 Ipm]. Perhaps an OE *Brȳningahalh* 'the HALH of *Brȳni*'s people', though the later development offers difficulties. Cf. LAGNESS.

Brigsley Li [*Briges-, Brigelai* DB, *Brigesla, Brighe(s)la* c 1115 LiS, *Briggele* 1202 Ass, *Brichelai* 1202 BM]. OE *brycg-lēah* 'LĒAH by a bridge', later Scandinavianized, *gs* being introduced for the palatal *cg*, which was unknown in Scandinavian.

Brigsteer We [*Brigster* 1265–75 Kendale]. '*Styr*'s bridge'; cf. ASPATRIA. *Styr* is a Scandinavian name.

Brigstock Np [*Bricstoc* DB, *-stoka* 1168 P, *Brichestoc* 1095–1100 AC, 1130 P, *Brigestok* 1221, *Brigstoke* 1226 Ep]. Probably OE *brycgstoc* 'STOC by the bridge', with unvoicing of *g* before *s*. *Brichestoc* &c. show intrusive *e*.

Brill Bk [*Bruhella* Hy 1 Abingd, c 1110 AC, *-hulla* 1156 P, *Brehull* 1198 Cur, *Brehilla* 1230 Ch]. An old British name identical with Welsh *bre* 'hill' (cf. BREDON), to which was added an explanatory OE *hyll* 'hill'.

Brilley He [*Brynlegh* 1259 PNHe, *Brunleg* 1267 Ipm, *Brunlege* 13 Misc]. First el. OE *bryne* 'burning'. The name means 'burnt clearing', 'place cleared by burning'.

Brimfield He [*Brume-, Bromefelde* DB, *Bremelfelda* 1123 PNHe, *Bromfeld* 1212 Fees]. OE *brēmel-feld* 'bramble field' with dissimilary loss of the first *l*.

Brimham YW [*Birnebeham* DB, *Bernebeam* 1135–40 YCh 65]. Second el. apparently OE *bēo-hamm* or *-hām* 'HAMM or HĀM where bees were found'. The first may be identical with BURN YW.

Brimington Db [*Brimintune* DB, *Brumington* 1183 P, *Bremiton* 1197 P, *Brimentone* 1230 FF], **Brimpsfield** Gl [*Brimesfelde* DB, *Brumesfeld* 1211–13 Fees, *Bremesfelde* 1221 Ass, *Brummesfelde* 1227 Flaxley]. 'The TŪN of *Brēme*'s people' and '*Brēme*'s FELD'.

Brimpton Brk [*æt Bryningtune* 944 BCS 802, *Brintone* DB, *Brinton* 1167 P]. 'The TŪN of *Brȳni*'s people.'

Brimstage Chs [*Brunstath* 1260 Court, *-e* 1275 Cl]. '*Brȳni*'s landing-place' (OE *stæþ*). Or the first el. may be OE *burna* 'stream'.

Brind YE [*Brende* 1188 ff. P, *Brenda* 1254 Val, *Le Brende* 1289 Cl]. A substantivized form of ME *brend* 'burnt'.

Brindle La [*Burnhull* 1206 P, 1246 Ass]. 'Hill by a stream' (OE *burna*).

Brindley Chs [*Birnedelegh, Burndelegh* 1288 Court, *Brundeley* 1347 BM]. 'Burnt wood'; cf. LĒAH.

Brineton St [*Brunitone* DB, *Brimengton* 1236 Fees, *Bruneton* 1272 Ass]. OE *Brȳningatūn* 'TŪN of *Brȳni*'s people'.

Bringhurst Le [*Bruninghyrst* 1188 (1332) Ch, *Brenningeherst* 1200 Cur, *Brinningehurste* 1229 Ep]. OE *Brȳninga-hyrst* 'the HYRST of *Brȳni*'s people'.

Bringsty He [*Brinkestye* c 1275, *Brenkesty* 1307 Ep]. '*Brynca*'s path.'

Brington Hu [*Brynintune* 974 BCS 1311, *Breninctune* DB, *Brinngton* 1254 Val], **B~** Np [*Brinintone, Brintone* DB, *Brinton* 1198 P, *Brunton* 1248 Ch, *Little Brynton* 1284 Ipm], **Briningham** Nf [*Bruninga-, Burningaham* DB, *Briningham* 1254 Val]. 'The TŪN and HĀM of *Brȳni*'s people.'

Brinkburn Nb [*Brinkeburn* 1188, 1196 f. P, 1253 Ch], **Brinkhill** Li [*Brincle* DB, 1200

Cur, *Brincla* c 1115 LiS, *Brinckell'* 1212 Fees, *Brenkel* 1314 Ipm], **Brinkley** Ca [*Brinkelai* 1177–94 Fr, *Brinkele* 1203 Ass, *Brunckele* 1291 Tax], **Brinklow** Wa [*Brichelawa* 1130, *Brinchelawa* 1174 P, *Brinkelawe* 1201 Cur], **Brinkworth** W [*Brinkewrða* 1065 KCD 817, *Brenchewrde* DB, *Brunkwurth* 1242 Fees, *Brenkewrth* 1275 Cl]. '*Brynca*'s BURNA (stream), LĒAH, HLĀW (burial-mound), WORP.' It is possible that we have also to reckon with an OE **brinc(e)* 'brink of a hill, steep slope', cognate with ON *brekka* &c. This word is out of the question in names that show forms in *u*, and it does not generally give a very good meaning.

Brinnington Chs [*Bruninton* 1248 Ipm, *Brinnington* 1290 Court]. OE *Brȳningatūn* 'the TŪN of *Brȳni*'s people'.

Brinscall La [*Brendescoles* c 1200 WhC]. 'Burnt huts'; cf. SKÁLI.

Brinsford St [*Brenes-, Brunesford* 996 Mon, *Brunesford* 1176 P, *Bruneford* 1227 Ass]. '*Brūn*'s ford.'

Brinsley Nt [*Brunesleia* DB, *-lega* 1197 P, *-leg* 1242 Fees]. '*Brūn*'s LĒAH.' One *Brun* was a tenant here t. Edw. Conf. (DB).

Brinsop He [*Hope* DB, *Bruneshopa* c 1130 Ordericus, *Brunhopa* 1178 P, *-hop* 1242 Fees]. Originally OE *hop* 'valley'. The later addition *Brunes-* &c. is very likely the Fr pers. n. *Brun*; cf. (BRIZE) NORTON.

Brinsworth YW [*Brinesford* DB, 1202 FF, *Brenesford* 1202 FF, *Brunisford* 1241 FF]. '*Brȳni*'s ford.'

Brinton Nf [*Bruntuna* DB, *Brinton* 1197 P, 1252 Ch, *Bryneton* 1291 Tax]. OE *Brȳningatūn*; cf. BRINGTON. The place is close to BRININGHAM.

Briscoe Cu [*Brethesco* 1203, *Brisco* 1204 Cur]. ON *Bretaskógr* 'wood of the Britons'. **Brisco** Cu [*Byrkscawe* 1231 P]. 'Birch shaw.' **Briscoe** YN is ON *birkiskógr* 'birch wood'.

Brisley Nf [*Bruselea* c 1105 BM, *Brisele* 1199 P, 1254 Val, *Brissele* 1270 Ch]. 'Gadfly-infested LĒAH.' First el. OE *brīosa* 'gad-fly'.

Brislington So [*Bristelton* 1194 P, 1243 Ass, *Bristleton* 1196 FF, *Brihthelmeston* 1199 P]. '*Beorhthelm*'s TŪN.'

Bristol Gl [*Brycgstow* 1063 ASC (D), *Bristou* DB, *Bricstou* 1169 P, *Bristoll* 1200 Cur]. 'The site of the bridge.' See STŌW.

Briston Nf [*Burstuna* DB, *Birston* 1191 P, *Birstone, Burstone* 1254 Val]. OE *Byrst-tūn* 'TŪN in the gap'. See BYRST. Burston is in a pass on the upper Bure.

Brit Do [*Bride* 1577 Harrison]. A back-formation from BRIDPORT [*Britport* 1426 AD].

Britain, Great B~ [*Brutaine* 1205 Layamon, *Breteyne, þe more Brutaine* 1297 Rob

Gl, *Bretayne, Bretaygne þe grete* 1338 Rob Br]. In its present form the name comes from OFr *Bretaigne* (cf. Fr *Grande Bretagne*). Great B~ in contradistinction to **Little** B~ [*Britannia Minor* c 1150 Monm, *þe lasse Brutayne* 1297 Rob Gl], an old name of BRITTANY, which was identical in form with Britain. The source is Brit-Lat *Britannia*. The modern form with *i* is due to influence from *Britannia*. There was also an OE *Breten* (*Breoton, Bryten*) ASC, OEBede, which was taken direct from *Britannia*. The latter is 'the land of the Britons'. The forms of the names of the country and the people in classical sources show a good deal of variation between *Britannia, Brittania, Britanni, Brittani, Brittones*. Greek sources also have forms with initial *P-*, as *Prettania* Diodorus, *Prettanoi* Strabo, Diodorus. The latter are probably the more original forms, *B-* being due to inexact rendering of the British form. If so, Britain is etymologically connected with OW *Priten*, MW *Pryden*, OIr *Cruithne* 'Picts', Welsh *Ynys Prydain* 'Britain'. The word for Picts seems to be derived from Welsh *pryd*, Ir *cruth* 'figure, picture', and refers to the custom of the ancient Britons to tattoo themselves.

Britford W [*Brytfordingea landscære* c 670, *Brutford* 826 BCS 27, 391, *Bretford, Bredford* DB, *Brideford* 1203 Cur, *Brudford* 1212 RBE]. OE *brȳdford* 'brides' ford'; cf. BRIDFORD.

Britwell Bk [*Bretewell* 1244 Cl, *Brutewelle* 13 AD], **B~ Prior** O [*Braðeuuelle* 997 KCD 697, *Brutuwylle* c 1050 ib. 896, *Brittewelle* 1316 Ch], **B~ Salome** O [*Brutewelle* 1235 Ep, *Brittewell* 1254 Val, *Brutewelle Solham* 1322 BM]. 'Bright spring'; cf. BRIGHTWELL. Salome (olim *Suleham*) is a family name, derived from SULHAM Brk. Aumaricus de Suleham held Britwell in 1236 (Fees).—B~ Prior belonged to the prior of Christchurch, Canterbury.

Brixham D [*Briseham* DB, *Brixeham* 1143 Totnes, *Brikesham* 1205 FF, *Brixham* 1242 Fees]. '*Beorhtsige*'s HĀM or HAMM.'

Brixton D [*Brisestone* DB, *Brixton* 1197 P, *Brikeston* 1200 Cur]. '*Beorhtsige*'s TŪN.'

Brixton Sr [*Brix(g)es stan* 1062 Th, *Brixiestan* DB, *Brissistan* 1230 P]. '*Beorhtsige*'s stone.'

Brixton Deverill W [*Devrel* DB (held by Brictric), *Brihtricheston* 1242 Fees]. '*Beorhtrīc*'s TŪN.' Cf. DEVERILL.

Brixworth Np [*Briclesworde* DB, *-wurde* c 1160 Add Ch 22011, *Brihteswrðe* 1198 P, *Brikelesworth* 1224 Ep]. '*Beorhtel*'s or *Beorhthelm*'s WORP.'

Broadbottom Chs [*Brodebothem* 1286 Court]. 'Broad valley'; cf. BOPM.

Broadcar Nf [*Bradeker* 1324 Ipm]. 'Wide marsh.' Cf. KERR.

Broadfield He [*Bradefelde* DB], **B~** Hrt [*Bradefella* DB, *-feld* 1254 Val]. 'Extensive FELD.'

Broadhembury D [*Hen-, Hanberia* DB, *Hembiri* 1227 Ch, *Brodehambur'* 1252 Fees, *Brodehembyri* 1274 Ep]. Originally *Hēaburg* (dat. *Hēanbyrig*) 'the high fort'. **Hembury Fort** is a large earthwork situated c 900 ft. above the sea. *Broad-* in the sense 'Great' for distinction from PAYHEMBURY.

Broadhempston. See HEMPSTON.

Broadholme Nt [*Brodeholm* DB, *Broddeholm* 1275 BM]. '*Broddi*'s island.' First el. the pers. n. ON *Broddi*, ODan, OSw *Brodde*.

Broadmayne, Broadnymet, Broadoak. See MAYNE, NYMET, BRADDOCK.

Broadstairs K [*Broadstayer* 1565 Hasted]. 'The broad steps.' A gateway to the sea is said to have been built at Broadstairs in 1440 (Arch Cant xii).

Broad Town W [*Bradeton* 1201 Cur, *la Bradetune* 1220 AD]. 'Broad (i.e. large) TŪN.'

Broadward He [*Bradeford* 1212 Fees, *Bradford* 1280 Ch]. 'Broad ford.'

Broadwas Wo [*Bradeuuesse*, obl. *-wassan* 779 BCS 233, *Bradewesham* DB]. 'Large fen'; see BRĀD and WÆSSE.

Broadwater Sx [*Bradewatre* DB, 1166 P]. 'Broad water', i.e. presumably 'broad stream', though there is no important stream here now.

Broadway Do, see WEY.

Broadway So [*Bradewei* DB, *-e* 1225 Ass], B~ Wo [(in) *Bradanuuege* 972 BCS 1282, *Bradeweia* DB]. 'Broad road.'

Broadwell Gl [*Bradewelle* 11 KCD 1367, DB], B~ O [*Bradewelle* DB], B~ Wa [*Bradewella* 1130 P]. 'Broad stream or spring.'

Broadwindsor. See WINDSOR.

Broadwood Kelly D [*Bradehode* DB, *Brawode* 1242 Fees, *Brawode Kelly* 1261 Ep], B~ Widger D [*Bradewode* DB, *Brodwode Wyger* 1310 Ep], B~ So [*Bradeuda* DB]. 'Large wood.'

B~ Kelly was held by William de Kelly in 1242 (Fees); cf. KELLY.—B~ Widger passed to the Wiger family in the 12th cent. Sir John Wyger held the manor in 1273 (Ep). *Wiger* is an OFr pers. n. of Germanic origin (OHG *Wigher*).

Brobury He [*Brocheberie* DB, *Brocbire* 1242 Fees, *-buř* 1249 ib.]. 'BURG on the stream' (the Wye).

OE **brōc** = OHG *bruoh* 'marsh', LG *brôk*, MDu *broek*. The old sense 'marsh' is still preserved in Kent and Sussex, where *brook* means 'a water-meadow, low marshy ground'. But the usual sense in English is 'brook, stream'. This sense probably arose in England. In pl. ns. the sense is usually 'brook', but the old sense may be reckoned with in the south-eastern counties. See BROCK, BROOK, -E, names in BROCK-, BROOK-,

BROBURY, BROC(K)TON, BROGDEN, BROUGHTON, BROXFIELD, &c. Common as a second el. in names of brooks and pl. ns. derived from such. Sometimes *brōc* is difficult to distinguish from OE *brocc*.

OE **brocc** 'badger' occurs occasionally as the first el. of pl. ns. Certain examples are BROCKHALL, BROCKHOLES; probable ones are BROCKHURST, BROXTED.

Brock R La [*Broc, Brok* 1228 Cl]. OE *brōc* 'brook'.

Brockdish Nf [*Brodise* DB, *Brochedisc* c 1095 Bury, *Brokedis* 1166 RBE]. 'The EDISC on the stream' (the Waveney).

Brockenhurst Ha [*Broceste* DB, *Brocheherst* 1158, *Brokenherst* 1182, 1184 P]. Apparently '*Broca*'s HYRST'; cf. *Brocan burna* BCS 787 (nr Clere Ha). OE *brocen* 'broken' does not go well with the early forms.

Brockford Sf [*Brocfort* DB, *Brocford* 1209 FF]. 'Ford over the brook.'

Brockhall Np [*Brocole* DB, *Brochole* 1220 Fees], Brockholes La [*Brochole* 1212 LaInq, *-holes* 1244 ib.]. OE *brocc-hol* 'badger hole or burrow'.

Brockham Green Sr [*Brocham* 1254 BM, *Brokham* 1241 FF]. Second el. HĀM or HAMM, first OE *brōc* 'brook' or *brocc* 'badger'.

Brockhampton Do [*Brochamton* c 1250 Glaston], B~ Gl nr Bishops Cleeve [*Brochamtone* 1221 Ass], B~ Gl nr Sevenhampton [*Brochampton* 1248 Ass], B~ Ha [*Brochematune* DB, *Brochampton* 1242 Fees], B~ He nr Bromyard [*Brockampton* 1251 Cl], B~ He nr Ross [?*Brochantone* 1166 RBE], Brockington Do [*Brochemtune* DB, *Brochampton* 1225 FF]. OE *Brōchæmatūn* 'the TŪN of the dwellers on the brook'. But cf. BROOKHAMPTON.

Brockholes. See BROCKHALL.

Brockhurst Wa [*Brochurst* 1200 Cur]. OE *brocc-hyrst* 'badger hill or wood'.

Brocklesby Li [*Brochelesbi* DB, *Broclosbi* c 1115 LiS, *Broclousebi* 12 DC]. 'The BY of *Broclos*.' This pers. n. is found in DB (Li) and is a Scandinavian nickname meaning 'breechless' (from ON *bróklauss*).

Brockley K [*Brocele* 1182–4 Reg Roff, *Brockele* 1328 Ch], B~ Sf [*Brocle, -lega* DB, *Brochelee* 1196 FF, *Brocleye* 1254 Val], B~ So [*Brochelie* DB, *Brockeleg* 1225 Ass]. B~ Sf is no doubt OE *brōc-lēah* 'LĒAH on the brook'. B~ K, So are perhaps more likely OE *Brocan lēah* '*Broca*'s LĒAH'; cf. BROCKENHURST.

Brockmanton He [*Brochem'ton* DB, *Brocmanetune* 1123 PNHe, *Brocmantona* 1242 Fees]. OE *brōcmanna tūn* 'the TŪN of the people on the brook'.

Brockton Sa nr Ironbridge [*Broctone* DB,

Brocton 1212 Fees, 1290 BM], **B~** Sa nr Lilleshall [*Brochetone* DB], **B~** Sa nr Patton [*Broctune* DB, *Brocton* 1222 FF], **B~** Sa nr Worthen [*Brockton* 1272 Eyton]. 'TŪN on a brook.'

Brockweir Gl [*Brocwere* 12 Fr, 1314 Ipm]. 'Weir by the brook.' B~ is on the Wye where it is joined by a brook.

Brockworth Gl [*Brocowardinge* DB, *Brocwurthin* 1183, *-wurdin* 1196 P, *Brochworth* 1220 Fees]. 'WORÞ(IGN) on a brook.'

Brocton St nr Stafford [*Broctone* DB, *Broeton* 1290 Cl], **B~** St nr Eccleshall [*Broctone* DB, *Brocton* 1327 Subs]. 'TŪN on a brook.'

Brodsworth YW [*Brodesworde* DB, *Broddeswrde* 1157 YCh 186, *-worth* 1222 FF]. Perhaps '*Brodd*'s WORÞ', the first el. being ON *Broddr*, pers. n. But more likely the first el. was originally OE *Brord*. In OE *Brordesworþ* one *r* would be apt to be lost owing to dissimilation.

Brogden YW [*Brokden* c 1307 Pudsay]. 'The valley of the brook.'

Brokenborough W [(in) *Brokeneberegge* 956 BCS 921, *Brocheneberge* DB]. 'The broken hill', perhaps in reference to the deep valley in which the Avon flows here.

OE **brŏm** 'broom' is sometimes used alone as a pl. n.; see BROME, BROOM, -E. It is rare as a second el.; cf. FALLIBROOME. Very frequent as a first el.; see e.g. BROM-, BROOM-, BRAM-, also BRANDON, BRENDON, BRUNDON, BRA(U)NTON, BRYMPTON, BREA-, BRYMORE, &c. The form *Bram-* arose through shortening. OE *brŏm* sometimes became *brom-*, but as a short *o* hardly ever occurred before a nasal in early OE, the dark *a* or open *o* that had developed from *a* before nasals, as in *mann* (*monn*), was often substituted, and this dark *a* later became ordinary ME *a* in most dialects.

Bromborough Chs [*Brombur* c 1160 Chester, *Brumbur* c 1220 ib.]. 'BURG where broom grew.'

Bromden Sa [*Bromdon* 1280 Misc]. 'Broom hill.'

Brome Sf [*Brom* DB, 1197 FF]. OE *brŏm* 'broom', here used collectively.

Bromeswell Sf [*Bromeswella*, *Brumeswella* DB, *Bromeswell* 1254 Val]. First el. OE BRŌM. The second el. may be OE *wella* 'a stream'. But the genitive form of the first el. would be peculiar. More likely the second el. is OE *swelle* 'swelling', here used in the sense 'hill'. Cf. SWELL.

Bromfield Cu [*Brumfeld* c 1145 WR, 1227 FF, *Bramfeld* c 1275 StB], **B~** Sa [*Brunfelde* DB, *Brumfeld* 1155 BM, *Bromfeld* 1228 Cl]. 'Broom-covered FELD.'

Bromhall Brk [*Brumhala* 1158 P, *Bromhal* 1212 Fees, 1228 Cl]. 'Broom-covered HALH.'

Bromham (-ōŏm-) Bd [*Brune-*, *Brimeham* DB, *Bromham* 1227 Ass, *Bruham* 1254 Val], **B~** W [*Bromham* DB, c 1090 BM]. 'HAMM or HĀM where broom grew.'

Bromholm Nf [*Bromholm* c 1150 Crawf, 1229 Ch]. 'Broom-covered island.'

Bromkinsthorpe Le [*Brunechinestorp* DB, *Bruneskinnestorp* 1233 Cl]. The first el. seems to be an OScand nickname *Brūnskinn* 'one with brown skin'.

Bromley Ess [*Brumleia*, *Brumbeleia* DB, *Brumlegh Magna*, *Brumleghe Parva* 1238 Subs], **B~** Hrt [*Bromeleys* 1456 BM], **B~** K [*Bromleag* 862 BCS 506, *Bronlei* DB], **Abbots B~** St [*Bromleage* 1002 Wills, *Bromlegh Abbatis* 1304 Ass], **Kings B~** St [*Bromelei* DB, *Bramlea Reg[is]* 1167 P]. OE *brōm-lēah* 'LĒAH where broom grew'. Abbots B~ belonged to Burton Abbey.

Bromley (-ŭ-) Mx [*Bræmbeleg* c 1000 CCC, *Brembeleg* 1200 Cur, 1219 Fees, *Brambele* 1235 Ass]. 'LĒAH where brambles grew.'

Bromlow Sa [*Bromlawe* 1255 RH]. 'Broom-covered hill'; cf. HLĀW.

Brompton (-ŭ-) Mx [*Bromton* 1309 FF], **B~** Sa nr Bishops Castle [*Brompton* 1249 Misc, 1272 Ipm], **B~** Sa nr Shrewsbury [*Brantune* DB], **Potter B~** YE [*Brunetone* DB, *Poterbrumton* 1285 FA], **B~** YN nr Ebberston [*Bruntun* DB, *Brumton* 1219 Ass], **B~** YN nr Northallerton [*Bromtun* c 1050 HSC, *-e* 1088 LVD, *Bruntone* DB], **Patrick B~** YN [*Brunton* DB, *Patricbruntun* 1157 YCh 354], **B~ on Swale** YN [*Brunton* DB, *Bramton* 1208 Cur, *Brumpton super Swale* 1316 FA]. OE *Brōmtūn* 'TŪN where broom grew'.

Patrick B~ from some early owner. **Potter B~** must have been named from potteries.

Brompton Ralph So [*Burnetone* DB, *Brumpton* c 1235 Dunster, *Bruneton* c 1250 ib., *Brompton Rad[ulph]i* 1274 RH], **B~ Regis** So [*Brunetone*, *Burnetone* DB, *Brompton Regis* 1291 Tax]. 'TŪN by BRENDON HILLS' (OE *Brūnan tūn*).

B~ Ralph was given c 1245 by Reginald de Mohun to Ralph fitz Urse (Dunster).

Bromsberrow Gl [*Brunmeberge* DB, *Bremeberga* 1166 P, *Bromesberga* c 1180 Fr, *-berg* 1220 Fees]. 'Broom-covered hill.' See BRŌM, BEORG.

Bromsgrove Wo [*Bremesgraf* c 822 BCS 308, *-grefan* 804 ib. 313, *Bremesgrave* DB]. '*Brēme*'s grove.'

Bromwich (brŭmïj), **West**, St [*Bromvic* DB, *Westbromwich* 1322 Ipm], **Castle B~** Wa [*Bromwich* 12 Dugdale, *Magna Bromwyce* 1285 QW], **Little B~** Wa, now **Ward End** [*Parva Bromvice* 1285 QW, *Wode-*, *Wodybromwych* 1292 Ipm]. 'WĪC where broom grew.'

Bromyard He [*Bromgeard* c 840 BCS 429, *Bromgerde* DB]. 'Broom enclosure'; see GEARD.

Brom-y-Clos He [*Broomy Close* 1650 PNHe]. 'Broomy close.'

Brondesbury Mx [*Bronnesburie* 1254 Val, *-bury* 1327 Ep, *Brondesbury* 1291 Tax, *-biri* 1328 Pp, *Brounesbury* 1322 Pat]. Apparently '*Brūn*'s (later *Brand*'s) manor'. One Brand was a canon of St. Paul's c 1180–1216.

Brook K [*Broca* 11 DM, *Broc* 1165 P], **B~** Wt [*Broc* DB, *Broke* 1346 FA], **Brooke** Nf [*Broc* DB, 1254 Val], **B~** Ru [*Broc* 1176 P, 1202 Ass]. OE *brōc* 'brook'.

Brookhampton O [*Hantone* DB, *Brochamton* 1279 RH], **B~** Wa [*Brochamton* 1203 Ass, 1231 Cl], **B~** Wo nr Astley [*Brokamton* 1275 Subs], **B~** Wo nr Ombersley [*Brokhampton* c 1220 PNWo]. **B~** O was originally *Hammtūn*, to which a distinguishing el. *Brook-* was added. The rest may be OE *Brōc-hāmtūn* 'HĀMTŪN on the brook' or identical with BROCKHAMPTON. **B~** Wa may be OE *brōchæmatūn* 'TŪN of the people of COMBROOK'.

Brookland K [*Broklandes* 13 Misc]. 'Land on the brook.'

Brooksby Le [*Brochesbi* DB, *Brokesbi* 1197 P]. 'BY on the brook' (the Wreak).

Brookthorpe Gl [*Brostorp* DB, *Brocthorp* 1226–8 Fees]. 'Thorp by the brook.'

Brookwood Sr [*Brocwod* 1274 Cl, *Brokewode* 1289 Cl]. 'Wood on the brook.'

Broom Bd [*Brume* DB, *Brom* 1203 AD], **B~** Du [*Brom* c 1170 Finchale], **B~** Wo [*Brom* 1168 f. P], **Broome** Nf [*Brom* DB, c 1095 Bury, 1190 P], **B~** Sa [*Brome* 13 Misc], **B~** Wa [*Brome* 710 BCS 127, DB, *Kingges Brome* 1279 Ipm]. OE *brōm* 'broom'. Cf. BROME.

Broomfield (-ŭm-) Ess [*Brumfelda* DB, *Bromfeud* 1254 Val], **B~** K nr Maidstone [*Brumfeld* 11 DM, *Bromfeld* 1242 Fees], **B~** So [*Brunfelle* DB, *Bromfeld* 1243 Ass]. See BROMFIELD.

Broomfleet YE [*Brungareflet* 1150–4 YCh 185, *Brungarflet* 1154–60 ib. 1825]. '*Brungar*'s FLĒOT or stream.' *Brungar* is the name of a moneyer t. Æthelred II.

Broomhall Chs [*Brunhala* DB, *Bromhale* c 1100 Chester, 1260 Court]. 'Broom-covered HALH.' Cf. BROMHALL.

Broomhaugh Nb [*Brunhalwe* 1242 Fees, *Bromehalwe* 1262 Ipm, *Bromhalgh* 1269 Ass]. 'Broom-covered haugh'; see HALH.

Broomhill K [*Bromhell* 1322 Ipm]. 'Broom-covered hill.'

Broomhill Sx [*Prumhel* c 1165 BM, *Prunhull* 1201, 1208 Cur]. 'Plum(tree) hill.' The first el. is an OE ***prūme** 'plum', corresponding to OHG *pfrūma*, MLG *prūme*, Du *pruim*. OE *plūme* 'plum' comes from Lat *prūnus* and shows a change of *r* to *l*.

Broomhope Nb [*Bromhop* 1242 Fees]. 'Broom valley.'

Broomley Nb [*Bromley* 1242 Fees]. See BROMLEY.

Broomsthorpe Nf [*Brunestorp* 1198, 1203 Cur]. '*Brūn*'s thorp.'

Broomy Holm Du [*Bromywhome* 1326 PNNb]. 'Broomy corner or valley'; see HWAMM.

Broseley Sa [*Burewardeslega* 1177 f., *Burgardeslega* 1194 P, *Borewardesleg* 1242 Fees]. '*Burgheard*'s or *Burgweard*'s LĒAH.' *Burgheard* is better evidenced than *Burgweard*.

Brothertoft Li, **Brotherton** Sf [*Brodertuna* DB, *Bretherton* 1207 Cur], **B~** YW [*Broðertun* c 1030, *Broðortun* c 1050 YCh 7, 9, *Broderton* 1193 ff. P], **Brotherwick** Nb [*B[r]otherwyc* 1242 Fees, *Brothirwike* 1252 Ipm]. 'The TOFT (TŪN, WIC) of *Brodor*.' *Brodor*, *Broder* is found as a pers. n. in DB. It is probably from ODan, OSw *Brodher*, lit. 'brother'. To some extent the source of the first el. may be the word *brother*. Cf. BRETHERTON.

Brotton YN [*Broctune* DB, *Brocton* 1272 Ipm]. 'TŪN on the brook.'

Brough (brŭf) Db [*Burgus* 1165 PNDb, *Burg* 1253 Ch], **B~** (brŭf) Nt [*Burgh* 1525 PNNt(S)], **B~** (brŭf) **under Stainmore** We [*Burc* 1198 P, *Burgh* 1228 Pat, *Burch under Steymor* 1279 Ass], **B~** YE on the Humber where it was crossed by Ermine Street [*Burg* c 1200 YCh 1829, *Burgh on Humber* 1239 Ch], **B~** YE at Skipsea [*Skipseburgh* 1321 f. Ipm], **B~** (brŏŏf) YN nr Catterick [*Burg* DB], **B~** (brŏŏf) YN nr Reeth [*Burg* 1218 FF]. OE *burg* 'fort'. The name refers in all cases to ancient camps, usually Roman ones. For **B~** Db cf. NOE.

Brougham (brŏŏm) We [*Bruham* 1228 Pat, *Burgham* 1362 Cl]. OE *Burg-hām* 'the HĀM by the BURG'. The *burg* is the Roman station of *Brocavum*.

Broughton is a common name. It has 3 distinct sources, which are sometimes not quite easy to keep apart:

1. OE *Brōc-tūn* 'TŪN on a brook': **Broughton** (-aw-) Bk at Bierton [*Brotone* DB, *Broctona* c 1155 PNBk], **B~** Bk nr Fenny Stratford [*Brotone* DB, *Broctone* 1237–40 Fees], **B~** Cu [*Brocton* 1208 FF, *Brochton* 1286 Ipm], **Church B~** Db [*Broctune* DB, *Brocton* 1208 FF], **West B~** Db [*Parva Broctona* W 2 Mon], **B~** Hu [*Broctune* DB], **B~** La nr Preston [*Broctun* DB], **B~** in Cartmel La [*Brocton* 1276 Ass, *Broghton* 1314 FF], **B~** in Furness La [*Brocton* 1196, 1235 FF], **B~** Astley Le [*Brohtone*, *Broctone* DB, *Broghton Astley* 1423 AD], **Nether B~** Le [*Broctone* DB, 1223 Ep, *Netherbroughton* 1394 Bodl], **Upper B~** Nt [*Brotone* DB, *Broctun* 1236 Fees], **B~** O [*Brohtune* DB, *Broctona* 1224 Ep], **B~ Poggs** O [*Brotone* DB, *Brocton*, *Brouhton* 1254 Val], **B~ Gifford** W [*Broctun* 1001 KCD 706, *Broctone* DB, *Brochton Giffard* 1415 FF], **B~** Wa [*Brocton* 1285 BM], **B~**

Hackett Wo [*Broctun* 972 BCS 1282, *-e* DB], **B~** YN nr Malton [*Broctun* DB], **Great & Little B~** YN in Kirby in Cleveland [*Broctun, Magna Broctun* DB], **B~** YW [*Broctune* DB, *Broghton* 1278 Ch].

2. OE *Burh-tūn* identical with BURTON. *Burh-* became *Bruh-* owing to metathesis: **Broughton** La nr Manchester [*Burton* 1177 P, *Burghton* 1323 FF], **Brant B~** (-ōō-) Li [*Burtune* DB, *Bructun* 1185 TpR, *-tone* 1209–19 Ep, *Brendebrocton* 1250 Pudsay], **B~** (-ow-) Np [*Burtone* DB, *Brohtune* 1125–8 LN, *Bructon* 1220 Ep, *Bruchton* 1220 Fees], **B~** Sa nr Bishops Castle [*Burhton* 13, *Burghton* 1366 BM], **B~** Sa nr Claverley [*Burgton* 1194 P, *Bureton* 1212 Fees], **B~** Sa nr Wem [*Burtune* DB, *Burhton* 1255 RH], **B~** St [*Burghton* 1281 Ass, *Borghton* 1327 Subs], **B~** Sx [*Bortone* DB, *Burghtone* 1265 Misc].

3. OE *Beorg-tūn* 'TŪN by a hill or barrow'; OE *eo* became ME *u* in some dialects, and *ur* became *ru* owing to metathesis: **Broughton** Ha [*Brestone* DB, *Berchton* 1176, *Brecton, Burchton* 1191 P, *Berghton* 1239 Ch], **B~** Li nr Brigg [*Bertone* DB, *-tuna* c 1115 LiS, *Berchtun* 1125–8 LN, *Berton* 1188 P, 1254 Val].

B~ Astley Le was held by Thomas de *Estleg* in 1203 (Cur).—**Brant B~** Li is on the BRANT, but the addition is the past part. of the verb *burn* and means 'burnt'. The place must have been burnt down at some early period.—**B~ Gifford** W was held by John Giffard in 1281 (Ch); see ASHTON GIFFORD.—**B~ Hackett** Wo was in the possession of the Hackett family from the 12th cent.; on the name see BEER HACKETT.—**B~ Poggs** O contains the family name *le Pugeis*, also found in STOKE POGES Bk. The family name is a byname of obscure origin, very likely denoting a man coming from a certain place.

Brown So. See BRENDON.

Brownedge La [*Browneegge* 1551 DL]. 'Brown hill.'

Browney R Du [*Brune* c 1190, *Brun* c 1195 Finchale]. OE *Brūn* or *Brūne* 'the brown one' with later addition of *-ey*, which may be OE *ēa* 'river'.

Brownsea or **Branksea Island** Do [*Brunkes'* 1235, *Brunkeseye* 1276 Cl, *Broncheshe* 1291 Tax]. **B~** is an island; the second el. is OE *īeg* 'island'. The first may be an unrecorded OE pers. n. *Brūnoc* derived from *Brūn*.

Brownsover. See OVER.

Brownston D in Modbury [*Brunardeston* 1219 FF, *Brunewardestuna* 1231 PND]. '*Brūnheard*'s or *Brūnweard*'s TUN.' *Brūnheard* is better evidenced than *-weard*.

Browston Sf [*Brockestuna* DB, *Brockeston* 1270 Ch, *Broxton* 1232 FF]. '*Brocc*'s TŪN.' OE *Brocc* seems to be evidenced in *Broccesham* 862 BCS 506 (now BROCKSHAM K), *Broccæs hlæw* 990 KCD 673. It may be OE *brocc* 'badger' used as a pers. n.

Broxa YN [*Brokesaye* 1090–6 YCh 855,

Brokesey 1316 FA]. First el. perhaps *Brocc* pers. n. The second may be OE *gehæg* 'enclosure'.

Broxbourne Hrt [*Brochesborne* DB, *Brocheburn* 1174 P, *Brokesburn* 1253 ff. Ch]. 'The stream of *Brocc* or of the badger' (OE *brocc*). Cf. BROWSTON.

Broxfield Nb [*Brokesfeud* 1256 Ass, *Brockesfeld* 1307 Ch]. 'FELD on the brook.'

Broxholme Li [*Broxholme* DB, *-holm* c 1115 LiS, 1254 Val, *Brokesholme* 1219 Ep]. 'Holm or island by the brook' (the Till).

Broxted Ess [*Brocheseued* c 1050 KCD 907, *Broccheshevot* DB, *Brockesheved* 1252 Ch]. OE *brocces hēafod* 'the badger's head' or 'badger hill'. Cf. HĒAFOD.

Broxton Chs [*Brosse* DB, *Broxun* 1259, *Broxne* 1287 Court, *Brexin* 1260 Court, *Brexen* 1278 Chester]. A variant of the common name BURWAINS, BORRANS &c. from OE **burgæsn* or the like. Broxton goes back to a side-form **borgæsn*, which became *Borgsne* > *Borhsne* and *Brohsne*, *Broxne* &c.

Broxtow Nt [*Brochelestou* DB, *Brocolvestov* (wap.) ib., *Broculvestowe* 1166 P, 1212 Fees]. '*Brocwulf*'s place.' *Brocwulf* pers. n. is not evidenced, but must have existed; *Brocheard* is recorded. There may have been a saint called *Brocwulf*. Cf. STŌW.

Brue R So [*Briuu* 681 BCS 61, *Bru* 744 ib. 168]. A British river-name related to Welsh *bryw* 'brisk, vigorous'. Cf. BREWHAM, BRUTON.

Bruer, Temple, Li [*la Bruere* Hy 2 TpR, *Bruer* 1201 Cur, *Brueria* 1202 Ass, (Templars of) *La Bruere* 1288 Ipm]. OFr *bruiere* 'heath'. Came to the Templars in 1169.

Bruera Chs [*Heeth* 12, *Bruera* c 1190 Chester]. Originally *Hæþ* 'the heath', later replaced by OFr *bruiere* or Lat *brueria* with about the same meaning.

Bruern O [*Bruer'* c 1175 AC, *Brueria* 1197 P, *Bruern* c 1200 Osney, *Brewernye* 1252 Misc]. OE *brēowærn* 'brewery' or identical with BRUER.

Bruisyard Sf [*Buresiart* DB, *-iard* 1191 P, 1203 Cur, *Buresgerd* 1204 FF]. Second el. OE *geard* 'yard'. The first may be OE *būr* (cf. BOWER, -S, BURES) or OE *gebūr* 'farmer'.

Brumby Li [*Brunebi* DB, *Brunneby* 1271 Ch, *Brouneby* 1340 Misc]. '*Brūni*'s BY.' First el. ODan *Brūni* pers. n.

Brumstead Nf [*Brumestada* DB, *Brunstede* c 1165 Holme, *Brum-, Bromstede* 1254 Val]. 'Broom-covered place.' See STEDE.

Brun R. See BURNLEY.

Brundall Nf [*Brundala* DB, *Brundhal* 1257 Ass, *Brundale* c 1180 Bodl]. Perhaps a compound of OE **brōmede* 'broomy' and HALH.

Brundish Sf [*Burnedich* 1177 P, *Burnedis*

1204 Cur, 1208 FF]. 'EDISC on the stream' (OE *burna*).

Brundon Sf [*Brandona* DB, *Brandune* 1178 AC]. OE *brōm-dūn* 'broom hill'.

Brunshaw La [*Brunschaghe* 1296 Lacy, *Brounshagh* 1311 Ipm]. 'Copse on R BRUN', which may have been originally OE *Brūne* 'brown river'; cf. BROWNEY.

Brunstock Cu [*Bruneskayth* 1253 P, -*stach* 1281 Ipm, *Brunskaith* 1332 Subs], **Brunskaith** Cu (lost) in Burgh by Sands [*Brunstayth* 1300, -*scaith* 1312 Ipm]. Second el. ON *skeiδ* 'race-course'. The first may be ON *bruni* 'burning' (area cleared by burning) or *brunnr* 'well, spring'.

Bruntingthorpe Le [*Brandinestor* DB, *Brantingestorp* 13 BM, 1254 Val, *Brentingestorp* 1199 FF]. '*Branting*'s (or *Brenting*'s) thorp.' *Brenting* pers. n. is found BCS 1130, *Branting* as the name of a moneyer t. Eadweard II &c.

Brunton Nb nr Embleton [*Burneton Batayll* 1242 Fees, *Brunton* 1296 Subs], **East, West, North, South B~** Nb [*Burneton* 1242 Fees, 1250 Ipm, *Burnton Est, Burneton West* 1296 Subs]. 'TŪN by the brook' (OE *burna*).

Brushfield Db [*Bricthicesfel* 12 BM, *Brittrichisfield* 13 Derby]. '*Beorhtric*'s FELD.'

Brushford D [*Brigeford, Brisforde* DB, *Brigheford* 1242 Fees], **B~** So [*Brigeford, Brucheford* DB, *Breggeford* 1195 P, *Bryschford* 1335 Ep]. 'Ford by or with a bridge.' Cf. BRIDGFORD.

Bruton So [*Briwetone* DB, *Briweton* c 1150 Fr]. 'TŪN on R BRUE.' **B~ W.** See BROUGHTON.

Bryanston Do [*Blaneford* DB, 1212 Fees, *B~ Brian* 1271 Ipm, *Brianeston* 1291 Tax, *Brieneston* 1295 Ch]. '*Brian*'s TŪN.' *Brian* is a Norman name of Breton origin. Brian de Insula de Blaneford is mentioned 1232 Pat. Cf. BLANDFORD.

OE **brycg** 'bridge'. See BRIDGE. As a first el. it takes various forms. See BRIDG(E)- (passim), BREIGHTON, BRIGHAM, BRIGSLEY, BRIGSTOCK, BRISTOL, BRUSHFORD. Common as a second el. DOVERIDGE is noteworthy. ON **bryggia** is the source of BRIGG, and is found in BRIGHOUSE, BRIGSTEER, WESTBRIGGS.

OE **brȳd** 'bride'. See BIRDFORTH, BIRDLIP, BRID-, BRITFORD.

Brymore So [*Brummore* 1299 BM]. 'Broom-covered moor.'

Brympton So [*Brunetone* DB, *Brimpton* 1264, *Bromton* 1331 Ep]. Identical with BROMPTON Mx &c.

Bryn Sa [*Bren* 1272 Ipm]. Welsh *bryn* 'hill, mound'.

Bryning La [*Birstaf Brinning* 1201 P, *Birstatbrunning* 1236 LaInq, *Brunigg* 1252,

Brining 1254 Ipm]. The original name may have been OE *Brȳningas* '*Brȳni*'s people', to which was prefixed ON *bȳiarstaδr* 'farmstead', the longer name meaning 'Bryning farm'. Less likely the name is a shortening of an original name consisting of ON *bȳiarstaδr* and OScand *Bryning* pers. n. and meaning '*Bryning*'s farm'. If so, ASPATRIA may be compared.

ON **bú** 'homestead' is found in BEWALDETH, BARNBOW.

Bubbenhall Wa [*Bubenhalle* DB, *Bubehull* 1230, *Bubbehull* 1231 Cl], **Bubnell** Db [*Bubenenle* DB, *Bobenhull* 1283 Ipm]. '*Bubba*'s hill.'

Bubwith YE [*Bubvid* DB, *Bobewid* 1226, *Bubwyht* 1246 FF, *Bubbewych* 1279–81 QW]. Probably OE *Bubban wīc* '*Bubba*'s wīc', later associated with OScand *viδr* 'wood'. Cf. COTTINGWITH, SKIPWITH.

OE **bucc** 'male deer' is the first el. of some names, as BUCKDEN, BUCKFAST. OE **bucca** 'he-goat' may also occur in pl. ns., but it is difficult to distinguish it from *Bucca* pers. n., which is generally a more probable source.

Buckby, Long, Np [*Buchebi* DB, 1190 P, *Buckeby* 1203 Cur]. '*Bucca*'s BY.' But probably a Scandinavianized OE *Buccan burg*; cf. BADBY. BUGBROOKE c 5 m. away may have been named from the same *Bucca*.

Buckden Hu [*Bugedene* DB, *Buge(n)den* 1185 P, *Buggenden* c 1156 RA, *Buccend[en]* 1150–60 Chron Rams, *Buggeden* 1227 Ch, *Bukeden* 1254 Val]. 'The valley of *Bucge* (fem.) or of *Bucca* or of the bucks.'

Buckden YW [*Buckedon* 1202 FF, *Bukeden* 1218, *Buckeden* 1235 FF, *Bugeden* 1227 FF]. 'The valley of the bucks' (OE *bucc* 'male deer').

Buckenham, New & Old, Nf [*Bucheham, Bucham* DB, *Bucheham* 1151 BM, *Nova Bukham* 1286 QW, *Vetus Bokenham* 1343 BM], **B~** Nf on the Yare [*Buc(h)anaham* DB, *Bokenham Ferye* 1451 BM], **B~ Tofts** or **Little B~** Nf [*Bokeham* Hy 2 BM, *Toftes* 1254 Val]. OE *Buccan hām* '*Bucca*'s HĀM'. See TOFT.

Buckenhill He [*Bokenhulle* 1328, 1350 Ep, 1407 BM]. First el. OE *bōcen* 'of beech'; cf. BOCKENFIELD.

Buckerell D [*Bucherel* 1166 P, *Bukerel* 1199 P, *Bokerel* 1278 Ep]. Unexplained. Cf. CHEVERELL, CHICKERELL.

Buckfast D [*Bucfæsten* 1046 KCD 1334, *Bucfestre* DB, *Bukfastmore* 1240 For], **Buckfastleigh** D [*Legh* 1286 Subs, *Leghe Bufestre* 1310 Ep, *Bucfastenlegh* 1353 Ch]. Buckfast is 'stronghold of the (male) deer' (OE *bucc*). Cf. FÆSTEN. Buckfastleigh is 'the LĒAH (probably forest) of Buckfast'.

Buckham Do [*Bochenham* DB, *Bukeham* 1285 FA]. '*Bucca*'s HĀM.'

Buckholt Ha [*Bocolt* DB, *Bocholte* 1231 Cl, (forest of) *La Boukholte* 1272 Ipm], B~ Sx [*Boccholte* c 1123 BM]. OE *bōc-holt* 'beech wood'.

Buckhurst Ess [*Bocherst* c 1135 BM], B~ Sx [*Bochirst* 1234 Cl]. OE *bōc-hyrst* 'beech grove'.

Buckingham Bk [*Buccingahamm* 918 ASC, *Buccingaham* c 1000 Saints, *Bochingheham* DB]. 'The HAMM of *Bucca*'s people.' **Buckinghamshire** is *Buccingahamscir* 1016 ASC.

Buckland, a common name, represents OE *bōcland* 'land held by charter' in contra-distinction to *folcland*. With the exception of one Buckland in Lincolnshire, the name is found only in the south. There are no examples in Wo, W, O, Ess. **Buck-land** Bk [*Bocheland* DB, *Buchelant* 1090 RA], B~ Brk [*æt Boclande* 957 BCS 1005, *Bocheland* DB], **East & West** B~ D [*Boche-landa, Bochelant* DB, *Estbokland, West Boc-launde* 1242 Fees], **North** B~ D [*Bochelant* DB, *Boclande* 1228 FF], B~ **Brewer** D [*Bochelanda* DB, *Boclande Bruere* 1290 Ch], **Egg** B~ D [*Bochelanda* DB, *Eckebokelond* 1221 Cl, *-bocland* 1242 Fees], B~ **Filleigh** D [*Bochelan* DB, *Bokelondfilleghe* 1333 Ep], B~ **Monachorum** D [*Bōcland* c 970 BCS 1247, *Bocsætena higweg* 1031 KCD 744, *Bochelanda* DB], B~ **in the Moor** D [*Bochelanda* DB, *Bokelaund in the More* 1318 Ch], B~ **Tout Saints** D [*Bochelanda* DB, *Bocland Touzseyns* 1303 FA], B~ **Newton** Do [*Boclond* 941 BCS 768, *Boche-lande* DB, *Niweton and Boclande* 13 Misc], B~ **Ripers** Do [*Bocheland* DB, *Boklond Ripers* 1359 FF], B~ Gl [*Bochelande* DB], B~ Ha [*Buclond* 1530 AD], B~ He [*Bouc-land* 1230 P], B~ Hrt [*Bochelande* DB, *Bocland* 1192 FF], B~ K nr Dover [*Boclond* 825 BM, *Bocheland* DB], B~ K nr Teynham [*Bocheland* DB, *Bocland* 1086 KInq, *Bok-lond next Tenham* 1309 FF], B~ Li (extinct) [*Bochelanda* c 1115 LiS, *Bocland* 1212 Fees], **West** B~ So [*Bocland* 904 BCS 610, 1065 Wells], B~ **Denham** So [*Boclande* 951 BCS 889, *Bochelande* DB, *Bokelonddynham* 1329 Ep], **Minchin** B~ or B~ **Sororum** So [*Bokeland* 12 Buckland, *Bocland* 1228 Cl], B~ **St. Mary** So [*Bochelande* DB, *Bokeland S. Marie* 1346 BM], B~ Sr [*Bochelant* DB, *Boclond* 1242 Fees].

B~ **Brewer** D was held by William Briwerre in 1219 (Ass). Brewer is an OFr byname and family name.—B~ **Denham** So was held by Oliver de Dinant in 1205 (Cur). Dinant from DINAN in Brittany.—**Egg** B~ D was held before the Conquest by *Hec* (or *Hecus, Heche*) vice-comes (DB), no doubt OE *Heca* pers. n.—B~ **Filleigh** D was held by Nicholas de Fyleleye in 1284–6 (FA). Cf. FILLEIGH.—**Minchin** B~ So was the site of B~ Priory. *Minchin* is OE *mynecen* 'nun'.—B~ **Monachorum** D was the seat of an abbey.—B~ **Ripers** Do was held by John de Ripirs in 1285 (FA). Ripers, in early sources often *de Ripariis*, is identical with Rivers from RIVIÈRE in Normandy.—B~ **St. Mary** So from the dedication of the church.—B~ **Tout**

Saints D was held by William de Tuz Seynts in 1242 (Fees). TOUSSAINT is a place in Seine-Inf. (France).

Bucklebury Brk [*Borgeldeberie* DB, *Burg-hildebiri* 1267 BM]. '*Burghild*'s BURG.' *Burghild* is a woman's name.

Bucklesham Sf [*Bukelesham* DB, *Bucles-ham* 1286 Ch]. '*Buccel*'s HĀM.' *Buccel is a derivative of *Bucca*.

Buckminster Le [*Bucheminstre* DB, *Bucce-menistre* 1180 P]. '*Bucca*'s minster or church.'

Bucknall Li [*Bokenhale* 806 BCS 325, *Buchehale* DB], B~ St [*Bucenhole* DB, *Buccenhal* 1227 Ch, *Bukenhal* 1230 P]. '*Bucca*'s HALH.' Cf. BUCC.

Bucknell O [*Buchehelle* DB, *Buckehulla* c 1130 Oxf, *Bukenhull* 1254 Val], B~ Sa [*Buchehal(l)e* DB, *Bukehill* 1175 P, *Buken-hull* 1270 Ch]. '*Bucca*'s hill.' Cf. BUCC.

Bucknowl Do [*Bubecnolle* 1285, *Bouknolle* 1412, *Boueknolle* 1431 FA]. '*Bubba*'s knoll.'

Bucks Cross D [*Bochewis* DB, *Bochiwis* 1168 P, *Buckish* 1325 Ipm]. '*Bucca*'s HĪ-WISC or homestead.'

Buckshaw Do [*Buggechage* 1194, *Bug(g)e-hag* 1195 ff. P, *Bukshawe* 1412 FA]. In spite of the early g-spellings probably 'buck shaw', 'grove of the deer'.

Buckton He [*Buctone* DB, *Buctun* 1252 Ch, *Buketon* 1292 QW], B~ Nb [*Buketun* 1208–10 Fees], B~ YE [*Bochetone* DB, *Buchetona* c 1130 BM, *Bucatona* c 1150 YCh 1154]. '*Bucca*'s TŪN.'

Buckworth Hu [*Buchesworde* DB, *Buckes-wrda* 1180 BM, *Buckewrth*' 1225 FF]. '*Bucc*'s or *Bucca*'s WORP.' Cf. BUXHALL.

Budbridge Wt [*Botebrigge* 1287–90 Fees, *-brigg* 1346 FA]. '*Bōta*'s bridge.'

Budbrooke Wa [*Budebroc* DB, 1190 P, *Buddebroc* 1236 Fees]. Identical with *Bud-dan broc* 978 KCD 620 (in boundaries of Tredington Wo). *Budda* is a well evidenced pers. n. Both names mean '*Budda*'s brook'.

Budby Nt [*Butebi* DB, *Buttebi* 1169 P]. '*Butti*'s BY.' *Butti* is an ODan, OSw, ON pers. n. Also ON *Buti* occurs.

Bude Co [*Bude* 1400 Ep, *Bedebay* 1468 Pat]. Bude is also the name of the stream at Bude [*the Bedewater* 1577 Harrison]. If originally a river-name, it may be identical with BOYD.

Būdle Nb [*Bolda* 1166 P, *Bodle* 1197 P, 1212 Fees, *Bodhill* 1242 Fees]. OE *bōpl* 'dwell-ing'; cf. BŌPL.

Budleigh, East, D [*Bodelie* DB, *Budelega* 1125–9 Fr, *Buddeleg* 1212 Fees]. '*Budda*'s LĒAH.' Cf. SALTERTON.

Budna Bd [*Budenhou* 1169 P, 1200 Cur, *Buddenho* 1195 Cur]. '*Budda*'s HŌH or spur of land.'

Budock Co [*ecclesia Sancti Budoci* 1208 Cl]. On St. Budock see ST. BUDEAUX.

Budworth, Great, Chs [*Budewrde* DB, *Buddewrtha* R 1 BM], **Little B~** Chs [*Bodeurde* DB, *-worth* 1291 Tax]. '*Bud(d)a*'s WORÞ.'

Buersill La [*Brideshull* 1228 Ass, *Berdeshull* 1292 QW, *Birdishill* 1324 Ct]. If the instance of 1228 belongs here, '*Bridd*'s hill'. Cf. BIRDSALL.

Buerton Chs nr Audlem [*Burtune* DB], **B~** Chs nr Chester [*Beurton* 1286 Court, *Buyrton* 1307 Ch]. Either OE *Byrh-tūn* (cf. BURTON (2) or OE *Bȳre-tūn* 'TŪN with a byre'. The first alternative is preferable.

Bugbrooke Np [*Buchebroc* DB, *Bukebroc* 1201 Cur, *Buckebrok* 1247 Ch]. 'The brook of the buck(s)' or '*Bucca*'s brook'.

Buglawton Chs [*Lauton* 1278 Ipm, *Buggelauton* 1287 Court]. Originally OE *Hlāwtūn* 'TŪN by the hill' (the Cloud); cf. LAWTON, LOWTON. The meaning of the addition *Bug-* is obscure. Possibly it is *bug* 'a bugbear, scarecrow'. Or it may be *Bugg*, the family name.

Bugley W [*Bugelighe, Bogelegh, Bokele* E 1 BM]. Perhaps '*Bucge*'s LĒAH.' OE *Bucge* is a woman's name.

Bugsworth Db [*Buggisworth* 1275 RH, *Bougesworth* 1285 For, *Buggesword* 1315 Ipm]. See WORÞ. The first el. is a pers. n., possibly an OE **Buduc*, a cognate of *Budeca* in BUTLEIGH.

Bugthorpe YE [*Buchetorp, Bughetorp* DB, *Buggatorp* 1157 YCh 354, *Buggethorp* 1219 FF]. '*Buggi*'s thorp.' ON *Buggi* is used as a byname and OSw *Bugge* as a Christian name.

Buildwas Sa [*Beldewes* DB, *Billewas* 1158 P, *Buldewas* 1169 P, c 1180 Fr, *Byldewas* 1248 FF]. Second el. OE *wæsse* 'swamp'. The first may be OE *bylda* 'builder' or *gebyldu* (= *gebytlu*) 'building'.

OE **bula** 'bull, bullock', corresponding to ON *boli*, ODan *bul*, MLG *bulle*, is not evidenced, but must have been in common use. It is the first el. of many pl. ns., as BOLNEY, BOOLEY, BOULMER, BULCAMP, BULLEY, BULMER, BULWICK. Some names may contain the Scand word, as BOWNESS We, BULBY. But there must also have been a pers. n. *Bula* (and also *Bulla*), from which it is not always easy to distinguish *bula*.

Bulbridge W [*Bolebrigge* c 1200 Salisbury, *-brygg* 1291 Tax, *Bulebrige* c 1200 Salisbury]. 'Bullock bridge.'

Bulby Li [*Bolebi* DB, 1190 P, *Bollebi* 1202 Ass]. First el. probably a pers. n., very likely *Bole* DB (Li), which may be a byname from OScand *buli* 'bull'. But Bulby may quite well be 'village where there was a bull'.

Bulcamp Sf [*Bulecampe* DB, *-camp* 13 BM]. 'Bull or bullock enclosure.' Cf. CAMP.

Bulcote Nt [*Bulecote* DB, 1236 Fees]. 'Shelter for bulls or bullocks.'

Bulford W [*Bultisford* 1178 BM, *Bultesforda* Hy 2 (1270) Ch, *Bultiford* 1199 Ch, *Bulteforde* 1291 Tax]. Perhaps 'ford where ragged robin grew'; cf. BOULTHAM. The first el. may be an adj. **bulutig* formed from *bulut* or an OE *Bulut-īeg* 'ragged robin island'. If so, the *s* of some early forms is intrusive.

Bulkeley (bo͝oklǐ) Chs [*Bulkelegh* 1259, *-lee* 1260 Court]. OE *bulluca-lēah* 'bullock pasture'.

Bŭlkington W [*Bulkinton* 1207 Cur, 1242 Fees, 1244 Cl], **B~** Wa [*Bochintone* DB, *Bulkintona* 12 DC, *Bulkinton* 1232 BM]. 'The TŪN of *Bulca*'s people.' OE *Bulca* is apparently evidenced in *Bulcan þyt* BCS 225.

Bulkworthy D [*Buchesworde* DB, *Bulkewurthi* 1228 FF, *-vurdh* 1230 P]. '*Bulca*'s WORÞ'; cf. BULKINGTON. Or 'bullock farm'.

Bulley Gl [*Bulelege* DB, *Bullega* 1169 P, *Bulleye* 1265 Ipm]. 'Pasture for bullocks.'

Bullingham (-nj-) or **Bullinghope** He [*Boniniope, Boninhope* DB, *Bullingehope* 1242 Fees, *Bolingehop* 1236 Ipm]. 'The valley (OE HOP) of **Bul(l)a*'s people.' The exchange of *-hop* for *-ham* is late.

Bullington Ha [*Bulandun* 1002 KCD 707, *Bolende* DB, *Bolyndon* 1316 FA]. '**Bula*'s DŪN.' Or the first el. may be BULA 'bull'.

Bullington Li [*Bolintone* DB, *Bulingtuna* c 1115 LiS, *-tona* 12 BM, *Bulington* 1195 P]. 'The TŪN of **Bula*'s people.'

OE **bulluc** 'bullock'. See BULKELEY, BULK-WORTHY.

Bulmer Ess [*Bulenemera* DB, *Bulemere* 1178 AC], **B~** YN [*Bolemere* DB, *Bulemer* 1130, 1156 P]. OE *bulena mere* 'bulls' lake'.

Bulphan Ess [*Bulgeuen* DB, *Bulewefen* 1238 Subs, *Bureghefen* 1244, *Burgefen* 1247 FF]. OE *burge-fen* 'fen belonging to the BURG' (Tilbury). The curious *l* for *r* is due to Norman influence.

Bulstrode Bk [*Burstroda* 1185 f., *Burestroda* 1193 f., *-strod* 1195 ff. P, *Bolestrode* 1195 Cur]. Identical in meaning with BULPHAN and with the same change of *r* to *l*. Second el. OE *strōd* 'marsh'.

OE **bulut** 'ragged robin'. See BOULTHAM, BULFORD.

Bulverhythe Sx [*Burewarehethe* 1229 Pat, *Bulewareheda* c 1150 Fr]. OE *burgwara hȳþ* 'the landing-place (HȲÞ) of the people of the BURG' (i.e. Hastings). For the change of *r* to *l* cf. BULPHAN, BULSTRODE. See -WARU.

Bulwell Nt [*Buleuuelle, Bulwelle* DB, *Bulewell* 1165, 1169 P]. Either '**Bula*'s stream' or 'bulls' stream'.

Bulwick Np [*Bolewyk* 12 NS, *Bulewic* 1163 P]. 'Bull(ock) farm.' Cf. COWICK, SHAPWICK &c.

Bulworthy D [*Bolewurði* 1168 P, *Buleworthy* 1233 Cl]. '*Bula's WORÞIG' or 'bullock WORÞIG'.

Bumpstead, Helion & Steeple, Ess [*Bum(m)esteda, Bunsteda* DB, *parua Bumsteda* 1166 P, *Bumpsted Helyun* 1238 Subs, *Bumstede ad Turrim* 1259 Ipm, *Stepilbumstede* 1261 FF]. Possibly contracted from *Bun-hāmstede* or the like. The first el. might then be OE *bune* 'reeds' or *Bune* a river-name; cf. *Bune*, the old name of **Claydon Brook** O, Bk [*Bunan* 995 KCD 1289].

Helion B~ was held by Tihel Britto or de Helion (from HELLÉAN in Brittany) in 1086 (DB). **Steeple B~** from the church steeple.

Bunbury Chs [*Boleberie* DB, *Bunnebury* 1259 Court, *Bunebury* 1279 Chester]. '*Buna*'s BURG.'

Buncton Sx [*Bongetune* DB, *Bungeton* 1207 Cur, 1242 Fees]. Very likely OE *Buningatūn* 'the TŪN of *Buna*'s people'.

Bungay Sf [*Bongeia, Bunghea* DB, *Bungheia* 1174, *Bungeia* 1175, 1191 P]. Probably OE *Buninga-ēg* 'the island of *Buna*'s people'.

Bunny Nt [*Bonei* DB, *-a* 1176 P, *Buneya* 1227 Ep]. 'Reed island' or 'island on the river *Bune*'. Cf. BUMPSTEAD. Bunny is on a stream.

Buntingford Hrt [*Buntingeford* 1185 TpR]. 'Ford haunted by buntings.' *Bunting*, the name of a bird, is first found in ME, but may very well be older.

Bunwell Nf [*Bunewell* 1198 FF, 1254 Val]. 'Reed stream' (first el. OE *bune* 'reed').

Bupton Db [*Bubandun* 1002 E, *Bubedune* DB, *Bubbendon* 1169, *Bubendona* 1197 P]. '*Bubba*'s hill.'

OE **būr** 'cottage' &c. See BOWER(S), BURES, BURCOT.

Burbage on the Wye Db [*Burebeche* 1172 P], **B~** Db in Padley [*Burbache* 1200–30 PNDb], **B~** Le [*Burhbeca* 1043 Th, *Burbece* DB, *Burbache* 1242 Fees], **B~** W [*Burhbece*, gen. *Burgbeces* 961 BCS 1067, *Burhbeces* (gen.) 968 ib. 1213, *Burhbec* c 1000 KCD 1312, *Burbetce* DB]. Burbage on the Wye, B~ in Padley and B~ W are undoubtedly OE *burg-bece* 'brook or valley of the BURG'; two of the OE examples actually denote a brook. Cf. BÆCE. Burbage Le is generally held to contain OE *bæc* 'hill, ridge'. The place is on the slope of a hill, but Burbage House is lower down near Soar Brook. The probability seems to be that this Burbage is identical in origin with the other three. If the second el. is OE *bæc*, the form is that of the old locative in *-i*, OE *-bece* from *-baki*.

Burcombe W [*Brydancumb* 937 BCS 714,

Bredecumbe DB, *Brudecumbe* 1242 Fees]. Cf. BRIDMORE W and next name.

Burcot O [*Bridicote* 1198 FF, *Bridecote* c 1225 Fridesw, *Brudecot* 1279 RH]. Possibly OE *brȳdecot* 'COT of the bride'. If Bridmore and Burcombe contain a pers. n. *Brȳda*, Burcot may be '*Brȳda*'s COT'.

Burcot Wo [*Bericote* DB, *Byrcote* 1275 Subs]. OE *byrig-cot* 'COT belonging to the BURG' (Bromsgrove, a royal manor).

Burcott Bk [*Burcote* 1184 P], **B~** Sa [*Burchota* 1176 P, *Burkot* 13 Misc], **B~** So nr Wells [*Burcotan* 1065 Wells, *Burecote* 1243 Ass]. The source is either OE *burg-cot* 'COT belonging to a BURG' or OE *būr-cot*, which might mean 'dwelling-place, cottage' in contradistinction to a sheep-cote &c. The first alternative is preferable in the case of Burcott Bk, which is near BIERTON. The second explanation holds good at any rate for *Burcote* 903 BCS 602 (Ha).

Burdale YE [*Bredhalle* DB, *Bredall* 1202 FF, *Breddal* 1246 FF]. OE *bred-hall* 'hall made of boards' (OE *bred*).

Burdon Du [*Byrdene* c 1050 HSC, *Bireden* 1196 P]. 'Valley with a byre' (OE *bȳre*).

Burdon, Great, Du [*Burdune* Hy 1 DST, *Burdon* 1195 (1335) Ch], **Burdon Head** YW [*Burghedurum* DB, *Burgedun* 1178 P]. 'Hill with a BURG', OE *burg(e)-dūn*. The same is the etymology of **Burden** YW at Harewood [*Burgedon, -dun* 1219 FF].

Bure Ha [*Bevra* 12 (1313) Ch, *Boure* 1316 FA]. Perhaps OE *be ōfre* '(the place) by the shore'. The place is near the sea-shore.

Bure R Nf [*Bure* 1577 Harrison]. A backformation from BRISTON (olim *Burston*) or BURGH.

Bures (būrz), **Mount** or **Little,** Ess [*Bura* DB, *Bures* 1199 Cur, *Bures Parva* 1254 Val, *Bures ad montem* 1290 BM], **B~** St. **Mary** Sf [*Adburam* DB, *Bure, Bura* ib., *St. Mary in Buri* c 1075 Fr, *Buren* c 1095 Bury, *Buras* c 1180 Bodl, *Bures Nostre Dame* 1282 Misc, *Bures Seinte Marie* 1359 BM]. OE BŪR in some sense, e.g. 'cottage'; cf. BOWER. The abnormal vowel of *Bures* is due to Norman influence.

Mount B~ and B~ St. Mary are opposite to each other on the Stour. Mount B~ is on the slope of rising land.

Burford O [*Beorgfeord* 752 ASC, *Bureford* DB, 1203 Ass]. OE *beorgford* 'ford by a hill or tumulus', if the first form belongs here. Otherwise identical with BURFORD Sa.

Burford Sa [*Bureford* DB, 1252 Fees, 1266 Ch]. Probably OE *burg-ford* 'ford by the BURG'. The place is c 1 m. from Tenbury.

OE **burg, burh** (gen. *burge, byrh,* dat. *byr(i)g, byrh*) is a very common element in pl. ns. The meaning is usually 'fortified place, fort'. Very often the reference is to a Roman or other pre-English fort; very likely this is often the meaning where it cannot at present be proved. Sometimes an Anglo-Saxon fort is referred to. In many cases *burg* probably denotes a fortified manor, and even a meaning 'manor' often occurs. Sometimes the meaning is 'town, borough'. It is impossible to decide in each case what is the exact meaning of the word.

Burg often occurs alone as a pl. n. See BROUGH, BURGH, BURROW, BURY, BERE (FERRERS), BERRY. As a second el. it is very common and appears variously as *-borough*, *-burgh* (from OE *-burh*) and *-berry*, *-bury* (from OE *-byrig*). As a first el. the word usually appears in the uninflected form *burg-* or the gen. form *burge-* whence BOUR-, BUR- (as in BOURTON, BURTON, BURWELL), BOROUGH- (as -BRIDGE), BURRA- (as -DON). Cf. also BURPHAM, BULPHAN &c. The gen. form *byrig-* (*byrh-*) occurs fairly often, usually a good deal changed in form, as in BERICOTE, BIRLEY, BIERTON, BERRINGTON, BURRINGTON, &c. Cf. BURTON (2), BOARHUNT.

It is doubtful if Engl *burrow* (for animals) is identical with *burg*. It is ME *borow*, but not recorded in OE. It occurs in some names, as MUSBURY La.

Burgate Ha [*Borgate* DB, *Burgat* 1227 Ch, 1242 Fees], B~ Sf [*Burgata* DB, *-gat* 1204 Cur, *-gate* 1254 Val], B~ Sr [*la Burgate* 1259 PNSr]. OE *burg-geat* 'gate of a BURG'.

Burgh, a common name, is **1.** generally OE *burg* 'fort': **Burgh** (bru̇f) **by Sands** Cu [*Burgh* c 1220 StB, *Burgh on the Sands* 1247 Ipm], B~ **on Bain** Li [*Burg* DB, *Burc* c 1115 LiS, *Burgus super Beyn* 1291 Tax], B~ (bŭru) **le Marsh** Li [*Burg*, *Burch* DB, *Burc* c 1115 LiS], B~ **next Aylsham** Nf [*Burc* DB], B~ **Parva** Nf [*Burg* 1254 Val], B~ **St. Margaret** or **Flegg Burgh** Nf [*Burc* DB, *Burg Sancte Margarete* 1254 Val, *Burgh in Fleg* 1342 BM], B~ **St. Peter** or **Wheatacre B~** Nf [*Qwetacre Sancti Petri* 1254 Val, *Whettaker Borowgh* 1515 BM], B~ (bŭru) Sf [*Burc* DB, *Burg* 1254 Val], B~ (bŭru) **Castle** Sf [*Burch* DB, *Burc* 1168 P, *Borough-Castell* 1281 Bodl], **Burghwallis** or **Burgh Waleys** YW [*Burg* DB, *Burghwaleys* 1283 Ch].

In some cases the name refers to a Roman fort, e.g. B~ **by Sands** Cu, B~ **Castle** Sf (by the old *Garianno*). In most cases the reason for the name is not obvious, and in some cases Burgh refers to a fortified manor. **Burgh St. Peter** seems originally to have been called WHEATACRE or WHEATACRE BURGH.—**Burghwallis** belonged to the Waleys family in the 12th cent. Henry Waleis of Burg occurs in 1170 ff. (P). *Waleys* means 'Welsh'.

2. OE *beorg* 'hill, mound': **Burgh Apton** Nf [*Berc* c 1050 KCD 907, (into) *Berhe* 1043, *Beorh, Apetune* c 1060 Wills, *Berc* DB, *Berc, Apetone* 1254 Val], **Burgh** Sr [*Berge* DB, *-s* 1196 FF, *Berga* 1199, *Berghes* 1206 Cur]. Cf. also SOUTHBURGH.

Burgh Apton is really Burgh and Apton. In 1321 Bodl are mentioned the rectory of *Bergh*

and chapel of *Apton*. Apton may be '*Api*'s TŪN'. ODan, OSw *Api* pers. n. is recorded. But a reduced form of Appleton is not impossible. A place called *Appelsco* is mentioned in connexion with Apton c 1060 Wills.

Burgham or **Burpham** Sr [*Borham* DB, *Burham* 1242 Fees, 1276 Cl]. OE *Burg-hām* or *-hamm* 'HĀM or HAMM by the BURG'.

Burghclere. See CLERE.

Burghfield Brk [*Beorhfeldinga gemaere* 946–51 BCS 888, *Borgefel* DB, *Bergefelda* 1167 P]. 'FELD by the hill.' See BEORG.

Burghill He [*Burgelle* DB, *Burchil* 1169 P, *Burghulle* 1212 RBE]. 'Hill of the fort.'

Burghley Np [*Burglea* DB, *Burgelai* 1163 P, *Burgele* 1227 Ch]. OE *burge-lēah* 'LĒAH (probably 'wood') belonging to the BURG' (i.e. Stamford).

Burham K [*Burhham* 10 BCS 1321 f., 995 KCD 688, *Burham* 1016–20 Th, *Borham* DB]. 'HĀM by the BURG' (perhaps Rochester).

Buriton (-ĕ-) Ha [*Buriton* 1227 Ch, 1263 Ipm, *Bergton* 1229 Cl]. OE *Byrg-tūn* 'TŪN by a BURG' or *Beorg-tūn* 'TŪN by a hill'?

Burland Chs [*Burlond* 1260 Court]. Presumably OE *burg-land* 'land belonging to the BURG'.

Burland YE [*Birland* DB, *-lande* 1193 P]. OE *byrh-land* 'land of the BURG'. Smith, PNER, derives the first el. from OE *bȳre* 'cowshed'. This is a possible alternative.

Burlescombe D [*Burewoldiscumbe* 1173–5 (1329) Ch, *Burlescumb* 1249 FF]. '*Burgweald*'s CUMB or valley.'

Burleston Do [*Burdalueston, Bordelestone* 939 BCS 738 f., *Burdeleston* 1212 Fees]. The first el. is no doubt a pers. n., but its original form is doubtful. The would-be OE forms are in a poor text.

Burley Hill Db [*Burleye* 1251 Ch], B~ Ha [*Burgelea* 1178 P, *Burle* 1251 Ch], B~ Ru [*Burgelai* DB, 1179 P], B~ Sa [*Burlegh* 1233 Cl, *Borle* 1292 QW], B~ YW nr Leeds [*Burchele* c 1200, *Burghlay* 1332 Kirkst], B~ **in Wharfedale** YW [*Burhleg* c 972 BCS 1278, (on) *Burhleage* c 1030 YCh 7, *Burghelai* DB, *Burghlay in Quervesdale* c 1300 BM]. OE *burg(e)lēah* 'LĒAH by or belonging to a BURG'.

Burleydam Chs [*Burley* c 1130 Mon, *Burle* 1253 Ch]. Identical with BURLEY. The addition is no doubt *dam* 'weir'.

Burlingham Nf [*Berlingaham, Sutberlingeham* DB, *Berlingeham* 1177 P, 1207 Cl, *Birlingham* 1198 FF, *Northbirlingham Sancti Andree, Sancti Petri* 1254 Val]. The first el. seems to be identical with BARLING, BARLINGS, BIRLING. Second el. HĀM.

Burlton He [*Burghelton, Burweltun* 1242 Fees], B~ Sa [*Burghelton* 1241 FF, *Burghulton* 1285 FA]. 'TŪN by a hill with a BURG' (an OE *Burg-hyll-tūn*). Burlton He is close to BURGHILL.

Burmarsh K [*Burwaramers* 616–18 BCS 837, *Burwaramersc* 1016–20 Th, *Burwarmaresc* DB]. 'The marsh of the *burgware* or people of the BURG or town' (i.e. Canterbury). Cf. -WARU.

Burmington Wa [*Burdintone* DB, *Breminton* 1170 P, *Burmenton* 1273 Cl]. OE *Beornmundingatūn* 'the TŪN of *Beornmund*'s people'.

Burn Hill Bk [*Burnhull* 1276 RH], **Burn** YW [*Byrne* c 1030 YCh 7, *Birne* 1279–81 QW]. OE *byrgen* 'tumulus, burial-mound'.

OE **burna (burne)** 'spring; brook, stream', now *bourne, burn*, corresponds to OHG *brunno*, OFris *burna* &c. 'spring, fountain'. In English pl. ns. the usual meaning is 'stream'. The original form *brunna* is still met with in pl. n. forms (see e.g. BOURNE Li). There is no reason to look upon these as due to Scandinavian influence. The word is often used alone as the name of brooks, e.g. **Bourne** Bk [*Burne* c 1540 Leland], Ess [*Burne* 1577 Harrison], **Burn** Y [*Brunne* 12 Fount] and sometimes as that of a place, as BOURN Ca, BOURNE Li. The dat. plur. *burnum* is the source of BURNHAM Li (two). *Burna* is common as the second el. of names of streams and as the first and second el. of names of villages and homesteads.

Burnage La [*Bronage, Bronnegge, Brownegg* 1322 LaInq]. Perhaps 'brown hedge'.

Burnaston Db [*Burnulfestune* DB, *Brunolviston* 1242 Fees, *Brunufystone* 13 Derby]. '*Brūnwulf*'s TŪN.' This name is not well evidenced, but is found also in BURNSIDE.

Burnby YE [*Brunebi* DB, -*by* 1201 FF, *Brunnebia* c 1155 Thurgarton Cart]. The place is near NUNBURNHOLME, and the probability is that the first el. of the name is OE *burna* 'stream'. Cf. Sw BRUNNBY 'BY by a spring'.

Burneston YN [*Brennigston* DB, *Brineston* 1246 Ipm]. '*Brȳning*'s TŪN.'

Burnett So [*Bernet* DB, 1107 (1300) Ch, *Burnet* 1227 FF, 1327 Subs]. OE *bærnet* 'burning, place cleared by burning'; cf. BARNET. The *u* may be due to influence from the verb *burn*.

Burnham Bk [*Burnham* c 880 BCS 553, *Burneham* DB, 1165 P, 1254 Val], **B~ on Crouch** Ess [*Burneham* DB, *Burnham* 1201 Cur, 1254 Val], **B~** Nf [*Brunham, Bruneham* DB, *Brunham* 1158, *Burnham* 1191 P, *Burneham* 1271 Ch]. OE *Burn(e)hām* 'HĀM on a stream'.

Burnham Nf consists of several villages, whose names are distinguished by various additions, **B~ Deepdale** [*Depedala* DB, *Depedale* 1381 BM] 'the deep valley', **B~ Market, B~ Norton** [*Norton* 1300 Ch, *Brunham Norton* 1457 AD], **B~ Overy** [*Brunham, Overia* 1457 AD] 'Burnham over the water', **B~ Sutton** [*Burnham Sutton* 1242 Fees], **B~ Thorpe** [*Brunhamtorp*

1199 P, *Burnhamtorpe* 1201 Cur], **B~ Westgate** [*Brunham Westgate* 1276 AD] 'the western gate'.

Burnham Li nr Barton upon Humber [*Brune* DB, *Brunum* c 1115 LiS, *Brunnum* 1157 YCh 354], **B~** Li in Axholme [*Brune* DB, *Brunhom, -ham* c 1200 DC]. OE *burnum* or *brunnum*, dat. plur. of BURNA, 'the streams or springs'. The latter meaning seems preferable for at least the second of the two names.

Burnham So [*Burnhamm* c 880 BCS 553, *Burneham* DB, *Burnham* 1170 P]. 'HAMM on the stream' (the Parret).

Burniston YN [*Brinnistun* DB, *Brinigstun* 1091–5 YCh 863, *Briningeston* 1219 Ass, *Brineston* 1234 FF]. '*Brȳning*'s TŪN.'

Burnley La [*Brunlaia* 1124, *Brunley* 1154 YCh 1486, 1475]. 'LĒAH on the burn or on the BRUN.' Brun may represent OE *Brūne* 'the brown one'; cf. BRUNSHAW. A place on the Brun is called BROWNSIDE.

Burnsall YW [*Brineshale* DB, *Brinneshale* c 1160 YCh 784, *Brunnishall* 1202 FF]. '*Brȳni*'s HALH.'

Burnside or **Burneside** We [*Brunolvesheved* c 1235 CWNS xxiv, *Brunolvishefd* c 1255 Kendale]. 'The headland or hill belonging to *Brūnwulf*.' Cf. BURNASTON.

Burntshiel. See ESPERSHIELDS.

Burntwood St [*Brendewode* 16 PNSt]. 'Burnt wood.'

Burpham Sx [*Burhham* c 920 Gale, *Bercheham* DB, *Bercham* 1121 AC, *Burcham* c 1140 AD]. OE *Burg-hām* 'HĀM by the BURG'. There is an ancient earthwork here.

Burradon Nb nr Newcastle [*Burgdon* Hy 2 Percy, *Buruedon* 1242 Fees], **B~** Nb nr Alwinton [*Burhedon* J Sc, *Burwedon* 1242 Fees]. OE *burg-dūn* 'hill with a BURG'.

Burrill YN [*Borel* DB, *Burell* 1316 FA]. OE *burg-hyll*, cf. BURGHILL.

Burringham Li [*Burringham*, 1199 P, *Burningham* 1218 Ass, *Burnygham* 1281 QW]. 'The HĀM of the dwellers on the stream.' B~ is on the Trent.

Burrington D [*Bernurtona, Bernintone* DB, *Burumtone* 1277, *Buringtone* 1284 Ep]. '*Beornwynn*'s TŪN.' *Beornwynn* is a woman's name.

Burrington He [*Boritune* DB, *Buriton* 13 BM], **B~** So [*Buringtune* R 1 Berk]. OE *Byrigtūn* 'TŪN by the BURG'.

Burrough Green Ca [*Burg* c 1044 Wills, 1254 Val, *Burch* DB]. OE *burg* 'fort'.

Burrow (or Burrough) on the Hill Le [*Burg* DB, 1254 Val, *Erthburgh* 1327 Subs], **B~** La nr Lancaster [*Burg* c 1200 CC], **B~ with Burrow** La [*Borch* DB]. OE *burg* 'fort'. Burrow with Burrow has remains of a Roman fort. The other Burrow

La is on a Roman road. At Burrow Le there must have been an earthwork.

Burrow So [*æt þam Beorge* 1065, *Bergh* 1325 Wells]. OE *beorg* 'hill'. There are several Burrows in D, which are derived in PND from OE *beorg*, though there are decisive early forms only for one or two.

Burscough La [*Burscogh* c 1190 LaCh]. 'Wood by the BURG.' Cf. SKÓGR. The fort is referred to as *Burgastud* 'the site of the old burg' c 1190 LaCh.

Burshill YE [*Bristhil* 1172 YCh 1391, *-hill* 1203 FF, *-hall* J Ass]. Perhaps 'hill with a landslip or gap' (OE *byrst*). Cf. BRISTON.

Bursledon Ha [*Brixenden* 12, *Burstlesden* 14 VH, *Norbursedone* 13, *Bollesdon* c 1270 Ep]. Very likely OE *Beorhtsiginga* DŪN. Cf. BRIXTON.

Eurslem St [*Barcardeslim* DB, *Borewardes-lyme* 1242 Fees, *Burewardeslime* 1252 Ch]. Cf. LYME. First el. OE *Burgweard* or *Burgheard* pers. n. *Burgheard* is better evidenced.

Burstall Sf [*Burgestala* DB, *Burcstal* 1194 P]. OE *burg-stall* 'site of a BURG' or simply 'BURG or fort'.

Burstall Garth YE [*Berestal* 1115 YCh 1304, *Bristall* 1160–2 ib. 1307, *Birstal* 1228 Ep]. OE *byrgstall*, identical in meaning with *burgstall*. Cf. BIRSTAL.

Burstead Ess [*Burgestede* c 1000 BCS 1306, *Burghesteda* DB, *Parva Burgested* 1204 Cur]. OE *burgstede* 'site of a BURG', perhaps 'site of an old fort'.

Burstock Do [*Burewinestoch* DB, *Burgestoche* 1179 P]. '*Burgwine*'s or *Burgwynn*'s STOC.' *Burgwynn* is a woman's name.

Burston Bk [*Briddesthorne* c 1215 PNBk, 1275 Ipm, *Bridelestorn* 1227 Ass]. '*Briddel*'s thornbush.' *Briddel*, unrecorded, is a derivative of *Bridd*.

Burston Nf [*Borstuna* DB, *Birston* 1196 FF, 1199 Cur, *Burston* 1212 Fees, *-e* 1254 Val]. Possibly 'TŪN by the landslip' (OE *byrst*). Cf. BRISTON.

Burston St [*Burouestone* DB, *Bureweston* 1242 Fees, 1255 Ass, *Burcheston* 1278 Ass]. The first el. is hardly OE BURG. It might be OE *Burgwine* or *Burgwulf*.

Burstow Sr [*Burestou* 1121 AC, *Burstowe* 1247 Ch, *Birstowe* 13 BM]. OE *burgstōw* or *byrgstōw* 'place by a BURG'.

Burstwick YE [*Brocstewic, Brostewic* DB, *Brustewic* c 1215 YCh 1398]. Burstwick is close to Burton Pidsea. In the light of one DB form, *Burst-* may well be derived from *burgsæta-* 'of the dwellers by the BURG'. The name would then mean 'the cattle-farm belonging to Burton Pidsea'. Or the first el. may be OE *brōc-sætan* 'dwellers on the brook'.

Burtholme Cu [*Burtholm* 1256 Lanercost]. 'Holm on R Burth.' **Burth** R [*Burth* 1169

Lanercost] might be a back-formation from a Welsh name containing Welsh *buarth* 'fold'. Cf. BIRDOSWALD.

Burton is in most cases **1.** OE *Burh-tūn* 'TŪN by a burg' or 'fortified manor': **B~ by Tarvin** Chs [*Burtone* DB, *Brunburton* 1282 Court], **B~** Chs in Wirral [*Burton in Wirhal* 1287 Court], **B~** Db [*Burtune* DB], **B~ Do** nr Dorchester [*Burton* 1212 Fees, 1231 Cl], **B~ Do** nr Stalbridge [*Buretune* c 1250 Glaston], **East & West B~** Do [*Bureton* 1212 Fees, *Estburton* 1280, *West Burton* 1279 Ch], **Long B~** Do [*Buryton* 1285, *Burton* 1428 FA], **B~ Ha** [*Bourton* 1316 FA], **B~ Lazars** Le [*Burtone* DB, *Burgtun* 1237 Cl, *Burton Sancti Lazar'* 1254 Val], **B~ Overy** Le [*Burtone* DB, *Burton Novereye* 1285 Ch], **B~ on the Wolds** Le [*Burtone* DB], **Gate B~** Li [*Bortone* DB, *Burtuna* c 1115 LiS, *Geiteburtone* 1219 Ep], **B~ by Lincoln** Li [*Burton* DB, *-tuna* c 1115 LiS], **B~ Pedwardine** Li [*Burtun* DB, *Burton Pedewardyn* 1402 FA], **B~ on Stather** Li [*Burtone* DB, *Burtonstather* 1275 RH], **B~** Nb [*Burton* 1242 Fees, 1257 Ch], **B~ Latimer** Np [*Burtone* DB, *-ton* 1228 Cl, *Burton Latymer* 1482 AD], **West B~** Nt [*Burtone* DB, *-ton* 1195 P], **B~** Sa nr Much Wenlock [*Burtune* DB], **West B~** Sx [*Westburgton* 1230 P], **B~ W** [*Burinton* 1204 Cur, *Burton* 1237 Cl], **B~ Hastings** Wa [*Burhtun* 1002 Wills, *Bortone* DB], **B~ Dassett** Wa (see DASSETT), **B~** We nr Warcop [*Burton* 13 Misc], **B~ in Kendal** We [*Bortun* DB, *Burtun* 1090–7 Kendale], **B~ Agnes** YE [*Burtone* DB, *Burton Agneys* 1231 Ass, *Anneysburton* 1257 Ch], **Bishop B~** YE [*Burtone* DB, *Bisshopburton* 1376 AD], **Brandesburton** YE [*Brantisburtone, Branzbortune* DB, *Brandesburton* 1219 FF], **Cherry B~** YE [*Burtone* DB, *Nordburtona* c 1200 YCh 1117], **B~ Constable** YE [*Santriburtone* DB, *Burton Constable* 1285 Ch], **B~ Fleming** or **North B~** YE [*Burtone* DB, *Burton Flemeng* 1234 FF, *Burtona Flandrensis* Hy 3 BM], **Hornsea B~** YE [*Burtune* DB, *Horneseburton* 1260 Ipm], **B~ Pidsea** YE [*Bortune* DB, *Pydese Burton* 1230 FC], **Constable B~** YN [*Bortone* DB, *Roald-Burton* 1270 Ipm, *Burton Constable* 1301 Subs], **B~ Dale** YN [*Bortun* DB], **B~ on Ure** YN [*Burtone* DB, *Burton upon Yor* 1254 AD], **West B~** YN [*Borton* DB], **B~ Hall** YW [*Burhtun* c 1030, c 1050 YCh 7, 9, *Burtone* DB], **B~ Leonard** YW [*Burtone* DB, *Burton St. Leonard* 1280 Ch], **B~ in Lonsdale** YW [*Borctune* DB, *Burtona de Lanesdala* 1130 P].

2. OE *Byrh-tūn* 'TŪN by or belonging to a BURG', the first el. being *byrh*, the gen. of *burg, burh*: **Burton Coggles** Li [*Bertune* DB, *Birton* 1208 Cur, 1254 Val], **B~ Joyce** Nt [*Bertone* DB, *Birtun, Burtun* 1236 Fees, *Birton Jorce* 1348 Misc], **B~ upon Trent** St [*Byrtun* 1002 Wills, *Bertone* DB, *Burton super Trente* 1234 Ep], **Kirkburton** YW [*Bertone* DB, *Birton* 1208 FF].

3. Burton Bradstock Do [*Bridetone* DB,

-ton 1157 Fr, *Briditonia* 1157 Fr]. 'TŪN on R Bredy or Bride' (see BREDY).

4. Burton Sx [*Botechitone* DB, *Budincatona* 1135–50 BM, *Bodeghetone* c 1300 Sele, *Boudeketon* 1314 Ipm]. '*Budeca*'s TŪN'; cf. BUTLEIGH.

5. Burton Salmon YW [*Breiðetun* c 1030 YCh 7, *Brettona* c 1160 YCh 36]. 'Broad TŪN', probably OE *Brāda-tūn* Scandinavianized (with ON *breiðr* for OE *brād*).

B~ Agnes YE from Agnes de Albemarle, married to William de Roumare; she is witness to a deed concerning B~ c 1175 (YCh 677).— **Bishop B~** YE belonged to the Abp. of York. **—B~ Bradstock** Do belonged to BRADENSTOKE.—**Brandesburton** YE must have been held by one *Brand* before the Norman Conquest. **—Cherry B~** YE may have been noted for its cherries.—**B~ Coggles** Li is said to contain the word *coggles* 'cobble-stones'.—**B~ Constable** YE must have been held by a Constable before 1285.—**Constable B~** YN was granted to Roald, Constable of Stephen, Earl of Richmond, c 1100 (VH).—**B~ Fleming** YE from the Fleming family.—**Gate B~** Li may be 'the Burton where goats were kept' (from OScand *geit* 'goat'). Or *Gait* as a family name as in HAMPTON GAY O.—**B~ Hastings** Wa was held by Henry de Hasteng in 1242 (Fees). The family name is from HASTINGS Sx.—**Hornsea B~** YE is near HORNSEA.—**B~ Joyce** Nt was held by Geoffrey de Jorz in 1236 (Fees). Jorz may be from JORT in Calvados.—**B~ Latimer** Np was held by William le Latymer in 1323 (Ipm). The family name *Latimer* means literally 'interpreter'.—**B~ Lazars** Le was the seat of a hospital for lepers.—**B~ Leonard** YW presumably from the dedication of the church.— **B~ Overy** Le was held by Robert de Novereia in 1229 (Ep). Cf. BURNHAM OVERY.—**B~ Pedwardine** Li came to a son of Thomas de Pedwardine by marriage c 1280 (Ipm). The family hailed from PEDWARDINE He.—**B~ Pidsea** YE took its surname from a now drained mere called Pidsea [*Piddese mere* 1260, *Pidesse* 1285 Ipm]. The second el. is OE *sǣ* 'lake'. The first seems to be related to PIDDLE.—**B~ Salmon** YW is obscure. Salmon may be a family name. **—B~ on Stather** Li is more correctly B~ Stather. *Stather* is ON *stǫðvar*, plur. of *stǫð* 'landing-place'.

Burtonwood La [*Burtoneswod* 1228 Cl]. 'Wood belonging to Burton', a lost place [*Burton* 1200 P].

Burwardsley Chs [*Burwardeslei* DB, *Berewardesleya* c 1110, *Burewardesleia* c 1150 Chester], **Burwarton** Sa [*Burertone* DB, *Burwardton* 1194 f. P, *Burwarton* 1199 FF]. '*Burgweard*'s LĒAH and TŪN.'

Burwash (bŭrĭsh) Sx [*Burgersa* 12 AD, *Burhercse* a 1170 BM]. OE *burg-ersc* 'ERSC by a BURG'.

Burwell Ca [*Burewelle* 969 Chron Rams, 1060 Th, *Buruuelle* c 1080 ICC, DB, *Burewell* 1203 Cur], **B~** Li [*Buruelle* DB, *Burewelle* c 1110 Fr, *Burwell* c 1115 LiS, *Burgwelle* 1292 BM]. 'Spring or stream by a fort.'

Bury Hu [*Byrig* 974 BCS 1311], **B~** La [*Biri* 1194 P, *Bury* c 1190 LaCh], **B~** (bĕrĭ) **St. Edmunds** Sf [*Sanctæ Eadmundes stow*

c 995 BCS 1288, (on) *Byrig* c 1035 Wills, *Sancte Eadmundes Byrig* 1038 BM, *Bery* Lydgate, Thebes], **B~** Sx [*Berie* DB, *Biri* 1200 FF]. OE *byrig*, dat. of BURG 'fort, town'.

Bury St. Edmunds was originally *æt Bæderices wirde* 945 BCS 808, *Beadriceswyrð* c 1000 Saints. The name means '*Beaduric*'s WORÞ'. St. Eadmund (d. 870) was buried at the place, which came to be known as St. Edmundsbury and finally as Bury St. Edmunds.

Burythorpe YE [*Bergetorp* DB, 1198 Cur, *Berkerthorp* 1199 YCh 624, *Berewethorp* 1239 Ep, *Bergertorp* 1242 Fees]. ON *Biargarþorp* '*Biǫrg*'s thorp'. ON *Biǫrg*, gen. *Biargar*, is a woman's name.

Busby YN [*Buschebi* DB, *Magna Buskebi*, *Parva Buskeby* c 1185 YCh 582]. First el. OScand *buski* 'shrub'.

Buscot Brk [*Boroardescote* DB, *Burwardescota* 1130 P]. '*Burgweard*'s COT.'

Bushbury St [*Byscopesbyri* 996 Mon, *Biscopesberie* DB]. 'The bishop's manor.'

Bushby Le [*Buszebia* 1175, *Bucebi* 1176 P, *Busseby* 1270 Ipm, 1327 Subs]. OScand *Buts bȳr* 'the BY of *Butr*'. *Butr* is an ON pers. n.

Bushey Hrt [*Bissei* DB, 1196 P, *Bichseyia* 1230 P, *Bisheye* 1230 Cl]. The name has been derived from OFr *boisseie* 'place covered with wood' and from OE **bysc* 'bush, thicket'. The place is near OXHEY and its name very likely has OE *gehæg* as second el. The first el. might be OE *byxe* (cf. BEXHILL &c.). *Byx-gehæg* might well have become *Bisheye* &c. owing to Norman influence. **Bushey** (or **Bushy**) **Park** Mx was named from BUSHEY Hrt.

Bushley Wo [*Bisclege* 11 Heming, *Biselege* DB, *Bisselega* 1159 P, *Busseleg* 1212 Fees]. *Bisc-* may be a reduction of *biscop*.

Bushmead Bd [*Bissop(es)med* 1227 Ass, 1231 FF], **Bushton** W [*Bissopeston* 1242 Fees, *Bisshoppeston* 1316 FA], **Bushwood** Wa nr Lapworth [*Bissopeswude* 1197 P]. 'The meadow, TŪN, wood of the bishop.'

Buslingthorpe Li [*Esetorp* DB, *Esatorp* c 1115 LiS, *Buslingthorpa* 12 Gilb, *Bisilingtorp* 1197 FF]. The original name means '*Esi*'s thorp'. *Esi* may be OE *Esi* or ODan *Esi*, *Æsi*. *Esatorp* was held c 1115 by one *Buselin* (LiS), whose name was added to *thorp* to form a new name. *Buselin* is an OFr name.

Buston, High & Low, Nb [*Buttesdune* 1166, *Uuerbuttesdun* 1186 P, *Butlisdon* (*Budlisdon*) *Superior, Inferior* 1242 Fees, *Butlesdon* 1249 Ipm]. Perhaps '**Buttel*'s DŪN'. A pers. n. stem *Butt-* seems to have been in use, to judge by BUTLEY, BUTSASH &c.

Butcombe So [*Budancumb* c 1000 Wills, *Budicome* DB, *Budecumb* 1225 Ass]. '*Buda*'s CUMB.'

Buteland Nb [*Boteland* 1242 Fees, *Botylaund* 1269, *Botelaund* 1279 Ass]. '*Bōta*'s land.'

Butleigh So [*Budecalech* 725 BCS 142, *Bodecunleighe* 801 ib. 300, *Budeclega* 971 ib. 1277, *Boduchelei* DB]. '**Budeca*'s or **Buduca*'s LĒAH.'

Butley Chs [*Bute-*, *Botelege* DB, *Butteleg* 1268 Chester], **B~** Sf [*Butelea* DB, *Butteule* 1195 FF, *Butele* 1198 FF]. Apparently '**Butta*'s LĒAH'; cf. BUSTON. *Buttinga graf* KCD 1369 (Wo) seems to contain a patronymic formed from *Butta* or *Butt*.

Butsash Ha [*Bottesasse* 1212 Fees, *Butesasshe* 1316 FA]. '*Butt*'s ash-tree.' Cf. BUSTON.

Butterby Du [*Beutroue* 1242 Ass, *Beautrove* 1296 Cl]. 'Beautiful find.' Second el. OFr *trueve*, *trove* 'find'.

Buttercrambe YN. See CRAMBE.

Butterlaw Nb [*Buterlawe* 1242 Fees, 1256 Ass], **Butterleigh** D [*Buterlei* DB, *-lea* 1188 P], **Butterley** Db [*Buterleg* 1276 RH, *-leye* 1330 QW], **B~** He [*Buterlei* DB, *-lega* 1138 AC]. 'Hill, LĒAH or pasture, which yielded plenty of butter.'

Buttermere (lake) Cu [*Butermere* 1230 Sc, *Bottermere* 1256 FF], **B~** W [*Butermere* 863, 931 BCS 508, 678, *Butremere* DB]. Literally 'butter mere', i.e. 'lake on whose shores were pastures that yielded plenty of butter'. B~ W is the name of a place, but was no doubt originally that of a lake. In the boundaries of *Butermere* in BCS 508 are mentioned *lilan mere* and *þrocmere*.

Butterton St nr Leek [*Buterdon* 1200 Cur, *Butterdon* 1223 FF, *Boterdon* 1236 FF]. 'Butter hill', i.e. 'hill with good pastures giving plenty of butter'. Cf. DŪN.

Butterton St nr Newcastle [*Butereton* 1182 P, *Botertun* 1208 FF]. 'Butter farm.'

Butterwick Du [*Boterwyk* 1131 FPD], **B~** Li nr Boston [*Butrvic* DB, *Buterwic* 1202 Ass], **East & West B~** Li [*Butreuuic* DB, *Buterwic* 1219 Ass], **B~** We [*Butterwyk* 1285 PNCu], **B~** YE [*Butruid* DB, *Buteruic* c 1160 YCh 1891], **B~** YN [*Butruic* DB, *Buterwic* 1227 FF]. 'Butter farm', 'dairy farm'.

Butterworth La [*Buterwrth* 1235 FF, *Butterwurth* 1246 Ass]. 'Butter farm.'

Buttington Gl [*Buttingtun* 894 ASC]. B~ is situated at a marked hill. It seems likely that the first el. is a name of the hill, derived from an OE **butt*, the source of ME *butt* 'thicker end' and related to OE *buttuc* 'end, piece of land'.

Buttsbury Ess [*Botolfvespirie* 1220 FF, *Botoluespirie* 1230 P]. '*Bōtwulf*'s pear-tree' (OE *pyrige* 'pear-tree'). The earlier name was *Ginge(s)*; cf. ING. *Ginge le viel defens* 1201 BM, *Ginges Joiberd* 1231, *Ginges Laundri* 1236 FF refer to Buttsbury.

Buxhall Sf [(æt) *Bucyshealæ* c 995 BCS 1289, *Boccheshale* 1050 Th, *Buckeshala*, *Bukessalla* DB, *Buchessala* 1165 P], **Buxlow** Sf [*Buckeslawe* 1250 Ipm, *Bukkeslowe* 1254 Val]. '*Bucc*'s HALH and HLĀW or tumulus.' OE *Bucc* pers. n. is not evidenced, but is presupposed by several pl. ns. It is not likely that OE *bucc* 'male deer' is the first el. in all these names.

Buxted Sx [*Boxted* 1199 Cur, *Boxstede* 1278 Pat, *Bocstede* 1230 FF]. OE *bōcstede* 'place where beeches grew', or *boxstede* 'place where box-trees grew'.

Buxton Db [*Buchestanes* c 1100 Mon v, *Bucstanes* 1230 P, 1251 Ch, *Bucstones* 1287 Court]. The name should be compared with **Buckstone** Gl, the name of a rocking stone, and with **the Buckstone** in Dixton par. Monmouth, a now destroyed loganstone (Bristol & Gl. Arch. Soc. ix). There were probably some logan-stones at Buxton. *Buckstone* 'logan-stone' possibly represents an OE **būg-stān* 'bowing stone'; cf. *buxom* from *būgsum*.

Buxton Nf [*Bukestuna*, *Buchestuna* DB, *Buxstone* 1254 Val]. '*Bucc*'s TŪN'; cf. BUXHALL.

OE **by** from ON *býr*, *bœr*, ODan, OSw *by* is common as a second el. in the parts of England where Scandinavians settled. As a first el. it is rare. Cf. however, BICKER, BYKER, BIERLOW (see BRAMPTON BIERLOW), BRYNING. OScand *by* denoted a village or a homestead. In English pl. ns. both these senses are to be reckoned with, but the exact sense cannot be determined in the actual instances. The first el. is mostly Scandinavian, usually a pers. n., but English and even Norman first elements also occur.

OE **byden** 'vessel, tub' must also, like *trog* 'trough', have been used in such a sense as 'shallow valley'. See BEEDON, BEDWELL, BIDWELL.

Byers Green Du [*Bires* 1345 Pat]. The plur. of OE *bȳre* 'cowhouse'.

Byfield Np [*Bifelde* DB, *Biffeld* 1199 P, *Bifeld* 1254 Val, 1260 BM]. B~ is in the bend of a river. The name may go back to OE *byge-feld* 'FELD in the *byge* or bend'.

Byfleet Sr [(æt) *Bifleote* 1062 KCD 812, *Biflete* 933 BCS 697, *Biflet* DB]. Originally OE *bī Flēote* '(the place) by the fleet or stream'.

Byford He [*Buiford* DB, 1242 Fees, *Buford* 1249 Fees, *Byford* 1249 Cl]. OE *byge-ford* 'ford by the bend' or 'ford where commerce took place'. The place is on the Wye, which does not make a very marked bend here. So the second alternative seems preferable. OE *byge* means 'traffic, commerce' and is related to the word *buy*.

OE **byge** 'bend' (of a river). See BYFIELD, BYFORD, BYTON, BYWELL.

Bygrave Hrt [(æt) *Bigrafan* 973, BCS 1297, (æt) *Biggrafan* 1015 Wills, *Bigrave* DB]. Originally OE *bī Grafan* '(the place) by the ditch'. Second el. probably an OE **grafa* 'ditch' corresponding to OHG *grabo*, OLG *gravo* 'ditch'.

OE **byht** 'bight, bend of a stream'. See NESBIT, SIDEBEET, BITTERNE.

Byker Nb [*Bikere* 1196 FF, *Byker* 1212 Fees, *Bychre* 1254 Val, *Bikerr* 1287 Ipm]. Identical with BICKER.

Byland YN [*Begeland* DB, *Beghlanda* c 1145 YCh 1827, *Beland* 1157 P, *Vetus Beland* 1209 FF]. 'The land of **Bēaga* or *Bēage*.' *Bēage* is a woman's name. Cf. BEAWORTHY.

Bylaugh Nf. See BELAUGH.

Byley Chs [*Bevelei* DB, *Magna, Parva Biueleg* 1250 Cl, *Biueleg* 1260 Court]. Apparently '*Bēofa*'s LĒAH'. Cf. BEVINGTON.

Byng Sf [*Benges, Benga* DB, *Beenges* 1242 P, *Benges* 1257 Ipm]. OE *Bēgingas* '**Bēaga*'s or *Bǣga*'s people'.

Byram YW [*Byrum* c 1030 ASCh, *Birum* c 1170 YCh 1634, *Burun* 1208 Cur, *Byrrom* 1316 FA]. OE *bȳrum*, dat. plur. of *bȳre* 'cowshed'.

OE **bȳre** 'shed, cowhouse', also in *cū-bȳre*, is rare in pl. ns. See BYERS GREEN, BYRAM, EDMONDBYERS. As a first el. it is difficult to distinguish *bȳre* from *byrg, byrh*, gen. of BURG. See BURDON Du, BIRLEY, BUERTON.

OE **byrgen** 'tumulus, burial-mound'. See BURN Bk, Y, BERNWOOD, BICESTER, HEBBURN, HEBRON, HEPBURN, WHITBURN.

OE **byrst** is only found in the sense 'loss, injury', but *eorþ-gebyrst* 'landslip' occurs. *Byrst* in a sense such as 'gap' or 'landslip' may be found in BRISTON, BURSHILL, BURSTON.

Bystock D [*Boystok* 1242 Fees]. '*Boia*'s STOC.'

Bytham, Castle & Little, Li [*Bytham* c 1067 Wills, *Bitham, Bintham, Westbitham* DB, *Biham* c 1100 Fr]. OE *Byþn-hām* 'HĀM in the valley'. OE *bytme, byþme, byþne* is rendered 'bottom, head of a valley'. It is a derivative of OE *boþm, botm* 'bottom'. Little Bytham is often *Bihamel* and the like in early sources [*Bihamel* 1212 Fees, *Byhamel* 1227 Cl]. The ending is the Fr diminutive suffix *-el*.

Bythorn Hu [*Bitherna* c 960 BCS 1061, *Bierne* DB, *Bitherne* 1248 FF]. '(The place) by the thorn-bush' (OE *þyrne* 'thorn-bush').

Byton He [*Boitune* DB, *Buton* 1287 Ipm, *Buyton* 1386 BM]. OE *Byge-tūn* 'TŪN by the bend' (OE *byge* 'bend'). The place is near a bend of the Lugg.

Bywell Nb [*Biguell* 1104–8 SD, *Biewell* 1195 (1335) Ch, *Biwel Petri, Bywell Andree* 1254 Val]. OE *byge-wella* 'spring in the bend'. The place is in a bend of the Tyne.

Byworth Sx [*Begworth, Byworthe* 1279 Ass]. '*Bēaga*'s or *Bēage*'s WORÞ.' Cf. BYLAND.

OE **byxe**, a derivative of *box* and meaning 'box grove' or 'box-tree' is not recorded, but is the source of BIX and the first el. of BEXHILL, BEXLEY, BIXLEY. BEXINGTON seems to contain an adj. *byxen* 'of box'.

C

Cābourn Li [*Caburne* DB, *Caburna* c 1115 LiS, *Kaburne* 1201 Cur]. 'Jackdaw stream' (ME *cā, cō* 'jackdaw' and OE *burna* 'stream'). The same el. is found in CAVILLE, CAWOOD, KABER.

Cābus La [*Kaibal* 1200–10 FC, *Cayballes* 1292 Ass]. OE *cǣg* 'key', here in an earlier sense such as 'peg', and **ball* 'a rounded hill'. The exact meaning of the name is not clear.

Cadbury D [*Cadebirie* DB, *-beria* 1093 Fr, 1188 P], **North & South** C~ So [*Cadanby*[*rig*] c 1000 Coins, *Cadeberie, Sudcadeberie* DB, *Northkadebir'* 1212 Fees]. '*Cada*'s BURG.' There are ancient camps at both places.

Caddington Bd [*Caddandun* c 1000 CCC, *Cadandune* c 1053 KCD 920, *Cadendone* DB, *Cadendona* 1145 BM]. '*Cada*'s DŪN.'

Cadeby Le [*Catebi* DB, *-by* 1228 Ch], **North** C~ Li [*Cadebi* DB, *Catebi* c 1115 LiS, c 1162 BM], **South** C~ Li [*Catebi* DB, c 1115 LiS], C~ YW [*Catebi* DB, *-by* 1201 FF]. '*Kāti*'s BY.' ODan *Kati*, ON *Kāti* is a well evidenced pers. n.

Cadeleigh D [*Cadelie* DB, *Cadeleghe* 1275 RH]. '*Cada*'s LĒAH.' Cadeleigh is near CADBURY.

Cadishead La [*Cadewalesate* 1212 Fees, *Cadewallessiete* 1226 LaInq]. Second el. OE (GE)SET 'fold, pasture'. The first may be the OE pers. n. *Ceadwalla*, or a stream-name *Cadan wælla* '*Cada*'s stream'.

Cadlands Ha [*Cadiland* 1198 FF, *Cadelande* 1291 Tax]. '*Cada*'s land.'

Cadmore End Bk [*Cademere* 1236 FF]. Probably '*Cada*'s mere or lake', though there does not seem to be any lake there now.

Cadnam Ha [*Cadenham* 1286 Ch], C~ W nr Chippenham [*Kadenham* 1242 Fees, *Cadenham* 1468 BM]. '*Cada*'s HĀM or HAMM.'

Cadney Li [*Catenai* DB, *Cadenai* c 1115 LiS, *Cadenaia* 1212 Fees]. 'Cada's island.'

Cadwell Li [*Cathadala* Hy 2 BM, *Cattedale*, *Catendale*, *Candale* 1202 Ass]. Probably, in spite of the forms in Ass, 'wild-cat valley'.

Cadwell O [*Cadewelle* DB, *Kadewalle* 1196 FF, *Kadewell* 1203 Cur]. 'Cada's spring.'

Caenby Li [*Couenebi* DB, *Casnabi* c 1115 LiS, *Cauenebi* 1191 ff. P, *Kauenbi* 1202, *Kauenebi* 1203 Ass, *Couenby* a 1223 RA]. The first el. may be identical with that of CAVENDISH, CAVENHAM. It is no doubt a pers. n. belonging to OE *cāf* 'active, bold', but it is hardly OE *Cāfa* (evidenced as *Caua* LVD). It seems to be a derivative of that name, e.g. an *OE *Cāfna*; cf. *Pægna* in PAIGNTON.

Caerlĕ·on, Welsh **Caerlleon ar Wysg**, Monm [*Iskalis* (for *Iska leg.*) c 150 Ptol, *Isca leg. II. Augusta* 4 IA; *Cair Legeion* (*Legion*) *guar Usic* c 800 HB, *Carleion*, *Carlion* DB, *Cairlion, civitas legionum* c 1150 LL, *Kaerleun*, i.e. *Legionum urbs* 1191 Gir, *Karliun bi Uske*, *Kairliun* 1205 Lay, *Cair llion ar Wysc* 13 Mab]. C~ is on the Usk and its name was originally identical with that of the river; see USK *infra*. Later arose the Latin name *Castra legionis* (or *legionum*) 'the camp of the (second) legion', and of this *Caerleon* is a Celticized form, Welsh *caer* 'castle, fort' having replaced Lat *castra*. The addition *ar Wysg* 'on the Usk' for distinction from Welsh *Caerlleon* 'Chester'.

OE **cærse, cerse, cresse** 'cress' is found several times as the first el. in pl. ns. The usual meaning is 'water-cress', as in CARSWELL, CASWELL, CRASSWALL, KERSWELL, CARSHALTON. But the word was also used of other kinds of cress; water-cress is in OE also known as *ēa-cerse*. OE *cærse* is the first el. also of CARSINGTON, CASSINGTON, KEARSLEY, KESGRAVE, where it is combined with the words TŪN, LĒAH, GRĀF 'grove'. Cf. also KERSAL.

Caerwent (kīre-) Monm [*Venta Silurum* 4 IA, *Cair Guent* c 800 HB, *Cairguent, urbs Guenti* c 1150 LL]. The original name is identical with the old name of WINCHESTER. Later Welsh *caer* 'castle, fort' was prefixed.

Cainham Sa [*Caiham* DB, *Kayham* 1265 Ch, *Caynham* 1291 Tax]. 'Cæga's HĀM or HAMM.' OE *Cæga* is presupposed also by CAINHOE, and the strong form *Cæg* or rather *Cægi* seems to be the first el. of CASSIOBURY. Cf. also KEYNSHAM.

Cainhoe Bd [*Chainehou, Cainou* DB, *Camhó* 1166 P]. '*Cæga's HŌH or spur of land.'

Caister next Yarmouth Nf [*Castra* 1044–7 KCD 785, *Castra* DB, *Castre* 1196 P], C~ **St. Edmunds** Nf [*Castre* c 1025 Wills, *Castrum* DB, *Castre Sancti Eadmundi* 1254

Val], **Caistor** Li [*Castre* DB, 12 DC]. OE *cæster, ceaster* 'Roman camp or fort'.

C~ **St. Edmunds** is held to be the Roman *Venta Icenorum*. The place belonged to Bury St. Edmunds.

Caistron Nb [*Cers* c 1160 YCh 1241, *Kerstirn* 1202 FF, *-thirn* 1244 Ch]. 'Thornbush by the marsh.' The first el. is ME *kers*, dial. *carse* 'marsh', a word probably related to ON *kiarr* (cf. KERR). Second el. OE *þyrne* or ON *þyrnir* 'thorn-bush'.

Cakeham Sx [*Cacham* 1235 Cl, 1248 AD], **Cakemore** Wo [*Cackemor, Cakemore* 1270 Ct]. Both names, like **Cakebole** Wo [*Kakebale* 1270 AD], appear to have as first el. a pers. n. This might be OE *Cæfca*, found in *Cæfcan græfan* Crawf, a derivative of *cāf* 'active, bold'. See HĀM, MŌR. Cakebole seems to have as second el. OE *ball in such a sense as 'round hill'; cf. CABUS.

Calbourne Wt [*Cawelburne* 826 BCS 392, *Cavborne* DB, *Cauelburn* 1181 P]. The name originally denoted the stream on which the place is, now **Caul Bourne** [(on) *Cavelburnan* BCS 392]. The first el. of the name may be identical with CALE.

OE **calc, cealc** 'chalk, limestone' is the first el. of several names, as CHALDEANS Hrt, CHALFORD Gl, O, CHALGRAVE Bd, CHALGROVE O, CHALTON Ha, CAWKWELL Li. The form *Chalk-* is the Saxon and Kentish *cealc*, while *Cawk-* is Anglian *calc*. CALKE Db, CHALK K, W apparently go back to a derivative *c(e)alce* 'lime-stone hill'. CHELSEA Mx, KELK, KELFIELD Y seem to contain a derivative with a suffix that caused *i*-mutation, e.g. an OE *c(i)elce* fem. or the like.

Calceby Li [*Calesbi* DB, c 1115 LiS, *Calseby* 1254 Val]. The first el. is identical with that of **Calceworth** wap. Li [*Calsvad* DB, *Calswat* c 1115 LiS, *Calsuad* 1194 P]. It is ON *Kálfr*, OSw, ODan *Kalf, Kalv*, a contracted form of *Kárulfr*. Calceby is OScand *Kalfs bȳr*, Calceworth being *Kalfs vað* 'Kalf's ford'. See VAÐ.

Calcethorpe Li [*Cheilestorp* c 1115 LiS, *Caillestorp* 1197 P, *Kaillestorp* 1212 Fees]. The place is near KELSTERN, and the two names have the same first el., which is probably an unrecorded pers. n. *Cægel*, a derivative of *Cægi* or *Cæga*; cf. CAINHAM, CASSIOBURY.

Calcot Brk nr Kintbury [*Colecote* DB, 1220, 1242 Fees], **Calcutt** W nr Cricklade [*Colecote* DB, *Colecote by Cheleworth* 1334 Ipm]. Perhaps 'Cola's COT'. But the name may mean 'shed where coal was kept'.

Calcott Brk nr Reading, **Calcott** Sa, **Calcutt** Bd [*Caldecote* 1224 FF], C~ **Wa** nr Southam [*Caldecote* DB, *-cot* 1242 Fees]. 'Cold COT'; see CALDECOTE &c.

OE **cald, ceald** 'cold' is a common first el. in pl. ns. See CALD-, COLD- &c. The original Saxon and Kentish form of the word was OE *ceald*, which should have given ME *chald, cheld, chold*, while the Anglian form was *cald*, which gave ME *cald, cold*. The Anglian form began at an early date to be introduced into Saxon and Kentish dialects (see e.g. CALCOTT, CALD-, CAUDLE, CAULCOTT), and the old Saxon and Kentish form is only occasionally found preserved in pl. ns., as in CHADWELL Ess, CHALFIELD W, CHALLACOMBE D, CHARLOCK Np, CHOLWELL So. *Cold* is often found as a distinguishing epithet before names, as in COLD ASTON. It no doubt refers to an exposed situation.

Caldbeck Cu [*Caldebek* c 1060 Gospatric's ch, *Caudebec* 1195 FF]. Really the name of the stream at the place, now **Cald Beck** [*Caldebec* 1225 Sc]. 'Cold brook.' See BECK.

Caldbergh YN [*Caldeber* DB, -*berg* 1270 Ipm]. 'Cold hill.'

Caldecote, Upper & Lower, Bd [*Caldecot* 1197 FF, *Magna et Parva Caldecote* 1234 Cl], C~ Bk [*Caldecote* DB], C~ Ca [*Caldecote* DB, -*cot* 1195 P], C~ Chs [*Caldecote* DB], C~ Hrt [*Caldecota* DB], C~ Hu nr Oundle [*Caldecote* DB], C~ Hu in Eynesbury Hardwicke [*Caldecota* 1194 P, -*cote* 1242 FF], C~ Nf [*Caldanchota* DB, *Caldecote* c 1080 Fr], C~ Np nr Towcester [*Caldecot* 1202 f. Ass], C~ Wa nr Nuneaton [*Caldecote* DB], (Chelveston cum) **Caldecott** Np [*Caldecote* DB], **Caldecott** Ru [*Caldecote* 1246 Ch]. OE *calde cot* 'cold hut' or *caldan cotu* 'cold huts'. The name often appears in the plural form. The reference may be to a hut or a shelter for animals in an exposed position, but the common occurrence of the name suggests that in many cases it has some technical meaning, probably the same as COLDHARBOUR, i.e. 'a place of shelter for wayfarers'.

Calder R Cu [*Kalder* c 1200 Mon, 1292 Ass], C~ R La, a trib. of the Ribble [*Caldre* c 1190 Whitaker, *Kelder* 1296 Lacy], C~ R La, a trib. of the Wyre [*Keldir* c 1200 Mon, *Caldre* 1228 Cl], C~ R YW [*Kelder* c 1170 YCh 1762, *Kaldre* 1279 Ass]. A British river-name identical with CALETTWR and CLETTWR in Wales and a compound of Welsh *caled* 'hard, violent' and *dwfr* (OBrit *dubro-*) 'water, stream'. On Calder Cu is **Calder in Copeland** [*Calder* 1179 P, *Kaldre* 1231 Ch].

Caldew R Cu [*Caldeu* 1201 Ch, *Caldew* 1242 Sc &c., *Calde* 1228 For]. Probably OE *cald-ēa* 'cold river', altered to *Caldew* owing to influence from OFr *ewe* 'water, stream'.

Caldicot Monm [*Caldecote* DB, 1291 Tax]. Identical with CALDECOTE.

Caldwell YN [*Caldeuuella* DB, -*well* 1208 FF]. 'Cold stream.' C~ is on Caldwell Brook.

Caldy, Great & Little, Chs [*Calders* DB, c 1100 Chester, *Caldeihundr.* 1183 P, (insula) *Caldei* c 1190 Gir, *Caldera* c 1245

Chester, *Caldey* 1285 Ch]. 'Cold island.' The early forms in -*r* represent a Scandinavian *Kald-eyiar* 'the (two) Caldys'.

Cale R So, Do [*Cawel, Wincawel* 956 BCS 923]. A pre-English river-name. *Win*- is no doubt Welsh *gwyn* 'white'. *Cawel* and *Wincawel* seem to have denoted different arms of the Cale. Cf. CALBOURNE.

Calehill K [*Calhull* 1327 Ch, *Calehell* 1410 BM]. 'Bare hill'; cf. CALU.

OE **calf, cealf** 'calf' is a common first element in pl. ns. The old Saxon and Kentish form *cealf* is preserved as *Chal*- in several pl. ns., e.g. CHALDON Do, Sr, CHALLACOMBE D, CHALLOCK K, CHALVEY Bk, CHAWLEIGH D, CHELVEY So, but the Anglian form is found in CALVERTON Bk, KELSTON So. In Anglian districts the form always has *c*-, e.g. CALLALY Nb, CALTON Db, St, Y, CALVELEY Chs, CALVER Db, CALWICH St, CAULDON St, CAWTON Y. The OE plur. form *calfru*, gen. *calfra* accounts for CALLERTON Nb, CALVERHALL Sa, CALVERLEY Y, CALVERTON Bk, Nt. The umlaut form *celf* is found in KELLOE, KELTON, KILPIN.

Calgarth We [*Calfgarth* 1437 Kendale]. 'Enclosure for calves.' See GARTH.

Calke Db [*Calc* 1132 BM, 1212 Fees, *Chalke* 1196 f. P]. See CALC.

Callaly Nb [*Calualea* 1161 P, *Calvelega* 1212 RBE, -*leya* 1212 Fees]. OE *calfa-lēah* 'pasture for calves'.

Callaughton Sa [*Calweton* 1251 Cl, 1291 Ch, *Caleweton* 1284 Misc]. The first el. is a derivative of CALU 'bare', either a pers. n. *Calwa* 'bald man' or more likely a hill-name *Calwe* (or *Calwa*) 'bare hill'; cf. CALLOW He.

Callerton Nb nr Ponteland [*Kaluerduna* Hy 2 (1271) Ch], **Black C~** Nb [*Calverdona* 1212, *Blackalverdon* 1242 Fees, *Calverdon, Blakecalverdon* 1256 Ass], **High C~** Nb [*Calverdon* 1242 Fees]. OE *calfra-dūn* 'hill where calves grazed'.

Callington Co [*Kelli wic ygkernyw* Mabinogion, *Cælling, Celling* 905 BCS 614, *Cællwic* 980–8 Crawf, *Caluuitona* DB, *Calwintona* 1188 P]. The OCo name was *Celliwic*, which may be identical with Welsh *celliwig* 'wood, forest' or a compound of Co *celli* 'grove' and *gwic* 'village', i.e. 'village by a grove'. The English added OE TŪN to the old name.

Callingwood St [(wood called) *Le Chaleng* 1247, *Calyngewode* 1280 Ass]. 'Debatable wood'; cf. THREAPWOOD. The first el. is ONFr *calenge* 'challenge'.

Callow Db nr Wirksworth [*Caldelauue* DB, -*lowe* 1299 FF]. 'Cold hill'; cf. HLĀW. The same is the origin of **Callow** nr Hathersage [*Caldelawe* 1208 FF]. **Callow** nr Mappleton is more doubtful [*Caldelawe* 1203 Cur, *Calwelawe* early 13 Derby, *Caldelowe* 1382 ib.]. It may have as first el. OE *calu* 'bare'.

Callow He [*Calua* 1180 P, *Calowe* 1292 QW], **C~ Hill** So nr Axbridge [(on ufewearde) *Calewen* 1068 E]. An OE *Calwe* or *Calwa*, derived from *calu* 'bald, bare' and meaning 'bare hill'.

Calmsden Gl [*Kalemundesdene, Calmundesdene* 852 BCS 466, *Calmundesden, Calmondesden* 1220 Fees]. 'The valley of *Calemund*.' This pers. n. is not otherwise evidenced. It seems to consist of OE *calu* 'bald' and the common el. *-mund*.

Calne (kahn) W [*Calne* 955 BCS (912), 978 ASC (E), DB, *et Calnæ* 997 KCD 698, *Cauna* DB]. Originally the name of Abberd Brook, on which the place stands. See COLNE La.

Cālow Db [*Calehale* DB, 1226 FF, *Calale* 1279 Ipm]. 'Bare HALH'; cf. CALU. OE *halh* may be here used in the sense 'spur of hill'.

Calshot Castle Ha in Fawley [(æt) *Celcesoran* 980 KCD 626, *Celceshord* 1011 Ann Wint, *Calchesorde* 1347 Misc, *Calshot castel* 1579 Saxton]. Second el. OE *ōra* 'firm foreshore or gravelly landing-place', 'hard'. First el. doubtless a pers. n., perhaps OE *Cælic* or an unrecorded *Cælci* (cf. *Brænci* in BRENCHLEY). The place is at the end of a spit of land extending into Southampton water. Hence OE *ord* 'point' replaced *ōra*, or *ord* was added to *Celcesōra*. Later the *r* of *-ord* was lost and *-d* became *-t*, perhaps before *Castle*.

Calstock Co [*Calestoch* DB, *-stoc* 1208 Cur, *Kallistok* 1291 Tax]. 'STOC belonging to CALLINGTON.' The first el. is the old name of Callington, OE *Cællwic*.

Calstone Wellington W [*Calestone* DB, 1166 RBE, *-tona* 1130 P]. If *Calveston* 1273 Ipm is for *Calneston* (PNW(S) xl) and G. de *Calneston*' (1333 Subs) in Lockeridge took his surname from C~, as suggested by Mr. N. Stridsberg of Stockholm, the name probably contains the village name CALNE, and since C~ is in and SE of Calne the name may even be *Calne Ēasttūn*. Calstone was held by Ralph de Wilinton in 1228 (Cl). The name may derive from WILLITON.

Calthorpe Nf [*Caleðorp* 1044–7 KCD 785, *Caletorp* DB, 1197 P]. '*Kali*'s thorp.' ON, ODan, OSw *Kali* is a well-evidenced pers. n.

Calthorpe O [*Cotthrop* 1279 RH, *Colthrop* 1285, 1354 PNO(S)]. The first el. may be OE *col* 'coal', *colt* 'colt', even *cāwel* 'cole'.

Calthwaite Cu [*Calvethweyt* 1285 For]. 'Thwaite where calves were kept.'

Calton Lees Db [*Calton* 1330 QW, 1431 FA], **Calton** St [*Calton* 1238 FF, 1292 Ch], **C~** YW [*Caltun* DB, *Calton* 1246 FF, *Calveton* 1304 Ch]. OE *Calf-tūn* 'TŪN where calves were reared'.

OE **calu** (gen. *calwes* &c.) 'bald, bare' occurs sometimes as the first el. of pl. ns., usually with a second el. meaning 'hill', e.g. CALEHILL, CALOW, CALUDON. See further CALVERLEIGH and CAWTHORN, also CALLAUGHTON and CALLOW.

Caludon Wa [*Canledon* 1265 Misc, *Caludon* 1275 Cl, *Calwedon* 1292 Ch]. 'Bare hill.' See CALU and DŪN.

Calveley (kahvlĭ) Chs [*Calueleg* c 1235 Chester, *Calveleye* 1287 Court]. 'Pasture for calves.' See LĒAH.

Calver (-ah-) Db [*Caluoure* DB, *Caluore* 1199 P, *Calfover* 1239 Derby], **Calverhall** Sa [*Cavrahalle* DB, *Chalvrehalle* 1219 Eyton, *Calverhale* 1315 Ch]. 'Ridge and HALH where calves grazed.' See OFER.

Calverleigh D [*Calodelie* DB, *Calewudelega* 1194 P, *Calwodeleghe* 1270 Ep]. 'LĒAH or clearing in the bare wood'; cf. CALU. Or the original name of the place may have been *Caluwudu*, and an explanatory LĒAH 'wood' was added.

Calverley YW [*Caverleia* DB, *Kalverlay* 12 BM, *Caluerlai* 1198 P]. 'Pasture for calves.' See LĒAH.

Calverton (-ăl-) Bk [*Calvretone* DB, *Calvreton* 1235 Cl], **C~** (-ahv-) Nt [*Caluretone* DB, *Caluertona* 12 DC]. 'TŪN where calves were kept' (OE *Calfra-tūn*).

Calvington Sa [*Calveton* 1198 Fees, *Keluiton* 1199 FF, *Kalvinton* 1209 Eyton]. OE *Calfa-tūn* 'TŪN where calves were kept'.

Calwich St [*Calowic* Hy 2 (1314) Ch, *Calewich* 1197 P]. OE *Calf-wīc* 'farm where calves were kept'.

Cam R Ca [*Camus* 1586 Camden, *Cam* 1610 Speed]. A back-formation from CAMBRIDGE.

Cam R Gl [*Cam* 1612 Drayton], **Cam** vil. Gl [*Camma* DB, 1194 P, *Camme, Ka(u)mne* 1221 Ass]. Cam is a British river-name, related to Welsh *cam* 'crooked' (OCelt **kambo-*). The base may be OBrit **cambanā*, perhaps the source also of CAMAN in Wales. Cf. CAMBRIDGE Gl.

Cam R So. A back-formation from CAMEL.

Cam Beck Cu [*Camboc* 1169 &c. Lanercost, *crooked Cambeck* 1622 Drayton]. An OBrit river-name derived from **kambo-* 'crooked' (cf. CAM Gl). The source is OBrit **cambāco-* meaning 'crooked stream' and corresponding to CAMOGUE in Ireland. From Cam Beck were named **Kirkcambeck** and **Little Cambeck** Cu [*Camboc* c 1160 YCh 175, *Cambok* 1292 QW].

Cam Brook R So [*Camelar, Cameler* 961 BCS 1073]. An OBrit river-name. Derivation from **kambo-* 'crooked' (cf. CAM Gl) or from *Camulos*, the name of a Celtic wargod, is possible. Cf. CAMELEY.

Cam Fell YW [*Camp* 1190 FC, *Camb* 13 (1307) Ch]. OE *camb* or OScand *kambr* 'comb', later also 'crest, ridge of a hill'. This el. is found in other names of hills in

Yorkshire. Thus **Cams House** in Aysgarth YN is *Camb* 1218 FF. Cf. CAMBO, CAMS-HEAD, also COMBRIDGE, COMBS.

Camberley Sr. A late name. The district was called *Cambridge Town* in 1862 after the Duke of Cambridge. The name was altered to Camberley 'for postal convenience' (VH).

Camberwell Sr [*Cambrewelle* DB, *Camerwella* 1175 P, *Camerewell* 1199 Cur, 1212 Fees, *Cambrewell* 1206 Cur]. The first el. of this, as of CAMERTON Cu, YE, CAMMERING-HAM, may be an OE *cranburna* or *cranmere* 'crane stream or lake' (with dissimilatory loss of the first *r*). Even an OE *cāmere* 'jackdaw mere' may be assumed. Probably the first el. is not the same in all the names. In Camerton YE, Cammeringham it would be a derivative in *-ingas* of a pl. n.

Camblesforth YW [*Camelesforde, Canbesford* DB, *Camforth* 12 Fr, *Cameleford* 1204 Obl, *Cameleg-, Camelesford* 1311 Ch]. The first el. might be a river-name corresponding to Welsh CAMLAIS 'crooked stream' (Welsh *cam* 'crooked' and *glais* 'stream'). If so, the *s* was sometimes misunderstood as a genitive ending.

Cambo Nb [*Camho* 1230 Sc, 1230 Ch, *Cambhogh, Cambhou* 1253 Pat]. OE *camb-hōh* 'spur of hill with a comb or crest'. Cf. CAM FELL.

Cambois (kămus) Nb [*Cammes* c 1050 HSC, *Kambus* Hy 2, *Cambus* 1204 FPD, *Cammus* 1195 (1335) Ch]. Identical with Welsh CEMMAES, KEMEYS and Ir *camus* 'a bay'. The name is a derivative of OCelt **kambo-* 'crooked', Welsh *cam*. Cambois is on Cambois Bay. It is British in origin, but probably influenced in form by Gaelic *cambus*.

Camborne Co [*Cambron* 1291 Tax, 1316 FF, *Camberoun, Cambron* 1309 Ep]. Co *cam* 'crooked' and possibly *bron* 'a round protuberance, the slope of a hill'.

Cāmbridge Ca [*Grantacaestir* c 730 Bede, *Grantacester* c 890 OEBede, *Grontabricc* c 745 Felix, *Grantebrycg* 875 ASC, *-e* c 894 Asser, *Grentebrige* DB, *Cantebruge* c 1125 Cambr Bor Ch, *Cambrugge* 15 Chaucer, Reves Tale(MS Lansd)]. Originally 'Roman fort on the GRANTA', later 'bridge over the GRANTA'. The loss of the first *r* and the change of *G-* to *C-* are due to Norman influence. **Cambridgeshire** is *Grantabrycgscir* 1010 ASC (E), *Cantebruggescir* 1142 N & Q, 8. S. viii. 314, *Cambruggeschire* c 1400 Rob Gl.

Cambridge Gl [*Cambrigga* 1200–10 Berk]. 'Bridge over R CAM.' See CAM Gl.

Cămel R Co [*Cambula* 1147 Monm, *Camble* c 1300 Rob Gl]. A Cornish river-name *Cambull*, identical with **Cambwll**, an old name of a stream in He [*Campull* c 1150 LL] and a compound of Co *cam* 'crooked' and *pul* (Welsh *pwll*) 'a stream'. On the Camel is **Camelford** [*Camelford* 1205 Layamon, *Camleford* 1256 FF]. 'Ford over R CAMEL.'

Camel, Queen & West, So [*Cantmæl* 995 Muchelney, *Camel, Camelle* DB, *Camel Regis* 1275 Ipm, *Camel Reginæ* 1280 FF, *Quene Cammell* 1431 FA, *Est-, Wescammel* 1291 Tax]. Camel is probably an old name of **Camel Hill,** a long ridge. The second el. is very likely identical with Welsh *moel* 'bare', also 'conical, bare hill'. The first may be Welsh *cant* 'rim'. From the same word may be derived the name QUANTOCK So.

Queen Camel from Eleanor, queen of Edward I, who gave it to her before 1280.

Cāmeley So [*Camelei* DB, *-leia* 1156 Wells, *Camele* 1186 P, 1201 Ass]. 'LĒAH on R *Cameler* or CAM BROOK.'

Camelford Co. See CAMEL R.

Cămerton Cu [*Camerton* c 1150 StB, *Camberton* c 1150 StB, 1290 Ch, *Camertona* c 1150 StB], **C~** YE [*Camerinton* DB, *Camerringtona* Hy 2 DC, *Kamerington* 1226 FF]. See CAMBERWELL.

Camerton So [*Camelertone* DB, *Camelarton* 1227 Ch]. 'TŪN on R *Cameler* or CAM BROOK.'

Cammeringham Li [*Came(s)lingeham* DB, *Camringham* c 1115 LiS, *Cambrigeham* 1126 Fr, *Cambringeham* 1192 Fr, *Cameringham* 1202 Ass]. With this may also be compared *Cameringcroft* 1257 FF (Mumby Li). See CAMBERWELL.

OE camp is only found in pl. ns. It comes from Lat *campus* 'field'. Like OFris and LG *kamp* it seems to have denoted rather 'an enclosed piece of land' than simply 'field', but the exact meaning is not clear. It is found in CAMPS, CAMPSEY, CAMPDEN, COMPTON Db, and as second el. in BARCOMBE, BULCAMP, MUSKHAM, RUSCOMBE, SACOMBE, WARNINGCAMP. In Ruscombe, Sacombe the first el. is a pers. n. In Barcombe it is OE *bere* 'barley'; in Bulcamp, *bula* 'bull or bullock'; in Muskham, OE *mūs* 'mouse'.

Campden, Broad & Chipping, Gl [*Campsætena gemære* 1005 KCD 714, *Campedene* DB, 1196 P, *Cheping Caumpedene* 1315 BM, *Brodecaumpedene* 13 AD]. OE *campa-denu* 'valley with camps or enclosures'. See BRĀD and CHIPPING. A correspondent suggests that *Broad* here means 'far, distant' (cf. *abroad*).

Camps, Castle & Shudy, Ca [*Canpas* DB, *Campes* c 1080 ICC, *Campecastel* 13 AD, *Sudekampes* 1219, 1230 FF, *Sudicampes* 1242 Fees, *Schudecamp* 1284–6 FA, *Caumpes magna* 1291 Tax]. OE *campas* 'fields or enclosures'.

Castle refers to a medieval castle. *Shudy* is probably ME *schudde*, EAngl dial. *shud* 'shed', which is apparently identical with OE *scydd* (in pl. ns.).

Campsall YW [*Cansale* DB, *Camsala* 1157

YCh 186, *Camsal* 1208 FF, *Cameshal* 1227, *Kemeshal* 1239 Ep]. The place is situated at a sharp bend of a stream. The probability is that the first el. of this name, like that of KEMPSTON Bd, is a Brit word for 'bend of a river', identical with the word for 'bay' mentioned under CAMBOIS. The OWelsh form of the word would be **cambeis* or the like. The second el. is OE HALH 'haugh'.

Campsey Ash Sf [*Campeseia* DB, *Campesse* 1211 Cur, *Campese* 1235 FF, *Campeseye Ass* 1254 Val]. OE *campes-ēg* 'island with a field or enclosure'. Cf. CAMP.

Ash seems originally to have been a separate place. In DB *Esce* is mentioned immediately after Campsey, and in 1249 the monks of *Campes* obtained free warren in *Eysse* (Ch).

Campton Bd [*Chambeltone* DB, *Camelton* c 1150 BM, 1193 ff. P, *Kamerton* c 1155 Oxf]. The first el. is no doubt a British stream-name identical with CAMEL Co.

Cams Hall Ha [*Kamays* 1242 Fees, *Cammeys* 1282 Ep, *Cams* 1412 FA]. The place is on Portsmouth Harbour. The name is no doubt the British name of the bay and identical with CAMBOIS.

Camshead YN [*Cambesheved* 1235 FF]. OE *camb* 'ridge' and *hēafod* 'hill'.

Cams House. See CAM FELL.

Candlesby Li [*Calnodesbi* DB, *Kandelesbi* 1202 Ass]. The first el. is identical with that of **Candleshoe** wap. Li [*Calnodeshou* DB, c 1115 LiS]. It is a pers. n., perhaps an OE **Calunōþ*; cf. CALMSDEN. First el. OE *calu* 'bald', second the common pers. name el. *-nōþ*. Candlesby and Candleshoe mean 'the BY and the burial-mound of **Calunōþ*'.

Candover, Brown, Chilton & Preston, Ha [*Cendefer* c 880 BCS 553, *Candeverre* 903 ib. 602, *Candevere, Candovre* DB, *Brunkardoure* 1296 Cl, *Chiltone Candevere, Prestecandevere* c 1270 Ep, *Preston Candeuer* 1291 Tax]. Candover was originally the name of the stream that runs past the places [*Cendefer* 701, c 830 BCS 102, 398]. This is a British river-name consisting of OW *cein*(Welsh *cain*) 'beautiful' and an *i*-mutated form of Welsh *dwfr* (OBrit **dubro-*) 'water, stream'.

Brown in **Brown C~** is a family name.—Chilton is an alternative name of Chilton C~ or the name of a part of the village; *Chiltone* occurs alone 1291 Tax, apparently as the name of Chilton C~. Cf. CHILTON.—**Preston C~** was originally *Prestecandevere* 'the Candover belonging to the priests'. In 1316 the manor belonged to the prior of Southwick. *Preston* is probably from *Prestene*, an alternative gen. form with the weak ending *-ene*, OE *-ena*.

Canewdon Ess [*Carenduna* DB, *Canuedon* 1181 P, *Kenewedone* 1228 Ch, *Canewedon* 1254 Val]. OE *Caninga-dūn* 'the DUN of *Cana*'s people'. Cf. CANFIELD &c., CANNINGS, and for the abnormal development of *-inga-* (> *-iga-* > *-ege-* > *ewe*) CONEY WESTON, DANBURY, MANUDEN, MONEWDEN.

Canfield, Great & Little, Ess [*Cane-, Chenefelda* DB, *utraque Canefelda* 1121 AC, *Canefeud magna, parva* 1291 Tax]. '*Cana*'s FELD.' *Cana* pers. n. occurs in DB and is also found in *Caneworth* 1275 Cl (Ess), CANFORD, CANLEY, CANWELL, CANWICK. A corresponding Cont. Germ. name is found in CANEGEM in Flanders [*Caningahem* 967].

Canford Magna, Little C~ Do [*Cheneford* DB, *Caneford* 1200 Ch, 1212 Fees], Canley Wa [*Canelea* 1180 f. P, *Caneleye* 1285 QW]. '*Cana*'s ford and LĒAH'; cf. CANFIELD.

Cann Do [*Canna* Hy 1 PNDo]. OE *canne* 'can', here used of a deep valley.

Cannings, All & Bishops, W [*Canegan mersc, Caningan mærsc* 1010 ASC (CDE), *Caninge, Cainingham* DB, *Caninges* 1091 Sarum, *Keninges* 1201 Cur, *Aldekanning* 1205 Obl, *Canyng Episcopi* 1294 Ch]. OE *Caningas* '*Cana*'s people'; cf. CANFIELD.

All C~ is 'Old Cannings'. Bishops C~ belonged to the bishop of Salisbury.

Cannington So [*Cantuctun* c 880 BCS 553, *Cantoctona, Cantetone* DB, *Cantinton* 1187 P, *Caninton* 1178 P]. 'TŪN by Quantock Hills.' See QUANTOCK.

Cannock St [*Canuc* 956 BCS 969, *Chenet* DB, *Cnot* 1156, *Canot* 1157 P, *Canoc* 1198 Fees, 1230 P]. A British hill-name, probably a Brit **Cunuc* identical with the first el. of CONSETT. Cf. CONOCK.

Canonbury Mx [*Canonesbury* 1373 PNMx]. 'The manor of the Canons', i.e. the Augustinian Canons at Smithfield.

Canonsleigh D [*Leiga* DB, *Leghe Canonicorum* 1283 Ed, *Canons Lee* 1286 Ch]. Originally LEIGH, from OE LĒAH, later Canonsleigh from the abbey for Canons founded c 1170.

Canterbury K [*Cantwaraburg* 754, *Contwaraburg* 851 ASC, *Cant-, Contwaraburg* c 890 OEBede, *Canterburie* 1086 KInq]. 'The BURG (fort or town) of the *Cantware* or people of Kent.' OE *Cantware* 'the people of Kent' is often found in ASC, e.g. 694. The OBrit name of Canterbury appears as *Darovernon* c 150 Ptolemy, *Durovernon* 4 IA, *Dorovernia* 605 BCS 5, *Doruuernis* c 730 Bede, *Dorubernia* c 894 Asser. The elements appear to be OBrit *duro-* 'fort' and *verno-* 'alders, swamp'. The meaning would be 'the swamp by the fort'.

Canterton Ha [*Cantortun* DB, *Kantarton* 1212 Fees]. OE *Cantwaratūn* 'the TŪN of the Kentishmen'. Cf. CANTERBURY. Some Kentishmen must have settled at Canterton in the OE period.

Cantley Nf [*Cantelai* DB, *Cantelea* 1196 P, *Kantele* 1212 Fees], C~ YW [*Canteleia* DB, c 1190 YCh 817, *Kantelai* 1212 FF]. '*Canta*'s LĒAH.' **Canta* is easily explained as a short form of names such as *Cantwine*. *Cantwine* itself is not well evidenced either, while *Centweald, Centwine* are. The non-umlauted *Cant-* has an exact parallel in OE *Cantware* 'Kentishmen' by the side of OE *Cent* 'Kent'.

Cantlop Sa [*Cantelop* DB, 1230 FF, *Cantelhope* c 1180 Eyton]. Second el. OE HOP 'valley'. The place is on COUND BROOK, and the first el. of the name might contain the name COUND. Cf. OE *Cameleac* ASC 918 from Welsh *Cyfeilliog* (with *a* from Welsh *y*).

Cantsfield La [*Cantesfelt* DB, *Canceveld* 1202 FF]. C~ is on **Cant Beck** [*Kant* 1202 FF]. If, as seems probable, Cantsfield means 'FELD on the CANT', *Cant* is an OBrit river-name, which may be related to **Caint** R (Wales) and KENN D, So. But Cantsfield might be 'the FELD of *Cant*'. If so, *Cant* is a pers. n. analogous to *Canta* in CANTLEY, and the river-name Cant is a back-formation.

Canvey Island Ess [*Caneveye* 1255, *Kaneweye* 1265, *Canefe* 1321 Ipm, *Caneve* 1325 FF]. Possibly OE *Caninga-ēg* 'the island of *Cana*'s people'; cf. CANEWDON. But the early forms with *v* instead of *w* are difficult to account for.

Canwell St [*Canewelle* 12 PNSt, *-well* 1209–35 Ep]. '*Cana*'s spring'; cf. CANFIELD. There is a spring called St. Modwen's Well at the place.

Canwick (kănĭk) Li [*Canewic, Canuic* DB, *Canewich* Hy 2 DC, *-wic* 1200 Cur]. '*Cana*'s WĪC'; cf. CANFIELD.

Cāpel K nr Tonbridge [*Capele* 1331 AD], **C~ le Ferne** K [*Capel ate Verne* 1377 FF, *Capell* 1431 FA], **C~ St. Andrew** Sf [*Capeles* DB, 1254 Val], **C~ St. Mary** Sf [*Capeles* 1254 Val, 1291 Tax], **C~ Sr** [*Capella* 1190 P]. 'The chapel', from ME *capel*, ONFr *capele* 'chapel'.

Capenhurst Chs [*Capeles* DB, *Capenhurst* 1278 Chester, c 1296 Court]. Second el. HYRST, here probably 'hill'. For the first cf. CAPTON.

Cāpernwray La [*Koupemoneswra* 1212 Fees, *Caupemanneswra* 1228 Cl]. 'The valley of the chapman.' The elements are ON *kaupmaðr* 'chapman' and VRÁ 'corner, remote valley'.

Capesthorne Chs [*Copestor* DB, *Capestorne* 1285, *-thorn* 1288 Court]. Second el. *þorn* 'thorn-bush'. The first is obscure.

Capheaton. See HEATON.

Cāple, How & Kings, He [*Capel, Cape* DB, *Caples* 1190 ff. P, *Huwe Capel* 1327 PNHe, *Houcaple* 1428 FA, *Kingescapoll* 1205 Cur]. ONFr *capele* 'chapel'. Cf. CAPEL. *How* is no doubt the pers. n. *Hugh*.

Capton D [*Capieton* 1278, *Capinton* 1285, *Capyat me* 1330 PND], **C~** So [*Capintone, -tona* DB]. With these may be compared

Capland So [*Capilande* DB, *-lond* 1243 Ass, *Cappilond* 1225 Ass]. It may be suggested that the first el. of these, as of CAPENHURST Chs, is an OE *cape* or the like 'lookout place', cognate with OHG *kapf* 'look-out place', MLG *kape* 'beacon' and with OE *capian* 'to look, peer'. Most of the places in question are situated high. Capland may be 'beacon island', the second el. being OE *iegland* 'island'.

Caradoc He [*Cayrcradoc* 1292 Ipm, *Cradoc* 1329 Ep]. Welsh *Caer Caradoc* 'the fort of Caradoc'. *Caradoc*, now *Caradog*, from OBrit *Caratacos*, is a well-known pers. n., sometimes shortened to *Cradoc*. *Caradoc* arose through ellipsis of *Caer*.

Carant or **Carrant** R Gl [*Carent* 778–9, *Cærent* 780 BCS 232, 236]. A British river-name, identical with CHARENTE in France (from *Carantonus* &c.] and derived from *carant-*, a pres. part. of *car-* 'to love'. The meaning may be 'pleasant stream'.

Carbrook YW [*Kerebroc* c 1210 YCh 1281], **Carbrooke** Nf [*Cherebroc, Weskerebroc* DB, *Kerebroc* 1195 FF, 1202 Cur]. The first el. may be a British river-name identical with KEER La.

Carburton Nt [*Carbertone* DB, *Karberton* 1169 P, *Carberton* 1174, 1197 P, 1228 Cl]. The place is nr Clumber at the foot of a considerable hill. The first el. of the name is probably OE *carr* 'rock'. The second may be OE *beretūn* (see BARTON). The name then means 'the grange by the hill'. Or the hill was known as *Carrbeorg* 'rocky hill', and the whole name means 'the TŪN by the rocky hill'.

Carcolston. See COLSTON.

Carcroft YW [*Kercroft* 12 Pont, *Kerecroft* 1204 FF, *Carecroft* 1197 P]. First el. perhaps as in KEARBY. Cf. CROFT.

Carden Chs [*Kawrdin* c 1235 Ormerod, *Cawardyn* 1302, 1304 Chamb]. Second el. OE WORÐIGN. The first may be OE *carr* 'rock', the first *r* being lost in *Carrworþign* (dissimilation). Higher C~ is on the slope of a marked round hill.

Cardeston Sa [*Cartistune* DB, *Cardistone* 1275, *Cardestone* 1277 Ep], **Cardington** Sa [*Cardintune* DB, *Cardinton* 1167 P, *Kardinton* 1190 P, *Cardytone* 1327 Subs]. The first has as first el. a pers. n. The second probably contains an *-ing*-derivative of the same name. There is an OE *Carda* in *Cardan ham(m)* and *Cardan hlæw* 949 BCS 877 (Berks). This probably goes back to an earlier *Cradda* or the like. Cardeston may have as first el. a cognate name *Cræddi* or the like. A definite etymology is impossible with the material available.

Car Dike or **Dyke**, an ancient ditch in Li and Np [*Karesdic* 12 DC, *Chardyk* 1261 FF]. Apparently 'the ditch of *Kāri* or *Kārr*'. Both are well-evidenced OScand pers. ns.

Cardington Bd [*Chernetone* DB, *Kerdingtone* 1229 Ep, *Kardinton* 1227 Ass]. OE *Cēnrēding(a)tūn* 'the TŪN of *Cēnrēd*'s people'.

Cardinham Co [*Cardinan* 1194 f. P, 1229 Fees, *Cardinam* 1251 FF]. Co *cer, car* 'town, castle' and a place-name *Dinan*, identical with DINAN in Brittany and derived from Co *dun, din* 'hill', Welsh *din* 'fort' &c., from OBrit *dūno-* 'hill fort'.

Cardurnock Cu [*Cardrunnoke* 1386, *-drunnok* 1468 FF]. A British name, whose first el. is Welsh *caer* 'fort' &c. The second may be a Brit word corresponding to Gaelic *dornach* 'pebbly, pebbly place', which is derived from Gael *dorn* 'fist' &c., corresponding to Welsh *dwrn*.

Careby Li [*Careby* 1199 (1332) Ch, *Karbi* 1202 Ass, *Kareby* 1219 Ep]. '*Kāri*'s BY.' ON *Kári*, ODan *Kare* is a pers. n., found as *Carig* on coins of Æthelred II and as *Cari* in DB.

Carey R Ɽ [*Kari* 1238 Ass], **Carey** vil. D [*Kari* DB, 1194 P]. Carey is a British river-name, identical with CARY So. Carey vil. is on the Carey.

Cargo Cu [*Kargho* 1195 P, *Kargou* 1255 P]. A hybrid name, the second el. being Scand *haugr* 'hill', the first being apparently Welsh *carreg* 'rock, stone ; cf. CARK.

Cargo Fleet YN [*Caldecotes* 12 Whitby]. See CALDECOTE &c. Fleet is OE *flēot* 'stream'.

Cărham Nb [*Carrum* c 1050 HSC, 1104–8 SD, *Karrum* 1252 Ch, *Karham* 1242 Fees]. OE *carrum*, dat. plur. of CARR 'rock'.

Carhampton So [(æt) *Carrum* 833, 840 ASC, *Carumtun* c 880 BCS 553, *Carentone* DB]. Originally *Carrum*, identical with CARHAM, but TŪN was added at an early period.

Carisbrooke Wt [*Karesbroc* a 1175, *Caresbroc* 12 BM, *Care-*, *Karebroc*, *Karisbroch* 1179 P, *Keresbroc* 1202 Cur]. The first el. is possibly a river-name related to CAREY.

Cark La [*Karke* 1491 PNLa], **High Cark** La [*Ouer Carke* 1606 PNLa]. OW *carrecc*, Welsh *carreg* 'rock, stone'.

Carkin YN [*Kerrecan* a 1175, *Karrecan* c 1200 YCh iv, *Kercan* c 1200 Pudsay]. Apparently OW *carrecc*, Welsh *carreg* (plur. *cerrig*), or OIr *carric* 'a rock' and a doubtful second el., possibly Welsh *can* 'white'.

Carlatton Cu [*Carlatun* 1186 P, *Karlatona* 1219 Fees, *Carlauton* 1242 Ch]. A hybrid name, consisting of an OE *Lād-tūn* 'TŪN on a stream' or *Lēac-tūn* 'TŪN where leeks were grown' and Welsh *caer* 'fort, city', here probably in the sense 'village'. The name thus means 'the village of Latton'.

Carlbury Du [*Carlesburi* 1198 (1271) Ch, *Carlebiry* 1313 RPD]. The elements seem to be OScand *karl* 'man, free peasant' or

Karl pers. n. and OE BURG. But very likely the name is a Scandinavianized form of OE *Ceorlaburg* 'the BURG of the *ceorls* or free peasants'.

Carlby Li [*Carlebi* DB, *Karlebi* 1202 Ass]. Identical with OSw *Karlaby*, which may be 'the BY of the free peasants' (OSw *karl*) or '*Karle*'s BY'. *Carle* pers. n., from ODan *Karle*, ON *Karli*, is found in DB.

Carlecotes YW [*Carlecotes* 1266 Misc, 1285 Ch]. See COT. First el. as in CARLBY.

Carlesmoor YW [*Carlesmore* DB, *-mor* 1274 Cl]. 'The moor of *Carl*' (ON, ODan *Karl*).

Carleton Cu nr Carlisle [*Karleton* 1212 Fees, 1290 Ch], **C~** Cu nr Drigg [*Karlton* c 1240 FC], **C~** Cu nr Penrith [*Karleton* 1250 Fees, *Carlatun* 1252 Cl], **C~** La [*Carlentun* DB, *Karleton* 1256 FF, *Magna, Parva Karlton* 1242 Fees], **East C~** Nf [*Karltun* 1046 Wills, *Carletuna* DB, *Est Karleton* 1311 BM], **C~ Forehoe** Nf [*Carletuna* DB, *Karleton Fourhowe* 1268 Ass], **C~ Rode** Nf [*Carletuna* DB, *Carleton Rode* 1201 FF], **C~ St. Peter** Nf [*Carletuna, Karlentona* DB], **C~** YW nr Pontefract [*Carleton* 1258 Ipm], **C~** YW nr Skipton [*Carlentone* DB, *Karleton* 1184 BM]. Identical with Carleton is **Carlton** Bd [*Carlentone* DB, *Carleton* 1198 FF], **C~** Ca [*Carletun* c 1000 BCS 1306, *Carletona* c 1080 ICC, *Carlentone, Carletone* DB], **C~** Du [*Carltun* c 1090 SD, *Carlentune* 1109 RPD], **C~** Le nr Market Bosworth [*Karletone* 1209–19 Ep, *Carleton juxta Boseworth* 1327 Subs], **C~ Curlieu** Le [*Carletone, Carlintone* DB, *Carleton Curly* 1273 Cl], **Castle, Great & Little C~** Li [*Carletone* DB, *Carletune, -tuna* c 1115 LiS, *Castre Karleton* 1253 Ep, *Maior Carleton* 1254 Val, *parva Karletona* 1209–19 Ep], **C~ le Moorland** Li [*Carlatun* c 1067 Wills, *Carletune* DB, 1202 Ass], **North & South C~** Li [*Carletune, Nortcarletone* DB, *Carletuna* c 1115 LiS], **C~ Scroop** Li [*Carletune* DB, *Carlentona* 1115 YCh 1304], **East C~** Np [*Carlintone* DB, *Carleton* 1199 P, 1254 Val], **C~** Nt nr Nottingham [*Karleton* 1182, *Carleton* 1197 P], **C~ in Lindrick** Nt [*Carletone* DB, *Carletuna* c 1150 DC, *Carleton in Lindric* 1212 Fees], **Little C~** Nt [*Karletun* 12 DC, *Carleton* 1242 Fees], **C~ on Trent** Nt [*Carletune* DB], **C~** Sf [*Carletuna* DB], **C~ Colville** Sf [*Carletuna* DB, *Carleton Colville* 1346 FA], **C~** YE [*Carlentun, Carletun* DB], **C~** YN in Helmsley [*Carletona* 1301 Subs], **C~** YN nr Rudby [*Carletun* DB], **C~** YN in Stanwick [*Cartun* DB, *Karleton* 1226 FF], **C~ in Coverdale** YN [*Carleton* DB, 1270 Ipm], **C~ Husthwaite** YN [*Carleton* DB], **C~ Miniott** YN [*Carleton* DB, *Carleton* 1207 Cur], **C~** YW nr Barnsley [*Carlentone, Carleton* DB], **C~** YW nr Guiseley [*Carletune* DB], **C~** YW nr Rothwell [*Carlentone* DB, *Carleton juxta Rothewell* 1402 FA], **C~** YW nr Snaith [*Carleton* DB, *Karleton in Balne* 127? Ipm, *Carleton juxta Snaith* 1293 QW]. Carleton and Carlton go back to OScand *Karlatūn*, which usually no doubt means 'the TŪN of the free men or peasants', but may in some cases represent OScand *Karla tūn* '*Karli*'s TŪN' (cf. CARLBY). The name *Karlatūn* is never found in Scandinavia, and very likely *Carl(e)ton* is in most cases due to Scandinavianization of OE *Ceorlatūn*; cf. CHARLTON, CHORLTON.

Castle Carlton Li is also called *Market Karleton* 1243 Ep. Castle or more correctly *Caster* seems to refer to a market town.—C~ **Colville** Sf was held by Robert de Colevill in 1230 (Cl). The rame is derived from COLLEVILLE in Normandy.—C~ **Curlieu** Le was held by Robert de Curly in the 13th cent. He died without male issue in 1274 (Nichols). The family name *de Curli* (*Curle*) occurs 1157 ff. Fr. It may be from CULLY in Normandy.—C~ **Forehoe** Nf is near Forehoe Hills, which gave its name to Forehoe hd. Forehoe means 'the four hills'.— C~ **Husthwaite** YN is near HUSTHWAITE.— C~ **Miniott** YN was held by John Mynot in 1346 (FA). The name Minot, which was borne also by a well-known medieval poet, is doubtless French.—C~ **le Moorland** Li means 'C~ in the moorland'.—C~ **Rode** Nf took its name from the local family. It was held by Walter de Rede in 1302, by Robert de Rode in 1346 (FA). From REDE Sf?—C~ **Scroop** Li was named from the Scrope family. Henry Scrop in Carlton is mentioned 1346 FA.

Carlingcott So [*Credelincote* DB, *Credlingcot* 1199 P, *Cridelincot* 1225 Ass]. 'The COT of *Cridel(a)*'s people.' Cf. CRIDDON.

Carling Howe YN [*Kerlinghou* 12 Guisb], **Carlinghow** YW [*Kerlinghowe* 1307 Wakef]. 'The hill of the old woman or hag.' Second el. OScand *haugr* 'hill'. The first is ON *kerling*, OSw *kærling* 'old woman' (Scotch *carline* 'woman, hag, witch'). Very likely the hill was one where witches were supposed to gather.

Carlisle Cu [*Luguvall(i)um* 4 IA, c 425 ND, *Lugubalia* c 730 Bede, *Luel* c 1050 HSC, c 1130 SD, *Cardeol* 1092 ASC (E), *Karlioli* (gen.) c 1100 WP., *Cærleoil* 1130 P; *caer liwelyd* Taliesin, Mod Welsh *Caer Liwelydd*]. The original name is generally explained as meaning 'the wall of the god *Lugus*'. The later forms do not quite agree with this etymology and Professor Jackson in PNCu(S) suggests derivation of the old British name from a pers. n. *Luguvalos*. To the old name was prefixed Welsh *caer* 'city'.

Carlswall Gl [*Crasowel* DB, *Kersewell* 1220 Fees, *Carswall* 1346 FA]. 'Cress spring or stream'; cf. CÆRSE.

Carlton. See CARLETON.

Carnaby YE [*Cherendebi* DB, *Kerendeby* 1155–7 YCh 1148, *Kernetebi* 1190 ff. P, *Kerneteby* 1267 Ipm]. The earliest forms suggest as first el. an OScand byname *Kærandi*, the pres. part. of *kæra* 'to prosecute at law'. The ON byname *Kærir* is held to mean 'the prosecutor', i.e. 'the litigious person'. But the later forms with *t* are difficult to explain from *Kæranda-bȳr*.

Carnforth La [*Chreneforde* DB, *Kerneford* 1246 Ass]. A form of CRANFORD 'cranes' ford'. For the form with *e* cf. *cren* 'crane' in Barbour.

Carperby YN [*Chirprebi* DB, *Carperbi* 1168 P, *Kerperby* 1218 FF]. The first el. has been identified with OIr *Cairpre* pers. n. (PNNR).

OE **carr** 'rock', found in Northumbrian texts, is a Celtic loan-word. The distribution of the element in pl. ns. suggests a British rather than an Irish or Gaelic source, but *carr* 'a rock' is only found in modern Irish and Gaelic and may be a shortened form of OIr *carrac* 'rock', which corresponds to OW *carrecc*, Welsh *carreg* 'rock'. We must assume that the old British language had a shorter form, which is the base of OE *carr*. See CARBURTON, CARDEN, CARHAM, CARHAMPTON, CARROW, PAINSHAW.

Carraw Nb [*Charrau* 12 BBH, *Karrawe* 1279 Ass, *Cadrere* 1280 Cl, 1296 Subs, *Carrawer* 1298 BBH]. The name may possibly represent an OBrit plur. of *carr* 'a rock' (OWelsh *carrou*) or a compound of *carr* and OE *rāw* 'row'. The later forms seem to have been influenced by ME *quarere* 'quarry' from OFr *quarriere*, MLat *quadraria*.

Carrington Chs [*Karinton* Hy 3 Pudsay, *Carington* 1294 Court]. The first el. offers difficulties. It can hardly be derived from OE *carr* 'rock', as the place is low on the Mersey.

Carrington Li, a late name, from Robert Smith, the banker, created Lord Carrington in 1796, who owned most of the land here.

Carrock Fell Cu [*Carroc* 1208 Sc, *Carrok* 1261 Ipm]. OW *carrecc* 'rock'.

Carrow Nf [*Charhó* 1158, *Carho* 1159 P, *Carhow* 1212 Fees]. Second el. OE HŌH 'spur of hill'. The first seems to be OE CARR 'rock'.

Carshalton Sr [*Æuueltone* 675 BCS 39, *Aweltun* c 880 BCS 553, *Avltone* DB, *Kersaulton* c 1150 BM, *Cressalton* 1275 Ch]. The original name was OE *Æwell-tūn* 'TŪN at the source of a stream'; cf. ALTON (2). *Cars-* is OE *cærse* 'cress'.

Carsington Db [*Ghersintune* DB, *Kercinton* 1251 Ch, *Kersinton* 1276 RH]. 'TŪN where cress grew.' The first el. appears to be OE *cærsen* 'of cress'.

Carswell Brk [*Chersvelle* DB, *Cressewell* 1191 ff. P]. 'Spring or stream where watercress grew.'

Carthorpe YN [*Caretorp* DB, *Carethorp* 1243 FF]. '*Kāri*'s thorp.' First el. ODan *Kare*, ON *Kári*; cf. CAREBY.

Cartington Nb [*Cretenden* 1220 Cur, *Kertindun* 1236, *-don* 1242 Fees]. 'The DŪN of *Certa*'s people.' *Certa* is a form with metathesis of *Cretta* (LVD); cf. CRATFIELD. The corresponding *Chrezzo* (*Chretzo*) is found in Old German.

Cartmel La [*Ceartmel, Cartmel* 12 SD, *Cartmel* 1177 ff. P, *Kertmel* 1188 BM]. An ON *kart-melr* 'sandbank by rocky ground' (ON **kartr,* Norw *kart* 'rocky ground' and *melr* 'sandbank'). *Cart-* may, however, be a Scandinavianized form of OE *ceart*; cf. CHART.

Cartworth YW [*Cheteruurde, Cheteuuorde* DB, *Cartewrth* 1274 Wakef]. 'The WORÞ of one *Certa*.' Cf. CARTINGTON.

Carwinley Cu [*Karwindelhov* 1202 FF, *Carwindelawe* 1267 Ch]. Probably a hybrid consisting of Welsh *caer* 'town' or rather 'village' and an English name, e.g. *Wendlan hōh* or *hlāw* 'the spur of land (or the hill or burial-mound) of *Wendla*'.

Cary R So [*Kari* 725 BCS 143, *Cari* 729 ib. 147]. From the stream are named **Castle C~, C~ Fitzpaine, C~ Lytes** and **Babcary** [*Caric, Cary* c 680 Ant Glast, *Cari* DB, (castellum de) *Cari* 1138 HHunt, *Castelkary* 1237 FF, *Stipelkari* 1225 Ass, *Lytilkary* 1439 BM, *Babba Cari* DB, *Babekary* 1212 Fees]. The original form may have been *Caric*, to judge by one early form. The name is a British river-name, which may be related to the river-names CHER, CHIERS in France (from Gaul *Carus*) and CAR in Wales and derived from the root *car-* 'to love' in Welsh *caru* &c. Cf. CARANT. The meaning of the name might be 'pleasant stream'.

Bab- in **Babcary** is the OE pers. n. *Babba*.— *Stipe[l]kary* was held in 1243 by Margery Fitz Payn (Ass). Fitzpaine means 'son of Pain'. *Pain* is an OFr pers. n., really a nickname meaning 'heathen'.—*Lytes* would appear to be really the adj. *little* (OE *lȳtel*), but perhaps it is here used as a byname or family name.

Casewick Li [*Casuic* DB, *Casewic* 1198 FF]. A Scandinavianized form of OE *Cēse-wic* 'cheese farm'. Cf. KESWICK and CHESWICK.

Cassington O [*Cersetone, Cersitone* DB, *Cressenton* 1103 Fr, *Kersinton* 1197 P, *Karssinton* 1236 Fees]. Identical with CARSINGTON.

Cassiobury, Cashio Hrt [*æt Caegesho* 793 BCS 267, *Caissou, Chaissou* DB, *Caysho* 1291 Tax]. '*Cǣge*'s HŌH or spur of hill'; cf. CAINHAM and KEYSOE. Cashio is now used only as the name of a hundred. The village is Cassiobury, in which *bury* 'manor' has been added. The hundred is sometimes called *Caysford* (13 Misc &c.). '*Cǣge*'s ford.'

Cassop Du [*Cazehope, Cassehopp* 1183 BoB]. OE *Cattes-hop* 'the valley of the (wild) cat(s)'. But the early forms may rather suggest a first el. *Cattesēa* or *Cattasæ* 'cat stream or lake'.

Castern St [*Cætespyrne* 1002 Wills, *Catesturn* 1227 Ass, *Casterne* 1327 Subs]. '*Catt's thorn-bush'? If so, the earliest form must be miswritten. Or possibly an OE pers. n. **Cætti,* derived from *Catt,* may have existed. Second el. OE *þyrne* 'thorn-bush'.

Casterton Ru [*Castretone* DB, *Magna Castretone* 1234 Ep, *Casterton Maior, Minor* 1254 Val], **C~ We** [*Castretune* DB, *Castreton* 1222 FF]. 'TŪN by a Roman fort'; cf. CEASTER. There are remains of a camp at Casterton Ru.

Casthorpe Li [*Caschingetorp, Chaschintorp* DB, *Caskingtorp* 1212 Fees]. '*Kaskin*'s thorp.' The pers. n. *Caschin* is found in DB, *Kaskin* c 1190 YCh 1576. Its etymology is obscure.

Castle Carrock Cu [*Castelcairoc* c 1160 WR, *-kairoc* 1209 P]. Welsh *Castell caerog* 'fortified castle'.

Castle Church St [*Castellum* 1293, *Castre* 1302 Ass]. 'The castle.' The place is close to a hill on which is Stafford Castle.

Castleford YW [*Ceasterford* 948 ASC (D), *Casterford* c 1130 SD, *Castreford, Castelforde* 1155–8 YCh 1451]. 'Ford by the Roman fort.' Ermine Street here crosses the Aire. The Roman station of *Legeolium* was at this place. See CEASTER.

Castle Howard YN. A late name that has displaced HINDERSKELFE. The modern mansion is named after the Howard family.

Castlerigg Cu [*Castelrig* 1256 Fount]. 'The ridge of or adjoining Derwentwater Castle.' There is also a stone circle (Castlerigg Circle) on the ridge.

Castlethorpe Li [*Castorp* DB, *Cheistorp* c 1115 LiS, *Keistorp* 1311 Ch]. Perhaps '*Cǣge*'s thorp'. Cf. CASSIOBURY. Or better, the first el. is an OScand pers. n., e.g. *Keikr* (gen. *Keiks*), found as an ON byname.

Castlethwaite We. 'The thwaite by the castle', i.e. Pendragon Castle.

Castleton Db [*castellum Willelmi peuerel* DB, *Castelton* 13 Derby], **C~ La** [*Castelton* 1246 Ass, *Villa Castelli de Racheham* 13 WhC], **C~ YN** [*Castelton* 1577 Saxton]. 'TŪN by the castle.' C~ YN was named from Danby Castle. For C~ Db and La see the examples.

Castlett Gl [*Cateslat* DB, *Catteslada* 1178 P, *-slade* 1220 Fees]. OE *catta* (or *cattes*) *slæd* 'the valley of the wild cats or cat'.

Castley YW [*Castelai* DB, *Castelea* 1166 P, *Castelay* c 1200 YCh 516, *Castellay* 1234 FF]. The first el. is OE *ceastel, cestel* found in *stancestil* &c. This is explained in BCS 282 as 'acervus lapidum', i.e. 'a heap of stones'. OE *ceastel* &c. is related to OSw *kaster,* ON *kǫstr* 'a heap' (of wood &c.). Second el. OE *lēah*.

Caston (-ah-) Nf [*Catestuna, Castestuna* DB, *Catestuna* 1121 AC, *Cattestun* 1191 FF, *-ton* 1194 ff. P]. '*Catt*'s or *Kāti*'s TŪN.' See CATT.

Castor Np [(be) *Cyneburge cæstre* 948 BCS 871, *Castra* 972 ib. 1281, *Castre* DB]. 'The Roman fort'; see CEASTER. Possibly identical with *Durobrivas* in IA route v.

Caswell Np [*Karswell* 1196 Cur], C~ O [*Cressewell* 1186 P, *Carsewelle* 1316 FA]. 'Spring or stream where water-cress grew.' Identical in origin are C~ Do and (presumably) So.

Catchburn Nb [*Cacheborn* 1279 Ass, *Cachebur* 1323 Ipm]. 'The stream of *Cæcca*'? OE *Cæcca* pers. n. is possibly evidenced in *Cæccam wæl* BCS 565, if *Cæccam* is miswritten for *Cæccan*.

Catcleugh Shin (a hill) Nb [*Cattechlow* 1279 Ass]. Catcleugh means 'the clough or ravine of wild cats'. Cf. CLŌH. *Shin* is a Scotch word for 'the sharp slope of a hill', really the ordinary word *shin* in a transferred sense.

Catcombe W [*Cadecoma* 1114 Fr, *Cadecumb* 1241 Cl]. '*Cada*'s coomb or valley.'

Catcott So [*Caldecote, Cadicote* DB, *Cadicot* 1225 Ass, *Katicote* 1243 Ass, *Caldecot* 1251 Cl]. '*Cada*'s COT.'

Cāterham Sr [*Catheham* 1179 ff. P, *Caterham* 1200 Cur, *Katerham* 1236 Fees, *Katreham* c 1270 Ep]. The first el. may be the same as that of CHADDERTON, CATTERTON &c. There is a prehistoric fort near the place.

Catesby Np [*Catesbi* DB, c 1200 BM, *-by* 1220 Fees, *Katteby* 1228 Ep, *Kateby* 13 BM]. '*Kāti*'s BY.' Cf. CADEBY.

Catfield Nf [*Catefelda* DB, *Cattfeld* 1197, *Catefeld* 1198 P]. 'FELD frequented by wild cats.'

Catford K [*Cateforde* 1311 Ipm, *Catford* 1331 Ch], **Catforth** La [*Catford* 1332 Subs]. 'Wild-cat ford.'

Catfoss YE [*Catefos(s)* DB, *Cathefossa* c 1160 YCh 1334]. '*Catta*'s FOSS or ditch.' See CATT. C~ is near CATWICK.

Cathanger So [*Cathangre* DB, 1197 P, *Kathangre* 1225 Ass]. 'Slope or wood frequented by wild cats.' Cf. HANGRA.

Catherington Ha [*Cateringatune* c 1015 BM, *Cateringeton* 1176 P, *Chaderinton* 1187 P, *Katerringeton* 1242 Fees]. See CHADDERTON. The place is on a hill.

Catherston Leweston Do [*Chartreston* 1268, *Cartreston* 1316 FF, *Carterestone* 1316 FA]. The first el. is manorial, viz. the family name *Chartray* (cf. BEER CHARTER). Leweston is probably the name of a local family, from LEWESTON Do. Walter de Leweston is mentioned along with John de Chartrey in 1256 FF in a document referring to Leweston.

Catherton Sa [*Carderton* 1316 FA]. Possibly identical with CARDESTON.

Catley He [*Catesley* 1242 Fees, *Cattelegh* 1251 Ch, *Catteley* 1279 Ipm], C~ Li [*Catelei* Hy 2 DC, *Catteleia* 1197, *Kattele* 1230 P]. 'Wild-cat wood.'

Catmore Brk [*Catmere, Catmeringa* (*-mæringa*) *gemære* 916, 931 BCS 633, 682, *Cat-* *meres gemære, Catbeorh* 951 ib. 892, *Catmere* DB]. 'Wild-cat lake.'

Caton D [*Cadetone* 1330 Subs]. '*Cada*'s TŪN.' C~ (-ā-) La [*Catun* DB, *Catton* 1186 P]. '*Kāti*'s TŪN.' On *Kāti* cf. CADEBY.

Catsfield Sx [*Cedesfeld* DB, *Catesfeld* 12 (1432) Pat, *Cattesfeld* 12 PNSx]. 'FELD frequented by wild cats.'

Catshall or **Catteshall** Sr [*Gateshela* 1130 P, *Catteshull* 1212, *Cateshell* 1219 Fees], **Catshill** Wo [*Catteshull* 1199 P, *-e* 1221 Ass]. 'Hill frequented by wild cats.'

Catsley Do [*Catesclive* DB, *Cattesclive* 1227 FF]. 'Cliff frequented by wild cats.'

Catsley Sa [*Gateschesleie* DB, *Cackesleg* 1242 Fees, *Catekesle* 1255 RH]. '*Catoc*'s LĒAH.' Welsh *Cadog*, MW *Cadoc, Catoc*, is a well-evidenced pers. n.

OE **catt** 'cat' in the sense 'wild cat' is probably the first el. of a good many pl. ns. But in all probability it was also used as a byname and pers. n. In pl. ns. where the second el. is a word such as LĒAH, SLÆD, or the like, the meaning 'wild cat' is as a rule to be assumed. There was also an OE pers. n. *C(e)atta*, which is evidenced once in Saints. It must be assumed in *Ceatian mære* and *Ceattan broc* 983 KCD 636 (W) and *Cattan ege* 966 BCS 1176 (O). There is also the OScand pers. n. *Kāti* (ON *Káti*, ODan *Kate*), which it is sometimes difficult to distinguish from *catt*.

Cattal YW [*Cathale, Catale* DB, *Parva Cathale* c 1200 YCh 536]. 'HALH frequented by wild cats.'

Cattawade Sf [*Cattiwad* 1247 Cl]. OE *cattgewæd* 'cats' ford'.

Cattenhall Chs [*Catenhale* c 1130 Chester, c 1300 ib.]. '*Catta*'s HALH.'

Catterall La [*Catrehala* DB, *Caterhale* 1212 Fees]. Doubtful. The name has been derived from OScand *kattar-hali* 'cat's tail', here used of a farm on account of the lengthened shape of its land. The pl. n. KATTERALL in Norway actually has this etymology. The situation of Catteral rather suggests a second el. HALH, but the first el. offers difficulties.

Catterick YN [*Katouraktónion* c 150 Ptolemy, *Cataractone* (abl.) 4 IA, *Cataracta*, (a) *uico Cataractone* c 730 Bede, *Cetreht*, (neah) *Cetrehtan, Cetrehttun* c 890 OEBede, *Catrice* DB, *Kateric* 1231 FF, MW *Cat(t)-raeth* Taliesin &c.]. Probably Lat *cataracta* 'waterfall', which suits the local conditions. If so, the name must have been changed by Britons, who substituted OBrit *catu-* 'war' for the original first el. and added a British suffix.

Catterick Moss Du [*Katerickesaltere* 1311 FPD] was probably named for some reason from CATTERICK. The el. *-saltere* is very likely identical with SALTER.

Catterlen Cu [*Kaderlenge* c 1165 WR, *Katerlen* 1201 P, *Katirlen* 1283 Ipm]. For the first el. see CHADDERTON. Second el. possibly OE *hlynn* 'a torrent' or, if the first form may be trusted, OE *hlinc* 'hill'. The place is on a stream and the map marks a camp close by.

Catterton YW [*Cadretune* DB, *Cadartona* 1157 YCh 186, *Cadreton* 1230 FF, *Catherton*, *Katerton* 1256 FC]. See CHADDERTON. C~ is near Bilbrough, which is on a hill, past which runs a Roman road.

Catthorpe Le [*Torp* DB, *Parva Thorp* 1254 Val, *Torpkat* 1276 RH, *Catthorp* 1316 FA]. Originally THORP. The additional *Cat-* is presumably the name of an owner, added for distinction from COUNTESTHORPE.

Cattishall Sf [*Catteshale* 1187 Bury, -*hal* 1238 Cl, *Catteshull* 1242 Cl, *Catteshill* 1269 Misc]. 'HALH or hill frequented by wild cats.'

Cattistock Do [*Cattesstok*, *Stoke* 939 BCS 738 f., *Stoche* DB, *Stok* 1212 Fees, *Cattestoke* 1291 Tax]. The form of 939 is really of no value. Originally *Stoc*. See STOC. The addition is probably a family name.

Catton Db [*Chetun* DB, *Catiton* 1208 Cur, *Catton* 1236 FF, 1242 Fees], C~ Nf [*Catetuna* DB, *Cattuna* ib., *Catton* 1212 Fees], **High & Low** C~ YE [*Cattune* DB, *Cattuna* 12 YCh 910, *Catton* 1200 FF], C~ YN [*Catune* DB, *Cattun* Hy 2 (1247) Ch]. '*Catta*'s or *Kāti*'s TŪN.' See CATT.

Catton Nb [*Catteden* 1229, -*dene* 13 BBH]. 'Wild-cat valley.'

Catwick YE [*Catingeuuic* DB, *Cattingewic* c 1130, *Cattewic* c 1125 YCh 1319, 1318, *Catewic* 1226 FF]. 'The wīc of *Catta*'s people.'

Catworth Hu [*Catteswyrð* 972–92 BCS 1130, *Cateworde* DB, *Catteswurda* 1163 P]. '*Catt*'s WORÞ.' Cf. CATT.

Caudle Green Gl [*Caldwella* c 1155 (1340) Ch, *Caldewell* 1270 Ipm]. 'Cold spring or stream.'

Caughall Chs [*Cochull* 1278 Chester]. 'Cock hill', i.e. 'hill frequented by fowls'.

Caughley Sa [*Cahing læg* 901 BCS 587, *Cacheleg* 1221 PNSa, *Kacheleg*, *Kakeleg* 1255 RH]. The original name may have been OE *Ceahhing* 'daw wood' (from OE *ceahhe* 'daw'), to which was added an explanatory LĒAH 'wood'.

COTE &c.
Caulcott O [*Caldecot* 1279 RH]. Cf. CALDE-

Cauldon St [*Celfdun* 1002 Wills, *Caldone* DB, *Caluedon* 1196 FF, *Calfdon* c 1200 Bodl]. OE *cælf-dūn* 'calves' hill'. *Cælf* is an *i*-mutated form of *calf*.

Cauldwell Bd [*Caudewell*' 1200 FF, *Chaldewell* 1224 Pat], **Caldwell** Db [*æt Caldewellen* 942 BCS 772, *Caldewelle* DB]. 'Cold spring.'

Caundle, Bishop's, Purse & Stourton, C~ **Marsh** Do [*Candel* DB, 1176 P, *Candele* DB, *Caundel Episcopi* 1285 FA, *Purscaundel* 1241 FF, *Pruscandel* 1275 RH, *Candelemers* 1245 FF]. The Caundles are situated some way apart near a chain of hills. Caundle is no doubt a British name of the chain of hills. Its etymology is obscure. Possibly there may be some connexion with *Cant-* in *Cantmæl* (now CAMEL So) and QUANTOCK HILLS. **Caundle Brook** was named from C~ Marsh and Bishop's Caundle.

Bishop's C~ belonged to the Bishop of Sarum, **Purse** C~ to Athelney Abbey. Purse may be OE *prēost* 'priest'. **Stourton** C~ was named from the Lords Stourton, who held the manor from t. Hy VI. It was known in the 13th cent. as C~ **Haddon** [*Caundel Haddone* 1276 FF]. Henry de Haddone got land in Caundle in 1202 (FF). The family took its name perhaps from HADDON LODGE in Stourton Caundle. Or the place was named from the family.

Caunton Nt [*Calnestune* DB, *Calnoðeston* 1167 P, *Kalnadatun* c 1155 DC, *Calfnadtun* 13 BM]. Perhaps '*Calunōþ*'s TŪN'; cf. CANDLESBY, CANDLESHOE Li. *Calfnadtun* may be due to popular etymology.

Cause Sa [*Alretone* DB, *Chaus* 1165 P, *Caos* 1200 BM, *Cauz* 1246 Ch, *Caures* 1255 RH]. The old name *Alretone* 'alder TŪN' was displaced by a Norman name, which is supposed to be from CAUX in Normandy. Roger Fitz Corbet, who held the manor in 1086, is said to have belonged to a family that hailed from Caux.

Causey Park Nb [*La Chauce* 1242 Fees], C~ **Pike** (hill) Cu [*Le Cauce* 1294 Cl]. ME *cauce*, *cause* from ONFr *cauciée*, Fr *chaussée* 'a paved way'. But a meaning 'embankment or dam' is also found in ME.

Cave, North & South, YE [*Cava*, *Cave* DB, *Cava* c 1120 YCh 1822, *Northkave* c 1150 YCh 1124, *Suthkave* 1246 FF]. Both Caves are near **Mires Beck**, which comes from the Wolds and must have a swift course at Cave. The name may be originally that of the stream and a derivative of OE *cāf* 'quick, prompt, nimble'. The OE form would be *Cāfe* fem.

Cavendish Sf [*Kauanadisc*, *Kanauadisc* DB, *Kavenedis*, *Kaftnedich* 1219, *Cavenedis* 1242 Fees, *Cauenedess* 1229 FF], **Cavenham** Sf [*Canauatham*, *Kanauaham* DB, *Cauenham* 1198 FF, *Caveham* 1210 FF, *Cavenham* 1291 Tax]. The first el. of the two names is no doubt a pers. n. derived from OE *cāf* 'bold, active'. It may be OE *Cāfa* (found as *Caua* LVD), but the earliest forms point rather to an OE *Cāfna; cf. CAENBY. Indeed, if *Kaftnedich* 1219 and *Canauatham* (evidently for *Cauanatham*) DB are not to be disregarded, we may have to start from an OE *Cāfnōþ*. The second el. of Cavendish is OE *edisc* 'enclosure, pasture'.

Caversfield O [*Cavrefelle* DB, *Kaveresfelde* 1225 Ep, *Caveresfeld* 1302 Ch], **Caversham** Brk [*Cavesham* DB, *Caueresham* 1174

P, *Caveresham* 1209–35 Ep], **Caverswall** St [*Cavreswelle* DB, *Cauereswell* 1167, *Chauereswella* 1185 ff. P, *Cavereswall* 1242 Fees]. These must be compared with OE *Caberes bec* 862 BCS 505 (Wittenham Brk), *Kaverash* R 1 (1227) Ch (not far from Caversham) and *Cauerswelle Broke*, *Cauershulle* (pratum) 1298 For (in boundaries of Wychwood For. O). The element common to all these must be a pers. n., probably connected with OE *cāf* in CAVENDISH &c. It may be an OE *Cāfhere* or a derivative with an *r*-suffix. The second el. of Caverswall is OE *wella* 'stream' in its West Midland form *walle* from OE *wælla*.

Cavick House Nf [*Cakwyc*, *Cakewyk* 1332 BM]. Perhaps '*Cæfca*'s wīc'; cf. CAKEHAM.

Caville or **Cavil** YE [*Cafeld* 959 YCh 4, *Cheuede* DB]. 'Jackdaw FELD.' First el. ME *cā* 'jackdaw'.

Cawkwell Li [*Calchewelle* DB, -*wella* c 1115 LiS, *Calkewell* 1206 Ass]. 'Chalk stream.'

Cawood La [*Kawode* c 1225, c 1250 CC], C~ YW [*Cawuda* c 972 BCS (1278), c 1030 YCh 7, *Cawude* 1184 P]. 'Jackdaw wood.' Cf. CABOURN, CAVILLE.

Cawston Nf [*Caupstuna*, *Caustuna* DB, *Causton* 1159, 1190 ff. P], C~ Wa [*Calvestone* DB, *Causton* 1200 Cur, 1283 Ch]. Both are probably '*Kalf*'s TŪN' the first el. being the OScand pers. n. *Kalfr* (cf. CALCEBY). The early loss of *l* is due to Norman influence.

Cawthorn YN [*Caltorn*, -*a* DB, *Calthorn* 1176 P, *Kaldthorn*, *Kalethorn* 1202 FF], **Cawthorne** YW [*Caltorn*, -*e* DB, *Calthorn* c 1125 YCh 1663]. One spelling points to the first el. of Cawthorn being OE *cald* 'cold', but 'cold thorn-bush' is a remarkable name, which one would not expect to find twice. The first el. at least of Cawthorne is rather CALU 'bare'.

Cawthorpe Li nr Bourne [*Caletorp* DB], C~ Li nr Covenham [*Caletorp* c 1115 LiS, *Caltorp* 1212 Fees], **Little C~** Li nr Louth [*Carletorp* 1205 Cur, *Calthorp* 1241 Ep, 1254 Val]. '*Kali*'s thorp.' Cf. CALTHORPE. Little C~ may be rather '*Karli*'s thorp'.

Cawton YN [*Calvetun* DB, *Calueton* 1163 P, 1226 FF]. OE *Calfa-tūn* 'TŪN where calves were kept'.

Caxton Ca [*Caustone* DB, *Kachestona* 12 Fr, *Cakeston* 1187, 1190 P, 1205 Cur]. The first el. may be a Scand. pers. n. or by-name *Kakkr*, found in KAKSRUD (Norway). This may be from Norw *kakk* 'nose' or ON *kǫkkr* 'lump', Norw *kakk* 'a knob' &c. This seems to be the most probable solution. Or it might be Scand *kax*, a side-form of *käx* 'umbelliferous plants', the source of Engl *kex*. If this should be right, later forms like *Kakeston* must be due to the first el. having been misunderstood as a pers. n.

Caynton Sa [*Caginton* 1180 P, *Kagintun*

1249 Ipm, *Caynton* 1327 Subs]. '*Cæga*'s TŪN' or 'the TŪN of *Cæga*'s people'. Cf. CAINHAM.

Caythorpe Li [*Catorp* DB, *Catetorp* 1203 Cur, *Cattorp* 1203 Ass], C~ Nt [*Cathorp* 1177, *Catthorp* 1179 P]. '*Kāti*'s thorp.' Cf. CADEBY and CATT.

Caythorpe YE [*Caretorp* DB, *Carthorp* 1100–15 YCh 1001, *Carethorp* c 1130 ib. 1063]. '*Kāri*'s thorp.' Cf. CAREBY.

Cayton YN [*Caitune*, *Caimtona* DB, *Keyton* 1243 FF], C~ YW [*Chetune* DB, *Caituna* 1146, *Caitona* 1150–3 YCh 79, 71]. '*Cæga*'s TŪN.' Cf. CAINHAM.

OE ceart. See CHART.

OE ceastel, cestel 'heap (of stones)'. See CASTLEY, CHASTLETON, CHESHAM.

OE ceaster, cæster, an early loan-word from Lat *castra*, means 'a city or walled town, originally one that had been a Roman station'. This is actually the meaning in many pl. ns., such as GLOUCESTER, MANCHESTER, CHESTER &c. But in many cases the meaning must have been 'prehistoric fort' generally. The Northumbrian names in -*chester*, for instance, cannot all denote old Roman stations. The usual form of the word in pl. ns. is *Chester*-, -*chester*. But *Caster*-, -*caster*, from OE *cæster*, is regular in some districts, viz. Y, NLa, Cu, We, Li, Ru, Nf, and occurs in Np. See CAISTER, CASTOR &c. *Caster*-, -*caster* is sometimes replaced by *castle*. Cf. BEWCASTLE, HORNCASTLE, CASTLEFORD. Owing to Norman influence *Chester*-, and especially -*chester*, often becomes *Cester*, -*cester* or even -(*c*)*eter*, as in GLOUCESTER, EXETER, WROXETER.

Cefn-y-Castell Sa means 'ridge with a castle'.

Ceiriog R Sa [*Ceirawc* 13 Mabinogion, *Keriok* 1577 Saxton]. **Chirk** is an Anglicized form of the name [*Chirc* 1165 P]. Ceiriog is a British river-name, perhaps related to CARY.

OE celde 'spring' is rare in OE and likewise in pl. ns. See BAPCHILD, HONEYCHILD. *Celde* is derived from OE *ceald* 'cold' and corresponds to OScand *kelda*. In Saxon the word would have appeared as OE **cielde*, **cilde*, and some names in *Chil*(*d*)- may contain it (e.g. CHILCOMBE Do). The Anglian form would have given ME *kelde* and could not be distinguished from OScand *kelda*. The el. *keld*, when found in Scandinavian England, is no doubt Scand *kelda*.

OE ceole 'throat' seems also to have been used in a transferred sense of a gorge or valley, perhaps also of a neck of land, but is difficult to distinguish as a first el. from *Cēola* pers. n. See CHALE, CHELL, CHILGROVE, CHOLLERFORD.

OE ceorl 'free peasant, villein'. See especially CHARLTON, also CHARL- (passim),

CHORLEY, CHORLTON, CHALTON Bd, CHUR-
WELL. Cf. CARL(E)TON.

OE ceosol, cisel 'gravel' is a fairly common
el. See CHELSING, CHESELBOURNE, CHESIL,
CHESELADE, CHISLE-, CHILLESFORD, CHISEL-
BOROUGH, also CHEESEBURN, CHISLET.

OE cēping 'market' &c. See CHIPPING.

Cerne R Do [Cerne 1244 Ass]. Like CHAR,
CHARN a derivative of Welsh carn 'rock,
stones'. The form to be expected is Charn
or Chern. Cerne is due to Norman influence.
From Cerne are derived Cerne Abbas,
Nether & Up Cerne Do [Cernel 10 Ælfric,
DB, 1114 ASC (H), Cerne DB, Cerna
1130 P, Upcerl(e) 1001–12 ASWrits, Obcerne
DB, Cerne Abbatis, Nithercerne 1291 Tax,
Cern Monachorum 1256 FF]. Cerne was
the site of an abbey. The form Cernel is
obscure. Possibly -el represents Welsh iâl
'fertile upland region'. Cf. DEVERILL.

Cerney, North & South, Cerney Wick
Gl [Cirnea, Cyrnea 852 BCS 466, æt Cyrne
999 KCD 703, Cernei DB, Northcerneye
1291 Tax, Suthcerney 1285 FA, Creneiewich
1206 Cur]. Really the name of the river on
which the places stand. See CHURN. Cerney
has been modified in form owing to Norman
influence. Wick is wīc, probably 'dairy-
farm'.

OE cēse, cīese 'cheese'. See CHEESEBURN,
CHESWARDINE, CHESWICK, CHISWICK, also
CASEWICK, KESWICK.

Cesterover. See OVER Wa.

Chaceley Wo [Ceatewesleah 972 BCS 1282,
Chedeslega 1167 P, Cheddeslega Hy 2 BM,
Chaseleia 1183 AC]. The first el. looks like
a pers. n. and has been compared with
Ceatwanberge 869 BCS 526 (a very poor
text). A pers. n. Ceatwe is very difficult to
explain, and it is preferable to connect
Ceatwes- with Welsh coed 'wood', Brit cēto-
from *kaito-, which appears as -ceat in
Penceat (see PENGE). The first el. of Chaceley may be identical with CHITTOE. Second
el. OE LĒAH, probably 'wood'.

Chackmore Bk [Chalkemere 1229 Cl, Chake-
more 1241 Cl, 1284–6 FA]. 'Ceacca's moor.'
Cf. CHECKENDON &c.

Chacombe or Chalcombe (-ā-) Np [Cewe-
cumbe DB, Chaucumba 1166, -cumbe 1195 P,
Chacombe 12 NS]. 'Ceawa's CUMB or valley.'
OE Ceawa pers. n. is found in Ceawan
hlæw 947 BCS 833 (now CHALLOW),
Ceawwanledge 854 ib. 476 (So), Ceawan or
Ceauuan hrycg 942 ib. 778 (Winkfield Brk),
also in CHAURETH.

Chadacre Sf [Chearteker 1046 Wills,
Chardeker 1275 RH, Chardacre 1303, -akre
1346 FA]. First el. OE ceart 'a rough common'; cf. CARTMEL, CHART. The change of
rt to rd is exemplified also in CHARD. The
loss of r is due to dissimilation. Chadacre
thus means 'the field by the rough common'.

Chadbury Wo [(on) Ceadweallan byrig c 860
KCD 289, Chadelburi (castle) 13 Chron
Eve]. 'Ceadwealla's BURG.'

Chaddenwicke W [Chedelwich DB, Cha-
delewic 1196 FF, Chadewich 1242 Fees,
Chadewych 1325 Pat]. 'Ceadela's WĪC.' Cf.
CHADDLEWORTH.

Chadderton La [Chaderton c 1200 WhC,
1224 Ass, Chaterton 1224 Pat]. Hanging
Chadder La not far from Chadderton
[Hengandechadre 1324, 1332 PNLa].

An early Welsh cader (cater) 'hill fort', identical
with OIr cathir 'town', Gael cathair 'circular
stone fort', was generally assumed by earlier
Celticists (as H. Pedersen, Thurneisen), and in
PNLa and DEPN (edd. 1–3) is taken to be the
first el. of CHADDERTON and many other names,
like CATERHAM, CATTERLEN, CATTERTON, CATHER-
INGTON. Later Celtic scholars do not reckon
with OW cader, and take hill-names like CADER
IDRIS to contain OW cateir (cadeir) 'chair' from
Lat cathedra. Some scholars, as Bruce Dickins
in PNCu(S) and Smith in EPN(S), substitute
this word for cader 'hill fort' in names of this
kind. If this is right, OW cateir must have been
used in transferred senses such as 'hill' or possibly 'hill fort'. The matter is not definitely
solved, and though Cader in CADER IDRIS &c. is
doubtless OW cateir 'chair', it is not impossible
there was also an OW cater (cader) corresponding
to OIr cathir.

Chaddesden Db [Cedesdene DB, Chades-
dena 1168 P, -dene 1236 FF, Chaddesdene
1258 FF]. 'Ceadd's or Ceaddi's valley.'
A pers. n. *Ceadd(i), a short form of Ceadwalla, must be assumed also for CHAD-
STONE.

Chaddesley (-ăj-) Corbett Wo [Ceadres-
leahge, aet Ceadresleage, Ceades-, Cedres-
leage 816 BCS 356 f., Cedeslai DB, Chaddes-
leye Corbett 1327 Subs]. The first el. looks
like a pers. n., and is in PNWo held to be
an OE *Ceadder. Such a name is difficult
to explain, and perhaps the first el. is identical with that of CHADDERTON. See LĒAH.

Corbett is an OFr nickname and pers. n., from
OFr corbet 'raven'. The Corbet family held
Chaddesley from the end of the 12th cent.

Chaddleworth Brk [Ceadelanwyrð 960 BCS
1055, Cedeledorde DB, Chadelwurda 1167
P]. 'Ceadela's WORP.' The pers. n. *Ceadela is presupposed also by CHADDENWICKE,
CHADLINGTON.

Chadkirk Chs [Chadkyrke 1534 VE]. 'The
church of St. Chad.'

Chadlington O [Cedelintone DB, Chedelin-
ton 1163 P, Chadelington 1196 FF, Chiade-
linton 1201 Cur]. 'The TŪN of Ceadela's
people.' Cf. CHADDLEWORTH.

Chadnor He [Chabenore DB, Chabbenour
c 1180 Fr, -ore 1242 Fees, Chabenhoure
1212–17 RBE]. 'The OFER or hill of Ceabba.'
A pers. n. Ceabba is presupposed by Ceabban dun 1033 KCD 752, Ceabban solo 796
BCS 282. It may be a short form of Cæd-
bæd or Cædbald. The b became d before n
by partial assimilation.

Chadshunt Wa [*Chadeleshunte* 1043 KCD 916, *Chaddeleshunt* c 1050 ib. 939, *Cedeleshunte* DB, *Chedelesfont* 1135 Ch]. 'Ceadel's spring.' *Ceadel* is a strong side-form of *Ceadela* in CHADDLEWORTH. See FUNTA.

Chadstone Np [*Cedestone* DB, *Chaddeston* 1220 Fees]. 'Ceadd(i)'s TŪN.' Cf. CHADDESDEN.

Chadwell St. Mary Ess [*Celdewella* DB, *Chaudewelle* 1210–12 RBE, *Chaldewell* 1238 Subs]. 'Cold spring.' The map marks St. Chad's Well close by.

Chadwell Le [*Caldeuuelle* DB, *-wella* 1177 ff., *Chaldewell* 1179 ff. P, *Caldewell* 1276 RH]. 'Cold stream.' The correct form in this Anglian district is *Caldwell*, not *Chaldwell*, and the earliest examples have *C*-. The place seems to be still called alternatively *Caldwell*. Bartholomew has both Caldwell and Chadwell. The earliest spellings with *Ch*- may denote a pronunciation *C*-; *ch* is a common early spelling for *k*. Perhaps the present Chadwell is due to spelling-pronunciation. *Chad*- for *Chald*-owing to dissimilation.

Chadwich Wo [*Celdvic* DB, *Chadeleswik* 1212 Fees]. Identical with CHADWICK Wa.

Chadwick La [*Chaddewyk* c 1180 WhC, *Chadewik* 1246 Ass], **C~** Wo [*Cheddewic* 1182 PNWo, *Chedewyke* 1327 Subs]. 'The wīc of Ceadda', perhaps St. Chad.

Chadwick Wa [*Chadeleswiz* J BM, 1242 Fees]. 'Ceadel's wīc.' Cf. CHADSHUNT.

Chaffcombe D [*Chefecoma* DB], **C~** So [*Caffecome* DB, *Chaffacombe* 1204 (1313) Ch, *Chaffecumbe* 1236 FF]. The same first el. is found in **Chafford** hd Ess [*Ceffeorda* DB, *Cheaffeworda* 1130 P]. It can hardly be anything else than a pers. n. *Ceaffa*, which might be a hypocoristic form of *cealf* 'calf' used as a pers. n.

Chagford D [*Chageford* DB, *Chaggesford* 1185 P, *Chaggeford* 1230 P]. 'Chag ford.' Dial. *chag* 'broom or gorse' comes from OE *ceacge*.

Chaigley (-āj-) La [*Chadelegh*, *Chaddesl'* 1246 Ass]. 'Ceadd(i)'s LĒAH.' Cf. CHADDESDEN.

Chailey Sx [*Cheagele*, *Chaglegh* W 2 PNSx, *Chageleye* 1256 FF]. 'Chag LĒAH.' Cf. CHAGFORD.

Chalbury Do nr Wimborne [*Cheoles burg* 946, *Cheoles byrig* 956 BCS 818, 958]. 'Cēol's fort.'

Chaldeans Hrt [*Celgdene* DB, *Chaldene* 1303 FA]. 'Chalk or limestone valley.'

Chaldon Herring or **East C~**, **West C~** Do [*Cealuaduna* DB, *Chaluedon* 1234 BM, *Chaluedon Hareng* 1243 FF, *Est-*, *Westchalvedon* 1269 Misc], **C~** Sr [*Cealuadune* 967 BCS 1198, *Cealfadune* 1062 KCD 812,

Chalvedon 1275 Ipm]. 'Hill where calves grazed.' See DŪN.

Herring is an OFr byname and family name, from *hareng* 'herring'. Therricus Harang held Chaldon in 1203 (Cur).

Chale Wt [*Cela* DB, *Chele* 1182, *Chale* 1168 P, *Chaledone* 1324 Misc]. Chale is on **Chale Bay**, where the map marks some 'chines' or ravines in a cliff. Probably the name is OE *ceole* 'throat', here used of a ravine. If so, a dialectal change of *eo* to *ea* has to be assumed.

Chalfield W [(at) *Chaldfelde* 1001 KCD 706, *Caldefelle* DB]. 'Cold FELD.'

Chalfont (-ahf-) **St. Giles** & **St. Peter** Bk [*Ceadeles funta* 949 BCS 883, *Celfunte* DB, *Chaufonte Sancti Egidii*, *Chaufunte Sancti Petri* 1242 Fees]. 'Ceadel's FUNTA or spring.' Cf. CHADSHUNT and CHADWICK Wa.

Chalford Gl [*Chalkforde* 1297 PNGl], **C~** O in Enstone [*Celford* DB, *Chauford* 1242 Fees], **C~** O in Aston Rowant [*Chalcford* 1185–6 Thame, 1279 RH]. 'Chalk ford', i.e. 'ford where limestone was carried across or where limestone was found'.

Chalgrave Bd [*Cealhgræfan* 926 BCS 659, *Celgrave* DB], **Chalgrove** O [*Celgrave* DB, *Chealgraue* c 1170 Oxf, *Chalcgrava* 1236 Fees]. OE *cealc-græf* or *-grafu* 'chalk-pit(s)'. *Cealhgræfan* 926 is probably miswritten for *-grafan*.

Chalk K [*Cealca*, (of) *Cealce* 10 BCS 1321 f., *Celca* DB, *Chalcha* 1165 AC, *Chelk* 1207 Cur], **Bower** & **Broad Chalk** W [*Cealcan gemere* 826 BCS 391, *æt Ceolcum* 955 BCS 917, *Cheolca*, (to) *Cheolcan* 974 ib. 1304, *Chelche* DB, *Chalche* 1174 P, *Burchalke* 1316 FA, *Brodechalke* 1415 AD]. An OE *cealce* 'limestone down', derived from OE *cealc* 'chalk, limestone'. The Wilts Chalks are in the **White Chalk** district. *Bower* seems to be BURG 'borough'. On *Broad* see BRĀD.

Chalk (shawk) **Beck** R Cu [*Shauk* c 1060 Gospatric's ch, *Schauk* 1285 For]. A British river-name *Scawōc*, derived from the base of Welsh *ysgaw* 'elder wood' and meaning 'river where eldertrees grew'. Cf. Breton *skavek* adj. 'abounding in elders'.

Chalk Farm Mx [*Chaldecote* 1253 Cl]. Identical with CALDECOTE &c.

Challacombe D nr Lynton [*Celdecomba* DB, *Chaudecumb* 1242 Fees]. 'Cold valley.' **C~** D in Combe Martin [*Chaluecumba* 1168 P]. 'Calves' valley.'

Challock (-ŏl-) K [(ad) *Cealfalocum* 824, (et) *Cealflocan* c 833 BCS 378, 412, *Cealueloca* 11 DM]. 'Enclosure for calves.' See LOCA.

Challow Brk [*Ceawan hlæw* 947 BCS 833, *Ceveslane* DB, *Cewehlewe* Hy 1 (1317) Ch, *Chawelawe* 1220 Fees]. 'Ceawa's burial-mound.' Cf. HLĀW and CHACOMBE.

Chalmington Do [*Chelmynton* 939 BCS 738, *Chelmeton* 1181 P, *Chelminton* 1212 Fees, *Chelmington* 1268 FF]. OE *Cēolmundingatūn* or *Cēolhelmingatūn* 'the TŪN of *Cēolmund*'s or *Cēolhelm*'s people'.

Chalton Bd in Moggerhanger [*Cerlentone* DB, *Cherleton* 1173 P, *Chauton* 1250 Ch]. OE *Ceorlatūn*. See CHARLTON. **C~** Bd in Toddington [*Chalfton* 1227 Ass]. 'Calf farm.'

Chalton Ha [*Cealhtun* 1015 Wills, *Ceptune* DB, *Chalkton* 1278 Ipm]. 'TŪN on or by the chalk down.' **Chalton Down** is near Chalton.

Chalvey (-ahv-) Bk [*Chalfheye* 1227 Ass, *Chalfeye* 1237–40, *Chalveye* 1242 Fees]. OE *cealf-īeg* 'calf island', or *cealf-gehæg* 'enclosure for calves'.

Chalvington (chahntn) Sx [*Calvintone* DB, *-ton* 12 BM, *Chalvintona* c 1150 Fr]. OE *Cealfingatūn* 'the TŪN of the people of *Cealf(a)*'. The pers. n. *Cealf* or *Cealfa* is not evidenced, but is explained as a nickname from *cealf* 'calf'.

Chancton Sx [*Cengeltune* DB, *Changhetona* 1150–69 Oxf, *Changeton* 1249 FF], **Chanctonbury Ring** [*Changebury* 1351 Ipm]. If the *l* of the DB form is inorganic, the first el. may be an OE folk-name *Cēaingas*, derived from *Ceawa* (see CHALLOW). The *w* would be dropped before *i*. But *Cengel-* might represent OE *Cēainga-hyll*.

Chanston He [*Chenestun, Cheineston* 1242 Fees]. A manorial name, the first el. being the family name *Cheney*. Cf. CHENIES.

Chapel en le Frith Db [*capella del Frith* 1332 Derby]. Frith is OE *fyrhþ* 'woodland'. 'The chapel in the woodland.'

Chapelthorpe YW [*Schapelthorpe* 1285, *Chapelthorp* 1316 Wakef]. Presumably 'thorp with a chapel'.

Chapmanslade W [*Chepmanesled*' 1245 PNW(S), *Chepmanslade* 1396 Ipm]. 'The valley of the chapmen.' Second el. OE *slæd* 'valley'.

Char R Do [*Cerne* c 1230 Wells, 1288 Ass]. Identical with CERNE. The Char gave its name to **Charmouth** Do [*Cernemude* DB, *-mue* 1212 Fees]: 'the mouth of the Char'.

Charborough Do [*Cereberie* DB, *Chereberge* 1212 Fees, *Chernebrug* 1219 Fees, *Cerebeg* 1253 FF]. The first el. may be an old name of the WINTERBORNE, identical with CERNE. Cf., however, CHARNWOOD. Second el. OE BEORG 'hill'.

Chard, South Chard So [*Cerdren* 1065 Wells, *Cerdre* DB, *Cerda* 1166 RBE, *Sutcherde* 1261 Wells]. OE *Ceart-renn* 'house in a chart or rough common', the first el. being OE *ceart* (cf. CHART) with the same change of *rt* to *rd* as in CHADACRE; the second OE *renn*, a side-form of *ærn* 'house'. *Cerdren* lost its final *-n* and the second *r*

was dropped owing to dissimilation. Near Chard is **Crimchard** [*Cynemerstun* 1065 Wells, *Kinemerscherd* 1196 P]. 'The part of Chard belonging to *Cynemær*.'

Chardstock D [*Cerdestoche* DB, *Cherdestokes* 1196 FF, *Cerdestok* 1200 Cur]. The place is less than 3 m. from CHARD. The name means 'STOC belonging to Chard'. See STOC.

Charfield Gl [*Cirvelde* DB, *Certfeld* c 1200 Bath]. 'FELD in a chart or rough common.' See CHART.

Charford, North & South, Ha [*Cerdicesford* 508 ASC, *Cerdeford* DB, 1185 P, *North-, Suthchardeforde* c 1270 Ep]. '*Cerdic*'s ford.'

The Chronicle states that the place was named after Cerdic, the West Saxon king, who won a battle here. Similarly *Cerdices ora* 495 ff. ASC and *Cerdices leaga* 527 ib. are no doubt understood to have been named from the king.

Charing K [*aet Ciornincge* 799 BCS 293, *Cerringges* 799 ib. 294, *Ciorringc* 799 BM, *Cheringes* DB, *Cyrringe* 11 DM, *Cherringis* 1121 AC]. The correct reading of the earliest example is no doubt *Ciorrincge*. This represents an OE *Ciorring* sing., a derivative of OE *Ceorra* (OKent *Ciorra*) pers. n., '*Ciorra*'s place or brook'. The place is on a stream.

Charing Cross Mx [*Cyrring* c 1000 Crispin, *Cherringe* 1197 FF, *la Charring* 1253 Cl]. OE *cierring* 'turning, turn' (from *cierran, cyrran* 'to turn'), referring to the great bend in the Thames near the place or possibly to a bend of the Roman road that ran west from London.

Charingworth Gl [*Chevringavrde* DB, *Chevringewrth* 1201 Cur, *Cheveringeworth* 1220 Fees]. Possibly 'the WORÞ of *Ceafor*'s people', *Ceafor* being a nickname from OE *ceafor* 'a beetle'.

Charlbury (-awl-) O [*Ceorling(c)burh* c 1000 Saints, *Cerlebiria* c 1160 RA, *Cherlebiria* 1234 Ep]. 'The BURG of *Ceorl*'s people.' *Ceorl* is well evidenced as a pers. n.

Charlcombe So [*Cerlecume, Cerlacuma* DB, *Cherlecumba* 1156 Wells, *-cumbe* 1225 Ass]. 'The coomb or valley of the *ceorls* or free peasants.'

Charlcote W [*Chedecotun* 1065 KCD 817, *Cherlecote* c 1300 PNW(S)]. See next name.

Charlcott Sa [*Cerlecote* DB, *Cherlecote* 1290 Ipm], **Charlecote** Wa [*Cerlecote* DB, *Cherlekote* 1196 FF]. 'The COT of the *ceorls* or free peasants.'

Charles D [*Carmes* DB, *Charnes* 1242 Fees, 1280 Ep, *Charles* 1244 Ass, 1291 Tax]. In PND explained as a compound of Co *carn* 'rock' and *lis, les* 'court, palace'. This may be right.

Charlesworth Db [*Cheuenesuurde* DB, *Chauelisworth* 1286 Court, *Chavelesworth* 1290 Ch]. Names in WORÞ usually have a pers. n. as first el. In this case a nickname derived from OE *ceafl* 'jaw' may be suggested. But the first el. may be *ceafl* used in a sense such as 'ravine'. The name was later influenced by the neighbouring **Charlestown** which is a late name, first recorded in 1843, according to PNDb(S).

Charleton D [*Cherletone* DB, *-ton* 1197 FF]. See CHARLTON.

Charley Le [*Cernelega* DB, *Cerneleia* 1130 P, *Cherlega* c 1125 LeS]. The place is nr CHARNWOOD FOREST and the name has the same first el., viz. Welsh *carn* 'a rock'. Cf. CHARNIE (forest) in Mayenne (France) [*Carneia* 989, *Carneta* 1109], which is described as a broken-up, rocky country. **Charley Knoll** is near Charley. The OE form would be *Cearn-lēah* 'the forest by the rock or rocky hill'.

Charlinch So [*Cerdeslinc, -ling* DB, *Cherdelinch* 1291 Tax, 1316 FA]. '*Cēolrēd*'s HLINC or hill.' Or the first el. may be the same as that of CHARD.

Charlock Np [*Chaldelacke* c 1250 BM, *Calde-, Scholdelak* 1291 Tax]. 'Cold stream.' Cf. LACU.

Charlton, a common name, is 1. usually OE *Ceorlatūn*. The same is the origin of CHARLETON, one CHALTON Bd, CHARLESTOWN and several CHORLTONS. Very likely CARL(E)-TON is at least to a great extent a Scandinavianized form of *Ceorlatūn*. OE *ceorl* means 'a freeman of the lowest rank, a free peasant'. But it is quite possible that already in OE times the word had come to be used also of a villein. Whether the name *Ceorlatūn* means 'TŪN of the free peasants' or 'TŪN of the villeins', it suggests that manorialism had made a good deal of advance in OE times, for even 'TŪN of the free peasants' presupposes that there were villages not held by freemen. In favour of the meaning 'TŪN of the villeins' may be adduced the fact that the Charltons are often found near important centres, as Charlton Kings nr Cheltenham, Charlton Ha nr Andover and so on.

Charlton Brk [*æt Ceorlatun* 956 BCS 925, *Cerletone* DB], **C~** Do nr Charminster [*Cherleton* 1242 Ch], **C~ Marshall** Do [*Cerletone* DB, *Cheorleton* 1187 Fr], **C~ Gl** nr Henbury [*Cherleton* 1204 Cur], **C~ Gl** nr Tetbury [*Chorlton* 1281 Ipm], **C~ Abbots** Gl [*Cerletone* DB, *Cherletone* 1221 Ass], **C~ Kings** Gl [*Cherleton* 1236 Fees, *Kynges* **C~** 1270 Ipm], **C~ Ha** [*Cherleton* 1192 P], **C~ Hrt** [*Cerletone* DB], **C~ K** nr Dover [*Cerlentone* DB, *Ceorletun* 11 DM], **C~ K** nr Greenwich [*Cerletone* DB], **North & South C~** Nb [*Charleton del North, Suth* 1242 Fees], **C~ Nb** nr Bellingham [*Carlton* 1195 (1335) Ch, *Charletona* 1279 Ass], **C~ Np** [*Cerlintone* DB, *Cherleton* 1220 Fees], **C~ on Otmoor** O [*Cerlentone* DB, *Cherleton upon Ottemour* 1314 Ipm], **C~ Sa** [*Cerlitone* DB, *Cherleton* 1212 Fees], **C~**

So nr Kilmersdon [*Cherelton* 1243 Ass], **C~** So nr Shepton Mallet [*Cerletone* DB], **C~ Adam** So [*Cerletune, Ceorlatona* DB, *Cherleton Adam* 13 Bruton], **C~ Horethorne** So [*Ceorlatun* c 950 Wills, *Cherleton Kanvill* 1225 Ass], **C~ Mackrell** So [*Cerletune* DB, *Cherletun Makerel* 1243 Ass], **C~ Musgrove** So [*Cerletone* DB, *Cherleton Mucegros* 1225 Ass], **Queen C~** So [*Cherleton* 1291 Tax], **C~ Sx** [*Cherleton* 1248 FF], **C~ W** nr Malmesbury [*Cherletune* 680 BCS 59, *Ceorlatun* c 965–71 ib. 1174, *Cerletone* DB], **C~ W** nr Pewsey [*Cherleton* 1203 Cur, 1225 Pat], **C~ W** nr Salisbury [*Cherleton* 1207 Cur], **C~ W** nr Shaftesbury [*Cherleton* 1216 Ch], **C~ Wo** nr Evesham [*Ceorletun* 780 BCS 235, DB], **C~ Wo** in Hartlebury [*Cherletona* 1182 PNWo].

C~ Abbots Gl belonged to Winchcomb abbey. —**C~ Adam** So was held by William fitz Adam in 1206 (FF).—**C~ Horethorne** So is in the old hundred of Horethorne (*Hareturna* 1184 GeldR). The name means 'grey thornbush' (OE *hāre þyrne*). It was held by Gerard de Camvile t. Stephen.—**C~ Mackrell** So took its name from the Makerel family. *Makerel* is a byname, no doubt meaning 'mackerel'.— **C~ Marshall** Do. Cf. STURMINSTER MARSHALL. —**C~ Musgrove** So was held by Richard de Mucegros in the time of King John. *Musgrove* is from MUSSEGROS in Normandy.—**Queen C~** So was given to Catherine Parr by Henry VIII.

2. **Charlton** Mx [*Cerdentone* DB, *Cherdinton* 1221–30 Fees]. 'The TŪN of *Cēolrēd*'s people.'

Charlwood Sr [*Cherlewod* 1199 Cur, *-wode* 13 BM]. 'The wood of the *ceorls* or peasants.'

Charminster Do [*Cerminstre* DB, *-ministr'* 1212 Fees, *Chernminstr'* 1291 Tax]. 'Minster or church on R CERNE.'

Charmouth. See CHAR.

Charn R Brk [*Cern* 958 BCS 1035]. Identical with CERNE. On the Charn or Ock is **Charney Basset** Brk [*Ceornei* 821 BCS 366, *Cearninga gemære* 958 ib. 1028, *æt, by Cern* 958 ib. 1035, *Cernei* DB]. The place was sometimes called *æt Cern* '(the place) on the Charn', sometimes *Cearn-ēa* or *Cearnīeg* 'Charn river' or 'island on the Charn'. *Cearningas* means 'the people of Charney'. *Basset* is stated in VH to be corrupt for *Basses*, a copyhold tenement in Charney.

Charndon Bk [*Credendone* DB, *Charendone* 1227 Ass, *Chardone* 1284–6, *Charndone* 1316 FA]. If the DB form is reliable, perhaps '*Cerda*'s DŪN', *Cerda* being a short form of *Cerdic*. But more likely the form is corrupt, and the first el. is the same as in CHARNWOOD.

Charnes St [*Ceruernest* DB, *Chauernese* 1197 P, *-nes* 1242 Fees, *Cauernessa* 1230 P]. OE *ceafor-næss* 'point of land where beetles abounded'.

Charney. See CHARN.

Charnock, Heath, and **C~ Richard** La [*Chernoc* a 1190 CC, *Hethechernoce* 1270 Ass; *Chernoch* 1194 P, *Chernok Ricard* 1288 Ipm]. A derivative of Welsh *carn* 'rock', either a river-name identical with CERNIOG in Wales or a district name meaning 'rocky district'.
Richard de Chernok is mentioned in 1246 Ass.

Charnwood Forest Le [*Charnewode* 1276 RH, 1288 Ipm]. First el. Welsh *carn* 'rock, stones'. Cf. CHARLEY, which may be an alternative name. The district is hilly, reaching 912 ft. at Bardon Hill.

Charsfield Sf [*Cerresfella, Ceresfelda, Caresfelda* DB, *Caresfeld* c 1150 Crawf, *Charesfeud* 1254 Val]. The forms point to OE *Cearesfeld*, whose first el. is very likely a river-name *Cear*, identical with CAR in Wales. Cf. CARY. The place is near a tributary of the Deben.

Chart, Great & Little, K [*Cert* 762 BCS 191, *Seleberhtes cert* 799 ib. 293, *Cert* 843, 858 ib. 442, 496, *Certh, Litelcert* DB, *Magna Chart* 13 BM], **Chart Sutton** or **next Sutton Valence** K [*Cært* (silva) 814 BCS 343, *Certh* DB, *Chert juxta Sutthon* 1280 Ep]. OE *ceart*, identical with dial. *chart* 'a rough common, overrun with gorse, broom, bracken' (K, Sr), and with Norw *kart* 'rough, rocky, sterile soil'. Also found in CHADACRE, CHARD, CHARFIELD, CHARTHAM, CHARTLEY, CHARTRIDGE, CHURT. Cf. also CARTMEL.

OE *ceart* had an *i*-mutated dative, WSax *ciert*, OKent *cert* (Sievers, Ae. Gr. § 284, note 7, Hallqvist 119 ff.). This explains OE *Cert* and later CHURT.

Charterhouse on Mendip So [(priory of) *Chartuse* 1243 Ass]. Charterhouse is an alteration by popular etymology of Fr *chartreuse* 'Carthusian house'. Cf. MENDIP.

Chartham K [*Certham* c 871 BCS 529, *Certaham* c 1050 KCD 896, *Certeham* DB]. 'HĀM in a chart or rough common.' Cf. CHART. The first el. may be the gen. plur. *cearta*.

Chartley St [*Certelie* DB, *Certelea* 1192 P, *Cerdel* 1232 Cl]. 'LĒAH in a rough common.' See CHART. The first el. seems to be in the plur. form.

Chartridge Bk [*Charderuge* 1191–4 PNBk, *Chardrugge* 1199 FF, *Chartrugge* 13 Misc]. First el. apparently OE *ceart* as in preceding names. Cf. CHARD.

Charwelton Np. See CHERWELL.

Chastleton O [*Ceasteltone* 777 BCS 222, *Cestretone* 1152–4 Eynsham, *Cesteltone* 1209–35 Ep, *Chasteltone* 1323 Eynsham]. 'TŪN by a CEASTEL or heap of stones'; cf. CASTLEY. The name may refer to the prehistoric camp marked in the vicinity. Later often influenced by CEASTER.

Chatburn La [*Chatteburn* 1251 Ch, 1258 Ipm]. '*Ceatta*'s stream.' *Ceatta* is mentioned in Saints and occurs in *Ceattan mære* 983 KCD 636. Cf. CATT.

Chatcull St [*Ceteruille* DB, *Chatculne* 1199 FF, 1327 Subs]. '*Ceatta*'s kiln' (OE *cylen*).

Chāter R Le, Ru [*Chatere* 1263 Ass]. Perhaps a Brit *cēto-dubron* 'forest stream'. Cf. CHATHAM, KETTON, CALDER.

Chatford Sa [*Chattefort* 1255, *-ford* 1274 RH]. '*Ceatta*'s ford.'

Chatham Green Ess [*Cetham* DB, *Chatham* 1303 FA, 1307 FF], **C~** K [*Ceðæma mearc* 995 KCD 688, *Cætham* 10 BCS 1321 f., *Ceteham* DB, *Chatham* 1195 P]. 'HĀM by the forest.' The first el. is Brit *cēto-* from OCelt *kaito-* 'forest' (Welsh *coed* &c.). Chatham Ess is in Great Waltham, whose name means 'HĀM by the forest' (OE *Weald-hām*). The same el. is found also in *Cæthærst* 946 BCS 1345, the name of a swine-pasture belonging to Swalecliffe K. *Ceðæma* for *Cet-hæma* means '(of) the people of Chatham'. This is probably the first el. also of **Chattenden** K [*Chatindone* 1281, *Chetyndone* 1287 Reg Roff]. The place is a few miles north of Chatham on the other side of the Medway. It was no doubt an outlying part of Chatham.

Chatley Ess [*Chatelee* 1199, *Chattel'* 1235 FF], **Chat Moss** La [*Catemosse* 1277 Ass, *Chatmos* 1322 LaInq]. '*Ceatta*'s LĒAH and moss.'

Chatsworth Db [*Chetesuorde* DB, *Chattesworth* 1276 Ass]. '*Ceatt*'s WORÞ.' A strong side-form of *Ceatta* is presupposed also by CHATTISHAM Sf.

Chattenden K. See CHATHAM.

Chatteris Ca [*Cæateric* 974 BCS 1311, *Chaterih* 1060 KCD 809, *Chateriz* c 1080 ICC, *Cetriz* DB, *Chatric* 1200, 1203 Cur]. This name has often been identified with CATTERICK YN, but there is really nothing to bear out this suggestion. The second el. may be RIC 'stream', the first being OE *Ceatta* pers. n. or better Brit *cēto-* 'forest'. See CHATHAM.

Chatterley St [*Chaderleg* 1212 Fees, *Chaterlyh* 1227 Ch, *Chaderley* 1252 Ch]. Second el. OE LĒAH. The first may well be that of CHADDERTON.

Chattisham Sf [*Cetessam* DB, *Chettesham* 1190 FF, *Chatesham* 1254 Val]. '*Ceatt*'s HĀM.' Cf. CHATSWORTH.

Chatton Nb [*Chetton* 1178 P, *Chatton* 1242 Fees, 1253 Ch]. '*Ceatta*'s TŪN.'

Chatwall Sa [*Chatewelle* 1185 TpR, *-walle* 1255 RH], **Chatwell** St [*Chatewall* 1327 Subs]. '*Ceatta*'s well or stream.' The second el. is OE *wella* in its West Midland form *walle*, from OE *wælla*.

Chaureth Ess [*Ceauride* DB, *Chaurea* 1185, *Chaurie* 1190 P, *Chaureth* 1303 FA]. OE *Ceawan riþ* '*Ceawa*'s stream.' Cf. CHALLOW.

Chawleigh D [*Calvelie* DB, *Cheluelega* c 1227 BM]. OE *cealfa-lēah* 'pasture for calves'.

Chawston Bd [*Calnestorne* DB, *Caluesterna* 1167, *Chaluesthorn* 1180 P]. '*Cealf*'s thornbush.' Cf. CHALVINGTON.

Chawton Ha [*Celtone* DB, *Chaltun* c 1195, *Chalvedone* c 1230 Selborne, *Chaueton* c 1272 AD]. OE *Cealfa-tūn* 'TŪN where calves were reared'.

Chaxhill Gl [*Chakeshull* 1220 Fees, *Cheakeshulle* 1227 Flaxley]. Perhaps '*Ceac*'s hill'. OE *Ceac* pers. n. is not evidenced, but *Cæc* occurs BCS 218. *Ceac* may be a nickname from OE *cēac* 'a pitcher', or related to the first el. of CHECKLEY. Or the first el. is OE *cēac* in some transferred sense.

Chazey O in Mapledurham [*Chawses* 1476–8] is a manorial name taken from the family of Walter de *Chauseia*, who is mentioned c 1180 (PNO(S)). The family seems to have come from CHOLSEY.

Cheadle (-ē-) **Bulkeley, C~ Moseley** or **Hulme** Chs [?*Cedde* DB, *Chedle* 1153–80 Ormerod, *Chedle* 1285 ff. Court, *Chedlee* 1326 Ipm], **C~** St [*Celle* DB, *Chedele* 1197 P, *Chedle* 1227 Ass, 1253 FF, *Chedlhe Basset* 1236 Fees]. Cheadle is a compound with OE *lēah* as second el. The first is probably Brit *cēto-* (Welsh *coed*) 'wood'; cf. CHEETHAM. Probably *lēah* is an explanatory addition, Cheadle meaning '*Chet* wood'; cf. CHETWODE.

Bulkeley and Moseley from local families. Richard de Bulkeley acquired C~ Bulkeley in the 14th cent. The Moseleys came into possession later (16th or 17th cent.).

Cheal Li [(æt) *Cegle* 852 BCS 464, *Ceila* DB, *Cheila* 1167 P]. Cheal is on a stream, referred to as *Cheylebecke* 13 FF. An OE **cegel*, corresponding to OHG *kegil* 'a peg, pole'. The meaning here may be 'pole' or 'plank bridge'. Cf. BĒAM in BENFLEET. Cf. CHELMARSH, CHEYLESMORE, also CHILMARK.

Cheam Sr [*Cegeham* 675 BCS 39, *Cegham* c 950 ib. 819, *Ceiham* DB, *Cheiham* 1199 Cur]. The first el. is an OE **ceg* related to *cegel* in CHEAL and to Norw *kage* 'a low shrub, a small tree with many branches', Swed dial. *kage* 'stumps'. The meaning may be 'stump'. If so, 'HĀM by the stumps'.

Chearsley Bk [*Cerles-, Cerdeslai* DB, *Cherdeslea* Hy 2 (1313) Ch]. '*Cēolrēd*'s LĒAH.'

Chebsey St [*Cebbesio* DB, *Chebbesee* 1222 FF, *Chebbeshey* 1236 Fees]. '*Cebbi*'s island or river land.' OE **Cebbi* is a normal formation from *Ceabba* or *Ceobba*. For *Ceabba* see CHADNOR.

Checkendon O [*Cecadene* DB, *Chakenden* 1236 Fees, *Chekenden* 1258 Ch], **Checkley** Chs [*Checkley* c 1130 Mon, *Chackileg* 1252 Ch, *Chackeleg* 1274 Ipm], **C~** He [*Chakkeleya* 1195, *Chakeleia* 1196 P, *Chakele* 1308 Ipm], **C~** St [*Cedla* DB,

Checkeleg 1196 FF, *Chekelee* R 1 Cur]. Apparently '*Ceacca*'s DŪN and LĒAH.' **Ceacca* pers. n. may be found in *Ceacca wylles heafde* 1012 KCD 1307, which occurs in boundaries of Whitchurch a few miles south of Checkendon. The same *Ceacca* would have given their names to Checkendon and *Ceacca wyll*, a stream. The fact that there are no less than three Checkleys may indicate, however, that the first el. is a common noun. A hill-name *ceacce* might possibly be referred to ON *kǫkkr* 'a lump' &c. Cf. CAXTON.

Chedburgh Sf [*Cedeberia* DB, *Cheddeberg* 1254 Val, *Chedeberwe* 1275 RH]. '*Cedda*'s BEORG or hill.'

Cheddar So [(æt) *Ceodre* c 880 BCS 553, *Ceod(d)rum* c 1000 Life of St. Dunstan, *Ceoddormynster* 1068 E, *Cedre* DB]. OE *Ceoder* is related to OE *cēod* 'a pouch', OHG *kiot* 'a purse'. It probably means a cave or a deep ravine and refers in reality to the deep gorge called Cheddar Gorge or the stalactite caverns at that place. **Cheddar Gorge** is referred to as *Ceoddercumb* 1068 E, and the caves very likely as *Chederhole* by Henry of Huntingdon (c 1130).

Cheddington Bk [*Cete(n)done* DB, *Chetendone* 1220 Fees]. '*Cetta*'s DŪN.' *Cetta* pers. n. is evidenced in *Cettantreo* BCS 210. But OE *cēte* 'a hut' might be thought of alternatively as first el.

Cheddington Do [*Chedinton* 1194 P, *Chedington, Chidinton* 1230 P, *Chedyndon* 1280 FF]. 'The TŪN of *Cedd*'s or *Cedda*'s people.'

Cheddleton St [*Celtetone* DB, *Chetilton* 1201 Cur, *Cheteltun* 1227 Ass]. 'TŪN in a CIETEL or narrow valley.'

Cheddon Fitzpaine, Upper C~ So [(of) *twam Cedenon* 11 KCD 897, *Ub-, Succedene* DB, *Chedene* 1182 P, *Cheddene* 1219 Fees]. The name is a compound containing as second el. OE *denu* 'valley'. The first is probably Brit *cēto-* 'wood', Welsh *coed*; cf. CHATHAM. But OE *cēte* 'hut' may also be thought of.

Fitzpaine means 'son of Pain'. Roger son of Pagan held Cheddon in 1226 (FF). On Pain see CARY FITZPAINE.

Chedglow (-ĕj-) W [*Chegeslei, Cheieslave* DB, *Cheggeslawa* 1168, *Cheggelewa* 1177 P, *Cheggelewe* 1242 Fees]. A still earlier reference is *Chegghemwllesbroke* 956 BCS 922, which means 'the brook of the Chedglow people'. The text is poor, but the first part of the name is an elliptical derivative with the suffix *-hǣme* 'dwellers' from the OE form of the name, which seems to have been **Cieggan hlǣw* or the like. The same first el. is found in the name of a neighbouring hill, *Cheggeberewe* c 1220 Malm. The original name of this hill was very likely OE *Ciegge* 'gorse-clad hill', derived from OE *ceacge* 'gorse', alternating with *Cieggan hlǣw* '*Ciegge* hill'. The latter was transferred to the neighbouring hamlet, and a new name for the hill was formed by adding OE *beorg* 'hill' to *Ciegge*. For full material see Anderson, *Hundred-names*, ii.

Chedgrave Nf [*Scatagraua* DB, *Chategrave* 1165–70 BM, *Chattegraua* 1158 P]. '*Ceatta*'s pit or grove.' The second el. may be OE *græf* 'grave, pit' or *grâf* 'grove'. The river-name **Chet** is a back-formation.

Chediston Sf [*Cidestan, -es, Cedestan* DB, *Chedestan* 12 BM, *Chedeston* 1203 Ass]. '*Cedd*'s stone.'

Chedworth Gl [(æt) *Ceddanwryde, -wyda* 872 BCS 535, *Cedeorde* DB, *Cheddewurda* 1194 P]. '*Cedda*'s WORÞ.'

Chedzoy So [*Chedesie* 729 BCS 147, *Cheddeseia* 1175 Wells, *Chedesia* 1194 f. P]. '*Cedd*'s island.'

Cheesden La [*Chesden* 1543 DL, 1546 FF]. 'Gravel valley.' Cf. CHISHALL.

Cheeseburn Nb [*Cheseburgh* 1286 Ch, 1293 QW, *Chesborne* c 1536 BBH]. The first el. appears to be the word *cheese*, but the combination with BURG is remarkable, and perhaps the name is identical with CHESEL-BOURNE or *Cisburne* BCS 356.

Cheetham La [*Chetam* 1212 Fees, *Chetham* 1226 LaInq]. Identical with CHATHAM, though Brit *cêto-* here appears as OE **cêt* instead of **cæt* in Chatham. Part of Cheetham is called **Cheetwood**, where an explanatory *wood* has been added to the original name of the wood, Brit *Cêt*.

Chelborough Do [*Celberge* DB, *Chalbergh* 12 Montacute, *Chauberge* 1236 Fees]. OE *cealc-beorg* 'chalk hill'.

Cheldon D [*Chadeledona* DB, *Chedeladon* 1185 Buckland, *Chedeldon* 1242 Fees]. '*Ceadela*'s DÛN.' Cf. CHADDLEWORTH.

Chelford Chs [*Celeford* DB, *Chelleford* 1245–50, *Cholleford* 1240–50 Chester]. '*Ceolla*'s ford.'

Chell St [*Chelle* 1227, 1252 Ch, *Ceolegh* 1313 PNSt]. Either '*Cêola*'s LÊAH', or the first el. is OE *ceole* 'throat', if that could be used of a ridge. C~ is on a long ridge.

Chellaston Db [*Celerdestune* DB, *Chelardeston* 1199 P]. '*Cêolheard*'s TÛN.'

Chellington Bd [*Chelewentone* 1219 Ep, *Chelinton* 1242 Fees, *Chelewynton* 1273 Cl]. '*Cêolwynn*'s TÛN.' *Cêolwynn* is a woman's name.

Chellow YW [*Celeslau* DB, *Chelleslawe* 1251 Ch, 1293 QW]. '*Cêol*'s HLÂW' (hill or tumulus).

Chelmarsh Sa [*Celmeres* DB, *Ceylmerys* 1252 Cl, *Cheylmerse* 1255 RH]. The first el. is identical with CHEAL Li, but the exact sense is doubtful. The place is on or near a long narrow ridge which may have been called 'the peg'. Second el. OE MERSC.

Chelmer. See CHELMSFORD.

Chelmick Sa [*Elmundewic* DB, *Chelmundewyk* 1241 FF], **Chelmondiston** Sf [*Chelmundeston* 1174 P, *Chelmondeston* 1219 Fees]. '*Cêolmund*'s WÎC and TÛN.'

Chelmorton Db [*Chelmaredon* 1196 Cur, *Chelemeredune* 1225 FF, *Chilmerdon* 1236 Fees, *Cheilmardon* 1265 Misc]. Not '*Cêolmær*'s DÛN'. It is not impossible that *Chel-mor-* is identical with CHILMARK. Second el. OE DÛN 'hill'.

Chelmscote Wa [*Chelmundescota* 1190 AC, *-cot* 1242 Fees]. '*Cêolmund*'s COT.'

Chelmsford (-ĕms-) Ess [*Celmeresfort* DB, *Chelmeresford* 1190 P]. '*Cêolmær*'s ford.' The river-name **Chelmer** [*Chelmer* 1576 Saxton] is a back-formation.

Chelsea Mx [*Cealchyþ* 785 ASC, *Celchyð* 785 BM, *Caelichyth* c 800 BCS 201, *Cealchithe* 1071–5 Reg, *Chelched* DB]. 'Landing-place for chalk or limestone.' Cf. CALC.

Chelsfield K [*Cillesfelle* DB, *Chilesfeld* 1086 KInq, *Chelesfeld* 1190 P, 1198 FF]. '*Cêol*'s FELD.'

Chelsham Sr [*Celesham* DB, *Chelesham* DB, 1177 ff. P]. '*Cêol*'s HÂM.'

Chelsing Hrt [*Cealsa* 1130 P, *Chelse* 1198 FF, 1212 Fees, *Chelsen* 1275 Cl]. OE *ceoslum*, dat. plur. of OE *ceosol* 'pebble, gravel', as suggested PNHrt(S). For metathesis of *sl* to *ls* CHILLESFORD and HALBERTON may be compared.

Chelston So [*Ceolfestun* 1065 Wells]. '*Cêolwulf*'s TÛN.'

Chelsworth Sf [*Ceorleswyrðe* 962 BCS 1082, *-weorð* 11 EHR 43, *Cæorlesweorþ* c 995 BCS 1288, *Cerleswrda* DB]. '*Ceorl*'s WORÞ.'

Cheltenham Gl [*Celtanhom, -homme* (dat.) 803 BCS 309, *Chinteneham* DB, *Chilteham* 1156 P]. Second el. HAMM. The situation of C~ at the foot of a high massif of hills suggests that the first el. is a hill-name, and this is borne out by the name *Cheltheved* ('Chelt Hill') 1248 Ass (PNEss 124), apparently nr C~. A hill-name *Celte* may be cognate with CHILTERN and of Brit origin, or possibly an old English word for 'hill' related by Ablaut to Norw *kult* 'lump, hillock' &c. The same el. appears to be found in CHILCOMB, CHILDERDITCH, CHILTINGTON.

Chelveston Np [*Celuestune* DB, *Chelveston* 1206 Cur]. '*Cêolwulf*'s TÛN.' Cf. CHELSTON.

Chelvey So [*Calviche, Caluica* DB, *Chalvy* 1285 FA]. OE *cealf-wîc* 'calf farm'.

Chelwood So [*Celeworde* DB, *Chelleworth* 1225 Ass, *Cheleworth* 1243 Ass]. '*Cêola*'s or *Ceolla*'s WORÞ.'

Chelworth W nr Crudwell [*Cellanwurd* c 890 BCS 569, *Cellewird* c 900 ib. 584, *Celeorde* DB]. '*Ceolla*'s WORÞ.' C~ W nr Cricklade [*Ceolæs wyrð* 965–71 BCS 1174,

Celewrde DB, *Celesworda* 1130 P]. '*Cēol*'s worþ.'

Cheney Longville Sa. See LONGVILLE.

Chenies (-ānĭ, -ēnĭ) Bk [*Isenhamstede* 1196 Cur, 1197 P, 1232 Ep, *Iselhamstede* 1232 Ep, *Iselhamstede Cheney* 1254 Val, *Cheynes* 1536 LP]. The old name was *Isenhamstede* rather than *Iselhamstede*. The elements may be a pers. n. **Isa* or a river-name (an old name of the Chess) and HĀMSTEDE. The manor was held by Alexander de Chednete in 1232 (Ep). *Chednete* is a form of the family name Cheney or Cheyne, which comes from one of the pl. ns. in France that go back to MLat *casnetum* 'oak grove' (cf. Fr *chêne* 'oak'), as CHESNOY, CHENOY, CHENAY.

Chepenhall Sf [*Cybenhale* c 1095, *Chebenhala* 1155–8 Bury, *Chebbehal* 1197 f. P]. '*Ceobba*'s HALH.'

Chepstow Monm [*Chepestowe* 1308, *Chepstowe* 1310, *Chipestowe* 1311 Pat]. OE *cēapstōw*, *cīepestōw* 'market-place'. The earlier name was **Strugull** or **Strigull** [*Strigoielg*, *Castellum de Estrighoiel* DB, *Strugull* c 1150, *Strigull* 1224 BM, *castrum Strigulense* 1191 Gir, *Striguil* 1193 P]. The Modern Welsh name is *Cas Gwent* (*Castellguent* c 1150 LL); cf. CAERWENT.

Cherhill W [*Ciriel* 1156 RBE, *Ceriel* 1156 ff. P, *Chiriel* 1215 Cl, *Chyriel* 1242 Fees]. No doubt a Brit name, whose second el. is Welsh *iâl* 'fertile upland region'; cf. DEVERILL. The first el. is possibly Welsh *caer* 'fort'. Oldbury Castle is on Cherhill Down. The first el. may also be Welsh *caer* in *caeriwrch* 'roebuck' (from *caper*-; cf. Lat *caper* and OIr *caera* 'sheep'). An exact analogy would then be offered by the partly Latinized *Caproialum* in Gaul (see Holder).

Cheristow D [*Chircstoua* 1168 P, *Cheristow* 1301 Cl]. 'Place of a church.' See STŌW.

Cheriton D in Brendon [*Ciretone* DB, *Chiriton* 1198, *Ceriton* 1205 FF], **C~ Bishop** D [*Ceritone* DB, *Cheritone* 1271 Ep, *Churiton* 1275 RH], **C~ Fitzpaine** D [*Cerintone* DB, *Churiton* 1242 Fees, *Chiriton* 1256 FF, *Cheriton Fitz Payn* 1335 Ipm], **C~ Ha** [*Cherinton* 1167 P, *Chiriton* 1284 Ch], **C~ K** [*Ciricetun* 11 DM, *Cheritun* 1158, *Ciriton* 1176 P], **North & South C~ So** [*Ciretona*, *Cherintone* DB, *Cheri-*, *Chirintone* 1198 FF, *Northchiriton* 1243 Ass, *Suthchuryton* 1329 Ep]. OE *Cyric-tūn* or *Cyr(i)ce-tūn* 'TŪN with or belonging to a church'. *Church* is OE *cirice*, *cyrice*. Particularly strong proof of the correctness of this etymology is the earliest form of Cheriton K.
C~ Bishop was held by the Bishop of Exeter. On Fitzpaine see CARY FITZPAINE. Roger son of Pagan held C~ Fitzpaine in 1256 (FF).

Cherrington Gl [*Cerintone* DB, *Chederintone* 1166 RBE, *Chiriton* 1196 P, *Chirinton* 1220 Fees]. Were it not for the form of 1166, this would obviously be identical

with CHERRINGTON Wa and CHERITON. If the 1166 form is trustworthy, the first el. may be a derivative with the suffix *-ingas* from **cēoder* in CHEDDAR.

Cherrington Sa [*Cerlintone* DB, *Chorrintona* 12 (1318) Ch, 1181 BM, *Cherington* 1230 P]. If the DB form is trustworthy, OE *Ceorlatūn* (or *Ceorlenatūn*, with analogical weak ending *-ena*), on which see CHARLTON, or OE *Ceorlingatūn* 'the TŪN of *Ceorl*'s people'. Otherwise OE *Ceorringatūn* 'the TŪN of *Ceorra*'s people'.

Cherrington Wa [*Chiriton* 1199 Rot Cur, 1242 Fees, *Cheriton* 1200 Cur, *Chirinton* 1203 ib., 1236 Fees]. Identical with CHERITON and CHURTON, though with a later intrusive *n* (*ng*).

Chertsey Sr [*Cerotaesei*, i.e. *Ceroti insula* c 730 Bede, *Ceortes eig* c 890 OEBede, *Ceorteseg* 871–89 BCS 558, *Certesy* DB]. '*Cerot*'s island.' *Cerot*, by *u*-mutation OE **Ceorot*, *Ceort*, is OBrit *Cerotus*, found in an inscription from London (see Holder). The same name is perhaps the first el. of *Ceortes beorg* 901 BCS 596 (Ha).

Cherwell (-ar-) R Np, O [*Ceruelle* 681 BCS 57, *Cearwellan* (obl.) 864, 929 ib. 509, 666, (to) *Cearwyllun* 944 ib. 792]. Second el. OE *welle* 'stream'. The first may be the same as that of CHARSFIELD, i.e. a Brit rivername identical with CAR in Wales. Or it may be an OE word corresponding to G *kar* 'kettle; hollow gorge', found probably in KARBACH, a name of rivers in Germany. On the Cherwell is **Charwelton** Np [*Cerweltone* DB, *Cerweltona* c 1110 NpCh].

Cheselade So [*Chesflod* 1201 Ass, *Cheselade*, *-lode* 1243 Ass]. OE *ceosol-flōde* 'gravel stream'.

Cheselbourne Do [*be Chiselburne* 869 BCS 525, *æt Ceosolburnan* 965 ib. 1165, *Ceoselburne* DB]. Really the name of the stream at the place [*Chiselburne* 869 BCS 525, (on) *Cyselburnan* 965 ib. 1165]. The name means 'gravelly stream'. First el. OE *ceosol*, *cisel* 'gravel, shingle'.

Chesham (-s-) Bk [*Cæstæleshamm* 966–75 Wills, *Cestreham* DB, *Chesham* 1247 Ass]. 'HAMM with or by a CEASTEL or heap of stones.' Cf. CASTLEY, CHASTLETON. The name was early associated with OE *ceaster* 'Roman fort'.
C~ Bois [*Chesham Boys* 1433 AD] may have as distinctive addition the family name Bois or (de) *Bosco*. Walter *le Bosch*' (for *de Bosch*') in Cestresham is mentioned in 1200 (FF). But it is possible a place in Chesham was called *Bois* or *Boscus* (from Fr *bois* or Lat *boscus* 'wood') and gave rise to the family name.

Cheshire [*Legeceasterscir* 980 ASC (C), *Cestrescire* DB]. Cf. CHESTER.

Cheshunt (-s-) Hrt [*Cestrehunt* DB, *Cestrehunte* 1197 FF, 1212 RBE, *Cesthunte* 1324 Ipm]. 'The huntsmen belonging to the chester.' Cf. HUNTA. The exact meaning of *chester* in this case is not clear. Cheshunt is on Ermine Street.

Chesil Bank Do [*the chisil* c 1540 Leland]. From OE *cisel* 'shingle'. Chesil Bank is a ridge of shingle or a bank of pebbles. **Chesilton** Do took its name from the bank.

Cheslyn Hay St [*Chistlin* 1236 Fees, *Chistelin* 1251, *Chisteling* 1252 Cl, *Haye of Chistelyn* 1293 Ass], **Chessel Down** Wt [*Chesthull* 1317 Abbr, *Chusthull* 1346 FA], **Chesthill** Sa, now lost [*Cesdille, Cestulle* DB, *Chesthull* 1212, 1236 Fees]. The same first el. is found also in **Chestham** Sx [*Chustham* 1305 Ass, *Chestham* 1313 FF]. The el. *Chest-* may be OE *ciest* 'a chest', here used of a coffin or coffins found at an old burial-place. At Chessel there is stated to be a heathen Anglo-Saxon cemetery. The second el. of Cheslyn may be OE HLINC 'a hill'.

Chessington Sr [*Cisendone, Cisedune* DB, *Chissindon* 1195 P, *Chissenden* 1196 P]. The numerous spellings with *ss* suggest that this is '*Cissa*'s DŪN' rather than a compound with an adj. *cisen* 'gravelly' as first el.

Chester Chs [*Dēoua* c 150 Ptol, *Deva* 4 IA, *ciuitas Legionum, Legacaestir*, Brit *Carlegion* c 730 Bede, *Legaceaster* c 890 OEBede, 894 ASC, *Ceaster* 1094 ASC (E), *Cestre* DB]. *Dēva*, the earliest name, is identical with the river-name DEE. Chester is on the Dee. Chester must also have been called Lat *Castra legionum*, which is the source of OE *Legacaestir*, and also, with substitution of Welsh *caer* 'fort, city' for *castra*, OW *Cair Legion* (or *urbs Legionis*) c 800 HB, Welsh *Caerlleon*. Later Chester supplanted the longer name *Legacaestir*.

Chester, Little, Db [*Cestre* DB, *Chestre* 1229 Ch, *Little Chester* 13 Derby]. Li. Chester is in Derby, which was a Roman station. See CEASTER.

Chester le Street Du [*Cunca-, Cunceceastre* c 1050 HSC, *Cestra* c 1160 Hexh, *Cestria in Strata* 1400 Surt. Soc. 9]. The place is on a Roman road. The el. *Cunca-, Cunce-* may represent a hill-name identical with CANNOCK, CONOCK and found also in CONSETT.

Chesterblade So [*Cesterbled* 1065 Wells, *Chestreblad* 1259 ib., *-blade* 1327 Subs]. The first el. is OE *ceaster* 'fort', which may refer to a camp marked in the map on a neighbouring hill. The second el. may be OE *blæd* 'blade, leaf' in some transferred sense such as 'ledge, terrace', or possibly OE *bledu* 'bowl, cup'. The place is in a hollow among hills.

Chesterfield Db [*ad Cesterfelda* 955 BCS 911, *Cestrefeld* DB, 1165 P], C~ St [*Cestrefeld Alani* 1167 P, *Cestrefeud* 1218 FF]. The first is on Ryknild Street; at the second Roman remains have been found. The name means 'FELD by the Roman station'.

Chesterford, Great & Little, Ess [*Ceasterford* 1004 HEl, *Cestreforda* DB, *Cestreford Magna, Parva* 1238 Subs]. 'Ford by the Roman station.' The place is on a Roman road.

Chesters Nb [*Scytlescester* 1104–8 SD]. OE *ceaster* 'fort'. *Scytles-* may be the gen. of a pers. n. *Scyttel*, but the *scyttels* found in SHUTTLEWORTH is more likely. The old fort may have been used as an enclosure for animals; cf. IRTHLINGBOROUGH.

Chesterton Ca [*Cestretone* DB, *-tun* 1156 P, *-ton* 1200 Cur], C~ Gl [*Cestertone* 1086 Glouc, *Cestreton* 1220 Fees], C~ Hu [*Ceastertuninga gemærie* 955 BCS 909, *Cestretune* DB], C~ O [*Cestertune* 1005 Eynsham, *Cestretone* DB, *-tune* 1212 Fees], C~ St [*Cestreton* 1214 FF, *Chesterton* 1276 Ipm], C~ Wa [*Cestretune* 1043 Th, *-tone* DB]. 'TŪN by a CEASTER or Roman station.'

C~ Ca and Gl are close to Cambridge (*Grantacaestir* in Bede) and Cirencester respectively. C~ Hu is held to be *Durobrivæ* in IA route 5; it is on Ermine Street. C~ O is on Akeman Street and near Bicester. C~ St had remains formerly of a walled fort. C~ Wa is near a Roman fort on Fosse Way.

Chestham Sx. See CHESLYN.

Cheswardine Sa [*Ciseworde* DB, *Chesewurda* 1160 f., *-wurða* 1169, *-wurdin* 1179, *Chessewurða* 1178 P, *Chesewurthin* 1212 Fees]. Perhaps 'cheese farm' in spite of numerous spellings with *ss*. Cf. WORP(IGN).

Cheswick (chĭzĭk) Nb [*Chesewic* 1208–10 Fees]. 'Cheese farm.' Cf. WĪC.

Chet R. See CHEDGRAVE.

Chetnole Do [*Chetenoll* 1242 P, *Chateknoll* 1316 FA]. '*Ceatta*'s knoll.'

Chettiscombe D [*Chetelescome* DB, *Chettiscome* 1284–6 FA]. First el. OE *cietel* in the sense 'a deep valley among hills'. The original name may have been *Cietel*, to which was added an explanatory *cumb* 'valley'.

Chettisham Ca [*Chetesham* c 1170, *Chedesham* 1221 PNCa(S), *Chetisham* c 1350 Rams]. Near by was *Chedes-* or *Chetesfeld* 1251 ff. PNCa(S). First el. of both perhaps OE *Cedd* pers. n.

Chettle Do [*Ceotel* DB, 1107 (1300) Ch, *Chetel* 1234 Cl]. OE *cietel* 'a deep valley among hills'.

Chetton Sa [*Catinton* DB, *Chatinton* 1167 P, *Chetinton* c 1210 BM, 1225 FF, *Chettynton* 1254 Ipm]. '*Ceatta*'s TŪN.'

Chetwode Bk [(ad) *Cētwuda* 949 BCS 883, *Ceteode* DB]. A British name of a wood, *Cēt* from *cēto-*, OCelt *kaito-* 'wood' (cf. Welsh *coed*), to which was added an explanatory OE *wudu*.

Chetwynd Sa [*Catewinde* DB, *Chetewind* 1242 Fees, *Chettewinde* 1233 Cl]. OE *Ceatta* pers. n. and OE *gewind* 'a winding ascent'.

The place is situated nr a hill called the Scar.

Cheveley Ca [(æt) *Cæafle* c 995 BCS 1289, *Cheaflea* 1022 KCD 734, *Chavelai* DB, (silva) *Ceauelai* 1086 IE, *Chafle* 1242 Fees]. The second el. is OE LĒAH 'wood'. The first is OE *ceaf* 'chaff', doubtless here used in a more general sense such as 'rubbish, fallen twigs'.

Cheveley Chs [*Ceofanlea* 958 BCS 1041, *Cavelea* DB, *Chevely* 1244 Ch]. '*Ceofa*'s LĒAH.' *Ceofa* (also in the form *Ciaba*) is a known name. C~ Nb. See CHEVINGTON.

Chēvening K [*Chivening* 1199, -*es* 1203 Cur, *Cheveninges* 1212 RBE]. As the place is at the foot of a considerable ridge, the probability is that the name is derived from an old name of the ridge. Welsh *cefn* 'back, ridge' is often used in pl. ns. (cf. CEFN-Y-CASTELL and CHEVIN). Very likely the OBrit name of the ridge was *Cefn*. Chevening would thus mean 'the dwellers at the ridge'.

Cheverell, Great & Little, W [*Chevrel* DB, *Capreolum* 1100–6 Fr, *Chiuerel* 1179 P, *Magna Chiverel, Cheverel* 1242 Fees, *Cheveroill* 1276 Cl, *Chiverel Magna, parva* 1291 Tax]. Unexplained. Cf. BUCKERELL. The form *Capreolum* shows that the name was sometimes supposed to be identical with Fr *chevreuil* 'roe-buck'. This also explains *Cheveroill*.

Chevet YW [*Cevet* DB, *Chivet* 1153–5 YCh (1497), 1230 Ep, *Chivot* 1233 Ep, *Chevet* 1244 Ipm]. The place is on a hill. The name is possibly identical with CHEVIOT.

Chevin, The, YW [*Scefinc* c 972 BCS 1278, (on) *Scefinge* c 1030 YCh 7]. The Chevin is a hill, and its name is Welsh *cefn* 'ridge'. Cf. CHEVENING. The initial *S-* in the OE forms may be a relic of an original Welsh *is* 'below', *Scefing* being really from Welsh *is cefn* 'below the ridge'. *Scefing* would then have been the name of a place at the foot of the Chevin.

Chevington, East & West, Nb [*Chiuingtona* 1212, *West Chivington* 1236, *Chivington del Est* 1242 Fees]. Apparently 'the TŪN of *Ceofa*'s or *Cifa*'s people'. The same name is found in **Cheveley** near Chevington [*Chiveleye* 1300 Ipm], which means '*Ceofa*'s or *Cifa*'s LĒAH'. For *Cifa* cf. CHIEVELEY.

Chevington Sf [*Ceuentuna* DB, *Cheventon* 1201 Cur, 1254 Val]. '*Ceofa*'s TŪN.'

Cheviot (-ĕ- or -ī-) Nb [*Chiuiet* 1182 P, *Chyvietismores* 1244 Ch, *Chyviot* 1250 Ipm, *Chivyet* 1251 Cl]. Probably a pre-English name. Etymology obscure.

Chevithorne D [*Cheuetorna* DB, -*thorn* 1198 FF]. '*Ceofa*'s thorn-bush.'

Chew Magna, C~ Stoke So [*Ciw* 1065 Wells, *Chiwe* DB, *Chiw* 1225 Ass, *Chiwestoch* DB], **Chewton Mendip** So [*Civtun* c 880 BCS 553, *Ciwetune* DB, *Cheuton by Menedep* 1313 Misc]. Chew and Chewton

are on the river Chew, but far apart. Chewton is at the source of the river. **Chew** is a Brit river-name, probably an ellipsis of a name such as *afon Cyw* 'the river of the chickens'. Welsh *cyw* means 'the young of an animal, a chicken' and seems to enter into the Welsh stream-names *foss ciu, pant ciu* c 1150 LL. Chew Stoke is 'STOC belonging to Chew'. Chewton is 'TŪN on the Chew'. For MENDIP see that name. There is a **Chewton Keynsham** near Keynsham (not on the Chew). Chewton may here be a family name.

Chewton Ha [*Chiventon* 12 (1313) Ch, *Cheveton* 1280 QW]. '*Cifa*'s TŪN.' Cf. CHIEVELEY.

Cheylesmore Wa nr Coventry [*Cheilesmore* Hy 3 AD, *Cheylesmore* 1275 Ipm, 1337 Ch]. First el. as in CHELMARSH. The place seems to be near a river. So a meaning 'plank bridge' is possible. Cf. CHEAL and CHILMARK.

Chich Ess, now **St. Osyth** [*Cicc* c 1000 Saints, *Cice* DB, *Chich* 1158 P]. The place is on a creek of the Colne. The name may well be a word related to Norw *kika* 'to bend', ON *keikr* adj. 'bent' and meaning 'a bend, a creek'.

Chicheley Bk [*Cicelai* DB, *Chichelei* 1151–4 Fr, *Chechel'* 1242 Fees]. Possibly '*Cičča*'s LĒAH'. OE *Cičča* is unrecorded, but it may be related to *Cic* (in *Cices weg* BCS 1045). OE *cīcen* 'chicken' might also be thought of as first el.

Chichester Sx [*Cisseceaster* 895 ASC, c 930 Laws, *Cicestre* DB]. '*Cissi*'s CEASTER.' *Cissi* is an unrecorded side-form of *Cissa*. Chichester was no doubt named from Cissa, son of Ælli. He must have been known also as *Cissi*.

Chickerell, East & West, Do [*Cicherelle* DB, *Chikerel* 1227 FF, *Estchykerel* 1285 FA, *Westchikerel* 1236, 1242 Fees]. Unexplained. Cf. BUCKERELL, CHEVERELL.

Chicklade W [*Cytlid* 901–24 BCS 591, *Ciclet* 1212 RBE, *Ciklet* 1242 Fees, *Chikelade* 1281 QW]. The OE form is in a good text and is authoritative. The later forms with *c* (*k*) are due to some special change. The second el. is probably OE *hlid* 'gate', the first being a form of Brit *cēto-* (Welsh *coed*) 'wood' (cf. CHEETHAM, CHITTOE). The name means 'gate leading to the wood'. C~ is in the downs north of Shaftesbury.

Chickney Ess [*Cicchenai* DB, *Chikenye* 1230 Ch, -*eye* 1233 Fees], **Chicksands** Bd [*Chichesane* DB, *Chichesant* 1156, -*sand* 1160, *Chikesant* 1161 P], **Chickward** He [*Cicwrdine* DB, *Chicwardin* 1267 Ipm]. The second el. of Chickney is OE *īeg* 'island', of Chicksands OE *sand* 'sandy soil', of Chickward OE WORÞIGN. The first el. may be OE *cīcen* 'chicken' in Chickney and Chickward, but this will hardly do for Chicksands. For this at least we have to assume an unrecorded OE pers. n. *Cica* or *Cicca*, related to *Cic*; cf. CHICHELEY.

Chidden Ha [*æt Cittandene* 956 BCS 976, *Chitteden* 1241 Ch, *Chidden* 1242 Cl]. '*Citta*'s valley.' *Citta* is no doubt a WSax side-form of OE *Cetta*. Cf. CHEDDINGTON Bk.

Chiddingfold Sr [*Chedelingefelt* c 1130 BM, *Chidingefald* 1200 P, 1206 Cur, *Chudinge-feld* 1287 Cl]. 'The fold of *Cidd*'s or *Cidda*'s people.' *Cidd* occurs in (on) *Ciddesbeara* 1033 KCD 1318 (Do), *Cidda* in the calendar of St. Willibrord, and the names are WSax side-forms of *Cedd* and *Cedda*.

Chiddingly (-lī) Sx [*Cetelingei* DB, *Chit-ingelehe* c 1263 Penshurst, *Cheddingeleg* 1247 Pat]. Identical with this is **Chiddingly Wood** in W. Hoathly Sx [*Citangaleahge* c 765 BCS 197]. The name may mean 'the LĒAH or wood of *Citta*'s people' (cf. CHID-DEN), but the base may just as well be Brit *cēto-* 'wood' with change of *cēt-* to OE *cīet*, *cīt*. Cf. CHICKLADE, CHITTOE.

Chiddingstone K [*Cidingstane* c 1110 Text Roff, *Chidingstan* 1263, *Chidingestane* 1280 Ch, *Chuddingestone* 1284 Ep]. Possibly 'the stone of *Cidd*'s people'. Cf. CHIDDINGFOLD.

Chideock (chĭdĭk) Do [*Cidihoc* DB, *Chidiok* 1316 FA]. A Brit name corresponding to Welsh *coediog* 'wooded' (from *coed*, Brit *cēto-* 'wood'). The name may be an old name of the stream at Chideock, now the **Chid**, or a name of the place itself.

Chidham Sx [*Chedeham* 1193 P, 1243 Ch, *Chideham* 1243 Pat, *Chudham* 1237 FF]. The first el. is OE *cēod* or *cēode* 'a bag', which refers here to one of the bays on which Chidham is situated. One of these, Bosham Channel, has a narrow opening and may well have been thought to resemble a bag. Hence 'the HĀM at the pouch-like bay'.

Chidlow Chs [*Chiddelowe* 1282 Court]. Apparently '*Cidda*'s hill or tumulus'.

Chieflowman. See LOMAN.

Chieveley Brk [*æt Cifanlea* 951 BCS 892, *Cifanlea* 960 ib. 1055, *Civelei* DB]. '*Cifa*'s LĒAH.' The pers. n. **Cifa* may be related to or even a side-form of *Ceofa* and *Ciaba*.

Chignall St. James, C~ Smealy Ess [*Cingehala*, *Cinguehella* DB, *Chikehala* 1180 P, *Chigehale* 1203 FF, *Chigenhal* 1230 Cl, *Chikenhale Iacob* 1254 Val]. Probably '*Cica*'s HALH'; cf. CHICKNEY &c. The change *k > g* is early, but may well be assumed. The name appears to have been associated with the word *chingle*, on which see CHINGFORD. St. James from the dedication of the church.— Smealy seems to have been a place in Chignall. It is referred to as *Smetheleye* 1254 Val and means 'smooth LĒAH'.

Chigwell Ess [*Cingheuuella* DB, *Chigge-well* 1187 P, *Chigwell* 1190 P, *Chichewell* 1200 Cur, *Chikewelle* 1254 Val]. Chigwell is not very far from Chingford and the name

may well have been influenced by CHING-FORD. The first el. seems to be identical with that of CHIGNALL. But if (apud) *Ceagewellam* 1095 RWrits 18 (Reg ii. 403) refers to Chigwell, the first el. appears to have been OE *ceacge* 'gorse', and later forms may be due to influence from CHIGNALL.

Chilbolton Ha [*Ceolboldingtun* 909 BCS 620, *Ceolbaldinctuna* 934 ib. 706, *Cilbode(n)tune* DB, *Chilbolton* 1284 Ch]. 'The TŪN of *Cēolbeald*'s people.'

Chilcomb Ha [*Ciltacumb* post 856 BCS 493, *Ciltancumb* 909 ib. 620, *Ciltecumbe* DB]. First el. as in CHELTENHAM. C~ is at Deacon Hill (471 ft.). The same first el. appears to be found also in **Chiltley** Ha at Bramshott [*Ciltelei* DB].

Chilcombe Do [*Ciltecome* DB, *Childecumb* 1198 P, *-cumbe* 1269 Misc]. 'The valley of the spring', the first el. being OE *cielde* 'a spring' (cf. CELDE), seems a very probable meaning. But *cilda*, gen. plur. of *cild*, is, of course, possible.

Chilcompton So [*Comtuna* DB, *Childe-cumpton* 1227 FF]. Originally COMPTON 'TŪN in a CUMB or valley'. *Chil-* is OE *cilda*, gen. plur. of *cild*. Cf. CHILTON.

Chilcote Le [*Cildecote* DB, *-cot* 1207 Cur], **C~** Np [*Cildecote* DB, *Childecote* 13 BM]. OE *cilda cot*. Cf. CHILTON.

Chilcott So [*Celicotan* 1065, *Cheolecote* 1157, *Chelechota* 1176 Wells]. '*Cēola*'s COTS.'

Childerditch Ess [*Celta* 695 BCS 87, *Cilten-dis*, *Ciltedic* DB, *Chiltendich* 1219 FF, 1219 Fees]. The identification of *Celta* is not certain, but it may well be an old name of Childerditch, which then means 'ditch belonging to *Celta*'. For *Celta* see CHELTEN-HAM. Childerditch is fairly high.

Childerley Ca [*Cilderlai*, *Cildrelai* DB, *Chil-derle* 1242 Fees]. 'The LĒAH of the children.'

Childrey Brk [(to) *Cillariðe* c 950 Wills, *Celrea* DB, 1220 Fees]. Really the name of **Childrey Brook** [*Cillariþ* 940, 944 BCS 761, 798]. The OE forms are in transcripts and represent OE *Cillan rīþ* or *Ciollan rīþ*. The name means '*Cilla*'s stream'; cf. RĪP. For *Cilla* see CHILHAM.

Childwall La [*Cildeuuelle* DB, *Childewalle* 1212 Fees]. 'The stream of the children.' The reason for the name is obscure. Alternatively the first el. might be the OE pers. n. *Cilda*, found in *Cildan spic* KCD 688. See WELLA.

Childwick Hrt [*Childwica* 1166 P, *Childe-wike* 1198 FF, *Child(e)wic* 1249 Ch]. OE *Cilda-wic*. Cf. CHILTON.

Chilfrome. See FROME.

Chilgrove Sx [*Chelegrave* 1200 FF, *Chule-grave* 1332 Subs]. Either '*Cēola*'s grove' or 'grove in a gorge', the first el. being OE *ceole* 'throat', perhaps also 'a gorge, gully'.

Chilham K [*Cilleham, Cylleham* 1032 Th, *Cilleham* DB, 11 DM]. '*Cilla*'s or *Ciolla*'s HĀM.' OE *Cilla* pers. n. occurs in *Kyllan rygc* 969 BCS 1242 (Wo) and probably in CHILDREY. *Cille* fem. is well evidenced. *Cille* 699 BCS 101 is identical with *Ceolswið* 688–90 ib. 74. *Cille* is a short form of *Cēolswið*. Similarly *Cilla* and *Ciolla* are short forms of names such as *Cēolbald, Cēolmund*. Chilham may well be *Ciollan hām*; cf. CHILLENDEN.

Chilhampton W [*Cildhantona* 1130 P, *Childhampton* 1242 Fees]. See CHILTON. Second el. HĀMTŪN.

Chillenden K [(an) *Ciollandene* c 833 BCS 412, *Cilledene* DB, *Cyllindaenne* 11 DM]. '*Ciolla*'s valley (OE DENU) or pasture (OE DENN).' *Ciolla* is a Kentish side-form of *Ceolla*.

Chillerton Wt [*Celertune* DB, *Chulierton* 1346 FA]. '*Cēolheard*'s TŪN.'

Chillesford Sf [*Cesefortda* DB, *Chiselford* 1211 FF]. OE *ceosol-ford* 'gravel ford'.

Chillingham Nb [*Cheulingeham* 1187 P, *Chevelingham* 1231 Cl, 1242 Fees]. 'The HĀM of *Ceofel*'s people.' **Ceofel* is a derivative of *Ceofa*.

Chillington D [*Cedelintone, -tona* DB, *Chedelington* 1200 Cur]. 'The TŪN of *Ceadela*'s people.' Cf. CHADDLEWORTH.

Chillington So [*Cherinton* 1231 Ch, *Cheleton* 1261 FF, *-tone* 1285 FA]. '*Cēola*'s TŪN.'

Chillington St [*Cillentone* DB, *Cildentona* 1130 P, *Chilinton* R 1 Cur, 1272 Ass]. '*Cilla*'s TŪN.' Cf. CHILHAM.

Chilmark W [*cigel marc, cigelmerc broc* 984 KCD 641, *Æt Chieldmearc* 929–40 BCS 745, *Chilmerc* DB, 1167 P]. The elements are OE *cegel, cigel*, found in CHEAL, and *mearc* 'mark', probably 'boundary mark'. OE *cegel* no doubt meant 'pole' and the like, and *cigel-mearc* would be 'boundary mark consisting of a pole'. OSw *rā* means 'a pole, a pole used as a boundary mark', and *rā-mark* means 'a boundary mark'. Cf. CHELMORTON.

Chilson O [*Childiston* 1236 Fees, *Childeston* 1448 BM], **Chilston** K [*Childeston* 1202 Cur, 1290 Ipm]. 'The TŪN of the *cild*' (probably in the sense of a young nobleman). Cf. CHILTON.

Chilsworthy D [*Chelesworde* DB, *Cheleswurth* 1246 Ipm]. '*Cēol*'s WORÞ.'

Chiltern Bk, O [*Cilternsætna* [land] 7 BCS 297, *Cilternes efes* 1006 KCD 715, *Ciltern* 1009 ASC (E)], **Chilthorne Domer** So [*Cilterne* DB, 1198 P, 1204 Cur, *Chilterne Dunmere* 1280 FF]. The Chiltern Hills are a well-known range of hills. Chilthorne Domer is by a hill. Both names contain a British hill-name, which may possibly be related to the word *Celt* (OCelt *Celtæ*), if that word, as is held by many scholars, is

related to Lat *celsus* and meant originally 'high'. From an OBrit adj. **celto-* 'high' a hill-name might have been formed. The suffix *-erno-* is well evidenced in Celtic.

Domer is a family name, perhaps from **Dimmer** nr Castle Cary [*Dunmere* 1241 Ass]. Henry de Dummere held *Cylterne Dumere* in 1276 (RH). Another portion was called **Chilthorne Vagg** [*Cilterne Fageth* Hy 1 Montacute, *Chylterne Fag* 1276 RH].

Chiltington, East, Sx [*Childetune* DB, *Chiltinton* 1285 BM], **West** C~ Sx [?*Cillingtúne* 969 Crawf, 1066 KCD 824, *Cilletone* DB, *Chyltinton* 1247 FF]. In spite of the variation in the early forms, the two names seem to be identical, the original form being OE *Ciltingatūn*. In West C~ was a place called *le Chilte* 1357 PNSx. The name seems to be derived from a hill-name identical with the first el. of CHELTENHAM. West C~ is near a marked hill of 250 ft., probably called *Chilte*. East C~ is at the foot of the Wolds, which may here have been called OE *Cilte*. Chiltington would thus seem to mean 'the TŪN of the dwellers at *Cilte* hill'. But after all West C~ may be 'the TŪN of *Cilla*'s people', the similarity to *Chilte* being accidental. If so, the later form *Chiltington* is due to influence from *Chilte* and East C~.

Chilton, a common name, usually, no doubt, represents 1. OE *Cilda-tūn*. This means literally 'the children's TŪN', but it is unlikely that this can always be the exact meaning, especially as *cilda-* gen. plur. is also found in several other names, as CHILCOMPTON, CHILCOTE, CHILDWICK, CHILHAMPTON. Also it is remarkable that these names do not show the normal plural *r* (OE *cildru*, gen. *cildra*). OE *cild* was also used as a title, of a youth of noble birth, sometimes synonymously with *æþeling* 'prince of the royal blood'. Probably in some of these names *cild* is used in this or some similar sense. Chilton may mean about the same thing as KNIGHTON. The name CHILCOTE does not go well with a meaning 'young nobleman'. Rather a meaning such as 'COT of the retainers' might be assumed. Childwick is stated in Gesta to have provided milk for young monks. To this group belong very likely or certainly: **Chilton** Bk [*Ciltone* DB], C~ Brk [*Cyldatun* 891 BCS 565, *Cildatun* 1052 KCD 796, *Cilletone* DB], **Great & West** C~ Du [*Ciltona* 1091, *Magna Chiltona* 1214 FPD], C~ **Candover** Ha [see CANDOVER], C~ K [*Chiltune, Chilton* 13 St Aug], C~ Sa [*Chylton* 1327 Subs], C~ Sf nr Clare [*Chilton* 1254 Val, 1316 FA], C~ Sf nr Stowmarket [*Ciltuna* DB, *Chilton* 1346 FA], C~ Sf nr Sudbury [*Ciltona* DB, *Chiltune* c 1180 Bodl], C~ **Cantelo** So [*Childeton* 1201 Cur, *Chiltone Cauntilo* 1361 Ep], C~ **Trinity** So [*Cilde-, Cilletone* DB, *Chileton* 1208 FF, *Chilton Sancte Trinitatis* 1431 FA], C~ **Foliat** W [*Cilletone* DB, *Chilton Foliot* 1221 Pat].

C~ **Cantelo** was held by Walter de Cantelu in 1201 (Cur). Cf. ASTON CANTLOW.—C~ **Foliat** was held by Sanson Foliot in 1236 (Fees). Foliot is an OFr nickname and family name, derived from OFr *foliot* 'trap'.—C~ **Trinity** presumably from the dedication of the church.

2. Chilton upon Polden So [*Ceptone* DB, *Cahalton* 1285, *Chauton* 1303 FA, *Cheltone* 1327 Subs]. OE *Cealc-tūn* 'TŪN on the limestone hill'. Cf. POLDEN.

3. Chilton Wt [*Celatune* DB, *Cheltona* 1173 P]. '*Cēola*'s TŪN.'

Chilvers Coton Wa [*Celverdestoche* DB, *Cheluerthescote* c 1155 DC, *Chelverdecote* c 1200 BM]. '*Cēolfriþ*'s COT.'

Chilwell Nt [*Cilleuuelle*, *Ciduuelle* DB, *Childewella* 1194 P]. Identical with CHILDWALL.

Chilworth Ha [*Celeorde* DB, *Cheleworth* 1230 Cl], C~ O [*Celelorde* DB, *Chelewrth* 1220 Fees], C~ Sr [*Celeorde* DB, *Cheleworth* 1232 Cl]. '*Cēola*'s WORP.'

Chimney O [(æt) *Ceommenige* c 1070 Ex, *Chymeney* 1316 FA]. '*Ceomma*'s island.' *Ceomma* is a short form of *Cēolmǣr*, *-mund*. It is found also in the name of a brook at Chimney [*Ceomina laca* 1005 KCD 714, *Ceoman lace* 1069 JAA 39], further in *Ceomman bricg* 985 KCD 652 and *Ceomman treow* 947 BCS 820.

Chineham Ha [*Chineham* DB, *Chinham* 1206 Cur, 1274 RH]. 'HĀM by a chine or ravine.' First el. OE *cinu* 'fissure, ravine'.

Chingford Ess [*Cingeford* c 1050 KCD 913, *-fort* DB, *Chagingeford* 1219 Bract, *Chingelford* 1242 Cl, 1243 FF]. 'Shingly ford.' First el. the word *shingle*, recorded in the form *chingle* from 1598.

Chinley Db [*Chineleia* c 1200 Derby, *Chinlegh* 1286 Court]. 'LĒAH by a ravine.' Cf. CHINEHAM.

Chinnock, East, Middle & West, So [*Cinnuc* c 950 Wills, *Cinioch* DB, *Cinnuc* c 1100 Montacute, *Estcinnok* 1243 Ass, *Westcinnok* 1241 Ass]. Possibly a derivative of OE *cinu* 'fissure, ravine'. The places are between two ridges. But the regular *nn* is remarkable. Formally OE *cinn* 'chin' (Goth *kinnus* 'cheek' &c.) would be preferable as the base. This word may have been used in a transferred sense of a hill of a certain shape. The ending *-uc*, *-ock* is diminutive as in *hillock*.

Chinnor O [*Chenore* DB, *Chennora* 1193, *Cennore* 1195 P, *Chenovere* 1236 Fees]. Second el. OE *ōra* or *ofer* 'edge'. The first may be a pers. n. OE *Ceonna*, a short form of *Cēolnōþ*. Cf. *Kiona* LVD. If so, '*Ceonna*'s hillside'. Chinnor is on the slope of the Chilterns.

Chipchase Nb [*Chipches* 1229 Pat, *Chipeches* 1256 Ass]. The elements may be OE *cipp* 'a beam, log, stock' and an OE *ceas* 'a heap', corresponding to ON *kǫs*, Norw *kas*, Sw *kase* 'a heap'. The meaning would be 'a heap or structure of logs', possibly a trap for animals made of logs.

Chipley Sf [*Chippeleye* 1254 Val, 1314 Ipm]. OE *cipp(a)lēah* 'wood where logs were got'. On OE *cipp* see CHIPCHASE. **Chipley** So [(oð) *Cyppan leage* ?854 CodWint, *Chippeleg* 1254 Ass]. Apparently '*Cippa*'s LĒAH'. Cf. CHIPPENHAM.

Chipnall Sa [*Ceppecanole* DB, *Chippeknol* 1260 Eyton]. '*Cippa*'s knoll.' Cf. CHIPPENHAM. Or the first el. may be as in CHIPLEY Sf.

Chippenhall Sf nr Cratfield [*Cibbehala*, *Cybenhalla*, *Cebbenhala*, *Cipbenhala* DB, *Chebenhale* 12 AD]. '*Ceobba*'s HALH.'

Chippenham Ca [*Cypeham* c 1080 ICC, *Chipeham* DB, *Chipenham* 1086 IE], C~ Gl nr Bishops Cleeve [*Cyppanhamm* 769–85 BCS 246], C~ W [*Cippanhamm* 878 ASC, c 880 BCS 553, *Cippanhomm* 901–24 ib. 591, *Chipeham* DB]. Identical with Chippenham are also **Cippenham** Bk and possibly SYDENHAM K. The name seems to mean '*Cippa*'s HAMM or HĀM'. The former is the meaning of the Gl and the W Chippenham. *Cippa* pers. n. is not recorded in independent use. Its occurrence in 4 or 5 names in *hām* or *hamm* is therefore remarkable. Probably the name of Chippenham W, which is an ancient place of importance, was transferred to some other places. Cf. CHIPNALL, CHIPSTABLE, which may contain the same pers. n.

Chipping La [*Chippin* 1203 Cur, *Chipping* 1242 LaInq], **Chippingdale** La [*Chipinden* DB, *Cepndela* 1102 LaCh]. Chipping is OE *cēping*, *cīeping* 'market, market town'. Chipping is often added before names of places that had a market, as C~ ONGAR Ess, C~ NORTON O, C~ SODBURY Gl.

Chippinghurst O [*Cibbaherste* DB, *Chibbenhurst* 1122 Fridesw]. '*Cibba*'s HYRST.' *Cibba* is a side-form of *Ceobba*. But *Ceobba* itself may be the first el.

Chipstable So [*Cipestaple* DB, *Chippestapel* 1251 Cl]. OE *Cippan stapol*. See CHIPPENHAM, STAPOL.

Chipstead K [*Chepsteda* 1191 f. P], C~ Sr [*Chepstede* 675, 933 BCS (39, 697), 1242 Fees]. OE *cēapstede* 'market-place'.

Chirbury Sa [(æt) *Cyricbyrig* 915 ASC (C), *Cireberie* DB, *Chiresbir* 1226–8 Fees]. 'BURG or fort with a church.'

Chirdon Nb [*Chirden* 1255 Ch, 1279 Ass, *Chyreden* 1256 FF], **Chirton** Nb [*Cheriton* 1203 Cur, *Chirton* 1256 FF, *Churton* 1293 QW]. Chirton is OE *Cyrictūn* 'church TŪN'. Cf. CHERITON. Chirdon may mean 'valley belonging to a church or with a church or chapel'. But the first el. might be a stream-name derived from OE *cierr* 'bend'. Chirdon is on a winding stream.

Chirton W [*Ceritone* DB, *Chiritun* 1221 Cl, *Churughton* 1316 FA]. OE *Cyric-tūn*; cf. CHIRTON Nb and CHERITON.

Chisbury W [*Cheseberie* DB, *Chisseburi* 1258 Ipm, *-bury* 1260, 1270 Ch]. '*Cissa*'s BURG.' There is an ancient camp near the place.

Chiselborough So [*Ceoselbergon* DB, *Ciselberg* c 1100 Montacute, *Chiselberge* 1253 Ch]. OE *ceosol-beorg* 'gravel hill'.

Chisenbury W [*Chesigeberie* DB, *Chisingburi* 1202 FF, *Chisingebur* 1227 Ch]. 'The BURG of *Cissa*'s people.' The place is near **Chisenbury Camp.**

Chishall Ess [*Cishella* DB, *Cheshull* 1199 Cur, *Parva Chishulle* 1212 RBE, *Chishell Magna* 1238 Subs]. 'Gravelly hill.' The first el. is an OE *cis* 'gravel', found in *Cisburne* 816 BCS 356, and corresponding to MHG *kis*, G *Kies*. OE *ceosol*, *cisel* 'gravel' is a derivative of the word.

Chisledon W [(æt) *Cyseldene* c 880 BCS 553, *Ciseldenu* 891 ib. 565, *Chiseldene* DB]. 'Gravel valley.' First el. OE *ceosol*, *cisel*, *cysel* 'gravel'.

Chislehampton O [*Hentone* DB, *Chiselentona* 1146 RA, *Chiselhamton* 1192 P]. Originally HAMPTON (from OE *Hēatūn*, dat. *Hēantūne* 'high TŪN'). Later *chisel* (from OE *cisel* 'gravel') was added for distinction from BROOKHAMPTON.

Chislehurst K [*Cyselhyrst* 973 BCS 1295, 998 KCD 700, *Chiselherst* 1159 P]. 'Gravel hill.' Cf. CHISLEDON and HYRST.

Chislet K [*Cistelet* 605 BCS 6, DB, 1175 P, *Chisteled* 1199, *Cislested* 1202 Cur, *Chistelet* 1242 Fees]. Very likely OE *cisel-stede* 'gravel place', as the form of 1202 suggests. *Cistelet* 605 is in a 15th-cent. transcript. Or possibly *cisel-flēot* 'gravelly stream'.

Chisnall Hall La [*Chysenhale* 1285 Ass, *Chisenhale* 1332 Subs]. 'Gravelly HALH', the first el. being OE **cisen* 'gravelly' from *cis*; cf. CHISHALL.

Chiswick Ess [*Ceseuuic* DB, C~ (-ĭzĭk) Mx [(of) *Cesuican* c 1000 CCC, *Chesewyc* 1230 P, *Cheswick* 1254 Val]. 'Cheese farm.'

Chisworth Db [*Chiseuurde* DB, *Chissewrde* 1197 FF]. '*Cissa*'s WORÞ.'

Chithurst Sx [*Titeherste* DB, *Chitesherst* 1248 Ass, *Chyteherst* 1279 FF, *Chutehurst* 1288 Ass]. See HYRST. First el. either OE *Citta* pers. n. (cf. CHIDDEN) or Brit *cēto-* 'forest'; cf. CHITTERNE, CHITTOE, CHUTE.

Chitterne All Saints, C~ St. Mary W [*Chetre* DB, *Cettra* 1167 P, 1232 Ch, *Chytterne* 1289 BM, *Cettre Beate Marie* 1291 Tax, *Chitterne Maiden* 1325 Pat]. A compound of Brit *cēto-* 'forest', a name of a forest (cf. CHUTE, CHITTOE), and OE ÆRN 'house'. The name thus means 'the house in the forest'. Chitterne is on Salisbury Plain, a highland tract.

Chittlehamholt D [*Chitelhamholt* 1288 Ass, *Chetelhampholt* (wood) 1314 Ipm], **Chittlehampton** D [*Citremetona* DB, *Citelhanton*

1177 P, *Chidelametun* 1218 Cl]. The two places are c 3 m. apart. The names mean 'the wood and the TŪN of the dwellers in the valley'. The first el. is OE *citelhǽme* 'dwellers in a CIETEL or valley among hills'. Chittlehampton is in a valley. Chittlehamholt was the wood belonging to the people at Chittlehampton.

Chittoe (-ŏŏ) W [*Chetewe* 1168 P, 1260 Ch, *Cuttewe* 1195 Cur, *Chutuwe* 1390 AD]. No doubt a derivative of or a compound containing Brit *cēto-*, Welsh *coed* 'wood'. It may be a derivative with the common suffix *-oviā*. Or it might be a compound with Welsh *yw* 'yew'. Welsh *coed yw* would mean 'yew wood'.

Chivelstone D [*Cheueletona* DB, *Chevelestuna* c 1135 Totnes]. OE *Ceofeles tūn*. On the pers. n. *Ceofel* see CHILLINGHAM.

Chivesfield Hrt [*Ciuesfeld* 1086 IE, *Chivelesfeld* 1200 Cur, *Chivesfeld* 1204 Cur, 1220 Fees]. '*Cifel*'s FELD.' *Cifel* has the same relation to *Cifa* (see CHIEVELEY) as *Ceofel* to *Ceofa* (cf. CHIVELSTONE).

Chobham Sr [*Chebeham*, *Chabbeham* a 675 BCS 34, *Chabbeham* c 1050 KCD 848, *Cebeham* DB, *Chabeham* 1254 Ipm]. '*Ceabba*'s HĀM.' *Ceabba* is found in *Ceabban dun* 1033 KCD 752 (Ha). Cf. CHADNOR.

Cholderton Ha [*Cerewartone* DB, *Chelewartona* 1175 P, *Chalwardtun* 12 Fr, *Chelewarton* 1200 Cur]. '*Cēolweard*'s TŪN.'

Cholderton W [*Celdrin-*, *Celdretone* DB, *Cheldrintona* 1175 P, *Cheldringet[on]* 1203 Cur]. 'The TŪN of *Cēolhere*'s or *Cēolrēd*'s people.'

Cholesbury Bk [*Chelwoldesbur* 1254 Val, *Chelewoldesbyr* 1262 Ass]. '*Cēolweald*'s BURG.'

Chollerford Nb, **Chollerton** Nb [*Choluerton* c 1175 PNNb, *Chelverton* 1242 Fees, *Cholverton* 1254 Val]. Chollerton may be '*Cēolferþ*'s TŪN'. But more likely it is 'the TŪN by *Cēolan ford* or *Ceolford*', an earlier name of Chollerford, which is not found in early sources. *Cēolan ford* would be '*Cēola*'s ford'. *Ceol-ford* would be 'ford in a *ceole* or gorge' (cf. CHILGROVE).

Cholmondeley (tshŭmlĭ) Chs [*Calmundelei* DB, *Chelmundeleg* 1287 Court], **Cholmondeston** (tshŭmsn) Chs [*Chelmundestone* DB]. '*Cēolmund*'s LĒAH and TŪN.'

Cholsey Brk [*Ceolesig* 891 BCS 565 &c., 1006 ASC (C), *Celsea* DB]. '*Cēol*'s island.'

Cholstrey He [*Cerlestreu* DB]. '*Ceorl*'s tree.'

Cholwell So [*Chaldewelle* 1201 FF, *Cheldewall* 1285 FA]. 'Cold spring.' Cf. CALD.

Cholwich D [*Caldeswyht* 13, *Choldeswych* 1411 BM, *Chaldeswych* 1249 FF]. Perhaps 'coldest WĪC'. Cf. CALD.

Choppington Nb [*Cebbington* c 1050 HSC,

Chabinton 1181, *Chabiton* 1182 P]. 'The TŪN of *Ceabba*'s people.' Cf. CHOBHAM.

Chopwell Du [*Cheppwell* c 1155 Newminster, *Cheppewell* 1279 Ass, *Chapwell* 1313 RPD]. Perhaps OE *cēap-wella* 'spring where commerce took place'.

Chorley Chs nr Nantwich [*Cerlere* DB], C~ Chs nr Macclesfield [*Chorlee, Cherleg* 1285 f. Court], C~ La [*Cherleg* 1246 Ass, *Cherle* 1252 FF], C~ St [*Cherlec* 1231 Cl]. The name is also found in Hrt (**Chorley Wood**) and Sa. OE *ceorla-lēah* 'the LĒAH of the *ceorls* or peasants'.

Chorlton Chs nr Nantwich [*Cerletune* DB], C~ Chs nr Malpas [*Cherlton* 1283 Ipm, *Chorleton* 1284 Ch], C~ **Hall** Chs nr Chester [*Cherleton* 1278, *Chorlton in Wirall* c 1300 Chester], C~ **upon Medlock** La [*Cherleton* 1177 P, 1196 FF], **Chapel** C~ St [*Cerletone* DB, *Cherleton* 1267 Ass]. Identical with CHARLTON.

Chorlton cum Hardy La [*Cholreton* 1258 Ass, *Cheluerton* 1259 Ass, *Chorleton* 1551 FF]. '*Cēolfrið*'s TŪN.'

Choseley Nf [*Cheseley* Hy 2 (1313) Ch, *Chusele* 1212 Fees, *Chosle* 1254 Val]. OE *ceosol-lēah* 'gravelly LĒAH'.

Choulton Sa [*Cautune* DB, *Cheleston* 1252 Fees, *Cheolton* 1291 Tax]. '*Cēol(a)*'s TŪN.'

Chowbent La [*Chollebynt, Shollebent* c 1350, *Cholle* 1385 VH]. The first el. seems to be *Cholale* 1323 LaInq, 1330 FF, the second being *bent* 'bent-land' (cf. BENTLEY &c.). *Cholale* has as second el. OE HALH. The first may be OE *Cēola* pers. n. or CEOLE in the sense 'gorge'. Cf. CHILGROVE.

Chowley Chs [*Celelea* DB, *Chelleye* 1290 Ipm]. '*Cēola*'s LĒAH.'

Chrishall Ess [*Cristeshala* 1068 EHR xi, DB, *-hal* 1198 FF, *-hale* 1200 FF]. Looks like 'Christ's HALH'. The meaning of such a name is not apparent. Perhaps the first el. is an early reduction of OE *cristelmæl* 'cross'. Cf. CHRISTLETON.

Christchurch Ha [*Cristescherche* 1177 P, *Cristechurch Twynham* 1242 Ch]. The original name was TWINHAM.

Christian Malford W [*Cristemaleford* 937, *At Cristemalford* 940 BCS 717, 752, *Cristemeleforde* DB]. 'Ford marked by a cross' (OE *cristelmæl*).

Christleton Chs [*Cristetone* DB, *Cristentune* c 1100, *Cristelton* c 1190 Chester, *Kirkecristelton* 1289 Court]. 'TŪN with a cross' (OE *cristelmæl*). Cf. LITTLETON, ROWTON Chs.

Christon So [*Cyrces gemæro* 1068 E, *Crucheston* 1197 Bruton, *Cricheston* 1204 Cur]. Originally *Cyrc* or *Cryc* from Brit *crūc*, Welsh *crug* 'hill'. The place is at the foot of Bleadon Hill. Later OE TŪN was added.

Christow D [*Cristinestowe* 1244 Ass, 1259 Ep]. 'Christian place.' The exact meaning of the name is not apparent.

Chudleigh D [*Cheddeleghe* 1259 Ep, *Chuddelegh* 1291 Tax]. '*Ciedda*'s LĒAH.' *Ciedda* is a normal WSax form of *Cedda*.

Chulmleigh (-ŭmlĭ) D [*Calmonleuge* DB, *Chulmelegh* 1276 RH]. '*Cēolmund*'s LĒAH.'

Chunal Db [*Ceolhal* DB, *Chelhala* 1185 P]. '*Cēola*'s HALH.'

Church La [*Chirche* 1202 FF, *Chyrche* 1284 Ass]. 'The church.'

Churcham Gl [*Hamme* DB, *Hamma* c 1145, *Chirchehamme* c 1233 Glouc]. Originally HAMM 'low-lying land on a river' (the Severn). Later *church* was added for distinction from HIGHNAM.

Churchdown Gl [*Circesdune* DB, *Kyrchesdon* 1190, *Kyrkesdon* 1191 P, *Chirchedon* 1221 Ep]. The place is at a high round hill, which was evidently called *Crūc*, from Brit CRŪC, Welsh *crug* 'hill', especially 'a round hill'. To this was added an explanatory OE DŪN 'hill'.

Churchfield Np [*Ciricfeld* 10 BCS 1129, *Chirchefeld* 1189 (1332) Ch]. 'FELD with a church.' There was a chapel here in the 12th cent.

Churchill D nr Barnstaple [*Cercelle* DB, *Churchehille* 1242 Fees], C~ D nr Loxbeare [*Chirchehill* 1238 Ass], C~ D nr Broad Clyst [(montem de) *Cherchull* 1281 Ass], C~ O [*Cercelle* DB, *Chirchehull* c 1175 Fridesw], C~ So [*Cherchille* 1201 FF, *Chyrchehull* 1243 Ass], C~ Wo nr Worcester [*Circehille* DB, *Cherchhull* 1209 Fees], C~ Wo nr Kidderminster [*Circhul* 11 Th, *Cercehalle* DB, *Chyrchull* 1275 Subs]. OE *cirichyll* 'church hill', either a hill with or near a church, or one belonging to a church. It is just possible that one or other of the Churchills in reality contains a Brit hillname *Crūc*, to which was added an explanatory OE HYLL (cf. CHURCHDOWN). If so, the name was associated at an early date with the word *church*. This etymology is to be assumed for **Church Hill**, the name of a hill in So [*Crichhulle* 705 BCS 112]. **Churchill** D in Malborough is *Curcheswille* 1201 FF, *Corcheswille* 1296 Ass. It consists of Brit *crūc* 'hill' and OE *wiella* 'well, stream'.

Churchover, -stanton. See OVER, STANTON.

Churchstow D [*Churechestowe* 1242 Fees]. 'Place of a church.'

Churn R. See CIRENCESTER.

Churnet R Chs, St [*Chirnet* 1250 Dieulacres, *Chernet* 1272 Ass]. A British river-name.

Churston Ferrers D [*Cercetone* DB, *Churechetone* 1242 Fees, *Churchtone* 1259 Ep]. Identical with CHERITON.

The manor was held by Hugh de Fereris in 1303 (FA). Cf. BERE FERRERS.

Churt Sr [*Cert* 688 BCS 72]. Identical with CHART.

Churton Chs [*Chirton* 1260 Court, 1290 Ipm, *Chyrchton* c 1334 Vale Royal]. OE *Cyric-tūn* 'church TŪN'. Cf. CHERITON.

Churwell YW [*Cherlewell* 1226 FF, *Chorlewelle* 1311 Ch]. 'The spring or stream of the *ceorls* or peasants.'

Chute Forest W [*Ceat* c 1110 RA, *Ceit* 1178 BM, *Cet* 1229 Cl, *Chut* 1259 Ipm]. Brit *cēto-*, Welsh *coed* 'forest', which became OE **Cīet*, **Cīt*, **Cȳt* owing to the influence of the initial palatal.

OE **cietel** 'kettle' was also used, like G *kessel*, of 'a deep valley surrounded by hills'. It is found in *Cytelwylle* 904 BCS 610, (to) *Cytelflodan* 931 ib. 682. See CHEDDLETON, CHETTISCOMBE, CHETTLE, CHITTELHAMHOLT, CHITTLEHAMPTON; cf. KETTLEWELL.

OE **cild** 'child' &c. See CHILTON.

Cinderford Gl [*Sinderford* 1258 Ch, *-e* 1258 Flaxley]. Identical with *Sinderford* 950 BCS 887 (in boundaries of Pucklechurch Gl). The first el. is OE *sinder* 'cinder, dross'. The exact meaning in the names is not clear.

Cinque Ports K, Sx [(de) *quinque portibus* 1191 OED, *the sink pors* 1297 Rob Gl]. OFr *cink porz* 'the five ports'.

OE **cinu** 'chine, ravine'. See CHINEHAM, CHINLEY.

OE **cipp** 'beam, log'. See CHIPCHASE, CHIPLEY.

Cippenham Bk [*Sippeham* 1163 P, *Chipeham* 1208 Cur, *Cippeham* 1250 Ep]. Identical with CHIPPENHAM. *C-* for *Ch-* is due to Norman influence.

Cirencester (sïsïter) Gl [*Korínion* c 150 Ptolemy, *Durocornovio* (abl.) 4 IA, *Cirenceaster* 577 ff. ASC, *Cirrenceastre* c 894 Asser, *Cirecestre* DB]. The place is on the river **Churn** [*Cyrnéa*, *Cirnea* c 800, 852 BCS 299, 466, *Cyrne* 999 KCD 703]. Cf. also CERNEY. If Ptolemy's *Korínion* is a mistake for *Kornion*, it may be explained as a shortened form of *(Duro)cornovium* in IA. The latter is a derivative of the tribal name *Cornovii*, whose territory was not far north of the Cirencester district. The name means 'the fort of the *Cornovii*'. The shortened form *Kornion* developed to OE **Ciern*, *Cyrn*, to which was added OE *ceaster* 'Roman station'. The river-name Churn may be a back-formation from *Cirencester* or an independent formation from the tribal name. In the latter case the meaning would be 'the river of the *Cornovii*'. The correct form of *Cirencester* would be *Chirenchester*, actually found (as *Chirenchestre*) in Layamon. The modern form is due to Norman influence.

OE **cir(i)ce**, **cyrice** 'church' is fairly common in pl. ns. See CHURCH, CHURCH-, CHERISTOW, CHERITON, CHERRINGTON, CHIR-

DON, CHIRTON, CHURSTON, CHURTON, also KIRTON. It is common as a second el.

OE ***cis** 'gravel', ***cisen** 'gravelly'. See CHISHALL, CHISNALL.

Clacton Ess [*Claccingtun* c 1000 CCC, *Clachintuna* DB, *Clachestona* 1130 P, *Claketon* 1202 FF, *Parva Claketon* 1254 Val, *Claketon Magna* 1291 Tax]. 'The TŪN of *Clacc*'s people.' OE *Clacc* is found in *Clacces wadlond* 774 BCS 216 (O).

OE **clæfre**, **clāfre** 'clover'. See CLAVER- (passim), CLOVERLEY, CLARBOROUGH, CLAREWOOD, also CLARENDON, CLARETON.

OE **clæg** 'clay' is a fairly common first el., as in CLAYHANGER, CLEHONGER, CLINGER, CLAYDON, CLAYTON, CLARE O &c. When used alone to form pl. ns., as in CLAY, CLEE, the meaning is 'clayey soil'. The adj. *clægig* or **clægen* often competes with *clæg* as a first el. See especially CLAYDON.

OE **clǣne** 'clean'. See CLANDON, CLANFIELD, CLANVILLE, CLENNELL.

Claife La [*Clayf* c 1275 PNLa, 1336 FC]. ON *kleif* 'a steep hill-side'.

Claines, North, Wo [*Cleinesse* 11 Heming, *Cleines* 1234 PNWo]. OE *clæg-næss* 'clayey point of land'.

Clandon, East & West, Sr [*Clenedone*, *altera Clendone* 675 BCS 39, *Clenedune* 1062 KCD 812, *Clanedun* DB]. OE (æt) *clænan dūne* '(at) the clean hill'. *Clean* refers to freedom from hurtful growth, thorn-bushes and the like.

Clanfield Ha [*Clanefeud* 1291 Tax], **C~** O [*Chenefelde* DB, *Clenefeld* 1195 P, *Clanefeld* 1226 BM]. 'Clean FELD.' Cf. CLANDON and *Clænefeld* 909 BCS 620 (nr Bp Waltham).

Clannaborough D [*Cloenesberge* DB, *Cloueneberge* 1239 FF]. 'Cloven hill.'

Clanville Ha [*Clavesfelle* DB, *Clanefelde* 1316 FA], **C~** So [*Clanefeld* 1219 Fees, 1225 Ass]. Identical with CLANFIELD, though *f* became *v* later.

Clapcot Brk nr Wallingford [*Clopecote* DB, 1230 P, *Clopcote* c 1180 Bodl], **Clapham** Bd [*Cloppam* 1060 KCD 809, *Clopeham* DB], **C~** Sr [*Cloppaham* 871–89 BCS 558, *Clopeham* DB, *Clopham* 1185 P], **C~** Sx [*Clopeham* 1073 Fr, DB, *Clopham* 1139–60 Oxf, 1225 FF], **Clapton** Brk [*Cloptona* 1167 P], **Clapton** or **Clopton** Ca [*Cloptona* c 1080 ICC, *Cloptune* DB, *Cloptum* 1196 Cur], **C~** Gl [*Cloptone* 1221 Ass], **C~** Mx [*Clopton* 1339 Lo Pleas], **C~** Np [*Clotone* DB, *Clopton* 1149 NpCh, 1177 P], **C~** So nr Maperton [*Cloppetona*, *Clopetone* DB, *Clopton* 1243 Ass], **C~** So nr Crewkerne [*Clopton* 1243 Ass, 1274 Ipm], **C~** in Gordano So [*Clotune* DB, *Clopton* 1225 Ass]; **Clophill** Bd [*Clopelle* DB, *Clophull* 1242 Fees]; **Clopton** Gl [*Cloptune* DB, *Clopton* 1251 Ch], **C~** Sf [*Clop-*, *Clopetuna* DB, *Clopton* 1186 P], **C~** Wa [*Cloptun*

988, 1016 KCD 666, 724], C~ Wo [*Cloptun* 985 KCD 649, *-tona* 1169 P]. The first el. of these is an OE *clop* not recorded in independent use, but found in *clopæcer*, *clophyrst* 972 BCS 1282, *clophangra* 863 ib. 508. The word is identical with MHG *klupf*, G dial. *klopf* 'rock', MDan *klop* 'block, lump'. G *klopf* is often found in pl. ns., as KLOPPENHEIM (one *Clopheim* 8). OE *clop* doubtless meant 'lump, hillock, hill'. Several of the places with names containing *clop* are situated on or near hills, some very prominent ones, as Clapton Gl, the So Claptons, Clopton Gl (at Meon Hill). In some cases the element seems to refer to a slight rise. Clophill is noteworthy. The element seems often to appear in the plural form. The word *clop* is found also as a byname: (John) *le Clop* 1274 Cl (Ha).

For **Clapton in Gordano** see EASTON IN GORDANO.

Clapdale YW [*Clapedale* 1190 FC], **Clapham** YW [*Clapeham* DB, c 1177 FC, *Clepeam* 1090–7, *Clapaham* c 1125 Kendale, *Claphaim* c 1170 FC, *Clapehamme* J Ass]. Clapdale and Clapham are on a stream. The first el. is probably an OE stream-name *Clæpe or the like, related to G KLAFFENBACH [*Chlaffintinpach* 11], whose first el. is cognate with OHG *chlaffôn* 'to sound'. The OE name would be related to OE *clappettan* 'to throb', ME *clappen* 'to clap' and mean 'brook where the water makes the stones in its bed clatter'.

Clapham Bd &c. See CLAPCOT.

Clappersgate We [*Clapper(s)gate* 1608 Kendale]. Dial. *clapper* means 'a rough or natural bridge across a stream, steppingstones'. *Gate* seems to be the word for a road, OScand *gata*.

Clapton. See CLAPCOT.

Clarborough Nt [*Claureburg* DB, *Clauerburg* 1185 P, *Clareburg* 1242 Fees]. 'Clover BURG.' The combination of *clover* and BURG is curious. Either *burg* here means simply 'manor' or the like, or the reference is to an old fort in a clover-field.

Clare O [*Cleyore* 1282 Ipm, 1284 Cl]. 'Clayey slope.' Cf. ŌRA.

Clare Sf [*Clara* DB, c 1145 BM, *Clare* 1198 FF]. Perhaps identical with CLERE.

Claremont Sr was so called by the Earl of Clare, who bought the property in 1714.

Clarendon W [*Clarendun* 1072 Round, *Feudal England* 304, *Clarendona* 1130, *-dun* 1157 P, *-don* 1165 BM]. Possibly OE *clæfren dūn* 'clover hill' in spite of the absence of forms with *v*.

Claret Hall Ess [*Clare* DB, *Clareta* 1165, *Claretta* 1194 P]. 'Little Clare.' C~ is opposite to CLARE Sf on the southern bank of the Stour. The diminutive ending *-et* is French.

Clareton YW [*Claretone* DB, *-ton* 1242 Fees, *Clarton* 1176 P]. C~ is in Claro wap. [*Clarehov* 1166, *Clarho* 1195 P]. Perhaps 'TŪN and hill where clover was grown'. Cf. CLARENDON.

Clarewood Nb [*Claver-*, *Clareworth* 1212 Cur, *Claverwrth* 1226–8 Fees]. 'Clover WORÞ.'

OE **clāte, clǣte** 'burdock'. See CLATFORD, CLATWORTHY, CLAVERTON So, CLOFORD, CLOTHALL, CLUDDLEY, CLEATHAM, CLEATLAM.

Clater Park He [*Cletera* 1166 RBE, *Clatere* 1269 Ipm]. Cf. CLATTERCOTT &c.

Clatford, Goodworth & Upper, Ha [*Cladford* DB, *Clatford* 1156 P, *Upclatford* 1316 FA, *Godorde* DB, *Godeworth* 1291 Tax], C~ W [*Clatford* DB, 1242 Fees]. 'Ford where burdock grew.' Cf. CLĀTE. Goodworth seems to have been the original name of Goodworth Clatford. The name means 'Gōda's WORÞ'.

Clattercott O [*Clatercota* 1167, *Clatrecote* 1199 P, *Clatercote* 1227 Ch], **Clatterford** Wt [*Clater-*, *Clatreford* 1287–90 Fees, *Claterford* 1346 FA]. The first el. of the names seems to be related to OE *clatrung* 'clattering, noise'. To this belongs *clatter* 'debris, loose stones', which may be the element sought for. The word has only been found in quite recent times, but may well be old. CLATER He may contain the word.

Clatworthy So [*Clateurde* DB, *-wurth* 1227 FF, *Clatewurthy* 1243 Ass]. 'WORÞ where burdock grew.' Cf. CLĀTE.

Claughton (-ăf-) Chs [*Clahton* 1260, *Clauhton* 1282 Court], C~ (-ī-) La in Garstang [*Clactune* DB, *Clacton* 1185 f. P, *Claghton* 1285 Ass], C~ (-ăf-) La nr Caton [*Clactun* DB, *Clahton* 1208 FF]. All three Claughtons are situated by hills. The first el. is ON *klakkr* 'a lump', Sw *klakk* 'a small hillock', ModIcel *klakkur* 'a rock'.

Claverdon Wa [*Clavendone* DB, *Claverdon* c 1155 Fr, *Claredon* 1316 FA]. 'Clover hill.' See DŪN.

Claverham So [*Claveham* DB, *Claverham* 1248 Ass]. 'Clover HAMM or HĀM.'

Clavering Ess [*Clæfring* 11 E, *Clauelinga* DB, *Clauering* 1159 P]. A derivative with the suffix *-ing* from OE *clǣfre* 'clover' and meaning 'clover field'.

Claverley Sa [*Claverlege* DB, *Clauerlai* 1163 P]. 'Clover LĒAH.'

Claverton Chs [*Claventone* DB, *Claverton* 1260 Court, 1285 Ch]. 'Clover TŪN.'

Claverton So [*Clatfordtun* c 1000 Wills, *Claftertone* DB, *Claferton* 1227 FF, *Claverton* 1212 Fees]. 'TŪN by *Clātford*.' Cf. CLATFORD.

Clawson, Long, Le [*Clachestone* DB, *Claxtun* 1236 Fees]. '*Clac*'s TŪN.' *Clac* c 980

BCS 1130 &c. is a Scand pers. n. (ODan *Klak*, OSw *Klakker*, ON *Klakkr*).

Clawthorpe We [*Clerkethorpe* 1277 Kendale]. 'The thorp of the clerks.'

Clawton D [*Clavetone* DB, *Clavatona* 1088 Totnes]. 'TŪN in a tongue of land.' The first el. is OE *clawu* 'a claw, cloven hoof', here 'the fork of a river'. The river-name **Claw** is a back-formation.

Claxby Li nr Alford [*Clachesbi* DB], C~ Li nr Market Rasen [*Cleaxbyg* c 1067 Wills, *Clachesbi* DB, c 1115 LiS, *Clakesbi* 1155–60 DC], C~ **Pluckacre** Li [*Clachesbi* DB, *Claxeby Pluc Acre, Claxby Pluk Acre* 1227 Ep]. '*Clac*'s BY.' Cf. CLAWSON.
Pluckacre is obscure. It consists of *pluck* vb. or sb. and *acre*. One might compare Sf dial. *plucky* 'heavy, clogging' (of clay &c.). Or the name may mean a field so poor that each ear had to be plucked.

Claxton Du [*Clachestona* 1091 FPD, *Claxtun* 1208–10 Fees], C~ Nf [*Clakestona* DB, *Claxtone* 1254 Val], C~ YN [*Claxtorp* DB, *Clakeston* 1176 P]. '*Clac*'s TŪN.' Cf. CLAWSON.

Claybrooke, Great & Little, Le [*Claibroc* DB, 1195 P, *Cleibrok* 1224 Ep]. Really the name of the brook at the place [*Clægbroc* 962 BCS 1096]. 'Clayey brook.'

Claycoton Np [*Cotes* 12 NS, *Cleycotes* 1284 FA, *Claycotone* 1329 QW]. Originally OE *cotu* 'the COTS'; *-coton* goes back to the dat. plur. *cotum*. Later *Clay* was added. The meaning is 'Coton in the clayey district'.

Claydon, Botolph (bŏtl), **East & Middle,** Bk [*Clai(n)done* DB, *Claindune* c 1130 Oxf, *Cleydon* 1220 Fees, *Botle Cleidun* 1224 Bodl, *Est Cleydon* 1247 Ass, *Middelcleydon* 1242 Fees], **Steeple** C~ Bk [*Claindone* DB, *Stepel Cleydon* 1209–35 Ep], C~ O [*Clæihæma broc* 956 BCS 947, *Claindona* c 1160 RA], C~ Sf [*Clainduna* DB, *Cleidun* 1198 FF]. 'Clayey hill.' The first el. is OE *clǣg* or *clǣgig* or **clǣgen* adj. 'clayey'.
Botolph is a popular etymology for OE *bŏtl* 'house, building', here perhaps 'manor'.— *Steeple* from the church steeple.

Claygate Sr [*Clæigate* 1065 BM, *Claigate* DB]. OE *clǣg* and *geat* 'gate'. 'Gate leading to the clayey district'?

Clayhanger Chs [*Clehongur* 1432 BM], C~ D [*Clehangre* DB], C~ St [*Cleyhungre* Hy 3 BM]. 'Clayey slope.' Cf. HANGRA and *Clǣighangra* 1016 ASC (C) in Mx.

Clayhidon D [*Hidone* DB, *Hydun* 1212, *Hidune* 1242 Fees, *Cleyhidon* 1485 Ct]. Originally HIDON, from OE *hīeg-dūn* 'hay hill, hill where hay was got'. *Clay* refers to clayey soil.

Claypole Li [*Claipol* DB, 1194 ff. P]. 'Clayey pool.'

Claythorpe Li [*Clactorp* DB, 1202 Ass]. '*Clac*'s thorp.' Cf. CLAWSON.

Clayton La nr Manchester [*Cleyton* c 1250 LaInq], C~ **le Dale** La [*Claiton* 1246 Ass, *Claiton in the Dale* 1327 Subs], C~ **le Moors** La [*Clayton* 1263, *Clayton super Moras* 1284 Ass], C~ **le Woods** La [*Claiton* c 1200 CC, 1227 Ass], C~ **Griffith** St [*Claitone* DB, *Cleyton* 1254 Ipm], C~ Sx [*Glaitone* 1073 Fr, *Claitune* DB], C~ YW nr Bradford [*Claitone* DB], C~ YW nr Mexborough [*Claitone* DB], **West** C~ YW [*Claitone* DB]. 'TŪN on clayey soil.'
C~ **Griffith** was held by Bertram Griffin till 1254 (Ipm). *Griffin* is a French family name, later supplanted by Welsh *Griffith*.

Clayworth Nt [*Clavorde* DB, *Claworth* c 1130 RA, *Clawurda* 1156, 1160, *-wurða* 1159, *Clauewurda* 1164, *Clauwurða* 1177 P]. 'WORÞ in a river-fork or tongue of land.' Cf. CLAWTON. Clayworth is situated at the junction of two streams.

Cleadon Du [*Clyvedon* 1280 Ch]. Identical with CLEVEDON.

Clearwell Gl [*Wellenton* 1220 Fees; *Clowerwall* 1444 Rudder, 1539 Bodl]. Second el. OE *wiella, wælla* 'spring'. The first cannot be explained without earlier forms. Possibly THE CLAWR Monm [*Clougur, Clour* c 1150 LL] may be compared.

Cleasby YN [*Clesbi* DB, *Clesebi* 1202 FF]. Perhaps OScand *Klepps-bȳr* '*Klepp*'s BY'. ON *Kleppr*, ODan *Klep*, OSw *Klæpper* are known pers. names. Cf. early forms of CLIXBY Li.

Cleatham Li [*Cletham* DB, c 1115 LiS, *Clatham* 1263 FF]. 'HĀM or HAMM where burdock grew.' First el. OE *clǣte*, a sideform of CLĀTE 'burdock'.

Cleatlam Du [*Cletlinga* c 1050 HSC, *Cletlum* 1271 FPD]. OE *clǣt(e)-lēah* 'LĒAH where burdock grew'. Cf. CLEATHAM. Cleatlam represents the dat. plur. *clǣt(e)-lēum*. The earliest form is really a derivative with *-ingas* from Cleatlam, meaning 'the people at Cleatlam'.

Cleator Cu [*Cletergh* c 1200 StB, 1294 Cl]. Second el. ERG 'a shieling'. The first may be ON *klettr* 'a rock, cliff'.

Cleckheaton. See HEATON.

Clee Li [*Cleia* DB, *Cle* c 1115 LiS, *Cleie* 1206 Ass]. OE *clǣg* 'clay, clayey soil'. Nr Clee is **Cleethorpes.**

Clee Hill, Brown Clee Hill, Titterstone Clee Hill, a massif of hills in Sa [*Clivas* 1232 FF], **Clee St. Margaret** and C~ **Stanton** Sa [*Cleie, Clee* DB, *Clyes* 1200 FF, *Clye Sancte Margarete* 1285 FA, *Cleo Staunton* 1290 Misc], **Cleeton** Sa [*Cleoton* 1241 FF, *Cletone* 1255 RH], **Cleobury** (-ĕ- or -ĭ-) **Mortimer** Sa [*Clai-, Cleberie* DB, *Claiberi* 1201 FF, *-bur* 1242 Fees, *-bir* 1266 Ch, *Clebury Mortimer* 1272 Misc], **North Cleobury** Sa [*Ufere Cleobyrig* Edw Conf. (Bowcock), *Cleberie* DB, *Cleybiri* 1241, *Northclaibiry* 1222 FF]. The Clees and

Cleoburys are situated near the Clee range of hills, Clee St. Margaret on the western slope of Brown Clee Hill, Clee Stanton on the slope of Titterstone Clee Hill, Cleeton on the northern slope of the latter, Cleobury Mortimer on the Rea near Clee Hill, North Cleobury near Brown Clee Hill. It is obvious that Clee, Cleo- represents the old name of the hill. *Clivas* 1232 is probably not the name of the hills, but OE *clifu* 'the cliffs', a generic term. Derivation of the hill-name from OE *clæg* 'clay', which some forms seem to indicate, is unlikely because the hills are famous for a hard rock (Dhu Stone), and also in view of the common spelling *Cleo-*. On the other hand *Clai-* is early and well attested. A common source of *Cleo-* and *Clai-* is difficult to think of. *Cleo-* may be an OE *clēo* (from *klewa-*) related to OE *clīewen* 'ball', OHG *kliuwi*, *kliuwa* the same, ON *klé* 'stone used as a weight in a loom' (from *klewan*). A hill-name may have been formed from such a word. CLEOBURY MORTIMER may be an independent formation with OE *clæg* or a derivative of it as first el. Later there was overlapping of the similar elements *Cleo-* and *Clai-*.

Cleobury Mortimer was held by Ralph de Mortemer in 1086 (DB) and by a namesake of his in 1236 (Fees). The name is derived from MORTEMER in France.

Cleeve, Bishops, Gl [*aet Clife* 769–85 BCS 246, *Clive* DB], **C~** So [*Clive* 1243 Ass, *Clyve* 1327 Subs], **Old C~** So [*Clive* DB, 1227 Ch], **C~ Prior** Wo [*Clive*, *Clyve* 11 Heming, *Clive* DB, *Clyve Prior*' 1291 Tax]. OE *clif* 'cliff, hill'. Cleeve represents the dat. sing. *clife*.

Bishops C~ stands at Cleeve Hill, called *Uuendlesclif* BCS 246 ('*Wendel*'s cliff'). It belonged to the Bishop of Worcester.—**C~ Prior** belonged to Worcester Priory.

Clegg La [*Clegg* c 1200 Whitaker, *Cleg* 1285 Ass]. The place stands at the foot of **Owl Hill**, which was no doubt once Clegg. The name is Scandinavian and related to ON *kleggi* 'haystack'.

Clehonger He [*Cleunge* DB, *Clahangra* 1184 P, *Clehungre* 1236 Ipm]. Identical with CLAYHANGER.

Clenchwarton Nf [*Ecleuuartuna* DB, *Clenchewarton* 1196 FF, *Clencwarton* 1205 Cur]. 'The TŪN of the *Clencware* or people at *Clenc*.' The first el. is identical with the name of an old hundred, *Clencware hundred* 11 EHR 43. *Clenc-* is no doubt identical with CLINCH W, but the exact meaning is obscure. Cf. -WARU.

Clennell Nb [*Clenhill* 1242 Fees, *-hil* 1290 Ch]. 'Hill free from hurtful growth.' Cf. CLANDON.

Clent Wo [*Clent* 11 Heming, DB, 1169 P]. The place is by a hill. The name is an old word for a hill, related to OSw *klinter*, ON *klettr* 'a hill, hillock'.

Cleobury. See CLEE HILL.

Clere Ha in **Burgh-, High-, & Kingsclere** [*Cleran* 749 BCS 179, (æt) *Clearan* c 880 BCS 553, *Clearas* 955 ib. 912, *Clere* DB, *Clara* c 1145 Fr, *Burclere* 1171 Ep, *Borcleare* 1176 P, *Alta Clera* c 1270 Ep, *Hauteclere* 1284 Ch, *Kyngeclera* Hy 1 (1270) Ch]. The OE form was *Clēare* or *Clēara*. This must be compared with *Cleara flod* 901, (æt) *Clearan floda* 909 BCS 596, 625 (nr North Waltham and far from the Cleres). In the latter we seem to have to do with a river-name. The Cleres are a good way apart from each other and not on the same stream. They are on a range of hills south of the river Enborne. *Clēare* may have been the old name of the ENBORNE, which gave its name to the district south of the river. The name may be derived from MW *clayar* 'gentle', Welsh *claear* 'lukewarm'. According to some scholars Welsh *claear* 'bright' is the same word. The latter would perhaps give a better sense. CLARE Sf might be identical in origin. Clare is on a tributary of the Stour.

Burghclere belonged to the Bishop of Winchester, who had a market here. The name means 'the borough of Clere'. The place is called *Novus Burgus de Clere* in 1218 (VH).—**Kingsclere** was an old royal manor. It belonged to King Alfred.

Clerkenwell Mx [*fons Clericorum* c 1100 Mon, *Clerkenwelle* 1182 P, *Clerekenewell* 1198 FF]. 'The spring of the clerks or clerics.' *Clerken-* is an analogical weak gen. plur. (ME *clerkene* from OE *-ena*).

Clevancy W [*æt Clife* 983 KCD 636, 638, *Clive* DB, *Clive Wancy* 1232 Ch]. Originally OE CLIF. Cf. CLIF. Wancy, a family name from WANCHY in Seine-Inf., was added for distinction from the neighbouring CLIFFE PYPARD. The OE example quoted may refer to the latter rather than to Clevancy, but very likely both manors are included.

Cleve He [*Clive* DB]. Identical with CLEEVE.

Clevedon So [*Clivedone* DB, *Cliuedon* 1172 P, *Clivedon* 1225 Ass, 1242 Fees]. OE *clifa-dūn* 'hill with cliffs'.

Cleveland YN [*Clivelanda* c 1110 YCh 932, *Kliflǫnd* Heimskringla]. 'The hilly district.' Cf. CLIF.

Cleveley La [*Cliueleye* c 1180 CC], **C~ O** [*Clivelai* c 1210, *Cliveleia* c 1235 ff. Winchc]. 'Cliff LĒAH.'

Clewer Brk [*Clivore* DB, *Clifwara* 1156 RBE, 1159 P, *Cliuewara* 1156 P, *-ware* 1198 FF], **C~ So** [*Cliveware* DB, *Clywar* 1276 RH]. OE *clif-ware* 'dwellers on a hill slope'; cf. -WARU, also CLIFFE K.

Cley (-i) Nf [*Claia* DB, *Claya* 1242 Fees], **Cockley Cley (-i)** Nf [*Cleia*, *Claia* DB, *Claia* 1199 Fees, *Cleye Omnium Sanctorum*, *Sancti Petri* 1254 Val, *Coclikleye* 1324 Ipm]. OE *clæg* 'clay, clayey soil'.

The additional Cockley is obscure. Perhaps it

is a pl. n., Cockley meaning 'cock wood, wood frequented by wild birds'.

Cliburn We [*Clibbrun* c 1150, *Clifburn* c 1250 WR]. 'Cliff stream.' The place is on the LEITH.

Cliddesden Ha [*Cleresden* DB, *Clereden* 1167 P, *Cledesdene* 1194, *Cludesdene* c 1250 Selborne, *Clidesdene* 1274 RH]. The first el. seems to be a derivative of OE *clūd* 'a rock'. The OE form may have been **clȳde* n. Second el. OE *denu* 'valley'.

OE **clif** 'cliff, rock, steep descent, promontory' is a common pl. n. el. The meaning varies. The most common one seems to be 'a slope' (not necessarily a steep one) or 'the bank of a river'. Cf. CLEVE, CLEEVE, CLIFF, CLIVE, CLYFFE, CLEADON, CLEVEDON, CLEVELAND, CLIFFORD, CLIFTON, CLIVEDEN, CLIVIGER &c. Common as a second el., usually in the form -*cliff(e)*, but cf. e.g. CATSLEY, GATLEY Chs, HECKLEY.

Cliff Ha nr Eling [*Sclive* DB], C~ Wa [*Cleve* 1392, *Clyve* 1405 AD], **North & South Cliff** YE [*Clive* DB, *North, Suth Clyf* 1307 Ch], **Cliffe** K nr Gravesend [*Clifwara gemære* 778 BCS 227, *Cliua*, (to) *Cliue* 10 ib. 1321 f., *Clive* DB], **West Cliffe** K [*æt Clife* 1042–4 BM, *Wesclive* DB, *Westcliua* 1173 P], **King's Cliffe** Np [*Clive* DB, *Cliua* a 1100 NpCh], **Cliffe** (-ēv) **Pypard** W [*Clive* DB, *Clive Pipart* 1231 Cl, *Cliva Pipard* 1242 Fees; cf. CLEVANCY], C~ YN [*Ileclif* c 1050 HSC, -*clife* c 1130 SD, *Cliue* DB]. OE **clif** 'cliff, slope'.

King's Cliffe was held by the king at the time of the Norman Conquest.—**Cliffe Pypard** was held by Richard Pipart in 1231 (Cl). Pypard is an OFr family name from OFr *pipart* 'piper'. —The first el. of *Ileclif*, the old name of **Cliffe** YN, may be *Illa* pers. n.; cf. ELEIGH.

Clifford Chambers Gl [*Clifford* 922, *æt Clifforda* 966 BCS 636, 1181, *Clifort* DB, *Clifford Chamberer* 1526 Glouc], C~ He [*Cliford* DB, *Clifford* 1230 P], C~ YW [*Cliford* DB, *Clifford* c 1170 YCh 1035]. 'Ford at a cliff or slope.'

C~ Chambers belonged to the Abbot of Gloucester and was administered by the Camerarius or Chamberlain.

Clifton Bd [*Cliftun* 944–6 BCS 812, *Cliftone* DB], C~ **Reynes** Bk [*Cliftone* DB], C~ Chs [*Clistune* DB, *Clifthona* c 1100 Chester], C~ Cu [*Clifton* 1204 P], C~ Db [*Cliptune* DB, *Clyfton* 1221–30 Fees], C~ **Maybank** Do [*Cliftun* 1002–14 KCD 708, *Clistone* DB, *Clifton Mabank* 1319 FF], C~ Gl [*æt Cliftune* 970 BCS 1257, *Clistone* DB], C~ La in Eccles [*Clifton* 1184 P], C~ La in Burnley [*Clifton* 1495 Ct], C~ **with Salwick** La [*Clistun* DB, *Clifton* 1257 FF], C~ Nb [*Clifton* 1242 Fees], C~ Nt nr Nottingham [*Cliftune* DB], **North & South** C~ Nt [*Cliftune* DB, *Southclyfton* 1327 Ipm], C~ O [*Cliftona* c 1170 Oxf], C~ **Hampden** O [*Cliftona* 1146 RA], C~ **Campville** St [*Clyfton* 942 BCS 771, *Cliftune* DB, *Clifton Caunvil* 1284 Ass], C~ **on Dunsmore**

Wa [*Cliptone* DB, *Cliftona* 1169 P], C~ We [*Clifton* 1291 Tax], C~ **on Teme** Wo [*Cliftun, Cliftun ultra Tamedam* 934 BCS 700, *Clistune* DB], C~ YN nr York [*Cliftune* DB], C~ **on Ure** YN [*Clifton* DB, C~ *upon Ure* 1317 Ch], C~ YW nr Brighouse [*Cliftone* DB], C~ YW nr Doncaster [*Cliftone* DB], C~ YW nr Otley [*Cliftun* c 1030 YCh 7, *Cliftun* DB]. 'TŪN on a hill or hill slope or the brink of a river.'

C~ **Campville** was held by Richard de Camvill in 1231 (Cl). Campville is a family name derived from CANAPPEVILLE or CANAPVILLE or CANVILLE in Normandy.—C~ **on Dunsmore**, see DUNSMORE.—Hampden in C~ **Hampden** is presumably a family name.—C~ **Maybank** was held by William *Malbeenc* in 1084 (GeldR). Maybank is an OFr byname, also appearing in the forms *Malbedeng, Malbanc* in DB. Cf. also NANTWICH.—C~ **Reynes** was held by Ralph de Reynes in 1303 (FA).

Climperwell Gl nr Brimpsfield [*Climperwelle* 1227 Flaxley, *Clymperwell* 1291 Tax]. First el. OE *clympre* 'lump of metal'. The exact sense in the name is not apparent.

Climping Sx [*Clepinges* DB, *Clenpinges* 1087, *Clinpingh[es]* c 1194 Fr]. '*Climp*'s people.' **Climp* is found in Clemsfold Sx [*Climpesfaude* 1285 Ass]. It is a nickname related to OE *clympre*, ON *kleppr* &c. 'a lump'.

Climsland Co, now **Stoke Climsland** [*Climeslande* 1217–20 BM, *Stoke* 1266 Ep]. The same first el. is found in **Climson** (in Stoke Climsland) [*Clymestun* c 970 BCS 1247, *Clismestone* DB, *Clemeston* 1177, *Climeston* 1194 P]. It looks like a pers. n., but very likely it is an old Cornish placename, perhaps of the hill by which the places are. The etymology is obscure.

Clinch or **Clench** W [*Cleynche* 1329, *Clench* 1355 Ipm]. The place is near a hill, and very likely C~ is really the name of the hill, related to Engl *clench*, *clunch* 'a lump, mass'. Possibly a related element is found in (on) *Clinca leáge* 941 BCS 765, (on) *Clincan leáge* 943 ib. 786 (near Tisted and Hinton Ampner Ha).

Clinger Gl nr Cam [*Claenhangare* DB, *Clehongre* 1287 QW]. Identical with CLAYHANGER.

Clint YW [*Clint* 1230 P, 1279–81 QW]. OSw *klinter*, Dan *klint*, ON *klettr* 'a hill'.

Clippesby Nf [*Clepesbei, Clipesby* DB, *Clipesbi* 1191 P]. '*Clip*'s BY.' *Clip*, the name of a moneyer 10 cent., is ON *Klyppr*.

Clipsham Ru [*Kilpesham* 1203 Cur, 1220, 1235 Ep, *Clyppesham* 1428 FA]. The first el. must be an OE el. *cylp* or the like, presumably a pers. n. **Cylp*, which may be related to Norw *kylp* 'a small sturdy fellow'. Second el. HĀM.

Clipston Np [*Clipestone* DB, -*tona* c 1155 DC, *Clipston* 1202 Ass], C~ Nt [*Clipstun* 1236 Fees], **Clipstone** Bd [*Clipeston* R 1

Cur], C~ Nf [*Clipestuna* DB, *Clipeston* 1199 Cur], C~ Nt [*Clipestune* DB, *Clipeston* 1196 Cur]. '*Clip*'s TŪN.' Cf. CLIPPESBY.

Clitheroe La [*Cliderhou* 1102 LaCh, 1176 P, *Clitherow* 1124 YCh 1486]. Second el. ON *haugr* 'hill'. The first may be OScand *kliðra* (Sw *klera*) 'song-thrush'.

Clive Chs [*Clive* DB], C~ Sa [*Cliua* 1176 P]. OE *clif* 'cliff' &c.

Cliveden Bk [*Cliueden* 1195, *Cliveden* 1200 Cur]. 'Valley among cliffs.'

Cliviger (-j-) La [*Cliuecher* 12 Kirkst, *Clyuacher* 1246 Ass]. 'Cliff acre.' The same palatalized form of *acre* is found here as in ALSAGER.

Clixby Li [*Clisbi* DB, *Clifsebi* c 1115 LiS, *Clipsebi* 1196, *Clessebi* 1193 P, *Clixeby* 1275 RH]. Identical with CLIPPESBY.

Clodock He [*ecclesia Sancti Clitauci* c 1150 LL, *Cladoc* 1266 Ewias]. '(The church of) St. Clydog.' Clydog (*Clitauc* c 1150 LL) is stated to have been a king and martyr.

Cloffocks Cu [*Fyt Cloffhou* 1610 Whellan]. OE *clōh-hōh* 'spur of land with a ravine'. *Fyt* is ON *fit* 'low-lying meadow'.

Cloford So [*Cladforda, Claford* DB, *Clatford* c 1150 Montacute, *Cloforde* 1327 Subs]. OE *clāt-ford* 'burdock ford'. Cf. CLATFORD.

OE *clōh, ME *cloʒ*, dial. *clough* 'a ravine'. See CATCLEUGH, DEADWIN CLOUGH, CLOFFOCKS, CLOTTON, CLOUGHTON.

Clophill, Clopton. See CLAPCOT &c.

Closworth So [*Clovewrde, Clovesuurda* DB, *Clouesword* c 1100 Montacute]. First el. the OE *clof* found in *Clofeshōh*, the name of an old meeting place, and meaning 'a crevice, valley'. Cf. ON *klof* 'a crevice'. See WORP.

Clothall Hrt [*Clatheala* c 1060 PNHrt(S), *Cladhele* DB, *Clothal* 1199 Cur, 1301 BM]. 'HALH where burdock grew.' Cf. CLĀTE.

Clotherholme YW [*Cludun* DB, *Cluthum* c 1160 YCh 72, *Clutherum* 1156 ib. 80, *Cluderun* 1195 P]. Looks like the dat. plur. of an OE *clŭder*, which may be related to OE *clŭd* 'a rock'. Cf. OE *stǣner* 'stony ground' from *stān*.

Clotton Chs [*Clotone* DB, c 1100 Chester, *Clottona* 12 Chester]. 'TŪN in a ravine.' Cf. CLŌH.

Cloud, Temple, So [*la Clude* 1199 P, *Clude, Cluda* 1204 f. Cur]. OE *clŭd* 'a rock', here in the sense 'a hill'. The Cloud is a prominent hill nr Buglawton Chs. Cf. CLUTTON. Temple C~ probably because the place belonged to the Templars.

Cloughton YN [*Cloctune* DB, *-ton* 1191 ff., *Clotton* 1195 P]. 'TŪN in a ravine.' Cf. CLŌH.

Clovelly (klōvĕ·lĭ) D [*Clovelie* DB, *Clovely*

1242 Fees, *Cloveli* 1276 Ipm, *Colf Ely* 1290 Ch]. Clovelly is only c 2 m. from Velly [*Felye* 1287, *Overefelye* 1301, *la Felye* 1333 PND] and is no doubt *Velly* with a prefixed OE *clōh* or *clof* 'ravine'. *Felye* (*Velly*) was apparently the name of the semicircular ridge on or by which the places are, so called on account of a fancied likeness to the felly (OE *felg*) of a wheel. Clo- in Clovelly refers to the ravine at whose bottom end the village is. *Overefelye* may be used in contrast to Clovelly, which may have been originally *Felye* too. OE *fealg* (*felg*) gives no good meaning and is not to be considered as a possible alternative.

Cloverley Sa [*Claverleg* 1255 RH]. 'Clover LĒAH.'

Clowne Db. See CLUMBER.

OE **clūd** 'rock', no doubt also 'hill'. See CLOUD, CLUTTON, also CLIDDESDEN, CLOTHERHOLME.

Cluddley Sa [*Clotleye* 1296 Ipm, *-legh* 1301 For]. 'LĒAH where burdock (OE CLĀTE) grew.'

Clumber Nt [*Clunbre* DB, *Clumbra* 1166 P, *Clumber* 1242 Fees], **Clowne** Db [*Clune* 1002 Wills, DB]. There was formerly a place **Clun** in Nt, apparently nr Carburton on the Poulter [*Clune* DB]. Clun seems to be an old river-name, identical with CLUN Sa. If so, it must be the old name of the Poulter. Clumber is on the Poulter at the foot of a hilly tract. The name seems to contain the river-name Clun and Welsh *bre* 'hill'. Clowne is on an arm of the Poulter.

Clun R Sa [*Colunus* 1572 Lhuyd, *Clune* 1577 Saxton], **Clun** (town) Sa [*Clune* DB, 1161 P, *Colunwy* 13 Brut], **Clunbury** Sa [*Cluneberie* DB], **Clungu·nford** (n-g) Sa [*Clone* DB, *Cloune Goneford* 1242 Fees], **Clunton** Sa [*Clutune* DB, *Clonton* 1267 Ipm]. The places mentioned are on the Clun and took their names from the river. Clun goes back to earlier *Colun-* and is identical with COLNE Ess. The name is British.

Clungunford was held by Gunward t. Edward the Confessor (DB). *Gunward* may be ON *Gunnvarðr* or OFr *Gundoard*, a name of OG origin.

Clutton Chs [*Clutone* DB, *Clutton* 1275 Ipm], C~ So [*Clutone* DB, *Clotton* 1205 Cur]. OE *Clūd-tūn* 'TŪN by a hill.' Cf. CLŪD. Clutton So is nr TEMPLE CLOUD.

Clyffe Do [*Clive* DB, 1212 Fees]. OE CLIF 'cliff' &c.

Clyst R D [*Clyst* 937, 963 BCS 721, 1103]. A British river-name related to Lat *cluo* 'to wash', OE *hlūttor* 'clean', the river-names CLYDE in Scotland and CLYDACH in Wales. The meaning is probably 'clean stream'. From the river-name are derived: **Broad Clyst** [*Glistun* 1001 ASC, *Clistun* c 1100 E, *Clistone* DB], C~ **Hydon** [*Clist* 1242 Fees, *Clist Hydone* 1268 Ep], C~ **St. George** [*Clyst Wicon* 963 BCS 1103, *Clyst Sancti*

Georgii 1334 Subs], **C~ St. Lawrence** [*Clist Sancti Laurencii* 1203 Cur], **C~ St. Mary** [*Clist Sancte Marie* 1242 Fees]. **Clyst William** [*Clistewelme* 1270 FF] means 'the source of the Clyst'. Second el. OE *æwielm* 'source of a river'.

C~ Hydon was held by Ricard de Hidune in 1242 (Fees). Cf. CLAYHIDON.

OE **cnæpp** 'top of a hill, hillock'. See KNAPP, KNEPP.

OE **cnēo** 'knee'. See KNAITH, KNEETON YN.

OE **cniht**. See KNIGHTON, KNIGHT- (passim).

OE **cnoll** 'a knoll'. See KNOLE, KNOWLE, KNOWLTON, BUCKNOWL, CHETNOLE, CHIP-NALL.

Coalbrookdale Sa [*Caldebrok* 1250 Eyton]. 'Cold brook valley.'

Coaley Gl [*Couelege* DB, c 1200 Berk, *-leg* 1220 Fees]. 'LĒAH in a cove or recess.' Cf. COFA.

Coalville Le. A late name. The town is in a coal district.

Coat, Coate, Coates, Coatham. See COT.

Cobb Do, a semicircular pier in Lyme Regis, dating from t. Edw I [*la Cobbe, Cobheye* 1295 Misc]. Identical with Engl *cob* 'roundish mass, lump' &c. It seems to presuppose an OE *cobb* or *cobbe*, related to Sw dial. *kobbe* 'round skerry' &c.

Cober R Co [*Coffar* 1284, 1286 Ass, *Chohor* 1336 Ch]. Unexplained.

Coberley (-ŭ-) Gl [*Culberlege* DB, 1221 Ass, *Cuthbertleia* c 1188 Bodl, *Cudbrigtlegh* 1230 Cl]. '*Cūþbeorht*'s LĒAH.'

Cobhall He [*Cobewelle* DB, *Cobbewell* 1242 Fees]. '*Cobba*'s stream or spring.' OE **Cobba* corresponds to OG *Cobbo* and appears as *Cobbo* (dat.) 1159 f., 1190 ff. P (Ha). It is probably the first el. of *Cobban dæne* 940 BCS 763, *Cobban lea* 956 ib. 974, *Cobban stan* 957 ib. 998 and of COBHAM. But the word mentioned under COBB may partly be the source.

Cobham K [*Cobba hamm* 939 BCS 741, *Cobbeham* 1195 P]. '**Cobba*'s HAMM.'

Cobham Sr [*Coveham* 675 BCS 39, *Couenham* 1062 KCD 812, *Covenham* DB, *Coveham* 1428 FA]. OE *Cofan-hām*, the first el. being either OE COFA (q.v.), here possibly referring to the bend of the Mole at the place, or rather OE *Cōfa* pers. n. Cf. COVENTRY. The change to Cobham is late.

Cock Beck YW [*Cock* 1293 Ass, *Koc* 1348 YD]. On the stream is **Cocksford** [*Cockesfort* c 1175 YCh 1569, *-ford* 1231 Cl, *Cokeford* 13 YD]. The river-name may be a back-formation from the pl. n., whose first el. may be OE *cocc* 'cock' the bird or the OE *cocc* found in names such as WITHCOTE, COCKHAMPSTEAD, COFTON D, COOKHAM, COUGHTON and in *haycock*, and which seems

to have meant 'a heap', 'a hillock', 'a clump of trees'. For OE *cocc* 'cock' as a first el. see COCKEY &c., COGDEAN, COQUET.

Cockbury Gl [*Coccanburh* 769–85 BCS 246, *Cocce-, Cokebiri* 1246 Ipm]. '*Cocca*'s BURG.' The pers. n. **Cocca* is presupposed by several other names. See the following names.

Cocken Du [*Coken* 1138–40 Finchale, *Cochena* Hy 2 FPD, *Cokene* 1195 (1335) Ch]. Probably OE *Coccan-ēa* '*Cocca*'s stream'. Cf. COCKBURY. For the loss of the second el. see WHITTON Li.

Cockenhatch Hrt [*Cochenac* DB, *Cochenach* Hy 1 BM, *Kokenhach* 1220 Fees]. '*Cocca*'s hatch or gate.' Cf. COCKBURY, HÆCC.

Cocker R Cu [*Coker* 1230 Sc, 1279 Ass]. On the river is **Cockermouth** [*Kokermue* 1195 FF, *Cokermuth* 1253 Pat] 'the mouth of the Cocker'. **Cocker** R La [*Cocur* 930 YCh 1, *Cokir* c 1155 LaCh]. On the river are **Cockerham** [*Cocreham* DB, *Kokerham* 1190 CC] 'HĀM on the Cocker', and **Cockersand** [*Kokersand* 1212 Fees, *Cocressand* 1215 P] 'the sandy bank of the Cocker'. **Cocker Beck** Du. On it is **Cockerton** [*Cocertun* c 1050 HSC]. **Cocker Beck** Nt [*Cokerbec* 1235 Ch]. Cocker is a Brit river-name, derived from OBrit **kukro-* 'crooked', which corresponds to early Ir *cúar* 'crooked, perverse'. The immediate base is the feminine form **cucrā*, which became **cocrā*, **cocr*. The meaning 'crooked river' suits all the Cockers.

Cockerington Li [*Cocrintone* DB, *Cocringtuna* c 1115 LiS, *Cokeringtune* 1197 FF]. It is possible that the LUD, at least in its lower course, was once known as Cocker. If so, the name means 'the TŪN of the dwellers on the Cocker'. North Cockerington is on the Lud.

Cockermouth, Cockersand, Cockerton. See COCKER.

Cockey Moor La [*Cokkaye Moor* 1545 DL]. OE *cocc-hege* 'enclosure for wild birds'.

Cockfield Du [*Cok(k)efeld* 1291 Tax]. '*Cocca*'s FELD.' Cf. COCKBURY.

Cockfield Sf [*Cokefeld* c 950 BCS 1012, (æt) *Cohhanfeldæa, Cochanfelde* c 995 BCS 1288 f., *Cochefelde* c 1095 Bury, *Cockefeld* 1196 FF]. '*Cohha*'s FELD.' *Cohha* pers. n. is found also in *Cohhanleh* 804 BCS 313.

Cockhampstead Hrt [*Cochamstede* 1004 KCD 1300, c 1010 Wills, *Cochehamestede* DB]. 'Homestead on a hill.' First el. OE *cocc* 'hill' (see COCK BECK). The place is on a hill.

Cocking Sx [*Cochinges* DB, *Cokinges* 1189 P]. '*Cocca*'s people.' Cf. COCKBURY.

Cockington D [*Cochintone* DB, *Cokinton* 1199, *Kokinton* 1230 P]. 'The TŪN of (the people of) *Cocca*.' Cf. COCKBURY.

Cocklaw Nb [*Coklau* 1479 BBH], **Cockle Park** Nb [*Cockhill* 1314 Ipm]. 'Hill frequented by wild birds', OE *cocc-hlāw* and *cocc-hyll*.

Cockley Cley. See CLEY.

Cocknage St nr Trentham [*Cokenache* 1195 ff., *Cokenach* 1198 P]. Identical with COCKENHATCH.

Cockthorpe Nf [*Torp* DB, *Coketorp* 1254 Val]. Originally THORP. The addition *Cock* is probably the name of an early owner. C~ O. See COKETHORPE.

Codbro Wa [*Coddebarwe* 1320, *Codbarue* 1363 AD]. '*Codda*'s grove.' Second el. OE BEARU. *Codda* pers. n. is not found in independent use, but is presupposed by many pl. ns. See esp. COTHERIDGE. *Coda* is found in *Codanford* (see CODFORD) and *Codan mæd* 956 BCS 942.

Coddenham Sf [*Codenham* DB, *Codeham* DB, 1242 Fees, *Codeneham* Hy 2 BM]. '*Cod(d)a*'s HĀM.' Cf. CODBRO.

Coddington Chs [*Cotintone* DB, *Cotintuna* c 1100, *Codinton* 12 Chester], C~ He [*Cotingtune* DB, *Cotinton* 1277 Cl], C~ Nt [*Cotintone* DB, *-tona* c 1175 Middleton]. 'The TŪN of *Cotta*'s people.'

Coddington Db [*Codintone* 1219 FF, *Codington* 1246 Darley]. '*Cod(d)a*'s TŪN.' Cf. CODBRO.

Codford St. Mary & St. Peter W [*Codanford* 901 BCS 595, *Codeford Sancti Petri, Sancte Marie* 1291 Tax]. '*Coda*'s ford.' See CODBRO.

Codham Ess [*Codanham* c 1000 CCC, *Coddeham* 1198 FF]. '*Coda*'s HĀM.' Cf. CODBRO.

Codicote Hrt [*æt Cuðeringcoton*, *Cuðingcoton* 1002 KCD 1297, *Codicote* DB, *Cudicote* 1198 (1301) Ch, 1272 Ch]. 'The COTS of *Cūþhere*'s people.' The original name was *Cūþhering(a)cotu*, but *Cūþhere* was also called by the short name *Cuda*, and alternatively the place was known as *Cuding(a)cotu*, whence Codicote.

Codnor Db [*Cotenovre* DB, *Codenoura* 1183 P, *Coddenovere* 1236 Fees, 1285 Derby]. '*Codda*'s ridge.' Cf. OFER and CODBRO.

Codrington Gl [*Cuderintuna* Hy 2 (1318) Ch, *Codrinton* 1287 QW]. 'The TŪN of *Cūþhere*'s people.'

Codsall St [*Codeshale* DB, 1271 Ass, *Coddeshal* 1167 P, 1248 Cl]. '*Cōd*'s HALH.' For OE *Cōd* see CUTSDEAN.

OE **cofa** is found in the senses 'cave, den; inner room'. In pl. ns. in the senses 'a recess with precipitous sides in the steep flank of a mountain' (esp. in the Lake district) and 'a small bay, a creek', senses found in dialects, are chiefly to be reckoned with. See COFTON, COVE, COVEN &c. But there was no doubt also a pers. n. *Cōfa*, related to *Coifi*,

which it is difficult sometimes to distinguish from *cofa*. See COBHAM Sr, COVENHAM &c.

Coffinswell D [*Willa, Welle* DB, *Welles* 1231 Cl, *Coffineswell* 1249 Ass]. Originally OE *Wiella* 'the stream'. *Willa* was held in 1185 by Hugh Coffin (Buckland). Coffin is a nickname and family name identical with OFr *coffin* 'basket, coffin'.

Cofton D [*Coctone* 1282 Ep, *Cofton* 1289 Cl]. The neighbouring **Cofford** appears as *Coccford* 1044 OSFacs. *Cocc-* is probably the word for 'a heap' &c. mentioned under COCK BECK, and refers to the hill at these places.

Cofton Hackett Wo [*æt Coftune* 780, *Coftun* 849 &c. BCS 234, 455 &c., *Costone* DB]. 'TŪN in a *cofa* or recess in a hill.' William Haket held C~ in 1166 (RBE). Cf. BEER HACKETT.

Cogdean Do [*Cocdene* hd 1084 GeldR, *Cocdene* 1212 Fees (hd), 1265 Misc]. 'Valley frequented by wild birds.' See COCK BECK.

Cogenhoe (kooknō) Np [*Cugenho* DB, 1176 P, 1202 Ass, 1220 Fees, *Cugeho* 12 NS, 1236 Fees]. '*Cugga*'s HŌH or spur of land.' *Cugga* pers. n. is otherwise found only in *Cuggan hyll* 974 BCS 1298. Cf. the related *Cycga* in *Cycgan stán* 969 BCS 1230.

Cogges O [*Coges* DB, 1103 Fr, 1200 Cur, *Cogas* 1166 Fridesw]. The place is at **Cogges Hill**. The name represents the plur. of an OE *cogg*, identical with ME and Mod *cog* 'cog of a wheel'. The meaning is here 'a hill'.

Coggeshall (kŏksl) Ess [*Kockeshale* c 1060 Wills, *Coghessala, Cogheshala* DB, *Coggeshal* 1168, 1191 P, *Parva Coggeshale* 1202 FF], **Cogshall** Chs [*Cocheshalle* DB, *Kogeshult* 1287, *Cogishull* 1289 Court]. '*Cogg*'s HALH and HYLL.' The related *Cogga* is found in *Coggan beam* 967 BCS 1200 (Ha), *Cocggan hyll* 931 ib. 670 (So). Both names are no doubt derived from OE *cogg* (see COGGES).

Coker, East, North & West, So [*Cocre* DB, 1195 P, *Est-, Nortkoker* 1243 Ass, *Westcocre* 1227 FF]. Really the name of the stream at Coker [*Coker water* c 1540 Leland]. The name is identical with COKER.

Cokethorpe O [*Coctorp* 1213 BM, *Cokthorp* 1254 Val, *Cocthrop* 1279 RH]. The early spellings suggest an OE *Cocc-þorp* 'thorp where cocks were reared'. Such a name in *thorp* is unusual. As C~ is not far from COGGES, the name might be 'thorp belonging to Cogges'. Before *th-* a *g* would be apt to become *c* (*k*).

OE **col** 'coal', generally no doubt 'charcoal', is common as the first el. of pl. ns., but is difficult to distinguish from *cōl* adj. 'cool' and from *Cola* pers. n. This is especially the case with original stream-names. Very likely a stream-name *Cole*, identical with

Kola in Norway, was formed from *col*, the meaning being 'black river'.

Colan Co [*Sanctus Colanus* 1205 Cur, (de) *Sancto Culano* 1262 FF, *Ecclesia Sancti Coelani* 1276 Ep]. '(The church of) St. Colan.' The forms point to the saint's name having been *Coelan* (cf. Welsh *coel* 'belief').

Colaton Raleigh D [*Colatun* c 1100 E, *Coletone* DB, *Coleton Ralegh* 1316 FA]. '*Cola*'s TŪN.' The manor was held by Wimundus de Ralegh in 1242 (Fees). Cf. RALEIGH D. The same is the origin of **Collaton St. Mary** D [*Coletone* 1261 Ep], C~ D in Malborough [*Coletona* DB], C~ D in Halwell [*Kolethon* 1242 Fees], **Colleton Barton** D [*Coleton* 1242 Fees]. Collaton in Malborough was held by *Cole* in the time of Edward the Confessor. *Cola* is a common pers. n. in late OE.

Colburn YN [*Corburne* DB, *Coleburn* 1198 Cur, *-brun* 1226 FF]. Perhaps 'cool stream'; cf. COL.

Colbury Ha [*Colebury* 1291 Tax, 1316 FA]. '*Cola*'s BURG.'

Colby Nf [*Colebei* DB, *Colebi* 1191 P], C~ We [*Coleby* c 1150, *Colleby* c 1170 WR, *Colebi* 1197 P]. '*Koli*'s BY.' ODan, OSw, ON *Koli* is a well-evidenced pers. n.

Colchester Ess [*Cair Colun* c 800 HB, *Colneceaster* 921 ASC, *Colenceaster* 931 BCS 674, *Colecestra* DB]. 'The Roman station on R COLNE.' The OBrit name was *Camulodunon*.

Coldcoats La [*Caldekotes* 1246 Ass, *Coldecotes* 1332 Subs], C~ Nb [*Caldecotes* 1242 Fees]. 'Cold COTS.' Cf. CALDECOTE &c.

Coldham Ca [*Koldam* 13 Fees, *Coldham* 1300 Ch]. 'Cold HĀM.'

Coldharbour. *Cold harbour* was formerly a common name for a place of shelter from the weather for wayfarers, constructed by the wayside. *Harbour* is here used in its old sense of 'shelter, lodging' (ME *hereberwe* &c., ON *herbergi* &c.). One in London is referred to as *Coldherberghe* 1317 Street-names 150.

Coldmeece. See MEECE.

Coldred K [*Colredinga gemercan* 944 BCS 797, *Colret* DB, *Colredan, -raedene, -red* 11 DM, *Colred* 1204 Cur]. First el. OE *col* 'coal'. The second may be an OE **rēod*, cognate with OHG *riuti* and meaning 'a clearing' (cf. ROTHEND, RIDDINGS &c.) or it may be OE *ryden* 'clearing' (cf. PNSr 364, PNEss 588).

Coldrey Ha [*Coleriche* c 1286 Ep, *Colrithe* 1323 BM]. Really the name of a stream, recorded as (to) *Colriðe* 909 BCS 627. 'Coal (black) brook.'

Coldridge D [*Colrige* DB, 1196 FF, *Colleruge* 1185 P]. 'Ridge where charcoal was made.'

Coldwaltham. See WALTHAM.

Coldwell Nb nr Kirkwhelpington [*Colewell* 1277 Ch]. Probably 'cool stream or spring'. **Coldwell** Nb nr Bavington [*Caldewell* 1325 Ipm], C~ Nb nr Stannington [*Caldewell* 1242 Fees, *Caldwell* 1346 FA]. 'Cold spring or stream.'

Cole R Brk, W. A back-formation from COLESHILL. Cf. LEINTHALL.

Cole R Wa [(in, on) *Colle* 849, 972 BCS 455, 1282]. **Colton Beck** La was formerly **Cole** [*Cole* 1247, *Colle* 1257 FC]. A Brit river-name identical with COOLE in France [*Cosla* 896, *Cola* 1239] and derived from the old Celtic word for 'hazel' (Welsh *coll* 'hazels'). Cf. COLESHILL.

Cole So [*Colna* 1212 Fees, *Colne* 1219 FF, *Kolle* 1285 FA]. Identical with COLNE R Ess.

Cole W [*Cusfalde* 1065 KCD 817, *Coufaud* 1283 Misc]. 'Cow-fold.'

Colebatch Sa [*Colebech* 1176 f. P, *-bache* 13 BM]. 'The valley of R Cole'? The place is on a stream, which may have been called *Cole*, from *col* 'coal' or *cōl* 'cool'. See BÆCE.

Colebrook D in Cullompton [*Colebroca* DB, 1176 P], C~ D in Plympton [*Colbroc* Hy 2 Ol, *-brok* 1328 Ch], **Colebrooke** D [*Colebroc* Hy 2 HMC iv, *-brok* 1241 FF]. Perhaps 'cool brook'.

Coleby Li in W. Halton [*Colebi* DB, 1202 Ass], C~ Li nr Lincoln [*Colebi* DB, 1212 Fees]. Identical with COLBY.

Coleford D [*Colbrukeforde* 1330 Ep]. The place is near COLEBROOKE. The present name is due to shortening.

Coleford Gl [*Colford* 1534 VE], C~ So nr Mells [*Culeford* 1234 FF, *Colford* 1291 Tax], C~ So nr Elworthy [*Colforde, Coleford* DB]. 'Charcoal ford', i.e. a ford over which charcoal was carried and where charcoal was therefore found. The isolated *Culeford* may be miswritten. But cf. COLERNE.

Colemere Sa [*Colesmere* DB, *Culemere* 1203 PNSa, *Colemere* 1274 Misc]. The place is on **Cole Mere**. The name may mean '*Cūla*'s mere'. Cf. COWLINGE, CULFORD &c.

Colerne (-ŭ-) W [*Colerne* DB, *Culerna* 1179, 1190, *Culerne* 1198 P, *Cullerne, Collern* 1270 Ipm]. The modern spelling suggests OE *colærn* 'house where charcoal was stored'. There would seem to have been an OE *cul* by the side of *col*. Cf. OSw, Dan *kul*.

Colesborne Gl [æt *Collesburnan, Colesburna* c 802 BCS 304, *Colesburnan ford* c 800 ib. 299, *Colesburne* DB]. The place is on the upper CHURN, which in part must have been known as *Colesburna*. The name means '*Col*'s stream'. *Col* pers. n. is found in *Colesleye* BCS 586, *Colesleie* ib. 922. It corresponds to ON *Kolr* and is related to *Cola*.

Coleshill Bk [*Coleshulle* 1279 PNBk], C~

Brk [*Colleshyll* c 950 Wills, *Coleselle, Coleshalle* DB, *Coleshull* 1220 Fees], **C~ Wa** [*Colles hyl* 799 BCS 295, *Coleshelle* DB, *-hell* 1162 P, *Colleshull* 1291 Tax]. C~ Bk, Brk seem to contain an unrecorded OE *coll* 'hill' corresponding to ON *kollr* 'head, top, hill', MLG *kol, kolle* 'head'. Note especially *Collhill* 817 BCS 361 (Wo). The hills were originally called *Coll*, and an explanatory *hyll* was added. C~ Wa is more difficult. The place is on the river COLE (OE *Coll*). But the gen. of the river-name ought not to appear as *Colles*, and possibly *Colles-* is here the gen. of a Brit pl. n. *Coll*, derived from Welsh *coll* 'hazels'. Of course, C~ Wa may be identical with the other Coleshills, and the river-name Cole might be a very early back-formation.

Colham Mx [*Colanhomm* 831 BM, *Coleham* DB, 1208 Cur, *Colnham* 1212 RBE]. '*Cola*'s HAMM.' C~ is near the COLNE, but cannot contain that river-name. *Cola* is found only in late OE texts, but it may well be old. Indeed it seems it must be assumed for several old pl. ns., as COLLINGHAM &c. Cf. also *Colungahrycg* 1015 Wills.

Colkirk Nf [*Colechirca, -kirka* DB, *-chirche* 1161, *-kerca* 1168 P]. '*Cola*'s or *Koli*'s church.' The first el. may be OE *Cola* or OScand *Koli* (cf. COLBY). The second el. varies somewhat in early records between OE *cirice* and OScand *kirkia*.

Collaton, Colleton. See COLATON.

Collingbourne Ducis & Kingston W [*Colengaburnam* 903 BCS 602, *at Colingburne*, (on) *Collengaburnan* 921 ib. 635, *Colinge-, Coleburne* DB, *Collingeburn* 1199 FF]. Really an old name of the upper BOURNE, whose lower course was in OE times called Winterbourne [*Winterburnan* 972 BCS 1286]. Possibly a still earlier name of the river was *Coll*, identical with COLE. If so, the name means 'the stream of the dwellers on the *Coll*'.

C~ Ducis was held by the Earls (later Dukes) of Lancaster.

Collingham Nt [*Colingeham* DB, 1194 P, *Colingham* c 1170 Middleton, *North-, Suthcolingham* 1291 Tax], **C~ YW** [*Collingeham* 1167, 1180 P, *Colingham* 1173 YCh 197, *Colingeham* 1180, 1191 ff. P], **Collington** He [*Col(l)intune* DB, *Colintun* 1242, 1249 Fees]. 'The HĀM and TŪN of *Cola*'s people.' Cf. COLHAM.

Collingtree Np [*Colestrev* (hd), *Colentrev* DB, *Colintrie* 1163 P, *-tr*[*e*] 1208 Cur, *Coluntre* 1251 Ipm]. '*Cola*'s tree.' Names in *-tree* generally have a pers. n. as first el. The same name is *Kolan treow* c 1000, *Colan treow* 1045 KCD 712, 780 (nr Hinton Ampner Ha).

Collow Li [*Caldecote* DB, *Caldecota* c 1115 LiS]. Identical with CALDECOTE &c.

Collyhurst La [*Colyhurst* 1322 LaInq]. 'Hill grimy with coal dust.' First el. *colly* adj.

Colmore Ha nr Alton [*Colemere* DB, 1180 P, *-mera* 1174 P, 1201 (1313) Ch, *Culemere* 1196 P]. 'Cool lake'; cf. COL. This etymology is suggested by the isolated early form with *u*, which may be from *ō*. There is no lake here now.

Colmworth Bd [*Colmeworde, Culmeuuorde* DB, *Colmwurda* 1167 P, *Colneworth* 1202 Ass]. It may be suggested that the first el. is identical with the river-name CULM. But the stream at the place is not winding. Or *Colm-* may represent an OE *Culhæma-*, the supposition being that there was once a place called *Culham* or the like near Colmworth. The meaning in the latter case would be 'the WORÞ of the Culham people'.

Coln R Gl [*Cunuglae* 721–43 BCS 166, *Colne* 1248 Ass]. On the Coln are **Coln Rogers, St. Aldwyn & St. Dennis** [(bi) *Cunelgan* 855 BCS 487, *Cunelgnan* 899 ib. 580, *Cungle* 962 ib. 1091, *Culne* DB, *Culna Rogeri* 1100, *Culna Sancti Elwyni* 1072, *Culna Sancti Aylwini* 1100 Glouc, *Colne Sancti Dionisij* 1291 Tax]. The river-name is unexplained. C~ **Rogers** was given to Gloucester Abbey by Roger de Gloucester (d. 1106).—C~ **St. Aldwyn** is said to have been named from a hermit (St. Ealdwine).—C~ **St. Dennis** belonged to the church of St. Denis at Paris in 1086 (DB).

Colnbrook Bk [*Colebroc* 1107 Abingd, 1190 P]. C~ is on an arm of the Colne, called *Colebrok* 1222 St Paul. The first el. of the name is very likely the river-name COLNE.

Colne (kōn) **R Ess** [*Colne* 1362 Pat]. On the river are **Colne Engaine, Earls, Wakes & White C~** [(at) *Colne* c 950 BCS 1012, *Colne* c 995 ib. 1289, *Colun* DB, *Colum* 1199, 1203 Cur, *Culn Quincy, de Ver, Vital'* 1238 Subs, *Colne Miblanc* 1225 Pat, *Colum Alba, Comitis, Engayn* 1254 Val, *Colne Wake* 1375 Cl]. The river-name is identical with COLNE Hrt and with CLUN. It is of British origin and had the form *Colun* originally; cf. CLUN and early forms of *Colchester*. The etymology is obscure.

C~ **Engaine** was acquired by Vital Engaigne in 1219 (FF). Engaine is a French byname, identical with or related to OFr *engaigne* 'ingenuity'.—**Earls C~** took its name from the family of de Vere, earls of Oxford. It belonged to Alberic de Vere in 1086 (DB).—Baldwin Wake (d. 1282) got land in Colne by marriage with a Quency. *Wake* is an English name.—**White C~** was held by Dimidius Blancus in 1086 (DB). *White* is due to wrong translation.

Colne (kōn) **R Hrt, Mx, Bk** [*Colenéa* 785 BCS 245, (be) *Colne* 894 ASC]. Identical in origin with prec. n. On this Colne is **Colney** Hrt [*Colneya* 1243 Ep, *Colneye* 1268 AD].

Colne Hu [*Colne* c 1050 KCD 907, DB]. Very likely named from a river, whose name is identical with COLNE Ess.

Colne (kōn) **La** [*Calna* 1124 Pont, *Caune* 1251 Ch, *Colne* 1296 Lacy]. Named from **Colne Water R** [*Coune* 1292 Ass]. **Colne R YW** [*Calne* c 1180, c 1200 YCh 1692,

1701, *Colne* 1344 Cor]. Colne is a Brit river-name, identical with CALNE W. It has been suggested that it belongs to the root of Lat *calare* 'to call', Welsh *ceiliog* 'cock' &c., and means 'roaring river'.

Colney Hrt. See COLNE. **Colney** (-ōn-) Nf [*Coleneia* DB, *Colneia* 1175 P, 1197 FF]. '*Cola*'s island.' The place is on the Yare.

Colsterdale YN [*Colserdale* 1301 Subs, *Costerdale* 1330 Ch], **Colsterworth** Li [*Colsteuorde* DB, *-worth* 1231 Ep, *Colstowurða* 1169 P, *Colsterworth* 1291 Tax, *Colstreworth* 1316 FA]. 'The valley and WORÞ of the charcoal-burners.' The first el. is an unrecorded OE *colestre* 'charcoal-burner', formed with the suffix *-estre* from *col* 'coal'.

Colston Basset Nt [*Coleston* DB, *-tun* 1160 BM, *Coleston Bassett* 1228 Ep], **Carcolston** Nt [*Colestone* DB, *Colistun* 1236 Fees, *Kyrcoluiston, Kerkolviston* 1242 Fees, *Kercolston* 13 BM]. '*Kol*'s TŪN.' First el. probably ON *Kolr* pers. n.

C~ Basset was held by Radulfus Basset c 1150 (Eynsham). See BERWICK BASSETT W.—The prefix *Car-* looks like *kirk* from OScand *kirkia*.

OE *colt* 'colt'. See COLTON St, COULTON.

Colthrop Brk [*Colsthorpe* Hy 1 (1317) Ch, *Colethrop* 1220, 1242 Fees]. '*Cola*'s thorp.'

Coltishall Nf [*Cokeres-, Coketeshala* DB, *Couteshal* 1200, 1207 Cur, *-hale* 1219 Misc, 1254 Val]. '*Cohhede*'s or *Coccede*'s HALH.' *Cohhede, *Coccede* are formations from *Cohha, Cocca,* analogous to *Luhhede, Lullede* from *Luhha, Lulla*.

Colton La [*Coleton* 1202 FF, *Colton* 1332 Subs]. 'TŪN on R COLE.'

Colton Nf [*Coletuna* DB, *-ton* 1199 FF], C~ YW nr Tadcaster [*Coletune* DB, *-tun* 1232 BM], C~ YW nr Whitkirk [*Cole-, Colletun* DB, *Choletuna* c 1160 YCh 1770]. '*Cola*'s or *Koli*'s TŪN.' Cf. COLBY.

Colton So [*Couleton, Chuleton* 1249 Misc]. '*Cūla*'s TŪN.' Cf. CULFORD &c.

Colton St [*Coltone* DB, *-ton* 1176 P, *-tun* 1227 Ass]. Perhaps OE *Colt-tūn* 'TŪN where colts were reared'. But '*Cola*'s TŪN' is also possible.

Colveston Nf [*Couestuna* DB, *Colveston* 1248 Ch, 1254 Val, *Colviston* 1316 FA]. Possibly '*Kolf*'s TŪN'. There may be an ON byname *Kolfr*. Or the first el. may be an unrecorded OScand pers. n. *Kolfastr*; cf. *Kolbeinn* &c.

Colwall He [*Colewelle* DB, *Colowella* 12 BM]. Perhaps 'cool stream or spring'. Cf. COL.

Colway Do [*Coleweye* 1242 Fees, *Calwehegh* 1346, 1428, *Calewey* 1431 FA, *Coleweheys* 1335 Ch]. Second el. OE *hege* or *gehæg* 'enclosure'. First el. very likely a hill-name *Calwe* or *Calwa* from *calu* 'bare'.

Cf. CALU. But the early *o*-forms offer difficulties.

Colwell D [*Colewille* DB, *-will* 1242 Fees]. The place is on a tributary of the COLY. The first el. may be the river-name.

Colwell Nb nr Chollerton [*Colewel* 1236, *-well* 1242 Fees, *Colwell* 1318 Ipm]. Either 'cool stream' or identical with *Coluullan broc* 958 BCS 1036 (O), which seems to have as first el. OE *col* 'coal'.

Colwich (kŏlij) St [*Colewich* 1240 Cl, *Colewyz, Colwich* 1247 Ass], **Colwick** (kŏlĭk) Nt [*Colewic, Colvi* DB, *Colewich* 1175 P]. Probably OE *col-wīc* 'WĪC where charcoal was got'.

Colwith Force We [*Colwith Bridge* 1712 Kendale]. Colwith Force is a waterfall, which was named from a forest. Colwith is ON *kol-viðr* 'forest where charcoal was burnt' or 'dark forest'. Force is ON *fors* 'waterfall, rapid'.

Colworth Bd [*Kaleworth* 1203 Ass, *Colingwurth* 1242 Fees]. Apparently originally OE *cāl-worþ* 'WORÞ where cole was grown'. *Coling-* from *Cālingas*, an elliptical form meaning 'the Colworth people'.

Colworth Sx [*æt Coleworð* 988 KCD 663, *Culewurth* 1230 P]. '*Cūla*'s WORÞ.' Cf. CULHAM.

Coly R D [*Cullig* 1005 KCD 1301], **Colyford** D [*Culiford* 1244 FF], **Colyton** D [*Culintona* 940–6 Laws, *Culi-, Colitone* DB]. Coly is a Brit river-name derived from Welsh *cul* 'narrow'. Colyford, Colyton are on the Coly.

Comb, Combe, Coombe (-ōō-) are common names. All three are here given together. Only some of the names are included. **North & South Coombe** D [*Coma* DB, *Cumb, Succumbe* 1206 Cur], **Combe Martin** D [*Comba* DB, *Cumbe Martini* 1265 Ch], **Combpyne** D [*Coma* DB, *Cumb* 1238 FF, *Combpyn* 1377 Subs], **Combe Raleigh** D [*Cumba* 1237 Cl], **Combeinteignhead** D [*Comba, Cumbe* DB, *Combe in Tenhide* 1227 Cl], **Combe Almer** Do [*Cumbe* 1244 FF], **Coombe Keynes** Do [*Cume* DB, *Cumb* 1212 Fees, *Cumbe Chaynes* 1284 Ipm], **Combe** Gl nr Chipping Campden [*Cumba* 1138 (1266) Ch], **Combe** Gl nr Wotton [*Combe* 1150–60 Berk], **Coombe** Ha [*Cumbe* DB, *Cumb* 1242 Fees], **Combe** O [*Cumbe* DB, *Cumba* 1156, 1194 P], **Combe** So in Huish Episcopi [*Cuma* 1065 Wells], **Abbas Combe** So [*Cumbe* DB, *Coumbe Abbatisse* 1327 Subs], **English Combe** So [*Engliscome, Ingeliscuma* DB, *Inglescumbe* 1227 FF], **Combe Florey** So [*Cumba* 1155–8 (1334) Ch, *Cumbeflori* 1291 Tax], **Combe Hay** So [*Come* DB, *Cumb(e) of Thomas de Ha(i)weie* 1225 Ass, *Cumbehawya* 1249 FF], **Monkton Combe** So [*Cume* DB, *Cumba* 1136 Bath], **Combe St. Nicholas** So [*Cume* 1070 Wells, *Cumbe* DB], **Combe Sydenham** So [*Come*

DB, *Cumbe* 1280 Dunster], **Temple Combe** So [*Come* DB, *Cumbe Templer* 1291 Tax, *Templecombe* 1387 Buckland], **Coombe** Sr [*Cumbe* DB, *Cumba* 1165 P], **Coombe Bissett** W [*Cumbe* DB, *Combebysset* 1385 Ipm], **Castle Combe** W [*Come* DB, *Castelcombe* 1315 Ch], **Combe** Wa [*Cumba* 1162 P, *Sancta Maria de Cumba* 1251 Ch]. Co(o)mbe is OE *cumb* 'a narrow valley'.

Abbas C~ So was held by the Abbess of Shaftesbury in 1086 (DB).—**C~ Almer** Do; see ALMER.—**C~ Bissett** W was held by Maneser Biset before 1186 (P). Bisset is OFr *biset* 'dark'.—**Castle C~** W derives its byname from a Norman castle.—**English C~** So seems to have had as first el. a pers. n. *Ingel*.—**C~ Florey** So was held by Hugh de Flury c 1155 (1334 Ch). Florey from FLEURY in France (several).—**C~ Hay** So was held by Thomas de Ha(i)weie in 1225 (Ass).—**C~ Keynes** Do belonged to William de Cahaignes in 1199 (P). Cf. ASHTON KEYNES.—**C~ Martin** D in 1265 was held by Nicholas son of Martin (Ch).—**Monkton C~** So belonged to the Bishop of Bath.—**Combpyne** D has its name from the Pyn family, who got it in the 13th cent. The name may be from LE PIN in Calvados (Normandy).—**C~ Raleigh** D was held by Henry de Ralegh in 1292 (Ch); cf. RALEIGH D.—**C~ St. Nicholas** So presumably belonged to the Priory of St. Nicholas in Exeter.—**C~ Sydenham** So was held by Johannes Sydenham in 1447 (Dunster).—**Combeinteignhead** D; see TEIGN-(HEAD).—**Temple C~** So came to the Templars before 1185.

Comberbach Chs [*Comberbeche* R 1, *Comburbach* 1333 Ormerod], **Comberford** St [*Cumbreford* 1187 f. P, *Cumberford* 1218 FF], **Combermere** Chs [*Combemare* c 1150 BM, *Cumbramara* 1157, *-mare* 1182, *Cumremara* 1186 P], **Comber Mere** Chs, on which Combermere stands [(mere of) *Cumbermare* 1286 Ch], **Comberton** Ca [*Cumbertone* DB, *Cumbertuna* 1156, *Cumbreton* 1190 P], **Comberton** Wo [*Cumbrincgtun* 972 BCS 1282, *Cumbrintune* DB]. The second el. of Comberbach is OE BÆCE. The first el. of the names is OE *Cumbra*, a well-evidenced pers. n., or in some cases possibly *Cumbra* gen. 'of Cumbrians', as in CUMBERLAND. Comberton Wo is 'the TŪN of *Cumbra*'s people'. *Cumbra* pers. n. is the first el. of **Cumberwood** Gl (in Chaceley) [*Cumbranweorþ* 972 BCS 1282].

Combpyne D. See COMB.

Combridge St [*Kanbrugge* 1246 Ch, *Combrugge* 1258 FF]. OE *camb-hrycg* 'ridge with a crest'. Cf. CAM FELL, COMBS. If the second el. is OE *brycg* 'bridge', the first is obscure.

Combrook Wa [*Cumbroc* 1233 Cl, *Cumbrok* 1316 FA]. 'Brook in a valley.' See CUMB.

Combs Db [*Comb* 1169 Pp, *Cambes* 1374 Gaunt Reg], **C~** Sf [*Cambas* DB, *Cambes* 1130 P, 1205 FF, *Combes* 1212 Fees, *Caumbes* 1230 P]. The plur. of OE CAMB 'comb, crest of a hill' &c. C~ Db is on the slope of **Black Edge** (a hill). Early forms such as *Coumbes* are due to the WMidl

change of *a* to *o* before nasals. The lengthened OE *o* became ME close *ō* and *ū*. C~ Sf seems to have been named from the spurs of hill at the place.

Combwell K [(to) *Cumwyllan* KCD 1363, *Cumbwell* c 1160 Arch Cant v]. 'Stream or spring in a valley.' Cf. CUMB.

Combwich (kŭmïj) So [*Comich, Commiz* DB, *Cumwiz* 1178 P]. 'wīc in a valley.' Cf. CUMB.

Comhampton Wo [*Cumbehampton* 1275 Ass]. First el. OE CUMB. Cf. HĀMTŪN.

Commondale YN [*Colemandale* 1272 Ipm]. '*Colman*'s valley.' *Colman* is OIr *Colmán*, from *Columbán*.

Compton (-ŭ- or -ŏ-), a common name, is **1.** usually OE *Cumb-tūn* 'TŪN in a CUMB or narrow valley': **Compton** Brk nr Streatley [*Comtun* Hy 2 BM, *Est-*, *Westcumpton* 1242 Fees], **C~ Beauchamp** Brk [*æt Cumtune* 955 BCS 908, *Contone* DB, *Cumton Beucamp* 1236 Fees, *Cumpton near le Witehors* 1273 Ipm], **C~ D** [*Contone* DB, *Cumpton* 1234 Fees], **C~ Abbas** Do [*Cumtun* c 871 BCS 531, *Contone* DB, *Cumpton Abbatisse* 1293 FF], **C~ Abbas West** Do [*Compton* 939 BCS 738, *Contone* DB, *Cumpton Abbatis* 1291 Tax], **Over & Nether C~** Do [*Cumtun* 946-51 BCS 894, *Cumbtun* 998 KCD 701, *Contone* DB], **C~ Valence** Do [*Contone* DB, *Cumpton Pundelarche* 1265 Misc, *Compton Valence* 1324 Ipm], **C~ Gl** nr Newent [*Cumpton* 1220 Fees], **C~ Abdale** Gl [*Cuntune* DB, *Cumpton* 1283 Ch], **C~ Greenfield & East C~** Gl [*Cumtún* 962 BCS 1089, *Contone* DB, *Compton Greneville* 1303 FA], **C~ Ha** [*Cuntune* DB, *Cumton* c 1195 BM], **East & West C~** So [*Coumpton* 1327 Subs], **C~ Bishop** So [*Cumbtune* 1067 Wells, (into) *Cumbtune* 1068 E, *Compton Episcopi* 1332 Ep], **C~ Dando** So [*Contone* DB, *Cumton* 1225 Ass, *C~ Daunon* 1256 Ass], **C~ Dundon** So [*Contone* DB, *Cumpton by Dunden* 1289 Ch], **C~ Durville** So [*Contone* DB, *Cumton Durevil* 1255 Ipm], **C~ Martin** So [*Comtone* DB, *Cumpton Martin* 1226-8 Fees], **C~ Pauncefoot** So [*Cuntone* DB, *Cumpton Paunceuot* 1291 Tax], **C~** Sr [*Comptone* 675 BCS 39, *Contone* DB], **C~ St** nr Tettenhall [*Contone* DB, *Cumpton* 1227 Ass], **C~ Sx** [*Cumtun* 1015 Wills, *Contone* DB], **C~ W** nr Enford [*Contone* DB, *Cumpton* 1242 Fees], **C~ Bassett** W [*Contone* DB, *Cumpton Basset* 1271 Ipm], **C~ Chamberlayne** W [*Contone* DB, *Compton Chamberleyne* 1316 FA], **Fenny C~** Wa [*Contone* DB, *Fennicumpton* 1242 Fees], **Long & Little C~** Wa [(in) *Litlan-Cumtune* 1005 KCD 714, *Cuntone, Contone parva* DB, *Long Compton* 1299 Ch], **C~ Scorpion** Wa [*Compton Scorfen* 1279 PNWa, *C~ Scorefen* 1316 FA], **C~ Verney** Wa [*Contone* DB, *Compton Murdak* 1323 AD], **C~ Wynyates** Wa [*Contone* DB, *Cumpton Wintace* 1242 Fees, *C~ Windgate* 1268 Ipm], **C~ Wt** [*Cantune* DB, *Coumpton* 1287-90 Fees].

2. Compton Db [*Campedene* J BM]. 'The valley of the fields.' Cf. CAMP, DENU.

C~ **Abbas** Do means 'the Abbess's Compton'. The place belonged to Shaftesbury Abbey from c 871.—C~ **Abbas West** Do means 'the Abbot's Compton'. It belonged to Milton Abbey. —C~ **Abdale** Gl is unexplained. Abdale looks like a pl. n.—C~ **Bassett** W was held by Fuke Basset in 1242 (Fees). Cf. BERWICK BASSETT.— C~ **Beauchamp** Brk was held by Walter de Bello Campo in 1220 (Fees). Cf. ACTON BEAUCHAMP.—C~ **Bishop** So belonged to the Bishop of Wells.—C~ **Chamberlayne** W was held by Galfridus Camerarius in 1234 (Fees).—C~ **Dando** So from the Dando or *de Alno* family. Alexander de Alno was in possession t. Hy 2. The name is written (de) *Alno, Auno, Alneto, Dauno* &c. The source is AUNOU in Normandy [*Alnetum* 12 Fr].—C~ **Dundon** So is nr DUNDON.—C~ **Durville** So was held by Eustachius de Dureuill in 1230 (P). Durville perhaps from DIERVILLE in Pas de Calais.—C~ **Greenfield** Gl was held by Richard de Greinvill in 1228 (Cl). The name is from one of the GRAINVILLES in France.—C~ **Martin** So seems to have been named from Martin de Tours, whose son Robert succeeded to the manor in the time of Hy 1.— C~ **Pauncefoot** So. Cf. BENTLEY PAUNCEFOTE. —C~ **Scorpion** Wa is due to popular etymology. The original form was *Scorfen*, possibly 'fen in a ravine', if the word *score* may be assumed for the Warwick dialect.—C~ **Valence** Do came to the Valence family t. Hy 3. Valence is a common pl. n. in France.—C~ **Verney** Wa was held by Richard Verney c 1450.—C~ **Wynyates** Wa was named from a pass. Cf. WINGATE.

Conder R La [*Kondover* c 1200 CC, *Candovere* 1246 Ass]. Identical with CAMDDWR in Wales, i.e. a compound of Welsh *cam* 'crooked' and *dwfr* 'stream'. Original *m* became *n* before *d*. Cf. the identical **Candor** Co [*Camdour* 1433].

Conderton Wo [*Cantuaretun* 875 BCS 541, *Canterton* 1201 Cur]. OE *Cantwaretūn* 'the TŪN of the Kentishmen'. Cf. CANTERBURY, CANTERTON. The name records a Kentish settlement.

Condicote Gl [*Condicote* DB, *Cundikote* 1128 Glouc, *-cota* 1193 P]. '*Cunda*'s COT.'

Condover (-ŭ-) Sa [*Conedoure* DB, *Cunedoura* 1130, *Cunedofre* 1169 P]. First el. the river-name COUND. The second may be Welsh *dwfr* 'stream' or OE *ōfer* 'bank'.

Coneysthorpe (-ŭ-) YN [*Coningestorp* DB, *Cuningestorp* 1167 P], **Coneythorpe** YW [*Conigthorp* 1275 Ep]. 'The king's thorp.' First el. OScand *konungr, kunungr*.

Coney Weston Sf [*Cunegestun* 11 EHR 43, *Cunegestuna* DB, *-tun* c 1095 Bury, *Cunewestone* 1254 Val], **Congerston** Le [*Cuningestone* DB, *Kinigston* 1209–35 Ep, *Cunigeston* 1247 Ass]. 'The king's TŪN.' Congerston, to judge by one ex., may well be a Scandinavianized form of OE *Cyningestūn*, and the same may be the case with Coney Weston. The change *-ing-* > *-ig-* > *-eg-* > *-ew-* is found in several names, as CANEWDON, MANUDEN, MONEWDEN.

Congham Nf [*Congre-, Conghe-, Concham* DB, *Cungheam* 1121 AC, *Congham* 1197 P, *Cangham* 1199 FF], **Congleton** Chs [*Cogeltone* DB, *Congelton* 1282 Ch, *Congilton* 1282 Court]. The first el. of the two names is obscure. It may be connected with the stem *kang-*, found in ON *kengr* (< *kangi-*) 'a bend', *kǫngull* 'a cluster (of grapes &c.)', but the meaning of the el. in the names cannot be determined. Congleton is at a bend of the Dane.

Congresbury So [(on) *Cungresbyri* c 894 Asser, *-byrig* c 1000 Saints, *Cungaresbyrig* 1065 Wells, *Cungresberie* DB]. 'Saint Congar's BURG.' According to Saints St. Congar was buried at the place. Congar is a Welsh name; cf. *Congur* LL, *Cyngar* Rees, *Congar* in Brittany. The local pronunciation is 'Coomsbury' (Hope). Cf. *Combebrey* 1612, *Coombesbury* 1758 Hill, PNSo, 20.

Congreve St [*Comegrave* DB, *Cumgrave* 1236 FF]. 'Grove in a valley.' Cf. CUMB, GRÆFE.

Conholt W [*Coueholt* 1242 Fees, *Covenholt* 1251 Cl]. 'Wood by or in a *cofa* or recess.' There are two deep valleys near the place.

Coningsby Li [*Cuningesbi* DB, *Coningesbi* c 1115 LiS]. OScand *Kunungsbȳr* 'the king's BY'.

Conington Ca [*Cunningtun* c 1000 BCS 1306, *Cunitone* DB], C~ (-ŭ-) Hu [*æt Cunictune* 957 BCS 1003, *Coninctune* DB]. 'The king's manor', probably Scandinavianized from OE *Cyningestūn*.

Conisbrough YW [*Cunugesburh* 1002 Wills, *Cuningesburg* DB, *-burch* 1121 AC]. 'The king's BURG', probably Scandinavianized from OE *Cyningesburg*.

Conisby Li [*Cunesbi* DB, *Cunigesbi* c 1115 LiS, *Nortkuningesby* 1219 Ep]. See CONINGSBY.

Coniscliffe (-kŭns-), **High,** Du [*Ciningesclif* 778 ASC (E), (in) *Cingcesclife* c 1050 HSC, *Cuniggesclive* 1202 FF]. 'The king's cliff.' The OE name has been Scandinavianized.

Conisford Nf in Norwich [*Cunegesford* 1165 P], **Conishead** (-ŭ-) La [*Cuningesheued* 1180–4 LaCh, 1235 FF], **Conisholme** Li [*Cuninggesolm* 1195 FF, *Cuningesholm* 1196 FF]. 'The king's ford, headland, and holm.' First el. OScand *kunungr*.

Coniston (-ŭ-), **Church,** La [*Coningeston* c 1160 LaCh, 1257 Ass], C~ **YE** [*Coningesbi* DB, *Cuningeston* 1190 YCh 1312], **Cold** C~ YW [*Cuningestone* DB, *Calde Cuningeston* 1202 FF], **Conistone** YW [*Cunestune* DB, *Conyston in Kettelwell* 1285 FA]. 'The king's manor.' Generally no doubt a Scandinavianized form of OE *Cyningestūn*.

Conksbury Db nr Bakewell [*Cranchesberie* DB, *Cankersburia* Hy 2 (1318) Ch, *Conkesburgh* 1339 DbAS xi]. OE *cranuces burg* 'BURG of the crane(s)'. Cf. CORNBROUGH, -BURY.

Conock W [*Cunet* 1212 RBE, *Kunek* 1242 Fees, *Coneke* 1316 FA]. A Brit hill-name identical with CANNOCK.

Cononley YW [*Cutnelai* DB, *Cunetlay* 1246 FF, *Conotlay*, *Cuniglay* 1254 Ipm, *Cuniglaye* 1273 Ep, *Conethelegh* 1277 Cl]. See LĒAH. The first el. seems to be identical with the river-name COUND. C~ is on a tributary of the Aire.

Consett Du [*Covekesheued* 1183 BoB, *Conekesheued* 1228 FPD]. Cf. CHESTER LE STREET. The first el. is a Brit hill-name **Cunuc*, identical with CANNOCK, CONOCK. The second is an explanatory OE *hēafod* 'head, hill'. Consett is on a prominent hill.

Constantine Co [*Sanctus Constantinus* DB]. '(The church of) St. Constantine.'

Cooden Sx [*Codiggis* 12 AD, *Codingg* 1230 P]. '**Cōda*'s people.' The same *Cōda* gave its name to *Codanclib* 772 BCS 208 (in the neighbourhood of Cooden).

Cookbury D [*Cukebyr'* 1242 Fees, *Cokebery* 1303 FA]. '*Cuca*'s BURG.' OE *Cuca* pers. n. may be found in *Cucan healas* 955–9 BCS 936. It is a short form of names in *Cwic-*.

Cookham Brk [*Coc(c)ham* 798, *Coccham* 965–71 BCS 291, 1174, *Cocheham* DB]. OE *cocc*, probably in the sense 'a hill', and HAMM. Cookham stands in a bend of the Thames at a ford called *Cocdun* 1220 Fees.
Cookham Dean took its surname from an early tenant. Osbert de la Dene held land in *Chocham* in 1220 (Fees).

Cook Hill Wo [*Cochilla* 1156 P, *Cochull* 1262 Ipm]. OE *cocc* 'hill' (see COCK BECK) with an explanatory OE HYLL.

Cookley Sf [*Cokelei* DB, *Kukeleia* Hy 2 (1268) Ch]. '*Cuca*'s LĒAH.' Cf. COOKBURY, CUCKNEY.

Cookley Wo [*Culnan clif* 964 BCS 1134, *Culla clife* 11 Heming, *Cuckele* 1281 Ct]. The first el. seems to be a pers. n. related to *Cūl(a)*. See COWLINGE.

Cookridge YW [*Cucheric* DB, *Cukeriz* 1192, 1198 Kirkst, *Cukeric* c 1190 YCh 1657]. The second el. seems to be OE *ric*, the word for a stream mentioned under CHATTERIS. The first may be the pers. n. *Cuca* (cf. COOKBURY) or rather a stream-name *Cuce*, derived from OE *cwicu*. Cf. CUCKMERE. A stream-name *Cuce* seems to occur 956 BCS 958 (*on Cucan*).

Cooksey Green Wo in Upton Warren [*Cochesei* DB, *Cokeseya* 1212 Fees], **Cooksland** St [*Cuchesland* DB]. '*Cucu*'s island and land.' **Cucu* seems to be a short form of names in *Cwic-*. *Cwic-* is from OE *cwicu* adj.

Coole Chs [*Couhull* c 1130 Mon, *Coule* 1316 Chamb]. OE *cū-hyll* 'cows' hill'.

Cooling K [*Culinga gemære* 778 BCS 227, *Culingas* 808 ib. 326, *Colinges* DB]. 'The people of *Cūl* or *Cūla*.' Cf. CULFORD &c.

Coombe. See COMB. **Coombes** Sx [*Cumbhæma gemæra* 956 BCS 961, *Cumbe* DB]. The plur. of OE CUMB 'valley'.

Cootham Sx [*Codeham* DB, *Coudham* 1296 Subs]. '*Cūda*'s HĀM or HAMM.'

Copdock Sf [*Coppedoc* 1195 P, *Coppedac*, *-oc* 1254 Val]. 'Copped oak', i.e. 'oak rising to a top'. OE *coppede* (from *copp* 'top') means 'provided with a top'.

Copeland Cu [*Couplanda* c 1125, *Caupuland*, *Caupalandia* 12 StB, *Coupland* 1228 Ch], **Coupland** Nb [*Coupland* 1242 Fees]. ON *kaupland* 'bought land'.

Copford Ess [*Coppanford* a 995 Wills, *Copeforda* DB], **Copgrove** YW [*Copegrave* DB, *Coppegraua* 1166 P, 1195 Cur, *-grave* 1220 FF]. '*Coppa*'s ford and grove.' *Coppa* is not evidenced in independent use.

Cople Bd [*Cochepol* DB, *Cogopol* c 1150 BM, *Coggepole* 1196 FF, *Coupol* 1254 Val]. '*Cocca*'s pool' or 'cocks' pool'. Cf. COCKBURY.

Copmanthorpe YW [*Copemantorp* DB, *Coupmanetorp* c 1200 YCh 554]. OScand *Kaupmanna-þorp* 'the thorp of the chapmen'.

Copnor Ha [*Copenore* DB, *Copponore* 12 BM]. '*Coppa*'s landing-place.' Cf. COPFORD and ŌRA.

OE **copp** 'top, summit' is the source of **Copp** La, the name of a hill. See COPPULL, COPSTON, PICKUP, SIDCUP, WARCOP. From *copp* is derived OE *coppede* 'provided with a top', found in COPDOCK, COPT HEWICK. **Copped Hall** Ess [*La Coppedehall* 1272 Ch] means 'hall with a high roof'.

Coppenhall, Church & Monks, Chs [*Copehale* DB, *Copenhale* 13 BM, *Chirchecopenhal* 1288 Court, *Munkescopenhale* 1295 Cl], **Coppenhall** St [*Copehale* DB, *Coppenhale* 1222 Ass, *Coppenhal* 1243 Cl]. '*Coppa*'s HALH.' See COPFORD.

Coppingford Hu [*Copemaneforde* DB]. 'The chapmen's ford.' First el. OScand *kaupmaðr*.

Copplestone D [(on) *Copelan stan* 974 BCS 1303]. 'The logan-stone, the rocking stone.' The same name seems to be *Copilleston* 1286 QW (nr Worlingham Sf).

Coppull La [*Cophill* 1218 Ass, *Cophull* 1243 LaInq]. 'Peaked hill.' Cf. COPP.

Copston Magna Wa [*Copstuna* Hy 1 (1251) Ch, *Copston* 1290 Ch]. The place is near a round hill. Hence perhaps 'TŪN by the COPP'. But the Anglo-Scand pers. n. *Copsi* (DB &c.), suggested in PNWa(S), is perhaps a more probable first el.

Coquet (kōkĭt) R Nb [*Cocwud(a)* c 1050 HSC, *Coqued* 1104–8 SD, *Coket* 1100–35 Brinkburn], **C~ Island** [*Insula Coket* 1135–54 Vita Oswini, *Coketeland* 1347 Percy],

Coquetdale [*Cokedale* c 1160 Newminster, *Choketdale* c 1160 FPD]. Coquet was originally the name of a forest. *Cocwudu* means 'forest frequented by cocks or wild birds'. The river-name is a back-formation, very likely from *Cocwud-dæl* 'Coquetdale', which became by normal development *Cocuddale* and *Cokeddale* and was understood to mean not 'the valley of the wood Coquet', but 'the valley of the river Coquet'.

Corbridge Nb [*Corebricg* c 1050 HSC, *Et-Corabrige* c 1130 SD, *Colebruge* 1100–7 YCh 457, *Corebrigge* 1158 P]. Corbridge is held to have its name from the Roman *Corstopitum* (IA), whose site was at Corchester nr Corbridge [*Colchestre* 1394 PN Nb]. If so, *Cor-* must represent a shortening of the old name, in which only the first syllable was preserved.

Corby Cu [*Chorkeby* c 1120 WR, *Corchebi* 1167 P, *Corcabi* R 1 (1308) Ch]. '*Corc*'s BY.' *Corc* is a well-evidenced Ir name.

Corby Li [*Corbi* DB, *Corebi* 1157 YCh 354, *-by* 1212 Fees], C~ Np [*Corebi* 1066–75 GeldR, 1167 P, *Corbei* DB, *Corbi* 1168 P, *Coreby* 12 NS]. '*Kori*'s BY.' *Kori* pers. n. is found in ON and OSw.

Coreley Sa [*Corna liþ* c 957 BCS 1007, *Cornelie* DB]. 'Slope frequented by cranes.' Cf. HLIÞ. The first el. is a metathesized form of OE *cran, cron*, found in the derivative *cornuc* for *cranuc*. In BCS 1007 are also mentioned *Corna broc* and *Corna wudu* (now **Corn Brook** Sa, **Cornwood** Wo) 'brook and wood frequented by cranes'.

Corfe Castle Do [*Corf* 955 BCS 910, 1162 P], **Corfe Mullen** Do [*Corf* DB, *Corf le Mulin* 1176 P, *Corfmulin* 1272 Ipm], Corfe So [*Corf* 1243 Ass], **Corfan** DB, *Corfham* 1160 P], **Corfton** Sa [*Cortune* DB, *Corfton* 1242 Fees], **Corton** Do [*Corfetone* DB, *Corfton* 1168 P], **Corton Denham** So [*Corfetone* DB, *Corftona* 1168 P, *Corftan Dynham* 1308 Misc], **Corve** R Sa [*Corue* 1256 Ass, *Corfe* 1272 Ass], **Coryates** Do [*Corfgetes westran cotan* 1024 KCD 741]. All these contain an unrecorded OE *corf* 'a pass', derived from OE *ceorfan* 'to cut' and meaning literally 'a cutting'. The places are situated near or in passes. Corfe is simply 'the pass'. Corton is 'TŪN at a pass'. Corton Do is near Coryates, which means 'the gate of the pass'. Corve Sa originally no doubt referred to the valley through which the river flows and was later transferred to the river. Corfham, Corfton Sa may mean 'HĀM and TŪN in the pass or on the river Corve'.

Corfe Mullen is the Corfe with a mill (OFr *mulin* 'mill').—**Corton Denham** was held by Hawis' de Dinan in 1204 (Obl). Cf. BUCKLAND DENHAM.

Corhampton Ha [*Cornhamton* 1201 Cur, c 1225 BM, *Cornhameton* 1242 Fees]. First el. OE *corn* 'corn'. See HĀMTŪN.

Corkickle Cu [*Corkekyll* 1200–13 StB]. Second el. the river-name KEEKLE. The first is obscure.

Corley Wa [*Cornelie* DB, *Cornlea* 1183 P]. OE *corna lēah* 'cranes' forest'. Cf. CORELEY.

Cornard Sf [*Corn(i)erda* DB, *Cornerde* c 1095 Bury, *Cornerth* 1196, *Cornherd* 1197 FF, *Corntherth Magna, Cornherth Parva* 1254 Val]. OE *corn-erþ* 'corn land'. Cf. ERÞ.

Cornbrook La. Really a name of a stream [*Le Cornebroke* 1322 LaInq], which means 'cranes' brook'. Cf. CORELEY.

Cornbrough YN [*Corlebroc* DB, *Cornburc* 1166, *Corneburc* 1167 P], **Cornbury Park** O [*Corneberie* DB, *Corneberia* (for.) 1159, 1190 P, *Cornebir'* 1247 Ass]. 'Cranes' BURG.' Cf. CORELEY. The meaning may be an old fort in which cranes had taken up their abode. Cf. OUTCHESTER. But cranes are wary birds, which live in remote places that are difficult of access. Their home might therefore well be called 'cranes' stronghold'.

Corndean Gl [*Querendon* 1207 Cur, *Corndena* 1181 ff. Winchc]. If the identification of the form of 1207 is correct, the name means 'mill hill' (OE *cweorn-dūn*). If not, the meaning may be 'valley where corn was grown'. The situation favours the first alternative.

Cornelly Co. '(The church of) St. Cornelius.'

Corney Cu [*Corneia, Cornea, Cornai* 12 StB], C~ Hrt [*Cornei* DB, *Corneia* 1198 AC]. 'Corn or crane island.'

Cornforth Du [*Corneford* 1196 P, 1208–10 Fees]. 'Cranes' ford.' Cf. CORELEY.

Cornhill Mx [*Cornehull* 1165, *Cornhell* 1188 AC]. 'Corn hill.' Stow says Cornhill ward was 'so called of a corne Market, time out of minde there holden'.

Cornhill Nb [*Cornehale* 12 DST, 1208–10 Fees]. 'Cranes' HALH.' Cf. CORELEY.

Cornsay Du [*Corneshowe* 1183 BoB]. 'Cranes' point of land.' Cf. HŌH and CORELEY.

Cornwall [*Cornubia* Vita Melori &c., MW *Cerniu*, Welsh *Cernyw*, Co *Kernow*, (on) *Cornwalum* 891 ASC, *-wealum* 997 ASC (E)]. The Brit name goes back to *Cornā-viā*, probably derived from the tribal name *Cornōvii*. OE *Cornwealas* means 'the Welsh in Cornwall'. This folk-name later became the name of the district. Cf. Introd. p. xiii.

Cornwell O [*Cornwelle* 777 BCS 222, *Cornewelle* DB], **Cornwood** D [*Cornehude* DB, *Cornwod, Curnwod* 1242 Fees]. 'Cranes' stream and wood.' Cf. CORELEY. The *u* of some forms of Cornwood is due to the OE lengthening of vowels before *rn*. Corn became OE *cōrn*, whence ME *cōrn* with close *ō*, which in some dialects developed to *ū* early. **Cornwood** Wo. See CORELEY.

Cornworthy D [*Corneorde* DB, *-worthi* 1205 FF, *-wurth* 1238 FF]. 'worþ(ig) where corn was grown.'

Corpusty Nf [*Corpestih, -stig* DB, *Corpesti* 1196 FF, 1203 Cur]. '*Corp*'s STIG', i.e. 'the path (or possibly the pigsty) of *Corp*'. *Corp* is ON *Korpr*, a byname from *korpr* 'raven'.

Corringham Ess [*Currincham* DB, *Curingeham* 1204 FF, *Curingham* 1206 Cur, *Currygeham* 1212 RBE]. 'The HĀM of *Curra*'s people.' *Curra* is also the base of the first el. of *Curringtun* 786 BCS 248, a house in Canterbury. It is not evidenced in independent use, but is easily explained as a short form of *Cūþrēd* &c.

Corringham Li [*Coringeham* DB, 1130, 1162 P, *Coringheham* c 1115 LiS, *Corincham* 1212 Fees]. Hardly identical with CORRINGHAM Ess, as no *u*-forms are recorded. First el. possibly from *Corra*, a side-form of *Curra*, or rather from an OE *Cora*, corresponding to OScand *Kori*. Cf. CORBY.

Corscombe Do [*Corigescumb* 1014, 1035 KCD 1309, 1322, *Cories-, Corscumbe* DB, *Coruscumb* 1244 Ass], **Croscombe** So [*Correges cumb* 705 BCS 113, *Coriscoma* DB]. Originally OE *corfweges cumb* 'the valley of the pass road'. Cf. CORFE &c. In the boundaries of Corscombe are mentioned *micla corf* in KCD 1309, *Corfget* in 1322. The *Corfestig* mentioned 1060–6 KCD 821 is an error for *Eofestig* (Tengstrand).

Corse Gl [*Cors* c 1165 Glouc, 1212 Fees, 1221 Ass]. Welsh *cors* 'bog, fen'.

Corsenside Nb [*Crossinset* 1254 Val, *Crossenset* 1291 Tax]. '*Crossan*'s shieling.' Cf. (GE)SET and SÆTR. First el. the Ir pers. n. *Crossán*.

Corsham (kŏs-) W [*Coseham* 1001 KCD 706, *Cosseham* DB, 1130 P]. '*Cos(s)a*'s HĀM.' *Cosa* (or *Cossa*) is not evidenced, but is a collateral form of *Cusa*. Cf. COSHAM, COSSINGTON &c.

Corsley W [*Corselie* DB, *-lea* 1167 P, *Corslee* 1206 Cur]. Welsh *cors* 'bog, fen' and OE LĒAH.

Corston Sa nr Clunbury [*Cozetune* DB, *Coston* 1272 Ipm]. '*Cott*'s TŪN.' Cf. COSSAL, COTTESBACH &c.

Corston So [*æt Corsantune* 941, *Corsantun* 972 BCS 767, 1287, *Corstune* DB]. 'TŪN on R *Corse*.' The river-name is *Corsan* (obl.) 941 BCS 767 &c. It is derived from MW *cors* 'reed', Welsh, Co *cors* 'bog'.

Corston W [(at) *Corsborne* 854 BCS 470, *Corstuna* 1065 KCD 817, *-tone* DB]. 'TŪN on Gauze Brook.' **Gauze Brook** is *Corsaburna* 701 BCS 103. Cf. CORSTON So.

Corton Do, C~ **Denham** So. See CORFE.

Corton Sf [*Karetuna* DB, *Corton* 1226, *Korton* 1235 FF]. '*Kāri*'s TŪN.' Cf. CAREBY.

Corton or Cortington W [*Cortitone* DB,

Cortyngton 1291 Tax, *Cortun* 1130–5 Sarum]. 'The TŪN of *Cort*'s people.' This name is found in *Cortes hamm* 955 BCS 917. It is related to the first el. of COSTOCK and to OE *Cyrtla* in KIRTLING &c. and derived from a lost adj. *cort* 'short', which corresponds to G *kurz* &c. and is the base of OE *cyrtel* 'kirtle'.

Corve, Coryates. See under CORFE.

Cŏryton D [*Cur(r)itun* c 970 BCS 1246 ff., *Coriton* DB]. The first el. is probably an old name of the river Lyd. See CURRY.

Cosby Le [*Cossebi* DB, *Cossibi* Hy 2 (1318) Ch, *Cosseby* 1236 Fees, *Cotesby* 1258 Ch]. First el. possibly the pers. n. *Cofsi, Copsi* (DB) from ON *Kupsi*, OSw *Kofse*. Or it may be *Cossa* as in COSHAM.

Coseley St [*Colseley* 1357 &c. PNSt]. Possibly identical with COWLERSLEY.

Cosford Sf nr Hadleigh [*Corsforde, Crosfort* DB, *Corsford* 1206 f. Cur]. First el. Welsh *cors* 'bog, fen'.

Cosford Wa [*Cosseford* 1246–9 BM, *Cosforde* 1272 Ipm]. '*Cossa*'s ford.' Cf. CORSHAM.

Cosgrove Np [*Covesgrave* DB, 12 NS, *Couesgraua* 1163, 1167 P]. '*Cōf*'s grove.' *Cōf* pers. n. is a side-form of *Cōfa* (cf. COFA) and related to OE *Coifi*.

Cosham Ha [*Cosham* 1015 ASC (E), *Cos(s)eham* DB, *Cosseham* 1175 P, *Cosham* 1170 P]. '*Cossa*'s HĀM.' Cf. CORSHAM.

Cossal Nt [*Coteshale* DB, *Cozale* c 1200 Middleton, *Cozhall* 1242 Fees]. '*Cott*'s HALH.' *Cott* pers. n. occurs in *Cottes hyrst* 962 BCS 1085. Cf. COTTESBACH &c.

Cossington K [*Cusintun* 10 BCS 1321 f., *Cusinton* 1230 P], C~ Le [*Cosintone* DB, *Cusintona* 1175, *-ton* 1185 P, *Cosintun* 1236 Fees], C~ So [*Cosingtone* 729 BCS 147, *Cosintone* DB, *Cusinton* 1196 P, 1225 Ass]. 'The TŪN of *Cusa*'s people.'

Costessey or Cossey Nf [*Costeseia* DB, c 1184 Fr, 1196 P, *Costesseia* c 1130 BM]. '*Cost*'s island.' *Cost* is found as a pers. n. 1160 ff. P (Db) and as a byname (*Harold Cost*) 1202 Ass (Li). *Kostr* is also found as a nickname in ON. The name may be Scandinavian. But if *Costesford* 675 ASC (E), probably **Cosford** Sa nr Albrighton [*Costeford* DB], is an old name, *Cost* must be English. It may belong to OE *cost* 'tried, excellent'. *Costic* in *Costices mylne* 961 BCS 1076 may be a derivative of *Cost*.

Costock Nt [*Cortingestoche(s)* DB, *-stoce* 1087–1100 Reg, *Kortlincstok* 1236 Fees, *Chirtlingastoca* 1158, *Curtlingestoke* 1211 PNNt(S)]. 'The STOC of *Cort*'s people.' Cf. CORTON W. *Cort* was apparently sometimes called by the hypocoristic names *Cortel* and *Cyrtla*.

Cŏston Le [*Castone* DB, *Caston* 1205 Cur, 1242 Fees, *Coston* 1227 Ch, 1254 Val]. The early ME form was *Cāstūn*. The source seems to be OScand *Kāts-tūn*, with ON *Kátr* pers. n. as first el.

Coston Nf [*Kar(e)ston* 1198 FF, *Corestone* 1254 Val, *Corston* 1291 Tax, 1316 FA]. First el. OScand *Kārr* pers. n.

OE *cot* n., also *cote* fem. 'cottage', ME *cot(e)* also 'a shelter, as for sheep'. OE *cotlif* also meant 'manor'. This may have been the meaning also of *cot* in some pl. ns. as BISCOTT, PRESCOT. The usual meanings in pl. ns. are no doubt 'cottage' and 'shelter for animals, esp. for sheep'. OE *cotsetla* meant 'a cottager', and a meaning 'cottage' is natural in names such as CHARLCOTT, SMETHCOTE. *Cot* often meant particularly 'a woodman's hut', as in the common names WOODCOTE, WOODMANCOTE. The meaning 'shelter for animals' is obvious in names such as BUL-, LAMCOTE. Cf. also HUR(D)-COT(T). In CALDECOTE &c. the meaning may be 'a shelter for wayfarers', and the same may be the meaning for some DRAYCOTTS. In compound names the first el. is often a pers. n. or some other designation for a person. Combinations with words such as *east, west* are common, as ASCOT, WESTCOT. The element often appears in the plur. form, regularly so when used alone to form a pl. n. But the plur. *cotu* became indistinguishable from the sing., unless -*s* was added, and even the dat. plur. *cotum* in many dialects became ME *cote*. When used alone as a pl. n. the word takes three chief forms:

1. (Without an ending indicating a plur. form): **Cote** O [*Cote* DB], **Coat** So [*Cotes, Kote* 1225 Ass], **Coate** W nr Devizes [*Cotes* 1283 Pat].

2. (With analogical -*s*): **Coates** Ca, **Coates** Gl nr Cirencester [*Cotes* 1221 Ass], **C~** Gl nr Winchcomb [*Cota* 1195, *Cotes* 1251 Winchc], **Cotes** Le nr Loughborough [*Cothes* 12, *Cotes* c 1200 DC], **Cotes de Val** Le [*Toniscote* DB, *Cotesdeyvill* 1285 FA], **Coates** Li nr Stow [*Cotes* DB, c 1115 LiS], **Great & Little Coates** Li [*Cotes, Sudcotes* DB, *Cotis, Cotun, Sut Cotun* c 1115 LiS, *Magna, Parva Cotes* 1242 Fees], **North Coates** Li [*Nordcotis* c 1115 LiS, *Northcotes* 1202 Ass], **Coates** Nt [*Cotes* 1316 FA], **Cotes** St nr Eccleshall [*Cota* DB, *Cotes* 1251 Ch], **Coates** Sx [*Kotes* c 1142 Fr].

3. (Names that preserve the OE dat. form): a. **Coton** Ca [*Cotes* R 1 Cur, 1203 Ass], **Cotton Abbotts** Chs [*Chotam Ordrici* c 1100 Chester, *Abbotescoten* 1288 Court], **Coton in the Elms** Db [*Cotune* DB, *Cotene* 1242 Fees], **Coton** Le [*Cotes* 1209–35 Ep, *Cotene* 1327 Subs], **Nun Coton** Li [*Cotes* DB, *Cotun* c 1115 LiS, *ecclesia Sancte Marie de Cottuna* 12 DC], **Coton** Np nr Guilsborough [*Cote* DB, *Cotes* 1195 P], **(Far) Cotton** Np [*Cotes* 1196 Cur, *Coten* 1324 Ipm], **Coton** O [*Cotes* 1316 FA], **Coton** Sa nr Wem [*Cote* DB, *Coten* 1285 FA], **Cotton upon Tern** Sa [?*Ludecote* DB, *Cota* 1121 Eyton], **Coton** St in Milwich

[*Cote* DB], **Coton** St NE. of Stafford [*Cote* DB, *Cotes* 1209 Cur, *Coton* 1285 FA], **Coton** St in Wigginton [*Coton* 1313 Ipm], **Coton** Clanford St [*Cote* DB, *Coton* 1291 Tax], **Coton** Wa [*Cotes* J BM, 1242 Fees]. b. **Coatham Mundeville** Du [*Cotum* R 1 FPD, *Cotum Maundevill* 1344 RPD], **Cottam** La [*Cotun* a 1230 CC, 1227 FF, *Cotum* 1246 Ass], **Cotham** Nt nr Newark [*Cotes, Cotune* DB, *Cotes* 1197 P, *Cotum* 1264 Ipm], **Cottam** Nt nr Retford [*Cotum* 1274 Ipm, 1303 FA], **Cottam** YE [*Cottun* DB, *Cotum* 1285 FA], **Coatham** YN [*Cotum* 1231 FF, 1272 Cl]. It will be noticed that the ending -*um* is preserved only in Du, La, Nt, Y. The form *Cot(t)on* from *Cotum* is found in the Midlands, Ca, Chs, Db, Le, Li, Np, Sa, St, O, Wa.

The surnames, so far as not self-explanatory, are generally family names. **Cotes de Val** (or **Deville**) from the Deville family, perhaps from DEVILLE in Normandy. **Coton Clanford** was named from a place of this name in Seighford. **Nun Coton** was the seat of a nunnery. **Coatham Mundeville** was held by Thomas de Amundevilla t. R 1 (1318 Cu), and sold by a namesake of his in 1274. The name comes from OMONVILLE or MONDEVILLE in Normandy, formerly *Amondeville* &c.

Cote, Cotes. See COT.

Cotgrave Nt [*Godegrave* DB, *Cotegrava* 1094 Fr, *Cottegraua* 1158, 1160 P, *Cotesgrava* Hy 2 (1316) Ch]. '*Cotta*'s grove', rather than OE *cota-grāf* 'grove with cots'.

Cotham. See under COT.

Cothelstone So [*Cothelestone* 1327 Subs, *Cuthelstone* 1333 Ep]. '*Cūþwulf*'s TŪN.'

Cothercott Sa [*Cotardicote* DB, *Cudardecote* 13 Eyton]. '*Cūþheard*'s COT.'

Cotheridge Wo [*æt Coddan hrycce*, (to) *Coddan hrycge* 963 BCS 1106, *Codrie* DB]. '*Codda*'s ridge.' Cf. CODFORD.

Cotherstone YN [*Codrestune* DB, *Cudereston* 1201 Ch, *Cudreston* 1226 FF]. '*Cūþhere*'s TŪN.'

Cotleigh D [*Coteleia* DB, *Cotteleg* 1150 Wells, -*lege* 1219 FF]. '*Cotta*'s LĒAH.'

Cotness YE [*Cotes* DB, *Cotenesse* 1199 FF, *Coutenesse* 1285 FA]. 'Headland with cots.'

Coton. See under COT.

Cotswolds, The, Gl [(montana de) *Codesuualt* 12 Gir, *Coteswaud* 1250 Pat, *Coddeswold* 1294 Cl]. '*Cōd*'s WALD or forest.' Cf. CUTSDEAN, which was named from the same *Cōd*. The forms of Cutsdean suggest that *Cod* had long ō. Cf. COODEN and CODSALL.

Cottam. See under COT.

Cottenham Ca [*Cotenham* c 1050 KCD 907, 1201 Cur, *Coteham* DB, 1130 P]. '*Cot(t)a*'s HĀM.'

Cottered Hrt [*Chodrei* DB, *Codruth* 1185 TpR, *Codreth* 1220 Fees, *Coudreya* 1239 Ep, *Coudray* 1258 Ch]. The absence of any

early forms with an *e* between *d* and *r* tells
against the first el. being *Cod(d)a* pers. n.
The spellings with *ou* may point to the *o*
having been long. No such el. as *cōd* is
known in English, but there is ON *kǒð*
'spawn of fish', which would go well with
the second el., which is OE *riþ* 'brook'.
Possibly *cod* in *codbait*, an earlier side-form
of *cadbait*, might be from an OE *cōd*.

Cotterstock Np [*Codestoche* DB, *-stoc*
1125–8 LN, *Copestoche* c 1175 Middleton,
Cotherstoke 12 NS, *-stock* 1254 Val, 1275
Ch]. OE *cope-stoc* 'hospital', the elements
being OE *copu* 'sickness, pestilence' and *stoc*
'place'. The *r* may be intrusive, or possibly
there was an OE *cop-ærn* 'hospital', which
was used alternatively with *copu* as the
first el.

Cottesbach Le [*Cotesbece* DB, *-bac* 1236
Fees, *Cottesbec* 1254 Val], **Cottesbrook**
Np [*Cotesbroc* DB, 1220 Fees, *Cottesbroc*
1220 Ep], **Cottesmore** Ru [*Cotesmore* DB,
1228 Ep, *Cottesmor* 1228 Cl, 1237 Fees].
'*Cott*'s valley, brook and moor.' Cf. COSSAL
and BÆCE.

Cottingham Np [*Cotingeham* DB, 1190 P,
Cotingham 1137 ASC (E), *Cottingeham* 1163
P], **C~** YE [*Cotingeham* DB, 1227 Ch,
Cotingham DB, 1201 Cur, *Cottingham* 1156
YCh 1388], **Cottingley** YW [*Cotingelai*
DB, *Cotingelegh, Cottingele* 1208 Cur,
Cottingle 1226, *-lay* 1240 FF]. 'The HĀM
and LĒAH of *Cott(a)*'s people.'

Cottingwith YE [*Coteuuid, Cotinwi* DB,
Cotingwith 1100–15 YCh 1001, *Cotingwic*
1195 Cur, *Cottingwic* 1157 YCh 354]. OE
Cottinga wīc 'the wīc of *Cott(a)*'s people',
later with *with* from OScand *viðr* 'wood'
substituted; cf. SKIPWITH. Cottingwith may
have been a dairy-farm belonging to Cottingham.

Cottisford O [*Cotesforde* DB, *-ford* 1180
BM, *Cottesford* 1242 Fees]. '*Cott*'s ford.'
Cf. COSSAL.

Cotton Chs &c. See under COT.

Cotton Db [*Codetune* DB, *Codinton* 1194
P], **C~** Sf [*Codetuna, Kodetun, Cotetuna,
Cottuna* DB, *Cotton* 1203 Cur, 1254 Val].
'*Cod(d)a*'s TŪN.' Cf. CODBRO.

Cotwalton St [*Cotewaltun* 1002 Wills, *Cote-
woldestune, Codewalton* 1176 P]. The
first el. is a pl. n. with OE WALD 'wood' or
a stream-name with OE WELLA 'stream' (in
its WMidl form *walle* from *wælla*) as second
el. The first may be OE *cot* 'cottage, hut'
or possibly *Cotta* pers. n. To this was
added TŪN.

Coughton (-ō-) He [*Cocton* Hy 3 BM, *Coctone* 1286 Ep], **C~** Wa [*Coctune* DB, *Coc-
tona* 1169 P, *Copton* 1242 Fees, *Cotton* 1241
Ch]. Both places are near prominent hills,
and the probability is that the first el. is OE
cocc 'heap, hill'. Cf. COOKHAM and COCK
BECK. C~ Wa is near the ridge on which
is COOK HILL.

Coulderton Cu nr Egremont [*Culdretun* 12,
Culdirton 13 StB, *Culdertone* 1294 Cl].
C~ is in a long tongue of land between the
Ehen and the sea. This may well have been
known by the Britons as *culdir*, a Welsh
word for 'narrow strip of land, isthmus'.
Cf. HOLME CULTRAM.

Coulsdon Sr [*Curedesdone* 675, *Cudredes-
done* 933 BCS 39, 697, *Cuðredesdune* 1062
KCD 812, *Colesdone* DB, *Culisdon* 1242
Fees]. '*Cūþrēd*'s DŪN.'

Coulston W [*Covelestone* DB, *Cuulestun*
1195 FF, *-ton* 1242 Fees]. '*Cufel*'s TŪN.'
OE *Cufel* pers. n. is a normal derivative
of *Cufa*. Cf. COWESFIELD.

Coulton YN [*Coltune* DB, 1167 P]. '*Cola*'s
TŪN' or *Col-tūn* 'TŪN where charcoal was
burnt' or *Colt-tūn* 'TŪN where colts were
bred'.

Cound Sa [*Cuneet* DB, *Conet, Conede* 1255
RH, *Cunet* 1256 Ass]. C~ is on **Cound
Brook** [*Cunette* c 1200 Gervase, *Cunethe*
c 1204, *Cunede* post 1236 Eyton]. A Brit
river-name identical with KENNET, KENT, the
base being Brit *Cunētiō*.

Coundon Du [*Cundun* 1196 P]. Perhaps
OE *cūna dūn* 'cows' hill'.

Coundon Wa [*Condelme* DB, *Cundelma* 1172
P, *Cundulme* 1257 Ch]. Second el. OE
ǣwylm 'source of a river'. The first is no
doubt a river-name identical with COUND.
Coundon is near the source of the Sherborne, which may have had the name
Cound. The OE base would be *Cuned-
ǣwylm*.

Countesthorpe Le [*Torp* 1209–35 Ep, *Cun-
tastorp* 1242 Fees, *Thorp Cuntasse* 1276
RH]. 'The countess's THORPE.'

Counthorpe Li [*Cudetorp* DB, *Cunnige-
torp* 1208 FF, *Cunetorp* 1219 Ass, *Coin-
thorp* 13 FF, 1288 Ipm, *Cunthorp* 1291 RA].
The DB form points to OE *Cūpan þorp*
'*Cūþa*'s thorp'. The later forms are explained from an alternative *Cūþing(a)þorp*,
in which medial *þ* was lost.

Countisbury D [*Contesberie* DB, *Cuntes-
beria* 1178 P, *-bir* 1200 Cur, *Cantebire* 1228
FF]. C~ has been rightly identified with
arx Cynuit c 894 Asser, stated to be a fort.
Cynuit is identical with COUND and KENNET
and with CYNWYD in Wales, but the name
here refers to a hill. Countisbury is thus
Brit *Cunēt*, to which was added an explanatory OE BURG. Brit *Cunēt* became normally
OW *Cynuit*, Welsh *Cynwyd*.

Coupland Nb. See COPELAND.

Courteenhall (kortn-) Np [*Cortenhale* DB,
12 NS, *Curtehala* c 1110 NpCh, *Cortenhal*
1196 P]. '*Curta*'s HALH.' *Curta* is closely
related to *Cort* in CORTON W, *Cortel* in
COSTOCK and *Cyrtla* in KIRTLING &c.

Cove D [*La Kove* 1242 Fees], **C~** Ha [*Coue*
DB, *Cove* 1261 Ch]. OE COFA. In Cove D

the meaning 'recess, valley running in among hills' is suitable. The exact meaning of Cove Ha is not so clear.

Cove, North, Sf [*Cove* 1204 Cur, 1235 Fees, 1254 Val], **South C~** Sf [*Coua* DB, *Cove* 1203 Cur, 1254 Val]. S. C~ is not far from the sea, and may have been named from a former cove or creek on the coast. N. C~ is far inland. Very likely it was an outlying part of S. C~ and was named from it. **Covehithe** is on the coast not far from S. Cove. The name means 'Cove harbour'. The old name was **North Hales** [*Northala*, *Nor-*, *Northals* DB, *Northales* 1254 Val]. It almost looks as if the original name was *North-hals* 'northern neck of land', referring to a lost spit of land.

Coven St [*Cove* DB, *Couena* 1175, *Couene* 1176 P, *Covene* 1242 Fees]. OE *cofan*, dat. and acc. of *cofa*, here probably in the sense 'valley that runs in among hills'. The form *Covene* has analogies in *Cotene* for COT(T)ON from OE *cotum*.

Cŏveney Ca [*Coveney* 1254 Val, *-e* 1291 Tax]. The first el. may be OE *cofa*, the second being *ēg* 'island'. The place is in old fen-land. There may have been a lake at the place, so that the meaning of *cofa* might be 'bay'.

Cŏvenham Li [*Covenham* DB, c 1095 YCh 855, *Coveham* c 1115 LiS]. 'HĀM at a COFA or recess.' There seems to be a valley that runs in among the Wolds here.

Covenhope He [*Camehop* DB, *Comenhop* 1292 QW, *-e* 1316 FA, *Kovenhop* 1242 Fees, *Covenhop* 1249 Fees]. The first el. can hardly be COFA. The DB form points to original *ā*. Apparently the first el. is as in CAENBY and perhaps CAVENDISH, CAVENHAM, though with *ā* preserved long, i.e. an OE pers. n. *Cāfna* or the like. Sometimes *fn* was assimilated into *m*.

Cŏventry Wa [(æt) *Couæntrēē* 1043–50 BM, *Cofantreo* 1053 ASC (C), *Cofentreium* c 1070 BM, *Coventrev* DB]. Names in *-tree* usually have as first el. a pers. n. Hence probably '*Cōfa*'s tree'. Cf. COFA.

Cŏver R YN [*Cobre* c 1130 SD (s.a. 797), *Couer* 1279 Ass]. A Brit river-name. The first el. may be a word corresponding to Welsh *cau* 'hollow' and *ceu-* in *ceunant* 'ravine', the second being the el. *bero-* found in Welsh *gofer* 'a rill', the river-name HYDFER &c. The meaning would be that of HOLBORN. From the **Cover** were named **Coverdale** [*Coverdal* 1202 FF], **Coverham** [*Covreham* DB, *Coverham* 1202 FF], **Coverhead** [*Coverhede* 1405 AD].

Coverack Co [*Porthcovrec* 1262 FF]. Perhaps elliptical from a name containing as second el. a pers. n. corresponding to Welsh *Cynfarch*, Bret *Convarch*.

Coverdale, Coverham. See COVER.

Covington Hu [*Covintune* DB, *Kuvintone*

1226 Ep, *Couyngton* 1260 Ass]. 'The TŪN of *Cufa*'s people.'

Cowarne, Much & Little, He [*Cuure* DB, *Couern* 1255 Ch, *Magna*, *Parva Coerna* 1242 Fees]. OE *cū-ærn* 'cow-house'.

Cowbridge Ess [*Cubrigea* DB], **C~** W [*Coubryge* 1409 FF]. Self-explanatory.

Cowdale Db nr Buxton [*Cudala* 1186 P]. 'Cows' valley.'

Cowden K [*Cudena* c 1100 Text Roff, *Couden* 1237 Cl, *Cudenne* 1254 Pat]. OE *cū-denn* 'pasture for cows'. Cf. DENN.

Cowden Nb [*Colden* 1286 Ch]. 'Valley where charcoal was burnt.'

Cowden YE [*Col(e)dun* DB, *Coldon* 1208 FF, *Parva Coldon* 1285, *Magna C~* 1428 FA]. 'Hill where charcoal was burnt.' See DŪN.

Cowdray Park Sx [*la Coudreye* 1279 Ass]. Fr *coudraie* 'hazel copse'.

Cowes (kowz) Wt [*le Estcove et Westcove* 1413 PNWt, *the Cowe*, *betwixt the Isle of Wight and England* 1512 LP, (roads called) *the Esturly or the Westerly Cowe*, (at) *the Cowe, in the Isle* 1539 ib.]. Originally no doubt a sand-bank off the coast of Wight, called 'the Cow', or perhaps two sand-banks so called. Later the name seems to have been transferred to two forts on each side of the mouth of the Medina.

Cowesby YN [*Cahosbi* DB, *Cousebi* 1200 Cur]. '*Kausi*'s BY.' *Kausi* is an ON nickname meaning literally 'tom-cat'.

Cowesfield W [*Colesfeld* DB, *Cuuelesfeld* 1167, 1197 P, 1243 Cl]. '*Cufel*'s FELD.' Cf. COULSTON.

Cowfold Sx [*Coufaud* 1232 Sele, *Cufaude* 1255 FF, *Coufold* 1336 Ipm]. Self-explanatory.

Cowhill Gl [*Couhull* 1327 Subs, *Cowhull* 1445 AD]. 'Cows' hill.'

Cowick Barton D [*Cuicland*, (on) *Cuike* c 1100 E, *Coic* DB, *Cuwike* 1228 FF], **Cowick** YW [*Cuwich* 1197 P, *Cuwic* 1223 FF]. 'Cow farm.' See WĪC.

Cowlam YE [*Colnun* DB, *Collum* c 1110 YCh 25, c 1155 ib. 830]. If the DB form is not to be taken literally, *kollum*, dat. plur. of ON *kollr* 'a hill'.

Cowlersley YW in Lockwood [*Colresleye* 1226 FF]. 'The charcoal-burner's wood.' See LĒAH. OE *colere* is not evidenced.

Cowley Bk [(to) *Cufanlea* 949 BCS 883, *Couele* 1198 FF], **C~** D nr Exeter [*Couelegh* 1237 Fees], **C~** D in Ashton [*Couelegh* 1333 Subs], **C~** (-ow-) & **Temple C~** O [*Couele* 1004 KCD 709, 1139 TpR, *Covelie* DB, *Cuueleia* 1199, *Templecoueleya* 13 Fridesw], **C~** St [*Covelav* DB, *Couele* 1314 Ch]. Partly no doubt OE *Cufan lēah* '*Cufa*'s LĒAH'. But it seems likely that some of the names at least contain a descriptive element liable to be combined with *lēah*. All the Cowleys are near hills, and there may have existed an OE word for a hill cognate with Norw *kuv* 'a round top', South G *chobel* 'a cliff'. Still better would be a word denoting something that was got from a wood, e.g. an OE *cufl* 'a block of wood, a log' or 'a stump'. Cf. Swed *kubb* 'a log'.

Cowley Db nr Winster [*Collei* DB], **C~** Db at Dronfield [*College* 1315 Derby]. OE *col-lēah* 'clearing or wood where charcoal was burnt'.

Cowley Gl [*Kulege* DB, *Culega* Hy 3 AD]. OE *cū-lēah* 'cows' pasture'.

Cowley Mx [(in) *Cofenlea* 959 BCS 1050, (in) *Cofanlea* 998 Th, *Covelie* DB]. '*Cōfa*'s LĒAH' or 'LĒAH in a COFA or valley'. Cf. COFA.

Cowling YN [*Torneton* DB, *Thornton Colling* 1202 FF, *Collyng* 1400 YInq]. Originally THORNTON. *Colling*, later *Cowling*, is a pers. n. denoting an early owner. Eventually the original name was lost altogether.

Cowling YW [*Collinghe* DB, *Collinge* 1279 Ch, *Cullyng* 1315 Ipm]. Cowling Hill [*Collinge* 1202 FF] is a hill of 1,000 feet. *Colling* may be the original name of the hill, or a derivative of a hill-name *Coll*, from OE *coll* or ON *kollr* 'a hill'. In the latter case the meaning is 'the place by *Coll* hill'. Cf. COLESHILL.

Cowlinge (kōōlinj) Sf [*Culinge* DB, *Culinges* 1195 P, *Cullinges* 1201 Cur, *Culing* 1203 Ass]. '*Cūl*'s or *Cūla*'s people', an OE *Cūlingas*. *Cūla* pers. n. is the first el. of CULHAM O and is found in *Culan fenn* 962 BCS 1082 (Chelsworth Sf). On *Cūl* see COOLING. It is the first el. of *Culeslea* DB, *-le* 1202, *Cullesle* 1228 FF (Sf).

Cowlishaw La [*Colle-*, *Cowleshawe* 1558 DL]. First el. perhaps as in COLLYHURST.

Cowm La [*Cumbe* 13 WhC]. OE *cumb* 'valley'. Cf. COMB.

Cowpe La [*Cuhope* c 1200 WhC, *Couhop* 1324 LaInq]. 'Cow valley.' Cf. HOP.

Cowpen (-ōō-) **Bewley** Du [*Cupum* c 1150 (1201) Ch, 1195 (1335) Ch], **Cowpen** (-ōō-) Nb [*Cupum* c 1175 Brinkburn, 1242 Fees, *Copun* 1198 (1271) Ch]. 'The coops.' ME *cūpe* (from OE *cȳpe*) means 'a basket, hencoop', also 'a wickerwork basket used in catching fish'. The latter is the meaning here. C~ Bewley belonged to Bewley manor.

Cowthorpe YW [*Coletorp* DB, *Coltorp* 1166, 1190 P, *Colethorp* 1246 FF]. First el. OE *Cola* or OScand *Koli* pers. n.

Cowton, East, North & South, YN [*Cudtone, -tun, Cotun* DB, *Nordcuton* 12 Pudsay]. '*Cūpa*'s or *Cūda*'s TŪN.' Were it not for the numerous DB spellings with *d*, the obvious etymology would be OE *Cū-tūn* 'cow farm'.

Coxford Nf [*Kokesford* 1203, *Coke-, Kokesford* 1207 f. Cur, *Cogesford* 1248 Cl], **Coxley** So [*Cokkesleghe* 1327 Subs], **Coxwell, Great & Little,** Brk [*Cocheswelle* DB, *Cokeswell* 1205 BM, 1242 Fees, *Cogeswell* 1220 Fees]. It is difficult to decide if the first el. is OE *cocc* 'wild bird' or *cocc* 'heap or hill' or an unrecorded pers. n. *Cocc* (cf. *Cocca*). Coxwell is near hills. Cf. COCK BECK. Second el. FORD, LĒAH, WELLA.

Coxwold YN [*Cuhawalda* 758 BCS 184, *Cucvalt* DB, *Cucuald* c 1110 RA, *Cucawald* 1157 YCh 354]. '*Cuha*'s forest.' Cf. WALD. *Cuha* pers. n. is not otherwise evidenced, but is a relative of *Cohha* in COCKFIELD Sf. Hope gives the pronunciation as *Cookwood*.

Crabhouse Nf, an old monastery [*Crabehus* 1245 Cl, *Crabbehus* 1254 Val]. The first el. seems to be ON, OSw *Krabbi* pers. n.

Crabwall Chs [*Crabbewell* 1199 (1265) Ch, *-walle* 1200–50 Chester]. Possibly 'stream in which crayfish were found'. OE *crabba* is only found in the sense 'crab', but no OE word for 'crayfish' is known.

Crackenthorpe We nr Appleby [*Crakintorp* 1202 FF, *Crakenthorp* 1292 QW], **C~** We in Beetham [*Crakintorp* 1255, *Crakangthorp* 1290 Kendale]. There is also *Cracanethorp* 13 CC (Caton La). The first el. is no doubt a pers. n. It may be a nickname identical with Norw *krakande* 'crawling, who walks with difficulty'.

Cracoe YW [*Cracho* 1202 P, *Crachou* 1257 Ch]. 'Spur of land frequented by crakes.' *Crake* is often held to be Scandinavian, but may well be native. See HŌH.

Craddock D [*Cradocumba* 1185 Buckland, *Cradok* 1249 FF]. Originally the name of a brook [*Craducc* 938 BCS 724]. The brook-name may be elliptical for a name with the pers. n. *Cradoc* from *Caradoc* (cf. CARADOC He) as second el. Or it might be an old stream-name derived from an adj. corresponding to Bret *karadek* 'amiable', a Welsh *caradog*. Cf. CARANT, CARY &c.

Cradley He [*Credelaie* DB, *Credelei* c 1195, *-le* 1241 Hereford, *Cradelea* 1170 P]. '*Creoda*'s LĒAH.'

Cradley Wo [*Cradeleie* DB, *-lega* 1180 ff., *-leia* 1197 f. P]. First el. probably a pers. n. **Crad(d)a**. OE *Carda, Cærda* seem to be due to metathesis of *Cradda* (*Crædda*). Cf. CARDESTON. C~ is held in PNWo to have perhaps as first el. OE *cradol* 'cradle'. This suggestion may be supported by the surname *atte Cradele* 1296 (Petworth Sx), adduced by Löfvenberg, who takes the word to have meant 'hurdle' or 'fence'.

Crafton Bk [*Croustone* DB, *Croftona* 1200 FF, *-ton* 1200 Cur, *Croxton* 1200 Cur]. OE *Croh-tūn* 'TŪN where wild saffron grew'.

Craiselound Li [*Lund* DB, *Craslund* c 1220 Bodl, Hy 3 BM]. Cf. *Crasegarth* Selby (in

Axholme, thus near Craiselound). *Crais-* might possibly be OE *cærse* 'cress'. Or it might be compared with Norw *kras* 'thicket'. See LUND.

Crake R La [*Crec* c 1160 LaCh, *Crayke* c 1160 FC, *Craic* 1196 FF]. 'Rocky stream', a derivative of OW *creic* 'rock'. Cf. CRAYKE.

Crakehall, Great & Little, YN [*Crachele* DB, *Crakehale* 1157 PNNR, (*Great*) *Crakehale* 1270 Ipm], **Crakehill** YN [*Crecala* DB, *Crakhale* 1301 Subs]. 'HALH frequented by water-crakes.'

Crakemarsh St [*Crachemers* DB, *Crakemers* 1242 Fees]. 'Marsh frequented by water-crakes.'

Crambe YN [*Crambom, Crambun* DB, *Crambum* c 1180 YCh 633, 1208 FF]. 'The bends', the dat. plur. of an OE **cramb* 'a hook or bend', related to OE *crumb* 'bent'. The name refers to bends of the Derwent. Near Crambe is **Buttercrambe** [*Butecrame* DB, *Butercram* 1208 FF]. 'The Crambe with rich pastures.'

Cramlington Nb [*Cramlingtuna* c 1130, *Cramlingatuna* c 1150 FPD, *Cramelington* 1242 Fees]. The first el. may be derived from OE *cranwella* 'cranes' spring'.

OE **cran** 'crane' is common as the first el. of pl. ns., especially in combination with words for brook, lake, ford, wood. See CRAN- (passim), CARNFORTH. There is no reason to assume any other meaning for the word than 'crane', such as 'heron'. The two birds are always kept well apart in early records. By metathesis *cran* or rather *cron* became *corn*, which is found in many pl. ns. See CORELEY, CORLEY, CORNBROOK &c. The derivative *cranuc* (*cornuc*) occurs in some names, as CONKSBURY, CRANFIELD.

Cranage Chs [*Croeneche* DB, *Craunach* c 1215, c 1274, *Craulach* c 1247, 1271, *Cranach* c 1290 Chester]. OE *crāwena-læcc* 'crows' stream'. Cf. LÆCC. *Crāwenalæcc* was simplified to *Crawlach* and *Crawnach*, the latter of which prevailed.

Cranborne Do [*Creneburna* DB, *Craneburna* 1163 P], **Cranbourne** Ha [*Cramburna* 901 BCS 596, *Craneburne* 1175 P], **Cranbrook** K [*Cranebroca* 11 DM, *Cranebroc* 13 BM]. 'Cranes' stream.'

Crandon So in Bawdrip [*Grenedone* DB, *Grandon* 1212 Fees, *Crendon* 1219 Fees, 1230 P, *Crandon* 1243 Ass]. It is doubtful if this is OE *grēne dūn* 'the green hill' later modified to *Crendon, Crandon,* or OE *crandūn* 'cranes' hill'.

Crane R Do, K, Mx. In all cases a back-formation (from CRANBORNE, CRANBROOK and CRANFORD respectively).

Cranfield Bd [*Cranfeldinga dic, Crancfeldinga* [gemære] 969 BCS 1229, *Crangfeldæ* 1060 Th, *Cranfelle* DB]. OE *cranuc-feld* 'FELD frequented by cranes'.

Cranford Mx [*Cranforde* DB, *Cranford* 1162 P], **C~ St. Andrew & St. John** Np [*Craneford* DB, *Cranford* 1167 P, *Craneford Sancti Andree, Sancti Iohannis* 1254 Val]. 'Cranes' ford.'

Cranham Ess [*Wocheduna, Craohv* DB, *Wokyndon Episcopi vel Crando* 1254 Val]. See OCKENDON. The later name is identical with CRANOE.

Cranham Gl [*Craneham* c 1160 Glouc, 1291 Tax]. 'HAMM frequented by cranes.'

Cranleigh Sr [*Cranlea* 1166, *Cranelega* 1167 P], **Cranley** Sf nr Eye [*Cranlea* DB, *Cranele* 1198 FF]. 'Cranes' wood.' Cf. LĒAH.

Cranmere Sa, **Cranmore** So [*Crenemere* 1084 GeldR, *-melle* DB, *Cranemere* 1196 P, 1241 BM]. 'Cranes' mere.'

Cranoe Le [*Craweho* DB, *Crawenho* 1198 P, *Craunhou* 1209–35 Ep]. OE *crāwena hōh* 'headland frequented by crows'.

Cransford Sf [*Craneforda* DB, *Cranesforda* ib., *Craneford* 1203 Cur, *Cranesford* 13 BM]. 'Cranes' ford.'

Cransley Np [*Cranslea bricg* 956 BCS 943, *Craneslea* DB, *-leia* 1202 Ass]. 'Cranes' wood.' *Crane* is here used collectively.

Cranswick YE [*Cransuuic, Cranzvic* DB, *Cranzwic* 1200–16 YCh 1265, *Crancewik* 1202 FF]. Apparently 'cranes' wīc'. The combination is somewhat curious, and possibly the original name was *Cransæ-wīc* 'wīc by the lake called *Cransæ*'. Cf., however, CRANWORTH.

Crantock Co [(Canonici) *S' Carentoch* DB, (ecclesia de) *Sancto Carentoco* Hy 1 (1270) Ch, *Seint Karentoc* 1234 FF]. '(The church of) Saint Carantoc.' The saint is called St. Carannog in Welsh. CARANTEC is a parish in Brittany named from the saint.

Cranwell Li [*Craneuuelle* DB]. 'Cranes' spring.'

Cranwich Nf [*Cranewisse* DB, *Crenewiz* 1200 FF, *Kernewiz* 1254 Val]. 'Cranes' meadow.' The second el. is OE *wisc* or **wisse* 'meadow'. Hope gives the pronunciation as *Cranice*. Cf. HAUTBOIS.

Cranworth Nf [*Cranaworda, Craneworda* DB, *Craneworth* 1211 FF]. 'Cranes' WORÞ', i.e. 'a WORÞ near which cranes were seen'.

Crasswall He [*Cressewell* 1231 Ch, *Crassewalle* 1255 RH]. 'Cress stream.'

Cräster Nb [*Craucestr'* 1242 Fees, *-cestre* 1245 Ipm]. 'Old fort inhabited by crows.'

Cratfield Sf [*Cratafelda* DB, *Cratefeld* 1165 BM, 1236, 1242 Fees]. '*Crǣta's* FELD.' *Crǣta* pers. n. is not evidenced, but is presupposed by CREETING Sf. The name may be related to OE *Cretta,* OHG *Chrezzo* &c. Cf. also the lost **Cratley** Nt [*Creilege* DB, *Cratela* c 1150 DC].

Crathorne YN [*Cratorne* DB, *-thorn* c 1170 YCh 688]. Perhaps identical in origin with *Crakethorn* 1218 FF (Ebberston), which is 'crakes' thorn-bush'.

Craven YW [*Crave* DB, *Cravena* c 1140 FC, *Crafna* c 1160 Hexh, *Crauene* 1166 P]. Possibly derived from Welsh *craf* 'garlic'. Craven is the name of a large district round the sources of the rivers Aire and Wharfe.

Crawcrook Du [*Crawecroca* 1130 P, *Kraukruke* 1242 Ass]. 'Bend frequented by crows.' The bend of a road seems to be referred to. For OE *crāwe* 'crow' as a first el. see CRAW-, CROW-, CRANAGE, CRANOE, CRASTER, CREACOMBE, CROMER, CROYDON Ca, So.

Crawley Bk [*Crauelai* DB, *little, great Craule* c 1195 Fr], **Crawley Bury** Ess [*Crauuelæa* DB], **Crawley** Ha [*Craweleainga mearc* 909, *Crawanlea* c 960 BCS 620, 1158, *Crawelie* DB], C~ O [*Craule* 1316 FA], C~ Sx [*Crauleia* 1203 PNSx]. 'Crows' wood.'

Crawley Nb [*Crawelawe* 1225 P, 1256 Ass]. 'Crows' hill.'

Crawshaw Booth La [*Croweshagh* 1324 LaInq, *Crawshaboth* 1507 Ct]. 'Crows' wood', with *booth* 'dairy-farm' added. See SCAGA.

Cray R K [*Cræges æuuelma* 798, (on) *Crægean* 814 BCS 291, 346, *Craie* c 1200 Gervase], **Cray Gill** R YW [*Creibecke* 1241 FF]. A Brit river-name identical with CRAI in Wales and derived from MW *crei* 'fresh, clean'. From Cray K are derived **Foots, North, St. Mary & St. Paul's Cray** [*Cræga* 988 BM, *Crægan* 965–93 KCD 1288, *Crai(e)* DB, *Sudcrai* DB, *Fotescraei, Northcraei* c 1100 Text Roff, *Creye sancte Marie* 1257 FF, *Creypaulin* 1291 Tax]. On Cray Gill is **Cray** [*Craie* 1190 P].

Foots Cray was held by *Godwine fot* in 1066 (DB). *Fot* is a nickname.—St. Paul's from the dedication of the church to St. Paulinus.

Crayford K [*Crecganford* 457 ASC, *Crainford* 1322 Ep]. 'Ford over R Cray.' The form in ASC is corrupt for *Cræganford*.

Crayke YN [*Crec* 685 BCS 66, *Creic* c 980 ib. 1255, DB, 1088 LVD, *Craic* 1176 P]. OW *creic*, Welsh *craig* 'a rock'. The place is on a ridge.

Creacombe D [*Crawecome* DB, *Creuecumbe* 1238 Ass, *Creucumbe* 1284 Ep]. 'Crows' valley.' First el. OE *crāwe* 'crow', which ought really to have had the form *crǣwe* in the nom., *crāwan* in oblique forms. Occasionally the nom. *crǣwe*, which is not evidenced, gave rise to an oblique form *crǣwan*.

Creake, North & South, Nf [*Creic, Creich, Suthcreich* DB, *Crech* 1190 P, *Cre(i)c* 1196 FF, *Northcrec* 1211 Cur]. Apparently identical with CRAYKE.

Creaton Np [*Cre-, Crep-, Craptone* DB,

Creton 1197 P, *Creiton* 1202 Ass]. First el. OW *creic*. Cf. CRAYKE.

Credenhill He [*Cradenhille, Credenelle* DB, *Creddehull* 1242 Fees, *Credenhull* 1249 Fees]. 'Creoda's hill.' Curiously enough the same name occurs in Wilts: *Creodan hyll* BCS 390, 566.

Crediton D [*Cridie* 739, *Cridiantun* 930 Crawf, *Cridiantun* 977 ASC (C), *Chritetona* DB]. 'TŪN on R Creedy.' **Creedy** R D [(on) *Crydian, Cridian* 739 Crawf, *Cridia* 1244 Ass]. A Brit river-name.

Creech Do [*Criz, Cric* DB, *Crihz* 1212 Fees, *Crihc* 1264 Ipm], **Creech St. Michael** So [*Crice* DB, *Criche* c 1100 Montacute, *Cruche* 1157 Fr]. Brit *crūc*, Welsh *crug* 'a hill'.

C~ Do is named from **Creech Barrow**, a hill of 655 feet. C~ **St. Michael** took its name from **Creechbarrow Hill**, which is referred to as 'collem qui dicitur britannica lingua *Cructan apud nos Crycbeorh*' 682 BCS 62. *Cructan* is 'the hill on R TONE'. C~ St. Michael is dedicated to St. Michael.

Creed Co [(rector) *Sancte Cride* 1275 Ep, *Cride* 1291 Tax, *Sancta Crida* 1310 Ep]. '(The church of) St. Creda or Crida.'

Creedy R. See CREDITON.

Creeksea or **Cricksea** Ess [*Criccheseia* DB, *Krekeset* 1198, *Crikeseie* 1200 Cur, *Krikesheth* 1240 FF]. 'The landing-place (OE *hȳþ*) at the creek'? If so, the creek must be the river Crouch.

Creeping Hall Ess nr Colchester [*Crepinges* DB, 1186 P, *Creppinges* 1195 ff. P]. 'Cryppa's people', an OE *Cryppingas*. *Cryppa* would be related to OE *cropp, Crypsa*.

Creeting St. Mary & St. Peter Sf [*Cratingas, Cratingis* DB, *Cretinges* 1199 P, *Cretinges, aliam Cretinges* 1212 Fees, *Creting Sancte Marie, Sancti Petri* 1254 Val]. '*Crǣta*'s people.' Cf. CRATFIELD.

Creeton Li [*Cretone* DB, *Cretun* 1212 Fees, *Creton, Cretton* 1202 Ass]. '*Crǣta*'s TŪN.' Cf. CREETING, CRATFIELD.

Creighton St nr Uttoxeter [*Crectone* 1166 RBE, *Creiton* 1222 Ass, *Cracton* 1242 Fees]. Identical with CREATON Np.

Crendon, Long, Bk [*Credendone* DB, *Creendon* 1169, *Crééndon* 1196 P]. 'Creoda's DŪN.'

Creslow Bk [*Cresselai* DB, *Kerselawa* 1176 P]. 'Hill where cress grew.' See HLĀW.

Cressage Sa [*Cristesache* DB, *Cristesech* 1185 P, *-eche* 1232 Ch]. 'Christ's oak.' The second el. is in the dat. form, OE *ǣc*. Possibly an oak at which the gospel was preached by missionaries.

Cressing Ess [*Kersiges* 1198, *Kersinges* 1204, *Kersing* 1235 FF]. 'Place where cress grew.' An OE *cærsing*.

Cressingham Nf [*Cressingaham, C~ Parva* DB, *Kersingeham* 1168 P, *Great, Little*

Kersingham 1264 Ipm]. The first el. looks like a derivative in *-ingas*. No pers. n. *Cressa* is known, but it might be a derivative with an *s*-suffix of *Cretta*. Or *Cressingas* might be elliptical and mean 'people from Cresswell' or the like.

Cresswell Db [*Cressewella* 1176 f. P], C~ Nb [*Kereswell* 1234 Cl, *Cressewell* 1242 Fees], C~ St [*Cressvale* DB, *Cressewella* 1190 P]. 'Stream where water-cress grew.'

Cretingham Sf [*Gretinga-, Gratingeham* DB, *Gretincgeham* 1086 IE, *Gretingham* 1195, 1198 FF]. The regular *Gr-* shows that this name cannot be related to CREET- ING. The first el. is a toponymic *Grēotingas* 'people living in a sandy or gravelly district', identical with the early tribal name *Greut- ungi*, the name of a Gothic tribe. The change of *Gr-* to *Cr-* may be due to in- fluence from Creeting.

Crewe Chs [*Crev* DB, *Cruue* 1288 Court, *Crue* 1346 BM], **Crewe Hall** Chs [*Crev- halle* DB, *Cryu* c 1100, c 1150, *Cruwe* c 1190 Chester]. Welsh *cryw* 'ford, stepping- stones'.

Crewkerne So [*Crucern* c 880 BCS 553, *Cruche* DB, *Cruke* 1225 Ass, *Crukerne* 1266 Ep]. 'The house at *Cruc.*' *Cruc* was the name of a spur of hill, derived from Brit *crūc*, Welsh *crug* 'a hill' (cf. CRICKET). Some- times Crewkerne was called *Cruc* alone. See ÆRN.

Crewood Hall Chs nr Frodsham [*Crewode* 1287 Court]. The first el. is probably a third Crewe (cf. CREWE), to which was added *wood*. Thus 'the wood at Crewe'.

Crewton Db was named from a family called Crewe.

Crīch Db [*Cryc* 1009 Hengwrt MS 150, *Crice* DB, *Cruc* 1166 RBE, *Cruch, Cruz* 1229 ff. Ch]. Brit *crūc* 'a hill'. The place is near a prominent hill called *Cruchill* 13 For.

Crichel, Long & More, Do [*Chircelford* 935 BCS 708, *Circel* DB, *Kerechel, Kerichel* 1202 FF, *Longcherchel* 1219 FF, *Mor Kerchel* 1212 Fees]. Brit *crūc* 'hill' (cf. prec. name), to which was added an explanatory OE *hyll. More* is OE *mōr* 'moor'.

Crick Np [*Crec* DB, 1220 Fees, *Kreic* 1201 Cur, *Creck* 1254 Val]. Either Brit *crūc*, as in CRICH &c., or OW *creic* 'rock'. Cf. CRAYKE.

Cricket Malherbie (măˈlerbĭ) & **St. Tho- mas** So [*Cruchet, Cruche* DB, *Cruket* 1201 Cur, 1242 Fees, *Cryket Malherbe* 1320 Ipm, *Cruk Thomas* 1291 Tax]. In DB *Cruchet* refers to C~ Malherbie, *Cruche* to C~ St. Thomas, and this distinction is still to be traced in Tax. The original name was *Cruc* from Brit *crūc* 'hill' (cf. prec. names). The ending *-et* is the French dim. suffix (cf. CLARET). Cricket thus means 'little

Cruc'. Later this name was transferred also to C~ St. Thomas.

Robert Malherbe held Cricket in 1228. Cf. on the name BOUGHTON MALHERBE.

Crickheath Sa [*Gruchet* 1272 Ipm]. Brit *crūc* 'hill' and OE *hǣþ* 'heath'.

Cricklade W [*Crecca gelad* 905 ASC, *Crac- gelad* c 975 Wills, *Crocgelad* 1008 KCD 1305, *Cricgelad, Cræcilad* 1016 ASC (D, E), *Crichelade* DB]. Apparently OE *crȳc-* or *crȳca-gelād* with Brit *crūc* 'hill' as first el. The reference would be to **Horsey Down** near the place. Second el. OE *gelād* 'passage'. The name would mean 'the passage over the Thames by the hill(s)'. The variation in the vowel of the first syl- lable would be due to different substitutions for Brit *u*. Or the first el. may be OW *creic*; cf. CRAYKE.

Cricklewood Mx [*le Crikeldwode* 1294, *Crikeledewod* 1321 PNMx(S)]. C~ is in Hendon, whose name means 'high hill'. *Crickle-* may be an earlier name of the hill (Brit *crūc* 'hill' and OE *hyll*; cf. CRICHEL). The middle element may be a reduced OE *hēafod*, 'hill' (cf. Pendle Hill for earlier Pendle) or OE *wudu* 'wood'.

Criddon Sa [*Critendone* 1166 RBE, *Cridelton* 1242 Fees, *Criddone* 1316 FA]. '*Cridela's* DŪN.' Cf. next name and CARLINGCOTT. **Cridela* is derived from *Crioda*.

Cridling Park, C~ Stubbs YW [*Cridelinc* 1157 YCh 186, *Cridlinc* 1173 ib. 197, *Cride- ling* 1202 FF, *Cridelinge* 1229 FF, *Crudeling* 1311 Ipm]. Perhaps rather a compound with OE *hlinc* 'hill' as second el. than a derivative in *-ing(as)* from *Cridela*. The first el. would seem to be the pers. n. *Crioda, Creoda.*

Stubbs presumably means 'the stubs' and refers to a clearing where stubs were left.

Crigglestone YW [*Crigestone* DB, *Criche- lest[on]* 1166 P, *Crikeleston* 1199, *Crigleston* 1202 FF]. 'TŪN *by Crȳc-hyll.*' *Crȳc-hyll* was no doubt an early name of the hill on which the place is. It is identical with *Crickle-* in CRICKLEWOOD.

Crimbles La [*Crimeles* DB, *Crumles* 1206 Cur, *Crimbles* 1207 FF]. OE *crȳmel* 'a small piece of land', a derivative of *crūma* 'crumb'. OE *crȳmel* is found in *Crymelhamm* 1005 KCD 1301.

Crimchard. See CHARD.

Crimple Beck YW [*Crempell* c 1185 YCh 515, *Crempel* 1293 Ass]. A Welsh *Crym- pwll* 'crooked stream', from Welsh *crwm* 'crooked' and *pwll* 'a stream'.

Crimplesham Nf [*Crepelesham* DB, *Crim- plesham* 1200 Cur, *Crimpelesham* 1203 Ass, *Crunplisham* 1291 Tax]. '*Crympel's* HĀM.' *Crympel* is an unrecorded pers. n. derived from OE *crump* 'crooked'.

Crimscote Wa [*Kynemarescote* 1232

PNWa(S), *Kirmarescote* 1370 AD, *Kirmiscote* 1417 BM]. '*Cynemær's* COT.' Cf. CRIMCHARD (under CHARD).

Cringleford Nf [*Cringelforð* 1043 or 1044 ASCh, *Kringelforda* DB, *Cringelford* 1191 P]. 'Ford by the round hill.' The place is on the Yare. The first el. may be Scandinavian. ON *kringla* means 'a circle'. It is found in **Cringleber** La, the name of a round hill.

Crockerton W [*Crokerton* 1350 FF, 1463 Ipm]. 'The potters' TŪN.' OE *croccere* 'potter' is not evidenced, but ME *crokkere* is found from c 1315.

OE **croft** 'a piece of enclosed land used for tillage or pasture, a small piece of arable land adjacent to a house' occurs occasionally, especially as the second el. of pl. ns.

Croft He [*Crofta* DB, *Croft* 1163 P], C~ La [*Croft* 1212 Fees], C~ Li [*Croft* DB, c 1115 LiS], C~ YN [*Croft* DB, 1202 FF]. 'The enclosure.'

Croft Le [*Craeft* 836 BCS 416, *Crebre* DB, *Creft* c 1160 AC, 1165 P, *Craft* 12 DC, 1209–19 Ep, 13 BM]. Identical with *Cræft* 931 BCS 678 (Brk or W). This must be OE *cræft* 'craft, a machine, engine'. A windmill or water-mill may be referred to.

Crofton Cu [*Croftotona* c 1150 StB, *Crofton* 1198 FF, 1201 P], C~ Ha [*Croftone* DB, *Crofton* 1242 Fees], C~ W [*Croftun* 1167 P, *Crofton* 1242 Fees], C~ YW [*Scroftune* DB, *Croftona* c 1125 YCh 1428, *Crofton* 1219 FF]. 'TŪN with a croft.'

Crofton K [*Croptun* 973 BCS 1295, 998 KCD 700, *Croftona* 1179 P]. OE *cropp* is found in various senses, as 'bunch, ear of corn, crop of a bird' &c. It could probably also be used of a mound or hill. There is a hill near Cropton. The probable meaning is 'TŪN by the hill'.

Crofton Li [*Crohcton* 1204 Cur, *Crocton* 13 FF]. Very likely OE *Croh-tūn* 'saffron TŪN'.

Croglin Cu [*Crokelyn* c 1140 WR, *Croclyn* 1274 Cl, *Crogline* c 1160 WR, *Crogelin* 1195 P]. The place is on **Croglin Water** [*Croglyng* 1341 ERN]. Probably a name of the stream, with OE *hlynn* 'a torrent' as second el. The first may be ME *crōk* 'a bend' (from OScand *krōkr*).

OE **croh** 'saffron', also perhaps in *colloncroh* 'water-lily', may be the first el. of some names, as CROFTON Li, CROUGHTON Chs, CROYDON Sr. But there must also have been an OE *crōh*, perhaps with the sense 'corner, valley', found in CROOM and very likely in some other names, as CROOKHAM Brk, CROWFIELD, CROWHURST Sx, and perhaps CROWLE Wo.

ME **crōk** 'crook, bend', probably from ON *krōkr*, OSw *krōker*, ODan *krōk*, is found in CROOK Du &c., CROOKES, CROOKHAM Nb, CROOKHOUSE, CRAWCROOK, perhaps CROGLIN.

Crōmer Nf [*Crowemere* 13 BM, *Crowmere* 1297 AD]. 'Crows' mere or lake.'

Cromford (-ŭ-) Db [*Crunforde* DB, *Crumford* 1204 Cur, *Crumbeford* 1251 Ch]. 'Ford by a bend' (of the Derwent). The first el. is a noun derived from OE *crumb* 'crooked'.

Cromhall Gl [*Cromhal*, *Cromale* DB, *Crumhala* 1190 P, *Cromhale* 1220 Fees], **Crompton** La [*Crumpton*, *Crompton* 1246 Ass]. 'HALH and TŪN in the bend of a river.' Cf. CROMFORD.

Cromwell (-ŭ-) Nt [*Crunwelle* DB, *Crumwella* 1186 P, *Crumbwell* 1230 P]. 'Winding stream.' First el. OE *crumb* 'crooked'.

Crondall (-ŭ-) Ha [(æt) *Crundellan* c 880 BCS 553, *Crundelas* 974 ib. (1307), 979 KCD 622, *Crundele* DB, *Crumdela* 1179 P]. OE *crundel*, *crundul*, which may mean 'a chalk-pit', 'a hollow'. It is rather doubtful if this can be a compound of OE *crumb* 'crooked' and *dell* or *dæl*.

Cronton La [*Crohinton* 1242 Fees, *Growynton* 1242 LaInq, *Crouington* 1246 Ass, *Crouwenton* 1333 FF]. Etymology doubtful, because it is not clear if the original vowel of the first syllable was *ā* or *o*. In the former case the base would be OE *crāwena tūn* 'crows' TŪN', 'TŪN where crows were common'. If the vowel was *o*, the first el. might be a derivative of OE *crōh*. See CROOM.

Crook (or **Crooke**) **Burnell** D in N. Tawton [*Cruc* DB, *Cruk* 1234 Fees, *Cruk Burnel* 1316 FA], **Crook Hill** Do [*Cruc* 1014, 1035 KCD 1309, 1322]. Brit *crūc* 'a hill'. Cf. CRŪC. Crook Hill is a conspicuous hill. The same is the first el. of **Crookbarrow Hill** Wo [*Cruchulle* 1182 PNWo).

Burnell is a family name. Robert Burnel held *Cruk* in 1234 (Fees). *Burnel* is from OFr *brunel* 'brown'.

Crook Du [*Cruketona* 1267 FPD, *Crok* 1304 Cl], C~ La in Shevington [*Crok* 1324 Ct], C~ We [*Croke* c 1175 Kendale, *Crok* 1297 ib.]. ME *crōk* 'a bend' (from ON *krōkr*, ODan *krōk*, OSw *krōker*). The reference is to a bend of a river.

Crookes YW [*Croche*, *-s* DB, *Crokis* 1297 Subs]. 'The bends.' Cf. CRŌK.

Crookham Brk [*Crohmham* 944 BCS 802, *Crocheham* DB, *Croukham* 1228 BM], C~ Ha [*Crocham* 1248, *Crokham* 1257, *Croukham* 1341 Crondal]. The first el. is very likely the unrecorded OE word *crōh* 'a corner' or 'a bend' or both, suggested as the source of CROOM. C~ Brk is on a ridge between the Kennet and the Enborne. The situation of C~ Ha is not so characteristic.

Crookham Nb [*Crucum* 1244 Ch, *Crukum* 1255 Ipm, *Crocum* 1279 Ass]. '(At) the bends (of the Till).' The dat. plur. of OScand *krōkr*; cf. CRŌK.

Crookhouse Nb [*Le Croukes* 1323 Ipm]. 'The bends.' Cf. CRŌK.

Croom YE [*Crogun* DB, *Croun* c 1110 YCh 25, *Crohum* J BM, 1235 FF, *Crouum* 1212 FF]. The base must be an OE *Crōgum or *Crōhum. The place is in a valley with several widenings or side-valleys. A word meaning 'valley' or the like has to be assumed, probably an OE *crōh* corresponding to ON *krá* 'a corner'. From this sense the transition would be easy to 'cul-de-sac' or 'narrow valley'. The original meaning of *krá* was 'bend'. The base is *kranhō, which is related to OE *cringan* lit. 'to twist'. See CROH.

Croome d'Abitot, Earls & Hill Croome Wo [*Cromman*, (æt) *Cromban* 969 BCS 1235, *Crumbe* DB, 1176 P, *Hylcromban* 1038 KCD 760, *Crombe Dabetoth* 1275 Subs, *Erlescrombe* 1495 Pat]. The places were named from the brook that runs past them, and whose old name was no doubt *Crombe* 'the winding stream'. The name comes from a Brit *Crombā < *Crumbā, identical with the nom. fem. of Welsh *crwm* 'crooked'. Another brook in Wo is referred to as *Crome* 972 BCS 1282.

The family name d'Abitot is derived from ABBETOT in France.—**Earls C~** belonged to the earls of Warwick.—**Hill C~** is on a hill.

Cropredy O [*Cropelie* DB, *Cropperia* 1109 Eynsham, *Croprithi* c 1275 Godstow]. Second el. OE *rīþig* 'brook'. The first may be OE *cropp* in some sense, possibly referring to some water-plants. For OE *cropp* see CROFTON K and the following names.

Cropston Le [*Cropeston* c 1125 LeS, *Cropston* 1196 P, 1252 Fees]. '*Cropp*'s TŪN.' This pers. n. is not otherwise evidenced, but it may be a nickname derived from OE *cropp* in such a sense as 'crop of a bird'.

Cropthorne Wo [*Cropponþorn* 780, *Croppanþorn* 841 BCS 235, 432, *Cropetorn* DB]. Near C~ was formerly a place called *Croppedune* WoC (PNWo). This suggests that the first el. is a hill-name *Croppe, derived from *cropp* (cf. CROFTON K). Also *Croppanhull* 705 BCS 112 (now **Crapnell** So) may contain a hill-name.

Cropton YN [*Croptune* DB, *-tun* 1167 P]. 'Hill TŪN.' Cf. CROFTON K.

Cropwell Bishop & Butler Nt [*Crophille*, *-helle* DB, *-hulla* 1177 f. P, *Croppehull Episcopi* 1316 FA, *Croppill Boteiller* 1265 Misc]. OE *cropp-hyll* 'hill with a crop or characteristic hump'. On *cropp* see CROFTON K.

C~ Bishop belonged to the Archbishop of York.—**C~ Butler** was held t. Hy 2 by Richard the Butler, probably son of Robert, the Butler of Ranulf Earl of Chester (Thoroton).

OE **cros**, ultimately from OIr *cross*, but through Scandinavian (ON *kross* &c.), occurs as the first and the second el. of pl. ns. Cf. CROSBY &c., HOAR CROSS.

Crosby Cu nr Maryport [*Crosseby* 1291 Tax, *Crosseby in Aldredale* 1316 Misc], **C~ upon Eden, High & Low C~** Cu

[*Crosseby* c 1265 WR, c 1275 StB], **Great & Little C~** La [*Crosebi* DB, *Crossebi* 1177 P, *magna Crossby* c 1190 LaCh, *Parva Crossby* 1242 Fees], **C~ Garrett** We [*Crosseby* 1200 Obl, *Crossebi Gerard* 1206 Cur], **C~ Ravensworth** We [*Crosseby Raveneswart* c 1160 YCh 175, *Crossebi* 1195 P]. OScand *Krossa-bȳr* 'BY with crosses'. It is noteworthy that in all the names the old form is *Crosse-*, which must represent the OScand gen. plur. *krossa-*.

Garrett is the French pers. n. *Gerard*, ultimately of German origin.—Ravensworth is the ON pers. n. *Hrafnsvartr*, lit. 'raven black'.

Crosby Li [*Cropesbi* DB, *Crochesbi* c 1115 LiS, *Crosseby* 1206 Ass], **C~ YN** [*Crox(e)bi* DB, *Crossebi* 1088 LVD, c 1155 YCh 952]. Identical with CROXBY.

Croscombe So. See CORSCOMBE.

Crosland, South, YW [*Cros-*, *Croisland* DB, *Crosland* c 1200 YCh 1701]. 'Land by a cross.'

Crosscanonby Cu [*Crosseby* 12 Holme C, *Crosseby Canoun* 1285, *Crosbycannonby* 1535, *Croscanonby* 1552 PNCu(S)]. C~ was originally a part of Crosby in Allerdale given to the canons of Carlisle.

Crosscrake We [*Croskrake* 1282–90 Kendale]. ON *kross* 'cross' and a pers. n. (ON *Kraki* or *Krákr* or the woman's name *Kráka*).

Crossdale Cu [*Crozedal* 1294 Cl]. 'Valley with crosses.'

Crossens La [*Crossenes* c 1250 PNLa, 1323 Ct]. ON *krossa-nes* 'headland with crosses'.

Crossrigg We [*Crosrig* 13 StB]. 'Ridge with a cross.' Second el. ON *hryggr* 'ridge'.

Crosthwaite Cu [*Crosthwayt* 1233 Ep, *-thweyt* 1246 FF], **C~ We** [*Crosthwait* c 1190 Kendale], **C~ YN** [*Crosthwait* 1201 Ch]. 'Clearing by a cross.' Cf. THWAITE.

Croston La [*Croston* 1094, c 1190 LaCh]. TŪN with a (market) cross.'

Crostwick (krŏsĭk) Nf [*Crostucit* DB, *Crosthweyt* 1302 Ch], **Crostwight** Nf [*Crostwit* DB, *Crostweit* 1211 FF]. Identical with CROSTHWAITE.

Crouch (-ow-) R Ess [*Crouch* 1576 Saxton]. A late name which may be an antiquarian's formation from early forms of CREEKSEA or a back-formation from a pl. n. containing OE *crūc* 'cross'. See next name.

Crouch End Mx [*Crouche* 1400, *Crouchend* 1465 AD]. Crouch is OE *crūc*, ME *cr(o)uche* 'cross', ultimately from Lat *crux*. The meaning of *End* is not quite clear. Crouch End may be 'the side where the cross stands'.

Croucheston W [*Cr(o)ucheston* 1328 Ipm, *Crucheston* 1340 FF]. The forms are too late for an etymology to be suggested. The earliest instance in PNW(S) (*Crocheston* 1249 Ass) is inconclusive. First el. perhaps a family name.

Croughton Chs [*Crostone* DB, *Croctona* c 1100, *-ton* c 1190 Chester]. The first el. may be OE *croh* 'saffron' or the *crōh* discussed under CROOM. C~ is at the end of a valley.

Croughton (-ō-) Np [*Crevel-*, *Criweltone* DB, *Crouelton* 12 NS, *Craulton*, *Crewelton* 1198 P, *Croulton* 1202 Ass]. 'TŪN in a river fork.' First el. an unrecorded OE *crēowel*, corresponding to OHG *crawil*, MLG, MDu *krouwel*, OFris *krawil* 'fork'. The base would be **krawila-*, whence OE *crēowel* in the same way as **mawilō* gave OE *mēowle*. The meaning of OE *crēowel* would be 'fork', here 'fork of a river'. C~ stands in a tongue of land between two streams.

Crowan Co [*Ecclesia Sancte Crawenne* 1238, *Sancta Crouwenna* (*Crewenna*) 1269 Ep, (Ecclesia de) *Crewenne* 1291 Tax]. '(The church of) St. Crevenna.' Crevenna is said to have been a virgin saint.

Crowborough St [*Crowbarwe* 13 PNSt], C~ Sx [*Cranbergh* 1292, *Crowbergh* 1390 PNSx]. 'Crow hill.' C~ St might also be 'crow grove' (OE *bearu*).

Crowcombe So [*Crauuancumb* 904, *Crawancumb* 968 BCS 612, 1219, *Crawecumbe* DB]. 'Crow valley.'

Crowdycote Db [*Crudecote* 1223, *-s* 1244 FF, *Croudecote* 1251 Ch, *Welleton Cruddecote* 1287 FF]. '*Crūda*'s COT.' OE *Crūda* pers. n. is evidenced in **Critchet Field** Sr [*Crudan scéat* 909 BCS 627], *Crudan scypsteal* 962 BCS 1085 (Mx). Cf. *Crūd* in *Crudes silba* 873 BCS 536 (now **Curlswood** K).

Crowell O [*Clawelle* DB, *Crauwelle* 1231 Ep]. 'Crows' spring or stream.'

Crowfield Sf [*Crofelda* DB, *Cropfeld* 1212 Fees, *Crosfeld* 1219 FF, *Croffeld* c 1230 Bodl]. First el. very likely *crōh* 'corner' &c. (cf. CROOM). The place is at the head of a valley.

Crowhurst Sr [*Craueherst* 1189 P, *Crawe-*, *Crowehurst* 1303, 1315 Ch]. 'Crow wood.'

Crowhurst Sx [*Croghyrst* 772 BCS 208, *Croherst* DB, *Cruherst* 1245 Ch]. First el. OE *crōh* as in CROWFIELD and CROOM. The form of 1245 points to an el. with OE *ō*, and the place is near a recess in the chain of hills. *Hurst* may be hill or wood.

Crowland (-ō-) Li [*Cruglond*, *Crugland*, *Cruuulond*, *Cruwland* c 745 Felix, *Cruwland* 10 Guthlac, *Crūland* c 1000 Saints, *Croiland* DB, 1202 Ass]. The first el. is an otherwise unknown word *crūw* (*crūg*), which may be cognate with Norw *kryl* 'a hump' (from **krūwila-*) and the first el. of CROUGHTON Np. Presumably *crūw* meant 'a bend' and refers to the bend of the Welland at Crow-

land. This bend is not now very marked but very likely was before the draining of the fens.

Crowle (-ōō-) Li [*Crull* c 1080 YCh 468, *Crul(e)* DB, *Crull* 1232 Ep]. Named from a river of the same name [*Crulla* c 1100, *Crull* 1352 Selby]. The name is derived from OE **crull*, ME *crull* 'curly' in a more general sense 'winding'. Owing to draining the river has now disappeared.

Crowle Wo [(æt) *Croglea* 836 BCS 416, *Crohlea* 840 ib. 428, DB, *Croelai* DB]. In BCS 428 is mentioned a stream at Crowle, called *Crohwælla*, which must be BOW BROOK or one of its head streams. Bow Brook is strongly winding. Possibly the first el. is the word *crōh* (cf. CROOM) in the sense 'bend'. *Crōhwælla* would be 'winding stream', Crowle 'LĒAH by the bends or by *Crōhwælla*'.

Crowmarsh Battle & Gifford, Preston C~ O [*Craumareis* c 1085 BM, *Cravmares* DB, *Croumerse* 1195 FF, *Craumershe Bataill*, *Cromershe Gifford* 1316 FA, *Prestecromerse* 1279 RH]. 'Marsh frequented by crows.'

C~ **Battle** and **Gifford** belonged to Battle Abbey and Walter Gifard respectively in 1086 (DB). On Gifford see ASHTON GIFFORD. **Preston** C~ is really 'the priests' Crowmarsh' (*Preston* from ME *prestene* 'of priests', with the weak gen. plur. in OE *-ena* instead of regular *-a*).

Crownthorpe Nf [*Congrethorp*, *Cronkethor* DB, *Crungelthorp* 1252 Cl, *Crungethorp* 1252 Cl, *Crunkelthorp* 1316 Ipm]. The first el. is very likely an OScand nickname belonging to Norw dial. *krungla* 'a crooked tree', *krunglutt* 'crooked' (of trees or limbs).

Crowton Chs [*Crouton* 1260 Court, 1308 Ipm]. Crewood is in Crowton. This may suggest that the first el. of the name is Welsh *cryw* 'ford', but the vowel is not what we expect.

Croxall St [*Crokeshalle* 942 BCS 773, *Crocheshalle* DB]. Apparently '*Cróc*'s hall'. *Croc* (moneyer t. Cnut, DB) is ON *Krókr*, ODan *Krók*, OSw *Króker*, originally a byname from *krókr* 'a hook'. The Scandinavian name here appears remarkably early.

Croxby Li [*Crocsbi*, *Crosbi* DB, *Crochesbi* c 1115 LiS, *Croxebi* 1202 Ass]. '*Cróc*'s BY.' Cf. CROXALL.

Croxdale Du [*Crokesteil* 1195 (1335) Ch, *Crokestail* c 1190 Godric]. '*Cróc*'s piece of land.' OE *tægl* 'tail' must have been used of a piece of land jutting out from a larger piece. This use is still found in Scotland.

Croxden St [*Crochesdene* DB, *Crokesdene* 1212 Fees], **Croxley** Hrt [*Crokesleya* 1166 RBE, *-lea* 1176 P, *Crochesle* 1198 (1301) Ch]. '*Cróc*'s valley and LĒAH.' Cf. CROXALL.

Croxteth La [*Crocstad* 1257, *Croxstath* 1297 LaInq]. '*Cróc*'s landing-place.' Second el.

ON *stǫð* or OE *stæþ* 'landing-place'. Cf. CROXALL.

Croxton Ca [*Crocestona* c 1080 ICC, *Crochestone* DB, *Crocstun* 1202 FF], C~ Chs [*Crostune* DB, *Croxton* 1260 Court], C~ (krǒsn) **Kerrial** Le [*Crohtone* DB, *Crocstona* c 1125 LeS, *Croxton Kyriel* 1247 Ass], **South** C~ (krǒsn) Le [*Crochestone* DB, *Sudcroxton* 1212 Cur], C~ Li [*Crochestune* DB, *Crochestuna* c 1115 LiS, *Croxton* 1196 P], C~ Nf nr Fakenham [?*Crochestune* c 1050 KCD 907, *Crokeston* 1242 Fees], C~ Nf nr Thetford [*Crokestuna* DB, *-tone* 1254 Val], C~ St [*Crochestone* DB, *Croxton* 1327 Subs]. '*Crōc*'s TŪN.' Cf. CROXALL.

C~ **Kerrial** was granted to Bertram de Cryoil in 1242 (Ch), and was held by Nicholas de Kyriel in 1247 (Ass). The family name was taken from CRIEL in Seine-Inf.

Croyde D [*Crideholde* DB, *-ho* 1242 Fees, *Cridenho* 1307 Ol, *Crude* 1276 Cl]. The earliest forms really belong to **Croyde Hoe**, a point of land. The probability is that Croyde is an old name of the headland. It would be an OE *crȳde*, formed from *crūdan* 'to press, make one's way'. Hoe is OE *hōh*, here in the sense 'headland'.

Croydon Ca [*Crauedena* c 1080 ICC, *Crauuedene* DB, *Craudene* R 1 Cur], C~ **Hill** So [*Craudon* 1243 Ass, *Croudon* 1331 Dunster]. C~ Ca is 'valley frequented by crows', C~ So 'hill frequented by crows'.

Croydon Sr [*Crogedena* 809, (æt) *Crogdene* c 871, (de) *Croindene* c 980 BCS 328, 529, 1133, *Croindene* DB, *-dena* 1168 P]. Perhaps 'valley where wild saffron grew'. The first el. appears in two different forms, with and without *n*. The form in *-n* may be an adj. in *-en* derived from *crog*. The OE forms would then be *Crog-denu* and *Crogen-denu*, both meaning 'saffron valley'. *Crogen* 'of saffron' is not found, but is a regular formation.

Brit, OW **crūc**, Welsh *crug* 'heap, barrow, hill, esp. a round hill or hillock', corresponding to Ir *crúach*, is very common in English pl. ns. The form of the element varies a good deal owing to different times of adoption and different substitutions of English for British sounds. The earliest Brit vowel in the word was *ū*. Hence Engl CROOK, CREWKERNE &c. At an early period *ū* passed into *ȳ*. Hence OE *Crȳc* (cf. CREECH So). Hence CRICKLEWOOD, CRIGGLESTONE, KIRKLEY, &c. Owing to sound-substitution OE *crȳc* became *crȳč* (with palatal *č*). Hence CREECH, CRICHEL, CRICH, CRUCHFIELD, CHRISTON &c. Finally, owing to association with OE *cirice* 'church' the metathesized form *cyrc* got palatal initial *č* too. Hence CHURCHDOWN, CHURCH HILL &c. An explanatory OE *hyll* or *dūn* 'hill' was often added to the original Brit *crūc*, as in CRICHEL, CHURCHDOWN. Brit *crūc* is the second el. of PENKRIDGE.

Cruchfield Brk [*Cruchesfeld* 1212 Fees,

Cruchefeld 1220 Fees, 1230 P]. First el. Brit *crūc* 'hill'. Cf. FELD.

Cruckmeole. See MEOLE.

Cruckton Sa [*Croctun* 1272 Ipm, *Crokton* 1308 Ipm]. 'TŪN by a hill.' First el. Brit CRŪC.

Crudgington Sa [*Crugetone* DB, *-ton* 1231 Cl, *Crugelton* 12, 13 Eyton]. The first el. may be an OE *crȳč-hyll* (cf. CRICHEL, CRUTCH), though with abnormal voicing of the *č*. The place is low on the Tern, but near a marked hill.

Crudwell W [*Criddanuille* 854, *Cruddewelle* 901 BCS 470, 586, *Creddewilla* 1065 KCD 817, *Credvelle* DB, *Credewella* 1181 P]. '*Creoda*'s stream.' The OE forms are in late transcripts.

OE **crumb** 'crooked'. See CROMWELL. CROMFORD, CROMHALL, CROMPTON seem rather to contain a word for 'bend' derived from *crumb*.

Crumbles, The, Sx [*Crumble* 1275 RH]. Identical with CRIMBLES La.

Crummock Water Cu [*Crombocwater* 1308 Ipm, *Crombokwatre* 1343 Cl]. The first el. is doubtless a river-name, denoting the upper Cocker, which flows through the lake. **Crummock Beck** is another stream in Cu, a tributary of the Waver [*Crumboc* 1201 Ch]. The name is British, derived from Welsh *crwm* (Brit *crumbo-*) 'crooked' and identical with CROMOGE, CRUMMOGE in Ireland. The Brit form would be *Crumbāco-*.

Crumpsall La [*Cormeshal* 1235 Ass, *Curmeshale* 1322 LaInq]. The first el. seems to be a pers n., which might be a derivative of OE *crumb* 'crooked'. Or it might be ON *Krumr*, lit. 'the crooked one'. Second el. HALH.

Crundale K [*Crundala* 11 DM, *Crumdal*, *Crundale* 1242 Fees]. Identical with CRONDALL Ha.

Crutch Wo [*Cruchia* Hy 2 (1285) Ch]. Brit *crūc* 'hill'. Cf. CRŪC.

Cruxton Do [*Froma* DB, *Fromma Johannis Croc* 1178 P, *Crocston* 1195 P]. Originally a part of FROME. Later Croxton from tenants bearing the name *Croc*. Willelmus Croc held the manor in 1195 (P).

Cryfield Wa [*Croilesfelda* 1204, *Crulefeld* 1284 Ch, *Crey-*, *Crewelfeilde* 1590 AD]. The place is situated at the junction of two streams. First el. as in CROUGHTON Np.

OE **cū** 'cow' is a common first el. See COW- (passim), COWES, COOLE, QUY. The gen. sing. *cȳ* is found in KEELE, KEYHAVEN, KEYMER, KYLOE, KYO, WHAW, the gen. plur. *cūna* perhaps in COUNDON Du.

Cubbington Wa [*Cobintone, Cubintone* DB, *Cubbintona* Hy 2 (1314) Ch, *Cubinton* 1201 FF]. 'The TŪN of *Cubba*'s people.' The name *Cubba* is not recorded in OE, but it is related to *Cybba* in *Cybban stan* 957 BCS 1002 (Brk) and *Cybbel* in *Cybles weorðig* 849 ib. 455. Cf. also KIBBLESWORTH and KIBWORTH.

Cubert Co [(Vicaria) *Sancti Cuberti* 1269 Ep, (Ecclesia) *Sancti Cutberti* 1291 Tax]. '(The church of) St. Cuthbert.'

Cubley Db [*Cobelei* DB, *Cubbeleg* 1232 FF, *Cubbelegh* 1255 Ch]. '*Cubba*'s LĒAH.' Cf. CUBBINGTON. *Cub* 'a young fox' might be thought of as first el., but has been first found in the 15th cent.

Cublington Bk [*Coblincote* DB, *Cubelintona* 1154 Eynsham]. 'The TŪN of *Cubbel*'s people.' *Cubbel* is a derivative of *Cubba*. Cf. CUBBINGTON.

Cuckamsley Knob Brk, a hill near Wantage [*Cwicchelmeshlæw* 1006 ASC (E)]. '*Cwichelm*'s burial-mound'; see HLĀW. It has been suggested that the Cwichelm who was buried at C~ was the West-Saxon king Cwichelm who died in 593 according to the Chronicle.

Cuckfield (-ōō-) Sx [*Kukefeld*, *Kukufeld* W 2 PNSx, *Cucufelda* 1121 AC, *Cokefeld* 1255 Ch]. This might be 'cuckoos' FELD', if *cuckoo* is an English word. But very likely it is French and at most the name has been associated with it. *Cuca* pers. n. or a form of OE *cwice* 'couch-grass' are more likely as the first el. Cf. COOKBURY.

Cucklington So [*Cocintone* DB, *Cukelingeton* 1212 Fees, 1274 RH]. 'The TŪN of *Cucol(a)*'s people.' *Cucola* is evidenced in CUXTON K. *Cucol* is found in *Cuceles hyll* KCD 741 (Do), in **Curscombe** D [*Cochalescome* DB, *Coklyscomb* 1333 Subs] and **Cuckoldscoomb** K [*Cukkelescumbe* 1226 Ass]. *Cucol* is a derivative of *Cuca, which itself is a short form of names in *Cwic-*.

Cuckmere (-ōō-) R Sx [*Cuckmer* 1577 Harrison]. A back-formation from **Cuckmere Haven** [*Cokemerehaven(e)* 1352 Ass, 1422 Ipm]. **Cuckmere** [*Cookemere* 1335 ERN] must have denoted a widened part at the mouth of the river, and its first element may be a river-name *Cuce* (cf. COOKRIDGE) or *Cuca* pers. n. or OE *cucu* (*cwicu*) 'living' in some special sense. Cf. *quicksand*.

Cuckney Nt [*Cuchenai* DB, *Cugeneia* 1187 P, *Cucheneia* c 1200, *Kuyekeney* c 1245 Bodl]. '*Cuca*'s island.' Cf. COOKBURY, COOKLEY Sf. *Cuca* is a short form of names in *Cwic-*.

Cuddesdon O [*æt Cupenesdune* 956 BCS 945, *Codesdone* DB, *Cudesdon* 1122 Fridesw]. '*Cūpen*'s DŪN.' *Cūpen* is a short form of names in *Cūp-*, as *Cūpwine*, formed with the diminutive suffix -*in*. The same name is found in *Cupænesford* 909 BCS 622.

Cuddington Bk [*Cudintuna* c 1120 Reg Roff, 1176 BM], C~ Chs nr Malpas [*Cun-*

titone DB, *Cudinton*, *Cudington* 1288 Court], C~ Chs nr Northwich [*Cudinton* 1278 Misc], C~ Sr [*Cotintone* 675, *Cudintone* 933 BCS 39, 697, *Codintone* DB, *Cudinton* 1198 P, 1201 Cur]. 'The TŪN of (the people of) *Cuda*.'

Cudham K [*Codeham* DB, 11 DM], **Cudworth** So [*Cudeworde* DB, *Cudewurth* 1243 Ass]. '*Cuda*'s HĀM and WORÞ.'

Cudworth YW [*Cudeuurdia* c 1185 YCh 1540, *Cutheworde* 12 ib. 1636, *Cudewrth* 1233, *Cutheworth* 1263 Ep]. '*Cūpa*'s WORÞ.'

Cuerdale La [*Kiuerdale* c 1190 LaCh, 1246 Ass, *Keuerdale* 1293 LaInq], **Cuerdley** La [*Kyuerlay*, *Kyuerdeleg*, *Cunercheleg* 1246 Ass, *Kyuerdelegh* 1275 LaInq]. Perhaps '*Cynferþ*'s valley and LĒAH'. For the loss of *n* before the labial cf. LYFORD.

Cuerden (-ūr-) La [*Kerden* c 1200 CC, 1246 Ass, *Kirden* 1212 LaInq]. Welsh *cerddin* 'ash-tree'.

Culbone So. A late name said to be from the saint's name *Columbanus*. The old name was **Kitnor** [*Chetenore* DB, *Kitenore* 1236 FF]. 'Hill slope frequented by kites' (OE *cyta*). Cf. ŌRA.

Culcheth La [*Culchet* 1201 P, *Kulcheth* 1246 Ass, *Culchit* 1258 Ass]. A Brit name identical with KILQUITE and COLQUITE Co, CILCOIT Monm and meaning 'back wood' or 'retreat in a wood'. The elements are Welsh *cil* 'back, corner, retreat' and *coed* 'wood'.

Culford Sf [*Culeford*, *Coleford* c 1025 BCS 1018 f., *Culeford* 11 EHR 43, -*forda* DB, -*ford* 1197 FF]. '*Cūla*'s ford.' Cf. COWLINGE, CULHAM O.

Culgaith (-ā·th) Cu [*Chulchet* c 1135 LaCh, *Culchet* 1203 P, *Culgait* c 1160 WR]. Welsh *cilgoed* 'back wood'. Cf. CULCHETH.

Culham, Upper, Brk [*Culnham* 1208 PWint, *Kilham* 1284 Ch, *Culham* 1402 FA]. 'HĀM or HAMM with a kiln.' Cf. CYLEN.

Culham O [*Culanham* 811, 815 BCS 352, 850, *Culanhom* 821 ib. 366, *Culham* 1242 Fees]. '*Cūla*'s HAMM.' Cf. COWLINGE. The place is in a bend of the Thames.

Culkerton Gl [*Culcorto(r)ne* DB, *Culkerton* 1204 Cur, 1220 Fees, *Culcretun* 1239 Ep]. The first el. may be a first pl. n., to which TŪN was added. It might be a Welsh *cilcawr* 'giant's retreat' or *cil-cor* 'dwarf's retreat'.

Cullercoats Nb [*Culvercoats* c 1600 PNNb]. 'Dove-cots' (first el. OE *culfre* 'pigeon, dove').

Cullingworth YW [*Colingauuorde* DB, *Culingeworth*, *Cullingewrthe* 1208 Cur, *Cullingwurth* 1235 FF]. 'The WORÞ of *Cūla*'s people.' Cf. COWLINGE, CULHAM O.

Culm R D [*Culum* 938 BCS 723, 1238 Ass]. 'Winding river.' Culm is derived from

Welsh *cwlwm* (Co *colm*, Bret *coulm*) 'a knot', no doubt also 'loop'. On the Culm are **Cullompton** [*Columtun* c 880 BCS 553, *Culumtun* a 1097 E, *Colunp* DB, *Culminton* 1165 P], **Culmstock** [*Culumstocc* 938 BCS 724, *Culmstok* c 1070 Ex, *Culmestoche* DB], **Uffculme** [*Offecoma* DB, *Uffe Culum* 1176 P]. Culmstock is 'STOC on the Culm' or 'the STOC belonging to Uffculme'. Uffculme is 'the Culm belonging to one *Uffa*'.

Culm Davy D [*Comba* DB, *Cumb* 1242 Fees, *Combe Davi* 1284–6 FA]. OE *cumb* 'valley'.

Davi from David de Wydeworth (1242 Fees). Culm Davy is near Culmstock, but not on the Culm.

Culmington Sa [*Comintone* DB, *Colmiton* 1160, *Colminton* 1161, *Culminton* 1197 P]. C~ is on the CORVE, which is a winding river and may well have been called Culm (cf. CULM). The probable meaning is 'the TŪN of the dwellers on the Culm'. But a possible alternative is OE *Cūphelmingatūn* 'the TŪN of *Cūphelm*'s people'.

Culpho Sf [*Culfole* DB, *Colfho* 1168, *Culfou* 1169, *Culfo* 1175 f., *Culfho* 1178 P, *Colvesho* 1250 Fees]. Probably '*Cūpwulf*'s HŌH or spur of land'. A later form of the pers. n. *Cūpwulf*, viz. *Cuulf*, appears in DB. The rare occurrence of gen. forms is somewhat remarkable, but not without analogies.

Culverthorpe Li [*Torp* DB, *Calewar-, Kalewarthorp* 1275 RH, *Kilwardthorp* 1338 Ch]. Originally THORP. The addition is the name of an owner, presumably *Chiluert* (DB) on which see KILLERBY, though numerous spellings with *a* in the first syllable are difficult to account for.

Culworth Np [*Culeorde* DB, *-wurðe* 1195 P, *-wurth* 1230 P, *-wrth* 1254 Val]. '*Cūla*'s WORÞ.' Cf. COWLINGE.

OE **cumb** 'a coomb, a deep hollow or valley' is an early loan from Celtic (MW, Welsh *cwm* 'a deep valley', Gaul *Cumba* pl. n., Bret *komb* 'a valley'). The base is *kumbā*, *kumbo-*. Welsh *cwm* is very common in pl. ns. Some names in *Cum-* in Cumberland are direct loans from the old Cumbrian language (e.g. CUMREW). English pl. ns. containing *cumb* are particularly common in the south-west (D, Do, So, see COMB), the reason being that narrow valleys of the coomb type are very common there. The absence or rare occurrence of *cumb* in some districts is due to topographical conditions, as in East Anglia, Li, or to Scandinavian influence, as in La, Y. In Scandinavian districts *gill* often designates a narrow valley. COMBE is found several times in Kent. COMPTON is widely distributed. In some cases *Combe* or *-combe* or the like has replaced some other el., as OE *camb* (in COMBS), *camp* (as SWANSCOMBE), or the dat. plur. ending *-um* (as ACOMB, ESCOMBE).

Cumberland [*Cumbraland* 945 ASC, *Cumerland* 1000 ASC (E), *Cumberland* ib. (D),

Cumberland c 1145 Facs. Nat. MSS of Scotland]. 'The land of the Cumbrians, i.e. Britons.' C~ originally denoted the British kingdom of Strathclyde, which embraced Cumberland. After C~ had been annexed by William Rufus, the name soon began to be used in its modern sense. *Cumbras* is from Welsh *Cymry* 'the Welsh'.

Cumberworth Li [*Combreuorde* DB, *Cumberworda* c 1115 LiS, *-worth* 1209–19 Ep], C~ YW [*Cumbreuurde* DB, *Cumberwrth* 1242 Fees]. '*Cumbra*'s WORÞ.'

Cumdi·vock Cu [*Combeðeyfoch* c 1060 Gospatric's ch, *Cumdeuoc* 1245 P]. A Brit name, consisting of Welsh *cwm* 'valley' and an OW *Dyfoc* derived from *du* (< *dubo-*) 'black', which may be an old name of CHALK BECK (cf. Ir DUVOG, a name of streams, and DEVOKE WATER), or a pers. n.

Cummersdale Cu [*Cumbredal* 1227 FF, *Cumberdale* 1292 QW, *Cumbresdall* 1293 Ipm]. 'The valley of the Cumbrians.' Cf. CUMBERLAND.

Cumnor Brk [*Colmonora* a 688 BCS 844, *Cumanora* 931 ib. 680, *Comenore* DB]. '*Cuma*'s hill slope.' One *Cumma* was abbot of Abingdon c 730 (BCS 155). The form *Colmonora* (like *Colmanora* 955 BCS 906, *Colmenora* 985 KCD 1283) is due to association with the Ir name *Colmán*. The spellings are in late transcripts.

Cumrew· Cu [*Comreu* 1202 FF, *Cumreu* 1211 P], **Cumwhi·nton** Cu [*Cumquintina* c 1200 WR, *-quintin* 1260 WR, *-quinton* 1227 FF], **Cumwhitton** Cu [*Cumquetinton* 1286 Ipm, *-quitington* 1294 ib.]. All three are old Cumbrian names, containing Welsh *cwm* 'valley'. The second el. of Cumrew is Welsh *rhiw* 'hill, ascent, slope' ('the valley by the hill', i.e. Cumrew Fell). Cumwhinton may be 'Quintin's valley'. Cumwhitton is a hybrid, the second el. being an OE pl. n. *Hwītingatūn*. The name means 'the valley by Whittington'.

Cundall YN [*Cundel, Goindel* DB, *Cundala* 1176 P, *Cumdal* 1235 FF]. Apparently the first el. is OE *cumb* 'valley'. The OE name was probably *Cumb*, to which Scandinavians added an explanatory *dalr* 'valley'.

Cunscough La [*Cunigescofh* a 1190 CC]. ON *kunungs-skógr* 'the king's wood'.

Curborough St [*Curborud* 1280 Ipm, *Curbur*' 1285 FA, *Curburgh* 1290 Ch, *Corbrun* 1291 Tax, *Curburn* 1428 FA]. OE *cweornburna* 'mill stream', with *burna* replaced by BURG. Cf. CWEORN.

Curbridge O [(æt) *Crydan brigce* 956 BCS 972, *Cradebrege* 1242 Fees, *Crudebrug* 1279 RH]. '*Creoda*'s bridge.'

Curdridge Ha [*Cuðredes hricgc* 901 BCS 596]. '*Cūprēd*'s ridge.'

Curdworth Wa [*Credeworde* DB, *Croddewurth* Hy 2 (1318) Ch, *Crudeworth* 1285 QW]. '*Creoda*'s WORÞ.'

Curland So [*Curiland* 1252 Ch]. First el. the pl. n. CURRY. But Curland is far from Curry, and probably the name means 'outlying land belonging to Curry'.

Curridge Brk [*Cusan hricg* 953 BCS 900, *Coserige* DB, *Cuserigge* 1157 P]. '*Cusa*'s ridge.'

Curry Mallet, C~ Rīvel, East & North C~, Curry Load So [*Curig* 854, 904 BCS 475, 612, *Churi, Curi, Nortcuri* DB, *Curiet, Nordcuri* 1156 P; *Curi Malet* 1225 Ass; *Curry Revel* 1225 Ass; *Byestcory* 1327 Subs; *Corilade* c 1157 Athelney, *Curilade road* 1233 Wells]. Curry is no doubt the old name of the stream that runs near the places. It is identical with CORY D (in CORYTON), but etymologically obscure.

C~ **Mallet** belonged to William Malet at least t. R 1. Malet is an OFr nickname, perhaps from OFr *malet* 'evil'. C~ **Rivel** was given to Richard Revel by R 1 (see 1194 P). Rivel or Revel is an OFr nickname (cf. OFr *revel* 'revolt'). **East** C~ is really 'east of R Curry'. **Curry Load** means 'the Curry road'.

Curthwaite Cu [*Kirkethuait* 1286 Ipm]. 'Thwaite by or belonging to a church.'

Cury Co [(Ecclesia de) *Sancto Corentino* 1284 Ass, (Villa) *Sancti Korentini* 1302 FF (Gover)]. Dedicated to St. Corentinus. The name CURY is late.

Cushat Law (hill) Nb [*Cousthotelau* c 1200 Newminster]. 'Wood-pigeons' hill.' OE *cúscote*, dial. *cushat*, means 'a wood-pigeon'. Cf. HLĀW.

Cusop He [*Cheweshope* DB, *Kiues Hop* 1196, *Kiweshop* R 1 Cur, *Kyweshop* 1291 Tax]. The first el. may be a Welsh stream-name *Cyw*, lit. 'the young of birds'; cf. CHEW. If the name was adopted late, initial *C-* would remain as *K-*.

Cusworth YW [*Cuzeuuorde* DB, *Cucewordh* 1208, *Cuzcewurth* 1237, *Cucewrth* 1240 FF]. '*Cūpsa*'s WORP.' The numerous spellings with *z, sc*, &c. show that we cannot start from *Cusa*. *Cūpsa* may be a derivative with an *s*-suffix from *Cūpa* or a short form of an unrecorded *Cūpsige*.

Cutcombe So [*Codecoma* (hd) 1084 Geld R, *Cudecumba* 1178 P]. '*Cuda*'s valley.'

Cutsdean Wo [*æt Codestune* 974 BCS 1299, 987 KCD 660, *-tune* DB]. '*Cōd*'s TŪN.' The same pers. n. is the first el. of COTSWOLDS and of *Codesuuella* 780, *-welle* 840 BCS 236, 430, apparently a place in Cutsdean.

Cutteslowe or **Cutslow** O [*Cuðueshlaye* 1004 Fridesw, *Codeslave* DB, *Cudeslawe* 1122 Fridesw]. '*Cūpen*'s burial-mound.' Cf. HLĀW, CUDDESDON.

Cuxham (-ōō-) O [*Cuceshæma gemære* 880 BCS 547, *Cuceshamm* 995 KCD 691, *Cuchesham* DB, c 1170 Oxf]. '*Cuc*'s HAMM.' *Cuc* or *Cucu* is a short form of names in *Cwic-*. Cf. COOKSEY.

Cuxton K [*Cucolanstan* 880 BCS 548, *Cucclestan* 10 ib. 1322, *Coclestane* DB]. '*Cucola*'s stone.' Cf. CUCKLINGTON.

Cuxwold Li [*Cucualt* DB, *Cucuwalt* c 1115 LiS, *Cucuwalg* c 1115 LiS, 12 DC]. '*Cuca*'s forest'; cf. WALD, COOKBURY. Or the name may be identical with COXWOLD YN.

OE **cweorn** 'a mill, esp. a handmill' no doubt also meant, like Goth *qairnus*, OHG *quirn*, ON *kvern*, 'a mill-stone'. This is the meaning in the OE *cweorn-dūn* that is the source of QUARLTON, QUARNDON, QUARRENDON, QUORLTON, perhaps CORNDEAN. The meaning is 'hill where mill-stones were quarried'. The same is the meaning in QUERNMORE, WHERNSIDE, perhaps QUARLEY Ha. OE *cweorn* must also have been used of a watermill or a windmill, as in OE *cwyrnburna* 962 BCS 1082, (æt) *cweornwelle* ib. 1129. The sense watermill is found certainly in CURBOROUGH. Cf. also QUARRINGTON, GORNAL. ON *kvern* is found in QUARMBY YW.

OE **cylen** 'kiln' is found in some pl. ns. KILHAM Nb, YE is OE *cylnum* dat. plur. *Cylen* is the first el. of some names beginning with *Kil-*, as KILBOURNE, KILNHURST, perhaps KILNSEA YE, also of CULHAM Brk. It is the second el. of CHATCULL, YARKHILL.

OE **cyning** 'king' is a common first el. in pl. ns. See KING- (passim). It is often found as a distinctive addition, as in KINGSCLERE, KING'S LYNN. OE *cyning* has often been replaced by the OScand equivalent. See KONUNGR.

D

Daccombe D [*Daccumba* 1178 P, *-cumbe* 1228 FF, *-cumb* 1242 Fees, *Daggecumba* 1193 Ol, *Doccuma* 1185 Buckland]. Derivation of the first el. from OE *Dæcca* (cf. DAGENHAM) leaves the *o*-form unexplained. Perhaps OE *dā-cumb* 'valley frequented by does'.

Dācre R Cu [*amnem Dacore* c 730 Bede, *Dakre* 1292 QW]. A Brit river-name derived from the base *dakru-* in Welsh *deigr*,

OBret *dacr* &c. 'a tear'. The meaning is 'trickling stream'. On the Dacre is **Dacre** vil. [*Dacor* c 1125 WMalm, *Dacre* 1211 P]. **Dacre** YW [*Dacre* DB, *Dacra* 12 Fount] was named from Darley Beck, whose name must once have been Dacre.

Dadford Bk [*Dodeforde* DB, *Doddeford* 1242 Fees]. '*Dodda*'s ford.'

Dadlington Le [*Dadelintona* c 1190 DC, *-ton* 1209-35 Ep]. The first el. is no doubt derived from an unrecorded OE pers. n. *Dæddel* or the like.

OE *dæl* 'valley' is not a common word, but must have been in use all over England, to judge by isolated names such as DALHAM K, Sf, DALWOOD D, DAWLEY Mx, DOVERDALE Wo. Names in *Dal-* and *-dale* are most frequent in the old Scandinavian districts and mostly contain ON *dalr*, ODan, OSw *dal* 'valley'.

Dagenham Ess [*Dæccanhaam* 692 BCS 81, *Dakeham* 1194 P, *Dakenham* 1254 Val]. '*Dæcca*'s HĀM.' *Dæcca may be a form with assimilation of *Dædca, a hypocoristic form of names in *Dǣd-* and corresponding to OHG *Tadica*.

Daglingworth Gl [*Dakelingwrth* 1200 Cur, *Daggelingewrth* 1220 Fees]. 'The WORÞ of *Dæccel*'s people.' *Dæccel is a diminutive of *Dæcca*. See prec. n.

Dagnall Bk [*Dagenhall* 1196 FF, 1228 Ep, *Dakenhale* 1308 Cal Wills (London)], **Dagworth** Sf [*Dagaworda* DB, *Daggewurða* 1166 P, *-wurthe* 1218 FF]. Perhaps '*Dæcca's HALH and WORÞ'. Or the first el. may be OE *Dægga, a short form of names in *Dæg-*, as *Dægheard*.

Dainton D in Ipplepen [*æt Doddintune*, (to) *Doddingtune* 956 BCS 952]. 'The TŪN of *Dodda*'s people.'

OScand *dal(r)*. See DÆL.

Dalbury Db [*Dellingberie* DB, *Dalebir*, *Dalenburi* 13 BM]. 'The BURG of *Dealla*'s people.' *Dealla* occurs as the name of a moneyer and corresponds to OHG *Tallo*. It is derived from OE *deall* 'proud, resplendent'. A woman's name *Cynedeall* is found in *Cynedealle rod* 990 KCD 673.

Dalby, Great & Little, Le [*Dalbi* DB, *Dalbia* c 1125 LeS, *parva Dalby* 1212 RBE, *Great Dalby* 1227 Ch], **Old D~ or D~ on the Wolds** Le [*Dalbi* DB, *Dalbia super Wald* c 1125 LeS], **D~ Li** [*Dalbi* DB, c 1115 LiS], **D~ YN** in Thornton Dale [*Dalbi* DB], **D~ YN** N. of York [*Dalbi* DB]. 'BY in a valley.' Cf. DALBY in Denmark and Sweden. **Old D~** means 'Dalby on the Wold'.

Dalch R D [*Doflisc* 739 Crawf]. A compound of Brit *dubo-* 'black' (Welsh *du*) and OW *gleis*, Welsh *glais* 'stream'. Hence 'black stream'. Identical with DULAIS, DULAS in Wales and with DAWLISH, DOUGLAS, DOWLISH and others in England.

Dalderby Li [*Dalderby* c 1115 LiS, 1221 Ep]. 'BY in a valley.' First el. ON *dæld* (gen. *-ar*), Dan *dæld*, Sw *däld* 'little valley'.

Dale Abbey Db [*Depedala* 1158 P, *La Dale* 1242 Fees], **Dale Town** YN [*Dal* DB, *Dale* 1201 FF]. 'The valley.' Cf. DÆL.

Dalemain Cu [*Dalman* 1254, 1317 Ipm]. Possibly a compound of the same type as ASPATRIA, the elements being ON *dalr* 'valley' and *Máni* pers. n.

Dalham K nr Rochester [*Dælham* 973 BCS 1296, *Delham* 1197 P], **D~ Sf** [*Dalham* DB, 1200 Cur]. 'HĀM in a valley.'

Dalling (-aw-), **Field, Nf** [*Dallinga* DB, *Dallenges* 1138, *Dalinges* c 1165 Fr, *Fildedalling* 1272 Ch], **Wood D~ Nf** [*Dallinga* DB, *Dallinge* 1242 Fees, *Wode Dallinges* 1198 FF]. OE *Dallingas* '*D(e)alla*'s people'. Cf. DALBURY.

Dallinghoo Sf [*Dal(l)inga-*, *Delingahou* DB, *Dalingeho* c 1150 Crawf]. 'The HŌH or spur of land of the *Dallingas*.' Cf. DALLING.

Dallington Np [*Dailintone* DB, *Daylington* 12 NS, 1227 Ch, *Dailinton* c 1145 Eynsham]. 'The TŪN of the *Dæglingas*.' *Dæglingas* is 'the people of *Dægel' (cf. DAYLESFORD) or 'of *Dægla' (cf. OHG *Dagilo*). The names are short forms of names in *Dæg-*.

Dallington (-ð-) Sx [*Dalintone* DB, *Dalington* 1201 Cur, *Dolinton* 1232 Cl]. Perhaps 'the TŪN of *Dealla*'s people'. Cf. DALBURY. Or the first el. may be derived from an OE *dā-lēah* 'doe wood'.

Dalston Cu [*Daleston* 1187, 1197 P, *Dalleston* 1190 ff. P, *Dalaston* 1201 Cur, 1219 Fees]. Hardly 'TŪN in the valley'. The first el. may be a pers. n. *Dall, cognate with *Dealla*. Cf. DALBURY. Some forms indicate that the name is OE *Ēast-tŭn*, with *dale* prefixed. There is an EASTON in Bowness c 8 miles away [*Estuna* Hy 2 WR].

Dalston Mx [*Derleston* 1294 PNMx(S), *Dorleston* 1388 FF]. '*Dēorlāf*'s TŪN.'

Dalton le Dale Du [*Daltun* c 720 Bede HAbb, c 1050 HSC], **D~ Piercy** Du [*Daltun* c 1150 Guisb, *Dalton Percy* 1370 AD], **D~ La** in Wigan [*Daltone* DB, *Dalton* 1212 Fees], **D~ in Furness** La [*Daltune* DB, *Dalton in fournais* 1332 Subs], **D~ Nb** nr Hexham [*Dalton* 1256 Ass], **D~ Nb** nr Stamfordham [*Dalton* 1201 FF, 1242 Fees], **D~ We** [*Daltun* DB, *Dalton* 1228 FF], **North D~ YE** [*Dalton* DB, *Northdaltona* c 1155 YCh 586], **South D~ YE** [*Delton* DB, *Suthdalton* 1260 Ass], **D~ YN** nr Richmond [*Daltun* DB], **D~ YN** nr Topcliffe [*Deltune* DB], **D~ upon Tees** YN [*Dalton* 1204 FF, *Dalton super Tese* c 1125 PNNR], **D~ YW** nr Kirkheaton [*Daltone* DB, 1235 FF], **D~ YW** nr Rotherham [*Daltone* DB, *Dauton* 1260 Ch]. 'TŪN in a valley.' Cf. DÆL.

D~ Piercy belonged to the Percy family till 1370. Cf. BOLTON PERCY.

Dalwood D [*Dalewude* 1195 ff. P, *Dalwde* 1201 Cur]. 'Wood in a valley.'

Damerham Ha [*Domra hamm* c 880 BCS 553, *at Domerham* 946 BCS 817, *Domarham* c 995 ib. 1288, *Dobreham* DB, *Dumbreham* 1156 P]. OE *Dōmera hamm* 'the HAMM of the judges'. D~ was an old royal manor.

Danbury Ess [*Danengeberia* DB, *Danegeberia* 1176 P, *Daningbery* 1254 Val, *Danwebiry* 1274 RH]. Rather 'the BURG of *Dene*'s people' than 'the BURG of the people from the valley'. D~ is in a high situation.

Danby YN nr Egton [*Danebi* DB], **D~ on Ure** YN [*Danebi* DB, *Daneby super Yore* 1316 FA], **D~ Wiske** YN [*Danebi* DB, *Daneby super Wiske* 13 BM]. 'The Danes' BY.' D~ Wiske is on R WISKE.

Dane R Chs [*Dauen* c 1220, *Dauene* 13 Chester, *Daan* 1416 AD]. A Brit river-name related to MW *dafn* 'a drop', *dafnu* 'to drop, trickle', Norw *dave* 'a pool' and meaning 'a trickling stream'. Cf. DAVENHAM, DAVENPORT.

Dane Court K nr Margate [*Dene* 1242 Fees], **D~ Court** K nr Tilmanstone [*Dane* 1310 BM]. OE DENU 'valley'.

Danehill Sx [*Denne* 1279 Ass, 1296 Subs]. OE DENN 'pasture'.

Danthorpe YE [*Danetorp* DB, 1190 P]. 'Danes' thorp.'

Darby Li [*Derbi* DB, 1212 Fees, *Dorby* 1316 FA]. Identical with DERBY.

Darent R K [*Diorente* 822 BCS 370, (of) *Dærentan* 983 KCD 640]. Identical with DERWENT. On the Darent is **Dărenth** [*Daerintan* 940, (æt) *Dæræntan* c 980 BCS 747, 1132, *Tarent* DB, *Derente* 1206 Cur]. Cf. DARTFORD.

Daresbury Chs [*Derisbury* c 1250 Chester, 1260 Court]. Perhaps '*Dēore*'s BURG'. But all the early forms noted have the spelling *Deris-*, which may point to OE *Dēoring* as the first el.

Darfield YW [*Dereuueld* DB, *Derfeld* c 1175 YCh (1638), 1208 FF]. 'FELD frequented by deer.'

Darlaston St nr Stone [*Deorlauestun* 956 BCS 954, *Deorlafestun* 1002 Wills, *Der-lavestone* DB], **D~** St nr Wednesbury [*Derlaveston* 1262 For, *Derlaston* 1316 FA]. '*Dēorlāf*'s TŪN.'

Darley Abbey Db nr Derby [*Derlega* 1199 P, *-leg* 1212 Fees, *Derley* 1230 P], **D~** Db nr Bakewell [*Dereleie* DB, *Derleia* c 1125, *-lega* 1155 RA]. OE *dēor-lēah* 'wood frequented by deer.'

Darlingscott Wo [*Derlingescot* 1210 Cur]. '*Dēorling*'s COT.'

Darlington Du [*Dearthingtun* c 1050 HSC, *Dearningtun* 1104–8 SD, *Derlinton* 1196 P]. 'The TŪN of *Dēornōþ*'s people.'

Darliston Sa [*Derloueston* 1199 FF, *Der-lawstun* 1249 Ipm]. '*Dēorlāf*'s TŪN.'

Darlton Nt [*Derluuetun* DB, *Derlintun* 1156 P, *Derleton* 1172 P]. '*Dēorlufu*'s TŪN.' OE *Dēorlufu*, a woman's name, is not recorded, but cf. *Herelufu*.

Darmsden Sf [*Dermodesduna* DB, *Der-mondesdoune* 1307 Ipm]. '*Dēormōd*'s DŪN.'

Darnall YW [*Darnehale* 13 BM, *Darnale*

1297 Subs]. 'Hidden HALH.' *Halh* may here mean 'a nook'. Cf. DIERNE.

Darnford Sf nr Needham Market [*Derneford* DB, 1200 Cur]. 'Hidden ford.' Cf. DIERNE.

Darnhall Chs [*Dernhal* 1240 Cl, *-hale* 1275 Ch, *Darnale* 1275 Misc]. Cf. DARNALL.

Darras Hall Nb [*Calverdon Araynis* 1242 Fees, *Calverdon Darreyne* 1346 FA, *Cal-verton Darrays* 1360 Pat]. See CALLER-TON. The surname is a family name derived from AIRAINES in Somme. Wydo de Araynis held the manor in 1242 (Fees). Finally Calverdon disappeared from the name.

Darrington YW [*Darni(n)tone* DB, *Dardin-tuna* 1148 YCh 179, *Dardinton* 1193 P, 1229 Ep, *Dardhinton* 1208 FF, *Darthingtone* 1205 FF]. 'The TŪN of *Dægheard*'s people.'

Darsham Sf [*Dersham*, *Diresham* DB, *Ders-ham* 1224 FF]. '*Dēor(e)*'s HĀM.'

Dart R D [(to) *Dertan* 10 BCS 1323, *Derte* 1249 Ass]. Identical with DERWENT. Hence **Dartington** D [?*Derentunehomm* 833 BCS 410, *Dertrintona* DB, *Dertinton* 1194 P] 'the TŪN of the dwellers on the Dart', **Dartmoor** D [*Dertemora* 1182 P], **Dart-mouth** D [*Dærentamuða*, *Dertamuða* 1049 ASC (C, D), *Dertemuðe* 1205 Layamon] 'the mouth of the Dart'.

Dartford K [*Tarentefort* DB, *-ford* 1159 P, *Derentef*[*ord*] 1194 P, *Derteford* 1194 P]. 'Ford over the DARENT.'

Dartington, Dartmoor, Dartmouth. See DART.

Darton YW [*Dertun* DB, *Dertona* c 1200 YCh 1716, *-ton* 1234 Ep]. OE *dēortūn* 'enclosure for deer'. The OE word is recorded in this sense.

Darwen (dărĕn) R La [*Derewente* 1227 FF]. Identical with DERWENT. On the river are **Lower** and **Over Darwen** [*Derewent* 1208 FF, 1246 Ass, *Superior Derwent* 1246 Ass, *Netherderwent* 1311 Ipm].

Dassett, Avon & Burton, Wa [*Derceto*, *Dercetone* DB, *Dercet* 1175 P, 1202 Ass, *Derchet* 1176 P, *Afnedereceth* 1185 TpR, *Avendercet*, *Magna Dercet* 1242 Fees, *le Cheping Derset* 1321 AD]. Probably an OE *dēor-cēte* 'shelter for deer', perhaps a roofed-over shelter where fodder was placed and deer could resort in severe weather. OE *cēte* (*cȳte*) 'hut' corresponds to Norw *køyta* 'a rough shelter made of pine-boughs and the like'. A Brit name containing the words for 'oak' and 'wood' (Welsh *derw*, *coed*), suggested in PNWa(S), is unlikely because the same name enters into DOSTHILL.

Avon must have been the old name of the arm of the Avon that has its source near **Avon D~.**— **Burton D~** was once a borough.

Datchet Bk [*Deccet* 10 KCD 693, *Daceta* DB, *Dachet* 1163 P]. Possibly an old Brit name, identical with *Decetia* Cæsar (now DECIZE in Nièvre), a derivative of the base *dek-* in Ir *dech* 'best', Lat *decus*.

Datchworth Hrt [*Dęcewrthe* 969 Crawf, *Dæcceuuyrthe* 1065 BM, *Tæccingawyrð* 11 E, *Daceuuorde* DB, *Tachewurth* 1240 Cl]. 'The WORÞ of **Dæčča* and his people.' Cf. *Dæcca* in DAGENHAM. For the change of *D*- to *T*- cf. TIDENHAM.

Dauntsey (-ah-) W [*Dometesis, Dameteseye* 850, (at) *Domeccesige* 854 BCS 457 f., 470, *Dometesig* 1065 KCD 817, *Dantesie* DB]. The OE forms are in poor texts, but the later development points to *t* being more trustworthy than *c*. Second el. OE *īeg* 'island'. The first may be an unrecorded OE pers. n. **Dōmgēat*.

Dăvenham Chs [*Deveneham* DB, *Davenham* 1278 Ipm], **Davenport** Chs [*Deneport* DB, *Devennport* c 1130 SD]. 'HĀM and town on R Dane.' See DANE and PORT.

Daventry (-ăn-) Np [*Daventrei* DB, *Dauintre* 1150–5 BM, 1199 FF]. The first el. might be a stream-name identical with DANE. But names in *-tree* usually have a pers. n. as first el. An OE **Dafa*, corresponding to OHG *Dabo, Tabo* and related to Goth *gadaban* 'to fit', OE *gedafen* 'fitting', may have existed. Cf. DAVINGTON.

Davidstow Co [*Dewestowe* 1313 Ep, *par. Sancti David* 1377 PT]. 'St. David's STŌW.' Cf. DEWCHURCH.

Davington K [*Dauinton* 1186 P, *Davynton* 1255 Ch, *Davinton* 1279 Ipm]. 'The TŪN of *Dafa*'s people.' Cf. DAVENTRY.

Davyhulme (dăvĭhōōm) La [*Hulme* 1276 Ass, *Defehulme* 1434 PNLa]. Originally *Hulm*; cf. HOLM. *Davy-* seems to be the adj. *deaf* in some sense, possibly the nickname of some owner.

Dawdon Du [*Daldene* c 1050 HSC, 1155 FPD]. OE *dæl* and *denu*, both 'valley'. Possibly elliptical for *Dalton dean*. Dawdon is near DALTON LE DALE.

Dawley Mx [*Dallega* DB, *Daulee* 1199 Cur]. 'LĒAH in a valley.'

Dawley Sa [*Dalelie* DB, *Dalilega* 1185 P, *-leg* 1242 Fees]. 'The LĒAH of *D(e)alla*'s people.' Cf. DALBURY.

Dawlish D [*Doflisc* 1044 OSFacs, c 1070 Ex, *Douelis* DB]. Originally the name of the stream at the place [*Doflisc ford* 1044 OSFacs], which is identical with DALCH.

Daylesford Wo [*Dæglesford* 718, 841, 875 BCS 139, 436, 540, *Deilesford* 777 ib. 222, *Degilesford* 979 KCD 623]. '**Dægel*'s ford.' Cf. DALLINGTON Np.

Deadwin Clough La [*Dedequenclogh* 1324 LaInq]. 'The clough of the dead woman.'

Deal K [*Addelam* DB, *Dela* 1158 ff. P, *Dale* 1275 RH]. OE *dæl*, OKent *del* 'valley'.

Dean, Upper & Lower, Bd [*Dene* DB], **D~** Cu [*Dene* c 1175 WR, *Dena* 1193 P], **D~ Prior** D [*Dena* DB, *Nitheredene* 1242 Fees, *Dene* 1261 Ep, *Dene Prioris* 1316 FA], **D~**

Forest, East, West & Little D~ Gl [*Dene* DB, *Dena* (for.) 1130 P, *Parva Dene* 1220 Fees], **East D~** Ha [*Dene* DB, *Estdena* 1167 P, *Dune* 1212 Fees], **Priors D~** Ha [*Dene* 1201 (1313) Ch, *Pryorsden* 1475 BM], **D~ O** [*Dene* DB, 1200 Cur], **East & West D~** Sx [*æt Dene* c 725 BCS 144, *Est-, Westdena* 1150 (1227) Ch], **East- & West-dean** Sx nr Beachy Head [(æt) *Dene* c 880 BCS 553, *Dene* c 894 Asser, *Dene, Esdene* DB], **West D~** W [(æt) *Deone* c 880 BCS 553, *Duene* DB, *Westdone* 1265 Misc, *Westdune* 1270 Ipm], **Deane** Ha [*Dene* DB, 1291 Tax], **D~** La [*Dene* 1292 QW, *Sayntemarien* 13 WhC]. OE DENU 'valley'.
D~ **Prior** belonged to Plympton priory.—**Priors D~** was given to Southwick Priory in 1201 (Ch).

Deanham Nb [*Danum* 1198 (1271) Ch, *Denum* 1242 Fees, 1254 Val, *Denhum* 1256 Ass]. OE *denum*, dat. plur. of DENU 'valley'.

Dearham Cu [*Derham* 12 StB, 1291 Tax]. OE *dēor-hamm* 'enclosure for deer'.

Dearnbrook YW [*Dernbroc* 12 Fount, *Dernebroc* 1206 ib.]. Really the name of the stream at the place [*Dernebroc* 1175 Fount]. 'Hidden stream.' Cf. DIERNE.

Dearne R YW [*Dirna* 1154 Pont, 1230 Ch, *Derna* 1157 Mon]. Probably derived from OE *dierne*: 'hidden stream'.

Dearnley La [*Dernylegh* 1324 Ct]. 'Hidden, solitary clearing.' Cf. DIERNE.

Debach (děbĭj) Sf [*Depebek, -becs, -bes, Debenbeis* DB, *Debech* 1201 Cur, *Debbeche* 1270 Ch, *Debach* 1250 FF], **Debenham** Sf [*Debham* c 1050 KCD 907, *Depham, Depheam, Depbenham* DB, *Debeham* 1168, 1172 P, *Debenham* 1226 FF]. Debach would seem obviously to be OE *dēopa bæce* 'deep valley', but the place is in a high situation. However, it is not very far from a tributary of the Deben, which runs in a deep valley. The probability is that Debenham also contains the adj. *deep* or a derivative of it, e.g. an OE river-name *Dēope* 'deep river'. The name then means 'HĀM at *Dēope*'. Debach may mean in reality 'the valley of *Dēope*', as the tributary may have had the same name as the main river. **Deben** (dēvn) R [*Deue* 1577 Harrison] is a late back-formation from Debenham.

Debden Ess nr Newport [*Deppedana* DB, *Depeden* 1227 FF]. 'Deep valley.'

Debden Green Ess nr Theydon Bois [*Tippedene* 1062 Th, *-dana* DB, *Tipeden* 1250 Cl]. The charter of 1062 mentions *Tippa-, Teppeburne*, a stream at Debden. Cf. TIP-TREE Ess. If the same el. is found in this and Debden, it seems to be a pers. n. *Tippa*. An earlier stream-name **Tippe* would suit Debden, but is not easy to explain.

Deben, Debenham. See DEBACH.

Deddington O [*Dædintun* c 1050 KCD 950, *Dadintone* DB, *-ton* 1190 P]. 'The TŪN of

Dǣda's people.' *Dǣda is a short form of names like *Dǣdhēah*. Cf. OHG *Dado*.

Dedham Ess [*Delham* DB, *Dedham* 1165, 1169 P, *Diham* 1202 Cur, *Didham* 1428 FA], **Dedworth** Brk [*Dideorde* DB, *Diddewurth* 1204 FF, *-worth* 1242 Fees]. '*Dydda*'s HĀM and WORP.' For OE *Dydda* cf. TIDENHAM.

Dee, Welsh **Dyfrdwy**, R Chs [*Dee* 1043 Th, *De* DB, *Dubr duiu* 10 Welsh Genealogies, *Deverdoeu* c 1214 Gir]. Brit *Dēvā* (cf. the earliest forms of CHESTER) 'the goddess' or 'the holy river'. The name is related to Lat *divus*. The river is in Welsh also called *Aerfen*, which means 'the war-goddess'. *Dyfrdwy* means 'the river Dee'; *Dyfr-* is Welsh *dwfr* 'water, river'.

Deene Np [*Den* 1065 BM, *Dene* DB, *Dena* 1163 P]. OE *denu* 'valley'. **Deenethorpe** Np [*Torp* DB, *Denetorp* 1169 P]. 'Thorp belonging to Deene.'

Deepdale, a common name, which means 'deep valley'. D~ La [*Dupedale* 1228 Cl], D~ YN [*Depedale* DB].

Deeping St. James & St. Nicholas, Market & West D~ Li, **Deeping Gate** Np [*Estdepinge, West Depinge* DB, *Deping Sancti Jacobi* 1209–35 Ep, *Estdeping Sancti Guthlaci* 1254 Val (Market D~), *Depynggate* 1390 Cl]. An OE *Dēoping* 'deep fen'. The district is still D~ Fen. D~ Gate is 'the road to Deeping'. Cf. GATA.

Deerhurst Gl [*Deorhyrst* 804 BCS 313, *Dorhirst* c 1050 KCD 830, *Derherste* DB]. 'Wood frequented by deer.'

Deerness R Du [*Diuerness* c 1200 ERN]. 'The river Ness.' First el. Welsh *dwfr* 'river' (cf. DEE). Second el. probably identical with NESS, the name of a river in Scotland.

Defford Wo [(in) *Deopanforda* 972 BCS 1282, *Depeforde* DB]. 'Deep ford.'

Deighton (-ē-) YE [*Distone* DB, *Dighton* 1285 FA], D~ (-ē-) YN [*Dictune* DB, *Distone* 1088 LVD, *Dicton* 1231 FF], D~ YW nr Huddersfield [*Dicton* 1297 Subs], **Kirk & North** D~ YW [*Distone* DB, *Northdicton* 1226 FF, *Suth Ditthon* 1285 FA]. OE *Dīc-tūn* 'TŪN by a ditch or dike or surrounded by a moat'.

Deightonby YW [*Dictenebi* DB, *Dicthenbi* 1486 Goodall]. An OE *Dīc-tūn*, to which was added Scand BY. Cf. DINSDALE.

De·lamere Chs [*foresta de Mara* 1248, *de la Mare* 1249 Cl]. OE *mere* 'lake'. *Dela-* is Fr *de la*. **Delapré Abbey** Np [(moniales de) *Prato* or *Sancte Marie de Prato*, (Abbatissa) *sancte Marie de Pratis* 1220 Ep]. Fr *de la pré* 'of the meadow'.

OE **delf**, **gedelf** 'digging, mine, quarry, ditch' occurs in some names, as **Delph** YW, **King's Delph** Hu (an old drain), STANDHILL. Cf. BIDDULPH, DILHORNE.

OE **dell** 'a dell, a deep hollow or vale' occurs sometimes in pl. ns., as in ARUNDEL.

Dembleby Li [*Dembelbi* DB, *-by* 1212 Fees, *Dembleby* 1242 Fees]. First el. perhaps dial. *dimble* 'a ravine with a watercourse through it'. The word may be of Scand origin; cf. Norw *dembel* 'a pool'.

Denaby YW [*Denegebi* DB, *Daningebi* 1195 P, *Deneby* 1219 FF]. 'The BY of the Danes.' First el. the OE gen. *Denigea* or *Dena* from *Dene* 'Danes'.

Denbury D [*Deveneberie* DB, *Devenebyr* 1242 Fees]. 'The BURG of the Devon people.' The *Defnas* may in this instance be the British aborigines, the *Dumnonii* (cf. DEVON). There is an old earthwork here.

Denby Db [*Denebi* DB, *Deneby* 1234 Fees], D~ YW nr Kirkheaton [*Denebi* DB, *Deneby* 1241 FF], D~ YW nr Penistone [*Denebi* DB]. 'The BY of the Danes.'

Denchworth Brk [*Denceswyrth, -wyrð* 811, 815, *æt Deneceswurþe* 947, *Deniceswyrð* 960 BCS 352, 850, 833, 1055, *Denchesworde* DB]. '*Denic*'s WORP.' *Denic is a normal derivative of *Dene* pers. n.

Dendron La [*Denrum, -run* 1269 Ass]. 'Clearing in a valley.' Cf. DENU, RŪM.

Denes, North & South, Nf [*Den* 1155–8 Holme]. OE *denu* 'valley'.

Denford Brk [*Denford* c 930 BCS 678, *Daneford* DB, *Deneford* 1220 Fees], D~ Np [*Deneforde* DB, *-ford* 1195 FF]. 'Ford in a valley.' See DENU.

Denge, D~ Marsh. See DUNGE MARSH.

Dengie Ess [*Deningei* 709–45 Bodley MS, *Denesig* c 950 Wills, *Daneseia* DB, *Danesy* 1212 Fees, *Danegeye* 1235 Ass, *Danengeye* 1276 FF]. OE *Denes-īeg* '*Dene*'s island', alternating with *Deninga-īeg* 'the island of *Dene*'s people'. The form with *a* for *e* is typically East Saxon. Cf. DENU.

Denham Bk nr Uxbridge [*Deneham* 1065 BM, 1195 Cur, *Daneham* DB, *Denham* 1163 P], D~ Sf nr Bury [*Denham* DB, 1254 Val], D~ Sf nr Eye [*Denham* DB, 1212 Fees]. 'HĀM in a valley.' Cf. DENU.

Denham Bk nr Quainton [*Dunindon* 1237–40, *Dunidon* 1242 Fees, *Duningdon* 1247 Ass]. 'The DŪN of *Dunn*'s people' or possibly OE *Dūninga-dūn* 'the hill of the people on the DŪN or hill'.

Denholme (dĕnum) YW [*Dennum* 1329, *Denum* 1332 Kirkst]. OE *denum*, dat. plur. of DENU 'valley'.

Denmead Ha [*Denemed* 1231, *-mede* 1292 Cl]. 'Meadow in a valley.'

OE **denn** n. 'a pasture, esp. a swine-pasture'. A more general sense 'pasture' is indicated by the name COWDEN K. *Denn* is common as the second el. of pl. ns. in the Kent and Sussex Weald district, and the numerous names in -*den* in Kent are generally names of old pastures. It is doubtful to what extent the word was used outside these districts. It is often difficult to distinguish names in -*denn* from those in -*denu*.

Dennington Sf [*Dingifetuna, Dingiuetuna* DB, *Dingieueton* 1169 P, *Dinniueton* 1190 P]. '*Denegifu*'s TŪN.' The woman's name *Denegifu* is not otherwise evidenced, but cf. *Denegȳþ* in *Denegiðegraf* 937 BCS 712. The change *e* > *i* in the first syllable is due to the palatalized *n* that followed it.

Denny Ca [*Daneya, Deneia* 1176 TpR, *Deneye* 1325 Misc]. Perhaps 'the Danes' island'.

Denshanger Np [*Dinneshangra* 937 BCS 712, *Duns-, Deneshanger* 13 BM]. '*Dynne*'s slope.' Cf. HANGRA.

Denston Sf [*Danardes-, Danerdestuna* DB, *Denardeston* 1220 FF]. '*Deneheard*'s TŪN.'

Denstone St [*Denestone* DB, -*ton* 1208 FF]. '*Dene*'s TŪN.'

Dent YW [*Denet* c 1200 Ass, 1231 FF, 1247 Ch, *Dent*, (dale of) *Dent* 1278 Kendale]. Probably a name of **Dent Crag**, a hill of 2,250 ft. Cf. **Dent** Cu, the name of a hill nr Cleator [*Dinet* c 1200 StB]. Dent may be derived from a Brit word corresponding to OIr *dinn, dind* 'a hill', ON *tindr* 'point, crag'. **Dent**, river-name, is a back-formation.

Denton Cu [*Denton* c 1180 WR, 1203 P], **D~** Du [*Denton* 1200 BM], **D~** Hu [*Dentun* 972–92 BCS 1130, -*tone* DB], **D~** K nr Dover [*Denetun* 799 BM, *Dane-tone* DB], **D~** K nr Gravesend [*Denetun, Danituna* 964–95, *Denituna, Denetun* 10 BCS 1132 f., 1321 f., *Danitone* DB], **D~** La nr Manchester [*Denton* 1255, 1278 FF], **D~** La nr Widnes [*Denton* 1246 Ass], **D~** Li [*Dentune* DB, *Denton* R 1 Cur], **D~** Nb [*Dentuna* Hy 2 AD, *Dentun* 1252 Ch], **D~** Nf [*Dentuna* DB, *Denton* 1199 P], **D~** O [*Denton* 1237 Ep, 1242 Fees, c 1265 Bodl], **D~** Sx [*Denton* 801, *Deanton* 825 BCS 302, 387], **D~** YW [*Dentun* c 972 BCS 1278, c 1030 YCh 7, DB]. 'TŪN in a valley'; cf. DENU.

Denton Np [*Dodintone* DB, *Dudinton* 1200 Cur, *Parva Dudinton* 1220 Fees]. 'The TŪN of *Dudda*'s people.'

OE **denu** 'a dene, a valley' is common as the first and second element of pl. ns.; it is also used alone as a pl. n. Cf. DEAN, -E, DEENE, DENES, DENHAM &c. As a first el. it is often difficult to distinguish from *Dene* 'Danes'. In East Saxon, where *i*-mutated *a* before nasals appears as OE *æ*, ME *a*, the form of the word is often *dane*. This form is met with also outside the East Saxon district in a stricter sense. EAST DEAN Ha and WEST DEAN W in early forms show a curious development, indicated by spellings such as *Duene, Done* &c. The OE base seems to be a form *deonu* with *u*-mutation, found in *Deone* BCS 553. Cf. DINDER.

Denver Nf [*Danefella, -faela* DB, *Denever* 1200 FF, -*e* 1254 Val]. Apparently OE *Dena fær* 'the passage of the Danes'.

Denwick (dĕnĭk) Nb [*Den(e)wyc* 1242 Fees, *Denewic* 1265 Misc]. 'WĪC in a valley.'

OE **dēop** 'deep' is the first el. of some names, as DEBDEN, DEEPDALE, DEPDEN, DEPTFORD, DIPTFORD, DIPPENHAM. DEEPING is a derivative of the word, as is perhaps the first el. of DEBACH, DEBENHAM. See DEOPHAM.

Deopham Nf [*Dep-, Diepham* DB, *Depham* 1227 Bodl]. D~ is nr **Seamere**, a lake, whose name was very likely OE *Dēop*, identical with OE *dēop* 'a deep place in the sea'. Second el. HĀM.

OE **dēor** 'animal' in pl. ns. no doubt means 'deer'. Cf. DARTON, DASSETT, DEAR-, DERE-, DYRHAM, DORDON, DORFOLD, DOSTHILL, DUR-BOROUGH, DURFORD, DURLEIGH. In DARBY, DERBY the first el. is the OScand equivalent (ODan, OSw) *diur*.

Depden Sf [*Depdana* DB, *Depedene* 1198 FF]. 'Deep valley.'

Deptford (dĕt-) K [*Depford* 1334 Ch, *Depe-ford* Chaucer], **D~** (dĕt-) W [*Depeford* DB]. 'Deep ford.'

Derby (-ar-) Db [*Norðworþig* c 1000 Saints, *Deoraby* 917 ASC (C), 942 ASC (A), *Deorby* 959–75 Coins, *Derby* DB], **West D~** La [*Derbei* DB, *Derbeia* 1153 BM, *Westderbi* 1177, -*derebi* 1201 ff. P]. Both are OScand *diurby* or *diuraby* 'BY where deer were seen', 'BY with a deer-park'. There is a DARLEY 2 m. N. of Derby. **Derbyshire** is *Deorby-scir* 1049, 1065 ASC (D), *Derbyscire* DB.

Dereham (-ēr-), **East**, Nf [*Derham* DB, 1254 Val, *Estderham* 1428 BM], **West D~** Nf [*Deorham* 798 ASC (F), *Der(e)ham* DB, *Derham* c 1095 Bury, 1193 DC, *Dierham* 1197 P, *Westderham* 1203 Cur]. Most likely OE *dēor-hamm* 'enclosure for ᵭeer'.

Derrington St [*Dodintone* DB, -*ton* 1242, *Dudington* 1236 Fees]. 'The TŪN of *Dudda*'s people.'

Derriton D [*Direton* 1238 Ass]. '*Dēora*'s TŪN.'

Derrythorpe Li [*Dodithorp* 1263 FF, *Dod-ingthorp* 1316 FA]. '*Dodding*'s thorp.' Cf. *Dodinc* pers. n. in DB.

Dersingham Nf [*Dersincham* DB, *Dersinge-ham* 1166 P, *Dersingham* 1203 Cur]. 'The HĀM of *Dēorsige*'s people.'

Derwent R Cu [*Deruuentionis fluuii* c 730 Bede, *Deorwentan stream* c 890 OEBede, *Derewent* 12 StB], **D~** R Db [(neah) *Deor-wentan* c 1000 Saints, *Derwenta* Hy 2 Rutland], **D~** R Du, Nb [*Dyrwente* c 1050 HSC, *Derwent* c 1155 Newminster], **D~** R Y [(amnem) *Deruuentionem* c 730 Bede, (be) *Deorwentan* c 890 OEBede, *Derewent* c 1145 YCh 373]. A Brit river-name *Derventiō*, found as the name of a place on the Yorkshire Derwent [*Derventione* (abl.) 4

IA], derived from Brit *dervā 'oak', Welsh derw &c. The name means 'river where oaks were common'. Cf. DARENT, DART, DARWEN, which all go back to Derventiŏ. From Derwent Cu are named **Derwent Fells** [Derewentfelles 1292 QW] and **Derwentwater** [Derwentewater 1210 FC, Derewentewater 1234 Cl].

Desborough Bk [Dustenberg DB, Dustebergahundredum 1190 P, Dusteleberg 1195 Cur, all denoting Desborough hd]. 'Hill where penny royal, OE (dweorge)dwostle, grew' (PNBk).

Desborough Np [Des-, Dereburg DB, Deresburc 1167 P, 1200 Cur, -burg 1208 BM, Desburc 1197 FF], **Desford** Le [Deresford, Diresford DB, Dersford 1209–35 Ep, 1257 Ch, Dessford 1253–8 Ep]. 'Dĕor(e)'s BURG and ford.' The loss of r is due to dissimilation.

Detchant Nb [Dichende 1166 RBE, Dichend 1242 Fees]. 'The end of the ditch or dike.'

Dethick Db [Dethek c 1200 Darley, 1275 RH, Dethic 1290 PNDb]. Perhaps OE dĕaþ-āc 'death oak', i.e. 'oak on which felons were hanged'. Dethick is not far from MATLOCK, which means 'oak at which moots were held'.

Detling K [Dytlinge 11 DM, Detlinges 1066–87 Reg Roff, 1197 FF, Detlingges 1230 P]. OE Dyttlingas 'Dyttel's people'. *Dyttel may be a short form of names in Dryht-.

Deuxhill Sa [Dehocsele DB, Dewkeshul 1255 RH, Deukeshulle 1277 Ep]. 'Deowuc's hill.' OE Deowuc pers. n. is found in Diuruces (Diowuces) þæþ 963 BCS 1119 (in boundaries of Aston nr Lilleshall Sa). Deuxhill and Aston are far from each other; so the common element of Deuxhill and Diuruces þæþ cannot be a topographical feature.

Deverill R W [Defereal, Deferael 968 Reg Wilt]. Either a Welsh dwfr iâl 'the river of the iâl or fertile upland region' or Brit Dubroialon 'the iâl on the stream'.

Cf. BRIXTON, HILL, KINGSTON, LONGBRIDGE, MONKTON DEVERILL, which were once all Deverill [Devrel DB, Defurel c 1140 Gaimar, Deurel 1165 P &c.]. Brixton, Hill &c. were originally distinctive additions, whereas nowadays it is no doubt Deverill that is looked upon as the distinctive element.

Devils Brook. See DEWLISH.

Devils Ditch Ca [dicum 905 ASC]. An ancient earthwork, which was ascribed to the agency of the devil. Devil's Ditch (Dyke) is the name also of other old earthworks.

Devils Water Nb [Diveles c 1230 Ep, 1269 Ass]. Identical with DALCH.

Devi·zes W [Divisas 1139 Ordericus, 1142 BM, Divise 1139 HHunt, (de) Divisis 1162 P]. Fr devises, Lat divisæ 'boundary'. An important boundary must once have run past Devizes.

Devoke Water Cu, a tarn [Duvokeswater c 1200 StB, Duuokwat' 1279 Ass]. First el. a Brit Dyfoc 'the little black one', as in CUMDIVOCK, here perhaps an old name of the tarn.

Dĕvon R Le, Nt [Dyvene 1252 Misc, Deven 1342 Pat]. Identical with DEVON, name of a river in Scotland, which is no doubt Brit Dubonā 'black river' (from Brit *dubo-, Welsh du 'black, dark').

Devon, -shire [(on) Defnum 894, (on) Defenum 897 ASC, Defenun 955 BCS 912, Defenascir 851 ASC, Defnascir 894 ASC]. Devon is identical with the tribal name Defnas 'men of Devon' (e.g. 823 ASC), which came to be used as a name of the territory (see Introd. p. xi). Defnas is Brit Dumnonii, the name of the Celtic aborigines, which was transferred to their Saxon conquerors. The Welsh name of Devon, Dyfnaint, goes back to Brit Dumnonia. It is Dibnenia Vita Gildæ, Domnonia c 894 Asser, Dyfneint Mab. **Devonport** is a late name.

Dewchurch, Much & Little, He [Lann Deui Ros Cerion c 1150 LL, Deuweschurche c 1225 Glouc, Deuschirch 1243 Ch]. 'The church of St. Dewi.' Dewi is the Welsh form of David.

Dewlish Do [Devenis DB, Deueliz 1194, Duuelis 1195, Deuelis 1196 P]. Named from **Devils Brook** [Deuelisc 869 BCS 525, Dovelz 1245 FF]. See DALCH.

Dewsall He [Dewyes Welle, Deuwewell 1242 Fees, (Ecclesia de) fonte David 1269 Ep, Deweswall 1291 Tax]. 'St. David's spring.' Cf. DEWCHURCH. Dewsall is little more than a mile from Dewchurch.

Dewsbury YW [Deusberia DB, Deubir 1202 FF, Deaubir 1230 P, Dewesbiri 1226 FF, Dewesbiry, Deaubir, Dyaubir 1267 Ep]. Possibly 'the BURG of Dewi or David' (cf. DEWCHURCH). But many forms point to an OE first el. with ēa. No doubt it is identical with OE Deaw in Deawes broc 972 BCS 1282. Deaw looks like a pers. n., but perhaps it is the common noun dĕaw 'dew' in some more original sense such as 'fluid, water'. It might be a stream-name.

Dexthorpe Li [Dr(e)istorp DB, Drextorp c 1180 Bury, 1212 Fees, Drexthorp 1206 Ass, 1242 Fees]. The first el. is no doubt a pers. n., perhaps a nickname formed from ON driúgr 'ample, large'. The first r was lost owing to dissimilation.

Dibberford Do [Dibberwurð 1002–14 KCD 708, Diberwrth 1252 Fees]. 'Dycgbeorht's WORÞ.'

Dibden Ha [Depedene DB, Diepedena 1165 P, Dupedene 1291 Tax]. 'Deep valley.'

OE dīc 'ditch, moat; dike, wall of earth, embankment', in pl. ns. often referring to prehistoric dikes. It is generally difficult to decide which is the sense in each individual case. The form varies between ditch and *dike*, both of which are native developments. See DITCH- (passim), DEIGHTON, DETCHANT, DICKLEY, DISHFORTH, DISS, DISSINGTON, DITTERIDGE, DITTON, CAR DIKE, DOWDYKE, WANSDYKE &c. OScand *dīki* perhaps in DIGBY.

Dicker Sx [*Dyker* 1261 FF, *Dikere* 1294 Misc]. ME *dyker* 'dicker, a number of ten' from Lat *decuria*, MLat *decora*, perhaps in allusion to a rent of a dicker of iron.

Dickleburgh Nf [*Dicclesburc* DB, *Dikel-, Dikleburg* 1254 Val]. It is possible that the first el. is a pers. n. *Dicel or *Dicla, but the vicinity of DISS suggests that it contains OE *dīc* or rather the pl. n. DISS. The name might go back to OE *Dīclēa-burg, Dīclēah* being 'the forest belonging to Diss'.

Dickley Ess in Mistley [*Dicheleia* DB, *Dikeleia* 1198 P, *-leg* 1203 Cur]. OE *dīc-lēah* 'LĒAH by a ditch'.

Didbrook Gl [*Didibroke* 1269 Pat, *Dyddebroke* 1316 FA]. '*Dydda*'s brook.' Cf. TIDENHAM.

Didcot Brk [*Dudecota* 1206, *-cot* 1207 Cur, *-cothe* 1212 Fees]. '*Dudda*'s COT.'

Didcote Gl [*Didicot* 1107 (1300) Ch]. '*Dydda*'s COT.' Cf. DIDBROOK.

Diddington Hu [*Dodinctun, Dodintone* DB, *Dudinton* 1220 Fees]. 'The TŪN of *Dudda*'s people.'

Diddlebury Sa [*Dudeneburia* 1147 LaCh, *Dudelebire* 1167, *-beri* 1193 P, *-bur* 1242 Fees]. '*Duddela*'s BURG.' *Duddela is a side-form of *Duddel*.

Didley He [*Dodelegie* DB, *Duddeleia* 1166 RBE, *Duddele* 1266 Misc]. '*Dudda*'s LĒAH.'

Didling Sx [*Dedelingis* Hy 2 (1361) Pat, *Dudelinges* 1237 Ch]. OE *Dyddelingas* '*Dyddel*'s people'.

Didlington Do [*æt Didelingtune, Dydelingtun* 946 BCS 818, *Dedilintone* DB]. 'The TŪN of *Dyddel*'s people.'

Didlington Nf [*Dudelingatuna* DB, *Dodelintona* 1086 IE, *Dudelington* 1254 Val]. 'The TŪN of *Duddel*'s people.'

Didmarton Gl [*Dydimeretun* 972 BCS 1282, *Dedmertone* DB, *Dudemerton* 1200 P, *Dodemarton* 1220 Fees]. The place is on the border between Gl and W. The probability is therefore that the name contains OE *gemǣre* 'boundary'. Perhaps *Dyddi*'s *gemǣrtūn* 'the boundary farm of *Dyddi*'. Cf. TORMARTON. But *Dudemær* pers. n. is recorded.

Didsbury La [*Dedesbiry* 1246 Ass, *Diddesbiry* 1276 Ass]. '*Dyddi*'s BURG.' *Dyddi is a normal derivative of *Dudd* or *Dudda*.

OE **dierne, derne** 'secret, hidden' is sometimes the first el. of names of brooks, fords, &c. It would refer to a brook &c. difficult to find. Cf. DARNFORD, DARN(H)ALL, DEARNBROOK, DORNFORD, DURNFORD, also DEARNE.

Dieulacres St [*Deulecresse* 1214–16 BM, 1228 Ch]. An old abbey. The name is an OFr *Dieu l'acreisse* 'may God increase it', a name analogous to DIEULOUARD 'Dieu le garde' and DIEU S'EN SOUVIENNE, names of monasteries in France.

Digby Li [*Dicbi* DB, *Diggebi* 1197 P]. 'BY at the ditch', probably an old drain.

Digley Chs. See DISLEY.

Digswell Hrt [*Dichelesuuelle* DB, *Diklenes-, Dikesneswell* 1198 Rot Cur, *Digeneswella* 1209–19 Ep]. The first el. may be an unrecorded pers. n. *Diccīn, related to the base of DITCHLING and the first el. of DISCOVE.

Dikler R Gl [*Thickeleure* c 1250 Eynsham]. A compound of OE *þicce* 'thick' and *lǣfer* 'rush'. It was originally a pl. n. Cf. *Tyckeleuere* (pratum) c 1250 Eynsham (i. 182). The meaning is 'clump of rushes' or 'place where rushes grew thickly'.

Dilham Nf [*Dilham, Dillam* DB, *Dilham* c 1150 Crawf]. 'HĀM or HAMM where dill was grown.' *Dill* is OE *dile*. The el. is found also in DILICAR, DILWORTH, DULWICH.

Dilhorne St [*Dulverne* DB, *-uerne* 1200 P, *Dilverne* 1236 Fees, *Delverne* 1281 Misc]. Perhaps 'house by a mine or quarry'. Cf. BIDDULPH and ÆRN.

Dilicar We [*Dilacre* c 1200 CC, 1208 Kendale]. 'Dill field.'

Dillington Hu [*Dilingtune* 974 BCS 1311, *Dellinctune* DB, *Dulintone* 1255 For], D~ So [*Dillington* 1243 Ass, *Dilyngton* 1275 Ipm]. 'The TŪN of *Dylla's or *Dylli's people.' A pers. n. stem *Dull*, derived from OE *dol*, Engl *dull* and corresponding to OHG *Doll-* in *Dolleo* &c., must be assumed for DULLINGHAM and other names.

Dilston Nb [*Deuelestune* 1172, *Diueliston* 1175 P]. 'TŪN on DEVILS WATER.'

Dilton W [*Dulintun* 1190 P, *Dultun* 1222 FineR, *Dulton* 1236 Fees]. '*Dylla*'s TŪN.' Cf. DILLINGTON.

Dilworth La [*Bileuurde* DB, *Dileworth* 1227 FF]. 'WORÞ where dill was grown.'

Dilwyn He [*Dilven, Dilge* DB, *Diliga* 1123 PNHe, *Dilun* 1138 AC, *Dilum* 1193 f. P, *Dilewe* 1277 Ep]. OE *dīglum*, dat. plur. of *dīgle* (*dīgol, dēgol*) 'concealment, a secret or shady place'. OE *diglum* once glosses 'recessibus'. The reference would be to a lonely place; cf. DARNALL and the like. OE *dīglum* sometimes became by metathesis *dīlgum*, whence most of the later forms. Cf. the change of *þl, sl* to *lþ(ld), ls* in bold from *bōþl* and *-gils* from *-gisl*.

Dimlington YE [*Dimelton* DB, c 1155 YCh 1352, *Dimbilton* 1260 Ipm]. First el. as in DEMBLEBY.

Dimmer So. See CHILTHORNE DOMER.

Dimsdale St [*Dulmesdene* DB, *Dimesdal* 1242 Fees, *Dymmesdale* 1281 Ass]. The first el. might be ME *dimple* 'dip in the ground', from which the first el. of **Dumplington** La [*Dumplinton* 1229 FF] is derived. If so, the *p* was lost early and *ml* became *lm*.

Dinchope Sa [*Dudingehope* c 1180 Eyton]. 'The HOP or narrow valley of *Dudda*'s people.'

Dinckley La [*Dunkythele, Dinkedelay* 1246 Ass, *Dinkedelegh* 1257 Ipm]. Second el. LĒAH. The first might be a Brit name of the place, e.g. a Welsh *Dincoed* 'fort by a wood'.

Dinder So [*Denrenn* 1065 Wells, *Dinre* 1174 P, *Dinra* 1176 P]. 'House in a valley', the elements being OE *denu* 'valley' and *renn*, an early form of ÆRN. Cf. CHARD.

Dinedor He [*Dunre* DB, 1176 P, 1242 Fees, *Dinra* 1170 P]. Probably a Welsh name, either *Din-fre* 'hill with a fort' or *Din-dre(f)* 'village by a hill'. Early Welsh *din* means 'a fortified hill, a fort'. Welsh *bre* (by lenition *fre*) is 'a hill'. Welsh *tre(f)* is 'a village, hamlet'. Dinedor is situated at Dinedor Hill.

Dingestow (-njĕ-) Monm [*Merthir Dincat* c 1150 LL, *Landinegath* 1191 Gir, *Landinegat* 1199 LL, *Dungestowe* 1405 Pat]. 'Church or holy place of St. Dingad.' OE *stōw* has replaced Welsh *llan* (cf. LANN). *Merthir* is Welsh *merthyr* 'saint', lit. 'martyr'. *Dingad* (OW *Dincat*, OBrit *Dunocatus*) was the name of two Welsh saints.

Dingle La [*Dingyll* 1246 Ass]. ME *dingle* 'a dingle, a deep dell or hollow'.

Dingley Np [*Dinglei* DB, *Dingelai* 1197 FF, *Dingelea* 1175 P]. 'LĒAH in a dingle.' Cf. prec. name.

Dinmore He [*Dunemore* 1189 Hereford, 1227 Ch, *-mor* 1212 Fees, *Dinemor* 1273 Cl]. Welsh *din mawr* 'great hill'. Cf. DINEDOR. The place is on Dinmore Hill.

Dinnington Nb [*Donigton* 1242 Fees, *Dunington* 1256 Ass]. 'The TŪN of *Dunn*'s people' or possibly 'of the people on the DŪN or hill'.

Dinnington So [*Dinnitone, Dunintone* DB, *Doniton* 1201, *Dunington* 1254 Ass], D~ YW [*Dunintone, Dunnitone* DB, *Dunington* 1191-3 Fr, *Dinigton, Dynington* 1268, 1271 Ep]. 'The TŪN of *Dynne*'s people.'

Dinsdale, Low, Du [*Ditneshall* c 1185 YCh 950, *Ditleshal* 1196 P, *Dictensale* 13 FPD], **Over D~** YN [*Dignes-, Dirneshale* DB, *Ditneshal* 1208-10 Fees]. OE *Dīctūnes-halh* 'the haugh belonging to Deighton'. DEIGHTON YN is not far away. Low and Over D~ are opposite to each other on the Tees.

Dinsley, Temple, Hrt [*Deneslai* DB, *Dyn-*

nesleya 1142 TpR, *Dineslea* 1166, 1192 P]. '*Dynne*'s LĒAH.' The place belonged to the Templars at least in 1142.

Dinting Db [*Dentinc* DB, *Duntinge* 1226 FF]. Perhaps derived with the suffix *-ing* from the hill-name discussed under DENT. Cf. **Dinthill** Sa nr Shrewsbury [*Dunthull* 1200 Cur].

Dinton Bk [*Danitone* DB, *Duninton* 1209 Fees, *Deniton* 1227 Ass], D~ W [*Domnitone* DB, *Dunnitone* 1198 P, *Dunneton* 1199 FF]. 'The TŪN of *Dunn*'s or *Dynne*'s people.' D~ W may be identical with *Duningland* (*Duningheland*) 860 BCS 499 (Forsberg, NoB 30, 153 f.).

Dippenhall Sr [*Dupehale* 1248, *Dupenhale* c 1307 Crondal]. 'Deep HALH.'

Diptford D [*Depeforde* DB, *Dippeford* 1230 P, *Dupeford* 1268 Ipm]. 'Deep ford.'

Dipton Du [*Depeden* 1339 PNNb], **D~** Nb [*Depeden* 1269 Ass]. 'Deep dean or valley.'

Discove So [*Dinescove, Digenescoua* DB, *Dichenes-, Dikenescova* 1166 RBE]. Second el. OE COFA; the place is in a valley. The first el. may be as in DIGSWELL.

Diseworth Le [*Digþeswyrþ* c 972 BCS 1283, *Diwort* DB, *Digaðeswrð, Digðeswrthia* c 1180 BM, *Digitheswurth* 1184 Berk]. '*Digoþ*'s WORþ.' The first el. appears to be a pers. n. of about the same formation as *Dogod* in DOWDESWELL. Cf. DISHLEY.

Dishforth YN [*Disforde* DB, *Disceford* 1202, *Diceford* 1208 FF]. OE *dīc-ford* 'ford by a dike or ditch'.

Dishley Le [*Dislea, Dexleia* DB, *Dixeleia* c 1125 LeS, *-leya* 1224 Ep]. '*Digoþ*'s LĒAH.' Cf. DISEWORTH. Dishley is only 4 or 5 m. from Diseworth.

Disley Chs [*Distislegh* 1285, *Distelee* 1288 Court, *Disteslegh* 1308 Ipm]. The forms suggest a pers. n. as first el., but no name is known that shows the form required. Near D~ is **Digley** [*Dyghleg(h)* grange 1287 Court]. This seems to be OE *dīc-lēah*. Disley may have been named from the same dyke. *Dis-* might be an OE *dīc-stīg*.

Diss Nf [*Dice* DB, *Dic* 1130 P, *Dize* 1158, *Disze* 1190, *Disce* 1191 P]. See DĪC. The change of *č* to *s* is due to Norman influence.

Dissington Nb [*Dicentona* Hy 2 (1271) Ch, *Dichintuna, Discintune* c 1190 Godric, *Dicheston* 1208 Cur]. It is possible that this is an OE *Dīchæmatūn*. Cf. DISS, DITCHAMPTON. The place is c 2 m. from the Roman Wall.

Distington Cu [*Dustinton* c 1230 StB, *Distington* 1256 FF, 1274 Cl]. Possibly the first el. is an OE *dýsten* 'dusty'. Cf. DUSTON. *Dýsting* might also have been the name of the short stream here, derived from *dūst* in a sense 'mist, spray': cf. G *Dunst* in such senses.

Ditcha·mpton W [æt *Dichæmatune* 1045 KCD 778, *Dechementune, Dicehantone* DB, *Dichamton* 1195 FF]. 'The TŪN of the dwellers by Grim's Ditch.'

Ditchburn Nb [*Dicheburn* 1236 Fees, 1252 P]. 'Stream by a ditch or dike.'

Ditcheat So [*Dichesgate* 842 BCS 438, *Dicesget* DB, *Dichesgete* 1196 P]. 'The gate in the dike.' Cf. *dices get* 739 Crawf (D). The dike is the Fosse Way, not far from which D~ is situated. In the boundaries of D~ in BCS 438 are mentioned *Dich, Dichforde, strete yate.*

Ditchford Np [*Dichesford* 1236 Fees, *Dicheford* 1330 FA], **Lower D~** Wa, **Upper D~** Wo [*Dicford* c 1050 KCD 804, DB, *Dichford* 1209 Fees, *Dicheford* 1230 Cl]. 'The dike ford.' At D~ Wa, Wo the Fosse Way crosses a stream.

Ditchingham Nf [*Dicingaham* DB, *Dichingeham* 1178, 1194, *Dikingeham* 1196 f. P]. 'The HĀM of the *Dicingas*', who may either be the people of a man named *Dic(c)a* or the like (cf. DITCHLING) or dwellers at a dike or ditch.

Ditchley O [*Dichelegh* 1227 FineR]. 'LĒAH by Grim's Dyke.'

Ditchling Sx [*Dicelinga* c 765 BCS 197, (æt) *Diccelingum* c 880 ib. 553, *Dicelinges* DB]. This cannot well be anything else than a derivative with -*ingas* of a pers. n. *Diccel, which may be the first el. of DIXTON. The etymology of the supposed name, like that of many other pers. ns., is obscure.

Ditteridge W [*Digeric* DB, *Dicherigga* 1168 P, -*rigge* 1242 Fees]. OE *dīc-hrycg* 'the ridge with the dike'. The Fosse Way runs along the ridge.

Dittisham D [*Didasham* DB, *Didisham* 1230 P, *Didesham* 1276 Ipm]. '*Dyddi*'s HĀM.' Cf. DIDSBURY.

Ditton Bk [*Ditone* DB, *Dittun* 1205 FF], **Fen D~** Ca [*Dictun* c 995 BCS 1288, *Dittona* 1254 Val, *Fen Dytton* 1286 FF], **Wood D~** Ca [*Dittona* c 1080 ICC, *Dictune* 1086 IE, *Wodedittone* 1228 FF], **D~ K** [*Dictun* 10 BCS 1321 f., *Dictune* DB], **D~ La** [*Ditton* 1194 P], **Long & Thames D~** Sr [*Dictun* 1005 KCD 714, *Ditune* DB, *Langedittone* c 1270 Ep, *Temes Ditton* 1235 FF]. OE *Dīc-tūn* 'TŪN by a dike or ditch'.

Fen D~ is on FLEAM DIKE, Wood D~ at DEVILS DITCH. D~ K is on the main road from Maidstone to London.

Ditton, Earls, Sa [*Dodentone* DB], **D~ Priors** Sa [*Dodintone* DB, *Dodintun* 1160 P, *Dudinton* 1212 Fees, *Dudington* 1230 P]. 'The TŪN of *Dudda*'s people.'

Earls D~ belonged to the earls of March, D~ Priors to Wenlock Priory.

Divelish R Do [*Deuelisch, Defelich* 968 BCS 1214]. Identical with DALCH.

Dixton Gl [*Dricledone* DB, *Diclisdon* 1169 P,

Dichelesdona 1175 Winchc, *Diclesdon* 1205 Cur]. Perhaps '*Diccel*'s DŪN'. Cf. DITCHLING. The place is by a hill, on which the map marks a camp. This suggests the alternative explanation of the first el. as *Dīc-hyll* 'hill with a dike'. But no early forms point to this base.

OE **docce** 'dock, water-lily', also in *ēa, sūr-, wududocce.* See next name, also DOCKING, DOCKLOW, DOGDYKE, DOGMERSFIELD.

Dockenfield Ha [*Docchenefeld* c 1150 (1341) Pat]. 'FELD where docks grew.' *Doccenaford* 909 BCS 627 must have been near D~.

Docker La [*Dokker* 1505 ff. FF], **D~** We [*Docherga* c 1180 Kendale, *Dochergebec* c 1210 NpCh]. 'Shieling in a valley.' Cf. ERG. First el. ON *dǫkk* 'a hollow, valley'.

Docking Nf [(et) *Doccyncge* c 1035 Wills, *Dochinga* DB, *Dokinges* 1166 RBE]. OE *Doccing* 'place where docks grew'. See DOCCE.

Docklow He [*Dockelawe* 1291 Tax]. 'Hill where docks grew.' Cf. DOCCE.

Dockray Cu [*Dochora* 1195 FF, *Dokwra* 1292 QW]. ON *dǫkk* 'a hollow, valley' or OE *docce* 'dock, sorrel' and (v)*rá* 'a corner'. Cf. VRÁ. Since the name occurs four times in Cu, the more probable first el. is OE *docce*.

Dodbrooke D [*Dodebroch* DB, *Doddebrok* 1242 Fees], **Dodcott** Chs [*Dodecote* c 1130 Mon, 1252 Ch, *Doddecote* 1260 Court]. '*Dodda*'s brook and COT.'

Dodd, Great, Cu, a hill of 2,807 ft. Identical with Dod in *Le Dod de Gillefinchor* 1230 Sc, the name of a hill near Loweswater (not Great Dodd). Cf. *dod* 'a rounded summit or eminence', found in Northern and Scotch dialects (1843 &c. OED).

Doddenham Wo [*Dodhæma pull* 779 BCS 233, *Dodeham* DB]. '*Dodda*'s HĀM.'

Dodderhill Wo [*Dudrenhull* 12 PNWo, *Duderhull* 1175 ib., 1221 Ass]. 'Hill where *dodder* or *Cuscuta* grew.' *Dodder*, corresponding to MLG *doder*, Sw *dodra, dudra* &c., is found from c 1265.

Doddershall Bk [*Dodereshell* 1167 P, *Dodreshill* 1207 Cur, *Dodhereshull* 1255 For]. The first el. looks like a pers. n., but no such name is known elsewhere.

Doddinghurst Ess [*Doddenhenc* DB, *Duddingeherst* 1218 FF, -*hurst* 1260 Ipm]. 'The wooded hill of *Dudda*'s people.' Cf. HYRST.

Doddington Ca [*Dodinton* DB, *Doddintona* 1086 IE, *Dudinton* 1230 FF, *Dudington* 1254 Val], **D~** Chs [*Dodynton* 1308 Ormerod, 1315 Misc], **D~ K** [*Duddingtun* 11 DM, *Dudinton* 1201 Cur, *Dudington* 1230 P], **D~ Li** W. of Lincoln [*Dodin(c)tune* DB, *Dodinton, Dudentun, Dudinget* 1205 Cur], **Dry D~** Li [*Dodintune* DB, *Dodinton* 1202 Ass, -*e* 1212 RBE], **D~** Nb [*Dodinton* 1207

Cur, *Dodington* 1242 Fees, *Dudington* 1256 Ass], **Great D~** Np [*Dudinton* 1174, c 1180 Fr, *Magna Dodington* 1309 BM]. OE *Duddingatūn* 'the TŪN of *Dudda*'s people'. In some instances the base may be OE *Dodda*. Sometimes the first el. may be derived from *dod* 'hill'. Cf. DODD. **D~** Nb is at **Dod Hill**.

Doddiscombsleigh D [*Leuga* DB, *Lega* 1263 Ipm, *Doddescumbeleghe* 1309 Ep]. Originally *Lēah*. The manor was held in 1259 by Ralph de Doddescumbe (Ep), who presumably hailed from **Doddiscombe** in Bampton. Doddiscombe is '*Dudd*'s valley'.

Dodford Np [*Doddanford* 944 BCS 792, *Dodeforde* DB], **D~** Wo [*Doddeford* 1232 Cl]. '*Dodda*'s ford.'

Dodington Gl [*Dodintone* DB, *Dudinton* 1226–8 Fees], **D~** Sa nr Cleobury Mortimer [*Dodington* 1285 FA], **D~** Sa nr Whitchurch [*Dodetune* DB, *Dodinton* 1261 Eyton, *Dudinton* 1284 Ipm], **D~** So [*Dodington* 1225 Ass, *Dodinton* n.d. Buckland]. 'The TŪN of *Dudda*'s or *Dodda*'s people.'

Dodleston Chs [*Dodestune* DB, *Dodeliston* c 1205 Chester, *Dudleston* 1330 Ch]. '*Duddel*'s TŪN.'

Dodnash Sf [*Dodenessa* 1188 P, *Dodeneis* 1254 Val, *Dudenessh* 1327 Ch]. '*Dudda*'s ash.'

Dodsley St [*Dedeslega* 1167 P, *Dadesleia* Hy 1 Burton, *Daddesleye* 1272 Ass]. The first el. looks like a pers. n. **Dæddi*. Cf. DADLINGTON.

Dodwell Wa [*Dodwell* 1312 Ipm]. '*Dodda*'s stream or spring.'

Dodworth YW [*Dodeswrde* DB, *Dudewurða* 1170 P, *Doddewurth* 1240 FF]. '*Dudda*'s WORÞ.'

Dogdyke Li [*Dockedic* 12 DC, 1256 BM]. 'Ditch where water-lilies grew.' Cf. DOCCE.

Dogmersfield Ha [*Ormeresfelt* DB, *Dochemeresfelda* 1106 Bath, *Docchemeresfeld* 1167 P, *Dogmeresfeld* 1198 P]. 'FELD by *Dogmere Lake.' The place is by a lake, not now called Dogmere, but once evidently *Doccemere* 'lake where water-lilies grew'. Cf. DOGDYKE.

Dogsthorpe Np [*Dodisthorp* 970 BCS 1258, *Doddesthorp* 972 ib. 1280, *Dodestorp* 1199 NpCh]. '*Dodd*'s thorp.'

Doiley Ha nr Finkley [*Digerlea* 1156, 1192 f., -*lege* 1195 ff. P, *Digerl*[*e*] (wood) 1233 Cl]. 'Thick wood.' First el. OE *diger, identical with Goth *digrs*, ON *digr* 'thick'. Cf. MLG *diger* adv. 'completely'. OE *diger* has not been found elsewhere.

Dolphenby Cu nr Edenhall [*Dolphinerbi* 1203 P, *Dolfanbi* 1282 Ipm], **Dolphinholme** La [*Dolphineholme* 1591 DL]. 'The BY and holm of *Dolfin*.' The name Dolfin is common in the north of England in the 11th cent. Possibly from ON *Dolgfinnr*.

Dolton D [*Duueltone* DB, *Dyvilton* 1235 Cl, *Dughelton* 1235 Pat, *Deweltone* 1279 Ep], **Dowland** D [*Duuelande* DB, *Duhelanda* 1173–5 (1329) Ch, *Dugheland* 1242 Fees]. The first el. may be identical with DUFFIELD, OE *Dūfe-feld*, though with normal Southern early change *f-* > *v-*. Final -*d* was lost before *t* in -*tūn*, and would easily disappear before *l* in -*land*. The early spelling *Dyvil-* may be for *Dvvil-*.

Don R Du [(ostium) *Doni amnis* 1104–8 SD, *Don* c 1140 Gaimar; *Donæmuþe* 757–8 BCS 184, *Done muþe* 794 ASC (E), an old name for JARROW], **Don** R YW [*Don* c 1200 YCh 1009, *Done* 1194–9 ib. 489]. On the Don is **Doncaster** YW [*Danum* 4 IA, *Cair Daun* c 800 HB, (æt) *Doneceastre* 1002 Wills, *Donecastre* DB, *Dunecast*' 1130 P]. Don is an old river-name, Brit *Dānā*, which is related to the name DANUBE and is really an old word for 'water', found in Sanskr *dānu* 'rain, moisture'. Doncaster was Brit *Dāno-*, a derivative of *Dānā*. Later OE CEASTER was added.

Donhead St. Andrew & St. Mary W [*Dunheued, Dunehefda* c 871, (to) *Dunheafdan* 955 BCS 531 f., 917, *Duneheve* DB]. OE *Dūn-hēafod* 'top of the down'.

Doniford So [*Duneford* 1196, *Duniford* 1197 P, *Donyford* 1268 Ch]. The first el. may be an OE *Dūn-ēa* 'hill stream', an old name of **Doniford Stream**.

Donington, Castle, Le [*Duni*(*n*)*tone* DB, *Doninton* c 1125 LeS, *Castel Donyngton* 1428 FA], **D~ le Heath** Le [*Duntone* DB, *Dunigton* 1254 Val, *Duninton* Hy 3, *Donygton super le heth* 1462 BM], **D~** Li nr Spalding [*Donninctune, Duninctune* DB, *Duningetona* R 1 (1290) Ch, *-ton, Dunninton* 1203 Cur], **D~ on Bain** Li [*Duninctune* DB, *Dunnington* 1202 Ass, *Donygton super Beyne* Hy 3 BM], **D~** Sa [*Dunnintun* 10 BM, *Donitone* DB, *Dunnincton* 1167 P]. 'The TŪN of *Dunn*(*a*)'s people.' There is the theoretical possibility that one or other name may contain OE *Dūningas 'dwellers on a hill'.

Donisthorpe Le [*Durandestorp* DB, 1242 Fees]. '*Durand*'s thorp.' *Durand* is a French name of German origin.

Donnington Brk [*Dunintona* 1167 P, *Doniton* 1236 Fees], **D~** Gl [?*Dunnestreatun* 779 BCS 229, *Dunnington* 1176 P, *Donington* 1262 Ipm], **D~** He [*Dunninctune* DB, *Donintone* 12 Glouc], **D~** Sa nr Wellington [*Dunniton* 1180 Eyton, *Duninton* 1201 FF, *Donynton* 1303 Misc]. Identical with DONINGTON. The identification of *Dunnestreatun* with Donnington Gl is very probable. *Dunnestreatun* was near Evenlode and the Fosse Way. This applies to Donnington. The place was then first called *Dunnan Strættūn* 'the TŪN on the Fosse Way belonging to *Dunna* or *Dunne*'. In the boundaries of the charter are mentioned *Dunnen dic, Dunnes slead, Dunnen cumb*. Later the name was changed to *Dunninga-tūn*. The eponym of *Dunnestreatun* was probably at least supposed to be *Dunne*, a nun to whom Æthelbald gave land in Withington Gl in 736 (BCS 156).

Donnington Sa nr Newport [*Derintune* DB, *Derinton* 1255 RH]. '*Dēora*'s TŪN' or 'the TŪN of *Dēora*'s people'.

Donnington (downtn) Sx [*Dunketone* 966 BCS 1191, *Dunketon* 1181 P]. '*Dunnuca*'s TŪN.' **Dunnuca* is a normal derivative of *Dunn*.

Donyatt So [(on) *Duuneȝete* 725 Muchelney, *Doniet* DB, *Duneiet* 1176 P, *Dunniete* 1212 Fees]. The form of 725 is in a late transcript and does not prevent us from deriving the name from *Dunnan geat* '*Dunna*'s gate'. In favour of this etymology tells the name *Dunnepool* found in the same charter. OE *Dūn-geat* does not give good sense, as the place is low on the river Isle.

Donyland, East, Ess [*Dunningland* c 995 BCS 1288, *Dunilanda* DB, *Doniland* 1158 P, *Est Dunilond* 1253 Ch]. '*Dunning*'s land.'

OE **dor** 'door' is sometimes found in pl. ns. in the sense 'a pass'. Cf. DORE Db.

Dorchester Do [*Durnonovaria* 4 IA, *Durngueir* c 894 Asser; *Dornuuarana ceaster* 847 BCS 451, *Dornwaraceaster* 864 BCS 510, *Dorecestre* DB]. The Brit name has been explained as meaning literally 'fist play' (Welsh *dwrn* 'fist' and *gwarae* 'play'), whence 'place where this is carried on'. The name would then have referred to the Roman amphitheatre. The real Dorset form was no doubt *Dorn-gweir*, with Co *dorn* corresponding to Welsh *dwrn*. By substitution the Brit name was adopted as OE *Dornwaru*, or the name was abbreviated to *Dorn*, from which was formed OE *Dornwaru* 'the Dorchester people'. This was combined with OE *ceaster* 'Roman station'. Cf. DORSET.

Dorchester O [*Dorcic, Dorciccaestræ* (gen.) c 730 Bede, *Dorceceaster* 635 ASC, *Dorchecestre* DB, *Dorkecestr'* 1190 P]. A Brit *Dorcic*, to which was added OE *ceaster* 'Roman station'. D~ is on a Roman road. *Dorcic* is derived from the root *derk-* in Bret *derch*, Welsh *drych* 'aspect', OIr *dercaim* 'I see', OE *torht* 'bright'. From this root are derived the river-name *Dorce* (see DORKING), Brit *Condercum* and CONDORCET in France [*Castrum Condorcense* 998]. *Dorcic* may mean 'bright or splendid place'.

Dordon Wa [*Derdon* Hy 2 (1398) Pat, 1285 BM]. OE *dēor-dūn* 'deer hill'.

Dore Db [*Dor* 942 ASC, *Dore* DB]. OE *dor* 'door, pass'. Dore is in a pass on the old boundary between Northumbria and Mercia, now between Yorkshire and Derbyshire.

Dore, Abbey, He [*Dore* 1147 PNHe, 1199 P, 1227 Ch, *Dora* a 1205 BM, *Doier* 1577 Saxton]. Named from **Dore**, the river [*Dour, Dor* c 1150 LL, *Dore* 1213 Cl].

Dore is identical with DOVER and goes back to Brit **dubrā*. OW *Dovr* became *Dour*, for which OE *Dor* was substituted. *Dour* would have given MW *Deur*, Welsh *Daur*. This form explains *Doier* in Saxton. See GOLDEN VALLEY.

Dorfold Chs [*Derfold* 1360 Chamb]. OE *dēor-falod* 'fold for deer'.

Dorking Sr [*Dorchinges* DB, *Dorkinges* 1180 f. P]. 'The dwellers on R **Dork*.' Presumably the Mole was once called *Dorce* 'bright river' (cf. DORCHESTER O). *Dorce* was the old name of a trib. of the COLE W [(innan) *Dorcan* c 1050 BCS 479].

Dormington He [*Dorminton* 1206 Cur, 1242 Fees, *Dormington* 1290 Ch]. 'The TŪN of *Dēormod*'s or *Dēormund*'s people.'

Dormston Wo [*Deormodesealdtun* 972 BCS 1282, *Dormestun* DB]. '*Dēormōd*'s TŪN.' The OE form means '*Dēormōd*'s old TŪN'.

Dorn Wo [*Dorene* 964 BCS 1135, *Dorne* 11 Heming, *Dorna* 1208–10 Fees]. Possibly identical with Gaul *Duronum* and a derivative of Brit *duro-* 'stronghold'. The place is on the Fosse Way and has Roman remains.

Dorney Bk [*Dornei* DB, *-a* 1186 P, *Dornee* 1245 Ep]. OE *dorena īeg* 'island of the humblebees' (OE *dora* 'humblebee').

Dornford O [*Deorneford* 777 BCS 222, *Darneford* c 1160 RA, *Derneford* 1194 P]. 'Hidden ford.' Cf. DIERNE and DARNFORD. The river-name **Dorn** is a back-formation.

Dorrington Li [*Derintone* DB, *-ton* 1170 P, *Dirinton* 1202 Ass, *Dirintone, Durinton* 1209–35, *Dirington* 1238 Ep], D~ Sa nr Woore [*Derintune* DB, *Derynton* 1285 FA, *Deorintone* 1327 Subs]. 'The TŪN of *Dēora*'s or *Dēore*'s people.'

Dorrington Sa nr Condover [*Dodinton* 1198 Fees, 1209, 1283 Eyton]. 'The TŪN of *Dodda*'s people.'

Dorset [*Thornsæta* c 894 Asser, (to) *Dorsæton* 955 BCS 912, (on) *Dorsætum* 978 ASC (C), *Dorseteschire* 940–6 BCS 817, *Dorsete* DB]. Really the name of the people of Dorset, *Dornsæte* 837, 845 ASC. Cf. Introd. p. xiii. *Dornsæte* is either formed from *Dorn*, an earlier form of DORCHESTER, or an elliptical formation from OE *Dornwaraceaster* 'Dorchester'. Cf. SOMERSET.

Dorsington Gl [*Dorsintone* 710 BCS 127, *Dorsitune* c 1050 KCD 964, *Dorsintune* DB, *Dersington* 1236 Fees]. 'The TŪN of *Dēorsige*'s people.'

Dorstone He [*Torchestone* DB; *Dorsintun* 1242 Fees, *Dorsington* 1309–24 BM; *Dorsutton* 1230 P, 1291 Tax]. The forms vary in a curious way. The DB form suggests a derivative of *Torhtsige* as the first el. The change to D- might be due to the influence of the river-name DORE. D~ is on the Dore. Some forms suggest *Dore Sutton*. The majority of forms point to OE *Dēorsiginga-tūn* 'the TŪN of *Dēorsige*'s people'.

Dorton (-oor-) Bk [*Dortone* DB, *-ton* c 1155 Oxf, *Durton* 1291 Tax, *Dourton* 1325 Cl]. 'TŪN in a pass.' First el. OE DURU 'door'. Cf. HAYDOR, LODORE.

Dosthill Wa [*Dercelai* DB, *Dertsechul* 1236, *Dercetehill* 1242 Fees, *Dersthull* 1316 Ipm]. 'Hill with a shelter for deer.' Cf. DASSETT.

Dotland Nb nr Hexham [*Dotoland* c 1160 Hexh, *Doteland* 1226 Ep]. '*Dot*'s land.' *Dot* may be ODan *Dota* (fem.) or OSw *Dote* (masc.).

Dotton D [*Dodingthon, Dodeton* 1242 Fees]. Identical with DODINGTON.

Doughton (-ŭf-) Gl [*æt Ductune* 775–8 BCS 226, *Dughtone* 1301 Ch], D~ Nf [*Doketon* 1196 FF, 1226–8, 1236 Fees]. 'Duck farm.'

Douglas R La [*Duglis* a 1220, *Dugeles* a 1232 CC]. Identical with DALCH and with DOUGLAS in Ireland and Man.

Doulting So [*Dulting* 725 BCS 142, 1125 WMalm, 1267 Ass, *Doltin* DB, *Duulting* 1267 Ass]. An old name of the Sheppey [*Doulting, Duluting* 705 BCS 112 f.]. A Brit river-name, to which was added OE *-ing*. *Dulut-* may be a compound of Welsh *du* 'black' and a word corresponding to OIr *loth* 'dirt', the meaning being 'dirty river'.

Dovaston Sa [*Douaneston* 1198 FF]. '*Dufan*'s TŪN.' *Dufan* is a Brit pers. n. derived from *dubo-* (Welsh *du*) 'black', Welsh *Dyfan*, Bret *Devan*, OIr *Dubán*.

Dove (-ŭ-) R Db [(an) *Dufan* 951 BCS (890), 1008 Burton Reg, *Duue* 1228 Ass], D~ R YN, a tributary of the Rye [*Duve* c 1110 YCh 352, *Duva* c 1157 ib. 355], D~ R YW, a tributary of the Dearne [*Duva* c 1145, c 1150 Riev]. A Brit river-name derived from Brit *dubo-* (Welsh *du*) 'black, dark'. **Dove** Sf is a late name. The valley of the Db Dove is **Dovedale** [*Duuedale* 1296 Abbr].

Dovenby Cu [*Duuaneby* 1230 FF, *Duuvaneby* 1286 Ci]. '*Dufan*'s BY.' ON *Dufan* pers. n. is a loan from OIr *Dubán*. Cf. DOVASTON.

Dover K [*Dubris* (abl.) 4 TP, c 425 ND, (portus) *Dubris* 4 IA, *Dofras* 696–716 BCS 91, *at Dobrum* 844 ib. 445, (on) *Doferum* c 1000 Saints]. Named from the stream at Dover, now **Dour** [*Doferware broc* c 1040 KCD 769, *Dour* 1577 Harrison]. The base is a Brit *Dobrā* from *Dubrā*, the old plur. of *dubro-* 'water' (Welsh *dwfr*). The name thus means 'the waters', i.e. 'the stream'.

Dover Beck Nt [*Douerbec* Hy 2 ERN, 1219 FF, *Doverbec* 1227 For]. The Brit river-name *Dover* from *Dubrā* (cf. DOVER) with an explanatory OScand *bekkr*.

Dovercourt Ess [(æt) *Douorcortæ* c 995 BCS 1289, *Druurecurt* DB, *Dovecourt* 1254 Val].

A Brit *Dover*, identical with DOVER, and OE *corte* (KCD 1363), a word of unknown meaning (PNEss). *Corte* may be cognate with OE *ceart* 'a rough common' and of similar meaning.

Doverdale (dor-) Wo [*Lunvredele* DB, *Duverdale* 1166 RBE]. Really the old name of **Elmley Brook** [*Douerdæl* 706, *Doferdæl* 817 BCS 116, 361], which consists of the old Brit name of the stream (identical with DOVER) and OE DÆL 'valley'.

Doverhay (-rĭ) So [*Doueri, Dovri* DB, *Duvreye* 1243 Ass]. Either a compound of a Brit river-name, identical with DOVER, and OE *īeg* 'island', or an old name of the stream at the place, identical with OE *Doferic*, the name of a stream in Wo, found 757–75, 962 BCS 219, 1087 &c. *Doferic* is a derivative of Brit *dubro-* 'water, stream'.

Doveridge (-ŭ-) Db [*Dubrige* DB, *Duvebrug* 1275 Ch, *-brigg* 1330 QW]. 'Bridge over R DOVE.'

Doward, Great & Little, He, the name of two hills [*Lann Dougarth* c 1150 LL, (mons) *cloartius* c 1140 Monm, *Cloard, Cloward* 1205 Lay]. 'The two hills', from OW *dou* 'two' and *garth* 'a hill'. The *Cl-* forms are miswritten.

Dowdeswell Gl [*æt Dogodeswellan* 781–98 BCS 283, *Dodesuuelle* DB, *Doudeswelle* 1221 Ass]. '*Dogod*'s stream.' *Dogod* is only recorded in this name. It is no doubt a derivative of OE *dugan* 'be of use, avail' with the same suffix as OE *metod* 'fate' from *metan* 'to measure'. Cf. DISEWORTH.

Dowdyke Li [*Duuedic* DB, 1202 Ass]. This could be 'ditch of the doves', but more likely the first el. is the pers. n. *Duve* found in Li in the 12th cent. (DC).

Dowland. See DOLTON.

Dowles Wo [*Dules* 1217 Pp, *Doules* 1292 Ass]. D~ is on Dowles Brook [*Doules* 1296 Eyton]. Dowles is a Brit river-name identical with DALCH.

Dowlish Wake, West D~ So [*Duuelis, Dovles* DB, *Duueliz* 1196 P, *Duueliz Wak* 1243 Ass, *Est-, Westdouelish* 1290 Ch]. Named from the stream at the place [*Douelish* 725 Muchelney, *Doueliz* 1243 Ass]. The name is identical with DALCH. Ralph Wac held the manor in 1189 (Wells). Cf. (Wakes) COLNE.

Down, East & West, D [*Duna* DB, *Estdoune* 1260, *Westdone* 1273 Ep], **Down St. Mary** D [*Done* DB, *Dune St. Mary* 1297 Pat], **Downe** K [*Doune* 1316 FA, *La Doune* 1368 BM]. OE DŪN 'hill'.

Downham Ca [*Duneham* DB, *Dunham* 1086 IE, 1203 Cur], D~ Ess [*Dunham* 1199 P, 1254 Val, *Dounham* 1428 FA], **D~ Market** Nf [(market æt) *Dunham* c 1050 Th, *Dunham* DB, (Forum de) *Dunham* c 1110, (Mercatus de) *Dunham* 1130 BM], **Santon D~**

Sf [*Dunham* DB, 1198 Cur, *Dounham* 1277 Ch]. OE *Dūn-hām* 'HĀM on a hill'.
Santon Downham is near SANTON in Nf.

Downham La [*Dunum* 1194 P, *Dounum* 1251 Ch], D~ Nb [*Dunum* 1186 P, 1251 Ch, *Dunhum* 1256 Ass]. OE *dūnum*, dat. plur. of DŪN 'hill'.

Downhead So [*Dunehefde* DB, *Duneheued* 1196 P, *-hefd* 1244 FF]. 'Top of the down.' Cf. DONHEAD.

Downholland La. See HOLLAND.

Downholme YN [*Dune* DB, *Dunum* 1231 Ass]. OE *dūnum*, dat. plur. of DŪN 'hill'.

Downton or **D~ on the Rock** He [*Duntune* DB, *Dunton* 13 Misc], **D~** Sa nr Stanton Lacy [*Dounton* 1291 Tax, 1320 Ch], **Clee D~** Sa [*Dounton* 1291 Ch], **D~** W [*Duntun* a 670, 826 &c. BCS 27, 391 &c., *Duntone* DB]. 'TŪN on or by a DŪN or hill.'
Clee D~ is near CLEE HILL.

Downwood He [*Dounewode* 1299 Misc]. 'Wood on a DŪN or hill.'

Dowsby Li [*Dusebi* DB, *-by* 1212 Fees, *Douseby* 1275 RH]. '*Dūsi*'s BY' (first el. ODan *Dusi* pers. n.).

Dowthorpe YE [*Duuestorp* DB, 1202 FF], **Dowthwaite** Cu [*Dowthwate* 1488 Ipm]. '*Duve*'s thorp and thwaite.' Cf. DOWDYKE.

Doxey St [*Dochesig* DB, *Dokeseia* 1168 P], **Doxford** Nb [*Dochesefford* Hy 2 FPD, *Dockesford* 1230 Pat, *Dux(e)ford* 1269 Ass]. '*Docc*'s ford.' The pers. n. *Docc* is not otherwise evidenced, but may be compared with *Ducc* in DUXFORD Ca.

Doynton Gl [*Didintone* DB, *Deinton* 1194 P, *Dedigtone* 1221 Ass, *Dointon* 1250 Cl]. 'The TŪN of (the people of) *Dydda*.' Cf. TIDENHAM.

OE dræg. See DRAX, DRAYCOTT, DRAYTON.

Drakelow Db [(æt) *Dracan hlawen* 942 BCS 772, *Drachelawe* DB, *Drakelawe* 1175 P]. OE *dracan hlāw* 'the dragon's mound'. Evidently a myth about a dragon was applied to the place.

Drascombe D in Drewsteignton [*Drosncumb* 739 Crawf, *Droscumb* 1212 Fees, 1230 P]. 'Dirty valley.' First el. OE *drōsn* 'dirt'.

Draughton (-aw-) Np [*Dractone* DB, *Drahton*, *Draiton* 1167 P, *Drachton* c 1170 BM, *Drayhton* 1291 Tax], **D~** (-âf-) YW [*Dractone* DB, *Drahton* 1275 Ep]. Either a form of OE *Drægtūn* 'Drayton' or a compound with OScand *drag* as first el. The meaning is in any case that of DRAYTON.

Drax YW [(æt) *Ealdedrege* 959 YCh 4, *Drac* DB, *Dracas* 1154 RA, *Drach* 1157, *Drachs* 1188, *Drax* 1190 f. P]. OE *dragu*, the plur. of DRÆG, here used in the sense 'a portage'. There was no doubt once a portage at Drax between the Ouse and the Aire, which meet c 4 m. east of DRAX. Cf. DRAYCOTT &c.

Drax, Long, YW nr Drax [*Langrak* 1208 Cur, *Langerak* 13 Misc]. Owing to association with DRAX the form was later changed to *Long Drax*. The second el. of the name *Langrak* is OE *racu* 'bed of a stream'. The meaning of *rak* in this name, as in LANGRICK Li, may be 'reach' or 'straight part of a river'. The meaning would then be 'long reach'.

Draycott Moor Brk [*Draicote* DB], **D~** Db [*Draicot* DB, 1230 P], **D~** O [*Draicote* DB], **D~** So nr Cheddar [*Draicote* DB, *Draycot* 1227 Ch], **D~** So in Limington [*Dregcota*, *Draicota* DB], **D~ in the Clay** St [*Draicote* DB, 1251 Ch], **D~ in the Moors** St [*Draicot* 1251 Ch, *Draycote* 1291 Tax], **Draycot Cerne** W [*Draicote* DB, *Draycote Cerne* 1402 FA], **D~ Fitz Payne** W [*Draicote* DB], **D~ Foliat** W [*Draicote* 1197 FF, *Dreykote Folyoht* 1307 Ipm], **Draycote** Wa [*Draicot* 1203 Cur], **Draycott** Wo nr Blockley [*Draicota* 1209 Fees], **D~** Wo nr Kempsey [*Draycote* 1275 Subs], **Drayford** D [*Draheford* DB, *Draiford* 1238 Ass];

Drayton Beauchamp Bk [*Draitone* DB, *Drayton Belcamp* 1239 Ep], **D~ Parslow** Bk [*Drai(n)tone* DB, *Drayton Passelewe* 1254 Val], **D~** Brk [æt *Draitune* 958, *Drægtun* 960 BCS 1032, 1058, *Draitone* DB], **Dry D~** Ca [*Draitone* DB, *Dreie Draiton* 1228 FF], **Fen D~** Ca [*Drægtun* 1012 PNCa(S), *Draitone* DB, *Fendreiton* 1188 P, 1202 FF], **D~** Ha in Barton Stacey [*Draitone* 903 BCS 602, *Drægtun* c 1000 Hyda, *Draitone* DB], **D~** Ha in Bighton [*Dregtun* 701, 956 BCS 102, 938, *Drayton* 1270 Ch], **D~** Ha in Farlington [*Drayton* 1242 Fees], **D~** Ha nr East Meon [*Drayton* 1376 Works], **D~** He [*Dreituna* 1123 PNHe], **D~** Le nr Rockingham [*Draiton* 1163 P], **Fenny D~** Le [*Draitone* DB, *Fenedrayton* 1465 AD], **D~** Li [*Draitone* DB], **West D~** Mx nr Uxbridge [*Drægtun* 939 BCS 737, *Draitone* DB], **D~ Green** Mx in West Ealing [*Drayton* 1387 Works], **D~** Nf [*Draituna* DB], **D~** Np at Daventry [*Dræghæma gemære* 1021-3 BM, *Drayton* 1220 Fees], **D~ Park** Np in Lowick [*Draiton* 1194 ff. P, *Drayton* 12 NS], **East D~** Nt [*Draitone* DB, *Est Draiton* 1276 Cl], **West D~** Nt [*Draitone* DB, *West Drayton* 1316 FA], **D~** O nr Banbury [*Draitone* DB, 1223 Ep], **D~** O nr Wallingford [*Drætona* 1146 RA], **D~ in Hales, Market & Little D~** Sa [*Draitune* DB, *Drayton en Hales* 1291 Tax, *Parua Drayton* 1327 Subs], **D~** Sa nr Shifnal [no early forms found], **D~** So nr Somerton [*Draitone*, *-tune* DB], **D~** So in S. Petherton [*Drayton* 1243 Ass, 1305 Ipm], **D~** St nr Penkridge [*Draitone* DB], **D~ Basset** St [*Draitone* DB, *Drayton Basset* 1301 Ass], **D~** Sx [*Draiton* 1199 FF, *Draitun* 1212 Fees], **D~** Wa [*Dræitun* 11 Th, *Draiton* 1195 P], **D~** Wo [*Dreiton* 1200 Mon, 1255 Ass].

All these contain OE *dræg*, a word never found in independent use, but often recorded in pl. ns. Besides with COT, FORD, and TŪN it is found combined with OE MERE in *Draymere* Hu, the name of a now drained mere [*Dreigmære* a 1022 KCD 733] with STĀN in *drægstan* 934 BCS 699 (nr DRAYCOT FITZ PAYNE), with DŪN in **Draydon** So [*Est-*, *Westdraydon* 1155–8 (1334) Ch]. It occurs alone in DRAX, and as a second el. in DUNDRY and perhaps DUNDRAW. The OE form was *dræg*, no doubt neuter, and corresponding to OScand *drag*. The latter is used in various senses, esp. 'a portage', i.e. a place where boats are dragged over a narrow piece of land or past an obstruction in the course of a river. Other senses are 'a narrow spit of land or island', 'a way along which timber can be dragged'. The general meaning is 'a place where something can or has to be dragged'. The word is derived from *draga* 'to draw'. Likewise *dræg* is derived from OE *dragan* 'to draw'. In English pl. ns. two senses of *dræg* can be distinguished. In many cases the meaning 'portage' is very probable. Here belong DRAX, DRAYTON Ha nr Farlington, which is in a tongue of land between two deep bays, E. and W. DRAYTON Nt, which seem to have been at the ends of a portage between the Trent and the Idle. In several cases the meaning seems to be 'a stiff hill, a steep slope or ascent where more than ordinary effort is required'; cf. *pull* in this sense. Here belong DRAYCOTT So nr Cheddar, DRAYCOTT Wo nr Kempsey, DRAYCOT FITZ PAYNE and others. Near DRAYTON Wa is a sharply rising hill, apparently called *Dray-hull* 1339 ff. [PNWa(S)]. Both the meanings 'portage' and 'stiff hill' shade off into the meaning 'pass', which is suitable in some cases, as for the St DRAYCOTTS. It is probable that Draycott in some cases refers to a house of shelter at the head of a pass or of a long hill.

The distinctive additions, so far as not self-explanatory, are family names. **Draycot Cerne** was held by Henry de Cerne in 1228 (Ch). Named perhaps from CERNE Do.—For **D~ Fitz Payne** cf. CARY FITZPAINE.—**D~ Foliat** was granted to Henry Foliot in 1209 (Berk). Cf. CHILTON FOLIAT.—**Drayton Bassett** belonged to the Bassets at least from c 1145. Cf. BERWICK BASSETT.—**D~ Parslow** was held by Ralf Passaquam in 1086 (DB). The name, OFr *Passelewe*, means 'pass the water'.

OE **dreng** 'a free tenant holding by tenure older than the Norman Conquest', from OScand *drengr* 'a young man, a lad, a servant', is found in some pl. ns., as DRINGHOE, DRINGHOUSES, DROINTON. The meaning of *dreng* may sometimes have been 'lad' or 'servant'.

Drewsteignton. See TEIGNTON.

Drewton YE [*Drowetone* DB, *Droutun* 1166 P, *Dreuton* 1206 FF]. First el. probably the OFr pers. n. *Dru* or *Dreu*, also *Drogo*, from OG *Drogo*.

Dreyton D [*Dreyton* 1285 Ass]. Identical with DRAYTON.

Drĭby Li [*Dribi* DB, *Driebi* 1130 P]. Apparently 'dry BY'. First el. OE *drȳge* 'dry'.

Driffield Gl [*Drifelle* DB, *Driffeld* 1190 P], **Great & Little D~** YE [(on) *Driffelda* 705 ASC (E), *Drifelt*, *-feld* DB, *Driffeld* 1100–8 YCh 426, *Dridfeld* 1165 P]. 'Dirty (manured) FELD.' First el. OE *drit* 'dirt'.

Drigg Cu [*Dreg* 12, c 1225 StB, *Dregg* 1294 Cl, *Dregge* 1300 Ipm]. Drigg is situated on the river Irt where it runs for two miles near the sea separated from it by a narrow tongue of land. A portage between the Irt and the sea might very well have been used just at Drigg. The name is very likely from OScand *drag* 'a portage'; cf. DRAYCOTT &c. The form with *e* may be explained from the dative **dregi*; cf. *degi*, dat. of *dagr* 'day'. But the Swed pl. n. *Dräg*, dealt with by Franzén, *Vikbolandets by- och gårdnamm* (1937), pp. 186 ff., appears to contain an *i*-mutated OSwed *drægh*, identical in meaning with ON *drag*. Dan dial. *dræg* occurs.

Drighlington YW [*Dreslin(g)tone* DB, *Drichtlington* 1202 FF, *Driclington* 1226 FF]. 'The TŪN of *Dryhthelm*'s or *Dryhtla*'s people.' *Dryhtla*, unrecorded, is a normal derivative of pers. ns. in *Dryht-*, and corresponds to OG *Truhtilo*.

Drimpton Do [*Dremintun*, *Driminton* 1250, *Dremeton* 1252 Fees]. '*Drēama*'s TŪN.' **Drēama*, a short form of *Drēamwulf*, is the first el. also of TREMWORTH K.

Dringhoe YE [*Dringolme* DB, *Drenghou* c 1165 YCh 1405], **Dringhouses** YW [*Drengus* 1234 FF, *Drenghus* 1252 Ch, *Drenghous* 1295 Ipm]. 'The hill or mound and the houses of the drengs.' The second el. of Dringhoe is OScand *haugr* 'mound or hill'.

Drinkstone Sf [*Drincestune* c 1050 KCD 907, *Drencestuna*, *Drincestona* DB, *Drencestun* c 1095 Bury, *Drencheston* 1192 P, *Drenchistone* 1254 Val]. '*Drēmic*'s TŪN.' **Drēmic* is a derivative of *Drēama* in DRIMPTON. The cognate *Dremca* is on record.

OE **drit** 'dirt'. See DRIFFIELD, DRYPOOL.

Drointon St [*Dregetone* DB, *Drengeton* 1199 P, 1284 Ch]. 'The TŪN of the drengs.'

Droitwich Wo [*Wiccium emptorium* 716 BCS 134, *Saltwic* 888 ib. 557, *Wich* DB, *Drightwich* 1347 Pat]. Originally *Wīc*; cf. WĪC. *Droit-* is OE *dryht* 'troop', also used as a first el. of compound words with a general laudatory sense, as in *dryhtsele* 'princely hall' &c.

Dromonby YN [*Dromundeby* c 1185 YCh 582, *-bi* 1185 P]. '*Dromund*'s BY.' First el. the ON byname *Drómundr*, really the name of a kind of ship.

Dronfield Db [*Dranefeld* DB, *Dronefeld* 13 Derby]. 'FELD frequented by drones' (OE *drān*).

Droxford Ha [*Drocenesford* 826, 939 BCS 393, 742, *Drocheneford* DB]. The first el. is related to OHG *trockan* 'dry', probably a noun derived from an OE adj. *drocen 'dry' and meaning 'dry place'.

Droylsden La [*Drilisden* c 1250 LaCh, *Drilesden* 1506 DL]. Possibly OE *drȳgewelles denu* 'the valley of the dry stream'.

Drumburgh Cu [*Drumbogh* 1170–5 (1332) CWNS iii, a 1240 Holme C, 1390 FF]. First el. Welsh *drum* 'ridge'. The second may be the Welsh word for 'small' found in ALLT BOUGH (see ALT). Or it might be OE *burh* with loss of *r* owing to dissimilation. Drumburgh is not far from BURGH LE SANDS. If so, 'ridge near Burgh'.

Druridge Nb [*Dririg* 1242 Fees], **Drybeck** We [*Dribeck* 1290 Ipm], **Drybrook** Gl [*Druybrok* 1282 For]. 'Dry ridge and stream.'

Drypool YE [*Dritpol* DB, *Dripol* 1226 FF]. 'Dirty pool', altered to Drypool to avoid unpleasant associations. First el. OE *drit* 'dirt'.

OE **dūce** 'duck'. See DOUGHTON, DUKINFIELD.

Duckington Chs [*Dochintone* DB, *Dukinton* Hy 3 BM], **Ducklington** O [*Duclingtun* 958 BCS 1036, *Ducelingdun* 1044 KCD 775, *Dochelintone* DB, *Dukelindona* c 1130 Oxf]. The first el. would seem to be derived from a pers. n. *Ducca and **Duc(c)el* : 'the TŪN of *Ducca*'s people' and 'the DŪN of *Ducel*'s people'. Neither name is recorded, but *Docc* may occur in DOXEY, and *Ducc* or the like must be assumed for DUXFORD Ca. It does not seem likely that Ducklington contains the word *duckling*. Possibly *Ducca* is recorded in *Duccenhull* BCS 923. The first el. of Duckington may also be derived from *Duduc*.

Duckmanton Db [*Ducemannestun* 1002 Wills, *Dochemanestun* DB, *Duchemanetun* c 1160 BM]. Looks like '*Duceman*'s TŪN'. *Duceman* is not evidenced, but it might have the same relation to *Ducca* as *Dudeman* to *Dudda*. Cf. prec. name.

Dudbridge Gl [*Dodebrugge* 1292 BM], **Duddenhoe** Ess [*Duddenho* n.d. AD, *Dodenho* 1251 Ch]. '*Dudda*'s bridge and HŌH or spur of land.'

Duddeston Wa [æt *Duddestone* 963 BCS 1100, *Dudeston* 1204 Cur]. '*Dudd*'s TŪN.'

Duddington Np [*Dodintone* DB, *Duditun* 1156 P, -*ton* 1206 Cur]. 'The TŪN of *Dudd(a)*'s people.'

Duddo Nb [*Dudehou* 1208–10 Fees, *Dudeho* 1228 FPD], **Duddoe** Nb [*Dudden* 1242 Fees]. '*Dudda*'s HŌH and DENU.'

Duddon Chs [*Duddon* 1288 Court]. '*Dudda*'s DŪN.'

Duddon R Cu, La [*Dudun* a 1140 LaCh,

Dudena c 1160 ib., *Duthen* 1196 FF]. Unexplained.

Dudleston Sa [*Dodeleston* 1267 Ipm]. '*Duddel*'s TŪN.'

Dudley Wo [*Dudelei* DB, *Duddele* 1221 Ass, *Doddeley* 1279 Cl]. '*Dudda*'s LĒAH.'

Dudmaston Sa [*Dodemanestun* Hy 1, *Dudemanneston* 1165 Eyton]. '*Dudeman*'s TŪN.'

Dudstone Sa [*Dudestune* DB, *Dudistone* 13 Misc]. '*Dudd*'s TŪN.'

Duffield Db [*Duuelle* DB, *Duffelda* Hy 2 DC, *Duffeld* 1236 FF], **North & South D~** YE [*Dufeld, Nortdufelt, Suddufeld* DB, *Duffeld* c 1185 YCh 992]. 'FELD frequented by doves.'

Dufton We [*Dufton* 1289 Ipm, 1291 Tax]. 'Dove farm.' Cf. Dovecot, a well-known pl. n.

Duggleby YE [*Difgelibi* DB, *Deuegelebi* 1190 P]. '*Dufgall*'s BY.' ON *Dufgall* is a pers. n. of Irish origin (OIr *Dubgall*, lit. 'black stranger').

Dukinfield (-ŭ-) Chs [*Dokenfeld* 12 Earwaker, *Dokinfeld, Dukenfeld* 1285 Court]. A possible etymology is OE *dūcena feld* 'FELD frequented by ducks'.

Dulas R He [*Duneleis* c 1135 Ewias, *Dunelays* 13 Misc, *Dyueleis* 1327 Ch]. Identical with DALCH. On the Dulas is **Dulas** vil. [*Dewlas* 1523 Glouc].

Dullingham Ca [*Dullingham* 1043–5 Wills, 1200 Cur, *Dullingeham* c 1080 ICC, DB]. 'The HĀM of *Dull*'s or *Dulla*'s people.' Cf. DILLINGTON.

Duloe Co [*Doulo* 1291 Tax, *Dulo* 1348, 1365 FF]. D~ stands on a ridge between the East and the West LOOE. The name means 'the two Looes'. Co *dow* means 'two'.

Dulverton So [*Dolvertune* DB, *Dulverton* 1212 Fees, *Dilvertone* 1225 Ass, *Delverton* 1291 Cl]. The first el. is very likely a name in *ford*, e.g. OE *dīegla ford* 'hidden ford'. Cf. DILWYN.

Dulwich (dŭlǐj) Sr [*Dilwihs* 967 BCS 1196, *Dilewisse* 1212 RBE, 1242 Fees, *Dilewyshe* 1279 QW]. OE *dile-wisse* or -*wisc* 'meadow where dill grew'.

Dumbleton Gl [*Dumeltuna,* (ad) *Dumoltan, Dumolatan* 930 BCS 667, *Dumbel-, Dumaltun* 995 KCD 692, (æt) *Dumeltan* 1002, *Dumeltun* 1004 Wills, *Dubentone* DB, *Dumbelton* 1206 Cur, 1230 BM]. The name probably denoted originally D~ Hill, a very prominent hill. Of the early forms those in -*tan* are probably most trustworthy. Possibly -*tan* is OE *tān* 'twig' &c., used in a transferred sense of a hill. The first el. may be a Brit name of the hill, containing Welsh *moel* 'bare hill' and *du* 'black' or *dwn* 'dark'. But OE (æt) *Dumeltan* may well be an oblique form of *Dumolta*, which may be analysed as *Dumol* and *tā* 'toe'. Dumble-ton Hill juts out from Alderton Hill and on the map looks rather like a big toe. *Dumol* might have been the old name of the whole massif of hills. Words for parts of the body are often used in transferred topographical senses.

Dummer Ha [*Dummere* DB, *Dunmere* 1196 P, 1198 FF]. 'Mere or lake on or by a DŪN or hill.' There is no lake here now.

OE dūn 'down, hill, mountain'. In pl. ns. the meaning varies from 'hill' (as in ASH-DOWN, BREDON, BRENDON, HAMBLEDON, SNOWDON) to 'a slight rise' (as in DOWNHAM Ca, SANTON DOWNHAM Sf, WATTON). HEN-DON 'the high DŪN' (Mx) rises to 280 ft. See ABINGDON. An important special sense is 'hill pasture', as in KINGSDOWN. See DOWN, -E, -HAM, DUNHAM, DUNTON &c. When *Down* is used as a distinguishing el. before names, the meaning is generally 'lower', from OE *of dūne*.

Dunchi·deock (-chĭdǐk) D [*Donsedoc* DB, *Dunsidioch* 1188 P, *Dunchidyok* 1291 Tax]. Early Welsh *din*, Co *dun* 'fort' &c. and Welsh *coediog* (earlier *coedioc*) 'wooded'. Cf. CHIDEOCK.

Dunchurch Wa [*Donecerce* DB, *Dunnes-chircch* 1200 Cur, *-chirch* 1236 Fees]. 'Dunn's church.' D~ is near DUNSMORE.

Dunclent Wo [*Dunclent* DB, *Dounclent* 1315 Ipm]. 'Down (i.e. lower) CLENT.' The place is on the slope of Clent Hills.

Duncote Np [*Doncote* 1276 BM, *Donecote* 1316 FA]. 'Dunna's COT.'

Duncton Sx [*Donechitone* DB, *Duneketon* 1181 P]. 'Dunnuca's TŪN.' Cf. DONNING-TON Sx.

Dundon So [*Dondeme* DB, *Dunden* 1236 Fees, 1243 Ass]. 'Valley by the DŪN.'

Dundraw Cu [*Drumdrahrigg* 1194, *Dun-drahe* c 1230 Holme C, *Dromdraw* 1308 Misc]. Welsh *drum* 'ridge', probably an old name of the ridge, and ON *drag*, per-haps a Scandinavianized form of OE *dræg*, the meaning being 'the steep ascent of Drum ridge'.

Dundry So [*Dundreg* 1065 Wells, *-drey* 1227 FF, *-dray* 1230 FF]. 'The steep ascent of the DŪN or ridge.' Cf. DŪN and DRAYCOTT. Dundry and East Dundry are on the slope of a long ridge, **Dundry Hill**.

Dunge (or **Denge**) **Marsh** K [*Dengemersc* 774 BCS 214, *-maris* 1071 Reg, *Dingemareis* 1225 Pat, *Dengemareys* 1278 QW]. 'Marsh belonging to DENGE.' Denge is doubtless **Denge** nr Chilham [*Denge* 1292 Ass], which may be explained as an OE *Den-gē* 'the valley district'. Cf. EASTRY and LYMINGE. Denge would then have been a district in OE times. Cf. BURMARSH nr Dunge Marsh, which is 'the marsh of the Canterbury people'. **Dunge Ness** was named from Dunge Marsh.

Dunham on the Hill Chs [*Doneham* DB, *Dunham* 1302 Chamb], D~ **Massey** Chs [*Doneham* DB, *Donham* c 1150 Mon, *Dun-ham Massy* c 1280 Misc], **Great & Little** D~ Nf [*Dunham* DB, *MagnaDunham* 1242 Fees]. 'HĀM on a DŪN or hill.'

D~ **Massey** belonged to Hamo de Masci in 1086 (DB). Masci from MASSY in France (Seine-Inf.).

Dunham Nt [*Duneham* DB, 1156 P, 1212 BM, *Donneham* 1291 Tax, *Dunham* 1157 f. P]. '*Dunna*'s HĀM.'

Dunhampstead Wo [*Dunhamstyde* 814 BCS 349, *Dunhæmstede* c 975 KCD 680]. 'Homestead on a hill.'

Dunhampton Wo [*Dunhampton* 1222 FF]. 'HĀMTŪN on a hill.'

Dunholme Li [*Duneham* DB, c 1115 LiS, 1202 Ass, *Dunham* c 1115 LiS, c 1155 BM]. '*Dunna*'s HĀM' rather than *Dūn-hām*, to judge by the situation.

Dunkenhalgh La [*Dunkansale* 1208–20 PNLa, *Dunkaneshalghe* 1285 Ass]. '*Dun-can*'s HALH.' Duncan is OIr *Donnchad*.

Dunkerton So [*Duncretone* DB, *Dunkerton* 1225 Ass], **Dunkery Hill** So [*Duncrey* 13 AD, *Dunnecray* 1298 Wells]. Dunkery is no doubt a Brit hill-name, perhaps from Welsh *din* 'hill fort', Co *dun*, and OW *creic* 'rock'. The loss of final *-c* is a difficulty, it is true. Dunkerton may contain an identical hill-name.

Dunkeswell D [*Doducheswelle* DB, *Donekes-well* 1228 FF]. '*Duduc*'s stream.'

Dunkeswick YW [*Chesuic* DB, *Dunkeswyk* c 1145 YCh 1862, *Dunkeswic* 1228 Cl]. 'Down Keswick.' Cf. KESWICK.

Dunmail Raise Cu [*Dunbalrase stones* 1610 Saxton]. *Raise* is ON *hreysi* 'cairn'. The first el. is stated to be *Dumnail*, the name of the last king of Cumbria.

Dunmow, Great & Little, Ess [(at) *Dune-mowe* c 950, *Dunmawe* 1043–5 Wills, (of) *Dunmæwan* c 1000 CCC, *Dommauua* DB, *Dunmawe Magna, Parva* 1238 Subs]. 'Hill meadow.' Second el. an OE **māwe* 'meadow' (from *māwan* 'to mow'). Cf. *mow* 'meadow' in Devon. First el. DŪN.

Dunnerdale La [*Dunerdale* 1293 LaInq]. 'The valley of R DUDDON.'

Dunnington YE nr Hornsea [*Dodintone* DB, *Dudinton* 1223 FF]. 'The TŪN of *Dudd(a)*'s people.'

Dunnington YE nr York [*Donniton* DB, *Duninton* 1200, *Dunnigton* 1202 FF], **Dun-ningworth** Sf [*Duniworda* DB, *Dunninge-wurða* 1177 P]. 'The TŪN and WORÞ of *Dunn(a)*'s people.'

Dunnockshaw La [*Dunnockschae* 1296 Lacy]. 'Dunnock or hedge-sparrow wood.'

Dunsby Li nr Bourne [*Dunesbi* DB, *Dunnes-by* 1242 Fees], D~ Li nr Sleaford [*Dunnesbi* DB, *Dunnysby* 1242 Fees]. '*Dunn*'s BY.'

Dunsden O [*Dunesdene* DB, *Denesden* 1231 Cl, 1242 Fees]. '*Dynne*'s valley.'

Dunsfold Sr [*Duntesfaude* 1259 AD, *-faud* 1272 FF, *-falde* 1291 Tax]. Apparently '*Dunt*'s fold'. *Dunt pers. n. seems to be presupposed also by DUNTISBORNE.

Dunsford D [*Dunesforda* DB, *Dunisford* 1237 Fees], **Dunsforth, Lower & Upper**, YW [*Dunesford* DB, 1202 FF, *Dunnesford* 1283 Ch], **Dunsley** YN [*Dunesle* DB, c 1165 YCh 899], **Dunsmore** Wa [*Dunesmore* R 1 Cur, *Dunnesmor* 1236 Fees]. '*Dunn*'s ford, LĒAH, moor.'

Dunstable Bd [(æt) *Dunestaple* 1123 ASC (E), *Dunestapla* 1130 P, *-stable* 1154 HHunt]. '*Dunn(a)*'s STAPOL.'

Dunstall Li [*Tonestale* DB, *Tunstal*, *Dunestal* c 1115 LiS], **D~ St** [*Tunstall* 13 BM, *Donestal* 1272 Ass]. OE *tūnst(e)all* 'homestead'.

Dunstan Nb [*Dunstan* 1242 Fees, 1256 Ass]. 'Stone on a DŪN or hill.'

Dunster So [*Torre* DB, *Dunestore* 1138 HHunt, *-torra* c 1150 Bath, *Dunestere* 1238 Ass, *Dunsterr* 1242 Fees]. Originally TORR 'a tor'. Later '*Dunn*'s Torr' from some early owner.

Dunston Db [*Dunstone* 1258 FF, *Dunestan* Hy 3 PNDb, *Doneston* 1292 Abbr]. '*Dunn*'s stone.'

Dunston Li [*Dunestune* DB, *Dunnestona* 1215 (1291) Ch, *-tun* 1264 Ipm], **D~ Nf** [*Dunestun* DB, *-ton* 1186 P], **D~ St** [*Dunestone* DB, *Doneston* 1242 Fees]. '*Dunn*'s TŪN.'

Dunstone D [*Dunestanetune* DB, *Dunstanestun* 1204 Cur]. '*Dunstān*'s TŪN.'

Dunterton D [*Dondritone* DB, *Duntertone* 1242 Fees, *-dune* ib.]. Perhaps a Brit *Dintref 'village by the castle' (PND), to which was added OE TŪN. There is an ancient castle here.

Dunthrop O [*Dunetorp* DB, *-trop* 1166 RBE, *Dunestorp* 1193 P]. '*Dunna*'s thorp.'

Duntisborne Abbotts & Rouse Gl [*Duntes-, Dantes-, Tantesborne* DB, *Duntesbourn Abbatis, -burn Militis* 1291 Tax, *Dontesborne Roue* 1303 FA, *Duntesbourn Rous* 1327 Subs]. Apparently '*Dunt*'s stream'. See DUNSFOLD.

D~ Abbotts belonged to Gloucester Abbey.—
D~ Rouse was held by Roger le Rus in 1285 (FA). Rouse is a nickname and family name from Fr *roux* 'red'.

Duntish Do [(on) *dounen tit* 941 BCS 768, *Dunhethis* 1249 FF, *Dunetisse, Dunedisse* 1264 Ipm, *Dunetys* 1280 Ch]. OE *dūnetisc* 'pasture or field on a hill'. The form of 941 is in a very poor text. OE *etisc corresponds to Goth *atisk* 'cornfield', OHG *ezzisch* 'a piece of land'. The exact meaning is doubtful.

Dunton Bd [*Donitone* DB, *Donton* 1202 Ass]. Possibly 'the TŪN of the people on the hill' (OE *Dūninga-tūn*).

Dunton Bk [*Dodintone* DB, *Dudinton* 1198 FF]. 'The TŪN of *Dudd(a)*'s people.'

Dunton Ess [*Dantuna* DB, *Dunton* 1206 Cur, *-e* 1254 Val], **D~ Bassett** Le [*Donitone* DB, *Dunton* 1199, 1230 P, *Dunton Basset* 1418 AD], **D~ Nf** [*Dontuna* DB, *Dunton* 1198, 1219, 1236 Fees], **D~ Wa** [*Dunton* 1241 Fees, *-tun* 1251 Ch]. 'TŪN on a DŪN or hill.'
D~ Bassett was held by Ralph Basset in 1242 (Fees). Cf. BERWICK BASSETT.

Dunwea·r So [*Dunewere* 1194 f. P, *Dunwer* 1236 FF, (mill of) *Were* 1232 Ch]. Second el. OE *wer* 'weir'. The low situation of the place suggests that the first el. is *Dunna* pers. n.

Dunwich (dŭnĭch) Sf [*Domnoc, Dommoc* c 730 Bede, *Domnoc* 636 ff. ASC (F), *Dommocceaster* c 890 OE Bede, *Dummucæ (Dammace) civitas* 803 BCS 312, *Duneuuic* DB]. A Brit name derived from Celt *dubno-* (Welsh *dwfn*) 'deep' and meaning perhaps 'port with deep water'. The later name is due to popular etymology or a contracted form of *Dumnuc-wīc*, OE *wīc* 'town' or 'port' having been added.

Dunwood Ha [*Dunewode* 1273 BM], **D~ St** [*Dunwode* 1278 Ipm]. 'Wood on a DŪN or hill.'

Durborough So [*Dereberge* DB, *-berg* 1238, 1243 Ass]. 'Deer hill' (OE *dēora-beorg*).

Durford Sx [*Dureford, Dereford* Hy 1 PNSx, 1204 Cur]. 'Deer ford.'

Durham Du [*Dunholm* c 1000 Saints, 1056 ASC (D), *Dunhelme* 1122 HHunt, *Donelme* 1191 FF, *Durealme* c 1170 Jordan Fantosme, *Duram* 1297 Rob Gl]. 'Holm or island with a hill.' D~ is nearly surrounded by the Wear and is built on a rocky hill. *Dun-* is OE *dūn*, while *holm* is OScand *holmr*. The later change is due to Norman influence. The county of Durham is comparatively late.

Durleigh So [*Derlege* DB, *Derleya* 13 BM, *Durlega* 1274 RH], **Durley** Ha [(to) *Deorleage* 901 BCS 596, *Derleie* DB]. Identical with DARLEY.

Durnford W [*Diarne-, Darneford* DB, *Derneford* 1166 RBE, 1190 P]. 'Hidden ford.' Cf. DIERNE.

Durrington Sx [*Derentune* DB, *Direnton* 1200, *Duringtone* 1219 FF], **D~ W** [*Derintone* DB, *Durentona* 1178 BM, *Durintone* 1212 RBE, *Diryngtone* 1291 Tax]. 'The TŪN of *Dēora* or of *Dēor(a)*'s or *Dēore*'s people.'

Dursley Gl [*Dersilege* DB, *Derseleie* 1195 f. P, *Dursleg* 1220 Fees]. '*Dēorsige*'s LĒAH.'

Durston So [*Derstona* DB, *Derston* 1181 P, *Durston, Dirston* c 1180 Buckland]. '*Dēor*'s TŪN.'

Durton or **Urton** La [*Overton* 1502 DL]. Originally *Overton* (q.v.). *D-* is the Fr prep. *de*.

OE duru 'door'. See DORTON, HAYDOR, LODORE. Cf. DOR.

Durweston Do [*Dervinestone* DB, *Dirwinestun* c 1100 Montacute]. '*Dēorwine*'s TŪN.'

Duston Np [*Dustone* DB, *Duston* 1178, 1190 P, 1202 Ass]. Perhaps simply 'dusty TŪN'.

Dutton Chs [*Duntune* DB, *Dottona* 12 Chester, *Dutton* 1288 Court], **D~** La [*Dotona* 1102 LaCh, *Duttun* 1182–5 YCh 199]. '*Dudda*'s TŪN.'

Duxbury La [*Deukesbiri* 1202 FF, *Dukesbiri* 1227 FF]. If the earliest form is reliable, the first el. is as in DEUXHILL. Otherwise as in DUXFORD Ca.

Duxford Brk [*Dudochesforde* DB, *Dodekelesford* 1316 FA]. '*Duduc*'s ford.'

Duxford Ca [*Dukeswrth* c 950 Wills, *Doches-urda* c 1080 ICC, *Dochesuuorde* DB, *Duckeswurthe* 1218 Ass, *Dukesword* 1230 P]. '*Duc*'s or *Ducc*'s WORÞ.' A pers. n. *Duc(c)* is not evidenced, but must be inferred from Duxford.

Dyke Li [*Dic* DB]. OE DĪC 'dike, ditch'.

Dymchurch K [*Deman circe* 11 DM, *Demecherche* 1243 StAug, 1291 Tax]. 'The church of the judge' (OE *dēma*) or '*Diuma*'s church'. One *Diuma* was bishop of Mercia in the 7th cent.

Dymock (-ĭ-) Gl [*Dimoch* DB, *Dimmok* 1156 RBE, *Dimmoch* 1156, 1190 P]. No doubt a Brit name, which might possibly be compared with MOCCAS and MOTTRAM and be a compound of Welsh *ty* 'house' and *moch* 'pigs'. The meaning would be 'pigsty'. *D-* would be due to lenition.

Dyrham Gl [*Deorham, -hamme* 950, 972 BCS 887, 1282, *Dirham* DB]. 'Enclosure for deer.' See HAMM.

E

OE **ēa** 'river', corresponding to OScand *ā*, OHG *aha* &c., is common in river-names. It is sometimes used alone; cf. EYE, RAY, REA, YEO. As the second el. it mostly appears now as *-ey*, as in WAVENEY, WISSEY. Some original river-names in *-ēa* have become pl. ns., as GRAVENEY. It often appears as a first el., as in EATON, ETON; cf. EAMONT. But some of these really seem to have as first el. OE *ēg* 'island'.

Each K [*Ece* DB, *Ecche, Heche* 11 DM]. OE *āēc*, dat. sg. of *āc* 'oak'.

Eachwick Nb [*Achewic* c 1160 Hexh, *Echewic* 1242 Fees]. The first el. seems to be a derivative of OE *āc* 'oak', either *āēcen* adj. 'of oaks' or a pers. n. **Æca*, a short form of names in *Āc-*.

Eagle Li [*Aclei, Aycle* DB, *Eicla* 1141 RA, *Aycle* 1254 Val]. OE *āc-lēah* 'oak wood', with substitution of OScand *eik* for OE *āc*.

Eaglesfield Cu [*Eglesfeld* 12, *Egclesfeld* c 1250, *Egglesfeld* 13 StB]. '*Ecgwulf*'s or *Ecgel*'s FELD.' **Ecgel* is a short form of names in *Ecg-*, perhaps found in *Ecgeles stiele* 1007 KCD 1303.

Eakring (ē-, *olim* ā-) Nt [*Ecringhe, Echeringhe* DB, *Aichringa* c 1150 DC, *Aikering* 12, 1229 BM]. OScand *Eik-hringr* 'oaks forming a circle'.

Ealand Li [*Aland* 1316 FA, 1372 Selby]. OE *ēaland* 'island, land by water'.

Ealing Mx [*Gillingas* c 700 Bodley MS, *Illing* 1130 P, *Gilling* 1243 Pat, 1292 QW, *Ylling* 1254 Val]. OE *Gillingas* '*Gilla*'s people'. Cf. YELLING Hu. OE **Gilla* may be a nickname belonging to *giellan* 'to scream' or a short form of names in *Gisl-*.

Eamont (ē-, yă-) R Cu, We [*Amoth* 12 Holme C, *Amot* 1285 For, *Emot* c 1235 CWNS x]. A back-formation from a pl. n. **Eamont** or from **Eamont Bridge** We [(æt) *Eamotum* 926 ASC (D), *Pons Amot* 1278 Ass, *Amotbrig* 1362 Test Karl]. OE *ēa-gemōt* means 'junction of streams'. The meeting of the brook from Dacre and the Eamont is probably referred to.

OE ***ēan** 'lamb'. See ENHAM, YEN HALL, also ENDON, ENFIELD.

OE **ēar** 'gravel'. See EARITH, ERITH, YARLET, YARMOUTH Wt.

Earby YW [*Eurebi* DB, *Euerby* 1260 Ipm]. The first el. may be ON *Iǫfurr*, OSw *Iauur* pers. n.

Eardington Sa [*Eardigtun* c 1030 Förster, *Themse*, p. 769, *Ardintone* DB, *Eardinton* 1203 Ass, *Erdinton* 1212 Fees]. 'The TŪN of *Ēanrēd*'s or *Ēadrēd*'s people.'

Eardisland He [*Lene* DB, *Erleslen* 1230 Pat, 1233 Cl, *-lan* 1230 Cl]. The first el. is OE *eorl* 'earl'. The second is the old district-name *Leon* (in *Leon-, Lionhi(e)na gemǣre* 958 BCS 1040), on which see LEOMINSTER. Cf. KINGSLAND, MONKLAND.

Eardisley He [*Herdeslege* DB, *Eierdesl'* 1249 Cl, *Eiardeleye* 1252 Ch, *Erdesleye* 1269 Ipm]. '*Ægheard*'s LĒAH.'

Eardiston Wo [*Eardulfestun* c 957 BCS 1007]. '*Eardwulf*'s TŪN.'

Earith Hu [*Herheth* 1244 Rams, *Earheth* 1260 Ass]. Identical with ERITH.

Earle (jerl) Nb [*Yherdhill* 1242 Fees, *Yerdhil* 1256 Ass, *Yerdill* 1289 Ipm]. 'Hill with an enclosure' (OE *geard* 'yard').

Earley Brk [*Erlei* DB, *Erlega* 1177 ff. P]. Very likely OE *earn-lēah* 'eagle forest' (cf. ARLEY &c.). But OE *ēar* 'gravel' is also a possible first el. Cf. ERITH.

Earlham Nf [*Erlham* DB, 1163 P, 1198 Fees, *Herlham* 1196 FF, 1242 Fees]. The first el. may be OE *eorl* 'earl' or a pers. n. *Herela* (cf. HARLING). In the latter case the loss of *H*- is due to dissimilation.

Earlstone Ha [*Erlestone* 1167 P, *Urleston* 1242 Fees]. 'The earl's TŪN.'

OE **earn** 'eagle' is common especially in the compound *earn-lēah* 'eagle wood', as in ARE-, AR-, EAR-, EARNLEY. See also e.g. ARN-, EARNWOOD, ARNOLD, ERIDGE, YARN- (passim).

Earnley Sx [*Earnaleach, Earneleagh, Earnelegh* 780, *Earneleia* 930 BCS 237, 669, 1334]. 'Eagles' wood.'

Earnshill So [*Erneshele* DB, *Erneshelle* 1194 P, 1225 Ass]. 'Hill on R Earn.' **Earn** is *Earn* 762, 966 Muchelney.

Earnstrey Sa [*Ernestreu* 1172 P, *-trie* 1199 P, 1200 FF]. OE *earnes trēo* 'the eagle's tree'.

Earnwood Sa [*Erne Wode* 1327 Subs, *Ernewode* 1333 Misc]. 'Eagles' wood.'

Earsdon Nb nr Newcastle [*Erdesdon* 1233 P], E~ Nb nr Felton [*Erdisduna* Hy 2, *Erdesdona* 1198 (1271) Ch, *Erdisdon* 1242 Fees]. '*Ēanrēd*'s or *Eorēd*'s DŪN.'

Earsham Nf [*Ersam* DB, *Earesham* (hd) c 1095 Bury, *Eresham* 1158 ff. P, 1212 Fees, *Erlsham* 1248 Cl]. The form of 1248, though isolated, probably indicates the etymology: 'the earl's HĀM'. Or the first el. may be as in EASTBURY Wo.

Earswick (er-) YN [*Edresuuic, Edrezuic* DB, *Ethericewyk* 13 PNNR]. '*Ēpelrīc*'s WĪC.'

Eartham Sx [*Ercheham* Hy 1 AD, *Ertham* 1279 Ass, *Urtham* 1279 QW]. First el. OE *erþ* 'ploughing, ploughed land'. Second el. HĀM or HAMM.

Earthcott Gl [*Herdicote* DB, *Herdecota* 1220 Fees, *Erthecote* 1289 Cl]. OE *eorþe-cot* 'earth hut'.

Easby Cu [*Essebi* (p) 1159 P (cf. CWNS xxvi. 288), *Ecchesby* 1363 PNCu(S), *Eseby* 1486 Ipm]. Perhaps OScand *Eskibȳr* 'BY near some ash-trees'. E~ YN nr Richmond [*Asebi* DB, *Eseby* 1208 FF], E~ YN nr Stokesley [*Esebi* DB]. '*Ese*'s BY.' *Ese* is ODan *Æse*, *Ese*, OSw *Æse*, ON *Æsi*.

Easebourne (ēz-) Sx [*Eseburne* DB, *-burna* 1166 P, *Isenburna* 1166 P], **Easenhall** Wa [*Esenhull* 1221 Ass, 1428 FA]. '*Ēsa*'s stream and hill.'

Eashing Sr [(æt) *Æscengum* c 880 BCS 553, (to) *Eschingum* c 910 ib. 1335, *Essinge* 1272 FF]. '*Æsc*'s people.'

Easington Bk [*Hesintone* DB, *-ton* c 1155 Oxf, *Esington* 1242 Fees]; E~ Du [*Esingtun* c 1050 HSC, *Esinton* 1196 P], E~ YE [*Hesinton, Esintone* DB, *Essintona* c 1100 YCh 1300, *Esington* 1227 FF], E~ YN [*Esingetun* DB, *Esington* 1208 FF], E~ YW [*Esintune* DB, *Esyngton* 1285 FA]. OE *Ēsingatūn* 'the TŪN of *Ēsa*'s or *Ēsi*'s people'.

Easington Nb [*Yesington* 1242 Fees, 1269 Ass]. The place is on a stream which may have been ME *Yese* (cf. OUSE BURN). If so, 'the TŪN of the dwellers on *Yese*'.

Easington O [*Esidone* DB, *Esendon* c 1150 &c. Godstow, *Esindone* 1209-19, 1223 Ep], **Easingwold** YN [*Eisincewald* DB, *Esingewald* 1169 P, 1208 FF]. 'The DŪN and wold of the *Ēsingas* or people of *Ēsi* or *Ēsa*.'

Easole K [æt *Oesewalum* 824, *Oesuualun* 832 BCS 378, 402, *Eswalt* DB, *Easole* 1242 Fees]. The second el. is OE *walu*, which is held to mean 'a ridge of earth or stone'. The first seems to be cognate with MLG *ōse* 'a ring-shaped handle', ON *æs* 'a hole in a shoe for the shoelace', from *ansiō. This would give OE *ǣs*, *ēs*. Easole is situated in a hollow between two ridges, which jut out so as to form a kind of fork. The name may mean 'ridges that form or look like a handle'.

OE **ēast** 'east' is common as the first el. of pl. ns. Often *ēa* was shortened and gave later *a*, as in ASCOT, ASTLEY, ASTON. But in some cases *East*- in a pl. n. goes back to OE *be ēastan* 'east of'. A name such as EASTBOURNE may well be OE *be ēastan burnan*, which is elliptical and means '(place) east of a stream'. An unrecorded OE *ēastor*, corresponding to OLG *ôstar*, OFris *âster*, ON *austr*, occurs in EASTREA, EASTRY; cf. also EASTRINGTON. OE *ēasterra* 'eastern' is found in EASTERGATE, EASTERTON.

Eastbourne Sx [*Burne* DB, 1227 Ch, *Estburn* 1279 Ass]. Originally *Burna* from the stream here. Later Eastbourne in contradistinction to WESTBOURNE.

Eastbridge K [*Eastbrige* 11 DM, *Heastbruge* c 1150 Fr, *Estbreg* 1219 Fees]. 'Eastern bridge.'

Eastburn YW [*Est-, Esebrune* DB, *Esteburn* n.d. Kirkst]. Probably OE *be ēastan burnan* 'east of the stream'. For **Eastburn** YE see KIRKBURN.

Eastbury Brk [(of) *Eastbury* c 1066 ASCh, *Estberi* 1165 P, *Estbir* 1208 FF]. 'Eastern BURG.'

Eastbury Wo at Hallow [*Earesbyrig* 11 Heming, *Eresbyrie* DB]. Perhaps '*Ēanhere*'s BURG'.

Eastby YW [*Estby* 1257 Ch]. 'Eastern BY.'

Eastchurch K [*Eastcyrce* 11 DM, *Estchirche* 1194 Fr]. 'Eastern church.'

Eastcote Np [*Edeweneskote* 1277 BM]. '*Ēadwine*'s COT.'

Eastcott Mx [*Estcote* 13 AD], E~ W nr Potterne [*Estcote* 1349 Ipm], E~ W nr Swindon [*Escot* 1490 Ipm], **Eastcotts** Bd [*Estcote* 1200 P, *-s* 1240 Ass], **Eastcourt** W [*Escote* 901, *Eastcotun* 974 BCS 586, 1301]. 'Eastern COTS.'

Eastdean. See DEAN.

Easter (ĕ-), **Good & High**, Ess [*Estre* c 1050 KCD 907, 1206 FF, *Estra* DB, *Godithestr'* 1200 P, *-estre* 1208 FF, *Heautestre* 1251 Ch, *Heyestre* 1254 Val]. Easter goes back to OE *ēstre*, which has the same relation to *eowestre* as OE *ēde* 'a flock' to *eowde*. The meaning is 'sheepfold'.

Good E~ from an early owner, a woman Godith (OE *Gōdgȳþ*). High E~ is on higher land than Good E~.

Eastergate Sx [*Gate* DB, *Estergat* 1263 FF]. 'Eastern gate.'

Easterton W [*Esterton* 1412 FA]. OE *ēasterra tūn* 'eastern TŪN'.

Eastham Chs [*Estham* DB, *Esteham* c 1100 Chester], E~ So nr Crewkerne [*Estham* DB, 1296 FF], E~ Wo [*Eastham* 11 Heming, *Estham* DB]. 'Eastern HĀM or HAMM.'

Easthampnett. See HAMPNETT.

Easthampstead Brk [*Yezhamesteda* 1167, *Yethamstede* 1176, 1180, *Yetzhamsteda* 1180 P, *Yashamsted* 1242 Fees]. Probably OE *geat(es)hāmstede* 'homestead by the gate'. The gate may be one that led to Windsor Forest.

Easthope Sa [(in) *Easthope* 901 BCS 587, *Esthop* 1242 Fees]. 'Eastern HOP or valley.'

Easthorpe Ess [*Estorp* DB, *Est Thorp* 1254 Val], E~ YE [*Estorp* DB]. 'Eastern thorp.'

Eastington D [*Estyngton* 1330 Subs], E~ Do [*Estington* 1259 FF, *Estinton* a 1280 Ep], E~ Gl nr Northleach [*Estintone* Hy 1, c 1275 Glouc, *-ton* 1231 Cl], E~ Wo [*Estinton* 1255 Ass]. OE *ēast in tūne* '(place) in the east of the village'. Cf. Introd. p. xviii.

Eastington Gl nr Stroud [*Esteueneston* 1220 Fees, *Estaneston* 1265 Misc, *Esteneston* 1275 Ipm]. '*Ēadstān*'s TŪN.'

Eastleach Martin & Turville Gl [*Lece* DB, *Lecha* Hy 1, *Est Lech* c 1145 BM, *Estlech Sancti Martini* 1291 Tax, *Estleche Roberti de Turville* 1221 Ass, *Lecheturvill* 1316 FA]. Originally LEACH from the river LEACH. Eastleach for distinction from NORTH-LEACH.

E~ **Martin** from the dedication of the church. —E~ **Turville** was held in part by Galiana de Turville in 1242 (Fees). Cf. ACTON TURVILLE.

Eastleigh. See LEIGH.

Eastling K [*Eslinges, Nordeslinge* DB, *Aeslinge* 11 DM, *Eslynges* 1242 Fees]. OE *Ēslingas* '*Ēsla*'s people'.

Eastlound Li [*Lund* DB, *Estlound* 1370 AD]. OScand *lundr* 'grove'. Eastlound for distinction from CRAISELOUND.

Eastney Ha [*Esteney* 1242 Fees, *-e* 1316 FA]. OE *be ēastan īege* 'east of the island', i.e. '(place) in the eastern part of Portsea Island'.

Eastnor He [*Astenofre* DB, *Estnover* 1166 RBE, *Estenoure* 1241 Ch]. OE *be ēastan ofre* '(the place) east of the ridge', i.e. **Eastnor Hill**.

Eastoft Li, YW [*Eschetofth* c 1170 YCh 487, *Esketoft* 13 Selby]. 'TOFT by an ash grove.' ON *eski* means 'ash-trees, ash grove'.

Easton mostly goes back to 1. OE *Ēast-tūn* 'eastern TŪN': **Easton** Brk [*Eston* 1221 Fees, 1231 Cl], E~ Cu [*Estuna* R 1 (1308) Ch], E~ Ha nr Winchester [*Eastun* 825, 961 BCS 389, 1076, *Estune* DB], **Crux** E~ Ha [?*Eastun* 796 BCS 282, *Estune* DB, *Eston Croc* 1242 Fees], E~ Hu [*Estone* DB], **Great** E~ Le [*Estone* DB], E~ Li [*Estone* DB, *Eston* 1202 Ass], E~ Nf [*Estone* 1044-7 KCD 785, *Estuna* DB], E~ **on the Hill** Np [*Estone* DB, *Eston* 1220 Fees], E~ **Maudit** Np [*Estone* DB, 1220 Ep], E~ Sf nr Framlingham [*Estuna* DB, *Eston* 1219 FF, 1254 Val], E~ **Bavents** Sf [*Estuna* DB, *Eston Bavent* 1330 Ch], E~ **in Gordano** So [*Estone* DB, *Eston in Gordon* 1293 FF, *E~ in Gorden* 1330 BM], **Stone** E~ So [*Estone* DB, *Stonieston* 1230 Ch], E~ W nr Devizes [*Eston* 1428 FA], E~ **Bassett** W [*Estun* 956 BCS 970], E~ **Grey** W [*Estone* DB, *Eston Grey* 1281 QW], E~ **Percy** W [*Estone* DB], E~ **Royal** W [*Eston* 1232 Ch], E~ YE [*Estone* DB, *Estona* c 1130 BM].

2. Various sources: **Easton** D in Cheriton Bishop [*Alvriketone* 1156, *Ailrichestone* 1159 RBE, *Ailrichestun* 1158 P]. '*Ǽþelrīc*'s or *Æþelrīc*'s TŪN.' E~ D in West Alvington [*Esteton* 1333 Subs], E~ Wt [*Estetone* 1244 FF]. OE *be ēastan tūne* '(place) east of the TŪN'. **Great & Little** E~ Ess [*E(i)stane* DB, *Eistane* 1166 RBE, *Eystan ad Tu rim* 1242 Fees (= Li. E~), *Eystan de Monte* 1235 Fees (= Gt E~), *Greater Estones* 1265 Misc]. OE *ēg-stān(as)* 'stone(s) by the island'. **Easton Neston** Np [*Estanestone, Adestanestone* DB, *Astaneston* 1224 Ep]. '*Ēadstān*'s TŪN.'

Easton Bassett W; cf. BERWICK BASSETT.—E~ **Bavents** Sf was held by Thomas de Bavent in 1316 (FA). BAVENT is a place nr Caen.—**Crux** E~ Ha was held by *Croch* the huntsman in 1086 (DB), by Elyas Croc in 1212 (Fees).—E~ **in Gordano** So has as distinctive addition the name of a district, which means 'triangular valley' (OE *gār-denu*; cf. 'dirty valley' (OE *gor-denu*). The first alternative suits the locality, but earlier material is wanted. Hill adduces a form *Gordeyne* of 1270 without reference. Cf. CLAPTON, WALTON, WESTON IN GORDANO.—E~ **Grey** W was held by Johannes Greiz in 1242 (Fees). The name has the form (John) *de Gray* 1236 Fees. *Greiz* is an adj. derived from Graye. The name is from GRAYE in Normandy.—E~ **Maudit** Np belonged to the Mauduit fee from the 12th cent. *Mauduit* (in Lat form *maledoctus*) is an OFr byname meaning 'badly educated'.—E~ **Percy** W [*Eton Peres* 13 PNW(S)] from an early owner named Peter or Piers. —E~ **Royal** W may have been a royal manor at the time of the Conquest.

Eastover So [*Estovore* 1323 AD]. 'Eastern bank.' See ŌFER.

Eastrea Ca [*Estrey* 966 BCS 1178, late copy]. 'Eastern island.'

Eastrington YE [*Eastringatun* 959 YCh 4, *Estrincton* DB, *Estrington* 1169 P, 1202 FF, *Aistrintun* 1199 FF]. OE *Ēastringas* may mean 'the people living east' (of some place, e.g. Howden), or it may mean 'the people of *Ēastra*' or the like. **Ēastra* would be a short form of names such as *Ēastorwine*.

Eastrip So nr Brewham [*Estrope* DB, *Estthrop* 1290 Cl], **Eastrop** Ha [*Estrope* DB, *Estropa* 1167 P], E~ W [*Esthrop* 13 BM]. 'Eastern thorp.'

Eastry K [*regio Eastrgena* 788, (to) *Eastorege* 805–31, (on) *Easterege*, *Eosterge* 811, (ad) *Eastrǣge* 824 BCS 254, 318, 332, 380, *Estrei* DB]. 'The eastern district.' The second el. is an OE *gē*, corresponding to G *Gau*. *Eastrgena* is the gen. (plur.) of a derivative meaning 'the Eastry people'. Cf. Goth *gauja* 'inhabitant of a district'. Eastry was no doubt once the name of a district, though it was later restricted to a village. Cf. LYMINGE, SURREY.

Eastwell K [*Estwelle* DB, *Eastwelle* 1267 Ipm], E~ Le [*Esteuuelle* DB, *Estwell* 1166 P, 1242 Fees]. 'Eastern spring or stream.'

Eastwick Hrt [*Esteuuiche* DB, *Estwic* 1138 NpCh], E~ YW nr Ripon [*East-wic* c 1030 YCh 7, *Estuuic* DB]. 'Eastern wīc.'

Eastwood Ess [*Estuuda* DB, -*wuda* 1181 P]. 'Eastern wood.'

Eastwood Nt [*Estewic* DB, *Estweit* 1165, *Est Twait* 1166 P]. 'Eastern thwaite.'

Eathorpe Wa [*Ethorp* c 1315 BM]. 'Thorp on the river (Leam).'

Eatington Wa [*Etendone*, *Etedone* DB, *Eatendon* 1174 P, *Etindon* 1236 Fees]. '*Ēata*'s DŪN' or more likely OE *eten-dūn* 'down used for grazing'. OE *eten* is derived from *ettan* 'to graze' and is found in *etenlǣs* 'common pasture land', also, with loss of -*n*, in *etelond* 'pasture land'.

Eaton, a common name, has two chief sources, which are not always easy to keep apart: **1.** OE *Ēa-tūn* 'TŪN on a river': E~ **Socon** Bd [*Etone* DB, *Eton* 1231 Ch, *Eaton cum Soca* 1645 PNBd], **Water** E~ Bk [*Etone* DB, *Eton* 1209 Fees], E~ Brk [*Eatun* 811 BCS 850, *Edtune* DB], E~ **Hastings** Brk [*Etton* 1190 P, *Eton Willelmi de Hasting'* 1220 Fees, E~ *Hastynges* 1298 Misc], E~ **Hall** Chs [*Eaton* c 1050 KCD 939, *Etone* DB], **Little** E~ Db [*Detton* DB, *Lytyll Eton* 1502 BM], E~ **He** nr Leominster [*Eatuna* 1123 PNHe, *Eatona* 1242 Fees, *Eton Leministre* 1278 Cl], E~ **Bishop** He [*Etune* DB, *Eton* 1241 Ch, E~ *Episcopi* 1316 FA], E~ **Tregose** He [*Edtune* DB, *Etuna*

1100 Glouc, *Eton Tregos* 1316 FA], E~ Nf [*Ettune*, *Ettuna* DB, *Etona* 1147–9 Holme, *Eton* 1232 Ch, -*e* 1254 Val], E~ Nt [*Etune* DB, *Etona* 1169 P, *Eton* 1242 Fees], **Water & Wood** E~ O [*Eatun* 864, 904, 929 BCS 509, 607, 666, *Etone* DB, *Wdeatone* c 1200 Eynsham, *Water Eton* 1268 Ch], E~ Sa nr Bishops Castle [*Eton* 1252 Fees, 1272 Ipm], E~ **Constantine** Sa [*Etune* DB, *Eton Costentyn* 1285 Ch], E~ **Mascott** Sa [*Etune* DB, *Eton* 1242 Fees, E~ *Marscot* 1255 RH], E~ **under Haywood** Sa [*Eton* 1227 Ch, 1255 RH], E~ **upon Tern** Sa [*Eton* c 1223, 1226 Eyton], **Water** E~ St [?*Eatun* 940, 949 BCS 746, 885, *Etone* DB, *Eton* 1262 For], **Castle** E~ W [*Ettone* DB, *Castell Eton* 1504 AD], **Water** E~ W [*Etone* DB, *Eatona* 12 (1316) Ch, *Watereton* 1372 Cl].

2. OE *Ēg-tūn* 'TŪN in an island or in land by a river'. Spellings such as *Eyton* may not always be conclusive, but may denote *Eton* or reflect some dialectal change of OE *ēa*, ME open *ē*. Here are given the names that are generally spelt with *ey*, *ei* &c.: **Eaton Bray** Bd [*Eitone* DB, *Eitona* 1130 P, *Eitun* 1156 P, *Eiton* 1209 Fees, *Eyton* 1227 Ch], E~ Chs nr Davenham [*Eyton* 1313, *Eayton* 1322 Ormerod], E~ Chs nr Tarporley [*Ayton*, *Eyton* 1304 Chamb], **Cold** E~ Db [*Eitune* DB, *Eyton* 1251 Ch, *Coldeyton* 1323 Ipm], E~ **Dovedale** Db [*Aitun* DB, *Eyton in Duuedale* 1296 Abbr, *Eyton super Douue* 1381 Derby], **Long** E~ Db [*Aitone* DB, *Long Eyton* 1288 FF, *Long* E~ 1322 BM], E~ Le [*Aitona* c 1125 LeS, *Eyton* 1236 Ep, 1254 Val], **Church & Wood** E~ St [*Eitone* DB, *Eiton* 1200 Cur, 1202 FF, *Eaiton* 1236 Fees, *Chirche Eyton* 1293 QW, *Wodeyton* 1284 Ass].

3. Eaton Chs nr Congleton [*Yeyton* 1285 Court, *Yeiton* 1290 Ipm, *Yayton* 1394 Ormerod]. Possibly a form of Eaton 2.

E~ **Bishop** He belonged to the Bishop of Hereford.—E~ **Bray** Bd came to Sir Reginald Bray in 1490 (VH).—E~ **Constantine** Sa was held by Thomas de Costentin in 1242 (Fees). CÒTENTIN is a district in Normandy.—E~ **Hastings** Brk was held by Ralph de Hastings in 1161 (RBE).—E~ **Mascott** Sa was granted to one Marescot between 1135 and 1160 and was held by William Marescot in 1242 (Fees).— *Socon* in E~ S~ Bd is the word *soke* (OE *sōcn*).— E~ **Tregose** He was held by Robert Tregoz in 1242 (Fees). Tregose (in early sources *Tresgoz*, *Tresgotz*, &c.) is from TROISGOTS in La Manche, Normandy.—*Water* means 'river'.

Eavestone YW [*Efestun* c 1030 YCh 7, *Euestone* DB]. First el. OE *efes* 'border'.

Ebberly D [*Edberleg* 1249 FF, *Ebberlegh* 1326 Ipm]. '*Ēadburg*'s LĒAH.' *Ēadburg* is a woman's name.

Ebberston YN [*Edbriztune* DB, *Edbrihteston* 1163 P, *Ebriston* 1202, *Edbrictiston* 1218 FF]. '*Ēadbeorht*'s TŪN.'

Ebbesborne Wake W [*Eblesburna* 826 BCS 391, *Eblesborne* DB, *Ebbeleburn Wak'* 1249 FF, *Ebbelesburne Episcopi*, E~ *Wak* 1291

Tax]. In reality the old name of the river **Ebble** [*Ebblesburna, Ybblesburna* a 672, *Eblesburna* 826 BCS 27, 391 &c.]. '*Ebbel*'s stream.' *Ebbel is a derivative of *Ebbe*; cf. *Ebbella* 736 BCS 154. A counterpart of the river-name is *Ebbleswell* Hy 2 (1251) Ch (Wa). This may suggest that *Ebbel* is a river-name, but such a name is difficult to explain. The modern river-name Ebble is a back-formation.
E~ **Wake** was held by Galfridus Wac in 1166 (RBE). Cf. (WAKES) COLNE.

Ebbsfleet K [*Ypwines fleot* 449 ASC (A), *Heopwines fleot* ib. (E), *Hyppelesfleot* 1308 Th, *Heppelesflete* 1280 Misc]. Probably *Heopwelles fleot* 'the river *Heopwell*'. *Heopwell* means 'stream where hips grew'. The forms from ASC are corrupt. *Heopwilles* in a source was misread as *Heopwines*.

Ebchester Du [*Ebbecestr* 1230 PNNb, *Hebcestr* 1291 Tax]. 'The Roman fort of *Ebba* or *Ebbe*.' *Ebba, Ebbe* are side-forms of *Æbba, Æbbe*, which are well evidenced. Cf. EBONY.

Ebernoe Sx [*Hyberneogh* 1262, *Iburnehew* 1271 Ass]. 'Spur of land by *Iburn*.' See HŌH. Iburn may be OE *īg-burna* 'stream by the island' or rather *īw-burna* 'yew stream'.

Ebony K [*Ebbanea* 833 BCS 408, *Ebbenea* 11 DM, *Ebbene* 1278 QW]. '*Ebba*'s or *Ybba*'s stream.'

Ebrington Gl [*Bristentune* DB, *Edbricton* 1200 Cur, *Eadbrithona* 1220 Fees]. 'The TŪN of *Eadbeorht*'s people.'

Ebury. See EYE Mx.

Ecchinswell Ha [*Eccleswelle* DB, *Egeneswell* 1176 P, *Echeneswelle* 1172 Ep, *-well* 1186 P, 1284 Ch]. Originally the name of the stream at the place, called *Ec(e)lesburna* 931 BCS 674. This name may be identical with **Ecclesbourne** R Db [*Ecclisborne* 1298 Ipm], and the first el. agrees with that of *Eclesbroc*, the OE name of the stream at ASHFORD Mx. Forms of Ashford show that the first el. had palatal *c*. *Ecclesbroc* was the name of a brook in Wo (c 975 KCD 682). Cf. also *Ecles cumb* 956 BCS 957 (So). Some of these cannot contain Brit *eclēs* 'church' (cf. ECCLES), and it is unlikely that the unrecorded pers. n. *Eccel*, which very likely once existed, can be assumed in all. The OE *ēcels* discussed under ETCHELLS is not a likely element in names of streams. It looks as if we have to assume an old river-name *Ecel, which might belong to ON *aka* 'to drive', Lat *ago* &c. An *l*-derivative actually occurs in Lat *agilis*, Sanskr *ajirás* 'swift, agile'. But the verb in question is unrecorded in West Germanic. If *Ecel* is a river-name, Ecchinswell means 'the river *Ecel*'.

Eccles K [*Aclesse*, (of) *Æcclesse* 10 BCS 1321 f., *Aiglessa* DB, *Ecclesse* 1166 RBE, *Eccles* 1208 FF]. Hardly identical with ECCLES La &c. The forms suggest a compound with OE *āc* (and *æc* gen. sg.) 'oak' as first el. The base may be OE *āc-lǣs*, *æc-lǣs* 'oak pasture'. Possibly *-lesse* developed from *-lǣswe*.

Eccles La [*Eccles* c 1200, *Ecclis* c 1250 CC, *Eckles* 1276 Ass], E~ Nf nr Attleborough [*Eccles* DB, 1212 Fees, *Ecclis* 1254 Val], E~ Nf nr Hickling [*Heccles* DB, *Eccles* DB, 1254 Val, *Ecles* 1272 Ch]. Brit *eclēs, OW *eccluys*, Welsh *eglwys*, OCo *eglos*, OIr *eclis* 'a church', from Lat *ecclesia*. Eccles Nf nr Hickling has no church now, but there is a ruin of an old church. It is likely that a good many names in ECCLES-have the Brit word as first el. If there is now no church in a place with such a name, the name may refer to a lost, even to a pre-English church. It is possible, however, that some contain a pers. n. *Eccel*, a derivative of *Ecca*.

Ecclesall Bierlow YW [*Eccleshale* c 1205 YCh 1295, *Eccheleshalla* c 1210 ib. 1279, *Eclishale* c 1200 DC]. Perhaps '*Eccel's HALH'. Cf. BRAMPTON BIERLOW.

Ecclesbourne R. See ECCHINSWELL.

Ecclesfield YW [*Eclesfeld* DB, c 1155 YCh 1266, *Ecclesfeld* 1190 P], **Eccleshall** St [*Ecleshelle* DB, *Eccleshale* 1227 Ass, *Ecclyshale* 1255 FF], **Eccleshill** La [*Eccleshull, -hil* 1246, *Eckeleshulle* 1276 Ass], **Eccleshill** YW [*Egleshil* DB, *Ecclesil* Hy 3, *Ecleshil* 1254 Calverley], **Eccleston** Chs [*Eclestone* DB, *Eccleston* 1285 Ch, 13 Chester], E~ La nr Chorley [*Aycleton* 1094 LaCh, *Ekeleston* 1203 FF], E~ La in Prescot [*Ecclistona* 1190 CC, *Ecleston* 1246 Ass], **Great & Little** E~ La [*Eglestun* DB, (in) *duobus Eccliston* 1212 Fees, *Great Ecleston* 1285 FF, *Parua Eccliston* 1261 Ass], **Eccleswall** He [*Egleswalle* 1274 Ipm, *Ecleswelle* 1275 RH, *-wall* 1292 QW]. The first el. is probably *eclēs* 'church'. See ECCLES (2). The second el. is FELD, HALH, HYLL, TŪN, WELLA (WÆLLA) 'a spring'.

Eccup YW in Adel [*Echope* DB, *Ecop* c 1215 Bodl]. '*Ecca*'s HOP.'

OE **ecg** 'edge' is used in ME and Mod dialects in senses such as 'the crest of a sharply pointed ridge, a ridge, steep hill or hillside'. See EDGE &c. Sometimes used as a second el., as in HARNAGE, HATHERSAGE, HEAGE.

Eckington Db [*Eccingtun* 1002 Wills, *Echintune* DB, *Ekinton* 1194 P], E~ Sx [*Achintone, Echentone* DB, *Achinctona* 1121 AC, *Eckentuna* 12 Fr], E~ Wo [*Eccyncgtun* 972 BCS 1282, *Aichintune* DB, *Akinton* 1197 P]. 'The TŪN of *Ecca* or his people.'

Ecton Np [*Echentone* DB, *Echeton* 1165 ff. P, *Eketon* 1221 Ep], E~ St [*Ekeyton* 1293 QW]. '*Ecca*'s TŪN.'

Edale Db [*Aidele* DB, *Eydale* 1305, *Edale* 1362 Ipm]. E~ is also used of the upper Noe. OE *ēg-dæl* 'the island valley'. OE *ēa-dæl* 'the valley of the river' is what would be expected. If *ēg-dæl* is the original form, it may be explained on the supposition that a place in the valley was called *Ēg* 'the island, the land on the stream'.

Edburton Sx [*Eadburgetun* c 1247 AD]. '*Ēadburg*'s TŪN.' *Ēadburg* is a woman's name.

Eddington Brk [*Eadgife gemære* 984 KCD (1282), 1050 E, *Eddevetone* DB, *Ediuetona* 1167 P]. '*Ēadgifu*'s TŪN.' *Ēadgifu* is a woman's name.

Eddisbury Chs [(æt) *Eadesbyrig* 914 ASC (C), *Edesberie* DB]. '*Ēad*'s BURG.' Cf. ADDISHAM.

Eddistone D [*Egereston* 1301 Ipm]. '*Ecghere*'s TŪN.'

Eddlethorpe YE [*Geduuales-, Guduuales-, Eduardestorp* DB, *Edelestorp* 1221 FF]. '*Ēadweald*'s thorp.'

Eden (ē-) R Cu, We [*Itouna* c 150 Ptolemy, *Edene, Eden* 1131 ff. WR], **Eden Burn** Du [*Hedene* c 1165 YCh 653, *Edeneburne* 1270 Ch], **Castle Eden** Du on Eden Burn [*Geodene, Iodene* c 1050 HSC, *Edene* 1195 (1335) Ch, c 1175 FPD]. Eden is a Brit river-name, identical with EDEN WATER in Roxburgh- and Berwickshire and with AFON EDEN in Merioneth [*Aberydon* 1370]. The name *Ituna* is very likely from **Pitunā*, which may be compared with various words belonging to the root *pi* 'to be full of sap', 'to gush forth', as Greek *pidúō* 'to gush forth', OIr *íath* 'a meadow'. Cf. also the river-name ESK. Brit *Ituna* became **Iduna*, whence OE **Idune* and with *u*-mutation *Iodune, Eodune* and later *Edene*.

Eden R Sx, K [*Eden* 1577 Harrison]. A back-formation from EDENBRIDGE.

Edenbridge K [*Eadelmesbrege* c 1100 Text Roff, *Edelmebrigg* 1213 Abbr, *Edoluesbrigg* 1199 Rot Cur, *Pons Edulfi* 1250 FF, *Edulve-brugge* 1292 Ch]. '*Ēadhelm*'s or *Ēadwulf*'s bridge.'

Edenfield (ē-) La [*Aytounfeld* 1324 LaInq]. OE *Ēg-tūn* 'TŪN in an island' (cf. EATON 2) and FELD.

Edenhall Cu [*Edenhal* 1159 f., 1223 P, -e 1290 Ch]. 'HALH by R EDEN.'

Edenham Li [*Edeham, Edeneham* DB, *Eden-ham* 1202 Ass, 1227 Ch]. '*Ēada*'s HĀM.'

Edenhope Sa [*Edenhope* 1272 Ipm, *Edene-hope* 1284 Cl]. '*Ēada*'s HOP.' Or the first el. might be OE *ēa-denu* 'valley of a river'.

Edensor (ĕn-) Db [*Ednesovre* DB, *-ouria* Hy 2 DC, *-ofre* 1196 FF]. '*Ēden*'s ŌFER or bank.' *Ēden* (from *Ēdīn*) is an unrecorded hypocoristic form of *Ēada*.

Edgbaston Wa [*Celboldestone* DB, *Ege-boldeston* 1221 Ass, *Egebaston* 1292 Ipm]. '*Ecgbald*'s TŪN.'

Edgbold Sa [*Edbaldinesham* DB, *Egbalden-ham* 1273 Ipm, *Egebaldham* 1327 Subs]. '*Ecgbald*'s HĀM.' The name shows the same curious shortening as EDGEMOND. The *-in-* of the DB form is remarkable. Perhaps the first el. is really a diminutive, *Ecgbaldīn*.

Edgcote Np [*Hocecote* DB, *Hochecote* 1159 BM, *-cot* 1224 Ep, *Echecott* 1223 Ep]. Perhaps OE *Hwicca cot* 'COT of the *Hwicce*' (cf. WYCHWOOD). Or the first el. may be a pers. n. *Hwicca*, lit. 'the Hwiccian'. For the sound-development cf. *such* from OE *swilc*.

Edgcott Bk [*Achecote* DB, *-cot* 1226 BM, *Echecota* 1163 P]. Possibly the first el. is OE *æcen* 'of oaks'. Or a pers. n. *Ecča*.

Edge Chs [*Eghe* DB, *Egge* 1260 Court], E~ Gl [*Egge* 1268 Glouc], E~ Sa [*Egge* 1276 Ipm]. OE ECG 'edge, hillside, hill'.

Edgefield Nf [*Edisfelda* DB, *Edichfeld* 1191 P, *Edesfeld* 12 BM, *Egesfeld* R 1 BM, 1197 FF], **Edgeley** Chs [*Edis(he)leg* 1287 Court, *Edisshelegh* 1304 Chamb], E~ Sa [*Edeslai* DB, *Edesleye* 1327 Subs]. First el. OE *edisc* 'park, pasture'. See FELD, LĒAH.

Edg(e)mond Sa [*Edmendune* DB, *Egmendon* 1165 ff. P, *Egmundon* 1227 Ch]. '*Ecgmund*'s DŪN.' Cf. EDGBOLD.

Edgerley Sa [*Eggredesl'* 1245 Cl, *Egardeleye* 1299, *Egardesley* 1308 Ipm]. '*Ecgheard*'s or *Ecgrēd*'s LĒAH.'

Edgeworth Gl [*Egesworde* DB, *Egewurd* 1138 AC, *Eggewrth* 1220, *-worth* 1236 Fees], E~ La [*Eggewrthe* 1212 Fees, *-worth* 1276 Ass]. 'WORþ by an edge or hillside.'

Edgton Sa [*Egedune* DB, *Egedon* 1237 FF, 1242 Fees]. 'Hill with an edge or brow.'

Edgware Mx [*Ægces wer* 972 BCS 1290, *Eggeswera* 1169 ff., *Egeswere* 1198 FF]. '*Ecgi*'s weir.'

Edingale St [*Ednunghal(l)e* DB, *Edeling(e)-hale* 1208 FF, *Ederingehal* 1191 P, *Edenyng-hale* 1272 ff. Ass]. 'The HALH of the people of *Ēden* (cf. EDENSOR) or *Ēadwine*.'

Edingley Nt [*Eddyngley* 1291 Tax, *Edingley* 1303 FA]. 'The LĒAH of *Eddi*'s people.'

Edingthorpe Nf [*Edmestorp* 1198 Fees, *Edinesthorp* 1254 Val, *Edmen-, Edyemsthorp* 1291 Tax]. ?'*Ēadhelm*'s thorp.'

Edington Nb [*Ydinton* 1196 f. P, *Idington* 1242 Fees]. 'The TŪN of *Ida*'s people.'

Edington So [*Eduuintone* DB, *Edinton, Edingtone* 1243 Ass]. '*Ēadwine*'s or *Ēad-wynn*'s TŪN.' *Ēadwynn* is a woman's name.

Edington W [(to) *Eþandune* 878 ASC, (æt) *Eōandune* c 880 BCS 553, *Edendone* DB]. 'Waste or uncultivated hill.' OE *ēþe* 'waste', corresponding to Goth *auþeis*, OHG *ōdi*, ON *auōr*, is found once in a poem (*ēōne ēōel* Daniel). The verb *ēþan, īþan* 'to lay waste' also occurs.

Edingworth So [*Lodenwrde, Iodena Wirda*

DB, *Hedeneworth* 1234, *Edenworth* 1274 Cl, *Edenewrthy* 1274 Ipm]. Formally the first el. might be an OE *ēow-denu* 'yew valley', but this hardly suits the locality. An OE *eow-denn* 'pasture for ewes' might be thought of.

OE **edisc** is usually rendered by 'enclosed pasture, park'. It is identified in a gloss with OE *deortuun* and *hortus cervorum*. A name such as BENDISH suggests a meaning 'tilled field'. Cf. EDGEFIELD, EDGELEY, BENDISH, BROCKISH, FARNDISH, GREATNESS, OXNEAD, STANDISH, THORNAGE.

Edlaston Db [*Duluestune* DB, *Edulveston* 1229 Pat], **Edlesborough** Bk [*Eddinberge* DB, *Eduluesberga* 1163, *Edulfesberga* 1168, *Aduluesberg* 1197 P]. '*Ēadwulf*'s TŪN and BEORG or barrow.'

Edleston Chs [*Edelaghston, Edelaston, Edelauston* 1288 Court]. '*Ēadlāc*'s TŪN.'

Edlingham (-nj-) Nb [*Eadwulfincham* c 1050 HSC, *Eadulfingham* 1104–8 SD]. 'The HĀM of *Ēadwulf*'s people.'

Edlington Li [*Ellingetone* DB, *Edlingtuna* c 1115 LiS, *Edlington* 1202 Ass], E~ YW [*Ellin-, Eilintone* DB, *Edlinton* J Ass, *Edlington* 1242 Fees]. 'The TŪN of *Ēadwulf*'s people' or the like. The immediate base may be a short form such as *Ēdla (cf. OG *Audila*).

Edmondbyers Du [*Edmundesbires* 1228 FPD]. '*Ēadmund*'s byres.'

Edmondsham Do [*Amedesham* DB, *Ædmodesham* 1176 P, *Edmodesham* 1196 P, 1205 Cur, 1226 FF, *Edmundesham* 1195 P, 1249 Ipm]. '*Ēadmōd*'s or *Ēadmund*'s HĀM.' *Ēadmōd* is unrecorded.

Edmondsley Du [*Edemennesleye* c 1190 Godric, *Edmanneslege* 1242 Ass]. Perhaps 'the LĒAH of the *ēdemann or shepherd'. OE *ēde* means 'a flock of sheep'. Or *Edeman* may be '*Ēadu*'s man'. Eda de E~ is mentioned c 1190 Godric.

Edmondthorpe Le [*Edmerestorp* DB, 1165 P, *Thorpe* c 1125 LeS, *Thorp Edm'* 1254 Val]. '*Ēadmǣr*'s thorp.'

Edmonton Mx [*Adelmetone* DB, *Edelmetona* 1130 P, *Edelmestun* 1236 Fees]. '*Ēadhelm*'s TŪN.'

Ednaston Db [*Ednodestun* DB, *Ednatheston* 1229 Pat]. '*Ēadnōþ*'s TŪN.'

Edstaston Sa [*Stanestune* DB, *Edestaneston* 1256 PNSa]. '*Ēadstān*'s TŪN.'

Edstone Wa [*Edricestone* DB, *Edricheston* 1195 BM]. '*Ēadrīc*'s TŪN.'

Edstone, Great & Little, YN [*Micheledestun, Parva Edestun* DB, *Edenston* 1167 P, *Edneston* 1231 Ass]. First el. as in EDENSOR.

Edvin Loach & Ralph He [*Gedeuen, Edevent* DB, *Gedesfenna* 1123 PNHe, *Yedefen* 1176 P, *Iadefen* 1212, *Yedefen Loges* 1242

Fees, *Yeddefenne Radh* 1291 Tax]. '*Gedda*'s fen.' OE *Gedda* is not recorded, but the cognate *Geddi* occurs. Cf. also YEADING.

E~ **Loach** was held by John de Loges in 1212 (RBE). Loges from one of the places of this name in France.—E~ **Ralph** was held by one Ralph in 1176 (P) and 1242 (Fees).

Edwalton Nt [*Edwoltone* DB, *Ædwaldton* 1183 P]. '*Ēadwald*'s TŪN.'

Edwardstone Sf [*Eduardestuna* DB]. '*Ēadweard*'s TŪN.'

Edwinstowe Nt [*Edenestou* DB, *Edenestowa* 1169, 1194 P, *-stowe* 1212 Fees, 1230 P]. '*Ēden*'s STŌW.' Cf. EDENSOR. *Eden* is in this case probably a hypocoristic form of *Eadwine* (St. Edwin). The chapel of St. Edwin in Edwinstowe is mentioned 1205 Cl, &c.; see PNNt(S).

Edworth Bd [*Edeuuorde* DB, *Eddewrþe* 1198 FF]. '*Edda*'s WORÞ.'

OE **efes** 'edge of a wood', later also 'brow of a hill'. See EAVESTONE, EUSTON, MEADS, BASHALL EAVES, HABERGHAM EAVES.

Effingham Sr [*Effingeham* 933 BCS (697), 1229 Pat, *Epingeham* DB]. 'The HĀM of *Effa*'s people.' *Effa* is a variant of *Æffa*.

Efford Co [*Ebbeford* 1184 P, 1236 Fees], E~ Ha [*Ebbeford* 1292 Cl]. 'Ford that can be used at ebb tide', as suggested in PND for three Effords in Devon.

OE **efn, emn** 'even, smooth, level' is found in EMBOROUGH, ENVILLE, EVENLEY, EVENWOOD, YANWATH, perhaps in EMNETH. Cf. NEMPNETT.

OE **ēg, īeg** 'island', also used of a piece of firm land in a fen and of land situated on a stream or between streams. Bede has a curious side-form *-eu*, as in *Heruteu* (see HART). The word is sometimes used alone as a pl. n., as in EYE, EYAM, RYE Sx. As a first el. *ēg* is often difficult to distinguish from *ēa*; cf. EATON. Often as a second el., as in CALDY, EASTREA, HILBRE, MERSEA. OScand *ey* would be merged in OE *ēg*. See e.g. BARROW La. OE **ēaland** and **ēgland** have the same senses as *ēg*. Cf. EALAND, ELAND, ELLAND, NAYLAND, SHOPLAND. The derivative OE **ēgoþ, īgoþ** 'ait' is found in MEDLEY.

Egbrough YW [*Ege-, Acheburg* DB, *Egburc* 1155–70 YCh 1502, *-burg* 1202 FF, *Eggeburg* 1240 FF]. '*Ecga*'s BURG.'

Egdean Sx [*Egedene* 1279 QW, *Eggedene* 1318 FF]. '*Ecga*'s dean.'

Egerton Chs [*Eggerton* 1260, 1282 Court]. '*Ecghere*'s or *Ecgheard*'s TŪN.'

Egerton (-j-) K [*Eardingtun* 11 DM, *Egardinton* 1203, *Ediardinton* 1206 FF, *Adgarinton* 1208 Cur]. 'The TŪN of *Ecgheard*'s people.'

Egford So [*Ecferdintone* DB, *Ehforton, Efferton* 1243 Ass, *Eggeforde* 1342 Misc]. The earliest forms mean 'the TŪN of (the people of) Egford'. Egford is '*Ecga*'s ford'.

Eggardon Do [*Giochresdone* 1084 GeldR, *Jekeresdon* 1204 Cur, *Ucresdon* 1219 FF, *Ecresdon* 1265 Misc, *Ekerdun* 1285 FA]. The first el. may be a hypothetical OE *Eohhere*, corresponding to ON *Ióarr*. Cf. EXBURY Ha.

Eggesford D [*Eggenesford* 1242 Fees, -*e* 1259 Ep, *Ekene-, Eggnesford* 1291 Tax]. 'The ford of **Ecgen* or **Eccen.*' The names would be derivatives of *Ecca* and *Ecga*.

Eggington Bd [*Ekendon* 1195 FF]. OE *æcen-dūn* 'oak hill' or '*Ecca*'s hill'.

Egginton Db [*Eghintune* DB, *Eggenton* 1242 Fees, *Eginton* 1228 Cl]. 'The TŪN of *Ecga* or his people.'

Egglescliffe Du [*Eggascliff* 1085 DST, *Eggescliua* 1163, *Eggles-, Ecclescliue* 1196 P]. First el. perhaps *eclēs* 'church'. Cf. ECCLES.

Eggleston Du [*Egleston* 1196 P]. The place is near where a stream called in its upper part **Eggles Hope** falls into the Tees. **Eggles Hope** is *Egleshope* 1161–7 YCh 562. The first el. of both names seems to be *Ecgwulf* or *Ecgel* pers. n. Cf. EAGLESFIELD.

Egglestone YN [*Eghistun* DB, *Egleston* 1198, *Eggleston* 1226 FF]. Identical with EGGLESTON.

Eggleton He [*Eglingtone* 1212 RBE, *Eglintune* 1219 Hereford]. 'The TŪN of *Ecgwulf*'s or *Ecgel*'s people.' Cf. EAGLESFIELD.

Egham Sr [*Egeham* a 675, 933 BCS 34, 697, DB, *Eggeham* c 1050 KCD (850), 1155–8 (1285) Ch]. '*Ecga*'s HĀM.'

Egleton Ru [*Egiltun, Egoluestun* 1218 For]. '*Ecgwulf*'s TŪN.'

Eglingham (-nj-) Nb [*Ecgwulfincham* c 1050 HSC, *Ecgwulfingham* 1104–8 SD]. 'The HĀM of *Ecgwulf*'s people.'

Egloshay·le Co [*Egglosheil* 1201 Ass, -*heyl* 1258 Pat]. 'Church on R HAYLE.' *Egglos* is Co *eglos* 'church'. Hayle is an old name of the Camel estuary.

Egloskerry Co [*Egloskery* 1291 Tax, 1377 PT]. Co *eglos* 'church' and probably a pl. n. identical with KERRY in Wales.

Egmanton Nt [*Agemuntone* DB, *Eggemonton* 1191–3 Fr]. '*Ecgmund*'s TŪN.'

Egmere Nf [*Eggemera* c 1035 Wills, 1165 P, *Egemere, Edgamera* DB, *Eggemere* 11 EHR 43, 1191 P]. '*Ecga*'s MERE' or '*Ecga*'s (GE)MÆRE or boundary'.

Egremont (ĕgrĭ-) Cu [*Egremont* c 1125 StB, *Egremunt* 1203 Cur, (de) *Acrimonte* c 1200 StB]. Apparently named from AIGREMONT in Normandy. E~ is on the EHEN, whose old name (*Egne*) may partly have suggested the name.

Egton (-k-) La [*Egetona* 1248, *Egeton* 1262 PNLa], E~ YN [*Egetune* DB, *Eggeton* 1187, 1195 P]. '*Ecga*'s TŪN.' The pronunciation with *k* tells against OE *ecg* 'edge' as first el.

Ehen (ān) R Cu [*Egre* c 1130, *Ehgena* c 1160 StB, *Egene* 1203 FF]. A Brit river-name. The early form *Egre* is due to influence from EGREMONT.

ON **eik** 'oak' is found in EYKE, GREENOAK, THONOCK, and some names such as AIGBURTH, AISKEW. In some cases *eik* has replaced OE ĀC, as in EAGLE Li. In AIKE *ai* is only a spelling for OE *ā*.

Eisey (īzĭ) W [*Eseg*, *æt Esig* 775–8 BCS 226, (*æt*) *Esege* 855 ib. 487, *Aisi* DB]. The first el. perhaps as in EASOLE K, though it is not clear what may have been the reason for the element here. Second el. ĒG 'island'.

Eland, Little, Nb [*Parva Elaund* 1242 Cl]. OE *ēaland* 'island'. Cf. PONTELAND.

Elberton Gl [?*Æwelburhehemediche* 986 KCD 654, *Eldbertone* DB, *Elbrihtona* 1167 P, *Eadbritthona* 1220 Fees, *Albricton* 1230 Cl]. If the form of 986 may be disregarded, apparently '*Ealdbeorht*'s TŪN'. But *Æwelburh* must have been very near Elberton. *Æwel-* would seem to be OE *æwiell* 'source of a river, spring'. But it may be for *Æþel-*, the original name having been *Æþelburgetūn* '*Æþelburg*'s TŪN'. Later forms point to a name in -*beorht* as first el.

Elborough So [*Illebera, Eleberie* DB, *Elleberwe* 1185 P, -*berewe* 1278 FF]. '*Ella*'s grove.' See BEARU. The name *Ella* sometimes appears in its true West Saxon form **Iella*; cf. ILFRACOMBE.

Elbridge K in Littlebourne [*þælbrycg* 948 BCS 869, *Thelebrigge* 1187 P]. OE *þelbrycg* 'plank bridge'. *Th-* was mistaken for the def. art. and dropped.

Elburton D [*Aliberton* 1254, *Aylberton* 1480 Pat]. '*Æþelbeorht*'s TŪN.'

Elcombe W [*Elecome* DB, *Ellecumba* 1168 P], **Elcot** Brk [*Ellecote* 1286 Ch], E~ W [*Elcot* 1237, *Ellecote* 1257 PNW(S)]. '*Ella*'s coomb and COT.'

Eldersfield Wo [*Yldresfeld* 972 BCS 1282, *Edresfelle* DB, *Ederesfeld* 1156, *Eld(e)res-, Eldredesfeld* 1195 P]. In BCS 1282 also occurs *Eldres ege* (not near Eldersfield). Very likely *Yldres-* is an inverse spelling for *Eldres-*. Perhaps both names contain a form, with early intrusion of *d*, of the tree-name *elder*, OE *ellern*.

Eldmire YN [*Elvetemer* 1236 Cl, -*e* 1246 Ch]. 'Swan mere.' See ELFETU.

Eldon Du [*Elledun* c 1050 HSC, 1104–8 SD]. '*Ella*'s hill.'

Eldwick YW [*Helguic, Heluuic* DB, *Helewike* 1273 YInq]. '*Helgi*'s WĪC.' *Helgi* is an OScand pers. n.

Eleigh, Brent & Monks, Sf [(*æt*) *Illanlege* c 995 BCS 1289, *Illeleia, Ilelega* DB, *Illeya Combusta, Illeya Monachorum* 1254 Val, *Illeya Arsa* 1260 Ch, *Brendeylleye* 1312,

Monekesillegh 1304 Ch]. '*Illa*'s LÊAH.' **Illa* is related to OG *Ilo, Illinc*. Cf. ILLINGTON, ELLISHAW.

Brent E~ must have been burnt down before 1254.—**Monks E~** belonged to St. Paul's in London.

OE **elfetu, ielfetu** 'swan' is the first el. of several names, as ALTHAM, ELDMIRE, ELVEDEN, ELVET, -HAM, ILTNEY. Cf. also ELKINGTON Np, ELTHAM. The corresponding ON *elpt* is found in ELTERWATER.

Elford Nb [*Eleford* 1256 Ass, 1250 Misc], E~ St [*Elleford* 1002 Wills, 1179 P, *Eleford* DB]. '*Ella*'s ford' or OE *ellern-ford* 'elder ford'.

Elham K [*Alham* DB, *Aelham* 11 DM, *Elham* 1182, *Eleham* 1189 BM, *Aleham* 1202 FF]. If the OE *Ulaham* 853, 964 BCS 467, 1126 does not belong here, as is usually assumed, probably identical with ALKHAM.

Eling Brk [*Elinge* DB, *Eling* 1220 Fees, 1240 Cl, *Yeling* 1246 Cl]. Derived with the suffix *-ingas* probably from some pers. n., e.g. OE *Eli*.

Eling Ha [*Edlinges* DB, *Eilling* 1130 P, *Elinges* 1181 ff. P, 1212 Fees]. Derived with the suffix *-ingas* from a pers. n. such as **Ēdla*, a hypocoristic form of *Ēada* (cf. OG *Audila*).

Elingdon W in Wroughton (lost) [*Ellendun* 823 ASC, (at) *Ellendune* 844, *Ællændun* 965–71 BCS 447, 1174, *Elendune* DB]. 'Elder-tree down.'

Elkesley Nt [*Elchesleig, -leie* DB, *Elkesle* 1227 Ep], **Elkington, North & South,** Li [*Alchinton* DB, *Helchingtuna* c 1115 LiS, *Northalkinton* 12 YCh 544, 1205 Ch, *Sudhelkinton* 1209–35 Ep], **Elkstone** Gl [*Elchestane* DB, *Elkestan* 1177 P, 1220 Fees], E~ St [*Elkesdon* 1227, 1253 Ass, *Elkesdun* 1251 Ch]. All these seem to contain the same element, which is probably a pers. n., very likely OE *Ēanlāc*, of which *Ēalāc* is a variant. Elkington is then 'the TŪN of the people of *Ēa(n)lāc*'. The rest are '*Ēa(n)lāc*'s LÊAH, STĀN, and DŪN'.

Elkington Np [*Eltetone* DB, *Heltedun* 1200 Cur, *Eltindon* 1283 Ch]. '*Elta*'s DŪN.' *Elta* pers. n. seems to be presupposed also by ELTHAM K. Cf. also ELTISLEY. But OE *elfetu, ilfetu* 'swan' may alternatively be assumed as the first el. of all except Eltisley.

Elkstone. See under ELKESLEY.

Ella, Kirk & West, YE [*Alvengi* DB, *Heluiglei* 1157 YCh 354, *Elvele* 1200 FF, *Westeluelle* 1305 YInq, *Esteluele* 1365 BM]. OE *Ælfingalēah*, perhaps alternatively *Ælfanlēah*, 'the LÊAH of *Ælfa* or his people'.

Elland YW [*Elant* DB, *Eiland* 1167 P, *Elande* 1202 FF]. OE *ēaland* 'island' or 'land by water'.

Ellastone St [*Edelachestone* DB, *Adelakeston* 1197 P, *Adlacston* 1236 Fees, *Athelaxton* 1242 Fees]. '*Ēadlāc*'s TŪN.'

Ellel La [*Ellhale* DB, *Elhale* c 1155 LaCh]. '*Ella*'s HALH.'

Ellen R Cu [*Alne* R 1, 1202 Ch, *Alen* 1279 Ass]. Brit *Alauna*. Cf. ALN. On the Ellen is **Ellenborough** Cu [*Alneburg* 12 StB, 1208 FF].

Ellenbrook La [*Elynbroke* 1544 LP]. 'Elder or alder brook.' Cf. ELLERN, ELRI.

Ellenhall St [*Linehalle* DB, *Ælinhale* c 1200 DC, *Elinhale* 1242 Fees, 1258 Ipm]. The regular *i* in the second syllable is noteworthy, and it may be the DB form is to be taken seriously. The original name may have been *Līn-halh* 'flax HALH', to which was prefixed OE *ēa* 'river'.

Ellenthorpe YN [*Adelingestorp* DB, *Ethelingetorp* 1228 Ep]. 'The thorp of the *æþeling*.'

Ellerbeck YN [*Elre-, Alrebec* DB, *Elrebek* 1243 FF], **Ellerburn** YN [*Elreburne* DB, *-burn* 1225 Ep, *Alreburne* c 1160 YCh 380]. 'Alder brook.' See ALOR, ELRI.

Ellerby YE [*Aluuardebi, Alverdebi* DB, *Elwardeby* 1286 Ch], E~ YN [*Aluuerdebi, Elwordebi* DB, *Elferby* 1252 Ass]. '*Ælfweard*'s BY.'

Ellerdine Sa [*Elleurdine* DB, *-wurth* 1196 P, *-wurthin* 1212 Fees]. '*Ella*'s WORÞIGN.'

Ellerker YE [*Alrecher* DB, *Elreker* 1202, 1226 FF]. 'Alder marsh.' Cf. ELRI, KERR.

OE **ellern** 'elder-tree' is often difficult to keep apart from ALOR, ELRI. See ELDERSFIELD, ELFORD, ELLENBROOK, ELSTEAD, ELSTOB.

Ellers, High & Low, YW [*Hegealres* 1185 P, *Heg(eh)alres* 1222 FF, *Heyhelleres* 1281 Ipm]. OE *alras*, the plur. of *alor* 'alder', later influenced by OScand *elri*.

Ellerton Sa [*Athelarton* 13 Eyton, *Ethelarton* 1285 FA]. '*Æþelheard*'s TŪN.'

Ellerton YE [*Elreton* DB, *Elretuna* c 1190 YCh 1173, *Alreton* 1206 Cur], E~ YN nr Bolton on Swale [*Alreton* DB, *Elreton* 1227 Ch], E~ **Abbey** YN [*Elreton* DB, 1219 FF]. 'Alder TŪN.' First el. ON *elri*, which may have replaced OE ALOR.

Ellesborough Bk [*Esenberge* DB, *Eselberge* 1195 Cur, 1196 FF]. '*Ēsla*'s hill.'

Ellesmere Sa [*Ellesmeles* DB, *-mera* 1172 P, *-mere* 1200 Ch, 1212 Fees]. '*Elli*'s mere.' Cf. ELSON Sa.

Ellingham Ha [*Adelingeham* DB, *Haslingueham* c 1165, *Alingeham* c 1170 Fr, *Elingeham* 1167 P]. 'The HĀM of *Ēdla*'s people.' Cf. ELING Ha.

Ellingham (-nj-) Nb [*Ellingeham* c 1130 FPD, 1254 Val, *Elingeham* Hy 2 FPD, 1254 Val], E~ Nf [*Elincham* DB, *Elingham* 1201 Cur], **Great & Little** E~ Nf [*Elincgham, Elingham* DB, *Hellingeham* 1198 FF, *Magna, Parva Elingham* 1242 Fees]. 'The HĀM of

Ella's people.' But some other name, such as *Édla* or *Eli*, may be the base.

Ellingstring YN [*Elingestrengge* 1198 Fount M, *Elyngstreng* 1282 Cl]. 'The "string" of the people of ELLINGTON.' Ellingstring is near Ellington. *Streng* may be dial. *string* 'a small vein of lead, a narrow vein of ore' from OScand *strengr* 'a string'.

Ellington Hu [*Elintune* DB, *-ton* 1163 P, *Alin-*, *Elinton* 1207 Cur], E~ Nb [*Elingtona* 1166 RBE, *Ellington* 1242 Fees, 1279 Ass], E~ YN [*Ellintone* DB, *Elinton* 1208 Cur, *Ellington* 1260 Ass]. 'The TŪN of *Ella*'s people.'

Ellington K [*Ealdingctuninga mearc* 943 BCS 784, *Elinton* 13 StAug]. 'The TŪN of *Ealda*'s people.'

Elliscales La [*Aylinescal*, *Alinscalis* c 1230 FC]. Perhaps '*Alein*'s huts'. Cf. SKÁLI.

Ellisfield Ha [*Esewelle* DB, *Elsefeld* 1167 P, *Ulsefeld* 1284, 1295 Ep]. The first el. may be OE *Ælfsige* or a short form of it. The vowel must then be the regular West Saxon *ie* (*i, y*), found in ILFRACOMBE. The OE base would be *Ielfsiges* (or *Ielfsan*) *feld*.

Ellishaw Nb [*Illishawe* 1254 Val, *Illeshawe* 1279 Ass]. '*Illa*'s shaw.' Cf. ELEIGH.

Ellonby Cu nr Skelton [*Alemby* 1267 Ch]. '*Alein*'s BY.' Alein is from OFr *Alain*.

Ellough Sf [*Elga* DB, *Elech* 1199 (1319) Ch, 14 BM, *Heleg* 1254 Val, *Elgh* 1291 Tax], **Elloughton** YE [*Elgendon* DB, *Elgedon* 1185, 1196 P, *Elege-*, *Helegedon*, *Elegeton* 1197 P, *Helgedon* 1216, *Elveton*, *Elvhetona* 1233 Ep]. Both names may contain ON *elgr* 'a heathen temple', which is found in many old Norwegian pl. ns., such as *Elgisætr*, *Elgi(ar)tún*, and in some Swedish pl. ns. Ellough would then be simply *Elgr* 'the temple', while Elloughton would be identical with ON *Elgitún*. It is true some early forms point rather to DŪN, but *dūn* and *tūn* are often mixed up with each other in pl. ns. The *n* in *Elgendon* would be intrusive.

Elm Ca [*Ælm* 656 ASC (E), *Eolum* 973 BCS 1297, *Elm* 1236 Ch], **Great & Little Elm** So [*Telma* DB, *Theaumes* 1247 Ipm, *Elme* 1327 Subs]. OE *elm* 'elm'. The *T-* of some early forms may be a relic of the prep. *æt*. *Eolum* is probably a dialectal variant of Elm. Cf. **Olmstead** Ca [*Olm(e)-*, *Elm(e)stede* &c. 13 PNCa(S)], in which the variation between *e* and *o* seems to point to earlier *eo*.

Elmbridge Wo [*Elmerige* DB, *Ammerugge* 1287 Cl], **Elmdon** Ess [*Elmerduna* DB, *Elmedon* 1199 P, *Elmedon* 1231 Cl], E~ Wa [*Elmedone* DB, *-don* 1242 Fees]. 'Elm ridge and DŪN.' The first el. is partly the adj. *elmen* 'of elms'.

Elmesthorpe Le [*Ailmerestorp* 1207 Cur, *Ailmerstorp* 1254 Val]. '*Æþelmǽr*'s thorp.'

Elmet YW, an old district; cf. BARWICK,

SHERBURN IN ELMET [(in) *silua Elmete* c 730 Bede, *Elmedsætna* [land] 7 BCS 297, *Elmet* c 800 HB, *Elmete* 1212 FF]. A Brit name identical with *Elfet*, the name of a cantred in Wales (Red Book of Hergest). A derivative is the pers. n. *Elmetiacos* in an ancient inscription in Carnarvon. The etymology of the name is obscure.

Elmham, North, Nf [*Ælmham* c 1035 Wills, *Elmenham* DB, *Elmham* 1167 P, *Northelmeham* 1252 Ch], **South** E~ Sf [*Almeham* DB, *Elmham* c 1105 BM, *Suthelmeham* 1252 Ch]. 'HĀM where elms grew.' First el. OE *elm* and *elmen* 'of elms'.

South E~ comprises several parishes, distinguished by additions such as St. James, St. Margaret &c. from the dedication of the churches. Elmham St. Cross is *Sancroft* 1254 Val, *Sandcroft* 1391 BM.

Elmington Np [*Elmintone* DB, *Elmenton* 12 NS, 1227 Ch]. First el. OE *elmen* adj. 'of elms'.

Elmley K [*Elmele* 1227 Ch, 1275 Ipm], E~ **Castle** Wo [(æt) *Elmlege* 780 BCS 235, *Elmelege* 11 Heming, (Castle of) *Elmeleye* 1312 AD], E~ **Lovett** Wo [*Elmesetene gemǽre* 817 BCS 361, *Ælmeleia* DB, *Almeleye Lovet* 1275 Subs, *Elmeleye Lovet* 1285 Cl]. 'Elm wood'; cf. LĒAH.

Lovett is an OFr byname and family name from OFr *lovet* 'wolf cub'.

Elmore Gl [*Elmour* 1176 P, *Elmoure* 1195 P, 1227 Ch]. 'Shore where elms grew.' See ŌFER.

Elmsall, North & South, YW [*Ermeshale* DB, *Elmeshale* 1242 Fees, *Suthelemeshal* 1230 Ep, *Northelmesale* 1350 BM]. 'HALH with an elm or with elms.'

Elmsett Sf [(æt) *Ylmesǽton*, *-sǽtun* c 995 BCS 1288 f., *Elmeseta* DB]. *Ylme-* is an inverse spelling for *Elme-*. It may be an OE **elme* n., a derivative of *elm* meaning 'elm-grove'. The second el. is either OE *-sǽtan* 'dwellers' or the plur. of OE *sǽte* 'a house' or (*ge*)*set* 'a fold' &c. In the text *æ* often stands for *e*.

Elmstead Ess [*Almesteda* DB, *Elmested* 1201 Cur, 1237 FF], E~ K nr Ashford [*Elman-*, *Elmesstede* 811 BCS 335 f., *Elmesteda* 1165 P], E~ K nr Bromley [*Elmsted* 1320 Hasted]. The first E~ K is probably from *Elm-hāmstede* 'homestead by the elms'. The same or OE *Elm-stede* 'place by the elms' may be the source of the other two.

Elmstone Hardwicke Gl [*Almundestan* DB, *Eilmundestan* 1221 Ass]. '*Ealhmund*'s stone.' Cf. HARDWICK. *Alchmundingtuun*, *Alhmundingtun* 889 BCS 559 f. is probably an old name of the place, meaning 'the TŪN of *Ealhmund*'s people'. This name seems to have been replaced by *Almundestan*, which presumably denoted a locality in *Alhmundingtun* and was named after the same Ealhmund.

Elmstone K [*Ailmereston* 1203 Cur, *Eylmereston* 1242 Fees]. '*Æþelmǽr*'s TŪN.'

Elmstree Gl [?*Æþelmodes treow* 962 BCS 1086, *Elmondestruo* 1201, *Ailmundestre* 1212–32 BM]. If the first ex. belongs here, '*Æþelmōd*'s tree'. If not, '*Æþelmund*'s tree'.

Elmswell Sf [*Elmeswella* 11 EHR 43, DB, -*well* 1200, 1203 Cur, 1254 Val]. 'Spring or stream where elms grew.'

Elmton Db [*Helmetune* DB, -*ton* 1176 P, *Elmeton* 1242 Fees, *Elmenton* 1276 RH]. First el. OE *elm* and *elmen* adj.

OScand **elri** 'alders, alder grove' is the first el. of several names, as ELLERBECK, ELLERKER. Cf. ELLERS, ELLERTON.

Elrington Nb [*Elrinton* 1229 Ep, *Elyrington*, *Elrington* 1256 Ass]. First el. OE *elren* 'of alders'.

Elsdon Nb [*Eledene* 1226 Ep, *Hellesden* 1236 Cl, *Elisden* 1242 Fees, *Ellesden* 1245 Ipm, 1254 Val, *Helvesden* 1325 Ipm]. Probably '*Elli*'s DENU or valley'.

Elsenham (-z-) Ess [*Elsen-, Alsenham* DB, *Elsenham* 1254 Val, *Elseneham* 1248 Ch]. '*Elesa*'s HĀM.'

Elsfield O [*Esefelde* DB, *Elsefeld* c 1130 Oxf, 1231 Cl, 1242 Fees]. '*Elesa*'s FELD.'

Elsham Li [*Elesham* DB, c 1115 LiS, *Ellesham* Hy 2 DC, 1223 Ep, *Elnes-, Eluesham* 1218 Ass]. '*Elli*'s HĀM.'

Elsing (-z-) Nf [*Helsinga* DB, *Alsinges* 1197 FF, 1203 Cur]. '*Elesa*'s people.' The OE base may really be a side-form *Ǽlesa*. But the *a*-forms may be due to Norman influence.

Elslack YW [*Eleslac* DB, *Elslac* 1219, *Elleslake* 1231 FF, *Elselak* 1240 FF]. '*Elli*'s or *Elesa*'s stream'. Second el. OE *lacu* 'stream'.

Elson Ha [*Æþelswiðetuninga lea* 948 BCS 865]. '*Æþelswiþ*'s TŪN.' *Æþelswiþ* is a woman's name.

Elson Sa [*Elleston* c 1247 Eyton, *Ellesdon* 1280 PNSa]. '*Elli*'s TŪN or DŪN.' The place is near ELLESMERE.

Elstead Sr [*Helestede* 1123 (1318) Mon (*Helstede* 1341 Pat), *Ellestede* 1197 P], **Elsted** Sx [*Halestede* DB, *Elnestede* 1212 Fees]. 'Place where elder (OE *ellern*) grew.'

Elsthorpe Li [*Aighelestorp* DB, *Eylestorp* 1212 Fees]. The first el. may be ODan *Egil* or OE *Ǽgel* pers. n. Cf. AYLESFORD &c.

Elstob Du [*Ellesstobbe, Ellestob* 1242 Ass]. 'Elder stub.' Cf. ELLERN.

Elston La [*Etheliston* 1212 Fees, 1259 Ass]. '*Eþelsige*'s TŪN.'

Elston Nt [*Elvestune* DB, *Eluestun* 1166 P *Ayleston* 1236 Cl, *Eyleston* 1252 Ch]. First el. probably an OScand pers. n., e.g. ODan

Ailaifr, ON *Eileifr* or ON *Eilífr* as in ALLITHWAITE.

Elston W [*Winterburn*' 1242 Fees, *Wynterborne Elistone* 1299 Ipm, *Eliston* 1316 FA]. Named from Elias Giffard, who held the manor in 1168 (P).

Elstow Bd [*Elnestou* DB, *Alnestoua* 1168 P, -*sto* 1178 BM]. '*Ǽllen*'s STŌW.' *Ǽllen (from *Ǽllīn*) is an earlier form of *Ǽlli*.

Elstree Hrt [*Tiðulfes treow* 785 BCS 245, *Ydolvestre* 1278 QW]. '*Tīdwulf*'s tree.' T- was lost owing to wrong division of æt *Tīdwulfes trēo*.

Elstronwick YE [*Astenewic* DB, *Alstineswich* 1190 P, *Elstanwik* 1297 Subs]. '*Ǽlfstān*'s WĪC.'

Elswick La [*Edelesuuic* DB, *Hedthelsiwic* c 1160 LaCh]. '*Eþelsige*'s WĪC.'

Elswick (ĕlsĭk) Nb [*Alsiswic* 1204 Ch, *Elsisseswich* 1210 Cur, *Elleswyke* 1254 Val]. '*Ǽlfsige*'s WĪC.'

Elsworth Ca [*Elesworð* 1060 KCD 809, -*worde* DB, *Ellesworth* 1219 FF, -*worde* 1230 P]. '*Elli*'s WORÞ.'

Elterwater (lake) La [*Elterwat[er]* c 1160 LaCh, *Helterwatra* 1196 FC]. An adaptation of an ON *Elptar-vatn* 'swan lake'. First el. ON *elpt* 'swan'.

Eltham K [*Elte-, Alteham* DB, *Healteham* 11 DM, *Elteham* 1242 Fees]. '*Elta*'s HĀM.' Cf., however, ELKINGTON Np.

Eltisley Ca [*Helteslay* 1202 FF, *Eltesle* 1228 FF, 1254 Val]. Apparently '*Elt*'s or *Elti*'s LĒAH'. The pers. n. *Elt(i)* is not found elsewhere, but cf. ELKINGTON Np.

Elton Brk [*Elphinton* 1220, *Elfreton* 1242 Fees]. The first el. is a pers. n. such as *Ǽlfhere* or *Ǽlfrēd*, or an *ing*-derivative of it.

Elton Chs nr Middlewich [*Helton* 1289 Ipm, *Elton* 1289 Cl], E~ Chs nr Chester [*Eltone* DB, *Elton* 1281 Court], E~ Db [*Eltune* DB, *Elton* 1282 FF], E~ Du [*Eltun* c 1090 SD, *Eligtune* c 1175 BM, *Elton* 1291 Tax], E~ He [*Elintune* DB, *Eleton* 1199 P], E~ La [*Elleton* 1246 Ass, *Elton* 1277 Ass], E~ Nt [*Ailetone* DB, *Eleton* 1242 Fees, *Elton* 1197 P], **Eltonhead** La [*Eltoneheued* a 1230 CC]. Most of these are no doubt '*Ella*'s TŪN', even if forms such as *Elton* are found somewhat early. Elton Du is perhaps rather *Ǽl-tūn* 'TŪN where eels were caught', but this will not do for some Eltons. Eltonhead is 'the hill at Elton'.

Elton Hu [*Æþeling-, Ǽilintun* 972–92 BCS 1130, *Adelintune* DB]. 'The TŪN of the *æþelings* or of *Æþelhēah*'s people.'

Eltringham (-nj-) Nb [*Heldringeham* c 1200 PNNb, *Eltrincham* 1242 Fees, *Helfryngham* 1346 FA]. 'The HĀM of *Ǽlfhere*'s people.'

Elvaston Db [*Aleuuoldestune* DB, *Elwadeston* c 1175 Fr, *Ailwaldestone* 1219 FF, -*ton*

1221–30 Fees]. '*Æþelwald*'s or *Ælfwald*'s TŪN.'

Elveden Sf [*Eluedena, Heluedana, -dona* DB, *Eluedene* c 1095 Bury, *-den* 1179 P, *Elveden* 1242 Fees]. 'Swan valley' from OE *elfetdenu*.

Elvet Hall Du [(æt) *Ælfetee* 762 ASC (E), *Æluet* c 1085 LVD, *Elvete(hale)* 1195 (1335) Ch, *Elvet* 1225 Ep]. OE *elfet-ēa* 'swan stream' or *elfet-ēu* 'swan island'. The exact situation of the original Elvet is not clear. For *-ēu* cf. ĒG. *Hall* from HALH.

Elvetham Ha [*Ylfethamm* 974 BCS 1307, *Elveteham* DB]. 'Swan HAMM.' Cf. ELFETU.

Elvington YE [*Alwintone* DB, *Eluinton* 1176 P, *Elvington* 1279–81 QW]. '*Ælfwynn*'s TŪN.' *Ælfwynn* is a woman's name.

Elwell Do [*Helewill* 1212 Fees, *-well* 1258, *-welle* 1285 Ch]. 'The wishing well.' First el. OE *hǣl* 'omen'. The spring at E~ is called The Wishing Well.

Elwick Du [*Ailewic* c 1150 YCh 650, *Ellewic* 1214 P, *-wick* 1239 Ep], E~ (ĕlĭk) Nb [*Ellewich* Hy 2 FPD, *-wic* 1195 (1335) Ch]. '*Ella*'s WĪC.'

Elworth Chs [*Ellewrdth* 1282 Court, *Helleworth* c 1300, *Elleworth* 1418 Ormerod], E~ Do [*Alevrde* DB, *Ellewrd* 1212 Fees, *Elleworth* 1281 Ipm]. '*Ella*'s WORÞ.'

Elworthy So [*Elwrde* DB, *Elleswurða* 1166 f. P, *Elleworthe* 1166 RBE, 1225 Ass]. '*Ella*'s or *Elli*'s WORÞ(IG).'

Ely (ē-) Ca [*Elge* c 730 Bede, *Elig* c 890 OEBede, 970 BCS 1267, *Ely* DB]. OE *ǣl-gē, ēl-gē* 'eel district'. Bede says that Ely got its name from the great number of eels caught in the fens there. *Ēlgē* was at an early period altered to *Ēlēg* 'eel island'. Cf. EASTRY.

Embaston. See AMBASTON.

Ember Court Sr. See IMBER Sr.

Emberton Bk [*Ambritone, Ambretone* DB, *Emberdestone* 1227 Ass]. '*Ēanbeorht*'s TŪN.' *Eanbeorht* occurs in the form *Æmbriht* in *Æmbrihtes gæt* BCS 1213.

Embleton Cu [*Emelton* 1195 FF, *Embleton* 1243 Cl, 1309 Ipm]. '*Ēanbald*'s TŪN.'

Embleton Du [*Elmedene* c 1190 Godric, 1208–10 Fees]. 'Elm valley.'

Embleton Nb [*Emlesdone* 1212 RBE, *Emelesdona* 1212 Fees, *Emildon* 1242 Fees, 1245 Ipm]. Possibly 'hill infested by caterpillars' OE *emel* 'caterpillar' is found. But '*Æmele*'s DŪN' is perhaps more probable.

Emborough So [*Amelberge* DB, *Emeneberge* 1200 Cur, *Eueneberia* 1194 P, *Emnebergh* 1238 Ass]. 'Smooth hill.' First el. OE *efn, emn* 'even'.

Embsay YW [*Embesie* DB, *Embeseie* c 1140 Fount]. '*Embe*'s island.'

Emington O [*Amintone* DB, 1224 Ep, *-ton* 1236 Fees, *Eminton* 1199, *Emigton* 1230 P]. 'The TŪN of *Eama*'s people.'

Emley YW [*Amelai, -leie* DB, *Emelaiebroc* c 1200 YCh 1688, *Emeleg, Emmesleg* 1203 Cur]. Most likely OE *elm-lēah* 'elm wood' with dissimilatory loss of one *l*.

Emmott La [*Emot* 1296 Lacy]. OE *ēagemōt* 'junction of streams'.

Emneth Nf [*Anemeða* 1170 P, *Enemeða* 1171 P, *-meth* 1203 Cur, *-methe* 1251 Ch]. The second el. may be OE *gemyþe* 'junction of streams, mouth of a stream'. If so, the first el. is very likely a river-name, an old name of the NAR. On this was formerly a place called **Emenhouse** [*Emenhus* 1250 Ass]. The original form of Emneth was then *Æmenan-gemyþe*. *Æmene* was the old name of the lower MOLE Sr [(*on*) *Æmenan* 1005 Eynsham]. The name is probably derived from an OE **ām*, corresponding to ON *eimr* 'mist'. But Emneth may have as second el. OE *mǣþ* 'meadow'. If so, the first el. is probably OE *efn, emn* 'smooth'.

Empingham Ru [*Epingeham* DB, *Empingeham* 1106–10 RA, 1166 P]. 'The HĀM of *Empa*'s people.' OE *Empa* is possibly found in *Empenbeorch* 956 BCS 970. It is related to OG *Ampho.*

Empshott Ha [*Hibesete* DB, *Ymbesete* 1242 Selborne, *Imbeschate* c 1270 Ep]. OE *imbescēat* 'bee grove', OE *imbe* (*ymbe*) 'swarm of bees' (= OHG *impi*) and SCĒAT.

Emscote Wa [*Edelmescote* 1236 Fees, *Edulmescote* 1284 Misc, *Edulvescote* 1314 Ipm]. '*Ēadhelm*'s COT.'

Emstrey Sa [*Eiminstre* DB, *Eministre* 1197 FF, *Eiminstre* 1256 (1332) Ch]. 'Church on the island or in land by the river.' Cf. MYNSTER. First el. OE ĒG.

Emswell YE [*Elmes-, Helmesuuelle* DB, *Helmeswella* 1157, *Elmeswella* c 1175 YCh 354, 441]. '*Helm*'s spring or stream' or 'elm spring or stream'.

Emsworth Ha [*Emeleswurth* 1224 Cl, 1239 Ch, *-whurth* 1244 Cl, *Emeleworth* 1231 Ch, *Elmeworth* 1231 Cl]. '*Æmele*'s WORÞ.' *Æmele* is found once. It is probably a derivative of the name-stem *Amal-*, which is so common in Continental names.

Enborne Brk [*Aneborne* DB, *Enedburn* 1220 Fees, *Enedeburne* 1292 Ch]. 'Duck stream.' First el. OE *ened* 'duck'. Enborne was originally the name of a tributary of the river **Enborne**. The latter was in OE *Aleburna* 749 &c., *Alorburna* 909 BCS 179, 624 &c. It is still called *Awborne* by Saxton in 1575. This name means 'alder stream'. The modern river-name is due to influence from ENBORNE vil.

OE **ende** 'end'. See INGATE, DETCHANT, HAZON, MILE END, WALLSEND.

Enderby Le [*Andretesbie, Endrebie* DB,

Endredeby 1226 Ep], **Bag E~** Li [*Andrebi, Adredebi* DB, *Endrebi* c 1115 LiS, *Bagenderby* 1291 Tax], **Mavis E~** Li [*Endrebi* DB, c 1115 LiS, *Enderby Malbys* 1302 BM], **Wood E~** Li [*Endrebi* DB, *Wodenderby* 1198 (1328) Ch]. '*Eindriði*'s BY.' Cf. AINDERBY.

The prefix *Bag* is obscure. Possibly it is a pers. n. **Mavis E~** was held by Wiliam Malebisse in 1202 (Ass). Cf. ACASTER MALBIS.

Endon St [*Enedun* DB, 1227 Ass, *-don* 1252 Ch]. Perhaps '*Ēana*'s DŪN.' *Eanandun* actually occurs 1003 KCD 1299 (Bengeworth). But the first el. is more likely as in ENHAM, YEN HALL.

OE ened, æned 'duck'. See ANDWELL, ANMER, ENBORNE, ENFORD, ENMORE.

Enfield Mx [*Enefelde* DB, *-feld* 1190 P, 1221–30 Fees, 1254 Val]. See FELD. First el. as in ENDON.

Enford W [*Enedford* 934 BCS 705 f., *-e* DB, *Eneford* 1223 Cl]. 'Duck ford.' Cf. ENBORNE.

England [*Englaland* c 890 OEBede, c 1000 Ælfric, c 1000 Saints, 1014 ASC (E)]. Originally 'the land of the Angles' (so in OEBede), later 'the land of the English'. OE *Engle* originally meant 'Anglians' (lit. 'people from ANGEL in Sleswig'), but began fairly early to be used in the sense 'Englishmen'. Eadweard is called *Engla þeoden* 942 ASC. *Englisc*, originally 'Anglian', is used in the sense 'English' 880 Laws, c 890 Alfred CP, 897 ASC. *Angelcynn* is used in the sense 'the English nation' or 'England' 880 Laws, c 890 Alfred CP. The Latin form *Angli* is used in the sense 'Englishmen' already by Gregory the Great.

Englebourne D [*Engleborne* DB, *-burn* 1251 Cl], **Englefield** Brk [*Engla feld* 871 ASC, *Englefel* DB, *-feld* 1196 FF]. 'The stream and FELD of the Anglians.' The names indicate Anglian settlements in Saxon territory. Cf. CONDERTON, EXTON Ha.

Englefield Sr [*Hingefelda* 967 BCS 1195, *Ingefeld* 1291 FF, *Ingelfeld* 1282 Ipm]. Perhaps '*Ingweald*'s FELD'. The form of 967 is in a late transcript.

Engleton St [*Engleton* 1242 Fees, 1285 FA, *Engelton* 1250 Ass]. 'The TŪN of the Anglians.' The Anglians may be East Anglian settlers in Mercia.

Enham, King's & Knight's, Ha [*Eanham* post 1008 Laws, *Etham* DB, *Enham* 1167 P, 1242 Fees, *Ennam Militis* 1316 FA]. 'HĀM or HAMM where lambs were bred.' First el. OE *ēan* 'lamb', identical with Lat *agnus* and the base of OE *geēan* 'with lamb' (cf. *gecielf* 'with calf') and *ēanian* 'to yean'. Cf. YEN HALL.

Enmore So [*Animere* DB, *Enemere* 1200 Cur, *Enedemere* 1315 Ipm]. 'Duck mere.' First el. OE *ened* 'duck'. There is a small lake at the place.

Ennerdale Cu [*Ananderdala* c 1195 StB; *Eghnerdale* 1321 Ipm, *Eynordale* 1322 Cl]. The earliest name means '*Anund*'s valley'. *Anander-* is *Anundar*, the gen. of the ON pers. n. *Anundr*. The later name is 'the valley of the EHEN'. *Eghner-* is an OScand gen. of the name Ehen (*Eghen* &c.). **Ennerdale Water** [*Eyneswater* 1338 Cl] was originally 'Ehen lake'. Cf. DERWENTWATER.

Enson St [*Hentone* DB, *Enstone* 1272 FF, *Eneston* 1275 Ass]. Possibly '*Ēansige*'s TŪN.'

Enstone O [*Henestan* DB, *Ennestan* 1185–7 Winchc, *Enneston* 1242 Fees]. '*Enna*'s stone.' Cf. (æt) *Ennan beorgum* BCS 932.

Entwisle La [*Hennetwisel* 1212 Fees, *Ennetwysel* 1276 Misc &c.]. Perhaps '*Enna*'s twisla'. OE *twisla* is 'a tongue of land in a river fork'. Other possibilities are *henna twisla* 'river fork frequented by water-hens' or *ened-twisla* (*ened* 'duck').

Enville St [*Efnefeld* DB, *Euenfeld* 1183 P, *Evenefeud* 1240 FF]. 'Smooth FELD.' First el. OE *efn* 'even, smooth'.

OE eofor, efer 'boar' is the first el. of some names. See ERISWELL, EVER- (passim), YAVERLAND.

OE eorþe 'earth'. See ARBURY, EARTHCOTT, YARBOROUGH.

OE eowestre 'sheepfold'. See AUSTERFIELD, OSBORNE, also EASTER.

Epney Gl [*Eppen*' 1252 PNGl, *Oppen* 1252 Misc]. '*Eoppa*'s island.'

Epperstone Nt [*Eprestone* DB, *Eperstona* c 1170 Middleton, *Eperston* 1242 Fees]. The first el. is no doubt a pers. n., e.g. OE *Eorphere*, suggested in PNNt(S).

Epping Ess [*Eppinges* DB, 1205 Cur, *Upping* 1227 FF]. 'The people on the upland.' A derivative *Yppingas* from OE *yppe* 'a raised place, a look-out place'. **Epping Upland** is *Eppynggehethe* 1287 AD, lit. 'Epping Heath'.

Eppleby YN [*Aplebi* DB, *Appelby* 1218 FF]. Identical with APPLEBY.

Eppleton Du [*Æpplingdene* c 1180 FPD, *Epplindena* 1180 Finchale, *Appelden* 1196 P]. 'Apple valley.' First el. OE *æplen* 'of apples'.

Epsom Sr [*Ebesham* 675, *Ebbesham* 973 BCS 39, 1296, *Evesham* DB, *Ebbesham* 1181 P]. '*Ebbe*'s HĀM.'

Epwell O [*Eoppanwyllan broc* 956 BCS 964, *Eppewell* 1206 Cur]. '*Eoppa*'s stream.'

Epworth Li [*Epeurde* DB, *Appe(l)wurda* 1179 P, *Eppeworth* 1233 Ep]. '*Eoppa*'s WORÞ.'

Ercall (ar-), High, Sa [*Archelov* DB, *Ercalewe* 1241 FF, *Ercalwe* 1256 Ass, *Magna Ercalewe* 1327 Subs], **Childs E~** Sa [*Arcalun* DB, *Parva Erkalawe* 1242 Fees,

Erkalwe parva 1291 Tax, *Childes Ercalewe* 1327 Subs]. High Ercall and Childs E~ are far apart on different sides of the Tern. Probably Ercall is an old Welsh name of the district. Cf. CHILTON.

Erdington Wa [*Hardintone* DB, *Erdinton* 1204 Cur, 1236 Fees, *Eredinton* 1285 QW]. 'The TŪN of *Eorēd*'s or *Ēanred*'s people.'

Eresby (ēr-) Li [*Iresbi* DB, *Eresbi* 12 DC, -*by* 1238 Ep, 1254 Val]. '*Iōar*'s BY.' First el. ON *Ióarr*, ODan, OSw *Iōar*.

Erewash (ĕrĭ-) R Db, Nt [*Yrewis* c 1175 Middleton, *Irewiz* 1229 For, -*wys* 1226–8 Fees]. Second el. an OE *wisc*(e) 'stream' as in WISKE. The first is OE *irre* 'wandering, winding'.

ON **erg, ærgi** 'a shieling, i.e. a hill pasture, a hut on a pasture', from MIr *airge* 'a dairy', Ir *airghe*, Gael *airidh* 'a shieling', is common in pl. ns. in Cu, La, We, Y. The el. is usually combined with Scand first elements, and the names must be looked upon as Scandinavian. The vowel of the word varies a good deal between *e* and *a*. This is due to various substitutions for the Ir sound. *Erg* is the second el. of several names, as ANGLEZARK, BATTRIX, BERRIER, CLEATOR, DOCKER, GRIMSARGH, KELLAMERGH, MEDLAR, SALTER, TIRRIL, TORVER, WINDER. It is often used alone as a pl. n., usually in the plural. An exception is ARROWE Chs. ARRAS YE is *erg* with an Engl plur. ending. The following go back to the dat. plur. *ergum, ærgum*: AIRYHOLME, ARGAM, ARK-HOLME, ARRAM, ERYHOLME.

Eridge Sx [*Ernerigg* 1202 FF]. 'Eagles' ridge.' See EARN.

Eriswell Sf [*Hereswella* DB, *Ereswell* 1183 P, 1242 Fees, *Evereswell* 1249 Ipm]. If the isolated form of 1249 is reliable, 'boar's stream'. See EOFOR.

Erith (ē-) K [*Earhyð* 695, *Earhið* c 960, *Earhetha*(*m*) c 960 BCS 87, 1097 f., *Erhede* DB]. 'Gravel harbour.' OE *ear* is only found as the name of a rune. It is identical with ON *aurr* 'gravel' and very likely had the same sense. *Ear-hȳþ* may mean 'gravelly landing-place' or 'harbour where gravel was exported or imported'. Cf. HȲþ, EARITH.

Erlestoke or **Earl Stoke** W [*Stoke* 1242 Fees, *Erlestoke* Hy 2 Montacute, -*stok* 1239 Ch]. 'The earl's STOC.'

Erme R D [*Irym* 1240 For, *Erm* 1280 AD, *Hyrm* 1282 Ass], **Ermington** D [*Ermentona* DB, *Ermintona* 1130 P, *Hermiton* 1201 FF, *Erminton* 1238 Ass, 1263 Ipm]. The etymology depends upon whether the river-name Erme is the older or a back-formation from Ermington. The latter could very well be from OE *Iermen-tūn* 'chief TŪN', the first el. being OE *iermen-*, *eormen-* in *eormencynn* 'mankind' &c. *Ier-men-* is a prefix meaning 'great'. The river-name would then be a back-formation.

If the river-name is the earlier formation, it might possibly be a Brit name related to ERMS in Germany, if that is a Celtic name.

Ermine Street. See ARRINGTON.

Erpingham (ar-) Nf [*Erpingham* 1044–7 KCD 785, *Erpingaham* DB]. 'The HĀM of *Eorp*'s people.' **Eorp* is a short form of names in *Eorp-*, as -*weald*, -*wine*.

Erring Burn R Nb [*Eriane* 1479 Hexh]. On the stream is **Errington** [*Erienton* 1202 FF]. The river-name is British and related to Welsh *arian* 'silver' (from *argant*). The meaning is 'bright stream'.

Erringden YW [*Ayrykedene* 1277 Wakef, *Ayrikedene* 1308 ib.]. The first el. looks like the ON pers. n. *Eiríkr*, but earlier forms are needed.

Errington. See ERRING BURN.

OE **ersc, ærsc**, found in *ersc-hen* 'quail' and in pl. ns., is identical with dial. *earsh*, *arrish* 'stubble field' and is generally held to have the same meaning. But the old meaning was probably rather 'ploughed field'. The word must be related to OE *erian* 'to plough', and a name such as RYARSH suggests a meaning 'tilled field'. Cf. BURWASH, OAKHURST, PEBMARSH, SUNDRIDGE, WONERSH, also GREATNESS.

OE **erþ, ierþ, yrþ** 'ploughing, ploughed land' is occasionally found in pl. ns., as EARTHAM, BRIGHTSIDE, CORNARD, ? FOXEARTH, HORNINGSHEATH.

Erwarton Sf [*Eurewardestuna* DB, *Euere-wardeston* 1196 FF]. '*Eoforweard*'s TŪN.'

Eryholme YN [*Argun* DB, *Ergum* c 1095 YCh 855]. The dat. plur. of ERG 'shieling'.

Escombe Du [*Ediscum* 10 BCS (1256), c 1050 HSC]. The dat. plur. of OE *edisc* 'park' &c.

Escot D [*Estcot* 1227 Ch]. 'Eastern COT.'

Escowbeck La [*Escouthebec* c 1240 CC]. *Escow-* is ON *Eskihofuð* 'ash-tree hill'.

Escrick YE [*Ascri* DB, *Ascric* 1157 YCh 354, *Escric* 1169 P, 1230 Cl, *Eskerick* 1227 FF]. OScand *eski* 'ash-trees' and the word *ric* 'stream' or 'ditch' discussed under CHATTERIS. Or the second el. may be ON *krikr* 'a bend, nook'.

Esh Du [*Esse* 12 Finchale, *Esshe* 1372 AD]. OE *æsc* 'ash-tree'.

Esher (ē-) Sr [(et) *Æscæron* 1005 Eynsham, *Esshere* 1062 KCD 812, *Aissele* DB, *Esshere* 1212 Fees]. Probably a compound of OE *æsc* 'ash-tree' and some word derived from OE *sceran* 'to cut'. OE *scearu* 'boundary' might be thought of. Or it may be OE *scear* '(plough)share'. The long ridge on which Esher is might have been likened to a ploughshare. Or there may have been an OE *scearu* with the same sense as ON *skǫr*, G *schar*, i.e. 'a border, a rim'.

Esholt YW [*Esseholt* c 1190 YCh 1785, *Escheholt* 1248 Ch]. 'Ash wood.'

Eshott (ĕ-) Nb [*Esseta* 1187 P, *Esset* 1242 Fees]. OE *æsc-scēat* 'ash grove'. Cf. SCĒAT.

Eshton YW [*Estune* DB, *Eston* 1207 Cur]. OE *Æsc-tūn* 'ash TŪN'.

Esk R Dumfr, Cu [*Ask* c 1200 Sc, *Eske* 1279 Ass], **South E~** Cu [*Esc* 12 StB, *Esk* 1242 FC], **E~** YN [*Esch* c 1110 Whitby, *Esc* c 1120 Guisb, *Esk* 1204 Ch]. A Brit river-name identical with EXE. **Eskdale** Cu on the northern Esk [*Eske dale* 1375 Barbour], **E~** Cu on the South Esk [*Eskedal* 1294 Cl], **E~** YN [*Eschedala* DB]. **Esk Hause** Cu [*Eskhals* 1242 FC]. Hause is OE *heals* or ON *hals* in the sense 'a pass'. Esk Hause is a pass at the source of the South Esk.

Eske YE [*Asch* DB, *Eske* 1297 Subs]. OScand *eski* 'ash-trees'. *Eski* is the first el. of EASTOFT, ESKETT, ESPRICK.

Eskett Cu [*Eskat* 1760 PNCu(S)]. Possibly identical with HESKET NEWMARKET.

Eslington Nb [*Eslinton* 1163, *Estlinton* 1170 P, *Eselinton* 1177 P, *Eslington* 1212 Fees]. 'The TŪN of *Ēsla*'s people.'

Espershields Nb [*Estberdesheles* 1230 PNNb]. The place is near **Burntshiel**, whose name means 'burnt hut'. Cf. SCHĒLE. Espershields is 'East Burntshiel'.

Espley Nb [*Espeley* 1242 Fees, *Aspele* 1252 P]. 'Aspen wood.'

Esprick La [*Eskebrec* c 1210 CC]. ON *eski-brekka* 'ash slope'.

Essendine Ru [*Esindone* DB, *Issendene* 1185 Rot dom, *Esenden* 1230 Cl]. '*Ēsa*'s DENU.'

Essendon Hrt [(into) *Eslingadene* 11 E, *Esindena* 1179, *Esenden* 1186 P]. 'The DENU or valley of *Ēsla*'s people.'

Essex [*East Seaxe* 894, *East Sexe* 904 &c. ASC, (on) *East Seaxum* 894 ib., *Exsessa* DB]. 'The East Saxons.' Cf. Introd. p. xi. OE *East Seaxe* 'the East Saxons' is found 604, 823 ASC.

Essington St [*Esingetun* 996 Mon, *Eseningetone* DB, *Esenington* 1227 Ass, 1240 FF]. 'The TŪN of *Esne*'s people.'

Esthwaite La [*Estwyth* 1539 FC]. Perhaps 'Eastern thwaite'. **E~ Water** [*Estwater* 1537 PNLa]. *Estwater* is probably from *Esthwaite Water*.

Eston YN [*Astun* DB, *Eston* 1229 FF]. Identical with ASTON from *Ēast-tūn*.

Etal (ē-) Nb [*Ethale* 1232 Cl, *Hethal* 1242 Fees]. '*Ēata*'s HALH' or *ete-halh* 'HALH used for grazing'. Cf. EATINGTON.

Etchells, Northenden & Stockport, Chs [*Echelis* 1248 Misc, *Echeles* 1286 Court]. An unrecorded OE *ēcels* 'addition', a derivative of (*ī*)*ēcan* 'to increase, add to'. The meaning is 'land added to a village or an estate'. Cf. NECHELLS.

Etchilhampton W [*Ec(h)esatingetone* DB, *Hechesetingeton* 1207 Ch, *Echeham[ton]* 1195 Cur, *Hechelhamt[on]* 1228 Cl, *Echelhampton* 1242 Fees]. A complicated case. There are two types, each with a folk-name derived from the original name as first el., one formed with -*sætan* (-*sætingas*), one with -*hǣme* 'dwellers'. For -*sǣtingas* cf. *Fromesetinga hagen* 'the haw of the Frome people' BCS 1127. The original name was possibly OE *Æchyll* 'oak hill' (*æc* being the gen. of *āc* 'oak'). The place is on the slope of a hill. The form *Eche-* may be elliptical like *Boc-* for *Bocland* in OE *Bochǣme* 'the Buckland people'. Both name-types mean 'the TŪN of the people at *Æchyll*'.

Etchingham Sx [*Hechingeham* 1158 P, *Echingehamme* 1176 Penshurst, -*ham* 1190 P]. 'The HAMM of *Ecci*'s people.'

Etherdwick YE [*Ethreduuic* J BM, *Edredewik* 1240 FF, *Etherdwyk* 1285 FA]. '*Æþelrēd*'s WĪC.'

Etherow R Chs, Db [*Ederhou* 1226 FF, *Ederou* 1285 For, *Edderowe* 1290 Ch]. A Brit river-name.

OE etisc. See DUNTISH, WRANTAGE.

Etloe Gl [*Ete(s)lau* DB, *Ettelawe* 1220 Fees]. '*Ēata*'s or *Etti*'s mound or hill.' See HLĀW.

Eton (-ē-) Bk [*Ettone* DB, *Eton* 1207 Cur, *Eitun* 1156 P]. OE *Ēa-tūn* 'TŪN on R Thames'.

Etruria St. Here was the earthenware manufactory erected by Josiah Wedgwood (d. 1795). His dwelling-house he named Etruria Hall and the village built for his workmen Etruria, clearly in allusion to Etrurian (Etruscan) pottery.

Etterby Cu [*Etardeby* 1246 Ipm]. '*Etard*'s BY.' In 1130 P is mentioned 'land which was Etard's'. *Etard* is a Fr name of German origin (OHG *Eidhart*).

Ettingshall St [*Ettingeshale* 996 Mon, *Etinghale* DB, *Ettingehal* 1175, *Etingehale* 1196 P, *Ettingeshale* 1261 FF]. Perhaps '*Etting*'s HALH'. *Etting* is unrecorded, but would belong to *Atta*, *Etti*. Possibly, however, the first el. is a verbal noun derived from OE *ettan* and meaning 'grazing'.

Etton Np [*Ettona* 1125–8 LN, 1175 P, *Ecton* 1189 (1332) Ch], **E~** YE [*Ettone* DB, *Ettona* c 1185 YCh 1098, *Etton* 1233 Ep, *Ecton* 1242 Fees, 1291 Tax]. Perhaps '*Ēata*'s TŪN'. The isolated spellings with *ct* are hardly to be taken seriously.

Etwall Db [*Etewelle* DB, -*well* 1185 P, 1242 Fees]. '*Ēata*'s stream.'

Eudon Burnell & George Sa [*Eldone* DB, *Eudon* 1183 P, *Eudon Burnel*, *E~ Jory* 1316 FA]. 'Yew hill.'

E~ Burnell was held by Robert Burnel in 1269

(Ch). Cf. ACTON BURNELL.—E~ George was held by William de Sancto Georgio in 1242 (Fees).

Euston Sf [*Euestuna* DB, *Euuestun* c 1095 Bury, *Euston* 1242 Fees]. '*Eof*'s TŪN' or OE *Efes-tūn*, if *efes* could be used in the sense 'bank of a river'.

Euxton (ĕks-) La [*Eueceston* 1187, *Euekeston* 1188 P]. '*Æfic*'s TŪN.'

Evedon Li [*Evedune* DB, *Euedon* 1196 FF, *Evedon* 1206 Cur]. '*Eafa*'s hill.'

Evelith Sa [*Ivelithe* 1200, *Ivelyth* 1292 Eyton]. 'Ivy slope' (OE *ifig* and HLIþ).

Evenley Np [*Evelai, Avelai* DB, *Evenle* 12 NS, 1220 Fees]. 'Smooth LĒAH', first el. OE *efn* 'even, smooth'.

Evenlode Wo [æt *Euulangeladæ* 772 BCS 209, *Eownilade* 779 ib. 229, *Evnilade* DB]. '*Eowla*'s passage or ferry.' **Eowla* corresponds to the Continental *Awila.* See (GE)LĀD. The river-name **Evenlode** is a back-formation from the pl. n. Cf. BLADON.

Evenwood Du [*Efenwuda* c 1050 HSC, 1104–8 SD, *Efnewda* 1131 FPD]. 'Level wood.' See EFN.

Evercreech So [*Evorcric* 1065 Wells, *Evrecriz* DB, *Euercriz* DB, 1176 P]. Originally no doubt CREECH, identical with CREECH (q.v.). *Ever-* is a distinctive addition, probably OE *eofor* 'boar'.

Everdon Np [*Eferdun* 944 BCS 792, *Eofordunenga gemære* 1021–3 BM, *Everdone* DB]. 'Boar hill.' First el. OE *eofor, efer* 'boar'.

Everingham YE [*Yferingaham* c 972 BCS 1279, *Evringham* DB, *Eueringeham* 1185, 1191 P]. 'The HĀM of *Eofor*'s people.' The *Y-* of the form of c 972 is probably an inverse spelling.

Everington Brk nr Yattenden [*Eurinton* 1176 P, *Euerinton* 1197 FF, *Yevrinton* 1220 Fees]. 'The TŪN of *Eofor*'s people.'

Everley W [*Eburleagh* 704 BCS 108, *Euerlai* 1173 P], E~ YN [*Evrelai* DB, *Everleg* 1240 FF]. 'Boar wood.' See EOFOR.

Eversden, Great & Little, Ca [*Eueresdona* c 1080 ICC, *Aueres-, Euresdone* DB, *Everesdon Magna, Parva* 1240 FF]. 'Boar's hill.'

Eversholt Bd [*Eureshot* DB, -*holt* 1185 P], **Evershot** Do [*Teversict* 1201 FF, *Evershet* 1286 Ch, *Euershut* 1293 FF], **Eversley** Ha [*Euereslea* c 1050 KCD 845, 1175 P, *Evreslei* DB]. 'Boar's wood.' First el. OE *eofor* 'boar', second OE HOLT, SCĒAT and LĒAH.

Everthorpe YE [*Euertorp* DB, *Everthorp* 1195 (1335) Ch, *Yver-, Iverthorp* c 1200 YCh 1128, 1130, *Jurethorp* 1252 Ch, *Yverthorp* 1285 FA]. The place is on high ground and the first el. of the name is probably OE *yferra* 'upper', varying with ON *efri* the same.

Everton Bd [*Euretone* DB, *Euerton* 1199 P], E~ La [*Evretona* 1094 LaCh, *Everton* 1201 P], E~ Nt [*Euretone* DB, *Euerton* 1185 P]. OE *Eofor(a)tūn* 'boar farm'.

Evesbatch He [*Sbech* DB, *Esebec* 1200, *Esebeche* 1201 Cur, -*bache* 1234 Cl]. '*Ēsa*'s valley.' Cf. BÆCE.

Evesham (ēv-, ēsham, ēsam) Wo [(æt) *Homme*, (into) *Eveshomme* 709 BCS 124, (on) *Eoueshamme* 1017–23 E, *Evesham* DB]. '*Eof*'s HAMM.'

Evington Gl [*Giuingtune* DB, *Yeninton* 1285 Ch, *Yivynton* 1303 FA]. 'The TŪN of *Geofa*'s people.' *Geofa* is found in *Geofanstig* 961 BCS 1074, *Geofandene* KCD 1355.

Evington Le [*Avintone* DB, *Evintona* c 1200 Fr, *Evington* 1254 Val]. 'The TŪN of *Eafa*'s or *Æfic*'s people.'

Ewanrigg Cu [*Ouenrig* c 1174 Holme C, *Wnering* 1187, *Ouerinc* 1190 ff. P]. Perhaps OE *eowena hrycg* 'ridge where ewes grazed'.

Ewart Nb [*Ewurthe* 1218 P, *Ewrth* 1242 Fees]. OE *ēa-worþ* 'WORþ on a river'.

Ewdness Sa [*Hendinas* 13 Misc, *Eudenas* 1360 AD]. A Welsh name, *Henddinas* 'old town'. *Hen-* seems to have been misread as *Heu-*.

Ewell, Temple, & E~ Minnis K [*Æwille* [mearc] c 772 BCS 207, *Ewelle* 959 ib. 1050, DB, *Templum de Ewell* 1213 RA], E~ Sr [*Euuelle* 675 BCS 39, *Eauuelle* 1065 BM, *Etwelle* DB, *Ewella* 1156 P]. OE *æwiell* 'spring, source of a river'.

Temple E~ is on the upper Dour. It was held by the Templars from the time of Hy 2. Minnis is OE *gemænness* 'community', here 'common'.

Ewelme O [*Lawelme* DB, *Eawelma* 1209–19 Ep], **Ewen** Gl [*at Awilme* 931, *Ewulm* 937 BCS 671, 719]. OE *æwielm* 'spring, source of a river'. Ewen is at the source of the Thames.

Ewerby Li [*Geresbi, Ieresbi* DB, *Iwarebi* Hy 2 DC, *Ywarebi* 1190 P]. '*Ivar*'s BY.' First el. ODan *Ivar*, ON *Ívarr*, pers. n. **Ewerby Thorpe** [*Oustorp* DB, 1212 Fees]. OScand *Aust-þorp* 'east thorp'.

Ewesley Nb [*Oseley* 1286 PNNb]. '*Ōsa*'s LĒAH', or better 'blackbird wood', the first el. being OE *ōsle* 'blackbird'.

Ewhurst Ha [*Ywyrstæ stigel* 1023 KCD 739, *Werste* DB, *Ywhurst* 1242 Fees], E~ Sr [*Iuherst* 1179 RA, *Ywehurst* 1208 Cur], E~ Sx [*Werste* DB, *Yuehurst* 1242 Fees]. 'Yew wood.' Cf. HYRST.

Ewood La in Blackburn [*Eywode, Euot* 1246 Ass], E~ (ē-) La in Haslingden [*Thewode* 1269, *Le Ewode* 1323 LaInq]. 'Wood on a river' (OE *ēa-wudu*). E~ Sr. See NEWDIGATE

Ewshott Ha [*Iweshete* 1305, -*schate* c 1307 Crondal]. OE *īw-scēat* 'yew grove'. See SCĒAT.

Ewyas Harold He [*Euias, Eugias, Euwias* c 1150 LL, *Euyas* Mab, *Ewias* DB, *Euuias Haraldi* 1177 P], **Ewyas Lacy**, now **Longtown**, He [*Ewias* DB, *Ewyas Lascy* Hy 3 Misc]. A Welsh name, derived from **ovi-* 'sheep' in Welsh *ewig* 'doe', Ir *ói* 'sheep', Lat *ovis*, with the suffix found in *dinas* 'town' from *din* &c. The name means 'sheep district'.

E~ Harold from Harold son of Earl Ralph, a nephew of the Confessor.—**E~ Lacy** belonged to the fee of Roger de Laci in 1086 (DB). Laci from LASSY in Normandy.

Exbourne D [*Echeburne* DB, *Yekesburne* 1242 Fees, *-bourn* 1292 Cl]. OE *gēaces-burna* 'cuckoo's stream'.

Exbury Ha [*Ykeresbirie* 1196, *-bir* 1197 P, *Ekeresbur'* 1212 Fees, *Hukeresbir* 1235 Cl]. '*Eohhere*'s BURG.' Cf. EGGARDON Do.

Exceat Sx [*Essete* DB, *Exeta* c 1150 Fr, *Esshetes* 1242 Fees]. OE *ǣc-scēat* 'oak grove'. Cf. SCĒAT. First el. OE *ǣc*, gen. of *āc* 'oak'.

Exe R So, D [*Iska* c 150 Ptolemy, *Uuisc* c 894 Asser; *Eaxan, Exan* (obl.) 739 BCS 1331 f., *Exe* 1238 Ass]. A Brit river-name, identical with AXE, ESK and with USK in Wales, also with ESK in Scotland and ISCH and others on the Continent. Brit *Iscā* became **Escā*, whence OE *Esce* and *Æsce*, which gave *Esk* and with metathesis *Exe* and *Axe*. The name is identical with OIr *esc*, Ir *easc* 'water' and probably comes from **pid-skā* or **pit-skā*, the root being *pi-* in Greek *pidúō* 'to gush forth' &c. Cf. EDEN.

On the Exe are **Up & Nether Exe** D [*Ulp-esse, Niresse* DB, *Nitherexe* 1196 FF, *Uppe Esse, Nytheresse* 1242 Fees], **Exebridge** D [*Exebrigge* 1255 FF], **Exeter** D [*Iska* c 150 Ptol, *Isca Dumnuniorum* 4 IA, *Cairuuisc* c 894 Asser, *Ad-Escancastre* c 750 Life of St. Boniface, *Escanceaster* 876, *Exanceaster* 877 ASC, *Execestre* DB], **Exford** So [*Aisse-ford* DB, *Exeford* 1243 Ass], **Exminster** D [*Exanmynster* c 880 BCS 553, *Esseminstre* DB], **Exmoor** D, So [*Exemora* 1204 Ch], **Exmouth** D [*Exanmuðan* 1001 ASC], **Exton** D [*Exton* 1242 Fees], **Exton** So [*Exton* 1216 Cl], **Exwick** D [*Essoic* DB]. Exeter is 'the Roman station on the Exe', or *Ex-* may even represent *Isca*, the old name of the station. For the later development see CEASTER. Exminster is 'the monastery on R Exe'. Exwick is 'wīc on R Exe'.

Exelby YN [*Aschilebi* DB, *Eskelby* 1252 Ass]. '*Eskil*'s BY.' ODan *Eskil* corresponds to ON *Áskell, Ásketill*.

Exford. See EXE.

Exhall Wa nr Coventry [(æt) *Ecclesbale* 1002 Wills, 1004 KCD 710, *Ekleshal, Eckeleshale* 1275 Ipm], **E~** Wa nr Stratford [*Ecclesbale* 710 BCS 127, *Eclesbelle* DB]. Identical with ECCLESHALL. *Ecclesbale* 1002, 1004 may rather be ECCLESHALL St.

Exminster, Exmoor, Exmouth. See EXE.

Exning Sf [*Esselinga* c 1080 ICC, *Essellinge* DB, *Exningis* 1158 P, *Ixninges* 1158 RBE, 1218 Cl]. OE **Gyxeningas* '*Gyxa*'s people'; cf. IXWORTH. The *-n-* may be the suffixal *n* of the *n*-stem *Gyxa*. Or the immediate base might be a derivative **Gyxīn*.

Exton D, So. See EXE.

Exton Ha [*æt East Seaxnatune* 940 BCS 758, *Essessentune* DB, *Exton* 1182 P]. 'The TŪN of the East Saxons.' The name records an East Saxon colony in Hants.

Exton Ru [*Exentune* DB, *Exton* 1185 P]. Possibly OE *Exna-tūn* 'ox farm', the first el. being a gen. plur. **exna* for *oxna* (cf. *exen* nom. acc. plur.) from *oxa* 'ox'.

Extwistle La [*Extwysle* a 1193 Whitaker, *-twisil* 1242 Fees]. Second el. OE *twisla* 'junction of streams'. The first may be as in prec. name.

Eyam (ēm) Db [*Aiune* DB, *Eyum* Hy 3 BM, 1236 Fees]. OE *ēgum*, dat. plur. of ĒG 'island' &c.

Eycote Gl [*Aicote* DB, *Aicota* 1209 Fees, *Eicote* 1221 Ass]. OE *ēg-cotu* 'COTS on the island.'

Eydon (ē-) Np [*Egedone* DB, *Eindon* 1202 Ass, *Eydon, Eyndon* 1254 Val]. '*Æga*'s DŪN.' Cf. AYNHO.

Eye (ī) R Le [*Eye* c 1540 Leland], **Eye Brook** Le [*Litelbe* 1218 For, 1227 Cl]. OE *ēa* 'river'. The spelling *Eye* reflects a dialectal sound-change.

Eye He [*Eia* c 1175 BM], **E~** Mx [*Eia* DB, 1204 Cur], **E~** Np [*Ege, Aege* 970 BCS 1258, *Ege* 972 ib. 1280 f., *Eya* 1199 NpCh], **E~** (and Dunsden) O [*Eye* 1294 Ch], **E~** (ī, ā) Sf [*Eia* DB, 1158 P, *Eye* 1190 P]. OE *ēg, īeg* 'island, land by water'. Eye Mx is lost, but the name is preserved in **Ebury** Sq. and Street in Westminster. Ebury means 'Eye manor'.

Eyford Gl [?*Æeoport* 872 BCS 535, *Aiforde* DB, *Heyford* 1220, *Heiford* 1236 Fees, *Ey-ford* 1303 FA]. Either *ēg-ford* 'ford by an island' or *hēg-ford* 'hay ford'.

Eyke Sf [*Eyk* 1270 Ipm, *Eyck* 1291 Tax]. OScand *eik* 'oak'.

Eynesbury Hu [(on) *Eanulfesbyrig* c 1000 Saints, *Enulesberia* c 1080 ICC, *Einulues-berie* DB]. '*Ēanwulf*'s BURG.' **E~ Hard-wicke** [*Herdwich* 1209 Abbr]. See HARDWICK.

Eynsford K [*Ænesford, Æinesford* c 960 BCS 1097 f., *Æinesford* 1130, *Einesford* 1156 P], **Eynsham** (ēnsh-) O [*Egonesbám* 571 ASC, *Egeneshomm* 864 BCS 509, *Egnesham* 1005 Eynsham, 1130 P, *Egbenes-ham* c 1137 Oxf, *Einegsham* c 1160 RA, *Egenesham* 1163, *Egeneisham* 1190 P, *Euenesham* 1160 P]. With these may be compared the lost **Eynsworth** Ess in Arkesden [*Einesuurda* DB, *Eynesworth* 13 FF] and AINSWORTH La. There is every reason to suppose that OE had a short form *Ægen* from names in *Ægen-*, though this stem is not well evidenced in English. Cf. OG *Agin*, *Agino* &c., short forms of names such as *Aganbold* &c. The hypothetical *Ægen* is no doubt the first el. of EYNSFORD, EYNSWORTH, AINSWORTH. But the earliest forms of EYNSHAM do not go well with such a base, and the numerous later spellings with *g* indicate that the form *Egoneshám* should be taken seriously. Here probably belongs *Iogneshomm* 825 BCS 384, described as of 300 hides and claimed by King Cenwulf of Mercia from Archbishop Wulfred. Eynsham belonged to the king of Mercia in 864 (BCS 509). *Iognes* is a Kentish form for *Eognes*, and *Egoneshám* 571 ASC is possibly miswritten for *Eogneshám*. *Eogen* is doubtless OW *Eugein*, an early form of Welsh *Owain* from OBrit **Esuganios* (Morris Jones 102). For the change of *esu* to OE *ēo* cf. OE *Eoccen* 'Ock' from W *eog* 'salmon' (**esōk-*). The second el. of Eynsham is OE *hamm*.

Eype (ēp) Do [*Estyep* 1300, 1319, *-e* 1372, *Yep* 1405 FF]. The place is on a stream, at whose mouth is Eype Mouth. Eype may be a river-name, derived from OE *gēap* adj. in some sense, such as 'crooked' or 'steep'. Perhaps Eype is not an original stream-name, but the name of a neighbouring hill. If so, YAPHAM may be compared. But *Estrhep* 1329 ADC 2808, which evidently refers to Eype, may indicate that the source is OE *hēap* 'heap'; cf. HEAP, SHAP. OE *ēa* in Do often became *ie*, as in YETMINSTER, and *Hiep* may well have lost its *h-*, especially in combination with *East-* and *West-*.

Eythorne K [*æt Heagyðeðorne* 805–31 BCS 318, *Ægyðeðorn* 824 ib. 381, *Egedorn* 11 DM, *Egethorn* 1212 RBE, *Egethorne*, *Heythorn* 1291 Tax]. '*Hēa(h)gȳþ*'s thorn-bush.' The woman's name *Hēa(h)gȳþ* is not otherwise evidenced.

Eythrope (ē-) Bk [*Edropa* 1167 P, *Ethrop* 1242 Fees]. OE *ēa-þorp* 'thorp on a river'.

Eyton Sa nr Wem [*Hetone* DB, *Eton juxta Bascherch* 1221 Eyton], **E~** Sa nr Ford [*Etune* DB, *Eton* 1242, 1252 Fees, *Eyton* 1274 Cl]. These seem to be OE *Ēa-tūn* 'TŪN on a river'.

Eyton on Severn Sa [*Aitone* DB, *Eyton* 1285 FA, *Eyton Abbatis* 1327 Subs], **E~ upon the Weald Moors** Sa [*Eyton* 1231, 1238 Cl, 1242 Fees, *-e* 1327 Subs, *Eyton super le Wildmore* 1344 PNSa]. OE *Ēg-tūn* 'TŪN on an island or in land by a river'. Weald Moors means 'wild, waste moor'.

Eyworth Bd [*Aieuuorde* DB, *Eiword* 1199 P, *Eyworth* 1232 Cl]. '* Æga*'s WORP.' Cf. EYDON, AYNHO.

F

Faccombe Ha [*Faccancumb* 863 BCS 508, c 950 Wills, *Facumbe* DB, *Faccumba* 1167 P]. '*Facca*'s CUMB or valley.' *Facca* is the first el. of (æt) *Faccanlea* 969 BCS 1232, (æt) *Fachanleage* 966 ib. 1182 (nr Stratford on Avon Wa), of FACKELEY, FAKENHAM. OG *Facco* occurs. *Facca* may be a modification of *Falca* (cf. FAWKHAM).

Faceby YN [*Feizbi*, *Foitesbi* DB, *Faicesby* 1208 FF]. OScand *Feits-bȳr*. First el. a nickname *Feitr* from OScand *feitr* 'fat'.

Făcit La [*Fagheside* 13 WhC]. 'Multi-coloured hill-side.' Cf. FĀG, SĪDE.

Fackeley Np [*Fakele* c 1170 BM, (boscus de) *Fakkeleya* 13 ib.]. '*Facca*'s wood.' Cf. FACCOMBE, LĒAH.

Faddiley Chs [*Faddelee* 1260 Court, *Fadile* 1271 AD, *Fadylegh* 1288 Ormerod], **Fadmoor** YN [*Fademora* DB, *-mor* 1204 Cur, *Faddemor* c 1150 Riev]. With these may be compared **Vaddicott** D [*Faddecote* 1212 Fees]. First el. possibly a pers. n. *Fadda*, related to OE *Fadol*, OG *Fato*, *Fadiko* &c. But Goth *faþa*, MHG *vade* 'fence' might also be compared.

OE *fæger* 'fair, beautiful'. See FAIR- (passim), FAREWELL, VERWOOD.

OE **fælging.** See FEALG.

OE **fær** 'passage' is found in some names, as HOLLINFARE, LAVER, probably DENVER, FARWAY.

OE **fæsten** 'stronghold'. See BUCKFAST, HOLDFAST, STOCKERSTON, VASTERNE.

OE **fāg** 'variegated, multicoloured' is the first el. of several pl. ns., as FACIT, FAWCETT, FAWSIDE (second el. OE *sīde* 'side, slope'), FAINTREE, FAWDON, FAWEATHER, FAWLER, FAWNLEES, FAWNS, VOWCHURCH, possibly FALSTONE, FOWNHOPE. In Facit and others the meaning may be 'flowery'. In Fawler a tessellated Roman floor is referred to.

Failsworth La [*Fayleswrthe* 1212 RBE, *Failesworth* 1246 Ass]. The first el. may be a derivative of OE *fēgan* 'to join', of about the same meaning as *scyttels* in SHUTTLEWORTH. A *fēgels* might possibly have been a hurdle.

Faintree Sa [*Faventrei* DB, *Fagentre* 1212 Fees, *Faentre* 1212 RBE, *Fayntre* 1274 Ipm]. OE *fāge trēo* 'multicoloured tree'.

Fairbourne K [*Fereburne* DB, *-borna* 1130 P, *Fareburna* 1165 P, *Farnburn* 1242 Fees], **Fairburn** YW [*Farenburne* c 1030 YCh 7,

Fareburne DB, *Farneburne* 1270 Bodl]. 'Stream by which ferns grew.'

Fairfield Db [*Fairfeld* 1230 P, *Fairefeld* c 1250 Derby], **F~** K [*Faierfeld* 1203 Cur, *Fairefeld* 1278 QW]. 'Beautiful FELD.'

Fairfield Wo [*Forfeld* 817 BCS 360, *Forfeud* 1255 For]. 'Hog FELD.' First el. OE *fōr* 'a hog, pig'.

Fairford Gl [(æt) *Fagranforda* 872 BCS 535, *Fareforde* DB, *Faireford* 1176, 1195 P]. 'Clear ford', the opposite of FULFORD. OE *fæger* also means 'untroubled, not disturbed' (of water).

Fairlee Wt [*Fayrelye* Hy 3 AD, *Fayreleg* Hy 3 BM]. 'Beautiful LĒAH.'

Fairley Sa [*Fernelege* DB], **Fairlight** Sx [*Farleghe* c 1175 Penshurst, *Farnleg* 1249 Ass]. 'Clearing overgrown with ferns.'

Fairstead Ess [*Fairsteda* DB, *Fairsted* 1229 Ch]. 'Beautiful place.'

Fakenham (-ā-) Nf [*Fachenham, Fagana, ham, Faganham* DB, *Fakenham* 1254 Val*Fakeham* 1212 Fees], **F~ Magna & Little F~** Sf [*Fakenham* c 1060 Th, *Fachenham, Litla Fachenham* DB, *Fakeham* 1242 Fees, *Fake(n)ham Magna, Parva* 1254 Val]. '*Facca*'s HĀM.' Cf. FACCOMBE.

Fal (-ǎ-) R Co [(to) *Fæle* 969 BCS 1231, (of) *Fæle* 1049 OSFacs, *Fale* c 1200 Gervase]. Etymology obscure. On the Fal is Falmouth [*Falemuth* 1235, -*mue* 1297 Cl].

Falcutt (-awk-) Np [*Faucot* 1220 Fees, *Faucot* 1284 FA]. Identical with **Fewcott** O [*Feaucote* 12 TpR, *Faucot* 1208 Cur, -*kote* 1209–35 Ep]. First el. OE *feawe* 'few', the meaning being 'few COTS' or 'humble COT'. Cf. OE *fēasceaft* 'miserable'.

Faldingworth Li [*Falding(e)urde* DB, *Faldinguorda* c 1115 LiS, -*wrd* 12 DC, -*worth* 1202 Ass]. 'WORÞ for folding animals.' OE *faldian* means 'to make sheep-folds, hurdle sheep'.

Falfield Gl [*Falefeld* 1327 Subs, 1347 Ipm]. Cf. FALLOWFIELD.

Falinge La [*Faleng* 13 WhC, *Falynge* 1323 Ct]. OE *fælging* 'newly cultivated land'. Cf. FEALG.

Falkenham Sf [*Faltenham* DB, 1291 Tax, 1331 BM, *Falcenham* 1254 Val]. The original form had *t*, not *k*. The first el. is obscure. It looks like a pers. n. **Falta* which may be derived from an OE **faltan* corresponding to OHG *falzan* 'to forge', and cognate with OE *anfielte*, OHG *anafalz* 'anvil'.

Fallibroome Chs [*Falinisbrom* Hy 3 Ipm, *Falingbrom* 1286 Court]. 'Broom thicket by a *fælging* or newly cultivated land.' Cf. FEALG.

Fallowdon Nb [*Falewedune* Hy 2 FPD, *Falewedon* 1233 P]. 'Fallow, i.e. yellow, DŪN or hill.'

Fallowfield La [*Fallufeld* 1317, *Falofeld* 1417 PNLa], **F~** Nb [*Faloufeld* 1296 Subs, *Falughfeld* 1355 Pat]. The forms are too late for a decision as to whether the first el. is OE *fealu* adj. 'fallow, yellowish' or *fealg* 'newly cultivated land'.

Fallowlees Nb [*Falalee* 1388 Ipm; *Fawleyburne* c 1235 Percy]. First el. OE *fealu* adj. or *fealg* sb. Cf. FALLOWFIELD, LĒAH. *Fawleyburne* is **Fallowlees Burn.**

Falmer (-ahm-) Sx [*Falemere* DB, -*mera* 1121 AC]. Possibly OE *fealwa mere* 'fallow mere', though we should expect to find traces of the *w* in early forms. Alternatively the first el. may be OE *fǣle* 'pleasant' with early shortening of *ǣ*.

Falmouth. See FAL.

OE **falod** 'fold', also 'enclosure for deer', is found alone as a pl. n. in FAULD and occasionally as a second el., e.g. in AL-, CHIDDING-, DOR-, DUNSFOLD, COLE W. Cf. especially STAT-, STOTFOLD. See also FALDINGWORTH, FAWDINGTON, FORTHERLEY.

Falsgrave YN [*Walesgrif* DB, -*graua* 1169 P, *Whallesgrave* 1241 FF, *Walegrive* c 1180 YCh 370, *Whalegrave* 1237 Cl]. 'Pit or hollow by a hill.' First el. OScand *hvǎll* 'hill', second OScand *gryfia* 'a pit, hollow'. ON *hvǎll* appears in the gen. sing. in names such as KVAALSAASEN, KVAALSVIG in Norway.

Falstone Nb [*Faleston* 1256 Ass, *Faustane* 1371 Sc]. The earliest example points to OE *fealu* adj. as first el. But OE *fāg* 'multicoloured' is more likely and indicated by the second example. Second el. OE STĀN 'stone'.

Fambridge, North & South, Ess [*Fanbruge* c 1050 KCD 907, *Fan-, Phenbruge* DB, *North Fambregg, Suthfambregg* 1291 Tax]. 'Bridge by a fen.' The form *fan* is the normal East Saxon form from OE *fæn*.

Fangdale YN [*Fangadala* c 1130 Riev, *Fangedale* c 1180 YCh 1845]. First el. OScand *Fangā* 'river for fishing' (ON *fang* 'hunting, fishing').

Fangfoss YE [*Frangefos* DB, *Fangefosse* 1120–9 YCh 449, 1219 Fees, *Fangelfosse* 1200 Cur]. Second el. *foss* 'a ditch' (cf. FOSS DYKE). Other names in -*foss* in the neighbourhood have a pers. n. as first el. (CATFOSS, WILBERFOSS). This suggests that *Fang-* goes back to a pers. n. too, but no name is known that suits the case.

Fanthorpe Li [*Fenthorp* 1202 Ass, *Falmettorp* 1212 Fees]. '*Feolumǣr*'s thorp.' Cf. FELMERSHAM. The first *r* was lost owing to dissimilation. Or the first el. may be a short form of *Feolumǣr*. Cf. FELMINGHAM.

Farcet Hu [*Faresheued*, (æt) *Farresheafde* 963–84 BCS 1128, *Fearresheafod* 973 ib. 1297]. 'Bull's head.' First el. OE *fearr* 'bull'. On the meaning of *hēafod*, which in this case cannot mean 'hill', see HĒAFOD.

Fardle D [*Ferdendelle* DB, *Ferthedel* 1204 Cur, 1242 Fees]. OE *feorþa dæl* 'fourth part'.

Fareham Ha [*Fernham* DB, *Fereham* c 1130 RA, *Ferham* 1136 AC]. 'Fern HĀM or HAMM.'

Farewell St [*Fagerwell* 1200 Ch, *Faierwell* 1200 P, *Faurewell* 1251 Cl]. 'Beautiful stream.'

Farforth Li [*Farforde* DB, *Fareford* 1170 P, *Forefort* c 1125 Fr, *Foresford* 1202 Ass]. If the variation between *a* and *o* in the first syllable is not due to dittography, the *o* of *ford* having been anticipated, the first el. might be OE *fær* 'passage', alternating with *fōr* 'journey', or else OE *fearh* 'pig', alternating with *fōr* 'pig'.

Faringdon Brk [*Fearndun* 924 ASC (C), *Færndun* c 970 Wills, *Ferendone* DB, *Cheping Farendon* 1242 Fees], **F~** Do [*Ferendone* 1084 GeldR, *Ferndone* 1178 P], **F~** Ha [*Ferendone* DB, *-don* 1186 P, *Farendone* c 1200 Ep], **Little F~** O [*Parua Ferendon* 1199 P, *Farendon* 1277 Cl]. 'Fern-clad hill.'

Farington La [*Farinton* a 1149 LaCh, 1212 Fees]. 'TŪN where ferns grew.'

Farlam Cu [*Farlam* 1169 PNCu, c 1210 WR, *Furlaham* (sic) 1183 Pp, *Farlaham* 1279 Ass, *Farlham* 1295 Cl]. Possibly OE *fearn-lēam* '(at) the fern clearings'. Or more likely *Fearnlēah-hām* 'HĀM by a fern clearing'. Cf., however, FARLETON.

Farleigh Wallop Ha [*Ferlege* DB, *Farnly* 1337 Ch, *Farle Mortymer* 1412 FA], **East & West F~** K [(on) *Fearnlege* 871–89, *-leag* 898 BCS 558, 576, *Ferlaga* DB, *East-, Westfarlegh* 1291 Tax], **F~ Hungerford** So [*Fearnlæh* 987 KCD 658, *Ferlege* DB, *Montford Farlegh* 1362, *Farlegh Hungerford* 1404 Ep], **F~** Sr [*Ferlega* DB, *Farnleg* 1264 BM], **F~** W [*Farlege* DB, *Fernelega* 1109–20 Sarum], **Monkton F~** W [*Farnleghe* 1001 KCD 706, *Farlege* DB, (Prior de) *Ferneleia* 1195 Cur, *Farley Monachorum* 1316 FA], **F~ Wick** W [*Farlegh Wyke* 1365 AD]. OE *fearn-lēah* 'fern-covered clearing'.

F~ Hungerford was acquired by Sir Thomas Hungerford in 1369.—**F~ Wallop** was owned by members of the Wallop family in the latter part of the 14th cent. Cf. WALLOP.—**Monkton F~** was the seat of a monastery. *Monkton* may be for *Monken*, gen. plur. of *monk* (ME *monkene*).—**F~ Wick** was originally WICK. The distinguishing el. from the proximity to Monkton Farleigh.

Farlesthorpe Li [*Farlestorp* 1190 P, 1202 Ass, *Fareslestorp* 1202 Ass]. Perhaps '*Farald*'s thorp'. ON *Faraldr* occurs as a fictitious name, and possibly in a pl. n.

Farleton La [*Fareltun* DB, Hy 2 BM, *Farlton* 1227 Ch, *Farleton* 1235 FF, ?*Farneton* 1208 Cur], **F~** We [*Fareltun* DB, *Farleton* 1199 Kendale]. Possibly 'the TŪN of *Faraldr* or *Farle*'. Both names are to some extent evidenced in ON. But it is

not impossible that the first el. is a compound containing OE *fearn* 'fern', e.g. *fearnhyll* or *fearnlēah*. Farleton We is near Farleton Fell.

Farley Hill Brk [*Farellei* DB, *Ferlega* 1190 P], **F~** Db [*Farleie* DB, *Farnleya* 13 BM], **F~ Chamberlayne** Ha [*Ferlege* DB, *Ferlega Camerarii* 1167 P, *Farle Chaumberleyn* 1346 FA], **F~** St [*Fernelege* DB, *Farleye* 1273 Ipm]. Cf. FARLEIGH. OE *fearn-lēah* 'fern-covered clearing'.

Chamberlayne is a family name, originally meaning 'chamberlain'. *Ferly* was held by Robertus Camerarius in 1212 (Fees).

Farlington Ha [?*Ferninduna, Ferningdon* 1168 f. P, *Ferlingeton* 1187 f. P, 1200 Cur]. If the first forms belong here, the original name seems to have been OE **fiernen dūn* 'fern-clad hill'. If not, the source may be OE *Fearnlēainga tūn* 'the TŪN of the dwellers at *Fearnlēah*'. Cf. FARLEY.

Farlington YN [*Ferlin-, Farlintun* DB, *Ferlinton* 1167 P]. OE *Fearnlēainga tūn*. Cf. prec. name.

Farlow Sa [*Fernelau* DB, *Ferlaue* 1206 Cur, *Ferlowe* c 1433 BM]. 'Fern-clad hill.' Cf. HLĀW.

Farmanby YN [*Farmanesbi* DB, *Farmanby* 1157 YCh 186, *Farmanneby* 1225 Ep]. '*Farman*'s BY.' *Farman* is a well-known pers. n., found on coins and in DB. It is ON *Farmann*, OSw *Farman*, lit. 'traveller', 'travelling trader'.

Farmborough So [*Fearnberngas* 901 BCS 589, *Ferenberge* DB]. 'Fern-clad hill.' Cf. BEORG.

Farmcote Gl [*Fernecote* DB, *Fernecota* 1220 Fees], **F~** Sa [*Farnecote* 1209 For]. 'COT(s) among the ferns.'

Farmington Gl [*Tormertona* 1182 Winchc, *Tormartona* 1220 Fees, *Thormerton* 1236 Fees]. First el. probably an OE *þorn-mere* 'mere where thorns grew'.

Farnborough Brk [(to) *Fearnbeorgan* 916, 931 BCS 633, 682, *Fermeberge* DB], **F~** Ha [*Ferneberga* DB, *Farnburge* 1243 Crondal], **F~** K [*Fearnbiorginga mearc* 862 BCS 506, *Ferenberga* 1180 P, *Farnberg* 1242 Fees], **F~** Wa [*Ferneberge* DB, *Farneberue* 1236 Fees]. 'Fern-clad hill(s).' Cf. BEORG.

Farncombe Sr [*Fernecome* DB, *Farncumbe* 1348 BM], **Farndale** YN [*Farnedale* c 1160 Riev, *Farendale* 1207 FF]. 'Valley where ferns grew.' See CUMB.

Farndish Bd [*Fernadis, Farnedis* DB, *Fernedis* 1194 P]. 'Fern-clad pasture.' Cf. EDISC.

Farndon Chs [*Ferentone* DB, *Farendun* c 1195 Chester], **East F~** Np [*Feren-, Faredone* DB, *Ferendon* c 1175 Bury, *Farendon* 12 NS], **West F~** Np [*Ferendone* DB, *Farendon* 12 NS, 1220 Fees], **F~** Nt [*Farendune* DB, *Ferendon* 1175 P]. OE *fearn-dūn* 'fern-clad DŪN'.

Farne Islands Nb [(in) *insula Farne, insula
... Farne* c 730 Bede, (locus ... cognomine)
Farne 8 Alcuin, Carmina, (on) *Fearne* 9
OET, *Farne, Farene* c 890 OEBede, *Farne-
heland* 1254 Val]. A derivative of OE *fearn*
or the word *fearn* itself, possibly because of
a supposed similarity of the group of islands
to a fern.

Farnham Royal Bk [*Ferneham* DB, *Farn-
ham* 1200 Cur], F~ Do [*Ferneham* DB,
Fernham 1201 Cur], F~ Ess [*Phernham* DB,
Fernham 1166 RBE, 1198 FF], F~ Sf
[*Farnham, Ferneham* DB, *Farnham* 1206
Cur], F~ Sr [(æt) *Fearnhamme* 894 ASC,
Fernham 688, 803–5 BCS 72, 324, *Fearna-
ham* 858 ib. 495, *Ferneham* DB], F~ YW
[*Farneham* DB, *Fernham* 1226 FF]. OE
Fearn-hām or *-hamm* 'HĀM or HAMM where
ferns grew'.

Farnham Nb [*Thirnum* 1242 Fees, 1307
Ipm, *Thernhamme* 1343 Percy]. OE *þyrnum*,
the dat. plur. of *þyrne* 'thorn-bush'. Less
likely OE *þyrn-homm* (*-hamm*), with early
change of *o* to *u* and loss of *h*.

Farnhill YW [*Fernehil* DB, *Farnhille* 1230
Ep]. 'Fern-clad hill.'

Farningham K [*Ferningeham* DB, 1198 FF,
Freningeham 11 DM, 1177, 1193 f. P, 1206
FF, *Freningham* 1206 Cur]. Possibly 'the
HĀM of the *Fearningas* or dwellers in a ferny
place'. But some pers. n. beginning with
Fr- would be preferable. We may compare
FRANJUM in Frisia (from *Franingaheim*),
which is held to be derived from a word
meaning 'lord' corresponding to OE *frēa*.

Farnley YW nr Leeds [*Fernelei* DB], F~
YW nr Otley [(on) *Fernleage* c 1030 YCh 7,
Fernelai DB], F~ Tyas YW [*Fereleia,
Ferlei* DB, *Farneleye Tyas* 1335 FF]. OE
fearn-lēah 'fern-covered clearing'.

F~ Tyas was held by Baldewyn le Tyeys in
1236 (FF). Tyas is OFr *tyeis* 'German'. The
name also appears as *Teutonicus*.

Farnsfield Nt [*Fearnesfeld* 958 YCh 2,
Farnesfeld DB, *Farnefeld* 1187 f. P]. 'Fern-
clad FELD.'

Farnworth La in Deane [*Farnewurd* 1185
P], F~ La in Prescot [*Farneword* 1324
WhC]. 'WORþ where ferns grew.'

Farringdon D [*Ferentone, Ferhendone* DB,
Ferndon 1242 Fees]. 'Fern-clad down.'
Farringdon in London, originally the ward
of Ludgate and Newgate, was renamed from
two successive aldermen, William de Farndon (1278–93) and Nicholas de F~ (1293–
1334).

Farrington Gurney So [*Ferentone* DB,
Ferenton 1225 FF]. 'TŪN where ferns grew.'
The manor was held by Robert de Gurnay in
1225 (FF). Cf. BARROW GURNEY.

Farsley YW [*Ferselleia* DB, *Ferselee* 1203
FF]. OE *fyrs-lēah* 'furze-covered clearing'.

Farthinghoe (farnĭgō) Np [*Ferningeho* DB,
1183 P, *-hou* 1196 P, *Faringho* 1220 Fees,

Farnighou 1239 Ep, *Faringho* 1291 Tax].
Perhaps simply OE (æt) *fearnigan hō* 'fern-
clad hill or spur of land'. More likely the
first el. is OE *Fearningas* 'people from(West)
FARNDON'. The change to *Farthing-* is due
to influence from FARTHINGSTONE.

Farthingstone Np [*Fordinestone* DB, *Fard-
ingestun* 1167, *-ton* 1177 P, 1206 Cur,
Fardeneston 1167 P, *Fordingeston* 1184 P,
Ferdingestone 1232 Ep, *Farthingeston* 1261
Ass]. The occasional *o* in the first syllable
is probably to be disregarded. The first el.
can hardly be the pers. n. *Færþegn* (Coins,
DB) from ON *Farþegn* in view of the
regular *d*. Possibly the original name was
Fearndūninga-tūn 'the TŪN of the FARNDON
people'. This would have given ME *Fard-
ingeton*, which might have been changed to
Fardingeston.

Farway D [*Farewei* DB, *-e* 1219 FF, *Fare-
weye* 1242 Fees]. OE **fær-weg* 'road'.

Fauld St [*Felede* DB, *Falede* 1236, 1242
Fees]. OE *falod* 'fold'.

Faulkbourne (-awb-) Ess [*Falcheburna* DB,
Falkeburn 1207 Cur, *Faukisburn* 1247 FF].
'*Fealca*'s stream' or 'falcons' stream'. OE
fealca is not evidenced.

Faulkland So [*Fouklande* 1243 f., *Falclond*
1243 Ass]. OE *folcland* 'land of the people',
'Crown land'. Cf. BUCKLAND. On *folcland*
and *bocland* see now also Stenton, *Anglo-
Saxon England*, pp. 306–8.

Faulstone W [*Fallerstone* 1275 RH, *Valleres-
tone* c 1286 Ep, *Fallardestone* 1328 Ipm].
'*Fallard*'s TŪN.' *Fallard* is an OFr pers. n.

Faversham K [*Fefreshám* 811 f., *Febresham*
815 BCS 335, 341, 353, *Fæfresham* c 935
Laws, *Favreshant* DB, *Fauersham* 1130 P].
Generally explained as 'the smith's HĀM',
the first el. being an early loan from Lat
faber. OE *fæfer* is not recorded elsewhere.

Fawcett Forest We [*Faxide* 1247 Ch,
Fausyde 1282 Kendale]. Identical with
FACIT.

Fawdington YN [*Faldingtun* 1247 Ch,
Faldington 1235 FF, *Faldintun* 1254 Misc].
OE *falding-tūn* 'TŪN where animals were
folded'. Cf. FALDINGWORTH.

Fawdon Nb nr Gosforth [*Faudon* 1242
Fees], F~ Nb nr Ingram [*Faudon* 1207 Cur,
1267 Ass, *Faundon* 1268 Ass]. 'Multicoloured hill.' Cf. FĀG, DŪN.

Faweather YW [*Faheddre* 1198 f. P, *Fag-
heder* 1235 FF]. Probably 'multicoloured
heath or heather'. Cf. FĀG, HEATHER.

Fawkham K [*Fealcnaham* 964–95, 973 BCS
1132, 1296, *Falchenham* 10 ib. 1322, *Faches-
ham* DB, *Fauke(n)ham* 1242 Fees]. '*Fealcna*'s
HĀM.' The pers. n. *Fealcna* occurs in *Wester-
falcna* in a version of a Northumbrian
genealogy in AŞC 560 (*-falca* in B, C).
Cf. also *Fealcnes ford* 898 BCS 576.

Fawler Brk [*Fageflor* Hy 2 Abingd, *Fagflur* 1178 P, *Fauflore* 1207 Cur], F~ O [*Fauflore* 1205 Cur, *Fauelore* 1220–2 Eynsham]. Identical with (to) *fagan floran* 904 BCS 607 (in boundaries of Eaton on the Cherwell). The name means 'variegated pavement' and refers to tessellated Roman pavements. One was actually discovered in a Roman villa nr Fawler O in 1865. Cf. FĀG, FLORDON.

Fawley Bk [*Falelie* DB, *Falle* 1199 FF, *Falele* 1234 Cl], F~ Ha [(to) *Faleðlea* 10 BCS 1161, *Falegia, Falelie* DB, *Falesleia* 1130, *Faleleia* 1194 P], F~ He nr Ross [*Filileia* 1142 PNHe, *Felileie* 1166 RBE, *Falleye* 1284 Ch], F~ He nr Weobley [*Fæliglæh* a 1036 Th, *Falle* 1303 FA]. F~ Ha must contain OE *filiþe, fæliþe*, which is held to mean 'hay' (see *filiþe*). The same is no doubt the first el. of F~ He nr Ross and probably that of F~ nr Weobley. These cannot well contain OE *fealg*. F~ Bk may be OE *fealg-lēah* 'LĒAH with a FEALG or clearing' or possibly a compound of OE *fealu* adj. and *lēah*, 'fallow-coloured LĒAH'.

Fawley Brk [*Faleslei* DB, *Faleslie* 1167, -*lega* 1190, *Faleweslega* 1177 P, *Falueley* 1230 P]. Identical with FAWSLEY Np.

Fawnlees Du [*Fawleys* 1359 PNNb], Fawns Nb [*Faunes* 1256 Ass], Fawside Du [*Fauside* 1365 Pudsay]. First el. OE FĀG 'multicoloured'. Second el. OE LĒAH, NÆSS 'headland', SĪDE 'slope'. Cf. FACIT.

Fawsley Np [(on) *Fealuwes lea* 944 BCS 792, *Faleuuesle, Faleuueslei, Felesleuue, Falelav* DB, *Fealeweslea* c 1110 NpCh, *Faleslea* 1167, *Faleweslea* 1190 P]. The name is identical with FAWLEY Brk. The first el. cannot well be a pers. n. derived from *fealu* adj., though such a name is in itself quite possible. It would be too remarkable to find such a name twice combined with LĒAH. The el. must be an OE noun *fealu*, a substantivized form of the adj. It might be a name of a forest, 'fallow-coloured wood'. Or, more likely, it is an animal's name and means 'fallow deer'. If so, it is here used collectively and the name means 'forest frequented by fallow deer'.

Faxfleet YE [*Faxflete* 1190 YCh 1312, -*flet* 1199, 1219 FF, *Faxeflet* 1228 Cl, 1230 P, *Flaxflet* 1185 TpR]. Probably '*Faxi*'s fleet' (cf. FLĒOT). OScand *Faxi* is evidenced as a pers. n. Alternatively the first el. might be ON *faxi* 'a horse'. The name would then have arisen owing to some incident in which a horse was concerned.

Faxton Np [*Fextone* DB, *Faxtona* 1167 P, *Faxton* 1220 Fees, *Fachestuna* 1121 AC]. First el. OE *feax* in a sense such as 'grass'. This is very likely the meaning of (to) *feaxum* 949 BCS 880. Cf. Norw *faks* 'coarse grass', G dial. *fachs* 'poor mountain grass'.

Fază·kerley La [*Fasacre*, -*legh* 1325 Ct]. *Fas*- may be OE *fæs* 'border, fringe'. *Fas-*

acre would then mean 'border strip'. Cf. LĒAH.

Fazeley St [*Faresleia, Farisleia* c 1142 Mon, *Faresleye* 1335 Ch]. 'Bull's LĒAH.' Cf. FARCET.

OE fealg, fealh, felh, ME *falwe* 'fallow, i.e. a piece of ploughed land, arable land; ground left uncropped for a year or more' corresponds to EFris *falge*, Bavarian *falg*. OE *fealgian* means 'to break up land'. In pl. ns. the meaning is 'land broken up, newly cultivated land'. It is difficult to distinguish *fealg* from OE FEALU adj. See FAL-, FALLOWFIELD, FAWLEY Bk. OE *felh* is found in FELPHAM, perhaps in FELLEY, FELLISCLIFFE. From *fielgan, a derivative of *fealg* that corresponds to G *felgen*, is formed OE *filging, fælging*, which means the same thing as *fealg*. See FALINGE, FALLIBROOME, FELLING, and cf. *Babban fæling* 849 BCS 455. There was also an OE *felg* 'felly, felloe, (part of) the rim of a wheel' which is probably found in CLOVELLY, VELLY and perhaps some other names.

OE fealu adj. 'fallow, of a pale brownish or reddish yellow colour, as withered grass or leaves' is found in some pl. ns., as FALLOWDON, perhaps FALMER. Cf. also FEALG and see FAWLEY Brk, FAWSLEY.

Fearby YN [*Federbi* DB, *Fetherby* 1184 PNNR, *Fegerbi* J Ass, *Fegtherby* 1301 Subs]. The first el. might be ON *fegrð* 'beauty', the old form being *Fegrðar-býr*.

OE fearn 'fern' is very common as the first el. of pl. ns. See FARN-, FEARN-, FERN-. But it takes several different forms in pl. ns. See FAIRBOURNE, FAIRLEY, FAIRLIGHT, FAREHAM, FARINGDON, FARRINGDON, FARINGTON, FARRINGTON, FARLEIGH, FARLEY, FARLOW, FARMCOTE, FARTHINGHOE, VERNHAM.

Fearnhead La [*Ferneheued* 1292 PNLa]. 'Fern-clad hill.' Cf. HĒAFOD.

OE fearr 'bull'. See FARCET, FAZELEY.

Featherstone Nb [*Fetherestanehalg* 1204 Cur, *Fetherstanhishalu* 1236 Fees, *Fetherstan* 1256 Ass], F~ St [*Feoþer(e)stan* 996 Mon, *Ferdestan* DB, *Federestan* 1186 P], F~ YW [*Fredestan, Ferestane* DB, *Federestana* 1122 Pont, *Fetherstan* 1166 P]. OE *feþerstān 'a tetralith', i.e. a cromlech, which consists of three upright stones and a headstone (Bradley). OE *fe(o)þer-, fiþer-* 'four' occurs in several compounds, such as -*fēte* 'four-footed', -*rīca* 'tetrarch'.

Feckenham Wo [*Feccanhom* 804, -*ham* c 960 BCS 313, 1006, *Fecheham* DB]. '*Fecca*'s HAMM.' *Fecca* is not evidenced elsewhere. It may be related to *Facca* in FACCOMBE &c. and *Fecc* in *Fecces wudu* 940 BCS 763.

Feering Ess [*Feringes* 1067 BM, 1206 Cur, *Feringas* DB, *Fering* 1196 FF, *Ferringes* 1206 Cur]. Possibly OE *Fēringas*, a derivative from a pers. n. formed from the adj.

fēre 'fit for service' or derived direct from *fēre*.

Feetham YN [*Fytun, Fyton* 1242 Fees, 1274 YInq]. Apparently the dat. plur. of ON *fit* 'river meadow' in spite of the early forms in *-un*. Cf. FELDOM.

Feizor YW [*Fegheserche* 1300, *Fegesargh* n.d. FC, *Fehhesherge* Fount]. Second el. ERG 'shieling'. The first is the pers. n. *Fech*, *Feg* DB (Yks), possibly from OIr *Fiach*.

Felbridge Sr [*Feltbruge* 12 Fr, *Feldbrigge* 1255 AD]. 'Bridge by a FELD.'

Felbrigg Nf [*Felebruge* DB, *Felebrigge* 1207 Cur]. OScand *fiol-bryggia* 'plank bridge'.

OE **feld** 'open country, land free from wood, plain' is common in pl. ns. The element is particularly common in old forest districts. In pl. ns. it is probably used in much the same sense as LĒAH. Probably *feld* denoted an open space of larger extent than *lēah*. Sometimes names in *-feld* denote large districts, as HATFIELD, ARCHENFIELD. The first el. is often a word such as *beonet, brōm, fyrs, hǣþ* (cf. BENT-, BROM-, FERS-, HAT-, HEATHFIELD) or an adj. such as *clǣne, scīr* (as CLAN-, SHERFIELD). Sometimes the el. takes the form *-ville*, the reason being that *f-* became *v-*, whereupon association with Fr *ville* took place: CAVILLE, CLANVILLE, ENVILLE, LONGVILLE. In the etymologies *feld* is generally left untranslated.

Feldom YN [*Feldun, Fildon* 1228 FF, *Feldom* 1301 Subs]. The dat. plur. of FELD.

Felhampton Sa [*Feldhampton* 1327 Subs]. 'HĀMTŪN in a FELD.'

Felixkirk YN [*Ecclesia S. Felicis* 1210 FF, *Felicekirke* a 1233 BM]. 'St. Felix's church.'

Felixstowe Sf [*Filchestou* 1254 Val, *-stowe* 1291 Tax, 1375 FF, *Fylthestowe* 1359 FF]. Probably 'St. Felix's place'. Cf. STŌW. Or possibly the first el. is *Filica* pers. n. in *Filican slæd* BCS 1093.

Felkington Nb [*Felkindon* 1208–10 Fees, *Felkendon* 1238 Pat]. Possibly 'the DŪN of *Feoluca*'s people'. *Feoluca would be a hypocoristic form of *Feolugeld* &c.

Felkirk YW [*Felekircha* c 1125 YCh 1428, *Felekirke* 13 BM]. 'Church made of boards.' First el. OScand *fiol* 'board'.

ON **fell, fiall** 'fell, mountain'. See e.g. BOW FELL, HAMPSFELL, WHINFELL.

Felley Priory Nt [*Feleleia* c 1145 Eynsham, *Felleya* Hy 2 (1316) Ch, *Falleg'* 1244 Cl]. Either OE *felh-lēah* 'LĒAH with newly cultivated land' (cf. FEALG) or *feliþ-lēah* 'LĒAH where hay was got' (cf. FILIÞE and FAWLEY).

Felling Du [*Fellyng* c 1220 FPD, *Le Felling* 1326 Misc]. ME *felling* 'clearing' (from *felle* 'to fell') or better OE *fælging* 'newly cultivated land'.

Felliscliffe YW [*Felgesclif* DB, *Fellesclive* 1230 Ep]. 'Cliff or slope with newly cultivated land.' See FEALG. We do not really expect OE *felg* to have had the gen. sg. *felges*, but the name may have arisen comparatively late.

Felmersham Bd [*Falmeresham* DB, *Felmeræsham* 1163 P, *Felmeresham* 1207 Cur]. '*Feolomǣr*'s HĀM.' *Feolomǣr* occurs in *Fiolomeresford* 963 BCS 1111 and in *Fealamæres broc* 709 ib. 124.

Felmingham Nf [*Felmincham* DB, *Felmingeham* 1175 ff. P]. 'The HĀM of *Feolma*'s people.' *Feolma is a normal short form of *Feolomǣr*.

Felpham Sx [*Felhhamm* c 880 BCS 553, *-ham* 953 ib. 898, *Falcheham* DB]. 'HAMM with newly cultivated land.' Cf. FEALG. Alternatively the first el. may be OE *felg* 'felly, rim of a wheel' in a transferred sense 'rim of the sea, coast'. F~ is on the sea.

Felsham Sf [*Fealsham* DB, *Fealsam, Felesham* c 1095 Bury, *Falesham* 1203 Cur, *Felsham* 1203 FF]. '*Fǣle*'s HĀM.' OE *Fǣle* occurs in *Fælesgræf* BCS 1282. It is derived from OE *fǣle* 'pleasant'.

Felsted Ess [*Feldestede* 1082 Fr, *-steda* 1177 P, *Felstede* DB, *Feltsted* 1238 Subs]. 'Place in a FELD.'

Feltham Mx [*Feltham* 969 Crawf, 1199 Fees, *Felteham* DB]. 'HĀM in a FELD.' Cf., however, FELTWELL.

Feltham So [*Fælet-, Fylethamm* 882 BCS 550, *Filetham* 1243 Ass, *Felethham* 1306 FF]. 'HAMM where hay was got.' Cf. FILIÞE.

Felthorpe Nf [*Felethorp, Faltorp* DB, *Feletorp* 12 BM, 1198 FF, *Felestorp* 1254 Val]. First el. very likely a pers. n. *Fæla*, derived from OE *fǣle* adj., on which see FELSHAM.

Felton He [*Feltone* DB, *Feltun* 1242 Fees], **F~** Nb [*Feltona* 1167 P, *Felton* 1242 Fees], **F~ Butler** Sa [*Feltone* DB, *-ton* 1176 P, *Felton Butiler* 1205–30 PNSa], **West F~** Sa [*Feltone* DB, *Felton* 1265 Ch], **F~** So in Winford [*Felton* 1243 Ass, 1285 ff. FA]. OE *Feld-tūn* 'TŪN in a FELD.'

F~ Butler was held by Hamo fitz Buteler c 1165 (Eyton), by Hamo Pincernator in 1242 (Fees).

Felton So, another name for Whitchurch [*Filton* 1243 Ass, 1316 FA, *Fylton* 1291 Tax]. OE *Filiþ-tūn* 'TŪN where hay was got'.

Felton Hill Nb [*Fyleton* 1245 Ipm, *Fileton* 1271 Ch]. '*Fygla*'s TŪN.' Cf. FIGHELDEAN. The first el. is hardly *filiþe* 'hay', for that word would not have had *i* in the North.

Feltwell Nf [*Feltwelle* c 1050 KCD 907, *Feltuuella, Fatwella* DB, *Falwella* 1121 AC, *Feltewell* 1169 P, 1196 FF, *Fautewelle* 1162 BM]. OE *felt* 'felt' occurs as the first el. of some plant-names, as *feltwurma* 'wild marjoram', *-wyrt* 'wild mullein'. There may well have been a plant-name derived from *felt*, e.g. *felte*. Such a name would give the best explanation of Feltwell. The second el. is OE WELLA 'spring'.

OE **fen**, ESax *fæn* 'fen, marsh' is fairly common in pl. ns. See, e.g., VANGE, FAM-BRIDGE, BULPHAN, SWINFEN, EDVIN, PINVIN.

Fenby Li [*Fendebi*, *Fenbi* DB, *Fembi* c 1115 LiS, *Fenby* 1231 Ch, 1242 Fees, *Fenneby* 1262 FF]. 'BY at the fens.'

Fenchurch Mx [*Fanchirche* 1292 AD]. Self-explanatory. *Fen-* in early records generally shows the ESax form *fan* from OE *fæn*.

Fencote He [*Fencote* DB], **Fencott** O [*Fencote* 1194 ff. P]. 'COT(s) in a fen.'

Fenham Nb nr Newcastle [*Fenhu*' 1256 FF, *Fenham* 1375 Cl], F~ Nb nr Holy Island [*Fennum* c 1085 LVD, Hy 2 LVD, *Fenham* 1254 Val]. OE *fennum*, dat. plur. of FEN, or OE *fen-homm* 'HAMM by a fen'.

Fenhampton He [*Fenhampton* 1354 AD]. 'HĀMTŪN by a fen.'

Feniton D [*Finetone* DB, *-tuna* 1185 Buck-land, *Feneton* 1169 P, *Vinetone* 1309 Ep]. 'TŪN by Vine Water.' **Vine** [(on) *Finan* 1061 ERN] may be derived from Welsh *ffin* 'boundary' and mean 'boundary stream'.

Fennymere Sa [*Finemer* DB, *Fennimare* 1226 Eyton, *Fennymer*' 1327 Subs]. Second el. OE *mere* 'lake'. The first is OE *fynig* 'mouldy, musty' from *fyne* 'mould'.

Fenrother Nb [*Finrode* 1189 P, *-rothre* 1232 Pat, *-rother* 1242 Fees, *Fenrother* 1256 Ass]. First el. OE *fin* 'a heap of wood'. The second might be OE *rop* 'a clearing' (cf. ROTHEND) with analogical addition of *-er*. Or there may have been an OE **roper* or **ruper* 'clearing'.

Fenstead Sf [*Finesteda* DB, *Finstede* 1195 ff. P]. OE *fin* 'heap of wood' and *stede* 'place'.

Fenton Cu [*Fenton* 1252 StB, 13 WR], F~ Hu [*Fentun* 1236 FF], F~ Li nr Newark [*Fentun* 1212 Fees, *Fenton* 1209–19 Ep], F~ Li in Kettlethorpe [*Fentuna* c 1115 LiS], F~ Nb [*Fenton* 1242 Fees], F~ Nt [*Fentone* DB, *-ton* 1200 Cur], F~ St [*Fentone* DB, *-ton* 1273 Ipm], **Church & Little** F~ YW [*Fentun* 963, *Fenntún* c 1030 YCh 6 f., *Fentun* DB]. 'TŪN by a fen.'

Fenwick Nb nr Kyloe [*Fenwic* 1208–10 Fees], F~ Nb nr Stamfordham [*Fenwic* 1242 Fees, 1250 Ipm], F~ YW [*Fenwic* 1166 P, 1206, 1208, 1226 FF]. 'WĪC by a fen.'

Feock Co [(ecclesia) *Sancte Feoce* 1264 Ep, (de) *Sancto Feoko* 1291 Tax]. '(The church of) St. Feoca or of St. Feoc.' There is some doubt as to whether the saint was a man or a woman. St. Fiac is a Breton saint. Förster, *Reliquienkultus*, pp. 105 ff., would derive the saint's name *Feoc* from a Cornish name corresponding to OW *Maioc*.

OScænd **feria** 'ferry'. See FERRIBY, FERRY-BRIDGE, KINNARD'S FERRY.

Fern Down Do [*Fyrne* 1321, *Ferne* 1358 FF]. OE *fiergen* 'wooded hill'. Cf. FERRY-HILL.

Fernham Brk [*Fernham* 821 BCS 366, Hy 1 Abingd]. 'HĀM among ferns.'

Fernhurst Sx [*Fernherst* c 1200 PNSx]. 'Ferny HYRST.'

Fernilee Db [*Ferneleia* a 1108 Mon]. 'Ferny LĒAH.'

Ferrensby YW [*Feresbi* DB, *Feringeby* 1239 FF, *Feringesby* 1316 FA]. This may be 'the BY of a man or men from the Faroe Islands'. ON *færeyingr* means 'inhabitant of the Faroe Islands'.

Ferriby, South, Li [*Ferebi* DB, c 1115 LiS, *Suthferebi* c 1130 BM], **North** F~ YE [*Ferebi* DB, 1190 P, *Feribi* c 1160 YCh 1895]. 'BY at the ferry.' The places are opposite each other on the Humber.

Ferring, East & West, Sx [*Ferring* 765 BCS 198, *Feringes* DB, *Ferringis* 1173 P]. Perhaps identical with FEERING.

Ferrybridge YW [*Ferie* DB, *Pons Ferie* 1226 FF, *Feribrige* 1393 AD]. 'Bridge by the ferry.' F~ is on the Aire near Ponte-fract.

Ferryhill Du [(æt) *Feregenne* 10 BCS 1256, *Ferie* Hy 2 FPD, *Ferigchan* 1354 Newcastle]. OE *fiergen*, *firgen* 'hill, wooded hill', identical with Goth *fairguni* 'hill'. Cf. FERN DOWN Do.

Fersfield Nf [*Fersafeld* c 1035 Wills, *Ferseuella* DB, *Fersfelde* 1212 Fees]. OE *fyrs-feld* 'furze-covered FELD'.

Fetcham Sr [*Fecham* 964–95 BCS 1132, *Feceham* DB, *Fecham* 1242 Fees, *Fecchenham* 1253 Ch]. Apparently OE *Feččan hām*, where **Fečča* would seem to be a derivative of *Facca* in FACCOMBE &c.

Fewcott O. See FALCUTT.

Fewston YW [*Fostune* DB, *Foteston* Fount, *Fosceton* 1280 Ch]. '*Fōt*'s TŪN.' The first el. may be OE *Fōt* (in *Fótes eige* 969 BCS 1229) or rather OScand *Fótr*. Cf. FOSTON. In Fewston *ō* retained its length.

Fiddington Gl [*Fittingtun* 1004 Wills, *Fitentone* DB, *Fitintona* 1220 Fees], F~ So [*Fitintone* DB, *-ton* 1236 Fees, 1243 Ass, *Fidington* 1304 Ch]. 'The TŪN of *Fita*'s people.' OE **Fita* corresponds to OHG *Fizo*. Cf. also FITZ.

Fiddleford Do [*Fitelford* 1243 Ass, 1315 FF]. '*Fitela*'s ford.' Cf. FITTLETON.

Field St [*Felda* 1130 P]. OE FELD.

OE **fiergen**, **firgen** 'wooded hill'. See FERN DOWN, FERRYHILL.

Fifehead Magdalen Do [*Fifhide* DB, *Fif-hidam* 1154 (1318) Ch, *Fyfyde Magdaleyne* 1408 Ep], F~ **Neville & St. Quintin** Do

[*Fifhide* DB, *Fifhid*' 1205 Cur, *Vyfhyde Nevyle* 1303 FA, *Fifhide Seynt Quyntyn* 1323 FF], **Fīfield** O nr Stow on the Wold [*Fifhide* DB, *-hid* 1220, 1236 Fees], F~ O nr Wallingford [*Fifide* 1316, *Fyfhide* 1346 FA], F~ W [*Fifide* 1230 Pat], F~ **Bavant** W [*Fifhide* DB, *Fiffide Escudemor* 1267 Ch]. OE *fīf hīde* 'five hides', 'an estate of five hides', the normal holding of a thegn.

F~ **Bavant** was held by Roger de Bavent in 1316 (FA). Cf. EASTON BAVENTS.—F~ **Magdalen** from the dedication of the church.—F~ **Neville** was held by William de Nevill in 1236 (Fees). Neville from NÉVILLE or NEUVILLE in Normandy.—Herebertus de Sancto Quintino in F~ **St. Quintin** is mentioned in 1205 (Cur). St. Quintin from ST. QUENTIN in Normandy or some other place with this name.

Figheldean (fīl-) W [*Fisgledene* DB, *Fugelden* 1203 Cur, *Fighelden* 1226 FF]. '*Fygla*'s valley.' *Fygla is a derivative of *Fugol* and is found in several pl. ns.

Filby Nf [*Filebey* DB, *Filebi* 1165 ff., *-bia* 1179 P]. The first el. is no doubt a pers. n., possibly ON, ODan *Fili*, which is not well evidenced, however.

Filey YE [*Fiuelac* DB, *Fivelai* 1125–30 YCh 1135, *Fifle* 1148 ib. 179, *-lea* 1195 P]. Apparently 'the five *lēah*'s or clearings'. Cf. SIXHILLS.

Filgrave Bk [*Filegrave* 1241 Ep, *Fillegrave* 1242 Fees]. '*Fygla*'s grave (pit) or grove'. Cf. FIGHELDEAN.

OE **filiþe, fileþe** occurs once in a gloss, where it renders Lat *foenam* 'hay', and several times in compound pl. ns., as *Filiðleage* 778 BCS 225 (W), *Fileþleage ford* 958 ib. 1027 (D), *Fileðcumb* 961 ib. 1067 (W), *Filedhamm* 956 ib. 923 (So). In one case it occurs alone in a charter: *up on fileþa* 943 BCS 780. If *fileþe* is a neuter noun, *fileþa* is the acc. plur., with *-a* for original *-u*. In this case the meaning of the word cannot be 'hay', but it may be 'hayfield'. FELTHAM So appears as *Fælet-, Fylethamm* 882 BCS 550. The form *Fælet-* indicates that *filiþe* goes back to an earlier form with *ie* from *ea*. If so, an *h* or a ჳ must have disappeared after *l*. The word may then be derived from OE *fealg* 'newly cultivated land'. The base would be **falჳiþia-*, where ჳ disappeared before *i*. *Filiþe* would mean something like 'hay growing on a fallow'. In non-Saxon dialects the form would be *feliþe or *fæliþe*. All the OE examples of the word are in West Saxon texts. The word is the first el. of FELTHAM So, some FAWLEYS, FILLEIGH, FELTON So, FILTON.

Filkins O [*Filching* 1174, *Filechinge* 1269 Eynsham, *Filking* 1185 TpR, *Filekinge* 1268 Val]. Perhaps 'the people of *Filica*' (in *Filican slæd* BCS 1093).

Filleigh D [*Filelei* DB, *Fillingeleg* 1199 P, *Fileleghe* 1281 Ep]. OE *filiþ-lēah* 'clearing where hay was got'. Cf. FILIÞE.

Fillingham Li [*Figelingeham* DB, *Figlinga-ham* c 1115 LiS, *Fugelingam, Figelingham* 1202 Ass]. 'The HĀM *of Fygla*'s people.' Cf. FIGHELDEAN.

Fillongley Wa [*Filunge-, Filingeli* DB, *Filungele(e), Filigele* 1206 Cur, *Filungeleg* 1236 Fees]. Possibly 'the LĒAH *of Fygla*'s people'. But the persistent *-u-* is remarkable, and it may be the name really consists of an original Longley (OE *Longan-lēa*) with a distinctive addition such as OE *fīn* 'heap of wood'. In West Midland dialects *long* often appears as *lung*.

Filton Gl [*Filton* 1187 P, 1220 Fees]. 'Hay farm.' See FILIÞE.

Fimber YE [*Fym(m)ara* 1121–37 YCh 456, 460, *Fimmare* 1208, *Fimmer* 1222 FF]. Perhaps OE *fīnmere* 'lake by a heap of wood'. See FĪN.

OE **fīn** 'heap (of wood)', also in *līm-, wudufīn*, is found in some names, e.g. FINDERN, FINDON, FYNHAM. Difficult to distinguish from OE *fīna* 'woodpecker'. See FINBOROUGH, FINMERE, FINSTOCK.

Finborough, Great & Little, Sf [*Fineberga* DB, *Parva Fineberg* 1226–8 Fees, F~ *Magna* 1254 Val], **Finburgh** Wa in Stoneleigh [*Fineberg* 1237 Cl]. 'Woodpecker's hill.' See FĪN.

Finchale (-k-) Du [?(æt) *Pincanheale* 788 ASC (E), *Finchale* c 1100 Finchale, *Finchhala* 1186 P, *Finkhal* 1230 Ep]. 'Haugh frequented by finches.' Cf. FINKLEY. If *Pincanheale* belongs here, the first el. may originally have been an OE *pinca*, the source of dial. *pink* 'chaffinch' and perhaps found in *pincanhamm* BCS 665.

Fincham Nf [*P(h)incham* DB, *Fincham* c 1095 Bury, *Fincheham* c 1150 Crawf]. 'HĀM frequented by finches.'

Finchampstead Brk [(æt) *Heamstede* 1103 ASC (E), *Finchamestede* DB, *Finchamstæde* 1098 ASC (E), *Finchemsted* 1220 Fees]. 'Homestead frequented by finches.'

Finchingfield Ess [*Fincingefelda* DB, *Finchingefeld* 1177, 1193, *Finchelesfeld* 1190, *Finchesfeld* 1194 P, *Finchefeld* 1203 Cur]. 'The FELD *of Finc* or his people.' *Finc* occurs as a byname in *Godric Finc* c 1050 KCD 923, and as a pers. n. in *Finces stapol* (*stapel*) 956, c 975 BCS 982, 1319.

Finchley Mx [*Finchelee, -leya* c 1210 St Pauls, *Fynchesl*' 1243 FF, *Finchesle* 1291 Tax]. 'Finch LĒAH.'

Findern Db [*Findre* DB, *Findena* 1188 P, *Finderne* 1204 Cur, 1242 Fees]. OE *fīn-renn* 'house for wood', the first el. being OE *fīn* 'heap of wood', the second OE *renn* 'house', an early side-form of ÆRN. Cf. CHARD, DINDER.

Findon Sx [*Fintona* 1073 Fr, *-tune, -dune* DB, *Findon* 1166 P, 1252 Ch]. 'Hill with a heap of wood.' Cf. FĪN.

Finedon (-ĭnd-) Np [*Tingdene* DB, *Thingdene* 12 NS, *-den* 1230 P]. 'Valley where things were held.' See ÞING. The change of *Þ-* to *F-* is late.

Fineshade Np [*Finesheved* 1227 Ch, *Finnesheved* 1234 Cl, 1254 Val]. '*Finn's* hill.' See HĒAFOD.

Fingest (-nj-) Bk [*Tingeherst* 1163 RA, 1163 (1329) Ch]. 'Thing hill.' See ÞING.

Finghall YN [*Finegala* DB, *Finyngale* 1157 PNNR, *Fingala* 1157 YCh 354]. 'The HALH of *Finn's* or *Fina's* people.' *Fina* may be a nickname from OE *fīna* 'woodpecker'.

Fingle Bridge D. Fingle is a stream-name [*Fengel* 938 BCS 724].

Finglesham K [*Đenglesham* c 832 BCS 403, *Fenglesham* 1072 BM, 1230 P]. 'The prince's HĀM.' OE has *fengel* and *þengel* 'prince'.

Fingringhoe Ess [(æt) *Fingringaho* c 995 BCS 1288 f., *Fingringho* 1202 Fr, 1254 Val], **Fingrith** Ess in Blackmore [*Phingeria* DB, *Fingrithe* 1203 FF, *Fingrith* 1212 Fees]. At first sight it would seem these two names have as first el. a pers. n. **Finger* or **Fingra*, a nickname that might quite well have been formed from *finger*. But Fingrith has as second el. OE *rīþe* 'brook'. The place is on or near a small stream. This may have been called *finger-rīþe* because of its smallness. Fingringhoe is on Roman River. This may have been *Finger-ēa*, and *Fingringas* may be 'the people on Roman River'. Alternatively, as suggested in PNEss, the *Fingringas* may have been named from the spur of land near the place, the supposition being that this spur was known as *Finger*. See HŌH.

Finkley Ha [*Finkel'*, *Finchel'* 1233 ff. Cl, *Fynkeleye* 1276 ib.]. 'Finch LĒAH.' Cf. FINCHALE.

Finmere O [*Finemere* DB, 1207 Cur, 1230 P, *Vinemere* 1237 Cl]. 'Mere frequented by woodpeckers' (cf. FINBOROUGH) or identical with FENNYMERE.

Finningham Sf [*Finingaham* DB, *Finegeham* c 1095 Bury, *Finingeham* Hy 2 (1268) Ch, 1191 P]. F~ is not very far from FINBOROUGH and possibly *Finingas* may be a derivative with ellipsis of that name meaning 'the Finborough people'. Or *Finingas* may be '*Fina's* people'. Cf. FINGHALL.

Finningley Nt [*Feniglei* DB, *Feningelay* 1176 P, *Finigleya* 1229 Ep]. Probably OE *fynig-lēah* with *fynig* 'a moist, marshy place' as first el. Or OE *fynige lēah* 'mouldy LĒAH' (cf. FENNYMERE). Finningley is in a low situation.

Finsbury Mx [*Finesbury* 1254 Val, *-bur'* 1275 RH]. '*Finn's* manor.'

Finsthwaite La [*Fynnesthwayt* 1336 FC]. '*Finn's* clearing.'

Finstock O [*Finestoches* c 1160, *Fines-stokes* c 1200 Eynsham]. 'STOC frequented by woodpeckers.' See FĪN.

Firbank We [*Frithebenk* 1230 Cl, *Frethebank*, *Frebanc* c 1240 CC]. 'Hill in a *frith* or woodland.' Cf. BANKE, FYRHþ. Firbank Fell reaches 1,040 ft.

Firbeck YW [*Fritebec* 1190, *Fridebech* 1197 f. P, *Frithebek* 1276 RH]. 'Beck in a *frith* or woodland.' Cf. FYRHþ.

Firby YE [*Friebia* DB, *Fridebi* 1202 FF], F~ YN [*Fredebi* DB, *Frytheby* c 1180 YCh 635]. '*Frithi's* BY.' First el. ODan *Frithi*.

Firle (fŭrel) Sx [*Firolaland* c 790 BCS 262, *Ferle* DB, *Fierles* 1201, *Estfirle* 1236, *Westferles* 1256 FF]. A derivative of a word for 'oak' or 'beech' found in OHG *fereh-eih*, Langobardic *fereha* 'oak'. OE **fierol* would be an adjective for 'covered with oaks', later used substantivally.

Firsby, East & West, Li [*Frisebi* DB, *Frisabi* c 1115 LiS, *Frisebi* 1212 Fees], **Firsby** Li nr Spilsby [*Frisebi* 1202 Ass, 1212 Fees, *-by* 1254 Val]. 'The Frisians' BY.'

Fishbourne Sx [*Fiseborne* DB, *Fissaburna* c 1090 Fr], **Fishburn** Du [*Fisseburne* c 1190 Godric, 1208–10 Fees]. 'Fish stream, stream with plenty of fish.'

Fisherton Anger W [*Fiscartone* DB, *Fissherton Ancher* 1412 FA], F~ **Delamere** W [*Fisertone* DB, *Fisserton* 1200 Cur, *Fyssherton Dalamare* 1412 FS]. 'Fishermen's TŪN.'

Richard son of Aucher held F~ **Anger** in 1242 (Fees). Aucher or Alcher the huntsman is mentioned c 1166 in Essex. *Aucher* was later misread as *Ancher*.—F~ **Delamere** was named from a local family. LA MARE is a common pl. n. in France.

Fisherwick St [*Fiscerwic* 1167 P, *Fischerewich* 1176 P]. 'Fishermen's WĪC.'

Fishlake YW [*Fiscelac* DB, *Fiskelak* 1230 P]. OE *fisc-lacu* 'fish stream'.

Fishley Nf [*Fiscele* DB, c 1130 Holme, *Fischel'* 1242 Fees]. Hardly 'LĒAH where fish were caught'. First el. perhaps an OE **fisca* 'fisherman' corresponding to Goth *fiskja*.

Fishmere Li [*Fiskermere* 1188, 1190 P]. 'Fishermen's mere.' Cf. FISKERTON.

Fishtoft Li [*Toft* DB, 1212 Fees]. See TOFT. *Fish* is possibly a family name.

Fishwick La [*Fiscuic* DB, *Fiskwic* 1202 FF]. 'WĪC where fish was sold.' Cf. WĪC.

Fiskerton Li [*Fiskertuna* 1060 KCD 808, *Fiscartone* DB], F~ Nt [*Fircertune* 958 YCh 2, *Fiscartune* DB]. OE *Fiscera tūn* 'fishermen's TŪN', Scandinavianized to *Fiskerton*.

Fitling YE [*Fidlinge*, *Fitlinge* DB, *Fiteling* c 1150 YCh 1345, *Fitling* 1194 P, *Fitlinges* 1207 Cur]. '*Fitela's* people.' Cf. FITTLETON.

Fittleton W [*Viteletone* DB, *Fitletone* 1212 RBE, *Fitelton* 1236 Cl], **Fittleworth** Sx [*Fitelwurða* 1168, *Fintleswrda* 1198 f. P, *Fitelewrth* 1199 FF]. '*Fitela*'s TŪN and worþ.' *Fitela* (= OHG *Fizzilo*) is found in *Fitelan slad* 934 BCS 705 (in boundaries of Enford, which is near Fittleton). No doubt the same *Fitela* gave their names to Fittleton and *Fitelan slæd*.

Fitz Sa [*Witesot* DB, *Fittesho* 1194 P, *Fittes* 1285 Ch]. This name shows the same curious shortening as EDGBOLD and EDGE-MOND. Second el. OE HŌH 'spur of land'. The first might be a pers. n. related to the hypothetical *Fita* in FIDDINGTON. Or it might be OE *fitt* n. 'fight'.

Fitzhead So [*Fifhida* 1065 Wells, *Fifida* 1178 ib., *Fyfhide* 1330 Ep], **Fivehead** So [*Fifhide* DB, 1225 Ass]. Identical with FIFEHEAD. The change to *Fitzhead* is remarkable.

Fixby YW [*Fechesbi* DB, *Fekesby* 1274 Wakef]. First el. perhaps as in FEIZOR.

Fladbury Wo [*Fledanburg* 692, (on) *Flaedan-byrg* 778–81 BCS 76, 238, *Fledebirie* DB]. '*Flǽde*'s BURG.' **Flǽde* (fem.) is a short form of names in -*flǽd*, as *Æþelflǽd*.

Flagg Db [*Flagun* DB, *Flagge* 1284 Derby, *Flagh* 1315 Ipm]. ME *flag* 'a sod, turf', which may be of Scand origin (cf. Icel *flag* 'spot where a turf has been cut away' &c.). A place where peat was got may be referred to.

Flamborough YE [*Flaneburg* DB, *Fleine-burhc* c 1130 BM, *Flamesburgh* c 1190 YCh 917]. '*Flein*'s BURG.' ON *Fleinn* is a known pers. name.

Flamstead Hrt [*Fleamstede* 1006 E, *Flame-stede* DB, *Fleme-*, *Flamested(e)* 1206 Cur]. OE *flēam-stede* 'refuge, sanctuary'. OE *flēam* means 'flight'.

Flamston W [*Flambertone* 1282 Ep, *Flam-bardeston* 1354 FF]. '*Flambard*'s TŪN.' *Flambard* is an OFr name of German origin (OG *Flanbert*). Walter, son of Robert Flambard held *Flambardeston* in 1227 (PNW(S)).

Flansham Sx [*Flennesham* 1221 FF, *Fleme-*, *Flomesham* 1279 Ass]. Unexplained.

Flasby YW [*Flatebi* DB, *Flatesbi*, *Flasceby* c 1160 FC]. OScand *Flats-býr*, the first el. being *Flat*, a byname from *flatr* 'flat'.

Flashbrook St [*Fletesbroc* DB, *Flocesbroc* 1242 Fees, *Flotesbroc* 13 Ronton]. First el. OE *flēot* 'stream'. *Flēot* may be the original name, to which was added an explanatory BRŌC.

Flass Hall Du [*Flaskes* 1313 RPD, *Le Flassh* 1382 Hatfield]. ME *flasshe*, *flask* 'pool'.

Flat Holme So, an island [*Flotholm* 1375 Misc, *Floteholmes* 1387 Pat]. Probably 'the island of the fleet' (OScand *floti*, OE *flota*).

The name refers to the fact that Viking fleets used the island as a base. A band of Vikings were starved out of Flat Holme in 918 (ASC). The old name was (æt) *Bradan Relice* 918 ASC, (into) *Bradan Reolice* 1067 ib. (D). *Relic* is OIr *reilic* 'cemetery' (from Lat *reliquiæ* 'relics').

Flaunden Hrt [*Flawenden* 13 AD, *Flauenden* 1248 Ep, *Flaunden* 1250 Cl]. The first el. may be the OE *flage* 'slab' postulated for FLAWFORTH, though perhaps rather in a sense such as 'ledge'.

Flawborough Nt [*Flodberge* DB, *Flouberewe* 1252 Cl, *Flaubergh* 1316 FA]. The first el. is very likely OE *flōh* 'fragment, a bit of stone', identical with OHG *fluoh* 'rock', ON *flō* 'layer, stratum'. The name may mean 'hill with slabs of stone' or 'flat hill'.

Flawforth Nt [*Flage-*, *Flaggeford* 1200 Cur]. 'Ford with flagstones.' *Flag* (or *flaw*) is probably an Engl word cognate with ON *flaga* 'slab', i.e. an OE *flage*.

Flawith YE [*Flagdthewat* c 1180 YCh 838, *Flathwath* c 1265, 1287 PNER], **F~** YN [*Flathwath* c 1190 YCh 796, *Flathewath* 1260 Ass, 1292 Misc]. The two Flawiths are evidently identical in origin. The second el. is OScand *vað* 'ford'. The first is very likely ON *flagð* 'a female troll, witch', Icel *flagð* the same, also 'a scold'. F~ may then be compared with names like SHOBROOKE, SHUCKBURGH, Icel *Tröllahals*, *Tröllaskógr* (in Landnáma).

Flaxby YW [*Flatesbi* DB, *Flaceby* c 1175 YCh 424]. Identical with FLASBY.

Flaxley Gl [*Flaxleg* 1160, -*lea* 1163, *Flexe-lega* 1179 P], **F~** YW [(on) *Fleaxlege* c 1030 YCh 7], **Flaxton** YN [*Flaxtune* DB, *Flax-ton* 1202, 1226 FF, 1228 Ep]. 'Clearing and TŪN where flax was grown.'

Fleckney Le [*Flechenie* DB, *Flekeneye* 1176 FF, *Fleckeneya* 1230 P], **Flecknoe** Wa [*Flechenho* DB, *Fleckeho* 1236 Fees]. If the first el. is not a pers. n. **Fleca* or the like, related to the base of FLETCHING, it may be an OE **fleca* 'hurdle', the source of *fleke*, a side-form of *flake* 'hurdle' (found from the 13th cent.). In OED it is suggested that *flake* (*fleke*) is a Scand loan-word.

Fledborough Nt [*Flatburche* 1060–6 KCD 818, *Fladburh* c 1080 Eynsham, *Fladeburg* DB, *Flatburch* 1090 RA, *Fletburg* 1242 Fees]. In spite of the early *a*-forms OE *Flēot-burg* 'BURG on a stream'.

Fleet Do [*Flete* DB, 1212 Fees], **F~** Ha [*Le Flete* 1506 Crondal], **F~** K [*Fleote* 798 BCS 291, *Fletes* DB], **F~** Li [*Fleot* DB, *Flet* DB, 1165 P], **F~** R Mx [*Flieta* 1159 TpR, *Flete* 1309 Pat], **F~ Prison** &c. Mx [*gaiola de ponte de Fliete* 1197 P]. OE *flēot* 'a stream, a creek' &c. Fleet Do is at **East** and **West Fleet**, a long narrow channel separated from the Channel by Chesil Bank.

Fleetham Nb [*Fletham* c 1180 FPD, 1254 Val]. 'HĀM by a fleet or stream.' Cf. KIRKBY FLEETHAM.

Fleetwood La. A late name. The town was named from Sir Peter Fleetwood (1836).

Flegg Nf [*Flec* (regio) 1014 StEdm, *Flec West, East Hundred de Flec, Eastflec* DB, *Flec* 1200 Cur, *Fleg* c 1155 Holme, 1193 P, *Fleeg* 1196 Cur, *Fleg, Flegg* 1254 Val]. Flegg is an old name of the low-lying district NW. of Yarmouth. The name may be derived from Dan *flæg*, Swed *flägg* 'flags or other water-plants'. Dan *flæg* is also used of a marsh where flags grow.

Flempton Sf [*Flemingtuna* DB, *Flameton* 1195 P, *Fleminton* 1197 FF]. Probably 'the TŪN of the Flemings'. The word *Fleming* occurs in *Flemingaland* 1075 ASC (D).

Flemworth Sf [*Flemewrda* 1164 BM, *Flimwurd* 1199 FF, *Flymworthe* 1237, *Flemewrth* 1271 FF]. First el. perhaps OE *fl(ī)ēma* 'fugitive, outlaw'.

OE **flēot** 'an estuary, an arm of the sea, a stream' is common in names of streams &c., some of which have become names of places. The usual meanings in pl. ns. are 'an estuary, a tidal stream, a creek or inlet, especially one in a tidal river'. The last meaning is probably that of names such as EBBSFLEET, FAX-, SWINE-, YOKEFLEET. The exact meaning is often doubtful.

Fletchamstead Wa [*Flichehamstede* 1189 TpR, *Flichamsted* 1200 Cur, *Flechampstede* 1288 Misc]. Perhaps 'homestead where flitches of bacon could be had'. *Flitch* is OE *flicce*. The *e*-forms do not go quite well with that base, but might be due to influence from OE *flǣsc* (*flǣc*) 'flesh'. Cf. FLITCHAM.

Fletching Sx [*Flescinge, -s* DB, *Flechinges* 1249 FF]. A derivative with the suffix *-ingas* from some pers. n., perhaps related to OG *Flaco*.

Fletton Hu [*Flettuna* 972 BCS (1280), 1125–8 LN, *Fletun* DB]. 'TŪN on a FLĒOT', i.e. the Nene.

Flimby Cu [*Flemingby* c 1174 Holme C, *Flemingeby* 1201 Ch, 1279 FF]. 'The BY of the Flemings.'

Flintham Nt [*Flintham* DB, 1185 P], **Flinton** YE [*Flintone, Flentun* DB, *Flinton* 1163–5 YCh 1347, 1260 Ass]. 'HĀM and TŪN where flints were found.'

Flitcham Nf [*Flicham* DB, 1203 Ass, 1227 Ch, *Flitcham* 1207 FF]. 'HĀM where flitches of bacon were produced.'

Flitton Bd [*Flictham* DB, *Flitte* 1166 P, *Flete* 1188 P, *Flitten* 1276 RH], **Flitwick** Bd [*Flicteuuiche* DB, *Fletwyk* 1242 Fees]. Flitton is the dat. plur. of OE FLĒOT 'fleet, stream' and Flitwick is 'WĪC by the streams'. The places are on the river Flitt and its tributaries. **Flitt** is a back-formation from

Flitton. OE *flēot* appears here in a dialectal form *flīet*, whence *flit* &c.

Flixborough Li [*Flichesburg* DB, *-burc* c 1115 LiS, *Flickesburc* 1202 Ass], **Flixton** La [*Flixton* 1177 ff. P, 1253 FF], F~ Sf nr Bungay [*Flixtuna* DB, *Flixton* 1254 Val], F~ Sf nr Lowestoft [*Flixtuna* DB, *Flixton* 1254 Val], F~ YE [*Fleustone* DB, *Flixtona* c 1170 YCh 1246, *Flixton* 1208 FF, *Flykeston* 1260 Ass]. '*Flīk*'s BURG and TŪN.' *Flīk* is ODan *Flic, Fliic* pers. n. Flixton represents ODan *Flīks-tūn*, with the OScand gen. form *Flīks* corresponding to an OE *Flīces*.

Flockthorpe Nf in Hardingham [*Flokethorp* DB, *Flochetorp* 1161, *Flochestorp* 1170 P], **Flockton** YW [*Flochetone* DB, *Floketon* 1201 FF]. '*Flōki*'s thorp and TŪN.' *Flōki* is an ON pers. n.

Flodden Nb [*Floddoun* 1517 f., *Flowdoun* 1521 Rot Scacc Scotland]. The name has not been met with in sources earlier than 1513, the year of the battle of Flodden. The name denotes Flodden Hill. In contemporary English sources the battle is referred to as the field of Branxton. The name Flodden contains OE *dūn* 'hill'. The first looks like OE *flōde* 'a channel, a stream', but possibly it is OE *flōh*; cf. FLAWBOROUGH.

OE **flōde** 'a channel', perhaps also 'an intermittent spring', is rare in pl. ns. Cf. INGLEWOOD Brk, CHESELADE, PRINCELET.

Flookburgh La [*Flokeburg* 1246 Ass]. '*Flōki*'s BURG.' Cf. FLOCKTHORPE.

Floore Np [*Flore* DB, 1190 P, *Flora* DB, 12 NS, 1156 P, 1220 Fees]. OE *flōr* 'floor, ground'. The exact sense is doubtful. OE *flōr* is used also in the sense 'threshing-floor', a sense quite possible here. MHG *vluor* means among other things 'a corn-field'. This sense may have occurred also in English.

Flordon Nf [*Florenduna* DB, *Florendone* 1254 Val, 13 Misc, *Flordone* 1291 Tax]. The first el. is OE *flōre* (gen. *flōran*), a derivative of *flōr*. *Flōre* occurs once in *upflōre* 'upper story' by the side of the common *upflōr*. Probably (to) *fagan floran* BCS 607 (see FAWLER) contains the dat. sing. of *flōre*, not the dat. plur. of *flōr*, and *Floraheafdo* 1069 JAA 39 is for *Floranheafdo*. The meaning of *flōre* is apparently much the same as that of *flōr*.

Flotmanby YE [*Flotemanebi* DB, 1205 FF, *Flotemanby* 1226 FF]. 'The BY of the Vikings.' OE *flotman* means 'Viking'. *Floteman* pers. n. occurs in DB.

Flotterton Nb [*Flotweyton* c 1160 YCh 1241, *Flotwaytun* 1236 Fees, *Flotewayton* 1256 Brinkburn]. First el. apparently an OE *flot-weg*, designating some kind of road, perhaps one partly made on floats, i.e. rafts or the like.

Flowton Sf [*Flochetuna* DB, *Floketon* 1201 Cur, *Floweton* 1503 BM]. '*Flōki*'s TŪN.' Cf. FLOCKTHORPE.

Flyford Flavell Wo [*Fleferth* 930, *Flæferth*, *Fleferð* 972 BCS 667, 1282, *æt Fleferht* 1002 KCD 1295, *Flavel* 1212 Fees, (wood in) *Flefrith* 1317 Pat]. Cf. GRAFTON FLYFORD. Flyford is an old forest name, whose second el. is a weakened form of OE FYRHþ 'frith'. There is some reason to believe that Flyford was also called *Ælflædetun* (BCS 1282), or rather a place in Flyford was so called. This suggests the possibility of the first el. of Flyford being a short form of *Ælflæd*, the original form being then *Flæde-fyrhþ* '*Flæd*'s woodland'. The OE forms are in transcripts, and *Fleferth* &c. may well be ME developments of *Flæde-fyrhþ*. *Flavell* is a Normanized form of Flyford, which was ultimately added to Flyford (Flavell) for distinction from Grafton Flyford.

Fobbing Ess [*Phobinge* DB, *Fobinges* 1125 Fr, *Fobbinges* Hy 2 (1227) Ch]. OE *Fobbingas* 'Fobba's people'. *Fobba*, a short form of *Folcheorht*, is found in *Fobban wyll* BCS 27, 863 and in FOVANT.

Fockbury. See FOLKINGTON.

Fockerby YW [*Fulcwardby* c 1170 YCh 487, *Folkardeby* 1242 FF, 1250 Fees]. '*Folk-varð*'s BY.' First el. ON *Folkvarðr*, ODan *Folcuard*.

Foddington So [*Fodindone, Fedintone* DB, *Fodindon* 1227 FF, 1243 Ass]. 'Hill used for grazing.' OE *fōding* or *fōdung* 'feeding, grazing' is not evidenced, but ME has *fōde* 'to feed', *fōdynge* 'feeding'.

Foggathorpe YE [*Fulcartorp* DB, *Folcware-thorp* 1157 YCh 354, *Folkerthorp* 1240 FF]. '*Folkvarð*'s thorp.' Cf. FOCKERBY.

OE **fola**, OScand **foli** 'foal'. See FOLLIFOOT, FOULRIDGE, FOWBERRY.

Foleshill Wa [*Focheshelle* DB, *Folkeshull* Hy 3 AD, *-hill* 1275 Ipm]. 'The hill of the people', perhaps because of the place being a meeting-place, or '*Folc*'s hill'. Cf. FOLKSWORTH.

Folke Do [*Folk* 1244 Ass, 1285 FA]. OE *folc* 'the people'. Cf. FREEFOLK.

Folkestone K [*Folcanstan* 696, c 833 BCS 91, 412, *æt Folcanstanæ* 824 ib. 378, *Stan* 993 ASC, *Fulchestan* DB]. '*Folca*'s stone.' **Folca* is a normal short form of names in *Folc-*.

Folkingham (-ŏk-) Li [*Folchinge-, Fulch-ingeham* DB, *Fulkingham* 1212 Fees, *Fuck-ingeham* 1218 Ass, *Folkingham* 1239 Ep]. 'The HĀM of *Folca*'s (*Fulca*'s) people.' Cf. FOLKESTONE.

Folkington (fōĭng-) Sx [*Fochintone* DB, *Fokintune* 1121 AD, *-ton* 1199 P, *Folkintone* c 1150 Fr]. 'The TŪN of *Folca*'s people.' Cf. FOLKESTONE. *Folca* was also called by

the pet name *Focca*. Cf. **Fockbury** Wo [*Fockebure, -bury* c 1200 &c. PNWo].

Folksworth Hu [*Folchesworde* DB, *Fulkes-wrthe* c 1155 Oxf, *Fucheswurd* 1180 P, *Fukesworth* 1207 Cur]. '*Folc*'s or *Fulc*'s WORþ.' **Folc* (*Fulc*) is a short form of names in *Folc-*.

Folkton YE [*Fulcheton* DB, *Folketun* c 1165 YCh 1250, *-ton* 1225 Ep, 1254 Ipm]. '*Folca*'s TŪN.' Cf. FOLKESTONE.

Follifoot YW [*Fulifet* 1167, *Folifeit* 1180 P, *Folifait* 1195 ff. P, 1206 FF, *Folifeit* 1204 FF], **Follithwaite** YW nr Wighill [*Folifayt* 1242 Fees, *Folyfayt* 1275 RH, 1285, 1303 FA, *Folitwait* 1316 FA]. The forms from P possibly belong to Follithwaite. The first el. is OE *fola* or ON *foli* 'foal'. The regular *-i-* or *-y-* may go back to a prefix *ge-*. The second el. may be OE *gefeoht* 'fight'. OE *folgefeoht* would mean 'horse-fight'. Horse-racing and horse-fighting, i.e. fights between horses, were common sports in ancient Scandinavia. Follifoot may have been a place where this kind of sport was carried on, and the name of the sport may have been transferred to the place.

Follingsby Du [*Foletebi* Hy 2, *-by* c 1180 FPD, *Folethebi* 1195 (1335) Ch], **Fonaby** Li [*Fuldenebi* DB, *Fulmedebia* 1177 P, *Felmetheby* 1226–8 Fees], **Fulletby** Li [*Fullobi* DB, *Fuledebi, Fuletebi* c 1115 LiS, *Fulletebi* Hy 2, *Fulotebi* 1163 DC, *Fulneteby* 1225 Ep], **Fulnetby** Li [*Fulnedebi* DB, *Fulnetebi* c 1115 LiS, *Fullethebi, Fulnathebi, Fulnotebi* 12 DC, *Fulnotesbi* 1200 Cur]. All these seem to have the same first el. No pers. n. is known that suits the case. The forms vary a good deal, but those with *ln* seem to be more original than those in *lm*, *ll*. Possibly the common el. is an OScand *full-nautr* 'one who has a full share'. No such word is known, but we may compare ON *iam-nautar* 'those who have an equal share'. The original form would be *Full-nautabȳr*. The early forms in *-e-* (*Fulneteby* &c.) might be due to influence from OE *genēat* 'companion', the equivalent of ON *nautr*.

Fŏnt R Nb [*Funt* c 1200 &c. Newminster, *Fount* 1208 Percy]. It is doubtful if Font can be identical with OE *funta*. It is probably a Brit river-name. Cf. next name.

Fonthill Bishop & Gifford W [*Funtgéall* 901, *Funtial* 901–24 BCS 590 f., *Fontel* DB, *Fontel Episcopi, Giffard* 1291 Tax]. The name is really that of **Fonthill Brook** [*Funtgeal, Funtal* 984 KCD 641]. The first el. is no doubt a Brit river-name identical with FONT. The second is Welsh *iâl* 'fertile upland region'. Cf. DEVERILL.

F~ **Bishop** belonged to the Bishop of Winchester from 901.—F~ **Gifford** was held by Berenger Gifard in 1086 (DB). Cf. ASHTON GIFFORD.

Fontley Ha [*Funtelei* DB, *Fonteleg* 1242 Fees]. 'LĒAH with a FUNTA or spring.'

Fŏntmell Do [*Funtemel* c 871, 932 BCS 531 f., 691, *Fontemale* DB]. Really the name of **Fontmell Brook** [*Funtamel* 704, *Funtemel*, *Funtmeales* (gen.) 939 BCS 107, 744]. 'Stream or spring by the bare hill.' First el. a Brit form of Lat *fontana* (cf. FUNTA), here perhaps in the sense 'stream'. The second is a name of **Fontmell Down**, derived from OCelt *mailo-* (Welsh *moel*) 'bare'. Welsh *moel* also means 'bare hill'.

Foolow Db [*La Foulowe* 1284, *Fuwelowe* 1338 Ipm]. 'Hill frequented by birds' (OE *fugol*). Cf. HLĀW.

Forcett YN [*Forset, Forsed* DB, *Forseta* 1157 YCh 354, *Fordseta* 1178 P]. 'Fold by a ford.' Cf. (GE)SET.

OE **ford** is very common in pl. ns., especially as a second el. As a first el. it is found e.g. in FORCETT, FORTON, FURTHO. As a second el. it sometimes takes the form *-forth*, which only appears in late sources. It is sometimes mixed up with WORÞ; cf. BROADWARD He. The first el. of names in *-ford* is often the name of an animal (as CRAN-, CRAW-, HART-, HORS-, OX-, SWINFORD, SHEFFORD) or of a person. Interesting types are BARFORD, HEYFORD. Compounds with *-ford* frequently form the first el. of pl. ns., as CLAVER-, DULVER-, HARVING-, MILVER-, WOTHERTON.

Ford He [*Forne* DB, *la Forda* 1127 AC, *Forda* 1242 Fees], **F~** Nb [*Forda* 1224 Pat], **F~** Sa [*Forde* DB, *Forda* 1161 P], **F~** So in Norton Fitzwarren [*Eford, Æford* DB, *Ford* c 1100 Montacute], **F~** So in Wiveliscombe [*Forda* 1065 Wells], **F~** Sx [*Fordes* c 1194 Fr], **Forde Abbey** Do [*Forda* 1158 P]. OE *ford* 'ford'. One Ford So is *E-*, *Æford* DB. This is OE *ēa-ford* 'ford over the river'.

Fordham Ca [*Fordham* c 975 ASCh, c 1080 ICC, *Fordeham* DB], **F~** Ess [*Forham* DB, *Fordham* 1181 P], **F~** Nf [*Forham, Fordham* DB, *Fordham* 1175–86 Holme]. 'HĀM by a ford.'

Fordingbridge Ha [*Forde* DB, 1242 Fees, *Fordingebrige* (hd) DB, *Fordingebrug* 1255 Ipm]. Originally *Ford*, later *Fordinga brycg* 'the bridge of the Ford people'.

Fordington Do [*Fortitone* DB, *Fortintun* 1156, *Fordintun* 1157 f. P, *Fordinget'* 1205 Cl]. 'The TŪN of the people by the ford.'

Fordington Li [*Fortintone* DB, *Forthintuna* c 1115 LiS, *Forthington* 1212 Fees, *Forþinton, Forþingeton* 13 BM; *Fordintun* c 1180 BM, *Fordington* 1230 Ep, 1254 Val]. Most likely identical with prec. name. If so, the *th*-forms are due to Scand influence. Alternatively the base might be a pers. n. **Forþ(a)*, a short form of names in *Forþ-*.

Fordley Sf [*Forle* DB, *Fordle* 1254 Val, 1265 Ch]. 'LĒAH by a ford.'

Fordon YE [*Fordun* DB, *-a* c 1150 YCh 1156]. OE *Ford-dūn* is perhaps not satisfactory. The place is at the foot of a long ridge along which runs an ancient road. The ridge may have been OE *Fōr-dūn* with *fōr* 'journey, way' or *fōr* 'pig' as first el. Cf. SWINDON.

Fordwich (-dĭch) K [*Fordeuuicum* 675, *Forduuic* 747 BCS 36, 173]. 'WĪC by a ford.'

Foreland, North, K [*Forland* 1326 Cl, *the Forland of Tenet* 1432 Pat]. *Foreland* 'cape, headland' is evidenced from 1580 (OED).

Foremark Db [*Fornewerche, -werk* 1242 Fees]. 'Old fort', from OScand *forn* 'old' and *verk* 'work, fort'.

Forest Hill O [*Fostel* DB, *Forsthulle* 1122 Fridesw, *-hell* 1192 P, *Forstella* 1164–6 Oxf]. 'Frost hill', i.e. 'hill often visited by frost'.

Forest Hill Sr. A late name containing the word *forest*.

Formby La [*Fornebei* DB, *Fornebia* 1177 P]. 'Old BY' (cf. FOREMARK) or '*Forni*'s BY'. *Forni* is a well-evidenced OScand name.

Forncett Nf [*Fornesseta* DB, *Fornesset* 1199 FF, *Fornesete* 1254 Val]. '*Forne*'s (GE)SET.' *Forne* from OScand *Forni* is found 970 BCS 1266.

Fornham All Saints, St. Genevieve & St. Martin Sf [*Fornham* 11 EHR 43, DB, c 1095 Bury, 1198 FF, *Genonefæforham* DB, *Geneuefes Fornham* c 1095 Bury, *Fornham Omnium Sanctorum, Sancte Genovefe, Sancti Martini* 1254 Val]. 'HĀM where trout were caught' or 'HĀM by the trout stream'. OE *forne* 'trout' corresponds to OHG *forhana*, OLG *forhna*. The Fornhams are on both sides of the LARK. It is possible that this river was once called *Fornēa* 'trout stream'.

Forrabury Co [*Forbyiri* 1291 Tax]. OE *foreburg* 'outwork'. F~ is nr Boscastle.

Forsbrook St [*Fotesbroc* DB, *-brock* 13 BM]. '*Fōt*'s brook.' Cf. FEWSTON, FOSTON.

OE **forsc** 'frog'. See FROSTENDEN, FROXFIELD.

Forscote So [*Fuscote* DB, *Foxcote* 12 Montacute, *Foxecote* 1243 Ass]. Cf. FOXCOTE.

Forston Do [*Fosardeston* 1236 Fees, *Forsardeston* 1285 FA]. '*Forsard*'s TŪN.' *Forsard* or *Fossard* is an OFr pers. n. and family name. William Forsard held the manor in 1285 (FA).

Forthampton Gl [*Forhelmentone, Fortemeltone* DB, *Fortelminton* 1167 P, *Forthelminton* 1220 Fees]. 'The TŪN of *Forþhelm*'s people.'

Fortherley Nb [*Falderle* 1208 Cur, *-leg* 1256 Ass]. 'The LĒAH of the sheep-folders.' *Folder*, OE **faldere*, is first found in OED in 1571, but as a surname in 1332 (Cu Subs).

Forton Ha nr Andover [*Forton* 1312 Ipm, 1316 FA], **F~** La [*Fortune* DB], **F~** Sa [*Fordune* DB, *Forton* 1246 Ch], **F~** St [*Forton* 1274 Ass]. 'TŪN by a ford.' F~

Ha is probably referred to as *Forde* 1046 KCD 1335 (Prof. O. Arngart).

Fosbury W in Tidcombe, by F~ **Camp** on a hill of 833 ft. [*Fostesberge* DB, *Forstebyri* 1268, *Forstesbery* 1270 PNW(S)], F~ W in West Overton [*Forstesbyria* Hy 2 (1270) Ch, *-beria* 1178 BM, *Forstebir'* 1225 Cl, *Forstesbur'* 1242 Fees]. The first el. has been derived by Tengstrand from an OE **forst*, a word related to OE *first* 'roof', here used of a hill. The double occurrence of the el. in the gen. form is an instance against this suggestion, and a personal designation would be preferable. OE *forwost* 'chief, chieftain', though found only in Northumbrian, would be suitable.

Foscote Np [*Foxcote* 1197 FF, 1200 Cur], F~ W nr Grittleton [(bi este) *foxcotone* 940 BCS 750], **Foscott** Bk [*Foxescote* DB, *Foxcota* 1167, 1192 P], F~ O [*Foxcote* DB]. See FOXCOTE.

Fosdyke Li [*Fotesdic* 1183, 1195 P, 1202 Ass]. '*Fōt*'s ditch.' *Fōt* is probably OScand *Fōtr* (gen. *Fōts*), originally a nickname.

Fosham YE [*Fos-, Fossham* DB, *Fosham* 1166 P]. 'HĀM by a *foss* or ditch.'

OE **foss** is not evidenced, but must have existed. The ultimate source is Lat *fossa*, but the proximate one is British (Welsh *ffos*, OBret, Co *fos* 'ditch, trench, dike'). Welsh *ffos* is common in pl. ns. OE *foss* no doubt meant 'a ditch', but very likely also 'a canalized stream'. See FOSHAM and the following names, also CATFOSS, FANGFOSS, WILBERFOSS.

Foss R YN [*Fossa* c 1210 YCh 321, *Fosse* 1220 For, 1226 Cl]. See prec. article.

Foss Dyke Li [*Fossedic* c 1155 DC, *Fossdic* 1281 Ass]. An ancient canal. The old name was no doubt *Foss* 'the ditch', to which was added an explanatory OE DĪC.

Fosse Way, an ancient road from Lincoln to near Axminster past Leicester, Stow on the Wold, Cirencester, Bath [*strata publica de Fosse* 956 BCS 922, *Fos* 978 KCD 620, *Fosse* 1235 Ep]. The road was named from the foss or ditch along it.

Foston Db [*Fostuna* 1158–9 P, *Foxtuna* 1159–60 ib., *Fostun* 13 BM], F~ Le [*Fostone* DB, *Fostuna* 1158, *Fosteston* 1169, 1196, *Fosceton* 1198 P], F~ Li [*Foxtun* DB, *Fotstun* 1212 Fees], F~ on the Wolds YE [*Fodstone* DB, *Fostun* 13 BM], F~ YN [*Fostun* DB, *Fotestun* 1231 Ass, *Fosceton* 1230 Ep]. '*Fōt*'s TŪN.' Probably OScand *Fōts-tūn*. First el. OScand *Fōtr*, gen. *Fōts*.

Fotherby Li [*Fodrebi* DB, *Fotrebi* c 1115 LiS, *Foterbi* Hy 2 DC, *-bia* 1212 Fees]. OScand *Fōtar-bȳr* '*Fōt*'s BY'. ON *Fótr* had the gen. *Fótar* by the side of *Fóts*.

Fotheringhay Np [*Frodigeia* 1075 PNNp, *Fodringeia* DB, 1163, 1166 P, 1202 Ass, *Foddringeia* 1176 BM, *Fodringeie* 1201 Cur]. The earliest form may suggest an original

Frōdinga-ēg 'the island of *Frōd(a)*'s people'. But the majority of forms indicate connexion with OE *fōdor* 'fodder'. The first el. is probably an OE **fōdring* 'foddering, grazing'. Cf. OE *fōdrere* 'pabulator' and FODDINGTON.

Fotherley St [*Fulwardlee* 12, *Fulverle* 13, *Fulfordleigh* 14 PNSt]. 'LĒAH by Fulford or the dirty ford.'

Foulden Nf [*Fugalduna* DB, *Fugeldona* 1166 P]. 'Hill frequented by birds.' See DŪN.

Fouldray Island. See FURNESS.

Foulness Ess [*Fughelnesse* a 1219 BM, *Fuelnesse* 1218 FF]. 'Headland frequented by birds.'

Foulness R YE [(on) *Fulanea* 959 BCS 1052, *Fulna* c 1175 Riev, *Fulne* 1350 Pat]. OE *fūle ēa* (dat. *fūlan ēa*) 'dirty river'.

Foulney (-ō-) **Island** La [*Fowley* 1537 PNLa, *Foulney* 1577 Saxton]. ON *Fugley* 'bird island'.

Foulridge (-ō-) La [*Folric* 1219 Ass, *-rigge* 1246 Ass]. 'Ridge where foals grazed.'

Foulsham Nf (fōlsam) [*Folsham* DB, *Folesham* 1156, 1168 ff. P, *Folisham* 1254 Val]. *Folesham* was a common surname in London in the 13th cent. It is not rarely written *Fulsham*, *Fullesham* and the like (SPML 55 f.). The name was undoubtedly taken from Foulsham, whose first el. was probably OE *Fugol*.

Fountains Abbey YW [*Fontes* 12 Fount M, *Sancta Maria de Fontibus* c 1132 Mon, Hy 2 BM]. The abbey was named from some springs found by the original settlers according to Matthew Paris. The later name is French.

Fourstones Nb [*Fourstanys* 1236 Fees, *Fourestanes* 1256 Ass]. 'Four stones', perhaps identical in meaning with FEATHERSTONE.

Fŏvant W [*Fobbanfuntan* [*boc*], *Fobbefunte* 901 BCS 588, æt *Fobbafuntan* 994 KCD 687, *Febefonte* DB, *Fofunte* 1242 Fees]. '*Fobba*'s spring.' Cf. FUNTA and FOBBING.

Fowberry Nb [*Folebir'* 1242 Fees, *-byr'* 1256 Ass]. 'BURG where foals were kept.' Cf. STŌD.

Fowey (foi) R Co [*Fawe* c 1200 Gervase, 1276 RH, *Fawy* 1241 Montacute]. Perhaps 'beech river', a derivative of the OCo equivalent of OBret *fau, fou*, Welsh *ffawydd* 'beeches' (from Lat *fagus*). *Faou* occurs as the name of a brook in Brittany. On the Fowey are **Fowey** port [*Fawy* 1255 FF, *Fawe* 1262 Ep] and **Fawton** [*Fauuitona*, *Fawintone* DB, *Fawyton* 1229 Fees].

Fowlmere Ca [*Fugelesmara, Fuglemære* DB, *Fulemere* c 1080 ICC]. 'Bird mere.'

Fownhope He [*Hope* DB, *Fauue Hope*, *Faghehop* 1242 Fees, *Fowehope* 1275 RH]. Originally HOP 'valley'. The addition was made for distinction from WOOLHOPE, which is 'Wulfgifu's Hope'. We expect Fownhope to contain a pers. n., but no such name as *Fāge* is known. (On) *faganstan* 980 KCD 627 may be 'the coloured stone'. Possibly Fownhope means 'the coloured Hope', the reference being to a painted building.

Foxcote Gl [*Fuscote* DB, *Foxcota* 1192 P], F~ Wa [*Foxcote* 1316 FA, 1370 AD], **Foxcott** Ha [*Fulsecote* DB, *Foxcote* 1146 Fr]. These should be compared with FORSCOTE and FOSCOTE, -COTT. It is improbable that these all mean 'fox-infested cottage'. Probably OE *foxcot* could be used in the sense 'foxes' burrow'.

Foxdenton La [*Denton* 1224 Ass, *Foxdenton* 1282 Ipm]. Cf. DENTON. *Fox-* perhaps because foxes were common at the place.

Foxearth Ess [*Focsearde* DB, *Foxherthe* 1198 FF, *Foxierth, Focsherde* 1202 FF, *Foxerde* 1249 BM]. Perhaps 'ploughed land (OE *erþ, earþ*) where foxes were common'. One would prefer 'fox's earth', but *earth* in this sense is late (16th cent. OED).

Foxhall Sf [*Foxehola* DB, *Foxhole* 1254 Val]. OE *fox-hol* 'foxes' burrow'.

Foxham W [*Foxham* 1065 KCD 817]. 'HĀM or HAMM where foxes were frequent.'

Foxholes La [*Foxholes* 1325 Ct], F~ YE [*Fox(o)hole* DB, *Foxholes* c 1130 YCh 1073]. Identical with FOXHALL.

Foxley Nf [*Foxle* DB, 1254 Val], F~ Np [*Uoxle* (hd) 1066–75 GeldR, *Foxeslea* DB], F~ W [*Foxelege* DB, *Foxleg* 1242 Fees]. OE *fox-lēah* 'fox wood'.

Foxt St [*Foxwiss* 1176 FF]. See FOXWIST.

Foxton Ca [*Foxtona* c 1080 ICC, *Foxetune* DB, *Foxton* 1202 FF], F~ Le [*Fox(es)tone* DB, *Foxton* 1159 P, 1254 Val], F~ YN [*Foustune* DB, *Foxtune* 1088 LVD]. 'TŪN where foxes abounded.' But it is quite possible that OE *Fox-dūn* is partly the source.

Foxton Du [*Foxedene* c 1170 Reg Dun, *Foxden* 1407 AD], F~ Nb [*Foxden* 1325 Ipm]. 'Fox valley.'

Foxwist Chs nr Whitegate [*Foxwist* 1260, 1286 Court]. 'Foxes' burrow.' OE *wist* 'dwelling', corresponding to ON *vist*, OHG *wist* 'staying, stay', is found in *hūswist* 'house'. Cf. also *nēahwist* 'neighbourhood'.

Foy He [*Lanntimoi, -tiuoi* c 1150 LL, *Foy* 1100 Glouc, *Foye* 1291 Tax]. *Lanntimoi* means 'the church of St. *Moi* or *Mwy*'. Cf. LANN. *Moi* corresponds to Bret *Moe* in *Moe-lan, Lanvoé*; *ti* is Welsh *dy* 'thy', placed before the name for hypocoristic purposes in accordance with Celtic usage. The Welsh form of Foy is said to be *Llandyffwy*.

Fradswell St [*Frodeswelle* DB, *-uella* 1155 BM]. '*Frōd*'s spring or stream.'

Fraisthorpe YE [*Frestintorp* DB, *Freistingthorp* c 1160 YCh 1361]. '*Freystein*'s thorp.' ON *Freysteinn*, OSw *Frøsten* is well evidenced.

Framfield Sx [*Framelle* (hd) DB, *Fremisfeld* 1223 PNSx, *Fremefeld* 1257 Sele]. '*Fremi*'s FELD.' Cf. FRENSHAM.

Framilode Gl [*Framilade* 1086, 1138 Glouc, *Fremelada* 1176 P]. OE *Frōm-gelād* 'passage (over the Severn) at (the mouth of) the Frome'.

Framingham Earl & Pigot Nf [*Framingaham* DB, *Framingeham* 1130, 1157, *Fremingham* 1198 P, *Frammingeham* 1157 BM, *Framelingham Comitis, Picot* 1254 Val]. Apparently 'the HĀM of *Fram*'s people'. *Fram* is found at least as the name of a moneyer, and is well evidenced on the Continent.

The Earl is the Earl of Norfolk.– Ralph Picot held F~ in 1235. Cf. ABINGTON PIGOTTS.

Framlingham Sf [*Framalingaham, Fram-(e)lingaham* DB, *Framillingeham* 1175 P], **Framlington, Longframlington** Nb [*Fremelintun* 1166 P, *Framlincton* 1196 FF, *Framelington* 1242 Fees]. 'The HĀM and TŪN of **Framela*'s people' or the like. *Framela* must be a derivative of *Fram*.

Frampton Do [*Frantone* DB, *Framton* 1157 Fr, *Fromton* 1212 Fees], F~ **Cotterell** Gl [*Frantone* DB, *Franton Ade Cotella* 1167 P, *Frompton* 1220, *Framton* 1236 Fees, *Frampton Cotell* 1257 Ch], F~ **Mansell** Gl [*Frantone* DB, *Frompton* 1212, *-a* 1220 Fees], F~ **on Severn** Gl [*Frantone* DB, *Frompton* 1220 Fees, *Fromton upon Severne* 1311 Ch]. 'TŪN on R FROME.' F~ Cotterell is on Frome (3). The other Gl Framptons are on Frome (4). The OE form was *Frōmtūn*. Here ō was shortened and for it was substituted the dark *a* or open *o* that developed from *a* before nasals.

F~ **Cotterell** was held by Adam Cotella in 1167 (P), by John Cotel in 1236 (Fees). Cotel is a Fr family name, perhaps identical with OFr *cotel*, a kind of trader.—F~ **Mansell** was held by John Maunsell in 1285 (FA). Mansel is a Fr byname, identical with MLat *mansellus* 'mansionarius' (Du Cange).

Frampton Gl nr Tewkesbury [*Freolintune* DB, *Freulintona* 1175 Winchc]. 'The TŪN of **Frēola*'s people.' *Frēola* is a normal derivative of *Frēo-* in pers. ns. and corresponds to OG *Frilo*.

Frampton Li [*Franetone, Frantune* DB, *Francton* 1183 BM, *Framtona* 1202 Ass, *-tun, Frantun, Francton* 1212 Fees, *Franketon* 1272 FF]. The different forms can be explained from an OE *Framecan tūn*. OE **Frameca* is a normal derivative of *Fram*.

Framsden Sf [*Framesdena* DB, *-den* 1213 FF, 1242 Fees, 1254 Val]. '*Fram*'s valley.' Cf. FRAMINGHAM.

Framwellgate Du [*Framwelgat* 1352 f. Durh Acc Rolls]. *Framwell* is a spring or was at the beginning of the 19th cent. The name may mean 'the strong spring' (OE *fram* 'vigorous'). *Gate* is OScand *gata* 'street'.

Franche Wo [*Frenesse* DB]. '*Frēa*'s ash-tree.' Cf. FRING.

Frankby Chs [*Frankeby* 1315 Ormerod]. '*Franki*'s BY.' ODan *Franki* (*Franco*), ON *Frakki* (from **Franki*) are known names.

Frankley Wo [*Franchelie* DB, *Frankeleg* 1212 Fees], **English & Welsh Frankton** Sa [*Franchetone* DB, *Fronchetone* 1166 RBE, *Franketon* 1242 Fees], **Frankton** Wa [*Francton* 1043 (1267) Ch, *Franchetone* DB, *Franketone* 1166 RBE, *-ton* Hy 2 (1235) Ch]. '*Franca*'s LĒAH and TŪN.'

Fransham Nf [*Frandesham, Frandeham* DB, *Franesham* 1198 FF, *Fransham* 1197 P, *Fransham Magna, Parva* 1254 Val]. It is unlikely that the first el. is ODan, OSw *Frændi*. More likely it is a name formed from OE *fræmde, fremede* 'strange'.

Frant Sx [(æt) *Fyrnþan* 956 BCS 961, *Fernet* 1177 BM, *Fernthe* 1296 Subs, *Frenthe* 1332 FF]. A derivative of OE *fearn* 'fern' meaning 'fern brake' or the like.

Frāting Ess [(at) *Fretinge* c 1060 Wills, *Fratinga, Fretinga* DB, *Fretenges* Hy 1 AD]. OE *Frætingas*, a derivative of OE *fræte* 'wanton, foul' or a noun corresponding to OHG *frāz* 'glutton'. The proximate base is no doubt a nickname *Frēt(a)*. Cf. FRET-TENHAM.

Fratton Ha [*Frodin(c)gtun* 982 Hyda, *Frodintone* DB, *Froditonia* c 1160 Oxf, *Frodyngton* 1307 Ch]. 'The TŪN of *Frōd(a)*'s people.'

Freckenham Sf [*Frekeham* 895 BCS 571, *Fracenham* 1071 Reg, *Frakenaham* DB, *Frechceham* 1161 BM, *Frekenham* 1225 FF]. '*Freca*'s HĀM.' *Freca* is found in *Frecanðorn* 904 BCS 610. Cf. also *Frecinghyr[s]t* 801, *Fraecinghyrst* 811 BCS 303, 339. *Freca* is derived from OE *frec, fræc* 'gluttonous, bold'.

Freckleton La [*Frecheltun* DB, *-a* c 1155 LaCh, *Frekenton* 1201 P]. First el. perhaps a pers. n. **Frecla* derived from *Freca*.

Freeby Le [*Fredebi* DB, *-bia* c 1125 LeS, *Fretheby* 1227 Ch, 1230 Cl]. '*Fræthi*'s BY.' First el. ODan *Fræthi, Frethi*.

Freefolk Ha [*Frigefolc* DB, *Frivolk* 1245 Ipm, *Frefork* 1271 Ch]. Perhaps 'the free people'. The name would refer to people who were freeholders. Or '*Frīg*'s people'. Cf. FROYLE, and see Introd. pp. xiv, xxx.

Freeford St [*Fraiforde* DB, *Freford* 1242 Fees, 1271 Ass]. 'Free ford', i.e. one that could be used without paying a toll.

Freemantle Ha [*Freitmantell* 1181, 1185 P, *Frigid Mantell* 1200 Obl, *Freidmantel* 1236 Ass]. The name is borrowed from France, where FROMENTEL is a common name. One

in Pas de Calais is *Frigidum Mantellum* 1233, *Froitmantel* 1279 &c. Freemantle was the name of a forest. The name means 'cold cloak'. It may be explained by the Swedish saying that 'the forest is the poor man's jacket'. The forest would at best be a cold jacket.

Freethorpe Nf [*Frietorp* DB, *Frethorp* 1254 Val]. '*Fræthi*'s thorp.' Cf. FREEBY.

Fremington D [*Framintone* DB, *Freminton* 1196 FF, 1206 Cur], **F~** YN [*Fremington* DB, *Fremmingeton* 1251 Ch]. 'The TŪN of *Fremi*'s people.' Cf. FRENSHAM.

Frenchay Gl [*Fromesham* (for *-shaw*) 1221 Ass, *Fromscawe* 1257 PNGl]. 'Wood on R FROME' (3). See SCAGA.

Frenchmoor Ha [*Freschemore* 1246 Ch, *Freynsemor* 1254, *la Frenshemore* 1309 Ipm]. 'The French moor.' The reason for the name is not apparent.

Frensham (-s-) Sr [*Fermesham* 10, *Fremesham* 967 BCS 1159, 1195, *Fermesham* 1190 ff. P]. '*Fremi*'s HĀM.' Cf. FRIMLEY, which is near Frensham. **Fremi* is a derivative of *Fram* (cf. FRAMINGHAM) or *fram* adj.

Frenze Nf [*Frense* DB, 1195 FF, 1254 Val, *Frenese* 1254 Val, (ripa de) *Frens*', (vill de) *Frenge* 1257 Ass]. Possibly a doublet of FRING, though with a palatal *g* and with substitution of *z* for *j* (*dzh*) owing to Norman influence.

Fresdon W [*Fersedon* 1263 Ipm], **Freseley** Wa [*Freselega* 1169 P, *-leg* 1236 Fees]. 'Furze-covered hill and clearing.' Cf. DŪN, LĒAH.

Freshford So [*Ferscesford* c 1000 Wills, *Ferseforð* 1001 KCD 706, *Fersshford* 1327 Subs]. 'Ford with fresh water.'

Freshwater Wt [*Frescewatre* DB, *Freschewatere* 1194 P]. Originally the name of the YARE. The name means 'river with fresh water'.

Fressingfield Sf [*Fessefelda* DB, *Frisingefeld* 1185 P, *Fresingefeld* 1197 P]. OE *fyrsen(e)feld* 'furze-covered FELD'. Cf. FERS-FIELD.

Freston Sf [*Fresantun* c 995 BCS 1289, *Fresetuna* DB]. 'The Frisian's TŪN.'

Fretherne Gl [*Fridorne* DB, *Frohorn* 1166 RBE, *Freorne* 1195 P, *Frethorn* 1236 Fees]. Second el. OE *þorn* 'thorn-bush'. The first may be OE *Frīg*, the name of the goddess. Cf. FROYLE.

Frettenham Nf [*Fretham* DB, *Freteham* 1174 P, *Fretenham* 1202 FF, 1267 Ch]. '*Frǣta*'s HĀM.' Cf. FRATING.

Frickley YW [*Frichelie* DB, *Frikeley* 1247 Ipm, *Frykelay* 1297 Subs]. '*Frica*'s LĒAH.' *Frica* occurs in *Fricanfenn* 904 BCS 610. OE *frec* has a side-form *fric*. Cf. FRECKEN-HAM.

Fridaythorpe YE [*Fridagstorp* DB, *Friday-thorp* c 1165 YCh 85, *Fridaithorp* 1221 FF]. '*Frigedæg*'s thorp.' OE *Frigedæg* is found in *Frigedæges tr[e]ow* BCS 1047 and *Frige-dæges east* ib. 197. Cf. OG *Frigdag*.

Friesthorpe Li [*Frisetorp* DB, *Frisatorp* c 1115 LiS, *Fristorp* Hy 2 (1329) Ch], **Fries-ton** Li nr Boston [*Fristune* DB, *-ton* 1198 FF, *Frestuna* Hy 2 DC, *Freston* 1254 Val], **F~** Li nr Caythorpe [*Fristun* DB, Hy 2 DC, *Freston* 1303 FA]. 'The thorp and TŪN of the Frisians.'

Frilford Brk [*Frileford* 965 BCS 1170, *Frieliford* DB, *Fridleford* 1220 Fees]. '*Fri-þela*'s ford.' *Friþela*, a derivative of *Friþu-* in many pers. ns., is found in *Friþela byrig* BCS 1002 (in boundaries of Hinksey not far from Frilford). Cf. Goth *Frithila*.

Frilsham Brk [*Frilesham* DB, *Fridlesham* 1174 P, 1220 Fees, *Fridelesham* 1190 P]. '*Friþel*'s HĀM.' *Friþel* is a strong side-form of *Friþela* in FRILFORD.

Frimley Sr [*Fremeley* 933, *Fremesleya* 967 BCS 697, 1195]. '*Fremi*'s LĒAH.' Cf. FRENSHAM, FREMINGTON.

Frindsbury K [*Freondesberia*, (of) *Frinon-desbyrig* 10 BCS 1321 f., *Frandesberie* DB, *Frendesberia* Hy 1 Reg Roff]. '*Frēomund*'s BURG.'

Fring Nf [*Frainghes*, *Frenge* DB, *Frainges* c 1140 BM, *Frenges* 1198 FF]. Perhaps '*Frēa*'s people'. *Frēa* would be a short form of names such as *Frēalāf*.

Fringford O [*Feringeford* DB, *Felinghefort* 1103 Fr, *Faringford* 1245 Ch, *Fyringford* 1266 BM, *Fringeford* 1205 Obl]. The first el. may be identical with FEERING. Or it might be OE *fēring* 'going, travelling'.

Frinsted K [*Fredenestede* DB, *Fridenastede* 11 DM, *Frethenestede* 1268 Ipm]. Cf. *Freoðene feld* 1062 Th. Both names contain an OE *freoþen*, a derivative from OE *friþian* 'to protect'. The meaning would be 'a fenced-in place, an enclosure'.

Frinton on Sea Ess [*Frie(n)tuna* DB, *Frien-ton* 1158 P, 1198 FF, *Frichtintone* 1212 RBE, *Frinton* 1199 Cur]. The first el. may be an OE **friþen*, identical with the *freoþen* found in FRINSTED.

Frisby by Galby Le [*Frisebi* DB, 1190 P], **F~ on the Wreak** Le [*Frisebie* DB, *-bia* c 1125 LeS, *Frisebi* 1200 Cur, c 1200 DC, *Friseby* 1254 Val]. 'The Frisians' BY.'

Friskney Li [*Frischenei* DB, *Freschena* c 1115 LiS, *Freschenei* c 1150 BM]. OE *Frescan ēa* 'river with fresh water'.

Frismarsh. See SUNK ISLAND.

Fristling Hall Ess in Margaretting [*Fest-inges* DB, *Ferstlinges* 1185 P, *Fristlingg* 1230 P]. Perhaps OE *fyrs-hlincas* 'furze-covered hills'.

Friston Sf [*Frisetuna* DB, *Freston* 1254 Val]. 'The TŪN of the Frisians.' Cf. FRIESTON.

Friston Sx [*Friston* 1200 Cur, 1243 FF, *Freston* 1262, *Fruston* 1347 FF]. This may be OE *Friges tūn* or the like. But perhaps rather OE *fyrs-dūn* 'furze-covered hill'. After *s* the *d* would be apt to become *t*.

Fritham Ha [*Friham* 1212 Fees, *Frytham* 1280, *Frythham* 1331 For (Gover)]. The name is probably analogous to FRITTON, and has as first el. OE *friþ*. *Ðruham* 749 BCS 180, *Tru(c)ham* DB, sometimes identified with F~, seem to belong to a lost place in Beaulieu.

Frithelstock D [*Fredelestoch*, *Fredeletestoc* DB, *Frithelakestoke* 1224, 1228 FF]. '**Friþu-lāc*'s STOC.'

Frithsden or **Friesden** Hrt [*le Fryth* (wood) 1285 Ch, *Frithesden* 1291, *-dene* 1293 Ch]. 'Dean or valley in woodland' (OE FYRHþ).

Frithville Li [*Le Frith* 1331 Ch]. The original name is OE *fyrhþ* 'woodland'. The addition *-ville* must be late.

Frittenden K [*Friðöingden* 804, 850 BCS 316, 459, *Frithindenne* 1243 StAug, *Freth-ingeden* 1279 Ep]. *Friðöingden* is mentioned in BCS 316 and 459 in connexion with *Friðesleah*, *-leas* (now FRIEZLEY). The latter is clearly '*Friþu*'s LĒAH', *Friþ(u)* being a short form of names in *Friþu-*, *-friþ*. Frit-tenden is the DENN or swine-pasture of the people of *Friþ(u)* or (with hypocoristic lengthening of *þ*) *Frippa*.

Fritton Nf [*Fride-*, *Fredetuna*, *Frithetuna* DB, *Fretone* 11 Holme, *Freton* 1199 FF], **F~** Sf [*Fridetuna* DB, *Freton* 1224 FF, *-e* 1254 Val]. *Friþetune* 1046 Wills is either Fritton Nf or Sf. Fritton very likely goes back to OE *friþ(u)-tūn* 'enclosed place, fenced-in TŪN'. Cf. OE *friþgeard* 'en-closure', G *Friedhof* 'churchyard', really 'enclosed place'.

Fritwell O [*Fert(e)welle* DB, *Frettewell* 1196 P, *Fretewell* 1203 Cur, *Fritewell* 1236 Fees]. 'Wishing-well.' First el. OE *freht*, *firht* 'augury'.

Frizinghall YW [*Frizinghale* 1265 Calver-ley, *Fresinghale* 1288 Ipm]. Perhaps 'furze-covered haugh'. Cf. FRESSINGFIELD, HALH.

Frizington Cu [*Frisingaton* 12 StB, *Fresin-ton* 1260 P]. Perhaps really OE *Frēsna tūn* 'the TŪN of the Frisians'. Or *Frēsa* may have been used as a pers. n. If so, 'the TŪN of *Frēsa*'s people'.

Frobury. See FROYLE.

Frocester (fröster) Gl [*Frowecestre* DB, *Froucestre* Hy 2 Glouc]. 'Roman station on R FROME' (4).

Frodesley Sa [*Frodeslege* DB, *-lega* 1167 P]. '*Frōd*'s LĒAH.'

Frodingham Li [*Frodingham* 1125-8 LN, 1254 Val, *Frodingeham* 1224 Ep, 1291 Tax], **North F~** YE [*Frotingham* DB, *Frohingham* c 1100 YCh 1300, *Frodigham* c 1145 ib. 1305, *North Frothyngham* 1297

Subs], **South F~** YE [*Sowth Frothingham* 1285 FA, *Suth Frothingham* 1297 Subs, *Frodyngham* 1301 Ch]. 'The HĀM of *Frōd(a)*'s people.' The forms with *-th-* are due to Scandinavian influence.

Frodsham Chs [*Frotesham* DB, *Frodesham* c 1100 Chester]. '*Frōd*'s HĀM.'

Frogmore Brk, D, Do, Ha, Hrt. The name usually means 'frog lake or pool'. This is the case with 5 Frogmores in Devon. **F~** Do in Toller Porcorum is *Frogmere* 1455 FF, **F~** Do in Handley is *Froggemere* 1244 Ass. **F~** Ha is *Frogmore* 1294 Ch. Forms in *-more* may well be misread or miswritten.

Frome (-ōō-) R (1) Do [*Frauu* c 894 Asser, (be) *Frome* 869, 966 BCS 525, 1186], **F~** R (2) So, W [*From*, *Frón* 701 BCS 105 f., *Frome* 1218 For], **F~** R (3) Gl, a tributary of the Avon [*Frome* 950 BCS (887), 1192 Glouc], **F~** R (4) Gl, a tributary of the Severn [*Frome* 1248 Ass, *Frome*, *Fraw* c 1540 Leland], **F~** R (5) He [*From* 840 BCS 429]. Frome is a Brit river-name, identical with FFRAW in Anglesey. Both are derived from the Welsh adj. *ffraw* 'fair, fine, brisk'. The base is Brit *frām-* (whence OW **frōm* and later **fraum*, **frauv*, *ffraw*). The ultimate base of Frome and Ffraw may be **sprām-* or **sprōm-*, which would be related to Lat *spargo* 'to sprinkle', Engl *sprinkle* &c. The later Welsh development *Frau* is recorded for the Do Frome by Asser and for Frome (4) by the pl. n. FROCESTER.

Frome St. Quintin, Vauchurch & Whitfield, Chilfrome Do [*Frome*, *Litelfrome* DB, *From* 1205 Cur, *Frome Quentyn* 1291 Tax, *Frome Voghechurche* 1297 Pat, *Fromesfoghechurche* 1319 FF, *Froma Witefeld* 1242 Fees, *Childefrom* 1206 Ch, *-frome* 1236 Fees]. Named from R FROME (1).

F~ St. Quintin was held by Herbert de Sancto Quintino in 1205 (Cur). Cf. FIFEHEAD ST. QUINTIN.—**F~ Whitfield** from a local family, resident here from c 1200.—*Chil-* from *Childe-*; cf. CHILTON.—*Vauchurch* means 'coloured church' (cf. FĀG).

Frome, Bishops, Canon, Castle, Halmonds & Priors, He [(æt) *Frome* a 1038 KCD 755, *Frome*, *Brismer-*, *Nerefrum* DB, *Froma* 1138 AC, *Frume al Evesk* 1252 Cl, *Froma Canonicorum*, *Froma Castri*, *Froma Heymund*, *From Prioris* 1242 Fees, *Frumhemund* 1252 Cl]. Named from R FROME (5). The bishop is the bishop of Hereford.—Canon refers to the canons of Lanthony, Priors to Hereford Priory.—**Castle F~** had a Norman Castle.—*Halmonds* is the gen. of a pers. or family name. It is spelt *Hamund* in 1242 (Fees).

Frome So [*Froom* 705 BCS 114, *Frome* 955 ASC, DB]. Named from R FROME (2).

Frostenden Sf [*Froxedena* DB, *Frosteden* 1242 Fees, *Frostendene* 1254 Val]. If DB may be trusted, OE *froxa-denu* 'frog valley'. Probably *Froxdene* became **Froxden* and **Froxten*, **Frosten*, whereupon a fresh *dene* was added.

Frosterley Du [*Forsterlegh* 1239 Cl]. 'The forester's clearing.' *Forester* is French.

Frowlesworth Le [*Frel(l)esworde* DB, *Fredlesuurða* 1175, *-wurða* 1176 P, *Frolleswurth* 1242 Fees, *-wrth* 1254 Val]. See WORÞ. The first el. is a pers. n., *Freopuwulf* or *Freopulāf* or a short form of such names, e.g. **Freopul*.

Froxfield Ha [(æt) *Froxafelda* 965–71 BCS 1174, *Froxfeld* 1316 FA], **F~** W [*Forscanfeld* 803–5 BCS 324, *Froxefeld* 1212 Cur, *Froxfeld* 1242 Fees]. 'FELD frequented by frogs.' First el. OE *forsc* 'frog'. But **F~** W is perhaps rather 'FELD by **Forsce* or the frog stream'. The stream would be that called *Forscaburna* 778 BCS 225.

Froyle Ha [*Froli* DB, *Froila* 1167 f., *Froile* 1185 P, *Frolia* 1196 FF, *Frohull* 1205 Obl, *Froille* 1230 P], **Frobury** Ha [*Frolebiri* 1185, *Frollebiri* 1186 P, *Frolebir'* 1212, *Froille*, *-byr'* 1236, *Frellesbur'* 1249 Fees]. The two places are a good way apart. Their names were once identical, apparently OE *Frēohyll*, which may be 'the hill of the goddess *Frīg*'. Cf. FREEFOLK.

Fryerning Ess [*Ginge Hospitalis* 1254 Val, *Fryer Inge* 1539 LP]. Cf. ING. *Fryern-* is ME *frērene* 'of the friars', i.e. the Knights Hospitallers.

Fryston, Ferry, Monk & Water, YW [(on) *Frypetune* 963 YCh 6, *Fristonam* c 1075 YCh 41, *Fristone* DB, *Friston and Feri* 1247 Ipm, *Fryston juxta aquam* 1428 FA]. '*Fripe*'s TŪN.' *Fripe* may be ODan *Frithi*.

Monk F~ belonged to Selby Abbey in 1086 (DB).—**Ferry** and **Water F~** are on the Aire.

Fryton YN [*Frideton*, *Fritun* DB, *Friton* 1224–30 Fees, *Fryton* 1239 FF]. Identical with FRITTON.

Fryup YN [*Frehope* 12 Guisb, *Frihop* c 1225 ib., 1234 FF]. First el. perhaps as in FROYLE. See HOP.

Fugglestone (fowlstn) **St. Peter** W [*Fugeleston* 1242 Fees, *Foleston* 1280 Cl]. '*Fugol*'s TŪN.'

OE *fugol* 'bird', i.e. 'wild bird', is the first el. of several names, especially such as contain words for lake, island, &c. See FOULNESS Ess, FOWLMERE, FULMER, FULBOURN, FOULDEN, FOULNEY. OE *Fugol* pers. n. is to be assumed in names containing such elements as COT, TŪN, STŌW. See FOULSHAM, FUGGLESTONE, FULSTONE, FULSCOT, FULSTOW.

OE *fūl* 'foul, putrid, rotten, dirty' is the first el. of some pl. ns., especially in combination with words for brook, ford, wood. See FOULNESS Y, FULBECK, FULBROOK, FULLEDGE, FULWELL, FULREADY, FULFORD, FULWOOD.

Fulbeck Li [*Fulebec* DB, 1130 P]. 'Foul or dirty brook.'

Fulbourn Ca [*Fuulburne* c 1050 KCD 907, *Fuleberne* DB, *Fugelburn* 1190 P, 1198 FF]. 'Brook frequented by birds.'

Fulbrook Bk [*Fulebroch* 1169, *-broc* 1191 ff. P], **F~** O [*Fulebroc* DB, *-broch* 1192 P], **F~** Wa [*Fulebroc* DB, 1198 FF, 1236 Fees]. 'Foul or dirty brook.'

Fulford D [*Foleford* DB, *Fuleford* 1242 Fees], **F~** So [*sordidum vadum* 854 BCS 476, *North-, Southfuleforde* 1327 Subs], **F~** St [*Fuleford* DB, 1167 P], **Gate & Water F~** YE [*Fuleford* DB, *Fuleforda, alia Fuleforda* R 1 (1308) Ch, *Waterfulforth* 1285 FA]. 'Dirty ford.' *Gate* is OScand *gata* 'road' and refers to the York–Doncaster road. Water F~ is on the Ouse.

Fulham Mx [*Fulanham* 704-5 St. Pauls, *Fullanhamm* 879, *-homm* 880 ASC, *Fullonham* c 894 Asser, *Fullenham* 957 BCS 1008, *Fuleham* DB]. '*Fulla*'s HAMM.' OE *Fulla* is not evidenced, but would correspond to OG *Vullo*. Cf. *Fullingadich* a 675 BCS 34 (Sr).

Fulking Sx [*Fochinges* DB, *Folkinges* c 1100 AD, 1260 FF]. '*Folca*'s people.' Cf. FOLKESTONE.

Fulledge La [*Fullach* 1510 Ct]. 'Dirty pool.' Cf. LÆCC.

Fullerton Ha [*Fugelerestune* DB, *Fughelerton* 1234 Cl]. 'The TŪN of the bird-catchers.' First el. OE *fuglere* 'bird-catcher'.

Fulletby. See FOLLINGSBY.

Fulmer Bk [*Fugelmere* 1198 FF, *Fughelemere* 1237-40 Fees]. 'Mere frequented by birds.'

Fulmodeston Nf [*Fulmotestuna* DB, *Fulmodeston* 1242 Fees, *-e* 1254 Val]. '*Fulcmod*'s TŪN.' *Fulcmod* is not found in OE and may be a Continental loan. OG *Folkmod* is well evidenced.

Fulnetby Li. See FOLLINGSBY.

Fulready Wa [*Fulrei* DB, *Fulrea* 1166 P, *Fulrithi* 1428 FA]. 'Foul or dirty brook.' See FŪL and RĪPIG.

Fulscot Brk [?*Follescote* DB, *Fugelescota* 1178 P, *Fughelescot* 1220 Fees]. '*Fugol*'s COT.'

Fulshaw Chs [*Fuleschawe* 1252 RBE, *Fulsawe* 1260, *Fulchauue* 1287 Court]. 'Foul wood' or 'wood frequented by birds'.

Fulstone YW [*Fugelestun* DB, *Fugeliston* 1274 Wakef]. '*Fugol*'s TŪN.'

Fulstow Li [*Fugelestou* DB, *Fuglestowa* c 1115 LiS]. '*Fugol*'s STŌW.', *Fugol* may have been a hermit, so that the name means '*Fugol*'s hermitage'.

Fulwell Du [*Fulewella* Hy 2, c 1200 FPD, *Fuleswell* 1195 (1335) Ch], **F~** O [*Fulewelle* DB, *-well* 1190 P]. 'Foul or dirty stream.'

Fulwood La [*Fulewude* 1228 Cl, *Fuluuode* 1252 Ch], **F~** Nt [*Folewode* 13 AD]. 'Foul or dirty wood.'

Fundenhall Nf [*Fundenhale kirke* c 1060 Wills, *Funde-, Fundahala* DB, *Fundenhal* 1254 Val]. '*Funda*'s HALH.' *Funda* is not evidenced, but may belong to OE *fundian* 'to depart, hasten', *fyndel* 'device' &c.

OE **funta** 'spring', perhaps also 'stream', is only found in pl. ns. See BEDFONT, BOARHUNT, CHADSHUNT, CHALFONT, FOVANT, HAVANT, MOTTISFONT, TEFFONT, TOLLESHUNT, URCHFONT, FONTLEY, FONTMELL. It is ultimately Lat *fontana*, but the immediate source is early Brit *funtōn (OBret *funton*, OCo *funten*, OW *finnaun*, Welsh *ffynnon*). The word is often found in Co and Welsh names, e.g. *fonton gén* 967 BCS 1197 (Co), *ffinnaun bechan* 'small spring' c 1150 LL.

Funtington Sx [*Fundintune* 12 PNSx, *Funtigton* 1252 Ch &c.]. The first el. might be OE FUNTA or a derivative of it.

Furness La [*Futhþernessa* c 1150 Hexh, *Fudernesium* 1127, *Furnesio* c 1155 LaCh]. Originally no doubt the name of the southernmost point of the Furness peninsula, RAMPSIDE POINT. The second el. is ON *nes* 'headland'. Outside Rampside is **Piel Island**, anciently **Fouldray** [*Fotherey* c 1327, *Fotheray* c 1400 FC]. This name is an ON *Fuðar-ey*, which means 'the island of *Fuð*'. The original name of Fouldray was *Fuð* (gen. *Fuðar*), which is identical with ON *fuð* 'podex'. ON *fuð* is found in names of skerries and small islands in Norway. The point opposite to Fouldray was called *Fuðar-nes*.

Furtho Np [*Forho* DB, *Fordho* 1220 Fees, *Fortho* 1254 Val]. 'HŌH by the ford.' Watling Street runs from Stratford straight up on the ridge by Furtho. For the loss of *d* in early forms cf. FORDHAM.

Fyfield Brk [*æt fif hidum* 956, *æt Fifhidan* 968 BCS 977, 1221, *Fivehide* DB], **F~** Ess [*Fifhida* DB, *Fifhid* 1202 Cur], **F~** Gl [*Fishyde* 12 Glouc], **F~** Ha [*æt Fifhidon* 975 BCS 1316, *Fifhide* DB], **F~** W [*Fiffhide* 1242 Fees]. See FIFEHEAD.

Fylde (-ī-), **The**, La, the W. part of Amounderness [*Filde* 1246 Ass]. OE *gefilde* 'plain'.

Fylingdales YN [*Figelinge, Nortfigelinge* DB, *Figelingam* c 1110 YCh 857]. '*Fygla*'s people.' Cf. FIGHELDEAN and DÆL.

Fynham Wa [*Fin-, Fynham* 1202 ff. PNWa(S), *Funham* 1221 Ass]. The place is near FINBURGH and its name will have the same first el. as that of the latter (OE *fina* 'woodpecker' rather than *fīn* 'a heap of wood'), the second being then perhaps *hamm* rather than *hām*.

OE **fynig** 'mouldy', 'marshy place'. See FENNYMERE, FINNINGLEY.

OE **fyrhþ** 'frith, wood, woodland' is found in some pl. ns. See CHAPEL EN LE FRITH, FRITHSDEN, FRITHVILLE, PIRBRIGHT.

OE **fyrs** 'furze' is the first el. of some names, as FARSLEY, FERSFIELD, FRESDON, FRESELEY, perhaps FRISTON Sx. The adj. *fyrsen* 'of furze' is found in FRESSINGFIELD, perhaps FRIZINGHALL.

G

Gable, Great, Cu [*Mykelgavel* 1338 Cl]. Gable is ME, dial. *gavel* 'gable' from ON *gafl*. Norw *gavl* is recorded in the sense 'a short mountain wall connecting two parallel mountain ridges'. The meaning 'mountain resembling a gable' given in EDD from Cumberland is possibly deduced from mountain names. ·

Gabwell D [*Gabewell* 1228 FF, *Nithergabewill* 1238 FF]. '*Gabba*'s spring.' *Gabba may be explained as a short form of names such as *Gārbeorht*. Cf. GAPTON.

Gaddesby Le [*Gadesbi* DB, *-by* c 1125 LeS, *Gaddesbia* 1177 P]. '*Gadd*'s BY.' First el. the OScand byname *Gaddr* (ON *Gaddr*, ODan *Gad*, OSw. *Gadd*) from *gaddr* 'a sting'.

Gaddesden, Great & Little, Hrt [*Gætesdene* 944–6 BCS 812, *Gatesdene* DB, *Parva Gatesdenn* 1205 Cur, *Magna, Parva Gatesden* 1254 Val]. Second el. DENU. The same first el. is found in Gatesbury Hrt nr Braughing [*Gatesbirie* 1197 FF, *-biry* 1212 RBE]. It is probably a pers. n. *Gæte, which may be a derivative of *gāt* 'goat', perhaps simply an OE **gæte(n)* 'kid' corresponding to Goth *gaitein*, and used as a nickname. The river-name Gade is a back-formation. An earlier name was *Gatesee* 1242 FF, an OE *Gætes-ēa* '*Gæte*'s stream'.

Gadshill K [*Godeshyll* 973 BCS 1296]. 'God's hill.'

OE gærs 'grass'. See GARSDALE, GARSDON, GARSTON, GRACECHURCH, GRASMERE, GREASBOROUGH, GRESHAM. Frequent spellings with *e* may partly be due to influence from OScand *gres*. An adj. *gærsen* 'grassy' seems to occur in GARSINGTON, perhaps GRASSINGTON. Cf. also GRASSENDALE, GRESSINGHAM.

Gagingwell (-ĕj-) O [*Gadelingewelle* 1193, *Gadelingwelle* Hy 3 Winchc]. OE *gædling* 'companion, kinsman' and OE *wella* 'spring'.

Gailey St [*Gageleage* 1002 Wills, 1004 KCD 710, *Gragelie* DB, *Gaeleg* 1267 Ch]. OE *gagol-lēah* 'LĒAH overgrown with bog myrtle'.

Gainford Du [*Geg(e)nford, Geagenforda* c 1050 HSC, *Et-Gegenforda* c 1130 SD, *Gainesford* c 1150 Crawf, *Gainefford* 1157 YCh 354, *Gaineford* 1196 P]. OE *gegn* 'direct' (of a road) and *ford*.

Gainsborough Li [*Gegnes-, Gæignesburh* 1013 ASC (E, D), *Gainesburg* DB, *Gleinesburc* c 1115 LiS]. '*Gegn*'s BURG.' *Gegn* is a short form of names such as *Gænbeald, Geanburh*, which contain OE *gægn* (in *ongægn* &c.). The *g* was preserved hard as in *again*.

Gainsthorpe Li nr Hibaldstow [*Gamelstorp* DB]. '*Gamel*'s thorp.' *Gamel* (DB &c.) is OScand *Gamall*.

Gaisgill We [*Gagesgylle, Gasegille, Gassegille* 1310 Whitby], G~ YW nr Barnoldswick [*Gasegile* 1182–5 YCh 199, R 1 Pudsay]. OScand *gāsa-gil* 'wild goose valley'. Cf. GIL. First el. OScand *gās* 'goose'. The first form of the We Gaisgill may be miswritten.

Galby Le [*Galbi* DB, *Gaubi* 1191 P, *Galby* 1232 Ep, 1254 Val, *Galeby* 1258 BM]. The absence of *e* between *l* and *b* in the earliest forms tells against the first el. being the ODan, OSw, ON *Galli* (a pers. n.). Possibly it is a noun *gald 'sterile soil' from OSw *galder* 'sterile', or OScand *galgi* 'gallows'.

OE galga, OScand **galgi** 'gallows'. See GALPHAY, GAWBER, GOWBARROW.

Galgate La [*Gawgett* 1605 CC]. Named from an ancient road running north past Kendal and called *Galwaithegate* c 1190 CC, *Galewethegate* c 1210 NpCh (nr Kendal). The name means 'the Galway road' and is said to refer to the road having been used by cattle drovers from Galway. *Gate* is OScand *gata* 'road'.

Galhampton So [*Galmeton* 1199 FF, *Galampton* 1303 FA], **Galmington** So [*Galameton* 1225 Ass, *Galampton* 1249 FF, *Galmetone* 1327 Subs], **Galmpton** (-ăm-) D nr Kingsbridge [*Walenimtona, Walementone* DB, *Galmeton* 1232 Cl, *Gaumeton* 1242 Fees], **Galmpton** (-ăm-) D in Churston Ferrers [*Galmentone* DB, *Galmeton* 1198 FF]. OE *gafolmanna tūn* 'TŪN inhabited by rent-paying peasants'. Cf. *gavelman* in OED.

Galphay YW nr Ripon [*Galghaga* Fount, *Galgagh* 1279–81 QW]. 'Enclosure where the gallows stood' (OE *galg-haga*).

Galsham D [*Gallecusham* 1189 Ol, *Galkysham* 1333 Subs]. 'HĀM or HAMM where wild comfrey grew.' OE *galluc* meant 'wild comfrey'.

Galsworthy D [*Galeshore* DB, 1244 Ass]. Second el. OE *ōra* 'bank, slope'. The first is possibly OE *gagol* 'bog myrtle, sweet gale'.

Galton Do [*Gavel-, Galtone* DB, *Gauton* 1236 Fees, *Gawelton* 1269 Ch, *Est-, Westgawelton* 1305 FF]. OE *Gagol-tūn* 'TŪN where bog myrtle or sweet gale grew', or *Gafol-tūn* 'TŪN subject to *gafol* or tax'.

Galtres Forest YN [*Galtris* 1171 ff., 1191 P]. 'Boar wood.' The elements are ME *galte* 'a boar or hog' (from ON *goltr*, OSw *galter*) and HRIS 'brushwood'.

Gamblesby Cu nr Melmerby [*Gamelesbi* 1177, 1197 P, *-by* 1212 Fees], G~ Cu in Aikton [*Gamelesby by Ayketon* 1305 Ipm, *Gamelsby* 1332 Subs]. '*Gamel*'s BY.' *Gamel* is OScand *Gamall*. Gamblesby nr

Melmerby is referred to as 'land of Gamel son of Bern' 1330 P.

Gamlingay Ca [*Gamelinge(i)* DB, *Gamelingeia* 1154 BMFacs, *Gamelingaye* 1201 FF]. 'The island of *Gamela*'s people.' *Gamela* is found in *Gamelanwyrþ* 946 BCS 813.

Gamston Nt nr E. Retford [*Gamelestune* DB, -*ton* 1211–13 Fees, 1229 Ep], G~ Nt nr Nottingham [*Gamelestune* DB, -*ton* 1275 RH]. '*Gamel*'s TŪN.' *Gamel* is OScand *Gamall*.

Ganarew He [*Genoreu* c 1150 Monm, *Guenerui* 1186 Fr, *Genoire* 1205 Lay]. Welsh *genau rhiw* 'pass of the hill'. G~ is situated between two hills. The elements are Welsh *genau* 'mouth, opening of a valley' (identical with GENEVA from *Genavā*) and *rhiw* 'hill, ascent, slope'. Cf. CUMREW.

Ganstead YE [*Gagenestad* DB, -*sted* 1196 FF, *Gaghenested* 1208 FF]. The first el. is apparently OScand *Gagni* pers. n. The second may be OScand too, either ON *stǫð* 'landing-place' or *staðr* (or rather plur. *staðir*) 'place, homestead', later Anglicized to -*stede*.

Ganthorpe Li [*Germuntorp* DB, *Germethorp* 1281 QW]. '*Germund*'s thorp.' *Germund* DB is ODan *Germund*.

Ganthorpe YN [*Gameltorp* DB, *Galmestorp* 1169 P, *Galmethorp* 1202 FF, *Gamelestorp* 1240 FF]. '*Gamel*'s thorp.' Cf. GAMSTON.

Ganton YE [*Galmeton* DB, 1206 FF, -*a* 1125–30 YCh 1135]. '*Galma*'s TŪN.' First el. OE *Galma* in *Galmanhó*, *Galmahó* 1055 ASC (C, D) in York. But the name might be identical with GALHAMPTON.

Gappah D [*Gatepade* DB, -*path* 1242 Fees]. OE *gāta-pæþ* 'goats' path'.

Gapton Sf nr Yarmouth [*Gabba-*, *Gabbetuna* DB, *Gapeton* 1198 FF]. '*Gabba*'s TŪN.' Cf. GABWELL.

OE **gāra** 'a gore, a triangular piece of land, a strip of land' occurs occasionally as the second el. of pl. ns.: e.g. BREDGAR, LANGAR, WALMSGATE, also alone (see GORE). See also AVERHAM, GARGRAVE, EASTON IN GORDANO.

Garboldisham (garblsm) Nf [*Gerboldesham* DB, 1233 Fees, *Garboldesham* 1254 Val]. '*Gǣrbald*'s HĀM.' OE *Gǣrbald* is not with certainty evidenced, but *Gǣr*- occurs in some names, as -*burh*, -*friþ*, -*weald*, -*wine*. It seems to be an *i*-mutated form of *Gār*-.

Gardham YE nr Market Weighton [*Gerdene* DB, *Gerthum* 1303 FA, *Gerthom* 1357 BM]. The dat. plur. of OScand *gerði* 'fence, enclosure'.

Gārendon Le [*Geroldon* c 1125 LeS, -*dun* 1156, 1166 P, *Gerewedon* 1173, *Gerolddon* 1178 P, *Gerewedon* 1193 P, 1202 Ass]. '*Gǣrwald*'s DŪN.' Cf. GARBOLDISHAM.

Garford Brk [*æt Garanforda* 940, *Garanford*

960 BCS 761, 1055, *Gareford* 1175 P, 1242 Fees]. '*Gāra*'s ford.' *Gāra is a short form of names in *Gār*- and -*gār*.

Garforth YW [*Gereford* DB, c 1090 YCh 350, 1226 FF]. Apparently '*Gǣra*'s ford'. *Gǣra would be a short form of names in *Gǣr*-. Cf. GARBOLDISHAM.

Gargrave YW [*Geregraue* DB, *Gairgrava* c 1160, *Gairegrave* 1214 FC, *Gargrave* 1182–5 YCh 199, c 1190 FC, *Garegrave* 1260 Ass]. Probably OE *gāran-grāf* 'grove in a gore', later partly Scandinavianized, ON *geiri* having replaced the synonymous OE *gāra*.

Garlinge K [*Groenling* c 824 BCS 851, *Grenling* 943 ib. 784, *Greneling* 13 StAug]. OE *grēn-hlinc* 'green hill'. See HLINC.

Garmondsway Du [*via Garmundi* 1104–8 SD, *Garmundeswai* 12 FPD]. '*Gārmund*'s road.'

Garmston Sa [*Garmundeston* 1301 For, *Garmeston* 1327 Subs]. '*Gārmund*'s TŪN.'

Garnstone Castle He [*Gernereston* 1294 Cl, *Gernestone* 1332 Ep]. '*Gerner*'s TŪN.' *Gerner* is an OFr pers. n. of OG origin (OG *Warinhari*, OFr *Guarnier* &c.).

Garren or **Garron** R He [*Garran* 1558 AD, 1577 Saxton]. Cf. LLANGARREN. A Brit river-name derived from OCo, Bret, Welsh *garan*, Gaul -*garanus* 'crane'.

Garrick Li [*Gerewic* Hy 2 (1316) Ch, -*wik* 1275 RH]. It cannot be decided if -*wīc* is OE WĪC or OScand VĪK 'a bay'. The first el. may be OScand *geiri*, ODan *gēri* 'a gore' or OE GĀRA or OE *Gāra* pers. n., later Scandinavianized.

Garrigill Cu [*Gerardegile* 1232 Ch, 1292 QW]. '*Gerard*'s valley.' See GIL. *Gerard* is an OFr pers. n. of OG origin.

Garrington K in Littlebourne [*Garwynnetun* 11 DM, *Garwintun* 1194 StAug, -*ton* 1200 Cur]. '*Gārwynn*'s TŪN.' *Gārwynn is an OE woman's name.

Garriston YN [*Gerdestone* DB, *Gertheston* 1184 PNNR, *Gerdeston* J Ass]. Perhaps '*Giarðar*'s TŪN'. ON *Giarðarr* is a known name. An *r* often disappears before *s*, especially if the name contains another *r* (dissimilation). Or the first el. may be OE *Gyrð*, *Georð* (*Guerd* DB) from ODan *Gyrdh*, *Gyurth*, OSw *Gyrdher*, *Giordher*.

Garrowby YE [*Ghervenzbi*, *Geruezbi* DB, *Gervordeby* 1281 Cl, *Gerwardby* 1285 Ipm]. '*Gerwarth*'s BY.' First el. the OScand name found in OSw as *Gervardh*. The DB forms may suggest ON *Geirviðr*, OSw *Gervidh* as the original first el.

Garsdale YW [*Garsedale* 1241 FF, 1279 Ass]. 'Grass valley.'

Garsdon W [*Gersdune* 701 BCS 103, *Gardone* DB, *Garsedon* 1228 Cl]. 'Grass hill.'

Garshall St [*Garnonshale* 1310 Ipm]. '*Garnon*'s hall.' *Garnon* is a family name from an OFr byname, identical with OFr *grenon*, *grenons*, *guernons* 'moustache'.

Garsington O [*Gersedune* DB, *Gersendona* 1130 P, *Garsindon* 1207 Cur, *Gersinton* 1195 P]. 'Grass-covered hill.' First el. OE **gærsen* 'of grass'.

Garstang La [*Cherestanc* DB, *Gairstang* c 1195 LaCh, 1247 Ipm &c.]. An OScand name, apparently containing ON *geirr* 'a spear' or *geiri* 'a gore' and *stong* 'a pole'. A boundary mark may be referred to.

Garston, East, Brk [*Esgareston* 1180 P, *Esegareston* 1220 Fees]. '*Esgar*'s TŪN.' The place was very likely named from *Esgar stallere*, who was a tenant in Lambourn Brk in 1066 (DB). *Esgar* is ODan *Esger* = ON *Ásgeirr*. Esgar stallere is also called *Asgar stalre*.

Garston Ha in Clere [*la Garston de Clere* 1251 Cl], G~ Hrt [*Garston* 1265 Misc]. OE *gærstūn* 'meadow'. The name may sometimes mean 'grazing-farm'.

Garston La [*Gerstan* 1094, 1142 LaCh, 1212 Fees, *Grestan* c 1155 LaCh, 1215 P]. 'Big stone', OE GRĒAT and STĀN.

Garswood La [*Grateswode* 1367 VH, *Gartiswode* 1479 FF]. The forms are too late for a definite etymology.

ME **garth** 'enclosed ground used as a yard, garden or paddock', dial. *garth* also 'a farm', from ON *garðr*, OSw *gardher* &c., occurs occasionally in pl. ns., as ARKENGARTHDALE, HAWSKER, PLUNGAR. Cf. also GARTON.

Garthorpe Le [*Garthorp* c 1125 LeS, 1199 FF, *Garetorp* 1180 P]. Perhaps '**Gāra*'s thorp', though the usual early form *Gar-* seems to point rather to GARTH as first el.

Garthorpe Li [*Gerulftorp* DB, *Geroldtorp* 1180 P]. '*Gerulf*'s thorp.' *Gerulf* is probably ODan *Gerulv*, ON *Geirulfr*.

Garton YE [*Gartun* DB, *Garton* 1190 YCh 1312, *-a* 1297 Subs], G~ **on the Wolds** YE [*Gartune* DB, c 1170 YCh 441, *Garton in Waldo* 1208 FF]. Probably OScand *Garðtūn*, analogous to OSw *Gardhby* 'BY with a fence'. In favour of derivation of *Gar-* from OScand *garðr* is the fact that GRIMSTON in Garton is called G~ GARTH.

Garveston Nf [*Gerolfes-, Girolfestuna* DB, *Gerolvestone* 1254 Val]. '*Gerulf*'s TŪN.' *Gerulf* may be ODan *Gerulv* (cf. GARTHORPE) or OE **Gǣrwulf* (cf. GARBOLDISHAM).

Garway He [?*Lann Guoruoe* c 1150 LL, *Garou* 1138 AC, *Langarewi* 1199, *Garewi* 1227 Ch]. If the ex. from LL belongs here, '*Guoruoe*'s church'. Cf. LANN. *Guoruoe* pers. n. occurs in LL.

Gastard W [*Gatesterta* 1155 RBE, *-stert* 1168 ff., 1186 P, 1230 Ch, *-herst* 1177 f. P]. OE *gāte-steort* from OE *gāt* 'goat' and *steort* 'a tongue of land'.

Gasthorpe Nf [*Gades-, Gatesthorp* DB, *Gaddesthorpe* c 1095 Bury, *Gatestorp* 1244 Cl, *Gadisthorp* 1275 BM]. '*Gadd*'s thorp.' Cf. GADDESBY.

OE **gāt** '(wild or tame) goat' is a common first el. in pl. ns., but is sometimes difficult to distinguish from GEAT 'gate'. See GAT(E)-, GOAT- (passim), GAPPAH, GASTARD, GAYHURST, GAYTON, GEDGRAVE, GOTHAM. The corresponding OScand **geit** is found in GATESGILL, and sometimes tends to replace OE *gāt*. See e.g. GATEFORD, -FORTH.

ON, OSw **gata** 'a road', ME **gate**, is found in names of roads and streets in the north and the Scandinavian Midlands, as in BOTCHER-, FRAMWELLGATE. Sometimes such names have become names of places, as CLAPPERSGATE, GALGATE, HARROGATE, HOLGATE.

Gatacre Sa [*Gatacra* 1160 f. P, *Gattacra* 1195 Cur, *Gatacre* 1208 FF]. 'Field by a gate' (OE GEAT), rather than 'goat field'.

Gatcombe Wt [*Gatecome* DB, *-cumbe* 1263 Ipm]. 'Valley frequented by (wild) goats.'

Găteacre La. Identical with GATACRE.

Gateford Nt [*Gaiteford* 1166 P, *Gayteford* 1316 Ch], **Gateforth** YW [*Gæiteford* c 1030 YCh 7, *Geiteford* 1166 P, *Gateford* 1316 FA]. OE *gāta-ford* 'ford of the goats', later Scandinavianized, OScand *geit* having replaced OE *gāt*.

Gatehampton O [*Gadindone* DB, *Gathanton* 1177 P, *Gathamptona* 1219 Fees]. 'HĀMTŪN by the gate' (OE GEAT).

Gateley Nf [*Gatelea* DB, *-leia* 1156 P, *Gotele* 1202 FF]. OE *gāta-lēah* 'clearing where goats were kept'.

Gatenby YN [*Ghetenesbi* DB, *Gaitenebi* 1208 Cur, *Geytenby* 1231 FF]. Explained in PNNR as '*Gaithan*'s BY', *Gaithan* being an OIr pers. n.

Gatesbury. See GADDESDEN.

Gatesgill Cu [*Geytescales* 1273 Cl, *Gaytsheles* 1337 WhC]. 'Shelter for goats.' Cf. SKÁLI. First el. OScand *geit* 'goat'.

Gateshead Du [*Ad Capræ Caput* c 730 Bede, *æt Rægeheafde* c 890 OEBede, *Gateshaphed* c 1170 Newcastle, *-heued* 1196 P]. 'Headland or hill frequented by (wild) goats.' Cf. HĒAFOD. The translator of Bede mistranslated *Capræ Caput*.

Gathurst La [*Gatehurst* a 1547 DL]. 'Goats' HYRST.' OE *geat* 'gate' is also a possible first element.

Gatley Chs [*Gateclyve, Gaticlyve* 1290 Court]. 'Goats' cliff.'

Gatley He [*Gatesleg, Gatleg* 1230 P, *Gatleye* 1275 Ep]. 'LĒAH by the pass.' Cf. GEAT. The place is in a pass. OE *gāt* 'goat' is also a possible first el.

Gatton Sr [*Gatatun* 871–89 BCS 558, *Gatone* DB, *Gatetuna* 1121 AC]. 'TŪN where goats were kept.'

Gaunless R Du [*Gauhenles* 12 FPD, *Gawenles* 1242 Ass, *Gaunles* 1291 RPD]. ME *gaghenles* 'useless' (from ON *gagnlauss*). The name may refer to scarcity of fish or the like.

Gautby Li [*Goutebi* 1196 FF, *Gauteby* 1212 Fees]. '*Gouti*'s BY.' *Gouti* DB &c. is ON *Gauti*, ODan *Gøti*.

Gauxholme La [*Gawkeholme* 1521 DL]. '*Gauk*'s holm or island.' *Gauk* is ON *Gaukr*, OSw *Gøker*, pers. n., really *gaukr* 'cuckoo'.

Gawber YW [*Galghbergh* 1304 Ipm]. 'Gallows hill.' First el. OE *galga* or OScand *galgi* 'gallows'. See BEORG.

Gawcott Bk [*Chauescote* DB, *Gauecota* 1090 RA, *Gavecote* 1255 RH, *Galcote* 1486 BM]. Possibly OE *gafol-cot*, the first el. being OE *gafol* 'tax, rent'. But the forms rather suggest an OE *Gafan cot*, *Gafa* being a pers. n. identical with OG *Gabo*.

Gawsworth Chs [*Govesurde* DB, *Gousewrdth* 1276 Ipm, *-wrthe* 1285, *Gowesworth* 1287 Court]. If the DB form is trustworthy, the first el. appears to be Welsh *gof* 'a smith', used as a pers. n. Welsh *gof* is the source of the well-known family name *Gough*.

Gawthorpe YW nr Dewsbury [*Goukethorpe* 1274 ff. Wakef], G~ YW nr Huddersfield [*Goutthorp* 1297 Subs, *Gawkethorp* 1324 Goodall]. '*Gauk*'s thorp.' Cf. GAUXHOLME. **Gawthorpe Hall** La [*Gouthorp* 1256, *Goukethorp* 1324 PNLa] was probably named from a family.

Gaydon Wa [*Gaidone* 1195 P, *Geydon* 1285 QW, 1316 FA]. Perhaps '*Gǣga*'s hill'. Cf. GAYTON.

Gayhurst Bk [*Gateherst* DB, *Gaherst* 1167 P]. 'Goats' HYRST', OE *gāta-hyrst*. The loss of *t* is due to Norman influence.

Gayles YN [*Aust-*, *Westgail* 12 YCh v, *Gayles* 1556 FF]. ON *geil* 'ravine'.

Gayton Chs [*Gaitone* DB, *Gayton* 1244 Misc, *Geyton* 1286 Court], G~ **le Marsh** Li [*Geiton* 1206 Ass, *Gayton* 1236 Ep], G~ **le Wold** Li [*Gedtune*, *Gettone* DB, *Gertuna* c 1115 LiS, *Gattunasoca* 1154 AC, *Gaitun* 1162 BM, *Gatton* 1200 Cur], G~ Nf [*Gaituna* DB, *Geitun* c 1150 Crawf, *-ton* 1198 Cur], G~ Np [*Gaiton* 1163, 1196 P, *Gainton* 1167 P, *Gauton* 12 NS], G~ St [*Gaitone* DB, *-ton* 1227 Ass, *Gayton* 1272 Ass], **Gaywood** Nf [*Gaiuude* DB, *Geywode* c 1105, *Gaiwde* c 1140 BM]. G~ le Wold is a Scandinavianized form of OE *Gāta-tūn* 'goat TŪN' (cf. GATTON, GATEFORD), and the same may well be the explanation of G~ le Marsh, possibly of G~ Chs. But this explanation is out of the question for G~ Nf, which must be taken together with

Gaywood, and improbable for Gayton Np, St. These must be compared with GAYDON and with GUIST Nf and GINGE Brk, for which a first el. or base **gǣg-* is to be assumed. This base may be identical with the first el. of *Gegan dene (lege)* 996 KCD 1292. A rivername *Gǣge related to OE *gǣgan* 'to turn aside' and the like (in *for-*, *ofergǣgan*) might explain some of the names, but Gayton Chs and Np are not on streams. The probability is therefore that the common element is an OE pers. n. *Gǣga related to OE *gǣgan*.

Gayton Thorpe Nf [*Torp* DB, *Thorp* 1302 FA, *Aylswiththorp* 1316 FA, *Aylswythorp* 1390 BM, *Geythorp* 1346, *Geytonthorp* 1401–2 FA]. The place was sometimes named from an owner called *Æþelswiþ* (a woman). *Geythorp* is probably shortened from *Geyton thorp*.

Gazeley Sf [*Gaysle* 1219 FF, *Gasel[e]* 1248 Ch, *Gaisle* 1254 Val]. '*Gǣgi's LĒAH.' The first el. is apparently a strong side-form of *Gǣga* in GAYTON &c.

OE **gēac** 'cuckoo' is the first el. of EXBOURNE and YAXLEY, but *Gēac* was very likely used as a pers. n. in OE, like *Gaukr* in Scandinavia, and is a possible first el. of YAXLEY and YAXHAM. See also GOXHILL.

OE **gear** 'weir, enclosure for catching fish', also in *mylengear* 'mill weir', is found in KEPIER, YARPOLE, YARWELL. YARM is the dat. plur. of OE *gear*.

OE **geard** 'a fence, hedge, an enclosure' is rare in pl. ns. See BROMYARD, BRUISYARD, RUDYARD, YARKHILL, YORTON, also PLUNGAR.

OE **geat, gæt** 'a gate' appears in names in the two forms *yate* (*-yatt*, *Yat-* &c.) and *gate*. The latter is due to the influence of the OE plur. *gatu*. Cf. YATE, YATELY &c., AYOT, BOYATT, DONYATT, LAMYATT, LEZIATE, MERRIOTT; GAT(E)ACRE, GATEHAMPTON, BURGATE, CLAYGATE, HUGGATE, PILSGATE, REIGATE, SKILGATE, SNARGATE, WYEGATE &c. The usual meaning is probably 'gate'. Sometimes the gate is one leading to a forest, as in NEWDIGATE, ? WOODYATES. In MARGATE, WESTGATE *gate* seems to refer to a natural opening in the sea wall. Sometimes OE *geat* is used of a gap in a chain of hills, as in *windgeat* BCS 1066; cf. WINGATE(S). OE *hlidgeat* 'a swing-gate' is found in LEADGATE, LIDGATE, LYDIATE. Cf. further BAGSHOT W, BOZEAT, DITCHEAT.

Gedding Sf [*Gedinga* DB, *Gedding* 1185, *-es* 1190 ff. P]. OE *Gyddingas* '*Gydda*'s people'. *Gydda* is found in *Gyddan dene* 943 BCS 789 and in GIDLEIGH. Cf. also GEDDINGE, GIDDING, GIFFORDS HALL.

Geddinge or **Giddinge** K [*Geddingge* 687 BCS 69, *Geddingc* 799 ib. 296, *Getinge* DB]. OE *Gydding* '*Gydda*'s place'. Cf. prec. name.

Geddington Np [*Geitentone*, *Gadintone* DB, *Gadintona* 1130 P, *Gaitintun* 1157, *-ton*

1167, 1194 P, *Gatinton* 1159 P, 1202 Cur, *Gattinton* 1163 P]. The probability seems to be that the name means 'the TŪN of *Gǣte*'s people'. Cf. GADDESDEN. Later the name was Scandinavianized, being associated with OScand *geit* 'goat' and *Geitir* pers. n.

Gedgrave Sf [*Gata-, Gategraua* DB, *Gategrave* 1275 RH]. OE *gāta-grǣf* 'goats' grove'.

Gedling Nt [*Ghellinge* DB, *Gedlinges* 1187 f., *Geddlinges* 1187 P, *Gedelinghes* 1249 Ep]. Probably '*Gēdel*'s or *Gēdla*'s people.' Cf. GILSTON Hrt. OE *Gēdla* would correspond to OG *Gōdila*. *Gēdel* and *Gēdla* are normal diminutives of *Gōda*.

Gedney Li [*Gadenai* DB, 1130 P, *Gedeneie* 13 DC, *Geddeney* 1226 Ep]. This may be '*Gydda*'s island' (cf. GEDDING) if we may assume that OE *y* sometimes became *e* in Lincs. Otherwise the first el. may be an OE *Gǣda*, a short form of names in *Gād-* (*Gadfrid* BCS 43) and related to OG *Gaido*.

ON geil 'a narrow ravine'. See GAYLES, HUGILL. OScand **geit** 'goat'. See GĀT.

Geldeston Nf [*Geldestun* 1242 Fees, *-tone* 1254 Val, *-ton* 1273 Cl]. '*Gyldi*'s TŪN.' Cf. GUILSBOROUGH.

Gelston Li [*Cheuelestune* DB, *Geueleston* 1202 Ass, *Geveleston* 1242 Fees, *Giveleston* 1272 FF]. The first el. appears to be a pers. n. of Scand origin derived from the verb *give*. Possibly a byname identical with ON *giǫfull* 'liberal'.

Gelt R Cu [*Gelt* c 1200 Lanercost, 1228 For]. A Celtic river-name derived from OIr *geilt* 'mad, wild' or its possible Brit equivalent *gwyllt*. Hence **Gelt Forest** [*Gelt* 1295 Ipm] and **Geltsdale** [*Geltesdal* 1295 Ipm].

Gembling YE [*Ghemelinge* DB, *Gemelinge* c 1185 YCh 984, *Gemeling* 1229 Ep, *Gamelinga* c 1170 YCh 1355]. '*Gamela*'s people.' Cf. GAMLINGAY.

Georgeha·m D [*Hama* DB, *Hamme* 1261 FF, *Ham Sancti Georgii* 1471 Ipm]. Originally *Hamm*; see HAMM. *George-* from the dedication of the church.

Germansweek D [*Wica* DB, *Wyk* 1242 Fees, *Wyke* 1270, *Wyke Germyn* 1458 FF]. See WĪC. *Germans-* from the dedication of the church.

Germoe Co [(parochia) *Sancti Germocii* 1377 PT]. '(The church of) St. Germocus.'

Gerrans Co [(ecclesia) *Sancti Gerendi* 1261 Ep, (de) *Sancto Gerendo* 1291 Tax]. '(The church of) St. Gerend or Gereint.'

Gestingthorpe Ess [*Gyrstlingaþorp* c 1000 BCS 1306, *Gristlyngthorp* c 1040 Wills, *Ghestlingetorp* DB, *Gestlingetorp* 1198 P]. 'The thorp of the *Gyrstlingas*.' *Gyrstlingas* appears also to be the source of GUESTLING

Sx. The origin of the folk-name *Gyrstlingas* is obscure.

Gibside Du [*Gippeset* 1339, *Gibset* 1375 PNNb], **Gibsmere** Nt [*Sypermere* (sic) 958 YCh 2, *Gipesmare* DB, *-mere* 1228 Cl]. '*Gyppi*'s (GE)SET or fold and mere or lake.' *Gyppi* is a pers. n. related to *Guppa* in GUPWORTHY and is the base of GIPPING.

Gidding, Great, Little & Steeple, Hu [*Gedelinge, Geddinge* DB, *Geddinges* 1168 P, *Guddinges* 1212 RBE, *Gyddinge, Magna Gidding* 1220 Fees, *Gydding Parva* E 1 BM, *Stepelgedding* 1260 Ass]. Identical with GEDDING. *Gedelinge* DB is miswritten for *Geddinge*. *Steeple* presumably from the church steeple.

Gidleigh D [*Gideleia* 1156, *Gedelega* 1158 P, *Giddelegh* 1212 Fees, *Gudeleghe* 1284 Ep]. '*Gydda*'s LĒAH.' Cf. GEDDING.

Giffords Hall Sf [*Giddin(c)gford* c 995 BCS 1289]. 'The ford of the *Gyddingas*.' Cf. GEDDING.

Giggleswick YW [*Ghigeleswic* DB, *Gicheleswik* c 1160, *Gekeleswik* 1221 FC, *Gicleswic* 1204 Cur]. '*Gikel*'s WĪC.' The pers. n. *Gichel* is found 1156 YCh 80. *Gikel* may be from an unrecorded OScand *Guðkell* (with *i*-mutation as in ILKETSHALL).

ON gil 'a ravine, narrow valley' is common in the north-west in names of valleys. Some of these have given names to homesteads or villages. Cf. GAISGILL, GARRIGILL, HOWGILL, IVEGILL, REAGILL, ROSGILL, SLEAGILL. Sometimes *-sgill* in modern names goes back to *-scale*, as in GATESGILL, SOSGILL.

Gilberdike YE [*Dyc* 1234, *Dyke* 1336 FF, *Gilberdyke* 1349, *Gilbertdike* 1376 FF]. OE DĪC. *Gilber-* must be the pers. n. *Gilbert*.

Gilby Li [*Gillebi* 1139 RA, *Gilby* 1316, *Kelbi* 1303, *Keleby* 1428 FA]. If the earliest form may be trusted, '*Gilli*'s BY'. Cf. GILSLAND.

Gilca·mbon Cu in Edenside [*Gilkamban* 1285 For, 1324 Ipm]. '*Kamban*'s valley.' See GIL. *Kamban* is an ON pers. n. of OIr origin (derived from *camm* 'crooked'). The order of the elements is Celtic. Cf. ASPATRIA.

Gilcrux (-ōoz) Cu [*Killecruce* 12 StB, *Gillecruz* 1230 FF, *Gillecruce* 1272 StB, *Gillecruice* 1308 Ipm]. The name is generally held to be a compound of Ir *cill* (the dat. form of OIr *cell* 'a church') and a word for *cross*. But the Ir word for cross is *cros*. Possibly the name represents a Welsh *cil crug* (OW *cil crūc*) 'retreat by a hill'. On *cil* see CULCHETH, on *crūc* see that word. *G-* would be due to Brit lenition as in GILLOW or to influence from OScand *gil* 'valley'. The form *-cruz* &c. would be due to Norman influence.

Gilderdale Forest Cu [*Gilderdale* 1279 PNCu(S), *Gildresdale* 1332 Subs]. First el.

ON *gildri* 'trap', found in **Gillerbeck** Cu, *Gylderbek* 1342 AD (We), and *Gilderschoh* 13 StB.

Gildersome YW [*Gildehusum* 1181 P, *Gildhus* 1226 FF, *Gyldusum* 1304 Ch]. '(Àt) the guild-houses.' *Gild-* is OScand *gildi* 'guild'.

Gildingwells YW [*Gildanwell* 1324 Ipm, *Gyldanwelles* 1345 FF]. G~ is close to WALLINGWELLS, whose name has a pres. part. *welland* as first el. *Gilding* is no doubt also a pres. part. in *-and*. It may belong to a verb meaning 'to gush' related to Norw *gyldra* 'a water-course in a ravine'.

Gillamoor YN [*Gedlingesmore* DB, *Gillingemora* c 1170 Middleton, *Gillingamor* 1207 Cur]. 'Waste belonging to Gilling in Ryedale.'

Gilling YN in Ryedale [*Ghellinge* DB, *Gillinga Ridale* 1157 YCh 354, *Gilling* 1208 Cur], G~ YN nr Richmond [*Ghelling(h)es* DB, *Gillinge* c 1090 YCh 350, *Gilling* 1220 Ep]. With Gilling nr Richmond is usually and no doubt correctly identified *Ingetlingum* c 730 Bede (*Inngetlingum* c 890 OEBede). The base must then be OE *Gētlingas* 'the people of *Gētla*', from *Gautilan*, a derivative of *Gēat* from **Gaut-*. The hard *G-* must be due to Scand influence. If *Ingetlingum* should not be identical with Gilling, the latter may go back to OE *Gyplingas*. Cf. GIVENDALE YW.

Gillingham Do [*Gillingahám* 1016 ASC(D), *Gelingeham* DB, *Gellingeham* 1130, 1155, *Gillingeham* 1156 P], G~ K [*Gillinge-*, *Gyllingeham* 10 BCS 1321 f., *Gelingeham* DB, *Gillingeham* 1212 RBE], G~ Nf [*Kildincham* DB, *Gelingeham* 1107–18 AC, *Gillingham* 1198 FF, 1275 RH, *Gilingham* 1254 Val]. 'The HĀM of *Gylla*'s people.' *Gylla* may be a hypocoristic form of **Gybla* (cf. GIVENDALE) or the base may be *Gybla* itself.

Gillow He nr Tretire [*Cil Luch* c 1150 LL, *Gilloch, Kilho* 1280 Ipm]. Welsh *cil* 'retreat' and *llwch* 'pool', i.e. 'retreat by the pool'.

Gilmonby YN [*Gillemaneby* c 1150 PNNR, *Gilmanby* 1301 Subs]. '*Gilman*'s BY.' First el. the pers. n. *Gillemon* found c 1217 YD, which is probably a compound of the pers. n. *Gille* (cf. GILSLAND) and *man*.

Gilmorton Le [*Mortone* DB, *Aurea Morton* 1249 Ep, *Gilden Morton* 1327 Subs]. The original name was MORTON. *Gil-* is OE *gylden* 'golden', an epithet often given to places, as in GUILDEN MORDEN. The meaning is 'rich' or 'splendid'. In ASC(E) 1052 it is stated that Abbot Leofric endowed Peterborough so that it was called *þa Gildene burh.*

Gilpin R We [*Gylpyne* 16 Kendale]. Probably named from the Gilpin family.

Gilsland Cu [*Gilleslandia* c 1185 WR, *-land* 1234 Ch, 1250 Ipm]. '*Gille*'s land.' Probably named from *Gille* son of Bueth who is mentioned in the Lanercost foundation charter (1169). In a document of 1155–7 he is called Gilbert son of Boet (WR). This suggests that *Gille* is a short form of the Fr name *Gilbert*. But more likely *Gille* is ON *Gilli* from OIr *gilla* 'servant' (or names in *Gilla-*). This name may have been taken to be a short form of *Gilbert*.

Gilson Wa [*Gudlesdone* 1232 Ass]. '*Gydel*'s DŪN.'

Gilston Hrt [*Gedeleston* 1197 f. FF, 1200 Cur, *Godeleston* 1199 Cur]. First el. perhaps as in prec. name. Or it may be '*Gēdel*'s TŪN'. Cf. GEDLING.

Gimingham Nf [*Gimingeham* DB, *Gemingheam* 1121 AC, *Gimmingeham* 1188, *Gummingeham* 1192 ff. P]. 'The HĀM of *Gymi*'s or *Gymma*'s people.' Neither name is recorded, but they would be normal derivatives of *Gum-* in *-beorht*, *-weald*.

Ginge, East & West, Brk [*Gainge* 815, *Gæging* 959 BCS 352, 1047, *Gainz* DB]. Ginge is really the old name of **Ginge Brook** [*Geenge* 726–37, *Gæingbroc* 959 BCS 155, 1047]. *Gæging* may be a derivative of *Gæga* pers. n. (cf. GAYTON) or of a noun belonging to OE *gægan* 'to turn aside' (cf. OFris *gēie* 'penalty', ON *geigr* 'damage').

Gipping Sf [*Gippinges* Hy 2 Waltham Cart, *Gippingneweton*, *Gypping* E 1 BM]. '*Gyppi*'s or *Gyppa*'s people.' Cf. GIBSIDE, GIPTON. **Gipping** R is a back-formation.

Gipton YW [*Chipetun* DB, *Gipetuna* c 1160, *Giptuna* 1159, c 1173 Pont]. '*Gyppa*'s TŪN.' **Gyppa* is a weak side-form of *Gyppi* in GIBSMERE.

Girlington YN [*Gerlinton* DB, *Girlington* 1251 YInq], G~ YW [*Gryllyngton* 1379 Goodall]. The first el. may be a derivative of a pers. n. related to ME *gurle*, *gerle* 'a youth', Mod *girl*. A pers. n. stem *Gur-* is possibly found in *Gyran torr* 938 BCS 724.

Girsby Li [*Grisebi* DB, c 1115 LiS], G~ YN [*Grisebi* c 1050 HSC, DB, *Grisibi* 1088 LVD]. The forms rather suggest OScand *Grīsa-bȳr* 'BY where pigs were reared' than '*Gris*'s BY'. *Gris* pers. n. is found in medieval English sources. It is ON *Griss*, ODan, OSw *Gris*, originally a nickname.

Girtford Bd [*Grutford* 1247 Ass, *Gretford* 1291 Tax]. OE *grēot-ford* 'gravelly ford'.

Girton Ca [*Gretton, Gryttune* 1060 Th, *Gretone* DB, *Gretton* 1206 Cur, *Grytton* 1291 Tax], G~ Nt [*Gretone* DB, *Gretona* 1163 RA, *Gretton* 1240 Cl]. OE *Grēot-tūn* 'TŪN on gravelly soil'.

Gisburn YW [*Ghiseburne* DB, *Giselburn* c 1195, *Gisleburn* c 1200 Pudsay, *Giseburn* 1218 FF]. G~ is near the Ribble and a tributary of it. The name may mean '*Gysla*'s stream' (cf. GISLEHAM) or the first el. may be an adj. **gysel* 'gushing', related to the first el. of GUSSAGE.

Gisleham Sf [*Gisleham* DB, 1254 Val, *Giselham* 1203 Ass, 1233 Cl], **Gislingham** Sf [*Gyselingham* c 1060 Th, *Gislingaham*, *Gissilincham* DB, *Gislingham* 1193 P, *Giselingham* R 1 Cur]. 'The HĀM of *Gys(e)la* and of his people.' OE *Gysla* is possibly evidenced in *Gyslan* (*Gislan*) *ford* 972 BCS 1282. *Gys(e)la* is related to the base of GISSING and of OG *Gusso*.

Gissing Nf [*Gessinga* DB, *Gissing* 1195, *Gessing* 1205, *Gissinges* 1210 FF, *Gissinge* 1242 Fees]. 'The people of **Gyssa* or **Gyssi*.' Cf. prec. name.

Gittisham D [*Gidesham* DB, *Giddesham* 1242 Fees]. '**Gyddi*'s HĀM.' Cf. GIDLEIGH, which contains a related weak *Gydda*.

Givendale, Great & Little, YE [*Ghiue-*, *Geuedale* DB, *Geveldala* c 1125 YCh 449, *Little Geveldale* 1227, *Gevendale*, *Giveldal* 1231 FF]. 'The valley of the river *Gifl*' (cf. IVEL). The hard *G-* is due to Scandinavian influence.

Givendale YW [(on) *Gyŏlingdale*, *Gyþinga deal* c 1030 YCh 7, *Gherindale* DB, *Gyvenedal* 1248 Ch]. 'The valley of **Gýþla*'s people.' Cf. GILLING. *Gýþla* is a hypocoristic form of names in *Gūþ-*. But *Gyŏingdale* is the reading in ASCh for *Gyŏlingdale* YCh 7. If that is right, the first el. is derived from some other short form of names in *Gūþ-*, e.g. **Gýþi*.

Gladley, Nares, Bd [*Gledelai* DB, *-leia* 1176 P, *Gladeleia* 1176 P]. Perhaps 'kite wood'. First el. OE *gleoda*, *glida* 'kite' with dialectal development of *eo* to *ea* as in ME *wale* for *weola*, *fale* for *feola* 'much' &c. Or the first el. may be OE *glæd* 'bright'.
Nares is apparently the name of an early owner.

Glaisdale YN [*Glasedale* 12 Guisb, *Glasedal* 1223, *Glasdale* 1228 FF]. 'The valley of R *Glas*.' Cf. GLAZEBROOK.

Glandford Nf [*Glam-*, *Glanforda* DB, *Snitesle Glaumford* 1254 Val, *Glamford* 1257 Ass, 1275 RH], **Glanford Brigg** Li now **Brigg** [*Glanford* 1183 P, *punt de G~* 1218 Ass, *G~ Brigg* 1235 Ch, *Glannford* 1256 FF, *Glaumford Bridge* 1294 Ipm]. First el. probably OE *gléam* 'merriment'. If so, the meaning is 'ford where sports were held'.
Glaven, river-name Nf, is a back-formation.

Glantlees Nb [*Glendeleya* 1201 Ch, *Glanteleye* 1256 Ass, *Glenteley* 1242 Fees, c 1250 BM], **Glanton** Nb [*Glentendon* 1186 P, *Glantedon* 1200 Ch, *Glentedun* 1212 Fees, *Glantendon* 1219 P]. Glantlees is at **Glantlees Hill,** while Glanton is on the slope of a hill where is **Glanton Pike.** The first el. is related to ME *glenten* 'to shine, look, move quickly', Norw *gletta* 'to peep, look' &c., G *Glanz, glänzen*. It may be suggested that the base is an OE **glente* 'look-out hill'. Cf. GLENTHAM, GLENTWORTH Li.

Glapthorn Np [*Glapthorn* 12 NS, 1202 Ass, *Glapethorn* 1189 (1332) Ch, 1229 Cl],

Glapton Nt [*Glapton* Hy 3 Ipm], **Glapwell** Db [*Glapewelle* DB, *Glapwelle* 1186 P, *Glapwell* 1242 Fees]. The material for Glapton is poor, but the probability is that it means '*Glappa*'s TŪN'. Glapwell is probably 'stream where buckbean grew'. Buckbean, *Menyanthes trifoliata*, is a waterplant. Its OE name was *glæppe* (gen. *glappan*). Glapthorn might be '*Glappa*'s thorn', but the early forms point rather to an OE *Glæp-þorn* than to *Glappan-þorn*. Probably OE *glæp-þorn* was the name of some shrub. OE *glæppe* is no doubt related to Swed *glappa* 'to be too wide', *glap* 'a fissure', Norw *glapa* 'to be open' &c., and there may well have been some OE word such as **glæp* which might be added to *þorn* to form the name of a shrub.

Glaramara Cu [*Gleuermerghe* 1211, *Houedgleuermerhe* 1210 FC]. G~ is a hill of 2,560 ft. *Houedgleuermerhe* is 'Glaramara Hill', *houed* being ON *hofuð* 'head'. For the order between the elements cf. ASPATRIA. The first el. of Glaramara is ON *gliúfr* 'an abrupt descent or chasm'. The rest of the name is obscure. Possibly *-erghe* is ERG 'a shieling'. The *m* would then be a relic of a middle element containing an *m*, e.g. *rūm* 'clearing'. In PNCu(S) the first el. is taken to be *gliúfrum*, the dat. plur. of ON *gliúfr*. This may be right, though it is somewhat difficult to believe that the dative should have been generalized so early. Incidentally it may be noted that a district in the Faroe Islands is called *á Glyvrum* (from ON *gliúfrum*) according to Matras, *Namn och bygd* 44, p. 56.

Glasbury He [*Glasebury* 1346 Ep, 1511 AD], **Glascote** Wa [*Glascote* Hy 2 Dugdale, 1321 BM, 1330 Misc]. First el. perhaps OE *glæs* 'glass'. Glascote might be 'glass-worker's hut', Glasbury 'BURG with glass windows'.

Glasney Co, old monastery [*Glasneye* 1282 Ep, *Glasneyth* 1291, 1306 Ass]. Originally a river-name containing Co *glas* 'blue, grey, green' and a river-name *Neth* identical with the old name of the STRAT. See STRATTON Co.

Glasson Cu [*Glassan* 1260 P, 1278 Cl]. Probably a Celtic name, but the exact etymology is obscure. *Glassan* is a well-evidenced Ir pers. n. Glasson might be an elliptical form of an Ir name consisting of a word for 'homestead' and *Glassan*.

Glasson La [*Glassene* c 1265 CC]. Probably identical with GLAZEN(WOOD).

Glassonby Cu [*Glassanebi* 1177, *Glassanesby* 1230 P]. '*Glassan*'s BY.' *Glassan* is an Ir pers. n.

Glassthorpehill Np [*Clachestorp* DB, *Clakestrop* 1198 Cur]. '*Clac*'s thorp.' Cf. CLAWSON.

Glaston Ru [*Gladestone* DB, *Glaston* 1225 Ep, 1254 Val, *Glaceton* 1286 QW]. The name appears as *Glathestun* in the forged charter BCS 22. This form suggests that the first el. is ON *Glaðr*, which seems to occur in some pl. ns., MDan *Glath*.

Glastonbury So [*Ineswytrin* 601 BCS 835, *Glastonia*, i.e. *Urbs Vitrea* Caradoc, Life of Gildas, *Glastingaea* 704, *Glastingei* 744 BCS 109, 169, *Glestingaburg* 732–55 Wiehtberht (Holder), (on) *Glæstingabyrig* c 1000 Saints, *Glæstingeberia* DB]. The original name was very likely something like *Glastonia*, a derivative of OCelt *glasto-*, Gaul *glastum* 'woad'. The meaning would be 'place where woad grew'. From the Celtic name was formed OE *Glæstingas* 'the people of Glastonia', to which was added OE *īeg* 'island' or BURG. The Welsh *Ineswytrin* (or *Ynisgustrin* Caradoc) is explained by Caradoc as meaning 'glassy island' (*ynis* 'island' and *gutrin*, Welsh *gwydrin*, 'vitrea', i.e. 'of glass'). It is probably a mistranslation of OE *Glæstinga ieg*. But it is worthy of notice that Lat *vitrum* also means 'woad'.

Glatton Hu [*Glædtuninga weg* 957 BCS 1003, *Glatune* DB, *Glattun* 1158 P]. Neither OE *Glæd-tūn* 'joyful TŪN' nor *Glæd-tūn* 'bright TŪN' is a likely formation, but OE *Glædatūn* might have been shortened to *Glædtun-* in a derivative with *-ingas*. *Glædtuningas* might thus be the people of *Glædatūn* 'bright TŪN'. Or more likely OE had a noun **glæd* corresponding to Sw *glad* 'an open place in a forest'.

Glaven. See GLANDFORD.

Glazebrook La [*Glasbroc* 1227 FF, *Glasebrok* 1246 Ass]. Named from **Glaze Brook**, a stream [*Glasebroc* c 1230 CC]. This is identical with **Glaze Brook** D [*Glas* 1240 For]. *Glas* is a Brit river-name derived from Welsh *glas* 'blue, green, grey'. **Glazebury** La is a late name formed from *Glaze-* in Glazebrook.

Glazeley Sa [*Gleslei* DB, *Gleseleia* 1194, *-leg* c 1230 Eyton, *Glaseleye* 1270 Ch]. Cf. GLEASTON La. Etymology doubtful. The first el. might be a river-name derived from OW *gleis*, Welsh *glais* 'stream', which forms the second el. of DOUGLAS, DALCH &c.

Glazenwood Ess [(on) *Glæsne* a 995 Wills, *Glasene* 1179 P, 1204 FF, *Glasnes* 1220, *Glasne* 1224 FF]. Probably identical with GLASSON La. The latter is on the Conder and the Lune and cannot have been named from a river. The two names may mean 'bright spot' and be related to OE *glisian* 'to glister', ON *glæsiligr* 'shining' &c., but their exact history is obscure.

Gleadless YW [*Gladeleys* 1277, *Gledeleys* 1300 Goodall, *Gleydlys* 1473 BM], **Gledhill** YW nr Halifax [*Gledehul* 1275 Wakef], **Gledholt** YW nr Huddersfield [*Gledeholt* 1297 Subs]. First el. OE *glida, gleoda* 'kite'. Second OE LĒAH, HYLL, HOLT.

Gleadthorpe Nt [*Gletorp* DB, *Gledetorp* 1275 RH, *-thorp* 1291 Ch]. First el. perhaps

OE *glida, gleoda* 'kite', though used as a pers. n.

Gleaston (-ē-) La [*Glassertun* DB, *Glestona* 13 StB, *Gleseton* 1269 Ass]. The first el. is possibly a river-name; cf. GLAZELEY. Or it may be related to GLAZEN(WOOD).

Gledhill, -holt. See GLEADLESS. **Gledhow** YW. See ALLERTON GLEDHOW.

Glemham, Great & Little, Sf [*Gl(i)emham, Glaimham* DB, *Glemmeham, Glamessam* 1086 IE, *Glemham* 1180 P, *Northglemham, Parva Glemham* 1254 Val], **Glemsford** Sf [*Glemesford* c 1050 KCD 907, c 1125 Bury, *Clamesforda* DB, *Glammesforda* 1086 IE, *Glamesford* c 1160 NpCh, *Glemeford* 1232 Cl]. The first el. of the names is possibly OE *glēam* 'merriment'. Cf. GLANDFORD. **Glem,** river-name, is no doubt a back-formation.

Glen Magna & Parva Le [*æt Glenne* 849 BCS 455, *Glen* DB, c 1200 Fr, *Glenne* 1199 FF, *Gleen* 1332 Misc, *Magna Glen* 1247 Ass, *Parva Glen* 1242 Fees]. Probably an old name of the Sence, which may be identical with GLEN Li, Nb or from Brit *glenno-* 'valley' (Welsh *glyn* &c.).

Glen R Li [*Glenye* 1276 RH, *Glen* 1365 Pat], **G~** R Nb [(fluuio) *Gleni* c 730 Bede, *Glene* c 890 OEBede, 1256 Ass]. A Brit river-name, derived from Brit *glano-* 'clean, holy, beautiful' (Welsh *glan* &c.) with a suffix *-iā* or *-io*.

Glencoyne Cu [*Glencaine* 1212, *Glenekone* 1255, *Glenkun* 1424 FF]. The elements are *glen* from Brit *glenno-* 'valley' (cf. GLEN Le) and an old name of Glencoyne Beck, possibly identical with Gaul *Cainos* and related to OIr *cáin* 'beautiful'. The meaning would be 'the valley of Glencoyne Beck'.

Glenderama·ckin Cu [*Glenermakan* 1278 CWNS xxiii], **Glenderate·rra** Cu [*Glenderterray* 1729 CWNS xviii, *Glendoweratera* 1789 PNCu(S)]. These names denote two streams which join to. form the Greta. *Glender-* is *Glunduuar* 1247 P and means 'the valley of the river' (a Welsh *Glyndwfr*, consisting of *glyn* 'valley' and *dwfr* 'stream'). The distinguishing elements *-mackin* and *-terra* are obscure. The *-a* of *Glendera* and the *-a* of *-terra* are probably ON *á* 'river'.

Glendon Np [*Clendone, Clenedune* DB, *Clendon* 1220 Fees, *Glendon* 1254 Val]. Identical with CLANDON.

Glendue· Nb [*Glendew, -e* 1239 Hexh]. 'Dark valley', from Welsh *glyn* 'valley' and *du* 'dark, black'.

Glenfield Le [*Clanefelde* DB, *Clenefeld* 1175 P, *Glenefeld* 1254 Val]. See CLANFIELD.

Glentham Li [*Gland-, Glant-, Glentham* DB, *Glentheim* c 1115 LiS, *Glentham* 1197 FF], **Glentworth** Li [*Glentewrde, -uurde* DB, *-worda* c 1115 LiS, *-wurða* 1166 P].

The places are not very far apart, but on opposite sides of a high ridge, near whose western edge Glentworth stands. Very likely the first el. of the names is the word for 'look-out place' (OE *glente) suggested under GLANTLEES Nb.

Glevering Sf [*Glereuinges* DB, *Gleringes* 1206 Cur, *Glering'* 1229 FF, *Glerthyngg* 1346 FA]. Perhaps '*Glēawfriþ*'s people'. OE *Glēawfriþ* may be compared with OG *Glauperaht, Glaumunt. Glēaw-* is OE *glēaw* 'wise'.

Glinton Np [*Clinton* 1060 KCD 809, *Glintone* DB, *-tona* 1121–3 RA, *Gluinton* 1227 Ch]. First el. probably identical with GLYNDE Sx.

Glodwick (glŏdĭk) La [*Glodic* 1190–8 PNLa, *Glothic* 1212 Fees, *Glodyght* 1474 VH]. Has been compared with GLODDAETH in Carnarvon on the supposition that the name is British (see PNLa). But the variation in the early forms is difficult to explain.

Glooston Le [*Glorstone* DB, *Gloreston* 1163, 1230 P]. '*Glōr*'s TŪN.' *Glor* pers. n. is found in HEl. Cf. also (tó) *Gloran ige* BCS 627.

Glossop Db [*Glosop* DB, *Glotsop* 1219 Fees, *Glossope* 1245 Ch, *Gloshop* 1290 Ch]. '*Glott*'s HOP or valley.' *Glott* is found in *Glottes wyll* 854 BCS 477 and is the base of **Glatting** Sx [*Clotinga* DB, *Clottinges* c 1145 Fr]. It is related to G *glotzen* 'to stare'.

Gloster Hill Nb [*Gloucestre* 12 Newminster]. No doubt named from GLOUCESTER.

Glosthorpe Nf nr Bawsey [*Glorestorp* DB, 1194 P]. '*Glōr*'s thorp.' Cf. GLOOSTON. *Glor* occurs as a surname 1202 FF. Roger and Richard Glor were litigants in connexion with a place in the neighbourhood of Glosthorpe, perhaps Glosthorpe itself.

Gloucester (glŏster) Gl [OBrit *Glevum* Holder, *Cair Gloui* c 800 HB, (ad) *Gleawecestre* 804 BCS 313, *Glowecestre* DB, *-ceastre* 1093 ASC (E), *Gleo-, Glouchæstre* 1205 Lay]. Brit *Glēvum* belongs to OW *gloiu*, Welsh *gloew, gloyw* (from *glēvo-*) 'bright' and means 'bright, splendid place' or the like. This was adopted at an early date in a form that gave OE *Gleaw-*, and later in a form that gave late OE *Glowe-*. To the original name was added OE *ceaster* 'Roman fort'. **Gloucestershire** is *Gleawcestrescir* 1016 ASC (D).

Glusburn YW [*Glusebrun* DB, *Glusebrunna* 1170 P, *-burna* 1182–5 YCh 199]. 'Bright stream.' The first el. belongs to MHG *glosen* 'shine, glimmer', ON *glys* 'gleam', Engl *gloʒe* 'to shine brightly' &c.

Glyme R O [*Glim* 958 BCS 1042]. 'Bright stream.' It is difficult to say if the name is English or Celtic. If English, it is related to *gleam*. On the Glyme is **Glympton**

[*Glimtune* 1049–52 KCD 950, *Glintone* DB, *-tona* 1143 Oxf, *Glimtun* 1236 Fees].

Glynch Brook Gl, He, Wo [*Glenc, -ing* 963 BCS 1109, *Glench* 13 Misc]. A Brit river-name *Glanic*, derived from *glano-* 'pure'. Cf. GLEN Li.

Glynde (-īn) Sx [*Glinda* 1165 P, *Glinde* 1210 FF, 1252 Ch], **Glynleigh** Sx in Westham [(to) *Glindlea* 947 BCS 821]. *Glind*, which is common in Sx pl. ns., may be identical with MLG *glinde* 'enclosure, fence'.

ON *gnípa* 'steep rock'. See KNIPE, KNIPTON.

Gnosall St [*Geneshale* DB, *Gnoweshalia* Hy 2 Berk, *Gnousale* c 1165 Fr, *Gnodwes-, Gnodeshall* 1199 PNSt, *Gnoushal'* 1221 Ass, *Gnoushale* 1222 Ass]. If the forms with *-d-* may be trusted, the first el. might be a nickname formed from OE *gnēaþ* 'niggardly'. The DB form rather tells in favour of such a base. The development would then have been *Gnēaþ > Gneaþ > Gnāþ > Gnōþ*. But the absence of early spellings with *a* is against the first el. being OE *gnēaþ*, and possibly we may postulate a formation from the same root with a suffix *þwa-*, something like Old Germanic *gnuþwa-*.

Goadby Le [*Goutebi* DB, 1182 BM], **G~ Marwood** Le [*Goltebi, Goutebi* DB, *Goutebia* c 1125 LeS, *-by* 1268 Ch]. '*Gouti*'s BY.' Cf. GAUTBY.

G~ Marwood was held in part by William Maureward in 1316 (FA). *Maureward* is an OFr nickname and family name meaning 'evil regard, evil eye'.

Goathill Do [*Gatelme* DB, *Gathulla* 1176 P, *Gothull* 1254–6 Ass]. 'Goat hill.'

Goathland (-ōd-) YN [*Godelandia* c 1110 YCh 396, *-land* 1205 Obl, *Gotheland* 1252 Ep]. '*Gōda*'s land' or possibly 'good land'. The name has been partly Scandinavianized (*th* for *d*).

Goathurst (-ōth-) So [*Gahers* DB, *Gothurste* 1292 FF]. 'Goat hill or wood.'

Godalming Sr [(æt) *Godelmingum* c 880 BCS 553, *Godelminge* DB, *-s* 1155 RBE, *Godhelming* 1173 P]. '*Godhelm*'s people.' *Godhelm* corresponds to OG *Godohelm*.

Goddington K [*Godinton* 1190 ff. P, 1197 FF]. 'The TŪN of *Gōda*'s people.'

Godington O [*Godendone* DB, *-dune* c 1160 Oxf, *Godindon* 1208 Cur]. '*Gōda*'s DŪN.'

Godley Chs [*Godel', Godelegh* 1285 f. Court]. '*Gōda*'s LĒAH.'

Godmanchester Hu [*Godmundcestre* DB, *Gudmundcestria* 1168 P]. '*Gōdmund*'s CEASTER or fort.' G~ was a Roman station.

Godmanstone Do [*Godemanestone* 1166 RBE, *Godmaneston* 1201 FF]. '*Godmann*'s TŪN.'

Godmersham K [*Godmeresham* 822, *Godmæreshám* 824 BCS 372, 378, *Gomersham* DB]. '*Godmǣr*'s HĀM.'

Godney So [*Godeneia* 971 BCS 1277, *Godnye* Hy 2 (1227) Ch]. '*Gōda*'s island.'

Godolphin Co [*Wulgholgan* 1194 P, *Gotholgan* 1345 AD]. Co *goth*, *gwyth* 'vein', here probably in the sense 'stream', and perhaps a river-name identical with OLCHON.

Godshill Ha [*Godesmanescamp* DB, *Godeshull* 1230 P, *Goddeshull* 1242 Fees], G~ Wt [*Godeshull* c 1270 Ep, 1340 BM]. Perhaps '*God*'s hill'. *God* may at least in the first instance be a short form of *Godmann*, if *Godes-* is miswritten for *God-*.

Godstone Sr [*Godeston* 1248, *Codeston* 1279 FF, *Coddestone* 1288 SrAS]. Either '*God*'s TŪN' or '*Cōd*'s TŪN'. Cf. CUTSDEAN. The old name was **Walkingstead** [*Wolcnæsstedæ* c 970, *Wolcnesstede* c 980 Wills]. This is '*Wolcen*'s place'. *Wolcen* corresponds to OG *Wolkan*.

Godstow O [*Godstow* c 1135 Godstow, *Godestou* 1156, 1190 P]. 'Place dedicated to the service of God.' G~ was a nunnery. See STŌW.

Godwick Nf [*Goduic* DB, *Godewic* c 1227 BM]. '*Gōda*'s WĪC.'

Gokewell Li [*Gaukewelle* 1163 BM, *Goukewell* 1212 Fees]. 'Cuckoo stream' or '*Gauk*'s stream'. Cf. GAUXHOLME.

Golborne Chs [*Colborne*, *-burne* DB, *Goldbur* 1260, *-burn* 1298 Court], G~ La [*Goldeburn* 1187 P, *-e* 1271 Ass, *Goldburne* 1203 P]. Identical with *Goldburna* 969 BCS 1240 (Wo). 'Stream where marsh marigold (OE *golde*) grew.' For OE *golde* see also GOLDHANGER, GOLDING, GOLDOR, GOLTHO, GOWDALL.

Golcar YW [*Gudlagesarc* DB, *Guthlacharwes* 1307 Wakef]. '*Guðlaug*'s ERG.' *Guðlaugr* is an ON pers. n.

Goldcliff Monm [*Goldclive*, i.e. *rupis aurea* 1191 Gir, *Goldclivia* 1291 Tax]. 'Golden cliff.' Giraldus Cambrensis says the place was named from a cliff which showed a golden colour when the sun shone upon it. G~ was the seat of a priory.

Golden Valley or **Vale** He [*Vallis Stradelie* (*Stratelie*) DB, *Estrateur*, *Istratour*, *Stratdour* c 1150 LL]. The Welsh name means 'the valley of the DORE' (Welsh *ystrad* 'valley'). OW *Istratour* was taken to mean 'valley of gold'; OW *our*, Welsh *aur* means 'gold'.

Golder O [*Goldhora* 987 KCD 661, *Goldor* 1236 Cl]. 'Slope where marigold (OE *golde*) grew.' See ŌRA.

Goldhanger Ess [*Goldhangra* DB, *-hangr* 1202 Cur, *-hangre* c 1230 Bodl], **Golding** Sa [*Goldene* DB, *Golden* 1222 FF]. 'Slope and valley where marigold (OE *golde*) grew.' See HANGRA, DENU.

Goldington Bd [*Goldentone* DB. *Goldinton* 1163, 1167 P]. '*Golda*'s TŪN.'

Goldsborough YN [*Golborg*, *Goldeburgh* DB, *Goldesburgh* 1303 FA], G~ YW [*God-(en)esburg* DB, *Goldesburc* 1166, *Godelesburc* 1170 P, *Godlesburc* c 1200 YCh 516, *Goldesburg* 1172 ib. 511]. G~ YW seems to be '*Godhelm*'s BURG'. Cf. GODALMING. G~ YN may be '*Golda*'s BURG'.

Goldshaw Booth La [*Goldiauebothis* 1324 LaInq]. '*Goldgeofu*'s booth or dairy-farm.' *Goldgeofu* is a woman's name.

Goldsoncott So [*Golsmithecote* AD]. 'The goldsmiths' cottage.'

Goldstone K [*Goldstaneston* 1202 FF]. '*Goldstān*'s TŪN.'

Goldstone Sa [*Goldestan* 1185 f. P]. '*Golda*'s stone.'

Goldthorpe YW [*Golde-*, *Guldetorp* DB, *Goldtorp* 1197 P]. '*Golda*'s thorp.'

Goltho Li [*Golthawe* 1209–35 Ep, *Goltehayt* 1275 f. RH, *Golthag* E 1 BM]. OE *gold-haga* 'enclosure where marigold (OE *golde*) grew'. *Goltehayt* looks like *Goltho thwaite*.

Gomeldon W [*Gumelesdon* 1230 FF, *Gomeledon* 1275 RH, 1311 Ipm]. '*Gumela*'s hill.'

Gomersal YW [*Gomershale* DB, *Gumereshal*, *Gumersale* 13, 14 BM]. '*Gōdmǣr*'s or *Gūþmǣr*'s HALH.'

Gomshall (-ŭ-) Sr [*Gomeselle* DB, *Gumeselva* 1168, *Gumesselua* 1174 P, *Gommeschulue* 1298 BM]. '*Guma*'s SCYLF or hill slope.' *Guma*, a short form of *Gumbeorht* &c., is found in *Gumanedisc* BCS 282.

Gonalston Nt [*Gunnuluestone* DB, *Gunnoluiston* 1175 P]. '*Gunnulf*'s TŪN.' *Gunnulf* (Coins &c.) is ON *Gunnolfr*, ODan *Gunnulf*.

Gonerby Li [*Gunfordebi*, *Gunnewordebi* DB, *Gunwardebi* 1190 P]. '*Gunward*'s BY.' *Gunward* (DB) is ON *Gunnvarðr*.

Goodameavy. See MEAVY.

Gooderstone Nf [*Godestuna* DB, *Gurreston* 1177 ff. P, *Gutherestone* 1254 Val, *Gutherstun* 1267 Misc]. '*Gūþhere*'s TŪN.'

Goodleigh D [*Godelege* DB, 1201 FF]. '*Gōda*'s LĒAH.'

Goodmanham YE [*Godmunddingaham* c 730 Bede, *Gudmundham* DB, *Guthmundham* c 1200 YCh 1122]. 'The HĀM of *Gōdmund*'s people.' The name has been influenced by OScand *Guðmundr*.

Goodnestone K nr Eastry [*Godwineston* 1196 FF, *Guodwinestone* 1279 Ep], G~ K nr Faversham [*Godwineston* 1208 Cl, 1242 Fees, 1291 Tax]. '*Gōdwine*'s TŪN.'

Goodrich He [*castellum Godric* 1102, *castellum Godrici* 1146 Fr, *Godrich* 1307 Ipm]. '*Gōdrīc*'s castle.' The later name is elliptical.

Goodrington D [*Godrintone* DB, *Godrington* 1198 FF]. 'The TŪN of *Gōdhere*'s people.'

Goodshaw Booth La [*Godeshagh, Godischaw* 1324 LaInq]. '*Gōda*'s or *Gōdgȳþ*'s SCAGA or wood.' *Gōdgȳþ* is a woman's name. See BŌÞE.

Goodwin Sands K [*Godewynsonde* 1371 Pat, *the Goodwyn* 1513 LP, *the Goodwins* Sh]. There is a tradition that the Goodwins were an island belonging to Earl Godwine that was washed away by the sea in 1097. But Goodwin may be a name of a dangerous shoal meaning literally 'good friend' and given for the same reason as the wolf is called *gullfot* (lit. 'goldenfooted') in some parts of Sweden.

Goodwood Sx [*Godivewod* 1225 Cl]. '*Gōdgifu*'s wood.' *Gōdgifu* is a woman's name.

Goodworth Ha. See CLATFORD.

Goole YW [*Gowle* 1553 Goodall]. Identical with *gool* 'a small stream, a ditch, a sluice' (1552 &c. OED). Goole is very likely referred to as *gulla . . . in Merskland* 1356 Selby ii. 49.

Goosey Brk [*Goseie* 815, *Goseig* 821 BCS 352, 366, *Gosei* DB]. 'Goose island.'

Goosnargh (gōōzner) La [*Gusansarghe* DB, *Gosenharegh* 1246 Ass]. '*Gōsan*'s ERG.' *Gosan, Gusan* is an OIr pers. n.

Goostrey Chs [*Gostrel* DB, *Gosetre* 1119, *-tro* c 1150, *Gorestre* c 1220, c 1255, *Gorstre* 1267 ff. Chester]. This can hardly be *Gōsan trēo*. *Gorst-trēo* 'gorse tree' gives no sense, and the modern form points to *ō*. Possibly '*Gōdhere*'s tree'.

Gopsall Le [*Gopeshille* DB, *Gopshull* 1242 Fees]. 'The serf's hill.' OE *gōp*, apparently 'servant', occurs in a Riddle. Here *gōp* may be used as a pers. n.

Gore Court K [*Gore* 1198 P], G~ (now St. Joan à Gore from a chapel) W [*Gare* DB, *Goren* 1242 Fees]. OE GĀRA.

Goring O [*Garinges* DB, c 1130 Oxf, 1209–19 Ep, *Garingies* 12 BM], G~ Sx [*Garinges* DB, 1203 Cur, *Garing* 1202 FF]. '*Gāra*'s people.' Cf. GARFORD.

Gorleston Nf [*Gorlestuna* DB, *Gurlestona* 1130 P, *-ton* 1235 Fees]. The first el. may be related to the word *girl*. Cf. GIRLINGTON. It is probably a pers. n.

Gornal St [*Gornhal* Hy 3, *Goronhale* 1375 BM, *Gwarnell, Guarnell* 15 PNSt]. Probably OE *cweorn-halh* 'mill HALH' with a change *c- > g-* analogous to that of *t- > d-* in DUNSTALL.

Gorran Co [(Ecclesia de) *Sancto Gorrono* 1270 Ep]. '(The church of) St. Goran or Goron.' The name is identical with Bret *Gouron* and Welsh *gwron* 'valiant'.

Gorsley Gl [*Gorstley* 1228 Cl]. 'Gorse-covered clearing.' *Gorse* is OE *gorst*, found also in GOSCOTE St.

Gorton La [*Gorton* 1282 Ipm, 1332 Subs]. 'Dirty TŪN.' **Gore Brook** [*Gorbroke* c 1250

LaCh] runs through the township. OE *gor* means 'dirt'.

Gosbeck Sf [*Gosebech* 1179 P, *-bec* 1203 Cur, 1212 Fees]. 'Goose stream.'

Gosberton (-z-) Li [*Gosebertechirche, Gozeberdechercha* DB, *Goseberdeschirche* 1167 P, *Gosberkirke* 1212 Fees]. '*Gosbeorht*'s church', later changed to Gosberton. *Gosbeorht* is probably a Continental name (OHG *Gauzpert, Gosbert* from *Gautberht*).

Goscote Le [*Gosecot* DB, *Gosecote* c 1125 LeS]. 'Hut for geese.'

Goscote St [*Gorstycote* 13 PNSt]. 'Hut among gorse.'

Gosfield Ess [*Gosfeld* 1198, 1202 FF, 13 BM, *Gosefeld* 1254 Val]. 'FELD frequented by (wild) geese.'

Gosford D [*Goseford* 1249 Ass], G~ O [*Goseforde* 1234 Osney], G~ Wa [*Gosseford* 1202 Ass], **Gosforth** Cu [*Goseford* 12 StB], G~ Nb [*Goseford* 1166 RBE, 1212 Fees]. 'Goose ford.'

Gosport Ha [*Goseport* 1250 Cl, 1285 Ch]. Perhaps 'market-place where geese were sold'.

Gossington Gl [*Gosintune* DB, *Gosinton* 1194 P, *Gosington* 1196 FF]. 'The TŪN of *Gōsa*'s people.' OE *Gōsa* is possibly evidenced in *Gosanwelle* BCS 754.

Goswick (gŏzĭk) Nb [*Gossewic* 1202 FF, *Gosewic* 1208–10 Fees]. 'Goose farm.'

Gotham (-ōt-) Nt [*Gatham* DB, c 1085 LVD, *Gataham* 1152 BM, *Gotham* 1291 Tax]. 'Homestead where goats were kept.'

Gotherington (gŭdhertn) Gl [*Godrinton* DB, *Gutherintona* 1209 Fees, *Goderinton* 1291 Tax]. 'The TŪN of *Gōdhere*'s people.'

Goudhurst (-ow-) K [*Guithyrste* 11 DM, *Gudherst* 1202 Cur]. '*Gūþa*'s HYRST.'

Goulceby Li [*Colchesbi* DB, *Colkesbi* 1193 P, *Golckesbi* 1185 TpR, *Golkesbi* 1212 Fees, *Golkesbi, Golcebi* 1202 Ass]. The correct form seems to be *Golkesby*. No pers. n. *Golk* is known. Possibly the first el. is ON *Guðleikr*, OSw *Gudhlek*. The common initial *C-* in early spellings may, however, indicate that the first el. is rather ON *Kolkr*, ODan *Kolk*, a byname.

Goverton Nt [*Sofertune* (sic) 958 YCh 2, *Guverton* 1287, 1302 Ass, *Goverton* 1302 FA]. The first el. is very likely the Anglo-Scand pers. n. *Gunuert, Gonuerd, Gunuer* (DB), from OScand *Gunnfrøðr*. For loss of *n* before a labial consonant cf. LYFORD, STOFORD and the like. OE *Godfrith*, suggested in PNNt(S), would not explain *-v-*.

Gowbarrow Cu [*Golbery* c 1250, *Golebergh* 1294 PNCu(S), *Gollebergh* 1294 Cl]. Perhaps 'windy hill', the elements being ON *gol* '(a gust of) wind' and *berg*.

Gowdall YW [*Goldale* 1220 Pont, *Goldhale* 1353 Goodall]. 'HALH overgrown with marigold' (OE *golde*).

Gowthorpe YE [*Gheuetorp* DB, *Gou(k)thorp* 1235 Ep, *Goutorp* 1235 FF]. See GAW-THORPE.

Goxhill Li [*Golse* DB, *Golsa* c 1115 LiS, *Gosla* 1194 P, *Gousele* 1212 Fees, *Gousle* 1218 Ass, 1232 Ep], G~ YE [*Golse* DB, *Gousla* c 1185 YCh 1310, *Gousle* 1246 FF]. OScand *Gauks-lā*, which may be 'the stream (OScand *lá*) of *Gaukr*' or a Scandinavianized form of OE *Gēaces-lēah*. Cf. YAXLEY.

Goyt R Chs, Db [*Guit* 1244 FF, *Gwid*, *Gwit*, *Goyt* 1285 For]. Welsh *gwyth* 'channel, conduit', identical with OCo *guid* 'a vein', MBret *goeth* 'a brook'.

OScand **grā(r)** 'grey'. See GRAYRIGG, GRAY-THWAITE.

Graby Li [*Greibi* DB, 12 DC, *Greyby* 1242 Fees, *Grayeby*, *Gratheby* 1275 f. RH]. The first el. may be ON *grey* 'bitch', possibly used as a nickname.

Gracechurch London [*Gerschereche* 1054 ASCh, *Gerscherche* 1190 P, *Garscherche* 1254 Val]. 'Grass church', so called perhaps because standing in a grass plot or having a roof of turves. Stow's statement that the name alludes to 'the Herbe market there kept' is not very likely to be correct.

Gracedieu Le [*la Gracedieu* 1243 Ep, *La Grace Deu* 1254 Val, *Gratia Dei* 1257 Misc]. 'God's grace.' Gracedieu was a monastery.

Grade Co [(Ecclesia) *Sancte Grade* 1291 Tax, *Grade* 1377 PT]. '(The church of) St. Grada.'

OE **græf** 'grave, trench, pit' and **grāf**, **græfe**, **grāfe** 'grove, brushwood, thicket' are difficult to keep apart, but the more common el. is doubtless the word for 'grove'. Even when the form rather suggests OE *græf*, the source is often *grāf* or *græfe*, as in GRAVELEY, GRAVESEND, GED-GRAVE, HARGRAVE &c. OE *græf* is certain in some cases, as CHALGRAVE, ORGRAVE, ORGREAVE Y, REDGRAVE. OE *grāf* is obviously the source in COSGROVE, GRAFHAM, -TON, GRAYSHOTT, OE *græfe* in CONGREVE, WHIT-GREAVE &c. For OE **grafa* 'ditch' (= OHG *grabo*) see BYGRAVE, GRAVENEY.

Graffham Sx [*Grafham* DB, *Grofham* 1292 Ch], **Grafham** Hu [*Grafham* DB, 1159 P, *Grofham* 1342 Cl]. OE *Grāfhām* 'the HĀM by the grove'.

Grafton Chs [*Grafton* 1358 Chamb], G~ Gl [*Graftone* Hy 3 Misc], G~ He [*Crafton* 1303, *Grafton* 1316 FA], G~ Regis Np [*Grastone* DB, *Grafton* 12 NS, *Grafton* 1204 Cur], G~ Underwood Np [*Grastone* DB, *Grafton* 12 NS, 1202 Ass], G~ O [*Graptone* DB, *Grafetona* c 1150 Fr, *Grafton* 1190 P], G~ Sa [*Grafton* 1291 Tax], **East & West** G~ W [*Graftone* DB, *Graftona* 1130

P, *Est-*, *Westgrafton* 1198 Fees], **Temple** G~ Wa [*Greftone* 962 BCS 1092, *Grastone* DB, *Grafton* 1182 P, 1189 BM], G~ **Fly-ford** Wo [*Graftun* 884, 972 BCS 552, 1282, *Garstune* DB], G~ **Manor** Wo [*Grastone* DB, *Grafton* 1212 Fees], G~ YW [*Graftune* DB, *Graftona* 1180–9 YCh 730]. OE *Grāf-tūn* 'TŪN in or by a grove'.

G~ **Flyford**. See FLYFORD.—**Temple** G~ was held by the Templars at least in 1189 (BM).—G~ **Underwood** means 'Grafton in the wood'.

Grain, Isle of, K [*Grean* c 1100 Text Roff, 1278 QW, 1291 Tax, *Grien* 1205 Ch, *Gren* 1232 Cl]. A derivative of an OE **grēon* 'sand, gravel' corresponding to MLG *grēn* 'sand on the sea-shore', MHG *grien* 'gravel, sandy shore', ON *grión* 'grits'.

Grainsby Li [*Grenesbi* DB, *Greinesbi* c 1115 LiS, c 1155 BM, *Grainesbia* 1212 Fees]. '*Grein*'s BY.' *Grein*, lit. 'a branch', is a common ON byname.

Grainthorpe Li [*Germund(s)torp* DB, *Ghermudtorp* c 1115 LiS]. '*Germund*'s thorp.' Cf. GANTHORPE.

Grampound Co [*Graundpont* 1373 AD, *Grauntpont* 1422 BM]. Fr *grand pont* 'great bridge.'

Granby Nt [*Granebi*, *Grenebi* DB, *Granebi* 1236 Fees, *-by* 1252 Ch, *Grenebi* c 1180 BM]. Apparently '*Grani*'s BY' (ON *Grani* pers. n.), but the common *Grenebi* is curious.

Grandborough Bk [*æt Grenebeorge* c 1060 KCD 962, *Grenesberga* DB], G~ Wa [*Greneburgan* 1043 Th, *-berge* DB, *-berg* 1198 Fees]. 'Green hill.' See BEORG.

Gransden, Little, Ca [*Grantandene* 973 PNCa(S), *Grentedene* c 1050 KCD 907, *-dena* c 1080 ICC, *Gratedene* DB, *Granten-dene* 1086 IE, *Granteden* 1194 P], **Great** G~ Hu [*Grantesdene* DB, *Grantendene* 1168 P]. '*Granta*'s or *Grenta*'s valley.' A person named *Grante* occurs in the Croyland Cart. A strong form *Grente* occurs in *Grentes mere* 1016 KCD 724. Cf. also GRANSMOOR. The names belong to Dan *grante* 'to complain', ON *grettast* 'grin, show one's teeth', South G *angranzen* 'to grumble' &c.

Gransmoor YE [*Grenzmore*, *Grentesmor* DB, *Grancemor* 1240 FF]. '*Grente*'s moor.' Cf. GRANSDEN.

Granta R Bd, Ess, Ca [*Gronte* c 745 Felix, *Grantan stream* c 890 OEBede, *Grante* 1286 Ass]. A Brit river-name related to Celtic-Latin *gronna*, *gromna* 'bog', ON *grunnr* 'shallow', Dan, Sw *grums* 'muddy deposit' &c. The meaning would be 'muddy river'. From the river-name is derived **Grant-chester** Ca [*Grenteseta* c 1080 ICC, *Grante-seta* DB, *-sete* 1242 Fees]. OE *Grante-sǣte* 'the dwellers on the Granta'. Cf. CAM-BRIDGE. The river is alternatively CAM.

Grantham Li [*Grantham*, *Granham*, *Grand-ham* DB, *Graham* DB, 1130 P, 1254 Val].

'*Granta*'s HĀM.' Cf. GRANSDEN. The loss of *nt* in many early forms is due to Norman influence. Or the first el. may be an OE *grand 'gravel', corresponding to LG *grand*. Cf. ON *grandi* 'sand-bank'.

Grantley YW [*Grantelege* c 1030 YCh 7, *Grentelai* DB, *Grantle* 1207 FF]. '*Granta*'s LĒAH.' Cf. GRANSDEN.

Grappenhall Chs [*Gropenhale* DB, 1291 Tax, *-hal* 1288 Court]. The first el. may be an OE *grōpe, related to OE *grōp*, *grēp*, *grēpe* 'a ditch, drain'. See HALH.

Grasby Li [*Gros(e)bi* DB, *Grossebi* c 1115 LiS, 1166 P, *Gressebi* 1165 P, 1202 Ass]. OScand *Griōtsbȳr* 'BY in a stony district'. Cf. GRÖSSBY in Sweden from *Griōtsbȳr*. First el. ON *griót* 'stones', OSw *grȳt* 'stony ground'.

Grasmere We [*Ceresmere* 1203 Cur, *Gresemere* 1246, *Gressemer* 1254 Kendale]. 'Grass mere', referring to grassy shores or to vegetation in the lake. The first el. is OScand *gres* 'grass'. The examples above refer to Grasmere vil. The lake is *Grysemere* 1374 Kendale. The early forms *Gresemere* &c. very likely indicate that the original name was OScand *Gressǣr* (*Gressiōr*), to which OE *mere* was added.

Grassendale La [*Gresyndale* 13 WhC, ?*Gresselond Dale* VH]. First el. ME *gresing* 'pasture' or *gresland* 'grass land'.

Grassington YW [*Ghersintone* DB, *Gersinton* 1212 FF, 1271 Ipm, *Gersigton* 1213 FF]. 'Grazing-farm.' First el. ME *gresing* 'grazing, pasture' or an OE *gærsen, *gersen 'of grass'. Cf. GARSINGTON.

Grassthorpe Nt [*Grestorp* DB, 1226 Ep, 1242 Fees]. First el. perhaps as in GRASBY. Or the first el. might be an OScand *Griōtr, a short form of *Griōtgarōr*; cf. ON *Griōti*. The first el. may alternatively be OEScand *gres* (*græs*) 'grass'.

Graston Do [*Gravstan* DB, *Grauestone* 1269 Ch, *Graveston* 1269 Misc]. OE *græf-stān 'gravestone' or *grafen stān* 'engraved stone'?

Grately Ha [*Greatteleiam* c 929 BCS 1341, (æt) *Greatanlea* c 935 Laws, *Greteleia* 1130 P]. 'Great LĒAH.'

Gratton D in High Bray [*Gretedone* DB, *Gratedene* 1242 Fees], G~ Db [*Gratune* DB, *Gratton* 1290 Cl]. G~ D is OE *grēate dūn* 'great hill'. The same may be the etymology of G~ Db.

Gratton D in Meavy [*Gropeton* 1242 Fees, 1303 FA]. Cf. GRAPPENHALL.

Gratwich St [*Crotewiche* DB, *Grotewic* 1176 P, *Grotewis* 1242 Fees, *Gretewyz* 1236, 1242 Fees]. The first el. is a derivative of OE *grēot* 'gravel', e.g. an OE *Grēote* (cf. GREET), but not a river-name, as G~ is nr the Blythe. The meaning may be 'gravelly place'. Second el. wīC.

Graveley Ca [*Græflea* 1060 Th, *Gravelei* DB], G~ Hrt [*Grauelai* DB, *Gravele* 1200 Cur]. 'LĒAH with brushwood in it' (cf. GRĀF, GRǢFE).

Graveney K [(æt) *Grafonaea*, *Grafoneah* 811 f. BCS 335, 341, *Grauenea* 11 DM]. Originally a name of the stream at G~ [*Grafon eah* 814 BCS 348]. The name corresponds to OG *Grabanowa* and contains an OE *grafa 'ditch' corresponding to OHG *grabo*, G *Graben*. Second el. OE ĒA.

Gravenhunger Sa [*Gravehungre* DB, *Gravinhunger* 1283 Ipm]. 'Slope covered with brushwood.' Cf. GRĀF, GRǢFE, and HANGRA.

Gravenhurst Bd [*Grauenhest* DB, *Gravenherst* 1206 FF]. 'Hill with a grove or brushwood.' Cf. GRĀF, GRǢFE, and HYRST.

Gravesend K [*Gravesham* DB, *Grauessend* 1157 P, *Graveshende* 1236 Fees]. Identical with **Gravesend** Np, an old hundred [*Grauesende* 1066-75 GeldR, DB]. The Np Gravesend may be referred to as *æt þæs grafes ende* 944 BCS 792. This is 'at the end of the grove'. Cf. GRĀF.

Grayingham Li [*Gra(i)ngeham* DB, *Greingheham* c 1115 LiS, *Grahingaham* 1157 Fr]. 'The HĀM of *Grǣg(a)*'s people.' *Grǣg(a)* is not evidenced, but is very likely the first el. of GREINTON. OG *Grawo* occurs. *Grǣg(a)* is derived from *grǣg* 'grey'.

Grayrigg We [*Grarigg* c 1165 Kendale, *-rig* c 1200 CC]. 'Grey ridge' (OScand *grā(r)* 'grey' and *hryggr* 'ridge').

Grayshott Ha [*Grauesseta* 1185 P, *Graveschete* c 1200 Ep]. OE *grāfes-scēat* 'strip of wood' or the like. Cf. GRĀF, SCĒAT.

Graythwaite La [*Graythwayt* 1336 FC]. See THWAITE. The first el. may be ON *grā(r)* or OE *grǣg* 'grey'.

Grazeley Brk [(on) *Grægsole burnan*, *hagan* 946-51 BCS 888, *Greshull* 1198 P]. Second el. OE *sol* 'mire, wallowing-place'. The first can hardly be OE *grǣg* 'grey'. It may be an OE noun *grǣg*, the base of ME, Mod *grey* 'badger'.

Greasbrough YW [*Gersebroc*, *Gres(s)eburg* DB, *Gresebroc* c 1160 YCh 175, 1195 P]. 'Grassy brook.' See GÆRS.

Greasby Chs [*Gravesberie* DB, *Grauesberi* c 1100, *-byri* c 1150, *Grauesbi* c 1155, *-by* c 1190 Chester]. OE *Grāfes-* or *Grǣfesburg* Scandinavianized to Grafesby. The name may mean 'BURG by a grove or by a trench or canal'.

Greasley (-ēz-) Nt [*Griseleia* DB, *Greseley* 1230 P, *Greselley* Hy 3 Ipm]. OE *grēosnlēah* 'gravelly LĒAH'. OE *grēosn* means 'gravel, pebble'.

OE **grēat** adj. 'thick, stout, big' is a rare first el. in pl. ns.; see GARSTON La, GRATELY, GRATTON. Names in *Great-*, *Gret-* usually contain OE GRĒOT 'gravel'. But *Great* is common as a distinguishing el.

Greatford Li [*Grite-*, *Greteford* DB, *Gretford* 1191, *Grafford* 1192 P]. 'Gravelly ford', OE *grēot-ford*. See GRĒOT.

Greatham (-ĕt-) Du [*Gretham* 1196 P, 1208–10 Fees], G~ Ha [*Greteham* DB, *Grietham* 1167 P, *Grutam* 1236 Fees], G~ (-ĭt-) Sx [*Gretham* DB, *Gretheam* 1121 AC, *Gruteham* 12 Fr]. OE *Grēot-hām* or *-hamm*. See GRĒOT.

Greatness K [*Gretaniarse*, *Greatnearse* 821, *Greotanedesces lond* 822 BCS 367, 370, *Greteness* 1206 Cur]. The first el. may be a stream-name *Grēote* 'gravelly stream' (cf. GREET). The second varies between OE EDISC and ERSC.

Greatworth (-ĕt-) Np [*Grentevorde* DB, *Gretteworth* 12 NS, *Gretewrd* 1200 Cur, *Gruttewrth* 1254 Val]. 'Gravelly WORÞ.' First el. OE **grēoten* 'gravelly'.

Grebby Li [*Gredbi*, *Greibi* DB, *Grebbi* 1212 Fees]. OScand *Griõt-bȳr* identical with GRYTTBY in Sweden, 'BY on stony ground'. Cf. GRASBY.

Greendale D [*Grendel* R 1, c 1200, *Grendil* c 1200 Torre, *Grendell* 1200 Ch, *Grindel* 1275 RH]. OE *Grēn-dæl* 'green valley'. The valley also gave name to **Greendale** or **Grindle Brook** [*Grendel* 963 BCS 1103].

Greenfield, common, is self-explanatory. G~ Li is *Grenefeld* c 1150 &c. BM.

Greenford Mx [*et Grenan forda* 845 BCS 448, *Greneforde* DB, *Greneford Magna* 1254 Val]. 'Green ford.' Cf. PERIVALE.

Greenhalgh (grēna) La [*Greneholf* DB, *Grenhole* 1212 Fees, 1216 Ch]. 'Green hollow.' Second el. OE HOLH 'hollow'.

Greenham Brk [*Greneham* DB, *Grenham* 1206 f. Cur]. 'Green HAMM.'

Greenham So in Stawley [*Grindeham* DB, 1201 Ass, *Gryndenham* 1327 Subs]. The first el. is probably a stream-name *Grinde*, derived from *grindan* 'to grind'. The meaning may be 'mill brook' or 'brook that grinds its bed, carries away gravel' or the like. Cf. *Grindanbroc* 877 BCS 544 (Ha).

Greenhaugh Nb [*le Grenehalgh* 1326 Ipm]. 'Green haugh.'

Greenhead Nb [*le Greneheued* 1290 Sc]. 'Green hill.' Cf. HĒAFOD.

Greenhill Wo. See GRIMLEY.

Greenhithe K [*Grenethe* 1264 Pat, 1277 Ipm, *Grenehethe* 1405 BM]. 'Green landing-place.' Cf. HȲÞ.

Greenhow YN [*Grenehou* c 1180 YCh 799, *Grenho* 1197 (1252) Ch], G~ YW [*Grenhou* 1269 Ch]. 'Green hill or mound.' Second el. OScand HAUGR or OE HŌH.

Greenoak YE [*Grenaic* 1199, 1202, *Greneic* 1199, 1208 FF]. 'Green oak.' Second el. OScand EIK.

Greenodd La [*Green Odd* 1774 map]. 'Green promontory.' Second el. ON ODDI 'promontory'.

Greenriggs We [*Grenerig* 1274 Kendale]. 'Green ridge.' Second el. ON HRYGGR.

Greenstead Ess nr Colchester [*Grenestede* c 958 BCS 1012, *Grenstede*, *Grænstydæ* c 995 ib. 1288 f.], **Greensted** (-ĭn-) Ess nr Chipping Ongar [*Gernesteda* DB, *Grenstede* 1254 Val]. 'Green place.'

Greenwich (grĭnij) K [*Gronewic* 918 BCS 661, *Grenewic* 964 Fr, *Grenawic* 1013 ASC (E, D), *Grenviz* DB, *East*, *West Grenewych* 1291 Tax]. 'Green WĪC.'

Greet R Nt [(andlang) *Greotan* 958 BCS 1029, *Girt* c 1540 Leland]. 'Gravelly stream.' OE *Grēote* is derived from OE *grēot* 'gravel'.

Greet Gl [*Greta* 12 Winchc, 13 BM], G~ Sa [*Grete* 1183 Eyton, 1204 Cur, *Groete* 1278 Misc, 1291 Tax], G~ Wo [*Grete* 1255 FF]. Either OE *Grēote*, river-name, identical with GREET Nt, or OE *grēot* 'gravel'.

Greetham Li [*Gretham* DB, 1259 Ipm, *Greham*, *Graham* 12 DC, *Greteham* 1233 Ep], G~ Ru [*Gretham* DB, 1202 Ass, 1238 Ep]. OE *Grēot-hām* or *-hamm*. Cf. GRĒOT.

Greetland YW [*Greland* DB, *Gretland* 1277 Wakef]. 'Gravelly land.' Cf. GRĒOT.

Greetwell Li [*Grentewelle* DB, *Gretwella* c 1115 LiS, *Gretewelle* 1120–2 YCh 467]. 'Gravelly stream.' Cf. GRĒOT.

Greinton So [*Graintone* DB, 1166 RBE, *Greinton* 1201 Ass, 1202 FF]. '*Grǣga*'s TŪN.' Cf. GRAYINGHAM.

Grendon Underwood Bk [*Grennedone* DB, *Grenedon* 1163, 1194 P], G~ Np [*Grendone* DB, *-don* 1186 P, 1220 Fees], G~ Wa [*Grendone* DB, *Grendon* 1236 Fees]. 'Green DŪN or hill.' Cf. GRAFTON UNDERWOOD.

Grendon Bishop & Warren He [*Grenedene* DB, *Grendene* 1242 Fees, *Grenden* 1241 Ch, *Grendone* 1241 Ep, *Grendon* 1249 Fees, *Grendene Waryn* 1291 Tax, *Grendone Episcopi* 1316 FA]. 'Green valley.' Forms in *-don* are incorrect; they are found for both Grendons.

G~ **Bishop** was held by the Bishop of Hereford.—Warinus de Grendene is mentioned c 1270 Glouc. He or a namesake of his was sheriff of Hereford early t. Hy 3. *Warin* is a Fr. pers. n. of OG origin.

OE **grēne** 'green' is a common first el. in pl. ns. See GREEN-, GREN- (passim), GRANDBOROUGH, GRINDALE, GRINDLEY, GRINGLEY, GARLINGE &c. Sometimes the source is rather OScand *grønn*. In cases such as WOOD GREEN Green is ME *grēne* 'green spot, village green, common'.

OE **grēosn** 'gravel, pebble'. See GREASLEY, GRESLEY, GRESSENHALL, GRISTON.

OE **grēot** 'gravel'. See GREET, GREAT-, GREET- (passim), GRET-, GIRTON, GIRTFORD

&c. The corresponding OScand word (ON *griót*, OSw *grȳt*) means 'stones, stony ground'. See GRETA, GRASBY, GREBBY.

Gresham Nf [*Gersam, Gressam* DB, *Gresham* 1194 P, 1242 Fees, *Gresseham* 1254 Val]. 'Grazing-farm', OE *Gærs-hām*. The form *Gres-* may be due to Scand influence.

Gresley (-ēz-), **Castle** & **Church**, Db [*Gresele* c 1125 LeS, *Griseleia* 1130 P, 1166 RBE, *Castelgresele* 1252 FF]. Identical with GREASLEY.

Gressenhall Nf [*Gressenhala* DB, *Gresenhal* 1203 Ass, *Gressinhale* 1254 Val, *Gressingehal* 1195, *Gressinghal* 1196 P, *Grossenhale* 1289 Bodl]. 'Gravelly HALH.' First el. OE *grēosn* 'gravel'.

Gressingham La [*Ghersinctune* DB, *Gersingeham* 1183, 1194 P, *Guersingueham* 12 Lancaster]. 'Grazing-farm.' First el. ME *gresing* 'grazing, pasture'.

Gresty Chs [*Greysty* 1312 Ormerod, 1400 BM, *Graysty* 1395 BM]. Possibly '*Grǣga*'s path' (OE STĪG). Cf. GRAYINGHAM. Or rather the first el. is as in GRAZELEY.

Greta (-ē-) R Cu [*Greta* 1278 CWNS xxiii], G~ R YW, La [*Gretagila* c 1215 CC, *Grythawe* 1307 YInq], G~ R YN [*Gretha* 1279 f. Ass]. ON *Griótá* 'stony stream'. Cf. GRĒOT. *Griótá* is a river in Iceland.

Gretton Gl [*Gretona* 1175 Winchc, *Greton* 1201 Cur, *Gretton* 1236 Fees], G~ Np [*Gretone* DB, *Gretton* 1163 P, 1220 Fees]. OE *Grēot-tūn* 'TŪN on gravelly soil'.

Gretton Sa [*Grotintune* DB, *-tun* c 1185 Eyton, *Grotington* 1195 Cur, *Gretinton* 1219 FF, *Greotytone* 1327 Subs]. OE *grēoten tūn* 'TŪN on gravelly soil'. First el. OE **grēoten* 'gravelly'.

Grewelthorpe YW [*Torp* DB, *Gruelthorp* 1279–81 QW, *Grewelthorpe* 1290 Misc, *Grouelthorp* 1303 FA]. Originally Thorp. The early forms of the first el. agree exactly with those of the word *gruel* 'fine flour' (from OF *gruel*), and it may be that word, possibly used as a nickname. A nickname may also have been formed from an early form of Fr *gruau* 'a young crane'.

Greysouthen Cu [*Craykesuthen* 1185–9 Holme C, *Creiksuthen* 1230 FF]. '*Suthan*'s cliff.' *Suthan* is an OIr pers. n., found also in English sources (*Sudan* DB, *Suthen* LVD). Cf. *Mælsuðan* (on OE coins) from OIr *Maelsuthain*. *Grey-* is MIr *craicc* 'crag, rock, cliff', identical with Welsh *craig*.

Greystoke (-ōk) Cu [*Creistock* 1167 P, *Craystok* 1292 QW, *Creystok* 1294 Ch]. 'STOC on R Cray.' Cray, identical with CRAY K, was probably the name of the tributary of the Petteril on which Greystoke is.

Greywell Ha [*Graiwella* 1167 P, *Greywell* 1235 Cl, *Greiwell* 1236 Ass]. Second el. OE *wella* 'a stream'. The first is very likely as in GRAZELEY.

Gribthorpe YE [*Gripetorp* DB, *Gripthorp* 1231 FF]. '*Grip*'s thorp.' First el. ODan *Grip*, ON *Grípr*.

Griff Wa [*Griva* Hy 2 BM, *la Griue* 1203 Ass, *La Greve* 1280 Ipm, 1285 QW]. ON *gryfia* 'hollow, pit , partly influenced by OE *grǣfe* 'grove'.

Griffe Db [*Grif* 1286 BM, 1294 Ch]. ON *gryfia* 'a hollow, pit'.

Grimblethorpe Li [*Grimchiltorp* c 1115 LiS, *Grimkiltorp* c 1162 DC, *Grimpilthorp* 1242 Fees]. '*Grimkell*'s thorp.' *Grimchell* DB is ON *Grímkell*.

Grimesthorpe YW [*Grimestorp* 1297 Subs]. '*Grīm*'s thorp.' *Grim* DB &c. is the common ODan *Grīm*, ON *Grímr*.

Grimley Wo [*Grimanlea, -leage* 851 BCS 462, *Grimanleh* DB]. The same first el. is found in **Greenhill** Wo [*Grimeshyll* 816, *Grimanhyll* 957 BCS 356, 993, *Gremanhil* DB]. 'Wood and hill haunted by a ghost or spectre.' OE *grīma* is found in these senses. Grimley and Greenhill are near each other.

Grimoldby Li [*Grimalbi, Grimoldbi* DB, *Grimolbi* c 1115 LiS, *Grimmoldibi* R 1 Cur]. '*Grīmaldi*'s BY.' *Grīmaldi* is found in ON and OSw.

Grimsargh (-zer) La [*Grimesarge* DB, *-argh* 1246 Ass]. '*Grīm*'s ERG.' Cf. ERG and GRIMESTHORPE.

Grimsbury O [*Grimberie* DB, *-beri* 1195, *-biri* 1197 P]. '*Grīm*'s BURG', but probably *Grīm* is not a man's name, but identical with *Grīm* in GRIM'S DITCH. Or else the original name was *grīman burg*. Cf. GRIMLEY.

Grimsby, Great, Li [*Grimesbi* DB, c 1115 LiS, 1130 P], **Little** G~ Li [*Grimesbi* DB, *Parva Grimesbia* c 1115 LiS, *Parva Grimesbi* 1212 Fees], **Grimscote** Np [*Grimescot* 1199 P, 1201 Cur, 1202 Ass]. '*Grīm*'s BY and COT.' Cf. GRIMESTHORPE.

Grim's Ditch W, an ancient earthwork on S. border of Wilts [*Grymmesdich, Grimesdiche* 1280 QW], **Grim's Dyke** or **Devil's Dyke** O, an ancient earthwork nr Wallingford [*Grimesdich* c 1220 AJ xxii, *Grymesdiche* 1298 Eynsham]. The same name is found in other places in early sources. *Grimes dic* 956 BCS (934, 985), 1045 KCD 778 is another ancient earthwork in South Wilts (S. of the Nadder). *Grymesdich* 1291, *Grimesdich* 1295 Ch was nr Berkhamstead Hrt. *Grimesdich* AD i was at Edgware Mx. The name is synonymous with WANSDYKE. In Old Norse *Grímr* is used as a byname of Oðinn. The name is identical with ON *grímr* 'a person who conceals his name', lit. 'a masked person', and related to OE *grīma* 'a mask'. It refers, like *Grímnir*, to Oðinn's well-known habit of appearing in disguise. No doubt the Saxons used *Grīm* in the same way.

Grimstead W [*Greme-, Gramestede* DB, *Gremesteda* 1165, *Grenested* 1161 P, *Grimested* 1242 Fees]. OE *Grēn-hǣmstyde* 'green homestead, homestead in green fields'. Cf. HĀMSTEDE.

Grimsthorpe Li [*Grimestorp* 1212 Fees]. Identical with GRIMESTHORPE.

Grimston Le [*Grimestone* DB, *-tona* c 1125 LeS], **G~** Nf [*Grimastun* c 1035 Wills, *Grimestuna* DB], **G~ Hill** Nt [*Grimeston* 1188 P, 1212 Fees], **G~** Sf [*Grimestuna* DB], **G~** YE in Garton [*Grimestone* DB], **G~** YE nr York [*Grimestone* DB, *-ton* 1221 FF], **Hanging G~** YE [*Grimeston* DB, *Hengandegrimeston* 1219 Ass], **North G~** YE [*Grimeston* DB, 1231 FF], **G~** YN [*Grimestone* DB], **G~** YW [*Grimestun* DB, *Grimestona* 1175 YCh 359], **Grimstone** Do [*Grimeston* 1212 Fees, 1285 FA]. 'GRĪM's TŪN.' Cf. GRIMESTHORPE. G~ YE in Garton is G~ GARTH in PNER [*Grymston Garth* 1618 FF]. Cf. GARTON.

Hanging means 'situated on a slope'.

Grindale YE [*Grendele* DB, *Grendala* 1125-30 YCh 1135, *-dal* 1207 Cur]. 'Green valley.'

Grindle Sa [*Grenhul* c 1190 Eyton]. 'Green hill.'

Grindleton YW [*Gretlintone* DB, *Grillington* 12 Pudsay, *Grenlington* 1251 Ch, *Grinlington* 1258 Ipm]. Perhaps 'the TŪN of *Grentel's* people'. ***Grentel** would be a derivative of *Granta*. See GRANSDEN. Or the first el. may be derived from an OE *Grēnlēah.* Cf. GRINDLEY.

Grindley St [*Grenleg* 1251 Ch], **Grindlow** Db [*Grenlawe* 1199 (1285) Ch]. 'Green LĒAH and hill.' Cf. HLĀW.

Grindon Du nr Stockton [*Grendune* 1208-10 Fees, *Gryndone* 1539 FPD], **G~** Du nr Sunderland [? *Grendune* c 1190 Godric, *Grendon* 1291 Tax], **G~** Nb nr Berwick on Tweed [*Grandon* 1208-10 Fees], **G~** St [*Grendone* DB, *Grenedun* 1236 Fees]. 'Green DŪN or hill.'

Gringley on the Hill Nt [*Gringeleia* DB, *-lay* 1184 P, *Gryngeleia* Hy 2 (1316) Ch, *Gringeleg* 1234 FF, *Gringele* 1252 Ch], **Little G~** Nt [*Grenleige, Greneleig* DB, *Grenley* 1242 Fees, *Grenlay* 1278 Ipm]. The two Gringleys are generally kept well apart in early sources. Little G~ is 'green LĒAH.' G~ on the Hill has as first el. a gen. plur. in *-inga. Gringe-* might possibly be from *Grēninga-, Grēningas* being 'people from Little Gringley' or 'dwellers on a green hill' or the like. But equally well the source may be OE *Grǣgingas.* Cf. GRAYINGHAM. A third possibility is OE *Gāringas*; cf. FRINGFORD O and the like. The ridge on which G~ is may have been called *Gār(a).* Eventually Little Gringley got the form originally belonging only to G~ on the Hill. The Gringleys are c 6 m. apart.

Grinsdale Cu [*Grennesdale* c 1180 Lanercost, *Gremesdale* 1279 Ass, *Grimesdal* 1281 Pat, *Grenesdale* 1200 FF, *Grinnisdal* 1271 Ipm]. It is probable, as suggested in PNCu(S), that the first el. is a pers. n., but that name cannot be an ON *Grennir* 'grinner', for such a word (or ON *grenna* 'grin') is not on record. ON *Grímnir* might be thought of.

Grinshill Sa [*Grivelesul* DB, *Grineleshul* 1242 Fees, *Greneleshull* 1320 Ch]. The original name may have been OE *Grēn-hyll* 'green hill', which was weakened to *Grēnel*, a fresh *hyll* being afterwards added. But the early *i* suggests that the first el. is rather OE *grin-hyll* 'hill with a trap'. OE *grin* means 'a snare, trap'.

Grinstead, East, Sx [*Grenesteda* 1121 AC, *Estgrenested* 1271 Ass], **West G~** Sx [*Grenstede* 1261 Ass, *Westgrenested* 1280 Ch]. 'Green place.'

Grinton YN [*Grinton* DB, *Grentone* c 1180 YCh 1140, *-ton* 1234 FF]. 'Green TŪN.'

Grisedale Pike Cu [*Grisedal* 1323 Ipm]. 'Pig's valley.' First el. OScand *gris* 'pig'.

Gristhorpe YN [*Grisetorp* DB, *Gristhorp* c 1180 YCh 370]. 'Gris's thorp.' *Gris* is ON *Gríss*, ODan *Gris* pers. n. from ON *griss* &c. 'pig'.

Griston Nf [*Grestuna, Gristuna* DB, *Gerdestuna* c 1150 Fr, *Greston* 1166 P, *Gristone* 1254 Val]. If the form *Gerdestuna* is reliable the first el. may be ODan *Gyrdh, Gyurth*; cf. GARRISTON. Otherwise it may be OE GRĒOSN.

Grittenham W [*Gruteham* 850 BCS 458, *Grutenham* 1065 KCD 817, *Gretenham* 1291 Tax]. Perhaps *Grēote* was an old name of Brinkworth Brook, on which G~ stands. Cf. GREET. Or the first el. might be OE **grēoten* 'of gravel'.

Grittleton W [*Grutelington* 940 BCS 750, *Gretelintone* DB, *Grutelington* 1242 Fees]. The first el. might be an OE *grēot-hlinc* 'gravel hill', or an OE *Grēotlingas* from a pl. n. such as *Grēot-lēah* 'gravel LĒAH'.

Grizebeck La [*Grisebek* 13 FC, 1292 Ass], **Grizedale** La nr Lancaster [*Grisedale* 1314 LaInq], **G~** La in Hawkshead [*Grysdale* 1336 FC]. 'Pigs' brook and valley.' Cf. GRISEDALE.

Groby Le [*Grobi* DB, 1180 P, c 1200 Fr, *Groubi* c 1140 BM, *Groebi* 1180 P]. G~ is c ½ m. from a tarn. This may have had the OScand name *gróf.* ON *gróf* means 'a torrent and the gully formed by it'. It is identical with Goth *gróba*, OHG *gruoba* 'pit, hollow'. G~ is pronounced *grōobi.*

Groombridge K [*Gromenebregge* 1239 Reg Roff, 1318 FF]. 'The grooms' bridge.' *Groom* is found in ME as *grome* 'boy, servant'.

Grŏsmont Monm [(de) *Grossomonte* 1187 P, *Grosmunt* 1193 P, (de) *Grosso Monte* 1291 Tax], G~ (-ŏm-) YN [*Grosmunt* 1226–8 Fees]. A French name meaning 'big hill'. G~ YN was a monastery, which was named from its mother priory GROSMONT in France.

Groton Sf [*Grotena* DB, *Grotene* 11 EHR 43, c 1095 Bury, 1201 Cur, 1254 Val]. OE *grot* means 'a particle'. OE *sandgrot* is 'a grain of sand', *meregrota* 'a sea-pebble, a pearl'. Groton is probably a name of the stream at Groton, an OE *Groten-ēa* 'sandy stream', with an adj. *groten as first el. For the loss of the final el. cf. WHITTON Li. OE *Grotena-ēa*, with the gen. plur. of OE *grota* as first el., is also possible.

Grove Bk [*Langraue* DB, *la Graue* 1197 FF, *Grava* 1222 Ep], G~ Brk [*La Graue* 1188 P, *Grove* 1316 FA], G~ Nt [*Graue* DB, *Graua* 1194 P]. OE GRĀF 'grove, thicket'.

Grovely Wood W [(on) *Grafan lea* 940 BCS 757, *Grauelea* 1168 P, *Grofle* 1317 Cl, *Gravelinges* DB, *Graueling* 1156 P, *Grauelinga* 1190 P]. It is possible that the first el. is OE *grafa* 'a ditch' (cf. GRAVENEY). Through Grovely Wood runs a Roman Road, called on the map 'Ditch' part of the way. This might have been called *grafa*. But the first el. may be OE *grāfe, grǣfe* 'grove, brushwood'. Cf. GRAVELEY. LĒAH is here 'wood'. *Gravelinges* means 'the dwellers at Grovely'.

Grundisburgh (grŭnzbrŭ) Sf [*Grundes-burch, -burc* DB, *-burg* 1235 FF, 1254 Val]. Very likely *Grund* was the old name of the place, and Grundisburgh means 'the BURG at Grund'. *Grund* may be OE *grund* 'foundation' from an old building-site. Cf. STANGROUND.

ON **gryfia** 'hole, pit', Engl dial. *griff* 'a deep, narrow glen' is found in some pl. ns., as GRIFF(E), FALS-, MULGRAVE, SKINNINGROVE.

Guestling Sx [*Gestelinges* DB, *Grestling* 1197 FF]. OE *Gyrstlingas*, identical with the first el. of GESTINGTHORPE.

Guestwick (-tĭk) Nf [*Geg(h)estueit* DB, *Geistweit* 1203 Ass, 1244 Ipm, *Geystweyt* 1242 Fees, *Geystethweyt* 1254 Val]. 'Thwaite belonging to GUIST.'

Guildford (-ĭlf-) Sr [*Gyldeford* c 880 BCS 553, *Gilde-, Geldeford* DB, *Guldeford* 1131–3 BM, *Geldeford* 1130, 1156, 1190 ff. P]. G~ is on the Wey where the river cuts through the long ridge called the **Hog's Back**. This was formerly **Guildown** [*Geldedon* 1190 ff. P, *Gildedon* 1251 Cl, *Mons Guldedonye* c 1282 Ep]. Guildford is probably 'ford where golden flowers grew'. The first el. would be OE *gylde, a derivative of *gold* and of the same meaning as *golde*, marigold. The word is very likely the first el. of *Gyldeburne* 843 BCS 442. In Guildford *gylde* would mean 'marsh

marigold'. The relation of Guildown to Guildford is not clear. Guildown may be elliptical for *Gyldeford-dūn*. Or it may mean 'ridge where golden flowers grew'.

Guilsborough Np [*Gildesburh* 1066–75 GeldR, *Gisleburg* DB, *Gildesburc* 12 BM, *-burg* 1225 Ep, 1254 Val]. '*Gyldi*'s BURG.' *Gyldi* is a short form of names in *Gold-*, as *Goldwine*.

Guisborough (gĭzbrŭ) YN [*Ghigesburg, Gighesborc* DB, *Gisburham* 1104–8 SD]. The first el. may be a pers. n. But the ON *Gigr* that has been suggested is of doubtful authenticity.

Guiseley (gĭz-) YW [*Gislicleh* c 972 BCS 1278, *Gisele* DB, *Giselai* c 1180 YCh 201, *Guyseley* 1291 Tax]. OE *Gislicleh* may well be a shortened form of *Gislica(n)-lēah* '*Gīslica*'s LĒAH'. *Gislica* is a normal hypocoristic form of names in *Gisl-*. If this is right, G- in the modern form is due to Scandinavian influence.

Guist (gĭst) Nf [(et) *Gæssǣte (Geysete)* c 1035 Wills, *Gegeseta* DB, *Geiste* 1200 Cur, 1254 Val]. First el. OE *Gǣga* or *Gǣgi*. Cf. GAYTON, GAZELEY. The second may be OE (GE)SET 'a fold' &c. or SǢTE 'a house'. Cf. ELMSETT. *Geistorp* 1199 Pp, *Geystorp* 1253 Ch is perhaps the present UPPER GUIST.

Guiting (gī-) **Power, Temple G~** Gl [(bi) *Gythinge* 814 BCS 351, *Getinge* DB, *Guttinges* 1221 Ass, *Gettinges Poer* 1220, *Guttinges Templ'* 1236 Fees]. Guiting seems to have been the old name of the upper Windrush. **Guiting** R is *Gytingbroc* 780, *Gytinc, -ges* 974 BCS 236, 1299. *Gyting* is a derivative of OE *gyte* 'flood'. The meaning is 'torrent'.

G~ **Power** from the local family. The name, which is (le) *Poer* in early sources, may be OFr *pohier, pouhier* 'of Poix, Picard'.—**Temple** G~ belonged to the Templars from c 1160.

Guldeford (gĭlf-), **East,** Sx [*Newguldford* 1508 PNSx]. Named from a Surrey family Guildford.

Gulval Co [*Ecclesia Sancte Welvele de Lanystly* 1328, *Sanctus Weluelinus* 1377 Ep]. '(The church of) St. *Gudwal*' according to Oliver (*Gudwal* or *Gulwal* Stanton), but the early forms hardly point to this. The dedication seems to be to St. Wolvela. An earlier name of the parish is *Lanestli* 1261 Ep.

Gumley Le [*Godmundeslaech* 749 BCS 178, *Godmundesleah* 779 BCS 230, KCD 1360, *Godmundelai, Gutmundeslea* DB, *Godmundeslee* 1197 FF, *Guthmandelai* 1147 BM, *Gumundeley* 1233 Ep]. '*Gōdmund*'s LĒAH', later influenced by OScand *Guðmundr*.

Gunby St. Nicholas Li [*Gunnebi* DB, *-by* 1212 Fees], G~ **St. Peter** Li [*Gunnebi* DB, 1212 Fees]. '*Gunni*'s BY.' *Gunni* DB &c. is ODan *Gunni*.

Gunby YE [*Gunelby* 1066–9, *Gundeby* 1070–83, *Gunneby* 1258 Selby]. If the first form is trustworthy, '*Gunhild*'s BY'. Cf. GUN-NERSBURY.

Gunnerby Li in Hatcliffe [*Gunresbi* DB, c 1115 LiS, *Gunnerby* 1242 Fees]. '*Gunner*'s BY.' *Gunner* DB &c. is ODan *Gunnar*, *Gunnær*, ON *Gunnarr*.

Gunnersbury Park Mx [*Gunnyldesbury* 1348 ff. FF]. '*Gunnhild*'s manor.' *Gunnhild* is ON *Gunnhildr*, OSw *Gunhild*, ODan *Gunild*, a woman's name.

Gunnerton Nb [*Gunwarton* 1170 P, 1242 Fees, *Gonewerton* 1268 Ipm]. '*Gunnward*'s or *Gunnware*'s TŪN.' *Gunward* would be ON *Gunnvarŏr*; *Gunware*, ON *Gunnvǫr*, a woman's name.

Gunness Li [*Gunnesse* 1199 P, 1202 Ass, 1250 Fees, *Gunnes* 1219 Ep]. '*Gunni*'s headland.' Cf. GUNBY. Second el. OScand *nes* 'ness, headland'. The ness seems to be a bend of the Trent.

Gunthorpe Li [*Gunetorp* c 1200 DC, -*thorp* Hy 3 BM], G~ Nf [*Gunes*-, *Gunatorp* DB], G~ Np [*Gunetorp* 1130, 1163 P, *Gunestorp* 1177 P], G~ Ru [*Gunetorp* 1200 Cur]. '*Gunni*'s thorp.' Cf. GUNBY.

Gunthorpe Nt [*Gulne*-, *Gunnetorp* DB, *Gunildethorp* n.d. Thoroton, *Gunnetorp* 1191–3 Fr]. '*Gunnhild*'s thorp.' Cf. GUN-NERSBURY.

Gunthwaite YW [*Gunhild*-, *Gunyldthwayt* 1284 Wakef]. '*Gunnhild*'s thwaite.' Cf. GUNNERSBURY.

Gunton Nf [*Gunetune* DB, *Gonetone* 1166 RBE], G~ Sf [*Guneton* 1198 FF, 1203 Cur]. '*Gunni*'s TŪN.' Cf. GUNBY.

Gunwalloe Co [(Eccl.) *Sancti Wynwoluy* 1291 Tax]. *Wynwoluy* is a saint's name, identical with OBret *Winwaloe*, Bret *Guenolé*.

Gupworthy So [*Guppewurþe* 1155–8 (1334) Ch]. '*Guppa*'s WORÞIG.' *Guppa* is a short form of OE *Guþbeorht* or the like. Cf. *Gyppa*, *Gyppi* in GIBSMERE, GIPTON &c. *Guppa* is also found in Guppy Do in Wootton Fitzpaine [*Guppe*-, *Gupehegh* 1254 Misc]. Second el. perhaps OE *gehæg* 'enclosure'.

Gusford Hall Sf nr Ipswich [*Gutthulues forda* DB]. '*Gūþwulf*'s ford.'

Gussage All Saints, St. Andrew & St. Michael Do [*Gissic*, *Gersicg* c 871 BCS 531 f., *Gyssic* 966–75 Wills, *Gessic* DB, (Church of All Saints) *Gersic* c 1100 Montacute, *Gersich* 1168 P, *Gessich* 1240 Ch, *Gessich Omnium Sanctorum* 1242 Fees, *Gissik St. Andrews* 1258 Ch, *Gissich Sancti Michaelis* 1285 FA]. OE *gyse, corresponding to OHG *gusi* 'water suddenly breaking forth', and *sic* 'water-course'. The common spelling with *rs* is probably erroneous, but it seems to have had a certain vogue.

Guston K [*Gocistone* DB, *Gutiestun* 11 DM, *Gutsieston* 1208 FF]. '*Gūþsige*'s TŪN.' *Gūþsige* is not recorded elsewhere.

Gutterby Cu in Egremont [*Godrickeby* 1235, *Goderickby* 1344 StB], G~ Cu in Whitbeck [*Godrikeby* 1209 FF]. '*Gōdrīc*'s BY.'

Guyhirn Ca [*la Gyerne* 1275, *le G(u)y(e)-herne* (-*hirne*) 13 &c. PNCa(S)]. The second el. is clearly OE *hyrne* 'corner, angle'. The first will be dial. *gye* 'a salt-water ditch', found only in Somerset dial. The word is doubtless French and may be the source of the river-name LA GUYE in Saône-et-Loire [*Guia* 1203, *Guie* 1238 Lebel p. 233 under *wiia* 'marécageuse']. According to PNCa(S) the tide came as far as Guyhirn.

Guyzance Nb [*Gynis* 1242 Fees, *Gysnes* 1254 Val]. A manorial name. *Guines* is a Norman family name derived from GUINES near Calais.

Gweek Co [*Wike* 1337, *Gwyk* 1358 FF]. Co *gwic* 'village' from Lat *vicus*.

Gwennap Co [(ecclesia) *Sancte Weneppe* 1269 Ep, (Ecclesia) *Sancte Wenep* 1291 Tax]. '(The church of) St. Wenep' (a woman saint).

Gwinear Co [(rector) *Sancti Wyneri* 1258 Ep, (de) *Sancto Wyniero* 1286 Ep]. '(The church of) St. Winnier.' *Winnier* corresponds to Bret *Guigner*, OW *Guinier* LL.

Gwithian Co [*parochia Sancti Goythiani* 1335 Subs]. Named from St. Gothianus (Oliver), whose name is identical with OBret *Gozien*, *Goezian*.

H

Habberley Sa [*Habberleg*, *Hatburleg* 1242 Fees, *Haburleye* 1346 FA], H~ Wo [*Harburgelei* DB, *Haberlega* 1184 P]. The first is '*Heaþuburg*'s LĒAH'. *Heaþuburg* (*Hæþburg*) is a woman's name. The second may be identical in origin, but OE *Hēahburg*, also a woman's name, is a possible alternative.

Habblesthorpe Nt [*Happelesthorp* 1154 YCh 155, *Harpeles*-, *Happelestorp* 1275 RH, *Harplesthorp* 1341 NI]. The first el. seems to be a pers. n., but its original form and history are obscure.

Habergham Eaves (häbergam) La [*Habringham* 1242 LaInq, *Habringeham* 1296 Lacy]. 'The HĀM of the dwellers at *Hēahbeorg*', *Hēahbeorg* being a hypothetical name of Horelaw, a prominent hill in the township. *Eaves* is here 'edge of a hill'.

Habertoft Li [*Halbertoft* 1166 P, 1317 Ipm, *Habertoft* 1389 Pat]. Probably '*Hagbarth*'s toft'. *Hagbarth* is an ODan name = ON *Hagbarðr*. Early spellings with *lb* probably stand for *bb*.

Häbrough Li [*Haburne* DB, *Haburc* c 1115 LiS, *Haburg* 1202 Ass, 1254 Val, *Hauburc* 1197 P]. OE *hēahburg* 'chief BURG', lit. 'high BURG'.

Habton YN [*Habetun, Abbetune* DB, *Parva Habeton* 1163–85 YCh 781, *Habbeton* 1208 Cur]. '*H(e)abba*'s TŪN.' **Heabba*, found also in HAPTON Nf, may be a short form of *Hēahbeorht* or the like.

OE *haca* 'hook'. See HACKFORD &c., HAWK-WELL Ess.

Haccombe D [*Hacome* DB, *Hakcumbe* c 1200 PND, *Haccumb* 1242 Fees, *Heccham* 1293 PND]. See CUMB. The first el. may be OE HÆCC 'a hatch' &c.

Haceby Li [*Hazebi* DB, *Hat(h)sebi* 1115 RA, *Hascebi* 1162 P, 1202 Ass]. OScand *Hadds bȳr*. First el. late OE *Hadd* pers. n. from ON *Haddr*.

Hacheston Sf [*Haces-, Hecestuna, Hecetuna* DB, *Hecetune* c 1095 Bury, *Hascheton* 1197 P, *Hacheston* 1292 Ch]. '*Hæcci*'s TŪN.'

Hackenthorpe Db [*Hakenthorp* 1327 Subs, *Hakunthorpe* 1423 Derby]. '*Hacun*'s thorp.' *Hacun* ASC (C) is ODan, OSw *Hākon*, ON *Hákon*.

Hackford Nf nr Reepham [*Hacforda* DB, 1196 FF, *Hakeford* 1204 Cur, 1242 Fees, 1372 BM], H~ Nf nr Wymondham [*Hakeforda* DB, -*ford* 1203 Cur, *Hacford* 1254 Val]. It has been suggested that the first el. is OE *hæcc* 'a hatch' &c. The numerous forms in *Hake-* rather suggest that it may be OE *haca* 'a hook' in the sense 'a bend'.

Hackforth YN [*Acheford* DB, *Hacford* 1205 FF, *Hakford* 1305 Ch]. Identical with HACKFORD.

Hackington K [*Hakinton* 1186, 1195 P, 1233 Cl, *Hakenton, Hakinton* 13 BM]. '*Ha(c)ca*'s TŪN.' Cf. *Hac(c)an pundfald* 961, 964 BCS 1080, 1144, and see HAGBOURNE.

Hackinsall La [*Hacunesho* c 1190 LaCh, 1221 Cl, -*hou* 1246 Ass]. '*Hacun*'s mound.' Cf. HACKENTHORPE and HAUGR.

Hackleton Np [*Hachelintone* DB, *Haclintona* 1155–8 (1329) Ch, *Hakelinton* 1202 Ass, 1220 Fees, *Hakelington* 12 NS]. 'The TŪN of *Hæccel*'s people.' **Hæccel* is a regular derivative of *Hacca*.

Hackness YN [*Hacanos* c 730 Bede, *Heacanos* c 890 OEBede, *Hagenesse* DB, *Hakanessa* 1092 YCh 862]. 'Hook-shaped headland.' Cf. HACKPEN. The elements are OE *haca* 'a hook' and *nōs*, corresponding to OSw, Dan, Norw *nōs* 'snout', here used to describe the prominent hill at the place. Later *nōs* was replaced by the cognate *næss* 'ness, headland'.

Hackney Mx [*Hakney* 1231 FF, *Hakeneye* 1242 Fees, 1294 QW]. Perhaps '*Haca*'s island'. Cf. HACKINGTON. Or the first el. may be *haca* as in HACKFORD.

Hackpen Hill D [*Hacapenn* 938 BCS 724, *Hakepen* 1249 FF], H~ Hill W [(an) *Hacan penne* 939 BCS 734]. Both hills are described as hook-shaped. The elements of the name are OE *haca* 'hook' and *pen* 'hill' from Welsh *pen* 'head, hill'.

Hackthorn Li [*Hagetorne* DB, -*thorn, Hakethorn* 1202 Ass, *Hacatorn, Hachethorna* c 1115 LiS, *Haggethorn* 1193 P]. OE *haguþorn* 'hawthorn'.

Hackthorpe We [*Hacatorp* c 1175 Kendale, *Hakethorp* c 1240 CWNS xxiv]. '*Haki*'s thorp.' First el. ON *Haki* pers. n.

Hackwood Ha [*Hagewod* 1228 Cl, *Hacwode* 1313 Misc]. First el. as in HAGLEY.

Haconby Li [*Hacunesbi* DB, *Haçunebi* 1164 P]. '*Hacun*'s BY.' Cf. HACKENTHORPE.

Haddenham Bk [*Hedreham* DB, *Hedenham* 1142–8 Reg Roff, *Hadenham* 1196 FF], H~ Ca [*Hædanham* 970 BCS 1268, *Hadreham* DB, *Hadenham* Hy 3 BM, 1282 Bodl]. '*Hæda*'s HĀM.' But the form *Hædanham* is suspicious. It may be for *Hæddan-* or *Headdanham*.

Haddington Li [*Hadinctune* DB, *Hadingtun* 1212 Fees]. 'The TŪN of *Headda*'s or *Hada*'s people.'

Haddiscoe Nf [*Hadescou* DB, -*sco* 1208 FF, 1236 Fees, *Haddesco* 1253 Cl]. '*Hadd*'s wood.' Cf. HACEBY. Second el. OScand *skōgr* 'wood'.

Haddlesey, Chapel & West, YW [*twa Haðelsæ, þridda Haðelsæ* c 1030 YCh 7, *Hedlesic* 1190 P, *Mediana Haþelsay* c 1200 YCh 497, *Westhathelsay* 1304 Ch, *Esthauʒelsay* c 1250 BM]. The second el. is OE *sæ* 'lake'. The first is obscure. Possibly it may be compared with that of the lost **Hathelton** in Bingley YW [*Hateltun* DB, *Hagelton* c 1166 YD, *Hadelton* c 1215 YD, 1234 FF]. The first el., *haðel*, is possibly cognate with Gk κοτύλη, κότυλος 'cavity, cup, bowl' and OE *haþo*, apparently 'arm-pit', in *haþolida* 'elbow' (cf. Holthausen). Gk κότυλος would correspond to an OE *haþol*. It is possible that *Haðel* was a name of the lake.

Haddon, Nether & Over, Db [*Hadun, -e* DB, *Uverehaddon* 1206 Cur, *Ufrehedon* 1230 P, *Netherhaddon* 1276 Ipm], H~ Lodge Do [*Haddone* (p) 1212 RBE], H~ Hu [*Haddedun* 951 PNHu, *Adone* DB, *Haddon* 1286 Ass], East & West H~ Np [*Eddone, Hadone* DB, *Haddon, Westhaddon* 12 NS, *Esthaddon* 1220 Fees]. OE *hǣþ-dūn* 'heather-covered hill'. H~ Hu is somewhat doubtful, however. It might be *Headdan dūn*. H~ Do may be manorial.

Hadfield Db [*Hetfelt* DB, *Haddefeld* 1185 f. P]. OE *hǣþ-feld* 'heather-covered FELD'.

Hadham, Much & Little, Hrt [*Haedham* c 960 Bodley MS, *Hedham* c 995 BCS 1288 f., *Hadham* c 1050 KCD 907, 1212 Fees, *Hadam* DB, *Heddeham* c 1175 BM]. Apparently identical with HADDENHAM.

Hadleigh Ess [(of) *Hæplege* c 1000 CCC, *Hadleg* 1199 P, *Hadleghe* 1238 Subs], H~ Sf [(into) *Hedlæge, Hædleage gemære* c 995 BCS 1288 f., *Hæõleh* c 1050 KCD 896, *Hetlega* DB, *Hadlega* 1183 P], **Hadley, Monken,** Hrt [*Hadley* 1248 Ch, *Haddeleye* 1254 Val, *Hadle, Hedle* 1291 Tax], H~ Sa [*Hatlege* DB, *Hadlega* 1191 P, *Hethlegh* 1238 Cl]. OE *hæp-lēah* 'heather-covered clearing'.

Monken H~ belonged to Saffron Walden from 1248 (Ch). *Monken* is 'of the monks'.

Hadley Wo [*Haddeley(e)* 13, 1327 PNWo]. '*Headda*'s LĒAH.'

Hadlow K [*Haslow* DB, *Haslo* 11 DM, *Hadlou, Haudlou* 1235 Cl, *Hadlo* 1241 Ep, 1242 Fees, *Haudlo* 1280 Cl]. OE *hæp-hlāw* 'heather-covered hill'. On *hæp* see HÆP. Wallenberg prefers OE *hēafod* 'head', here 'chief', as first el. This would explain forms like *Haudlou* better.

Hadlow Down Sx [*Hadleg* 1254 Pat, *Haddele(gh)* 1279, 1296 PNSx]. Cf. HADLEIGH.

Hadnall Sa [*Hadehelle* DB, *Hedenhola* 1167 P, *Hadenhale* 1242 Fees]. '*Headda*'s HALH.'

Hadstock Ess [*Hadestoc* c 1050 KCD 907, 1166 P, *-stok* 1197 (1233) Ch]. '*Hada*'s STOC.'

Hadstone Nb [*Hadeston* 1189 P, *Hadistona* 1236 Fees, *Haddeston* 1251 Ch]. '*Hæddi*'s TŪN.'

Hadzor Wo [*Headdesofre* 11 Heming, *Hadesore* DB]. '*Headd*'s hill or slope.' **Headd* is a short form of names in *Heard-*. Cf. OFER.

OE **hæčč** 'hatch, i.e. a gate or wicket, a floodgate or sluice, a grating used to catch fish at a weir' is found in some pl. ns. The usual meaning is probably 'a gate, esp. one leading to a forest'. See HATCH, STEVENAGE, also HECK, HEACHAM, HACKFORD. An OE *hæcce, hecce*, possibly 'a fence of rails', is postulated by B-T Suppl on the strength of (on lang) *heccan* (*hæccan*) BCS 963.

OE **hæfen** 'haven'. See KEY-, WHITEHAVEN.

OE **hæfer,** ON *hafr* 'he-goat' and ON *hafri,* OE **hæfera* 'oats'. See (Market) HARBOROUGH, HAVER- (passim).

OE **(ge)hæg** 'hay, enclosed piece of land, meadow', ME *hay* also 'forest fenced off for hunting'. See HAY, CHALVEY, HARPURHEY, HARTHAY, OXHEY, STREETHAY, WOODHAY. As a first el. it is difficult to distinguish this word from *hay* 'dried grass'. Cf. HAYDON, HAYTON, HAYWOOD.

OE **hæs, -e** is only found in pl. ns. It corresponds to LG *hees, hēse* 'brushwood, underwood' and is related to MHG *heister*, MLG *hēster* 'young oak or beech'. An early form is seen in *Silva Cæsia* = Heserwald in Tacitus. See HAYES, HEYSHAM, HEST, HESTON. A full discussion of the wordgroup is found in V. Ekenvall, *De svenska ortnamnen på hester* (Lund, 1942), pp. 152 ff. The meaning 'beech or oak wood' assumed by the writer in PNLa is perhaps to be preferred to that of 'brushwood'.

OE **hæsel** 'hazel' is a common first el. in pl. ns. See HASLE-, HAZEL- &c., also HASWELL, HESSAY, HESWALL and others. Second el. in BADSADDLE. OScand *hesli* 'hazels' in HASLAND, HESSLE. The adj. *hæslen* 'of hazels' in HASLINGDEN &c.

OE **hæp** means 'heather and other plants or shrubs found upon heaths' and 'a tract of uncultivated land'. The first meaning is found also in OHG. Both meanings are found in pl. ns. As a first el. *hæp* no doubt generally means 'heather' &c., as in HADDON, HADLEY, HATFIELD, HAYFIELD, HEATHFIELD, HETHEL. When used alone or as a second el. *hæp* means 'heath', as in HETHE, BLACKHEATH. Forms such as HADLEY, HATFIELD are due to special developments of *p* before *l* and *f*. A side-form *hāp* is shown by pl. ns. to have been used in K and Sx. See HADLOW, HOATH, -LY.

OE **hafoc** 'hawk'. See HAUXLEY &c., HAWK- (passim), HAWRIDGE.

OE **haga** 'fence, fenced enclosure', also 'enclosed dwelling in a town' and OScand *hagi* 'enclosure' are found in pl. ns. as a first and second el. and are also used alone, as in HAIGH, HAUGH, HAW, HOUGH Li. It is the first el. of HAUGHMOND, HAWCOAT, HOUGHAM Li and some other names. It is the second el. of BELAUGH, BYLAUGH &c., BREARY, GOLTHO, LOCKO, STODDAY, THORNEY Nt, THORNHAUGH, WELLOW Nt &c. Identical in form is OE *haga* 'haw, berry of the hawthorn'. See HAGLEY, HAUGHLEY. OE *haguporn, hægporn* 'hawthorn' is usually the source of HATHER- in pl. ns., as HATHERLEIGH &c. Cf. HACKTHORN.

Hagbourne, East & West, Brk [*Hacca-, Hacceburna* 891 BCS 565, *æt Hacceburnan* 990-4 BM, *Hacheborne* DB]. Hagbourne is the old name of a stream by the villages. Its OE form was really *Haccanburna*, as shown by the side-form *Haccanbroc* 944 BCS 801, referring to the stream at Hagbourne. In BCS 1143 in boundaries of Aston Tirrold, which is close to Hagbourne, is mentioned, besides *Hacce-, Hæccebroc*, also *Hæcceleas dic*, which represents an OE *Haccan lēah*. The common element in the names is doubtless a pers. n. **Hacca*, which is not recorded in independent use. The related *Hæcci* is on record. Cf. also HACKINGTON.

Haggerston Mx [*Hergotestane* DB, *Heregodeston* 1221-30, *Haregodeston* 1242 Fees]. '*Heregod*'s stone.' *Heregod* is possibly a genuine OE name, but its late appearance [*Hargod* 1004 KCD 1300, *Haregod* mon.) suggests the possibility of loan from the Continent.

Haggerston Nb [*Agardeston* 1196 P, *Hagardestun* 1208–10 Fees, *Hagardestone* 1228 FPD]. The first el. is probably a Fr family name derived from OFr *hagard* 'wild, strange'.

Hagley Sa nr Clunbury [*Haggele* 1272 Ipm, *-leye* 1341 NI], H~ So nr Wiveliscombe [*Haggelegh* 1243 Ass, *-lee* 1276 RH], H~ St [*Hageleia* 1130, *-lega* 1169 P, *Haggleges* 1166 RBE, *Haggele* 1242 Fees], H~ Wo [*Hageleia* DB], **Haglow** or **-loe** Gl [*Haggelow* 1437 PNGl]. The first el. cannot well be a pers. n. It is no doubt the OE form of dial. *hag* 'haw, fruit of the hawthorn', which is found all over England. It represents an OE **hacga*, which may be looked upon as a kind of hypocoristic form of *haga*. Cf. OE *twigge* by the side of *twig*, plur. *twigu*. Hagley is then 'wood where haws were found'. Hagloe is 'hill where haws grew'.

Hagnaby Li nr Spilsby [*Hagenebi* DB, *Hagenesbia, Hahnebia* 1142 NpCh], H~ Li in Hannah par. [*Haghnebi* R 1 BM, *Hagnebi* 1202 Ass, 1212 Fees, *-by* 1228 Ep]. '*Hagne*'s BY.' First el. ODan *Haghne*, OSw *Hagne*, ON *Hǫgne*.

Hagworthingham Li [*Haberdingham, Hacberding(e)ham* DB, *Hagwordingheheim* c 1115 LiS, *Aburdingeham* 1167 P, *Hacwrdhingham* 1197 FF, *Hagwrðingham* 1198 P, *Hagworthingham* 1202 Ass]. The first el. is apparently a derivative of a pl. n. in *-worþ*, e.g. OE **Hæcg-worþ*, the first el. being the word for *haw* (OE **hacga*) suggested under HAGLEY. The name would then mean 'the HĀM of the Hagworth people'. The second el. sometimes shows the OScand form *-heim*.

Haigh La [*Hage* 1194 P, *Hagh* 1298 FF], H~ YW in Elland [*Hagh* 1198 Fount], H~ YW nr Barnsley [*Hagh* 1379 PT]. OE *haga* or OScand *hagi* 'enclosure'.

Haighton La [*Halctun* DB, *Halechton* 1226 LaInq, *Halghton* 1327 Subs]. 'TŪN in a haugh.' See HALH.

Hailes Gl [*Heile* DB, *Heilis* 1114 Fr, *Hailes* 1173 P, 1221 Ass]. Very likely from an old name of the stream at Hailes, called *Haylebrok* 1256 Winchc. Cf. (HAIL) WESTON.

Hailey O [*Hayle, Hyle* 1279 RH, *Haylle* 1316 FA]. 'Hay clearing.'

Haileybury or **Hailey** Hrt [*Hailet* DB, *Heilet* Hy 1, *Heyle* 1374 BM]. Identical with HAILEY O, if the early forms in *-t* are to be disregarded. If they are trustworthy, the name may be identical with **Haylot** La [*Hailett* 1584 PNLa], which means 'hay lot', 'allotment for grazing'. The el. *-bury* means 'manor'.

Hailsham (-ĕls-) Sx [*Hamelesham* DB,

Heilesham 1198 FF, 1230 P]. '*Hægel*'s HĀM.' **Hægel* pers. n. corresponds to ON *Hagall* and is related to OG *Hagilo*. It is found in other pl. ns., as HAZELEIGH and HAYLING.

Hainford Nf [*Hemfordham* c 1060 Wills, *Han-, Hamforda* DB, *Heinford* 12 BM, 1199 FF, *Henford* 1206 Cur], **Hainton** Li [*Haintone* DB, *Heintuna* c 1115 LiS, *-ton* 1193, *Hainton* 1197 P]. First el. an OE **hægen* 'enclosure' or the like, corresponding to OHG, OLG *hagan* 'a kind of thorn-bush', MLG *hagen* 'hedge', G *Hain* 'grove'. Layamon has *hain* in the sense 'enclosure, park'.

Hainworth YW [*Hageneuuorde* DB, *Haghenwrde* 1230 YD]. '*Hagena*'s WORþ.'

Haisthorpe YE [*Ascheltorp, Aschiltorp, Haschetorp* DB, *Hascheltorp* 1190 YCh 1312, *Hasthorp* 1260 Ipm]. 'The thorp of *Hǫskuldr*.' *Hǫskuldr* is an ON pers. n.

Hālam Nt [*Healum* 958 YCh 2, *Halum* 1198 FF, 1331 BM]. The dat. plur. of OE HALH 'a corner' &c. The place is in a valley.

Halberton D [*Halsbretona* 1184 GeldR, *Halsbretone* DB, *Hauberton* 1188 P, 1247 Ch, *Halbertone* 1269 Ep]. Probably OE *Hæselbearu-tūn* 'TŪN by a hazel grove'.

Halden, High, K [*Hadinwoldungdenne* 11 DM, *Hathwoldindanna* 1157 StAug, *Hadewoldineden* 1185 P, *Hathewolding* 1215 FF, 13 StAug]. Apparently an OE *Hapuwealding* '*Hapuweald*'s land' (cf. -ING), to which was added OE *denn* 'swine-pasture'. Or else 'the DENN of *Hapuweald*'s people'.

Haldenby YW [*Haldaneby* 1100–8 YCh 470, *Haldenebi* 1190 P, *Haldaneby* 1226 FF]. '*Halfdan*'s BY.' *Halfdan* is ON, OSw *Halfdan*, ODan *Haldan*.

Hale Chs [*Hale* DB, 1260 Court], H~ Cu [*Hale* 1227 P, 1291 Tax], H~ Ha [*Hala* 1161 P, *Hale* 1219 Fees, *La Hale* 1242 Fees], H~ La [*Halas* 1094 LaCh, *Hales* 1227 Ch, *Hale* 1201 P], **Great & Little** H~ Li [*Hale* DB, *Magna Hale* 1204, *Hales, Halh* 1205 Cur, *Hal, Parva Hal* 1212 Fees], H~ Sr [*Hale* 1222 PNSr], H~ We [*Hale* c 1185 CC, 1266 Pat]. OE HALH, dat. *hale* 'nook, haugh' &c.

Hales Nf [*Hals* DB, c 1095 Bury, *Hales* 1236 Fees], **Sheriff** H~ Sa [(æt) *Halen* 1002 Wills, 1004 KCD 710, *Halas* DB, *Hales* 12 BM, *Little* H~ 1222 FF, H~ *upon Lousyerd* 1283 Ass, *Shiruehales* 1301 For], H~ St [*Hales* 1291 Tax], **Halesowen** Wo [*Hala* DB, *Hales* 1195 ff. P, H~ *Ouweyn* 1276 Misc]. The plur. of OE HALH 'a corner' &c. The reference is generally to a remote valley or a recess in a hill. The identification of *Halen* 1002, 1004 is doubtful.

Sheriff H~ was held by Rainald Bailgiole, sheriff of Salop, in 1086 (DB).—**Halesowen** got its surname from Owen, son of David, a Welsh prince who married a sister of Henry 2. Owen became lord of Hales in 1204.

Halesworth Sf [*Healesuurda* DB, *Halesuuorda* DB, *-wurde* 1195 P]. The first el. is probably identical with that of HALSALL, i.e. a pers. n. *Hæle or the like. See WORP.

Halewood La [*Halewode* c 1200 CC]. 'Wood by HALE.'

Halford D [*Halford* 1275 RH, 1284–6 FA], H~ Wa [*Haleford* 1176 Fr, *Halecford* 1190 P]. 'Ford in a HALH', i.e. 'nook, narrow valley'. Cf. *Halhford* BCS 966 (Tadmarton O).

Halford Sa [*Hauerford* 1155 BM, *Hawkeford* 1535 VE]. 'Hawkers' ford.'

OE **halh, healh** 'a corner, angle, a retired or secret place, cave, closet, recess' is very common in pl. ns., both alone and as a first or second el. The meaning is generally difficult to establish in each case. In the South and Midlands the usual meaning seems to be 'a nook, recess, remote valley'. In the North *halh* developed a curious special meaning, viz. 'haugh, a piece of flat alluvial land by the side of a river'. The intermediate sense is 'land in a corner formed by a bend'. The sense 'haugh' is that usually found in pl. ns. in the North. A meaning 'spur of hill' is possibly sometimes found, as in CALOW. The form of the el. also varies a good deal. The Southern form was *healh*. Hence HELE D, So. The *h* was lost in inflected forms. Hence the variation between forms such as HALGH and HALE. The plur. form is found in HALES and HAL(L)AM. *Halh* is the first el. of HALTON, HALLATON, HALLOUGHTON, HOLTON (2), WESTHOUGHTON, HAIGHTON and others. As a second el. it varies particularly. Cf. e.g. MIDGEHALL, CRUMPSALL, ORDSALL, ELLEL. Most names in -HALL contain *halh*.

Halifax YW [*Feslei* DB, *Haliflex* c 1175 AD, *Halifax* 1268 Ep]. 'Holy flax field.' The loss of the second *l* is due to dissimilation.

OE **hālig** 'holy' is the first el. of HALIFAX, HAL(L)IWELL, HOLYWELL and others. The inflected forms usually lost the vowel before *g* (*hālge* plur. &c.). Hence HALLATROW (*hālge trēo*), HALSTOW, HALWELL, HOLWELL &c.

Haliwell Mx [*Haliwell* 1201 Cur, 1230 P]. 'Holy spring.'

OE **hall, heall** 'hall', also 'residence, manorhouse', and 'a building for worship', 'a building for legal purposes, a court of law'. The word is not very common in pl. ns., at least in names of villages. WOODHALL is a common name. It may mean 'building where forest courts were held'.

OE **hall, heall** 'rock, stone' (= ON *hallr*, Goth *hallus*) may be found in HALLAM YW, HAWLEY Ha.

Hallam, Kirk & West, Db [*Burhhalum* 1006 Hengwrt MS 150, *Halun* DB, *Kirkehalum* 1242 Fees, *Westhalum* 1230 P.]

Identical with HALAM. OE *halh* appears to mean 'nook, remote valley' in this name.

Hallam YW [*Hallun* DB, *Hallum* 1297 Subs, *Halumsira* 1161 YCh 1268, *-shire* 1276 RH]. Possibly identical with HALLAM Db, but the common *ll* tells against this. Probably it is the dat. plur. of OE *heall* (*hall*) 'a rock, stone' (B-T Suppl). Cf. SCĪR.

Hallaton Le [*Alctone* DB, *Halecton* 1167 P, *Halc-, Halechtone* 1229 Ep]. 'TŪN in a HALH or narrow valley.'

Hallatrow So [*Helgetrev* DB, *Halghetre* 1259 Wells]. 'Holy tree.'

Halliford, Lower & Upper, Mx [(to) *Halgan forde* 962 BCS 1085, *Halgeford* 969 Crawf, 1196 P]. 'Holy ford.'

Hallikeld YN [*Halikeld* 1226 FF]. 'Holy spring.' Cf. **Hallikeld Spring** [*fontes de Halikeld* 1202 FF], which gave its name to **Hallikeld** Wapentake.

Halling (-aw-) K [*Hallingas* 765–91 BCS 260, *Heallingwara mearc* 880 ib. 548, (of) *Heallingan* 10 BCS 1322, *Hallinges* DB]. '*Heall*'s people.' The same *Heall* gave its name to a lake at Halling [*Hallesmeri, -mere* BCS 260, 548]. The pers. n. *Heall*, which may be related to OG *Halo* &c., is unrecorded in independent use, but may be found also in *Hallesborge* BCS 125. Cf. next name.

Hallingbury (-ŏ-), **Great & Little,** Ess [*Hallinge-, Halingheberia* DB, *Hallingeberia* 1130 P, *Hallingeber' Parva* 1238 Subs, *Hallingber' magna* 1291 Tax]. 'The BURG of *Heall*'s people.' Cf. HALLING.

Hallington Li [*Halington* 806 BCS 325, *Halintun* DB, *Haligtune* c 1115 LiS, *Halington* 1202 Ass, 1212 Fees, *Hallinton* 1209–35 Ep]. The first el. may be a derivative with *-ingas* of the pers. n. found in HALESWORTH, HALSALL.

Hallington Nb [*Halidene* 1247 Ep, *-den* 1256 Ass]. 'Holy valley.'

Halliwell La [*Haliwelle* c 1200 CC, *-well* 1246 Ass]. 'Holy spring.'

Halloughton (hawtn) Nt [*Healhtune* 958 YCh 2, *Halton* 1291 Tax], H~ Wa [*Halughton* 1367, 1411 AD]. 'TŪN by a HALH.' H~ Nt is situated at a narrow valley called **Halloughton Dumble.**

Hallow (hŏ-) Wo [(of) *Halhe(o)gan*, (de) *Heallingan* 816, *Hallege* 964 BCS 356, 1135, *Halhegan* DB]. 'Enclosures (OE *hagan*) in or by a HALH.' H~ is in a tongue of land between two streams. Cf. HAWLING.

Halnaby (awn-) YN [*Halnatheby, -bi* 12 PNNR]. '*Halnath*'s BY.' Halnath de Halnatheby is mentioned c 1200 Marrick. The etymology of the name is obscure.

Halnaker (hän-) Sx [*Helnache* DB, *Halfnakere* 1316 FA]. OE *healfne æcer* (acc.) 'half an acre'.

OE **hals, heals,** ON **hals** 'neck' must have been used in various transferred senses, which are found in pl. ns. ON *hals* also meant 'projecting part of something, a narrow piece of land'. ME *hals* is used of 'a narrow neck of land or channel of water'. Dial. *halse* is recorded in the senses 'a defile, narrow passage between mountains' and (usually in the form *hause*) 'a narrower or lower neck or connecting ridge between two heights or summits, a col'. MHG *hals* also meant 'a long ridge'. In English pl. ns. the chief senses are 'col', 'promontory or projecting piece of land or headland'. In HALSWAY the meaning seems to be 'defile'. See HALSE &c., HAWES.

Halsall La [*Heleshale, Herleshala* DB, *Halsale* 1212 Fees, *Haleshal* 1246 Ass]. Perhaps '*Hæle*'s haugh'. *Hæle may be OE *hæle* 'hero' used as a pers. n. Cf. HALESWORTH.

Halse (hawz) Np [*Hasou* DB, *Halsho* 1198 P, 1220 Fees, *Halsou* R 1 BM, *Hals* 1284 FA]. OE *hals-hōh* 'necklike point of land'. The place is between two valleys.

Halse So (*Halse* DB, 1243 Ass, *Hause* 1152 Buckland]. OE *heals* 'neck', referring to a neck of land.

Halsham YE [*Halsaham* 1033 YCh 8, *Halsham* DB, *Hausham* 1212 FF, *Est-*, *Westhalsam* 1260 Ass]. 'HĀM on the neck of land', here referring to the Holderness peninsula.

Halsnead La [*Grewinton Halfsnede* 12 VH, *Halsnade* 1246 Ass]. 'The half part' (of CRONTON). Cf. SNÆD.

Halstead Ess [*Haltesteda* DB, *Haudested* 1202 FF, *Haldstede* 1218 FF], H~ K [*Halsted* 1201 FF, *Aldested* 1212 RBE, *Haltested* 1272 Ipm], H~ Le [*Elstede* DB, *Hallested* c 1125 LeS, *Hald-*, *Hautsted* 1230 P]. OE *h(e)ald-stede*. The first el. is hardly *heald* adj. 'sloping', as this would have been in the weak form. Very likely it is *hold* 'a place of refuge, shelter, or temporary abode' (Lay &c.) from OE (ge)*heald*. The meaning would be 'a place of shelter for cattle'.

Halstock Do [(in) *Halganstoke* 998 KCD 701, *Haleghestok* 1279 Cl]. 'Holy place.' Cf. STOC.

Halston Sa [*Halstune* DB, *Hallestan* 1221, *Halstan* c 1338 PNSa]. Has been explained as 'holy stone', but the forms do not suggest that. Rather 'stone in a HALH or recess'.

Halstow, High, K [*Halgesto* c 1100 Text Roff, *-stowe* 1274 Reg Roff], **Lower H~** K [*Halgastaw* 11 DM, *Halegestowe* 1200 FF, *-sto* 1219 Fees]. 'Holy place.' Cf. STŌW.

Halsway So [(æt) *Healswege* c 1080 ASCh, *Halsuueia* DB, *Halseweie* 1166 RBE, *Hausweie* 1176 P]. 'The pass road.' Cf. HALS. The place is in a pass between hills.

Haltcliff Cu [*Halteclo* 1208 FF, *Alteclo* 1236 P, *Hauteclo* 1252 P, *Hawtecliffe* 1523

PNCu(S)]. 'The angular (winding) ravine.' First el. OE *h(e)alhiht* 'angulosus'; cf. CLŌH. H~ is in the valley of the Caldew, which is here extremely sinuous. For the early reduction of *halhiht* to *halt* cf. SALTLEY, which is *Sautlega* as early as c 1170. Fr *haut* 'high' is first recorded in English c 1450 and is evidently not to be considered as the first el.

Haltemprice YE [*Hautenprise* 1324 Pat, *Hautempris* 1340 FF]. A monastery founded in 1322. The name is French and means 'high enterprise' (Fr *haut* and *emprise*).

Haltham Li [*Holtham* DB, 1255 Ch, 1254 Val, *Holteim* c 1115 LiS]. 'HĀM by a wood.' See HOLT. Sometimes Scandinavianized, OScand *heimr* replacing OE *hām*.

Halton Bk [*Healtun* c 1033, c 1050 KCD 1321, 1336, *Haltone* DB], H~ Chs [*Heletune* DB, *Halton* 1259 Court], H~ La [*Haltune* DB, *Halghton* 1246-51 LaInq], H~ Hole-gate Li [*Haltun* DB, *Haltona* c 1150 DC, *Halton juxta Stephing* 13 BM], **East H~** Li [*Haltune* DB, *-tun* c 1115 LiS, *Halton* 1254 Val], **West H~** Li [*Haltone* DB, *Halton* c 1115 LiS, *Halghton* 1219 Fees], **Lady & Priors H~** Sa [*Halghton* 1327 Subs], H~ YW nr Leeds [*Halletune* DB, *Halghton* 1235 FF], **East H~** YW [*Haltone* DB, *Esthalton* 1314 Ipm], **West H~** YW [*Halctun* 12 Pudsay, *Halton* 1259 Ipm]. OE *H(e)alhtūn* 'TŪN in or by a HALH'. The meaning 'haugh' is suitable for the La Halton and for E. & W. H~ Li.

Holegate is stated to refer to rocks of green sandstone that overhang the road to Halton. If this is right, *gate* is OScand *gata* 'road'.— **Lady H~** from the dedication of the church.— **Priors H~** belonged to Bromfield Priory.

Halton Nb [*Haulton* 1161 P, *Hawelton* 1212, *Hawiltona* 1236, *-ton* 1242 Fees]. H~ is on the slope of a hill on whose top is **Halton Shields.** The old name of the hill was very likely *hāw-hyll* 'look-out hill', *hāw being related to OE *hāwian* 'to gaze on, survey'.

Haltwhistle Nb [*Hautwisel* 1240 Sc, *-tvysel* 1254 Val, *Hawtewysill* 1279 Ass]. 'The junction of streams by the hill'; cf. TWISLA. First el. OE HĒAFOD, here 'hill'.

Halvergate Nf [*Halfriate* DB, *Haluergiata* 1158, *-iet* 1177, *-gata* 1182 P]. Second el. OE *geat* 'gate'. The first seems to be connected with the word *half*, but the form is not clear. The name remains unexplained. It is not certain that its second el. is OE *geat* 'gate'. It could be a combination of *half* and *heriot*, OE *heregeatu*. The name would then mean 'land for which a half heriot was paid' or the like and may be compared with *heregeatland* 1002 Wills and names derived from OE *morgengifu*, on which see EPN(S).

Halwell D [(to) *Halganwille* 10 BCS 1335, *Halgwelle* 1259 Ep], **Halwill** D [*Halgewilla* DB, *Haliwill* 1228 FF]. 'Holy spring.'

OE **hām** 'village, estate, manor, homestead'

is one of the most common elements in pl. ns. The most common meaning is probably 'village'. It is never used alone and rarely as the first el. except in *hāmstede*, *hāmtūn* (q.v.). It is often difficult to distinguish *hām* from HAM(M). See Introduction ii. 2. (*b*).

OE **ham(m)**, **hom(m)** 'meadow, esp. a flat low-lying meadow on a stream', also 'an enclosed plot, a close' is a very common el. in pl. ns. It is frequently used alone to form pl. ns., and it occurs both as a first and as a second el. It is difficult to distinguish it from HĀM, unless early spellings with *mm* or *o* occur. The original meaning of *hamm* is generally held to be 'enclosure', but in pl. ns. it is so often used to refer to flat land on a river or even in the bend of a river, that 'water-meadow' must be assumed to be one of the chief meanings of the word. The corresponding EFris *hamm* means 'a pasture or meadow surrounded with a ditch', LG *hamm* 'a piece of enclosed land, meadow'. OE *hamm* is the source of: **Ham, East & West**, Ess [*Hamme* 958 BCS (1037), 969 Crawf, *Hame* DB, *Estham* 1206 Cl, *-hamme* 1219 Fees, *Westhamma* 1186 BM, *West Hamm* 1198 FF], **H~** Gl at Berkeley [*Hamma* 1194 P, *Hamme* 1195 P], **H~** Ha [*Hamme* 1282 Ep], **H~** K [?*æt Hamme* 875 BCS 539, *Hama* DB, *Hamme* 11 DM], **H~** So at Creech St. Michael [*Hamm* c 1100 Montacute], **High & Low H~** So [*Hamme* 973 BCS 1294, *Hame* DB, *Nitherhamm* 1264 Ipm, *Heyghe Hamme* 1330 Ch], **H~** Sr [*Hamma* c 1150 Crawf, 1168 P, *Hamme* 1194 ff. P], **H~** W [*æt Hamme* 931 BCS 677 f., *Hame* DB], **Hamp** So [*Hame* DB, *Hamme* 1225 Ass]. Most Hams are on low-lying land. It is worthy of notice that all the Hams here discussed are in the south of England.

Hamble R Ha [*Homelea* c 730 Bede, (into) *Hamele* 901 BCS 596, *Hamel* 1369 Pat]. The name is derived from the OE adj. *hamel* (in pl. ns.) which seems to have meant 'maimed', but very likely meant originally 'crooked'. On the Hamble is **Hamble le Rice** [*Amle* 1147 Fr, *Hamele* 1270 Arch 50, *Prioratus de hamele in the Rys* 1404 ib.]. *Rice* is OE *hris* 'brushwood'.

Hambleden Bk [(æt) *Hamelan dene* 1015 Wills, *Hanbledene* DB]. Perhaps '*Hamela*'s valley'. *Hamela* pers. n. is not evidenced, but seems to occur in some pl. ns., as HAMBLETON La.

Hambledon Ha [(to) *Hamelandune* 956 BCS 976, *Hamledune* DB, *Hameledon* 1192 P], **H~** Sr [*Hameledone* DB, *-don* 1203 Cur, 1242 Fees], **H~ Hill** Do [(on) *Hamelendune* 932 BCS 691], **Upper Hambleton** Ru [*Hameleduna* 1067 BM, *Hameldun*, *Hameldune Cherchesoch* DB, *Hameldon* 1202 Ass], **Hambleton Hill** YN [*Hameldune* 13 Guisb], **Black H~** YN [*Hameldon* 1290

Mon], **Great Hameldon** (hill) La [*Hameldon* a 1194 Kirkst]. All these contain OE *hamel* adj. and DŪN and denote or originally denoted hills. Hambledon Ha, Sr and Upper Hambleton denote villages situated near hills. OE *hamel*, like OHG *hamal*, ON *hamall*, no doubt meant 'maimed'. In the hill-name the meaning may be 'bare, treeless' or 'cut-off', i.e. 'level'. Cf. HAMBLE.

Hambleton La [*Hameltune* DB, *-ton* 1177 P], **H~** YW [*Hameltun* DB, *-ton* 1087 Selby]. Apparently '*Hamela*'s TŪN'. Cf. HAMBLEDEN. It should be noticed, however, that in Hambleton Y is a richly wooded hill called **Hambleton Hough** or **Haugh** [*hoga de Hamelton* Selby i. 287].

Hambrook Gl [*Hanbroc* DB, *Hambroke* 1327 Subs]. First el. OE *hān* 'rock, stone'.

Hamdon Hill So [*Hamedone* c 1100 Montacute, *Homedon* 1244 Ass, *Hamedon* 1284 Ipm]. OE *hamma-dūn* 'hill with or among *hamms*'. See HAM(M).

OE **hamel** adj. See HAMBLE &c., HAMILTON, HUMBLEDON. **Hameldon.** See HAMBLEDON.

Hamer La [*Hamer* 1572 PNLa]. OE *hamor* 'rock, cliff'. OE *hamor* is not evidenced in this sense, which applies to OScand *hamarr* and OG *hamar*.

Hameringham Li [*Hameringam* DB, *Hamringheheim* c 1115 LiS, *Hameringeham* 1190 P]. 'The HĀM of the dwellers at the *hamor* or hill.' Cf. HAMER.

Hamerton Hu [*Hambertune* DB, *Hamertun* 1153 BM]. Possibly 'TŪN by a hill', though it is a mile to the nearest hills. Perhaps rather 'hammer TŪN', i.e. 'place where there was a hammersmithy'. Cf. HAMMERSMITH. The place is on Alconbury Brook.

Hamilton Le [*Hamelton* c 1125 LeS, 1242 Fees, *Hameldon* 1220–35 Ep, *Hameld'* 1236 Fees]. The original form is doubtful. The name may be identical with HAMBLEDON or with HAMBLETON La.

OE **hamm**. See HAM(M) after HĀM.

Hammersmith Mx [*Hameresmythe*, *-smithe* 1312 Selden Soc 33]. 'The hammersmithy', 'the hammersmith's smithy'. Cf. G *Hammerschmiede*, Sw *hammarsmedja*. Engl *hammersmithy* is not in OED, but no doubt once existed.

Hammerton YW in Slaidburn [*Hamereton* DB, *Hamerton* 1168 P], **Green & Kirk H~** YW [*Hanbretone*, *Ambretone* DB, *Hamerton* c 1150 YCh 535, *Grenhamerton* 1176 P, *Kyrkehamerton* 1226 FF], **Hammerwich** St [*Humeruuich* DB, *Hamerwich* 1191 P, *-wic* 1220 Ass]. 'TŪN and WĪC by a *hamor* or hill.'

Hammill K [*Hamolde* DB, *Hammolde* 11 DM, *Hemwold* 1200 FF, *Hamelewold* 1232 Cl, *Hammewolde* 1240 Ch]. '*Hamela*'s WALD or wood.' Cf. HAMBLEDEN.

Hammoo·n Do [*Archethamm* 939 BCS 744, *Hame* DB, *Ham Galfridi de Moion* 1194 P, *Hame Mohun* 1282–4 Dunster]. Originally *Hamm*. Cf. HAM(M).

Hammoon belonged to William de Moion in 1086 (DB). *Moon* (*Moyon*) from MOYON in the Côtentin, Normandy. Cf. ORCHARD Do.

Hamnish He [*Hamenes* DB, *Hamenessce* 1123 PNHe, *Hamenes* 1242 Fees]. Apparently OE *Hāman-æsc* '*Hāma*'s ash-tree'.

OE hamor 'hammer' &c. See HAMER &c.

Hamp So. See HAM(M).

Hampden Bk [*Hamdena* DB, *Hampdene* c 1200 PNBk]. Perhaps 'valley with a HĀM or village'.

Hampen Gl [*Hagenepene* DB, -*penne* 12 Fr, 1234 Cl]. '*Hagena*'s pen or enclosure.'

Hamphall Stubbs. See STUBBS.

Hampnett Gl [*Hantone* DB, *Hamtona* c 1130 Oxf, *Hamtonett* 1211–13, *Hamptonet* 1220 Fees], East H~ Sx [*terram Heantunensem* 680 BCS 50, *Antone* DB, *Esthamtonette* 1275 RH], Westhampnett Sx [*Hentone* DB, *Hantonet* 1187 Fr, *Westhamtonet* 1317 Ipm]. OE *Hēa-tūn*, dat. *Hēantūne* 'high TŪN' with addition of the Fr diminutive ending -*et*.

Hampole YW [*Hanepol, Honepol* DB, *Hanepol* c 1160 YCh (1502), 1230 Ep]. '*Hana*'s pool' or 'cocks' pool'. See HANA.

Hampreston Do [*Hame* DB, *Hamma* 1107 (1300) Ch, *Hammes & Prestinton* 1203, *Hamme Preston* 1283 FF]. A combination of two names HAM and PRESTON. Cf. HAM(M).

Hamps R St [*Hanespe* c 1200 Burton, *Hanse* 1577 Saxton]. A river-name identical with HAFHESP in Wales. *Hafhesp*, from earlier *Hamhesp*, means 'summer-dry' and refers to a stream that goes dry in summer. This is a characteristic of the Hamps.

Hampsfell La, a hill which gave its name to Hampsfield [*Hamesfell* 1292–9 FC, 1314 FF]. The name means '*Ham*'s fell or hill'. *Hamr* is an ON pers. n.

Hampshire [*Hamtunscir* 755 &c. ASC, 1060–6 KCD 820, *Suðhamptonscir* c 1050 ib. 845, *Hamtesira* c 1155 Laws, *Hamptessira* 1297 Rob Gl]. Named from *Hamtūn*, the old name of SOUTHAMPTON. An abbreviated form is Hants [*Hantescire* DB, -*scira* 1156 P].

Hampstead Marshall Brk [*Hamestede* DB, (terra comitis Marescalli de Spenes et) *Hamsted* 1220 Fees], H~ Norris Brk [*Hanstede* DB, *Hamsted* 1220 Fees, *Hampstedeferrerys* 1409 BM], H~ Mx [*Hamstede* 959, 978 BCS 1351, 1309, *Hemstede* 959 BCS 1351, *Hamestede* DB], H~ Wt [*Hamstede* DB]. OE *hāmstede* 'homestead, manor'.

H~ Marshall belonged to the Lord (now Earl) Marshal of England.—H~ Norris was sold to John Norreys a 1450 (VH).

Hampsthwaite YW [*Hamethwayt* c 1180 YCh 510, *Hameleswaith* 1208 Cur, *Hamesthwait* 1225 FF, -*thueit* 1230 Ep]. 'The thwaite of *Hamr* or of *Hamall*.' *Hamr* and *Hamall* are ON pers. ns.

Hampton has three distinct sources:

1. OE HĀMTŪN: Meysey H~ Gl [*Hantone* DB, *Hamtone Rogeri de Meisi* 1221 Ass, *Meiseishampton* 1287 QW], H~ Gay & Poyle O [*Hantone* DB, *Hampton* 1248 Ep, *Hamtona Gaitorum* c 1130 Oxf, *Geithamton* 1203 Ass, *Hampton Poile* 1428 FA], H~ Lovett Wo [*Hamtona* 716 BCS 134, *Hamtune* DB, *Hampton Lovet* 1291 Tax].

2. OE *Hamm-tūn* 'TŪN in a HAMM': Hampton He nr Bodenham [*Homtona* 1242 Fees], H~ Bishop He [*Hantune* DB, *Homptone*, 1240, 1270 Ep], H~ Wafer He [*Hantone* DB, *Hampton Waffre* 1286 Ch], H~ Mx [*Hamntone* DB, *Henton*, *Hampton* c 1130 Oxf, *Hamton* 1202 Cur], H~ Lucy Wa [*Homtun* 781 BCS 239, *Hantone* DB, *Hampton Episcopi* 1285 QW]. Cf. SOUTHAMPTON.

3. OE *Hēa-tūn*, dat. *Hēan-tūne* 'high TŪN': Hampton Chs [*Hantone* DB, *Hanton* 13 BM], H~ Sa on the Severn [*Hempton* 1391 Ipm], Welsh H~ Sa [*Hantone* DB, *Hampton Howell* 1292 QW], H~ in Arden Wa [*Hantone* DB, *Hantuna in Ardena* Hy 2 BM, *Hamton in Ardern* 1201 Cur, *Hampton in Arden* 1242 Fees], H~ on the Hill or H~ Curli Wa [*Hampton Curly* 1316 FA], Great & Little H~ Wo [*Hamtona* 709, (æt) *Heantune* 780 BCS 125, 235, *Heamtun*, *Hamtun* 988 KCD 662, *Hantun* DB]. Cf. MINCHINHAMPTON.

H~ in Arden, see ARDEN.—Gay in H~ Gay is a Norman family name, probably from OFr *gaite*, *guaite* 'guard'.—H~ Lovett was held by Henry Luvet in 1242 (Fees). Lovett is from OFr *lovet* 'wolf cub'.—H~ Lucy came to Thomas Lucy of Charlecote c 1550. Lucy is a family name, perhaps from LUCÉ or LOUCÉ in Normandy.—Meysey H~ was held by Robert de Meisi before 1185 (TpR). Meysey is a family name from MAISY in Normandy.—H~ Poyle came to Walter de la Puile in 1268 (Ipm). *La Puile* is a French family name.—H~ Wafer from a local family. Simon le Wafre(r) is mentioned 1212 ff. Fees (He). *Wafre* may be OFr *wafre* 'wafer' or *wafrer* 'waferer'.

Hamptworth W [*Hanteworth* 1232 Cl, *Hamteworthe* c 1270 Ep]. The place is near the Hants boundary. The first el. may be *Hamtūn*, elliptical for *Hamtūnscīr*. See WORÞ.

Hamsey Sx [(æt) *Hamme wiþ Læwe* c 961 BCS 1064, *Hame* DB, *Hammes Say* 1322 Ipm]. OE HAMM.

Say is the name of the family that held the place from the early 13th cent. (from SAI in Normandy).

Hamstead St [*Hamsted* 1227, 1293 Ass]. OE *hāmstede* 'homestead'.

OE hāmstede 'homestead' and very likely 'manor' is the source of HAMPSTEAD,

HAMSTEAD, and is the second el. of several names, as BERKHAMSTEAD, EASTHAMPSTEAD, HEMPSTEAD Gl, NETTLESTEAD, SANDERSTEAD, SWINSTEAD. A side-form of *hămstede* is *hæmstede*, found e.g. in *Hæmstedes* (*Hemstedes*) *geat* 979 KCD 622, *Netelhæmstyde*, *Sondemstyde* 871–89 BCS 558, *Dunhæmstede* post 972 KCD 680, *Stanhæmstede* 990 ib. 673. Cf. also FINCHAMSTEAD and *Hanchemstede* 692 BCS 81. This *hæmstede* is the source of some HEMPSTEADS. *Hæmstede* is due to *i*-mutation, which took place in the form *hămstyde*, which became *hæmstyde*, whence often *hæmstede* owing to the dialectal change *y* > *e* or to weakening of *y* in the unstressed syllable. The same kind of *i*-mutation is found in OE *Hæmgisl* from *Hămgisl*, *Ēdgȳþ* from *Ēadgȳþ*.

Hamsteels Du [*Hamstele*, *-stelis* 1242 Ass]. The elements are OE *hăm* 'home' &c. and *stigol*, here in the dialectal sense 'steep ridge' found in *steel*.

Hamsterley Du [*Hamsteleie* c 1190 Godric, *Hamsterlege* 1242 Ass, *-le* 1307 RPD]. See LĒAH. The first el. is no doubt an OE **ham(e)stra*, corresponding to OHG *hamastro*, OLG *hamstra* 'corn-weevil' and of similar sense.

OE **hămtūn** is only found in pl. ns. It is on the whole a rare element. The meaning is no doubt much the same as that of BOLTON, i.e. 'the village proper' in contradistinction to outlying parts, or even 'the chief manor of a large estate'. Some *Hămtūns* achieved great importance, as NORTHAMPTON. Cf. Introd. p. xvi f. Modern HAMPTON only rarely represents OE *hămtūn*. The same is true of *-hampton* as a second el. Sometimes the source is OE *-hēantūn* 'high TŪN'. But in the majority of cases *-hampton* represents OE *-hæmatūn*. A name such as BROCKHAMPTON goes back to OE *Brōchæmatūn* 'the TŪN of the *Brōchæme* or dwellers on the brook', *-hæme* 'dwellers' being a derivative of OE *hăm*. Of this type are DITCH-, POOL-, SEVENHAMPTON and others.

Hamworthy Do [*Hamme* 1236 Fees, *Hamme juxta la Pole* 1285 FF, *Hamwurthy* 1535 VE]. First el. OE HAMM. The second is presumably OE WORÞIG.

OE **hăn** 'stone, rock'. See HAMBROOK, HANFORD Do, HANHAM, HENFIELD, HONING, HONLEY.

OE **hana** 'cock, wild bird'. See HAMPOLE, HANDBOROUGH, HANDFORTH, HAN- (passim), HONICKNOWLE. Cf. HANKFORD.

Hanbeck Li [*Handebek* 1242 Fees, 1275 RH]. '*Handi*'s beck.' *Handi* is an ON byname.

Hanbury St [*Hamb*[*ury*] c 1185 Fr, *Hambur* 1251 Ch], H~ Wo [*Heanburh* c 765, (in) *Heanbyrg* 836 BCS 220, 416, *Hambyrie* DB]. 'High BURG.'

Hanby Li nr Welton le Marsh [*Hundebi*, *Hunbia*, *Humbi* DB, *Humbi* c 1115 LiS,

Hambi 1212 Fees, *-by* 1221 Ep, 1242 Fees]. The earliest forms point to OScand *Hundabȳr* '*Hundi*'s BY'. *Hundi* is an ON byname and pers. n. The name may have been altered owing to its unpleasant associations.

Hanby Li nr Folkingham [*Handebec* 1212, 1242 Fees]. Identical with HANBECK.

Hanchurch St [*Hancese* DB, *-churche* 1212 Fees]. 'High church.' See HĒAH.

Handborough, Church & Long, O [*Haneberge* DB, 1195 P, *-berga* 1143 Oxf, *-berg* 1230 P, ?*Hageneberga* 1156 P]. If the form of 1156 is reliable, '*Hagena*'s hill'. Otherwise '*Hana*'s hill' or even 'cocks' hill'.

Handbridge Chs [*Bruge* DB, *Honebrugge* 1260 Court, 13 BM, *Hunebrugge* 1289 Court]. '*Hana*'s bridge.' *Hana* appears in the WMidl form *Hone*, which sometimes even passed into *Hune*. Cf. HANDSACRE, HANKELOW.

Handforth Chs [*Haneford* 1158–81 Chester, *Honeford* 1260 Court]. '*Hana*'s ford' or 'cocks' ford'. The fact that there are three Hanafords in Devon tells in favour of the latter alternative. Cf. HANA, HANFORD St.

Handley Chs [*Hanlei* DB, *-legh* c 1175, *-leg* c 1200 Chester], H~ Db [*Henlege* DB], **Mid, Nether & West** H~ Db [*Henleie* DB, *Hanleg* 1230 Cl], H~ Do [(at) *Hanlee*, (in) *Henlee* c 871 BCS 531 f., *Hanlege* DB, *Henleg* 1212 Fees], H~ Np [*Haunleg*, *Hanlegh*, *Henle* 1234 Cl]. OE *Hēa-lēah*, obl. *Hēanlēa*, 'high LĒAH'. Handley Do is probably an old name of Cranborne Chase. The place is on the eastern side of the Chase, but in BCS 970 (at) *Heanlegen* occurs in the boundary of Compton Abbas west of the Chase.

Handsacre St [*Hadesacre* DB, *Hendesacra* 1167, *Hundesacra* 1176, *Handesacra* 1196 P, *Hondesacr'* 1242 Fees]. The first el. appears to be a pers. n. **Hand*, originally a nickname from *hand* 'manus'; cf. *Fōt*. If so, the name *Hand* did not keep the inflexion of the word *hand*, but formed its genitive with *-es*. For the forms with *o*, *u* cf. HANDBRIDGE.

Handsworth St [*Honesworde* DB, *Huneswordne* 1209 FF, *-wurth* 1222 Ass, *Honesworthe* 1242 Fees]. '*Hūn*'s WORÞ.' The late form *Hands-* may be due to influence from HANDSACRE.

Handsworth YW [*Handeswrde* DB, *-wrth* 12 BM, *-wrda* c 1185 YCh 1274]. '*Hand*'s WORÞ.' Cf. HANDSACRE.

Hanford Do [*Hanford* DB, *Hámford* 1194 P, *Hamford* 1197 FF, *Haunford* 1228 Cl]. OE *hăn-ford*, the first el. being OE *hăn* 'stone, rock'.

Hanford St [*Heneford* DB, *Honeford* 1212 Fees, *Hondford* 1327 Subs]. Apparently identical with HANDFORTH.

Hangleton Sx nr Hove [*Hangetone* DB,

Hangeltuna c 1115, *-tona* 1121 AC, *Hengelton* 1248 Ass], H~ Sx nr Ferring [*Hangleton* 1380 PNSx]. The first el. can hardly be OE *hangra*, as no spellings with *r* occur. It is probably another derivative of the verb *hang* with an *l*-suffix, meaning 'slope'. Cf. OE *hangelle* 'a hanging thing'.

OE **hangra** is a derivative of the verb *hang* and must originally have denoted 'a slope'. Cf. *hanging* 'steep, situated on a steep slope'. Nowadays *hanger* means 'wood on the side of a steep hill or bank'. This sense may have developed already in OE times, but the old sense 'slope' probably lived on, to judge by the common name CLAYHANGER (CLE-HONGER, CLINGER &c.), which must mean 'clayey slope'. The first el. is often the name of a tree, as OAKHANGER. See also BARNACLE, BINEGAR, BRAMINGHAM, GOLD-HANGER, HARTANGER, RISHANGLES.

Hanham Gl [*Hanun* DB, *Hanum* 1153, c 1155 Berk, 13 BM, *Hanam* a 1173 Berk]. This looks like the dat. plur. of *hān* 'stone'. But OE *hānum* would have become *Hanen*, *Hane* in Gl. Probably the name is a compound with OE *hamm*, *homm* as second el. The first el. may be OE *hana* 'cock' or *Hana* pers. n. or *hān* 'stone'.

Hankelow Chs [*Honkyloue* 1260, *Hun(e)kelowe* 1281, *Honekelow* 1282 Court]. '*Haneca*'s hill.' Cf. HANDBRIDGE and the following names, esp. HANKHAM.

Hankerton W [*Hanekyntone* 681, 901 BCS 59, 589, *Honekynton* 1065 KCD 817, *Hanckinton* 1242 Fees]. '*Haneca*'s TŪN.'

Hankford D [*Hanecheforda*, *Hancheford* DB], **Hankham** Sx [*æt Hanecan hamme* 947 BCS 821, *Henecham* DB]. '*Haneca*'s ford and HAMM.' **Haneca* is found in several names. It is derived from *Hana*, which is evidenced as the name of a moneyer and occurs in several names (e.g. *Hanan wurð* 901 BCS 588). Like OScand *Hani*, OG *Han(n)o* it is a nickname identical with *hana* 'cock'.

Hanley St [*Henle* 1212 Fees, *Hanlih* 1227 Ch], **H~ Castle** Wo [*Hanlege* DB, *Heanlega* 1182 P], **H~ Child & William** Wo [*Hanlege* DB, *Hanleg* 1198 FF, *Chuldrenehanle* 1265 Misc, *Williames Henle* 1275 Ass]. Identical with HANDLEY.

H~ William from William de la Mare, who held Hanley in 1212 (RBE) or the William de la Mare whose son Thomas held H~ in 1198 (FF).—For **Child Hanley** see CHILTON.

Hanlith YW [*Hangelif* DB, *Hahgenlid* 12 Fount, *Hahenelid* 1219 FF, *Haunlith* 1260 YInq]. '*Hagena*'s HLIP or slope.' Or the elements are OScand *Hagne* (cf. HAGNABY) and *hlīð* 'a slope'.

Hannah Li [*Hanai*, *Hanei(e)* 12 DC, *Haneye* 1228 Ep]. '*Hana*'s island' or as next name.

Hanney, East & West, Brk [*Hannige* 956, *Haniges hamm* 958, *æt Hanige, Hannige* 968 BCS 949, 1035, 1224, *Hannei* DB, *Est-*

henneya, Westhenn' 1220 Fees]. Apparently OE *hanena īeg* 'island frequented by (wild) cocks'.

Hanningfield, East, South & West, Ess [*Haneghe-, Haningefelda* DB, *Westhanegefeld* 1208 FF, *Haningefeld* 1212 Fees, *Est-, Sut-, Westhaningefeld* 1254 Val], **Hannington** Ha [*Hanning-, Hanitun* 1023 KCD 739, *Hanitune* DB, *Hanincton* 1185 BM], **H~** Np [*Hani(n)tone* DB, *Haninton* 1195 FF, 1224 Ep]. 'The FELD and TŪN of *Hana*'s people.' Cf. HANKHAM.

Hannington W [*Hanindone* DB, *Hanedone* 1212 RBE, *Hanendon* 1242 Fees]. '*Hana*'s hill' or 'cocks' hill'. See HANA.

Hanslope Bk [*Hammescle, Anslepe, Hamslape* DB, *Hames(c)lape* 1104–6 RA, *Hamesclapa* 1159, *Hamslope* 1198 P]. The first el. may be *Hāma* pers. n. The second is the el. *slæp* found in several pl. ns., as ISLIP &c. It is generally held to mean 'a slippery, miry place'. In Hanslope 'slope' would be more suitable. Such a sense may well have developed from 'slippery place'.

Hanthorpe Li [*Hermodestorp* DB]. '*Heremōd*'s thorp.'

Hanwell Mx [*Hanewelle* 959 BCS 1050, *Hanawella* 998 Th, *-uuelle* 1065 BM, *Hanewelle* DB]. '*Hana*'s spring' or 'cocks' spring'. See HANA.

Hanwell O [*Hanewege* DB, *-weie* 1220, *-wey* 1242 Fees, *Haneuell* 1236 Fees]. '*Hana*'s road', later changed to *-well*.

Hanwood Sa [*Hanewde* DB, *-wude* 1180 P]. 'Cock wood.' See HANA.

Hanworth, Cold, Li [*Haneurde* DB, *-worda* c 1115 LiS, *Calthaneworth* 1322 Ipm], **Potter H~** Li [*Haneworde* DB, *-wrda* 1157 YCh 354, *-wrth* 1206 Ass], **H~** Mx [*Haneworde* DB, *-wrth* 1212 RBE, *Hanesworth* 1254 Val]. '*Hana*'s WORP.'
Potter must refer to potteries.

Hanworth Nf [*Haganaworda* DB, *Hanewrth* 1270 Ipm]. '*Hagena*'s WORP.'

Happisburgh (hāzbru) Nf [*Hapesburc* DB, *Apesburga* c 1150 Fr, *Hapesburg* 1272 Ch]. '*Hæp*'s BURG.' The same *Hæp* gave its name to **Happing** hundred, in which Happisburgh is [*Hapinga* DB, *Hapingeh'* 1156 P]. Happing means '*Hæp*'s people'. The pers. n. **Hæp* is also found in HAPSFORD. It belongs to OE *gehæp* 'fit'.

Hapsford Chs [*Hapisford* 1288 Court, *Hapesford* 13 Chester]. '*Hæp*'s ford.' Cf. HAPPISBURGH.

Hapton La [*Apton* 1242 Fees, *Hapton* 1246 Ass]. 'TŪN by a *heap* or hill.' Cf. HEAP.

Hapton Nf [*Habetuna, Habituna* DB, *Habeton* 1198 FF, *Hapetun* 1242 Fees]. '*H(e)abba*'s TŪN.' Cf. HABTON.

OE **hār** adj. 'hoary, grey' is a fairly common first el. in pl. ns., though by no means so frequent as has been sometimes assumed. It often occurs combined with stone, as in HARSTON, HOARSTONE, HORSTON, and *hoarstone*, lit. 'a grey lichen-covered stone', came to be a technical term for such a stone used as a boundary mark. It is with certainty only combined with words for objects that are or sometimes are naturally grey, as cross (HOAR CROSS), oak (HARROCK), withy (HOARWITHY), hill (HARLOW), wood (HAREWOOD Ha). Cf. HARWELL Brk, which is a special case. It is often stated that OE *hār* had developed the meaning 'boundary', 'boundary-defining'. This theory is not well founded. It is unlikely, or at any rate it has not been proved, that the first el. of names beginning in *Har(e)-*, as HARDEN, HAREWOOD, is generally the adj. *hār*. There is every probability that some names such as HAREWOOD, HARGRAVE, HARROP contain the word *hara* 'hare'. We must also reckon with an element *hær* or the like that has only recently been discovered. It is certainly found in HAROME YN, HERNE Bd and may be suspected to enter into some other names, such as HARROLD, HARNAGE, HARNHILL &c. The exact meaning and OE form of the word are unknown. It is related to Sw *har* n. 'stony ground', a LG and Du *har*, *hare* that is found in many pl. ns., as HAAR, HAREN (in early sources *Hare*, *Harun* &c.) and which is stated to mean 'height', 'ridge', 'height covered with wood'. The OE form may have been *hær* n. The words are related to Ir, Welsh *carn* 'cairn' and a derivative is very likely OE *hearg* 'heathen temple', ON *horg* 'heap of stones, altar' (originally meaning 'stone altar'). The meaning of OE *hær* may have been 'stone, stony ground'.

Harberton D [*Herburnat'* 1108 Sarum, *Hurbertun* 1212 Fees, *Hirbirton* 1276 Cl]. 'TŪN on R Harbourne.' **Harbourne** is *Hurburn* 1244 Ass, *-e* 1315 Totnes. On it is **Harbourneford** [*Herberneforda* DB, *Hurberneford* 1242 Fees]. The name seems to consist of OE *hēore*, *hȳre* 'gentle, pleasant' and BURNA.

Harbledown K [*Herebolddune* 1175, *Herboldon* 1179, *Herebaldon* 1196, *boscus Hereboldi* 1200 P]. 'Herebeald's DŪN or hill.'

Harborne St [*Horeborne* DB, *Horeburn* 1221 Ass]. OE *horu-burna* 'dirty stream'. OE *horh*, *horu* means 'dirt'.

Harborough, Market, Le [*Hauerberga* 1177, 1190 P, *Haverberge* 1237 Ch, *Mercat Heburgh* 1312 BM]. The first el. may be OE *hæfer* 'he-goat'. But more likely it is an unrecorded OE **hæfera* 'oats' corresponding to OScand *hafri*, OHG *habaro*, OLG *havoro*, or even OScand *hafri*. 'Hill where oats were grown.'

Harborough Magna & Parva Wa [*Herdeberge* DB, *-berwe* 1274 Ipm, *Herdeborough Magna* 1316 FA, *Little Herdebergh* 1305 Ch]. OE *heord* 'herd, flock' and *beorg* 'hill'.

Harbottle Nb [*Hirbotle* c 1220 Sc, *Hyrbotle* 1245 Ipm, *Hirebotel* 1279 Ass, *Herbotle* 1291 Tax]. 'The dwelling of the hireling(s).' First el. OE *hȳra* 'hireling'. Cf. BŌÞL.

Harbourne. See HARBERTON.

Harbridge Ha [*Herdebrige* DB, *Hardebriggs* c 1270 Ep, *-brygg* 1316 FA]. Perhaps 'Hearda's bridge'. **Hearda* is a normal short form of names in *Heard-*, *-heard*. 'Hard bridge' does not seem plausible.

Harbury Wa [(æt) *Hereburgebyrig* 1002 Wills, 1004 KCD 710, *Erburgeberie* DB, *Herburberi* 1200 Cur]. 'Hereburg's BURG or manor.' *Hereburg* is a woman's name.

Harby Le [*Herdebi* DB, 12 DC, *-by* 1242 Fees], **H~** Nt [*Herdebi*, *Herdrebi* DB, *Hertheby* Hy 2 (1316) Ch, 1291 Misc]. The first el. might be OScand *hiǫrð* 'herd, flock' (OScand *Hiarðar-bȳr*). More likely it is a pers. n., viz. ON *Herrøðr* (gen. *Herruðar*), ODan *Heroth* (*Herothus* Saxo).

Harcourt Sa nr Cleobury Mortimer [*Havretescote* DB, *Havekercot* 1255, *-e* 1274 RH]. 'The hawker's cottage.' OE *hafocere* 'hawker, falconer' is on record.

Harcourt Sa nr Wem [*Harpecote* DB, *-cot* 1191 P]. Cf. HARPENDEN &c.

Harden St [*Haworthyn*, *-werthyn* 14 PNSt]. Identical with **Hawarden** Flint [*Haordine* DB, *Haurdina* c 1100 Chester, *Haworthyn* 1275 Ipm]. Possibly OE *Hēa-worþign* 'high WORÞIGN'.

Harden YW nr Bingley [*Hareden* c 1166, *Harden* c 1215 YD, *Hareden* 1234 FF]. Perhaps 'hare valley'.

Hardenhuish (harnish) W [*Hardenhus* DB, *Hardehiwis* 1178 P, *Herdenehywys* 1258 Ipm]. It is doubtful if *Heregeardingc Hiwisce* 854 BCS 469 belongs here. If not, *Harden-* may be identical with HARDEN YW. *Heregeard* is an OE pers. n. See HĪWISC.

Hardham Sx [*Heriedeham* DB, *Heringham* 1189 P]. 'Heregȳþ's HĀM or HAMM.' *Heregȳþ* is a woman's name.

Hardhorn La [*Hordern* 1298 WhC, 1327 Subs]. OE *hordern* 'store house'.

Hardingham Nf [*Hardingeham* 1161 P, *Hardingham* 1242 Fees, 1275 RH]. 'The HĀM of Heardrēd's or **Hearda's* people.' Cf. HARBRIDGE.

Hardingstone Np [*Hardingestorp*, *-tone* DB, *-torn* 12 NS, 1224 Ep]. 'Hearding's thorn-bush.' *Heardinc* (*Harding*) pers. n. is recorded in the 11th cent.

Hardington So nr Radstock [*Hardintone* DB, *Hardington* 1225 Ass], **H~ Mandeville** So [*Hardintone* DB, *Herdintone* 1166 RBE, *Hardin(g)ton*, *Herdinton* 1243 Ass]. 'The TŪN of Heardrēd's or Hēahrēd's or **Hearda's* people.' Cf. HARBRIDGE.

H~ Mandeville was held by Galfridus de Mondeville in 1166 (RBE). Mandeville, earlier *Magneville*, is from MANDEVILLE in Normandy.

Hardisworthy D [*Herdesworth* 1284–6 FA, -*e* 1326 Ipm]. '*Hererēd*'s WORÞIG.'

Hard Knott Cu, a hill [*Hardecnuut* c 1210, *Ardechnut* 1242 FC]. Second el. ON *knútr* 'knot', Norw *knut* 'peak'. In PNCu(S) a doublet of the name in the neighbourhood of Loweswater is pointed out (*Hardecnut* 1230 FF). This tells in favour of the first el. being the word *hard*, originally ON *harŏr*, though rather in such a sense as 'difficult, inaccessible' than 'craggy'.

Hardley Ha [*Hardelie* DB, *Hardel'* 1212 Fees], H~ Wt [*Hardelei* DB]. 'Hard clearing', perhaps referring to hard soil.

Hardley Nf [*Hardale* DB, 1286 QW, *Hardele* c 1115 Holme, 13 BM, *Hardeleygh* 1268 Ch]. Apparently identical with preceding name, though some forms suggest a second el. HALH.

Hardmead Bk [*Herould*-, *Herulf*-, *Horel-mede* DB, *Harewemede* 1194 Rot Cur, *Harle-mede* 1223 Ep, *Hardmede* 1284–6 FA]. Possibly '*Heardwulf*'s or *Heoruweald*'s meadow'. But H~ is not very far from Harrold, and the name might mean 'meadow belonging to HARROLD'.

Hardres (hardz), **Lower & Upper**, K [(in) *Haredum* 785, (in) *Haraðum* 786 BCS 247 f., *Hardes* DB, *Hardan* 11 DM, *Hardres* 1191 P, *Heghardres* 1242 Fees, *Netherhardres* 1247 Ch]. The plur. of an OE *haraþ* (*harad*), corresponding to and identical in meaning with OHG *hart* 'wood' (the source of the name HARZ).

Hardstoft Db [*Hertestaf* DB, *Hertistoft* 1257 FF]. '*Heort*'s toft.' *Heor(o)t* pers. n. is not recorded, but cf. HARTING and *Heortla* in HARTLEBURY, ON *Hiǫrtr*, OHG *Hiruz*. *Heorot* is identical with OE *heorot* 'hart'.

Hardwell Brk in Compton Beauchamp [(on) *Hordwyllæ* 856, (æt) *Hordwelle* 903 BCS 491, 601, *Hordewelle* 1220 Fees]. 'Treasure spring', probably referring to a spring into which coins or other articles of value were thrown for sacrificial purposes.

Hardwick (-dĭk) Bk [*Harduich* DB, *Herde-wyc* 1209 Fees], H~ Ca [*Hardwic* c 1050 KCD 907, -*uic* DB, *Herdewyk* 1250 Fees], H~ Du [*Herdwich* Hy 2 FPD, *Herdewich* 1195 (1335) Ch], H~ Le [*Herdwyk* 1252 Cl], H~ Li [*Hardwic* DB], H~ Nf nr Bungay [*Herdeuuic, Hierduic* DB, *Herdwik* 1254 Val], H~ Nf nr King's Lynn [*Herdwic* 1242 Fees], H~ Np [*Heordewican* c 1067 Wills, *Herde*-, *Hardewiche* DB, *Herdewic* 1164 BM], H~ Nt [*Herdewic* Hy 2 (1316) Ch, *Herthewyk* 1286 Ch], H~ O nr Banbury [*Herdewyke* 1316 FA], H~ O nr Bicester [*Hardewich* DB, *Herdewic* c 1130 Oxf], H~ O nr Yelford [*Herdwich* 1200 Cur, -*wick* 1245 Ch], H~ Ru [*Herdewyk* 1316 Ipm], H~ Sa nr Bishops Castle [*Hordewik* 1237 FF, *Nor-bur' Herdewyke* 1255 RH], H~ Sa nr Shrewsbury [*Herdewica* 1155–8 PNSa, *Herdewyke* 1291 Tax], H~ Sa nr Wem [*Herdewyk*

1284, 1320 Ch], H~ Sf [*Herdewic* c 1130, c 1150 Bury], **Priors** H~ Wa [*Herdewyk* 1043 Th, *Herdewiche* DB, *Herde-wych Priour* 1331 AD], H~ Wo [*Herdwicke* 1299 PNWo], H~ **Green** Wo [*Herdwich* 1183 f. P], **East & West** H~ YW [*Harduic* DB, *Herdewica* 1120–2 YCh 1430], **Hard-wicke** Gl nr Stroud [*Herdewike* c 1200 Glouc], H~ Gl nr Tewkesbury [*Herdeuuic* DB], H~ He [*la Herdewyk* 1309–24 BM]. OE *heord(e)wic* 'WĪC for the flock', i.e. 'sheep farm'.

Priors H~ from the Prior of Coventry.

Hardy La [*Hardey* 1555 FF]. Second el. ĒG 'island'. The first is doubtful.

Hareby Li [*Harebi* DB, 1154–60 RA, 1202 Ass]. '*Hari*'s BY.' ON *Hari* is a nickname, literally 'the hare'.

Harefield Mx [*Herefelle* DB, -*feld* 1201 Cur, -*feud* 1236 Fees, *Herrefeld* 1177 P, *Harefeld* 1242 Fees, *Heresfeld* 1200 Cur]. This looks like OE *Herefeld* 'field of the army or of the people', but the meaning of such a name is not apparent. Cf. HARLOW.

Harehope Nb [*Harop* 1185 P, 1236 Fees, *Harhop* 1242 Fees]. 'Hares' valley.'

Harescombe Gl [*Hersecome* DB, -*cumbe* c 1160 Glouc, *Harescombe* 1287 QW], **Haresfield** Gl [*Hersefel* DB, -*feld* 1211–13, 1220 Fees, *Harsefelde* c 1160 Glouc, *Hares-feld* 1287 QW]. The common first el. would seem to be an unrecorded pers. n. *Hersa* or *Heresa* corresponding to OScand *Hariso*, apparently the first el. of Hescott D [*Hersecote* 1168 P]. But the places are on opposite sides of **Haresfield Hill**, and *Herse*- might possibly represent an old name of the hill, identical with ON *hiarsi*, OSw *hiæsse* 'top of the head'. Such a word is not evidenced in English, however. See CUMB, FELD.

Harewood Ha [*Harwode* 1198 (1260) Ch, *Horwud* 1238 Cl, *Harewood* 1246 Cl], H~ He [*Harewuda* 1138 AC, 1188 P, -*wod* 1252 Fees], H~ (har-) YW [(æt) *Harawuda* 10 Rushw MS, *Hareuuode* DB, -*wod* 1209 FF]. H~ Ha has as first el. OE *hār* 'grey'. The name means 'grey wood'. The same may be the etymology of the other two, but 'hares' wood' is more probable. H~ YW may even have as first el. the OE *hær* 'stony ground' or the like mentioned under HĀR. The place is on a high ridge.

Harford D [*Hereford* DB, *Herford* 1291 Tax]. See HEREFORD and HEREPÆÞ.

Harford Gl [*Heortford* 743 BCS 165 (the ford), *æt Heortford* 963 ib. 1105, *Hurford* DB, *Hertford* 1220 Fees]. 'Stags' ford.'

Hargham (harf-) Nf [*Herkeham* DB, *Herc-ham* DB, 1166 RBE, *Herceam* 1121 AC]. '*Hereca*'s HĀM.'

Hargrave Chs [*Haregrave* 1287 Court], H~ Np [*Haregrave* DB, 12 NS, 1220 Fees], H~ Sf [*Haragraua* DB, *Haregraue* c 1150

Bury, *Hargrave* 1254 Val], **Hargreave Hall** Chs in Li. Neston [*Haregrave* DB]. 'Hares' grove' or 'grey grove'. It is unlikely that the latter is the meaning in all cases. Cf. GRÁF, GRÆFE.

Harkstead Sf [*Herchesteda* DB, *Herkestede* 1198 FF, 1254 Val]. Names in STEDE rarely have a pers. n. as first el. but in this case an exception will have to be admitted. The name means '*Hereca*'s place'.

Harlaston St [*Heorelfestun* 1002 E, *Horulvestone* DB, *Herlaueston* 1165 P]. '*Heoruwulf*'s TŪN.' For the pers. n. *Heoruwulf* or *Heorulf* see also HARSTON Ca.

Harlaxton Li [*Herlavestune* DB, *-tona* 1180–3 Middleton, *-ton* 1234 Ep, *Herlakiston* 1276 RH]. '*Heorulāf*'s or *Herelāf*'s TŪN.' Neither name is on record, but cf. HARLESCOTT. *Heorulāf* corresponds to OScand *Hiǫrleifr*.

Harle, Little, & Kirkharle Nb [*Herle* 1177 P, 1242 Fees, *-lee* 1196 FF, *Kyrkeherl*' 1242 Fees, *Parva Herle* 1279 Ass]. Second el. OE LĒAH. The first might be **Herela* pers. n. Or it may be OE *herg*, a late form of *hearg* 'heathen temple'.

Harlescott Sa [*Herlaveschot* 1160–5 PNSa, *Erlauescote* 1199 FF]. First el. as in HARLAXTON.

Harlesden Mx [*Herulvestune* DB, *Herleston* 1241 Ep, 1254 Val, *Herlesdon* 1291 Tax]. See HARLESTON.

Harleston D [*Harliston* 1252 FF], **H~** Nf [*Heroluestuna* DB, *Heroluestun* c 1095 Bury, *Harolveston* 1228 Ch], **H~** Sf [*?Heorulfestun* 1015 Wills, *Heroluestuna* DB, *Heroluestun* c 1095 Bury, *Herleston* 1197 P], **Harlestone** Np [*Herolvestune* DB, *Herleston* 1170 P]. '*Heoruwulf*'s or *Herewulf*'s TŪN.'

Harley Sa [*Harlege* DB, *-leg* 1229 Cl, *-le* 1242 Fees], **H~** YW nr Wentworth [*Harlay* 1297 Subs], **H~ Wood** YW nr Todmorden [*Harley* 1379 PT]. Probably 'hares' wood'.

Harling, East & West, Nf [*Herlingham* 1046, (at) *Herlinge* c 1060 Wills, *Herlinga* DB, *Herlinge* c 1095 Bury, *Est-*, *Westherling* 1242 Fees]. '*Herela*'s people.' Cf. OG *Herilo* and the next names, also HARLTON. **Herela* is a normal short form of names in *Here-*, *-here*.

Harlington Bd [*Herlingdone* DB, *Herlingedon* 1190 P], **H~** YW [*Herlyngton* 1345 FF]. 'The DŪN and the TŪN of *Herela*'s people.' Cf. HARLING.

Harlington Mx [*Hygeredingtun* 831 BCS 400, *Herdintone* DB, *-ton* 1206 Cur]. 'The TŪN of *Hygerēd*'s people.'

Harlow Ess [(at) *Herlawe* 1043–5 Wills, *Herlaua* DB, 1190 P, *-laue* 1202 FF]. OE *here-hlāw* (or rather *her-hlāw* with loss of *e* as in *herpæþ*) 'the mound of the people'. A hundred meeting-place is referred to. Harlow is also a hundred.

Harlow Hill Nb [*Hirlawe* 1242 Fees, 1269 Ass, *Hyrlawe* 1245 Ipm, *Herlauwe* 1254 Val]. Possibly identical with prec. name, though the early *i* is remarkable.

Harlow YW. See HARROGATE.

Harlsey, East & West, YN [*Herlesege*, *Herelsaie* DB, *Herleseie* 1088 LVD, *Herleseia* 1170–6 YCh 728]. Perhaps '*Herel*'s island', the first el. being a strong variant of *Herela* in HARLING &c., apparently found in *Herelesho* 1130 YCh 7. But '*Heregils*'s island' is quite possible. Cf. ĒG.

Harlthorpe YE [*Herlesthorpia* 1150–60 PNER, *Herlethorp* c 1200 YCh 1133]. First el. probably an OScand pers. n. such as ODan *Herlef* or *Herlugh*.

Harlton Ca [*Harletona* c 1080 ICC, *-ton* 1242 Fees, *Herletone* DB, *Herlenton* c 1150 Fr, *Herleton* 1203 Cur]. '*Herela*'s TŪN.' Cf. HARLING.

Harmby YN [*Hernebi* DB, *-by* 1219 FF, 1252 Ass]. '*Hiarni*'s BY.' ODan *Hiarni*, OSw *Hiærne* are known names or bynames.

Harmondsworth Mx [*Hermodesworde* DB, *-worth* 1233 Cl, *Hermondesworth* 1316 FA], **Harmston** Li [*Hermodestune*, *Hermestune* DB, *Hermedeston* 1202 Ass]. '*Heremōd*'s WORÞ and TŪN.'

Harnage Sa in Cound [*Harenegga* 1167 P, *Hernegie* 1232 Ch, *Hernegg* 1234 Cl, *Harnegge* 1327 Subs]. Second el. OE *ecg* 'edge, steep ridge'. Neither *hār* adj. nor *hara* 'hare' seems possible for the first el. It looks like a derivative of OE **hær* 'stone' (see HĀR), i.e. an OE **hæren* or **heren* 'stony, rocky'.

Harnham Nb [*Harnaham* 1242 Fees, *Hernham* 1272 Ipm, *Herneham* 1285 Pat], **East & West** H~ W [*Hareham* 1130 P, 1212 RBE, *Estharnham* Hy 3 BM, *Westharham* 1277 Fine]. Here the adj. **hæren* or **heren* suggested under HARNAGE may also be thought of. Especially for Harnham Nb *hār* adj. or *hara* is out of the question, and *hyrne* 'corner' is ruled out by the absence of spellings with *i*.

Harnhill Gl [*Harehille* DB, *Harnhilla* 1177 P, *-hull* 1220 Fees, 1331 BM]. OE *hāra hyll* 'grey hill' or 'hares' hill' may be suggested. Cf. HĀR.

Harome YN [*Harun*, *Harum* DB, *Harum* c 1170 Riev]. The dat. plur. of OE **hær* 'stone' or the like. Cf. HĀR.

Harpenden Hrt [*Harpendene* 1196 P, *Harpendena* 1262 Ch], **Harpham** YE [*Harpein*, *Arpen* DB, *Harpenna* 1130 P, *Harpham* c 1160 YCh (1064), 1237 FF], **Harpley** Nf [*Herpelai* DB, 1121 AC, *Harpelai* DB, *Harpele* 1206 FF, 1254 Val], **H~** Wo [*Hoppeleya* 1222 FF, *Harpel*' 1275 Subs], **Harpsden** O [*Harpendene* DB, 1219 Ep, *Harpedene* 1212 Fees, *Harpisden* 1236

Fees], **Harpswell** Li [*Herpeswelle* DB, 13 BM, *Harpeswella* c 1115 LiS, 1185–7 DC, -*well* 1212 Fees], **Harpton** He [*Herton* 1308 Ipm], **Harptree, East & West,** So [*Harpeðreu*, *Herpe-*, *Harpetreu* DB, *Harpetre* 1172 P, *Est*, *West Herpetre* c 1185 BM]. No doubt the first el. of these goes back to more sources than one. For Harpswell and perhaps Harpton OE *hearpere* 'harper' is a probable first el. One *r* would be apt to disappear owing to dissimilation. If so, 'the harper's spring and TŪN'. Harpenden and Harpsden should be compared with *hearpdene* 966 BCS 1176 (at Newnham Murren O). Harpsden, as shown in PNO(S), was actually named from that valley. Here we probably have OE *hearpe* 'salt-harp' (cf. SALTHROP), and this el. suits Harpham and Harpley Nf, which are not far from the sea, and Harpley Wo. Frequent *e*-spellings are relics of OE *ea*. For Harptree a pers. n. would be preferable, but one DB form possibly suggests as first el. OE *herepæþ*.

Harperley Du [*Harperleia* 1183 BoB], **Harper's Brook** R Np [*Harperesbroc* c 1200 Gervase, *le Harperisbrok* 1299 For]. 'The harper's clearing and brook.'

Harpford D [*Harpeford* 1168, 1230 P, *Herpeford* 1212 Fees], H~ So at Wellington [*Herpoþford* 904 BCS 610, *Herpoðford* 1065 Wells]. OE *herepæþ-ford* 'ford over which a main road led'.

Harpham, Harpley. See HARPENDEN.

Harpole Np [*Horpol* DB, 1254 Val, *Horepol* 12 NS, 1202 Ass]. OE *horh-pōl* 'dirty pool'.

Harpsden, Harpswell, Harpton, Harptree. See HARPENDEN.

Harpurhey La [*Harpourhey* 1320 Mamecestre]. 'Harper's hay or enclosure.' Named from William Harpour, who received land here before 1322.

Harras Cu nr Whitehaven [*Harrais* c 1220 StB, *Harreys* c 1225 Holme C]. Second el. OScand *hreysi* (Northern dial. *raise*) 'a cairn'. The first may be ON *horgr* 'altar'.

Harraton D in Aveton Gifford [*Harvedetone* 1274, *Harwodetone* 1285 Ep]. 'TŪN by *Harwood.' Cf. HAREWOOD.

Harraton Du on the Wear [*Hervertune* c 1190 Godric, -*ton* 1297 Pp]. Probably OE *Hereford-tūn* 'TŪN by the main ford'. H~ is some way east of a Roman road.

Harrietsham K [*Hæri-*, *Herigeardes hamm* 964–95 BCS 1132, *Hergeardesham* 1043 Th, *Hariardesham* DB]. 'Heregeard's HAMM.'

Harringay. See HORNSEY.

Harrington Cu [*Halfringtuna*, *Haverinton* c 1160, *Hafrincton* 12, *Haveringtuna* c 1200 StB]. 'The TŪN of *Hæfer's* people.' For *Hæfer* see HAVERSHAM.

Harrington Li [*Haringtona*, *Harintun* 12 DC, *Harinton*, *Harminton* 1202 Ass, *Haring-*

ton 1212 Fees]. Possibly the first el. might be a derivative in -*ingas* of OE **hær* 'stony ground'. See HĀR. Cf. HARRINGWORTH.

Harrington Np [*Arintone* DB, *Hetheringtone* 12 NS, -*ton* 1228 Ch, 1249 Ep, *Hezerinton* c 1236 BM]. Identical with **Hetherington** Nb [*Hetherinton* 1288 Ipm]. Usually taken to have a first el. derived from OE **Hæþhere* pers. n. But it is doubtful if a pers. n. stem *hæþ*- exists. *Hæþred* and the like may contain an el. *hæþ*, cognate with OE *heaþu*. Perhaps the first el. is a derivative of the word *heather* (cf. HEATHER). If so, the names mean 'the TŪN of the dwellers on a heath'.

Harringworth Np [*Haringeworde* DB, -*wurða* 1167 P, *Haringworthe* 12 NS, -*wurth* 1226 Ep]. Cf. HARRINGTON Li and WORÞ.

Harrock Hall La [*Harakiskar* c 1260 CC]. 'Grey oak.' See HĀR; -*kar* is carr 'marsh'.

Harrogate YW [*Harrogate* 1512, *Harlogate* 1522, -*gait* 1605 Knaresborough Wills]. The first el. is **Harlow**, the name of a neighbouring hill, meaning 'grey hill'. The second el. is probably *gate* from OScand *gata* 'road', used in the north country sense 'right of pasturage for cattle, pasturage'.

Harrold Bd [*Hareuuelle* DB, -*wolda* 1163 P, -*uuald* 1253 Ch]. Very likely OE **hær* 'stone' &c. (cf. HĀR) and WALD 'wold'.

Harrop YW [*Harrop* 1274, *Harehoppe* 1307 Wakef]. Identical with HAREHOPE.

Harrow on the Hill Mx [*Gumeninga hergae* (dat.) 767, *æt Hearge* 825 BCS 201, 384, *Hergas* 832 ib. 402, *Herges* DB]. OE *hearg* 'heathen temple'. *Gumeningas* must have been the name of the early inhabitants, meaning '*Guma*'s people'. **Harrow Weald** [*Waldis in Harwes* 1303, *Welde* 1382 AD]. Cf. WALD.

Harrowby Li [*Herigerbi* DB, *Herierebi* 1202 Ass, 1212 Fees, *Herierbi* a 1241 Berk]. '*Herger's* BY', the first el. being OSw *Hærger*, ON *Hergeirr*.

Harrowden Bd [*Herghetone*, *Hergentone* DB, *Harewedon* 1166 P], **Great & Little** H~ Np [*Hargindone*, *Hargedone* DB, *Harhgeduna* 1155–8 (1329) Ch, *Harewedon* 1202 Ass, *Maior*, *Parva Harewedon* 1220 Fees]. OE *hearga-dūn* 'hill with heathen temples'.

Harrowsley Sr [*Herewoldesleg* 1242 Fees, -*le* Hy 3 BM]. '*Hereweald*'s LĒAH.'

Harston Ca [?*Heorulfestun* 1015 Wills, *Herlestona* c 1080 ICC, -*tone* DB, *Herleston* 1230 P]. '*Heoruwulf*'s TŪN.' The ex. of 1015 may belong to HARLESTON Nf or Sf.

Harston Le [*Herstan* DB, *Harestan* c 1125 LeS, 1191 f. P, *Hareston* 1180 P]. OE *hāra stān* 'grey stone', 'grey boundary stone'.

Harswell YE [*Erseuuelle* DB, *Hersewella* 1130 P, -*well* 1260 Ass]. '*Her(e)sa*'s stream.' Cf. HARESCOMBE.

Hart Du [*Heruteu, -ei, -eig* c 730 Bede, *Heorotea* c 890 OEBede, *Hert* 1130–5 YCh 671, *Herte, Hert* 1242 Ass]. *Heruteu* is rendered by Bede 'insula cervi'. This is correct, the name 'stag island' referring no doubt to the headland or peninsula on which Hartlepool stands. **Hartlepool** (-lĭ-) Du [*Herterpol* c 1180 YCh 673, *-pul* c 1190 Godric, *Hertelpol* 1195 (1335) Ch, *-e* 1242 Ass, *Hertepol* 1242 Ass, 1254 Val] means 'the pool by Hart', the reference being perhaps to the bay south of the peninsula. The original name was very likely *Heruteu*, this name including both the present Hart and Hartlepool. The latter was later distinguished from the former by the addition *pool*. Hartlepool is thus from *Heorotēg-pōl*, which became *Herte-pōl* and, owing to influence from *Herterness* (now Hartness, the old name of the district of Hart and Hartlepool), *Herterpol* and by dissimilation *Hertelpol*, Hartlepool. **Hartness** [*Heorternesse* c 1050 HSC, *Heorternysse* c 1130 SD, *Herternes* c 1125 Guisb] is OE *Heorotēg-hērnyss* 'the district subject to Hart', which was reduced to *Heorthernyss* and *Heorternyss* &c. OE *hērnyss, hȳrnyss* originally meant 'obedience, subjection'. The shortening of OE *Heorotēg* (Bede's *Heruteu*) to early ME *Hert* is no doubt due to influence from *Herthernyss*. Hartness was probably the lordship of the old monastery of Heruteu or Hartlepool. Bede's *Heruteu* contains a variant form *-eu* of OE *ēg* 'island'. *Herut-* is normal OE *heorot* 'stag'.

Hartanger K in Barfreston [*Hertange* DB, *Herthangre* 1242 Fees]. 'Stag slope.' Cf. HANGRA.

Hartburn, East, Du [*Herteburna* c 1190 Godric, *-burn* 1208–10 Fees], **H~** Nb [*Herteburne* 1198 (1271), 1204 Ch]. 'Stag stream.' East H~ is on Hartburn Beck, H~ Nb on Hart Burn.

Harter Fell Cu [*Herterfel* c 1210, *Herterfelbek* c 1210, 1242 FC]. 'Stag fell', ON *Hiartar-fell*, the first el. being the gen. of ON *hiǫrtr*.

Hartest Sf [*Hertest* c 1050 KCD 907, DB, *Herterst* DB, *Herthyrst* c 1095 Bury, *Hertherst* 1200 Cur]. 'Stag hill or wood.' The loss of the second *r* is due to dissimilation. Cf. HYRST.

Hartfield Sx [*Hertevel* DB, *-feld* 12 Fr]. 'Open land frequented by stags.'

Hartford Chs [*Herford* DB, *Hertford* 1278 Misc], **East & West H~** Nb [*Hertford* 1198 (1271) Ch], **Hartforth** YN [*Herford* DB, *Hertford* 1234 FF]. 'Stag ford.'

Hartford Hu [*Hereforde* DB, *Herford* 1147 BM]. Identical with HEREFORD.

Hartham W [*He(o)rtham* DB, *Hertham* 1182 P]. Probably *heorot-hamm* 'enclosure for deer'. Identical in meaning is **Harthay** Hu [*Herteia* 1215 RA]. Cf. (GE)HÆG.

Harthill Chs [*Herthil, Harthil* 1259 Court],

H~ Db [*Hortil* DB, *Herthil* 1176 P], **H~** YW [*Hertil* DB, *Herthull* 1202 FF]. 'Hill frequented by stags.'

Harting Sx [*Heartingas, Hertingas* 970 BCS 1265 ff., *Hertinges* DB, *Herting* 1130 P, 1196 FF]. '*Heorot*'s people.' On the pers. n. *Heor(o)t* see HARDSTOFT.

Hartington Db [*Hortedun* DB, *Hertendon* 1200–25 Derby, *Hertindon* 1251 Ch]. OE *heorta-dūn* 'stags' hill'. The *-n-, -ng-* is intrusive.

Hartington Nb [*Hertweitun* 1171 P, *Hertwayton* 1242 Fees]. 'TŪN by the stags' path.' The path must have been one used by stags.

Hartland D [*Heortigtun* c 880 BCS 553, *Hertitone* DB, *Hertilanda* 1130, 1168, *Hertti-ilaund* 1230 P]. The original name was *Heorot-īeg* 'stag island', identical with HART and referring to the Hartland peninsula. To this was added TŪN and later LAND.

Hartlebury Wo [(to) *Heortlabyrig* 817, *Heortlanbyrig* 10 BCS 361, 1320, *Huerteberie* DB]. '*Heortla*'s BURG.' The name *Heortla* is also found in *Heortlaford* 985 KCD 653 (in bounds of Hartlebury). *Heortla* is not evidenced in independent use. But cf. HARTLINGTON and OHG *Hirzil*, also *Heort* in HARDSTOFT &c.

Hartlepool. See HART.

Hartley Dummer Brk [*Hurlei* DB, *Hurtlea* 1167 P, *Hertleg* 1198 P, 1242 Fees, *Hurtleye Dommere* 1361 BM], **H~** Do [*Herleg* 1212 Fees, *Hertlegh* 1229, 1231 Cl], **H~ Mauditt** Ha [*Herlege* DB, *Hertlie* 1212 Fees, *Hertlye Maudut* 1306 BM], **H~ Westpall** Ha [*Harlei* DB, *Hertlegh Waspayl* c 1270 Ep], **H~ Wintney** Ha [*Hurtlege* 12 (1337) Ch, *Hurtle Monialium* c 1270 Ep, *Hertleye Wynteneye* 13 VH], **H~** K nr Gravesend [*Erclei* DB, *Hertle* 1253 Ch, *Hertlegh* 1278 QW], **H~** K in Cranbrook [*Heoratleag* 843 BCS 442]. OE *heorot-lēah* 'stag wood or clearing'.

H~ Dummer was held by Richard de Dunmere in 1242 (Fees). Cf. DUMMER Ha.—**H~ Mauditt** was held by William Malduith as early as 1086 (DB). Cf. EASTON MAUDIT.—**H~ Westpall** from the Waspail family. Waspail is a nickname meaning probably 'waster' and related to Fr *gaspiller* 'to waste, squander', which is held to consist of the stem of Fr *gâter* (< *wast-*) and *paille* 'straw'.—On **H~ Wintney** see WINTNEY. The prioress of Wintney held the manor in 1228 (Cl).

Hartley Nb [*Hertelawa* 1167 P, *-lawe* 1242 Fees]. 'Stag hill.' Cf. HLĀW.

Hartley We [*Hartecla* 1285 For, *Hardecla* c 1285 StB, *Hardcla, Hartcla* c 1290 WR]. The first el. does not seem to be OE *heorot* 'stag'. It may be OE *harað* 'wood'; cf. HARDRES. The second is possibly OE *clā* 'claw', a side-form of *clēa, clawu* and used in the same sense as in CLAWTON. The place is near the junction of two brooks.

Hartleyburn Nb is on **Hartley Burn** [*Hertlingburne* c 1195, *Hertleburn* c 1170 Lanercost]. 'Hartley stream.' The form *Hertlingburne* means 'the stream of the Hartley people'.

Hartlington YW [*Herlintun* DB, *Hertlington* 1219 FF, 1280 Ipm]. 'The TŪN of *Heortla*'s people.' Cf. HARTLEBURY.

Hartlip K [*Heordlyp* 11 DM, *Hartlep* 1219 Fees, *Hertlepe* 1273 Ch]. OE *heort-hliep* 'leap-gate for stags'. Cf. HINDLIP.

Hartoft YN [*Haretoft* 1316 FA]. See TOFT. First el. as in HAREBY.

Harton Du [*Heortedun* 1104–8 SD, *Herteduna* Hy 2 FPD]. 'Stags' DŪN or hill.'

Harton YN [*Heretune* DB, *Haretona* 1157 YCh 354, R 1 (1308) Ch]. The first el. is possibly OE *hær 'stone'. See HĀR.

Hartpury Gl [*Hardepirer*' c 1155 Eynsham, *-pirie* 1167 P, *-piria* Hy 2 Glouc]. 'Hard pear-tree', i.e. 'pear-tree with hard pears'. Cf. *suran apoldran* BCS 610 'apple-tree with sour apples'. In the first ex. *-pirer* shows French *poirier* (OFr *perier*) instead of OE *pirige* 'pear-tree'.

Hartshead La [*Hertesheued* 1200 P], **H~** YW [*Hortesheue* DB, *Hertes Heved* 1206 FF]. 'Stag's hill.' Cf. HĒAFOD.

Hartshill Wa [*Ardreshille* DB, *Hardredeshella* 1152 BM, *-hill* 1204 Cur]. '*Heardrēd*'s hill.'

Hartshorne Db [*Heorteshorne* DB, *Herteshorn* 1196 FF]. 'Stag's headland.' See HORN.

Hartwell Bk [*Herdewelle* DB, *Hertwell* 1205 Obl], **H~** Np [*Hertewelle* DB, *Hertwella* c 1155 NpCh], **H~** St [*Hertwalle* 1361 PNSt]. 'Stags' spring or stream.'

Hartwith YW [*Hertwith* 1535 Fount M]. 'Stag wood.' Second el. OScand *viðr* 'wood. forest'.

Harty, Isle of, K [*Heortege* 11 DM, *Hertei* 1086 KInq, *Herteye* 1242 Fees]. 'Stag island.' See ĒG.

Harvington Wo in Chaddesley Corbett [*Herewinton* 1275 Subs]. '*Herewynn*'s TŪN.' *Herewynn* is a woman's name.

Harvington Wo nr Evesham [*Hereford* 799, *-a* 802 BCS 295, 307, *Herverton* 709 ib. 125, *Herefordtun* 964 ib. 1135]. 'TŪN by the army ford.' Cf. HEREFORD.

Harwell Brk [*æt Haranwylle* 956, *Harawille* 973 BCS 1183, 1292, *Harwelle* DB]. In BCS 1183 (on) *Harandúne* is mentioned in the boundaries. This is **Horn Down** nr Harwell. The hill-name means 'grey hill'. Its original name was no doubt *Hāra* or *Hāre* 'the grey one' (cf. BROWN). Harwell is 'stream by or coming from Horn Hill'.

Harwell Nt [*Hereuuelle* DB, *-well* 1242 Fees]. 'Pleasant stream', the first el. being

OE *hēore*, *hȳre* 'pleasant'. Cf. HARBOURNE D (under HARBERTON).

Harwich Ess [*Herwyz* 1238 Subs, *Herewyk* 1253 Ch, *-wyz* E 1 BM]. OE *herewic* 'camp'.

Harwood La in Bolton [*Harewode* 1212 Fees, 1241 FF], **Great & Little H~** La [*majori Harewuda* a 1123 Whitaker, *Magna, Parva Harwode* 1327 Subs], **H~** Nb nr Rothbury [*Harewuda* c 1155 BM, *-wud* 1236 Cl], **H~ Shiel** Nb [*Harewode* a 1214 Mon], **H~ Dale** YN [*Harwood* 1301 Subs]. Identical with HAREWOOD.

Harworth Nt (hărŭth) [*Hareworde* DB, *-wrthe* 1191–3 Fr, *-wurth* 1242 Fees]. See WORþ. The first el. may be the OE *hær 'stone, stony ground' discussed under HĀR.

Hasbury Wo [*Haselbury* 1270, *Heselbure* 1272 Ct]. OE *hæsel-burg* 'hazel BURG'.

Hascombe Sr [*Hescumb* 1232 Ch, *-cumbe* 1243 Misc, *Hassecumbe* 1266 FF, *Hascoumbe* 1307 Ch]. 'Witch's valley.' Cf. HESCOMBE. First el. OE *hætse*, *hægtesse* 'witch'. See CUMB.

Haselbech (-ĭch) Np [*Esbece* DB, *Haselbech* 12 NS, 1202 Ass]. 'Hazel valley.' See BÆCE.

Haselbury Bryan Do [*Haseberg* 1201 Cur, *Haselber* 1237 FF, *-bere* 1298 FF], **H~ Plucknett** So [*Halberge* DB, *Heselberge* 1176 P, *Haselbere* 1201 Cur, 1270 FF, 1327 Subs, *Haselbare Ploukenet* 1431 FA]. OE *hæsel-bearu* 'hazel grove'.

Guy de Bryene held *Haselbere* (**H~ Bryan**) in 1361 FF. The family took its name from BRIENNE in France.—**H~ Plucknett** was held by Alan de Plugenet in 1268 (Ch).

Haseley O [*Hæseleia*, (æt) *Hæsellea* 1002 KCD 1296, *Haselie* DB], **H~** Wa [*Haseleia* DB, *Haselea* 1194 P], **H~** Wt [*Haselie* DB]. OE *hæsel-lēah* 'hazel wood'.

Haselor Wa [*Haselov[r]e* DB, *-overe* 1236 Fees], **Haselour** St [*Haselovre* 1242 Fees, *-ouere* 1293 Ass]. 'Hazel slope.' Cf. OFER.

Hasfield Gl [*Hasfelde* DB, *-feld* 1167 P, 13 BM]. 'Open land where hazels grew.'

Hasketon Sf [*Haschetuna* DB, *Hasketun* 1253 Ipm, *-tone* 1254 Val]. '*Haseca*'s TŪN.' OE *Haseca (from *hasu* 'grey'; cf. also HASSINGHAM) is not directly evidenced, but is found in *Heasecan* (*Heahsecan*) *berh* BCS 513 f.

Hasland Db [*Haselund* c 1200 RA, *Heselund* Hy 3 Derby, *Haseland* 1276 QW]. 'Hazel grove.' The elements are OScand *hesli* 'hazels' and *lundr* 'grove'.

Haslemere Sr [*Heselmere* 1221 Cl, *Haselemere* 1316 FA, *Haselmere* 1435 BM]. 'Hazel mere.'

Haslingden La [*Heselingedon* 1241 Cl, *Haselen-*, *Heselindene* 1246 Ass]. 'Hazel valley.' First el. OE *hæslen* 'of hazel'.

Haslingfield Ca [*Heslingefelda* c 1080 ICC, *Haslingefeld* DB, *Heselingafeld* 1157 YCh 354, *Haselingefeld* 1202 FF]. Apparently 'the FELD of the *Hæselingas*'. These may be 'the dwellers at a place called Haseley', though no such place seems to be recorded in the neighbourhood.

Haslington Chs [*Hasillinton* 13, *Haselinton* Hy 3 BM]. 'TŪN among hazels.' First el. OE *hæslen* 'of hazel'.

Hassall Chs [*Eteshale* DB, *Hattesale* 13 BM, *Hassale* 1288 Court]. Looks like '*Hætt*'s HALH'. *Hætt* would be a byname formed from OE *hætt* 'hat' and corresponding to ON *Hǫttr*. Or the first el. might be OE *hætse* 'witch'. Cf. HASCOMBE.

Hassingham Nf [*Hasingeham* DB, *Hasingham* 1254 Val]. Apparently 'the HĀM of the *Hasingas* or people of *Hasu*'. *Hasu* would be a pers. n. formed from OE *hasu* 'grey' and cognate with OG *Haso*.

Hassop Db [*Hetesope* DB, *Hatsope* 1236 Fees, *Hashop* 1229 Ch]. See HOP. First el. as in HASSALL.

Hasthorpe Li [*Haroldestorp* DB]. '*Harald*'s thorp.' *Harald* is ON *Haraldr*, ODan *Harald*.

Hastingleigh K [(of) *Hæstingalege* 993 Hyda, *Hastingelai* DB, *Haestingelege* 11 DM], **Hástings** Sx [*Hastingas* 790 BCS 259, *Hæstingaceaster* 1050 ASC (D), c 1100 Laws, *Hastinges* DB]. The *Hæstingas* were an ancient tribe, referred to by Symeon of Durham under the year 771 as *Hestingorum gens*. The name is derived from a pers. n., probably *Hæsta*; cf. *Hæstan dic* 985 KCD 647 (K). *Hæsta* is derived from OE *hæst* 'violence', *hæst* 'violent', cognate with Goth *haifsts* 'fight', OHG *haist* 'violent'. OHG *haist* is found in pers. ns., as *Haistulf*, *Heistilo* &c. Hastingleigh is no doubt the 'LĒAH of the *Hæstingas*', some of the tribe having settled in Kent.

Haston Sa [*Haueston* 1241 FF, *Hauston* 1242 Fees, *Hastan* 1327 Subs]. Seems to be OE *hēafod-stān*, whose meaning is not apparent.

Haswell Du [*Hessewella* 1131, *-welle* 1155 FPD, *Hessewell* 1253 Ch], **H~** So nr Goathurst [*Hasewelle* DB, *Halswell* 1243 FF]. OE *hæsel-wella* 'hazel spring or stream'.

Hatch Bd [*la Hache* 1232 FF, 1247 Ass], **H~** Ha nr Basing [*Heche* DB, *Heccha* 1167 P, *Hacche* 1212 Fees], **H~ Beauchamp** So [*Hache* DB, *Hach* 1212 Fees, *Hache Beauchampe* 1243 Ass], **West H~** So [*Hache* 1201 Cur, *Westhache* 1243 Ass], **H~** W [*Hache* 1200 Cur, *Heche* 1201 Cur, *Hacche*, *Westhacch* 1242 Fees]. OE *hæcc* 'a hatch'. The meaning is generally 'a gate, esp. in or leading to a forest'. But **H~** Bd may refer to a floodgate or sluice.
Robert de Bello Campo held **H~ Beauchamp** before 1212 (Fees). Cf. ACTON BEAUCHAMP.

Hatcham Sr [*Hacheham* DB, *Hachesham*

1235 Ch, 1242 Fees, *Hacchesham* 1285 BM]. '*Hæcci*'s HĀM.'

Hatcliffe Li [*Hadecliue* DB, *-cliua* c 1115 LiS, *Haddecliua* 1219 Ep]. '*Headda*'s cliff or slope.'

Hatfield Broad Oak Ess [*Hadfelda* DB, *Hatfelda* 1127 BM, *Hatfeld Brodehoke* c 1130 PNEss, *Hadfeld Regis* 1188 P], **H~ Peverel** Ess [*Hæpfeld* c 995 BCS 1289, *Hadfelda* DB, *Hadfeld Peuerell'* 1166 P], **H~** He [*Hetfelde* DB, *Parva Hethfeud* 1242 Fees], **H~** Hrt [*Haethfelth*, *-feld* c 730 Bede, *Hæðfeld* c 890 OEBede, *Hetfelle* DB, *Hatfeld* 1254 Val], **H~** Nt [*Haytfeld* 1275 RH, *Hethfeld* 1332 For], **H~** Wo [? *Australis Hepfeld* 892 BCS 570, *Hathfeld* 1275 Subs], **Great & Little H~** YE [*Haiefelt*, *Haifelt* DB, *Haitefelde* c 1155 YCh 1346, *Est Hattfeld* 1226 FF], **H~** YW [*Haethfelth* c 730 Bede, *Hæðfeld* c 890 OEBede, *Hedfeld* DB, *Haitfeld* c 1185 YCh 815, *Hadfeld* 1199 P, *Haytefeld* 1297 Subs]. OE *hæpfeld* 'FELD or open land where heather and similar shrubs grew'. The Nt and Y forms show influence from OScand *heiðr* 'heath'.

H~ Broad Oak from a wide-spreading oak.—**H~ Peverel** was held by Ranulf Piperell in 1086 (DB). Cf. BARTON PEVEREL.

Hatford Brk [*Hevaford* DB, *Hauetford* 1176 P, *Hauedford* 1220 Fees, *Hautford* 1291 Tax]. 'Ford by a hill.' First el. OE HĒAFOD 'head, headland'.

Hatherden Ha [*Hetherdene* 1324 AD, *Hetherden Militis* 1355 BM]. 'Hawthorn valley.' First el. OE *hagu-*, *hægþorn* 'hawthorn'.

Hatherleigh D [*Hadreleia* DB, *Hatherlega* 1193 Ol, *-leg* 1228 FF], **Down & Up Hatherley** Gl [*Hegberleo* 1022 KCD 1317, *Athelai* DB, *Haiderleia* c 1150 BM, *Dunheytherleye* 1273 Cl, *Hupheberleg* 1220 Fees, *Uphatherleya* c 1275 Glouc]. 'Hawthorn wood.' Cf. HATHERDEN. In the original *hægþorn-lēah* n was lost early between the surrounding consonants, and the vowel *o* was weakened to *e*.

Hathern (-ădh-) Le [*Avederne* DB, *Hacthurne* 1230 Ep, *Hawethurn* 1254 Val, *Hauthirne* 1255 Ipm]. 'The hawthorn.' The early spellings suggest an OE base *hagupyrne, which is not otherwise evidenced. Cf. *pyrne* 'thorn-bush' from *porn*.

Hatherop Gl [*Etherope* DB, *Hadrop* 1086 Glouc, *Heythrop* 1211-13, *-trop* 1220 Fees, *Heʒthtrop* 1307 Winchc]. Perhaps 'high thorp', but some forms may point to the first el. being OE *hiehpu* 'height'.

Hathersage Db [*Hereseige* DB, *Haueresheg* 1200 P, *Haureshegg* 1230 P, *Haveresheʒh*, *Haversech* 1242 Fees, *Hathersegge* 1264 Ipm]. **H~** is in a valley by a steep ridge, Millstone Edge. It is no doubt the old name of this; hence the second el. of the name is OE *ecg* 'edge, steep ridge'. The first el. is probably OE *hæfer* 'he-goat'.

Hatherton Chs [*Haretone* DB, *Hatherton* c 1300 Chester], H~ St [*Hagenþorndun* 996 Mon, *Hargedone* DB, *Hatherdon* 1262 For]. 'Hawthorn TŪN and DŪN or hill.' Cf. HATHERDEN.

Hatley, Cockayne, Bd [*Hattenleia* c 960, (æt) *Hættanlea* c 1000 BCS 1062, 1306, *Hatelai* DB], **East** H~ and H~ **St. George** Ca [*Hateleia* c 1080 ICC, *Hatelai* DB, *Esthatteleia* 1199 P]. The three Hatleys are close together on a piece of elevated land. The name may mean '*Hætta*'s LĒAH', though *Hætta* is not recorded. It would be a derivative of *hætt* 'hat'. Cf. HASSALL. But it is possible that **Hætte* was a name of the hill. Cf. HATTINGLEY.

Cockayne H~ came to the Cockayne family in 1417 (VH). H~ **St. George** was held by Sir Baldwin de Sancto Georgio in 1282 (Ipm).

Hattersley Chs [*Hattirsleg* 1248 Ipm, *Hattersleg* 1260 Court]. Possibly 'deer wood', the first el. being OE *hēahdēor* 'deer'. The *d* might have been unvoiced after *h*.

Hattingley Ha nr Medstead [*Hattingele* 1204 Cur, *-lega* 1225, *-lige* 1240 Selborne, *-lig* 1242 Fees]. Possibly 'the *lēah* of *Hætt*'s people'. Cf. HASSALL. But H~ is on a hill and the hill may well have been called *Hætt* (lit. 'the hat'). Cf. **Hatt** Ha in Mottisfont [*Hatte* 1206 Cur]. Hatt is at Hatt Hill. Cf. also (boscus de) *Hat* 1198 FF (Np). Hattingley may thus be 'the LĒAH of the dwellers at HÆTT'.

Hatton Chs nr Runcorn [*Hatton* c 1230 Ormerod, 1293 AD], H~ **Hall** Chs nr Chester [*Hetone* c 1100 Chester, *Hatton* 1259 Court], H~ Db [*Hatune* DB, *Hetton* Hy 2 BM, *Hatton* 1230 FF], H~ Li [*Hatune* DB, *Hattuna* c 1115 LiS, *Hattune* 12 DC], H~ Mx [*Hatone* DB, *Hatton* 1230 P], H~ Sa in Shifnal [*Hatton* 1212 Fees, 1275 Cl], **Cold & High** H~ Sa [*Hatune, Hetune* DB, *Hatton* 1242 Fees, *Hatton on Hineheth* 1268 Ch, *Colde Hatton* 1233 Cl, *Heye Hatton* 1327 Subs], H~ St [*Hatton* 1205 Cur, *Hadton* 1227 Ass], H~ Wa [*Hattona* 1163 BM, *Hatton* 1242 Fees]. OE *Hæþ-tūn* 'TŪN on a heath'.

Cold and High H~ Sa are on **Hine Heath**. This name may mean 'the heath of the monks' (OE *hiwan*).

Haugh Li [*Hage* DB, *Hag(h)a, Hah* 12 DC, *Haghe* 1204 Cur]. OE *haga* 'enclosure'.

Haugham Li [*Hecham* DB, *Hacham, Hecham* 12 DC, *Hagham* 1212 Fees]. OE *Hēah-hām* 'high HĀM'.

Haughley Sf [*Hagele* c 1040 Wills, *Hagala* DB, *Haggle* 1247 (1326) Ch, *Haule* 1254 Val, *Hagenet* 1165 P, 1219 Fees]. Identical in the main with HAGLEY, the first el. being OE *haga* 'haw, fruit of the hawthorn'. Hence 'haw wood'. *Hagenet*, a common early form, seems to be due to Norman influence.

Haughmond Sa [*Haghmon* c 1135, 1141 PNSa, *Hageman* 1156 ff., *Hagheman* 1196, 1230 P, *Hagemam* c 1200 Gir, *Haweman* 1255 RH, *Shortehagheman* (alnetum) 1232 Cl]. The place is near a prominent hill, **Haughmond Hill,** to which it is very probable that the name was originally applied. The form in Gir may suggest that the second el. was Welsh *mam* 'breast', also used of a hill (cf. MANSFIELD). If so, the first el. is obscure. It seems more likely, however, that the correct form was *-man* (*-mon*). This may be the word *man*, which is used in the north of England of a cairn, a pile of stones marking a summit. The first el. might then be OE *haga* 'enclosure' or even *haga* 'haw'. The name *Shortehagheman* suggests an English name. Possibly *man* is used rather of the hill than of a cairn on it, so that the meaning might be 'hill where haws grew'.

Haughton Chs [*Halecton* Hy 3 BM, *Halghton* 1311 Ormerod], H~ **le Skerne** Du [*Halhtun* c 1050 HSC, *Haloughton* 1291 Tax], H~ La [*Halghton* 1307 FF, 1322 LaInq], H~ Nb [*Haluton* 1177, *Halghton* 1284 PNNb, *Haluton* 1293 QW], H~ Sa nr Haughmond [*Halekton* 1242 Fees, *Halton subtus Haghmon* 1291 Tax], H~ Sa nr Oswestry [*Halchton* 1285 FA], H~ Sa nr Shifnal [*Halghton* 1281, *Halgtone* 1384 BM], H~ St [*Haltone* DB, *Halcton* 1242 Fees]. OE *Halh-tūn* 'TŪN in or by a HALH'. In the Du, La, Nb exx. *halh* is 'haugh'. H~ Sa nr Shifnal is in a remote valley. For the rest the exact meaning of *halh* is not particularly evident.

Haughton Nt [*Hoctun* DB, *Hocton* 1191–3 Fr, *Hoctune* 1200–3 BM]. OE *Hōh-tūn* 'TŪN on a HŌH or ridge'.

OScand (ON) **haugr,** OSw *høgher,* ODan *haugr, høg* 'heap, mound, hillock, hill' is common in pl. ns. of the Scandinavian districts in England, but it is not always easy to distinguish it from OE HŌH. Where the first el. is a Scand word, the source is probably the Scand word *haugr*. The element is sometimes used alone to form pl. ns., as in HOLME ON THE WOLDS YE, HOON, and as a first el. in HUBY, HUGILL, but mostly it appears as the second member of names. The form varies a good deal. Cf. BECCONSALL, CLITHEROE, DRINGHOE, HACKINSALL, ULPHA Cu. In names of hundreds the meaning is often 'grave-mound', as in ASLACOE, CANDLESHOE, HAVERSTOE Li, or 'moot hill', as in GREENHOE Nf, THINGOE Sf.

Haulgh La [*Halgh* 1332 FF]. OE HALH 'haugh'.

Haunton St [*Hagnatun* 942 BCS 771, *Hagheneton* 1249 FF, *Hauneton* 1271 Ass]. '*Hagona*'s TŪN.'

Hautbois (hŏbǐs), **Great & Little,** Nf [*Hobbesse* 1044–7 KCD 785, c 1140 Holme, *Hobuisse, Ohbouuessa* DB, *Hobbossa* 1183 Holme, *Hobissa* 1191 P, *Hobwiss, -wise* 1200 Cur, *Hau(t)boys* 1242 Fees, (de) *Haltobosco*

1200 Cur, *Hauboys Maior, Minor* 1254 Val].
The original form seems to be *Hob-wisse*,
the second el. being an OE **wisse* 'meadow',
cognate with WISSEY. The first should be
compared with EFris *hobbe* 'a hummock',
a mudbank rising like a head over water',
Sw dial. *hobb* 'fertile spot in a field or
meadow, where corn or grass grows thicker',
hobbe 'a tuft of thick grass'. OE **hobb* may
mean 'a hummock' or 'a tussock' or the like.
Hob-wisse might be 'meadow with tussocks
or by a hummock'. The place is on an
elevation. Owing to folk-etymology the
name was associated with Fr *haut* 'high'
and *bois* 'wood'.

Hauxley Nb [*Hauekeslaw* 1204 Ch, *Hauekis-
lawe* 1242 Fees]. 'The hawk's or **Hafoc's*
mound.' See HLĀW.

Hauxton Ca [*Hafucestun* c 975 ASCh,
Hauekestune c 1050 KCD 907, *Hauextona*
c 1080 ICC, *Havochestun* DB]. '*Hafoc's*
TŪN.' *Hafoc* pers. n., not evidenced in
independent use, must be assumed for
several pl. ns., as HAWKESBURY, HAWKS-
WORTH. Cf. ON *Haukr* &c. (common).

Hauxwell YN [*Hauocheswelle* DB, *Hauekeswell* 1197 P, 1219 FF]. '**Hafoc's* or the
hawk's spring.'

Hăvant Ha [*Hamanfunta* 935 BCS (707), 980
KCD 624, *Havehunte* DB, *Hafhunte* 1256
Ass]. '*Hāma's* spring.' Cf. FUNTA. For the
loss of *m* before *f* cf. STOFORD, LYFORD.

Haven Street Wt [*la Hethene Stret* 1339,
la Hethenestret 1345 AD]. 'The heathen
street', 'road built by heathens'.

Haverah Park YW [*Heywra* 1227 Cl,
Haywra 1310 Ch]. 'Remote valley where
hay grew.' See VRĀ.

Haverbrack We [*Halfrebrek* 1090-7, *Hafre-
brec* 1120-30, *Haverbrec* 1205 Kendale].
OScand *hafra-brekka* 'hill or slope where
oats grew'. Cf. BREKKA. OScand *hafri* is
'oats'.

Havercroft YW [*Hauerecroft* 1191 ff. P].
'Oat croft.' Cf. prec. name.

Haverhill (hāvrĭl) Sf [*Hauerhella, -hol* DB,
-hell 1190 P, *Haverhell* 1158 P]. 'Hill where
oats were grown.' Cf. MARKET HARBOROUGH.

Haverholme Li [*Haversholm* 12 BM, *Hauer-
holm* 1171 P, 1206 Ass, *Haverholm* 1212
Fees]. 'Holm where oats were grown.'
First el. OScand *hafri* 'oats'.

Haverigg Cu [*Haverig* c 1185 FC, *Haverigg*
c 1190 StB]. 'Ridge where oats were grown',
or 'he-goat's ridge'. Second el. OScand
hryggr. The first may be OScand *hafri*
'oats' or *hafr* 'he-goat'.

Havering atte Bower Ess [*Haueringas* DB,
1187 Oxf, *Haueringes* 1192 ff. P, *Haverynge
atte Bure* 1305 Misc]. OE *Hæferingas*
'*Hæfer's* people'. Cf. HAVERSHAM.
Bower House is close to Havering.

Haveringland Nf [*Heueringalanda* DB,
Haueringlond 12 BM, *Heueringeland* 1203
Ass]. 'The land of *Hæfer's* people.' Cf.
HAVERING.

Haversham Bk [*Hæfæresham* 966-75 Wills,
Havresham DB]. '*Hæfer's* HĀM.' **Hæfer*,
which is the base of HAVERING and HAVER-
INGLAND, is no doubt a nickname from OE
hæfer 'he-goat'. Cf. OScand *Hafr*, a common name.

Haverthwaite La [*Haverthwayt* 1336 FC].
'Clearing where oats (OScand *hafri*) were
grown.'

Haw Gl [*Haga* 1169 P, *Hawes* Hy 3 Misc,
Hawe 1327 Subs]. OE *haga* 'enclosure'.

Hawarden. See HARDEN St.

Hawcoat La [*Hawcote* 1537 PNLa]. 'COT in
a *haga* or enclosure.'

Hawerby Li [*Hauuardebi* DB, *Hawardabi*
c 1115 LiS, *Hawardeby* 1254 Val]. '*Hawarð's* BY.' *Hawarð* in Kirkdale runic
inscr., *Hauuard* DB is ODan *Hawarth*, ON
Hávarðr. The same name is found in
Haverstoe, the name of the hundred where
Hawerby is [*Hawardeshou* DB, c 1115 LiS].
This is OScand *Hávarðar-haugr* '*Hávarð's*
burial-mound'. Cf. HAUGR.

Hawes YN [*Hawes* 1614 PNNR]. No doubt
hals 'hause'. Cf. HALS.

Hawes Water We [*Havereswater* 1199 Ch].
'*Hæfer's* or *Hafr's* lake.' Cf. HAVERSHAM.

Hawick Nb [*Hawic* 1242 Fees, *Hawik* 1285
Ipm, *Hauwyk* 1346 FA]. Possibly OE *Hēa-
wīc* 'high wīc'. The place is in a high
situation. But the first el. might be the
word *hāw* 'look-out (place)' suggested for
HALTON Nb.

Hawkchurch D [*Hauekech(i)erch* 1196 ff. P,
Hauekescherich c 1201 Salisbury]. '*Hafoc's*
church.' Cf. HAUXTON.

Hawkedon Sf [*Hauochenduna, Hauokeduna*
DB, *Hafkindun* 1195 Cur, *Hauekedon* 1242
Fees]. 'Hawks' hill.' Some forms rather
suggest '*Hafoca's* hill'. *Hafoca* is not
recorded.

Hawkesbury Gl [*Havochesberie* DB, *Hauo-
chesburia* 1183 AC]. '*Hafoc's* BURG.' Cf.
HAUXTON.

Hawkhill Nb [*Hauechil* 1178 P, *Haukhill*
1314 Ipm], **Hawkhurst** K [*Hauochesten*
DB, *Hauekherst* 1291 Tax]. 'Hawk hill and
hurst.' Cf. HYRST.

Hawkinge K [*Hauekinge* 1204 Pp, *Hauekyng*
1242 Fees]. A derivative of *hafoc* 'hawk'
(OE *Hafocing* 'hawk wood') or of *Hafoc*
pers. n. (OE *Hafocing* '*Hafoc's* place' or
Hafocingas '*Hafoc's* people'). Cf. HAUXTON.

Hawkley Ha [*Haveclige* 1234, *-lye* c 1255
Selborne]. OE *hafoc-lēah* 'hawk wood'.

Hawkridge Brk [*Heafochrycg* 956 Abingd, *Hauechrugge* 1185 f. P], H~ So [*Hauekerega* 1194, *-regg* 1195 P, *Haweckrig* 1225 Ass]. 'Hawk ridge.'

Hawkshead La [*Hovkesete* c 1200 LaCh, *Haukesset* c 1220 FC]. '*Hauk*'s shieling.' First el. OScand *Haukr* pers. n. See sÆTR.

Hawkstone Sa [*Hauekestan* 1185 P, *Hauekeston* 1276 Ipm]. OE *hafoces stān* 'hawk's stone'.

Hawkswick YW [*Hochesuuic* DB, *Houkeswyk* 1226 FF, *Haukeswyk* 1285 FA]. '*Hauk*'s wīc.' Cf. HAWKSHEAD.

Hawksworth Nt [*Hochesuorde* DB, *Houkeswrda* 1179, *-wurda* 1188 P, *Hokiswrh* 1236 Fees]. Either '*Hōc*'s worþ' or '*Hauk*'s worþ', the latter a Scandinavianized form of *Hafoces-worþ*.

Hawksworth YW [(on) *Hafecesweorðe* c 1030 YCh 7, *Hauocesorde* DB, *Hauekeswrth* 1226 FF]. '*Hafoc*'s worþ.' Cf. HAUXTON.

Hawkwell Ess [*Hacheuuella, Hechuuella, Hacuuella* DB, *Hakewell* 1202 FF, 1236 Fees]. 'Crooked stream', the first el. being OE *haca* 'hook'. The place is on a fairly winding stream.

Hawkwell Nb [*Hauekeswell* 1242 Fees, *Haukewell* 1260 Ipm, 1269 FF], H~ So in Dulverton [*Havekeuuelle* DB, *Hauecwell* 1225 Ass]. 'Hawks' stream or spring.'

Hawley Ha [*Hallee* 1248, *Hallely* 1281, *Hallie* 1287 Crondal]. The first el. appears to be OE *heall* 'hall'. Second el. LĒAH. 'Wood or clearing with a hall.' Cf. WOODHALL. Or the first el. may be OE *heall* 'stone'.

Hawley K [*Hagelei* DB, *-leg* 11 DM, *Halgeleg* 1203 FF]. 'Holy lēah.'

Hawling Gl [(to) *Halhagan, Hallinga homm* 816 BCS 356, *Hallinge* DB, *-s* 1221 Ass, *Hallinghis* 1174 Fr]. Cf. HALLOW. The forms of 816 are given in a charter dealing with Hallow, but must refer to Hawling, as Turkdean is mentioned in connexion with them. Hawling is no doubt a colony from, or an outlying district belonging to, Hallow, and the name means 'the Hallow people'. The name of the mother village (*Halhagan*) seems sometimes to have been applied also to Hawling.

Hawnby YN [*Halm(e)bi* DB, *Halmbi* c 1160 Riev, *Halmeby* c 1170 YCh 1838]. '*Halmi*'s or *Hialmi*'s BY.' Both names are found in OScand.

Hawne Wo [*Hale* 1270, *-n* 1274, *Halin* 1271 Ct]. OE *healum*, the dat. plur. of *healh*. See HALH. The place is in HALESOWEN.

Hawold YE nr Huggate [*Holde* DB, *Howald* 1157, *Houwald* c 1157 YCh 354, 1095]. OE *hōh-wald* 'forest on a ridge'.

Haworth YW [*Hauewrth* 1209 FF, *Hawrthe* c 1246 Pont, *Hawurth* 1252 Ass, *Hauwarth*

1311 Ipm]. See WORÞ. The first el. may be OE *haga* 'enclosure' or rather *haga* 'haw'. Cf. HAGWORTHINGHAM.

Hawridge (-ð-) Bk [*Haurege* c 1130 Oxf, *Hauecrugge* 1227 Ass]. 'Hawk ridge.'

Hawsker YN [*Houkesgarth* c 1125 YCh 838, *-gard* 1167, *Haukesgarð* 1176 P]. '*Hauk*'s enclosure.' Cf. HAWKSHEAD and GARTH. The name is Scandinavian.

Hawstead Sf [*Haldsteda* DB, 1180 P, *Halsteda* 1181 P, *Haustede* 1242 Fees]. See HALSTEAD.

Hawthorn Du [*Hagethorn* 1155 FPD, *Hagethhorn* c 1190 Godric]. 'The hawthorn.'

Hawthorne Abbey Chs [*Harethorne* J Ormerod]. 'Grey thorn-bush.' Cf. HĀR.

Hawthorpe Li [*Awartorp* DB, *Hawrthorp* c 1160 DC, *Awardetorp, Haldwardtorp* 1218 f. Ass]. First el. rather ODan *Hawarth* than *Hawar* (ON *Hávarr*).

Hawton Nt [*Holtone, Houtone* DB, *Houton* 1228 Ep, *Hautone* 1270 BM]. OE *Hol-tūn* 'TŪN in a hollow'. The early loss of *l* is due to Norman influence.

Haxby YN [*Haxebi* DB, *-by* 1228 YCh 785]. '*Hāk*'s BY.' *Hákr* is an ON pers. n.

Haxey Li [*Acheseia* DB, *Haxei* 1212 Fees, *Haxay* c 1220 Bodl]. Perhaps '*Hāk*'s island', OScand *Hāks-ey*, the first el. being ON *Hákr* pers. n.

Haxton or **Hacklestone** W [*Hakenestan* 1173 P, *-ston* 1239 Ch, *Hakeneston, Hakeleston* 1277 Misc, *Hakeneston by Nether Avene* 1318 AD, *Acleston* 1287 Cl]. Haxton, on the Avon, is sometimes called HACKLESTONE, and the form Haxton is then restricted to **Haxton Down** east of the place. Sometimes Haxton and Hacklestone are treated as two distinct names. It is evident, however, that they are identical, the earlier form of both being *Hakenestan*. The name means '*Hæccīn*'s stone', *Hæccīn* being a pet-form of *Hacca*. The later form with *l* is due to dissimilation.

Hay He [*La Haye* 1259 (1309) Ipm], H~ We [*the Hay* 1260 Kendale]. OE *gehæg* 'enclosure'.

Haycrust Sa [*Hauekehurst* 1232, *Hauekeshurst* 1241 Cl]. 'Hawk wood.' H~ is the name of a forest.

Haydock La [*Hedoc* 1169, *Heddoch* 1170 P, *Haidoc* 1212 Fees]. A Welsh *Heiddiog* 'barley place, corn farm', from Welsh *haidd* 'barley'. Cf. CEIRCHIOG in Anglesey, derived from *ceirch* 'oats'.

Haydon Do [*Heidon* 1201 FF, *Hedon* 1204 Cur, *Haidon* 1412 FA], H~ So [?*Hægdun* 1046 KCD 1334, *Heidun* 1225 Ass], H~ W [*Haydon* 1242 Fees, *Heidon* 1322 Misc]. Probably 'hay down', though the first el.

might be OE *hege* or *gehæg* 'hedge, enclosure'.

Haydon Bridge Nb [*Hayden* 1236, 1242 Fees, *-e* 1271 Ch]. 'Hay valley.'

Haydor Li [*Hai-, Heidure* DB, *Heidure* 1202 Ass, *Hedure* 1205 Cur]. 'The high door.' The second el. is OE *duru* 'door', here used in the sense 'pass'. The name refers to the pass in the ridge to the west of Haydor. Cf. *Hegedure* FC, the name of a pass near LODORE Cu. Lodore means 'the low door or gap'.

Hayes D, Do, a common name of minor places, is the plur. of OE *gehæg* or *hege* 'enclosure'.

Hayes K [*Hese* 1168 P, 1391 BM], H~ Mx [*on Lingahæse* 793 BCS 265, (to) *Hæse* 831 ib. 400, *Hesa* DB]. OE *hæs* 'wood' &c. See HÆS. *Hese* 838 BCS 418 (K) seems to refer to a place other than Hayes vil.

Hayfield Db [*Hedfeld* DB, *Hayfeld* 1307 FF, *Great Hayfield* 1338 Derby]. 'Open land where grass grew', or a Scandinavianized form of HATFIELD.

Hayle R Co [*Heylpenword* 1260 FF, *Heyl* c 1450 ERN]. Identical with this is **Hayle**, the old name of the Camel estuary [*Hehil, Heil* c 954 Ann Cambr, *Hægelmuða* c 1000 Saints]. An OBrit *Saliā*, identical with SHIEL in Scotland and derived from the stem *sal-* 'salt' in Welsh *heli* 'brine', *halen* 'salt'. The meaning is 'salt river', 'estuary'. On the Hayle is **Hayle** port Co [*Heyl* 1265, *Heyll* 1318 Glasney]. The name of the other Hayle is preserved in EGLOSHAYLE.

Hayling, North & South, Hayling Island Ha [*Hallinges* 1215 Cl, *Helynge* 1253 BM; *Heglingaig*, (to) *Hæglingaiggæ* 956, (æt) *Heilincigæ* 956 BCS 979 f., *Halingei, Helingey* DB, *Hailinges island* c 1140 Fr]. Hayling is 'Hægel's people'. Cf. HAILSHAM. H~ Island is 'the island of the *Hæglingas* or *Hægel*'s people'.

Haylot. See HAILEYBURY.

Hayne D is a common name of minor places. It appears in early sources in forms such as *La Heghe* 1242 Fees, *la Hachen* 1275 RH,. *La Heghen* FA &c. It is really identical with HAYES D, except that *Hayne* represents the dat. plur., *Hayes* the nom. acc. plur. It is doubtful if Hayne can be the dat. plur. of OE *gehæg* only, for that would have been OE *gehagum*. At least partly the source seems to be OE *hegum*, the dat. plur. of *hege* 'hedge'. The change *-um* > *-en* is normal. Cf. also next name and HAINFORD.

Haynes Bd [*Hagenes* DB, *Hagnes* c 1150 BM, *Hawenes* 1202 Ass]. Has been explained as *Hagonan-næss* 'Hagona's næss or headland' or as *haga-næss* 'land on which stood a haw'. Rather the name is the plur. of an Engl word corresponding to OG *hagan* 'hedge' &c. Cf. HAINFORD. In Hain-

ford this word appears as *hain* from OE *hægen. But OE may also have had a form *hagen; cf. *bragen* by the side of *brægen* 'brain'. Haynes would then mean 'the enclosures'.

Hayton Cu nr Aspatria [*Hayton* 1277 Ch, 1292 QW], H~ Cu nr Brampton [*Heiton* c 1200, *Haiton* c 1240 WR], H~ Nt [*Heiton* 1176 P, *Haythona* 13 BM, *Heyton* 1291 Tax], **Lower & Upper** H~ Sa [*Heyton* 1233 Cl, 1242 Fees], H~ YE [*Haiton* DB, *Hayton* 1228 FF]. OE *Hēg-tūn* 'hay farm'.

Haytor D [*Eofede torr* BCS 1323, *Idetordoune* 1566 PND]. See TORR. *Eofede* seems to be an adj. in *-ede* (from *-ōdi*), formed from the base or first el. of OE *ifig* 'ivy'. Cf. OHG *ebahewi* &c. 'ivy'.

Haywards Heath Sx [*Heyworth* 1261 Ass, *Hayworthe* 1276 FF]. The first el. may be OE *hēg* 'hay' or *hege* 'hedge'. See WORÞ.

Haywood He [*Haywode* 1276 Ep], H~ Nt [*Heywod* 1232, *-wud* 1237 Cl], H~ Sa [*Heywode* 1250 Fees], **Great & Little** H~ St [*Haiwode* DB, 1176 P, *Heywode* 1279 Ass]. 'Enclosed wood.' First el. OE *gehæg* or *hege* 'enclosure'.

Hazeleigh Ess [*Halesleia* DB, *Magna Hailesle* 1212 Fees, *Heylesley* 1218 FF]. '*Hægel*'s LĒAH.' Cf. HAILSHAM.

Hazeley Ha [*Heishulla* 1167 P, *-hull* 1203 Ch, *Hesulle* 1212 Fees, *Heysole* 1274 RH, *Haysull* 1317 Ch]. The second el. appears to be OE *hyll* 'hill', the first being perhaps OE *gehæg* 'enclosure'. But the form of 1274 suggests that H~ may have as second el. OE *sol* 'mire' (cf. GRAZELEY). If so, the name might be identical with (to, of) *Higsolon* 985 KCD 652 (in Michelmersh and far from Hazeley). *Higsolon* presumably has as first el. OE *hīg, hēg* 'hay' and means 'wet hayland'.

Hazelmere Bk [*Heselmere* 13 AD]. 'Hazel mere.'

Hazlebadge Db [*Heselebec* DB, *Haselbech* 1252 FF]. 'Hazel valley.' Cf. BÆCE.

Hazleton Gl [*Hasedene* DB, *Heseldene* c 1130 Oxf, *-tona* c 1162 Winchc]. 'TŪN among hazels.' The earliest forms point to original *hæseldenu* 'hazel valley'.

Hazlewood Db [*Haselwode* 1327 Ipm], H~ Sf [*Haselewod* 1254 Val], H~ YW [*Heseleuuode* DB, *Heselwode* 1188 (1271) Ch, *-wod* 1242 Fees]. 'Hazel wood.' The YW Hazlewood appears to have been influenced by OScand *hesli* 'hazels'.

Hazon Nb [*Heisende* 1170 P, *Heysandan, Haysand* 1242 Fees, *Haysand* 1267 Ipm]. Perhaps OE *heges-ende* 'the end of the hedge'. But the *a*-forms are really too numerous, and OE *sand* 'sand' is perhaps a more probable second el. The first el. would then be OE (GE)HÆG or HEGE.

Heacham (-ĕ-) Nf [*Hecham* DB, 1203 Ass, *Hecgham* 1191 FF, *Hecham, Heccham, Hek-*

ham 1254 Val]. Apparently identical with HEIGHAM Nf.

Headcorn K [*Hedekaruna* 11 DM, *Hede-crune* 1240 Ch, *-crone* 1248 BM]. The second el. is OE *hruna*, cognate with Icel *hrun* 'fall', *hruni* 'landslip, debris at the bottom of a hill', OHG *rono* 'a fallen tree-trunk'. OE *hruna* is found in (on) *þone ealdan hrunan* 956 BCS 955. The meaning of OE *hruna* was very likely 'a fallen tree' or 'fallen trees'. The first el. may be an OE pers. n. *Hydeca, a derivative of *Huda*.

Headingley YW [*Hedingeleia* DB, *Est, West Hadigleia* c 1135 Bodl, *Est-, West-haddinge-leia* c 1170 YCh 1558]. 'The LĒAH of *Hedde*'s people.'

Headington O [*Hedenedune, -done* 1004 Fridesw, *Hedintone* DB, *Heddendona* 1114–16 RA, *Hedendun* 1156 P, *-don* 1219 Fees]. The earliest forms point to OE *Hedenan dūn*, whose first el. is identical with that of *Hedenan mós* 975 BCS 1312. The common first el. seems to be an unrecorded pers. n. *Hedena*, which may be related to *Headda* and the like.

Headlam Du [*Hedlum* c 1190 Godric, c 1220 Pudsay, *Hedlem* 1316 Cl]. OE *hæþ-lēam* the dat. plur. of *hæþ-lēah* 'clearing overgrown with heather'.

Headley Ha [*Hallege* DB, *Hedlegh* 1248 Cl, *-e* c 1255 Selborne], H~ Sr [*Hallega* DB, *Hetlega* 1188, *Hedlega* 1190 P, *Hethleg* 1253 Ch], H~ Heath Wo [*Hæðleage sceagan* 849 BCS 455], H~ YW [*Hethleia* c 1180 YCh 1742]. OE *hæþ-lēah* 'clearing overgrown with heather'.

Headon (-ē-) Nt [*Hedune* DB, *Heddon* 1176 ff. P, *Headun* 1247 Ep], H~ Hill Wt [*Het-done* 1324 Misc]. OE *hæþ-dūn* 'DŪN or hill covered with heather'. H~ Nt may also be OE *hēa-dūn* 'high ridge'.

Headstone Mx [*Hegeton* 1348, *Heggestone* 1367, *Heggedon* 1438 PNMx(S)]. The material is not very conclusive, but the forms are best explained from OE *Hecg-tūn* 'TŪN with a hedge'. *Hedgton* seems to have become *Hedston* early owing to sound-substitution, but the old spelling *Heggeton* was often retained traditionally.

OE **hēafod** 'head' was used in various transferred senses such as 'headland, summit, upper end, source of a stream'. In pl. ns. the meaning varies between 'promontory', as in BEACHY HEAD, LINDETH; 'hill', as in GATESHEAD, HARTSHEAD, READ; 'upper end', as in SHIRESHEAD, **Waterhead** La (at the N. end of Coniston Water); 'source', as in COVERHEAD, RIBBLE HEAD. The meaning 'head' must be assumed in some cases, as in RAMPSIDE, lit. 'ram's head', owing to likeness to one. It has been suggested by Dr Bradley that some names such as FARCET, SHEPSHED, SWINESHEAD, whose first el. is the name of an animal, refer to an ancient custom of setting up an animal's head on

a pole to mark a hundred meeting place. This is merely a hypothesis, in support of which nothing has been adduced, and probably it will prove possible to account for names of this kind without resorting to it.

Heage or **High Edge** (hēj) Db [*Heyheg* 1251 Ch, *Heghegge* 1330 FA, *Heege* 1471, *Heegge* 1485 BM]. OE *hēa-ecg* 'high edge or ridge'.

OE **hēah** 'high' is very common as the first el. of pl. ns., mostly referring to high situation, but sometimes used in the sense 'high, tall', as in combination with words for hill or mound. The element usually appears in the weak form *hēa*, dat. *hēan*, as in *Hēatūn, Hēantūne*. Only rarely does it appear in the uninflected form *hēah*, chiefly in combination with HĀM (hence HIGHAM). Otherwise the form *High-* is usually a sign of late formation, as in HIGHWORTH, which was originally WORTH. The analogical weak form *hēaga, -n* may sometimes occur. In the north the *-n* of the weak form disappeared early, the OE dat. forms being *Hēalēge, Hēatūne* (as against *Hēanlēage, Hēantūne* in the Midlands and South). Hence Northern names such as HEALEY, HEELEY, HEATON &c., which are rarely found outside the Northern area. The *ēa* in forms such as *Hēanlēage, Hēantūn* was shortened, the result being later HANLEY, HENLEY, HENTON and HAMPTON, HINTON &c.

Healaugh (ē-) YN [*Helagh in Swaledal* 1200 Ch], H~ YW [*Hailaga* DB, *Helage* DB, c 1185 YCh 538, *Helagh* 1224–30 Fees], **Healey** La in Spotland [*Hayleg* 1260 FF, *Heghlegh* 1332 Subs], H~ La in Chorley [*Helei* 1215 P, *Helegh* 1314 LaInq], H~ Nb nr Hexham [*Heley* 1268 Ipm], H~ Nb nr Morpeth [*Heley* c 1235 Newminster], H~ Nb nr Rothbury [*Heley* Hy 1 Brinkburn], H~ YN [*Helagh* 1279–81 QW], H~ YW nr Dewsbury [*Helay* 1348 YD]. OE *hēa-lēah* 'high clearing or wood'. The form *-lagh* (*-laugh*) may be due to Scandinavian influence.

Healing Li [*Heg*(h)*elinge* DB, *Heghelinga* c 1115 LiS, *Hailinges* 1180, *Heilinge* 1194 P]. Identical with HAYLING.

Heanor Db [*Hainoure* DB, *Henovere* 1236 Fees, 1258 FF]. OE (æt) *hēan-ofre* '(at) the high ridge'. Cf. OFER.

Heanton (-ā-) **Punchardon** D [*Hantone* DB, *Hainton* 1214 FF, *Hyaunton* 1242 Fees, *Heauntone Punchardone* 1320 Ep], H~ **Satchville** D [*Hantone* DB, *Heannton Sechevill* 1284–6 FA], **Kingsheanton** D [*Hagintone* DB, *Kyngesheighampton* 1387 Fine]. OE *Hēa-tūn*, dat. *Hēan-tūne* 'high TŪN'. *Hagintone* may represent OE *Hēagan-tūn* (cf. HĒAH).

H~ **Punchardon** was held by Robert de Ponte Cardonis in 1086 (DB). The family name is from *Pontchardon* ('thistle bridge') in Normandy.—H~ **Satchville** was held by John de Sicca Villa in 1242 (Fees). SECQUEVILLE is the name of two places in Normandy.

Heap Bridge La [*Hep* 1226 Bardsley, *Hepe* 1278 VH]. OE *hēap* 'heap', here perhaps in the sense 'hill'. OE *hēap* is also found in HAPTON and SHAP.

Heapey La [*Hepeie* 1219, *Hepay* 1246, *Hephay* 1248 Ass]. OE *hēop-hege* or *-gehæg* 'hip hedge or enclosure'.

Heapham Li [*Iopeham* DB, *Iopheim* c 1115 LiS, *Hepham* 1202 Ass, 1212 Fees]. 'HĀM where hips grew.' Cf. HĒOPE.

OE **heard** 'hard' is rare in pl. ns. See e.g. HARDLEY Ha, HARTPURY.

OE **hearg** 'a heathen place of worship, a temple, sacred grove, idol' is identical with OHG *haruc* 'grove, holy place', ON *horgr* 'stone altar, cairn'. The original meaning seems to be 'stone altar'. Cf. *hær* under HĀR. See HARROW, -DEN, PEPER HARROW, HARLE.

Heath Bd [*la Hethe* 1276 Ass], H~ Db [*Heth* 1257 FF], H~ He [*Hed* DB, *Hethe* 1242 Fees], H~ Sa [*Hethe* 1237 FF, *La Hethe* 1267 Ipm], H~ YW nr Wakefield [*Heth* 1121 Mon]. OE Hᴁᴘ 'heath'.

Heathcote Db [*Hedcote* DB, *Hethcote* 1244 FF], H~ Wa [*Hethcot* 1196 FF]. 'COT on a heath.'

Heathencote Np [*Heymindecot* 1220, *Hekemundecot*, *Hemundescot* 1236 Fees, *Heghmundecotes* 1307 Ch]. 'Hēahmund's COT(s).'

Heather (-ē-) Le [*Hadre* DB, *Hethere* 1209–35 Ep, *Hethere* 1276 RH]. Apparently 'heather'. The word *heather*, found from 1335 in OED, chiefly in the north, is rare in pl. ns. Cf. UTTOXETER. HEATHER would seem to mean 'a heath' rather than 'heather'.

Heatherslaw Nb [*Hedereslawa* 1176 P, *-lau* 1255, *Hederislaw* 1291 Ipm]. 'Deer hill.' First el. OE *hēahdēor* 'stag, deer'. See HLĀW.

Heathery Clough Du [*Hethereclogh* 1432 PNNb]. 'Heathery valley or ravine.'

Heathfield (hĕfl) So [*Hafella* DB, *Haðfeld* 1159 P, *Hethefeld* 1199 Cur], H~ Sx [*Hatfeld* 1230 P, *La Hethfeld* 1275 RH]. 'Open land overgrown with heather.'

Heathpool (-ĕth-) Nb [*Hetpol* 1242 Fees, *Hethpol* 1250 Ipm]. Professor Mawer points out that the hill at H~ is Hetha and renders the name by 'pool under Hetha'. If so, the name is elliptical.

Heathwaite La [*Heittheuuot* 1273 Pat], H~ or **Haythwaite** YN [*Haithwait* c 1175 PNNR]. 'Clearing where hay was got.'

Heatley Chs [*Hethileg* 1286 Court]. 'Heathy clearing.'

Heaton Norris La [*Hetton* 1196 FF, *Heton Norays* 1282 Ipm], H~ **under Horwich** La [*Heton* 1227 FF, *Heton under Horewich* 1332 FF], H~ **with Oxcliffe** La [*Hietune* DB, *Hetun* c 1160 LaCh], **Great & Little**

H~ La [*Heton* c 1200 CC, *Little Heton* 1235 FF], H~ Nb nr Newcastle [*Heton* 1256 Ass], H~ Nb nr Norham [*Heton* 1183 BoB, *-a* 1208–10 Fees], **Capheaton, Kirkheaton** Nb [*Magna, Parva Heton* 1242 Fees, *Little Heton* 1232 Ch, *Cappitheton* 1454 Pat], **Heaton** YW nr Bradford [*Heton* c 1166 YD], **Cleckheaton** YW [*Hetun* DB, *Claketon* 1285 FA, *Clakheton* 1317 Ch], **Earls** H~ YW [*Etone* DB, *Heton Comitis* 1286, *Erlesheeton* 1308 Wakef], **Hanging** H~ YW [*Etun* DB, *Hingandeheton* 1266 Misc], **Kirkheaton** YW [*Heptone* DB, *Hetun* c 1190 YCh 1694]. OE Hēa-tūn 'TŪN situated on high land'.

Cap- in **Capheaton** is said to mean 'chief' (from Lat *caput*).—*Cleck-* in **Cleckheaton** is probably OScand *klakkr* 'hill'; cf. CLAUGHTON. —**Earls Heaton** from the Earl of Warren and Surrey (1297 Wakef).—*Hanging* in **Hanging** H~ means 'sloping, situated on a slope'.—For H~ **under Horwich** see HORWICH.—H~ **Norris** was held by the Norris family from the 12th cent. Norris is Fr *norrois* 'Norwegian'.

Heavitree D [(on) *Hefatriwe* c 1130 E, *Hevetrove* DB, *Hevetre* 1201 FF, 1280 Ep, *Hevedtre* 1270 Ep]. 'Hefa's tree.'

Hebburn Du [*Heabyrm* 1104–8 SD, *Heberine* Hy 2 FPD, *Heberne* 1195 (1335) Ch]. OE *hēa-byrgen* 'high tumulus'.

Hebden YW [*Hebedene* DB, *-den* c 1225 FC, *Hebbedenna* 12 Fount, *-den* 1228 FF], H~ **Bridge** YW [(aqua de) *Heppedene* 1279 Ass, *Hepden Bridge* 1508 Goodall]. 'Hip valley.' OE *hēope-denu*. For the early change *p > b* cf. **Skibeden** YW [*Scipeden, Schibeden* DB], which means 'sheep valley'.

Hebron Nb [*Heburn* 1242 Fees, 1251 Ch, *Heborin* 1262 Ipm]. Very likely identical with HEBBURN. **Hebron Hill** reaches 424 ft.

OE **hecg** 'hedge'. See e.g. HEDGECOURT, HEACHAM, HEIGHAM Nf, HECKFIELD, HESSETT.

Heck YW [*Hecca* 1157 YCh 186, 1195 P, *Heck* 1226 FF, *Hecke* 1242 Fees]. Northern *heck*, corresponding to *hatch*. See HᴁCC. A gate may be referred to.

Heckfield Ha [*Hizfeld* 1194 Selborne, *Hecfeld* 1208 Cur, 1242 Fees, *Hechfeld* 1207 Cur, *Heggefeld* 1280 QW]. Perhaps OE *hēah-feld* 'high plain'. OE *hecg* 'hedge' might also be thought of as first el.

Heckingham Nf [*Hechingheam* DB, *Huchingaham* 1163 BM, *Hechingham* 1198 FF, *Hekigham* 1203 Ass, *Hekingeham* 1245 Ch], **Heckington** Li [*Hechintune* DB, *Hegkington* 1195 Cur, *Hekinton* 1202 Ass, *Heckingtun* 1212 Fees]. 'The HĀM and TŪN of Heca's people.'

Heckley Nb [*Hecclive* 1242 Fees, *-clif* 1284 Percy, *Heckelive* 1307 Ch, *Hetcliffe* 1354 Percy]. OE *hēa-clif* 'high cliff' or *hæp-clif* 'cliff where heather grew'.

Heckmondwike YW [*Hedmundewic* 1166 P, *Hecmundewik* Hy 3 BM, *Hecmundeswyk* 1261 YInq]. 'Hēahmund's wīc.'

Heddington W [*Edintone* DB, *Hedinton* 1201 Cur, *Hedington* 1316 FA]. 'The TŪN of *Hedde*'s people.'

Heddon on the Wall Nb [*Hedun* 1175 P, *Heddun* 1262 Ipm, *Hedon super murum* 1242 Fees], **Black H~** Nb [*Hedon* 1271 Ipm, *Nigra Heddon* 1242 Fees]. OE *hǣp-dūn* 'hill where heather grew'. H~ on the Wall is on the Roman Wall.

Heddon, East & West, Nb [*Hidewine* 1178, *Hiddewin* 1187 P, *Hydewin del Est, H~ del West* 1242 Fees]. '*Hidda*'s pasture.' The second el. is an OE **winn* 'pasture', corresponding to Goth *vinja*, ON *vin*, OHG *winne* 'meadow, pasture'. The word is the first el. of WIMBORNE Do.

Hedenham Nf [*Hedenaham* DB, *Hedenham* 1180 P, *Heddenham* 1296 BM]. '*Hedena*'s HĀM.' Cf. HEADINGTON.

Hedgecourt Sr [*Hegecurt* 1302 Ch, *Le Heggecourt, Le Heycourt* 1314 Ipm]. Presumably the original name was HEDGE from OE *hecg* 'hedge', to which was added Fr *court* 'manor'.

Hedgeley (-ĭj-) Nb [*Hiddesleie* c 1150 Percy, *Hiddisley* 1236 Fees, *Higgeley* 1335 Percy]. '*Hiddi*'s LĒAH.'

Hedgerley Bk [*Huggeleg* 1195 P, *-legh* 1237– 40 Fees, *Huchele* 1242 Fees]. Apparently '*Hycga*'s LĒAH', **Hycga* being a short form of names in *Hyge-* as *-beald*.

Hedingham (hĭnĭnggam), **Castle & Sible,** Ess [*Heding-, Hiding-, Hainḡheham* DB, *Hethingaham* Hy 1 Abingd, *Hidingeham* 1199 FF, *Heingeham* 1194 P, *Heyngham Sibille* 1231 FF, *Henigeham Sibille* E 1 BM, *Hengham ad castrum* 1254 Val]. The same first el. is found in **Hinckford,** the name of the hundred where Hedingham is [*Hidingaforda* DB, *Heingeford* 1190 P]. The OE base is probably *Hȳpinga hām (ford),* the *Hȳpingas* being a tribe named either from a landing-place (*hȳp*) on the Colne or from a man *Hȳp(a).* **Hȳp(a)* is a normal short form of *Hyp-* in *Hypwalda.*

Sible is the woman's name *Sibilla (Sibyl).* A lady of this name must have held a manor in Hedingham before 1231.

Hedley Du in Lamesley [*Hedley* 1382 Hatfield], **H~** Du in Lanchester [*Hedley* 1183 BoB], **H~** on the Hill Nb [*Hedley* 1242 Fees, 1256 Ass]. OE *hǣp-lēah* 'clearing overgrown with heather'.

Hednesford St [*Hedenesford, Edenesford* 13, 14 PNSt]. Perhaps '*Heddin*'s ford', **Heddin* being a diminutive form of *Headda* pers. n. Cf. *Hednesbroc* 1192–1219 BM (Sx), *Hedenesdene* 877 BCS 544 (Ha), also HENSALL YW.

Hĕdon YE [*Heldone* 1115 YCh 1304, *Heddone* 1160-2 ib. 1307, *Haddun* 1158 P, *Heddon* 1208 FF]. OE *hǣp-dūn* 'DŪN overgrown with heather'. The elevation is very slight.

Hedsor Bk [*Heddesore* 1196 Cur, 1196 P, *-our* 1208 Fees, *Hedleshore* 1195 Cur]. '*Hedde*'s bank.' Cf. ŌFER. H~ is situated on the Thames on a high bank.

Heeley YW [*Heghlegh* 1366 Goodall, *Helay* 1379 PT]. See HEALEY.

Heene Sx [*Hene* DB, *Henam* (acc.) 1121–5 BM, *Hen* 1195 P, *Hyen* 1193 P, 1219 FF &c., *Hean* 1271 Ass]. The derivation from an OE **hiwun* 'family' adopted in PNSx(S) and that from OE **hægen* 'enclosure' suggested in EPN(S) for the name must be abandoned. The common spelling *Hyen* points to OE *Hēan.* This base agrees with that for HAYNE K (an old hundred name), which has been taken to be connected with OE *hēah* 'high'. But Heene is on low ground near the sea and cannot have a name derived from *hēah.* Possibly the name is identical with HIEN in Belgium [*Hauinum* 814, *Hehun* 997 NGN iii. 136], which may be derived from a word for 'hay' (OE *hēg* from **hawia-* or ON *há* 'aftermath').

OE **hēg, hīeg, hīg** 'hay' is a fairly common first el., but is difficult to distinguish from (GE)HÆG, HEGE. See HAILEY(BURY), HAY-, HEY- (passim), CLAYHIDON, HIGHWAY. ON **hey** 'hay' is found in HEATHWAITE.

OE **hege** 'hedge' probably occurs sometimes in pl. ns. See HAYNE, HAYWOOD, HEXTELLS, THORNESS &c. But it is generally impossible to distinguish it from (GE)HÆG, which appears to be more common.

Heigham Nf in Norwich [*Hecham* DB, 1163 Holme, 1291 Tax, *Heigham, Hecgham* 1254 Val], **Potter H~** Nf [*Echam* DB, *Hecham* c 1160 Holme, *Hegham Pottere* 1182 BM, *Potteres Hecham* 1254 Val]. The name might be supposed to be 'high HĀM', but topographical considerations and the early forms forbid this. The same name is HEACHAM and very likely HITCHAM Sf. Some forms point to OE *Hecg-hām* 'HĀM by or with a hedge'. If so, *čǧ* became unvoiced before *h.* Or it may be *Hecc-hām, hecc* being a side-form of *hæcc* (B-T Suppl). See further HÆCC. There must have been potteries at Potter H~.

Heighington (hī-) Du [*Heghyngtona* 1183 BoB, *Hekenton* 1195 (1335) Ch, *Hekinton* 1227 Ep, *Hehingtone* 1228 FPD]. 'The TŪN of *Heca*'s people.' In this name medial *c (k)* seems to have become *g* and a spirant.

Heighington (hā-) Li [*Hickinton, Hictinton* 1242 Fees, *Hutington* 1285, *Hiccinton* 1316 FA]. Apparently identical with HEIGHTINGTON.

Heighley Castle St [*Heolla* DB, *Helyh* (Castle) 1227 Ch, *Heleye* 1273 Ipm]. Presumably identical with HEALEY in spite of the DB spelling.

Heightington Wo [*Hutinton* 1275, *Hutdynton* 1332 Subs]. 'The TŪN of *Hyht*'s people.' Cf. HIXON.

Heighton (hā-) **Street** Sx [*Hiectona* c 1150 AD, *Heghton* 1262 Ass], **South H~** Sx [*Hectone* DB, *Hezton* 1296, *Sutheghton* 1327 Subs]. Possibly OE *Hēah-tūn* 'high TŪN' with *hēah* in the uninflected form, but the surname *atte Heghe* (*High*) borne by a tenant in Heighton Street in 1327 and 1332 (Subs; cf. Löfvenberg) indicates that the Downs that separate H~ Street and South H~ were once called OE *Hēah* 'the height' (a substantivized use of OE *hēah* 'high'). Heighton is thus OE *Hēah-tūn* 'TŪN by *Hēah*'.

Hel R. See HELSTON.

Helbeck or **Hillbeck** We [*Hellebek* 1231 FF, 1279 Ass], **Helbeck** YN [*Hellebec* 1199 P]. According to EDD *hell-beck* is 'a rivulet, esp. one issuing from a cave-like recess'. First el. ON *hellir* 'a cave'.

OE **helde, hi(e)lde** 'slope' is found in AKELD, LEARCHILD, STOCKELD, REDHILL Sr, TYLER-HILL, perhaps HELTON, HILCOTT Gl. There was also an OE *helde* 'tansy'.

Hele So [(æt) *Hele* 11 KCD 897, *Hela* DB, *Hele* 1201 Ass]. OE *hēale*, the dat. of OE *healh* 'a corner' &c. See HALH. The name HELE is common in D.

Helford Co [*Helleford* 1230 Cl, *Hayleford* 1318 Pat]. 'Ford over R Hayle.' Hayle is the old name of **Helford** river [*Haill* 1602 Carew]. See HAYLE.

Helhoughton Nf [*Helga-, Hælgatuna* DB, *Helgetun* c 1150 BM, *Helewetone* 1157 RBE]. '*Helgi*'s TŪN.' *Helgi, Helga* &c. (DB &c.) is ON *Helgi*, ODan, OSw *Hælghi*.

Hellaby YW [*Helgebi* DB, *Helghby* 1303 FA]. '*Helgi*'s BY.' Cf. HELHOUGHTON.

Helland Co [*Hellaund* 1303 FA]. Identical with Welsh *hen-llan* 'old church' from Welsh *hen*, 'old' (= Co *hen*) and *llan* 'enclosure, church' (= Co *lan*). **Helland** in Probus [*Henlant* DB] is the same name. Cf. also HENTLAND.

Hellesdon Nf [*Hægelisdun, Haglesdun* c 985 St. Edm (Abbo's *Passio Sancti Eadmundi*), *Hailesduna* DB, *-don* 1180 P, 1196 FF, *Heilesdon* 1199 FF]. '*Hægel*'s DŪN or hill.' Cf. HAILSHAM. H~ was the place where St. Edmund suffered martyrdom. *Hægelisdun* has been identified with Hoxne and with Hollesley, but is obviously Hellesdon.

Hellidon Np [*Eliden* 12 NS, *Helidon* 1193 P, *-den* 1220 Fees, *Haliden* 1246 Cl]. Perhaps 'holy valley', the first el. being OE *hælig*, a rare side-form of *hālig*.

Hellifield YW [*Hælge-, Helgefeld* DB, *Helgefeld* 1203 FF, *Helwefeld* 1233 Cl]. '*Helgi*'s field.' Cf. HELHOUGHTON. It is just possible, however, that the name might be a Scandinavianization of OE *Hālga-feld* 'holy field'.

Hellingly (-lĭ·) Sx [*Helingeam* 1121 AC, *Hellingeleghe* Hy 3 AD, *Hillinggelige* 1279

Ass]. 'The LĒAH of the *Hiellingas*.' *Hiellingas* might be a derivative of an OE **Hielle*, a side-form of *Heall*; cf. HALLING. Or it might be a derivative of OE *healh* 'corner' &c. The place is in a tongue of land between two streams. *Hiellingas* (or more correctly *Hielingas*) would mean 'the dwellers in the tongue of land'.

Hellington Nf [*Halgatuna* DB, *Hel(e)getone, Helegheton* 1254 Val]. See HELHOUGHTON.

ME **helm** (from OE *helm* or ON *hialmr*) 'helmet' is also used in the sense 'a roofed shelter for cattle, a shed' (first ex. in OED 1501). This sense is probably Scandinavian. Norw *hjelm* means 'a haystack with a primitive roof', Dan dial. *hjelm* 'a kind of barn'. Hence *helm* in names of minor places in the north. Cf. BENSHAM and HELMSHORE.

Helmdon Np [*Elmedene* DB, *Helmendene* 12 NS, *Halmeden* 1163 P, *Helmedene* 1249 Ep]. '*Helma*'s valley.' *Helma* is a short form of names in *Helm-, -helm*. It seems to occur in *Helman hyrst* 838 BCS 418, now ELMHURST K.

Helmingham Nf nr Norwich [*Helmingeham* DB, 1192 P], **H~** Sf [*Helmingheham* DB, *Helmingueham* c 1160 (1331) Ch]. 'The HĀM of *Helm*'s people.' *Helm* is found in Widsith and in *Helmes treow* 968 BCS 1213.

Helmshore La [*Hellshour* 1510 Ct]. 'Steep cliff with a cattle-shed.' Cf. HELM and SCORA.

Helmsley (hĕmz-) YN [*Elmeslac* DB, *Helmesley* 12 Whitby, *-lay* 1226 FF]. '*Helm*'s LĒAH.' Cf. HELMINGHAM.

Helmsley (hĕmz-), **Gate & Upper**, YN [*Hamelsec(h)* DB, *Homeleseya* c 1130, *Hemelseya* c 1160 YCh 169, 175, *Over Hemelsey* 1301 Subs]. '*Hemele*'s island' (OE ĒG).
Gate refers to a Roman road. It is OScand *gata* 'road'.

Helperby YN [*He(o)lperby* c 972 BCS 1278 f., *Helprebi* DB, *Hilprebi, Ilprebi* DB, *Elperby* 1219 FF], **Helperthorpe** YE [*Elpetorp* DB, *Helprethorp* c 1110 YCh 25, *Helperthorp* c 1180 Middleton]. Apparently '*Hialp*'s BY and thorp', the first el. being ON *Hialp* (gen. *Hialpar*), a woman's name, though it is somewhat surprising to find this uncommon name in two Engl pl. ns. Cf. HILPERTON, which may contain an OE pers. n. derived from *help* with an *r*-element.

Helpringham Li [*Helperincham, Helpericham* DB, *Helpringham* 1138 NpCh, 1212 Fees, *Helpringeham* 1218 Ass]. '*Helpric*'s HĀM' or possibly *Helpricinga-hām* 'the HĀM of *Helpric*'s people'.

Helpston Np [*Hylpestun* 948, 972–92 BCS 871, 1130, *Helpeston* 1163 P, c 1185 NpCh, 1198 FF, 1202 Ass]. '*Help*'s TŪN.' **Help* is a normal short form of *Helpric*. The *y* in the OE forms may be an inverse spelling. Cf., however, HILPERTON.

Helsby Chs [*Helesbe* DB, *Ellesbi* 1186 P, *Hellesby* 1216, 1241 Chester]. The first el. may be OScand *hellir* 'a cave' or *hiallr* in such a sense as 'a ledge on the side of a hill' (a sense found in Norw pl. ns.). Helsby is nr a steep little hill, **Helsby Hill.**

Helsington We [*Helsingetune* DB, *Helsinton* 1246 Ipm]. H~ is on the southern slope of **Helsington Barrows,** a long high ridge. This was no doubt called *Hals* (cf. HALS), and the dwellers on the ridge were called *Helsingas*. The name thus means 'the TŪN of the dwellers on *Hals* ridge'.

Helston Co [*Henlistone* DB, (burgus de) *Helleston* 1186 P, *Helleston in Kerrier* 1310 Ch], **Helstone** Co [*Henliston* DB, *Helleston in Trigg* 1310 Ch]. A hybrid name, consisting of Co *henlis* 'old court' (Co *hen* 'old' and *lis* 'court, hall') and OE TŪN. The river-name **Hel** is a back-formation.

Helton We [*Helton* 1196 P, *Helton Flechan* 1314 Ipm]. Identical with these were originally HILTON We and Do, both *Helton* in early sources. The first el. might be OE *helde* 'slope' or *helde* 'tansy'. For the We names ON *hiallr* might be thought of. Cf. HELSBY.

Helve·llyn Cu [*Helvillon* 1577 Saxton, *Lauuellin* 1600 Camden]. The forms are too late for an etymology to be suggested.

Helwith YN [*Helwathe* 1282 YInq]. 'Ford paved with flat stones' (ON *hella* 'flat stone' and *vað* 'ford'). Identical with **Helwath** (Beck) YN [*Hellewath* 1230 Whitby] and *Hellawath* c 1120, *Hellewath* c 1200 Guisb, the name of a locality near Guisborough.

Hem, The, Sa nr Shifnal [*Hemma* 1182 P, *Hemme* 1310, 1322 Ipm]. Identical with HEM Montg [*Heme* DB]. OE *hemm* 'hem, border'. The exact meaning of the name is not clear. Cf. OFris *hemm* or *hemme* 'place enclosed for a single combat'.

Hemblington Nf [*Hemelingetun* DB, *Hemelington* 1252 Ch]. 'The TŪN of *Hemele*'s people.'

Hemel Hempstead. See HEMPSTEAD.

Hemerdon D [*Hainemardun* DB, *Henemerdona* Hy 2 Ol, *Hennemerdon* 1284–6 FA]. The situation does not really suit derivation from OE *hennamere-dūn* 'hill by the hen pool'. The second el. may be OE *mǣrdūn* 'boundary hill', the first being OE *higna* gen. plur. 'of monks'. See HĪWAN.

Hemingbrough YE [*Hamiburg* DB, *Hemyngburgh* 1080–6 YCh 974, *Hemingaburg* Hy 2 FPD, *Hemmingeburch* 1153–60 ib. 937, *Hemingaborg* 1026 Ottar svarti in Knytlinga saga]. It has been suggested that this is 'the BURG of *Heming*', *Heming* being possibly identical with the Hemingr jarl who operated in the north of England in the early 11th cent. If this is right, the earliest form *Hemingaborg* would be corrupt. More likely

Hemingbrough is 'the burg of the *Hemingas* or *Hemmingas*'. Cf. HEMINGFORD &c.

Hemingby Li [*Hamingebi* DB, *Hemmingebi* 1173 P, *Hemmingkebi* 1212 Fees]. Perhaps '*Heming*'s BY'. *Heming* is ODan *Heming*, ON *Hemingr*. The *-e-* in the earliest forms may be intrusive.

Hemingfield YW [*Himlingfeld* 1276 RH, *Hymelingfeld* 1303 YInq]. Perhaps 'the FELD of *Hymel*'s people'. Cf. HEMLINGTON, HEMSWORTH YW. But the first el. might be an OE **hymelen*, derived from *hymele* 'hop plant': 'FELD where hops grew'.

Hemingford Abbots & Grey Hu [*Hemmingeford* 974 BCS 1310, *Hemminggeford* c 1000 HEl, *Hemmingaford* 1012 Soc Ant iii, *Emingeforde* DB, *Hemmingford Abbatis* 1276 RH, *Hemingford Grey* 1316 FA]. 'The ford of *Hemma*'s or *Hemmi*'s people.'
H~ **Abbots** was held by the Abbot of Ramsey. H~ **Grey** from the Gray family (from GRAYE in Normandy). The manor was held by Reginald de Grey in 1276 (RH).

Hemingstone Sf [*Hamingestuna* DB, *Hemingeston* 1201 Cur, 1212 Fees, *Hemmingeston* 1206 Cur]. '*Heming*'s TŪN.' Cf. HEMINGBY.

Hemington Le [*Aminton* c 1125 LeS, *Hemingeton* 1204 Cur, *Hemminton* c 1200 BM], H~ Np [*Hemmingtune* 1077 Chron Rams, *Hemintone* DB, *Heminctona* 1149 NpCh], H~ So [*Hammingtona* DB, *Heminton* 1176 P, *Hemington* 1212 Fees]. 'The TŪN of *Hemma*'s or *Hemmi*'s people.'

Hemley Sf [*Helmela, Helmelea* DB, *Helmele* 1219 FF, 1254 Val]. '*Helma*'s LĒAH.' Cf. HELMDON.

Hemlington YN [*Himelige-, Himelintun* DB, *Hamelinton* 1206 Obl, *Hemelington* 1253 Ch]. 'The TŪN of *Hymel*'s people'; cf. HEMSWORTH YW. Or possibly 'the TŪN of *Hemele*'s people'.

Hempholme YE [*Henepeholm, Hempholm* n.d. Bridlington]. 'Holm where hemp grew.'

Hempnall Nf [*Hemen-, Hamehala* DB, *Hemehal* 1199 P, 1203 Cur, 1242 Fees]. '*Hemma*'s HALH.' The place is in a valley.

Hempshill Nt [*Hamessel* DB, *Hemdeshil* c 1200 Middleton, *-hil* 1242 Fees, *Hindishul* 1239 Ep]. '*Hemede*'s hill.' *Hemede* (or *Hemmede*) is not evidenced in independent use, but it seems to occur in HEMSWORTH Do and is a derivative of *Hemma* analogous to *Luhhede* from *Luhha*, *Lullede* from *Lulla*.

Hempstead Ess [*Ham(e)steda* DB, *Hamsted* 1203 Cur, *Hemsted* 1235 Cl], **Hemel Hempstead** Hrt [*Hamelamstede* DB, *Hemelhamsteda* 1173 P, *Hamelhamsted* 1228 Cl], H~ Nf nr N. Walsham [*Hemsteda* DB, *Hemsted* 1212 Fees, *Hemstede* 1254 Val, *Hempstede* 1302 FA]. These are probably OE *hāmstede* 'homestead'. On *e* for *a* see HĀMSTEDE.

Hemel in **Hemel Hempstead** is an old name

of a district [(paga) *Haemele* c 705 Bodley MS]. In this case *Hempstead* may be due to weak stress: *Hemelhamsted* > *-hemsted*.

Hempstead Gl [*Hechanestede* DB, *Heihamp-stud* 1287 QW, *Heyhamstede* 1292 Ch]. OE *hēah-hāmstede* or *hēa-hāmstede* 'high home-stead'.

Hempstead Nf nr Holt [*Henep-*, *Hemesteda* DB, *Hempstede* c 1130 BM, *Hemstede* 1242 Fees]. 'Place where hemp (OE *henep*) was grown.'

Hempston, Little, & Broadhempston D [*Hamistone* DB, *Hemmeston* 1221 FF, 1242 Fees, *Great Hemmeston* 1232 FF, *Brode-hempstone* 1362 Ep, *Parua Hæmeston* 1176 P]. '*Hemme*'s or *Hǣmgils*'s TŪN.' On *Broad-* see BRĀD. **Hems** R [*Hemese* 1287 PND] is '*Hemme*'s or *Hǣmgils*'s stream'.

Hempton Gl [*Hempton* 1327 Subs], **H~** O [*Hentone* DB, *Hyantona* c 1225, *Hentone* 1254 Eynsham]. OE *Hēa-tūn*, dat. *Hēan-tūne*, 'high TŪN'.

Hempton Nf [*Hamatuna* DB, *Hemton* 1242 Fees, *Hempton* 1254 Val]. '*Hemma*'s TŪN.'

Hemsby Nf [*Hemesbei*, *Heimesbei* DB, *Hemesby* 1103–6 BM, *Hammesbi*, *Hemmesbi* 1177 P, 1203 Ass]. First el. perhaps ODan **Hēmer*, OSw *Hēmer*, ON *Heimir* (gen. *Heimis*).

Hemsted K nr Cranbrook [*?Hǣmstede*, *Hamstede* 993 Hyda, *Empsted* 1254 Ass, *Hemstede* 1292 Ass], **H~** K nr Lyminge [*Empestede* 1240 Ass, *Hemstede* 1275 RH, *Hempstede* 1278 Ass]. If the identification of the forms from Hyda is correct, this Hemsted is OE *hāmstede* (or *hǣmstyde*, cf. HĀMSTEDE). The second is OE *henep-stede*. Cf. HEMPSTEAD Nf nr Holt.

Hemswell Li [*Helmeswelle* DB, *Helmes-*, *Halmeswella* c 1115 LiS, *Helmeswell* 1202 Ass]. '*Helm*'s spring.' Cf. HELMINGHAM. The map marks a chalybeate spring close by.

Hemsworth, East & West, Do nr Wimborne [*Hemedesworde* DB, *-wurth* 1243 FF, *Hemmesdeswurda* 1194, *Hemmedeswrda* 1195 P, *Hendesworth* 1236 Fees; *Hemeleswurth* 1224 Cl, *Hameleswrth* 1257 FF, *Esthemeles-worth* 1304, *Westhameleswerth* 1303 Ch]. The probability seems to be that the two Hemsworths once had different names, *Hemedesworþ* and *Hemelesworþ*. *Hemeles-wurth* 1224 Cl appears to be West H~. About 1300 *Hemelesworth* had come to be applied to both. But it may be the common name was *Hemedeswerþ* and that *Hemeles-worþ* is due to influence from the well-known name *Hemele*. *Hemedesworþ* is '*Hemede*'s WORÞ'. Cf. HEMPSHILL.

Hemsworth YW [*Hamelesuurde*, *Hilmeuuord* DB, *Hymeleswrde* c 1170 YCh 1548, *Himeles-wurðe* 1191, *-wurda* 1192 P, *Himlesword* c 1200 YCh 1594]. This can hardly be anything else than '*Hymel*'s WORÞ'. *Hymel*

pers. n. is unrecorded, but the stem *Hum-* appears in *Hymma* BCS 519 (and the probably corrupt *Hymora* BCS 148). The pers. n. stem *Hum-* may have arisen in names such as *Hūnbeald*, *Hūnbeorht*. Cf. HEMINGFIELD, HEMLINGTON.

Hemyock D [*Hamihoc* DB, *Hemmiac* 1212 Fees, *Hemihoc* 1228 FF, *Hemiok* 1238 Ass, 1254 FF]. Has been explained as a Brit stream-name **Samiāco-*, derived from *samo-* (Welsh *haf* 'summer'). But an Engl *Hem-man hōc* or *Hemman āc* is possible. *Hemma* is a pers. n. OE *hōc* means 'hook, bend', also 'enclosure'; *āc* is 'oak'. Cf. HENNOCK.

Henbury Chs [*Hamede-*, *Hameteberie* DB, *Hemdebury*, *Hendebiry* 1289 f. Court]. The first el. may be *Hemede* pers. n. (cf. HEMS-WORTH Do) or rather a weak side-form **Hemeda*.

Henbury Do [*Hennebyr* 1244 FF, *Hembyr*, *Hymbur* 1249 ib., *Hymburi* 13 BM]. Apparently OE *Hīgna-burg* 'the BURG of the *hīwan* or monks'.

Henbury Gl [*Heanburg* 692, *æt Heanbyrig* 791–6 BCS 75, 273, *Henberie* DB]. 'High BURG.' It is not quite clear if high situation or a high building is meant.

Hendon Du [*Hynden* 1382 Hatfield]. 'Valley frequented by hinds', OE *hind-denu*.

Hendon Mx [*Hendun* 959, *Heandunes ge-mære* 972 BCS 1050, 1290, *Handone* DB, *Hendon* 1199 Cur]. OE *Hēa-dūn*, dat. *Hēan-dūne*, 'high DŪN or hill'.

Hendred, East & West, Brk [*Hennarið* 956, *æt Henne riðe* 962 BCS 975, 1095, *Henret* DB, *Esthenred* 1200 Cur, *Est*, *West Henrede* 1220 Fees]. 'Stream frequented by water-fowl.' Cf. HENN, RIþ. The name originally denoted a stream [*Henna rið* 984 KCD 1281].

Henfield Sx [*Hanefeld* 770 BCS (206), 1166 RBE, *Hamfelde* DB, *-feld* 1230 P, *Henfeld* 1275 RH]. The forms do not go well with OE *hēan-feld* 'high FELD'. Rather they point to OE *hān(a)-feld* 'FELD with rocks' (OE *hān* 'rock, stone').

OE **hengest** 'stallion'. See HENSTRIDGE, HINKSEY, HINXHILL &c. Often difficult to distinguish from the pers. n. *Hengest*.

Hengrave Sf [*Hemegretham* DB, *-grede* c 1095 Bury, *Hemmegredhe* 1157–80 Bodl, *Hemegrede* 1198 FF, *-grave* 1242 Fees, 1264 Ipm]. '*Hemma*'s meadow.' The second el. is an OE **grēd*, corresponding to OFris *grēd* 'pasture land, water meadow', EFris *grēde* 'pasture land', and related to the verb *grow* &c. The base is **grōði-* or the like. Later *-grave* is due to popular etymology.

Henham Ess [*Henham* 1043–5 Wills, DB, 1254 Val], **H~** Sf [*Henham* DB, 1207 Cur]. OE *Hēa-hām*, dat. *Hēan-hām*, 'high HĀM'.

Henheads La [*Henhades* 1464 Whitaker]. 'Hen hills, hills frequented by wild birds.' See HĒAFOD.

Henhull Chs [*Henulle* 1301 BM, *Henhull* 1304 Chamb]. 'High hill' or 'hen hill'. Cf. HĒAH, HENN.

Henhurst K [*Hænhersta, Hennhyst* 10 BCS 1321 f., *Hanehest* DB, *Henherst* 1186 P]. 'Wood frequented by wild birds.' See HENN, HYRST.

Henley on Thames O [*Heanlea* 1186 ff. P, *Hanlea* 1192 P, *Henleg* 1219 Fees, *Hanleya* 1224 Ep], H~ Sf [*Henleia* DB, *Hanley* 1219 FF, *Henleye* 1242 Fees], H~ So nr Wearne [*Henleighe* 973 BCS 1294, *Henlegh* 1243 Ass], H~ So nr Crewkerne [*Henley* c 1300 BM], H~ Sr [*Henlea* 675 BCS 39, 1062 KCD 812, *Henlei* DB], H~ in Arden Wa [*Henlea* Hy 2 Fr, *Hanleye* 1285 QW, *Henleye in Ardern* 1378 AD]. OE *hēa-lēah*, dat. *hēan-lēa(ge)*, 'high LĒAH (wood or clearing)'. The exx. from So and Sr might be 'hen LĒAH', but the situation favours 'high LĒAH'.

Henley Sa [*Haneleu* DB, *Hennele* 1242 Fees, *-leg* 1255 RH]. 'Wood frequented by wild birds.' See HENN, LĒAH.

Henlow Bd [*Haneslauue* DB, *Hennelawe* 1207 FF, *Henlawe* 1207 Cur, *Hen-, Hanlawe* 1232 f. Cl]. OE *henna-hlāw* 'hill frequented by wild birds'. The DB *-s-* may be a spelling for *h(l)*.

Henmarsh Gl [*Hennemerse* 1236 Ipm]. 'Marsh frequented by wild birds.' Cf. MORETON-IN-MARSH.

OE **henn** 'hen' is the first el. of some names, as HENDRED, HENHEADS, HENLEY Sa &c. The reference is normally to wild birds, as moorhens, waterhens, partridges. In HENTON it is 'domestic fowl'.

Hennock D [*Hainoc, Hanoch* DB, *Henoc* 1234 Fees, *Hyanac* 1242 Fees]. 'High oak.' Cf. HĒAH, ĀC.

Hennor He [*Heanoura* 1123 PNHe]. Identical with HEANOR.

Henny, Great & Little, Ess [*Henies, Hanies, Heni* DB, *Heny* 1202 FF, *Little Hennye* 1248 FF, *Magna Heneye* 1254 Val]. OE *henn-īeg* 'island or river land frequented by wild birds'.

Hensall YW [*Edeshale* DB, *Hedenessale* c 1190 YCh 498, *Hedneshale* 1198 P, *Hechersale* 1226 FF, *Hethensale* 1280 Ipm], **Henshaw** Nb [*Hedeneshalch* 12 BBH, *Hetheneshalgh* 1298 ib., *Ethensalch* c 1300 AD]. Perhaps '*Hepīn*'s HALH or haugh', *Hepīn* being a diminutive of *-īn* formed from *Hæp-* in *Hæpbeorht, -red* &c. Hensall, however, may contain a name *Heddīn*; cf. HEDNESFORD.

Hensingham Cu [*Hensingham, Hunsingham* 12 StB, *Ensingham* 1276 FF]. 'The HĀM of *Hȳnsige*'s people.' *Hȳnsige,* not otherwise evidenced, is a form with *i*-mutation of *Hūnsige.*

Hensington O [*Hansitone* DB, *Hensintona* c 1130 Oxf, *Hencinton* 1196 Cur, 1234 Cl, *Hensinton* 1232 Cl, 1242 Fees]. 'The TŪN by *Hensing*.' *Hensing* may be the name of a wood or stream derived from OE *hēns* 'hens', found in *hensbroc* 770 BCS 204 (Wo). The same first el. is found in *Hensinglade* 1004 Fridesw (*Hensislade* KCD 709) and *Hense-, Hessingrave* 1276 Cl, both names of localities close to Hensington.

Henstead Sf [*Henestede* DB, *Henstede* 1254 Val, 1272 Ch]. The place is in a low situation. The first el. might be OE *henn* 'hen'.

Henstridge So [*Hengstesrig* 956 BCS 923, *Hengest(e)rich* DB, *Heynstrugge* 1243 Ass]. 'Ridge where stallions were kept' (first el. OE *hengest* 'stallion'). The place is near HORSINGTON, and in the boundary of BCS 923 *Horspol* is mentioned.

Henthorn La [*Hennethyrn* 1258 Ipm, *-thyrne* 1276 Ass]. OE *henn-þyrne* 'spinney where wild birds were found'. Cf. HENN, ÞYRNE.

Hentland He [*Hennlann Dibric* c 1150 LL, *Hentlan, Henlande* 1291 Tax]. Welsh *henllan* 'old church'. Cf. HELLAND. HENLLAN is a common name in Wales.

Henton O [*Hyenton* 1220, *Henton* 1236 Fees]. OE *Hēa-tūn*, dat. *Hēan-tūne*, 'high TŪN'.

Henton So [*Hentun* 1065 Wells]. Apparently OE *Henn-tūn* 'TŪN where hens were kept'.

Henwick Np lost [*Hyne-, Henewyk* 12 NS, *Hane-, Henewyc* 1230 P, *-wic* 1248 Ipm], H~ Wo [*Higna gemære* 851 BCS 462, *Henewic* 1182 P]. OE *Hīgna-wīc* 'the WĪC of the *hīwan* or monks'.

Henwood Wa, a nunnery [*Hinewude* 1200 P, 1246 Cl, *Henewode* 1334, *Hynewode* 1369 AD]. 'The nuns' wood.' Cf. HĪWAN.

OE **hēope** 'hip, the fruit of the wild rose', perhaps also *hēopa* 'dog-rose' (cf. OLG *hiupo*, OHG *hiufo* the same). See EBBSFLEET, HEAPEY, HEAPHAM, HEBDEN, HEPPLE &c., HETTON Du, SHIPTON Y.

OE **heord(e)wīc** 'WĪC for a herd or flock of domestic animals', i.e. as a rule 'a sheepfarm'. See HARDWICK.

OE **heorot** 'hart, stag' is a common first el. in pl. ns. See e.g. HART, HART-, HERT-, HURT- (passim), HARFORD Gl, HURSTLEY.

Hepburn (-ĕb-) Nb [(montem) *Hybberndune* c 1050 HSC, *Hibburn* 1242 Fees, 1428 FA]. Perhaps identical in the main with HEBRON, though the source would be OE *hēah-byrgen* 'high tumulus'. *Hēah* would give *hēh-* and perhaps late OE *hīh-, hī-*.

Hepple Nb [*Hephal* 1205 Cur, *-e* 1212 Fees, *Hyephal* 1229 Pat]. OE *hēop-halh* 'haugh where hips grew'.

Hepscott Nb [*Hebscot* 1242 Fees, *Hebbescotes* 1289 Ipm, *Heppescotes* 1257 Ch]. '*Hebbi*'s COTS.' **Hebbi* would be derived from *H(e)abba* in HABTON &c.

Heptonstall YW [*Heptonstall* 1274 Wakef, 1316 FA]. 'The STALL or stable or the like in Hebden.' HEBDEN (q.v.) is OE *Hēope-denu*. This became **Hepten* and, owing to association with TŪN, *Hepton*.

Hepworth Sf [*Hepworda* DB, *Hepewurde* 1193 P, *Hepwrthe* 1196 FF], **H~** YW [*Heppeuuord* DB, -*wrth* 1274 Wakef]. 'Hip WORÞ.' Possibly 'hedge where hips grew'.

OE **here** 'army, host, multitude' in pl. ns. chiefly occurs in HEREFORD and HEREPÆÞ (q.v.), but also in HARLOW, HARWICH, and perhaps HAREFIELD. OScand *herr* also meant 'the whole people', as in *allsheriar þing* 'the national assembly'. OE *here* very likely has this sense in HARLOW and perhaps HARE-FIELD.

Hereford He [*Hereford* 958 BCS 1040, DB], **Little H~** He [*Lutelonhereford* DB, *Herefordia parva* 1122 BM]. 'Army ford.' If this is the meaning, the reference would be to a ford where a marching column could pass in closed order. Cf. HERFORD in Germany. Another possibility would be to take H~ to be elliptical for *herepæþford* (cf. HEREPÆÞ). Cf. also HARFORD, HARTFORD, HARVINGTON. *Hereford* is not used of a ford where a Roman road crosses a river. This was *strǣtford*. But a *hereford* seems to have been an important ford.

Herefordshire is *Herefordscir* a 1038, a 1056 KCD (755, 802), 1048 ASC (E).

OE **herepæþ** 'army road, road large enough to march soldiers upon', 'through road'. Distinct from *strǣt*, the name for a Roman road. See HARPFORD and the discussion under HARPENDEN. **Harford** D in Crediton is *Herepaðford* 739 Crawf.

Hergest He [*Hergest*(*h*) DB, *Heregast* 1251 Ch, *Hergast* 1278 Ep]. **H~** Ridge reaches 1,389 ft. A Welsh name of obscure etymology.

Herne Bd in Toddington [*Hara* 1183 P, *Hare* 1211 Cur, 1237–40 Fees, *Haren* 1276 Ass]. OE **harum*, dat. plur. of **hær* 'stone' or the like. Cf. HĀR.

Herne or **Heron** or **Hurn** Ha [*Herne* DB, *Hurna* 1150 (1313) Ch, *Hurne* 1242 Fees], **H~** K [(monasterium) *aethyrnan* 11 DM, *Hierne* 1389 BM]. OE *hyrne* 'corner, angle'. Herne Ha is at a wide bend of the Stour. H~ K seems to be in a curving valley.

Hernhill K [*Haranhylle* 11 DM, *Harehull* 1237 Cl, *Harenhull* 1247 Ch]. Apparently 'grey hill' (cf. HĀR) with early shortening of *ā*.

Herriard Ha [*Henerd* DB, *Herierda* a 1162 Oxf, *Heriet* 1167 P, *Herierd* 1251 Ipm, *Hereyerd* 1252 FF]. Second el. OE *geard* 'enclosure'. The first might be OE *hearg* 'heathen place of worship'. Cf. early forms of HARROW with *e*, also MARDEN W.

Herringby Nf [*Haringebei* DB, -*bi* 1177 P, *Heringebi* 1196 FF, *Haringbi* 1254 Val]. '*Hæring*'s BY.' First el. ON *Hæringr* pers. n.

Herringfleet Sf [*Herlingaflet* DB, *Herlingefleth* 1202 FF, *Herlingflet* 1254 Val, *Heringflete* c 1255 Bodl]. 'The stream of *Herela*'s people.' Cf. HARLING and FLĒOT.

Herringswell Sf [*Hyrningcwylle*, *Herningwelle* 11 EHR 43, *Hyrningwella*, *Herningawella* DB, *Haringwell* 1242 Fees, *Heringwell* 1254 Val]. 'The spring of the *Hyrningas*.' The *Hyrningas* would be 'the people dwelling at the corner or angle'. *Hyrne* may have been the name of the horseshoe-shaped ridge near the place.

Herrington Du [*Erinton* 1196 P, *Heringtone* c 1250 FPD]. Perhaps 'the TŪN of *Here*'s people'. **Here* would be a short form of names such as *Herefrið*. Cf. HERSTON. *Hering* occurs ASC (E).

Hersham Sr [*Hauerichesham* 1175 P, *Heverichesham* 1231 Cl]. The first el. seems to be a pers. n., possibly **Hæferic*. Cf. HAVERSHAM.

Herstmonceu·x (-sōō) Sx [*Herst* DB, *Hurst quod fuit Willelmi de Munceus* 1243 Cl, *Herstmonceus* 1287 Ep]. OE HYRST 'wooded hill'.

The Monceux family was in possession from the 12th cent. Monceux from MONCEAUX (one in Calvados, Normandy).

Herston Do [*Herestona* DB, *Herston* 1318 FF]. '*Here*'s TŪN.' Cf. HERRINGTON. One *Her* held part of the manor in 1066 (DB).

Hertford (harf-) Hrt [*Herut*-, *Heorutford* c 730 Bede, *Heorotford* 673, 913 ASC, *Heortford* 1130 P]. 'Stag ford.' **Hertford-shire** is *Heortfordscir* 1011 ASC (E), c 1050 KCD 866.

Hertingfordbury Hrt [*Herefordingberie* DB, *Hertfordingebur*' 1240 Ep]. 'The BURG (manor?) of the Hertford people.' The metathesis of the *ing*-element is remarkable.

Hescombe So in Odcombe [*Hascecomba* DB, *Hececumb* c 1100, *Hetsecumb* c 1150, *Hatsecumbe* c 1155 Montacute]. Identical with HASCOMBE.

Hesket, High & Low, Cu [*Heskgeth* 1330 Pat, *Heskaith* 1337 Orig, *Hescath* 1346 Pat], **Hesketh** La [*Heschath* 1288 LaInq, *Heskayth* 1298 FF], **Hesketh Grange** YN [*Hesteskeith* c 1155 Riev]. OScand *hesta-skeið* 'race-course'. Horse-racing was a favourite sport of the old Scandinavians. Cf. also WICKHAM SKEITH.

Hesket Newmarket Cu [*Eskeheued* 1227 FF]. OScand *eski* 'ash-trees' and OE *hēafod* 'head, hill'. The name was associated with HESKET. There was a market here.

Heskin La [*Heskyn* 1257 Ass, 1301 FF]. A Welsh name. Welsh *hesgen* (OW *hescenn*) means 'sedge, rush'. The meaning in pl. ns. may be that of the cognate OIr *sescenn*, viz. 'marsh'.

Hesleden, High & Monk, Du [*Heseldene* c 1050 HSC, Hy 2 FPD, *Hæseldene* c 1085

LVD, *Munkhesilden* 1324 FPD]. 'Hazel valley.' The first el. seems to be OE **hesel*. The monks were those of Durham.

Heslerton, East & West, YE [*Heslerton, -e* DB, *Heslertona* c 1165 YCh 803, *Heselerdton* 1246 FF, *West Heslardtona, Est, West Heslartona* 1297 Subs]. The first el. seems to be a compound containing OE *hæsel* 'hazel', e.g. *hæsel-geard* 'hazel enclosure'.

Heslington YE [*Haslinton* DB, *Heselington* c 1190 YCh (320), 1190 P]. Identical with HASLINGTON.

Hessay YW [*Hesdesai* DB, *Heselseia* 1169 P, *Hesleshai* c 1150 YCh 528, *Hessai* c 1090 YCh 350]. OE *hæsel-sǽ* or OScand *heslisǽr* 'lake where hazels grew'.

Hessett Sf [*Heteseta* DB, *Hecesete* c 1095 Bury, *Hecheset* 1203 Ass, *Heggeset* 1225 FF, *Haggesete, Hegessete* 1254 Val]. 'Fold by a hedge.' Cf. (GE)SET, HECG.

Hessle YE [*Hase* DB, *Hesla* Hy 2 DC, *Hesel* 1157 YCh 354, 1242 Fees], **H~** YW [*Hasele* DB]. OScand *hesli* 'hazel grove'. **Hessleskew** YE [*Heselescof* 1202 FF]. 'Hazel wood'; *-scof* is OScand *skōgr* 'wood'.

Hest La [*Hest* 1177 ff. P]. An OE **hǽst*, a derivative of *hǽs* 'beech or oak wood' corresponding to OHG *Haist*.

Hestercombe So [*Hegsteldescumb* 854 BCS 476, *Hestercumba* 1155–8 (1334) Ch]. 'The valley of the *hægsteald*.' OE *hægsteald* or *hagusteald* means 'warrior, bachelor', but must once have been used in the same sense as OHG *hagustalt*, i.e. 'owner of a *haga* or enclosure', a younger son who had no share in the village, but had to take up a holding for himself outside. The OG *hagustalt* formed a definite class in the community. Cf. HEXHAM.

Heston Mx [*Hestone* c 1130 Oxf, *Heston* 1254 Val]. OE *Hǽs-tūn* 'TŪN in the HǼS or wood'. Heston is c 3 m. south of HAYES.

Heswall Chs [*Estvelle* DB, *Haselewelle* 1252 RBE, *Haselwell* 1287 Court]. 'Hazel spring.' Second el. OE *wælla*, a form of WELLA.

Hethe O [*Hedha* DB, *Heða* 1176 P, *Hethe* 1206 Cur, 1220 Fees]. OE HǼÞ 'heath'.

Hethel (-ĕth-, -ĕth-) Nf [*Hethella, Hathelle* DB, *Hethill* 1250 Cl]. OE *hǽþ-hyll* 'hill overgrown with heather'.

Hethersett Nf [*Hederseta* DB, 1254 Val, *-sete* 1252, 1276 Ipm]. Perhaps 'fold for deer'. First el. OE *hēahdēor* 'stag'. See (GE)SET.

Hethfelton Do [*Hafeltone* DB, *Hethfelton* 1280 Ch]. 'TŪN in open land overgrown with heather.'

Hett Du [*Het* c 1168 FPD, *Hett* 1335 Ch]. OE *hætt* 'hat', here used of a marked hill. Cf. HATTINGLEY.

Hetton le Hill & le Hole Du [*Heppedun* 1180 Finchale, *Hepedon* c 1230 FPD]. OE *hēope-dūn* 'DŪN or hill where hips grew'. **H~ le Hole** is at the foot of the hill.

Hetton Nb [*Hetton* 1163 P, *Hethton* 1289 Cl], **H~** YW [*Hetune* DB, c 1200 FC]. 'TŪN on a heath' (OE *Hǽþ-tūn*).

Heugh Du [*le Hough* 1411 PNNb], **H~** Nb [*Hou* 1279 Ass]. OE HŌH 'ridge, spur of land'.

Heveningham Sf [*Heueniggeham* DB, *Eueningeham* 1193 P, *Heveningham* 1200 Cur, 1254 Val]. 'The HĀM of *Hefa*'s people.' The *-n-* would be the *n* of the stem of *Hefa* (gen. *Hefan*). Or the immediate base may be a diminutive **Hefīn*.

Hever K [*Heanyfre* 814 BCS 346, *Hevre* 1242 Fees, *Heuere* 1279 Ass]. 'High edge.' See HĒAH, YFER.

Heversham We [*Hefresham* c 1050 HSC, *Eureshaim* DB, *Hevresham* 1157 YCh 354, R 1 (1308) Ch]. Possibly '*Hēahfriþ*'s HĀM'. *Hēahfriþ* would give late OE *He(a)ferþ*. Anyhow the first el. is a pers. n. *Hæfer* does not explain the regular *e*.

Hevingham Nf [*Heuincham* DB, *Hevingham* 1242 Fees, 1252 Ch]. 'The HĀM of *Hefa*'s people.'

Hewelsfield Gl [*Hiwoldestone* DB, *Hiwaldestun* 12 Fr; *Hualdesfeld* 1140–50 Fr, *Huwaldesfeld* 1227 Ch]. '*Hygeweald*'s TŪN and FELD.'

Hewick, Bridge & Copt, YW [*Heawic, oþer Heawic* c 972 BCS 1278, *Suthewic, Hauuic* DB, *Hewyk ad pontem* 1290 Ipm, *Copedhewike* 1297 Subs]. 'High WĪC.'
Copt is OE *coppede* 'provided with a *copp*, peaked'.

Hewish So in Yatton [*Hiwis* 1198 P, 1223 FF], **H~** So nr Crewkerne [*Hywys* 1327 Subs]. See HĪWISC.

Heworth Du [*Hewarde* 1091, *Hewrtha* Hy 2 FPD, *Hewrthe* Hy 1 (1300) Ch], **H~** YN [*Heworde* DB, *Hewud* 1219 FF]. 'High WORÞ.' **H~** YN is in a low situation, so 'high' refers to the nature of the WORÞ ('high fence'?).

Hexham Nb [*Hagustaldes ea* 681, *Hagustaldes ham* 685 ASC (E), *Hagustaldensis ecclesia* c 730 Bede, *Heagostealdes ea* c 890 OEBede, *Hextoldesham* 1188 P]. The original name was *Hagustaldes ēa*, later refashioned to *Hagustaldes hām*. On *Hagustald* see HESTERCOMBE. The original name referred to the stream at Hexham: 'the *hagustald*'s stream'.

Hextells or **Extall** St [*Hegstal* 1176 P, *Hehstall* 1227 Ass, *Hegestall* 1272 Ass]. OE *hegesteall* 840 BCS 428, rendered by B-T (Suppl) 'site of a hedge, place with a hedge (?)'.

Hexthorpe YW [*Hestorp* DB, *Hexthorp* 1279

Ipm]. *'Hegg*'s thorp.' ON *Heggr* pers. n. is found in Landnáma.

Hexton Hrt [*Hegæstanestone* DB, *Hecstaneston* 1198 (1301) Ch, *Hecstonstun* 1219 Pp]. *'Hēahstān*'s TŪN.'

Heybridge Ess [*Hebrege* 1222 St Paul, *Habrugg* 1236 Fees, *Hebrugg* 1254 Val]. 'High bridge.' The old name was *Tidweldington* c 950, *Tidwoldingtun* c 995 BCS 1012, 1289, *Tydwoldyngton* 1316 Ch ('the TŪN of *Tidweald*'s people').

Heydon Ca [*Haidena, Haindena* DB, *Heiden* 1200 Cur, 1236 Fees]. 'Hay valley.'

Heydon Nf [*Heidon* 1196 FF, *Heydon* 1242 Fees, 1253 Ipm]. 'Hay DŪN or hill.'

Heyford, Nether & Upper, Np [*Heiforde, Haiford* DB, *Heyford, Little Heyford* 12 NS, *Inferior, Superior Heyford* 1220 Fees], **Lower & Upper H~** O [*Hegford* 995 KCD 1289, DB, *Haiforde* DB, *-ford* c 1130 Oxf, *Heyford Magna* 1242 Fees]. 'Hay ford', probably 'ford used at the time of the hay harvest'.

Heynings Li, old priory [*Hening* 1220, *Heninges, Heyings* 1237 Ep, *Heyninges* 1268 Ch]. ME *haining* 'enclosure', from *hain* 'to enclose', perhaps from OScand *hegna*.

Heyrod La [*Heyerode* 1246 Ass, *Heghrode* 1422 PNLa]. 'High clearing.' Cf. ROD.

Heysham (-ĕsh-) La [*Hessam* DB, *Hesheim, Hesham* c 1190 LaCh]. OE *Hǣs-hām* 'HĀM in a wood'. Cf. HǢS.

Heyshott (hē-) Sx [*Hesset* 1244 Cl, *Hetheshete* 1288 Ass]. 'SCĒAT where heather grew.'

Heytesbury W [*Hestrebe* DB, *Hehtredeberia* c 1115 Salisbury, *Hegtredebiri* c 1115 Sarum, *Hehtredesberi* 1156 P]. *'Hēahþrȳþ*'s BURG.' *Hēahþrȳþ*, a woman's name, is not recorded.

Heythrop (-ē-) O [*Hetrop* 1209–19, *Hethrope* 1224 Ep, *Hetrop* 1242 Fees]. 'High thorp.'

Heywood La [*Hewude, Heghwode* 1246 Ass]. 'High wood.'

Heywood W [*Heiwode* 1224 FF, *Hewode* 1412 FA]. 'Enclosed wood.' First el. OE *hege* 'hedge' or *gehæg* 'enclosure'.

Hibaldstow Li [*Hiboldestou* DB, *Hibaldestoua* c 1115 LiS, *Hibaldestowa, Huboldestou* 1088 RA]. *'Hygebald*'s STŌW or burial place.' The old name was *Cecesēg 'Cec*'s island'. In Saints it is stated that St. *Higebold* was buried in Lindsey *on Cecesége* near the river Ancholme.

Hickleton YW [*Chicheltone, Icheltone* DB, *Hikalton* c 1175 YCh 584, *Hikelton* 1200 Cur]. *'Hicela*'s TŪN.' *Hicela* is not recorded, but *Hicel* occurs in *Hiceles wyrþ* (*wyrð*) BCS 27, 862.

Hickling Nf [*Hikelinga* DB, *Hikelinges* 1191 P, *Hikeling* 1206 Cur], **H~** Nt [*Hikelinge* 11 Th, *Hechelinge* DB, *Hikelinga* 1185 P,

Hicalinga R 1 (1308) Ch]. *'Hicel(a)*'s people.' Cf. HICKLETON.

OE **hīd, hīgid** 'hide, land adequate for the support of one free family, as much land as could be tilled with one plough in one year'. The original meaning was 'household', as indicated by the rendering *familia* in Bede and early charters. Hence the pl. n. HYDE, which may mean 'homestead consisting of one hide'. The word is often used as a second el. with a numeral prefixed, as FIFEHEAD (FIFIELD, FIVEHEAD), NYNEHEAD. Cf. TEIGNHEAD, PIDDLETRENTHIDE. A pers. n. is the first el. of TILSHEAD.

Hidcote Bartrim & Boyce Gl [*Hudicota* 716 BCS 134, *Hidi-, Hedecote* DB, *Hudicota* 1175 Winchc, 1190 ff. P, *Hudycote Bartram, Boys* 1327 Subs]. Perhaps *'Hydica*'s COT'. Cf. HEADCORN.
Bertrannus de *Hudicota* is mentioned 1190 ff. P.—Ernolf de Bosco was tenant of *Hidecot* in 1200 (Cur).

Hidden Brk [(on) *Hyddene* 984 KCD (1282), 1050 E, *Hudden* c 1170 Fridesw, *Hudden* 1242 Fees]. OE *hȳþ-denu* 'valley with a landing-place'.

Hidon. See CLAYHIDON.

Hiendley YW [*Hindelei* DB, *-lay* 1297 Subs]. 'Wood frequented by hinds' (OE *hind* 'female of the hart').

Higford Sa [*Huchefor* DB, *Hugeford* 1206 Cur, *Huggeford* 1242 Fees]. Apparently OE *Huggan ford*, the first el. being a pers. n. **Hugga*, which is found also in HIGHLEY Sa. *Hugga* might be an early short form of names in *Hyge-* (from *Hugi-*) or possibly a development of *Hudeca*.

Higham Gobion Bd [*Echam* DB, *Heham* 1166 P, *Heygham Gobyon* 1341 NI], **H~** Db [*Heyham* 1284 FF, *Hegham* 1330 QW], **H~ Ess** [*Hecham* DB], **H~ K** nr Canterbury [*Hegham* 1242 Fees, 1346 FA], **H~ Upshire** K [*Heahhaam* c 765, *Hehham* 774 BCS 199, 213, *Hecham* DB], **H~ La** [*Hegham* 1296 Lacy], **H~ on the Hill** Le [*Hecham* 1220–35 Ep, *Heyham* 1230 P, 1254 Val], **Cold H~** Np [*Hecham* DB, 1198 Fees, 1254 Val], **H~ Ferrers** Np [*Hehham* 1066–75 GeldR, *Hecham* DB, *Heccham Ferrar'* 1279 Cl], **H~ Sf** nr Manningtree [*Hecham* c 1050 KCD 907, *Heihham* DB, *Hegham* 12 BM], **H~ Sf** nr Newmarket [*Heyham* 1275 RH, *Hegham* 1303 FA], **H~ YW** [*Hegham* 1297 Goodall]. OE *Hēah-hām* 'high HĀM'.
H~ Ferrers was held by Comes de Ferariis in 1166 (RBE).—Cf. BERE FERRERS.—**H~ Gobion** from the Gobion family, resident here from the 12th cent. Gobion is a nickname identical with *gudgeon* (Lat *gobio*).

Higham Sx, an old name of Winchelsea [*Iham* 1200 FF, *Ihomme* 1205 Cl]. OE *ieg-hamm* 'HAMM in an island or forming an island'.

Highampton D [*Hantona* DB, *Hyauntone* 1242 Fees, *Hegheheauntone* 1308 Ep]. OE

Hēa-tūn, dat. *Hēan-tūne* 'high TŪN', with addition of a fresh *high*.

Highbridge So [*Highbridge* 1324 Misc, (juxta) *Altum Pontem* 1327 Subs]. 'High bridge.'

Highbury Mx [*Heybury* c 1370 Gesta]. 'High BURG.'

Highclere. See CLERE.

Highgate Mx [*Heygate* 1391 FF, *Higate* 1466 AD]. 'High (toll) gate.'

Highhead Cu [*le Heghheved* 1323 Ipm]. 'High hill.' Cf. HĒAFOD.

Highleadon Gl. See LEADON.

Highley Sa [*Hugelei* DB, *-leg* 1242 Fees, *Huggel*[e] 1246 Ipm, *Huggeleye* 1291 Tax]. '*Hugga*'s LĒAH.' Cf. HIGFORD.

Highlow Db [*Heghlawe* Hy 3 Derby, *Heyelawe* 1242 FF, *Heelowe* 1265 Misc]. 'High hill.' See HLĀW.

Highnam Gl [*Hamme* DB, *Hynehamme* 1100 Glouc, 1316 FA]. 'The monks' Hamm.' Cf. CHURCHAM. The place belonged to Gloucester Abbey. Cf. HĪWAN.

Highway W [*Hiwei* DB, *Hyweie* 1232 Ch, *Hiweie* 1242 Fees]. 'Road for carrying hay.' First el. OE *hēg*, *hīeg* 'hay'. The form is here WSax *hīeg*, *hīg*.

Highweek D [*Teinnewic* 1205 Layamon, *Hegewyk* 1281 QW]. Originally 'wīc on R TEIGN', later 'high wīc'.

Highworth W [*Wrde* DB, *Wurða* 1156, *Wurda* 1190 P, *Hegworth* 1232 Cl]. Originally WORTH (see WORÞ), with later addition of *high*.

Hilborough Nf [*Hildeburhwella* DB, *Hildeburwrthe* 1242 Fees, *Hilburgwrth* 1254 Val], **Hilbre Island** Chs [*Hildeburghey* c 1235 Chester, *Hildeburweye* c 1300 ib.]. '*Hildeburg*'s WORÞ and island.' *Hildeburg* is a woman's name.

Hilcott Gl [*Willecote* DB, *Huldicota* 1209 Fees, *Hildecote* 1303 FA]. Perhaps 'COT on a slope', the first el. being OE *helde*, *hielde* 'a slope'.

Hilcott W [*Hulcote* 1195 Cur, *Hulecot* 1237 Cl, *Hulkot* 1242 Fees]. Perhaps 'COT on a hill'. But the absence of spellings with *i* suggests identity with HULCOTE.

Hildenborough K [*Hildenne* 1291 Tax, 1314 Ipm, *Hildenborough* 1389 Pat]. '*Hilda*'s DENN', with late addition of *-borough*.

Hildenley YN [*Hildingeslei* DB]. '*Hilding*'s LĒAH.' *Hilding* is unrecorded in OE.

Hildersham Ca [*Hildricesham* DB, *Hildrichesham* c 1080 ICC, 1242 Fees]. '*Hildrīc*'s HĀM.'

Hilderstone St [*Hildulvestune* DB, *-ton* 1227 Ass]. '*Hildwulf*'s TŪN.'

Hilderthorpe YE [*Hilgertorp* DB, *Hilderthorp* 1125–30 YCh 1135, *-thorp* 1246 FF]. '*Hildiger*'s thorp.' *Hildiger* is an ODan pers. n.

Hilfield Do [*Hylfeld* 939 BCS 738, *Hulfeld* 1212 Fees]. 'Open land on a hill.'

Hilgay Nf [*Hillingeiæ* 974 BCS 1311, (æt) *Hyllingyge* 11 Thorney Fragm, *Huling-, Hidlingheia* DB, *Helingeia* 1103–6 BM, *Helegeye* 1254 Val]. Cf. HILLINGTON, whose first el. is the same as that of Hilgay. Hilgay may be 'the island of *Hȳpla*'s people'. *Hȳpla is a short form of names such as *Hȳphere, -walda*. But *Hydla, a derivative of *Huda*, is also possible.

Hill End Brk [*Hulle* 1263 Ipm], **North & South Hill** Co [*Northehull* 1291 Tax, *Suthhulle* 1270 Ep], **Hill** Gl [*Hilla* DB, *Hulle* 1220 Fees], **H~** Ha nr Nursling [*Hulla* 1167 P, *La Hull* 1236 Cl], **H~ Farrance** So [*Hyll* 11 KCD 897, *Hulla* DB, *Hull* 1182 P, *Hull Ferun* 1253 Ch], **H~ Chorlton** St [*Hylle* 1194 P], **H~ Deverill** W [*Devrel* DB, *Hull* 1130–5 Sarum, *Hulle Deverel* 1242 Fees], **H~ Wo** in Halesowen [*Hulle* 1270 ff. Ct], **H~ and Moor Wo** [*Hylle* c 1050 KCD 923, *More et Hylle* DB]. OE *hyll* 'hill'.

H~ Chorlton is nr CHAPEL CHORLTON.—H~ Deverill was originally DEVERILL (q.v.). *Hill* is really a distinctive addition.—Hill Farrance was held by Robert Furon in 1182 (P). *Furon* is an OFr byname, identical with Fr *furon* 'ferret', lit. 'pilferer'.

Hillam YW nr Pontefract [(on) *Hillum* 963, *Hyllum* c 1030, *Hillum* c 1050 YCh 6 f., 9], **H~** YW nr Aberford [*Hullum* 1236 Ch, *Hillum* 1303 FA]. The dat. plur. of OE *hyll* 'hill': '(at) the hills'.

Hillbeck. See HELBECK.

Hillborough Wa [*Hildeburhwrthe* 710 BCS 127, *Hildeborde* DB, *Hildeburworth* 1202 Ass]. Identical with HILBOROUGH.

Hillesden Bk [*Hildesdun* 949 BCS 883, *-don* 1163, 1185 P, 1230 Cl]. '*Hild*'s DŪN.' *Hild* is found in *Hildes hlæw* KCD 621.

Hillhampton Wo [*Hilhamatone* DB]. OE *Hyllhǣma-tūn* 'the TŪN of the dwellers on the hill'. Cf. HĀMTŪN.

Hillingdon Mx [*Hildedun* 1078–85, *Hildendune* 1160–91 ChronEve, *Hillendone* DB, *Hillendon* 1229 FF, *Hilledon* 1238 FF, *Hilenden* 1254 Val]. '*Hilda*'s or *Hilla*'s DŪN.' Both are known OE pers. names.

Hillington Nf [*Helingetuna, Nidlinghetuna, Idlinghetuna* DB, *Hillingeton* 1177, 1181, *Hellingeton* 1185 P]. First el. as in HILGAY.

Hillmorton Wa [*Mortone* DB, *Hulle, Morton* 1252 Fees, *Hullemorton* 1265 Ch]. Originally two places HILL and MORTON, later thrown into one called Hillmorton.

Hillsea Ha [*Hulesey* 1281 Cl, *-e* 1316 FA, *Hulsea* 17 VH]. Second el. ĒG 'island'. The first may be an OE *hyles*, corresponding to OHG *hulis* 'holly'. The place is in a low situation.

Hillsley Gl [*Hildeslei* DB, *-leg* 1220 Fees, *-lege* 1221 Ass]. '*Hild*'s LĒAH.' Cf. HILLES-DEN.

Hilmarton W [*Helmerdingtun* 962 BCS 1081, *Helmerintone* DB, *Helmerton* 1198 Fees]. 'The TŪN of *Helmheard*'s people.' *Helmheard* is identical with OHG *Helmhart*.

Hilperton W [*Help(e)rintone, Helperitune* DB, *Hulprinton* 1242 Fees]. The same first el. is found in *Hulpryngmor* 964 BCS 1127 (in boundaries of Steeple Ashton nr Hilperton, late copy). Hilperton may mean 'the TŪN of *Hylpric*'s people'. *Hylpric* is a sideform of *Helpric* and shows the same vowel as OHG *hulfa*, OLG *hulpa* 'help'. OE *hylp* 'help' must have existed. It is found as *hilp, hylp* in the ME Ferumbras.

Hilston YE [*Heldovestun* DB, *Hildulueston* 1166 P, *Hildolueston* 1240 FF]. '*Hildwulf*'s or (OScand) *Hildulf*'s TŪN.' Cf. HILDER-STONE, HINDOLVESTON.

Hilton Db [*Hiltune* DB, *Hilton* 1208 Cur, *Helton* 1197 P, *Hulton* 1208 FF], H~ Hu [*Hilton* 1196 FF, *Hulton* 1227 Ass], H~ St [*Hylton* 996 Mon, *Iltone* DB, *Hulton* 1262 For], H~ YN [*Hiltune* DB, *Hilton* 1218 FF]. OE *Hyll-tūn* 'TŪN on a hill'.

Hilton Do [*Heltona* DB, *Helton* 1212 Fees, 1227 FF], H~ We [*Helton* 1291 CWNS xxi, *Helton Bacon* 1314 Ipm]. Cf. HELTON We.

Himbleton Wo [*Hymeltun* 816, 884 BCS 356, 552, *Himeltun* DB]. 'TŪN where *hymele* grew.' *Hymele* may refer to the hop plant or to some similar plant. The brook at Himbleton, now Bow Brook, was formerly *Hymelbroc* 840 BCS 428 &c.

Himley St [*Himelei* DB, *Humelilega* 1185 P, *Humelele* 1242 Fees]. 'LĒAH where *hymele* grew.' Cf. HIMBLETON.

Hincaster We [*Hennecastre* DB, 1190–5 LaCh, *Hine-, Henecastre* 1237 Kendale, *Hanecaster* 1260 FF]. Second el. OE CEASTER 'Roman fort'. The first is doubtful. Some forms suggest OE *higna ceaster* 'the CEASTER of the monks', but the earliest forms tell against this. OE *henn* 'hen, wild bird' may not seem very likely, but it is possible that wild birds might have taken up their abode in a deserted fort.

Hinchingbrooke Hu [*Hychelingbrok* 1260 Ass, *Inchinbrok* 1378, *Fynchyngbroke* 1462 BM]. Etymology doubtful.

Hinchwick Gl [*Hunchewic* 1189 (1372) Ch, 1218 AD, *Hinkewik* 1205 Ch]. Apparently '*Hynca*'s WĪC'. The form with *ch* may be due to misreading of early *ch*, meant for *k*. Or there may have been a form *Hynci* used by the side of *Hynca*.

Hinckley Le [*Hinchelie* DB, *Hinkelai* 1176 P, *-lai* Hy 2 DC]. '*Hynca*'s LĒAH.'

OE **hind** 'hind, female of the hart'. See HIND- (passim), HENDON Du, HIENDLEY.

Hindburn (hin-) R La [*Hyndborn* 1577 Saxton]. 'Hind stream' (OE *hind* 'female of the hart').

Hinderclay Sf [*Hildericlea* c 1000 (c 1300) ASCh, *Hildercle* 10 BCS 1013, *-clea* DB, *-cle* 1254 Val, *Hyldreclea* c 1095 Bury]. 'Tongue of land where elder grew.' The first el. is ME *hilder* (*hyldyr, hildertre* &c.) 'elder', a word related to Dan *hyld*, Sw *hyll*, OHG *holuntar* 'elder', and probably native in English. It is evidenced in East Anglia. Cf. ILDERTON. The second el. appears to be OE *clēa* 'claw', here used in the sense 'tongue of land in a river fork' (cf. CLAWTON). A still better etymology would perhaps be obtained if there was an OE derivative of *hilder* with a *k*-suffix. If so, the second el. would be OE LĒAH. *Hinder-* for *Hilder-* owing to dissimilation. Cf. next names.

Hinderskelfe YN [*Hildreschelf* DB, *Hilderscelf* 1170–85 YCh 633]. OScand *Hildar skialf* '*Hild*'s ledge of land'. Cf. SCYLF. *Hildr* is an OScand woman's name. H~ is now usually CASTLE HOWARD.

Hinderwell YN [*Hildrewelle* DB, *Hilderwella* c 1140 YCh 906, *-well* 1204 Cur, 1226 Ep]. Usually explained as '*Hild*'s spring', *Hild* being St. Hilda of *Streoneshalh*. *Hilder-* would then be due to Scandinavianization. The Scand gen. *Hildar* having replaced OE *Hilde*. But the first el. might be as in HINDERCLAY.

Hindhead Sr [*Hyndehed* 1571 PNSr]. 'Hill frequented by hinds.' Cf. HĒAFOD.

Hindley La [*Hindele* 1212 Fees, *-leye* 1259 Ass]. Identical with HIENDLEY.

Hindlip Wo [*Hindehlep* 966 BCS 1180, *Hindelep* DB]. OE *hinde-hlīep* 'a leap-gate for hinds'. A deer-leap is a lower place in a hedge or fence where deer may leap. Cf. HLĪEP.

Hindolveston Nf [*Hildolueston* 11 Wills, *Hidolfestuna* DB, *Hildolveston* 1206 Cur, *Hindoluestone* 1254 Val]. Identical with HILDERSTONE St. Cf. HINDERCLAY.

Hindon W [*Hynedon* 1275 RH, 1284 Ch]. OE *higna dūn* 'the DŪN of the monks or nuns'. Cf. HĪWAN. The place is near Shaftesbury.

Hindringham Nf [*Hindringham* 11 Wills, 1204 Cur, *Hindringaham* DB, *Hindringeham* 1203 Ass]. Possibly the first el. *Hindringas* is derived from OE *hinder* 'behind' and means 'the people dwelling behind', perhaps behind the hills near which the place stands. Second el. HĀM.

Hine Heath. See HATTON Sa.

Hingham Nf [*Ahincham, Hincham* DB, *Heingeham* 1173, 1190 P, *Heingham* 1167 f. P, *Hengham* 1158 P]. Probably 'the HĀM of *Hega*'s people'. But the base of the first el. might be OE *hēah* 'high'. The place is by a hill.

Hinksey, North & South, Brk [*Hengestesie(g)* 821 BCS 366, *Henxtesia superior* c 1222 Fridesw, *Nort-, Suthhenctesey* 1242 Fees]. '*Hengest*'s island' or 'the island of the stallion' (OE *hengest*).

Hinstock Sa [*Stoche* DB, *Hinestok* 1242 Fees, *Hynestok* 1281 Ipm]. 'The STOC of the monks.' Cf. HĪWAN.

Hintlesham Sf [*Hintlesham* c 1040 Wills, DB, 1158 P, *Huntlesham* 1168 P, 1235 Fees]. '*Hyntel*'s HĀM.' **Hyntel* is a derivative of *Hunta*. Cf. *Hintleswode* 1180 P (Nf).

Hinton, a common name, has at least two sources:

1. OE *Hēa-tūn*, dat. *Hēan-tūne*, 'TŪN situated on high land': **Hinton Waldrist** Brk [*Heantunninga gemære* 958 BCS 1028, *Hentone* DB, *Henton* 1192, *Hanton* 1193 P], **H~ St. Mary** Do [*Haintone* DB, *Henton* 1212 Fees], **H~** Gl nr Bristol [*Heanton* Hy 3, *Henton* 1266 BM], **H~ Admiral** Ha [*Hentune* DB, *Henton* 1242 Fees, *H~ Damarle* 1412 FA], **H~ Ampner** Ha [*Heantun* 1045 KCD 780, *Hentune* DB, *Hinton Amner* 13 VH], **H~ Daubney** Ha [*Henton* 1316, *H~ Daubeney* 1412 FA], **H~ He** in Eardisland [*Hentun* 1190 PNHe], **H~ Blewett** So [*Hantone* DB, *Hentun Bluet* 1246 Cl, 1268 FF], **Bower H~** So [*Hanton Mertoc* 1225, *Burehenton* 1243 Ass], **H~ Charterhouse** So [*Hantone* DB, *Henton* 1212 Fees, *H~ Charterus* 1273 FF], **H~ St. George** So [*Hantone* DB, *Heanton* 1219 Bath, *Hentun Sancti Georgii* 1246 Cl], **Broad H~** W [*Hen-, Hantone* DB, *Henton* 1203 Cur, 1232 Ch, *Brodehenton* 1333 FF], **Great H~** W [*Henton* 1316, 1412 FA].

2. OE *Higna-tūn* 'the monks' or nuns' TŪN'; but cf. HĪWAN: **Cherry Hinton** Ca [*Hintona* c 1080 ICC, *Hintone* DB, *Hingtone* 13 PNCa(S), *Hyneton* 1299 Pat], **H~ Martell** or **Magna & H~ Parva** Do [*þare hina gemære* 946 BCS 818, *Hinetone* DB, *Hineton* 1151–7 Fr, *H~ Martel* 1226 FF, *Parva Hyneton* 1285 FA], **H~** Gl nr Berkeley [*Hineton* 1220 Fees, *Hinton* Hy 3 Berk], **H~ on the Green** Gl [*Hinhæma gemæru* 1042 KCD 764, *Hinetune* DB, *Hyneton* 1316 FA], **H~ He** nr Hereford [*Hinetone* 1290 Ep], **H~ He** nr Peterchurch [*Hinetune* DB, *Hyniton in Straddel* 1372 AD], **H~ Np** in Woodford Halse [*Hintone* DB, *Hinton* 12 NS, *Hyneton* 1199 FF], **H~ in the Hedges** Np [*Hintone* DB, *Hinton* 1202 Ass, *Hineton* 1254 Val], **H~** Sa nr Cleobury Mortimer [*Hinetone* W 1 or 2 Eyton], **H~ Hall** Sf [*Hinetuna* DB], **H~** So nr Mudford [*Hyneton* 1303 FA, *-e* 1319 AD], **Little H~** W [*Hynyton, Hyneton* 854 BCS 477 f., *Hineton* 1242 Fees].

Some of these Hintons belonged to monastic establishments, H~ on the Green to St. Peter's, Gloucester, Little H~ to St. Swithun, Winchester.

H~ Admiral Ha was held by Reginald de

Albamara in 1242 (Fees) and by William de Fortibus count of Aumâle in the 12th cent. (1313 Ch). *Admiral* is due to popular etymology. The family name is from AUMÂLE in France.—**H~ Ampner** Ha is 'the almoner's Hinton'. He was the almoner of the priory of St. Swithun, Winchester.—*Blewett* in **H~** **Blewett** So is a Fr nickname and family name. OFr *bleuet* means 'bluish'.—*Bower* in **B~** **Hinton** So is presumably OE BŪR (q.v.).—**Broad H~** W; see BRĀD.—**H~ Charterhouse** So was a priory for Carthusian monks founded in 1232.—**Cherry H~** Ca. *Cherry* is simply *cherry* the fruit.—**H~ Daubney** Ha was held by Johannes Aubeny in 1316 (FA). The family name is perhaps from AUBIGNY nr Falaise (Normandy).—**H~ in the Hedges** Np is obscure.—**H~ Martell** Do was held by Eudo Martel in 1212 (Cur). *Martell* is a Fr nickname (OFr *martel* 'hammer').—**H~ St. George** So from the dedication of the church.—**H~ St. Mary** Do belonged to the abbey of St. Mary, Shaftesbury.—**H~ Waldrist** Brk was held by Thomas de S. Walerico in 1192 (P). The name is from ST. VALERY in France (several).

Hints Sa nr Coreley [*Hintes* 1242 Fees, *Hyntes* 1292 QW], **H~ St** [*Hintes* DB, 1199 FF, 1220 Ass]. Welsh *hynt* means 'road'. Hints is an English plural of this. Hints St is on Watling Street.

Hinwick (hinik) Bd [*Hene-, Haneuuic* DB, *Henewich* 1166, *Hennewic* 1167, *-wich* 1175 P]. OE *henna-wīc* 'wīc where hens were kept'.

Hinxhill K [*Haenostesyle* 11 DM, *Henxhille* 1288 Ep, *-helle* 1291 Tax]. '*Hengest*'s hill' or 'the hill of the stallion'. Cf. HENSTRIDGE.

Hinxton Ca [*Hestitone, Histetone* DB, *Hengstetton* 1202 FF, *Henxtenton* 1203 Cur, *Henxton* 1242 Fees]. Hardly simply *Hengestes tūn* '*Hengest*'s TŪN' or *hengestes-* or *hengesta-tūn* 'TŪN with the stallion(s)'. Perhaps *Hengestinga-tūn* 'the TŪN of *Hengest*'s people'.

Hinxworth Hrt [*Hain(ge)steuuorde* DB, *Heingstewurde* 1176 P, *Hengstewrd* 1199 FF]. Either '*Hengest*'s WORÞ' or OE *hengesta-worþ* 'enclosure for stallions'. Cf. HENSTRIDGE.

Hippenscombe W [*Huppingescumbe* 1259 Ipm, *Hippingescumbe* 1292 Cl]. The first el. may be *hippings* 'stepping-stones', though that word is only found in northern dialects.

Hipper R Db [*Hipere* 13, 1350 Derby, *Hypir* 1276 RH], **Hipperholme** YW [*Huperun* DB, *Yperum* 1202 FF, *Hiperum* 1231 FF]. Hipperholme is the dat. plur. of the OE word (**hyper* or the like) that is the base of dial. *hipper* 'osier'. The river-name Hipper may be a derivative of this word or a shortening of earlier *Hyper-ēa* or *-brōc*.

Hipswell YN [*Hiplewelle* DB, *Hipleswell* 1203 Cur, *Hepleswell* 1228 FF]. 'Stream with stepping-stones'? The first el. may be an OE **hyppels* 'stepping-stones', a derivative of **hyppan* 'to hop' from which is formed dial. *hippings* 'stepping-stones'.

Hirst Nb [*Hirst* 1242 Fees], **H~ Courtney, Temple H~** YW [*twa Hyrst* c 1030 YCh 7, *Est-, Westhyrst* 1235 FF, *Hirst Courtenay* 1303, *Templehurst* 1316 FA]. OE *hyrst* 'wooded hill, wood, hill'.

John de Courtney and the Master of the Knights of the Temple held the manors in 1235 (FF). Courtney from COURTENAY in France (several). The Templars got land in Hirst c 1175.

Histon Ca [*Histonona* (sic) 1086 IE, *Histone* DB, *Histon* 1201 Cur, *Heston* 1166 P, 1203 Ass, *Huston* 1188 P, 1206 Cur, 1224, 1250 FF]. The first syllable must have had the vowel *y*. Possibly an OE *Hȳþsæta-tūn* 'the TŪN of the dwellers at the *hȳþ* or landing-place'. H~ is at Beach Dyke.

Hitcham Bk [*Hucheham* DB, 1231 Ch, *Huccham* 1179 RA, *Hucham* 1220 Fees, *Hiccheham* 1382 Pat]. Perhaps '*Hyčča*'s HĀM'. *Hycca* would be related to *Hucca* and the like in HUCKNALL &c. Or possibly *Hwicca-hām* 'the HĀM of the *Hwicce*'. We must then assume that a number of *Hwicce* had settled in Bucks. For the change *Hwičče* > *Hwučče* > *Hučče* cf. such from OE *swilc*, and EDGCOTE.

Hitcham Sf [*Hetcham, Hecham* DB, *Heccham* 1198 FF, *Hecham* 1254 Val]. Identical with HEACHAM, if that comes from *Heccham*.

Hitchin Hrt [(ad) *Hiccam* 944–6 BCS 812, *Hicche* 1062 KCD 813, *Hiz* DB, *Hichene* 1147 TpR, *Hiche* 1197 FF, 1212 Fees, *Hycche* 1230 P]. *Hiccam* BCS 812 no doubt stands for *Hiccum*. This is the dative of the tribal name *Hicce*, which is found (in the gen. form *Hicca*) in the Tribal Hidage (7 BCS 297). Cf. Introd. ii. 1. *Hicce* might possibly be derived from an old name of the river **Hiz**, which may belong to Welsh *sych* 'dry', a word common in stream-names, as SYCHNANT in Wales. The present river-name Hiz is no doubt a late back-formation.

Hittisleigh D [*Hiteneslei* DB, *Huttenesl[e]gh* 1242 Fees, 1275 RH]. '*Hyttīn*'s LĒAH.' *Hyttīn* is a diminutive of *Hyht*, on which see HIXON.

Hive YE [*Hyðe* 959 YCh 4, *Hidon* DB, *Hithe* 1231 FF]. OE *hȳþ* 'landing-place'.

OE hīwan plur. 'members of a household or of a religious house' frequently occurs in pl. ns. in the gen. (in the gen. *higna, hiona*, as HENWICK, HINDON, HINSTOCK, HINTON (2). The usual meaning is no doubt 'members of a religious house', but 'domestics' is sometimes possible. The gen. plur. *hiona* explains the common early form *Hene-* in pl. ns.

OE hīwisc 'family, household; a family-holding of land, hide' is the source of the pl. ns. HEWISH, HUISH. The common occurrence of this name is noteworthy. The meaning may be that of HYDE, i.e. 'homestead consisting of one hide'. *Hīwisc* sometimes occurs as the second el., as in BUCKS. The element is only found in D, So, W.

Hixon St [*Hustedone* DB, *Huchtesdona* 1130 P, *Huhtesdon* 1239 Ass, *Huyhtesdon* 1289 Ass]. '*Hyht*'s DŪN or hill.' *Hyht*, which belongs to OE *hyht* 'hope, joy, pleasure', is found in *Hihtes gehæg* 963 BCS 1106.

Hiz. See HITCHIN.

OScand **hlaða** (ON *hlaða*, OSw *laþa*) 'barn'. See LATHAM, LATHOM, LAYTHAM, ALDOTH, SILLOTH.

OE **hlāw, hlǣw** 'low, hill, mound'. In pl. ns. the meaning ranges from 'mound, burial-mound' to 'hill, mountain'. In names with a pers. n. as first el. the meaning is generally 'burial-mound'. Sometimes names in *-low* refer to mounds used as meeting-places, as in SPELLOW, MUTLOW. The meaning 'hill, mountain' is obvious in names of hills, such as **Horelaw, Pike Law**. The form *hlǣw* is found e.g. in LEW, LEWES. OE *hlāw* gives later *low* or *law*, the latter especially in the north. See LAW-, LOW- (passim). As a second el. the form varies between *-low*, *-loe*, *-law*, and occasional *-ley*, as in KEARSLEY, KIRKLEY Nb. See also WHARLES.

OE **hlēo, hlēow** means 'shelter, refuge', but was also used in a concrete sense. A meaning 'shelter, hut' is obvious in *turfhlēo* 967 BCS 1201, which means 'hut made of turves'. See e.g. LEEFORD, LIBBERY, LYDDEN K. A side-form *hlīeg, hlīg* is found in LAYHAM, perhaps LEYBURN. Another derivative occurs in OE charters, viz. OE *hlywe* or *hlywa* in (on) *ðes cyninges hlywan* KCD 713, (on) *Upicenes hlywan* 1046 KCD 783.

OE **hlid** 'gate'. See HLIÞ. OE **hlidgeat**. See GEAT.

OE **hlīep, hlȳp, hlēp** 'leap' also occurs in concrete senses such as 'leaping-place, a place to be jumped over', perhaps also 'steep slope, abyss' or 'a precipitous fall of a river', but some examples that show senses like these rather belong to OE *hlīepe, hlȳpe* fem., which means 'a place to jump over' and perhaps 'precipitous fall in a river'. See BIRDLIP, LEPTON, LIPWOOD. The sense 'place to jump over, place where a fence can be leapt' is obvious in HARTLIP, HINDLIP. OE *hlīepgeat* 'a leap-gate' is the source of LYPIATT.

OE **hlinc** 'a bank separating strips of arable land on a slope, a rising ground, ridge' is fairly common in pl. ns. But there must also have been a form *hlinč* with palatal *č*, whence dial. *linch* 'rising ground, a ledge' &c. See e.g. LINCH, LINTZ, LINSLADE, MOOR-LINCH, SHANKLIN, STANDLYNCH. There must also have been an OE *hlenc* (from *hlanki-*) with about the same senses as *hlinc*. It is the source of dial. *lench* 'a shelf of rock' and is found in LENCH. A change *nc > ng* is often to be noted, as in LING-WOOD, LYNG, SWARLING, SYDLING, GARLINGE. Cf. also LINKENHOLT, LINTON.

OE **hliþ** n. (plur. *hliþu, hleoþu*) 'a slope, hillside, declivity, hill' sometimes occurs in pl.

ns. The dat. plur. *hliþum, hleoþum* is the source of KIRK-, UPLEATHAM, LYTHAM. AD-GARLEY has lost the final *þ*. There must have been an OE side-form *hlid*, found in LYDD, LYDBURY NORTH, LYDHAM Sa, perhaps YARLET. But there was also an OE **hlid** 'gate'. See CHICKLADE, LIDSTONE. OScand **hliŏ** (ON *hliŏ*, gen. *-ar*, OSw *liþ*) is not uncommon in the North. See LYTH, LYTHE, LITHERLAND, LITHERSKEW, AINSTABLE, KEL-LET, KELLETH.

OE **hlōse** 'a pigsty' is found in LOOSE, LISCOMBE So, LOOSLEY, LOSELEY, LOSCOMBE Do, perhaps LOSTOCK. Cf. Introd. p. xxxi.

OE **hlot** 'lot, share' may be the second el. of HAILEY Hrt, HAYLOT, and the first el. of LADBROOKE; cf. KINLET, SHIRLET.

OE **hlyn** 'maple'. See LINEAL, LINFORD.

OE **hlynn** 'a torrent'. See LYN, -TON, LOWLYNN.

OE **hnecca** 'neck' seems to be used in the sense 'neck of land' in NECTON.

OE **hnoc** 'wether sheep'. See NOCTON, NOTTON.

OE **hnutu** 'nut'. See NUT- (passim), NOTLEY, NURSLING, NURSTEAD.

Hoar Cross St [*Horcros* 1230 P, *Horecros* 1251 Ch, *Harecros* 1242 Fees]. 'Grey cross.' Cf. HĀR.

Hoarstone Wo [*Horeston* 1221 Ass, *Horstan* 1240 WoP]. Identical with HARSTON Le.

Hoarwithy He [*La Horewythy* E 1 BM]. 'The whitebeam.' *Hoar-withy* is still used in some dialects in this sense. The whitebeam or *Pyrus Aria* is 'a small tree having large leaves with white silky hairs on the under side' (OED). The tree is often mentioned in boundaries in OE charters (e.g. *þone haran wiðig* 875, 961 BCS 542, 1066). This has often been held to mean 'boundary willow'.

Hoath K [*la Hathe* 13 StAug, *Hothe* 1422 BM]. OE *hāþ* 'heath', a side-form of HÆP.

Hoathly, East, Sx [*Hodlegh* 1290 Ipm], **West** H~ Sx [*Hadlega* 1121 AC, *Hodlega* 1155 PNSx]. OE *hāþ-lēah* 'heather-covered clearing'. Cf. HOATH.

Hoborough K [*æt Holanbeorge, Holan beorges burna* 838, *Holanbeorges tuun* 841 BCS 418, 437]. Apparently 'hollow mound, mound with a hole in it'.

Hoby Le [*Houcbig* c 1067 Wills, *Hobie* DB, *Houbia* c 1125 LeS, *Hobi* 1183 P, *Houby* 1254 Val]. '*Hauk*'s BY' with early loss of the gen. *s*; cf. HAWKSHEAD.

OE **hōc** 'hook' must have been used also in senses such as 'bend', 'projecting corner' or even 'spur of hill'. Cf. HOOK, HOOKE, WITHYHOOK. A meaning 'enclosure' is found in ME *inhoc, -hok* (see OED) and possibly in some pl. ns., e.g. LIPHOOK. There

may also have been an OE *hūc*, cognate with ON *húka*, MHG *hûchen* 'to crouch'. Cf. HOOK YW.

OE **hocc** 'hock, mallow'. See HOCK- (passim).

Hockenhull Chs [*Hokenhul* 1271, *Hokenul* 1279 Chester]. '*Hoc(c)a*'s hill.'

Hockerill Hrt [*Hokerhulle* Hy 3, *-hille* 1427 PNHrt], **Hockering** Nf [*Hokelinka* DB, *Hokeringhes* 12 RBE, *Hokering* 1205 f. FF, *-e* 1254 Val], **Hockerton** Nt [*Hocretune* DB, *Hochertun* c 1155, *Hocretona* Hy 2 DC, *Hokerinton* 1242 Fees], **Hockerwood** Nt [*Hocer wuda* 958 YCh 2, *Hockerwod* 1329 QW]. Hockerwood and Hockerton are near a marked ridge. The probability is that *Hocker-* represents an old word for 'hill' or 'hump', an OE *hocer*, cognate with G *Höcker*, MHG *hocker, hogger* 'a knob, hump'. Hockerill is then 'hill with a hump', Hockering 'the people at the hill'.

Hockham Nf [*Hocham* DB, 1160 P, 1254 Val, *Hougham* 1204 Cur]. Most likely OE *Hocc-hām* 'HĀM where hocks or mallows grew'. But '*Hocca*'s HĀM' may be thought of.

Hockleton Sa [*Hokelton, Hukelton* 1242 Fees, *Hokeleton* 1332 Misc]. '*Hucela*'s TŪN.' **Hucela* is a derivative of *Hucca* in HUCK-NALL and is found also in HUCCLECOTE.

Hockley Ess [*Hocheleia* DB, *Hockele* 1198 FF, *Little Hocklegh* 1232 FF, *Little Hokkele* 1234 Ch]. '*Hocca*'s LĒAH.' But OE *hocc* 'hock, mallow' is a possible first el.

Hockley Wa [*Hokelowe* 1345, *Huckeloweheth* n.d. AD]. Perhaps '*Hucca*'s mound or hill'. Cf. HUCKNALL, HLĀW.

Hockliffe Bd [*Hocgan clif* 1015 Wills, *Hoc-cliua* 1190 P]. Perhaps '*Hocga*'s cliff', though *Hocga* is unrecorded. If the word *hog* partly goes back to OE *hocga*, it is a more probable first el.

Hockwold Nf [*Hocuuella* DB, *Hocwood* 1198 FF, *-wolde* 1242 Fees]. 'WALD where hocks or mallows grew.'

Hockworthy D [*Hocoorde, Hochaorda* DB, *Hockeworthe* 1274 Ep]. '*Hocca*'s WORÞ(IG).'

Hodcott Brk [*Hodicote* DB, 1242 Fees, *-cot* 1220 Fees]. '*Hoda*'s COT.' *Hoda* is found in *Hodan hlæw* BCS 899, 1121.

Hodder R YW, La [*Hodder* 930 BCS 1344, *Hoder* 1226 FF]. An OW *Hōŏ-ŏufr* 'pleasant stream', the elements being Welsh *hawdd* 'easy', originally 'pleasant, peaceful' and related to Welsh *hedd* 'peace', and *dwfr* 'water, stream'. Cf. HODNET.

Hoddesdon Hrt [*Hodesdone* DB, *Hoddesdone* 1166 RBE, *-don* 1195 P, *Hodesdon* 1212 Fees]. '*Hod*'s DŪN.' Cf. HODSOCK and next name.

Hoddington Ha [*Hoddingatun* 1046 BM, *Odingetone* DB, *Hodingetona* 1219 Fees].

'The TŪN of *Hod*'s or *Hoda*'s people.' For *Hoda* see HODCOTT. *Hod(d)* is found in *Hodes hlæw* BCS 687, *Hodeshliŏ* ib. 1041, *Hodes mære* ib. 1199, *Hoddes stocc* BCS 756.

Hoddlesden La [*Hoddesdene* 1296 Lacy, *-den* 1311 Ipm, *Hodelesdon* 1324 Abbr]. '*Hod*'s valley.' The *l* seems to be intrusive.

Hodnell Wa [*Hodenhelle* DB, *-hull* 1196 P, 1236 Fees]. '*Hoda*'s hill.' Cf. HODCOTT.

Hodnet Sa [*Hodenet, Odenet* DB, *Hodnet, Hodenet* 1230 P]. Identical with early Welsh (*Glyn*) *Hodnant* Rees 108. Hodnant consists of OW *hōŏ* 'pleasant, peaceful' (cf. HODDER) and *nant* 'valley, stream'. OW *hōŏ* (Welsh *hawdd*) corresponds to Co *hueth* 'tranquil', which is the first el. of *Hennon* Co (St. Breward) [*Hethnant* 1400 AD]. The second *n* of *Hodnant* was lost owing to dissimilation as in *Sekenet* 1256 Ass, identical with *Sechenent* 1169 ff. Lanercost, the lost name of a stream in Cu. *Sekenet* corresponds to SYCHNANT 'dry brook' in Wales.

Hodsock Nt [*Odesach* DB, *Hodeshac* 1188 P, *Hoddeshac* 1242 Fees]. '*Hod*'s oak'; cf. HODDESDON. Curiously enough the same name is found in Wo (*Hodes āc* BCS 1282).

Hodson W [*Hodeston* 1223 FF, 1315 Pat]. '*Hod*'s TŪN.' Cf. HODDESDON.

Hoe, East, Ha [*Hou* DB, 1167 P, *Ho* 1242 Fees], **West Hoe** Ha [*How* 1236 VH], **H~** Nf [*Hou* DB, *Ho* 1165, *Hó* 1166 P, *la Hoge* 1200 Cur]. OE HŌH 'spur of hill, ridge'.

Hoff We [*Hofes* c 1160 YCh 175, *Hof* c 1200 (1294) Ch]. OE or ON *hof* 'house, temple'. Quite possibly a Scandinavian heathen temple is referred to. Hof is a very common pl. n. in Scandinavia, but *hof* hardly occurs in native English names.

Hofflet Li [*Holflet* 1175, *-fliet* 1197 P]. 'Hollow or deep stream.'

ON *hǫfuŏ*, OSw *huvup*, Dan *hoved*, ON *hǫfŏi* 'head' are also used in the senses 'promontory, projecting hill or ridge'. The el. is found in Engl pl. ns. See e.g. ESCOWBECK, PREESALL, HOLLETH, WHITEHAVEN, and cf. HOWDEN.

Hoggeston Bk [?*Hocgestun* c 1000 Wills, *Hochestone* DB, *Hoggeston* 1200 Cur]. First el. OE *hogg* 'hog' or *Hogg*, an unrecorded pers. n. The identification of the OE form is doubtful, as the charter is a Somerset one, but at any rate it is etymologically identical with Hoggeston.

Hoghton (-aw-) La [*Hoctonam* c 1160 LaCh, *Houton* 1227 Ass]. OE *Hōh-tūn*. See HOUGHTON (1).

Hognaston Db [*Ochenauestun* DB, *Hokenaston* 1241 RA, 1446 BM]. A curious name. A possible solution is an OE *Hoccan æfesn-tūn*. OE *æfesn* means 'pasturage'. *Hocca* is a pers. n. The meaning would be '*Hocca*'s grazing-farm'.

Hog's Back. See GUILDFORD.

Hogshaw Bk [*Hocsaga* DB, *-shaŏe* 1166 P, *Hoggeshag* 1199 Cur, *-shawe* 1199 P]. 'Copse where hogs (or wild boars) were found.' Cf. HOGGESTON.

Hogsthorpe Li [*Hocges-, Hoggestorp* 12 DC, *Hoggestorp* 1195 P, *-thorp* 1242 Fees]. '*Hogg*'s thorp.' Cf. HOGGESTON.

OE **hōh** 'heel; projecting ridge of land', dial. *hoe, heugh* 'crag, cliff, precipice, a height ending abruptly'. In pl. ns. the meaning varies from 'steep ridge' to 'slight rise'. The OE inflexion was *hōh*, gen. *hōs*, dat. *hō*, plur. *hōs*, gen. *hō*, dat. *hōm*. Later were formed gen. *hōges*, dat. *hōge*, plur. *hōas, hōgas* &c. The word is common as a first and a second el. and it occurs alone, as HOUGH (from *hōh* nom.), HOE, HOO(E) (from *hō* dat.), HOSE (from *hōs* or *hōgas* plur.). As a first el. it appears in the stem-form *hōh* in HOGHTON, HOUGHTON, in the form *hō* in HOOTON, HUTTON (perhaps from *hō* gen. plur.), HOLLAND, HOYLAND &c. As a second el. it mostly appears as *-hoe*, as FINGRINGHOE (from *hō* dat.), sometimes as (*h*)*ow*, as RAINOW, SCOTTOW. Cf. also CORNSAY, KEW, LUBBENHAM, STENIGOT. It is sometimes difficult to distinguish *hōh* from OScand *haugr*.

Holbeach (hōlbēch) Li [*Holebech* DB, *-e* 1170 P]. 'Hollow, i.e. deep, brook.' Cf. BÆCE.

Holbeam. See HOL(H) adj.

Holbeck Nt [*Holebek* c 1180, *Holbek* 1332 PNNt(S)]. 'Hollow, i.e. deep, brook.' The brook, to which the name was originally applied, was OE (on) *holan broc* 958 YCh 2. The OE name has been Scandinavianized.

Holbeck Nt, vil. [*Hollebec, Holebec* 1227 PNNt(S)]. Identical with the other Holbeck, which is in Southwell.

Holbeton (hŏb-) D [*Holbouton* 1229 FF, *Holbogatone* 1256 Ep]. 'TŪN in or by a hollow bow.' A bend may be referred to.

Holborn (-ōb-) Mx [*Holeburne* DB, (pons de) *Holeburn* 1191 P], **Holbrook** Db [*Holebroc* DB, (aqua de) *Holebrok* 1280 Ass], **H~** Do [(on) *Holambrok* 968 BCS 1214, *Holebrok* 1412 FA], **H~** Sf [*Holebroc* DB, 1177 P]. 'Hollow brook', i.e. 'brook running in a deep ravine'. Some of the names are also recorded as the name of the brook at the place, e.g. *Holeburne* 959 BCS 1351 (Mx).

Holbury Ha in Fawley [*Holeberi* 1187 ff. P, *Holebury* 1316 FA], **H~** Ha in E. Tytherley [*Holebury* 1245 Ch, *-biry* 1270 Ch]. 'Low-lying BURG' does not seem very probable. Possibly 'old fort with breaches in its walls'. A possible first el. is a pers. n. *Hōla*. Cf. HOL(H) adj.

Holcombe D in Dawlish [(æt) *Holacumbe* c 1070 Ex, *Holecomma* DB], **H~** (hōkʊm) **Burnell** D [*Holecumba* DB, *Holecumbe*

Bernard 1263 Ep], **H~ Rogus** D [*Holancumbes landscare* 958 BCS 1027, *Holecoma* DB, *Holcombe Roges* 1281 Ass], **H~** Do [*Holancumb* 998 KCD 701, (æt) *Holancumbe* 1002–14, c. 1006 ib. 708, 1302], **H~** Gl [*Holecumbemed* 13 Berk], **H~** La [*Holecumba* a 1236 Whitaker, *Holcoumbe* 1296 Lacy], **H~** O [*Holecumba* Hy 2 (1320) Ch, -*cumbe* 1231 Ch], **H~** So nr Kilmersdon [*Holecumbe* 1243 Ass, 1276 RH]. 'Hollow, i.e. deep, coomb or ravine.' Burnell is a late modification of the pers. n. Bernard. Ralph son of Bernard held **H~ Burnell** in 1242 (Fees).—**H~ Rogus** was held by *Rogo* in 1086 (DB). *Rogo*, also *Rogus*, is an OFr pers. n. of German origin (OG *Roggo*).

Holcot (-ŭk-) Bd [*Holacotan* 969 BCS 1229, *Holecote* DB], **H~** Np [*Holecote* DB, 12 NS, 1195 P]. Apparently 'COT in the hollows', *hola* being the gen. plur. of OE *hol*.

Holcroft La [*Holecroft* 1246 Ass]. 'Croft in the hollows.' Cf. HOLCOT.

Holden YW [*Holedene* DB, *Holden* 13 Pudsay]. 'Hollow, i.e. deep, valley.'

Holdenby Np [*Aldenesbi* DB, *Haldenebi* 1170, 1190 P]. Identical with HALDENBY.

Holdenhurst Ha [*Holehest* DB, *Holehurst* 1172 P]. 'Holly wood.' Cf. HOLEGN.

Holderness YE [*Heldernesse* DB, -*neis* 1130, -*nes* 1170 P, *Heoldernessa* 1160–2 YCh 1307, *Hildernessa* c 1130 BM, *Heuderness* 1228 Ep; *Holdernessa* 1166 P, -*ness* 1208 FF]. Has been explained as 'the ness of the hold'. A *hold* was a man of high rank in the Danelaw. The word is ON *hǫldr*. This etymology is very likely correct, but the variation of the early forms offers difficulties.

Holdfast Wo [*æt Holenfesten, æt Holanfæstene* 967 BCS 1204 f., *Holefæst* 11 Heming, -*fest* DB]. The second el. is OE *fæsten* 'a stronghold'. The first looks like a form of OE *holh* 'hollow', but this does not give a likely meaning. Either it is OE *holegn* 'holly' (*Holan-* is then corrupt) or less likely it is a pers. n. *Hōla*, corresponding to OHG *Huolo*.

Holdgate Sa [*Stantune* DB, *castellum Holo-goti* 1185 TpR, *Castrum Holegot* 1242 Fees, *Castrum de Holegot* 1277 Ep, *Holgod* 1327 Subs]. *Stantune* was held by *Helgot* (DB). *Helgot* is OFr *Helgot* from OG *Helgaud*, *Hildegaud*. *Castrum Holegot* or *Castel Holegod* 'Holegot's castle' was taken to mean 'Holegot castle', and the pl. n. *Holgod*, Holdgate was evolved.

Holdingham Li [*Haldinge-, Holingham* 1202 Ass, *Haldingham* ib., 1276 RH, 1316 FA]. 'The HĀM of *Hald*'s people.' Cf. HOLSWORTHY.

Holditch Do [*Holedich* 1242 Fees, 1247 Misc]. 'Deep ditch.' Cf. HOL(H).

Holdsworth YW [*Haldewrth* 1276 RH], **Holdworth** YW [*Haldewrde* DB, -*wrth* 1297 Subs]. '*Halda*'s WORP.' *Halda* is an unrecorded pers. n. derived from OE *h(e)ald* 'bent'. Cf. HOLDINGHAM, HOLSWORTHY.

OE **holegn** 'holly, holm-oak' is found alone as a pl. n., as HOLME Do, YW, HOLNE D, HULNE Nb, and fairly often as a first el., e.g. in HOLDENHURST, HOLDFAST, HOLLINFARE, HOLLINGTON, HOLLINGWORTH, HOLMFRITH, HOLMWOOD, HOLNEST.

Holford So [(æt) *twam Holaforda* 11 KCD 897, *Holeforde* DB, -*ford* 1176 P]. Cf. (on) *holan ford* 956 BCS 945 (O), evidently identical in origin. The name means 'hollow ford', i.e. perhaps 'ford in a deep valley'.

Holgate YW [*Holegate* 1200 P, *Holgate* 1218 Ep]. 'Hollow road.' Second el. OScand *gata* 'road'.

OE **hol(h)** adj. 'hollow' is common in pl. ns., but rarely in its original sense, as in **Holbeam** D [*Holebema* DB] 'hollow tree', HOBOROUGH K (also H~ HILL Np) 'hollow mound or hill', HOLBURY. In pl. ns. the usual sense is 'sunken, deep', in names of streams 'running in a deep ravine'. It is mostly combined with words for a stream or valley, as in HOLBORN, -BROOK, -WELL, -COMBE, -DEN, -DITCH, or a lake, as in HOLMER, or a road, as in HOLLOWAY, HOLWAY. Cf. also HOLFORD. It is doubtful if it could be used in the sense 'low-lying' of a homestead or hill or the like, as in HOLTON, HOLWORTH. It is quite possible that for such names we have to reckon with a pers. n. *Hōla*, corresponding to OHG *Hōlo*, *Huolo*.

OE **hol(h)** sb. 'hole, deep place in water; cave, burrow' must also have been used in the sense 'hollow, depression in the ground'; cf. Scotch *howe* 'hollow place or depression'. The sense 'burrow' is found in BROCKHALL &c., FOXHALL, FOXHOLES. 'Hollow' is the sense in GREENHALGH, INGOL, TOCKHOLES and others. Partly the source may here be OScand *hol*. *Holh* is the first el. of HOLCOT, perhaps HOLCROFT. Cf. also HOOLE, HULSE Chs. On ON *hóll* 'hill', which may sometimes be suspected to be the source of names in *Hol-*, see HVĀLL.

Holker (-ōk-) La [*Holkerre* 1276 Ass, -*ker* 1342 FF]. 'Hollow marsh', 'marsh with hollows or depressions'. See KERR.

Holkham Nf [*Holcham* DB, 1159, 1162 P, 1203 Ass]. First el. OE *holc* 'hollow, cavity'. A lake in the park may be referred to. Second el. HĀM.

Hollacombe D [*Holecome* DB, *Holecoumb* 1276 RH]. Identical with HOLCOMBE.

Holland Ess [*Holand* c 1000 CCC, *Hoilanda*, *Holanda* DB, *Parva Hoilande* 1212 RBE, *Hoylande Magna* 1238 Subs], **Downholland** La [*Holand* DB, *Dunholand* 1298 LaInq], **Upholland** La [*Hoiland* DB, *Upholand* 1226 La Inq], **Holland** Li [*Hoiland* DB, 1156, 1190 P, -*e* 1130 P]. OE *hō(h)-land* 'land on or by a HŌH or spur of hill'.

Hollesley (hōzlĭ) Sf [*Holeslea* DB, 1177 ff.
P, *-lega* 1186 P, *-le* 1254 Val]. First el.
possibly OE HOL(H) 'a hollow'. But the
forms rather suggest an OE pers. n. **Hōl.*
Cf. HOLH, where a pers. n. *Hōla* is conjectured.

Holleth La [*Holout* 1242, *Holauth* 1320 CC].
OScand *hol-hǫfuð* 'hill with a hollow'.
There are some ponds on the hill.

Hollinfare La [*Le Fery del Holyns* 1352 VH,
Hollynfare 1556 FF]. 'The ferry by the
hollies.' Second el. OE *fær* 'passage', here
'ferry'.

Hollingbourne K [*Holingeburna*, (to) *Hol-
inganburnan* 10 BCS 1321 f., (æt) *Holunga-
burnan* 1015 Wills, *Hoilingeborde* DB]. 'The
stream of *Hōla*'s people' (cf. HOLH adj.), or
possibly 'the stream of the people dwelling
in the HOLH or hollow'.

Hollington Db [*Holintune* DB, *-ton* 1252
Ch], **H~** St [*Holyngton* 13 PNSt], **H~** Sx
Holintun, Horintone DB, *Holintuna* 12
AD]. OE *Holegn-tūn* 'TŪN among hollies'.

Hollingworth Chs [*Holisurde* DB, *Holin-
wrth* 1285 Court], **H~** La [*Holyenworth* 1278
FF]. 'Holly WORP.'

Holloway Mx [*Holwey* 1480 PNMx]. 'Hollow or sunken road.' Cf. HOLH adj.

Hollowell Np [*Holewelle* DB, 12 NS, *-well*
1242 Fees]. 'Deep stream.' Cf. HOLH adj.

Hollym YE [*Holam, Holun* DB, *Holume* 1260
Ipm, *Holaym* 1297 Subs, 1316 FA, *Holm*
1292 Ch]. The place is close to **Holmpton**
[*Holmetune, Ulmetun* DB, *Holmetona* 1160–2
YCh 1307]. This appears to be OE *Hol-
hǽma-tūn* 'the TŪN of the Hollym people'.
Hollym would then be OE *Hol-hām* or
rather *Hol-hamm* 'HĀM or HAMM in a hollow'. *Holhamm* or *-homm* might have become *Holum* early.

Hollytreeholme or **Hallytreeholme** YE
[*Halitreholm* c 1180 YCh 1410, 1290 Ch].
'Island with a holy tree.'

Late OE **holm** from ON *holmr*, OSw *holmber*,
Dan *holm*, also OScand *holmi* 'small island',
'a piece of dry land in a fen, a piece of
land partly surrounded by streams or by
a stream'. It is common as the second el.
of pl. ns. (as AXHOLME) and is often used
alone (see HOLME). In modern forms it is
sometimes exchanged for *-ham*. Conversely
Holme sometimes appears for original HAMM
(as HOLME LACY He) and *-holm* for *-ham*,
as BLOXHOLM, DUNHOLME. Instead of
Holme, -holme sometimes appears *Hulme,
-hulme*. This is the ODan, OSw by-form
hulm. See HULME, LEVENSHULME.

Holme Bd [*Holme* DB, (in) *Hulmo* 1179 P],
Holme St. Cuthbert Cu [*Sanct Cuthbert
Chappell* 1538 PNCu(S)], **H~** Db nr
Bakewell [*Holun* DB, *Hulm* 1278 Misc,
Holm by Bauquell 14 Derby], **H~** Db nr
Brampton [*Holun* DB, *Hulme* 1258 FF,

Hulm Hy 3 BM], **H~** Hu [*Hulmus* 1217 Pat,
Holme 1252 Ch], **H~** La [*Holme* 1305 Lacy,
Holm 1311 LaInq], **H~** Li nr Brigg [*Holm*
DB, c 1115 LiS], **H~** Spinney Li [*Holm*
DB, *Holma* c 1130, *Holmum* 1139 RA], **H~**
Nf nr Norwich [*Holm* c 1025 ff. Wills, *Hul-
mus* 1158 P], **H~** Hale Nf [*Holm* 902 ASC
(C), c 961 BCS 1064, DB, *Hale, Holm* 1254
Val, *Holmhel* 1267 Misc], **H~** next Runc-
ton Nf [*Holm* 1254 Val], **H~** next the
Sea Nf [*Holm* c 1035 Wills, DB, *Hulmum*
Hy 3 BM], **H~** Nt [*Olm* 12 BM, *Holme*
1316 FA], **H~** Pierrepont Nt [*Holmo* DB,
Holm 1211–13 Fees, *Holme Peyrpointe* 1571
BM], **H~** We [*Holme* DB, *Holm* c 1190
Kendale], **H~** YE [*Holm* DB, *Holm by
Pagle* 1285 Ch], **H~** upon Spalding Moor
YE [*Holme* DB, *Spaldiggeholm* Hy 2 BM,
Holm in Spaldingmor 1293 QW], **H~** YN
in Pickhill [*Hulme* DB, *Holm* 1088 LVD],
North H~ YN [*Holme, Holm* DB, *Northolm*
1208 FF], **South H~** YN [*Holm, -e* DB].
OScand *holmr* 'island' &c. Cf. HOLM.

Holme Hale Nf seems to be due to amalgama-
tion of two places, Holme and Hale.—**H~
Pierrepont** was held by Annora de Perpunt
in 1303 (FA). Her husband Sir Henry Pierpont
seems to have acquired the manor. The name
is from PIERREPONT in France (several).

Holme Cultram Cu [*Culterham* c 1130 SD
(s.a. 854), *Holmcultran* 1153 SD, *Holme-
coltrame* c 1220 FC, *Holmcoltran* 1290 Cl].
'The holm belonging to *Culterham*.' The
order of the elements is Celtic; cf. ASPATRIA.
The OE name was *Culterhām*. First el.
very likely Welsh *cul-dir* (from *-tir*) 'narrow
strip of land, isthmus'. Cf. COULDERTON.
The place is on a long low ridge.

Holme, East & West, Do [*Holna* DB,
c 1107 BM, *Holne* 1242 Fees, *Westholme*
1316 FF], **H~** YW nr Holmfirth [*Holne*
DB, 1274 Wakef]. OE *holegn* 'holly'. The
change *n > m* in Holme YW took place in
HOLMFIRTH (q.v.).

Holme Lacy He [*Hamme* DB, *Hamme
Hugonis de Laci* 1167 P, *Hamma* 1190 P,
Homme Lacy 1221 Hereford, *Hamme Lacy*
1242 Fees]. OE HAMM 'low-lying meadow'
&c.

Roger de Laci held the manor in 1086 (DB).
Cf. EWYAS LACY.

Holme on the Wolds YE [*Hougon* DB,
Hogum 1100 YCh 965, *Haum* c 1135 ib.
970, *Howum* 1202 FF, *Holme super Wolde*
1578 BM]. OScand *haugum*, dat. plur. of
haugr 'hill'.

Holmer (*-ōm-*) Bk [*Holemere* 1208 Cur,
Holmere 1311 Ipm], **H~** He [*Holemere* DB,
Holemare 1273 Misc]. 'Mere in a hollow.'
Cf. HOLH adj.

Holmescales We [*Eschales* 1201 CC, *Holme
Scales* 1297 Kendale]. 'The shielings belonging to HOLME.' Cf. SKÁLI.

Holmesfield Db [*Holmesfelt* DB, *-feld* c
1160 BM]. 'FELD belonging to HOLME', i.e.

Holme nr Brampton, which is only a few miles distant.

Holmfirth YW [*Holnefrith* 1274 Wakef, *Holmfrithes* 1328 AD]. 'The woodland belonging to Holme.' Cf. HOLME YW and FYRHÞ. Before the labial *f n* became *m*.

Holmpton YE. See HOLLYM.

Holmwood Sr [*Homwud* 1241 Cl, *-wude* 1243 Misc]. 'Wood in a HAMM or low-lying river land.' Other Holmwoods are probably OE *holegn-wudu* 'holly wood'.

Holne (hōl) D [*Holle* DB, *Holna* 1178 P, 1198 FF]. OE *holegn* 'holly'.

Holnest Do [*Holeherst* 1185, 1194 P, *Holnehurste* 1279 For]. 'Holly wood.' Cf. HYRST.

Holsworthy (hōlzerï) D [*Haldeword*, *-urdi* DB, *Haldewurth* 1228 FF, *Halleswrthia* 12 BM, *Haldesworth* 1242 Fees, *Holdesworthe* 1308 Ep]. '*Heald*'s WORÞIG.' *Heald* is an unrecorded pers. n. derived from OE *heald* 'bent', found also in **Halsworthy** D [*Haldeswurthy* 1249 Ass].

OE **holt** 'wood' as a second el. is mostly combined with a tree-name, as ACOL, OC-COLD, KNOCKHOLT, AISHOLT, BIRCHOLT, BUCKHOLT. HOLT is a fairly common name: **Holt** Do [*foresta de Winburne* DB, *Winburneholt* 1185 P, *Holte* 1313 Ch], H~ Ha [*Holt* 1167 P], **Nevill** H~ Le [*Holt* 1166 RBE, 1220–35 Ep], H~ Nf [*Holt* DB, 1242 Fees], H~ So [*Holt* 1225 Ass], H~ St [*Hout* 1247 Cl], H~ W [*Holte* 1242 Fees, *Holt* 1252 Ch], H~ Wa [*Holtto* c 1200 Middleton], H~ Wo [*Holte* DB]. R. de Nevill was patron of Nevill H~ in 1220–35 (Ep). Cf. FIFEHEAD NEVILLE. See WIMBORNE.

Holtby YN nr York [*Boltebi* DB, *Holtebi* c 1125 YCh 936], H~ YN in Ainderby Mires [*Holtebi* DB, *-by* 1231 FF]. '*Holti*'s BY.' *Holti* is a common OScand pers. n.

Holton has four or five distinct sources: 1. OE *Hōh-tūn* 'TŪN on a spur of land': H~ Li nr Beckering [*Houtune* DB, *-tuna* c 1115 LiS, *-ton* 1202 Ass], H~ le Clay Li [*Holtun* DB, *Houtuna* c 1115 LiS, *Hocton* 1202 Ass], H~ le Moor Li [*Hoctune* DB, *Houtuna* c 1115 LiS, *Houton* 1202 Ass, H~ *in la More* 1331 Ch]. 2. OE *H(e)alh-tūn* 'TŪN in or by a HALH or remote valley': H~ O [*Healhtun* 956 BCS 945, *Halcton* 1192 P, *Halton* 1223 Misc], H~ So [*Healhtun* c 1000 Wills, *Halton* 1219 FF]. 3. OE *Holh-tūn* 'TŪN by a hollow' or *Holt-tūn* 'TŪN by a holt': H~ Do [*Holtone* DB]. H~ is near H~ Mere. 4. H~ Sf nr Halesworth [*Holetuna* DB, *-tun* 12 BM, *-ton* 1254 Val], H~ **St. Mary** Sf [*Holetuna* DB, *-ton* 1254 Val, 1258 Ch, *-tun* 1270 Ipm]. 'TŪN in hollows' hardly suits H~ St. Mary, and the two names very likely mean '*Hōla*'s TŪN'. Cf. HOLH adj.

Holverston Nf [*Huluestone kirke* c 1060 Wills, *Holuestuna* DB]. '*Holmfast*'s TŪN.' First el. ON *Holmfastr*, OSw *Holmfast*, *Holuaster* pers. n.

Holway So [*Holeweie* 1225 Ass, *-weye* 1247 FF]. Identical with HOLLOWAY.

Holwell Do nr Cranborne [*Holewella* 1194 P], H~ Hrt [*Holewelle* 969 Crawf, (in) *Holewelle* 1066 KCD 824, *Holewell* 1236 Fees], H~ Le [*Holewelle* DB, *-well* c 1125 LeS, 1200 Cur]. 'Stream in a deep valley.' Cf. HOLH adj.

Holwell Do nr Broadway [*Halegewelle* DB, *Brodewaye Hallewolle* 1285 FA, *Halghwell* 1307 FF], H~ O [*Haliwelle* DB, 1209–35 Ep]. 'Holy spring or stream.'

Holwell Do nr Sherborne [*Holewala* 1188 P, *-wal* 1194 P, *-wale* 1196 P, 1212 Fees, *Holewale in Blakemor* 1251 Misc]. The second el. is the OE *walu* mentioned under EASOLE K. The word is held to have meant 'a ridge of earth or stone'. It is found in the compounds *dicwalu* and *stanwalu* (B-T Suppl). There is good reason to believe that the word was also used of a fence or enclosure. Holwell may be explained as 'sunk hedge, ha-ha'.

Holwick YN [*Holewyk* 1235 FF, *-wic* 1251 Ch]. Second el. OE WĪC 'dairy-farm' or the like. The first may be *hol* sb. 'hollow' or *holegn* 'holly'.

Hŏlworth Do [(at) *Holewertþe* 939 BCS 738, *Holverde* DB, *Holewrdhe* 1212 Fees]. The OE ex. is in a very poor text. Names in *-worth* mostly have a pers. n. as first el. Hence very likely '*Hōla*'s WORÞ'. Cf. HOLH adj. But *holh* 'hollow' is a possible first el.

Hŏlybourne Ha [*Haliborne* DB, *-burna* 1167 P]. 'Holy stream.'

Holy Island Nb [*Healand* c 1150 SD, *Halieland* 1195 (1335) Ch]. Self-explanatory. There was a famous abbey on the island. Cf. LINDISFARNE.

Holy Oakes Le [*Haliach* DB, 1187 P, *Halyok* 1396 BM]. 'Holy oak.'

Holyport Brk [*Horipord* 1220 Fees]. 'Dirty market town.' The change to *Holy-* is probably intentional.

Holystone Nb [*Halistan* 1242 Fees, *Halystane* 1254 Val]. 'Holy stone.' The place was the site of a Benedictine abbey. But the abbey was probably built here because of a stone with religious associations, perhaps one at which the gospel had been preached. Cf. BOSTON.

Hŏlywell Hu [*Haliewelle* DB], H~ K [*Haliwelle* 13 StAug], **East & West** H~ Nb [*Halewell* 1218 P, *Haliwell* 1242 Fees]. 'Holy spring.'

Holywell Li [*Helewelle* 1190 P, *-well* 1212 Fees, 1256 Gilb]. 'The wishing-well.' First el. OE *hǽl* 'omen'. Cf. ELWELL

Homersfield (-ŭ-) Sf [*Humbresfelda* DB, *Humrefeld*, *Humbresfeld* c 1130 BM, *Humeresfeud* 1254 Val]. '*Hūnbeorht*'s FELD.'

Homerton Mx [*Humburton* 1343 PNMx]. '*Hūnburg*'s TŪN.' *Hūnburg* is a woman's name.

Homington W [*Hummingtun* 956 BCS 962, *Humitone* DB, *Humintona* 1130, *Huminton* 1190 P]. 'The TŪN of *Humma*'s people.' **Humma* may be a short form of *Humbeald* for *Hūnbeald* &c.

Honeyborne, Cow, Gl [*Hunniburne* 1221 Ass, *Huniburn* 1251 Ch], **Church H~** Wo [*Huniburna* 709 BCS 125, *-burne* DB]. The two Honeybornes are close together. They are named from a stream called *Hunigburna* KCD 1368. 'Honey stream, stream on whose banks honey could be gathered.' Cf. *Hunigburna* 840 BCS 428, the lost name of a stream near Crowle Wo. The two Honeybornes are referred to as '*æt Hunigburnan twegen Weorþias*' 817 BCS 361. *Cow* is no doubt *cow* the animal.

Honeychild K [*Hunechild* c 1150 Fr, 1168 P, *Honi-*, *Hunichild* 1227 ff. Ch]. '*Hūna*'s spring.' Second el. OE *celde*. Cf. BAPCHILD.

Honeychurch D [*Honechercha* DB, *Hunichurche* 1242 Fees]. '*Hūna*'s church.'

Honeywick So nr Bruton [*Hunewic* 12 Bruton, *-wica* c 1155 Fr, *-wic* 1207 Cur]. '*Hūna*'s WĪC.'

Honicknowle D [*Hanechelole* DB, *Hanecnolle* 1242 Fees]. '*Hana*'s knoll' or 'knoll frequented by wild cocks'. See HANA.

Honiley Wa [*Hunileg* 1207 Cur, *-legh* 13 BM]. 'Wood where honey was got.'

Hōning Nf [(aecclesia de) *Hanninge* 1044-7 KCD 785 (*Haninge* Holme), *Haninga* DB, *Haninges* c 1150 Crawf, 1231 Ch]. OE *Hāningas* 'the people at the HĀN or rock'. Very likely *hān* here means 'hill' and refers to the small hill at the place.

Honingham (-ŭ-) Nf [*Hunincham* DB, *Huningeham* c 1184 Fr, *Huningham* 1202 FF, 1227 Ch]. 'The HĀM of *Hūn(a)*'s people.'

Honington Li [*Hundintone* DB, *Hundingtune* 1172 BM, *-ton* 1212 Fees]. 'The TŪN of *Hund*'s people.' Cf. HOUNSLOW.

Honington Sf [*Hunegetuna* DB, *-ton* 1254 Val, *-tune* Hy 3 BM, *Honeweton* 1305 BM]. 'The TŪN of *Hūn(a)*'s people.' For the change of *Hūninga-* to *Hūnege-* &c. cf. CANEWDON. The early forms forbid derivation of the first el. from OE *hunig* 'honey'.

Honington Wa [*Hunitona* 1043 Th, *Honington* c 1050 KCD 939, *Hunitone* DB, *Honyton* 1257 Ch], **Honiton** (-ŭ-) D in S. Molton [*Hunitona* DB]. OE *Hunig-tūn* 'honey TŪN', 'homestead where honey was produced'.

Honiton (-ŭ-) D par. [*Honetone* DB, *Huneton* 1211 FF, 1230 P]. '*Hūna*'s TŪN.'

Honiton, Clyst, D [*Hinatun* c 1100 E, *Hinetun* 1219 FF, *Clisthineton* 1291 Tax]. OE *Hīgna-tūn* 'the TŪN of the monks'. The place belonged to Exeter Cathedral.

Honley YW [*Haneleia* DB, *-lay* 1242 Fees, *Honeley* 1252 Ep, 1274 Wakef]. OE *hāna-lēah* 'LĒAH with stones or rocks' (OE *hān* 'stone, rock').

Hoo, St. Mary's H~, H~ St. Werburgh K [*Hogh*, (in) *Hoge* c 700, Æt *Hoe* 697, (æt) *Hó* 964-95 BCS 89, 91, 1132, *How hundr.* DB, *Ho* 1161 P; *Ho St. Mary* 1272 Ipm, *Hoo St. Werburga* 1314 FF], **H~** Sf [*Ho* c 1050 KCD 907, *Hou* DB, *Ho* 1198 (1253) Ch]. OE *hōh*, dat. *hō(e)* 'spur of land'.

Hood D [*Hode* 1242 Fees], **H~** YN [*Hode* 1143 Mon, *Hod* 1218 Cl, 1239 FF, *Houd* 1235 FF]. OE **hōd*, identical with OFris *hōde*, OHG *huota*, G *Hut* 'care, protection', and the base of OE *hēdan* 'to heed'. OE *hōd* must have meant 'protection', but also 'shelter'. Cf. HOTHAM.

Hooe D [*Ho* DB, 1237 Cl, *Hoo* 1199 Cur], **H~** Sx [*Hou* DB, *Ho* W 1 (1312) Ch]. Identical with HOO.

Hook Ha nr Titchfield [*Houch* DB, *la Hoke* 1167 P, *Hok* 1242 Fees], **H~** Ha nr Basingstoke [*Hoc* (wood) 1223 AD], **H~** Sr [*Hoke* 1227 ff. PNSr], **H~** W [*la Hok* 1238 Cl, *La Hoke* 1316 AD]. OE *hōc* 'hook'. The meaning of the word varies. In Hook nr Titchfield *hōc* seems to mean 'headland'. Hook Sr, W are on hills. The meaning of the other Hook Ha is not quite clear. Cf. HOOKE.

Hook YW [*Húc* Hy 2 DC, *Huuc, Huck* n.d. Pont, *Huc* 1208 FF, 13 BM, *Huuc* 1294 Ch]. Possibly OE *hōc*, but more likely OE **hūc*, cognate with ON *húka* 'to crouch'. Hook is in a pointed piece of land formed by a bend of the Ouse.

Hooke Do [*Lahoc* DB, *Hoc* c 1100 Montacute, 1200 Cur, *Hoka* 1230 P]. OE *HŌC*; cf. HOOK. Hooke seems to have been named from a bend of the river Hooke and its valley. **Hooke** R is a back-formation. Cf. TOLLER.

Hook Norton O [*Hocneratun* 917 ASC, *Hocceneretun* ib. D, *Hochenartone* DB, *Hokenarton* Hy 2 (1267) Ch, *-a* c 1130 Oxf, *-e* 1225 Ep, *Hoke Norton* 1291 Tax]. The first el. is clearly the genitive of a folk-name. It may be suggested that the original name was *Hoccan ōra* '*Hocca*'s hill slope', from which was formed *Hoccanēre* 'the people at *Hoccanōra*'. Hook Norton would then mean 'the TŪN of the people at *Hoccanōra*'. The same first el. is found in *Hokernesse* c 1260, c 1270 Osney. The locality was in Hook Norton and apparently an eminence (land upon and under *Hokernesse*). The spur of hill by Hook Norton may well be meant. The el. *-nesse* is OE *næss* 'headland'.

Hoole Chs [*Hole* 1119, c 1150, 1268 Chester, 1288 Court]. OE *holh* 'hollow'. The immediate source is the dat. *hōle*, in which *o* had been lengthened owing to loss of *h*.

Hoole, Much & Little, La [*Hull, -e* 1204 FF, *Hole* 1212 Fees, 1246 Ass, *Litlehola* c 1200, *Magna Hole* c 1235 CC, *Little Hoole* 1423 FF]. OE *hulu* 'husk' gives ME *hule*, which is recorded also in the sense 'hut, hovel'. This is the source of Hoole.

Hoon Db [*Hougen* DB, *Howyn, Howene* 1275 RH, *Howen, Hawen* 1280 Ass]. OScand *haugum*, dat. plur. of *haugr* 'hill, mound'.

Hooton Chs [*Hotone* DB, *Hoton* c 1260 Chester], **H~ Levett** YW [*Hotone* DB, *Hoton Livet* 1242 Fees], **H~ Pagnell** YW [*Hotun* DB, *Hotton Painel* 1192, *Hoton Painell* 1195 P], **H~ Roberts** YW [*Hotun* DB, *Hoton Robert* 1285 FA]. OE *Hō-tūn*. See HUTTON.

Custancia Livet held **H~ Levett** in 1242 (Fees). Livet is a Fr family name. LIVET is a common pl. n. in France.—**H~ Pagnell** belonged to the fee of William Painel in 1204 (FF), and Ralph Paganel, who founded Holy Trinity, York, in 1089, gave among other land also the church of *Hotona* to the priory. Cf. BOOTHBY PAGNELL. —Robert de Hooton held **H~ Roberts** in 1285, but the name was borne by at least one of his predecessors.

OE **hop** is stated to mean 'a piece of enclosed land in the midst of fens', but words such as *fen-, morhopu* in Beowulf suggest some more general meaning, such as 'dry land in a fen'. In pl. ns. the usual meaning is that of *hope* in dialects, viz. 'a small enclosed valley, a smaller opening branching out from the main dale, a blind valley'. In MEATHOP, MYTHOP *hop* has its OE sense. The meaning 'valley' is found in BACUP, BRINSOP, COWPE, the HOPES &c. It is possible there was an OE *hopu* 'privet'. If so, it would be a probable source of *Hop-* in HOPWAS, HOPWOOD, but there is some doubt about the authenticity of the word (see B-T Suppl). Cf. STANFORD LE HOPE Ess.

Hope Green Chs [*Hope* 1282 Court], **H~ D** [*la Hope* 1281 Ass], **H~ Db** [*at Hope* 926 BCS 658, *Hope* DB], **H~ under Dinmore** He [*Hope* DB, *Hope sub' Dinnemor* 1291 Tax], **H~ Mansel** He [*Hope* DB, *Hoppe Maloisel* 12 Fr, *Hope Mal Oysel* 1242 Fees], **Sollers H~** He [*Hope* DB, *Hope Solers* 1242 Fees], **H~ All Saints** K [*Hope* 1240 FF], **H~ La** [*le Hope* 13 WhC], **H~** Sa nr Shelve [*Hope* 1242 Fees], **H~ Baggot** Sa [*Hop* 1242 Fees, *Hope Bagard* 1241 FF], **H~ Bowdler** Sa [*Fordritishope* DB, *Hop* 1201 Cur, *Hope* 1201 FF, *Hopebulers* 1273 Ipm, *-bolers* 1275 Ep], **Hopesay** Sa [*Hope* DB, *Hope de Say* 1255 RH, *Hope Say* 1280 Ep], **Hope** YN [*Hope* 13 PNNR]. OE HOP 'valley'. All the places are in valleys, usually of the type generally denoted by *hope*.

H~ Baggot was held by Robert Bagard in 1242

(Fees). Baggard is evidently a French family name.—Robert de Bullers held **H~ Bowdler** in 1201 (Cur), and the de Bulers family was here in the time of Hy 1. Cf. ASHFORD BOWDLER.—**H~ under Dinmore.** See DINMORE.— **H~ Mansel** from the local family. Mansel, olim *Maloisel*, is a nickname meaning 'ill bird'.— **Hopesay** was held by Picot de Say in 1086 (DB). Cf. HAMSEY.—**Sollers H~** was held by Walter de Solar[iis], son of James de Solar[iis], in 1242 (Fees). Cf. BRIDGE SOLLERS.

Hoppen Nb [*Hopum* 1242 Fees, 1256 Ass]. The dat. plur. of HOP 'valley'.

Hopperton YW [*Hopretone* DB, *Hopertona* c 1130 (1394) YCh 169, *-ton* 1168 P, 1204 FF]. 'The hoopers' TŪN.' *Hooper* 'maker of hoops' is evidenced in OED in the 16th cent. Cf. however *R. le Hopere* 1245 Cl (Ha).

Hopton Db [*Opetune* DB, *Hopton* 1251 Ch, 1299 FF], **H~ Sollers** He [*Hopetune* DB, *Hoptun* 1242, *Hoptun Solers* 1249 Fees], **H~ Sa** at Great Ness [*Hopton* 1285 FA], **H~ Sa** nr Hodnet [*Hotune* DB, *Hopton* 1242 Fees, 1285 FA], **H~ Cangeford** Sa [*Hopton* 1242 Fees, *Hopton Cangefot* 1315 Ipm], **H~ Castle** Sa [*Opetune* DB, *Hopton* 1242 Fees], **Monkhopton** Sa [*Hopton* c 1180 Eyton, *-tone* 1327 Subs], **H~ Wafers** Sa [*Hoptone* DB, *Hopton Wafre* 1236 Brecon, *Hopton Waffre* 1278 Ep], **H~ Sf** nr Lowestoft [*Hoppetuna, Opituna* DB, *Hopeton* 1242 Fees], **H~ Sf** nr Thetford [*Hopetuna* DB, *Hopeton* 1156–60 Bury, *-tone* 1254 Val], **H~ St** [*Hoptuna* 1167 P, *-ton* 1204 Cur, 1236 Fees], **H~ YW** [*Hoptun* DB, *Hopetune* c 1110 Fr]. OE *Hop-tūn* 'TŪN in a valley'. This is accurate for most Hoptons. **H~ Sf** nr Lowestoft seems to show *hop* in the sense 'enclosure in the midst of fens'. On the possibility of an OE *hopu* 'privet', which might sometimes be the first el., see HOP.

Herbert Cangefoot is mentioned in connexion with Clee St. Margaret in 1199. Cangefot is no doubt an OFr nickname.—**Monkhopton** belonged to Wenlock Priory.—**H~ Sollers.** See BRIDGE SOLLERS.—**H~ Wafers.** See HAMPTON WAFER.

Hopwas (hŏpŭs) St [*Opewas* DB, *Hopwas* 1166, 1185 P, *Hopewas* 1256 Misc]. 'Marsh with a HOP or enclosure.' Cf. HOP, WÆSSE.

Hopwell Db [*Opeuuelle* DB, *Hopewell* 1197 P, *Hopwell* 1242 Fees]. 'Stream in a valley.' See HOP.

Hopwood La [*Hopwode* 1278 Ass, *Hopewode* 1285 Ass], **H~ Wo** [(in) *Hopwuda, Hopwudeswic* 849 BCS 455, (in) *Hopwuda* 934 ib. 701]. **H~ La** is in a wooded valley called **Hopwood Clough.** *Hop-* is here *hop* 'valley'. In **H~ Wo** the meaning of the first el. is not clear, and OE *hopu* 'privet', if it existed, might be seriously thought of.

Horbling Li [*Horbelinge* DB, *-beling, -bulling* 1195, *Horblinges* 1208 Cur, *-belling* 1212 Fees]. The place is close to BILLINGBOROUGH. The name is OE *Hor-Billingas* 'the Billing on muddy land'. Cf. HORH. Billing is identical with BILLING Np.

Horbury YW [*Horberie* DB, *-biri* c 1125 YCh 1663, *-beria* 1176 P, *-bir'* 1206 FF]. Apparently OE *Horh-burg* 'BURG on dirty or muddy land'. This does not seem a probable etymology and is hardly suggested by the situation of the place. Perhaps the source is OE *hordburg*, lit. 'treasure-burg', found in poetry.

OE **hord** 'hoard, treasure'. See HORDLE &c., HARDHORN, HARDWELL, HURDLOW, perhaps HORBURY.

Horden Du [*Horedene* c 1050 HSC, *Horden* 1260 Pat]. 'Dirty valley.' Cf. HORH.

Hordle Ha [*Herdel* DB, *Hordhull* 1242 Fees, 1263 Ipm]. Probably OE *hord-hyll* 'treasure mound'.

Hordley Sa [*Hordelei* DB, *-leg* 1237 FF, *Hordileg* 1255 RH]. Hardly OE *hord-lēah* in view of early *Horde*-. OE *hordern-lēah* 'LĒAH with a storehouse' may be suggested. Cf. HARDHORN. OE *hordern* is the source of **Hordron** YW nr Penistone [*Horderne* 1290 Ch].

Horfield Gl [*Horefelle* DB, *-feld* 1287 QW]. 'Dirty FELD.' OE *horu-feld*; cf. HORH.

OE **horh, horu** 'filth, dirt' is fairly common as the first el. of pl. ns.; see HOR- (passim), HARPOLE, HOLYPORT. The meaning is no doubt 'mud'.

Horham (-ŏr-) Sf [*Horham* c 950 BCS 1008, c 1095 Bury, c 1150 Crawf]. 'Dirty or muddy HĀM.' See HORH.

Horkesley, Great & Little, Ess [*Horchesleia* c 1130 Bodl, *Horkesle* 1212 RBE, *Vetus Horkeleg* 1219 Fees, *Horkele* Hy 3 BM, *Horkeleghe Parva* 1238 Subs, *Horkesle Monachorum, Horkele Maior* 1254 Val, *Horskeleye parva* 1298 BM], **Horkstow** Li [*Horchetou* DB, *Horchestou* c 1115 LiS, *Horkestoue* c 1140 DC, *-stow* 1202 Ass]. The first el. may be an OE **horc*, cognate with dial. *hurk* 'a temporary shelter for young lambs, formed of hurdles wattled with straw' (Np, Wa). The word belongs to *hurk* vb. 'to crouch, cower', ME *hurkle* 'to cower'. Second el. LĒAH, STŌW.

Horley O [*Hornelie* DB, *-leia* c 1115 RA, *Hornlea* 1190, *Horlega* 1197 P]. 'LĒAH in a HORNA or tongue of land.' Cf. HORNTON on the same stream somewhat higher up. Horley is in a tongue of land formed by two streams.

Horley Sr [*Horle* 12 BM, *Hornly* c 1270 Ep, *Horneleye* 1279 QW]. H~ is 3 m. west of HORNE and the name probably means 'wood belonging to Horne'. The OE form was *Horn-lēah*.

Hormead, Great & Little, Hrt [*Horemede* DB, *Hormad* 1204–12, 1212 Fees, *Magna, Parva Hormed* 1254 Val]. OE *horh-* or *horu-mǣd* 'muddy meadow'. Cf. HORH.

OE **horn** 'horn' was used in various transferred senses, such as 'horn-like projection, gable (with horn-like ornament)' and the like. German *Horn* is also used of a spit of land or a peak. These senses are to be reckoned with also for Engl *horn*. See HORNE, HORNDON, ASHORNE, IMBERHORNE, WOODHORN. But there was also an OE *horna*, found with certainty in HORNE Ru, WAREHORNE and probably in HORLEY O, HORNCASTLE. OE *horna* may have had the same senses as *horn*, but it seems also to have been used in the same sense as *hyrne*, i.e. 'corner, bend'. From this easily developed the sense 'tongue of land' (in a river fork &c.). *Horn* and *horna* cannot be kept apart in all cases. HORNCHURCH has as first el. *hornede* adj. 'provided with horns'.

Horn Down. See HARWELL Brk.

Hornblotton So [*Horblawetone* DB, *Hornblauton* 1236 FF, *Hornblaneton* 1276 RH, *Hornbloutone* 1327 Subs]. Probably OE *hornblāwera tūn* 'the TŪN of the hornblowers'. OE *hornblāwere* is on record. The loss of *r* is due to dissimilation. Alternatively the first el. might be an OE **hornblāwa* 'hornblower'.

Hornby La [*Hornebi* DB, *-by* 1227 FF], H~ We [*Horneby* 1365 CWNS xxiv], H~ YN nr Bedale [*Hornebi* DB, *-bia* 1132 WR]. '*Horni*'s BY.' *Horni* DB is perhaps OSw *Horni* pers. n.

Hornby YN in Great Smeaton [*Horenbodebi* DB, *Hornbotebi* 1088 LVD, *Horneby* 1243 FF]. '*Hornbođi*'s BY.' ON *Hornbođi* is only recorded as a variant of *Holdbođi*.

Horncastle Li [*Hornecastre* DB, *-castra* 1130 P, 12 DC]. 'The Roman station in the *horna* or tongue of land' between the rivers Bain and Waring. Cf. HORN, CEASTER.

Hornchurch Ess [*Monasterium Cornutum* 1228 Cl, *Hornedecherche* 1311 AD]. 'Horned church', i.e. 'church with horn-like gables'. See HORN.

Horncliffe Nb [*Hornecliff* 1208–10 Fees]. 'Cliff in a *horna* or tongue of land.' See HORN.

Horndon, East & West, Ess [*Torninduna* DB, *Thorendon* 1185 P, *Thorindon Magna, Parva* 1254 Val, *Esthorndone* 1275, *Westorenden* 1274 RH]. OE (æt) *þornigan-dūne* 'hill where thorns grew'. *Est-, West-þornendun* became *Estornendun, Westornendun*, which was understood to mean *East, West Hornendun*, partly owing to influence from HORNDON ON THE HILL.

Horndon on the Hill Ess [*Horninduna* DB, *-done* 1212 RBE, *-don* 1227 Ch, *Horningdon* 1277 Ch, *Horndon* 1254 Val], **Horrington, East & West,** So [*Hornningdun, ođer Horningdun* 1065 KCD 816 (*Hornningdune* 1065 Wells), *Horningdon* 1243 Ass, *Esthorningedon* 1268 FF]. The first el. is probably a hill-name *Horning*, derived from *horn* and meaning 'horn-like hill, peak'. To this was added an explanatory DŪN.

Horne Ru [(to ðæm ham on) *Hornan* (ðæm wuda) 852 BCS 464, *Horne* DB, *Horn* 1229 Ep]. OE *horna* 'corner, bend'. H~ is on a stream which makes a sharp bend north of the village.

Horne Sr [*Horne* 1208 Cur, 1229 FF]. OE *horn*, probably in the sense 'hill, projecting spur of hill'. Cf. HORN.

Horning Nf [*Horningga*, (at) *Horninggen* a 1020, *Horningge* 1044–7 Holme, *Horninga* DB, *Horning* 1254 Val]. OE *Horningas* 'the people at the HORNA or bend'. H~ is at a sharp bend of the Bure. Cf. HORN.

Horninghold Le [*Horniwale* DB, *Horninuald* 1106–23 (1333) Ch, *Horningewald* 1163 P]. 'The WALD or woodland of the *Horningas*.' H~ is in a winding valley, and *Horningas* may be 'the dwellers in the HORNA or bend'. Cf. HORN.

Horninglow St [*Horninglow* Hy 1 Burton, -*e* 1327 Subs]. First el. very likely a hill-name *Horning* (cf. HORNDON ON THE HILL), to which was added OE *hlāw* 'hill'.

Horningsea Ca [(æt) *Horninges ige* c 975 ASCh, *Horninggeseie* c 1050 KCD 907, *Horningeseie* c 1080 ICC, -*ie* DB, -*eye* 1251 Ch]. Perhaps '*Horning*'s island'. *Horning* may be a nickname identical with OE *hornung* 'bastard'.

Horningsham W [*Horning(es)ham* DB, *Horningesham* 1242 Fees]. Probably 'the HĀM at the hill called *Horning*' (cf. HORNDON ON THE HILL). H~ is at a marked spur of hill.

Horningsheath & Horringer Sf [*Horning(g)esh'de*, -*hæð* c 950 BCS 1008, *Horningasearðe*, *Horningeseorðe* 11 EHR 43, *Horningeserda*, -*worda* DB, -*eorda* c 1095 Bury, -*hearde* 1166 RBE, *Horning(g)esherth Magna, Parva* 1254 Val]. Horningsheath and Horringer are probably identical in origin, though it is possible the first was originally -ERþ, the second -WORþ. The places are in a bend of the Linnet. Horning may be an OE *horning* 'bend' or it may be an old name of the Linnet meaning 'winding stream'. Second el. OE *erþ, ierþ, earþ* 'ploughed land'.

Horningtoft Nf [*Horninghetoft* DB, *Horningetoft* 1203 FF, 1270 Ch]. Probably 'the toft belonging to HORNING or people from Horning'. Horning and Horningtoft are far apart.

Hornington YW [*Horninctune, Hornitone* DB, *Hornington* 1190 P, *Horningeton* 1208 FF]. 'The TŪN of the *Horningas* or dwellers in the HORNA or tongue of land.' H~ is in a tongue of land between the Wharfe and the Foss.

Hornsby Cu [*Ormesby* c 1210 WR, 1371 FF]. '*Orm*'s BY.' *Orm* is a well-known pers. n. from ON *Ormr* &c

Hornsea YE [*Hornesse(i)* DB, (mara de) *Hornesse* c 1145, *Hornesse* c 1155 YCh 1302, 1320, (mara de) *Hornese* 1208 FF, 1231 Ep]. H~ is situated on **Hornsea Mere**, and Hornsea is the old name of the lake. The name must be compared with HORNSJÖN, HORNTJÄRN and the like in Sweden, which mean 'lake with horns or corners, angular lake'. Hornsea may be OE *hornasǣ*. Hornsea Mere has an irregular shape.

Hornsey Mx [*Haringeie* 1201 Cur, *Haringue* 1200 FF, *Harengheye* 1232 FF, *Haringesheye* 1243 Cl, *Harnesey* 1543, *Hornsey* 1564 FF]. The original name **Harringay** continued to be applied to the manor house till it was demolished about 1870 and is now the name of a district in London. The change from *Harringay* to *Hornsey* is difficult to explain. The original name may have been OE *Hǣring-gehæg*, the second el. being OE *gehæg* 'hay, enclosure'. The first may be the name of a wood, derived from *hār* 'grey' and meaning 'grey wood'.

Hornton O [*Hornigeton* 1194 Rot Cur, *Horningtun* c 1195 BM, *Hornintton* 1242 Fees]. OE *Horningatūn*. As H~ is only 1½ m. from HORLEY, the probability is that *Horningas* means 'the people from Horley' and Hornton 'the TŪN of the Horley people'.

Horrabridge D [*Horebrigge* 1345 Ass, *Le Horebrugg* 1348 BM]. 'Grey bridge.' See HĀR.

Horringer. See HORNINGSHEATH.

Horringford Wt [*Horningeford*, -*e* 1235, 1255 PNWt, *Horingeford* 13 AD, 1287–90 Fees, 1316 FA]. The first el. may be a derivative of OE *horna* 'a river fork'. The name then means 'the ford of the dwellers by a river fork'.

Horrington So. See HORNDON ON THE HILL.

Horse Eye Sx [*Horsig* 947 BCS 821, *Horsie* 1197 FF]. 'Horse island.'

Horseheath Ca [*Horseda* c 1080 ICC, *Horsei* DB, *Horesathe* 1198 AC, *Horseth* 1245 FF, *Horsheth* 1283 Pat]. OE *hors-hǣþ* 'heath where horses were kept'.

Horsell Sr [*Horsele, Horisell* Hy 3 PNSr, *Horeshull* c 1270 Ep, *Horishull, Horshull* 1279 QW]. OE *hor-gesella* 'shelter for animals in a muddy place'. See HORH, (GE)SELL. The name was at an early date associated with the word *hill*.

Horsenden Bk [*Horsedene, -dune* DB, *Horsendon* 1176 P, *Horsin(g)done* 1221 Ep], **Horsendon Hill** Mx in Sudbury [*Horsendun* 1203 FF]. OE *horsa-dūn* 'hill where horses were kept', with inorganic *n*.

Horsepath O [*Horspadan* DB, -*pade* c 1130 Oxf, -*pathe* 1192 P]. 'Horse path.'

Horsey Nf [*Hors(h)eia, Horseia* 1202 FF, *Horseye* 1254 Val], **H~ Pignes** So nr Bridgwater [*Hursi* DB, *Horsye* 1227 FF,

Peghenes DB, *Pagenesse, Pegeness* 1201 Cur, *Horsy, Pegenesse* 1327 Subs]. 'Horse island.'

Pignes is no doubt the name of a place near Horsey, which was joined with it. Pignes may be '*Pæcga*'s NÆSS'. Cf. PAGHAM.

Horsford Nf [*Hosforda* DB, *Horsford* 1254 Val], **Horsforth** YW [*Horseford* DB, *Horsford* c 1200 YCh 731]. 'Horse ford.'

Horsham St. Faith Nf [*Horsham* DB, *Ecclesia S. Fidis de Horsham* 1163 BM], H~ Sx [*Horsham* 947, 963 BCS 834, 1125]. 'HĀM or HAMM where horses were kept.'

Horsington Li [*Horsintone* DB, *-tun* c 1140 DC, *-ton* 1202 Ass, *Horsington* 1254 Val]. 'The TŪN of *Horsa*'s people.'

Horsington So [*Horstenetone* DB, *Horsinton* 1179 P, 1225 Ass]. OE *horspegna tūn* 'the TŪN of the horsekeepers or grooms'. H~ is near HENSTRIDGE.

Horsley Db [*Horselei* DB, *-lee* 1212 Fees], H~ Gl [*Horselei* DB, *-lega* 1176 P], H~ Nb [*Horseley* 1242 Fees], **Long H~** Nb [*Horsleg* 1196, *Horslega* 1197 P], **East & West H~** Sr [(on) *Horsalege* 871–89 BCS 558, *Horslege* c 1050 KCD 896, *Horslei* DB], H~ St [*Horselega* 1165, *Horslega* 1167 P]. OE *horsa-lēah* 'pasture for horses'. See LĒAH.

Horsmonden K [*Horsbundenne* c 1100 Text Roff, *Horsburdenne* 1212 RBE, 1285 Ch, *Horsmonden* 1263 Ipm, *Horsmundenn* 1278 QW, *-denn'* 1312 Ep]. The present name appears to mean 'the DENN or pasture of the horsekeepers', though *horseman* in this sense is hardly evidenced. The old name may have as first el. an earlier name of the stream at the place, viz. *Horsburna*.

Horstead Nf [*Horsteda* DB, 1166 P, *Horsstud* R 1 Cur], **Horsted** K [*Horsum stydæ* 860–2, *Horstede* 10 BCS 502, 1321, *Horstede* 1254 Ass], **Horsted Keynes** Sx [*Horstede* DB, *Horsestud* 1195 Cur, *Horsted Kaynes* 1307 Misc], **Little H~** Sx [*Horstede* DB, *Little Horstede* 1307 Misc]. 'Place where horses were kept, horse-farm.' The form of 860–2 for the Kent H~ is no doubt corrupt.

H~ Keynes was held by Wills de Cahainges in 1086 (DB). Cf. ASHTON KEYNES.

Horston Db [*Harestan* 1205 P, 1230 Cl]. Identical with HARSTON Le.

Horton Bk nr Datchet [*Hortune* DB], H~ Bk at Ivinghoe [*Hortone* DB], H~ **by Malpas** Chs [*Hortone* 1289 Court], H~ Do [*Horetuninge gemære* 946 BCS 818, *Hortun* 1033 KCD 1318, *Hortune* DB], H~ K nr Canterbury [*Horatun*, *-e* 874 BCS 538, *Hortone* DB], **H~ Kirby** K [*Hortune* DB, 11 DM, *Horton Kyrkeby* 1346 FA], **Monks H~** K [*Hortune* 1035 BM, 11 DM, *-tun* DB], H~ Nb nr Blyth [*Horton* 1242 Fees, *H~ Shireve* 1270 Ch], H~ Nb nr Dodding-ton [*Horton Turbervill* 1242 Fees], H~ Nb nr Ponteland [*Horton* 1242 Fees, *Hortun*

1252 Ch], H~ Np [*Hortone* DB, *Horton* 1220 Fees], H~ O [*Hortone* DB, *-tun* 1220 Fees], H~ Sa nr Wem [*Hortune* DB, 1327 Subs], H~ Sa nr Wellington [*Hortune* DB, *Horton* 1327 Subs], H~ So [*Horton* 1242 Ass], H~ Sr [*Horton* 1178 PNSr], H~ St [*Horton* 1239 Ass, 1252 Ch], H~ W [*Horton* 1191 P], H~ YW nr Bradford [*Hortona* c 1195 YCh 1795, *-ton* 1246 Ipm], H~ YW nr Gisburn [*Hortun* DB, *Horton* 1226 FF], **H~ in Ribblesdale** YW [*Hortune* DB, *Hortuna de Ribblesdala* 1150–5 LaCh]. All these are OE *Horh-tūn* or *Horu-tūn* 'TŪN on muddy land'. Cf. HORH.

H~ Kirby was held by Gilbert de Kirkeby in 1254 (Kn Fees).—**Monks H~** was a priory.

Horton Gl [*Horedone* DB, *Horton* 1200 P, *Heorton* 1291 Tax, *Herton* 1303 FA]. OE *heorta-dūn* 'hill frequented by stags'. The place is on the slope of a ridge.

Horwich (hŏrij) La [*Horewych for.* 1254 Misc, *Harewych* 1277 VH, *Horewiche* 1282 Ipm]. OE (æt) *hāran wicum* 'the grey wych elms'. OE *wice* means 'wych elm'.

Horwood Bk [*Horwudu* 792 BCS 264, *-wuda* c 1155 Oxf, 1167 P]. 'Muddy wood.' Cf. HORH.

Horwood D [*Horewode, Hareoda* DB, *Horwude* 1198, 1202 FF, *Harewde* 1219 FF]. Probably identical with prec. name. Forms with *a* may be due to the wish to avoid unpleasant associations. If the *a*-forms are trustworthy, the name means 'grey wood'.

Hose Le [*Hoches, Howes* DB, *Houwes* c 1125 LeS, *Houes* 1236 Fees]. OE *hōhas* or *hōgas*, plur. of HŌH 'hill, spur of land'.

Hotham YE [*Hode* 963 YCh 5, *Hode, Hod-hum* DB, *Hothum* c 1160 YCh 971, *Hoðum* 1167 P, *Houthum* 1203 Cur]. OE *hŏðum*, dat. plur. of *hŏd* 'a shelter'. See HOOD. The name has been Scandinavianized, *Hōðum* being adapted as OScand *Hōðum*.

Hothersall (-ŏdh-) La [*Hudereshal* 1199 Ch, 1201 P, *-e* 1257 Ipm, *Huddeshal* 1206 P]. See HALH. The first el. seems to be a pers. n. such as OE **Huder*, which is also found in HUDDERSFIELD.

Hothfield K [*Hathfelde* 11 DM, *Hatfelde, Hedfield* 1212 RBE, *Hothfelde* Hy 3 Ipm]. OE *hǣþ-feld* 'FELD overgrown with heather'. Cf. HǢp and HOATH.

Hothorpe Np [*Udetorp* DB, *Hudtorp* 1236 Fees, *Huttorp* c 1155 DC, 1202 Ass, *Huthorp* 1220 Fees]. '*Hūda*'s thorp.'

Hōton Le [*Hole-, Hohtone* DB, *Holtuna* 1158 Fr, *Houtona* c 1200 DC, *Houton* 1198 Fees, 1220–35 Ep]. OE *Hōh-tūn* 'TŪN on a spur of hill'.

Hough Chs at Wybunbury [*Houcht* 1287, *le Hogh* 1287 f. Court], H~ Chs nr Knuts-ford [*Hoh* 1176–1208 Chester], H~ Db [*Hoge* DB, *Hogh* 1285 FF]. OE HŌH 'spur of hill'.

Hough on the Hill Li [*Hach, -e, Hag* DB, *Hac, Hag* 1100–35 RA]. OE *haga* 'enclosure'. **Hougham** Li nr Hough on the Hill [*Hacham, Hacam* DB, *Acham* 1090 RA, *Hauham* 1212 Fees]. 'The HAMM belonging to HOUGH.' *Hough(am)* for *Haugh(am)* is late and probably an inverse spelling due to the common change of *ōu* to *au* (aw) in names like HOUGHTON from *Hōhtūn*, BROUGHTON from *Brōctūn*.

Houghall (-ŏf-) Du [*Hocchale* 1228 FPD, *Howhale* 1292 Ch]. 'HALH or haugh at a spur of hill.' See HŌH.

Hougham (-ŭf-) K [*Hucham, Huham* DB, *Hucham* 1086 KInq, *Huhcham* 11 DM, *Hugham* 1178 P, *Hougham* 1271 Ipm]. Possibly OE *Hōh-hām* 'HĀM on a spur of hill', though the change *ō > ū* appears remarkably early. But more likely '*Huhha*'s HĀM' (OE *Huhhanhām*) with the same early reduction as in OFFHAM and the like. For *Huhha* see HUGHENDEN.

Hougham Li. See HOUGH ON THE HILL.

Houghton, a common name, has several sources:

1. Usually OE *Hōh-tūn* 'TŪN on a spur of hill': H~ **Conquest** Bd [*Houstone* DB, *Hocton* 1202 Ass, *Houghton Conquest* 1316 FA], H~ **Regis** Bd [*Houstone* DB, *Hohtun* 1156 P, *Houghton Regis* 1323 Ch], H~ **Cu** [*Hochton* 1321 Ipm], **New & Stoney H~** Db [*Holtune* DB, *Hochtone* 1280 Derby, *Hoghthon* 1289 FF], **H~ le Side** Du [*Hoctona* 1200 BM], **H~ le Spring** Du [*Hoctun* (endorsed *Hoghton Springes*) c 1220 Pudsay, *Hoghton* 1291 Tax], H~ **Ha** [*Hohtuninga mearc* 982 KCD 633, *Houstun* DB, *Hochton* 1185 P], H~ **Hu** [*Hoctune* DB], H~ **La** in Winwick [*Houton* 1263 Ass, *Hoghton* 1327 Subs], **H~ on the Hill** Le [*Hohtone* DB, *Hoghtone* 1220 Ep, *Hocton* 1254 Val], H~ **Li** [*Hoctune, Hochtune* DB, *Hocton* 1202 Ass], H~ **Nb** nr Heddon on the Wall [*Houcton* 1242 Fees, *Hocton* 1256 Ass], **Little & Long H~** Nb [*Houcton Magna, Parva* 1242 Fees, *Magna Houton* 1281 Percy], H~ **Nf** nr W. Rudham [*Houton* 1254 Val, -e 1291 Tax], **H~ on the Hill** Nf [*Houton* 1212 Fees, 1254 Val], **H~ St. Giles** Nf [*Hohttune, Houtuna* DB, *Hocton* 1212 Fees], **Great & Little H~** Np [*Hohtone* DB, *Magna Houtona* 1199 FF, *Parva Houtone* 1233 Ep], **Hanging H~** Np [*Hohtone* DB, *Hangadehouton* 1230 Cl, *Hangandehouton* 1275 RH], H~ **Sx** [*Hohtun* 683, *Hocton* 957 BCS 64, 997, *Hocton* 1227 Ch], **Glass H~** YW [*Hoctun* DB, *Hoghton* 1316 FA].

H~ Conquest from the Conquest family, mentioned in connexion with the manor from 1223 (FF).—**Glass H~** apparently from a glass manufactory.—**Hanging H~** means 'H~ on a slope'.—**H~ St. Giles** from the dedication of the church.—**Side** in **H~ le Side** is OE *side* 'hill slope'.—Henry Spring in **H~ le Spring** is mentioned c 1220 (Pudsay). *Spring* is no doubt *spring* 'a sprig, a sapling' (c 1300 &c.), also 'a young man' (1559 &c.), used as a byname.

2. OE *Halh-tūn* 'TŪN in a haugh': **Little H~** La [*Halughton* 1253 FF, *parva Halghton* 1310 WhC], **Westhoughton** La [*Halcton* c 1210 CC, 1258 Ass, *Westhalcton* c 1240 CC], **Great H~** YW [*Haltun* DB, *Halcton* 1297 Subs, *Magna Halghton* 1303 FA], **Little H~** YW [*Haltone* DB, *Parva Halghton* 1303 FA].

3. **Houghton** YE [*Houetun* DB, *Hoveton* c 1200 YCh 1128, *Houeton* 1241 FF]. '*Hofa*'s TŪN' (cf. HOVETON, HOVINGHAM) rather than 'TŪN where *hōfe* or hove (a plant) grew'.

4. **Houghton** D. See HOWFIELD.

Hound Ha [*Hune* DB, 1176 P, 1251 Ch, *Hona* 1242 Fees]. OE *hūne* 'hoarhound'.

Houndsditch Mx [*Hundes-, Hondesdich* 1275 RH]. The probable meaning is 'ditch into which dead dogs were thrown'. The ditch is stated to date from the time of King John. This rules out the OE pers. n. *Hund* as first el.

Houndstone So [*Hundestone* DB, -ton 1201 Ass], **Houndstreet** So [*Hundesterte* 1243 Ass, 1277 FF, *Houndesterte* 1327 Subs, *Hundestret* 1316 FA]. '*Hund*'s TŪN and piece of land.' Cf. STEORT. But Houndstreet may be 'dog's tail', owing to a fancied likeness of a piece of land to one.

Hounslow Mx [*Honeslaw* DB, *Hundeslauwe* 1242 Fees, -lowe, -lawe 1252 f. Cl]. '*Hund*'s barrow.' *Hund* is found in *Hundeshlæw* BCS 687 (Brk), *Hundesgeat* ib. 887 (Gl).

Hove (hōov) Sx [*Houue* 1296 Subs, *Houve* 1302 Cl, *Huve* 1341 NI]. Apparently OE *hūfe* 'hood', here used in the sense 'shed, shelter' or the like.

Hŏveringham Nt [*Horingeham* DB, *Houeringeham* 1167 P, *Hoveringham* 1235 Cl]. The place is in a low situation. The first el. may be derived from a nickname derived from OE *hofer* 'hump' (cf. OE *hoferede* 'hunchbacked'). If so 'the HĀM of *Hofera*'s people'. In PNNt(S) the first el. is taken to be a folk-name *Hoferingas* derived from a hump of ground called *Hofer*. This is preferable if the topography admits it. There is a considerable hill near Hoveringham, but on the opposite side of the Trent at Kneeton. Possibly the *Hoferingas* came from there.

Hoveton Nf [*Houeton* 1044–7 Holme, *Houetuna* DB, *Houeton* 1186 P], H~ **YN**, lost [*Houetune* DB, -tona 1197 (1252) Ch], **Hŏvingham** YN [*Hovingham* DB, *Houingeham* c 1110 RA, *Hovingeham* 1157 YCh 354]. Probably Hoveton and HOUGHTON YE mean '*Hofa*'s TŪN' and Hovingham 'the HĀM of *Hofa*'s people'. OE *Hofa* is evidenced as *Hova* in Aldhelm's poem *Æthilwaldus ad Hovam comitem*.

Howbury K [*Hov* DB, *Litelhou* 1242 Fees, *Hobury* 1379 BM]. OE *HŌH* 'spur of hill' with *bury* 'manor' added.

Howden Nb [*Holden* c 1290 Percy], Howden Clough YW [*Holeden* 1202 FF]. Identical with HOLDEN.

Howden YE [*Æt Heafuddæne, Heofoddene* 959 YCh 4, *Hovedene* DB, *-den* 1080–6 YCh 974]. The OE name means 'head valley', possibly 'chief valley'. Later the OScand equivalent (ON *hǫfuð,* OSw *huvuþ*) was substituted for OE HĒAFOD.

Howe Nf [*Hou, Howa* DB, *Howe* 1254 Val], H~ YN [*Hou* DB, 1246 FF]. OScand *haugr* 'hill, barrow'.

Howell Li [*Welle, Huuelle* DB, *Huwella* 1165, *Huwelle* 1190 P, *Huwell* 1202 Ass]. The situation forbids OE HŌH 'spur of hill', which is rendered unlikely also by the regular early *Hu-.* A possible etymology is OE *hūn-wella* 'the stream of the cubs' with loss of *n* as in STOWFORD, LYFORD.

Howfield K [*Hughefeld* 13, 1390, *Huggifeld* c 1215, *Huggefeld* 1284 BM, *Hugefeld* 1291 Tax]. The first el. may be a pers. n. *Huhha* or the like, found also in Houghton D in Bigbury [*Hugheton* 1242 Fees] and in Highweek [*Hugeton* 1238, *Huggeton* 1249 Ass].

Howgill YW nr Gisburn [*Holegile* c 1240 Pudsay, *Holgill* 1285 FA]. 'Deep valley.' Cf. HOLH adj. and GIL.

Howgrave YN [*Hograve, Hogram* DB, *Hograue* 1088 LVD, *Hougraue* 1196 FF]. 'Grove on a HŌH or spur of hill.' See GRĀF.

Howick La [*Hocwike* a 1096 Pont, *Hokewike* a 1122 LaCh, *Houwyk* 1246, *Hoghwyk* 1276 Ass]. OE *Hōc-wīc* or *Hōh-wīc* 'wīc at a spit of land'. See HŌC, HŌH.

Howick (-ō-) Nb [*Hewic* c 1100 PNNb, *Hawic* 1230 Pat, *Hawick* 1279 Ass, *Howyc* 1242 Fees]. The earliest forms suggest OE *Hēa-wīc* 'high wīc' (cf. HEWICK). The later forms seem due to a refashioning of the name under the influence of OE HŌH.

Howle Hill He [*Hulla* DB, *Hule Cnolle* 1286 Ep, *Hule* 1305 Ep], H~ Sa [*Hugle* DB, *Hulam* 1253 Ch]. H~ Hill is on the slope of a hill of 659 ft. H~ Sa is near a hill. Clearly the name originally denoted a hill. It is related to G *Hügel* 'hill' and *Hugl,* the ON name of a high island in Norway. There must have been an OE *hugl* or *hugol* 'hill'.

Howsell Wo [*Howeshulle* c 1230, *-hell* n.d. AD iii]. The place is on the northern slope of Malvern Hill. This may have been known as OE HŌH 'the hill' (gen. *hōs, hōges*). To this early name in the gen. form was added an explanatory *hill.*

Howsham Li [*Usun* DB, *Husum* c 1115 LiS, 1177 P], H~ YE [*Huson* DB, *Husum* 1227 Ch, *Housom* 1297 Subs]. The dat. plur. of OE or OScand *hūs* 'house'. HUSUM is a common name in Denmark.

Howtel Nb [*Holthal* 1202 FF, 1242 Fees, *-hale* 1226 P]. 'Wooded haugh.' See HOLT, HALH.

Howthorpe YN [*Holtorp* DB, 1167 P, 1234 FF, *Holetorp* 1166 P]. '*Holti*'s thorp.' Cf. HOLTBY.

Howton He [*Hutun* 1242 Fees, *Hugetun, Hueton* 1249 Fees]. Probably 'Hugh's TŪN', Hugh being the OFr name *Hue* from OG *Hugo.*

Hoxne (hŏksn) Sf [*Hoxne* c 950 BCS 1008, c 1035 Wills, *Hoxana, Hoxa* DB, *Hoxe* 1121–4 BM, *Hoxna* Hy 1 (1232) Ch]. The place is on a spur of land between the Waveney and one of its tributaries. The name is probably OE *hōhsinu* 'heel-sinew', which, to judge by the later *hockshin, hough,* was probably used also in the sense 'hough'. The place was named from the similarity of the spur of land to the hough of a horse.

Hoxton Mx [*Hochestone* DB, *Hocston* 1221 FF, *Hoxtone* 1254 Val]. '*Hōc*'s TŪN.'

Hoylake Chs. No early forms found. Really *Hoyle Lake.* Outside H~ is East Hoyle Bank. Hoyle may be OE *holh* 'hollow'. Cf. HOOLE Chs. If so, the name originally denoted the pool later known as Hoyle Lake.

Hoyland, High, YW [*Holand* DB, *Heghholonde* 1329 FF], Nether & Upper H~ YW [*Holand, Hoiland* DB, *Holand* 1240 FF], H~ Swaine YW [*Holant* DB, *Hoiland* 1200 Cur, *Holandeswayn* 1266 Misc, *Swaynholand* Hy 3 BM]. OE *hōh-land* 'land on or by a spur of hill'.
Suanus de Hoiland is mentioned 1194–1211 YCh 1686. He died in 1129 (YCh 1664 with pedigree). The correct form of the name is *Swayn* from ON *Sveinn,* ODan, OSw *Sven.*

OE **hrace, hræce** 'throat' in a transferred sense such as 'pass' may be the source of RAKE and of the first el. of RACTON, RAGDALE.

OE **hræfn** 'raven' is the first el. of some names, as RAINOW, RAVELEY, RAVENDALE, RENSCOMBE, but is often difficult to distinguish from the pers. n. *Hræfn,* which must be postulated (see e.g. RAMSBURY) and from OE *hramsa* 'garlic' and *ramm* 'ram'. OScand **hrafn** (ON *hrafn* &c.) is found in some names, as RANSKILL, RAVENSCAR, RAVENSTONEDALE.

OE **hrāgra** 'heron' is the first el. of RAWRETH.

OE **hramsa** (plur. *-n*) 'ramson, wild garlic, *Allium ursinum*' (= MLG *ramese,* Norw *rams*) is a common first el. of pl. ns. See RAMS- (passim), ROMSLEY. In compounds the word often appears as OE *hrames-,* as in *hramæs hangra* 987 KCD 658, *hrameslea* 944 BCS 801.

OE **hrēod** 'reed' is a very common pl. n. el., mostly found as the first member. See, however, ROWDE. The form varies a good deal owing to different development of *ēo* in dialects; thus *ēo* often becomes *ō*. *Hrēod* is sometimes difficult to distinguish from OE *rēad* 'red'. It may be due to influence from the latter, when OE *hrēod* appears as Mod *Rad-*, as in RADBOURN(E), RADIPOLE. See further RED-, REED- (passim), RIDGE-WELL, RIDLEY, RINGMORE, RODBOURNE, ROD-LEY, RUDFORD. OE *hrēodmere* 'reedy lake' is the first el. of RADMANTHWAITE, REDMAR-LEY, REDMARSHALL, RODMARTON.

ON **hreysi**, **hreysar** plur. 'cairn' is found in DUNMAIL RAISE, HARRAS, ROSEACRE.

OE **hring** 'ring, circle' is found in RING-BOROUGH, RINGMER, RINGSHALL, RINGSTEAD. The exact meaning of the word is not clear. It probably varied. See also EAKRING.

OE **hrīs**, OScand *hris* (ON *hrís*, OSw *ris*) 'brushwood' is a fairly common el. of pl. ns. RISE, RISEHOLME, RYSOME go back to *hrīsum*, dat. plur. *Hrīs* is the second el. of ACRISE, GALTRES. As a first el. it sometimes shows influence from OE *risc* 'rush', as in RISHANGLES, RUSTON Nf, RUSWARP. See RIS- (passim), RESTON, RYSTON. There must have been an OE *hrīsen* adj. 'of brushwood'. See RISBOROUGH, RISELEY, RISLEY, RISSING-TON.

OE **hrōc** 'rook'. See ROCK-, ROOK-, ROX-(passim), also RUCKINGE, RUCKLAND, RUCK-LEY.

ON **hross** 'horse' is found in ROSEDALE, ROSGILL. The corresponding OE word is *hors*, developed from earlier *hross*. It seems very likely that OE sometimes had the form *hross*. See ROSLEY, ROSS HALL, ROSSALL.

OE **hrucge** 'woodcock'. See RUGLEY, RUG-MERE.

OE **hruna** 'a fallen tree, a log'. See HEAD-CORN, RUMBRIDGE, RUNFOLD, RUNHALL.

OE **hrung** 'a pole'. See ROUNTON, RUNCTON Nf, and cf. RANGEWORTHY.

OE **hrycg** 'back; ridge' is common in pl. ns. See e.g. FOULRIDGE, HENSTRIDGE, TAND-RIDGE, RIDGE, RUDGE, RIDGEACRE, RIDG-WARDINE, RUDGWICK, RUGELEY, REIGHTON. OScand **hryggr** (ON *hryggr*, OSw *rygg*) is found in some names, as RIBBY, RIG- (passim), CROSS-, GRAY-, LAMBRIGG, MANSRIGGS. Sometimes the OScand word has replaced the English one, as in MARRICK, RIGTON.

OE **hrȳper** 'ox, cattle'. See ROTHER- (passim), RYTHER.

Hubberholme YW [*Huburgheham* DB]. Perhaps '*Hūnburg*'s HĀM'. *Hūnburg* is a woman's name. For the loss of *n* cf. STOW-FORD, LYFORD.

Hubbridge or **Howbridge Hall** Ess in Witham [*Hobruge* DB, 1205 FF, *Hobregge* 1197 FF]. 'Bridge by a HŌH or ridge.'

Huby YN [*Hobi* DB, 1167 P, *Hoby* 1181 YCh 420], H~ YW [*?Hobi* 1167 P, *Hugby*

c 1285 Bodl]. This may be 'BY at a HŌH or spur of hill'. But more probably it is identical with Sw HÖGBY, OScand *Haugbȳr* 'BY at a barrow or hill'. Cf. HAUGR.

Hucclecote Gl [*Hochilicote* DB, *Hukeling-cote* 1221 Ass]. 'The COT of *Hucel*'s people' or the like. A pers. n. *Hucel* or *Hucela* must be assumed for HOCKLETON, HUGGLESCOTE. Cf. HUCKNALL &c., which contain a pers. n. *Huca* or *Hucca*.

Hucking K [*Hugginges* 1195 Cur, *Hoking* 1215 Cl, *Hukinge* 1246 Ipm]. '*Hucca*'s or *Hōc*'s people.' Cf. HUCKNALL.

Hucklow Db [*Hochelai* DB, *Parva Hoke-lawe* 1253–8 Derby, *Hukelowe* 1265 Ch, *Magna Huckelowe* 1301 BM]. '*Hucca*'s mound or hill.' Cf. HUCKNALL, HLĀW.

Hucknall, Ault, Db [*Hokenhale* 1291 Tax, *Hukenalle* 1428 FA], H~ **Torkard** Nt [*Hochenale*, *Hochehale* DB, *Huccenhal* 1163 f. P, *Hukenhal* 1198 Fees, *Huckenhale* 1198 FF, *Hukenale Torcard* 1288 Misc]. The two Hucknalls are c 11 m. apart, but about midway between them is HUTHWAITE or HUCKNALL UNDER HUTHWAITE. This suggests that Hucknall was once the name of a large district which comprised both Hucknalls. Hucknall means '*Hucca*'s HALH', *halh* referring here to a valley, whether that at Ault H~ or H~ Torkard. The pers. n. **Hucca* is found also in HUCKLOW and perhaps in HUCKING. It might be a develop-ment of a *k*-derivative of *Hūda* (OE *Hūd(e)ca*).

Ault has been explained as Fr *haut* 'high'.— H~ **Torkard** was held by Geoffrey Torchard in 1195 (P). Torkard is a Fr family name, probably originally a pers. n. of G origin (e.g. OG *Droctard*, *Truhthard*).

Huddersfield YW [*Oderesfelt* DB, *Hudres-feld* 1121–7 YCh 1428, *Huderesfeld* 1297 Subs]. '*Huder*'s FELD.' Cf. HOTHERSALL.

Huddington Wo [*Hudigtuna gemæra* a 840 BCS 428, *Hudintune* DB]. 'The TŪN of *Hūda*'s people.'

Huddleston YW [*Hudelestun* c 1030 YCh 7, *Hudlestona*, *Hudelestuna* c 1175 YCh 1719 f., *Hudeleston* 1223 FF], **Hudswell** YN [*Hudreswelle* DB, *Hud(e)leswell* 12 Mon, *Hudeswell* 1199 P, 1202, 1226 FF]. '*Hūdel*'s TŪN and well.' **Hūdel* is a derivative of *Hūda*.

Huggate YE [*Hughete* DB, *Hugat* c 1150 YCh 1238 f., *Hugate* 1176 P, 1219 FF]. Second el. OE *geat*, presumably in the sense 'pass'. The first cannot well be OE HŌH, but it might be OE *hūc* 'a point of land'. Cf. HOOK YW.

Hugglescote Le [*Hukelescot* 1227 Ch, 1233 Cl, *Hoclescot* 1236 Fees]. '*Hucel*'s COT.' Cf. HUCCLECOTE.

Hughenden Bk [*Huchedene* DB, *Huggenden* 1186 P, *Hugenden* 1195 Abbr, *Huchendenn* 1203 Cur]. '*Hucca*'s or *Huhha*'s valley.'

Cf. HUCKNALL, HOWFIELD. A change of intervocalic *c* > *g* > *ȝ* has analogies in other names in the district; cf. COPLE, MOGGERHANGER Bd.

Hughley Sa [*Lega* 1169–76 Eyton, *Hugh Leghe* 1327 Subs]. Originally LEY from OE lēah. *Hugh* from Sir Hugh de Lega, who is mentioned c 1170 (Eyton).

Hugh Town in the Scilly Isles was named from the hill above it on which Star Castle was built in 1593 [(the) *Hew Hill* 1593, 1595 State Papers]. *Hew* may be a form of OE *hōh* 'ridge, hill'. See HŌH.

Hugill We [*Hogayl* 1255 FF, *Hagayl* 1274 Ipm, *Hugayl* 1274 Kendale]. Second el. OScand *geil* 'a ravine'. The first may be OScand *haugr* 'hill, barrow'. See HUBY.

Huish D nr Torrington [*Iwis* DB, *Hywis* 1242 Fees], **North H~** D [*Hewis* DB, *Northywys* 1284–6 FA], **South H~** D [*Heuis* DB, *Hywis* 1242 Fees], **H~ So** in Burnham [*Hiwis* DB, *Hywys* 1280 Wells], **Beggearn H~** So nr Nettlecombe [*Hewis* DB, *Beggerhywys* 1276 RH], **H~ Champflower** So [*Hiwis* DB, *Hywis* 1212 Fees, *Hywys Champflur* 1274 RH], **H~ Episcopi** So [*Hiwissh* 973 BCS 1294, *Hiwisc* 1065 Wells], **H~ W** [*Iwis* DB, *Hiwis* 1163 P, 1198 Fees]. OE *hīwisc* 'household, hide'.

Beggearn is the gen. plur. of ME *beggere* 'beggar', here no doubt in the sense 'friar'.— **H~ Champflower** was held by Thomas de Chanflurs (or Campo-florido) in 1212 (Fees). CHAMPFLEURY [*Campus Floridus* 1264] is a place in Normandy.—**H~ Episcopi** belonged to the see of Wells.

Hulam Du [*Holum* c 1050 HSC, c 1200 FPD]. OE *hōlum*, dat. plur. of HOLH 'hollow'. The *o* was lengthened as in HOOLE Chs.

Hulcote Np [*Hulecote* DB, -*cot* 1202 Ass, 1220 Fees, *Hullecot* 1237 Cl]. First el. OE *hulu* in such a sense as 'shed, hovel'. Cf. HOOLE La. The exact meaning of the name is not clear. Possibly 'hut for shelter'.

Hulcott (hŭk-) Bk [*Hoccot* 1200 FF, *Huccot* 1237–40 Fees, -*e* 1239 Ep, *Hulecot* 1228 Pat]. Early forms with *l* are rare and perhaps to be disregarded. If so, '*Hucca*'s COT'. Cf. HUCKNALL.

Hull Chs [*Hulle* 1283 Ipm, 1384 AD]. Identical with HOOLE La.

Hull, Bishops, So [*Hylle* 1033 KCD 750, *Hilla* DB, 1155–8 (1334) Ch, *Hulle* 1225 Ass, *H~ Episcopi* 1327 Subs]. OE *hyll* 'hill'. The manor belonged to the bishop of Winchester.

Hull R YE [*Hull* c 1000 Saints, *Hul* 1156 YCh 1388]. A British river-name. The river gave their names to Hull port [*portus de Hull* 1276 RH] and Hull priory and chapel [*capella de Hulle* 1291 Ep], also to Hull town, which, however, was formerly and is still alternatively KINGSTON UPON HULL (q.v.).

Hulland Db [*Hoilant* DB, *Holond* 1249 Ch, -*land* 1262 BM]. Identical with HOYLAND.

Hullasey Gl nr Kemble [*Hunlafesed* DB, *Hunlaweshyde* 1169 P, *Hunlaueseta* Hy 2 (1268) Ch, *Hunlauesheda* 1192 P]. '*Hūnlāf*'s HȲP or landing-place.'

Hullavington (hŭlingtn) W [*Hunlavintone* DB, *Hunlauinton* 1190 P]. 'The TŪN of *Hūnlāf*'s people.'

Hulme (hūm), **Cheadle** (-ē-), Chs [*Hulm* 1363, *Chedulholme* 1528 Ormerod], **Church H~** Chs [*Churche Hulm* c 1292 Ormerod], **Hulme Walfield** Chs [*Walefeld*, *Wallefilde* and *Hulm* 1290 Ipm, *Hulm juxta Wallefeld* 1308, *Hulm Wallefeld* 1338 Ormerod], **H~** (hōom, hūm) La in Manchester [*Hulm* 1246 Ass, *Over-*, *Netherhulm* 1324 LaInq], **H~** La in Winwick [*Hulm* 1246, 1276 Ass], **H~ Hall** La in Manchester [*Hulme* 1343 VH], **H~ St** [*Hulme* 1227 FF, *Hulm*, *Holm under Kevermund* 1293 Ass]. ODan *hulm* 'holm, small island, piece of land on a stream'. Cf. HOLM.

For **Cheadle H~** see CHEADLE.—**Hulme Walfield** contains a pl. n. **Walfield**, whose first el. seems to be OE *wella* (*wælla*) 'stream' &c.

Hulne (hōol) Nb [*Holme* (wood) 1248 Ipm, *Holyn* 1265 Misc, (boscus de) *Hulm* 1279 Ass, *Holne* 1284 Percy]. OE HOLEGN 'holly'.

Hulse Chs [*Holes*, *Holis*, *Holys* c 1250 ff. Chester]. The plur. of OE HOLH 'hollow'. The OE form was *Hōl(u)* (from *Holhu). To this was later added the ending -*es*.

Hulton La [*Hilton* 1200 P, *Hulton* 1212 Fees], **H~ Abbey** St [*Hulton* 1235 Cl]. 'TŪN on a hill.' See HYLL.

OE **hulu** 'husk', ME *hule* 'hut or hovel'. See HOOLE La, HULL Chs, HULCOTE.

Hulverstone Wt [*Hunfredeston* 13 VH, *Humfrideston* 1346 FA]. '*Hūnfriþ*'s TŪN.'

Humber He [*Humbre* DB, *Humbra* Hy 1 Brecon, *Humbre* 1242 Fees]. Originally the name of **Humber Brook**, which is identical with HUMBER R.

Humber R [*Humbri* (fluminis) c 720 Bede HAbb, c 730 Bede, *Humbrae* (fluminis) c 730 Bede, (fluminis) *Humbrę* 832 OET, *Humbre* 827 &c. ASC, c 890 Alfred CP, *Humber* 1147 Monm]. A Brit river-name, which may be a compound of the prefix *su*- (OW *hu*-, *hy*-) 'good, well' and the word for river found in AMBER. Humber was formerly a common name of streams in England.

Humberstone Le [*Humerstane* DB, *Humberstan* c 1150 BM, Hy 2 (1318) Ch, *Humbristona* c 1200 Fr]. '*Hūnbeorht*'s stone.'

Humberstone Li [*Humbrestone* DB, *Humberstein* c 1115 LiS, *Humbrestan* Hy 2 DC, *Humberstain* 1228 BM]. The place is near the Humber. The name means 'the Humber stone'. The second el. is sometimes

Scandinavianized (-*stain* for OE -*stān*). The stone from which the place was named is referred to by Gervase Holles in *Lincolnshire Church Notes* (1634–42) as 'a great Boundry blew Stone (lying) just at the place where Humber looseth himselfe in the German Ocean'. The passage is quoted in *Li Record Soc.* i. 13, as pointed out in a private communication to the author by Mr. H. V. Thompson of Stoke-on-Trent.

Humbledon Hill Du [*Hameldone, Homeldun* 12 FPD, *Hameldon* 1382 Hatfield], **Humbleton Hill** Nb [*Hameldun* 1170 P, *Hamildon* 1242 Fees, *Hameldon* 1256 Ass]. Identical with HAMBLEDON. The *u*-forms are phonetic variants going back to OE *homel* by the side of *hamel*.

Humbleton YE [*Humeltone* DB, *Humbleton* 1190 YCh 1312, *Humbiltona* 1297 Subs]. Probably 'TŪN where hops grew'. First el. ON *humli* 'the hop plant', which may have replaced OE *hymele* (cf. HIMBLETON).

Humburton YN [*Burton* DB, 1276 Ipm, *Hundesburton* 1224–30 Fees]. OE *Burh-tūn*, see BURTON. *Hund* must have been an early owner.

Humby Li [*Humbi* DB, *Humby* 1242 Fees]. Apparently '*Hundi*'s BY'. Cf. HANBY.

Humshaugh Nb [*Hounshale* 1279 Ass, *Homeshalk* 1318 Ipm]. The earliest form suggests '*Hūn*'s haugh'. The later *m*-form is difficult to explain.

Huncoat La [*Hunnicot* DB, *Hun(n)ecotes* 1241 FF, *Hunecote* 1296 WhC], **Huncote** Le [*Hunecote* DB, -*cot* 1253 Ch, *Honecote* 1270 Ipm]. '*Hūna*'s COT(s).'

Hundersfield La [*Hunnordesfeld* 1202, *Hunewrthefeld* 1235 FF]. 'The FELD of Hunworth.' Hunworth, from *Hūnan worþ*, must be an old name of the place.

Hunderthwaite YN [*Hundredestoit* DB, *Hunderthuait* J Ass]. Possibly '*Hūnrøð*'s thwaite'. *Hūnrøðr* is an ON pers. n.

Hundleby Li [*Hundelbi* DB, 1190 P, *Hundelby* 1209–35 Ep]. '*Hundulf*'s BY.' *Hundulf* is ON *Hundolfr*.

Hundon Li [*Hunidune, Humendone* DB, *Huneduna* c 1115 LiS, *Hunedon* 1193 P], **H~** Sf [*Hunen-, Hunedana* DB, *Huneden* 1219 FF, *Hunedene* 1263 Ipm]. One is '*Hūna*'s DŪN', the other is '*Hūna*'s DENU or valley'.

Hungerford Brk [*Hungreford* 1101–18 Fr, *Hungerford* c 1148 Fridesw, 1219 Fees]. Literally 'hunger ford', 'ford where people had to starve'.

Hungerstone He [*Hunegarestun* 1242 Fees, *Huniegarestun* 1249 Fees]. Probably OE *Hūnan gærstūn* '*Hūna*'s pasture land'.

Hungerton Le [*Hungretone* DB, *Hungerton* c 1125 LeS, 1191 P], **H~** Li [*Hungretune* DB, *Hungertuna* 1106–23 (1333) Ch].

'TŪN with poor soil, where people had to starve.'

OE **hunig** 'honey'. See HONEYBORNE, HONILEY, HONINGTON, HONITON, HUNNINGTON. The el. refers to a wood or other place where wild honey could be found or to a homestead where bees were kept. Cf. BEWICK, BICKERTON and the like.

Hunmanby YE [*Hundemanebi* DB, 1205 FF, *Hundmannebi* 1196 P]. Perhaps 'the BY of the dog-keepers'. OScand *hundamann* might be synonymous with *hundasveinn*, but is not recorded.

Hunningham Wa [*Huningeham* DB, *Huningham* 1236, *Hunningham* 1242 Fees]. 'The HĀM of *Hūna*'s people.'

Hunnington Wo [*Honinton(e)* 1270 ff., *Honewynton* 1276 Ct]. OE *Hunig-tūn* 'TŪN where honey was produced'.

Hunscote Wa [*Unestonescota* 1176 P, *Hunstanescot* 1206 f. Cur]. '*Hūnstān*'s COT.'

Hunsdon Hrt [*Honesdone* DB, *Hunesdone* 1220 Fees]. '*Hūn*'s DŪN.'

Hunshelf YW [*Hunescelf* DB, *Huneself* 1227 BM]. '*Hūn*'s SCYLF or ledge of land.' Or the first el. may be OE *hūn* 'a cub'.

Hunsingore YW [*Hulsingovre* DB, *Hunsinghour* 1195 Cur, *Hunsigour* 1208 FF, -*ouer* 1241 Misc]. 'The OFER or ridge of *Hūnsige*'s people.'

Hunslet YW [*Hunslet* DB, *Hunesflet* c 1180 YCh (1620), 1202 FF]. '*Hūn*'s FLĒOT or stream.'

Hunsley, High & Low, YE [*Hund(r)eslege* DB, *Hundeslai* 1100 YCh 1894, -*leie* c 1115 ib. 966]. '*Hund*'s LĒAH.' Cf. HOUNSLOW.

Hunsonby Cu [*Hunswanby* 1292 QW]. 'The BY of the dog-keepers.' First el. ON *hundasveinn* 'dog-keeper'.

Hunstanton (hŭnstn) Nf [*Hunstanestun* c 1035 Wills, *Hunestanestuna* DB, *Hunstanestun* c 1150 Crawf], **Hunstanworth** Du [*Hunstanwortha* 1183 BoB, -*worth* 1291 Tax]. '*Hūnstān*'s TŪN and WORÞ.'

Hunsterson Chs [*Hunsterton* 1260 Court, E 1 BM, *Honstretton* post 1292 BM]. The name may be a combination of a name identical with HOUNDSTREET So and TŪN. Or it might be a STRETTON with a pers. n. *Hūn* or OE *hunta* prefixed.

Hunston Sf [*Hunterstuna* DB, -*ton* 1197 FF, *Hunterestun* c 1095 Bury]. OE *hunteres tūn* 'the huntsman's TŪN'. See HUNTA.

Hunston Sx [*Hunestan* DB, 1230 Cl]. '*Hūn*'s stone.'

Hunsworth YW [*Hunddeswrth* 1195 YCh 1767, *Hundeswurth* 1226 FF]. '*Hund*'s WORÞ.' Cf. HOUNSLOW.

OE **hunta** 'huntsman' is a common first el., esp. in the gen. plur. form *huntena*, as in

HUNTINGDON, HUNTINGFORD, HUNTINGTON, HUNTON, HUNTWICK, HUNTWORTH. But there was also an OE pers. n. *Hunta*. An interesting group is formed by BONHUNT, CHESHUNT, which seem to have OE *huntan* as second el. Cheshunt means 'chester huntsmen', later 'the homestead or village occupied by them'. An OE **hunte* or the like 'hunting' or 'hunting-ground' is unrecorded. Names in *-hunt* of the kind here suggested have many analogies in Germany, as MÜNCHEN 'Munich', lit. 'the monks', FORSTERN (OHG *Forstarûn*) 'the foresters', ZEILARN (OHG *Zîdalarin*) 'the bee-keepers'. Cf. Introd. ii. 1. *Hunter* is first evidenced in OED from c 1250, but must have existed in OE times, as shown by HUNSTON Sf (in DB). Cf. also HUNTERCOMBE. *Cranhunterestone* 801 BCS 300 very likely represents *Cranhuntena stān* 'the crane-hunters' stone' in the original. The text is a late transcript.

Huntercombe O in Nuffield [*Huntercumbe* 1231, *-cume* 1237 Cl]. 'The huntsmen's valley.'

Huntingdon Hu [*Huntandun* 921 ASC, *Huntedun* DB]. 'The huntsman's hill' or possibly *Hunta's* hill'. **Huntingdonshire** is *Huntadunscir* 1011 ASC (E).

Huntingfield Sf [*Huntingafelde* DB, *Huntingefeld* c 1180 Bodl]. 'The FELD of *Hunta's* people.'

Huntingford Do [*Hunteneford* 1279 For], H~ Gl [*Huntenaford* 940 BCS 764, *Hunteneford* 1228 Ch]. 'The huntsmen's ford.'

Huntington Chs [*Huntingdun* 958 BCS 1041, *Hunditone* DB, *Huntinthona* c 1100 Chester, *Huntindun* 1233–7 Chester, *-don* 1244 Ch], H~ St [*Huntendon* 1198, 1236 Fees, *Huntedon* 1247 Ass, *Huntingdon* 1262, *Hontindon* 1271 For], H~ YN [*Huntindune* DB, *Huntingedon* 1188 P, *Huntington* 1202 FF]. OE *huntena dūn* 'the DŪN of the huntsmen'. H~ Chs may be OE *huntingdūn* 'hill for hunting'.

Huntington He nr Hereford [?*Huntena tun* 757–75, 796 BCS 218, 277, *Huntenetune* DB], H~ He nr Kington [*Hantinetune* DB, *Huntinton* 1228 Pat, 1230 Cl, *Huntiton* 1267 Ipm], H~ Sa nr Little Wenlock [*Hantenetune* DB, *Huntiton* 1255 RH, 1317 Ch]. OE *huntena tūn* 'the TŪN of the huntsmen'.

Huntley Gl [*Huntelei* DB, *-leia* 1146 Fr]. 'The wood of the huntsman.' Cf. LĒAH.

Hunton Ha [*Hundatún* c 909 BCS 629, *Hundinton* 1167 P]. OE *Hunda-tūn* 'TŪN where the hounds were kept'.

Hunton K [*Huntindune* 11 DM, *-tone* 1212 RBE, *-ton* 1257 Ch], H~ YN [*Huntone* DB, *Hunton* 1231 FF, 1236–8 Pudsay]. 'The TŪN of the huntsman or huntsmen.'

Huntroyde La [*Huntrode* 1412 VH]. 'The ROD or clearing of the huntsman or of *Hunta*.'

Huntsham D [*Honesham* DB, 1242 Fees]. '*Hūn's* HĀM.'

Huntsham He [*Hondsum* c 1200 Glouc, *Hunsum* 1233 Cl]. The name looks like OE *hondsum* 'handsome'. If so, it was probably used first as a byname and surname of a person and was later transferred to the place, a very unusual phenomenon. But *Honsom* is actually used as a surname 1298 Glouc (Laur. *Honsom*, not *de Honsom*, which occurs, however, ib. 1301).

Huntshaw D [*Huneseve* DB, *Hunshaue* 1242 Fees], **Huntspill** So [*Hunespille* 1084 Geld R, *Honspil* DB, *Hunespil* 1170, *-pille* 1177 P]. '*Hūn's* shaw and pool.' Cf. PYLL. Or Huntshaw is OE *hūn-scaga* 'cub shaw'.

Huntstile So [*Hustille* DB, *Hunestille* 1212 RBE, *Hunstille* 1285 FA]. '*Hūn's* or *Hunta's* stile' or 'the stile of the huntsman'.

Huntwick YW [*Huntewich* 1202 FF, *Huntewykes* 1280 Ch, *Huntwyk* 1402 FA], **Huntworth** So [*Hunteworde* DB, *-worth* 1225 Ass]. 'The WĪC and WORÞ of the huntsman or of *Hunta*.'

Hunwick Du [*Hunewic* c 1050 HSC, 1104–8 SD], **Hunworth** Nf [*Huneuurda*, *Hunaworda* DB, *Hunewrth*, *Hunesworth* 1211 f. Cur, 1254 Val]. '*Hūna's* WĪC and WORÞ.'

Hurcot So nr Somerton [*Herdecot* 1212 Fees, *-cote* 1274 RH, *Hurdekote* 1276 RH], **Hurcott** So nr Seavington [*Herdecote* 1260 Wells, 1291 Tax], **Hurdcott** W nr Wilton [*Hardicote* DB, *Hurdecote* 1270 Ipm], H~ W nr Salisbury [*Herdicote* DB]. More likely *hierda cot* 'the hut of the herdsmen' than *heorde-cot* 'COT for the flock of sheep'.

Hurdlow Db [*Hordlawe* 1244 FF, *Hordlowe* 1251 Ch]. OE *hord-hlāw* 'treasure mound'. First el. OE *hord* 'hoard, treasure'.

Hurdsfield Chs [*Hirdisfeld*, *Herdisfeld* 1285 Court, *Hurdisfeld* a 1303 BM]. '*Hygerēd's* FELD.'

Hurleston Chs [*Hurdleston* 1278 Ipm, 1316 BM, *Hurdeleston* 1311 Ipm, *Hurdlaston* 1325 BM]. Hardly 'TŪN with a hurdle'. But H~ is near an Aston, and the name may be ASTON with *hurdle* prefixed for distinction. Note the form of 1325.

Hurley Brk [*Herlei* DB, *Hurlea* 1167 P, *Hurnlye* 1220, *Hurnle* 1242 Fees], H~ Wa [*Hurlega*, *Hurnlee* c 1180 Middleton, *Hurleg* 1207 Cur]. OE *hyrn-lēah* 'LĒAH in a HYRNE or corner'. H~ Brk is near a big bend of the Thames. H~ Wa is by a marked hill with a peculiar recess in it.

Hurn. See HERNE.

Hursley Ha [*Hurselye* 1171 Ep, *Hurselege* c 1255 Selborne, *-leye* c 1270 Ep]. Possibly OE *horsa-lēah* 'pasture for horses'. The *u*-form is remarkable; cf., however, forms of HORSEY So. The first el. may also be an OE *hyrse* 'mare' (cf. ON *hryssa*), also found in **Hursley Bottom** W (*on hyrsleage* 939 BCS 734). See SPN, p. 65, PNW(S).

Hurst Brk [*La Hurste* 1242 Fees, *Hurst* 1252 Cl], H~ K SE. of Ashford [*Herst* 1219 Fees, *Herste* 1227 Ass], H~ So [*Hurst* 1285 FA], H~ Wa [*Hurst* 1285 QW]. OE *hyrst* 'hill, wood, wooded hill'.

Hurstbourne Priors & Tarrant Ha [(juxta) *Hissaburnam* a 790, (æt) *Hysseburnan*, (æt) þam *nyðeran Hysseburnan* c 880 BCS 258, 553, *Hysseburna* 961 ib. 1080, *Esseborne* DB, *Huphusseburn* 1242 Fees, *Husseburne Prior'*, *Regis* 1291 Tax]. Really the name of the stream at the place [(ofer) *Hysseburnan* 901 BCS 594]. An identical name is HUSBORNE Bd. The first el. is OE *hysse* 'a tendril, vine-shoot'. The reference may be to winding water-plants.

H~ **Priors** belonged to the monks of Winchester, H~ **Tarrant** to Tarrant Abbey.

Hurstley He [*Hurtesleg* 1242 Fees, *-leye* 1282 Ep]. 'The LĒAH (clearing or wood) of the stag.'

Hurstpierpont Sx [*Herst* DB, *Herst Perepunt* 1279 Ass]. Originally *Hyrst* (see HYRST).

Robert [de Pierpoint] held the manor in 1086 (DB). Cf. HOLME PIERREPONT.

Hurtmore Sr [*Harmera* DB, *Hertemere* 1242 Fees, *Hurtemere* c 1270 Ep]. 'Stags' lake.'

Hurworth Bryan Du [*Hordeworðhe* 1211, *Hurtheworth* 1212 Cur, *Hurworth Bryan* 1438 PNNb], **H~ on Tees** Du [*Hurdewurda* 1158 YCh 400, *-wurd* 1196 P]. 'Enclosure made from hurdles.' See WORÞ. First el. an unrecorded OE *hurþ*, corresponding to ON *hurð*, Goth *haurds*, OLG *hurth* and meaning very likely 'hurdle'. H~ **Bryan** was held in 1211 by Brian, son of Alan Earl of Richmond.

OE, OScand **hūs** 'house' is mostly found in Scandinavian England and chiefly in the (dat.) plur., as in HOWSHAM, NEWSHAM, NEWSHOLME &c., AYRESOM, GILDERSOME, LOFTHOUSE &c., WOTHERSOME. Other exx. are ONE-, SALT-, STONEHOUSE, HUSTHWAITE.

Husborne (-z-) Crawley Bd [*Crawelai* DB, *Crawelye et Husseburne* 1237-40, *Craule Husseburn* 1242 Fees, *Hussebourne Craule* 1331 Misc]. Husborne was originally a river-name [(of) *Hysseburnan* 969 BCS 1229], on which see HURSTBOURNE. Crawley was originally a distinct manor. Eventually it and Husborne were joined into one, which became known as Husborne Crawley.

Husthwaite YN [*Hustwait* 1167 P]. 'Clearing with a house on it.' See THWAITE.

Hüthwaite Nt [*Hothweit* 1208 Obl, *Hothweyt* 1288 Ipm, *Hokenhale Houthwayt* 1330 PNNt(S), *Hucknoll Howthwaite* 1611 BM], H~ YW [*Hothweit* 1219 FF]. 'Clearing on a HŌH or spur of land.' Cf. THWAITE.

Huttoft Li [*Hotoft* DB, c 1115 LiS, *Hottoft* 1202 Ass]. 'Toft on a HŌH or spur of land.'

Hutton John Cu [*Hoton* 1291 Tax, *H~ Johan* 1296 Cl], **H~ Scough** or **H~ in the Forest** Cu [*Hoton* 1212 Fees, *H~ in foresta* 1248 Cl], **H~ Roof** Cu [*Hotunerof* 1278 Ass, *Hoton Rouf* 1287 Cl], **H~ Henry** Du [*Hotun* c 1050 HSC], H~ **Ess** [*Atahov* DB, *Hotone* 1254 Val, *-ton* 1291 Tax], **H~ La** in Penwortham [*Hotun* a 1180 LaCh, *Hoton* c 1200 CC], **H~ La** in Quernmore [*Hotun* DB], **H~ La** in Wennington [*Hoton* a 1227 CC], **Priest H~ La** [*Hotune* DB, *Prest Hoton* 1280 Kendale], **H~ So** [*Hotune*, *Hutone* DB, *Hocton* 1243 Ass, *Hutton* 1291 Tax], **H~ Roof** We [*Hotun* DB, *Hotunariof* 1157 YCh 354, *Hoton Rofh* c 1175, *Hotonriwe* c 1190 Kendale], **New & Old H~**, **H~ in the Hay** We [*Hotun* DB, *Hoton* 1170-80, 1274, *Old H~* 1297, *H~ Hay* 1297 Kendale, *H~ in Laya* 1283 Ipm], **H~ Cranswick** YE [(*Cranzvic* and) *Hotone* DB, *Hoton by Crauncewyk* 1310 Ch], **H~ Bonville** YN [*Hotune* DB, *Hoton Benevill* 1285 FA], **H~ Bushell** YN [*Hotun* DB, *Hoton Buscel* 1253 Ep], **H~ Conyers** YN [*Hotone* DB, *Hotonconyers* 1198 FountM], **H~ Hang** YN [*Hotun* DB, *Hoton Hange* 1290 Ch], **High & Low H~**, **H~ Hill** YN [*Hotun* DB, *Hoton* 1202 FF], **H~ le Hole** YN [*Hotun* DB, *Hegehoton* J Ass, *Hoton Underheg* 1285 FA], **H~ Lowcross** YN [*Hotun* DB], **H~ Magna, Little H~** YN [*Hotune* DB], **H~ Mulgrave** YN [*Hotune* DB, *Hoton juxta Mulegref* 1303 FA], **H~ Rudby** YN [*Hotun* DB, *Hoton by Ruddeby* 1310 Ch], **Sand H~** YN nr York [*Hotone* DB, *Sandhouton* 1219 FF], **Sand H~** YN nr Thirsk [*Hotune* DB, *Sandhoton* 12 Fount], **H~ Sessay** YN [*Hotun* 1252 Ch], **Sheriff H~** YN [*Hotone* DB, *Shirefhoton* c 1200 YCh 1054], **H~ Wandesley** YW [?*Hotun* c 1030 YCh 7, *Wendesle*, *Hoton* 1226, *Hotun Wandelay* 1253 FF]. OE *Hō-tūn*. First el. OE HŌH 'a spur of hill'. H~ **Ess** was originally *æt þam hō* 'at the HŌH'.

H~ **Bonville** YN was held by Robert de Boneville t. Hy 3 (from BONNEVILLE in Normandy).—H~ **Bushell** YN was held by Alan Buscel c 1140 (YCh 372). Buscel is a byname, derived no doubt from OFr *bucel* 'small barrel'. —H~ **Conyers** YN. Conyers may be from COIGNIÈRES in Seine-et-Oise or COGNERS in Sarthe.—H~ **Cranswick** YE is near CRANSWICK.—H~ **Hang** YN is in Hang wapentake.— H~ **in the Hay** We was in the Hay of Kendal.— H~ **Henry** Du was held by Henry de Essh c 1380 (Hatfield).—H~ **le Hole** YN is said to be 'H~ in the hollow', while the epithet *Underheg* refers to a hay for hunting (PNNR).— H~ **John** Cu from some early owner.—H~ **Lowcross** YN from a place close by [*Loucros* 12 Guisb]. *Low-* may be identical with the first el. of LOWTHORPE YE.—H~ **Mulgrave** and **Rudby** YN are near MULGRAVE and RUDBY respectively.—**Priest H~** La was held by the rector of Warton.—H~ **Roof** Cu from a Norse or Norman owner *Rolf* (ON *Hrólfr*).—H~ **Roof** We from some early owner. *Roof* may be OFr *Riulf* from OG *Ricwulf*.—**Sand H~** YN from sandy soil.—H~ **Scough** Cu was in Inglewood Forest. *Scough* is OScand *skógr* 'wood'.— H~ **Sessay** YN is near SESSAY.—**Sheriff H~** YN was held by the sheriff of York.—H~

Wandesley YW from a lost place [*Wandeslage* DB]; cf. WENSLEY.

Huxham D [*Hochesham* DB, *Hokesham* 1212 Fees, 1230 P]. '*Hōc*'s HĀM.'

Huxley Chs [*Huxeleg* 1260 Court, *Huxleg* 1271, *Hoxeleg* 1279 Chester, *Huxelegh* 1285 Ch, *Huckysley* 1284 Ipm]. Perhaps '*Hōc*'s LĒAH', but earlier forms are needed.

Huyton (-ī-) La [*Hitune* DB, *Hutona* 1189–96 LaCh]. OE *Hȳþ-tūn* 'TŪN by a landing-place'. Cf. HȲþ.

ON hváll, hóll 'round hill'. See WHALE, FALSGRAVE, STAFFIELD, WARTHALL.

OE hwǣte 'wheat'. See WHEAT- (passim), WADDINGTON Sr, WHADDON, WHATBOROUGH &c.

OE hwamm meant 'corner' (of a house or room). It was probably used also in the same sense as the cognate ON *hvammr*, i.e. 'small valley' or probably, more exactly, 'nook, valley surrounded by high hills', which is the meaning of Norw *kvam*. See BROOMY HOLM, ULGHAM, ULWHAM, WHITWHAM.

OE hwearf, hwerf 'wharf, embankment, shore' is the first el. of WHARTON We, WHERSTEAD. The exact meaning of the element in the pl. ns. is not apparent.

OE hwēol 'wheel, circle' is the first el. of several names, as WELBATCH, WHEELTON, WHELNETHAM, WHILTON. The OE side-form *hweowol* is found in WELLSBOROUGH. The exact meaning of the word is probably not the same in all the names. Sometimes the meaning may be 'water-wheel'. Sometimes a round object is probably referred to, e.g. a stone circle. An illustrative instance is the following from StB (Cu): 'a quibusdam circulis qui vocantur *le Wheles* juxta Harashowe'.

OE hwer 'kettle, cauldron'. See WHARRAM, WHERWELL.

OE hwerfel, hwyrfel is the source of later *whirl, whorl* 'fly-wheel of a spindle, curl, spiral' &c., and corresponds to MLG *wervel* 'circle' &c., ON *hvirfill* 'circle, crown of the head' &c. It is only found in charters, as (on) *þone hwyrfel* 938 BCS 724, *hwerfel dic* 1001 KCD 705. The meaning may be 'circle' or 'whirlpool'. See QUARLES, WHARLES, WHORLTON Nb.

OE hwīt 'white' is a common first el. in pl. ns., as in WHISTON, WHIT-, WHITE- (passim). There is hardly ever any reason to assume any other meaning for the word in pl. ns. than 'white, light-coloured'.

Hyde Bd [*la Hide* 1197 FF], **H~** Chs [*Hyde* 1285, *Hide* 1288 Court], **H~ Abbey** Ha [*Hida* Hy 1 Hyde, *Hyda* 1190 P], **H~ Park** Mx [*Hida* 1204 FF]. OE HĪD 'hide' (of land).

Hykeham Li [*Hicham* DB, 1138 NpCh, 1202 Ass, *Hiccham* 1160–5 NpCh, *Hikham* 1195 FF, *Northicam* DB, *Nort-*, *Suthicham* 1212 Fees]. The first el. may be OE *hīce*, the name of a bird, perhaps identical with *hīcemāse* 'blue titmouse'. Second el. HAMM or HĀM.

OE hyll 'hill' is common alone and as a first and second el. of pl. ns. Cf. HILL, HULL, HILLAM, HILTON, HULTON &c. As a second el. it varies a good deal; cf. e.g. COPPULL, HETHEL, PENDLE, WHITTLE, SHELFIELD, COOLE.

Hylton Du [*Helton* 1195 (1335) Ch, *Hilton* 1291 Tax, 1312 RPD]. It is doubtful if this is OE *Hyll-tūn*. Cf. HELTON We.

OE hyrne 'corner', also no doubt 'bend', is sometimes found in pl. ns. See HERNE, HURLEY.

OE hyrst, ME hurst &c. 'hillock, knoll, esp. one of a sandy nature, copse, wood, wooded eminence'. The original meaning was very likely 'brushwood'; cf. the cognate Welsh *prys* 'brushwood'. OG *hurst* is rendered by 'brushwood'. The exact meaning cannot always be determined. The probable meaning is 'wood' in names such as ASHURST, NUTHURST. See HERST, HIRST, HURST &c., HARTEST, HOLNEST.

Hythe (hīdh) K [(on) *Hyþe* 1052 ASC, *Hede* DB, *Hythe*, *Hethe* 11 DM, *Heða* 1177 P]. OE *hȳþ* 'landing-place'. OE **hȳþ** is fairly common in pl. ns. See HIVE, HIDDEN, HUYTON, HYTON, also HEDINGHAM. It often has as first el. a word denoting the product imported or exported, as in CHELSEA, LAMBETH, ROTHERHITHE, probably EARITH, ERITH, but sometimes a pers. n., as in HULLASEY, MAIDENHEAD, PUTNEY, STEPNEY. Cf. also ALDRETH, STOCKWITH.

Hyton Cu nr Bootle [*Hytun* 1220–30, c 1220 StB]. Identical with HUYTON.

I

Ibberton Do [*Abristentona* DB, *Ebrictinton* 1211 FF, *Edbrichton* 1245 Cl]. 'The TŪN of *Ēadbeorht*'s people.'

Ible Db [*Ibeholon* DB, *Ibole* 1288 Ipm, 1308 FF]. OE *Ibban holu* '*Ibba*'s hollows or valley'. Cf. HOLH. *Ibba* is found as the name of a moneyer and in *Ibban stan* 951 BCS 892 (Brk). Cf. also IBSTOCK &c.

Ibsley Ha [*Tibeslei* DB, *-a* 1166 RBE, *Ibeslehe* 13 AD, *Ibbeslig* 1242 Fees]. '*Ibbi*'s or *Tibbi*'s LĒAH.' **Ibbi* and **Tibbi* are normal side-forms of *Ibba* and *Tibba*. If the original form was *Tibbeslēah* the loss of *T-* is due to wrong division of *æt Tibbeslēa*. If *Ibbeslēah* is the old form, *T-* was transferred from the prep. *æt* (*æt Ibbeslēa*).

Ibstock Le [*Ibestoche* DB, *-stok* 1209–35 Ep, *Ebbestoka* Hy 2 DC, *Ybestock* 1254 Val]. '*Ibba*'s STOC.' Cf. IBLE.

Ibstone Bk [*Hibestanes, Ebestan* DB, *Ibbastana* c 1160 Oxf]. '*Ibba*'s stone.' Cf. IBLE.

Iburndale YN [*Ybrun* c 1180, *Yburne* 1308 Whitby]. 'The valley of R Iburn.' **Iburn**, the old name of the river at I~ [*Ybrun* c 1180 Whitby], means 'yew stream' (OE *iw-burna*).

Iccomb Gl [*Icancumb* 781, *Iccacumb* 964 BCS 240, 1135, *Iccumbe* DB]. '**Ica*'s valley'; cf. CUMB. In the boundaries in BCS 240 is mentioned *Icangæt* '*Ica*'s gate'.

Ickburgh Nf [*Iccheburna, Ic(c)heburc* DB, *Ykeburc* 1193 P, *Ikeburc* 1199 FF]. '*Ica*'s BURG.' Cf. ICCOMB.

Ickenham Mx [*Ticheham* DB, 1163 P, *Tikenham* 1203 Cur, *Ikeham* 1203 Cur, *Ikenham* 1252 Ch]. '*Ica*'s or *Tica*'s HĀM.' Cf. IBSLEY. For *Tica* see TICKENHAM.

Ickford Bk [*Iforde* DB, *Ycford* 1199 P, *Ikeforde* 1226 Ep]. '*Ica*'s ford.' Cf. ICCOMB.

Ickham K [*Ioccham* 785, *Iocham* 786, *Geocham* a 958 BCS 247 f., 1010, *Gecham* DB, *Ieocham* 11 DM, *Icham* 1233 BM]. 'HĀM comprising a yoke of land.' OE *geoc* denoted one-fourth of a suling or 50 to 60 acres. Later *yoke* came to be used of a small manor. Cf. YOCKLETON.

Ickleford Hrt [*Ikelineford* Hy 2 AD iii, *Ikelesforde* 1219 Ep, *Icleford* 1220 Fees, *Hiclingford* Hy 3 BM, *Ikelingford* 1303 FA]. OE *Iceles-ford* '*Icel*'s ford' and *Icelinga-ford* 'the ford of *Icel*'s people'.

Icklesham Sx [*Ikelesham*, (to) *Icoleshamme* 772 BCS 208, *Ichelsham* 1161, *Yclesham* 1190 P]. '*Icel*'s HAMM.'

Ickleton Ca [*Icelingtun* c 1000 BCS 1306, *Ichelintone* DB, *Yclinton* 1194 P]. 'The TŪN of *Icel*'s people.'

Icklingham Sf [*Ecclingaham, Etclingaham* DB, *Echelincgham* 1086 IE, *Ikelingeham* 1242 Fees, 1254 Val]. Probably 'the HĀM of *Yccel*'s people'. OE **Yccel* is a normal derivative of *Ycca* and *Ucca* (cf. UCKFIELD &c.). The *cc* of *Yccel* would be palatalized, but in the derivative *Ycclingas* it would remain stopped before the consonant.

Icknield Way, an ancient British road from Norfolk to Dorset [*Iccenhilde weg, Icenhylte* 903, *Icenhilde weg* 1043–53 BCS 601, 603, 479, *Ikenildestreta* 1185 P, *Ykenildestret* 1227 FF, 1297 Rob Gl], **Icknield** or **Ryknild Street** [*Ykenilde Strete* 1275 Ass]. The etymology of the name has not been found. ICKLEFORD and ICKLETON are on Icknield Way, which is sometimes called Ickleton Street. Icknield Street seems to have been transferred from Icknield Way.

Ickornshaw YW [*Icornsawe* 1279 Ch]. 'Squirrel wood.' Cf. SCAGA. First el. ON *ikorni* 'squirrel'.

Ickwell Bd [*Gikewell* 1195 Cur, *-e* 1202 Ass, *Jekewelle* 1240 Ass]. First el. possibly OE *gēoc* 'help'. The name would then be analogous to BOTWELL Mx. OE *ēo* often becomes *i* in Beds.

Ickworth Sf [*Ikewrth* c 950 Wills, *Iccawurð* 1047–65 BM, *Icceuuorde* c 1095 Bury, *Hikewrd* 1196 FF, *Ikewrthe* 1254 Val]. '*Ica*'s (or *Icca*'s) WORÞ.' Cf. ICCOMB.

Idbury O [*Ideberie* DB, *-bir'* 1236 Fees]. '*Ida*'s BURG.'

Iddesleigh D [*Edeslege* DB, *Edwislega* 1107 (1300) Ch, *-lege* 1219 FF]. '*Ēadwīg*'s LĒAH.'

Iddinshall Chs [*Etingehalle* DB, *Edinchale* c 1100, c 1150, *Idinchale* c 1190, *Idinghale* c 1235 Chester, *Idingehale* 1287 Court]. 'The HALH of *Ida*'s people.'

Ide (ēd) D [(*æt*) *Ide* c 1070 Ex, *Ide* DB, *Yde* 1291 Tax]. Possibly an old rivername.

Ideford (idĭ-) D [*Yudaforda* DB, *Yddeford* 1281 QW, *Yuddeforde* 1309 Ep, *Giddeforde* 1315 Ep]. Possibly '*Giedda*'s ford', **Giedda* being a side-form of *Geddi*. The gen. plur. of OE *giedd* 'song, speech' is also a possible first el. Cf. GLANDFORD, GLEMSFORD.

Iden (ī-) Sx [*Idene* DB, *Idenne* 1295 Misc]. OE *īg-denn* 'pasture in a piece of land in a marsh'.

Idle (ī-) R Nt, Li [*Idlae* c 730 Bede, *Idle* c 890 OEBede, *Iddel* 958 BCS 1044]. A derivative of OE *īdel* 'idle', perhaps in the sense 'slow'. But *īdel* is held to have had the original meaning 'shining, bright'. That might be the sense here.

Idle YW [*Idla* c 1190 YCh 1785, *Hidel* 1212 f. FF, *Idel* 1271 Ipm]. Perhaps a derivative of OE *īdel* 'idle' in the sense 'uncultivated land'.

Idlicote Wa [*Etelincote* DB, *Iteli-*, *Utelicota* Hy 2 (1314) Ch, *Utlicote* 1291 Ch]. 'The COT of *Yttel*'s people.' *Yttel* is a diminutive of *Utta*. *Uttel* is recorded.

Idmiston W [*at Idemeston(e)* 947, 970 (late copies) BCS 829, 1259, *Ydemeston* 1190 P, *Idemereston* 1268 FF, *Edemeston* 1305 PNW(S)]. First el. no doubt a pers. n., e.g. *Idhelm* or *Idmær*, preferred in PNW(S), or even *Ēadmær* with early change of *ēa* to *īe* and *i*. Regular *Ide-* goes best with *Idhelm*.

Idridgehay Db [*Edrichesei* 1230 P, *Iriggehay* 1252 Ch, *Eddricheshey*, *Iddurshey* 1484 Derby]. '*Ēadrīc*'s hay.' See (GE)HÆG.

Idsall Sa, now Shifnal [(æt) *Iddeshale* 836 BCS 416, *Iteshale* DB]. '*Iddi*'s HALH or valley.'

Idstone Brk [*Edwineston* 1199 FF, c 1235 Fridesw]. '*Ēadwine*'s TŪN.'

Idsworth Ha [*Iddesworth* 1238 Cl, *-worth* 1257 Ch, *Lidesworth* 1242 Fees]. '*Iddi*'s WORÞ.'

Iffley O [(to) *Gifetelea* 1004 Fridesw, *Givetelei* DB, *Iuittelai* 1165 P, *Ghyftele* 1234 Cl, *Yveteleg* 1236 Fees]. The first el. may be an old word for 'plover' or some similar bird cognate with MHG *gîbitz*, G *Kiebitz* 'plover'. Cf. Engl *peewit*, *tewhit* 'lapwing'. Second el. OE LĒAH.

Ifield (ī-) K [*Yfeld* 1198 P, *Yffeld* 1203 Cur, *Ifelde* 1212 RBE], **I~** Sx [*Ifelt* DB, *Yfeld* 1212 RBE]. OE *īw-feld* 'open land where yew grew'.

OE **īfig** 'ivy'. See IVY-, also ST. IVES Ha, HAYTOR.

Ifold Sx [*Ifold* 1296 Subs]. OE *īeg-falod* 'fold in river land'. Cf. ĒG.

Iford (ī-) Sx [*Niworde* DB, *Yford* 1219 FF], **I~** W [*Igford* 987 KCD 658]. The last is 'ford by an island'. The first may be identical in origin or 'yew ford'.

Ifton Heath Sa [*Iftone* 1272 Ipm]. Earlier material is wanted.

Ightenhill (īt-) La [*Ightenhill* 1242 LaInq, *Ichtenhille* 1296 Lacy], **Ightfield** Sa [*Istefelt* DB, *Hichtefeld* 1175 f., *Ihttefeld* 1230 P, *Ihtenefeld* 1260 Court]. The first el. may be Welsh *eithin* 'furze' (< *ektin* < *aktin*).

Ightham (īt-) K [*Ehteham* c 1100 Text Roff, *Eitham* 1232 Pat, *Heitcham* 13 BM, *Eychtham* 1291 Tax, *Eghteham* 1322 Ipm]. '*Ehta*'s HĀM.' *Ehta* may be a side-form of *Ohta*.

Iken (ī-) Sf [*Ykene* 1212 RBE, 1212 Fees, *Ikene* 1254 Val]. Possibly an old rivername related to ITCHEN. But it may be simply OE *Ican ēa* '*Ica*'s stream'.

Ilam St [*Hilum* 1002 Wills, 1004 KCD (710), 1227 Ass, *Ylum* Hy 1 Burton, 1208 FF]. Apparently derived from an old name of the Manifold identical with HYLE Ess (see ILFORD). The plural form would be analogous to OE *Liminum* 'Lympne' from *Limen* (river).

Ilchester So [*Givelcestre* DB, *Giuelcestr'* 1156, *Iuelcestr'* 1157 P]. 'Roman fort on R Yeo' (formerly *Gifl*). Cf. YEOVIL, YEO.

Ilderton Nb [*Ildretona* c 1125, *Hildreton* Hy 2 (1336) Ch, *Hildirton* 1242 Fees, *Hillerton* 1346 FA]. 'Elder TŪN.' Cf. HINDERCLAY.

Ilford Ess [*Ilefort* DB, *-ford* 1167, *Yleford* 1171 P, *Hileford* 1234 FF]. 'Ford over the Roding' (formerly Hyle). Hyle is (innán) *Hile*, (andlang) *ealdan Hilæ* 958 BCS 1037, *Hyle* 1250 Waltham Cart. It is a Brit name related to Ir *silim* 'to distil', Welsh *hil* 'seed' and meaning 'trickling stream'.

Ilford So [*Ileford* 1260 Wells, 1269 Ass]. 'Ford over R ISLE.'

Ilfracombe D [*Alfreincome* DB, *Alferdingcoma* 1168 P, *Alfredescumbe* 1249 FF, *Ilfridecumbe* 1279 Ch]. 'The coomb of *Ælfrēd*'s people.' *Ælfrēd* here appears in its correct WSax form *Ielfrēd, *Ilfrēd. Its first el. is OE *ielf* 'elf' (Angl *ælf*).

Ilkerton D [*Incrintona* DB, *Hilcrinton* 1242 Fees]. 'The TŪN of *Ælfgār*'s people.' *Ælfgār* appears in its correct WSax form *I(e)lfgār. Cf. prec. name.

Ilkeston Db [*Tilchestune* DB, *Elchesdona* 1155–7 YCh 1148, *Hilkesdon* 1236, *Ilkesdon* 1242 Fees, *Elkesdon* 1252 Ch, *Ilkesdon* E 1 BM]. Apparently identical with ELKSTONE St.

Ilketshall Sf [*Ilcheteleshala*, *Elcheteshala*, *Ulkesala* DB, *Ilketeleshal* 1186 P, *Hulketeleshal* 1228, *Ilketeleshal* 1248 Ch]. '*Ulfketill*'s HALH.' *Ulfketill* is a well evidenced OScand pers. n., which is often found also in England. *Ulfcytel* (ASC 1004) was alderman of East Anglia. The immediate source of the first el. is a side-form of *Ulfketill* that is not evidenced in Scandinavia, viz. *Ylfketill*, which shows the same *i*-mutation as OSw *Æskil*, ODan *Eskil* by the side of ON *Ásketill*, OSw *Yskil* from *Osketill*, OSw, ODan *Thyrkil*, *Thørkil* from *þorketill* &c.

Ilkley YW [*Hillicleg* c 972 BCS 1278, (on) *Yllic-leage* c 1030 YCh 7, *Illicleia* DB, *Illeclay* 1245 FF, *Ylkelay* 1234 FF, *Ilkelay* 1259 Ipm]. Probably '*Illica*'s LĒAH', *Illica being a derivative of *Illa* in ELEIGH, ILLINGTON. Cf. ILTON, GUISELEY. The identification of *Olikana* Ptol with Ilkley is very doubtful, and it is difficult to explain OE *Hillic-*, *Yllic-* from *Olikana*.

Illey Wo [*Hillely* J, *Hilleleye* Hy 3 PNWo, *Illeleya* 1271, *Hilleye* 1276 Ct]. '*Hilla*'s or *Illa*'s LĒAH.'

Illington Nf [*Illynton* c 950 Wills, *Illinketuna* DB, *Illingeton* 1160 P], **Illingworth** YW [*Hilling-*, *Yllingwrth* 1297 Wakef]. 'The TŪN and WORþ of *Illa*'s people.' Cf. ELEIGH.

Illo·gan (-ŭ-) Co [(Ecclesia) *Sancti Illogany* 1291 Tax, (rector) *Sancti Elugani*, *Yllugani* 1308 Ep]. '(The church of) St. Illogan.' The name is apparently a diminutive of Bret *Illec*, Welsh *Illog*.

Illston on the Hill Le [*Elvestone* DB, *Eluestun* 1166, *Iluestona* 1176 P, *Ilveston* 1231 Ch, 1242 Fees]. '*Iolf*'s TŪN.' ON *Iolfr*, *Iulfr* (from *Iō-olfr*) might well have given ME **Elf*, **Ilf*. The name appears to be the source of *Iaulf* KCD 806, *Ialf* DB.

Ilmer Bk [*Imere* DB, *Ilmere* 1161–3 Reg Roff, 1229 Pat]. OE *igil-mere* 'hedgehog mere'. Cf. *ilmere* BCS 1037.

Ilmington Wa [*Ylmandun* 978 KCD 620, 10 Ælfric, *Ilmedone* DB, *-dona* 1169 f. P, *Ylmindon* 1272 Ch]. The first el. is a derivative of *elm* the tree-name, an OE **ilme* 'elm' (cf. *birce* by the side of *beorc* &c.). Second el. DŪN. I~ is by a hill.

Ilminster So [*Illemynister* 995 Muchelney, *Ileminstre* DB]. 'Minster on R ISLE.'

Ilsington D [*Ilestintona* DB, *Ilstingtun* c 1200 Buckland, *Ilstinthon* 1242 Fees]. 'The TŪN of the people of *Ilfgiest* or *Ilfstān*.' *Ilfstān* would be a WSax form of *Ælfstān* (cf. ILFRACOMBE). *Ælfgiest* (*Ilfgiest*) is unrecorded, but cf. OG *Albgast*.

Ilsington Do [*Elsangtone* DB, *Ilsington* 1257, *Elsinton* 1260 FF]. 'The TŪN of *Ælfsige*'s people.' For the form with *i* cf. ILFRACOMBE.

Ilsley, East & West, Brk [*Hildeslei* DB, *Illeslai* 1130 P, *Hyldesle* 1241 Ch, *Est Hyldeslye*, *Westyldesl'* 1220 Fees]. '*Hild*'s LĒAH'; cf. HILLESDEN. The loss of *H-* took place in the combinations *East*, *West Hildesley*.

Iltney Ess in Mundon [*Altenai*, *Eltenai* DB, *Elteneye* 1233 Cl]. OE *ielfetan* (or *ielfetna*) *ēg* 'swan island'. Cf. ELFETU.

Ilton So [*Atiltone* DB, *Ilton* 1243 Ass]. 'TŪN on R ISLE.'

Ilton YN [*Ilcheton* DB, *Ilketon* 1196 ff. P, 1226 FF]. '*Illica*'s TŪN.' Cf. ILKLEY.

Imber Sr [*Limeurde* DB, *Immewurth* 1223 Fine, *-worth* 1280 Ipm]. '*Imma*'s WORþ.' The DB form might point to an OE **Gimma* corresponding to OG *Gimmo*.

Imber W [*Inemerie* DB, *Imbmeram* 1166 RBE, *Immemer* 1198 Fees, *-mere* 1212 RBE]. '*Imma*'s lake.' *Imma* is the first el. of *Ymmanedene* 968 BCS 1215 (in bounds of the neighbouring Edington), and of *Imendone* 1161 RBE (a lost place in Imber). See PNW(S).

Imberhorne Sx [*Hymberhorn(e)* c 1100 &c.

PNSx, *Hinberhorn* 1229 Cl, *Himberhorne* a 1290 AD]. 'Raspberry hill' (OE *hindberie* 'raspberry' and HORN).

Immingham Li [*Imungeham* DB, *Imminge-ham* c 1115 LiS, *Emmingham* 1090–6 YCh 855]. 'The HĀM of *Imma*'s people.'

Impington Ca [*Impintune* c 1050 KCD 907, *Empintona* c 1086 IE, *Epintone* DB, *Emping-ton* 1201 Cur, *Impinton* 1201 FF]. 'The TŪN of *Empa*'s people'; cf. EMPINGHAM. The early change of *e* to *i* is somewhat surprising, but the form of c 1050 is in a late transcript.

Impney Wo at Dodderhill [*Ymeneia* 1176 P, *Imenea* 1212 Fees]. '*Imma*'s island.'

Ince Chs [*Inise* DB, *Ynes* c 1100, c 1150 Chester], **I~ Blundell** La [*Hinne* DB, *Ines* 1212 Fees, *Ins Blundell* 1332 Subs], **I~ in Makerfield** La [*Ines* 1202 P, *Ynes* 1206 P, *Ins in Makerfeld* 1332 Subs]. Welsh *ynys* 'island, water meadow'. The name is very apt for Ince in Chs, which forms (with Elton) an island in the low-lying country on the Mersey.

I~ Blundell passed to the Blundell family c 1200. Blundell is a Fr nickname and family name meaning 'the blond one'. See MAKERFIELD.

Ing Ess [*Ginga*, *Inga* DB, *Ginges* R 1 BM, *Gynges* 1230 P, *Mounteneye Giginge* 1363 Cl]. An OE **Gigingas*, which may mean 'the people of *Giga*'. OE **Giga* corresponds to OG *Gigo* in GINKHOVEN (olim *Giginc-hova*). *Ing* is no longer used alone, but is found in INGATESTONE, INGRAVE, FRYERNING, MARGARETTING, MOUNTNESSING. Cf. also BUTTSBURY. All these places form a group.

-ing is a common ending of pl. ns. It goes back to several sources:

1. A derivative suffix *-ing* sing., (a) in old pl. ns., especially names of streams (as GUITING, LEEMING, WENNING), hills (as *Riv-ing*; see RIVINGTON and cf. HORNDON), but also names of inhabited places (as BOCKING, CLAVERING, KEMSING, LAWLING, NEDGING); (b) in common nouns that have become pl. ns., as CHARING Mx, CHIPPING, FALINGE &c. See also BIGGING, RYDING (under RYDDAN).

2. OE *-ingas* (dat. *-ingum*) plur. Names of this kind are derivatives of pers. ns., as BARKING, BARLING, COOLING, HALLING, HASTINGS, HAVERING, POYNINGS, or of the name of a river or other topographical word, as AVENING, CHEVENING, DORKING, EPPING, NAZEING, ULTING. These names originally denoted the inhabitants of the places. Derivatives of pers. ns. may mean 'the sons (or descendants) of —' or 'the dependants (men) of —'. In the etymologies the usual explanation given is '—'s people', which thus leaves the exact sense open. The derivatives from pl. ns. mean 'the dwellers at —'. AVENING (OE *Æfeningas*) is thus 'the dwellers at the river called Avon', later 'the village of these *Æfeningas*'. Cf. Introd. ii. 1.

3. In not a few cases -*ing* has replaced an earlier ending of different form, as CHELSING from *Chelsen* (and *Chelse*), WORTHING from OE *worþign*. Names in -*ling* are often old compounds in -*hlinc*, in which *c* became *g*, as in SWARLING, SYDLING. Cf. also EAKRING.

-**ing**- is very common in combination with elements like -HĀM, -HAMM, -TŪN, -FELD, -LĒAH, as in BEDDINGHAM, EFFINGHAM, LASTINGHAM; BEDDINGTON, WASHINGTON, WHITTINGTON; BEDINGFIELD, KNOTTINGLEY &c. As a rule -*ingham* is OE -*ingahām* or -*ingahamm* (as GILLINGHAM, OE *Gillingahám*, BUCKINGHAM, OE *Buccingahamm*), early ME -*ingeham*, while -*ington* is mostly OE -*ingtūn*, early ME -*ington* (as BEDDINGTON, OE *Beaddingtun*). The -*inghams* are clearly as a rule 'the HĀM of the -*ingas*' (*Beddingham* 'the HĀM of the *Bēadingas* or *Bēada*'s people'). The meaning of -*ingtūn* has been much discussed. Some scholars hold that in this case -*ing*- is a kind of adjectival suffix, *Beddington* meaning '*Beadda*'s TŪN'. But the probability is that in most cases -*ingtūn* is a shortening of -*ingatūn*, the *a* being lost in the inflected -*ingatūne*. It should be noticed that -*hām* was unchanged in the dative.

But in a great many cases -*ing*-, especially in -*ington*, is due to a change of some other element. Sometimes it represents a weak ending -*an*, as in ABINGDON, BULLINGTON Ha, HUNTINGDON (OE *Abban dūn*, *Bulan dūn*, *Huntan dūn*) or in the common NEWINGTON from OE *nīwan tūn* 'the new TŪN'. Sometimes -*ing*- has developed from an *n* of different origin, as in HOLLINGWORTH (*holegn-worþ*) or WRIGHTINGTON (*wyrhtena tūn*), or even from an intrusive *n*, as in BERRINGTON, CHERRINGTON, HONINGTON (from *Byrig*-, *Cyric*-, *Hunigtūn*), or CANNINGTON from *Cantuctūn*, DENNINGTON from *Denegifetūn* &c.

Ingardine Sa [*Ingewyrð* 10 BCS 1317, *Ingurdine* DB, *Yngewurthe* 1188 P]. See WORþ. The first el. is possibly the word for 'hill' discussed under INGON.

Ingarsby Le [*Inuuaresbie* DB, *Inguarebi* 1177 P, *Hingwardeby* 1209-35 Ep]. '*Ingwar*'s BY.' *Ingvar* (*Inguarus* DB &c.) is ON *Ingvarr*, OSw *Ingvar*, ODan *Ingwar*.

Ingate Sf at Beccles [*Endegat* 1201 Cur, *Endegate* (church) 1208 FF, *Beccles Endegate* 1334 FF]. Presumably 'the gate at the end (of the territory)'.

Ingatestone Ess [*Ginges ad Petram* 1254 Val, *Gynges Atteston* 1283 Cl]. 'ING at the stone.' Cf. ING. It has been suggested that the stone was a Roman milestone, but *petra* rather suggests a rock or an erratic block or the like.

Ingbirchworth YW [*Bercewordе* DB, *Yngebyrcheworth* 1424 YD]. 'Birch WORþ.' The later *Ing*- may be ON *eng* 'meadow'.

Ingerthorpe YW [*Ingeridtorp* 1162 YCh 120, *Yngridetorp* 1190 P, *Ingritorp* 1201 FF]. '*Ingrið*'s thorp.' *Ingrið* (*Ingrede* DB &c.) is ON *Ingiríðr*, OSw *Ingerith*, *Ingridh*, a woman's name.

Ingestre St [*In Gestreon* (for *Ingestreon*) DB, *Ingerstrent* 1242 Cl, *Ingestret*, *Higestront* 1242 Fees, *Ingestraund* 1250 Ass, *Ingestre* 1236 Fees]. It looks as if the second el. is OE *gestrēon* 'gain, property', here used in some special topographical sense. See STRENSALL. The first might be the word for 'hill' discussed under INGON. There is a marked hill at the place.

Ingham Li [*Ingeham* DB, *Ingheham* c 1115 LiS, *Ingaham* 1163 RA, *Ingham* 1202 Ass], I~ Nf [*Hincham* DB, *Ingham* c 1165 Holme, 1208 FF, *Ingeham* 1248 Ch], I~ Sf [*Ingham* DB, c 1095 Bury, 1251 Ch, *Hingham* 1121-35 Bury]. '*Inga*'s HĀM.' *Inga* is only found as the name of a moneyer, and genuine names in *Ing*- are rare in OE. But at least *Inguburh* and *Ingweald* are well evidenced.

Ingleborough Hill YW [*Ingelbrrc* c 1170, -*burgh* c 1185, 1220 FC, 1293 QW]. The first el. is identical with that of INGLETON YW. The second is OE BURG, which may refer to **Ingleborough Cave**, a remarkable cave with stalactites &c. If so, *burg* is here used in the sense of modern *burrow*. But there is an ancient camp on the top of the hill.

Ingleby Db [*Englaby* 1009 Hengwrt MS 150, *Englebi* DB, -*by* 1228 BM, 1242 Fees], I~ Li [*Englebi* DB, *Englabi* c 1115 LiS, *Engelby* 1251 Ch], I~ Arncliffe YN [*Englebi* DB, *Engleby juxta Arneclif* 1285 FA], I~ Barwick YN [*Englebi* DB, *Caldengilbi* 1279 YInq], I~ Greenhow YN [*Englebi* DB, *Engelby* 1150-5 YCh 570, *Engilby juxta Grenehowe* 1285 FA]. OScand *Englabȳr* 'the BY of the English'.

I~ Arncliffe &c. from the neighbouring ARNCLIFFE [*Erneclive* DB], BARWICK [*Berewic* DB], GREENHOW.

Inglesham W [*Inggeneshamm* c 950 Wills, *Incgenæs ham* 965-71 BCS 1174, *Inglesham* 1161 P, *Ingelesham* 1212 Fees]. '*Ingīn*'s HĀM or HAMM.' *Ingīn* is a derivative of *Inga*. Cf. INGHAM.

Inglethorpe Nf [*Yngelesthorpe* Hy 3 RBE, *Hingultorp* Hy 3 Misc]. '*Ingulf*'s thorp.' *Ingulf* may be OE *Ingwulf* (not well evidenced) or OScand *Ingulfr*.

Ingleton Du [*Ingeltun* c 1050 HSC, 1104-8 SD]. The first el. is a pers. n., possibly OE *Ingeld* or *Ingwald*. OScand *Ingialdr* is also possible.

Ingleton YW [*Inglestune* DB, *Ingilton* 1235, -*e* 1245 FF, *Ingelton* 1240 FF, 1247 Ch]. Ingleton is at the foot of INGLEBOROUGH HILL. The two names Ingleborough and Ingleton must be explained together. The common first el. might be a pers. n. such as OE *Ingeld* or OScand *Ingialdr* or *Ingolfr*, but neither goes quite well with the early forms. If there was an OE el. *ing* 'hill' or the like (cf. INKPEN and INGON), the best explanation is to take *Ingle* to represent an OE *Ing-hyll*. Also a hill-name *Inga(n) hyll* '*Inga*'s hill' might be thought of.

Inglewood Brk [*Ingleflot, Ingheflot* DB, *Ingafloda* Hy 2 (1270) Ch, *Ingeflod* 1220, 1242 Fees, 1252 Ch]. The place is near INKPEN, and must contain the same el. *Ing-* as the latter (first el. perhaps OE *Inghyll*). Second el. OE FLŌDE.

Inglewood Forest Cu [*Engleswod* 1150, *Englewode* 1158, 1189 Holme C, *Englewud* 1227 Ch]. OE *Engla wudu* 'the wood of the English'. The name tells of an English settlement in Welsh territory.

Ingmanthorpe YW [*In Gemunstorp* (for *Ingemunstorp*) DB, *Ingmanthorp* 1285 FA]. '*Ingimund*'s thorp.' *Ingimundr* is a Scand name.

Ingoe Nb [*Hinghou* 1229 Pat, *Inghou* 1242 Fees, *Ingou* 1279 Ass]. Perhaps '*Inga*'s hill'; but cf. INGON.

Ingol La [*Ingole* 1200 Ch, *Ingol* 1246 Ass]. '*Inga*'s hollow or valley.' **Ingolhead** [*Ingolheued* 1310 LaInq] adjoins Ingol and its name may mean 'the upper end of Ingol'.

Ingoldisthorpe Nf [*Torp* DB, *Ingaldestorp* 1203 Cur, 1254 Val, *-thorp* 1242 Fees]. '*Ingiald*'s thorp.' First el. ON *Ingialdr*, OSw *Ingiæld* &c. The river-name *Ingol* is a back-formation.

Ingoldmells Li [*in Guldelsmere* DB, *Ingoluesmera* 1095–1100 AC, *Ingoldesmeles* 1180 P, *Ingaldemoles* 1212 Fees]. '*Ingiald*'s sand-banks.' First el. as in prec. name. Second el. OScand *melr* 'sand-bank'.

Ingoldsby Li [*Ingoldesbi* DB, *Yngoldebi* 1202 Ass]. First el. as in INGOLDMELLS.

Ingon Wa [*Ingin* 704–9 BCS 122, *Hynge* 12 Mon, *Inge* 1199 FF, 1209 Fees, *Ingewrthe* 1200 Cur, *Inggene* 1315 Ass]. The form of 704 is in a late transcript, and may well stand for original *Ingum*, which the later forms seem to presuppose. If so, the name must be the dat. plur. of a word such as *ing*, which may be identical with the first el. of INKPEN and of *Incghæma gemære* 880 BCS 547, *Ingham* 1049–52 KCD 950 (a lost place in the Chilterns). It has been suggested that the pers. n. el. *Ing-* is identical with Gk *énkhos* 'lance'. If that is right, an OE *ing* might have existed with the sense 'a peak, hill'. Such a sense would be very suitable for Inkpen and the lost Ingham, as well as for Ingleborough Hill. Ingon is on the slope of a hill, but not one that could be called a peak.

Ingram Nb [*Angerham* 1242 Fees, *Angreham* 1254 Val]. OE *angr* 'grassland' (cf. ANGERTON, ANGRAM, ONGAR) and HĀM or HAMM.

Ingrave Ess [*Gynges Rad'* 1238 Subs, *Gingeraufe* 1276 Abbr]. 'Ralph's ING.' See ING. One Ralf was tenant in 1086 (DB).

Ingthorpe Ru [*Torp* c 1125 LN, *Ingeðorp* BCS 22 (late copy), *Ingelthorp* 1189 (1332) Ch, *Ingetorp* 1203 Ass]. Probably '*Ingi*'s ÞORP'. *Ingi* is an OScand pers. n.

Ingworth Nf [*Inghewurda* DB, *Ingewrde* c 1145 Holme, *-wrthe* 1242 Fees]. '*Inga*'s WORÞ.'

Inkberrow Wo [*Intanbeorgas* 789, *-beorgum* c 822 BCS 256, 308, *Inteberge* DB]. '*Inta*'s hills or mounds.'

Inkersall, West, Db [*Hinkershil* 1242 Fees, *-hill* c 1290 BM, *Hinckreshill* 1264 Ipm]. The name originally began in *H-*, which was lost owing to dissimilation. It may be suggested that the first el. is an OE *hīgna æcer* 'the field of the HĪWAN or monks'.

Inkersall Nt is late for **Winkerfield** [*Wirchenefeld* DB, *Werkenefeld* 1229 Cl, *Wircnesfeud* c 1180 DC]. The first el. is taken in PNNt(S) to be a pers. n. **Wyrcen*. It is perhaps rather OE *wyrhtena* 'of the wrights'. Cf. RIGBOLT Li (*Wirchebald* 1238 Ep), BECKENHAM K, and the like.

Inkpen Brk [*Ingepenn* 931 BCS 678, *Ingepene* DB, 1167 P, *Ingepenn* 1220 Fees, *-penne* 1227 Ch, *Yngelpenne* 1236 Fees, *Inkepenne* 1291 Tax]. Inkpen is near **Inkpen Beacon** (1,011 ft). The second el. is presumably Brit *penn* (Welsh *pen*) 'head, hill'. The first cannot be a pers. n. *Inga*, as the form of 931 is in an original charter. The hypothetical OE *ing* 'hill' would be very apt here. Cf. INGON.

Inny R Co [*Æni* 1044 KCD 770, *Eny* 1229 Ol]. A doublet of ONNY, though with British *i*-mutation.

Inskip La [*Inscip* DB, *Inscype* 1246 Ass, *Inscyp* 1272 AD]. The first el. may be Welsh *ynys* 'island' &c. (cf. INCE). The second might be OE *cȳpe* 'osier basket', later also 'osier basket for catching fish'.

Instow D [*Johannesto* DB, *Jonestowe* 1242 Fees, *Yenestowe* 1291 Tax]. 'St. John's STŌW or church.'

Intwood Nf [*Intewda, -wida* DB, *-wude* 1207 Cur]. '*Inta*'s wood.'

Inwardleigh D [*Lega* DB, *Inwardlegh* 1235 FF]. '*Inwar*'s LĒAH.' The manor was held at the time of the Conquest by one *In(e)war*. *Inwær* is a common OE form of OScand *Ivarr* (< **Inhu-haria-*).

Inworth Ess [*Inewerth* 1206 Cur, 1235, 1270 FF]. '*Ina*'s WORÞ.'

Iping (ī-) Sx [*Epinges* DB, *Ipinges* 1212 AD]. '*Ipa*'s people.' *Ipa* is found in *Ipan lea* 984 KCD 1281.

Ipley Ha in Fawley [*Yppeleigh* 1212 Fees, *Eppel'* 1235 Cl, *Ippele* 1316 FA]. Either '*Ippa*'s LĒAH' (cf. *Ippan beorg* 955 BCS 917) or OE *yppe-lēah* 'wood by a hill'. Ipley is in a low situation, but close to a ridge.

Ipplepen D [*Iplanpen*, (to) *Ipelanpænne* 956 BCS 952, *Iplepene* DB]. '*Ipela*'s PEN or fold.' *Ipela is a derivative of *Ipa*.

Ippollitts Hrt [*S. Ypollitus* 1283, *Polytes* 1412 BM]. 'St. Hippolytus's church.'

Ipsden O [*Yppesdene* DB, *Ipesden* 1195, *Ypeden* 1204 Cur, *Ippeden* 1233 Cl, *Ippisden* 1236 Fees]. Ipsden is on the lower slope of the Chiltern Hills. I~ **Heath** is high. First el. of the name OE *yppe* 'hill' &c., referring to the Chiltern Hills.

Ipsley Wa [*Epeslei* DB, *Ypeslea* 1190 P, *Ypelai* 1192 P, *Yppesleg* 1236 Fees, *Uppesleg* 13 AD]. The place is on a small round hill, which was no doubt known as OE *Yppe*. Cf. YPPE. Second el. OE LĒAH.

Ipstones St [*Yppestan* 1175 P, *Ipestanes* 1206 Cur, *Ipstone* 1220 Ass, *Ippestanes* 1244 FF]. The absence of spellings with *u* points to '*Ippa*'s stone' rather than *yppe-stān*. Cf. YPPE.

Ipswich Sf [*Gipeswíc* 993 ASC, 1010 ASC (E), *-wic*, *-wiz* DB, *Gepeswiz* DB, *-wíc* 1130 P]. Apparently '*Gip*'s or *Gipe*'s wíc'. OE *Gip(e)* is unrecorded. It might belong to OE *gipian* 'to yawn', LG *gīpen* the same &c. But it is possible that *Gip-* is a common noun *gip* belonging to *gipian* and meaning 'gap', 'opening'. Cf. Swed *gipa* 'corner of the mouth', Norw *gipa* 'gaping wound'. *Gip* would then have been an old name of the broad estuary of the Orwell.

Irby Chs [*Erberia* c 1100, c 1150, *Irrebi* c 1190 Chester], I~ **on Humber** Li [*Iribi* DB, *Irebi* DB, c 1115 LiS, *Yrebi* 1202 Ass], I~ **in the Marsh** Li [*Irebi* c 1115 LiS, 1212 Fees, *Yreby* 1257 Ch], I~ YN [*Irebi* DB, 1088 LVD]. 'The BY of the Irish', OScand *Írabȳr*.

Irchester Np [*Yranceaster* 973 BCS 1297, *Irencestre* DB, *Irecestr'* 1168 P]. '*Ira*'s or *Yra's CEASTER or Roman fort.' *Ira* is found as the name of a moneyer.

Ireby Cu [*Yrebi* 1185 P, *Ireby* 1236 Cl], I~ La [*Irebi* DB, *Yrebi* 1215 P]. Identical with IRBY.

Ireleth (īr-) La [*Irlid* 1190, *Ireleyth* c 1200, *Irelith* 1292 FC]. 'The hill-slope of the Irish', an ON *Íra hlíð*.

Ireton, Kirk & Little, I~ Wood Db [*Hiretune*, *Iretune* DB, *Little Ireton* 1315 FF]. 'The TŪN of the Irish.'

Iridge Sx [*Yrugge* 1248 Ass, *Iwrugge* 1316 FA]. 'Yew ridge.' Cf. IW.

Irk R La [*Irk*, *-e* 1322 LaInq]. Etymology obscure.

Irlam La [*Urwil-*, *Urwelham* c 1190 CC, *Irwelham* 1259 Ass]. 'HĀM on R IRWELL.'

Irmingland Nf [*Erminc-*, *Urminclanda* DB, *Ermingland* 1196 FF, *Irmingeland* 1207 Cur].

'The land of *Eorma*'s (*Irma*'s) people.' *Eorma (*Irma) is probably a short form of names in *Eormen-*, *Irmen-*.

Irnham Li [*Gerneham* DB, *Erneham* 1100–8 Fr, *Yrneham* 1191 ff. P, *Yrenham* 1202 Ass]. '*Georna*'s HĀM.' *Georna is a short form of names like *Friþu-*, *Heregeorn*.

Ironbridge Sa, **Ironville** Db. Late names. Self-explanatory.

Irstead Nf [*Irstede* c 1140 Holme, 1242 Fees, 1254 Val]. The place is in low fen country. *Ir-* may well be OE *gyr* 'mud'; cf. JARROW. See STEDE.

Irt R Cu [*Irt* 1279, 1292 Ass]. On the Irt is **Irton** Cu [*Yirreton* 1228 Ep, *Iryton* 1292 QW, *Irtona* 13 StB]. *Irt* is obscure. It might be derived from OE *gyr* 'mud'.

Irthing (-dh-) R Cu, Nb [*Irthin*, *Irthing* 1169, *Erthina* c 1200 Lanercost, *Erthing* 1279 Ass]. A Brit river-name. On the river is **Irthington** Cu [*Irthinton* 1169 Lanercost, *Irthington* 1295 Ipm, *Erthington* 1279 Ass].

Irthlingborough (artlburu) Np [*Yrtlinga burg* 780 BCS 1334, *Erdi(n)burne* DB, *Hyrtlingberi* 1137 ASC (E), *Urtlingburch* 1179 P, *Hertlingburc* 1199 FF, *Ertlingeburc* 1203 Ass]. OE *yrþlinga burg* 'the BURG of the ploughmen'. The name is best explained if it may be assumed that I~ was an old fort which was used for the purpose of keeping oxen. Cf. SALMONSBURY (under BOURTON) and STŌD. The change of *yrþling* to *yrtling* has analogies in *Hatfield* from *Hǽþfeld*, OE *bōtl* from *bōþl* &c.

Irton Cu. See IRT. **Irton** YN [*Iretune* DB, *Irton* 1228 FF]. 'The TŪN of the Irish.'

Irwell R La [*Urwel*, *Urwil* 12 CC, *Irewel* c 1200 WhC, *Irrewelle* 1277 Ass]. 'Winding stream.' First el. OE *irre*, *eorre* 'angry', originally 'straying, wandering'.

Isbourne R Gl, Wo [(in) *Esenburnen* 709, *Esegburna* 777, *Esingburnan* 930 BCS 125, 223, 667 f., *Eseburne* 988 KCD 662]. '*Ēsa*'s stream' and 'the stream of *Ēsa*'s people'.

Ise R Np [(andlang) *Ysan* 956 BCS 943, *Ise* 1247 Ass]. A derivative of OUSE, the base being *ūsiōn.

Isell (ī-) Cu [*Ysala* 1195 P, *Yshale* 1271 Ipm, *Issal*, *Isale* 1291 Tax], **Isfield** Sx [*Isefeld* 1214 FF]. '*Isa*'s HALH and FELD.' *Isa is identical with OG *Isa*, *Iso*.

Isham (īs-) Np [*Ysham* 974, 1060 BCS 1310, KCD 809, *Isham* 974 BCS 1311, DB, 1203 Ass]. 'HĀM on R ISE.'

Isis (īsis) R, the Thames at Oxford [*Isa* c 1350 Higden, *Ise* 1347 Pat, *Isis* 1577 Harrison]. An artificial formation from *Tamise* (*Tamesis*), early forms of THAMES, which was supposed to be a combination of *Thame* (the name of one of the head-streams of the Thames) and a hypothetical *Ise* (*Isis*).

Isle R So [*Yle* 693 ff., 13 Muchelney, *Ile* 1280 Ass]. A Brit river-name very likely identical with *Ila* Ptol (now ILIDH in Scotland), which has been held to be a derivative of the root *pi* in Greek *pino* 'to drink', or of the root of Norw *ila* 'a spring', G *eilen* 'to hurry'. On the Isle are **Isle Abbotts** So [*Yli* 966 Muchelney, *Ile* DB, *Ile Abbatis* 1291 Tax] and **I~ Brewers** So [*Isle* DB, *Ile Brywer* 1275 Ipm].

I~ Abbotts belonged to Muchelney Abbey.—
I~ Brewers was held by Richard Briwer in 1212 (Fees). Cf. BUCKLAND BREWER.

Islebeck YN [*Iselbec* DB, *Yselbec* c 1200 YD, *Iserbec* 1208 Cur]. Named from **Isle Beck** [*Yserbec* c 1200 YD], which may be '*Isolf*'s beck'. *Isolf* is ON *Ísolfr*.

Isleham (ĭzlăm) Ca [*Yselham* 895 BCS 571, *Gisleham* DB, *Iselham* 1232 FF]. '*Gīsla*'s HĀM.' *Gīsla* is a short form of names in *Gīsl-*.

Isleworth (ĭzl-) Mx [*Gislheresuuyrth* 695 BCS 87, *Gistelesworde* DB, *Ysteleswurde* 1180 P, *Istelesworth* 1221–30 Fees]. '*Gīslhere*'s WORÞ.' The loss of *r* is due to dissimilation. The *t* in early forms is an intrusive consonant that developed between *s* and *l*.

Islington (ĭz-) Mx [*Gislandun* c 1000 CCC, *Isel-*, *Isendone* DB, *Iseldon* 1197 FF, *-e* 1242 Fees, 1254 Val]. '*Gīsla*'s DŪN'; cf. ISLEHAM. The interchange between *l* and *n* is due to the form *Gīslan dūn*.

Islington Nf [*Elsington* 11 EHR 43, *Ilsingha-*, *Ilsinghetuna* DB, *Ilsingtune* c 1095 Bury, *Hiselingeton* 1166 P]. 'The TŪN of *Elesa*'s people.'

Islip (ĭz-) Np [(æt) *Isslepe, Hyslepe* 972–92 BCS 1130, *Islep* DB, *Yslep* 1202 Ass], **I~** (īs-) O [*Giðslepe* c 1050 KCD 862, *Gihtslepe* 1065 BM, *Letelape* DB, *Ichteslep* 1242 Fees]. 'SLÆP at the rivers *Ise* and *Ight*.' Islip Np is on the Nene some way below its junction with the Ise. Evidently this part of the Nene was once known as Ise. *Ight* is an old name of the river RAY, a tributary of the Cherwell [*Geht* c 848 BCS 452, *Giht* 983 KCD 1279, *Ychte* 1185 ff. P]. Ight is very likely a cognate of IEITHON in Wales, a derivative of Welsh *iaith* 'language'. *Slæp* is usually held to be 'slippery place'. But Islip O may be referred to KCD 1279 as '(of) ðan ealdan *slæpe* up andlang *Giht*'. This rather suggests a meaning such as 'place where things are dragged, portage' (cf. DRAYCOTT). *Slæp* is related to MLG *slēpen*, OHG *sleifen* 'to drag'. See further SPN, p. 184 ff., where names in *slæp* are dealt with.

Ismere House Wo [*Husmeræ* 736, (provincia) *Usmerorum* 757–75, *Usmere* 964 BCS 154, 220, 1134]. Second el. OE *mere* 'lake'. The first is identical with the river-name OUSE and may be the old name of the stream that runs through the lake. *Usmerorum* is a Latin genitive of a tribal name *Usmere* 'the people at *Usmere*'.

Isombridge Sa [*Asnebruge* DB, *Esnebrugg* 1249 Ipm]. 'The bridge of the servants.' First el. OE *esne* 'servant'.

Itchel Ha [*Ticelle* DB, *Hichelle* 1165, *Ichulle* 1279 Crondall]. An old river-name, which may have denoted the Hart and the Whitewater [*Icæles* (æwilmas) 973–4 BCS 1307]. Perhaps related to ITCHEN.

Itchen R Ha [*Icene* 701, (on) *Ycænan* 825, *Iccene* c 830 BCS 102, 389, 398, *Ichene* 1256 Ass], **I~** R Wa [(on) *Ycenan* 998 Crawf, (in) *Ycenan* 1001 BM, *Huchene*, *Ichene* 1262 Ass]. Cf. ITCHINGTON Gl. Itchen is a Brit river-name, perhaps related to the tribal name *Iceni*. On the Itchen Ha are **Itchen Abbas** Ha [*Icene* DB, *Ichene Monialium* 1167 P] and **Itchen Stoke** Ha [*Stoche* DB, *Ichenestok* 1291 Tax].

I~ Abbas belonged to the Abbey of St. Mary, Winchester. *Abbas* is for *Abbess*. See STOC.

Itchenor, West, Sx [*Iccanore* 683 BCS 64, *Icenore* DB, *Westigenore* 1243 FF]. '*Ycca*'s ŌRA or landing-place.'

Itchingfield Sx [*Ec(c)hingefeld* 1222, 1256 FF, *Hechingefelde* c 1235 Sele]. 'The FELD of *Ecci*'s people.'

Itchington Gl [æt *Icenantune* 967 BCS 1206, *Icenantun* 991 KCD 677, *Icetune* DB]. TŪN on R *Icene*. The brook on which the place stands must have been once called *Icene*. Cf. ITCHEN.

Itchington, Bishops, Wa [*Icetone* DB, *Ichinton Episcopi* 1291 Tax], **Long I~** Wa [(æt) *Yceantune* 1001 BM, *Icentone* DB, *Longa Ichenton* 1262 Ass]. 'TŪN on R ITCHEN.'

Bishops I~ belonged to the Bishop of Lichfield.

Itterby Li in Clee [*Itrebi* DB, *Yterby* 1212 RBE]. OScand *ytri býr* 'outer BY.'

Itteringham Nf [*U(l)trincham* DB, *Itringham* 1203 Cur, 1242 Fees, *Iteringham* 1202 FF]. Either 'the HĀM of *Ytra*'s or *Ytri*'s people' or 'the HĀM of the people dwelling outside'. *Ytra (*Ytri) might be derived from *Otr* (cf. OTTERDEN). In the second alternative the first el. would be derived from OE *ȳterra* 'outer'.

Ive R Cu [*Yue* 1285 For, *Ive* 1307 Pat]. ON *Ifa*, probably derived from the stem *iwa-* 'yew' (ON *ýr* &c.). Here is **Ivegill** 'the valley of the IVE'.

Ivel (ĭ-) R Bd, Hrt [*Givle* c 1180 Warden Cart., *Giuele* 1232 FF, *Yivele* 1294 Ipm]. The river-name is identical with **Yeo** So [*Gifl* 946–51 BCS 894]; cf. also GIVENDALE. The name means 'forked river' and is related to OBret *gablau*, Welsh *gafl* 'fork'. On the Ivel are **Northill** Bd [*Nortgiuele* DB, *Norgivel* 1221 Cl] and **Southill** Bd [*Sudgiuele* DB, *Suthgivel* 1214 FF]. The second el. of these is the tribal name OE *Gifle* (*Gifla* gen. plur. 7 Trib Hid) 'the dwellers on the IVEL'. Cf. Introd. ii. 1.

Iver (ĭ-) Bk [*Evreham* DB, *Eura* c 1130 Oxf, 1163 P, *Eure* 1196 P]. OE *yfer* 'edge, steep slope'.

Iveston Du [*Ivestan* 1183 BoB, *Yvestan* 1297 Pp]. '*Ifa*'s stone.' *Ifa* is found as the name of a moneyer.

Ivinghoe Bk [*Evingehov* DB, *Iuingeho* 1195 Cur]. 'The HŌH or spur of land of *Ifa*'s people.'

Ivington (ĭ-) He [*Ivintune* DB, *Iuentonia* c 1145 Oxf]. 'The TŪN of *Ifa*'s people.'

Ivonbrook Db [*Winbroc* DB, *Ivelbrok* 1269 Ass, *Yuenbroc* 13 Derby]. '*Ifa*'s brook.'

Ivybridge D [*Ponte Ederoso* 1280 AD], **Ivychurch** K [*Iue circe* 11 DM, *Ivechirch* 1242 Pat], **I~** W [*Capella Ederosa* 1155 RBE, *monasterium Ederosum* 1156 P, *Ivy-church* 1247 Pat]. 'Ivy-covered bridge and church.'

OE *īw* 'yew'. See EWHURST, EWSHOTT, IFIELD, IRIDGE, IWADE, IWODE, ULEY. Sometimes difficult to distinguish from OE ĒG (*īeg, īg*) 'island'; cf. IFORD.

Iwade K [*Ætwangeraede* 11 DM, *Ywada* 1179 P, *Ywad* 1208 Cur, *Iwade* 1257 Ch]. The first form may be corrupt for *æt Iwangewæde*. Anyhow *Iwade* is probably 'yew ford', the elements being OE *īw* 'yew' (or a derivative of it) and *gewæd* 'ford'.

Iwerne (ū-) R Do [*Iwern broc* 958 BCS 1033]. A Brit river-name identical with *Iérnos* Ptolemy (an old name of the KEN-MORE in Ireland) and derived from *īvo-* 'yew' (Welsh *yw* &c.). The river gave their names to **Iwerne Minster** and **I~ Court-ney** or **Shroton** Do [*Ywern, Hywerna* c 871, *Iwern* 956 BCS 531 f., 970, *Iwerne, Werne, Euneminstre* DB, *Ywerne Curtenay* 1261 Ch, *Iwerne Munstre* 1278 QW].

I~ Courtney belonged to Hawis de Curtenei in 1212 (Fees). Cf. HIRST COURTNEY.—Minster is probably OE *mynster* in the sense 'church'.—**Shroton** [*Shereueton* 1374 FF] is 'sheriff's TŪN'. Cf. also STEEPLETON IWERNE.

Iwode Ha [*Iwuda* 1167 P, *Ywode* 1231 Sel-borne, *Iwode* 1250 Cl]. 'Yew wood.'

Ixworth Sf [(æt) *Gyxeweorde* c 1025 BCS 1018, (of) *Ixewyrðe* 11 EHR 43, *Giswortha, Icsewrda* DB, *Ixewurða* 1168 P]. '*Gicsa*'s or *Gycsa*'s WORÞ.' Cf. EXNING. *Gicsa (*Gixa, *Gycsa, *Gyxa) may be a nickname belonging to OE *gesca, geocsa, gihsa* 'hic-cough'. **Ixworth Thorpe** is *Torp* DB, *Ixeworth thorp* 1305 BM.

J

Jacobstow Co [*Jacobestowe* 1270 Ep], **Jacobstowe** D [*Jacopstoue* 1331 Ep, *Jacobestauwe* 1349 Ass]. 'Church dedicated to St. James.'

Jarrow Du [*In Gyruum* c 730 Bede, *Gyruum, Girwe* 1104–8 SD, *Jaruum* 1158 YCh 400, *Jarwe* 1228 FPD]. 'The *Gyrwe*.' The name is really that of a tribe, which became a place-name in the same way as *Cornwealas* became CORNWALL. Cf. Introd. ii. 1. The tribe of the *Gyrwe* is *Gyruii* c 730 Bede, *Gyrwa* (gen. plur.) c 890 OEBede. The name is derived from an old word for 'mud' or 'fen' found in OE *gyr*, ON *giọr* 'mud'. The base is *gerwō fem. *J-* instead of *Y-* is due to Norman influence. The *Gyruii* mentioned by Bede dwelt in the fen districts round Peterborough. The *Gyrwe* of Durham may have migrated from the Peterborough district or they may have got their name independently from a fen.

Jervaulx (jervŏ, jarvĭs) YN [(de) *Jorvalle* c 1145 Mon v, *Girevalle* c 1200 BM,

Gereuall' 1195 FF]. 'The URE valley.' The name is French and perhaps a translation of Engl *Ure-dale*. See URE, YORDALE.

Jesmond Nb [*Gesemue* 1205 P, 1216 Cl, *-muthe* 1275 RH]. 'The mouth of OUSE BURN' (q.v.). The initial *J-* and the form *-mond* are due to Norman influence.

Jevington Sx [*Lovingetone* DB, *Govingetona* 1189 PNSx, *Gyvingerton* 1248 Cl]. 'The TŪN of *Geofa*'s people.' Cf. YEAVELEY. *J-* is due to Norman influence.

Johnby Cu [*Ionesbi* 12 CWNS xxxii, *Jo-hannebi* 1200 P, *Johanbi* c 1205 WR]. 'John's BY.'

Johnson Hall St [*Johannestun* 1227 Ass]. 'John's TŪN.'

Jolby YN in Croft [*Joheleby* c 1195 PNNR, *Joeleby* 1219 FF]. '*Johel*'s BY.' *Johel* (*Joel*) is a French name, no doubt of Breton origin (= Welsh *Ithel*, OW *Iudhail*). A *Joel* lived at Jolby c 1170.

K

Kaber We [*Kaberge* a 1195, c 1250 WR, 1200 FF]. 'Jackdaw hill.' Cf. CABOURN.

OScand **karl** 'freeman'. See CARL- (passim).

OScand **kaupmaðr** 'chapman'. See CAPERN-WRAY, COPMANTHORPE, COPPINGFORD.

Kea Co [*Ecclesia Sancte Kee* 1451 FF (Gover)]. Identical with ST. QUAY in Brittany, which means 'the church of St. *Ke*'. *Ke* is identical with *Kei* in Mab and goes back to earlier *Cai*. A variant form of Kea is **Landegea** (in Kea) [*Landighe* DB, *Landegei* 1185 P, *Landegeye* 1235 Cl]. It contains Co *lan* 'church' and a hypocoristic form of *Ke* meaning 'thy Ke'. Cf. the identical LANDKEY and LANDEWEDNACK.

Keadby Li [*Ketebi* 1185 TpR, 1199 P, *Keteby* 1275 RH]. '*Keti*'s BY.' First el. ODan *Kæti, Keti.*

Keal, East & West, Li [*Cale, Estre-, Westrecale* DB, *Cal', Oustcal'* c 1115 LiS, *Cales* c 1135, *Keles* 12 DC]. ON *kiǫlr* (gen. *kialar*) 'keel, ridge'.

Kearby YW [*Cherebi* DB, *Kerebi* 1193 P, *-by* 1242 Fees]. '*Kærir*'s BY.' First el. ODan *Kærer,* OSw *Kærir.*

Kearsley La [*Cherselawe* 1187, *-lawa* 1188 P, *Kersleie* c 1220 CC]. 'Cress hill or clearing.' See HLĀW, LĒAH.

Kearsley Nb [*Kerneslawe* 1245 Ipm, 1279 Ass, *Kereslaw* 1346 FA]. '*Cynehere*'s or *Cēnhere*'s hill or mound.' Cf. KERSALL, HLĀW.

Kearsney K [*La Kersuner*' 1242 Fees, *Kersonere* 1286 Ipm]. Fr *cressonnière* 'place where cress grows'.

Kearstwick We [*Kestwhait, -what* 1576 Kendale]. Possibly a modification of KESWICK. Or the elements may be ON *kióss* 'valley' and *þveit* (see THWAITE).

Kearton YN [*Karretan* Hy 3 Misc, *Kirton* 1298 YInq, *Kerton* 1301 Subs]. First el. as in KEARBY.

Keckwick or **Kekewick** Chs [*Kecwyc, Kequik* 1287 f. Court, *Keckwyk* 1295 Cl]. Perhaps '*Cæfca*'s WĪC'. *Cæfca* is found in *Cæfcan græfan* 739 Crawf.

Keddington Li [*Cadi(n)ton* DB, *Chedingtuna* c 1115 LiS], *Kedingtuna* 12 DC, *Kidington* 1257 FF], **Kedington** Sf [*Kydington* 1043–5 Wills, *Kidituna* DB, *Kedintune* Hy 2 BM, *-ton* 1200 Cur]. 'The TŪN of *Cyd(d)a*'s people.'

Kedleston Db [*Chetelestune* DB, *Ketleston* 1206 Cur]. '*Ketel*'s TŪN.' *Ketel* (*Chetel* DB &c.) is ON *Ketill,* ODan *Ketil,* OSw *Kætil.*

Keekle Beck Cu [*Chechel* 1120–35, *Kekel* c 1230, *Kikil* c 1450 StB]. Perhaps identical

with the Norw river-name *Kykla* or derived from an adj. *kikall* 'winding' that seems to occur in Norw. pl. ns. Cf. CORKICKLE.

Keelby Li [*Chelebi* DB, c 1115 LiS, *Kelesbi* 1202 Ass, *Kelebi* 1203 ib., 13 BM]. OScand *Kiǫlar-bȳr* 'BY at a *kiǫlr* or ridge' (cf. KEAL).

Keele St [*Kiel* 1169 ff., *Kyel* 1230 P, *Kel* 1211 FF]. OE *cȳ-hyll* 'cow hill'.

Keer R We, La [*Kere* 1262 CC, *Keere* c 1350 For]. A Brit river-name derived from Brit **cēro-,* identical with Ir *cíar* 'dark'.

Keevil W [*Kefle* (gen.) 964 BCS 1127, *Chivele* DB, *Kyvelegh* 1240 Cl, *Cuvel* 1242 Fees, *Cuvele* 1327 Ch]. Second el. OE LĒAH. The first may be an OE pers. n. *Cyfa* (cf. *Cufa* and see KILWORTH). Or it might be OE *cȳf* 'tub, vessel', the name meaning 'wood where material for tubs was got'.

Kegworth Le [*Cacheuuorde, Cogeworde* DB, *Caggworth* c 1125 LeS, *Kagwrth* 1196 FF, *Kegworth* 1209–35 Ep]. See WORÞ. The first el. is perhaps a pers. n., but its history is obscure. The ON byname *Kaggi* (also *Kaggr*) might be compared.

Keighley (kēthli) YW [*Chichelai* DB, *Kikeleia* 1170–9 YCh 1872, *Kikhele* 1244 Ep, *Kye Leya* c 1231 Pudsay, *Kyghele* 1246 FF]. '*Cyhha*'s LĒAH.' **Cyhha* is related to OE *Cohha.*

Keinton Mandeville So [*Chintune* DB, *Kynton* 1243 Ass, *Kyngton Maundevill* 1280 FF]. OE *cyne-tūn* or *cyning-tūn* 'royal manor'.

The manor was held by Geoffrey de Mandevill in 1243 (Ass) and by William de Mandevill before him. Cf. HARDINGTON MANDEVILLE.

Keisby Li [*Chisebi* DB, Hy 2 DC, *Kisebi* Hy 2 DC, 1202 Ass]. '*Kisi*'s BY.' ON *Kisi,* lit. 'cat', occurs as a byname.

Keisley We [*Kesclif(f)* 1317, 1323 Ipm]. ON *kióss* 'valley' and *klif* 'cliff'.

Kelbrooke YW [*Chelbroc, Cheuebroc* DB, *Kelebrok* 1240 FF, 1300 Ch, *Kellebrok* 1260 Ipm]. The first el. may be a stream-name *Cēle* from *cōl* 'cool'. Or it might be an unrecorded pers. n. **Cēnla,* derived from *Cēne* and names in *Cēn-*.

Kelby Li [*Chelebi, Chillebi* DB, *Kellebi* Hy 2 DC, 1202 Ass, *Kelleby* 1242 Fees]. Perhaps identical with KEELBY.

OScand **kelda** (ON *kelda,* OSw *kælda,* early Dan *kælde*) 'spring' is fairly common in Scandinavian England. Cf. KELLET, KEL-LETH, KELSICK, TRINKELD. See also CELDE.

Keldholme YN [*Keldeholm* 1201 Ch, 1204 Cur]. 'Holm with a spring.' Cf. KELDA.

Kelfield Li [*Kelke-, Calkefeld* Hy 2 DC, *Kelkefeud* Hy 2, *-feld* 13 BM], K~ YE [*Chelchefeld* DB, *Calcefeld, Kelkfelda* c 1150 Selby, *Kelkfeld* Hy 3 BM]. 'Chalk FELD.' First el. identical with KELK.

Kelham Nt [*Calun* DB, *Chelum* c 1155 DC, *Chelun* 1166 P, *Kelum* 12 DC]. '(At) the ridges.' The dat. plur. of OScand *kiǫlr* 'ridge'. See KEAL.

Kelk, Great & Little, YE [*Chelche* DB, *Kelka* c 1170 YCh 1356, *Mangna Kelck*' 1297 Subs, *Little Kelk* 1290 Ch]. A derivative of OE *calc* 'chalk'. Cf. CHELSEA, which has as first el. a derivative of *cealc*, for instance an OE **celce* (< **kalkiōn*). The non-palatalization of the second *k* in *Kelk* may be due to Scand influence. An OE *cælce* is also the first el. of *Kælcacæstir* (or *Calcaria ciuitas*) c 730 Bede, which is usually identified with Tadcaster, but might be supposed to be KELK.

Kellamergh La [*Kelfgrimeshereg* 1201 P, *Kelgrimisarhe* 1236 Fees]. '*Kelgrim*'s ERG or shieling.' *Kelgrim* has been derived from an ON **Ketilgrímr*.

Kellaways W. A shortened form of TYTHERTON K~ or KELWAYS. Kelway is a family name, which appears as (de) *Chailewai* 1165 P (Gl), (Elyas de) *Kaylewe* 1255, (Thomas) *Caylewey* 1275 RH, (Johannes) *Kalewaye* 1286 Malm. It is possibly derived from CAILLOUET in Eure, Normandy [*Cailloel* 1157].

Kellet, Nether & Over, La [*Chellet* DB, *Kellet* 1194 P, *Kelleth* 1212 Fees, *Overkellet* 1277 Ass, *Netherkellet* 1299 FF], **Kelleth** We [*Keldelith* early 13 CWNS xi]. ON *keld(u)-hlíð* 'slope with a spring'. Cf. KELDA, HLÍÐ.

Kelleythorpe YE [*Calgestorp* DB, *Kellingtorp* 1190 P, *Kelingtorp* 1226 FF]. Perhaps originally '*Kiallak*'s thorp'. ON *Kiallakr* is from Ir *Ceallach*. The later forms point to a pers. n. *Keling* or *Kelling*, doubtless the byname *Keling* from ME *keling* 'cod' (probably a Scand word) borne by a former tenant in York (*Grim Chelyng* 1150–61 YCh 224).

Kelling Nf [*Chillinge, Killinge* c 970 HEl, *Kellinga, Challinga* DB, *Kellinges* 1177, 1191 P]. OE *Cyllingas* 'the people of *Cylla*' (in *Cyllanhricg* KCD 1369, *Cyllan beorg, wyll* 772 BCS 208).

Kellington YW [*Chel(l)inctone* DB, *Kelington* 1190, *Killington* 1191 P]. 'The TŪN of *Cylla*'s people.' Cf. KELLING.

Kelloe Du [*Kelflau* c 1170 Reg Dun, *Kellawe* 1225 Ep]. OE *celf-hlāw* 'calf hill'.

Kelly D [*Chenleie* DB, *Chelli* 1166 RBE, *Kelli* 1194 P]. If the DB form is reliable, the name is identical with KENLEY. If not, the source is Welsh, Co *celli* 'grove'.

Kelmarsh Np [*Keilmerse, Cailmarc* DB, *Keilmers* 1199 FF, 1201 Cur, *Chailesmers* c 1155, *Keilmers* 12 BM]. Second el. OE *mersc* 'marsh'. The first is derived in some way or other from OE *cǽg* 'a key', originally 'a peg' or the like. It might be a pers. n. **Cǽgel* or **Cǽgla* or a common noun *cǽgel*, e.g. a plant-name.

Kelmscott O [*Kelmescote* 1279 RH, 1316 Ipm]. '*Cēnhelm*'s COT.'

Kelsale Sf [*Keleshala* DB, *Kelleshalle* 1228 Ch, *Keleshale* 1254 Val], **Kelsall** Chs [*Kelsale* 1260, *Kelishal* 1291, *Keleshale* 1297 Court], **Kelsey, North & South,** Li [*Colesi, Chelsi, Northchelesei* DB, *Calisei* 1094 Fr, *Cheleseia, Nordcheleseia* c 1115 LiS, *Nordchelsei* c 1140 RA, *Suthkelleseye* 1262 Ipm], **Kelshall** Hrt [*Keleshelle* c 1050 KCD 907, *Cheleselle* DB, *Chyllessella* 1086 IE, *Keleshele* 1198 FF, *-hull* 1212 Fees]. The first el. of these seems to be a pers. n. **Cēl(i)* or **Cǽl(i)*, but no such name is known. OE *Cēol* with K- owing to Scand influence will not do for all the names, and *Cylli* only for Kelshall Hrt. Possibly *Keles-* represents original *Cēnles-*, the gen. of an OE **Cēnel*, which would be a regular derivative of *Cēn-* (in *Cēnhelm* &c.). *Cēnles-* would give *Kēles-*. The second el. is OE HALH, ĒG, HYLL respectively.

Kelsick Cu [*Keldsyke* 13 StB]. OScand *keldu-sík* 'rivulet from a spring'. See KELDA.

Kelstern Li [*Che(i)lestorne* DB, *Chelestuna* c 1115 LiS, *Kaylsterne* 1209 (1252) Ch, *Kaillesterna* 1212 Fees]. '*Cǽgel*'s thornbush.' The elements are a pers. n. **Cǽgel*, related to *Cǽge* in CASSIOBURY &c., and OE ÞYRNE 'thorn-bush'. K~ is near CALCETHORPE, which see.

Kelston So [*Calvestona* Hy 1 Bath, *Calveston* 1178 Wells, *Kelveston* 1260 Bath]. 'TŪN where calves were reared.' We expect *Chalveston* or *Chelveston*, but K- is due to Midland influence.

Kelthorpe Ru [*Ketelesthorp* 1296 Subs, *Kettelthorpe* 1545 LP]. '*Ketil*'s thorp.' Cf. KEDLESTON.

Kelton Cu [*Keltona* c 1160, c 1200 StB]. OE *Celf-tūn* 'calf TŪN'.

Kelvedon Ess [(æt) *Cynlauedyne* 998 Crawf, *Kyn(e)levedene* 1066 Th, *Chelleuedana* DB, *Keuleveden* R 1 Cur], **Kelvedon Hatch** Ess [*Kyleuuedun* 1065 BM, *Keluenduna* DB, *Kelewedon* 1218 FF, *Kelvedon* 1291 Tax]. The first is '*Cynelāf*'s DENU or valley'. *Cynelāf* is an unrecorded woman's name. The el. *lāf* is rare in OE women's names, but *Brithlave* occurs in HEl and *Oslava* in Flor. K~ **Hatch,** as suggested by Smith in PNEss, has as first el. OE *cylu* 'spotted' and means 'multicoloured hill'.

Hatch may refer to a forest gate. The place is near Brentwood.

Kemberton Sa [*Chenbritone* DB, *Kembricton* 1242 Fees]. '*Cēnbeorht*'s TŪN.'

Kemble Gl [*Kemele* 682, 854, *Cemele* 688 BCS 63, 70, 470, *Kemeleshage* 956 ib. 922, *Chemele* DB]. Possibly a Brit name derived from *Camulos*, the name of a Celtic god.

Kemerton Gl [?*Cyneburgingctun* 840 BCS 430, *Chene-*, *Chinemertune* DB, *Kenemerton* 1190 ff. P, *Kenemarton* 1220 Fees]. 'The TŪN of *Cyneburg*'s people', if the first example belongs here. Otherwise '*Cynemǣr*'s TŪN'.

Kempley Gl [*Chenepelei* DB, *Kempelea* 1195 P, *Kenepeleg* 1220, 1236 Fees, *-lege* Hy 3 Misc]. First el. OE *cenep* 'a moustache; a bit of a bridle', identical with OFris *kenep* 'moustache', ON *kanpr* 'a projecting part of a wall', lit. 'a beard'. The exact meaning of *cenep* in the pl. n. is obscure. Probably the word was used of some plant. Cf. KEMPSHOT and *Cenepesmor* 772 BCS 210 (Wo).

Kempsey Wo [*Kemesei* 799 BCS 295, *Cymesig* 977 KCD 612, *Kymesei*, *Kemesige* 11 Heming, *Chemesege* DB]. '*Cymi*'s island.' *Cymi* is also found in KEMPSTON Nf and is really identical with OE *Cymen*.

Kempsford Gl [*Cynemæres ford* 800 ASC, *Chenemeresforde* DB, *Kynesmersford* c 1200 BM]. '*Cynemǣr*'s ford.'

Kempshot Ha [*Campessete* DB, *Campeseta* c 1125 Oxf, *Kempeschete* 1274 RH]. First el. as in KEMPLEY. The second is OE SCĒAT. Names in *-scēat* often have the name of a tree or plant as first el. This tells in favour of *cenep* being a plant-name.

Kempston Bd [*Kemestan* 1060 KCD (809), 1199 Cur, *Coembestune* c 1050 Rams, *Kembestone* 1047 ChronRams, *Camestone* DB, *Kembeston* 1176 P]. 'TŪN by the bend.' The first el. is identical with CAMBOIS, CAMS. The place is situated at a sharp bend of the Ouse.

Kempston Nf [*Kemestuna* DB, *Chemest.* Hy 2 BM, *Kemston* 1291 Tax]. '*Cymi*'s TŪN.' Cf. KEMPSEY.

Kempton Park Mx [*Chenetone* DB, *Keniton* 1221–30 Fees, 1230 P, *Keninton* 1228 Ch]. '*Cēna*'s TŪN.'

Kempton Sa nr Clun [*Chenpitune* DB, *Kempeton* 1256 Ass]. '*Cempa*'s TŪN.' *Cempa* is a byname from OE *cempa* 'a warrior'.

Kemsing K [*Cymesinc*, *Cymesinges* cert 822 BCS 370, æt *Cymesing* c 958 BCS 1031, *Chemesing* 1156 P, *Kemesinges* 1166 RBE]. Another example of the name is *Cymesing* 944 BCS 797 (in boundaries of Sibertswold, thus not referring to Kemsing). Apparently '*Cymesa*'s place'. See -ING. *Cymesa* is unrecorded, but may belong to OE *cȳme* adj. 'comely'.

Kenardington K [*Kynardingtune* 11 DM, *Kenardintona* 1175 P, *Kynardinton* 1242 Fees]. 'The TŪN of *Cyneheard*'s people.'

Kenchester He [*Chenecestre* DB, *Kenecestre* 1166 RBE, *Kenecestr'* 1235 Cl]. '*Cēna*'s CEASTER or Roman fort.'

Kencott O [*Chenicota* c 1130 Oxf, *Kenigco* 1228 Ep, *Kenecot* 1229 Cl]. '*Cēna*'s COT.'

Kendal We, originally **Kirkby Kendal** [*Cherchebi* DB, *Cherkaby Kendale* 1090–7 Kendale, *Kirkeby in Kendale* c 1240 FC, *Kendal'* 1190 P]. Cf. KIRKBY. Kendal is 'the valley of R KENT'.

Kenderchurch He [*Lanncinitir* c 1150 LL *Ecclesia Sancti Kenedr'* 1291 Tax, *Kendurchirche* 1428 FA]. 'St. *Cynidr*'s church.' On St. Cynidr see Rees, p. 340.

Keni·djack Co [*Kynygiek*, *Kynysiek* E 3 Ass]. In the exx. in Ass the name of the river at the place. Kenidjack goes back to Brit *Cunētiāco-*, a derivative of the river-name *Cunētiō* (KENNET). Very likely the river was *Cunētiō*, the place on it *Cunetiāco-*. See KENNET.

Kenilworth Wa [*Chinewrde* DB, *Kenildewurda* 1165, *Kinildewurða* 1173, *Kenillewurd* 1190 P]. '*Cynehild*'s WORÞ.' *Cynehild* is a woman's name.

Kenley Sa [*Chenelie* DB, *Kenelee* (wood) 1203, *Kenele* 1219, *Keneleg* 1228 FF], K~ Sr [*Kenele(e)* 1255 &c. PNSr]. '*Cēna*'s LĒAH.'

Kenn D [*Chent* DB, *Ken* 1168 P, 1274 Ipm], K~ So [*Chent* DB, 1157 P, *Chen* DB, *Kenne* 1200 FF]. Both places are on streams called KENN. The name is an original river-name *Kent*, identical with CAINT Anglesey [*Ceint* 13 Red Book], Gaul *Cantia*. The loss of *-t* at least partly took place in the compounds KENNFORD, KENTON. **Kennford** D [*Keneford* 1300 Ch]. See KENTON.

Kennerleigh D [*Kenewarlegh* 1219, *Kinwardelegh* 1244 Ass]. '*Cyneweard*'s LĒAH.'

Kennet R W, Brk [*Cynetan* (obl.) c 849 Asser, 939 BCS 734, *Cynete* 984 KCD 1282, *Kenete* 1221 Pat]. A Brit river-name *Cunētiō*, found as the name of a place on the Kennet (apparently now Mildenhall) in the abl. form *Cunetione* 4 IA, and identical with CYNWYD (the name of a place in Merioneth). The name denotes a hill in COUNTISBURY. Identical with Kennet are KENNETT, KENT, COUND Wa. Cf. also COUNDON, KENTWELL. On the Kennet are **East & West Kennett** W [æt *Cynetan* 939 BCS 734, *Cynetan* 972 BCS 1285, *Chenete* DB] and KINTBURY. An OCelt *kuno-* 'high' was generally held by earlier Celticists (e.g. Holder and Stokes) to be the first el. of OBrit and Gaul pers. ns. like *Cunotamos*, *Cunobarros*, and to be the base of OBrit *Cunētiō*. Later Celtic scholars do not reckon with such a word, and *Cuno-* in pers. ns. has at least by some been identified with Welsh *cwn* 'dogs' (sing. *ci*). It is not clear if this word can be the base of all such names as OBrit *Cunētiō*, the first el. of KINVER, CONOCK (CANOCK, the first el. of

CONSETT). Some of these appear to be named from hills, and one would suppose that they contain a pre-English (British) hill-name or a word for 'hill' or the like. For the ultimate history of these names we have to look to Celtic scholars.

Kennett R Sf, Ca [no early forms found]. Cf. KENNET. On the Kennett is Kennett Ca [*Chenet* DB, *Kenet* c 1080 ICC, 1161 BM, 1230 P]. Cf. also KENTFORD.

Kennford. See KENN.

Kenninghall Nf [*Keninchala, Cheninkehala* DB, *Keninghale* 1212 Fees, 1254 Val], **Kenningham Hall** Nf [*Kenincham* DB, *Kenigham, Kimingham* 1254 Val, *Kyningham* 1275 RH]. 'The HALH and HĀM of *Cyna*'s people.' Kenninghall may also be 'the HALH of *Cēna*'s people'.

Kennington Brk [*Chenitun* 821, *Cenigtun*, (æt) *Cenintune* 956 BCS 366, 971 f., *Chenitun* DB, *Keninton* 1242 Fees], K~ Sr [*Chenintune* DB, *Kenigton* 1275 Ipm]. 'The TŪN of *Cēna*'s people.'

Kennington K [*Chintun* 1072 BM, *Kynigtune* 11 DM, *Chenetone* DB, *Kenintuna* 1157 StAug, *Kenitton* 1270 Ch]. OE *cyne-tūn* 'royal manor'. In Kennington is **Conningbrook** [*Cuningbrok* 13 StAug, *Cunebrok* 1270 Ch]. The non-mutated vowel of the first syllable is remarkable.

Kennythorpe YE [*Cheretorp* DB, *Kinnerthorp* c 1180 (1464) Pat, *Keneringthorp* 1268 FF, *Kenerthorp* 1285 FA, 1297 Subs]. First el. perhaps OE *Cēnhere* or *Cynehere*; certainly not ON *kennari* 'teacher' (PNER).

Kensal Green Mx [*Kingisholte* 1253 FF, *Kingesholt* 1290 Ipm, *Kynsale Grene* 1550 Pat]. 'The king's wood.' Cf. WORMWOOD SCRUBBS, which is close by.

Kensey R Co [*Kensi* 13 Ol, *Kyensy* 1272, *Kensy* 1306 Launceston]. A river-name related to KENN and CAINT (from *Cantiā*). A change of *nt* to *ns* is common in Cornish.

Kensington Mx [*Chenesitun* DB, *Kensiton* 1221–30 Fees, *Kensington* 1235 FF]. 'The TŪN of *Cynesige*'s people.'

Kenstone Sa [*Kentenesdene* c 1190–4, 1257 Eyton, *Kentenisdena* 1228 BM]. Perhaps '*Centwine*'s valley'.

Kenswick Wo [*Checinwiche* DB, *Kekingwik* 1208 Fees, *Kekingewic* 1242 P]. See WĪC. The first el. is derived with the suffix *-ingas* from the pers. n. found in KECKWICK.

Kensworth Bd [*Ceagnesworthe* 975 HEl, *Caneswurde* DB, *Keneswurda* 1168 P, *-worth* 1221 Ep]. '*Cægin*'s WORÞ.' Cf. *Keneswey* 1291 Ch, the name of a road at Berkhamstead Hrt. Cf. KEYNSHAM.

Kent, the county [*Cantium* 51 B.C. Cæsar, *Kántion* Diodorus, Strabo, *Kántion ákron* Ptol, *Cantia* c 730 Bede, *Cent* 568 &c. ASC, 835 OET, *Cænt* 871–89 BCS 558]. Prob-

ably derived from Celt *canto-* (Welsh *cant*), 'rim', border': 'border land' or 'coast district'.

Kent R We, La [*Kent* c 1175 &c. Kendale, *Kenet* 1246, 1256 Ass, *Keent* 1278 Ass]. Identical with KENNET.

Kentchurch He [*Lan Cein* c 1150 LL, *Ecclesia Sancte Keyne* 1205 PNHe, *ecclesia de Sancta Kayna* 1277 Ep]. 'St. Ceina's church.' St. Ceina was a woman saint (Rees, p. 607). The name presumably belongs to Welsh *cain* 'bright, beautiful'.

Kentford Sf [*Cheneteforde* W 1 (1318) Ch, *-fort* 1109 BM, *Keneteford* 1203 Ass]. 'Ford over R Kennett.'

Kentisbeare D [*Chentesbere* DB, *Kentelesbere* 1242 Fees, *-bire* 1252 FF], **Kentisbury** D [*Chentesberie* DB, *Kentelesberi* 1260, *Kentesbyri* 1275 Ep]. The first el. appears to be an unrecorded OE *Centel* pers. n., a derivative of *Cent-* in *Centwine* &c. The same first el. is found in *Kentelisbroch*, *Kentelesmore* c 1200 Coll (in Kentisbeare) and in *Kenteleswurth* 1236 FF, *-worthe* 1431 FA (in Do). The second el. of Kentisbeare is OE BEARU 'grove', that of Kentisbury OE BURG 'fort'.

Kentish Town Mx [*Kentisston* 1207 FF, *Kentissetune* 1254 Val]. 'The TŪN of the Kentishmen.'

Kentmere We [*Kenetemere* 1272 Ipm, *Kentemere* 1274 Kendale]. The place is on the upper Kent by a now drained mere, which was known as **Kent Mere** 'mere formed by R Kent'.

Kenton D [*Chentone* DB, *Chentun* 1156 P]. 'TŪN on R KENN.'

Kenton Mx [*Keninton* 1232 FF]. Identical with KEMPTON Mx.

Kenton Nb [*Kinton* 1242 Fees, *Kynton*, *Quenton* 1256 Ass, *Kyn(g)ton* 1346 FA]. OE *cyne-tūn* 'royal manor'.

Kenton Sf [*Chenetuna, Kenetuna* DB, *Kenetona* 1179, *Cheniton* 1181 P, *Kingeston* 1252 Ch]. OE *cyne-tūn* 'royal manor' or '*Cēna*'s or *Cyna*'s TŪN'.

Kentwell Sf [*Kanewella* DB, *Kenetwelle* 1162 RBE, *Kenetewell* 1168, 1176 P, *Kentewelle* 1156–80 Bury]. Kentwell is not far from the GLEM and may be an old name of the river. If so, the original name was Kennet (OE *Cynete*), to which was added an explanatory OE *wella* 'stream'. Cf. KENNET.

Kentwood or **Kent Wood** Brk [*Kenetewuda* 1188, *Kenetwude* 1198 P]. 'Wood on R KENNET.'

Kenwick Sa [*Kenewic* 1203 Ass, *-wike* 1205–10 Eyton]. '*Cēna*'s WĪC.'

Ken Wood Mx. The earliest certain forms in PNMx(S) (*Canewood* 1543, *Cane Wood*

1558-79) are too late for an etymology to be suggested.

Kenwyn Co [*Keynwen* 1259 Ep, *-wyn* 1363 BM, *Kenwen* 1265 Ep]. Co *keyn* 'ridge' and *gwyn, gwen* 'white'.

Kenyon La [*Kenien* 1212 Fees, 1269 Ass, *Kenian* 1242 Fees, 1246 Ass, *Kynian* 1276 Ass]. Possibly from an OW **crūc Enion* 'Einion's mound', the name being misunderstood as *cruc Cenion*. See CRŪC. *Einion*, MW *Enniawn*, is a common name. In favour of the etymology suggested it may be pointed out that there was formerly a Bronze Age barrow (116 feet above Ordnance datum) at the place. See *Proceedings of La and Chs Antiquarian Society* 21 (1903). Communicated by Mr. A. C. B. Mercer of Woodford (Chs).

Kepier Du [*Kypier* 12 Mon, 1244 Ep, *Kyppyere* 1248 Ep, *Kippiard* 1237 Cl]. 'Weir with a contrivance for catching fish.' First el. OE *cȳpe* 'basket', later 'an osier basket used for catching fish'. Second el. OE *gear* 'weir'.

Kepwick (kĕpĭk) YN [*Cap-, Chipuic* DB, *Chepewic* 1166, *Kepwic* 1198 P]. A Scandinavianized form of an OE *cēap-* or *cēpe-wīc* 'market-place'. Cf. OE *cēap-, cīepestōw* 'market-place', *cēapstrǣt* 'trade street' &c.

Kerdiston Nf [*Kerdestuna* DB, *Kerdeston* 1200 P, 1267 Ch, *Kertheston* 1242 Fees]. '*Cēnrēd*'s TŪN.'

Keresforth YW nr Barnsley [*Crevesford* DB, *Keueris-, Keueresforth, Keueresford* 13, 14 BM]. '*Cēnfriþ*'s ford.' *Cheure* DB may be a form of *Cēnfriþ*. For the loss of *n* cf. STOFORD.

Keresley (karzli) Wa [*Kereslega* 1180 P, *Keresleye* 1275 Ipm]. Perhaps 'cress LĒAH', or better '*Cēnhere*'s LĒAH'. Cf. KEARSLEY Nb.

Kermincham Chs [*Cerdingham* DB, *Kerthyngham* 1275, *Cherdingham* 1278 Ipm, *Kerthingham* 1288, *Kermincham* 1286 Court]. 'The HĀM of *Cēnfriþ*'s people.' OE *Cēnfriþingahām* explains the variant forms.

Kerne Wt [*Lacherne* DB, *La Kerm* 1202 Cur, *Kurne* 1287-90 Fees]. OE *cweorn, cwyrn* 'quern, mill'. **Kerne Bridge** He [*Kernebrigges* 1272 Ep] may contain the same word.

ME **kerr** 'bog, fen, esp. one grown up with low bushes &c., a boggy or fenny copse', from ON *kiarr* 'brushwood', Norw *kjerr* 'wet ground, esp. where brushwood grows', Swed *kärr* 'fen, marsh' &c., is fairly common in pl. ns. See e.g. ALTCAR, BICKER, BYKER, ELLERKER, HOLKER.

Kersal La [*Kereshala* 1142, *Kershala* c 1175 LaCh]. 'HALH where cress grew.'

Kersall Nt [*Cherueshale* DB, *Kyrneshale* 1196 P, *Kyrueshal* 1197 P, *Kirneshall* 1264 Ipm]. '*Cynehere*'s HALH or valley.' Cf. KEARSLEY Nb.

Kersey Sf [*Cæresige* (gemære) c 995 BCS 1289, *Careseia* DB, *Karsee* 1220, *Kerseye* 1235 FF]. Probably 'cress island', the first el. being OE *cærse* 'cress'.

Kershope Cu [*Creshop* 1201 Cur]. 'Cress valley.' See HOP.

Kersoe Wo in Elmley [(æt) *Criddesho* 780 BCS 235]. '*Criddi*'s HŌH or spur of land.' **Criddi* is related to *Creoda*.

Kerswell D in Broad Clyst [*Carswill* 1212 Fees, 1315 Ipm], **Abbotskerswell** D [æt *Cærswylle, Cærswellan landscore* 956 BCS 952, *Carsuella* DB, *Kareswill* 1242 Fees, *Karswill Abbatis* 1284-6 FA], **Kingskerswell** D [*Carsewelle* DB, *Kyngescharsewell* 1270 FF]. 'Cress spring.'
Abbotskerswell was held in 1086 by the abbot of Horton, **Kingskerswell** by the King.

Kesgrave Sf [*Gressegraua* DB, *Kersigrave* 1231 Cl, *Kersse-, Kessegrave* 1254 Val]. 'Ditch or grove where cress grew.' Cf. GRÆF. The loss of *r* is due to dissimilation.

Kessingland Sf [*Kessingalanda* DB, *Kessingeland* 1219 FF, *Kessingland* 1242 Fees, *Cassingeland* 1251 Ch]. 'The land of *Cyssi*'s people.' Cf. KESTON.

Kestē·ven Li [*Ceoftefne* c 1000 Ethelwerd, *Chetsteven* DB, *Ketsteuene* 1185, 1194 P]. The first el. is probably an old district name derived from Brit *cēto-*, Welsh *coed* 'wood'. The second is OScand *stefna* 'a meeting', here in the transferred sense 'district with a common meeting-place, an administrative district'.

Keston K [*Cystaninga mearc* 862, *Cysse stanes gemæru* 973 BCS 506, 1295, *Chestan* DB, *Kestan* 1205 Cur]. '*Cyssi*'s stone.' **Cyssi* is a normal derivative of OE *Cussa*.

Keswick (kĕzĭk) Cu [*Kesewik* 1276 Ch], K~ Nf nr Norwich [*Chese-, Kesewic* DB, *Kesewic* Hy 3 BM], K~ Nf nr N. Walsham [*Casewic* c 1150 Crawf, 1254 Val, *Kesewike* 1316 FA], **East** K~ YW [*Chesuic* DB, *Chesewich* 1197 P, *Estekeswyke* c 1145 YCh 1862]. Cf. DUNKESWICK. A Scandinavianized form of OE *cēsewīc* 'cheese farm'. Cf. CHESWICK, CHISWICK.

Ketford Gl nr Dymock [*Chitiford* DB, *Kettford* Hy 3 BM]. 'Kite ford.' Identical with *Cytanford* 1005 KCD 714 (Ditton Sr).

Ketley Sa [*Cattelega, Kettelea* 1177 P, *Keteleg* 1262 Eyton, *Ketteleye* 1327 Subs]. 'Wild cat wood', OE *catta lēah*.

Ketsby Li [*Chetelesbi* DB, *Chetlesbi* c 1115 LiS, *Ketellesbi* 1212 Fees]. '*Ketil*'s BY.' Cf. KEDLESTON.

Kettering Np [æt *Cytringan* 956, *Keteiringan* 963-84, *Kyteringas* 972 BCS 943, 1128, 1280, *Cateringe* DB, *Ketering* c 1200 NpCh], **Ketteringham** Nf [*Keteringham* c 1060 Wills, 1242 Fees, *Keterincham, Kitrincham* DB, *Ketteringham* 1263 Ipm]. OE

Cytringas and *Cytringa hām*. The folk-name *Cytringas* is difficult to explain. It might be derived from a short name evolved from OE *Cūþfriþ* (whence *Cutfriþ*).

Kettlebaston Sf [*Kitelbeornastuna* DB, *Kytel*-, *Chethelbernestun* 1095 Bury, *Ketelberneston* 1208 Cur]. 'Ketelbern's TŪN.' *Ketelbern* (DB, *Cytelbearn* 963 BCS 1113) is ON *Ketilbiǫrn*, ODan *Ketilbiørn*, OSw *Kætelbiorn*.

Kettleburgh Sf [*Chetelbiria*, *Chettlebiriga*, *Kettleberga* DB, *Keteleberga* 1188 P, *Ketelberwe* 1235 Fees]. 'Ketil's hill.' Cf. KETTLEBY. It is just possible that the name is a Scandinavianized form of an OE *cetelbeorg* 'hill by a narrow valley'. Cf. CIETEL.

Kettleby, Ab, Le [*Chetelbi* DB, *Ketelbia* c 1160 DC, *Abeketlebi* 1236, *Abbe Ketlebi* 1237 Ep], **Eye K~** Le [*Chitebie* DB, *Chetelbia* c 1125 LeS, *Kedlesby* 1236 Fees, *Eketilby* 1529 ERN], **K~** Li [*Kitlebig*, *Kytlebi* c 1067 Wills, *Chetelbi* DB, *Chetlebi* c 1115 LiS], **K~ Thorpe** Li [*Torp* DB, c 1115 LiS]. 'Ketil's BY.' Cf. KEDLESTON. The almost total absence of the gen. *s* is somewhat remarkable, but has many analogies. See e.g. THIRKLEBY, THIRTLEBY, THORGANBY, THURLBY, THURGARTON.

Ab (OE *Abba*) from an early owner. Cf. *Gaufridus Abbe* 1199 Cur (Le).—*Eye* is OE *ēa* 'river'. **Eye Kettleby** is on a tributary of the river EYE.

Kettleshulme Chs [*Keteleshulm* 1285 Court]. 'Ketil's holm or island.' Cf. HOLM.

Kettlesing YW [*Ketilstringe* 1446–58 Fount M, *Kettyllsynge* 1546 Knaresborough Wills]. 'Ketil's string.' Cf. ELLINGSTRING and KEDLESTON.

Kettlestone Nf [*Ketlestuna* DB, *Ketleston* 1200 Cur], **Kettlethorpe** Li [*Ketel(s)torp* 1220 Ep, *Ketelestorp* 1249 Ep], **K~** YE [*Torp* DB, *Ketelestorp* 1227 FF]. 'Ketil's TŪN and thorp.' Cf. KEDLESTON and KETTLEBY.

Kettlewell YW [*Chetelewelle* DB, *Keteluella* 1173 YCh 197, *Ketelwell* 1222 FF]. A Scandinavianized form of OE *cetel-wella* 'stream in a narrow valley'. Cf. CIETEL.

Ketton Du [*Cattun* c 1085 LVD, *Cathona* 1091, *Chettune* Hy 2 FPD, *Ketton* 1195 (1335) Ch]. Identical with CATTON (1).

Ketton Ru [*Chetene* DB, *Chetena* 1146, *Chetenea* 1163 RA, *Ketene* 1174 Fr, 1199 FF]. Really an old name of the CHATER, the second el. being OE *ēa* 'river'. The first may be a tribal name derived from *Ket-* in KESTEVEN and meaning 'the Kesteven people' (an OE *Cētan* plur.). For the loss of the final element cf. WHITTON Li.

Keverstone Du [*Kevreston* 1306, *-e* 1317 Pat, *Keverstone* 1361 AD]. 'Cēnfriþ's TŪN.' Cf. KERESFORTH.

Kew Sr [*Caiho* (surname) 1202 P, *Cayho* 1327, *Kayho* 1330 Ha Rec. Soc. 6, *Keew*

1538, *Kewe*, *Keyo* 1592 BM]. OE *Cǣg-hōh*, the elements being OE *cǣg* in some sense (cf. CABUS) and HŌH, probably in the sense 'projecting piece of land'. Kew is in a sharp bend of the Thames.

John de Caiho was one of the sheriffs of London in 1201–2.

Kewstoke So [*Stoke super mare* 1265 Ep, *Kiustok* 1274 Ipm]. Originally STOKE. Kew appears to be the saint's name *Kew*.

Kex Beck YW, a trib. of the Wharfe [*Kexegilbec* 1227 FF, *Kexbec* 1244 Bridl]. 'Brook where *kex* grew.' *Kex*, a Scand word (cf. Sw *hundkäx*, Dan *hundekjæks*), denotes large hollow-stemmed umbelliferæ.

Kex Beck YW, a trib. of the Laver [*Kesebec* 12 YCh 83, *-bek* 1268 Ass]. First el. ON *kióss* 'narrow valley', Sw *kjusa* the same.

Kexbrough YW [*Ceze-*, *Chizeburg* DB, *Kesceburg* c 1170 YCh 1681, *Keseburc* 1194 P, *Kexeburg* 1284 YInq], **Kexby** Li [*Cheftesbi*, *Chestesbi* DB, *Chezbi*, *Chetesbi* c 1115 LiS, *Keftesby* 1202 Ass, *Kestesbi* 1212 Fees]. 'Kept's BURG and BY.' ON *Keptr* is used as a byname.

Kexby YE [*Ketelesby* c 1160 YCh 85, *Kexebi* c 1175 ib. 444, *Kexeby* 1278 Ch]. 'Ketil's BY.' But the identification of the first form is not quite certain. If the form *Ketelesby* c 1160 does not belong here, the first el. may be an ODan *Kēk* (MDan *Kegh*) corresponding to the ON byname *Keikr* (from ON *keikr* 'bent backwards'), as alternatively suggested in PNER.

Kexmoor YW [*Chetesmor*, *Cotesmore* DB, *Ketelmora* 1224–30 Fees]. 'Ketil's moor.'

Keyford So [*Caivel*, *Chaivert* DB, *Keyferz*, *Kayvel* Hy 3 BM, *Caiver* 1303 FA]. Apparently an OE *Cǣg-fyrhþ*, the first el. being OE *cǣg* in some sense (cf. CABUS), the second OE *fyrhþ* 'frith, wood'.

Keyham Le [*Caiham* DB, *Cahiham* c 1125 LeS, *Kaiham* 1199 P, *Cayham* 1209–35 Ep]. Either 'Cǣga's HĀM' (cf. CAINHAM) or *Cǣg-hām* (cf. CABUS).

Keyhaven Ha [*Kihavene* c 1170 Fr, *Kyhaven* 1228 FF]. OE *cȳ-hæfen* 'harbour where cows were shipped'.

Keyingham YE [*Caingeham* DB, 1115 YCh 1304, *Kaingham* 1190 P]. 'The HĀM of *Cǣga*'s people.' Cf. CAINHAM.

Keymer (-ī-) Sx [*Chemere*, *Chemele* DB, *Kiemela* 1107–18 AC]. OE *cȳ-mere* 'cow mere'.

Keynsham So [*Cægineshamme* c 1000 Ethelwerd, *Cainesham* DB, *Keinesham* 1170 P]. 'Cǣgin's HAMM.' *Cǣgin* is a derivative of *Cǣga* (cf. CAINHAM).

Keysoe (kāsō) Bd [*Chaisot*, *Caissot* DB, *Kaiesho* Hy 2 (1317) Ch, *Kaisho* 1195 Cur, *Kaysho* 1237–40 Fees]. Identical with CASSIO[BURY].

Keyston (-ĕ-) Hu [*Chetelestan* DB, *Ketelestan* 1165 P]. '*Ketil*'s stone.'

Keythorpe Le [*Caitorp* DB, *Keythorp* 1316 FA]. '*Cǣga*'s thorp.' Cf. CAINHAM.

Keyworth (kŭ-) Nt [*Caworde* DB, *Kaworda* 1178 PNNt(S), *Kewurda* Hy 2 DC, *Kieword* 1201 Ch, *Kewurth* 1242 Fees]. Possibly OE *cǣg-worp* 'enclosure made of poles' (cf. CABUS).

Kibblesworth Du [*Kibleswrthe* 1185 FPD]. '*Cybbel*'s WORP.' *Cybbel* is found in *Cybles weorðig* 849 BCS 455. Cf. CUBLINGTON Bk.

Kibworth Beauchamp & Harcourt Le [*Chiburde* DB, c 1125 LeS, *Kibewrda* c 1160 DC, *Cubworth* 1200 Cur, *Kybeworth Beauchamp* 1315 Ipm, *Kibbeworth Harecourt* 13 Fees]. '**Cybba*'s WORP.' Cf. *Cybban stan* BCS 1002.

K~ Beauchamp was held by Walter de Bellocampo c 1125 (LeS). Cf. ACTON BEAUCHAMP. K~ Harcourt was held by Robert de Harewecurt in 1202 (Ass). Harcourt is from HARCOURT in Normandy.

Kidbrooke K [*Ketebroc* 1202 FF, 1207 BM], **K~ Park** Sx [*Ketebrokebregge* 1438 Ct]. 'Kite brook.' Cf. *Cytanbroc* 932 BCS 692.

Kiddal YW [*Chidale* DB, *Kidall* 1303 FA]. OE *cȳ-dæl* 'cow valley'. See CŪ.

Kidderminster Wo [*Chideminstre* DB, *Kedeleministra* 1155 RBE, *Kidministra* 1167, *Kedemenistra* 1190, *Kydeministr'* 1194 P, *Kidelministr'* 1212 Fees, *Kideministre* 1227 Ch, *Kyderemunstre* 13 BM]. '*Cydda*'s or *Cydela*'s minster.' There was formerly a monastery at K~. Forms with *-r-* are rare in early sources. The *r* is intrusive and a kind of anticipation of the final *-er*. For *Cydela* cf. KIDLINGTON.

Kiddington O [*Chidintone* DB, *Cudintona* 1209–19 Ep, *Nethercudinton*, *Kudinton Superior* 1242 Fees]. '*Cydda*'s TŪN' or 'the TŪN of *Cydda*'s people'.

Kidland Nb [*Kideland* 1271 Ch, 1244 Cl]. '*Cydda*'s land.'

Kidlington O [*Chedelintone* DB, *Kedelintona* c 1130 Oxf, *Cudelinton* 1170 P, 1242 Fees]. 'The TŪN of *Cydela*'s people.' **Cydela* is a derivative of *Cuda*, *Cyd(d)a*.

Kidsley Db [*Kidesleage* 1009 Hengwrt MS. 150, *Chiteslei* DB, *Kideslea* 1176 P, *-leia* c 1200 BM]. '*Cyddi*'s LĒAH.' *Cyddi* is found in *Cyddesig* 968 BCS 1221.

Kielder Nb [*Keilder* 1326 Ipm, *Kailder* 1330 Fine, *Keldre* 1370 Cl]. Really the old name of **Kielder Burn** [*Keylder* 1542], which is identical with CALDER. Note *Kelder* and the like for Calder in early sources.

Kigbeare D [*Cacheberga* DB, *Kadekeber* 1256 Ass, *Cadekebere* 1303 FA, *Cadekbear* 1391 Ipm]. The second el. is OE *bearu* 'grove'. The first is identical with that of *Cadaca hryge* 843 BCS 442, *Kadekeregge* 1232 Subs (K), and perhaps with Cad R So

[*Caducburne* 725 Muchelney]. OE *cadac*, of which *Cadaca* (*hrygc*) is a gen. plur., may be identical in meaning with *caddow* (*cadaw* 1440) 'jackdaw'. Kigbeare would then mean 'jackdaw grove'. OE *cadac* would be a compound of OE **cā* 'jackdaw' and an OE **dāc* derived with a *k*-suffix (cf. *cornuc* from *cran*, *styrc* &c.) from the base of *daw* (cf. OHG *tāha*, MHG *tāhe*). *Caddow* is a compound of *cā* and *daw*, both of which mean 'jackdaw'.

Kilbourne Db [*Killebrun* 1200 P, *-burn* 1236 Ch, E 1 BM, *Kileburn* 1236, 1242 Fees], **Kilburn** Mx [*Cune-*, *Keneburne* c 1150 Mon, *Keleburne* 1207 FF, *Keleburn* 1236 FF], **Kilburn** YN [*Chileburne* DB, *Killebrunna* 12 Riev, *-brun* 1209 FF]. OE *cylenburna* 'stream by a kiln'.

Kilby Le [*Cilebi* DB, *Kilebi* 1165 P, 1202 Ass, *Kildebi* 1195 ff. P, *Kyldeby* 1209–19, *Kildeby* 1219 Ep]. Probably a Scandinavianized form of OE *Cilda-tūn*. Cf. CHILTON, KILDWICK.

Kilcot Gl nr Newent [*Chilecot* DB, *Killicote* 1221 Ass, *Killicot* 1254 Ipm], **Kilcott** Gl nr Charfield [(on) *Cyllincgcotan* 972 BCS 1282, *Chillecota* 1169 P]. 'The COT of *Cylla*'s people.' Cf. KELLING.

Kildale YN [*Childale* DB, *Kildalam* c 1180 YCh 659]. The first el. may be ON *kill* 'a narrow bay', here in the sense 'a narrow valley'.

Kildwick YW [*Childeuuic* DB, *Kildewike* 1267 Ep]. A Scandinavianized form of OE *Cilda-wīc*. Cf. CHILDWICK.

Kilham Nb [*Killum* 1177 P, 1242 Fees, *Kylnom* 1323 Ipm], **K~** YE [*Chillun* DB, *Killum* 1100–8 YCh 426, 1206 FF]. OE *cylnum*, dat. plur. of *cylen* 'kiln'.

Kilkhampton Co [*Chilchetone* DB, *Kilkamton* 1195 Cur, *Kilcanton* 1202 FF, *Kylchampton* 1238 FF]. No doubt a hybrid name, consisting of an old Cornish name and OE *-hǣmatūn* (cf. HĀMTŪN). The original name may have been a combination of Co *cil* 'recess' and e.g. *loch* 'pool'. Cf. MW *cil luch* c 1150 LL.

Killamarsh Db [*Chinewoldemaresc* DB, *Kinewaldesmers* 1249 Ch]. '*Cynewald*'s marsh.'

Killerby Du [*Culuerdebi* 1091 FPD, 1196 P, *Kiluerdebi* 1207 FPD], **K~** YN nr Catterick [*Chiluordebi* DB], **K~** YN in Cayton [*Chiluertesby* DB]. '*Kilvert*'s BY.' This pers. n. (*Chiluert* DB) is also found in CULVERTHORPE, KILVERSTONE, KILWARDBY, and in the lost **Killerwick** La [*Chiluestreuic* DB, *Kilverdiswic* 1190 FC]. The name has not been explained. It might be suggested that it is a byname, viz. an ON *kylfu-vǫrðr* 'one who defends the prow of the ship'. ON *kylfa* is used of the beak on a ship's stem, while *vǫrðr* means 'defender, guard'.

Killinghall YW [*Chenihalle, Kilingala* DB, *Chilingehal* 1165 P, *Killingehal* 1206 Obl]. 'The HALH or haugh of *Cylla*'s people.' Cf. KELLING.

Killingholme Li [*Chelvingeholm* DB, *Chiluingheholm* c 1115 LiS, *Kiluingeholm* 1144 BMFacs]. The original name may have been an OE *Cylfingas* (from *Cynwulfingas*), to which was added OScand *holm*. Or it may be simply 'the HOLM of the *Cynwulfingas*'.

Killington We [*Killintona* 1175, *-ton* 1176, *Killington* 1193, *Kellinton* 1195 P, *Kylington* 1247 Ipm]. 'The TŪN of *Cylla*'s people.' Cf. KELLING.

Killingwoldgraves YE [*Kynewaldgrave* 1169 (1327) YCh 86, *Kinewaldesgraue* 1197 P]. '*Cynewald*'s grave or grove.'

Killingworth Nb [*Killingwrth* 1242 Fees, *Kelingwrth* 1292 Cl]. 'The WORÞ of *Cylla*'s people.' Cf. KELLING.

Kilmersdon So [*Kunemersdon* 951 BCS 889, *Chenemeresdone* DB, *Kinemeresdon* 1176 P]. '*Cynemǣr*'s DŪN.'

Kilmeston Ha [*Cenelmestun* 961 BCS 1077, *Cylmestuna* BCS 1160, *Chelmestune* DB, *Culmestone* 1282 Ep]. '*Cynehelm*'s TŪN.'

Kilmington D [*Chenemetone* DB, *Culmiton* 1194 P, *Kelminton* 1272 Ipm], K~ W [*Chelme-, Cilemetone* DB, *Kelmeton, Culmetone* 1289 Ipm, *Culminton* 1251 Cl]. 'The TŪN of *Cynehelm*'s people.'

Kilnhurst YW [*Kilnehirst* 1379 PT]. 'Kiln hill or wood.'

Kilnsea YE [*Chilnesse* DB, *Chinlesei* 1115, 1160–2 YCh 1304, 1307, *Kilneseia* 1222 FF, *Kilnese* 1228, *Kelneseia* 1232 Ep], **Kilnsey** YW [*Chileseie* DB, *Kilneseiam* 1150–3, *Kilnesey* 1162 YCh 68, 81, *Kylneshei* c 1205 FC]. The first may be OE *cylen-sǣ* 'lake with a kiln', but this hardly suits the second. The first el. may be an OE *Cynel*, a derivative of *Cyna*, or even *Cynehelm*, the second being OE ĒG or (GE)HÆG.

Kilnwick YE [*Chileuuit, -wid* DB, *Kilnewic* 1226 FF]. OE *cylenwic* 'WĪC with a kiln'.

Kilnwick Percy YE [*Chelingewic* DB, *Killingwych* 1160–5 YCh 749, *Killingewic* 1218 FF, *Killingwik Perci* 1303 FA]. 'The WĪC of *Cylla*'s people.' Cf. KELLING.

The manor was held by Robert de Percy in 1160–5 (YCh). It came to Ernald de Percy t. Hy 1. Cf. BOLTON PERCY.

Kilpeck He [*Chipeete* DB, *Cilpedec, Lann Degui Cilpedec* c 1150 LL, *Kilpeec* 1167, *-pedet* 1176, *-pech* 1193 P]. A Welsh name. First el. Welsh *cil* 'corner, retreat'. The second is obscure. It may be compared with (*nant*) *pedecou* c 1150 LL (Monm).

Kilpin YE [*Celpene* 959 YCh 4, *Chelpin* DB, *Kilpin* 1199 FF, c 1200 YCh 1130]. OE *celf-penn* 'pen for calves'.

Kilsall Hall Sa [(æt) *Cylleshale* 10 BCS 1317]. '*Cylli*'s HALH or valley.' Cf. *Cylles ege* 964 BCS 1129.

Kilsby Np [*Kyldesby* c 1050 KCD 939, *Kildesbig* 1043 Th, *Chidesbi* DB, *Kildebi* 1139 RA, *Kildesbi* 1155–62 ib.]. The first el. is a Scandinavianized form of OE *cild* 'young nobleman'.

Kilton Nt [*Kileton* Hy 3 DbAS xiv, *Kilton* 13 Duk, *Kelton* 1301 Thoroton]. TŪN with a kiln.' **Kilton** So [*Cylfantun* c 880 BCS 553, *Chilvetune* DB]. See KILVE. **Kilton** YN [*Chiltun* DB]. Probably a Scandinavianized form of OE *Cilda-tūn*. Cf. CHILTON.

Kilve So [*Clive, Cliua* DB, *Kelua* 1186 P, *Kylve* 1243 Ass, *Culue* 1329 Ep]. Kilve is near **Kilton** [OE *Cylfantūn*]. Apparently Kilve is OE *Cylfe* and Kilton contains the same element. *Cylfe* may be an OE *cylfe* identical with ON *kylfa* 'a club' and used of an eminence. There is a hill at Kilve.

Kilverstone Nf [*Culuertestuna* DB, *Kilverdestun* 1202 FF, *Kelewerdestone* 1254 Val]. Cf. KILLERBY.

Kilvington Nt [*Cheluintone, Chelvinctune* DB, *Kilvintun* 1236 Fees], **North K~** YN [*Cheluintun* DB, *Keluintune* 1088 LVD, *Kilvinton* 1200 FF, *Northkilvington* 1240 FF], **South K~** YN [*Chelvinctune* DB, *Kiluinton* c 1190, *Suth Kiluingtona* Hy 3 BM]. 'The TŪN of *Cynewulf*'s people.'

Kilwardby Le [*Culvertebi* c 1125 LeS, *Culverdeby* 1270 Ipm]. See KILLERBY.

Kilworth, North & South, Le [*Chivelesworde, Cleveliorde* DB, *Kiuelewurd* 1177, *Kiuelivvurd* 1185, *Kiuelingwurda* 1191, *Kiuelingewurðe, Cuuelingwurd* 1195 P, *Nortkeueligworth, Suth Kiuiligwrth* 13 BM]. 'The WORÞ of *Cyfel*'s people.' *Cyfel* is a derivative of *Cufa*.

Kimberley Nf [*Chineburlai* DB, 1161 P, *Cheneburlai* 1162 P, *Kineburle* 1254 Val]. '*Cyneburg*'s LĒAH.' *Cyneburg* is a woman's name. **K~** Nt [*Chinemarelie* DB, *Kinemarle* c 1200 Middleton]. '*Cynemǣr*'s LĒAH.' **K~** Wa in Kingsbury [*Kynebaldeleye* 1311 BM]. '*Cynebald*'s LĒAH.'

Kimberworth YW [*Chibereworde* DB, *Kimberwurth* 1222, 1226 FF]. '*Cyneburg*'s WORÞ.' *Cyneburg* is a woman's name.

Kimble, Great & Little, Bk [*Cynebellinga gemǣre* 903 BCS 603, *Chenebella, Chenebelle Parva* DB, *Kinebelle* 1196 FF, *Magna Kenebell* 1254 Val]. OE *cyne-belle* 'royal hill'. Cf. BELCHALWELL.

Kimblesworth Du [*Kymliswrth, Kimleswrthe* Hy 3 BM]. '*Cynehelm*'s WORÞ.'

Kimbolton He [*Kimbalton* Hy 3 BM], **K~** Hu [*Chenebaltone* DB, *-boltona* 1130 P, *Kinebalton* 1232 Cl]. '*Cynebald*'s TŪN.'

Kimcote Le [*Chenemundescote* DB, *Kinemundescot* 1167 P]. '*Cynemund*'s COT.'

Kimmeridge Do [*Cameric* DB, *Kimerich* 1212 Fees, *Kemerich* 1230 P]. Second el. perhaps OE *RIC 'stream'. The first may be OE *cȳme* 'convenient' &c.

Kimmerston Nb [*Kynemereston* 1244 Ch, *Kenemeriston* 1255 Ipm]. '*Cynemǣr*'s TŪN.'

Kimpton Ha [*Chementune* DB, *Keminton* 1167 P, *Cumeton* 1304 Ep], **K~** Hrt [*Kamintone* DB, *Cuminton* 1198 FF, *Kymeton* 1236 Ep]. '*Cyma*'s TŪN.'

Kimsbury Gl [*Kynemeresburia* 1121 Glouc]. '*Cynemǣr*'s BURG.' There is an ancient camp here.

Kinder Db [*Chendre* DB, *Kynder* 1285 For, 1293 Ipm]. **Kinder Scout** is a prominent hill (2,088 ft.), the highest peak in the Peak district. The probability is that Kinder is an old hill-name. If so, it may be a Brit name consisting of Brit *Cunētiō* (cf. COUNTISBURY) and *brigā* (Welsh *bre*, mutated *fre*) 'hill'. The Welsh form would be *Cynwydfre*. For the loss of *f* cf. MELLOR. *Scout* is the north country *scout* 'a high rock or hill' from ON *skúti* 'overhanging rock'.

Kinderton Chs [*Cinbretune* DB, *Kindreton* 1240 Cl, *Kindirton* 1289 Court]. *Cinbretune* DB is no doubt for *Cindretune*. '*Cynrēd*'s TŪN.'

Kineton Gl [*Kinton* 1191 P, 1252 Ch, *Kyngton* 1330 Ch], **K~** Wa [*Cyngtun* 969 BCS 1234, *Quintone* DB, *Kincton* 1230 Ch]. OE *cyne-tūn* or *cyning-tūn* 'royal manor'. OE *cyne-* 'royal' is common, as in *cynebotl* 'palace', *-hām* 'royal manor'.

King Water R Cu [*King* 1169 ff. Lanercost, *King*, *Keeng* 1292 Ass]. Probably elliptical for *cyninges-burna* or the like. 'The king's stream.' **Kingwater** is on the King.

Kingcombe Do in Toller Porcorum [*Chimedecome* DB, *Kendecumb* 1212 Fees, *Kemthecumb* 1226 FF, *Kentecumba* 1236 Fees]. 'Valley where wall-germander (OE *cymed*) grew.'

Kingerby Li [*Chenebi* DB, *Chimerebi* c 1115 LiS, *Kinierbi* 1163–5 BM, *-bia* 1212 Fees, *Cunehereby* 1208 FF]. '*Cynehere*'s BY.'

Kingham O [*Caningeham* DB, *Keingham* 1236, *Kaingeham* 1220, 1242 Fees]. 'The HĀM of *Cǣga*'s people.' Cf. KEYINGHAM. The DB form may be for *Cahingeham*.

Kingsbridge D [*Cinges bricg* 962 PND, *Kingesbrig* 1230 P]. Self-explanatory.

Kingsbury Mx [*Kynges byrig* 1044–6 BM, *Chingesberie* DB], **K~ Episcopi** So [*Cyncgesbyrig* 1065 Wells, *Chingesberie* DB], **K~ Regis** So at Milborne Port [*Kingesberi* 1200 Cur]. 'The king's BURG.'

K~ Episcopi from the Bishop of Bath (DB).

Kingsbury Wa [*Chinesberie* DB, *Kinesburi* Hy 1 BM, 1222 FF, *Kineberia* 1190 P]. '*Cyne*'s BURG.'

Kingsclere. See CLERE.

Kingscote Gl [*Chingescote* DB, *Kingescota* 1191 P]. 'The king's COT.'

Kingsdon So [*Kingesdon* 1194 P, 1201 Ass], **Kingsdown** K nr Deal [*Kingesd.* 1177, *Kingesdon* 1371 BM], **K~** K nr Gravesend [*Kingesdone* 1166 RBE, *-dun* 1171 P], **K~** K nr Sittingbourne [*Cyningesdun* 850 BCS 459, *Kyngesdon* 1229 Ch]. 'The king's DŪN or hill pasture.'

Kingsettle, hill So [*Kingessettle* 1251 Misc]. OE *cyninges setl* 'the king's seat'. There is a monument with a statue of King Alfred on the hill.

Kingsey Bk [*Eya* 1174 RA, *Kingesie* 1197 FF, *Kyngeseya* 1232 Ep, *Eye* 1236 Fees]. 'The king's Eye.' Cf. TOWERSEY, ĒG.

Kingsford Wa [*Kingesford* 1187 P]. 'The king's ford.' **Kingsford** Wo [*Cenungaford* 964 BCS 1134, *Keningeford* 1265 Misc]. 'The ford of *Cēna*'s people.'

Kingsheanton. See HEANTON.

Kingsholm Gl [*Kingesham* 1211–13 Fees, *Aula regis* Glouc, *Kyngeshamme* 1267 Glouc, *Kingeshome* Hy 3 Ipm]. 'The king's HAMM', though the form *Aula regis* may suggest 'the king's HĀM'.

Kingside Cu [*Kyngesetemire* 1292 Holme C]. 'The king's shieling.' See (GE)SET, SÆTR, MÝRR.

Kingsland He [*Lene* DB, *Kingeslan* 1213 Cl, *-len* 1230 Pat]. Cf. LEOMINSTER and EARDISLAND, MONKLAND, LYONSHALL. *Lene* is an old district name. Kingsland is the part belonging to the king.

Kingsley Chs [*Chingeslie* DB, *Kingisleg* 1260 Court], **K~** Ha [*Kyngesly* 1210–15 Selborne, *Kyngeslye* 1293 BM], **K~** St [*Chingeslei* DB, *Kingeslegh* 1227 Ass]. 'The king's LĒAH.'

Kingslow Sa [*Kynsedel* 1215, *-leg* 1226 Eyton]. OE *cyne-seþl* 'the king's seat'.

Kingsnorth K [*Kingesnade* Hy 3 BM, *Kyngesnode* 1278 QW]. Cf. *Cyningessnade* 850 BCS 459 (K, but not Kingsnorth). 'The king's *snād* or wood.' See SNĀD, SNÆD.

Kingsteignton. See TEIGNTON.

Kingsthorpe Np [*Torp* DB, *Kingestorp* 1190 P], **K~ Lodge** Np [*Chingestorp* DB, *Kynesthorp* 12 NS]. 'The king's thorp.'

Kingston, a common name, is usually **1.** OE *Cyninges-tūn* 'the king's TŪN, royal manor': **K~ Bagpuize** Brk [*Cinghæma gemære* 958, *Cingtuninga gemære* 959, *Cingestun* 970 BCS 1028, 1047, 1260, *Cyngestun* c 977 E, *Kyngeston Baggepus* 1291 Tax], **K~ Lisle** Brk [*Kingeston* 1220 Fees, *Kyngeston Lisle* 1322 Ch, *Kyngeston del Isle* 1336 Ch], **K~** Ca [*Kingestona* c 1080 ICC, *Chingestone* DB], **K~** D nr Modbury [*Kingeston* 1242 Fees], **K~** D nr Sidmouth [*Kingeston* 1249 Ass], **K~** Do nr Corfe [*Chingestone* DB, *Kyngeston Abbatisse* 1297 FF], **K~** Do nr Dorchester [*Kingeston* 1247

FF, *Kyngeston Marlevard* 1280 FF], K~
Lacy Do [*Kingeston* 1191 P, 1234 Cl,
Kyngeston Lacy 1335 Ipm], K~ Russel Do
[*Kyngeston* 1212 Fees, *Kyngeston Russel* 1284
Ch], K~ Gl [*Kingston* 1243–5 Berk], K~
Cross Ha [*Kingeston* 1194 ff. P], K~ K
[*Kyngestun* 11 DM, *Kyngeston* 1279 Ep],
K~ Blount O [*Chingestone* DB, *Kingeston*
1200 Cur], K~ Sf [*Kingestun* c 1050 KCD
907, *Kyngestuna* DB], K~ So nr Taunton
[*Kyngestona* 1155–8 (1334) Ch, *Kyngestone*
1327 Subs], K~ Pitney or K~ juxta Yeovil
So [*Kingeston* 1285 FA], K~ Seymour So
[*Chingestone* DB, *Kingeston Milonis de
Sancto Mauro* 1196 P, *Kyngeston Saymor*
1327 Subs], K~ on Thames Sr [*Cyninges
tūn* 838 BCS (421), 979 ASC (E), *Cyngestun*
972 BCS 1290, *Chingestune* DB], K~ St
[*Kingeston* 1166 P, 1227 Ass], K~ near
Lewes Sx [*Chingestona* 1121 AC], K~
by Sea Sx [*Chingestune* DB, *Kyngeston
Bouci* 1315 Pat, *Kingeston Bouscy* 1317
Misc], K~ Deverill W [*Kingesdeverell* 1206
Cl, *Kingeston Deverel* 1250 Cl], K~ Wa
[*Chingestune* DB, *Kingeston* 1242 Fees],
K~ Wt [*Chingestune* DB, *Kingeston* 1229
Ch], K~ upon Hull YE [*Burgus super
Humbre* 1239 Ep, *Kyngeston super Hul* 1299
BM]. K~ upon Hull YE was originally Wike
[*Wyk*' 1160–80 PNER] or Hull (q.v.). The
manor came to Edward I in 1292 and was
given the name Kingston. The old name
Wike is OE *wīc* in the sense 'dairy-farm' or
'dwelling-place', hardly OScand *vīk* 'bay',
as suggested in PNER.

2. Kingston upon Soar Nt [*Chinestan*
DB, *Kinestan* W 2 Reg, *Kenestan* 1238
Cl]. OE *cyne-stān* 'royal stone'.

K~ Bagpuize Brk was held in 1086 by Ralph
[de Bachepuz], in 1242 (Fees) by William de
Bakepuz. The name is from BACQUEPUIS in
Normandy.—K~ Blount O was held by Hugo
le Blund in 1279 (RH). Blount or *Blund* is
a family name, originally a byname (OFr *blund*
'blond').—K~ Deverill W was originally DEVE-
RILL (q.v.). In this case Kingston (or King's)
is the distinguishing addition.—K~ Lacy Do
was held by John de Lasey in 1230 (FF). Cf.
EWYAS LACY.—K~ Lisle Brk was named from
the family *del Isle* or *de Insula*, lit. 'of the isle'.
L'ISLE is a common pl. n. in France.—K~
Pitney So. See PITNEY.—K~ Russel Do was
held by John Russel in 1212 (Fees). Russel is
OFr *roussel* 'red'.—K~ by Sea Sx is corrupt
for K~ *Busci*. Robert de Busci held the manor
in 1199 (FF). Busci is said to be from BOUCÉ
in Normandy.—K~ Seymour So. Seymour is
a family name from one of the places called
ST. MAUR in France.

Kingstone Winslow Brk [*Wend(e)lesclive*
1242 Fees, *Kyngeston* 1316 FA], K~ He nr
Hereford [*Chingestone* DB, *Kyngeston* 1198,
Kingestun 1242 Fees], K~ He nr Ross
[*Chingestune* DB]. 'The king's TŪN.'

On Winslow (*Wendlesclif*) see CLEEVE (Bishops).

Kingstone So [*Chingestana* DB, *Kingestan*
1194 f. P, 1212 Fees]. 'The king's stone.'

Kingswear D [*Kingeswere* 1170–96 Totnes].
'The king's weir.'

Kingswinford. See SWINFORD.

Kingswood Gl nr Wotton [*Kingeswodam*
1166 RBE], K~ Gl nr Bristol [*Kingesuuode*
1252 Cl], K~ Sr [*Kingeswod* 1202 Cur,
-wode 1212 Fees], K~ Wa [*Kyngeswode* 1407
AD]. 'The king's wood.'

Kingthorpe Li [*Chinetorp* DB, *Chin(e)torp*
c 1115 LiS, *Kinctorp* 1212 Fees, *Kuningkes-
torp* 1202 Ass]. 'The king's thorp.'

Kingthorpe YN [*Chinetorp* DB, *Kinthorp*
1198 Fees]. Perhaps '*Cyna*'s thorp'.

Kington Magna & Little K~ Do [*Chintone*
DB, *Kinton* 1203 Cur, *Magna, Parva King-
ton* 1242 Fees], K~ He [*Chingtune* DB,
Kinton 1187 P], K~ St. Michael W [*at
Kingtone* 934 BCS 704, *Chinctuna* c 1185
BM, *Kynton Mich[aelis]* 1281 QW], West
K~ W [*Westkinton* 1194 P, 1202 Cur,
Westkyngton 1249 Ass], K~ Grange Wa
[*Cintone* DB, *Kington* 1236 Fees, *Kyngton*
1313 Misc], K~ Wo [*Chintune* DB, *Kington*
1236 Fees]. Identical with KINETON. The
original form may have been OE *cyne-tūn*
'royal manor', but a change to *cyning-tūn*
must have taken place early.

Kingwater. See KING WATER.

Kingweston So [*Chinwardestune* DB, *Kyne-
wardeston* 1243 Ass]. '*Cyneweard*'s TŪN.'

Kinlet Sa [*Chinlete* DB, *Kinleet* 1185 TpR,
Kinlet 1201, 1211 FF]. Perhaps an OE
cyne-hlīet 'royal share'. Cf. SHIRLET. The
manor was held at the Conquest by Queen
Edith. OE *hlīet* is found in the sense 'lot',
but probably also in the sense 'portion'.

Kinnard's Ferry Li [*Kinerdefere* 1185 TpR,
Kinardesferi 1219 Ass]. '*Cyneheard*'s ferry.'

Kinnerley Sa [*Chenardelei* DB, *Kinardeslegh* 1223 Cl], Kinnersley He [*Cyrdes leah*
a 1038 KCD 755, *Curdeslege* DB, *Kyn-
ardesle* 1242 Fees, *Kinardeslegh* 1252 ib.],
Kinnersley Wo [*Kinardeslege* 1221 Ass,
-le 1232 Ch, *Kynarsleie* 1314 Ipm]. '*Cyneheard*'s LĒAH.' The earliest examples of
Kinnersley He show a contracted OE form
(*Cyrdes* gen. from *Cyneheardes*).

Kinnersley Sa [*Chinardeseie* DB, *Kynardes-
heye* 1256 (1332) Ch, *Kinardeseie* 1291 Tax].
'*Cyneheard*'s island.'

Kinnersley Sr [*Kynwardeleg, Kynewardel*'
1253 Abbr]. '*Cyneweard*'s LĒAH.'

Kinnerton Chs [*Kynarton* 1240 Cl, *Kin-
narton* 1260 Court]. '*Cyneheard*'s TŪN.'

Kinniside Cu [*Kynisheved* 1322 Ipm]. See
HĒAFOD. The first el. may be OE *Cyne*.

Kinoulton Nt [*Kinildetune* 11 KCD 971,
Chineltune DB, *Cheneldestona* 1152 BM,
Kinelton 1211–13 Fees]. '*Cynehild*'s TŪN.'
Cynehild is a woman's name.

Kinsham Wo [*Kelmesham* 1209 Fees,
Kilmesham 1275 Ass]. '*Cynehelm*'s HĀM.'
Cf. *Cylmes gemære* KCD 618, where *Cylmes*
is a shortened form of *Cynehelmes*.

Kinsley YW [*Chineslai* DB, *Kyneslay* 1245 Ipm]. '*Cyne*'s LĒAH.'

Kinson Do [*Chinestanestone* DB, *Kenstaneston* 1230 Cl]. '*Cynestān*'s TŪN.'

Kintbury Brk [(æt) *Cynetan byrig* 931 BCS 678, *Cheneteberie* DB]. 'BURG on R KENNET.'

Kinvaston St [*Kinwaldestun, Kineuoldestun* 996 Mon, *Chenwardestone* DB, *Kyneswaldestan* 1227 Ass]. '*Cynewald*'s TŪN.'

Kinver St [*Cynibre* 736 BCS 154, *Cynefaresstan* 964 ib. 1134, *Chenevare* DB, *-fara* 1130 P]. Second el. Welsh *bre* 'hill', in the mutated form *fre*. For the first cf. KENNET. Possibly *Cynibre* is an adaptation of a Welsh *Cynfre*, the first el. having been associated with OE *cyne-* 'royal'.

Kinwalsey Wa [*Kinewoldeseye* 1276 Ipm, *Kynewaldeshey* 1292 Ch]. '*Cynewald*'s hay.' See (GE)HÆG.

Kinwarton (kǐnertn) Wa [*Kineuuarton* 714 BCS 130, *Chenevertone* DB, *Kinewarton* 1169 P]. '*Cyneweard*'s TŪN.'

Kiplin YN [*Chipeling* DB, *Kepling* 1205 Obl, *Kipeling* 1208 Cur]. Possibly '*Cyppel*'s people', *Cyppel* being a diminutive of *Cuppa*.

Kipling Cotes YE [*Climbicote* DB, *Kibblincotes* 1190 P, *Kiblingecotes* 1279 Ipm]. 'The COTS of *Cybbel*'s people.' Cf. KIBBLESWORTH.

Kippax YW [*Chipesch* DB, *Kippeys* 1155-8 YCh 1451, *Kipais* 1190 P, *Kypask, -ax* 1293 Ass], **Kipton** Nf [*Chiptena* DB, *Kipton* 1280 Ipm, 1302 FA]. The first el. may be a pers. n. *Cyppa*, related to *Cuppa*. Kippax has as second el. OE *æsc* 'ash-tree', partly Scandinavianized to *-ask* (whence *-ax*).

Kirby, West, Chs [*Kirchebi* 1154-81 Chester, *Kirkebi* 1205 BM, *Westkirkeby* 1289 Court], **K~ le Soken** Ess [*Kyrkebi* 1181 StPaul, *Kyrkeby* 1254 Val], **K~ Bellars** Le [*Chirchebi* DB, *Kirkeby super Wreic* 1242 Fees, *Kirkeby Belers* 1428 FA], **K~ Bedon** Nf [*Kerkebei* DB, *Kirkeby Bydon* 1291 Tax], **K~ Cane** Nf [*Kerkeby* DB, *Kyrkeby* c 1095 Bury, *K~ Cam* 1282 Cl, *Kirkebycaam* 1375 BM], **K~ Hall** Np [*Chercheberie* DB, *Chirchebi* 1163 P], **Monks K~** Wa [*Chircheberie* DB, *Kirkebi* Hy 2 BM, *Kirkeby Moynes* or *Monachorum* 1305 Ch], **K~ Grindalythe** YE [*Chirchebi* DB, *Kirkebi in Krandale* 1180-95 YCh 1077, *Kirkeby Crandala* c 1190 ib. 1080], **K~ Underdale** YE [*Cherchebi* DB, *Kircabi in Hundolvesdala* 1157 YCh 354, *Kirkeby Hundoldale* 1254 Ep], **K~ in Cleveland** YN [*Cherchebi* DB], **K~ Hill** YN nr Boroughbridge [*Chirchebi* DB, *Kirkeby in Mora* 1224-30 Fees], **K~ Hill** YN nr Richmond [*Kirkebi* c 1160 Marrick, *Kyrby Hylle* AD i], **K~ Knowle** YN [*Chirchebi* DB, *Kirkeby subtus Knoll* 1230 Ep, *K~ Knol* 1279 Cl], **K~ Misperton** YN [*Chirchebi* DB, *Mispertona Kirkeby* c 1090, *Kircabimispertun* 1157 YCh 350, 354], **K~ Ravens-**

worth YN [*Kirkeby Ravenswathe* 1280 YInq], **K~ Sigston** YN [*Kirchebi* 1088 LVD], **K~ Wiske** YN [*Chirchebi* DB, *Kirkeby super Wisc* c 1180 YCh 673, *K~ Wisch* 1212 Cur]. ON, OSw *kirkiubȳr* 'church village, village with a church'.

K~ Bedon Nf was given to Hadenald de Bidun t. Hy 1 and was held by John de Bidon before 1212 (Fees). The family took its name from BIDON (several in France).—Hamo Beler held **K~ Bellars** Le c 1166 (DC). Beler is a nickname (Fr *bélier* 'ram').—**K~ Cane** Nf was held by Walter de Cadamo in 1205 (Cur), by Maria de Cham in 1242 (Fees). Cane (*Cham*) is a family name derived from CAEN in France.—**K~ Grindalythe** YE was originally **K~ Crandale** 'crane valley', to which was added ON *hlið* 'slope'.— **K~ Hill** YN from high situation.—**K~ Knowle** YN is by **Knowle Hill** (OE *cnoll* 'knoll').— **K~ Misperton** YN is really Kirby and Misperton. The latter is *Mispeton* DB, *Mispertona* c 1090 YCh 350. *Misper-* may be an OE *mistbeorg* 'foggy hill' or 'dung hill'.—**Monks K~** Wa was a priory.—**K~ Ravensworth** YN. See RAVENSWORTH.—**K~ Sigston** YN is near SIGSTON.—**K~ le Soken** Ess. Soken is an early form of *soke*.—**K~ Underdale** YE was originally **K~ Hundolvesdale** '*Hundulf*'s valley'. *Hundulf* (BCS 1130) is ON *Hundolfr*, OSw *Hundulf*.

Kirby Muxloe Le [*Carbi* DB, *Carobi* c 1200 Fr, *Kereby* 1236 Fees, *Kerby Muckless* 1799 BM], **Cold K~** YN [*Carebi* DB, *Kerebi* c 1170 Riev, *Kareby, Kereby* 1209 FF]. Identical with KEARBY.

Muxloe seems to be a modification of *muckless*.

Kirdford Sx [*Kinredeford* 1229 Cl, *Kenredeford* 1241 FF]. '*Cynerēd*'s ford.'

Kirkandrews Cu nr Penrith [*Kirkandreas* a 1147, c 1160 WR], **K~ Cu** on the Esk [*Kirchandr.* c 1165 CWNS xxix], **K~ upon Eden** Cu [*Kirkandres* c 1210 Marrick, 1332 Subs]. 'St. Andrew's church.' The order between the elements is Celtic. Cf. ASPATRIA.

Kirkbampton. See BAMPTON.

Kirkbride Cu [*Chirchebrid* 1163 P, *Kirkebride* 1189 P]. 'St. Bride's church.' Cf. BRIDEKIRK and KIRKANDREWS.

Kirkburn YE. Close to this are **East-, Southburn**. These were all once *Burn* [*Burnous* DB, *Burnus* c 1180 YCh 659, *Aust-, West-, Sudburne* DB, *Estbrunne, Kirkebrun* 1272 Ipm, *Kirkebrunnon* 1272 Cl]. *Westburne* DB is Kirkburn. OE *burna* 'stream'. *Burnous* DB &c. seems to be an alternative name *Burn-hūs* 'house on a stream'.

Kirkburton. See BURTON.

Kirkby (kerbǐ) La [*Cherchebi* DB, *Kierkeby* 1207 P], **K~ Ireleth** La [*Kirkebi* c 1195 FC, *-by* 1227 FF, *Kirkeby Irelith* 1278 Ass], **K~ Mallory** Le [*Cherchebi* DB, *Kyrkeby Malure* 1285 FA], **K~ on Bain** Li [*Chirchebi* DB, c 1115 LiS, *Kyrkeby super Bein* 1226 Ep], **East K~** Li [*Cherchebi* DB, *Circhebia* 1142 NpCh], **K~ Green** Li [*Cherchebi* DB

Kirkebi 1202 Ass], K~ **Laythorpe** Li [*Chirchebi* DB, *Kirkebi et Leitorp* 1206 Cur, *Kirkeby Leylthorp* 1316 FA], K~ **cum Osgodby** Li [*Kyrchebeia* 1146 RA, *Kirkeby* 1254 Val], K~ **Underwood** Li [*Cherchebi* DB, *Kyrkeby* 1242 Fees], K~ **in Ashfield** Nt [*Chirchebi* DB, *Kyrkeby in Essefeld* 1237 Ep], K~ **Lonsdale** We [*Cherchebi* DB, *Kircabilauenesdala* 1090–7 (1307) Ch, *Cherkeby Lonnesdale* 1090–7 Kendale], K~ **Stephen** We [*Cherkaby Stephan* 1090–7 Kendale, *Kircabi Stephan* 1157 YCh 354], K~ **Thore** We [*Kirkebythore* 1179 Holme C, *Kirkebithore* 1223 Pat], K~ **Fleetham** YN [*Chirchebi* DB, *Kirkby et Fleteham* 1287 FA, *Kyrkbyfletham* 1291 Tax], K~ **Moorside** YN [*Chirchebi* DB, *Kirkeby Moresheved* 1282 Cl], K~ **Malham** YW [*Chirchebi* DB, *Kirkeby Malgam* 1250 Ep], K~ **Malzeard** YW [*Chirchebi* DB, *Malassart* 1155–95 YCh 83, *Kirkeby Malesard* 1242 P], K~ **Overblow** YW [*Cherchebi* DB, *Kirkeby Oreblowere* 1211 Cur, K~ *Orbelawer* 1242 Ep, *Kirkby Ferers* 1291 Tax], **South** K~ YW [*Suthkyrkeby* 1226 FF], K~ **Wharfe** YW [*Chirchebi* DB, *Kyrkeby upon Werf* 1254 Ipm]. Identical with KIRBY.

K~ in Ashfield Nt. Ashfield 'FELD with ash-trees' must be a district name.—**K~ Fleetham** YN is K~ and Fleetham. The latter (*Fleteham* DB) is 'HĀM on a *flēot* or stream'.—**K~ Green** Li may contain ME *grene* 'village green'.—**K~ Ireleth** La from vicinity to IRELETH.—**K~ Laythorpe** Li is K~ and Laythorpe. The latter [*Ledulvetorp* DB, *Layltorp* 1196 FF, *Leithorp* 1202 Ass] is '*Leiðulf*'s thorp'. First el. ON *Leiðulfr* pers. n.—**K~ Lonsdale** We. See LONSDALE.—**K~ Malham** YW. See MALHAM.—**K~ Mallory** Le was held by Richard Mallor' c 1225 (Ep) and belonged to the Mallorys in the 12th cent. Mallory is a Fr family name, originally a nickname, perhaps OFr *maleuré* 'malheureux'.—Malzeard (**K~ M~** YW) is OFr *mal assart* 'poor clearing'.—**K~ Moorside** YN. Moorside is *Moresheved* 'top of the moor'.—**K~ Overblow** YW means 'the Kirkby of the smelters'. The addition is an OE *ōrblāwere* 'smelter'.—**K~ Stephen** We apparently from an early owner, though it was given by Ivo Taillebois to St. Mary's Abbey, York, and Stephanus its abbot (WR, p. 412).—**K~ Thore** We from an early owner. *Thore* is ON *þórir* &c.—**K~ Underwood** Li means 'K~ in the forest'. **K~ Wharfe** YW is on the WHARFE.

Kirkcambeck. See CAM BECK.

Kirkdale La [*Chirchedele* DB, *Kirkedale* 1185 P], K~ YN [?*Kirkedale* 1202 FF]. 'Valley with a church.' The church at K~ Y is called *Sanctus Gregorius minster* in the Kirkdale runic inscription of c 1060.

Kirkham La [*Chicheham* DB, *Chercheham* 1094 LaCh], K~ YE [*Chercham*, *Chirchan* DB, *Chercheham* c 1125 (1336) Ch, *Kirk-(h)aham* c 1200 YCh 1079]. A Scandinavianized form of OE *Ciric-hām* 'church village'.

Kirkharle. See HARLE.

Kirkhaugh Nb [*Kyrkhalwe* 1254 Val, *Kirkehalghe* 1279 Ass]. 'Haugh with a church.' See HALH.

Kirkheaton. See HEATON.

OScand **kirkia** 'church' is common in pl. ns. in Scandinavian England. See KIRBY, KIRKBY &c. It is the second el. of some names, as FELKIRK, ORMSKIRK, PEAKIRK. Where *Kirk-* is combined with English elements, it is no doubt often a Scandinavianized form of OE *cirice*, as in KIRKHAM, KIRKLEES, KIR(K)STEAD, KIRTON.

Kirkland Cu nr Penrith [*Kerkelanda* 1194 P, *Kirkeland* 1229 P], K~ Cu at Blennerhasset [*Kirkeland* 1290 Ch, 1332 Subs]. 'Land belonging to a church.'

Kirkland La [*Kirkelund* c 1230 CC, *Kirkelund wood* 1247 Ipm]. 'Church wood.' Cf. LUND.

Kirkleatham YN [*Westlidum*, *Weslide* DB, *Lithum* 1130–5 YCh 671, *Kyrkelidun* 1180 P]. OE *hliþum*, dat. plur. of HLIÞ 'slope'. *Kirk-* for distinction from UPLEATHAM.

Kirklees YW [*Kyrkelegh* 1246 FF, *Kyrkeleis* 1242 Cl], **Kirkley** Sf [*Kirkelea* DB, *-lee* 1200 Cur]. 'LĒAH with or belonging to a church.' The original first el. was no doubt OE *cirice*. There was a nunnery at Kirklees.

Kirkley Nb [*Crikelawa* 1176 P, *Crekellawe* 1267 Ch]. The original name was Brit *crūc* 'hill' (see CRŪC), to which was added OE HYLL 'hill'. Later, when the compound became obscured, a further OE HLĀW 'hill' was added.

Kirklington (kĭt-) Nt [*Cyrlinstune* 958 YCh 2, *Cherlinton* DB, *Kirtlingtun* Hy 2 DC], K~ YN [*Cherdinton* DB, *Kirtelington* 1198 FountM, *Kertlingeton* 1220 FF]. 'The TŪN of *Cyrtla*'s people.' OE *Cyrtla* is found in *Cyrtlan geat* 739 Crawf.

Kirklinton. See LYNE R Cu.

Kirknewton. See NEWTON.

Kirkoswald Cu [*Karcoswald* 1167 P, *Kirkoswald* 1235 FF]. 'St. Oswald's church.' Cf. KIRKANDREWS.

Kirksanton Cu [*Santacherche* DB, *Kirkesantan* c 1185 FC]. 'St. Sanctan's church.' *Sanctan* is the name of several Irish saints.

Kirkstall YW [*Kirkestal* 1153 Kirkst, *Kyrkestal* 1185 P]. 'Site of a church.' Cf. STALL.

Kirkstead Li [*Chirchesteda* 1157 f. P, *Kirkested* 1202 Ass]. A partly Scandinavianized form of OE *ciricstede* 'site of a church'.

Kirkstone Pass We [*Kirkestain* a 1184 CWNS xxiv, *the Rayse of Kyrkestone* 16 Kendale]. A Scand name meaning 'church stone'. Some prominent stone must have been so called owing to a fancied resemblance to a church.

Kirkwhelpington. See WHELPINGTON.

Kirmington Li [*Chernitone* DB, *Chirnigtuna*, *Cherlingtuna* c 1115 LiS, *Kirmiton*

1202 Ass, *Kirmington* 1225 Ep, *Kurnington* 1233 Ep]. 'The TŪN of *Cynemǣr*'s people.'

Kirmond le Mire Li [*Chevremont* DB, *Chesfremund* c 1115 LiS, *Keuermunt* c 1152 DC, *Kuermunt* c 1150 BM]. A French name, probably transferred from France, where CHÈVREMONT is common and QUÈVREMONT also occurs. The name means 'goat hill'. The Latinized form (de) *Caprimonte* occurs as a pers. n. 1100–5 YCh 856, (de) *Capramonte* Hy 2 Gilb. The place is in a valley among hills. *Mire* refers to wet ground.

Kirstead Nf [*Kerkestede* c 1095 Bury, *Kirkestede* 1206 Cur]. See KIRKSTEAD.

Kirtling Ca [*Curtelinge* c 1080 ICC, *Chertelinge* DB, *Kertlinges* 1177 P], **Kirtlington** O [*Cyrtlinctune* 944–6 BCS 812, *Kyrtlingtun* 977 ASC (C), *Certelintone*, *Cortelintone* DB, *Kertlinton* 1190 P]. '*Cyrtla*'s place' and 'the TŪN of *Cyrtla*'s people'. Cf. KIRKLINGTON. Kirtling was formerly pronounced 'Catlidge'. This probably indicates that it is a singular name in *-ing* meaning '*Cyrtla*'s place'. *Cyrtlan hlinc*, suggested in PNCa(S), is improbable, since *hlinc* is not with certainty combined with a pers. n. in early pl. ns.

Kirton Li nr Boston [*Chirchetune* DB, *-ton* 1159 P, *Cerchetone* 1130 P], **K~ in Lindsey** Li [*Chirchetone* 1070–87 RA, DB, *Kirketune* 1155–60 DC], **K~** Nt [*Circeton* DB, *Kyrketona* 12 DC], **K~** Sf [*Kirketuna* DB, *Kirketon* 1285 BM]. 'Church village.' Probably a Scandinavianized form of OE *Ciric-tūn* or *Circe-tūn*. Cf. CHERITON.

Kislingbury Np [*Ceselingeberie*, *Cifelingeberie* DB, *Cheselingebiri* 1167, *Kiselingeberia* 1176 P]. 'The BURG of *Cysel(a)*'s people.' **Cysel(a)* is a derivative of *Cusa*.

Kitley D [*Kitelhey* 1309 Ipm]. 'Kite wood', OE *cȳtan-lēah*. **Kitnor** So. See CULBONE.

Kittisford So [*Chedesford* DB, *Kedeford* 1236 Fees, *Kideford* 1257 Ass, *Kydesford* 1327 Subs]. '*Cyddi*'s ford.'

Kiverknoll He [*Kynernoc*, *Kinernoc* 1230 P, *Kyuernou* 1299 Ipm]. Probably a Welsh name, to which was added OE *hōh* 'ridge'. The original name may have contained Welsh *bryn* 'hill'. Cf. MALVERN.

Kiveton YW [*Ciuetone* DB, *Keueton* 1297 Subs, *Kyveton* 1324 Ipm]. The situation of the place on a prominent hill may suggest that the first el. is OE *cȳf* 'a tub', here used in a transferred sense of the hill.

OScand **kleif**. See CLAIFE.

Knaith Li [*Cheneide* DB, *Kneia* 1199 Cur, *Keneya*, *Cneie* 1225 Ep, *Cnaythes* 1254 Val]. OE *cnēohȳþ* 'landing-place by the knee or bend'. The place is at a bend of the Trent.

Knapp Ha [*Chenep* DB, *Cnapp* 1242 Fees]. OE *cnæpp* 'top, mountain top'.

Knapthorpe Nt [*Chenapetorp* DB, *Knapetorp* R 1 (1308) Ch], **Knaptoft** Le [*Cnapetot* DB, *-toft* 1209–35 Ep], **Knapton** Nf [*Kanapatone* DB, *Gnapenton* 1193 P, *Cnapeton* 1254 Val], **K~** YE [*Cnapetone* DB, *-ton* 1191 P, *Knapetona* 1157 YCh 354], **K~** YW [*Cnapetone* DB, *Cnapton* c 1180 YCh 464], **Knapwell** Ca [*Cnapwelle* 1043–5 Wills, *Cnapenwelle* 1060 Th, *Chenepewelle* DB, *Cnapwella* 1190 P]. '*Cnapa*'s thorp, toft, TŪN, stream.' OE *Cnapa* occurs as the name of a moneyer. It may be OE *cnapa* 'boy, servant' used as a byname Sometimes the first el. may be OE *cnapa* 'boy' itself.

Knaresborough YW [*Chenaresburg* DB, *Chenardesburg* 1130 P, *Canardesburc* 1157, *Cnardesburc* 1159 P, *Knaresburg* 1230 P]. Skelden nr Knaresborough was formerly **Knaresford** [*Cnearresweorð* c 1030 YCh 7, *Kenaresforde*, *Chenaresford* DB]. The probability is that the first el. of both is OE *Cēnheard* and that *Cnearresweorð* is corrupt. But it is possible that Knaresborough was *Cēnheardes-burg*, the other *Cnearres-ford* (or *-worþ*). If so, the early forms without *d* of Knaresborough may be due to influence from the latter. *Cnearres-* is then identical with the first el. of KNARESDALE.

Knaresdale (knarz-) Nb [*Knaresdal* 1254 Val, *-dale* 1291 Tax]. In Knaresdale is a place called **Knar** [*Knarre* 13 AD, 1326 Ipm]. The latter is ME *knar* 'a rugged rock', also 'a knot in wood', related to LG *knorre*, Du *knar* 'a stump, knob'. Knaresdale is 'the valley by Knar'.

Knayton YN [*Chenevetone*, *Cheniueton* DB, *Cneveton* 1233 Cl]. '*Cēngifu*'s TŪN.' *Cēngifu* is a woman's name. For the loss of the first vowel cf. KNARESBOROUGH, KNEETON, KNOWSLEY &c.

Knebworth Hrt [*Chenepeworde* DB, *Knebbewrth* 1220 Fees, *-wrthe* 1292 BM]. '*Cnebba*'s WORÞ.'

Knedlington YE [*Cnyllingatun* (no doubt for *Cnytlingatun*) 959 YCh 4, *Cledinton* DB, *Knedlington* 1285 FA]. 'The TŪN of *Cnytel*'s people.' *Cnytel* is found as the name of a moneyer.

Kneesall (-s-) Nt [*Cheneshale* DB, *Cneshala* 1176 P, *Keneshale* 1230 BM, *Kneshal* 1226 Ep], **Kneesworth** Ca [*Knesewrth* 1251 Cl, *Knesworthe* Hy 3 Ipm, *-worth* 1276 RH]. '*Cynehēah*'s HALH and WORÞ.' For the loss of the first vowel cf. KNAYTON. By the side of *Knesworth* there were 13th-cent. side-forms *Kenes-*, *Kynesworth*, developed from *Cynehēasworþ*, and also occasional *Knenes-*, *Knynesworth*, a blend of *Knesworth* and *Kenes-*, *Kynesworth*.

Kneeton Nt [*Cheniueton* DB, *Knivetun* 1236 Fees, *Kenyueton* 1291 Tax]. Identical with KNAYTON.

Kneeton YN [*Naton* DB, *Cneton* 12 PNNR]. OE *Cnēo-tūn* 'TŪN at a knee or bend of a road'.

Knepp Castle Sx [*Knepp* c 1145 Sele, *Lacneppe* Hy 2 (1361) Pat, *Cnapp* 1209 Cur]. OE *cnæpp* 'top, mountain top'.

Knettishall Sf [*Ghenetessala, Gnedeshalla* DB, *Gnedeshale* c 1095 Bury, *Gnatteshale* 1188 P, *Gnateshale* 1190 P]. First el. OE *gnætt* 'gnat', perhaps used as a nickname. The second is OE HALH.

Knightcote Wa [*Knittecot* 1242 Fees, *Knyghtcote* 1404 AD], **Knightley** St [*Chenistelei* DB, *Knihtele* 1227 Ass]. 'The COT and LĒAH of the knights.' Cf. KNIGHTON.

Knighton Brk [*Nisteton* DB, *Knicteton* c 1155 Fridesw, *Knighteton* 1220 Fees], **Chudleigh K~** D [*Chenistetone* DB, *Knytteton* 1282 Cl], **K~** Do nr Beer Hackett [*Knyghton* 1362 FF, 1431 FA], **K~** Do nr Durweston [*Knicteton* 1212 Fees, *Knyghteton* 1303 FA], **East K~** Do [*Knysteton* 1285 FA, *Knyghteton* 1313 Ch], **West K~** Do [*Chenistetone* DB, *Cnititon* 1208 Cur], **K~** Le [*Cnihtetone* DB, *Cnichtingtuna* 1146 RA, *Knicteton* 1204 RA], **K~** So [*Knytteton* 1372 Wells], **K~** St nr Adbaston [*Chnitestone* DB, *Knichton* 1222 Ass], **K~** St nr Mucklestone [*Chenistetone* DB], **K~** W [*Knichteton* 1200 FF, *Cnicteton* 1200 Cur], **K~ on Teme** Wo [*Cnihtatun* c 957 BCS 1007, *Cnistetone* DB], **K~** Wt [*Chenistone* DB, *Knighteton* 1302 Ep]. OE *Cnihta-tūn* 'the TŪN of the knights'. OE *cniht* in pl. ns. probably refers to a household'servant of a lord, a knight. **K~ W** is referred to as *þare cnihtaland* 955 BCS 917 (in boundaries of Chalk).

Knightsbridge Mx [(in) *Cnihtebricge* 1042–66 PNMx(S), *Cnichtebrugge* Hy 3 BM, *Knichtebrugg* 1270 Misc]. 'The bridge of the knights.' Cf. KNIGHTON.

Knightstone D [*Cnizteston* 1284–6 FA]. 'The knight's TŪN.'

Knightwick Wo [*Cnihtawice* 964 BCS 1135, *Cnihtewic* DB]. 'The WĪC of the knights.' Cf. KNIGHTON.

Knill He [*Chenille* DB, *Cnulla* 1242, *Knulle* 1249 Fees]. An OE *cnylle* 'knoll', derived from *cnoll* 'knoll'.

Knipe We nr Bampton [*Gnype* 1314 Ipm, 1360 Kendale]. ON *gnípa* 'steep overhanging rock'.

Knipton Le [*Cnipetone, Cniptone* DB, *Knipton* c 1125 LeS, *Gnipeton* 1180 P, *Gnipton* 1206 Cur, 1242 Fees]. If *Gn-* is original, the first el. is ON *gnípa* (see KNIPE). If *Kn-* is to be preferred, the first el. may be Norw *knip* 'narrow place'. Either etymology would do. **K~** is in a narrow valley with high hills at its sides.

Knitsley Du [*Knyhtheley* 1303 RPD]. 'The LĒAH of the knights.' Cf. KNIGHTON.

Kniveton (nĭftn) Db [*Cheniuetun* DB, *Kniueton* 1169 P]. Identical with KNAYTON, KNEETON Nt.

Knock We [*Chonoc-salchild* 1150–62 YCh 1241, *Knok* 1323 Ipm]. OIr *cnocc*, Ir *cnoc* 'a hillock'. **Knock Pike** reaches 1,306 ft.

Knockholt K nr Sevenoaks [*Ocholt, Nord Ocholte* 1197 FF, *Sudacholt* 1203 FF, *Okholte* 1285 Ch, *Nocholt* 1353 FF], **K~** or **Knockhall** K nr Gravesend [*Okolte* 1260 Ipm]. OE *āc-holt* 'oak wood'. The *N-* has been carried over from the def. art. (OE *þǣm* > *þen*): *æt þǣm ācholte*.

Knockin Sa [*Cnochin* 1165, *Cnukin* 1196 P, *Knokyn* 1197, *Knukin* 1198 FF]. Welsh *cnycyn* 'bump, small hillock'. Cf. *Knukyn* (monticulus) 1307–23 Chester (Chs).

Knoddishall Sf [*Cnotesheala* DB, *Knodeshal* 1234 FF, *Knoteshal* 1275 RH]. '*Cnott*'s HALH.' *Cnott* is found in *Cnottis rode* KCD 1364. Cf. KNOTTING.

Knole K [? *æt Cnollam* 985 KCD 647, *Cnolle* 1327 Subs]. OE CNOLL 'knoll'.

Knook (-ōō-) W [*Cunuche* DB, *Cnuke* 1212 RBE, *Knuc* 1234, *Cnuk* 1242 Fees, *Knuch* 1249 BM]. Welsh *cnwc* 'bump, hillock'. Or possibly identical with CANNOCK. **Knook Barrow** rises to 621 ft.

Knossington Le [*Nossitone, Closintone* DB, *Knossinton* c 1125 LeS, *Cnossintona* a 1160 DC, *Cnossington* 1254 Val]. It is possible that the first el. is derived from an OE *cnoss* 'hill', related to ON *knauss* 'rounded hill', Sw *knös* the same, MLG *knust*, G dial. *chnûs* 'a knot', Engl *knot* &c. But a pers. n. may also have developed from this stem. Cf. OSw *knös* 'a goblin, a terrible person', Norw, Dan *knøs* 'a proud, over-bearing person'.

Knostrop YW [*Knousthorp* 1335 Whitaker]. '*Cnūt*'s thorp.' *Cnūt* is a Scand name.

Knott End La [*Hacunshou Cnote* c 1265 CC]. ME *knot* 'a hill', from OE *cnotta* 'knot'.

Knotting Bd [*Chenotinga* DB, *Cnotting* 1163 P, *Cnotinges* 1224 Bract]. The situation of the place at a hill of 311 ft. may suggest that the name means 'the dwellers at the hill' (cf. prec. name). But as *knot* 'hill' is only found in Northern dialects, the base is perhaps rather OE *Cnotta* pers. n., originally a nickname from OE *cnotta* 'a knot'. This name is no doubt the base of the first el. of *Cnottinga hamm* 952 BCS 895 (Brk) and of KNOTTINGLEY.

Knottingley YW [*Notingeleia* DB, *Chodtingalaia* 1148 YCh 179, *Cnottingaleie* 1155–70 ib. 1502, *Cnottingeleg* 1226 FF]. 'The LĒAH of *Cnotta*'s people.' Cf. KNOTTING.

Knowle Hill Do [*Cnolle* 1212 Fees, *La Cnolle* 1228 FF], **Church K~** Do [*Chenolle* DB, *Cnolle* DB, 1285 FA], **K~** So nr Bedminster [*Canole* DB, *Cnolle* 1196 P], **K~**

So nr Chew [*Knolle* 1327 Subs], K~ So nr Long Sutton [*Knolle* 1341 BM], K~ So nr Wincanton [*Chenolle* DB, *Cnolle* 1254 Ipm], **K~ St. Giles** So [*Knolle* 1189 Wells, *Cnolle* 1285 FA], K~ Wa [*La Cnolle* 1251 Ch]. OE *cnoll* 'knoll, hillock'.

Knowlton Do [*Chenoltune* DB, *Cnolton* 1168 P, 1212 Fees], K~ K [*Chenoltone* DB, *Cnoltun* 1070–82 StAug, *Cnoltune* 11 DM]. 'TŪN by a knoll.'

Knowsley (nōzlĭ) La [*Chenulueslei* DB, *Knuvesle* 1199 FF, *Knouwesley* 1246 Ass]. '*Cēnwulf*'s or *Cynewulf*'s LĒAH.' Cf. KNAYTON.

Knowstone (now-) D [*Chenutdestana, Chenudestane* DB, *Cnutstan* 1220 FF]. '*Cnūt*'s stone.' *Cnūt* is a Scand name (ON *Knútr* &c.).

Knoyle, East & West, W [æt *Cnugel* 948, *Cnugel* c 956 BCS 870, 956, (on) *Cnugellege* 984 KCD 641, *Chenvel* DB, *Cnoel* 1188 P, 1200 Cur, *Childe Knoel* 1202 FF, *Stepelknoel* 1228 Cl]. An OE *cnugel* or *cnuwel* related to ON *knúi*, OSw *knōe* 'knuckle', OE *cnuwian* 'to crush'. The ridge by which the Knoyles stand must have been called *cnugel* 'the knuckle' owing to a fancied resemblance.

Knuston Np [*Cnutestone* DB, *Cnoteston* 1220 Fees, *Cnoston* 1236 Fees]. '*Cnūt*'s TŪN.'

Knutsford Chs [*Cunetesford* DB, *Knottisford* 1282 Court]. Apparently '*Cnūt*'s ford'.

Knutton St [*Clotone* DB, *Cnoton* 1212 Fees, *Cnutton* 1227 Ch, *Knotton* 1256 Ch]. Perhaps '*Cnūt*'s TŪN'.

Knuzden La [*Knuzdenbroke* 1200–8 PNLa, *Knowesden* n.d. WhC]. Etymology doubtful.

ON **konungr**, OSw **konunger, kununger,** ODan **konung** 'king' is a common first el. See CONEYSTHORPE, CONINGSBY &c., CUNSCOUGH. The names normally have *u* in the first syllable (from OScand *kunungr*). But in several cases the Scand word has evidently replaced OE *cyning*. See e.g. CONINGTON, CONISBROUGH, CONISCLIFFE.

ON **kringla** 'circle'. See CRINGLEFORD.

Kyle R YN [*Kijl* 1220, *Kil* 1228 For]. A Brit river-name derived from Welsh *cul* 'narrow'.

Kyloe Nb [*Culeia* 1195 (1335) Ch, *Kylei* 1208–10 Fees, *Kyley* 1254 Val]. OE *cȳ-lēah* 'cow-pasture'.

Kym R Hu. See KIMBOLTON and (HAIL) WESTON. Kym is a back-formation.

Kyme, North & South, Li [*Chime, Nortchime* DB, *Kyma* c 1115 LiS, *Chimba* 1130 P, *Chimb'* 1159 P, *Kimbe* 1202 Ass]. Probably an OE *cymbe*, a derivative of OE *cumb* 'a vessel, tub' and denoting a depression in the ground.

Kynaston He in Hentland [*Kyneuarstone* 1336 Ipm, *Kynastone* 1334 Ep], K~ He in Much Marcle [*Kynewardestone* 1294 BM]. '*Cyneweard*'s TŪN.'

Kynaston Sa [*Chimerestun* DB, *Kineuerdeston* 1198 FF]. '*Cynefriþ*'s TŪN.'

Kyo Du [*Kyhow* c 1240 Finchale]. OE *cȳ-hōh* 'cow hill'.

Kyre (-ē-) Magna & Parva Wo [*Cyr* 11 Heming, *Chure, Cuer* DB, *Cura* 1212 Fees], **Kyre Brook** R [*Cura* 13 PNWo]. A Brit river-name related to CURRY. On Kyre Brook are **Kyrebach** He [*Curebache* in an early source] and **Kyrewood** Wo [*Curewod* 1275 Subs]. See BÆCE.

L

Laceby Li [*Levesbi* DB, *Leyseby* c 1115 LiS, *Laifsebi* Hy 2 DC, *Leusebi* 1156, *Leissebi* 1168 P]. OScand *Leifs-bȳr* '*Leif*'s BY'. First el. ON *Leifr*, ODan *Lev* pers. n.

Lach Dennis Chs [*Lece* DB, *Lache Deneys* 1260 Court, *Lache Maubanc* 1288 Chester], **Lache** Chs [*Leche* DB, c 1100, c 1150 Chester, *Lache* 1285 Ch]. See LÆCC. Dennis must be a pers. n. or family name.

Lackenby YN [*Lache(ne)bi* DB, *Lackenbi* 1202 FF, *Lachaneby* 1231 Ass, *Lakeneby* 1234 FF]. The first el. may be a pers. n. OIr *Lochán* has been suggested. The first el. might better be an ON byname *Hlakkandi* from *hlakka* 'to cry, shout' or an ODan *Lakkande* from *lakke* 'to walk slowly'. For early loss of *d* cf. CRACKENTHORPE.

Lackford Sf [*Lec-, Lacford* 11 EHR 43, *Le(a)cforda, Lacforda* DB, *Leacforde* c 1095 Bury, *Lacford* 1253 Ch]. OE *lēac-ford* 'ford where leeks grew'.

Lackham W [*Lacham* DB, 1242 Fees, 1252 BM]. The first el. looks like OE *lacu*, but since the place is in Lacock in a bend of the Avon the name is more probably a contracted OE *Lacuc-hamm* 'HAMM belonging to LACOCK'.

Lackington, White, So [*Wyslagentona* DB, *Withlachinton* 12 Wells, *Wichtlakington* 1225 FF]. 'The TŪN of *Wihtlāc*'s people.'

Lācock W [*Lacok* 854 BCS 470, *Lacoc* DB, *La coc* 1100–10 RA]. An OE *lacuc* 'streamlet', a derivative of LACU. Cf. LAYCOCK.

Lacon Sa nr Wem [*Lach* DB, *Lak* 1228 FF,

Laken 1285 FA]. OE *lace*, dat. *lacum*, the plur. of LACU 'stream'.

OE **lacu** 'stream, water-course'. See LACON, LAKE, LACKHAM, further BABLOCK HYTHE, CHARLOCK Np, FISHLAKE, MEDLOCK, MORTLAKE, SHIP-, STANDLAKE. Cf. LACOCK, LAYCOCK.

OE **lād, gelād** 'road, path, water-course'. Cf. ODu, LG *lēde* 'water-course, conduit'. Except in CURRY LOAD, OE *lād* seems to mean 'water-course' when used in pl. ns., as in LOAD, LODE, SHIPLATE, WHAPLODE. Cf. LAYTON La. OE *gelād* appears to mean 'passage over a river' in CRICKLADE, EVEN-, FRAMILODE, LECH-, LINSLADE. In AQUALATE it is perhaps 'water-course'.

Ladbrooke Wa [*Hlodbroc* 998 Crawf, *Lodbroc* DB, 1226–8 Fees, *Lotbroc* 1236 Fees]. Originally the name of the stream at L~ (*Hlodbroc* 998 Crawf). The first el. may be OE *hlot, hlod* 'lot'. The meaning would be 'a stream used for the purpose of drawing lots, of divining the future'.

Ladhill YN [*Laddedale* c 1160 Riev, *Laddale* 1201 FF]. Perhaps '*Ladda*'s valley'. OE *Ladda* is found as a byname (*Godric Ladda* KCD 1351). *Ladda* is probably *lad* 'boy'.

Ladock Co [*Ecclesia Sancte Ladoce* 1268 Ep, *Sancta Ladoca* 1291 Tax, *Seynt Ladok* 1359 FF]. '(The church of) St. Ladoca', a woman saint.

OE **læcc, lecc,** ME *lache, leche* 'a stream flowing through boggy land, a bog' is found in some pl. ns. See LACH, -E, LEACH, LATCHFORD, LASHBROOK, LECHLADE, CRANAGE, FULLEDGE, SHOCKLACH.

OE **læfer** 'rush, yellow iris, flag'. See LAVERTON So, LEARMOUTH, LEVER, -TON Li.

OE **læge** 'fallow'. See LEYLAND.

OE **læl** 'twig, withe'. See LALEHAM, LEALHOLM.

OE **læs** (gen. *læswe*) 'leasow, pasture' is a rare el. in pl. ns. It is the source of **Leasowe** Chs. It may be the first el. of LEZIATE, LISSETT, and the second el. of ECCLES K, BECCLES, BRECKLES.

OE **(ge)lætu** 'junction of roads' (in *wega gelætu*) must also have been used in other senses. OE *wæter-gelæt* 'water-conduit' (= OHG *wazzar gilâz*) occurs. OE *gelæt* is the source of *leat* 'an open water-course to conduct water for household purposes, mills &c.' (1590 &c. OED). Some such sense is no doubt to be assumed for LONGLEAT, and perhaps LEDBURN, LETWELL.

OE **læwerce, lāferce** 'lark, laverock'. See LARKBEARE &c., LAVERSTOCK, -STOKE &c.

Lagness Sx [*Langan ersc* 680 BCS 50, *Langeners* 1179 P, *Lageners* 1242 Fees]. 'Long pasture.' Cf. ERSC. The loss of *n* is due to dissimilation. Cf. BRIGNALL.

Laindon Ess [(of) *Ligeandune* c 1000 CCC, *Legen-, Leienduna* DB, *Leindon* 1199 FF, *Leyndon* 1260 Ch]. The first el. seems to be a river-name identical with LEA.

Lainston Ha [*Lewyneston* 1280 Ass, *Leynestone* c 1270, *Lenistone* c 1286, *Levestone* c 1294 Ep]. '*Lēofwine*'s TŪN.'

Lake W [*Lake* 1316 FA, 1325 Pat]. OE LACU 'stream'. The same is no doubt the origin of **Lake** in other counties.

Lakenham, Old & New, Nf [*Lakemham* DB, *Lakeham* 1212 Fees, *Lakenham* 1211 FF, 1247 Ch]. Probably '*Lāca*'s HĀM'. **Lāca* (= OHG *Laico*) is a short form of names in -*lāc*.

Lakenheath Sf [*æt Lacingahiδ* 945 BCS 809, *Lakingheδe* 1020–3 KCD 735, *Lakingahethe* DB, *Lachingeia, Lachingahutha* c 1120, c 1150 BM]. 'The landing-place of *Lāca*'s people'; cf. LAKENHAM and HȲÞ. Or the first el. may be OE *Lacingas* 'people at a LACU or stream'.

Laleham Mx [*Laelham* 1062 KCD 812, *Leleham* DB, *Lelham* a 1134 Fr, *Lalham* 1206 Cur]. First el. OE *læl* 'twig, withe, whip'. Cf. LEALHOLM. The meaning in pl. ns. is probably 'withy, willow'.

OE **lām** 'loam' is probably the first el. of some names, as LAMARSH, LAMAS &c., but it cannot be distinguished from LAMB.

Lămarsh Ess [*Lamers* DB, *Lammers* 1233 Fees, *Lammerssh* 1327 Ch], **Lămas** Nf [*Lamers* DB, *Lammesse* 1044–7 KCD 785, c 1150 Bodl, *Lammasse* 1186 P]. OE *lām-* or *lamb-mersc* 'loam marsh' or 'marsh where lambs were kept'. The first alternative seems preferable.

OE **lamb** 'lamb' is the first el. of several pl. ns., but cannot always be distinguished from LĀM 'loam'. See LAMBETH &c., LAMCOTE, LAMESLEY. The OE plur. of *lamb* was *lambru* (gen. *lambra*). Hence perhaps LAMBERHURST and LAMMERMOOR in Scotland.

Lamberhead Green La [*Londmerhede* 1519 FF]. 'Boundary hill' (OE *landgemǣre* 'boundary' and *hēafod* 'hill').

Lamberhurst K [*Lamburherste* c 1100 Text Roff, *-herst* 1205 Cur, *Lamberherste* 12 (1285) Ch]. 'Lambs' hill or wood.' Cf. LAMB. But *Lamber-* may be from OE *lām-* or *lamb-burna*. Cf. LAMERTON.

Lambeth Sr [*Lámbhyð* 1041 ASC (E), (into) *Lambehyδe* 1062 KCD 813, *Lamhytha* 1089 BM, *Lamheda* 1188 ff. P]. 'Harbour where lambs were shipped.' Cf. HȲÞ.

Lambley Nb [*Lambeleya* 1201 Ch, *-leye* 1256 Ass], L~ Nt [*Lambeleia* DB, *Lameleya* 1191–3 Fr, *-leia* 1212 Fees]. 'Pasture for lambs.' Cf. LĒAH.

Lambourn R Brk [*Lamburna* 943, (on) *Lámburnan* 949 BCS 789, 877]. On the river are **Lambourn** and **Up Lambourn**

[(æt) *Lambburnan* c 880 BCS 553, *Lamburninga mærce* 1050 E, *Lamborne* DB, *Uplamburn* 1190 P, *Chepinglamburn* 1227 Ch, 1242 Fees]. OE *lamb-burna* 'stream where lambs are washed'. OE *lām-burna* is a possible alternative. Lambourn and Up Lambourn are referred to respectively as (on) *byrihæmetune* and (on) *vphæmetoune* c 1066 ASCh.

Lambourne Ess [*Lamburna* DB, *-burne* 1198 FF]. Identical with LAMBOURN.

Lambrigg We [*Lambrig* c 1190 Kendale, *Lamberig* c 1210 NpCh]. 'Ridge where lambs grazed.' Second el. OScand *hryggr* 'ridge'.

Lambrook So [*Landbroc* 1065 Wells, *Lambrok* 1201 Ass, *Lanbroc* 1227, *Estlambrok* 1268 FF]. First el. OE *land*. The meaning may be 'boundary brook' (Dr. Grundy).

Lambton Du [*Lambton* 1421 FPD], **Lamcote** Nt [*Lanbecote*, *La[m]becotes* DB, *Lambecote* 1198 FF]. 'TŪN and COTS where lambs were kept.'

Lamellan Co [*Lanmaylwen*, *Lamaylwyn* 1303, *Lamaylwen* 1306 FA]. 'Maylwen's church.' *Maylwen* is identical with MBret *Melguen*. See LANN.

Lamerton D [(æt) *Lamburnan* c 970 BCS 1247, *Lambretone* DB, *Lamerton* 1242 Fees]. 'TŪN on Lumburn Water.' **Lumburn** was originally *Lamburna* (*Lambre* Hy 2 Buckland is a back-formation from *Lambretone*). See LAMBOURN.

Lamesley Du [*Lamelay* 1297 Pp, *Lamesley* 1291 Tax]. OE *lamba-lēah* or *lambes-lēah* 'pasture for lambs'. For early loss of *b* cf. LAMBETH.

Lamonby Cu [*Lambeneby* 1267 Ch, *Lambenby*, *Lambingby* 1277 Ipm]. 'Lambin's BY.' *Lambin* is a French short form of *Lambert*.

Lamo·rran Co [*Lanmoren* 969 BCS 1231, *Lammoren* 1194 P, *Lanmoren* 1268 Ep]. 'Moren's church.' See LANN. *Moren* is identical with Bret *Moran* and with *Moryn* in Gospatric's charter (c 1060, Cu); *-morin* is the second el. of several OBret pers. ns. (*Hael-*, *Iudmorin* &c.).

Lamplugh Cu [*Lamplou* c 1150, *-plogh* 12, *Lanploch* c 1210 StB, *Lanplo* 1181 P, *Landplo* 1241 Cl]. An old Cumbrian name, corresponding to a Welsh *llan plwy* 'the church of the parish'. Welsh *plwy*, *plwyf* (= OBret *pluiu*, *ploi*, OCo *plui*, MCo *plu* 'parish') is a loan from Lat *plēbs*. The early Engl *-plo* is an adaptation of OW *plui* or the like. See LANN.

Lamport Bk [*Land-*, *Lanport* DB, *Langeporte* c 1150 Fr], **L~** Np [*Langeport* DB, 1196 FF, 1202 Ass], **L~** Sx [*Langeport* a 1107 PNSx, *Lamport* 1173 P, *Langepord* 1197 FF], **Old Langport** K [*Lan-*, *Lamport* DB, *Langeport* 11 DM, *-porte* 1198 FF], **L~** So [*Longport* c 930 Coins, *Lanport* DB,

Langeport DB, 1225 Ass]. There was formerly another Langport in K nr Canterbury [*Lanport* DB, *Langeport* 1291 Tax], and a *Langeport* Hrt is mentioned in DB. OE examples of the name are *lang port* 680 BCS 50 (at Pagham), *Langport* 956 BCS 982 (Meon, Ha). OE *lang-port* means 'long town' or rather 'long market-place' and no doubt referred to a market-place consisting of a row of booths along the road and consequently of considerable length. Before *þ* the *ng* was assimilated to *m*. Hence often *Lamport*.

Lampton Mx [*Lamptonfeld* 1375 Cl]. OE *Lamb-tūn* 'TŪN where lambs were reared'.

Lamyatt So [*Lamieta* DB, *Lamiete* 1185 TpR, *Lamiet* 1238, *Lamiette* 1249 FF]. Second el. OE *geat* 'gate'. Neither 'lamb gate' nor 'loam gate' seems very satisfactory, and it may be suggested that the source of the name is an OE *hlamm-geat* 'swinggate', the first el. being related to OE *hlemman* 'to clash, dash'. Cf. Goth *hlamma* 'a trap' and ON *hlemmr* the same, Swed *läm* 'a hatch' &c. Cf. the surname *atte Lamesete* (sic) 1315, *atte Lomezate* 1327 (Wo) Löfvenberg 124.

Lancashire [*honor de Lancastre* 1140 LaCh, *Comitatus de Lancastra* 1169 P, *Lancastreshire* 14 Higden]. **Lancaster** [*Loncastre* DB, *Lanecastrum* 1094 Lancaster, *Loncastra* 1127 Ch]. 'Roman fort on R Lune.' Lancaster is the county town of Lancashire.

Lancaut Gl [*Landcawet* 11 BCS 928, *Lann Ceuid*, *podum Ceuid* c 1150 LL, *Langcaut* 1221 Ass, *Lancaut* 1291 Tax]. 'The church of St. Cewydd.' Cf. LANN. St. Cewydd was a Welsh saint.

Lanchester Du [*Langecestr'* 1196 P, 1238 Cl, *-cestria* 1237 Cl]. 'Long CEASTER or Roman fort.'

Lancing Sx [*Lancinges* DB, 1196 Cur]. A Normanized form of OE *Wlencingas* 'the people of *Wlenca* or *Wlanc*'. Cf. LINCHMERE, LONGSLOW. *Wlencing*, son of Ælle, is mentioned in ASC (s.a. 477).

OE, ON **land** 'land' is a common second el. of pl. ns. The exact meaning is not always apparent. Usually it seems to mean 'estate, landed property', as in the numerous cases where the first el. is a pers. n., e.g. DOTLAND, GILSLAND, GOATHLAND. Sometimes a very large estate is referred to, as in RUTLAND. Some names in *-land* denote a district, as CLEVELAND, HARTLAND, HOLLAND. Interesting names are COPELAND, BUCKLAND, SUNDERLAND. In some cases *-land* refers to a portion of a village or estate, as in NEWLAND. In names such as GREETLAND, SWARLAND the first el. refers to the nature of the soil, and in LITHERLAND the situation is indicated by the first member. *Land* is rare as a first el. See LAND- (passim), LAMBROOK.

Landbeach Ca [*Beche* 1242 Fees, *Londbech* 1235 Cl, *Inbeche* 1276 Val]. Landbeach is near WATERBEACH and probably both were once *Bece, Beche*, i.e. OE BÆCE, BECE 'brook, valley'. They were later distinguished as Waterbeach 'Beche on R Ouse' and Landbeach 'Beche inland'.

Landcross D [*Lanchers* DB, *Lancarse* 1242 Fees, 1265 Ep, *Lancras* 1318 Ch]. Etymology obscure.

Landermere Ess in Thorpe le Soken [*Landimer* 1211 FF]. OE *landgemǽre* 'boundary'.

Landewe·dnack Co [*Ecclesia Sancti Wynewali de Landewenesek* 1268 Ep, *Landewinnek* 1305 AD, *Sancti Wynwolayi de Lanwynnocke* 1310 Ep]. 'The church of St. Gwennock or *Winnoc*.' The name of the saint was really *Wynwalo*, identical with Bret *Guenolé*, earlier *Win-waloe* (cf. GUNWALLOE), but he was also called by the short form *Winnoc* or, with the possessive pronoun for 'thy' prefixed, *Te-winnoc* and the like. *Thy* is Co *te, de*. The form with *dn* is due to a late Co change of *nn*. See LANN.

Landford W [*Langeford* DB, *Laneford* 1242 Fees, 1291 Tax, 1316 FA]. Apparently 'lane ford' with change to *Langeford* and *Landford* owing to popular etymology.

Landican Chs [*Landechene* DB, *Landekan* 1240–9 Chester, *Landecan* 1281 Court]. 'The church of St. *Tecan*.' Cf. LANN. OW *Tecan* is the name of a saint.

Landkey D [*Landechei* 1166 RBE, *Landege* 1225 Ep, *Londekey* 1284–6 FA]. 'The church of St. *Cai*.' See LANN. Cf. the identical LANDEGEA Co (under KEA) and LANDEWEDNACK.

Landmoth YN [*Landemot* DB, *-e* 1088 LVD]. OScand *landamōt* 'meeting of lands, boundary'. *Landamót* is mentioned in Landnáma as the name of a place in Iceland.

Landrake Co [*Landerhtun* 1018 KCD 728, *Landrei* DB, *Lanrak* 1291 Tax]. Co *lanherch* 'an open place in a wood, a glade' (= Welsh *llannerch*). In the OE example OE *tūn* has been added.

Land's End Co [*the Londis end* 14 OED]. Said to be a translation of Co *Pen an Wlas* 'end of the land'. For an earlier name cf. PENWITH.

Landulph Co [*Landylp* 1280, *-hylp* 1311 Ep, *Landilp* 1312 Ch]. First el. Co *lan* 'church'. The second is no doubt a saint's name.

Landwade Ca [*Parua Landwathe* 1195 FF, *Landwath* 1212 RBE, *Landwade* 1282 Ipm]. Probably OE *land-gewæd*, sometimes with substitution of OScand *vað* for the OE word. The exact meaning of the name ('land ford') is not clear. Possibly 'chief ford' or 'boundary ford' (PNCa(S)).

Laneast Co [*Lanerst* 1291 Tax, *Laneyst* 1428 FA]. Co *lan* 'church' and an obscure saint's name.

Laneham Nt [*Lanun* DB, 1194 P, *Lanum* 1186 P, 1212 Fees]. OE *lanum* '(at) the lanes'. Cf. LANU.

Lanercost Cu [*Lanercost* 1169 WR, *Lanrecost* 1195 P]. The first el. is Welsh *llannerch* 'glade'; cf. LANDRAKE. The second el. is obscure.

OE **lang** 'long' is a common first el. referring to the length of a piece of land or the like or to the height of a tree or stone &c. See LANG-, LONG-, also LAGNESS, LAMPORT, LANCHESTER, LANTON, LAUNTON.

Langar Nt [*Langare* DB, 1212 Fees, *Langar* 1163 P]. 'Long gore.' Cf. GĀRA. L~ is on a long ridge, and it is possible *gāra* refers to it.

Langcliffe YW [*Lanclif* DB, *Langecliff* 1270 Ch]. 'Long cliff', i.e. Langcliffe Scar.

Langdale We [*Langedenelittle* c 1160 LaCh, *Langedena, -dala* 1179 P, *Langedal* 1252 Ch]. 'Long valley.'

Langdon D in Werrington [*Langedon* 1273 Cl], L~ Do [*Langedon* 1285 FA], L~ **Hills** Ess [*Langenduna* DB, *Langedun* 1169 P], **East & West** L~ K [(to) *Langandune* 861 BCS 855, *Estlangedoun, Westlangedone* 1291 Tax], L~ Wa [*Langedone* DB, *-don* 1253 Ch]. 'Long DŪN or hill.'

Langenhoe Ess [*Langhou* DB, *-hó* 1167 P, *Langenho* 1254 Val]. 'Long HŌH or ridge.'

Langford Bd [*Longaford* 944–6 BCS 812, *Langeford* DB], L~ D in Cullompton [*Langeforde* DB], L~ Ess [*Langheforda* DB, *Langeford* 1176 P], L~ Nf [*Langaforda* DB, *Langeford* 1254 Val], L~ O [*Langefort* DB, *-ford* c 1140 RA], L~ **Budville** So [*Langeford* 1212 Fees, L~ *Budevill* 1305 FF], **Lower & Upper** L~ So [*Langeford* DB], **Hanging, Little & Steeple** L~ W [(æt) *Langanforda* 943 BCS 783, *Langeford* DB, *Hangindelangeford* 1242 Fees, *Langeforde parva, Stupelangeforde* 1291 Tax]. 'Long ford.'

L~ **Budville** was held by Richard de Buddevill before 1212 (Fees). The name comes from some place in France (? Boutteville in Normandy).—*Hanging* means 'sloping, situated on a slope'.—*Steeple* from the church steeple.

Langford Nt [*Landeforde* DB, *Landeford* 1201 Cur]. Possibly '*Landa*'s ford'. **Landa* (= OG *Lando*) is a short form of *Landfrib, Landwine* &c. But an OE *Landaford* 'boundary ford' is equally possible and perhaps preferable. L~ is not far from the Lincs border.

Langhale Nf nr Kirstead [*Langahala* DB, *Langenhal* 1179 P]. 'Long HALH.'

Langham Do nr Gillingham [*Langeham* 1157 P], L~ Nf [*Langham* 1047–70 Wills, DB, *Langaham* DB], L~ Ru [*Langham* 1202 Ass, 1269 For], L~ Sf [*Langham* DB, c 1095 Bury, *Langeham* 1205 FF]. 'Long HĀM (village or homestead).'

Langham Ess [*Laingaham* DB, *La Winge-ham* 1130 P, *Lavigahan* 1138 Fr, *Leingeham* 1190 f. P, *Lawingeham* 1198 FF]. 'The HĀM of *Lāwa*'s people.' Cf. LONGHAM Nf.

Langham Row Li [*Langholm* 1219 Ass, 1317 Ipm]. 'Long holm.'

Langho La [*Langale* 13 WhC]. 'Long haugh.' See HALH.

Langley, L~ Marish Bk [*Langel'* 1163 P, *Langeley* 1208 Fees, *Langele Marais* 1316 AD], **L~** Brk [*Lonchelei* DB, *Langelea* 1167 P], **L~** Db in Heanor [*Langeleie* DB], **Kirk & Meynell L~** Db [*Langelei* DB, *Chircehelongeley* 1273 Ipm, *Langelle Meynill* 1284–6 FA], **L~** Du in Lanchester [*Lange-leye* 1232 Ch], **L~** Ess [*Langelega* 1166 RBE, *Langleie* 1205 FF], **L~** Ha [*Langelie* DB, *Langelega* 1165 P], **L~** Hrt nr Steven-age [*Langeleya* 1220 Fees, *-leye* a 1292 BM], **Abbots & Kings L~** Hrt [(æt) *Langalege* c 1050 KCD 962, *Langelai* DB, *Langel' Regis* 1254 Val, *Langley Abbots* 1302 AD], **L~** K nr Maidstone [*Longanleag* (obl.) 814 BCS 343, *Langvelei* DB], **L~** Le [*Langleya* 1209–19 Ep, *Langel'* 1254 Val], **L~** Nb [*Langeleya* 1212 Fees, *-ley* 1256 Ass], **L~** Nf [*Langale* DB, *Langeleg* 1201 FF, *-le* 1254 Val], **L~** O [*Langeleya* 1230 P, *-leg* 1231 Cl], **L~** Sa [*Langvelege* DB, *Langeleg* 1226–8 Fees], **L~** So [*Langele* 1065 Wells], **L~ Burrell** W [*Langelegh* 940 BCS 751, *Lange-fel* DB, *Langele Burel* 1309 AD], **Kington L~** W [*Langhelei* DB], **L~** Wa [*Longelei* DB, *Langelleie* 12 Fr], **L~** Wo [*Longeley* 1270 Ct]. 'Long LĒAH (wood or clearing).'
Abbots L~ Hrt was held by the Abbot of St. Albans.—**L~ Burrell** W was held by Petrus Burel in 1242 (Fees). *Burel* is a French nick-name and family name. OFr *burel* means 'a coarse woollen cloth' and is held to have been the source of ME *borel* 'lay, unlearned'.—**Kington L~** adjoins Kington St. Michael W.—**L~ Marish** Bk from the Mareis family. It was held by Christiana de Mariscis in 1285 (Cl). The name comes from MARAIS in France (from *marais* 'marsh').—**Meynell L~** Db was held by Robertus de Maisnell t. Hy 1 (DbAS ix. 45). Meynell is a family name taken from one of the MESNILS in France. OFr *mesnil* (Lat *mansionile*) means 'village'.

Langley Park Cu [*Langliferga* Mon vi, 556, *Langelyve Erghe* c 1250 FC]. '*Langlif*'s ERG or shieling.' See ERG. *Langlif* is an ON woman's name.

Langney Sx [*Langelie* DB, *Langania* 1121 AC]. 'Long island.'

Langport. See LAMPORT.

Langrick, Langriville Li [*Langrak* both 1243 Cl, *Langrake* 1260 FF]. See DRAX, LONG. Langriville is a late name, formed by addition of Fr *ville* to Langrick.

Langridge So [*Lancheris* DB, *Langerig* 1225 Ass, *Langerigge* 1276 RH], **Langrigg** Cu [*Langrug* 1189, *-rig* 1195 P]. 'Long ridge.' Langrigg has as second el. OScand *hryggr* 'ridge'.

Langrish Ha [*Langerisse* 1273 Cl, *Langrixe* c 1285 Selborne, *Langeryshe* 1316 FA]. 'Long rush-bed' or 'tall rushes'. Second el. OE *risc* 'rush' or a derivative of it.

Langsett YW [*Langeside* 1200–14 YCh 1793, *Langgesid* 1208 ib. 1798]. 'Long slope.' Cf. SĪDE.

Langstone D [*Langeston* 1324 Ipm], **L~** Ha [*Langeston* 1289 Misc]. 'The long-stone', i.e. 'menhir'.

Langstroth Dale YW [*Langestrode* 1201 Cur, 1202 FF]. 'Long marsh.' Cf. STRŌD.

Langthorne YN [*Langetorp* DB, *Langethorn* 1246 FF]. 'Tall thorn-bush.'

Langthorpe YN [*Torp* DB, *Langliuetorp* 12 PNNR, *Langelisthorp* 1228 Ep]. '*Langlif*'s thorp.' Cf. LANGLEY PARK Cu.

Langthwaite La [*Langethwayte* n.d. Lan-caster], **L~** YN [*Langethwait* 1167 P], **L~** YW [*Langetovet* DB, *-thweit* 1219 FF]. 'Long thwaite or clearing.'

Langtoft Li [*Langetof* DB, *Langetoft* 1167 f. P], **L~** YE [*Langetou* DB, *Langetoft(h)* c 1165 YCh 161, 1251]. 'Long toft.'

Langton, a common name, is **1.** OE *Langa-tūn* 'long village or homestead': **L~ Her-ring** Do [*Langetone* DB, *Langtona* Hy 1 BM, *Langeton Heryng* 1384 FF], **L~ Long Blandford** Do [see BLANDFORD], **L~ Ma-travers** Do [*Langeton* 1206 Cur, *L~ Mau-trevers* 1420 FF], **Church, East & West L~** Le [*Lang(e)tone* DB, *Langeton* c 1125 LeS, *Chirch L~* 1316 FA, *Estlangeton, Langeton West* 1327 Subs], **L~** Li nr Horn-castle [*Langetone & Torp* DB, *Langhetuna* c 1115 LiS], **L~ by Wragby** Li [*Langetone* DB, *-tuna* c 1115 LiS], **L~ juxta Partney** Li [*Langetune* DB, *Langhetuna* c 1115 LiS], **L~** We [*Langeton* 1314 Ipm], **L~** YE [*Lanton* DB, *Langatuna* 1157 YCh 354, *Langeton* 1168 P], **Great & Little L~** YN [*Langeton* DB, *Great L~* 1223 FF].

2. Langton Du [*Langadun* c 1050 HSC, 1104–8 SD, *Langeton* 1313 RPD]. 'Long DŪN or hill.'

3. Tur Langton Le [*Terlintone* DB, *Tir-linton* 1166 P, *Tirlingeton* 1206 Cur, *Turlin-ton* 1165 P, 1205 Cur]. 'The TŪN of *Tyrhtel*'s or **Tyrli*'s people.' Cf. TERLING. OE *Turla* is found in *Turlan homm* 940 BCS 764. *Tyrli* is a regular derivative of it. The name was associated with that of the neighbouring (East &c.) Langton.

Langton Herring was held by Philip Harang in 1268 (FF). Cf. CHALDON HERRING.—**L~ Matravers** was held by John Mautravers in 1281 (FF). Matravers is a French family name, originally a nickname containing Fr *mal* and *travers*, perhaps in the sense 'obstacle, trouble' or 'mishap, misfortune'.

Langtree D [*Langetreu* DB, *-tre* 1228 FF], **L~** La [*Longetre* c 1190 LaCh, *Langetre* 1206 FF]. 'Tall tree.'

Languard Wt [*Langred* 1287–90 Fees, 1397 BM]. 'Long reed-bed.' Cf. HRĒOD.

Langwathby (längenbï) Cu [*Langwadebi* 1159, *Langwathebi* 1228 P, *Languadeby* 1242 Ch]. 'BY at the long ford.' First el. OE *lang-gewæd* 'long ford'.

Langwith, Upper, Db [*Langwath* 1208 FF, 1270 Ch], **Nether L~** Nt [*Langwad* 1194 P, *-wath* 1291 Ch], **L~** YE [*Languelt* DB, *Langwat* 1234 Cl, *-wath* 1276 Ch], **Langworth** Li in Coningsby par. [*Langwath* 1209, 1252 Ch], **East & West Langworth** Li nr Wragby [*Langwath* Hy 2 (1291) Ch, (pons de) *Langwath* 1202 Ass]. OScand *langa vað* 'long ford', later associated with OScand *viðr* 'wood' and with OE worþ.

Lanhy·drock (-ĭ-) Co [*Lanydret* 1291 Tax, *-hedrek* 1366 FF]. 'The church of St. Hydroc.' Cf. LANN. *Hydroc* may be derived from the adj. found in Welsh as *hydr* 'strong'.

Lani·vet (-ĭ-) Co [*Lannived* 1268 Ep, *Lannyvet* 1283 FF]. Identical with LANNEVET in Brittany. 'The church of St. Nivet.' Cf. *-nimet, -niuet* in pers. ns. such as *Eid-, Gurniuet* c 1150 LL, OBret *Iudnimet*. *Nivet, Nimet* belongs to Gaul *nemeton* 'holy place', Welsh *nyfed* 'shrine'.

Lanlivery Co [*Lanlyveri* 1291 Tax, *-levery* 1323 Ep, *-livery* 1428 FA]. Identical with LANLIVRY in Brittany. Cf. LANN. The second el. is a saint's name.

OW, MW lann, Welsh *llan,* Co *lan* 'enclosure, yard, church' is common in pl. ns. in Wales and Cornwall, and occurs fairly often in He, Cu. The meaning in pl. ns. is usually 'church'. See LAN-, LLAN- (passim), also LAMELLAN, LAMORRAN, LEWANNICK and others. The original form *land* is still found in some names, at least in early forms. The corresponding word is found in OIr *land* and in Gaul *landa,* which is the source of Fr *lande* (cf. LAUNDE).

Lanrea·th (-ĕth) Co [*Lanredoch* DB, *Lanreydhou* 1260 FF, *Lanreython* 1263 Ep, *-reitho* 1283 Ep, *Landreyth* 1377 FF]. Perhaps 'court of justice', the elements being Co *lan* (see LANN) and the plur. of Co *reith* 'justice', corresponding to OBret *reith,* Welsh *rhaith* (plur. *rheithiau*) 'justice, oath'.

Lansallos Co [*Lansaluus, -salhus* DB, *-salewys* 1291 Tax, *-celewys* 1283 Ep]. See LANN. The second el. is no doubt a saint's name, perhaps corresponding to Bret *Salot* in LANSALOT.

Lansdown So [*Lantesdune ecge* c 1067 Bath, *Lantesdon* 1228, *Launtesdon* 1230 Cl, *Lawntesdon* 1228 Ch]. L~ is the name of a ridge, at which is LANGRIDGE. The first el. is doubtless a derivative of the adj. *long* (an OE **langet* analogous to OE *efnet, þiccet* from *efn* 'even' and *þicce* 'thick') or a compound containing the word. OE *langet* survives as dial. *langet* 'a long strip of

ground; a long, narrow wood; a neck of land' (Gl, He, Db). The last seems to be the meaning here.

Lante·glos by Camelford Co [*Nanseglos* 1309 Ep], **L~ by Fowey** Co [*Lanteglos* 1249 Ass, 1283 Ch]. Both were probably originally *nant eglos* 'valley of the church'. *Lan-, Llan-* frequently occurs for original *nant.* Cf. LANTHONY.

Lanthony Priory Gl [*Lantoeni* 1130 P, *-toni* 1199 Cur]. Named from the mother abbey of LLANTHONY in Monm [*Lanthotheni,* more correctly *Nanthotheni* 12 Gir]. The name means 'the valley of the HONDDU' [*Hodni* c 1150 LL]. Original *nant* has been replaced by *llan.*

Lanton Nb [*Langeton* 1242 Fees, 1256 Ass]. 'Long TŪN.'

OE lanu 'lane' is rare in pl. ns. See LANEHAM, LANDFORD W, LENWADE. A special sense occurs in **Asland,** the name of the lower Douglas La, viz. that of Scotch dial. *lane* 'the hollow course of a large rivulet in meadow-land; a brook whose movement is hardly perceptible; the smooth slowly moving part of a river'. See also WATENDLATH.

Lapal (-ă-) Wo [*Lappol* 1220 FF, 1270 Ct, *Laphole* 1272, *Lappehol* 1276 Ct], **Lapford** D [*Eslapaforda, Slapeford* DB, *Lapeford* 1107 (1300) Ch]. '*Hlappa*'s hollow and ford.' In the DB forms *Sl-* is used for OE *Hl-.* Cf. LAPWORTH.

Lapley St [*Lepelie* DB, *Lapeleia* 1130, *Lappeleia* 1200 P]. Perhaps '*Læppa*'s LĒAH'. **Læppa* might be a side-form of *Leppa.* Or the first el. may be OE *læppa* 'tag, end', also 'a district'.

Lapworth Wa [*Hlappawurþin* 816 BCS 356, *Lappawurðin* 11 Th, *Lapeforde* DB, *Lappewurðe* 1197 P]. '*Hlappa*'s WORPIGN.' *Hlappa* is not evidenced in independent use.

Lark R Sf, Ca [*Lark* 1735]. A back-formation from LACKFORD.

Larkbeare D [*Laurochebere* DB, *Lauerkeberia* 1199 P, *-beare* 1272 Ipm], **Larkfield** K [*Lauercefeld* 11 DM], **Lark Stoke** Gl [*Stok* 1220, *Lavirkestok* 1236 Fees], **Larkton** Chs [*Lavorchedone* DB, *Laverketon* 1282 Court]. 'BEARU or grove, FELD, STOC, and DŪN frequented by larks.' Cf. LÆWERCE. In Larkton *-dūn* became *-tūn* by assimilation.

Larling Nf [*Lurlinga* DB, *Lurlinges* 1180 P, *Lirlinge* 1242 Fees, 1254 Val]. The forms point to OE *Lyrlingas,* which would seem to mean '*Lyrel*'s people'. **Lyrel* might be related to ME *lorel* 'worthless person', OE *lyre* 'loss'.

Lartington YN [*Lyrtingtun* c 1050 HSC, *Lertinton* DB]. 'The TŪN of *Lyrti*'s people.' **Lyrti* may be a derivative of *Lorta* in *Lortan hlæw* 934 BCS 705. Cf. OE *belyrtan* 'to cheat'.

Larton Chs [*Layrton* 1291 Tax, *Lairton* 1515 Ormerod]. OScand *Leir-tūn* 'TŪN on clayey soil'.

Lasborough Gl [*Lesseberge* DB, *Lasseberga* c 1150 (1318) Ch, *Lesseberg* 1242 Fees], **Lasham** Ha [*Esseham* DB, *Lasham* 1175 P, *Lesham* 1200 Cur, *Lesseham* 1206 Obl], **Lassington** Gl [*Lessedune* DB, *-don* 1220 Fees, *Lessendon* 1265 Ipm, *Lassendone* 13 Glouc]. The first el. might be OE *lǣssa* 'smaller'. But it is noteworthy that Lasborough and Lassington are not very far apart, and it would be remarkable if *lǣssa* should be used twice in the district. More likely the common el. is OE *Leaxa* pers. n., either a Normanized form (cf. forms of LAXTON, LEXDEN) or a form with hypocoristic assimilation of *x* to *ss* (cf. *Seassa, Sessa* for *Seaxa* in SESSINGHAM). The second el. is OE BEORG, HĀM, DŪN.

Lashbrook O nr Shiplake [*Lachebroc* DB, *Lechebroc* R 1 BM]. First el. OE LÆCC.

Laskill Pasture YN [*Lauescales* c 1170 Riev, 1218–19 FF, *-schales* 1201 FF]. Second el. ON *skáli* 'hut'. The first is possibly Norw *lav*, Sw *lav* 'lichen'.

Lassington Gl. See LASBOROUGH.

Lastingham YN [*Laestingaeu, Lǣstingæ* c 730 Bede, *Lǣstinga ea* c 890 OEBede, *Lestingeham* DB, *Lestingham* c 1090 YCh 350]. Originally 'the island of the *Lǣstingas*', later 'the HĀM of the *Lǣstingas*'. *Lǣstingas* would seem to mean '*Lǣsta*'s people'. *Lǣsta might be compared with ON *Leistr*.

Latchford Chs [*Lacheford, Lachisford* 1288 f. Court], L~ O [*Lacford* 1236 Fees, *Lacheford* 1279 RH]. 'Ford over a LÆCC or stream.'

Latchingdon Ess [*Laecedune* 1065 Th, *Lacen-, Lachen-, Lessenduna* DB, *Lechendon* 1200 Cur]. It is not impossible that there may have been an OE *lǣcce related to OE *lǣccan* 'to catch' and meaning 'a trap'. ME *latch* is recorded in the sense 'gin, snare'. It is probably a native word.

Latham YW. Probably identical with LATHOM, LAYTHAM.

Lathbury Bk [*Late(s)berie* DB, *Lateberi* 1163, *-a* 1167 P, *Lathebur* 1254 Val]. 'BURG made of laths.' *Lath* is OE *lætt*, ME *latt, lappe.*

Lathkil R Db [*Lathkell* 1308 PNDb(S), 1577 Saxton]. Etymology obscure.

Lathom (lādham) La [*Latune* DB, *Lathum* 1201 ff. P]. The dat. plur. of OScand *hlaða* 'lathe, barn'.

Latimer Bk [*Isenhampstede Latymer* 1389 Ipm]. Originally *Isenhamstede*; cf. CHENIES. Later *Isenhamstede Latymer*, and finally Latimer alone. William Latymer got the manor in 1330. Cf. BURTON LATIMER.

Latteridge Gl [*Laderugga* 1176 P, *Ladderuge* 1221 Ass]. Perhaps OE *lāde-hrycg* 'ridge with a road or stream'. Cf. LĀD.

Lattiford So [*Lodreford, Lodereforda* DB, *Loderford* 1243 Ass]. 'The beggars' ford.' OE *loddere* means 'beggar, vagabond'.

Latton Ess [*Lattuna* DB, *Latton* 1197 P, *Lactone* 1212 RBE], L~ W [*Latone* DB, *Latton* 1242 Fees, *Lacton* 1251 AD, 1281 QW]. OE *Lēac-tūn* 'TŪN where leeks were grown'. Cf. LAUGHTON, LEIGHTON.

Laughern Brook R Wo [*Lawern* 757–75, 816 BCS 219, 357, *Lawerna* 1253 WoP]. A Brit river-name, derived from the word for 'fox' (Welsh *llywarn*, OBret, OCo *louuern*). On the river are **Laughern** (lorn) [*æt Lawern* 963 BCS 1108, *Lavre* DB] and **Temple Laughern** [*Lauuarne* 1252 Ch, held by the master of the Temple].

Laughterton Li [?*Leugttricdun* c 680 BCS 840, *Lactertun* 1227 Ep, *Lachterton* 1253 Cl, *Laghterton* 1316 FA]. The form of c 680 may be misspelt for *Leagttricdun*. If so, it may belong here, and the OE form was *Leahtric-dūn* 'hill where lettuce (OE *leahtric*) grew'. Only it would be remarkable to find this Lat loan-word (Lat *lactuca*) so early in a pl. n. Cf. LEIGHTERTON.

Laughton Le [*Lachestone* DB, *Lacton* 1200 Cur, *Lectone* 1219 Ep, *Lethton* 1233 Cl], L~ Li nr Gainsborough [*Lactone, Lastone* DB, *Lactuna* c 1115 LiS, *Lactun* 1212 Fees, *Lecton* 1209–35 Ep], L~ Sx [*Lestun* DB, *Lacton* 1229 Pat, *Lehton, Lechton, Lecton* 1240 ff. Ch], L~ en le Morthen YW [*Lastone* DB, *Lacton* 1228 Cl, *Latton in Morthing* 1230 FF, *Lacton Imorthing* 1256 Ch]. OE *Lēac-tūn* 'TŪN where leeks were grown'. OE *lēactūn* 'kitchen-garden' occurs. See MORTHING.

Laughton Li nr Folkingham [*Lohtun* c 1067 Wills, *Loctone* DB, *Lohcton, Locton* 1204 Cur]. OE *Loc-tūn* 'enclosed TŪN'. Cf. LOCA.

ON **laukr** 'leek'. See LAWKLAND, LOUGHRIGG.

Launcells Co [*Landsev* DB, *Launceles* 1238, 1269 FF, *Lanceles* 1244 FF]. This might be OE *Land-selas* 'country halls'. Cf. SELE. English names are common in the district.

Launceston (lahnstn) Co [*Lanscavetone* DB, *Lanstauaton* c 1180 BM, *Lanzaueton* 1184 P, *Lanceuetona, Lanstaueton* early 13 Ol, *Lancaveton* 1228 Cl]. A hybrid name, OE *tūn* having been added to an Old Cornish name. An OCo *Lan Stephen* has been suggested, but it is doubtful if this is compatible with the early forms.

ME **launde** from OFr *lande* 'glade, pasture', now *laund, lawn*, is found in some pl. ns. It is sometimes difficult to distinguish from LAND. BLANCHLAND, **Old & New Laund Booth** La [*Oldeland, Newland* 1462 Whitaker], **Launde** Le [*Landa* 1163, *La*

Landa 1180 P, *la Launde* 1202 Ass] contain this word.

Launton O [*Langtune* c 1050 KCD 865, *Langtun* 1065 BM, *Lantone* DB]. 'Long TŪN.'

Lăvant R Sx [*la Lovente* 1225 Cl]. A Brit river-name identical with LOVAT Bk, LOVAT in Scotland, LAVANT in Carinthia [*Labanta* 890] and derived from the root of Lat *lābor* 'to glide' &c. *Lavant* is used in some Southern dialects in the sense 'a landspring breaking out on the downs, a brook that is dry at some seasons'. No doubt this is the river-name Lavant that has developed into a common noun. Cf. *pharos* 'a lighthouse' from Pharos. On the Lavant are **East** and **Mid Lavant** [*Loventone* DB, 11 DM, *Louentona* 1121 AC, *Lavent* 1227 Ch, *Estlovent* 14 BM, *Midlouente* 1288 Ass, *Westlovente* 1289 Ep].

Lavendon (lahn-) Bk [*Lauendene* DB, 1201 Obl], **Lăvenham** Sf [*Lauanham* c 995 BCS 1288 f., *Lauenham* DB, *Lave-*, *Lavenham* 1254 Val]. '*Lāfa*'s valley and HĀM.'

Lăver, High, Little & Magdalen, Ess [(at) *Lagefare* c 1010 Wills, 1004 KCD 1300, *Lagafara*, *Laghefara* DB, *High Laufare* 1247 FF, *Alta L~* 1291 Tax, *Parva Lagefare*, *L~ Magna* 1212 Fees, *Laufar la Magdelene* 1263 FF]. An OE *lagufær* or *-faru* 'ford', consisting of OE *lagu* 'sea, flood, water' and *fær* or *faru* 'passage'. Cf. DENVER.
Magdalen from the dedication of the church.

Laver R YW [*Lauer* 12 Fount, *Laure* 1307 YInq]. A Brit river-name, identical with *Læfer* 949 BCS 879, Gaul *Labara*, Welsh LLAFAR, all of which are derived from the adjective found as Welsh *llafar* 'vocal, resounding', OIr *labar* 'talkative'. The name means 'babbling brook'.

Laversdale Cu [*Lefredal* c 1200 WR, *Leversdal* c 1225 WR, *Leveresdale* 1296 Ipm]. '*Lēofhere*'s valley.'

Lăverstock W [*Lawrecestohes* DB, *Laverkestok* 1221 Pat], **Laverstoke** Ha [*Lavrochestoche* DB, *Lauerchestoch* 1158 P]. 'STOC frequented by larks.' Cf. LǢWERCE.

Laverton Gl [*Lawertune* 1220–43 PNGl]. Perhaps 'TŪN frequented by larks'. Cf. LARKTON.

Laverton So [*Lavretone* DB, *Laurton* 12 BM, 1196 P, *Lawerton* 1238 Ass, *Lauwerton* 1243 Ass]. Quite possibly OE *Lǣwerc(e)tūn* 'TŪN frequented by larks'. The interchange of *v* and *w* would go well with such a base. But the first el. may be OE *lǣfer* 'rush, iris' or a river-name identical with LAVER YW.

Laverton YW [*Lavretone* DB, *Lavertun* 1294 Ch]. 'TŪN on R LAVER.'

Lavington Li. See LENTON.

Lavington, East, or **Woolavington** Sx [*Levitone* DB, *Lovinton* 1219 Fees, *Wellauenton* 1208 FF, *Wullavinton* 1230 FF, *Est-*

leuyngton 1288 Ass], **Barlavington** Sx [*Berleventone* DB, *Berlavinton* 1242 Ch]. Here belongs *Lavingtunes dic* c 725 BCS 144, but there is a lacuna before the name, and a syllable may have been lost. The two places were once *Lāfingatūn* 'the TŪN of *Lāfa*'s people', but they were early distinguished by the additions *wella* (*wylla*) 'stream' and *beorg* 'hill' or *bere* 'barley'.

Lavington, Bishop's (or **West**) & **Market** (or **East**), W [*Laventone* DB, *Lauinton* 1186 P, *Lavinton Episcopi* 1233 Cl, *Stupellavintona* 1242 Fees]. Identical with preceding name.
Bishop's L~ belonged to the Bishop of Sarum.

Lawford Ess [*Lalleford* c 1042 Wills, *Lele-*, *Laleforda* DB, *Laleford* 1158 P], **Church & Long L~** Wa [*Leileforde*, *Lille-*, *Lelleford* DB, *Ledleford* 1086–94 Fr, *Lalleford* Hy 2, *Longa Lalefort* 12 BM, *Churche*, *Long Lalleford* 1235 Ch]. '*Lealla*'s ford.' *Lealla* corresponds to OG *Lallo*.

Lawhi·tton Co [*Land Withan* 905 BCS 614, *Landwiþan* 980–8 Crawf, *Langvitetone* DB, *Lawyteton* 1291 Tax]. The present name seems to be a hybrid, OE TŪN having been added to OCo *Landwiþan*. The latter probably means 'the church of *Wiþan*', but the history of the pers. name is not clear.

Lawkland YW [*Laukeland* c 1200, c 1375 Pudsay]. ON *laukaland* 'land where leeks were grown'.

Lawley Sa [*Lavelei*, *-lie* DB, *Laueleye* 1285 FA]. '*Lāfa*'s LĒAH.'

Lawling Ess [(æt) *Lellinge* c 995 BCS 1289, *Lælling* 1006 KCD 715, *Lalinge* DB]. '*Lealla*'s land.' Cf. -ING and LAWFORD.

Lawshall Sf [*Lawessela* DB, *Laueshel* c 1095 Bury, *Laweshell* 1194, 1196 P, *Laugesale*, *Laugetsille* 1253 BM]. OE *hlāw-gesella* 'shelter or hut on a hill'. Cf. HLĀW, (GE)SELL.

Lawton Chs [*Lavtune* DB, *Lautona* 1119, c 1150 Chester], **L~ He** [*Lavtone*, *Lavtune* DB, *Lauton* 1249 Fees]. 'TŪN on a hill.' See HLĀW.

Laxfield Sf [*Laxefelda*, *Lessefelda* DB, *Lexfelde* c 1095 Bury, *Lexefelde* 1168 P]. '*Leaxa*'s FELD.' *Leaxa* is found in *Leaxan oc* 942 BCS 775.

Laxton Np [*Lastone* DB, *Laxetona* 1130 P, *-ton* 1198 Fees, *Laxinton* 12 NS], **L~ Nt** [*Laxintune*, *Leston* DB, *Laxintona* c 1200 DC, *Lexinton* 1212 Fees], **L~ YE** [*Laxinton* DB, *Laxingetun* 1199 FF]. '*Leaxa*'s TŪN' or 'the TŪN of *Leaxa*'s people'. Cf. LAXFIELD.

Laycock YW nr Keighley [*Lacoc* DB, *Lackoc* 1285 FA]. Identical with LACOCK. L~ is on Laycock Beck.

Layer Breton, de la Haye & Marney Ess [*Legra* DB, *Leigre* 1212 RBE, *Leghere* 1255 Ass, *Legra de Haya* 1236 Fees, *Legere Britonis, de la Haye* 1238 Subs, *Leyre*

Bretoun, Marnu 1254 Val]. Probably an old name of Layer Brook (PNEss). Cf. LEIRE.

Breton means 'Breton'. Lewis Brito gave land in L~ **Breton** to St. John's Abbey, Colchester, in the 12th cent. (Wright).—Iuliana de Haia held *Leiren* (L~ **de la Haye**) before 1185 (TpR). The surname is from LA HAYE in Normandy.—L~ **Marney**. Hugh de Marinni held *Legre* in 1207 (FF). The name is taken from MARIGNY in Manche.

Layham Sf [*Hligham* c 995 BCS 1289, *Leiham* DB, *Laiham* 1207 FF]. Skeat's suggestion that the first el. is identical with ON *hlý*, OFris *hlī* 'shelter' is probably correct. The usual form of the OE word corresponding to ON *hlý* is *hlēow*, but a side-form **hlīeg* is possible.

Laysters He [*Last* DB, *Lastes* 1228 FF, 1242 Fees, *Lastres* 1242 Fees, 1257 Ch]. Unexplained.

Laysthorpe YN [*Lechestorp* DB, *Laysthorp* 1239 FF]. '*Leik*'s thorp.' *Leikr* is an ON pers. n.

Layston Hrt [*Lefstanchirche* c 1140 BM, *Leostanecherche* 1198 AC, *Lefstonchirche* 1313 BM]. '*Lēofstān*'s church.' The name was later misunderstood as containing a place-name, and *church* was dropped. An earlier name was *Ichetone* DB, *Ykinton* 1212 Fees, which is '*Ica*'s TŪN'.

Laytham YE [*Ladone, Ladon* DB, *Lathom, Lathum* 1225 f. FF]. Identical with LATHOM.

Layton La [*Latun* DB, *Latona* c 1140 LaCh]. OE *Lād-tūn* 'TŪN on a stream'. See LĀD.

Layton, East & West, YN [*Lastun, Latton* DB, *Laton* 1199 P, 1228 Ep, *Westlaton* 1270 Ipm]. OE *Lēac-tūn* 'TŪN where leeks were grown'. See LĒAC, LAUGHTON.

Lazenby YN in Ormesby [*Laisinbia, Le(i)singebi* DB, *Laisingby* 1237 FF], L~ YN in Northallerton [*Leisinghi* DB, *Laisingbi* 1088 LVD, *Leysingeby* 1204 FF], **Lazonby** Cu [*Leisingebi* DB, 1166 P, *Laysingby* 1247 Ipm]. '*Leysing*'s BY' or 'the freedman's (freedmen's) BY'. ON *leysingr* (*leysingi*) means 'freedman' and was also used as a byname.

le placed before a distinctive addition, as in BOLTON LE SANDS, HAUGHTON LE SKERNE, is the French definite article, which has replaced an English one. In early sources *le* is used more widely, and also *la* occurs. See, e.g., LEA, LODE. See also LEAFIELD.

Lea R Bd, Hrt, Ess, Mx [(on) *Ligan, Ligean* 880 Laws, (on) *Lygan, Lygean* 895 ASC (A, D), *Luye* 1228 Ass, *Leye* 1274 Ass, *La Lye* 1354 Pat, *Lea* 1576 Saxton]. The form varies between ESax *Leye*, WSax *Luye*, and Midland *Lye*. A Brit river-name, which may be derived from the base *lug-* 'light' in Gaul *Lugu-*, OIr *Lug*, Welsh *Lleu*, the name of a deity, Welsh *go-leu* 'light' &c. The name may even mean 'the river of the god *Lugus*'.

Lea by Backford Chs [*Wisdelea* DB, *Lee* c 1230, *la Lee iuxta Bacford* c 1275 Chester], **L~ cum Newbold** Chs [*Lai* DB, *Lay* c 1100, *Leey* c 1150 Chester], L~ Db [*Lede* DB, *Lee* 1326 Ipm], L~ He [*la Lega* 1195 PNHe, *la Le* 1228 Cl], L~ La [*Lea* DB, *Legh* 1246 Ass], L~ Li [*Lea* DB, *Le* c 1115 LiS, *Lee* 1212 Fees], L~ W [*La Le* 1242 Fees]. OE *lēa*, dat. of OE *lēah* 'wood or clearing'.

OE **lēac** 'leek' is the first el. of LACKFORD, LECKHAMPSTEAD, LECKHAMPTON, also of *Lēac-tūn*, which is the source of LAUGHTON, LATTON, LEIGHTON, LAYTON YN, LETTON. This is hardly OE *lēactūn* 'kitchen-garden', but 'TŪN where leeks were grown'.

Leach R Gl [*Lec* 721–43 BCS 166, *Leche* 1577 Saxton]. OE LÆCC, LECC 'stream'. The stream gave their names to EAST- and NORTHLEACH, early forms of which are *Lecche* 872 BCS 535, *Laeceæ* 1070–87 Hereford, *Lechia, Lichia* 1127 AC.

Lead YW [*Lede, Lied* DB, *Lede* 1193 P, *Leade, Ledewudheved* c 1200 YCh 1615, *Leddewdeheued* 1208 FF]. Perhaps OE *hlēo-wudu*, contracted to *hlēodu*. The meaning would be 'wood with a shelter'.

Leadenham (lĕ-) Li [*Ledeneham* DB, *Ledenham* 12 DC, 1191 P, 1202 Ass, *Ledeham* 1194 P]. Perhaps '*Lēoda*'s HĀM'. **Lēoda* would be a short form of *Lēodweald* &c. OE *lēaden* 'of lead' or a pl. n. identical with LYDDEN K do not seem likely first elements. But a derivative of the OE *lēod* which seems to be the source of Li dial. *leed* 'the reed meadow grass, *Glyceria aquatica*', is possible.

Leadgate Du [*Lidgate* 1590 PNNb]. OE *hlidgeat* 'swing-gate'.

Leadon (lĕdn) R Gl, Wo, He [(of) *Ledene* 972 BCS (1282), 978 KCD 619, *Leden* 1248 Ass]. A Brit river-name derived from OBrit *litano-* 'broad' (Welsh *llydan* &c.). From the river were named **High-** and **Upleadon** Gl [*Hyneledene* 1267 Glouc, *Hymeleden* 1291 Tax; *Ledene* DB, *Upledene* c 1275 Glouc], **Leadon** He [*Ledene, Lede* DB, *Ledene* 1242 Fees], **Upleadon** He [*Upleden* 1212, *-e* 1242 Fees].

High- and **Upleadon** Gl are not far apart and were no doubt both once *Ledene*. Highleadon is 'the Leadon of the *hīwan* or monks' (of Gloucester Abbey). Upleadon is higher up the Leadon. **Upleadon** He is on a trib. of the upper Leadon.

Leafield O [*la Felde* 1213 Cur, *la Feld* n.d. Eynsham ii. 95]. 'The field.' *Lea-* is the French definite article (*le, la*).

Leagram La [*Lathegrim* 1282 VH, *-grym* 1425, *Laythgryme* 1349 PNLa]. Perhaps an OScand *leið-grīma* 'a blaze to indicate a road' (OScand *leið* 'road' and *grīma* 'a mark on a tree to indicate a boundary').

Leagrave Bd [*Littegraue* 1224 Cl, *Lihte-, Littlegraue* 1227 Ass]. 'Light grove', probably one with the trees far apart. The first el. was early associated with the word *little*.

OE **lēah** masc. (dat. *lēa, lēage*) and *lēah* fem. (dat. *lēa, lēage, līeg*) is a very common pl. n. el. It corresponds to OHG *lōh* 'grove', LG *lōh* 'thin wood', Du *-loo* (in WATERLOO &c.), ON *ló* 'low-lying meadow' and Lat *lūcus* 'grove'. The original meaning was 'an open place in a wood, a part in a wood with the trees scattered so that grass can grow'. In English pl. ns. two senses are to be reckoned with. The more common one is 'open place in a wood, glade', probably not really a cleared place, but a naturally open space. If the rendering 'clearing' is used, it should be taken in the sense 'glade'. This sense 'open land' is obvious in names such as BENTLEY, FARSLEY. It appears specialized to 'meadow, pasture-land' in names like CALVERLEY, LAMBLEY, STUDLEY. A meaning 'open land used as arable' is obvious in RAYLEIGH, WHEATLEY, LINLEY, FLAXLEY and the like. The other main sense is 'wood, forest'. The great forest of WEALD in K and Sx is called *Andredesleage* 477 ASC. *Wulleleah* is called a wood 817 BCS 361. *Weogorena leag* BCS 357 is WYRE FOREST. The meaning 'wood' is probable in names such as ASHLEY, HASELEY, OAKLEY or CATLEY, ROCKLEY, YAXLEY (with an animal's name as first el.), or STOCKLEIGH, STAVELEY, YARDLEY (where the first el. denotes a product from a wood). Cf. also BADDESLEY, which appears to be an old name of the New Forest. *Lēah* is common in names denoting places for heathen worship, as THUNDERSLEY. See WĒOH. The meaning may here be 'grove' or 'glade'. Names in *-lēah* are naturally most common in old woodland districts. As the exact meaning of *lēah* is generally doubtful in pl. ns., it is mostly left untranslated in etymologies.

Lēah often occurs alone as a pl. n. See LEA, LEE (generally from the dat. *lēa*), LEIGH (from the uninflected *lēah* or from the dat. *lēage*). The plural form is seen in LEECE, LEES, LEESE, LEIGHS, LEAM. *Lēah* is rare as a first el. Possible cases are LEISTON, LEYSDOWN, LYHAM. It is extremely common as the second el., where it generally appears as -LEY (BRADLEY &c.) or -LEIGH (as HADLEIGH, STOCKLEIGH). Occasional forms are seen in ACLE, BALE, EAGLE, MARCLE, OCLE, SALL, NOSTELL, SIXHILLS, ELLA, BARLOW Db. Scandinavianization accounts for HEALAUGH, SKIRLAUGH. The dat. plur. *-lēam* is found in ACKLAM, CLEATLAM and others.

Leake Li [*Leche* DB, *Lech* c 1185 NpCh, *Leke* 1212 Fees], **East & West L~** Nt [*Leche* DB, *Lec* 1204 Cur, *Lek, Westerlek* 1242 Fees, *Estirlek* 1291 Tax], **L~** YN [*Lece, Leche* DB, *Leche* 1088 LVD, *Leke* 1231 FF]. All the places are on streams, and the name originally denoted the stream. The source is an OE **lece*, derived from an OE **lecan* 'to drip, leak', corresponding to ON *leka*, OHG *lechan* and cognate with OE *læcc, lecc*. OE *lece* may well have been

a word for 'brook', but Leake (OE *Lece*) may also have been a river-name.

Lealholm (lēlum) YN [*Lelun* DB, *Lelum* 1272 Ipm, *Lelhom* 1272 Cl]. OE *lǣlum*, dat. plur. of *lǣl* 'a twig, withe', probably here in the sense 'withy, willow'. Cf. LALEHAM.

Leam (lĕm) R Np, Wa [(on) *Limenan* 956 BCS 978, (on) *Leomenan*, (of) *Leomanan* 1033 KCD 751, *Lemene* 1232 Ass, *Leme* c 1540 Leland]. A Brit river-name identical with LYMN.

Leam Nb in Redesdale [*Leum* 1176 P, *Maior, Parva Lem* 1242 Fees, *Lower Lem* 1298 Ipm], **The Leam** Du [*Lem* c 1200 FPD]. OE *lēam* or *lēum*, dat. plur. of LĒAH. At Leam Du is **Leamside** [*le Lemside* 1380 PNNb].

Leamington (-ĕ-) **Hastings** Wa [*Lunintone* DB, *Leminton* 1198 Fees, *Lementon* 1242-9 BM, *Lymyngton* 1280 Ch], **L~ Priors** Wa [*Lamintone* DB, *Lamminton* Hy 2 (1314) Ch, *Leminton* 1242 Fees, *Lemynton Prioris* 1327 PNWa]. 'TŪN on R LEAM.' **L~ Hastings** was held by Aytropius, son of Humfrey Hastings, in 1280 (Ch). **L~ Priors** belonged to Kenilworth Priory from 1122 (VH).

Learchild Nb [*Levericheheld* 1242 Fees, *Leverilcheld, Leverichesbille* 1247 Sc]. '*Lēofrīc*'s slope.' Second el. OE *helde* 'slope'.

Learmouth Nb [*Leuremue* 1177, *Livermue* 1227 P, *Levermue* 1251 Ch]. 'The mouth of R **Lever**' [*Leuer* 1293 Ass]. Lever is derived from OE *lǣfer* 'rush' or 'iris'.

Leasam Sx [*Leuesham* 1200, *Lieuesham* 1206 FF, *Leuelesham* 1279, *-hamme* 1288 Ass]. '*Lēofel*'s HAMM.' **Lēofel* (**Līefel*) is a normal derivative of *Lēofa*.

Leasingham Li [*Leuesingham, Lessingham* DB, *Lefsingham* 1202 Ass, *Levesingeham* 1221 Ep]. 'The HĀM of *Lēofsige*'s people.'

Leasowe Chs. See LǢS.

Leatherhead Sr [(æt) *Leodridan* c 880 BCS 553, *Leret* DB, *Lereda* 1156, *Ledreda* 1160, *Leddrede* 1195 P]. The elements appear to be OE *lēode* 'people' and **rida* or **ride* 'riding-path' or 'ford over which it was possible to ride'. OE *rida* (or *ride*) would be a formation from *rīdan* analogous to *stiga* or *stige* in *ānstiga* (*-stige*) from *stīgan* (cf. ANSTEY). The name probably means 'the public ford'. Leatherhead is on the Mole, where it is crossed by an important road.

Leathley YW [*Ledelai* DB, *Leeleia* 1166 P, 12 Pudsay, *Lethelaye* 1291 Tax]. Perhaps *hleoþa-lēah* 'LĒAH on the slopes' (first el. OE *hliþ*, plur. *hleoþu*, 'slope').

Leaton Sa [*Letone* DB, *Leton* 1212 Fees]. The first el. may be OE HLĒO 'shelter', or (GE)LǢT 'water-course'.

Leaveland K [*Levelant* DB, *Liofeland* 11 DM, *Livelande* c 1180 Fr, *Leueland* 1230 P]. '*Lēofa*'s land.'

Leavenheath Sf [(heath of) *Levynhey* 1292 AD, *Levenesheth* 1351 Copinger]. '*Lēofwine*'s heath.'

Leavening YE [*Ledlinghe* DB, *Leyingges*, *Levingg* 1242 Fees, *Levenyng* 1281 FF, *Leguingge* 1297 Subs]. The forms vary too much for a definite solution to be possible. OE *Lēofhēahingas* '*Lēofhēah*'s people' may be suggested tentatively.

Leaventhorpe YW [*Leventhorp* c 1300 Whitaker]. '*Lēofwine*'s thorp.'

Leavington, Castle, YN [*Leuetona* DB, *Levinton* 1230 Cl, *Castellevinton* 1219 Fees], **Kirk L~** YN [*Leuetona, Lentune* DB]. 'TŪN on R LEVEN.'

Lebberston YN [*Ledbeztun, Ledbestun* DB, *Ledbrizton* 1206 FF]. '*Lēodbriht*'s TŪN.'

Lechlade Gl [*Lecelade* DB, *Lechelad* 1211–13 Fees, *Lichelad* c 1194, 1200–5 RA]. 'The passage (over the Thames) near R LEACH.' Cf. LEACH and (GE)LĀD.

Leck La [*Lech* DB, *Leec* 1196 CC, *Lec* 1251 Ipm]. Identical with LEAKE. Leck is on Leck Beck.

Leckby YN [*Ledebi* DB, *Letteby, Lecceby* 1301 Subs]. '*Liót*'s or, *Liōti*'s BY.' *Liót* is an ON woman's name. *Lióti* is unrecorded, but ODan *Liuti* may occur in pl. ns. ON *liótr* means 'ugly'.

Leckford Ha [*Legh-, Leaht-, Legford* 947 BCS 824 ff., *Lechtford, Lecford* DB, *Legford* c 1270 Ep]. Possibly OE *lēah-ford*. Cf. LĒAH. The OE forms are in poor transcripts, but *Leahtford* and DB *Lechtford* rather suggest a first el. OE **leaht*, which might possibly be a derivative of the stem in OE *læccan* 'to catch'. Cf. LATCHINGDON. Bradley has suggested that *leaht* is a derivative of the stem in OE *lecan* and meant 'irrigation channel', and Forsberg (p. 68) draws attention to Continental *lecht* (found in the Belgian name ANDERLECHT), whose etymology is not sufficiently clear. If *leaht* belongs to *lecan*, the probable meaning here would be 'channel, side-channel'. The Test runs in two arms at Leckford.

Leckhampstead Bk [*Lechamstede* DB], **L~** Brk [*Lechamstede* 815, 821 BCS 352, 366, *æt Leachamstede* 943 ib. 789, *Lecanestede* DB], **Leckhampton** Gl [*Lechantone, Lechametone* DB, *Lechamton* 1211–13 Fees]. 'HĀMSTEDE and HĀMTŪN where leeks grew.' Cf. LĒAC.

Leconfield YE [*Lachinfeld* DB, *Lecingfeld* 1130–8 YCh 970, *Lekingefeld* 1218 FF]. First el. OE *Lecingas* 'people at the *lece* or brook'. Cf. LEAKE.

Ledburn Bk [*Leteburn* 1212 Cur, 1288 Orig, 1299 Ipm]. First el. *leat* 'water-conduit'. See (GE)LÆTU, BURNA.

Ledbury He [*Liedeberge* DB, *Ledburia* c 1140 Hereford, *Lindeberia* 1167, *Lideberia* 1169

P, *Ledebur* 1241 Ch]. L~ is on the LEADON and presumably the first el. of the name is the river-name, though the rarity of forms with *n* is remarkable. Cf. LYDBURY.

Ledsham Chs [*Levetesham* DB, *Leuedesham* c 1100 Chester]. '*Lēofgēat*'s or *Lēofede*'s HĀM.' Cf. LIDSTONE D.

Ledsham YW [*Ledesham* c 1030 YCh 7, DB, 1155–8 YCh 1451], **Ledston** YW [*Ledestune* DB, *-tun* c 1090 Pont, *Ledistona* 1155–8 YCh 1451]. First el. LEEDS, originally the name of a district.

Ledwell O [*Lede-, Ludewelle* DB, *Lydewell* 1270 Ch, *Ledewelle* 1226 Ep]. The first el. is identical with LYD. Second el. OE *wella*. The original name may have been OE *Hlȳde*, to which was added an explanatory *wella*.

Ledwyche, Lower & Upper, Sa [*Ledewic* DB, *-wich* 1155 BM, *Ledwic* 1203 Ass, *-wiz* 1242 Fees]. The places are on **Ledwyche Brook.** Second el. OE WĪC. The first may be as in LETCOMBE. Or it may be an OE **Lēoda* pers. n. or an old name of the stream.

Lee Bk [*Lega* 1182 P], **West L~** Ess [*Lea* DB, *Westlee* 1291 Tax], **L~** Ha in Romsey [*Ly* 1236 Ipm, *Lye* 1280 Ass], **L~ on the Solent** Ha [*Lie* 1212 Fees, *Lee* 1281 Cl], **L~ K** [*Lee* DB, *Lega* 1206 Cur], **L~ Priory** K [*La Lee* 1240 Ass], **L~** Sa nr Pontesbury [*Lee* 1327 Subs], **Leebotwood** Sa [*Lege* DB, *Leg de Bottewud* 1212 Fees], **Lee Brockhurst** Sa [*Lege* DB, *Leye under Brochurst* 1285 Ipm]. The dat. of OE LĒAH.

Botwood [*Botewde* DB] may mean '*Bōta*'s wood', but see BOTLEY.—L~ **Brockhurst** from an adjoining place [*Brokhurst* 1290 Ipm].

Leece La [*Lies* DB, *Lees* 1269 Ass]. OE *lēas*, plur. of LĒAH.

Leeds K [*Hlyda, Hledes* 11 DM, *Esledes* DB, *Ledes* 1186 P, *Lhedes* 1235 Cl]. L~ is on a stream, which must have been called *Hlȳde* 'loud brook'. Cf. LYD.

Leeds YW [*Loidis* c 730 Bede, c 890 OEBede, *Ledes* DB, 1190 P, *Leedes* c 1185 YCh 1746 f., *Liedes* 1181–9 BM, *Leddes* 1100–8 Fr]. *Loidis* in Bede is the name of a district (*regio*), but the name was later restricted to the chief place in it. The name is British and formed with the same suffix (*-iss-*) as *Lindis* (LINDSEY Li). The original vowel of the first syllable must have been ō, which was umlauted to æ, whence ē. Possibly the base is **plōd-*, related to Gk *plōtós* 'flowing', OE *flōd*, Goth *flōdus* 'river' and derived from the verb for 'flow' found in Gk *plōō*, OE *flōwan* &c. Leeds would then be 'district on the river (Aire)'.

Leeford D [*Leoford* 1200 Cur, 1209 Ol]. OE *hlēo-ford* 'ford with a shelter'.

Leegomery Sa [*Lega* DB, 1199 P, *Lega que*

fuit Johannis de Cumbrai 1200 P, *Lega Cumbr'* 1235 FineR]. OE LĒAH.

The manor was held by Alfred de Cambrai in 1167 (Eyton). Cambrai (Cumbrai) from CAMBRAI in France (dep. Nord).

Leek St [*Lec* DB, *Lech* c 1100 Chester, 1188 P, *Leke* 1247 Ass]. Identical with LEAKE.

Leeming Beck YN [*Leminc, Leming* Hy 2 (1348) Ch, *Liemwic* c 1200 Gervase, *Lemyng* 1293 Ass]. A derivative of OE *lēoma* 'ray, radiance'; cf. ME *leeming* 'shining'. On the stream is **Leeming** [*Leming* Hy 2 (1348) Ch, *Lemming* 1202 FF, *Lemyng* 1251 Fount].

Leen R Nt [*Liene* c 1200 Middleton, *Lene* 1218 For, 1227 Cl]. Identical with *Lēon* He (see LEOMINSTER) and derived from the root **lei-* 'to flow' in Welsh *lliant* 'stream'.

Lees La [*the Leese* 1604 PNLa], **Leese** Chs [*Leyes* 1208–29, 1267, *Leghes, Leys* 1244 Chester]. The plural of OE LĒAH.

Leesthorpe Le [*Luvestorp* DB, *Luiestorp* 1229 Cl, *Leves-, Lyvestorp* 1276 RH]. '*Lēof*'s or *Lēofhēah*'s thorp.' *Lēof* is unrecorded.

Leftwich Chs [*Wice* DB, *Leftetewych* 1278, *Leftedewich* 1311 Ipm]. '*Lēoftǣt*'s wīc.' *Lēoftǣt* (Th 299) is a woman's name. Cf. also *Leoftǣta* or *Leoftǣte* (*Leoftǣtan* gen.) c 990 ASCh.

Legbourne Li [*Lecheburne* DB, *-burna* c 1115 LiS, *Lecceburne* 1158 Fr]. 'Trickling stream.' First el. identical with LEAKE.

Legsby Li [*Lagesbi* DB, *Leggesbi* 1202 Ass, 1212 Fees]. '*Legg*'s BY.' ON *Leggr*, lit. 'leg', is a byname.

Leicester (lĕster) Le [*Legorensis civitas* 803 BCS 312, *Ligera ceaster* 917, *Ligora ceaster* 942 ASC, *Ledecestre* DB, *Legrecestra* 1130 P, *Leirchestre* 1205 Lay]. William of Malmesbury, Gesta Pontificum, says L~ was named 'a *Legra* fluvio'. Leicester is on the Soar, but *Legra* may have been an alternative name or rather the name of the tributary on which LEIRE stands. The old name of this river might then have been identical with LOIRE in France (Gaul *Ligeris*). But Leicester cannot be 'Roman fort on R *Legra*'. The early forms suggest as first el. a tribal name in the gen. plur., an OE **Ligore* or the like, which may mean 'dwellers on R *Legra*'. The OE form of the river-name may have been *Ligor* or *Legor*. Leicestershire is *Lægreceastrescir* 1087, *Lepecæstrescir* 1124 ASC (E), *Ledecestrescire* DB.

Leigh, Bessels, Brk [*Leia, Leoie* 965 BCS 1170, *Leie* DB], **High L~** Chs [*Lege* DB, *Legh* 1286 Court], **Little L~** Chs [*Lege* DB, *Legh* 1295 Cl], **Northleigh & Southleigh** D [*Lege* DB, *North-, Suthleghe* 1291 Tax], **Westleigh** D [*Weslega* DB, *Westlegh* 1242 Fees], **L~** (lī) Do nr Sherborne [*Lega* 1228 FF], **L~** (lī) Do nr Wimborne [*Lege* DB,

12 Fr], **L~** (lē) **on Sea** Ess [*Leye* 1254 Val, *La Leye* 1267 Ch], **L~ Gl** [*Lalege* DB, *Leia* 13, *Leghe* 1412 BM], **L~, East L~** Ha nr Havant [*Lega* 1203 Cur, *Estle* 1272 Ch, *Esteleyghe* 1316 FA], **Eastleigh** Ha in S. Stoneham [*Estleie* DB, *Estleg* 1242 Fees], **L~ K** [*Lega* c 1220, *Legh* c 1240 Bodl], **L~ La** [*Leeche* 1276 CC, *Legh* 1276 Misc], **Westleigh** La [*Westlegh* 1238 Ass], **North & South L~** O [*Lege* DB, *Lega* 1192 P, *Northleg* 1225 Ep, *Suthleye* 1291 Tax], **L~ Sa** [*Lege* 1199 PNSa], **L~** So nr Winsham [*Lege* DB, *Lega* 1176 Wells], **Abbots L~** So [*Lege* DB, *Legh of the Abbot of St. Augustin* 1243 Ass], **L~ upon Mendip** So [? (æt) *Leage* c 1000 Wills, ?*Legh* 1243 Ass], **L~** (lī) Sr [*Leghe* Hy 2 BM, *Leya* 1230 P, *La Legh* 1298 BM], **Church L~** St [*Lege* 1002 Wills, 1004 KCD 710, DB], **L~** W nr Westbury [*Lia* 1242 Fees, *Leye* 1318 Ch], **L~ Delamere** W [*Lega* 1242 Fees], **L~** (lī) Wo [*Beornoðesleah* 972 BCS 1282, *Lege* DB]. OE LĒAH. The immediate base is mostly the dat. form *lēage*. The pronunciation varies between (lē) and (lī).

Abbots L~ So belonged to the Abbot of St. Augustine's, Bristol.—*Beornoðesleah* (**L~** Wo) is '*Beornnōp*'s LĒAH'.—The manor of **Bessels Leigh** Brk was held by Petrus Besyles in 1412 (FA).—**L~ Delamere** W was held by Adam de la Mare in 1242 (Fees). Cf. FISHERTON DELAMERE.

Leighs (lēz), **Great & Little,** Ess [*Lega* DB, 1171 P, *Leyes* 1251 Ch, *Magna, Parva Lega* 1254 Val]. Identical with LEIGH. Great and Little Leigh were called Leighs, and this form was transferred to the individual Leighs.

Leighterton Gl [*Lettrintone, Letthrintone* c 1140, *Lettrentone* c 1215 Glouc, *Lechtintone* 1221 Ass]. Apparently identical with LAUGHTERTON.

Leighton Buzzard Bd [*Lestone* DB, *Lectona* c 1140, *Lechtona* 1163 RA, *Letton Busard* 1254 Val], **L~ Chs** nr Nantwich [*Lecton* Hy 3 BM, *Leghton* 1289 Court], **L~ Chs** nr Neston [*Lestone* DB, *Leychtona* 1240–9 Chester], **L~ Bromswold** Hu [*Lestona* 1070–87, *-tuna* 1090 RA, *Lectone* DB, *Letton super Bruneswald* 1254 Val], **L~ La** [*Lecton* 1255 Ipm, *Leghton* 1301 FF], **L~ Sa** [*Lestone* DB, *Leocton* 1188 P, *Lecton* 1198 FF]. OE *Lēac-tūn* 'TŪN where leeks were grown'. Cf. LĒAC, LAUGHTON.

L~ Buzzard from a family of the name. OFr *busard* means 'a buzzard'; it is here used as a nickname. **Bromswold** [*Bruneswald* 1168 P] means '*Brūn*'s wold'. It seems to have been originally a separate vill.

Leighton (-ī-), **Green,** Nb [*Lytedon* 1242 Fees, *Lyhtedon* 1272 Ipm]. 'Bright hill.' First el. OE *lēoht* 'light'.

Leinthall Earls & Starkes He [*Lentehale* DB, *Lintehale* ib., *Leintall Comites* 1275 Ep, *Leinth. Sterk.* Hy 3, *Leinthale Starkare* 13 BM]. 'HALH on R LENT.' L~ is on a tributary of the Teme, which was no doubt once Lent, OE *Lēonte*. Cf. *Lente* 854, 931 BCS 477, 675 (an old name of the COLE Brk, W), (on) *Leontan*, (in) *Liontan* 704–9 BCS 123 (an old name of a brook in Wo). *Lēonte* is identical with or related to Welsh *lliant* 'a torrent, stream'.

Starkes is the gen. of a pers. n. (*Sterker* or *Starker*), which may be identical with *Stercher* DB from ON *Styrkárr*, ODan *Styrkar*, or rather with *Estarcher* DB, a Fr name from OG *Starchari*.

Leintwardine (-ĕ-, -īn) He [*Lenteurde* DB, *Lenttuwurda* 1180 P]. 'WORÞ(IGN) on R LENT.' Cf. LEINTHALL. *Lēonte* may have been an alternative name of the lower Clun.

Leire Le [*Legre* DB, *Leire* 1227 Ep, *Leyre* 1242 Fees]. Very likely an old river-name. Cf. LEICESTER.

Leiston (-ā-) Sf [*Ledes-*, *Leistuna* DB, *Legestona* 1168 P, *Leeston* 1179 P]. Derivation from OE *Lēagestūn* 'TŪN in a LĒAH' is not very satisfactory. L~ is near the coast, and a first el. OE *līeg*, *lēg* 'fire' might be thought of. Cf. LEYSDOWN.

Leith (lēth) R We [*Leeth* 1777 Nicolson & Burn]. Probably a back-formation from a pl. n. containing the word HLIÞ 'a slope'.

Leith Hill Sr [*Lalida* 1167, *La Lida* 1168 P]. OE HLIÞ 'slope'.

Lela·nt or **Uny Lelant** Co [*Lananta* c 1200 Ol, *c* 1260 Ep, *Parochia Sancti Eunini de Lananta* 1327 Subs (Gover)]. *Lananta* is held to mean 'the church of St. Anta'. There was then a later dedication to another saint, according to Loth St. Ewinus (Bret *Ewin*). Hence *Uny*.

Lelley YE [*Lelle* 1284 Pat, 1297 Subs]. Cf. LÆL, LĒAH.

Lemington Gl [*Limentone*, *Leminingtune* DB, *Lemeninton* 1221 Ass, *Lemynton* 1287 QW]. L~ is near Knee Brook, which may have been called *Limen* (cf. LYMN, LEAM). If so, the name means 'TŪN on the *Limen*' and 'the TŪN of the dwellers on the *Limen*'.

Lemmington Nb [*Lemetun* 1158 P, *Lemechton* 1186 P, *Lemocton* 1201 Ch, 1242 Fees]. 'Brook-lime TŪN.' OE *hleomoc* means 'brook-lime' (*Veronica Beccabunga*).

Lemon R D [*Lymenstream* 10 BCS 1323, *Limene* 1244 Ass]. Identical with LYMN.

Len R K. See LENHAM.

Lench, Abbots, Wo [*Abeleng* DB, *Abbelench* 1227 FF], **Atch** L~ Wo [*Achelenz* DB, *Aches Lenche* 1262 For], **Church** L~ Wo [*æt Lench* 860–5 BCS 511], **Rous** L~ Wo [*æt Lenc* 983 KCD 637, *Lenc* 11 Heming, *Lelenz Rand'* 1167 P, *Lench Rondulph* 1291 Tax], **Sheriffs** L~ Wo [*Lench Alnod* 716 BCS 134, *Scherreuelenche* 1221 Ass]. All these were once OE *Hlenc*. The OE forms are in late transcripts. OE **hlenc*,

a side-form of *hlinc* (from **hlanki-*), probably meant 'a hill'.

The distinctive additions are OE *Abba* (Abbots L~), *Æcci* (Atch L~), Norman *Randulf* (Rous L~) and *sheriff* (OE *scīrgerēfa*).

Lenham K [*Leanham*, *East Leanaham* 850, *Leanaham* 858 BCS 459, 496, *Leanham* 11 DM]. '*Lēana*'s HĀM.' **Lēana* is related to OG *Launus*, *Launobaudus* &c. The river-name **Len** is a back-formation.

Lenton or **Lavington** Li [*Lofintun* c 1067 Wills, *Lavintone* DB, *Launton* 1093–1100 YCh 13, *Lenton* 1202 Ass]. Perhaps '*Lēofa*'s TŪN'.

Lenton Nt [*Lentone* DB, *Lenton* 1164 P]. 'TŪN on R LEEN.'

Lentworth La [*Lenteworth* 1324 LaInq]. Probably 'WORÞ on R LENT'. The stream at L~ may have been so called. Cf. LEINTHALL.

Lenwade Nf [*Langewade* (= *Lan-gewade*) 1198–9 FF, *Londe-*, *Lonewade* 1257 Ass, *Lonwade* c 1330 Blomefield]. The elements may be OE LANU 'lane' and (GE)WÆD 'ford'.

OE *lēoht* 'light, bright' is sometimes found in pl. ns. See LIGHTHORNE, LEAGRAVE, LEIGHTON Nb. The meaning is probably as a rule 'light-coloured', but 'thin, with trees far apart' is probable in LEAGRAVE.

Leominster (lĕmster) He [*Leomynster* 10 BCS 1317, c 1000 Saints, 1046 ASC (C), -*minstre* DB]. The Welsh form is or was *Llanllieni*. The first el. is OE *Lēon*, the old name of a district on the Arrow and Lugg, preserved also in EARDIS-, KINGS-, MONKLAND and LYONSHALL. It is *Lionhina* (*Leonhiena*) *gemære* 958 BCS 1040. *Lēon* (*Līon*) represents an OW *lion* or *lian*, of which *llieni* (in *Llanllieni*) is a plur. form. It is identical with the river-name LEEN and belongs to the root **lei-* 'to flow' in Welsh *lliant* 'stream'. Probably we have to assume a Welsh word *llion* 'stream' (Pughe actually gives *llion* 'floods'), and *Llanllieni* means 'the church on the streams' or 'in the district of the streams (Arrow and Lugg)'. Leominster may be a translation of the Welsh *Llanllieni*.

Leppington YE [*Lepinton* DB, *Lepenton* 1196 FF, *Leppington* 1279–81 QW]. 'The TŪN of *Leppa*'s people.'

Lepton YW [*Leptone* DB, -*tuna* c 1170 YCh 1681, *Lepton* 1246 FF]. Hardly '*Leppa*'s TŪN'. Probably the first el. is OE *hlīep*, *hlēp* 'leap' as in BIRDLIP, perhaps in the sense 'abyss'. Lepton is on a steep hill.

Lesbury Nb [*Lechesbiri* c 1190 Godric, *Lescebr'* 1228 FPD, *Lecebir'* 1242 Fees, *Lescebyry* 1254 Val]. 'The BURG of the leech.' OE *lǽce* is 'leech, physician'.

Lesnew·th Co [*Lisnewic* 1233 Cl, *Lysnewyth* 1238 FF]. 'New hall' (Co *lis* 'court, hall, palace' and *newydh* 'new').

Lessingham Nf [*Losincham* DB, *Lesingham*

1254 Val, 1275 RH]. 'The HĀM of *Lēofsige*'s people.'

Lessness K [*Leosne* 1042–50 OSFacs, 1065 BM, *Hlosnes* 11 DM, *Lesneis, Loisnes* DB, *Liesenes* 1086 KInq, *Liesnes* c 1150 BM, 1195 P, *Hliesnes* c ?188 Gervase, *Lesnes* 1194 P, 1202 Curj. R. Forsberg rightly objects to my earlier suggestion (eds. 1–3) that the name contains OE *næss* and takes it to have had initial *Hl-*, but he gives no etymology of his own. If the OE name was a simplex *Hlēosn(e)*, it must be a derivative with the suffix *-asnō (-isnō)* in Goth *hlaiwasnōs* (pl.) 'tomb', OE *lyfesn* 'charm'. Two stems might then be thought of as the base. One is *hlaiw-* in Goth *hlaiwasnōs*, OE *hlāw*, *hlǣw*. A base *hlaiwisnō* might yield OE *hlēosn*. Early *Hlēosne* would be the plur. form, which was later changed to *Hlēosnes*. Lessness would then be a counterpart of the Goth word for 'tomb'. The alternative would be the stem of OE *hlēo* 'shelter' (from **hlewa-*). *Hlewasnō* would probably become OE *hlēosn*, whose meaning might be 'shelter, hut'. Since the place is on a ridge, the later form *Leosnes* might have been associated with the word *næss* 'headland' and have come to be felt to be a compound with this word as second el.

Letchworth Hrt [*Leceworde* DB, *Lechewrde* 1198 Cur, *Luchewrthia* c 1200, *-wrth* Hy 3 BM]. Second el. OE WORÞ. The first had OE *y*. It is probably an OE **lycce* 'enclosure' or the like, related to OE *loc* and corresponding to ON, Norw *lykkia*, Sw *lycka* 'a piece of enclosed land', OHG *luccha*, MHG *lücke* 'a gap'. The element seems to be found in *lychaga* 1014 KCD 1309 (Do). The same first el. is found in LISCOMBE Bk, LITCHAM, LITCHBOROUGH.

Letcombe Basset & Regis Brk [*Ledecumbe* DB, 1212 Fees, *Hledecumba* Hy 2 Abingd, *Ledecumba* 1136 Fr, 1156 P, *Ledecumbe Basset, Regis* 1291 Tax]. The first el. seems to have begun with *Hl-*, to judge by the form from Abingd. Possibly it is OE *hlēda* 'seat, bench', in such a sense as 'ledge'. See CUMB.

L~ **Basset** was held by Richard Basset c 1158 (Abingd). See BERWICK BASSETT.

Letheringham Sf [*Ledringa-, Letheringaham* DB, *Letheringham* 1235 FF, 1254 Val], **Letheringsett** Nf [*Leringa-, Laringaseta* DB, *Letheringsete* 1254 Val]. Perhaps 'the TŪN and (GE)SET of *Lēodhere*'s people'. But OE *Lēodhere* is not with certainty evidenced. Both places are on streams, which may have been called **Hlēoþre* (from OE *hlēoþor* 'sound, melody').

Letton He nr Ludlow [*Lectune* DB, *-thona* 12 BM, *Lecton* 1242 Fees], L~ He nr Weobley [*Letune* DB, *Lettun* 1242 Fees, *Lecton* 1291 Tax], L~ Nf [*Let(e)tuna* DB, *Lecetuna* 1086 IE, *Lecton* 1200 Cur]. The first two are no doubt OE *Lēac-tūn* (cf. LEIGHTON). The first el. of L~ Nf is perhaps rather OE **lece* 'brook'. Cf. LEAKE.

Letwell YW [*Lettewelle* c 1150 DC, c 1175 BM, 1190 ff. P]. The first el. may be as in LEDBURN. The second is OE WELLA 'stream'.

Leven R La [*Leuena* c 1160 LaCh, *Levena* c 1160 ff. FC, *Levene* 1246 Ass], L~ R YN [*Leuene* 1268 Ass, *Leven* 1293 Ipm]. A Brit river-name identical with *Libníos* c 150 Ptol (in Ireland) and LLYFNI, LLYNFI [*Lyfni* c 1150 LL] in Wales. The name may be derived from the adj. for 'smooth' found in Welsh *llyfn*.

Leven YE [*Leuene* DB, *Levene* 1260 Ass, 1297 Subs]. Probably originally the name of the stream at the place. See prec. name.

Lĕvens We [*Lefuenes* DB, *Levenes* 1187 ff. Kendale, 1241 FF, *Lewenes* 1196 FF, *Lefnes* 1170–81 Kendale]. The second el. is OE NÆSS, here in the sense 'headland, projecting ridge'. Some forms suggest as the first el. OE *Lēofwynn*, a woman's name. But it might be OE *Lēofa* or OE *lēaf* 'leaf' (or **lēafig* 'leafy').

Levenshulme (lĕvnzōōm) La [*Lewyneshulm* 1246 Ass]. '*Lēofwine*'s holm.' Cf. HOLM.

Lever, Darcy, Great & Little, La [*Parua Lefre* 1212 Fees, *Leoure* 1227 FF, *Leure, Lever* 1246 Ass, *Magna Leure* 1285 Ass, *Darcye Lever* 1590 Bolton Reg]. The plur. of OE *lǣfer* 'rush' or 'iris' or a derivative of it meaning 'a rush bed'. It is also possible that the old name of the Croal, on which the places are, was **Lǣfre* 'rush stream'. Cf. LEARMOUTH.

Darcy L~ came to Sir Thomas D'Arcy c 1500. Darcy is a family name (from ARCY in France).

Leverington Ca [*Leverinton* 1210 Cur, *Leveringtun* 1239 FF, *-a* 1254 Val]. 'The TŪN of *Lēofhere*'s people.'

Leverton Brk nr Hungerford [*Leofwartun* 1050 Abingd, *Lewartone* DB, *-ton* 1220 Fees]. '*Lēofwaru*'s TŪN.' *Lēofwaru* is a woman's name.

Leverton Li [*Leuretune* DB, *Leuerton* 1167, *Lefrinton* 1180 P, *Leuerton* 1212 Fees]. Perhaps OE *Lǣfer-tūn* 'TŪN where rushes grew'; cf. LǢFER.

Leverton, North & South, Nt [*Cledretone, Legretone* DB, *Legretuna* 1146, *Leertona* 1163 RA, *Leirton* 1166 P, *Legerton* 1212 Fees, *Leuertona* 1175 P, *North-, Suthleverton* 1291 Tax]. *Cledretone* DB indicates that the name began with *Hl-*. First el. perhaps OE **Hlēogār* pers. n. Or it may be an OE *hlēo-gāra* 'gore with a shelter' (cf. GĀRA). Cf. AVERHAM, which shows a similar sound-development.

Levington Sf [*Leuetuna, Leuentona* DB, *Leuington* 1254 Val]. '*Lēofa*'s TŪN.'

Levisham YN [*Leu(u)ecen* DB, *Levezham* 13 Ch, *Levesham* 1234 FF]. '*Lēofgēat*'s HĀM.'

Lew (lōō) R D [*Lyu* 1282 Ass], **L~ Water** R D [*Lywe* 1565 ERN]. Identical with LLIW in Wales and no doubt derived from an adj. *lliw* 'brilliant', related to Welsh *lliw* 'colour'. On the Lew is **North Lew** [*Leuia* DB, *Liw* 1228 FF, *Northlyu* 1282 Ass]. On Lew Water is **Lew Trenchard** [*Lewe* DB, *Lyu*, *Lywe* 1242 Fees, *Liw Trenchard* 1274 Ep].

Lew Trenchard was held by William Trenchard in 1242 (Fees). Trenchard is a byname, related to Fr *trancher* 'to carve'. Cf. also LIFTON.

Lew O [*æt Hlǣwe* 984 Hengwrt MS 150, *Lewa* DB, *Lewes* 1198 FF]. OE HLǢW 'hill'.

Lewa·nnick Co [*Lanwenuc* c 1120 EHR xiv, *Lanwenech* 1261 Ep]. See LANN. The second el. is a saint's name, perhaps identical with Bret *Guenoc*.

Lewell (lōō-) Do [*Lewelle* DB, *Liwella* 1194 f. P, *Liwelle* 1202 FF]. Perhaps 'spring with a shelter', the first el. being OE *hlēo* 'shelter'.

Lewes (lōōis) Sx [(wiþ) *Lǣwe*, (juxta) *Laewes* c 961 BCS 1064 f., *Leuuas* 1081–5 BM, *Lewes* DB]. The plur. of OE *hlǣw* 'hill'.

Leweston Do [*Leweston* Hy 3 BM, 1256 FF]. '*Lēofwīg*'s TŪN.'

Lewisham K [*Liofshema* (mearc) 862, *Lievesham* 918 BCS 506, 661, *Leofsnhǣma* [mearc] 987 BM, *Liofesham* 11 DM, *Levesham* DB, *Leueseham* 1081 Ep, 1203 FF, 1275 Cl]. '*Lēofsa*'s HĀM.' **Lēofsa* is a short form of *Lēofsige*.

Lewknor O [(æt) *Leofecanoran* c 994 KCD 693, *Levec(h)anole* DB, *Leovechenora* Hy 2 Abingd]. '*Lēofeca*'s ŌRA or slope.'

Lexden Ess [(æt) *Læxadyne* c 995 BCS 1289, *Lessendena*, *Lassendena*, *Laxendena* DB, *Lexedone* 1254 Val]. '*Leaxa*'s valley.' See DENU and LAXFIELD.

Lexham, East & West, Nf [*Lecesham* DB, *Lechesham* 1158, 1196 P, 1197 FF, *Est-*, *Westlechesham* 1242 Fees]. 'The leech's HĀM.' Cf. LESBURY.

Leybourne K [*Lilleburna*, *Lillanburna* 10 BCS 1321 f., *Leleburne* DB, *Leiburn* 1193 P]. Originally the name of the stream at the place [*Lylleburna* 942–6 BCS 779]. '*Lylla*'s stream.' **Lylla* is a side-form of *Lulla*. The second *l* was lost owing to dissimilation.

Leyburn YN [*Leborne* DB, *Laibrunn* 1208 Ass, *Layburn* 1246 FF]. Second el. OE BURNA 'stream'. The first may be as in LAYHAM.

Leyland La [*Lailand* DB, *Leilandia* c 1160 LaCh, *Leylond* 1246 Ass]. 'Fallow or untilled land.' First el. OE *lǣge* 'fallow'.

Leysdown K [*Legesdun* 11 DM, *Leesdona* 1175 P, *Leisdon* 1230 P]. Possibly OE *lēages dūn* 'hill at a LĒAH'. But the situation

of the place on low ground by a hill of 73 ft., where there is now a coastguard station, suggests that the first el. of the name is OE *lēg* 'fire', here in the sense 'beacon fire'. The hill would be an excellent place for a beacon, since it is in the Isle of Sheppey on the Thames estuary.

Leyton Ess [*Lygetun* 1065 BM, *Lei(n)tuna* DB, *Luiton* 1201 Cur]. 'TŪN on R LEA.'

Lez·ant Co [*Lansant* 1276 Ep, 1291 Tax]. 'The church of the saint' (Co *sant*, *sans*).

Leziate (lĕjĕt) Nf [*Lesiet* DB, *Lesgate* 1197 P, *Les-*, *Lisegate* 1254 Val]. 'The gate of a LǢS or meadow.'

Libbery Wo [(into) *Hleobyri* 872 BCS 1282]. OE *hlēo* means 'shelter'. *Hlēoburg* would be 'sheltering BURG', 'stronghold'. OE *hlēoburg* actually occurs in this sense in Beowulf.

Lichfield St [*Letoceto* (abl.) 4 IA, *Lyccidfelth*, *Liccidfeld* c 730 Bede, *Licced-*, *Liccetfeld* c 890 OEBede, *Lichesfeld* 1130 P]. Brit *Lētocēton* means 'grey wood' (cf. Welsh *llwyd* 'grey' and *coed* 'wood'). This became OE *Licced*, to which was added OE FELD. The name means 'open land in *Licced* forest'.

Lickle R La [*Licul* a 1140, c 1180 LaCh]. Unexplained.

Liddel R Cu [*Lydel* 12 Sc, *Lidel* c 1165 CWNS xxix]. The river gave its name to **Liddel** Cu [*Lidel* c 1165 CWNS xxix, 1219 Cl, *Liddel* 1267 Ch]. Liddel is an OE *Hlȳdan-dæl* 'the valley of R *Hlȳde*'. See LYD.

Lidden R Do [(bi) *Lidenan* 968 BCS 1214, *Lidenne* 1244, *Ludene* 1288 Ass]. Identical with LEADON.

Liddington W [at *Lidentune* 940 BCS 754, *Ledentone* DB, *Ludinton* 1242 Fees]. 'TŪN on R *Hlȳde*.' The river-name appears as (andlang) *Hlydan* 1043–53 BCS 479. Cf. LYD. **Liddington** Ru [*Lidentone* DB, *Lidinton* 1167 P, 1202 Ass] probably has the same origin.

Lidgate Sf [*Litgata* DB, *Lidgate* 1254 Val]. OE *hlidgeat* 'swing-gate'.

Lidlington Bd [*Litincletone* DB, *Littlingeton* 1180 P]. 'The TŪN of *Lȳtel*'s people.' OE *Lȳtel* 'the little one' is not evidenced in independent use.

Lidstone D [*Lyuedeston* 14 BM, *Lydeston* 1335 Ch]. '*Lēofede*'s TŪN.' Cf. *Leouede* E, p. 262.

Lidstone O [*Lidenestan* c 1235 Winchc, *-ston* 1261 Ipm]. L~ is near ENSTONE, and the name is Enstone with a distinguishing element, e.g. *Lida* pers. n., or OE *hlid* 'gate'.

Lifton D [*Liwtun* c 880, c 970 BCS 553, 1247, *Listone* DB, *Leftun* 1156, *Liftuna* 1157 P]. 'TŪN on R Lew.' Lew was once the name also of the river LYD, on which Lifton is, but came to be restricted to the arm at Lew Trenchard. The change $w > f$ is abnormal.

Lighthorne Wa [*Listecorne* DB, *Litthethurne* 1236 Fees, *Lychtethirn* 1252 Ch]. 'Light-coloured thorn-bush' (OE PYRNE).

Lilbourne Np [*Lilleburne* DB, *-burna* 12 DC], **Lilburn** Nb [*Lilleburn* 1170 P, *Parva Lilleburn* 1201 Cur, *West Lilleburn* 1256 Ass], **Lilford** Np [*Lilleforde* DB, *-ford* 1230 P, *Lillingford* 1205 Cur]. '*Lilla*'s stream and ford.'

Lillechurch K [*Lillecheriche* 1176 BMFacs, *-cherche* 1183, 1190 ff. P]. '*Lilla*'s church.'

Lillesdon So [*Lillesdon* 1225 Ass, 1252 Cl], **Lilleshall** Sa [*Lilsætna gemære* 963 BCS 1119, *Linleshelle* DB, *Lilleshull* 1162 P, *-hell* 1200 Cur]. '*Lill*'s DŪN and hill.' OE *Lill* is found in *Lilles ham* BCS 479.

Lilley Hrt [*Linleia* DB, *-lege* 1204–12, *-lee* 1212 Fees]. OE *lin-lēah* 'LĒAH where flax was grown'.

Lilling, East & West, YN [*Lilinge, Lilinga* DB, *Lillinga* c 1130 YCh 456]. OE *Lillingas* '*Lilla*'s people'.

Lillingstone Dayrell & Lovell Bk [*Lill-ingestan* DB, 1130 P, *Litlingestan Daireli* 1167 P, *Lullingeston, Lillingstan* 1236 Fees]. 'The stone of *Lȳtel*'s people' (cf. LIDLING-TON).

Dayrell is a family name derived from AIRELLE in Normandy (*Dayrell* is really *d' Airelle*).—The Lovells were in L~ from the 13th cent. Lovell is a Fr byname and family name derived from OFr *lovel* 'wolf cub'.

Lillington Do [*Lilletone* 1166 RBE, *Lilli(n)ton* 1180 P, *Lullinton* 1200 Cur, *Lillingtone* 1260 FF]. 'The TŪN of *Lilla* (or **Lylla*) or of his people.' Cf. LEYBOURNE.

Lillington Wa [*Lillintone* DB, *Lillinton* 1203 Cur, 1236 Fees]. 'The TŪN of *Lilla*'s people.'

Lilstock So [*Lulestoch* DB, *Lullinstoke* 1204 Pp, *Lillingstok* 1285 FF]. 'The STOC of *Lylla* and of his people.' Cf. LEYBOURNE.

Limber, Great & Little, Li [*Lindbeorhge* c 1067 Wills, *Lim-, Linberge* DB, *mangna Limberga* 1202 Ass, *Parva Linberga* c 1115 LiS]. 'Lime-tree hill.' Cf. LIND.

Limbury Bd [*Lygeanburg* 571 ASC]. 'BURG on R LEA.'

Limebrook. See LINGEN.

Limehouse Mx [*les lymostes* 1367 Cor, *-hostes* 1380 AD]. 'The lime-oasts', i.e. lime-kilns.

Limehurst La [*Lymehirst* 1379 Bardsley]. First el. the forest name Lyme. See LYME.

Limington So [*Limin(g)tone* DB, *Limintone* c 1200 Montacute, *-ton* 1235, *Liming-, Lemington* 1243 Ass]. The first el. may be a river-name identical with LYMN, denoting a tributary of the Yeo.

Limpenhoe Nf [*Limpeho, Linpeho* DB, *Limpenho* 1193 ff. P]. '*Limpa*'s HŌH or hill.' **Limpa* may belong to the verb *limp*. Cf. OE *lempihealt*.

Limpole Nt [*Lympol* 1311 Ipm]. Possibly 'lime-tree pool'.

Limpsfield Sr [*Limenesfelde* DB, W 1 (1312) Ch, *-feld* 1121 BM, *Linesfeld* 1082–7 BM]. The first el. is very likely a Brit name corresponding to Gaul *Lemonum* and derived from the word for elm (Welsh *llwyf* &c.). Cf. LYMN. Limpsfield would then mean 'open land in an elm wood'.

OE, OScand *līn* 'flax' is the first el. of some names, as LILLEY, LINACRE, LINDLEY, LIN(E)-THWAITE, LINLEY &c., LYNEHAM, LYFORD, but it is often impossible to distinguish it from LIND 'lime-tree', as in LINTON.

Linacre Ca [*Linacra* c 1080 ICC], **L~** La [*Linacre* 1212 Fees]. 'Flax field.'

Linby Nt [*Lidebi* DB, *Lindebi* 1164 ff. P, *-by* 1212 Fees]. OScand *Linda-bȳr* 'lime-tree BY'.

Linch Sx [*Lince* DB, *Linces* 1194 P, *Linche* 1244 Ipm]. OE HLINC 'hill'.

Linchmere Sx [*Wlenchemera* 1187 P, *-mere* 1228 Cl]. '*Wlenca*'s lake.' **Wlenca* is derived from OE *wlanc* 'proud'. Cf. LANC-ING.

Lincoln (lĭngkŭn) Li [*Lindon* c 150 Ptol, *Lindo* (abl.) 4 IA, *Lindum colonia* c 650 Rav, *Lindocolina* c 730 Bede, *Lindcyl(e)ne, -colne* c 890 OEBede, *Lindcylene* 942 ASC, *Lincolia* DB]. *Lindon* is identical with Welsh *llyn* 'a lake' and refers to a widening of the Witham, still partly preserved as **Brayford Mere**. The place was first called *Lindon*, later *Lindon colonia*, whence Lincoln. OE *-cylene* has arisen through English *i*-muta-tion. **Lincolnshire** is *Lincolnescire* 1016 ASC (D, E).

OE, OScand **lind** 'lime-tree' is the first el. of several names. See e.g. LIND-, LYND-(passim), LIMBER, LINBY, LINSTEAD, LIN-WOOD. It is the source of LYNE Sr. Cf. LIN.

Lindal La [*Lindale* c 1220 FC], **Lindale** La [*Lindale* 1246 Ass]. 'Lime-tree valley.'

Lindeth La [*Lyndeheved* 1344 Orig]. 'Lime-tree hill.' Cf. HĒAFOD.

Lindfield Sx [*Lindefeldia, Lendenfelda* c 765 BCS 197, *Lindefeld* 12 AD]. 'Lime-tree FELD.' The first el. is OE *linden* 'of lime-trees'.

Lindisfarne Nb [*insula Lindisfarnensis, ecclesia Lindisfaronensis* c 730 Bede, *Lindis-farena* c 890 OEBede, *Lindisfarna ee* 779 &c. ASC (E)]. 'The island of the *Lindes-faran* or Lindsey people.' Bede calls the Lindsey people *gens Lindisfarorum*. Lindis-farne would then be a colony from Lindsey. But it is possible that *Lindisfaran* in Lindisfarne has the meaning 'people who have been to or regularly go to Lindsey'. Cf. ON *Jorsalafari* 'one who has been to Jerusalem' &c. If so, the name indicates close intercourse between Lindisfarne and Lindsey. In any case the first el. of Lindisfarne is OE *Lindisfaran*, consisting of *Lindis*, the old name of North Lincolnshire, and *faran* 'travellers'. The second el. is ĒG 'island'.

Lindley Le [*Lindle* 1209–35 Ep, 1236 Fees, *Linle* 1242 Fees, 1276 RH], L~ YW nr Huddersfield [*Lillai* DB, *Linley* 1297 Subs], **Old L~** YW [*Linleie* DB]. OE *lin-lēah* 'LĒAH where flax was grown'. L~ Le might also be *lind-lēah*.

Lindley YW nr Otley [*Lindeleh* c 972 BCS 1278, (on) *Linde-leage* c 1030 YCh 7]. 'Lime-wood.'

Lindrick YW [*Lindric* 1225 FF]. Identical with this is **Lindrick** Nt, which occurs in CARLTON IN LINDRICK [(fossatum de) *Lindric* c 1150 DC, (boscus de) *Lindric* 1199 (1232) Ch, *Carleton in Lindric* 1212 Fees]. 'Lime-tree stream.' Cf. RIC.

Lindridge Wo [*Lynderycge* 11 Heming]. 'Lime-tree ridge.'

Lindsell Ess [*Lindesela, -seles* DB, *-sel'* c 1130 Oxf]. OE *lind-gesella* 'huts among lime-trees'. Cf. (GE)SELL.

Lindsey Li [*prouincia Lindissi, Lindissae prouincia* c 730 Bede, (in) *Lindesse* c 890 OEBede, *Lindissa* Alcuin, (on) *Lindesse* 838, 873 ASC, *Lindesig* c 894 Asser, *Lindesi* DB]. A Brit derivative of *Lindon*, the old name of Lincoln, to which was added OE *ēg* 'island'. The district was practically an island, before the fens on the Witham were drained.

Lindsey Sf [*Lealeseia* c 1095 Bury, *Leleseia* 1191 FF, *Lelleseye* 1233 FF]. '*Lelli*'s island.' **Lelli* is a derivative of *Lealla* in LAWFORD.

Lineal Sa [*Lunehal* 1222 FF, *Lunyhal* 1221, *Lunyal* 1280 PNSa]. OE *hlyn* 'maple' and HALH.

Linethwaite Cu [*Linthwait* 1331 StB]. 'Flax clearing.' Cf. LĪN, THWAITE.

Linford, Great & Little, Bk [*Linforde* DB, *-ford* 1176 P, *Lindford* 1175 P, *Lunforde* 1220–34 Ep, 1227, 1241 Ass (PNBk), *parua Linford* 1166 P, *Magna Linford* 1242 Fees]. The first el. is perhaps OE *hlyn* 'maple', rather than OE *lin* 'flax' or *lind* 'lime-tree'.

Lingen He [*Lingham* DB, *Lingen* c 1150 Hereford, *Lingein* 1178, 1183, 1190 P, *Lingeyne* 1237 FF]. Near Lingen is **Limebrook** [*Lingebrok* 1221 Ass, *-broc* 1226 Cl]. Both places are on a brook, which was probably once *Lingen, Lingetn*. This may be a Welsh *llyn-gain* (Welsh *llyn* 'liquid, water' and *cain* 'clear, beautiful'), i.e. '(brook) with clear water'.

Lingfield Sr [*Leangafeld* 871–89 BCS 558, *Lingefeld* 1168 P]. 'The FELD of the *Lēangas*.' *Lēangas* is for *Lēah-ingas* 'people in a LĒAH'. Cf. LINKFIELD.

Lingwood Nf [*Lingewode* 1199 FF, *Lingwude* 1254 Val]. OE *hlinc-wudu* 'wood by a hill'. Cf. LYNG.

Linkenholt Ha [*Linchehou* DB, *Lynkeholte* c 1145 Glouc, *Linkeholt* 1242 Fees, *Lynkenold* 1289 Ep]. OE *hlinca-holt* 'wood on the hills'. But very likely there was an OE **hlince* by the side of *hlinc* and with the same sense.

Linkfield Sr in Reigate [*Lencanfeld* 871–89 BCS 558, *Linkefeld* c 1180 PNSr]. *Lenca* may be a later form of *Lendca*, a pers. n. corresponding to OG *Landico*.

Linkinhorne Co [*Lankinehorn* 1235 FF, *Lankynheorn* 1291 Tax]. 'The church of St. *Cynheorn*.' Cf. Welsh *Cynhaearn*, Bret *Conhoiarn*.

Linley Sa nr Bridgnorth [*Linléé* c 1166 NpCh, *Linley* 1272 Ipm], L~ Sa nr Lydbury [*Linlega* c 1150 PNSa, *Linleg* 1255 RH]. OE *linlēah* 'LĒAH where flax was grown'.

Linmouth Nb [*Lynemuve* 1242 Fees, *-muth* 1268 Ipm]. 'The mouth of R LYNE.'

Linsheeles Nb [*Lynsheles* 1292 QW, *Lyndesele* 1314 Pat]. 'Shieling among lime-trees.'

Linslade Bk [*Hlincgelad* 966 BCS 1189, 966–75 Wills, *Lincelada* DB, 1163 P]. 'The passage by the hill.' Cf. HLINC, (GE)LĀD. L~ is on the Ouzel below a hill.

Linstead Magna & Parva Sf [*Linestede* DB, *Magna, Parva Linstede* 1254 Val]. 'Place where flax was grown.'

Linsted K [*Lindested* 1247 StAug, *-stede* 1291 Tax]. 'Place where lime-trees grew.'

Linstock Cu [*Linstoc* 1212 Fees, *-stoke* 1254 P]. 'STOC where flax was grown.'

Linthorpe YN [*Levingthorp* 12 Whitby]. '*Lēofing*'s thorp.'

Linthwaite YW [*Linthwait* 1208 FF]. See LINETHWAITE.

Linton Ca [*et Lintune* 970 BCS 1268, (et) *twam Lintunum* 11 KCD 725, *Linton* DB], L~ Db [*Linton* 942 BCS 772, *Linctune* DB, *Linton* 1242 Fees], L~ He nr Ross [*Lintune* DB, *-tun* 1156 P], L~ **Grange** YE [*Linton* DB], **West L~** YE [*Lynton* 1316 Misc], L~ **upon Ouse** YN [*Luctone* DB, *Linton* 1176 P], L~ YW nr Skipton [*Lipton* DB, *Linton* 1225 FF], L~ YW nr Wetherby [*Lintone* DB, ?*Lingtona* 1167 P, *Lenton* 1201 Cur, *Linton* 1208 FF]. Most of these are no doubt OE *Lin-* or *Lind-tūn* 'flax or lime-tree TŪN'. For L~ nr Skipton OE *hlynn* 'torrent' would suit admirably. L~ upon Ouse and L~ nr Wetherby may be OE *Hlinc-tūn* 'TŪN by a hill'. See HLINC.

Linton K [*Lilintuna* c 1100 Text Roff, *Lillington* 1226 Ass]. 'The TŪN of *Lilla*'s people.'

Linton Nb [*Linttuna* 1137 Newminster, *Lynton* 1242 Fees]. 'TŪN on R LYNE.'

Lintz Du [*Lince* c 1155 Newminster]. OE HLINC 'hill'. **Lintzford** Du [*vadum de Lince* c 1155, *Lynchesforde* c 1300 Newminster].

Linwood Ha [*Lindwude* 1200 P, *Lindewode* 1271 Ch], **L~** Li nr Market Rasen [*Lindude*, DB, *Lindwda* c 1115 LiS], **L~** Li nr Tattershall [*Lyndwde* 13 FF]. 'Lime-wood.'

Liphook Ha [*la Leephook* 14, *Liephok* 15 VHHa]. OE *hlīep* 'leap, leaping-place' and *hōc* perhaps 'enclosure' as in *inhōc*. The meaning would be 'enclosure with a leap for deer'. The place is on the lower slope of Hindhead.

Lipwood Nb [*Lipwude* 1176 P, *-wode* 1256 Ass]. Possibly the first el. is OE *hlīep, hlēp* 'leap', here in the sense 'steep slope, abyss'.

Liscard Chs [*Lisecark, Lisenecark* 1260 Court]. 'Hall on a cliff', the elements being Welsh *llys* 'hall' and *carreg* 'cliff'. The *en* is probably the OW definite article.

Liscombe Bk [*Lichecumbe* 1207 f. Cur, *Lychescumb* 1251 Cl, *Liscumbe* 1276 RH]. See CUMB. First el. as in LETCHWORTH.

Liscombe So nr Winsford [*Loscumb* 1251 Ass]. 'Valley with a pigsty.' See HLŌSE, CUMB.

Liskeard (-kard) Co [*Lyscerruyt* 11 Th, *Liscarret* DB, *Liscaret* 1194 AC, *Leskered* 1229 Fees]. The first el. is Co *lis, les* (= Welsh *llys*) 'court, hall'. The second may well be a pl. n. containing Co *caer* 'town, castle' and possibly Co *ruid* 'free' or a pers. n. derived from it (OW *Ruid*).

Liss Ha [*Lis* DB, 1198 FF, *Lissa* 1174 P]. Welsh *llys* 'court, hall'.

Lissett YE [*Lessete* DB, *Leset* Hy 2 DC]. OE *lǣs-geset* 'fold in a meadow'. See LǢS, (GE)SET.

Lissington Li [*Lessintone* DB, *Lissigtuna* c 1115 LiS, *Lissingtona* c 1200 DC, *Leusinton* 1202, *Lissinton* 1203 Cur, *Linsinton* 1242 Fees]. 'The TŪN of *Lēofsige*'s people.'

Lisson Mx [*Lilestone* DB, *Lilleston* 1198 Fees]. '*Lill*'s TŪN.' Cf. LILLESDON.

Liston Ess [*Lissingtun* c 995 BCS 1289, *Listuna* DB, *Liston* 1176 P, *Leston* 1219 Fees]. Cf. LISSINGTON. Liston might contain a hypocoristic form **Lissa* or the like from *Lēofsige* (*Līofsige*).

Litcham Nf [*Licham, Lecham, Leccham* DB, *Lucham* 1197 FF, *Litcham, Lucham* 1254 Val]. The first el. must have contained a *y*. It is probably OE **lycce* 'enclosure'; cf. LETCHWORTH. Second el. HĀM.

Litchborough Np [*Liceberge* DB, *Lichebarue* 12 NS, 1199 P, *-berw*' 1202 Ass]. 'Hill with an enclosure.' Cf. prec. name and BEORG.

Litchfield Ha [*Liveselle* DB, *Lieueselua* 1168 P, *Liuesulve* 1212, *Lidescelve* 1219, *-sulfe* 1242 Fees]. Second el. OE *scylf* 'hill' or 'ledge'. The first was originally *Live-*, which may be identified with OE *hlif* in *Hlifgesella* 843 BCS 442. This is cognate with OE *hlifian* 'to tower', ON *hlifa* 'protect', *hlif* 'protection'. It may be an OE word meaning 'shelter'. The change to *Lide-* is difficult to explain. Possibly the name was associated with OE *hliþ, hlid* 'slope'.

Litchurch Db [*Ludecerce* DB, *Litlecherche* 1197 P, *Lutchurch* 1212 Fees]. 'Small church.'

Litherland or **Down L~** La [*Liderlant* DB, *Litherlande* 1202 FF], **Uplitherland** La [*Literland* DB, *Uplittherland* 1207 Ch], **Litherskew** YN [*Litherskewe* 1606 PNNR]. ON *Hliðarland* and *Hliðarskógr* 'land and wood on a slope'. ON *hlið* 'slope' had the gen. *hliðar*.

Litlington Ca [*Litlingetona* c 1080 ICC, *Lidlin(g)tone* DB, *Litlington* 1242 Fees]. 'The TŪN of *Lȳtel*'s people.' Cf. LIDLINGTON.

Litlington Sx [*Litlinton* 1191 P, *Litleton* 1199 Cur]. 'Small TŪN.'

Little Beck YN [*Lithebech* c 1110 ff. Whitby]. ON *Hliða(r)bekkr* 'brook coming from the hill side(s)'. Cf. LITHERLAND.

Littleborough La [*Littlebrough* 1577 Harrison], **L~** Nt [*Litelburg* DB, *Lutilburg* 1242 Fees]. 'Small fort or borough.'

Littlebourne K [*Littelburne* 696 BCS 90, *Liteburne* DB]. 'Small stream.'

Littlebury Ess [*Lytlanbyrig* c 1000 BCS 1306, *Litelbyria* DB]. 'Small fort.'

Littlecote Bk [*Litecote* DB, *Litlecot* 1198 P], **L~** W nr Hilmarton [*æt Lytla codun* 962 BCS 1081, *Litlecote* DB], **L~** W nr Hungerford [*Litlecote* 1412 AD]. 'Small COTS.'

Littleham D nr Bideford [*Liteham* DB, *Litleham* 1219 FF], **L~** D on the Exe [*Littleham*, (æt) *Lytlanhamme* 1042 KCD 1332, *Liteham* DB]. The second is 'small HAMM'. The first may be 'small HĀM'.

Littlehampton Sx [*Hantone* DB, *Hamton* 1229 Cl, *Lyttelhampton* 1482 Ipm]. Originally HĀMTŪN. *Little* is a late addition.

Littlemore O [*Litemora* 1177, *Litlemora* 1191 P, *-mor* 1236 Fees]. 'Small moor.'

Littleover. See OVER.

Littleport Ca [*Litelport* DB]. 'Small town.'

Littleton Chs [*Parua Cristentona* c 1150, *Parua Christleton* a 1250 Chester]. Originally Little Christleton (see CHRISTLETON), later Littleton.

Littleton Do [*Liteltone* DB], **L~ upon Severn** Gl [*Lytletun* 986 KCD 654, *Liteltone* DB], **West L~** Gl [*Litentune* DB,

Litleton 1240 Cl], L~ Ha in Kimpton [*Litel-tone* DB], L~ Ha nr Winchester [*Lithleton* 1285 Ch, *Littleton* 1291 Tax], L~ Mx [*Litleton* 1185 P, *Litlingeton* 1201 Cur, *Litlinton* 1242 Fees], L~ So nr Somerton [*Liteltone* DB], High L~ So [*Liteltone* DB, *Heghelitleton* 1324 Wells], Stony L~ So [*Liteltone* DB], L~ Drew W [*Litletun* 1065 KCD 817, *Liteltone* DB, *Litleton Drewe* 1316 FA], L~ Pannell W [*Liteltone* DB, *Lutleton Paynel* 1317 Ipm], North, Middle & South L~ Wo [*Litletona* 709, *Lytletun*, *alia Litletun* 714 BCS 125, 130, *pry lytlen tunes* c 860 KCD 289, *Liteltune* DB, *Middleton*, *Northlitleton*, *Sutlitinton* 1251 Ch]. 'Small TŪN.'

L~ Drew was held by Walter Drew in 1242 (Fees). Drew is an OFr pers. n. of OG origin (OG *Drogo*, OFr *Dru*).—L~ Pannell was held by Willelmus Painel in 1253 (Cl). Cf. BOOTHBY PAGNELL.

Littlewick Brk [*Lidlegewic* c 1050 KCD 844]. 'WĪC belonging to Littley', a lost place. Forsberg draws attention to *hildleage* (for *hlid*-) 940 BCS 762, which shows that the first el. is OE *hlid* 'gate' or *hlid* 'slope'.

Littleworth Bk [*Litlengeworth* 1227 Ch]. 'The WORÞ of *Lȳtel*'s people.' Cf. LIDLINGTON.

Littleworth Brk [*Weorþe* 955, (æt) *Wyrðæ* 965–71 BCS 906, 1174, *Wurda* 1195 P, *Parva Wurth* 1242 Fees]. Originally WORTH. *Little*- for distinction from LONGWORTH.

Litton Db [*Litun* DB, *Litton* 1273 Ipm, *Lutton* 1302 FA], L~ Cheney Do [*Lideton* 1204 Cur, 1212 Fees, *Ludeton* 1204 Cur, 1236 FF, *Ludinton* 1236 Fees], L~ So [*Hlytton* c 1050, *Hlittun* 1065 Wells, *Litune* DB, *Lidtona* 1176, *Lutton* 1245 Wells], L~ YW [*Litone* DB, *Littuna* 1148 YCh 179, *Lictona* 1182–5 YCh 199, *Lyttona* c 1210 FC]. OE *Hlȳdan-tūn* 'TŪN on R *Hlȳde*' or 'TŪN on a torrent'. See LYD. *Lictone* YCh 199 is probably to be disregarded. If not, this Litton is OE *lictūn* 'burial-ground'.

L~ Cheney was held by Ralph Cheyne in the late 14th cent. Cf. CHENIES.

Livermere Sf [*Leuuremer* c 1050 KCD 907, *Liuermera* DB, -*mere* c 1095 Bury, *Litla Liuermera* DB, *Maius Liuremere* c 1095 Bury, *Liuremere* 12 BM]. The place is situated on a lake. The almost regular *i* of the early forms tells against OE *læfer-mere* 'lake where rush or iris grew'. Possibly the lake was thought to resemble a liver in form. Or the first el. may be as in LIVERPOOL.

Liverpool La [*Liuerpul* a 1194 LaCh, -*pol* 1211 P, *Liverpol* 1246 Ass, *Litherpol* 1222–6 LaInq]. L~ was originally the name of the Pool, a tidal creek, now filled up. *Liver*- is to be compared with OE *lifrig*, ME *livered* 'coagulated, clotted' (as in *þe liuerede se* Rob Gl 'the Red Sea'). Cf. *liver-sea* 16 OED, G *Lebermeer* 'the Red Sea'. The

name may mean 'pool with thick water'. Or Liver may be the old name of one of the streams that fell into the pool. If so, it is identical with *Lifra* in Norway ('stream with thick water').

Liversedge YW [*Livresec* DB, *Liversegge* 1198 FountM, *Luvereseg* 1212 Cur]. '*Lēofhere*'s ECG or ridge.'

Liverton YN [*Liuretun* DB, *Livertun* c 1170 YCh 891]. The first el. is probably a stream-name *Lifra*. Cf. LIVERPOOL.

Livesey La [*Liveseye* 1227 FF, *Liveshey* 1243 LaInq]. 'Island with a shelter.' Cf. LITCHFIELD.

Liza R Cu [*Lesagh* 1294, 1322 Cl]. 'Bright river' (ON *lióss* 'light' and *á* 'river'). Identical with *Ljósá* in Norway and Iceland.

Lizard Co [*Lisart*, *Lusart* DB, *Lesard* 1302 FF]. 'High court' (Co *lis* 'court, hall' and *ard* 'high').

Lizard Hill Sa [*Lusgerde* 664, *Lusgeard* 680 BCS 22, 49 (late texts), *Lusegarde* 1199 (1285) Ch, *Lus*-, *Lisgarde* 1291 Tax]. The name may be a Welsh *llys garth* 'hall by a hill'. Cf. DOWARD and prec. name.

Llancillo He [*ecclesia Sancti Sulbiu*, *Lann Sulbiu* (*Suluiu*) c 1150 LL]. '*Sulbiu*'s church.' Cf. LANN.

Llancloudy He [*Lann Loudeu* c 1150 LL, *Lontlendi* 1266 Ipm]. '*Loudeu*'s church.' MW *Loudeu* is a pers. n. whose first el. is found also in *Loubran* &c.

Llandinabo He [*Lann Iunabui* c 1150 LL, *Landinabo* 1279 Ep]. '*Junabui*'s church.' The second el. is the OW pers. n. *Iunapui* (*Junabui*) with hypocoristic *di* 'thy' prefixed. Cf. LANDEWEDNACK.

Llangarren He [*Lann Garan* c 1150 LL, *Langaran* 1291 Tax]. 'Church on R GARREN.' But originally probably *Nant Garan* 'the valley of the Garren'.

Llangrove He [*Longe grove* 1372 Ipm]. 'Long grove.'

Llangstone Monm [*Langeston* 13 LL, 1297 BM]. 'The long stone', perhaps 'the menhir'. The spelling with *Ll*- is due to Welsh influence.

Llanrothal He [*Lann Ridol* c 1150 LL, *Lanrethal* 1277 Ep, *Lanrothal* 1291 Tax]. See LANN. The second el. is probably a saint's name.

Llanthony Monm. See LANTHONY.

Llanvair Waterdine Sa [*Watredene* DB, *Waterdene* 1278 Ep, *Thlanveyr* 1284 Cl, *Llanver* 1560 BM]. Llanvair is 'the church of St. Mary' (Welsh *Mair*, mutated *Fair*). Waterdine is 'the valley of the river (Teme)'. It is probably a distinguishing addition, Welsh *Llanfair* being a very common name.

Llanwarne He [*Ladgvern* DB, *Lann Guern* c 1150 LL, *Lanwaran* 1291 Tax]. 'Church by the swamp or alders.' Welsh *gwern* means 'swamp, alder grove'.

Llanyblodwell Sa. 'Church by Blodwell.' Blodwell is the old name of a tributary of the Tanat [*Blodwelle* c 1200 Gervase]. Blodwell (hamlet) is *Blodvol* 1254 Val, *Blodowauham, Bloduorvaur* 1272, *Bledewelle Vaghan, Vaur* 1302 Ipm]. The etymology of the name is obscure.

Llanymynech Sa [*Llanemeneych* 1254 Val, *Llanymeneich* 1282 Eyton]. 'The church of the monks.' Welsh *mynach* 'monk' formerly had the plur. *myneich*.

Load, Long, So [*La Lade* 1285 FA, 1292 Misc]. OE LĀD 'water-course'.

Lobthorpe Li [*Lopintorp* DB, *Loupingtorp* 1212 Fees]. '*Louping*'s thorp' or 'the thorp of the fugitive'. First el. ON *hlaupingr* 'fugitive', probably used as a pers. n.

OE **loca** 'enclosure' is the second el. of CHALLOCK K, PORLOCK, and the first el. of some names in LOCK-. OE *loc* 'enclosure' also occurs.

Lockeridge W [*Locherige* DB, *Lokeruga* 1141–3, *Locrugge* 1185 TpR, *Lokerigg* 1242 Fees]. Perhaps 'ridge with a *loca* or enclosure'.

Lockerley Ha [*Locherlega, Locherslei* DB, *Lokerlay* 1194 f. P, -*le* 1203 Cur]. Cf. *Lokeresleag* 994 KCD 687, *loceres weg* 948 BCS 866. *Locere* is possibly identical with ME *lōkere* 'keeper, shepherd' (1340 &c.).

Locking So [*Lockin* 1212 Fees, *Lokkinges* 1249 FF, 1264 Ep]. '*Locc*'s people.' OE *Locc* occurs as a byname Th 636 and is found in several pl. ns.

Lockinge, East & West, Brk [*Lakinge* 868, *Lacing* 956 BCS 522, 935, *Lachinges* DB, *Lakinges, Westlaking* 1220 Fees]. *Lācing* was originally the name of the stream at L~ [*Lakincg* 868, *Lacing* 956, 958 BCS 523, 935, 1032]. The stream-name *Lācing* masc. may be a derivative of OE *lāc* 'play', *lācan* 'to play' (cf. Norw LEIKEBÆK) or of a pers. n. *Lāc*, a short form of names in -*lāc* (cf. ON *Leikr*, OG *Laico*).

Lockington Le [?*Lochamtona* 11 KCD 971, *Lokinton* c 1125 LeS, *Lokintone* 1223 Ep, *Lokington* 1254 Val], L~ YE [*Locheton* DB, *Lokintona* 1154–60 YCh 1118, *Lokinton* 1228 Ep, *Lukinton* 1226 FF]. If *Lochamtona* belongs to Lockington Le, it must have as first el. OE *loc(a)*, the second being *hāmtūn* or -*hǣmatūn*, 'homestead by an enclosure' or 'the TŪN of the people by the enclosure'. If not, the name may mean 'the TŪN of *Locc*'s people'. Cf. LOCKING. Lockington YE, if *Lukinton* 1226 may be trusted, would seem to mean 'the TŪN of *Luca*'s people'. OE *Luca* is found in *Lucan beorh* 961 BCS 1066, *Lucan weorðig* ib. 1343.

Locko Db [*Lokhaye* 1258 FF, *Lochay* c 1261 Derby, *Lokhawe* 1276 RH]. OE *loc-haga* 'enclosure'. The two elements have much the same meaning.

Lockton YN [*Lochetun* DB, *Loketon* 1167 P, 1241 FF, *Lokinton* 1198 Fees]. The first el. may be OE *loca* 'enclosure', but the *n*-forms are difficult. OE *locen* 'closed, enclosed' might also be thought of.

Lockwood YW [*Loc(k)wode* 1275 Wakef, *Locwode* 1297 Subs]. OE *loc-wudu* 'enclosed wood'.

Lodden R Do [*Lyden* 1236 Cl, *Lydene* 1279 For]. Identical with LEADON.

Loddington Le [*Ludintone* DB, -*ton* c 1125 LeS, *Ludington* 1248 Ch, *Lodington* 1209–35 Ep], L~ Np [*Lodintone* DB, *Ludinton* 1199 FF, *Lodinton* 1220 Fees]. 'The TŪN of *Luda*'s people.'

Loddiswell D [*Lodesville* DB, *Lodiswill* 1212 Fees, *Lodeswell* 1230 P]. '*Lod*'s spring.' A pers. n. *Lod* is presupposed also by LODSWORTH.

Loddon R Brk, Ha [*Lodena* c 1215 Gir, *Loden* 1250 Cl, *Lodene* 1279 For]. A Brit *Lutnā* 'muddy river', derived from the base *lutā* 'mud' in OIr *loth* 'mud' and in the Gaul river-names *Luteva, Lutosa*. A base *lutno-* is found in Gaelic *lòn* 'marsh, mud'. *Lutnā* regularly gave *Lodnā* and *Lodn*.

Loddon Nf [(into) *Lodne* 1043 Wills, *Lotna, Lothna, Lodnes* DB, *Lodne* c 1095 Bury, *Lodne* 1198 FF]. Loddon is an old name of the river CHET, identical with LODDON Brk. **Loddon** hd [*Lothninga* DB, *Lodninge Hundret* c 1095 Bury] is 'the dwellers on the Loddon'.

Lode Ca [*la Lade* 1242 P, *Lada* n.d. AD]. OE LĀD 'water-course'.

Loders, Uploders Do [*Lodre* DB, c 1100 Montacute, *Lodres* DB, 1212 Fees]. *Lodre* is very likely an old name of the stream at the place. It may be a compound with Brit *dubro-* as second el. (cf. ANDOVER, CALDER &c.). The first el. might be identical with LOOE.

Lodore Cu [*Laghedure* 1210 f. FC]. 'The low door', i.e. the lower gap in the ridge between Watendlath and Borrowdale. There was also an upper gap called *Heg(h)edure* 'the high door' 1210 f. FC. The latter is no doubt the present **High Lodore**. *Laghe-* is the word *low* (OScand *lāgr*), while *dure* is OE *duru* 'door'. There is a famous waterfall at Lodore.

Lodsworth Sx [*Lodesorde* DB, -*wurða* 1166 P]. '*Lod*'s WORP.' Cf. LODDISWELL.

Lofthouse YW nr Ardsley [*Locthuse, Loftose* DB, *Lofthuse* 1242 Fees], L~ YW nr Harewood [*Lofthuse* DB, *Lofthusum* c 1145 YCh 1862], L~ **Hill** or **Loftus Hill** YW [*Locthusun* DB, *Lofthus* 1219, 1233 FF],

Loftsome YE [*Lofthus* 1208 FF], **Loftus** YN [*Loctehusum* DB, *Lofthus* 12 Guisb]. ON *lopthús* 'a house with an upper floor'. The name usually appears in the plur. form (dat. *-húsum*).

Lolworth Ca [*Lolesuuorde* DB, *Lulleswrõe* 1199 P, *Lolleworth* 1242 Fees, *-wrth* 1251 Ch]. '*Lull*'s WORÞ.'

Loman R D [*Lomund Water* 1577 Harrison, *Leman* 1612 Drayton]. Probably identical with LEAM, LYMN. One of the places named from the Loman is *Lemene* 1297 FF. On the Loman are **Chieflowman** [*Lonmine* DB, *Childelumene* c 1166 Montacute], **Craze Loman** [*Lonmele* DB, *Luminee* 12 (1329) Ch, *Lomene Clavile* 1284–6 FA], **Uplowman** [*Oppaluma* DB, *Uplomene* 1303 FA].

Chieflowman is 'the Loman of the children or knights'.—**Craze L~** is 'Clavile's Loman'. The manor was held by Walter de Clavill in the 12th cent. (Ch), by William de Clavill in 1242 (Fees). Clavile is a French family name (from CLAVILLE in Normandy).

Lomax. See LUMB.

Londesborough YE [*Lodenesburg* DB, *Landenesburgh* c 1110 YCh 25, *Lonesburgh* 1136–9 ib. 31]. '*Lothen*'s BURG.' *Lothen* (*Loðen* 1046 ASC) is ON *Loðinn*, ODan *Lothæn*, a nickname meaning 'hairy'.

London [*Londinium* 115–17 Tacitus, 4 IA, *Londinion* c 150 Ptolemy, *Lundin(i)um* Ammianus Marcellinus, *Lundonia* c 730 Bede, *Lundenburg* 457 ff. ASC, (on) *Lundenne* 839, (on) *Lundene* 962 ASC, (in) *Lundenne*, *Lundenceaster* c 890 OEBede, *Lundres* 12 Fantosme, *Lundin* 1205 Lay]. *Londinium* is no doubt a derivative of a stem *londo- 'wild, bold', found in OIr *lond* 'wild'. The immediate base may be a pers. n. *Londinos* or a tribal name formed from the adjective. **London Bridge** is (æt) *Lundene brigce* 10 BCS 1131.

Londonthorpe Li [*Lunde(r)torp* DB, *Lundretorp* 1202 Ass, *Londenetorp* 1180–3 Middleton]. OScand *Lundar-þorp* 'thorp by a grove'. OScand *lundr* was *lundar* in the genitive.

Longbenton. See BENTON.

Longborough Gl [*Langeberge* DB, *-berga* 1193 ff. P, *-birge* 1221 Ass]. 'Long hill.'

Longbridge Deverill W [(in) *Longo Ponte Deverell*, *Deverel Lungpunte* 1252 Misc, *Deverellangebrigge* 1330 Ch]. The original name was DEVERILL (q.v.). Longbridge 'the long bridge' (in early records usually in the Fr form) is a distinctive addition.

Longcot Brk [*Cotes* 1316 FA, *Longcote* 1332 Ipm]. 'Long COTS.'

Longden Sa [*Langedune* DB, *Longedun* 1236 Fees]. 'Long hill.' See DŪN.

Longdendale Chs, Db [*Langedenedele* DB, *-dala* 1158 P]. 'The valley of Longden', which itself means 'the long valley'.

Longdon upon Tern Sa [*Langvedune* DB], **L~ St** [(æt) *Langandune* 1002 Wills, *Langedun* 1158, *-don* 1195 P], **L~ Wo** at Tredington [æt *Longandune* 969 BCS 1243, *Longedun* DB], **L~ Hill** Wo in Bengeworth [(into) *Langandune* 972 BCS 1282]. 'Long hill.' See DŪN.

Longfield K [(æt) *Langanfelda* 964–95 BCS 1132, *Langafel* DB, *Langefeld* 11 DM]. 'Long FELD.'

Longfleet Do [*Langeflete* 1230 P]. 'Long FLĒOT or channel.'

Longford Db [*Langeford* 1197 P], **L~ Gl** [*Langeford* 1107 (1300) Ch, 1200 Cur], **L~ He** [*Longeford* 1256 Ipm], **L~ Mx** [*Langeford* 1327 FF], **L~ W** [*Langeford* DB, c 1195 Cur]. 'Long ford.'

Longford Sa [*La[n]ganford* 1002 Wills, *Langeford* DB, 1191 ff. P]. This Longford is not on a stream, but it is near a road, **The Longford**, which runs from Watling Street through Newport and farther west. Near this road is another **Longford** at Hodnet [*Langeford* 13 Eyton]. Longford is also, according to Duignan, the name of a part of Watling Street between Church Bridge and Four Crosses (in Staffordshire). At least in the name of the road (The Longford) it is tempting to assume as second el. Welsh *ffordd* 'road'.

Longframlington. See FRAMLINGTON.

Longham Nf [*Lawingham* DB, 1200 Cur, *Lowingeh'* 1199–1200 FF, *Laingeh'* 1208 Obl, *Langham* 1254 Val]. Longham is in **Launditch** hd [*Lawendic* DB, *Lawendichhdr*. 1190 P, *Lawendich*, *Lowedich* 1202 FF]. The latter seems to be OE *Lāwan dic*, the first el. being a pers. n. *Lāwa*. Longham is 'the HĀM of *Lāwa*'s people'. Cf. LANGHAM Ess. *Lāwa* is not recorded.

Longhirst Nb [*Langherst* 1200 Cur, 1242 Fees]. 'Long HYRST.'

Longhope Gl [*Hope* DB, *Hop* 1206 Cur, *Longehope* Hy 3 Misc]. 'Long HOP or valley.'

Longleat W [*Langelete* 1235 Salisbury, *La Langhelete* 13 BM]. 'Long stream or conduit.' Cf. (GE)LÆTU.

Longmynd (-i-) Sa [*Longameneda* 12 Haughmond Cart, *Longa foresta* 1199 P, (foresta de) *Longa Muneta* 1212 Fees, *Longemynede* 1275 Cl]. The second el. is Welsh *mynydd* 'hill'. The Longmynd is a long and broad ridge.

Longner Sa [*Langvenare* DB, *Langenhalre* 1223 Ass], **Longnor** Sa [*Longenalra* c 1170 Eyton, *Longenolre* 1333 Ch], **L~ St** nr Penkridge [*Longenalre* DB, *Longenolre* 1327 Subs]. 'Tall alder(s)' or 'long alder copse'. Second el. OE *alor* 'alder'.

Longney Gl [(in) *Longanege* 972 BCS 1282, *Langenei* DB]. 'Long island.'

Longnor St nr Buxton [*Langenoure* 1227 FF, *Longenovere* 1277 Misc]. 'Long OFER or ridge.' See also L~ under LONGNER.

Longparish Ha. 'The long parish.' A late name, which has supplanted **Middleton** [*Middeltune* DB, *Midelton* 1304 Ep].

Longridge La [*Langrig* 1246 FF], L~ St [*Langrige* 1199 FF, *Langerugge* 1236 FF]. 'Long ridge.'

Longsdon St [*Longesdon* 1242 Fees, 1252 Ch]. Cf. LONGSTONE Db, which has the same early forms. Both are by long ridges. Perhaps in both cases the ridge was called LONG, to which was added an explanatory DŪN.

Longslow Sa [*Walanceslav* DB, *Wlaunkeslawe* 1230 P, *Wlonkeslawe* 1242 Fees]. '*Wlanc*'s burial-mound.' Cf. HLĀW. **Wlanc* is a pers. n. derived from OE *wlanc* 'proud'. Cf. LANCING and LINCHMERE.

Longstock Ha [*æt Stoce* 982 KCD 633, *Stoches* DB, *Langestok* 1233 Cl]. 'Long STOC.'

Longstone Db [*Langesdune* DB, *Langsdune* 1225, *Langesdone* 1258 FF]. See LONGSDON.

Longton La [*Lange-*, *Longetuna* 1153–60 LaCh], L~ St [*Longeton* 1212 Fees, *-a* Hy 3 Misc], **Longtown** Cu [*Longeton* 1267 Ch], L~ He [*Longa villa* 1540 PNHe]. 'Long TŪN.'

Longtown He was formerly EWYAS LACY.

Longville in the Dale Sa [*Longefewd* 1255 RH, *Longfeld* 1291 Tax], **Cheney L~** Sa [*Langvefelle* DB, *Langefeud* 1242 Fees]. 'Long FELD.' The *f* was voiced to *v*, and the name was associated with Fr *ville*.
Cheney L~ was held by the Cheney family at least from the early 14th cent. (FA). Cf. CHENIES.

Longworth Brk [*æt Wurðe* 958 BCS 1028, *Wurth* 1242 Fees, *Langwrth* 1291 Tax], L~ La [*Langeworthe* c 1210 CC]. 'Long WORÞ.' Cf. LITTLEWORTH.

Longworth He [*Langeford* 1242 Fees, 1281 Ch]. 'Long ford.'

Lonsdale La, We [*Lanesdale* DB, *-dala* 1130 P, *Lonesdale* 1169 P]. 'The valley of the LUNE.'

Lonton YN [*Lontune* DB]. 'TŪN on R LUNE' (YN).

Looe R Co [*Loo* 1301 Ipm, 1365 FF]. Co *lo* 'an inlet of water, a pool' (= Welsh *llwch*, Ir *loch*). The name was presumably at first restricted to the mouth of the river. On the Looe are **East & West Looe** [*Lo* 1237 BM, 1297 Cl, *Lohe* 1244 FF].

Loose (lōoz) K [*Hlose* 11 DM, *Losa* 11,0 P, *Lose* c 1195 BM]. OE HLŌSE 'pigsty'.

Loosley Row Bk [*Losle* 1241 Ass]. 'LĒAH with a pigsty.' Cf. HLŌSE.

Lopen So [*Lopen*, *-e* DB, *Lopena* 1166 RBE, *Luppena* n.d. Bruton, *Lopen* 1244 Ass]. Perhaps '*Lufa*'s PEN or fold'.

Lopham Nf [*Lopham* DB, 1177 P, *Loppham* 1198 FF]. '*Loppa*'s HĀM.' *Loppa* is found in *Loppancomb* BCS 828, *Loppandyne* BCS 1289.

Loppington Sa [*Lopitone* DB, *Lopinton* 1199 P, *Lopington* 1230 P]. 'The TŪN of *Loppa*'s people.' Cf. LOPHAM.

Lorbottle Nb [*Luuerbotle* 1178, *Leuerbotle* 1179 P, *Loverbothill* 1236, *Liverbothill* 1244 Fees]. Possibly '*Lēofhere*'s *bōtl* or homestead'. But the woman's name *Lēofwaru* would go better with the absence of gen. *s*.

Lordington Sx [*Harditone* DB, *Herdinton* 1196 P, *Lerdingetuna* 1219 FF, *Lurdyngton* 14 BM]. 'The TŪN of *Lēofrēd*'s people.'

Lorton Cu [*Lorenton* 1195 FF, *Loretona* 12 StB, *Lortone* 1197 P]. The first el. is either a brook-name *Hlōra*, identical with LORA in Norway, or a pers. n. *Hlōra*. *Hlóra* fem. is a mythical ON pers. n. Both belong to the base of OE *hlōwan* and mean 'the roaring one'. The name is no doubt Scandinavian. Low L~ is on the Cocker, High L~ on Whitbeck. Probably *Hlōra* was the old name of Whitbeck, whose name indicates that it is a swift stream with foamy water.

Loscoe Db [*Loscowe* 1277 BM, *Loschowe* 1401 Derby], L~ YW in Ackton [*Loft Scoh* c 1200 Kirkst]. OScand *Loftskógr* 'wood on a hill'. Cf. LOTHWAITE.

Loscombe Do [*Loscumbe* 1268 FF]. Identical with *hloscumb* 933 BCS 695 (Bradford Abbas Do), which means 'valley with a pigsty'. Cf. HLŌSE.

Loseley Sr [*Losele* DB, 1206 Cur]. 'LĒAH with a pigsty.' Cf. HLŌSE.

Losenham K [*Hlossanham* 724 BCS 141, *Lossenham* 1205 FF, *Losham* 1265 Ch]. '*Hlossa*'s HĀM.' **Hlossa* is derived with an *s*-suffix from *Hlōþ-* in *Hlōþhere*.

Lostford Sa [*Lokefford* 1199 FF, *Loskesford* 1212 Fees, *Lockesford* 1241 FF]. Perhaps OE *loxes ford* 'the lynx's ford' with metathesis.

Lostock Gralam Chs [*Lostoch* c 1100 Chester, *Lostoc* c 1200 CC, *le Lostoke Graliam* 1288 Court], L~ La in Bolton le Moors [*Lostok* 1205 FF], L~ La in Eccles [*Lostoke* 1322 LaInq], L~ R La [*Lostoc* c 1200 CC, *Lostok* 13 WhC]. If Lostock is an old name of an inhabited place, it is easily explained as OE *hlōs-stoc* 'STOC with a pigsty'. But the river-name is not easily explained as a back-formation, and the Lostocks are on streams, which may have been called *Lostoc*. If Lostock is an old river-name it is no doubt derived in some way from Welsh *llost* 'tail'. There may have been a Welsh *llostog* 'beaver', a substantival use of *llostog* 'provided with a tail'. Lostoc might then be elliptical for a Welsh name meaning 'beaver stream'. One *Gralamus* held Lostock Gralam c 1200 (CC). Gralam is no doubt a French pers. n.

Lostwi·thiel (-ĭth-) Co [*Lostwetell* 1194 P, *Lostudiel* 1195 P, *Lostwhidiel*, *-wydiel* 1269 FF]. The name goes with WITHIEL SW. of Bodmin, which is some miles away on the other side of an upland district. This district may have been called Withiel, and Lostwithiel would be 'the end (lit. the tail) of Withiel'. Co *lost* means 'a tail'.

Lothersdale YW [*Lodresdene* DB, *Lodderesden* 1202 FF, *Lothereston* 1285 FA]. '*Hlōphere*'s valley' or rather 'the vagabond's valley'. OE *loddere* means 'a beggar, vagabond'.

Lotherton YW [*Luttringtun* 963, *Lutering(a)-tun* c 1030 YCh 6 f., *Luterington* c 1190 YCh 1613, 1225 FF]. The original form may well have begun in *Hl*-; cf. *Rypum* for *Hrypum* in YCh 7. If so, the first el. may be an OE *Hlūtringas* 'the people at a stream or spring called *Hlūtre*'. *Hlūtre* would mean 'the clean one' (from OE *hlūttor* 'clean'). Cf. LUTTERWORTH.

Lothing, Lake, Sf takes its name from the old **Lothing** (now Mutford) hd [*Ludinga* DB], which adjoins the hundred of **Lothingland** Sf [*Luthinglond* c 950 BCS 1008, *Ludingalanda* DB, *Luðingeland* 1198 P]. Lothing may be OE *Hlūdingas* or *Ludingas* 'the people of *Hlūd* or *Luda*', and Lothingland 'the (outlying) land of these'. If so, *th* must be due to Scandinavian influence.

Lothwaite or **Lowthwaite** Cu in Uldale [*Louthweit* 1230, *Lofthweit* 1245 P, *Loftthwayt*, *Loftethayt* 1398 Ipm], L~ Cu in Wythop [*Loftweic*, *-wic* 1195 FF]. An ON *Loft-þveit* 'clearing on a hill'. ON *loft* is used in pl. ns. in the sense 'hill'.

Loton Sa [*Luchetune* DB]. '*Luca*'s TŪN.' On OE *Luca* see LOCKINGTON.

Lottisham So [*Lottisham* 744, *Lottis-*, *Lotthesham* 842 BCS 168, 438, *Lotesham* 1238 FF]. '*Lott*'s HĀM.' OE **Lott* no doubt belongs to OE *lot* 'deceit, wile'.

Loud R La [*Lude* 1246 FF, *Loude* 1350 For]. OE *Hlūde* 'the loud one', derived from OE *hlūd* 'loud'.

Loudham Sf [*Ludham*, *Ludeham* DB]. '*Hlūda*'s HĀM.' **Hlūda* is a derivative of *hlūd* 'loud'.

Loudwater Bk [*la Ludewatere* 1241 Ass]. The place is on a stream, called *Ludewatir* c 1310 Godstow. 'Loud water.' Cf. LOUD.

Loughborough (lŭf-) Le [*Lucteburne* DB, *Lucteburga* Hy 2 DC, *Luchteburc* 1186 P]. '*Luhhede*'s BURG.'

Loughrigg (lŭf-) We [*Loukrig* c 1270, *Loghrygg* 1274 Kendale, *Loucrigg* 1275 Cl].

OScand *lauk-hryggr* 'ridge where leeks grew'. Leek is ON *laukr*.

Loughton (low-) Bk [*Lochintone* DB, *Lufton* c 1155 Oxf, *Parva Lughtone* 1219 Ep, *Lughton* 1237-40 Fees]. '*Luhha*'s TŪN.'

Loughton (low-) Ess [*Lukintone* 1062 Th, *Lochintuna* DB, *Luketon* 1225 Cl]. '*Luca*'s TŪN.' Cf. LOCKINGTON.

Loughton Sa [*Luchton* c 1138, *Luhtune* c 1225 Eyton]. Possibly '*Luhha*'s TŪN'.

Lound Li [*Lund* DB, 12 DC], L~ Nt [*Lund* DB, 1166 P], L~ Sf [*Lunda* DB, *Lund* 1254 Val]. OScand *lundr* 'grove'.

Loundthwaite Cu [*Lontwayt* 1255 P, *Lounethweyt* 1316 Ipm]. First el. dial. *lown* 'calm, quiet; shelter, sheltered place', from ON *logn* 'calm' &c. See THWAITE.

Louth (lowth) Li [*Hludensis monasterium* 790 ASC (F), *Lude*, *-s* DB, *Luda* 1093 RA, c 1115 LiS]. Named from the river LUD, whose name is identical with LOUD La.

Lovat R Bk, Bd [*Lovente* c 1200 Gervase, *Louente* 1262 Ass]. See LAVANT.

Loventor D [*Lovenetorre* DB, *Lavonatora* 1166 RBE]. '*Lēofwynn*'s tor.' *Lēofwynn* is a woman's name. See TORR.

Loversall YW [*Loures-*, *Geureshale* DB, *Luvereshale* 1207 FF, *Liureshal* 1198 P, *Liveressall* 1234 FF]. '*Lēofhere*'s HALH.'

Lovington So [*Lovintune* DB, *Louinton* 1187 f. P]. '*Lufa*'s TŪN.'

Low R Nb. See LOWICK.

Lowdham Nt [*Ludham* DB, *Ludeham* 1166 P, 1227 Ch, *Loudam* 1258 Ipm]. Identical with LOUDHAM. But the upper Cocker Beck might alternatively have been called LOUD.

Lowesby (-ōz-) Le [*Glowesbi* DB, *Lousebia* c 1125 LeS, *-by* 1220 Ep, 1254 Val]. The first el. is very likely OScand *lausa* 'slope', found (as *-lösa*, *-löse*) in many Scand pl. ns.

Lowestoft (-ōs-) Sf [*Lothu Wistoft* DB, *Lothewistoft* 1212 Fees, *Lowistoft* 1219 Fees]. '*Hloðvér*'s toft.' *Hloðvér* is an ON pers. n.

Loweswater (-ōz-) lake Cu [*Lawes-*, *Lausewatre* c 1203 StB, *Loweswatre* 1230 FF]. The lake gave its name to a place [*Lowswater* 12 StB, *Laweswator* 1188, *Laueswat'* 1190 P]. *Lowes-* probably represents OScand *Laufsær*, identical with LÖVSJÖN, a common name of lakes in Sweden, and meaning 'leafy lake'. To this was added *water* 'lake'.

Lowick (-ō-) La [*Lofwik* 1202 FF, *Laufwik* n.d. FC]. ON *Lauf-vik* 'leafy bay'. First el. ON *lauf* 'leaf'.

Lowick (-ō-) Nb [*Lowich* 1181 P, *Lowyc* 1242 Fees]. 'WĪC on R Low.' **Low** [*Low* c 1540 Leland] is dial. *low* 'a shallow pool left in the sand by the receding tide'. The word is used of several tidal streams in Nb. From the sense 'tidal pool' developed 'tidal stream'. The source is Ir, Gael *loch* 'lake, arm of the sea'. On the Low is Lowlynn [*Loulinne* 1208–10 Fees]. Second el. OE *hlynn*, dial. *linn* 'waterfall'.

Lowick (lŏĭk, lŭfĭk) Np [*Luhwic* DB, *Lofwyc* 12 NS, *Luffewich* 1167 P]. '*Luhha*'s or *Luffa*'s WĪC.' For OE *Luffa* see LUFFENHALL.

Lowlynn. See LOWICK Nb.

Lowman. See LOMAN.

Lowther (-owdh-) R We [*Lauther* c 1160 FC, *Louther* 1278 Ass]. Possibly a Brit river-name identical with LAUDER, the name of a place in Scotland. This is an old word for 'bath' (Gaul *lautro*-; cf. Brit *Lavatres*). OIr *lóthar* means 'a canal'. But Lowther may be a derivative of ON *lauðr* 'froth' and mean 'foaming river'. On the river is Lowther [*Lauder* c 1180, *Louther* 1195 Kendale, *Loudre* 1195 P].

Lowthorpe YE [*Log(h)etorp* DB, *Loutorp* 1234 FF]. '*Logi*'s thorp.' First el. ON *Logi*, OSw *Loghe* pers. n.

Lowthwaite Cu. See LOTHWAITE.

Lowton La [*Lauton* 1202 ff. P, 1212 Fees]. 'TŪN on a hill.' See HLĀW.

Lox Yeo R So [(on) *Loxan*, *Loxs* 1068 E]. Identical with Lox So, a lost name of a trib. of the Avon [(innan) *Loxan* 931 BCS 670]. A Brit river-name of doubtful etymology. On the Lox Yeo is Loxwood [*Loxanwuda* 956 BCS 959].

Loxbeare D [*Lochesbere* DB, *Lokeberga* 1196, *Lockesbere* 1205 FF], Loxhore D [*Lochesore* DB, *Lokesore* 1256 FF], Loxley St [*Locheslei* DB, *Lockesley* 1236 Fees], L~ Wa [*Locsetena gemære* 985 KCD 651, *Lockeslea* 11 Th, *Locheslei* DB]. '*Locc*'s BEARU or wood, ŌRA or bank and LĒAH.'

Loxton So [*Lochestone* DB, *Lokestone* 1212 RBE, *Loxton* 1259 FF]. As L~ is on LOX YEO, the name must be 'TŪN on Lox Yeo' in spite of the earliest forms.

Loynton St [*Levintone* DB, *Levynton* 1281 Ass]. '*Lēofa*'s TŪN.'

Lubbenham Le [*Lubanham*, *Lubeham*, *Lo-benho* DB, *Lubeho* 1147 BM, *Lubenho* 1208 BM, 1254 Val, *Lubenham* 1291 Tax]. '*Lubba*'s HŌH or hill.' The variant *-ham* is probably from a dat. plur. side-form *-hōm*, whence *-hŏm*, *-ham*. *Lubba* is a geminated form of *Lufa* (from *Luƀa*).

Lubbesthorpe Le [*Lupestorp* DB, *Lubestorp* 1229 Cl, *Lubbestorp* 1251 Cl]. '*Lubb*'s thorp.' *Lubb* is a strong side-form of *Lubba* in LUBBENHAM.

Luccombe So [*Locumbe* DB, *Loucumba* 1183 P, *Luuecumbe* c 1271 Dunster], L~ Wt [*Lovecombe* DB, *Louecumba* 1155 BM]. Perhaps '*Lufa*'s CUMB'. But the first el.

may be OE *lufu* 'love': 'valley where courting was done'.

Lucker Nb [*Lucre* 1170 P, *Lukre* 1242 Fees]. OScand *lō-kiarr* 'marsh frequented by sandpipers'. Cf. KERR. ON *ló* means 'sandpiper'.

Luckington So [*Lochintone* DB, *Lokintone* 1166 RBE, *-ton* 1201 Ass], L~ W [*Lochintone* DB, *Luchinton* 1195 Cur, *Lukintona* 1242 Fees]. 'The TŪN of *Luca* or his people.' Cf. LOCKINGTON.

Lucton He [*Lugton* 1185, 1193 P]. 'TŪN on R Lugg.'

Lud R Li [*Ludhena*, *Ludeney* 12 (1314) Ch, *Lude* c 1540 Leland]. OE **Hlūde* 'the loud one'; cf. LOUD, LOUTH. Alternatively the river was called *Hlūdan-ēa* 'the river Loud'. This form is preserved in Ludney [*Ludena* c 1115 LiS, *Ludenho* 1202 Ass]. On the Lud is also Ludborough [*Ludeburg* DB, *-burc* c 1115 LiS, Hy 2 Gilb]. 'BURG on R Lud' or 'BURG belonging to LOUTH'.

Ludbrook D [*Ludebroch* DB, *-broc* 1204 Cur]. Originally a stream-name: 'loud brook'.

Luddenham K [*Luddenham* 11 DM, *Lude-ham* 1212 RBE, *Lodenham* 1242 Fees]. '*Luda*'s HĀM.'

Luddesdown K [*Hludesduna*, *-dun* 10 BCS 1321 f., *Ledesdune* DB, *Ludesdon* 1186 P]. '*Hlūd*'s DŪN.' *Hludesbeorh* 939 BCS 741 may be an alternative name of the hill from which Luddesdown took its name, or it means '*Hlūd*'s burial-mound'. **Hlūd* is from OE *hlūd* 'loud'.

Luddington Li [*Ludintone* DB, *Ludingeton* 1229 Ep], L~ Wa [(æt) *Ludintune* c 1000 BCS 1318, *Ludingtun* 11 Th]. 'The TŪN of *Luda*'s people.' Cf. LODDINGTON.

Luddington in the Brook Np [*Lullingtun* 972–92 BCS 1130, *Lullintone* DB]. 'The TŪN of *Lulla*'s people.' The name was later influenced by LUTTON Np.

Ludford Li [*Lude(s)forde* DB, *Ludesfort*, *Ludeforda* c 1115 LiS]. 'Ford on the way to LOUTH.' The place is on the Bain west of Louth.

Ludford Sa nr Ludlow [*Ludeford* DB, Hy 3 BM]. *Lude* DB is held to refer to Ludlow. The place is on the Teme. *Lude* is probably OE **Hlūde* 'the loud one', referring to a rapid in the Teme. Ludford is 'ford by the rapid'.

Ludgarshall Gl nr Newington Bagpath [*Lutegareshale* 1220, *Letegareshale* 1280 PNGl], Ludgershall (lurg-) Bk [*Lutte-gersahala* Hy 2 (1285) Ch], L~ (lūg-) W [(æt) *Lutegaresheale* 1015 Wills, *Litlegarsele* DB, *Lotegarsal* Hy 1 (1268) Ch, *Lutegareshala* 1190 P, *-hal* 1203 Cur, *Lutegreshal* 1228 Cl, *Lutele-*, *Lutegrashale* 1281 QW], Lurgashall Sx [*Lutesgareshale* 12 Fr, *Lute-gareshal(e)* 1224 FF]. There is also a *Lutle-*, *Lutegreshale* 1281 Ass (Highworth W).

'Small *gærs-healh.*' OE *gærshealh* 'grass HALH' would mean 'grass corner or hollow, grazing-ground'. The second *l* was often lost owing to dissimilation. Sir Allen Mawer, *Studia neophilologica* 14, pp. 92 ff., gives good reasons for identifying *Lutegaresheale* 1015 Wills with Ludgershall W. The etymology of this pl. n. is much disputed. The editors of EPS take the first el. to be a pers. n. *Lutegar*. Tengstrand, pp. 219 ff., suggests an unrecorded OE *lūtegār* 'trap-spear'.

Ludgvan or **Ludjan** Co [*Luduha[n]* DB, *Luduhanum* 1087–91 Fr, *Luduon, -von* 1260 FF, *Ludgwon* 1291 Tax]. Possibly '(the church of) St. Ludwan or Ludowanus'.

Ludham Nf [*Ludham* 1021–4, 1044–7 KCD 740, 785, DB, 1253 Ch, *Ludeham* 1170 P]. '*Luda*'s HĀM.'

Ludlow Sa [*Ludelaue* 1138 HHunt, *-lawa* 1177 P]. 'Hill by the rapid.' See LUDFORD and HLĀW. Ludlow is on the Teme near Ludford.

Ludney Li. See LUD.

Ludstone Sa [*Luddesdon* 1250 Cl]. Identical with LUDDESDOWN.

Ludwell Db [*Lodouuelle* DB], L~ O in Wootton [*Ludewelle* DB, c 1130 Oxf], L~ W [*Ludewell* 1195 FF]. 'Loud stream.' Cf. LOUD.

Ludworth Db [*Lodeuorde* DB, *Ludewurda* 1185 P], L~ Du [*Ludeuurthe* 12 FPD]. '*Luda*'s WORÞ.'

Luffenhall Hrt [*Luffenheðe, -hale* 939 BCS 737, *Lufenelle, Lufenhate* DB], **North &** **South Luffenham** Ru [*Lufenham* DB, *Luffenham* 1167 P, Hy 2 DC, *Norlufeham* 1179 P, *Suthluffenhama* 1209–19 Ep]. '*Luffa*'s HALH and HĀM.' *Luffa is a geminated form of *Lufa*.

Luffield Bk [*Luffeld* 1180–3 FF, 1200 Cur]. '*Lufa*'s or *Luffa*'s FELD.'

Luffincott D [*Lughyngecot* 1242 Fees, *Loghingecote* 1284–6 FA, *Luffingecote* 1275 Ep]. 'The COT of *Luhha*'s people.'

Lufton So [*Lochetone* DB, *Luketun* 1227 FF, *-ton* 1340 BM]. '*Luca*'s TŪN.' Cf. LOCKING-TON.

Lugg, Welsh **Llugwy** R Sa, He [*Lucge* c 1000 Saints, *Lugge* 1231–3 Hereford, *Lhygwy* 1572 Lhuyd]. Identical with LLUGWY and LLIGWY, names of rivers in Wales. The name is derived from the base *leuk-* or *louk-* of Welsh *llug* 'light', Gk *leukós* 'white', and means 'bright stream'. **Lugwardine** He [*Lucvordine* DB, *Lugwurðin* 1168 P]. 'WORÞIGN on the Lugg.'

Lullingstone K [*Lolingestone* DB, *Lullingestan* 1200, *-ton* 1208 Cur]. '*Lulling*'s TŪN.'

Lullington Db [*Lullitune* DB, *Lullingtone* 1254 FF], L~ So [*Loligtone* DB, *Lullyngton*

1272 FF], L~ Sx [*Lullinton* 1192 P]. 'The TŪN of *Lulla* or his people.'

Lulsley Wo [*Lolleseie, Lulleseia* 12 VH]. '*Lull*'s island or river land.'

Lulworth Do [*Luluorde* DB, *Lullewurda* 1194 P, *Estlulleworth* 1285 FA, *Westlullewrth* 1258 FF]. '*Lulla*'s WORÞ.'

Lumb in Rossendale La [*Le Lome* 1534 Ct]. Dial. *lum* 'a well for the collection of water in a mine'. In the pl. n. the meaning is no douþt 'pool'. The same el. is found in **Lomax** La [*Lumhalghs* 1324 Ct].

Lumburn. See LAMERTON.

Lumby YW [*Lundby* 963, c 1030 YCh 6 f., *Lumby* c 1110 YCh 43]. 'BY at a grove'. See LUND. LUNDBY is a common name in Denmark and Sweden.

Lumley Du [*Lummalea* c 1050 HSC, *Lummelei, Lummesleie* c 1190 Godric, *Lumeleia* 12 Finchale]. 'LĒAH by the pool(s).' Cf. LUMB.

OE **lund** (in pl. ns.) from ON *lundr*, OSw *lunder* 'grove, copse' is the source of LOUND, LUND, LUNT. It is the first el. of LONDON-THORPE, LUMBY. As a second el. it is now always -LAND, as in BOY-, KIRK-, ROCK-, RUCK-, SNEL-, SWAN-, SWITH-, TIMBER-, TOSE-, UPSLAND. **Lund** La [*le Lund* a 1268 CC], L~ YE [*Lont* DB, *Lunde* c 1170 YCh 991], L~ YW [*le Lund* 13 Selby], **Lunt** La [*Lund* 1251 CC].

Lundy Island D [(Insula de) *Lundeia* 1189 TpR, 1199 Ch, (Isle of) *Lunday* 1281 Ch]. 'Puffin island.' ON *lundi* means 'puffin'. The name is Scandinavian. It is recorded as *Lundey* c 1145 Orkneyinga saga.

Lune R La, We [*Loin* c 1160 LaCh, *Lon* c 1160 Kendale, *Loon* 1186–90 CC, *Lone* 1202 FF], L~ R YN [*Loon* 1201 Ch, *Lon* 1235 FF]. A Brit river-name cognate with SLANEY in Ireland [olim *Sláine*], which is derived from OIr *slán* 'healthy, sound'. OIr *slán* is often found in names of springs and then means 'health-giving'. This is probably the meaning of the river-name, which is derived from an unrecorded Brit word corresponding to OIr *slán*.

Lunt La. See LUND.

Luppitt D [*Lovapit* DB, *Louepette* 1267 Ep]. '*Lufa*'s pit or hollow.'

Lupridge D [*Luperige, Kluperiga* DB], **Lupton** We [*Lupetun* DB, *Luppeton* 1199 Kendale]. Apparently '*Hluppa*'s ridge and TŪN'. The name *Hluppa is unexplained.

Lurgashall. See LUDGARSHALL.

Lusby Li [*Luzebi* DB, *Lucebi* c 1142 RA, 1176 P]. OScand *Lūts-bȳr. Lútr* is an ON pers. n.

Lushill W [*Lusteshull* 1166 RBE, *-e* 1242 Fees, *Lustreshell* 1240 Ch, *-hull* 1268 Pat]. The first el. is probably a reduced form of OE *lūsþorn* 'spindle-tree'.

Lustleigh D [*Leuestelegh* 1242 Fees, *Leuiste-, Luuestelegh* 1276 Ipm]. '*Lēofgiest*'s LEAH.' *Lēofgiest* is not evidenced.

Luston He [*Lustone* DB, *Luston* 1230 P]. Etymology doubtful.

Lutley St [*Luctelega* 1167 P, *Luteleg* 1199 P]. 'Small LĒAH.' OE *lȳtla lēah.*

Luton (-ōō-) Bd [*Lygetun* 792 BCS 264, *Lygtun* 917 ASC, *Loitone* DB, *Luituin* 1156 P]. 'TŪN on R LEA.'

Luton D in Bishopsteignton [*Leueton* 1238 Ass, *Luneveton* 1303 FA]. '*Lēofgifu*'s TŪN.' *Lēofgifu* is a woman's name. **L~** D in Broadhembury [*Levinton* 1227 Ch, *Liuetone* 1269 FF], **L~** K [*Leueton* 1275 Reg Roff, *Lyeueton* 1313 Ass]. '*Lēofa*'s TŪN.'

Lutterworth Le [*Lutresurde* DB, *Lutreworth* 1202 Ass, *Lutterwurth* 1242 Fees]. The first el. may be a river-name **Hlūtre*, derived from OE *hlūttor* 'clean, pure'. If so, *Hlūtre* was an old name of the SWIFT.

Lutton Li [*Luctone* DB, *Lochtona* c 1175 Middleton, *Lutton* 1236 Fees]. 'TŪN by a pool' (OE *luh* from Welsh *llwch* 'pool').

Lutton Np [*Lundingtun* 972–92 BCS 1130, *Ludintune* c 1060 BM, *Luditone* DB]. 'The TŪN of *Luda*'s people.'

Lutton, East & West, YE [*Ludton* DB, (cum) *duabus Luttunis* c 1110 YCh 25, *Lutton* 1166 P]. 'TŪN on R **Hlūde*.' See LOUD. The places are on a stream.

Lutwyche Sa [*Loteis* DB, *Lotwych* 1292 Ass]. Identical with *Lootwic* 717 BCS 137 (Wo). The first el. is no doubt identical with Du *loete*, LG *lōte* 'a shovel used to remove mud from ditches and canals'. The exact meaning of OE *lōt* cannot be determined. See WĪC.

Luxborough So [*Lolochesberie* DB, *Lollokesbourgh* c 1240 Dunster, *Lochesberge* 1150–61 Fr]. '*Lulluc*'s hill.' See BEORG.

Luxulian (lŭksi·lyan) Co [*Lauxsolian, -silyan* 1329 AD]. '*Sulian*'s monastery.' The first el. is identical with Welsh *lloc* 'monastery'. The second is a pers. n. identical with Bret *Sulian, Sulien,* OW *Sulgen.*

Lyd R D [*Lide* 1249 Ass]. OE *Hlȳde,* a river-name derived from *hlūd* 'loud' and meaning 'roaring stream, torrent'. *Hlȳde* must have been a very common name (see e.g. LYDE, LIDDINGTON, LITTON), and very likely OE *hlȳde* was a common noun for 'torrent'. See also LYDFORD.

Lydbrook Gl [*Ludebrok* 1282 For]. 'The river *Hlȳde*.' Cf. LYD.

Lydbury North Sa [*Lideberie* DB, *Leddebur*' 1212 Fees, *Lindeberinort* 1167 P, *Liddebiry North* 1223 Cl]. OE *Hlidaburg* 'BURG on the downs or slopes'. L~ is near LYDHAM (q.v.). The first el. is OE *hlid,* a sideform of HLIP 'slope'.

North for distinction from LEDBURY He.

Lydd K [*ad Hlidum* 774 BCS 214, *Hlide* 11 DM, *Lhida* Hy 2 (1313) Ch]. The dat. plur. of OE *hlid* 'slope' (see HLIP).

Lydden K nr Dover [*Hleodaena* 11 DM, *Liedenne* 1176 BMFacs, *Liedon* 1205 Cur], **L~** K in Thanet [*Ledene* 13 StAug], **L~ Valley** K nr Sandwich [*Lydene* 1278, *Lhydene* 1313 Ass]. OE *hlēo-denn* 'pasture with a HLĒO or shelter'.

Lyde He [*Leode, Lude* DB, *Luda* 1173 Hereford], **L~** So [*Eslide* DB, *Lude* 1236 Fees]. Both are named from streams which must have been called *Hlȳde.* See LYD.

Lydeard St. Lawrence & Bishop's L~ So [*Lid(e)geard* 854, 904 BCS 476, 610, *Lidegeard* 1065 Wells, *Lidigerd* 11 KCD 897, *Lidiard, Lediart* DB, *Lydiard Sancti Laurencii,* **L~** *Episcopi* 1291 Tax], **Lydiard Millicent & Tregoze** W [*Lidgeard, æt Lidgerd, Lidegæard* 901 BCS 590, *Lidgeard* 901–24 ib. 591, *Lidiarde* DB, *Lydyerd Milsent* 1291 Tax, *Lidiard Tregoz* 1196 Ewias]. Lydeard and Lydiard are situated by prominent hills, to which the name was no doubt originally applied. The same name also occurs in *Lidgeardes beorg* BCS 834, 1125 (Sx), which means 'Lidgeard hill'. The name is British and identical with *Litgart (-garth)* c 1150 LL (in Wales), whose second el. is Welsh *garth* 'hill'. The first el. is not clear.

Bishop's Lydeard belonged to the Bishop of Wells.—**L~ St. Lawrence** from the dedication of the church.—**Lydiard Millicent** from a lady named Millicent.—**L~ Tregoze** came to Robert Tregoz by marriage before 1194 (Ewias). See EATON TREGOSE.

Lydford D [(to) *Hlidan* c 910 ASCh (Burghal Hidage), *Hlydanford* 997 ASC (C, D), *Hlidaford* ib. (E), 1018 Crawf]. 'Ford over the LYD.' The place appears originally to have been (æt) *Hlȳdan* from the Lyd.

Lydford, East & West, So [*Lideford* DB, 1194 P, *Ludeford* 1194 f. P, *Estludeford* 1291 Tax, *Westludeford* 1243 Ass]. 'Ford over the torrent.' Cf. LYD. L~ is on the Brue.

Lydham Sa [*Lidum* DB, *Lidun* 1267 Ch, 1272 Ipm, *Lideham* 1250 Ipm]. The dat. plur. of OE *hlid* 'a slope'. Cf. HLIP and LYDBURY NORTH.

Lydiard W. See LYDEARD.

Lydiate La [*Leiate* DB, *Liddigate* 1202 FF, *Lidiate* 1212 Fees]. OE *hlidgeat* 'swing-gate'.

Lydley Heys Sa [*Litlega* DB, *Lidlegee* 1185 TpR]. 'Small LĒAH.' See (GE)HÆG.

Lydlinch Do [*Litelinge* 1166 RBE, *Lidelinz* 1182 P, *-linch* 1285 FA]. 'Hill on R LIDDEN.' See HLINC.

Lydney Gl [*Lidaneg* 972 BCS 1282, *Ledenei* DB, *Lideneie* 1221 Ass]. '*Lida*'s island.' **Lida* is identical with OE *lida* 'sailor'. But Lydney may be 'the sailor's island'.

Lye He [*Lecwe*, *Lege* DB, *Lege* 1242 Fees], L~ Wo [*Lega* 1275 Subs]. Identical with LEIGH.

Lyford Brk [*Linfordinga gemære* 940, *æt Linforda* 944 BCS 761, 798, *Linford* DB, *Liford* W 1 Abingd]. 'Ford where flax grew.' See LĬN. For the loss of *n* cf. STOFORD.

Lyham Nb [*Leum* 1242 Fees, *Lyum* 1256 Ass, *Leyham* 1269 FF, *Leyum* 1279 Ass, *Lyhum* 1289 Ipm]. OE *lēah-hamm* 'HAMM by a LĒAH' or OE *lēagum*, dat. plur. of LĒAH.

Lyme Hall, L~ Handley Chs [*Lyme* 1313 Ipm]. Lyme is the old name of a large forest district which is often found as a distinctive addition to pl. ns., as in ASHTON UNDER LYNE La, NEWCASTLE UNDER LYME St. It is the second el. of AUDLEM Chs, BURSLEM St, and the parts of the Honour of Lancaster outside Lancashire are often referred to in early sources as the honour *extra limam*, i.e. 'beyond the Lyme'. The forest of Lyme included Macclesfield Forest. Lyme is a Brit name derived from the word for 'elm' found in Welsh *llwyf* &c. Cf. LYMN.

Lyme R D, Do [*Lim* 774 BCS 224, *Lym* 938 ib. 728]. A Brit river-name, very likely identical with Welsh *llif*, Co *lif* 'flood, stream' (probably from *lim*) and related to LEEN (see also LEINTHALL). On the Lyme are Lyme Regis Do [*Lim* 774, *at Lym* 938 BCS 224, 728, *Lime* DB, *Lyme Regis* 1285 (1321) Ch], and Uplyme D [*Lim* DB, *Uplim* 1238 Ass].

Lyminge (līmĭnj) K [*Liminge* 689, *Limingae* 697, *Liminiaeae* 740, *aet Liming(g)e* 798 BCS 73, 97, 160, 289, *Leminges* DB]. OE *Limengē* 'the district on R *Limen*' (see LYMPNE). Lyminge once denoted a large district, but was restricted to its chief place. Lyminge is not on the river. Cf. EASTRY.

Lymington (lĭ-) Ha [*Lentune* DB, *Limington* 1186, *Liminton* 1196 P, *Lemneton* Hy 3 Ipm]. The stream here was probably once *Limen*. Cf. LYMN.

Lyminster (līmster) Sx [*Lullyngmynster* c 880 BCS 553, *Lolinminstre* DB, *Limenistr'* 1202 Cur]. 'The church or monastery of *Lulling* or of *Lulla*'s people.'

Lymm Chs [*Lime* DB, *Limme* 1260 Court]. Lymm is on a stream, which may have been OE **Hlimme* 'roaring brook', formed from *hlimman* 'to resound, roar'.

Lymn (līm) R Li [*Limene*, *Lime* 12 (1331) Ch, *Limine* 1276 RH], Lympne, an old name of the East Rother Sx, K, which formerly fell into the Channel at Rye, and whose old course is marked by the Royal Military Canal [*Liminaea* 697, *Liminæa* 740, *Limenæ*, *Limen* 724 BCS 98, 160, 141, *Lymene* 1241 Ass]. A Brit river-name, identical with LEAM and LEMON, also with LEVEN (Gael *Leamhain*) in Scotland. The name is de-

rived from the Celtic word for 'elm' found in OIr *lem*, Welsh *llwyf*. From this are derived Gaul *Lemonum*, (lacus) *Lemannus* 'Lake Leman' and others. The OBrit base was *Lemanā* (cf. the forms for Lympne town), which became OE *Limene* or (with velar mutation) *Leomene* (in LEAM, LEMON). From the river-name Lympne is derived the name of Lympne (līm) town K [*portus Lemanis* 4 IA, *Lemannis* c 425 ND, (of) *Liminum* 805–10, *et Liminum* 811 BCS 330, 332]. This name is in the plural form (*Lemanæ*, dat. *Lemanis* &c.).

Lympsham So [*Linpelesham* 1225 Ass, *Limpelesham* 1254 Val]. The first el. might be a pers. n. (cf. LIMPENHOE) or a compound such as *lind-pyll* 'pool where lime-trees grew'.

Lympstone D [*Levestone* DB, *Leveneston* 1238 Ass, *Lumeneston* 1291 Tax]. '*Lēofwine*'s TŪN.' The name shows a change of *vn* to *mn* and *m*.

Lyn R D, So [*Lyn* 1282 Ass]. OE *hlynn* 'torrent'. The river has two arms, the East and the West Lyn, from which are named East and West Lyn D [*Line* DB, *Lyn* 1242 Fees, *Est-*, *Westlyn* 1303 FA]. From the Lyn are also named Lynmouth D [*Lynmouth* 1330 Subs] and Lynton D [*Lintone* DB].

Lynch, East, So [*Linz* 1259, *Estlinche* n.d. Buckland, *Lynche* 1325 Ipm]. OE HLINC 'hill'.

Lyndhurst Ha [*Linhest* DB, *Lindeherst* 1165 P]. 'Lime-wood.' See HYRST.

Lyndon Ru [*Lindon* 1167, 1197 P, *Lindone* 1230 Ep]. 'Lime-tree hill.' Cf. LIND, DŪN.

Lyne, Black & White, R Cu [*Leuen, -e* 1292 Ass, *Levyn* 1383 Pat]. Identical with LEVEN. On the river are Kirklinton and Westlinton [*Leuentona* 1188 P, *Kirklevyngton* 1332 Subs, *Westleventon* 1250 Ipm].

Lyne R Nb [*Lina* c 1050 HSC, 1137 ff. Newminster]. A Brit river-name perhaps related to LEEN.

Lyne Sr in Chertsey [*la Linde* 1208, *la Lynde* 1306 FF], L~ Sr in Newdigate [*La Linde* 1185 P]. OE *lind* 'lime-tree'.

Lyneham O [*Lineham* DB, *Linham* 1236 Fees], L~ W [*Linham* 1285 Ch, *Lynham* 1291 Tax]. 'HĀM or HAMM where flax was grown.' Cf. LĬN.

Lynesack Du [*Lynesak* 1307 RPD]. If, as seems probable, the second el. is OE *āc* 'oak', the first is no doubt a pers. n., e.g. *Lēofwine*.

Lynford Nf [*Lineforda* DB, *Lineford* 1197 P, *Linford* 1252 Cl]. 'Flax ford' or 'ford on the road to Lynn'. The place is south-east of Lynn.

Lyng Nf [*Ling* DB, c 1160 Holme, *Lins* 1157 YCh 354, *Ling* 1254 Val]. OE HLINC 'hill' with voicing of *c* as in LYNG So.

Lyng So [(to) *Lengen* c 910 ASCh (Burghal Hidage), *Relengen* 937 BCS 715, *Lege* DB, *Lenga* c 1180 Buckland, *Leng* 1225 Ass]. OE *hlenc* 'hill' (cf. LENCH) with voicing of *c*. The form of 937 is in a poor copy and evidently corrupt.

Lynher (līner) R Co [*Linar* 1018 JAA xxxix, *Liner* 1125 WMalm], **Lynor** R D [*Linor* 958 BCS (1027), c 1160 ERN]. A Brit river-name, related to LYNE Nb.

Lynmouth. See LYN.

Lynn, King's, North, South, & West, Nf [*Lynware* (hd) 11 EHR 43, *Lena, Lun* DB, *Lynna* c 1105, *Linna* c 1140 BM, *Lenn* c 1095 Bury, 1196 FF, *Luna* 1121 AC, *Lenna* 1160 P, *Nordlen* 1199 FF]. Very likely identical with *Lindon*, the original name of LINCOLN (Brit *lindo-*, Welsh *llyn* 'lake'), though in a later form. The name would then refer to a pool at the mouth of the Ouse. *Lynware* means 'Lynn people'.

Lynor R. See LYNHER.

Lynton D. See LYN.

Lyonshall He [*Lenehalle* DB, *Lenhal,* *-es* 1227 ff. Ch]. First el. as in LEOMINSTER. Second el. HALH, here 'valley'. Lyonshall is a long way from Leominster. It is not far from the Arrow.

Lypiatt Gl [*Lippegat* 1220 Fees, *Lipegate* 1287 Ipm]. Cf. *hlypgeat* 972 BCS 1282. The name is identical with *leapgate* 'a low gate in a fence that can be leaped by deer, while keeping sheep from straying'. The same is the origin of **Lypiate** So [*Lupiat, la Lypiat* 1242 P].

Lyscombe Do nr Milton Abbas [(at) *Liscombe* 939 BCS 738, *Liscome* DB, *Liscumb* 1212 Fees]. The first el. is the *lisc* found in *Liscbroc* 1019 KCD 730, the name of a stream near Lyscombe. OE *lisc* is identical with OHG *lisca*, Du *lisch* 'reeds' and the like. See CUMB.

Lytchett Matravers & Minster Do [*Lichet* DB, *Litsed* 1236, *Lischet* 1242 Fees, *Lyceministr'* 1253 FF, *Lechet Ministre* 1269 FF, *Luchet Mautravers* 1291 Tax]. Probably identical with the first el. of LICHFIELD, i.e. Brit *Lētocēto-* 'grey wood'.

L~ **Matravers** was held by Hugh Maltrauers in 1086 (DB). Cf. LANGTON MATRAVERS.— Minster may mean 'church'.

OE **lȳtel** 'small' is a common first el. in pl. ns. Cf. LITTLE- (passim). Sometimes the second *l* was lost owing to dissimilation. Cf. LITCHURCH, LUDGARSHALL &c.

Lyth (lēth, līth) We [*Le Lyth* 1247, *Lith* 1301 Kendale]. ON *hliδ* 'slope'.

Lytham (-ĭdh-) La [*Lidun* DB, *Lithum* 1201 P]. The dat. plur. of OE *hliþ* 'slope'.

Lythe YN [*Lid* DB, 1195 P, *Lith* 1194 P, 1225 Ep]. ON *hliδ* 'slope'.

Lythwood Sa [*Lia* 1199 P, *Lythewod* 1250, *La Lithewode* 1280 Cl]. 'Wood on a slope' (OE *hliþ*).

Lyveden Np [*Lieueden, Luuedene* 1178 P, *Liveden* 1220 Fees, *Liueden* 1230 P]. '*Lēofa's* valley.'

Lyve·nnet R We [*Leveneth* 13 CWNS xi, *Lyuened, Leuenyd* 1292 Ass]. A Brit river-name.

M

Mabe Co. Probably a saint's name.

Mablethorpe Li [*Malbertorp* DB, 1209–19 Ep, *Maltorp* DB, c 1115 LiS, *Malbretorp* 12 DC]. '*Malbert's* thorp.' *Malbert* is an OFr pers. n. of OG origin.

Macclesfield Chs [*Maclesfeld* DB, c 1150 Chester, *Macclesfeld* c 1100 Chester, *Makelesfeld* 1183 P]. M~ was a great forest. The first el. may be an old name of the forest, identical with **Mackley** Db [*Makelai* c 1150 Mon, *Mackele* 1210 Cur, *-leg* 1252 FF], which may mean '*Macca's* forest'. Macclesfield would then be 'open land in Mackley forest'. But the two places are too far apart for it to be probable that the same forest gave both their names.

Macefen Chs [*Masefen* 1260 Court, E 1 Ormerod]. First el. perhaps Welsh *maes* 'field'. Or it may be OE *māse* 'titmouse'. See FEN.

Mackley. See MACCLESFIELD.

Mackney Brk [*Maccaneig* 891, *-ig* 945 BCS 565, 810, *Makeni* 1196 FF], **Mackworth** Db [*Macheuorde* DB, *Mackeworth* 1211 Cur]. '*Macca's* island and worp.' Cf. *Macan broc* 949 BCS 880 and OG *Maco.*

Maddington or **Maiden Winterbourne** W [*Maidenewinterburn* 1205 Ch, *Medinton* 1198 Fees, *Madintone* 1212 RBE]. Originally WINTERBOURNE. M~ is 'the TŪN of the maidens'. First el. OE *mægden* (*mæden*) 'maiden', here 'nun' (of Amesbury).

Madehurst (-ăd-) Sx [*Medhurst* 1255 FF, *Madhurst* 1279 Ass]. OE *mǣdhyrst* 'hurst by or with a meadow'. Forms such as *Meslirs* c 1150 Fr, *Medliers* 1188 P are due to association with OFr *medlier* 'medlar'.

Mādeley Sa [*Madelie* DB, *Madelega Prioris* 1167 P], M~ St nr Newcastle [*Madanlieg* 975 BCS 1312, *Madelie* DB], M~ St nr Uttoxeter [*Madelie* DB, *-leye* 1176 FF]. '*Mada's* LĒAH.' **Mada* may be a nickname formed from OE *mād* 'foolish'. M~ Sa is on **Mad Brook** [*Madebroc* 13 ERN]. The two Madeleys St are c 15 m. apart on opposite sides of an upland district. Very likely this was *Mādanlēah* '*Māda*'s forest'.

Madely Gl nr Gloucester [*Methlegh* 1234 Cl]. OE *mǣþ-lēah* 'glade where mowing was done'.

Madingley Ca [*Matingeleia* c 1080 ICC, *Mading(e)lei* DB, *Maddingelea* 1193 P]. 'The LĒAH of *Māda*'s people.' Cf. MADELEY.

Madjeston Do [*Malgereston* 1206 Cl, *Maugereston* 1266 FF]. '*Malger*'s TŪN or manor.' *Malger* is an OFr pers. n. of OG origin.

Madley He [*Medelagie* DB, *Matle* ('hoc est bonus locus') c 1150 LL, *Maddeleia*, *Madele* c 1200 Hereford]. OW *matle* means 'good place' (first el. OW *mat*, Welsh *mad* 'good, beneficial', second Welsh *lle* 'place'). The English forms show association with OE LĒAH.

Mădresfield Wo [*Madresfeld* c 1086 PNWo, *Metheresfeld* 1192 P]. OE *mǣþeresfeld* 'the mower's FELD'.

Madron (mădren) Co [(Ecclesia) *Sancti Maderi* 1205 Cur, (ecclesia) *Sancti Maderni* 1276 Ep]. '(The church of) St. Madernus.'

OE **mǣd** (*mēd*), dat. -*we* 'meadow' is not a common pl. n. el. It usually appears without the *w*, as in BREIGHTMET, BUSH-, HARDMEAD, RUNNYMEDE, MEDBOURNE, MED-LOCK, METFIELD, MADEHURST. An exception is SHIPMEADOW.

OE **mǣddre**, ON **maðra** 'madder'. See MAYFIELD St, MATTERDALE.

OE **mǣgden**, **mǣg(e)þ** 'maiden, girl'. See MADDINGTON, MAID-, MAIDEN-. The el. is sometimes difficult to distinguish from **mǣgþe** 'mayweed', found in MAYFIELD Sx, MAYTHAM. See MAIDSTONE, MAYFORD.

OE **mǣl** 'mark, cross' occurs in MALDEN, MALDON, MAULDEN, MELDON. A related adj. *mǣl* or *mǣle* is the first el. of MELBURY, MILLBARROW and perhaps some other names.

OE **(ge)mǣne** (obl. also **gemāna**) 'common'. See e.g. MAN(A)TON, MANGREEN, MENWITH. OE **(ge)mǣnness** 'community'. See EWELL (MINNIS). OE **(ge)mǣnscipe**. See MINSKIP.

Maer St [*Mere* DB, 1242 Fees]. OE *mere* 'lake'. The place is near a tarn.

OE **(ge)mǣre** 'boundary' is the first el. of MAESBROOK, -BURY, MARPLE, MARRICK, MARWELL, MEARLEY, MERRIDGE, MERSEY, MEERS-BROOK and some other names. It is difficult to distinguish it from *mere* 'lake', but the latter is far more common. Second el. in LANDERMERE, UDIMORE and some others. Cf. DID-, TORMARTON Gl.

Maesbrook Sa [*Meresbroc* DB], **Maesbury** Sa [*Meresberie* DB, *Mersburi* 1272 Ipm]. OE *Mǣres-brōc* and -*burg* 'brook and BURG

on the boundary'. Both are nr OFFA'S DYKE and in the old *Mersete* hundred [*Mersete* DB = OE *Mǣrsǣte* 'boundary people').

Maesbury Camp. See MARKSBURY.

OE **mǣþ** 'mowing' is found in MADELY, METHAM.

Maghull (magŭ·l, *olim* māl) La [*Magele* DB, *Maghele* a 1190 CC, *Maghal* 1219 Ass]. Second el. HALH. The first is probably OE *mǣgþe* 'mayweed'. Cf. early forms of MAYFIELD Sx.

Maidenhead Brk [*Maydehuth* 1248 Ch, *Maydenhythe* 1428 FA]. 'The maidens' landing-place.' Cf. HȲÞ.

Maidenwell Li [*Welle* DB, *Maidenwell* 1212 Fees], **Maidford** Np [*Merdeford* DB, *Maideneford* 1167 P, *Meideford* 1200 Cur]. 'The maidens' spring and ford.'

Maidstone K [*Mǣidesstana*, *Mǣgþan stan* 10 BCS 1321 f., *Meddestane* DB, *Maegdestane* 11 DM]. Probably 'the maidens' stone'. One OE form seems to suggest the word *mǣgþe* as the first el., but 'mayweed stone' gives no good meaning. Probably the original form was *mǣgþa-stān*, which came to be misunderstood. See MÆGDEN.

Maidwell Np [*Medewelle* DB, *Maidewell* 1198 P]. 'The maidens' spring or stream.'

Mainsforth Du [*Maynesford* 1183 BoB, 1304 Cl, *Maineford* 1208-10 Fees]. The first el. may be a pers. n. *Mǣgen*, a short form of *Mǣgenfriþ* &c.

Mainstone Ha in Romsey [*Maihiweston* 1242 Fees, *Mayhueston* 1346 FA]. '*Mayhew*'s TŪN.' *Mayhew* is from a French form of *Matthew*.

Mainstone He [*Maineston* 1206 Cur, *Mayneston* 1242 Fees]. The first el. is very likely a Fr family name.

Mainstone Sa [*Meyneston* 1284 Cl]. OE *mǣgenstān* 'big rock'.

Maisemore Gl [*Mayesmora* 1138 Glouc, *Maismora* 1167 P, *Meismore* 1221 Ass]. First el. Welsh *maes* 'plain, field'. The second may be Welsh *mawr* 'large' ('large field') or OE *mōr* 'moor'.

Makeney Db [*Machenie* DB, *Makeneye* 1236 Fees]. '*Mac(c)a*'s island.' Cf. MACKNEY.

Maker Co [*Macretone* DB, *Macre* 1202 FF]. Identical with Welsh *magwyr*, OW *macyrou* plur., OBret *macoer* 'wall, ruin', from Lat *macēries* 'wall'. *Macretone* may denote a place in Maker. The name may refer to ruins of an ancient building.

Makerfield La, an old district; cf. ASHTON, INCE, NEWTON IN M~ [*Macrefeld* 1121 La Ch, *Makerefeld* 1213 P]. 'Open land by a wall or ruin.' Cf. MAKER.

Malborough (mawl-) D [*Malleberge* 1270 FF, *Merleberg* 1275 RH]. Identical with **Marlborough** (mawl-) W [*Merleberge* DB,

Mærle beorg 1110 ASC (E), *Merleberga* 1130 P]. Possibly '*Mærla*'s hill', *Mærla being a short form of names such as *Mærheard*. But the double occurrence of the name suggests a significant first el., which may be OE *meargealla*, *mergelle* 'gentian'. The same may be the first el. of **Marwell** D [*Merlewill* 1259 Ipm].

Malden Sr [*Meldone* DB, *Maldone* 1279 QW], **Maldon** Ess [*Mældun* 913 ASC, *Mal-*, *Melduna* DB, *Mealdona* 1130, *Maldon* 1160 P]. 'Hill with a *mēl* or monument or cross.'

Malham YW [*Malgun* DB, *Malghum* 1208 FF, *Malgum* 1257 Ch]. The dat. plur. of a Scand word related to the Swedish lake-name MALJEN (from *Malghe*), ON *mǫl* 'gravelly soil', *melr* 'sandbank' &c. The exact Scand base is not clear (an OScand *malg-* or an adj. *maligr*, in a def. form *malgi*?). The name means something like 'stony or gravelly place'. At Malham are **M~ Moor** [*Malghemore* Fount] and **M~ Tarn** [*Malge-*, *Malhewater* ib.].

Malins Lee Sa [*Malineleg* 1262 For, *-lee* 1300 BM]. Originally no doubt LEE. The place is nr LEEGOMERY. Malin was an early owner, a lady.

Mallerstang We [*Malrestang* 1223 FF, *Malverstang* 1228 Pat]. The second el. is ON *stǫng* 'a pole', probably a boundary mark. The first is identical with MELLOR and Welsh *Moelfre* 'bare hill'. Cf. MALVERN.

Malling (-aw-), **East & West**, K [*Meallingas*, *East Meallinga gemære* 942–6 BCS 779, (of) *Meallingan* 10 ib. 1322, *Mellingetes* DB], **South M~** Sx [(æt) *Mallingum*, *Meallinges in Suthsexan* 838 BCS 421 f., *Mellinges* DB, *Mallinges* 1212 RBE, *Suthmelling* 1232 Pat]. '*Mealla*'s people.' *Mealla is related to OG *Malo*, *Mello*, *Mallobaudes* &c.

Malmesbury (-ahmz-) W [*Maildufi urbs* c 730 Bede, *Maldulfes burgh* c 890 OEBede, *M(e)aldumesburg* 675 BCS 37, (on) *Ealdelmesbirig* c 1000 Saints, *Mealdelmes byrig* 1015 ASC (E), *Malmesberie* DB]. Many other forms are found in copies of OE charters. The original form was no doubt *Maldufesburg*. Cf. *Maildubiensis æcclesia* 892 BCS 569. *Maildu(l)f*, a Scot, is stated by William of Malmesbury to have founded the monastery. The correct form would be OIr *Maeldub* or *Maelduib* (*b* pronounced as *v*). Later the name was modified in various ways. Sometimes the place was called *Ealdhelmesburg* after *Aldhelm*, who is said to have been abbot of Malmesbury. Malmesbury may be a compromise between *Maldulfesburg* and *Aldhelmesburg*.

Malpas (-awl-) Chs [*Malpas* c 1125 Chester, (ecclesia de) *Malo passu* 1291 Tax]. A French name meaning 'bad (difficult) passage'. The same name is found in Co and Monm.

Maltby Li nr Louth [*Maltebi* DB, c 1115

LiS], **M~ le Marsh** Li [*Maltebi* DB, c 1115 LiS], **M~ YN** [*Maltebi* DB], **M~ YW** [*Maltebi* DB]. '*Malti*'s BY.' ODan *Malti* is a common name.

Malton YN [*Maltune* DB, *Maaltun* c 1150 SD, *Mealton*, *Vetus M~* 1190 P]. OE *Middeltūn*, Scandinavianized to *Meðaltūn* and *Miaðaltūn*.

Malvern, Great, Wo [(in, ondlang) *Mælfern* c 1030 Förster, *Themse*, p. 769, *Malferna* DB, *Maluernia* 1130 P, *Magna Malverna* 1228 Cl], **Little M~** [*Parva Malvernia* 1232 Cl]. Named from **Malvern Hill**, which had a Brit name meaning 'bare hill', a Welsh *Moel-fryn* (*moel* 'bare' and *bryn* 'hill'). *Moel* was OW *mēl* from **mailo-*. Engl *Mal-* represents some intermediate form between *mailo-* and *moel*.

Mamble Wo [*Momela gemæra* c 957 BCS 1007, *Mamele* DB, 1232 FF]. *Momela* in the ex. of c 957 appears to be the gen. of a folk-name, e.g. *Momele*, which is probably derived from a Brit name of the place. This name is very likely connected in some way with the word *mam* 'hill' discussed under MAMHEAD.

Mamhead D [*Mammeheve* DB, *-havede* 1242 Fees]. The second el. is OE HĒAFOD 'hill'. The first is the el. *Mam-* found in several hill-names, as **Mam Tor** Db. See also MANSFIELD. *Mam* is common in Irish and Scotch names of hills, and the source is here Ir *mamm* 'breast'. Very likely Welsh *mam* ('mother, womb') was used in the same sense. OE *mamme* 'a teat' is a Latin loan-word, and it is doubtful if it was a popular word.

Mana·ccan Co [(Ecclesia) *Sancte Manace in Menstre* 1309 Ep]. The saint's name implied is of obscure history.

Manaton D [*Magnetone*, *Manitone* DB, *Manneton* 1200 P, *Maneton* 1269 FF]. '*Manna*'s TŪN' or OE *gemæna* (*gemāna*) *tūn* 'common TŪN'.

Manby Li nr Louth [*Mannebi* DB, *Magnebi* 1212 Fees], **M~** Li nr Brigg [*Mannebi* DB, c 1115 LiS, *Manneby* 1257 Ch]. '*Manni*'s BY.' First el. ODan, OSw *Manni*.

Mancetter Wa [*Manduessedo* (abl.) 4 IA, *Manacestre* 1196 FF, *Manecestre* 1236 Fees]. The first el. is a reduced form of OBrit *Manduessedon*, to which was added OE CEASTER.

Manchester La [*Mamucio* (abl.) 4 IA, *Mameceaster* 923 ASC, *Mamecestre* DB, *Manchestre* 1330 LaInq]. OBrit *Mamucion*, to which was added OE CEASTER.

Mānea Ca [*Moneia* 1177, *Maneia* 1178 f., *Moneya* 1178 P]. Second el. OE *ēg* 'island'. The first may be OE *gemǣne* (*gemāna*) 'common'. Formally OE *manu* (*monu*) 'mane' would be more suitable, but it is not easy to see what it would mean here. Cf., however, FAXTON.

Manfield YN [*Mannefelt* DB, *-feld* 1202 FF, *Manefeld* 1228 FF]. '*Manna*'s FELD' or, as indicated by numerous early spellings with single *n*, as *Manafeld* 1159–71, *Manefeld* 1173–86 YCh iv, more likely OE *gemāna feld* 'common field'.

Mangerton Do [*Mangerton* 1207 Cur, 1318 FF]. OE *mangera tūn* 'the traders' TŪN'.

Mangotsfield Gl [*Manegodesfelle* DB, *-feld* 1167 P, *Mangodesfeld* 1231 Ch]. '*Mangod*'s FELD.' *Mangoda* is found BCS 1309, *Manegot* DB. The name may be Continental.

Mangreen Nf [*Manegrena* DB, *Mangrene* 1395 AD]. 'The common green.' First el. OE *gemǣne* 'common'.

Manifold R St [*Manifold* c 1540 Leland]. 'Winding river' (lit. 'with many folds'). For OE **manig** see also MANNINGTREE, MONYASH.

Manley Chs [*Menlie* DB, *Manleye* 1283 Ipm]. 'Common wood.' Cf. MANGREEN, LĒAH.

Manningford Abbots, Bohun & Bruce W [*Maningaford* 987 Hyda, *Maniford* DB, *Manningeford* 1212 Cur, M~ *Abbatis* 1291 Tax, *Manyngford Brewose* 1297 Pat, *Maningford Boun* 1316 FA]. 'The ford of *Mann(a)*'s people.'
M~ **Abbots** belonged to the Abbey of St. Peter, Winchester.—*Bohun* and *Bruce* are family names derived from BOHON and BRIOUZE in Normandy. Henry de Boun held M~ Bohun in 1212 (Cur).

Manningham YW [*Maningeham* 1249 Ch, *Maynigham* 1298 Wakef, *Mayningham* c 1304 Calverley]. 'The HĀM of *Mægen*'s people.' Cf. MAINSFORTH.

Mannington Do [*Manitone* DB, *Maniton* 1242 Ch], M~ Nf [*Manninctuna* DB, *Manington* 1254 Val]. 'The TŪN of *Mann(a)*'s people.'

Manningtree Ess [*Manitre* 1274 RH, *Manyngtre* 14 BM]. Apparently 'many trees.' Cf. MONYASH. Or '*Manning*'s tree' or 'the tree of *Mann(a)*'s people'.

Mansell Gamage & Lacy He [*Malueshyll* a 1056 KCD 802, *Malveselle, Malveshille* DB, *Maushil* 1169, *Mauneshulla* 1194, *Maweshull* 1198 P, *Maumeshull Gamages, Lacy* 1242 Fees]. The first el., like that of MAWSLEY, is an OE *malu (gen. malwes) 'sand, gravel, gravel hill' or the like, related to ON *møl* 'gravelly soil', Sw *mal* 'stones, gravel'. OE *Malwes-hyll* means 'gravel hill'. *Malwes-* became *Malmes-* owing to a kind of assimilation or to association with OE *mealm* 'sand' or the like.
M~ **Gamage** was held by Matthew de Gamagis in 1194 (P); *Gamagis* from GAMACHES in Normandy.—M~ **Lacy** was held by Roger de Lacy in 1086 (DB). Cf. EWYAS LACY.

Mansergh (-*zer*) We [*Manzserge* DB, *Mannissergh* 1206 FF]. '*Man*'s ERG or shieling.' First el. OScand *Man* (gen. *Mans*).

Mansfield Nt [*Mamesfeld* DB, *Mammesfelt* 1093 RA, *Mamefeld* 1130 P, *Mammefeld* 1212 Fees]. The first el. is no doubt the name of a hill; cf. *Mammesheved* 1232 For, which means 'the top of *Mam* hill' or '*Mam* hill'. Cf. MAMHEAD. Mansfield is 'open land by *Mam*'. The river-name **Maun** is a back-formation. **Mansfield Woodhouse** [*Wodehuse* 1230 P, *Maunsfeld Wodehus* 1289 Ipm].

Mansriggs La [*Manslarig* 1520 VH, *-rigges* 1539 FC]. The first el. may be ON *manslagari* or OE *manslaga* 'homicide'. Second el. OScand *hryggr* 'ridge'.

Manston Do [*Manestone* DB, *Manneston* 1236, 1242 Fees], M~ K [*Mannestone* 1285 Misc], M~ YW [*Manestune, Mainestune* DB, *Manston* 1285 FA]. '*Mann*'s TŪN.'

Manthorpe Li nr Witham on the Hill [*Mannetorp* DB], M~ Li nr Grantham [*Mannetorp* 1185 TpR, 1212 Fees]. One of these is *Mannethorp* c 1067 Wills. '*Manni*'s thorp.' Cf. MANBY.

Manton Li [*Malmetun* 1060–6 KCD 819, *-tune, Mameltune* DB, *Malmetuna* c 1115 LiS, *-tona* c 1145 DC]. First el. OE *malm* (*mealm*) 'sand' or 'chalky earth'.

Manton Nt [*Mennetune* DB, *Manton* Hy 2 (1316) Ch], M~ Ru [*Manatona* 1130–3 Fr, *Maneton* 1202 Ass, 1223 Ep], M~ W [*Manetune* DB, *Manynton* 1235 Cl]. Either '*Manna*'s TŪN' or rather 'common TŪN' (OE *gemǣna tūn*).

Manuden Ess [*Magghedana, Menghedana* DB, *Manegedan*' c 1130 Oxf, *Manegadenna* c 1150 BM, *Maneweden* 1254 Val]. 'The valley of *Mann(a)*'s people.' For the change of *Manninga-* to *Manewe-* cf. CANEWDON.

Maperton So [*Malpertone* DB, *Maperton* 1219 Fees]. 'Maple TŪN.' First el. OE *mapuldor* 'maple'.

Maplebeck Nt [*Mapelbec* DB, 1166 P]. 'Maple brook.' First el. OE *mapul* 'maple' in *mapultrēow*.

Mapledurham Ha [*Malpedresham* DB, *Mapelderesham* 1190 P], M~ O [*Mapeldreham* DB, *Mapeldoreham* 1195 P], **Mapledurwell** Ha [*Mapledrewelle* DB, *Mappedreuuella* c 1125 Oxf, *Mapeldurewelle* 1183 P]. 'Maple HĀM and stream.' Cf. MAPERTON.

Maplescombe K [*Mapeldrescamp* 11 DM, *Mapledescam* DB, *Mapeldurescamp* 1195 Fr]. 'Maple field.' Second el. OE CAMP, altered to *combe*. Cf. MAPERTON.

Maplestead, Great & Little, Ess [*Mapulderstede* 1065 BM, *Magna Mapeldonestede, Parva Mapestede* 1254 Val]. 'Place where maples grew.' Cf. MAPERTON.

Mapperley Db [*Maperlie* DB], **Mapperton** Do nr Winterborne Zelstone [*at Mapeldertune* 943 BCS 781, *Mapledretone* DB], M~ Do nr Beaminster [*Malperetone* DB,

Mapeldoreton 1236 Fees]. 'Maple LĒAH and TŪN.' Cf. MAPERTON.

Mappleborough Wa [*Mapeldosbeordi* 714 BCS 130, *Mapelberge* DB]. 'Maple hill.' OE *mapuldor-beorg*.

Mappleton Db [*Mapletune* DB], M~ YE [*Mapletone* DB]. 'Maple TŪN.' Cf. MAPLE-BECK.

Mappowder Do [*Mapledre* DB, *Mapeldrea* 1121 AC, *Mapodre* 1236 Fees]. OE *mapuldor* 'maple'.

Marazion (mărazī·n) Co [*Marghasbigan* c 1200 Ol, (in) *Parvo Foro* or *Marhasbean* 1311 AD, *Marghasbighan* 1359 AD]. 'Little market' (Co *marchas* 'market' and *bichan, bechan* 'small'). There was formerly by M~ a place called **Market Jew** [*Marchadyou* c 1200 Ol, *Marcasiow* 1313, *Marghasdiow* 1358 AD], whose second el. may be Co *dyow* 'south'. Market Jew is given by Bartholomew as an alternative name of Marazion. The article is largely based on information from Mr. Gover.

Marbury Chs nr Whitchurch [*Merberie* DB, *Merebury* 1260 Court], M~ Chs nr Northwich [*Merebiria* Hy 3 Ormerod]. 'BURG by a lake.' Both are on lakes.

March Ca [*Mercc* 1086 IE, *Merche* DB]. Perhaps OE *mearc* 'boundary' with palatal *c* from a locative form in *-i*.

Marcham Brk [*Mercham* 835, *æt Merchámme* 901, *Mercham*, dat. *-hamme* 965 BCS 413, 592, 1169, *Merceham* DB]. 'HAMM where smallage (OE *merece*) grew.'

Marchamley Sa [*Marcemeslei* DB, *Merchemeslega* 1185 P, *Merchemelee* 1206 FF]. '*Merchelm*'s LĒAH.' *Merchelm* no doubt has as first el. the folk-name *Mierce, Merce* 'Mercians'. The loss of *l* is due to dissimilation.

Marchington St [*æt Mærcham* 951 BCS 890, *Mærchamtun* 1002 Wills, *Merchametone* DB, *Mercinton* Hy 2 Derby, *Mercington* 1230 Ass]. The original name was *Mercham*, which is identical with MARCHAM rather than from *Merca-hām* 'the HĀM of the Mercians'. Marchington is *Merchæmatūn* 'the TŪN of the *Mercham* people'.

Marchwood Ha [*Merceode* DB, *Merchewude* 1254 Ipm]. 'Smallage wood.' Cf. MARCHAM.

Marcle, Much & Little, He [*Merchelai* DB, *Parva Markelay* 1208 Cur, *Mercley, magna Merkel'* 1236 Cl, *Mangna Markele* 1242 Fees]. OE *mearc-lēah* 'boundary wood'.

Mardale Green We [*Merdale* 1278 FF]. 'Lake valley.' M~ is south of HAWES WATER.

Marden He [*Maurdine* DB, *Mauordine* 1138 AC, *Maurdin* 1169 P, *Magewurdin* 1177 P]. If the last form is trustworthy, the first el. is identical with MAUND, which is c 3 m. away. See WORÞIGN.

Marden K [*Maeredaen* 11 DM, *Meredenna* 1166 P, *-den* c 1220 Bodl]. 'Pasture for mares' (OE *miere, mere*). The second el. is OE DENN.

Marden, East, North & Up, Sx [*Upmerdone* 931–40 BCS 640, *Meredone* DB, *Normerdon* 1207 Cur]. 'Boundary hill.' Cf. (GE)MÆRE, DŪN.

Marden W [*Mercdene* 941, *Merhdæne* 963 BCS 769, 1118, *Meresdene* DB, *Mergdena* 1168 P]. Perhaps OE *mearc-denu* 'boundary valley'.

Marefield Le [*Merdefelde* DB, *Merðefeld* 1169, 1177 P, *Mardefeud* 1247 Ass]. 'FELD frequented by martens.' Marten is OE *mearþ*.

Mareham le Fen, M~ on the Hill Li [*Mæringe* 1060 Th, *Marun, Meringhe* DB, *Maring* 1202 Ass, 1254 Val, *Maringes* 1237 Ep, *Marum* c 1200 NpCh, 1241 Ep]. The form *Marum* seems generally to refer to M~ le Fen, *Meringhe* &c. to M~ on the Hill in early sources. Later the form in *-ing* disappeared. Mareham is the dat. plur. of OE *mere* 'lake' (the original *i*-stem dat. plur. *marim was exchanged for *marum* at an early date). *Meringe* is OE *Meringas* 'the dwellers by the lakes'.

Maresfield Sx [*Mersfeld* 1234 Cl, *Meresfeld* 1248 Ass, *Meresefeld* 1293 Cl]. OE *meriscfeld* 'FELD by a marsh'.

Marfleet YE [*Mereflet* DB, *-fleit* 1166 P, *Merflet* 1246 FF]. The first el. may be OE *mere* 'lake' or *gemǽre* 'boundary'. Second el. OE FLĒOT 'stream'.

Margaretting Ess [*Ginga* DB, *Gynge Margarete* 1291 Tax, *Margretyinge* 1408 Pat]. 'The ING of St. Margaret.' See ING.

Margate K [*Meregate* 1254, *Margate* 1258 FF, *Margate* 1293 RBE]. 'Gate leading to the sea.' See MERE.

Mărham Nf [*Merham* c 1050 KCD 907, 1292 Ch, *Marham* DB, 1199 Cur, 1230 P]. OE *Mær-hām* 'HĀM by a mere'. See MERE. There is now no lake at M~.

Marhamchurch Co [*Maronechirche* DB, *Marwencherche* 1275 Ep]. Perhaps 'the church of *St. Mærwynn*' (a woman saint).

Marholm (mărŭm) Np [*Marham* c 1060 BM, 1167 P, 1200 Cur]. Identical with MARHAM.

Mariansleigh D [*Lege* DB, *Marinelegh, Seyntemarilegh* 1242 Fees]. 'St. Mary's LEIGH.'

Marishes YN. 'The marshes.' Cf. MERSC. In DB and later sources the parish consisted of several manors with names in *-mersc*, as *Aschilesmares, Odulfesmares*, (in) *paruo Mersc* &c. DB ('*Askell*'s, *Auðulf*'s, Little marsh'). *Askell, Auðulfr* are ON pers. ns.

Mark So [*Mercern* 1065, *Merkerun* c 1070, *Merker* 1164 Wells, *Merke* 1201 FF, *Merk*

1225 Ass]. OE *mearc-ærn* 'boundary house'. The same first el. is found in *Merkemere* 973 BCS 1291 (late copy), the name no doubt of a lake near Mark.

Markby Li [*Marche-, Marchesbi* DB, *Marchebi* c 1115 LiS, 12 DC, *Markebi* 12 DC]. '*Marki*'s BY.' *Marki* is an ODan pers. n.

Markeaton (martn) Db [*Marchetone* DB, 1226 FF, *Merchetune* DB, *Marketon* 1236, 1242 Fees, *Markenton* 1251 Ch]. Very likely OE *Mearcēa-tūn* 'TŪN on the boundary river'.

Markfield Le [*Merchenefeld* DB, *Merkenefelda* 1209–19 Ep, *Merkinfeld* 1254 Val]. 'The open land of the Mercians' (OE *Mercna feld*).

Markham Clinton, East M~ Nt [*Marcham* DB, 1169 P, *Westmarcham* DB, *Estmarcham* 1191–3 Fr]. OE *Mearc-hām* 'HĀM on the boundary'.

On Clinton see ASTON CLINTON.

Markingfield YW [*Merchefeld* DB, *Merchingfeld* c 1140 YCh 64, *Merkenfeld* 1297 Subs], **Markington** YW [*Mercingatun* c 1030 YCh 7, *Merchinton* DB, *Merkington* 1297 Subs]. Probably 'the FELD and the TŪN of the Mercians' (OE *Mercna feld* and *tūn*). If so, the form *Mercingatun* is corrupt or due to a later change. It is possible that Markingfield is *Mercna feld*, and *Mercingatūn* contains a derivative *Mercingas* 'the people of Markingfield'.

Marksbury So [*at Merkesburi* 936, *Mercesburh* 941 BCS 709, 767, *Mercesberie* DB]. Identical in origin is **Maesbury** So [*Merkesburi* 705 BCS 112]. Maesbury Camp is an ancient camp NW. of Doulting. See also MASBROUGH. Marksbury is nr **Winsbury Hill** [*Wineces burug* 963 BCS 1099]. This must be '*Winec*'s BURG'. Marksbury seems to be '*Mǣrec*'s BURG', **Mǣrec* being a short form of *Mǣrheard* &c., analogous to OG *Maricus, Maricho*. The treble occurrence of the name is noteworthy. Cf. BADBURY.

Markshall Ess [(æt) *Mearcyncg seollan* 998 Crawf, *Mercheshala* DB, *Markeshale* 1232 Ch]. Probably OE *mearc-gesella* 'huts at the boundary'. *Mearcgesella* became *Mearcigsella* and an excrescent *n* was introduced. Probably *-seollan* is only a poor spelling for *-sellan*. See (GE)SELL.

Markshall Nf [*Merkeshalle, -hala, Markeshalla* DB, *Markes-, Merkeshala* 1254 Val]. Hardly identical with M~ Ess. First el. perhaps as in MARKSBURY. Second el. HALH or HALL.

Markyate Hrt [*Boscus* 12 Mon, *Markeyate* 1239 Ep, 1254 Val]. 'Boundary gate.' See MEARC. The old name may have been *Wood*.

Marland, Peters, D [*Merland* DB, 1238 FF, *-landa* 1185 P, *Petermerland* 1242 Fees]. 'Land on a lake' (OE *mere*).

The church is dedicated to St. Peter.

Marlborough. See MALBOROUGH.

Marlbrook Sa nr Neen Sollars [*Marebroc* 1195 P, *Merebroc* 1200 Cur]. Named from a brook [*Mærabroc* c 957 BCS 1007], whose name seems to mean 'boundary brook'. Cf. (GE)MǢRE.

Marlcliff Wa, Wo [(æt) *Mearnan clyfe, Marnan clive* 872–4 BCS 537, (to) *Marana cliue* 1005 KCD 714]. The first el., *Mearna*, is found also in MARNHAM, MARNHULL. It looks like a pers. n., which may have an analogy in OG *Marningum*.

Marldon D [*Mergheldone, Merledone* 1308 Ep]. 'Hill overgrown with gentian.' Cf. MALBOROUGH.

Marlesford Sf [*Marles-, Merlesforda* DB, *Marle(s)ford* 1235 FF]. M~ is only c 2 miles from **Martley** [*Martele, Mertlega* DB]. Martley means 'marten wood' (OE *mearþ-lēah* with change *þ* > *t* as in OE *bōtl* from *bōþl*). Marlesford may well be 'Martley ford'. Cf. MARLINGFORD, MARTLESHAM.

Marley D [*Merlegh* 1242 Fees], **M~** Du [*Merleia* 1183 BoB]. OE *gemǣr-lēah* 'boundary LĒAH'.

Marley K [*Merille* 1254, *Merile* 1292 Ass]. 'Pleasant LĒAH.' See MYRGE.

Marley YW nr Bingley [*Mardelei, Merdelai* DB, *Martheley* 1311 Ch]. 'Marten wood.' Cf. MARLESFORD.

Marlingford Nf [*Mardingforð, Marþingforð* c 1000 Wills, *Merlinge-, Marthingeforda* DB, *Mearthingforde* c 1095 Bury, *Merlingeford* 1197 FF]. The OE form must have been *Mearþlinga ford*. It is possible that *Mearþlingas* means 'the people of *Mearþ* or *Mearþel*', **Mearþ* being OE *mearþ* 'marten' used as a pers. n. (cf. ON *Mǫrðr*) and **Mearþel* a derivative of it. But there may have been a *Mearþ-lēah* 'marten wood' also in Nf. Cf. MARLESFORD.

Marlow Bk [(æt) *Mereláfan* 1015 Wills, *Merlave* DB, *Merlaua* 1184 P, *Parva Merlaue* 1204 Cur, *Magna Merlaue* 1237–40 Fees]. The first el. is OE *mere* 'lake'. The second appears to be OE *lāfe*, plur. of *lāf* 'remnant, remains' (in the OE ex. in the dat. plur. *lāfum*, later *lāfan*). The meaning of the name is not clear.

Marlston Brk [*Marteleston* 1242 Fees]. '*Martel*'s manor.' Galfridus Martel held the manor in 1242. Cf. HINTON MARTELL.

Marlston Chs [*Merlestone* DB, *Marleston* 1247 Cl]. Etymology doubtful.

Marnham, High & Low, Nt [*Marneham, alia Marneham* DB, *Marnaham* c 1175 Middleton], **Marnhull** Do [*Marnhulle* 1254 Val, 1316 FA, *Marenhull* 1274 Pat, 1308 FF]. See MARLCLIFF.

Marple Chs [*Merpel* 1248 Ipm, *Merphull* 1285 Court, *Merpil* 1288 Court]. The place is on a ridge by the upper Mersey, which forms a county boundary. The obvious etymology is OE *mǽrhop-hyll* 'hill by the boundary valley'.

Marr YW [*Marra* DB, 1100–15 YCh 1001, *Mar* 1196 P, *Mara* 1200 Cur]. Possibly OScand *marr* in the sense 'marsh, fen'. The OScand word is rare and mostly used of the sea, but *mar* is found in Sw dialects in the sense 'fen, bog', in Faroese in the sense 'mud'.

Marrick YN [*Marige* DB, *Marrich* c 1190 Godric, *Marrig* 1240 FF]. A Scandinavianized form of OE *(ge)mǽr-hrycg* 'boundary ridge'.

Marrington Sa [*Meritune* DB, *Merinton* 1242 Fees]. M~ is at Marrington Dingle nr the Welsh border. OE *Mǽringa tūn* 'the TŪN of the borderers'. *Mǽringas* from OE *gemǽre* 'boundary'.

Marron R Cu [*Meran* 1282 StB]. **Mockerkin Tarn**, not far from the Marron, is *Ternmeran* 1343 Cl. The name seems to be '*Meriaun*'s tarn'. For the order of the elements cf. ASPATRIA. *Meriaun* is the OW form of Welsh *Meirion*. The river-name may be elliptical for *beck Meran* or the like '*Meriaun*'s brook'.

Marsden, Great & Little, La [*Merkesden* 1195 ff. P, *Merkelesden* 1246 Ass, *Merclesden major, Little Merkelstene* 1242 LaInq], M~ YW [*Marchesden* 1274, -*dene* 1277 Wakef, *Marchedene* 1293 Ass]. First el. OE *mercels* 'mark' &c., very likely in the sense 'boundary'. Second el. OE *denu* 'valley'.

Marsh Gibbon Bk [*Merse* DB, *Mersh Gibwyne* 1292 Ipm], **Marsh** Sa [*Mersse* DB]. OE *mersc* 'marsh'.
Gibbon is a family name, originally a Fr pers. n. *Giboin* from OG *Gebawin*.

Marsham Nf [*Marsam* DB, *Marsham* 1252 Ch], **Marshfield** Gl [*Meresfelde* DB, *Maresfeld* 1221 Pp, *Mersschefeld* 1452 BM], **Marshwood** Do [*Mersoda* a 1174 Fr, *Merswude* 1188 P, *Merschwode* 1329 BM]. 'HĀM, FELD and wood by a marsh.' See MERSC.

Marske YN nr Saltburn [*Mersc* DB, c 1180 YCh 659, *Merscum* 1104–8 SD], M~ YN nr Richmond [*Mersche* DB, *Mersk* 1234 FF]. 'The marshes.' The form with *sk* probably comes from the dat. plur. *merscum*, but may also be due to Scand influence.

Marston Moretaine Bd [*Mercstuninga* (gemǽre) 969 BCS 1229, *Merestone* DB], **Fleet M~** Bk [*Merstone* DB, *Fletemerstone* 1223 Ep], **North M~** Bk [*Merstone* DB, *Nordhmerston* 1237–40 Fees], **M~** Chs [*Merston* 1304, *Merschton* 1316 Chamb], **M~ Montgomery** Db [*Merston* 1242 Fees, *Marston Mountegomery* 1350 Derby], **M~ upon Dove** Db [*Merstun* DB], **Broad M~ & Long M~** or **M~ Sicca** Gl [*Merstuna* 1043 Th, *Merestone* DB, *Brodemershtone* 1361 AD, *Drye Merston* 1250 Winchc, *Merston Sicca* 1291 Tax, *Longa Merston*

1285 FA], **M~ He** [*Merstune* DB, *Merstun* 1242 Fees], **M~ Stannett** He [*Merstun* DB, 1242 Fees], **Long M~** Hrt [*Merston* 1194 P, *Long M~* 1325 AD], **Potters M~ Le** [*Mersitone* DB, *Poteresmerston* 1043 (1267) Ch], **M~ Li** [*Merestune* DB, *Merstona* a 1167 DC], **M~ St. Lawrence** Np [*Merestone* DB, *Mersshton Sancti Laurencii* 1329 QW], **M~ Trussell** Np [*Mersitone* DB, *Merston Trussel* 1236 Fees], **M~ O** [*Merston* 1122 Fridesw, *Mershton* 1316 FA], **M~ Bigot** So [*Mersitone* DB, *Merston Bygod* 1348 BM], **M~ Magna** So [*Merstone* DB, *Merscetun* c 1100 Montacute, *Great Merston* 1248 Ch], **M~ St** nr Stafford [*Merstone* DB, *Mershton* 1316 FA], **M~ St** nr Penkridge [*Mersetone* DB, *Mershton* 1316 FA], **M~ W** [*Merstone* 1309 BM], **M~ Maisey** W [*Merston* 1199 FF, *Mershtone Meysi* 1302 AD], **South M~** W [*Merston* 1242 Fees, *Suthmershton* 1330 FF], **M~ Wa** nr Wolston [*Merston iuxta Auonam* c 1050 KCD 939, *Merston* 1237 Cl], **Butlers M~** Wa [*Mersetone* DB, *Merston Le Botiler* 1176 Fr], **M~ Culey**, now **M~ Green** Wa [*Merston Quilly* 1316 FA], **M~ Jabbett** Wa [*Merstone* DB, *Merston Jabet* 1242 Fees], **Lea M~** Wa [*Merstone* DB, *La Le, Merston* 1253 Ch, *Merston juxta La Lee* 1428 FA], **Priors M~** Wa [*Merston* 1236 Fees, *M~ Priors* 1316 AD], **Long M~** YW [*Mersetone* DB, *Merston* 1190 f. P]. OE *Mersc-tūn* 'TŪN by a marsh'.

M~ Bigot So was held by Richard le Bigod before 1195 (P). Bigot is a Fr nickname and family name, probably identical with *bigot*.—**Broad M~** Gl. See BRĀD.—**Butlers M~** Wa from the local family. Ralph Boteler was tenant t. Steph (Dugdale).—**M~ Culey** Wa was held by the Culeys from the 13th cent. Culey from CULEY in Normandy.—**Fleet M~** Bk from a stream.—**M~ Jabbett** Wa was held by Henry Jabet before 1242 (Fees).—**Lea M~** Wa seems to have been Lea and Marston.—Roger de Meysi held **M~ Maisey** W in 1212 (RBE). Cf. MEYSEY HAMPTON.—**M~ Montgomery** Db was held by William de Mungumeri in 1242 (Fees). The name is from MONTGOMMERY in Normandy.—The family name Moretaine (**M~ Moretaine** Bd) is from MORTAIN in Normandy. —**Potters M~** Le is 'the potters' M~'. There must have been potteries here.—**Priors M~** Wa from the prior of Coventry.—**M~ St. Lawrence** Np from the dedication of the church.—**M~ Stannett** He. Surname obscure. —**M~ Trussell** Np belonged to the Trussells from the time of Hy 2 (Bridges). Cf. ACTON TRUSSELL.

Marstow He [*Lann Martin* c 1150 LL, *Martinestowe* 1291 Tax]. 'St. Martin's church.' See STŌW.

Marsworth Bk [*Mæssanwyrð* 966–75 Wills, *Missevorde* DB, *Messewurda* 1163 P]. '*Mæssa*'s WORP.' Cf. MASSINGHAM. OE *Mæssa* is not recorded in independent use. Cf. OG *Maso, Masso*.

Marten W [*Mertone* DB, *Mereton* 1200 P, 1227 Ch]. Identical with MARTIN, MARTON.

Marthall Chs [*Marthall* 1507 AD]. Possibly 'market hall'.

Martham (-th-) Nf [*Martham* DB, 1191 P, 1254 Val]. 'HĀM or HAMM frequented by martens.' See MEARÞ.

Martin Ha [*Mertone* 946 BCS 817, *Meretun* 12 VH], M~ K [*Meretum* 861 BCS 855, *Mereton* 1162 P], M~ La in Burscough [*Merretun* DB, *Mertona* c 1190 LaCh], M~ La in Dalton in Furness [*Meretun* DB], M~ Li nr Horncastle [*Mærtune* 1060 Th, *Martone* DB, *Martuna* c 1115 LiS], M~ Li nr Timberland [*Martona* 12 DC, *Marton* 1212 Fees], M~ Nt [*Martune* DB, *Marton* 1217 BM], M~ Hussingtree Wo [*Meretun* 972 BCS 1282]. OE *Mere-tūn* or *Mær-tūn* 'TŪN by a lake'. Cf. MARTON. The lake has in many cases disappeared. One M~ La is on the now drained Martin Mere. M~ Nt is near the Yorkshire border, and a meaning 'boundary TŪN' (OE *gemærtūn*) is perhaps preferable.

M~ Hussingtree is M~ and Hussingtree [*Husantreo* 972 BCS 1282]. *Husantreo* is '*Hūsa*'s tree'.

Martindale We [*Martindale* 1246 Ipm]. '*Martin*'s valley.'

Martinhoe D [*Matingeho* DB, 1196 FF, *Mattingeho* 1228 FF]. 'The HŌH or spur of land of *Matta*'s people.' Cf. MATTINGLEY. *Matta* goes with OG *Mazo*, *Matto*, *Mezzi*.

Martinscroft La [*Martinescroft* 1332 Subs], **Martinsthorpe** Ru [*Martinestorp* 1206 Cur]. '*Martin*'s croft and thorp.'

Martlesham Sf [*Merlesham* DB, *Martlesham* 1254 Val, Hy 3 Misc]. Cf. MARLESFORD and MARLINGFORD. Martlesham is some 8 miles from Marlesford and Martley. A connexion is not absolutely impossible.

Martley Wo [*Mertlega* 11 Heming, (in) *Mærtleages ecge* c 1030 Förster, *Themse*, p. 769, *Mertelai* DB, *Mardelege* 1156 RBE, *Martheleg* 1255 Fees]. '*Marten* wood', OE *mearþ-lēah*. Cf. MARTLEY Sf under MARLESFORD.

Martock So [*Mertoch* DB, *Mertoc* 1176 P, 1230 P, *Meretoc* 1209 Pp, *Merttoke* 1243, *Merkestok* 1265 Ass]. Probably OE *Mere-stoc* 'STOC by a lake'. The loss of *s* is due to Norman influence. *Merkestok*, which is identified with Martock by the editor of Ass, shows that the second el. is *stoc*. It might suggest that the first is OE *mearc*, but no doubt it is miswritten.

Marton Chs nr Macclesfield [*Merutune*, *Meretone* DB, *Merton* 1248 Ipm], M~ Chs nr Northwich [*Merton* 1276, 1312 Ch], M~ La [*Meretun* DB, *Mertona* 1176 P], M~ Li [*Martone* DB, *Martuna* c 1115 LiS], M~ Sa nr Chirbury [*Mertune* DB, *Merton* 1242 Fees], M~ Wa [*Merton* c 1155 Fr, *Mereton* 1206 Cur], Long M~ We [*Merton* 13 Misc], M~ YE in Bridlington [*Martone* DB, *Marton* 1235 FF], M~ YE nr Swine [*Meretone* DB, *Martona* 1155-7 YCh 1148], M~ YN nr Middlesbrough [*Martune* DB, *Marton* 1206 FF], M~ YN in Sinning-ton [*Martun* DB, *Marton* 1167 P], M~ in the Forest YN [*Martun* DB, *Martona* 1181 YCh 420], M~ le Moor YN [*Marton* 1198 FountM, *Marton on the Moor* 1292 Ch], M~ YW nr Boroughbridge [*Martone* DB, *Marton in Burgesir* 1219 FF], East & West M~ YW [*Martun* DB, *-ton* 1147-50 YCh 1471]. OE *Mere-tūn* or *Mær-tūn* 'TŪN by a lake'. Cf. MERE. Some of the Martons are still by lakes, or lakes are known to have been formerly there. M~ Chs nr Macclesfield was named from the now drained **Marton Mere**. M~ La is at **Marton Mere**. M~ Sa is on **Marton Pool**. For M~ We cf. 'magnum (parvum) vivarium de Merton' 13 Misc. But OE *Mær-tūn* 'TŪN on a boundary' is sometimes a possible alternative.

Marvell Wt in Carisbrooke [*Miryfeld* 1359 VH]. 'Pleasant FELD.' Cf. MYRGE.

Marwell Ha [*Merewelle* 1194 Selborne, 1284 Ep, *Merewell* 1255 Ch]. 'Boundary stream', OE (*ge*)*mær-wella*. M~ D. See MALBOROUGH.

Marwood D [*Merode*, *Mereude* DB, *Merewde* 1219 FF]. Perhaps 'boundary wood'. Cf. (GE)MÆRE.

Marwood Du [*Marawuda* c 1050 HSC, *Marwode* 1316 Ipm]. Apparently OE *māra wudu* 'the greater wood'. A possible analogy is offered by LASBOROUGH &c.

Marylebone Mx [*Tyborne* al. *Maryborne* 1490 FF]. Originally TYBURN, later altered to *Maryborne* from a church dedicated to St. Mary, and by popular etymology to *Marylebone* (as if 'Mary the good').

Maryport Cu. A late name. The port is old, but was called *Ellnesfoote* 1566 CWNS xxi and later. The old name means 'the mouth of R ELLEN'. The new harbour made here about 1750-60 was named, as pointed out in CWOS iii, p. 303, after the founder's wife Mary.

Marystow D [*Ecclesia Sancte Marie Stou* 1266 Ep, *Marystowe* 1334 Ep]. 'St. Mary's church.' See STŌW.

Marytavy. See TAVY.

Masbrough (-ăz-) YW [*Merkisburg* 1202 FF, *Merkesburc* 1206 Cur]. Apparently identical with MARKSBURY and MAESBURY So.

Masham (-ăs-) YN [*Massan* DB, *Mesham* 1200 Obl, *Masseham* 1251 Ch], **Mashbury** Ess [*Massebirig*, *Masceberia* DB, *Messebir* 1203 Cur, 1212 Fees]. '*Mæssa*'s HĀM and BURG.' Cf. MARSWORTH. Mashbury might also be *Mæccan burg*. Cf. MESSING.

Mason Nb [*Merdisfen* 1242 Fees, *Merdesfen* 1260-3 Newcastle, *Merdefen* 1242 Cl]. '*Mærheard*'s fen.'

Massingham, Great & Little, Nf [*Masinge-*, *Masincham* DB, *Massingeham* 1202 FF, *Magna Massingham*, M~ *Parva* 1254

Val]. 'The HĀM of *Mæssa*'s people.' Cf. MARSWORTH.

Matching Ess [*Matcinga, Metcinga* DB, *Macinges* c 1130 Oxf, *Maching* 1232 FF]. OE *Mæccingas* '*Mæcca*'s people'. *Mæcca* is found in *Mæccanfer* 854 BCS 476. Cf. MACKNEY.

Matfen Nb [*Matefen* 1159 P, 1213 Ch, *Matesfen* c 1190 Godric, *Matfen* 1236 Fees, *Mathfen* 1291 Ch]. Perhaps '*Matta*'s fen'. Cf. MATTINGLEY, MARTINHOE.

Matfield K [*Mattefeld* c 1230 BM, 1328 Ch, *Mettefeld* 1275 RH]. Perhaps '*Matta*'s FELD'. Cf. prec. name.

Mathon (-ādh-) Wo, He [*Matme* DB, *Mademe* 1242 Fees, *Matheme* 13 AD]. OE *māþm* 'treasure, gift'. The reason for the name is not apparent.

Matlask Nf [*Matelasc, -esc* DB, *-aske* 1198 FF, *Matolask* 1179 RA]. OE *mæþl-æsc* 'ash where a moot was held'. For the change *þl > tl* cf. OE *bōtl* from *bōþl*. The form *-ask* is due to Scand influence.

Matlock Db [*Meslach* DB, *Matlac* 1196 Cur, *Mathlac* 1233 Derby, *Matloc* 1204 Cur]. OE *mæþl-āc* 'oak where a moot was held'. Cf. MATLASK.

Matra·vers Do [*Lodre* DB, *Lodres Luttone* 1285 FA, *Lodres Mautravers* 1356 FF]. Originally LODERS. *Lutton* from LITTON CHENEY. The manor was held by John Mautravers in 1303 (FA). Cf. LANGTON MATRAVERS. Matravers is elliptical.

Matson Gl [*Matresdone* 1121, *Mattresdone* c 1170 Glouc, [*Duntisburn*] *Matrisdon* 1236 Fees]. Second el. DŪN. The first is obscure. It may be a Brit name of the hill.

Matterdale Cu [*Matherdal* 1323 Ipm]. 'Valley where madder grew.' The first el. is OE *mæddre* or ON *maðra* 'madder'. Cf. *Moðruvellir* in Iceland (Landnáma).

Mattersey Nt [*Madressei* DB, *Mareseia* c 1200 DC, *Mathersay* 1254 Val]. Second el. ĒG 'island'. The first seems to be a pers. n., possibly an OE *Mæþelhere.

Mattingley Ha [*Matingelege* DB, *-lega* 1167 P, *Mattingely* 1251 Ipm]. 'The LĒAH of *Matta*'s people.' Cf. MARTINHOE.

Mattishall Nf [*Mateshala* DB, *-hal* 1200 Cur, *Matteshala* 1193 f. P]. Second el. OE HALH. The first el. looks like a strong side-form of *Matta* in prec. name. **M~ Burgh** [*Berk* 1204 FF, *Parva Berg* 1254 Val]. OE *beorg* 'hill'.

Maugersbury Gl [*Meilgaresbyri* 714, (æt) *Mæþelgares byrig* 949 BCS 130, 882, *Malgeresberiæ* DB]. '*Mæþelgār*'s BURG.'

Maughonby (mǎfnbǐ) Cu nr Kirkoswald [*Merghanby* 1288 Cl]. '*Merchiaun*'s BY.' OW *Merchiaun* is a pers. n. corresponding to Welsh *Meirchion*.

Maulden Bd [*Meldone* DB, *Mealdon* 1180 P]. Identical with MALDON Ess.

Maun R Nt [*Mome* Hy 3, *Mone* 1335 PNNt(S), *Man* 1622 Drayton]. A backformation from MANSFIELD.

Maunby YN [*Mannebi* DB, *Magnebi* 1166 P, 1202 FF, *Magneby* 1240 FF]. '*Magni*'s BY.' First el. ON *Magni*.

Maund Bryan, Rose M~ He [*Mage, Magene* DB, *Magena* 1161 P, *Magene* 1187 Hereford, *Magen* 1212 Fees, *Magene Brian* 1242 Fees, *Rons Maune* 1433 PNHe]. Maund is an old name of a district [(on) *Magonsetum* 811, (in pago) *Magesætna* 958 BCS 332, 1040]. The inhabitants are called *Magesæte* 1016 ASC (E). *Magen* is from an early form of Welsh *maen* 'stone', used in a more original sense 'plain'. Cf. Ir *magen* 'place'. *Maen* is derived from Welsh *ma* (< *mago-*) 'field'.

Brian de Maghene lived in the 12th cent. (Hereford). Cf. BRAMPTON BRYAN.—Rose in **Rose M~** may be identical with *Rose* in ROSE ASH or *Rous* in (ROUS) LENCH.

Mautby Nf [*Malteby* DB, *-bi* 1168 P]. Identical with MALTBY.

Mawdesley La [*Madesle* 1219, *Moudesley* 1269 Ass]. '*Maud*'s LĒAH.' *Maud* is a Fr pers. n. of OG origin (OG *Mahthild*).

Mawgan in Meneage, M~ in Pyder Co [(in) *Sancto Maugan*, (villa) *Sancti Ma(l)gani* 1206 Cur (1), (Ecclesia) *Sancti Maugani* 1291 Tax (2), (Ecclesia) *Sancti Maugani in Kerier* 1308, *S~ M~ in Pyderschire* 1309 Ep]. '(Church of) St. Maugan.' *Maugan* is identical with Bret *Maugan* (perhaps from *Mal-cant*).

Meneage is an old name of a part of the Lizard peninsula. For Pyder see PETHERWIN.

Mawnan Co [(Ecclesia) *Sancti Maunani* 1281 Ep]. The saint's name implied is obscure.

Mawsley Np [*Malesle* 1066–75 GeldR, *-lea* DB, *Maleuesle* 1202 Ass]. First el. as in MANSELL He. Mawsley is OE *Malwes-lēah* 'wood by a *malu* or gravel ridge'. Cf. WYTHEMAIL.

Mawthorpe Li [*Mauthorp* 1251 Cl, *Malthorpe* 1311 Ipm, *Malthorp* 1316 FA]. Probably '*Malti*'s thorp'. Cf. MALTBY.

Maxey Np [(of) *Macuseige* 963–84, (æt) *Macusie* 972–92 BCS 1128, 1130, *Makeseia* 1199 FF]. '*Maccus*'s island.' *Maccus* (Battle of Maldon) is an Irish-Scandinavian form of *Magnus*.

Maxstoke Wa [*Machitone* DB, *Makestoka* 1170 P]. '*Mac(c)a*'s STOC.' Cf. MACKNEY, MAKENEY.

Mayfield St [*Medevelde* DB, *Matherfeld* c 1180 Mon, 1252 Ch, 1269 Ass]. 'Madder FELD.' See MÆDDRE.

Mayfield Sx [*Magefeud* c 1200, 1248, *Megthefeud* 1279 PNSx]. 'FELD where MÆGÞE or mayweed grew.'

Mayford Sr [*Maiford* 1212 Fees, *Maynford* 1230 P, *Mayford* 1236 Fees]. No doubt identical with *Mægþeford* 955 BCS 906 (Abingdon, Brk), *Maȝþeford* 931 ib. 672 (Norton, Gl). This may be 'maidens' ford' (OE *mægþ*) or 'ford where mayweed grew' (OE *mægþe*). Cf. MÆGDEN.

Mayland Ess [*La Mailanda* 1185 ff. P]. OE *æt þǣm ēglande* 'at the island', with the *m* of the def. art. carried over to the name.

Mayne, Little, & Broadmayne Do [*Maine* DB, *Maynes* c 1100 Chester, *Maena, Magene* 1186 f. P, *Brademaene, Parva Maene* 1202 FF]. Welsh *maen* 'stone'. Cf. MAUND. The name may refer to the stone circle at Broadmayne. For *Broad-* see BRĀD.

Maytham K [*Maiham* c 1185 Penshurst, *Meyhamme* 1242 Fees, *Matham* 1314 Pat]. 'HAMM overgrown with MÆGÞE or mayweed.'

Meaburn, King's & Mauld's, We [*Maiburne* c 1115, *Meabrun* c 1125 WR, *Medbrunne* 13 CWNS xi, *Meabruna Gerardi* c 1150 WR, *Meburnemaud* 1278 Ipm, *Meburn-Regis* c 1290 WR]. If the form *Medbrunne* is correct, OE *mēd-burna*, the first el. being MÆD 'meadow'. Cf. MEDBOURNE, MEDLOCK.
Mauld's M~ from Maud, married to William de Veteriponte (t. Hy 2).

Meads Sx [*Mades* 1196 FF, *Medese* 1296 Subs]. OE *mǣd-efes* 'the edge of the meadow-land'. OE *efes* means 'edge, border' (of a wood &c.).

Meaford (-ĕ-) St [*Meþ-, Metford* DB, *Medford* 1175 P]. As M~ is on the Trent where it is joined by a tributary, the name must mean 'ford at the junction of streams' (see (GE)MȲÞE), in spite of the regular *e*-form.

Mealrigg Cu [*Midelrig* 1189 Holme C]. 'Middle ridge.' Second el. OScand *hryggr*.

OE **mearc** 'mark, boundary mark, boundary, border, border district'. The last sense is found in OE *Mierce* 'Mercians', lit. 'borderers'. In pl. ns. the usual meaning is no doubt 'boundary mark, boundary'. See MARCH, MARK, MARCLE, MARDEN W, MARKHAM, MARKYATE, CHILMARK.

Meare So [*Mere* DB, 1225 Ass]. OE *mere* 'lake'. In DB 10 fishermen and 3 fisheries are mentioned under Mere.

Mearley La [*Merlay* 1241 FF]. 'Boundary LĒAH.' See (GE)MǢRE.

OE **mearþ** 'marten'. See MAREFIELD, MARLESFORD, MARLEY, MARLINGFORD, MARTHAM, MARTLESHAM, MARTLEY, MERSTHAM.

Mease (mēs) R Le, Db, St [*Meys* 1247, 1272 Ass]. Probably derived from OE *mēos* 'moss'. On the river is **Measham** Le

[*Messeham* DB, *Meisham* 1182 P]. 'HĀM on R Mease.'

Meathop We [*Midhop* c 1185, *Mithehop* c 1200 CC]. 'Enclosure or piece of firm land in fens.' See HOP. OE *mid-* has been replaced by OScand *miðr* 'middle'.

Meaux (mūs) YE [*Melse* DB, 1154 YCh 1385, *Mealsa* 1162, *Mealse* 1197 P]. Probably OScand *mel-sǽr* 'lake with sandy shores'. Cf. MEL(R). Identical with MELSJÖN in Sweden. The name was later associated with MEAUX in France.

Meavy (-ā- or -ē-) R D [*Mǽwi* 1031 KCD 744, *Mewy* 1291 Pat]. It has been suggested that *Mavia* in Rav (*Maina* in some MSS) refers to the Meavy. This may well be right. If so, Meavy is a Brit name, derived from the adj. found in MBret *mau* 'agile, gaillard'. The name would mean 'merry brook'. On the river are **Meavy, Goodameavy, Hoo M~** [*Mǽwi* 1031 KCD 744, *Mewi* DB, 1194 P, *Gode-, Hughemewy* 1242 Fees]. The additions are OE *Gōda* and Fr *Hugh*, pers. ns.

OScand **meðal** 'middle' is found in some names, as MALTON YN, MEDLAR, MELBOURNE YE, MELTON (some), MELWOOD. As a rule *meðal* has replaced an OE MIDDEL.

Medbourne Le [*Medburne* DB, *Med-, Metburna* c 1115 (1333) Ch, *Medburna* 1165 P], M~ W [*Medeburne* 940 BCS 754, *-bourne* 955 ib. 904]. OE *mǣd-burna* 'meadow stream, stream with meadows on its banks'. Cf. MEDLOCK.

Meddon D [*Madone* DB, *Meddon* 1234 Fees, *Mededon* 1242 Fees]. 'Meadow hill.' See DŪN.

Mēden R Nt [*Medine* 1227 Cl, 1227 For, *Medme* 1300 For, *Modome* 1338 Pat], **Medina** R Wt [*Medine* 1196 HMC, *Medme* 1280 Ass, *Medeme* 13 AD]. The correct form of both is *Medme* (*Medeme*). *Medine* is probably often a misreading of *Medme*. The name means 'middle one' and is formed from OE *meoduma, medema* 'middle'. The Meden is the middle one of three rivers that join to form the IDLE. The Medina divides the Isle of Wight into two roughly equal parts.

Medlar La [*Midelarge* 1215 CC, *-ergh* 1235 FF]. 'Middle ERG or shieling.'

Medley O [*Middeleya* 1147 Os, *Middileit* c 1145 Fridesw]. 'Middle island.' The second el. varies between OE *ēg* and *ēgoþ* (*ait*); cf. ĒG.

Medlock R La [*Medlak* 1292 Ass, *Medelake, -loke* 1322 LaInq]. OE *mǣd-lacu* 'meadow stream'. Cf. MEDBOURNE.

Medmenham Bk [*Medemeham* DB, *Medmenham* 1200 P, 1210 Cur], **Medmerry** Sx at Selsey [*Medemenige* 683 BCS 64]. 'Middle HAMM and island.' First el. OE *medema* 'middle'.

Medomsley Du [*Madmeslei* c 1190 Godric, *Medomesley* 1183 BoB]. 'Middlemost LĒAH.' First el. OE *medumest* 'middle'. Cf. *Westmæstun* BCS 197 'westernmost TŪN'.

Medstead Ha [*Medestede* c 1235 Selborne, *Medested* 1282 Ep]. 'Place in a meadow.' See STEDE.

Medway R K, Sx [*Meduuuæian* 764, *Medeuuæge* 765–91, *Miodowæge* 880 BCS 195, 260, 548, *Medwæg* c 894 Asser, *Medeweye* 1227 Ass]. A compound of the river-name WEY (identical with WYE) and perhaps Celt *medu* 'mead' (Gaul, OCo *medu*, Welsh *medd*), the first el. referring to the colour of the water.

Meece St in **Cold-** & **Millmeece** [*Mess* DB, *Mes* 1208 Cur, *Coldemes* 1272 Ass, *Mulnemes* 1289 Cl]. OE *mēos* 'moss'.

Meerbrook St [*Merebroke* 1338 Misc]. 'Boundary brook.' See (GE)MÆRE. The brook is referred to as *Merebroc* 1330 Ch.

Meering Nt [*Meringe* DB, *-s* 1242 Fees, *Meringa* c 1163 RA]. OE *Meringas*, identical with MAREHAM ON THE HILL. The place, as pointed out in PNNt(S), is situated by a pool.

Meersbrook Db [*Meresbroc* 12 Beauchief Cart]. The brook at M~ is called *Merebroc* 1155–8, *Meresbroch* 1154 YCh 1451, 1475. The name means 'boundary brook'. Cf. (GE)MÆRE and MERSEY. The brook forms the boundary between Derby and Yorkshire.

Meertown or **Meretown** St [*Mera* DB, 1167 P, *Mere* 1198 P]. The original name means 'the lake'. The place is on AQUALATE MERE.

Meesden Hrt [*Mesdone* DB, 1254 Val, *Misedon* 1253 AD]. 'Mossy hill.' Cf. MĒOS.

Meese Brook St, Sa [*Mees* 1266 Ch]. Identical with MEASE. On the stream is **Meeson** Sa [*Mestun* 1249 Ipm]. 'TŪN on Meese Brook.'

Meeth D [*Meda* DB, *Meðe* 1176, *Meða* 1178 P, *la Methe* 1259 Ep], **Meethe** D in S. Molton [*la Methe* 1249 Ass]. The situation of the places suggests that the name is OE *gemȳþe* 'junction of streams'.

ON mel(r) 'sandhill, sandbank', Engl dial. *meal*, *meol* the same is found in MEOLS, CARTMEL, INGOLDMELLS, RATHMELL, TRANMERE, MEAUX, also AMBLESIDE.

Melbourn Ca [*Meldeburna* 970 BCS (1265 ff.), c 1080 ICC, *Melleburne* DB]. 'Stream on whose banks milds (OE *melde*) grew.' *Milds* is a name for various species of *Atriplex* and *Chenopodium*.

Melbourne Db [*Mileburne* DB *Meleburn* 1164 P, 1219 FF]. 'Mill stream.' Mills in M~ are mentioned t. Hy 2 Derby.

Melbourne YE [*Middelburne* DB, *Medelbornn* 1285 Ipm]. 'Middle stream.' OE *middel* has been replaced by OScand *meðal*.

Melbury Abbas Do [*Meleburge imare*, *Mealeburg* 956 BCS 970, *Meleberie* DB, *Melbury Abbatisse* 1291 Tax], **M~ Bubb**, Osmond & Sampford Do [*Mele(s)berie* DB, *Mellebir* 1202 Cur, *Melebir* 1212 Fees, *Melebury Osmund* 1243 BM, *Bub Melebur* 1280, *Melebury Saunford* 1313 FF]. The places are by hills. **Melbury Hill** [*Meleberig dun* 956 BCS 970] reaches 863 ft. It is possible that the second el. is in reality OE *beorg* 'hill', with change of *eo* to *u*. If so, Melbury is identical with **Millbarrow Down** Ha nr Bishops Waltham [(on) *Mælan beorh* 909, (neah) *Mælan beorge* 961 BCS 622, 1077]. The last-mentioned name means 'multicoloured hill', the first el. being OE *mæl(e)*, found in *unmæle* 'spotless'.

M~ **Abbas** belonged to the Abbess of Shaftesbury. *Abbas* is thus *Abbess.*—M~ **Bubb** from an early owner. William Bubbe held the manor in 1212 (Fees). Bubb is OE *Bubba. Bubbancumb* nr Melbury is mentioned c 1010 KCD 708.—M~ **Osmond** from the dedication of the church to St. Osmund.—M~ **Sampford**. Sampford is a family name (from Sampford D or So?).

Melchbourne Bd [*Melceburne* DB, *Melcheburn* 1163 P]. First el. ME *mielch*, *milch* 'giving milk', probably referring to good pastures where cows gave plenty of milk.

Melchet Park Ha [*Milchete* DB, *Milset* 1222, *Milcet* 1236, *Mulset*, *Melkecet* 1244 f. Cl, *Melchet* 1279 For]. An old forest-name. Second el. Brit *cēt* (Welsh *coed*) 'wood'. The first may be Welsh *moel* 'bare'.

Melcombe Horsey Do [*Melcome* DB, *Melecumbe* 1198 Cur, *-cumb* 1212 Fees, *Melcombe Horsey* 1535 VE], **M~ Regis** Do [*Melecumb* 1238 Cl, *-cumbe* 1280 Ch, *Melcombe Regis* 1391 FF]. OE *meoluc-cumb* 'valley where milk was got, fertile valley'. OE *meoluccumb* is found BCS 620 (Ha).

The Horsey family were in possession of M~ Horsey in the 16th cent.

Meldon D [*Meledon* 1176 P]. 'Multicoloured hill.' Cf. MELBURY.

Meldon Nb [*Meldon* 1242 Fees, 1254 Val]. Identical with MALDON.

Meldreth Ca [*Melreda* c 1080 ICC, *-rede* DB, *Milree* 1201 Cur, *Mulri* 1238 Cl, *Melreth* 1261 FF, *Melrith* 1263 Ipm]. OE *myln-rīþe* 'mill stream'.

Melford, Long, Sf [*Melaforda* DB, *Meleforde* c 1095 Bury, *-ford* 1235 Ch]. OE *mylen-ford* 'mill ford'.

Melkinthorpe We [*Melkamestorp* 1195 FF, *Melcanetorp* c 1215 CWNS xxiv]. First el. the Ir pers. n. *Maelchon*.

Melkridge Nb [*Melkrige* 1279 Ass]. 'Milk ridge.' Cf. MEOLUC.

Melksham W [*Melchesham* DB, 1144 AC, 1156, 1190 P, *Mulcheham* 1194 Rot Cur, *Melkesham* 1198 FF]. Possibly 'HAMM where cows gave plenty of milk', though the gen. in -es is remarkable. Cf. MEOLUC.

Mell Fell. See WATERMILLOCK.

Melling La in Halsall [*Melinge* DB, *Mellinges* 1194 P], M~ La in Lonsdale hd [*Mellinge* DB, *Mellynges* 1094 LaCh]. An OE folk-name *Mellingas*, which may be 'the people of *Moll* or of *Malla*' (cf. MALLING).

Mellis Sf [*Melles, Mellels* DB, *Melles* 1198 FF, 1254 Val], **Mells** Sf [*Mealla* DB, *Melne* c 1160 Harl Ch, *Melnes* 12 Blythburg Cart, *Melles* 1254 Val], **Mells** So [*at Milne* 942 BCS 776, *Mulle* DB, *Melnes* 1196 P, *Melles* 1225 Ass]. 'The mills' (OE *mylen*).

Mellor Ch [*Melner* 1330 QW], M~ La [*Malver* c 1130 Whitaker, *Meluer* 1246 Ass]. Identical with Welsh *Moelfre* 'bare hill' (Welsh *moel* 'bare', from *mailo-*, and *bre* 'hill'). Both places are on the slope of prominent hills.

Mells. See MELLIS.

Melmerby (mĕlerbĭ) Cu [*Melmorby* Hy 3 Ipm, 1291 Tax], M~ YN nr Coverham [*Melmerbi* DB, *Melmorbi* 1202 FF], M~ YN nr Ripon [*Malmerbi* DB, *Melmorby* 1301 Subs]. '*Melmor*'s BY.' *Melmor* (11 Gospatric's charter) is OIr *Mael-muire* 'St. Mary's servant'.

Melplash Do [*Melpleys* c 1155 Salisbury, *Muleples* 1242 Fees, *Est Meleplessch* 1333, *Milplassh* 1449 BM]. 'Mill pool.' Cf. PLÆSC.

Melsonby YN [*Malsenebi* DB, *Melsanebi* 1202 FF]. Perhaps '*Maelsuithan*'s BY'. *Maelsuithan* is an Ir pers. n.

Meltham YW [*Meltham* DB, 1297 Subs, *Muletham* 1316 FA]. The form of 1316 suggests as first el. an OE *mylen-gelæt* 'mill stream'. Cf. (GE)LÆTU.

Melton Mowbray Le [*Medeltone* DB, *Melton* 1200, *Miauton* 1201 Cur, *Melton Moubray* 1284 Cl], M~ **Ross** Li [*Medeltone* DB, *Meltuna* c 1115 LiS, *Melton Roos* 1402 FA], **Great & Little** M~ Nf [*Middilton, Methelton, Lithle Meddeltone* c 1060 Wills, *Meltuna, Parua Meltuna* DB, *Magna Melton* 1242 Fees], M~ YE [*Methelton'* 1219 Ass, *Melton* 1316 FA], **High** M~ YW [*Middeltun, Medeltone* DB, *Melton le Heyg* 1285 FA], **West** M~ YW [*Middeltun, Medeltone* DB]. OE *Middel-tūn*, Scandinavianized to *Meðaltūn*, whence *Melton*. **Melton Constable** Nf [*Maeltuna* DB, *Meutone* 1212 RBE, *Melton Constable* 1320 Ch], M~ Sf [*Meltune* c 1050 KCD 907, *Meltuna* DB] may be identical in origin, but decisive forms have not been found. OE *Mæl-tūn* (cf. MALDON) or even *Mylen-tūn* 'mill TUN' may be suggested.

Melton Constable (Constabularius de Melton 1197 P) was held by the constable of the bishop

of Norwich.—M~ **Mowbray** was held by Roger de Moubray c 1125 (LeS). Mowbray from MONTBRAY in Normandy.—M~ **Ross** was held by William de Ros in 1303 (FA). The family took its name from ROTS in Calvados, Normandy (olim *Ros, Roos*).

Melverley Sa [*Melevrlei* DB, *Milverlegh* 1311 Cl, 1311 Ipm]. 'LĒAH by the mill ford', OE *Mylenford-lēah.*

Melwood Li [*Methelwode* 12 DC]. OE *Middelwudu* 'middle wood', Scandinavianized. Cf. MELTON.

Membland D [*Mimidlande* DB, *Mimilaunde* 1242 Fees]. First el. perhaps a streamname identical with MINT.

Membury D [*Maneberie, Manberia* DB, *Menbir* 1204 Cur, *Membir'* 1212 Fees]. See BURG. Welsh *maen* 'stone' has been suggested as first el., the reference being to an old 'castle'. But OE *gemæne* 'common' may also be thought of.

Membury W nr Ramsbury [*Minbiry* c 1066 ASCh, *Mimbir'* 1196 FF, *Mimmebir'* 1242 Cl, *Mymbury* 1291 Tax]. Derivation of the first el. from Welsh *min* 'edge, brink, border', suggested in PNW(S), is possible, but an English first el. is preferable, and OE *myne* in the sense 'love' or 'memorial' may be better. In the first alternative the common street name LOVE LANE might be compared.

Mendham Sf [*Myndham, Mendham* c 950 Wills, *Mendham* DB, *Mendeham* 1168 P, 1196 Fr]. '*Mynda*'s HĀM.' **Mynda* is derived from *Munda* in MUNDFORD &c.

Mendip Hills So [*Menedepe* 1185 TpR, *Mendep, Menedup* 1225 Ass, *Minedepe* 1236 FF, *Munedep* 1235 Cl]. The first el. is Welsh *mynydd* 'hill'. The second el. may be OE HOP 'valley'. If so, Mendip at first denoted a valley, probably that which cuts Mendip Hills into two parts, and was later transferred to the hills.

Mendlesham Sf [*Melnes-, Mundlesham* DB, *Mendlesham* 1165 P, 1198 FF]. '*Myndel*'s HĀM.' **Myndel* is derived from *Munda.* See MUNDFORD &c.

Menethorpe YE [*Mennistorpe* DB, *Meny(g)-thorp* 1297 Subs, *Menythorp* 1303 FA]. '*Menning*'s thorp.' *Menning*, perhaps a Scand name, is found c 1050 YCh 9.

Menheniot Co [*Mahiniet* 1260 Ep, *Manhunghet* 1291 Tax]. Co *maen* 'stone' and a pers. n. corresponding to OW *Huniat* in the 9th cent. Book of St. Chad.

Menston YW [*Mensinctun* c 972 BCS 1278, *Mensingtun* c 1030 YCh 7, *Mersintone* DB, *Mensinton* 1190 P]. Apparently 'the TŪN of *Mensa*'s people'. *Mensa* might be related to *Menta* (in MENTMORE). Or there may have been an OE **Mensige*, ə side-form with *i*-mutation of *Mansige* (moneyer t. Cnut).

Menthorp YE [*Menethorp* 1166 P, *-torp* 1219 FF]. First el. perhaps OE *gemǽne* 'common'.

Mentmore Bk [*Mentemore* DB, 1179 P, *Mentemor* 1203 Cur, *Mantemor* 1276 RH]. '*Menta's* moor.' The pers. n. *Menta* is not found, but corresponds to OG *Mantio*, *Manzo*.

Menwith YW [*Menwit* 1230 Ep, *Menewyth* 1318 Misc]. 'Common wood.' An OE (*ge*)*mǽna wudu* with OScand *viðr* introduced for the synonymous *wudu*.

Meole (-ē-) **Brace** Sa [*Melam* DB, *Mole* 1203 Cur, *-s* 1210 FF, *Meeles* 1242 Fees, *Melesbracy* 1273 Ipm]. M~ is on **Meole Brook** [*Mola*, *Meola* c 1130 Ordericus, *Mele* c 1200 Gervase]. On this is also **Cruckmeole** [*Meole* 1327 Subs] higher upstream. The river-name is very likely an early back-formation from CRUCKMEOLE, which it is most natural to explain as 'the bare hill' (Welsh *crug* 'hill' and *moel* 'bare'). Close to Cruckmeole is CRUCKTON. Cruckmeole may have been misunderstood as 'the hill on the Meole'.

M~ **Brace** was held by Aldolf de Bracy in 1206 (Cur). The family name may come from BRASSY nr Amiens.

Meols (mĕls), **Great & Little**, Chs [*Melas* DB, *Parua Moeles* 1200–45 Chester, *Mangna Molles* 1287 Court, *Litlemolis* 1283 Ipm], **North M~** (mēlz) La [*Otegrimele* DB, *Moles* a 1149, *Moeles* 1153–60 LaCh, *Nor Muelis* 1229 Ass], **Ravensmeols** La [*Mele* DB, *Ravenesmeles* 1190–4 LaCh]. ON *melr* 'sand-bank, sand dune'.

Otegrimele DB means '*Auðgrim's* Meols'. *Auðgrimr* is an ON pers. n. Ravensmeols is '*Hrafn's* Meols'. *Hrafn* is an ON pers. n.

OE meoluc 'milk'. See MELK-, MILK- (passim), MELCOMBE, MULBARTON. The element usually indicates that the place had good pastures, where cows gave much milk.

Meon R Ha [*Meonea* a 790, (ofer) *Meóne* 824, (innán) *Méone* 932 BCS 258, 377, 689]. A Brit river-name, perhaps related to Gaul *Moenus* 'Main'. On the Meon are **East & West Meon** [*Meanuarorum prouincia* c 730 Bede, *Meanware mægð* c 890 OEBede, (æt) *Meone* c 880, *æt Meóne* 956 BCS 553, 982, *Mene*, *Estmeone* DB, *Mienes* 1156 P, *Westmenes* 1284 Ch] and **Meonstoke** [*Menestoche* DB]. The last is 'STOC belonging to Meon' or 'STOC on the Meon'.

Meon Hill Gl, a prominent hill of 637 ft. [*Mene* DB, 1236 Fees, *Muna* 1159, *Mina* 1190, *Muena* 1191, *Muene* 1196 P]. The forms refer to a place by the hill. Probably Meon was originally a brook-name identical with MEON Ha. The hill is between two arms of a brook.

Meopham (-ĕp-) K [*Meapaham* 788, 964–95 BCS 253, 1132 f., *Mepeham* DB]. '*Mēapa's* HĀM.' The same name is found in MEPAL. **Mēapa* may be related to Engl *mope*.

OE **mēos** 'moss', identical with OHG *mios*, is found in MEASE, MEECE, MEESE, MEESDEN and perhaps some other names.

Mēpal Ca [*Mephal* J FF, *-hale* 1254 Val]. '*Mēapa's* HALH.' Cf. MEOPHAM.

Meppershall Bd [*Malpertesselle*, *Maperteshale* DB, *Maperteshala* 1176 P, *Meperteshale* 1200, 1205 FF]. 'HALH with a maple or maples.' First el. OE *mapuldor* or *mapultrēo* 'maple'. The forms with *e* may represent OE **meapul*. Cf. *Mepelesbarwe* 848 BCS 453.

Mercaston Db [*Merchenestune* DB, *Murkelistone* 1245, *Murkamstone* 1252 FF, *Murcaston* 1278 Derby, *Murcaneston* 1281 FF, *Murkanston* 1297 Ipm]. Forms like *Murkelis-*, *Murkamstone* are probably to be disregarded. The normal forms may be explained from a combination of OE *myrce* 'dark' and *Ēast-tūn*, either OE *Myrcan Ēast-tūn* 'Easton (Aston) on R *Myrce*' or *Myrca Ēast-tūn* 'dark Easton (Aston)'. *Myrce* 'dark river' may have been the old name of Cutler Brook, on which M~ stands. There are two Astons in Db, both at a considerable distance from Mercaston, however. Cf. BATHEASTON, DALSTON Cu.

OE **mere** 'lake, mere' is a common pl. n. el. A side-form *mǽre* occurs, though rarely. A form *mǽr-*, analogous to *bǽr-* for *bere* 'barley' (in OE *bǽrlic* &c.) must have been common as a first el., as in MARHAM and several MARTONS. Cf. MAREHAM. OE *mere* is found in several names of lakes, as BUTTER-, GRAS-, WINDERMERE, but many original names of lakes were transferred to places on the lake. The word is used alone in MAER, MEARE, MERE, DELAMERE. As a second el. it appears as *-mere*, *-mer* (as BULMER, STURMER Ess, BARMER, CROMER Nf). Frequently *-mere* has been replaced by *-more*, as in CRANMORE, MONMORE, PEASEMORE, rarely by *-mire*, as in ELDMIRE, REDMIRE. As a first el. it appears as *Mar-* (MARBURY, MARLAND, MARTIN, MARTON &c.) or *Mer-* (MERTON &c.). The meaning 'sea' is rare, but is found in MARGATE, MERSEA.

Mere Chs [*Mera* DB], **M~** Li [*Mere* 1202 Ass, 1212 Fees], **M~** (mēr) W [*Mere* DB, 1166 RBE]. OE MERE 'lake'. M~ Chs is by a lake. M~ W is not far from a tarn.

OE **merece** 'smallage'. See MARCHAM, MARCHINGTON, MARCHWOOD.

Meretown. See MEERTOWN.

Merevale (mĕrĭ-) Wa [*Mireual'* 1157 P, *-vallis* 1189 (1292) Ch, *Mirival* c 1190 BM, *Murivall*, *Mirivalle* Hy 3 BM]. A monastery founded in 1148. The original name may have been *Mira vallis*, transferred from the Continent. But the name must at least have been associated with OE *myrge* 'merry', and it is possible that Merevale is a part translation of an OE *myrge-denu* 'pleasant valley'. Cf. MERIDEN.

Mereworth K [*Meran worð* 843 BCS 442, *Mæreweorð* c 960 ib. 1097, *Mæranwyrþ* 10 ib. 1322, *Marovrde* DB]. '*Mæra*'s WORÞ.' **Mæra* is a short form of names like *Mærheard*.

Meriden Wa [*Myrydene* 1441 BM, *Miryden* 1443 AD]. OE *myrge-denu* 'pleasant valley'. The place is in early sources called **Alspath** [*Ailespede* DB, *Allespathe* 1221 Pp, *-path* 1236 Fees]. '*Ælli*'s path.'

Merridge So [*Malrige* DB, *Merige* 1201 FF, *Merigge* 1327 Subs]. 'Boundary ridge.' Cf. (GE)MÆRE.

Merrington, Kirk, Du [*Mærintun* c 1085 LVD, *Merintona* Hy 2, *Kyrke Merington* 1331 FPD]. 'The TŪN of *Mæra*'s people.' Cf. MEREWORTH.

Merrington Sa [*Muridon* 1254 Eyton, 1327 Subs]. 'Pleasant hill.' Cf. MEREVALE, MERIDEN. The place is called *Gellidone* DB, *Gulidon* 1245 FF.

Merriott So [*Meriet* DB, 1194 P, 1201, 1225 Ass, *Muriet* 1329 Ep]. Probably OE (ge)-*mærgeat* 'boundary gate', in spite of the isolated *Muriet*. Cf. (GE)MÆRE. Or the first el. may be OE *miere* 'mare'.

Merrow Sr [*Marewe* 1185 BM, *Merewe* 1187, 1190 f. P, 1212 Fees, *Merwe* 1200 Cur]. The place is on the slope of **Merrow Downs**, which may have been called OE *mær-ræw* 'boundary row (i.e. ridge)'. Cf. (GE)MÆRE, RÆW. Or the base may be OE *mære wēoh* 'famous temple'.

OE **mersc, merisc** 'marsh' is found alone as a pl. n. in MARISHES, MARSH, MARSKE. As a first el. it occurs in MARSHAM &c., MARSTON (common), MARESFIELD, MERSTON and others. It is fairly common as a second el., as in HENMARSH, LAMARSH, LAMAS.

Mersea (mahzĭ) **Island, East & West M~** Ess [*Meresig* 895 ASC, *Myresig* c 995 BCS 1288, *Meresai* DB, *Estmereseia* 1196 P, *Westmeresheye* 1238 Subs]. 'The island in the sea.' In ASC *Meresig* is called '*igland ... ute on þære sǽ*'. See ĒG, MERE.

Mersey (-z-) R Chs, La [*Mærse* 1002 Wills, *Mersham* (acc.) DB, *Merse* 1142 LaCh, *Merese* 1228 Cl]. OE *Mæres-ēa* 'boundary river'. Cf. (GE)MÆRE. The Mersey is the boundary between Cheshire and Lancashire, and was once that between Mercia and Northumbria. The lower Mersey seems also to have been called WEAVER. See WERVIN.

Mersham K [*Mersaham* 858, 863 BCS 496, 507, *Merseham* DB]. '*Mæsra*'s HĀM.' **Mæsra* is a pers. n., which may have been formed with an *s*-suffix from the stem *Mær-* in *Mærheard* &c.

Merstham Sr [*Mestham* 675 BCS 39, *æt Mearsætham* 947 ib. 820, *Mersetham* c 1050 KCD 896, *Merstan* DB, *Merstham* 1202 Cur]. Perhaps OE *mearþsæt-hamm* 'HAMM

by a trap for martens'. The elements would be OE *mearþ* 'marten', *sæt* 'ambush' and *hamm*. A *þ* would be lost early between *r* and *s*.

Merston K [*Mersctun* 774 BM, *Mer(i)ston* 1242 Fees], M~ Sx [*Mersitone* DB, *Merschtone* 1304 Ipm], M~ Wt [*Merestone* DB, *Merston* 1287–90 Fees]. OE *Mersc-tūn* 'TŪN by a marsh'.

Merther Co nr Truro. M~ nr Sithney is *Merthersithun* 1230 FF. Merther is identical with MERTHYR in Wales, which is Welsh *merthyr* 'martyr': '(the church of) the martyr'.

Merton D [*Mertone* DB, *Merton* 1246 Ipm], M~ Nf [*Meretuna* DB, *Mertuna* 1121 AC], M~ O [*Meretone* DB, *Meriton* 1227 Ch]. OE *Mere-tūn* 'TŪN by a lake'.

Merton Sr [(on) *Merantune* 755 ASC, *at Mertone* 967 BCS 1196, *Meretone* DB, *-tune* 1152 BM, *-ton* 1159 P]. Probably '*Mæra*'s TŪN', in spite of the spelling with *e* in the earliest example. Cf. MEREWORTH. OE *mere* 'mare' ought to have been *miere* or *myre* (*Mierantun*) in ASC.

Meshaw D [*Mavessart* DB, *Malessart* 1176 P, *Mausard* 1242 Fees, *Meusawe* 1249 Ass, *Meushagh* 1316 FA]. Evidently the name was by Normans made into *mal assart* 'poor clearing'. But this could not have given Meshaw. The old name probably had as second el. OE *scaga* 'wood', possibly interchanging with *sceard* 'gap'. The first el. may have been a stream-name identical with MEAVY.

Messing Ess [*Metcinges* DB, *Mecinges* 1166 RBE, *Medsinges* 1199 FF]. A doublet of MATCHING, though with *ts* (> *ss*) for *č* owing to Norman influence.

Messingham Li [*Mæssingaham* c 1067 Wills, *Messingeham* DB, 1181 P, *Massingeham* c 1115 LiS, *Massingham* 1265 Ch]. Identical with MASSINGHAM.

Metfield Sf [*Medefeld* 1214, 1229 FF]. 'Meadow FELD.' Cf. MÆD.

Metham YE [*Metham* 1312 Ipm]. was *mæþ-hamm* 'HAMM where mowing OE done'. Cf. MÆÞ.

Metheringham Li [*Medric(h)esham* DB, *Methricham* 1185 TpR, *Mederingeham* 1193 f. P, *Madringeham* 1219 Ass, *Metheringham* 1231 Ep]. The first el. looks like a pers. n. But it ma. be an OE *mæd-ric* (*mēd-ric*) 'meadow stream'. Cf. RIC.

Methley YW [*Medelai* DB, *Metheleia* c 1160 YCh 1452, *Medelay* 1226 FF]. OE *Middel-lēah* 'middle LĒAH', Scandinavianized. Cf. next name.

Methwold Nf [*Medelwolde* c 1050 KCD 907, *Methelwalde* DB, *-wolda* 1171 P]. M~ is between HOCKWOLD and NORTHWOLD. The name means 'middle wold' and is OE *Middel-wald* with *middel* replaced by

OScand *meðal* 'middle'. **M~ Hythe** [*Meþel-woldehyþe* 1277 Ely Cart, *Otringheia* DB, *Oteringhithe* 1316 FA]. Cf. OTTERDEN and HŸP.

Mettingham Sf [*Metingaham* DB, *Meting-ham* 1230 P, 1235 Ch]. 'The HĀM of *Metti*'s people', *Metti* being a derivative of *Matta* in MARTINHOE &c. Cf. OG *Mezzi*.

Metton Nf [*Metune* DB, *Metton* 1197 FF, 1206 Cur]. The first el. may be OE MÆD 'meadow' or MÆÞ 'mowing'.

Mevagi·ssey Co [*Mavagisi* 1410 AD]. '(The church of) SS. *Mewa* and *Ida*', Co *Mew ag Ida* 'Mew and Ida'. Co *d* often becomes *s*.

Mexborough YW [*Mechesburg* DB, *Mekes-burg* c 1150 YCh 1664, *Mekeburc* 1196 P]. '*Mēoc*'s BURG.' *Meoc* may be ODan *Miuk*, a nickname from *miuk* 'meek'. But OE *Mēoc* seems to occur in *Meocesdun* 944 BCS 801 (Brk).

OE **micel, mycel** 'large, big' is found in several names, as MICHELMERSH, MIT-CHAM, MITCHELDEAN, MUCHELNEY, MICK-FIELD, MICKLEBRING, MICKLEFIELD &c., MIDDLETON (2). In some names the source is rather OScand *mikill*. *Much* when used as a distinctive addition means 'great'.

Michaelchurch He nr Tretire [*Lann mihac-gel cil luch* c 1150 LL], **M~ Escley** He [*Michaeleschirche* c 1275 Ewias]. 'Church dedicated to St. Michael' (Welsh *Mihangel*). Escley is the name of a stream. For *cil luch* cf. GILLOW.

Michaelston y Vedw Monm [*ecclesia sancti Michaelis* c 1348 LL]. 'St. Michael's.' *Vedw* will be Welsh *bedw* 'birch'.

Michaelstow Co [*Mighelestowe* 1302 FF]. 'St. Michael's church.' See STŌW.

Micheldever Ha [*Mycendefr* 862, *Mycel-defer* 901 BCS 504 f., 596, *Miceldevre* DB]. Originally a name of the stream at M~ [*Myceldefer* 901, 904 BCS 596, 604]. The second el. is a form of Brit *dubro-* 'water, river'; cf. ANDOVER, CANDOVER. The first el. seems to be OE *micel* 'great', which may be a rendering into English of a Brit word for great (OW *mor*, Welsh *mawr*). Or the Brit name may have had as first el. Welsh *mign* (from **micn*) 'bog', which by popular etymology was made into OE *micel*. Note the earliest OE form.

Michel Grove Sx [*Muchelegraua* 1193 P], **Michelmersh** Ha [*æt Miclamersce* 985 KCD 652, *Muchelemareis* 1167 P], **Mick-field** Sf [*Mucelfelda* DB, *Miclefeld* c 1095 Bury, *Mikelefeld* 1242 Fees]. 'Large grove, marsh and FELD.'

Micklebring YW [*Mikelebrinc* 1206 Cur, *Mikelbrink* J Ass, *Mykelbring* 1254 Ep]. 'Large brink or hill.' It is doubtful if *brink* is a native or a Scand word.

Mickleby YN [*Michelbi* DB, *Miclebi* 1185–90 YCh 1046]. OScand *Miklibȳr* 'large village'. The name is common in Scandinavia.

Micklefield YW [(on) *Miclanfelda* 963 YCh 6, *Miclafeld* c 1030 ib. 7], **Mickleham** Sr [*Micleham* DB, *Mikelham* 1242 Fees], **Micklethwaite** YW [*Muceltuoit* DB, *Mikelthwait* 1208 FF], **Mickleton** Gl [(to) *Mycclantune* 1005 KCD 714, *Mucletona* 1183 AC], **M~ YN** [*Micleton* DB, *Mickilton* 1251 Ch], **Mickley** Nb [*Michelleie* c 1190 Godric, *Mickeley* 1242 Fees]. 'Large FELD, HĀM or HAMM, THWAITE, TŪN, LĒAH.'

Mickleover. See OVER.

OE **mid** 'middle' is found in MEATHOP, MIDDOP, MIDHOPE, MIDHURST, MYTHOP.

OE **middel** is a common first el. It is usually uninflected and forms a compound with the second el., as OE *Middeltūn*. It refers to the situation of a place between (two) other places. See MIDDLE- (passim), MIL-TON, MEALRIGG, MILCOMBE &c. OE *middel* is often replaced by OScand *meðal, miaðal*, as in MALTON, MELTON, MELWOOD, METH-WOLD.

Middle Sa [*Mulleht* DB, *Muthla* 1121 Eyton, *Mhutle, Mudle* 1242 Fees]. OE (*ge*)*mȳþ-lēah* 'LĒAH by the junction of streams'. As the junction of streams is some way off, LĒAH must mean 'wood'. OE *gemyðleag* occurs BCS 164 (Gl).

Middleham, Bishop, Du [*Middelham* Hy 2 FPD, *Midelham* 1195 (1335) Ch, 1208–10 Fees], **M~ YN** [*Medelai* DB, *Midelhaym* 1240 FF]. 'Middle HĀM.'

Middleney So [*Midelneia, Midelenie* DB]. 'Middle island.' OE *middel* is inflected.

Middlesb(o)rough (-brōof) YN [*Mid(e)les-burc* c 1165 YCh 709, *Middelburg* 1272 Ipm]. 'Middlemost burg' (OE *midleste burg*). Cf. MIDDLEWICH.

Middlesex [(provincia) *Middelseaxan* 704 BCS 111, (in) *Middil Saexum* 767 ib. 201, *Middelseaxe* 1011 ASC (E), (in) *Middelsexan* 1071–5 Reg, *Midelsexe* DB]. A tribal name 'the Middle Saxons', later used of their territory. Cf. Introd. ii. 1.

Middlestone Du [*Malder-, Melderstayn* 1366 PNNb]. The material is too scanty. One might guess that the first el. is identical with MELLOR and the first el. of MALLER-STANG. The place is near a prominent hill. Second el. ON *steinn* 'stone'.

Middlestown YW [*Midle Shitlington* 1325, *Middleton* 1523 Goodall]. Originally 'Middle SHITLINGTON'.

Middlethorpe YW [*Midelthorp* 1297 Selby]. Self-explanatory.

Middleton is usually 1. OE *Middel-tūn* 'middle TŪN': **M~** Db nr Winster [*Middel-tune* DB], **M~ by Wirksworth** Db [*Middel-tune* DB, *Midelton juxta Wyrkesworth* 1297

FF], **Stoney M~** Db [*Middeltone* DB, *Middleton juxta Heyum* 1283 FF, *Middilton juxta Eyum* 1347 BM], **M~** Du nr Auckland [*Middeltun* 1104-8 SD], **M~ St. George** Du [*Middlinton* 1238 Cl, *Middleton Sancti Georgii* 1291 Tax], **M~ in Teesdale** Du [*Middeltun* 1161-7 YCh 562, *Middelton super Teisam* 1198 (1271) Ch], **M~** Ess [*Middeltun* c 1050 KCD 896, *Mildeltuna* DB], **M~** La in West Derby hd [*Midelton* 1212 Fees], **M~** La in Salford hd [*Middelton* 1194 P], **M~** La in Lancaster par. [*Middeltun* DB], **M~** Nb nr Belford [*Middelton* 1242 Fees], **M~** Nb nr Hartburn [*Middelton Morel* 1242 Fees], **M~** Nb nr Ilderton [*tres Middelton* 1201 Cur, *le Midlest, North, Suth Middiltun* 1236 Fees], **M~** Nf [*Mideltuna* DB], **M~** Np nr Rockingham [*Middelton* 1197 FF], **M~ Cheney** Np [*Mideltone* DB, *Middelton Cheyndut* 1342 Cl], **M~ Stoney** O [*Middeltone* DB, *Mudelingtona* 1209-19 Ep, *Mudelinton* 1242 Fees, *Middelington* 1251 Ep], **M~** Sa nr Chirbury [*Mildetune* DB, *Myddeltone* 1327 Subs], **M~** Sa nr Ludlow [*Middeltone* DB], **M~** Sa nr Oswestry [*Middleton* 1272 Ipm], **M~ Scriven** Sa [*Middeltone* DB, *-ton* 1327 Subs], **M~** Sf [*Mideltuna* DB, *-ton* 1203 Cur], **M~** Sx [*Middeltone* DB], **M~** Wa [*Mideltone* DB], **M~** We [*Middeltun* DB, *Medilton in Lonesdale* c 1160 StB], **M~** Wt [*Middelton* 1280 Ch], **M~ on the Wolds** YE [*Middeltun* DB, *Mideltona* 1297 Subs], **M~** YN nr Pickering [*Middeltun* DB], **M~ Quernhow** YN [*Middeltun* DB, *Middelton Quenerowe* 1329 FF], **M~ Tyas** YN [*Middeltun* DB, *Midilton Tyas* 14 PNNR], **M~ upon Leven** YN [*Middeltun* DB, *Midleton in Cliveland* 1204 Ch], **M~** YW nr Ilkley [*Middeltun* c 972 BCS 1278, *Meðeltun* c 1030 YCh 7, *Middeltune* DB], **M~** YW nr Rothwell [*Milde(n)tone* DB, *Middelton* 1209 FF]. M~ Stoney O shows inflected forms of the first el. The *u*-forms seem due to the rounding influence of *m*.

2. Middleton on the Hill He [*Miceltune* DB, *Miclatuna* 1123 Leominster Cart, *Mitletona* 1242 Fees]. OE *Micla-tūn* 'large TŪN'.

3. Middleton Baggot & Priors Sa [*Mittilton* c 1200 Eyton, *Mittelington* 1222 FF, *Mitletone* 1291 Tax, *Muttulton* 1349 Eyton]. The first el. seems to be OE *gemyþlēah* 'LĒAH at the junction of streams'. Cf. MIDDLE Sa. For the change *þl > tl* cf. MARTLEY.

Middleton Baggot Sa. See HOPE BAGGOT.— M~ Cheney Np was held by Simon Chendut in the 12th cent. (NS). Cf. CHENIES.—M~ Priors Sa belonged to Wenlock Priory.—M~ Quernhow YN. Cf. AINDERBY QUERNHOW.— M~ St. George Du from the dedication of the church.—M~ Scriven Sa is 'the scrivener's M~' (ME *scrivein* from OFr *escrivain*), but Scriven may here be a family name.—Stoney M~ Db no doubt from stony soil.—M~ Stoney O is on stony soil.—M~ Tyas YN. Cf. FARNLEY TYAS.

Middlewich Chs [*Mildestvic* (hd), *Wich* DB, *Middelwich* 1185 P, *Medius Vicus* 1240 Cl].

'The middle wich' (see wīc). M~ is between Nantwich and Northwich. The DB form shows the OE superlative *midlest*.

Middlewood He [*Midewde* DB, *Middelwde* 13 AD]. Self-explanatory.

Middlezoy So [*Soweie, Sowy* 725, 971 BCS 142 f., 1274, 1277, *Sowi* DB, *Middlesowy* 1227 FF]. Cf. WESTON ZOYLAND, formerly *Westsowi*. *Sowi* contains a stream-name identical with SOW, SOWE, *-i* being OE *ēg, īeg, īg* 'island'.

Middop YW nr Colne [*Mithope* DB, *Midhop* c 1150, *Midhope* 1182-5 YCh 641, 199]. Formally identical with MEATHOP, but *hope* seems to mean 'valley'.

Middridge Du [*Midrige* 1183 BoB]. 'Middle ridge.'

Midford So [*Mitford* 1001 KCD 706, 1296 FF]. OE (ge)*mȳþ-ford* 'ford at the junction of streams' (Cam Brook and Wellow). Cf. (GE)MȲÞE.

Midge Hall La [*Miggehalgh* 1390 FF], **Midgehall** W nr Wootton Bassett [*Micghæma gemæra* 983 KCD 636, 638], **Midgham** Brk [*Migeham* DB, *Migham* 1198 FF, *Mighala* 1156, *Miggehal* 1190 ff. P]. OE *mycg-healh* 'midge-infested nook'. The form *-ham* for *-healh* in the Brk name may be due to a derivative *Mycghǣme* 'the people at *Mycghealh*', as seen in the OE example of Midgehall.

Midgley YW nr Halifax [*Micleie* DB, *Miggelay* 1238 Cl], **M~** YW nr Barnsley [*Migelaia* 1160-75 YCh 1730, *Miggeley* 1234 Ep]. 'Midge-infested LĒAH.'

Midhope YW nr Ecclesfield [*Midhop* 13 BM]. Identical with MIDDOP.

Midhurst Sx [*Middeherst* 1186, 1190 P]. 'Middle hurst.'

Midley K [*Midelea* DB, *Middelea* 11 DM]. OE *Middel-ēa* 'middle stream' or OE *Middel-lēah* 'middle LĒAH'.

Midloe Hu [*Middelho* 1135-60 Rams, 1198 (1286) Ch]. 'Middle HŌH or spur of hill.'

Migley Du in Lanchester [*Miggeleye* 1232 Ch]. A variant of MIDGLEY.

Milborne St. Andrew & Stileham Do [*Muleburne* 939 BCS 738, *Meleburne, -borne* DB, *Muleburne St. Andrew* 1294 FF, *Milborn Stylam* 1431 FA], **M~ Port** So [(æt) *Mylenburnan* c 880 BCS 553, *Mele-, Mileburne* DB, (Burgh of) *Mileburn* 1225 Ass, *Milleburnport* 1249 FF], **Milborne** Nb [*Meleburna* 1158 f. P, *Mulneburn* 1201 FF], **M~** W at Malmesbury [*Milburn* 1315 Orig], **Milburn** We [*Milnebrunn* 1200 FF, *Meleburn* 1247 Cl]. OE *mylen-burna* 'mill stream'. The stream at M~ Port is *Mylenburna* 933, *-burnna* 946 BCS 695, 894.

Port in M~ Port is OE *port* 'town'. The place was a borough.—M~ Stileham is obscure.

Milby YN [*Mildebi* DB, 1166 P, *Mildeby* 1228 Ep, 1246 FF]. '*Mildi*'s BY.' *Mildi* is a common ON byname.

Milcombe O [*Midelcumbe* DB, *-cumba* c 1160 RA]. 'Middle valley.' See CUMB.

Milcote Wa [*Mulecote* 710 BCS 127, *Melecote* DB, *Mylekote* 11 Th]. 'Mill cottage.'

Milden Sf [*Mellinga* DB, *Meldinges* c 1130 Bury, *Meldingg* 1254 Val]. Perhaps '*Melda*'s people'. OE *Melda* is possibly found in (on) *Meldanige* BCS 810. But more likely OE *Melding* 'place where orach (OE *melde*) grew'. Cf. CRESSING, WRATTING.

Mildenhall Sf [(at) *Mildenhale* c 1050 KCD 832, *Mitdenehalla*, *Mudenehalla* DB, *Middelhala* 1130, *Mildehal* 1158, *Middehala* 1162 P, *Mildenhale* c 1200 Bodl], **M~** (mīnawl) W [*Mildanhald* 803–5 BCS 324, *Mildenhalle* DB, *Mildehale* 1241 Ch, *-hal* 1261 Ipm]. The last is clearly '*Milda*'s HALH or nook'. *Milda* is a pers. n. formed from OE *milde* 'mild'. The first seems to be OE (æt) *middelan hale* '(at) the middle HALH'.

Mile End Ess [*la Milende* 1200 P, 1256 FF], **M~** Mx [*la Mileende* 1441 AD iii]. 'The end of the mile.'
M~ Ess is c 1 m. from the ford north of Colchester, **M~** Mx 1 m. from Aldgate.

Mileham Nf [*Meleham*, *Muleham* DB, *Meleham* a 1122 Fr, 1160 P]. OE *Mylen-hām* 'HĀM with a mill'.

Milford Db [*Muleforde* DB], **M~** Ha [*Melleford* DB, *Melneford* a 1189 BM], **M~** W [*Meleford* DB, 1198 Fees, *Muleford* Hy 3 Ipm], **North M~** YW [*Mileford* DB, *Meleford* 1166 P, *Northmilford* 12 Selby], **South M~** YW [*Mysenford* 963, *Myleford* c 1030 YCh 6 f., *Mileford* 1234 FF]. 'Mill ford, ford by a mill.' Other Milfords are no doubt identical in origin.

Millbarrow Ha. See MELBURY.

Millbeck We. See UNDERMILLBECK.

Millbrook Bd [*Melebroc* DB, *Mulebrok* 1220 Subs], **M~** Ha [æt *Melebroce* 956 BCS 926, *Mylebroces ford* 1045 KCD 781, *Melebroc* DB]. 'Mill brook.'

Millichope Sa [*Melicope* DB, *Millinghope* 1199 P, *Millingehope* 1249 Eyton, *Myllynchop* 1327 Subs]. Apparently OE *mylenhlinč-hop* 'valley by the mill hill'. Cf. HLINC, HOP. A windmill must be referred to.

Millington Chs [*Mulintune* DB, *Mulneton* 1259 Court, *Millyngton* 1278 Chester], **M~** YE [*Milleton* DB, *Milingtona* c 1155 YCh 1242, *Milinton* 1254 Ipm]. OE *Mylentūn* 'TŪN with a mill'.

Millmeece See MEECE.

Millom Cu [*Mulnum* c 1190 LaCh, *Mullum* 1206 P, *Milnum* 1287 StB]. OE *Mylnum*, dat. plur. of *mylen* 'mill'.

Millow Bd [*Melnho* 1062 Th, *Melehou* DB, *Milneho* 1200 Cur]. 'Mill hill.' See HŌH.

Millthorpe Li [*Milnetorp* 1202 Ass], **Milnethorpe** Nt [*Multhorp* 1284 Misc], **Milnthorpe** We [*Milnethorpp* 1348 CWNS xiv], **M~** YW. 'Thorp with a mill.'

Milnrow La [*Milnehuses* 13 WhC, *Mylnerowe* 1554 DL]. 'Mill row.' Dial. *row* means 'a row of houses'.

Milnthorpe. See MILLTHORPE.

Milson Sa [*Mulstone* DB, *Mulston*, *Muleston* 1242 Fees]. '*Myndel*'s TŪN.' Cf. MUNSLOW and MENDLESHAM. *Myndles-tūn* would be apt to become *Mynles-* and *Myllestūn*.

Milsted K [*Milstede* 11 DM, *Milsted* 1219 Fees, *Middelstede* 1226 Ass, *Mildestede* 1285 Ep]. Not OE *mylen-stede* 'place of a mill', but OE *middel-stede* 'middle place', as indicated by the regular *i*.

Milston W [*Mildestone* DB, *Mildistona* 1178 BM, *Middestone* 1212 RBE, *Midleston* 1242 Fees]. 'Middlemost TŪN', OE *midlesta tūn*.

Milton has two distinct sources, 1. OE *Middel-tūn* 'middle TŪN': **M~** Bryant Bd [*Middeltone* DB, *-ton* 1179 RA, *Mideltone Brian* 1303 FA], **M~** Ernest Bd [*Middeltone* DB, *Middelton Orneys* 1330 QW], **M~** Keynes Bk [*Middeltone* DB, *M~ Kaynes* 1227 FF], **M~** Brk [*Middeltun* 956 BCS 935, *-tune* DB], **M~** Ca [*Mideltune* c 1050 KCD 907, *Middelton* DB], **M~** Abbot D [*Middeltone* DB, *-ton* c 1180 BM, *M~ Abbot*'s 1297 Pat], **M~** Damerel D [*Mideltone* DB, *Middelton Aubemarle* 1301 Pat, *M~ Albemarl* 1314 Ch], **South M~** D [*Mideltone* DB, *Middelton* 1219 FF], **M~** Db nr Chapel en le Frith [*Middeltune* DB], **M~** Abbas Do [*Middeltun* 964 ASC, *Mideltune* DB, *Middelton Abbatis* 1298 FF], **M~** on Stour Do [*Mideltone* DB, *Middelton* 1236 Fees], **West M~** Do [*Mideltone* DB, *-ton* 1212 Fees], **M~** Ha nr Portsmouth [*Middelton in Portesia* 1188 P, *Middeltona* 1219 Fees], **M~** Ha nr Christchurch [*Mildeltune* DB, *Middelton* 1242 Fees], **M~** next Sittingbourne K [*Middeltun* 893 ASC, *-tone* DB], **M~** or **Middleton Malzor** Np [*Mideltone* DB, *-ton* c 1200 NpCh], **M~** O nr Deddington [*Middelton* 1240 Ch, *Middelton* 1291 Tax], **Great & Little M~** O [*Mideltone* DB, *-tona* 1139 RA], **M~** under Wychwood O [*Middelton* 1199 P, *Midelinton* 1219 Fees, *Midelton* 1316 FA], **M~** So nr Martock [*Midelton* 1285 FA], **M~** So nr Wells [*Middeltun* 1065 Wells], **M~** So in Worle [*Middeltone* DB], **M~** Clevedon So [*Mideltune* DB, *Middelton* 1201 FF, *Milton Clyvedon* 1408 Ep], **Podimore M~** So [see PODIMORE], **M~** Sr [*Mildetone* DB, *Middeltone* 1212 RBE], **M~** W nr Hindon [*Middelton* 1281 PNW(S)], **M~** Lilbourne W [*Middelton* 1198 Fees, *M~ Lillebon* 1282 Ipm].

2. OE *Mylen-tūn* 'mill TŪN': **M~** Cu [*Milneton* 1285 PNCu(S)], **M~** Db nr Repton [*Melton* 1228 BM, 1260 FF], **M~**

K nr Gravesend [*Meletune,* (to) *Melantune* 10 BCS 1321 f., *Meletune* DB], **M~** next Canterbury K [*Melentun* 1044 Th, *Meletone* 1249 Ipm], **M~** Nb [*Mulliton* 1204 Ch], **M~** Park Np [*Mylatun, Myletun* 972–92 BCS 1130, *Meletone* DB], **M~** Nt [*Miletune* DB, *Mulneton* 1203 Cur], **M~** St [*Mulneton* 1227 Ch], **M~** We nr Milnthorpe [no early forms found].

M~ Abbas Do was an abbey.—**M~** Abbot D belonged to the Abbey of Tavistock.—One Robert son of Bryan held **M~** Bryant Bd t. Hy 2 (VH). Cf. BRAMPTON BRYAN.—**M~** Clevedon So from a local family. William de Clyvedon held the manor c 1200 (Bruton).—**M~** Damerel D was held by Robert de Albemarle in 1086 (DB). Cf. HINTON ADMIRAL.—Ernisus de Middelton (**M~** Ernest Bd) is mentioned 1193 P. Ernest, from *Erneis,* is OFr *Erneis* from OG *Arnegis.*—**M~** Keynes Bk from the Keynes family. Lucas de Kaynes held the manor in 1221 (Ep). Cf. ASHTON KEYNES.—**M~** Lilbourne W was held by Walter de Lillebon in 1242 (Fees). The family name is from LILLEBONNE in Seine-Inf. (France).—**M~** Malzor Np was held by Henricus Mala opera c 1200 (NpCh), by William Malesoures in 1202 (Ass). The name means 'ill works'.

Milverton So [*Milferton* 11 KCD 917, *Milvertone* DB, *Melverton* 1253 FF], **M~** Wa [*Malvertone* DB, *Mulvertun* 1200 Cur, *Milverton* 1236 Fees]. 'TŪN by the mill ford.' Cf. MELVERLEY.

Milwich St [*Melewich, Mulewiche* DB, *Millewyz* 1236 Fees]. 'Mill WĪC.'

Mimms, North, Hrt [*Mimmine* DB, *Mimmes* 1212 Fees, *North Mimmes* 1254 Val], **South M~** Mx [*Mimes* DB, *Mimmes* 1236 Fees, 1254 FF]. Skeat's suggestion that **M~** is a folk-name *Mimmas* may be correct, but the etymology of such a name is obscure.

Mimram R Hrt [*Memeran* (obl.) 913 ASC, *Mimeram* c 1130 HHunt, *Méran, Meran, Mæran* 913 ASC (B, C, D)]. The correct OE form was doubtless *Memere,* which probably has an analogy in the first el. of MORVILLE Sa. The name may be related to MINT and mean 'babbling brook'. The stem may well be that of Norw *mimra,* Dan *mimre* 'to babble', but the name must be English or British. The form *Mimram* is not the normal development of OE *Memere,* but must have been taken from an old source, e.g. Henry of Huntingdon.

Minchinhampton Gl [*Hantone* DB, 1180–7 Fr, *Minchenhamtone* 1221 Ass]. Originally OE *Hēatūn* (dat. *Hēantūne*) 'high TŪN'. The addition is OE *mynecen* 'nun'. **M~** belonged to the nunnery of Caen. The name means 'the nuns' Hampton'.

Mindrum Nb [*Minethrum* c 1050 HSC, 1177 P, *Mindrum* 1227 Ch, *Mundrum* 1251 Ch]. A Brit name, containing Welsh *mynydd* 'mountain' and *trum* or *drum* 'ridge'. Cf. MYNYDD DRYMMAU in Glamorgan. The meaning would be 'mountain with a ridge'.

Minehead So [(æt) *Mynheafdon* 1046 KCD 1334, *Maneheve* DB, *Menehewed* 1225 Ass].

M~ is by a hill, on which are **E. & W. Myne** (-ē-) [*Mene* DB]. The hill is *Menedun* 1225 Ass. Probably the hill was Welsh *Mynydd,* which was adopted as OE *Myned* (cf. LONG-MYND). To this was added OE *dūn* 'hill', and *Myned-dūn,* whence *Mynedūn,* was taken to mean *Myne-dūn* 'Myne Hill'. Hence MYNE and MINEHEAD. The latter thus means 'Myne Hill'. Cf. HĒAFOD.

Minety W [*Mintig, Minti(h), Minty(g)* 844 BCS 444, 447, *Minti* 1185, 1190 P]. OE *mintēg* 'mint island' or *mintēa* (dat. *mintie*) 'mint stream'. Mint is OE *minte* 'mint' (the plant).

Miningsby Li [*Melingesbi* DB, *Mithingesbia, Minigesbia* 1142 NpCh, *Mithingesbi* 12 DC, c 1200 NpCh]. '*Miðiung*'s BY.' First el. OSw *Midhiung,* ON *Miðiungr* pers. n.

Minley Ha [*Mindeslei* DB]. First el. perhaps as in next name. See LĒAH.

Minsden Hrt in Hitchin [*Menlesdene* DB, *Mendlesdenn* 1203 Cur]. '*Myndel*'s valley.' Cf. MENDLESHAM.

Minshull, Church, Chs [*Maneshale* DB, *Chirchemunshull* 1289 Court, *Chirche Munsulf* 1295 BM, *Munchulf* 1331 Ch], **M~ Vernon** Chs [*Manessele* DB, *Munshull Vernoun* 1309 AD]. The Minshulls are on opposite sides of the Weaver. The second el. is OE SCYLF 'ledge, bank'. The first, as shown by the DB forms, had OE *a(o),* which became ME *o, u,* and late *i.* Minshull is OE *Monnes scylf* '*Monn*'s ledge of land'.

Vernon is a family name (from VERNON in France).

Minskip YW [*Minescip* DB, *Menescipe* 1166, *Mæn-, Manschipe* 1168 P, *Mineskip* 1225 FF]. A Scandinavianized form of OE *gemænscipe* 'community of goods', here 'common'.

Minsmere Sf [*Milsemere, Mensemara* DB, *Mennesmer* 1265 Ch, *Amynnesmere* 1452 Pat]. OScand *mynni* 'mouth of a river' (sometimes the synonymous *āmynni*) and OE *mere* 'lake', the last referring to a widening of the river. The river-name **Minsmere** is a back-formation.

Minstead Ha [*Mintestede* DB], **M~** Sx [*Mintestede* 1170 P]. 'Place where mint grew.'

Minster K in Thanet [*Menstre* 694 BCS 86, 1239 Ch], **M~** K in Sheppey [*Menstr'* 1203 Cur, *Menstre* Hy 3 BM], **M~ Lovell** O [*Minstre* DB, *Ministre* 1206 Cur, *Munster Lovell* 1291 Tax]. OE *mynster* 'monastery'

For **M~** Lovell see LILLINGSTONE LOVELL. William Luvel (or Lupellus) was tenant in 1206 (Cur).

Minsterley Sa [*Menistrelie* DB, *Munstreleg* 1246 Ch], **Minsterworth** Gl [*Mynsterworðig* (-*worþig*) c 1030 Förster, *Themse.* p. 769, *Minstredwrð* 1154 Flaxley,

Minsterwrde 1221 Ass, *Menstreworth* 1231 Ch]. 'The LĒAH and WORÞ of the monastery.' Minsterworth belonged to St. Peter's, Gloucester.

Mint R We [*Mymid* c 1180, *Mimed* c 1210 (1294) Ch, *Mimmet* c 1200 NpCh]. **Mint House** is *Mimet* DB. Mint is an old rivername derived from the root mim- 'to make a sound' in Sanskrit *mimāti* 'bleats, cries' &c. Cf. MIMRAM. The name is British.

Minterne Do [*Minterne* 987 KCD 656, *Mintra* 1165 P, *Mynterne* 1291 Tax]. Apparently OE *minte* 'mint' (the plant) and *ærn* 'house'. The meaning of such a name is not clear. OE *minte* 'mint' is also found in MINETY, MINSTEAD.

Minting Li [*Mentinges* DB, *Mintingis* c 1115 LiS, *Mentinges* c 1125 (1336) Ch, *Muntinges* 1219 Ass]. '*Mynta*'s people.' For *Mynta cf. OHG *Munizo*.

Mintlyn Nf [*Meltinga* DB, *Mintlinge* c 1140 BM, *Myntlinge* 1254 Val]. Perhaps '*Myntel*'s people'. *Myntel (or *Myntla) would be derived from *Mynta* in MINTING.

Minton Sa [*Munetune* DB, -*ton* 1212 Fees]. 'TŪN by the mountain', i.e. LONGMYND. First el. Welsh *mynydd* 'mountain'.

Minworth Wa [*Meneworde* DB, *Munnewrth* c 1200 Middleton, -*worth* 1346 Misc]. '*Mynna*'s WORÞ.' *Mynna corresponds to OG *Munio* and is a short form of pers. ns. in *Myne-*, corresponding to OG *Muni-* in *Munifrid* &c. OE names in *Myne-* are unrecorded. OE *myne* 'love' &c. corresponds to Goth *muns*, ON *munr* &c.

Mirfield YW [*Mirefelt* DB, c 1180 YCh 1692, *Mirifeld* 1246 Ass]. 'Pleasant FELD.' First el. OE *myrge* 'merry' &c.

Misbourne R Bk [*Misseburne* 1407, *Messeborne* 1475 ERN], **Missenden** Bk on the river [*Missedene* DB, *Mesendena* 1154 AC, *Messendena* 1163 RA], **Miswell** Hrt [*Misseuuelle* DB, *Messewell* 1204–12 Fees, *Mossewell* 1231 Ch]. The first el. of these is probably a derivative of OE *mos* 'moss', an OE *mysse, which may be identical and synonymous with Dan *mysse* 'water arum' (*Calla palustris*), from a base *musjōn. Dan *mysse* is related to Sw *missne* 'water arum' or 'buckbean' (*Menyanthes trifoliata*), which goes back to earlier *mysne*. The first el. of Misbourne and Missenden may also be a stream-name derived from *mysse* and meaning 'stream where water arum grew'.

Miserden Gl [*Grenhamstede* DB, 1221 Ass, *Musardera* 1187, *la Musardiere* 1191 P]. *Grenhamstede* was held in 1086 by Hascoit Musard. *Musardere* is a derivative of *Musard*: 'Musard's manor'. Musard :3 a nickname (OFr *musard* 'dreamer').

Missenden. See MISBOURNE.

Misson (-z-) Nt [*Misne* DB, 1197 P, *Misun* R 1 Cur, *Misene* 1228 Ep, *Miseneya* 1247

Ep]. Identical with MUIZEN in Belgium [two, one *Musena* c 1150, the other *Musinium* c 680, *Musinne* 10] and MIJSEN in Holland [*Misen* 11], which represent an OG *musin-* cognate with *mos* and the first el. of MISBOURNE &c. The meaning would be 'marsh' or the like, which suits the situation. Cf. Mansion, *Vlaamsche Plaatsnamen*, Schönfeld, *Mededelingen* 33. 126 f., SPN 114.

Misterton Le [*Minstretone* DB, *Munesterton* 1209–35, *Mustertone* 1222 Ep], **M~** Nt [*Ministretone* DB, *Mistertona* 1166 P], **M~** So [*Mintreston* 1199 P, *Musterton* 1316 FA]. 'The TŪN of the monastery' or 'TŪN with a church'. Cf. MYNSTER.

Mistley Ess [*Mitteslea* DB, *Misteleg* 1225 FF, -*leye* 1254 Val]. 'Wood where mistletoe (OE *mistel*) grew.'

Miswell. See MISBOURNE.

Mitcham Sr [*Micham* 675 BCS 39, *Michelham* DB, *Micheleham* 1177 P]. OE *Micelhām* 'great HĀM'.

Mitcheldean Gl [*Dena* 1220 Fees, *Magna Dene* 1282 For, *Micheldeane* 1316 FA]. 'Great Dean.' Cf. DEAN.

Mitchell Co [*Meideshol* 1239 Ch, *la Medissole* 1277, *la Medeshole* 1321 AD iv]. Perhaps '*Mēde*'s HOLH or valley'. Cf. *Medeshamstede* under PETERBOROUGH.

Mite R Cu [*Mighet* 1209 FF, *Mite* 1292 Ass]. A Brit river-name. **Miterdale** [*Myterdale* 1322 Ipm] 'the valley of the Mite' shows a Scand gen. form (OScand -*ar*) of the river-name.

Mitford Nb [*Midford* 1196 P, 1201 Cur, *Mitford* 1254 Val, *Mithford* 1280 AD]. Identical with MIDFORD. The place is at the junction of the Font and the Wansbeck.

Mitton, Little, La [*Parva Mitton* 1242 Fees, *Little Mutton* 1283 FF], **M~** Wo nr Bredon [*Myttun* 841 BCS 433, *Muttone* 11 Heming, *Mitune* DB], **Upper & Lower M~** Wo [*Mettune* DB, *Mutton* 1227 Ch, *Ouermitton* 1221 Ass], **Great M~** YW [*Mitune* DB, *Mangna Mitton* 1241 Cl]. OE (*ge*)*mȳþ-tūn* 'TŪN at the junction of streams'. See (GE)MȲÞE. Great and Little Mitton are at the junction of the Hodder and the Ribble.

Mixbury O [*Misseberie* DB, *Mixeburia* c 1130 Oxf, *Mixeberia* 1190 P, *Mixseberi* 1242 Fees]. 'Dunghill BURG.' First el. OE *mixen* 'dunghill'. The combination is curious.

Mixon St [*Mixne* 1219 FF, 1227 Ch]. OE *mixen* 'dunghill'. Cf. OLDMIXTON.

Mobberley Chs [*Motburlege* DB, *Modberleg* 1260 Court]. 'LĒAH or glade with a *gemōtbeorg* or assembly mound.' Cf. *gemotbeorh* BCS 392. Or the first el. may be identical with MODBURY D.

Moccas He [*Mochros, locus porcorum* c 1150 LL, *Moches* DB, *Mocras* 1202 Cur, *Mocros* 1291 Ch]. Welsh *Mochros* means 'moor for swine' (*moch* 'swine, pigs' and *rhos* 'moor').

Mockerkin Tarn. See MARRON.

Mocktree. See MOTTRAM.

Modbury D [*Motheria* DB, *Modberia* 1182 P, *-byre* 1242 Fees, *Motbury* 1291 Tax]. OE *gemōt-burg* 'BURG where moots were held'. Possibly originally *gemōt-beorg* (cf. MOBBERLEY). This is the source of the hundred name **Modbury** Do [*Modberg* 1207 Ch, *Motberge* 1265 Misc].

Moddershall St [*Modredeshale* DB]. 'Mōd-rēd's HALH.'

Modney Nf in Hilgay [*Modmeneya* 1283 Bodl, *Medmeneye* 1291 Tax]. 'Middle island.' Cf. MEDMERRY. First el. OE *meoduma* 'middle' with change *eo* > *o*.

Moggerhanger (mŏr-) Bd [*Mogarhangr'* 1216 Cl, *Mogerhanger* 1240 FF, *Mokerhanger* 1276 Ass, *-hangre* 1290 Cl, *Mouerhanguer* 1289 Ipm]. Second el. OE *hangra* 'slope'. First el. obscure. Its form was apparently early ME *moker*.

Mōlash K [*Molesse* 1212 Cur, *Malesse* 1240 Ass, *Molesshe* 1294 Cl, *Mollesh* 1315 Ch]. Perhaps identical in meaning with MATLASK, but with late OE *māl* 'action at law' (from OScand *māl*) as first el. Or the first el. might be OE *māl* 'mark'. See ÆSC.

Mole R D [*Moll* 1553 Pat]. A back-formation from MOLTON. Formerly *Nymet*.

Mole R Sx, Sr [*Moule* 1577 Harrison]. A back-formation from MOLESEY. Cf. DORKING, EMNETH.

Molescroft YE [*Molescroft* DB, 1203 FF], **Molesdon** Nb [*Molliston* 1242 Fees, *Moleston* 1256 Ass, *Molesdon* 1273 Cl]. '*Moll*'s or *Mūl*'s croft and TŪN.'

Molesey Sr [*Muleseg* a 675 BCS 34, *-eige* 933 ib. 697, *Molesham* DB, *Mulesee* 1212 Fees], **Molesworth** Hu [*Molesworde* DB, *Mulesworth* 1234 Cl]. '*Mūl*'s island and WORÞ.'

Molland D [*Mollande* DB, *Mollanda* 1100–3 (1332) Ch, 1204 Ch, *Mouland* 1202 FF, *Modland* 1205 FF], **North & South Molton** D [*Nort-, Sudmoltone* DB]. As Molland is a good way from the Moltons, it is probable that *Mol* was the name of (part of) the chain of hills (Exmoor Forest), on whose southern slope Molland and North Molton are. A hill-name *Mol* might possibly be derived from Welsh *moel* 'bare hill', if the word was adopted comparatively late. *Gloucester* from OBrit *Glēvum* appears as *Glowecestre* in DB. The vowel of *Glēvum* was the same as that of Welsh *moel* (from *mēl* < *mailo-*). *Glēvum* is from **Glaivo-*. If this is right, Molland and North Molton mean 'land and TŪN by *Moel* or the bare hill'. South Molton would be best explained as a later extension of North Molton.

Mollington Chs [*Molintone* DB, *Molyngton Banastre* 1287 Court], M~ O [*Mollintun* 1015 Wills, *Mol(l)itone* DB, *Mulinton* 1220 Fees, *Mollington* 1230 P]. 'The TŪN of *Moll*'s people.'

Molton D. See MOLLAND.

Mondrum Chs [*Mondrem* 1311 Ipm, 1316 Chamb, 1320 BM]. OE *mandrēam* 'joyous life among men, joyous noise'. Such a name might conceivably have been given to a place where village sports were carried on or feasts were held.

Monewden Sf [*Munega-, Mungadena* DB, *Munegeden* 1194 P, *Monewedene* 1254 Val]. OE *Mundinga-denu* 'the valley of *Munda*'s people' (cf. MUNDHAM), whence *Mundiga-, Mundga-* and *Mungadenu*. For the change *-inga-* > *-ewe-* cf. CANEWDON. But *nd* may have become *n* owing to dissimilation. Cf. also next names.

Mongeham (-ŭnj-) K [*Mundelingeham* 761, *Mundlingham* 833 BCS 190, 405, *Mundingeham* DB, *Mundingham* 11 DM, *Muningeham* 1195 P, *Monigeham* 1251 Ch]. 'The HĀM of *Mundel*'s people.' **Mundel* is a derivative of *Munda* (cf. MUNDHAM).

Mongewell (mŭnjel) O [(æt) *Mundingwillæ* 966–75 Wills, *Mongewel* DB, *Mungewell* 1242 Fees, 1281 Ch]. 'The spring or stream of the *Mundingas*.' Cf. MONEWDEN.

Monkhopton. See HOPTON.

Monkland He [*Leine* DB, *Munkelen* c 1180 Fr, *Monekeslane* 1193 P]. 'The part of *Lene* belonging to the monks' (i.e. the abbey of CONCHES in Normandy). Cf. EARDISLAND, KINGSLAND and LEOMINSTER. *Lene* is the name of an old district.

Monkleigh D [*Lega* DB, *Monckeleghe* 1266 Ep]. 'The LEIGH belonging to the monks' (of Montacute). See LEIGH.

Monkseaton. See SEATON.

Monksilver So [(æt) *Sulfhere* 11 KCD 897, *Selvere, Selvre* DB, *Siluria* Hy 2 (1290) Ch, *Monkesilver* 1249 FF]. Silver is very likely a stream-name derived from OE *seolfor* 'silver'. Cf. **Silver Beck** Cu [*Siluerbeck* 1285 For]. The meaning would be 'clear stream'. The monks were those of Goldcliff in Monmouthshire.

Monkton D [*Muneketon* 1244 FF], M~ **Wyld** Do [see WILD], M~ Du [*Munecatun* 1104–8 SD], M~ K [*Munccetun* c 960 BCS 1065], **West** M~ So [*Monechetone* DB, *Westmonketon* 1397 Buckland], M~ **Deverill** W [*Devrel* DB, *Deverel* 1254 Val, *Deverel Monketon* 1275 RH, *Deverel Monachor'* 1291 Tax], M~ **Farleigh** W [see FARLEIGH], **Bishop** M~ YW [*Munecatun* c 1030 YCh 7, *Monucheton* DB], **Moor & Nun** M~ YW [*Monechetune* DB, *Munketon on the Moor* 1300 Ch, *Nun Monketon* 1303 FA]. 'The TŪN of the monks.' The exact reason for the name is not always apparent, but places called Monkton must once have belonged to a monastery. M~ Du belonged to Jarrow from the 11th cent., **West M~** So and **M~ Deverill** W to Glastonbury, **Bishop M~** to the Archbishop of York.—M~ Deverill was originally DEVERILL (q.v.) and Monkton (or Monks') is a distinctive addition.—Nun M~ was Monkton long before the nunnery there was founded.

Monmore St [*Monnemere* 1291 Tax]. '*Manna*'s lake' or 'the men's lake'. There is a lake at the place.

Monmouth (mŭn-), Welsh **Trefynwy**, Monm [*Munuwi muða* 11 ERN, *Monemude* DB, *Munemuta* 1191 Gir]. 'The mouth of the MONNOW.' M~ is at the confluence of the Monnow with the Wye. The place is called *Aper Mynuy* 'the mouth of the Monnow', *Castell Mingui, Castellum de Mingui* 'the castle on the Monnow' c 1150 LL. Monmouth may be a translation of early Welsh *Aper Mynuy*.

Monnington in Straddel He [*Mane-, Manitune* DB, *Monintun* 1242 Fees, *Monyton Straddel* 1316 FA], **M~ on Wye** He [*Manitune* DB, *Moninton* 1237 Cl, *Monynton supra Wyam* 1418 BM]. '*Manna*'s TŪN' or 'the TŪN of *Mann(a)*'s people'. Cf. STRADDLE.

Monnow R He [*Mingui, Mynui* c 1150 LL, *Munuwi muða* 11 ERN]. 'Little Wye.' The Monnow is a tributary of the Wye. The first el. is a word for 'small' found in Co *minow, menow*.

Monsal Db [*Morleshal* 1200 P, *Mornes(h)ale* 13 Derby]. Second el. OE *halh* 'valley'. The first cannot be determined with the material available.

Montacute So [*Montagud* DB, (de) *Monteacuto* 1156 f. P]. A French name identical with MONTAIGU in France. One is nr Caen. The OE name was *Biscopestun* (*Biscopestone* DB). Montacute means 'pointed hill'. The modern form comes from the Latinized *Mons acutus*.

Montford Sa [*Maneford* DB, 1241 Cl, *Moneford* 1255 RH]. '*Manna*'s ford' or 'the men's ford' (perhaps in contradistinction to a maidens' ford).

Monxton Ha [*Anna de Becco* c 1270 Ep, *Monkestone* 15 VH]. Originally ANN. Monxton may in reality be 'Monks' Ann'. Cf. SARSON under ANN. M~ belonged to the abbey of Bec in Normandy.

Monyash Db [*Maneis* DB, *Moniasse* 1200 P, -*ass* 1242 Fees, *Moniasche* 1316 BM]. OE *manig æsc* 'many an ash', in contradistinction to the neighbouring **One Ash** [*Aneisc* DB], 'single ash'. Cf. **Moneylaws** Nb in Carham [*Manilawe* 1242 Fees]. 'Many hills or mounds.'

Moorby Li [*Morebi* DB, -*by* 1254 Val]. OScand *Mōra-bȳr* 'BY at the moors or fens'.

Moore Chs [*Mor* 1311 Ipm]. OE *mōr* 'moor, fen'.

Moorlinch So [*Mirieling* 971 BCS 1277, *Merielinz* 1196 P, *Murieling* 1202 FF, *Mirielinch* 1256 FF]. 'Pleasant hill.' Cf. HLINC. First el. OE *myrge* 'merry'.

Moorsholm, Great & Little, YN [*Morehusum* DB, *Great Moresum* 1242 FF, *Grant, Petite Moresum* 1273 Cl]. '(At) the moorhouses.' Moor is here 'waste upland'.

Moorsley Du [*Moreslau* Hy 2, *Morueslaue, Moreslawe* 12 FPD, *Moreslave* c 1190 Godric]. If the form *Morueslaue* is to be considered, the first el. may be that of MONSAL. Otherwise it is the word *moor*.

Moorton Gl [*Morton* 1301 BM], **M~** O [*Morton* 1208 Cur]. 'TŪN by a fen.' M~ O is near NORTHMORE.

OE **mōr** 'moor, waste upland; fen' is common in pl. ns. The usual meaning is 'fen'. 'Waste upland' is seen in DART-, EXMOOR and the like, MORCOTT, MOORSHOLM and others. See e.g. MOORE, MORE, MOORTON, MOR-, MORE- (passim), MURCOTT, MURTON. Cf. MERE.

Morborne Hu [*Morburne* DB, -*burna* 1158 P]. 'Fen stream.'

Morchard Bishop D [*Morchet* DB, 1166 P, *Morcherd* 1226 FF, *Morchet Episcopi* 1207 Ch], **Cruwys** (krōōz) **Morchard** D [*Morchet* DB, *Morceth* 1242 Fees, *Morcherde* 1262 Ep, *Morcestr' Crues* 1281 QW]. 'Great wood.' The name is British, the elements being Welsh *mawr* (OW *mor*) 'great' and *coed* (from *cēt*) 'wood'. The Morchards are c 5 miles apart, but were doubtless named from the same wood. The change from *Morchet* to *Morchard* is due to association with the word *orchard*.

M~ Bishop from the Bishop of Exeter.— **Cruwys M~** was held by Alexander de Crues before 1242 (Fees). The name may be from CRUYS-STRAËTE in Dép. Nord.

Morcombelake Do [*Mortecumbe* 1240 Wells]. Identical with *Mortan cumb* 1043–53 BCS 479 (W). *Morta* may be a pers. n. (cf. MORTLAKE). But if dial. *mort* 'a young salmon' is an old word, it is to be preferred in these two names. The fish-name is related to Norw *murt*, Swed *mört* 'roach', Icel *murti* 'young trout'. See LACU.

Morcott Ru [*Morcote* DB, 1177 P]. 'Cottage on the moor.'

Morda R Sa [*Mordaf* 1295 Ch]. 'Great *Taf*.' See TAME. The first el. is OW *mor* 'great' (Welsh *mawr*). The form -*daf* is due to lenition. **Morda** vil. is on the river.

Morden, Guilden & Steeple, Ca [*Mórdun* 1015 Wills, *Morduna* c 1080 ICC, *Gildene Mordon* 1204 Cur, *Stepelmordun* 1242 Fees], **M~** Do [*Mordone* DB, *Mordun* 1182 P], **M~** Sr [*Mordúne* 969 Crawf, -*dune* 1065 BM]. OE *mōr-dūn* 'hill in fens'.

On *Guilden* see GILMORTON.—*Steeple* from the church steeple.

Mordiford He [*Mordiforde* c 1230 Hereford, *Mordeford* 1242 Fees]. The first el. may be a Welsh *mor-dy* 'great house' (Welsh *mawr* 'great' and *ty* 'house').

Mordon Du [*Mordun* c 1050 HSC, 1104–8 SD, *Mordon* 1196 P]. Identical with MORDEN.

More Sa [*la Mora* 1181 P, *la More* 1198 FF]. OE *mōr* 'moor' or 'fen'.

Morebath D [*Morbade* DB, *Morbathe* 1259 Ep]. 'Bath in a moor or fen.' Bath refers to some chalybeate springs.

Moreby YE [*Morebi* DB, 1190 P]. Identical with MOORBY.

Morecambe La. A late name. The town is on Morecambe Bay, which came to be so called because of the identification of the bay with Ptolemy's *Morikámbē* suggested in 1771 by Whitaker.

Moredon W [*æt Mordune* 943 BCS 788, *Mordone* DB]. Identical with MORDEN.

Moreleigh D [*Morlei* DB, *Morlegh* 1242 Fees]. 'LĒAH in a moor.'

Moresby (mŏrĭsbĭ) Cu [*Moriceb*[*y*] Hy 2 StB, -*bi* 1195 P]. '*Maurice*'s BY.' Maurice is a French name.

Morestead Ha [*Morstede* 1172 Ep, 1291 Tax]. 'Place by a moor or fen.'

Moreton, Maids, Bk [*Mortone* DB], **North & South** M~ Brk [*Mortun* 891 BCS 565, *Sud-*, *Northmorton* 1220 Fees], M~ Chs nr Birkenhead [*Morton* 1291 Tax], M~ Chs nr Congleton [*Great Morton sub Lyme* 1289 Court], M~ **Hampstead** D [*Mortone* DB, *Morton Hampsted* 1493 Ipm], M~ Do [*Mortune* DB], M~ Ess [*Mortuna* DB], M~-**in-Marsh** Gl [*Morton* 714 BCS 130, *Morton in Hennemersh* 1253 Ch], M~ **Valence** Gl [*Mortune* DB, *Morton* 1220 Fees], M~ **Jeffreys** He [*Mortune* DB, *Morton Jeffrey* 1273 PNHe], M~ **on Lugg** He [*Mortune* DB, *Morton juxta Logge* 1291 Tax], M~ **Pinkney** Np [*Mortone* DB, *Geldenemortone* 1219, *Guldenemorton* 1226 Ep], M~ O [*Morton* 1291 Tax], M~ Sa nr Oswestry [*Mortune* DB], M~ **Corbet** Sa [*Mortone* DB, *Morton Corbet* 1284 Ch], M~ **Say** Sa [*Mortune* DB, *Morton de Say* 1255 RH], M~ St nr Colwich [*Mortone* DB], M~ St nr Gnosall [*Mortone* DB], M~ St nr Hanbury [*Mortune* DB], M~ **Morrell** Wa [*Mortone* DB, *Morton et Merehull* 1316 FA, M~ *Merehul* 1336 AD]. OE *Mōr-tūn* 'TŪN by a fen'.

M~ **Corbet** Sa was held by Richard Corbet c 1200. Cf. CHADDESLEY CORBETT.—M~ **Jeffreys** He from some early owner.—**Maids** M~ Bk is not sufficiently clear.—M~-**in-Marsh** Gl is a corruption of M~ *Henmarsh* (see HENMARSH).—M~ **Morrell** Wa was originally M~ and Morrell, the latter being *Merhull* 'boundary hill'. See (GE)MÆRE.—M~ **Pinkney** Np was held by Henry de Pinkeny in 1236 (Fees). The family name is from PICQUIGNY in Picardy.—M~ **Say** Sa from the Say family. It was held

by Hugh de Sai in 1199 (FF). Cf. HAMSEY.—M~ **Valence** Gl came to William de Valencia t. Hy 3. Cf. COMPTON VALENCE.

Morfe St [*Moerheb, Moreb* 736 BCS 154, *Morve* DB, *Morf, Morue* 1166 P]. A Brit name. Possibly a reduction of an OW *mor-dref* 'big village'.

Morland We [*Morlund* c 1140 ff. WR, *Murlund* J BM]. 'Grove by a moor.' Cf. LUND.

Morley Db [*Morlege* 1002 Wills, *Morelei* DB], M~ Du nr Evenwood [*Morley* 1312 RPD], M~ Nf [*Morlea* DB, *Morleg* 1201 FF], M~ YW [*Moreleia* DB, *Morlai* 1202 FF]. 'LĒAH by a fen or moor.'

Morningthorpe Nf. See THORPE, MORNING.

Morpeth Nb [*Morthpath* c 1200 Hexh, 1256 Ass, *Morpeth* 1200 Ch, *Morpath* 1257 Ch]. Apparently OE *morþ-pæþ* 'murder path'.

Morston Nf [*Merstona* DB, *Marston* 1185 P, *Merston* 1252 Ch]. Identical with MARSTON.

Mort(e)hoe D [*Morteho* DB, 1168 P, 1204 Cur]. Near by is **Morte Point**, a promontory. No doubt the first el. of Morthoe is the name of the promontory. Morte is related to dial. *murt* 'small person', MHG *murz* 'stump', the fish-name *mort* (see MORCOMBELAKE). The name may mean 'the stump'. Second el. HŌH.

Mortham YN [*Mortham* DB, 1270 Ipm]. '*Morta*'s HĀM.' Cf. MORTLAKE. A pers. n. *Morta* would be related to the words mentioned under MORCOMBELAKE and MORTEHOE.

Morthing or **Morthen** YW, a district [*Morthinges* J Ass, *Mordhingg* 1230 P, *Morthing* 1297 Subs]. Probably *Mōr-þing*, lit. 'assembly of (the people of) the moors', but used in the sense 'the moor district'.

Mortimer Brk. See STRATFIELD MORTIMER.

Mortlake Sr [*Mortelaga, -lage* DB, -*lace* 11 DM, *Murtelac* c 1120 Eadmer, *Mortelak* 1228 Cl]. If the second el. is OE *lacu* 'stream', the first is very likely the fish-name *mort* 'young salmon' (see MORCOMBELAKE). The DB forms may suggest that the second el. is rather dial. *lag* 'long, narrow, marshy meadow' (cf. *lacgeburnan* 757–75 BCS 219). If so, the first el. is rather the pers. n. *Morta*. See MORTHAM.

Morton Db [*Mortun* 1002 Wills, -*e* DB], M~ Du nr Houghton le Spring [*Mortona* 1183 BoB], M~ **Palms** Du [*Mortona* 1208–10 Fees], M~ **Tinmouth** Du [*Mortun* c 1050 HSC, 1104–8 SD], M~ Li nr Gainsborough [*Mortune* DB], M~ **by Bourne** Li [*Mortun* DB], M~ **by Lincoln** Li [*Morton* 1242 Fees], M~ **on the Hill** Nf [*Morton* 1196 Cur, 1219 Fees], M~ Nt nr Retford [*Nortmorton* DB, *Mortona* 12 DC], M~ Nt nr Southwell [*Mortune* 958 YCh 2, DB], M~ **Bagot** Wa [*Mortona* DB, *Morton Bagot* 1291 Tax], **Abbots** M~ Wo [*Mortun* 708

BCS 120, -e DB], **Birtsmorton & Castle M~** Wo [*Mortun* 1235 Ch, *Brittesmoretone* 1204 Cur, *Castel Morton* 1346 FA], **M~ Wt** [*La Morton* 1287-90 Fees], **M~** YN [*Mortun* DB], **M~ upon Swale** YN [*Mortun* DB], **East & West M~** YW [*Mortune* DB, *Est-*, *Westmorton* 1231 FF]. See MORETON.

Abbots M~ Wo belonged to the Abbot of Evesham.—**M~ Bagot** Wa came to William Bagod t. Hy 2. Bagot is a Fr pers. n., originally a nickname.—**Birtsmorton** Wo belonged to the family of *le Bret* from the 12th cent. *Bret* means 'Breton'.—**M~ Palms** was held by Bryan Palmes till 1569.—**M~ Tinmouth** Du belonged to the monastery of Tynemouth.

Morvah Co [*Morueth* 1377 PT]. '(The church of) St. Morwetha.'

Morval Co [*Morval* 1238 FF, *Morvalle* 1309 Ep]. Etymology obscure.

Morville Sa [*Membrefelde* DB, *Momerfeld* 1200 P, *Mainerfeld*, *Menneresfeld* 1235 Cl, *Momerefeld* 1291 Tax]. **M~** is on **Mor Brook**, which was probably once *Memere* or *Meomere*. Cf. MIMRAM, FELD.

Morwenstow Co [*Morwennestohe* 1273 Ep, *Morewynstouwe* 1291 Tax]. 'The church of St. Morwenna.'

Morwick (mŏrĭk) Nb [*Morewic* 1161 P, *Morwic* 1166 RBE]. 'wīc in a fen.'

OE **mos**, ON *mosi* 'bog, swamp, morass' is found in MOSS, MOZE, CHAT MOSS, MOSEDALE, MOSSER, MOSTON, MOZERGH and others.

Mosb(o)rough Db [*Moresburh* 1002 Wills, *-burg* DB, *Moresbur.* Hy 3 BM]. 'Fort in a moor.'

Mosedale Cu [*Mosdale* 1300, *Mosedal* 1308 Ipm]. 'Valley with a moss or peat bog.'

Moseley St [*Moleslei* DB, *Mollesleg* 1227 Ass, *-le* 1242 Fees]. 'Moll's LĒAH.'

Moseley Wo [*Museleie* DB, *Moseleia* 1195, *-lege* 1197 P]. 'LĒAH infested by mice.'

Moss YW [*Mosse* 1476 FF]. 'The moss or morass.'

Mosser Cu [*Moserg* 1203 Cur, *Mosergh* 1321 Ipm]. 'ERG or shieling in a moss.'

Mosterton Do [*Mortestorne* DB, *-torn* 1196, *-thorn* 1210 FF]. '*Mort*'s thorn-bush.' **Mort* is a pers. n. related to *Morta* in MORTHAM. Or possibly *mōteres þorn.* Cf. MOTTISFONT.

Moston Chs nr Chester [*Morcetone* c 1125, *Morsetona* c 1150, *-ton* 1208-26, *Morston* c 1305 Chester]. Very likely OE *Mōrsæ-tūn* 'TŪN by a lake *Mōrsæ* or lake by a moor'. The place is in a low situation.

Moston Chs nr Middlewich [*Moston* 1286 Court, 1289 Ipm], **M~ La** [*Moston* 1195 FF, 1235 Ass], **M~ Sa** [*Mostune* DB, *Mostone* 1327 Subs]. 'TŪN by a moss.'

OE **(ge)mōt** 'meeting' is found in pl. ns. in two senses: (1) 'junction of streams', as in EAMONT, EMMOTT, (2) 'meeting, assembly, moot', as in MOBBERLEY, MODBURY, MOTCOMBE, SKIRMETT. OScand *mōt* 'meeting, junction of streams' is found in BECKERMET, BECKERMONDS, LANDMOTH.

Motcombe Do [*Motcumbe* 1311 Ipm]. OE *gemōt-cumb* 'valley where moots were held'.

Motherby Cu [*Mothersby* 1317, *Motherby* 1323 Ipm]. '*Mothir*'s BY.' ODan *Mothir*, OSw *Moðor* is a pers. n., from *mothir* 'mother'.

Mottenden K [*Modinden* 1236 Misc, *-denn* 1275 RH, *Motinden* 1251 FF], **Mottingham** K [*Modingahema mearc* 862, *Modingahammes gemæro* 973 BCS 506, 1295, *Modingeham* 1044 Th, 1081 Fr]. 'The DENN (swine-pasture) and HAMM (or HĀM) of *Mōda*'s people.' **Mōd(a)* is a short form of names in *Mōd-*, *-mōd*. Cf. OHG *Muato*.

Mottisfont Ha [*Mortesfunde*, *Mortelhunte* DB, *Motesfont* 1167 f., *-funt* 1170 P, *Motefunt* 1203 Cur]. Second el. OE FUNTA 'spring'. The first is probably OE *mōtere* as in next name. The place might have been called alternatively *Mōtesfunta* 'spring where moots were held'.

Mottistone Wt [*Modrestan* DB, *Motereston* 1291 Ep, *Motestan* 1176 ff. P]. OE *mōteres stān* or *mōtera stān* 'the stone of the speaker(s) or pleader(s)'. The name must refer to a stone at a meeting-place from which the judge spoke or a person pleaded his cause. The place was clearly named from the Mottistone, a large menhir on the hill above it.

Mottram St. Andrew Chs [*Motre* DB, *Mottrum* 1248 Ipm, 1285 Court, ?*Motern* 1304 Chamb], **M~ in Longdendale** Chs [certain early forms not found]. Has been derived from OE *mōt-ærn* 'court house'; cf. (GE)MŌT, ÆRN. But in **M~** St. Andrew is **Mottershead** [*Moctresheved* 1287 Court, *Mottresheved* 1304 Chamb]. The first el. of this seems to be identical with **Mocktree** in Bromfield Sa [*Moctre* Hy 2 (1235) Ch, *Moctro*, *Mouhtre* 1235 Glouc] and with MOCHDRE in Wales, which means 'pig farm' (Welsh *moch* 'pigs' and *tref* 'homestead, hamlet'), the second being OE *hēafod* 'hill'. Mottram may be identical with the first el. of Mottershead. If so, it is the dat. plur. of the Welsh word.

Mouldsworth Chs [*Moldeworthe* 1153-81 Chester, *Molde(s)w(o)rth* 1260 Court]. **M~** is at the foot of a considerable hill. The first el. probably refers to the hill and is OE *molda* 'top of the head' used in a transferred sense. *Moldi* is a common name of hills in Norway. See WORP.

Moulsecoombe (mows-) Sx [*Muliscumba* c 1110, *Molescumba* 1121 AC], **Moulsford** Brk nr Wallingford [*Muleford* c 1110 Bodl, *Mullesford* 1130-5 Eynsham, *Muleford* 1207 Cur, *Molesford* 1220 Fees], **Moulsham** (-ōō-) Ess [*Mulesham* 1065 BM,

1202 FF, *Molesham* DB], **Moulsoe** (-ŭls-) Bk [*Moleshou* DB, *-ho* c 1155 Oxf, *Mulesho* 1189 P]. '*Mūl*'s coomb, ford, HĀM, HŌH.' *Mūl* is a known name. It is found also in *Muleshamstede* 891 BCS 565, which must have been nr Moulsford. It is just possible that OE *mūl* 'mule' may be the first el. of one or two of the names. The pers. n. *Mūl* might be a nickname from *mūl* 'mule', but more likely it belongs, like OE **Mūla*, to the old word for 'muzzle' found in G *Maul*, OHG *mûla*, MLG *mûle* fem., OFris *mûla*, ON *mûli* masc. No doubt this word was once known to the Anglo-Saxons.

Moulton Chs [*Moletune* DB, *Multon* 1260 Court], M~ Li [*Multune* DB, *Muleton* 1165 P, *-a* Hy 2 DC, 1209-19 Ep], M~ St. **Michael, Little** M~ Nf [*Mulantun* c 1035 Wills, *Muletuna* DB, 1183 Holme, *Muleton Maior, Minor* 1254 Val], M~ Np [*Multun* 1066-75 GeldR, *Multone* DB, *Multon* 1200 Cur, *Muleton* 1202 Ass, *Molentun* 1205 Pp], M~ Sf [*Muletuna* DB, *-ton* 1198 P, 1235 FF, *-tun* 1242 Fees], M~ YN [*Moltun* DB, *Muleton* 1176 ff. P]. M~ Nf is '*Mūla*'s TŪN'. On *Mūla*, which is identical with the ON byname *Mūli*, see MOULSECOOMBE. The other Moultons probably have the same etymology, though OE *Mūl(a)tūn* 'TŪN where there were mules' is at least a possible alternative. M~ Np (*Multun* GeldR) is the most likely case.

Moulton Nf nr Yarmouth [*Modetuna, Mothetuna* DB, *Modeton* R 1 Cur, *Mothetun* 1202 FF]. '*Mōda*'s TŪN.' Cf. MOTTENDEN. Or the first el. may be ON *Mōði*, OSw *Modhi*.

Mountfield Sx [*Montifelle* DB, *Mundifeld* 12 (1432) Pat]. '*Munda*'s FELD.' Cf. MUND-FORD.

Mountnessing Ess [*Ginga* DB, *Gynges Munteny* 1237 FF, *Mounteneysynge* 1467 AD]. See ING. The name means 'Mounteney's ING', Mounteney being the name of the local family, derived from MONTENAY or MONTIGNY in Normandy.

Mounton Monm [*Monketowne* 1535 VE]. 'The TŪN of the monks (of Chepstow Priory).'

Mountsorrel Le [*Munt Sorel* 1152 BM, *Muntsorell* 1190 P]. The place had a strong Norman castle, whose name may be a transplantation of MONTSOREAU nr Saumur or MONT-SOREL nr Rennes. Mountsorrel is on the SOAR, and it has been suggested that *-sorrel* is 'Soar hill'.

Mousen Nb [*Mulefen* 1167, *Mulesfen* 1186 P, 1212 Fees, *Mullesfen* 1219 Fees]. '*Mūl*'s fen.'

Mow (mow) **Cop** Chs [*Rocha de Mowa* 1280, *Mouhul* 1317 BM]. The place is on a hill of 977 ft. at the county boundary. *Mow* is OE *mūga* 'heap' and probably refers to a boundary cairn. *Cop* is OE *copp* 'hill'.

Mowsley (mowzlï) Le [*Muselai* DB, *Muslai* Hy 2 (1318) Ch, *Musele* 1200 Cur]. 'Mouse-infested LĒAH.'

Mowthorpe YN [*Muletorp* DB, 1167 P, *Multhorp* 12 YCh 34]. '*Mūli*'s thorp.' Cf. MOULTON.

Moxby YN [*Molzbi, Molscebi* DB, *Molesbi* 1158, *Molsebi* 1190 ff. P]. '*Mōðolf*'s BY.' ON *Mōðolfr* is a pers. n. Cf. ON *Hrólfr* from *Hrōð-wulfr*.

Moxhull Wa [*Moxhul*' c 1200 Middleton, *Moxshulf* 1428 FA]. Second el. OE SCYLF 'ledge, hill'. The first may be OE *Mocca* pers. n. or OE *mox* (or *Mox*) in *Moxes dun* 825 BCS 390 (cf. ib. 566, 1071).

Moze Ess [*Mosa* DB, 1254 Val, *Mois* 1236 Cl, *Moese* 1270 FF]. OE *mos* 'marsh, moss'.

Mozergh We [*Moserga* 1196 FF]. Identical with MOSSER.

Muchelney So [(of) *Miclanige* c 990 ASCh, *Muceleneia* 1084 GeldR, *Micelenye* DB, *Mucheleneia* 1160 P]. 'Large island.'

Mucking Ess [*Muc(h)inga* DB, *Muckinges* 1199 FF, 1203 Cur]. '*Muc(c)a*'s people.'

Mucklestone St [*Moclestone* DB, *Mukleston* 1221 FF]. '*Mucel*'s TŪN.'

Muckton Li [*Machetone* DB, *Munchetune* c 1115 LiS, *Muketun* c 1110 Fr, Hy 2 DC]. '*Muca*'s TŪN.'

Mudford So [*Mudiford, Mundiford* DB, *Mudiford* 1176 P, 1201 Ass, *Mudeford* 1201 FF]. 'Muddy ford.' *Muddy* is found first in 1413, *mud* in the 14th cent. (OED). Evidently the words go back to OE times.

Mudgley So [*Mudesle* 1157, *-liegh* 1164 Wells, *Modeslega* 1176 ib.]. The elements may be *mud* (cf. MUDFORD) and LĒAH, though the gen. form is against this.

Mugginton Db [*Mogintun* DB, *Mugginton* 1242 Fees, *Mokyncton* 1330 FA]. 'The TŪN of *Muca*'s people.' Or there may have been an OE pers. n. *Mucga*.

Muggleswick Du [*Muclingwic* c 1170, *-wik* 13 FPD, *Mukeleswyk* 1291 Tax]. '*Mucel*'s wīc' and 'the wīc of *Mucel*'s people'.

Muker YN [*Meuhaker* 1274 YInq]. 'Narrow field.' The elements are ON *miór* 'narrow' and *akr* 'field'.

Mulbarton Nf [*Molkebertuna, Molkebertestuna* DB, *Mulkebertun* 1250 Ass, *-ton* 1254 Val]. OE *Meolc-beretūn* 'outlying dairy-farm'. Cf. MEOLUC, BERETŪN. For the change *eo* > *o, u* cf. MODNEY.

Mulgrave YN [*Grif* DB, *Mulegrif* c 1160 Whitby, *Mulegreve* 1251 Ass]. Originally *Grif* from ON *gryfia* 'pit, hollow'. The addition may be ON *Mūli* pers. n. or rather ON *mūli* 'headland, crag'. This would suit the situation.

Mullion Co [(Rector) *Sancti Melani* 1262 Ep, *Seynt Melan* 1284 FF]. '(The church of) St. Melan.' Identical with **St. Mellons** Monm [(ecclesia de) *sancto Melano* 14 LL].

Mulwith YW [*Mulewath* c 1175 YCh 124, 1241 Ep]. '*Mūli*'s ford.' Cf. MOULTON and VAÐ.

Mumby Li [*Mundebi* DB, *Mumbi* c 1115 LiS, *Munbi* 1160 f. P]. '*Mundi*'s BY.' First el. OScand *Mundi*.

Muncaster Cu [*Mulcaster* c 1150, *Molecastre* c 1190 StB, *Mulecastr*' 1236 P]. '*Mūla*'s CEASTER or Roman fort.' Cf. MOULSECOOMBE. Or the first el. is ON *mūli* 'crag'. Cf. MULGRAVE.

Munden, Great & Little, Hrt [*Mundene* 944–6 BCS 812, *Munddene* 11 E, *Mundene* DB, *parva Mundena* 1209–19 Ep, *Munden Magna* 1254 Val], **Mundon** Ess [*Munduna* DB, *-donia* 1166 RBE, *-dun* 1212 Fees]. '*Munda*'s valley and hill.' Cf. MUNDFORD. The early occurrence of Munden without a trace of the original ending *-an* may, however, suggest that the first el. at least of Munden is a common noun *mund*, but the meaning of such an element is obscure. Possibly early Mod *mound* 'hedge' goes back to OE *mund*.

Mundesley Nf [*Muleslai* DB, *-le* c 1150 Crawf, *Munesle* 1208 Cur, 1254 Val]. '*Mundel*'s LĒAH.' Cf. MUNSLEY, MUNSLOW, also MONEWDEN, MONGEHAM. *Mundles-* became *Munles-*, whence *Mules-* and *Munes-*.

Mundford Nf [*Mundefort* c 1050 KCD 907, *-forda* DB, *-ford* 1242 Fees], **Mundham** Nf [*Mundaham* DB, *Mundham* 1158 P, *Mundeham* 1197 FF, 1264 Ch], **North & South Mundham** Sx [*se northra, other Mundan ham* 680 BCS 50, *Mundreham* DB]. '*Munda*'s ford and HĀM.' *Munda* is not recorded in independent use, but is a regular short form of names in *-mund* and corresponds to OG *Mundo*, ON *Mundi*.

Mundon Ess. See MUNDEN.

Mungri·sedale (mŭn-) Cu [*G[r]isdale* 1254 Ipm, *Mounge Gricesdell* 1600 CWOS vi, *Mung Grisedale* 1647 CWNS xxv]. 'Pig valley.' Cf. GRISEDALE. The church appears to be dedicated to St. Mungo, and the addition to the old pl. n. is presumably the saint's name, as pointed out PNCu(S).

Munsley He [*Munes-, Moneslai, Muleslage* DB, *Muneslega* c 1175 Hereford], **Munslow** Sa [*Mulslaye* 1110–15 Eyton, *Mulselawahundr*. 1187 P, *Munceloue* 1261 Pat, *Munsselawe* 1256 Ass]. '*Mundel*'s LĒAH and HLĀW or tumulus.' Cf. MUNDESLEY.

Murcot O [*Morkote* 1149 Osney], **Murcott** W [*Morcotun* 1065 KCD 817]. 'COT(S) in a fen.'

Murrow Ca [*Morrowe* 1376, 1383 PNCa(S)]. The elements are OE *mōr* 'marsh' and *rāw* 'row (of houses)'.

Mursley Bk [*Muselai* DB, *Murselai* c 1155 Oxf, *Muresle* 1195 Cur, *Meresle* 1203 Cur]. Looks like '*Myrsa*'s LĒAH', but *Myrsa* pers. n. is otherwise unknown.

Murston K [*Morieston* 1165 f., *Morestona* 1169 P, *Moriston* 1199 P, 1212 StAug]. 'TŪN by a fen.' But the earliest forms perhaps rather suggest 'Maurice's TŪN'. Cf. MORESBY.

Murton Du nr Dalton le Dale [*Mortun* 1155 FPD], M~ Du nr Sedgefield [*Westmorton* 1432 PNNb], M~ Nb [*Morton* 1204 Ch], M~ We [*Morton* 1288 FF], M~ YN at York [*Mortune* DB], M~ YN nr Hawnby [*Mortun* DB]. Identical with MORETON.

Musbury D [*Musberie* DB, *-biri* 1219 Fees], M~ La [*Musbiri* 1311 LaInq]. 'Old fort inhabited by mice' or 'mouse burrow'. The first alternative is probable for M~ D, where there is an ancient camp.

Muscoates YN [*Musecote* c 1160 Riev, *-s* 1227 FF], **Muscott** Np [*Misecote* DB, *Musecote* 12 NS, *-cot* 1220, *-cotis* 1236 Fees]. 'Mouse-infested huts.'

Musgrave, Great & Little, We [*Musegrave* c 1215, c 1240 CWNS xxiv, *Magna, Parva Musegrave* 1292 QW]. 'Grove frequented by mice.'

Muskham, North & South, Nt [*Muscham, Nordmuscham* DB, *Muscampe* 1155 Fr, *-camp* 1166 RBE, 1187 f. P, 1212 Fees, *Sutmuscham* 1242 Fees]. OE *mūs-camp* 'mouse-infested field'.

Muston Le [*Moston* c 1125 LeS, *Mustun* 1106–23 (1333) Ch, *Museton* 12 DC, *Muston* 1200 Cur], M~ YE [*Mustone* DB, *Mussetuna* c 1170 YCh 1174, *Museton* 1201 Cur, 1240 FF]. The last may well be '*Músi*'s TŪN', with the ON byname *Músi* as first el. M~ Le may be OE *Mūs-tūn* 'mouse-infested TŪN', or *Must-tūn*, the first el. being an el. *must* 'muddy stream or place' or the like found in old records (e.g. *Must* 972 BCS 1280, used of a fen stream).

Muswell Hill Mx [*Mosewella* 1152–60 BM]. 'Mossy well.'

Mutford Sf [*Mutford* DB, 1157 P, 1198 FF, 1212 Fees, *Muthford* 1264 Ipm]. Apparently OE *(ge)mōtford* 'ford at which moots were held', in spite of the early *u*-forms. M~ is also the name of a hundred.

OE **mūþa** 'mouth of a river' is often the second el. of pl. ns., the first being a river-name, as CHAR-, DART-, STOUR-, WEYMOUTH, JESMOND. Cf. PORTSMOUTH.

Mutlow Chs [*Motlowe* 1354 Chamb]. OE *(ge)mōt-hlāw* 'mound where moots were held'.

Muxton Sa nr Wellington [*Mukeleston* 1186 Eyton]. '*Mucel*'s TŪN.'

Myerscough La [*Mirscoh* 1246 Ass].

'Marshy wood' (ON *mýrr* 'marsh, bog' and *skógr* 'wood').

OE **mylen** 'mill' (from Lat *molina*) is a common pl. n. el. See MELLIS, MELLS, MILLOM, MIL-, MILL- (passim), MELBOURNE Db, MELDRETH, MELFORD, MELPLASH, MELVERLEY, MILEHAM. Sometimes used as a second el., as in SHOTTER-, WESTMILL.

Mylor Co [(Ecclesia) *Sancti Melori* 1258 Ep, (de) *Sancto Meloro* 1291 Tax]. '(The church of) St. Melor.' The saint was also known in Brittany (St. Melor or Méloir).

Mynde He [*la Munede* 1299 Ipm]. Welsh *mynydd* 'mountain'.

Myndtown Sa [*Munete* DB, *Muneta* 1166 RBE, *la Munede* 1181 P]. Originally MYND (*Munede* &c.) from Welsh *mynydd* 'mountain', *town* being a late addition. The place is by LONGMYND. Cf. MINTON.

Myne So. See MINEHEAD.

OScand **mynni** 'mouth of a river', often in pl. ns., as Sw ÅMINNE 'mouth of the river', is found in MINSMERE, AIRMYN, STALMINE.

OE **mynster** 'minster' (from Lat *monasterium*) was used in OE in the senses 'monastery' and 'church of a monastery', but the word must also have been used of a parish church. On *mynster* in the sense *matrix ecclesia* see Stenton, *Anglo-Saxon England*, pp. 148 ff. See MINSTER, MINSTER-, MISTER-

TON, CHAR-, EX-, LEO-, LY-, STURMINSTER &c., EMSTREY.

OE **myrge** 'merry' in p l. ns. no doubt means 'pleasant'. See MARLEY K, MARVELL, MEREVALE, MERIDEN, MERRINGTON Sa, MIRFIELD, MOORLINCH.

ON **mýrr**, OSw *myr*, Dan *myr*, ME *mire* 'wet, swampy ground, a boggy place' is sometimes found in pl. ns., as MYERSCOUGH, AINDERBY MIRES. But *-mire* in pl. ns. sometimes represents OE *mere*.

OE **(ge)mýþe** 'waters' meet, junction of streams' is fairly common in pl. ns. See e.g. MEAFORD, MEETH, -E, MIDDLE, MIDFORD, MITFORD, MITTON, MYTON, MYTTON. It is found alone also in **The Mythe** Gl [*Mitha* 1291 Tax], the name of a place at the junction of the Avon and the Severn; in the dat. plur. in **Mytham Bridge** Db [*Mithomford, Mythomstede* 1285 For]. **Mytholmroyd** (mīdhumroid) YW [*Mithomrode* 1307 Wakef] is 'the clearing at Mytholm' (OE *gemýþum*). See ROD.

Mythop La [*Midehope* DB, *Mithop* 1212 Fees]. Identical with MEATHOP.

Myton Wa [*Mytun* 1033 E, *Muitone* DB], M~ YE (lost) [*Mitun* DB, *Miton* 1196 FF], M~ (-i-) on Swale YN [*Nyŏtun, Mytun* c 972 BCS 1278 f., *Mitune* DB], **Mytton** Sa [*Mutone* DB]. OE (ge)mýþ-tūn 'TŪN at the junction of streams'.

N

Naburn YE [*Naborne* DB, *Naburn* 1167, -a 1169, -e 1230 P, *Neiburn* 1200 Cur]. N~ was originally the name of a stream. First el. perhaps OE *nafu* 'nave' (here 'bend'?). Cf. (to) *nafæ* 909 BCS 624. Derivation from an OE *Narwburna* 'narrow stream' is alternatively suggested by the writer in *Studia neophilologica*, x, p. 105 f.

Nackington K [*Natyngdun* 993 Hyda, *Latintone* DB, *Natindune* 11 DM, *-don* 1200 FF]. An OE *næt* 'wet' must be assumed on the strength of NATELY, NETLEY. This may be the first el. also of Nackington, which would mean 'wet hill'. Cf. WATTON YE. The change of *t* to *k* is late.

Nacton Sf [*Nachetuna* DB, *-ton* 1165 P, *Naketun* 1233 Fees]. The first el. is probably a pers. n., e.g. ON *Hnaki* or *Nǫkkvi*. Scand pers. ns. are common as the first el. of pl. ns. in the district.

Nadder R W [*Noodr* 705, (on) *Nodre* 860, 901, (be) *Noddre* 958 BCS 114, 499, 588, 1030]. The base is a Brit *Nŏtr*, derived from the root *snā* 'to flow' in Welsh *nawf*, OIr *snám* 'to swim', Ir *snuadh* 'stream' &c. Cf. NOE.

OE **næddre** 'adder'. See NETHERFIELD.

OE **næss** 'headland, cape', ON *nes*, OSw *næs* the same are found in several pl. ns. There must also have been an OE sideform *ness*, seen in NESS Sa and others, and NAZE Ess presupposes a form *næs* (with short *s*). The Engl word also meant 'headland, projecting ridge', as in NESS Sa, YN, NAZEING, NASSINGTON, NESWICK, CLAINES, TOTNES, while the sense 'cape' is seen in NASS, NAZE, SHARPNESS, WIDNES and others. OScand *nes* is the second el. of AMOUNDERNESS, FURNESS, HOLDERNESS, SKEGNESS and others, while NESS Chs and some others are doubtful.

Nafferton Nb [*Natferton* 1187 P, *Natfertun* 1236, *-ton* 1242 Fees], N~ YE [*Nadfartone* DB, *Natferton* c 1185 YCh 623]. '*Nattfari*'s TŪN.' ON *Náttfari*, ODan *Natfari* pers. n., lit. 'wanderer by night', is on record.

Nailsbourne So [*Nailesburn* 1200 FF], **Nailsea** So [*Nailsi* 1196 P, *Naylesye* 13 or 14 BM], **Nailstone** Le [*Neylliston* 1209–35, *Naylestone* 1225 Ep], **Nailsworth** Gl [*Nailleswurd* 1196, *-wurŏe* 1197 P, *Naylesworth* 1247 Berk]. '*Nægl*'s stream, island, TŪN and WORÞ.' OE **Nægl* is a byname from *nægl* 'nail', corresponding to OG

Nagal, ON *Nagli*. The same pers. n. is the first el. of *Neglescumb* 854 BCS 476 (nr Nailsbourne), and of *Neglesleag* 716–43, (on) *Nægleslege* 896 BCS 164, 574 (nr Nailsworth).

Nantwich Chs [*Wich* DB, *Wicus* Hy 2, *Nametwihc* 1194, *Wicus Malbanc* Hy 3 BM, *Nantwich* 1281 Court, *le Namtewyche Maban* 1302 Pat]. OE wīc 'town'. *Nant-* is from *named* 'famous'.

The manor was held by Willelmus Malbedeng in 1086 (DB). Hence sometimes Wich Malbank. Cf. CLIFTON MAYBANK.

Nappa YN [*Nappay* 1251 Ch, 1331 FF], N~ YW [*Napars* DB, *Nappai* 1182–5 YCh 199, *Naphay* 1226 Ep], **Napton on the Hill** Wa [*Neptone* DB, *Nepton* 1170 P, *Napton* 1174, 1176 P, *Napton, Cnapton* 1236 Fees]. The first el. of all three is OE *hnæpp* 'bowl', used in a transferred sense of a hill resembling an inverted bowl. The second el. is OE *gehæg* 'enclosure' and TŪN.

Nar R. See NARBOROUGH Nf.

Narborough Le [*Norburg* 1209–35 Ep, *Northburg* 1254 Val]. 'North fort.'

Narborough Nf [*Nereburh* DB, *-burg* c 1150 Crawf, 1254 Val], **Narford** Nf [*Nereforda* DB, *-ford* 1166 P, 1254 Val]. The places are not far apart on the river NAR. But the river-name is clearly a late back-formation. *Nere-* may be a derivative of OE *nearu* 'narrow' (an OE *neru from *narwīn) meaning 'narrow place, pass'. The names would mean 'BURG and ford at a pass'.

Nare Point Co. See PENARE.

Narford. See NARBOROUGH Nf.

Naseby (-āz-) Np [*Navesberie* DB, *Navzebe[r]ia* 1094 Fr, *Nauesbi* 1167 P]. '*Hnæf*'s BURG.' *Hnæf* is a figure of heroic saga (Beowulf, Widsith), and very likely the prehistoric fort to which the name originally referred was held to have been built by him. OE BURG was replaced by OScand BY.

Nash Bk [*Esse* 1231 Cl], N~ He [*Nasse* 1291 Tax, *Asshe* 1308 Ipm], N~ Sa [*Eshse* 13 BM, *Assh* 1308 Ipm]. 'The ash-tree(s).' OE *æt þæm æsce* or *æscum* became ME *atten ashe* and by wrong division *atte Nashe*.

Nass Gl [*Nest* DB, *Nasse* 1327 Subs]. OE *næss* 'cape'.

Nassington Np [*Nassintone* DB, *-tona* 1168, *Nessinton* 1177, 1191 P]. 'The TŪN of the *Næssingas* or dwellers on the NÆSS.' N~ is on a broad headland, which may have been called *Næss*. But the place is just outside the old **Nassaburgh** hd [(de) *Nasso* 12 NS, *Nes de Burc* 1180 P, *Nassus burgi* 1227 BM]. This name means 'the ness of Peterborough'. The hundred forms a promontory between the Welland and the Nene and is referred to in BCS 1128 as 'þam nesse þe Medeshamstede (i.e. Peterborough) onstent'. Probably the *Næssingas* of Nassington were settlers coming from that ness. Cf. NÆSS.

Nateby La [*Nateby* 1204 FF], N~ We [*Nateby* 1246 Ass]. '*Nate*'s BY.' There are some traces of an OScand pers. n. *Nate*. ON *nata* 'nettle' might also be thought of as first el.

Nately Scures, Up N~ Ha [*Nataleie* DB, *Natelega* c 1195, *-leye* 1234 Selborne, *Nateligh Scures, Opnatelegh* 1291 Tax, *Upnateley* 1274 RH]. 'Wet LĒAH.' The first el. is no doubt an unrecorded OE *næt* 'wet' corresponding to OHG *naz*, OLG *nat*. Cf. NETLEY MARSH, NOTGROVE.

N~ Scures was held by Roger de Scures in 1220 (Cur). Scures from ESCURES in Normandy.

Natland We [*Natalund* c 1175, *-lunt* 1246 Kendale, *Natelund* 1246 ib.]. Perhaps '*Nate*'s grove'. Cf. NATEBY and LUND.

Natton Gl [*Natone* DB, *Nacton* 1291 Tax, *Natton* 1327 Subs]. OE *Nēat-tūn* 'cattle farm'.

Naughton Sf [*Nawelton* c 1150 Bury, *Nauelton* 1191 FF, *Navelton* 1254 Val]. Either '*Nagli*'s TŪN', the first el. being ON *Nagli* (cf. NAILSBOURNE), or OE *Nafol-tūn*, the first el. being OE *nafola* 'navel' used in some transferred sense. The absence of spellings with *gh* or *g* tells against the first alternative.

Naunton Gl on the Windrush [*Niwetone* DB, *Newenton* 1287 QW, 1291 Tax], N~ Gl NW. of Winchcomb [?(æt) *Niwantune* 1004 Wills, *Niwetone* DB, *Newenton* 1235 Cl], N~ Wo in Ripple [*Newentone* c 1120 PNWo], N~ **Beauchamp** Wo [(in) *Niuuantune* 972 BCS 1282, *Newentune* DB, *Newenton Beauchamp* 1370 BM]. A doublet of NEWNTON, OE (æt) *Nēowantūne, Nīwantūne*. The sound-development is not clear, possibly it was *Nēowan- > Neōwan- > Nōwantūne*, whence *Naunton*, but none of the early forms point to it.

N~ Beauchamp was held by William de Bellocampo in 1167 (P). See ACTON BEAUCHAMP.

Nāvenby Li [*Navenebi* DB, *Nauenebi* 1170, 1190 P]. '*Nafni*'s BY.' *Nafni* (LVD) is ODan, ON *Nafni*.

Navestock Ess [*Nasingestok* 967 BCS 1210, *Naesingstoc* c 970 Bodley MS, *Nas(s)estoca, Nessetocha* DB, *Navestoke* 1283 FF]. 'STOC belonging to NAZEING or to the Nazeing people.' The change to *Nave-* is not easy to explain.

Naworth Castle Cu [(peel of) *Naworthe* 1323 Sc, *Naward* 1325–50 Map, 1335 Sc, *Naward castle* 1375 Sc). Ranulf Dacre got permission in 1335 to fortify and kernel his manor of Naward. The second el. is OE *weard*, here perhaps 'fort', 'stronghold'. The first may be OE *nearu* in such a sense as 'narrow place' or 'strait, difficulty'. Alternatively it might be OE *nearra* 'nearer'. In either case an *r* was lost owing to dissimilation.

Nawton YN [*Nageltone, Nagletune* DB, *Nawelton* 1202 FF]. '*Nagli*'s TŪN.' Cf. NAUGHTON.

Nayland Sf [*Eilanda* DB, *Eiland* 1167 P, *Leiland* 1234 Cl, *Neiland* 1227 Ch]. OE *ēgland* 'island, river land'. N~ is on the Stour. The *N-* has been carried over from the dat. of the def. art. (*atten Eilande* from OE *æt þǣm ēglande* became *atte Neilande*). The change may have taken place particularly in the name STOKE BY NAYLAND [*Stoke atte Neylaunde* 1303 Ch].

Naze, The, Ess [(to) *Eadulfes næsse* 1052 ASC (F), (to) *Ealdulfes næse* 1052 ib. (E)]. '*Eadwulf*'s ness.' The second el. is OE NÆS(S). The name was formerly used also of WALTON ON THE NAZE [*Eduluesnæsa* 939 BCS 737, *Eduluesnase* 1181, *Edolvesnase* 12 StPaul]. The first el. was later lost.

Nazeing Ess [*Nesingan*, (into) *Nassingan* 1062 Th, *Nasinga* DB, *Nesinges* 1199 Pp, *Nasinges* 1205 Cur]. 'The dwellers on the NÆSS or spur of land.' Cf. NÆSS.

Neadon D. See Introd. p. xviii.

Neasden Mx [*Neasdune* 939 BCS 737, *Nesdone* 1254 Val], **Neasham** (-ē-) Du [*Nes(s)ham* 1158 YCh 400, *Neshan* 1202 FF]. The first el. is rather ME *nese* 'nose', also 'ness, headland', than NÆSS. Neasden is 'nose-shaped hill or ridge'. See DŪN. Neasham is 'HĀM or HAMM by the nose-shaped bend'. The Tees makes a sharp bend here.

OE **nēat** 'cattle'. See NATTON, NEATHAM, NESFIELD, NETTON, NOTTON Do, W.

Neatham Ha [*Neteham* DB, *Nietham* 1147 BM, *Netham* 1156 ff. P]. 'Cattle farm.' See NĒAT.

Neatishead Nf [*Netheshird* 1021–4, *Netheshirda* 1044–7 Holme, *Snateshirda* DB, *Neteshirde* c 1100 BM]. '*Snǣt*'s *hīred* or household.' Cf. SNETTISHAM. The forms from Holme are in late transcripts. The loss of *S-* is due to Norman influence. Cf. NOTTINGHAM.

Nechells Wa [*Echeles* c 1180 Middleton, *Le Echeles* 1290, *Le Necheles* 1322 Ipm]. Identical with ETCHELLS. The *N-* has been carried over from the dat. of the def. art. Cf. NAYLAND.

Necton Nf [*Nechetuna, Neketuna* DB, *Neketona* 1168 P]. 'TŪN by a neck of land.' OE *hnecca* 'neck' was probably used also of a neck of land. The place is situated at the foot of a ridge.

Nedging Sf [(æt) *Hnyddinge* c 995 BCS 1289, *Neddinge* c 1050 KCD 907, *Niedinga* DB, *Nedding* 1235 FF]. A singular name in *-ing*. See -ING. The name is probably derived from a pers. n. *Hnydda or *Hnyddi, which is the base also of the first el. of NEEDINGWORTH. It may be related to Engl *nod*, Norw *nudd* 'a small nail' and the like. The name ought to appear as *Neddinge*, but the *g̃* of the final syllable was passed back into the first.

Needham, High, Db nr Longnor [*Nedham* 1244 FF, 1251 Ch], N~ Nf [*Nedham* 1352 f. BM, 1428 FA], N~ **Market** Sf [*Nedham* 13 BM, *Neidham* 1331 Misc, *Nedeham Markett* 1511 AD], N~ **Street** Sf [*Nedham* c 1185 Bodl, 1219, 1235 FF], N~ Sf in Yaxley [*Nedeham* Copinger], **Needwood Forest** St [*Nedwode* 1248 BM, *Neydwode* c 1265 BM]. The first el. of these is no doubt OE *nīed, nȳd, nēod* 'need'. Needham is probably analogous to HUNGERTON and means 'needy, poor homestead or village'. OE *nīed* meant among other things 'distress, hardship', no doubt also 'poverty'. Needwood may mean 'wood resorted to in an emergency', i.e. a wood where outlaws found a refuge or the like.

Needingworth Hu [*Neddingewurda* 1163, 1167, *-wurða* 1170 P, *Niddingeworth* 13 AD]. 'The WORÞ of *Hnydda*'s people.' Cf. NEDGING.

Needwood. See NEEDHAM.

Neen Savage & Sollars Sa [*Nene* DB, 1242 Fees, *Nena* 1193, *Niene* 1195 P, *Nenesauvage* Hy 3 Misc, *Nen Solers* 1274 RH]. Neen is the old name of the river REA, on which the places are [*Nen* c 957 BCS 1007]. Neen is a Brit river-name, identical with NENE.

N~ **Savage** belonged to Adam le Savage in the early 13th cent. Savage is a nickname from Fr *sauvage* 'savage'.—N~ **Sollars** was held by Roger de Solariis c 1195 (Eyton). Cf. BRIDGE SOLLERS.

Neenton Sa [*Newentone* DB, *Nenton* 1242 Fees, *Neynton* 1255 RH]. N~ is on the Rea, formerly NEEN, and the name may mean 'TŪN on R Neen'. If so, the DB form is due to popular etymology. But the original name may have been OE (æt) *Nēowan-tūne* 'new TŪN', which was changed to *Neenton* owing to association with the river-name.

Neepsend YW [*Nipisend* 1297 Subs]. 'The end of the hill or brae.' The first el. is dial. *nip* 'a steep ascent of a road, a hill' from Sw, Norw *nipa* 'a crag, steep river bank', a word related to ON *hnípa* 'to overhang', OE *hnipian* 'to droop'.

Neithrop O [*Ethrop* 1224 FF, *Nethrop* 1278–9 RH, *Nethropp* 1316 FA]. 'Nether thorp.'

Nelson La. A late name derived from the Lord Nelson Inn.

Nempnett Thrubwell So [*Emnet* c 1200 Flaxley, 1208 Cur, *Empnete* 1242 Wells]. OE *emnet* 'plain', with *N-* carried over from the dat. of the def. art. Cf. NAYLAND. Thrubwell is *Trubewel* 1201 Ass, *Tribnelle* 1227 Flaxley, *Trubewelle* 1239, *Threbwell* 1299 FF. It seems to have been a place close to Nempnett. Its first el. may be related to the word *throb*, so that the name means 'gushing spring'.

Nene (-ĕ-, -ĕ-) R [*Nyn* 948, 964 BCS 871, 1129, *Nén* 972 BCS (1280), c 1000 Saints, *Nene* c 1200 Gervase, *Neen* 1281 QW]. See NEEN.

Nent R Cu [*Nent* 1314 Ipm, *Vent* 1576 Saxton]. A Brit name derived from Welsh *nant* 'valley, brook'. From the Nent were named **Nenthead** and **Nenthall**, the latter being probably represented by *Nentesbire* 1230 Ep ('byre on R Nent').

OE **neoþerra, niþerra** 'nether, lower'. See NEITHROP, NETHER-.

OE **nēowe, nīewe, nīwe** 'new' is an extremely frequent pl. n. el. See NEW- (passim). Special developments are seen in NOBOLD, NOBOTTLE, NOWTON, where *ēow* became *ēōw* and *ōw*, and in NAUNTON (q.v.). There was an OE side-form *nīge*, found in NINHAM, NITON, NYETIMBER, perhaps NINFIELD. It is restricted to Sr, Sx and Wt.

Neroche Forest So [*Nerechich* 1236 ff. Cl, *-e* 1252 Ch, *Neracchich* 1298 Wells, *Errechich* 1237, *Recchiche* 1241, *Rachiche* 1244 Cl, *Rechich* 1243 Ass]. Neroche seems to be 'Nearer Rechich' (first el. OE *nierra* 'nearer'). *Rechich* or *Rachich* may be OE *ræcc-wīc* 'wīc where hunting-dogs were kept', with early loss of *w*. Cf. WINCH.

Nesbit Nb in Doddington [*Nesebit* 1242 Fees, *-bith* 1262 Ipm], **Nesbitt** Du [*Nesebite* c 1220 FPD, *Nesbitt* 1311 RPD], N~ Nb in Stamfordham [*Nesebite* 1242 Fees, *-bith* 1298 BBH]. The name looks like the North country word *nesebit* 'the iron that passes across the nose of a horse and joins the branks together'. If the places were named from a nesebit, the reason would be some sort of similarity between them and a nesebit. More likely the second el. is OE *byht* 'bight, bend', so that the name means 'noselike bend'. Spellings with *-th* are reminiscences of OE *-ht*. For the first el., which is ME *nese* 'nose', see NEASDEN.

Nesfield YW [*Nacefeld* DB, *Nethes-*, *Nesthesfeld* 1212 f. FF, *Netfeld* 1212 Cur, *Nessefeld* 1271 Ipm]. OE *nēates-feld* 'open land where cattle were kept'. Cf. NĒAT.

Ness Chs [*Nesse* DB, *Nessa* c 1100, c 1150 Chester]. ON *nes* or OE *ness* 'headland, cape'. Ness is in the innermost part of the Wirral peninsula, and it may be a relic of the name NESS for the peninsula. But there is a short ridge close by, which may be referred to. **Neston** [*Nestone* DB, *-tona* c 1100, *-tuna* c 1150 Chester] is close to Ness.

Ness, Great & Little, Sa [*Nessham, Nesse* DB, *Nesse* 1160 ff., 1190 ff. P], **East & West N~** YN [*Neisse, Nesse* DB, *Nesse, Westnes* 1202 FF]. Both places are on or by projecting ridges. The source of the name is OE *ness*. See NÆSS.

Neston Chs. See NESS Chs.

Neswick YE [*Nesseuuic* DB, *Nessingwyk*

1285 FA]. 'wīc by a ness or projecting ridge.' Cf. NÆSS.

OE **netele** 'nettle'. See NETLEY, NETTLE- (passim).

Netheravon. See AVON.

Netherbury Do [*Niderberie* DB, *Nitherbury* 1285 FA], **Nethercot** Np [*Nethercote* 1345 AD], **Nethercote** O [*Nethercot* 1208 Cur, *Nuthercot* 1220 Fees]. 'Nether BURG and COT.'

Netherfield Sx [*Nedrefelle* DB, *-felde* 1121-5 BM, *Nadrefeud* 1271 Ch]. 'FELD infested by adders.' Adder is OE *næddre*.

Netherhampton W [*Otherhampton* 1208 Cur, *Noþerha[m]pton* 1242 Fees], **Netherton** Nb [*Nedertun* c 1050 HSC, *Neterton* 1207 Cur, *Nodirton* 1307 Ipm], N~ Wo nr Bredon Hill [*Neoþeretun* 780 BCS 235, *Neotheretune* DB]. 'Lower HĀMTŪN and TŪN.' The forms with *o* are due to OE *eo*.

Netherthong. See THONG.

Netley Ha SE. of Southampton [(æt) *Lætanlia* 955-8 Wills, *Latelie* DB, *Leteleye* 1241 Ch, *Lattel*' 1246 Cl, *Letele* 1329 BM]. The change to *Netley* is late. The first el. must have had long *æ*. The only OE word *læt* known is *læt* 'freedman' in the Kentish laws. If there was also a form *læta*, Netley might mean 'the LĒAH of the freedman'. But more likely we have to postulate an OE *læte* adj., derived from *lætan* and meaning 'left alone' and the like. OE *ǣlæte* means 'desert, desolate'. OE *læte* in the present name would mean 'deserted' or rather 'left fallow, untilled'.

Netley Marsh Ha [*Natan leaga* 508 ASC, *Nutlei* DB, *Natale* 1316 FA]. In ASC it is stated that a Brit king called *Natan leod* was slain and that afterwards the land was called *Natan leaga* as far as Charford. Netley is a long way from Charford at the other end of the New Forest, but the entry seems to suggest that *Natan leaga* was the name of a district. For the etymology see NATELY.

Netley Sa [*Netelie* DB, *Netelegh* 1209 Eyton]. 'Nettle LĒAH.'

Netteswell (nĕts-) Ess [*Nethleswelle* 1062 Th, *Netlesuell*' 1196 FF, *Nedliswell* 1248 FF], **Nettlesworth** Du [*Netles-, Nettelworth* 1312 RPD]. '*Nēþel*'s stream or spring and WORÞ.' *Nēþel* is a derivative of *Nōþ-, -nōþ* in pers. ns. Cf. OG *Nandilo*. But Nettlesworth may be identical with NETTLEWORTH.

Nettlebed O [*Netelbedde* 1252 Ipm, *La Netelbedde* 1276 Ipm]. 'Nettle-bed, place overgrown with nettles.' Cf. OE *hrēod-bedd* 'reed bed, clump of reeds'.

Nettlecombe Do [*Netelcome* DB, *-cumb* 1212 Fees], N~ So [*Netelcumbe* DB, *Netelcumbe* 1243 Ass], N~ Wt [*Netelcumbe* 1316 FA], **Nettleham** Li [*Netelham* DB, c 1100

RA, 1166 P], **Nettlestead** K [*Netlasteda, Netlestede* 10 BCS 1321 f., *Netlestede* 1242 Fees], N~ Sf [*Netlesteda* DB, *Netlesteda* 1166 P], N~ Sr at Chelsham [*Netelam-, Netelhæmstyde* 871–89 BCS 558, *Netlested* 1197 FF]. 'CUMB or valley, HĀM, HĀMSTEDE, STEDE or place where nettles grew.' The identification of *Netelamstyde* BCS 558 is not certain. It is usually identified with Nettlestead K.

Nettleston Wt [*Hoteleston* DB, *Noteleston* 1329 BM, *Nutleston* 1431 FA]. The first el. seems to be a pers. n. Cf. OG *Neozzo*, perhaps *Niuzilo*, from the base in OE *nēotan* 'to enjoy'.

Nettlesworth. See NETTESWELL.

Nettleton Li [*Neteltone* DB, *-tuna* c 1115 LiS], N~ W [*at Netelin(g)tone* 944, 956 BCS 800, 933, *Niteletone* DB, *Netelton* 1242 Fees], **Nettleworth** Nt [*Net(t)leswurda* 1195 f., *Netlewurða* 1197, *-wrða* 1198 P]. 'TŪN and WORP where nettles grew.' The first el. of Nettleton W is OE *netele* and **netlen* adj. 'of nettles'.

Netton W at Durnford [*Netetun* 1242 Fees, *Netton* 1309 Ipm]. OE *Nēata-tūn* 'cattle farm'.

Nevendon Ess [*Nezendena* DB, *Neuendene* 1218 FF, *-den* 1270 FF]. Probably 'flat valley' (OE *efn* 'even, flat' and *denu* 'valley'), the *N-* having been carried over from a prep. *in* or the dat. of the def. art. (cf. NAYLAND).

Newark Np [*Newerc* 1227 Ch], N~ Nt [*Newercha* 1054–7, *Niweweorce* 1075–92 Eynsham, *Newerche* DB], N~ Sr [*Novus Locus* 1205 Cur, *Newerk* 1414 BM]. 'New work', i.e. 'new fort' in the case of the first two. N~ Sr was a priory, and *work* seems to mean 'building'. The second el. is OE *(ge)weorc*.

Newbald, North & South, YE [*Niubotle* 963 BCS 1113, (æt) *neowe boldan* c 972 BCS 1279, *Niwebolt, Niuuebold* DB]. 'New building.' See BŌPL.

Newball Li [*Nevberie* DB, *Neobole* c 1115 LiS, *Neubele* c 1175 DC]. OScand *nÿbøle* 'new homestead'. The elements are OScand *nÿ(r)* 'new' and *bøle* 'homestead, dwelling'. *Nÿbøle* is the source of the common Sw pl. ns. NYBBLE, NIBBLE.

Newbegin YN [*Neubiggin* 1310 Ch], **Newbiggin** (nibikan) Cu [*Neubigon* 1198 Cur, *-bigging* 1290 Ch], N~ Du in Bishopton [*Neubigin* 1208–10 Fees], N~Du nr Middleton in Teesdale [*Neubigging* 1316 Ipm], N~ Nb in Blanchland [*Neubyggyng* 1378 Cl], N~ Nb in Newburn [*Neubiging* 1242 Fees], N~ Nb in Norham [*Neubiging* 1208–10 Fees], N~ **by the Sea** Nb [*Niwebiginga* 1187 P, *Neubigging* 1242 Fees], N~ We nr Appleby [*Neubigging* 1223 Pat], N~ YN [*Neubigginge* c 1240 BM]. 'New building or house.' Cf. BIGGIN. There are several other Newbiggins in Du, Nb, Y and Cu.

Newbold Astbury Chs [*Neubold* DB, *Neobold* c 1100 Chester], N~ Db [*Newebold* DB, *Neubaude* 1226 FF], N~ La [*Neubolt* c 1200 WhC, *Newbold* c 1300 ib.], N~ **Saucy** Le [*Neobold[i]a* Hy 2 DC, *Neobolt* 1242 Fees, *Neobold Sauce* 1327 Subs], N~ **Verdon** Le [*Newebold* DB, *Neubold Verdon* 1324 AD], N~ Np [*Neubold* 1203 Ass, *Newebold* 1220 Fees, *Neubo* 1228 Ep], N~ Nt [*Neubold* DB, *Niwebold* c 1150 Eynsham, *Niwebote* 1158 P], N~ **on Avon** Wa [*Newebold* DB, *Neubolde on Avene* 1347 Misc], N~ **Pacey** Wa [*Niwebold* DB, *Neubold Pacy* 1236 Fees], N~ **on Stour** Wo [(æt) *Nioweboldan* 991 KCD 676, *Newbold-on-Stoure* 1383 AD]. 'New building.' See NEWBALD. The local distribution of the name is noteworthy.

N~ **Astbury** is nr ASTBURY.—N~ **Pacey** was held by Adam de Pasci t. John. Pacey is from PACY in Normandy.—N~ **Saucy** was held by Robert de la Sauce 1242 Fees. Saucy is a family name derived perhaps from SAUSSEY in Manche. Cf. SALCEY.—N~ **Verdon** came to Bertram de Verdon t. Stephen. The name is from VERDUN in France (one in Normandy).

Newborough St [*Neuboreg* 1280 Ass, *Neuburgh* 1327 Subs]. 'New BURG.'

Newbottle Du [*Neubotl* 1196 P], N~ Np [*Neubote, Niwebotle* DB, *Neobotha* c 1125 Fr], N~ **Bridge** Np [*Neubotle* Hy 3, 1404 BM]. 'New building.' See BŌPL. Newbottle is a variant of NEWBOLD.

Newbourn Sf [*Neubrunna* DB, *Neubrounia* c 1160 (1331) Ch, *Neubrunn* 1254 Val]. 'New stream.' See BURNA. The stream must have changed its course.

Newbrough Nb [*Nieweburc* 1203 P, *Neweburgh* 1256 Ass], **Newburgh** La [*Neweburgh* 1431 PNLa], N~ YN [*Nouo Burgo* 1199 Pp, *Newburgh* 1247 Ch]. 'New fort or borough.' The last is the meaning of the La name.

Newburn Nb [*Neuburna* 1121–9 YCh 458, *Nieweburn* 1165 P, *Neuburne* 1212 Fees, *-beri* 1201 P]. Identical with NEWBOURN.

Newbury Brk [*Neuberie* c 1080 Fr, *Niweberia* 1103–7 RA, (castrum) *Neubiriæ* 1152 HHunt]. 'New BURG or castle.'

Newby Cu [*Neubi* c 1175 WR], N~ We [*Neuby* c 1160 YCh 175, *Neweby* 1200 FF], N~ YN in Scalby [*Neuby* 13 Percy], N~ YN in Stokesley [*Neubie* c 1236 PNNR], N~ **Wiske** YN [*Neuby* 1252 Ch, *Neuby super Wysk* 1285 FA], N~ YW nr Clapham [*Neubie, Neuby* c 1170 FC], N~ YW nr Ripon [*Neubi* 1170–80 YCh 124, *-by* 1231 FF], N~ YW nr Stainborn [*Neuby* a 1190 Fount]. 'New BY.' **Newby Bridge** La [*New bridge* 1577 Saxton] seems to be really 'new bridge'.

Newcastle upon Tyne Nb [*Novum Castellum* 1130 P, *Nouum Castellum super Tinam*

1168 P, *Novum Castrum* 1254 Val], N~ Sa [*Novum castrum* 1284 Cl], **N~ under Lyme** St [*Nouum Oppidum sub Lima* 1168 P, *Novum castellum subtus Lymam* 1173 LaCh, *Novum Castrum super Are* (*Nef Chastel sus Are*) 1305 Chr. & Mem. 98, (de) *Newcastle super Are* 1316–17 Letter Books of the City of London]. 'New castle.' The castles are post-Conquest ones. See LYME Chs. *Are* seems to be an old name of Lyme Brook, perhaps identical with Ayr in Scotland.

Newchurch K [*Nevcerce* (hd) DB, *Niwancirc* 11 DM, *Newechirche* 1198 FF], N~ Wt [*Niwecherch* 1291 Tax]. 'New church.' There are other examples of the name.

Newdigate Sr [*Niudegate* Hy 2 BM, *Neudegate* 1201 Cur, 1229 FF, *Newdegate* c 1270 Ep, *Niwodegate* 1312 Pat]. 'Gate to the new wood.' There is in N~ a place, formerly a wood, called **Ewood** [*Iwode* 1312 FF]. Newdigate might be 'the gate leading to Ewood'. If so *N-* must have been carried over from a preceding prep. *in* or from the dat. of the def. art. Ewood is OE *īw-wudu* 'yew wood'. The first alternative seems preferable.

Newenden K [*Newedene* DB, *Niwendenne* 11 DM, *Newendenna* 1157 StAug]. 'New DENN or pasture.'

Newent Gl [*Noent* DB, *Nowent* 1167 P, *Newent* 1221 Ass, *Neuwent* 1253 Ch]. A Brit name corresponding to Gaul *Noviantum* or *Novientum* (now NOGENT, a common name), which means 'new place'.

Newerne Gl [*Niware* DB, *Newern* 1282 For]. 'New house.' Cf. ÆRN.

New Forest Ha [*Nova Foresta* DB, *Noveforest* 1154 HHunt]. The forest was created by William I.

Newhall Chs [*La Nouehall* 1252 Ch, *Nova Aula* 1275 Cl], N~ Db [*Le Newehale* 1284–6 FA, *Nova Aula* 1302 FA]. 'New hall.' See HALL.

Newham Nb nr Bamborough [*Neuham* 1242 Fees, 1252 Ch], N~ Nb nr Morpeth [*Neuham* 1242 Fees, *Neweham* 1256 Ass], N~ YN [*Neuham* DB, *Newenham* 1206 FF]. 'New HĀM.'

Newhaven Sx. A late name. The old name was *Mechingas* 1121 AC, *Mecinges* 1204 Cur.

Newholm YN [*Neueham* DB, *Neuham* 1090–6 YCh 855]. 'New HĀM.'

Newhouse. See NEWSHAM.

Newick Sx [*Niwicha* 1121 AC, *Newike* 12:y FF]. 'New wīc.'

Newington Bagpath Gl [*Neueton* DB, *Newentone Baggepathe* 1327 Subs], N~ K nr Hythe [*Neventone* DB, *Niwantun* 11 DM, *Neuton juxta Heth* 1285 Ch], N~ K nr Sittingbourne [*Newetone* DB, *Niwantune* 11 DM], **Stoke N~** Mx [*Neutone* DB,

Stokene Neuton 1294 QW], N~ O nr Wallingford [*Newintune* 997, *Niwantun* c 1050 KCD 697, 896], **North N~** O [*Newinton* 1200 Cur, *North Newenton* 1299 Ch], **South N~** O [*Neuinton* 12 Berk, *Suthnewenton* 1281 Misc], N~ Sr [*Niwetun* 1212 Fees, *-tune* 13 BM]. 'New TŪN', a variant of NEWTON.

Bagpath is a place nr **N~ Bagpath** [*Baghepathe* 1221 Ass, *Baggapath* c 1238 Berk]. The name means '*Bagga*'s path' or else the first el. is the word of uncertain meaning suggested under BAGLEY. An animal's name would go well with *path*.—*Stoke* in **Stoke N~** is OE *stoccen* 'of stocks'.

Newland Gl [*Newelond* 1251 Ch, *Nova terra* 1221 Fees], N~ La [*Neulande* 1276 FC], N~ Wo [*la Newelande* 1221 Ass, *Newelond* 1327 Subs], N~ YW [*Neuland* 1234 FF, 1299 Ipm], **Newlands** Cu [*Neuland* 1323 Ipm], N~ Nb [*Neuland* 1343 Pat]. 'Newly-cleared land' or 'newly-acquired land'.

Newlyn, East, Co [(Ecclesia) *Sancte Neuline* 1264 Ep, *Eglosnyulyn* 1415 AD iv]. '(The church of) St. Newelina.' **Newlyn** nr Penzance is *Lu(e)lyn* in early sources.

Newmarket Sf [*Novum Forum* 1200 Cur, Hy 3 BM, *Novum Mercatum* 1219 FF, *la Newmarket* 1418 AD]. 'New market-place.'

Newminster Nb [(abbathia) *Novi Monasterii* 1137 Newminster]. 'New monastery.'

Newnham Bd [*Neweham* 1198 FF], N~ Ca [*Neuham* 1195, *Newenham* 1202 FF], N~ Gl [*Nevneham* DB, *Niweham* 1130 P], N~ Ha [*Neoham* c 1125 Oxf, *Neweham* 1198 FF, *Niuenham* 1212 Fees], N~ Hrt [*Nevham* DB, *Nieweham* 1198 FF, *Newenham* 1291 Tax], N~ K [*Newenham* 1177 Reg Roff, 1230 P], N~ Np [(æt) *Niwanham* 1021–3 BM, *Niwenham* 1202 Ass], **N~ Murren** O [*Niwanham*, dat. *-hamme* 966 BCS 1176, (æt) *Niwanhám* 966–75 Wills, *Neuham Morin* 1236 Fees], N~ Wa in Aston Cantlow [*Neweham* DB, *Newenham* 1316 FA], **Kings N~** Wa [*Neowenham* 1043 Th, *Newenham* c 1050 KCD 939, *Newenham Kynge* 1285 QW], **N~ Paddox** Wa [*Niweham* DB, *Parva Newenham* 1305 Ch], N~ Wo [*Neowanham* c 957 BCS 1007]. 'New HĀM', from OE *nēowa hām*. This is no doubt the etymology also of N~ Murren O in spite of the dat. *-hamme*.

N~ Murren was held by Richard Morin in 1231 (Ch). Morin may be a nickname from OFr *morin*, *mourin* 'mortal, pale'.—**N~ Paddox** seems to have its surname from a paddock or small park.

Newnton, Long, W [*Niuentun* 681 BCS 58, *Newentuna* 1065 KCD 817, *Long Newenton* 1331 FF], **North N~** W [*Norþniwetune* 892 BCS 567, *Nywantun* 934 ib. 699, *Newetone* DB]. 'New TŪN', a doublet of NEWTON.

Newport Monm [*Novus Burgus* 1191 Gir, 1291 Tax, *Niweport* c 1218 Gir]. 'New town.' The Welsh name is **Cas-newydd**

[*Castell newyd ar Wysc* c 1400 Brut y Tywys-ogion], which has the same meaning.

Newport Pagnell Bk [*Neuport* DB, 1151–4 Fr, *Neuport Paynell* 1220 Cl], N~ D [*Neuport* 1295 Ch], N~ Ess [*Neueport* c 1080 ICC, *Neuport* DB, *Niweport* 1141 BM], N~ Sa [*Niweport* c 1050 Coins, *Nouus Burgus* 1174 P, *Neweburg* 1232 Cl, *Newport* 1237 FF], N~ Wt [*Novus Burgus* Hy 3 BM, *Neuport* 1287–90 Fees]. 'New town.' See PORT. N~ **Pagnell** was held by Fulc Paganellus 1151–4 (Fr). Cf. BOOTHBY PAGNELL.

Newsells Hrt in Barkway [*Nevsela* DB, *Newesel* 1212 Fees, *Neuseles* 1251 Ch]. Second el. OE *sele* 'hall, dwelling, house'.

Newsham Du nr Yarm [*Neusum* 12 FPD, *Neusom on Teyse* 1316 Misc], N~ La in Preston par. [*Neuhuse* DB, *Newesum* 1246 Ass], N~ La in Walton par. [*Neuhusum* 1212–17 RBE, *Neusum* 1200 Ch], N~ or **Newhouse** Li [*Neuhuse* DB, *-hus* 12 DC, *Neosum* c 1115 LiS], N~ Nb [*Neuhusum* 1207 Cur, *Neusum* 1242 Fees], N~ YN in Appleton [*Neuhuse*, *Newehusum* DB, *Neusum* 1231 Ass], N~ YN in Kirby Ravensworth [*Neuhuson* DB], N~ YN in Kirby Wiske [*Newehusum* DB, 1088 LVD, *Newesum* 1202 FF], N~ or **Temple Newsam** YW [*Neuhusum* DB, *Temple Neusom* 1334 Misc], **Newsholme** YE nr Howden [*Nevhusam* DB, *Newesum* 1303 FA], N~ YE nr Brid-lington [*Neusom* 1285 FA], N~ YW nr Gisburn [*Neuhuse* DB, *Neusum* 1226 Ep], N~ YW nr Keighley [*Neuhuse* DB, *Neusum* 1285 FA]. 'New houses', from OE (æt) *nēowan hūsum*, dat. plur. of *nēowe hūs* 'new house'. There are other examples of the name in the northern counties. Cf. HŪS.

Newstead Li nr Brigg [*Nouus Locus* 1202 Ass, *Newsted* 1227 Ch], N~ Li nr Stamford [*Novus Locus juxta Stanford* 1246 Fees], N~ Nb [*Newstede* 1377 Ipm], N~ **Abbey** Nt [*Nouus locus* 1169 P, *Nouus Locus in Schirewod* 1230 P], N~ YN [*Newestede* 1301 Subs]. 'New place.' The Newsteads in Li and Nt were monasteries, and here the name means 'new monastery'. The others may mean 'new farmstead or cattle-farm'. Cf. STEDE.

Newthorpe Nt [*Nevtorp* DB, *Newetorp* 1169 P], N~ YW [(on) *Niwan-þorp* c 1030 YCh 7, *Neutorp* 1231 FF]. 'New thorp.'

Newtimber Sx [*Nitimbre* 960 BCS 1055, *Nivembre* DB]. 'New timbered house.' The second el. is OE *timber* or rather *getimbre* 'building, house'. Cf. NYETIMBER.

Newton, probably the most common Eng-lish pl. n., means 'new homestead or village' and represents OE *nēowa tūn*, dat. *nēowan tūne* (or in the North *nēowa tūne*). It is identical in origin with NEWTOWN, NEWING-TON, NAUNTON. The name is found in all the northern counties (except We), in the north Midland counties (except Ru), in the south Midland counties (Sa, He, Wa, Np,

Nf, Sf, Ca, also Ess), regularly in the south-western counties (D, So, Do, Ha), in Bd, Bk, Hu. In O and W Newton occurs side by side with other forms (1 Newton O, 3 W). In other southern counties Newton is absent. Its place is taken by forms with preserved *n* or by the form NITON. Gl has 2 Naunton, 1 Newington, Wo 3 Naunton, Sr 1 Newington, Mx 1 Newington, Kent 3 Newington, Wt 1 Niton. Further New-ington occurs thrice in O, Newnton twice in W. Sussex may have 1 Nyton, identical with Niton. There are no examples in Brk. A full list of all the Newtons is hardly necessary, and here are given chiefly New-tons with surnames and a few select ex-amples of other kinds, especially such as are found in OE sources.

Newton Blossomville Bk [*Newenton* 1202 FF, N~ *Blosmevill* 1254 Val], N~ **Longville** Bk [*Nevtone* DB, *Niwentona* 1106–9 Fr, *Newenton Longevile* 1254 Val], N~ **in the Isle** Ca [*Neweton* 1233 Fees, *Neutona* 1254 Val], N~ **by Daresbury** Chs [*Neuton near Dersbury* 1423 AD], N~ **Arlosh** Cu [*Arlosk* 1185, 1304 Holme C], N~ **Reigny** Cu [*Niweton* 1185 P, *Neuton-reynye* 1275 Ipm], N~ **Abbot** D [*Nova villa* 12 Ol, *Nyweton Abbatis* 1270 Ass], N~ **Bushel** D [*Nyweton juxta Teng* 1281 QW], N~ **Ferrers** D [*Niwetone* DB, *Neweton Ferers* 1303 FA], N~ **Poppleford** D [*Poplesford* 1226 Cl, *Neutone Popleforde* 1331 Ep], N~ **St. Cyres** D [(æt) *Niwantune* c 1070 Ex, *Nywetone Sancti Ciricii* 1338 Ep], N~ **St. Petrock** D [æt *Nywantune* 938 BCS 725], N~ **Tracy** D [*Newentone* DB], **King's** N~ Db [*Newetun* DB, *Kynges-neuton* 1352 Pat], N~ **Solney** Db [æt *Niwantune* 956 BCS 944, *Neuton Suloni* 1305 FF], **Maiden** N~ Do [*Newetone* DB, *Maydene Neweton* 1316 BM], N~ **Bewley** Du [*Neuton* 1195 (1335) Ch, N~ *Belu* c 1350 DST], N~ **Cap** Du [*Newatun* c 1050 HSC, *Newton capp* 1382 Hatfield], **Long** N~ Du [*Lang Newton* 1260 FPD], N~ **Valence** Ha [*Newentone* DB, *Nyweton Valence* 1346 FA], **Welsh** N~ He [*Neuton* 1341 NI], **Water** N~ Hu [*Niwantun* 973 BCS 1297], N~ **in Makerfield** La [*Neweton* DB, *Neuton Macreffeld* 1257 Ch], N~ **Burgo-land** Le [*Neutone* DB, *Neuton Burgilon* 1390 AD], **Cold** N~ Le [*Neutun* 1236 Fees, *Coldenewton* 1428 FA], N~ **Harcourt** Le [*Niwetone* DB, *Neuton Harecurt* 1284 AD], N~ **by Toft** Li [*Neutone* DB, *Toft, Neuton* 1242 Fees], N~ **le Wold** Li [*Neutone* DB, *Waldneweton* 1236 Ep], N~ **by the Sea** Nb [*Neuton super Mare* 1242 Fees], N~ **on the Moor** Nb [*Neuton super Moram* 1242 Fees], **Kirk-** & **Westnewton** Nb [*Niwe-tona* 1123–8 (1336) Ch, *Neuton del Est*, *West* 1242 Fees], N~ **Underwood** Nb [*Neuton* 1242 Fees, *Newton under Wood* 1296 Subs], N~ **Flotman** Nf [*Niwetuna* DB, *Neuton Floteman* 1291 Tax], N~ **Bromswold** Np [*Niwetone* DB], **Woodnewton** Np [*Niwe-tone* DB, *Niwentona* c 1120 RA, *Wodeneuton*

1274 Misc], N~ Nt [*Neutone* DB], N~ Purcell O [*Neuuenton* 1180 BM, *Niweton* 1198 Fees, *Neuton Purcel* 1245 Ch], Old N~ Sf [*Neweton* 1196 FF, *Vetus Neuton* E 1, *Eldneuton* 1418 BM], North N~ So [*Newetune* DB], N~ St. Loe So [*Niwetone* DB, *Nywetonseyntlou* 1336 Ep], N~ Toney W [*Newentone* DB, *Nywetone Tony* 1315 Ipm], South N~ Without W [(in) *Niwantune* 943 BCS 782, *Newentone* DB], N~ Regis Wa [*Niweton* 1179 P, *Kyngesneweton* 1285 Ipm], Out N~ YE [*Niuuetone* DB, *Utneuton* 1285 FA, *Out Neutona* 1297 Subs], Wold N~ YE [*Neuton* DB, *Neuethon in Waldo* 1154–60 YCh 880, *Neuton super Waldam* 1214 FF], N~ Morrell YN [*Neuton* DB, *Neuton Morel* 1280 Ipm], N~ Mulgrave YN [*Neuuetune* DB, *Neweton juxta Mulgrave* 1285 FA], N~ on Ouse YN [*Newetone* DB, *Neuton super Usam* 1230 Cl], N~ le Willows YN [*Neuton* DB], Bank N~ YW [*Neuton in Cravene* 1239 FF], N~ Kyme YW [*Niuueton* DB, *Neuton Kyme* 1285 FA], Potter N~ YW [*Snitertun* DB, *Potters Neuton* 1285 FA].

N~ Abbot D from the Abbot of Torre.—N~ Arlosh Cu. The old name was Arlosk, which is doubtless British and is explained in PNCu(S) as Welsh *ar* 'on' and *llosg* 'burning', possibly referring to an area cleared by fire. The low situation of the place near Morecambe Bay tells against a meaning 'land cleared by fire', but the elements of the name may well be those suggested. The name might indicate a place that had been burnt down. Cf. BRENT PELHAM and the like. Pughe records a Welsh *arlosg* 'singed'. —Bank N~ YW from a family of the name. John del Bank got the manor by marriage c 1400.—N~ Bewley Du belonged to the manor of BEWLEY.—N~ Blossomville Bk from the family of (de) Blosseville, mentioned in connexion with the manor from c 1200. The name is from BLOSSEVILLE in Normandy.—N~ Bromswold Np was named from its situation nr Bromswold. Cf. LEIGHTON BROMSWOLD. —N~ Burgoland Le from the Burgilon family. The name means 'Burgundian'.—N~ Bushel D was held by Theobald Bussel in 1281 (QW). Cf. HUTTON BUSHELL.—N~ Cap Du may be 'N~ by the *cap* or hill.'—N~ Ferrers D was held by William de Ferers in 1242 (Fees). Cf. BERE FERRERS.—N~ Flotman Nf is said to be so named from a ferryman, but *Floteman* occurs as a pers. n. in DB. Cf. 'molendinum de Floteman' c 1195 Holme. OE *flotman* means 'sailor, pirate'.—N~ Harcourt Le came to Richard de Harcourt c 1240. See KIBWORTH HARCOURT.—N~ Kyme YW was held in part by Simon de Kimbe in 1242 (Fees). The name may be from KYME Li.—N~ Longville Bk belonged to the church of St. Faith of Longueville in France from the 12th cent.— Maiden N~ Do is obscure.—N~ in Makerfield La. See MAKERFIELD.—N~ Morrell YN was claimed by Petronel of Neuton, daughter of William Morell, in 1231 (FF). Morell is a nickname from OFr *morel* 'brown, dusky'.— N~ Mulgrave YN from its situation nr this place.—Out N~ YE must be 'outer N~'.— N~ Poppleford D was originally POPPLEFORD 'pebbly ford' (OE *popel* 'pebble').—Potter N~ YW from potteries.—N~ Purcell O was held by Ralph Purcell in 1198 (Fees). OFr *porcel* means 'pig'.—N~ Reigny Cu was held by William de Reigny in 1185 (P). Cf. ASHREIGNY.

—N~ St. Cyres D was dedicated to St. Ciricius.—N~ St. Loe So from the family of St. Lo. Roger de Sancto Laudo held the manor in 1122 (Bath). ST. LO is a place in Manche (France).—N~ St. Petrock D belonged to the monastery of St. Petrock, Bodmin.—N~ Solney Db was acquired by Alfred de Suleini (*Solenneio*) in 1204 (Derby). The name is from SUBLIGNY nr Avranches.—N~ by Toft Li from the neighbouring TOFT.—N~ Toney W from the Tony or Toeni family, which took its name from TOSNY in Normandy.—N~ Tracy D was held by Henry de Tracy in 1242 (Fees). Cf. BOVEY TRACY.—N~ Underwood Nb is 'N~ in the wood'.—N~ Valence Ha was held by Willelmus de Valencia in 1249 (Cl). Cf. COMPTON VALENCE.—Water N~ Hu is on the Nene.—Welsh N~ He was presumably in Welsh hands.—N~ le Willows YN must be 'N~ by the willows'.—South N~ Without W means 'N~ without or outside Wilton'.—Wold N~ YE is 'N~ on the Wold'.

Newtown is a later form of NEWTON and is also fairly common. The meaning is probably as a rule 'new hamlet'. N~ Wt, however, was a borough from the 13th cent. Examples are N~ Ha [*Nova villa* 1284 Ch], N~ Nb [*Nova Villa* 1242 Fees, *Le Neuton* 1310 Ipm], N~ Wt [*Newetone* c 1270 Ep].

Nibley, North, Gl [(to) *Hnibban lege* 940 BCS 764, *Nubbeleia* c 1200 Berk, *Nibbelege* 1221 Ass]. There is another Nibley nr Chipping Sodbury and a third nr Blakeney Gl. In BCS 764 is also mentioned *Nybban beorh*. No doubt *Hnybbe* or *Hnybba* was the name of a neighbouring hill, and the name is identical with Engl *nib* (1585 &c. OED) 'point, peak' &c., from an OE *hnybbe* or *hnybba* 'point, tip' or the like. The word is related to Norw, Sw *nubb* 'small nail', ON *hnúfa* 'stump nose' &c.

Nibthwaite La [*Neubethayt* 1246 Ass, *Neburthwait* 1336 FC]. 'Clearing by the new booth.' First el. probably ON *nýr* 'new' and *búð* 'booth'.

Nidd R YW [*Nid* c 715 Eddi, *Nidd* c 730 Bede, *Nide stream* c 890 OEBede]. A Brit river-name, cognate with NEDD or NEATH in Wales [*Ned* c 1150 LL], NIDDA [*Nida* 2], NIED [*Nita* 1018] in Germany. Probably derived from the root *nei*- 'to be brilliant' in Lat *nídeo*, OIr *niamde* 'brilliant' &c. On the river is Nidd vil. [*Nith*, *Nit* DB, *Nid* 1165 P]. The valley is Nidderdale [*Niderdale* 1155 YCh 76], which shows the river-name in a Scand gen. form (OScand -*ar*).

Ninebanks Nb [*Ninebenkes* 1228 Ep, *Nine bankes* 1296 Subs]. 'The nine banks or hills.'

Ninfield Sx [*Nerewelle* DB, *Niuenefeld* 1230 P, *Nimenefeld* 1255 FF]. Perhaps OE (æt) *níwan gemǣnan felda* 'the new common field'. Or better (æt) *niwnumenan felda* 'the newly reclaimed (open) field (Anderson, *Hundred-names*, iii. 103).

Ninham Wt [*Newenham* 1310 Ipm], Niton Wt [*Neeton* DB, *Nyweton* c 1270 Ep, *Nyton* 1412 FA]. 'New HĀM and TŪN.' The word *new* appears in the OE form *nige*.

Nobold Np [*Newbolt* Hy 3 BM, *Neubold* 1284 FA], **Nobottle** Np [*Nevbote* DB, *Neubottle* 12 NS, *Neuwebotle* 1202 FF]. 'New building.' See BŌPL and cf. NEWBOLD, NEWBOTTLE. OE *nēowe* shows the development to *nōwe* mentioned under NĒOWE.

Nocton Li [*Nochetune* DB, *Noketona* Hy 2 DC, *-ton* 1202 Ass]. The first el. may be OE *hnoc* 'wether sheep'.

Noctorum Chs [*Chenoterie* DB, *Cnoctirum* 1119, *-tyrum* c 1150 Chester]. First el. Ir *cnoc* 'hill'. Cf. KNOCK. The second is possibly OIr *tirim* 'dry'.

Noe R Db [*Noue* a 1300 AD]. A Brit rivername related to NAAB [olim *Naha*] and NAHE [*Nava* Tacitus, *Nawa* 8] in Germany and derived from the root *snāu-* 'to flow' found in Gk *naō* 'to swim' &c. Cf. also NADDER. From the river-name is derived the old name of BROUGH Db [*Nauione* (abl.) 6 Rav].

Noke O [*Ac(h)am* DB, *Ake* 1209–19 Ep]. OE (æt) *þǣm ācum* '(at) the oaks'. Cf. NASH.

Nonington K [*Nunningitun* 11 DM, *Nonynton* 1291 Tax]. 'The TŪN of *Nunna*'s people.'

Norbiton Sr [*Norberton* 1205 Cur, *Northeton*, *Norbeton* 1272 Ch, *Norbeton* 1303 ff. AD]. 'North barton.' Cf. BARTON.

Norbreck La [*Northbrek* 1267 Ass]. 'Northern hill.' Second el. ON *brekka* 'hill, slope'.

Norbury Chs nr Whitchurch [*Norberie* DB, *Nortbury* 1289 Court], N~ Chs nr Stockport [*Nordberie* DB, *Northbury* 1248 Ipm], N~ Db [*Nordberie* DB, *Norburi* Hy 3 BM], N~ Sa [*Norbir'* 1237 FF, *Northbur'* 1242 Cl], N~ Sr [*Le Northbury* 1314 Ipm], N~ St [*Nortberie* DB, *Nordbiri* 1198 P]. 'Northern BURG.'

Norfolk [*Norfolk* 1043–5 Wills, *Norðfolc* 1075 ASC (E), *Nordfolc* DB]. 'The northern people', in contradistinction to SUFFOLK, the southern part of East Anglia. In the OE Bede *Norþfolc* is used to denote the people north of the Humber.

Norham (-ŏr-) Nb [*Northham* c 1050 HSC, *Norham* c 1085 LVD, 1130, 1196 P]. 'Northern HĀM.' An earlier name was *Ubbanford* c 1000 Saints, which means '*Ubba*'s ford'.

Norland YW [*Northland* 1274 Wakef], **Norley** Chs [*Nortleg* 1259, *Northle* 1288 Court]. 'Northern LAND and LĒAH.'

Normanby Li [*Normanebi* DB, *Nordmanabi*, *Normanebi* c 1115 LiS], N~ **by Spital** Li [*Normanebi* DB, *Nordmanabi* c 1115 LiS], N~ **by Stow** Li [*Normanebi* DB, *Normannæbi* c 1115 LiS, *Normannebia iuxta Stou* 1146 RA], N~ **le Wold** Li [*Normane(s)bi* DB, *Nordmanabi* c 1115 LiS], N~ YN nr Middlesbrough [*Northmannabi* c 1050 HSC, *Normanebi* DB], N~ YN nr Pickering [*Normanebi* DB], N~ YN nr Whitby [*Nor-*

manebi DB]. OScand *Norðmannabȳr* 'the BY of the Northmen or Norwegians'.

Normansburgh Nf [*Normanesberht* Hy 2 BM, *Normannesberg* 1203 Ass]. 'The Northman's hill or mound.' But ON *Norðmaðr* was used as a pers. n. and *Norðman* (*Norman*) was also used in Old English.

Normanton Db by Derby [*Normantune* DB], **South N~** Db [*Normentune* DB], **Temple N~** Db [*Normantune* DB], N~ Le nr Bottesford [*Normantona* 1209–19 Ep, *-ton* 1242 Fees], N~ **le Heath** Le [*Normenton* 1209–35 Ep, *Normanton super le Heth* 1327 Subs], N~ **Turville** Le [*Normanton* Hy 3 BM, N~ *T'uill* 1327 Subs], N~ Li [*Normenton* DB, *Normanton* 1202 Ass], N~ Nt nr Southwell [*Normantun*, *Nordmantune* 958 YCh 2, *Normantun* DB], N~ **upon Soar** Nt [*Normantun* DB, *Normanton super Sore* 1225 Ep], N~ **upon Trent** Nt [*Normentune* DB], N~ **on the Wolds** Nt [*Normantun* DB], N~ Ru [*Normanton* 1183 P, *-tone* 1227 Ep], N~ YW [*Normantone* DB, *-ton* 1208 Cur]. 'The TŪN of the Northmen or Norwegians.'

Temple N~ belonged to the Templars in 1185 (TpR).—For **N~ Turville** see ACTON TURVILLE.

Norrington W [*Northintone* 1212 RBE, *Norpinton* 1242 Fees]. OE *norþ in tūne* '(the part) north in the village, the northern part of the village'. Cf. EASTINGTON.

OE **norþ** 'north' is a common first el. of pl. ns.; see NORTH-, NOR- (passim), NARBOROUGH, NUTBOURNE Sx. In some cases NORTH- may go back to OE *be norþan* 'north of'; cf. NORTHWOOD.

Northa·llerton YN [*Alure-, Aluertune* DB, *Aluertuna* 1088 LVD, *North Alverton* 1293 Cl]. '*Ælfhere*'s or *Ælfrēd*'s TŪN.'

Northam D [*Northam* DB, 1157 Fr]. 'Northern HĀM.' **Northam** Ha [*Northam* DB, *Norham* 1151 BM, *Homme* 1291 Tax]. 'Northern HAMM.'

Northa·mpton Np [*Hamtun* 917 &c. ASC, *Hamton* 972–92 BCS 1130, *Norðhamtun* 1065 ASC (C), *Hantone*, *Northantone* DB]. OE HĀMTŪN (q.v.). North- for distinction from SOUTHAMPTON. **Northamptonshire** is *Hamtunscir* 1011 ASC (C, D), *Northantonescire* DB.

Northaw or **Northall** Hrt [*Nort Haga* (silva) 1077–93 Gesta, *Northawe* 1198 (1301) Ch, *Norhaghe* 1200 Cur]. 'Northern HAGA or enclosure.'

Northborough Np [*Norðburh* 656 ASC (E), *Norburg* 1202 Ass]. 'Northern BURG.'

Northbourne K [*Nortburne* 618 BCS 13, *Norborne* DB]. 'Northern stream.'

Northenden Chs [*Norwordine* DB, *Norwrdina* 1119, c 1150, *-wrthin* c 1220 Chester, *Northerden* 1360 Chamb]. 'Northern WORÞIGN.'

Northfield Wo [*Nordfeld* DB], **Northfleet** K [*Northfleta, Flyot* 10 BCS 1321 f., *Norfluet* DB]. 'Northern FELD and FLĒOT.'

Northiam (norjam) Sx [*Hiham* DB, *North Hyham, Nordhyam* c 1210 Penshurst, *North Ihamme* 1302 FF]. 'North Higham.' Higham seems to be OE *hīeg-hamm* 'HAMM where hay was got'.

Northill Bd. See IVEL. **N~** Co. See HILL.

Northington Ha [*Northametone* 903 BCS 602, *Norhameton* 1167 P]. OE *Norþhǣmatūn* 'the TŪN of the dwellers to the north' (of Winchester).

Northleach Gl [*Lecce* DB, *Northleche* 1215 Glouc, *-lecche* 1227 BM]. 'North LEACH.' The place is on the river LEACH and was at first *Leche. North* was added for distinction from EASTLEACH. See LEACH, EASTLEACH.

Northleigh. See LEIGH.

Northmore O [*la Mora* 1195 P, *Mora* 1208 Cur, 1229 Ep]. Self-explanatory.

Northolt Mx [(æt) *Norðhealum* 960-2 BCS 1063, *Northala* DB, *Norhale* 1236 Fees]. 'Northern HALHS.' The meaning of HALH is not clear.

Northorpe Li nr Gainsborough [*Torp* DB, c 1115 LiS, 1139 RA, *Nortorp* 1202 Ass]. 'Northern thorp.'

Northover So [*Nordoure* 1180 P, *Northovere* 1242 Fees]. 'Northern bank.' N~ is on the northern bank of the Yeo. See ŌFER.

Northowram, Northrepps. See OWRAM, REPPS.

Northumberland was originally the land north of the river Humber, including the Anglian parts of Scotland, as in *Norþhymbre* 867 ASC, *Norðhymbralond* 895 ASC OE. *Norþhymbre* originally meant 'dwellers north of the Humber', but was also used of the territory. After the land north of the Tweed had been ceded to Scotland (finally in 1018 as a result of the battle of Carham), the name was restricted to the part of England north of the Humber. But Cumberland and the rest of western Northumbria at an early date got special names of their own, and in 1065 (ASC (D)) Yorkshire (*Eoforwicscir*) and *Norðhymbraland* are mentioned side by side. Here *Norðhymbraland* seems to be used only of the present Northumberland and Durham. Durham became a palatinate under the Bishop of Durham and was no longer reckoned as part of Northumberland. In its modern sense Northumberland is used after c 1100 (*Norhumberland* 1130 P).

Northwich Chs [*Wich, Norwich* DB, *Northwich* c 1150 Chester]. OE wīc 'town'. *North-* in contradistinction to NANTWICH and MIDDLEWICH.

Northwick Gl [(to) *Norðwican* 955-9 BCS 936], **N~** Wo [*Norðwica* 964 BCS 1135]. 'Northern wīc or dairy farm.'

Northwold Nf [*Northuuold* 970 BM, *Nortwalde* DB]. 'North Wold.' N~ is north of HOCKWOLD and METHWOLD. The whole territory seems to have been OE *Wald*, i.e. *wald* 'forest'.

Northwood Mx [*Northwode* 1438 Pat], **N~** Wt [*Northwode* 1287-90 Fees, 1316 FA]. 'North wood.' But one Northwood K is *bi Northanuude* 727 BCS 846. This is '(the place) north of the wood', and the same may be the origin of some other Northwoods. Another N~ K is *Northuuda* 832 BCS 402. This is 'north wood'.

Nortoft Np [*Nortot* DB, *Northtoft* 1176 P]. 'Northern toft.'

Norton is a very common name, which means 'north TŪN', 'a homestead or village north of another'. N~ Chs [*Nortune* DB], **N~** Db [*Nortune* DB], **N~** Du [*Norðtun* c 1000 BCS 1256], **Cold N~** Ess [*Nortuna* DB, *Coldenorton* 1350 Ipm], **N~ Mandeville** Ess [*Nortuna* DB, *Norton de Mandevill* 1238 Subs], **N~** Gl nr Gloucester [*Nortune* DB], **N~** Ha [*Nortune* 903 BCS 602, DB], **N~ Canon** He [*Nortune* DB, *Norton Canons* 1327 Ipm], **N~** Hrt [*Norðtun* 1007 Crawf], **N~** K [*Nordtone* DB, *Northtune* 11 DM], **East N~** Le [*Nortone* DB, *Est Norton* 1327 Subs], **King's N~** Le [*Nortone* DB, *Kyngesnortona* R 1 (1253) Ch], **N~ juxta Twycross** Le [*Nortone* DB], **Bishop N~** Li [*Nortune* DB, *Nordtuna* c 1115 LiS, *Bischopnorton* 1402 FA], **N~ Disney** Li [*Nortune* DB, *Norton Isny* 1331 Ch], **N~ Folgate** Mx [*Nortonfolyot* 1433 FF], **Blo N~** Nf [*Nortuna* DB, *Blonoron* 1291 Tax], **Pudding N~** Nf [*Nortuna* DB, *Pudding Norton* 1276 Cl], **N~ Subcourse** Nf [*Nortuna* 1044-7 KCD 785, DB, *Norton Supecors* (*Subcors*) 1282 Cl, *N~ Soupecors* 1326 Bodl], **Wood N~** Nf [*Nortuna* DB, *Wudnorton* 1199 Pp], **N~** Np [*Nortone* DB, *Norton juxta Davintre* 1242 Ipm], **Greens N~** Np [*Nortone* DB, *Grenesnorton* 1465 FF], **N~** Nt [*Norton* 1194 P], **Brize N~** O [*Nortona* c 1130 Oxf, *-e* 1222 Ep, *Northone Brun* c 1267 Eynsham], **Chipping & Over** (*olim* **Cold**) **N~** O [*Norton Mercatoria* 1246 Ch, *Chepyngnorton* 1280 Ep, *Caldenorthon* 1229 Ch], **N~ in Hales** Sa [*Norðtun* 1002 Wills, *Nortune* DB, *Norton in Hales* 1291 Tax], **N~** Sf [*Nortuna* DB], **N~ Fitzwarren** So [*Nortone* DB, *Nortun by Tantone* c 1100 Montacute], **N~ under Hamdon** So [*Nortone* DB, *Norton under Hamedon* 1246 Wells], **N~ Hawkfield & Malreward** So [*Nortone* DB, *Norton Hautevill & Malreward* 1238 Ass], **Midsomer N~** So [*Midsomeres Norton* 1248 Ch, *Midsummernorton* 1269 FF], **N~ St. Philip** So [*Nortune* DB, *Norton Sancti Phillipi* 1316 FA], **N~ Canes** St [æt *Norðtune* 951 BCS 891, *Nortone* DB, *Norton-super-le-Canok* 1289 Ass], **Cold N~** St [*Calde Norton* 1227 Ch], **N~ in the Moors** St [*Nortone* DB, *Norton super le Mores* 1285 FA, *N~ under Keuremunt* 1227

Ch], N~ W nr Malmesbury [*Nortun* 937 BCS 719, *Nortone* DB], N~ Bavant W [*Nortone* DB, *Norton Bavant* 1412 FA], N~ Lindsey & Nether N~ Wa [*Nortun* 12 Fr, *Norton Lindeseye* 1316 FA, *Nether Nortone* 1377 AD], N~ Wo nr Evesham [*Nortona* 709 BCS 125, *-tune* DB], Bredons N~ Wo [*Nortune* DB], N~ juxta Kempsey Wo [*Norŏtun* 989 KCD 671, *Norton juxta Kemeseye* 1346 FA], King's N~ Wo [*Nortune* DB, *Kinges Norton* 1221 Ass], N~ YE [*Nortone* DB], N~ le Clay YN [*Nortone* DB, *Norton in le Drit* 1301 Subs], N~ Conyers YN [*Nortone* DB, *-tune* 1088 LVD, *Norton Coniers* 1316 FA], N~ YW [*Nortone* DB].

N~ Bavant W. See EASTON BAVENTS. The manor was held by Roger Bavent in 1344 (Misc).—Bishop N~ Li belonged to the Bishop of Lincoln.—Blo N~ Nf has as addition ME *blo* from OScand *blā(r)* 'blue, dark', possibly in the sense 'bleak', or the word that is the first el. of BLOFIELD.—Bredons N~ Wo is nr BREDON HILL.—Brize N~ O was held by William Brun in 1200 (Cur). *Brun* is OFr *brun* 'brown' used as a byname.—N~ Canes St possibly from CANNOCK.—N~ Canon He was held by the Dean and Chapter of Hereford.—Chipping N~ O. See CHIPPING.—N~ le Clay YN from clayey soil. Drit is 'dirt'.—N~ Conyers YN was held by Roger de Koiners in 1196 (FF). Cf. HUTTON CONYERS.—N~ Disney Li from the de Isney family from ISIGNY in Normandy. William de Ysini is mentioned in connexion with Norton c 1150 (DC).—N~ Fitzwarren So from a family of the name. Fitzwarren is 'son of Warin' (OFr *Warin, Guarin* from OG *Warin*).—N~ Folgate Mx for N~ Foliot; cf. CHILTON FOLIAT.—Greens N~ Np from the Green family (14th cent.).—N~ in Hales Sa is nr HALES St.—N~ under Hamdon So. See HAMDON.—N~ Hawkfield St is corrupt for N~ Hauteville. Reginald de Alta Villa held N~ before 1219 (Fees). Hauteville perhaps from HAUTTEVILLE nr Coutances.—N~ Lindsey Wa is corrupt for N~ Limesi. The Limesi family held the manor from the 12th cent. The name is from LIMESY nr Rouen.—N~ Malreward So was held by William Malreward in 1238 (Ass). Cf. GOADBY MARWOOD.—N~ Mandeville Ess from the Mandeville family, resident here from c 1200. Cf. HARDINGTON MANDEVILLE.—Midsomer N~ So is said to be so called in allusion to the festival held at midsummer on the day of St. John, the patron saint.—Pudding N~ Nf is unexplained.—N~ St. Philip So presumably from the dedication of the church.—N~ Subcourse Nf probably from a family name. Hermannus *Sorlecors* (*Surlecors*) is mentioned 1177 P (Nf). *Surlecors* seems to mean 'on the river' (cf. Fr *cours d'eau* 'stream').—N~ juxta Twycross Le. See TWYCROSS.

Norwell Nt [*Nortwelle* DB, *Norwell* 1167 P]. 'North stream or spring.' The place is NE. of SOUTHWELL.

Norwich (nŏrǐj) Nf [*Norŏwic* c 930 Coins, 1004 ASC (E), c 1035 Wills, *Noruic* DB]. 'North town.' See wīc.

Norwood Mx [*Northwode* 1294 FF], N~ Sr [*Norwude* 1176 P, *Northwode* 1284 Ch]. Identical with NORTHWOOD.

Noseley Le [*Noveslei* DB, *-lai* c 1135 Ordericus, *-leia* 1221 Ep, *Nouesle* 1251 Ch]. '*Nōþwulf*'s LĒAH.' Cf. ROLLESTON Le, ROUSHAM, which show similar early reduction of a long name.

Nostell YW [*Osele, Osle* DB, *Nostlay* 1121–7 YCh 1428, *Nostla* 1135–40 YCh 1015, *Noslay* 1227, *Nostle* 1228 Ch]. OE *ōsle* 'blackbird' and *lēah* 'wood' or 'glade'. *N-* is explained as in NAYLAND, NEVENDON.

Nosterfield Ca in Shudy Camps [*Nostresfelda* c 1080 ICC, *Ostresfeld* 1179, *Ostrefeld* 1180 P, *Nostrefeld* J BM], N~ YN [*Nostrefeld* J Ass, *Nosterfeld* 1257 YD]. The *N-* is inorganic as in NOSTELL. An el. *oster* is found in Osterland K [*Osterland* c 961 BCS 1064], *Ostercumb* 909 BCS 622, OSTERLEY PARK Mx. The word may be related to OE *ōst* 'a knot, knob'. Its meaning may be 'hillock' or the like.

Notgrove Gl [*ad Natangrafum* 716–43 BCS 165, *Nategraue* DB, *-grava* 1209 Fees, *-grave* 1349 BM]. Cf. *Nataleahes æsc* c 800 BCS 299 (in bounds of Withington c 3 m. from Notgrove). *Nataleah* seems to be a variant form of *Natangrafum*. The name probably means 'wet wood'. See NATELY, NETLEY MARSH. The *o* of Notgrove apparently points to OE *Nātan-*, but the 13th and 14th cent. forms have *a*. This indicates that *Not-* is due to a late change. See GRĀF.

Notley Bk [*Nutlee* 1200 Cur, *Nutele* 1204 FF], Black & White N~ Ess [(into) *Hnutlea* 998 Crawf, *Nutlea* DB, *Blanche Nutele* 1232, *Black, White Nuteleye* 1240 FF]. 'Nut wood.' See LĒAH.

Nottingham Nt [*Snotengaham* 868, *Snotingaham* 922 ff. ASC, *Snotingeham* DB, *Notingham* 1087–1100 Reg]. 'The HĀM of *Snot*'s people.' *Snot* pers. n. is found in DB. The loss of *S-* is due to Norman influence. Nottinghamshire is *Snoting(a)hamscir* 1016 ASC (D, E), *Snotingehamscyre* DB.

Notting Hill Mx [*Knottynghull'* 1356 Works, *Knottyngwode* 1376 Cl, *Knottinge Bernes* 1476 PNMx]. Possibly named from KNOTTING Bd, Notting being really a family name.

Nottington Do [*Notinton* 1212 Fees, *Notingeton* 1234 BM]. 'The TŪN of *Hnotta*'s people.' *Hnotta* is found in *Hnottan forð* BCS 1217, *Hnottan mæræ* ib. 491 and is derived from OE *hnot* 'bald-headed'.

Notton Do [*Natton* 1350 FF, *Neton* 1370 FF], N~ W [*Natton* 1345 Misc, 1412 AD, *Netton* n.d. AD]. OE *Nēat-tūn* 'cattle farm'.

Notton YW [*Notone, Norton* DB, *Nocton* 1186 P, 1260 Ass, *Nottone* 1201 FF, *Nottun* 1226 FF]. OE *Hnoc-tūn*; first el. OE *hnoc* 'a wether sheep'.

Nowton (-ō-) Sf [(at) *Newetune* c 950 Wills, *Neotuna* DB, *Neutune* c 1095 Bury, *Neuton,*

Nouton 1254 Val]. A variant of NEWTON, with change of *ēow* to *ōw* as in NOBOLD, NOBOTTLE.

Nuffield or **Tuffield** O [*Togfelde* 1209–19, *Toufeld* 1209–35, *Tofelde* 1229 Ep, *Todfeld* 1254 Val, *Tuffeld* 1428 FA]. *N-* and *T-* must be later additions, as in NAYLAND, *Ticelle* DB (see ITCHEL). The original name may have been *hōh-feld* 'FELD by a spur of hill'.

Nunburnholme YE [*Brunham* DB, *Brunnum* 1206 FF, 1231 FF, *Brunhum* 1291 Tax]. OE *Burnhamm* (*-homm*) 'HAMM by a stream'. There was a nunnery at the place.

Nunea·ton Wa [*Etone* DB, *Eaton* 1155–9 Fr, Hy 2 DC, *Nun Eton* 1290 Cl]. OE *Ēatūn* 'TŪN on the river (Anker)'. N~ was a priory for nuns.

Nuneham Courtenay O [*Niwanhæminga londgemære* 940 BCS 760, *Newham* 1227 Ep]. Identical with NEWNHAM. Cf. IWERNE COURTNEY.

N~ **Courtenay** belonged to the Courtneys from 1214.

Nunkeeling YE [*Chelinge, Chiling(h)e* DB, *Killing* 1200 Cur, 1222 FF]. Identical with KELLING Nf (OE *Cyllingas* 'the people of *Cylla*'). N~ was a priory for nuns.

OE **nunne** 'nun' is a late addition in NUN-BURNHOLME, NUNEATON, NUNKEELING. In names such as NUNNEY, NUNWICK &c. it is difficult to decide whether we have this word or the OE pers. n. *Nunna*.

Nunney So [*Nonin* DB, *Nony* 1219 FF, *Nuni* 1243 Ass]. 'The island of the nuns or of *Nunna*.'

Nunnington YN [*Noning-, Nunnigetune* DB, *Nunintun* 1167 P]. 'The TŪN of *Nunna*'s people.'

Nunthorpe YN [*Torp* DB, *Nunnethorp* 1240 FF]. 'The Thorp of the nuns.' There was a nunnery here.

Nunton Np nr Stamford [*Nunnetun* 963–84 BCS 1128]. '*Nunna*'s TŪN.'

Nunwell Park Wt [*Nonoelle* DB, *Nunewill* 1199 FF]. '*Nunna*'s or the nuns' spring.'

Nunwick (nŭnĭk) Nb [*Nunnewic* 1166 f. P], N~ YW [*Nunnewic* c 1030 YCh 7, *Nonnewic* DB]. Cf. NUNNE, WĪC.

Nursling Ha [*Nhutscelle* 8 Life of St. Boniface, æt *Nutscillinge* 877, *Hnutscillingc* 909 BCS 544, 620 f., *Notesselinge* DB]. The earliest form of the name means 'nutshell'. The later name is a derivative of this. Apparently the name is jocular, referring to a tiny place.

Nurstead K [*Notestede* DB, *Nutstede* 1242 Fees], N~ Sf [*Hnutstede* c 995 BCS 1289]. 'Place where nuts grew.'

Nutbourne Sx nr Pulborough [*Nordborne* DB, *-burne* 1195 P, *Nuteburn* 1275 Ipm]. 'North stream.' **Nutbourne** Sx nr Chich-

ester [*Notburna* Hy 2 (1387) Pat]. 'Stream on whose banks nuts grew.'

Nutfield Sr [*Notfelle* DB, *Nutfelda* 1170 P], **Nutford** Do [*Notforda* DB, *Nutford* 1242 Fees], **Nuthall** Nt [*Nutehale* DB, *Notehal* 1194 P], **Nuthampstead** Hrt [*Nuthamstede* 1212 Fees], **Nuthurst** Sx [*Nothurst* 1228 Pat], N~ Wa nr Stratford [*Hnuthyrst* 704–9 BCS 123], **Nutley** Ha [*Noclei* DB, *Nutlie* 1212 Fees], N~ Sx [*Nutleg* 1249 FF], **Nutwell** D [(of) *Hnutwille* 1072–1103 E]. 'FELD, FORD, HALH, HĀM-STEDE, HYRST, LĒAH, WELLA where nuts grew.'

Nyetimber Sx in W. Chiltington [*Nitinbreha(m)* DB, *Niutimbre* 1283 Ipm], N~ Sx in Pagham [*Niwetimbr'* 1279 Ass, *Nitimbre* 1275 Cl]. See NEWTIMBER and NĒOWE.

Nyland (-ī-) Do [*Iland* DB, *Liland* 1212 Fees]. OE *īegland* 'island' with addition of *N-* as in NAYLAND. *L-* in *Liland* is the French def. art. Identical in origin is **Nyland** So [*Ederedeseie* 725 BCS 142, (insula de) *Adredesia* 1344 BM]. The old name means '*Ēadrēd*'s island'.

Nymet (-ī-) **Rowland, N~ Tracy, Broadnymet** D [*Nymed* 974 BCS 1303, *Limet* DB, *Nimet Rollandi, Bradenimet* 1242 Fees, *Nemethe Tracy* 1270 Ep]. Named from *Nymet*, the old name of the river YEO, a tributary of the Taw [*Nimed, Nymed* 739 BCS 1331 f.]. The name is British and related to Gaul *nemeton* 'holy place', OW *nimet* in pers. ns., Welsh *nyfed* 'shrine'. An adj. *nemed* is found in OIr *brátha nemed* 'the dooms of the nobles'. The river-name may be derived from an old adj. *nemeto-* 'holy', or it may mean 'river at a holy place or grove'. Cf. NYMPTON.

N~ **Rowland** was held by one Rolandus in 1166 (RBE), N~ **Tracy** (now BOW) by Oliver de Trascy in 1212 (Fees). See BOVEY TRACY. *Broad-* means 'Great'. See BRĀD.

Nympsfield Gl [*Nymdesfeld* 872 BCS 535, *Nimdesfelle* DB, *Nimedesfeld* 1236 Fees]. 'FELD by a holy grove or place.' No doubt the Brit name of the place was *Nemeto-* 'holy place'. Cf. NYMET.

Nympton, Bishop's, D [*Nimetone* DB, *Nemetone Episcopi* 1269 Ep], **George N~** D [*Nimet* DB, 1249 Ipm, *Nymet St. George* 1274 Ipm, *Nymeton Sancti Georgij* 1291 Tax], **King's N~** D [*Nimetone* DB, *Nuneton Regis* 1230 P]. The places are on the MOLE, formerly *Nymet* [*Nimet* 1238 Ass]. See NYMET.

Bishop's N~ belonged to the Bishop of Exeter. The early forms show much vacillation between the forms *Nimet* and *Nimeton*.

Nynehead So [(of) *Nigon Hidon* 11 KCD 897, *Nichehede* DB, *Nigenid* c 1100 Montacute]. '(Estate of) nine hides.'

Nyton Sx [*Nyton* 1327 Subs]. Perhaps identical with NITON. In PNSx the name is derived from OE *Īeg-tūn* 'TŪN in the island or river-land' with *N-* prefixed as in NAYLAND.

O

Oadby Le [*Oldebi* DB, *Outheby* 1199 FF, *Oudeby* 1204 Cur]. '*Auði*'s BY.' First el. ON *Auði*, ODan *Øthi* pers. n.

Oake So [(æt) *Acon* 11 KCD 897, *Acha* DB], **Oaken** St [*Ache* DB, *Oken* 1327 Subs]. OE *ācum*, dat. plur. of *āc* 'oak'.

Oakengates Sa [*Lee Okynyate* 1535 VE]. 'Oaken gate.'

Oakenshaw YW [*Akanescale* 1254 Ipm, *Okenschagh* 1402 FA]. 'Oak wood.' OE *ācen* 'of oaks' and SCAGA 'wood, copse'.

Oakerthorpe Db [*Ulchilthorp* 1229 Ch, *Hulkerthorpe* 1237 FF]. '*Ulfkell*'s thorp.' The first el. is ON *Ulfkell, -ketill*, ODan *Ulkel*, OSw *Ulfkil*, a name commonly found in England (*Ulfketel, Ulchel* DB &c.). Cf. Robert son of Ulkel 1224 Pat (at Horsley nr Oakerthorpe).

Oakford D [*Alforda* DB, *Acford* 1166 RBE, *Ocford* 1224 Cl]. 'Ford by the oaks.'

Oakham (ōō-) Ru [*Ocham* 1067 BM, *Ocheham Cherchesoch* DB, *Ocham* 1202 Ass, 1229 Ch]. '*Oc(c)a*'s HĀM.' *Oca* is found in *Ocan lea* (see OCKLEY), *Occa* in *Occan slǽw* 969 BCS 1230. The addition *Cherchesoch* in DB is OScand *kirkiusōkn* 'parish.'

Oakhanger Ha [*Acangre* DB, *Achangra* 1175 P, *Akehanger* 1250 Ch]. 'Oak slope.' See HANGRA.

Oakhurst Hrt in Shenley [*Acersc* 944–6 BCS 812, *Okersh* 1287 BM]. 'Oak park'; see ERSC.

Oakington Ca [*Hokintona* c 1086 IE, *Hochinton* DB, *-a* 1130 P, *Hokinton* 1200 Cur]. '*Hoca*'s or *Hocca*'s TŪN.' *Hocca* is well evidenced. *Hoca* is found in *Hocan edisc* BCS 1123.

Oakle Street Gl [*Acle* 1310 BM], **Oakleigh** K nr Higham [*Acleah* 774, *Hacleah* 805 BM], **Oakley** Bd [*Acleia* c 1060 KCD 919, *Achelei* DB], **O~** Bk [*Achelei* DB, *Aclai* 1197 P], **O~** Do [*Aclaeh* 805, æt *Aeclea* 844 BCS 321, 445, (æt) *áclee on westsæxum* c 975 Rit Dun], **Great & Little O~** Ess [*Accleia* DB, *Ocle Magna, Parva* 1238 Subs], **O~** Park Gl [*Achelie* DB], **Church O~** Ha [(to) *Aclea* 10 BCS 1161, *Aclei* DB, *Acle, Chirchocle* 1206 Cur], **North O~** Ha [*Acle* 1206 Cur, *Northacle* 1280 QW], **Great & Little O~** Np [*Achelau* DB, *Accle* 1176 P, *Maior, Parva Acle* 1220 Fees], **O~** O [*Aklye* 1220 Fees, *Ocle* 1451 BM], **O~** Sf [*Acle* DB, c 1095 Bury], **O~** So [*Achilei* 1084 GeldR, *Achelai* DB, *Akele* 1205 Cur], **O~** St nr Lichfield [(æt) *Ácclea* 1002 Wills, *Acle* DB]. OE *āc-lēah* 'oak wood' or 'glade where oaks grew'. The examples under O~ Do refer to the place where the synod was held. There is good reason to identify it with Oakley Do.

Oakmere Chs [*Okmere* (stew) 1348 Vale Royal]. 'Oak lake.' There is a lake at the place.

Oaksey W [*Wochesie* DB, *Wokesai* 1195 Cur, *Wokesia* 1197 FF]. '*Wocc*'s island.' *Wocc* is found in *Wocces geat* BCS 594, 1080 (Ha). Cf. WOKING.

Oakthorpe Le [*Achetorp* DB]. '*Áki*'s thorp.' *Áki* is a common OScand name.

Oakworth YW [*Acurde* DB, *Ocwurde* 1246, *Akeworthe* 1255 YInq]. 'Oak WORÞ.'

Oare Brk [*Orhæma gemære* 951, æt *Oran* 968 BCS 892, 1225, *Ore* 1242 Fees], **O~** K [*Oran* 11 DM, *Ore* DB, *Ora* 1162 P], **O~** W [*Æt Motenes oran*, (æt þam) *oran* 934 BCS 699, *Ore* 1232 Ch, 1242 Fees]. OE *ōra* 'border, edge' &c. In Oare K *ōra* means 'shore'. In the others it means 'hill-side, slope, ridge'. See ŌRA.

Oare So [*Are* DB, *Ar* 1194 P, *Hor* 1212 Fees]. Named from **Oare Water** [*Ar* 1279, 1301, *Ore* 1298 For]. The river-name is identical with AYR in Scotland [*Ar* 12], AHR in Germany &c. It is a Brit river-name.

Oborne Do [(æt) *Womburnam* 974 BCS 1308, *Wonburna* 998 KCD 701, *Wocburne* DB, *Wuburn* 1227 FF]. 'Crooked, winding stream.' First el. OE *wōh* 'crooked'.

Oby Nf [*Othebei, Oebei* DB, *Oubi* 1196 FF]. Identical with OADBY, though with loss of *ð*.

Occlestone Chs [*Aculvestune* DB, *Aculuiston* 1210–20 Chester]. '*Ácwulf*'s TŪN.'

Occold Sf [*Acholt* c 1050 KCD 907, *Acolt* DB, *Achold* 1201 Cur, *Ocolt* 1254 Val]. 'Oak copse.'

Ock R Brk [*Eoccen* 931, 940 BCS 684, 761, *Ocenne wyllas* 953–5 ib. 902, *Ocke* Hy 1 Abingd]. A Brit river-name derived from the word for salmon, OCo *ehoc*, MW *ehawc*, Welsh *eog*.

Ockbrook Db [*Ochebroc* DB, 1166 P, *Okebroc* 1185 P]. '*Oc(c)a*'s brook.' Cf. OAKHAM.

Ockendon Ess [*Wokendune* 1067 BM, *Wochen-, Wochaduna* DB, *Wokindon* 1230 P]. '*Wocca*'s DŪN or hill.' Cf. WOKEFIELD.

Ockham Sr [*Bocheham* DB, *Hocham* 1170 P, *Ocham* 1291 Tax]. Identical with OAKHAM.

Ockley Common Sr nr Farnham [(to) *Ocan léa* 909 BCS 627], **O~** Sr nr Horsham [*Hoclei* DB, *Okelee* 1203 Cur, *Ockele* 1242 Fees, *Okele* 1279 QW]. '*Oca*'s LĒAH.' The two Ockleys are a good way apart, but if LĒAH means 'wood' they may have been named from the same wood.

Ocle (ō-) **Pychard** He [*Acle* DB, *Acle Pichard* 1242, *Aclepihard* 1249 Fees]. Identical with OAKLEY.

Pychard from its Norman owners. Roger Pichard held the manor in 1242 (Fees). Pychard is OFr *pichard* 'green woodpecker'. There was also a **Lyre Ocle** [*Acle Lyre* 1242 Fees], held by the Abbey of Lyre in Normandy.

Octon YE [*Ocheton* DB, *Ocatuna* 1170–85 YCh 1065, *Oketon* 1222 FF]. '*Oca*'s TŪN.' Cf. OAKHAM.

Odcombe So [*Udecome* DB, *Odecumba* 1157 ff. P, *Odecumbe* 1201 FF]. '*Uda*'s CUMB or valley.' Or perhaps rather OE *wudu-cumb* 'wood coomb'. Cf. ODIHAM.

OScand **odd(r), oddi** 'point, cape'. See GREENODD, RAVENSER ODD.

Oddingley Wo [*Oddingalea* 816, 840 BCS 356, 428, *Odduncalea* 963 ib. 1108, *Oddunclei* DB]. 'The LĒAH of *Odda*'s people.'

Oddington Gl [*Otintone* 872 BCS 535, *Otintone* DB, -*tona* 1066–87 Glouc], O~ O [*Otendone* DB, *Otindon* 1242 Fees, -*dun* 1246 Ch]. '*Ota*'s TŪN and DŪN.' *Ota is found in *Otanhyrst* 811 BCS 339.

Odd Rode. See RODE.

Odell (ō-) or **Woodhill** (wŭdl) Bd [*Wadehelle* DB, *Wahella* 1163, *Wahull* 1193 P]. OE *wād-hyll* 'woad hill'.

Odiham (ō-) Ha [*Wudiham* 1116 ASC (E), *Odiham* DB, 1130 P, c 1140 RA]. OE *wudiga hamm* 'wooded HAMM'. The loss of *W-* must be due to Norman influence.

Odsey Ca [*Oddeseth* 1198 (1286), *Oddeseye* 1252 Ch]. The place gave its name to a hundred [*Odesei hund*' DB, *Oddesech*'-*hundredum* 1191 P, *Oddeseth* 1248 Ass]. '*Odda*'s SĒAP.'

Odstock W [*Odestoche* DB, -*stocha* 1174 P, -*stoka* 1199 FF]. '*Oda*'s STOC.'

Odstone Brk nr Ashbury [*Ordegeston* DB, *Ordeiston* 1220 Fees]. '*Ordhēah*'s TŪN.'

Odstone Le [*Odestone* DB, *Oddestuna* 12 DC]. '*Odd*'s TŪN.' *Odd* is ON *Oddr*, OSw *Odder*.

OE **ōfer** (= MLG *ōver*, G *Ufer*, Du *oever*) means 'river bank', 'border, margin'. The sense 'river bank' is found in some pl. ns., as OVER Ca, NORTH-, WESTOVER So, TANSOR Np, some ORTONS and OVERTONS. But *ofer* often refers to a steep slope or even a hill or ridge. A typical case is EASTNOR He, which means 'east of the *ofer*', i.e. Eastnor Hill. A sense-development from 'border, margin' to 'edge, brae of a hill' and even 'hill, ridge' is possible. Cf. ŌRA. But there is also a form *ufer* (in OVER Chs, Db, OWRAM &c.), which is related to YFER, *over* adv. &c. Very probably we have to reckon with an OE **ofer** (with short *o*), *ufer* meaning 'hill, steep slope' and the like. This kind of sense is found in ASHOVER, OKEOVER, OVER Chs, Db, WELLINGORE, WENTNOR, and

in some ORTONS and OVERTONS. In some names OVER- goes back to OE *uferra*, ME *overe* 'upper', e.g. in OVERBURY Wo, OVERTON Ha, W. The OE prep. *ofer* 'over' is rarely to be assumed as a pl. n. element. Cf., however, OVERY.

Offa's Dyke, the old boundary between England and Wales [*Offedich* 1184 P (He), -*e* 13 AD, *claud offa* Red Book]. The dyke is said to have been built by King Offa. Cf. *Offan dīc* 854 BCS 475 (So). *Claud* is Welsh *clawdd* 'ditch'.

Offchurch Wa [*Ofechirch* 1043 (1267) Ch, *Offecherche* 1197 P]. '*Offa*'s church.'

Offcote Db [*Ophidecotes* DB, *Offidecot* 1251 Ch, Hy 3 BM, -*cote* 1272 FF, *Offedecote* 1265 Abbr]. Possibly '*Offede*'s COTS', *Offede* being a derivative of *Offa* analogous to *Lullede* from *Lulla*. The normal -*i*- in the second syllable in early forms tells against this. Probably the name consists of *Offa* and a compound word, e.g. an OE *wida-cot* 'wood-cot' (for pl. ns. in *Wood-* with the original OE *i* preserved see WUDU, WITTON). The name would then mean '*Offa*'s wood-cot'. Offcote and Underwood form a parish.

Offenham Wo [*Offeham* 709, *Uffaham* 714 BCS 125, 130, *Offenham* DB]. '*Uffa*'s or *Offa*'s HAMM.'

Offerton Chs [*Offerton* 1248 Ipm, 1289 Court], O~ Db [*Offretune* DB, *Offerton* 1200 P, Hy 3 Derby]. The first el. may be a name in -*ford*, e.g. *Offan ford* '*Offa*'s ford'. The pers. ns. *Oftfōr* and *Ōsfrið* might also be thought of.

Offerton Du [*Uffertun* c 1050 HSC, -*ton* c 1180 FPD]. The first el. may be an OE *up-ford* 'upper ford'. Cf. OFFORD.

Offham K [*Offahames gemǣre* 942–6, *Offaham* 10 BCS (779, 1322), 11 DM], '*Offa*'s HĀM.'

Offham (ō-) Sx [*Wocham* c 1092 PNSx, 1199 FF, *Wogham* 1296 Subs]. '*Weohha*'s HĀM.' *Weohha*, a short form of names in *Wēoh-*, is on record.

Offington Sx [*Ofintune* DB, *Offentun* c 1200, *Offingetone* 13 BM]. 'The TŪN of *Offa*'s people.'

Offley, Great & Little, Hrt [*Offanlege* 944–6 BCS 812, *Offelei, altera* O~ DB, *Parva Offeleg* R 1 Cur], **Bishops & High O~** St [*Offeleia, -lie* DB, *Bissopstoffeleg* 1285 FA, *Alta Offyleye* 1316 FA]. '*Offa*'s LĒAH.' **Bishops O~** from the Bishop of Lichfield.

Offord Cluny & Darcy Hu [*Ope-, Upeforde* DB, *Uppeford* 1195 BM, *Upford* 1210 Cur, *Offord Willelmi Daci* 1220 Fees, *Offorde le Daneis* 1225 Ep, *Offord Daci* 1254 Val, *Offord Cluny* 1252 Ep]. 'Upper ford.' O~ **Cluny** belonged to the Abbey of Cluny in France.—*Darcy* is a corruption of *Daci*, the gen. of *Dacus* 'Dane', itself a Latinized form of OFr *Daneis*, here used as a family name.

Offton Sf [*Offetuna* DB, *Offintone* 1166 RBE, *parua Offigetun* 1198 FF]. '*Offa*'s TŪN' and 'the TŪN of *Offa*'s people'.

Offwell D [*Offewille* DB, *Uffewill* 1230 P, -*wille* 1242 Fees]. '*Uffa*'s stream.' The place is on Offwell Brook.

Ogbourne St. Andrew & St. George W [*Ocheburne* DB, *Occheburna* 1133 Fr, *Okeburn Sancti Andree, Sancti Georgii* 1428 FA]. '*Occa*'s stream.' See OAKHAM.
The surnames are the names of the patron saints.

Ogle Nb [*Hoggel* 1170, *Ogle* 1181 P, *Oggill* 1242 Fees]. '*Ocga*'s hill.'

Oglethorpe YW [*Oceles-, Oglestorp* DB, *Occlistorp* c 1180 YCh 1026, *Okelesthorp* 1240 FF]. '*Odkell*'s thorp.' *Odkell* (*Otkell*) is a well-evidenced ON pers. n.

Ogley Hay St [*Oggele* 1231 Cl, (Hay of) *Uggeleye* 1292 Cl]. '*Ocga*'s LĒAH.'

Ogston Db [*Oggodestun* 1002 Wills, *Oggedestun* 1004 KCD 710, *Oughedestune* DB]. '*Oggod*'s TŪN.' *Oggod* pers. n. is found in BCS 1130. It may be of Continental Germanic origin.

Ogwell (ō-), **East & West,** D [*Wogwel, Woge-, Ogewille* DB, *East-, Westwogewelle* 1278 Ipm]. Originally a name of the stream at the place [(to) *Wogganwylle* 956 BCS 952, *Woggawill* 10 ib. 1323]. The first el. may be a pers. n. related to *Wocc(a)*. Cf. OAKSEY, WOKING.

Okeford, Child, O~ Fitzpaine, Shilling O~ or Shillingstone Do [*Acford* DB, *Acheford* 1180 P, *Chiltacford* 1212 Fees, *Childacford* 1227 FF, *Ocford Fitz Payn* 1321 FF, *Acford of Robert Eskylling* c 1155 Montacute, *Acford Eskelling* 1215 Cl]. 'Oak ford.'
Child may mean 'knight' or the like; cf. CHILTON.—O~ Fitzpaine came to Robert son of Payn in 1264 (Ipm). Cf. CARY FITZPAINE.— Shilling O~ was held by *Schelin* in 1086 (DB), by Robert son of Scilling c 1155 (Montacute). The pers. n. is doubtless French, but ultimately of OG origin.

Okehampton D [*Ocmundtun, Ocmondtun* c 970 BCS 1245 ff., *Ochenemitona* DB], **Monk O~** D [*Monacochamentona* DB, *Munekeokementon* 1242 Fees]. 'TŪN on R OKEMENT.' Monk O~ once belonged to Glastonbury Abbey and is referred to as *Ocemund, Occemund* 851 ff. Ant Glast. **Okement** [*Okem* 1244, *Okemund* 1282 Ass] is a Brit river-name. Its first el. is the same as that of OGMORE [*Ocmur* c 1150 LL] and OGWEN in Wales, viz. an old adj. for 'swift' found in OW *diauc*, Welsh *diog* 'lazy' and corresponding to Lat *ōc(ior)*. The second el. is difficult to explain definitely.

Okeover St [*Ácofre* 1002 Wills, 1004 KCD 710, *Acoure* DB]. 'Slope where oaks grew.' See OFER.

Olantigh K [*Olenteye* 1270 Ass, *Holmthege, Holitege* E 1 PNK]. Apparently OE *holegntēag* 'holly enclosure'. See HOLEGN, TĒAG.

Olchon R He [*Elchon* c 1150 LL, *Holzham* 13 AD]. A Welsh river-name derived from an adj. corresponding to OIr *olc, elc* 'bad, evil'. The base would be *Ulconā*.

Old or **Wold** Np [*Walda* DB, 1167 P, *Wolde* 1291 Tax]. OE WALD 'wold, wood'

Oldberrow Wa [*Aet Ulenbeorge* 709, *Ulanbearh* 963 BCS 124, 1111, *Oleberge* DB]. '*Ul(l)a*'s barrow.' The place is c 1 m. from ULLENHALL (q.v.). *Ulla* pers. n. is found in *Ullan crypel* 909 BCS 624.

Oldbury on the Hill Gl [(on) *Ealdanbyri* 972 BCS 1282, *Audeburia* 1220 Fees], **O~ upon Severn** Gl [*Aldeberi* 1208 Cur], **O~** K [*Ealdebery* 1303 Ipm], **O~** Sa [*Aldeberie* DB, *Aldebur*' 1242 Fees], **O~** Wa [*Aldeberie* 12 PNWa, *Oldebury* 1278 Misc], **O~** Wo [*Oldebure* 1270 Ct]. 'Old fort.' A pre-English fort is no doubt generally referred to.

Oldcastle Chs [*Le Veu Chastel* 1284 Ch, *Oldecastell* 1289 Court]. 'Old castle.'

Oldcoates Nt [*Ulecotes* 1199 (1232) Ch, -*cote* 1212 RBE, -*cot* 1230 P]. 'COTS inhabited by owls.'

Oldham La [*Aldholm* 1226–8 Fees, -*hulm* 1227 Ass]. 'Old holm.' See HOLM.

Old Hurst Hu [*Waldhirst* 1227 Ass, *Hirst* 1228 FF]. The place is situated near **Woodhurst** Hu [*Wdeherst* 1209 FF]. Both must have been once called *Hyrst* 'wood' or 'hill', and were later distinguished by the elements *wold* and *wood*.

Oldland Gl [*Aldelande* DB, *Oldelonde* 1327 Subs]. 'Old land.'

Oldmixton So [*Almixton* 1200, *Alde-, Eldemixne* 1202 Cur]. 'Old dunghill.' Second el. OE *mixen* 'dunghill'. See MIXON.

Ollerton Chs [*Alretone* DB, *Olreton* 1288 Court], **O~** Nt [*Alretun* DB, -*ton* 1176 P]. 'TŪN among alders.' OE *alor* 'alder' is *owler* in many dialects.

Olney (ōnĭ) Bk [*Ollaneg* 979 KCD 621, *Olnei* DB]. '*Olla*'s island.' The same name is found in Gl (ALNEY). **Olla* is related to *Ulla* (see OLDBERROW) and is found in ALTHORP Np.

Olney Np [*Anelegh* c 1220 For]. OE *āna lēah* 'lonely glade'.

Olton Wa [*Alton* 1221 Ass, *Olton* 1325 AD iii]. 'Old TŪN'.

Olveston Gl [*Ælves-, Alfestun* 955–9 BCS 936, *Alvestone* DB, *Olueston* 1167 P]. '*Ælf*'s TŪN.' *Ælf* is a short form of names in *Ælf*-.

Ombersley Wo [*Ambreslege* 706, *Ambresleie* 714, *Ombersetene gemære* 817 BCS 116, 130, 361]. '*Ambr*'s LĒAH.' Cf. AMESBURY. The same first el. appears in *Ombreswelle* BCS 116 '*Ambr*'s spring'.

Ompton Nt [*Almuntone, Almentune* DB, *Almeton* 1182 P]. '*Alhmund*'s TŪN.'

One Ash Db. See MONYASH.

Onecote (ŏn-) St [*Anecote* 1199 FF, *-cot* 1203 Ass, *Onecote* 1272 Ass], **Onehouse** Sf [*Anhus* c 1060 Wills, *Amuhus* DB, *Anhus* DB, 1240 BM, *Onhus* 1275 RH]. 'Lonely COTS and house.' Cf. ANCOATS.

Ongar, Chipping & High, Ess [(at) *Aungre* 1043–5 Wills, *Angra* DB, 1160 P, *Aungre ad Castra, Alta A~* 1254 Val, *Chepyng Hangre* 1388 Pat, *High Angre* 1240 FF]. OE *angr* 'grazing land'. Cf. ANGERTON. Chipping refers to a market.

Onibury Sa [*Aneberie* DB, *Onyber'* 1243 FF]. 'BURG on R ONNY.'

Onley Np [*Onlee* 1273 Ipm, *Onle* 1345 Cl, *Oneley* 1412 FA]. Identical with OLNEY Np.

Onn, High & Little, St nr Penkridge [*Otne, Anne* DB, *Onna, Othna* c 1130 Ordericus, *Onne* 1221 FF, *Onna* 1230 Cl]. Possibly Welsh *odyn* 'kiln' (< from *otn).

Onneley St [*Anelege* DB, *Oneleia* 1185 TpR, *Onilegh* 1211 Cur, *Onyleye* 1293 Ass]. '*Anna*'s (*Onna*'s) LĒAH' or identical with OLNEY Np.

Onny R Sa [*Onye* 1236 FF, *Oneye* 1301 For]. 'River on whose banks ashes grew.' The name is derived from Welsh *on* 'ashes'. The final element may be OE *ēa* 'river'.

Onslow Sa [*Andrelav, Andreslave* DB, *Hundreslawe* 1190 Eyton, *Ondeslawe* 1203 Ass, 1237 FF, *Hundeslawe* 1272 Eyton]. '*Andhere*'s burial-mound.' *Andhere corresponds to OG *Andhari*. For the *u*-forms cf. HANDBRIDGE.

Onston Chs [*Aneston* 1183 f., *Oneston* 1185, *Honeston* 1186 f. P]. Perhaps OE *āna stān* 'lonely or single stone'.

Openshaw La [*Opinschawe* 1282 Ipm]. 'Open, i.e. unenclosed, wood.'

OE **ōra** 'border, margin, bank' has much the same meanings as ŌFER, and the two words are often difficult to keep apart. The meaning 'river bank', 'shore' is found in OARE K, ORFORD, WINDSOR and others. A special meaning 'firm fore-shore or gravelly landing-place' is found in ITCHENOR, ROWNER Ha and perhaps other names. Senses such as 'edge of a hill, steep slope' or even 'hill, ridge' are found in OARE Brk, W, ORE Sx, BICKNOR K, ORCOP.

OE **ōra** 'ore'. See ORGRAVE, ORGREAVE, ORRELL, ORSETT.

Orby Li [*Heresbi* DB, *Orreby* c 1115 LiS, *Orrebi* 1202 Ass]. '*Orri*'s BY.' ON *Orri* is used as a byname. *Orri* means 'black cock'.

Orchard D [*Orcherd* 1242 Fees], **O~ Portman** So [*Orceard* 854, *Orcerdford* 882 BCS 475, 550, *Orchyard* 1225 Ass], **O~ Wyndham** So [*Orchard* 1424 Dunster]. OE *ortgeard, orceard* 'orchard'.

O~ **Portman** came to Walter Portman t. Hy 6. Johannes Wyndeham de Orchard is mentioned c 1619 (Dunster).

Orchard Do [*at Archet, Archet hamm* 939, *at Archet* 963 BCS 744, 1115, *Orchet* 1330 Cl]. Identical with ARGOED in Wales (several), which is Welsh *argoed* 'wood, shelter of wood'. The name was later associated with *orchard*.

Orchardleigh So [*Orcerdleia* DB, *Orchardesleg* 1219 Fees]. 'Glade with an orchard.'

Orcheston St. George & St. Mary W [*Orc(h)estone* DB, *Ordrycheston* 1314 Ipm, *Orchestone Georgii* 1291 Tax]. '*Ordrīc*'s TŪN.'

The surnames from the dedication of the churches.

Orcop He [*Orcop* 1138 AC, 1168 P, *Orecop* 1173 P]. OE *ōra* 'ridge' and COPP 'top'. Orcop Hill is more than a mile north of the village.

Ord Nb [*Horde* 1196 P, *Orde* 1208 Ch]. OE *ord* 'point, sword'. The name may refer to a long ridge, on which **Middle Ord** is. Cf. ORWELL Ca.

Ordsall La [*Ordeshala* 1177, *-hal* 1201 f. P], **O~** Nt [*Ordeshale* DB, *-hal* 1197 P, *Ordishall* 1242 Fees]. '*Ordhēah*'s or *Ord*'s HALH or haugh.' *Ord would be a short form of names in *Ord-*. OE *ord* does not seem to give a good meaning.

Ore R Sf [*Orus* 1577 Harrison]. A back-formation from ORFORD.

Ore Sx [*Ora* 1121–5 BM, *Ore* 1230 P]. OE *ōra* 'slope, ridge'.

Orford La [*Orford* 1332 FF, *Overforthe* 1465 PNLa]. 'Upper ford.' Cf. OFER.

Orford Li in Stainton le Vale [*Erforde* DB, *Iraforda, Ireforde* c 1115 LiS, *Ireford* 1218–19 Ass]. OE *Íra ford* 'the ford of the Irish' or *Yran ford* '*Yra*'s ford' (cf. IRCHESTER). There was a nunnery here called *Urford* or *Irford* by Tanner in *Notitia Monastica* (1695).

Orford Sf [*Oreford* 1164 P, 12 BM, 1212 Fees]. 'Ford at the sea-shore.' Cf. ŌRA.

Orgarswick K [*Ordgarescirce, Ordgaresuuice* 11 DM, *Orgareswyke* 1254 KnFees]. '*Ordgār*'s WĪC.'

Orgrave La [*Ouregraue* DB, *Oregraua* c 1160 LaCh], **Orgreave** YW [*Nortgrave* DB, *Orgrave* 1357 Goodall]. 'Ore-pit.' See ŌRA, GRÆF.

Orgreave St [*Ordgraue* 1195 f. P, *Ordegrave* 13 PNSt]. The second el. is OE *græfe* 'grove'. The first is doubtful. OE *ord* does not seem suitable, unless it could be used of the bank of a river. O~ is near the Trent. A pers. n. *Orda* may have existed.

Orlestone K [*Orlavestone* DB, *Ordlauestone* 11 DM, *-ton* 1208 FF]. '*Ordlāf*'s TŪN.'

Orleton He [*Alretune* DB, *-tun* 1242 Fees, *Olretton* 1249 Fees], O~ Wo [*Ealretun* 1023 KCD 738, *Alretune* DB]. Identical with OLLERTON.

Orleton Sa [*Erleton* c 1150 Eyton, *-tun* 1249 Ipm]. OE *Eorla-tūn* 'TŪN of the earls'.

Orlingbury Np [*Ordlingbære* 1066–75 Geld R, *Ordinbaro* DB, *Orlinberga* 1130 P, *Ordelinberg* 1202 Ass, *Ordinbere* 1207 Cur, *Ordlingber'* 1220 Fees, *-berg* 1254 Val]. 'The hill of *Ordla*'s people.' *Ordla is a short form of names in *Ord-*. The second el. is not clear. The earliest forms point to OE *bær* 'pasture' or *bearu* 'grove' rather than to *beorg*, but the forms in GeldR and DB really refer to the hundred.

Ormerod La [*Ormerode* 1305 Lacy, 1311 LaInq]. '*Orm*'s or *Ormarr*'s clearing.' See ROD. *Ormr* and *Ormarr* are OScand names.

Ormesby Nf [*Ormisby* c 1025 Wills, *Ormesbei* DB], O~ YN [*Ormesbi* DB, *-by* 1218 FF], **North Ormsby** Li [*Urmesbyg* c 1067 Wills, *Ormesbi* DB, c 1115 LiS, 1202 Ass], **South O~** Li [*Ormesbi* DB, *Ormesbi*, *Ormeresbi* c 1115 LiS]. '*Orm*'s BY.' *Orm*, *Urm*, from ON *Ormr*, ODan *Orm*, is common in England. For the form *Urm* see URMSTON. The form *Ormeresbi* LiS may suggest that South Ormsby has as first el. ON *Ormarr*, ODan *Ormær*.

Ormside We [*Ormesheved* 1256 Kendale]. '*Orm*'s hill.' Cf. HĒAFOD, ORMESBY.

Ormskirk La [*Ormeschirche* a 1196 LaCh, *Ormeskierk* 1203 P]. '*Orm*'s church.' Cf. ORMESBY. Ormskirk was held by one *Orm* in 1203 (P).

Orpington K [*Orpedingtun* a 1011, 1032 Th, *Orpinton* DB]. 'The TŪN of *Orped*'s people.' *Orped is a pers. n. formed from OE *orped* 'active, energetic'. *Orped* is used as a byname or family name in William Orped 1228 Cl (Ha).

Orrell La in Wigan [*Horhill* 1202 P, *Orhille* 1206 P], O~ La in Sefton [*Orhul* 1299 PNLa]. 'Ore hill.' Cf. ŌRA.

Orsett Ess [*Aetorseaþan* 957 Bodley MS, (of) *Orseaþum* c 1000 CCC, *Orseda* DB, *Orsathe*, *Orset* 1231 FF]. 'Ore-pits', 'pits where bog-ore was obtained', or 'chalybeate springs'. Cf. ŌRA, SĒAP.

Orston Nt [*Oschintone* DB, *-tona* 1093, *-tuna* 1146 RA, *Oskinton* 1198 P, *Orston* 1254, 1272 Ep, 1276 RH]. Identical with OSSINGTON. Both seem to mean 'the TŪN of *Ōsrīc*'s people'. The differentiation of the names was probably partly intentional.

Orton Cu [*Orreton* 1227 P, 1292 QW]. '*Orri*'s TŪN.' Cf. ORBY.

Orton Longueville & Waterville Hu [*Ofertuninga gemære* 955, *æt Ofertune* 958 BCS 909, 1043, *Ovretune* DB, *Uuertun* 1158 P, *Overton Henrici de Longa Villa* 1220 Fees, *Ouerton Longavill* 1247 Ep, *Ouerton Wateruile* 1248 FF], **Cole (-ŏ-) O~** Le [*Ovretone* DB, *Overton* c 1125 LeS, *Cole Orton* 1571 BM], **O~ on the Hill** Le [*Wortone* DB, *Overton* 1209–19 Ep, 1254 Val], **O~** Np [*Overtone* DB, *-ton* 1263 Ipm], **Water O~** Wa [*Overton* E 1 BM, 1285 QW], **O~** We [*Overton* 1278 Ch, 1291 Tax]. O~ Hu and Water O~ are OE *Ōfer-tūn* 'TŪN on a river bank'. The others are either OE *Ofer-tūn* 'TŪN on a slope' or *ufera tūn* 'upper TŪN'. It is impossible, without OE examples, to decide between these alternatives.

O~ **Longueville** was held by Henricus de Longauilla before 1185 (TpR). The name is perhaps from LONGUEVILLE nr Bayeux.—Wido de Waltervilla held land in *Overtone* in 1125–8 (LN). Walterville is presumably from a place in France, the name meaning 'Walter's manor'. —**Cole O~** has coal-mines.—**Water O~** is on the Thame.

Orwell Ca [*Oreuuella* c 1080 ICC, *Ore-, Ord(e)-, Oreduuelle* DB, *Oruuella* 1087 Fr]. The original form of the name was probably *Ord-wella*, where *d* was dropped between the two consonants. O~ is at the foot of a long ridge, which was probably called *Ord*; cf. ORD. The meaning of the name is 'spring at Ord ridge'.

Orwell R Sf [(into) *Arewan, Arwan* 1016 ASC (D, E), *Orewell* 1341 Pat]. The identification of the ASC forms is doubtful, and even if it is correct, the later form Orwell may be of independent origin. OE *Ar(e)we* is identical with ARROW Wa. Orwell, if *Or-* is not a later development of OE *Arwe*, means 'river by the shore'. See ŌRA. **Orwell Haven** is *Orewell* 1216, 1223 Cl &c., *Erewell(e)* 1214 Cl, 1215 ff. Pat. The last form tells in favour of *Or-* coming from OE *Earwe* (*Arwe*).

Osbaldeston (ŏzbaldĕstn, locally awbĭstn) La [*Osbaldeston* 1246 Ass, 1292 FF], **Osbaldwick** YN [*Osboldewic* DB]. '*Osbald*'s TŪN and WĪC.'

Osbaston (ŏz-) Le [*Sbernestun* DB, *Osberneston* 1200 Cur, 1230 Ch], **Osberton** Nt [*Osbernestune* DB, *Osberton* 1242 Fees]. '*Osbeorn*'s TŪN.' Late OE *Osbeorn* is an Anglicized form of ON *Ásbiǫrn*, ODan *Asbiorn*.

Osborne (ŏz-) Wt [*Austeburn* 1316 Ipm, *-bourne* 1346, *-borne* 1431 FA, *Austerborne* 14 PNWt]. Second el. BURNA. The first may be identical with AUST or OE *eowestre* 'sheepfold'.

Osbournby Li [*Osbernebi, Esbernebi* DB, *Osbernesbi* 1206 Ass]. '*Asbiorn*'s BY.' Cf. OSBASTON. The DB form *Esbernebi* contains the ODan side-form *Esbern*. DB has both *Osbern* and *Esbern*.

Osea Island Ess [*Uveseia* DB, *Oveseye* 1303 FA]. '*Ufic*'s island.'

Oseby Li [*Asedebi* DB, 1202 Ass, *Asedeby* 1242 Fees]. The first el. may be the ODan pers. n. *Aswith* (*Asede* in 1408).

Osgathorpe Le [*Osgodtorp* DB, *Osgodestorp* c 1125 LeS, *Angodesthorpe* 1225 Ep], **Osgodby** Li in Bardney [*Osgotebi* DB, c 1115 LiS, *Ansgotebi* c 1115 LiS, *Ansgotabi* c 1155 DC, *Angoteby* 1212 Fees], **O~** Li nr Corby [*Osgotbi* c 1060 KCD 908, *Osgotebi* DB, *Angotesby* 1189 (1332) Ch], **O~** Li nr Market Rasen [*Osgote(s)bi* DB, *Osgotabi* c 1115 LiS, *Angotebi* 1202 Ass], **O~** YE [*Ansgote(s)bi* DB, *Osgetebi* 1200 Cur], **O~** YN [*Asgozbi* DB, *Osgoteby* 1234 FF], **Osgoodby** YN [*Ansgotebi* DB, *Angoteby* 1235 FF]. The first el. is ON *Asgautr*, ODan *Asgot* pers. n., Anglicized to *Osgot*, *Osgod*. OScand *Ās-* comes from *Ans-*; this form explains early spellings such as *Ansgotebi*.

Osleston Db [*Oslavestune* DB, *Oslaueston* 13 Derby]. '*Ōslāf*'s TŪN.'

Osmaston Db nr Derby [*Osmundestune* DB, *-ton* 1206 Cur, *Osemundestun* 1226 FF], **O~** Db nr Ashbourne [*Osmundestune* DB, *-ton* 1219 FF, *Osemondeston* 1285 FF]. '*Ōsmund*'s TŪN.'

Osmington Do [*Osmingtone* 939 BCS 739, *Osmentone* DB, *Osminton* 1212 Fees]. 'The TŪN of *Ōsmund*'s people' (OE *Ōsmundinga-tūn*).

Osmondiston Nf [*Osmundestuna* DB, *-ton* 1174 P]. '*Ōsmund*'s TŪN.'

Osmondthorpe Nt [*Oswitorp* DB, *Osmundthorp* 1331 BM], **O~** YW [*Osmundestorp* 1185 TpR, 1197 P]. '*Ōsmund*'s or *Āsmund*'s thorp.' Cf. next name.

Osmotherley La [*Asemunderlawe* 1246 Ass, *Osmoundrelawe* 1332 Subs], **O~** YN [*Asmundrelac* DB, *Osmunderle* 1088 LVD]. '*Āsmund*'s HLĀW or hill and LĒAH.' The first el. is ON *Āsmundr*, ODan *Asmund*, here in the gen. form *Asmundar*.

Osney O [*Osanig* 1004 Wills, *Osineia* 1156, *Oseneia* 1157 P]. '*Ōsa*'s island.'

Ospringe K [*Ospringes* DB, *Ospringe* 11 DM, *Ospring* 1164 P, *Ofspringe* 1168 P, *-s* 1197 FF, *Ofsprung* 1240 Fees]. Probably from an OE **or-spryng* 'spring', closely related to OHG *ursprinc* 'spring', a word that has given rise to several pl. ns. An OE **ofspryng* with the same sense is also possible.

Ossett YW [*Osleset* DB, *Oselset* 1275 Wakef]. '*Ōsla*'s (GE)SET or fold.' Cf. OZLEWORTH.

Ossington Nt [*Oschintone* DB, *Oscinton* 1167 P, 1208 f. Cur, *Oscington* 1275 RH]. Probably identical with ORSTON.

Osterley Park Mx [*Osterlye* 1294 FF, *Osturle* 1375 Cl]. See NOSTERFIELD, LĒAH.

Oswaldkirk YN [*Oswaldescherca* DB, *-kirke* 1243 FF]. 'Church dedicated to St. Oswald.'

Oswaldtwistle La [*Oswaldestwisel* 1246 Ass]. 'Tongue of land belonging to *Ōswald*.' See TWISLA.

Oswestry (ŏz-) Sa [*Osewaldstreu* c 1190 PNSa, *Oswaldestre* 1272 Ipm, *Croesoswald* 1254 Val]. 'St. Oswald's tree.' *Tree* possibly in the sense 'wooden cross'. The Welsh form *Croesoswald* means 'Oswald's cross'. The place is called *Blancmuster* 1233 Cl, *-mostre* (alternatively) 1272 Ipm: 'white minster'.

Otby Li [*Ote(s)bi* DB, *Ottebi* c 1115 LiS, *Octhebi* 1154 BM, *Otteby* 1242 Fees]. '*Otti*'s BY.' First el. ON *Ótti*, ODan *Otti*.

Oteley Sa [*Otley* 1280 Eyton]. 'LĒAH where oats were grown.' Oats is OE *ātan*.

Otford K [*Otteford* 832 BCS 402, 1161 P, *Otefort* DB, *-ford* 1170 P]. Identical with *Ottanford* 909 BCS 627 (Sr): '*Otta*'s ford.' **Otta*, a short form of names in *Oht-* and related to *Ohta*, is found also in OTHAM, OTLEY, OTMOOR.

Otham K [*Oteham* DB, *Otham* 1130 P, *Otteham* 11 DM, 1242 Fees], **O~** Sx [*Otteham* c 1207 Penshurst]. '*Ot(t)a*'s HĀM.' Cf. ODDINGTON, OTFORD.

Otherton St [*Orretone* DB, *Oðerton* 1167 P, *Otherton* 1242 Fees], **O~** Wo [*Othertun* 1240 WoP], **Othery** So [*Othri* 1225 Ass, *Otheri* 1263 FF]. Apparently 'the other TŪN and island'.

Othorpe Le [*Actorp* DB, c 1125 LeS, *Aketorp* 1209-35 Ep, 13 NpCh]. Identical with OAKTHORPE.

Otley Sf [*Otelega* DB, *Oteleia* 1198 FF, *Otteleia* ib., *-le* 1212 RBE], **O~** YW [(on) *Ottanlege* c 1030 YCh 7, *Ot(h)elai* DB, *Ottelea* 1195 P], **Otmoor** O [*Ottemore* 1340 Misc]. '*Otta*'s LĒAH and MŌR or fen.' Cf. OTFORD.

Otter R So, D [*Othery* 963 BCS 1104, *Otrig* 1061 ERN, *Oteri* 1238 Ass]. OE *oter-ēa*, dat. *oter-īe* 'otter river'. From the river are named **Otterford** So [*Oteriforð* 854 BCS 476, *Otriford* 1225 Ass], **Otterton** D [*Otritone* DB, *Otrintonam* 1157 Fr], **Mohun's Ottery** D [*Otri* DB, *Otery* 1242 Fees, *Otery Moun* 1276 Ipm], **Ottery St. Mary** D [*Otri*, *Otrei* DB, *Otery Sancte Marie* 1242 Fees], **Upottery** D [*Upoteri* 1005 KCD 714, *Otri* DB], **Venn Ottery** D [*Fenotri* 1156 ff. P, *Fenoteri* 1212 Fees].

Mohun's O~ was held by Reginald de Mohun in 1242 (Fees). Cf. HAMMOON.—**O~ St. Mary** belonged to the church of St. Mary in Rouen in 1086 (DB).—**Upottery** is '(settlement) up the Otter'.—*Venn* is OE FEN 'fen, marsh'.

Otterbourne Ha [*Oterburna* c 960 BCS 1158, *Otreburne* DB], **Otterburn** Nb [*Oterburn* 1217 Pat], **O~** YW [*Otreburne* DB, *Oterburn* 1226 FF]. 'Otter stream.'

Otterden K [*Otringedene* DB, *Ottringedene*, *Ottrindaenne* 11 DM, *Otringeden* 1182 P, 1242 Fees]. OE *Otringa-denn*; cf. DENN. The *Otringas* may be 'people of *Oter*' or 'the people at a place called Ottery' or the like. *Oter* would probably be OE *oter* 'otter' used as a byname; cf. ON *Otr*. The same uncertainty attaches to OTTERINGTON, OTTRINGHAM, and the old name of METHWOLD HYTHE.

Otterford So. See OTTER.

Otterham Co [*Otrham* DB, *Otterham* 1234 Cl]. Probably OE *oter-hamm* 'HAMM frequented by otters'. Cf. OTTER.

Otterhampton So [*Otramestone*, *Otremetone* DB, *Oterhanton* 1180 P]. OE *Oterhǽma-tūn* 'the TŪN of the *Oterhǽme*', i.e. 'the dwellers at *Oterham* or *Oterburna*' or the like. No place of such a name is known in the neighbourhood.

Otterington, North, YN [*Otrinctun* DB, *-tuna* 1088 LVD, *Otheringeton* 1208 Ch, *Northoterington* 1241 FF], **South O~** YN [*Ostrinctune*, *Otrintona* DB, *Sonotrinctune* 1088 LVD, *Oteryngeton* 1233 FF]. 'The TŪN of the *Otringas*.' See OTTERDEN.

Ottershaw Sr [*Otershaghe* c 890 BCS 563]. 'Otter wood.' See SCAGA.

Otterton, Ottery St. Mary &c. See OTTER.

Ottery D nr Tavistock [*Odetrev* DB]. '*Odda*'s tree.'

Ottery R D, Co [*Otery*, *Oter* 1284 Ass]. Identical with OTTER D.

Ottringham YE [*Otrege*, *Otringeham* DB, *Oteringeham* 1155–7 YCh 1148]. If *Otrege* DB belongs here, the original name of the place was OE *Oterēg* 'otter island', and Ottringham is 'the HĀM of the people of *Oterēg*'. Cf. OTTERDEN.

Oughterby Cu [*Uchtredebi* 1192 P, *Uttredeby* 1292 QW]. '*Uhtrēd*'s BY.'

Oughterside (owt-) Cu [*Uchtredsete* 1260 PNCu(S), *Ughtredsate* 1344 Cl]. '*Uhtrēd*'s shieling.' See SǢTR, (GE)SET.

Oughtibridge YW [*Uhtinabrigga* 1161 YCh 1268]. The first el. appears to be a pers. n. in *Uht-*.

Oughtrington (ōōt-) Chs [*Uttrington* 1296 Rep, *Hughtrington* 1345, *Ughtryngton* 1399 Ormerod]. 'The TŪN of *Uhtrēd*'s people.' Cf. OUGHTERSIDE.

Oulston YN [*Uluestun* DB, 1231 FF]. '*Ulf*'s or *Wulf*'s TŪN.' The first el. may be OScand *Ulfr* or OE *Wulf* (with loss of *W* owing to Scand influence).

Oulton Chs in Over [*Altetone* DB, *Oldeton*, *Oaldeton* 1287 Court], **O~** St nr Stone [*Oldeton* 1251 Ass]. OE *alda tūn* 'old TŪN'.

Oulton Cu [*Ulveton* 1286 Ipm]. '*Wulfa*'s TŪN' with loss of *W-* owing to Scand influence.

Oulton (ō-) Nf [*Oulstuna* DB, *Oueltune* 1199 FF, *Oulton* 1219 Fees, *Owelton* 1253 Cl]. '*Ouðulf*'s TŪN.' *Ouðulf* (Coins &c.) is from ON *Auðulfr*. For loss of *ð* cf. OWMBY.

Oulton (ō-) Sf [*Aleton* 1203 Cur, *Alton*, *Oltun* 1275 RH], **O~** YW [*Aleton* 1180 P, *Altun* 1175 YCh 1873, *Olton* 1251 Ch]. '*Āli*'s TŪN.' Cf. AILBY.

Oundle (ow-) Np [(in) *Undolum* c 715 Eddi, (in prouincia) *Undalum*, *Inundalum* c 730 Bede, *Inundalum*, *Undalana mægð* c 890 OEBede, (into) *Undelum* 972–92 BCS 1130, *Undola* c 1000 Saints, *Undele* DB]. An old tribal name. Cf. Introd. p. xiii f. The fact that the medial vowel is never dropped and the interchange between *a* and *o* suggest that the original vowel was *ā*. OE *undāl* would be an adj. formed from *un-* and *dāl* 'share' or 'division'. The meaning would be 'that has no share' or 'undivided'. From such an adj. a tribal name might have been derived.

Ousby (ōōzbi) Cu [*Uluesbi* 1195 P, *-by* 1224 P, *Ulvesbi* 1214 P]. '*Ulf*'s BY.' Cf. OULSTON.

Ousden (owz-) Sf [*Uuesdana* DB, *-dene* 1198 FF]. 'Owl valley.' First el. OE *ūf* 'owl'.

Ouse (ōōz), **Great,** R [(on) *Usan* 880 Laws, c 1000 Saints, *Use* 937 BCS 712, c 1200 Gervase, *Ouse* 1279 RH], **O~** Y [*Usa* 780–2 Alcuin, (on, andlang) *Usan* 959, 963 BCS 1052, 1352, *Usa* c 1130 SD, *Use* 1226 Cl, *Ouse* 1268 Ass]. A Brit river-name derived from the root *ved-*, *ud-* 'water' found in Sanskrit *udán-* 'water', *udrá-* 'a water animal', Engl *otter*, OIr *usce* 'water' &c. The base is IG *udso-*, found in Sanskrit *útsa-* 'a well'. *Udso-* became *Usso-*, *Uss*, whence *Ūs* and OE *Ūse*.

Ouse R Sx [*Isis* 1577 Harrison, *Ouse* 1612 Drayton]. A late and artificial formation, possibly from LEWES.

Ouse Burn R Nb [(in) *Jhesam* 13 Newcastle, *Yese* 1293 Ass]. Either derived from an OE **gēosan* 'to gush', identical with ON *giósa* (cf. GUSSAGE), i.e. an OE *Gēose* 'gushing river'; or a derivative of the Brit word found in Welsh *ias* 'boiling, seething', cognate with OHG *gesan* 'to ferment', Sanskrit *yásati* 'boils'. Cf. JESMOND.

Ouseburn, Great & Little, YW [*Usebruna*, *-burne* DB, *Kirkeby juxta Useburn* c 1150 YCh 535, *Granthusebarne* 1226 FF]. The place is on a tributary of the OUSE Y. The name means 'stream that falls into the Ouse'.

Ousefleet YW [*Useflete* 1100–8, *-fleoth* c 1170 YCh 470, 487]. The place is on the Ouse. The meaning of the name is not clear. Possibly there was originally a creek of the Ouse here, called Ouse fleet. Cf. FLĒOT.

Ouston Du [*Ulkestan* 1328 Cl, *Ulleston* 1382 Hatfield], O~ Nb nr Stamfordham [*Hulkeleston* 1201 FF, *Ulkilleston* 1242 Fees]. '*Ulf-kell*'s TŪN.' Cf. OAKERTHORPE.

Ouston Nb nr Whitfield [*Ulvestona* 1279 Ass]. '*Ulf*'s or *Wulf*'s TŪN.' Cf. OULSTON.

Outchester Nb [*Ulecestr'* 1206 Cur, 1236 Cl]. 'Roman fort inhabited by owls' (OE *ūle* 'owl').

Outwell Ca, Nf [(æt) *Uuyllam* 970 BM, *Wella, Utuuella* DB, *Utwell* 1202 FF, 1256 BM]. Outwell and UPWELL are close together and once formed a whole, the name being *Wella* 'the spring'. They were later distinguished as Out- and Upwell. Outwell may be the later settlement that sprang up outside the old village.

Ouzel R Bk [*Ousel* 1847 Lipscomb]. A late name for earlier LOVAT. Another form of the name seems to be WHIZZLE BROOK, which means 'weasel brook'. The change to Ouzel is probably due to influence from OUSE. The Ouzel falls into the Ouse.

Ovenden YW [*Ovinden* 1246 FF, *Ovendene* 1266 Misc]. '*Ofa*'s valley.'

Over Ca [*Ouer* 1060 KCD 809, *Oura* c 1080 ICC, *Ovre* DB, *Ouere* 1189 BM]. OE *ōfer* 'river bank'.

Over Chs [*Ōvre* DB, *Ufre* c 1150 Chester, *Huuere* 1246 Ch, *Overe* 1291 Tax], **Little-& Mickleover** Db [*Ufre, Parua Ufre* DB, *Magna Oufra* a 1113 Burton, *Magna, Parua Uure* 1226, *Magna Oura* 1233–9 BM], O~ Gl [(30 hides) *under Ofre* 804 BCS 313, *Overam* Hy 2 Glouc, *Overe* 1242 Ipm]. OE *ufer, ofer* 'slope, ridge' or the like. See OFER. O~ Chs and Db are on ridges or slopes of ridges. O~ Gl is on the bank of the Severn, but at the foot of a hill. The name originally meant '(place) at the foot of the hill'.

Over Wa in **Browns-, Cester-, & Church-over** [*Wavre, Wavra, Gavra* DB, *Wafre* Hy 2 DC, *Bruneswavere* 1236, *Cetrestwaver* 1242 Fees, *Thestrewaure* 1305 Ch, *Chirche-wouere* 1291 Tax]. Apparently an old name of the Swift, identical with WAVER Cu [*Wovere Watir* 14]. See PNWa(S). The meaning is 'winding stream', an accurate name.

Brownsover from *Bruno*, who had land in *Gavra* in 1086 (DB). Cf. BRIZE NORTON. Cester- may be for *Chester-*.

Overbury Wo [*Uferebreodun vel Uferebiri* 875 BCS 541, *Ovreberie* DB]. Cf. BREDON. The present name means 'upper BURG'. An old earthwork may be referred to.

Overpool Chs [*Pol* DB, *Pulla* c 1165, *Huuerpulle, Uuerpulle* c 1200 Chester]. 'Upper Pool.'

Oversley Wa [*Oveslei* DB, *Ouresleia* c 1140 BM, *Oversleie* 1154–8 (1340) Ch]. Perhaps OE *ōferes-lēah* 'LĒAH by a river bank'. Cf. ŌFER.

Overstone Np [*Oveston* 12 NS, 1236 Fees, *Uviston* 1220 Fees, *Oviston* 1221 Ep]. '*Ufic*'s TŪN.'

Overstrand Nf [*Othestranda* DB, *Over-strand* 1231 Cl, *Ovestronde, Ouerstronde* 1254 Val, *Ovirstrond, Ovestronde* 1275 RH]. If the DB form is correct, perhaps 'the other shore'. Cf. OTHERTON. Sidestrand is not far away. But more likely the first el. is OE *ofer*, here in the sense 'edge, margin'. The name would mean 'shore with a steep edge, narrow shore', as against SIDESTRAND 'broad shore'.

Overton Chs nr Frodsham [*Overton* 1284 Cl, 1300 Misc], O~ **by Malpas** Chs [*Ovretone* DB, *Overton* 1293 Court], O~ Db [*Nether Overton* 1281 FF], O~ Ha [*Uferantun* 909 BCS 625 f., *Ovretune* DB], O~ (ð-, ō-) La [*Ouretun* DB, *Ouerton* 1177 P], **Cold** O~ Le [*Ofertun* c 1067 Wills, *Ovretone* DB, *Caleverton* (hd) c 1125 LeS, *Kald-overton* 1203 Cur], **Market** O~ Ru [*Overtune* DB, *Marketesoverton* 1200 Cur, *Market Overton* 1286 QW], O~ Sa nr Ludlow [*Overton* 1199 FF, 1242 Fees], **West** O~ W [(to) *Uferan tune, Oferan tunes* 939, *æt Ofærtune* 949 BCS 734, 875, *Ovretone* DB, *Westovertone* 1275 RH], O~ YN [*Ouertune, Ovretun* DB, *Overtona* c 1090 YCh 350]. O~ Ha and W are OE *ufera tūn* 'upper TŪN'. O~ La, YN, from the situation of the places, are OE *Ōfer-tūn* 'TŪN on a river bank'. The rest are on slopes or ridges, and the name may here be either OE *ufera tūn* or *Ofer-tūn*, *ofer* being used in the sense 'slope' or 'ridge' (see OFER). Very likely most are *ufera tūn*.

Overy O [*Overeye* 1545 Mon]. O~ is nr Dorchester on the other side of the Thame. The name means 'across the river' (OE *ofer-īe*). Cf. BURNHAM OVERY.

Oving (ōō-) Bk [*Olvonge* DB, *Uvinges* 1237–40 Fees], O~ (ōō-) Sx [*Uuinges* 956 BCS (930), 1183 P, *Ouingges* 1230 P]. '*Ufa*'s people.'

Ovingdean (ōō-) Sx [*Hovingedene* DB, *Uuingeden* 1199 FF]. 'The valley of *Ufa*'s or *Ofa*'s people.'

Ovingham (ð-, -nj-) Nb [*Ovingeham* 1238 Ep, *Ovingham* 1245 Ipm], **Ovington** (ð-) Nb [*Oventhuna* Hy 2 (1271) Ch, *Ovinton* 1201 Ch, *Ovigton* 1242 Fees]. 'The HĀM and TŪN of *Ofa*'s people.' The places are close together.

Ovington Ess [*Ouituna* DB, *Uvinton* 1227 FF], O~ Ha [*æt Ufinctune* c 960 BCS 1153, *Ovinton* 1189 BM], O~ Nf [*Uvinton* 1202 FF, *Uvington* 1254 Val, *Oviton* 1263 Ch]. 'The TŪN of *Ufa*'s people.'

Ovington YN [*Ulfeton* DB, *Ulvington* 1251

Ipm]. 'The TŪN of *Wulfa* or of *Wulfa's* people.' Cf. OULTON Cu.

Ower Ha nr Eling [*Hore* DB, 1212 Fees, *Ore* 1284 QW, *Ores* 1177 P]. OE *ōra* 'bank, shore'.

Owermoigne (ōr-) Do [*Ogre* DB, *Oweres* 1212, *Our'* 1219 Fees, *Ogres* 1244 Fees, 1269 Ch, *Oares* 1285 FA, *Oure Moigne* 1314 FF]. Probably OE *ofer* in the sense 'slope, ridge'. The common spelling *Ogres* is curious.

The manor was held by Radulfus Monachus in 1212 (Fees). The name is often written *le Moyne, le Moyngne*. It is a family name derived from Fr *moine* 'monk'.

Owersby, North & South, Li [*Aresbi, Oresbi* DB, *Ouresbi* c 1115 LiS, 1163–5 BM, *Oueresbi* 1207 Cur]. Possibly '*Ávar's* BY'. ON *Ávarr*, ODan, OSw *Awair* is a pers. n.

Owlerton WR, town [*Allerton* 1219 Ass]. Identical with OLLERTON.

Owlpen Gl [*Olepenne* c 1200 Berk, 1220 Fees, *Olpenne* 1287 QW]. '*Olla's* pen or enclosure.' Cf. OLNEY Bk.

Owmby Li nr Caistor [*Odenebi* DB, *Ounebi* c 1115 LiS, *Oudenbi* 1155–8 RA, *Outhenebi* 1155–60 DC], **O~ by Spital** Li [*Ovne(s)bi* DB, *Ounabi* c 1115 LiS, *Ounebi* 1202 Ass]. '*Auðun's* or *Aun's* BY.' Cf. AUNBY.

Owram YW in Northowram [*Ufrun* DB, *Northuuerum* 1202 FF] and **Southowram** [*Overe, Oure* DB, *Uuerune* 1242 Fees, *Southouerum* 1286 Wakef]. The dat. plur. of OE *ofer, ufer* 'slope, ridge'. The places are not far apart, but on different ridges or hills.

Owslebury (ūzl-) Ha [*Oselbyrig* c 960 BCS 1158, *Oselebury* 1316 FA]. '*Ösla's* BURG.' The form of c 960 is in a late transcript. Cf. OZLEWORTH.

Owsthorpe or **Ousethorpe** YE in Pocklington [*Uluestorp* 1203 FF, *Ulvestorp* 1219 Fees]. '*Ulf's* thorp.' Cf. OULSTON. **Owsthorpe** YE in Eastrington [*Ousthorpe* 1316 FA] is OScand *Austþorp* 'east thorp'.

Owston Le [*Osulvestone* DB, *Osolvestona* Hy 2 (1253) Ch, *Osuluestan* 1185 P]. '*Oswulf's* TŪN.'

Owston Li [*Ostone* DB, *Oustuna* Hy 2 DC, *Ouston* 12 BM], **O~** YW [*Austhun, Austun* DB, *Oustona* c 1150 Crawf, *Oustun* c 1190 YCh 1585]. OScand *Aust-tūn* 'eastern TŪN', but very likely a Scandinavianized OE *Éast-tūn*.

Owstwick YE [*Osteuuic* DB, *Austwich* 1177 P, *Ostwik* 1202 FF]. Probably a Scandinavianized form of an OE *Éast-wíc* 'eastern wÍc'. Cf. OWSTON Li.

Owthorne YE [*Torne* DB, *Uttethorne* 1285 FA, *Out Thoren* 1297 Subs]. Originally THORN 'the thorn-bush'. Later Out Thorne 'outer Thorne', for distinction from a part of the village farther inland. Owthorne is on the sea.

Owthorpe Nt [*Ovetorp* DB, *Uuetorp* 1194, *-thorp* 1230 P]. '*Ufa's* or *Ufi's* thorp.' *Ufi* is ON *Úfi*, OSw *Ufi* pers. n.

Oxborough Nf [*Oxenburch* DB, *Oxeburg* 1194 P], **Oxcliffe** La [*Oxeneclif* DB, *Oxeclive* 1201 ff. P], **Oxcombe** Li [*Oxecumbe* DB, 1203 Cur, *Ox(e)cum* c 1115 LiS], **Oxenbold** Sa [*Oxibola* DB, *Oxebald* 1194 P, *Oxenebold* 1205 Cur], **Great & Little Oxendon** Np [*Oxendone* DB, *-don* 1176 P, *Maior, Minor Oxendon* 1220 Fees], **Oxenhall** Gl [*Horsenehal* DB, *Oxenhale* 1221 Ass], **Oxenholme** We [*Oxinholme* 1274 Kendale], **Oxenhope** YW [*Oxope* 1191 P, *Oxenhop* 1280 Ch], **Oxenton** Gl [*Oxnadunes cnol* 977 KCD 617, *Oxendone* DB, *Oxsendon* 1176 P]. 'BURG, CLIF, CUMB, BOLD, DŪN, HALH, HOLM, HOP, DŪN where oxen were kept.'

Oxford O [*Oxnaford* 912 ASC, *Oxenaford* c 1000 Saints, *Oxeneford* DB]. 'Ford for oxen.' **Oxfordshire** [*Oxnafordscir, Oxenafordscir* 1010 f. ASC (E)].

Oxhey Hrt [*æt Oxangehæge* 1007 Crawf, *Oxehei* 1165 P]. 'Enclosure for oxen.' Cf. (GE)HÆG.

Oxhill Wa [*Octeselve* DB, *Ohteselua* c 1150, *Octhesselua* 1157 BM, *Ofte Schelua* 1187 P, *Ocsulve* 1236 Fees]. OE *Ohtan scylf* '*Ohta's* ledge of land or hill'. The place is on the slope of a slight ridge. *Ohta* is found written *Octa*.

Oxley St [*Oxelie* DB, *Oxeleg* 1236 Fees]. 'Pasture for oxen.' See LÉAH.

Oxnead Nf [*Oxenedes* DB, *Oxenedich* 1254 Val, *Oxenedish* 1302 Misc]. OE *oxna edisc* 'pasture for oxen'. Cf. EDISC.

Oxney K nr Deal [*Oxena gehæg* 1042–4 BM, *Oxenia* 1242 Fees]. 'Enclosure for oxen.' Cf. (GE)HÆG.

Oxney, Isle of, K [(on) *Oxnaiea* 724 BCS 141, *Oxeneya* 1212 StAug], **O~** Np nr Peterborough [*Oxanege, -ige* 972–92 BCS 1130, *Oxeney* 1249 Ch]. 'Island where oxen were kept.'

Oxshott Sr [*Okesseta* 1180 P, *Occasate, Oggeschate* J BM, *Hokeset* 1202 Cur, *Oggessot* 1235, *Hoggesete* 1281 BM, *Ockeschete* 1265 Misc]. '*Ocga's* SCÉAT.'

Oxspring YW [*Ospring, Osprinc* DB, *Oxspring* 1154–9 YCh 1665, *Oxprig* 1260 Ass, *Ospring* 1305 YInq]. This name may be identical in the main with OSPRINGE K, the source being an OE *or-spring*. But the first el. may be OE *oxa* 'ox'. If so, the second el. is probably ME *spring* 'a copse, grove'. If the first alternative is correct, the later Oxspring is due to popular etymology.

Oxted Sr [*Acstede* DB, *Akested* 1177 P, *Ocsted* 1225 Ass]. 'Place where oaks grew.'

Oxton Chs [*Oxtone* 1275 Ipm, *Oxton* 1282 Court], **O~** Nt [*Oxetune* DB], **O~** YW [*Oxetone* DB, *Oxton* 1201 Cur], **Oxwick** Nf [*Ossuic* DB, *Oxewic* 1242 Fees]. 'Farm where oxen were kept.'

Ozengell K [*Osinghelle, Osingehelle* n.d. StAug]. 'The hill of *Ōsa*'s people.'

Ozleworth Gl [*Oslanwyrþ* 940 BCS 764, *Osleuuorde* DB, *Oseleworth* 1220 Fees]. '*Ōsla*'s WORÞ.' *Ōsla* pers. n. is not evidenced in independent use. It is a formation from *Ōsa, Ōsbeald* &c. OE *ōsle* 'blackbird' is an unlikely first el. in this case.

P

Packington Le [*Pakinton* 1043 Th, 1188 P *Pakyngton* c 1050 KCD 939, *Pachintone* DB, *Pakenton* 1201 Cur], **P~** St [*Pagintone* DB, *Pakentone* 1166 RBE, *Pakinton* 1167 P, *Pakigton* 1230 P], **Great & Little P~** Wa [*Patitone* DB, *Pakinton* 1236 Fees, 1268 Ipm], **Packwood** Wa [*Paggewod* 1043 (1267) Ch, *Pachawud* 1195 Cur, *Pa(c)kwode* 1196 FF]. Apparently '*Pac(c)a*'s TŪN' or 'the TŪN of *Pac(c)a*'s people' and '*Pac(c)a*'s wood'. Cf. PAKEFIELD, PAKENHAM, also PAGLESHAM, which seem to contain the same or a related pers. n. No common noun is known in English that might be the first el. of all these names. A pers. n. *Pac(c)a* is not known either. ON *Pakki* occurs as a byname.

Padbury Bk [*Pateberie* DB, *Padeberi* 1163, *Paddeberi* 1167 P]. '*Padda*'s BURG.'

Paddington Mx [*Padintun* 959 BCS 1050, *Paddington* 998 Th, *Padinton* 1168 f. P]. 'The TŪN of *Padda*'s people.'

Paddington Sr in Abinger [*Padendene* DB, *Patinden* 1185, 1190, *Pattesden* 1191 P, *Padinden* 1212 Fees, 1230 P, *Patenden* 1230 Cl]. '*Padda*'s or *Patta*'s valley.' Cf. *Pattan dene* 973–4 BCS 1307 (Crondall, Ha).

Paddlesworth K nr Dover [*Peadleswurthe* 11 DM, *Padelesworth* 1341 Pat], **P~** K nr Snodland [*Petelesuurthe, Peadleswyrþ* 10 BCS 1321 f., *Pellesorde* DB, *Padleswrth* 1242 Fees]. The first el. seems to be a pers. n. *Pættel or *Pæddel related to *Peatta* or *Pead(d)a*. If the original form had *t*, we may compare PATSHULL and *Pætla* in Petteridge K [*Pætlanhryge* 747, *Pætlan hrycg* 942–6 BCS 175, 779]. See WORÞ.

Paddock Wood K [*Parrok* 1346 FA]. OE *pearroc* 'paddock, enclosure'.

Padfield Db [*Padefeld* DB, *Paddefeld* 1185 f. P]. '*Padda*'s FELD' or 'FELD frequented by toads or frogs'. Cf. FROXFIELD. *Pade* 'toad' is found 1137 ASC.

Padiham La [*Padiham* 1251 Ch, *Padingham* 1292 QW]. 'The HĀM of *Padda*'s people.'

Padley Db [*Paddeley* 1220–30 PNDb]. See LĒAH. First el. as in PADFIELD.

Padstow Co [*Sancte Petroces stow* 981 ASC (C), *Petrokestowe* 1297 Cl, *Padristowe*

1351, *Padestou* 1361 FF]. 'St. Petroc's church.' St. Petroc is the patron saint, and the neighbouring LITTLE PETHERICK or ST. PETROC MINOR is so called for distinction from Padstow. Padstow is identical with *Petrocys stow, Pætrocysstow* 11 E, an alternative name of BODMIN.

Padworth Brk [*æt Peadanwurðe* 956 BCS 984, *Peteorde* DB, *Pedewrtha* a 1162 Oxf, *Padeworth* 1220 Fees]. '*Peada*'s WORÞ.'

OE **pæþ** 'path' is the second el. of some names, as ALSPATH (see MERIDEN), GAPPAH, HORSEPATH, MORPETH. First el. in PATELEY, perhaps PATTON We.

Pagham Sx [*Pecgan ham* 680, *Pacgan hamm* 10 BCS 50, *Pageham* DB]. '*Pæcga*'s HAMM.' *Pæcga is a form with gemination of *g* of *Pæga*.

Paglesham Ess [*Paclesham* 1065, 1221 BM, *Pachesham* DB, *Pakelesham* 1203 Cur]. '*Pæccel*'s HĀM.' Cf. PACKINGTON &c. *Pæccel* is found in Patchway Sx [*Petteleswig* c 765 BCS 197, *Peccheleswia* 12 PNSx]. '*Pæccel*'s temple.' See WĒOH.

Paignton D [*Peintone* DB, *Peintona* 1159 Buckfast, *Painton* 1230 P]. Possibly '*Pæga*'s TŪN', but rather '*Pægna*'s TŪN'. *Pægna* is found in *Pægnalaech* Bede.

Pailton Wa [*Pailintona* 13, *Paylintona* 1302 BM, *Paylyngton, Palyngton* 1322 Ipm]. 'The TŪN of *Pægel*'s people.' *Pægel is a diminutive of *Pæga*.

Painley YW nr Paythorne [*Padehale, Paghenhale* DB, *Pathenhal, Pathanal, Panhale* c 1200 Pudsay, *Pathenhale* 1226 FF]. Cf. PANBOROUGH. Second el. HALH.

Painshaw (-ĕ-) Du [*Pencher* 1183 BoB, 1305 BM, *-chare* 1472 BM]. Identical with *Pencer* in *Pencer setna gemære* 849 BCS 455 (in bounds of Wast Hills Wo). Probably a Brit name consisting of Welsh *pen* 'top' and an *i*-mutated plur. of *carr 'a rock' (see CARR), i.e. an OCo, OW *cerr* 'rocks'. This form may well be the second el. of *pollicerr* 977, *pollcerr* 1059 E (now Polkerth Co) and *ryt y cerr* c 1150 LL (in Wales), quoted PNW(S) under Cherhill. These names would then mean 'pool and ford of the rocks'.

Painsthorpe YE [*Thorf* DB, *Paines Thorp* c 1090 YCh 350]. 'Pain's thorp.' *Pain* is a Fr pers. n., identical with *paiën* 'heathen'.

Painswick Gl [*Wiche* DB, *Payneswik* 1265 Misc]. Originally *Wīc*. The surname from Pain Fitzjohn (d. 1137). Cf. prec. name.

Pakefield Sf [*Paggefella* DB, *Pagefeld* 1198 FF, *Pakefeld* 1228 FF, *-feud* 1254 Val], **Pakenham** Sf [*Pakenham* c 950 BCS 1008, *Paccenham* 11 EHR 43, *Pachenham* DB, c 1095 Bury]. '*Pacca*'s FELD and HĀM.' Cf. PACKINGTON.

Palgrave, Great & Little, Nf [*Pag(g)raua* DB, *little Pagrave* 1157 Fr, *Paggrave* 1202 FF, *Great, Little Pograve* 1278 Cl]. Both the first and the second el. are doubtful. The second may be OE *græf* 'grave' or *grāf* 'grove'. The first may be a pers. n., e.g. *Paga* or **Pacca* as in PAKENHAM.

Palgrave Sf [*at Palegrave* 962 BCS 1084, *Pallegrafe* 11 EHR 43, *Palegraue* c 1035 Wills, *-graua* DB]. Doubtful like P~ Nf, but etymologically distinct from it. Possibly *pāla-grāf* 'grove where poles were got'. But one OE form rather suggests as first el. an OE **Palla* pers. n., cognate with *Pælli*, or OE *pall* (see PAULTON).

Palling (-aw-) Nf [*Pallinga, Palinga* DB, *Pallenges* 1199 P, *Pallinge* 1254 Val]. '*Pælli*'s people.' *Pælli* LVD is a derivative of the adj. for 'firm, stiff' found in LG, Fris *pal, pall* (an OE *pall, peall*).

Pallingham Sx [*Palingham* 1199 Cur, 1233 Cl, *Pallingeham* 1244 Cl]. 'The HĀM of *Pælli*'s people.'

Pallion Du [*le Pavylion* 1328 PNNb]. 'The pavilion.' A French name.

Palstre K [(æt) *Palstre* 1032 Th, *Palestrei* DB, *Palstrege* 11 DM, *Palstre* 1207 Cur]. OE *palestre, palstr* 'spike' in some unrecorded sense, as 'point of land', and OE *ēg* 'island'.

Palterton Db [*Paltertun* 1002 Wills, *Paltertune* DB]. The first el. may be an old name of the ridge at which P~ is. The river POULTER comes from the other side of the ridge. A hill-name *Palter* may be explained as a compound of *pall* 'ledge' (cf. PAULTON) and OE *torr*. DUNSTER may be compared.

Pamber Ha [*Penberga* 1165 ff. P, *-berg* 1204 Cur, *Penbere* 1253 Ipm]. 'Hill with a *pen* or enclosure.' Brit *pen* 'hill' is improbable, as there is no very prominent hill here. Some forms may point alternatively to OE *bearo* 'grove' or *bǽr* 'pasture'.

Pamington Gl [*Pæuintun* 977 KCD 617, *Pamintonie* DB, *Pannentona* 1107 (1300) Ch, *Pamynton* 1287 QW]. It looks as if *m* goes back to earlier *vn*. If so, the first el. may be related to that of PEVENSEY &c.

Pamphill Do [*Pamphilla* 1168 P]. An el. *pamp* is found in several names, as PANTON, PONTON. It must be related to ME

pampe 'to pamper', ON *pampi* a byname, Sw *pamp*, Dan *pamper*, G dial. *pfampf* used of thick-set people, Sw *pampen* 'swollen'. The word may have been used also of objects, e.g. a hillock or mound.

Pampisford Ca [*Pampeswrda* c 1080 ICC, *-uuorde* DB, *-wrth* 1254 Val]. '*Pamp*'s WORÞ.' **Pamp* is a pers. n. belonging to the group of words discussed under PAMPHILL.

Pan, Great & Little, Wt [*Lepene* DB, *La Penne* 1263 Ipm, *Panne* 1489 BM]. The early forms point to OE *pen* 'enclosure' rather than to *panne*, which would mean 'salt pan'.

Panborough So [*at Patheneberghe* 956, *Papeneberga* 971 BCS 920, 1277, *Padenaberia* DB, *Patheneberg* Hy 2 (1227) Ch]. Second el. OE *beorg* 'hill'. The first may be the same as that of PAINLEY, PATHLOW. It seems to belong in some way to OE *pæþ* 'path'. Possibly we may postulate an OE **paþa* 'wayfarer'. Panborough would then be an OE *paþena beorg*, which suits the early forms.

Pancrasweek D [*Pancradeswike* 1198 FF, *Pankardeswik* 1242 Fees]. Originally no doubt *Wīc*. The church is dedicated to St. Pancras.

Panfield Ess [*Penfelda* DB, *Panfeld* 1254 Val, *Pantfeld* 1428 FF]. 'FELD on R PANT.'

Pangbourne Brk [(at) *Peginga burnan*, *Pægeinga burnan* 843 BCS 443, *Pangeborne* DB, *Pangeburne* 1166 RBE]. Really the old name of the river **Pang** [*Panganburna* 956 BCS 919]. 'The stream of *Pæga*'s people.' *Pang* is a late back-formation.

Pangdean Sx [*Pinwedene, Pinhedene* DB, *Pengedene* c 1100, *Penkedena* 13 PNSx]. Second el. DENU. The first may be as in PANGBOURNE.

Pannal YW [*Panhal* 1170 P, *-e* 1291 Tax, *Panehale* 1280 Ch]. Second el. OE *halh* 'haugh'. The first seems to be OE *panne* 'pan', here used of a rounded valley.

Panshanger Hrt near Hertford [*Paleshangre* 1197 FF, *Palleshangre* 1203, *Paneshangr'* 1204, *Peneshangr'* 1206 Cur]. The second el. is OE *hangra* 'slope' &c. The first may be *pall* 'ledge' as in PAULTON or OE *penn* (ESax *pænn*) 'enclosure'. Change of *n* to *l* owing to Norman influence is not rare if a following syllable contains an *n* (*ng*); cf. e.g. *Belingehou, Belintone* DB for BENGEO, BENNINGTON Hrt. The place is near Hertford, thus in a district where Norman influence would easily make itself felt.

Pant R Ess [*Pentæ* (*Paente*) *amnis* (gen.) c 730 Bede, *Pente* (*Pante*) *stæð* c 890 OEBede, *Pantan stream* c 1000 Battle of Maldon]. A derivative of Welsh *pant* 'a valley'.

Panton Li [*Pantone* DB, *Pantuna* c 1115 LiS, *Pamtun, Pantun* 12 DC]. OE *Pamp-tūn*, on which see PAMPHILL. The place is on a ridge.

Panxworth Nf [*Pankesford, Pancforda* DB, *Pankesford* c 1165 Bodl, 1202 FF, *Pangeford* 1254 Val]. The first |el. is very likely a pers. n. or nickname identical with the surname *Panke* 1221 Ass (Robert Panke of Mancetter Wa). Such a name may be derived from an animal's name cognate with OSw *panka* 'a kind of fish', used as a byname, Sw *panka* 'a young bream', Sw dial. *panke* 'a young pig', *panker* 'a small boy' and the like. Second el. FORD.

Papcastle Cu [*Papecastre* 1267 Misc, 1286 Ipm]. Second el. OE *cæster, ceaster* 'Roman fort'. The first may well be ON *papa, papi* 'hermit'.

Papplewick Nt [*Papleuuic* DB, *Papelwic* 1212 Fees, *-wyc* 1230 P]. 'WĪC on pebbly soil.' First el. OE *papol* 'pebble' in *papolstān*, also *popel*, in *popelstān*, found in NEWTON POPPLEFORD, POPPLETON. Cf. also PEOPLETON, PEPLOW.

Papworth Everard & St. Agnes Ca [*Papeworde* DB, *Papewurda* 1160 P, *Pappewrth* 1228 FF, *Anneys Papwrth* 1241 FF, *Pappewrth Agnetis, Everard* 1254 Val]. Near P~ is **Papley Grove** [*Pappele* 1334 BM]. '*Papa*'s WORP.' *Papa* is found in *Papanholt* 901 BCS 596, *Pappenholt* 1228, *Papenholt* 1253 Cl (Ha). Cf. also PAVENHAM. *Papa* is not found in independent use, but has Continental analogies.

Everard and St. Agnes from early owners. Agnes de Papewurda is mentioned 1160 P.

Parbold La [*Perebold* 1200 FF, *Perbold* 1212 Fees]. 'Homestead where pears grew.' Cf. BŌPL.

Pardshaw Cu [*Perdishaw* c 1203 StB, *Perdyshowe* 1397 FF]. Possibly '*Preed*'s barrow'. OE *Preed* pers. n. is found in LVD. If the vowel was shortened early, metathesis might take place. Second el. ON *haugr*.

Părham Sf [*Perreham* DB, *Pereham* 1206 Cur, *Perham* 1254 Val], **P~** Sx [*Perham* 959 BCS 1050, DB, 1207 Cur]. 'HĀM where pears grew.'

Parkham D [*Perceham* DB, *Parkeham* 1242 Fees, *Parcham* 1254 FF]. OE *pearrocahamm* 'HAMM with paddocks'.

Parley, West, Do [*Perlai* DB, *Perlea* 1187 P, *Westperele* 1305 FF], **East P~** Ha [*Perle* 1242 Fees, *Estperle* 1346 FA]. OE *per-lēah* 'LĒAH where pears grew'.

Parlington YW [*Pertilin(c)tun* DB, *Parlinton* c 1215 BM]. 'The TŪN of *Pertel*'s people.' **Pertel* is a diminutive of *Pearta*. Cf. PARTNEY.

Parndon Ess [*Peren-, Perinduna* DB, *Perendon* 1230 P, *Perendon Magna, Parva* 1254 Val], **Parnham** Do [*Perham* 1228 FF, *Parnham* 1431 FA]. 'DŪN and HĀM where

pears grew.' The first el. may be partly an OE **peren* 'of pears'.

Parr La [*Par* 1246 Ass, *Parre* 1298 LaInq]. An OE **pearr* 'enclosure', corresponding to OHG *pharra* 'parish', originally 'district' or the like. From *pearr* are derived ME *parren* 'to enclose, fold' and *pearroc* 'paddock'.

Parracombe D [*Pedrecumbe, Pedracomba* DB, *Parcumb* 1240 Cl, *Piarecomb* 1303 FA]. OE *peddera cumb* 'the pedlars' valley'. OE *peddere* is not recorded, but no doubt existed. The word is found c 1225, as a byname 1166 P. For the loss of *d* cf. PARRET.

Parret R Do, So [*Pedredistrem* 725 BCS 143, (*op*) *Pedridan* 658, (be eastan) *Pedredan* 894 ASC, *Pedret* c 1200 Gervase, *Peret* 1233 ff. Wells]. Unexplained. The same name occurs in early sources used of a stream in Gl or Wo [(of) *Pedredan* 988, (innan) *Pederedan* 1003 KCD 662, 1299)].

Parson Drove Ca [*Personesdroue* 1324 PNCa(S), (way called) *Parsonsdrove* 1509 AD]. *Drove* (OE *drāf*) is used locally in the Fen District of a road along which horses or cattle are driven. *Parson* may be *parson* or the word used as a family name.

Partington Chs [*Partinton* 1260 Court]. 'The TŪN of *Pearta*'s people.' Cf. PARTNEY.

Partney Li [*Peartaneu* c 730 Bede, *Peortanea* c 890 OEBede, *Partene* DB, *Partenay* DB, 1208 Cur]. '*Pearta*'s island or river land.' The place is on the Lymn. OE *Pearta* is not recorded in independent use. It is found also in PARTINGTON; cf. also *Peartingawyrth* BCS 262 (Sx), PARLINGTON, PERTENHALL, PERTWOOD.

Parwich (părĭch) Db [*Peuerwich* 966 BCS 1175, *Pevrewic* DB, *Peuerwiz* 1236 FF]. 'WĪC on R *Pever*.' Cf. PEOVER. The stream at P~ must have been called *Pever*.

Passenham Np [*Passanhamm* 921 ASC, *Passon-, Passeham* DB]. '*Passa*'s HAMM.'

Paston (-aw-) Nb [*Palestun* 1176 P, *Paloxton* 1227 Ch, *Palwiston* 1242 Fees, *Palxton* 1256 Ass]. '*Palloc*'s TŪN.' **Palloc* is related to *Pælli*.

Paston (-ah-) Nf [*Pastuna* DB, *-tun* c 1150 Crawf, *Paxton* 1194 ff. P], **P~** (-ah-) Np [*Pastune, -tun* 972 BCS 1280 f., *Paston* 1167 P, *Paxtona* 1199 FF]. Cf. also *Paskeden* 1195 Cur (Bk). The first el. seems to be an OE **pæsc*, or **pæsce*, which might be compared with MDu *pasch* 'pastureland' and the first el. of PASSCHENDALE in Holland (one is *Pascandala* 9). But MDu *pasch* may be from Lat *pascuum*, and it is doubtful if this can be the source of *Pascan-* in *Pascandala* or OE *pæsc(e)*. More likely OE *pæsc(e)* is the source of dial. *pash* 'a soft mass, a puddle' and had about the same meaning. It may have been formed from the verb *pat* (ME *patte*) in the same way as OE *plæsc* 'puddle' from OE *plættan* 'to strike'.

Patcham Sx [*Piceham* DB, *Peccham* a 1100 PNSx, *Petcham* 1316 FA], **Patching** Ess [*Pacingas* DB, *Pachinges* 1207, *Pecinges* 1208 Cur], **P~** Sx [*Pettinges* 947 BCS 823, *Pæccingas* 960 ib. 1055, *Petchinges* DB]. '*Pæcca*'s HĀM' and '*Pæcca*'s people'. ***Pæcca** is related to *Pacca* in PACKINGTON &c.

Pateley Bridge YW [*Patleiagate* 1175 Fount, *Patheleybrigge* 1320 Ch]. OE *pæþlēah* 'LĒAH by a path'. For the change *þ* > *t* cf. *bōtl* from *bōþl* &c.

Pathlow Wa [*Patelav* (hd) DB, *Pathelawa* 1174 P, *-lawe* 1232 Fees]. The second el. is OE *hlāw* 'barrow, tumulus'. The first may be as in PANBOROUGH.

Patmore Hrt [*Patemere* DB, 1203 Cur, *Pattemera* 1166 P]. '*Patta*'s or *Peatta*'s lake.' For *Peatta* cf. PATNEY. *Patta* is found in *Pattan dene* 973-4 BCS 1307 (Ha).

Patney W [*æt Peatanige* 963 BCS 1118, (æt) *Peattanigge* c 1050 ib. 390, *Pateneia* 1205 Pp]. '*Peatta*'s island.' *Peatta* might be a form with assimilation of *Pearta*.

Patrington YE [*Æt Patringtune*, *Pateringatun* 1033 YCh 8, *Patrictone* DB, *Patrinton* 1187, 1190, *Paterington* 1194 P]. The material does not render it probable that the first el. is the name *Patric*. A definite suggestion cannot be made.

Patrixbourne K [*Burna* 11 DM, *Borne* DB, *Patingesburn* 1203 Cur, *Patrickeburn* 1228 Cl]. Originally *Burna* 'the stream'. The manor was held by William Patrick t. Stephen.

Patshull St [*Pecleshella* DB, *Patleshull* 1200 Cur, *-hul* 1242 Fees]. '*Pættel*'s hill.' Cf. PADDLESWORTH.

Patterdale We [*Patrichesdale* a 1184 CWNS xxiv, *Patricdale* 1246 Ipm]. '*Patric*'s valley.' Perhaps named from the Patric son of Bernard who witnessed a deed regarding Docker and Grayrigg We c 1200 or earlier (CWNS xxiv).

Pattesley Nf [*Patesleia* DB, 1199 P, *Pattesle* 1203 Cur]. Cf. LĒAH. First el. as in PATSHULL or PATTISHALL.

Pattingham St [*Patingham* DB, *Pattingeham* 1158, *Patingeham* 1169 P]. 'The HĀM of *Peatta*'s people.' Cf. PATNEY.

Pattishall Np [*Pascelle* DB, *Pateshill* 12 NS, *-hell* 1190 P], **Pattiswick** Ess [*Pateswyck* 1246 Ch, *-wyk* 1265 Ch]. '*Pætti*'s (or *Pættel*'s) hill and WĪC.' Cf. PADDLESWORTH. ***Pætti** would be a derivative of *P(e)atta*.

Patton Sa nr Easthope [*Peat(t)ingtun* 901 BCS 587, *Patintune* DB]. 'The TŪN of *Peatta*'s people.' Cf. PATNEY.

Patton We [*Patun* DB, *Pattun* 1170-84 Kendale]. '*Patta*'s TŪN' or OE *Pæþ-tūn* 'TŪN by the path'.

Paul Co [*St. Paulinus* 1266 Ch]. '(The church of) St. Paulinus.'

Paulerspury Np [*Pirie* DB, *West Pyria* 12 NS, *Pirye Pavely* 13 AD iv, *Paulesperie* 1412 FA]. Originally *Pirige* 'the pear-tree'. The addition *Paulers-* for distinction from POTTERSPURY.

Paulerspury was held by Robert (de Pavely) in 1086 (DB), by Robert de Paueilli before 1194 (P). The name is from PAVILLY in Seine Inférieure.

Paull YE [*Paghel*, *Pagele* DB, *Pagla* 1115, *Pagela* c 1165 YCh 1304, 1309, *Paghel* 1208 FF, *Pagula* (ferry) 1223, (passagium de) *Pavel* 1226 FF, *Pawelflet* 1260 Ipm]. The source seems to be an OE *pagol, which may be a side-form of OE *pægel* 'a winemeasure', originally 'a peg', Du *pegel* 'a peg'. Paull is on the Humber, and a meaning 'foot-bridge' is out of the question. It may have been 'pole, stake', e.g. one put up to mark the place of a passage or ferry.

Paulton So [*Palton* 1171 P, 1201 Cur, *Pealton* 1194 P, *Peanton* 1225 ff. Ass]. *Peanton* evidently for *Peauton*. The source is an OE *Peall-tūn*, whose first el. *peall may be a word meaning 'ledge' corresponding to ON *pallr*, Sw *pall*, Dan *pald* 'raised place, footstool, terrace' &c. **P~** is high up on a hill slope. Löfvenberg adduces the surname *de la Palle* (13, 14), borne by persons mentioned in connexion with Cameley c 2½ m. from Paulton. There was evidently a place called *Pall* near Paulton.

Pauntley Gl [*Pantelie* DB, *-le* 1206 Cur, *-leg* 1220 Fees]. 'LĒAH by a valley.' First el. Welsh *pant* 'valley'.

Pauperhaugh Nb [*Papwirthhalgh* Hy 1 Brinkburn]. 'The haugh of Papworth.' Cf. PAPWORTH Ca.

Pavenham Bd [*Pabeneham* DB, *Papenham* 1195 Cur, *Papeham* 1240 Ass, *Pabenham* 1242 Fees, *Pavenham* 1491 Ipm]. '*Papa*'s HĀM.' Cf. PAPWORTH. The change *p* > *b* > *v* has analogies in Bd and neighbouring counties.

Pawlett So [*Pavelet*, *Paulet* DB, *Poulet* 1186 Buckland, 1194 P, *Poolet* 1212 Fees]. Second el. OE *flēot* 'stream'. The first may be OE *pāl* 'pole'. OE *pāl-flēot* would easily become *pā-flēot* owing to dissimilation. The meaning would be 'fleet with stakes, stream obstructed by stakes'.

Paxton Hu [*Parchestune*, *Pachstone* DB, *Pachestona*, *-tonia* 1161 RA, *Pacstonia* 1164 BM, *magna Paxton* 1163, 1190 P, *Parva P~* 1220 Fees, *Magna Praxton* 1245 For]. The first el. is found also in *Paches-*, *Paccheswell(e)* c 1230 ff. RA, the name of a spring or stream in Gt Paxton, and is doubtless a pers. n. *Pæcc* cognate with *Pacca* in PACKINGTON, *Pæccel* in PAGLESHAM, as suggested already by Skeat in PNHu.

Payhembury D [*Hanberie* DB, *Paihember* 1236 BM, *Payhaumbir'* 1242 Fees]. Cf. BROADHEMBURY. The original name was *Hēaburg*, dat. *Hēanbyrig* 'high fort'. *Pay-* from an early owner, perhaps a Saxon named *Pǣga*.

Paythorne YW [*Pathorme, Pathorp* DB, *Paththorn* 12 Pudsay, *Pathorn* 1197 P, *Pattorp* 1187 ff. P]. Perhaps OE *pæþ-þorn* (or *-þorp*) 'thorn-bush or thorp by a path'. But it is worthy of notice that PAINLEY is only about one mile away.

Peak Db [*Pecsætna* (land) 7 BCS 297, *Peaclond* 924 ASC, (mons) *Pec* c 1130 HHunt, *Pech* 1157, 1159 P]. **Peak Cavern** is *Pechesers* DB, *ers* being OE *ears* 'podex'. Peak is an OE **pēac* 'hill, peak', a word cognate with Du *pôk* 'dagger', Sw *påk* 'cudgel', Sw dial. *pjuk* 'point, hillock', Norw *pauk* 'a stick', OE *pūca* &c. The same el. is found in PECKFORTON, PEGSDON.

Peakirk Np [*æt Pegecyrcan* 1016 KCD 726, *Peychirche* 12 NS, *Peikirke* 1198 FF, *Peichirche* c 1202 NpCh]. 'St. Pega's church.' St. Pega was a sister of St. Guthlac. OE *cirice* 'church' has been replaced by Scand *kirk*.

Peamore D [*Peumera* DB, *Paumera* 1194 P]. Has been explained as 'peacock's mere' (PND). OE *pēa, pāwa* 'peacock' appears as *peaw* 1719 in Somerset dialect.

OE **pearroc** 'paddock, enclosure'. See PADDOCK, PARKHAM.

Peasemore Brk [*Pesemere* 1166 RBE, 1167 P, 1212 Fees], **Peasenhall** Sf [*Pesehala, Pisehalla* DB, *Pesenhal* 1228 FF], **Peasmarsh** Sx [*Pisemerse* 12 AD, *Pesemershe* 13 PNSx]. 'Lake, HALH and marsh where peas grew.' Cf. PISU. In Peasenhall the first el. is partly OE *pisen* 'of peas'. At least in Peasemore and Peasmarsh the reference is probably to some wild plant resembling a pea, e.g. marsh-trefoil, also called buckbean.

Peatling Magna & Parva Le [*Petlinge* DB, *Pellinguis* 1190–1204 Fr, *Pedlinges* 1196 P, *Magna Petling, Parva Pedling* 1242 Fees]. '*Pēotla*'s people.' **Pēotla* is a diminutive of OE *Pēot (Pīot)* pers. n.

Pebmarsh Ess [*Pebeners* DB, 1238 Subs, *-herse* 1202 FF, *-ershe* 1296 Ch]. '*Pybba*'s ERSC or pasture land.'

Pebsbury So [*Pibbesbyrig* 1065 Wells]. '*Pybbi*'s BURG.' **Pybbi* is a side-form of *Pybba*.

Pebworth Gl [*Pebewrthe* 848 BCS 453, *Pebbewurðy* c 1012 KCD 898, *Pebeworde* DB]. Probably '*Pybba*'s WORÞ'. Cf. PEDMORE.

Peckforton Chs [*Pevretone* DB, *Pecfortuna* c 1100, c 1150 Chester]. 'TŪN by *Pēacford*.' *Pēacford* (*Pecford* 1288 Court) was the name of a ford here, named from **Peckforton**

Hills, which must have been called *Pēac*. Cf. PEAK.

Peckham, East & West, K [(de) *duobus Peccham,* (to) *þam twam Peccham* 10 BCS 1321 f., *Pecheham* DB, *Peccham* 1198 Fees, *Est Pecham* 1293 QW, *West Pekeham* 1202 FF], **P~** Sr [*Pecheham* DB, *Pecham* 1178 PNSr, 1200 Cur]. Perhaps both are OE *Pēac-hām* 'HĀM by a hill'. Cf. PEAK. **P~ Rye** is *Peckham Rye* 1589 AD. Rye is OE *rīþe* 'brook'.

Peckleton Le [*Pechintone* DB, *Petlington* 1180 P, *Peyhtelton* Hy 3 Misc, *Peghtelton* 1292 Cl]. '*Peohtla*'s TŪN' or 'the TŪN of *Peohtel*'s people'. **Peohtel* (**Peohtla*) from names in *Peoht-.*

Pedmore Wo [*Pevemore* DB, *Pubemora* 1176 P, *Pebbemore* 1292 Ipm]. '*Pybba*'s moor.'

Pedwardine He [*Pedewrde* DB, *Magna, Parva Pedewardin* 1292 QW]. '*Pēoda*'s (*Piuda*'s) WORÞIGN.'

Pedwell So [*Pedewelle* DB, *-well* 1201, 1243 Ass]. '*Pēoda*'s stream or spring.'

Peelings Sx [*Pellinges, Palinges* DB, *Pedlinga* 1186 ff. P]. Probably identical with PEATLING.

Pegsdon Bd [(æt) *Peácesdele* 1015 Wills, *Pechesdone* DB, *Pekesdene* 1228 FF]. 'Valley by a hill.' Cf. PEAK. Second el. originally OE *dæl*, later *denu*, both 'valley'.

Pegswood Nb [*Peggiswrth* 1242 Fees, *Peggeswurthe* 1259 Sc]. '*Pecg*'s WORÞ.' *Pecg* is found in *Pecgesford* 958 BCS 1023. It may be a nickname formed from *peg*, which is not found until ME, however.

Pēlaw Du [*Pellowe* 1242 Ass, *Pelawe* 1297 Pp]. See PELTON.

Peldon Ess [(at) *Piltendone* c 950, (æt) *Peltandune* c 995 BCS 1012, 1288 f., *Peltenduna* DB, *Pultindone* 1212 RBE]. The OE form must have been *Pyltan dūn*. *Pylta* may be related to the first el. of POULTNEY and to OE **pyltan*, the base of ME *pilte, pulte* 'to thrust, put'. Perhaps a pers. n. *Pylta*.

Pelham, Brent, Furneux & Stocking, Hrt [*Peleham* DB, 1186 P, 1203 Ass, *Peldeham* 1177 P, *Pelham Combusta* 1210 Cur, *Barndepelham* 1230 Pat, *P~ Forneaus* 1287 QW, *Stokenepelham* 1243 Cl]. If *Peldeham* belongs here, probably '*Pēotla*'s HĀM'. Cf. PEATLING. Otherwise '**Pēola*'s HĀM'. Cf. *Pēol* in PELSALL.

Brent P~ must have been burnt down before 1210.—**Furneux P~** was held by Radulfus de Furnellis in 1212 (RBE). Furneux is from FOURNEAUX (lit. 'the furnaces') in Normandy.—**Stocking P~** means 'the P~ built of logs' (OE *stoccen* adj. 'of logs, wooden').

Pelsall St [*Peoleshale* 996 Mon, *Peleshale* DB, *-hala* 1167 P]. '*Pēol*'s HALH.' *Halh* may here mean 'land between two streams'.

OE *Pēol* is found also in *Pioles clifan* BCS 1282.

Pelton Du [*Pelton* 1312 RPD]. P~ and Pelaw are near each other. Pelaw has as second el. OE *hlāw*, here 'barrow, mound'. The probability is that both names have as first el. a pers. n., e.g. *Pēola*; cf. PELHAM.

Pely·nt Co [*Plunent* DB, *Plenint* 1229, 1236 Fees, *Plenent* 1275 Ep]. Has been explained by Henderson as *plou Nent* 'parish of St. Nonna', *Nent* being a hypocoristic form. *Plenint* seems to have become by assimilation *Plelint* and by dissimilation *Pelynt*.

Pemberton La [*Penberton* 1201 P, *-breton* 1202 FF, *Pemberton* 1212 Fees]. 'BARTON by the hill.' See PEN.

Pembridge He [*Penebruge* DB, *-brigia* 12 BM, *-brigg* 1230 P]. Possibly 'bridge by the pens or enclosures'. The forms point rather to a first el. with short *n*, possibly OE *Pægna*; cf. PAIGNTON.

Pembury K [*Pepingebir'* 1205 Cur, *Papingbyr* 1257 Ch, *Pepingeburi* 1262 Ipm, *Peapyngeberi* 1309 BM]. The first el. is a derivative in *-ingas* of a pers. n., whose form, however, is doubtful.

OE **pen (penn)** 'pen, enclosure' and Brit pen(n) in Welsh *pen*, Co *pen* 'head, top, summit, end', also 'point, promontory', Welsh *pen* 'chief' (adj.) are often found in pl. ns., but are not always easy to keep apart. As a second el. OE *penn* 'pen' is no doubt as a rule to be assumed, as in HAMPEN, IPPLEPEN, KILPIN, OWLPEN. But INKPEN may well contain the Brit word. Probably Brit *pen* was to some extent used by the Anglo-Saxons in the sense 'hill'. The situation suggests *pen* 'enclosure' for PAMBER. Situation on or near a hill points to the Brit word, though not necessarily, as for PENN Bk, St. The Brit word is certainly or probably the first el. of PEMBERTON, PENWORTHAM, hill-names such as PENDLE, PENYGHENT, or PENSELWOOD, PINHOE. Several names containing the el. *pen* are Brit names altogether, as PENCOYD, PENCRAIG, PENDOCK, PENGE, PENGETHLY, PENKETH, PENKRIDGE, PENNARD, PENRITH, PENYARD, and several Cornish names. **Pen Hill** So [*æt þam Pænne* 1065 Wells] is Brit *pen*, and **Pen Cross** D [*peon mynet* 938 BCS 724] is Welsh *pen mynydd* 'top of the hill'. In the last Brit *pen* appears in a form *peon*, which is found also in other names, as in PENDOCK, PENSELWOOD, PINHOE, and is not easy to explain.

Penare Co at Nare Point [*Pennarð* 967 BCS 1197] and P~ Co nr Mevagissey [*Penhard* 1303 FA, *Penarth* 1309 Ipm] are both named from promontories. The base is Co *pen ard* or *arth* 'high headland' or a Co *penard* corresponding to Welsh *penardd* 'promontory'. **Nare (Point)** is due to wrong division of *Pennard*.

Penbury Gl [*Penneberie* DB, *Pendeberiam* 1192 P, *Penthebery* 1192 Fr]. '*Penda*'s BURG.'

Pencombe He [*Pencumbe* 12 Glouc, 1206 Cur, 1233 Cl]. 'Valley with a pen or enclosure.'

Pencoyd He [*Pencoyt* 1291 Tax, 1301 Misc]. Welsh *pen coed* 'end of the wood'. Welsh *coed* is Brit *cēt*, OW *coit*.

Pencraig He [*Penncreic* c 1150 LL, *Pencrek* 1347 Ep]. 'The top of the crag.' Second el. Welsh *craig* 'crag, rock', from OW *creic*. See PEN.

Pendeford St nr Wolverhampton [*Pendeford* DB, 1282 Ch], **Pendeley** Hrt nr Aldbury [*Pentlai* DB, *Pendelleg* 1230 P]. '*Penda*'s ford and LĒAH.'

Pendle (Hill) La [*Pennul* 1258 Ipm, *Penhul* 1305 Lacy]. Welsh *pen* 'top, hill' with an explanatory OE *hyll* 'hill' added. Cf. *Penhyll* 11 Th (now **Penn Hall** Wo).

Pendlebury La [*Penelbiri* 1202, *Penlebire* 1206 P, *Penhilbyry* 1284 Ass]. 'BURG by *Penhill*.' This must have been the name of the ridge on which Pendlebury is. Cf. PENDLE. **Pendleton** La nr Pendlebury [*Penelton* 1200 f. P, *Penhulton* 13 WhC]. 'TŪN on *Penhill*.' **Pendleton** La nr Pendle Hill [*Peniltune* DB, *Penhulton* 1272 Ass]. 'TŪN by Pendle Hill.'

Pendock Wo [*Pe(o)nedoc* 875 BCS 541 f., DB, *Penedoc* 967 BCS 1208]. A Welsh name consisting of Welsh *pen* and **heiddiog* (earlier *heiddioc*) 'of barley' or 'barley field' (cf. HAYDOCK La). The name would mean 'hill where barley was grown' or 'the end of the barley field'. P~ is on the lower slope of Malvern Hill.

Pendō·mer So [*Penne* DB, *Penna* 1180 P, *Penne dommere* 1311 Ep]. The original name was PENN, which may be OE *penn* 'enclosure' or Brit *pen* 'hill'. The place is at the foot of a considerable hill. *Domer* is a family name, on which see CHILTHORNE DOMER.

Pendra·gon We [*Pendragon* (castle) 1314 Ipm]. Held to have been named from Uther Pendragon, father of Arthur.

Penge Sr [*Penceat* 1067 BM, *Pænge* (wood) 957 BCS 994, *Peenge* 1203 Cur]. A Brit name consisting of *pen* (Welsh *pen* 'end' or rather 'chief') and Brit *cēt* (Welsh *coed*) 'wood'. For the loss of *-t* cf. TRUNCH.

Pengethly He [*Penketlin* 1275 RH, *Penketly* 1332 Ep]. Welsh *pen* 'end' &c. and *celli* 'wood'. 'The end of the wood' or 'chief wood'.

Penhill (hill) YN [*Pennel* 1155–84 TpR, *Penle* 1202 FF, *Penhill* 1577 Saxton]. Identical with PENDLE.

Penhurst Sx [*Penehest* DB, *-hurst* 1238 FF, *Penherste* 1197 P, *-hurst* 1249 Fees]. First el. as in PEMBRIDGE. See HYRST.

OE **pening** 'penny' is no doubt the first el. of PENNINGTON Ha, La (1), PENTON Ha. The meaning would be 'TŪN that had to pay a penny geld' or the like. Names such as PENISTONE, PENSTHORPE contain an el. *pening* too, but hardly in the sense 'penny'.

Penistone YW [*Pengeston(e)*, *Pangeston* DB, *Peningeston* 1199 P, *-a* c 1190 YCh 1677]. Near P~ was formerly **Penisale** [*Penigheshal* c 1200, *Penigeshalg* c 1215 YCh 1803, 1805, *Peningeshalge* 1209 (1252) Ch]. If the first el. is a pers. n., it is probably OE *pening* used as a byname. We may compare the family name *Penny*, found at least from 1273. ON *Peningr* is used as a byname. But Penistone is situated at a high ridge, which may have been called *Penning*, the name being derived from Brit *pen*. A hill called *Penningstein howe*, *Penigstain hou* in Kirkby Lonsdale We is mentioned c 1200 CC. *Penning* may also here be a hill-name.

Penk R. See PENKRIDGE.

Penketh La [*Penket* 1242 Fees, *-keth* 1259 Ass]. Identical with PENCOYD.

Penkhull St [*Pinchetel* DB, *Pencul* 1169 P, *Penkhil* 1230 P]. The DB form suggests a compound consisting of Brit *Pencēt* (identical with PENKETH) and OE *hyll*. We must then suppose that a Brit place on the hill was called *Pencēt*.

Penkridge St [*Pennocrucio* (abl.) 4 IA, *Pencric* 958, 10 BCS 1041, 1317, *Pancriz* DB, *Pencrich* 1156, *Peinchriz* 1158 P]. A Brit name consisting of Brit *penno-* (Welsh *pen*) and a derivative of OBrit *crūcā* (Welsh *crug*, see CRŪC) 'hill, mound'. The Roman station of *Pennocrucium* appears to have been on rising ground near the river Penk. Hence a meaning 'mound on a hill' or 'hill summit' is suitable. **Penk,** the name of the river at P~ [*Penk* 1577 Saxton], is a back-formation. In early times a different back-formation was used [*Penchrich* 996 Mon, *Pencriz* 1300 For].

Penn Bk [*Penna de Tapeslawa* 1188 P, *Lapenne* 1197 FF]. OE *penn* 'enclosure' or Brit *pen* 'hill'. The place is on a hill.

Penn, Lower & Upper, St [*Penne* DB, 1176 P, *Netherpenne* 1271 For, *Overpenne* 1318 Ch]. Probably Brit *pen* 'hill'.

Penn Hall Wo. See PENSAX.

Pennard, East & West, So [*Pengerd* 681, *Pennard* 705, *Pengeard mynster* 955 BCS 61, 112, 903, *Pennarminstre* DB, *East Pennard* 1243 Ass]. A derivative of the pl. n. is (*Eanulf*) *Penearding* 901-24 BCS 591, which means 'of Pennard'. The places are at **Pennard Hill** (395 ft.), which the name originally designated. Pennard is Welsh *pen ardd* 'high hill' (cf. PEN; Welsh *ardd* means 'high'). This became OE *Pen-eard*, and sometimes *Pen-geard* owing to association with OE *geard*.

Pennington Ha [*Penintune*, *Penigtone* 13 VH, *Penyton* 1316 FA], P~ La nr Ulverston [*Pennigetun* DB, *Penigtona* c 1160 LaCh, *Peninton* 1187 P]. See PENING.

Pennington La in Leigh [*Pinington* 1246 FF, *Pynington* 1246 Ass]. Perhaps 'the TŪN of *Pinna*'s people'. *Pinna* occurs in *Pinnan rod* KCD 767.

Penny Bridge La. Named from a local family *Penny*.

Penrith Cu [*Penrith* c 1100 WR, *Penred* 1167 P, 1242 Ch]. 'The chief ford' (Welsh *pen* 'chief' and *rhyd*, OW *rit* 'ford').

Penruddock Cu [*Pendredoch* 1276 Cl, *Penreddok* 1285 CWNS x, *Penruddoc* 1292 Sc]. Evidently a Brit name with Welsh *pen* as first el. The second is obscure.

Penry·n Co [*Penrin* 1259 Ch, *Penryn* 1275 Ep]. Co *penryn* (= Welsh *penrhyn*) 'promontory, cape'.

Pensax Wo [*Pensaxan* 11 Heming, *Pensex* 1231 PNWo]. Probably a Welsh *Pen Saeson* 'the hill of the Saxons'. In P~ is **Penn Hall** [*Penhyll* 11 Th, *Penhull* 1221 Ass]. Cf. PENDLE.

Pensby Chs [*Penesby* 1261-3 Chester, *Pennesby* 1309 Ormerod, 1316 Misc]. The place is on a marked hill. This may have been called *Pen*, and *Pensby* is 'BY on Pen hill'.

Pense·lwood So [(æt) *Peonnum* 658 ASC, (æt) *Peonnan* 1016 ib. (D, E), *Penne* DB, (boscus de) *la Penne* 1274 RH, *Penne in Selewode* 1345 Ep]. OE *Peonnum* is the dat. plur. of OE **penn* (**peonn*), which must be from Brit *pen* 'hill'. The place is on a long ridge. The addition *-selwood* for distinction from PENDOMER. See SELWOOD.

Pensfold. See PEVENSEY.

Pensford So [*Pensford* 1400 AD, *Penesford* 1412 FA]. First el. perhaps OE *penn* 'pen'.

Pensham Wo [(in) *Pedneshamme* 972 BCS 1282, *Pendesham* DB]. '*Peden*'s HAMM.' **Peden* is a derivative of *Peada*.

Penshurst K [*Pensherst* 1072 BM, *Peneshurst* 1203 Cur, *Peveneshurste* 1252-72 Ep]. '*Pefen*'s hurst.' See PEVENSEY.

Pensnett St [*Pensnet* 1244 Cl, *Pensned* 1322 Ipm, *Peninak* 1247 FF, *Pennak* 1273 Ipm, *Penynak* 1292 Misc]. Second el. OE *snǣd* 'a piece of woodland.' P~ was a chase. The first el. may well be Brit *pen* 'hill', to judge by the situation. Hence 'wood on Pen hill'. The form *Peninak* is due to Norman influence. *Pensnǣd* became by sound-substitution *Peninet*, which came to be read as *Peninec*. The change to *Peninak* may be due to the influence of French names in *-ac*.

Pensthorpe Nf [*Penestorpa* DB, *Penetorp* 1195 Cur, *Pengestorp* 1254 Val]. Apparently identical with a lost **Penthorp** YE [*Pene-*

gestorp, Peningestorp 1200 FF]. '*Pening*'s thorp.' On ON *Peningr* see PENISTONE.

Pentire Point Co [*Pentir* 1201 FF, 1208–10 Fees]. The forms quoted may not refer to Pentire Point on the Camel estuary, but to places with the same name. The meaning is 'end of the land' (Co *pen* 'end' and *tir* 'land').

Pentlow Ess [(at) *Pentelawe* 1043–5 Wills, *Pentelauua* DB, -*lau* 1166 RBE], **Pentney** Nf [*Penteleiet* DB, *Pentenay* 1200 Cur, *Penteneya* 1254 Val]. Pentlow is on the Stour. The second el. is OE *hlāw* 'barrow, tumulus'. Hence no doubt '*Penta*'s barrow'. **Penta* is a derivative of *Pant*, found also in *Panteshede* DB (a lost place in Banwell So). Pentney may well be '*Penta*'s island or river land', but a river-name identical with PANT is a possible first el.

Penton Grafton or **Weyhill** Ha [*Penitone* DB, *Peninton Abbatis* 1167 P, *Penyton Croftyn* 1316 FA], **P~ Mewsey** Ha [*Penitone* DB, *Penintona Roberti* 1167 P, *Penitune Meysi* 1264 AD]. See PENING. The two Pentons are close together.

P~ **Grafton** belonged to the Abbey of Grestayn in Normandy. *Croftyn, Grafton* are corruptions of *Grestayn*.—P~ **Mewsey** was held by Robert de Meisy in 1212 (Fees), perhaps identical with the Robert of 1167 (P). For *Mewsey* cf. MEYSEY HAMPTON.

Pentrich Db [*Pentric* DB, *Pentrich* 1229, *Pentriz* 1251 Ch], **Pentridge** Do [*Pentric* 762 Muchelney, *Pentringtone* 940–6 BCS 817, *Pentric* DB, 1107 (1300) Ch, *Pencriz* 1187 P]. Pentrich is on a marked hill. Pentridge is by **Pentridge Hill** (600 ft.). The names are identical and represent a Brit hill-name. It may be identical with **Pentyrch** Glam [*Penntirch* 12 Life of St. Cadoc]. Pentyrch is on a marked hill, and the name has as first el. Welsh *pen* 'hill'. The second el. is perhaps **tyrch*, an old gen. sg. of *twrch* 'boar': 'boar's hill'.

Penwith Co, an old name of Land's End [*Penwihtsteort* 997 ASC (E), *Penwæðsteort* ib. (D), *Penwiðsteort* 1052 ib. (C, D), *Penwid* 1186, -*wed* 1194 P]. It is a moot point if *Pengwaed* in the Welsh Mabinogion belongs here. Penwith has been explained as 'pointe en vue' (Loth), i.e. 'point seen from afar' or the like. *Pen* is 'promontory'; *-with* would be identical with OBret *uuid*-in pers. ns. and the like, and related to Welsh *gwedd* 'aspect' (< *vidā*) and *gwydd* 'presence' (< *veid*-), Lat *videre* &c. Welsh *gwydd* 'wood', Co *gwydh* 'trees' would suit formally, but hardly topographically.

Penwortham (-dh-) La [*Peneverdant* DB, *Penuertham* a 1149 LaCh, 1212 Fees, *Penwertham* 1205 P]. A hybrid name, probably consisting of Welsh *pen* 'hill', an old name of the eminence on which P~ stands, and OE *Worþ-hamm* or -*hām* 'enclosed homestead' or the like. Cf. WORTHAM.

Penyard He [*Penyerd Regis* 1227 Ch,

Peniard, Penierd 1228 Cl]. Identical with PENNARD. **Penyard Hill** is a conspicuous hill.

Penyghent, hill, YW [*Penegent* 1307 YInq, *Penaygent* R 2 Whitaker, Craven]. A Welsh *pen y gaint*, perhaps 'the hill of the border country' (cf. KENT); *y* is the Welsh definite article. A Welsh *caint* 'plain, open country' has been assumed, but its existence is disputed by Jackson.

Penza·nce Co [*Pensans* 1332 Ch, 1367 AD, *Pensant* 1367 AD, *Pensaunce* 1552 BM]. 'Holyhead', the elements being Co *pen* 'cape' and *sans*, earlier *sant*, 'holy'.

Peopleton (pĭp-) Wo [*Piplincgtun* 972 BCS 1282, *Piplintune* DB, *Puplinton* 1254 WoP]. The first el. has been compared with that of **Pepper Wood** in Belbroughton Wo [*Pupperode* 1230 P, *Pipperod* 1262 For], which apparently contains a pers. n. *Pyppa*. Peopleton would be 'the TŪN of *Pyppel*'s people'. But the first el. of Peopleton might be a stream-name derived from OE **pyppel* 'pebble'. Cf. *pyppelriðig* 955 BCS 906, PEPLOW.

Peover R Chs [*Peuerhee* 13, *Peuere* 1277 Ancestor ii]. A Brit river-name *Pefr*, identical with PEFFER in Scotland and derived from Welsh *pefr* 'radiant, bright'. To this was added OE *ēa* 'river'. Cf. PERRY R Sa, PARWICH. On the Peover are **Nether & Over Peover** (-ē-) [*Pevre* DB, *Pevere* 1278 Ipm, *Peverhe* 1260, *Over Pevre*, *Netherepevre* 1287 f. Court].

Pĕper Harrow Sr [*Pipereherge* DB, *Pyperhargh* 1291 Tax]. OE *pīpera hearg* 'the HEARG of the pipers'. But probably Harrow is here a pl. n. identical with HARROW, to which was added *pīpera* 'of the pipers'. Cf. HEARG, HARROW.

Peplow Sa [*Papelav* DB, *Peppelawe* 1232 FF, *Pippelawe* 1256 Ass]. Perhaps 'pebble hill'. OE *pyppel* 'pebble' appears as *puble* c 1290, *pible* 1542 (OED). Cf. PEOPLETON.

Perdiswell Wo [*Perdeswell* 1182 PNWo]. '*Preed*'s spring.' Cf. PARDSHAW.

Pĕrivale Mx [*Pyryvale* 1508, *Peryvale* 1564 FF]. 'Pear-tree valley.' The name is late and has replaced LITTLE GREENFORD. Second el. *vale* 'valley', a French word.

Perlethorpe Nt [*Torp* DB, *Peureltorp* 1159, *Peuerelestorp* 1167 P]. 'The THORP belonging to the *Peverels*.' The manor no doubt once belonged to a Peverel. Cf. BARTON PEVERELL.

Perranarworthal Co [*Harewithel* 1187, *Arwrthel* 1198 P, *Arwoethel* 1303 FA, *Arwythel* 1337 FF], **Perran Uthnoe** (ūthnō) Co [*Peran Uthnoe* 1202 Pp, *Odenol* DB, *Hithenho* 1214 FF, *Hutheno* 1229 Fees], **Perranzabuloe** (-zăbūlō) Co [*Canonici S. Pierani* DB, *Sanctus Piran* 1195 FF].

Perran is the saint's name *Peran (Piran)*, Bret *Pieran*, Ir *Cíaran*.

Arworthal must be an earlier name of the place and has been explained by Loth, *Revue celtique*, 37, pp. 179, 303, as containing a Co *gothal* corresponding to Bret *gwazell* 'wet or marshy ground' and related to MBret *goeth* 'brook' &c. (see GOYT). *Ar* seems to be the Co prep. *ar* 'on'.— *Uthnoe* is likewise an earlier name of the place. It is possibly an elliptical use of a pers. n. corresponding to OW *Iudnou* (c 1150 LL); *Iud-* is Mod Welsh *udd* 'lord'.—**Perranzabuloe** is 'Perran in the sand' (Lat *in sabulo*).

Perrott, South, Do [*Pedret* DB, *Superete* 1218 Salisbury], **North Perrott** So [(æt) *Peddredan* c 1050 KCD 839, *Peret* DB, 1291 Tax]. Named from the river PARRET.

Perry Hu [*Pirie* DB, *Peri* c 1110 RA], **P~ Court** K [*Perie* DB, *Pirie* 11 DM], **Waterperry & Woodperry** O [*Pereio, Pereiun* DB, *Periet* c 1130 Oxf, *Waterpiria* 1209–35 Ep, *Wdeperie* 1220 Fees], **Perry Barr** St [*Pirio* DB, *Piri* 1176 P, *Pirie* 1242 Fees]. OE *pirige* 'pear-tree', sometimes in the plur. (*pirigan*, dat. *pirigum*).

Waterperry is on the Thame. **P~ Barr** is nr BARR.

Perry R Sa [*Peueree* Hy 2 ERN, *Pevereye* c 1250 Eyton]. See PEOVER.

Pershore Wo [*Perscoran* 972 BCS 1282, (on) *Persceoran* c 1055 KCD 804, (on) *Prescoran* c 1035 E, *Persore* DB]. Dial. *persh* 'osier' (ME *persche* 'twig') has been taken to be the first el., the second being OE *ōra* 'bank'. If *persh* is an old word, this is probably the correct etymology. One might also suggest as first el. the word found in PRISTON.

Pertenhall Bd [*Partenhale* DB, *Pertinhala* 1179 P], **Pertwood** W [*Perteworde* DB, *-wurda* 1166 P, *Perteswrth* 1200 FF]. '*Pearta*'s HALH and WORÞ.' Cf. PARTNEY.

OE **peru** 'pear' is found in several names, as PARBOLD, PARHAM, PARLEY, PRESHAW, PRESTED, SPURSHOT. OE **peren* 'of pears' perhaps in PARNDON, PARNHAM, PRINSTED.

Peterborough Np [*Medeshamstedi* c 730 Bede, *-stede now Burchus* 972 BM, *Burh* 972–92 BCS 1130, *Burg* DB, *Burgus sancti Petri* 1225 RA, *Petreburgh* 1333 Cl]. The old monastery, which was destroyed by the Danes, was called Medeshamstede: '*Mēde*'s homestead'. **Mēde*, a derivative of *Mōd-* in pers. ns., is also found in *Medeswæl* 654 ASC (E) '*Mēde*'s pool' (OE *wæl* 'weel, pool'), evidently a pool in the Nene at Peterborough. The new monastery became known as *Burg* 'the town or borough', later Peterborough from the dedication of the abbey.

Peterchurch He [*Petruschirche* 1428 FA]. 'Church dedicated to St. Peter.'

Peterley Bk [*Piterleia* c 1150 PNBk, *-lee* 1196 FF, *Puterle* 1291 Tax, *Peterleye* 1302 Ch]. Perhaps 'pear-tree LĒAH', with dis-

similatory loss of the first *r*. *Pertre* is found c 1300, and OE *per-trēo(w)* or *pirge-trēo(w)* may well have existed.

Petersfield Ha [*Peteresfeld* Hy 2 VH, 1182 P, *Peterfeld* 1230 P]. 'St. Peter's FELD.' The church is dedicated to St. Peter.

Petersham Do [*Petrishesham, Pitrichesham* DB, *Pidrischesham* 1219, *Pitrichesham* 1242 Fees, *Petrichesham* 1259 FF, *Piterichesham* 1264 Ipm], **P~** Sr [*Piterichesham* 675, 933 BCS 39, 697, *Patricesham* DB]. '*Peohtric*'s HĀM or HAMM.' *Peohtric* is unrecorded, but cf. *Peohtgils, -hūn, -helm* &c.

Peterstow He [*Lann petyr* c 1150 LL, *Peterestow* 1207 Cur, *-e* 1277 Ep]. 'St. Peter's church.' *Lann* is Welsh *llan* 'church'.

Petertavy. See TAVY.

Petham K [*Pettham* c 961 BCS 1065, *Piteham* DB, *Pytham* 11 DM, *Petham, -hom* 1203 FF]. 'HAMM by the pit or hollow.' The form *Pethom* points to *hamm*.

Petherick, Little, or **St. Petrock Minor** Co [(ecclesia) *Sancti Petroci* 1371 Ep]. 'The church of St. Petroc.' Cf. PADSTOW.

Petherton, North, So [*Nordperet, Nort Peret, Peretune* DB, *Norpereton* 1212 Fees], **South P~** So [*Sudperet, Sutpetret, Sudperetone* DB]. 'TŪN on R PARRET.'

Petherwin, South, Co [*Suthpydrewyn* 1269, *Pitherwyne* 1275 Ep], **North P~** D [*Pidrewin* c 1145, *Nordpydrewyn* 1269 Ep]. The two Petherwins are several miles apart and cannot have formed a whole. The first el. appears to be identical with the old name of the hundred of **Pyder** [*Piderscire* 1130, 1186 P]. If *-win* is Co *gwin* 'white', as suggested in PND, 'White Pyder' must have been the name of the district in which the places are.

Petrockstow D [*Petrochestov* DB, *Petrochestona* 1150 Fr]. Identical with PADSTOW.

Petsoe Bk [*Petrosho* 1151–4 Fr, *Pottesho* 1197 FF, *Petesho* 1303 FA]. '*Pēot*'s HŌH or spur of land.'

Pett K [*Pytte* 11 DM, *Pette* 1325 AD], **P~** Sx [*Pette* 1196 FF]. OE *pytt* 'pit, hollow'.

Pettaugh Sf [*Petehaga, Pettehaga* DB, *Pethag* 1219 FF, *Petteshaghe* 1275 Cl]. '*Pēota*'s *haga* or enclosure.'

Petteril R Cu [*Peterel* 1268 For, *Peyterel* 1285 For, *Peterell* 1338 Pat]. Unexplained.

Pettistree Sf [*Petrestre* 1253 Ch, 1254 Val, *Petristre* 1291 Tax]. '*Peohtrēd*'s tree.'

Petton D [*Petetona* c 1150 Bath, *Peatetone* 1242 Fees, *Pyaton* 1303 FA]. '*Peatta*'s TŪN.' Cf. PATNEY.

Petton Sa [*Pectone* DB, *-ton* 1155 ff. Eyton]. The nearest hill of any importance is a couple of miles away. Still the name may be OE *Pēac-tūn* 'hill TŪN'. Cf. PEAK.

Petworth Sx [*Peteorde* DB, *-wurða* 1168 P, *Peteswurda* c 1150 Fr, *Puetewurth* 1181 P]. '*Pēota*'s worþ.'

Pevensey Sx [*Pefenesea* 947 BCS 822, *Peueneséa* 1050 ASC (D), *Pevenesel* DB]. '**Pefen*'s river.' The same pers. n. is found in **Pensfold** Sx [*Peuenesfeld* 1301 PNSx], in PENSHURST, and with loss of *-n* in PEWSEY, PEWSHAM. The etymology of the pers. n., whose base seems to be *Paf-*, is obscure.

Pevington K [*Piventone* DB, *Piuen-, Piuingtune* 11 DM, *Pevinton* 1242 Fees]. 'The TŪN of *Pēofa*'s (*Peufa*'s) people.'

Pewsey (-z-) W [(æt) *Pefesigge* c 880, *Pevesige* 940 BCS 553, 748, *Pevesie* DB, *Peuesia* 1156 P], **Pewsham** W [*Peusham* 1238 Cl, *Peuseham* 1245 Cl, *Pewesham* 1284 Cl]. '**Pefe*'s island or river land and HĀM.' Cf. PEVENSEY.

Pexall Chs [*Pexul* 1285 Court, *Pexhille* 1296, 1305 Lacy]. OE *Pēaces-hyll* '*Pēac* hill.' Cf. PEAK. The original name of the hill was no doubt *Pēac*.

Peyton Ess [*Pachenhou* DB, *Pakenho* 1255 FF, *Pakenho* alias *Patenhall* t. Eliz PNEss]. '*Pac(c)a*'s HŌH or spur of land.' Cf. PACKINGTON. The change of *kn* to *tn* is analogous to that in WARTNABY and that of *kl* to *tl* in WATLING STREET and the like.

Peyton Sf [*Peituna* DB, *Peiton* 1242 Fees]. '*Pǣga*'s TŪN.'

Phepson Wo [*Fepsetnatun* 956 BCS 937, *Fepsetenatun* DB]. 'The TŪN of the *Fepsǣtan*.' *Fepsǣtan* must be connected with the tribal name *Infeppingum* c 730 Bede, *on Feppingum* c 890 OEBede, which denoted a tribe in Middle Anglia. *Fepsǣtan* might mean 'members of the *Feppingas* tribe'.

Phillack Co [(ecclesia) *Sancte Felicitatis* 1259, 1282 Ep]. '(The church of) St. Felicitas.'

Philleigh Co [*Sancti Filii de Eglosros* 1312 Ep]. '(The church of) *Sanctus Filius*.' The earlier name was **Eglosros** [*Eglossos* DB, *Eglosros* 1279 Ep]: 'church in the moor' (Co *eglos* 'church' and *ros* 'moor').

Piall. See PYON.

OE **pīc** 'point, pike' is not a common element in pl. ns., but certain or possible examples are PICKHILL, PICKUP, PICKWELL. The meaning 'peak, pointed hill' is not evidenced in OE, but may well have occurred. *Pike*, the name of a fish, is not evidenced in OED until 1314, but may well be of OE date. See PICKBURN, PICKMERE. Some names may contain a pers. n. *Pīca*, which is not evidenced. *Pūchil* occurs, however, in LVD, and *Pic* Th 617.

Pickburn YW [*Picheburne* DB, *Pikeburne* c 1190 YCh 1585, *-burn* 1202 FF]. 'Stream in which pike were found' or '**Pīca*'s stream'.

Pickenham Nf [*Pichenham, Pi(n)kenham* DB, *Pikenham* 1198 FF, *Nortpykenham* 1291 Tax, *Sutpikeham* 1242 Fees]. It is possible that the original name was *Pincan-hām* '*Pinca*'s HĀM' with dissimilatory loss of the first *n* (cf. BRIGNALL, LAGNESS). But more likely it was *Pīcan-hām* '**Pīca*'s HĀM'.

Pickering YN [*Picheringa* DB, *Pikeringes* c 1110 PNNR, *Pinchering* 1130 P, *Pikering* 1208 Cl]. Apparently an old tribal name *Piceringas*, for which a final solution has not been found. The el. *picere* found in **Pixham** Wo [*Picresham* c 1086, *Pikeresham* 1221 Ass], *Piceres homm* KCD 1368 (Wa), *Pikiresford* 1202 FF (Huddersfield Y) may give the clue, but its meaning is doubtful. Or Pickering might be 'people at *Pīcōra*' ('edge of the *pīc* or hill'), Dickering wap. YE being 'people at *Dīcōra*' ('dyke edge').

Pickhill YN [*Picala* DB, *Pichala* c 1160 YCh 175, *Pikehal* 1208 FF, *Pikehale* 1270 Ipm]. '**Pīca*'s HALH or nook.' OE *Picahalh* 'HALH by the hills' would also do. There are some small hills at Pickhill.

Picklescott Sa [*Pikelescote* a 1231 Eyton, *Piclescot* 1255 RH]. '*Pīcel*'s COT.' Cf. *Pūchil* in LVD.

Pickmere Chs [*Pikemere* 1274 Ipm, *Pyckmere* 1283 Ipm]. The place is by **Pike Mere**, whose name no doubt means 'lake where pike were found'.

Pickthorne Sa nr Stottesdon [*Pichetorne* DB, *Piketorn* 1194 P]. 'Spiky thorn', the first el. being OE *pīc* 'point'. *Pike* in the sense 'prickle, thorn' is noted from c 1305 (OED).

Pickup Bank La [*Pycoppe* 1296 Lacy, *Pickope Bank* 1595 PNLa]. OE *pīc-copp* 'hill with a PĪC or sharp point'. *Bank* means 'hill'.

Pickwell Le [*Pichewell* DB, *Picwell* c 1125 LeS, 1236 Fees, *Pikewella* 1209-19 Ep]. OE *pīc-wella* 'stream or spring by the peak(s)'. The place is in a high situation near a couple of high hills.

Pickworth Li [*Picheuuorde* DB, *-wurða* 1170 P, *-worth* 1202 Ass], **P~** Ru [*Pikesworth* 1203 Ass, *Pikeworda* 1209-19 Ep, *Pickewurth* 1226 Ep], **Picton** Chs [*Pichetone* DB, c 1100, c 1150 Chester], **P~** YN [*Piketon* 1200 P, c 1200 BM, *Picton* 1251 FF]. '*Pīca*'s worþ and TŪN.' Cf. PĪC.

Piddinghoe· (-hōō) Sx [*Pidingeho* 1204 Cur, *Pedingeho* 1224 FF, *Pudingehou* 1291 Tax], **Piddington** Np [*Pidentone* DB, *-ton* 12 NS, *Pedinton* 1167 P, *Pudinton* 1298 Ipm], **P~** O [*Petintone* DB, *Pydentona* c 1160 Fridesw, *Pidinton* 1187 P, *Pedinton* 1212 BM]. 'The HŌH or spur of land and the TŪN of *Piuda* or *Pydda* or his people.' *Piuda* is recorded. **Pydda* would be related to *Puda*.

Piddle or **Puddle** R Do [*Pidelen stream* 966 BCS 1186, *Pidele* 1229 Ch, *Pudeie* 1325

Abbr]. A river-name of Germanic origin, cognate with MDu *pedel* 'low land, fen land, marsh'. The same is the origin of **Piddle Brook** R Wo [*Pidele* 708 BCS 120, *Pidelan stream* 972 ib. 1282, *Pidwuella* 930 ib. 667, *Pidele* 1229 Ch]. One example shows a shorter form *pid-* of the same meaning.

Several places were named from the PIDDLE Do; the names vary between *Piddle* and *Puddle*. P~ **Hinton** [*Pidele* DB, *Pidele* called *Hinctune* 1082–4 Fr, *Hine Pudele* 1285 FA], **Piddletrenthide** [*Pidrie* DB, *Pidele Trentehydes* 1212 Fees, *Pudele thrittyhide* 1314 FF], **Puddletown** [*at Uppidelen* 966 BCS 1186, *Pitretone* DB, *Pideltona* Hy 2 BM], **Affpuddle** [*rus iuxta Pydelan* 987 KCD 656, *Affapidele* DB, *Effepidel* 1212 Fees], **Bryants** P~ [*Pidele* DB, *Pidel Turbervill* 1238 Cl], **Tolpuddle** [*Pidele* DB, *Tolepidele* 1212 Fees], **Turners** P~ [*Pidele* DB, *Tunerepidel* 1242 Salisbury]. Cf. also ATHELHAMPTON, WATERSTON.

Affpuddle from a Saxon owner named *Æffa*, identical with Ælfriðus, who was owner in 987 (KCD 656).—**Bryants** P~ from Brianus de Turbervill, who is mentioned as holding *Pidele Turbervyll* in 1316 (FA). Cf. BRAMPTON BRYAN.—P~ **Hinton** is 'the Piddle of the *hiwan* or monks' (of Mortain).—**Tolpuddle** was given c 1050 by Tola, widow of Urc, to Abbotsbury (see KCD 841 and Fees, p. 92). *Tola* is a Scand name.—**Piddletrenthide** is 'Piddle of thirty hides' (*trente* is French for thirty). Curiously enough the Piddle is alternatively Trent and is called *Tarente* by Florence of Worcester.— **Turners** P~ was held by Walter Tonitruus in 1084 (GeldR), by Henry Tonere in 1280 (Ch). *Tonere* is Fr *tonnerre* 'thunder' used as a by-name.

Piddle, North, Wo [*Pidelet* DB, *Pydele* 1234 FF], **Wyre** P~ Wo [*Pidele* DB, 1209 Fees]. Both are on Piddle Brook, on which see *supra*.

Wyre is connected in some way with Wyre Forest, which may have extended as far as Wyre Piddle. See WYRE FOREST and WORCESTER.

Pidley Hu [*Pydele* 1228 Ch, *Pudele* 1319 Fine]. '*Pydda*'s LĒAH.' Cf. PIDDINGHOE. The place is on a ridge, and the first el. cannot well be the word *pid-* mentioned under PIDDLE (in *Pidwuella*).

Piel or **Peel Island** La. Named from a peel castle. Cf. FURNESS.

Piercebridge Du [*Persebrigc* c 1050 HSC, *-brige* 1104–8 SD, 1207 FPD]. The ex. in SD is found in an entry for the years 820–5. If the form is taken from an old source, the first el. cannot be the Fr pers. n. *Piers*. It may be the word *persh* 'osier', ME *pershe* 'twig, withe' mentioned under PERSHORE.

Pigdon Nb [*Pikedenn* 1205 Cur, *Pikeden* 1226 P, *Pykeden* 1242 Fees]. Most probably '*Pica*'s DENN or pasture'. OE *pic* 'peak' might be thought of, but it would then be used in the plur. (*pica denn*), and 'the DENN by the peaks' hardly suits the situation.

OE **pil** 'pile, pointed stick, stake' possibly

occurs in some names. But a pers. n. el. *Pil-* occurs in *Pilheard*, and *Pil* or *Pila* would be normal short forms of this. ON *pill*, Sw *pil* 'willow' is a possible el. in Scand parts of England.

Pilham Li [*Pileham* DB, 1202 Ass, *Phileham* 1139 RA]. Probably '*Pila*'s HĀM. *Pilahām* or *-hamm* 'HĀM or HAMM with piles' gives no good sense. Cf. PIL.

Pilkington La [*Pulkinton* 1202, *Pilkenton* 1204 P, *Pilkington* 1246 Ass]. 'The TŪN of **Pileca*'s people.' Cf. PILTON Np.

Pillaton Co [*Pilatona*, *Piletone* DB, *Pileton* 1291 Tax]. '*Pila*'s TŪN.' Cf. PIL.

Pillaton Hall St nr Penkridge [*Pilatehala* a 1113 Burton, *Pilatonhall* 1271 For, *Pilotenhale* 1300 For]. 'HALH or nook where pilled oats grew.' Pilled oats or *pillotes* (1551 OED) is the name of a kind of oats, in which the grain is free from the husk or glumes. The OE form must have been *pil-ātan*. The word is found in the pl. n. *Pilate Croft* a 1186 BM (Denby YW). The earlier name of Pillaton was **Bedington** [*Bedintun* 1002 E, *Beddintone* DB]. 'The TŪN of *Bēda*'s people.'

Pillerton Hersey & Priors Wa [*Pilardetune*, *Pilardintone* DB, little *Pilardentona* c 1125 Fr, *Pilardeston* 1170 P, *Pylardington Hercii* 1316 Ipm]. 'The TŪN of *Pilheard*'s people.' P~ **Hersey** was held by Hugo de Hersy in 1206 (Cur) and came to him t. R 1. Hersi from HERCÉ in France.—P~ **Priors** belonged to the monks of Ware, later to the Prior of Sheen.

Pilley Ha [*Piste(s)lei* DB, *Pyleleye* 1316 FA], P~ YW [*Pillei* DB, *Pillay* 1194 f. P]. The last is OE *pil-lēah*, very likely 'wood where piles were got'. The first may be identical in origin.

Pilling R La [*Pylin* 1246 CC]. Possibly a diminutive of Welsh *pyll* 'pool, creek'. P~ vil. is *Pylin* c 1195, 1201 CC, 1270 Ass, *Pelyn* 1320 CC.

Pilsbury Db [*Pilesberie* DB]. '*Pil*'s BURG.'

Pilsdon Do [*Pilesdone* DB, *Pillesdun* 1168 P, *Pulesdune* 1185 TpR, *Pulesdon* 1200 Cur, *Pyulesdon* 1269 Ch]. Cf. PILSON.

Pilsgate Np [*Pilesget* 963–84, *-geat* 972–92 BCS 1128, 1130, *Pillesgete* DB, *Pilesgate* 1198 P]. '*Pil*'s gate.' Cf. PIL.

Pilsley Db SE. of Chesterfield [*Pilleslege* 1002 Wills, *Pinneslei* DB, *Pileslea* 1170 P, *Pillesleg* 1226 FF]. Perhaps '*Pinnel*'s LĒAH'. *Pinnel* may be found in *Pinnelesfeld* 796 BCS 280 (now **Pinesfield** Hrt).

Pilsley Db nr Edensor [*Pirelaie* DB, *Pilisleg* 1205 Obl]. '*Pil*'s LĒAH.'

Pilson Sa nr Chetwynd [*Plivesdone* DB, *Pivelesdon* 1200 Cur, *Piwelesdon* 1248 Cl, *Piuelisdon* 1288 Court]. Cf. PILSDON Do. The first el. may be a pers. n. *Pēofel*, a diminutive of *Pēof(a)*, found as *Peuf*, *-a* LVD.

Pilton D [*Pilton* 10 BCS 1335, *Pilton* 1121 Fr], **P~** So [*Piltune* 725 BCS 142, *-tone* DB, *Pulton* 1243 Ass]. 'TŪN by a pill or creek.' Cf. PYLL. **P~** So is nr PYLLE.

Pilton Np [*Pilchetone* DB, *Pilkenton* 1189 (1332) Ch, 1254 Val]. '*Pileca*'s TŪN.' Cf. PILKINGTON.

Pilton Ru [*Pilton* 1202 Ass, 1225 Ep, *-a* 13 NpCh, *Piletone* 1227 Ep]. Perhaps 'TŪN by a pill or creek'. If so, the Gwash must be meant.

Pimperne R Do [*Pimpern, -welle* 935 BCS 708]. If the name is British, the first el. may be OW *pimp* 'five'. The second might be Welsh *pren* 'tree', and the name would mean 'five trees'. If so, it would be a back-formation from the name of **P~** vil. [*Pinpre* DB, 1178 P, *Pimperne* 1271 Ipm].

Pinchbeck Li [*Pyncebek* 1051 KCD 795, *Pincebec* DB, 1227 Ch, *Pinchebech* 1183 BM]. The second el. is no doubt, or was originally, OE *bæce, bece* 'stream', as in HOLBEACH, WATERBEACH, WISBECH; *-beck* seems due to influence from OScand *bekkr*. It is possible that the present name Pinch-beck is really due to a kind of metathesis of *Pinkbeach. If so, the first el. may be either OE *pinca* 'a finch' (cf. FINCHALE) or possibly *pink* 'a minnow' (1490 &c.). But the latter is *penk* in early sources and perhaps not to be thought of, though 'minnow stream' is a very probable meaning. Possibly there was an OE *pinč* by the side of *pinca*.

Pinchinthorpe YN [*Torp* DB, *Pinzunthorp* c 1200 YCh 753, *Pynchunthorp* 1336 FF]. **P~** was held c 1200 by Willelmus Pinzun. The family name is an OFr nickname, taken from OFr *pinçon, pinchon* 'pincers, forceps'. The old name was simply THORP.

Pinden K [*Pinindene* 10 BCS 1321 f., *Pinnedene* 11 DM, DB]. Either 'the DENN of *Pinna*'s people' (cf. PILSLEY) or 'DENN with an enclosure', the first el. being an OE *pinning* formed from *pinnian* 'to confine' (possibly found in *Pinninge* 1035 BM).

Pinesfield. See PILSLEY.

Pinge or **Punge Wood** Brk [*Punge* Mon iv. 36] Etymology obscure. Possibly *Punningstoce* 811, 815 BCS 352, 850 belongs here.

Pinhoe D [(æt) *Peonhó* 1001 ASC, *Pinho* 1238 Ass; *Pynnoc* c 1100 E, *Pinnoch* DB]. A hybrid name, consisting of Brit *pen* 'hill' and OE *hōh* 'spur of hill'. The place is by a short ridge. The side-form *Pynnoc* is curious; it might represent a diminutive form identical with PINNOCK Gl.

Pinley Wa [*Pinneleya* Hy 2 (1229) Ch, *-lei* R 1 BM, *-leye* 1326 Ch], **Pinner** Mx [*Pinora* 1232 FF, *Pinnora* 1232 Ch]. The first el. is very likely OE *pinn* 'peg, pin'. Both places are near narrow ridges, and *pinn* may be used to designate them. See LĒAH, ŌRA. The latter means 'slope'.

Pinnock Gl [*Pignoscire* DB, *Pinnokesscr'* 1194 P, *Pinnok* 1248 Cl]. Apparently the old name was *Pinnok*, though *-shire* is generally added in early sources. The meaning of *scir* is not apparent. *Pinnok* may be explained as a diminutive *pennuc* or *pennoc* formed from Brit *pen* 'hill' either in British or in English. Cf. PINHOE.

Pinvin Wo [*Pendefen* 1187, 1190 P, 1275 Subs]. '*Penda*'s fen.'

Pinxton Db [*Penkeston* 1208 Cur, *Penekeston* 1236 Fees, *Penkiston* 1244 Ipm]. Possibly an old name identical with PENKETH, to which was added OE *tūn*. The base would be OE *Pencētes-tūn*.

Pipe He [*Pipe* DB, *la Pipe* 1272 Hereford]. OE *pīpe* 'water-pipe, water-course', really referring to the brook here. **Pipe** St nr Lichfield [*Magna, Parua Pipa* 1167 P] is often mentioned in early sources. The name is preserved in **Pipehill**. Pipe is identical in origin with Pipe He.

Pipewell Np [*Pipewelle* DB, *-well* 1157 ff. P, 1197 FF, *Pippewell* 12 NS]. First el. probably OE *pīpe* as in PIPE, the reference being to a small stream.

Pirbright Sr [*Perifrith* 1166 RBE, *Perifrið* 1173 P, *Pyrifright* c 1270 Ep, *Pirbrigth* 1316 FA]. OE *pirig-fyrhþ* 'pear-tree wood'. Cf. PIRIGE, FYRHþ. The change of *f* to *b* is anomalous.

OE **pirige, pyrige** 'pear-tree' is found alone and as a first and a second el. in pl. ns. See PERRY, BUTTSBURY, HARTPURY, PAULERS-, POTTERSPURY, PERIVALE, PIRTON, PURITON, PURTON, PYRTON, PIRBRIGHT, PURLEY, PYR-FORD.

Pirton Hrt [*Peritone* DB, *-ton* 1197 FF, *Piriton* 1283 BM], **P~** Wo [*Pyritun* 972 BCS 1282, *Peritune* DB]. 'TŪN where pear-trees grew.'

Pishill O [*Pushulle* 1219 f., *-hull* 1247 Fees], **Pishiobury** Hrt [*Pyssoubur* 1294 Cl, *Pissho* 1310 Ipm]. 'Hill and HŌH where peas grew.' Cf. PISU. The meaning of *-bury* is 'manor'.

OE **pisu, piosu, peosu** 'pea' (< Lat *pisum*) is found in several pl. ns., as PEASEMORE, PEASENHALL, PISHILL, PISHIOBURY, PUSEY. The form *Pus-* comes from *piosu*. Probably *pisu* in some cases refers to a wild plant; cf. PEASEMORE.

Pitchcombe Gl [*Pichenecumbe* 1211–13 Fees, 1221 Ass, 1230 Ch, *Pichenescumbe* 1226–8 Fees]. 'Valley where pitch was obtained.' First el. OE *picen* 'of pitch'. The *s* of one early form is probably intrusive.

Pitchcott Bk [*Pichecote* 1176 P, *-cot* 1220 Fees]. 'COT where pitch was made.' Pitch is OE *pic*. *Pichecot* for *Pichcot* owing to the difficulty of pronouncing the group of consonants.

Pitchford Sa [*Piceforde* DB, *Picheford* 1194,

1196 P, *Pichford* 1176 P, *Picford* 1242 Fees].
First el. OE *pic* 'pitch', here used of mineral
pitch. There is still a bituminous well in
existence or was in Eyton's time.

Pitcombe So [*Pidecombe* DB, 12 Bruton,
Pidecumba c 1155 Fr]. See CUMB. First
el. the *pid-* found in *Pidwuella*. See
PIDDLE.

Pitminster So [(to) *Pipingmynstre* 938, *æt
Pippingmynstre* 941 BCS 729, 770, *Pipe-
minstre* DB, *Pupmunstre* 1330 Ep, *Put-
mynstre* 1327 Subs]. 'The MYNSTER or
church of *Pippa*'s or *Pyppa*'s people.' For
Pyppa cf. PEOPLETON. The change *p > t*
before *m* is remarkable.

Pitney So [*Petenie* DB, *Puttenaya* 1225 Ass,
Petteney 1230 Ch]. '*Pytta*'s island.' *Pytta
is a side-form of *Putta*. '*Pēota*'s island' is
a possible alternative.

Pitsea Ess [*Piceseia* DB, *Pichesheye* 1198
(1252) Ch, *-eye* 1238 Subs, *Petceseye, Pet-
cheseye* 1285 QW]. Perhaps '*Pīce*'s island'.
*Pīce would be a derivative of *Pic*.

Pitsford Np [*Pides-, Pitesford* DB, *Pictes-
ford* 1236 Fees, *Pithisford* 1270 Ipm].
'*Peoht*'s ford.' *Peoht is a short form of
names in *Peoht-*.

Pitstone Bk [*Pinceles-, Pincenestorne* DB,
Pichelesporne 1220 Fees, *Pikelesthorn* 1248
Cl]. '*Pīcel*'s thorn-bush.'

Pitt Ha [*Pette* 1286 Ch, *la Putte* 1316 FA].
OE *pytt* 'pit, hollow'.

Pittington Du [*duo Pittindunas* c 1085 LVD,
Pitinduna Hy 2 FPD, *Pitingdun* c 1190
LaCh]. Presumably 'the hill of *Pytta*'s
people'. Cf. PITNEY.

Pittleworth Ha [*Puteleorde* DB, *Puttteles-
word* 1212, *Putlesworth* 1242 Fees]. '*Pyttel*'s
WORÞ.'

Pitton W [*Putenton* 1167 P, *Petton* 1194 P,
Putton 1198 Fees]. The earliest form suggests '*Pytta*'s TŪN'. Cf. PITNEY.

Pixham. See PICKERING.

Pixley He [*Picheslei* DB, *Pictele* 1206 Cur,
Pikesl' 1242 Fees]. '*Peoht*'s LĒAH.' Cf.
PITSFORD.

OE **plæsc**, Mod *plash* 'a shallow piece of
standing water, a marshy pool, a puddle',
corresponding to MDu *plasch* 'pool', is
found in PLAISH, PLASH, MELPLASH. Cf. also
PLUSH.

Plaish Sa nr Cardington [*Plesha* DB, *Plassh*
1327 Subs]. See PLÆSC. Another place of
the same name at Aston nr Lilleshall may be
referred to by *Plesc, Plæsc* 963 BCS 1119.

Plaistow Db nr Crich [*Plaustowe, Plage-
stoue* c 1200 Darley], P~ (-ah-) Ess [*Play-
stowe* 1414 Pat]. OE *plegstōw* 'play-
ground'. The same is the origin of **Plaistow** D, K, Sx. In Plaistow Db the first el.
is the OE *plaga*, ME *plawe*, that occurs as
a side-form of *plega* 'play'.

Plaitford Ha [*Pleiteford* DB, *Pleitesford* 1234
AD]. 'Play ford.' First el. probably an
OE *pleget 'playing' derived from *plegian*.
Cf. OE *bærnet* 'burning' from *bærnan* 'to
burn'.

Plash So nr Elworthy [*Plesse* c 1245 Dunster,
Plasshe 1428 FA]. See PLÆSC.

Platt Bridge La [*Platte* c 1225 CC]. Dial.
plat 'a foot-bridge'.

Plawsworth Du [*Plauworth* 1297 Pp,
Plawesworth 1345 RPD]. The first el. is
the form of OE *plega* 'play' discussed under
PLAISTOW, but it is difficult to say if it is
simply this word, the name meaning 'enclosure for sports', or a pers. n., a short
form of names in *Pleg-*, as *Plegmund*. The
gen. form suggests the latter.

Plaxtol K [*Plextole* 1386 Cl]. Identical with
PLAISTOW. The *-l* is excrescent.

Playden Sx [*Pleidena* DB, *-dene* 1107 Fr,
Pleindenne 1225 Penshurst]. If *den* is DENN
'pasture', the first el. is an OE *Plega rather
than *plega* 'play'. But *denn* may mean 'a
hollow'. If so, the name may mean 'DENN
where deer played'. Cf. PLEGA.

Playford Sf [*Playford* 11 KCD 978, *Plege-
forda* DB, *Pleiforda* 1130 P]. OE *pleg-ford*
'ford where sports were held'.

Plealey Sa [*Pleyleye* 1308 Ipm, 1327 Subs].
'Play glade.' Here very likely of a place
where deer played. Cf. PLEGA, LĒAH.

Pleasington (-ĕz-) La [*Plesigtuna* 1196 YCh
1524, *Plesington* 1267 Ass]. 'The TŪN of
Plēsa's people.'

Pleasley (-ĕz-) Db [*Pleseleia* 1166 RBE,
1208 Cur, *-leg* 1221-30 Fees]. '*Plēsa*'s LĒAH.'
P~ **Hill** Nt [*Hulle juxta Pleseleye* 1285
PNNt(S)]. Originally *Hill*. The two places
are opposite to each other on different banks
of the Meden.

Pledgdon Ess [*Plicedana* DB, *Plycheden*
1238 Subs, *Plicheden* 1251 Ch, *Plechenden*
1272 FF]. '*Plycca*'s valley.' *Plycca is
a side-form of *Plucca*.

OE **plega, plæga, plaga** 'play, sport, game'
is found in PLAISTOW, PLAXTOL, PLAYFORD,
PLEALEY, PLOWDEN, perhaps PLAWSWORTH,
PLAYDEN. Cf. PLAITFORD. The meaning
may sometimes be 'place where deer play',
as is certainly the case in **Deerplay** La [*Der-
plaghe* 1296 Lacy].

Plemstall or **Plemonstall** Chs [*Pleymunde-
stowe* 1291 Tax, 1297 Chester]. '*Plegmund*'s
STŌW or hermitage.' Plegmund, archbishop
of Canterbury 890-914, is said to have lived
as a hermit at Plemstall in Cheshire.

Plenmeller Nb [*Plenmenewre, Playsmale-
vere* 1256 Ass, *Playnmelor* 1279 Ass]. A
Welsh *blaen Moelfre* 'the top of *Moelfre* or
the bare hill'. Cf. MELLOR, BLENCARN. P~
Moor is a mountain. The substitution of
Engl *P-* for Welsh *B-* is not without
analogies.

Pleshey Ess [*Plaisseiz* (castle) 1143 HHunt, *Plessetum* 1228 Ch], **Plessey** Nb [*Pleisiz*, *Pleisetum* 1203 Cur, *Plesset* 1257 Ch]. OFr *plaisseis* or *plaissiet* 'an enclosure, park or forest, formed by a plashed hedge, i.e. one with bent and interwoven branches'. From the first is formed the common Fr pl. n. PLESSIS, from the other Fr PLESSÉ. The names Pleshey and Plessey may have been transferred from France.

Plompton. See PLUMPTON.

Plowden Sa [*Plaueden, Pleweden* 1252 Eyton, *Plowedene* 1286 Ep]. 'Valley where sports were held, or where deer played.' Cf. PLEGA.

Pluckley K [*Pluchelei* DB, *Plucelea* 11 DM, *Plukele* 1207 Cur]. '*Plucca*'s LĒAH.'

Plumbland Cu [*Plumlund* 12 StB, 1229 Pat]. 'Plum-tree grove.' See LUND.

OE **plūme** (Lat *prunus*) meant both 'plum' and 'plum-tree'. It is found in many pl. ns., no doubt in the sense 'plum-tree'. See PLUMBLAND, PLUMLEY, PLUMPTON, &c., PLUNGAR. See also BROOMHILL. Another OE word for plum-tree was *plȳme*. See PLYMPTON, PLYMSTOCK, PLYMTREE.

Plumley Chs [*Plumleia* 1119 Chester]. 'Plum-tree LĒAH or wood.'

Plumpton Wall Cu [*Haia de Plumton* 1212 RBE, *Plumton* 1247 Ipm], **P~** or **Field-plumpton** La [*Pluntun* DB, *Fildeplumpton* 1323 LaInq], **Woodplumpton** La [*Pluntun* DB, *Wodeplumpton* 1327 Subs], **P~** Np [*Pluntune* DB], **P~ End** Np [*Plumpton* 1220 Fees], **P~** Sx [*Pluntune* DB], **P~** or **Plompton** YW [*Plontone* DB, *Plumton* 1190 P]. 'TŪN where plum-trees grew.'

Plumstead K [*Plumstede* 961–9 BCS 1173, *Plum(e)stede* DB], **P~** Nf nr Aylsham [*Plumestede* DB, *Plumstede* 1254 Val], **Great & Little P~** Nf [*Plumestede* DB, *Parva Plumbsted, Grimene Plumsted* 1254 Val, *Parva, Grimere Plumstede* 1302 FA]. 'Place where plum-trees grew.'

The addition *Grimere* for Great P~ seems to be the OScand pers. n. *Grimarr*.

Plumtre Nt nr Nottingham [*Pluntre* DB, *Plumtr[e]* 1206 Cl], **P~** Nt nr Bawtry [*Plumptre* 1265 Misc, *Plumtre by Bautre* 1300 Pat]. OE *plūm-trēow* 'plum-tree'.

Plungar Le [*Plungar* c 1125 LeS, *Plumgar* 1242 Fees, *Plungard* 1186 P, 1236 Fees, *Plumgarth* 1291 Tax]. OE *plūm-gāra* 'piece of land where plum-trees grew', with occasional substitution of *garth* from OScand *garðr* 'enclosure' for *gāra*.

Plush Do [*Plyssche, Plisshe* 891 BCS 564, *ad Plussh'* 941 ib. 768, *Plys* 12 Montacute, *Plys, aqua de Plys* 1268 Ass]. Apparently an OE **plysc*, cognate with PLÆSC and of a similar meaning.

Plym R D [*Plyme* 1238, *Plime* 1244 Ass]. An early back-formation from PLYMPTON.

From the river was named **Plymouth** [*Plimmue* 1231, *Plummuth* 1235 Cl] 'the mouth of the Plym'. The old name of Plymouth was SUTTON [*Sutona* DB]. **Plympton** [*Plymentun* 904 BCS 610, *Plimtun* c 1135 E, *Plintone* DB] is OE *Plȳm(an)tūn* 'plum-tree TŪN'. **Plymstock** D [*Plemestocha* DB, *Plumstok* 1228 FF] is 'STOC where plum-trees grew' or elliptical for *Plympton stoc*.

Plymtree D [*Plumtrei* DB, *Plimtree* 1199 Cur]. OE **plȳm-trēow* 'plum-tree'.

Pockley YN [*Pochelac* DB, *Pokelai* c 1190 Riev, *Pockele* 1232 FF], **Pocklington** YE [*Poclinton* 1100–8 YCh 426, *Pochelinton* 1169, 1190 P], **Pockthorpe** YE [*Pochetorp* DB, *Poketorp* 1195 P, *-thorp* 1227 FF]. The last is clearly '*Poca*'s thorp', or possibly '*Pohha*'s thorp'. Pocklington is 'the TŪN of *Pocel*'s (*Pohhel*'s) people'. Pockley seems to be '*Poca*'s (*Pohha*'s) LĒAH'. *Pohha* is evidenced, but not *Poca, Pocel, Pohhel*. If there was an OE *poc(c)e* 'frog' (cf. POLEBROOK) it would suit Pockley.

Podimore or **Podimore Milton** So [*Middeltone* 966 BCS 1188, *Mideltone* DB]. 'Middle TŪN.' Podimore must have denoted some locality nr Milton. It appears to be identical with **Podmore** St NW. of Eccleshall [*Podemore* DB, *Podemor* 1288 Ass], which probably means 'frog moor'. ME *pode* 'frog' is found c 1250.

Podington Bd [*Podintone* DB, *Pudinton* 1163 P, *Puddington* 1231 Ep]. 'The TŪN of *Puda*'s people.'

Pointon Li [*Pochinton, Podintone* DB, *Poinnton* 1165 P, 1198 FF, *-tun* 1212 Fees]. 'The TŪN of *Pohha*'s people.'

OE **pōl** 'pool, deep place in a river', also 'tidal stream' (cf. LG *pôl*, G *Pfuhl*) is a common pl. n. element. See e.g. POOL, POOLE, POOLHAM, POULTON, HAMPOLE, LIVERPOOL. The meaning varies between 'pool' and 'stream'. The OE side-form *pull*, which often interchanges with *pōl* also in later forms, is Welsh *pwll* (= Bret *poull*, Co *pol*), which had the same meanings as OE *pōl*. A form *poll*, which occurs in ME stream-names, as in WAMPOOL (q.v.), and means 'stream', is Welsh *poll*, Co *pol*. It is the source of *pow* 'a slow-moving stream', found in Scotland. The stream-names **Pow**, **Pow Beck**, **Powmaughan** Cu contain this word. Pow is *Pol* c 1170 StB. Powmaughan [*Polmergham* 1486 Ipm] means '*Merchiaun*'s stream' (cf. MAUGHONBY).

Polden Hill So [*Pouldon* 1241 Cl, *Poweldun* 1235–52 Glaston]. The second el. is OE *dūn* 'hill'. The first is a lost pl. n. [*Pouelt* 705, *Poelt, Poholt* 725, *Pouholt* 729 BCS 113, 142 f., 147]. The correct form is *Po-* or *Pouholt*, the second el. being OE *holt* 'wood'. The place comprised 20 hides and must have been near Polden Hill. The el. *Po(u)-* is obscure. Welsh *pau* (OW *pou*) 'country' from Lat *pagus* has been suggested. The would-be OE forms are in late transcripts, and it is not absolutely impossible that *Po(u)holt* may be from *Pōlholt* 'wood by the pool' with dissimilatory loss of the first *l*. Or POUGHILL, POUGHLEY may be compared.

Polders, Poldhurst. See POWDERHAM.

Polebrook Np [*Pochebroc* DB, 1166 P, *Pokebroc* 12 NS, 1200 Cur, 1227 Ch, *Pockebroc* 1203 Ass]. The first el. may possibly be an OE **poc(c)e* 'frog', related to MLG, MDu *pogge* the same. 'Frog brook' yields good sense. Cf. POCKLEY.

Polesden Sr [*Polesdene* 1198 FF, *Palesden* 1202, *-denn* 1204 Cur]. OE *Pāles-denu*, the first el. being OE *pāl* 'pole' or a pers. n. **Pāl* derived from it.

Polesworth Wa [*Polleswyrð* c 1000 Saints, *Poleswurth* 1200 Cur]. *'Poll's* WORÞ' or 'WORÞ by the hill'. ME *poll* 'head' (c 1290 &c.) presupposes OE *poll*. This might have been used in the sense 'hill' too. There is a hill nr Polesworth. But *poll* would easily give rise to a nickname, and the gen. form points to a pers. n. as first el.

Poling Sx [*Palinge* Hy 2 (1361) Pat, *-s* 1257 Sele, 1299 Ch, *Polyng* 1306 Ipm]. OE *Pālingas*, a tribal name, as suggested also by *Palinga Schittas* 953 BCS 898 (a lost place in Petworth Sx) 'the sheds of the Poling people'. *Pālingas* might possibly be 'people by a pole' (OE *pāl*), but more likely it is 'the people of *Pāl'*. Cf. POLESDEN.

Pollicott Bk [*Policote* DB, *-cota* 1130 P, *-cote* c 1155 Oxf, *Pulicote* 1241 Ass]. If *Poli-* goes back to *Pōling(a)-*, as seems probable, the name may mean 'COT of the people from *Pōl'*. The latter must then be a lost pl. n.

Pollington YW [*Polingtonia, Pouilgleton* 1160, *Pouelington* c 1185, *-tona* c 1200 YCh 484 f., 495 f., *Poulinton* 1197 P]. See POOL YW.

Polpe·rro Co [*Portpira* 1303 Pat, *Porpira* 1379 AD iii]. The first el. is Co *porth* 'port'. The change to *Pol-* is due to dissimilation. The second el. may be a streamname.

Polruan Co [*Porthruan* 1284 Ass (Gover), *Polruan* 1292 Ch, *Polruwan* 1335 FF]. The first el. seems to have been originally *porth* 'harbour'. The second is obscure. It is not identical with RUAN.

Polscoe Co [*Polscat* DB, *-scad* 1198 P, *-scoth* 1359 FF]. 'Boat pool.' Co *scath* means 'boat'.

Polsham So [*Paulesham* 1065, *Pauleshamesmede* 1361 Wells]. '*Paul's* HAMM.'

Polsloe D [*Poleslevge* DB, *Poleslawa* 1178 P, *Polslo* 1230 P]. OE *pōl-slōh* 'marsh by a pool'.

Polstead Sf [*Polstede, Polstyde, Polestede* c 995 BCS 1288 f., *Polesteda* DB]. 'Place by a pool.'

Poltimore D [*Ponti-, Pultimore* DB, *Pultimor* 1242 Fees]. The first el. may be compared with that of PELDON and POULTNEY. Second el. OE *mōr* 'moor'.

Ponsonby (-ŭ-) Cu [*Puncuneby, Puncunesbi* 12 StB]. '*Puncun's* BY.' John son of Puncun, mentioned 1177 P (Cu), was owner of P~. *Puncun* is an OFr nickname and pers. n. identical with OFr *poinçon*, Engl *puncheon* 'awl, punch' &c.

Pont R Nb [*Ponte* 1269 Ass, *Pont* 1479 Hexh]. Identical with PANT. Cf. PONTOP.

Pontefract YW [*Fracti-pontis* (gen.) 1069 (1141) Ordericus, *Pontefracto* (dat.) 1100–2 YCh 1418, *Pontfreit* 1177 P]. A Latin and French name meaning 'broken bridge'. The pronunciation is (pŏntĭfrăkt), locally (pŭmfrĭt).

Pontë·land Nb [*Eland* 1242 Fees, *Punteland* 1203 Cur]. OE *ēgland* or *ēaland* 'island, land on a river'. *Pont-* is the river-name PONT.

Pontesbury Sa [*Pantesberie*, *-bury* 1203 Ass, *Pontesbiri* 1236, *Pantebur'* 1242 Fees], **Pontesford** Sa [*Pontesford* 1308 Ipm, 1327 Subs]. Either 'BURG and ford on (over) R **Pant*' (cf. PANT) or '*Pant's* BURG and ford'. The gen. form in *-es* is in favour of the second alternative. *Pant* pers. n. is on record; cf. also PENTLOW.

Ponton, Great & Little, Li [*Pamptune*, *Pamtone, Great Panptune, Magna, Parva Pantone* DB, *Pampton* 1245 Ep, *Ponton* 1227 Ch]. Identical with PANTON. Ponton Heath is an elevated district.

Pontop Du [*Pontehope* c 1245 FPD]. 'The HOP or valley of Pont Burn.' **Pont Burn** [*Pont* 1153–9 Newminster]. Cf. PONT.

Pool YW [(on) *Pofle* c 1030 YCh 7, *Povele* DB, *Pouela* 1166 f. P, *Poule* 1191 P]. Clearly not OE *pōl* 'pool'. The place is on the Wharfe. The OE form was *Pofl* or *Pofel*. From such a base is derived the first el. of POLLINGTON, which denotes a place in a low-lying situation, but not on a stream. The meaning of the word **pofl* is unknown. One might compare dial. *poffle* 'a small piece of land', found in **Maxpoffle** Roxb [*Max-poffil* 1317 OED] and *Prestpofill*, *-pofle*, *-pofyll* 1479 BBH, but it is doubtful if there can be a connexion.

Pool, South, D [*Pole* DB, *Suthpole* 1284–6 FA], **Poole** Chs [*Pol* DB], **P~** Do [*Pole* 1194 P, *La Pole* (port) 1235 Cl], **P~ Keynes** Gl [*Pole* 931 BCS 673, 1241 Cl]. 'The pool.' On *Keynes*, see ASHTON KEYNES.

Pooley Wa [*Povele* 13 Mon, *Poueleye* 1285 BM, *Powelee, Pouelee* 1259 Ipm, *Poleye* 1285 Ipm]. First el. perhaps identical with POOL YW.

Pooley Bridge We [*Pulhoue* 1252 FF,

Pulhou 1291 lpm]. 'Hill or mound by the pool.' Second el. OScand *haugr*. The place is at the lower end of Ullswater.

Poolham Li in Woodhall [*Polum* Hy 2 DC, 1212 Fees]. The dat. plur. of OE *pōl* 'pool'.

Poolhampton Ha [*æt Polhæmatunæ* 940, *æt Polehametune* 956 BCS 763, 974, *Polemetune* DB]. 'The TŪN of the dwellers by the pool.' See HĀMTŪN.

Poorton, North & South, Do [*Pover-, Pourtone* DB, *Poertona* 1168 P, *Pourton* 1212 Fees, *Supereporthon* 1229 BM], **Powerstock** Do [*Pourestoca, Povrestoch* DB, *Pourstoke* 1195 P, *-stok* 1205 Cl]. A shorter form *Power* is found in the 12th cent. MS Cott Faust A ii in a grant by Egbert. It is doubtful whether Poorton or Powerstock is meant. The two places are on the two arms that join to form the Mangerton river. *Power* might possibly be an old name of the river, but its etymology is obscure. Powerstock may be 'STOC on R Power' or 'STOC belonging to the old village of Power'.

Popham Ha [*Popham* 903 BCS 602, *Popeham* DB, *Poppham* 1212 Fees]. Doubtful. The name seems to contain an OE word *pop(p)* of unknown meaning, possibly a shorter form of *popel* 'pebble'.

Poplar Mx [*Popeler* 1350 FF, *Le Popeler* 1412 FA]. 'The poplar.' Poplar is a Fr word.

Poppleton, Nether & Upper, YW [*Popeltun* c 972 BCS 1278, (in) *duabus Popletunis, Popletone, -tune* DB, *Popelton* 1190 P]. First el. OE *popel* 'pebble' (in *popelstān*).

Porchester Ha [*Porceastra* 904, *Porteceaster* c 960 BCS 613, 1157, *Portcestre* DB]. 'The Roman fort by the port or harbour.' No doubt *Port* was once the name of Portsmouth Harbour.

Poringland, East & West, Nf [*Porringa-, Porringhelanda* DB, *Porringeland* c 1095 Bury, *Poringlond Maior, Minor* 1254 Val]. Unexplained. Derivation of the first el. from OE *porr* 'leek' (< Lat *porrus*) does not seem probable.

Porkington Sa [*Porchinton* 1161 ff. P, *Porkintun* 1236 (1295) Ch]. Cf. **Portley** Sr [*Porkele* 1225 ff. PNSr], which probably contains an OE *Porca*, a nickname related to G *pfurch* 'small person', ON *purka* 'sow'. Porkington is then 'the TŪN of *Porca*'s people'.

Porlock So [*Portloca* 918 ASC, *Portloc* DB]. 'Enclosure by the harbour.' See PORT and LOCA. P~ is on Porlock Bay.

OE **port** 'harbour, town' (from Lat *portus*) and **port** 'gate' (from Lat *porta*) are found in pl. ns. *Port* 'harbour' is found e.g. in PORLOCK, PORTSMOUTH (&c.), PORTISHEAD, PORTLAND. *Port* 'town', esp. 'market town', is probably as a rule the meaning in names

such as ALPORT, BRIDPORT, NEWPORT, WEST-PORT. Cf. also LAMPORT. *Port* 'gate' is certain in PORTGATE.

Portbury So [*Porberie* DB, *Portberi* 1159 P, *-buri* 1196 FF], **Portishead** So [*Portesheve* DB, *-heved* 1200 Cur, 1225 Ass]. Portishead is by a long ridge along the Severn estuary, while Portbury is a little way inland. The names mean 'ridge and BURG by the harbour'.

Portgate Nb [*Portyate* 1269 Ass]. P~ is at a gap in the Roman Wall where Watling Street runs through it. The old name was *Port* 'the gate', to which was added an explanatory OE *geat* 'gate'.

Porthallow Co [*pord alaw, perd alau* 967 BCS 1197, *Porthaleu* 1333 AD]. 'Port on R Allow.' First el. Co *porth* 'harbour'. The river-name is identical with ALAW R in Anglesey and related to Welsh *alaw* 'music'.

Portington YE [*Portiton* DB, *Portinton* DB, 1234 FF, *Portington* 1285 FA]. 'The TŪN of the towns-people' (an OE *Portingas*). York might be referred to.

Portinscale Cu [*Portquenescales* n.d. CWNS xxi]. See SKÁLI. First el. OE *portcwēn* 'prostitute'.

Portisham Do [*æt Porteshamme* 1024 KCD 741, *Portesham* DB]. 'HAMM belonging to the port or town.' The town may be Abbotsbury.

Portishead So. See PORTBURY.

Portland Do [*Port* 837 ASC, *On Portlande* 872 BCS 535, *Porland insula* DB]. The old name was *Port* 'the harbour', referring to Portland Harbour. Later *land* was added.

Portlemouth D [*Porlamuta* DB, *-mue* 1219 FF, *Portlemue* 1262 FF]. Probably OE *Portwellan-mūþa* 'the mouth of *Portwella* or the harbour stream'. E. P~ is at the mouth of the river that runs past Kingsbridge. W. P~ is west of the river a little farther inland.

Portley. See PORKINGTON.

Porton W [*Poertone, Portone* DB, *Pourton* 1161, 1194 P, *Powertone* 1212 RBE]. Apparently identical with POORTON Do. If the first el. is a river-name, it must have denoted the Bourne.

Portsdown Ha [*Portesdon* DB, 1161 P], **Portsea Island** Ha [*Porteseia* c 1125 Oxf, 1167 P, *-ia* 1168 P], **Portsmouth** Ha [*Portesmuþa* 501 ASC, *-muða* 1101 ib. (E), *-muda* 1123 AC, 1194 P]. Portsmouth Harbour was no doubt once *Port* 'the harbour'. Portsmouth is at the entrance to the harbour and means 'the mouth of Port harbour'. Portsea Island bounds the Harbour on the east: 'Harbour Island'. Portsdown is a ridge north of Portsmouth Harbour.

Portskewett Monm [*Portascihð* 1065 ASC (C, D), *Poteschiuet* DB, *Porthskywet* 1131

Mon, *Porteskywet* 1291 Tax, *Porth Isceuin* c 1150 LL, *portus Eskewin* 1191 Gir, *Porth Ysgewydd* Myvyrian Arch. of Wales 749]. A Welsh name, whose first el. is Welsh *porth* 'harbour', the second being possibly derived from Welsh *ysgaw* 'elder wood'; cf. BOSCAWEN, CHALK BECK. Pughe gives Welsh *ysgewin* adj. 'being of elder wood'.

Portslade Sx [*Porteslage* DB, *Portes Ladda* 1080–1108 Fr, *Porteslad'* 1179–86 AC]. Apparently OE *Portes-lād* 'the stream by the port or harbour'. The reference may be to the creek south of the place.

Portsmouth. See PORTSDOWN.

Portswood Ha [(on) *Portes wuda* 1045 KCD 776, *Porteswuda* 1167, -*wude* 1197 P]. 'The wood belonging to the town' (i.e. Southampton, in which P~ is).

Posenhall Sa [*Posenhall* 1226, *Pesenhale* 1256 PNSa]. 'HALH where peas grew.' The first el. is OE *pisu, piosu* 'pea' or an adj. derived from it.

Poslingford Sf [*Poslingeorda* DB, -*uuorde* c 1095 Bury, *Poselingwrtha* 1195 FF], **Possingworth** Sx in Waldron [*Posingeworde* 12 PNSx, *Poselingewurth* 1238 Cl]. 'The WORþ of the *Poslingas*'; cf. POSTLING.

Postern Db [*Posterne* 1300 QW]. OE *post-ærn* 'house made of posts or timber'.

Postling K [*Postingas* DB, *Postlinges* 1212 RBE]. OE *Poslingas* 'Possel's people'; cf. POSLINGFORD. *Possel* is not directly evidenced, but *Poss* is found in *Posses hlæw* 940 BCS 756, *Possa* in **Poston** Sa [*Possetorne* DB, -*thorn* 1194 P], POSTWICK.

Postlip Gl [*Poteslepe* DB, 1221 Ass, 1236 Fees, *Pottesleppe* 1220 Fees]. The second el. appears to be OE SLÆP (q.v.). First el. possibly OE *pott* 'pool'.

Poston He [*Poscetenetune* DB, *Postone* 1100 Glouc, *Puttestun* 1242 Fees]. The first el. is a tribal name in -*sætan* 'dwellers', which was formed from a name of the place where the tribe lived, very likely a name of the ridge on which Poston is, e.g. an OE *Puttandūn*; cf. PUTFORD. The OE form seems to have been something like *Putsætnatūn*.

Poston Sa. See POSTLING.

Postwick (pŏsĭk) Nf [*Possuic* DB, *Posswyc, Poswyk* c 1147 Holme, *Possewik* 1175–86 Holme, 1254 Val]. '*Poss(a)*'s wīc.' Cf. POSTLING.

Potcote Np [*Potcote* 1202 Ass, -*cot* 1220 Fees], **Potlock** Db nr Twyford [*Potlac* DB, 1176 P, -*lok* 1304 FF]. The first el. is OE *pott* 'pot', very likely in a transferred sense such as 'hole, pit'. Potlock is 'stream in a hollow' or 'stream with deep pools'. See COT, LACU.

Potsgrove Bd [*Potesgrave* DB, 1212 Cur, *Putesgraue* 1200 P, *Pottessgrave* 1247 Ass, *Portesgrave* 1242 Fees, 1428 FA]. Perhaps

OE *pottes-grāf* 'grove by a pool' (cf. POTT). There is a small lake nr P~. But some forms may suggest OE *potteres-grāf* 'the potter's grove'. The first *r* would easily be lost (dissimilation).

Pott Hall YN [*Pott* 12 PNNR, *Pot* 1301 Subs]. OE *pott* 'pot', here in the sense 'a pool' or 'tarn'. There are two tarns here. **Pott Beck** is *Pozebec* c 1200 Fount (*Poze-* from *Pott-sǣ* 'Pott lake').

Potterne W [*Poterne* DB, 1195 P, *Poterna* 1165 P, 1195 FF]. 'House where pots were made, pottery', or 'potters' house', i.e. OE *pottera ærn*.

Potterspury Np [*Perie* DB, *Espirie* 1229 Cl, *Potterespyrie* 1315 Ipm]. Originally *Pirige* 'the pear-tree'. *Potters-* for distinction from PAULERSPURY. There were potteries here.

Potterton YW [*Potertun* DB, -*ton* 1195 P]. OE *pottera tūn* 'the potters' TŪN'.

Potto YN [*Potho* 1202, -*howe* 1208, -*hou* 1218 FF], **Potton** Bd [*Pottune* c 960, *Pottun* c 1000 BCS 1062, 1306, *Potone* DB]. The first el. is OE *pott* 'pot'. Potto, whose second el. is OScand *haugr* 'mound', may mean 'mound where pots had been found', Potton 'TŪN where pots were made', but *pott* may mean 'hollow' in both.

Poughill (-ŭf-) Co [*Pochehelle* DB, *Pocwell* 1247 FF, *Pohewille* 1269 Ep, *Poghewille* 1314 FF], **P~** (powel) D [*Pochehille* DB, *Pohgehille* 1198, *Pokehill* 1219 FF, *Poghhill* 1242 Fees], **Poughley** Brk [*Pohanlech* (-*læh*) 821 BCS 366, *Poghele* 1232 (1329) Ch, 1291 Tax]. Perhaps '*Pohha*'s spring, hill and LĒAH'. OE *pohha* 'pouch' cannot be considered, but there may have been some transferred use of the word that might occur in pl. ns.

Poulshot (-ō-) W [*Paulesholt* 1186 P, *Paulisholt* 1242 Fees]. 'Paul's wood.'

Poulter R Nt [*Paltr'* Hy 6 Newstead Cart]. Perhaps a back-formation from PALTERTON. An old name may have been CLUN.

Poultney Le [*Pontenei* DB, *Pulteney* 1209–35, -*eia* 1228 Ep, -*eye* 1258 FF]. Apparently '*Pulta*'s island'. Cf. PELDON Ess, POLTIMORE D. A pers. n. *Pulta* is unrecorded.

Poulton Chs nr Pulford [*Pontone* DB, *Puntona* c 1150 BM], **P~ cum Seacombe** Chs [*Pulton* 1260, 1288 Court], **P~ cum Spital** Chs [*Pontone* DB, *Pulton Launcelyn* 1286 Court], **P~** Gl [*Pultun* 855 BCS 487, *Pulton* 1242 Ipm], **P~** K [*Poltone* DB, -*tun* 1235 Ch], **P~ with Fearnhead** La [*Poltona* 1094, *Pultona* 1142 LaCh], **P~ le Fylde** La [*Poltun* DB, *Pultona* 1094 LaCh], **P~ le Sands** La [*Poltune* DB, *Pulton* 1201 P]. 'TŪN by a pool.' Cf. PŌL.

Poundisford So [*Punderford* 1225, *Punderesford* 1243 Ass]. 'The pinder's ford.'

Pounder 'pinder' is recorded in OED from 1622. The surname *le pundere* is found 1176 P.

Poundon Bk [*Paundon* 1255 For, *Pondon* 1291 Tax, *Powendone* 1316 FA]. First el. possibly as in PAMPHILL. See DŪN.

Poundstock Co [*Pondestoch* DB, *Pondestok* 1291 Tax]. OE *pund-stoc* 'STOC with a pound', very likely 'pinfold' (cf. OE *pundfald*).

Pŏvington Do [*Povintone* DB, *-ton* 1212 Fees, *Pieuynton* 1316 FA]. '*Pēofa*'s TŪN.'

Pow. See PŌL.

Powderham D [*Poldraham* 1050–73 E, *Poldreham* DB, *Puderham* 1219 FF, 1230 P]. The first el. is identical with MDu *polre, polder* 'low-lying land reclaimed from the sea'. This el. is found also in **Poldhurst** K [*Polre* 1194 P, *Polres* 1246 Ch], **Po(u)lders** K [*Polr* 1220 Cl]. Cf. *Polre mariscus* 1252 StAug (at Chislet).

Powerstock Do. See POORTON.

Powick (-ō-) Wo [*Poincguuic* 972 BCS 1282, *Poiwic* DB]. 'The wīc of *Pohha*'s people.'

Powmaughan. See PŌL.

Pownall Chs [*Pounhale* 1276 Ipm, 1287 Court]. Cf. POYNINGS.

Poxwell Do [*Poceswylle* 987 KCD 656, *Pocheswelle* DB, *Pokeswll* 1212 Fees]. Probably the name should be analysed as *Poceswylle*, the second el. being cognate with OE *swelle* 'hill'. Cf. SWELL. For the first el. cf. POCKLEY, POLEBROOK.

Poynings (pŭnĭnz) Sx [*Puningas* 960 BCS 1055, *Poninges* DB, *Punninges* 1230 P]. OE *Pūningas* is a tribal name, which is derived from a pers. n. **Pūn* or **Pūna*, a derivative of which is found in *Puneces wurði* BCS 1323 (nr Ashburton D). An identical name is PÜNING in Münster, Germany [*Puningun* 890 &c.]. Cf. POYNTON Chs. The pers. n. stem *pūn-* may be related to OE *pūnian* 'to pound', if that is derived from a word meaning 'peg, pestle' or the like. POWNALL Chs may contain an OE *pūn* in such a sense.

Poyntington Do [*Ponditone* DB, *Puntintuna* Hy 1 (1270) Ch, *Pontington* 1250 FF, 1265 Ep]. 'The TŪN of *Punt*'s people.' *Punt* is found in *Puntes stan* BCS 934.

Poynton Chs [*Poninton* 1248 Ipm, 1276 Chester]. First el. identical with POYNINGS.

Poynton Sa nr Shawbury [*Peventone* DB, *Pevinton* 1255 RH]. '*Pēofa*'s TŪN.'

Prawle D [*Prenla* DB, *Prahulle* 1204 Cur, *Praulle* 1242 Fees]. An OE **præ(w)hyll* or **prā(w)hyll* 'look-out hill', the first el. being related to OE *bepriwan* 'to wink', Mod Engl *pry*.

Preen, Church & Holt, Sa [*Prene* DB, *Prena* 1194 P, *Preone* 1245 FF, *Prune* 1255 RH, *Holprena* 1234 FF, *Chirche Prene* 1301

For]. OE *prēon* 'pin, brooch', used in a transferred sense of the characteristically shaped hill on which the places are. Holt is HOLT 'wood'.

Prees Sa [*Pres* DB, 1255 RH, *Prees* 1291 Tax], **Preese** La [*Pres* DB, *Prees* c 1200 CC, 1259 Ass]. Welsh *pres, prys* 'brushwood, covert'. The same el. is found in **Preesall** La [*Pressouede* DB, *Preshoued* c 1190 LaCh, *-hou* ib., 1246 FF], whose second el. is or was originally ON *hofuð* 'head, headland'.

Prendwick Nb [*Prendewic, Prendwyc* 1242 Fees, *Prenderwyk* 1256, *Prandewick* 1279 Ass]. The first el. seems to be OE *Prend* recorded as a variant of the pers. n. *Præn* in Ethelwerd. PRINCETHORPE Wa is *Prondestorp* (for *Prendes-* ?) 1221 Ass.

Prenton Chs [*Prestune* DB, *Prestona* c 1100, c 1150 Chester, *Prenton* 1260 Court]. The earliest forms suggest identity with PRESTON. If so, Prenton is difficult to explain. Perhaps the original name was *Prænes-tūn* '*Præn*'s TŪN'. On OE *Præn* pers. n. see PRENDWICK.

OE **prēost** 'priest, parson' is found in several names, as PRESCOT, PRESTON, PURSTON &c. It is impossible always to decide whether the meaning of such names is 'village with a priest', 'parsonage' or 'place belonging to a priest or a college of priests'.

Prescot La [*Prestecota* 1178 P, *-cote* c 1190 LaCh], **P~** O [*Prestecote* 1220, *-cot* 1231 Ep, *-kot* 1236 Fees], **Prescott** Gl [*Prestecote* 1287 QW, *Prescote* 1291 Tax]. 'The priests' cottage, parsonage.'

Preshaw Ha [*Presshagh* 1291 Tax, *Pershawe* 1412 FA]. 'Pear(-tree) wood.' See SCAGA.

Preshute (prĕshut) W [*Prestcheta* 1186 P, *Preschete* 1223, *Preschut* 1252, *Preshut* 1254 Salisbury]. The second el. may be identical with CHUTE. If so, 'the Chute belonging to the priest'. But there was an OE *ciete, cēte* 'cottage', which would suit the name. Cf. CHEDDON, DASSETT.

Pressen Nb [*Prestfen* 1177 P, 1242 Fees]. 'The priest's fen.'

Prestbury Chs [*Presteb*[*uria*] c 1175, *-buri* 1221 Chester], **P~** Gl [(into) *Preosda byrig* 889 BCS 560, *Presteberie* DB]. 'The priests' manor.'

Prested Ess in Feering [*Peresteda* DB, *Perstede* 1206 FF]. 'Place where pears grew.'

Presteigne (-ēn) Sa, Radnor [*Presthemede* 1252 Hereford, *-hemed* 1291 Tax]. 'Household of priests.' Second el. OE *hǣmed*, which is recorded only in the sense 'marriage, sexual intercourse', but must also have meant 'household'.

Presthope Sa [*Presthope* 1167 P]. 'Priests' valley.'

Preston is OE *Prēosta-tūn* 'the TŪN of the priests' except in P~ Candover (see CAND-OVER) and P~ Crowmarsh (see CROWMARSH).
Preston Bisset Bk [*Prestone* DB, *Prestinton* 1163 P, *Prestona Manass[eri]* 1167 P], **P~ on the Hill** Chs [*Prestona* 1157–94 Chester], **P~ Quarter** Cu [*Prestona* c 1130 StB], **P~ Do** nr Weymouth [*Prestun* 1228 Pat, *Prestone* 1285 FA]; cf. HAMPRESTON, **P~ le Skerne** Du [*Prestetona* 1091 FPD, *Preston super Skiryn* 1384 BM], **P~ on Tees** Du [*Prestuna* Hy 2 FPD, *Preston upon Teas* 1402 PNNb], **P~ Gl** nr Cirencester [*Prestitune* DB], **P~ Gl** nr Ledbury [*Prestone* c 1160 Glouc], **P~ upon Stour** Gl [*Præston* DB, *Preston super Sturham* 1291 Tax], **P~ Candover** Ha [see CANDOVER], **P~ on Wye** He [*Prestetune* DB, *Prestone super Weye* 1221 Hereford], **P~ Wynne** He [*Prestetune* DB], **P~ Hrt** [*Prestun* 1185 TpR], **P~ next Faversham** K [*Preostantun* 941 BCS 766, *Prestetone* DB], **P~ near Wingham** K [*Prestetune* DB, *Preston* 1200 Cur], **P~ La** [*Prestune* DB, *Prestona* 1094 LaCh], **P~ Mx** [*Prestone* 1212 RBE], **P~ Nb** nr Ellingham [*Preston* 1242 Fees], **P~ Nb** nr Tynemouth [*Prestona* 1198 (1271) Ch], **P~ Capes & Little P~** Np [*Pres(te)-tone* DB, *Great Preston* 1256 Ipm, *P~ Capes* 1335 Ch, *Parva P~* 1220 Fees], **P~ Deanery** Np [*Prestone* DB, *Decanus de Preston* 1199 BM], **P~ Crowmarsh** O [see CROWMARSH], **P~ Ru** [*Prestetona* 1130 P, *Preston* 1240 Ep], **P~ Brockhurst** Sa [*Preston* DB], **P~ Gubbals** Sa [*Prestone* DB, *Preston Gobald* 1292 QW], **P~ Montford** Sa [*Prestune* DB, *Preston juxta Moneford* 1199 Ch], **P~ upon the Weald Moors** Sa [*Prestune* DB, *Preston in Wyldmore* 1262 Eyton], **P~ Sf** [*Preston* c 1060 Wills, *Prestetona* DB], **P~ Plucknett** So [*Prestetone* DB, *Preston Plukenet* 1285 FA], **P~ Sx** nr Brighton [*Prestetone* DB], **East & West P~** Sx [*Prestetune* DB, *Estpreston* 14 BM, *Westprestone* 1339 Misc], **P~ Bagot** Wa [*Prestetone* DB, *Preston Bagot* 1345 AD ii], **P~ Patrick & Richard** We [*Prestun* DB, *Preston Patrick* 1235, *P~ Richard* 1301 Kendale], **P~ YE** [*Prestone* DB, *Prestitonia* c 1100 YCh 1300], **P~ under Scar** YN [*Prestun* DB], **Great & Little P~** YW [*Prestun, -e* DB], **Long P~** YW [*Prestune* DB, *Prestona in Cravana* 1175 YCh 359].

For additions such as (le) Skerne, (upon) Stour, (on) Tees, Wye, see these names.—**P~ Bagot** Wa was held by Symon Bagoth in 1236 (Fees); see MORTON BAGOT.—**P~ Bisset** Bk was held by Manasser Biset in 1167 (P), by Ansellus Biset in 1208 (Cur); see COMBE BISSETT.—**P~ Brockhurst** Sa. See LEE BROCKHURST.—**P~ Capes** Np was held by Hugh de Capes in 1255 (Fees), called Hugh de *Capes* 1244 Fees. Capes is a Fr family name, perhaps from CAPELLES-LES-GRANDS in Normandy.—**P~ Deanery** Np gave its name to a deanery.—**P~ Gubbals** Sa was held in 1086 by Godebold the priest, apparently an Englishman.—**P~ Montford** Sa is nr MONTFORD.—**P~ Patrick** We after Patrick grandson of Gospatric de Workington (early 13th cent. Kendale).—On **P~ Plucknett** So see HASELBURY PLUCKNETT.—

P~ Richard We after the Richard son of Uhtred who confirmed his father's gift to Cockersand c 1215 (Kendale).—**P~ under Scar** YN is at the foot of a steep hill. Scar is *scar* 'rock, crag, precipice'.—**P~ upon the Weald Moors** Sa. See EYTON Sa.—**P~ Wynne** He was held by *Dionisia la Wyne* in 1303 (FA). Wynne may be of Welsh origin (Welsh *gwyn* 'white').

Prestwich La [*Prestwich* 1194 P], **Prestwick** Nb [*Prestwic* 1242 Fees]. 'The priests' WĪC or parsonage.'

Prestwold Le [*Prestwolde, -uuald* DB, *-walde* 1229 Ep]. 'The priests' wood.'

Priddy So [*Pridi* c 1180, *Pridia* c 1185 BM, *Pridie* 1219 FF]. A derivative of Welsh *pridd* 'earth, soil'. The place is on Mendip Hills.

Priestcliffe Db [*Prestecliue* DB, 1200 P]. 'The priests' cliff.'

Primethorpe Le [*Torp* DB, *Prymesthorp* 1316 FA]. '*Prim*'s Thorp.' *Prim* is the name of a moneyer t. Eadmund I.

Princelet Wt [*Prymesfloude* 1316, *-flode* 1346 FA]. '*Prim*'s FLŌDE or stream.' See prec. name.

Princethorpe Wa [*Prenestorp, Pernesthorp* 1262 FF, *Prensthorp* 1428 FA]. '*Præn*'s thorp.' Cf. PRENDWICK, PRENTON.

Princetown D. Named from the Prince Regent. The prison was built in 1808.

Prinknash Gl [*Prinkenesse* 1121 Glouc]. Second el. OE *æsc* 'ash-tree'. The first is obscure.

Prinsted Sx [*Pernestede* 1253 FF, *-sted* Hy 3 Misc]. 'Place where pears grew.' Cf. PERU.

Priston So [*Prisctun, Pristun* 931 BCS 670, *Prisctun* a 1087 E, *Prisctone* DB, *Prisshtone* 1327 Subs]. First el. Welsh *prysg, prysgl* 'copse, thicket'.

Prittlewell Ess [*Pritteuuella* DB, *Pritelewell* 1166, *-wella* 1194 P]. 'Babbling brook.' First el. an adj. **pritol* belonging to OE *pritigian* 'to chirp'.

Privett Ha nr Petersfield [(æt) *Pryfetes flodan* 755 ASC (*Pryftes* ib. E), *Pruuet* c 1245 Selborne, *Prevet* 1329 Ipm]. 'Privet copse' (*Ligustrum vulgare*).

Probus Co [*Sanctus Probus* DB, (Ecclesia) *Sancti Probi* 1269 Ep]. '(The church of) St. Probus.' *Probus* from Lat *probus* 'honest'.

Prudhoe Nb [*Prudho* 1173 P, *Prudehou* 1212, 1242 Fees]. '*Prūda*'s HŌH or spur of land.' *Prūda* from *prūd* adj. 'proud'. *Prud* occurs BCS 1250.

Publow So [*Pubelawe* 1219 Ass, *Pubbelowe* 1259 FF, *Puppelawe* 1262 Ipm]. '*Pubba*'s HLĀW or barrow.' **Pubba* is related to *Pybba*.

Puckeridge Hrt [*Pokerich* 1314, 1327 Ch, 1343 BM]. 'The stream of the goblin or watersprite.' First el. OE *pūca* 'goblin'. Second OE **ric* 'stream'.

Puckington So [*Pokintuna* DB, *Pukinton* 1201, 1244 FF]. '*Pūca*'s TŪN' or 'the TŪN of *Pūca*'s people'. OE **Pūca* is a nickname from *pūca* 'goblin'. Cf. the OSw byname *Puke*.

Pucklechurch Gl [*Puclancyrce* 946 ASC (D), *Pucelancyrcan* 950 BCS 887, *Pulcrecerce* DB]. '*Pūcela*'s church.' **Pūcela* is a diminutive of *Pūca* in prec. name.

Puddington Chs [*Potitone* DB, *Potinton*, *Podinton* 1260 Court], P~ D [*Potitone* DB, *Putingthon* 1242 Fees]. 'The TŪN of *Puta*'s or *Putta*'s people.'

Puddle. See PIDDLE.

Pudleston He *Pillesdune* DB, *Putlesdone* 1212 RBE, *Puttlesdune* 1242 Fees, *Pudlesdun* 1249 Fees]. 'The hill of the mouse-hawk' or '*Pyttel*'s hill'. OE *pyttel* occurs in *bleri pittel*, *blerea pyttel*.

Pudlicott O [*Pudelicote* 1176 P, *-cota* 1181 Eynsham, *-cot* 1242 Fees]. 'The COT of *Pudel*'s people' rather than 'COT by the puddle'. **Pudel* would be a diminutive of *Puda*. *Puddle* is recorded from c 1330.

Pudsey (-s-) YW [*Podechesaie* DB, *Pudekeshee* 1203, *-hay* 1219 FF]. '*Pudoc*'s island or river land.' **Pudoc* may be a diminutive of *Puda* or OE *pudoc* 'wen, wart' used as a nickname.

Pulborough Sx [*Poleberge* DB, *Polemberg* 1166 RBE, *Puleberga* 1168 P]. 'Hill or barrow by the pools.' *Polemberg* may represent OE *Pōlhǽma-beorg*.

Pulford Chs [*Pulford* DB, c 1100 Chester], **Pulham** Do [*Poleham* DB, *Puleham* 1130 P], P~ Nf [*Polleham* c 1050 KCD 907, *Pulham*, *Pullaham* DB, *Pulham* 1251 Ch]. 'Ford and HĀM or HAMM by the pool or pools.' The forms vary between *pōl* and *pull*.

Pulloxhill Bd [*Polochessele* DB, *Pullokeshull* 1196 f. P]. Apparently '*Pulloc*'s hill'. *Pulloc* is unrecorded.

Pulverbatch, Castle & Church, Sa [*Polrebec* DB, *Purlebech* 1196 P, *Pulrebeche* 1212 Fees, *Pulverebach* 1291 Tax, *Castel-*, *Chirchpolrebache* 1301 For]. Second el. OE *bæce*, *bece* 'valley, stream'. The first is a stream-name related to Norw *puldra* 'to gush', Sw *porla*, *pollra*, Norw *purla* 'to purl', also to dial. *prill* 'a rill', also *purl* (earlier *pirle*, *perle*, &c.).

Puncknowle (pŭnel) Do [*Pomacanole* DB, *Pumernolle* 1202 FF, *Pomecnolle* 1291 Tax]. OE *plūm-cnoll* 'plum-tree knoll', with dissimilatory loss of the first *l*.

Purbeck, Isle of, Do [(tellus) *Purbicinga* 948 BCS 868, *Porbiche* (hd), *Porbi* DB,

Porbica 1107 (1300) Ch, *Purebic* 1240 Cl]. The first el. may be OE *pūr* 'bittern'. The second is possibly an old word meaning 'headland' related to OE *becca* 'pick-axe'. But a better etymology is probably reached if *-beck* is taken to be an old word for 'bill, beak', the name meaning 'the bill of the bittern'. An OE *bica*, literally 'pecker', might have been used in the senses 'beak' and 'woodpecker'; for such a use of *bica* cf. BICKLEIGH. The name would be analogous to RAMPSIDE.

Purbrook Ha [*Pukebrok* 1248 Ass, 1255 FF]. 'Brook of the watersprite'; cf. PUCKERIDGE.

Purfleet Ess [*Purteflyete* 1285 PNEss, *Pourteflet* 1312 Cl]. ?'*Purta*'s stream.' Cf. FLĒOT. *Purta* may occur in *Purtan ig* 962 BCS 1093.

Puriton So [*Peritone* DB, *Piriton* 1212 Fees]. 'Pear-tree TŪN.' First el. OE PIRIGE.

Purleigh Ess [(on) *Purlea* 998 Crawf, *Purlai* DB, *Purle* 1212 Fees], **Purley** Brk [*Porlei* DB, *Purlye* 1220, *-le* 1242 Fees]. 'Bittern LĒAH.' Cf. PURBECK.

Purley Sr [*Pirlee* 1200 FF, *Pirelea* c 1220 Hyde]. 'Pear-tree LĒAH.' First el. OE *pirige* 'pear-tree'.

Purslow Sa [*Posselav* DB, *Pusselawe* (hd) 1226–8 Fees]. '*Pussa*'s barrow.' **Pussa* is a side-form of *Pusa*. The *r* is a late addition.

Purston Np [*Prestetone* DB, *Purston* 1220 Fees], P~ **Jaglin** YW [*Preston* DB, *Preston Jakelyn* 1334 FF]. A variant of PRESTON. *Jakelin* is a Fr diminutive of *Jacques*.

Purton Gl nr Berkeley [*Peritone* DB], P~ Gl in Lydney [*Periton* 1190 P, *Piriton by Lydeneye* 1327 Misc], P~ or **Perton** St [*Pertona* 1167 P, *Periton* 1193 P], P~ W [*et Pirigean* 796 BCS 279 (evidently an earlier name), *Puritone* 796, *Piritune* 854 BCS 279, 470, *Piritone* DB]. 'Pear-tree TŪN.' First el. OE *pirige* 'pear-tree'.

Pusey Brk [*Pesei*, *Peise* DB, *Pusie* W 1 Abingd, *Pesee* 1180 P]. 'Pea island.' First el. OE *pisu*, *piosu* 'pea'.

Putford D [*Podiford*, *Potiforde*, *Pudeforde* DB, *Pudiford* 1199 P, *Putteford*, *Churiputteford* 1242 Fees, *Westpoteford* 1284–6 FA]. '*Putta*'s ford.' But there may well have been an OE **putta* 'kite', to judge by the diminutive *pyttel* (in *bleri pittel* 'mousehawk') and ME *puttok* 'kite'. Such a first el. would be suitable here.

Putley He [*Poteslepe* DB, *Putelega* c 1180 Hereford, *-leg* 1206 Cur]. 'Kite wood' or '*Putta*'s LĒAH'. Cf. prec. name.

Putney Sr [*Putelei* DB, *Potenhipe* c 1327 Beves of Hamtoune]. OE *Puttan-hȳþ* '*Putta*'s landing-place'.

Puttenham Hrt [*Puteham* DB, *Putteham* 1204–12, *Putenham* 1212 Fees], P~ Sr

[*Puteham* 1199 FF, *Poteham* 1291 Tax].
'*Putta*'s HĀM.'

Putton Do [*Podinton* 1237 FF, *Podintone* 1285 FA]. '*Puda*'s TŪN.'

Puxton So [*Pukereleston* 1212 Fees, 1227 FF]. '*Pukerel*'s TŪN.' *Pukerel* is no doubt a Fr family name. Robert Pukerel(l) is mentioned 1158–9 RBE, 1176 P (So).

Pyder. See PETHERWIN.

Pyecombe Sx [*Picumba* W 2, *Piccumbe* c 1100 PNSx]; OE *pie-cumb* 'valley infested by gnats'.

OE **pyll** 'pill, tidal creek, stream' from OW *pill* is found in some names, as PYLLE, HUNTSPILL, UPHILL, PILTON.

Pylle So [*Pil* 705, *þæt pyl* 955 BCS 112, 903, *Pille* DB, *Pulle* 1276 RH]. 'The creek.' In the first examples *pyl* (*pil*) is used as a common noun.

Pyon, Canon & Kings, He [*Pionie, Pevne* DB, *Peuna* Hy 2 Marden, *Pionia* c 1200, 1219–31 Hereford, *Pyonia* 1242 Fees, *Pyone*

canonicorum 1221 Hereford, *King's Pyon* 1285 Ipm]. An OE *pēona ēg* 'gnat island'. First el. OE *pēo, pie* 'insect'. Cf. PYWORTHY. The same el. is found in **Pymore** Do [*Pimore* 1236 Fees], and the river-name **Piall** D [*Piall, Pial* 13 PND]: OE *pie-halh* 'gnat-infested valley'.

Canon P~ belonged to the cathedral of Hereford, **Kings Pyon** to Edward the Confessor in 1066 (DB).

Pyrford Sr [*æt Pyrianforda* 956 BCS 955, *Piriford* 1067 BM]. 'Pear-tree ford.'

Pyrton O [*Pirigtun* 766 BCS 221, *Peritone* DB]. Identical with PURTON.

Pytchley (pīchli) Np [*Pihteslea ford* 956 BCS 943, *Pihteslea* DB, *-le* 1201 Cur]. '*Pēoht*'s LĒAH.' Cf. PITSFORD.

OE **pytt** 'pit, hole, cavity' is a rare el. in pl. ns. See e.g. PETT, WOOLPIT and cf. BEAUMONT Ess.

Pyworthy D [*Paorde* DB, *Peworthy* 1239 Ch, *-wrthe* 1262 Ep, *Piworthi* 1285 Origl]. See WORPIG. The first el. is OE *pēo, pie* 'insect', here very likely used as a nickname.

Q

Quadring Li [*Quadheueringe, Quedhaveringe* DB, *Quadhaueringe* 1170, 1197 P, 1202 Ass]. A name analogous to HORBLING and consisting of OE *cwēad* 'dirt' (here no doubt 'mud') and a tribal name *Hæferingas* identical with HAVERING Ess.

Quainton Bk [*Chentone* DB, *Quentona* 1167 P, *Queinton* 1236 Fees]. OE *Cwēne-tūn* 'the queen's manor'.

Quantock Hills So [*Cantucuudu* 682 BCS 62, *Cantok* (for.) 1274 RH, *mons de Cantok* 1314 BM]. *Cantoche* DB refers to some place nr the hill. *Cantuc* is a Brit name of the ridge or chain of hills. It may be a derivative of Celtic *canto-* 'circle, rim' (Gaul *cantus*, Welsh *cant* 'rim of a circle, tyre'). Cf. CAMEL. The ridge must alternatively have been called OE *Cantuc(es)-hēafod* 'Quantock hill' (cf. HĒAFOD). This name lives on in **East & West Quantoxhead** So (at the northern end of the ridge) [*Cantocheve* DB, *Cantokesheued* 1185 P, *-heved* 1212 Fees, Est-, *Westcantokeshende* 1327 Subs]. The ridge also gave its name to CANNINGTON [*Cantuctun* c 880 BCS 553].

Quarles Nf [*Hucrueles* DB, *Warfles* 1175 P, *Quarueles* 1199 Cur]. OE *hwerflas*, plur. of OE *hwerfel* 'circle'. Cf. WHARLES La, WHORLTON Nb, YN. Some stone circles may be referred to.

Quarley Ha [*Cornelea* 1167 P, *Querli* c 1270 Ep, *-leye* 1291 Tax]. 'LĒAH or glade with a mill or where millstones were got.' See CWEORN.

Quarlton La [*Querendon* 1246 Ass, *Querdon* 1304 Ch]. OE *cweorndūn* 'hill where millstones were got'. See CWEORN.

Quarmby YW [*Cornebi* DB, *Querneby* 1237 Cl, 1274 Wakef]. OScand *Kvernbȳ(r)* 'BY with a mill'. OScand *kvern* corresponds to OE *cweorn.*

Quarndon Db [*Cornun* DB, *Querendon* c 1200 Chester]. See QUARLTON and CWEORN.

Quarnford St [*Querneford* 1227 Ass]. 'Ford by a mill.' See CWEORN.

Quarr Wt [*Quarraria* a 1155, *Quadraria* Hy 2 BM, *Quarrer* 1289 Bodl]. ME *quarrere* 'quarry', from OFr *quarriere*, MLat *quarraria, quadraria.*

Quarrendon Bk [*Querendone* DB, *Querendona* c 1140 RA], **Quarrington** Du [*Querendune* c 1190 Godric]. Identical with QUARLTON.

Quarrington Li [*Cuernintune* 1060 Th, *Corninctun* DB, *Querinton* 1202 Ass]. The first el. must be a derivative of OE *cweorn* 'mill'. A windmill must be referred to. An OE *cweorning* 'mill stream' is hardly to be thought of in this low-lying district. Probably the name means 'the TŪN of the *Cweorningas* or millers'.

Quatford Sa [*Quattford* c 1030 Förster, *Themse*, p. 769, *Qvatford* DB, *Catford* c 1100 Fr, *Quatford* c 1130 Ordericus], **Quatt** Sa [*Quatone* DB, *Quatte* 1212 Fees, 1291 Tax]. A bridge over the Severn, probably at Quatford, is *Cwatbrycg* 896 ASC, *Quatbricg* a 1118 Flor. QUATT is a shortening of *Quatton* analogous to ALBRIGHT, EDGBOLD, EDGE-MOND from *Albrighton* &c. *Cwatt* occurs as a byname in *Leofwine Cwatt* 1016 E 226 (gen. *Cwattes*). The names thus seem to mean 'Cwatt's ford, TŪN and bridge'. But we should expect the gen. form *Cwattes* in these names. Were it not for OE *Cwatbrycg* (and the byname *Cwatt*), the obvious history of the names would be as follows. Quatt was originally OE *Cwēadtūn*, Quatford OE *Cwēadford* 'TŪN and ford in a muddy place' (cf. QUADRING). *Cwēad*- would have its *ēa* shortened to *ea* (whence ME *a*) and become *Cweat* before the *f* of *ford* (hence ME *Quatford*). The bridge that was built at Quatford was naturally named *Cweatbrycg*. It is quite possible this is the correct solution after all, though the similarity to the byname *Cwatt* must then be accidental, and the OE form *Cwatbrycg* must have developed from *Cweatbrycg*. A change of *ea* to *a* after *w* at least occurs in Old Northumbrian. A third possibility is that OE had a word *cwatt*, which could be used in pl. ns. and could also give rise to a byname. The OE byname *Cwatt* has been identified with Mod E *quat* 'a pimple' (1579 &c.). It is, of course, possible that the pl. ns. contain the same word used of hill or knoll near the ford, but this is doubtful.

Quedgeley Gl [*Quedesleya* c 1145 ff. Glouc, *Quedelee* 1201 Cur]. The first el. appears to be OE *cwēad* 'dirt'. See LĒAH.

Queenborough K [*Queneburgh* 1376 St Aug]. Q~ became a borough in 1367. The name was given in honour of Queen Philippa.

Queenhill Wo [*Cynhylle*, *Cumhille* 11 Heming, *Cunhille* DB, *Queinhull* 1209 Fees, *Kin-*, *Kunhull* 1221 Ass]. Probably OE *cyne-hyll* 'royal hill', whence *Kinhill* and the like. This was sometimes written *Quinhill*, which was misread (with *qu-* as in *queen*).

Queensbury YW is a late name, which seems to have been given in 1863.

Quemerford (kŭm-) W [*Camerford* 1204 Obl, *Kemerford* 1226-8 Fees, *Cameresford* 1292 Cl, *Quemerford* 1240-5 Salisbury, *-e* 1294 Ipm]. OE *Cynemǣres-ford* 'Cynemǣr's ford', with the same development as in QUEENHILL.

Quenby Le [*Qveneberie* DB, *-bia* c 1125 LeS, *-by* 1242 Fees]. An OE *Cwēne-burg* 'the queen's manor', whose second el. was replaced by OScand BY.

Quendon Ess [*Kuenadana* DB, *Quendene* 1254 Ipm]. The first el. may be OE *cwēn* or rather *cwene* 'woman' (*Cwenena-denu* 'the women's valley').

Queniborough Le [*Cuinburg* DB, *Quen-*

burg c 1125 LeS, *Queningburc* 1236, *Queniburg* 1242 Fees]. Probably OE *Cwēne-burg* 'the queen's manor'.

Quenington Gl [*Qvenintone* DB, *-tona* 1138 Glouc, *Quentona* 1169 P]. OE *Cwenena-tūn* 'the women's TŪN'.

Quernmore La [*Quernemor* 1228 Cl, *-more* 1278 FC]. 'Moor where millstones were got.' See CWEORN.

Quethiock Co [*Quedoc* 1201, *Queidike church* 1230 FF, *Quedik* 1291 Tax]. No Cornish word can begin in QU-. Some substitution must have taken place. Either the name originally began in *gw-*, and *qu-* is due to English substitution, as in OE *Cwæspatrik* BCS 1254 for *Gwas Patric* 'Gospatric'. The name may then be a saint's name corresponding to Bret *Gouezec* (OBret *Wedoc*). Or the name may be identical with CHIDE-OCK, i.e. an adj. for 'wooded' derived from Co *coit* 'wood' and corresponding to Welsh *coediog*. Cf. MBret *coadyc*, *koedig* 'little wood'. In this case substitution of *qu-* for *cu-* or *co-* would have taken place.

Quickbury. See QUY.

Quidenham Nf [*Cuidenham* DB, *Quideham* 1177 P]. '*Cwida*'s HĀM.' *Cwida corresponds to OHG *Quito*.

Quidhampton Ha [*Quidhampton* 1316, 1412 FA], Q~ W [*Quidhampton* 1242 Fees, *-hamton* 1287 Ipm]. The same name is found in Wt [*Quedhampton* 1287-90 Fees, *Quydhampton* 1346 FA]. The first el. is probably, as suggested in PNW(S), OE *cwēad* 'dirt, dung'. For *i* (< *ie*) from *ēa* cf. BINCKNOLL, BINSTEAD.

Quinton Gl [*Quentone* 848 BCS 453, *Quentona* 1183 AC], Q~ Np [*Quintone* DB, *Quenton* 12 NS, 1176 P, 1220 Fees], Q~ Wo [*Quenton* 1221 Ass, *Quinton* 1275 Subs]. 'The queen's manor.'

Quixhill St [*Quikeshull* 1272 FF, 1279 Ass]. '*Cwic*'s hill.' *Cwic is a short form of names in *Cwic-*, as *Cwichelm*.

Quob Ha in Titchfield [*la Qvabbe* 1198 FF, *Quabbe* 1243 Cl, *la Quabbe* 1282 Ep, 1311 ff. Ipm]. 'Marshy place or bog.' *Quab* in this sense is recorded in OED from 1617, but it is found in a late transcript of an OE charter: (on) *Heahstanes quabben* 968 BCS 1218 (Do). Du *kwabbe* is identical in meaning.

Quorndon or **Quorn** Le [*Querendon* 1209-35 Ep, *Querondon* 13 BM]. Identical with QUARLTON.

Quy Ca [*Choeie* c 1080 ICC, *Coeia* DB, *Cueye* 1212 RBE, *Coueye* 1273 Ipm, *Queye* 1261 FF]. OE *cū-ēg* 'cow island'. For the sound-development cf. **Quickbury** Ess in Sheering [*Cuica* DB, *Cuwyk* 1258 FF], from OE *cū-wic*.

R

OE **rā** (*rāha*) 'roe-deer, roebuck', in compounds also *rāh-*, as in *rāhdēor* 'roe-deer', *rāhhege* 'fence or enclosure for roe-deer', is the first el. in some names, as RODDEN, ROEL or ROWELL, ROGATE. OE *rāhhege* is the source of ROFFEY. OScand *rā* 'roe-deer' is the first el. of RASKELF and ROWLAND. OE *rǣge* 'the female of the roe' is found in some names, as READ, ROEBURN La, REIGATE Sr.

ON **rá**, OSw **rā** 'a pole', OSw also 'a boundary mark' is the first el. of RABY, ROBEY, ROBY, and probably of RAUGHTON.

Rabley Hrt [*Wrobele* 1235 AD, *Wrobbele* 1274 AD, 1311 Ipm]. Identical with **Robley** Ha [*Wrobban léa æfisc* 909 BCS 625]. The first el. may be related to ME *wrobbe* 'to blab', *wrobber* 'an informer'. If so, it is no doubt a nickname. Cf. WRABNESS, WRIBBENHALL.

Raby Chs [*Rabie* DB, *Rabi* c 1100 Chester, -*by* 1260 Court], **R~** Du [*Raby* c 1050 HSC, 1334 Misc]. Identical with Dan RAABY, OSw RABY, Sw RÅBY, which consist of OSw *rā*, Dan *raa* 'a boundary mark' and BY. The meaning may be 'BY situated near a boundary mark' (a hundred boundary or the like) or 'BY with boundary marks of a certain kind'.

Rackenford D [*Rachenefode, Litelrachene-ford* DB, *Racherneford, Racarneforde* c 1150 Buckland, *Rakerneford* 1238 f. FF]. OE *racu* 'bed of a stream' and OE *ærneford* 'ford that can be passed on horseback' (cf. OE *ærneweg* 'road fit to ride on'). Cf. RACU.

Rackham Sx [*Recham* 1166 RBE, 1196 Cur, *Rakham* 1295 Ch]. OE *Hrēac-hām*. OE *hrēac* means 'a (hay) rick', but is here used of **Rackham Hill** (625 ft.), which must have been held to resemble a hayrick. The meaning is 'HĀM at Rackham Hill'.

Rackheath Nf [*Racheitha* DB, *Racheia* DB, 1153–68 Holme, 1197 FF, *Racheth* 1252 Ch]. The second el. is apparently OE *hȳþ* 'landing-place'. The place is a couple of miles from the Bure, but there may have been a stream at Rackheath in earlier days, or the name originally denoted a landing-place on the Bure belonging to the village. The first el. may be OE *racu* 'bed of a stream' &c.

Racton Sx [*Rachetone* DB, *Rakentune* 1121 PNSx, *Rakintona* c 1150 Fr]. 'TŪN in the pass.' The first el. is identical with RAKE, i.e. OE *hrace* or *hræce* 'throat', here in transferred use 'a pass'.

OE **racu** only occurs in the compounds *ēa-*, *strēam-racu*, which are held to mean 'bed of a stream, water-course'. The simple word *racu* was evidently used in the same sense; see RACKENFORD, RACKHEATH, RAGDALE. The original meaning of the word was about the same as that of the corresponding Du *rak*, i.e. 'a stretch'. Du *rak* is used particularly of a stretch of road. A more original meaning than 'watercourse' is probably found in *Langrake* (see LONG DRAX and LANGRICK), where *rake* seems to mean 'reach, straight part of a river'.

OE **rād** 'riding', but also 'road' (as in *hronrād, swanrād* 'the sea', lit. 'the road of the whale and swan'). The meaning 'riding' is seen in RADFORD, RADWAY, RODWAY. The meaning 'road' is evidenced in RADSTOCK. On OE **rǣde* adj. (as in *rǣdehere* 'cavalry') as a possible el. in pl. ns. see RADFORD.

Radbourn Wa [*Hreodburna* 998 Crawf, *Redborne* DB, *Rodburn* 1268 Ipm]. Originally the name of the stream at the place, 'stream where reeds grew'. *Hrēod-* became ME *Rod-*, later *Rad-*; cf. LADBROOKE Wa.

Radbourne Db [*Radeburne, Rabburne* DB, *Rodburn* 1242 Fees, -*burne* 13 BM, *Redburna* 1171 P]. In spite of the DB *a*-forms most likely identical with RADBOURN.

Radcliffe La [*Radecliue* DB, 1200 P], **R~ on Trent** Nt [*Radeclive* DB, 1226–8 Fees, *Radeclyf super Trent* 1291 Tax], **Radclive** Bk [*Radeclive* DB, *Redeclive* 1314 Ch]. OE *rēade clif* 'red cliff'.

Radcot O [*Redcota* 1163 P, *Radcote, Retkot* 1236, *Redcot, Rethcot* 1242 Fees]. Either 'red COT' or rather 'reed COT', i.e. 'cottage with roof made of reeds'.

Raddington So [*Radingtone* 891 BCS 564, *Radingetune* DB, -*ton* 1198 Cur, 1225 Ass]. 'The TŪN of the *Rǣdingas*.' The *Rǣdingas* would be 'the people of **Rǣd(a)*', the latter being a short form of names in *Rǣd-*, -*rǣd*.

Radford D in Plymstock [*Reddeford* 1249 Ass, *Radeford* 1275 RH], **R~** Nt nr Nottingham [*Redeford* DB, *Radeford* Hy 2 (1316) Ch, 1212 Fees], **R~** Nt nr Worksop [*Radeford* Hy 2 (1316) Mon], **R~** O [*Radeford* DB, c 1280 Winchc, *Rodeford* 1316 FA], **R~** Wa nr Coventry [*Raddeford* 1354 AD, *Radford* 1411 ff. Coventry Leet Bk], **R~ Semele** Wa [*Redeford* DB, *Radeford* 1202 Ass, 1242 Fees, *Radeford Semely* 1325 Misc], **R~** Wo in Rous Lench [*Radeford* 1230 PNWo], **R~** Wo in Alvechurch [*Radeford* 1182 PNWo]. Radford probably means in most cases 'red ford'. For some of the places it is stated that the soil is red in or near the ford, as for R~ nr Nottingham, R~ Semele, R~ in Alvechurch. But R~ O appears to be OE *rādeford* 'ford that can be passed on horseback', and sometimes Radford may represent an OE *rǣdeford* with the same meaning. See RĀD.

R~ Semele was held by Galfrid de Simily in 1242 (Fees) and by an earlier member of the family already t. Hy 1 (Dugdale). Semele is a Fr family name from SEMILLY in Manche (Normandy).

Radipole Do [*Retpole* DB, *Redpole* 1166 RBE, 1194 P, *Radepol, Radipol, Retpol* 1237 Cl]. 'Reedy pool' (OE *hrēod-pōl*).

Radley Brk [*Radelega* 1176 P, *Radelege, Redelea* c 1225 Abingd, *Radeley* 1242 Fees]. 'Red LĒAH.'

Radmanthwaite Nt [*Redmareswerc* 1197 f. P, *Rodmarthweyt* Hy 3, 1288 Misc]. 'Thwaite or clearing by a reedy lake.' First el. OE *hrēodmere*.

Radmore St [*Redamora* 1157, *Rademora* 1156, 1158 P, *Radmore* 1227 Ch]. 'Red moor.'

Radnage Bk [*Radenhech* 1161 f., *Radenach* 1175, *Radenech* 1176 P, *Radenache* 1200 BM]. OE *rēade āc*, dat. *rēadan ǣc*, 'the red oak'.

Radstock So [*Stoche* DB, *Stokes Elie de Clifton* 1198 Cur, *Radestok* 1221 FF, *-stoke* 1225 Ass, *Rodestoke* 1276 RH]. Originally STOKE; see STOC. *Rad-* is OE *rād* 'road'. The place is on the Fosse Way.

Radstone Np [*Rodestone* DB, *Rodestona* 1163, *-tun* 1167 P, *Rudstan* 1198 P, *Rodestan* 1201 Cur]. OE *rōde-stān* 'rood stone', i.e. 'stone with a cross' or the like. Cf. RUDSTON. The second el. was at an early period associated with TŪN.

Radway Wa [*Radwei, Rodewei* DB, *Radewey* 1198 Fees]. OE *rādweg* 'roadway', i.e. 'road fit to ride on'. The topographical information in PNWa(S) may indicate alternatively the meaning 'red road'.

Radwell Bd [*Radeuuelle* DB], **R~** Hrt [*Radeuuelle* DB, *-wella* 1167, *-welle* 1195 f. P]. 'Red spring or stream.' Radwell Hrt was named from the stream or spring referred to as *Readan wylles heafdan* 1007 Crawf.

Radwinter Ess [*Redeuuintra* DB, *Radewintre* 1200 FF, *-winter* 1212 RBE]. The second el. is probably OE *trēo* 'tree'. If so, the first el. is probably an unrecorded OE woman's name *Rǣdwynn*.

OE **ræcc** 'a dog that hunts by scent' is found in NEROCHE, ROCHFORD.

OE **rǣge**. See OE RĀ.

OE **rǣw, rāw** 'row, hedgerow' is found in REW(E), BAGGROW, BAGRAW, MILNROW, WOOD-ROW, perhaps MERROW. In later periods the word *row (raw)* is used particularly of 'a row of houses, a street, a hamlet'. These are probably often the senses in pl. ns.

Ragdale Le [*Ragendele* DB, *Rachedal* c 1125 LeS, *Rakedal* 1242 Fees, 1254 Val]. First

el. probably as in RACTON, though the meaning 'pass' is not so obviously suitable in this case.

Ragley Hall Wa [*Rageleia* 710 BCS 127, *-lega* 1180 P, *Raggeleia* 1154–8 (1340) Ch, 1176 P]. 'Wood where lichen grew.' OE *ragu* means 'lichen', and *rag* 'a kind of moss' is evidenced from 1758 (OED).

Ragnall Nt [*Ragenehil* DB, *Raghenehull, Rawenhell* 1230 P, *Ragenhil* 1242 Fees]. '*Ragni's* hill.' First el. ON *Ragni*, OSw *Ragne* pers. n.

Rainford La [*Raineford* a 1198 LaCh, *Reineford* 1202 FF], **Rainham** Ess nr Dagenham [*Raineham, Renaham* DB, *Renham* 1192 P]. '*Regna's* ford and HĀM.' **Regna* is a short form of names in *Regn-* as *Regengār, Regnhēah, Regnhere*. But *Ricingahaam* 695 BCS 87, mentioned with Dagenham, may be Rainham. If so, 'the HĀM of *Rica's* people'.

Rainham K [*Roegingahám* 811 BCS 335, *Raenham* 11 DM, *Renham* 1130, 1165 P]. The first el. is a tribal name *Roegingas*, whose etymology is obscure.

Rainhill La [*Reynhull, -hill* 1246 Ass]. '*Regna's* hill.' Cf. RAINFORD.

Rainow Chs [*Ravenhoh* 1288, *Ravenouh* 1290 Court]. 'Raven hill.' Cf. HRÆFN, HŌH.

Rainton Du [*Reiningtone* c 1170, 1228 FPD], **R~** YN [*Reineton* DB, *Rennington* 1202 FF, *Reynington* 1231 FF]. 'The TŪN of *Regna's* people.' Cf. RAINFORD, RENNING-TON.

Rainworth Nt [*Rayngwath* 1280 Cl, *Reynewathford* 1300 For]. A Scand name meaning 'clean ford' (cf. SHEREFORD). The elements are OScand *hreinn* 'clean' and *vað* 'ford'.

Raisthorpe YE [*Redrestorp* DB, *Reidestorp* 1163 P]. '*Hreiðar's* thorp.' First el. ON *Hreiðarr*, ODan *Rether*, OSw *Redhar* pers. n. (*Reider, Reder* DB).

Raithby Li nr Louth [*Radresbi* DB, *Reithebi* 12 DC, 1202 Ass]. '*Hreiðar's* BY.' Cf. RAISTHORPE.

Raithby Li nr Spilsby [*Radebi* DB, Hy 2 DC, *Radabi* c 1150 DC, *Radthebi, Rathebi* 1202 Ass]. '*Hraði's* BY.' First el. ON *Hraði*, ODan *Rathi* pers. n.

Rake, The, Sx [(ate) *Rake* 1327 Subs]. OE *hrace* or *hræce* 'throat', here in the sense 'pass'. Cf. RACTON. Rake is in a pass on the boundary between Ha and Sx.

Raleigh (-aw-) D [*Radeleia* DB, *Radlea* 1175, *-lega* 1176, *Raelega* 1162, *Ralega* 1161 P]. The original form must have had a first el. containing a *d* or *þ*. The name may mean 'red LĒAH' (OE *rēada lēah*).

Rame Co [*Rame* DB, 1229 Fees, 1263 Ep].
Rame Hill is *Ramhill* 1324 FF. Rame is
nr the Devon border, where many pl. ns.
are of English origin. Rame may thus be
an English name and identical with OHG
rama (from *hrama*) 'a post, frame, barrier'
or the like. OE *hremman* 'to hinder' is
derived from a corresponding OE **hramu*.

OE **ramm** 'ram' is difficult to distinguish
from *hræfn* 'raven' and *hramsa* 'wild garlic'.
It is doubtless the first el. of RAMPSIDE,
RAMPTON, RAMSHORN, and perhaps of RAN-
TON.

Rampisham Do [*Ramesham* DB, 1236
Fees]. The first el. may be OE *ramm* 'ram'
or rather a pers. n. *Ram* (found in DB),
a byname formed from *ramm* 'ram'. Less
likely seems derivation from OE *hramsa*
'wild garlic'.

Rampside La [*Rameshede* 1292, *-heved* 1336
FC]. 'Ram's head.' R~ was originally the
name of a promontory, which may well have
been thought to resemble a ram's head.

Rampton Ca [*Ramtona* c 1080 ICC, *Rantone* DB], R~ Nt [*Rametone* DB, *Ramton*
1198, 1201 Cur, *Rampton* Hy 2 (1316) Ch,
1242 Fees]. OE *Ramm-tūn* 'TŪN with a
ram or where rams were reared'.

Ramsbottom La [*Romesbothum* 1324 Ct].
'Wild garlic valley.' Cf. HRAMSA, BOÞM.

Ramsbury W [*æcclesia Corvinensis* 905 BCS
614, *Rammesburi* 947 BCS 828, (tó) *Hrem-
nesbyrig* 980–8 Crawf, *Ramesberie* DB, *Rem-
nesbery* 1281 QW]. OE *Hræfnesburg*, which
may mean '*Hræfn*'s BURG' or 'BURG inhabited
by ravens'. The rendering *æcclesia Corvinensis* is not definite proof of the latter alternative. OE *Hræfn* may safely be assumed to
have been used; cf. OG *Hraban*, ON *Hrafn*.

Ramsdale Ha [*Ramesdela* 1170 Oxf], R~
YN [*Ram(m)esdal* 1240 FF], **Ramsden
Bellhouse, Crays & Heath** Ess [*Rames-
dana* DB, *-den* 1158 BM, 1208 Cur,
Ramesden Belhous, Gray 1254 Val, *Rames-
den Crei* 1274 RH], R~ O [*Ramesdon* 1179
RA, *Rammesden* 1279 RH, 1316 FA]. 'Wild
garlic valley' or possibly 'ram valley'. See
HRAMSA, RAMM, DENU.
Ramsden Bellhouse was held by Ricardus de
Belhus in 1208 (Cur). Bellhouse means 'belfry'.
—R~ **Crays** was held by Simon de Craye in
1252 (Cl). *Craye* is from CRAY or CRAYE in
France.

Ramsey Ess [*Rameseia* DB, *Rammesye* 1224
FF], R~ Hu [(into) *Hramesege* c 1000 BCS
1306, (æt) *Hramesige, Ramesige* c 1000
Saints, (on) *Ramesige* 1011 Byrhtferth].
'Wild garlic island.' See HRAMSA, ÉG.

Ramsgate K [*Ramisgate, Remisgate, Rem-
mesgate* n.d. StAug]. '*Hræfn*'s gate.' Cf.
RAMSBURY. The gate is one which leads to
the sea through the chalk cliffs.

Ramsgill YW [*Ramesgile* 1198 Fount],
Ramsgreave La [*Romesgreve* 1296 Lacy],

Ramsholt Sf [*Ramesholt* DB, 1166 P],
Ramshope Nb [*Rammeshope* c 1230
PNNb]. 'Gill or valley, grove, wood, hope
or valley, where wild garlic grew.' See
HRAMSA. OE *ramm* 'ram' is a possible
alternative.

Ramshorn St [*Rumesoura* 1197 P, *Romes-
overe* 13 PNSt, *Romesor* 1309 Ipm]. 'The
ram's slope or hill-side', rather than 'wild
garlic slope'. The second el. is OE OFER
'slope, hill-side'.

Ranby Li [*Randebi* DB, c 1115 LiS, *-by*
a 1166 BM]. '*Randi*'s BY.' First el. OScand
Randi pers. n. (OSw *Rande*).

Ranby Nt [*Ranebi, Ranesbi* DB, *Raneby*
1247 FF]. '*Hrani*'s BY.' First el. ON *Hrani*,
ODan *Rani* pers. n. (*Hrani* KCD 739, 743
&c.).

Rand Li [*Rande* DB, *Randa* c 1115 LiS,
Rande 1165 DC, 1206 Ass], R~ **Grange**
YN [*Randes* 12 PNNR, *Rand* 1251 Ass].
OE *rand* 'brink, edge, margin, shore'. The
el. is also found in RAUNDS, perhaps RAN-
WORTH. In East Anglia *rond* means 'a
marshy, reed-covered strip of land lying
between the natural river-bank and the
artificial embankment'. This sense may be
that in Rand Li.

Randwick Gl [*Rendewiche* 1121, *-wike* c
1145 Glouc, *Rindewyk* 1220 Fees]. 'WĪC on
a ridge.' The first el. is very likely an OE
rind- 'hill, ridge', cognate with Norw *rinde*
'ridge', Crimean Gothic *rintsch* 'hill'. OE
rind- is found in several pl. ns., as *Rind-
burna* 759 BCS 187 (nr Andoversford Gl),
Rinda crundel 958 ib. 1022 (Brk). For the
interchange of *i* and *e* cf. RENDCOMBE.

Rangeworthy Gl [*Rengeswurda* 1167 P,
Ryngeworth 1303 FA, *Rungeworthe* 1349
Subs]. The first el. looks like a derivative
of OE *hrung* 'rung, pole'. It may be an
OE **hrynge* or better an adj. **hryngen* 'made
of poles'. If so, the name means 'enclosure
made of poles or stakes'.

Ranskill Nt [*Ravenschel* DB, *Ravenskelf*
1275 RH]. 'The raven's hill' or '*Hrafn*'s
hill'. First el. OScand *hrafn* 'raven' or
Hrafn pers. n., second OScand *skialf* (see
SCYLF).

Ranston Do [*Iwerne* DB, *Randelfstone* 1274
Ipm, *Randolvestone* 1277 Ch]. '*Randulf*'s
TŪN.' *Randulf* is a Norman name of Scand
origin. See IWERNE.

Ranton or **Ronton** St [*Rantone* DB, *Ramton*
1208 Cur, *Rontun* 1236 Fees]. Very likely
identical with RAMPTON. Or it may be OE
Rand-tūn 'TŪN on a bank'. Cf. RAND.

Ranworth Nf [*Randwðe* 1044–7 KCD
785, *Randuorda* DB, *Randewrtha* c 1158
Holme, *Randeworth* 1203 Ass, *Randeswrth*
1242 Fees]. See WORÞ. The first el. may
be OE *rand* 'border, margin' &c. or the
OScand pers. n. *Randi* (cf. RANBY).

OE **råp** 'rope'. See ROPE, STYRRUP.

Rapton. See WRABNESS.

Rasen, Market, Middle & West, Li [*Rase, Rasa, Resne* DB, *Rasa, Media, Parua Rasa* c 1115 LiS, *Rasne* Hy 2 BM, *Magna Rasna* c 1150 BM, *Rasen, Westrasen* 1202 Ass, *Est, Media, West Rasne* 1242 Fees]. Market Rasen was formerly East R∼. OE *ræsn* 'plank', here probably in the sense 'plank bridge'. The river-name **Rase** is a back-formation.

Raskelf YN [*Raschel* DB, *Raskelf* 1242 Fees]. OScand *rā-skialf* 'roe-deer headland'. See RĀ, SCYLF.

Rastrick YW [*Rastric* DB, *Rastrik* 1274 Wakef]. Probably the name of a brook; cf. *Rastrikebroc* c 1200 Fount. The second el. is OE *ric 'stream'. The first is obscure. An OE *ræsn-ric* 'stream with a plank bridge' (cf. RASEN) might have become ME *Rastric*.

Ratby Le [*Rotebie* DB, *Rotebi* c 1200 Fr, *-by* 1209–35 Ep]. Identical with *Rotaby* 1021–3 KCD 736 (nr Newnham Np). '*Rōta*'s BY.' Cf. RUTLAND. The *Rōta* whose name is preserved in Rutland must have been a great land-owner, and he may have owned both Ratby and *Rotaby*.

Ratchwood Nb [*Wrethewode* 1279 PNNb]. 'Outlaw's wood.' First el. OE *wrecca* 'outlaw'.

Ratcliff Mx [*Radclif* 1422 FF], **Ratcliffe Culey** Le [*Redeclive* DB, *Radeclive* 1209–35 Ep], **R∼ on the Wreak** Le [*Radeclive* DB, 1209–35 Ep], **R∼ upon Soar** Nt [*Radeclive* DB, *Radecliva super Soram* R 1 BM]. OE *rēade clif* 'red cliff'.

R∼ **Culey** was held by Hugo de Culy in 1285 (FA). Culey from CULEY in Normandy.

Ratham. See ROTTINGDEAN.

Rathmell YW [*Rodemele* DB, *Routhemele* 1235 FF]. Identical with *Rauðamelr* in Iceland. The name means 'red sandbank'. Cf. RAUÐR, MEL(R).

Ratley Wa [*Rotelei* DB, *Rotteleia* Hy 2 BM, *-leg* 1205 Misc]. Perhaps '*Rōta*'s LĒAH'. OE *Rōta* is found as a byname (*Æþelstan Rota* BCS 917) and *Roting* occurs. Both are derived from OE *rōt* 'merry'.

Ratling K in Nonington [*Rytlinge* 11 DM, *Rethlinge* 1176 BMFacs, *Retlinge* 1212 RBE]. OE *rȳt-hlinc* 'hill with rough growth, rough shrubs &c.'. Cf. RȲT. For the change *nc* > *ng* see HLINC.

Ratlinghope Sa [*Rotelingehope* DB, *Rotelinghop* 1255 RH]. 'The valley of *Rōtel*'s people.' *Rōtel is a diminutive of *Rōta*; cf. RATLEY.

Rattery D [*Ratreu* DB, *Radetre* c 1240 PND]. 'The red tree.'

Rattlesden Sf [*Rattesdene* 11 KCD 907, *Ratesdana, Ratlesdena* DB, *Retlesden* 1198 P, *Ratlesden* 1198 FF, 1200 Cur]. Second el. OE *denu* 'valley'. The first is obscure.

Rauceby Li [*Rosbi, Roscebi* DB, *Roucebi* 1146 RA, 1170 P, *Raucebi* 1202 Ass, *Nord-, Sutrouceby* 1242 Fees]. OScand *Rauðs-býr* '*Rauð*'s BY'. First el. ON *Rauðr*, ODan *Røth* pers. n., lit. 'the red one'.

ON *rauðr*, OSw *røþer*, Dan *rød* 'red' is found several times in combination with *clif*: RAWCLIFFE, ROECLIFFE, ROCKCLIFF. In all probability all these are Scandinavianized forms of OE *rēade-clif*. See further RATH-MELL, RAWMARSH, also RAWTHEY. There was also an OScand pers. n. *Rauðr* 'the red one'. See RAUCEBY.

Raughton (-ahf-) Cu [*Ragton* 1182, *Rachton* 1186 P, *Rahton* 1202 FF]. 'TŪN on Roe Beck.' **Roe Beck** [*Ranhe* for *Rauhe* 1272, *Rache* 1285 For] is OScand *rā-ā* (from *rā-ah*) 'boundary stream'. Cf. ON RĀ.

Raunds (rahns) Np [(æt) *Randan* 972–92 BCS 1130, *Rande* DB, *Raundes* 12 NS]. The plur. of OE *rand* 'border' &c. See RAND.

Raveley Hu [*Ræflea* 974 BCS 1311, *Rauelai* 1163 P]. OE *hræfn-lēah* 'raven wood'.

Ravendale Li [*Ravenedal* DB, *Ravendala* c 1115 LiS, *Estravendal* 1254 Val, *Westravendale* 1219 Ep]. 'Raven valley.'

Ravenfield YW [*Rauenesfeld* DB, *Ravenesfeld* 1154 YCh 1475, *Ragenefeld* 1188 ff. P, *Ravenefeld* 1246 FF]. '*Hræfn*'s FELD' (cf. RAMSBURY) or possibly 'ravens' FELD'.

Ravenglass Cu [*Rengles* c 1170 CWNS xxix, *-glas* 1208 P, *Reynglas* c 1250 StB, *Ravenglas* 1297 Cl]. A Goidelic name containing OIr, Gael *rann* 'part, share' and *Glas* pers. n.: '*Glas*'s share'.

Raveningham Nf [*Rauenicham* DB, *Rafningeham* 1177 P, *Raueningham* 1203 Ass]. 'The HĀM of *Hræfn*'s people.' Cf. RAMSBURY.

Ravensbourne K [*Randesbourne* 1360, *Rendesburne* 1372 Ipm]. The forms are too late for a definite etymology. The first el. might be as in RENDLESHAM.

Ravenscar YN (no early forms found). 'The ravens' rock.' Second el. OScand SKER.

Ravenscroft Chs [*Ravenescroft* n.d. AD]. '*Hræfn*'s croft.' See RAMSBURY.

Ravensdale Db [*Rauenes . . .* DB, *Ravenesdale* 1251 Ch]. The first el. may be OE *Hræfn or OScand *Hrafn* pers. n. or OE *hræfn* 'raven'.

Ravensden (rahnz-) Bd [*Rauenesden* 1180, 1190 ff. P]. '*Hræfn's or ravens' valley.'

Ravenser Odd YE [(af) *Hrafnseyri* c 1145 Orkneyinga saga, *Ravenser* 1240 FF, *R∼ Hodde* 1260 Ipm, *Ravenserod* 1299 Ch]. '*Hrafn*'s gravel bank.' Second el. ON *eyrr* 'gravel bank'. *Odd* is ON *oddr, oddi* 'point of land'. Ravenser and R~ Odd were submerged by the sea c 1400. Cf. SPURN HEAD.

Ravensmeols. See MEOLS.

Ravensthorpe Np [*Ravenestorp* DB, 1199 FF], R~ YN [*Rauenestorp* DB]. '*Hrafn*'s thorp.' *Hrafn* is an OScand pers. n.

Ravenstone Bk [*Raveneston* DB, *Raueneston* 1163 P, *Ravenestone* 1225 Ep], R~ Le [*Ravenestun* DB, c 1125 LeS]. '*Hræfn*'s or *Hrafn*'s TŪN.' Cf. RAMSBURY.

Ravenstonedale We [*Rauenstaindal* 1223 FF, *Ravenstandal* 1251 Ch]. 'The valley with the ravens' stone.' The first el. is apparently OScand *hrafn(a)steinn* 'ravens' stone'.

Ravensworth Du [*Rӕveneswurthe* 1104–8 SD, *Raveneswrd* c 1180 Newcastle]. '*Hræfn*'s WORÞ.' Cf. RAMSBURY.

Ravensworth YN [*Raveneswet* DB, *-wad* 1157 YCh 354, *Ravenswath* 1227 Ch]. '*Hrafn*'s ford' (OScand *Hrafn* pers. n. and *vað* 'ford').

Raventhorpe Li [*Ragenal-, Rageneltorp* DB, *Ragheniltorp* c 1115 LiS, *Ragnildthorp* 1228 Ep]. '*Ragnhild*'s thorp.' First el. ON, ODan *Ragnhildr*, a woman's name (*Ragenild* DB &c.).

Rawcliffe, Out & Upper, La [*Rodeclif* DB, *Outroutheclif* 1324 LaInq, *Uproucheclive* 1246 Ass], R~ YN nr York [*Roudeclif, -e* DB, *Roudacliva* R 1 (1308) Ch], R~ **Bank** YN [(in) *Readeclive* 1104–8 SD, *Roudeclif* DB, *Roucheclive* 1242 FF], R~ YW [*Roupeclif* 1070–85 YCh 468, *Routheclive* 1238 Cl]. 'Red cliff.' R~ Bank is OE *rēade clif*, which was later Scandinavianized, OScand *rauðr* taking the place of OE *rēad*. The same is no doubt the history of the other Rawcliffes.

Rawdon YW [*Roudun, Rodum* DB, *Roudon* c 1200 YCh 1874, *Raudon* 1202 FF]. Possibly OE *rēade dūn* 'red hill' with substitution of OScand *rauðr* for OE *rēad*.

Rawmarsh YW [*Rodemesc* DB, *Rumareis* 1204 Cur, 1206 FF, (de) *Rubeo Marisco* 1240 FF, *Routhemersk* 1293 QW]. 'Red marsh.' The second el. is OE *mersc* 'marsh'. The first was doubtless once OE *rēad*, replaced later by OScand *rauðr*.

Rawreth Ess [*Raggerea* 1177, *Ragerugge* 1183 P, *Ragherethe* 1240, *-eth* 1242 FF, *Raureth* 1267 Ch]. 'Herons' stream.' First el. OE *hrāgra* 'heron', second OE *rīþ* 'stream'.

Rawtenstall La [*Routonstall* 1324 LaInq]. 'Roaring pool.' First el. ME *routand* 'roaring'. Second el. STALL 'pool'.

Rawthey R YW, We [*Routha, Roudha* 1235–55 CC], **Rothay** R We [*Routha* 1275 Ch, *Rawthaw* 1390–4 Kendale]. Perhaps 'trout stream'. The second el. is OScand *ā* 'river'. The first cannot well be OScand *rauðr* 'red',

but it may be an OScand **rauði* 'red one', a name for a trout.

Ray R Bk, O [*la Ree* 1363 Pat], R~ R W [*the Rey* 1576 Saxton]. A variant of REA. For earlier names of the Rays see ISLIP, WROUGHTON.

Raydon Sf [*Reindune, Rienduna* DB, *Reidunia* 12, *Reindun* c 1200 BM]. 'Rye hill.' First el. OE *ryge* 'rye' and *rygen* 'of rye'.

Rayleigh Ess [*Rageneia, Ragheleia* DB, *Reilee, Rielie* 1181 P, *Reileia* 1200 P, *Reylegh* 1219 Fees]. 'LĒAH where rye was grown.' Cf. prec. name. The name may alternatively have as first el. OE *rӕge* '(female) roe'.

Rayne Ess [(æt) *Hrӕgenan* c 1000 BCS 1306, (æt) *Rӕgene* a 995 Wills, *Raines* DB, *Reine* 1065 BM, *Reines* 1194 P]. R~ might be related to OE *oferhrӕgan*, found once and possibly meaning 'to tower' or 'to cover over', MHG *ragen* 'to tower, jut'. An OE **hrӕgene* (from *hraginōn*) might have meant 'shelter, hut' or 'eminence'.

Raynham Nf [*Reineham, Sutreineham* DB, *Reinham* 1199 FF]. '*Regna*'s HĀM.' Cf. RAINFORD.

Rea R Ca [*le Ee* 1447, *le Ree* 1455 ERN], R~ R Sa, Wo [*in þære éa Nen* c 957 BCS 1007, *Ree* 1310 PNWo], R~ R Wa [*Rhée* 1577 Harrison]. OE *æt þære ēa* 'at the river' became ME *atter ē, atterē*, which was wrongly divided as *atte rē*. *Re* was supposed to be a river-name. Cf. NEEN. On REA Wa see also TAME *infra*.

Reach Bd [*Reche, Rache* 1276 Ass], R~ Ca [*Reche* 1086 IE, *Recher* Hy 1 BM, *Reche* 1276 RH]. Engl *reach* 'portion of a river' is found in 1536 (OED). It is probably a derivative of OE *rӕcan* 'to reach', &c. The pl. n. Reach presupposes an OE *rӕc* (< *raikiō*), cognate with OE *rӕcan* and with ON *reik* 'parting of the hair', Sw *rēk* 'a stripe' &c. (also in pl. ns.). OE *rӕc* probably meant something like 'a strip'.

OE *rēad* 'red' is a common first el., but often difficult to distinguish from OE *hrēod* 'reed'. The latter ought never to give ME *Rad-*, whereas *rēad-* was often shortened to *rĕad-*, whence ME *Rad-*. But forms like *Rad-* are occasionally found even where the source is demonstrably OE *hrēod*, and are not decisive proof of OE *rēad*. OE *rēad* is found in REDE R, RADCLIFFE, RATCLIFF, REDCLIFF &c., RADFORD, RETFORD, RADNAGE, RADWELL, REDMILE, RODMELL and others.

Read La [*Revet* 1202 P, *Reved* 1246 Ass, *Rieheved* 1418 Whitaker]. 'Roe headland.' Cf. RӔGE (under OE RĀ), HĒAFOD.

Reading (-ĕ-) Brk [(to) *Readingum* 871, (from) *Readingum* 872 ASC, *Reddinges* DB]. OE *Rēadingas* 'the people of *Rēad(a)*'. OE **Rēad(a)* is a byname formed from *rēad* 'red' and corresponds to ON *Rauðr* (a common name) and *Rauði* (a nickname).

Reagill We [*Reuegile* 1176 P, *Revegyll* 1260 Kendale]. 'Fox valley' (ON *refr* 'fox' and *gil* 'valley').

Rearsby Le [*Redresbi, Reresbi* DB, *Reresby* 1236 Fees]. Identical with RAITHBY (1).

Reasby Li [*Reresbi* DB, c 1115 LiS, *-by* 1203 Ass]. Probably identical with REARSBY, but ON *Hrørekr*, ODan *Rørik* pers. n. is a possible alternative first el.

Reaveley Nb [*Reueley* 1242 Fees, *Reveley* 1269 Ipm]. Possibly 'the reeve's LĒAH' (see (GE)RĒFA), but the first el. may also be OE *hrēof* 'rough'.

Recu·lver K [*Regulbium* c 425 ND, *Reculf mynster* 669 ASC, *Racuulfe* c 730 Bede, *Reaculfe* c 890 OEBede, *Ricuulfi* c 765 BCS 199, *aet Ræculfo* 825 BCS 384, *Roculf* DB, *Raculvre* 1276 Ch]. Many more OE forms are on record. The name has been derived from an OBrit word for 'beak, bill' (found in Welsh *gylf* and cognate with OIr *gulba* 'beak, point'; OBrit *gulbā* or *gulbiā*) with a prefix corresponding to Lat *præ* or *pro*, the name meaning 'promontory'. IG *p* always disappears in Celtic. This is probably in the main correct, though the curious fact that *g* appears as OE *c* remains unexplained. Possibly the prefix was one that ended in *k* (cf. Welsh *rag*, OW *rac* 'before, against') or *s* (cf. Greek *pres-*). After *k* or *s*, Brit *g* would become *k*. The final *-er* is a late addition. Jackson takes the name to contain the prefix *ro-* (from *pro-*) in the sense 'great', the meaning being 'great headland'.

Redbourn Hrt [(æt) *Reodburne* c 1060 KCD 962, *Redborne* DB], **Redbourne** Li [*Radburne, Reburne* DB, (in) *Ratburno* 1090 RA, *Redburna* c 1115 LiS, *Rodburn* 1224 Ep]. OE *Hrēodburna* 'reedy stream'. Redbourn is on the VER, which is *Redburne* 1284 ERN.

Redbridge Ha [*Rodbrige* DB, *Redbregg* 1250 Fees]. The original name was *Hreutford* ('id est, uadum harundinis') c 730 Bede, *Hreodford* c 890 OEBede: 'reed ford, ford where reeds grew'. The bridge built at the ford got the name *Hreodbrycg* 956 BCS 926 (*Hreodbricg* 1045 KCD 781), which really seems to mean 'the bridge at *Hrēodford*' (elliptical).

Redcar YN [*Redker* c 1170 YCh (768), 1231 FF, *Rideker* 1272 Ipm]. 'Reedy marsh' (OE *hrēod* 'reed' and ON *kiarr* 'marsh').

Redcastle Sa at Weston under Redcastle [*Radecliffe* (rock) 1227 Ch, *-clif* 1228 BM, (Castrum de) *Radeclive* 1229 Cl, *Rubeum Castrum* 1276 Ipm]. The old name is identical with RADCLIFFE and means 'red cliff'. Redcastle appears to be elliptical for Redcliff castle.

Redcliff So at Bristol [*Radeclive* c 1180 Wells, *Redecliva*, (rubeam rupem called) *Cliva* n.d. Buckland]. 'Red cliff.' See RADCLIFFE.

Reddish La [*Rediche* 1212 RBE, *Redich* 1212 Fees]. 'Reed ditch' (OE *hrēod-dīc*).

Redditch Wo [(de) *Rubeo Fossato* c 1200 Madox, *la Rededich* 1247 FF]. Very likely identical with REDDISH. The translation 'rubeum fossatum' may quite well be due to popular etymology. But 'red ditch' is of course possible.

Rede R Nb [*Rede* c 1200 ERN, 1279 Ass]. OE *Rēade* 'the red one', a derivative of *rēad* 'red'. The valley is Redesdale [*Redesdale* 1075 ERN, 1212 Fees].

Rede Sf [*Reoda, Reda, Riete* DB, *Reode* c 1095 Bury, *Rede* 1254 Val, *Wrede* 1269 Ipm]. The isolated form *Wrede* may be disregarded. The source may be OE *hrēod* 'reed', here used in a collective sense: 'the reeds, the reed bed'. Another possibility is an OE **rēod* 'clearing'. See RĒOD.

Rĕdenhall Nf [*Radahalla, Radanahalla, Redanahalla* DB, *Redehal* 1166, *Redhala* 1186 P, *Redenhal* 1199 Cur]. Probably 'reedy HALH' in spite of the *a*-forms. First el. OE *hrēoden* 'reedy'.

Redgrave Sf [(on) *Redgrafe* 11 EHR 43, *-graue* c 1095 Bury, *Redegraue* 1179 P]. 'Reed ditch.' The elements are OE *hrēod* 'reed' and *græf* 'grave', here 'ditch'.

Redhill, Regilbury So [*Ragiol* DB, *Ragel* 1193, *Ragelbiri, Rachelburi* c 1200 Flaxley, *Rachel* 1254 Ass, *Raggel* 1289 FF]. The first el. may be OE *rā-ecg* 'roe hill', the second being OE *hol* 'hollow' or *hyll* 'hill'.

Redhill Sr [*Redehelde* 1301 Pat]. 'Red slope.' See HELDE.

Redisham Sf [*Redesham* DB, 1202 FF, 1267 Ch]. The regular gen. form suggests a pers. n. as first el. rather than OE *hrēod* 'reed'. It may be OE *Rēad*, on which see READING.

Redland Gl [*Thriddeland* 1209 Fees, *Yriddelond* 1285, *Thriddelond* 1346 FA, *Theriddelond* 1349 Subs]. 'The cleared land', OE *þæt rydde land*. See RYDDAN.

Redlingfield Sf [*Radinghefelda* DB, *Radelingfeud* Hy 2 (1285) Ch, *Radlingefeld* 1166 ff. P, *Redlingefeld* 1203 Ass, *Ridelingefeud* 1254 Val]. 'The FELD of *Rædel*'s or *Rædla*'s people.' *Rædel* is found as the name of a moneyer. **Rædla* would correspond to OHG *Ratilo*. Both are short forms of names in *Rǣd-*.

Redlynch So [*Redlisc* DB, *Redlis* 1219 FF, 1225, 1243 Ass, *Redlinch* 1225 Ass]. OE *hrēod-lisc* 'reed marsh'. On OE *lisc*, which probably meant 'reeds', but also 'marsh', see LYSCOMBE.

Redmain Cu [*Redeman* 1184 P, 1229 Pat, *Rademan* 1202 P]. Derivation from Welsh *rhyd maen* 'stone ford', suggested in PNCu(S), is not to be thought of, since the place is not on a stream. The first el. looks like OE *rēad* 'red', and *-main* may be dial. *man* 'cairn'. Cf. HAUGHMOND.

Redmarley Wo [*Ridmerlege, Redmerleie* DB, *Rudmerlega* c 1150 Surv], **R~** (rĭd-) **d'Abitot** Wo [*Reodemæreleage* 963 BCS 1109, æt *Rydemæreleage* 978 KCD 619, *Ridmerlege* DB, *Rudmarleye Dabetot* 1324 Ch]. 'LĒAH with a reedy lake' (OE *hrēodmere*).
R~ d'Abitot was held by Urse d'Abitot in 1086 (DB). Cf. CROOME D'ABITOT.

Redmarshall Du [*Rodmerehil* 1208–10 Fees, *Redmerhill* 1260 Pat]. 'Hill by a reedy lake.' Cf. prec. name.

Redmile Le [*Redmelde* DB, *Redmilde* 1202 Ass, *Remilde* 1221 Ep]. '(Place with) red soil.' The second el. is an OE *mylde*, derived from *molde* 'earth, soil'. The same word is Sw *mylla* 'mould, loose earth'. Cf. RODMELL.

Redmire YN [*Ridemare* DB, *Ridemere* 1166 P, 1204 FF, *Redmera* 1167 P]. OE *hrēodmere* 'reedy lake'.

Rednal Wo [(æt, on) *Wreodanhale* 780, 849, 934 BCS 234, 455, 701, *Weredeshale* DB]. As *a*-mutation is rare in West Saxon, the *eo* of the first el. was probably long, and connexion with OE *gewrid* 'thicket' is unlikely. An OE *wrēode* might be a derivative of OE *wrēon* 'to cover' and mean 'shed'. Second el. HALH.

Redruth (-rōo·th) Co [*Ridruthe* 1259, *Rudruth* 1283 Ep, *Riddruth* 1291 Tax]. 'Red ford', the elements being Co *rid* (= Welsh *rhyd*) 'ford' and *rudh* (= Welsh *rhudd*) 'red'.

Redwick Gl [(to) *Hreodwican* 955–9 BCS 936, *Redeuuiche* DB, *Radewic* 1230 Cl]. 'WĪC where reeds grew.'

Redworth Du [*Redwortha* 1183 BoB]. 'WORÞ where reeds grew.'

Reed Hrt [*Retth, Rete* DB, *Ruith* c 1150 Fr, *Red, Rud* 1204 Cur, *Ruth* 1212, 1219 Fees, *Rued* 1254 Val]. This can be neither OE *hrēod* 'reed' nor **rēod* 'clearing', though very likely the name was associated with *reed*. The source is OE RỸT, RỸHT 'rough growth' or 'rough common' or the like. See RỸT. There may quite well have been a side-form *rŷhþ* or *rŷþ* of this word.

Reedham Nf [*Redham* 1044–7 KCD 785, *Redeham* DB, *Redham* 1158 P], **Reedley Hallows** La [*Redelegh Halowez* 1464 Whitaker], **Reedness** YW [*Rednesse* c 1170 YCh 487, 1200 FF]. 'Reedy HĀM or HAMM, LĒAH and ness.' Reedley Hallows means 'haughs belonging to Reedley'.

Reepham (-f-) Li [*Refam, Refaim* DB, *Refham* c 1115 LiS, 1202 Ass], **R~** (-f-) Nf [*Refham* DB, 1203 Cur, 1254 Val]. 'Manor held or run by a reeve' (OE *gerēfa*). The OE form must have been (*ge*)*rēfhām*, analogous to OE *gerēfærn* 'court-house', *gerēfmǣd* 'meadow under the supervision of a reeve'.

Reeth YN [*Rie* DB, *Ryth* 1224, *Rithe* 1226 FF]. OE *riþ* 'stream'. R~ is on Arkle Beck.

OE (ge)rēfa. See REEPHAM, REAVELEY.

ON refr, OSw **ræver** 'fox'. See REAGILL.

Reigate (rī-) Sr [*Reig.* Hy 2 BM, *Regata* 1185 P, *Reigat'* 1199 FF, *Regate* 1203 Cur, *Reigate* 1212 Fees]. 'Roe gate'; cf. ROGATE. First el. OE *rǣge* 'female of the roe'. *Gate* might well mean 'pass' in this case, but 'gate' is perhaps more likely.

Reighton (-ē-) YE [*Rictone* DB, *Ricton* 1201 Cur, *Richtona* 1125–30 YCh 1135, *Richton* 1231 FF]. OE *Hrycg-tūn* 'TŪN by a ridge'. Cf. DEIGHTON.

Remenham Brk [*Rameham* DB, *Remeham* 1167 P, *Remenham* 1242 Ipm, *Rumeham* 1242 Fees]. Perhaps OE *Rioman-hām* 'HĀM by the rim or bank'. OE *rima* would be *rioma* in some dialects. For the change *io > u* cf. PUSEY.

Rempstone Nt [*Rampestune, Repestone* DB, *Rampestona* Hy 2 (1316) Ch, *Rempeston* 1231 Ep]. '*Hrempi*'s TŪN.' *Hrempi* is unrecorded, but cf. *Hrempingwiic* 798 BCS 289 (K) and OG *Rampo, Hremfing* pers. n. The name belongs to OE *gehrumpen* 'wrinkled' (past part. of **hrimpan*), *hrympel* 'wrinkle', Norw *ramp* 'lean person' and the like.

Rendcombe Gl [*Rindecumbe* DB, *Uuer Rindecumb* 1175 AC, *Rendecumb* 1242 Cl, *Renden Cumb* 1262 Ipm]. 'The valley of R *Hrinde*.' The river-name [*Hrindan broc* 852 BCS 466] is a derivative of OE *hrindan* 'to push, thrust' and means 'the torrent', lit. 'the thruster'.

Rendham Sf [*Rimdham, Rindham, Rindeham* DB, *Rindham* 1203 Cur, *Rendham* 1254 Val, 1268 Ch]. If the form *Rimdham* DB is trustworthy, the first el. may be OE *rŷmed* 'cleared' from *rŷman*.

Rendlesham Sf [*Rendlæsham* i.e. *mansio Rendili* c 730 Bede, *Rendlesham* c 890 OE Bede, DB]. '*Rendel*'s HĀM.' **Rendel* must be a short form of names in *Rand-*; such names are not well evidenced in OE, but are common in Scandinavia and on the Continent.

Renhold Bd [*Ranhale* 1220 Subs, 1229 Ep, *Ronhale* 1239 Ep, 1252 Ch, 1274 Cl, *Runhale* 1247 Ass]. The forms with *a* and *o* seem to be the most trustworthy. Isolated *e*-forms are probably mistakes for *o*. The rare *u*-forms may partly be mistakes for *a*. The first el. is thus probably a word with *ā*, which later became *ō*. It may be OE *rā* 'roe-deer' in the gen. sg. or plur. Thus 'the nook of the roe-deer'. See HALH.

Rennington Nb [*Reiningtun* 1104–8 SD, *Renninton* 1176 P, *Renigton* 1242 Fees]. Identical with RAINTON. Symeon of Durham tells us that *Reiningtun* was named from a certain *Reingualdus*. Clearly this man was also called *Regna* for short and his people were *Regningas*.

Renscombe Do [(in) *Hreminescumbe* 987 KCD 656, *Romescumbe* DB, *Rembescumb* 1212 Fees]. OE *hremnescumb* 'raven valley'. OE *hremn* is a side-form of *hræfn* 'raven'.

Renwick Cu [*Rauenwich* 1178, *Raueneswich* 1190 P]. Either '*Hrafn*'s (or **Hræfn*'s) wīc' or 'wīc on R RAVEN'. The river-name is *Raven* 12 Lanercost, 13 WhC. It may be a back-formation from Renwick, or it may be an independent formation, possibly OE *hræfn* 'raven' used in a transferred sense of a river with dark water.

OE **rēod*, cognate with OHG *riuti* 'cleared land' and used in the same sense, is assumed in PNSx as the second el. of COLDRED and some other names. Most of the names adduced may be satisfactorily explained from OE *hrēod* 'reed'. Coldred is *Colredinga* in 944. If belonging here, it ought to have appeared as *Colreodinga*, but the language of the charter is not above suspicion, and *ēo* may have become *e* in an unstressed position. If OE *rēod* 'clearing' existed, it is a possible source of REDE Sf.

Repps Nf [*Repes* DB, *Reppes* c 1150 Fr, 1171 P, *Repples* 1191 P, *Reples* 1203 Ass], **Northrepps** Nf [*Norrepes* DB, *Nordrepples* 1185 P, *Northreppes* 1254 Val], **Southrepps** Nf [*Sutrepes* DB, *-repples* 1209 FF, *Suthreppes* 1254 Val]. The original form was evidently early ME *Repples*. This is related to OE *ripel* 'a strip', esp. 'a strip of wood', found in RIPPLE K, Wo and in some OE examples. The form (be) *repple* occurs 1033 KCD 752. The same word is Norw *ripel*, *repel* 'a strip', also of a strip of land or wood. OE *ripel* cannot have given *Repples*, but we may assume a side-form **ripul*, whence **riopul*, **reopul*, ME *repel*. The loss of *l* is possibly due to Norman influence. The name Repples would mean 'the strips' and refer to strips of land in a fen or the like that could be tilled.

Repton Db [*Hrypadun* c 745 Felix, 848 BCS 454, (on) *Hreopandune* 755 ASC (F), *Rapendune* DB, *Rep"endon* 1197 FF, *Repedon* 1236 Fees]. 'The hill of the *Hrype* tribe.' The same tribe gave its name to RIPON and probably some places near Ripon (see RIBSTON, RIPLEY). The etymology of the tribal name is obscure.

Rere Cross. See REY CROSS.

Reston, North & South, Li [*Ristone* DB, *Ristuna* c 1115 LiS, *Riston* 1170 P, 1202 Ass, *Rustun* 1193 P, *North Riston* 1274 Ipm]. OE *Hris-tūn* 'TŪN by brushwood'.

Reston We [*Rispeton* 1272 ff., *Respeton* 1297 Kendale]. The first el. is an unrecorded OE word cognate with OHG *hrispahi*, G *Rispe*, MLG *rispe* 'brushwood'.

Restormel Co [*Rostormel* 1310, *Restormel* 1331 Ch]. 'Moor at the bare hill.' The elements are Co *ros* 'moor', *tor* 'mountain' and *moel* 'bare'.

Restro·nguet (-ngg-) **Creek** Co [*Restrangret* 1234 FF, *Restronget* 1322 Misc]. 'Ford by *Tronget*.' First el. Co *rid*, *res* 'ford'. The second is a pl. n. *Tronget* or the like, consisting of Co *tron*, *trein* (= Welsh *trwyn*) 'nose, promontory' and *coid* (from *cēt*) 'wood' and meaning 'wood on the promontory'.

Retford, East & West, Nt [*Redforde* DB, *Rat-*, *Retford* 1230 P, *West Retford* 1278 Ipm, *Estretford* 1375 BM]. 'Red ford.' Cf. RADFORD.

Rettendon Ess [*Rettendun* c 995 BCS 1289, *Ratendune* c 1050 KCD 907, *-duna* 1086, *Retendon* 1254 Val]. The earliest form tells against an original form *Rettan dūn*. The first el. is rather an adj. which might be a derivative of OE *rætt* 'rat' (OE **rætten* 'infested with rats').

Revelstoke D [*Rawelestok* 1219 Ass, *Rewelstoke* 1417 AD]. See STOC. The original name was STOKE, *Revel* being a Norman family name borne by some early owner(s). Cf. CURRY RIVEL.

Revesby Li [*Resuesbi* DB, *Reuesbia* 1142 NpCh, *-bi* 1154 BM]. '*Ref*'s BY.' First el. ON *Refr*, ODan *Ræf*, pers. n., originally a nickname 'the fox'.

Rew Wt [*Rewe* 1287–90 Fees], **Rewe** D [*Rewe* DB, 1242 Fees]. OE *ræw* 'row', here in the sense 'a row of houses'.

Rey Cross YN [*Rerecros on Stanmore* Hy 2 (1348) Ch, *Rere Crosse* c 1275 StB, *Reir croiz de Staynmore* 1280 CWNS xxvii]. Perhaps OSw *rør* 'cairn' and *cross*, i.e. 'cross in a cairn'. Rey Cross is on the boundary between Yorkshire and Westmorland and may be a boundary mark.

Reydon Sf [*Rienduna* DB, *Reydone* 1254 Val]. 'Rye hill.' Cf. RAYDON.

Reymerston (rĕ-) Nf [*Raimerestuna* DB, *Reimerestona* 1168 P]. '*Raimer*'s TŪN.' *Raimar* (DB) may be a Continental name, but OE *Regenmær* may quite well have existed.

Rhiston Sa [*Ristune* DB, *Russeton*, *Riston* 1242 Fees, *Ruston* 1318 Misc]. First el. OE *risc*, *rysc* 'rush'.

Rib R Hrt [*Ribbe* 13 AD]. Probably a back-formation from a pl. n. containing OE *ribbe* 'hound's-tongue, ribwort'. If an original river-name it may be derived from *ribbe* in the sense 'water-cress' (found in E. Anglian dial.).

Ribbesford Wo [*Ribbedford* 1023 KCD 738, *Ribetford* 11 Heming, *Ribeford* DB]. Either 'ford by or with a bed of *ribbe*' or 'ford where *ribbe* grew' (cf. RIB). The first el. may be an OE *ribb-bed* or an adj. **ribbede* 'overgrown with *ribbe*'.

Ribble R YW, La [*Rippel* c 715 Eddi, *Ripam* DB, *Ribbel* 930 YCh 1, 1002 Wills, *Ribble* c 1130 SD]. If *ribyll* in the Myvyrian

Archaeology 143 refers to the Ribble, the name is probably British, and its etymology is obscure. If the name is English, it might be an adj. *ripel* 'tearing', derived from OE *ripan* 'to reap' (originally 'to tear'). On the Ribble is **Ribbleton** La [*Ribleton* 1201 P].

Ribby La [*Rigbi* DB, 1169 P]. 'BY on a ridge' (ON *hryggr*).

Ribchester La [*Ribelcastre* DB, *Ribbelcestre* 1215 P]. 'Roman fort on R Ribble.'

Ribston YW [*Ripestain, -stan* DB, *Ribestan* 1173 YCh 197, *Ribbestain* 1202 FF]. No doubt OE *Hrypa stān* 'the stone of the *Hrype*' (cf. RIPON). The stone may have been a boundary stone marking the territory of the *Hrype* tribe, or a stone at the meeting-place of the tribe. An intervocalic change *p > b* is found early in other names, e.g. in HEBDEN.

Ribton Cu [*Ribbeton* 12, 13 StB]. 'TŪN where *ribbe* or hound's-tongue grew.'

Riby Li [*Ribi* DB, c 1115 LiS, *Riebi* 1159 P, *-by* 1202 Ass]. Possibly 'BY where rye was grown'. First el. OE *ryge* 'rye'. If so, Riby is probably a Scandinavianized form of OE *Rygetūn*.

OE **ric** 'stream, ditch' is unrecorded, but must be postulated for several names, as the stream-name SKITTERICK (see ERN), LINDRICK, RASTRICK, WHELDRAKE; cf. CHATTERIS. It appears to be found in **Glynde Reach** Sx [*Ritche* (sewer) 1544 PNSx, perhaps *Riche* 1332 Subs]. The extinct vil. of **Riche** Li [*Riche* DB, *Rike, Richehundred* 1200 Cur] may have been called *Ric* in OE from a stream. OE *ric* is related to MHG *ric* a narrow road', Sw dial. *raik*, Norw *reik* 'a stripe', OE *rǣcan* 'to reach' &c.

Riccal R YN [*Rycaluegr[eines]* 1252 Riev, *Ricolvegraines* 1332 Pat], **Riccal House** (on the stream) [*Ricalf* DB, 1293 QW, *Ricalue* 1257 Ch]. The name means 'the calf of the Rye, little Rye'. Riccal is a trib. of the Rye. *Calf* is used here in about the same way as when a small island near a larger one is called the calf of the latter (e.g. the CALF OF MAN). The el. *-graines* (ON *grein*) means 'fork or branch of a river'.

Riccall YE [*Richale* DB, c 1130 SD, 1227 FF, *Richehale* 1190 P]. 'Rica's haugh.' *Rica* is found in *Rican ford* KCD 713 and is a short form of names such as *Ricsige*.

Richards Castle He [(baronia) *Castri Ricardi* 1212 Fees]. The castle is stated to have been built by Richard son of Scrob, a Frenchman who came into England in the time of Edward the Confessor.

Richborough K [*Routoupiai* c 150 Ptolemy, *Rutupina litora* Lucanus, *Rutupino* (abl.) Juvenal, *Ritupis, portus Ritupis* 4 IA, *Rutupis* c 425 ND, *Rutubi portus, Reptacæstir* c 730 Bede, *Raette* 11 DM, *Ratteburg*

1197 FF, *Retesbrough* 14 VHK ii. 4]. The forms point to a first syllable with *ū*, which later became *i* (cf. Welsh *din* from early *dūno-*). The stem is thus *rūt*, which may belong to the root *reu* 'to tear out, dig' &c. in Lat *ruo, rutrum* 'a spade' &c. and possibly mean 'ditch, trench'. The el. *-up-* is probably a suffix. The exact meaning of the name cannot be determined. The Brit name was taken over into English, where *t-p* underwent metathesis to *pt*; hence *Repta-* in Bede and with assimilation of *pt* to *tt Raette* DM. Richborough evidently arose by *burg* being added to the genitive of ME *Rette*. The etymology of the Brit name suggested is of course tentative, but nothing better has been proposed.

Richmond Sr [*Richemount* 1502 AD]. The earlier name was SHEEN, which was replaced by Richmond after the accession (1485) of Henry 7, previously earl of Richmond. The title was taken from **Richmond** YN [*Richemund* 1108–14 YCh 25, *-munt* 1167 P]. The latter was named from one of the Richemonts in France.

Rickerby Cu [*Ricardeby* 1247 Ipm]. 'Richard's BY.' Richard is a Norman name, also found in **Rickergate** Cu, a street in Carlisle [*vicus Ricardi* c 1206 Holme C], and **Rickerscote** St [*Ricardescote* DB].

Rickinghall Inferior & Superior Sf [(at) *Rikinghale* 10 BCS 1013, *Ricynga-,Rikingehale* 11 EHR 43, *Rikingahala, Richingehal(l)a* DB, *Uprichingehale* c 1095 Bury, *Rikinghale Inferior* Hy 3 BM]. 'The HALH of *Rica*'s people.' Cf. RICCALL.

Rickling Ess [*Richelinga* DB, *Richelinges* 1185 P, *Riclinges* 1214 Cl]. '*Ricel(a)*'s people.' A pers. n. *Ricel(a)* has Continental cognates. Or '*Ricola*'s people'. *Ricola* was queen of Essex in the 6th cent.

Rickmansworth Hrt [*Prichemareworde* DB, *Rikemaresworth* 1198 (1301) Ch]. '*Ricmǣr*'s WORÞ.' *Ricmar* is well evidenced on the Continent.

Riddings Db [*Rydynges* 1296 FF]. OE *ryding* 'clearing'.

Riddlesden YW [*Redelesden* DB, *Redlesden* c 1180 YCh (1867), 1226 FF], **Riddlesworth** Nf [*Redelefuuorda* DB, *Redleswrth* 1242 Fees, *Redeleswrth* 1256 Ipm, *Rydeleswrth* 1274 Ipm]. '*Hrēþel*'s valley and WORÞ.' *Hrēþel* is found in Beowulf and in *Hredlesstede* BCS 741.

Ridge Hrt [*la Rugge* 1275 AD]. OE *hrycg* 'ridge'. **Ridgeacre** Wo [*Rugacre* 1271 Ct]. 'Field on a ridge.'

Ridgewell (rĕj-) Ess [*Rideuuella* DB, *Redeswell* R 1 Fr, *Rodewell* 1245 Ch, *Radeswella* 1163 P, *Redeswell* 1274 RH]. 'Reedy stream.' First el. OE *hrēod* 'reed'.

Ridgmont Bd [*Rugemund* 1227 Ass, (de) *Rubeomonte* 1349 Cl], **R~** YE [*Rugemunt* 1166 f. P, *Rugeomont* 1260 Ipm]. A Fr name meaning 'red hill', probably a Fr *Rougemont* transplanted into England.

Ridgwardine Sa [*Ruggewurd* 1188, *Rugwrthin* c 1203 Eyton]. 'WORÐIGN on or by a ridge.'

Riding Nb [*Ryding* 1262 Ipm]. OE *ryding* 'clearing'.

Riding, East, North & West, Y [*Estreding, Oustredinc, Est* (*Nort, West*) *Treding, Westreding* DB, *Nortrithing* 1198 Fees]. The same division was formerly made in Lincolnshire, where Lindsey was divided into three Ridings [*Nort-, Sudtreding, Westreding, Nort* (*Sud, West*) *Treding* DB, *Nortriding, Suttriding, West Triding* c 1115 LiS]. Riding is ON *þriðiungr*, OSw *thrithiunger* 'third part', whose initial consonant was lost in combination with *east, west* &c. The names are Scandinavian, as was the division itself, and East Riding was originally OScand *Austþriðiungr* (cf. one DB form).

Ridley Chs [*Riddeleg* c 1255 Chester, *Ridleg* 1260 Court], R~ Nb [*Ryddeley* 1268, *Rydeley* 1271 Ipm]. OE *rydde lēah* 'cleared LĒAH'. See RYDDAN.

Ridley Ess in Terling [*Retleia* DB, *Redleigh* 1385 BM], R~ K [*Redlege* DB, 11 DM, *Riddelee* 1198 FF, *Rodlegh* 1291 Tax]. OE *hrēod-lēah* 'reedy LĒAH'.

Ridlington Nf [*Ridlinketuna* DB, *Ridelington* 1254 Val, *Redlington* 1199 P, 1267 Ch], R~ Ru [*Redlinctune Cherchesoch* DB, *Ridelinton* 1167 P, *Rodlinton* 1202 Ass, *Ridlingtona* 1209–19 Ep], **Rillington** YE [*Redlintone* DB, *Rillington* 1190 P, *Ridlinton* 1229 Ep]. The first el. of these may be derived from a pers. n., e.g. *Hrēþel*, or from a pl. n. identical with RIDLEY (1 or 2).

Ridware, Hamstall, Mavesyn & Pipe, St [*Rideware* 1004 PNSt, DB, *-wara* 1169 P, *Hamstede Ridewale* 1236, *Hamstal, Media Ridewar* 1242 Fees, *Ridewale Mauvaisin* 1236 Fees, *Pipe Ridware* 14 PNSt]. The second el. is very likely OE *-waru* 'dwellers' as in CLEWER. The first is possibly Welsh *rhyd* 'ford' or OE **ride* as in LEATHERHEAD. The meaning would in either case be 'dwellers by the ford'.
Hamstall is OE *hāmsteall* 'homestead, residence'. The meaning would be 'demesne farm' or the like. **Mavesyn** R~ from the Malveisin family, resident here at least from t. Hy 1. Cf. BERWICK MAVISTON. **Pipe** R~ from the Pipe family. Robert de Pipe (from PIPE St) got the manor c 1285.

Rievaulx (rēvō, locally rīverz) YN [*Rieuall.* 1148–50 BM, *Rievalle* 1157 YCh 401]. A translation of Engl *Ryedale* 'the valley of the Rye'.

Rigbolt or **Rightbolt** Li [*Writtebaud* 13 BM, *Writebaud* 1251 Ep]. OE *wyrhtan bold* 'the wright's dwelling'.

Rigmaiden We [*Rigmaiden* 1255, 1302 LaInq]. 'The maiden's ridge'? Cf. ASPATRIA.

Rigsby Li [*Rig(h)esbi, Richesbi* DB, *Rigesbi* c 1115 LiS, *Riggesbi* 1193 P, 1202 Ass]. 'BY on a ridge' (OScand *hryggr*) or '*Hrygg*'s BY' (ON *Hryggr* pers. n.).

Rigton YW nr Leeds [*Ritun* DB, *Ricton* 1200 Kirkst, *Rigton* 1237 FF], R~ YW nr Kirkby Overblow [*Riston* DB, *Rigton* 1588 FF]. 'TŪN on a ridge' (OE *hrycg*, ON *hryggr*).

Rillington. See RIDLINGTON.

OE **rima** 'rim, border, bank, coast'. See RYME, REMENHAM, RIMPTON, RIMSWELL, also RIMINGTON, RINGWOOD.

Rimington YW [*Renitone* DB, *Rimingtona* 1182–5 YCh 199, *-ton* 12 Pudsay, *Rymmigton* 1244 FF], **Rimington Brook** [*Rimingden* c 1280 Sawley Cart]. *Riming* may be an old name of the brook, or of the ridge on which Rimington is, in either case derived from OE RIMA.

Rimpton So [(*æt Rimtune* 938, *Rimtun* c 975 BCS 730, 931, *Rintone* DB]. 'TŪN on the RIMA or border.' R~ is on the Dorset border.

Rimswell YE [*Rimeswelle* DB, *-well* Hy 2 BM, 1208 FF]. R~ is on a slight elevation in low-lying surroundings and near the sea. The probable base is *rim-swelle* or *riman-swelle* 'hillock near the coast'. Cf. SWELL.

Ringborough YE [*Ringheburg, -borg* DB, *Ringeburg* 1285 FA]. Apparently an OE *Hringa-burg*, the first el. being OE *hring* 'circle' in the gen. plur. Some stone circles or circular entrenchments might be referred to.

Ringland Nf [*Remingaland* DB, *Ringeland* 1206 Cur, *-lond* 1219 Fees]. The first el. is a tribal name in *-ingas*, but the original form of the element is obscure. One might think of an OE *Rȳmingas*, derived from an unrecorded *Rȳmi* (a short form of names in *Rūm-*) or *Rimingas* 'people on the border' or the like (cf. RIMA).

Ringleton K [*Ringuentun* 1070–82 StAug, *Ringetone* DB, *Ringleton* 1242 Fees, *Ringelton* 1265 Misc]. R~ is nr **Ringlemere** [*Ryngwynemere* 1278 Ass]. The latter might be supposed to contain a name **Hringel* 'circular lake'. But the earliest forms point to an OE **Hringwynn* pers. n. (fem.) as the first el. of both names.

Ringmer Sx [*Ryngemere* 1276, *Ryngmere* 1289 FF]. OE *Hringmere* 'round lake'.

Ringmore D nr Modbury [*Reimore* DB, *Red-, Ridmore* 1242 Fees], R~ D nr Teignmouth [*Rumor* DB, *Redmor* 1275 RH, 1284–6 FA]. OE *hrēod-mōr* 'reedy moor'.

Ringsfield Sf [*Ringesfelda* DB, *-feld* 1267 Ch, *Ringefeld* 1235 FF, 1264 Ch]. '*Hring*'s FELD' or 'FELD with a circle'. See HRING. **Hring* is a short form of names in *Hring-*. Cf. ON *Hringr* pers. n.

Ringshall Sf [*Ringhesehla, Ringeshala* DB, *Ringeshale* 1198 FF, *-hal* 1203 Cur,

Renggeselle 1355 BM]. OE *Hring-gesella*, the first el. referring to a stone circle or some other round object. The second is OE (*ge*)*sell* 'shelter for cattle'.

Ringstead Do [*Ring(h)estede* DB, *Ringstede* 1227 FF, 1264 Ipm]. The first el. is very likely OE *hringe* 'a salt-pan'. The place is on the sea. Cf. STEDE.

Ringstead Nf [*Ringstyde* c 1050, -*stede* 1060 Th, *Rincsteda* DB, *Ringestede* 1186 P], R~ Np [*Ringstede* 12 NS, 1220 Fees, -*sted* 1227 Ch, *Ringestede* 1203 Cur]. First el. OE *hring* 'circle'. A stone circle may be referred to, or a circular enclosure or the like. Cf. STEDE.

Ringwood Ha [(to) *Rimucwuda* 955, *Rimecuda* 961 BCS 917, 1066, *Rincvede* DB, *Ringwode* 1199 P, 1219 Fees]. *Rimuc* may be a derivative of OE *rima* 'border' &c., perhaps in the sense 'boundary wood'.

Ringwould K [*Roedligwealda* 861 BCS 855, *Ridlingwalde* 1275 RH, *Rudelingewealde* Hy 3 Ipm, *Ringwald* 1185 P]. 'The weald of *Hrēþel*'s people.' See WALD.

Ripe Sx [*Ripe* DB, c 1150 Fr, *Rip*, *Ryp* 1240 ff. Ch, *Ripp* 1288 Ass]. Identical with *Ripp* 741 BCS 160, a wood in Kent. OE *ripp* is very likely related to *ripel* 'a strip' and had about the same meaning, i.e. 'a strip of wood' &c.

OE **ripel** 'a strip, a strip of wood' is the source of RIPPLE K, Wo and probably the first el. of RIPLEY and perhaps some other names. A side-form **riopul* appears to be the base of REPPS. OE examples of *ripel* are (andlang) *riple* 968 BCS 1218, *Myntleage riple* BCS 624, *Suggariple* c 1050 HSC. The last has OE *sucga*, the name of a bird, as first el. OE *ripel* corresponds to Norw *ripel* 'a strip' (*skogarripel* 'a strip of wood'). The word is derived from OE *ripan* 'to reap', lit. 'to tear, pluck' and means 'a piece torn off'. It is recorded in the sense 'coppice, thicket' in He dial.

Ripley Db [*Ripelie* DB, *Rippelega* 1176 P, *Rippelle* 1240 FF], R~ Ha [*Riple* DB, *Ripela* 1167 P], R~ Sr [*Ripele* 1220 Cl, *Reppele* 1240 Pat, *Ryppeleye* 1279 QW]. OE *ripel-lēah* 'LĒAH that has the shape of a strip'. Cf. RIPEL.

Ripley YW [*Ripeleia* DB, *Ripelai* 1165 P, *Rippeleg* 1202 FF]. Possibly identical with the other Ripleys. But this Ripley is near Ripon and may be suspected to be 'the LĒAH or wood of the *Hrype* or Ripon people'.

Riplingham YE [*Ripingham* DB, *Riplingham* c 1180 YCh (1126), 1202 FF, *Ripplingeham* 1180 P], **Riplington** Nb [*Riplingdon* 1242 Fees, -*tone* 1251 Sc]. 'The HĀM and TŪN of the *Riplingas*', i.e. perhaps 'people at a strip (of wood or land)'. Riplingham and Riplington are on or by long ridges, which may have been called *Ripel*.

Ripon YW [(in) *Hrypis*, *Hripis* c 715 Eddi,

Inhrypum (monastery) c 730 Bede, c 890 OEBede, *Hrypsætna* (cirican), *Onhripum* c 890 OEBede, (Æt) *Rypum* c 1030 YCh 7, (æt) *Hryopan* c 1000 Saints, *Ripum*, *Ripun* DB]. Ripon is the dat. plur. of OE *Hrype*, a tribal name, found also in REPTON Db. Cf. Introd. p. xiii f.

Rippingale Li [?*Hrepingas* 675 ASC (E), (in) *Repingale* 806 BCS 325, *Repinghale* DB, c 1160 DC, *Repingehal* 1166 P, *Reppingehal* 1221 Ep]. *Hrepingas* 675 need not refer just to Rippingale, but evidently it is identical with its first el. The latter is 'the HALH of the *Hrepingas*'. *Hrepingas* is derived from a pers. n. cognate with the pers. n. stem found in OG *Raffo*, *Rafold* &c., ON *Hrappr*. The last is identical with *hrappr* 'active', which belongs to ON *hrapa* 'to hurry'.

Ripple K [*Ryple* 1086 KInq, *Ripple* 1235 Cl, *Riple* 1275 RH], R~ Wo [*Rippell* 680 BCS 51, *Rippel* DB]. OE *ripel* 'a strip'. R~ Wo may have been named from the tongue of land in which the place stands.

Ripponden YW [*Ryburne-*, *Riburnedene* 1307 f. Wakef, *Rybunden* 1566 YD]. 'The valley of the RYBURN.' The latter is *Riburn* 1308 Wakef. Its first el. may be identical with the river-name RYE, or it may be OE *hrife* 'fierce'.

Ripton, Abbot's & King's, Hu [*Riptone* c 960 BCS 1061, *Riptona*, *Ripptune* 974 BCS 1310 f., *Riptune* DB, *Riptona Abbatis*, *Ripton Regis* 1163 P]. 'TŪN by a strip of wood.' Cf. RIPE.

Abbot's R~ belonged to Ramsey Abbey.

Risborough, Monks & Princes, Bk [(æt) *þǣm éasteran Hrisanbyrge* 903 BCS 603, *Risenbeorgas* 1004 Wills, *Riseberge* DB, *Monks Ryseberge* 1290 AD iii, *Pryns Rysburgh* 1433 Pat]. 'Hill(s) covered with brushwood.' First el. OE *hrisen* adj.

Monks R~ belonged to Christchurch, Canterbury.—Princes R~ was held by the Black Prince.

Risbury He [*Riseberie* DB, -*bur*' 1212 Fees]. 'BURG in brushwood' (OE *hris*).

Risby Li nr Market Rasen [*Risebi* DB, 1154 BM, *Risabi* c 1115 LiS], R~ Li nr Roxby [*Risebi* DB, c 1115 LiS, 1196 P, *Risabi* c 1115 LiS, *Risby* 1254 Val], R~ Sf [*Rysebi* 11 EHR 43, *Resebi*, *Risebi* DB, *Resebi* 1179 P, *Rissebi* 1166 RBE, *Risseby* c 1265 Bodl], R~ YE [*Risbi* DB, *Rizebi* 1167 P, *Risceby* 1229 Ep, *Resceby* 1297 Subs]. The Li Risbys are probably OScand *Hris(a)bý* 'BY in brushwood'. Cf. RISBY in Denmark (several). R~ Sf and YE are more likely identical with Sw RYSSBY, OSw *Rytzby*, which is *Rýðs-býr* 'BY at a clearing'.

OE **risc, rysc** 'rush' is a fairly common first el. and it is used alone as a pl. n. in RUSHOLME (OE *ryscum* dat. plur.), as a second el. in LANGRISH. As a first el. it is sometimes difficult to distinguish from *hris* 'brushwood'. See RHIS-, RISH-, RUSH- (passim), ROSEDEN, RUISHTON, RUISLIP.

Rise YE [*Risun* DB, *Risa* c 1165 YCh 1361, *Rise* 1251 FF, *Ryse* 1297 Subs], **Riseholme** Li [*Risun* DB, c 1115 LiS, *Risum* 1254 Val]. The plur. of OE *hris* 'brushwood', partly in the dat. form *hrisum.*

Riseley Bd [*Riselai* DB, 1156 P, *Risle* 1199 P, *Risele* 1202 FF], **R~** Brk [*Riselee* n.d. AD]. OE *Hris-lēah* or *Hrisen-lēah* 'brushwood LĒAH'.

Rishangles Sf [*Risangra* DB, *Rishangr'* 1203 Ass, *Rissangeles* 1254 Val]. 'Brushwood slope.' See HRĪS, HANGRA.

Rishton La [*Riston* 1200–8 PNLa, *Ruston* 1242 Fees, *Russhton* 1332 Subs], **Rishworth** YW [*Rissheworthe* 1276 YInq]. 'TŪN and WORÞ where rushes grew.'

Rising (-īz-), **Castle,** Nf [*Risinga* DB, *Risinges* 1190 P, *Castel Rising* 1254 Val], **Wood R~** Nf [*Risinga* DB, *Resinges* 1121 AC, 1206 FF, *Risinges* 1185 P, *Woderisingg* 1291 Tax]. Either OE *Hrising* or *Hrisingas* 'place by brushwood' and 'people at the brushwood' or OE *Risingas* '*Risa*'s people'. OE *Risa* would correspond to OG *Riso*. In the absence of OE forms it cannot be decided if the names began in *Hr-* or *R-.*

Risley Db [*Riselei* DB, *Riseleg* 1236 Fees], **R~** La [*Ryselegh* 1284, *Risseley* 1285 Ass]. Identical with RISELEY.

Rissington, Great & Little, Wyck R~ Gl [*Rise(n)dune* DB, *Risendona* c 1130 Oxf, *-don* 1200 Cur, *Braderisendon* 1220, *Wik Risindon* 1236 Fees]. OE *Hrisen-dūn* 'hill covered with brushwood'.

Riston, Long, YE [*Ristune* DB, *Riston* 1200 FF, *Restona* 1297 Subs]. 'TŪN in brushwood.' The place is near RISE.

OE **rīþ, -e, -ig** 'a small stream', corresponding to OLG *ritha, rithe,* Fris *riede,* is still used in some dialects in the forms *rithe, rife.* It is found in several names of streams, some of which have become names of places. *Rīþ* alone is the source of REETH, RYDE. As a second el. the form varies. See e.g. CHAURETH, MELDRETH, RAWRETH, SHEPRETH, FINGRITH, TINGRITH, HENDRED, CHILDREY, SHOTTERY, PECKHAM RYE. The longer form *rīþig* is found in CROPREDY, FULREADY. *Rīþ* is the first el. of RITTON.

Ritton Nb [*Rittona* c 1145 Percy, *Rittun* 1236 Fees]. 'TŪN on a *rīþ* or stream.'

Rivelin Bridge YW. *Rivelin* is the name of a stream [*Riveling water* 1637 ERN]. *Rivelindale* is *Rivelingdene* 1300 ib. The source is *riveling* 'rivulet', found from 1615, but evidently a much older word.

Rivenhall Ess [*Reuen-, Ruuenhala* 1068 EHR xi, *Ruenhale, Ruuuenhala, Riuuehala* DB, *Riuenhale, Riwehal* 1195 FF, *Rewenhale* 1254 Val]. This cannot well be OE (æt) *rūwan heale* 'the rough HALH', unless

owing to Norman influence *ū* became Fr *ü.* Possibly the first el. is a stream-name *Rȳwe* 'the rough one', derived from *rūh* 'rough'.

River K [*Rip'ia* 1199 FF, *Riveria* 1199 Ch, *Ripera* 1219 Fees, *La Ryvere* 1228 Ch]. Apparently OFr *rivere* 'river'.

River Sx [*Euere* 1279 Ass, *Rivere* 1396 Ipm]. Identical with **Rivar** W nr Ham [*on ǒa yfre* 931 BCS 677]. OE *yfer* 'edge, brow of a hill', the *R-* being a relic of the dat. of the def. art. (OE *æt þǣre yfre*), whose *r* was carried over to the noun. See YFER.

Riverhead K [*Reddride* 1278 Ass]. Identical with ROTHERHITHE.

Rivington La [*Revington, Rowinton* 1202 FF, *Ruhwinton* 1212 Fees]. 'TŪN by Rivington Pike' (a hill). The latter is *Roving* 1325 LaInq, apparently a derivative of OE *hrēof* 'rough': 'rough hill'.

Rixton La [*Rixton* 1201 ff. P, *Richeston* 1260 Ass]. '*Ric*'s or *Ricsige*'s TŪN.' *Ric* is not evidenced, but a normal short form of names in *Ric-.*

Roach R Ess. A back-formation from ROCHFORD.

Road So [*Rode* DB, 1201 FF, *Roda* 1201 Ass, *la Rode* 1230 Ch], **Roade** Np [*Rode* DB, 12 BM]. OE *rod* 'clearing'.

Robertsbridge Sx [*Pons Roberti* 1199 Cur]. Named from Robert de St. Martin, founder of Robertsbridge Abbey (1176).

Robey Du [*Raby* 1359 AD]. Identical with RABY.

Robley. See RABLEY.

Roborough D [*Raweberge* DB, *Ruaberga* 1166 RBE, *Rugheberg* 1242 Fees]. 'Rough hill.' See RŪH, BEORG.

Roby La [*Rabil* DB, *Rabi* 1185 P, *Roby* 1304 Ch]. Identical with RABY.

Rocester (rōster) St [*Rowcestre* DB, *Rouecestre* 1208 FF, *Rovecestre* 1225 FF, *Rocestre* 1246 Ch]. '*Hrōþwulf*'s *ceaster* or Roman fort.' Cf. early forms of ROUSHAM O.

Rŏch R La [*Rached.* 13 WhC, *Rachet* 1292 Ass, *Rach* 12 BM, *Rache* c 1200 WhC], **Rochdale** La [*Recedham* DB, *Rachedham* a 1193 WhC; *Rachedal* c 1195 PNLa, 1246 Ass]. Rochdale is 'the valley of the Roch'. Roch was originally *Rached,* but *Racheddale* became *Rachedale,* and a new rivername *Rache* was formed by back-formation. *Rached* itself may be a back-formation from *Rachedham,* the early name of Rochdale, if it has as first el. OE *reced, ræced* 'hall, house' ('HĀM with a hall'). But it is possible that *Rached-* is a Brit name consisting of OW *rac* 'against' and *coet* 'wood'. It would then be a name of the district or possibly a river-name ('district or river opposite to the forest').

Roche Co [*la Roche* 1233 FF, 1258 Ep], **R~** YW [(Abbatia de) *Rupe* 1199 (1232) Ch, *La Roche* 1251 Ch]. Fr *roche* 'rock'.

Rochester K [*Hrofaescaestre* c 730 Bede, *Hrofesceaster* c 700 Laws, 839 ASC, *Hrofescester* 811 BCS 339, *Rovecestre* DB]. 'The Roman fort *Hrofi*.' *Hrofi* (*ciuitas Hrofi* c 730 Bede, *civitas Hrobi* 842 BCS 439) is a clipped form of the British name [*Durobrivis* (dat. plur.) 4 IA, (in) *Dorubreui* c 730 Bede, *Dorobrevi* 844 BCS 445], which means 'the bridges of the stronghold' (Brit *duro-* 'stronghold' and *brivā* 'bridge'). *Durobrivæ* must have been accented on the *o* (this kind of stress is common in Gaulish), and *b* became *v* by lenition. When the name was taken over by the English, the unstressed initial syllable was lost, and the *v* (*w*), which had become final, was lost as in DEE from *Dēva*. For some reason (association with OE *hrōf* 'roof'?) the name got an initial *H-*. Hence a form *Hrofri* would develop, in which the second *r* was lost by dissimilation, and early OE *Hrofi* arose. To this was added OE *ceaster*.

Rochester Nb [*Roff'* 1208 Cur, *Rucestr* 1242 Fees, *Rouschestre* 1325 Ipm]. Possibly named from ROCHESTER K. Or the first el. may be OE *hrōc* 'rook'.

Rochford Ess [*Roche(s)fort* DB, *Rocheford* 1195 Cur, *Rochesford* 1197 FF, *Racheford* 1200 P], **R~** Wo [*Ræccesford* 11 Heming, *Recesford* DB, *Rocheford* 1242 Fees]. OE *ræcces-ford* 'the ford of the hunting-dog' (see RÆCC). The change to *Roch-* is analogous to that in ROCH, but it may have been helped by the influence of the Fr pl. n. *Rochefort*.

Rock Nb [*Rok* 1242 Fees]. OE *rocc* 'rock' (an early Romance loan-word).

Rock Wo [(del) *Ak* 1224, *Roke* 1259 FF]. OE (æt) *þǣre āce* '(at) the oak'. Cf. RIVER Sx.

Rockbeare D [*Rochebere* DB, *Rokebere* 1196 FF, -*bear* 1275 RH], **Rockbourne** Ha [*Rocheborne* DB, -*burna* 1157 f. P, *Rokeburn'* 1201 Cur]. 'Rook wood and stream.' See HRŌC, BEARU, BURNA.

Rockcliff Cu [*Redeclive* 1202 FF, *Radeclive* 1203 Cur, *Roudecliua* 1185 P, *Routheclive* 1203 Ass]. Identical with RAWCLIFFE. Originally OE *rēade clif* 'red cliff'.

Rockhampton Gl [*Rochemtune* DB, *Rochamton* 1220 Fees]. OE *Hrōchǣma-tūn* 'TŪN of the dwellers at Rookham' (or a place with some other name containing OE HRŌC 'rook'). Cf. HĀMTŪN.

Rockingham Np [*Rochingeham* DB, 1104–6 RA, 1130 P, *Roginggham* 1137 ASC (E), *Rokingeham* 1197 FF]. 'The HĀM of *Hrōc(a)*'s people.' OE *Hrōc* is found in *Hroces seað* (see ROXETH), *Hrōca* in *Hrocan leah* BCS 1047, and *Hrōc* corresponds to OHG *Hroch*, ON *Hrókr*. Rockingham has an exact parallel in ODu *Hrokingahem* 815–44.

Rockland St. Mary Nf [*Rokelund* DB, *Roclund* 1254 Val], **R~ St. Peter, R~ All Saints & St. Andrew's** Nf [*Rokelund* DB, *Roclund Toftes, Omnium Sanctorum, Sancti Andree* 1254 Val, *Rokelund Sancti Petri* 1291 Tax]. 'Rook wood.' Second el. OScand *lundr* 'grove'.

Rockley W [*Rochelie* DB, *Roclee* 1185 TpR, *Rokeleg* 1230 P]. OE *hrōc-lēah* 'rook wood'.

Rockmoor. See THROCKENHOLT.

OE **rod**, dial. *royd* 'a clearing in a forest' (= OHG *rod*, ON *ruð*) is the source of ROAD, -E, RODD, RODDAM, RODE, and the second el. of some names, as BLACKROD, ORMEROD, HUNTROYDE.

OE **rōd** 'rood, cross' is the first el. of RAD-STONE, RUDSTON.

Rodbaston St S. of Penkridge [*Redbaldes-tone* DB, -*ton* 1198 P, *Rodbaldeston* 1221 Ass, 1236 Fees]. '*Rēdbald*'s TŪN.' The change of *ē* to *o* is remarkable.

Rodborough Gl [*Roddanbeorg* 716–43, 896 BCS 164, 574, *Rodberghe* 1294 Cl]. OE *rodd* 'rod', ME *rodde* also 'branch, stick' cannot be the first el., but there might have been an OE **rodde*, corresponding to ON *rudda* 'club', Norw *rodda* 'a raised pole' and meaning 'boundary pole' or the like. A pers. n. *Rodda* might easily have been formed from OE *rodd*. Second el. OE *beorg* 'hill'.

Rodbourne W [*Reodburna* 701, 758 BCS 103, 185, *Redburn* 1232 Cl], **R~ Cheney** W [*Redborne* DB, *Rodebourn Chanu* 1438 AD iii]. OE *hrēod-burna* 'reedy stream'. The stream at R~ Cheney is *Hreodburna* 943 BCS 788.

On Cheney see CHENIES. Ralph le Chanu held the manor in 1242 (Fees).

Rodd He [*La Rode* 1356 Ipm]. OE *rod* 'clearing'. **Roddam** Nb [*Rodun* 1201 Cur, *Roden* 1203 P, *Rodum* 1236 Fees]. OE *rodum*, dat. plur. of ROD 'clearing'.

Rodden So [*Reddene* DB, *Radena* 1166 P, *Reddona, Raddona* 1166 RBE, *Raden* 1238 Ass, *Rodene* 1297 FF]. OE *rā-denu* 'roe valley'. The *e*-forms may point to an alternative *ræg-denu* with OE *ræge* 'female of the roe' as first el.

Roddlesworth La [*Rodtholfeswrtha* c 1160 LaCh, *Rotholueswurth* 1246 Ass]. '*Hrōþ-wulf*'s WORÞ.'

Rode Heath Chs [*Rodeheze* 1280 BM], **Odd Rode** Chs nr R~ Heath [*Rode* DB, *Odderode* 1368 Ormerod], **North R~** Chs [*Rodo* DB]. OE *rod* 'clearing'. *Odd* is a pers. n. (OE *Odda*).

Roden R Sa [*Roden* 1256 Ass]. On the river is **Roden** hamlet [?*Rutunio* (abl.) 4 IA, *Rodene* 1242 Fees]. The river-name is very likely MW *Trydonwy* (for *Rydonwy*) in Marwnad Cynddylan. It goes back to Brit **Rutunā*, which seems to belong to the root *reu* in Welsh *rhuthr* 'rush, attack', Lat *ruo* &c. The meaning would be 'swift river'.

Roding R Ess [*Rodon* 1576 Saxton] is a back-formation from Roding villages. For an old name see ILFORD Ess.

Roding (rōōdhǐng), **Abbess, Aythorpe** (äthrŏp), **Barwick, Beauchamp, Berners, High, Leaden** (-ē-), **Margaret & White,** Ess [*duae Rodinges* c 1050 KCD 907, *Rodinges, Rodingis* DB, *Roinges* DB, 1196 P, *Royng' Alba, Berners, Roynges Abbatisse, Beuchamp, Sancte Margar'* 1238 Subs, *Rothingg Abbatisse* 1254 Val, *Roeng Aytrop* 1248 FF, *High Roinges* 1225 FF, *Ledeineroing* 1248 FF, *Rothing plumbi* 1291 Tax, *White Roeng* 1248 FF]. OE *Hrōþingas* 'the people of *Hrōþ(a)'; cf. OE *Hroda,* OG *Hrōdo.*

Abbess R~ belonged to the Abbess of Barking.—**Aythorpe R~** was held by William son of Ailtrop c 1200 (BM). The name is written *Aitrop, Eitrop, Eutropius* 1200 ff. Cur and is a Fr form of *Eutropius.*—Barwick is OE BERE-WIC.—**Beauchamp R~** was held by John de Bello campo in 1233 (Ch). See ACTON BEAUCHAMP.—*Berners* is a family name taken from BERNIÈRES in Normandy. Hugo de Berneris held the manor in 1086 (DB).—**Leaden R~** from a church with a leaden roof [*Ledenechirche* c 1100 Mon].—**Margaret R~** from the dedication of the church.—**White R~** from the colour of the church.

Rodington Sa [*Rodintone* DB, *-ton* 1203 Cur]. 'TŪN on R Roden.'

Rodley Gl [*Rodele* DB, *Redlega* 1157 P, *-leg* 1220 Fees, *-lege* 1227 Flaxley]. OE *hrēod-lēah* 'reed LEAH'. **Rodley** YW [*Rothelaye* Hy 3 BM]. If the isolated spelling is trustworthy, identical with ROTHLEY.

Rodmarton Gl [*Redmertone* DB, *Rodmarton* 1220 Fees]. 'TŪN by a reedy lake' (OE *hrēodmere*).

Rodmell Sx [*Redmelle* DB, *Radmelde* 1202 FF]. Identical with REDMILE.

Rodmersham K [*Rodmaeresham* 11 DM, *Rodmaresham* 1197 FF, *Rodmeresham* 1204 Ch]. '*Hrōþmǣr*'s HĀM.' **Hrōþmǣr* is identical with OHG *Hrōtmār.*

Rodsley Db [*Redlesleie, Redeslei* DB, *Roddeslea* 1183 P, *Redisleye* 1277 BM]. Apparently OE *Hrēodlēah* (cf. RIDLEY 2), to which was added another LEAH.

Rodway So nr Cannington [*Radeweye* 1241 BM, *Rodweye* 1233 Wells]. Identical with RADWAY.

Roe Hrt. See ROTHEND. **Roe Beck.** See RAUGHTON.

Roeburn La [*Reborn* 1292 Ass, *Roburn* 1577 Saxton]. OE *rǣgan-burna* 'the stream of the female of the roe'. **Roeburndale** is *Reburndale* 1285 Ipm, *Rebrun-, Reynbrundale* 1301 FC.

Roecliffe YW [*Routhecliua* 1170 P, *-cliue* 1208 FF]. Identical with RAWCLIFFE.

Roel or **Rowell** Gl [*Rawelle* DB, *Rawell* 1174 Fr, *Rowell* Hy 3 Misc]. OE *rā-wella* 'roe stream'.

Roffey or **Roughey** Sx [*La Rogheye* 1281, 1331 AD vi, *Rozghee* 1296 Subs]. OE *rāhhege* 'deer-fence, enclosure for roe-deer' (BCS 932 &c.). See OE RĀ. *Rāhhege* became ME *Rŏghey, Rŏughey,* and this was associated with the adj. *rough.* Hence occasional spellings such as *Rugheye.*

Rofford O nr Stadhampton [*Roppanford* 1002 KCD 1296, *Ropeford* DB, 1196 FF]. The same first el. is found in the name of the brook at R~ [*Hroppan broc* 774 BCS 216, *Roppanbroc* KCD 1296]. It is no doubt a pers. n. *Hroppa,* which may be explained as a short form of names like *Hrōþbeorht* (not evidenced, but *Hrēþbeorht* is).

Rogate Sx [*la Rogate* 1196, *Rogate* 1203 FF, *la Ragat* 1229 Ch]. 'Gate for roe-deer.' Cf. REIGATE.

Roke O [*Rokes* c 1252 BM]. Identical with ROCK Wo.

Rokeby YN [*Rochebi* DB, *Rokeby* 1204 FF]. The same name is **Rookby** We [*Rochebi* 1178 P, *Rokebi* 1201 Cur]. Either 'BY where rooks were plentiful' or 'the BY of some man called *Hroca*' (or the like).

Rollesby Nf [*Rotholfuesbei, -by,* *Roluesbi* DB, *Roluesbi* 1196 FF, *Rollesbi* 1193 f. P]. '*Hrōlf*'s BY.' ON *Hrólfr,* ODan *Rolf* comes from *Hrōðulfr,* and the original form is still found in early examples of Rollesby.

Rolleston (rōlstn) Le [*Rovestoṇe* DB, *Rolueston* 1170 P, *Rolvestona* Hy 2 (1318) Ch], R~ St [*Roðulfeston* 942 BCS 771, *Rólfestun* 1002 Wills, *Rolvestune* DB, *Rolleston* 1291 Tax], **Rollestone** W [*Rolveston* 1242 Fees, *-tone* 1291 Tax]. '*Hrōþwulf*'s or *Hrōlf*'s TŪN.' Cf. ROLLESBY.

Rolleston (rōlstn) Nt [*Roldestun, Rollestone, Rolvetune* DB, *Roldeston* 1219 Fees, *Rollestun* 12 DC]. '*Hrōald*'s TŪN.' First el. ON *Hróaldr* pers. n.

Rollright, Great & Little, O [*Rollendri, parua Rollandri, Rollandri major* DB, *Rollendriz* 1090 RA, *Rollendricht* 1091 Eynsham, 1192 P, *Magna Rollindricht* 1234 Ep, *Roulandrith* 1247 Ass]. The numerous spellings with *-a-* suggest that the middle el. is *land.* The name may contain OE *landriht* 'privileges belonging to the owner of land' (Beowulf &c.), here synonymous with *landār* 'property'. If so, the first el. is no doubt a pers. n., e.g. **Hrolla,* a short form of *Hrōþlāf* or the like. *Rollandun* (*Wrollendun*) 944 BCS 795 might possibly contain this name.

Rolvenden (-ŏl-) K [*Rovindene* DB, *Ruluindaenne* 11 DM, *-den* 1185 P, *Rodelindenn* 1275 RH]. 'The DENN or pasture of *Hrōþwulf*'s people.'

Romaldkirk YN [*Rumoldescherce* DB]. 'St. Rūmwald's church.'

Romanby YN [*Romundebi, Romundrebi* DB, *Romundabi* 1088 LVD, *Romundeby* 1219 FF]. '*Hrōmund*'s BY.' First el. ON *Hrómundr*, OSw *Romunder* pers. n.

Romansleigh (rŭmzli) D [*Liega* DB, *Reymundesle* 1228 FF, *Romundeslegh* 1242 Fees]. Stated to be 'the Leigh of St. Rumon'. The early forms suggest a pers. n. *Romund*.

Romford (-ŭ-) Ess [*Rumford* 1200 Ch, 1212 RBE, 1247 Cl, 1271 Ch]. Perhaps 'broad (roomy) ford', the first el. being OE *rūm* 'roomy'. But we expect the form to be early ME *Rumeford*, if this is right; and it is quite possible *Rum-* goes back to earlier *Run-*, which may be OE *rūn* 'council, discussion' (cf. RUNNYMEDE) or *hruna*, as in HEADCORN. **Rom** R is a back-formation.

Romiley Chs [*Rumelie* DB, *Romilee* 1285 Court]. 'Spacious LĒAH' (first el. OE *rūm* 'roomy' &c.).

Romney (-ŭ-) K [*Rumenea* 1052 ASC (E), 11 DM, *Romenel* DB, *Rumenel* 1130 P, *Rumenal* 1247 Ch]. Originally the name of a river [*Rumenea* 895, *Rumenesea* 914 BCS 572, 638], whose second el. is OE ĒA 'river'. The first el. seems to be derived from OE *rūm* 'spacious', but its formation and meaning are obscure. *Rumen* may be an old name of **Romney Marsh**. The latter is referred to as *Merscuuare* 774 BCS 214, *Merscware* 796 ASC (E), 838 ib. (A), *regio Merscuuariorum* 811 BM. *Merscware* means 'the marsh-dwellers'. A derivative of Romney is found in *Ruminingseta* 697 BCS 98 'the fold of the Romney people'; cf. (GE)SET.

Romsey (-ŭ-) Ha [*Romes(e)ye* 966 BCS 1187, *Rumesig* 971 ASC, c 1000 Saints, *Romesy* DB, *Rumeseia* 1167 P]. '*Rūm*'s island.' **Rūm* is a short form of names in *Rūm-*.

Romsley Sa [(æt) *Hremesleage* 1002 Wills, *Rameslege* DB, *Rameslea* 1167 P, *-leye* 1212 RBE, *Rommesleye* 1287 Ipm], **R~** Wo [*Romesle* 1270 Ct, *-leye* 1291 Tax]. 'Wild garlic LĒAH.' See HRAMSA.

Ronton St. See RANTON.

Rookby We. See ROKEBY.

Rookhope Du [*Rochop* 1242 Ass], **Rookley** Wt [*Rokle* 1287–90 Fees, *Rouklye* 1316 FA]. 'Valley (HOP) and wood (LĒAH) frequented by rooks.'

Rookwith YN [*Rocvid* DB, *Rokewik* 1342 Misc]. 'Rook wood.' Second el. ON *viðr* 'wood'.

Roos YE [*Rosse* DB, *Rossa* 1135–40 YCh 1152, *Russa* 1161 P, *Russe* 1202 FF], **Roose** La [*Rosse* DB, *Ros* 1155 LaCh, *Roos* 1336 FC]. Welsh *rhôs* 'moor, heath, plain', identical with Bret *ros* 'hillock, usually one where heather grows', Ir *ros* 'promontory'. In Roos the meaning may be 'moor' or 'promontory', in Roose it is 'moor, heath'.

Roothing. See RODING.

Rooting K in Pluckley [*Rotinge* DB]. '*Rōt*'s or *Rōta*'s people.' For *Rōta* see RATLEY, for *Rōt* see RUSCOMBE. The names are derived from OE *rōt* 'merry'.

Rope Chs [*Rap* Hy 2, *Rop* Hy 3 BM]. OE *rāp* 'rope'. The meaning may be that of *rape* in Sx, the name of a division of land, originally perhaps used of a place for the assembly fenced off with stakes and ropes.

Ropley Ha [*Ropeleia* 1198 FF, *Roppeley* 1240–50 Selborne]. First el. as in ROFFORD.

Ropsley Li [*Ropeslai* DB, *Roppeslea* 1170 P, *-le* Hy 2 DC, *Roppele* 1212, 1242 Fees]. The first el. may be a strong side-form of *Hroppa* in ROFFORD. See LĒAH.

Rorrington Sa [*Roritune* DB, *Roriton* 1316 FA]. 'The TŪN of *Hrōr*'s people.' **Hrōr* is a byname formed from OE *hrōr* 'vigorous, strong'. Cf. ROYSTON YW.

Rose Ash D [*Aissa* DB, *Esse* 1242 Fees, *Rowesassche* 1400, *Aysch Raff* 1404 Ep]. Originally ASH 'ash-tree'. *Rose* is '*Ralph*'s'. One Ralph de *Ese* in Rose Ash is mentioned in 1198 (FF), one Ralph de *Esse* ib. 1261 (Ep).

Rose Castle Cu [*Rosa* 1275 WR, *La Rose* 1288 Cl, (manor of) *Rose* 1291 Ch]. Rose Castle is the residence of the Bishop of Carlisle. *Rose* is no doubt the word *rose* used in some special sense, e.g. as a general epithet for excellence or beauty or in allusion to the rose as the emblem of the Virgin Mary.

Roseacre La [*Raysacre* 1283 FF, *-aker* 1286 Ipm]. 'Field with a cairn' (ON *hreysi*).

Roseberry Topping YN, a hill [*Othenesberg* 1119 Guisb]. '*Oðinn*'s hill.' The present name may have developed from the old one, the initial *R-* having been carried over from a preceding preposition *under*.

Rosedale YN [*Russedal* 1130–58 (1201) Ch, 1165 P, *Rossedale* c 1190 YCh 694, 1244 Fees]. ON *hrossa-dalr* 'horse valley'.

Roseden Nb [*Russeden* 1242 Fees, *Russhden* 1346 FA]. 'Valley where rushes grew.'

Rosgill We [*Rosgyl* c 1250 CWNS xxiv]. ON *hross(a)-gil* 'horse valley'. See GIL.

Rosley Cu [*Rosseleye* 1285 PNCu, *Rosseley* 1317 Holme C]. There may have been an OE *hross* by the side of *hors* 'horse'. If so, 'LĒAH where horses grazed'.

Rosliston Db [*Redlavestun* DB, *Restlavestune* 1226 FF, *Rostlavestona* R 1 Derby, *-ton* 1236, *Roustloviston* 1242 Fees, *Rostlaweston, Roxlaueston* Hy 3 Derby]. Apparently OE *Hrōþlāfes tūn* '*Hrōþlāf*'s TŪN'. Abnormal *s* for *þ* is then due to Norman sound-substitution.

Ross He [*Rosse* DB, *Ros* 1199 P, 1242 Fees, *Roos* 1291 Tax], **R~** Nb [*Rosse* 1208–10 Fees, *Ross* 1250 Ipm]. Welsh *rhôs*; see ROOS. Ross He is on a steep hill; the meaning is here 'hill'. The Welsh name is *Rhossan ar Wy*. Ross Nb is on a promontory. Ross may here mean 'promontory'.

Ross Hall Sa nr Shrewsbury [*Rosela* DB, *Roshala* 1170 P, *-hale* 1242 Fees], **Rossall** La [*Rushale* DB, *Rossale* 1216, *Roshal* 1222 Cl]. Probably 'HALH where horses grazed'. Cf. ROSLEY, HROSS.

Rossendale La [*Rocendal* 1241 Cl, 1242 LaInq, *Rossendale* 1292 QW, *Roscindale* 1296 Lacy]. Earlier material is wanted for a definite etymology.

Rossington YW [*Rosington* c 1190 YCh (817), 1222 FF, *Rosenton* 1207 FF, *Rosingtun* 1249 Ep]. Possibly 'the TŪN of the people on the moor', the first el. being derived from Welsh *rhôs* (cf. ROOS).

Rostherne Chs [*Rodestorne* DB, *Roudes-, Routhestorn* 1226–8 Chester]. '*Rauð*'s thorn-bush.' First el. the ON pers. n. *Rauðr*. See RAUCEBY.

Roston Db [*Roschintone* DB, *Rocinton* 1252 Ch, *Rossynton* Hy 3 BM]. Perhaps 'the TŪN of *Hrōþsige*'s people'.

OE *roþ* or *roþu* 'clearing'. See ROTHEND, ROTHLEY, ROTHWELL, also RODLEY YW.

Rothay. See RAWTHEY.

Rothbury (rŏth-) Nb [*Routhebiria* c 1125 Hexh, *Rodebir* Hy 2 (1271) Ch, *Rothebyri* c 1190 Godric, *-buri* 1295 Ch, *Routhebyr* 1291 Tax]. '*Hrōþa*'s BURG.' Cf. RODING.

Rothend Ess in Ashdon [*Roda* DB, *Rothe* 1279 FF]. OE *roþ(u)* 'clearing'. The word is not evidenced in OE, but has an exact equivalent in OFris *rothe* 'clearing'. It is found also in ROTHLEY, ROTHWELL. **Roe** Hrt is *Roðe* 939 BCS 737.

Rother R Db, YW [*Roder* c 1170 YCh 1480, 1276 RH, *Rodur* 1388 Derby]. A Brit name, which may consist of the intensifying prefix ro- (Welsh *rhy-* in *rhylaw* 'heavy rain, &c.) and Welsh *dwfr* 'water, river'. The meaning would be 'chief river'. On the river is **Rotherham** YW [*Rodreham* DB, *Roderham* 1200 Cur, 1228 BM].

Rother R Sx, K [(flumen) *Rothori* Mon v, *Rother* 1575 Saxton]. A back-formation from ROTHERFIELD. For the old name see LYMN. **Rother** Ha, Sx. A back-formation from **Rotherbridge** (hd) [*Redrebrige* DB]: 'bridge for oxen' (see HRȲPER). The old name of this Rother was **Shire** [*Scir* 956 BCS 982, *Sire* c 1200 Gervase], which means 'bright river' (from OE *scir* 'bright, clear').

Rotherby Le [*Redebi* DB, *Rederbia* c 1125 LeS, *Rederebia* 1181 P, *Reytherby* 1254 Val]. Identical with RAITHBY Li (1).

Rotherfield Ha [*Hryðerafeld* 1015 Wills, *Reðeresfeld* 1167 P, *Rutherefeld* c 1235 Selborne], **R~Greys & Peppard** O [*Redrefeld* DB, *Reðeresfeld* 1194 P, *Rutthereffeld*

1246 Ch, *Retherfeld Grey* 1313 AD, *Retheresfelde Pipard* 1233 Ep], R~ Sx [*Hryðeranfeld* c 880 BCS 553, *Reredfelle* DB]. 'FELD or open land where cattle grazed.' See HRȲPER.

R~ Greys was held by Robert de Gray in 1242 (Ep). He was a nephew of Archbishop Gray. See EASTON GREY.—*Peppard* is a Fr family name, probably from OFr *pipart* 'piper'.

Rotherham YW. See ROTHER R (1).

Rotherhithe (*olim* rĕdrīf) Sr [*Rederheia* 1100–7 (1330) Ch, *Ruerhee* 1199 FF, *Rutherhee* 1204 Cur, *-heth* 1268 BM]. 'Landing-place where cattle were shipped.' Cf. HRȲPER, HȲÞ.

Rothersthorpe Np [*Torp* DB, *Trop que fuit aduocati de Bethun* 1196 P, *Trop Advocati* 1220 Fees, *Retherestorp* 1231 Ch, *-trop* 1247 Misc]. 'The advocate's Thorp.' OE *rædere* must have meant 'adviser, counsellor', a meaning given for ME *redere*. But the word must also have been used in the sense 'legal adviser, advocate'. The name has been influenced by the word *rother* 'cattle' (OE *hrȳþer*). R~ was held in the 12th cent. by the advocate of Béthune.

Rotherwas He [*Retrowas* DB, *Rudrewas* 1242 Fees, *Retherwas* 1322 Ipm], **Rother-wick** Ha [*Retherwyk* 1194 Selborne, *Rutherwyc* 1235 Cl]. 'Cattle swamp and farm.' See HRȲPER and WÆSSE, WĪC.

Rothley (rōthlī) Le [*Rodolei* DB, *Rodeleia* c 1125 LeS, *Roelay* 1153 BM, *Rothele* 1254 Val], **R~** Nb [*Ruelea* 1195 f. P, *Rotheley* 1233 P, 1242 Fees], **Rothwell** Li [*Rodewelle* DB, *Rod(e)wella* c 1115 LiS, *Rothewell* 1288 Ipm, 1291 Tax], **R~** (rŏðĕl) Np [*Roðewelle* 1066–75 GeldR, *Rodewelle* DB, *Rowell* 1156 P], **R~** YW [*Rodewelle, Rodouuelle* DB, *Rothenwella* 1121–7 YCh 1428, *Rothewell* 1291 Tax]. 'LĒAH with, and spring or stream by, a clearing.' See ROTHEND.

Rotsea YE [*Rotesse* DB, *Rottese* J Ass, 1260 Ch, *-see* 1239 FF]. The second el. is OE *sǣ* 'lake'. The first may be OE *Rōt* or *Rōta* pers. n. (see ROOTING). OE *hrot, rot* 'thick fluid, scum' does not suit the early forms so well.

Rottingdean Sx [*Rotingedene* DB, *Rotingesdena* 1121 AS]. 'The valley of *Rōt(a)*'s people.' For the pers. n. *Rōt(a)* see RATLEY, ROOTING, RUSCOMBE, RUTLAND. *Rōta* is the first el. of **Ratham** Sx [*Roteham* 1279 &c. PNSx]. Cf. also **Ruttingham** Sx [*Rottingeham, Rotingehamme* 1200 Cur].

Rottington Cu [*Rodintona, Rotingtona* c 1125 StB, *Rotington* 1211 P]. First el. as in ROTTINGDEAN.

Roudham Nf [*Rudham* DB, 1199 FF, 1254 Val]. The first el. may be OE *rūde* 'rue' (the plant). Hence 'HĀM where rue was grown'.

Rougham (-ŭf-) Nf [*Ruhham* DB, 1203 Cur, *Rugham* 1182 P, *Rucham* 1198 FF], **R~** Sf [*Rucham* c 950 BCS 1008, 1013,

Ruhham 11 EHR 43, DB, *Rugham* 1254 Val]. The first el. seems to be OE *rūh* 'rough', but if so, probably a noun *rūh* 'rough ground', though this sense is not evidenced in OED for *rough* until c 1480. The sense 'roughness, rough surface' is found in Ancren Riwle. Second el. HĀM.

Roughbirchworth YW [*Bercewrde* DB]. Cf. INGBIRCHWORTH. *Rough-* is no doubt *rough* adj.

Roughey. See ROFFEY.

Roughton (-ōō-) Li [*Rocstune* DB, *Ructuna* c 1115 LiS, *Ruchtuna* 1163 Bodl, *Ructon* 1202 Ass, 1232 Ep], R~(-ow-) Nf [*Rustuna*, *Rostuna*, *Rugutune* DB, *Rocton* 1196 FF, *Ruchton* 1254 Val], R~ Sa in Worfield [*Roughton* 1316, *Rowton* 1318 Ipm]. First el. as in ROUGHAM. It is possible, however, that the Li and Nf Roughton may be a Scandinavianized form of OE *Ryge-tūn* 'rye farm', OScand *rugr* having replaced OE *ryge*.

Roundhay YW [*La Rundehaia* c 1180 YCh 1509, *Rotunda Haia* 1201 Cur, *Rundehaye* 1294 Ch]. 'Round enclosure.' The same name occurs in Yardley Np [*le Rundehai* 1325 Ipm]. A Fr name (*ronde haie*).

Roundthwaite We [*Rounerthwayt* 1294 Cl, *Rounthwayt* 1338 Ch]. 'Clearing with mountain ash.' See THWAITE. First el. ON *raun* (gen. *raunar*) 'rowan, mountain ash'.

Roundway W [*Rindweia* 1149 Sarum, *Ryndewey* 1316 FA, *Ryndway* 1337 FF, *Rundewey* 1419 Ipm]. 'Cleared road', the first el. being OE *rȳmed* 'cleared'. Cf. RENDHAM.

Rounton, East & West, YN [*Runtune*, *Rontun* DB, *Rungtune* c 1130 YCh 944, *Rongetona* 1168 P, *Rungheton* 1208–10 Fees, *Rungeton* 1218 FF]. Identical with RUNCTON Nf. The first el. is OE *hrung* 'a rung, pole'. The reference is very likely to a primitive bridge over marshy ground formed by poles placed close together at right angles to the direction of the road. Bridges of this kind (called *kavelbro*) are still seen in Sweden. It is noteworthy that North and South Runcton Nf are more than 4 m. apart.

Rousdon (-owz-) or **Down Ralph** D [*Done* DB, *Rawesdon* 1284–6 FA]. OE *dūn* 'hill'. *Rous-* (*Ralph*) from Radulfus de Duna (1155–7 Fr).

Rousham O [*Rowes-*, *Rovesham* DB, *Rodulveshama* Hy 2 (1267) Ch, *Rodolvesham* c 1200 Bodl, *Rowulvesham* 1212 Fees, *Rolesham* c 1200 Bodl]. '*Hrōþwulf*'s HĀM.'

Routh (-ōō-) YE [*Rute*, *Rutha* DB, *Rudhe* c 1150 YCh 1380, *Routh* 1297 Subs]. The vowel must have been long. Formally the base might be a word corresponding to OHG *rûda*, OLG *hrûtho* 'scab', ON *hrúðr* 'scurf'. This word might have been used in a transferred sense of rough ground.

Rowberrow So [*Rugebera* 1177, *Ruberga* 1194 P, *Rugeberg* 1227 Ch], **Rowborough** Wt in Brading [*Rodeberge* DB, *Rouvebergh* 1284 BM], R~ Wt in Carisbrooke [*Rougheberg* 1282 Pat]. 'Rough hill.' See RŪH, BEORG.

Rowde (-ō-) W [*Rode* DB, *Rudes* 1187, 1190 P, *la Rode* 1230 Cl]. Perhaps OE *hrēod* 'reed, reed bed'.

Rowden He [*Ruedene* DB, *Rugedun* 1242, 1249 Fees]. 'Rough hill.' See RŪH, DŪN.

Rowell. See ROEL.

Rowfant Sx [*Rowfraunte* 1574 PNSx]. OE *rūh* 'rough' and OE *fyrnþa* 'fern brake'. See FRANT.

Rowhedge Ess [*Rouhegy* 1346, *-hegge* 1494 PNEss]. 'Rough hedge.'

Rowington Wa [*Rochintone* DB, *Rokintun* 1157 BM, *-ton* 1206 Cur, *Rouhwinton* 1291 Tax]. OE *Hrōcingatūn* 'the TŪN of *Hrōc(a)*'s people'. Cf. ROCKINGHAM. An unusual change of *k* has taken place.

Rowland Db [*Ralunt* DB, *Raalund* 1169 P]. OScand *rá-lundr* 'roe wood'.

Rowley D [*Rodeleia* DB, *Rughelegh* 1242 Fees], R~ Du [*Ruley* 1229 FPD, *Rowley* 1372 AD ii], R~ **Regis** St [*Roelea* 1173, *Ruelega* 1174 P, *Ruleye* 1272 Ass], R~ YE [*Ruley* 1227 Ep, *Roule* 1276 Ipm], R~ YW [*Ruley* 1246 FF]. 'Rough LĒAH.' See RŪH.

Rowlston YE [*Roolfestone*, *Roluestun* DB, *Rolleston* 1246 FF], **Rowlstone** He [*Rolveston* 1276 Cl, *Roulestone* 1280 Ep]. Identical with ROLLESTON (1).

Rownall St [*Rugehala* DB, *Ruanhall*, *Ruhenhal'* 1221 Ass]. 'Rough HALH.'

Rowner (-ow-) Ha [*Ruwanoringa gemæro* 948 BCS 865, *Ruenore* DB, *Rugenore* 1114 ASC]. 'Rough landing-place.' See RŪH, ŌRA.

Rowney Hrt [*Ruweney* 1239 Ep, *Rouneya* 1254 Val, *Ruenheye* Hy 3 BM]. 'Rough enclosure.' See RŪH, (GE)HÆG.

Rowrah (-ōō-) Cu [*Rucwrabek* 1248, *Rukwra* 14 StB]. OScand *rug-vrá* 'nook where rye was grown'.

Rowsham (-ow-) Bk [*Roduluesham* c 1130 Oxf, *Rollesham* 1170 P]. '*Hrōþwulf*'s HĀM.'

Rowsley (-ōz-) Db [*Reuslege* DB, *Rolvesle* 1204 Cur]. '*Hrōþwulf*'s LĒAH.'

Rowston (-ow-) Li [*Rouestune* DB, *Rolveston* 1209–35 Ep]. Identical with ROLLESTON (1).

Rowthorn Db [*Rugetorn* DB, *Ruethorn* 1242 Fees]. 'Rough thorn-bush.' See RŪH, ÞORN.

Rowton (-ow-) Chs [*Rowecristelton* 12 Ormerod, *Roghe Cristelton* 1287 Court]. A shortened form of ROUGH CHRISTLETON.

Rowton Sa W. of Shrewsbury [*Rutune* DB, *Ruton* 1233 Cl, 1273 Ipm], R~ YE [*Rughe-*

ton DB, *Ruton* 1241 FF]. OE *Rūh-tūn* 'TŪN with rough soil' or the like. See ROUGHTON.

Rowton Sa nr Ercall Magna [*Routone* DB, *Rowelton* 1195 P, *Ruelton* 1212 Fees, *Roulton* 1233 Cl]. Perhaps 'TŪN at a rough hill'.

Roxby Li [*Roxebi, Roscebi* DB, *Rochesbi* c 1115 LiS, *Rokesbi* 12 DC], **R~** YN in Pickhill [*Rokesby* 1235, 1251 FF]. OScand *Hrōks bȳr,* the first el. being ON *Hrókr* pers. n.

Roxby YN in Hinderwell [*Roscebi, Rozebi* DB, *Raucebi* 1145–8 Whitby], **R~** YN in Thornton Dale [*Rozebi, Rosebi* DB, *Rouceby* 1242 FF]. Identical with RAUCEBY.

Roxeth Mx [*et Hroces seaðum* 845 BCS 448, *Roxhethe* 1422 FF]. '*Hrōc*'s pits or lakes.' Cf. ROCKINGHAM. OE *sēaþ* means 'pit, well, lake'. Some salt-pits or watering-ponds may be referred to.

Roxham or **Roxholm** Li [*Rochesham* DB, *Rokesham* 1206 Cur], **Roxton** Bd [*Rochesdone* DB, *Rokesduna* 1209–10 Ep], **R~** Li [*Roxton* 1212, 1242 Fees]. '*Hrōc*'s HĀM, DŪN and TŪN.' Cf. ROCKINGHAM. Roxton Bd may also be 'rook hill'.

Roxhill. See WRAXALL.

Roxwell Ess [*Rokeswelle* 1291 Tax]. '*Hrōc*'s stream' or 'rook stream'.

Roydon Ess [*Ruindune* DB, *Reidona* c 1130 Bodl, *Reindon* 1204, *Roindon* 1208 Cur], **R~** Nf nr Diss [(et) *Rygedune* c 1035 Wills, *Regadona, Ragheduna* DB, *Reydon* 1242 Fees], **R~** Nf nr Lynn [*Reiduna* DB, *Ridone* 1254 Val], **R~** Drift Sf nr Long Melford [*Rigen-, Rigindun* c 995 BCS 1289]. 'Rye hill.' Cf. RAYDON.

Royston Hrt [*Crux Roaisie* 1184 BM, *Crux Roheis* 1209–19 Ep, *Croyroys* 1262 Ipm; *Reyston* 1280 FF, *Roiston* 1286 Misc]. The original name means '*Roese*'s cross', referring to a cross set up by a certain Lady Roese or Roheis. From this was named a priory founded near the spot t. Hy 2. The later name Royston may be 'TŪN by Crux Roys', the first el. having been dropped, or a translation of the old name into 'Roese's stone'.

Royston YW [*Rorestone, -tun* DB, *-tune* 1155–9 YCh 1168]. '*Hrōr*'s TŪN.' Cf. RORRINGTON.

Royton La [*Ritton* 1226 LaInq, *Ryton* 1260 FF, *Ruyton* 1327 Subs]. 'Rye farm.' Cf. RYGE.

Ruan Lanihorne Co [*Lanrihorn* 1318, *-hoern* 1329 FF, *Sti Rumoni de Lanryhorn* 1350 Subs], **Ruan Major & Minor** Co [(ecclesia de) *Sancto Rumono Parvo* 1277, (rector) *Sancti Rumoni Magni* 1315 Ep]. Ruan is the saint's name *Rumon* (Lat *Romanus*). Lanihorne has as second el. an OCo pers. n. corresponding to or containing MBret *Haiarn,*

Hoiarn. The first may be Co *lan* 'church, enclosure' or *lanherch* 'glade'.

Ruardean Gl [*Rwirdin* DB, *Reworthin* 1200 Cur, *Roworthin* 1220 Fees]. See WORÐIGN. The first el. may be OE *ryge* 'rye' or rather Welsh *rhiw* 'hill, ascent'. The place is on the slope of a prominent hill.

Ruborough Hill So NW. of Taunton [(in) *Rugan beorh* 854, (to) *Ruwanbeorge* 904 BCS 476, 610]. See ROBOROUGH.

Ruckcroft Cu [*Rucroft* 1211 P, 1231–5 WR]. 'Rye croft.' First el. OScand *rugr.*

Ruckinge K [*Hroching* 786, *Hrocing* 805 BCS 248, 1336, *Rochinges* DB, *Roking* 1202 Cur]. OE *hrōcing* 'rook wood'.

Ruckland Li [*Rocheland* DB, *Roclund* 12 DC], **Ruckley** Sa nr Tong [*Rochelai* 1139 Eyton], **R~ and Langley** Sa [*Rocle* 1253 Ch]. See ROCKLAND, ROCKLEY.

Rudby YN [*Rodebi* DB, *Ruddebi* c 1165 YCh 713, *-by* 1228 Ep]. First el. very likely an OScand pers. n. ON *Rudda*, a woman's name, occurs.

Rudchester Nb [*Rodecastre* 1251 Pat, *Rucestre* 1256 Ass]. '*Rudda*'s CEASTER or Roman fort.'

Ruddington Nt [*Rodintun, Roddintone* DB, *Rudinton* 1182 f. P, *Rutington* 1231 Ep]. 'The TŪN of *Rudda*'s people.'

Rudford Gl [*Rudeford* DB, 1138 Glouc, *-e* 1221 Ass, *Rodeforde* c 1160 Glouc]. First el. possibly OE *hrēod* or *hrēoden* (*hrieden*) 'reed, reedy' or as in RUDHEATH.

Rudge Gl [*la Rugge* 1112, *Rugge* c 1120 Glouc], **R~** Sa [*Rigge* DB, *Rugge* 1188 P]. OE *hrycg* 'ridge'.

Rudgwick Sx [*Regwic* 1210 FF, *Rugewik* 1240 FF]. 'WĪC on a ridge.'

Rudham, East & West, Nf [*Rudeham* DB, 1147 BM, *Ruddaham* 1163 BM, *Est Rudham, Westrudham* 1254 Val]. '*Rudda*'s HĀM.'

Rudheath Chs [*Ruddheth* 1271, *Rudeheth* 1277 Chester, *Rudheth* 1288 Court]. '*Rudda*'s heath', or 'marigold heath', if *rud* 'marigold' (14th cent. OED) is an old word.

Rudston YE [*Rodestan, -stain* DB, *Ruddestan* 1100–22 YCh 452, *Rudestan* 1231 FF]. OE *rōde-stān*, lit. 'rood stone'; cf. RADSTONE. The place was named from a monolith near the church.

Rudyard St [*Rudegeard* 1002 Wills, *Rudierd* DB, *Rudeyard* 1330 Ch]. Perhaps 'yard or garden where rue (OE *rūde*) was grown'. But R~ is on Rudyard Lake, and *yard* might refer to an enclosure for fish. If *rudd*, the name of a fish, is an old word (first ex. in OED 1606), Rudyard might mean 'pond where rudds were kept'.

Rufford La [*Ruchford* 1212 Fees, *Rughford* 1327 Subs], **R~** Nt [*Rugforde* DB, *Ruchford* c 1150 DC, *Ruford* 1185 P], **Rufforth** YW [*Ruford* DB, *Rucheford* c 1110 Fr, *Ruhford* c 1190 YCh 556]. 'Rough ford.' See RŪH.

ON rug(r), OSw **rugher, rogher**, Dan **rug** 'rye'. See ROUGHTON, ROWRAH, RUCKCROFT.

Rugby Wa [*Rocheberie* DB, *Rokebi* 1200 Cur, *-by* 1236 Fees]. '*Hrōca*'s BURG' rather than 'BURG inhabited by rooks'. Cf. ROCK-INGHAM. OE *burg* was replaced by OScand BY.

Rugeley (-ŭj-) St [*Rugelie* DB, *-lega* 1157, *Ruggelega* 1156, 1190 P]. 'LĒAH on a ridge' (OE *hrycg*).

Rugley Nb [*Ruggele* 1210 Cur, *-ley* 1242 Fees, *Rogeley* 1256 Ass], **Rugmere** Mx [*Rugemere* DB, *Ruggemere* 1207 Cur]. 'Woodcock LĒAH or wood and lake.' *Rug-* is identical with OE *hrucge* (*rugge*) in *Hrucggan broc* 704–9 BCS 123, *Hrucgan cumb* 739 ib. 1331, *Ruggan sloh* 988 KCD 667. OE *hrucge* is probably related to and identical in meaning with Norw *rugda* 'woodcock'. At any rate the fact that OE *hrucge* is combined with words for brook, lake, swamp, wood suggests that it denoted a bird like the woodcock. *Hrucge* and *rugda* are derived from the stem *hruh, hrug* 'to emit hoarse sounds' in Icel *hrygla* 'death rattle', Norw *rugla* 'to rattle in the throat' &c.

OE **rūh** 'rough, uncultivated, knotty' is fairly common as a first el. The form of the el. varies to some extent owing to the fact that in OE inflected forms *h* was often exchanged for *w* (gen. *rūwes*, weak *rūwa, -n* &c.). See e.g. ROUGH- (passim), RO-, RUBOROUGH, ROWBERROW, ROWNER, RUFFORD, RUSPER. Very likely there was also an OE *rūh* sb. 'rough or uncultivated ground', found in ROUGHAM, ROUGHTON, ROWTON.

Ruishton So [*Risctun* 854, 880 BCS 475, 549]. 'TŪN where rushes grew.'

Ruislip (rīslip) Mx [*Rislepe* DB, *Rislep* 1230 FF, *Risselep* 1252 Ch, *Russelep* 1254 Val, *Rushlep* 1315 FF]. The elements are OE *rysc* 'rush' and *slæp*, here probably 'wet place'. R~ is derived in PNMx(S) from OE *risc* 'rush' and *hlype* (better *hliepe*) 'leap'. This is perhaps formally better, but the meaning of such a name is not clear.

OE **rūm**, ON **rúm**, OSw **rūm** 'room', probably also 'clearing', is found in DENDRON La and probably some other names. OE **rūm** adj. 'roomy, spacious' is the first el. of ROMILEY, RUNCORN and perhaps some other names, as ROMFORD, RUMBURGH, RUMWORTH, but derivation from *rūm* adj. often offers difficulties.

Rumbles Moor YW [*Rumbesmore* 1235 FF]. '*Rumbald*'s moor.'

Rumbridge Ha [*Runbrigga* 1180 P, *Rumbrigge* 1320 Ipm, 1345 Misc], **Rumburgh** Sf [*Romburch* 1047–64 Holme, *Rumburg*

c 1130, *Romburch* 12 BM, *Romburg* 1154 YCh 354, 1207 Cur]. The first may be 'broad bridge' (cf. RŪM), but the first el. is more likely OE *hruna* 'a fallen tree, log'. For the second also, the regular early form *Rum-* (*Rom-*) instead of *Rume-* tells against the adj. *rūm*. OE *rūn* 'deliberation, council' or *hruna*, just mentioned, might be thought of.

Rumwell So [*Runwille* 1327 Subs]. See RUNWELL.

Rumworth La [*Rumwrth* 1205 FF, *-worth* 1278 Ass]. Perhaps 'spacious WORþ' (cf. RŪM), but the monosyllabic form of the first el. may suggest rather *rūm* 'cleared place'.

OE **rūn** 'secret, council, secret discussion' &c. is the first el. of RUNNYMEDE, and may be that of some other names, as RUMBURGH, RUNWELL.

Runcorn Chs [*Rumcofa* 915 ASC (C), *Runcoua* 1154–60 (1329) Ch, *Runcore* 1259 Court]. 'Wide bay.' The name refers to the broadening of the Mersey below Runcorn. *Rūm* adj. is here used in its uninflected form. The change of *-cofa* to *-core* may be due to misreading of *v* as *r*.

Runcton, North & South, R~ Holme Nf [*Run(c)getun* 11 EHR 43, *Runghetuna* DB, *Runget* 1158 P, *Norþrungetone* 1276 Ipm, *Suthrungetone* 1291 Tax, *Rungeton Holm* 1276 Cl]. See ROUNTON.

Runcton Sx [*Rochintone* DB, *Rogentona* 1110–17, 1155 Fr]. Identical with ROWING-TON.

Runfold Sr [*hrunigfealles wæt* 974 BCS 1307, *Runifall(e)* 1210 ff. PNSr]. OE *hrunigfeall* for *hrungefeall* 'falling of trees, place where trees have fallen', *hrun-* being OE *hruna* 'fallen tree' (see HEADCORN); *wæt* is OE *wæt* 'wet' in substantival use ('wet place').

Runhall Nf [*Runhal* DB, 1206 Cur, *-e* 1254 Val], **Runham** Nf [*Rom-*, *Ronham* DB, *Runnaham* 1163 BM, *Runham* 1165 P, 1196 Cur]. *Romham* DB is no doubt to be disregarded. The first el. may be OE *rūn* 'council' or *hruna* 'a fallen tree, log', but for Runham a pers. n. *Rūna* corresponding to OG *Rūno*, ON *Rúni*, OSw *Rune* is perhaps more probable. See HALH, HĀM.

Runnington So [*Runetone* DB, *Ronneton* 1202 Cur, *Runneton* 1233 FF, *Runnyngton* 1306 FF]. Possibly 'the councillor's TŪN', the first el. being OE *gerūna* 'councillor', or '*Rūna*'s TŪN'.

Runnymede Sr [*pratum . . . Ronimede* (var. *Runingmeð*) 1215 Magna Carta, *Runimede* 1215 (1318) Ch, *Rumened* 1244 Cl]. 'The meadow in council island.' The first el. is OE *Rūnīeg* 'council island, assembly island'. The name seems to indicate that Runny-mede was an ancient meeting-place.

Runswick YN [*Renneswyc* 1272 Cl, *Rumeswyk* 1293 QW, *Remmeswyk* 1327 BM]. The forms are too late and conflicting for an etymology to be possible.

Runton, East & West, Nf [*Runetune* DB, *-tona* 1175–86 Holme, *-ton* 1185 P, 1209 FF]. '*Rūni*'s or *Rūna*'s TŪN.' See RUNHALL.

Runwell Ess [*Runweolla*, *Runewelle* 939 BCS 737, *Runewella* DB, *-well* 1203 Cur]. Cf. RUMWELL. The first el. is very likely OE *rūn* 'secret, council' &c. The name may refer to a spring or stream at which a meeting-place was, or rather to a wishing-well. Cf. FRITWELL.

Ruscombe Brk [*Rothescamp* 1091, 1220 Sarum, *Rotescamp* 1167 P, c 1209 Salisbury]. '*Rōt*'s pasture-ground.' See CAMP and ROOTING.

Rushall Nf [*Riuessala* DB, *Riuishale* 1175 P, *Riveshale* 1242 Fees, 1254 Val, *Reueshall* 1264 Ch]. Second el. HALH. The first might conceivably be OE *hrif* 'womb' used in some transferred sense, but is more likely a pers. n. formed from OE *rif* or *hrife* 'fierce'.

Rushall St [*Rischale* DB, *Rushale* 1195 P, 1242 Fees]. 'HALH overgrown with rushes.'

Rushall W [*Rusteselve*, *-selle* DB, *Rusteshala* 1161 P, *-hale* 1242 Fees]. Second el. OE *scylf* 'ledge, bank'. The first is obscure. *Rustingdenn* 850 BCS 459 (K) may be compared, but is itself unexplained. Possibly we may assume an OE *rust* 'rest', corresponding to MHG *rust*, MLG *ruste*. The name would then mean 'bank where wayfarers took a rest'.

Rushbrooke Sf [*Ryssebroc* c 950 Wills, *Ryscebroc* DB], **Rushbury** Sa [*Riseberie* DB, *Rissiberia* 1180 For, *Russhebur* 1283 Ch], **Rushden** Hrt [*Risendene* DB, *Russenden* 1190 P, *Ressenden* 1212 Fees], **Rushden** Np [*Risdene* DB, *Ressendene*, *Rissendene* 1230 Ep], **Rushford** Nf [*Rissewrth* c 1060 Wills, *Rusceuuorda* DB, *Rischewrthe* 1242 Fees], **Rushmere** Sf nr Ipswich [*Risce-*, *Ryscemara* DB], **Rushmere** Sf nr Lowestoft [*Risce-*, *Ryscemara* DB]. The first el. is OE *risc*, *rysc* 'rush', partly *riscen* adj. 'rushy'. The second is BRŌC, BURG, DENU, WORÞ, MERE.

Rushock Wo [*Russococ* DB, *Rossoc* 1166 RBE, *Roisoc* 1167 P]. Probably identical with **Rushwick** Wo, named from *Rixuc*, a stream mentioned 963 BCS 1106, which appears to mean 'rushy brook or place'. The DB form *Russococ* may be miswritten for *Russoc*, but it is possible that it contains the OE *cocc* found in WITHCOTE. Rushock might even be from OE *rysc-cocc* 'clump of rushes'. **Rushock** He [*Ruiscop* DB, *Russhoc* 1330 Ep] may be identical in origin.

Rusholme La [*Russum* 1235 FF, *Reshum* 1417 PNLa]. OE *ryscum*, dat. plur. of *rysc* 'rush'.

Rushton Chs [*Rusitone* DB, *Ruston* 1240 Cl], **R~** Np [*Ris(e)tone* DB, *Riston* 1163 P, *Ruston* 1199 FF, *Rishton* 1327 BM], **R~** St [*Risetone* DB, *Ruston* 1278 Ipm]. 'Rush TŪN.'

Rushwick. See RUSHOCK.

Ruskington Li [*Rischintone*, *Reschintone* DB, *Rischinton* 1166 P, *Riskinton* 1202 Ass, *Resketon* 1267 Ch]. Probably OE *riscen*, *ryscen* 'rushy' and TŪN, *sk* being due to Scand influence.

Rusland (-z-) La [*Rolesland* 1336 FC]. '*Hrōald*'s or *Hrōlf*'s land.' Cf ROLLESTON.

Rusper Sx [*Rusparre* 1219, *Rugesparre* 1238 FF, *Rughesparre* 1278 Cl]. 'Rough enclosure.' See RŪH, SPEARR.

Rusthall K [*Rusteuuellæ* 765–91 BCS 260, *Rustuwelle* c 1180 Arch Cant vi]. 'Rust-coloured spring.'

Rustington Sx [*Rustitona* Hy 2 (1361) Pat, *Rustinton* 1180, *Rustincton* 1186 P, *Rustington* 1255 FF]. Doubtful. A pers. n. *Rusta corresponding to OG *Rusto* has been suggested as the base. *Rustingdenn* 850 BCS 459 (K) may be compared.

Ruston, East, Nf [*Ristuna* DB, *Ristone* 1129 Holme, *Riston* 1198, 1208 Cur, *Estriston* 1405 BM], **Sco R~** Nf [*Ristuna* DB, *Soriston* 1346, *Ryston* 1402 FA, *Skouriston* 1425 AD]. OE *Hris-tūn* 'TŪN in brushwood'. *Sco* is OScand *skōgr* 'wood'.

Ruston Parva YE [*Roreston* DB, *Roston* 1227 FF]. '*Hrōr*'s TŪN.' Cf. RORRINGTON, ROYSTON YW.

Ruston YN [*Rostune* DB, *Rostona* c 1195 YCh 381, *Ruston* 1167 P, 1234 FF]. The first el. may be OE *hrōst* 'roost, perch', but here used in a sense similar to that of OLG *hrōst*, which means 'rafters of a roof'. A similar sense is found in OE *hrōstbēag* 'woodwork of a roof'. The name would refer to a building with a special kind of roof.

Ruswarp (rŏŏzup) YN [*Risewarp* 1145–8 YCh 872]. OE *hris* 'brushwood' and *geweorp* 'throwing' &c., dial. *warp*, 'silt, silt land'.

Rutland [*Roteland* c 1060 KCD 863, DB, *-lande* 1080–7 Reg, *Rotland sokene* 1377 Langland]. '*Rōta*'s land.' Cf. ROOTING, ROTTINGDEAN, RATLEY &c. In DB Rutland is partly assessed under Lincolnshire and Northamptonshire. In its origin Rutland was a large soke. It was administered in OE times and later for successive queens of England.

Ruttingham. See ROTTINGDEAN.

Ruyton (-ī-) Sa [*Ruitone* DB, *Ruton* 1276 Ipm]. OE *Ryge-tūn* 'rye farm'.

Ryal Nb [*Ryhill* 1242 Fees], **Ryhill** YE [*Rihull* 1219 FF], **R~** YW [*Rihella* DB],

Ryle Nb [*Parva Rihull* 1212 Fees, *Ryhull* 1254 Ipm, *Ryel* 1256 Ass]. 'Rye hill.'

Ryarsh K [*Riesce* DB, *Rierssh* 1242 Ep]. 'Rye field.' See ERSC.

Ryburgh, Great & Little, Nf [*Reieburh, Parva Reienburh* DB, *Rieburc* 1165 P, *Riburg Magna* 1291 Tax, (in) *Parvo Riburc* 1198 FF]. 'BURG where rye was grown.'

Ryburn R. See RIPPONDEN.

Rycote O [*Reicote* DB, *Ruicota* 1177 P]. 'Rye cottage.'

Rydal We [*Ridale* 1240, 1274 Ipm]. Probably 'valley where rye was grown'.

OE *ryddan* 'to clear land' is not evidenced, but OE *āryddan* 'to strip, plunder' occurs. *Ryddan* corresponds to ON *ryðia*, OSw *ryþia* &c. The past part. of the verb, OE *ryded, rydd*, is the first el. of some names, as REDLAND, RIDLEY (1). From RYDDAN was formed OE *ryding*, evidenced once in the faulty form *hryding*, 'clearing, cleared land'. This is the source of *ridding* 'clearing', still common in dialects and found in several pl. ns., as RIDING Nb. For an OE *ryden* 'clearing' cf. COLDRED.

Ryde Wt [*la Ride* 1257 PNWt, *la Ryde* 1324 Misc]. 'The stream.' Dial. *ride* (Ha, Wt) is a variant of OE *riþ* 'a small stream'.

Rye Sx [*Ria* 1130 P, c 1197 Penshurst, *la Rye* 1247 Ch]. OE *æt þære iege* 'at the island', which became *atter ie, atterie* and by wrong division *atte Rie*. Rye is really on an island.

Rye R YN [*Ria* 1132 ff. Riev, 1201 FF, *Rie* 1279 Ass]. A Brit river-name, which may be identical with *Rhiw* in Wales. For the loss of *w* cf. DEE. The name may be cognate with Lat *rivus* 'stream', or it may be formed from Welsh *rhiw* 'hill, ascent'.

OE *ryge* 'rye' is found in several pl. ns. See RY- (passim), RAYLEIGH, REYDON, ROYTON, RUYTON &c. An adj. *rygen* 'of rye' is sometimes the first el., as in RAYDON, ROYDON.

Ryhall Ru [*Rihala* 963 ASC (E), *Righale* 1066–9 KCD 927, *Riehale* DB, *Rihale* 1179 P]. 'HALH where rye was grown.'

Ryhill. See RYAL.

Ryhope Du [*Reofhoppas* c 1050 HSC, *Refhope* c 1190 Godric, *Riefhope* 1196 P]. Second el. HOP 'valley'. The first seems to be OE *hrēof* 'rough', but its exact meaning is not clear.

Ryknild Street. See ICKNIELD WAY.

Ryle. See RYAL.

Rylstone YW [*Rilestun, Rilistune* DB, *Rillestun* 1166 P, *-ton* 1200–30 FC, 1251 Ass]. Perhaps OE *Ryneles-tūn* 'TŪN by the brook'. OE *rynel* 'brook' is on record. Assimilation of *nl* to *ll* is well evidenced.

Ryme Do [*Rima* c 1160 Salisbury, *Rym* 1280 FF, 1298 Ch, *Ryme* 1229 Pat, 1293 FF]. OE *rima* 'border, rim'. The place is on the slope of a ridge.

Sometimes **R~ Intrinseca** 'inner R~' in contradistinction to the lost **R~ Extrinseca**.

OE **rysc.** See RISC, RYSC.

Rysome YE [*Rison* DB, *Rysun* 1240 FF, *Risum* 1285 FA]. The dat. plur. of OE *hris* 'brushwood'. Cf. RISE.

Ryston Nf [*Ristuna* DB, 1121 AC]. OE *Hris-tūn* 'TŪN in brushwood'.

OE **rȳt, rȳht** is found once in a law. The context does not give a clear indication as to the meaning. The passage is: *Gif fyr sie ontended ryht* (var. *ryt*) *to bærnenne* 'if a fire is kindled in order to burn *ryht*'. *Ryht* is rendered by Liebermann 'brushwood that has been pulled up', by B-T (Suppl) 'rough growth on land'. The word is clearly a derivative of OE *rūh* 'rough', either with the suffix *-iþō* (OE *rȳhþ*, whence easily *rȳht*, or *rȳþ*) or with the suffix *-itia-*, as in OE *efnett* 'plain' (early OE **rȳhet*, whence *rȳht*, *rȳt*). The meaning of the word would be 'roughness' in the first alternative, 'rough place' or the like in the second. 'Rough growth' or 'rough common' would be a very likely meaning and suits the OE context. OE *rȳ(h)t* is found in REED, and very likely in RATLING, possibly in some other names.

Ryther YW [*Ridre, Rie* DB, *Rie* 1212 FF, *Ryther* 1257 Ch, *Rither* 1291 Tax]. The name has been derived from OE *geryþeru*, held to mean 'clearing'. But the existence of such a word is doubtful. Ryther may be an OE *hrȳþer-ēa* or *-ēg* 'cattle stream or island'.

Ryton Du [*Ritona* 1183 BoB, *Ryton* 1291 Tax], **R~** Sa nr Beckbury [*Ruitone* DB, *Ruton* 1242 Fees], **R~** Sa nr Condover [*Rutton* 1250 Eyton, *Ritton* 1260 Ipm], **R~** Wa [*Ruyton* E 1 BM], **R~** on Dunsmore Wa [*Rietone* DB, *Ruyton* 1316 FA]. OE *Ryge-tūn* 'rye farm'.

For R~ on Dunsmore see DUNSMORE.

Ryton R Nt. A back-formation from a pl. n. *Ryton* (now Rayton). The old name was *Blithe*. See BLYTH.

Ryton YN [*Ritun* DB, *Riton* 1163 P, 1240 FF, *Ryton in Rydall* 1403 AD]. 'TŪN on R RYE.'

S

Sabden La [*Sapeden* c 1140 LaCh, *Sappeden* 1377 Ct]. 'Spruce valley.' See SÆPPE.

Sacombe Hrt [*Seuechampe, Sueuecampe* DB, *Sauecampe* 1197 FF, *Swauecampe* 1310 BM]. '*Swǣfa*'s CAMP or field.' **Swǣfa* is a short form of names like *Swǣfheard*.

Sacriston Heugh Du [*Segrysteynhogh* 1312 RPD]. 'The HŌH or spur of land of the sacristan' (of Durham). First el. the word *sexton* in an earlier form and meaning. The source is OFr *segrestein, secrestein* from MLat *sacristanus*.

Sadberge Du [*Satberga* Hy 2 Finchale, *Sadberga* 1169, *Sethberga* 1177 P, *Sedberge* 1198 Fees]. ON *setberg* 'flat-topped hill'. *Setberg* occurs as a name in Iceland and Norway.

Saddington Le [*Sadintone, Setintone* DB, *Satintone* 1176, 1181 P, *Seddinton* 1199 FF, *Sadinton* 1228 Ch]. Etymology obscure. A possible etymology is 'the TŪN of *Sǣgēat*'s people'. *Sǣgēat* is a known name. It is true we should expect the *g* to have left a trace; cf. *Saiet* DB. But cf. BREADSALL, BREASTON Db, TEAN St. Norman influence might account for monophthongization of *ai*.

Saddle Bow Nf in Wiggenhall St. Mary [*Sadelboge* 1198 FF, *Satelbowe* 1314 Ipm]. 'The saddle-bow', possibly referring originally to an arched bridge or to a piece of land similar in shape to a saddle-bow. The name is used of an area in 1198.

Saddleworth YW [*Sadelword, -worth* c 1230 WhC]. 'WORÞ on a saddle or saddle-like ridge.'

OE **sǣ** 'lake; the sea' is a fairly common pl. n. el. A non-umlauted side-form *sā* (cf. OE *Sabeorht* by the side of *Sæbeorht* &c.) is found in SAHAM, SOHAM. As a second el. *sǣ* means 'lake' except in the late name SOUTHSEA. See e.g. HADDLESEY, HORNSEA, KILNSEA, MEAUX, SKIPSEA, WITHERNSEA. As a first el. it usually means 'the sea', as in SEACOMBE, -FORD, -HAM, -SCALE, -TON(most), SILLOTH. It means 'lake' in SA-, SOHAM, SEACROFT, SE(A)MER, SEATHWAITE, SEATON YE. OScand (ON) *siór, sær* 'lake' may partly be the source in some names.

OE **sǣppe** 'spruce fir'. See SABDEN, SAPLEY. A correspondent points out that the spruce is hardly indigenous in England, and that *sæppe* is not a likely el. in pl. ns. This is right, but *sæppe* may have been used of some other conifer, as the Scotch fir.

OE **sǣt** 'trap'. See MERSTHAM.

OE **-sǣte, -sǣtan** 'dwellers' is found in OE in *Arosǣtan, Cilternsǣtan* 'dwellers on the Arrow (in the Chilterns)' and the like. Some folk-names of this kind have passed into pl. ns. See DORSET, SOMERSET, GRANTCHESTER and Introd. ii. 1. Names in *-sett* in Nf and Sf, as BRICETT, HESSETT, LETHERINGSETT probably do not belong here. See (GE)SET, SÆTE. See also BURSTWICK, HISTON, POSTON.

OE **sǣte** 'a house' may be found in some names, as BRENZETT, ELMSETT, GUIST.

ON **sætr** 'a shieling, hill pasture' is doubtless the second el. of several names of places whose situation suggests that they were originally shielings. In early sources the el. appears as *-set, -sat*, but the modern form is generally *-side* or *-head*. See e.g. AMBLE-, ANNA-, OUGHTER-, SELSIDE, HAWKSHEAD, also BLENNERHASSET. ON *sætr* is the first el. of SATTERTHWAITE, SETMURTHY. It is sometimes difficult to distinguish *sætr* from OE (GE)SET 'a fold'.

Saham Toney Nf [*Saham* DB, 1168 P, *Seham* 1130 P, *Saham Tony* 1498 AD iii]. 'HĀM by the lake.' Cf. SǢ.
The manor was held by Roger de Toni in 1199 (P). See NEWTON TONEY.

Saighton (-ā-) Chs [*Saltone* DB, *Salhtona* c 1100, *Salighton* 1249–65 Chester]. OE *Salh-tūn* 'TŪN by sallows'. Cf. SALH.

St. Albans Hrt [*æcclesia sancti Albani* 792 BCS 264, *sancte Albanes stow* 1007 Crawf, *Villa S. Albani* DB, (æt) *Sancte Albane* 1116 ASC (E)]. Self-explanatory. The Brit name of the place was *Verulamium* Tac, &c., whence *Uerlamacæstir* c 730 Bede, *Werlameceaster* c 890 OEBede. Another early name was *Uaeclingacæstir* Bede, *Wæclingaceaster* OEBede, *Wætlingaceaster* 990 KCD 672. This name means 'the Roman fort of the *Wæclingas*' (or '*Wacol*'s people'). Cf. WATLING STREET, WATLINGTON.

St. Allen Co [(ecclesia) *Sancti Alluni* 1261, *Sancti Aluni* 1284 Ep]. *Allen* is a pers. n. from OCo *Alun*, cognate with OBret *Alunoc*. The Brit base is *Alauno-*; cf. the rivername ALN.

St. Anthony in Meneage Co [*Sancti Antonini in Manahec* 1269 Ep], **St. A~ in Roseland** Co [(ecclesia de) *Sancto Anton' in Roslond* 1291 Tax]. Meneage and Roseland are old names of districts.

St. Austell Co [(ecclesia de) *Austol* 1138–55 Ep, *St. Austol* 1251 FF]. '(Church of) St. *Austol*.'

St. Bees (bēz) Cu [*Cherchebi* c 1125, *Sancta Bega* 12, *Kirkebibeccoch* c 1195 StB, (ecclesia) *Sancte Bege* 1291 Tax]. The priory, founded c 1125, was dedicated to Sancta Bega, a virgin saint mentioned by Bede (*Begu*) and said to be of Irish descent.

St. Breock Co [*Sancti Brioci de Nansent*

1309 Ep, *Nanssent* 1335 FF]. St. Brioc, Bret *Brieuc*, Welsh *Briog* in LLANFRIOG, represents OBrit *Brigāco-* (from *brig-, OIr *brig*, Welsh *bri* 'dignity, rank'). *Nansent* means 'the valley of the saints' (Co *sant*, *sans*, plur. *syns*).

St. Breward Co [(ecclesia de) *Sancto Bruereto, Sancti Brueredi* 1272 Ep]. *Breward* is presumably a Cornish saint.

St. Briavels (-ĕ-) Gl [(Castellum de) *Sancto Briauel* 1130 P, *Sanctus Brievellus* 1207 Cur]. *Briavel*, MW *Briauail* c 1150 LL, seems to go back to OCelt *Brigomaglos*.

St. Budeaux D [*Bucheside* DB, *Buddekeshid* 1242 Fees]. 'St. *Budoc*'s *higid* or house-hold.' The forms represent an earlier name of the parish, now preserved as **Budshead** in St. Budeaux. *Budoc* is a pers. n. derived from OCelt **boudi-* 'victory' (in OIr *búaid*, Welsh *budd* &c., also in *Boadicea*, *Boudicca* Tac).

St. Buryan Co [(ecclesia) *sancte Beriane* 943 BCS 785, (Terra) *Sancte Berrione Virginis* DB]. Cf. Bret *Berrien* in LAN-VERRIEN &c. Nothing appears to be known about the saint.

St. Cleer Co [(ecclesia de) *Sancto Claro* 1212 Cur, *Seintcler* 1230 Cl]. *Sanctus Clarus*, an unknown saint.

St. Clether Co [*Seyncleder* 1249 FF, (ec-clesia) *Sancti Clederi* 1261 Ep]. St. Cleder, said to have been a brother of St. Nechtan, gave its name also to CLEDER in Brittany.

St. Columb Major & Minor Co [(ecclesia) *Sancte Columbe* 1266 Ep, (ecclesia) *Sancte Columbe majoris* 1291 Tax, (capella) *Sancte Columbe Minoris* 1284 Ep]. Named from St. Columba the Virgin.

St. Decumans So [(ecclesia) *Sancti Decimi* 1203 Cur, (church of) *St. Decuman* 1243 Ass]. St. Decuman is said to be a Welsh saint who died as a hermit at Dunster.

St. Denys Ha [(ecclesia) *Sancti Dionisii* 1192 AC]. Identical with SAINT DENIS in France.

St. Devereux He [(ecclesia) *Sancti Dubricii* 1291 Tax]. St. Dyfrig, OW *Dubric*, is a well-known Welsh saint.

St. Ende·llion Co [(ecclesia) *Sancte Ende-liente* 1260 Ep]. An unknown saint.

St. Enoder Co [*Heglosenuder* DB, (vicaria) *Sancti Enodri* 1271 Ep]. An unknown saint. *Heglos-* is Co *eglos* 'church'.

St. Erme Co [*St. Ermes* 1266 Ch, *Sancti Ermetis* 1283 Ep]. **St. Erney** Co [*S. Erne* 1377 PT]. Unknown saints.

St. Erth Co [(vicaria) *Sancti Ercii* 1257–80 Ep]. Said to be named from St. Ercus, bishop of Slane, Ireland. An earlier name is *Lannutheno* 1233 FF, (ecclesia de) *Lan-uthno* 1269 Ep, which means 'the church of Uthno'. Cf. PERRAN UTHNOE.

St. Ervan Co [(rector) *Sancti Hermetis* 1258 Ep]. Apparently a doublet of ST. ERME.

St. Eval Co [(ecclesia) *Sancti Uvelis* 1260 Ep, (de) *Sancto Uvelo* 1291 Tax]. Cf. *Uuel*, the name of a Breton saint. The name may come from Lat *humilis* 'humble'.

St. Ewe Co [*Sancta Ewa* 1282 Ep]. The name has been compared with the Breton saint's name *Eo*, which denoted a male saint.

St. Frideswide O [(Canonici) *S' Fride-suidæ* DB, *Sancta Frethesuith* 1195 FF]. The monastery was founded in honour of St. Frideswide, who was buried at Oxford according to Saints. The saint is called *Sancte Fryðesweoð* in Saints. The correct OE form would be *Friþuswiþ*.

St. Gennys Co [*St. Ginnes* 1244, (in) *Sancto Ginasio* 1246 FF, (ecclesia) *Sancti Gignasii* 1269 Ep]. According to Oliver St. Genesius.

St. Germans Co [(æcclesia) *S' Germani* DB]. Named from St. German of Auxerre.

St. Giles in the Heath D [(capella) *Sancti Egidii* 1202 Launceston], **St. G~ in the Wood** D [*Stow St. Giles* 1330 PND]. St. Egidius. *Giles* is a Fr form of the name.

St. Gluvias Co [(ecclesia de) *Sancto Gluviaco* 1291 Tax, *St. Glywyatus* 1333 AD iii]. Apparently a local saint.

St. Helens Wt [(Prior de) *S. Elena* Hy 2 BM]. The church is dedicated to St. Helen.

Saint Hill. See SIGNET.

St. Issey Co [(in) *Sancto Ydi* 1195 P, *Seint Idde* R 1 Cur, (ecclesia) *Sancte Ide* 1257–80 Ep, *St. Jidgey* E 1 BM]. Cf. MEVAGISSEY. There seems to have been some doubt whether the saint was a man or a woman. It is true the forms *Sancto Ydi* and *Seint Idde* are stated to refer to Gidgey in St. Issey, but no doubt the two are doublets of the same name.

St. Ive (ēv) Co [(ecclesia) *Sancti Ivonis* 1291 Tax], **St. Ives** Hu [*S. Ivo de Selepe* 1110 Rams, *S. Ivo de Slepe* 1130 BM]. Both are no doubt named from the same Ivo. Cf. SLÆP.

St. Ives Co [*Sancte Ya* 1377 PT, (Capella de) *Porthye* 1331 Ep, *Porthia* 1335 AD]. *Ia* corresponds to Bret *Ie* in PLOUYÉ [*Ploehie* 1337] 'the parish of St. *Ie*'. *Porthia* is 'the port of *Ia*'.

St. Ives Ha [(des) *Iuez* 1167, *Yuez* 1187, 1190 P, *Yvez*, *Yvetis* 1212 Fees, *les Yvez in Nova Foresta* 1244 Cl]. This is not a saint's name. It appears to be a derivative of OE *ifig* 'ivy' (cf. the shorter form *if-* in *ifiht* 'clad with ivy'), i.e. an OE *ifet 'clump of ivy' or the like.

St. Just in Penwith Co [(ecclesia) *Sancti Justi in Penwithe* 1334 Ep], **St. J~ in Rose-land** Co [(ecclesia de) *Sancto Justo* 1202 FF, *Sancti Justi in Roslonde* 1282 Ep]. *Just* (from Lat *Justus*) is also a Breton saint. Cf. PENWITH, ST. ANTHONY.

St. Keverne Co [*Sanctus Achebrannus* DB, (in) *Sancto Akaverano* 1201 FF, *St. Kaveran*

1236 FF]. The name is said to be Ir *Aed Cobhran.*

St. Kew Co [*Lannoho, Lanehoc* DB, *Landeho* 1261 Ep, *Lanewe* or *Kewe* 1576 BM]. The modern name is a saint's name identical with *Kew* in Brittany, earliest form *Caio.* The old name is etymologically different, and means 'church of St. *Docco* (*Dochou*)'.

St. Keyne Co [(ecclesia) *Sancte Keyne* 1291 Tax]. *Keyne* is a woman's name derived from Welsh *cain* 'beautiful'.

St. Mābyn Co [(ecclesia de) *Sancto Malbano* 1234 FF, (rector) *Sancte Mabene* 1266 Ep]. *Mabon* is a well-known Welsh and Breton name. Here there is doubt about the sex of the saint.

St. Mary Church D [*See Maria Circea* c 1070 Ex]. Self-explanatory.

St. Mawes (mawz) Co [*Sanctus Maudetus* 1345, *Seynt Mausa* 1467 AD iv]. The saint's name is identical with Bret *Maudez,* earlier *Maudeð, Mautith.*

St. Mellion Co [*Sanctus Melanus* 1198 P, 1280 Ep]. Identical with MULLION.

St. Mellons Monm. See MULLION.

St. Merryn Co [(rector) *Sancte Marine* 1259 Ep]. Merryn is identical with Bret *Merin* and Welsh *Merin* in BODFERIN, LLANFERIN.

St. Mewan Co [*Sanctus Mewanus* 1305 AD iv, 1318 Ep]. Cf. the Bret saint's name *Meven.*

St. Michael Caerhays Co [*S. Mich. Karyheys* 1400 BM]. Caerhays is *Karihays* c 1300 Ep, *Caryhays* 1329 AD iv. Etymology doubtful.

St. Michael Penkevil Co [*Sancti Michaelis de Penkevel* 1264 Ep]. *Penkevil* is 'hill of the horse' (Co *pen* 'summit' and *cevil,* Welsh *ceffyl* 'horse').

St. Michael on Wyre La [*Michelescherche* DB, (eccl.) *Sancti Michaelis Super Wirum* c 1195 Ch].

St. Michael's Mount Co [*Sanctus Michael* DB, *Mons Sancti Michaelis* 1169 P, *Mihælesmunt* 1205 Layamon]. 'St. Michael's hill.'

St. Minver Co [*St. Menefrede* 1256 FF, *Sancta Mynfreda* 1291 Tax]. St. Menefreda according to Oliver.

St. Neot Co [*Nietestov, Neotestov,* (Clerici) *S. Neoti* DB, *Sennet* c 1100 Montacute], **St. Neots** (-ē-) Hu [*S' Neod* 1132 ASC (E), (villa) *S. Neoti* 1203 FF]. According to Saints *sancte Neót* (martyr or priest) was buried at Eynesbury, is close to St. Neots.

St. Nicholas at Wade K [(Villa) *Sancti Nicholai* 1254 KnFees, *St. Nicholas by Wade* 1456 AD]. *Wade* is OE *gewæd* 'ford', and the ford may be the *gewæd* mentioned 943 BCS 780.

St. Nighton Co [*St. Nictan* 1284 FF]. *Nechtan* is shown by its form to be an Ir name. St. Nechtan is a famous saint.

St. Osyth (tōozĭ) Ess [*Sancta Ositha* 1130 P]. The old name of the place was CICC, but a priory was built here and dedicated to St. Osyth, and the name St. Osyth soon displaced *Cicc. Osgȳþ (sancte Osgið),* granddaughter of Penda, was buried at *Cicc* according to Saints.

St. Paul's London [*Monasterium . . . quod dedicatum est in nomine Sancti Pauli Apostoli* 678–81 BCS 55, *Paulus byrig æt Lundænæ* 10 BM].

St. Pinnock Co [(ecclesia) *Sancti Pynnoch* 1291 Tax]. Oliver gives the saint's name as *Pynocus.*

St. Teath (tĕth) Co [*Tethe* 1259, (ecclesia) *Sancte Thetthe* 1266 Ep, *Sancta Thetha* 1278 FF]. St. Tetha according to Oliver.

St. Tudy Co [*Ecglostudic* DB, *St. Tudy* 1258 FF]. Cf. Bret SAINT TUDY &c. The name is British, derived from Welsh *tud,* Co *tus* 'people'. *Ecglos* is Co *eglos* 'church'.

St. Veep Co [*St. Veep* 1284 FF, (ecclesia) *Sancti Vepi* 1351 FF]. Sanctus Vepus or Sancta Vepa according to Oliver.

St. Wenn Co [(vicar) *Sancte Wenne* 1260 Ep]. Cf. Bret *sainte Guen.* The name is derived from Co *gwen* 'white'.

St. Weonards (-ě-) He [*Lann Sant Guainerth* c 1150 LL, (eccl.) *Sancti Waynard* 1291 Tax]. The history of the saint, who is called *St. Gwennarth,* is obscure. Cf. LANN.

St. Winnow Co [*Sanwinvec* DB, *Sanctus Winnocus* 1166 RBE]. The name is identical with Bret *Winnoc,* Welsh *Gwynnog* and no doubt derived from Welsh, Co *gwyn* 'white'.

Saintbury Gl [?*Svineberie* DB, *Seinesberia* 1186 ff. P, *-bir'* 1203 Cur, *Seineburia* 1220 Fees]. Perhaps '*Sæwine*'s BURG'. Or the first el. may be **Sæwynn,* a woman's name.

Salcey Forest Np [*Boscus de Salceto* 1212 Cl, *Salcet* 1274 Ipm]. A Fr name derived from Lat *salicetum* 'willow wood' and identical with Fr SAUSSAY, SAULCET, SAULCY, and others.

Salcombe D nr Kingsbridge [*Saltecumbe* 1244 Ass, *Saltcomb* 1303 FA], **S~ Regis** D [*Sealtcumbe* c 1070 Ex, *Selcome* DB, *Saltcumb* 1242 Fees]. 'Salt valley.' See CUMB. The first contains *salt* adj., the name referring to a creek with salt water. The second contains *salt* sb. The name means 'valley where salt was made' or the like.

Salcott Ess [*Saltcot* J BM, *Salcot* 1231 FF]. 'COT where salt was made or stored.'

Salden Bk [*Scelfdun* 792 BCS 264, *Sceldene* DB, *Schaldene* 1176 P]. It is possible that *Scelfdun* and later *Sceldene* &c. are distinct names, though *Scelfdun* must be sought in the immediate vicinity of Salden. *Salden* may be from *Scelfdūn-denu.* Such a name must have been reduced early. *Scelfdun* is 'hill with a SCYLF or ledge'. Salden is on a hill.

Sale Chs [*Sale* 1260, *la Sale* 1285 Court]. OE *salh*, dat. *sale*, 'sallow'.

Saleby Li [*Salebi* DB, 12 BM, *Salesbi* 1166 P, -*by* 1209–35 Ep]. '*Sali*'s BY' (first el. ODan *Sali*, OSw *Sale* pers. n.).

Salehurst Sx [*Salhert* DB, -*hirst* 1205–16 BM]. 'Sallow wood.' Cf. SALH.

Salesbury (sălzbrī) La [*Sale(s)byry* 1246 Ass, *Salebiry* 1258 Ipml, 'BURG by Sale Wheel.' **Sale Wheel** [*Salewelle* 1296 Lacy] is a pool in the Ribble. The name means 'pool where sallows grew'. See WÆL.

Salford (sahf-) Bd [*Saleford* DB, 1156 P], **S~** (sŏl-) La [*Salford* DB, 1177 P]. 'Sallow ford.' See SALH.

Salford O [*Saltford* 777 BCS 222, 1208 Cur], **Abbots S~**, **S~ Priors** Wa [*Saltforde major et minor* 714 BCS 130, *Salford* DB, *Saltford Abbotes, Prior* 1316 FA]. 'Salt ford', i.e. 'ford over which salt was carried'.

Abbots S~ from the Abbot of Evesham, **S~ Prior** from the Prior of Kenilworth.

OE **salh, sealh** 'sallow'. *Sallow* is chiefly used of certain species of the genus *Salix* of a low-growing or shrubby habit (esp. *Salix caprea*), as distinct from 'osier' and 'willow'. *Salh* alone is used as a pl. n. in SALE, ZEAL, -S, and probably SALHOUSE. It is a fairly common first el. See SAIGHTON, SALEHURST, SALESBURY, SALFORD, SALKELD, SALL, SAUL, SALPERTON, SALTON, SALWICK, SAUGHALL, SAWBRIDGE, SAWDON, SAWLEY, SELBORNE, SELWOOD. Cf. also SALTLEY, SELABY &c., SILCHESTER.

Salhouse Nf [*Salhus* 1291 Tax, *Sallowes* 1543 Blomefield]. 'The sallows.' See SALH.

Saling Ess [*Salinges* DB, 1212 RBE, *Salingues* 1192 Fr]. The name may be derived from SALH 'sallow' or 'willow' and mean 'dwellers at the willows' or else 'willow copses'; cf. THURNING &c. **S~** is known for its cricket-bat willows.

Salisbury (sawlzbrī), **Old Sarum** W [*Sorviodunum, Sorbiodonum* 4 IA, (æt) *Searobyrg* 552 ASC, *Sarisberie* DB, *Sarum* 1091 Sarum, *Salesbir'* 1206 Ch, -*buri* 1205 Layamon, *Vetus Sarum* 1195, *Sarum, Nova Sarisberia* 1227 Salisbury]. The earliest forms refer to Old Sarum. When the people of old Sarum removed to the present town, the new town took over the name of the old. The Brit name consists of *dūnon* 'fort' and an el. *Sorvio-* of doubtful etymology. The Saxons dropped the second el. and, apparently owing to association with OE *searu* 'armour' &c., made the remaining part into OE *Searu*, to which OE *burg* was added. Later the first *r* was exchanged for *l* (dissimilation) in Norman pronunciation.

Salkeld Cu [*Salchild* c 1100 WR, *Olde Salehhild* 1164 P, *Saleghill* 1180 P, *Salighild* 1242 Ch, *Salkhull* 1230, *Salochild* 1236 Cl]. Great & Little **S~** are on opposite sides of the Eden. The first el. of the name is SALH 'sallow'. The second may be OE *hylte* 'wood' (in *hēah*-, *scōmhylte*), *lt* having become *ld* in the same way as *rt* became *rd* in CHARD or *nc* > *ng* in LYNG &c. Later the el. was associated with *hill*.

Sall (sawl) Nf [*Salla* DB, *Salle* 1196 FF, 1212 Fees, *Saulle* 1197 FF]. OE *Salhlēah* 'sallow wood'.

Salmonby (-ăm-) Li [*Salmundebi* DB, c 1150 DC, 1206 Ass]. '*Salmund*'s BY.' First el. ODan *Salmund*, ON *Sǫlmundr*.

Salperton Gl [*Salpretune* DB, *Salpertona* 1169 P, -*tone* 1221 Ass]. Probably OE *Sealhburn-tūn* 'TŪN by a stream where sallows grew'. Cf. LAMERTON for similar early reduction. *Sealhburna* would give *Salpurne*. Cf. SALH.

OE **salt, sealt** sb., adj. 'salt'. It is not always easy to say if *salt* as the first el. is the substantive or the adjective. The adj. would generally be inflected, and early forms in *Salte-* point to the adj. If the early forms are regularly *Salt-*, the first el. is probably the subst. See SALCOMBE, SALCOTT, SALFORD, SALT- (passim). OE *saltærn* 'salt-house, salt-works or house for storing salt' is found in SEASALTER, SALTERTON (Budleigh). OE *saltere* 'salt-worker' or 'salt-seller' is frequent in pl. ns. See SALTER- (passim), SALTRAM, SAWTRY.

Salt St [*Selte* DB, *Salt* 1167 P, *Saute* 1236 Fees]. OE **selte* 'salt-pit' or the like. In West Midland OE **selte* (derived from *salt* with a *jŏn*-suffix) would appear as *sælte*, ME *salte*. There are salt-works within 2 m. of Salt.

Saltash Co [*Aysh* 1284 FF, *Esse* 1316 Ipm, *Saltesh* 1337 Ch]. Originally ASH 'the ash-tree'. *Salt* either because of some salt-works or owing to the situation of the place on the Tamar estuary, where the water would be salt.

Saltburn YN [*Salteburnam* c 1185 YCh 767]. 'Salt stream.' The place is near the sea.

Saltby Le [*Saltebi* DB, -*bia* c 1125 LeS]. Probably '*Salte*'s BY.' OSw *Salte* may exist, and ON *Salt*, OSw *Salter* pers. n. occurs.

Salter Cu [*Salterghe* 12, *Saltherge* c 1195 StB]. 'Salt shieling.' *Salt* may refer to a salt-hole in a marsh.

Salterford Nt [*Saltreford* DB, *Salterford* 1241 Cl], **Salterforth** YW [*Salterford* Hy 3 Kirkst]. 'Salt-sellers' ford.' Cf. SALT.

Salterhebble YW is 'the salt-sellers' foot-bridge'. *Hebble* is a dialectal word for 'foot-bridge'.

Salterton, Budleigh, D [*Saltre* 1210 (1326) Pat, *Salterne* 1405 PND]. OE *sealt-ærn* 'salt-works'.

Salterton, Woodbury, D [*Salterton* 1250

FF, 1306 Ass], **S~** W [*Saltertun* 1198 P]. OE *sealtera tūn* 'the salt-workers' or salt-sellers' TŪN'.

Saltfleet Haven Li [*Salfluet* DB, *Saltflet-haven* 1346 Misc], **Saltfleetby** Li [*Salflatebi* DB, *Salfletebi* c 1115 LiS, *Saltfleteby* 1202 Ass, *Sauflet Omnium Sanctorum*, *Sauflet Sancti Clementis*, *Saltfletbi Sancti Petri* 1254 Val]. *Saltflēot* 'salt creek or stream' was evidently the OE name of Long Eau, on which the villages are. Here OE *salt* adj. seems to be used in its uninflected form. The stream falls into the sea and was no doubt salt at least at high tide. Saltfleetby is a Scand name formed by adding BY to the stream-name.

Saltford So [*Sanford* DB, *Saltford* 1291 Tax]. 'Salt (tidal) ford.' The Avon was formerly tidal at Saltford. Cf. FRESHFORD higher up the river.

Salthouse Nf [*Salthus* DB, 1242 Fees]. 'House for storing salt.'

Salthrop W [*Salteharpe* DB, *Sauteharp* 1198 FF, *Saltharp* 1241 Cl]. Identical with *Saltherpe* 956 BCS 922 (in Wilts, but not referring to Salthrop), i.e. an OE *sealt-hearpe* 'an apparatus for sifting salt'. The word *harp* is used in modern English of various implements used for winnowing grain or the like.

Saltley Wa [*Sautlega* c 1170 Middleton, *Salughtley* 13 PNWa, *Salutleye* 1285 QW]. 'LĒAH overgrown with sallows.' First el. an OE adj. *saleht*, derived from SALH.

Saltmarsh He [*Saltemers* 1130, *-mareis* 1167 P, (de) *Salso Marisco* 1347 Ep], **Saltmarshe** YE [*Saltmersc* DB, *-mareis* 1194 P]. 'Salt marsh.' The first is inland, and some salt springs must have given rise to the name. The second is on the tidal Ouse.

Salton YN [*Saletun* DB, *-ton* 1176 P]. 'TŪN among sallows.' See SALH.

Saltram D [*Salterham* 1249 Ass]. 'The salt-workers' HĀM.'

Saltwood K [(æt) *Sealtwuda* a 1011 Th, *Salteode* DB, *Sealtwude* 11 DM]. 'Salt wood', perhaps one where salt was made. The place is near Hythe. Cf. SALTSKOG in Sweden on the Baltic.

Salvington Sx [*Saluinton* 1248, *Saluington* 1250 FF]. 'The TŪN of *Sǣwulf*'s people.'

Salwarpe R Wo [*Saluuerpe* 692, *Saleuuearpe* 770, *Saloworpe* 817 BCS 77, 204, 362]. 'Sallow winding stream', from OE *salu* 'sallow, brownish yellow' and an OE *weorpe*, formed from *weorpan* 'to throw', lit. 'to twist'. On the river is **Salwarpe** vil. Wo [*Salouuarpe*, (into) *Salewarpan* 817 BCS 360, 362, *Salewarpe* DB].

Salwick La [*Saleuuic* DB, *-wic* 1201 P]. 'WĪC among sallows.' See SALH.

Sambourn Wa [*Samburne* 714 BCS 130, *Sandburne* DB], **Sambrook** Sa [*Semebre* DB, *Sambrok* 1285 FA]. 'Sandy stream.'

Samlesbury La [*Samerisberia* 1179, *Sameles-bure* 1188 f. P, *-bur* 1212 Fees, *Schameles-biry* 1246 Ass, *Scamelesbyry* 1277 Ass]. Etymology obscure. If the name originally began in *Sh-*, the first el. may be OE *sceamol* 'bench' &c. in some topographical sense such as 'ledge'.

Sampford Courtney D [*Sanford* DB, *Sand-fort* 1093 Fr, *Saunforde Curtenay* 1262 Ep], **S~ Peverell** D [*Sanforda* DB, *Saunford Peverel* 1275 RH], **S~ Spiney** D [*Sanforda* DB, *Saunford Gerardi de Spineto* 1234 Cl, *Sandford Spynee* 1303 FA], **Great & Little S~** Ess [*Sanforda* DB, *Samford* 1130 P, *Sanford Magna, Parva* 1238 Subs], **S~ Arundel** So [*Sanford* DB, *Samford Arundel* 1240 Ass], **S~ Brett** So [*Sanford* DB, *Saunford Bret* 1306 Ch]. OE *Sand-ford* 'sandy ford.'

S~ Arundel was held by Roger Arundel in 1086 (DB). Arundel is a nickname from OFr *arondel* 'swallow'.—**S~ Brett** was held by Simon Bret t. Hy 1 (Collinson). *Bret* means 'the Breton'.—**S~ Courtney** was held by the Courtneys before 1242 (Cl). Cf. HIRST COURT-NEY.—Matilda Peverel held **S~ Peverell** c 1150 (Montacute). Cf. BARTON PEVEREL.—**S~ Spiney.** Spiney is a family name derived from a Fr pl. n. identical with Engl *spinney* (Lat *spinetum* 'thorn-bush thicket').

Sancreed Co [(ecclesia) *Sancti Sancredi* 1291 Tax]. According to Oliver St. San-credus. The name looks more like a Cornish form of St. Faith (Co *cred*, Welsh *cred* 'faith').

Sancton YE [*Santun* DB, Hy 2 BM, *Sand-tona* 1175, *Santon* 1195 P, *Sancton* c 1200 YCh (1130), 1237 Ep]. Cf. SANTON Li. OE *Sand-tūn*, which became *Santun* and, owing to association with Lat *sanctus*, was some-times written *Sancton*.

OE **sand**, ON **sandr**, OSw **sander** 'sand' is a common first el. of pl. ns., and then means simply 'sand'. See SAMBOURN, SAM-BROOK, SAMPFORD, SANCTON, SAND- (passim), SENLAC. In SOUND and in names such as CHICKSANDS, COCKERSAND, WASSAND *sand* means 'a sandbank' or 'sandy soil' or the like. Cf. SEND.

Sandal Magna YW [*Sandala* DB, *Sandale* 1175 P, *Sandal Major* 1247 Ep, *Le Sande-hale* 1318 Misc], **Kirk & Long Sandall** YW [*Sandala, Sandela* DB, *Sandhala* 1148 YCh 179]. 'Sandy haugh.' See HALH.

Sandbach Chs [*Sanbec* DB, *Sondbache* 1260 Court]. 'Sandy stream or valley.' See BÆCE.

Sandbeck YW [*Sandbec* 1148 YCh 179, 1222 FF]. 'Sandy brook.' See BECK.

Sanderstead Sr [*Sondemstyde, Sondenstede* 871–89 BCS 558, *Sandestede* DB, *Sandre-stede* 1291 Tax]. 'Sandy homestead.' See

HĀMSTEDE. The OE form is here *hǣm-styde*.

Sandford, Dry, Brk [(ad) *Sandforda* 811, *Sandford* 821 BCS 850, 366, *Sandeford* 1167 P], **S~ D** [*æt Sandforda* 930, 1008–12 Crawf], **S~ Orcas** Do [*Sanford* DB, *Sandford* 1243 Ass, *Sandford-Orskuys* 1348 Ep], **S~ St. Martin** O [*Sanford* DB, *Saunforda* 1209–19 Ep], **S~ on Thames** O [*æt Sandforda* 1050 KCD 793, *Sanford* DB], **S~ Sa** nr Prees [*Sanford* DB, *Sontford* 1236 Fees], **S~ We** [*Sanfford* Hy 3 Misc, *Sandford* 1314 Ipm]. 'Sandy ford.'

Orcas is a Fr family name, written (Helye) *Oriescuilz* 1177 P, (Ricardus) *Dorescuilz* 1195 P.

Sandgate K [*Sandgate* 1257 Ipm]. 'Sand gate', perhaps 'gate leading to the sandy shore'.

Sandhoe Nb [*Sandho* 1225 Ep, 1232 Ch]. 'Sandy HŌH or spur of hill.'

Sandhurst Brk [*Sandherst* 1175 P, *-hurst* 1316 FA], **S~ Gl** [*Sanher* DB, *Sandherst* 1167, 1195 P, *-hurst* 1211–13 Fees], **S~ K** nr Cranbrook [*Sandhyrste* 11 DM]. 'Sandy hurst or hill.'

Sandiacre Db [*Sandiacre* DB, *-acra* 1158 P, *Sendiacre* 1201 FF]. 'Sandy field.'

Sandleford Brk [*Sandraford* 1180 P, *Sandelford* 1182 P, 1242 Ch, *Sandeliford* 1183 f., *Sanlesford* 1185 P]. The first el. appears to be a compound containing *sand*, possibly an OE *sandwell* 'sandy stream'. The place is on the Enborne.

Sandon Brk nr Hungerford [*Sandon* 1220, 1242 Fees], **S~ Ess** [*Saundon* 1274 RH, *Sandon* 1303 FA], **S~ Hrt** [*Sandona* 939 BCS 737, *Sandone* DB], **S~ St** [*Sandone* DB, *Sandun* 1236 Fees], **Sandown Sr** [*Sandon* 1246 Cl]. OE *sand-dūn* 'sand hill'. Sandon Ess is *Bedenesteda* DB, 1190 P. *Bedene-* is OE *byden* 'valley'. Cf. BEEDON.

Sandown Wt [*Sande* DB, *Sandham* 1287–90 Fees, 1316 FA]. Apparently OE *Sand-hamm*.

Sandridge D [*Sandruge* 1212 RBE], **S~** (sahn-) Hrt [*Sandrige* DB, *Sanrigg* 1204 Cur]. 'Sand ridge.'

Sandringham Nf [*Santdersincham* DB, *Sandringham* 1195 Cur, *Sandringeham* 1254 Val]. S~ is close to Dersingham, and its name is *Dersingham* with a distinctive addition *sand*: 'sandy Dersingham, D~ on sandy soil'.

Sandtoft Li [*Sandtofte* 1157 YCh 354]. 'Sandy TOFT.'

Sandwich K [*Sondwic* 851, *Sandwic* 993 ASC, *Sandwice* DB, *Sandwiz* 1165 P]. 'wīc on sandy soil.' The meaning of *wīc* is probably 'market town'.

Sandwith Cu [*Sandwath* c 1280 StB]. 'Sandy ford', the second el. being ON *vað* 'ford'.

Sandy Bd [*Sandeia* DB, 1185 P, *Sandee* 1203 Cur]. OE *Sand-ēg* 'sandy island'.

Sankey R La [*Sanki* 1202 FF, *Sonky* 1228 WhC]. A Brit river-name. On the Sankey are **Great & Little S~** [*Sonchi* c 1180 LaCh, *Sanki* 1212, *Sonky* 1242 Fees].

Sansaw Sa [*Sondsawe* 1327 Subs, *-shawe* 1347 Eyton]. 'Sandy wood.' See SCAGA.

Santon Cu [*Santon* 13 StB], **S~ Li** [*Sanctone*, *Santone* DB, *Santuna* c 1115 LiS, *Santun* 1212 Fees], **S~ Nf** [*Santuna* DB, *-tona* 1121 AC]. OE *Sand-tūn* 'TŪN on sandy soil'. Cf. DOWNHAM.

Sapcote Le [*Sape-*, *Scepecote* DB, *Sapecota* c 1200 Fr, *Scapecotes* 1230 P]. OE *scēap-cot* 'shelter for sheep'. For *a* we may compare SCOPWICK. The change of OE *sc* to *s* may be due to Norman or Scand influence.

Sapey, Lower, or **S~ Pichard** Wo [*at Sapian* 781 BCS 240, *Sapie* DB, 1195 ff. P, *Sapi Pichard* 1242 Fees], **Upper S~** He [*Sapy* 1291 Tax]. The two Sapeys are c 3 miles apart on Sapey Brook, which may have been originally *Sapey*. The base seems to be an OE *Sapige* or rather *Sæpige*. The latter may be from OE *sæpig* adj. 'sappy, full of sap'.

Miles Pichard held a fee in Sapey in 1212 (RBE). Cf. OCLE PYCHARD.

Sapiston Sf [*Sapestuna* DB, *-tune* c 1095 Bury, *-ton* 1204 Cur, 1254 Val]. Possibly OE *Sāperes-tūn* 'the soap-maker's TŪN', though the absence of *r* in early spellings is irregular. An *r* often disappears before an *s*, but the early forms generally preserve it occasionally. Cf. SAPPERTON.

Sapley Hu [*Sappel'*, *Sapele* 1232 Cl]. 'Spruce wood.' Cf. SABDEN, SÆPPE.

Sapperton Db [*Sapertune* DB, *Sapirton* 1242 Fees], **S~ Gl** [*æt Saperetún* 969 BCS 1239, *Sapletorne* DB, *Saperton* 1211–13 Fees], **S~ Li** [*Sapretone* DB, *-ton* 1269 Ch], **S~ Sx** [*Sabertona* 1210 PNSx, *Saperton* 1247 Ch]. OE *Sāpera-tūn* 'the TŪN of the soap-makers'. OE *sāpere* is not recorded, but *sopere* occurs c 1225 in the Ancren Riwle in the sense 'soap-seller'. A still earlier example is (Will.) *le Sopere* 1195 f. P.

Saredon, Great & Little, St [*Sardone*, *Seresdone* DB, *Sardonia* 1166 RBE, *Little Sardon* 1251 Ass, *Magna Sardon* 1316 FA]. The places are on or near a brook formerly called **Sarebrook** [*Searesbroc* 996 Mon, *Sarebrok* 1290 Ch]. The first el. of both may be a pers. n. cognate with Goth *Sarus*, ON *Sǫrr*. OE *Searu* seems to be evidenced. Saredon would then mean 'Searu's hill'. Another possibility is that the hill had a name containing OE *sēar* 'sere, withered' and that the brook was named from the hill.

Sark R Cu [*Serke* c 1200 Sc]. A Brit river-name.

Sarnesfield He [*Sarnesfelde* DB, *-feld* 1183 P, 1242 Fees, *Sarnefella* 1127 AC]. 'FELD by the road.' Welsh *sarn* means 'road'.

Sarratt Hrt [*Syreth* 1077, *Syret* 1119–46 Gesta, *Siret* 1166 RBE, *Seret, Serret* 1198 (1301) Ch, *Sarot* 1291 Tax]. Neither of the suggestions in PNHrt(S), a ME *seret*, &c. 'dry place' from OE *sēar* 'withered' or OE *sieret* &c. 'place of ambush or for snaring' from OE *searu* 'trick', is convincing. The common early spelling with *-th* may indicate that the name contains OE *hǽþ*. The first el. might then be an OE **sȳre* corresponding to Scand *syra* 'sorrel', or better OE *sweora, swyra* 'neck, col', referring to the prominent short ridge at the place. For loss of *w* cf. e.g. KERNE, SACOMBE, SHEARSBY, TIVERTON D. OE *y* often becomes *e* in Herts. Forms with *a* as well as final *-t* are then due to Norman influence.

Sarre K [*ad Serræ, Seorre* 761 BCS 189, *Syrran* 11 DM, *Serra* Hy 2 (1313) Ch, *Serres* 1204 FF]. Possibly an old Brit rivername. Gervase of Canterbury mentions *Serres* among rivers, and *water of Serre* occurs 1392 Pat. The name might be a derivative of the root **serp-* 'to crawl, slip' found in Lat *serpo*.

Sarsden O [*Secendene* DB, *Sercesd.* c 1160, *Cerchesdene* c 1190, *Cerchesden* c 1225 Eynsham, *Cercendenne* 1206 Cur, *Cercedene* 1225 Ep, *-den* 1242 Fees]. Perhaps 'church valley', the OE base being *Circandenu*. In any case the name has been strongly modified owing to Norman influence. The forms in *-s-* (*Cerchesdene* &c.) are abnormal, but may be due to a late change.

Sarson. See ANN. **Sarum, Old.** See SALISBURY.

Satley Du [*Sateley* 1228 FPD, 1291 Tax]. Second el. LĒAH. The first is possibly OE (*ge*)*seotu* from (GE)SET 'fold' with change *eo* > *ea*. Cf. forms of MEDOMSLEY.

Satterleigh D [*Saterlei* DB, *-leye* 1277 FF]. Has been explained as 'the LĒAH of the robbers', the first el. being OE *sætere* 'robber' (PND).

Satterthwaite La [*Saterthwayt* 1336 FC]. 'Clearing by a shieling.' See THWAITE, SÆTR.

Saughall Massie Chs [*Salghale* 13, *Salghale Mascy* 1322 Ormerod], **Great & Little S~** Chs [*Salhale* DB, *Salchale* c 1100, *Parua Salighale* a 1271 Chester]. OE *Salh-halh* 'HALH where sallows grew'.

On S~ **Massie** see DUNHAM MASSEY.

Saul Gl [*Sallege* 1221 Ass]. Identical with SALL.

Saundby Nt [*Sandebi* DB, 1165, 1176 P]. 'BY on sandy soil.'

Saunderton Bk [*Santesdune* DB, *Santresdon* 1195 P, *Sauntredon* 1196 FF, *Santre*(*s*)*don* 1197 f. P, *Sandresdon* 1200 Cur], **Saunderton Lee** Bk [*Santerley* 1227 Ass, *Saunterle* 1365 Ch]. The first el. of these is very

likely OE **sængde trēo* 'burnt tree, tree blasted by lightning'. OE *sengan* (*sængan*) means 'to singe, burn'. The names would mean 'hill and LĒAH with a blasted tree'.

ON saurr 'mud, muddy place'. See SAWREY, SOSGILL, SOWERBY.

Sausthorpe Li [*Saustorp* 1175, *Sauztorp* 1189 P, *Saucetorp* 1222 Ep]. OScand *Sauðs-þorp* '*Sauð*'s thorp'. *Sauðr*, lit. 'sheep', is a common byname in ON.

Săvernăke Forest W [*Safernoc* 934 BCS 699, *Savernac* 1156 P, *Savernak* 1275 Cl]. In all probability derived from a river-name identical with SEVERN. The name may have denoted the Bedwyn river or some river near by. Severn often appears in early sources with *a* in the first syllable. The suffix *-āco-* is common in British.

Savick Brook La [*Savoch* c 1200, *Safok* a 1268 CC, *Savok* 1252 Ch]. A Brit rivername.

Sawbridge Wa [*Salwebrige* DB, *Salebrugia* 1100–30 NpCh]. 'Bridge by the sallows.' Cf. SALH.

Sawbridgeworth Hrt [*Sabrixteworde* DB, *Sabrichteworda* 1130 P, *Sabrichtesuuorde* 1163 Bury, *Sebrichewurth* 1245 Ch]. '*Sæbeorht*'s WORÞ.'

Sawdon YN [*Saldene* 1289 Cl]. 'Sallow valley.' See SALH, DENU.

Sawley Db [*Salle* DB, *Sallawa* 1166 P, *Sallowe* 1242 Fees], **S~** YW nr Ripon [*Sallege* c 1030 YCh 7, *Sallai* DB], **S~** YW nr Clitheroe [*Sallea* 1162 P, *Salleie, Salieleie* c 1195 FC]. The first is 'sallow hill', the last two are 'sallow LĒAH'. Cf. SALH, HLĀW.

Sawrey La [*Sourer* 1336, 1400 FC]. ON *saurar*, plur. of *saurr* 'mud, muddy place'.

Sawston Ca [*Salsingetune, Selsingetona* 970 ChronRams, *Salsintona* c 1080 ICC, *Salsiton* DB]. 'The TŪN of *Salsa*'s people.' OE **Salsa* corresponds to OSw *Salsi*, ON *Sǫlsi* and is a short form of names in *Sele-*. *Sele-* is OE *sele, salor* &c., an old *s*-stem.

Sawtry Hu [*Saltreiam* 974 BCS 1310, *Saltrede* DB, *Saltreda* 1183 P]. OE *Sealtera hyþ* 'the landing-place of the salt-sellers'.

Saxby Le [*Saxebi* DB, *-bia* 1175 P, *Sessebia* c 1125 LeS], **S~** Li nr Lincoln [*Sassebi* DB, *Saxsabi, Saxsebi* c 1115 LiS, *Saxeby* 1206 Cur], **S~ All Saints** Li [*Saxebi* DB, *Saxeby* 1221 Ep]. '*Saxi*'s BY.' ODan, OSw, ON *Saxi* is a well-known name.

Saxelby Le [*Saxelbie* DB, *-by* 1219, *Saxeleby* 1220–35 Ep], **Saxilby** Li [*Saxlabi, Saxlebi* c 1115 LiS, *Saxelebi, Saxolebi* c 1150 DC]. '*Saxulf*'s BY.' First el. ODan *Saxulv*, ON *Sǫxulfr*.

Saxham, Great & Little, Sf [*Saxham, Sexham* DB, *Saxham* 1197 FF, *Saxham*

Magna, Parva 1254 Val]. Probably 'the HĀM of the Saxons', an OE *Seax-hām*. Another possibility is that OE *seax* had preserved the old sense 'stone, rock', found in some Germanic languages. If so, the name means 'HĀM by the rock', referring to the hill at the places.

Saxilby. See SAXELBY.

Saxlingham Nf nr Holt [*Saxel-, Sexelingaham* DB, *Saxlingham* 1199 FF], S~ **Nethergate & Thorpe** Nf [*Seaxlingaham* 1046 Th, *Saiselingaham* DB, *Saxlingaham* 1163 BM, *Saxlinghamtorp* 1254 Val, *Saxlyngham Neyergate, Thorp* 1291 Tax]. The first el. is a derivative in -*ingas* of a pers. n. such as **Seaxel* or possibly *Seaxhelm* or the like.

Saxmu·ndham Sf [*Sasmunde(s)ham* DB, *Saxmundham* 1213 FF]. '*Seaxmund*'s HĀM.' OE *Seaxmund* is not recorded.

Saxon Street. See SAXTON.

Saxondale Nt [*Saxeden* DB, *Saxendal* 1221–30, -*dall* 1242 Fees]. 'The valley of the Saxons' (OE *Seaxna denu* or *dæl*).

Saxtead Sf [*Saxsteda* DB, *Saxstede* 1202 FF]. It is doubtful if OE *seax* 'rock' would give good sense, even if it existed. Perhaps '*Seaxa*'s place'.

Saxthorpe Nf [*Saxthorp* DB, 1254 Val]. '*Saxi*'s thorp.' Cf. SAXBY.

Saxton Ca [*Sextuna* c 1080 ICC, -*tone* DB, *Sexton* 1219 FF], S~ YW [*Saxtun* DB, -*tona* 1175 YCh 359, *Sextona* Hy 2 (1294) Ch]. OE *Seax-tūn* 'TŪN of the Saxons'. Cf., however, SAXHAM. S~ Ca is now preserved in SAXON STREET (HALL).

Scackleton YN [*Scacheldene* DB, *Scakelden* c 1090 YCh 350]. Probably a Scandinavianized form of OE *Scacol-denu* or the like. The place is on a small steep hill. Probably we have to assume an OE *scacol* or the like, cognate with OHG *scahho* 'a strip, tongue or point of land', ON *skekill* 'tongue of land'. The el. may be identical with *scacel* in *scacalwic* BCS 834, *scaceluuic* ib. 1125 (Sx).

Scafell (skaw fĕl) Cu. Probably ON *Skálafell* 'hill with a shieling'. Cf. SCAWDALE.

Scaftworth Nt [*Scafteorde* DB, *Seftewurd* 1185 P, *Skaftwurth* 1341 NI]. '*Skapti*'s WORÞ.' *Skapti* is a well-evidenced OScand pers. n. But the name may be a Scandinavianized form of an OE *Sceaftan worþ*.

OE **scaga** 'shaw, thicket, small wood, grove' is the source of SHAW and occurs as the first and the second el. of names. See SHAW-(passim), FUL-, OAKEN-, STRUMP-, WISHAW and others, FRENCHAY, SANSAW.

Scagglethorpe YE [*Scachetorp* DB, *Scakelthorp* 1297 Subs], S~ YW [*Scachertorp* DB, *Scakeltorp* 1227 FF, *Skakelthorpe* 1253 Ch]. '*Skakli*'s thorp.' OSw *Skakli* pers. n. occurs, and ON *Skǫkull* is a byname.

Scaitcliffe La nr Todmorden [*Scatecliffe* 1575 DL], S~ La in Accrington [*Sclateclyff* 1527 Ct]. 'Slate cliff.' *Slate* is often *sclate* in early sources.

Scalby YE [*Scalleby* 1230 P], S~ YN [*Scallebi* DB, 1190 ff. P]. '*Skalli*'s BY.' First el. ON *Skalli*, OSw *Skalle*, really a byname from *skalle* 'skull'.

OE **scald, sceald** 'shallow' is the first el. of some names as SHALBOURNE, SHALFORD, SHADWELL. In SCALDWELL, SCALFORD, *sc*- for *sh*- is due to Scand influence. SHELFORD may contain a derivative of *scald* (an OE **sceldu* 'shallow place').

Scaldwell Np [*Scaldewelle* DB, 1220 Fees]. 'Shallow stream.' Cf. SCALD.

Scaleby Cu [*Scaleby* 1247 Ipm, 1291 Tax]. 'BY with huts.' See SKÁLI.

Scales. See SKÁLI.

Scalford (-awf-) Le [*Scaldeford* DB, c 1130 BM]. 'Shallow ford.' See SCALD.

Scamblesby Li [*Scamelesbi* DB, *Scamelbi* 1146 RA, *Scamelesby* 1163 RA, 1202 Ass]. Perhaps '*Skammlaus*'s BY'. *Skammlaus* would be a nickname formed from OScand *skammlauss* adj. 'shameless' &c. The first el. may also be the OSw pers. n. *Skammhals*, lit. 'short-neck', found in a runic inscription.

Scammonden YW [*Scambayndene* n.d. AD, *Scambanden* 1275 Wakef]. '*Skammbein*'s valley.' *Skammbeinn*, lit. 'short-legged', is a known ON byname.

Scampston YE [*Scameston* DB, *Scamastuna* c 1130 YCh 620, *Scamestun* 1157 ib. 354, *Scameliston* 1202 FF], **Scampton** Li [*Scantone* DB, *Scantuna* c 1115 LiS, *Scamtona* 12 DC]. Both seem to have a first el. derived from ON *skammr* 'short'. Scampton may be 'short TŪN' or '*Skammi*'s TŪN', *Skammi* being a nickname formed from the adj. Scampston would be '*Skamm*'s TŪN', *Skammr* being an unrecorded nickname from the same adj.

Scarborough YN [*Escardeburg* c 1160 YCh 364, *Scardeburc* 1158 ff. P, *Skarðaborg* Kormak's saga]. According to the ON Kormak's saga, *Skarðaborg* was built by Þorgils Skarði, the date being c 965. S~ thus means '*Skarði*'s BURG'. *Skarði* is a nickname, meaning 'hare-lipped'.

Scarcliff Db [*Scardeclif* DB, -*clive* 1226 FF, 1242 Fees, *Scartheclive* 1235 FF]. 'Cliff with a scar or gap.' The first el. is probably OE *sceard* adj. 'notched, gashed', later Scandinavianized (*sc*- for *sh*-, *th* for *d*).

Scarcroft YW [*Scardecroft* 1166, 1197 P, 1246 FF, *Scarthecroft* 1304 Ch]. See CROFT. First el. as in SCARCLIFF or SCARBOROUGH.

Scargill YN [*Seacreghil, Scracreghil* DB, *Scakregill* 1172 f. P]. 'Merganser valley.' See GIL. First el. Sw *skrake* 'merganser', with metathesis of *r*. The word seems to have been used in Norway, to judge by the nickname *Skraki*.

Scarisbrick (-ārz-) La [*Scharisbrec* c 1200 PNLa, *Skaresbrek* 1238 FF]. '*Skar*'s hillside or slope.' See BREKKA. First el. ODan *Skar* pers. n. (in pl. ns.).

Scarle, North, Li [*Scarle, Scaruell* 1202 Ass, *Parva Scarle* 1230 Ep, *Northscarle* 1240 Cl], **South S~** Nt [*Scornelei* DB, *Scarlai* 1147, *Scarla* 1163 RA, *Suthscarl'* 1240 Cl]. OE *Scearn-lēah* 'dirty LĒAH'. See SCEARN. *Sc-* for *Sh-* owing to Scand influence.

Scarning Nf [*Scerninga* DB, *Scerninges* 1199 Pp, *Skerning* 1253 Ch]. A derivative of OE *scearn* 'dirt', perhaps originally the name of a stream. If so, the meaning is 'dirty brook'. For *Sc-* see prec. n.

Scarrington Nt [*Scarintone* DB, *Shernintona* 1166, *Scherninton* 1167 P]. OE *scearnig* 'dirty' and TŪN. For *Sc-* see SCARLE.

Scarthingwell YW [*Scardingwell* 1202, 1225 FF, *Scarthingwell* 1333 FF]. Second el. OE *wella* 'stream'. The first may be a Scandinavianized form of an OE *Scearding(as)*, a derivative of OE *sceard* 'a gap' &c.

Scartho Li [*Scarhou* DB, *Scartho* 1190 P, *Scarthou* 1202 Ass]. Second el. O Scand *haugr* 'mound, hill'. The first may be OScand *Skarði* pers. n. (cf. SCARBOROUGH) or *skarð* 'gap, notch' &c.

Scawby Li [*Scal(l)ebi* DB, *Scallabi, Scallebi* c 1115 LiS]. Identical with SCALBY.

Scawdale Fell Cu [*Houedscaldale* 1210 FC]. ON *Skála-dalr* 'valley with a hut'. See SKÁLI. *Houed* is ON *hofuð* 'head'. *Houedscaldale* means 'the top of Scawdale'. For the order of the elements cf. ASPATRIA.

Scawton YN [*Scaltun* DB, *-a* 1154–60 YCh 1831]. First el. ON *skáli* 'hut, shed'.

OE **scēad, scēaþ** 'boundary'. See SHADINGFIELD, SHADWELL, SHATTON, SHAWELL, SHEAF.

OE **sceaft** 'shaft, pole'. See SHAFTOE, SHEBBEAR.

OE **scēap, scēp, scīp** 'sheep' is a common first el. of pl. ns., especially in combination with COT, FORD, WĪC. See SHAPWICK, SHEEP-, SHEP-, SHIP- (passim), SHEFFIELD, SHEFFORD, SHIFFORD, SAPCOTE. A Scandinavianized form is seen in some names, as SCOPWICK, SKIPTON, SKIPWITH.

OE **sceard** 'notch, gap', adj. 'notched'. See SHARDLOW, SHARSTONE, MESHAW, also SCARCLIFF &c. Cf. SHURDINGTON.

OE **scearn** 'dung, filth, mud' is the first el. of SHARNBROOK, SHARNFORD, SHERNBORNE, SHERRINGTON, SHORNCOTE, also, with *Sc-* owing to Scand influence, of SCARLE. Cf. also SCARNING, SCARRINGTON.

OE **scearp** 'sharp, rugged', doubtless also 'steep' (cf. OHG *scarph* 'steep'). See SHARP- (passim).

OE **scearu** 'boundary' in *land-scearu*. See SHAROW, WALDERSHARE.

OE **scēat** is recorded in senses such as 'piece of cloth, quarter of the earth, corner'. In pl. ns. a meaning such as 'strip of land' is to be assumed. But as names in -*scēat* mostly have as first el. a tree- or plantname (e.g. ALDER-, BRAM-, EW-, HEY-, KEMP-, SPURSHOT(T); cf. also GRAYSHOTT with OE *grāf* 'grove' as first el.), it seems probable that the word had developed meanings such as 'piece of land left untilled and overgrown with trees or plants' or even 'piece of wood, park'. BAGSHOT Sr was the name of a wood. In EMPSHOTT the first el. is OE *imbe* 'swarm of bees', in EVERSHOT OE *eofor*. The word is used alone in SHEAT, SHEET, SHUTE. The form of the element varies in ME sources between -*shet*, -*shat*, and -*shute*, -*shite*. The latter forms go back to OE **sciete*, probably a locative form of *scēat* (in early OE -*i* < -*i*).

OE **scēne, scīene** 'beautiful, bright'. See SHEINTON, SHEN- (passim).

ME **schēle**, dial. *sheel, shiel* 'hut, shed' (cognate with ON *skiól* 'shed, hut' or *skáli*) is not a common pl. n. el. See SHIELDS, SHELDON Db, AXWELL, ESPERSHIELDS, LINSHEELES. See also SKELBROOKE.

Scholar Green Chs [*Scholehalc, Scolal* E 1 BM]. 'HALH with a hut.' Cf. HALH, SKÁLI.

Scholes. See SKÁLI.

OE **scīd** 'shide'. See SHIDE, SHEDFIELD.

Scilly Islands [*Sylinancim* (acc.) c 400 Sulpicius Severus, *Sully* Hy 1 Ol, *Sullia* 1186 P, *Syllingar* Heimskringla]. A preEnglish name of doubtful etymology.

OE **scinna** 'spectre'. See SHINCLIFFE, SKINBURNESS.

OE **scipen** 'shippon, byre, cattle-shed'. See SHIPPEN, SHIPPON, SKIPLAM.

OE **scīr** 'shire, district' was used in OE times not only in the modern sense, but also of a smaller district. The exact meaning is not clear in such cases as PINNOCK(SHIRE), WILPSHIRE. The ordinary meaning 'shire' is seen in names of counties, as BERKSHIRE, and in names like SHIRESHEAD, SHIREOAKS, probably SHERWOOD. In SKIRMETT *Sk-* is due to a later change. Names such as SHERFIELD, SHIRLAND, SHIRLET, SHIRLEY very likely contain *scīr* 'shire', but *scīr* adj. may possibly be thought of in some cases.

OE **scīr** adj. 'clear, bright' is the first el. of SHERBO(U)RNE, -BURN, SHER(E)FORD, SHIRBURN, SHIREBROOK, SHIRWELL. See also SHERE, SHEERNESS.

OE **scīrgerēfa** 'sheriff'. See SCREVETON, SHIRENEWTON, SHREWLEY, SHREWTON, SHROTON, SHURTON.

Scofton Nt [*Scofton* 13 Duk]. Possibly '*Skopti*'s TŪN'. ON *Skopti* is a pers. n.

Scole Nf. See SKÁLI.

Scopwick Li [*Scap(e)uic* DB, *Scapewic* c 1150 DC, *Scapwic* 1170 P]. OE *scēap-wīc* 'sheep farm', Scandinavianized.

OE **scora**, found in *Waldmeresscora* 824 BCS 381, is the source of Mod *shore* 'bank', found from the 14th cent., and dial. *shore* 'a steep rock'. It is related to OE *scorian* 'to project' (of rocks). The meaning 'steep rock or hill' is the meaning of *shore* in SHOREHAM, SHORESWORTH, SHORWELL, HELM-SHORE, while 'bank' is the meaning in SHOLING, SHOREDITCH. A cognate word *scorra*, corresponding to OHG *scorro* 'steep declivity', is the first el. of SHERSTON.

Scorbrough YE [*Scogerbud* DB, *Scoure-burgh* 1291 Tax]. OScand *skōgar-búð* 'booth in a wood'. See BŌÞE, SKŌGR. *Scoger-* from OScand *skōgar* gen.

Scoreby YE [*Scornesbi* DB, *Scorreby* 1246 FF, 1315 Ipm]. '*Skorri*'s BY' (first el. ON, ODan *Skorri*).

Scorton La [*Scourton* c 1550 DL], S~ YN [*Scortone* DB, *-ton* 1231 FF]. 'TŪN in or by a ravine.' First el. ON *skor* 'a rift in a rock or a ravine'.

Scosthrop YW [*Scotorp* DB, *Scotthorp* 1226 FF]. '*Skotte*'s thorp.' *Skotte* pers. n. is found in the Sw pl. n. SKOTTORP.

Scotby Cu [*Scottebi* 1167 P, *Scotebi* 1197 P]. OScand *Skottabȳr* 'BY of the Scots'.

Scotforth La [*Scozforde* DB, *Scoteford* 1204 FF, *Scotford* 1212 Fees]. 'Ford of the Scot or Scots.' The place is near GALGATE (q.v.).

Scothern (-ŏth-) Li [*Scotstorne, Scotorne* DB, *Scotstorna* c 1115 LiS, *Scottorna* 1146, *Scosthorne* 1203–6 RA]. 'The thorn-bush of the Scot or Scots.'

OE **Scott**, plur. *Scottas*, in early OE meant 'Irishman', but was later applied to the Gaels in Scotland, to whom it soon became restricted. Several pl. ns. in SCOT-, SHOT-seem to contain the word *Scott*. When found in the South or Midlands these names no doubt indicate settlements of Irishmen. Those found in the North presumably contain *Scott* in the sense 'Scotsman'. But it is not always easy to distinguish *Scott* from other elements.

Scotter Li [*Scottere* 1060–6 KCD 819, *Scotere* DB, 1098 Reg, *Scotre* DB, *Scotra* c 1115 LiS]. The place is close to Scotton. *Scotter* may be *Scotta trēo* 'the tree of the Scots'.

Scotterthorpe Li [*Scaltorp* DB, 1212 Fees, *Saltorp* c 1115 LiS]. Perhaps '*Skalli*'s thorp'. Cf. SCALBY. S~ is in Scotter, which accounts for the change to *Scotterthorpe*.

Scottlethorpe Li [*Scotlattorp, Scodlotorp, Scodlouestorp, Scodlestorp* 1202 Ass, *Scot-lathorp* J BM, *Scotlautorp*' 1212 Fees]. The first el. is apparently a pl. n. *Scothlāw*.

Scotton Li [*Scottun* 1060–6 KCD (819), 1212 Fees, *Scotone* DB, *Scottuna* c 1115 LiS], S~ YN [*Scottune* DB, *Scottuna* 1157 YCh 354], S~ YW [*Scotone* DB, *Scottun* 1167 P, *Scottona* c 1180 YCh 513]. Apparently 'TŪN of the Scots'. Cf. SHOTTON, SCOTT.

Scottow Nf [*Scoteho* 1044–7 KCD 785, *Scotohou* DB, *Scothowe* 1117 Holme, *Scot-houe* 1177 P]. OE *Scothōh* or *Scottahōh* 'HŌH or spur of land of the Scots', to which was added OScand *haugr* 'hill'.

Scoulton Nf [*Sculetuna* DB, *-ton* 1198, 1212 Fees]. '*Skúli*'s TŪN.' First el. ON *Skúli*, ODan *Skule* pers. n., found in the form *Scula* DB &c.

OE **scræf** 'cave, den, hovel'. In pl. ns. the meaning is probably 'hollow, ravine' and the like. See SCRAFTON, SCRIVEN, SHARES-HILL, SHARLSTON, SHRAWARDINE, SHRAWLEY. *Sc-* for *Sh-* is due to Scandinavian influence.

Scrafield Li [*Screidefeld* 1184 Bury, *Scraide-feld* Hy 2 DC, *Screthefeld* 1206 Cur, *Scraidhesfeld* 1212 Fees]. Apparently a combination of ON *skreið* 'landslip' and OE FELD.

Scrafton YN [*Scraftun, -ton* 1260 Ass]. OE *Scræf-tūn* 'TŪN in a ravine' (cf. SCRÆF.)

Scrainwood (skahn-) Nb [*Scravenwod* 1242 Fees, *Scrawenewude* 1256 Ass, *Scranewode* 1289 Cl]. OE *Scrēawena wudu* 'wood of shrewmice or of villains'. OE *scrēawa* means 'shrewmouse', but ME *schrewe* also means 'rascal, villain'.

Scrane Li [*Vetus Screinga, Scrainga* Hy 2 BM, *Scrainges* 1197 P, *Screhinges, Scrahing* 1202 Ass]. Probably a Scandinavianized form of an OE name. Scrane may be an OE **scræging*, derived from an OE **scraga*, corresponding to MHG, MLG, MDu *schrage* 'trestle'. SCRHAMBACH in Germany (*Scraginpach* c 1140) is held to mean 'brook with poles placed slantwise'. OE *scræging* might mean 'structure made of poles'. Alternatively Scrane might be identical with the first el. of SCRAYINGHAM.

Scraptoft Le [*Scraptofte* c 1050 KCD 939, *Scrapentot* DB, *Screpetoft* 1191 f. P, *Scrape-toft* 1205 FF]. See TOFT. The first el. is no doubt an OScand pers. n. ON *Skrápi* and *Skrápr* occur.

Scratby Nf [*Scroutebi* c 1025 BCS 1017, *Scroutebei, Scroteby* DB]. '*Skrauti*'s BY.' ON *Skrauti* occurs as a byname (lit. 'person given to display').

Scrayingham YE [*Screngham, Escraingham* DB, *Scraingeham* 1157 YCh 354, 1208 FF]. Probably OE *Scirhēahinga-hām* 'the HĀM of the people of *Scirhēah*', with shift of stress as in KNOWSLEY and the like and *Sc-* for *Sh-* owing to Scandinavian influence. *Scirhēah* is unrecorded, but a regular formation.

Scredington Li [*Scredinctun, Scredintune* DB, *Scredinton* 1196 P, 1202 Ass]. A Scandinavianized form of some OE name. The first el. may be connected in some way with OE *scréad* 'shred'. Or it might be OE *Scirheardingas* 'Scirheard's people'.

Scremby Li [*Screnbi* DB, 1170 P, *Scrembi* 1191 ff. P, *Screinbi* 1202 Ass], **Scremthorpe** Li [*Scremtorp* 1212 Fees]. First el. a byname cognate with ON *Skræmir* (from *skræma* 'to frighten'), perhaps an OScand *Skræma*.

Scremerston Nb [*Schermereton* 1196 P, *Scremerstun* 1208–10 Fees]. A manorial name, the first el. being a family name identical with *escrimer* 'fencer' (from Fr *escrimeur*), Scotch *Scrymgeour*. Alexander Skirmissarius is mentioned FPD (12th cent.).

Screveton (skrētn) Nt [*Screuintone, Screvetone* DB, *Screveton* 1201 Cur]. A Scandinavianized form of OE *scirgeréfan tūn* 'the sheriff's manor'.

Scrivelsby Li [*Scrivelesbi* DB, *Scriflebi* c 1115 LiS, *Scriuelesbi* 1202 Ass]. The first el. seems to be a pers. n., perhaps a nickname identical with ON *skrifli* 'a piece', Icel *skrifli* 'a poor fellow'.

Scriven YW [*Scrauing(h)e* DB, *Scrauin* 1167 P, *Screving* 1208 FF]. A derivative of OE SCRÆF 'cave, hollow' &c. with the suffix -*ing* ('place at a *scræf*').

Scrooby Nt [*Scroppen þorp* 958 YCh 3, *Scrobi* DB, *Scroby* 1225 Ep, *Scruby* 1242 FF]. '*Skroppa*'s thorp and BY.' ON *Skroppa* pers. n. occurs.

Scropton Db [*Scrotun, Scroftun* DB, *Screpton* 1251 Ch, *Scropton* 1275 RH]. First el. probably as in SHROPHAM, though in a Scandinavianized form.

Scruton YN [*Scurvetone* DB, -*ton* 1210 Cur, *Scuruentune* 1185 TpR]. '*Scurfa*'s TŪN.' *Scurfa* occurs in ASC as the name of a Scandinavian jarl. It is a byname from ON *skurfa* 'scurf'.

OE scucca 'demon, goblin' is the first el. of SCUGDALE, SHOBROOKE, SHOCKLACH, SHUCKBURGH &c.

Scugdale YN in Guisborough [*Scuggedale* c 1190 YCh 696], S~ YN in Whorlton [*Schugedale* 1228 Cl]. 'Goblins' valley.' First el. OE *scucca*, Scandinavianized.

Sculcoates YE [*Sculekotes* 1197, -*cotes* 1209, 1223 FF], **Sculthorpe** Nf [*Sculatorpa* DB, *Sculetorp* 1174–80 BM]. '*Skúli*'s COTS and thorp.' Cf. SCOULTON.

Scunthorpe Li [*Escumetorp* DB, *Scunptorp* 1245 FF]. '*Skúma*'s thorp.' First el. ON *Skúma* pers. n.

OE scylf, scelf 'rock, crag', no doubt also 'ledge' and 'bank of a river', is a common el. in pl. ns. There was also OE **scylfe, scilfe** 'ledge, shelf', but probably used in other senses too. The two are not always easy to keep apart. The exact meaning of the elements in pl. ns. is often difficult to determine. *Scylf* (*scelf*) is the source of SHELF, SHELL, SHELVE. It is the first el. of SALDEN, SELLY OAK, SHELDON Wa, SHELFANGER, SHELFIELD, SHELLAND, SHELLEY, SHELTON, SHILTON (some), SHILVINGHAMPTON, SHULBREDE, also of SKELTON (Scandinavianized). As a second el. it appears as -*shelf* in HUN-, TAN-, TIB-, WADSHELF, but generally it takes some other form, as in OXHILL, SHARESHILL, MINSHULL, BASHALL, BRAMSHALL, GOMSHALL, RUSHALL W, LITCHFIELD. OE *scylfe, scilfe* is the first el. of SHILDON, SHILSTONE, SHILTON (some), SHILVINGTON, SILPHO, SILTON, SILVINGTON. The corresponding OScand **skialf** is found in some names, as HINDERSKELFE, RASKELF, RANSKILL, SKUTTERSKELFE, ULLESKELF.

OE scyttel, scyttels. See SHUTLANGER, SHUTTLEWORTH.

Seaborough Do [*Seveberge* DB, *Seveberugh* 1256, *Sevenbergh* 1306 FF]. 'Seven hills' (OE *seofon beorgas*).

Seabrook Bk [*Sebroc* 1203 Cur, *Seybroc* 1250 Fees]. 'Slow-moving brook.' The first el. seems to be an OE *sæge* adj. related to *sigan* 'descend, move' and found in OE *onsæge* 'assailing'. *Sæge* would mean 'trickling, slow-moving'.

Seacombe Chs [*Secumbe* 13 AD i]. 'Valley by the sea.'

Seacourt Brk nr Oxford [*æt Seofecanwyrþe, Seofocanwyrð* 957 BCS 1002, *Seuacoorde* DB]. '*Seofeca*'s WORÞ.' *Seofeca* is closely related to OG *Sibico*.

Seacroft YW [*Sacroft* DB, *Secroft* c 1200 YCh 1553]. 'Croft by a lake.' There is no lake here now.

Seaford Sx [*Saford* c 1150 Fr, *Seford* 1180 P]. 'Ford by the sea.'

Seagrave Le [*Satgrave, Setgraue* DB, *Satgraua, Sedgraue* 12 DC, *Sethgravia* Hy 2 Berk, *Segraue* 1197 FF]. Second el. OE *gráf* 'grove' or *græf* 'ditch'. The first may be OE (*ge*)*set* 'fold' or *séaþ* 'pit, pool'.

Seagry W [*Segrie, Segrete* DB, *Segrea* 1190 P, *Segrey* 1221 Ass, *Segre* 1242 Fees, *Segreth* 1399 PNW(S)]. OE *secg-riþ* 'sedge brook'.

Seaham Du [*Sæham* c 1050 HSC, *Seham* 1155 FPD]. 'HĀM by the sea.'

Seal K [*La Sela* DB, *Sele* 1233 Ch, *La Seele* 1317 BM]. OE *sele* 'hall'.

Seal, Nether & Over, Db [*Scella, Scela, Sela* DB, *Scegla* c 1125 LeS, *Seile* 1198 Cur, *Sceile* 13 Fees, *Scheleg, Parva Scheyl* 1242 ib.]. An OE *scegel*, a diminutive of *scaga* 'shaw, wood'. S- for Sh- must be due to Norman influence.

Seale Sr [*Sela* 1210, *Sele* 1218 PNSr]. OE *sele* 'hall' or the dat. of OE *sealh* 'sallow'.

Seamer YN nr Scarborough [*Semær* DB, *Semer* DB, -*e* 1235 Ep], S~ YN nr Stokesley [*Semer* DB, -*e* 1227 Ep, *Samare* c 1095 YCh 855]. The original name was OE *Sǽ* 'the lake'. After OE *sǽ* had ceased to be in common use in the sense 'lake', an explanatory *mere* was added.

Searby Li [*Seurebi* DB, c 1115 LiS, *Safrebi* LiS, *Sauerbi, Seuerbi* 1155–8 RA, *Seure-(de)bi* 1196 P]. '*Sæfari*'s BY.' *Sæfari* is an ON pers. n. It is the source of *Savari* LVD.

Seasalter K [*sealtern et Fefresham* 858 BCS 496, *Seseltre* DB, *Saesealtre* 11 DM]. OE *sæ-sealtærn* 'salt-house or salt-works on the sea'.

Seascale Cu [*Sescales* c 1165 WR]. 'Hut on the sea.' See SKÁLI.

OE **sēaþ** 'pit, pond'. See ORSETT, ROXETH.

Seathwaite La [*Seathwhot* 1592 PNLa]. 'Clearing on the lake' (Seathwaite Tarn).

Seaton R Co [*Seythyn* 1302 Ass, *Seythen* 1441 Ct]. Possibly derived from OCo *seit*, Welsh *saith* 'a pot, cauldron' ('stream with pot-holes'?).

Seaton Cu nr Bootle [*Seton* c 1250 StB], S~ Cu nr Maryport [*Seton* c 1174, 1201–12 Holme C], S~ D [*Seton* 1244 FF], S~ Du nr Seaham [*Sætun* c 1050 HSC, *Setun* 1155 FPD], S~ **Carew·** Du [*Setona* R 1 (1318) Ch, *Seton Carrowe* 1345 RPD], S~ **Delaval, Monkseaton** Nb [*Setuna* Hy 2 (1271) Ch, *Seton de la Val* 1270 Ch, *Seton Monachorum* 1380 Ipm], **North** S~ Nb [*Seton* 1242 Fees, 1258 Ipm], S~ YE [*Settun* DB, *Setone* 1297 Subs], S~ **Ross** YE [*Seton* DB, 1226 FF], S~ YN [*Scetun* DB, *Seton* 1306 Ch]. OE *Sǽ-tūn*, which means 'TŪN on the sea' in all the cases except the Seatons in YE. One of these is on Hornsea Mere. Cf. SǼ.

S~ **Carew** was held by Petrus Carou t. R 1 (Ch). *Carou* appears to be a byname, to judge by the absence of the prep. *de*.—S~ **Delaval** was held by the de la Val family, who took their name from LE VAL in Normandy.—**Monkseaton** from the monks of Tynemouth.—S~ **Ross** belonged to the Ross fee. Cf. MELTON ROSS.

Seaton Ru [*Seieton, Segen-, Segestone* DB, *Saitona* 1130 P, *Saieton* 1167 P, *Segeton* 1178 P, *Seinton* 1197 P]. First el. perhaps a brook-name *Sǽge*, identical with the first el. of SEABROOK. A pers. n. *Sǽga* is unrecorded, but might have arisen as a hypocoristic form of names like *Sǽgēat, Sǽgār.*

Seat Sandal (hill) We [*Satsondolf* 1274 Ipm]. '*Sandulf*'s SǼTR.' First el. ON *Sǫndulfr* pers. n.

Seavington (-ĕ-) **Denis, St. Mary or Vaux & St. Michael** So [*Seofenempton* c 1025 Athelney, *Seovenamentone, Sevenemetone, Sevenehantune* DB, *Sevenhampton Abbatis, Deneys* 1276 RH, S~ *Michaelis*

1291 Tax, *Sevenhamtone Vaus* 1327 Subs]. See SEVENHAMPTON.

S~ **Denis** was held by Adam le Daneys ('the Dane') in 1253 (Ch).—*Vaux* is a family name. Hubert de Vallibus was tenant in 1257 (FF). VAUX is a common pl. n. in France.

Seawell Np in Blakesley [*Sewelle* DB, *Sewewell* 12 NS, *Seuowell* 1220 Fees]. 'The seven wells.' A tradition of seven springs is common. Cf. *Seofenwyllas* BCS 165, and see SEWELL, SYWELL.

Sebergham (sĕbrum) Cu [*Setburgheham* 1224 P, *Seburgham* 1270 Sc, *Sedburgham* 1285 PNCu, *Saburgham* 1228 Cl]. Perhaps 'shieling belonging to Burgham', the elements being ON *sætr* and a lost pl. n. *Burgham*. For the order of the elements cf. ASPATRIA. The suggestion in PNCu(S) that the name means 'BURG of *Sǽburg* (a lady)' may well be correct, however.

OE **secg** 'sedge'. See SEDGE- (passim), SESSAY, SEAGRY, SETCHEY.

Seckington Wa [*Seccandun* 755 ASC, *Sechintone* DB, *Sechendon* 1175 P, 1204 Cur]. '*Secca*'s DŪN.' *Secca* pers. n. occurs in Widsith and is the first el. of *Seccanham* KCD 898 and **Seckford** Sf at Bealings [*Sekeforda* DB, *Secheford* 1206 Cur].

Sedbergh (-ber) YW [*Sedbergt* DB, -*berch* a 1177 FC, *Satberg* 1257 Ch], **Sedbury** YN [*Sadberge* 1157 PNNR]. Identical with SADBERGE.

Sedgeberrow Wo [*æt Segcesbearuue* ·777, *Secgesbearuwe* 964 BCS 223, 1135, *Seggesbarve* DB]. '*Secg*'s grove.' *Secg is OE *secg* 'warrior' used as a pers. n. Cf. SEDGLEY.

Sedgebrook Li [*Sechebroc* DB, *Segebroc* 1195 FF, *Seggebroc* 1230 P]. 'Sedge brook.'

Sedgefield Du [*Ceddesfeld* c 1050 HSC, *Segesfeld* c 1190 Godric, 1208–10 Fees, 1229 Ep]. If the first form belongs here, '*Cedd*'s FELD'. If not, cf. SEDGEBERROW.

Sedgeford Nf [*Seces-, Sexforda* DB, *Sicheford* 1166 RBE, *Secheford* ib., 1212 Fees, *Sechesforde* c 1140 BM, *Sekeford* 1190 P]. Hardly 'sedge ford'. A pers. n. *Secci, a side-form of *Secca* in SECKINGTON, might be thought of, or an OE *sæce, derived from *sican* 'to trickle' and cognate with OHG *seich* 'urine'. OE *sæce* would mean 'stream, rill'.

Sedgehill W [*Seghulle* Hy 1, *Segeshull* 1249, *Segehull* 1279 PNW(S), *Seggehull* 1398 Ipm]. OE *secg* 'sedge' is not an element one expects to find in combination with hill. Possibly '*Secga*'s hill'.

Sedgemoor So [*Seggemore* 1263 FF], **Sedgewick** Sx [*Segwike* 1222 FF]. 'Moor and WĪC where sedge grew.'

Sedgley St [*Secgesleage* (gemæra) 985 KCD 650, *Segleslei* DB, *Seggeslegh* a 1211 BM]. '*Secg*'s LĒAH.' See SEDGEBERROW.

Sedgwick We [*Sigghiswic* c 1185, *Siggeswic* 1190, c 1200 Kendale]. '*Siggi*'s wīc.' *Siggi* is probably an OScand name.

Sedlescombe Sx [*Sales-, Selescome* DB, *Sedelescumbe* 1205–16 BM]. 'Coomb with a homestead.' First el. OE *sedl, sedel* 'residence, abode'.

Sedsall Db [*Segessale* DB, *Seggeshal* 1275 RH]. '*Secg*'s HALH.' See SEDGEBERROW.

Seend W [*Sinda* 1190, *Seinde* 1194 P, *Sendes* 1203 Ch, *Sende* 1212 RBE], **Seend Head** [*Sendeneheued* 1279 For], **Seend Row** [*Senderowe* 1268, *Sendenerewe* 1281 Ass]. Seend looks like a doublet of SEND. But the place is nr a stream formerly called SEMNET (see SEMINGTON), and apparently Seend is identical with the river-name, which probably was alternatively *Semned*.

Seer Green Bk [*la Sere* 1223 Bract, *La Cere* 1273, *le Shere* 1309 Ipm]. A derivative of OE *sēar* 'withered, dry'.

Seething Nf [*Sithinges* DB, *Seinges* 1181 P]. Probably '*Sīþ(a)*'s people'. OE **Sīþa* would correspond to OG *Sindo*.

Sefton La [*Sextone* DB, *Sefftun* a 1222 CC, *Ceffton* 1236 Cl]. 'TŪN where rushes grew.' First el. ON *sef* 'rush'.

Seghill Nb [*Sihala* Hy 2, *Syghal* 1198 (1271) Ch, *Seyhale* 1296 Subs, *Seikhale* 1318 Misc]. The first el. may be a stream-name **Sige*, derived from OE *sīgan* (cf. SEABROOK). Second el. HALH.

Seifton. See SIEFTON.

Seighford (sī-) St [*Cesteforde* DB, *Cesterford* n.d. Ronton, *Seteford* 1208 Cur]. A Normanized form of *Chesterford*, with loss of the first *r* owing to dissimilation.

Seisdon (-ēz-) St [*Sais-, Seisdone* DB, *Saiesdona* 1130 P, *Seyxdun* 1236, *Seisdon* 1242 Fees]. Very likely OE *Seax-dūn* or *Seaxesdūn* 'the hill of the Saxons or of *Seax*'. **Seax* is a normal short form of names in *Seax-*. The change to *Sais-* may be due to Norman influence.

Selaby Du [?*Selebi* 1196 P, *Seletby* 1317 Cl]. Possibly a hybrid name, the first el. being an OE *selet* 'sallow copse', derived from SALH.

Selattyn Sa [*Sulatun* 1254 Val, *Sulatton* 1420 Ipm]. *Acton* with a distinguishing addition, e.g. *sulh* 'gully'. The name has been subject to Welsh influence.

Selborne Ha [*Seleborne* 903 BCS 602, *Selesburne* DB, *Seleburn* 1197 BM]. Originally the name of Oakhanger Stream [(water of) *Seleburne* 1233 Selborne]. First el. the plur. of OE *sealh* 'sallow' or a derivative of it (an OE *sele*). Cf. next name.

Selby YW [*Seleby* c 1030 YCh 7, *-bi* c 1050 HSC, c 1155 DC, 1190 P, *Saleby* 1093 RA, *-bi* 1218 Pp]. *Selby* may well be a Scandinavianized form of OE *Seletūn*, and the

place of this name mentioned 779 ASC (E), c 1050 HSC, may be Selby. First el. very likely a derivative see OE *salh* 'sallow', meaning 'sallow copse'. Cf. SELBORNE.

Sele Priory Sx [*Sela* 1080 Sele, *Sele* c 1096 Fr]. OE *sele* 'hall' or identical with SALE. For OE *sele* 'hall' see also SEAL(E), NEWSELLS, perhaps LAUNCELLS.

Selham Sx [*Seleham* DB, *Seltham* c 1200 PNSx, *Suleham* 1209 FF]. 'HĀM by a sallow copse.' First el. an OE **sele, *siele*, alternating with **selet, *sielet* (cf. OE *þyrnet* 'thorn brake' &c.). See SELBY, SELABY, SILCHESTER.

Selker Cu [*Selekere* c 1250 StB]. 'Sallow marsh.' First el. ON *selia* 'sallow', found also in SELSIDE, SILECROFT.

OE **(ge)sell**, a derivative of *sele*, is found in names of swine-pastures, as *bocgeselle* BCS 197, *hlifgesella* ib. 442, *Rindigsel* ib. 194. The first el. of *bocgeselle* is OE *bōc* 'beech'. For *hlifgesella* see LITCHFIELD. *Gesell* seems to mean 'shelter for animals' or 'herdsmen's hut'. See BOARZELL, BRAMSHILL, HORSELL, LAWSHALL, LINDSELL, MARKSHALL Ess, RINGSHALL, SPILSILL, STRADISHALL. The el. often appears in the plur. form (*gesella, geselle*).

Sellack He [*Lann Suluc* c 1150 LL, *Selak* 1301 Misc]. '*Suluc*'s church.' *Suluc* is a hypocoristic form of *Suliau* or Tysilio, to whom the church is dedicated. Tysilio is another hypocoristic form of Suliau, formed by prefixing the pronoun *ty* 'thy'.

Selling K [*Setlinges* DB, *Sedling* 11 DM, *Sellinge* 1086 KInq, *Selling* 1206 FF], **Sellinge** K [*Sedlinges* DB, *Sedling, Sellinge* 11 DM, *Sellinges* 1226 Ass]. These two seem to be identical in origin with OFris *Sedlingi* and with the first el. of ZEDELGEM nr Bruges [*Sedelgem* 1167], from *Sedelingahem*. The base is in some way OE *seþ(e)l* 'residence, abode' (= OLG *sethal* &c.), but the exact history of the names Selling and Sellinge &c. is difficult to determine.

Selly Oak Wo [*Escelie* DB, *Selvele* 1204 Cur, *Selleg* 1236 Fees]. A Normanized form of SHELLEY. Oak is a late addition.

Selmeston (sīmsn) Sx [*Sielmestone* DB, *Sihalmeston* 1252 Ch]. '*Sigehelm*'s TŪN.'

Selsdon Sr [*Selesdun* 871–89 BCS 558]. First el. OE **Sele*, a short form of names in *Sele-*, or OE *sele* 'hall' or **sele* 'sallow copse' (cf. SELBORNE).

Selsey Sx [(in) *Seolesiae* c 715 Eddi, *-ig* c 890 OEBede, *Selaeseu, insula uituli marini* c 730 Bede, *Siolesaei* 780 BCS 1334]. 'Seal island.'

Selside We [*Selesat* 1196 FF, *-e* c 1195 Kendale], S~ YW [*Selesat* DB, *-sete* 1190 FC]. ON *Seliu-sætr* 'sallow shieling'. *Seljusetr* (*-sætr*) is a common pl. n. in Norway. Cf. SELKER, SÆTR.

Selston Nt [*Salestune* DB, *Selestun* 1249 Ep, *Selveston* 1277 Cl]. '*Sæwulf*'s TŪN', if the ex. of 1277 belongs here. Otherwise the first el. may be as in SELSDON.

Selwood So [*Sealwyda* 878, *-wuda* 894 ASC, *Seluudu* c 894 Asser, *Selevuda* 1168 P]. 'Sallow wood.' Cf. SALH. Asser says the British name was *Coit Maur* 'sylva magna' (Welsh *coed* 'wood', *mawr* 'great').

Selworthy So [*Seleuurde* DB, *-worh* 1243 Ass, *Syleworth* 1291 Tax]. 'WORÐIGN by a sallow copse' (OE *siele*, cf. SELHAM).

Sem R. See SEMLEY.

Semer Sf [*Seamera* DB, *Semere* c 1095 Bury, 1208 FF, 1254 Val], **S~** Nf nr Dickleburgh [*Semere* DB, c 1095 Bury, (fishery of) *Semere* 1265 Misc], **Sĕmer Water** YN [*Semerwater* 1153 Mon]. Cf. SEAMER. Semer Water was originally **Sæ*, to which was added an explanatory *mere*. The process was repeated, when *water* was added to *Semere*.

Semington W [*Sempletone* E 3 For, *Semington* 1470 BM]. 'TŪN on R *Semnet*.' **Semington Brook** is *Semnit* 964 BCS 1127, *Semnet* 1228 Cl. The name is related to SEM in SEMLEY and corresponds to early Fr *Sumeneta*.

Semley W [(on) *Semeleage* 955 BCS 917, *Semele* c 1190 Salisbury, *-leg* 1242 Fees]. 'LĒAH on river Sem' [*Semene* 984 KCD 641, 1278 QW]. *Semene* corresponds to SUMÈNE in France [*Simina* 12, *Sumena* 1585], SOMME [*Sumina* Gregory of Tours, *Sumena* Rav]. A related name is possibly SYFYNWY in Wales. The name is derived from the root *su-*, *seu-* in Sanskr *sŏma* 'beverage' &c.

Sempringham Li [*Sempingaham* 852 BCS 464, c 1067 Wills, *Sepingeham* DB, *Sempringham* 1150–3 YCh 1108, *Sanpingeham* 1162, 1165 P, *Semplingam* 1195 FF]. The original form seems to be *Sempingaham* (without *r*). The first el. seems to be identical with that of SENGKOFEN in Germany [*Sempinchovun* c 900], for which Förstemann suggests derivation from a pers. n.

Sence R Le, a trib. of the Anker [*Sheynch* 1307 Cl]. OE *scenc* 'a cup, drinking-can' (cf. SHENTON). The name may refer to a river with plenty of good drinking-water. **Sence** Le, a trib. of the Soar, no doubt has the same origin. *Sence* for *Shench* is due to Norman influence.

Send Sr [(æt) *Sendan* 960–2 BCS 1063, *Sande* DB, *Sende* 1291 Tax]. A derivative of OE *sand* 'sand', e.g. an OE **sende* fem. 'sand dune, sandy place'.

Senlac Sx [*Senlac* 12 Ordericus, *Sandlake* 1296 Subs]. OE *sand-lacu* 'sandy brook'. *Senlac* is a Normanized form.

Sennen Co [(parochia) *Sancte Semane* 1377 PT]. A saint's name. Cf. Bret *Saint Senan*.

Seph R YN [*Sef* 1170–85 Riev, 1201 FF]. 'Slow stream.' Cf. Sw *Såveån* from OSw *sæver* 'calm, slow'.

Serlby Nt [*Serlebi* DB, *-by* 1191–3 Fr]. '*Serle*'s BY.' First el. ON *Sørli* &c. (*Serlo* DB).

Sessay YN [*Segege* 1088 LVD, *Sezai* DB, *Secey* 1182 P, *Scezzay* 1246 FF]. OE *secgēg* 'sedge island'. The abnormal development is due to Scandinavian influence. The unfamiliar sound-combination (in *sedge*) was replaced by *ds*, whence *ts*, *ss*.

Sessingham Sx [*Sesinge-*, *Sasingham* DB]. 'The HĀM of *Se(a)ssa*'s people.' **Se(a)ssa* is a hypocoristic form of names in *Seax-*. But possibly *Sessingham* is a Normanized form of *Seaxingahām* 'the HĀM of *Seaxa*'s people.'

OE **(ge)set**, plur. *(ge)setu*, *-seotu*, means 'dwelling, place of residence', also 'place where animals are kept, fold' and the like. This word is doubtless the source of *-set(t)* in many pl. ns., but it is often difficult to distinguish from SÆTE and SÆTR, with which it often coincides in meaning. The exact meaning is not always easy to determine. Probably it was often 'fold', but especially in names of old villages it may rather be 'homestead' or even 'village'. The el. is particularly common in East Anglia, as in FORNCETT, HETHER-, LETHERING-, STRAD-, TATTER-, WHISSONSETT Nf, BRICETT, ELM-, HES-, WETHERING-, WISSETT Sf. Cf. also LISSETT YE, OSSETT YW, TARSET Nb. It sometimes appears as *-side*, as in SIMONSIDE Nb. As a first el. it may occur in SATLEY.

Setchey Nf [*Seche*, *Siecche* 1202 FF, *Sechiche* 1242 Fees, *Sechithe* Hy 3 BM, *Sechyth* 1291 Tax]. Second el. OE *hȳþ* 'landing-place'. The first might be as in SEDGEFORD. But Setchey is nr Winch. If the latter is from *Winn-wic*, *Setch-* might be from *Sæ-wic* 'wīc on the lake'. Setchey is in a low situation, and there may well have been a lake here. But Setchey may after all be OE *Secg-hȳþ* 'sedge-covered landing-place' with change of *dg* to *ch* before *h*.

OE **setl** 'seat, abode'. See SETTLE, KINGSETTLE, also ANSLEY, WAST HILLS.

Setmurthy Cu [*Satmurdac* 1195 FF]. '*Murdac*'s shieling.' See SÆTR. *Murdac* or *Murdoch* (*Murdac*, *-oc* DB) is OIr *Muiredach*, Gael *Murdoch*.

Settle YW [*Setel* DB, c 1200 Pudsay, 1249 Ch]. OE *setl* 'seat', here 'dwelling, abode'.

Settling Stones Nb [*Sadelingstan*, *Sadelestanes* 1255 Ass, *Sadlystanes* 1295, *Sadelyngstanes* 1347 Misc]. Probably 'the saddlingstones', the name indicating the place on Stanegate where wayfarers got into the saddle after the ascent of a steep hill (Grindon Hill).

Settrington YE [*Sendriton* DB, *Seteringetune* c 1090 SD, *Setteringtona* c 1130 YCh 1073, *Setrinton* 1177 P]. Doubtful. The first el. might possibly be a derivative of OE *seohtre* 'drain, ditch'. *Aqueductus de Setrington* is mentioned 1190–1220 YCh 626.

Sĕven R YN [*Sivena* 1100–13 YCh 352, *Sivene* 1204 Ch]. A doublet of SEM, though we must then assume that Brit *Suminā* had become **Syfen* owing to lenition.

Sevenhampton Gl [*Sevenhamtone* DB], S~ W [*Suvenhantone* 1212 RBE, *Sevehampton* 1251 Cl], **Sevington** W [*æt Seofonhæmtune* 1043 KCD 767]. OE *Seofonhǽma-tūn* 'the TŪN of the *Seofonhǽme*' or dwellers at a locality called 'Seven wells' or the like (cf. SEAWELL). Sevenhampton Gl may have been named from *Seofenwyllas*, a locality mentioned BCS 165 in the bounds of Aston Blank 3 or 4 m. from Sevenhampton. **Sevington** Ha at Tichborne is probably from *Sevenhampton*. It was named from the (on) *syfan wyllan* mentioned 938 BCS 731 in the bounds of Tichborne. Cf. SEAV-INGTON.

Sevenoaks K [*Sevenac* (var. *-acher*) 1200 Cur, *Sevenak* 1218 Ep, *Seuenok* 1230 RA]. 'Seven oaks.' Cf. SIEBENEICH in Germany (at least six known).

Sĕvern R [*Sabrina* 115–17 Tac, 6 Gildas, c 730 Bede; *Habren* c 800 HB, *Hafren* c 1150 LL; *Sæferne* (obl.) 757–75, 816 BCS 219, 356, *Sæfern* 896 ASC, *Sauerna* DB, *Saverne* c 1140 Gaimar, *Seuerne* 1205 Lay]. Identical with the old name of a stream at Bedford [*Seuerne* 13 ERN] and *Sabrann*, the old name of the Lee in Ireland. The etymology of this ancient river-name is not clear. Cf. SEVERN STOKE under STOKE.

Sevington K [*Seivetone* DB, *Seyueton'* 1314 FF]. '*Sǽgifu*'s TŪN.' *Sǽgifu* is a woman's name. S~ Ha, W. See SEVENHAMPTON.

Sewardsley Np [*Sewardeslege* 12 AD ii, *-lega* 1180 P, *-leia* 13 BM]. '*Sǽweard*'s LĒAH.'

Sewardstone Ess [*Siwardeston* 1178 ff. P, 1212 Fees]. '*Sigeweard*'s TŪN.'

Sewell Bd in Houghton Regis [*Sewelle* DB, *Seuewell* 1193 P], S~ or **Showell** O [*Seve-, Sivewelle* DB, *Sefewella* c 1160 Eynsham]. Identical with SEAWELL.

Sewerby YE [*Siuuarbi, Siwardbi* DB, *Siwardebi* Hy 2 BM]. '*Sigeweard*'s BY.'

Sewstern Le [*Sewesten* DB, *Seustern* c 1125 LeS, *Sewesterna, Seuesterre* 1166 P, *Seuesterne* 1199 FF]. The second el. seems to be the same as that of SYDERSTONE, TANSTERNE, i.e. an OE *sterne* or the like. Possibly it might be a form with metathesis of *gestrēon*. See INGESTRE. The first el. of Sewstern is also doubtful. It might be OE *seofon* 'seven' or possibly OE *Sǽwīg*, pers. n.

Sexhow YN [*Sexhou* c 1160–80 YCh 692]. Possibly '*Sekk*'s mound', an OScand *Sekks-haugr*. *Sekkr* is an ON byname. But 'six hills' is a possible alternative.

Sezincote Gl [*Ch(i)esnecote* DB, *Sesnecot* 1205 Cur, *Scesnecot* 1236 Fees]. The first el. is an OE **cisen* 'gravelly', belonging to OE *cis-, ceosol*. *S-* for *Ch-* is due to Norman influence. See COT.

Shabbington Bk [*Sobintone* DB, *-ton* 1200 P, *Shobinton* 1231 Cl, *Shoppinton* 1167 P]. '*Sceobba*'s TŪN.' OE **Sc(e)obba* is found in several pl. ns., as *Scobban ora* 956 BCS 932, *Scobbanwirth* 744 ib. 168 and *Scobban byrygels* 990 KCD 673. The last means '*Scobba*'s burial-place'.

Shackerley La [*Shakerlegh* 1332 Subs], **Shakerley** La [*Shakerlee* c 1210 CC, *-ley* 1284 Ass], **Shackerstone** Le [*Sacrestone* DB, *Scaceston* 1236 Fees, *Shakerston* 1327 Subs]. 'LĒAH (wood) and TŪN of the robber(s).' Cf. *Scakeresdalehefd* c 1190 LaCh (Ormskirk La). The first el. is OE *scēacere* 'robber', corresponding to OHG *scâhhari*. It is true we should expect the OE word to be *scēcere* in Anglian dialects, but the word is regularly *scéacere* in Old Northumbrian texts.

Shadforth Du [*Shaldeford, Shaldeforth* 1183 BoB]. 'Shallow ford.' See SCALD.

Shadingfield Sf [*Scadenafella* DB, *Shadenefeld* Hy 2 (1268) Ch, *Schadenesfeld* 1190 P, *Schadenefeld* 1250 FF]. The first el. may be OE *scāden* 'separated' or *scēad-denu* 'boundary valley'. Cf. SCĒAD. The place is near Hundred River.

Shadoxhurst K [*Schettokesherst* 1239 FF, *Sadhokesherst* 1267 Ipm, *Shattokesherst* 1271 BM]. The first el. is obscure. See HYRST.

Shadwell Mx [*Shadewell* 1223 FF, *Shaldewell* 1316 Ch]. 'Shallow stream.' Cf. SCALD.

Shadwell Nf [*Shadewell* 1314 Ipm], S~ YW [*Scadeuuelle* DB, *-well* 1166 P, *Schadwelle* c 1200 YCh 1587]. 'Boundary stream.' Cf. SCĒAD and *Sceadwellan* BCS 1282.

Shaftesbury Do [*Sceaftesburi* c 871, 956 BCS 531, 970, (to) *Sceafnesbirig* 955 ib. 912, *Sceftesburg* c 894 Asser, (to) *Sceaftenesbyrig* 1015 Wills, (on) *Sceaftsbirig* c 1000 Saints, *Scepteseberie* DB]. '*Sceaft*'s BURG', alternating with '*Sceaften*'s BURG'. *Sceaften* is a hypocoristic form of *Sceaft*, which is itself a short form of names in *Sceaft-*, as *-here, -wine*. *Sceaft* occurs in *Sceaftes hangra* BCS 629.

Shaftoe Nb [*Shatfho* 1231 Cl, *Schafhou* 1242 Fees, *Shafthou* 1256 Ass], **Shafton** YW [*Sceptun* DB, *Scaftona* 1155–70 YCh 1533, *Scafton* 1230 P]. The first el. seems to be OE *sceaft* 'shaft, pole'. A boundary mark may be meant, at least in Shaftoe. Norw *skapt* is used of a lower ridge projecting from a hill and the like. Such a sense would do for these names, but the sense has not been evidenced in England. See HŌH, TŪN.

Shakerley. See SHACKERLEY.

Shalbourne W [*Scealdeburnan* 955 BCS 912, *Scaldeburne* DB], **Shalcombe** Wt [*Eseldecome* DB, *Shaldecumbe* 1284 BM], **Shalden** Ha [*Scealdedeninga gemære* 1046 KCD 783, *Seldene* DB, *Scaldeden* 1167 P], **Shalfleet** Wt [*æt Scealdan fleote* 838 BCS 423, *Seldeflet* DB], **Shalford** (-ahf-) Ess [*Scaldefort* DB, *-ford* 1254 Val], **S~** Sr [*Scaldefor* DB, *-ford* 1199 Cur]. 'Shallow stream (BURNA), valley (CUMB, DENU), stream (FLĒOT), ford.' Cf. SCALD. The stream that gave its name to Shalfleet is (on) *Scealdan fleot* 949 BCS 879.

Shalstone Bk [*Celdestone* DB, *Scaldestona* c 1130 Oxf, *Saldestuna* c 1160 NpCh]. Possibly *Scealdwelles tūn* 'TŪN at the shallow stream'. Cf. SCALD, SHELSWELL.

Shangton Le [*Santone, Sanctone* DB, *Scanketon* c 1125 LeS, *Schanketon* 1206 Cur]. The first el. is OE *scanca* 'shank, leg'. This word might well have given rise to a nickname (cf. *Longshanks*), but more likely the word is here used in a transferred sense of the long narrow spur of hill at Shankton. The name thus means 'TŪN at the spur of hill'.

Shanklin Wt [*Sencliz* 1287–90 Fees, 1306 Ch, *Schentling* 1287–90 Fees, 1306 Ch, *Schencling* 1324 Misc]. The second el. is OE *hlinc* 'hill'. The first is OE *scenc* 'cup' (cf. SENCE). The description of the place in PNWt shows that a meaning 'hill with a cup (or, better, a can)', i.e. a waterspout' is apposite.

Shap We [*Hepe* 1228 Cl, *Yhep* 1241 PNCu, *Heppe* 1291 Tax, *Sheppe* 1300 Cl]. OE *hēap* 'heap'. The name refers to **Shap Stones**, the ruins of a prehistoric stone circle. The change of *Hēap* to *Hiap* and *Shap* has analogies in *Shetland* from ON *Hialtland* and *Shapinsay* from *Hialpandisey*. Cf. also SHIPTON YE, YN.

Shapwick Do [*Scapeuuic* DB, *Sepwik* 1236 Fees, *Scepwyk* 1275 RH], **S~** So [*Sapwic* 725 BCS 142, *Sapeswich* DB, *Schepwich* 1173 P]. OE *scēap-wic* 'sheep farm'.

Shardlow Db [*Serdelau* DB, *Serdelaw* 1202 FF, *Sherdelawe* 1240 FF, *Schardelow* 1242 Fees]. 'Notched mound, mound with a notch or indentation.' First el. OE *sceard* adj. 'notched' &c.

Shareshill St [*Servesed* DB, *Sarneshull* 1213 FF, *Sarsculf* 1227 Ass, *Shareweshulf* 1252 Cl]. OE *Scræf-scylf* 'hill by a narrow valley'. See SCRÆF, SCYLF. *Scræf* here appears with metathesis of *r*. The place is on a marked hill.

Sharlston YW [*Scharvestona* 1180–5 YCh 1542, *Sarneston* 1242 Fees, *Sharweston* 1297 Subs]. 'TŪN by a SCRÆF or narrow valley.' Cf. SHARESHILL.

Sharnbrook Bd [*Scernebroc* DB, *Shernebroc* 1163, *Sharnebroc* 1167 P], **Sharnford** Le [*Scearnford* 1002 Wills, 1004 KCD 710, *Scerneforde* DB]. 'Muddy brook and ford.' Cf. SCEARN.

Sharow YW [*Sharho* 1239 Ep, *Sharehow* 1297 Subs]. 'Boundary hill.' First el. OE *scearu* (in *landscearu*) 'boundary'. Second el. OE HŌH.

Sharpenhoe Bd [*Scarpeho, Serpenho* 1197 FF, *Scarpenho* 1253 Ch]. 'Steep spur of land.' See SCEARP, HŌH.

Sharperton Nb [*Scharberton* 1242 Fees, *Scharperton* 1296 Subs]. The place is at **Sharperton Edge**, a steep little hill. This was no doubt called *scearpa beorg* 'steep hill'. Cf. prec. name.

Sharples Hall La [*Charples* 1212 Fees, *Scharples* 1246 Ass]. A derivative of OE *scearp* in the sense 'steep' (OE *scearpol* or the like 'steep place') or a compound with *scearp*, e.g. *scearp-læs* 'steep meadow'.

Sharpness Gl [*Nesse* DB, *Sharpenesse* 1349 PNGl]. 'Abrupt headland.' Cf. NÆSS.

Sharrington Nf [*Scarnetuna* DB, *-tune* 12 BM, *Sharnetone* 1254 Val]. OE *Scearn-tūn* 'muddy TŪN'. Cf. SCEARN.

Sharstone Chs [*Sharston* 1248 Ipm]. Presumably OE *scearda stān* 'notched stone'.

Shatton Db [*Scetune* DB, *Scatton* 1230 P]. Either OE *Scēad-tūn* 'TŪN on the boundary' (cf. SCĒAD) or *Scēat-tūn* 'TŪN in a corner of land'. S~ is in a tongue of land between the Derwent and a brook.

Shaugh D [*Scage* DB, *Saghe* 1242 Fees]. OE *scaga* 'shaw, copse'.

Shavington Chs [*Shawynton* 1260, *Shavinton, Shaventon* 1287 Court], **S~** Sa [*Savintune* DB, *Schauinton* 1227 FF, *Sauigton* 1230 P]. 'The TŪN of *Scēafa*'s people.'

Shaw Brk [*Sagas* c 1080 Fr, *Essages* DB, *Shage* 1167 P], **S~** La [*Shaghe* 1555 BM], **S~** W [*Schaga* 1167 P, *Schage* 1200 Cur]. OE *scaga* 'shaw, copse'.

Shawbury Sa [*Sawesberie* DB, *Schageberia* 1183 P], **Shawdon** Nb [*Schaheden* 1232 P, *Schauden* 1242 Fees]. 'BURG and valley by a shaw or copse.'

Shawell Le [*Savelle* DB, *Schadewelle* 1224 Ep, *Schathewell* 1276 RH, *Shathewell* 1316 FA]. 'Boundary stream' (OE *scēap-wella*). Cf. SHEAF.

Sheaf R Db, YW [*Scheve* 1183, *Scheth* 14 ERN]. OE *scēap, scǣp* 'boundary', identical with OFris *skēth* 'distinction', OHG *skeida* 'boundary'. The word is related to OE SCĒAD. The Sheaf forms the boundary between Derby and Yorkshire. For the change *ð > v* cf. GIVENDALE (2).

Shearsby Le [*Svevesbi, Sevesbi* DB, *Senesbi* 1196 P, *Schevesby* 1276 RH]. '*Swǣf*'s BY.'

Cf. SACOMBE Hrt. *Swǽf* is found in (to) *Swæfes heale* 940 BCS 762.

Shearston So [*Siredestone* DB, *Siredeston* 1254 Ass]. '*Sigerēd*'s TŪN.' The manor was held by *Sired* in 1066 (DB).

Sheat Wt [*Essvete* DB, *La Schete* 1287–90 Fees]. OE SCĒAT, perhaps in the sense 'park'.

Shebbear D [(of) *Sceftbeara* 1050–73 E, *Sepesberie* DB, *Seftberia* 1168, *Schaftberege* 1195 P]. OE *sceaftbearu* 'grove where poles were got'.

Shebdon St [*Schebbedon* 1267 Ch]. '*Sceobba*'s DŪN.' Cf. SHABBINGTON.

Shedfield Ha [(to) *scida felda* 956 BCS 953, *Schidefeld* c 1270, c 1285 Ep]. First el. OE *scīd* 'piece of wood split thin', possibly in the sense 'foot-bridge'. Cf. SHIDE.

Sheen St [(æt) *Sceon* 1002 Wills, *Sceon* DB, *Shene* 1265 Ass], **Sheen** Sr, the old name of Richmond [(on) *Sceon* c 950 BCS 1008, *Sceanes* 1130 P, *Scenes* Hy 3 BM, *Sienes* 1204 Cur]. The original form of the name seems to have been *Scēo*, gen. *Scēon*. This may be compared with Norw *skjaa* 'shed, kiln' (from *skewō*), a word derived from the root *sku-* in ON *skiól* 'shelter' and the like. OE *scēo* may have meant 'shelter, shed'.

Sheepstor D [*Sitelestorra* 1168, *Schetelestorre* 1182 P, *Sytelestorre* 1242 Fees]. *Sheeps-* is a late form due to association with *sheep*. The original first el. may have been OE *scitels* 'dung', referring to the dung of sheep. Second el. *tor* 'hill'. See TORR.

Sheepwash D [*Sepewais* 1184 P, *Shepwasse* 1249 FF], S~ Nb [*Sepewas* 1178 P]. OE *scēapwæsce* 'place for washing sheep'.

Sheepy Magna & Parva Le [*Scepehe, Scepa* DB, *Scepeia* 1190 P, (*Parva*) *Shepe* 1209–35 Ep, *Magna Shepeye* 1327 Subs]. OE *scēap-ēa* or *scēap-ēg* 'sheep river or island'.

Sheering Ess [*Sceringa* DB, *Seringes* 1212 RBE]. A tribal name, perhaps derived from a pers. n. cognate with OG *Scarius, Scering* &c., which are derived from OHG *scara* 'army'. Cf. SHELLINGFORD.

Sheerness K [*Shernesse* 1221 Cl, *Sirnesse* 1329 Ch]. 'Bright headland.' First el. OE SCĪR adj. or the base of Mod *sheer* (OE *scǽre*).

Sheet Ha [*Syeta* a 1210, *la Syete* c 1255, *la Shyte* 1266 Selborne]. See SHEAT.

Sheffield Brk nr Reading [*Sewelle* DB, *Scheaffelda* 1167 P, *Scefeld, Sceofeld, Suefeld* c 1202 BM]. The first el. is probably OE *scēo* 'shelter' as in SHEEN. Cf. FELD.

Sheffield Sx [*Sifelle* DB, *Shipfeud* 1275 RH]. 'Sheep FELD.' OE *scip* is a common form of *scēap* 'sheep'.

Sheffield YW [*Scafeld* DB, *Sed-, Sadfeld* 1184 f. P]. 'FELD on R SHEAF.'

Shefford Bd [*Sepford* 1220 Subs, *Shepford* 1247 FF], **East & West** S~ Brk [*Siford* DB, *Schipforda* 1167 P, *Est-, Westsipf'* 1220 Fees]. 'Sheep ford.'

Sheinton Sa [*Scentune* DB, *Seinton* 1197 P, *Shenton* 1242 Fees]. OE *Scēna-tūn* 'beautiful TŪN'.

Sheldon D [*Sildene* DB, *Schildene* 1185 Buckland, 1291 Tax]. OE *scylf-denu* 'valley with steep sides'. See SCYLF.

Sheldon Db [*Scelhadun* DB, *Schelehaddon* 1230 P]. The place is near HADDON. Sheldon consists of ME *schēle* 'shed, hut' and Haddon (Haddon with a shed or sheds).

Sheldon Wa [*Scheldon* 1190 P, *Sheldon* 1236 Fees]. OE *scylf-dūn* 'hill with a SCYLF or flat top' or the like.

Sheldwich K [*Scilduuic* 784 BCS 243, *Sceldwik* 1198 FF]. 'WĪC with a shelter.' First el. OE *scild, sceld* 'shield', here perhaps in the recorded sense 'protection'.

Shelf YW [*Scelf* DB, *Shelf* 1311 Ch]. OE SCYLF 'hill' &c.

Shelfanger Nf [*Sceluangra* DB, *Scelfhanger* c 1095 Bury]. OE *scylf* 'hill' &c. and *hangra* 'slope'. But there is not much of a hill here. Perhaps *scylf* means 'plateau'.

Shelfield St [*Scelfeld* DB, *Schelfhul* 1271 For, *Shelfhull* 1300 For], S~ Wa [*Shelfhull* 1315 Ipm, *Scelefhull* 1328 Ch]. OE *scylf-hyll*, perhaps 'hill with a plateau'.

Shelford Ca [*Scelford* c 1050 KCD 907, *-a* c 1080 ICC, *Escelforde* DB, *Scheldford* 1190 P, *Shelford parva* 1228 FF, *Schelford Magna* 1254 Val], S~ Nt [*Scelforde* DB, *Sceldford* c 1155 DC, *Scelford* 1232 Ch, *Scheldford* 1276 Ipm]. The first el. may be OE *sceld* in the sense 'shelter' (cf. SHELDWICH) or an OE *sceldu* 'shallowness, shallow place'.

Shell Wo [*Scylfweg* 956 BCS 937, *Scelves* DB]. OE *scylf* 'bank'.

Shelland Sf [*Sellanda* DB, *Sevelond* 1219 Fees, *Shevelond* 1234 FF]. OE *Scylf-land* 'land on a slope'.

Shelley Ess [*Senleia* DB, *Schelveleye* 1276 FF, *Schelflee* 1278 Ipm], S~ Sf [(to) *Scelfleage* c 995 BCS 1289, *Sceueleia* DB, *Schelfleye* 1254 Val], S~ YW [*Scelneleie, Sciuelei* DB, *Skelflay* 1242 Fees, *Scheflay* 1297 Subs]. OE *Scylf-lēah* 'LĒAH on a slope or ledge'.

Shellingford Brk [*Scaringaford* 931 BCS 683, *Serengeford* DB, *Sceringeford* Hy 1 Abingd, *Schalingef'* 1220 Fees]. The first el. seems to be identical with SHEERING Ess.

Shellow Bowells Ess [*Scelga, Scelda* DB, *Scelléé* 1198 FF, *Selewes* 1238 Subs, *Sheleghes* 1244 FF, *Scheuele Boueles* 1303 FA]. The name is correctly explained in PNEss as originally an alternative name of the Roding. It is an OE *Sceolge* (obl. *Sceolgan*), a derivative of OE *sceolh* 'wry, oblique', here 'winding'. There was formerly also a S~ Jocelyn; hence the plur. form in early sources.

The manor was held by John and Ralph de Büeles in 1249 (FF). Bowells perhaps from BOUELLES in Seine-Inf.

Shelsley Beauchamp Wo [*Celdeslai* DB, *Sceldeslega* c 1150 Surv, *Scheldeslegh Beauchampe* 1255 Ass], S~ **Walsh** Wo [*Sceldeslæhge* 11 Heming, *Caldeslei* DB, *Seldesleg le Waleis* 1242 Fees]. The two places are opposite to each other on the Teme. The name means '*Sceld*'s LĒAH'. *Sceld* occurs in (on) *Sceldes heafda* 1016 KCD 724, *Scyldestreow* 955 BCS 917.

S~ **Beauchamp** was held by the Beauchamps from the 12th cent. Cf. ACTON BEAUCHAMP.— S~ **Walsh** was held by Johannes Walensis in 1212 (RBE). The name means Welsh.

Shelswell O [*Scildeswelle* DB, *Saldewelle* 1209–19 Ep, *-well* 1242 Fees, *Shaldewell* 1289 Ch, *Schelde(s)well* 1255 RH]. 'Shallow stream.' See SCALD. The *s* is intrusive.

Shelton Bd [*Eseltone* DB, *Sheltune* 1197 FF, *Schylton* 1276 Ass], S~ Nf [*Sceltuna* DB, *Scelton* 1203 Ass], S~ Nt [*Sceltun* DB, *Scelton* J BM], S~ Sa [*Saltone* DB, *Shelfton* 1221 Eyton], S~ St [*Scelfitone* DB, *Schelton* 1190 P]. OE *Scylf-tūn* 'TŪN on a bank or ledge'.

Shelve Sa [*Schelfe* 1180, *Schylve* 1249 Eyton]. OE SCYLF 'hill, ledge' &c. The place is in a high situation.

Shelwick He [*Scelwiche* DB, *Sheldwik* 1241 Ch]. Identical with SHELDWICH K.

Shenfield Ess [*Scenefelda* DB, *Shenefeld* 1165 P]. 'Beautiful FELD.' See SCĒNE.

Shengay or **Shingay** Ca [*Sceningeie* c 1080 ICC, *Scelgei* DB, *Senegaia* 1087–93 Fr, *Schenegeia* 1196 P]. 'The island of the *Scēningas* or *Scēne*'s people.' Cf. SHINGHAM. OE *Scēne* (*Sciene*) pers. n. is found in *Scynes weorþ* 947 BCS 820 and is related to OHG *Scōnea*. It is derived from OE *scēne* 'beautiful' or a short form of names containing this el. *Scenwulf* occurs LVD.

Shenington O [*Senendone* DB, *Senedon* 1195 P, *Schenindon* 13 BM], **Shenley** Bk [*Selelai* DB, *Schenlega* 1183 P], S~ Hrt [*Scenlai* DB, *Shenlee* 1205 Cur]. 'Beautiful DŪN and LĒAH.' First el. OE *scēne* 'bright, beautiful'.

Shenstone St [*Scenstan* 11 PNSt, *Seneste* DB, *Scenestan* c 1130 Oxf, *Shenestan* 1168 P]. '*Scēne*'s stone' or 'beautiful stone'. Cf. SHENGAY, SHENFIELD.

Shenton Le [*Scenctun* 1002 Wills, 1004 KCD 710, *Scentone* DB]. 'TŪN on R SENCE.'

Shephall Hrt [*Escepehale* DB, *Sepehal* 1199 P, *Scepehale* 1219 Pp], **Shepley** YW [*Scipeleia*, *Seppeleie* DB, *Schepelay* 1242 Fees]. 'HALH and LĒAH where sheep were kept.'

Shepperton Mx [*Scepertun* 959 BCS 1050, 1065 BM, *-tone* DB]. Perhaps OE *scēaphierda tūn* 'TŪN of the shepherds'.

Sheppey, Isle of, K [*Scepeig* 696 BCS 91, *Sceapig* 832, 855 ASC, *Scape* DB]. 'Island where sheep were kept.'

Shepreth Ca [*Esceprid* DB, *Shepereth* 1232 FF, 1291 Tax]. OE *scēap-riþ* 'brook in which sheep were washed'.

Shepscombe Gl [*Sebbescumbe* 1263 Ipm]. The first el. may be an OE *Scebbi*, a sideform of *Sceobba*. Cf. SHABBINGTON.

Shepshed Le [*Scepe(s)hefde* DB, *Shepesheued* 1167 P, *Schepeheued* 1191 P]. The place is fairly high, and the name may mean 'hill where sheep grazed'. Cf. HĒAFOD.

Shepton Beauchamp So [*Sceptone* DB, *Septon Belli campi* 1266 Ep], S~ **Mallet** So [*Sepetone* DB, *Scheopton Malet* 1226–8 Fees], S~ **Montague** So [*Sceptone* DB, *Shepton* c 1150 Bruton, *Schuptone Montagu* 1285 FA]. OE *Scēap-tūn* 'sheep farm'.

S~ **Beauchamp** was held before 1212 by Robert de Bellocampo (Fees). See ACTON BEAUCHAMP.—S~ **Mallet** was held by Robert Malet t. Hy 1 (Collinson). Cf. CURRY MALLET.— S~ **Montague** was held by Drogo de Montacute in 1086 (DB). The name is from MONTAIGU in Normandy.

Sheraton Du [*Scurufatun* c 1050 HSC, *Scurvertune* c 1190 Godric, *Schuruetone* 12 FPD]. Perhaps identical with SCRUTON.

Sherborne Do [(æt) *Scireburnan* 864, c 880 BCS (510, 553), 910 ASC, *Scireburne* DB], S~ Gl [*Scireburne* DB, *Schireburn* 1193 P], S~ **St. John & Monk** S~ Ha [*Sireburne*, *-borne* DB, *Sireburna* c 1125 Oxf, *Shireburna Johannis* 1167 P, *Schireburne Monachorum* c 1270 Ep], S~ Wa [*Scireburne* DB, *Shireburn* 1248 Ch], **Sherbourne** R Wa [*Schireburn* 1310, *Shirburn* 1352 AD], **Sherburn** Du [*Scireburne* c 1170 FPD, *Shyreburn* 1237 Cl], S~ YE [*Schireburn*, *Sciresburne* DB, *Schireburn* c 1150 YCh 1154, *Sciraburna* a 1182 BM], S~ **in Elmet** YW [(to) *Scireburnan* 963, *Scyreburna* c 1030 YCh 6 f., *Scireburne* DB]. 'Bright stream.' First el. OE *scīr* 'bright, pure'.

Sherborne St. John was held by Robert de Sancto Johanne in 1245 (Ch).—**Monk** S~ was a priory.—See ELMET.

Shere Sr [*Es Sira* DB, *Shyre* 1242 Fees, *Shyr* 1251 Ch]. The place is on a stream, which was probably once known as *Scir* 'bright one'. Cf. ROTHER K, Sx, formerly *Scir*. See SCĪR adj.

Shereford Nf [*Sciraforda* DB, *Shireford* 1206 FF]. 'Clear ford.' Cf. SCĪR adj.

Sherfield English Ha [*Sirefelle* DB, *Schirefeld* 1291 Tax], S~ **upon Loddon** Ha [*Schirefeld* 1179 RA, *Scirefeld* 1212 Fees]. 'FELD belonging to the shire' or 'bright or clean FELD'. If the first el. is the adj. *scir*, the meaning may be that of *clean* in CLAN-

FIELD, i.e. 'free from noxious growth' or the like.

S~ English was held by Richard Lengleis ('the Englishman') in 1303 (Ep).—Cf. LODDON.

Sherford D [*Scireford* c 1050 KCD 926, *Sirefort* DB], S~ So [*Shireforde* 1327 Subs]. Identical with SHEREFORD.

Shĕringham Nf [*Silingeham* DB, *Siringeham* 1174 Fr, *Scheringham* 1242 Fees, *Schyringham* 1291 Tax], **Sherington** Bk [*Serintone* DB, *Schirincton* 1172, *Schirinton* 1180 P]. 'The HĀM and TŪN of *Scira*'s people.'

Shermanbury Sx [*Salmonesberie* DB, *Sirmannesburi* 1245 Sele, *Shyremannebyr'* 1249 FF]. Perhaps 'the manor of the *scirman* or sheriff'.

Shernborne Nf [*Scernebrune* DB, *Scarnebrune* 1254 Val], **Sherrington** W [*Schearntune* 968 Reg Wilt, *Scarentone* DB, *Sherinton* 1167 P]. 'Muddy brook and TŪN.' Cf. SCEARN.

Sherston W [*Scorranstan* 896 BCS 574, *Sceorstan* 1016 ASC (E), *Sorestone* DB, *Scorestan* 1168 P]. 'Stone on a steep ridge.' First el. an OE **scorra*, corresponding to OHG *scorro* 'steep declivity'. Cf. SCORA.

Sherwood Nt [*Sciryuda* (for *-wuda*) 958 YCh 3, *Shirewuda* 1164 P]. 'Wood belonging to the shire.'

Shevington La [*Shefinton* c 1225 CC, *Schevinton* 1288 Ipm]. 'TŪN by **Shevin* hill.' Localities called *Shevynlegh* and *Shevynhulldiche* are mentioned 1329 and 1362 nr S~, which stands on the slope of Shevington Moor. The hill-name may well be identical with the old form of the CHEVIN.

Sheviock Co [*Savioch* DB, *Sevioc* 1229 Fees, *Sheviok* 1306 FA]. No doubt identical with YSCEIFIOG in Flint, (Llanfihangel) ES-GEIFIOG in Anglesea [*Eskeyuyok* 1352 Rec Carn] and with *isceuiauc* c 1150 LL. *Tref ir isceiauc* ib. is translated into Latin as 'villa proclivii'. *Isceviauc* (Sheviock) would then mean 'slope' or 'sloping'.

Shide Wt [*Side* DB, *Schyde* 1287–90 Fees, *Schidhambrigge* 1324 Misc]. OE *scid* 'shide, piece of wood split thin'. The name may refer to a foot-bridge, as suggested by the ex. of 1324. The place is on the Medina.

Shields, South, Du [*Scheles* 1235 FPD], **North S~** Nb [*Chelis* 1268 Ipm, *Nortscheles* 1275 RH]. ME *schēle* 'a temporary building, a shepherd's summer hut, a shed'.

Shifford O [*Scip-*, *Scypford* 1005 KCD 714, *Scipford* DB]. 'Sheep ford.'

Shifnal Sa [*Scuffanhalch* 664 BCS 22, 675 ASC (E), *Shuffenhale* 1315 Ch]. '*Scuffa*'s HALH or valley.' *Scuffa* is unrecorded, but seems to have cognates in OG. The example of 664 is in a forged charter, but the form seems genuine.

Shilbottle Nb [*Siplibotle* 1228 FPD, *Shimplingbot'* 1238 Cl, *Schipplingbothill* 1242 Fees]. Second el. OE *bōtl* 'dwelling' &c. The first el. is probably OE *Sciplēaingas* 'the Shipley people'. There is a Shipley a few miles from Shilbottle.

Shildon Du [*Sciluedon* 1214 P, *Schilvedon* 1291 Tax]. OE *Scylfe-dūn* 'hill with a plateau or peak'. Cf. SCYLFE.

Shillingford D nr Exeter [*Selingeforde* DB, *Sullingford* 1234, *Sillingeford* 1242 Fees, *Syllingesford* 1267 Ass], S~ D in Bampton [*Sellingeford* 1180 P], S~ O [*Scillingeford* 12 (1316) Ch, *Scillyngforde* c 1200 Godstow]. A difficult name. It might be suggested that *Shilling-* is a stream-name derived from OE *sciell* 'resounding', but S~ O is on the Thames. OE *scilling* 'shilling' is not to be seriously considered. The first el. seems to be a tribal name *Scillingas*, which may be derived from OE *sciell* 'resounding'. *Scilling*, a fairly common pers. n., seems to be derived from this adj. or rather from a pers. n. **Sciell(a)* formed from *sciell*. But the triple occurrence of the name Shillingford calls for some special explanation. Possibly the Devon fords were named from the Oxford one.

Shillingstone. See OKEFORD.

Shillingthorpe Li [*Scheldintorp* 1193 P, *Scheldingthorp* 1276 RH, *Seldigtorp* 1212 Fees]. '*Skelding*'s thorp.' *Skeldyng* occurs among freemen of York. The name is no doubt Scandinavian. Cf. OSw *Skioldung*.

Shillington Bd [*Scytlingedune* 1060 Th, *Sethlindone* DB, *Scetlingedon* 1202 P]. The first el. seems to be a tribal *Scytlingas*, which may be derived from an unrecorded pers. n. **Scytla* or **Scyttel*. OE *Scytta* may be the first el. of *Scyttan mere (dun)* 774 BCS 216.

Shilstone D nr Modbury [*Silfestana* DB, *Silvestane* 1242 Fees], S~ D in Drewsteignton [*Selvestan* DB, 1238 Fees, *Shilston* 1263 Ipm]. Explained in PND as an OE *scylfe-stān* 'cromlech', *scylf* referring to the flat stone on top. The first el. is rather *scylfe* than *scylf*.

Shilton Brk [*Sculfton* 1205 BM, *Shilftun* 1242 Fees], **Earl S~** Le [*Sceltone* DB, *Sulton* 1209–35 Ep], S~ O [*Scylftun* 1044 KCD 775, *Schulton* 1254 Val], S~ Wa [*Scelftone* DB, *Selton* 1169, *Sulton* 1180 P]. The first el. is SCYLF or SCYLFE. Cf. SHELTON.

Shilvinghampton Do in Portisham [*Scilfemetune* DB, *Selfameton* 1212 Fees]. OE *Scylfhǣma-tūn* 'the TŪN of the people dwelling at a SCYLF' (slope or bank).

Shilvington Nb [*Schilington, -don* 1242 Fees, *Schilvyngton* 1346 FA]. 'The TŪN of the people at a SCYLFE.'

Shimpling Sf [*Simplingham* c 1035 Wills, *Simplinga, -ham* DB, *Scimplinge, Scimplingeham* c 1095 Bury], S~ Sf [*Simplinga* DB,

Simpling 1236 Fees, *Scimpling* 1275 RH].
'The people of *Scimpel*.' This unrecorded
pers. n. is a nickname formed from a word
corresponding to OHG *scimph* 'joke'.

Shincliffe Du [*Scinneclif* c 1085 LVD, *Sine-clive* 1195 (1335) Ch]. 'The cliff of the
spectre or demon, haunted cliff.' First el.
OE *scinna* 'spectre' &c.

Shinfield Brk [*Soanesfelt* DB, *Schiningefeld*
1167 P, *Sinningefelde* 1269 Hereford].
'*Scēne*'s FELD' and 'the FELD of *Scēne*'s
people.' Cf. SHENGAY.

Shingay. See SHENGAY.

Shingham Nf [*Siengham* 1198 FF, *Scing-ham* DB, *Sengham* 1207 Cur, *Shengham*
1254 Val]. 'The HĀM of *Scēne*'s people.' Cf.
SHENGAY.

Shingle Hall Ess nr Epping [*Chingledehall*
1272 Ch]. 'Hall with shingled roof.' *Shingle*
'wooden tile' is found in ME as *scincle*,
schyngle, *singel*. See SINGLETON.

Shinglewell K [*la Chingledewell* 1240 Ass,
Shingledewell 1327 FF]. 'Pebbly stream.'
Shingle 'pebbles' is found in the forms
chingle, *shingle* 16th cent. The word is
found in CHINGFORD, SINGLEBOROUGH, perhaps SINGLETON La. As suggested in PNK,
the name may alternatively mean 'a well
covered with a shingled roof'.

Shipbourne K [*Siburne* 11 DM, *Scipburn*
1198 FF, *Sipburne* 1242 Fees]. 'Stream
where sheep were washed.'

Shipden Nf [*Scepedane* DB, *Schipden* 1252
Cl]. 'Valley where sheep were kept.'

Shipdham Nf [*Scipdham*, *Scipedeham* DB,
Sipedham 1200 Cur, *Schipedham* 1254 Val].
The first el. might be a derivative of OE
scēap 'sheep' analogous to OE *eowde* from
eowu 'ewe'. If so, 'HĀM with a sheep-cote
or flock of sheep'.

Shipham So [*Sipeham* DB, *Schepham* 1291
Tax]. 'Sheep farm.'

Shiplake O [*Sciplak* 1236 Fees, *Schipelak*
1291 Tax], **Shiplate** So nr Bleadon Hill
[*Scypeladæspyll* 956 BCS 959, *Siplade* 1203
Cur]. 'Stream where sheep were washed.'
Second el. OE LACU and LĀD 'stream'.

Shipley Db [*Scipelie* DB, *Schippelæa* 1177
P], S~ Du [*Shepley* 1349 PNNb], S~ Nb
[*Schepley* 1236, *Schipley* 1242 Fees], S~
Sa [(æt) *Sciplea* 1002 Wills], S~ Sx [*Scape-leia* 1073 Fr, *Sepelei* DB], S~ YW [*Scipeleia*
DB, *Shepele* 1225 FF]. 'Pasture for sheep.'
Cf. LĒAH.

Shipmeadow Sf [*Scipmedu* DB, *Shipmedwe*
1254 Val]. 'Meadow for sheep.'

Shippen YW [*Scipene* DB, *Skipen* 1236 Cl],
Shippon Brk [*Sipene* DB, *Scipena* W 2
Abingd]. OE *scipen* 'cow-shed'.

Shipston on Stour Wa [*Scepuuæisctune*
764–75 BCS 205, *Scepwæsctun* 964 ib. 1135,

Scepwestun DB]. 'TŪN at a sheepwash.' In
BCS 205 is mentioned 'the ford of *Scepes-uuasce*'.

Shipton Lee Bk [*Sibdone* DB, *Scipdon* 1207
FF]. 'DŪN where sheep were kept.'

Shipton Gorge Do [*Sepetone* DB, *Sipton*
1236 Fees], S~ **Moyne** Gl [*Sciptone* DB,
Schipton Moine 1287 QW], S~ **Oliffe** &
Sollars Gl [*Scipetune*, *Sciptune* DB, *Schip-ton* 1303 FA], S~ **Bellinger** Ha [*Scep-tone* DB, *Shupton* 1270 Ch, *Shupton Beren-ger* 14 VH], S~ **on Cherwell** O [*Sceaptun*
1005 KCD 714, *Sciptune* DB, *Siptune super
Charewelle* 1213–28 Eynsham], S~ **under
Wychwood** O [?*Siptone* 777 BCS 222,
Sciptone DB, *Shupton under Wycchewode*
1391 AD], S~ **Sa** [*Scipetune* DB, *Shipton*
1291 Ch]. 'Sheep farm.' Cf. SCĒAP.

S~ **Bellinger** from the Berenger family.
Ingram Berenger held the manor in 1296 (VH).
Berenger is a Fr pers. n. of OG origin.—S~
Gorge was held by Thomas de Gorges in 1285
(FA). Gorges perhaps from GORGES nr Cou-
tances.—S~ **Moyne** was held by William le
Moygne in 1221 (Ruding). Cf. OWERMOIGNE.—
S~ **Oliffe & Sollars.** Thomas Olyve in
Shipton is mentioned in 1347 (Glouc), and
William de Solers in 1303 (FA). Cf. BRIDGE
SOLLERS.

Shipton YE [*Epton* DB, *Hyepton* 1176 P,
Sipton 1219 FF, *Shupton* 1234 FF], S~
YN [*Hipton* DB, *Hiepetuna* 1157 YCh 354,
Schupton 1180–7 ib. 550]. OE *Hēop-tūn*
'TŪN where hips or briars grew'. The sound-
development is similar to that in SHAP.

Shirburn O [*Scireburne* DB, *Sireburn* 1242
Fees]. See SHERBORNE.

Shirebrook Db [*Scirebroc* 1202 FF, 1230
P]. 'Bright stream.' Cf. SCĪR adj.

Shirenewton Monm [*Shirevesneuton* 1323
Pat]. 'The sheriff's Newton.'

Shireoaks Nt [*Scirakes* Hy 2 (1316),
Shirakes 1286 Ch]. 'The shire oaks,' referring to the shire meeting-place or some
oaks at its boundary. Cf. **Shire Oak** St.

Shireshead La [*Shireshead* 1577 Saxton].
'The upper end of the shire.' The place is
near the northern boundary of Amounderness hundred.

Shirland Db [*Sirelunt* DB, *Sirlund* 1199 P,
Schirlund 1226 FF]. 'Grove of the shire',
possibly an old meeting-place.

Shirlet Forest Sa [*Schirlet* 1172 P, *Sirlet*
1190 Eyton, *Shirlet* 1235, 1241 Cl]. 'The
share of the shire.' An OE *scirlett* actually
occurs in *bisceopes scirlett* 1038 E 239. Here
scir would mean 'office' rather than 'shire'.
Cf. KINLET.

Shirley Db [*Sirelei* DB, *Schyrelayg* 1230 P,
Schirleg 1247 Ch], S~ Ha nr Southampton
[*Sirelei* DB, *Schirle* 1316 FA], S~ Sr
[*Shyrley* 1461 FF], S~ Wa [*Shireleye* 1369,
1371 AD i]. 'LĒAH belonging to the shire'
or 'LĒAH where the shire moot was held'.

Shirwell D [*Sirewelle* DB, *Shirewill* 1242 Fees]. 'Clear spring.' Cf. SCĪR adj.

Shitlington YW [*Scellintone* DB, *Schitlingtona, Schetlintona* 1155 Pont, *Schitelington* 1145–60 YCh 1721]. First el. as in SHILLINGTON Bd.

Shobdon He [*Scepedune* DB, *Scobbedun* 1242 Fees, *Shobdon* 1265 Ch]. '*Sccobba*'s DŪN.' Cf. SHABBINGTON.

Shobrooke D nr Crediton [*Sotebroch* DB, *Sokebroc* 1215 Cl, *Schokebrocke* 1259 Ep]. Originally the name of a brook [*Sceocabroc* 938 BCS 726]. 'Goblin brook'; first el. OE *scucca* (*sceocca*) 'goblin'.

Shoby Le [*Seoldesberie* DB, *Siwaldebia* c 1125 LeS, *Siwaldeby* 1209–35 Ep]. '*Sigewald*'s BURG', with *burg* replaced by OScand BY, or '*Sigvaldi*'s BY'. *Sigvaldi* is an OScand pers. n.

Shockerwick So [*Sokerwicha* 1166 RBE, *Shokerwyk* 1243 Ass]. 'The WĪC of the shockers or people who pile sheaves into shocks.' *Shocker* is first evidenced in OED from 1820, but occurs as a surname c 1300 and is evidently an old word.

Shocklach Chs [*Socheliche* DB, *Schoclache* 1260 Court]. 'Goblin stream.' Cf. LÆCC, SHOBROOKE.

Shoebury (shōō-), **-ness** Ess [(to) *Sceobyrig* 894 ASC, *Soberia* DB, *Magna, Parua Schoberia* 1167 P]. First el. identical with SHEEN. The meaning seems to be 'sheltering, protecting BURG'.

Sholden K [*Shoueldune* 1176 BMFacs, *Seueldon* 1198 FF, *Schoweldon* 1242 Fees, *Shoueldon* 1251 Ch]. Second el. OE *dūn* 'hill'. The first may be OE *scofl* 'shovel' used in some transferred sense or referring to some fancied likeness of the hill to a shovel.

Sholing Ha [*Sorlinga* 1167 ff. P, *Sholling* 1251 Ch]. OE *scor-hlinc* 'hill on the shore'. Cf. SCORA, HLINC.

Shopland Ess [*Scopingland* c 1000 CCC, *Scopelanda* DB, *Scopiland* 1208 FF]. 'Island with a shed.' First el. OE *sceoppa*, ME *schoppe* 'shop, shed'. OE *sceoppa* is found once, apparently in the sense 'treasury'.

Shopwyke Sx [*Sepewica* c 1150 Fr, *Schepwich* 1188 P, *Shapewyk* 1318 Ipm]. OE *scēap-wīc* 'sheep farm'.

Shoreditch Mx [*Soredich* c 1148 AD, *Soresdic* 1183–4, *Sordig*', *Soredich* 1204 St Pauls, *Schoresdich* 1221 FF, *Soresdich* 1242 Fees, *Schoredich* 1235 FF]. 'Ditch leading to the shore' (of the Thames).

Shoreham K [*Scorham* 822 BCS 370, *Shorham* 1275 RH], **Old S~, S~ by the Sea** Sx [*Sorham* 1073 Fr, c 1100 Oxf, *Sore*(s)*ham* DB, *Shorham* 1156, *Sorreham* 1169 P]. 'HĀM at a rock or steep slope.' See SCORA.

Shoreston Nb [*Schoteston* 1177 P, *Schouteston* 1196 P, *Schotiston, Shetston* 1249 Misc]. The first el. looks like an OE *Scēot*, which may be a pers. n. formed from OE *scēot* 'quick'. The form with *r* is due to association with the word *shore*. The place is near the sea.

Shoresworth Nb [*Scoreswurthin* c 1085 LVD, *-wrthe* c 1170 FPD, *-worthe* 1195 (1335) Ch]. 'WORÞ(IGN) on a SCORA or steep slope.' Names in WORÞ mostly have pers. ns. as first el. This may have caused the first el. of S~ to have been taken for a pers. n. and a genitival *s* to be introduced.

Shorncote W [*Schernecote* DB, 1221 Ass, *Sernecote* 1235 Cl, *Scerncote* 1242 Fees]. 'COT in a muddy place.' Cf. SCEARN.

Shorne K [*Scorene* c 1100 Text Roff, *Shorna* 1159 P, *Shornes* 1176 BMFacs, *Schorene* 1193 Ep]. Connected with OE SCORA 'rock, steep slope' &c. The place is on the slope of a steep hill. An OE *scoren* (fem.), derived from OE *scorian* 'to project' and meaning 'projection', is possible.

Shortflatt Nb [*Le Scortflat* 1284, *Shortflat* 1324 Ipm], **Shortgrove** Ess [*Scortegraua* DB, *Shortegrava* 1208 Cur], **Shorthampton** O [*Sorthampton* 1242 Fees, *Schorthampton* 1291 Tax]. 'Short flat or furlong, grove, HĀMTŪN.'

Shorwell Wt [*Sorewelle* DB, *Schorewelle* 1287–90 Fees]. 'Hill stream.' Cf. SCORA. The stream at the place comes from uplands.

Shotesham Nf [*Shotesham* 1044–7 KCD 785, *Scotessam, Scotesham* DB, *Schotesham Omnium Sanctorum, Sancte Marie* 1254 Val]. '*Scott*'s HĀM.' *Scott* pers. n. (lit. 'the Scot or the Irishman') is found in *Scottes healh* 958 BCS 1036.

Shotford Nf [*Scotoford* DB], **S~** Sf [*Scotforð* c 950 Wills, *Shotford* 1291 Tax]. The places are on opposite sides of the Waveney. The name may mean 'ford of Scots'. But OE *gescot* 'contribution, payment' is perhaps preferable as first el. The name would refer to a ford where a toll was taken.

Shotley Nb [*Schotley* 1242 Fees, *Scoteley* 1256 Ass, *Schotley* 1262 Ipm], **S~** Sf [*Scoteleia* DB, *Soteleg* 1212, *Schottele* 1242 Fees]. S~ Nb is perhaps OE *Scotta lēah* 'the LĒAH of the Scots'. OE *cūsc*(*e*)*ote* means 'wood pigeon'. This suggests that there may have been a simplex *sc*(*e*)*ote* in a similar sense. This would give a good etymology: 'pigeon wood'. **Shotley Bridge** Du is near S~ Nb.

Shotover O [*Scotorne* DB, *Sotora* 1130, *Sottour*' 1230 P, *Shotovr*' 1231 Cl]. Second el. OE *ofer* 'hill, slope'. S~ is at Shotover Hill. First el. as in SHOTTLE.

Shottermill Sr [*Shottover* 1537, *Schotouermyll* 1607 PNSr]. *Shotover* is probably a family name.

Shottery Wa [*Scotta rið* 704–9 BCS 123, *Scotriðes gemæro* 1016 KCD 724]. 'The brook of the Scots or Irishmen.' Cf. RĪÞ. The brook is called *Scotbroc* KCD 724, and in the same charter is mentioned *Scothomm*.

Shottesbrook Brk [*Sotesbroc* DB, *-broch* 1167 P, *Sc(h)otebroc* 1190 P, *Sotebroc* 1230 P]. OE *sceota* means 'trout' (cf. Engl *shoat*). Shottesbrook may well mean 'trout stream'. The common early genitival *s* may be intrusive.

Shotteswell Wa [*Soteswalle, -well* c 1135 Fr, *Shoteswell* 1166, *Schotewella* 1188, *-well* 1190 P]. Second el. OE *wella* 'spring'. The first may be OE *Scott* pers. n. Cf. SHOTES-HAM.

Shottisham Sf [*Scotesham* DB, *Sotesham* 1186 P, *Schatesham, Snetesham* 1254 Val, *Shettisham* 1313 Ipm]. Hardly identical with SHOTESHAM. More likely the first el. is as in SHORESTON.

Shottle Db [*Sothelle* DB, *Schethell* 1191 ff. P]. In OHG pl. ns. occurs an el. *sciez* 'slope, steep place'. The word belongs to the word for 'shoot' and means literally 'place that shoots down or up'. The base is *skeut-*, which would give OE *scēot*. This word probably existed in OE and is found in Shottle, which thus means 'hill with a steep slope, steep hill', in SHOTOVER and some of the next names.

Shotton Du nr Easington [*Sottun* c 1165 YCh 653], **S~** Du nr Grindon [*Sceottun* c 1050 HSC, *Siotona* 1183 BoB, *Shotton* 1249 Cl], **West S~** Du [*Scottun* c 1050 HSC], **S~** Nb [*Sothune* 1196 P, *Schotton* 1242 Fees], **S~ in Glendale** Nb [*Scotadun* c 1050 HSC, *Schotton* 1291 Tax]. The last is 'hill of the Scots', the name referring perhaps to some incident in Border warfare. The fact that *Shotton* is only found in Du and Nb may suggest that the other names should be interpreted as 'TŪN of the Scots', and this is perhaps the correct explanation in some cases. But Shotton nr Stockton is rather *Scēot-tūn* or even *Scēot-dūn* 'TŪN on the slope', or 'steep hill'. A similar explanation is possible for the rest except S~ in Glendale. Cf. SHOTTLE.

Shotwick Chs [*Sotowiche* DB, *Sotewica* c 1100, c 1150 Chester, *Rowheschetewyk* Hy 2, *Rowe Shetewyke* Hy 3 BM, *Schetowyca* c 1235 Chester, *Schotewic* 1214–23 Chester]. The spellings *Soto-* DB, *Scheto-* Chester suggest that the first el. is a compound containing OE *hōh* 'ridge' &c. Shotwick is at the foot of the Wirral ridge. The name would then be 'wīc at *Scēothōh*', the latter meaning 'steep ridge'. Cf. SHOTTLE.

Shouldham Nf [*Sculham* 1043–5 Wills, *Sculdeham* DB, *Schuldham* 1177 P, *Shuldham* 1251 Ch]. The first el. may be a non-mutated form of OE *scyld* 'debt, due'. *Scultheta* actually occurs once for *scyldhæta*. If this is right, Shouldham might mean

'gavelland' or the like. It might be compared with YELDHAM.

Shoulton Wo [*Selgeton, Scolegeton* c 1220 PNWo]. The first el. looks like OE *scēolēge* 'squinting', used as a nickname, but may also be an alternative name of LAWERN BROOK, which is very winding, identical with SHELLOW.

Showell O. See SEWELL.

Shrawardine Sa [*Saleurdine* DB, *Shrewardin* 1165, *Shrawurdin* 1166 P, *Srawurthin* 1212 Fees]. OE *Scræf-worþign* 'WORÞIGN at a SCRÆF or hollow'. The *scræf* may be **Shrawardine Pool**.

Shrawley Wo [*Scræf-, Screfleh* 804 BCS 313, *Escreueleia* c 1150 Surv, *Schraveleg* 1212 Fees]. 'LĒAH by a SCRÆF', *scræf* referring perhaps to the recess in the hill close by.

Shrewley Wa [*Servelei* DB, *Shreveleg* 1198 Fees, *Shreueleg* Hy 3 Ipm, *Shrewele* 1285 QW]. 'The LĒAH of the sheriff' (OE *scirgerēfa*). Forms like *Scraueleia* 1150, *Scravele* 12 PNWa(S) may indicate original identity with SHRAWLEY. If so, the name seems later to have been understood as 'the sheriff's LĒAH'.

Shrewsbury (-ōz-) Sa [*civitas Scrobbensis* 901 BCS 587, (at) *Scropesbyri* 1006 ASC(F), (into, on) *Scrobbesbyrig* 1016 ib. (D, E), 1102 ib. (E), *Sciropesberie* DB, *Salopesberia* 1094–8 Fr]. '*Scrobb*'s BURG.' *Scrob* is evidenced as the name of the father of the Richard who built Richards Castle in Herefordshire c 1050. This Scrob is often said to have been a Norman, but if so, *Scrob* is very likely an English nickname given him. *Scrob* is related to Fris *scrob* 'brushwood', Norw *skrubb* 'a gruff person' &c., and probably meant about the same thing as the Norw word. It is usual to derive the first el. of Shrewsbury from a word related to Engl *shrub* and meaning 'brushwood'. It has even been suggested that the name is a translation of the Welsh name of the place, viz. *Pengwern*. But that name means 'the end of the swamp'. *Scrobbesburg* seems to have become *Shrovesbury*, whence *Shrows-bury*. The spelling *Shrewsbury* arose on the analogy of words like *shrew, shrewd*, which were formerly often pronounced alternatively as *shrow, shrowd*.

Shrewton W [*Winterburn Shyreveton* 1236 Ch, *Schereneton, Schreveton* 1281 QW]. 'The sheriff's manor.' The original name was WINTERBOURNE.

Shrigley, Pott, Chs [*Schriggel* 1285, *Shriggeleg* 1288 Court]. 'Wood frequented by shrikes' (OE *scric* 'shrike', a bird). *Pott* is a family name.

Shrivenham Brk [*Scriuenham, Scrivenan-hom* 821 BCS 366, (to) *Scrifenanhamme* c 950 Wills, *Seriveham* DB, *Scriveham* 1212 Fees]. Second el. HAMM. The first appears to be the past part. of OE *scrifan* 'decree, allot, adjudge (sentence), impose (penance)'. The exact meaning of the el. is obscure.

Shropham Nf [*Scerpham, Scerepham, Screpham* DB, *Schrepham* 1166 f. P, *Shropham* 1231 FF, *Scropham* 1242 Fees, *Shorpham* 1283 Ipm]. The variation in the forms seems to point to a first el. with the diphthong *eo* or *ēo*, an OE *scrēop* or *sceorp*. OE *sceorp* means 'dress', which does not seem to be suitable in a pl. n. OE *scrēop* is not recorded. Nothing definite can be suggested.

Shropshire [*Scrobbesbyrigscir* 1006 ASC (E), (into) *Scrobsæton* 1016 ASC (C), *Scropscir* 11 Th, *Sciropescire* DB, *Scrobscyr* 1087 ASC (E), *Salopescira* 1094–8 Fr, 1156 P]. An elliptical form of *Scrobbesbyrigscir* 'the shire with Shrewsbury as its head'. *Salopescira* (whence *Salop*) is a Normanized form.

Shroton. See IWERNE.

Shrubland Sf [*le Scrublond* n.d. AD, *Shrubblund* 1301 FF, *Shribland* 1557 BM]. OE *scrybb* 'shrub, shrubs' and OScand *lundr* 'grove'.

Shuckburgh Wa [*Socheberge* DB, *Succheberga* 1163 BM, *Schuckeberwe* 1242 Fees], **Shucknall** He [*Shokenhulle* 1377 Ep]. 'Goblin hill, haunted hill.' First el. OE *scucca* 'demon, goblin'. See BEORG.

Shudy Camps. See CAMPS.

Shulbrede Priory Sx [*Shelebred* 1212 Cur, *Schulbrede* 1261, *Scheluebred* 1316 PNSx]. 'Strip of land on a slope.' Cf. SCYLF, BREDE.

Shunner Howe YN [*Senerhou* 1223 FF, *Shonerhowes* 15 Whitby]. ON *Siónar-haugr* 'look-out hill'. ON *sión* is 'view'. ON *si* (*sj*) became Engl *sh*.

Shurdington Gl [*Surditona* c 1150 BM, *Schurdentone* 1148, *Scerdintona* c 1170 Glouc]. The first el. is a derivative of OE *sceard* 'gap' or *sceard* adj. 'notched', e.g. an OE *scierde* 'gap, pass' or the like. S~ is nr S~ **Hill**, which is separated from a larger hill by a pass.

Shurton So [*Shur(r)eveton* 1219 f., *Shireveton* 1228 Dunster]. 'The sheriff's manor' (OE *scirgerēfa*).

Shustoke Wa [*Scotescote* DB, *Shitestok* 1241 Cl, *Sustock* 1242 Fees]. Perhaps '*Scēot*'s STOC'. Cf. SHORESTON.

Shute D [*Schieta* c 1200 HMC Rep 4, *Schete* 1228 FF, *La Shete* 1242 Fees]. See SHEAT.

Shutford O [*Schiteford* 1148–66 RA, *Shutteforde* 13 AD, *Shetteford* 1283 Ch]. Perhaps 'the archer's ford' (OE *scytta* 'archer') or '*Scytta*'s ford'. Cf. SHILLINGTON Bd.

Shutlanger Np [*Shitelhanger* 1163 P, 1220 Fees, *Sutelhangra* 1186 P, *Scetelhangre* 1197 P]. OE *scytel* 'shuttle, arrow, bar, bolt' and *hangra* 'slope, wood', perhaps 'wood where shuttles were got'.

Shuttington Wa [*Cetitone* DB, *Schetynton* 13, *Schutinton* 1327 PNWa]. Perhaps 'the TŪN of *Scēot*'s people'. Cf. SHUSTOKE.

Shuttleworth La in Bury [*Suttelesworth* 1227 FF, *Shitleswurth* 1246 Ass]. There are two more Shuttleworths in La, and one in YW. The name consists of OE *scyttels* 'bar, bolt' and WORÞ. The name seems to mean 'enclosure made of bars of a certain kind'.

Sibbertoft Np [*Sibertod* DB, *Sibertoft* 1198 Fees, Hy 3 BM]. '*Sigbiorn*'s TOFT.' First el. ODan *Sigbiorn*, OSw *Sighbiorn*, ON *Sigbiǫrn* pers. n.

Sibdon Carwood Sa [*Sibetune* DB, *Sibbetone* 1166 RBE]. '*Sibba*'s TŪN.' Carwood is near S~.

Sibertswold K [*æt Swyðbrihteswealde* 940, at *Sibrighteswealde* BCS 755, 797, *Sibertesuuald* DB]. '*Swiþbeorht*'s WALD or wood.' The loss of *w* is due to dissimilation. Cf. also SACOMBE.

Sibford Ferris & Gower O [*Sibbeford* 1153 TpR, 1231 Cl, *Parva Sibeford* 1242 Fees, *Sibbeford Goyer* 1220 Fees]. '*Sibba*'s ford.' S~ **Ferris** belonged to the Ferrers fee before 1153 (TpR). Cf. BERE FERRERS.—S~ **Gower** was held by Thomas Guher before 1231 (Cl). Guher is a Fr form of OG *Guother*.

Sibsey Li [*Sibolci* DB, *Cibeceia, Sybeceia* 1151–3 DC, *Cybezay* 1233 Ch]. '*Sigebald*'s island.' See ĒG.

Sibson Hu [*Sibestune* DB, *Sibston* 1233 FF]. '*Sibbi*'s TŪN.'

Sibson or **Sibstone** Le [*Sibetesdone* DB, *Sibbedesdone* 1220, *Sybbedesdone* 1230 Ep]. '*Sigebed*'s DŪN.'

Sibthorpe Nt [*Sibetorp* DB, *Sibbetorp* J BM]. '*Sibba*'s or *Sibbi*'s thorp.' First el. OE *Sibba* or ODan *Sibbi*.

Sibton Sf [*Sibbetuna* DB, *-tun* 1156 P]. '*Sibba*'s TŪN.'

OE **sic.** See GUSSAGE, SYKEHOUSE. Cf. KELSICK.

Sicklinghall YW [*Sidingale* DB, *Sicclinhala* c 1150 Crawf, *Sikelingehal* 1220 FF]. 'The HALH of *Sicel(a)*'s people.' OE *Sicel(a)* is related to *Sica* in *Sicanburh* BCS 1023.

OE **sīd** adj. 'broad, spacious' is the first el. of SIDEBEET, SIDESTRAND, ?SINFIN, SYDENHAM, SYDERSTONE, SYDLING.

Sid R D [*Side* c 1250 Ol, *Syde* 1284 BM]. A derivative of OE *sīd* adj., possibly in the sense 'broad', but rather in such a sense as 'low-lying, running in a deep valley'. Hence **Sidbury** D, named from an old earthwork [(æt) *Sidebirig* c 1070 Ex, *Sideberia* DB], **Sidford** D [*Sideford* 1238 Ass], **Sidmouth** D [*Sidemuða* 1072–1103 E, *Sedemude* DB].

Sidbury Sa [*Sudberie* DB, *-beri* 1176 P]. OE *Sūþburh* 'southern BURG'.

Sidbury Wo. See sŪp.

Sidcup K [*Cetecopp* 1254 Ass, *Setecoppe* 1301 Subs]. Second el. COPP. The first may be OE *set* 'camp'.

Siddington Chs [*Sudendune* DB, *Sudingdone* 1286 Court, *-ton* ib.]. OE *be sūpan dūne* '(place) south of the hill'. An elliptical name. Cf. Introd. p. xviii.

Siddington Gl [*Sudi(n)tone* DB, *Suentune* W 1, *Suthintuna* 1146 Fr]. OE *sūp in tūne* 'south in the village, the southern part of the village'. Cf. SINTON, SODINGTON, and Introd. p. xviii.

OE **sīde** 'side', ME *side* also 'slope of a hill, esp. one extending for a considerable distance'. See SYDE, FACIT, FAWCETT, LANGSETT, WHERNSIDE, SITTINGBOURNE.

Sidebeet La in Rishton [*Sydebiht* 1278 FF]. 'Wide bend.' See sĪD, BYHT.

Sidestrand Nf [*Sistran* DB, *Sidestrande* R 1 Cur]. 'Broad shore.' See sĪD.

Sidford. See SID.

Sidlesham (-s-) Sx [*Sidelesham* 683, *-stede* 714 BCS 64, 132, *Sydelesham* 1227 Ch]. '*Sidel*'s HĀM.' *Sidel* is a hypocoristic form of names in *Side-*, as *-man*, *-wine*. Cf. OG *Sito*, *Situli*, *Sidimund* &c.

Sidlow Bridge Sr [*Sideluue melne* 12 BM]. First el. an unrecorded OE woman's name *Sidelufu*.

Sidmouth. See SID.

Siefton or **Seifton** (sēfn) Sa [*Sireton* DB, *Ciraton* 1086 Eyton, *Siveton* 1257 Ch, *Syueton* E 1 BM]. If *r* in the first two forms is misread for *v* or *f*, the first el. may be OE *Sigegifu* or even *Sægifu*, both women's names.

Sigglesthorne YE [*Siglestorne* DB, *Sighelesthorn* 1246 FF]. '*Sighulf*'s thorn-bush.' First el. OSw *Sighulf*, ON *Sigolfr*.

Signet O [*Senech* 1285 Ass, *Seynat*, *Saynet* 1289 Ipm, *Seynet* 1307 Ipm]. OE *senget* 'place cleared by burning' is suggested in PNO(S), but it is not quite clear how this could have become *Seynet*. Cf. however **Saint Hill** D [*Sengethill* 1249 PND].

Sigston YN [*Sighestun* DB, *Siggestune* 1088 LVD]. '*Sigge*'s TŪN.' First el. ON *Siggi*, OSw *Sigge*.

Silchester Ha [*Silcestre* DB, *-cestra* 1167 P, *Sele-*, *Silechæstre* 1205 Layamon, *Cilcestr'* 1233 Cl, *Cylchestre* 1278 Pat]. The first el. may be a derivative of OE *sealh* 'sallow', OE **siele*, **sele* 'sallow copse'. Cf. SELBY, SELHAM. Silchester was a Roman station, the Brit name being *Kalēoua* Ptol, *Calleva* IA. This name is related to Welsh *celli* 'wood'. It is possible that the first el. is really a worn-down form of Brit *Callēva*. This might have become PrimE *Calliw*, with change of *ē* to *i* as in KENNET from

Brit *Cunētiō*, and the *w* might have been lost as in DEE from Brit *Dēva*. The further development would have been from *Calli* to *Cealli* (with fracture of *a* before *ll*) and from that to *Cielli* (owing to *i*-mutation), whence OE *Ciell* and later *Cill*. Spellings such as *Sele-* might be due to association with OE *sele* 'hall' or 'sallow copse'.

Sileby Le [*Siglesbie*, *Siglebi*, *Seglebi* DB, *Siglebia* c 1125 LeS]. '*Sighulf*'s BY.' Cf. SIGGLESTHORNE.

Silecroft Cu [*Selecrotf* c 1200 StB, *-croft* 1213 P]. 'Croft where sallows grew.' First el. ON *selia* 'sallow'.

Sil Howe YN, a hill nr Goathland [*Sylehou* 1108–14, Hy 2 Whitby]. The first el. may be ON *sigla* 'a mast', here used of a boundary mark or a landmark. Cf. names such as *Siglunes*, *-vík* in Iceland. Second el. ON *haugr* 'hill'.

Silk Stream. See SULH.

Silkby Li, lost, cf. SILK WILLOUGHBY [*Silkebi* 1212, *-by* 1242 Fees, *Selkeby* 13 FF]. '*Silki*'s (*Selki*'s) BY.' *Selki*, *Silki* is an ON byname from *selki* 'young seal'.

Silkmore St [*Selchemore* DB, *Selkemor* 1230 Cl], **Silkstead** Ha [*Silkested* 1243 Pp, *Selkstede* 1243 Ep, *Selkestede* 1285 Ch, *Sulkestede* 1316 FA]. An el. *selk-*, *silk-* is found in several other names. Near Silkstead is mentioned *Silkeley* (wood, pasture) 1243 Pp, *Selkeleye* 1243 Ep. **Selkley** (hd) W is *Selkelai* 1170 P. *Selkeden* 1198 FF, *Selkleg* 1228 FF were in Sf. The el. is probably found in OE *siolucham(m)* 990 KCD 673 (Ha). *Sioluc* may be a derivative of an OE **sēol*, **siol* (< *sihula-*), a sideform of EFris, MLG *sīl* 'a drain, canal' (< *sihila-*). The words belong to OE *sēon* 'to filter, flow'. *Sioluc* may have meant 'a small drain, rill'.

Silkstone YW [*Silchestone* DB, *Silcheston* 1167 P, *Silkiston* 1195 P], **Silksworth** Du [*Sylceswurthe* c 1050 HSC, *Silkeswrthe* c 1180 FPD]. '*Sigelāc*'s TŪN and WORP.'

Silloth Cu [*Selathe* 1299 Sc]. 'Barn on the sea.' Cf. HLAÐA.

Silpho YN [*Sifthou* 1145–8, *Silfhou* 12, 1230 Whitby, *Silfho* 1231 Ass]. Most probably an OE *Scylf-hōh* 'ridge with a peak or with a plateau', Scandinavianized. As the sound *sh* was unknown to early Scandinavian, an *s* might be substituted for it. The first el. is really OE *scylfe*. See SCYLF.

Silsden YW [*Siglesdene* DB, *-den* 1314 Ipm, *Sighelesden* 1303 FA]. The first el. might be a brook-name (OE *Sigol*) derived from OE *sigan* 'to sink, move'. Second el. DENU.

Silsoe Bd [*Siuuilessou*, *Sewilessov* DB, *Siuelisho* 1175 P, *Sivelesho* 1199 FF], **Silsworth** Np nr Watford [*Sivelesworth* 1220 Fees, 1329 Ch]. Both seem to contain an unrecorded pers. n. **Sifel*, cognate with OHG *Sibilo*. See HŌH, WORP.

Silton Do [*Seltone* DB, *Selton* 1194, 1230 P, *Salton* 1291 Tax, *Sylton* 1412 FA]. First el. probably as in SELBY, SILCHESTER, i.e. an OE *siele 'sallow grove'.

Silton, Nether, YN [*Silftune* DB], **Over S~** YN [*Silftune, Siluetune* DB, *Parva Silton* 1301 Subs]. OE *Scylf-tūn* (cf. SHELTON, SHILTON) 'TŪN on a SCYLFE or hill'. S- for Sh- owing to Norman influence. One of the Siltons is probably referred to as *Shilton* 1231 FF.

Silverdale La [*Selredal* 1199 Ch, *Celverdale* 1292 QW, *Silverdale* 1382 FF]. 'Silver valley', the name referring to the silver-grey rocks in the place.

Silverley Ca [*Seuerlai* c 1080 ICC, *Seiluerleia* 1086 IE, *Severlai* DB, *Selverleia* Hy 2 BM, *Silverleg* 1228 FF]. First el. doubtless the word *silver*. Many names of plants and trees contain the word, as *silver-weed, -wort*. Silverley might be elliptical for a name containing such a word. See LĒAH.

Silverstone Np [*Sulueston* 942 BCS 773, *Silvestone, Selvestone* DB, *Seluestona* 1130, *-tuna* 1156 P, *Silveston* 1200 (1260) Ch]. '*Sæwulf*'s or *Sigewulf*'s TŪN.'

Silverton D [*Sulfretone* DB, *Seluerton* 1180 P, *Sulfertone* 1242 Fees, *Syluerton* 1249 FF]. The numerous forms with *f* suggest that the first el. is OE *sulhford* or *sylhford* 'ford in a gully'. *Sulford* actually occurs as the name of a locality not far from Silverton in BCS 723, but it is not certain that S~ was named from just this Sulford.

Silvington Sa [*Silvintone* a 1118 Eyton, *Silviton* 1154 Mon, *Silvynton* 1291 Tax]. A Normanized form of OE *Scylfe-tūn* or *Scylfinga-tūn*, 'TŪN on a slope' or 'the TŪN of the dwellers on the slope'.

Simene R. See SYMONDSBURY.

Simonburn Nb [*Simundeburn* 1229 Ep, *Symundesburn* 1291 Tax], **Simonside** Du in Monk Wearmouth [*Simundset* 12, *Symondset* 1276 FPD], **S~** (-ī-) Nb nr Rothbury [*Simundessete* 1279 Ass], **Simonstone** La [*Simondestan* 1278 Ass], **S~** YN [*Simundestan* 1195 ff. P], **Simonswood** La [*Simundeswude* 1207 P]. '*Sigemund*'s stream, (GE)SET or fold, stone and wood.'

Simpson Bk [*Siwinestone* DB, 1230 Ep, *-ton* 1237–40 Fees]. '*Sigewine*'s TŪN.'

Sinderby YN [*Senerebi* DB, *Sinderbi* 12 Pudsay, *-by* 1231 FF]. Either '*Sindri*'s BY' or rather 'southern BY', the first el. being Dan *søndre*, Norw dial. *syndre* 'southern'. S~ is south of Pickhill. *Sindri* is a well-evidenced ON name.

Sinderland Chs [*Sundreland* DB]. See SUNDERLAND.

Sindlesham Brk [*Scindlesham* 1242 Fees, *Syndlesham* 1365 AD]. Possibly the first el. is an OE *Synnel, derived from *Sunna* pers.

n. *Synnles* would become *Syndles*. Second el. HĀM.

Sinfin Db [*Sydenefen* 1251 Ch, *Seden-, Sudenfen* 1297 f. Ipm, *Sidenfen* 1322 FF]. ?'Broad fen.' Cf. SĪD.

Singleborough (sĭngkl-) Bk [*Sincleberia* DB, *Singleberghe* 1106–9 Fr, *-berge* c 1155 Oxf]. 'Shingle hill.' Cf. SHINGLEWELL.

Singleton La [*Singletun* DB, *Syngelton* c 1190 LaCh, *Schingelton* 1169 ff. P]. Either 'TŪN with shingled roof' or 'TŪN on shingly soil'. Cf. SHINGLE HALL, SHINGLEWELL.

Singleton Sx [*Silletone* DB, *Sengelton* 1185 P, *Schingelton* 1181 P, *Sangelton* 1327 Ipm]. *Sængelwicos* BCS 144 seems to have some connexion with *Singleton*. The first el. is taken in PNSx to be dial. *sangle* 'a handful of ears of corn, sheaf', identical with LG *sangele*. The meaning would be 'brushwood' or the like. But such a sense is not evidenced, and the derivation of the first el. from a word *sengel* 'burnt clearing' suggested by Anderson, *Hundred-names* iii. 70 f., is doubtless correct. *Sengel*, found also in the old name of COWDRAY PK Sx [*La Cengle* 1273, *La Sengle* 1284 Ipm], is related to OE *sengan* 'to burn'.

Sinnington YN [*Siuenintun, Sevenictun* DB, *Sivelington* 1201 Ch]. 'The TŪN of the dwellers on R SEVEN.'

Sinton, Leigh, Wo [*Sothyntone* (in) *Lega* 12 AD]. See SIDDINGTON. S~ is in the south end of LEIGH par.

Sipson Mx [*Sibwineston* c 1150 AD 3236, *Sibeston* 1341 FF]. '*Sibwine*'s TŪN.'

Sisland (sīz-) Nf [*Sislanda* DB, *Sigeland* 1206 Cur, *Sislond* 1254 Val]. '*Sigehēah*'s or *Sige*'s land.' *Sige* would be a short form of names in *Sige-, -sige*.

Sissinghurst K [*Saxingherste* c 1180 Arch Cant vi, *Saxingeherste* 1206 FF]. 'The hurst of *Seaxa*'s people.' The change of *x* to *ss* may be due to Norman influence.

Siston Gl [*Sistone* DB, *Sixtune* 1240 PNGl, *Syston* 1247 Ch]. First el. as in SISLAND.

Sithney Co [*St. Sythyn* 1230 FF, *Sanctus Sydnius* 1270 Ep]. A saint's name, identical with Bret *Sezny*.

Sittingbourne K [*Sidingeburn* 1200 FF, 1227 Fees, *-burne* 13 BM, *Sidingburn* 1279, *Sythynggeburn* 1296 Ep]. The first el., OE *Sidingas*, may be a derivative of OE *side* 'slope'. S~ is on the lower slope of a long ridge. Second el. OE *burna* 'stream'.

Sixhills Li [*Sisse* DB, *Sixla* c 1115 LiS, *Sixlei* 1196 Cur, *Sixele* 1212 Fees]. 'Six clearings or glades' (OE LĒAH). Cf. FILEY.

Sizergh (sīzer) We [*Sigredeshergh* 12 PNCu, *Sigaritherge* c 1175 Kendale]. 'The ERG of *Sigríðr*.' ON *Sigríðr* is a woman's name.

Sizewell Sf [*Syreswell* 1240 Ass (Reaney), *Syswell* 1280 Bodl]. First el. probably OE *Sigehere*; cf. SYERSTON.

ON **skáli** 'a hut, shed, temporary building', ME *scale* 'a hut or temporary shelter, a shed' is common in the pl. ns. of Scandinavian England, esp. the North. It mostly appears in the plural form. *Skáli* alone is often used as a pl. n., as **Scales** Cu (several), S~ La nr Aldingham [*Scales* 1269 Ass], S~ La in Ribbleton [*Ribelton Scales* 1252 Ch], **Newton with S~** La [*Skalys* 1501 CC], **Scole** Nf [*Escales* 1191 P], **Scholes** YW in Barwick in Elmet [*Scales* 1258 Ipm]. First el. in SCAFELL, SCAWDALE, SCAWTON, SCHOLAR GREEN. Second el. in BOWSCALE, BRINSCALL, ELLISCALES, PORTINSCALE, WINSCALES, WINSKILL, GATESGILL, SOSGILL and many others.

ON **skarð** 'cleft, mountain pass' is found in some names, as AYSGARTH, SCARTHO. In cases like SCARCLIFF, SCARCROFT, SCARTHINGWELL, ON *skarð* may have replaced an OE *sceard*.

Skeckling YE [*Scachelinge* DB, (in) *Scachelingis* c 1100 YCh 1300, *Eschechilinga* 1115 ib. 1304, *Skekling* 1228 Ep]. Very likely a Scandinavianized form of an OE **Scæclingas* or the like, which is possibly derived from the OE *scacol* postulated for SCACKLETON. *Scacol* would here mean 'promontory'. Skeckling is in the eastern part of the Holderness peninsula.

Skeeby YN [*Schirebi* DB, *Schittebi* 1187 P, *Skidebi* 1208 Cur]. Identical with SKIDBY.

Skeffington Le [*Sciftitone* DB, *Sceftinton* c 1125 LeS, 1165, 1192 P]. A Scandinavianized form of OE *Sce(a)ftinga tūn* 'the TŪN of *Sceaft*'s people'. Cf. SHAFTESBURY.

Skeffling YE [*Sckeftling, Sceftlinga* 1150–76 YCh 1399, *Sheftling* 1154–60 YCh 1825, *Sceflinges* 1204 FF, *Sheftelyng* 1301 AD i, 1338 Ch]. An OE *Sceftlingas* 'the people of **Sceftel* or **Sceftla*'. Cf. *Sceaftel* in *Sceaftles ora* 956 BCS 982. The base is OE *Sceaft* (see SHAFTESBURY).

Skegby Nt nr Mansfield [*Schegebi* DB, *Skegeby* 1227 Ch], S~ Nt nr Normanton upon Trent [*Scachebi* DB, *Skeggeby* 1230 P], **Skegness** Li [*Schegenesse* 1166 P, *Skegenes* 1256 BM]. '*Skeggi*'s BY and ness.' First el. ODan, ON *Skeggi* pers. n.

ON **skeið** 'a race-course'. See BRUNSTOCK, HESKET, HESKETH, WICKHAM SKEITH. SCAITCLIFFE does not contain this word.

Skelbrooke YW [*Scalebro* DB, -*broc* 1163 P, *Scelebroch* c 1170 YCh 1548, *Schelebrok* 1230, *Skelebrok* 1253 Ep], **Skellow** YW [*Scanhalle* DB, *Scalehale* 1180–95 YCh 1585, *Skelehall* 1204, *Skelhal* 1243 FF]. The two places are near each other. The first el. may be ME *schéle* (OE **scél* or **sceol*) 'shieling, hut', which was Scandinavianized to *skél-* and, owing to association

with ON *skáli*, to *scale-*. Second el. BRŌC, HALL or HALH.

Skell R YW [*Scel* 12, *Skell* 13 Fount]. A derivative of ON *skiallr* 'resounding'. Cf. SKELWITH. **Skeldale** [*Scheldale* c 1207 Fount], **Skelden** [*Scheldene* 1175 Fount] mean 'the valley of the Skell'.

Skellingthorpe Li [*Scheldinchope, Schellingop* DB, *Scellinghou* 1138, *Eschellingho* 1160–5 NpCh, *Scheldinghop* 1141 RA, *Scheldinghou* 1202 Ass, *Skeldinghop* 1238 Ep]. The original second el. was HOP in the sense 'dry land in a fen'. The first is probably an OE *Sceldingas* '*Sceld*'s people', Scandinavianized. Cf. SHELSLEY.

Skellow. See SKELBROOKE.

Skelmanthorpe YW [*Scelmertorp* DB, 1195 P, *Skelmertorp* 1242 Fees], **Skelmersdale** La [*Schelmeresdele* DB, *Skelmersdale* 1202 FF], **Skelsmergh** We [*Scelmeresherhe* c 1190, *Skelmeresherge* Hy 3 Kendale]. '*Skelmer*'s thorp, valley, and ERG or shieling.' *Skelmer* may go back to an OScand *Skialdmarr*.

Skelton Cu [*Sheltone* c 1160 YCh 175, *Schelton* 1187 P], S~ YE [*Scilton, Schilton* DB, *Skeltun* 1199 FF], S~ YN in Marske [*Scelton* 12 PNNR, 1272 Ipm], S~ YN nr Saltburn [*Scheltun, Sceltun* DB, *Sceltona* c 1175 YCh 660], S~ YN nr York [*Sc(h)eltun* DB, *Skelton* 1181–4 YCh 423], S~ YW nr Boroughbridge [*Scheltone* DB, *Sceltona* c 1175 YCh 124], S~ YW nr Leeds [*Sceltun* DB, *Scheltuna* c 1160 YCh 1770]. A Scandinavianized form of OE *Scylftūn, Scelftūn*. Cf. SHELTON, SHILTON. The meaning of the first el. varies between 'hill', 'bank' and the like.

Skelwith (skĕlĭth) La [*Schelwath* 1246 Ass, *Skelwath* 1332 Subs]. 'Ford by Skelwith Force.' Second el. ON *vað* 'ford'. The first is an old name of the waterfall, an ON *Skiallr*, derived from *skiallr* 'resounding, roaring'. Cf. SKELL.

Skendleby Li [*Scheueldebi* DB, *Scendelbi* Hy 2, *Schendelebia* c 1150 DC]. The DB form may stand for *Scheneldebi*, which might point to a first el. OE *Scéne-helde* 'beautiful slope'. If so, *Sk-* is due to Scand influence.

Skenfrith Monm [*ynys Gynwreid* c 1400 Brut y Tywyssogion, *Ysgynfraith* 16 Owen's Pembrokeshire, iii, 291, *Kenefrid* 1190, 1193 P, *Skynefrith* 1291 Tax]. The name means 'the island of Cynwraidd', the elements being W *ynys* 'island' and *Cynwraidd* pers. n. The stress fell on the first syllable of the second el., and the initial syllable of *ynys* appears to have been misunderstood as the prep. *yn* 'in'. S~ is on the Monnow.

ON **sker** 'rock, reef', Norw *sker* also 'rocky hill', Engl *scar* 'rock, crag, bed of rough gravel' &c. See SKERTON, RAVENSCAR, PRESTON UNDER SCAR.

Skerne R Du [*Schyrna* c 1190, *Scirne* 13 FPD], **Skerne Beck** YE [*Shyrne* 1240 Cl, *Skiren* 1260 Ass]. Probably a derivative of OE *scir* 'bright, clear' with *Sk-* owing to Scand influence. On Skerne Du is **Skirningham** [*Skirningheim* c 1090 SD, *Schirningaham* Hy 2 FPD]. 'The HĀM of the dwellers by the SKERNE.' On Skerne Beck is **Skerne** YE [*Schirne* DB, *Skirne* 1222 FF].

Skerton La [*Schertune* DB, *Skerton* 1201 P]. 'TŪN by the reefs', the reference being to some sand-banks in the Lune. Cf. SKER.

Skewsby YN [*Scoxebi* DB, *Scousby* 1226 FF]. OScand *Skōgs-bȳr* 'BY at a wood'. Cf. SKÓGR.

Skeyton (-ī-) Nf [*Scegutuna* DB, *Sceketuna* c 1150 Fr, *Scegeton* 1191 P]. '*Skeggi*'s TŪN.' Cf. SKEGBY.

ON skialf. See SCYLF.

Skibeden. See HEBDEN.

Skidbrook Li [*Sc(h)itebroc* DB, c 1115 LiS, *Scitebroc* 1230 P, *Skydbrok* 1328 BM]. OE *Scite-brōc* 'dirty brook', Scandinavianized.

Skidby YE [*Scyteby* c 972 BCS 1279, *Schitebi* DB]. '*Skyti*'s BY.' *Skyti*, lit. 'archer', occurs as an ON byname.

Skiddaw· Cu [*Skithoc* 1230 FF, *Skythou* c 1252, *Skiddehawe* 1256 CWNS xxi]. Second el. ON *haugr* 'hill'. The first might be ON *skyti* 'archer', but is perhaps rather an ON word related to Norw *skut* 'projecting crag', *skuta* the same, Sw *skuta* in names of hills. No word with *i*-mutation belonging to this group is known, however.

Skilgate So [*Scheligate, Schiligata* DB, *Schillegat* 1195 P, *Skilegate* 1243 Ass]. 'Boundary gate.' First el. *skill* vb. 'to separate', a Scand word found early in English (*scylian* 1049 ASC, MS C). *Scylget* is found 956 BCS 970 (Do).

Skillington Li [*Scillintun* c 1067 Wills, *Schillintune* DB, *Schillingetona* 1146 RA]. A Scandinavianized form of OE *Scillinga-tūn*. For the first el. cf. SHILLINGFORD.

Skinburne·ss Cu [*Skyneburg* 1175, *Schineburgh* 1185 Holme C, *Skinburness* 1319 Cl]. The name *Skyneburg* may be a Scandinavianized form of an OE *Scinnan burg* 'the BURG of the demon, haunted castle'. Cf. SHINCLIFFE. Second el. OScand *nes* 'headland'.

Skinnand Li [*Schinende* DB, *Schinande* 1185 TpR, *Skinand* 1230 P, *Skynende* 1242 Fees]. OScand *skinandi*, pres. part. of *skina* 'to shine', i.e. 'the bright one'. *Skinandi* may have been a name of the Brant.

Skinningrove YN [*Scineregrive* c 1175 YCh 893, *Skinergrive, Scinergreve* 1272 Ipm]. 'The skinners' pit.' *Skinner* and *gryfia* 'pit' are Scand words.

Skiplam YN [*Scipnum, Skipenum* 12 Riev]. A Scandinavianized form of the dat. plur. of OE *scipen* 'shippon, cow-shed'.

Skippool La [*Skippull* 1593 PNLa]. Named from a stream [*Skippoles* 1330 PNLa]: 'ship stream'. Skippool was formerly a harbour. *Skip* is OScand *skip* 'ship'.

Skipsea YE [*Skipse* c 1170 YCh 1356, *Scipse* 1226 Ep, *Skipse* (mere) 1260 Ipm, *Shipse* 1294 Ch]. 'Lake for ships, harbour for ships.' Either a Scandinavianized OE *Scip-sǣ* or an OScand *Skip-siōr*.

Skipton on Swale YN [*Schipetune* DB, *Skipton super Swale* 1242 Fees], **S~** YW [*Scipton* DB, 1190 P, *-tun* 1151 SD, *Skipton* 1260 Ipm]. A Scandinavianized OE *Sciptūn* 'sheep farm'. Skipton YW is still famous for its sheep markets.

Skipwith YE [*Schipewic* DB, *Scipewiz* 1166 P, *Scippewic* 1200 FF, *Skipwith* 1291 Tax]. OE *scip-wīc* 'sheep farm', Scandinavianized. OScand *viðr* 'wood' has replaced OE wīc.

Skirbeck Li [*Schirebec* DB, *Scirebec* 1157 YCh 354, *Scirebech* 1168 P]. OScand *skiri bekkr* 'bright brook'.

Skirfare R YW [*Scirphare* n.d. Fount 350]. 'Bright stream.' The elements are OScand *skirr* 'bright, clear' and a word for 'brook' derived from OScand *fara* 'to go'. Cf. the ON river-name *Fara*.

Skirlaugh YE [*Scherle, Scir(e)lai, Schireslai* DB, *Scirlaga* c 1155 YCh 1346, *Skirlagh* 1240 FF]. A Scandinavianized form of SHIRLEY.

Skirlington YE [*Schereltune* DB, *Skirling-tona* c 1150 YCh 1306, *Skirlinton* 1232 FF]. Probably a Scandinavianized form of OE *Scirlēainga-tūn* 'TŪN of the SHIRLEY people'.

Skirmett Bk [*la Skiremote* c 1307 PNBk]. A modified form of OE *scīrgemōt* 'shire moot'.

Skirningham. See SKERNE.

Skirpenbeck YE [*Scarpenbec* DB, *Scarpin-bec* 1157, *Scerpingbec* c 1165, *Scirpincbec* c 1170 YCh 354, 831, 834]. First el. an OScand *skerping*, a derivative of *skarpr* 'dry, barren' meaning 'barren land'. The word is the source of the common Norw pl. ns. *Skjerping, -en*. Second el. OScand *bekkr* 'brook'.

Skirwith Cu [*Skirewit* 1205 FF, *Skirwyth* 1304 Cl]. A Scandinavianized form of OE *Scir-wudu* (cf. SHERWOOD). OScand *viðr* is equivalent to OE *wudu*.

ON skógr, OSw skōgher, ME *scogh* 'wood' is the first el. of SCORBROUGH, SKEWSBY, and the second el. of some names, as AISKEW, BRISCOE, BURSCOUGH, HADDISCOE, MYER-SCOUGH, SWINSCOE, THURNSCOE.

Skutterskelfe YN [*Codreschelf* DB, *Scuðer-schelf* 1176 P, *Skotherskelfe* 1301 Subs].

The first el. is probably a stream-name *Skvaðra* 'chattering brook', a derivative of *skvaðra* 'to chatter'. But *Skvaðra* (*Skoðra*) also occurred as a byname in ON and is a possible source. The second el. is ON *skialf* in such a sense as 'ledge, bank'.

OE **slā(h)** 'sloe' is found in SLOLEY, SLAUGHAM. The corresponding OScand word (OSw *slān*, Dan *slaa*) is found in SLAITHWAITE. OE *slāhþorn* 'black-thorn' is the first el. of SLAUGHTERFORD W, perhaps SLAIDBURN.

OE **slæd** 'valley, dell'. See BAGSLATE, CASTLETT, WEEDSLADE, SLEDDALE, SLEDMERE.

OE **slæp** is only found in charters. It is related to OE *slipor* 'slippery', OHG *slîf(f)an* 'to glide' &c., ON *sleipr* 'slippery' (whence Engl dial. *slape*), MLG *slêpen*, OHG *sleifen* 'to drag, lug'. *Slæp* is usually held to mean 'slippery or miry place'. This suits some of the places with names containing the el., as RUISLIP and SLEPE, the old name of ST. IVES Hu [*Slepa* 974 BCS 1310, *Slepe* DB], but for some, as the ISLIPS, a meaning 'portage, place where boats or other objects are dragged' would be more suitable. Cf. ISLIP and DRAYCOTT &c. Such a sense would easily develop from 'slippery place'. In HANSLOPE 'slope' would be a suitable meaning. See also POSTLIP, SLAPTON. Löfvenberg concludes from the surname *de la Slape*, *atte Slape* 1276 ff. (So) that there was an OE *slæp* 'mud, mire, marsh'. Such a word may alternatively be the first el. of SLAPTON.

Slaggyford Nb [*Chaggeford* 1218 P, *Slaggiford* n.d. AD i, *Slaggingford* 1267 PNNb]. 'Muddy ford.' The first el. is an adj. *slaggy* derived from ME *slag* 'slippery with mud, muddy'.

Slaidburn YW [*Slateborne* DB, *Sleiteburna* 1154 YCh 1475, *Slaiteburn* 1229 Ep, *Slaghteburne* 1294 Ch]. Possibly OE *slāhþorn-burna*, whence *Slahterburna* and, with dissimilatory loss of the first *r*, *Slaghteburne*. See SLĀ(H). A better etymology may be obtained, however, if the SW. and WMidl word *slait* 'a level pasture, a down, a sheepwalk' (OE *slæget*) can be assumed for the early Yorkshire dialect. See Löfvenberg, *Studier i modern språkvetenskap* (Stockholm) xvii. 87 ff.

Slaithwaite YW [*Sladweit* 1178, *Slathwait* 1191 P, *Sclagtwayt* 1277, *Slaghthayth* 1286 Wakef]. 'Clearing where sloes grew.' First el. OE *slāh* 'sloe' or its OScand equivalent. Cf. THWAITE.

ON **slakki** 'shallow dell or valley' is found in some names as WITHERSLACK.

Slaley Nb [*Slaveleia* 1166 RBE, *Slaule* 1170 P, *Slaueley* 1242 Fees]. 'Muddy LĒAH.' The first el. is an OE *slæf* or the like, corresponding to Dan *slaf* 'mud' and related to Engl *slabber* &c. It is found in **Sladen** La [*Slaueden* 13 WhC] and *Slauilache* CC.

Slapton Bk [*Slapetone* DB, *Slaptone* 1222 Ep], S~ D [*Sladone* DB, *Slapton* 1244 Ass], S~ Np [*Slaptone* DB, *-ton* 12 NS, 1236 Fees]. 'TŪN by a SLÆP', perhaps 'miry place'. Cf. SLÆP.

Slaugham (-ăf-) Sx [*Slacham* a 1100 PNSx, *Slagham* 1248 Ass, *Slaucham* 1272 FF]. 'HĀM or HAMM where sloes grew.' Cf. SLĀ(H).

Slaughter, Lower & Upper, Gl [*Sclostre* DB, *Slochtra* 1169, *Sloc(h)tres* 1190 P, *Sloughtre superior* 1291 Tax]. Near S~ is **Slaughterford** [*Slohtranford* 779, *Slohterword* 949 BCS 230, 882]. OE **slōhtre* 'slough, muddy place'. The word seems only to be found in English in these names, but the corresponding word is found in German names, as SCHLÜCHTERN [*Sluohterin* 1025].

Slaughterford W [*Slachtoneford* 1176 P, *Slahtreford* 1291 Tax, *Slaghteneford* 1298 Cl]. OE *slāhþorn-ford* 'ford by the blackthorn(s)'.

Slawston Le [*Slagestone, Slachestone* DB, *Slaghestuna* 1106-23 (1333) Ch, *Slaeston* 1163 P, *Slaweston* 1242 Fees]. 'Slag's TŪN.' ON *Slagr* is found as a byname.

Sleaford Li [*Slioford* 852 BCS 464, *Sliowaford* 852 ASC (E), *Eslaforde* DB, *Sliforde* c 1200 Gervase]. 'Ford over R Slea.' **Slea** (OE **Sliow* or the like) is closely related to SCHLEI in Germany [*Slia* in early sources], also the first el. of SCHLESWIG [*Sliaswic* Saxo]. The names are related to Norw *sli* 'slime', OE *slēow, sliw* 'tench' and mean 'slimy, muddy stream'.

Sleagill We [*Slegill* 1208 FF, *Slegile* 1279 Ass, *Sleuegile* 1294 Ch]. Cf. *Slævdal* in Norway, which is held to have as first el. ON *slefa* 'saliva'. *Sleagill* may have as first el. a stream-name derived from this word ('trickling stream'). Cf. GIL.

Sleddale We nr Kendal [*Sleddale* 1229, 1246 Kendale], S~ We nr Shap [*Sleddall* 1249 CWNS xiv]. First el. OE *slæd* 'valley' (in Yks now *sled*). The old name of the valleys was no doubt *Slæd* (*Sled*) and an explanatory *dale* was added by Scandinavians.

Sledmere YE [*Slidemare* DB, *Sledemere* 1166 P, c 1165 YCh 1084]. 'Lake in a valley.' Cf. prec. name.

Sleekburn Nb [*Sliceburne* c 1050 HSC, *Sclikeburn* 1182 P]. 'Muddy stream.' First el. dial. *sleech, slitch, sleek* 'mud' (*slike* 1375).

Sleightholme (-ĕt-) YN [*Slethholm* 1234 FF, 1254 Pat], **Sleights** (-ĕt-) YN [*Slechetes, Sleghtes* c 1223 Whitby]. *Sleight* is ON *slétta* 'a level field', dial. Engl *sleet* 'a flat meadow' &c. *Slétta* is from **slehta*. This early form is preserved in the Engl names. Cf. HOLM.

Sleningford YW [*Sleaninga ford* c 1030 YCh 7, *Scleneforde, Sclenneford* DB, *Sclein-*

ingeford 1204 FF]. The first el. is a folk-name *Slēaningas*, which is of doubtful etymology. It might possibly be a derivative of an OE *slēa* (gen. *slēan*), cognate with Norw *slaa* 'grass-grown slope' (from an earlier *slah-*). Such a sense suits the locality.

Slimbridge Gl [*Heslinbruge* DB, *Slimbrugia* 1153 Berk, *-brug* 1220 Fees, *-brig* 1225 FF]. 'Bridge in a muddy place.' First el. OE *slim* 'slime, mud'.

Slindon St [*Slindone* DB, *-don* 1242 Fees], **S~** Sx [*Eslindone* DB, *Slindon* 1188 P]. The first el. may be identical with SLYNE or with the first el. of SLINFOLD. Second el. OE DŪN 'hill'. The meaning may be 'sloping hill'.

Slinfold Sx [*Stindefald* 1166 P, *Slindefold* 1226 FF, 1251 Cl]. 'Fold on a slope.' The first el. seems to be an old word corresponding to OSw *slind* 'side'. The meaning would be 'side of a hill, slope'. A word *slind* is actually evidenced in Sussex in the surname *de Slinde* 1332 Subs (Bexhill).

Slingley Du [*Slingelawe* 1155 FPD]. 'Hill where animals were snared.' First el. Engl *sling* 'loop, noose or snare', corresponding to G *Schlinge*. See HLĀW.

Slingsby YN [*Selungesbi, Eslingesbi* DB, *Slengesby* 12 Whitby]. '*Sleng*'s BY.' First el. an ON byname *Slengr*; cf. Norw *sleng* 'idler'.

Slipton Np [*Sliptone* DB, *-ton* 1167 P]. First el. perhaps OE *slyppe* 'paste, slime'. If so, 'muddy TŪN'.

OE **slōh** 'slough, mire'. See SLOUGH, POLSLOE. A derivative *slōhtre*, no doubt with the same meaning, is found in SLAUGHTER.

Sloley Nf [*Slaleia* DB, *Slalee* 1207 Cur, *Sloleye* 1254 Val], **S~** Wa [*Slalea* 1174 P, *Sloleye* 1408 AD]. OE *Slā-lēah* 'LĒAH where sloes grew'.

Sloothby Li [*Slodebi* DB, 1202 Ass, *Slothebi* 12 DC]. '*Slōði*'s BY.' *Slōði* is an ON pers. n., originally a nickname. Cf. Norw *slode* 'a clumsy person', Icel *slóði* 'a good-for-nothing'.

Slough (-ow) Bk [*Slo* 1196 Cur, (del) *Slo* 1196 P]. OE *slōh* 'slough, mire'.

Slyne (-ī-) La [*Sline* DB, *Slina* 1177 P, *Sline* 1203 P]. An OE *slinu* 'slope', related to Norw *slein* 'gently sloping ground' and *slind* in SLINFOLD.

Smales Nb [*Smale* 1279 Ass], **Smalesmouth** Nb. Both are on Smales Burn, which must have been called *Smale* 'the narrow one', from OE *smæl* 'narrow'.

Smallburgh Nf [*Smaleberga* DB, 1177 P, *Smalberge* 1147–9 Holme]. 'BEORG on river *Smale*.' *Smale* is an old name of the ANT [*Smale, Smallee* 1363 Works]. Cf. prec. name. OE *beorg* may here mean 'bank'.

Smalley Db [*Smalleage* 1009 Hengwrt MS 150, *Smalei* DB, *Smalleg* 1226 FF, 1242 Fees], **Smallwood** Chs [*Smaleuuod* 1252 Ch]. 'Narrow LĒAH and wood.' First el. OE *smæl* 'narrow'.

Smardale We [*Smeredal* 1197 f. P, *-dale* 1202 FF], **Smarden** K [*Smeredaenne* 11 DM, *Smeredenne* 1332 Ch]. 'Butter valley and pasture', i.e. 'valley and pasture where there was good grazing'. First el. OE *smeoru* 'fat, grease', also 'butter', as in *smeorumangestre* 'butter-woman'. See DÆL, DENN.

Smaws YW nr Tadcaster [*Smahus* 1200 P, 1234 FF, *Smawes* 1260 Ipm]. OScand *smā-hūs* 'small house(s)'. First el. OScand *smār* (ON *smár*, OSw *smār*) 'small, little'.

Smeaton, Great, YN [*Smiþatun* 966–92 BCS 1255, *Smidetune* DB], **Little S~** YN [*Smidetune* DB, *litle Smithetune* 1088 LVD], **Kirk & Little S~** YW [*Smedetone* DB, *Smydetona* c 1150 YCh 1494, *Smetheton* 1286 Ipm]. 'The smiths' TŪN.' The form *Smethe-* is partly due to the OE gen. plur. *smeoþa*, where *eo* arose from *i* owing to velar mutation.

Smeeth K [*Smiða* 1018 BM, *Smeth* 1246 Ipm, *Smethe* 1279 Ep]. OE *smiþþe* 'smithy', also *smeðe* c 1030, *smeþe* 15 OED. The form with *e* seems to come from the inflected form *smiþþan*, which could become *smeoþþan*.

Smeetham Hall Ess [*Smedetuna* DB, *Smetheton* 1199 P, 1230 Ch], **Smeeton Westerby** Le [*Smitetone* DB, *Smitheton* 1208 BM, *Smytheton Westerby* 1316 FA]. 'The smiths' TŪN.' Cf. SMEATON, WESTERBY.

OE **smeoru** 'butter'. See SMARDALE. It is found also in **Smercote** Wa [*Smerecote* DB, 1285 Ch] and **Smerrill** Db [*Smerehull* 1272 Rutland]. 'Butter COT and hill.'

Smestow Brook St [*Smethestall* 1300 For, *Smestall* 1577 Saxton]. *Smethestall* must be the old name of a pool in the TRYSULL. The name means 'smooth (i.e. still, stagnant) pool'. Second el. OE *stall* in *wætersteal* 'stagnant water, smooth place in a river'. First el. OE *smēþe* 'smooth'.

Smethcote Sa nr Hadnall [*Smethecot* 1242 Fees, *Smedecote* 1256 Ass], **Smethcott** Sa [*Smerecote* DB, *Smethecot* 1327 Subs]. 'The smiths' cottage.' Cf. SMEATON.

OE **smēþe** 'smooth'. See SMESTOW, SMITHDOWN, SMITHFIELD, SMITHILLS.

Smethwick Chs [*Smethewyk* 1331 Ormerod], **S~** (smĕdhĭk) St [*Smedeuuich* DB, *Smethewic* 1221 Ass]. 'The smiths' dwelling.' Cf. SMEATON.

Smisby Db [*Smidesbi* DB, *Smiðesbi* 1166 P]. 'The smith's BY.'

Smite R Le, Nt [*Smyte* 1280 Ass, 1316 Pat], **S~ Brook** Wa. *Smite* is a common old river-name. One that gave its name to **Smite** in Wo is (to) *Smitan* 978 KCD 618.

Another in Wilts is (of) *Smitan* 854 BCS 477. The name is derived from OE *smitan* in a non-recorded sense 'to glide, slip'. Cf. ME *smite* 'to move rapidly', Sw *smita* 'to slip away' and the like. The name means 'gliding stream'. On Smite Brook Wa was Smite hamlet [*Smitha* DB, *Smite* 1251 Ch].

OE **smiþ** 'smith' is a common first el. of pl. ns. Cf. SMEATON, SMEETON, SMETH- (passim). SMISBY contains the corresponding ON *smiðr*, OSw *smiþer* &c. OE *smiþþe* 'smithy' is found in SMEETH, HAMMERSMITH.

Smithdown La [*Esmedune* DB, *Smededon* 1185 P, *Smethedon* 1202 P], **Smithfield** Mx [*Smethefelda* c 1145 AC, *-feld* 1197 FF], **Smithills** La [*Smythell* 1322 LaInq, *Smethehill* 1506 DL]. 'Smooth DŪN, FELD and hill.' First el. OE *smēþe* 'smooth'.

Smockington Wa [*Snochantone* DB]. If the isolated form is reliable, the name means 'homestead at the end of the ridge', the first el. being ME *snōke* 'a projecting point or piece of land, a promontory', which is found as *snoca* 956 BCS 959. S~ stands at the end of a prominent ridge. See HĀMTŪN.

OE **snæd, snåd** is only found in charters, and its meaning is variously given as 'piece of land', 'clearing', 'a piece of woodland'. The last sense is based on a passage in BCS 442 (A.D. 843): 'unus singularis silva . . . quem nos theodoice *snad* nominamus'. The exact meaning of this is hardly quite clear, but it may well point to the meaning given. The el. is found in SNEYD, SNEATON, HAL-SNEAD, KINGSNORTH, PENSNETT, WHIPSNADE. Cf. SNAITH.

Snailwell Ca [*Sneillewelle* c 1050 KCD 907, *Sneileuuelle* c 1080 ICC, *Snegeluuella* 1086 IE, *Sneilewella* 1193 P]. 'Stream frequented by snails.' First el. OE *snægl, snegl* 'snail'.

Snainton YN [*Snechintune* DB, *Snechint'* 1166 P, *Snekinton* 1168 P, *Snegintona* 1158 YCh 402, *Sneintun* 1247 Ch]. Perhaps OE *Snocinga-tūn* 'TŪN of *Snoc*'s people'. Cf. SNORSCOMB and SNEINTON.

Snaith YW [*Esneid, Esnoid* DB, *Sneid* 1169 P, *Snaith* c 1110 YCh 43, 1249 Ch]. ON *sneið* 'a piece'. But Snaith may be a Scandinavianized form of OE SNĀD, SNÆD.

Snape Sf [*Snapes* DB, 1254 Val], **S~** YN [*Snapa* 1154 YCh v]. An OE *snæp* of doubtful meaning is found BCS 1124 in the phrase '*andlang dun and snæp*'. This el. is found in several minor pl. ns. in the south (D, Sx). One ex. ('*unam snappam terre*' c 1200) quoted in PNSx suggests a meaning such as 'piece of land', and the plural form of Snape Sf in early sources would go well with such a meaning. In the north *snape* is common in pl. ns., e.g. in La, and the probable meaning is here 'pasture', perhaps 'inferior pasture, winter pasture'. This has been derived from ON *snap* 'scanty grass for sheep to nibble at in snow-covered

fields' or 'poor grazing'. This is probably the source of SNAPE YN, BLACKSNAPE.

Snarehill Nf [*Snareshul* DB, *-hel* c 1095 Bury, *-huelle* 1168 P]. Apparently '*Snear*'s hill'. **Snear* may be a pers. n. derived from OE **snear* 'swift' (cf. OE *snierian* 'to hasten'). Cf. ON *Snari* from the equivalent ON *snarr*.

Snarestone Le [*Snarchetone* DB, *Snarkeston* 1188 P, 1242 Fees, *Snargeston* 1196 P]. The first el. is clearly a pers. n., which may be a derivative of *Snear* in SNAREHILL, an OE *Snaroc*. Cf. OG *Snaracho, Snarung*.

Snarford Li [*Snardesforde, Snerteforde* DB, *Snarteforde* c 1115 LiS, *Narteford* 1163-5 BM, *Snarteford* Hy 3 BM, 1231 Ep]. '*Snort*'s ford.' First el. ON *Snǫrtr*, gen. *Snartar*.

Snargate K [*Snergathe* c 1197 Penshurst, *Sneregate* 1242 Ch, Hy 3 BM, *Snere-, Snaregate* 1290 Ep]. First el. probably OE *sneare* 'snare'. If so, 'gate where snares for animals were placed'.

Snave K [*Snaues* 1182 P, *Snathes* 1202 FF, *Snaves* 1235 Cl, 1242 Fees]. Possibly an OE *snafa* or the like, cognate with OHG *snabul*, MLG *snavel* 'bill, beak' &c. and meaning 'spit of land, narrow strip'. Norw *snav* means 'spit of land'.

Sneachill. See SNETTISHAM.

Sneaton YN [*Snetune* DB, *Snetona* 1100-15 YCh 857, *Snetton* 1163 ff. P]. OE *Snæd-tūn* 'TŪN by a piece of woodland'. Cf. SNÆD.

Sneinton (-ĕ-) Nt [*Notintone* DB, *Snotinton* 1166, 1169, 1197 P, *Snainton, Sneynton* 1230 P]. OE *Snotinga-tūn* 'the TŪN of *Snot*'s people'. Cf. NOTTINGHAM. The place is close to Nottingham.

Snelland Li [*Sneleslunt* DB, *Snelleslund* c 1115 LiS, 1202 Ass], **Snelshall** Bk [*Snelleshal* 1226 Pat, 1228 Ch], **Snelson** Chs [*Senelestune* DB, *Snelleston* 1271-4 Chester], **Snelston** Db [*Snellestune* DB, *-ton* 1177 P]. '*Snell*'s LUND or grove, HALH, and TŪN.' The first el. of *Snelland* is very likely ON *Sniallr*. That of the rest is more likely the corresponding OE *Snell*, a derivative of *snell* 'quick, active' and found in independent use and in pl. ns., as *Snellescumb* 854 BCS 476, *Snelles hamm* 932 ib. 691.

Snetterton Nf [*Snetretuna* DB, *Sneterton* 1192 ff. P, *Snitertona* 1195 FF, *-ton* 1254 Val]. '*Snytra*'s TŪN.' **Snytra*, also found in SNITTERTON and in SNITTERLEY, the old name of Blakeney Nf, is a derivative of OE *snottor, snytre* 'wise'.

Snettisham Nf [*Snet(t)esham* DB, *Snetesham* 1161 P, 12 BM, 1196 FF]. '*Snæt*'s HĀM.' OE *Snæt* is found in (to) *Fnætes wyllan* (for *Snætes wyllan*, now **Sneachill** Wo) 977 KCD 612 and (with loss of *S-*) in NEATISHEAD. It is related to OG *Snato, Snazi*. But the quantity of the vowel is doubtful, and it is possible we have to assume the OE forms *Snæt (in Sneachill) and *Sneti (in Snettisham, Neatishead).

Sneyd (-ē-) St [Sned 1256 Ch]. See SNÆD.

Snibston Le [Suipestona c 1125 LeS, Snipeston 1201 Cur, 1209–35 Ep]. 'Snip's TŪN.' ON Snípr occurs as a byname.

Snilesworth YN [Snigleswath c 1160 YCh 1846, Snileswath 1230 FF]. The elements are ON snigill, Swed snigel 'snail', here very likely used as a byname, and OScand vað 'ford'.

OE **snīte** 'snipe'. See SNITTERFIELD, SNYDALE.

Snitter Nb [Snitere 1176, Snittera 1177 P, Snitter 1242 Fees]. The place is on a narrow ridge in an exposed situation. The name may belong to ME sniteren 'to snow', dial. snitter 'a biting blast', but the formation of the name is not clear. Cf. BLAND.

Snitterby Li [Snetrebi, Esnetrebi DB, Snitrebi c 1115 LiS, Sniterbi 1212 Fees]. 'Snytra's BY.' Cf. SNETTERTON.

Snitterfield Wa [Snitefeld DB, Esnitevele c 1135 Fr, Snytenefeud 1257 Ch]. OE Snitena-feld 'FELD frequented by snipes'. See SNĪTE.

Snitterton Db [Sinitretone DB, Sniterton 1232 Cl, Smuterton Hy 3 BM]. Identical with SNETTERTON.

Snoddington Ha in Shipton Bellinger [Snodintone DB, Snodington 1242 Fees]. 'The TŪN of Snodd's people.' Snodd is found in SNODLAND, UPTON SNODSBURY and Snodeswic 1002 Wills. The name is derived from snod 'smooth, sleek, even', though this word is not found until the 15th cent. (OED).

Snodhill He [Snauthil 1195, Snathil 1196 P, Snodehull c 1225 BM, Snodhull 1242 Fees]. The earliest form suggests 'snowy hill', the first el. being an OE *snāwede 'snowy'. S~ is on the slope of a high hill.

Snodland K [Snoddingland 838 BCS 418, Snodingcland 964–95 ib. 1132, Esnoiland DB]. 'The land of Snodd's people.' Cf. SNODDINGTON.

Snoreham Ess [Snorham 1245 FF, 1254 Val]. The first el. seems to be identical with **Snore Hall** Nf [Snora DB, 1212 Fees]. The names presuppose an OE *snōr, for which no definite etymology can be suggested. Cf. SNOW HILL Street-names 180 [Snore Hylle Hy 3].

Snoring, Great & Little, Nf [Snaringes DB, 1166 RBE, Snarringes 1200 Cur, Parva Naringes 1242 Fees, Snoring 1314 Ch]. 'Snear's people.' Cf. SNAREHILL. Or the Stiffkey, on which the places are, may have been OE Sneare (from *snear 'swift'). OE snear comes from *snarha-. The vowel would be lengthened alternatively when the h was lost.

Snorscomb Np [Snoces cumb 944 BCS 792, Snochescumbe DB, Snokescumb 1220 Fees, Snotescumbe 1275 Ipm]. Probably '*Snoc's CUMB or valley'. Alternatively the first el. might be an OE snōc, corresponding to OSw snōkr, Dan snog 'snake'. ME snoke 'projecting piece of land' (see SMOCKINGTON) was OE snōca (gen. snōcan).

Snowford Wa [Snawford 1001 BM]. First el. OE snāw 'snow'.

Snowshill Gl [Snawesille DB, Snoweshull 1251 Ch]. The gen. form of the first el. tells against a meaning 'snow hill', but is not definitely against it. S~ is by a hill of 921 ft. OE Snāw pers. n. seems to be recorded in **Snosmeres** Sr [Snawes mere 956 BCS 955]. Snahard LVD seems to contain OE snāw 'snow'. In Scand languages the word for 'snow' occurs in pers. ns., as ON Snæbiǫrn, Snæulfr, and even ON Snær, ODan Snio are recorded.

Snydale YW [Snitehale DB, Snithal 1219 FF]. 'Haugh frequented by snipes.' Cf. HALH and SNITTERFIELD.

Soar R [Sora 1147 Monm, Hy 2 DC, Sore 1247 Ass]. A Brit river-name, probably identical with SAAR and SERRE on the Continent [Sara in early sources]. The name belongs to the root ser- 'to flow' in Lat serum 'fluid' &c.

Soberton Ha [Sudbertune DB, Suberton 1167, 1190 P, Soberton 1291 Tax]. 'South barton.' Cf. BARTON, SURBITON.

Sock Dennis So [Soca, Soche DB, Soc c 1180, Sok Hy 3 BM, Sok Deneys 1257 FF], **Old Sock** So in Mudford [Vetus Stoke 1316 FA, Oldesok 1359 BM]. The two places are a good three miles apart on a stream. The name is OE soc 'sucking', found in þas soces seað 932 BCS 691 in a topographical sense. Soc belongs to OE sūcan 'to suck', socian 'to soak' and may well have been used in the sense 'marsh, quagmire', lit. 'a marsh that sucks up things'. Probably the reference is here rather to a marsh than to the stream.

Sock Dennis was held by Johannes Dacus in 1236 (Fees). Dacus is a Latinized form of OFr deneis 'Danish'.

Sockbridge We [Sokebrec a 1184 CWNS xxiv, Socabret c 1180 Kendale, Sochebred 1279 Ipm]. Probably OE soca-bred 'a board serving as a footbridge over marshy spots'. Cf. SOCK. Second el. OE bred 'board'.

Sockburn Du [(æt) Soccabyrig 780 ASC (E), Socceburg c 1050 HSC, 1104–8 SD, Sockburn c 1130 YCh 944]. 'Socca's BURG.' *Socca is a derivative of OE socc 'sock'. Cf. ON Sokki.

Sodbury, Chipping & Old, Gl [(æt) Soppan byrg 872–915 BCS 582, Sopeberie DB, Sodbury Mercata 1316, Olde Sobbury 1346 FA]. 'Soppa's BURG.' *Soppa, also found in SOPWORTH Gl and in SOPLEY, SOPWELL, corresponds to OG Suppo.

Sodington Hall Wo [*Supintuna gemæru* c 957 BCS 1007, *Sudtone* DB]. OE *sūþ in tūne* 'south in the town (village), the southern part of the village', an elliptical name. Cf. SIDDINGTON. *Supintuna* in the OE ex. is the gen. of *Sūpintūne*, a folk-name meaning 'the dwellers south in the village'.

Softley Du [*Softelawe* c 1200, *-leie* 13, *-ley* 1242 Finchale], **S~** Nb [*Softeley* 1277 PNNb]. 'Soft LĒAH.' The epithet would refer to soft, spongy soil.

Soham Ca [*Sægham* c 995 BCS 1289, *Saham* c 1080 ICC, DB, 1156, 1190 P, *Seham* 1260 Ch], **Earl & Monk S~** Sf [*Saham* 11 EHR 43, DB, 1195 Cur, *Earl, Monks Saham* 1235 FF, *Saham Comitis* 1254 Val]. OE *Sǣ-hām* or *Sā-hām* 'HĀM by the lake'. On *Sā-* see SǢ. There are no lakes at the places now, but in DB the lake (*lacus*) of Soham Ca is mentioned. The form *Sægham* is curious. Possibly the *g* is a glide developed between *æ* and *ā*.

Earl S~ was held by the Earl of Norfolk, **Monk S~** by Bury St. Edmunds.

OE sol 'muddy place, wallowing-place for animals'. The latter sense is obvious in *heortsol* BCS 204. The word is found in **Soles** K in Nonington [*Soles* DB, 1201 Cur]. See BRADSOLE, GRAZELEY, SOLWAY MOSS, SOYLAND.

Solberge YN [*Solberge* DB]. OScand *sōlberg* 'sunny hill'. ON *sól* means 'sun'.

Solent Ha [*Soluente* c 730 Bede, c 890 OEBede, *Solente* c 890 OEBede, (on) *Solentan* 948 BCS 865, *le Soland* 1395 Pat]. A Brit name, but the etymology is not clear.

Solihull Wa [*Sulihull* 1242 Ch, 1243 Cl, 1291 Tax, *Solyhulle* 1315 BM]. The first el. is identical with that of *sulig cumb* 972 BCS 1282, *sulig graf* 963 ib. 1108, and with OHG *sulag*, which means 'pigsty' and is found in pl. ns. like SULGEN (OHG *Sulaga* &c.).

Sollom La [*Solaynpull* c 1200 CC, *Solame* 1372 FF]. Perhaps ON *Sól-hlein* 'sunny slope', from ON *sól* 'sun' and *hlein* (Norw *lein* 'slope').

Solport Cu [*Solpert* 1246 Sc, 1302 Cl, *-perd* 1295 Ipm]. The second el. is Welsh *perth* (from *pert*) 'bush, brake, hedge'. The first el. is obscure.

Solway (-ŏl-) **Firth** Cu [*Sulewad* 1229 P, *Sulwath* 1292 Ass]. Second el. ON *vað* 'ford'. The first may be ON *súl* 'pillar' or *súla* 'Solan goose'. **S~** according to PNCu(S) was named after Lochmaben Stone, a granite boulder which marked the Scottish end of the ford. This is very probable, but a correspondent testifies to the common occurrence of solan geese in the district.

Solway Moss Cu [*Solum* 1246, *Solom* 1282 Ipm]. OE *solum*, dat. plur. of SOL.

Somborne, King's, Little & Upper, Ha [*Sumburne* DB, *Sunburna* 1159 P, *Opsun-*

burna 1167 P, *Parva Sunburn* 1242 Fees, *Kyngessumburne* 1256 Ass]. Named from the brook at the place, which is *Swinburna* c 909 BCS 629. 'Wild boar stream.'

Somerby Le [*Sumerlidebie, Summerdebi* DB, *Someredebia* c 1125 LeS], **S~** Li nr Brigg [*Sumertebi* DB, *Sumerdebi* c 1115 LiS], **S~** Li nr Corringham [*Sumerdebi* DB, c 1115 LiS], **S~** Li nr Grantham [*Sum(m)erdebie* DB, *Sumerdebi* 1138 NpCh]. '*Sumarliði*'s BY.' First el. ON *Sumarliði* pers. n., also found as *Sumerlida* DB &c. The name means 'summer warrior'.

Somercotes Db [*Somercotes* 1276 Ass], **S~** (-ŭ-) Li [*Summercotes* DB, *Sumercotis* c 1115 LiS]. 'Huts used in summer.'

Somerford Chs [*Sumreford* DB, *Sumerford* 1288 Court], **S~ Keynes** Gl [*Sumerford* 683, 931 BCS (65, 671), 1130 P, *Somerford Keynes* 1291 Tax], **Great & Little S~** W [*Sumerford* 939 BCS 719, *Somerford* 956 BCS 922, *Sumreford* DB, *Somerford Magna, Parva S~* 1291 Tax]. 'Summer ford, ford available in summer only.'

S~ Keynes was held by William de Kaines in 1212 (Fees). Cf. ASHTON KEYNES.

Somerleyton (-ŭ-) Sf [*Sumerledetuna* DB, *Sumerletun* c 1185 Bodl]. '*Sumarliði*'s TŪN.' Cf. SOMERBY.

Somersby Li [*Summerdebi* DB, *Sumerdebi* c 1115 LiS]. Identical with SOMERBY.

Somerset [*Summurtunensis paga* c 894 Asser, (on) *Sumærsæton* 1015 ASC (E), (earl over) *Sumersæton* 1048 ib., *Sumersetescir* 1122 ib.)]. OE *Sumorsǣte* or *-sætan* (*Sumursætum* dat. 845 ASC, *Sumorsæte* nom., *Sumursætna* gen. 878 ib.) originally meant 'the Somerset people' and later became the name of the district; cf. DORSET. *Sumorsǣte* is elliptical for *Sumortūn-sǣte* 'the dwellers at Somerton, people dependent on Somerton'.

Somers(h)all, Church, & S~ Herbert Db [*Summersale* DB, *Sumereshala* 1179 P, *Chirchesomereshal* 1278 FF, *Somersale Herbert* E 1 Derby]. '*Sumor*'s HALH or valley.' **Sumor* is a pers. name formed from *sumor* 'summer'. Cf. OHG *Sumar*, ON *Sumarr*, ODan, OSw *Somar*.

William son of Herbert in Somershall is mentioned 1206 Cur.

Somersham (-ŭ-) Hu [*Summeresham* c 1000 HEl, *Sumeresham* c 1050 KCD (907), 1163 P, *Sumersham* DB]. Probably '*Sumor*'s HĀM.' Cf. prec. name. But if the *Suðmere* mentioned in the *banlieu* of Ramsey in KCD 1364 was at Somersham, the name may mean 'the HĀM at the south mere'.

Somersham Sf [*Sumersham* DB, *Sumeresham* 1242 Fees]. '*Sumor*'s HĀM.' Cf. SOMERSHALL.

Somerton Li [*Summertune* DB, *Sumerton* 1242 Fees], **East & West S~** Nf [*Sumerton* 1044–7 Holme, *Somertuna* DB, *Est-, Westsumertone* 1254 Val], **S~** O [*Sumertone* DB,

1222 Ep], S~ So [*Sumurtún* 733 ASC, *Sumortun* 901–24 BCS 591, *Summertone* DB]. OE *sumortún* 'summer dwelling, TŪN used only in summer'. Originally the name would have denoted a place to which the cattle were taken and the people removed during summer.

Somerton Sf [*Sumerledetun* 1046, *Somerledeton* c 1060 Wills, *Sumerledetuna* DB, *Sumerledestun* c 1095 Bury]. Identical with SOMERLEYTON.

Sompting (sowntǐng) Sx [*Suntinga gemǽre* 956 BCS 961, *Sultinges* DB, *Suntinges* 1186 Fr, *Sumptinges* 1242 Fees]. 'The dwellers at the marsh.' The base is an OE **sumpt*, **sunt* 'marsh' or 'pool', identical with OHG *sunft* 'fen, marsh', MLG *sumpt* the same, and the source of **Sunt, Sunte** Sx. The word is related to Engl *sump* (found in *Brunes sumpe* 1240 FF Ess), *swamp* and the like. It is found in the form *sumpth* in the sense 'pit to collect water' 1616 CWNS xxviii. 13.

Sonning (-ǔ-) Brk [*Soninges* DB, *Sunningas* W 2 Abingd, *Sunninges* 1167 P]. OE *Sunningas* 'Sunna's people'. The folk-name is found also in SUNNINGDALE, -HILL, -WELL Brk. *Sunna* is found in SUNBURY and corresponds to OG *Sunno*.

Sookholme (sǔkm) Nt [*Sulcholm* 1189, *Sulkesholm* 1200 P, *Sulgholm* 1230 Cl, 1280 Ch]. 'HOLM by a SULH or gully.'

Soothill YW [*Sothull* 1219 FF, -*hill* 1266 Misc]. The first el. is presumably OE *sōt* 'soot', perhaps referring to charcoal-burning.

Sopley Ha [*Sopelie* DB, *Sappeleia* c 1130, *Soppeleia* c 1160 (1332) Ch, *Soppele* 1263 Ipm], **Sopwell** Hrt [*Sopewell* 1223 Ep, *Soppewell* 1291 Tax], **Sopworth** W [*Sopeworde* DB, *Sopesworth* 1206 Cur, *Soppewrth* 1291 Tax]. 'Soppa's LĒAH, WELLA, WORÞ.' Cf. SODBURY.

Sosgill Cu [*Saurescalls* 1208 FF]. 'Huts at a muddy place.' Cf. SAURR, SKÁLI.

Sotby Li [*Sotebi* DB, *Soteby* Hy 2 BM, 1225 Ep, *Sottebi* 1155 BM, 1206 Ass]. '*Sōti*'s BY.' ODan, ON *Sóti* is a well-known pers. n., and is found as *Sota* in DB &c.

Sotherton Sf *Sudretuna* DB, *Sutherton* 1229 Ch]. OE *Sūþerra tún* 'southern TŪN'.

Sotterley Sf [*Soterlega* DB, *Soterle* 1188 ff. P, 1242 Fees]. See LĒAH. The first el. is obscure. Possibly it may be that of SOTTERUM in Frisia [*Sotrenheim* 10].

Sotwell Brk [*æt Suttanwille* 945, (æt) *Suttanvvlle* 948 BCS 810, 864, *Sotwelle* DB, *Sothuwella, Sotewlle* a 1162 Oxf, *Sottewell* 1242 Fees]. The same first el. is apparently found in SUTCOMBE D, and both names have been held to contain an unrecorded pers. n. *Sutta*. But in the heading of BCS 864 Sotwell is called *Suttunæ wylle*. If this form is trustworthy, *Suttan-* in the above forms

is a weakened variant of *Suttūn* 'Sutton', which must then have been the old name of Sotwell. The charters BCS 810, 864 are not originals.

Soudley Gl [*Suthleg* 1258 Ch]. 'Southern LĒAH.'

Soulbury (sǔl-) Bk [*Soleberie* DB, *Suleberi* 1151–4 Fr, *Sulebire* 1198 FF]. 'BURG by a SULH or gully.'

Soulby Cu [*Suleby* 1225 P, *Soulebi* 1286 Ipm], S~ We [*Sulebi* 1195 P, *Suleby* 1278 FF]. A third ex. of the name is SULBY Np. It is possible that the first el. of all three is OE *sulh* 'gully', but it would be remarkable to find three names in -BY containing this element. It is probably partly ODan *Sula* pers. n. or ON *súl* 'post'.

Souldern O [*Suleþorna* c 1160, -*þorne* c 1190 Eynsham, *Sulthorna* 1209–19 Ep], **Souldrop** Bd [*Sultrop* 1196 P, -*drop* 1202 FF]. 'Thorn-bush and thorp in a SULH or gully.'

Sound Chs [*Sonde* 1274 Ipm, 1282 Court]. OE *sand* 'sand, sandy soil'. OE *sand* must have become *sond* and *sund*. Cf. *lung* for *long* in WMidl.

Sourton D [*Swuran tun* c 970 BCS 1247, *Surintone* DB, *Surethon* 1242 Fees]. 'TŪN by a col.' First el. OE *swira, swēora* 'neck, col'.

Southall (sowthawl) Mx [*Sudhalle* 1206 Cur, *Sudhale* 1212 RBE], **Southam** Gl [*Surham* DB, *Sutham* 1286 Ipm], S~ Wa [*Suðham* 965 BCS 1166, *Suþham* 998 Crawf, 1001 BM, *Sucham* DB]. 'Southern HALH and HĀM.'

Southa·mpton Ha [*Homtun* 825 BCS 390, *Hamtun* 837 ASC, 1045 KCD 781, *Homwic* c 1130 SD, *Hantune* DB, *Hanton* 1190 P, *Suðhamtun* 962 BCS 1094, *Suhantune* 1158 Oxf, *Suðhamtune* 1205 Lay]. OE *Hamm-tūn* 'TŪN in a HAMM or river land'. Southampton for distinction from NORTHAMPTON.

Southborough K [*la South Burgh in Tunbrigge* 1450 Pat]. 'Southern borough or fort.'

Southbroom W [*Suthbrome* 1227 Salisbury, -*brum* 1231 Pat]. 'Southern broom-brake.' Cf. BRŌM.

Southburgh Nf [*Berc* DB, *Berg Maior* 1254 Val, *Suthberg* 1291 Tax]. OE *beorg* 'hill'. *South* perhaps for distinction from MATTISHALL BURGH.

Southburn. See KIRKBURN.

Southchurch Ess [*Sudcyrcean* c 1050 KCD 896, *Sudcerca* DB]. 'Southern church.'

Southcoates YE [*Sotecote*, -*s* DB, *Sotcote* 1235 Cl]. '*Sōti*'s COTS.' Cf. SOTBY.

Southcot Brk [*Sudcote* DB, *Suthcote* 1242 Fees]. 'Southern COT.'

Southease Sx [*Sueise* 966 BCS 1191, *Suthese* 1291 Tax]. OE *Sūþ-hǣs* 'southern wood'. See HÆS.

Southe·nd on Sea Ess [*Sowthende* 1481 PNEss]. Originally part of PRITTLEWELL, of which it was the southern end. S~ began to be a resort for holiday-makers about 1800.

Southery (-ŭdh-) Nf [*Suthereye* 942, *Suðereye* c 950 BCS 774, 1008, *Sutreia* DB, *Suthereie* c 1095 Bury]. 'Southern island' (OE *sūþer-* or *sūþerra* and ĒG).

Southfleet K [*Suthfleotes*, *Fliot* 10 BCS 1321 f., *Sudfleta* DB], **Southgate** Mx [*Suthgate* 1370 AD]. 'Southern FLĒOT or stream and gate.'

Southill Bd. See IVEL. **S~ Co.** See HILL.

Southington Ha [*Suthampton* 1346 FA, *Sothyngton* 1412 FA]. Perhaps OE *Sūþhǣma-tūn* 'TŪN of the people to the south'. Cf. the neighbouring POOL-, QUIDHAMPTON.

Southleigh. See LEIGH.

Southmere Nf [*Sutmere* DB, *Sudmere* 1198 FF]. 'Southern lake.' There is no lake here now.

Southminster Ess [*Suðmynster* c 1000 ASCh, *Sudmunstra* DB, *Suthministr'* 1275 RH]. 'Southern church.' Cf. MYNSTER.

Southoe Hu [*Sutham* DB, *Sudho* 1186 P, 1220 Fees], **Southolt** Sf [*Sudholda* DB, *Sutholt* 1291 Tax]. 'Southern HŌH or spur of hill and wood.'

Southorpe Li in Edenham [*Sudtorp* DB], S~ Li nr Gainsborough [*Torp* DB, *Suthorp* 1254 Val], S~ Np [*Sudtorp* DB]. 'Southern thorp.'

Southover Sx [*Suthoure* 1121 AC, *Suthenouere* 1255 FF]. OE *be sūþan ōfre* '(the place) south of the bank or shore'.

Southowram. See OWRAM.

Southport La. A late place and name.

Southrepps. See REPPS.

Southrey (-ŭdh-) Li [*Sutreie* DB, *Suderei* 1115 LiS, *Surrea* 1163 RA]. Identical with SOUTHERY.

Southrop Gl [*Sudthropa* c 1145 BM, *Suthrop* 1211–13 Fees], **Southrope** O [*Suthrop* 1316 FA]. 'Southern thorp.'

Southsea Ha [*Southsea Castle* 1652 First Dutch War in Navy Rec Soc]. Doubtless referred to as *le South Castell of Portesmouth* 1545 LP. Southsea Castle was built in 1540.

Southwaite Cu [*Thougthuayth* 1380, *Touthwaite* 1461 CWNS xxiii]. The first el. may be OE *þōh* (*þohæ*) 'clay'. Cf. THWAITE. The first el. was later associated with the word *south*.

Southwark (sŭdhŭk) Sr [*Suðgeweork* 1023, *Suþ*(*ge*)*weorc* 1052 ASC (D), *Sudwerche* DB, *Sudhiwerke* 1197 FF]. 'Southern fort.' OE *geweorc* means 'fortification'. An earlier form *Suþriganaweorc* 10 Burghal Hidage means 'the fort of the Surrey people', *Suþrige* being a derivative of SURREY.

Southwell (sŭdhl) Nt [*Æt Suðwellan* 958 YCh 2, (æt) *Suðwillum* c 1000 Saints, *Sudwelle* DB]. 'Southern spring.' Named from Lady Well at the church. *South-* for distinction from NORWELL.

Southwick Du [*Suthewich* Hy 2 FPD, *-wic* 1195 (1335) Ch], S~ Gl [*Sudwicha* DB], S~ Ha [*Sudwic* 12 BM, 1212 Fees], S~ Np [(æt) *Suthwycan* 972–92 BCS 1130, *Sudwic* 1130 P], S~ Sx [*Sudewic* 1073 Fr, *Sudwica* c 1100 Oxf], S~ W [*Sothewyke* 1322 Ch]. 'Southern wīc.'

Southwold Sf [*Sudwolda* DB, *Sudwald* 1227 Ch], **Southwood** Nf [*Suthuuide*, *Sudwda* DB, *Suthwode* 1254 Val]. 'Southern wood.' Cf. WALD.

Southworth La [*Suthewrthe* 1212 Fees]. 'Southern WORþ.'

Sow R St [*Sowa* a 1118 Flor, *Sowe* 1272 Ass], **Sowe** R Wa [*Sowe*, *Souwe* 13 &c. PNWa(S), *Sowe* c 1540 Leland]. On the Sowe is **Sowe** vil. Wa [*Sowe* 1043 Th, *Sowa* DB, *Sowe* 1177 P]. *Sow*(*e*) is a Brit rivername cognate with Gaul *Savus*, *Sava* and derived from the root *seu-* 'to flow, liquid' in OE *sēaw* 'juice', Welsh *sug* the same, OIr *suth* 'milk'.

Sowber. See SOLBERGE.

Sowerby, Castle, Cu [*Sourebi* (castellum) 1186 P, *Soureby* 1219 Fees], S~ La nr Inskip [*Sorbi* DB, *Soureby* 1246 Ass], S~ **Hall** La in Dalton [*Sourebi* DB], **Brough** S~ We [*Soureby by Burgh* 1314 Ipm], **Temple** S~ We [*Saureby* c 1225 WR, *Templessoureby* 1292 QW], S~ YN nr Thirsk [*Sorebi* DB, *Soureby* 1228 FF], S~ **under Cotcliffe** YN [*Sourebi* DB, *Suleby sub Koteclyf* 1285 FA], S~, S~ **Bridge** YW [*Sorebi* DB, *Soureby* 1252 Ch, S~ *Brygge* 1478 Pat]. Identical with *Saurbœr* in Iceland, so called according to Landnáma because the district was boggy. First el. ON *saurr* 'mud, dirt'.

Brough & Cotcliffe S~ because near these places.—**Temple** S~ belonged to the Templars.

Sowton D [*Southton* 1420 PND]. 'Southern TŪN.' The place was formerly called CLIST (or CLYST) FOMISON [*Clis* DB, *Clist Fomicon* 1242 Fees]. Cf. CLYST. The addition is a family name.

Soyland Moor YW [*Soland* 1274 ff. Wakef]. OE *sol-land*, the first el. being OE *sol* 'muddy place'.

Spalding Li [*Spaldyng* 1051 KCD 795, *Spallinge* DB, *Spaldingis* c 1115 LiS, *Spaldinges* 1199 (1330) Ch], **Spaldington** YE [*Spellinton* DB, *Spaldiggetun* Hy 2 BM, *Spaldington* c 1200 YCh 445], **Spaldwick** Hu [*Spaldwic* c 1050 KCD 907, *-vice* DB, *Spaldewic* 1086 IE, 1163 P, *-wica* 1155–8,

Espalde(s)wic 1126, 1139 RA], **Spalford** Nt [*Spaldesforde* DB, *Spaldeford* 1183 P, 1265 Misc, *Spaldyngfordvvath* 1329 QW]. To these should be added (**Holme upon**) **Spalding Moor** YE [*Spaldinghemore* 1172 YCh 1391]. In the 7th cent. Trib Hid (BCS 297) is mentioned a tribe *Spaldas* or *Spalde* (in *Spalda* [*land*]), no doubt to be located in the fens of Hunts, Np, Lincs. Spaldwick is very likely the WĪC of some members of this tribe. Spalford Nt may commemorate some other group of it. The *Spaldingas* who gave their names to Spaldington and Spalding Moor very likely came from the Spalding district. Spalding itself probably goes back to OE *Spaldingas* 'descendants of the *Spaldas*' or 'members of the tribe of *Spaldas*'. But the tribal name *Spaldas* is etymologically obscure. The tribe may quite well have migrated from the Continent under the name *Spaldas*, and the name may have been taken from some place on the Continent. Spaldas is presumably derived from *spaldan* 'to cleave' (not recorded in OE, but cf. OHG *spaltan*). From this was formed a noun meaning 'cleft, ravine' or the like, which gave rise to pl. ns. SPAUWEN in Holland [*Spalden* 1096] and L'ESPAIX in France [*Spalt* 11] are derived from such a word. A place *Spald* may, of course, have been found also in England.

Spanby Li [*Spane(s)bi* DB, *Spanebi* 1138 NpCh, *Spannebi* 1170 P]. The first el. is ON *spánn*, OSw *spān* 'chip, shingle for roofing', and the name may mean 'homestead with shingled roof' or 'place where shingles were made'. Cf. SPAUNTON.

Spårham Nf [*Sparham* c 1060 Wills, DB, 1191 P, 1254 Val, *Sperham* 1196 FF, 1264 Ipm]. 'HĀM or HAMM with or by an enclosure.' See SPEARR.

Sparket Cu [*Sperkeheved* 1254 Ipm], **Sparkford** Ha nr Winchester [*Sparkeford* 1212 Fees, 1311 Ipm], **S~** So [*Spercheford, Sparkeforda* DB, *Sperkeford* 1242 Fees], **Sparkhays** So in Porlock [*Sperkheys* 1419 Dunster], **Sparkwell** D [(æt) *Spearcan wille* c 1070 Ex, *Sperchewelle* DB, *Sparke-vill* 1242 Fees]. A surname *atte Sperke* 1333 Subs (Do) is mentioned in PND. There must have been an OE *spearca* or *spearce* with some sense that rendered its liable to enter into pl. ns. Such a word would belong to OE *spræc* 'shoot, twig', *spracen* 'Rhamnus frangula', Norw *sprake* 'juniper', ON *sprek* 'dry twig'. The probability is that the word had a meaning such as 'brushwood', or else denoted some particular tree or shrub.

Sparsholt Brk [æt *Speresholte, Speresholt* 963 BCS 1121 f., *Spersolt* DB, *Speresholt* 1156 P], **S~** Ha [*Speoresholt* 901 BCS 594, *Spæresholt* 1060-6 KCD 820, *Speresholt* 1167 P]. Probably 'wood where (shafts for) spears were obtained'.

Spaunton YN [*Spantun* DB, *-ton* 1207 FF]. 'TŪN with shingled roof.' Cf. SPANBY.

Spaxton So [*Spachestone* DB, *Spaxton* 1227 FF]. '*Spak*'s TŪN.' *Spakr* is an ON byname. The name may well contain a Scand pers. n.

OE **spearr**, found in the pl. n. *Wynburge-spær* 947, *Wynburgespear* 963 BCS 834, 1125, seems to be cognate with OHG, OLG *sparro*, ON *sparri* 'rafter, beam', ME *sparre*, *sperre* 'spar'. The OE form would seem to have been *spearr* rather than *spær* or *spear*. In BCS 834 occur spellings such as *hleaw* for *hlæw*. *Spearr* very likely meant 'enclosure in a wood'. See SPARHAM, RUSPER, perhaps SPORLE.

Speckington So [*Speketon* 1253 Ass, *Spekynton* 1281 FF, *Spekington* 1285 FA]. The first el. may be related to SPEKE.

Speen Brk [*Spinis* (abl.) 4 IA, (silva) *Spene* 821 BCS 366, *Spone* DB, *Spenes* 1167 P, *Wodespene* 1275 Cl]. The identification of Speen and *Spinæ* must be upheld, in spite of the phonological difficulty of deriving *Speen* from Lat *Spinæ* (with long *i*). Skeat preferred to derive Speen from OE *spōn* 'shaving, chip'. The difficulty may be overcome by assuming that the Brit name of the place was a word related to OIr *scé*, Welsh *ysbyddad* 'hawthorn', Co *spedhes* 'brambles, briars', an OW **Spian* or the like, which was Latinized to *Spinæ* 'the thorn-bushes'. OW *Spian* would give OE *Spēon*, ME *Spēne*.

Speeton YE [*Spretone, Specton, Spetton* DB, *Spe(c)tone* 1166 YCh 1139, *Spetona* c 1180 ib. 1141]. If *Spretone* DB is reliable, the OE form was *Sprēot-tūn*, which was changed to *Spēot-tūn* owing to substitution of OScand *spiót*. *Sprēot* means 'a pole'. Possibly the long narrow ridge at the place was called *Sprēot*. If the forms in *ct* are to be taken seriously and not as spellings for *tt*, the first el. may be identical with that of SPETTISBURY, SPEXHALL. Or it may be the same as in SPETCHLEY.

Speke La [*Spec* DB, *Speke* 1252 Ipm, *Speek* 1332 Subs]. OE *spæc* (gen. plur. *spāca*) occurs in the sense 'twig' or the like. The word is related to OHG *spah(ho)* 'dry brushwood', MLG *spaken* 'dry twigs'. The meaning of the word in Speke may well be 'brushwood'.

Speldhurst K [*Speldhirst* 765-91 BCS 260]. First el. OE *speld* 'splinter, piece of wood'. Second el. OE *hyrst* 'hill' or 'wood'. The meaning may be 'hill or wood where chips of wood were found'. OE *speld* is the first el. of **Spilsill** K [*Speldgisella* 814 BCS 343], which means 'wooden shed'.

Spellow La [*Spellowe* 1306 FF]. OE *spell-hlāw* 'moot hill'. Cf. **Spelhoe** (hd) Np [*Spelhoh* 1066-75 GeldR], which means 'moot ridge'. OE *spell* means 'speech, discourse'.

Spelsbury O [(æt) *Speoles byrig* 10 BCS 1320, *Spelesberie* DB, *-beri* c 1090 Eynsham].

The first el. is no doubt derived from an OE (lost) equivalent of OHG *spehan* 'to spy, watch'. It may be an OE *spēol* adj. 'watchful', whence a pers. n. *Spēol*. Less likely seems a noun *spēol*, corresponding to Lat *speculum* and meaning 'look-out place'.

Spen, High, Du [*le Spen* n.d. Newminster], **Spen Valley** YW. The name Spen is often found in records from the Northern counties. *Spenne* occurs at least thrice in CC (La), *Munckehayespen* 1175–90 YCh 1026, *Brathewaitspen* c 1260 Selby, *le Spen* Newminster 25 (Nb). The name is derived from the word *spenne* found in Sir Gawain (*spene* Wars of Alexander). The meaning of the word has not been found. In Sir Gawain a fox is said twice to jump 'over a *spenne*'. This suggests a meaning such as 'hedge', and the word may be a derivative of OE *spannan* 'to clasp, fasten'.

Spennithorne YN [*Speningetorp* DB, *Spinithorn* c 1190 Godric, *Spenyngthorne* 1317 Ch], **Spennymoor** Du [*Spendingmor* c 1336 Ep, *Spennyngmore* 1381 Pat]. The probability is that *spenning* is a word equivalent to *spen*.

Spernall Wa [*Spernore* DB, *Sperenoura* 1176 f. P, *Spernovere* 1328 Ch]. OE *spæren* 'of chalk' and *ōfer* 'bank'. S~ is on the Arrow.

Spetchley Wo [*Spæcleahtun* 816, (æt) *Speclea*, *Spæclea* 967 BCS 356, 1204 f., *Speclea* DB]. 'Speech LĒAH', i.e. 'glade where moots were held'. Low Hill, the meeting-place of Oswaldslow hd, is nr Spetchley.

Spettisbury Do [*Spehtesberie* DB, *Spectesb'i* 1162 P, *Spectebury* 1291 Tax], **Spexhall** Sf [*Spectesbale* 1197, *Spictesbale* 1198 FF, *Spectishal* 1305 BM]. The first el. is OE *speoht*, *speht* 'woodpecker', in Spettisbury more likely used as a pers. n. The word *speight* is not found until 1450 (OED), but is no doubt an old word. See HALH.

OE spic. See SPITCHWICK, SPIXWORTH, WANSBECK.

Spilsby Li [*Spilesbi* DB, 12 DC, *Spillesby* 1209–35 Ep]. '*Spilli*'s BY.' The first el. is a nickname derived from ON *spillir* 'waster'.

Spilsill. See SPELDHURST.

Spindleston Nb [*Spilestan* 1166, *Spindlestan* 1187 P, *Spinlestan* 1212 Fees]. 'Spindle rock.' The place was named from a pillar of whinstone, probably owing to a fancied likeness to a spindle (OE *spinele*).

Spinney Abbey Ca [*Spinetum* 1254 Val]. 'The spinney' (OFr *espinei*, MLat *spinetum*).

Spital in the Street Li [*Hospitale* 1204 Cur, *Spitelenthestrete* 1322 BM]. 'The hospital' (on Ermine Street).

Spitchwick D [*Spicewite* DB, *Spichwic* 1169, *Spichewic* 1199 P, *Spychewik* 1242 Fees]. Perhaps 'bacon farm', the first el.

being OE *spic* 'bacon'. But the first el. might be the OE word *spic* in WANSBECK.

Spithead Ha [*Spithead* 1653 First Dutch War in Navy Rec Soc]. Named from a sandbank, **Spit Sand.** *Spit* in the sense 'reef, point of land' is first evidenced in OED from 1673.

Spixworth Nf [*Spikesuurda* DB, *Spicasurda* 1163, *Spicheswrtha* 12 BM]. '*Spic*'s WORP', the first el. being a nickname, perhaps derived from *spic* 'bacon', or simply 'bacon farm'.

Spodden. See SPOTLAND.

Spofforth YW [*Spoford* DB, *Spotford* 1218 FF, 1230 Cl, *Espodeford* 1294 Percy]. The first el. may be as in SPOTLAND, i.e. the word *spot*, but the meaning of the word here is not apparent. Second el. OE *ford* 'ford'.

Spondon (-ōō-) Db [*Spondune* DB, 1167 P, *Spondon* 1170 P, *Spandon* 1177 P, 1233 Ch]. First el. OE *spōn* 'shaving, chip', sometimes replaced by the ON equivalent *spánn*. Perhaps 'hill where shingles were made'.

Sporle Nf [*Sparle(a)*, *Esparlea* DB, *Esparlaium* 1146 Fr, *Sperly* c 1195 Sele, *Sporle* 1254 Val]. Perhaps 'LĒAH with an enclosure' (see SPEARR), though the later *o* offers difficulty.

Spotland La [*Spotlond* c 1180 WhC, *-land* 1285 Ass]. Here is **Spodden Brook** [*Spotbrok* 13 WhC]. The first el. may be OE *splott*, later *spot* 'spot, plot of land'. One *l* would easily be lost in *Splottland*.

Spratton Np [*Spretone*, *Sprotone* DB, *Sprotton* 12 NS, 1236 Ep]. OE *Sprēot-tūn*, the first el. being OE *sprēot* 'pole'. S~ is on a projecting ridge, which may well have been called *Sprēot*. Cf. SPEETON.

Spreyton D [*Spreitone* DB, *Spreiton* 1242 Fees]. OE *Spræg-tūn* 'TŪN in brushwood'. OE *spræg* is the source of ME, Mod *spray* 'twig, fine brushwood' &c. The word is common in names of minor places in Devon.

Spridlington Li [*Spredelintone*, *Sperlinctone* DB, *Spridlinc-*, *Spritlingtuna* c 1115 LiS, *Spridlinctun* 1212 Fees]. Apparently OE *Sprytlinga-tūn*, the *Sprytlingas* being '*Sprytel*'s people'. Cf. *Sprot* pers. n. (mon. Edw. Conf., DB), also the pers. n. *Sperling* in ELPN 64.

OE spring, spryng 'spring, well' is found in OSPRINGE K. ME *spring* also meant 'a plantation of young trees, esp. one for rearing or harbouring game, a spinney; a copse'. This may be the source of spring in OXSPRING, WOODSPRING. See next names.

Springfield Ess [*Springafelda*, *Springhefelda* DB, *Springefeld* 1190 P, 1198 FF]. Apparently OE *Springinga-feld*. *Springingas* may be derived from OE *spring* 'spring' or from *spring* in some other sense. A meaning 'projection' would be suitable, and OE *spring* might have had such a sense.

Springthorpe Li [*Springetorp* DB, *Spring(e)-torp* 1224 Ep]. 'Thorp by a *spring* or copse.' Cf. SPRING.

Sprint R We [*Spritt*, *Spret* c 1195, *Sprit* c 1235 Kendale]. A Scand river-name derived from ON *spretta* 'to jump, start', Sw *spritta* also 'to spirt'. Cf. *Sprintaaen* in Norway. *Spretta, spritta* come from *sprenta, sprinta*. The change to *Sprint*, however, may be due to the influence of the neighbouring MINT.

Sproatley YE [*Sprotele*, *Sprotelai* DB, *Sprotleya* 1135–40 YCh 1152, *Sprottelea* 1196 P]. 'LĒAH or wood consisting of young trees' or the like. First el. OE *sprota, sprot* 'shoot, sprout, twig'.

Sproston Chs [*Sprostune* DB, *Sprouston* 1271–4 Chester], **Sproughton** Sf [*Sproeston* 1191 ff., *Sprouton* 1198 FF, *Sproustone* 1254 Val], **Sprowston** Nf [*Sprowestuna, Sprotuna* DB, *Sprouston* c 1125 BM]. '*Sprow*'s TŪN.'

Sprotb(o)rough YW [*Sproteburg* DB, *Sprotteburg* 1250 Ep, *Sprotburgh* 1291 Tax]. The first el. may be OE *sprot(a)* 'shoot, twig' (cf. SPROATLEY), but looks more like a pers. n. OE *Sprot* exists (cf. SPRIDLINGTON). *Sprota* may also have been used.

Sproughton, Sprowston. See SPROSTON.

Sproxton Le [*Sprotone* DB, *Sproxcheston* c 1125 LeS, *Sproxton* 1166 P, *Sprokeston* 1236 Fees], S~ YN [*Sprostune* DB, *Sproxtona* 1157 YCh 354, *Sprocston* 1186 P]. '*Sprok*'s TŪN.' OSw *Sprok* seems to occur.

Spurn Head YE [*Ravenserespourne* 1399 Pat]. Henry IV landed at this place, called variously *Ravenspurn* and *Ravensburgh* (so Sh R II). The former is the correct form. *Ravens-* is RAVENSER (q.v.). The second el. is *spurn*, found in senses such as 'beak of a war-galley', 'a sharp projection or edge on a horse-shoe' (16th cent. &c.). A meaning 'point of land' is not given in OED, but must have occurred. The word comes from *spurn* vb. 'to spurn, kick'.

Spurshot Ha nr Romsey [*Purisute* 1242 Fees, *Purschite* 1245 Ch]. OE *pirig-scēat* 'pear-tree SCĒAT'.

Spurstow Chs [*Spuretone* DB, *Sporstowe* 1260 Court, *Spurstowe* 1289 Chester]. The first el. seems to be OE *spor* 'track, footprint'. Possibly the word is here used in the sense 'trackway'. Second el. OE *stōw*, perhaps in the sense 'hermitage'.

Stadhampton O [*Stodeham* 1146 RA, *Stodham* 1316 FA]. OE *stōd-hamm* 'river meadow where horses were kept'. The el. *-ton* seems to be a late addition.

ON **staðr**, OSw **stadher** 'place, town' &c. is common in pl. ns. in Scandinavia, but it is doubtful if any Engl pl. ns. contain the element. Cf., however, BIRSTWITH, BRYNING, GANSTEAD.

OE **stæf** 'staff, stick'. See STALYBRIDGE, STAVELEY. Cf. STAFFIELD.

OE **stæfer** 'stake', ME *staver* 'rung of a ladder, stake for a hedge' (= OSw *staver*, Dan *staver*). See STARETON, STAVERTON Np, Sf, STARBOTTON.

OE **stæner** 'stony ground'. See STAINDROP, STANDERWICK, STERNDALE.

OE **stæþ** 'landing-place' (= ON *stǫð*). See STAITHES, STATHAM, STATHE, STAFFORD St, BICKERSTAFFE, BIRSTWITH, BRIMSTAGE.

Staffield Cu [*Stafhole* c 1225 WR, *Stafole* 1246 FF, *Staffold* 1274 Cl]. ON *staf-hóll* 'hill with a staff'. Second el. ON *hóll* 'round hill'. *Staf-* refers to a boundary mark.

Stafford, East & West, Do [*Stanford* DB, 1321 Ipm, *Stafford* 1265 BM, *Weststaforde* 1316 Ipm]. OE *Stān-ford* 'stony ford'. Loss of *n* is common before *f*. Cf. STOFORD.

Stafford St [*Stæþ* 10 Coins, *Stæfford* 913 ASC (C), *Stadford* DB, *Statford* 1130 Pj. OE *Stæp-ford* 'ford by a STÆP or landing-place'. **Staffordshire** is *Stæffordscir* 1016 ASC (D, E).

Stagenhoe Hrt [*Stagnehou* DB, *Stagho* 1253 Ch]. 'Stags' HŌH or spur of land.' First el. OE *stagga* (gen. plur. *staggena*) 'stag'.

Stagsden Bd [*Stachedene* DB, *-den* 1199 P, *Staggeden* 1183 P, *Stachesden* 1196 FF]. 'Stake valley.' First el. OE *staca* 'stake'. A boundary mark may be referred to.

Stain Li [*Stein* c 1115 LiS, 1202 Ass]. OScand *steinn* 'stone'.

Stainall or **Staynall** La [*Staynole* a 1190, *Stainhole* a 1220 FC]. OE *stān-holh* 'stony hollow', Scandinavianized.

Stainborough YW [*Stan-, Stainburg* DB, *Steinburgh* c 1165 YCh 1507]. OE *Stān-burg* 'stone fort', Scandinavianized.

Stainburn Cu [*Steinburna* c 1125 StB], S~ YW [*Stanburne* c 972 BCS 1278, DB, *Stainburne* DB]. OE *Stān-burna* 'stony stream', Scandinavianized.

Stainby Li [*Stigandebi* DB, 1195 P, *Stighendebi* 1206 Ass]. '*Stigandi*'s BY.' First el. ON, ODan *Stigandi* pers. n.

Staincross YW [*Staincros* DB]. OScand *stein-kross* 'stone cross'.

Staindrop Du [*Standropa* c 1050 HSC, 1125–8 YCh (934), 1131 FPD, *Staindrop* 1195 (1335) Ch]. OE *Stæner-hop* 'valley with stony ground'. See STÆNER. *Stænerhop* would become *Stanrop* and *Standrop*. S~ is in a valley.

Staines Mx [*Stána* 969 Crawf, (æt) *Stane* 1009 ASC (E), *Stáne* 1050–65 BM, *Stana* 1051–65 E, *Stanes* DB]. OE *stān*, dat.

stāne 'stone'. Perhaps named from a milestone on the Roman road that runs past Staines. The preservation of *ā* in the name may be due to Norman influence. Cf., however, STEANE. OE *ǣ* became ESax *a*.

Stainfield Li nr Haconby [*Stentvith* DB, *Steynthweyt* 1268 Ch]. OScand *Stein-þveit* 'stony clearing'.

Stainfield Li nr Lincoln [*Steinfelde* DB, *-felda* c 1115 LiS], **Stainforth** YW nr Thorne [*Steinforde* DB, *-ford* 1199 (1232) Ch], S~ YW nr Settle [*Stainforde* DB, *Stainford* 1226 FF]. OE *Stān-feld*, *-ford* 'stony FELD and ford', Scandinavianized.

Staining La [*Staininghe* DB, *Stanynggas*, *Steyninges* 1211–40 WhC]. Probably a Scandinavianized OE *Stāningas*, which may be '*Stān*'s people' or 'dwellers at a stone or in a stony district'. OE *Stān* is not evidenced, but *Steinn* is common in Scandinavia, and *Stān* may be a hypocoristic form of names in *-stān*.

Stainland YW [*Stanland* DB, *Staynland* 1326 Ipm], **North Stainley** YW [*Stanleh* c 972 BCS 1278, (on) *Norð Stanlege* c 1030 YCh 7, *Nordstanlaia* DB, *Northstainle* 1246 FF], **South** S~ YW [(on) *Nyrran Stanlege* c 1030 YCh 7, *Stanlai* DB, *Staynleya* c 1180 YCh 509], **Stainmore** YN, We [*Stanmoir* c 990 CWNS xxvii, *Stanmore* Hy 2 (1348) Ch, *Staynmor* 1292 QW]. OE *Stān-land*, *-lēah*, *-mōr* 'stony land, LĒAH, moor', Scandinavianized.

Nyrra is OE *nierra* 'nearer'.

Stainsacre YN [*Stainsaker* 1090–6 YCh 855, *Steinesacr'* 1176 P]. '*Stein*'s field.' First el. ON *Steinn* pers. n. See ÆCER.

Stainsby Db [*Steinesbi* DB, 1170 P]. '*Stein*'s BY.' See prec. name.

Stainsby Li [*Stafnebi* 1196, *Stafnesbi* 1197 P, *Stavenesby* 1226 Ep]. '*Stafn*'s BY.' ON *Stafn* is used as a byname.

Stainton Cu on the Eamont [*Stainton* 1166 P, 1254 Ipm], S~ Du nr Barnard Castle [*Staynton* c 1150 YCh 566], S~ La [*Steintun* DB, *Steynton* 1246 Ass], S~ Li nr Waddingham [*Stain-*, *Stantone* DB, *Steintuna* c 1115 LiS], S~ by **Langworth** Li [*Staintune* DB, *Steintuna iusta Languat* c 1115 LiS, *Stantone* c 1125 Fr], **Market** S~ Li [*Staintone* DB, *Steintuna* c 1115 LiS, *Steynton Market* 1286 Ipm], S~ **le Vale** Li [*Staintone* DB, *Steintuna* c 1115 LiS], S~ We [*Steintun* DB, *Stainton* c 1175 Kendale], S~ YN nr Middlesbrough [*Steintun* DB, *-ton* 1198 P], S~ YN in Downholme [*Staynton* 1343 FF], S~ **Dale** YN [*Steintun* DB, S~ YW nr Doncaster [*Stantone*, *Staintone* DB, *Steinton* 1202 FF], S~ YW nr Gargrave [*Staintone* DB, *Stayntona in Cravana* 1154 YCh 480]. A Scandinavianized form of STANTON, OE *Stāntūn*.

Stainton, Great & Little, Du [*Staninctona* 1091, *Steinintune* c 1250 FPD]. S~ is on

an ancient Roman cross-road. The name may be OE *Stānweg-tūn* 'TŪN on the paved road'. Cf. STANNINGTON.

Staithes YN [*Setonstathes* 1415 YInq]. OE *stæþ* 'landing-place'. S~ is nr Seaton, for which it was the landing-place.

Stakesby YN [*Staxebi* DB, *Stachesby* c 1095 YCh 855]. '*Staki*'s BY.' *Staki* is an ON byname.

Stalbridge Do [(in) *Stapulbreicge* 998 KCD 701, *Staplebrige* DB, *Stalbriggh* 1346 FA]. 'Pile bridge, bridge built on piles.' OE *stapul* means 'post, pillar, pile'.

Stalham Nf [*Stalham* 1044–7 KCD 785, DB, 1166 RBE]. 'HĀM by a STALL or pool.'

Stalisfield K [*Stanefelle* DB, *Stealesfelde* 11 DM, *Stalisfeld* 1173 P, *Stalefeld* 1200 Cur, *Stallefeld* 1202 FF]. Identical with *Stealles feld* c 765 BCS 197 (Sx). 'FELD with a *steall* or stable.'

OE **stall, steall** 'place; stable, stǎll; pool in a river' is found in pl. ns. The meaning varies. The meaning 'place' is found in BIRSTAL(L), BURSTALL, BORSTAL, HEXTELLS; cf. HAMSTALL RIDWARE, TUNSTALL. 'Stable' is apparently the meaning in STALISFIELD, HEPTONSTALL, 'pool' that in STALHAM, STALMINE, RAWTENSTALL, TROUTS DALE.

Stallingborough Li [*Stalingeburg* DB, *Stalinburc* c 1115 LiS, *Stalingeburc* c 1130 BM, *Stalingburg* 1233 Ep, 1254 Val], **Stallington** St [*Stalenton* 1251 Ch, *Stalington* 1265 Ass, 1293 QW]. The first el. of Stallingborough is a folk-name, probably *Stælingas*, to judge by the persistent single *l*, derived perhaps from OE *stæl* 'place' in the pregnant sense seen in *stælwierþe*, the source of Mod *stalwart*. Stallington may alternatively have as first el. an OE *stān-hlinc* 'stony hill'.

Stalling Busk YN [*Stalunesbusc* 1218 FF]. 'The stallion's bush.' Second el. OScand *buski* 'bush'. 'Stallion' is OFr *estalon*, Fr *étalon*.

Stallington. See STALLINGBOROUGH.

Stalmine La [*Stalmine* DB, *-min* 1206 P, *-myn* 1262 FF]. 'The mouth (OScand *mynni*) of the pool.' Second el. OE *stall* 'pool', here in a sense 'slow stream'.

Stalybridge Chs [*Stavel'* 1285 Court, *Stauelegh* 1369 BM]. The original name of the place was STAVELEY. Cf. that name.

Stambourne Ess [*Stanburna* DB, *-burn* 1196 P], **Great & Little Stambridge** Ess [*Stanbruge* DB, 1195 P, *Stanbreg Magna*, *Parva* 1238 Subs]. OE *Stān-burna* 'stony stream' and *Stān-brycg* 'stone bridge'.

Stamford Li [*Steanford* 922, *Stanford* 942 ASC, *Stanford* DB], S~ Nb [*Staunford* 1242 Fees, *Stanford* 1257 Ch], S~ **Bridge** YE [*Stanford* c 730 Bede, c 890 OEBede, *Stanfordbrycg* 1066 ASC (C), *Pons Belli*

c 1155 YCh 830, *Pundelabataille* 1251 Cl].
'Stony ford.'

Stamford Bridge is often called *Pons Belli* and
the like in early sources in reference to the
famous battle.

Stamford Hill Mx [*Sanford* 1255 Misc,
Staneford 1321 FF]. Originally *Sandford*.

Stamfordham Nb [*Stanfordham* 1188 P,
1254 Val]. 'HĀM at the stony ford.'

OE **stān** 'stone, stones' is a very common
pl. n. el. It is used alone as a pl. n. in
STAINES, STEANE, STONE, where a Roman
milestone or some prominent stone of
another kind may be referred to. As a
second el. it generally appears as -*ston*(*e*),
and if sufficiently early forms are wanting
it is sometimes difficult to distinguish it
from -*tūn* with a first el. ending in -*s*.
Sometimes the name refers to a memorial
stone or a stone monument, as in FEATHER-
STONE, RADSTONE, RUDSTON, SHILSTONE, or
to a natural stone or stones remarkable in
some way, as in BUXTON, COPPLESTONE, GAR-
STON La, SPINDLESTON, SYSTON. 'Boundary
stone' is the meaning of HARSTON Le and of
-*stān* in most names with a pers. n. as first
el., e.g. CHEDISTON, KESTON, probably also
such as CHIDDINGSTONE, LILLINGSTONE,
where the first el. is a folk-name in -*ingas*. A
meaning 'meeting-place stone' is obvious in
names of hundreds, such as FOLKESTONE K,
GUTHLAXTON Le, where the first el. is a
pers. n. (cf. *scirgemot æt Ægelnoðes stane*
a 1036 Th), LEIGHTONSTONE Hu, MORLESTON
Db, whose first el. is a pl. n., or HURSTING-
STONE Hu, where it is a folk-name in -*ingas*.
The exact meaning is often obscure, as in
BOSTON, BRIXTON Sr and others. KINGSTON(E)
is sometimes 'the king's stone'.

As a first el. *stān* refers sometimes to
building material, as in STAMBRIDGE, STAN-
BURY, STANION, STANWARDINE, STONEHENGE,
STONEHOUSE, also in STANWAY, STOWEY &c.,
and in STANDHILL, STONEGRAVE, STONY-
DELPH, which mean 'quarry', to a stony
bottom in STAMBOURNE, STANFORD &c.,
STANDLAKE, to stony soil in STANDISH, STAN-
LEY, STANNEY and the like. The exact
meaning of *stān* in STANTON, STAUNTON,
STON(E)HAM and the like is difficult to
determine. It may have referred to stony
soil, to some prominent stone(s), possibly
to an enclosure or building made of stone.
In the first syllable *stān* generally appears
as *Stan-*, more rarely as *Stone-* or even
Ston- (STONTON). The *n* was often lost
before a labial, as in STAFFORD Do, STAWARD,
STAWELL, STO(W)FORD, STOWELL, STOWEY,
STAVERTON D, also in STORRIDGE, STAWLEY.
OE *stān* has often been replaced by
OScand STEINN. See STAIN- (passim).

Stanbridge Bd [*Stanbrugge* 1165, -*brigge*
1196 P], S~ Ha [*Stanbrig* 1242 Fees, -*brigg*
1252 Misc]. 'Stone bridge.'

Stanbury YW [*Stainburg* 1235 FF, *Stanbiri*
1249 Ch]. 'Stone fort.'

Standen Brk [*Standen* 1203 Cur, 1242 Fees],
S~ La [*Standen* 1258 LaInq], S~ W
[*Standene* DB, -*den* 1167 P]. 'Stony valley.'
See DENU.

Standen Wt [*Standone* DB, c 1270 Ep,
Staundon 1283 Ep]. 'Stony hill.' See DŪN.

Standerwick So [*Stalrewiche* DB, *Stanre-
wic* 1231 Cl, -*wike* 1334 Ep]. OE Stǽner-
wic 'WĪC on stony soil'. Cf. STÆNER.

Standhill O nr Watlington [*Stangedelf*
1002 KCD 1296, *Stanidelf* 1220 Ep]. 'The
quarry' (OE *stān* and *gedelf* 'digging, pit').

Standish Gl [*Stanedis* 872 BCS 535, DB,
Stanedisch 1291 Tax], S~ La [*Stanesdis*
1178 P, *Stanedis* 1207 P]. OE *Stān-edisc*
'stony pasture'.

Standlake O [*Stanlache* c 1155 Eynsham,
-*lac* 1194 P], **Standlynch** W [*Staninges*
(no doubt for *Stanlinges*) DB, *Staulinc* 1198
Fees, *Stanlinch* 1319 Bodl]. 'Stony stream
and hill.' See LACU, HLINC. A *d* was
developed between *n* and *l*.

Standon Hrt [*Standune* 944–6 BCS 812,
-*done* DB], S~ St [*Stantone* DB, *Standon*
1190 P, 1248 FF]. 'Stony DŪN or hill.'

Stanesgate Priory Ess [*Stanesgata* DB,
Stanagata 1121 AC]. Second el. OE *geat*
'gate'. The first may be STĀN or *Stān*
pers. n.

Stanfield Nf [*Stanfelda* DB]. 'Stony FELD.'

Stanford Bd [*Stanford* DB], S~ **Dingley**
Brk [*Stanford* DB, 1220 Fees, 1267 BM],
S~ **in the Vale** Brk [*Stanford* 1174 P,
1248 Ch], S~ **le Hope** Ess [*Stanford* 1068
EHR xi, 1200 Cur], S~ **Rivers** Ess [*Stan-
fort* DB, *Stanforde Ryveres* Hy 3 BM], S~
Bishop He [*Stanford* DB, S~ *Episcopi* 1316
FA], S~ **Regis** He [*Stanford* DB, *Kinge-
stanforde* 1242 Fees, *Stanford Regis* 1272
Ipm], S~ **K** [*Stanford* 1035 BM, 11 DM],
S~ **Nf** [*Stanforda* DB], S~ **on Avon** Np
[*Stanford* DB, S~ *super Hauene* 13 Selby],
S~ **upon Soar** Nt [*Stanford* DB, S~ *super
Sora* 1222 FF], S~ **on Teme** Wo [*Stan-
fordesbrycg* c 1030 Förster, *Themse*, p. 769,
Stanford DB, S~ *on Temede* 1317 Cl].
'Stony ford.'

S~ **Bishop** belonged to the Bishop of Hereford.—S~ **Dingley** was held by Robert Dyngley
in 1428 (FA).—S~ **le Hope** is nr **Broad Hope**,
where the Thames makes a wide bend (**Lower
Hope Reach**). Hope is presumably OE *hop* in
one of its senses.—S~ **Rivers** was held by
Richard de Ripar[iis] in 1218 (Fees) and came
to the Rivers family in 1213 (Wright). Cf.
BUCKLAND RIPERS.

Stanghow YN' [*Stanghou* 1272 Ipm, 1280
Ch]. ON *stong* 'pole' (probably a boundary
mark) and *haugr* 'hill'.

Stanground Hu [*Stangrund* c 1000 PNHu,
-*grun* DB]. Either 'stony ground' (cf. names
like CLAY, GREET) or 'stone foundation',
referring to the foundation of an ancient
building. Cf. GRUNDISBURGH.

Stanhoe Nf [*Stanhou* DB, *Stanho* DB, 1173 P], **Stanhope** Du [*Stanhopa* 1183 BoB, *-hop* 1228 Ep]. 'Stony HŌH or ridge and HOP or valley.'

Stanion Np [*Stanere* DB, *Stanerna* 1163 P, *Stanern* 1196 Cur]. OE *stān-ærn* 'stone house'. An ancient cromlech may be referred to.

Stanlawe or **Stanlow** Chs [*Stanlawe* 1178–90 Chester, *Stanlawa* J BM]. 'Stony hill' (OE HLĀW).

Stanley Db [*Stanlei* DB, *-lega* 1169 P], S~ Du nr Consett [*Stanley* 1297 Pp], S~ Du W. of Durham [*Stanlegh* 1241 BM], **King's & Leonard** S~ Gl [*Stanlege* DB, *ecclesia Sancti Leonardi de Stanleya* 1138 Glouc, *Stanllegh Leonardi* 1285 FA, *Kingestanleg* 1220 Fees], S~ **Pontlarge** Gl [*Stanlege* DB, *Stanleye Poundelarge* 1324 Misc], S~ St [*Stanlega* 1130 P], S~ W [*Stanlege* DB, *Stanleia* 1189 BM], S~ YW [*Stanlei* DB, *Stanlay* 1297 Subs]. OE *Stān-lēah* 'stony LĒAH'.

S~ **Pontlarge** was held by William de Ponte Arche in 1246 (Ipm). The name is derived from PONT DE L'ARCHE in Normandy.

Stanmer Sx [*Stanmere* 765 BCS 197, DB], **Stanmore** Mx [*Stanmere* 793 BCS 267, DB]. 'Stony mere or lake.'

Stanney Chs [*Stanei* DB, *-a* c 1150 Chester]. 'Stony island' (OE *Stān-ēg*).

Stanningfield Sf [*Stanfelda* DB, *Stanefeld* 1197 FF, 1242 Fees]. 'Stony FELD.' The first el. is OE *stān*, perhaps alternating with *stānen* 'stony'.

Stanninghall Nf [*Staningehalla* DB, *Staningehal* 1166 P]. The first el. is identical with STAINING. See HALH.

Stannington Nb [*Stanigton* 1242 Fees, *Stanington* 1254 Val]. OE *Stānweg-tūn* 'TŪN on the paved road' (the Great North Road). Cf. STAINTCN Du.

Stansfield Sf [*Stanesfelda*, *Stenesfelda* DB, *Stanesfelde* c 1095 Bury, *Stanefeld* 1196 FF, 1204 Cur, 1253 Ch], S~ YW [*Stanesfelt* DB, *-feld* 1246 FF], **Stanshope** St [*Stanesope* DB, *Staneshop* 1227 Ch]. Perhaps '*Stān*'s FELD and HOP or valley'. Cf. STAINING. But the first el. may be OE *stān* 'stone'.

Stanstead Abbots & St. Margaret Hrt [*Stanestede* DB, *Stanstede Abbatis de Wautham* 1247 BM, *Stanstede Abbatis* 1254 Val], S~ Sf [*Stanesteda* DB, *Stanstede* 1197 FF], S~ Sx [*Stansted* 1203 Cur], **Stansted Mountfitchet** Ess [*Stanesteda* DB, *-stede* Hy 2 BM], S~ K [*Stansted* 1231 FF, 1254 Ass]. 'Stony place.' Cf. STEDE.

S~ **Abbots** belonged to Waltham Abbey.—S~ **Mountfitchet** was held by Richard de Muntfichet in 1190 (P). The name is derived from MONTFIQUET in Normandy.—S~ **St. Margaret** Hrt was called (Pons de) Thele till the 16th cent. [*Pons de Thele* R 1, *Pons Tegule* 1200,

Thele 1296 PNHrt(S)]. The name means 'tile bridge', but was associated with *thele* 'plank'.

Stanthorne Chs [*Stanthurl* 1278 Ipm, 1304 Chamb, 1420 BM]. OE *stān-þyr(e)l* 'stone door'. Second el. OE *þyr(e)l* 'hole, aperture'. The reason for the name is not apparent.

Stanton, Long, Ca [*Stantona* c 1080 ICC, *-tone*, *-tune* DB, *Long Stanton* 1282 Ipm], **Churchstanton** D [*Stantone* DB, *Cheristontone* 1258 Ep], S~ Db nr Burton on Trent [*Stantun* 900 f., 968 BCS 583, 587, 1211], S~ **by Bridge** Db [*Stantun* DB, *Staunton* 1323 Ch], S~ **by Dale** Db [*Stantone* DB, *Stanton juxta Dale* 1428 FA], S~ **in Peak** Db [*Stantune* DB, *Stanton in alto pecco* 1372 Derby], S~ **St. Gabriel** Do [*Stantone* DB, *Staunton Gabriell* 1434 FF], S~ Gl [*Stantone* DB, *-tona* 1175 Winchc], **Fen** S~ Hu [*Stantun* 1012 PNHu, *-tone* DB, *Fenstanton* 1260 Ass], **Stoney** S~ Le [*Stantone* DB, *Stonystaunton* 1363 BM], S~ **under Bardon** Le [*Stantone* DB, *Stanton subtus Berdon* 1285 FA], S~ Nb [*Stantuna* 1201 Ch], S~ **on the Wolds** Nt [*Stantun* DB, *Staunton on Seggeswold* 1286 AD], S~ **Harcourt** O [*Stantone* DB, S~ *Harecurt* 1268–81 Eynsham], S~ **St. John** O [*Stantone* DB, *Stanton* 1197 P, *Stantona Iohannis de Sancto Iohanne* 1155–61 Eynsham], S~ **upon Hine Heath** Sa [*Stantune* DB, *Staunton super Hyne Heth* 1327 Subs], S~ **Lacy** Sa [*Stantone* DB, *Stauntone Lacy* 1277 Ep], **Long** S~ Sa [*Stantune* DB, *Longa Stanton* 1245 FF], S~ Sf [*Stantun* 11 EHR 43, *-a* DB], S~ **Drew** So [*Stantone* DB, *Stanton Drogonis* 1253 Cl, *Standondru* 1291 Bath], S~ **Prior** So [*æt Stantune* 963, *Stantun* 965 BCS 1099, 1164, *-tone* DB, *Staunton Prior's* 1276 Ch], S~ St [*Stantone* DB], S~ **Fitzwarren** W [*Stantone* DB, *-ton* 1196 Cur], S~ **St. Bernard** W [*Stantun* 903 BCS 600, *-tone* DB], S~ **St. Quintin** W [*Stantone* DB, *Staunton St. Quintin* 1283 Misc]. OE *Stān-tūn*, which probably as a rule means 'TŪN on stony ground'. Sometimes a Stanton was named from some prominent stone or stones, as S~ Harcourt O.

S~ **under Bardon** Le is on the slope of Bardon Hill. Bardon is probably OE *Beorgdūn* 'barrow hill'.—S~ **by Dale** Db is nr Dale Abbey.—S~ **Drew** So was held by one Drogo in 1225 (Ass); cf. LITTLETON DREW.—S~ **Fitzwarren** W was held in part by Fulco filius Warini in 1196 (Cur); cf. NORTON FITZWARREN. —S~ **Harcourt** O was held by Ricard fil. Will. de Harecurt in 1166 (RBE). Cf. KIBWORTH HARCOURT.—S~ **upon Hine Heath** Sa. See HATTON.—S~ **Lacy** Sa was held by Roger de Laci in 1086 (DB). Cf. EWYAS LACY.—S~ **Prior** So belonged to the Prior of Bath.—S~ **St. Bernard** W, **St. John** O, & **St. Quintin** W from local families. Emmingus de Vico Sancti Johannis held a fee in S~ in 1246 (Ch). The name apparently means 'St. John Street' (in Oxford). Herbert de Sancto Quintino held S~ in 1212 (RBE). Cf. FIFEHEAD ST. QUINTIN.—S~ **on the Wolds** Nt. *Seggeswold* means '*Secg's* wold'. Cf. SEDGEBERROW.

Stantonbury Bk [*Stantone* DB, *Staunton Barry* 1300 Ipm]. See STANTON. The addition is a family name *Barri* (of French origin). Radulfus Barri held a fee in S~ in 1236 (Fees).

Stantway Gl nr Westbury on Severn [*Stauncteweie* 1221 Ass]. 'Stony road.' First el. OE *stāniht* 'stony'.

Stanwardine Sa nr Baschurch [*Sturdine* DB, *Stanwardin* 1193 P]. Perhaps 'enclosure made of stones'. Cf. WORþIGN.

Stanway Ess [(æt) *Stanwægun* c 995 BCS 1289, *Stanwega* DB], S~ Gl [*Stanwege* DB, *-weya* 1220 Fees], S~ He [*Stanewei* DB], S~ Sa [*Staneweie* DB, *-wey* 1242 Fees]. 'Paved road.' At least Stanway Ess and He are on Roman roads.

Stanwell Mx [*Stanwelle* DB, 1254 Val]. 'Stony spring or stream.'

Stanwick (stănĭk) Np [*Stanwigga* 10 PNNp, 1125-8 LN, *Stanwige* DB, *Stanewigge* 12 NS, *-wig* 1199 FF, *Stanewica* 1209-19 Ep]. Very likely 'the logan-stone', the second el. being an OE *wicga*, identical in form with *wicga* 'a beetle'. The original meaning of the latter is 'wriggler', the word being cognate with *wiggle* (13 &c.), MLG *wiggen* 'to move' &c. *Stānwicga* means 'stone wiggler, a wiggling stone'.

Stanwick YN [*Steinuuege*, *duæ Stenueges* DB, *Stainwegges* 1233 Ep], **Stanwix** Cu [*Steynweuga* c 1160 YCh 175, *Stainwegges* 1197 P]. 'Stone walls', the plur. of ON *stein-veggr* 'stone wall', also 'stone building'. The latter is on Hadrian's Wall. The former refers to some other ancient wall or walls.

Stapeley Chs [*Steple* DB, *Stapeleg* 1260 Court, *Stapelehe* 1272 Chester], **Stapely** Ha in Odiham [*Stapeleg* 1185, *Stapelea* 1190 P]. OE *stapol-lēah*, which may well mean 'wood where posts were got'.

Stapenhill St [*Stapenh'* 942 BCS 773, *Stapenhille* DB, 1200-10 BM]. Perhaps OE (æt) *stēapan hylle* 'steep hill'. Shortening of *ēa* to *ĕa*, whence ME *a*, may well have taken place. The form of 942 is in a late transcript.

Staple K [*Staples* 1205, *Staple* 1240, *Stapele* 1247 Ch], S~ **Fitzpaine** So [*Staple* DB, 1212 Fees]. OE *stapol* 'post, pillar'. The *staple* in the latter case was probably the big sarsen stone still found near the church (suggested by Mrs. Sixsmith of Thurlbear, Taunton).

S~ **Fitzpaine** was held by Robert Fitzpaine t. E 3. Cf. CARY FITZPAINE.

Stapleford Ca [*Stapelforda* 956 BCS 1346, *Stapleford* c 1080 ICC], S~ Chs [*Stapleford* DB], S~ **Abbots & Tawney** Ess [*Staplefort* DB, *Staplford Abbatis Sancti Edmundi* 1255 FF, *Stapilford Thany* 1291 Tax], S~ Hrt [*Stapeleford* 1198 FF, *Stapleford* 1254 Val], S~ Le [*Stapeford* DB, *Stapleford* 1199 FF], S~ Li [*Stapleforde* DB, *Stapleford*

1212 Fees], S~ (-ǎ-) Nt [*Stapleford* DB, *Stapelford* 1194 P], S~ W [*Stapleford* DB, *Stapelford* c 1115 Salisbury]. OE *stapol-ford* 'ford marked by a post'. Some forms suggest an alternative OE *stapola-ford* 'ford marked by posts'.

S~ **Abbots** belonged to the Abbot of Bury.— S~ **Tawney** was held by Richard de Tauny in 1253 (Cl). The name (*de Tania* 1190 P) is from LE TANNEY in Normandy.

Staplegrove So [*Stapilgrove* 1327 Subs], **Staplehurst** K [*Stapelhurste* 1242 Fees]. 'Grove and hurst where posts were got.'

Stapleton Cu [*Stapelton* 1190 P, *Stapleton* 1250 Ipm], S~ Gl [*Stapelton* 1208 Cur, 1250 Cl], S~ He [*Stepeltone* 1286 Ep, *-ton* 1308 Ipm], S~ Le [*Stapelone* DB, *Stapelton* 1185 P], S~ Sa [*Stepleton* Steph Eyton, 1242 Fees, *Stapelton* 1203 Ass], S~ So [*Stapelton* 1212 Fees, *Stapleton* 1236 Ass], S~ YN [*Staple(n)dun* DB, *Stapeltun* 1166 P], S~ YW [*Stapleton* DB, *Stapeltona* 1155-8 (1230) Ch]. 'TŪN by a post.' The exact meaning of the name is not clear. S~ He seems to have been originally *Stēpeltūn* 'TŪN with a steeple', and S~ Sa may have the same origin. But the *e*-forms are not so regular here, and possibly they represent OE *steapol*, a side-form of *stapol* with velar mutation.

Staploe Bd [*Stapelho* 1203 Cur, a 1228 BM]. 'HŌH or ridge with a post or pillar.'

OE *stapol* 'post, pillar' is a common pl. n. el., especially in combination with FORD, TŪN. In STAPLEFORD, STALBRIDGE the meaning of the word is clear, and STAPELEY, STAPELY, STAPLEGROVE, STAPLEHURST, STAPLOE offer no difficulties. Less clear is STAPLETON. *Staple* is also found as a second el. See e.g. BARNSTAPLE, CHIP-, DUN-, WHITSTABLE.

Starbotton YW [*Stamphotne* DB, *Stauerbotle* R 1 Cur, *Starbotene* 1268 Fount]. 'Valley where stakes were obtained.' The elements are OE *stæfer* or rather OScand *staver* 'stake' and OScand *botn* 'the innermost part of a valley'.

Stareton (-ar-) Wa [*Stauertun* 1157, *Stauerton* 1160, 1190 P]. Identical with STAVERTON Np.

Starston Nf [*Sterestuna* DB, *-tun* c 1095 Bury, *Stirston* 1205 Cur]. 'Styr's TŪN.' First el. ON *Styrr*, ODan *Styr* pers. n. (*Ster* DB).

Start Point D [*the Sterte* c 1540 Leland]. OE *steort* 'tail', also 'promontory'. The point gave its name to **Start Farm** [*La Sterte* 1309 Ch].

Startforth YN [*Stretford* c 1050 HSC, *Stradford* DB, *Stredford* 1104-8 SD]. A variant of STRETFORD.

Statfold St [*Stotfold* 1291 Tax, *Stottesfeld* 1293 Ass, *Stotfolt* 1327 Subs]. Most probably OE *stōd-falod* 'stud-fold'. But the first el. may be late OE *stott* 'a horse' or ME *stott* 'a young castrated ox, a steer'.

Statham Chs [*Stathum* 1285 Ormerod]. The dat. plur. of OE *stæþ* 'landing-place'.

Stathe So [*Stathe* 1233 Wells, 1234 FF]. OE *stæþ* 'landing-place'.

Stathern (-ăth-) Le [*Stachedirne* DB, *Stacthirn* c 1125 LeS, *Stakethurne* J Berk, *-thirne* 13 Fees]. Second el. OE *þyrne* 'thorn-bush'. The first is OE *staca* 'stake'. The meaning of the compound is not clear. Possibly a special kind of thorn-bush was called *stacanþyrne* or *stæcþyrne*.

Staughton, Little, Bd [*Estone* DB, *Stoctuna* 1167 P, *Parva Stocton* 1242 Fees], **Great S~** Hu [*Stoctun* c 1000 BCS 1306, *Stottun* 1163 P, *Stoctun* c 1198 BM]. A variant of STOCKTON.

Staunton Gl [*Stanton* 1146 Fr, 1220 Fees], **S~ on Arrow** He [*Stántún* 958 BCS 1040, *Stantune* DB], **S~ Harold** Le [*Stantone* DB, *Stanton Haraut* 1242 Fees, *S~ Harald* 1327 Subs], **S~** Nt [*Stantun* DB, *Stanton* 1177 P], **White S~** So [*Stantune* DB, *Whitestaunton* 1337 Ep], **S~** Wo [*Stantun* 972 BCS (1282), 978 KCD 619]. A variant of STANTON.

S~ Harold was held by Harold de *Leec* (cf. LEAKE) 12th cent. (TpR).—**White S~** perhaps from the colour of the church.

Staunton on Wye He [*Standune* DB, *Standon* 1194 P]. 'Stony hill.' See DŪN.

Staveley Db [*Stavelie* DB, *Staveleia* 1212 Fees], **S~** La [*Stavelay* 1282 FC], **S~** We [*Stavele* 1212 RBE, *-ley* 1274 Kendale], **S~** YW [*Stanlei* DB, *Staflea* 1167 P, *Staueleie* 13 NpCh]. OE *stæf-lēah* 'wood where staves were got'.

Staverton D [*Stofordtun, Stafortuna* c 1070 Ex, *Stouretona* DB, *Staverthon* 1242 Fees]. OE *Stānford-tūn* 'TŪN by a stony ford'. For the loss of *n* cf. STAFFORD Do. The change *ā* > *ō* appears remarkably early in this name.

Staverton Gl [*Staruenton* DB, *Stawerton* 1249, *Staverton* 1250 Cl]. Possibly identical with prec. name. But the first el. may be OE *stæfer* (cf. next name) alternating with an adj. **stæfren* 'of stakes'.

Staverton (stăr-) Np [*Stæfertun* 944 BCS 792, *Stavertone* DB], **S~** Sf [*Stauertuna, Stauretona* DB, *Stauerton* 1190 P]. OE *stæfer-tūn* 'TŪN made of stakes'. In this case OE TŪN in the sense 'fence' gives the best sense: 'enclosure made of stakes'.

Stăverton W [*Stavretone* DB, *Staverton* 1212 Cur]. Very likely OE *Stānford-tūn*. Cf. S~ D. The place is on the Avon.

Stavordale So [*Staverden* 1218 For, 1243 Ass, *-dal* 1243 Ass, *-dale* 1272 Wells]. Perhaps OE *Stānford-denu* 'valley with a stony ford'. But the first el. may be OE *stæfer* 'stake'.

Staward Nb [*Staworth* 1215–55 Ep, *-e* 1272 Sc, *Stannord* 1291 Ipm]. OE *stān-worþ* 'enclosure made of stone'. Cf. next name.

Stawell So [*Stawelle* DB, *Stanwelle* 1276 RH]. 'Stony stream or spring.' An *n* was often lost before *w*. Cf. STOWEY.

Stawley So [*Stawei* DB, *Stauleyg* 1236 Fees, *Stanlegh* 1243 Ass]. Apparently a variant of STANLEY. The loss of *n* is remarkable.

Staxton YE [*Stac(s)tone, Staxtun* DB, *Staxtona* c 1190 YCh 1221]. '*Stakk*'s TŪN.' ON *Stakkr* occurs as a byname and in pl. ns.

Staythorpe Nt [*Startorp* DB, *Starctorp* 1195, *Staretorp* 1196 P, *Starestorp* 1247 Ch]. Apparently '*Stari*'s thorp'. *Stari* is an ON pers. n.

Steane Np [*Stane* DB, *Stanes* 12 NS, 1220 Fees, *Stenes* 1250 Ep, 1254 Val, *Stene* 1293 BM]. OE *stān* 'stone'. The form with *i*-mutation is most easily explained from an OE locative form in *-i*. Cf. SCĒAT with its by-form *sciete*.

OE **stēap** 'steep'. See STEEP- (passim), STAPENHILL.

Stearsby YN [*Estiresbi, Stirsbi* DB, *Steresbi* 1167 P]. '*Styr*'s BY.' Cf. STARSTON.

Stebbing Ess [*Stabinga, Stibinga* DB, *Stubbinges* 1183 P, *Stebinges* 1212 RBE]. Either '*Stybba*'s people' or OE *stybbing* 'clearing', identical with ME *stubbing* (cf. STUBBINS). OE *Stybba* pers. n. is found in *Stybban snad* 960 BCS 1054.

Stechford Wo [*Stichesford* 1267 Ipm, *Stigford* 1275 Ass, *Stichford* 1296 AD]. OE *sticce* 'sticky' occurs (in *þæt sticce* 'the sticky matter'). The probability is that the adj. had palatal *c*. If so, it may be the first el. of the name, which would mean 'sticky, slimy ford'. The *s* of the first ex. is then inorganic.

OE **stede, styde** 'place, site of a building'. The sense 'site' is found in such names as BURSTEAD, HALSTEAD, KIRSTEAD, MILSTED, TUNSTEAD. Cf. HĀMSTEDE. In many such cases *stede* loses its meaning, and *hāmstede* means much the same thing as *hām*. CHIPSTEAD is 'market-place'. In STIDD La the word means 'place of worship', and this is probably the meaning of *stead* in NEWSTEAD, when the name denoted a monastery. In the numerous other names in *-stede* the simple sense 'place' is hardly satisfactory. When *stede* is combined with names of trees, as ASHTEAD, BOXTED, ELMSTEAD, MAPLESTEAD, the meaning may be 'a group of oaks, an oak copse' &c. But BANSTEAD, BINSTEAD, NETTLESTEAD can hardly mean simply 'place where beans grow' &c. This kind of rendering is frequently used in etymologies in this book, but only because the exact meaning is obscure. In northern counties *stead* means 'farm'. In early sources *stede* is clearly used in the sense 'vaccary, dairy-farm'. **Abbeystead** La is *vaccary del Abbey*

1323 LaInq. It seems probable that such a sense occurred also in southern counties. HORSTE(A)D, TISTEAD are best explained as '(outlying) pasture for horses or kids'. *Haldstede* may well have meant 'pasture with a shelter or hut'. See HAL-, HAWSTEAD. In some cases the first el. of names in *stede* is some topographical word, as in FEL-, MORE-, STANSTE(A)D, or an adj., as FAIR-, GREENSTEAD. Pers. ns. are rarely found combined with *stede*, but there is no reason to doubt that there are some exceptions. **Cowstead** K [*Cudesteda* 1194 P] contains *Cūpa* or *Cūda* pers. n. *Hredles stede* BCS 741, *Scelces stede* ib. 469 can hardly be anything else than '*Hrēpel*'s and *Scealc*'s (or the servant's) STEDE'. See also HARKSTEAD, WINESTEAD.

Stedham Sx [*Steddanham* 960 BCS 1055, *Stedeham* DB, 1162 P, *Stodeham* 1188 P]. 'The stallion's HAMM, HAMM where stallions grazed.' **Stedda* is a hypocoristic form of *stēda*, analogous to ON *stedda* 'mare'.

Steel Nb [*Le Stele* 1269 Ass]. Identical with dial. *steel* 'ridge, precipice', *stile* 'a steep path up a ridge'. The source is OE *stigol* 'stile', also no doubt 'steep ascent', lit. 'place where one has to climb'. **Steel Fell** We contains the same word. **Steel** Sa nr Prees [*Stile* DB, *Style* 1327 Subs] contains OE *stigol*, but perhaps in the ordinary sense 'stile'. *Steel* is a regular development of OE *stigol* in many dialects.

Steep Ha [*la Stiepe* 1230–5, *Stype* 1234, *la Stupe* c 1275 Selborne]. A derivative of OE *stēap* 'steep' meaning 'steep place', perhaps an OE **stiepe* fem.

Steephill Castle Wt [*Stupele* 1316 FA]. Apparently OE *stēpel*, *stiepel* 'steeple'.

Steep Holme So [(æt) *Steapan Re(o)lice* 915 ASC (B, C, D), *Stepholm* R 1 Berk, *Stupeholm* Hy 3 Berk, *Stepelholme* 1331 Ep]. 'Steep island.' For the OE name cf. FLAT HOLME.

Steeping Li [*Stepinge* DB, *Stepping* 1205 Cur, *Steppinges* 1209–35 Ep, *Parua Steping* 1199 P]. '*Stēapa*'s people.' *Steapa* occurs c 975 HEl. The river-name **Steeping** is a back-formation. Cf. LYMN.

Steeple Do [*Stiple* DB, *Stupel* 1212 Fees], S~ Ess [*Stepla* DB, *Stieples* 1163–70 AC]. OE *stēpel*, *stiepel* 'steeple'.

Steepleton Iwerne Do [*Iwerna* DB, *Stepeltone* 1212 RBE, *Stipel-*, *Stupelton* 1291 Tax]. 'TŪN with a church steeple.' Cf. IWERNE.

Steers Pool R La [*Styrespol* 1235 FF, *Sterespol* 1292 Ass]. '*Styr*'s stream.' Cf. PŌL and STARSTON.

Steetley Db in Whitwell [*Stiveleia* Hy 2 (1316) Ch, *Stiueclea* 1166 RBE, *Stukeley* 1572 BM]. Identical with STEWKLEY.

Steeton YW nr Bolton Percy [*Stiuetun* 963, *Styfetun*, *Styfingtun* c 1030 YCh 6 f., *Stive-*

tune DB], S~ YW nr Keighley [*Stiuetune* DB, *Stiveton* 12 Pudsay]. OE *Styfic-tūn* 'stub TŪN'. Cf. STYFIC. A *c* in this position was dropped early.

ON **steinn**, ODan, OSw stĕn 'stone' is the source of STAIN and the first el. of STAINFIELD (1), STANWICK YN, -WIX, STENWITH. In many cases it has replaced OE STĀN· Cf. STAIN- (passim).

Stella Du [*Stelyngleye* 1183 BoB], **Stelling** Nb [*Stelling* 1242 Fees]. The last is dial. *stelling* 'cattle-fold, place where cattle take shelter from the sun'. The first is 'pasture with a stelling'. See LĒAH.

Stelling K [*Stellinges* DB, *Steallinge* 11 DM, *Stellinge* 1294 Misc]. Perhaps '*Steall(a)*'s people'. OE *Stealla* is not evidenced, but cf. OG *Stallo*.

Stembury Wt [*Staneberie* DB, *Stevenebir*' 1287–90 Fees]. 'Stone fort' (OE *stænen* 'of stone' and BURG).

Stenigot Li [*Stangehou* DB, *Steninghog* 1199 (1330) Ch, *Staningeho* 1202 Ass, *Stainigot* 1212 Fees, *Stanigot* 1263, *Stenynghod* 1267, *Stanigod* 1272 FF]. The earliest forms point to OE *Stāninga-hōh*, the second el. being OE HŌH 'spur of hill'. *Stāningas* may be 'the people at a stone or rock'. But the common final *d* or *t* is remarkable. Possibly the OE **hōd* 'shelter' found in HOTHAM &c. is the second el.

Stenson Db [*Steintune* DB, *Steineston* 1206, *-tun* 1208 Cur]. '*Stein*'s TŪN.' Cf. STAINSBY.

Stenwith Li [*Stanuuald* DB, *Steinwath* Hy 2 (1316) Ch, 1212 Fees]. OScand *Stein-vað* 'stony ford'.

OE **steort** 'tail, promontory, tongue of land'. See START, STERT, HOUNDSTREET.

OE **stēpel**, **stiepel** 'steeple'. See STEEPHILL, STEEPLE, -TON, also STAPLETON He, Sa. As a distinctive addition *Steeple* refers to a church steeple (s~ MORDEN &c.).

Stepney Mx [*Stybbanhyp* c 1000 CCC, *Stibenhede* DB, *Stubbehuða* 1173, *Stebbehede* 1190 P, *Stebenheth* 1242 Fees]. '*Stybba*'s HȲp or landing-place.' Cf. STEBBING.

Steppingley Bd [*Stepigelai* DB, *Stepingelea* 1167 P]. 'The LĒAH of *Stēapa*'s people.' Cf. STEEPING.

Sterndale Db [*Sternedale* 1251 Ch, *Stenredal* 1263 Ch, *Stenerdale* 1288 Cl]. 'Stony valley.' First el. OE *stæner* 'stony ground'.

Sternfield Sf [*Sternesfelda* DB, *-feud* 1254 Val, *Sternefeld* 1235 FF]. The first el. might be an OE pers. n. **Sterne* derived from OE *sterne* 'stern'. Or SEWSTERN &c. might be compared.

Stert So [*Esturt* DB, *Sterta* 1166 RBE, 1212 Fees], S~ W [*Sterte* DB, *Stertes* 1197 P, *Sterte* 1270 Ipm]. OE *steort* 'promontory' &c. The reference is to spurs of hill.

Stetchworth Ca [*Steuicheswrðe* c 1050, *Steuecheworde* c 1060 KCD 907, 932, *Stiuechesuurda* c 1080 ICC, *Stiuicesuuorde* DB, *Stivecheswrthe* 1235 FF]. Perhaps 'stub WORÞ'. *Styfic-worþ* may have been altered to *Styficesworþ*, because names in WORÞ mostly have pers. ns. as first el. Cf. STYFIC. Or '*Styfic*'s WORÞ'.

Stĕvenage Hrt [*Stithenæce* 1065 BM, *Stigenace* DB, *Stitheneach* 1209–19 Ep, *Stivenach* 1203 Cur, *-e* 1254 Val]. OE (æt) *stiþanhæcce* 'the strong hatch or gate'. But '*Stiþa*'s hatch' is possible, as *Stiþ-* occurs in *-beorht*, *-wulf* &c., and perhaps preferable.

Steventon Brk [*Stifingehaeme gemæra, Stifingchæma gemære* 964 BCS 1142, *Stivetune* DB, *Stivinton* 1220 Fees], S∼ Ha [*Stivetune* DB, *Stiuintona* 1167 P, *Stiventon* 1231 Cl], **Stevington** (stĕfn) Bd [*Stiuentone* DB, *Stiuiton* 1196 P, *Steventon* 1284–6 FA], S∼ (-ĕv-) Ess in Ashdon [*Stauintuna* DB, *Stiuenton* 1166 P, *Steuinton* 1197 FF]. Probably identical with STEETON. It is true we should expect to find some forms with *u*, if the first el. is *styfic* 'stump'. Possibly we may postulate an OE pers. n. *Stif*, derived from *stif* 'stiff'. Partly the meaning may be 'the TŪN of *Stīf*'s people'.

Stewkley Bk [*Stiuelai* DB, *Stiuecelea* 1183 P]. OE *Styfic-lēah* 'stump LĒAH'.

Stewton Li [*Stivetone* DB, *Stiuetuna* c 1115 LiS, *-tun* 1199 FF]. Identical with STEETON.

Steyning (-ĕn-) Sx [(æt) *Stæningum* c 880 BCS (553), c 1000 Saints, *Estaninges* 1085 Fr, *Staninges* DB]. OE *Stæningas*, which may mean '*Stān*'s people' or 'dwellers at a stone'. The derivative in this case shows *i*-mutation. Of course it is possible that the immediate base had *i*-mutation. Cf. STAINING.

Stibba·rd Nf [*Estanbyrda, Stabyrda, Stabrige* DB, *Stiberde* 1202 FF, 1242 Fees, 1291 Tax, *Stibyrd* 1270 Ch, *Stiburde* 1316 FA]. The curious vacillation in the early forms is possibly due to the existence of two variant names, *Stigbyrde* and *Stānbrycg* or *Stānbyrde*. The first el. of the former is OE *stig* 'path'. The second el. is an OE *byrde*, corresponding to Fris *bird, berd* 'bank of a river or road', which is found in pl. ns. and is derived from *bord* 'border' &c. OE *byrde* would mean 'bank'. Stibbard is some way from the Wensum. The name may mean 'road-side'. Cf. BARDFIELD.

Stibbington Hu [*Stebintune* DB, *Stibbinctuna* c 1150 PNHu, *Stibbingtona* 1209–19 Ep]. First el. identical with STEBBING.

Stickford Li [*Stichesforde* DB, *Sticceforda* 1142 NpCh, *Esticheford* 12 DC, *Stikeforde* 1209–19 Ep], **Stickney** Li [*Stichenai* DB, *Sticcenaia* 1142 NpCh, *Stikeneia* 12 DC, *Stikenay* 1202 Ass]. The two places are c 2 m. apart. Stickney is between two streams, which run almost parallel for a long way and form a kind of island. Stickford

is higher up between the same streams. The peculiar situation suggests that the 'island' was called *Sticca* 'the stick' or *Sticcan ēg*, and that STICKFORD was *Sticcan ford* 'the ford in the narrow island called *Sticca*'.

Sticklepath D [*Stikelepethe* 1280 Ep]. 'Steep path.' OE *sticol* means 'steep'.

Stickney. See STICKFORD.

Stidd La [*Stede* 1276 Ass, (Camera Sancti Salvatoris vocata) *Le Stede* 1338 Whitaker]. OE STEDE 'place of worship'. Stidd is an old chapelry.

Stiffkey (-ŭk-) Nf [*Stiuekai* DB, *Stiuekeia* 1203 Ass, *Stivekeye* 1242 Fees]. OE *Styficēg* 'stump island, island with stumps of trees'. The river-name S∼ is a backformation. Cf. SNORING.

Stifford Ess [(on) *stiþforde* c 1090 Hickes *Diss. Epist.* (Reaney), *Stiforda* DB, *Stiford* 1177 BM, *Stifford* 1199 P]. First el. apparently not OE *stig* 'path' but the OE plantname *stiþe* (cf. STISTED).

OE **stig** 'path' is found in STYFORD, perhaps STYAL, BRANSTY (see BRASSINGTON), CORPUSTY, GRESTY. OE **stig** 'pigsty' may be found in some of these and in STISTED. Cf. ANSTEY, ANSTY.

OE **stigol** 'stile' &c. See STEEL, STILTON, HAMSTEELS.

Stildon Wo [*Stilladun* c 957 BCS 1007, *Stilledune* DB, *Stillindon* 1275 Subs]. *Stilladun* can hardly be for *Stillandun*. In all other names in the text *n* is preserved in a similar position. *Stilla-* is the gen. plur. of OE *stiell, stæll* 'place for catching fish'. The form *still* is found in *Bykenstill, still* 937 BCS 715 (late language). The original meaning was no doubt 'trap'; cf. Du *stel* 'trap for wild animals'. Stildon is 'hill where traps for animals were placed'. The OE *still* may alternatively have had the sense of dial. *stell* (Nb, Cu &c.), viz. 'enclosure for giving shelter to sheep or cattle'. An early ex. of this is *Cowestel* 1225 Ep (Y). See SPN, p. 156.

Stillingfleet YE [*Steflingefled* DB, *Steuelingeflet* 1190 P, *Stiuelingflet* 1211 FF], **Stillington** Du [*Stilligtune* c 1190 Godric, *Stilyngton* 1408 AD], S∼ YN [*Stivelinctun* DB, *Stiuelintone* 1170 P]. 'FLĒOT or stream and TŪN of *Stȳfel*'s people.' *Stȳfel* is a diminutive of *Stūf*. It is not certain, however, that Stillington Du is identical with S∼ YN, as no spellings with *v* are recorded.

Stilton Hu [*Stichiltone* DB, *Stichelton* 1167 P, *Stigelton* 1227 Cl]. 'TŪN at a stile.' But as the place is at the foot of a hill, *stigol* may here have the sense 'ascent'. Cf. STEEL.

Stinchcombe Gl [*Stintescombe* c 1155 Berk, *-cumbe* 1256 Ipm], **Stinsford** Do [*Stincteford* DB, *Stinteford* 1236 Fees, *Stintesford* Hy 3 Ipm, *Styntesford* 1303 FA]. The same first el. is found in *Stintesford* 892, 934 BCS 567, 699 (N. Newton W). *Stint* is dial. *stint* 'sand-piper, esp. the dunlin' (found from 1486 in OED).

Stirchley Sa [*Styrcleage* 1002 Wills, 1004 KCD 710, *Stirchelega* 1167 P]. 'Pasture for young bullocks or heifers.' Cf. STYRC.

Stirton YW [*Strettuna* 1159 Pont, *Stretton* 1226 FF]. A variant of STRETTON.

Stīsted Ess [*Stistede* 1046 Wills, c 1095 BM, *Stiesteda* DB, *Stisteda* 1183 P, *Stidsted* 1198, -*stede* 1204 FF]. Were it not for the last two examples, the etymology would be undoubtedly OE *stig-stede* 'place with a pigsty' (OE *stig*). If these forms are trustworthy, the first el. is OE *stiþe* 'lamb's cress' or 'nettle'. Cf. STIFFORD.

Stitchcombe W [*Stotecome* DB, -*cumba* 1167 P, *Stutescumb* 1228 Cl, 1242 Fees]. 'Valley infested by gnats.' First el. OE *stūt* 'gnat'.

Stithians Co [(rector) *Sancte Stethyane* 1268 Ep, (ecclesia) *Sancte Stediane* 1291 Tax]. A saint's name, according to Oliver *St. Stedianus*. The forms point to a woman saint.

Stittenham YN [*Stidmum* DB, *Stitlum* 1185 P, *Stiklum* 1260 PNNR]. *Stitlum* should be read as *Sticlum*. This is the dat. plur. of a word derived from OE *sticol* 'steep' and meaning 'steep place, slope, hill'.

Stivichall (stīchal) Wa [*Stiuethal* 1183 P, *Stiuechal* 1202 Ass, *Stivechale* 1274 Ipm]. 'HALH with stumps of trees.' Cf. STYFIC, HALH.

Stixwould Li [*Stigeswalde* DB, *Sticheswald* c 1115 LiS, -*walda* 1130 P, *Stikeswald* 1212 Fees]. '*Stig*'s wold or wood.' First el. ON *Stigr*, ODan *Stig* pers. n.

Stoberry Park So [*Stabergh*, -*burgh* 13 Wells], **Stoborough** Do nr Wareham [*Stanberge* DB, 1284 Cl, *Stobargh* 1431 FA]. OE *Stān-beorg* 'stony hill'.

Stobswood Nb [*Stobbeswod* 1250 Cl, -*wude* 1256 Ass]. First el. *stob* 'stump of a tree', a variant of *stub*.

OE **stoc** is found in the sense 'monastery, cell'. The original meaning 'place' is recorded in Orm c 1200 (*i faderr stoke* 'in a father's stead' &c.), and Symeon of Durham renders *Wdestok* by 'silvarum locus'. OE also had *stocweard* 'townsman' and *stocwic* as a variant of *stoc*. The meaning 'monastery, cell' is a specialization of the meaning 'place'. Cf. STŌW. *Stoc* is etymologically related to *stōw*, *styde*. In pl. ns. a meaning such as 'holy place, monastery' is obvious in HALSTOCK, and it is probably found in some other names. BRADENSTOKE was a monastery, and there was a monastery, founded in the 10th cent., at TAVISTOCK. KEWSTOKE seems to contain a saint's name. BINDON Abbey Do is called *Bindonestok* 1236 Ch. A meaning 'meeting-place' is

plausible in the hundred-names **Redbornstoke** Bd, perhaps **Winterstoke** So. But these meanings cannot be assumed for the bulk of cases. Nor can *stoc* mean simply 'place', except in some special cases, as TOSTOCK, if that means 'look-out place'. STOKE alone is a very common pl. n., and many names now consisting of *stoc* and some other el. were originally Stoke, as ALVER-, BISHOP-, REVELSTOKE, HIN-, LONG-STOCK, also, of course, such as STOKE DOYLE, STOGUMBER, STOGURSEY, STOKENHAM. The fact that STOKE is such a common name indicates that the places so named were once dependent on some village or manor (see Introd. p. xvi f.). This is corroborated by the fact that a good many names in -*stoc* have as first el. the name of a neighbouring village. CALSTOCK and CHARDSTOCK are c 3 m. distant from CALLINGTON and CHARD respectively. The names mean 'STOC belonging to Callington (Chard)'. BASINGSTOKE is near BASING, MEONSTOKE near MEON, PURTON STOKE near PURTON. NAVESTOCK is c 9 m. from NAZEING, but may quite well have been an outlying farm belonging to Nazeing. The exact meaning of *stoc* may have varied, but the probability is that it was generally 'cattle-farm, dairy-farm'. This is indicated by the name POUNDSTOCK, which has as first el. the word *pound* 'fold'. STOKENHAM was formerly also *Hurdestoke* (1198 Cur) 'STOC of the flock or of the cowherds'.

The first el. of names in -*stoc* is often a pers. n., as AD-, FRITHEL-, HADSTOCK, sometimes a tribal name, as COSTOCK, or a common noun, as BRIG-, LAVERSTOCK, or a river-name or some other pl. n., as GREY-STOKE, TAWSTOCK.

As a first el. *stoc* is probably often to be assumed rather than *stocc*. Many STOCK-TONS (STAUGHTONS, STOUGHTONS) are probably *Stoc-tūn*. Stoughton Sr is opposite to Stoke. *Stoc* is certainly the first el. of STOCKLAND, STOCKLINCH, STOCKSFIELD, STOCKWOOD, STOKESLEY. See further STOCK-, STOKE- (passim).

The form of the element is mostly Stoke, -stoke. But the uninflected form *stoc* would give Stock, -stock, a form sometimes found. In early sources Stoke often appears in the plural form Stokes, showing that the OE form was in many cases *stocu* plur. STOKE-HAM represents the dat. plur. OE *stocum*.

OE **stocc** 'stock, trunk of a tree' is a far rarer pl. n. el. than *stoc*, and can as a rule easily be distinguished from it. As a second el. it is found in WARSTOCK. STOCK (from *stocc*) alone occurs as a pl. n., perhaps in the sense 'foot-bridge'. As a first el. *stocc* is fairly common, and also a derivative *stoccen* 'made of stocks' occurs. See STOCK- (passim), STOKENCHURCH, (Stoke) NEWINGTON, (Stocking) PELHAM.

Stock Gaylard Do [*Stoches* DB, *Stoke Coilard* 1304 Ipm, *Stokk Coillard* 1335 BM], S~ Ess [*Herewardestoc* 1239 Ch, -*stok* 1254 Val], S~ So [*Stoke* 1303 FA], S~ YW

[*Stoche* DB, *Stock* 1147–50 YCh 1471, *Stok* 1246 FF]. A variant of STOKE, from OE STOC.

Gaylard, formerly *Coilard*, is presumably a family name.

Stock and Bradley Wo [*Stokke and Bradeleye* 1376 Pat, *Stoke Bradley* 1418 PNWo]. Near this is **Stock Wood** [*la Stolke* 1271 For]. *Stock* is OE *stocc* 'stock'.

Stockbridge Ha [*Stocbrigge* 1227 Ch, *Stokbregg* 1258 Ipm]. 'Stock bridge.'

Stockbury K [*Stochingeberge* DB, *Stocingabere* 11 DM]. 'The BÆR or swine-pasture of the people at Stoke.' First el. OE *Stocingas*. STOKE is a village not very far from Stockbury.

Stockeld YW [*Stochilde* 1166 P, *Stokelde* 12, *Stockelde* c 1200 Pudsay, *Stokild*, *Stokeld* 1208 FF, *Stokeheld* 1257 Ch]. 'Slope with tree-trunks or stocks.' Second el. OE *helde* 'slope'. The first might also be *stoc* 'place' &c.

Stockerston Le [*Stoctone* DB, *Stocfatestona* 1167 P, *Stocfaston* 1209–35 Ep, *Stokefaston* 1254 Val]. OE *stocc-fæsten* 'stronghold built of tree-trunks, block-house', changed into *Stokfaston*, or OE *Stoccfæsten-tūn*.

Stockham Chs [*Stoccum* c 1173 Ormerod (from Arley Charters), *Stoccum*, *Stockum* 1288 Court]. 'The stocks', dat. plur. of OE *stocc*.

Stockingford Wa [*Stoccingford* 1155–9 Fr, *Stockiford* Hy 2 (1318) Ch]. OE *Stoccenford* 'ford with a stock or tree-trunk'.

Stockland D [*Stocland* 1201 Abbr, *Stokeland* 1212 Fees], **S~ Bristol** So [*Stocheland* DB, *Stoclande* 1166 RBE]. OE *Stoc-land* 'land with or belonging to a STOC'.

S~ **Bristol** belongs to the chamber of Bristol.

Stockleigh English D [*Stochelie* DB, *Stockelegh* 1242 Fees, *Stokeley Engles* 1268 Ep], **S~ Pomeroy** D [*Stochelie* DB, *Stockele Pomeray* 1266 Ep], **S~ Du** [*Stocaleia* R 1 (1308) Ch, *Stocheleya* 12 BM], **S~ St** nr Rolleston [*Stochilea* c 1170 Fr, *Stokelee* 1330 FA]. Either OE *Stocc-lēah* 'wood from which stocks were got' or OE *Stoc-lēah* 'LĒAH with or belonging to a STOC'.

S~ **English** was held by Gilebertus Anglicus in 1242 (Fees), by Sir Hugh le Engles in 1268 (Ep).—S~ **Pomeroy** was held by Radulf de Pomerei in 1086 (DB), by Henry de Lapumerai in 1200 (Cur). Cf. BERRY POMEROY.

Stocklinch Magdalen & Ottersay So [*Stoche* DB, *Stok* 1243 Ass, *Stokelinges* 1196 P, *Stokelinz* 1201 Ass, *Stokelynche Magdalene* 1349 Ep, *Stokel[inz] Ostricer* 1257 Misc, *Stoke Ostrizer* 1285 FA]. Originally STOKE (cf. STOC), to which was added OE *hlinc* 'hill'.

S~ **Magdalen** from the dedication of the church.—*Ostricer* in S~ **Ottersay** is a family name, originally ME *ostreger*, *ostringer* from OFr *ostruchier*, MLat *austurcarius* 'keeper of goshawks'. Willelmus Austurcarius in So is mentioned 1194 P. John le Ostricer was tenant in 1243 (Ass).

Stockport Chs [*Stokeport* 1188 P, *-porte* c 1190 LaCh, 1260 Court, *Stockford* 1283–8 Chester]. Apparently 'stock PORT or town', but perhaps originally *Stockford*. The place is on the Mersey.

Stocksfield Nb [*Stokesfeld* 1242 Fees, 1256 Ass]. 'FELD belonging to the STOC' (very likely Hexham). *Stoc* here means 'holy place, monastery'.

Stockton Chs nr Malpas [*Stocton* 1306 Ormerod], **S~ Heath** Chs [*Stocton* 1287 Court], **S~ on Tees** Du [*Stocton* 1196 P, *Stoketun* 1208–10 Fees, *-tone* 1228 FPD], **S~ He** [*Stoctune* DB], **S~ Nf** [*Stoutuna* DB, *Stocton* 1180 ff. P], **S~ Sa** nr Ironbridge [*Stochetone* DB, *Stocton* 1200 P], **S~ Sa** nr Welshpool [*Stocton* 1242 Fees], **S~ W** [*Stottune* DB, *Stocton* 1190 P], **S~ Wa** [*Stocton* 1249 Ass, 1291 Tax], **S~ on Teme** Wo [*Stoctun* c 957 BCS 1007, *Stotune* DB], **S~ on the Forest** YN [*Stochetun*, *Stocthun* DB, *Stocatuna* 1148 YCh 179], **S~ YW** [*Stochetun* DB, *Stocton* c 1140 YCh 1862]. OE *Stoc-tūn* 'TŪN with or belonging to a STOC', or OE *Stocc-tūn* 'homestead built of logs'. Most Stocktons probably belong to the first category.

Stockwell Sr [*Stokewell* 1197 FF, *Stokwelle* 1310 Ch]. 'Stream with a foot-bridge consisting of a tree-trunk.'

Stockwith, East, Li, West S~ Nt [*Stochithe* 12 Subs, *Stokhede* 1188 P, *Stochith'* 1226 Cl]. OE *stocc-hȳþ* 'landing-place made of stocks'. The two Stockwiths are opposite to each other on the Trent.

Stockwood Do [*Stocwode* 1221–3 Montacute, *Stokes sancti Edwoldi* 1238 Pat]. Originally STOKE (see STOC). Later *wood* was added.

OE **stōd** 'stud, herd of horses' is a common first el. of pl. ns. See STADHAMPTON, STOD-, STOOD-, STUD- (passim). OE *stōd-falod* 'stud-fold' is the source of STATFOLD, STOTFOLD. *Stōdfalod* is the source of **Studfold**, **Stotfold** &c., common names of old (Roman or other) enclosures. The names indicate that the Anglo-Saxons often used such old enclosures for horse-folds.

ON **stǫð** (plur. *stǫðvar*) 'landing-place' is found in BURTON ON STATHER, CROXTETH, TOXTETH.

Stodday La [*Stodhae* c 1200 CC, *Stodaye* 1246 LaInq, *Stodhag* 1262 Ass]. OE *stōd-haga* 'enclosure for horses'. See STŌD, HAGA.

Stodmarsh K [*Stodmerch* 675, *-mersche* 686 BCS 36, 67]. 'Marsh where horses grazed.'

Stody (-ŭ-) Nf [*Estodeia* DB, *Stodheye* 1254 Val]. OE *stōd-gehæg* 'enclosure for horses'.

Stoford So nr Yeovil [*Stafford* 1225 Ass, *Stoford* 1274 Ipm], **S~ So** nr Halse [*Stau-*

ford n.d. Buckland, *Stoford* 1281-90 ib.], S~ W [*Stoford* 1284 Ipm, *Stouford* 1352 Cl]. OE *Stānford* 'stony ford'. Cf. STAF-FORD Do.

Stogu·mber So [*Stoke Gunner* 1225, 1248, *Stok Gomer* 1249 Ass, *Stokegumer* 1249 FF]. Originally STOKE. See STOC. The addition is the name of an owner. The variation between *Gunner* and *Gumer* may point to original *Gunmer* from OG *Guntmar*.

Stogu·rsey So [*Stoche* DB, *Stok Curcy* 1212 Fees, *Stoke Curcy* 1241 BM]. Originally STOKE. Cf. STOC. The manor was held by William de Curci t. Hy 1. The name is from COURCY in Normandy.

Stoke, a very common name, represents OE *stoc* (or *stocu* plur.) in various special senses. See STOC. In a great many cases a distinctive addition has been made in post-Conquest times. S~ **Goldington** Bk [*Stoches* DB, *Stoch Petri de Gold'* 1167 P, *Stokegoldington* 1275 RH]. Goldington from G~ in Bd.—S~ **Hammond** Bk [*Stoches* DB, *Stokes Hamund* 1242 Fees]. Held by one Hamon 12th cent. *Hamon* is an OFr pers. n. of OG origin.—S~ **Mandeville** Bk [*Stoches* DB, *Stoke Mandeville* 1284-6 FA]. Cf. HARDINGTON MANDEVILLE.—S~ **Poges** Bk [*Stoches* DB, *Stokepogeis* 1292 Ipm]. Held by Hubert le Pugeis in 1255 (RH). Cf. BROUGHTON POGGS.—S~ Chs nr Chester [*Stoke* 1260 Court].—S~ Chs nr Nantwich [*Stoke* E 3 Ormerod].—S~ **Climsland** Co [see CLIMSLAND].—S~ D in Hartland [*Nistenestoch* DB, *Nectanestoke* 1189 Ol]. Dedicated to St. Nectan. Cf. ST. NIGHTON.—S~ **Canon** D [*Hrocastóc* 938 BCS 723, *Stoctun* c 970 ib. 1244, *Stoche* DB, *Stoke Canonicorum* 1316 FA]. Held by Exeter monastery.—S~ **Damarel** D [*Stoches* DB, *Stok Aubemarl* 1281 Ass]. Held by Robert de Albamarla in 1086 (DB). Cf. HINTON ADMIRAL.—S~ **Fleming** D [*Stoc* DB, *Stokes* 1218 FF, *Stoke Flandrensis* 1261 Ep, S~ *Flemmeng* 1275 Ipm]. Held by William le Flemeng (Flandrensis) in 1219 (FF). Cf. BRATTON FLEMING.—S~ **Gabriel** D [*Stoke-Gabriel* 1309 Ep]. Dedicated to St. Gabriel. —S~ **Rivers** D [*Stoche* DB, *Stoke Ryvers* 1284-6 FA]. Held by the Rivers family from the 12th cent. From RÉVIERS (Norm.). —S~ Db [*Stoche* DB, *Stoke* 1265 Misc]. — S~ **Abbott** Do [*Stoche* DB, *Stoke Abbots* 1275 FF]. Held by the Abbot of Sherborne. —**East** S~ Do [*Stoches* DB, *Stokes* Hy 3 Ipm].—S~ **Wake** Do [*Stoche* DB, *Stoke Wake* 1285 FA]. Cf. WAKES COLNE.—S~ **Bishop** Gl [(æt) *Stoce* 804, 883 BCS 313, 551, *Stoche* DB, *æt Bisceopes stoce* 984 KCD 646]. Held by the Bishop of Worcester.— S~ **Gifford** Gl [*Stoche* DB, *Stoke Elye Giffardi* 1221 Ass, *Stokes Giffard* 1243 Ass]. Held by Osbern Gifard in 1086 (DB). Cf. ASHTON GIFFORD.—S~ **Orchard** Gl [*Stoches* DB, *Stoke Archer* 1287 QW]. Held by Johannes le Archer in 1244 (Fees). Archer is a family name, originally 'archer'.—S~

Charity Ha [*Stoches* DB, *Stokecharite* c 1270 Ep]. Held by Henry de la Charite in the 13th cent. Charity is thus a family name.—S~ **Edith** He [*Stoches* DB, *Edithe-stoc* c 1180 Fr, *Stoke Edithe* 1242 Fees]. Named from Queen Edith, who held the manor at the time of the Conquest.—S~ **Lacy** He [*Stoke Lacy* 1242 Fees]. Cf. EWYAS LACY.—S~ **Prior** He [?(æt) *Stoce* a 1038 KCD 755, *Stoca* DB]. Belonged to the Prior of Leominster.—S~ K [*Stokes* 738, *Stoc* 10 BCS 159, 1322, *Estoches* DB].— S~ **Golding** Le [*Stokes* 1200 Cur, *Stok* 1209-35 Ep]. Held by Petrus de Goldinton in 1200 (Cur).—S~ **Rochford, North & South** S~ Li [*Stoc* c 1067 Wills, *Stoche(s)*, *Nort-*, *Sudstoches* DB]. Ralph de Rocheford had land near S~ in 1303 (FA).—[S~ **Newington** Mx. See NEWINGTON.]—S~ **Ferry** Nf [*Stoches* DB, *Stokeferie* 1248 Ch]. Named from a ferry over the Wissey.— S~ **Holy Cross** Nf [*Stoches* DB, *Crouche-stoke* c 1150 Bodl]. Presumably from the dedication of the church. *Crouch* is an old form of *cross*.—S~ **Albany** Np [*Stoche* DB, *Stok Aubeney* 1254 Val]. Albany is a family name taken from one of the AUBIGNYS in France. Willelmus de Albinni held *Stoch* in 1156 (P).—S~ **Bruern** Np [*Stoches* DB, *Stokbruer* 1254 Val]. Cf. BUCKLAND BREWER. William Briwere held Stoke in 1212 (RBE). —S~ **Doyle** Np [*Stoche* DB, *Stoke Doyle* 1428 FA]. Henricus de Oilli occurs in 1189 (1332 Ch) in a document where *Stokes* is mentioned. Cf. ASCOT D'OILLY.—S~ **Bardolph** Nt [*Stoches* DB, *Stokes Doun Bardulf* 1195 P, *Stokebardolf* 1329 QW]. Bardulf is a family name, originally a pers. n., here probably of Norman origin.—**East** S~ Nt [*Stoches* DB, *Stokes* 1163 RA].—S~ **Lyne** O [*Stoches* DB, *Stoke Insule* 1254 Val, S~ *del Ile* 1317 Ch]. *Lyne* from *Ile*. Otue[l] de Insula was tenant in 1167 (P), 1198 (FF). Cf. KINGSTON LISLE.—**North & South** S~ O [*Stoch* DB, *Stoches* c 1160 RA].—S~ **Talmage** O [*Stoches* DB, *Stokes Talemasche* 1219 Ep]. Held by Petrus Talemasche in 1207 (Cur). Talmage is a by-name from OFr *talemache* 'knapsack'.—S~ **Dry** Ru [*Stoche* DB, *Drie Stoke* 1220 Ep]. 'Dry Stoke.'—S~ Sa nr Greet [*Stok* 1203 Cur].—S~ **St. Milborough** Sa [*Godestoch* DB, *Stoke St. Milburg* 1291 Ch]. *Mildburga abbatissa* is mentioned in a Salop charter 901 BCS 587.—S~ **upon Tern** Sa [*Stoche* DB, *Stoke super Tyrne* 1316 FA]. See TERN.—S~ **Ash** Sf [*Stoches* DB, *Stoche* c 1095 Bury]. *Ash* may be the name of the tree.—S~ **by Clare** Sf [*Stoches* DB, *Stokes near Clare* 1287 Cl]. Cf. CLARE.— S~ **by Nayland** Sf [*Stoke* c 950, *Stoc* 970, c 995 BCS 1012, 1269, 1288 f., *Stokeneylond* 1272 FF]. Cf. NAYLAND. There was a Saxon monastery here.—**East** S~ So [*Est Stoke* 1350 BM].—S~ **Lane** or **St. Michael** So [*Mikelstok* 1243 Ass, *Stoke Michaelis* 1428 FA]. *Lane* is obscure.—**North** S~ So [*Norþstoc* 808 BCS 327].—S~ **Pēro** So [*Stoche* DB, *Stoke Pyro* 1326 Ep]. Held by

William de Pyrhou in 1243 (Ass). *Pero* is a family name derived from PIROU in Normandy.—**Rodney S~** So [*Stoches* DB, *Stokes Giffard* 1243 Ass]. Held by Richard de Rodene in 1303 (FA). He got it by marriage with a Giffard.—**S~ St. Gregory** So [*Stokes* 1225 Ass, (chapel of) *St. Gregory of Stoke at Northcuri* 1233 Wells].—**S~ St. Mary** So [*Stoc* 854, *Æs Stoce* 882 BCS 475, 550, *Stocha* DB]. Dedicated to St. Mary.— **South S~** So [*Tottanstoc* 961 BCS 1073, *Sudstoca* 1156 Wells]. The OE form means '*Totta*'s STOC'.—**Stoney S~** So [*Stoche* DB]. —**S~ sub Hamdon** So [*Stoca* DB, *Stokes under Hamden* 1248 Ch]. Cf. HAMDON.— **S~ Trister** So [*Stoche* DB, *Tristrestok* 1265 Ep, *Stoketristre* 1304 Ch]. *Trister* is obscure. It has been derived from the family name *del Estre*.—**S~ d'Abernon** Sr [*Stoche* DB, *Stokes de Abernun* 1253 Ch]. Gilbert de Abbernun held the manor in 1236 (Cl). Abernon is a Fr family name from ABENON in Normandy.—**S~ next Guildford** Sr [*Stochæ* DB, *Stok juxta Gildeford* 1205 Ch].—**S~ upon Trent** St [*Stoche* DB, *Stoke* 1232 Ch].—**North S~** Sx [*Stoches* DB, *Northstok* 1230 P].—**South S~** Sx [*Stoches* DB, *Sudstok* 1242 Fees].—**West S~** Sx [*Stokes* 1206 Cur].—**Beeching S~** W [*Stoke* 941 BCS 769, *Bichenestoch* DB]. Beeching perhaps OE *biccena* 'of bitches'.— **Earl S~** W. See ERLESTOKE.—**S~ Farthing** W [(be eastan) *Stoke*, *Stochæmaland* 955 BCS 917, *Stoke Verdon* 1412 FA]. The manor was held by Rois de Verdun in 1242 (Fees). Cf. NEWBOLD VERDON.—**S~** Wa [*Stokes* 1235 Ch].—**S~ Bliss** Wo [*Stoch* DB, *Stoke de Blez* 1242 Fees]. Bliss is the family name of Blez or Bledis (so 1242 Fees), apparently from BLAY in Normandy. —**S~ Prior** Wo [*Stoke* 770 BCS 204, *Stoche* DB, *Stok Prior* 1291 Tax]. Belonged to Worcester Priory.—**Severn S~** Wo [*Stoc* 972 BCS 1282, *Stoche* DB, *Savernestok* 1212 Fees]. Situated on the Severn.

Stokeham Nt [*Stokum* 1242 Fees, *Stocum* 1303 FA]. The dat. plur. of STOC (q.v.).

Stokeinteignhead D [*Stoches* DB, *Stokes in Tynhide* 1279 Ep]. Originally STOKE. The addition means '(manor of) ten hides'. It was associated with the river-name TEIGN.

Stŏkenchurch Bk [*Stockenechurch* c 1200 PNBk]. OE *stoccene cyrice* 'church built of stocks, timber church'. OE *stoccen* means 'of stocks'.

Stokenha·m D [*Stokes* 1242 Fees, *Stok in Hamme* 1276 RH]. Originally STOKE. The addition *Hamm* is the name of a district. The name means 'low-lying river land'.

Stokes Bay Ha [*Stokes juxta mare* 1174 Fr]. The plur. of STOC.

Stokesay Sa [*Stoches* DB, *Stoksay* 1256 Ass]. Originally STOKE. The manor was held by Hugh de Sei in 1195 (Cur). Cf. HAMSEY.

Stokesby Nf [*Stokesbei* DB, *Stokebi* 1155 Fr, 1168 P, *Stokesbi* 1194 P]. Very likely OE *Stoc*, to which was added OScand BY.

Stokesley YN [*Stocheslage* DB, *Stokesley* c 1120 YCh 559, *Stocaleia* 1157 ib. 354, *Stoclai* 1197 P]. 'LĒAH belonging to or with a STOC.' The place may have belonged to Stockton on Tees.

Stŏnar K [*Stanora* 1178 P, *Stanores* 1243 StAug, *Stonore* 1293 RBE]. OE *Stān-ōra* 'stony shore or landing-place'.

Stondon, Lower & Upper, Bd [*Standone* DB, *-don* 1200 FF], **S~ (-ō-) Massey** Ess [*Staundun* 1062 Th, *Standon de Marcy* 1238 Subs]. OE *Stān-dūn* 'stony hill'. **S~ Massey** from the Marci family (13th cent.). MARCY is the name of a place in Calvados, Normandy.

Stone Bk [*Stanes* DB, c 1145 Oxf], **S~** Gl [*Stane* 1250 Cl, *Stone* E 1 Berk], **S~** Ha [*Ad Lapidem* c 730 Bede, *Æt Stane* c 890 OEBede, *Stone* 1324 Ipm], **S~** K nr Dartford [*Stanes*, (of) *Stane* 10 BCS 1321 f., *Estanes* DB], **S~ cum Ebony** K [*Stane* c 1185 Penshurst], **S~ next Faversham** K [*Stane* 11 DM, *Stone* 1316 FA], **S~** So in Mudford [*Stane* DB, *la Stane* 1243 Ass], **S~** So nr Bristol [*Stone* 1327 Subs], **S~** St [*Stanes* 1187 P, 1201 Cur], **S~** Wo [*Stanes* DB, 1212 Fees]. OE *stān* or *stānas* 'the stone or stones'. The exact meaning is generally obscure. Stone Ha was very likely named from a stone marking the passage from Hampshire to Wight. Stone So in Mudford gave its name to Stone Hundred. The stone from which it took its name, called 'The Hundred Stone', is still *in situ* (Anderson, *Hundred-names*, ii. 58).

Stoneaston So. See EASTON.

Stonebury Hrt [*Stanes* DB, *Stoneberi* 1220 Fees]. Originally STONE, *bury* 'manor' having been added.

Stonegrave YN [*Staningagrave* 757–8 BCS 184, *Stane-*, *Stainegrif* DB, *Steingreua* 1163 P, *Stenegreve* 1218 FF]. Originally no doubt OE *Stān-græf* 'quarry', later remodelled to *Stāninga-græf* 'the quarry of the *Stāngræf* people', eventually to OScand *Steingryfia* 'stone pit, quarry'. Cf. GRYFIA.

Stoneham, North & South, Ha [*Stanham* 925–41, *æt Stanham* 932 BCS 649 f., 692, *Stanham* DB, *Stonham* 1281 BM]. 'HĀM by a stone or with stony ground.'

Stonehenge W [*Stanenges* c 1130 HHunt, *Stanhenge* 1205 Lay, *Stonheng* 1297 Rob Gl]. 'Stone monument.' The second el. is derived from the verb *hang*. It may be the OE **hencg* that is the source of *hinge*. More likely it is OE *hengen* 'hanging, gibbet' &c., the meaning being here 'that which is hung up'. The reference would be particularly to the horizontal stones resting on pillars.

Stonehouse D [*Stanehus* DB], **S~** Gl [*Stanhus* DB, *Stonhus* Hy 3 Ipm]. 'Stone house.'

Stoneleigh Wa [*Stanlei* DB, *-leia* 1153 BM, *Stonle* 1285 QW], **Stonely** Hu [*Stanlegh* 1260 Ass]. A variant of STANLEY.

Stonesby Le [*Stovenebi* DB, *Stonesbia* c 1125 LeS, *Stovenesbi* 1204 Cur]. The first el. may be ON *stofn* 'stump of a tree', OE *stofn* the same. The normal gen. form suggests a pers. n. *Stofn, but the OE name of the place may have been *Stofn* (cf. STOVEN), from which was formed *Stofnesby*.

Stonesfield O [*Stuntesfeld* DB, 1167, 1194 P, *Stuntefelda* 1130, *-feld* 1195, 1230 P]. '*Stunt*'s or *Stunta*'s FELD', *Stunt(*a*) being a nickname formed from OE *stunt* 'foolish'.

Stoneton Wa [*Stantone* DB, *Stonton* 1316 FA]. A variant of STANTON.

ON **stǫng** (gen. *stangar*), OSw *stang* 'pole'. See STANGHOW, GAR-, MALLERSTANG.

Stonham Aspall, Earl & Little S~ Sf [*Stonham* c 1040 Wills, *Stanham* DB, 1190 P, *Stanham Comitis* 1254 Val, *Parva Stonham* 1219 FF]. Identical with STONHAM. **S~ Aspall** was held by Roger de Aspale in 1292 (FF). Cf. ASPALL.—Comes Rogerus (Bigod) in *Stanham* (**Earl S~**) is mentioned 1212 RBE.

Ston(n)al St [*Stanahala* 1143 Oxf, *Stanhala* 1167 P, *Stonhal* Hy 3 BM]. 'Stony HALH.'

Stonor O [(in) *Stanora lege* 774 BCS 216, *Stonor* 1279 RH]. OE *Stān-ōra* 'stony ridge or slope'. Cf. ŌRA.

Stonton Wyville Le [*Stantone* DB, *Staunton Wyvile* 1265 Misc]. A variant of STANTON.

The manor was held by Robert de Wivill 1209–35 (Ep). The family name appears as *de Widuill* 1152–67 AC, *de Wiuilla* Hy 2 DC &c. It comes from GOUVILLE in Normandy (*Wiwilla* 1233).

Stonydelph Wa [*Stanidelf* 1202 FF]. OE *stān-gedelf* 'quarry'. Cf. STANDHILL.

Stonyhurst La [*Stanyhurst* 1358 FF]. 'Stony hill.' Cf. HYRST.

Stoodleigh D [*Stodlei* DB, *Stodleg* 1205 FF]. 'Horse pasture.' Cf. STŌD, LĒAH.

Stopham Sx [*Stopeham* DB, *Stopham* 1234 Cl], **Stopsley** Bd [*Stoppelee* 1198 FF, *Stopesleia* 1199 P, *Stoppeleg* 1202 Ass, *Stoppislee* 1262 BM]. The names must be compared with *Stoppingas* 723–37 BCS 157, the name of a district in Wa, also with **Stop Street** in Fonthill Gifford W [*Stoppe* 1372 AD], which shows that there was a topographical word *stoppe*. This might be identical with OE *stoppa* 'pail, bucket', but the topographical sense is obscure. The situation of the places does not seem to give a definite clue. A sense 'hollow, pit' might be conjectured. Second el. HĀM, LĒAH.

ON **storð** 'brushwood' occurs in some minor names in northern counties, e.g. **Storrs** La [*Stordis* 1242 Fees, *Storthes* c 1350 LaCh], **Storth** We [*Storthes* 1349 Cl].

Storeton Chs [*Stortone* DB, Hy 2 Chester, *-tuna* c 1150 Chester]. 'Big TŪN', the first el. being ON *stórr* 'big'.

Storridge He [*Storugge* Hy 3 Misc]. Probably OE *Stān-hrycg* 'stony ridge'.

Storrington Sx [*Storgetune, Estorchetone* DB, *Storkinton* 1185 P, *Storgeton* 1242 Fees, 1263 FF, *Storketon* 1255 Sele]. Has been explained as OE *Storca-tūn* 'homestead with storks'. This may be right.

Storrs. See STORÐ.

Stort R Ess, Hrt [*Stort* 1586 Camden]. A back-formation from **Bishop's Stortford** Hrt [*Storteford* DB, 1178 AC, 1200 Cur, *Sterteford* 1199 Cur, 1278 Ass]. Apparently OE *Steorta-ford* (cf. STEORT) or *Steortan ford*. In the latter case we may compare *Steortan leag* 938 BCS 731, which may contain a pers. n. *Steorta. In the former case the meaning would be 'ford by the tongues of land', which does not seem suitable. A nickname *Steorta* would easily be formed from OE *steort* 'tail'. Cf. the ON byname *stiartr*.

Storth. See STORÐ.

Stotfold Bd [*Stodfald* 1007 Crawf, *Stotfalt* DB], **S~** YW [*Stodfald, Stotfalde* DB, *Stodefald* 1252 Ch]. 'The stud-fold.'

Stottesdon Sa [*Stodesdone* DB, *-don* 1162, 1167 P, *Stottesdun* 1160, *-don* 1161, 1194 P]. OE *Stōdesdūn* 'the hill of the herd of horses' or *Stottesdūn* 'hill of the horse or bullock'. The first alternative seems preferable.

Stoughton (-ō-) Le [*Stoctone* DB, *-tona* 1174 BM], **S~** (-ow-) Sr [*Stocton'* 1225 Cl], **S~** (-aw-) Sx [*Estone* DB, *Stoctona* 1121 AC]. A variant of STOCKTON. **S~** Sr is on the Wey opposite to STOKE. It is clearly *Stoc-tūn*, not *Stocc-tūn*.

Stoulton (stōtn) Wo [*Stoltun* 840 BCS 430, DB]. **S~** is nr Low Hill, the meeting-place of the hundred. It is suggested in PNWo that *stōl* in the name refers to some seat of authority. The name would mean 'TŪN with the judge's chair'.

Stour (-ow-, -ōō-) R (1) K [*Stur* 686, 814 BCS 67, 344, *Sture* (gen.) 811 ib. (335), 1035 BM], **S~** (-ow-) R (2) Ca, Ess, Sf [*Sture* (fluminis) c 894 Asser, (into) *Sture* c 995 BCS 1289, *Stura* c 1200 Gervase], **S~**(-ow-) R (3) W, Do, Ha [(on) *Sture* 944, *Stoure* 968 BCS 793, 1214], **S~**(-ow-) R (4) O, Gl, Wo, Wa [*Stuur* c 757, *Stur* 764–75 BCS 183, 205], **S~**(-ow-) R (5) St, Wo [*Stur* 736, (on) *Sture* 866 BCS 154, 513, *Stoure* 1300 For]. A Brit river-name, identical with STURA in Italy (*Stura* Pliny) and derived from the root steu- in Sanskrit *sthāvará-* 'firm', Lat *stauro*, ON *staurr* 'a pole' &c. The name probably means 'strong, powerful river'.

Stour Provost, East & West S~, Stourpaine Do [*Stur, -e* DB, *Stures Pratellorum*

1243 Cl, *Sturprewes* 1307 FF, *Stures Paen* 1242 Fees, *Stureweston* 1290 Ch]. Named from the river Stour (3).

Provost is a corruption of *Prewes*, which represents PRÉAUX in Normandy. The manor belonged to Préaux Abbey.—**Stourpaine** was held by Pagan son of William in 1226 (FF). Cf. BOOTHBY PAGNELL.

Stourbridge (-er-) Wo [*Sturbrug* 1255 Ass]. 'Bridge over R Stour (5).'

Stourmouth K [*Sturmutha* 1089 BM, -*mude* 11 DM, *Sturemuda* 1190 P]. 'Mouth of R Stour (1).'

Stourport Wo. A late name. The place is on R Stour (5).

Stourton St [*Sturton* 1227 PNSt], S~ (-er-) W [*Stortone* DB, *Sturton* 1182 P], S~ Wa [*Sturton* 1206 Cur, 1229 Cl]. 'TŪN on R Stour' (5, 3, 4 respectively).

Stoven Sf [*Stoune, Stouone* DB, *Stovene* 1201 Cur, *Stofne* 1254 Val]. OE *stofn* 'stem, tree-stump'.

OE **stōw** is recorded in senses such as 'place', 'inhabited place', 'holy place, hermitage, monastery', probably 'church'. The word is found in many compounds, e.g. *cēap-, pleg-, wicstōw*, where the meaning is 'place'. *Stōw* has meanings in common with *stoc*, of which it is a cognate, but there is no reason to assume for the word the sense 'dairy-farm'. STOW alone is not nearly so common as a pl. n. as STOKE. When used alone, *stōw* certainly meant in some cases 'holy place, hermitage, monastery'. See e.g. STOW ON THE WOLD, ST. MARY STOW. A meaning 'meeting-place' is plausible for STOWMARKET (originally STOW). The exact meaning of the other Stows is not apparent. It may sometimes be 'church'.

Stōw hardly occurs as a first el. except in late combinations such as STOWLANGTOFT. STOWELL, STOWFORD and the like contain OE *stān*. As a second el. *stōw* means 'place' in such names as BURSTOW, CHURCHSTOW, PLAISTOW, WISTOW, probably BRISTOL. Most names in -*stow(e)* have as first el. a saint's name and mean 'place dedicated to a saint, church', as BRIDESTOWE, EDWINSTOWE, FELIXSTOWE, INSTOW, JACOBSTOW(E), MARSTOW, MARYSTOW, PETROCKSTOW; cf. also GODSTOW. HALSTOW is 'holy place'. In PLEMSTALL *stōw* means 'hermitage', in HIBALDSTOW 'burial-place', but the place was very likely dedicated to St. Hygebold. The exact meaning of -*stow* is not quite clear in BROXTOW, FULSTOW, WALTHAMSTOW, WISTANSTOW, WISTOW Le, whose first el. is a pers. n. Yet Wistanstow, Wistow may well be named from St. Wigstan. In the others 'hermitage' is a probable meaning. Very difficult etymologically are HORK-, SPURSTOW.

Stow cum Quy Ca [*Stoua* 1086 ICC, *Stowe* 1202 FF, *Stow* 1254 Val, *Coweye and Stowe* 1271 FF], S~ **Maries** Ess [*Stowe* 1230 FF, S~ *Mareys* 1420 FA], S~ **on the Wold** Gl [*Edwardestowe* c 1107 BM, *Stowe*

Sancti Edwardi 1330 Ch, *Stowe* 1221 Ass], **Long** S~ Hu [*Estou* 1086 IE, *Stou* 1163 RA, *Long Stowe* 1380 AD], **St. Mary** S~ Li [*Sce Marian stow* c 1067 Wills, *S' Maria de Stou* DB, (ecclesia) *sancte Marie de Stou* 1090 RA], S~ **Bardolph** Nf [*Stou* DB, *Stowe* 1244 Ch], S~ **Bedon** Nf [*Stou* DB, *Stouwebidun* 1287 Misc], S~ Sa [*Stowe* 1291 Tax], **West** S~ Sf [*Stowa* DB, *Westowe* 1254 Val], **Stowe** Bk [*Stov* DB], **Long** S~ Ca [*Stou* DB, *Longstowe* 1291 Tax], S~ Li [*Estou* DB, *Stoue* 1212 Fees, *Stowe* 1254 Val], S~ Np [*æt Stowe* 956 BCS 986, *Stowe* DB], S~ St [*Stowe* 1242 Fees, 1251 Ch]. OE *stōw* in various senses. Cf. STŌW.

Stow Bardolph was held by William Bardulf in 1244 (Ch). Cf. STOKE BARDOLPH Nt.—**Stow Bedon** was held by John de Bidun in 1212 (Fees). Cf. KIRBY BEDON.—**Stow Maries** was named from a local family. Robert de *Marisc'* held the manor in 1250 (FF). Cf. LANGLEY MARISH.

Stowell Gl [*Stanuuelle* DB, *Stawell* 1220 Fees, -*e* 1221 Ass], S~ So [*Stanwelle* DB, *Stawell* 1225, 1243 Ass], S~ W [*Stowelle* 1300 Ch]. OE *Stān-wella* 'stony stream'. Cf. STAWELL.

Stowey So S. of Bristol [*Staweie* 1246 Wells, *Staweye* 1327 Subs], **Nether & Over** S~ So [*Stawei* DB, *Stawaye* 1243 Ass, *Overstaweie* 1220 FF, *Nutherestoweye* 1276 RH]. OE *Stān-weg* 'paved road'.

Stowford (-ō-) D nr Lifton [*Staford* DB, *Stafford* 1242 Fees, *Stouford* 1303 FA], S~ D in Colaton Raleigh [*Stauford* c 1200 Torre, *Staford* 1242 Fees], **East & West** S~ D in W. Down [*Staveford* DB, *Stouford* 1289 Ass], S~ O [*Stauuorde* DB, *Staford* 1254–5 RH], S~ W [*Stanford* 987 KCD 658]. OE *Stān-ford* 'stony ford'. Cf. STOFORD, STAFFORD Do.

Stowlangtoft Sf [*Stou* DB, *Stowe* 1206 Cur, *Stowelangetot* 13 BM]. Originally *Stowe*. The manor was held by Richard de Langetot in 1206 (Cur).

The family name is Norman. LANGUETOT (from Scand *Langatoft*) is a common pl. n. in Normandy.

Stowmarket Sf [(ecclesia de) *Stou* DB, (forum de) *la Stowe* 1253 Cl, *Stowmarket* 1268 Ch, *Stowemarket* 1269 FF]. The hundred is called **Stow** hd [*Stou* DB]. Perhaps *stōw* is here 'meeting-place'. *Market* was added because the place had a market.

An earlier name of the place was **Thorney** [*Torneia* DB, *Thorneye* Hy 3 BM] 'thorn island'. **Stowupland** is near S~ on higher land.

Stowood O [*Stawode* c 1142 Fridesw, *Stawud, Stowud* 1235 f. Cl]. OE *Stān-wudu* 'stony wood'.

Stowting K [*Stuting* 1044 Th, *Stotinges* DB, *Stutinge(s)* 11 DM]. A derivative in -*ing* (sing.), perhaps from the OE *stūt* 'hill' found in some Devon pl. ns.

Stradbroke Sf [*State-*, *Stetebroc* DB, *Stradebroc* 1168, 1177 P]. OE *Stræte-brōc* 'brook by a (? Roman) road'.

Straddle He, an old name of the Golden Valley [*Stradel, vallis Stradelei, Stratelie* DB, *Vallis de Strada* 1169 P]. OE *Straddæl*, consisting of a shortened form of the Welsh name of the valley (*Stratdour* &c. c 1150 LL 'the valley of the Dore'), whose first el. is Welsh *ystrad* 'valley', and OE *dæl* 'valley'. Cf. GOLDEN VALLEY, MONNINGTON.

Stradishall Sf [*Stratesella* DB, *Strateshell* 1203 Ass, *Stratezell* 1228 FF, *Stradesele* 1254 Val]. OE *Stræt-gesell* 'shelter on the road'. Cf. GESELL.

Stradsett Nf [*Strateseta* DB, *-sete* 1254 Val, *Stradesete* 1242 Fees]. 'Place on the Roman road.' Cf. (GE)SET. The place is on Fen Road, a Roman road.

OE stræt, strēt 'street, Roman road', as in WATLING STREET. In some cases the word may well have been used of a paved road of other origin than Roman. See STREAT, STREET, STRETE, STRAD-, STRAT-, STREAT-, STREET-, STRET- (passim), STARTFORTH, STREFFORD, STRELLEY, STURTON, TRAFFORD.

Stragglethorpe Li [*Tragertorp* 1212, *-thorp* 1242 Fees, *Stragerthorp* 1242 ib.]. The first el. is clearly a pers. n. It may be *streaker* 'a kind of hound', from OFr *stracur, strakur*, which is used as a term of abuse for a person by Dunbar and as a byname (Robert Stracour) 1332 Subs (Cu). S~ Nt may be identical in origin.

Stramshall St [*Stagrigesholle* DB, *Strangricheshall* 1221 FF, *-hull* 1227 Ass, *Strongeshulf* 1269 Ass]. Perhaps '*Stronglic*'s hill'. OE *Stronglic* pers. n. is evidenced. Or an OE pers. n. *Strongric may be postulated.

OE strand 'shore'. See OVER-, SIDESTRAND, STRANTON. The word is the source of **the Strand** in London [*Stronde* 1185 TpR].

Strangeways La [*Strangwas* 1322 LaInq, *Strangways* 1326 Ct]. OE *strang* 'strong' ·and *gewæsc* 'washing up or overflow of water'. The name would mean 'strong current' or the like.

Stranton Du [*Stranton* 1159 P, *-a* c 1180 YCh 659]. 'TŪN on the shore.' See STRAND.

Stratfield Mortimer Brk, now usually **Mortimer** [*Stradfeld* DB, *Stratfeld Hug. de Mortem'* 1167 P, S~ *Mortymer* 1412 FA], **S~ Turgis** Ha [*Stradfelle* DB, *Stratfeld* 1158 P, S~ *Say* 1277 Ipm, S~ *Turgys* 1289 Cl]. 'FELD on the Roman road' (from Silchester to London). — S~ **Mortimer** was held by Radulf de Mortemer in 1086 (DB). Cf. CLEOBURY MORTIMER.— S~ **Saye** was held by Robert de Say in 1227 (Ch). Cf. HAMSEY.—S~ **Turgis** was held by the Turgis family at least from c 1270. Turgis is a Norman pers. n. of Scand origin.

Stratford Bd [*Stretford* 1312, *Stratford* 1312, 1325 Ipm], **Fenny S~** Bk [*Fenni Stratford* 1252 Ch], **Stony S~** Bk [*Stani Stratford* 1202 FF], **Water S~** Bk [*Stradford* DB], **S~ Langthorne** Ess [*Strætforde* 1067 BM], **S~ le Bow** Mx [see BOW], **Old S~** Np [*Forstratford* 1330 FA, *Old Stratford* 1498 AD], **S~ St. Andrew** Sf [*Straffort* DB, *Strafford* 1254 Val], **S~ St. Mary** Sf [*Strætford, Stredford* c 995 BCS 1288 f., *Strætford* c 1000 ib. 1306, *Stratfort* DB], **S~ sub Castle** W [*Stratford* 1091 Sarum, *Stratford under the Castle of Old Sarum* 1353 AD], **S~ Toney** W [*Stretford* a 672, 826, c 932 BCS 27, 391, 690, *Stradford* DB], **S~ on Avon** Wa [*Æt-stretfordæ* 691-2, *Ufera Stretford* 845 BCS 76, 450, *Stradforde* DB, *Strafford on Avon* 1255 Ch]. OE *Stræt-ford* 'ford by which a Roman road crossed a river'. All the Stratfords are on Roman roads.

S~ **Langthorne** from a tall thorn-bush. Cf. *Langethorn* 1199 FF (Ess).—*Forstratford* (under **Old S~** Np) contains OScand *forn* 'old'.—S~ **Toney** was held by Ralph de Touny before 1242 (Fees). Cf. NEWTON TONEY.

Stratton Bd [*Stratone* DB], **S~** Do [*Stratton* 1212 Fees, 1275 RH], **S~** Gl [*Stratone* DB, *Stretton* 1220 Fees], **East & West S~** Ha [*Strattone* 903 BCS 602, *Stratune* DB, *Strattona* 1167 P], **Long S~** or **S~ St. Mary & St. Michael** Nf [*Estratuna, Stratuna, Stretuna* DB, *Long Stratton* 1275 Cl, *Strattone sancti Michaelis* 1254 Val, *Stratton Sancte Marie* 1291 Tax], **S~ Strawless** Nf [*Stratuna* DB, *Stratton Streles* 1446 AD], **S~ Audley** O [*Stratone* DB, *Strettun* 1182 BM, *Stratton Audeley* 1491 AD], **S~ Hall** Sf [*Strattuna* DB], **S~ on the Fosse** So [*Stratone* DB, *Stratton super la Fosse* 1347 Ep], **Over S~** So [*Stratone* DB], **Stoney S~** So [*Strettun* 1065, *Stratton* 1262 Wells], **S~** Sr [*Strættun* 964-95 BCS 1132], **S~ St. Margaret** W [*Stratone* DB, *Stretton Sanct' Margaret'* 1427 Ch]. OE *Stræt-tūn* 'TŪN on a Roman road'. Nearly all the Strattons are on known Roman roads.

S~ **Audley** was held by James de Alditheleg in 1252 Ch. See AUDLEY St.—S~ **on the Fosse** is on the FOSSE WAY.—S~ **St. Margaret, Mary, Michael** from the dedication of the churches.—S~ **Strawless** probably means what it seems to mean.

Stratton Co [*Strætneat on Triconscire* c 880 BCS 553, *Stratone* DB, *Stretton* 1249 FF]. *Strætneat* means 'the valley of the river Strat' or 'the river Strat', whose old name was **Neth** [*Neth, Neet, Neht* 13 ERN]. *Neth* is probably identical with OIr *necht* 'clean'. *Stræt* is Welsh *ystrad* 'valley' or OCo *stret* 'a stream'. Later the river was taken to be called *Strat* or *Stræt*, and the name *Stratton* was formed. Or Stratton is a worn-down form of *Strætneat-tūn*. On *Triconscir* see TRIGG.

Streat Sx [*Estrat* DB, *Strete* 1272 Ipm]. OE STRÆT 'Roman road'. The place is on a Roman road.

Streatham (-ĕ-) Sr [*Stretham* 675, 933 BCS 39, 697, *Stratham* 1062 KCD 812]. 'HĀM on the Roman road.' The place is on Stane Street.

Streatlam Du [*Stretlea* c 1050 HSC, *Stretelam* 1316, *Stretlem* 1317 Cl]. OE *Strēt-lēah* 'LĒAH on a Roman road'. *Streatlam* is from the dat. plur. -*lēam*.

Streatley Bd [*Strætlea* c 1050 KCD 920, *Stradlei* DB], S~ (-ē-) Brk [*Stretlee* 699 BCS 100, *Estralei* DB, *Stretleg* 1242 Fees]. 'LĒAH on the (Roman) road.' It is not certain that the road at S~ Bd was Roman.

Street He [*Strete* DB, 1242 Fees, *Strate* 1196 P], S~ K [*Stræt* 1016–20 KCD 732, *Straeta* 11 DM], S~ So nr Glastonbury [*Stret* 725, 971 BCS 142, 1274, *Strete juxta Glastone* 1330 BM], S~ So nr Winsham [*Strate* DB, *Strete* 1254 Val]. OE STRĀT 'Roman road'. The places are on or near Roman roads.

Streethay St [*Stretheye* 1262 For]. 'Enclosure on the Roman road' (Ryknild Street). Cf. (GE)HÆG.

Streetley (now Littlebury) Green Ess [*æt Stretle, Stratlai* 11 KCD 725, 907], Streetly Ca nr Linton [*Stradleia, Stratleie* 1086 IE], S~ Wa [(on) *Strætléa*,(into) *Strétlie* 957 BCS 987]. 'LĒAH on the Roman road.' The places are on such.

Streetthorpe YW [*Stirestorp* DB, c 1175 Middleton]. '*Styr*'s thorp.' Cf. STARSTON.

Strefford Sa [*Straford* DB, *Streford* 1255 RH]. 'Ford where a Roman road crosses a river.' Cf. STRATFORD. The place is near Stretton on Watling Street.

Strelley Nt [*Straleia* DB, *Stratlega* 1167 P, *Stretleg* 1212 Fees]. OE *Strētlēah*, identical with STREATLEY. It is not known that there was a Roman road here.

Strensall YN [*Strenshale* DB, *Strenehal* 1167 P, *Strensale* 1228 YCh 785, *Streneshal* 1251 Ass]. S~ must be compared with *Streoneshalh*, said to be the old name of Whitby [*Streaneshalh, Strenæshalc, Streonaeshalch* c 730 Bede, *Streoneshealh* c 890 OEBede], *Streoneshalh* KCD 1358 (nr Bengeworth Wo), *Streon halh, be Streonen halæ* BCS 1139 (nr Wick Episcopi Wo). Even if *Streoneshalh* in Bede should be identified with STRENSALL, as has been suggested, it is difficult to believe that *Strēon* can here be a pers. n. The combination with HALH only would be too remarkable. A pers. n. *Strēon* is in itself quite plausible, as *Streonberct, -uulf* occur in LVD and *Streona* is found as a byname. Presumably the first el. is OE *gestrēon* in some sense. The word means 'gain, profit, wealth' and might have been used of land won by draining or reclaimed in some other way. Cf. INGESTRE.

Strensham (-s-) Wo [(in) *Strengesho* 972 BCS 1282, *Strengesham* 1212 Fees]. '*Strenge*'s HĀM', originally '*Strenge*'s HŌH'.

Cf. STRINGSTON and *Strengesburieles* BCS 458. The name is derived from OE *strenge* 'strong'.

Strete D in Blackawton [*Streta* 1194 P, *Strete* 1270 FF], S~ Raleigh D [*Estreta* DB, *Strete Ralegh* 1303 FA]. OE *stræt* 'Roman road'.

Henry de Ralegh was tenant of S~ Raleigh in 1242 (Fees). Cf. RALEIGH.

Stretford He [*Stratford* DB, *Stretford* 1316 FA], S~ La [*Stretford* 1212 Fees, 1325 FF]. OE *Strēt-ford*, a variant of *Strǣt-ford* (cf. STRATFORD). Both Stretfords are on Roman roads.

Strethall Ess [*Strathala* DB, *Strethale* 1212 RBE], Stretham Ca [*Strætham* c 975 ASCh, *Stratham* 1086 IE, *Stradham* DB]. 'HALH and HĀM on a Roman road.' Cf. STRÆT.

Strettington Sx [*Stratone* DB, *Estretementona* 1100–3 (1332) Ch, *Strethamton* 1212 Cur]. Originally *Strǣttūn* 'TŪN on Stane Street', later changed into *Strǣthǣma-tūn* 'TŪN of the *Strǣttūn* people'. Cf. HĀMTŪN.

Stretton Chs nr Malpas [*Stretton* 1282 Court, *Strecton*(bis) 1287 ib.]. It is doubtful if this can be 'TŪN on a Roman road'. The place does not seem to be with certainty on such a road. The first el. might be a metathesized form of OE *steorc* 'young bull or heifer'.

Stretton Chs nr Runcorn [*Strettona* Hy 2 Ormerod, *Stretton* 1260 Court], S~ Db [*Strǣttun* 1002 Wills, *Stratune* DB], S~ Grandison He [*Stratune* DB, *Strettona* c 1180 BM, *Stretton Graundison* 1350 Ep], S~ Sugwas He [*Stratone* DB, *Strattone by Sugwas* 1334 Ep], S~ en le Field Le [*Stretone* DB], S~ Magna & Parva Le [*Stratone* DB, *Great Stretton* c 1275, *Little S~* 1290 Bodl], S~ Ru [*Stratone* DB, *Stretton* 1254 Val], All, Church & Little S~ Sa [*Stratun* DB, *Alured Stretton* 1262 Eyton, *Aluethestret*', *Parva Stretton* 1327 Subs, *Strattonedal* 1228 Ch, *Chirchestretton* 1337 Ch], S~ St nr Burton on Trent [*Stretton* 942 BCS 771, *Strǣttun* 1002 Wills, *Stratone* DB], S~ St nr Penkridge [*Estretone* DB, *Stretton* 1242 Fees], S~ Baskerville Wa [*Stratone* DB, *Stretton Bakervill* 1285 QW], S~ on Dunsmore Wa [*Stratone* DB, *-tona* 1133 BMFacs, *Stratton upon Dunnesmore* 1262 FF], S~ on the Fosse Wa [*Stratone* DB, *Stretton super Fosse* 1316 FA], S~ under Fosse Wa [*Stretton* 1291 Tax, 1409 BM]. OE *Strǣt-tūn* or *Strēt-tūn* 'TŪN on a Roman road'. All the places are on Roman roads.

All S~ from an early owner, but it is not clear if the name was OE *Ælfrēd* or the woman's name *Ælfgўþ*.—S~ Baskerville was held by William de Baskervill t. Hy 1 (Dugdale). The name is from BACQUEVILLE in France.—For Dunsmore, Fosse, Sugwas see these names. —S~ Grandison was held by William de Grande Sono in 1303(FA). Grandison is a well-known family name.

Strickland, Great & Little, We [*Styrke-land* c 1235, *magna Stirkeland* 1292 WR, *Little S~* 1274 Kendale], **S~ Ketel & Roger** We [*Stercaland* DB, *Stirkelandes Ketel* 1278, *Stirkeland Ketel* 1280, *S~ Roger* 1310 Kendale]. 'Land (pasture) for young bullocks or heifers.' First el. OE *styrc, steorc.*

Uchtred son of Ketel had land in S~ c 1190 (Kendale). Ketel is ON *Ketill* pers. n.—**S~ Roger** perhaps from Roger de Brounolfes-hefed, who had land in Strickland in 1340 (Kendale).

Stringston So [*Strangestona* 1084 GeldR, *Strengestune* DB, *Strengestone* 1166 RBE]. '*Strenge*'s TŪN.' Cf. STRENSHAM.

Strixton Np [*Strixton* 12 NS, 1220 Fees, *Stricston* 1202 Ass]. '*Stric*'s TŪN.' *Stric* is mentioned as father of one of the festermen at Medeshamstede in BCS 1130, and one *Stric* held Wollaston nr Strixton in 1066 (DB). *Stric* may be the ON byname *Stríkr.*

OE strōd, strōþ 'marshy land overgrown with brushwood' is the source of STROOD, STROUD. Cf. STROXTON, BULSTRODE, LANG-STROTH DALE. A derivative *strother* with the same or a similar meaning is also found in pl. ns.

Strood K [*Strod* 889 BCS 562, *Stroda* 1159 P], **Stroud** (-ow-) Gl [*La Strode* 1221 Ass, *Strode* E 1 BM], **Stroud Green** Mx [*Strodegrene* 1562 FF]. Cf. STRŌD.

Stroxton (strawsn) Li [*Stroðistun* c 1067 Wills, *Stroustune* DB, *Strouueston* 1254 Val]. Perhaps OE *Strōpes-tūn* 'TŪN in marshy land'.

Strubby Li nr Alford [*Strobi* DB, *Strubbi* Hy 2 BM], **S~** Li in Langton by Wragby [*Strubi* DB, *Strutebi* c 1115 LiS, *Strubbi* 1202 Ass]. Apparently '*Strūt*'s BY'. *Strútr* is an ON byname. ON *strútr* means 'a cone-like ornament on a head-dress or cap'. It was no doubt used of other cone-like objects, e.g. of a cone-like hill. But such a sense is out of the question in the case of the first Strubby, and not probable for the second.

Strumpshaw Nf [*Stromessaga* DB, *Trume-shah* 1204 Cur, *Strumeshag* 1212 Fees, *Strumpsawe* 1291 Tax, *Strumpeshache* 1295 Ipm]. First el. an OE *strump*, corresponding to MLG *strump*, MHG *strumpf* 'stump'. The meaning would be 'stump wood'. See SCAGA.

Stubbington Ha [*Stubitone* DB, *Stubbinton* 1242 Fees]. Perhaps '*Stubba*'s TŪN', **Stubba* being a side-form of *Stybba* (cf. STEBBING). Cf., however, next name.

Stubbins La [*Stubbyng* 1563 Ct]. ME *stub-bing* 'cleared land'.

Stubbs, Walden, YW [*Eistop, Istop* DB, *Stubbis* c 1180 YCh 1555, *Stubbeswaldyng*

1327 FF]. 'The stubs or tree-stumps', the plur. of OE *stubb* 'stub'.

William son of Walding witnessed the document of c 1180 (supra). *Walding* is very likely for *Waldin*, an OFr name of OG origin. There is also a **Hamphall Stubbs** nr Hampole YW [*Stubbes* 1230 Ep, *Stubbes Lacy* 1285 FA]. *Hamphall* is a form of HAMPOLE.

Stubton Li [*Stubetune* DB, *-tun* 1212 Fees, *Stubenton* 1243 Ep]. Either '*Stubba*'s TŪN' (cf. STUBBINGTON) or OE *Stubba-tūn* 'TŪN where tree-stumps were found'.

Stuchbury or **Stutsbury** Np [*Stoteberie* DB, *-byr* 12 NS, *Stutesbiria* 1155–8 (1329) Ch, *Stotesbur* c 1230 BM, *Stuttebyri* 1228 Ep]. '*Stūt*'s BURG.' Cf. STUSTON. The correct form is OE *Stūtes-burg*, but the second *s* was sometimes lost owing to dissimilation. **Stūt* may be a nickname from *stūt* 'gnat'.

Studdal K [*Est-, Weststodwolde, Stodwolde* n.d. StAug]. OE *Stōd-weald* 'wood where horses were kept'.

Studfold. See STŌD.

Studham Bd [*æt Stodham* c 1060 KCD 945, *Estodham* DB]. OE *Stōd-hām* or *-hamm* 'homestead or enclosure where horses were bred'.

Studland Do [*Stollant* DB, *Stodland* 1236, *-londe* 1242 Fees]. 'Land where horses were kept.' See STŌD.

Studley O [*Stodlege* 1005 Eynsham, *-leg* 1230 P], **S~** W [*Stodlega* 1168 P, *-leg* 1232 Ch], **S~** Wa [*Stodlei* DB, *-leia* 1130 P], **S~** YW [(on) *Stodlege* c 1030 YCh 7, *Stollei, -lai* DB, *Stodlee* 1202 FF]. OE *Stōd-lēah* 'pasture for horses'. *Stodlege* 1005 Eynsham is perhaps rather STUDLEY Wa.

Stukeley, Great & Little, Hu [*Stivecleia, Styneclea* 974 BCS 1310 f., *Stivecle* DB]. Identical with STEWKLEY.

Stuntney Ca [*Stuntenei* DB, *-eye* 1252 Ch, *Stonteneia* 1086 IE]. '*Stunta*'s island.' Cf. STONESFIELD. **Stunta* is formed from OE *stunt* 'foolish'. Alternatively the first el. may be OE *stunt* adj. in the Mod dial. sense 'steep', as suggested PNCa(S).

Sturmer Ess [*Sturmere* c 1000 Battle of Maldon, *Sturemere* 1193 P]. 'Lake formed by R Stour.' The place is on the upper Stour (2).

Sturminster Marshall Do [*Stureminster* c 880 BCS 553, *Sturminstre* DB, *Stur-menystr' Mareschal* 1280 FF]. 'Minster or church on R Stour' (3). Held by Comes Marescallus in 1212 (Fees).

Sturminster Newton Do [*at Stoure* (*Nywetone* heading) 968 BCS 1214, *Newen-tone* DB, *Sturministr' Nyweton* 1291 Tax, *Sturmunstre juxta Newton Castel* 1327 FF]. Identical with prec. name. Sturminster Newton and Newton are on opposite sides of the Stour.

Sturry K [*Sturigao* 605, *Sturrie, Sturige* 675 BCS 6, 35, 41, *Esturai* DB]. OE *Stūr-gē* 'the Stour district'. Cf. EASTRY. S~ is on R Stour (1).

Sturston Db [*Sturstone* c 1200 Derby, *Stirstone* 1226 FF], **S~** Nf [*Esterestuna* DB, *Stirstun* 1254 Val]. Identical with STARSTON.

Sturton by Scawby Li [*Straitone, Stratone* DB, *Strettun* 1212 Fees], **Great & Little S~** Li [*Stratone* DB, *Strettuna* c 1115 LiS, *Stratton* 1209 (1252) Ch], **S~ by Stow** Li [*Stratone* DB, *Strettuna* c 1115 LiS], **S~ Grange** Nb [*Stretton* 1242 Fees, 1290 Ch], **S~ le Steeple** Nt [*Estretone* DB, *Strettun* 1236 Fees], **S~ Grange** YW [*Stretun* DB, *Straton* 1100–8'Fr]. A variant of STRETTON (2). With the exception of S~ Grange Nb the places are on known Roman roads.

Stuston Sf [*Stutestuna* DB, *-ton* R 1 Cur]. '*Stūt*'s TŪN.' Cf. STUCHBURY.

Stutton Sf [*Stottuna, Stuttuna* DB, *Stutton* 1220 FF, *Stuttone* 1254 Val], **S~** YW [*Stouetun, Stutune, Stutone* DB, *Stutton* 1230 FF, 1242 Fees]. The first el. may be OE *stūt* 'gnat' or ON *stútr*, OSw *stūter* 'bullock' or even an OE **stūt*, which seems to mean 'hill'. Cf. STOWTING.

Styal Chs [*Styhale* c 1200, *Stiale* 1331 Earwaker, *Stihal* 1286 Court]. 'HALH with a pigsty' or 'HALH by a path'. First el. OE STIG or STĪG.

Styche Sa [*Stucha, Stuche* c 1200 ff. Eyton]. OE *stycce* 'a piece' or *styfic* 'a stump'.

OE *styfic* 'stump of a tree' is a fairly common first el. in pl. ns. See e.g. STEETLEY, STEETON, STEVENTON &c., STEWKLEY, STEWTON, STIFFKEY, STUKELEY. The common occurrence of the pl. n. *Styfic-tūn* suggests that OE *styfictūn* had some special technical sense, e.g. denoted an enclosure of a certain type.

Styford Nb [*Styfford* 1212 RBE, *Stiford* 1262 Ipm]. OE *Stig-ford*, first el. OE *stig* 'path'.

OE **styrc, styric, steorc** 'young bullock or heifer'. See STIRCHLEY, STRICKLAND, perhaps STRETTON Chs (1).

Styrrup Nt [*Estirape* DB, *Stirap* 1200 P, 1236 Fees, *Stirop* 1242 Fees]. OE *stigrāp* 'stirrup'. The place stands by a hill whose shape certainly resembles a stirrup as seen in the map. If OE *stigrāp* meant 'stirrup' and not 'stirrup-leather', as it must once have done, it is very probable that Styrrup was so called from the ridge, whose name must have meant 'the stirrup'.

Subberthwaite La [*Sulbythwayt* 1284 Ass]. 'Clearing belonging to or at SULBY.' Sulby must be a lost place.

Suckley Wo [*Suchelei* DB, *Succhelege* 1156, *-leia* 1174 P, *Suggelega* 1180 ff. P]. OE *sucga* is the name of a bird. This word occurs also in *hægsugga, hegesugge* 'hedge

sparrow', which in ME appears as *heysoke*, in modern dialects as *haysuck*. There was apparently an OE *succa* by the side of *sucga*, probably with the same meaning. Suckley is then OE *Succan-lēah* (or *Sucgan-lēah*) 'wood where these birds were found'. OE *succa* may be found also in *Succan scylf* BCS 1071, *Succan pyt* ib. 1234. OE **sucga** is found in SUDBROOKE (2), SUGNALL, SUGWAS.

Sudborough Np [*Suthburhc* 1065 BM, *Sutburg* DB, *Sudburg* 1230 P]. 'Southern BURG.'

Sudbourne Sf [*Sutborne* c 1050 KCD 907, *Sutburna* DB, *Suthburna* 12 BM], **Sudbrooke** Li nr Lincoln [*Sutbroc* DB, *Sudbroc* 1202 Ass, *Suthbroca* 1209–19 Ep]. 'Southern stream.'

Sudbrooke Li in Ancaster [*Suggebroch* 1168 P, *-broc* R 1 BM]. First el. OE *sucga*, the name of a bird.

Sudbury Db [*Sudberie* DB], **S~** Mx [*Sudbery* 1294 Ass, *Subyry* 1294 FF], **S~** Sf [*Sudberi* 798 ASC (F), (into) *Suðbyrig* c 995 BCS 1289, *Sutberia* DB]. 'Southern BURG.'

Sudeley Gl [*Sudlege* DB, *Suthleia* 1175 Winchc]. 'Southern LĒAH.'

Suffield Nf [*Sudfelda* DB, *-feld* 1168, 1191 P], **S~** YN [*Sudfelt, -feld* DB]. 'Southern FELD.'

Suffolk [(pagus) *Suthfolchi* 895 BCS 571, *Suðfolc* c 1055 BM, 1075 ASC (E), *Sudfulc* DB, *Sudfolka* c 1095 Bury]. 'The southern folk.' Cf. NORFOLK.

Sufton He [*Shuffton* 1332 BM, *Suffton* 1390 Ipm]. First el. perhaps as in SHIFNAL.

Sugnall St [*Sotehelle* DB, *Sugenhulle* 1222 Ass, *Sogenhul* 1242 Fees]. The first el. is OE *sucga* a bird or *Sucga* pers. n. See HYLL.

Sugwas (sŭgas) He [*Sucwessen* DB, *Sugwas* 1251 Misc]. The elements are OE *sucga* (or *succa*; cf. SUCKLEY) the name of a bird, and WÆSSE 'swamp'.

Sugworth Brk [*Sogoorde* DB, *Suggewurth* 1242 Fees]. '*Sucga*'s WORÞ.'

Sulber YW [*Solberhc, -berc* 1190 FC]. OScand *sōlberg* 'sunny hill'. Cf. SOLBERGE.

Sulby Hall Np [*Solebi* DB, *Sulebi* 1158 P, 1202 Ass, *Sulehby* 1243 PNNp]. See SOULBY. The form *Sulehby* tells in favour of OE *sulh* as the first el. of Sulby.

Sulgrave Np [*Sulgrave* DB, *Solegreue* Hy 3 BM]. 'Grove in a gully.' Cf. SULH.

OE **sulh** (gen. *sylh*) is used in several pl. ns. OE *sulh* means 'a plough', but it may have been used in senses such as 'furrow' and 'gully, narrow valley'. It seems to be the source of the river-name Silk Stream Mx [*Sulh, Sulc* 957 BCS 994, *Sulh* 972 ib. 1290]. See further SOULBURY, SOULBY, SOULDERN, SOULDROP, SULBY, SULGRAVE, SULHAM, SULHAMPSTEAD, also SILVERTON D.

Sulham Brk [*Soleham* DB, *Sulham* Hy 3 AD iii, *Suleham* 1291 Tax]. 'HĀM in a SULH or narrow valley.'

Sulhampstead Abbots & Bannister Brk [*Silamested* 1198 AC, *Silhamstede* 1202 FF, -*sted* 1220 Fees, *Silhamstede Abbatis* 1291 Tax, *S~ Banastre* 1292 Ch]. 'Homestead in a SULH or narrow valley.' *Sulh* here appears in the gen. form *sylh*. The places are on opposite sides of a narrow valley.
S~ Abbots belonged to Reading Abbey.—**S~ Bannister** was held by William Banastre in 1198 (AC). Banastre is an OFr family name.

Sullington Sx [*Sillinctune* 959 BCS 1050, *Sillintone* DB, *Selinton* 1166, *Silingeton* 1176 P, *Shellyngton* 1266 FF, *Sullyngtone* 1291 Tax]. The first el. is identical with the (atte) *Sullingg* that occurs as a surname 1327 Subs (in Storrington, which adjoins Sullington). *Sullingg* probably represents an earlier *sielling* or *sieling*. OE *sieling* may be a derivative of OE *sealh* 'sallow' ('sallow copse').

Summergil or **Somergil** R He [*Somergil* c 1540 Leland]. Cf. *Somergelde* 1394 Cl (Sa). The second el. is derived from OE *gelde* 'barren' and the name means '(river) dry in summer'. It is very likely a translation of Welsh *Hafhesp* (see HAMPS).

Summerhouse Du [*Sumerhusum* 1200 FF], **Summerseat** La [*Sumersett* 1556 Ct]. 'Houses, SÆTR or shieling used in summer.'

Sunbury Mx [(æt) *Sunnanbyrg* 960–2, æt *Sunnanbyrig* 962 BCS 1063, 1085, *Sunneberie* DB]. '*Sunna*'s BURG.' Cf. SONNING.

Sunderland Cu [*Sunderland* 1332 Subs], **S~** Du [*Sunderland* c 1168 FPD], **S~** La [? *Sunderlond* 1262 Ass]. OE *sundorland* 'separate land'. The exact meaning may have varied between 'land separated from the main estate' and 'private land'. It has been pointed out that OEBede says Bede was born in the *sundurlond* of the Abbey of Jarrow.

Sunderland, North, Nb [*Suðlanda* 1177 P, *Sutherlannland* 12 PNNb, *Sunderland* 1187 P]. Originally OE *Sūþ-land* and *Sūþerre land* 'southern land'.

Sunderlandwick YE [*Sundre(s)lanwic* DB, *Sundarlandawic* 1157 YCh 354]. 'WĪC belonging to Sunderland' or 'WĪC that was *sundorland*'. Cf. SUNDERLAND.

Sundon Bd [*Sunnandune* c 1050 KCD 920, *Sonedone* DB]. '*Sunna*'s DŪN.' Cf. SONNING. Hardly 'sunny hill'.

Sundorne Sa [*Sundra* a 1157, *Sundrene* 1240 Eyton, *Sondene* 1291 Tax]. OE *sundor-ærn* 'separate house' or *sunor-ærn* 'house for the *sunor* or herd of swine'.

Sundridge K [*Sunderhirse* 1072 BM, -*hersce* 11 DM, *Sondresse* DB, *Sundresse* 1203 Cur]. Second el. OE *ersc* 'park, pasture'. First el. as in SUNDORNE.

Sunk Island YE [*Frisamersc* 1122–37 YCh 310, *Frisemareis* 1130 P, (villa de) *Frisemarasco, Frismareis* c 1200 YCh 1402, 1404]. The old name means 'marsh of the Frisians'. The name is purely English and no doubt pre-Conquest. The village has been devoured by the sea.

Sunningdale Brk [no early forms found], **Sunninghill** Brk [*Sunningehull* 1190 P, *Suninguehull* 1220 Fees], **Sunningwell** Brk [(ad) *Sunnigwellan* 811, 815 BCS 850, 352, *Sunningauuille* 821 ib. 366, *Soningeuel* DB]. 'Valley, hill and spring of *Sunna*'s people.' Cf. SONNING. The *Sunningas* at the three places no doubt belonged to the same tribe.

Sunt. See SOMPTING.

Surbiton Sr [*Subertone* 1203 FF, *Surbeton* 1351 FF, -*tone* 1265 Misc]. 'South barton.' Cf. NORBITON.

Surfleet Li [*Sverefelt* DB, *Surfliet* 1167 P, -*flet* 1212 Fees, *Sudflet* 1182, -*fliete* 1195 P]. 'Sour stream.' Cf. FLĒOT. *Sudflet* is probably a conscious change due to a wish to avoid unpleasant associations.

Surlingham Nf [? *Herlingaham* 1046 Th, *Sutherlingaham* DB, *Surlingeham* 1250 Ass]. Probably OE *Herlingahām* 'the HĀM of the *Herelingas*', with *sūþ* 'south' prefixed. Cf. HARLING.

Surrey [(on) *Suþrige* 722 ASC, *Suðrig* 1011 ASC (E), (in) *Suþregum* 871–89 BCS 558, *Sudrie* DB]. OE *Sūþer-gē* 'southern district'; cf. EASTRY. But partly the name goes back to an OE word for the people. These are called *Suþrige* 823 ASC, and Bede calls Surrey *regio Sudergeona* (*Suðrig(e)na land* OEBede). The latter name has as second el. a derivative of OE *gē*, corresponding to Goth *gauja* 'inhabitant of a *gawi* or district'.

Sussex [(on) *Suþ Seaxe* 722 ASC, (upon) *Suð Seaxum* 895 ib., *Suþ Seaxna lond* 773 ib., *Suth-Seaxa* c 894 Asser, *Sudsexe* DB]. Originally the tribal name South Saxons [*Suð Seaxe* 607 ASC] used later also of their country. Cf. ESSEX, MIDDLESEX.

Sustead Nf [*Surstede, Sutstede* DB, *Suthstede* 1101–7 Holme, 1254 Val]. 'Southern place.' Cf. STEDE.

Sutcombe D [*Sutecome* DB, *Suttecumb* 1242 Fees, *Suthtecumbe* 1269 Ep]. Perhaps originally *Sūþ-cumb* 'southern valley', whence *Sutcumb* and with an intrusive *e Sutecumb*. Cf. SOTWELL.

OE **sūtere,** ON **sútari** 'shoemaker'. See SUTTERBY &c.

OE **sūp** 'south' is a very common first el. in pl. ns. See e.g. SOUTH-, SUD- (passim), SIDDINGTON, SINTON, SOBERTON, SODINGTON, SOTWELL, SOUDLEY, SOWTON, SUDELEY, SUFFIELD, SUFFOLK, SURBITON, SURLINGHAM, SUSSEX, SUSTEAD, SUTTON. OE *be sūþan* 'south of' is found in SIDDINGTON Chs and perhaps in some other names. **Sidbury** Wo is

se haga be suðan byrig 963 BCS 1108. OE
sūper- (cf. OLG *sūthar-*, OFris *sūther*, ON
suðr) is the first el. of SURREY, and *sūperra*
'southern' that of SOTHERTON and of SOUTH-
ERY, SOUTHREY, if it is not *sūper-*.

Sutterby Li [*Sutrebi* DB, c 1115 LiS,
Suterbi 1202 Ass], **Sutterton** Li [*Suterton*
1200 Cur, 1202 Ass, 1242 Fees, *Sutterton*
1254 Val]. 'The shoemakers' BY and TŪN.'
First el. OE *sūtere* or OScand *sūtari* 'shoemaker'.

Sutton is a very common name, which goes
back to OE *Sūþ-tūn* 'southern TŪN'. It is
possible that in isolated cases Sutton may
represent OE *be sūpan tūne* '(the place) south
of the village', but no certain cases are on
record.

Sutton Bd [*Sudtone* DB], **S~ Courtenay**
Brk [*Suðtun* 892–901 BCS 581, *Sudtone* DB,
Sutton Courtenay 1340 Ep], **S~ Ca** [*Sud-
tone* DB], **S~ Chs** in Wirral [*Sudtone* DB],
S~ by Macclesfield Chs [*Sutton* 1245–50
Chester], **S~ by Middlewich** Chs [*Sud-
tune* DB], **Guilden S~** Chs [*Sudtone* DB,
Guldenesutton c 1209 Ormerod], **S~ D**, now
Plymouth [*Sudtone* DB], **S~ on the Hill**
Db [*Suttun* 1002 Wills, *Sudtun* DB], **S~
Scarsdale** Db [*Suðtun* 1002 Wills, *Sud-
tune* DB, *Sutton in Dal* 1242 Fees], **S~
Waldron** Do [*Suttun* 932 BCS 691, *Sutton
Waleraund* 1297 Cl], **S~ Ess** [*Suttuna* DB],
Bishop's S~ Ha [*Sudtone* DB, *Suttona
Episcopi* 1167 P], **Long S~** Ha [*æt Suðtune*
979 KCD 622, *Sudtune* DB, *Longesuttone*
c 1220 Crondal], **S~ Scotney** Ha [*Sudtune*
DB, *Sutton Scoteneye* 1346 FA], **S~ St.
Michael & St. Nicholas** He [*Su(d)tune*
DB, *Suttune* 1242 Fees], **S~ K** nr Deal
[*Suttone* 1155 RBE], **East S~ & S~
Valence** K [*Suðtun* 814 BCS 343, *Sudtone*
DB, *East Sutton* 1265 Ch, *S~ Valence* 1316
FA], **S~ at Hone** K [*Sudtone* DB, *Suttone
atte hone* 1281 BM], **S~ La** [*Sutton* 1200
Abbr, 1246 Ass], **S~ Cheney** Le [*Sutone
DB, Sutton* 1220–35 Ep], **S~ in the Elms**
Le [*Sutone* DB, *Sutton* 1220–35 Ep], **S~ Li**
nr Newark [*Suttun* 1212 Fees], **Long S~**
Li [*Sudtone* DB, *Sutton* 1202 Ass], **S~ in
the Marsh** Li [*Sudtune* DB, *Suttuna* c 1115
LiS], **S~ Nf** [*Suttuna* DB], **S~ Np** nr
Peterborough [*Suðtun* 948 BCS 871, *Sut-
tona* 1199 NpCh], **S~ Bassett** Np [*Sutone*
DB, *Sutton* 1220 Fees], **King's S~** Np
[*Sudtone* DB, *Suttun Regis* 1252 Ch], **S~
Nt** nr Bingham [*Suttun* 1236 Fees], **S~ Nt**
nr Retford [*Æt Suttune* 958 YCh 3, *Sud-
tone* DB], **S~ in Ashfield** Nt [*Sutone* DB,
Sutton in Assefeld 1288 Ipm], **S~ Bon-
nington** Nt [*Sudtone, Bonnitone* DB, *Sutton
upon Sore and Bonyngton* 1288 Ipm, *Sutton
Bonynton* 1340 Ch, *Bunningtun* c 1085 LVD,
Buningatuna Hy 2 FPD], **S~ on Trent** Nt
[*Sudtone* DB, *Sutton super Trente* 1221–30
Fees], **S~ O** [*Sutton* 1207 f. Cur], **S~ Sa** nr
Shrewsbury [*Sudtone* DB], **S~ Maddock**
Sa [*Suðtun* 1002 Wills, *Sudtone* DB, *Suthona
Gerueröi Coch* 1168 P, *Sutton Madok* 1276

Ipm], **S~ Sf** [*Suthtuna* DB], **S~ Bingham**
So [*Sutone* DB], **Long S~** So [*Sudton* 878
BCS 545, *Sutune* DB, *Langesutton* 1312
Ipm], **S~ Mallet** So [*Sutone* DB, *Sutton
Malet* 1280 FF], **S~ Montis** So [*Sutone*
DB, *Sutton Mountagu* 1335 Ep], **S~ Sr**
[*Suptone* 675 BCS 39, *Suðtone* 1062 KCD
812, *Sudtone* DB], **S~ Sx** [*Suðtun* c 880
BCS 553, *Sudtone* DB], **S~ Benger** W
[(at) *Suttune* 854 BCS 470, *Sudeton Berenger*
1377 Pat], **S~ Mandeville** W [*Sudtone* DB,
Sutton Maundevyle 1275 Ipm], **S~ Veny**
W [*Sudtone* DB, *Fenni Sutton* 1291 Tax],
S~ under Brailes Wa [*Sutton* 1203 Cur],
S~ Coldfield Wa [*Sutone* DB, *Sutton in
Colefeud* 1269 Ch, *S~ in le Colfeld* 1289
Misc], **S~ Wo** [*Sudtune* DB, *Sutton* 1212
Fees], **Full S~** YE [*Fulesutton* 1234 Ep],
S~ on Hull YE [*Sudtone* DB, *Sutune juxta
Hul* 1172 YCh 1391], **S~ upon Derwent**
YE [*Sudtone* DB, *Sutton super Derwent* 1233
Ep], **S~ on the Forest** YN [*Sudtune* DB,
Sutton sub Galtris 1242 Ep], **S~ Howgrave**
YN [*Suptun* 966–92 BCS 1255, *Sudtone* DB,
Sutton Hograve 1249 Cl], **S~ under White-
stone Cliffe** YN [*Sudtune* DB, *Sutton under
Whitstanclif* 1292 Cl], **S~ YW** nr Campsall
[*Sutone* DB], **S~ YW** in Kildwick [*Sutun*
DB], **S~ YW** nr Ripon [*Suðtun* c 1030
YCh 7, *Sudton* DB], (**Byram cum**) **S~ YW**
[*Sutton* 1193 ff. P].

S~ in Ashfield Nt. Cf. KIRKBY IN ASHFIELD.—
S~ Bassett Np was held by Richard Basset
in the 12th cent. (NS). Cf. BERWICK BASSETT.—
S~ Benger W. One *Berenger* was an under-
tenant here before 1066 (DB). Cf. SHIPTON
BELLINGER.—**S~ Bingham** So was held by
John de Bingham t. Hy 1.—**Bishop's S~** Ha
belonged to the Bishop of Winchester.—**S~
Bonnington** Nt was originally two manors,
Sutton and Bonnington. Bonnington is iden-
tical with BONNINGTON K.—**S~ under Brailes**
Wa. See BRAILES.—**S~ Cheney** Le. Cf.
CHENIES.—**S~ Coldfield** Wa means S~ in Cold-
field. The latter is OE *Colfeld* 'FELD where
charcoal was burnt'.—**S~ Courtenay** Brk
was held by Reginald de Curtenai in 1161 (P).
Cf. HIRST COURTNEY.—**Full S~** YE is really
Foul S~ (OE *fūl* 'dirty').—**Guilden S~** Chs.
See GILMORTON.—**S~ at Hone** K was named
from some stone, perhaps a boundary stone
(OE *hān* 'stone').—**S~ Howgrave** YN. See
HOWGRAVE.—**S~ Maddock** Sa was held by
Madoc son of Iorwerth Coch, a Welshman, in
1188 (P). Iorwerth Coch ('the red') was son of
Meredith, prince of Powys.—**S~ Mallet** So
was held by Ralph Malet in 1200 (Cur). Cf.
CURRY MALLET.—**S~ Mandeville** W was held
by Robert de Mandevill in 1236 (Fees). Cf.
HARDINGTON MANDEVILLE.—**S~ Montis** So was
held by Drogo [de Monteacuto] in 1086 (DB).
Cf. SHEPTON MONTAGUE.—**S~ Scarsdale** Db
from the Earls of Scarsdale. **Scarsdale** hd
[*Scarvedele, Scaruesdele* DB] is probably
'Skarf's valley', the first el. being ON *Skarfr*
pers. n. (from *skarfr* 'cormorant').— **S~
Scotney** Ha from the Scotney family resident here at least from 1236 (VHHa). The
name is French (from ESCOTIGNI in Normandy).
—**S~ Valence** K was held by William de
Valenc' in 1275 (RH). Cf. COMPTON VALENCE.—
S~ Veny W is 'Fenny S~'.—**S~ Waldron** Do
contains the OFr pers. n. *Waleran* (*Galerant*)
from OG *Walahram*. The manor was held by

Walerannus in 1086 (DB), by the heirs of Walter Walerant in 1212 (Fees).—S~ **under** White-stone Cliffe YN is at the foot of Whitestone Cliffe, a hill of 1,053 ft.

Swāby Li [*Suabi* DB, c 1115 LiS, *Suauebi* 12 DC]. '*Svāfi*'s BY.' OScand *Svāfi* is not well evidenced, but seems to have existed. *Swaue* occurs in DB and *Swafa* on coins.

Swadlincote Db [*Sivardingescotes* DB, *Suartlincot* 1208 FF, *Swardlincote* 1309 Ch, *Swartlyngcote* 1330 FA]. The first el. seems to be an OE *Sweartling* or OScand *Svartlingr* pers. n.

Swaffham (-ŏ-) **Bulbeck & Prior** Ca [*Extra Suafham* c 1050 KCD 907, *Suafham* c 1080 ICC, DB, *Swafham* 1196 FF, *S~ Monialium* 1254 Val, *S~ Prior* 1261 FF, *Suafham Bolebek* 1267 Misc], S~ Nf [*Suafham* DB, *Suaffham* c 1130 BM, *Swafham* 1230 P]. OE *Swāf-hām* (or *Swǣf-hām*), whose first el. is the tribal name *Swǣfas* in its stem form. The name means 'the HĀM of the *Swǣfas*'. *Swǣfas* is 'Swabians', *Suevi*, *Suebi* &c. in early sources. Members of this tribe must have been among the early invaders of Britain. The el. *Swǣf-* in OE pers. ns. is the tribal name.

Hugo de Bolebech (from BOLBEC nr Le Havre) had land in S~ c 1080 (ICC).—S~ **Prior** belonged to the Prior of Ely.

Swāfield Nf [*Suafelda* DB, *Suathefeld* c 1150 Crawf, *Swathfeld* 1197 FF]. First el. OE *swæþ*, *swaþu* 'track', here used in the later sense 'a longitudinal division of a field'.

Swainby YN in Pickhill [*Suanebi* DB, *Suenebi* 1111–22 PNNR], S~ YN in Whorl-ton [*Swaynsby* 13 BM, *Swa(y)neby* 1314 Pat]. '*Svein*'s BY' or 'the BY of the swains'. ON *Sveinn*, ODan *Sven* is a common pers. n. ON *sveinn*, OSw *sven* means 'a young man, servitor'. It occurs in Sw SVENNEBY from OSw *Svenaby*.

Swainsthorpe Nf [*Sueinestorp* DB, *Sweinestorp* 1196 FF], **Swainston** Wt [*Sweyneston* 1255 Ch, -e 1284 BM], **Swainswick** So [*Sweyneswyk* 1291 Tax, 1302 FF]. '*Svein*'s thorp, TŪN and WĪC.' Cf. SWAINBY.

Swalcliffe (-āk-) O [*Swalewecliua* c 1190 RA, -*cliue* 1190–5, *Swalecliue* 1196 DC]. 'Swallow cliff, cliff where swallows nested.' First el. OE *swealwe* 'swallow'.

Swale R K [*Suuealuue fluminis* 812, *Sualuæ* 815 BCS 341, 353], S~ R YN [*Sualua* c 730 Bede, *Swalwan stream* c 890 OEBede, *Swale* 1268 Ass], S~ R Brk, now Blackwater [*Swalewe* 1272 Ass, 1300 Cl]. Identical with SCHWALB [*Swalawa* c 802], SCHWALE [*Suala* 12] in Germany. The name is related to *swallow* (the bird) and belongs to the root *svel-* 'to move, plash' &c. in MHG *swalm* 'whirlpool', OE *swillan* 'to wash' &c. The meaning of the name seems to be 'whirling, rushing river'. **Swaledale** YN is *Sualadala* c 1130 BM, *Swaledal* 1159 P.

Swalecliffe K [*æt Swalewanclife* 949 BCS

874, *Soaneclive* DB, *Swalesclive* 1242 Fees]. Either identical with SWALCLIFFE or 'cliff by the SWALE'. The place is not far from the mouth of the Swale.

Swallow Li [*Sualun* DB, *Sualwa* c 1115 LiS, *Swalwe* 1163 RA, *Sualewe* 1175 P]. Very likely an old river-name identical with SWALE. The place is in a well-defined valley.

Swallowcliffe W [*Swealewanclif*, *rupis irundinis* 940 BCS 756, *Swaloclive* DB]. Identical with SWALCLIFFE.

Swallowfield Brk [*Svalefelle* DB, *Sualewesfeld* a 1162 Oxf, *Sualewefeld* 1167 P]. 'FELD on R Swale.' See SWALE Brk.

Swalwell Du [*Sualwels* 1183 BoB]. 'Swallow spring or stream.'

OE **swan** 'swan' and **swān** 'herd, esp. swineherd', but originally 'young man, servant', are not easy to keep apart in pl. ns. *Swan*, the bird, is the probable first el. of SWANBOURNE, SWANMORE, while *swān* is the probable source when the second el. is a word such as TŪN, WĪC, ÞORP. Cf. SWANAGE, SWANTHORPE &c.

Swanage Do [*Swanawic* 877 ASC, *Suuanuuic* c 894 Asser, -*wic* DB, *Swanewiz* 1183 P]. 'The WĪC of the herds.' Swanage is on **Swanage Bay**, and a meaning 'swan bay' may seem tempting. But no OE word *wic* 'bay' is known. A meaning 'swannery' is possible, however, and perhaps preferable.

Swanbourne Bk [*Suanaburna* 792 BCS 264, *Sveneborne* DB]. 'Stream frequented by swans.'

Swanland YE [*Suenelund* 1189 P, *Swaneslund* 1237 Cl, c 1265 Bodl, *Swainesland* 1300 Ipm]. '*Svein*'s grove' or 'the grove of the swains'. Cf. SWAINBY, LUND.

Swanley K [*Swanleg* 1203 FF]. See LĒAH. First el. OE *swan* 'swan' or *swān* 'herd'.

Swanmore Ha [*Suanemere* 1205 Cur, *Swanemere* c 1245 Selborne]. 'Lake frequented by swans.'

Swannington Le [*Sueniton* 1207 Cur, *Swaninton* 1242 Fees, *Swaniton* 13 ib.], S~ Nf [*Sueningatuna* DB, *Suaneton* 1191, *Sueiningeton* 1192 P, *Swaningeton* 1202 FF]. Apparently 'the TŪN of *Svein*'s people'. Cf. SWAINBY.

Swanscombe K [*Suanescamp* 695 BCS 87, *Svinescamp* DB, *Suanescomp* 1199 Cur]. 'The CAMP or pasture of the SWĀN or swineherd.'

Swanthorpe Ha [*Swanethorp* 1233 Ch, *Swandrop* 1248, -*thrope* 1334 Crondal]. 'The thorp of the swineherds.' Cf. SWĀN.

Swanton K nr Mereworth [*Suuanatuna*, *Swanatun* 10 BCS 1321 f.], S~ **Abbott** Nf [*Swaneton* 1044–7 Holme, *Suanetuna* DB, *Abbot Swanton* 1451 AD], S~ **Morley** Nf [*Suanetuna* DB, *Swaneton* 1212 Fees], S~

Novers Nf [*Suaneton* 1047–70 Wills, *-tuna* DB, *-ton* 1200 Cur]. OE *Swāna-tūn* 'TŪN of the (swine)herds'.

S~ Abbott belonged to Holme Abbey.—S~ Morley was held by Robert de Morle in 1346 (FA). He got it by marriage with Hawise Marshall. In 1316 the manor was held by Johannes le Mareschal (FA).—S~ Novers was held by Milo de Nuiers in 1200 (Cur). The name is from NOYERS-BOCAGE in Normandy.

Swanwick (-ŏnĭk) Db [*Swanwyk* c 1278 Beauchief Cart], S~ Ha [*Swanewic* 1231 Ch, *Swannewyk* 1242 Fees]. 'wĪc of the (swine)herds.' Cf. SWĀN.

Swarby Li [*Svarrebi* DB, *Svarrebi* 1199 P, *-by* 1233 Ep]. '*Svarri*'s BY.' *Svarri* is an ON byname.

Swardeston (sworstn) Nf [*Suerdestuna* DB, *Swerdeston* 1202 FF, *-tone* 1254 Val, *Suerdesdon* 1230 P]. '*Sweord*'s TŪN.' *Sweord* occurs in *Sweordes stan* 883 BCS 551.

Swarkeston Db [*Suerchestune* DB, *Swerkeston* 1230 P]. The first el. may be ON *Suærkuir* (*Sørkvir*) or ODan *Swerkir*, OSw *Swerker* pers. n.

Swarland Nb [*Swarland* 1242 Fees, *Swareland* 1256 Ass]. 'Heavy land', 'land heavy to plough', the first el. being OE *swǣr, swār* 'heavy'.

Swarling K [*Sueordhlincas* 805, 812 BCS 321, 341, *Swerlinges* 1205 Obl]. The elements are OE *sweord* 'sword' and *hlinc* 'hill'. The reason for the name is obscure.

Swarraton Ha [*Swerwetone* 903 BCS 602, *Serveton* c 1150 (1341) Pat, *Sarweton* 1242 Fees]. The first el. might be a compound of OE *swǣr* 'heavy' and *wǣd* 'ford', the name meaning 'TŪN by a heavy ford'. OE *Swǣrwǣdtūn* must have given *Swǣrwetūn*.

Swarthmoor La [*Swartemore* 1537 PNLa]. 'Black moor.' First el. OE *sweart* or ON *svartr* 'black'.

Swathling (-ādh-) Ha [*Swæðeling* 909, *Swæðelingford, Swæpelingeforde* 932 BCS 620, 692]. Apparently connected with OE *swæþ, swaþu* 'swathe, track', but the formation and meaning are obscure.

Swaton Li [*Suavintone, Svavetone* DB, *Suavetona* 1126 Fr, *Suaueton* 1190 P]. '*Swāfa*'s TŪN.' *Swāfa* may be of OScand origin (cf. SWABY). But *Swāfa* is a possible side-form of OE *Swǣfa*.

Swavesey Ca [*Suauesheda* c 1080 ICC, *Svavesye* DB, *Suaueseia* c 1155 BM, *-heða* 1172 P, *Suaveshide* 1203 Cur]. '*Swǣf*'s landing-place.' Cf HȲP. *Swǣf* occurs in (to) *Swæfeshcale* BCS 762 and is a normal short form of names in *Swǣf-*. OE *ǣ* often appears in East Saxon and sometimes in Ca as *ā*.

Sway Ha [*Sveia* DB, c 1150 Rutland, *Sweye* 1263 Ipm]. Perhaps a river-name, derived from OE *swēge* 'sounding' or from the OE

swegan that is the source of ME *sweie* 'to go, move'.

Swayfield Li [*Suafeld* DB, *Swafeld* 1198 FF, *Suathefeld* 1206 Ass]. Identical with SWAFIELD.

Sweeney Hall Sa [*Swyne* 1272 Ipm]. OE *Swin-ēa* 'river frequented by wild boars'.

Sweethope Nb [*Swethop* 1215–55 Ep, 1279 Ass]. Apparently 'sweet (pleasant) valley'. See HOP.

Swefling Sf [*Sueflinga, Suestlingan* DB, *Sueftlinges* c 1150 Crawf, *Swiftling* 1222 FF, *Sweftling* 1254 Val]. '*Swiftel*'s people.' OE *Swift* pers. n. is recorded. For *e* instead of *i* cf. WHEPSTEAD.

Swell, Lower & Upper, Gl [*Swelle* 706, *Suella major* 714 BCS 118, 130, æt *Suuelle* 1055 KCD 801, *Svelle* DB, *Netherswell* 1287 QW, *Ouereswell* 1336 BM], S~ So [*Sewelle* DB, *Swella* c 1080 Reg, *Swell* 1212 Fees, 1219 FF]. An OE *swelle* 'swelling', here used of a hill or ridge. The places are by or on hills or ridges. Other examples of the name Swell are (on) *þe suellen* 801 BCS 300 (at Butleigh So) and *La Swell* between Bruyton and Upton 1243 Ass (at Bruton and Upton Noble So, between which there is a hill). *Swelle* is derived from OE *swellan* 'to swell'.

Swepstone Le [*Scopestone* DB, *Swepeston* c 1125 LeS, *Swepston* 1220–35 Ep]. '*Sweppi*'s TŪN.' *Sweppi* is a side-form of *Swæppa*.

Swere R O [*Swere* 1577 Harrison]. A back-formation from Swerford O [*Surford* DB, *Swereford* 1200 Cur, 1220 Fees]. 'Ford by a col.' Cf. SWĪRA.

Swettenham Chs [*Suetenhala* 1183 P, *Swetenham* 1259, 1288 Court]. '*Swēta*'s HĀM.'

Swift R Le, Wa [*Swift* 1577 Harrison]. A late name meaning literally 'swift river'. But the river is winding and slow-moving, and possibly *swift* is here used in an earlier sense 'winding'. A former name seems to have been *Waver*. Cf. OVER Wa.

Swilgate R Gl [*Suliet* c 1540 Leland, *Swilyate* 1577 Saxton]. No doubt a back-formation from a pl. n. meaning 'flood-gate' or the like.

Swilland Sf [*Suinlanda* DB, *Suinelanda* 1185 P]. 'Land where pigs were kept.'

Swillington YW [*Suilligtune* DB, *Swinlentona* c 1150 Crawf, *Swinlingtune* c 1185 YCh 1637]. 'The TŪN of the dwellers at Swinley or Swinwell' or the like. An earlier name of the place may have been *Swinlēah* 'pig wood' or *Swinwella* 'wild boar stream'.

Swimbridge D [*Birige* DB, *Svimbrige* 1225 Ep]. '*Sǣwine*'s bridge.' The manor was held by *Sawin* in 1066 (DB).

Swin, The, Swin Channel outside the Essex coast [*the Swyn* 1365 Cl]. An OE *swin 'creek, channel', identical with Du zwin the same. This word is possibly found in some other names, as SWINE, -FLEET.

OE swīn 'swine, pig, wild boar' is a common first el. in pl. ns. Cf. SWILLAND, SWIN-(passim), SOMBORNE. It is generally impossible to decide whether wild or domestic swine are referred to. SWINTON is no doubt 'pig farm'.

Swinbrook O [Svinbroc DB, Swinbroc 1197 FF], Swinburn Nb [Swineburn 1236, 1242 Fees]. 'Pig brook.'

Swinden Gl [Svindone DB, Swindon 1220 Fees], S~ YW nr Skipton [Suindene DB, Suinden 1190 P, 1226 Ep]. 'Pig hill and valley.' See DŪN, DENU.

Swinderby Li [Sunderby, Suindrebi DB, Sunderby, Swinderby 1209-35 Ep]. 'Southern BY', the first el. being OScand sundri 'southern', found in the common SØNDERBY in Denmark, SÖNNERBY in Sweden. The change to Swinderby may be due to association with the word swine.

Swindon St [Swineduna 1167 P, Suindun 1236 Fees], S~ W [Svindune DB, Swinedon 1205 Cl]. 'Pig hill.' See DŪN.

Swine YE [Swine DB, Suine c 1150 YCh 1360]. Perhaps OE swin 'creek'. Cf. SWIN and next name.

Swinefleet YW [Swinefleth, Swyneflet c 1200 YCh 492 f.]. Either 'pig stream' or the first el. is OE swin 'creek'. Cf. prec. name and FLEOT. It is possible that the Humber was sometimes called Swin and that Swinefleet means 'the Swin creek'.

Swineshead Bd [Suineshefet DB, Suinesheued 1198 P]. 'Pig hill.' The name seems to refer to one of the ridges in the neighbourhood. See HĒAFOD.

Swineshead Li [Swines hæfed 675 ASC (E), æt Suinesheabde 786-96 BCS 271, Swynesheued 1230 P]. The place is in a low situation. Near it was formerly Swinefleet [Swireflet 1202 Ass]. Swineshead may mean 'the source of the Swin', Swin being an early name of Swinefleet. Cf. SWIN.

Swinethorpe Li SW. of Lincoln [Sueinestorp 1196 FF, Sweynesthorp 1263 FF]. 'Svein's thorp.' Cf. SWAINBY.

Swinfen St [Swyneffen 1232 Ass, Swynefen 1252 FF], Swinford Brk [Swynford 931 BCS 680, Swyneford 1242 Fees], S~ Le [Svin(e)ford, Svinesford DB, Suinford 1190 P], Kingswinford St [Svinesford DB, Kyngesswynford 1322 Ipm], Old S~ Wo [Swinford c 950 BCS 1023, Svineforde DB]. 'Pig fen and ford.'

Swingfield K [Swinesfeld 1202 Cur, 1242 Fees]. 'Pig FELD.'

Swinhoe Nb [Swinhou 1242 Fees, Swyneho 1280 Ch], Swinhope Li [Suinhope DB,

Suinahopa c 1115 LiS]. 'Pig hill and valley.' Cf. HŌH, HOP.

Swinithwaite YN [Synningthwait 1172 YCh 200, Swiningethwait 1202 FF, Swiningwait 1203 Cur, Sinythethweyt 1251 Ass]. See THWAITE. First el. perhaps ON sviðningr 'place cleared by burning'.

Swinscoe St [Swyneskow 1248 FF]. 'Pig wood.' Second el. OScand skōgr 'wood.'

Swinside Cu in Lorton [Swynesheued 1260 PNCu(S)]. 'Pig headland.' S~ Cu in Millom [Swynesat 1242 FF]. 'sÆTR where pigs were kept.'

Swinstead Li [Suinham, Suinhamstede DB, Swinehamsted 1203 Cur]. 'Homestead where pigs were reared.'

Swinthorpe Li nr Wragby [Sonetorp DB, Sunetorp c 1115 LiS, Suinetorp 1175 P, Sunthorp 1229 Ch]. 'Suni's thorp.' ODan Suni is a common pers. n. For the change to Swin- cf. SWINDERBY.

Swinton La [Suinton 1258 Ass], S~ YN nr Malton [Suintune DB], S~ YN nr Masham [Suinton DB], S~ YW [Suintone DB]. 'Pig farm.'

OE swīra, swēora 'neck' was also used in the sense 'hause, col'. See SWYRE, SOURTON, SWERFORD. Cf. also BOULSWORTH, which may contain the cognate OScand sviri.

Swithland Le [Swithellund 1209-19, Swithelunde 1224 Ep, Swythlund 1236 Cl]. Second el. OScand lundr 'grove'. The first may be OScand sviðinn 'burnt' or else a cognate subst., e.g. ON sviða, OSw sviþa 'land cleared by burning'. This is the sense of dial. Engl swidden, which is the source of the pl. n. Sweden, often found in northern counties. The first el. may also be this swidden from ME swithen, on which see Man, lv. 135 f.

Swyncombe O [Svinecumbe DB, Swinecombe 1227 Ch]. 'Pig valley.'

Swynnerton St [Sulvertone DB, Suinnerton 1242 Fees, Swynaferton, Swynforton 1272 Ass]. OE Swinford-tūn 'TŪN by the pig ford'.

Swyre Do [Svere DB, Swere 1196 P, Suure 1275 RH]. OE SWĪRA, SWĒORA 'col'.

Syde Gl [Side DB, 1242 Fees]. OE side 'slope'.

Sydenham Damarel D [Sidelham DB, Sideham 1184 P, Sydenham Albemarlie 1297 Pat], S~ O [Sidreham DB, Sideham 1194 P, Sidenham 1285 QW], S~ So nr Bridgwater [Sideham DB, Sidenham 1243 Ass]. 'Wide HAMM.' Cf. SĪD.

S~ Damarel was held by Johannes de Alba Mara in 1242 (Fees). Cf. HINTON ADMIRAL.

Sydenham K [Chipeham 1206 Cur, Cypenham n.d. Reg Roff, Shippenham 1315 FF]. Identical with CHIPPENHAM, S- for Ch-being due to Norman influence and -*d*- being late for *þ*.

Syderstone Nf [*Cidesterna, Scidesterna* DB, *Sidesterne* 1198 FF, 1203 Cur]. First el. apparently OE *sīd* 'broad'. The second seems to be as in SEWSTERN.

Sydling Do [*Sidelyng* 939 BCS 739, *Sidelince* DB, -*linz* 1200 Cur]. 'Broad hill', the elements being OE *sid* 'broad' and HLINC.

Sydmonton Ha [*Sidemanestone* DB, -*ton* 1200 P, 1204 Cur, *Sidemanton* 1169 P]. '*Sideman*'s TŪN.'

Syerscote St [*Siricescotan* 1100 PNSt, *Fricescote* (for *Sirices*-) DB, *Sirescot* 1242 Fees]. '*Sigeric*'s COTS.'

Syerston Nt [*Sirestune* DB, *Siristun* 1236 Fees]. '*Sigehere*'s TŪN.'

Sykehouse YW [*Sykhowse* 1555 FF]. 'House on a sīc or stream.'

Syleham Sf [*Silham* c 950 BCS 1008, *Seilam* DB, *Seleham* 1156 BM, *Sileham* 1174 P]. Either 'HĀM by a *sylu* or miry place' or 'HĀM in a *sulh* or gully'. Cf. SULH.

Symondsbury (-ĭ-) Do [*Simondesberge* DB, *Symundesberg* 1237 FF]. '*Sigemund*'s barrow or hill.' See BEORG. The river-name **Simene** is a back-formation.

Syndercombe So [*Sindercome* DB, -*combe* n.d. Buckland]. First el. OE *sinder* 'cinder, dross'.

Syon House Mx [*Syon* 1415 Ch]. Sion Abbey was founded in 1414-15. It was named from the Biblical Sion.

Syresham (sīs-) Np [*Sigresham* DB, *Sigeresham* 1151 BM, *Sigheresham* 1221 Ep]. '*Sigehere*'s HĀM.'

Sysonby Le [*Sistenebi* DB, *Sixtenebi* DB, -*bia* c 1125 LeS, *Sixtenesby* 1236 Fees]. '*Sigstein*'s BY.' First el. ON *Sigsteinn*, OSw *Sigsten* (*Sistain* DB).

Syston (-ĭ-) Le [*Sitestone* DB, *Sithestun* 1201 Cur, *Sidhestone* 1219 f., *Sidestone* 1220 Ep, *Sithestan* 1254 Val, *Sydestan* 1291 Tax], **S~** Li [*Sidestan* DB, 1205 Cur, *Sithestan* 1212 Fees]. The second is OE *sida stān* 'broad stone'. The first may be so too, but forms in -*tūn* are earlier than those in -*stān*. Possibly it is '*Sigehæþ*'s TŪN'. If it is *sida stān*, the numerous forms in -*th*- are due to Scand influence.

Sywell (sīel) Np [*Snewelle* DB, *Siwell* 12 NS, -*a* 1209-19 Ep, *Seuewell* 1236 Fees]. 'Seven wells.' Cf. SEAWELL. First el. OE *syfan*, a known form of the numeral *seofon* 'seven'.

T

Tablehurst. See TEALBY.

Tabley Chs [*Stab(e)lei* DB, *Thabbelewe* c 1160 Chester, *Tabbeleg* 1260, -*leye* 1281, *Over Tabbele* 1289 Court]. '*Tæbba*'s LĒAH.'

Tachbrook, Bishops, & T~ Mallory, Wa [*Taschebroc, Tacesbroc* DB, *Tachesbroch* 1177 P, *Tachelesbroc* 1200 Cur, *Tachebrok Mallore* 1291 Tax]. Originally the name of the brook at the place [*Tæceles bróc* 1033 E]. Another stream called *Tæcles bróc* in Wo is mentioned 969 BCS 1242. The first el. of both appears to be an OE **tæcels*, derived from OE *tæcean* 'to teach', originally 'to show', and meaning perhaps 'something that marks a boundary'. The brook at T~ was an important boundary.
Bishops T~ belonged to the Bishop of Chester (DB). **T~ Mallory** was held by Henry Mallore in 1200 (Cur). Cf. KIRKBY MALLORY.

Tackley O [*Tachelie* DB, *Takalege* 1158 Fr, *Takkelea* 1176 P, *T(h)ackele* 1209-19 Ep], **Takeley** Ess [*Tacheleia* DB, c 1130 Oxf, *Takelea* 1176 P, -*leia* 1194 BM]. The same first el. is found in **Tackbear** D [*Tacabeara* DB, *Takebeare* 1336 Misc, *Tekebeare* 1447, *Tegebere* 1491 Ipm], **Acton** Do [*Tacatone* DB, *Tachetona* 1109 Mon]. In the last *T*- was lost as in ELSTREE. It may be an earlier form of *teg, tag* 'a young sheep' (found from the 16th cent.), which is no doubt related to Sw *tacka* (OSw *takka*) 'ewe'. The second

el. of Tackbear is OE *bearu* 'grove'. Tackley, Takeley would mean 'pasture for tegs'.

Tacolneston (tăklstn) Nf [*Tacoluestuna* DB, *Takolueston* 1185 P, *Tacolneston* 1203 Cur]. '*Tātwulf*'s TŪN.' The change of *Tāt*- to *Tac*- is dissimilatory, while -*n*- for -*v*- is due to misreading of -*u*- in early forms.

Tadcaster YW [(to) *Táda* 1066 ASC (C), *Tatecastre* DB, 1227 FF, *Tathecastre* c 1150 BM, *Tadecastre* 1212 RBE, 1218 FF]. Probably '*Tāta*'s CEASTER or Roman fort'. The form *Táda* in ASC is difficult to judge. It may indicate that the first el. contained a *d*, not a *t*. But probably it is due to misreading of a form in some source used, perhaps an abbreviated form that looked like *Táda*. Cf. KELK.

OE **tadde, tādige** 'toad', or rather an unrecorded form **tāde*, the source of ME *tode*, is probably found in some pl. ns., as TADDIFORD, TADMARTON, TATHWELL. Very likely the meaning was sometimes 'frog' rather than 'toad'. The sense 'frog' is found in ME.

Taddiford House Ha at Hordle [*Tadeford* 1311 AD]. 'Toad or frog ford.' Cf. TADDE.

Taddington Db [*Tadintune* DB, -*ton* 1235 Ch, *Tatinton* 1200 P, 1263 Ch, *Tatingtone* 1275 RH], **T~** Gl [? *Tateringctun* 840 BCS 430, *Tatintone* DB, *Tadinton, Tatinton* 1227

Ch]. 'The TŪN of *Tāta*'s people.' If *Tateringctun* 840 belongs here, *Tātingas* in Taddington Gl is elliptical for *Tātheringas* or *Tāta* was a short form of *Tāthere*.

Tadley Ha [(æt) *Tadanleáge, Taddanleáge* 909 BCS 625 f., *Tadel* 1243 Cl], **Tadlow** Ca [*Tadeslaue* c 1080 ICC, *Tadelai* DB, *Tadelawe* 1197 FF, *Taddelawe* 1242 Fees]. The first el. might be OE *tadde, *tāde* 'toad'. But the probability is that Tadlow contains OE *hlāw* 'tumulus' and then no doubt has a pers. name as first el. An OE *Tāda* is unrecorded, but may have developed from *Tāta*.

Tadmarton O [*Tademertun, Tademærtun, Tadmertún, Tadmærton* 956 BCS 964 ff., *Tademertone* DB]. The first el. is an OE *Tāde-mere* 'frog pool'. Cf. *Thademere* c 1245, *Taddemere* c 1265 Selborne (in Thedden nr Selborne Ha).

Tadworth Sr [*Þeddewurþe* 675, *-uuerþe* 933 BCS 39, 697, *Ðæddeuurðe* 1062 KCD 812, *Tadeorde* DB, *Taddewurth* 1242 Fees]. The OE forms are not in good texts, and Ð may quite well be misread for D. If so, the name means '*Dæda*'s WORÞ'. The later change of *D-* to *T-* would be as in TIDEN-HAM. For *Dæda* cf. DEDDINGTON.

OE **tægl** lit. 'tail'. See CROXDALE.

Takeley. See TACKLEY.

Tale R D [(on) *Tælen* 1061 ERN, *Tala* 1185 Buckland]. A derivative of OE *getæl* 'quick, active' (= OFris *tel* 'swift', OLG *gital* 'quick') meaning 'the swift one'. From the river-name are derived **Talaton** [*Taletone* DB] and **Tale** [*Tale* DB, 1237 FF].

Talke St [*Talc* DB, *Talk* 1252 Ch, *Talke* 1276 Ipm]. Unexplained. The place is on a prominent ridge, of which Talke may have been the name. One might compare with it **Talkin** Cu [*Talcan* c 1200, *Talkan* c 1215 WR, *Talkaneterne* 1294 Ipm]. This may contain Welsh *tal* 'forehead, front, end'. The second el. is not clear.

Talland Co [*Tallan* 1264 Ep, *Talland* 1291 Tax]. Possibly a saint's name. OCo *Talan, Telent* pers. n. is found in Bodmin manumissions Th 626 f.

Tallentire Cu [*Talentir* Hy 2, *-tire* 1200–25 StB, *Talghentir* 1208 FF, *Talentyr* 1234 Cl]. A British name, the elements being Welsh *tal* 'front, end' and *tir* 'land' and *-en-* the def. art. (OBret, OCo *en*, OW *hin*). The name would mean 'end of the land'.

Tallington Li [*Talintune* DB, 1230 P, *Talingtun* 1212 Fees]. 'The TŪN of *Tæl*'s or *Tala*'s people.' *Tæl (*Tala) would be derived from OE *getæl*. Cf. TALE.

Talton Wo [*Tætlintun* 991 KCD 676, *Tadlinton* 1175 P, *Tatlinton* 1209 Fees]. 'The TŪN of *Tætel*'s people.' For *Tætel* see ADLESTROP and TATSFIELD.

Tamar R Co, D [*Tamáros* c 150 Ptol,

Tamaris c 650 Rav, *Tamur* 980–8 Crawf, *Tamer* 1018 ERN, *Tambra* c 1125 WMalm]. A Brit river-name identical with TAMBRE in Spain [*Tamaris* Mela, *Tamára* c 150 Ptol] and related to TAME. Cf. TAMERTON.

Tame R St, Wa [*Tame* c 1000 Saints, 1228 Ass; cf. *Tomsetna gemære* 849 BCS 455 'the boundary of the dwellers on the Rea, a tributary of the Tame'], T~ R YW, La, Chs [*Tome* 1292 Ass, *Tame* 1322 LaInq], T~ R YN [*Tame* 12, 13 Guisb]. Cf. THAME, TEAM. A Brit river-name identical with TAFF and TAF in Wales [*Tam, Taf* c 1150 LL] and meaning perhaps 'dark river'. If so, it is related to OIr *temen* 'dark', Sanskrit *támas* 'darkness'. Förster, *Themse*, is inclined to derive TAME, THAME, THAMES and related names from the root *tā* 'to melt' (in Engl *thaw* &c.), the meaning of the names thus being 'fluid, water'. As derivatives from the root with a suffix *m* that might have given OCelt *tam-* have nowhere been found, the suggestion is a not very likely hypothesis.

Tamerton, North, Co [*Tamerton* 1180 P, 1235 Ch], T~ **Foliot & King's** T~ D [*Tamáre* c 150 Ptol, *Tambretone, Tanbretona* DB, *Kyngestamerton* 1281 Ass, *Tamereton Foliot* 1263 Ipm]. 'TŪN on R TAMAR.' The OBrit name was a direct derivative of the river-name.

T~ **Foliot** was held by Robert Folioth in 1242 (Fees). Cf. CHILTON FOLIAT.—**King's** T~ was held by the king in 1086 (DB).

Tamhorn St [*Tamahore* DB, *Tamehorn* 1167 P, *Tamenhorn* 1166 RBE]. 'The bend of R TAME (1).' Cf. HORN.

Tamworth St [*Tamouuorði, Tamouuorthig* 781, *Tomeworðig* 799 BCS 239 f., 293, *Tameworþig* 922 ASC, *Tamuuorde* DB]. 'WORÞIG on R TAME.'

Tanat R Sa [*Tanad* 1263 Brut, *Tanat* 15 WWorc]. A Brit river-name identical with OW *tanet* in pers. ns. and meaning 'brilliant river'. OW *tanet* is derived from *tan* 'fire'.

Tandridge Sr [*Tenhric* c 960 BCS 1155, *Tenrige* DB, *-hrigg* 1212 Fees, *Tanrigges* c 1270 Ep, *Tanerigg* 1279 QW]. The original form was very likely OE *Dennhrycg* 'ridge with *denns* or swine-pastures'. For the change *D-* > *T-* cf. TADWORTH. The *a* of the first syllable represents OE *æ*, which occurs in Sr for *e* before nasals.

Tanfield Du [*Tamefeld* 1179 Hexh]. 'FELD on R TEAM.'

Tanfield, East & West, YN [*Tane-, Danefeld* DB, *Tanefeld* J Ass, 1240 FF]. The first el. is perhaps OE *tān* 'twig, sprout, shoot, branch'. Du *teen* also means 'osier'. This might be the meaning here.

OE **tang** 'tongs' must also have been used of the fork of a river and land in such a fork. But there must also have been an OE ***twang** in the same sense, related to OHG *zwange* 'tongs' and the base of OE *twengan* 'to tweak'. The two words are not always easy to keep apart. See TANG, TONG(E), TANGLEY, TANGMERE, TONGHAM.

Tang Hall YN in Murton [*Tanga* 1167 P, *Tonge* 1221 FF]. See TANG.

Tangley Ha [*Tangelea* 1175 P, *-lie* 1212 Fees], T~ Sr [*Tangelee* 1315 Ipm]. 'LĒAH in a tongue of land.' See TANG.

Tangmere Sx [*Tangmere* 680 BCS 50, *Tangemere* 11 DM, DB]. Second el. OE *mere* 'lake'. The lake has disappeared, and the exact meaning of TANG cannot be determined. The lake may have had a shape that resembled a pair of tongs.

Tankersley YW [*Tancresleia* DB, *Thankerleia* c 1150 Crawf, *Tancredeslay* 1194 f. P]. '*Þancrēd*'s LĒAH.'

Tankerton K [*Tangrenton* 1242 Fees, *Tangreton* 1258 Ch, *Tankerton* 1271 Ipm]. 'The TŪN of *Þancrēd*'s people.' OE **Þanchere* (cf. OG *Thancheri*) would be a still better base.

Tannington Sf [*Tatintuna* DB, *Tatingetona* 1168 P, *-tun* 1199 FF, *Tatingtone* 1254 Val]. 'The TŪN of *Tāta*'s people.' The change *t > n* is late.

Tanshelf YW [*Taddenesscylf* 947 ASC (D), c 1130 SD, *Tatesselle* DB, *Tanessolf* c 1170 YCh 1598]. The second el. is OE *scylf*, here 'hill' or 'slope'. The first is very difficult to determine owing to the variation of the early forms. It seems to be a pers. n.

Tansley Db [*Tanes-*, *Teneslege* DB, *Taneslea* 1175 ff. P], **Tansor** Np [*Tanesovre* DB, *-oura* 1110–23 RA, *-our'* 1198 P]. The first el. appears to be OE *tān* 'branch', here used in a transferred sense of a valley branching off from the main dale (in TANSLEY) and of a branch of a river (in TANSOR, which is on a branch of the Nene). See LĒAH, ŌFER. The latter is here 'shore, bank'.

Tansterne YE [*Tansterne* DB, *Tanestorn* 1198 P, *-terne* 1240 FF, *Tanstern* 1233 Cl]. The elements may be OE *tān* 'twig' &c. and the el. *sterne* suggested under SEWSTERN.

Tanton YN [*Tametun* DB, *-ton* 1170 P]. 'TŪN on R TAME (3).'

Tanworth Wa [*Tanewrth* 1201 Cur, 1242 Cl, *-e* 1206 Cur, *Tonewrth* 1316 FA]. The first el. might be OE *tān* 'branch' &c. ('enclosure made of branches') or an OE *Tāna* pers. n. derived from it.

Tapeley D [*Tapeleia* DB, *Tappeleg* 1178 P]. Tapeley may be identical with *Tæppe leag* 901 BCS 596. The first el. may be OE *tæppa* 'tap' in an earlier sense 'peg' (? 'wood where pegs were got').

Taplow Bk [*Thapeslav* DB, *Tapeslawe* 1187, *Tappelawe* 1196 P, 1208 Cur]. Taplow was named from a barrow in the old churchyard. The name probably has as first el. a pers. n. *Tæppa* occurs in **Tapners** K

[*Teppanhyse* 765–91 BCS 260; *-hyse* is identical with HAYES K], *T(e)appa* in *-n treow* 942 BCS 778. Cf. HLĀW.

Tappington K [*Tapinton* 1212 Fees, 1245 FF], **Tapton** Db [*Tapetune* DB, *Tappetona* Hy 3 BM, *-tone* 1266 FF]. '*Tæppa*'s TŪN.' Cf. TAPLOW.

Tarbock La [*Torboc* DB, *-bok* 1257 Ch, *Thorboc* 1242 Fees, *Torbroke* 1311 LaInq, *Thornebrooke* 1232–56 CC]. OE *þorn-brōc* 'thorn brook' changed to *Torboc* owing to dissimilation (loss of *r*) and Norman influence (*T-* for *Th-*).

Tardebigge Wo [(æt) *Tærdebicgan* 10 BCS 1317, *Terde-*, *Tyrdebicgan* 11 Heming, *Terdeberie* DB, *-bigge* 1156 P]. Unexplained.

Tarleton Gl [*Tornentone*, *Torentune* DB, *Torleton* 1291 Tax]. A variant of THORNTON. *T-* for *Th-* is due to Norman influence. The change of *n* to *l* is a case of dissimilation.

Tarleton La [*Tarleton* c 1200 CC, 1298 FF]. '*Þarald*'s TŪN.' ON *Þaraldr* is a variant of *Þóraldr*.

Tarnacre La [*Tranaker* c 1210 CC]. 'Field frequented by cranes.' First el. ON *trani*, *trana* 'crane'.

Tarporley Chs [*Torpelei* DB, *Torperleg* 1282, *-le* 1287, *Thorperle* 1287 Court, *Torperley* a 1293 Chester]. The place stands by a prominent hill, which may well have been called *Torr* (see TORR). The rest of the name might then be OE *per-lēah* 'pear wood or glade'. The name would mean 'Perley by the hill called Torr'.

Tarrant R Do [*Terrente* 935, *Terente dene* 956 BCS 708, 970, *Tarente* 1253 FF]. Identical with the old name of the ARUN Sx [*Trisántōnos* (gen.) c 150 Ptol, *Tarente* c 725 BCS 145, 1263 Ass]. The name is a variant of TRENT. From the Tarrant were named several places, all, except T~ Crawford, originally called **Tarrant** [*Terente* c 871 BCS 531, *ad Terentam* 935 BCS 708 (= T~ Hinton), *Tarente*, *Terente* DB, *Tarenta* 1165 P].

T~ Crawford [? *Crawan ford* 956 BCS 958, *Craveford* DB, *Craweford* 1242 Fees]. Crawford 'crow ford' was the old name of the place.—**T~ Gunville** [*Tarente Gundevill* 1233 Ch, T~ *Gundevil* 1242 Fees]. The manor was held by Robert de Gundeuill' in 1181 (P). The family name is from GONNEVILLE in Normandy. —**T~ Hinton** [*Tarente Hyneton* 1285 FA]. Hinton is 'the manor of the *hīwan* or inmates of Shaftesbury Abbey'.—**T~ Keynston** [*Tarente Kahaines* 1225 Sarum, *Tarrent Kahaynes* 1237 Ch]. The manor was held by William de Cahaignes in 1199 (P). Cf. ASHTON KEYNES.— **T~ Launceston** [*Tarente de Lowyneston* 1285 FA, T~ *Lowyneston* 1288 FF]. *Lowyneston* may be '*Lēofwine*'s TŪN.'—**T~ Monkton** [*Tarente Monachorum* 1291 Tax]. The manor was held by the church of Tewkesbury in 1107 (1300 Ch), by that church and the Abbess of Caen in 1291 (Tax).—**T~ Rawston** [*Tarente Willelmi de Antioche* 1242 Fees, *Tarrant Rawston* 1535

VE]. Rawston refers to an owner, apparently one Ralph.—T~ **Rushton** [*Tarente Petri de Russell* 1242 Fees, *Tarente Russeaux, Russeauston* 1314 Ipm]. *Russeaux* from ROUSSEAUX (several in France). The manor seems to have been granted to Peter de Rusceaus in 1216 (Cl).

Tarring Neville Sx [*Toringes* DB, *Torringes* 1194 P, *Terringes* 1291 Cl], **West T~** Sx [*Terringges* 941, *Teorringas* 946 BCS 766, 811, *Terringes* DB]. 'Teorra's people.' *Teorra* may be a short form of names in *Tīr-*.
The Neville family was in possession in 1254 (Ipm). Cf. FIFEHEAD NEVILLE.

Tarrington He [*Tatintune* DB, *-tonia* 1144 Fr, *Tadintune* 1146 Fr]. 'Tāta's TŪN.'

Tarset Nb [*Tyreset* 1244 Cl, *Tyrsete* 1267 Pat]. Perhaps 'Tīr's fold'. Cf. (GE)SET. For *Tīr* see TERRINGTON Nf. Or the first el. may be as in TIRRIL.

Tarvin Chs [*Terve* DB, *Terven* 1286 Court, 1297 AD]. Named from Tarvin R [*Teruen* 1209 ff. WhC, *Tervin* 1209 ff. Chester]. The source is Welsh *terfyn* 'boundary'. The river presumably once formed an important boundary.

Tasburgh (tāz-) Nf [*Taseburc* DB, 1197 P, *-burg* 1202 FF, *-burgh* 1242 Fees, *Tasseburc* 1199 FF]. Perhaps 'Tǣsa's BURG'. OE *Tǣsa* pers. n. would be derived from OE *getǣse* 'convenient, pleasant' and may be found in *Tǣsan mǣd* 825 BCS 390. The river-name **Tas** [*Tas* 1577 Harrison] is a back-formation.

Tasley Sa [*Tasselegh* 1230 Cl, *-leg* 1242 Fees, *Thasseleg* 1233 Cl]. 'LĒAH overgrown with teasels.' OE *tǣsel* 'teasel' appears as *tassyll, tassel* 16, 17 (OED).

Tatenhill St [*Tatenhyll* 942 BCS 771, *Tattenhull* 1251 Ch], **Tatham** (-ā-) La [*Tathaim* DB, *Tateham* 1202 FF]. 'Tāta's hill and HĀM.'

Tathwell Li [(æt) *Taðawyllan* 1002 Wills, *Taðawillan* 1004 KCD 710, *Tadewelle* DB, 1168 P, *-wella* c 1115 LiS, *Taddewell* 1156 P]. 'Frog stream'; cf. TADDE. The form with *th* is due to Scand influence. The identification of the first two examples is somewhat doubtful.

Tatsfield Sr [*Tatelefelle* DB, *Tatelesfeld* 1253 Ch, *Tatlesfeld* 1229 FF]. 'Tætel's or Tǣtel's FELD.' *Tatel* is the name of a moneyer. For *Tætel* cf. TALTON.

Tattenhall Chs [*Tatenale* DB, *-hala* c 1100 Chester]. 'Tāta's HALH.'

Tattenhoe Bk [*Taddenhó* 1167, *Tatenho* 1180 P, *Toternho* 1237–40 Fees]. 'Tāta's HŌH.' The spelling *Toternho* is due to confusion with TOTTERNHOE Bd.

Tatterford Nf [*Taterforda* DB, *-ford* 1203 Ass, *Tateresford* 1207 FF, *Tatersford* 1254 Val]. **Tattersett** Nf [*Tatessete* DB, *Tatersete* 1199 P, 1203 Cur, 1254 Val]. 'Tāthere's ford and (GE)SET.'

Tattershall Li [*Tateshale* DB, *Tatesala* c 1115 LiS, *Tatersala, Tatrehalla* 12 DC, *Tatersall* 1212 Fees]. 'Tāthere's HALH or valley.'

Tattingstone Sf [*Tatistuna* DB, *Tatingeston* 1219, *-tun* 1226–8 Fees]. 'Tāting's TŪN.' *Tāting* is a derivative of *Tāta*.

Tatton Chs [*Tatune* DB, *Tatton* Hy 3 Pudsay], **T~** Do nr Portisham [*Tatentone, Tatetun* DB, *Tattun* 1212 Fees], **Tatworth** So [*Tattewurthe* 1254 Misc, *Tateworth* 1315 Ipm]. 'Tāta's TŪN and WORÞ.'

Taunton So [*Tantun* 722 ASC, 737, 854 BCS 158, 475]. 'TŪN on R TONE.'

Taverham Nf [*Taverham, Tauresham* DB, *Tauerham* 1168, 1191 P]. The first el. is OE *tēafor* 'red pigment, vermilion', found also in *Tǣafersceat* 966–75 Wills. OE *tēafor* is here probably used of red earth.

Tāvy R D [*Taui* 1125 WMalm, 1238 Ass]. A Brit river-name, probably related to TAME and TAMAR but with later change of intervocalic *m* to *v*. The base would be an OBrit *Tamio-* or the like. From the river were named **Marytavy** [*Tavi* DB, *Ecclesia Sancte Marie de Tavi* 1270, *Tavymarie* 1413 Ep], **Petertavy** [*Tawi* DB, *Peterestavi* 1276 Ep], where the additions are taken from the dedication of the churches, and **Tavistock** [*at Tauistoce* 981 KCD 629, *Tæfingstoc* 997 ASC (C, D), *Tæfistoc* c 1000 Saints, *Tavestoc* DB]. Tavistock is probably 'the STOC belonging to Tavy (Mary- and Peter-tavy)'.

Taw R D [*Táwmuða* 1086 ASC (D), *Tavus* a 1118 Flor, *Tau* 1244 Ass]. Identical with TAY in Scotland [*Tavus* Tacitus]. The name is generally held to be related to Welsh *taw* 'silent', the meaning being 'the silent river'. But more likely it belongs to the root *tevā, teu* 'to swell' in Sanskr *tavás* 'strong' (= 'powerful river'). **Tawstock** D [*Tauestoca* DB, *Taustoche* 1157–60 Fr, *Toustok* 1227 Ch], **Bishop's Tawton** D [*Tautona* DB, *Tautone Episcopi* 1304 Ep], **North T~** D [*Tawetone* DB, *Chepintauton* 1199 Obl, *Nortauton* 1264 Ipm], **South T~** D [*Tavetone* DB, *Suthtaut[on]* 1212 Fees]. 'STOC and TŪN on R Taw.'
Bishop's T~ belonged to the Bishop of Exeter.

Taxal Chs [*Tackishalch* 1273 Ipm, *Tatkeshal, Tackesal* 1285 Court]. 'Tātuc's HALH or valley.' *Tātuc* is a derivative of *Tāta*.

Taynton Gl [*Tetinton, Tatintone* DB, *Thetintone* Hy 2, *Tepingtone* 12 Glouc, *Teintona* 1167 P, *-ton* 1220 Fees]. 'The TŪN of Tǣta's people.'

Taynton O [*Tengctun* c 1053–7 AS Writs (also R Writs), *Teigtone, Tentone* DB, *Teinton* 1163 P, 1242 Fees, *-e* 1229 Ep]. First el. an old name of Hazelford Brook, identical with the river-name TEIGN.

OE **tĕafor** 'red pigment, vermilion'. See TAVERHAM, TIVERTON Chs, and cf. TERRINGTON YN, TEVERSALL &c.

OE **tĕag** 'enclosure, close, a common pasture', dial. *tye*, is the base of TEIGH, and the second el. of OLANTIGH, TILTY. Cf. also TEY.

Tealby Li [*Tavelesbi, Tauelebi* DB, *Tablesbeia* 1094 Fr, *Teflesbi* c 1115 LiS, *Tevelesby* 1209–35 Ep, *Thevelbe* 1252 Ch], **Tellisford** So NE. of Frome [*Tefleforð* 1001 KCD 706, *Tablesford* DB, *Tevellsford, Teveleford* J Berk], **Thelsford** Wa [*Theuelisford* 1200–12 BM, *Tevelesford* 1209–35 Ep, *Teflesford* 1232 AD]. With these must also be compared **Tablehurst** Sx in Forest Row [*Tauelhurst* c 1200, *Tavelhurst* 1296 PNSx]. The first el. seems to be an OE **tæfli, *tefli*, derived from an OE *tæfl* 'chess-board' and probably meaning 'a plateau'. We may compare TAFELBERG in Holland [*Tafalbergon* 11], *Zabelstein* 12 (Förstemann). But THELSFORD Wa often has *Th-* in early records and may have begun in *Th-* originally. It might then contain the OE pers. n. *Theabul* 697 BCS 97.

Team R Du [*Tomemuthe* 1104–8 SD, *Thama* c 1190, *Tame* 13 Newcastle]. See TAME.

Tean R St [*Tayne* 1577 Saxton]. Apparently identical with TEIGN. On the Tean is **Upper Tean** St [*Tene* DB, 1204 Cur, *Teyne* 13 PNSt].

Tĕbay We [*Tibeia* 1179 P, 1201 Cur, *Tibbay* c 1200 (1294) Ch, *Tibbeie* 1224 P]. '*Tibba*'s island.' See ĒG.

Tebworth Bd [*Teobbanwyrþe* 926 BCS 659, *Tebbewurþ* 1227 Ass]. '*Teobba*'s WORÞ.' OE **Teobba* is easily explained as a short form of *Þeodbald* and the like.

Tedburn St. Mary D [*Tettaborna* c 1120 E, *Teteborne* DB]. Originally the name of the stream at the place [(on) *Tettan burnan* 739 Crawf]. '*Tette*'s or **Tetta*'s stream.' OE *Tette*, a woman's name, is well evidenced, while *Tetta* is not.

Teddesley St [*Teddesl'* 1236 Fees, *Tudeslegh* 1246, *Teddesleg* (Hay) 1252, *Tedeslegh, Tidesleye* 1275 ff. Cl]. '*Tydi*'s LĒAH.'

Teddington Mx [*Tudintún* 969 Crawf, *Tudincgatun* c 970 BCS 1174, *Tudinton* 1197 FF]. 'The TŪN of *Tud(d)a*'s people.'

Teddington Wo [*Teottingtun* 780, *Teotintun* 964, *Teottincgtun* 969 BCS 236, 1135, 1233, *Teotintune* DB]. 'The TŪN of *Teotta*'s people.' OE *Teotta* is not recorded. Cf. TETTENHALL.

Tedsmore Sa [*Teddesmere* c 1205 Eyton]. The first el. is identical with that of the next name. The second seems to be OE *gemǽre* 'boundary'.

Tedstone Delamere & Wafer He [*Tedesthorne* DB, *Tedethorn, Thedesthorne, Thoddesthorne la Mare* 1242 Fees, *Teddesthorn la Mare, Teddesthorne Wafre* 1249 Fees]. The

second el. is OE *þorn* 'thorn-bush'. The first is a pers. n. related to OE *Teodec*, and probably a short form of names in *Þeod-*.

T~ Delamare was held by Thomas and Jordan de la Mare in 1200 (Cur). Cf. FISHERTON DELAMERE.—Robert le Wafre in **T~ Wafer** is mentioned in 1242 (Fees). Cf. HAMPTON WAFER.

Tedworth, South, Ha [*Tudanwyrð* c 975 Wills, *Todeorde, Tedorde* DB, *Tudewrth* 1203 Cur, *-worth* 1236 Fees], **North T~** W [*Todeworde* DB, *Thudewrda* 1178 BM, *Tudewrth* 1242 Fees]. Cf. also *Teodeorda* c 1150 Fr. '*Tuda*'s WORÞ.' The two Tedworths are close together, though in different counties.

Tees (tēz) R [*Tesa* 1026 Knytlinga saga, *Tese* c 1050 HSC, *Tesa* 1104–8 SD, *Teisa* c 1090 Reg, 1104–8 SD, *Taise* c 1130 SD]. A Brit river-name, related to Welsh *tes* 'heat, sunshine', Ir *teas* 'heat'. The name may mean 'boiling, surging river'. **Teesdale** is *Tesedale* c 1130 SD.

Teeton Np [*Teche* DB, *Theche* Hy 2 NpCh, *Teacne* 1196 P, *Tekne* 1220 Fees]. A derivative of OE *tācn* 'token, sign', e.g. an OE **tǽcne* fem. 'beacon'. The place is on the end of a ridge.

Teffont Ewyas & Magna W [*be Tefunte* 860, *Teofuntinga gemǽre* 940, *at Teofunten* 964 BCS 500, 757, 1138, *Tefonte* DB, *Teffunt Ewyas* 1275 RH]. 'Boundary spring or stream.' Cf. FUNTA. The first el., like that of TYBURN, is an OE **tēo*, corresponding to OFris *tia* 'boundary' or 'boundary line'. The word is derived from OE *tēon* 'to draw'.

T~ Ewyas belonged to Ewyas Barony (He).

Teigh (tē) Ru [*Tie* DB, *Ti* 1202 Ass, *Ty* 1254 Val]. OE *TĒAG* 'enclosure'.

Teign (tīn, tēn) R D [*Teng* 739 Crawf, *Teine* 1205 Layamon, *Teyng* 1244 Ass, *Teygne* 1282 Ass]. The original form was *Tegn* (as in *Tegntun* 1001 ASC). The name is a Brit river-name, related to Welsh *taen* 'sprinkling' (from **tagnā*), Lat *stagnum*, OBret *staer* 'river'. The meaning is simply 'stream'. Several places are named from the river. **Teigngrace** (tĭng-) [*Taigne* DB, *Teyngegras* 1331 Ep]. The addition is a family name. Geoffrey Gras held the manor in 1352 (Pat). Gras is Fr *gras* 'fat'.—**Teignmouth** (tin-) [*Tengemuða* 1044 OSFacs, *Teingnemuth* 1253 Ch].—**Bishopsteignton** [*Taintona* DB, *Teygtone Episcopi* 1262 Ep]. Held by the Bishop of Exeter.—**Drewsteignton** [*Taintone* DB, *Teyngton Drue* 1275 RH]. Held by Drogo 1210–12 (RBE). Cf. LITTLETON DREW.—**Kingsteignton** [*Tegntun* 1001 ASC, *Teintone Regis* 1259 Ep, *Kingestentone* 1273 Ipm]. Held by the king in 1086 (DB). **Teignhead** (tīnĭd) in STOKEINTEIGNHEAD &c. has nothing to do with the river-name. It means 'ten hides'.

Teise R K [*Theise* 1577 Harrison]. A backformation from TICEHURST, called *Theise Hirst* by Harrison.

Tellisford So. See TEALBY.

Telscombe Sx [*Titelescumbe* 966 BCS 1191, *Tetelescombe* 1275 RH]. The first el. appears to be a pers. n. **Titel*, found in *Titlesham* 765 BCS 198 (Sx).

Teme, Welsh **Tefaidd, Dyffryn Tefeidiad**, R Sa, Wo, He [*Temede* 757–75, *Temede stream* c 779, 963 BCS 219, 233, 1107, (in) *Temedan* 816 ib. 357, *Temede* 1256 Ass]. A Brit river-name related to TAME. The Welsh name Tefaidd is apparently not now known in Wales (Owen's *Pembrokeshire* iii. 333 f.), but cf. the same work i. 202 (footnote 2).

Temple in pl. ns. indicates that the place once belonged to the Knights of the Temple. See TEMPLE NEWSAM and the like. **Templeton** Brk nr Kintbury is *Templeton Templariorum* 1220 Fees. **Templeton** D [*Templum* 1206 Cur, *Templeton* 1334 Buckland] is also called **Combe Temple**.

Tempsford Bd [*Tæmeseford* 921 ASC, *Temesanford* 1010 ib. (E), *Tamiseforde* DB]. 'Ford on the road to the Thames' (i.e. to London).

Tenbury Wo [*Temedebyrig* 11 Heming, *Tamedeberie* DB]. 'BURG on R TEME.'

Tendring Ess [*Tendringa* DB, *Tendringes* c 1145 Colchester, *Tenring* 1195 FF, *Thendring* 1200 Cur, *Tendringg* 1254 Val]. As in Essex an *i*-mutated *a* before nasals often appears as *a* in ME forms, we should expect frequent forms such as *Tandring*. Very few seem to occur. This suggests that the base of the name had OE *y*, which became in Essex *e*. If so, the name may be compared with TÜNDERN in Germany on the Weser [*Tundiriun* 1004, *Tundirin* 1025]. The etymology is obscure. The name may be derived from the stem *tundr*- in OE *tynder*, OHG *zuntirra* &c. 'tinder', but the exact meaning is obscure. One might think of a meaning 'beacon'. Tendring would then be 'the people at the beacon'. But the *Tendringas* may be 'people from Tündern'.

Tenterden K [*Tentwardene* 1179, *-den* 1180 P]. 'The DENN or swine-pasture of the Thanet people' (first el. OE *Tenetwaru*; cf. CANTERBURY).

T~ belonged to Minster in Thanet (StAug 29). *Tenetwara brocas* 968 BCS 1212 seems to have been near Tenterden ('the brooks or fens of the Thanet people').

Terling (-ar-) Ess [*Terlinges* c 1050 KCD 907, *Terlingas* DB, *Tertlinces* 1086 IE, *Terdlinge* 1237 FF, *Tyrlinge* 1338 Ipm]. OE *Tyrhtlingas* '*Tyrhtel*'s people'. The river-name Ter is a back-formation.

Tern R St, Sa [*Tren* 12 Taliesin, *Tirne* c 1200 Gervase, 1256 Ass, *Terne* c 1200 Sa Deeds]. A Brit river-name derived from Welsh *tren* 'strong, powerful'.

Terrington Nf [*Tilinghetuna* DB, *Terintona* 1121 AC, *Tirintuna* 1103–31, 1133–69 BM, *Tiringet'* 1205 Cur]. 'The TŪN of *Tīr(a)*'s people.' **Tir(a)* is a short form of names in *Tir*-.

Terrington YN [*Teurinctune* DB, *Tiuerinton* 1175 P, *Thiverington* 1202 FF, *Tiverington* 1226 FF]. The first el. is connected with OE *tiefran* 'to paint' and might be OE *tiefrung* 'painting'. If so, a mural painting or the like might be referred to. Cf., however, TEVERSALL &c., where it is suggested that OE *tiefran* may have meant also 'to practise sorcery'.

Terwick (tĕrĭk) Sx [*Tortewyk* 1271 Ch, *Turdewyk* 1291 PNSx]. Perhaps most likely '*Torhta*'s WĪC'. **Torhta* is a normal short form of names in *Torht*-. OE *tord* 'dung' might be the first el., but the regular *Torte*-, *Turde*- makes difficulties.

Test R Ha [(on) *Terstan* 877, 901 BCS 544, 594, *Tærstan stream* 1045 BM, *Terste* 1234 Cl, *Test* 1425 Pat]. A Brit river-name related to Welsh *tres* 'toil, labour', *tren* 'strong', *treio* 'to ebb'. The meaning may be 'running water, stream'. **Testwood** is *Lesteorde* DB, *Therstewode* 1242 Fees.

Testerton Nf [*Estretona* DB, *Testertun* 1242 Fees, *-ton* 1254 Val]. Dr. Schram aptly suggests that the first el. is identical with that of TESTERBANT in Holland, which is an old word for 'southern' (lit. 'right') related to Goth *taihswa* 'right hand'. But the DB form *Estretona* may indicate that the name is OE *Ēasterra tūn* 'eastern TŪN' and that *T*- is a remnant of a prep. *æt* (Introd. p. xviii).

Teston (tēsn) K [*Terstan*, *Cærstan* 10 BCS 1321 f., *Testan* DB, *Terstan* 1263 Ipm, *Tearstan* 1285 Ch]. The second el. is OE *stān* 'stone'. The first seems to be OE *tær* adj. 'gaping, cleft' or *taru* sb. 'rent, gap' (from *teran* 'to tear'). The meaning would be 'stone with a cleft or hole'.

Tetbury Gl [*Tettan monasterium* 681 BCS 58, (to) *Tettan byrg* 872–915, *Tettanbyrig* 10 BCS 582, 1320, *Teteberie* DB]. '*Tette*'s BURG or manor.' *Tette*, a sister of Ine, was abbess of Wimborne. There must once have been a monastery at Tetbury.

Tetchwick (tĕchĭk) Bk [*Tochingeuuiche* DB, *Totingwich* 1166, 1197 f. P, *Thochewik* 1237–40 Fees]. 'The WĪC of *Tota*'s people.'

Tetcott D [*Tetecote* DB, *Tetticot* 1242 Fees]. '*Tette*'s or **Tetta*'s COT.' Cf. TEDBURN.

Tetford Li [*Tedforde* DB, *-forda* c 1115 LiS, *Thetford* 12 DC]. Identical with THETFORD.

Tetney Li [*Tatenai* DB, c 1115 LiS, *Tataneina* Hy 2 BM, *Teteneia*, *Thateneia* Hy 2 DC]. '*Tæta*'s island.'

Tetsworth O [*Tetelesworth* 1241 Abbr, *Tetlesworth* 1339 Bodl, *Tettesvrda* c 1175 Bodl, *-wrthe* 1279 RH]. '*Tætel*'s WORþ.' Cf. TALTON.

Tettenhall St [(æt) *Teotanheale* 910 ASC (C, E), (æt) *Totanheale* ib. (D), *Totehala*,

Totenhale DB, *Tettenhala* 1169 P]. '*Teota's or *Teotta's HALH or valley.' Cf. TEDDING-TON Wo.

Tetton Chs [*Tadetune* DB, *Tetton* 1287 Court]. The first el. is a pers. n., e.g. *Tǣta*.

Tetworth Hu [*Tethewurða, Tettewrda* c 1150 BM, *Tetteworth* 1209 For]. '*Tetta*'s worþ.' Cf. TETCOTT.

Teversall Nt [*Tevreshalt* DB, *Tivresholt* 1204 Cur, *Thiversold* 1275, *Tyversolde* 1280 Ep, *Tyversald, Tyversalt* 1291 Tax], **Teversham** Ca [*Teuresham* c 1050 KCD 907, DB, *Teuersham* c 1080 ICC, DB, *Taversham* 1130 P, *Tevresham* 1198 Fees]. The first el. may be an OE *tiefrere* (*tēfrere*) 'painter', perhaps in the sense 'one who marks sheep'. Cf. TERRINGTON YN. But OE *tiefran* corresponds to G *zaubern*, Du *tooveren* 'to practise sorcery', and OE *tēafor* 'red pigment' to OHG *zoubar*, OFris *tāver*, ON *taufr* 'sorcery'. OE *tiefran* may well have been used in the sense 'to practise sorcery', and *tiefrere* in the sense 'sorcerer'. Such a sense would be possible in pl. ns. The second el. of Teversham is HĀM. That of Teversall is not so clear. It may be HOLT, but the early forms may point to a word with OE *a*. It might be OE (*ge*)*heald* 'shelter', for which see HALSTEAD.

Tew, Great & Little, Duns T~ O [(æt) *Tiwan* 1004 Wills, *Tewe, Teowe* DB, *Tiw, -e* 1130 P, *Tiwa Magna* 1165 P, *Parva Tiwe* 1207 Cur, *Donestiva* c 1200 Bodl, *Dunnestywa* 1232 Ep]. The name seems to be cognate with the el. *-tǣwe, -tiewe, -tēowe* found in *æltǣwe* 'in good health, excellent', *manigtiewe* 'skilful'. This el. appears to be related to OE *teohh* 'race, generation, troop', MHG *zeche* 'row, order'. The meaning of the OE word (? *tiewe*) may have been 'row', whence 'lengthy object'. Tew may then have been the name of the long ridge at which the places are. Duns Tew is c 4 m. from the other Tews. Duns Tew is 'the Tew belonging to *Dunn*'.

Tewin Hrt [*Tiwingum* 944-6 BCS 812, (æt) *Tywingan* 1015 Wills, *Theunge* DB, *Tiwinge* 1166 RBE]. Perhaps 'the people of *Tiwa*', *Tiwa being a short form of names in *Tiw-*, as *Tēoweald, Tēowulf*.

Tewkesbury Gl [*Teodechesberie, Teodekesberie* DB, *Theokesbiria* 1107 (1300) Ch, *Teokesberia* 1168 P]. '*Tēodec*'s BURG.' *Tēodec* is found in *Teodeces leage* 963 BCS 1111 (Wa). It is a hypocoristic form of names in *Þēod-*.

Tey, Great & Little, Marks T~ Ess [(at) *Tygan* c 950 BCS 1012, *Tigan* c 995 ib. 1289, *Teia* DB, 1165 P, *Teie* 1196 FF, *Theya Magna* 1238 Subs]. A derivative of OE *tēag* 'enclosure', probably OE *tiege fem.
Marks Tey is *Teye Mandevill* 1254 Val and often. It was held by the Merk family under the Mandevilles. Merk is from MARCK near Calais.

Teynham (-ĕn-) K [*Tena-, Teneham* 798, *Tenham* 801 BCS 291, 301, *Taen(e)ham* 11 DM, *Tenham* c 1140 BM]. The first el. is a pers. n., perhaps *Tēna (< *Tȳna*), a short form of names in *Tūn-*. From the same name is derived the first el. of **Timbold Hill** K (not far from Teynham) [(to) *Teninge faledun* 850 BCS 459, *Tenegefeld* 1204 Ch]. The name means 'the fold belonging to the Teynham people'.

OE þæc 'roof, thatch' is found in THATCHAM, THAXTED, where it means 'thatch'. There was also an OE *þaca* 'roof' (and 'thatch'?), which is perhaps found in THAKEHAM.

Thakeham (-ăk-) Sx [*Tacaham* 1073 Fr, *Taceham* DB, *Tacheham* 1167 P]. Perhaps 'thatched homestead', the first el. being OE *þaca* 'roof'. Cf. þÆC.

Thame (tām) R Bk, O [(on) *Tame* 956 BCS 945, *Tame strem* 1004 Fridesw, *Tame* 1241 Ass]. Identical with TAME. On the Thame is **Thame** town O [*Tamu* 675 BCS 39, (æt) *Tame* 971 ASC (B), *Tame* DB].

Thames (temz) R [*Tamesis* 51 B.C. Cæsar, c 894 Asser, *Tamesa* 115-17 Tac, *Tamensis* 417 Orosius, c 730 Bede, *Tamisa* 681 BCS 56; *Temis* 683 BCS 65, *Temes* 843 ib. 443, c 893 Alfred Or, *Temse* 1387 Trev]. The name is a Brit river-name, cognate with Sanskr *Tamasā*, the name of a tributary to the Ganges, *tamasá-* 'dark', Lat *tenebrae* &c. Cf. TAME. The name means 'dark river'.

Thanet (thăn-) K [*Tanatus* 3 Solinus, *Tanatos* c 730 Bede, *Tenid* 679, *Tænett* 949 BM, *Tenet* c 890 OEBede, 943 BCS 780, *Tanet* DB, *Tænate* 1205 Lay]. The name is identical or cognate with the river-name TANAT. It may mean 'bright island' or 'fire island' (from a beacon or lighthouse).

Thanington K [*Tan(n)ingtun* 833 BCS 407 f., *Taniton* 1202 FF, *Tenitune* 11 DM, *Teninton* 1254 Ass]. The first el. is possibly derived from THANET, the supposition being that the place had some connexion with the Isle of Thanet. It might be simply *Tænet-tūn*, or else *Tæninga-tūn, Tæningas* being an elliptical formation from *Tænet*.

Tharston Nf [*Therstuna, Sterstuna* DB, *Therestone* 1254 Val, *Therston* 1286 QW]. The first el. is clearly a pers. n., perhaps ODan *Therir*, for which there is some evidence.

Thatcham Brk [*Þæcham* c 970 BCS 1174, *Taceham* DB, *Tacheham* 1167 P, *Thacham* 1212 Fees]. 'Thatched homestead.' Cf. þÆC.

Thatto La [*Thetwall* 12 VH, *Thotewell* 1246 Ass]. OE *þēote* 'water-pipe' and *wella* 'spring' or 'stream'.

Thaxted Ess [*Tachesteda* DB, *Takesteda* 1176 P, *Thacsted* 1291 Tax]. OE *þæc-stede* 'place where thatch was got'.

Theakston YN [*Eston* DB, *Thekeston* 1157 PNNR, *Textone* c 1160 YCh 175]. First el.

perhaps OE *Þēodec, a short form of names in Þēod-. Cf. TEWKESBURY.

Thealby Li [*Tedulfbi* DB, *Tedolfbi* c 1115 LiS, *Tethelby* 1209–19 Ep]. 'Þiŏŏulf's BY'; first el. ON Þiŏŏulfr.

Theale Brk [*Thele* 1220 Fees, c 1230 BM, *La Thele* 1291 Tax], T~ So [*Thela* 1176, *la Thele* 1310 Wells]. OE þelu, plur. of þel 'plank', the name referring to a bridge or path formed by planks.

Thearne YE [*Thoren* 1297 Subs, *Thorne* 1309 f. Ep, *Thurne* 1573 BM]. OE þorn 'thorn-bush'.

Theberton Sf [*Tiberton* 1178, 1186 ff. P, *Tiburton* 1198 FF, *Teberton* 1200 FF, *Thebertun* Hy 3 BM]. 'Þēodbeorht's TŪN.'

Thedden Ha [*Tedena* 1168 P, *Thetdene* 1234, *Thutdene* c 1270 Selborne]. OE þēote 'water-pipe', perhaps also 'stream', and DENU 'valley'.

Theddingworth Le [*Tediworde, Tedinges-worde, Tevlingorde* DB, *Theingurda* c 1140 BM, *Tedingewrth* 1206 Cur]. 'The WORÞ of Þēoda's people.' *Þēoda is a short form of names in Þēod-.

Theddlethorpe Li [*Tedlagestorp* DB, *Te-dolftorp, Dedloncstorp* c 1115 LiS, *Tedlaue-, Thedlactorp* 12 DC, *Tedlauetorp* 1204 Cur]. 'Þēodlāc's thorp.' OE Þēodlāc is not evidenced, but cf. OG *Theodilacus, Theut-leich.*

OE þel 'plank' is found in THEALE, THELNET-HAM, THELWALL. OE þelbrycg 'plank bridge' is the source of **Thelbridge** D [*Talebrige* DB, *Thelebrig* 1242 Fees] and ELBRIDGE.

Thelnetham Sf [*Theluetcham, Teluettteham, Teolftham* DB, *Thelfet-, Theluetham* c 1095 Bury, *Teluedham* 1196 FF, *Elnetham* 1202 FF, *Thelnetham* 1254 Val]. *Thelnetham* is due to misreading of *Theluetham*. The name must be explained in connexion with WHELNETHAM Sf. Both were no doubt once *Elfethamm* (cf. ELVETHAM) 'HAMM frequented by swans' (OE elfetu), later distinguished by the addition of þel 'plank' and whēol 'wheel' respectively.

Thelsford. See TEALBY.

Thelveton Nf [*Teluetuna, Teluetaham* DB, *Thelueston* 1183 P, *Telvetune* 1198 FF]. 'Þialfi's TŪN.' ODan *Thialvi* as well as ON Þialfi occurs, and OSw *Thiælvi* is a common name. *Teluetaham* DB is no doubt due to influence from THELNETHAM.

Thelwall Chs [*Þelwæl* 923 ASC, *Thelewell* 1241 Cl, *-wall* 1259 Court]. 'Pool by a plank bridge.' Cf. þEL. The second el. is OE wæl 'a weel, a deep pool, a deep still part of a river'. Cf. SALESBURY.

Themelthorpe Nf [*Timeltorp* 1203 Cur, *-thorp* 1219 Fees, *Thymelthorpe* 1248 Ch]. '*Þymel's or *Þymli's thorp.' The first el. is a nickname formed from OE þymel 'thimble'

or ON þumall 'thumb'. ON Þumli occurs as a byname. Cf. THIMBLEBY.

Thenford Np [*Taneford* DB, *Tanford* c 1130 Oxf, *Thayniford* 12 NS, *Teinford* 1186 P, *Teneford* 1185 P]. OE þegna-ford 'ford of the thanes'. The exact meaning of OE þegn here is obscure.

Theobald Street Hrt [*Titeberst* DB, 1204 Cur]. The second el. appears to be OE byrst, probably in the sense 'landslip'. Cf. BURSTON Nf. The first as in TIDCOMBE. The present form is due to popular etymology.

Therfield Hrt [*Ðerefeld* 1060 KCD 809, *Furreuuelde* DB, *Thirefeld* 1198 (1301) Ch]. 'Dry FELD.' First el. OE þyrre 'dry'.

Thetford, Little, Ca [(æt) Þiutforda c 975 ASCh, *Liteltedford* DB, *Littleteodford* 1086 IE, *Theford* 1337 BM], T~ Nf [*Þeodford* 870 ASC, *Þeotford* 952 ASC (D), *Tedfort* DB]. 'The people's ford', i.e. 'chief ford' or the like. The river-name **Thet** [*Thet* 1586 Camden] is a back-formation.

Theydon Bois, Garnon & Mount Ess [*Petdene gemære* 1062 Th, *Teidana, Tain-dena* DB, *Taiden* 1163 P, 1200 Cur, *Theyden* 1236 Fees, *Thayden de Bosco*, T~ *Gernun* 1238 Subs, *Theyden de Monte* 1254 Val, *Theydone Boys* 1399 BM]. Þetdene Th appears as þecdene KCD 813. The latter is probably the better reading and þecdene may represent OE þæcdene 'valley where thatch was obtained', as suggested PNW(S) xxxvi f. The change of þæc- to Thay- has an analogy in **Braydon Hook** W [*Bræcdene* 968 BCS 1213].

T~ **Bois** was held by Hugh de Bossco in 1240 (FF), but the family name seems to be of local origin and derived from a wood in Theydon. *Boscus de Taiden* is mentioned 1190 P.—T~ **Garnon** was held by Radulfus Gernun in 1200 (Cur). *Gernun* is OFr grenon 'a moustache', here used as a nickname.—T~ **Mount** is on a hill.

Thicket YE [*Thickeheved* 1219, *Tikeheved* 1231 FF]. OE þicce 'thick' and hēafod 'head, headland'. The place is low on the Derwent, which here forms a blunt bend. Possibly 'headland with thick vegetation'.

Thickley Du [*Thiccelea* c 1050 HSC]. 'Dense wood.' Cf. THICKET and LĒAH.

Thimbleby Li [*Stimblebi* DB, *Timlebi* c 1115 LiS, *Thymelby* 1219 Ep], T~ YN [*Timbelbi* DB, *Themelebi* 1088 LVD, *Thi-milisby* 1208 FF]. 'Þymli's BY.' Cf. THEMEL-THORPE.

OE þing 'meeting, court of justice', OScand þing 'public meeting or assembly, parliament' is the first el. of some pl. ns. See next names and FINEDON, FINGEST, TING-RITH, also MORTHING. The el. is Scandinavian in THINGWALL and no doubt also in some other cases, as **Thingoe** hundred Sf [*Ðinghowe* 1042–66 Th, *Thingehov* DB] from OScand Þinghaugr 'assembly mound', but there is no reason to take it to be always of Scand origin.

Thinghill He in Withington [*Tingehele* DB, *Thinghull* 1242 Fees]. 'Assembly hill.'

Thingwall Chs [*Tinguelle* DB, *Thyngwall* Hy 3 BM], T~ La [*Tingwella* 1177 P, *Thingwalle* 1212 Fees]. ON *þingvǫllr* 'assembly field'. Cf. *þingvellir* in Iceland, the place of the Althing, and TYNWALD in the Isle of Man.

Thirkleby YE [*Turgislebi* DB, *-by* c 1110 YCh 25]. '*Þurgils*'s BY.' First el. ON *Þorgils*, OSw *Thorgisl*, ODan *Thurgils*.

Thirkleby YN [*Turchilebi* DB, *Thurkeleby* 1237 FF]. '*Þurkil*'s BY.' First el. ON *Þorkell*, OSw *Þurkil*, ODan *Thurkil* (*Þurcil* ASC).

Thirlby YN [*Trillebia* 1187 Riev, *-bi* 1226 FF, *Therelby* 1208 FF]. OScand *Præla-bȳr* 'BY of the thralls'. *Thrall* is often *thrill* in ME northern texts.

Thirlmere Cu [*Thyrlemere* 1574 Collingwood, Lake District History]. Etymology obscure. First el. possibly OE *þyrel* 'hollow'.

Thirlwall Nb [*Thurlewall* 1256 Ass]. Called 'murus perforatus' by Fordun. The place is on the Roman Wall, which must have had a gap here. First el. OE *þyrel* 'perforated'.

Thirn YN [*Thirne* DB, 1270 Ipm]. OE *þyrne* 'thorn-bush'.

Thirsk YN [*Tresch* DB, *Tresc* DB, 1130 P, *Trescs* c 1150 Crawf, *Thresca* 1148 YCh 179]. OSw *thræsk*, Sw *träsk* 'lake, fen'.

Thirston (thrŭstn) Nb [*Thrasfriston*, *Th(r)afriston* 1242 Fees, *Thrastereston* 1258 Ipm]. '*Præsfriþ*'s TŪN.' OE *Præsfriþ* is not evidenced. The el. *þras-* is common in OG pers. ns. and belongs to Goth *þras* (in *þrasabalþei*), ON *þrasa* 'to threaten' &c.

Thirtleby YE [*Torchilebi* DB, *Turkillebi* 1202 FF]. Identical with THIRKLEBY YN.

Thistleton La [*Thistilton* 1212 Fees], T~ Ru [*Tisteltune* DB, *Thisteltun* 1212 Fees, *-tone* 1226 Ep]. 'TŪN where thistles abounded.'

Thixendale YE [*Sixte(n)dale*, *Xistendale* DB, *Sixtenedale* 1157 YCh 354, *Sixendale* 1297 Subs]. '*Sigstein*'s valley.' Cf. SYSONBY.

Thoby Priory Ess [*Ginges* c 1185 Bodl, (prior of) *Ginges Tobye* 1242 FF]. *Thoby* is the pers. n. *Toby*. The first prior was called *Tobias*. Cf. ING.

Thockrington Nb [*Thokerinton* 1223 Ep, 1256 Ass, *Thokerington* 1254 Val]. The first el. seems to be related to OE *þocerian* 'to move to and fro'. This is no doubt derived from a noun or adj. *þocor* of unknown meaning, perhaps 'unsteady'.

From this a pers. n. or some topographical term may have been derived.

Tholthorpe YN [*Þorp* c 972 BCS 1279, *Turulfestorp* DB]. '*Þurulf*'s thorp.' First el. ON *Þórolfr*, OSw *Þorulver*, ODan *Þurulfr*. Tholthorpe was no doubt a thorp belonging to the *Þurulfestune* mentioned with *Þorp* BCS 1279.

Thomley (-ŭ-) O [*Tumbeleia* DB, *Thumeleya* 1124–30 Fridesw]. The first el. is OE *þūma* 'thumb', here used as a nickname or in a transferred sense, such as 'dwarf, pigmy' (cf. Tom Thumb). *Thumb* in such senses is found late, but might have been used early. If so, Thomley might possibly mean 'LĒAH or wood haunted by dwarfs or fairies'.

Thompson Nf [*Tomestuna* DB, *Tomestun* 1191 FF, *Tumestone* 1242 Fees]. '*Tumi*'s TŪN.' *Tumi* (DB) is ODan *Tumi* pers. n.

Thong K [*Thuange* c 1200 Reg Roff], T~ YW in Nether- & Upperthong [*Thwong* 1274, *-e* 1277, *Overthong* 1286 Wakef]. With these may be compared *Ðwangtun* c 1060 KCD 962 (given to St. Albans Hrt) and the old name for CAISTOR Li [*Þwong-Chastre* 1205 Lay, *Thwangcastre* 1322 Ipm]. Formally the el. is identical with OE *þwang* 'thong', which may here be used in a transferred sense. But the el. may be an independent derivative from an OE *þwingan* 'to force' &c., corresponding to OLG *thwingan*, OFris *thwinga*, OHG *dwingan*, G *zwingen*. MLG *dwenge* means 'a trap'. The meaning of the el. in Engl pl. ns. cannot be settled at present.

Thonock Li [*Tunec* DB, *Tuneic* c 1115 LiS, *Thunneck* 1276 RH, *Tunnayk* 1281 QW]. 'Thin oak', from OScand *þunnr* 'thin' and *eik* 'oak'.

Thoralby YN [*Turoldesbi*, *Toroldesbi* DB, *Thoroldeby* 1230 FF]. '*Þurold*'s BY.' First el. ON *Þóraldr*, ODan, OSw *Thorald*, OE *Þurold* (from Scand).

Thoresby, North, Li [*Toresbi* DB, c 1115 LiS, *Thorisby* 1242 Fees], **South** T~ Li [*Toresbi* DB, 1212 Fees], T~ Nt [*Yuresby* 958 YCh 3, *Turesbi* DB], T~ YN [*Toresbi* DB]. '*Þori*'s (*Þuri*'s) BY.' First el. ON *Þórir*, ODan *Þurir*, *Thuri*, OSw *Thore*, *Thure*.

Thoresthorpe Li [*Thuorstorp* DB, *Thoresthorp* 1242 Fees], **Thoresway** Li [*Toreswe* DB, *Toresweia* c 1115 LiS, *Thoreswaia* Hy 2 BM]. '*Þori*'s thorp and road.' Cf. THORESBY.

Thorganby Li [*Turgrimbi*, *Torgrembi* DB, *Torgrim(e)bi* c 1115 LiS], T~ YE [*Turgisbi* DB, *Turgrimesbi* 1194 ff. P]. '*Þorgrim*'s BY.' First el. ON *Þorgrímr*, ODan *Thorgrim* (*Turgrim* DB).

Thorington Sf [*Toren-*, *Tornintuna* DB, *Turritune* J BM, *Thurintone* 1254 Val]. OE *Þorn-tūn*, apparently varying with *Þyrne-tūn* 'TŪN where thorn-bushes grew'.

Thorlby YW [*Torederebi, Toreilderebi* DB, *Thorledby* 1315 Ipm]. '*Þorald*'s BY.' Cf. THORALBY. The name represents OScand *Þoraldar bȳr*, with the first el. in the OScand gen. form *Þoraldar*.

Thorley Hrt [*Torlei* DB, *Thorneley* 1212 Fees], T~ Wt [*Torlei* DB, *Thornlega* 1186 P]. 'Thorny LĒAH'; cf. THORNLEY.

Thormanby YN [*Tor-, Turmozbi* DB, *Thurmodeby* 1234 FF]. '*Þormōð*'s BY.' First el. ON *Þormóðr*, OSw *Þormōþer*, ODan *Thormoth*.

OE **þorn** 'thorn-bush' is a very common pl. n. element, and also occurs alone as a pl. n. See THORNE &c. It is often found as the first el. Here it may sometimes alternate with derivatives such as *þorniht, þornen* 'thorny'. See THORN- (passim), FARMINGTON, HORNDON, TARBOCK, TARLETON Gl, THO(R)RINGTON, THORLEY, THORVERTON. It is also a common second el., where it mostly appears as *-thorn(e)*, as in GLAPTHORN, but also in other forms, as in BISTERNE, FRETHERNE, ROSTHERNE, SCOTHERN, SOULDERN, MOSTERTON. The first el. of names in *-thorn* varies a good deal in meaning.

Thornaby on Tees YN [*Tor-, Turmozbi* DB, *Thormodebi* 1202 FF]. Identical with THORMANBY.

Thornage Nf [*Tornedis* DB, 1166 P, *Thornedisch* 1254 Val, *Thornege* 1291 Tax]. 'Pasture where thorns grew.' See EDISC.

Thornborough Bk [*Torneberge* DB, *-berga* 1167 P, *Thornberge* 1246 Ch], **Thornbrough** YN in S. Kilvington [*Thornebergh* c 1190 PNNR, *-berg* 1233 FF], T~ YN in W. Tanfield [*Thornbergh* 1198 Fount M]. 'Thorn hill.' See BEORG.

Thornbrough Nb [*Thorneburg* 1242 Fees, 1256 Ass], T~ YW [*Torneburg* 1242 Fees, *Thornburg* 1246 FF], **Thornbury** D [*Torneberie* DB, *Thornbir'* 1242 Fees], T~ Gl [(to) *Þornbyrig* 896 BCS 574, *Turneberie* DB], T~ He [(æt) *Þornbyrig* 10 BCS 1317, *Torneberie* DB]. 'BURG where thorns grew' or 'BURG protected by a thorn hedge'.

Thornby Np [*Torneberie* DB, *Thirnebi, Thurnebi* R 1 BM, *Turneby* 1220 Fees]. Originally OE *Þornburg*, later Scandinavianized to *Þyrneby* with OScand *þyrnir* 'thornbush' as first el.

Thorncombe Do nr Blandford [*Tornecome* DB, *Turnecumb* 1234 Cl], T~ Do N. of Lyme [*Tornecoma* DB, *-cumba* c 1140 BM, *Thorncumbe* 1291 Tax], **Thorncote** Bd [*Thornecote* 1206 FF], **Thorndon** Sf [*Tornduna* DB, *Thorndune* c 1095 Bury]. 'Valley, COT, hill where thorns grew.' See CUMB, COT, DŪN.

Thorne So nr Yeovil [*Torne* DB, *Thorn* c 1100 Montacute], T~ **Falcon** So [*Torne* DB, *Thorn fagun* 1265 Ep, *Thornfagun* 1268 FF], T~ **St. Margaret** So [*Torne* DB,

Thorn St. Margaret 1251 Wells], T~ YW [*Torne* DB, *Thorn* 1242 Ch]. OE ÞORN 'thorn-bush'.

Falcon from *Fagun* is no doubt a Norman family name.

Thorner YW [*Tornoure* DB, *Turnofra* 1170 P, *Thornouer* c 1180 YCh 509]. 'Slope overgrown with thorn-bushes.' Cf. OFER.

Thorness Wt [*Thornheye* 1324 Misc]. 'Thorn hedge.' Second el. OE *hege*.

Thorney Ca [*Þornig* c 960, 973 BCS 1131, 1297, (of) *Þorneie* 1066 ASC (E)], T~ Mx, the site of Westminster Abbey [*Torneia* 785, *Thorney* 969 BCS 245, 1228], T~ Sf (see STOWMARKET), T~ So [*Torelie* DB, *Thorne* 1316 FA], **West** T~ Sx [*Þorneg* 1052 ASC (D), *Tornei* DB]. 'Island overgrown with thorn-bushes.'

Thorney Nt [*Torneshaie* DB, *Thornehawe* Hy 3 BM, *Thornhaghe* 1282 Ch]. OE *þornhaga* 'enclosure formed by thorn-bushes'.

Thornford Do [*Thornford* 946–51 BCS 894, *Torneford* DB]. 'Thorn-bush ford.'

Thorngrafton Nb [*Thorgraveston* c 1150 PNNb, *Thoringraston* 1176 P, *Thorngrafton* 1365 Pudsay]. OE *Þorngrǣf-tūn* 'TŪN by a thorn brake'. Cf. GRĀF.

Thorngumbald YE [*Torne* DB, *Thoren Gumbaud* 1297 Subs]. Originally THORN. *Gumbald* is a family name derived from the OFr pers. n. *Gumbaud* from OG *Gundobald*.

Thornham K [*Turneham* DB, 1156 P, *Thornham* 11 DM], T~ La [*Thornham* 1246 Ass], T~ Nf [*Tornham* DB, c 1140 BM, 1197 FF], T~ **Magna & Parva** Sf [*Thornham, Marthorham,* (in) *paruo Thornham* DB, *Magna Thornham* 1235 FF]. 'HĀM where thorn-bushes grew.'

Marthorham is OE *māra Þornhām* 'greater T~'. T~ **Magna** is often called **Pilcock** or T~ **Pilcock** [*Pilcok* Hy 3, *Thornham Pilekoc* 1306 BM]. Pilcock is possibly 'willow copse or hill', the elements being OScand *pill* 'willow' and OE *cocc* 'heap' &c. Cf. COCK BECK.

Thornhaugh Np [*Thornhawe* 1189 (1332) Ch, 1230 P]. Identical with THORNEY Nt.

Thornhill Db [*Thornhull* 1230 P], T~ Do nr Stalbridge [*Thornhulle* 1377 FF], T~ Do nr Wimborne [*Tornehelle* DB, *Thornhill* 1212 Fees], T~ W [*Thornhulle* 1291 Tax], T~ YW [*Tornil* DB, *Tornhill* 1190 P]. 'Hill overgrown with thorn-bushes.'

Thornholm YE [*Thirnon* DB, *Thyrnom* 1266 Ipm]. OE *þyrnum*, dat. plur. of *þyrne* 'thorn-bush'.

Thornley Du [*Thornley* 1382 Hatfield], T~ La [*Thorenteleg* 1202 FF, *Thornideley* 1246 Ass]. 'Thorny LĒAH or glade.' T~ La has as first el. OE *þornihte* or **þornede* 'thorny'.

Thornley Du in Kelloe [(æt) *Ðornhlawa* 1071–80 ASCh, *Tornalau* 1104–8 SD]. 'Thorn hill or mound.'

Thornsett Db [*Tornesete* DB, *Thorneset* 1285 For]. 'Fold by thorn-bushes.' Cf. (GE)SET.

Thornship We [*Fornhep* 1226, 1231 FF]: 'Old Shap.' Cf. SHAP. First el. ON *forn* 'old'.

Thornthorpe YE [*Torgrimestorp* DB, *Thorgrimthorp* c 1185 YCh 33]. '*Þorgrim*'s thorp.' Cf. THORGANBY.

Thornthwaite Cu [*Thornthwayt* 1254 Ipm], T~ We [*Thornthwait* 1329 Ipm], T~ YW [*Tornthueit* 1230 Ep]. 'Thwaite where thorn-bushes grew.'

Thornton Bk [*Territone* DB, *Thornton* 1209 Fees], **Childer** T~ Chs [*Childrethornton* 1305 Chester], T~ **Hough** or **Mayow** Chs [*Toritone* DB, *Matheue Thornton* 1287 Court, *Thorneton Maheu* 1307 Ormerod], T~ **le Moors** Chs [*Torentune* DB], T~ **Do** in Marnhull [*æt Þorntune* 958 BCS 1033, *Torentone* DB], T~ **La** in Poulton le Fylde [*Torentun* DB, *Thorneton* 1246 Ass], T~ **La** in Sefton [*Torentun* DB, *Thorneton* 1246 Ass], T~ **Le** [*Torrenton* 1209–35 Ep, *Thornigton* 1254 Val], T~ **Curtis** Li [*Torentone* DB, -*tuna* c 1115 LiS], T~ **le Fen** Li [*Thorenton* 1218 Ass], T~ **by Horncastle** Li [*Torintune* DB, *Torentuna* c 1115 LiS], T~ **le Moor** Li [*Torentun* DB, *Torntuna* c 1115 LiS], T~ **Nb** nr Berwick [*Thornetona* 1208–10 Fees], **East & West** T~ Nb [*Torinton* 1203 Cur, *Thorneton* 1242 Fees], T~ **YE** [*Tornetun* DB], T~ **le Beans** YN [*Gristorentun* DB], T~ **Bridge** YN [*Torentone* DB, *Thorenton on Swale* 1275 Misc], T~ **le Clay** YN [*Torentun* DB, *Thornton* c 1180 YCh 1051], T~ **Dale** YN [*Torentune* DB, *Thornetone Cundale* Hy 3 Misc], T~ **on the Hill** YN [*Torenton* 1167 P, *Thorenton on the hill* 1275 Misc], T~ **le Moor** YN [*Torentune* DB, *Thornton in the Moor* 1310 Ch], T~ **Riseborough** YN [*Tornentun* DB, *Thorneton de Riseberg* 1285 FA], T~ **Rust** YN [*Torenton* DB, *Thorneton Ruske* c 1156 Mon, *Thornton Rust* 1198 Fount M, *Thorentonrust* 1260 Ass], T~ **Steward** YN [*Tornentune* DB, *Thornton Stiward* 1280 Ipm], T~ **le Street** YN [*Torentun* DB, *Tornton in Via* 1208–10 Fees, *Thorneton in Strata* 1285 FA], T~ **Watlass** YN [*Torretun* DB, *Thorneton Watlous* 1270 Ipm], T~ **YW** nr Bradford [*Torentone* DB], T~ **YW** nr Skipton [*Torentun* DB, *Torneton* 1260 Ipm], **Bishop** T~ **YW** [*Þorntun* c 1030 YCh 7, *Torentune* DB], T~ **in Lonsdale** YW [*Tornetun* DB, *Thorenton* 1297 Subs]. 'TŪN where thorn-bushes grew.'

T~ **le Beans** YN is 'the T~ where beans were grown'. *Gristorentun* DB is 'the T~ belonging to one *Gris*'. *Griss* is an ON nickname, lit. 'pig'.—**Bishop** T~ YW belonged to the Archbishop of York.—**Childer** T~ Chs is 'the T~ of the children'.—T~ **le Clay** YN from clayey soil. Cf. CLÆG.—T~ **Curtis** Li from a family name (from OFr *curteis* 'courteous').—T~ **Hough** Chs must be 'T~ by the ridge' (OE *hōh*), while *Mayow* is a form of *Matthew*; cf. MAINSTONE Ha. Christopher del Hogh in Thornton Mayowe is mentioned in 1420

(Ormerod).—T~ **Riseborough** YN is near Riseborough Hill. Wood of Torenton called *Risebergh* is mentioned t. John in 1310 Ch. *Riseberge* is 'brushwood hill'.—T~ **Rust** YN is obscure.—T~ **Steward** YN was held c 1100 (Mon iii) by Wymar steward to the Earl of Richmond.—T~ **le Street** YN is on a supposed Roman road.—T~ **Watlass** YN is T~ and Watlass. The latter [*Wadles* DB, *Watlos* 1205 Cur] is held to represent ON *vatnlauss* 'waterless'.

Thoroton Nt [*Toruertune* DB, *Turuerton* 1177, 1194, 1230 P, *Thururerton* 1242 Ipm]. '*Þurferð*'s TŪN.' *Þurferð* (ASC &c.) is ON *Þorrøðr* (from -*frøðr*) pers. n.

OE **þorp, þrop** is a rare word, and its meaning is doubtful. It was certainly used in the sense 'farm', possibly in the sense 'hamlet'. There is no reason to suppose that it meant 'village'. The places with names containing *þrop* are as a rule insignificant. The probability is that a *þrop* was a dependent farm, an outlying dairy-farm belonging to a village or manor. See Introd. pp. xvi f. Native names in *þrop* very often have a first el. meaning 'east, west, south' &c. (ASTROP, EASTRIP, WESTRIP, SOUTH-ROP &c.). Native names generally have the form *þrop*, whence THROOP(E), THRUP(P), THROPHILL &c., the second el. of HATHEROP, NEITHROP, SOULDROP, WILLIAMSTRIP &c. But *þorp* also occurs, as in GESTINGTHORPE Ess, SWANTHORPE Ha. The element is not common in purely English districts, but a fair number of instances occur in Gl, O. In some counties it is unknown, as D, K, Mx.

OScand **þorp** is a common pl. n. element in Scandinavia, especially in Denmark and Sweden. It is comparatively rare in Norway and absent in Iceland. It is very common in the Danelaw, but very rare in the north-western counties, where Norwegians settled. Thorpes are a sign of Danish settlement. ODan *thorp* means 'a smaller village, due to colonization from a larger one'. The latter was *adelby* 'the mother village'. OSw *þorp* means 'a farm, a new settlement', more rarely 'a village', and in later Swedish *torp* has come to mean 'croft'. A *þorp* was a settlement of far less importance than a *by*. The original meaning of *þorp* was 'newly reclaimed land, new settlement'. It should not be rendered by 'village', but rather by 'farm'. In origin the Danelaw thorps were evidently as a rule outlying, dependent farms belonging to a village. This is indicated partly by the fact that THORPE alone is a very common place-name. A thorp belonged to a mother village and was often simply called 'the thorp'. It is also indicated by the fact that a great many places with names containing *thorp* were named from a neighbouring village. Examples are BURNHAM and B~ THORPE, SAXLINGHAM and S~ THORPE Nf, BARKBY and BARKBY THORPE Le. BURNHAM THORPE was clearly a farm or hamlet dependent on Burnham. The first el. is frequently a pers. name, often of Scand origin. In many cases a distinguishing first el. has been added to an original THORP. This is often an English or a Norman pers. n.

Thorp Arch YW [*Torp* DB, *Thorp de Arches* 1272 Cl]. The manor was held by William de Arches c 1150 (YCh 535). *Arches* is from ARQUES in Normandy.—T~ **Perrow** YN [*Torp* DB, *Thorp Pirrowe* 1285 FA]. The manor was held by Thomas of Pirhou in 1219 (FF). Perrow is in PNNR derived from **Pirhou** Nf [*Pirenhou* DB, *Pirnho* 1188 P], which means 'pear-tree hill' (OE PIRIGE and HŌH), but cf. STOKE PERO So.

Thorpe, Castle, Bk [*Throp* 1255 For, *Castelthorpe* 1252 Ep]. This is OE PROP.—T~ Db [*Torp* DB].—T~ **Du** nr Easington [*Thorep* c 1050 HSC]. Probably OE PORP.—T~ **Bulmer** Du [*Thorpebulmer* 1312 RPD]. Belonged to Ralph de Bulmer from early 14th cent.—T~ **Thewles** Du [*Thorpp Thewles* 1265 Finchale]. *Thewles* is ME *thewles* 'immoral'.—T~ **le Soken** Ess [*Torp* 1181, 1226 StPaul]. Cf. KIRBY LE SOKEN.—T~ **Acre** Le [*Torp* DB, *Thorp Haueker* 1319 BM]. *Haueker* is OE *hafocere* 'hawker'.—T~ **Arnold** Le [*Torp* DB, *Thorp Ernad* 1239 Ep, *Torp Ernald* 1253–8 ib.]. Held by Ernald de Bosco in 1130 (P) and later. *Ernald* is from OFr *Ernaut*, a pers. n. of OG origin (OG *Arnald*).—T~ **Langton** Le [*Torp* DB, *Thorp Langton* 1428 FA]. Belonged to Langton.—T~ **Parva** or **Little** T~ Le [*Torp* DB, *Parva Thorpe* 1285 FA].—T~ **Satchville** Le [*Thorp* c 1125 LeS, *Thorp Secheville* 1316 FA]. Held by Radulfus de Secheville in 1210–12 (RBE). The name is from SECQUEVILLE in Normandy.—T~ **in the Fallows** Li [*Torp* DB, c 1115 LiS].—T~ **on the Hill** Li [*Torp* DB, *T~ sur le Tertre* 1281 Ch].—T~ **Latimer** Li [*Thorp* 1212 Fees]. Held by Thomas le Latimer in 1212. *Latimer* means 'interpreter'.—T~ **St. Peter** Li [*Torp* DB]. Dedicated to St. Peter.—T~ **Tilney** Li [*Torp* 1203 Cur]. Tilney [*Tileneia* 1185 TpR; cf. TILNEY Nf] is an extinct hamlet in T~.—T~ **le Vale** Li nr Ludford [*Thorp iuxta Lodeford* 1335 BM].—T~ **Abbots** Nf [*Thorp* DB, *Torp Abbatis* 1254 Val]. Belonged to Bury St. Edmunds.—T~ **next Haddiscoe** Nf [*Torpe* 1254 Val, *Thorp cum Hadesco* 1316 FA]. See HADDISCOE.—T~ **Market** Nf [*Torp* DB, *Torpmarket* 1251 Cl]. 'T~ with a market.'—**Morning** T~ Nf [*Torp, Maringatorp* DB, *Meringetorp* 1198 FF]. Probably 'T~ belonging to *Mǣringas* or *Meringas*', a lost place with a name identical with MEERING or MAREHAM.—T~ **next Norwich** Nf [*Torp* DB, *Thorp juxta Norwycum* 1302 FA].—T~ **Parva** Nf [*Torp* DB, *T~ Parva* 1254 Val].—T~ **Achurch** Np [see ACHURCH].—T~ **Hall** Np [*Porp* 972–92 BCS 1130, *Torp juxta Burch* 1179 P].—T~ **Lubenham** Np [*Torp* DB, *Thorp juxta Lubenho* 1220 Fees]. Situated nr LUBBENHAM Le.—T~ **Malsor** Np [*Alidetorp* DB, *Thorp*

Malesoures 1220 Fees]. Held by Fucher Malesoures in the 12th cent. (NS). Cf. MILTON MALZOR. — T~ **Mandeville** Np [*Torp* DB, *Trop* 1220 Fees, *Throp Mundevill* 1306 Ch]. OE PROP. The manor was held by Richard de Amundevill in 1252 (Ch). Cf. COATHAM MUNDEVILLE.—T~ **Underwood** Np [*Thorp* 12 NS, *Torp sub bosco* 1248, *Torp Underwode* 1255 For]. *Underwood* means 'in the forest'.—T~ **Waterville** Np [*Torp* 1125–8 LN, *T~ Watervile* 1265 Misc]. Held by Wido de Waltervilla in 1125–8 (LN). Cf. ORTON WATERVILLE.—T~ Nt nr Newark [*Torp* DB, c 1150 BM].—T~ **in the Glebe** or **Bochart** Nt [*Torp* DB, *Bochardistorp* 1236, *Thorp Bossard* 1242 Fees]. Held by Johannes Bochard in 1236. *Bochard* is a Fr. pers. n. and family name. Cf. BOTCHERBY.—T~ **by Water** Ru [*Torp* DB, *Thorpbythewatir* 1459 AD]. On the Welland.—T~ **Common** Sf [*Torpa* DB, *Torp* 1202 FF].—T~ **Morieux** Sf [*Porp* c 995 BCS 1288, *Guvetorp* 1201 Cur, *Thorp Morieux* 1330 FA]. Held by Roger de Murious in 1201 (Cur). *Morieux* from MORIEUX in Côtes-du-Nord. *Guvetorp* from *Gua*, mother of Roger de Murious, mentioned 1201 Cur.—T~ **Sr** [*Torpe* a 675 BCS 34, *Ðorpe* c 1050 KCD 848, *Torp* DB].—T~ **Constantine** St [*Torp* DB, *Thorp Costentin* c 1245 Cl]. Galfrid de Costetin held a fee in Thorp in 1212 (Fees). Cf. EATON CONSTANTINE.—T~ **Bassett** YE [*Torp* DB, *Thorp Basset* 1236 Ep]. Held by William Basset in 1204 (FF). Cf. BERWICK BASSETT.—T~ **le Street** YE [*Rud(e)torp* DB, *Thorp* 1226 FF]. On a Roman road.—T~ **under Stone** YN [*Torp* 1188 P, *Thorp Understane* 1314 Ch]. The place is at the foot of a hill.—T~ **le Willows** YN [*Torp* DB]. 'T~ by the willows.' *Le* is the Fr definite article.—T~ **Audlin** YW [*Torp* DB, *Thorp Audelyn* 1379 PT]. Held by William son of Aldelin in 1190 (YCh 1641). Aldelinus de Aldefelde is mentioned in 1156 (YCh 80). *Audlin* is a Fr pers. n. derived from an OG name.—T~ **in Balne** YW. Cf. BALNE.—T~ **on the Hill** YW nr Leeds [*Torp* DB, *Thorp othe Hull* 1309 Ch].—T~ **sub Montem** YW [*Torp* DB]. 'T~ at the foot of the hill.'—T~ **Salvin** YW [*Torp* DB, *Thorp Salvayn* 1309 Ch]. Held by Henricus Selvein t. R 1 (Cur). *Salvin* is an OFr nickname derived from OFr *salvagin* 'savage'.—T~ **Stapleton** YW [*Torp* DB, *Thorp Stapelton* 1303 FA]. The Stapleton family was in possession from the time of Hy 3 (1240 FF).—T~ **Underwoods** YW [*Torp* DB, *Thorp under Wood* 1292 Ch]. Cf. T~ UNDERWOOD Np.—T~ **Willoughby** YW [*Torp* DB, *Thorp Wyleby* 1303 FA]. Robert de Willeby in Thorpe is mentioned 1237 ff. (Selby).

Thorpland Nf [*Torpaland* DB, *Torpeland* 1177 P, *Thorpland* 1242 Fees]. 'Land belonging to Thorpe.' The first el. is in the plur. form. Cf. OScand *Þorpar* plur., a common pl. n.

Thorrington Ess [*Torinduna* DB, *Thurinton* 1253 Ch, *Thoritone* 1291 Tax]. First el. OE ÞORN or ÞYRNE. Second TŪN.

Thorverton D [*Toruerton* 1182 P, *Torverton* 1201 Cur, 1212 Fees, *Thurfurton* n.d. Ol]. OE *Þornford-tūn*, 'TŪN by a ford marked by a thorn-bush'.

OScand *þræll* 'thrall, serf'. See THIRLBY, THRELKELD.

Thrandeston Sf [*Thrandeston* c 1035 Wills, *Thrandestuna* DB, *Throndestun* c 1095 Bury]. '*Þrand*'s TŪN.' First el. ON *Þrándr*, *Þróndr*, ODan *Thrond* pers. n.

Thrapston Np [*Trapestone* DB, -*tona* 1138 NpCh, -*tun* 1160–5 ib., *Thrapeston* 1285 BM]. The suggestion in PNNp that the first el. is a pers. n. cognate with Old Germanic *Trapsta*, *Trafstila* (v. Förstemann) and Goth *þrafstjan* 'to comfort' is very likely correct.

Threapland Cu [*Threpeland* 1326 Ipm], **Threapwood** Chs. 'Debatable land, wood.' First el. ME *threpen* (OE *þrēapian*) 'to contend, dispute'.

Threckingham (-ē-) Li [*Trichingeham* DB, *Triccingeham* a 1118 Flor (s.a. 675), *Thrikingeham* 1178–84 YCh 1460]. The first el. is a tribal name of doubtful etymology. The base may be a pers. n. derived from the stem of OE *þræc*, ON *þrekr* 'force, courage' &c. or from a word identical with G *Dreck*, ON *þrekkr* 'dirt', here 'mud'. In either case the vowel *i* is abnormal. Perhaps we may adduce the lost pl. n. *Tric* DB (an earlier name of Skegness Li). Cf. also THREXTON Nf.

Threlkeld Cu [*Trellekell* 1197 P, *Threlekelde* a 1247 CWNS xxiii]. 'The spring of the thralls.' Cf. KELDA, ÞRÆLL.

Thremhall Ess [*Tramhale* 1194, *Tremhall* c 1200, *Trimhall* 13 BM, *Thremhale* 1310 Ipm, *Tremenhale* 1295 Pat]. If the name began in *þ*-, the first el. belongs to OE *þrymm* 'strength, glory, army' (*Þrymma* pers. n.; cf. OE *þrymma* 'warrior'). If it began in *T*-, the first el. is *Trymma* as in TRIMLEY.

Threshfield YW [*Freschefelt* DB, *Treskefeld* 1193 P, 1231 FF]. The name seems to mean 'place where corn was threshed'.

Threxton Nf [*Trectuna*, *Trestuna* DB, *Threkeston* 1242 Fees, 1254 Val, *Trikestun* 1208 FF]. Cf. THRECKINGHAM. A pers. n. *Þrec* seems a more likely first el. than an OE *þrec* 'dirt'.

Thrigby Nf [*Trukebei*, *Trikebei* DB, *Trikebi* 1177 P, *Thryckeby* 1291 Tax]. '*Þrykki*'s BY.' *Þrykki* is a short form of ON *Þrýðríkr*, which appears in forms such as *Thridicke* 1426, *Trycke* 1531.

Thrimby We [*Tirneby* 1200, *Thirneby* 1241 FF]. 'BY at a thorn-bush' (ON *þyrnir*) or

'*Thyrne*'s BY'. *Thyrne* is an OScand pers. n. recorded as *Pirne* c 1050 YCh 9. Identical with Thrimby is the lost **Thirnby** La [*Tiernebi* DB, *Thirneby* a 1219 CC].

Thringstone Le [*Trangesbi* DB, *Trengeston* c 1200 BM, *Threngeston* 1276 RH, 1276 Cl]. The first el. is identical with that of *Threingesthorp* 1276 RH (Le). It is probably an unrecorded OScand *Præingr*, derived from ON *þrár* 'obstinate'. Second el. originally BY.

Thrintoft YN [*Tirnetofte* DB, -*toft* 1170 P]. See TOFT. First el. as in THRIMBY.

Thriplow Ca [*Tripelan* c 1050 KCD 907, *Trippelaue* c 1080 ICC, *Trepeslau* DB, *Treppelawe* 1206 Cur, *Trippelawa* 1177 P, -*lawe* 1228 FF]. Second el. OE *hlāw*, which may mean 'hill' or 'tumulus'. The first must have had the vowel *y*. It may be an OE pers. n. *Tryppa*, a short form of names like *Þrypbeorht*, or a nickname *Tryppa*, belonging to OE *treppan* 'to tread'.

Thrislington Du [*Tursteintun* 1208–10 Fees]. '*Þorstein*'s TŪN.' First el. ON *Þorsteinn*, OSw *Þorsten*, ODan *Thursten*.

Throckenholt Ca [*Prokonholt* 656 ASC (E), *Trokenholt* 1240 Pp], **Throcking** Hrt [*Trochinge* DB, *Throcking* 1198 FF, *Trockinga* 1209–19 Ep, *Trokyng* 1291 Tax], **Throckley** Nb [*Trocchelai* 1161, *Trokelawa* 1177 P, *Throckelaue* 1212, *Throkelawe* 1242 Fees], **Throckmorton** Wo [*Throcmortune* 11 Hickes, *Trochemerton* 1176 P]. These must be compared with **Drockbridge** Ha [*Procbrigg* 826 BCS 393, -*byrg* 939 ib. 742], **Rockmoor Pond** Ha [*Procmere* 863 BCS 508, *P(o)rocmere* 961 ib. 1080], also *Trocketon* 1287–90 Fees, *Troketon* 1346 FA (Wt). OE *proc* is found in the senses 'a piece of timber on which the ploughshare is fixed (= dial. *throck*); table' (of the tables used by the moneychangers in the Temple). *Proc* is cognate with ON *þrekr*, OE *þracu* 'force' &c., and its original sense was probably 'prop, support'. The sense 'table' may have developed from 'trestle'. The sense 'prop' is very suitable in Drockbridge. Throckmorton no doubt has as first el. *procmere*, whence also Rockmoor, and this may well mean 'lake with trestles for the support of a bridge, e.g. one for washing'. Throckenholt seems to contain an adj. *procen*, and the name may mean 'wood where throcks were got'. Throckley contains OE *hlāw*, very likely in the sense 'tumulus'. If so, the first el. is probably a pers. n. *Proca*, which may well have been formed from *proc*, and such a name would be suitable as the first el. of Troketon and the base of Throcking, but all three may contain or be derived from *proc*, though the exact sense of the word must then remain doubtful.

Throop Ha [*La Thrope* 1274 (1313) Ch], **Throope** W [*Trope* 1185 P, *Throp* 1202 FF]. OE *prop* 'farm' &c. See ÞORP.

Throphill Nb [*Trophil* 1166 RBE, *Throphill* 1242 Fees], Thropton Nb [*Tropton* 1177 P, *Thropton* 1242 Fees]. First el. OE *þrop*, which may mean 'farm'. A better etymology for Thropton is given by the meaning 'cross-roads' evidenced for *þrop*.

Throston Du [*Thorston* 1344 Ipm]. '*Þori*'s TŪN.' Cf. THORESBY.

Througham (-ŭf-) Gl nr Stroud [*Troham* DB, *Truham* 1190 ff. P], Throwleigh(-ōō-) D [*Trule* DB, *Throulegh* 1242 Fees], Throwley K [*Trevelai* DB, *Trulege* 11 DM, *Thrulege* 1163–5 Fr, *Triwele* 1235 Cl], T~ St [*Truele* 1208 FF, *Truleg* 1227 Ass]. The first el. is OE *þrūh* 'water-pipe, conduit', originally 'a hollowed-out tree-trunk', identical with ON *þró*, OHG *drūh* 'trough' &c. In pl. ns. it probably refers to a deep valley. For another instance see FRITHAM. Second el. HĀM, LĒAH.

Throxenby (-ŏs-) YN [*Thurstanebi* 1167 P, *-by* 1276 Percy]. '*Þorstein*'s BY.' Cf. THRISLINGTON.

Thrumpton Nt nr Nottingham [*Turmodestun* DB, *Thurmundeston* 1226–8, *Turmodistun* 1236 Fees]. '*Þormōð*'s TŪN.' Cf. THORMANBY. T~ Nt nr Retford [*Thurmeston* 1278 Ipm, 1327 Subs]. Perhaps identical with the other Thrumpton.

Thrup O [*Trop* DB, *Tropa* Hy 2 (1267) Ch, *Thrupp* 1394 BM], Thrupp Brk [*Thrope* 1316, *Le Throp* 1402 FA], T~ Gl [*Thrope* 1359 BM], T~ Np [*Torp* DB, *Trop* 1207 Cur]. OE *þrop* 'farm'. See þORP.

Thruscross YW [*Thorecros* c 1180, *Thorescros* c 1210 YCh 513 f.]. '*Þori*'s cross.' Cf. THORESBY.

Thrushel R D [*Frischel* 1244 Ass, *Thrusshel* 1575 Saxton], Thrushelton D [*Tresetone* DB, *Thrisselton*, *Thrysselthon* 1242 Fees]. Thrushelton is 'TŪN frequented by thrushes'. The river-name is a back-formation. OE *þryscele* 'thrush' is not evidenced, but *thrishel* is used in Devon dialects, and *thruschyl* is found in the 15th cent.

Thrussington Le [*Turstanestone* DB, *Tursteineston* 1175 P, *-tuna* c 1200 DC, *Thurstington* 1344–6 BM]. Identical with THRISLINGTON.

Thruxton Ha [*Turkilleston* 1167 P, *Turcleston* 1236 Fees, *Trokeleston* 1316 FA], T~ He [*Thurclestun* 1249 Fees, *-ton* 1291 Tax]. '*Þurkil*'s TŪN.' Cf. THIRKLEBY YN.

Thrybergh YW [*Triberge* DB, 1204 FF, *Thriberg* 1297 Subs]. 'Three hills.' See BEORG.

Thulston Db [*Turuluestun, Torulfestune* DB, *Turleston* 1221–30 Fees]. '*Þurulf*'s TŪN.' Cf. THOLTHORPE.

Thunderfield Sr [*Þunresfeld* c 880 BCS 553, *-felda* 933 ib. 697], Thunderley Hall Ess [*Tondreleia* 1143, *Tunderleia* c 1143 BM, *Tunrele* 1199 FF], Thundersley Ess [*Thun-reslea* DB, *Tunderle* 1203 Cur], Thundridge Hrt [*Tonrinch* DB, *Tozriche* c 1200 Fr]. 'FELD, LĒAH, HRYCG or ridge dedicated to the god *Þunor*.' Cf. THURSLEY.

Thurcaston Le [*Turchitelestone, Turchilestone* DB, *Thurketlestone* c 1125 LeS]. '*Þorketil*'s TŪN.' *Þorketill* is the older form of *Þorkell*, on which see THIRKLEBY (2). It is frequent in early English sources, as *Þurcetel* in charters, *Turketel* DB.

Thurgarton Nf [*Ðurgartun* 1044–7 KCD 785, *Turgartuna* DB], T~ Nt [*Turgarstune* DB, *Turgarton* 1175 P], Thurgoland YW [*Turgesland* DB, *Turgarland* 1202 FF]. '*Þorgeir*'s TŪN and LAND.' First el. ON *Þorgeirr*, OSw, ODan *Thorger*, Anglicized to *Þurgar* KCD 789.

Thurland La [*Thurland* 1465 Pat]. If *Thorolfland* 1247 CC belongs here, '*Þorolf*'s land'. Cf. THOLTHORPE.

Thurlaston Le [*Turlauestona* 1166 P, *-ton* 1200 FF, 1230 P, *Thurleston* 1254 Val], T~ Wa [*Torlavestone* DB, *Thurlaveston* Hy 2 (1235) Ch, *Thurlauestona* 1229 BM]. '*Þorleif*'s TŪN.' First el. ON *Þorleifr*, OSw *Þorleuer*.

Thurlbear So [*Tierleberge* 1084 GeldR, *Torlaberie* DB, *Turelberiz* 1219 Fees]. OE *þyrel* 'having a hole' and *beorg* 'hill', the name meaning 'hill with a hollow'.

Thurlby Li nr Alford [*Toruluesbi* DB, *Turlebi* 1202 Ass], T~ Li nr Bourne [*Turolvebi* DB, *Turlebi* 1202 Ass], T~ Li nr Lincoln [*Turolfbi, Turolue(s)bi* DB, *Torlebi* 1141 RA]. '*Þurulf*'s BY.' Cf. THOLTHORPE.

Thurlei·gh (-ī) Bd [*La Lega* DB, *La Leya* 1227 Ep, *Thyrleye* 1372 Ipm]. OE (æt) *þǣre lēage* '(at) the glade or wood'.

Thurlestone D [(from) *ðyrelan stane* 847 BCS 451, *Torlestan* DB, *Therlestane* 1242 Fees]. 'Stone with a hole.' First el. OE þYREL. Named from T~ Rock, a rock with a natural hole.

Thurlow, Great & Little, Sf [*Tritlawa, Tridlauua* DB, *Thrillauue* c 1095 Bury, *Trillawe Magna, Parva* 1254 Val]. Perhaps OE *þrȳþ-hlǣw* 'famous tumulus' or 'assembly hill', the first el. being OE *þrȳþ* 'might, troop, host', often used in compounds as a laudatory epithet, as *-ærn, -bearn* &c. Or the first el. may be an OE *þride 'deliberation', belonging to OE *þridian* 'to deliberate'.

Thurloxton So [*Turlakeston* 1195 Buckland, *Thurlokestone* 1285 FA]. '*Þurlak*'s TŪN.' First el. ON *Þorleikr, -lákr*, ODan *Þurlakr* &c. *Þurlac* is found BCS 1130.

Thurlston or Thurston Sf nr Ipswich [*Toroluestuna, Turoluestuna* DB], T~ YW [*Turulfestune, Turolueston* DB]. '*Þurulf*'s TŪN.' Cf. THOLTHORPE.

Thurlton Nf [*Thuruertuna* DB, *Thurvertone* 1254 Val]. Identical with THOROTON.

Thurmaston Le [*Turmodestone* DB, *Thurmodeston* 1191 P]. '*Þormōð*'s TŪN.' Cf. THORMANBY.

Thurnby Le [*Turnebi* Hy 2 (1318) Ch, *Thirneby* 1239 Ch]. Identical with THRIMBY.

Thurne Nf [*Thirne*, *Thyrne* 1044–7 Holme, *Thirne* 1198 FF]. OE *þyrne* 'thorn-bush'. The river-name **Thurne** is a back-formation.

Thurnham K [see THORNHAM]. **T~** La [*Tiernun* DB, *Thurnum* a 1160 CC]. OE *þyrnum*, dat. plur. of *þyrne* 'thorn-bush'.

Thurning Nf [*Tyrninga*, *Turninga* DB, *Tiringes* 1203 Cur, *Therning* 1211 FF], **T~** Np [*Torninge* DB, *Turninges* 1187 NpCh, 1207 Cur]. A derivative of OE *þyrne* 'thorn-bush', either OE *þyrning* 'place where thorn-bushes grew' or OE *þyrningas* 'people at the thorn-bush(es)'.

Thurnscoe YW [*Ternusche* DB, *Tyrnesco* 1190 P, *Thirnesco* 1233 Ep]. OScand *þyrniskógr* 'thorn-bush wood'.

Thurrock, Grays, Little & West, Ess [*Turruc(ca)*, *Turoc(ha)*, *Thurrucca* DB, *Turroc* 1130 P, *Parva Turroch* 1201, *West Turroc* 1205 Cur, *Turrok de Grey* 1238 Subs]. OE *þurruc* means 'bottom of a ship, where dirt collects'. The Thurrocks are at a bend of the Thames, and this bend may have been called *þurruc* owing to a fancied similarity to a ship's bottom. But dial. *thurrock* is also used in the sense 'drain'. This may be the meaning here.

Grays T~ was granted to Henry de Grai by Richard I. Cf. EASTON GREY.

Thursby Cu [*Thoresby* c 1165 WR, *Toresbi* 1183 P]. Identical with THORESBY.

Thursford Nf [*Tureforde*, *Turesfort* DB, *Turesford* 1231 Cl, *Thirsford* 1291 Tax]. The first el. seems to be OE *þyrs* 'giant, demon'. If so, the name must be due to some local legend.

Thursley Sr [*Thoresle* 1296 BM, *Thursle* 1329 PNSr]. Identical with THUNDERSLEY. For the loss of *n* cf. *Thursday* from OE *þunresdæg*. **T~** is nr TUESLEY.

Thurstaston Chs [*Turstanetone* DB, *Thurstanestona* c 1125, c 1150 Chester], **Thurston** Sf in Hawkedon [*Thurstanestuna* DB, *Turstaneston* c 1145 Bury]. Identical with THRISLINGTON.

Thurston Sf nr Bury [*Thurstuna*, *Torstuna* DB, *Thurstune* c 1095 Bury, *Thurston* 1226–8 Fees]. '*Þori*'s (*Þuri*'s) TŪN.' Cf. THORESBY and see THURLSTON.

Thurstonfield Cu [*Turstanfeld* c 1234 Holme C, *Thurstanfeld* 1334 Ipm], **Thurstonland** YW [*Tostenland* DB, *Tursteinland* 1191 f. P]. '*Þorstein*'s FELD and LAND.' Cf. THRISLINGTON.

Thurton Nf [*Tortuna* DB, *Thermtona* c 1150 Fr, *Thuriton* 1248 Ch, *Thurnton* 1302 BM]. First el. OE *þyrne* 'thorn-bush'.

Thurvaston Db [*Turverdestune* DB, *Turuerdeston* 1188 P]. Identical with THOROTON.

Thuxton Nf [*Turstanestuna* DB]. Identical with THRISLINGTON.

Thwaite Nf nr Aylsham [*Ðweyt* 1044–7 KCD 785, *Tuit* DB, *Thweit* 1254 Val], **T~ St. Mary** Nf [*Thweit* 1254 Val], **T~** Sf [*Theyt* 1228 FF, *Thueyt* Hy 3 BM]. 'The thwaite.' The meanings of *thwaite* (from OScand *þveit*) vary a good deal, and it is impossible to say what the exact sense is in each name. ON *þveit* means 'a meadow, a piece of land', Norw *tveit* 'a piece of meadow in a wood, a cleared meadow, a clearing'. Engl dial *thwaite* means 'a forest clearing, a piece of land fenced off or enclosed, a low meadow', &c. The element is a common second member of names in La, Cu, We, YN, less common in other Scandinavian counties. Cf. CROSTWICK, CROSTWIGHT, GUESTWICK Nf, STAINFIELD Li. It is possible that a variant form OE **þwit* or OScand **þvit* occurs in some names, as **Inglewhite** La, TREWHITT Nb.

OE **þwang**. See THONG.

Thwing YE [*Twenc* DB, *Thueng* c 1200 YCh 761, *Twenge* 1206 Ass]. Apparently OScand *þvengr* 'a shoe-lace', a word related to OE *þwang*. Thwing is at a long ridge, which may have been called 'the Thong'.

OE **þyrel** adj. 'perforated, having a hole' is found in THIRLWALL, THURLBEAR, THURLESTONE, perhaps THIRLMERE. OE *þyrel* 'a hole' is found in STANTHORNE Chs.

OE **þyrne**, OScand *þyrnir* 'a thorn-bush' is not always easy to distinguish from *þorn*. It is used alone in THORNHOLM, THURNHAM (from the dat. plur. *þyrnum*), THIRN, THURNE. It is the first el. of some names, as THURNSCOE, THURTON, but is difficult to distinguish from the pers. n. *Þyrni* (cf. THRIMBY, THURNBY, THRINTOFT). THURNING is a derivative of OE *þyrne*. *Þyrne* is the second el. of several names, as BYTHORN, HENTHORN, LIGHTHORNE, STATHERN, CAISTRON, CASTERN, WINSTER Db.

OE **þyrre** 'dry'. See THERFIELD, TURVILLE.

OE **þyrs** 'giant, demon'. See THURSFORD, TUSMORE.

Tibberton Gl [*Tebriston* DB, *Tibristone* W 1, *Tribricthuna* 1146 Fr, *Tiberton* 1211–13 Fees], **T~** Sa [*Tetbristone* DB, *Tibrinton* 1180 P, *Tibbrihtonia* 1181 BM], **T~** Wo [*Tidbrihtingctun* 978–92 KCD 683, *Tidbertun* DB]. 'The TŪN of *Tidbeorht*'s people.'

Tibenham Nf [*Tybenham* 1044–7 KCD 785, *Tibenham* DB, *Tibeham* 1242 Fees], **Tibshelf** Db [*Tibecel* DB, *Tibbeshelf* 1179 P, *-schelf* 1226 FF], **Tibthorpe** YE [*Tibetorp* DB, *Tipetorp* DB, 1166 P, *Tibethorp* 1272 Ipm]. '*Tibba*'s HĀM, SCYLF or slope, thorp.'

OE **ticcen** 'kid' is the first el. of some pl. ns., as TICEHURST, TISTED, &c. The *cc* was palatal and became ME *ch*, owing to Norman influence sometimes *s*, as in TICEHURST. But in forms such as *ticcnes* (gen.) the *cc* would remain hard, and such forms might give rise to a ME *ticken* (cf. TICKENCOTE). The *-n* of the word was no doubt sometimes lost in OE, so that OE **ticce* arose. But there were also the OE pers. ns. *Tica* and *Ticcea*, from which it is sometimes difficult to distinguish *ticcen*.

Ticehurst Sx [*Tycheherst* 1248 Ass, *Thichesherst* 1263 FF], **Tichborne** Ha [(be) *Ticceburnan* 909, 938 BCS 622, 731]. 'Kid hurst and stream.' The stream at Tichborne is (inon) *Ticceburnan* 701 BCS 102.

Tickencote Ru [*Tichecote* DB, *Tichencota* 12 DC, *Tikencot* 1199 FF]. 'Cote for kids.' Cf. TICCEN.

Tickenham So [*Ticaham, Ticheham* DB, *Tiche(s)ham* 1201 Ass]. '*Tica*'s HĀM.'

Tickenhurst K [*Tikenherst* 1070–82 StAug, *Tikeneherst* n.d. ib.]. OE *ticcna-hyrst* 'kid hurst'.

Tickford Bk [*Ticheforde* DB]. 'Kid ford' or '*Tica*'s ford'.

Tickhill YW [*Tichehilla* c 1150 RA, *-hill* 1156 P, *Ticahil* 1157 YCh 186]. '*Tica*'s hill.'

Ticknall Db [*Ticenheale* 1002 Wills, *-healle* 1004 KCD 710, *Tichenhalle* DB, *Tikenhala* 1177 P]. 'Kid HALH.' Cf. TICCEN.

Tickton YE [*Tichetone* DB, *Tiketona* 1297 Subs]. '*Tica*'s TŪN.'

Tidbury Ring Ha S. of Whitchurch [(of) *Tudanbyrig* 1019 Hyde]. '*Tuda*'s BURG.'

Tidcombe W [*Titicome* DB, *Titecumba* 1197 P, *Tydecumb* 1220 Fees, *Titecumbe* 1242 Fees]. The first el. appears to be identical with that of *Tittandun* 930, 972 BCS 667, 1282 (Wo). It may be an OE **Titta*, which is easily explained as a short form of names like *Tidfriþ, -stān*, where *d* became *t*. But it may also be a common noun. One might think of ME *tite-* in *titmouse* (*titemose* c 1325). A bird-name would be suitable in *Titegraue, -graua* 1176 ff. P (Ha).

Tiddington O [*Titendone* DB, *Tatin-, Teten-, Totin-, Tudendon* 1208 Cur, *Tetindon* 1242 Fees]. '*Tytta*'s DŪN.' **Tytta* is a side-form of *Tutta*.

Tiddington Wa [*æt Tidinctune* 969 BCS 1232, *Tidantun* 985, *Tidingtun* 1016 KCD 651, 724]. '*Tida*'s TŪN' and 'the TŪN of *Tida*'s people.'

Tiddy R Co [*Tudi* 1018 ERN]. A derivative of a Welsh *tud* 'good', corresponding to OIr *túath* 'left', lit. 'good', and related to Lat *tutus*. Hence **Tideford** [*Tuddeford* 1284 Ass].

Tidenham Gl [*æt Dyddanhame, Dyddanhamm* 956 BCS 927 ff., *Dyddenhamm* 1060–6

KCD 822, *Tide-, Tedeham* DB]. '*Dydda*'s HAMM or river land.' **Dydda* is a side-form of *Dudda*. T- from *æt D-*.

Tideswell (tĭdzel) Db [*Tidesuuelle* DB, *Tiddeswell* 1230 P]. '*Tidi*'s stream.'

Tidmarsh Brk [*Tedmerse* 1196 f. P, *Thudmers* 1300 Ipm, *Tydemershe* 1428 FA]. Perhaps '*Tydda*'s marsh'. **Tydda* is a side-form of *Tudda*.

Tidmington Wa [*æt Tidelminctune* 977 KCD 614, *Tidelmintun* DB]. 'The TŪN of *Tidhelm*'s people.'

Tidworth. See TEDWORTH.

Tiffield Np [*Tifelde* DB, *-feld* 1163 P, *Tiffeld* 12 NS, 1193 f. P, 1202 Ass]. The first el. is obscure. It might be the word *ti* found in OSw *tybast*, Sw *tibast*, G *zeibast* 'daphne' (the shrub), which seems to be related to OHG *zîdal* 'swarm of bees'. The word may mean 'bee' or 'swarm of bees'. OE *tig* in *forþtig* 'porch' &c. (cf. OHG *zich* 'village meeting-place') is hardly a word that would be combined with FELD. But OE *Ti(w)*, the name of the god, is a possible first el.

OE **tigel** 'tile'. See TILEHURST, TILEY, TYLEY, also TYLERHILL.

Tilberthwaite La [*Tildesburgthwait* 1196 FF]. 'Thwaite at *Tillesburg*.' The latter is *Tillesburc* 1157–63 LaCh. It means '*Tilli*'s or *Tilhere*'s BURG'.

Tilbrook Hu [*Tilebroc* DB, 1202 Ass, *Tillebroc* 1206 Ass]. '*Tila*'s stream.' The river-name Til is a back-formation.

Tilbury, East & West, Ess [*Tila-, Tillaburg* c 730 Bede, *Ti(i)laburh* c 890 OEBede, *Tiliberia* DB, *Estillebery* 1199, *Westtillebire* 1203 FF], **T~ juxta Clare** Ess [*Tiliberia* DB, *Tillebere* 1197 FF, *Tillebiria juxta Clare* 1212 RBE, *Tyllebery* 1254 Val]. '*Tila*'s BURG.'

Tilehurst Brk [*Tigelherst* 1167 P, *Tiyelhurste* 1242 Fees], **Tiley** Do [*Tileye* 1264 Ipm, *Tilee* 1314 FF]. 'Hill and LĒAH where tiles were made or where tiles were found.' Cf. *Tigelhyrst* 1062 Th (Ess), *Tihelleah* 956 BCS 982 (Ha), TYLEY.

Tilford Sr [*Tileford* c 1140 Mon]. First el. OE *til* 'convenient' or *Tila* pers. n.

Till R Li [*Til* c 1190 ERN], **T~** R Nb [*Till* c 1050 HSC, *Tille, Tilne* 1256 Ass]. Perhaps identical with TILLE R in France [*Tyla* 7, *Tila* 830]. The name may be cognate with Welsh *tail*, MBret *teil* 'stercus, fimus' and other words belonging to the root *tei, ti* 'to dissolve, flow'. Cf. TYNE. The name may mean 'stream'. On the Till Li is **Till Bridge** [*Tilbrigge* 1357 Works].

Tilley Sa nr Wem [*Tyleweleye* 1327 Subs]. First el. OE *telga* 'branch, bough', dial. *tillow, tellow*.

Tillingdown Sr nr Tandridge [*Tellingedone* DB, *Tillingeden, Tillingdon* 1290 Ch]. The place is called *Tilmundesdoune* 1302 Ch. The original name seems to have been *Tilmundes dūn* '*Tilmund*'s hill'. This was exchanged for *Tillinga dūn*, where *Tillingas* is formed from a short form of *Tilmund*, OE *Tilla*.

Tillingham Ess [*Tillingeham* c 610 BCS 8, *Tillingham* c 950 Wills, DB, *Thillingeham* 1163–70 AC]. 'The HĀM of *Tilli*'s people.'

Tillington He [*Tillinton* 1188 P, *Tullinton* 1235 Cl, 1242 Fees]. 'The TŪN of **Tylla*'s or **Tylli*'s people.' OE *Tulla* is evidenced.

Tillington St [*Tillintone* DB, -*ton* 1242, *Titlingeston* 1236 Fees]. 'The TŪN of *Titel*'s people.' Cf. TELSCOMBE.

Tillington Sx [*Tullingtun* 960 BCS 1055, *Tulintona* c 1150 Fr, *Tolletun* 1198 FF]. 'The TŪN of *Tulla*'s people.'

Tilmanstone K [*Tilemanestun* 1072 BM, -*tone* DB]. '*Tilman*'s TŪN.'

Tilmouth Nb [*Tyllemuthe* c 1050 HSC, *Tillemuthe* 1104–8 SD]. 'The mouth of R TILL.'

Tiln Nt [*Tilne* DB, *Tilnea* 1194 P], **Tilney** Nf [*Tilnea* 1170, 1190 P, *Tillenee* 1197 FF, *Tilneie* 1207 Cur, *Tilneye* 1242 Fees]. '*Tila*'s river or island.' See ĒA, ĒG.

Tilshead W [*Theodulveside, Tidulfhide* DB, *Tidolfeshida* 1168 P, *Tidulveshida* 1198 Fees]. '*Tidwulf*'s hide.' See HĪD.

Tilstock Sa [*Tildestok* 1211 Cur, *Tyldestok* 1327 Subs]. '*Tidhild*'s STOC.' *Tidhild* is a woman's name.

Tilston Chs nr Malpas [*Tillestone* DB, -*ton* 1291 Tax]. '*Tilli*'s TŪN.'

Tilston Fearnall Chs [*Tidulstane* DB, *Tideluestan* c 1100, c 1150, *Tiduluestan* c 1190 Chester]. '*Tidwulf*'s stone.'
Fearnall is doubtless 'ferny HALH'. Ferny Lees is not far away.

Tilsworth Bd [*Pileworde* DB, *Thuleswrthe* 1202 FF, *Twylesworth* 1242 Fees, *Tyules-worth* 1250 Cl]. Apparently '*Tyfel*'s WORÞ'. **Tyfel* might be derived from *Tuf* BCS 1130.

Tilton Le [*Tile-, Tillintone* DB, *Tilton* 1163 P], **Tilty** Ess [*Tileteia* DB, 1156 P, *Tyleteye* 1155 RBE]. '*Tila*'s TŪN and tye.' Cf. TĒAG.

OE **timber**, OScand **timbr** 'timber, wood' is the first el. of TIMBERLAND &c., TIMPER-LEIGH, TIMSBURY. OE *timber* or *getimbre* 'building, timbered house' is found in NEW-, NYETIMBER.

Timberland Li [*Timberlunt* DB, -*lund* 1155 BM]. 'Grove where timber was got.' Cf. LUND.

Timberscombe So [*Timbrecumbe* DB, *Timbrescumba* 1176 P, *Timbercumbe* 1227 FF]. 'Valley where timber was got.'

Timble YW [*Timmel* c 972 ASCh (*Tun mel* BCS 1278), *Timbel* c 1030 YCh 7, *Timble* DB, *Tinbel* 1173–85 YCh 513]. Timble is in a high situation. The second el. may well be Welsh *moel* 'bare, bare hill' (from earlier *mēl*). The first may be Welsh *din* 'hill fort', with change to *tin* as in TINTAGEL &c. (so-called provection). The name would mean 'fort on the bare hill'.

Timbold. See TEYNHAM.

Timperleigh Chs [*Timperley* Hy 3 Pudsay, *Tympirleg* 1285 Court]. Very likely from *Timber-lēah* 'timber wood'. As regards *þ* for *b* cf. *timperon* 'timber building' (Cu), presumably from *timber-ærn*.

Timsbury Ha [*Timbreberie* DB, *Timberebir* 1227 Ch]. 'Timbered fort.'

Timsbury So [*Timesberua, -berie* DB, *Timberbarewe* 1200 FF, *Timberesberwe* 1233 FF]. 'Timber grove.' See BEARU.

Timworth Sf [*Timeworda* DB, *Timuuorde* c 1095 Bury, *Timeworthe* 1166 RBE]. '*Tima*'s WORÞ.' *Tima* seems to be found in *Timan* (*Tyman*) *hyll* BCS 1111.

Tincleton Do [*Tincladene* DB, *Tingledon* 1202 FF, *Uptincleden* 1257, *Holetincleden* 1260 FF]. First el. perhaps OE **tȳnincel*, a side-form of *tūnincel* 'small farmstead'. Derivatives in -*incel* do not generally have *i*-mutation, but cf. OHG *gensinklî, eninklî* from *gans, ano*.

Tindale Cu [*Tindale* 12 Lanercost, *Tiniel-side* 12 ib., *Tynyelfell* 1486 Ipm], **Tinnel** Co in Landulph [*Tinieltun* 1018 KCD 728, *Tiniel* 1291 Tax]. *Tiniel* may have as second el. Welsh *iâl* 'fertile upland region'. The first might be *din* 'fort' as in TINTAGEL. But Tindale Cu is on a tributary of the South Tyne, which may have been called *Tyne*. If so, *Tindale* means 'the Tyne valley' and *Tiniel* may be a Brit *Tinoialon* 'upland on the Tyne'.

Tingewick (tĭnjĭk) Bk [*Tedinwiche* DB, *Tingwich* 1163 P, *Tengewicha* 1167 P]. 'The wĪC of *Tida*'s or *Tēoda*'s people.'

Tingrith Bd [*Tingrei* DB, *Tingrith* 1209–19 Ep]. OE *þing-riþ* 'assembly stream'.

Tinhead W nr Westbury [*Tunheda* 1190 P, *Tyn-, Tunhide* 1240–5 Salisbury]. OE *tȳn hide* 'ten hides'.

Tinnel. See TINDALE.

Tinsley YW [*Tineslawe* DB, 1196 P, 1230 Ep, *Tunneslowe* 1292 YInq]. '*Tynne*'s barrow.' **Tynne* is a side-form of *Tunna*.

Tinta·gel (-ăj-) Co [*Tintagol* c 1145 Monm, *Tintaieol, Tintageolestun* 1205 Lay, *Tintagel* 1212 RBE, *Tinthagel* 1229 Fees]. The local form is said to be *Dundadgel*. The first el. is Co. *din, dun* 'hill, fort', with provection to *tin* as in TINTERN, TINDAETHWY, TENBY in Wales. (Tenby is Welsh *Dinbych*.) The second el. is obscure.

Tintern Monm [*Dindyrn, Dindirn, Tindirn, Tindyrn* c 1150 LL, *Tinterna, Tynterna* 1131 Mon, *Tynterne* 1268 Misc]. The first el. is W *din* 'hill, fort' with provection to *tin* as in TINTAGEL &c. The second has been identified with W *teyrn* 'king' from *tigerno-*. The form *Dinteyrn* is actually given from an early source in Geirfa Barddoniaeth Gynnar Gymraeg (under *din*).

Tintinhull So [*Tintehalle, Tintenella* DB, *Tintenhille* 1168 P, *Tintenhull* 1219 Fees]. Second el. OE *hyll* 'hill'. The first is obscure. OHG *Zinzo* pers. n. might possibly be compared. Cf. (Ailwinus) *Tint* 1176 P (Ca, Hu).

Tintwistle Chs [*Tengestvisie* DB, *Tenge-, Tyengetwisell, Tyngetwisel* 1286 Court]. Second el. OE *twisla* 'fork of a river'. The first may be a river-name identical with TEIGN.

Tinwell Ru [*Tedinwelle* DB, *Tineguella* 1125–8 LN, *Tinewell* 1189 (1332) Ch, 1220 Ep]. Perhaps 'the stream of *Tida*'s people'. An OE **Tidna* or **Tidin* would be better from a formal point of view.

Tipalt Burn Nb [*Typwolde fote* 1542 ERN]. Originally a pl. n. with OE *wald* 'wood' as second el. The first might be OE *yppe* 'hill', with *t-* from a prep. *æt* (*æt Yppewalde*).

Tipton St [*Tibintone* DB, *-ton* 1242 Fees]. '*Tibba*'s TŪN.'

Tiptree Ess [*Typpetre* c 1225 BM, *Tipetre* 1236 Fees, *Tippetre* 1291 Tax]. Possibly '*Tippa*'s tree.' *Tippa* is not evidenced, but cf. DEBDEN Ess, also *Tipemere* 1190 P (Ess).

Tirle Brook Gl [*Tyrl* 780, 769–85 BCS 236, 246]. An Engl river-name cognate with ME *tirle* 'to turn, make a rattling noise', *trille* 'to roll, purl'.

Tirley Gl [*Trinleie* DB, *-lege* 1221 Ass, *Trilleg* 1196 P]. Identical with (on) *Trindlea* 901 BCS 595 (W), 932 ib. 689 (Ha), *Trinlech* 821 BCS 366 (Brk). Cf. also (on) *Trindellea* 956 BCS 959 (So). The meaning is 'round glade', the first el. being perhaps an adj. **trind* 'round' or **trindel* 'circle'. Cf. OE *trinde* 'round lump', *trendel* 'ring, circle'.

Tirril We [*Tyrerhge* c 1189 CWNS x, *Tyrergh* 1257 P, *Tyrel* 1292 QW]. ON *tyri* 'dry resinous wood' and ERG 'shieling'. The change *r* > *l* is due to dissimilation.

Tisbury W [*Tyssesburg* 7 Letter of St. Boniface, *Tissebiri* 759 BCS 186, (to) *Tyssebyrig* 901–24 BCS 591, *Tisseberie* DB]. '**Tyssi*'s or **Tissi*'s BURG.' *Tissi* may be a short form of *Tidsige*.

Tissington Db [*Tizinctun* DB, *Tiscintona* c 1141 Mon iii, *Ticintona* Hy 2 DC, *Tyscin-*

ton 1242 Fees]. 'The TŪN of *Tidsige*'s people.'

Tisted, East & West, Ha [*Ticces stede* 932, *æt Ticcestede* 941 BCS 689, 765, *Tistede* DB, *Esttistede* 1291 Tax, *Westistude* 1234–6 Selborne], **Titchfield** Ha [*Ticefelle* DB, *Tichesfeld* 1168, 1194 P, *Tichefeld* 1219 Fees]. 'Place and FELD where kids were kept.' OE *ticcen* here appears in the shortened form *ticce*. Cf. *Ticcenesfeld, Ticcefeld* c 909 BCS 629 (Crawley Ha).

Titchmarsh Np [*Tut(e)an Mersc* 973 BCS 1297, *Ticceanmersc* c 975 PNNp, *Ticemerse* DB, *Tychemeris* c 1180 NpCh]. '*Ticcea*'s marsh.'

Titchwell Nf [(et) *Ticeswelle* c 1035 Wills, *Tigeswella, Tigeuuella* DB, *Tichewell* 1206 Cur]. 'Kid spring.' Cf. TICCEN.

Titley He [*Titel(l)ege* DB, *Titelea* 1194 P]. See LĒAH. First el. as in TIDCOMBE.

Titlington Nb [*Tedlintona* 1123–8, *Titlingtona* 1154–81 (1336) Ch, *Tidlington* 1167, *Titlinton* 1197 P]. Apparently 'the TŪN of *Titel*'s people'. Cf. TELSCOMBE.

Titsey Sr [*Tydices eg* 964–95 BCS 1132, *Ticesei* DB]. '*Tydic*'s island.' **Tydic* is cognate with *Tuda, Tydi*.

Tittenhanger Hrt [*Tidenhangra* 1198 (1301) Ch, *Tyndenhangr'* 1234 Cl]. '*Tida*'s slope.' Cf. HANGRA.

Tittenley Sa [*Titesle* DB, *Tutenlegh* 1304 Chamb, *Titenlegh* 1347 Ormerod]. '*Tytta*'s LĒAH.' Cf. TIDDINGTON O.

Tittensor St [*Titesovre* DB, *Titneshovere* 1236, *-overe* 1242 Fees]. '*Titten*'s *ofer* or slope.' **Titten* is a derivative of *Titta* (cf. TIDCOMBE).

Tittleshall Nf [*Titeshala* DB, *Titleshal* 1200, *Tetles-, Titleshal* 1205 f. Cur, *Tutleshal* 1275 Cl]. '*Tyttel*'s HALH.' *Tyttel* is cognate with *Tutta, Tyttla*.

Tiverton Chs [*Tevretone* DB, *Teverton* 1260 Court]. First el. OE *tēafor* 'red pigment, vermilion'. Cf. TAVERHAM.

Tiverton D [(æt) *Twyfyrde* c 880 BCS 553, *Tovretone* DB, *Tuiverton* c 1150 Fr, *Little Twuuertona* 1168 P, *Teverton, Tuverton* 1205 f. Cur]. Originally *Twifyrde* 'double ford'. Later TŪN at the double ford'. Cf. TWYFORD.

Tivetshall Nf [(of) *Tifteshale* 11 EHR 43, *Teuetessalla, Tiuetessala, Teuetesshala* DB, *Tiueteshale* c 1095 Bury, *Tiftes-, Tivetshale* 1254 Val]. The first el. may be a form of *tewhit, tewit* 'lapwing', which appears in forms such as *tewfet, tufit* &c. See HALH.

Tixall St [*Ticheshale* DB, *Tikeshala* 1167 P, *-hale* 1242 Fees], **Tixover** Ru [*Tichesovre* DB, *-oure* 1104–6 RA, *-ora* 1130–3 Fr, *-oura* 1166 P, *Tikesoura* 1163 RA]. 'Kid's HALH and bank.' OE *ticcen* here appears with hard *c(k)*.

TOCKENHAM [476] TOLLESBURY

Tockenham W [*Tockenham* 854 BCS 481, *Tocheham* DB]. '*Toc(c)a*'s HĀM.' *Toca* is found in *Tocan stan* 983 KCD 636, 638 (in bounds of Cliffe Pypard nr Tockenham). The same *Toca* gave their names to Tockenham and *Tocan stan*. *Tocca* is found in *Toccan sceaga* 755-7 BCS 181.

Tocketts YN [*Theostcota* 1104-8 SD, *Toscutun*, *Tocstune* DB, *Tofcotes* 1187 P]. OE *þēos cotu* 'the servant's huts'. OE *þēow* was sometimes *þēos* in the genitive. *Tof-* 1187 probably for *Tos-*. *T-* for *Th-* is due to Norman influence.

Tockholes La [*Tocholis* c 1200 CC, *-holes* 1246 Ass]. '*Tocca*'s or *Tōki*'s hollow.' Cf. TOCKENHAM, TOXTETH, HOLH.

Tockington Gl [*Tochintune* DB, *Tokinton* 1199 P, 1220 Fees]. 'The TŪN of *Toc(c)a*'s people.' Cf. TOCKENHAM.

Tockwith YW [*Tocvi* DB, *Tockwic*, *-with* 1121-7, *Tocwic* 1120-2 YCh 1428, 1430, *Tocwyz* 1249 Ch, *Tockewyht* 1280 Ch]. The second el. was originally OE wīc, later exchanged for OScand *viθ(r)* 'wood'. The first seems to be OE *Toc(c)a* pers. n. Cf. TOCKENHAM.

Todber Do [*Todeberie* DB, *Toteberga* 1177 P, *-bera* 1194 P, *Toddebir* 1228, *Todeberwe* 1268 FF]. '*Tota*'s hill or grove.' Cf. BEARU, BEORG.

Todber YW [*Toddebergh* 12, *Thodeberc* 13 Pudsay]. Perhaps 'fox hill', the first el. being ME *tod* 'fox' (1170 &c. OED).

Toddington Bd [*Totingedone* DB, *Tudingedon* 1166 P, *Tudingdon* 1238 Cl], T~ Gl [*Todintun* DB, *Tudintone* 1221 Ass, *Tutington* 1236 Fees]. 'The DŪN and TŪN of *Tuda*'s people.'

Todenham Gl [*Todanhom* 804 BCS 313, *Toteham*, *Teodeham* DB, *Todenham* 1291 Tax]. The identification of some of the above forms is doubtful. Perhaps '*Tēoda*'s HAMM or HĀM'.

Todmorden YW [*Tottemerden*, *Totmardene* 1246 Ass, *Todmarden* c 1300 WhC]. '*Totta*'s boundary valley.' Second el. OE (ge)*mǣrdenu*.

Todridge Nb [*Todrige* 1479 BBH], 'Fox ridge.' Cf. TODBER YW.

Todwick YW [*Tatewic* DB, *-wik* 1233 Ep, *Totewyk* 1300 Ch]. '*Tāta*'s wīc.'

OScand toft, topt (ON *topt*, OSw *toft*, *tompt*, Dan, Norw *toft*, Dan, Sw *tomt*) originally meant 'site of a house and its outbuildings, house site', and this meaning is still in common use. Sw *tomt* means 'a plot'. From this sense developed such senses as 'field near a house' or 'messuage, homestead'. The latter sense is recorded for ON *topt*. In Engl pl. ns. the meaning is either 'site of a house' &c. or 'deserted site' (as in ALTOFTS) or 'messuage, homestead'. Names containing *toft* are chiefly found in the East Midlands and in Yorkshire. The first el. of names in *-toft* is mostly a pers. n. (as in BROTHER-, LOWES-, SIBBER-, WIBTOFT) or an adj. (as in BLACK-, BRA-, LANGTOFT). The simple word *toft* is sometimes used as a pl. n. Toft Ca [*Tosta* c 1080 ICC, *Tofth* DB, *Toft* 1242 Fees], T~ Li nr Bourne [*Toftlund* DB, *Toft* 1212 Fees], T~ next Newton Li [*Tofte* DB, *Toft* c 1115 LiS], T~ Monks Nf [*Toft* DB, *Toft monachorum* 1386 BM], T~ Wa [*Toft* 1291 Tax], West Tofts Nf [*Stofftam* DB, *Toftes* 1199 P, *Westtoftes* 1291 Tax].

T~ Monks belonged to the Abbey of Préaux in Normandy (1199 Fr).

Toftrees Nf [*Toftes* DB, 1254 Val]. Apparently 'the tofts'.

Togstone Nb [*Toggesdena* 1130, 1177 P, *Toggisden* 1236, 1242 Fees]. '*Tocga*'s valley.' Or there may have been a strong side-form of *Tocga*.

Tolethorpe Ru [*Toltorp* DB, *Toletorþ* 1202 Ass, *Tolthorp* 1273 Ipm]. '*Tōli*'s thorp.' First el. OSw, ODan *Toli* (*Toli* DB).

Tolland So [*Tádland* 11 KCD 897, *Talanda* DB, *Taland* 1266 Ep, *Tolonde* 1327 Subs]. Originally OE *Tān-land* 'land on R TONE'. Tolland is on a tributary of the Tone, which must have been called Tone too.

Tollard Farnham Do [*Tollard* 1202 Cur, 1204 FF, *Toullard* 1204 Cur], T~ Royal W [*Tollard* DB, 1167 P, 1195 Cur]. The elements are Welsh *toll* 'having holes, pierced' and *ardd* 'hill'. The meaning may be 'hill intersected by valleys'. The two Tollards are near each other.

T~ Farnham is near Farnham.—T~ Royal is stated to have belonged to King John.

Tŏller Fratrum & Porcorum Do [*Tolre* DB, c 1100 Montacute, 1195 P, *Tolre Fratrum*, *Porcorum* 1341 NI, *Suynestholre* 1259 FF, *Swyntolre* 1288 Ass]. An old name of the river HOOKE. Near the source of the Hooke is Toller Whelme [(on) *Tollor æwylman* 1035 KCD 1322, *Tolreewelme* 1334 FF], whose name means 'the source of the *Tollor*'. The river-name may go back to an early Welsh *toll-ŏur* 'hollow stream, stream with deep holes or running in a deep valley'. For *toll* cf. TOLLARD. The second el. is Welsh *dwfr*, *dwr* 'stream'.

T~ Fratrum belonged to Forde Abbey.—T~ Porcorum must have been famous for its pigs.

Tollerton Nt [*Troclavestune* DB, *Torlauetun* 12 DC, *Turlaueston* 1183 P]. '*Þorleif*'s TŪN.' Cf. THURLASTON.

Tŏllerton YN [*Tolentun*, *Tolletune* DB, *Tolereton* 1167 P, *Tolnertona* 1293 PNNR]. 'The tollers' or tax-gatherers' TŪN' (OE *Tolnera-tūn*). The earliest forms may point to an original name *Toln-tūn* 'TŪN where taxes were paid'.

Tŏllesbury Ess [*Tolesberia* DB, *-bir'* 1218 Fees], Tolleshunt (tŏlznt) d'Arcy,

Knights & Major Ess [(of) *Tollesfuntan* c 1000 CCC, *Tolesfunte* 1068 EHR xi, -*hunte* DB, *Toleshuntetregoz* 1239 FF, *Tholeshunte Militis* 1238 Subs, *T~ Malgeri* 1257 Ch, *Tolleshunte Chyvaler* 1272 FF, *Toleshunte Mauger* 1254 FF]. '*Toll*'s BURG or manor' and '*Toll*'s spring' (cf. FUNTA). *Toll* is an unrecorded pers. name. Tolleshunt is near Tollesbury.

Tolleshunt d'Arcy was held by the Tregoz family till the beginning of the 15th cent. Cf. EATON TREGOSE. Robert Darcy got land here in 1441 (Pat). The Darcy family took its name from ARCY in Normandy.—**T~ Knights** was presumably held by knight's service.—**T~ Major** was held by *Malger* in 1086 (DB). *Major* is a corruption of this name, for which see MADJESTON.

Tolpuddle. See PIDDLE.

Tolworth Sr [*Taleorde* DB, -*worda* 1130, -*wurda* 1161 P, -*worth* 1241 FF]. '*Tala*'s WORÞ'; cf. TALLINGTON. The change to *Tol-* seems to be late. *Tolesworth*' occurs 1179 RA, but in a copy of 1345.

Tonbridge (-ŭ-) K [*Tonebrige* DB, *Tonebricg* 1087 ASC (E), *Thunnebrigge* 1230 P]. '*Tunna*'s bridge.'

Tone Nb [*Tolland* 12, 13 Newminster, 1296 Subs]. Perhaps OE *toln-land* 'land on which toll is paid'.

Tone R So [*Tan* 682, 705, *Táán*, *Tán* 854 BCS 62, 113, 475 f., *Thon* 1243 Ass]. A Brit river-name, perhaps related to Gaul *Tanarus*, the name of a river in Italy and a byname of Jupiter, Lat *tonare* &c. If so, the name means 'roaring stream'. Or the name may be related to Welsh *tan* 'fire'. Creech Hill on the Tone was in British called *Cructan* (BCS 62). This may well mean 'fire hill', i.e. 'beacon hill'. The river-name might be an early back-formation from this.

Tong Sa [(into) *Tweongan* 10 BCS 1317, ? (æt) *Twongan* 1002 Wills, *Tvange* DB, *Twanga* 1167 P, *Tange* 1176 P], **T~ YW** [*Tuinc* DB, *Tange* 1176 P, *Tange* 1203 FF], **Tonge** K [*Tangas* DB, *Tanga* 11 DM, 1161 P, *Twhonge* 1465 BM], **T~ La** in Prestwich [*Tange* 1212 Fees, *Twannge* 1212 RBE], **T~ with Haulgh** La [*Tonge* 1323 LaInq, 1332 Ass]. OE *twang* or partly *tang* 'tongs, fork of a river'. See TANG.

Tonge Le [*Tunge* DB, *Tunga* c 1125 LeS, Hy 2 BM]. OE *tunge* 'tongue', here 'tongue of land'.

Tongham Sr [*Tuangham* R 1 Mon v, *Twangham* 1244 FF, 1272 Ipm, *Tangham* 1251 Cl]. 'HĀM in a tongue of land.' Cf. TANG.

Tooley Le [*Tolawe* Hy 3 BM, 1278 Misc]. Possibly OE *tōt-hlāw* 'look-out hill'.

Tooting Graveney, Upper T~ Sr [*Totinge* 675 BCS 39, *Tottingas* 1067 BM, *Totinges*

DB, 1197 FF, *Toting Gravenee* 1314 Ipm]. '*Tōta*'s people.'

T~ Graveney was held by the Gravenel family (from Graveney in Kent?) from the 12th cent.

Topcliffe YN [*Topeclive* DB, -*cliue* 1166 P, *Toppeclive* 1218 FF]. 'Cliff with tops or peaks' does not give good sense, as the place is on the Swale in no high situation. '*Toppa*'s river bank' would be more suitable. *Toppa* might be a weak side-form of *Topp* (in TOPSHAM).

Topcroft Nf [*Topecroft* DB, c 1095 Bury, *Topescroft* c 1095 Bury, *Toppecroft* 1206 Cur]. *Tope* pers. n. is found in HEl and DB. It is from Dan *Topi*. Topcroft seems to be '*Topi*'s croft'.

Toppesfield Ess [*Topesfelda* DB, *Toppesfeld* 1197 FF, 1204 Cur], **Topsham** D [*Toppesham*, -*hamme* (dat.) 937 BCS 721, (æt) *Toppeshamme* c 1070 Ex, *Topeshant* DB]. '*Topp*'s FELD and HAMM or river land.' *Topp* is also found in *Toppes ora* ('*Topp*'s landing-place') BCS 721. It is not found in independent use.

Torbryan D [*Torre* DB, *Torre Briane* 1238 Ass, *Torbriane* 1270 FF]. OE *torr* 'hill'.

The manor was held by Wydo de Brianne or Brionne in 1242 (Fees). Brionne is in Eure.

Torkington Chs [*Torkinton* 1182 P, 1248 Ipm], **Torksey** Li [*Turecesieg* 873 ASC, *Turcesig* ib. (D, E), *Torchesey* DB, -*eia* 1153 BM]. 'The TŪN of *Turec*'s people' and '*Turec*'s island'. The pers. n. *Turec* (or rather *Turoc*) is not evidenced in independent use. It may be derived from the root of Goth *gatarhjan* 'to distinguish'.

Tormarton Gl [*Tormentone* DB, *Tormertona* 1183 AC, -*ton* 1209 Cur]. The place is on the Wilts border. The original name was no doubt OE *Mǣrtūn* 'TŪN on the boundary'. Later OE *torr* 'hill' was added for distinction from DIDMARTON.

Tormoham D [*Torre* DB, *Torre Brywere* c 1200 Torre, *Torre Moun* 1279 Cl]. OE *torr* 'hill'.

For the additions cf. BUCKLAND BREWER, HAMMOON. T~ was held by William de Mohun in 1242 (Fees).

Torne R YW [*Thorn* c 1160 Kirkst]. Apparently a back-formation from a lost pl. n. *Thornwath*, which means 'ford on the road to THORNE'. Cf. VAD.

Torpenhow Cu [*Torpennoc* 1163 P, *Thorpennou* 1212 Fees, *Torpenno* 1224 P]. The name contains the elements *torr* 'hill', *pen* 'hill' (Welsh *pen*), and OE *hōh* 'ridge, spur of land', or else *torr* and OW *pennou*, the plur. of *pen*, or *torr* and a name identical with PINHOE D. In the first alternative the Brit name would have been *Torr pen*, to which was added OE *hōh*. *Torr pen* would be analogous to OW *tormeneth* 'top or breast of the hill'. The meaning would be about the same in the other alternatives.

Torquay· D [*Torrekay* 1591 PND]. A late name meaning 'the quay at Tor(moham)'.

OE **torr** 'high rock, rocky peak, hill', dial. *tor* (Co, D, So, Db &c.), is a loanword from Co *tor* 'prominence, womb, mountain', Welsh *tor* 'bulge, belly, boss', Gael *torr* 'a tor, hill'. See TOR- passim, DUNSTER, HAYTOR &c. **Torre Abbey** D at Torbryan and Tormoham is *Torre* R 1 Torre, *Torr* 1199 FF.

Torridge R D [*Toric* 938 BCS 725, *Torix* 1238 Ass, *Torighe* 1371 Cl], **Tory Brook** D [*Torygg* 13 PND]. A Brit river-name identical with TERIG in Flint and derived from Welsh *terig* 'rough'.

Torrington, Black, D [*Torintona* DB, *Blaketorrintun* 1219 Fees], **Great T~** D [*Tori(n)tona* DB, *Chipping Toriton* 1296 Misc], **Little T~** D [*Toritona* DB, *Parva Toriton* 1242 Fees]. 'TŪN on R TORRIDGE.'

Torrington, East & West, Li [*Terintone* DB, *Tiringtuna* c 1115 LiS, *-tun* 1165, *-tona* 12 DC, *Est Tyrington* 1232, *West Tirinton* 1209–35 Ep]. Identical with TERRINGTON Nf.

Torrisholme La [*Toredholme* DB, *Toroldesham* 1201 P]. '*Þorald*'s holm.' Cf. THORLBY.

Tortington Sx [*Tortinton* DB], **Torton** Wo in Hartlebury [*Tortintuna* 1182 PNWo, *Torchinton* 1229 Ch]. 'The TŪN of *Torhta*'s people.' Cf. next name.

Tortworth Gl [*Torteword* DB, *-wurð* 1178 P, *-worth* 1220 Fees]. '*Torhta*'s WORÞ.' *Torhta is a short form of names in *Torht-*.

Torver La [*Thoruergh* 1190–9 LaCh, *Torvergh* 1246 Ass]. 'Peat shieling' (ON *torf* 'turf, peat' and ERG). Or the meaning may be 'hut made of sods'.

Torworth Nt [*Turdeworde* DB, *Thordworth* 1199 (1232) Ch, *Thordeswrð* 1200 P, *Thorchewurh* 1275 RH]. Probably '*Þorð*'s WORÞ'. First el. ON *Þórðr* (gen. *Þórðar*), ODan *Thorth* pers. n.

Tory Brook. See TORRIDGE.

Toseland Hu [*Toleslund* DB, *Touleslund* 1220 Fees]. '*Tóli*'s grove.' Cf. TOLETHORPE and LUND.

Tosson Nb [*Tosse, Thosse* 1150–62 YCh 1241, *Thosan* 1203 P, *Tossan* 1205 Cur, *Tossin* 1236, *Tossen* 1242 Fees]. Tosson Hill reaches 1,447 ft. The source may be OE *tōt-stān* 'look-out stone'.

Tostock Sf [*Totestoc, Totstocha* DB, *Totstoche* c 1095 Bury, *Totestok* 1226–8 Fees]. OE *tōt-stoc* 'look-out place'. The place is on a prominent hill.

OE ***tōtærn*** 'look-out house, watch-tower' is the first el. of TOTTERNHOE, TOTTERTON. The el. *tōt-* seems to be derived from OE *tōtian* 'to peep out, protrude'. *Toot* 'look-out hill' is found from 1387 and is held in OED to be perhaps short for *toothill*, which

is found in 1250. As a first el. *tōt-* is found in several other names, as TOSSON, TOSTOCK, TOTHAM, TOTHILL.

Totham, Great & Little, Ess [*Totham* c 950 Wills, c 995 BCS 1289, *Tot(e)ham* DB, *Thotham Magna, Parva* 1238 Subs]. 'Look-out HĀM.' Great T~ is on the slope of a hill, near Beacon Hill.

Tothill Li [*Totele* DB, 1158 Fr, 1242 Fees, 1255 Ch], T~ Mx [*Tothulle, -hell, -hill* 12 BM]. 'Look-out hill.' The forms of Tothill Li rather suggest OE *Totan lēah*, but the map has a Toot Hill close by.

Totley Db [*Totingelei* DB, *Totenleg* 1221–30 Fees]. 'The LĒAH of *Tota*'s people.'

Totmonslow St [*Tatemaneslav* DB, *-manneslawa* 1175 P]. '*Tātmann*'s HLĀW or barrow.'

Totnes D [*Totanæs* 979–1016 Coins, *Tottaness* 11 Crawf, *Totenais* DB, *Tottenas* 1205 Layamon]. '*Totta*'s NÆSS or headland.'

Toton Nt [*Tovetune* DB, *Toueton* 1230 P, *-tun* 1236 Fees]. '*Tófi*'s TŪN.' First el. ON *Tófi*, OSw *Tove*, ODan *Tovi*. The form *Tolvestone* DB possibly indicates that the first el. was originally OScand *Þólfr* (from *Þórolfr*), of which *Tófi* was a short-form.

Tottenham Mx [*Toteham* DB, c 1130 BM, *Totenham* 1265 FF], **Tottenham Court** Mx [(of) *Þottanheale* c 1000 ASCh, *Totehele* DB, *-hale* R 1 BM, *Totenhale* 1254 Val, *Totten-Court* Ben Jonson]. '*Tota*'s or *Totta*'s HĀM and HALH.' The initial *Þ-* of *Þottanheale* is possibly wrong for T.

Tottenhill Nf [*Tottenhella* DB, *Totehill* 1251 Ch]. '*Totta*'s hill.'

Totteridge Hrt [*Taterugg* 1248, *Tatterigg* 1251 Ch]. '*Tāta*'s ridge.'

Totternhoe Bd [*Totenehou* DB, *Toterhou* 1176 P, *Toternho* 1207 Cur]. 'Ridge with a look-out house.' Cf. TŌTÆRN.

Totterton Sa [*Toterton* 1180 P, *Toderton* 1327 Subs]. Probably 'DŪN or hill with a look-out house'. Cf. prec. name.

Tottington K nr Maidstone [*Totintune* DB, 11 DM], T~ La [*Totinton* 1212 Fees, *Totington* 1233 FF], T~ Nf [*Totingtonne* 1044–7 KCD 785, *Totintuna* DB, *Totingeton* 1193 ff. P], **Totton** Ha [*Totintone* DB, *-ton* 1212 Fees]. 'The TŪN of *Tota*'s people.'

Toulston YW [*Toglestun* DB, *Touleston* 1185, 1190 P, c 1200 YCh 533]. '*Toglos*'s TŪN.' First el. *Toglos* pers. n. A Danish jarl so called was slain at Tempsford in 921 (ASC).

Tove R. See TOWCESTER.

Tow Law Du [*Tollawe* 1423 PNNb]. Perhaps OE *tōt-hlāw* 'look-out hill'. Cf. TOOLEY.

Towcester (towst*er*, tōst*er*) Np [*Tofeceaster* 921 ASC, *Tovecestre* DB, *-cestr'* 12 NS]. 'Roman fort on R TOVE.' The Roman fort here was *Lactodoron*. **Tove** [*Toue* 1221 Cl] is derived from an adj. *tōf 'slow, dilatory', cognate with MDu *toeven*, MLG *toven* 'to linger'.

Towe·dnack Co [(parochia) *Sancti Tewynoti* 1377 PT]. A saint's name, identical with that found in LANDEWEDNACK. *To-* is the pronoun for 'thy', here used for hypocoristic purposes.

Towersey Bk [*Eie* DB, *Turrisey* 1237–40 Fees, *Tureseye* 1252 Cl]. 'The island.' The manor was held by Richard de Turs in 1252. *Turs* from TOURS in France. Cf. KINGSEY.

Towneley La [*Tunleia* c 1200 Whitaker, *-ley* 1242 Fees]. 'LĒAH belonging to the TŪN' (i.e. Burnley).

Towthorpe YE [*Touetorp* DB, *-thorp* 1231 FF], T~ YN [*Touetorp* DB]. '*Tōfi*'s thorp.' Cf. TOTON.

Towton YW [*Touetun* DB, *-ton* 1206 Cur]. Identical with TOTON.

Toxteth La [*Stochestede* DB, *Tokestath* 1212 Fees]. '*Tōki*'s landing-place.' First el. ON *Tóki*, ODan, OSw *Toki*. Second el. ON *stǫð* 'landing-place'.

Toynton, High & Low, Li [*Tedin-, Todintune* DB, *Tidinton* 1166 P, *Teinton* 1199, *-tune* 1230 P, *Tynton Superior, Toynton Inferior* 1254 Val]. 'The TŪN of *Tēoda*'s people.'

Toynton All Saints & St. Peter Li [*Totintun(e)* DB, *Totingtuna, Totintona* 12 DC, *Thoynton Omnium Sanctorum, Sancti Petri* 1254 Val]. 'The TŪN of *Tota*'s people.'

OE **træppe** 'trap'. See BAWDRIP, TRAFFORD Np.

Trafford, Bridge, Mickle & Wimbolds, Chs [*Tro(s)ford, Traford* DB, *Trochford* c 1100, *Trocford* c 1190 Chester, *Wimbaldesthrofford* 1288, *Great Trogthforde* 1290 Court]. 'Ford in a valley.' Cf. TROG. *Wimbold* may be the *Winebald* (*Wynebaud*) vicecomes mentioned 1121–c 1150 Chester.

Trafford La [*Stratford* 1206 P, *Straforde* 1212 RBE, *Trafford* c 1200 LaCh, 1212 Fees]. A Normanized form of OE *Strētford*. Trafford is close to STRETFORD.

Trafford Np in Chipping Warden [*Trapeford* DB, *Trapesford* 12 NS]. 'Ford by a trap' (OE *træppe*). A trap for otters or the like may be referred to.

Tranby YE [*Tranebi* J Ass, *-by* 1221 FF]. '*Trani*'s BY.' *Trani* is an OScand nickname identical with *trani* 'crane'.

Tranmere Chs [*Tranemor* 1260, *-mol* 1282, 1287, *-moll* 1288 Court, *-mel* 1290 Ipm]. 'Cranes' sandbank', the elements being ON *trani* 'crane' and *melr* 'sandbank'.

Tranwell Nb [*Trennewell* 1268 Ipm, *Tranewell* 1289 Ipm]. 'Cranes' stream.' Cf. TRANMERE.

Trawden La [*Trochdene* 1296, *Troudene* 1305 Lacy]. OE *trog-denu* 'flat valley'. Cf. TROG.

Treales (-ālz) La [*Treueles* DB, 1206 P, *Trivel* 1249 Ipm]. Identical with TREFLYS Carnarvon, MBret *Trefles* &c., the elements being Welsh *tref* 'village' and *llys* 'court'. The name seems to mean 'township of the court'.

Treborough So [*Traberge* DB, *Trebergh* 1225 Ass]. 'Hill where trees grew.'

Tredington Gl [*Trotintune* DB, *-ton* 1185, *Tretinton* 1195 f. P]. 'The TŪN of *Trota*'s people.' *Trota* is found as the name of a moneyer. Cf. TROTTISCLIFFE.

Tredington Wa [*Tredingctun* 757, *Tredingtun* 10 BCS 183, 1320, *Tyrdintune* 964 ib. 1135, *Tredinctun* 978 KCD 620, DB]. 'The TŪN of *Tyrdda*'s people.' *Tyrdda comes* had held T~ before 757 (BCS 183). *Tyrdda* is no doubt from *Trydda*. Cf. (on) *Tryddingleage* 863 BCS 508. *Trydda* (*Tredda*) is derived from OE *tredan* 'to tread'.

Treeton YW [*Tretone, Trectone* DB, *Tretona* c 1130 Oxf, c 1195 YCh 1276, *-ton* 1204 FF]. 'TŪN by the tree(s).'

Welsh **tref, tre**, Co *trev, tre* 'homestead, village, town' is a common first el. in pl. ns. of Wales and Cornwall, and examples occur also in Herefordshire and Lancashire. See TREALES and the following names. Common Cornish names are e.g. TREGAIR, TREGEAR 'hamlet of the *caer* or fort', TREGARN 'hamlet of the *carn* or rock', TREMAINE 'hamlet of the *maen* or stone'.

Tregate He in Llanrothal [*Treget* 1131–44, *-ket* 1144, *-jet* 1146 Fr]. Welsh *tre goed* 'hamlet of the wood' (Welsh *tre* and *coed*).

Tregavethan Co [*Treganmedan* DB, *Tregemadan* 1221 Cl]. See TREF. The second el. is possibly a personal name.

Tregony Co [*Trefhrigoni* 1049 KCD 787, *Treguni* 1229 Fees, *-goni* 1260 Ep]. First el. TREF. The second is obscure.

Trelleck Monm [*Trilecc, Trilec, Trylec* c 1150 LL, *Trillek* 1131 Mon, *Trellek* 1291 Tax]. Apparently 'the three stones' from the three stones still standing in the place, the elements being W *tri* masc. 'three' and W *llech* 'stone', or perhaps rather a cognate of the word, as *llech* is now a fem. word.

Tremaine Co. See TREF.

Trematon Co [*Trefmeutun* c 970 BCS 1247, *Tremetone* DB]. A Cornish name of doubtful etymology, to which was added OE *tūn*. *Tre* is OCo *tref*, Co *trev, tre* 'homestead'.

Tremworth K [*Dreaman uuyrð* 824 BCS 378, *Dreamwurthe* 11 DM, *Tremeworth* 1263 Ipm]. '*Drēama*'s WORÞ.' Cf. DRIMPTON, DRINKSTONE.

OE **trendel** 'circle'. See TRENTISHOE, TRULL.

Trene·glos Co [*Treneglos* 1269 Ep, 1291 Tax]. 'The church village.' The elements are Co *tre* 'hamlet' and *eglos* 'church', *n* being a relic of the def. art. (Co *an*).

Trenholme YN [*Traneholm* 1176 P]. 'Crane island.' Cf. TRANMERE.

Trenowth Co nr Truro [*Trefneweð, -næwð* 969 BCS 1231]. 'New hamlet or homestead.' Cf. TREF. Second el. Co *newydh* 'new'.

Trent R St, Db &c. [*Trisantona* 115–17 Tacitus, *Treanta, Treenta* c 730 Bede, *Treontan* (obl.) c 890 OEBede, 924 ASC, *Trente* DB; *Trahannoni fluminis* c 800 HB, *Taranhon* 12 Taliesin], T~ Do, another name of the Piddle [*Terente* a 1118 Flor, *Trent* c 1540 Leland]. Cf. also TARRANT, which is identical in origin. A Brit rivername *Trisantōn*, consisting of *tri-* 'through, across' and *santōn*, a word related to Welsh *hynt* 'road', OIr *sét* 'journey'. The name seems to be mean 'trespasser' and would be used of a river liable to floods. Cf. OBret *Treanton* pers. n.

Trent Do [*Trente* DB, 1225 Ass, *Trenta* 1163 P]. Originally a name of the stream at the place, identical with the river-name TRENT.

Trentham St [*Trenham* DB, *Trentham* 1156 P]. 'HĀM on R TRENT.'

Trentishoe D [*Trendesholt* DB, *Trenlesho* 1203 Cur, *Trendelesho* 1242 Fees]. 'HŌH or spur with a circular top.' First el. OE *trendel* 'ring, circle'.

OE **trēo(w)** 'tree' is a common second el. of pl. ns. Names of this kind refer to some prominent tree, sometimes one with religious associations, as in HALLATROW, sometimes one remarkable for its size, as in LANGTREE. The first el. is frequently a pers. n. In these cases the tree was probably often one marking a meeting-place, and the first el. may well be the name of a lawman. Several hundred-names have *tree* as a second element. As a first el. the word is not so common. See e.g. TREBOROUGH, TREETON, TREWICK, TREYFORD, TRING. The form of the element is mostly -*tree*, *Tree-* (as in ELMSTREE, COLLINGTREE, FAINTREE), but other forms occur, as in AUSTREY, AYMESTREY, GOOSTREY; COVENTRY, DAVENTRY; SCOTTER, WARTER; BISHOPS-, WANSTROW, TROWBRIDGE, TROWELL, TROWSE.

Trescott. See TRYSULL.

Treswell Nt [*Tireswelle* DB, *Tireswell* Hy 2 (1316) Ch, *Tyriswell* 1242 Fees]. '*Tir*'s spring or stream.' *Tir* is a short form of names in *Tir-*.

Tretire He [*Rythir* 1212 RBE, *Ryttyr* 1265 Ipm]. Welsh *rhyd hir* 'long ford'. The elements are Welsh *rhyd*, OW *rit* 'ford' and *hir* 'long'. T- is due to association with Welsh *tre* 'homestead' &c.

Treville He [*Triueline* DB, *Triuel* 1159 ff. P, *Trivel(broc)* 1227 Ch]. Perhaps 'hamlet with a mill'. The second el. is Welsh *melin* (mutated *felin*) 'mill'. The first may be Welsh *tre* 'hamlet' &c., but the regular *i* offers some difficulty, unless the Welsh name was *Tre-y-felin*.

Trewhitt Nb [*Tirwit* 1150–62 YCh 1241, *Tyrwit* 1236 Fees, *Tyrewit* 1240 FF, *Thyrewhyt* 1269 Misc]. ON *tyri* 'dry resinous wood' and *þvit* (see THWAITE).

Trewick Nb [*Trewyc* 1242 Fees, *Trowyk* 1269 Ass]. 'Tree WĪC.'

Treyford (-ē-) Sx [*Treverde* DB, *Treferd* 1256 Ch]. 'Tree ford, ford marked by a tree or provided with a tree-trunk to assist in crossing.' Second el. *fyrde*, found often in TWYFORD.

Triermain Cu [*Trewermain, Treverman* c 1200 WR]. Perhaps Welsh *tref yr maen* 'homestead at the stone'.

Trigg Co, an old district [*pagum . . . Tricurrium* 9 Life of St. Sampson (LL 19), *Triconscir* c 880 BCS 553, *Trigerscire* 1130 P, *Tregesir* 1211 FF, *Trigge Major, Minor Triggeshire* 1291 Tax]. Identical with TRÉGUIER in Brittany. Cf. Gaul *Tricorii*, a tribal name meaning 'those with three armies' (*corio-* 'army'). *Pagus Tricurius* would be 'a district consisting of three divisions'.

Trimdon Du [*Tremeldon* 1196 P, *Tremedon* 1262 BM]. The elements may be OE *trēomæl* 'wooden monument, cross' and DŪN 'hill'.

Trimingham Nf [*Trimingeham* 1185 P, *Tremingham* 1276 Misc]. 'The HĀM of *Trymma*'s people.' *Trymma* would be a short form of names in *Trum-*.

Trimley St. Mary & St. Martin Sf [*Tremlega, Tremelaia* DB, *Tremle Beate Marie, Sancti Martini* 1254 Val]. '*Trymma*'s LĒAH.' Cf. prec. name.

Trimpley Wo [*Trinpelei* DB, *Trimpelcge* 1221 Ass]. '*Trympa's LĒAH.' Cf. TRUMPINGTON.

OE **trind(e).** See TIRLEY.

Tring Hrt [*Tredunga, Treunge* DB, *Trawinge* 1176 P, *Treange* 1207 FF, *Trahing* 1212 Fees, *Trehenge* Hy 3, *Trehanger* 1265 Misc]. The last form gives the clue to the etymology. It suggests OE *trēo-hangra* 'slope where trees grew'. The second *r* was lost owing to dissimilation. Anderson, *Hundred-names* iii. 29, adds several exx. which confirm the etymology, e.g. *Trehangr'* 1199 Cur, *Triangre* J Mon.

Tritlington Nb [*Turthlyngton* c 1170 Newcastle, *Tirtlington* 1242 Fees]. 'The TŪN of *Tyrhtel*'s people.'

OE **trog** 'trough', later also 'hollow or valley resembling a trough, bed or channel of a stream'. Cf. **Trough of Bowland** in La [*Trogh* c 1350 LaCh]. See TRAFFORD Chs, TRAWDEN.

Troston Sf [*Trostingtun* c 1000 BCS 1306, *Trostuna* DB]. 'The TŪN of **Trost(a)*'s people.' Cf. OG *Trostila*, *Trostheri* &c. The name is related to Engl *trust*.

Trottiscliffe K [*Trottesclib* 788, *Trotescliua*, (of) *Trotescliue* 10 BCS 253, 1321 f., *Totesclive* DB, *Trottesclive* 1268 Ch]. '*Trott's cliff.' The name, which is also found in **Trottsworth** Sr [*Trotteswurth* 1242 Fees, 1243 Cl], is related to MLG *trot*, G *trotz* 'defiance'.

Trotton Sx [*Traitone* DB, *Tratinton* Hy 1 PNSx, 1252 Ch, *Traditona* 12 Ordericus, *-ton* 1230 FF]. Possibly 'TŪN of *Trott*'s people'; cf. prec. name. Or the base may be an OE *Trætt, cognate with MHG *traz* 'defiance'.

Trough of Bowland. See TROG.

Troughend Nb [*Trocquen* 1242 Fees, *Trequenne* 1279 Ass, *Trehquen*, *Troghwen* 1293 QW]. Unexplained. Possibly a Brit name.

Troutbeck Cu [*Troutbek* 1332 Subs], T~ We [*Trutebeck* 1272 Ipm]. 'Trout stream.'

Trouts Dale YN [*Truzstal* DB, *Trucedale* 1314 Ipm]. OE *truht-stall* 'trout pool'. Cf. STALL. Trout is OE *truht*.

Trowbridge W [*Trobrigge* 1184 P, *Troubrug* 1212 Pat, *Trebrigg* 1311 Ipm]. 'Wooden bridge.' Cf. TRĒO.

Trowell Nt [*Trowalle* DB, *Trowella* 1166 P]. 'Tree stream', perhaps one with a bridge formed by a tree-trunk.

Trowse (trōs) **Newton** Nf [*Treussa*, *Treus*, *Newotona* DB, *Trous* 1254 Val, *Trowes cum Newtone* 1316 FA]. OE *trēo-hūs* or OScand *trē-hūs* 'wooden house'.

Trull So [*Trendle* 1225 Ass, 1314 Ep, *Trull* 1483 AD]. OE *trendel* 'ring, circle'.

Trumpington Ca [*Trumpintune* c 1050 KCD 907, *-tona* c 1080 ICC, *-tone* DB]. 'The TŪN of *Trump(a)*'s people.' **Trump(a)* may belong to Goth *trimpan*, Engl *tramp*, Sw *trumpen* 'surly', *trumpe* 'surly person' &c. Cf. TRIMPLEY.

Trunch Nf [*Trunchet* DB, 11 Mon v, 49, *Truch* 1203 Cur, *Trunch* 1254 Val]. Perhaps a name transferred from France. LE TRONCHET in Ille-et-Vilaine is *Trunchetum* Hy 2 (1291) Ch, *Tronchetum* 1159–78 Fr. The Abbey had possessions in Norfolk. If Trunch is indigenous in England, it is no doubt of Celtic origin, and may be identical with the second el. of RESTRONGUET Co. For the loss of *-t* cf. PENGE.

Truro Co [*Triuereu*, *Triureu* 12 (1285) Ch, *Triueru* 1195 P]. According to Henderson the first el. is probably Co *tri* 'three', but the rest of the name is obscure.

Trusham (-is-) D [*Trisma* DB, 1291 Tax, *Trisme* 1260 Ep]. Perhaps a derivative of OE *trus*, on which see next name.

Trusley Db [*Trusselai* 1166 RBE, *-lea* 1177, *-lega* 1179 P, *-leia* 12 BM]. 'Wood with fallen leaves and rubbish.' First el. OE *trus* 'fallen leaves' &c. There is some doubt as to the quantity of the *u* of OE *trus*. Mod *trouse* 'brushwood' suggests *ū*, at least as an alternative.

Trusthorpe Li [*Dr(e)uistorp* DB, *Struttorp* 1196 FF, 1202 Ass, *Strustorp* 1231 Ep, *Trustorp* R 1 Cur, 1212 Fees]. '*Strūt*'s thorp.' Cf. STRUBBY. The loss of *S-* is due to dissimilation.

Trym R Gl. See WESTBURY ON TRYM.

Trysull (trēzl) St [*Treslei* DB, *Tresel* 1176 P, *Trisel* 1236 Fees]. An old name of Smestow Brook [*Tresel* 985 KCD (650), 996 Mon vi]. The name is cognate with TEST, Welsh *tres* 'toil, labour'. The exact meaning is obscure. On Smestow Brook is **Trescott** St [(æt) *Treselcotum* 985 KCD 650].

Tubney Brk [*Tubbeneia* W 1 Abingd, 1166 RBE, *Tobenie* DB]. The same first el. is found in *Tubbanford* 942, *Tubbaford* 965 BCS 777, 1169, which must have been near Tubney. The common el. is probably OE *Tubba pers. n., which may be a short form of *Tūnbeorht*. Second el. OE *ēg* 'island'.

Tuckerton So [*Tukerton*, *Tokerton* n.d. Buckland, *Tokertone* 1285 FA]. 'TŪN of the tuckers or fullers.'

Tuddenham, East & North, Nf [*East*, *Nord Tudenham*, *Toddenham* DB, *Tudenham* 1198 FF, *Tuddeham* 1199 P], T~ Sf nr Mildenhall [*Todenham* DB, *Tudeham* Hy 2 BM, *Tudenham* 1235 FF], T~ **St. Martin** Sf nr Ipswich [*Tudenham*, *Todenham* DB, *Tudenham* 1280 FF]. '*Tudda*'s HĀM.'

Tudeley K [*Tivedele* DB, *Tiuedele* 11 DM, *Teudele* 1238 Ep, *-lee* c 1265 Bodl, *Tudely* 1238 Ep]. Second el. OE LĒAH. The first looks like an OE *ifede 'ivy-covered' with *T-* from the prep. *æt*.

Tudhoe Du [*Tudhow* 1279 PNNb], **Tudworth** YW [*Tudeworde* DB, *-worth* n.d. AD]. '*Tudda*'s HŌH or spur of land and WORP.'

Tuesley Sr [*Tiwesle* DB]. 'LĒAH dedicated to the god *Tiw*.'

Tuffley Gl [*Tuffelege* DB, *-leye* 1100, *-leya* 1154–79 Glouc]. '*Tuffa*'s LĒAH.' *Tuffa* is found in DB and may be a short form of *Tūnfrip*.

Tufton Ha [*Tochiton* DB, *Tokinton* 1198 (1260) Ch, *Tokington* c 1270 Ep]. 'The TŪN of *Tucca*'s or *To(c)a*'s people.' Cf. TOCKENHAM. *Tuccingeweg* occurs 901 BCS 596 in bounds of Cranbourne a few miles from Tufton.

Tugby Le [*Tochebi* DB, 1167 P, *Tokebi* 1190 ff. P]. '*Tōki*'s BY.' Cf TOXTETH.

Tugford Sa [*Tuga-*, *Tuggeford* c 1138 Eyton, 1277 Ep, *Tukeford* 1237 FF], **Tughall** Nb [*Tughala* 1104–8, *Tuggahala* c 1150 SD, *Tughal* 1242 Fees]. Apparently '*Tucga*'s ford and HALH'. *Tucga might have developed from **Tud(e)ca*. Cf. BAGINTON, BAGNALL.

Tumby Li [*Tunbi* DB, *Tumbi* c 1115 LiS]. OScand *Tūn-bȳr* 'BY with a *tūn* or fence'.

OE **tūn** originally denoted 'fence' (cf. G *Zaun*) or 'enclosure', but must at an early date have developed the meaning 'enclosure round a house, toft', whence 'homestead', 'village' and 'town'. The meanings 'homestead' and 'village' must have arisen at an early date, as shown among other things by the numerous names in *-tūn*, even *-ingatūn*, found in Normandy and the Boulogne district and apparently due to early Saxon colonization in the district. Saxons are mentioned here by Gregory of Tours in the 6th cent. Some of the names are recorded early, e.g. BAINCTHUN (*Bagingatun* 811), TODINCTHUN (*Totingetun* 807), WADENTHUN (*Wadingatun* 1084). The three names, by the way, have exact counterparts in Engl BAINTON YE, TOTTINGTON, WADDINGTON. GODINCTHUN corresponds to Engl GODDINGTON. There is no reason to doubt that many Engl names in *-tūn* are very old. But *tūn* continued to be in living use as a pl. n. el. till post-Conquest times, and on the whole names in *-tūn* are later than names in *-hām*. On the relations between these two elements see Introd. pp. xiv ff.

Tūn is never found alone as a pl. n., and very rarely as a first el., except in TUNSTALL, TUNSTEAD, and the like, but it is the most common second element. The meaning is doubtless as a rule 'homestead' or 'village'. Many names in *-tūn* denote villages, but many of these may have developed from homesteads. A meaning 'enclosure' is probably found in GARSTON (OE *gærstūn* 'paddock'), DARTON, STAVERTON Np, perhaps partly in LEIGHTON. Cf. STYFIC. In names such as BARTON, SHEPTON, SWINTON the original meaning of *-tūn* may well have been 'outlying, dependent farm, dairy-farm' or the like. Such a meaning is also possible in names like BRINTON, SNEINTON, WINTERTON Li; the places are situated near BRININGHAM, NOTTINGHAM, WINTERINGHAM. The first el. varies a great deal in meaning. It is often a pers. n., and frequently a folkname (especially one in *-ingas*; cf. -ING). It is often a river-name or a topographical word denoting position (as in BROUGHTON, BURTON, EATON, SEATON) or a descriptive common noun (as in ACTON, ASHTON, THISTLETON, THORNTON), or a word denoting a product (as in RYTON, HONINGTON, PLUMPTON), or an adjective or adverb (as HEATON, NEWTON, LITTLETON, UPTON). The

first el. is often a Scand word or pers. n. (as in SKERTON, GRIMSTON, THURGARTON; *-tūn* may here sometimes be OScand *tūn* 'homestead') or even a French word or name (as in CASTLETON, MADJESTON, WATERSTON).

The later form *-town* is rare in pl. ns. (as in NEWTOWN). Such names are late.

Tunbridge Wells K. Named from TONBRIDGE. The medicinal springs are said to have been discovered in the time of James I.

Tunstall Du [*Dunstall* 1196 P, *Tunestel* 1208–10 Fees], T~ K [*Tunestelle* DB, *Tunsteal* 11 DM], T~ La [*Tunstalle* DB, *Tunstall* 1235 FF], T~ Nf [*Tunestalle* DB, *Tunstal* 1196 FF], T~ Sa [*Tunstal* 1327 Subs], T~ Sf [*Tunestal* DB, *Tunstall* 1242 Fees], T~ St, town [*Tunstal* 1212 Fees, *-stall* 1227 Ch], T~ St nr Adbaston [*Tunestal* DB, *Tunstall* 1267 Ch], T~ YE [*Tunestal* DB, *Donestal* c 1100 YCh 1300], T~ YN in Catterick [*Tunestale* DB], T~ YN nr Stokesley [*Ton(n)estale* DB]. OE *tūnstall*, *-steall* 'site of a farm, farmstead'. Identical with DUNSTALL.

Tunstead Db [*Tounstede* 1200–50 Darley], T~ La [*Tunstede* 1324 LaInq], T~ Nf [*Tunstede* 1044–7 KCD 785, *Tunesteda* DB]. OE *tūnstede* 'farmstead'.

Tunworth Ha [*Tuneworde* DB, *-wurda* 1177 P, *Tunnewrthe* 1194 Selborne]. Perhaps '*Tunna*'s WORÞ'. But cf. æt *Tuneweorðe*, *Tunwæorðinga gemære* 957 BCS 994 (Mx), and **Townworth** La [*Tunneworthe* c 1550 WhC], which seem to have as first el. OE *tūn*. *Tūnworþ* might be 'farm with a fence'. If Tunworth is from *Tūn-worþ*, the *-e-* of the early forms is intrusive.

Tupholme Li [*Tupeholm* c 1175, *Topeholm* 12 DC, *Tupholm* c 1175 BM, *Tupeholma* 1209–19 Ep]. 'Rams' island' or '*Tupi*'s island'. *Tup* 'ram' is found from the 13th cent. *Tupi* is an ODan pers. n.

Tupsley He [*Topeslage* DB, *-le* 1241 Hereford]. 'Pasture for rams.' Cf. TUPHOLME.

Tupton Db [*Top(e)tune* DB, *Tuppeton* 1199 P]. First el. as in TUPHOLME.

Turkdean Gl [(on) *Turcandene* 716–43, *Turcadenu* 779 BCS 165, 230, *Turchedene* DB]. 'The valley of the river *Turce*.' The river is referred to as *Turcanwyllas heafod* BCS 165. *Turce* is identical with TWRCH in Wales [*Turc*, *Turch* c 1150 LL], which literally means 'boar'. The name is said to refer to rivers which form deep channels or holes in which they sink into the earth and are lost for a distance.

Turnastone He [*Thurneistun* 1242 Fees, *Turneyston* 1250 Ipm, *Thurneston* 1252 Fees]. The first el. may be OE *þornisc*, perhaps 'thorn brake'; see BCS 1343. The spellings with *ei*, *ey*, however, are curious and may point to OE *þornhege* as first el.·

Turnworth Do [*Torneworde* DB, *Turne-*

worda 1204 (1313) Ch, *Turnewurth* 1234 Cl, 1237 FF, *Thorneworthe* 1316 FA]. OE *þyrne* or *þorn* 'thorn-bush' and worþ. The meaning may be 'enclosure formed by thorn-bushes'.

Turton La [*Turton* 1212 Fees, *Thurton* 1257 Ch]. '*Þori*'s or *Þuri*'s TŪN.' Cf. THORESBY.

Turvey Bd [*Torueie* DB, *Turueia* 1165, *Turfeia* 1176 P]. 'Turf island', i.e. 'island with good grass'.

Turville Bk [*Þyrefeld* 796 BCS 281, *Tirefeld* 1176 P]. Identical with THERFIELD.

Turweston (terstn) Bk [*Turvestone* DB, *Thurveston* 1254 Val]. Identical with THOROTON. Or first el. ODan *Purfastr.*

Tushingham Chs [*Tusigeham* DB, *Tussinhgham* 1260, *Tussingham* 1288 Court]. Possibly 'the HĀM of *Tūnsige*'s people'.

Tusmore O [*Toresmere* DB, *Turesmere* c 1130 Oxf, *Tursmere* 1237 Ep, *Thuresmere* 1242 Fees]. OE *þyrs-mere* 'lake haunted by a giant or demon'.

Tutbury St [*Toteberia* DB, *Totesbery* 1140–50, *-berie* 1141, *Stutesberia* 1139–60 Fr, *Stuteberia* 1176 P, *Tuttebury* 1200 FF]. '*Tutta*'s BURG' or '*Stūt*'s BURG'. Cf. STUCH-BURY. It is difficult to decide if the name originally began in *T-* or *St-*. In the latter case the loss of *S-* is due to Norman influence.

Tutnall Wo [*Tothehel* DB, *Tottenhull* 1262 For]. '*Tutta*'s hill.'

Tuttington Nf [*Totington, Tutintune* 1044–7 Holme, *Tutincghetuna* DB, *Tuttington* 1198 FF, *Tutingeton* 1200 Cur]. 'The TŪN of *Tutta*'s people.'

Tuxford Nt [*Tuxfarne* DB, *Tukesford* Hy 2 (1291) Ch, 1212 Fees, *Tuxford* Hy 2 (1316) Ch, *Tuxeford* 1227 Ep]. The first el. seems to be an early form (*tux*) of *tusk* 'a tuft of rushes' &c. (1530 &c.).

OE twang. See TANG.

Tweed R [*Tuidi fluminis* c 730 Bede, *Tuidon* (*Tuéode, Tuede*) *stream* c 890 OEBede, *Twtode* c 1000 Saints, *Tuida, Tweoda* c 1050 HSC, *Tweda* 1104–8 SD]. Very likely cognate with *Touésis* Ptol, an old name for the SPEY in Scotland or according to some referring to the Tweed itself. This name belongs to the root *tevā* 'to swell, be powerful' in Sanskr *tavás* 'powerful'. **Tweedmouth** (-ĕd-) is *Tuedemue* 1208–10 Fees.

Twemlow Chs [*Tuamlawe* 13 BM, *Twemelawe* 1259 Court, *Twamlawe* c 1210, *Tuamlowe* 1283–8 Chester]. OE *be twǣm hlāwum* 'by two hills'.

Twerton So [*Twertone* DB, *Twyuerton* 1225 Ass, *Twiverton* 1236 FF]. Identical with TIVERTON D.

Twickenham Mx [*Tuican hom, Tuiccanham* 704 BCS 111, *Tuicanham,* (in) *Tuican-*

hamme 793 ib. 265, *Tuuiccanham* 941 BCS 766; in the 17th cent. and later often *Twittenham*]. Generally explained as '*Twicca*'s HAMM'. *Twicca* is unrecorded, but may well have existed. But the situation of the place in a tongue of land between the Thames and the Crane suggests that the first el. may be an unrecorded OE *twicce, related to *twicen* 'fork of roads' and meaning 'river fork'. *Twicen* presumably comes from an adj. *twic* 'double', related to MHG *zwic* 'a peg'.

Twigmore Li [*Twigemor, Twiggemore* 1202 Ass], **Twigworth** Gl [*Twigeworth* 1220 Fees, *Twiggeworth* 1251 Ch]. The first el. may be OE *twigge* 'twig, branch' in some sense. Twigworth might be 'enclosure made of twigs'. But OE *Twicga* pers. n. occurs.

Twineham Sx [*Tuineam* W 2 PNSx, *Tuynhe* 1226 FF, *Twynem* 1242 Fees]. OE *betwēon ēam* '(the place) between the streams'. The name is elliptical. Identical in origin with **Twinham,** the old name of Christchurch Ha [(æt) *Tweoxneam* 901 ASC, (at) *Twynham* 939 BCS 738, *Thuinam* DB].

Twinstead Ess [*Tumesteda* DB, *Tuinested* 1201 Cur, *Twinsted* 1203 FF]. The first el. appears to be OE *twinn* 'double', but the exact meaning of the name is not clear. See STEDE and cf. TYTHROP.

OE twisla 'fork of a river, land in such a fork' is found in some names. See TWISTLETON &c. Second el. in ENTWISLE, EXTWISTLE, HALTWHISTLE, OSWALD-, TINTWISTLE.

Twistleton YW [*Thwisilton* 1208 FF, *Tweselton* 1297 Subs], **Twiston** La [*Tuisleton* 1102, *Twisleton* c 1140 LaCh]. TŪN in the fork of a river.' Cf. TWISLA.

Twitchen D [*Twechon* 1442 Pat, *Twycchyn* 1524 Subs]. OE *twicen* 'fork of a road'.

Twizel Castle Nb [*Tuisele* 1208–10 Fees], **Twizell** Nb [*Twisle* c 1050 HSC, *Tuysil* 1242 Fees]. OE TWISLA 'fork of a river'. Twizel is in a tongue of land formed by the Till and the Tweed.

Twycross Le [*Tvicros* DB, *Tuicros* 12 DC]. 'Double cross', perhaps one with four arms, showing the way at a cross-roads.

Twyford Bk [*Tveverde, Tuiforde* DB, *Tuiford* 1163 P], **T~** Brk [*Tuiford* 1170 P], **T~** Db [*Tviforde* DB, *Tuiford* 1206 Cur], **T~** Ha [*Tuifyrde* c 960 BCS 1158, *Tviforde* DB], **T~** Le [*Tuiuorde* DB, *Tuiford* 1190 P], **T~** Li [*Tuiforde* DB, *Twiford* 1206 Ass], **T~** Mx [*Tveverde* DB, *Tuiferde* 12 StPaul], **T~** Nf [*Twyford* 1254 Val]. 'Double ford', either one over a river that had two arms or perhaps a place where there were two fords side by side in the same river. The base is in several cases OE *twifyrde*, the earliest example of which is Bede's *Adtuifyrdi*, rendered 'ad duplex vadum'. Here *-fyrdi* may be a neuter *ja*-derivative from *ford* or a locative in OE *-i*.

Twyning (-ĭ-) Gl [*Bituinæum* 814 BCS 350, *Tuninge, Tveninge* DB]. The original name was identical with TWINEHAM. The later form is a derivative with the suffix *-ingas*, meaning 'the people of *Bituinæum*'.

Twywell Np [*Twiwel* 1013, *-well* c 1025 KCD 1308, 1329, *Teowelle, Tviwella* DB, *Twiwell* 1154–69 NpCh]. 'Double stream.'

Tyburn (-ĭ-) Mx [*Tiburne* DB, *-burn* 1275 RH]. Originally a stream-name [*Teobernan, -burnan* (obl.) 959 BCS 1048, 1351]. 'Boundary stream.' Cf. TEFFONT.

Tyby Nf nr Guestwick [*Tytheby* DB]. Cf. TIBY in Sweden [*Tidhæby* 1309]. '*Tidhe*'s BY.'

Tydd St. Giles Ca [*Tit* c 1165 NpCh, *Tid* 13 Fees, *Tyde* 1268 Ch], **T~ St. Mary** Li [*Tite, Tid* DB, *Tit* 1094 Fr, *Tid* 1168 P, 1202 Ass, 1212 Fees, 1257 Ch]. OE *titt* 'a teat', here used in a transferred sense of a slight hill. The Tydds are in low-lying country, but there is a slight rise nr Tydd St. Mary. Cf. *tid* 'a small cock of hay' (Li), *tid, tit* 'teat, udder' (Li, Np, Y &c.).

Tyldesley (tĭldzlĭ) La [*Tildesleia* c 1210 CC, *-le* 1212 Fees]. '*Tilwald*'s LĒAH.'

Tylerhill K [*Teghelerehelde* 1363 BM]. 'The tilers' slope' (OE *helde*).

Tyley Bottom Gl [(on) *tigel leage* 940 BCS 764]. Cf. TILEY.

Tyne R Nb, Du &c. [*Tina* c 150 Ptol, *Tina, Tinus* c 730 Bede, 1104–8 SD, *Tine* 875 ASC, c 890 OEBede, c 894 Asser, *Tinam Australem* c 1130 SD]. Identical with TYNE in Scotland. The river-name probably means 'river' and is derived from the root *ti-* 'to dissolve, flow' in TILL, OE *þinan* 'to dissolve'. It is evidently British. **Tynedale** is *Tindala* 1158 P. Cf. TINDALE.

Tyneham Do [*Tigeham, Tingeham* DB, *Tigeham* 1185, *Tiham* 1194 P, *Tynham* 1280 Ch]. The first el. might be an OE **tige* 'goat', corresponding to OHG *ziga*, G *Ziege*, and related to OE *ticcen*.

Tynemouth (tĭn-) Nb [(æt) *Tinan muþe* 792, (towardes) *Tine muðan* 1095 ASC (E)]. 'The mouth of R TYNE.'

Tyringham (tĭ-) Bk [*Telingham, Tedlincham* DB, *Tiringeham* 1186 P]. 'The HĀM of *Tidhere*'s or *Tir(a)*'s people.' Cf. TERRINGTON Nf.

Tyrley St [*Tirelire* DB, *Tyrle* (wood) 1247 Ass, *Tyrlegh* 1283 Cl]. 'LĒAH on R TERN.'

Tysoe Wa [*Tiheshoche* DB, *Tiesoch* Hy 1, *Thiesho* 1131–40 BM, *Tisho* 1201 Cur]. OE *Tiges hōh* 'spur of land dedicated to the god *Tig* (*Tīw*)'. OE *Tīw* sometimes has the gen. *Tiges*. The same may be the first el. of *Tyesmere* 849 BCS 455 (Wo) and (on) *Tislea* 1023 KCD 739 (Ha). The etymology suggested is rendered likely by the fact that at Tysoe was a cut figure of a horse, after which the Vale of the Red Horse was named. The horse may have been a monument to a victory won by the Anglo-Saxons dedicated to the war-god.

Tythby (tĭdhbĭ) Nt [*Tiedebi* DB, *Titheby* c 1190 Middleton, *Tytheby* 1242 Fees]. Identical with TYBY.

Tytherington Chs [*Tederinton* c 1250, 1258–91, *Tiderton* 1250–88 Chester, *Tyderington* 1285, *Tuderyngton* 1288 Court], **T~** Gl [*Tidrentune* DB, *Tiderinton* 1193 P, *Tidrinton* 1220 Fees], **T~** W nr Heytesbury [*Tuderinton* 1242 Fees, 1282 Cl], **Tytherton Kelways & Lucas** W [*Tedrintone, Tedelintone* DB, *Tidrinton* 1195 Cur, *Tuderington* 1202, *Tiderington* 1227 FF, *Tuderinton* 1242 Fees, *Tuderyngton Caylewey, Lucas* 1428 FA]. Partly 'the TŪN of *Tidhere*'s people', but this cannot well be the origin of all the names. Some forms point to a base with OE *y* in the first syllable. Possibly there was an OE nickname derived from OE *tiedre* 'fragile, weak' or from OE *tŭddor* 'progeny'. Tytherton Kelways is now KELLAWAYS (q.v.).

Tytherton Kelways was held by Elyas de Cailleway before 1227 (FF). Cf. KELLAWAYS.—**T~ Lucas** was held by Richard Lucas in 1202 (FF).

Tytherleigh D in Chardstock [*Tiderlege* Hy 2 PND, *Tuderlege* c 1201 Salisbury, *Tyderlegh* 1255 FF], **East & West Tytherley** Ha [*Tiderlege, -lei, Tederleg* DB, *Tederlea* 1168 P, *Estuderlegh* 1291 Tax, *Westiderlega* 1219 Fees]. 'Young wood.' First el. OE *tiedre* 'weak, fragile'. Cf. LĒAH.

Tytherton. See TYTHERINGTON.

Tythrop O [*Duchitorp* DB, *Twytrop* 1242 Fees, *-throp* 1255 RH, *Tythrop* 1384 AD]. OE *twi-þrop* 'double homestead'.

Tywardrea·th (-ĕth) Co [*Tiwardrai* DB, *Tywardrait* 1138–55 Ep, *Tywardraith* 1235 Ch]. 'House on the sands.' The elements are Co *ti* 'house', *war* 'on', and *traith* (= Welsh *traeth*) 'the sandy beach of the sea, sands'.

U

Ubbeston Sf [*Upbestuna* DB, *Ubbestun* Hy 2 BM, *-ton* 1206 FF]. '*Ubbi*'s TŪN.' First el. ON, ODan *Ubbi*, OSw *Ubbe* pers. n.

Ubley So [*Tumbeli* DB, *Ubbele* 1213, *-leia* 1223 FF]. '*Ubba*'s LĒAH.'

Uckerby YN [*Ukerby* 1198 Fount M, *Huckerby* 1219 FF]. The first el. is no doubt an OScand pers. n., e.g. an unrecorded *Ūkyrri* from *úkyrr* 'restless' (cf. ON *Kyrri* from *kyrr* 'quiet') or *Ūt-Kári* (from *út* 'out' and *Kári* pers. n.), analogous to ON *Ūt-Steinn.*

Uckfield Sx [*Uckefeld* 1220 P, *Ukkefeld* ib., 1248 Misc]. '*Ucca*'s FELD.' *Ucca* is found as a byname BCS 1132 f. and corresponds to OG *Ucco.*

Uckington Gl [*Hochinton* DB, *Uchintone* 1221 Ass], U~ Sa nr Wroxeter [*Uchintune* DB, *Ukington* 1199 (1285) Ch]. 'The TŪN of *Ucca*'s people.'

Uddens Do [*æt Udding(c*) 956 BCS 958, *Uddyng* 1331 Misc]. Possibly '*Udda*'s place'. OE *Udda* is found in *Uddanhom* 843 BM.

Udimore (ŭ-) Sx [*Dodimere* DB, *Hudimere* 1197 FF, *Odimere* 1249 FF]. Either '*Uda*'s boundary' or 'the boundary of the wood'. Second el. OE *gemǽre* 'boundary'. If the first is OE *wudu* 'wood', ODIHAM may be compared.

OE **uferra** 'upper'. See OFER.

Uffculme. See CULM.

Uffington Brk [*Uffentun, Offentona* c 931 BCS 687, *Offentone* DB], U~ Li [*Offintone* DB, *-tona* 1114–16 RA, *Uffinton* 1225 Ep, *Offington* 1254 Val], U~ Sa [*Ofitone* DB, *Uffitun* 1177 P, *Offinton* 1255 RH]. '*Uffa*'s TŪN' or 'the TŪN of *Uffa*'s people'.

Ufford Np [*Uffawyrða gemǽre* 948 BCS 871, *Uffewrð* 1199 FF, *-wurth* 1202 Ass], U~ Sf [*Uffeworda* DB, *Ufford* 1195 P]. '*Uffa*'s WORÞ.'

Ufton Nervet Brk [*Offetune* DB, *Uffeton* 1199 FF]. '*Uffa*'s TŪN.'

Nervet is a family name, originally a nickname, appearing as *Neirenuit, Neire Nuit* 1207 Cur, *Nigranox* 1236 Fees. Richer Neyrnut held the manor in 1242 (Fees). *Nervet* is due to misreading of *n* as *u.* The name means 'black night'.

Ufton Wa [*Hulhtune* 1043 Th, *Olufton* c 1050 KCD 939, *Ulfton* 1043 (1267) Ch, *Ulchetone* DB, *Hulugton* 1221 Pp, *Oloughton* 1291 Tax]. The first el. is no doubt correctly identified in PNWa(S) with that of *wulluht graf* 1001 KCD 705, which apparently represents the present Ufton Wood. OE *wulluht* appears to be a derivative with the suffix *-uht* (cf. ON *-óttr*, OG *-oht*) of

OE *wulle* 'wool'. Possibly the word was used as the name of certain plants such as cotton-grass. *Woolgrass, -thistle, -weed* are found in later English. *Wulluht* would then mean 'abounding with woolgrass' or the like.

Ugborough D [*Ulgeberge* DB, *Uggabergh* 1242 Fees, *Oggeberg* 1263 Ipm], **Ugford** W [*Ucganford* 958 BCS 1030, *Uggafordinga landscore* 1045 KCD 778, *Ogeford* DB, *Uggeford* 1195 FF], **Uggaton** Wt [*Ugelton* 1201 Cur, *Uggeton* 1287–90 Fees]. '*Ugga*'s hill, ford, and TŪN.' OE *Ugga (Ucga)* pers. n. is not recorded, but must have existed.

Uggeshall Sf [*Uggiceheala, Ugghecala, Wggessala* DB, *Uggecala* c 1095 Bury, *Huggechale* 1242 Fees, *Ugechale, Ugeshale* 1254 Val]. '*Uggeca*'s HALH.' *Uggeca* is a derivative of *Ugga.* See UGBOROUGH.

Ugglebarnby YN [*Ugleberdesbi* DB, *Ugelbardeby* 1100–15 YCh 857]. '*Uglubarði*'s BY.' *Uglubarði* is an unrecorded ON byname, composed of *ugla* 'owl' and *Barði* pers. n.

Ughill YW [*Ughil* DB, Hy 3 BM], **Ugley** Ess [*Ugghelea* DB, *Uggelegh* 1238 Subs, *-le* 1274 FF]. '*Ugga*'s hill and LĒAH.' Cf. UGBOROUGH.

Ugthorpe YN [*Ughetorp, Ugetorp* DB, *Uggathorp* 1157 YCh 186]. '*Ugga*'s or *Uggi*'s thorp.' Cf. UGBOROUGH. *Uggi* is a known ON name.

Ulceby Li nr Barton [*Ulvesbi* DB, *Ulesbi* c 1115 LiS, *Ulsebi* 12 DC], U~ Li nr Well [*Ulesbi* DB, c 1115 LiS, *Ulseby* 1201 Cur]. OScand *Ulfs-býr* '*Ulf*'s BY'. First el. ON *Ulfr*, OSw *Ulver*, ODan *Ulf.*

Ulcombe K [*Ulancumbe* 946 BCS 811, *Olecumbe* DB, *Ulecumbe* 1212 RBE]. 'The valley of the owl' (OE *ūle*).

Uldale Cu [*Ulvedale* 1230 Sc, 1300 Ipm]. OScand *ulfa-dalr* 'wolves' valley'.

OE **ūle** 'owl' is the first el. of OLDCOATES Nt, OUTCHESTER, ULCOMBE, ULGHAM, ULWHAM.

Uley Gl [*Euuelege* DB, *Iwele* c 1180 Berk, *Iwelega* 1232 Ch]. 'Yew wood' (OE *iw-lēah*).

OScand **ulfr** (ON *ulfr*, OSw *ulver*, ODan *ulv*) 'wolf' is the first el. of some names, as ULDALE, ULLOCK, ULPHA Cu. It is sometimes difficult to distinguish from *Ulfr* pers. n. Sometimes *ulf* has replaced the synonymous OE *wulf.* Cf. ULLEY.

Ulgham (ŭf-) Nb [*Wlacam* 1139 Newminster, *Ulweham* 1242 Fees, *Ulcham* 1251, *Ulgham* 1290 Ch, *Howltham* 1323 Ipm], **Ulwham** Nb in Featherstone [*Ulgheham* 1479 BBH]. 'Owl valley or nook.' The elements are OE *ūle* 'owl' and *hwamm* 'corner, angle'.

Ullenhall Wa [*Holehale* DB, *Ulenhala* 1187 P, *Olenhal* 1242 Fees]. '*Ul(l)a*'s HALH or valley.' Cf. OLDBERROW.

Ulleskelf YW [*Oleschel* DB, *Ulfscelf* 1226 FF, *-skelf* 1235 FF]. '*Ulf*'s bank.' Cf. ULCEBY, SKIALF (under SCYLF).

Ullesthorpe Le [*Ulestorp* DB, *Olestorp* 1190 P, *Olvestorp* 1278 Ipm]. OScand *Ulfs-þorp* '*Ulf*'s thorp.' Cf. ULCEBY.

Ulley YW [*Ollei* DB, *Ulflay* 1242 Fees]. OE *wulf-lēah* 'wolf wood' with loss of *w*-owing to Scand influence. Cf. WOOLLEY.

Ullingswick He [*Ullingwic* DB, *Olinge-wiche* 1127 AC, *Willyngwyke* 1167, *Wylyng-wyche* c 1200 Glouc]. 'The wīc of *Willa*'s people.'

Ullock Cu nr Keswick [*Great Ulfelayth* 1235 FF, *Ullaik* 1332 Subs], U~ Cu SE. of Work-ington [*Ulnelayke* 1248, *Ulvelayk* 1272 StB]. ON *ulfa-leikr* 'wolves' play, place where wolves play'. Cf. DEERPLAY (under PLEGA). See ULFR. Second el. ON *leikr* 'play'.

Ullswater, lake, Cu, We [*Ulneswater* 1292 QW, *Ulveswatre* 1323 Ipm]. '*Ulf*'s lake.' Cf. ULCEBY.

Ulnes Walton. See WALTON.

Ulpha Cu [*Ulfhou* 1337 Ipm]. OScand *ulf-haugr* 'wolf hill'. Cf. ULFR, HAUGR. *Wolfhou* 1279 Ass is probably anglicized from OScand *Ulfhaugr*.

Ulpha We [*Ulvay*, *Uluay* 1420 Kendale]. Perhaps OE *wulf-hege* 'enclosure for trap-ping wolves', with loss of *w* owing to Scand influence.

Ulrome YE [*Ulfram*, *Ulreham* DB, *Ulram* c 1170 YCh (826), 1226 FF]. '*Wulfhere*'s or rather *Wulfwaru*'s HĀM' with loss of *W*-owing to Scand influence. *Wulfwaru* is a woman's name.

Ulting Ess [*Ultinga* DB, *Hultinges* 1166 RBE, *Ultinges* ib., 1226 Cl]. 'The people on the river **Ult*.' Ult may be an old name of the Chelmer, identical with L'OUST in Brittany [*Ult* 834].

Ulverscroft Le [*Ulvescroft* 1174 BM, *Ulues-croft* 1196 FF]. '*Ulf*'s croft.' Cf. ULCEBY.

Ulverston (ōōstn) La [*Ulurestun* DB, *Ul-verston* 1180–4 LaCh]. '*Wulfhere*'s TŪN' with loss of *W*- owing to Scand influence, or '*Ulfar*'s TŪN', the first el. being ON *Ulfarr* pers. n.

Ulwham. See ULGHAM.

Umberleigh D [*Umberlei* DB, *Womberlegh* 1284–6 FA, 1322 Misc]. 'LĒAH on R **Win-burna*.' The stream at the place seems to have had a name identical with WIMBORNE.

Uncleby YE [*Unchelsbi*, *Unglesbi* DB, *Hun-kelbi* c 1090, *Hunchilebi* 1157 YCh 350, 354]. The first el. is an OScand pers. n., identical with *Hunchil* DB (perhaps from an unrecorded OScand **Hūnketill*), or *Un-*

ketel t Hy 2 (IPN 185), RH (? a form of *Ásketill*, in Engl sources also *Anketill* &c.; cf. OE *Unlaf* by the side of *Anlaf*, from ON *Ólāfr*).

OE under adv., prep. 'under, below' is found in several elliptical names, as UNDER-BARROW, generally denoting a place situated at the foot of a hill. In UNDERWOOD *under* means 'within' (a forest), the literal meaning being probably 'below the trees of a forest'. Both senses are often found where *under* is used in distinctive additions, as ASHTON UNDER HILL, THORPE UNDER STONE, ASCOT UNDER WYCHWOOD, HEATON UNDER HOR-WICH. In UNDERMILLBECK *under* seems to mean 'south of'.

Underbarrow We [*Underbarroe* 1517 Ken-dale]. '(Place) at the foot of Helsington Barrows' (a high hill, called *Le Bergh* 1332 Kendale; cf. BEORG).

Underley Hall We [*Underlai* 1282 Ken-dale]. Possibly identical with Underly He in Wolferlow.[*Hunderlithe* 1242 Fees, *Unde-lich* 1316 FA]: '(place) at the foot of the slope' (OE HLIÞ).

Undermillbeck We [*Undermylnebek* 1390–4, *Under Milnbek* 1442 Kendale]. '(Place) below, i.e. south of Millbeck.' Millbeck R [*Mulnebec* 1220–46, *Milnebek* 1442 Ken-dale] means 'mill stream'.

Underriver K [*sub le Ryver* (pers. n.) 1477 Will]. OE *under yfre* 'below the hill'. Cf. RIVER Sx. The place is below River Hill.

Underskiddaw Cu. '(Place at the foot of SKIDDAW.'

Underwood Db [*Hunderwude* J Derby, *Underwode* E 1 BM, 1287 FF], U~ Nt [*Underwood* 1490 AD i]. '(Place) within the wood.' Cf. UNDER.

Unstone Db [*Hones-*, *Onestune* DB, *Oneston* Hy 3 BM, 1263 FF, *Onistone*, *Honeston* E 1 Derby]. Perhaps '*Ōn*'s TŪN'. *Ōn* would be a strong side-form of *Ōna*.

Unsworth La [*Hundeswrth* 1291 Ch, 1292 QW, *Undesworth* 1322 LaInq]. '*Hund*'s WORÞ.' Cf. HOUNSLOW.

Unthank Cu nr Dalston [*Unthank* 1332 Subs], U~ Cu nr Gamblesby [*Unthanke* 1332 Subs], U~ Nb nr Alnham [*Unthanc* 1207 Cur, *Unthank* 1242 Fees], U~ Nb nr Haltwhistle [*Unthanc* c 1200 Abbr]. There are other examples of the name in Cu, Nb and YN. The name belongs to OE *unþances* 'without leave' and refers to a squatter's farm.

Uny Lelant. See LELANT.

OE up, upp(e) adv. 'up, above'. In UPTON and the like *Up-* means 'upper' and indi-cates a situation higher than neighbouring places or sometimes perhaps higher up stream. Names such as UPAVON, UPLEADON, where the second el. is a river-name, are elliptical and mean '(place) higher up the

AVON (LEADON &c.)'. In UPHILL the first el. may be OE *uppan* prep. 'above'.

Upavon. See AVON.

Upchurch K [*Upcyrcean* 11 DM, *Upchirche* 1241 Fees], **Upham** Ha [*Upham* 1201 Cur, 1284 Ch, *Uppham* 1291 Tax]. 'Upper church and HĀM.'

Uphill So [*Opopille* DB, *Uppepull* 1197 Bruton, *Uppehill* 1176 P]. Apparently OE *uppan pylle* '(place) above the pill or creek'. The place stands on the lower Axe.

Upholland, Upleadon. See HOLLAND La, LEADON.

Upleatham YN [*Upelider* DB, *Uplithum* c 1150 Whitby, 1272 Ipm]. 'Upper slopes.' Cf. KIRKLEATHAM. U~ is higher than Kirkleatham. *Upelider* DB seems to be a Scandinavianized form, ON *Upphliŏir*.

Uplitherland, Uplowman, Uplyme. See LITHERLAND, LOMAN, LYME.

Upminster Ess [*Upmynster* 1062 Th, *Upmunstra* DB]. 'Upper church.' Cf. MYNSTER.

Upottery. See OTTER.

Upperby Cu [*Hobrihtebi* 1164 P, *Hobriteby* 1200 Fr]. '*Hūnbeorht*'s BY' (with loss of *n* before *b*) or '*Hubert*'s BY', the OFr name having been anglicized.

Upperthong YW. See THONG.

Upperthorpe Li [*Hubaldestorp* DB]. '*Hūnbald*'s thorp.'

Uppingham Ru [*Yppingeham* 1067 BM, *Uppingeham* 1080–7 Reg, 1167 P]. 'HĀM of the people on the hill.' First el. OE *Yppingas* or *Uppingas*, derived from UPP or YPPE.

Uppington Sa [*Uppinghǽma gemǽra* 975 BCS 1315, *Opetone* DB, *Upton* 1251 Cl, *Oppinton* 1195 P, *Uppinton* 1237, 1245 FF]. Originally OE *Upp-tūn* 'upper TŪN', later changed to *Upping(a)tūn* 'TŪN of the *Uppingas* or Upton people'.

Upsall YN in S. Kilvington [*Upsale* DB], U~ YN nr Middlesbrough [*Upesale* DB, *Upsale* c 1185 YCh 758]. OScand *Uppsalir* 'higher homestead', the source of the common Norwegian pl. n. OPSAL. The second el. is OScand *salr* (plur. *salir*) 'hall, homestead, dwelling'.

Upsland YN [*Upsale* DB, *Upselun* 1218 FF, *Upsalund* 13 Fount]. The original name was UPSALL, the addition being OScand *lundr* 'grove'.

Upton, a common name, nearly always goes back to **1.** OE *Upp-tūn* 'higher TŪN': U~ Bk [*Upetone* DB], U~ Brk [*Optone* DB, *Upton* 1220 Fees], U~ Chs nr Macclesfield [*Opton* 1285 Court, *Upton Superior* 1315 BM], U~ **by Birkenhead** Chs [*Optone* DB, *Uptone in Wyrhale* 1307 Ipm], U~ **by Chester** Chs [*Huptun* 958 BCS 1041, *Optone* DB, *Uptuna* c 1125 Chester], U~ D

nr Kingsbridge [*Uppeton* 1242 Fees], U~ **Hellions** D [*Uppetone Hyliun* 1270 Ep], U~ **Pyne** D [*Opetone* 1264, *Uppetone Pyn* 1283 Ep], U~ or Hawkesbury U~ Gl nr Chipping Sodbury [*Uptun* 972 BCS 1282], U~ Gl nr Tetbury [*Opton* 1236 Fees], U~ **Cheney** Gl [*Upton* 1208 Cur, 1313 Ch], U~ **St. Leonards** Gl [*Optune* DB, *Uptone Sancti Leonardi* 1310 Glouc], U~ Ha N. of Andover [*Optune* DB], U~ **Grey** Ha [*Upton Grey* 1281 Cl], U~ He in Brimfield [*Upetone* DB], U~ **Bishop** He [*Uptune* DB, *Opton Episcopi* 1291 Tax], U~ Hu [*Opetune* DB], U~ La [*Upton* 1251 Ch], U~ Le [*Upton* 1196 P, 1209–35 Ep], U~ Li [*Opetune* DB, *Uppetune, Uptuna* c 1115 LiS], U~ Nf [*Uptune* DB, *Uppeton* 1165 P], U~ Np nr Ailsworth [*Uptun* 948 BCS 871, *Upton* DB], U~ Np nr Northampton [*Optone* DB, *Oppetona* c 1125 Oxf, *Uppetona* 1175 BM], U~ Nt nr E. Retford [*Upetun* DB], U~ Nt nr Southwell [*Uptun* 958 YCh 2, *Opetone* DB], U~ O [*Optone* DB], U~ **Cressett** Sa [*Ultone* BD, *Upton* 1242 Fees], U~ **Magna** Sa [*Uptune* DB, *Upton Magna* 1291 Tax], **Waters** U~ Sa [*Uptone* DB, *Upton Waters* 1346 FA], U~ So nr Dulverton [*Upton* 1225 Ass], U~ **Noble** So [*Opetone* DB, *Upton le Noble* 1291 FF], U~ W nr Hindon [*Uppeton* 1242 Fees], U~ **Scudamore** W [*Uptun* c 990 Wills, *Opetone* DB, *Uppton* 1242 Fees, *Upton Escudemor* 1267 Ch], U~ Wa nr Alcester [*Optone* DB, *Upton* 1428 FA], U~ Wa nr Kineton [? *Optone* DB, *Upton* 1236 Fees], U~ **on Severn** Wo [*In Uptune* 897, *Uptún* 962 BCS 575, 1088, *Upton* DB], U~ **Snodsbury** Wo [(in) *Snoddesbyri* 972 BCS 1282, *Snodesbyrie* DB, *Upton* 1221 Ass, *Snodisbur' Upton* 1287 Misc], U~ **Warren** Wo [*Uptona* 716 BCS 134, *Uptune* DB, *Upton Waryn* 1291 Tax], U~ YE [*Uptun* DB], U~ YW [*Uptone* DB, *Opton* 1218 FF].

2. Upton Ess in W. Ham [*Hupinton* 1203 FF, *Uptoun* 1496 BM]. OE *upp in tūne* 'up in the village, in the upper part of the village'. Cf. SIDDINGTON Gl and Introd. p. xviii.

3. Upton Lovell W [æt *Ubbantune* 957 BCS 992, *Ubeton* 1200 Cur, *Ubbedon Lovell* 1476 Ipm]. '*Ubba*'s TŪN.'

Upton Bishop He belonged to the Bishop of Hereford.—U~ **Cheney** Gl. Cf. CHENIES.— U~ **Cressett** Sa came to Thomas Cressett by marriage (13th cent.). Cressett is a French family name, originally a nickname, perhaps derived from OFr *crais* 'fat'.—U~ **Grey** Ha was acquired by John de Grey (d. 1271) and was held by Richard de Grey in 1334 (Ch). Cf. EASTON GREY.—U~ **Hellions** D was held by William de Helihun in 1242 (Fees). Cf. BUMPSTEAD HELION.—U~ **Lovell** W came to the Lovells about 1 R 2. Cf. LILLINGSTONE LOVELL.— U~ **Noble** So from a local family.—U~ **Pyne** D was held by Herbert de Pyn in 1264 (Ep). Cf. COMBPYNE.—U~ **St. Leonards** Gl from the dedication of the church.—U~ **Scudamore** W was given to Godfrey Escudamore shortly after 1150. Scudamore is a French family name.—U~ **Snodsbury** Wo was originally

Snoddesburg 'Snodd's BURG'; cf. SNODDINGTON. The pers. n. Snodd is found in Snoddeslea 840 BCS 428 (in boundaries of Crowle, which adjoins Upton Snodsbury).—U~ **Warren** Wo was held by William son of Warin in 1242 (Fees). Cf. GRENDON WARREN.—**Waters** U~ Sa was named from Walter Fitz-John, who died in 1201 (Eyton).

Upway Do. See WEY.

Upwell Ca, Nf [æt Uuyllan 970 BM, Wella 1147 BM, Upwell 1251 Ch, 1254 Val]. 'Upper Well.' Cf. OUTWELL.

Upwood Hu [Upwude 974 BCS (1311), 1253 BM]. 'Upper wood.'

Urchfont W [Ierchesfonte DB, Erchesfonta 1176, Archesfunte 1180 P, Urichesfunte 1242 Fees]. Second el. OE FUNTA 'spring'. The first is a pers. n., perhaps OE *Eohrīc, corresponding to ON Iórekr.

Ure R YN [Earp c 1000 Saints, Jor c 1140 Pat, c 1180 BM, Yor c 1190 YCh 797, Yore 1276 RH, Ure c 1540 Leland]. Probably an OBrit Isurā, identical with Gaul Isura, now ISAR, a trib. of the Danube, and related to Isarā (cf. AIRE). From Isurā was derived Isurion, the Brit name of Aldborough, a Roman fort on the Ure [Isoúrion c 150 Ptol, Isurium 4 IA]. Intervocalic s was dropped in later British, and Isurā would give OE Ior, Eor, whence ME Yōr. Earp in Saints is probably due to a mistake. See ERN.

Urishay He [Haya Hurri 1242 Fees, Hay Urry 1325 AD iv]. 'Urri's forest enclosure.' Urri is identical with Urricius (Roger fil. Urricii) 1150–4 Hereford. One Urri the Engineer is also called Wlricus Balistarius 1233 (Fees). See (GE)HÆG.

Urmston La [Urmeston 1212 Fees, Ormeston 1284 Ass]. 'Urm's TŪN.' Urm is ODan Urm, a side-form of ON Ormr, OSw Ormer.

Urpeth Du [Urpathe 1297 Pp]. 'Bison path.' First el. OE ūr 'bison'. Cf. URSWICK.

Urswick, Great & Little, La [Ursewica c 1150 FC, -wic 1194 BM, -wik 1269 Ass, Magna Urswic c 1185 FC, Parva Urswik 1257 Ass]. The place stands at Urswick Tarn, which was no doubt called OE Ūrsǽ 'bison lake'. The name means 'wīc by Ūrsǽ'. OE ūr would here be used of wild cattle.

Ushaw Du [Ulveskahe 12 Finchale, Uues- shawe 1312 RPD]. 'Wolves' shaw or wood', OE Wulf(a)-scaga, with loss of W- owing to Scand influence.

Usk, Welsh **Wysg,** R Monm [Uisc, Uysc, Huisc c 1150 LL, Wysc 13 Mab, (on) Wylisce Axa 1050 ASC (D), Huscha DB, Uske 1205 Lay]. Identical with ESK, EXE, &c. The OBrit form of the name, Isca, is found denoting CAERLEON (q.v.). Iscā became Escā owing to ā-affection and e- was lengthened, becoming regularly later Welsh wy [ui]. Engl Usk is due to substitution of u for wy. From the river was named the priory and town of Usk [Usk 1131 Mon, Usca, Uska c 1150 BM].

Usselby Li [Osoluabi, Osoluebi c 1115 LiS, Osolfby 1209–35 Ep]. 'Ōswulf's BY', but very likely Oswulf is here an anglicized form of ODan Asulf, ON Ásulfr.

Usworth Du [Useworth 1183 BoB, Oswrde c 1190 Godric]. 'Ōsa's WORÞ.'

OE **ūt** 'out', **ūterra** 'outer'. Cf. OUTWELL, OWTHORNE, UTTERBY.

Utkinton Chs [Utkynton 1303 ff. Chamb, 1358 Ormerod]. 'The TŪN of Uttoc's people'; cf. Uttokishal 1289 Court (Chs). *Uttoc is a diminutive of Utta.

Utterby Li [Uttrebi 1197 P, -by 1209–35 Ep]. 'Outer BY.' Originally no doubt OScand Ytri-býr, with OE ŪTERRA substituted for the OScand word.

Uttoxeter (ŭksĕter, ŭtshĕter) St [Wotoches- hede DB, Uttokishedere 1175 P, Wittokes- hather 1242 Fees, Uittokesather, Huttokes- ather 1251 Ch]. The first el. is a pers. n. *Wittuc (cf. WIXHILL Sa), a derivative of Witta. There seems also to have been a side-form Wuttuc (from Wiuttuc < Wittuc). The second el. appears to be the word heather, here used in the sense 'heath'. Cf. HEATHER. But possibly the second el. was originally hǽþærn 'house on the heath'. For loss of final -n, cf. SEASALTER.

Uxbridge Mx [Uxebregg 1200 P, Woxebruge 1219, Wyxebrigge 1220 FF, Wuxebrug 1242 Cl, Wixebrige 13 BM], **Uxendon** Mx nr Harrow [Woxindon 1258 FF, Woxedone 1290 Ep]. The first el. of these is the tribal name Wixan 7 Trib Hid (BCS 297). This comes from Wihsan and is related to Goth weihs 'village', the OHG pl. n. Wihsa (now WIECHS &c.), Lat vicus &c. Wihsan in the Middlesex dialect became Wyhsan, Wuhsan and the like. Cf. WSax wuht from wiht. In Wuxebrigg, Wuxendon initial W- was eventually dropped. The two names testify to a settlement of Wixan in Middlesex. Cf. WHITSUN BROOK, DŪN.

V

OScand **vað** (ON *vað*, OSw *vaþ*, Dan *vad*) 'ford', identical with OE *gewæd*, is found in several Engl pl. ns., but it is sometimes difficult to distinguish it from *þveit*, *viðr* 'wood' and even worþ. See WAITHE, WATH, WASSAND, WATFORD, WINDERWATH, LANG-WITH, MULWITH, SANDWITH, SKELWITH, STENWITH, LANG-, RAIN-, RAVENSWORTH, SOLWAY.

Vaddicott. See FADDILEY.

Vale Royal Chs [*Vallis Regalis* 1307 BM]. Self-explanatory.

Vange Ess [*æt Fengge*, (to) *Fænge* 963 BCS 1101, *Phenge* DB, *Fange* 1203 Cur, *Vahnge* Hy 3 BM]. OE *fen-gē* 'fen district'. Cf. ELY, LYMINGE, SURREY. The name shows normal ESax *a* from OE *æ* for *i*-mutated *a* before nasals and change of initial *f* to *v*.

ON **varða, varði** 'cairn'. See WARBOYS, WARBRECK, WARCOP, WARTHALL.

Vasterne W nr Wootton Bassett [*Festerne* 1233 Cl, *Fasterne* 1235 Ch, 1272 FF, *La Fasterne* 1281 Ipm]. OE *fæsten* 'fortress', changed to *fæstern* owing to association with *ærn* 'house'.

OScand **vatn** 'water, lake'. See WASDALE, WASTWATER, WATENDLATH.

Vaudey Abbey Li [*Vallis Dei* 1157 P, Hy 2 DC]. 'Valley dedicated to God.' The modern form of the name is French.

Vauxhall (vŏ-) Sr [*Faukeshale* 1279 Ass, *-halle* 1308 Pat]. Named from Falkes de Breauté, who married the heiress of this land c 1220 (VH). *Falkes* is an OFr pers. n., probably of OG origin. The second el. is OE *heall* 'hall'.

ON **veggr** 'wall'. See STANWICK YN, STAN-WIX.

Velly D. See CLOVELLY.

Ventnor Wt [(farm of) *Vintner* 1617]. The name is manorial, *Vintener* being a surname derived from *vintener* or *vintainer* 'officer of a vintaine or 20 men' (PNWt). The old name was *Holewey* 13, 14 VH ('hollow road').

Ver R Hrt [*Wærlame* c 1000 Saints, *Verus* 1572 Lluyd]. Back-formations from the old names of St. Albans [Brit *Verulamium*, *Uerlamacæstir* Bede]. For a genuine old name of the stream see REDBOURN.

Vernham's Dean Ha [*Ferneham* 1219 Fees, *Fernham* 1232 Cl]. 'HĀM or HAMM among ferns.' Change of *f*- to *v*- as in VANGE.

Verwood Do [*Fairwod* 1329, *Fayrwod* 1436 FF]. 'Fair wood.' For *V*- cf. prec. name.

Veryan Co [(parochia) *Sci Simphoriani* 1278 Ass]. 'Church of St. Symphorian.' An old name was *Elerchi* DB, *Elerky* 1231 FF.

Vexford So [*Fescheforde* DB]. No doubt identical with FRESHFORD.

OScand **viðr** (ON *viðr*, OSw *viþer*, Dan *ved*) 'wood, forest', identical with OE *wudu*. See e.g. ASKWITH, BLAWITH, ROOKWITH. In some cases the OScand word has replaced OE *wudu*, as in BECKWITH, SKIRWITH, YAN-WATH, or even OE wīc, as in BUB-, COTTING-, SKIP-, TOCKWITH.

OScand **vík** 'bay' is a rare el. in Engl pl. ns. See e.g. BLOWICK, LOWICK La, WIGTOFT.

Vine, R. See FENITON.

Virginstow D [*Virginestowe* c 1180 BM, *Virgenestowe* 1278 Ep]. 'Church dedicated to St. Bridget the Virgin.'

Virley Ess [*Salcota* DB, *Salcote Verly* 1291 PNEss, 1428 FA]. The original name was SALCOTT (q.v.). The manor was held by Robert de Verli in 1086 (DB). Robert de Werley held *Saltecote* in 1276 (Cl). The family name de Verli is from VERLY in AISNE (France).

Vobster So [*Fobbestor* 1234 FF, *-ter* 1243 Ass]. '*Fobb*'s tor.' Cf. FOBBING, TORR. *Fobb is a strong side-form of *Fobba*.

Vowchurch He [*Fowchirche* 1291 Tax, *Fowechirch* 1316 FA]. 'Multicoloured church.' First el. OE *fāg* 'multicoloured'. Cf. FROME VAUCHURCH.

OScand **vrā** (ON *rá*, OSw, ODan *vrā*) 'corner, nook' is used in pl. ns. in Scandinavia, denoting places with a remote or secluded situation, as surrounded by hills or merely isolated from other homesteads. The element may be rendered by 'remote valley' or 'isolated place'. See e.g. WRAY, WREA, WREAY, WRAYTON, CAPERN-WRAY, DOCKRAY, HAVERAH.

Vyrnwy R Wales, Sa [*Y Vyrnwy* 1201 Arch Cambr xiii, *Efyrnwy* 13 Mab]. The correct form is *Efyrnwy*, which goes back to earlier *Hefyrnwy. The river is one of the head-streams of the Severn and the name is a derivative of the name SEVERN, Welsh *Hafren* (from *Sabrinā*). Initial *H*- was lost in the unstressed first syllable, and *e* for OBrit *a* is due to *i*-affection.

W

Waberthwaite Cu [*Waythebutwayth* c 1200 StB, *-thwait* 13 FC]. ON *veiðibúð* 'hunting or fishing shed' and *þveit* (see THWAITE). ON *veiðr, veiði* mean 'hunting, fishing'.

Wackerfield Du [*Wacarfeld* c 1050 HSC, *Wakerfeld* 1268 Pat]. The first el. might be an OE **wācor* 'wicker, willow', an ablaut form of *wicker*, ME *wiker*, from OSw *viker, vikor* 'Salix pentandra'. The word is cognate with OE *wāc* 'weak', *wican* 'to yield'.

Wacton He [*Wakintun* 1242, 1249 Fees, *Waketon* 1303 FA], W~ Nf [*Waketuna* DB, *Waketone* 1101–7 Holme, *-tun* 1198 FF]. '*Wac(c)a*'s TŪN.' OE *Wacca* is recorded in *Waccan hám* BCS 1319.

OE **wād** 'woad' is found in some names, but is difficult to distinguish from *gewǽd* 'ford' and *Wada* pers. n. See ODELL, WADBOROUGH, WADDEN HALL, WADDON, WADHURST, WADLEY.

Wadborough Wo [*Wadbeorgas*, (in) *Uuadbeorhan* 972 BCS 1282, *Wadberge* DB]. 'Woad hills.' See WĀD.

Wadden Hall K in Waltham [*Wadenhal* 1181 P, *-hale* 1212 RBE, 1324 Ep, *Wodenhale* 1279 Ep]. 'HALH where woad grew.' First el. OE **wāden* 'of woad'.

Waddesdon Bk [*Votesdone* DB, *-dun* 12 BM, *Wottesdona* 1168 P, *-don* 1185 P, 1220 Fees, *Wettesdon'* 1167 P, *-dun* 1225 Cl]. The first el. is identical with that of *Wotesbroc* 1004 Fridesw, which denotes a stream close to Waddesdon. Anderson (Arngart), *Hundred-names* iii. 4 f., suggests that the first el. had a form with *eo*, without making a further suggestion. It is evidently an OE *Weott*, a hypocoristic form of pers. ns. like *Weohtgār*, which became later *Wett* and *Wott*. The place and the brook were named from the same man.

Waddingham Li [*Wadingeham* DB, 1168 P, *-heim*, *Wadingham* c 1115 LiS], **Waddington** Li [*Wadintune* DB, *-tonia* c 1150 DC, *Wadington* 1254 Val]. 'The HĀM and TŪN of *Wada*'s people.'

Waddington Sr [*Whatindone* 675, *Hwætedun* 871–89 BCS 39, 558]. 'Wheat hill.'

Waddington YW [*Widitun* DB, *Wadington* 1241 Cl, *Wadyngton* 1336 FF]. The name goes with **Waddow** close by [? *Waddaw* 1136–52 YCh 1004, *Wadhowhey* 1438 YD]. Waddow is on the Ribble and might have as first el. OE *gewǽd* 'ford'. Waddington might be 'the TŪN of the Waddow people'. Or better Waddow is '*Wada*'s HŌH or spur of land', Waddington being 'the TŪN of *Wada*'s people'.

Waddingworth Li [*Watlinworð* 1060 Th, *Wadingurde* DB, *Wadigworda* c 1115 LiS, *Wadingworda* 1170–5 DC]. 'The WORÞ of

Wada's people.' *Watlinworð* is for *Wadinworð*.

Waddon Do [*Wadone* DB, *Waddon* 1207 Cl, 1212 Fees], W~ Sr [*Waddona* 1107–29 (1330) Ch]. OE *wād-dūn* 'woad hill'.

Waddow. See WADDINGTON YW.

Wade Sf in N. Cove [*Wada* 1165, 1194, *Waða* 1167, *La Wada* 1188 P]. OE *(ge)wǽd* 'ford'.

Wadebridge Co [*Wade* 1382 Dexter]. Perhaps OE *(ge)wǽd* 'ford', with later addition of *bridge*.

Wadenhoe (wŏdnō) Np [*Wadenho* DB, 1167 P, 1254 Val, *Wadeho* 1186 P, 1249 Ch]. '*Wada*'s HŌH or spur of hill.'

Wadhurst Sx [*Wadehurst* 1253 Ch, *Wodhurst* 15 PNSx]. The form with *o* suggests as first el. OE *wād* 'woad', but the usual early form *Wade-* tells against this. Perhaps '*Wada*'s hurst'.

Wadley Brk [*Wadele* 1242 Fees, 1402 FA, *Wadeleye* 1291 Tax]. The first el. may be OE *Wada* pers. n. or **wāden* 'of woad'. See LĒAH.

Wadshelf Db nr Brampton [*Wadescel* DB, *Wadescelf* J BM]. '*Wada*'s hill.' Cf. SCYLF. In this and some other names the mythical hero *Wada* may be referred to.

Wadsley YW [*Wadesleia, Wadelei* DB, *Wadeslei* 1200 P, *Wadeley* 1279–81 QW], **Wadswick** W [*Wadeswica* 12 BM, *-wyke* 1226 Ch], **Wadsworth** YW [*Wadesuurde* DB, *Wadeswurth* 1246 FF]. The pers. n. *Wada* can hardly be the first el. of all these. OE *gewǽd* 'ford' may be suggested, but is out of the question for Wadswick, and the genitival form rather tells against the word. Perhaps an OE **Wæddi*, derived from *Wadda*, would meet the case best. See LĒAH, WĪC, WORÞ.

Wadworth YW [*Wadewrde* DB, *-wurðe* 1166 P, *Wade Wrthe* 1191–3 Fr, *Waddewurth* 1202 FF]. '*Wada*'s WORÞ.'

OE **(ge)wǽd** 'ford' is a rare el. in pl. ns. It is the source of WADE, the second el. of CATTAWADE, IWADE, LANDWADE, LENWADE, and perhaps the first el. of WADEBRIDGE, WATFORD Np.

OE **wægn** 'wain, wagon'. Cf. WAINFLEET, WANBOROUGH W, WANGFORD, WANSBECK, WANSTEAD, WONFORD (2).

OE **wæl** 'a weel, a deep pool, deep water of a stream', dial. *weel* 'a whirlpool, a deep, still part of a river' &c., is the source of WEEL and the second el. of THELWALL. See also SALESBURY, PETERBOROUGH. The **Weel** is a pool in the Tees at Cauldron Snout.

OE (ge)wæsc 'wash, the washing of the waves upon the shore, surging movement of the sea or other water' is found in STRANGE-WAYS, WASHINGBOROUGH, perhaps WASH-FIELD, WASHFORD D. See THE WASH, where a later sense of the word is found. OE wæsce 'washing, place for washing' is the first el. of WASHBOURNE D, WASHBROOK, and the second el. of SHEEPWASH. Cf. SHIPSTON.

OE wæsse is only found in pl. ns. The meaning is probably 'wet place, swamp'. See WASH river, WASS, ALLERWASH, ALRE-, BROAD-, BUILD-, HOP-, ROTHER-, SUGWAS, BOLAS, WASHINGLEY &c.

OE wæt, wēt 'wet' is the first el. of some names, as WEDDINGTON, WEETWOOD, WET-TENHALL &c. Cf. WEETING. In WATTON YE OE wēt has been replaced by OScand vātr 'wet'.

OE wæter 'water, river, lake'. See WATER-(passim). As a second el. wæter means 'river' in BLACK-, BROAD-, FRESHWATER and others, 'lake' in DERWENT-, ELTERWATER &c. Sometimes water in pl. ns. represents the pers. n. Walter, as in BRIDGWATER, WATERS UPTON.

Wainfleet Li [Wenflet DB, Weinflet c 1115 LiS, Waineflet c 1165 Bury, Weynflet Omnium Sanctorum 1229 Ep, Weynfled Beate Marie 1254 Val]. OE wægn-flēot 'stream that can be crossed by a wagon'. The place is on the Steeping.

Waitby We [Wateby 1247, 1257 P]. Possibly 'wet homestead', the first el. being ON vátr 'wet'. It is true the place is in a high situation.

Waithe Li [Wade DB, Wada c 1115 LiS, Wath 1202 Ass, Wathe 1254 Val]. OScand vAÐ 'ford'.

Wakefield Np nr Stony Stratford [Wacafeld DB, Wakefeld 12 NS, Wachefeld 1159 P], W~ YW [Wachefeld DB, -felda 1121 AC, Wakefeld 1219 FF, Wacfeld c 1180 YCh 1713]. The name has been explained as 'Waca's FELD'. Waca is not evidenced in independent use, but no doubt once existed; cf. the byname (Hereward) the Wake. But the occurrence of two Wakefields is noteworthy, and the name is rather 'field where the wake or annual festival was held, where wake-plays were given'. Wake in this sense is found from c 1225 and is no doubt an old word. The name would be particularly suitable for Wakefield YW, the home of the famous Towneley Plays.

Wakering Ess [Wacheringa DB, Wacrinense (monasterium) c 1130 SD, Wakeringes 1197 FF, Parva Wakering 1233 Fees, Wakeringe Magna 1254 Val]. 'Wacer's people', OE Waceringas.

Wākerley Np [Wacherlei DB, -lai 1163 P, Wakerle 1209 Fees]. Cf. WACKERFIELD, LĒAH.

Walberswick Sf [Walberdeswike 1199 (1319) Ch, Walbereswic 1235 Cl, Walberteswyk 1275 RH, 1286 QW]. 'Waldberht's or Walhberht's wīc.' Neither name is with certainty evidenced in England, but cf. OG Waldiberht, Walhberht.

Walberton (-awb-) Sx [Walburgetone DB, Walburgheton 1230 FF]. 'Wealdburg's or Wealhburg's TŪN.' These women's names are not with certainty evidenced in England. But cf. OG Waldburg.

Walbrook R Mx [Walebroc 1114-30 Chron Rams, -broch 1119 Colchester]. OE Weala brōc 'brook of the Welsh or of the serfs'. Cf. WALH and Walemerse 1212 Cur (Stepney).

Walburn YN [Walebrune 12 PNNR, -brun 1222 FF]. 'Brook of the Welsh.' Cf. WALH.

Walby Cu nr Carlisle [Walleby 1292 QW, Walby 1354 Cl]. 'BY on the Roman Wall.'

Walcot Brk [? Wealcotes leah 968 BCS 1225], W~ Li nr Billingham [Walecote DB, 1212 Fees, -cot 1202 Ass], W~ Li nr Folkingham [Walecote DB, -cota 1153-6 BM, -cot 1202 Ass], W~ Li nr W. Halton [Walecote 1051-60 KCD 806, DB, -cot 1202 Ass], W~ Np nr Barnack [Walecot 1125-8 LN, 1189 (1332) Ch], W~ Lodge Np in Fotheringhay [Walecote 1261 Ass], W~ O nr Charlbury [Walecot 1220, -kot 1242 Fees], W~ Sa nr Wellington [Walecota 1160 P, -cote 1230 P], W~ Sa nr Lydbury North [Walcote 1316 FA], W~ W nr Swindon [Walecote DB, -cot 1208 Cur], W~ Wa at Grandborough [Walecote DB, -cot 1236 Fees], Walcote Le [Walecote DB, 1176 FF], Walcott Nf [Walecota DB, Wallekote 12 BM], W~ Wo [Walecote c 1150 Surv, 1265 Misc]. OE Walacot, Wealacot 'cottage of the serfs or of the Welsh'. The former alternative seems on the whole preferable. See WALH.

W~ Brk is now usually called Wawcott. Wealcotes leah BCS 1225 is mentioned in bounds of Oare, which is some way from Wawcott, but the name may mean 'woodland belonging to Wawcott'. The form Wealcot may seem to tell against this being OE Wealacot, but -a- may have been dropped in the inflected form -cotes. Besides, the charter is in a 13th-cent. copy.

OE wald, weald corresponds to G Wald 'wood' and originally meant 'forest, woodland'. But it came to be applied especially to high forest land; when the forest had been felled the name remained, and wald developed the sense 'open upland ground, waste ground'. In pl. ns. wald probably referred in the first instance to woodland. The Anglian form was wald, whence ME wōld, wald (in the north). The Saxon-Kentish form was wēald, whence ME wēld, Mod weald. But in pl. ns. the ea often remained short. Hence the word appears as wald often in the South, as in WALDER-SHARE K, WALDRON Sx. Owing to influence from Standard English the form wold has often replaced original weald in later times,

as in RINGWOULD, SIBERTSWOLD, WOMENS-
WOLD K. OE *w(e)ald* alone has given rise
to WEALD, WIELD, OLD, WOLDS. For the
word as a first el. see WALD- (passim),
WALGRAVE, WALTHAM, WALTON, WAULDBY
and others. Examples of names with
w(e)ald as a second el. are COTSWOLDS, CUX-,
METH-, NORTHWOLD, STIXWOULD, HAMMILL,
HARROLD, HORNINGHOLD.

Walden, Saffron, Ess [*Waledana* DB, *Wale-
dena* 1141 BM, *-dene* 1198 FF, *Safforne-
walden* 1582 AD], **King's & St. Paul's
W~** Hrt [*Waledene* 888 BCS 557, (on)
Wealadene 11 E 276, *Waldene* DB, *Waleden*
1158 P, *Waldan Regis* 1190 BM], **W~** YN
[*Waldene* 1270 YInq, *Waledene* 1321 Mon].
'The valley of the Britons.' See WALH.

Saffron is apparently the word 'saffron'. It is
stated that saffron was extensively grown at the
place.—St. Paul's W~ belonged to St. Paul's,
London.

Walden Stubbs. See STUBBS.

Waldershare K [*Walwalesere* DB, *Weald-
warescare* 11 DM, *Walwarssare* 1262 RBE].
'The boundary (? or share) of the Weald
people.' The first el. is OE *Wealdwaru*
'people dwelling in the weald or wood'.
The second is OE *scearu*, perhaps in the
sense 'boundary' (found in *landscearu*) or
'share, territory'.

Walderton Sx [*Walderton* 1168 P, *Waldriton*
1291 Tax, *Waldryngton* 1331 Ch]. 'The
TŪN of *Wealdhere*'s people.'

Waldingfield, Great & Little, Sf [*Wæald-
inga fæld*, *Wealdinga feld* c 995 BCS 1288 f.,
Waldingefelda DB, *Waldingfeud Magna*,
Waudingefeud Parva 1254 Val]. 'The FELD
of the dwellers by the *wald* or wood.'

Walditch Do nr Bridport [*Waldic* DB, *-e*
1212 Fees]. OE *weall-dic* or *weald-dic*
'ditch by a wall or wood'.

Waldridge Bk [*Wealdan hrigc* 903 BCS 603,
Waldruge DB]. '*Wealda*'s ridge.'

Waldridge Du [*Walrigge* 1297 Pp]. Apparently 'ridge with or by a wall'.

Waldringfield Sf [*Waldringfeld* c 950 BCS
1008, *Waldringafelda* DB]. 'The FELD of
Waldhere's people.'

Waldron Sx [*Waldrene* DB, *Waldrena* 1121
AC, *Walderne* 1197 P]. 'House in a wood',
OE *Weald-ærn* or *Weald-renn*. Cf. ÆRN.

Wales YW [*Wales*, *Walis* DB, *Wales* 1291
Tax]. OE *Walas* 'the Welsh'. The name
is identical with WALES the country. Here
is **Waleswood** [*Waleswode* 1293 Cl].

Walesby Li [*Walesbi* DB, c 1115 LiS, 1204
Cur], **W~** Nt [*Walesbi* DB, a 1184 DC].
'*Val*'s BY.' First el. ON *Valr*, ODan *Val*
pers. n. Walesby Li is in **Walshcroft**
wapentake [*Walescros* DB, *Walescroft* c 1115
LiS]. '*Val*'s cross.'

Walford Do in Wimborne Minster [*Walte-*

ford DB, 1307, 1326 FF, *Waltesford* c 1140
BM]. First el. OE *wealt* 'shaky, unsteady'.
The name refers to a ford difficult to cross.

Walford He nr Ross [*Waleeford* DB, *Walford* 1166, 1212 RBE]. OE *Wealh-ford*
'Welsh ford, ford of the Welsh', the word
Wealh being here used in the uninflected
form.

Walford He nr Leintwardine [*Waliford(e)*
DB, *Welleford* 1242 Fees, *Walleford* 1347
BM], **W~** Sa [*Waleford* DB, *Walleford* 1292
QW, *Wallesford* 1241 FF]. OE *Wælle-ford*
'ford over the river'. OE *wella* appears in
its WMidland form *wælla*.

Walgherton Chs [*Walcretune* DB, *Walcer-
ton* 1260 Court, *Walgherton*, *Walquerton* E 1
BM]. '*W(e)alhhere*'s TŪN.'

Walgrave Np [*Waldgrave* DB, *Waldegrave*
12 NS, 1202 Ass]. OE *wald* 'wood' and
grāf 'grove' or *græf* 'grave, ditch'. As
pointed out in PNNp, W~ adjoins OLD (olim
Wald, *Wold*), and the probable meaning is
'grove belonging to Old'.

OE walh, wealh (plur. *walas*, *wealas*) meant
in the first instance 'Briton', but was also
used in the sense 'serf'. It is impossible to
say which sense should be assumed in each
of the pl. ns. containing the word. But as
most serfs in early OE times would be
Britons, the difference between the two
senses would not be very marked. The
word is used alone in WALES. It is a common first el., especially in WALCOT, WALTON
(WALLINGTON). See also WALDEN, WALLASEY,
WALMER, WALPOLE Sf, WALWORTH. One
WALFORD He seems to contain the uninflected form *wealh*.

Walham Green Mx [*Wendene* 1274, *Wan-
den* 1276, *Wanam Grene* 1545 FF]. Named
after a family from Wenden Ess (PNMx(S)).

Walhampton Ha [*Wolnetune* DB, *Welhamp-
ton* 1285 Ch]. First el. OE *wiell*, *wyll*
'spring, stream'. Cf. HĀMTŪN.

Walkden La [*Walkeden* 1325 Ct]. The first
el. may be a stream-name *Walce* (cf. WALK-
HAM) or a pers. n. *Wealaca* (cf. WALKING-
HAM). Second el. OE *denu* 'valley'.

Walker Nb [*Waucre* 1242 Fees, *Walkyr*
1268 Ipm]. 'Marsh by the Roman wall.'
Second el. ME *kerr* from OScand *kiarr*.

Walkeringham Nt [*Wacheringeham* DB,
Walcringham Hy 2 (1316) Ch, 1212 Fees,
Walcringeham 1247 Ch], **Walkerith** Li
[*Walkerez* 13 BM, *Walcreth* 1316 FA].
The places are near each other on and near
the Trent. Walkerith seems to be OE
w(e)alcera hӯþ 'the landing-place of the
fullers'. But Walkeringham can hardly
contain a derivative of *wealcere*. If the
two names belong together etymologically,
Walkeringham was originally OE *Walcera-
hӯþ-hamm*. But the similarity may be
accidental, and Walkeringham may mean
'the HĀM of *Walhhere*'s people'.

Walkern Hrt [*Walchra* DB, *Walkern* 1222 StPaul, 1241 Ep]. OE *wealc-ærn* or *weal-cera ærn* 'fulling-mill' or 'fullers' house'. OE *wealc-ærn* is not recorded.

Walkham R D [*Walkam(p)*, *Walkham* 13 (1408) Pat], **Walkhampton** D [*Walchentone* 1084 GeldR, *Wachetone* DB, *Walchintun* 1158 P, *Walkamtone* 1259 Ep]. Walkham is a back-formation from Walkhampton, but the latter very likely contains an old name of the river, viz. *Wealce* 'the rolling one', a derivative of OE *wealcan*. Walkhampton may be OE *Wealchæma tūn* 'the TŪN of the dwellers on R *Wealce*' or OE *Wealcan tūn* 'TŪN on R *Wealce*'.

Walkingham YW [*Walchingeham* DB, *Walkingeham* 1226 FF], **Walkington** YE [*Walchinton*, *Walcheton* DB, *Walketuna* c 1115 YCh 966, *Walkinton* 1202 FF]. 'The HĀM and TŪN of *W(e)al(a)ca*'s people.' OE *Wealaca* occurs in *Wealacan dic* BCS 475 and corresponds to ODu *Waloco*, OLG *Waliko*.

Walkley YW [*Walkelay* E 1 BM, *Walkley* 1361 Hall, Sheffield]. Perhaps '*W(e)al(a)ca*'s LĒAH'.

Walkwood Wo in Feckenham [*la Wercwode* 1221 Ass, *Wercwude* 1230 P, *-wud* 1237 Cl, *Werkewod* 1230 P]. OE *Weorcwudu*, which may be analogous to OE *weorcstān* 'stone for building' and mean 'wood where building material was got'.

OE **wall, weall** 'wall' is found in some pl. ns., mostly referring to ancient forts, especially Roman forts or walls. Several places were named from the Great Wall. See WALL, WALBY, WALKER, WALLBOTTLE, WALLSEND, WALPOLE, WALSOKEN, WALTON, WALWICK, further BESTWALL, THIRLWALL.

Wall Nb on the Roman Wall [*Wal* 1166 P], W~ St nr Lichfield [*Walla* 1167 P, *La Wal* 1242 Fees], W~ **Grange** St nr Leek [*Wal juxta Lek* 1293 QW]. OE *wall* 'wall'. The first obviously belongs here. So does the second, which is held to be the site of the Roman station of *Letocetum* (cf. LICHFIELD). The last is less certain.

Wall under Haywood Sa [*Walle sub Eywode* 1255 RH], **East W~** Sa [*Walle, Welle* 1200 Cur], **Walltown** Sa nr Neen Savage [*Walle* DB, *La Walle* 1248 Eyton]. OE *wælla* 'stream'. The first two are near each other on the two arms of Byne Brook. The last is far from the others near a stream. OE *wella* here appears in its WMidland form *walle* (OE *wælla*). Cf. HAYWOOD.

Walla Brook D, a trib. of the Dart [*Walebrok* 1240 For], **Walla** or **Western Wella Brook** D [*Westerewalbroke* 1240 For, *Huttere Welebroc* c 1235 Buckfast]. OE *Weala brōc* 'the brook of the Welsh'. The forms vary between *Wĕala brōc* (whence *Walebroc*) and *Wēala brōc* (whence *Welebroc*).

Wallasey Chs [*Walea* DB, *Waley* c 1100, *Waleie* c 1150 Chester, *Waylayesegh* 1362 Ipm]. Originally OE *Wala-ēg* 'island of the Welsh'. Later a second *ēg* (ME *ey*) was added to the ME gen. form of the name. The name means 'Waley island'.

Wallbottle Nb [*Walbotle* 1176 P, *Walbothill* 1242 Fees]. 'Homestead on the Roman Wall.' See BŌPL. The place is probably referred to as *Ad Murum* c 730 Bede, *æt Walle* c 890 OEBede.

Wallerscote Chs nr Onston [*Walrescota* 1186, *-cote* 1187 P]. 'The salt-boiler's cottage.' First el. OE *wyllere* (in *wylleres seaðon* 995 KCD 691), Mercian *wællere* 'salt-boiler', Chs dial. *waller* the same.

Wallingford (wŏ-) Brk [*Weli-, Wælingford* 821, *Welingford* c 891 BCS 366, 565, *Welengaford* c 893 Alfred Or, *Wealingaford* 1004 Wills, *Walingeford* DB, 1197 FF]. 'The ford of *Wealh*'s people.'

Wallington Brk [*Waleton* 1195 FF], W~ Ha [*Walintone* 1288 BM], W~ Sr [*Waletona* 1076–84 Reg, *-tone* DB, *-ton* c 1180 BM]. OE *Weala-tūn* 'the TŪN of the Welsh'.

Wallington Hrt [*Wallingtone* DB, *Wenlingeton* 1198 FF, *Waudlington* 1212, *Wandlington* 1236 Fees]. 'The TŪN of *Wændel*'s people.'

Wallington Nb [*Walington* 1242 Fees, 1262 Ipm, *Warlington* 1256 Ass]. 'The TŪN of *W(e)alh*'s people.' Cf. WALLINGFORD.

Wallington Nf [*Wal(l)inghetuna* DB, *Wallingtone* Hy 3 RBE]. Possibly 'the TŪN of the people by the wall'. The place is near the Ouse. An embankment might be referred to.

Wallingwells Nt [*Wallendewell* 1227 Ep, *Wellandewell* 1240 FF, *Wallandewelles* 13 BM]. 'Welling, gushing springs.' In the foundation charter of Wallingwells Priory (c 1150 DC) the site is called 'locum . . . iuxta fontes et riuum fontium'.

Wallop, Nether & Over, Ha [*Wallope* DB, *Wallop* 1130, 1162 f., *Walhope* 1230 P, *Wollop inferior* c 1270, *Wallop superior* 1283 Ep]. OE *wiell-hop, wæll-hop* 'valley of the stream'. Perhaps the common form in *a* is due to an OE *weall-hop*, with non-mutated vowel as in *Cantwaru* by the side of *Cent*.

Wallsend Nb [*Wallesende* c 1085 LVD]. 'The end of the Roman Wall.'

Walltown Nb nr Haltwhistle [*Waltona* 1279 Ass]. 'TŪN on the Roman Wall.' W~ Sa. See WALL.

Walmer K [*Wealemere* 11 DM, *Walemer* 1242 Fees]. 'Mere of the Welsh.'

Walmersley La [*Walmeresley* 1262 Ass, *-legh* 1332 Subs]. The first el. may be an OE *wald-mere* or *wald-gemǣre* 'lake by the wood' or 'boundary of the wood'. See LĒAH.

Walmesley La [*Walmesley* 1577 Harrison] is held to have been named from an early owner.

Walmley, W~ Ash Wa [*Warmelegh* 1232 FF]. Apparently 'warm LĒAH' (glade or wood).

Walmsgate Li [*Walmesgar* DB, *-gare* c 1110 Fr, 1202 Ass, *Walmeresgara* c 1115 LiS, *Walmeresgare* 1193 P]. The first el. might be as in WALMERSLEY La. Or an OE pers. n. *Waldmǽr* or *Walhmǽr* may have existed. Cf. OG *Waldomar*, Goth *Walahmar*. Second el. OE *gāra* 'gore, triangular piece of land'.

Walney La [*Wagneia* 1127, c 1130, *Wageneia* 1155 LaCh, *Waghenay* 1336 FC]. 'Grampus island.' First el. ON *vǫgn* 'grampus'. The name is an ON *Vǫgney*.

Walpole Nf [*Walepol* c 1050 KCD 907, *Walpola* DB, *-pole* 1121 AC, *-pol* 1198 P, 1200 Cur]. 'Pool by the wall.' A Roman bank is referred to.

Walpole Sf [*Walepola* DB, *Walepol* 1254 Val, 1265 Ch]. 'Pool of the Welsh.' See WALH.

Walsall (-s-) St [? (æt) *Waleshó* 1002 Wills, *Waleshale* 1163, *-hala* 1169 P, *-hal* 1201 Cur]. '*W(e)alh*'s HALH or valley.'

Walsden La [*Walseden* 1235 FF]. Perhaps '*Walsa*'s *denu* or valley'. *Walsa might be related to the first el. of WALSINGHAM.

Walsgrave on Sowe Wa is a later variant of SOWE Wa. It is *Woldegrove* 1411 PNWa(S). The elements are OE WALD, GRĀF.

Walsham, North, Nf [*Norðwalsham* 1044–7 KCD 785, *Walsam* DB, *Norwalesham* 1169 P, *Waleshann* 1203 Cur], **South W~** Nf [*Suðwalsham* 1044–7 KCD 785, *Walessam*, *Walesham*, *Walsam* DB, *Walesham* 1190 ff. P], **Walsham le Willows** Sf [*Wal(e)sam* DB, *Walesham* c 1095 Bury, 1203 Cur]. Perhaps '*Walh*'s HĀM'. But the early forms of the type *Walsham* may suggest that the first el. is rather OE *Wæls* (cf. WALSING-HAM). If so, the original form was *Wælses-hām*.

Walshford YW [*Walesford* 1227 Ch, 1232 FF, 1240 Cl]. 'The ford of the Welshman or of *W(e)alh*.' Cf. WALH.

Walsingham (-s-), **Great & Little,** Nf [*Walsingaham* c 1035 Wills, DB, *Walsinge-ham magnum* DB, *Magna Walsingham* 1200 FF, *Little Walsingham* 1263 Ipm], **W~** Nf in East Carleton, lost [*Walsingham* 1046, c 1060 Wills, *Walsincham* DB]. 'The HĀM of *Wæls*'s people.' *Wæls* is found in Beowulf.

Walsoken Nf [*Walsocna*, *-socne* 974 BCS 1310 f., *Wallsocne* 1060 Th, *Walsoca* DB]. 'The soke by the wall.' The place is near WALPOLE Nf.

Walsworth Gl [*Waleswurthe* 1221 Ass, *Waleworth* 1244 Fees, *Walesworth* 1271 Ch]. '*Wealh*'s WORÞ' or 'the WORÞ of the Welshman or Welshmen'.

Walterstone He [*Walterestun* 1249 Fees]. 'Walter's manor.' Named from Walter de Lacy, who held the manor soon after the Norman Conquest.

Waltham St. Lawrence Brk [*Waltham* DB, 1203 Cur, *Wautham* 1212 Fees, *Waltham Sancti Laurentii* 1291 Tax], **White W~** Brk [*Waltham* 675 BCS 39, *æt Wuealtham*, *Wealtham* 940 BCS 762, *Waltham* DB, *Wytewaltham* 1242 Fees], **W~ Holy Cross** Ess [*Waltham* DB, 1205 Cur], **Great & Little W~** Ess [*Waltham*, *Waldham* DB, *Uualtham* c 1095 Bury, *Waltham* 1212 Fees, *parua Waltam* 1197 FF, *Waltham Magna* 1238 Subs], **Bishops W~** Ha [*Walthám* 904, *Wealtham* 956 BCS 613, 976, *Waltham* DB], **North W~** Ha [*Wealtham* 909 BCS 625, *Northe Wautham* 1289 Ep], **W~ K** [*Wealtham* 11 DM, *Waltham* 1291 Tax], **W~ on the Wolds** Le [*Waltham* DB, 1194 Fr], **W~** Li [*Waltham* DB, c 1115 LiS, 1202 Ass], **Coldwaltham** Sx [*Uualdham* 683 BCS 64, *Waltham* 957 ib. 997, *Cold Waltham* 1341 NI], **Up Waltham** Sx [*Waltham* DB, *Up Waltham* 1371 Pens-hurst]. OE *W(e)ald-hām* 'HĀM at a wood'. Note especially Waltham on the Wolds. OE *weald* became *wealt* before the *h* of the second syllable.

Bishops W~ belonged to the Bishop of Win-chester.—W~ **Holy Cross** was an Abbey. DB mentions *Canonici sancte Crucis de Walt-ham*.—W~ **St. Lawrence** from the dedication of the church.—**White W~** perhaps from the colour of some building(s).

Walthamstow Ess [*Wilcumestouue* 1067 BM, *-stou* DB, *Welcomstowe* 1107–27 BM]. '*Wilcume*'s STŌW or holy place.' *Wilcume* was the name of an abbess and queen.

Walton, a common name, has at least three different sources: (1) OE *W(e)ala-tūn* 'the TŪN of the Britons or of the (British) serfs', (2) OE *W(e)ald-tūn* 'TŪN in a wood or on a wold', (3) OE *W(e)all-tūn* 'TŪN by a wall' or less probably 'with a wall'. A pos-sible fourth source in the West Midlands is OE *Wælle-tūn* 'TŪN by a stream'. It is difficult to keep the various sources apart. Early forms such as *Waletone* (e.g. in DB) point to *W(e)alatūn*. Early forms such as *Waltone* are not absolutely conclusive, as they may go back to OE *W(e)alatūn*, but may on the whole be taken to indicate OE *W(e)aldtūn* or *W(e)alltūn*. The last two sources are particularly difficult to keep apart. It is really only the situation of places that gives a hint as to the etymology. On the whole *W(e)aldtūn* is a more likely source than *W(e)alltūn*, except when the situation points definitely to the latter.

1. Walton Inferior & Superior Chs [*Waletona* 1154–60 (1329) Ch, *Netherwalton*

1295 Cl], W~ Db nr Chesterfield [*Waletune*
DB, *-tone* 1208 FF, *-ton* 1236 Fees], W~
upon Trent Db [*Waletun, -ton* 942 BCS
772 f., *-tune* DB, *Waleton on Trent* 1289
Cl], W~ **on the Naze** Ess [*Waletuna* 12,
-ton 1222 StPaul, *Edulvesnasse by Waleton*
1320 Misc], W~ K nr Folkestone [*Waltun*
11 DM, *Waleton* 1204 Pp, *-tune* 1263 Ipm],
W~ **le Dale** La [*Waletune* DB, *-ton* 1241
Cl, *Walton in La Dale* 1304 FF], W~ **Hall**
La [*Walletun* DB, *Waletona* 1190 CC], W~
on the Hill La [*Waletone* DB, *-ton* 1177
P], **Ulnes** W~ La [*Waleton* 1203 FF, *Ulnes-
walton* 1285 Ass], W~ **on the Wolds** Le
[*Waletone* DB, 1222 Ep, *-tona* Hy 2 DC,
-ton 1209 Cur, *Walton on the Wald* 1285 Cl],
W~ Sf [*Waletuna* DB, *-ton* 1159 P, 1228
Ch, 1254 Val], W~ **on Thames** Sr [*Wale-
tone* DB, *-tona* 1168, *-ton* 1190 ff. P, *Waleton
super Thamis'* 1279 QW], W~ St nr Bas-
wich [*Waletone* DB], W~ St nr Eccleshall
[*Waletone* DB, *-ton* 1242 Fees, 1285 FA],
W~ St nr Stone [*Waletone* DB, *-tona* c
1130 BM, *Waleton juxta Stanes* 1285 FA],
W~ Sx in Bosham [*Waleton* 1227 Pat, 1229
FF], W~ YN in Kirkdale, lost [*Waletun*
DB], W~ YW nr Wakefield [*Waleton* DB,
1242 Fees, *-tuna* 1159–80 YCh 1681], W~
YW nr Wetherby [*Wale-, Walitone* DB,
Waletona 1141–4 YCh 358].

2. *W(e)aldtūn* or less probably *W(e)alltūn*:
W~ Bk in Aylesbury [*Waltona* 1090, 1139,
-tuna 1146 RA], W~ Bk nr Newport Pagnell [*Waldone* 1219, *Waltona* 1225, *Waltone*
1231 Ep], **High & Low** W~ Cu [*Walton*
c 1150, c 1175 StB], W~ **Cardiff & Deerhurst** W~ Gl [*Waltone, Valton* DB, *Walton*
1194 P, 1220 Fees, *W~ Kerdef* 1303 FA],
Wood W~ Hu [*Waltune* DB, *Waltona* 1155,
Wodewalton 1300 BM], W~ Le [*Waltone*
DB, *Waleton* 1220–35 Ep, *Wauton* 1242
Fees], **Isley** W~ Le [*Walton* 1220–35 Ep,
Islywalton 1327 Subs], W~ Li [*Waltuna*
1146, *-tona* 1163 RA, *Waleton* 1212, *Walton*
1242 Fees], **East** W~ Nf [*Waltuna* DB,
Est Waleton 1252 Cl], W~ Np nr Peter-
borough [*Waltun* 972, 972–92 BCS 1130,
1280 f., *Wealtun* 1016 KCD 726, *Waletone*
DB], W~ **Grounds** Np [*Waltone, Waletone*
DB, *Walton* 12 NS, 1220 Fees], W~ Sa nr
Onibury [*Walton* 1243 FF, 1285 FA], W~
Sa nr Worthen [*Waleton* 1199 FF, *Walton*
1292 QW], W~ **or Bridge** W~ Sa nr Mor-
ville [*Walton* 1233 f. Cl], **Wenlock** W~ Sa
[*Walton* 1262 Ch], W~ So nr Glastonbury
[*Waltone* DB, *Walton* 1196 ff. P], W~ So
nr Kilmersdon [*Waltune* DB, *Wauton* 1243
Ass], W~ **in Gordano** So [*Waltona* DB,
Wauton 1252 Ch, *Waltone in Gordano* 1333
Ep], W~ **on the Hill** Sr [*Waltone* DB,
Wauton 1268 Ch, 1279 QW], W~ **Grange**
St [*Waltone* DB, *Walton* 1291 Tax, 1292
Ch], W~ **d'Eiville & Mauduit** Wa
[*Waltone* DB, *-tona* R 1 BM, *Walton
Deyvill* 1236 Fees, *Wauton Maudut* 1285
Cl], **Little** W~ Wa in Monks Kirby [*Wal-
ton* 1305 Ch, 1428 FA], W~ **Head** YW
[*Waltone* DB, *Walton* 1303 FA]. Some of

these may represent OE *W(e)alatūn* or
Wælletūn.

3. *W(e)alltūn*: **Walton** Cu on the Roman
Wall [*Waltun* c 1175 WR, *Walton* 1291
Tax], **West** W~ Nf [*Waltuna* DB, *-tona*
1081–7 BM, 1121 AC, *Westwaletone* 1254
Val], W~ O [*Waltone* DB, *-tona* c 1130 Oxf].
West Walton is nr WALPOLE Nf (q.v.). W~
O is near the City wall of Oxford and was
no doubt named from it.

W~ **Cardiff** Gl was held by William de Kaerdif
in 1263 (Ipm) and a namesake of his held land
in Gl in 1166 (RBE). The family took its name
from Cardiff in Wales.—W~ **d'Eiville** Wa
came to the d'Eyville family t. Hy 1 (Dugdale).
See COTES DE VAL.—W~ **in Gordano** So. See
EASTON IN GORDANO.—**Isley** W~ Le presumably
from a family name.—W~ **Mauduit** Wa was
acquired by William Mauduit in 1208 (Dugdale). Cf. EASTON MAUDIT.—W~ **on the Naze**
Ess. See NAZE.—**Ulnes** W~ La means 'Ulf's
Walton' (cf. ULCEBY). The early spelling *Ulues*
must have been misread as *Ulnes*.

Walwick (wŏlĭk) Nb [*Wallewik* 1262 Ch,
-wyk 1297 Misc]. 'wɪC on the Roman
Wall.'

Walworth Du [*Walewrth* 1207 FPD], W~
Sr [*Wealawyrð* 1006, *-wurð* c 1050 KCD
715, 896, *Waleorde* DB]. 'woRÞ of the
Britons.' See WALH.

Wambrook So [*Wambrok* 1280 FF, 1306
FF, *Wrambrok* 1291 FF]. If the last ex. is
trustworthy, the first el. would seem to be
ME *wrang* 'wrong', i.e. 'crooked'. But more
likely the first el. is OE *wōh* 'crooked', the
name going back to OE (æt) *wōn brōce* '(at)
the winding brook'.

Wampool R Cu [*poll Waðoen* c 1060 Gospatric's ch, *Wathenpol* 1292 Ass, *Wathelpol*
R 1 (1301), 1201 Ch]. The first el. seems to
be ON *vaðill* 'ford', the second being POLL
'pool, stream'. In the first ex. the order
between the elements is of the Celtic type.

Wanborough Sr [*Weneberge* DB, *Wane-
berga* Hy 1, *Wenebergia* 1147 BM, *Waneberg*
1231 Cl]. '*Wenna*'s hill.' *Wenna* is found
in *Wennan stan* 854 BCS 476. The first el.
shows vacillation between *e* and ESax *a*
from OE *æ* for *i*-mutated *a* before the nasal.

Wanborough W [(æt) *Wenbeorgan, Wænbeorgon* 854, (æt) *Wenbeorgan* c 1050 BCS
477 ff., *Wemberge* DB, *Wamberga* 1178 P,
Wanberge 1205 Cur]. The first el. is obscure. The most trustworthy form is (æt)
Wænbeorgon BCS 478, which is in an
original charter. *Wæn-* looks like a form of
OE *wægn* 'wain, wagon', but it is not easy
to see what a compound *wægnbeorg* could
mean. Cf., however, *Wagenberg* 1137 in
Germany. The second el. is OE *beorg* 'hill'.
The place is situated on Ermine Street at
a considerable hill, called *Wenbeorg* 854
BCS 477.

Wandle R. See WANDSWORTH.

Wandon End Hrt. See WAVENDON.

Wandsworth Sr [*Wendles wurð* 693 BCS 82, *Wendlesuurthe* 1067 BM, *Wand(el)esorde*, *Wendelesorde* DB, *Wendleswurda* 1185, *Wandleswurde* 1195 P]. 'Wendel's worp.' For the interchange of *e* and *a* cf. WAN-BOROUGH Sr. The river-name **Wandle** is a back-formation [*Vandalis riuulus* 1586 Camden].

Wangford Sf nr Southwold [*Wankeforda* DB, *Wangeford* 1238 Cl, 1242 Fees, 1254 Val]. 'Ford by the open fields.' First el. OE *wang* 'open field'. Skeat points out that Wangford Green was all open common till 1817.

Wangford Sf nr Thetford [*Wamforda* DB, *Waineford* 1190, *Wainford* 1197 P, 1242 Fees, 1254 Val]. OE *wægn-ford* 'ford that can be passed by a wagon'. The same is the origin of **Wangford** hd Sf [*Waine-, Wanneforda* DB, *Weinforde* c 1095 Bury, *Weineford* 1172 P]. The ford was at **Wainford Mills** on the Waveney nr Bungay [*Waynforth* 1491 BM].

Wanlip Le [*An(e)lepe* DB, *Anlepia* c 1125 LeS, *Anelep* 1208 Cur, *Onlep* 1316 FA]. OE *ānliepe* 'isolated, single'. The place was by a swamp, and the name might have referred to a narrow footbridge or some stepping-stones, which could only be crossed in single file. The change of *o* to *wa* is similar to that in *one* from OE *ān*.

Wannerton Wo [*Wenuertun* DB, *Wenfertone* 1275 Subs]. The first el. is an old name of the stream at the place [*Wenferð, -ferþ* 866 BCS 513 f.]. The etymology of the stream-name is obscure.

Wansbeck R Nb [*Wenspic* 1137, *-spik*, *Wanspic* 12, *Wanspik* 13 Newminster, *Wanespik* 1256 Ass]. The second el. is very likely the word *spic* found in **Poles Pitch** Sx [*Spolspiche* 1316 Pat], *Holanspic* 747 BCS 175 &c. The meaning of *spic* is not known, but LG *spike* 'brushwood cause-way', Du *spik* 'bridge made of tree-trunks or brushwood' (in pl. ns.) may be compared. If *spic* was used in a similar sense, the first el. of Wansbeck might be OE *wægn* 'wagon' and the name would mean 'bridge that could be crossed by a wagon'. The river-name would be a back-formation.

Wansdyke Ha, W, So, an ancient dyke from Andover to Portishead [*Wodnes dic* 903, 934, 960 f. BCS 600, 699, 1053, 1073]. 'Wōden's dyke', i.e. 'dyke supposed to have been built by the heathen god Woden'.

Wansford Np [*Wylmesford* 972–92, *Welmesford* 970 BCS 1130, 1258, *Walmesford* Hy 2 NpCh, 1176 P]. First el. OE *wielm*, *wylm*, *wælm* 'flowing, bursting forth'. The place stands on the Nene where it is joined by a tributary. Probably there was a whirlpool here, and the name means 'ford by the whirlpool'.

Wansford YE [*Wandeford* 1218, *Wandes-ford* 1235 FF, 1259 Ipm]. Apparently 'the ford of the moles'. OE *wand* means 'a mole'.

Wansley D [*Wanteslegh* 1242 Fees, 1326 Ipm], **W~** or **Wantsley** Do [*Wanteslegh* 1252 Fees]. 'Want's LĒAH.'

Wanstead Ess [*Wænstede* 1066 KCD 824, *Wenstede* 1065 BM, *Wenesteda* DB, *Wanstede* 1197 FF], **W~** Ha in Southwick [*Wansted* 1212, *-e* 1219 Fees, *Wenstede* 1212 RBE, 1250 Fees]. The first el. appears to be OE *wægn*, *wæn* 'wain, wagon', but the exact meaning of the compound is not clear.

Wanstrow So [*Wandestreow* 1065 Wells, *-treu* DB, *-tre* 1182 P, 1201 FF, *Wandelestr'* 1225 Ass]. 'Wændel's tree.'

Wantage Brk [*Waneting* c 880, 955 BCS 553, 912, *Uuanating* c 894 Asser, *Wanetinz* DB]. An old name of the stream at the place [*Wanotingc broc* 956, *Wanetincg*, *Waneting* 958 BCS 949, 1032], probably derived from OE *wanian* 'to decrease' and meaning 'intermittent stream'.

Wantisden Sf [*Wantesdena, -dana* DB, *-dene* 1254 Val]. '*Want*'s valley.'

Wantsum R K [*Uantsumu* c 730 Bede, *Wantsumo* c 890 OEBede, (on) *Wantsume* 944 BCS 791]. An OE adj. **wandsum*, **wendsum* 'winding'.

Wapley Gl [*Wapelei, Wapelie* DB, *Wapelai* 1165 P, *Wappeleia* 1189 BM]. OE *wapol-lēah*, the first el. being OE *wapol* 'bubble, froth', very likely also 'pool, mire'. Cf. OFris *wapul*, *wepel* 'pool, mire'.

Wapley House YN [*Wapelhawe* 1199 P, *Walepol* 1226–8 Fees, *Walplehous* 1231 Ass]. Wapley is identical with WALPOLE Sf. There is a small lake at the place. House is really *Howes*, the plur. of *how* from OScand HAUGR.

Waplington YE [*Waplinton* DB, 1198 Fees, *Waplingtona* c 1200 YCh 445, *Wapelinton* 1200 FF]. 'The TŪN of the people at a pool or mire.' Cf. WAPLEY Gl. East and West Moor are near.

Wappenbury Wa [*Wapeberie* DB, *Wapenbiria* c 1200 BM, *Wappenbiri* c 1200 DC, *Wappebury* 1236 Fees], **Wappenham** Np [*Wapeham* DB, 1163 P, *Wappenham* 12 NS, *Wappeham* 1220 Fees]. A pers. n. *Wappa* or *Wæppa* is unrecorded, but must be assumed on the strength of these two names. Second el. BURG, HĀM or HAMM.

Wappenshall Sa nr Wellington [*Whatmundeshal* 1228 Eyton, *-hall, -hal* 1230 FF]. 'Hwætmund's HALH or HALL.'

Wapping Mx [*Wapping* (mill) 1231 FF, *Wappingge atte Wose* 1345 AD]. Very likely derived from a word for 'marsh' related to OE *wapol* (cf. WAPLEY Gl). *Wose* is OE *wāse* 'mud', here used of a marsh.

Warbleton Sx [*Warborgetone* DB, 1187 P, *Warblinton* 1242 Fees]. '*Wǣrburg*'s TŪN' and 'the TŪN of *Wǣrburg*'s people'. *Wǣrburg* is a woman's name.

Warblington Ha [*Warblitetone* DB, *Werblinton* 1186 P, 1203 Cur, *Warblingetun* 1269 Misc]. If the DB form may be trusted, the first el. looks like an unrecorded woman's name *Wǣrblīþ*, but more likely the name means 'the TŪN of *Wǣrbeald*'s people'.

Warborough O [*Warberge* 1200 Cur, *Wareberg* 1231 Cl, *Wareberwe* 1278 Ch]. 'Watch hill.' The first el. is OE *waru* 'defence, guard, care'. The place is close to Town Hill. Forms like *Wardeberg* 1200 Cur, *Werdeberge* 1206 FF may point to an original *Weardbeorg*, but may equally well be explained as due to modification of *Warebeorg* owing to association with OE *weard*.

Warboys Hu [*Wardebusc*, *Weardebusc* 974 BCS 1310 f., *Wǣrdebusc* 1077 ChronRams, *Wardebusc* DB]. OScand *varði* 'beacon' and *buski* 'bush(es)'.

Warbreck La [*Wardebrecca* 1147 LaCh, *Warthebrek* 1324 LaInq]. 'Beacon hill.' See VARÐI, BREKKA.

Warbstow Co [*Capella Sancte Werburge* c 1180 Ol, *Warbestow* 1377 PT]. 'St. Werburg's STŌW or church.'

Warburton Chs [*Wareburgetune*, *Warburgetone* DB, *Werburgtuna* c 1150 Chester]. '*Wǣrburg*'s TŪN.' *Wǣrburg* is a woman's name.

Warcop We [*Warthecop* 1201 FF, *Wardecop* 1197 f. P]. 'Beacon hill.' The elements are ON *varði* 'beacon' and OE *copp* 'hill'. **Warcop Fell** reaches 2,106 ft.

Warden, Old, Bd [*Wardone* DB, *Guardona* 1163 RA], **W~ Law** Du [*Wardona* 1183 BoB], **W~** K [*Wardon* 1207 Pat, 1219 Fees, 1228 Ch], **W~** Nb [*Waredun* c 1175 Hexh, *Wardon* 1205 P, *Wardun* 1236 Fees], **Chipping W~** Np [*Werdun* 1066–75 Geld R, *Waredone* DB, *Wardon* 12 NS, *Chepyngwardon* 1387 BM]. OE *weard-dūn* 'watch hill'.

Chipping means 'market'. A market in Chipping Warden was abolished in 1227 (Ch). *Law* is OE *hlāw* 'hill'.

Wardington O [*Wardinton* 1279 RH, E 1 BM]. A pers. n. *Wearda* has been assumed on the strength of *Weardan hyll* BCS 663 (Ha), but this may contain an OE *wearde* or *wearda* corresponding to ON *varða*, *varði* 'beacon', 'cairn'. This would be a suitable first el. of Wardington.

Wardle Chs [*Warhelle* DB, *Wardhul* 1278 Ipm, *Wordhull* 1286 Court], **W~** La [*Wardhill* 1218 Ass]. OE *weard-hyll* 'watch hill'. The form *Wordhull* shows the same development as ME *yord* for *yard*.

Wardleworth La [*Wordelword* c 1200 WhC]. 'WORÞ belonging to WUERDLE.'

Wardley Ru [*Werlea* 1067 BM, *Warleia* c 1160 DC, *-leg* 1202 Ass, *-leya* 1223 f. Ep, *Wardele* 1269 For, *-legh* 1284 Cl]. Identical with WARLEY Ess, unless the late forms with *rd* are to be trusted. If so, the first el. is OE *weard* 'watch'.

Wardlow Db [*Wardelawe* 1258 FF, *Wardlowe* 1275 RH], **Wardour** W [*Weardora* 901–24 BCS 591, *Werdore* DB, *Werdor* 1200 Cur]. 'Watch hill and slope.' Cf. HLĀW, ŌRA, WEARD.

Ware Hrt [*Waras* DB, *Wares* 1173 P, 1198 FF, 1212 Fees, *Wara* 1191 P, *Ware* 1254 Val]. OE *wer*, *wær* 'weir'. Cf. WER. Ware is on the Lea.

Wareham Do [*Werham* 784 ASC, c 894 Asser, *Wærham* c 930 Laws, *Warham* DB, 1130 P]. 'HĀM by a weir.' Cf. WER.

Warehorne K [*Werahorna* 830 BCS 396, *Werhorna* 1032 Th, 11 DM, *Werahorne* DB]. 'Bend by the weirs.' Cf. WER, HORN. The place was on the river LYMPNE, and is now on the Royal Military Canal.

Warenford Nb [*Warneford* 1256 Ass], **Warenton** Nb [*Warnetham* 1209 P, 1212 Fees, *Warentham* 1256 Ass]. 'Ford over and HĀM on WARREN BURN.'

Waresley Hu [*Wedreslei*, *Wederesle* DB, *Wereslea* 1169 P, *-le* c 1195 BM]. 'Pasture for wethers.' See LĒAH. W~ Wo. See WARRINGTON La.

Warfield Brk [*Warwelt* DB, *Warefeld* 1171, 1176 P, *Werrefeld* 1228 Cl]. W~ is near Winkfield. In the boundaries of the latter in BCS 778 is mentioned *wernanuuellæ* (*wylle*). This suggests that Warfield is really OE *Wærnanwell(an)-feld* 'FELD by *Wærnanwella*' or 'wren's stream'. Cf. WRÆNNA. But perhaps 'FELD on a weir'.

Warford Chs [*Wareford* DB, *Wereford* 1260 Court, *Vetus Werford* 1271–4 Chester]. 'Ford by a weir.' First el. OE *wer*, *waru* 'weir'.

Wargrave Brk [*Weregrauæ* 1061–5 BM, *-grave* DB, 1212 Fees, *-graua* 1130, 1156, *Werregraua* 1162 P]. 'Grove or grave by the weirs.' See GRÆF, GRĀF, WER. The place is on the Thames.

Warham He nr Breinton [*Werham* DB, 1322 Ipm, *Warrham* c 1170 Hereford], **W~** Nf [*Warham*, *Guarham* DB, *Warham* 1175 P, 1200 Cur, 1242 Fees, *Warham Beate Marie, Omnium Sanctorum* 1254 Val]. 'HĀM by a weir.' W~ He is on the Wye, W~ Nf on the Stiffkey. See WER.

Wark Nb nr Bellingham [*Werke* 1279 Ass, *Werk in Tyndale* 1294 Ch], **W~** Nb nr Cornhill [*Werch* 1158 P, *Werke* 1212 Fees]. OE (*ge*)*weorc* 'fort'.

Warkleigh D [*Warocle* 1100–7 Fr, *-leia* 1204 Cl, *Wauerkelegh* 1242 Fees], **Warkworth** Np [*Wauercuurt* 1153, *-curt* 1155–8 RA, *Wavercurt* 1208 Cur, *Wauerkeworthe* 1220 Ep, *Warcwrth* 1257 Ipm]. See LĒAH, WORÞ. It has been suggested that the first el. might be an OE *wæferce* 'spider' (cf. OE *gangewæfre* 'spider' and OE *læwerce* 'lark'). This is, of course, very doubtful. The first el. looks more like an OE *wæferuc*.

Warkton Np [*Werkentune* 11 ASCh, *Werchintone* DB, *Werketon* 12 NS, *Wercheton* 1163, *Werkinton* 1177 P, *Werkeneton* 1228 Cl], **Warkworth** Nb [*Werceworthe* c 1050 HSC, *-worde* 1104–8 SD, *Werkewurda* 1182, 1194 P, *-wrth* 1242 Fees]. '*Werce*'s or *Werca*'s TŪN and WORÞ.' One *Werce* (*Werca*) was abbess of Tynemouth in the 7th cent.

Warkworth Np. See WARKLEIGH.

Warlaby YN [*Warlavesbi, Werlegesbi* DB, *Warlageby* 1208, *Warlaweby* 1212 Cur]. The first el. is no doubt OE *wærloga* 'traitor' (*warlau* Cursor Mundi), used as a nickname.

Warle·ggon Co [*Worlegan* 1334 Ep, 1355 FF, *Warlaygan* 1377 PT]. The first el. seems to be Co *war* 'on'. The second is obscure.

Warleigh D [*Ward(es)legh* 1242 Fees]. Perhaps 'LĒAH on the river bank'. First el. OE *waroþ* 'shore'. The place is on the Plym.

Warleigh So [*Werlegh* 1001 KCD 706, *Herlei* DB], **Warley** Ess [(at) *Werle* c 1040 Wills, *War(e)leia* DB, *Warle* 1212 Fees]. OE *wer-lēah* 'LĒAH by a weir'.

Warley Salop & Wigorn Wo [*Werwelie* DB, *Weruesleg* 1212 Fees, *Worveleg* 1236 Fees, *Weruele* 1270 Ct]. 'Pasture for cattle.' Cf. WEORF, LĒAH.

W~ **Salop** was formerly in Salop, while W~ **Wigorn** was in Worcestershire.

Warlingham Sr [*Warlyngham* 1155–8 (1330) Ch, *Warlingham* 1197 FF, 1201 Cur, *Warlingeham* c 1270 Ep, *Werlingham* 1274 Cl]. 'The HĀM of *Wærla*'s people.' *Wærla* would be a hypocoristic form of names in *Wær-*. Or the base might be OE *Wærlāf* or the like.

Warmfield YW [*Warnesfeld* DB, *Warnefeld* 1121–7 YCh 1428, 1252 Ep, *Wernefeld* 1201 Cur]. 'FELD of the stallions or wrens.' Cf. WARNFORD, WRÆNNA.

Warmingham Chs [*Wermingham* 1260, 1287 Court, *Wernyngeham* 1289 Ipm]. 'The HĀM of *Wærmund*'s people.' Cf. *Warmundestrou* (hd) DB, in or near which W~ was.

Warminghurst Sx [*Wurmincgehurste* 12 PNSx, *Wurmyngeherst* 1296 Subs], **Warmington** Np [*Wyrmingtun, Wermingtun* 972–92, *Wermingtun* 972 BCS 1130, 1280, *Wermintone* DB, *Wirminton* 1202 Ass].

'HYRST and TŪN of *Wyrm*'s people.' *Wyrm* is not recorded in independent use, but cf. ON *Ormr*, OG *Wurm*, also OE *Wurmhere*, OG *Wurmger, -hart* &c. Cf. also WORMEGAY.

Warmington Wa [*Warmintone* DB, *-tona* 1123–46 Fr, *-ton* 1206 Cur]. 'The TŪN of *Wærmund*'s people.' Cf. WARMINGHAM.

Warminster W [*Worgemynster* 901–24 BCS 591, *Guerminstre* DB, *Werminister* c 1115 Sarum]. '*Mynster* or church on R Were.' The river-name is identical with WORF Sa.

Warmley Gl [*Wurmelegh* 1309 PNGl]. 'LĒAH infested by reptiles' (OE *wyrm*).

Warmsworth YW [*Wermesford* DB, *Wermesworth* 1100–15 YCh, *Wermundesworth* 1267 Ep]. '*Wærmund*'s WORÞ.'

Warmwell Do [*Warm(e)welle* DB, *Wermewull* 1242 Fees]. 'Warm spring.'

Warnborough, North & South, Ha [*Weargeburninga gemæra* 1046 KCD 783, *Wergeborne* DB, *Waregeburna* 1167 P, *Warweburn* 1236 Fees, *Warneburne* 1183 BM]. Originally the name of the WHITEWATER, on which N. W~ is [(æt) *Weargeburnan* 973–4 BCS 1307]. The name means 'felon stream', i.e. 'stream in which felons were drowned'. First el. OE *wearg* 'felon'.

Warndon Wo [*Wermedun* DB, *Warmendone* c 1086 PNWo]. '*Wærma*'s hill.' *Wærma* may be a hypocoristic form of *Wærmund*. Or 'the DŪN of *Wærmund*'s people.'

Warne D. See WAWNE.

Warnford Ha [*Wernæford* c 1053 KCD 1337, *Warneford* DB, 1198 FF, 1230 P], **Warnham** Sx [*Werneham* 1166 P, *Warenham* 1219 Cl, *Wernham* 1256 Ass], **Warningcamp** Sx [*Warnecham* DB, *Warnekomp, Warnescamp* 1242 Fees]. Cf. also *wærnan hyll* 958 BCS 1028 (Brk). Several explanations are possible for the first el. of these, and it may not be the same in all three. The pers. ns. *Wærna* (a short form of *Wærnoþ*) and *Werna* (a short form of names in *Wern-*) may well have existed. OE *wærna, wrænna* means 'wren', and an OE *wræna* 'stallion' may well have existed. The last alternative seems preferable for Warningcamp ('stallion enclosure', cf. CAMP) and Warnham, which is near Horsham. 'Stallion ford' also gives a good meaning.

OE *waroþ* 'shore, bank'. See WARLEIGH D, WARWICK Cu.

Warpsgrove O [*Werplesgrave* DB, 1242 Fees, *Werpesgrave* 1205 Cur]. 'Grove by the stepping-stones or bridleway.' Cf. WORPLESDON Sr, **Wirples Moss** La [*Wirplesmos* c 1190 LaCh]. First el. an OE *werpels*, a derivative of *weorpan* 'to throw' and the source of dial. *wapple, worple, worples* 'bridleway'. Cf. also *Wyrpleswey* (road) AD ii. 412 (Np).

Warren Burn Nb [*Pharned* c 1050 HSC, *Warned* 12 SD, *Warnet* 1157 Percy, 1212 Fees]. A Brit river-name derived from Brit *verno-* (Welsh *gwern*) 'alders'. Cf. BERNÈDE in France [*Vernedus* 960]. The name means 'alder stream'.

Warrington Bk [*Wardintone* c 1175 PNBk, *Wardington* 1294 Ch]. 'The TŪN of *Wǣr-heard*'s people.'

Warrington La [*Walintune* DB, *Werington* 1246 Ass, *Werinton* 1259 Ass]. 'The TŪN of *Wǣr*'s people' or 'TŪN at a weir'. Cf. WARWICK. OE **Wǣr*, a short form of names in *Wǣr-*, is the first el. of **Waresley** Wo in Hartlebury [(to) *Wǣresleage* 817 BCS 361].

Warsash Ha in Hook has been supposed to have taken its name from the la Warr family. Hook was held by Thomas West, Lord de la Warr, in 1488 (VH). Cf. ÆSC. Mr. Gover, however, points out that the name is *Weresasse* 1272 Ass. The situation of the place at the mouth of the Hamble would suit OE *wer* 'weir' as first el., but OE *Wǣr* as in WARESLEY Wo is also possible.

Warsill YW [*Warteshale* 1146 YCh 79, *Warthsala* 1150–3 ib. 71, *Warzhale* 1215–55 Ep]. Perhaps an OE *weard-gesell* 'watch-tower, guard-place'. Cf. OE *weardseld* in these senses. An OE *weard-sæl* would suit better formally.

Warslow St [*Wereslei* DB, *Werselow* 1300 PNSt, *-e* 1327 Subs]. OE *weardsetl-hlāw* 'hill with a watch-tower' may be suggested.

Warsop, Market, Nt [*War(e)sope*, *Wareshope* DB, *Warsopa* 1180 P, *Waresop*, *Warshop* 1233 ff. Ch]. Second el. OE *hop* 'valley'. The first may be a pers. n. *Wǣr* derived from OE *wǣr* 'cautious'. OE *wearg* 'outlaw' would be a suitable first el., but does not go quite well with the early forms.

Warstock Wo [*Le Horestok* 1331 Misc]. 'Grey tree-trunk.' See HĀR, STOCC.

Warter YE [*Wartre* DB, 1156 YCh 1388, *Wartra* 1166 P, *Wartria* 1162–5 YCh 1120, c 1200 BM]. OE *weargtrēo* 'gallows, gibbet'.

Warthall Cu [*Warthehol* c 1220, *Warthole* 13 StB]. 'Hill with a beacon or cairn' (ON *varði* 'beacon, cairn' and *hóll* 'hill').

Warthermarske YN [*Wardonmersk* 1198 FountM]. 'Marsh at a watch hill.' Cf. WARDEN. Second el. OE *mersc* with *sk* owing to Scand influence.

Warthill YN [*Wardhilla*, *-hille* DB, *Warthill* 1221 FF]. OE *weardhyll* 'watch hill'. The change of *d* to *t* is the same as in WALTHAM.

Wartling Sx [*Werlinges* DB, *Wertlingis* 12 AD, *Wortling* 1275 RH]. '*Wyrtel*'s people.' *Wyrtel* is found in **Worsham** Sx [*Wyrtles-ham* 772 BCS 208]. The two places are not far apart.

Wartnaby Le [*Worcnodebie* DB, *Wartnadeby* c 1125 LeS, *Warcnatebi* 1169 P, *Warcnodbi*

c 1200 BM]. The place is in a high situation. The first el. may be a word meaning 'watch hill' or the like, e.g. an OE *weard-cnotta* (cf. KNOTT END) or an OScand *varð-knǫttr* 'hill with a cairn'. But the first el. is perhaps better explained as an OE pers. n. *Weorcnōþ*, *Worcnōþ* (cf. WORKINGTON), though the usual *a* in the first syllable offers difficulty.

Warton La in Kirkham [*Wartun* DB, *Warton* 1207 Cur], **W~ with Lindeth** La [*Wartun* DB, *Warton* 1246 Ass], **W~** Nb [*Wartun* 1236, *Warton* 1242 Fees]. OE *Weard-tūn* 'watch place, look-out place'.

Warton Wa [*Waverton* 1212 Cur, *Wauertone* 1285 BM]. See WAVERTON Chs.

OE *-waru* fem. sing., later *-ware* plur. 'inhabitants' is often used in compounds such as *ceasterwaru* 'the townspeople', *Lundenwaru* 'the Londoners'. CLEWER goes back to OE *clifwaru* 'people at a cliff or on a slope'. See further BURMARSH, CANTERBURY, CANTERTON, CONDERTON, TENTERDEN, WALDERSHARE, which have as first el. a word in *-waru*. Cf. Introd. p. xiii.

Warwick Cu [*Warthwic* 1132, *Warthewic* c 1140 WR, *Warthwik* 1258 P]. 'WĪC on the bank' (of the Eden). First el. OE *waroþ* 'shore, bank'.

Warwick (wŏrĭk) Wa [(into) *Wǣrincgwican* 723–37 BCS 157, (æt) *Wǣringwicum*, *Wǣrincwic* 914 f. ASC (C, D), (in) *Wǣrincwicum* 1001 KCD 705, *Warwic* DB]. See WĪC. The first el. might be an OE *Wǣringas* '*Wǣr*'s people' (cf. WARRINGTON La). But perhaps it is simply an OE **wǣring*, a side-form of OE *wering* 'weir, dam'. **Warwick-shire** is *Wǣrincwicscir* 1016 ASC (D), *Wǣringscir* 1016 ASC (E), *Wǣringwicscir* 1062 BM.

Wasdale, W~ Head Cu [*Netherwacedal*, *Wastedaleheved* 1338 Cl], **W~** We [*Wascedal*, *Wacedalbec*, *-terne* 1235 CWNS xiv, *Wascedale* 1282 Kendale]. ON *Vat(n)sdalr* 'valley of the stream or lake' (see VATN). Identical with the common Norwegian pl. n. VASDAL and VATSDALR (Landnáma) in Iceland. Cf. WASTWATER.

Wash, The, the estuary of the rivers Ouse, Nene, Welland, Witham [*the wasshes* a 1548, *these Lincolne-Washes* Sh King John, *the Washes* 1617 OED]. The Washes were two fordable portions of the estuary between Lincolnshire and Norfolk. *Wash* is here used in the sense 'a sand-bank or tract of land alternately covered and exposed by the sea, a portion of an estuary admitting of being forded or crossed on foot at low tide'. The source is OE *gewæsc* 'wash, washing movement of the water' &c., later used also of the land washed by the sea.

Wash R Ru, Le, Li [*Wasse* 1198 FF, 1275 RH, *Wass* 1269 For, *Washe* c 1540 Leland]. OE *wæsse* 'swamp', here probably used in a derived sense 'stream'.

Washbourne D [*Waseborne* DB, *Wasseburn* 1230 P, *Waysseburn* 1276 Ipm]. 'Stream where washing (of sheep or clothes) was done.' First el. OE *wæsce* 'washing'.

Washbourne Gl [*aet Uuassanburnan* 780, *Uuassanburna* 840 BCS 236, 430, *Waseborne* DB]. First el. OE *wæsse* 'wet place, swamp'. Perhaps it is identical with WASH R, so that the name means 'the river *Wæsse*'.

Washbrook Sf [*Wasebroc* 1198 FF, *Wasse-broc* 1254 Val, *Waschebrok* 1338 AD]. Identical in meaning with WASHBOURNE D.

Washburn YW [*Walke(s)burn, -a* 1173–85, 1203–15 YCh 513 f., *Walshburn* 1307 YInq]. Possibly OE *walceres* (*walcera*) *burna* 'the fuller's or fullers' stream'. The first *r* would easily be lost owing to dissimilation.

Washfield D [*Wasfelte* DB, *-feld* 1165 P, *Wasshfeld* 1334 Ep]. The first el. seems to be OE *gewæsc* 'wash, surging of water'. The name would mean 'FELD by the rapids or whirlpool'.

Washford Pyne D [*Wasforde* DB, *Wasse-ford* 1242 Fees, *Wayshford* 1316 FA]. First el. as in WASHFIELD.

Herbert de Pinu held the manor in 1219 (FF). Cf. COMBPYNE.

Washford So [*Wecetford* c 960 BCS 1149, *Wecheford* 1243 Ass, *Wachetford* 1367 BM]. The place is near WATCHET. The name means 'ford on the road to Watchet'.

Washingborough Li [*Washingeburg* DB, *Wassingeburc* 1170 P, *Wassingburc* 1193 P]. 'BURG of the people at the whirlpool' (see (GE)WÆSC). The place stands on the Witham where it is joined by a tributary.

Washingley Hu [*Wasingelei* DB, *Wassinge-lai* 1163, *-lega* 1167 P]. 'LĒAH of the people at the WÆSSE or swamp.' There is a lake close by.

Washington Du [*Wassyngtona* 1183 BoB, *Waissenton* c 1190 Newcastle, *Wessinton* 1196 P, *Wassinton* 1211 P], W~ Sx [*Wess-ingatun* 946–55, *æt Wassingatune* 947 BCS 819, 834, *Wassingetone* 1073 Fr, *-tona* c 1100 Oxf, *Wasingetune* DB]. Neither OE WÆSSE nor (GE)WÆSC seems to be a suitable base. The name seems to mean 'the TŪN of *Wassa*'s people'. Cf. WASHINGHAM Sr (lost) [*Wátsingahám* 693 BCS 82, *Wassingeham* c 1225 BM], which may have a first el. derived from an OE **Wāpsige* (*Wāp-* = OE *wāp* 'hunting') or a short name **Wāpsa*. Cf. OG *Waido, Weidheri* &c.

Wasing Brk [*Walsince* DB, *Wawesing* 1186 P, *Waghesing* 1220, *Wawesenge* 1236, *Wahes-inge* 1242 Fees]. The OE base seems to be *Wagesingas* or the like. OHG *waganso*, ON *vangsni* mean 'ploughshare'. The base is held to be *wagansan-*. This ought to give OE **wagōsa, *wagusa* or the like. The place is in a tongue of land between the Kennet and the Enborne. This might have been called *Wagusa*, lit. 'the ploughshare'.

Wasperton Wa [*Waspertune* 1043 Th, *Was-mertone* DB, *Wasperton* 1196 FF, 1242 Fees]. It may be suggested that the elements are OE *wæsse* 'swamp' and *beretūn* 'barton'. Cf. ASHPRINGTON for a similar change of *sb* to *sp*.

Wass YN [*Wasse* 1541 Mon]. OE *wæsse* 'swamp'.

Wassand YE [*Wadsande* DB, *Watsand* 1122–37 YCh 1302, *Wathsand* c 1155 DC, *-sonde* c 1155 YCh 1345, *-sand* 1292 Ch]. OScand *vað-sandr* 'sand-bank by the ford'.

Wast Hills Wo in Alvechurch [*æt Wærset-felda, Wearsetfeld* 780, *Werstfeld, Wærsethyll* 934 BCS 234, 701]. OE *weardsetl-feld* and *-hyll* 'FELD and hill with a watch-tower'.

Wastwater Cu [*Wassewater* 1294, *Was-water* 1322, *Wastwater* 1338 Cl]. The name of the lake is connected with WAS-DALE Cu, also called *Wastdale*. Wasdale is 'the valley of Wastwater', the lake having been called at one time simply *Vatn* 'the lake'. Probably Wastwater is a reduced form of *Wasdale-water* 'the lake in Wasdale'.

Watchet So [*Wæced* 918 ASC, *Wecedport* 987 ib. (E), *Wæcet* 962 BCS 1094, *Wacet* DB, *Wechet* 1243 Ass]. Identical with Gaul *vo-cēto-* 'lower wood' (in *mons Vocetius* Tacitus, in Switzerland). Gaul *vo-* corresponds to Welsh *go-* (from *gwo-*) 'under', but in some cases to Welsh *gwa-*, as in *gwas* 'servant', *gwastad* 'level'. The second el. is Brit *cēto-*, Welsh *coed* 'wood'.

Watchfield Brk [*Wacenesfel* 726–37, *Uuac-enes-, Wæthenesfeld* 821 BCS 155, 366, *Wachenesfeld* DB, *Wechenesfeld* 1220 Fees]. '*Wæccin*'s FELD.' **Wæccin* is a diminutive of *Wacca*. See WACTON.

Watchingwell Wt [(to) *Hwætincgle* 968 Reg Wilt, *Watingewelle* DB, *Whatingewelle* 1287–90 Fees, *Whatlyngwelle* 1316 FA]. 'The stream of the *Hwætingas* or people of *Hwæt*.' An OE *Hwæt* may be a short form of names in *Hwæt-* or a byname formed from OE *hwæt* 'brisk, bold'.

Watcombe D [*Whatecomb* 1414 PND]. 'Wheat valley.'

Watendlath Cu at Watendlath Tarn [*Watt-endlane* R 1, *Wattendelan* J PNCu(S), *Watt-intundelau* 1210, *Wathenthendelau* 1211 FC]. *Watend-* is no doubt an ON *vatn-endi* 'lake end', a side-form of *vatns-endi*, whence the common Norw pl. n. VASSENDEN. The final el., according to PNCu(S), is *-lan* rather than *-lau*. This is doubtless right, and *-lan* is identical with the second el. of ASLAND, i.e. dial. *lane* 'a slowly moving part of a river' and the like (see LANU). W~ is situated on Watendlath Tarn, where Wat-endlath Beck issues from it, and the lane will be the upper part of the beck. Later *-lane* was replaced by *lathe* from ON *hlaða* 'barn'.

Waterbeach Ca [*Bechia* a 1086 YCh, *Vtbech* DB, *Beche* 1206 Cur, *Waterbech* 1238 FF, 1242 Fees]. Cf. LANDBEACH. Originally *Beche*, from OE *bæce, bece* 'stream, valley'. Later *Waterbech* for distinction from Landbeach. The place is on the Cam. *Vtbech* is 'outer *Beche*'.

Watercombe Do [*Watrecome* DB, *Watercombe* Hy 1 BM, *-cumbe* 1242 Fees], **Waterden** Nf [*Waterdenna* DB, *-dene* 1188, *-dena* 1191 P]. 'Valley with a stream or lake.' Waterden is not far from Egmere.

Waterfall St [*Waterfal* 1201 Cur, *-fall* 1272 Ass]. OE *wætergefeall* 'waterfall', but in this case evidently referring to the place where the river (the Hamps) disappears into the ground.

Waterhead. See HĒAFOD.

Wateringbury K [*Woðringabyras*, (of) *Woðringaberan, Uuotryngebyri* 964–95, *Oteringaberiga*, (to) *Wohringabyran* 10 BCS 1132 f., 1321 f., *Otringeberge* DB, *Ottringeberia* 11 DM, *Wotringeberi* 1242 Fees]. As *ō* often becomes *uo* in Kent, the first el. might possibly be derived from *Ōhthere* pers. n. The meaning would be 'the BURG (or BÆR or BȲRE) of *Ōhthere*'s people'.

Watermillock Cu [*Weþermeloc* 1200–30 CWNS xxiv, *Wethermelok* 1254 Ipm]. The place stands at **Little Mell Fell** (1,657 ft.). **Great Mell Fell** (1,760 ft.) is near by. *Mell* is Welsh *moel* 'bare hill'. *Meloc* in Watermillock is a diminutive of *moel*, meaning 'Little Moel', and identical with MELLOCK, the name of a hill in Scotland. Originally Great Mell Fell was *Mell* (Welsh *Moel*), Little Mell Fell being *Meloc* (Welsh **Moelog*). *Wether-* (later *Water-*) is probably *wether* 'sheep'.

Waterperry. See PERRY.

Waterstock O [*Stoch* DB, *Waterstokes* 1209–19, *-stoke* 1236 Ep]. Originally *Stoc* (see STOC). The place is on the Thame.

Waterston Do [*Pidere* DB, *Pidela Walteri* 1212 Fees, *Walterton* 1226–8 Fees]. Originally PIDDLE. Walter was the name of an early owner.

Watford Hrt [*Watford* 944–6 BCS 812, 1007 Crawf, 1230 P, *Wathford* c 1190 Gesta]. First el. apparently OE *wāþ* 'hunting'. Cf. HUNTINGFORD.

Watford Np [*Watford* DB, 12 NS, 1209 Fees, *Wadford* DB, 1177 P, *Wathford* 1239 Ep, 1290 Ch]. Probably identical with WATFORD Hrt. But it may also be suggested that the original name was OScand *vað* (or OE *gewæd*, Scandinavianized) 'ford', to which was added an explanatory *ford*. For the change *þ* > *t* see HATFIELD.

Wath (wăth) YN in Hovingham [*Wad* DB, *Wath* 1224–30 Fees], **W~** YN nr Ripon [*Wat* DB, 1239 Ep], **W~ upon Dearne**

YW [*Wade, Wat* DB, *Wath* 1156 Fr, *Wath super Dyrne* 14 BM]. OScand *vað* 'ford'.

Watling Street [*Wætlingastræt* 880 Laws, 956 BCS 986, *Wæclingastræt* 926, 944 BCS 659, 792]. The correct form is OE *Wæclingastræt*, the first el. being identical with that of an early name of ST. ALBANS. The name very likely meant originally 'the road to St. Albans', the presupposition being that it was first applied to the part between London and St. Albans.

Watlington Nf [*Watlingetun* 11 EHR 43, *-tone* 1166 RBE, *Watlingtone* 1254 Val], **W~** O [*Wæclinctun* 880 BCS 547, *Watelintone* DB, *Watlintuna* c 1135 BM]. The last is certainly 'the TŪN of *Wacol*'s or *Wæcel*'s people', the first el. being identical with that of WATLING STREET. A pers. n. *Wæcel* or *Wacol* is not recorded, but is clearly to be assumed. It is found also in WATTLESBOROUGH. It is OE *wacol* 'watchful' used as a pers. n. A side-form **wæcel* may well have existed. W~ Nf is very likely identical with W~ O, though no forms with *c* (*k*) are on record.

Watnall Nt [*Watenot* DB, *Wattenho* 1200 Obl, *-hou* c 1200 Middleton, *Watenho* 1205 Pp]. '*Wata*'s HŌH or spur of land.' OE **Wata* pers. n. seems to be vouched for by *Watancumb* BCS 246. Cf. OHG *Wazo, Wezilo* &c.

Wattisfield Sf [*Watlesfelda* DB, *-feld* c 1150 Bury, 1197 FF, *Uueatlesfeld* c 1095 Bury]. Very likely '*Wacol*'s FELD'; cf. WATLINGTON. OE *watol* 'wattle' is also a possible first el.

Wattisham Sf [*Weeesham* DB, *Wechesham* 1182, *Wachesham* 1184 P]. '*Wæcci*'s HĀM.' **Wæcci* is a derivative of OE *Wacca* and in reality identical with **Wæccin* in WATCHFIELD.

Wattlesborough Sa [*Wetesburg* DB, *Wetlesborc, Watelesbur* 1242 Fees, *Waklesburg* 1257 Eyton]. '*Wacol*'s or *Wæcel*'s BURG.' Cf. WATLINGTON.

Watton Hrt [*Wadtun* 11 E, *Wattúne* 969 Crawf, *Watone* DB, *Watton atte Stone* 1311 Misc]. OE *Wād-tūn* 'TŪN where woad grew'.

Watton Nf [*Wadetuna* DB, *-ton* 1203 Ass, *Waditone* 1254 Val]. '*Wada*'s TŪN.'

Watton YE [*Uetadun* c 730 Bede, *Wetadun, Wǽtadún* 969 OEBede, *Wattune, Watun* DB]. 'Wet DŪN', originally OE *wēta dūn*, later with OScand *vātr* 'wet' substituted for the OE word. There is no hill at Watton, and *dūn* must refer to a very slight rise.

Wauldby YE [*Walbi* DB, *Waldebi* 1190 P, *Waldbi* 1208 Cur]. 'BY on the Wold.' The place is on the slope of Cave Wold.

Wavendon (wŏndn) Bk [*Wáfanduninga* (gemǽru) 969 BCS 1229, *Wauendone* DB], **Wandon End** Hrt at King's Walden

[*Wavedene* DB, *Wavendun* 1203 Cur, *Wawenden* 1207 Cur, 1236 Fees]. It has been suggested that the first el. is an OE *Wafa pers. n., identical with *Waba*, a name given by Piper. It might also be an OE *Wāfa cognate with OG *Waibilo* &c., OE *Wǣba*. OE *Wafa* appears as *Waua* DB (K). Second el. DŪN and DENU.

Waveney R Sf, Nf *Wahenhe* 1275 RH, *Wagenho* 1286 Ass, *Wawneye* 1485 BM]. The first el. is identical with WAWNE. The second is OE *ēa* 'river'.

Waver R Cu [*Wafyr* c 1060 Gospatric's ch, *Waura* 12 Holme C, *Waver* R 1 (1307) Ch, *Wauere* 1279 Ass]. 'Winding stream.' A derivative of OE *wæfre* 'wandering, flickering'. On the Waver is **Waverton** Cu [*Wauerton* 1186 ff. P, *Waverton* 1227 FF].

Waverley Sr [*Wauerleia* 1147 BM, -*le* 1156, -*lay* 1196 P], **Waverton** Chs [*Wavretone* DB, *Wauertone* c 1100, *Waueretone* c 1150 Chester, *Waverton* 1260 Court], **Wavertree** La [*Wauretreu* DB, *Wavertrea* 1177, *Wavertre* 1196 P]. The first el. of these must be compared with that of WARTON Wa, WHARTON He, WOORE (olim *Waver* &c.). The meaning of the el. *waver* is unknown. It is doubtful if it can be identified with dial. Engl *waver* 'a common pond', or 'a young tree left standing when the surrounding wood is felled, a twig shooting from a fallen tree'. These words have not been found in early sources. LG *waver* 'soft spongy ground' has been adduced, but such a sense hardly suits the names containing the el. *waver*. Now *waver* is well evidenced in Continental names, apparently associated with woodland. Waverley has an exact counterpart in *Waverlo* Holland (1200), the name of a brushwood. *Waverwald* is found in Holland. *Wabra silva* (now WOËVRE in France) is mentioned by Gregory of Tours. The probability seems to be therefore that the Engl *waver* meant 'brushwood' or something similar. Wavertree may contain OE *wæfre* 'flickering' and mean 'the shaking tree'.

Wawcott Brk. See WALCOT.

Wawne YE [*Wagene, Waghene* DB, *Wagna* 1150–3, 1151 YCh 1381, 1383, *Waghen* 1150–3 ib. 40, *Waune* 1228 Ep]. An OE *wagen, a derivative of OE *wagian* 'to wag' and meaning 'quaking bog'. A related word is found in **Warne** D [*Wagefen* 1194 P, *Waghefenne* 1242 Fees], which means 'quaking fen'.

Waxham Nf [*Waxtonesham* 1044–7 KCD 785, *Wacstanesham, Wacstenesham* DB, *Waxstonesham* c 1150 BM, *Waxstanesham* R 1 Cur, *Wextonesham* 1248 Ch]. Possibly '*Wǣgstān's HĀM'. But the name may contain a word related to OE *wacian* 'to keep watch', e.g. OE *wæcce* 'watch'. If so, the name would mean 'HĀM by the stone where watch was held'. The place is close to the sea.

Waxholme YE [*Washam* DB, *Waxham* DB, 1297 Subs]. 'Homestead where wax was produced.'

Wayford So [*Waiford* 1206 Cur, 1207 FF, 1225 Ass]. 'Way ford.'

Weald, South, Ess [*Weld* 1062 Th, *Welda* DB, *Suthwelde* E 1 FF], **North Weald Bassett** Ess [*Walla* DB, *Walda* 1130 P, *Northwolde* 1244 FF]. OE *weald* 'woodland'. At least in N. Weald the name refers to Epping Forest.

N. W~ **Bassett** was held by Philip Basset in 1260 (Ch). Cf. BERWICK BASSETT.

Weald, The, K, Sx, Ha [*Waldum* 1185 TpR, *Wald* 1235 Cl, 1275 RH, *þe Welde* 1297 Rob Gl]. OE *weald* 'woodland'. The old name was ANDRED.

Weald O [*Walda* 1188 P, *Welde* 1216, *Walde* 1229 Cl]. Identical with prec. names.

Weald Moors. See WILDE.

Wear (-*ēr*) R Du [*Wirus* c 720 Bede HAbb, *Uiurus* ib., c 730 Bede, *Wiire þære ea* c 890 OEBede, *Werra, Weorra, Weor* c 1050 HSC, *Wer* c 1175 Finchale]. The real OE form was *Wior, Wēor*, which comes from Brit *Visur-*, a name related to Gaul *Visurgis* (now WESER in Germany), *Visera*, Welsh *gwyar* 'blood'. The name means 'water, river'. **Weardale** is *Werredal* 1227 Cl, *Weredal* 1242 Ass.

OE *weard* 'watch' is the first el. of some pl. ns. See WARD- (passim), WARSILL, WARSLOW, WARTHERMARSKE, WARTHILL, WARTON, WAST HILLS. Special senses are found in WESTWARD, WOLFORD. On a possible OE *wearda or *wearde 'beacon' see WARDINGTON.

Weardley YW [*Wartle* DB, *Wiuerhelayes* 1138–50 YCh 1862, *Wiverdeleia* R 1 Cur, *Wyveresdesley* 1303 FA]. '*Wigferþ*'s LĒAH.'

Weare (-*ēr*) **Gifford** D [*Were* DB, *Weregiffarde* 1328 Ep], **W~** So [*Werre* DB, *Wera* 1169 P, *Were* 1242 Fees]. OE *wer* 'weir'.

W~ **Gifford** was held by a Giffard in 1219 (Ass). Cf. ASHTON GIFFORD.

OE *wearg* 'outlaw, felon, criminal' is found in some names referring to a stream or place where felons were executed. See WARNBOROUGH, WEYBOURNE, WREIGH BURN, WREIGHILL, WRELTON. OE *weargrōd, weargtrēo* 'gallows' are the source of WORGRET and WARTER. An interesting name is *Wargemere* 1189–98 St Pauls, p. 218 (Mx), probably 'lake in which felons were drowned'.

Wearmouth, Bishop & Monk, Du [*Uuiuraemuda, Uuiremuða* c 730 Bede, *Wiramuþa* c 890 OEBede, *Weremutha* 1195 (1335) Ch, *Wermuth Episcopi, Monachorum* 1291 Tax]. 'The mouth of the Wear.'

The manors belonged to the Bishop and the monks of Durham.

Wearne So [*Warne* DB, *Werne* 1219 Fees, 1225 Ass]. Originally a name of the stream at the place [*Werne, Wernanstrem, Wernanford* 973 BCS 1294]. The name means 'alder stream' and is derived from Brit *verno-*, Welsh *gwern* 'alders'. Cf. Gaul *Vernodubron* 'alder stream' (now VERDOUBLE in France).

Weasenham Nf [*Wesenham* DB, 1199 P, 1242 Fees, *Weseham* 1205 Cur, *Wesinham Omnium Sanctorum, Wesinhamthorp* 1291 Tax]. The first el. is obscure. The place is fairly high, and connexion with the words mentioned under WISSEY is not probable. OE *wēse* 'moist' is hardly to be thought of. The first el. may be a pers. n. related to *wisi-, wisu-* 'good' in *Visigothæ* 'West Goths' and OG pers. ns. OE **Wisa* would have given **Weosa*, later *Wese-*.

Weaste La. A form of *waste* 'uncultivated land, common'.

Weaver R Chs [*Weever* 1133 Mon, *Wevere* 1276 Ch, *Wiure* c 1284 Vale Royal]. A Brit river-name, probably identical with WIPPER in Germany and derived from the root of Lat *vibrare* &c. The name would mean 'winding river', a very apt name. On the Weaver are **Weaver Hall** [*Wevre* DB, *Weuere* c 1300 BM], **Weavercote** [*Wyvercote* 1289, *Wevercote* 1290 Ipm], **Weaverham** [*Wivreham* DB, *Weueresham* c 1100 Chester, *Weuerham* c 1150 ib.]. Cf. also WERVIN, which shows that the name Weaver must have been applied formerly also to the Mersey below its junction with the Weaver.

Weaverthorpe YE [*Wifretorp* DB, *Wiveretorp* c 1110, *Wivertorp* 1153 YCh 25, 28, *Wiverestorp* 1150–75 YCh 85]. The first el. is a pers. n., e.g. an OScand *Vidfari* 'traveller' (cf. *Widfare* Coins, *Wifare, Wiuara* DB) or OE *Wigferþ*.

OE weax 'wax'. See WAX-, WEX-.

Webbery D [*Wibeberia* DB, *Wibbebyria* 1235 Bract]. '*Wibba*'s BURG.' *Wibba* is found in *Wibbandun* 568 ASC.

Webton. See WEOBLEY.

Weddiker Cu [*Wedaker* 12 StB, *-akre* 1322 Ipm]. OE *wēod-æcer* 'weedy field'.

Weddington Wa [*Watitune* DB, *Wetinton* 1236, 1242 Fees, 1265 Misc, *Weddington, Waddinton* 1285 QW]. OE *Wæta-tūn* 'wet TŪN'.

Wedmore So [*Weþmor* 878 ASC, *Wedmor* 878 ASC (E), c 880 BCS 553, *-more* DB, *Wædmor* c 894 Asser, *Weddmor* 1065 Wells, *Wadmor* 1201 Cur]. The testimony of the early forms is conflicting. On the whole the probability seems to be that the first el. is an OE **wǣþ*, a side-form of *wāþ* 'hunting'. Cf. WEMBDON. The change *þ > d* before *m* is regular. The only difficulties are the forms with *e* 878, c 880. W~ would mean 'moor for hunting'.

Wednesbury (wĕj-, wĕnzbrĭ) St [*Wadnesberie* DB, *Wodnesberia* 1166, 1190 P, *Wednesbiri* 1227 Ass], **Wednesfield** (wĕj-, wĕns-) St [*Wodnesfeld* 996 Mon, 1227 Ass, *-felde* DB, *Wednesfeld* 1251 Cl]. '*Wōden*'s BURG and FELD.' The change of *ō* to *e* is the same as in *Wednesday*, OE *Wōdnesdæg*.

Weedon Bk [*Weodun* 944–6 BCS (812), 1066 KCD 824, *Wedonhull* 1328 Misc], **W~ Beck** Np [*Weoduninga gemære* 944 BCS 792, *Wedone* DB, *Wedon Beke* 1379 Cl], **W~ Lois** Np [*Wedone* DB, *Wedona* 12 NS, *Wedon Pinkeny* 1241 Ep, *Leyes Weedon* 1475 PNNp, *Loyeswedon* 1535 BM]. OE *wēodūn* 'hill with a temple'. First el. OE *wēoh* 'heathen temple'.

W~ Beck belonged to the abbey of Bec Helloün in Normandy at least from the 12th cent. (1167 P).—**W~ Lois** was held by Gilo [de Pinkeni] in 1086 (DB). Cf. MORETON PINKNEY. The history of the addition *Lois* is not quite clear.

Weedslade Nb [*Wideslade* 1197 P, *Nortwitheslod* 1203 Cur, *Wydeslad* 1242 Fees]. 'Withy valley.' Cf. SLÆD, WĪÞIG.

Weeford St [*Weforde* DB, *Weford* 1200 P, 1242 Fees, *Weoford* 1291 Tax]. 'Ford by a heathen temple.' Cf. WEEDON.

Week St. Mary Co [*Wich* DB, *Wyk S. Marie* 1291 Tax], **W~ Ha** nr Alton [*Wyk* 1282 Ep, *Wyke juxta Bynstede* 1333 BM], **W~ So** nr Stockland Bristol [*Wyke* 1274 RH, 1285 FA], **W~ Wt** in Godshill [*Wica* DB, 1142–55 Fr], **Weeke** Ha nr Winchester [*Wyke* c 1270 Ep, 1316 FA]. OE wĪC, probably in the sense 'dairy-farm' or the like.

Weekley Np [*Wicleaford* 956 BCS 943, *Wiclei* DB, *Wichelai* 1094 Fr, *Wichelea* 1167 P, *Wicle* 1199 FF]. 'Wych elm wood.' First el. OE WICE 'wych elm'.

Weel YE [*Wela* DB, *Wele* 1346 FA]. OE *wæl* 'pool, deep still part of a river'. The place is on the Hull. **The Weel.** See WÆL.

Weeley Ess [*Wilgelea* 11 St Pauls, *Wileia* DB, *Welega* 1166 P, *Wyley* 1254 Val]. The earliest example shows that the name means 'willow LĒAH' (cf. WELIG) and is not OE *wēolēah*, as later forms suggest.

Weelsby Li [*Wivelesbi* DB, *Uiflesbi* c 1115 LiS]. First el. ON *Vifill*, ODan *Wiwil* pers. n.

Weethley Wa [*Withelea* 714 BCS 130, *Widelega* 1176 P]. 'Willow wood.' Cf. WĪÞIG, LĒAH.

Weeting Nf [*Watinge* c 1050 KCD 907, *Wetinge* DB]. 'Wet district.' See WÆT, -ING.

Weeton La [*Widetun* DB, *Witheton* 1249 Ipm], **W~** YE [*Wideton* DB, *Wytheton* 1314 Ch], **W~** YW [*Widetune, Widitun* DB, *Withiton* 1226 FF]. OE *Wīþig-tūn* 'TŪN among willows'.

Weetwood Nb at Chatton [*Wetewude* 1197 f. P]. 'Wet wood.' Cf. WÆT.

OE **weg** 'way, road' is the first el. of WAYFORD, WHALEY Chs, and the second el. of several names, as BROADWAY, HOL-, HOLLOWAY, RAD-, RODWAY, STANWAY, STOWEY, THORESWAY. Cf. also FLOTTERTON, HARTINGTON Nb, STANNINGTON, STAINTON Du.

Weighton, Little, YE [*Wideton* DB, *Witheton* 1276 Ch]. Identical with WEETON.

Weighton (-ē-), Market, YE [*Wicstun* DB, *Wichtona* 1133 YCh 132, -*ton* 1165 P, *Wihtun* 1156 P, *Wicton* 1204 Cur]. OE *wīc-tūn* 'homestead, dwelling'.

Welbatch Sa nr Bayston [*Huelbec* DB, *Whelbache* c 1275 Ep]. OE *Hwēol-bæce* 'valley with a wheel or circle'.

Welbeck Nt [*Wellebec* 1185 P, 1193 DC, 1212 Fees, -*bek* 1243 BM]. Originally the name of a stream [*Wellebec* 1179 (1328) Ch]. The stream was no doubt OE *Wella* 'the stream', to which was added an explanatory OScand *bekkr*.

Welborne Nf [*Walebruna* DB, *Welebrun* 1203 Ass, *Wellebrunne* 1254 Val], **Welbourn** Li [*Wellebrune* DB, -*burn* 1196, 1198 P], **Welburn** YN nr Crambe [*Wellebrune* DB], **W~** YN nr Kirkdale [*Wellebrune* DB, -*burna* 1160–5 YCh 164]. OE *welle-burna* 'brook coming from a spring' or 'welling spring'. Cf. BURNA.

Welbury YN [*Welleberge* DB, -*berg* 1226 FF]. 'Spring hill.' There is a holy well here.

Welby Le [*Alebi(e)* DB, -*bia* c 1125 LeS, *Oleby* 1242 Fees]. '*Ali*'s BY.' First el. ON *Áli*, ODan *Ali* pers. n. Cf. AILBY, WANLIP.

Welby Li [*Wellebi* DB, Hy 2 DC, 1202 Ass]. 'BY by a spring.' Cf. WELLA.

Welcombe D [*Walcome* DB, *Wellecombe* 1301 Cl]. 'Valley with a spring or stream.'

Weldon, Great & Little, Np [*parua Weledene, Wale(s)done* DB, *Welledon* 1163 P, 1220 Fees, *Magna W~* 1186 P]. 'Hill with a spring or by a stream.' See DŪN.

Welford Brk [*æt Weligforda* 949, 956 BCS 877, 963, *Waliford* DB]. 'Ford by the willow(s).'

Welford on Avon Gl [*Welleford* 1187 P, 1203 Cur, *Welneford* 1215, 1229 AD]. OE *Wellanford* or ?*Wellnaford* 'ford over the stream(s) or by the spring(s)'.

Welford Np [*Wellesford* DB, *Welleford* Hy 2 DC, 12 NS]. 'Ford over the stream.'

Welham Le [*Wale-, Wele-, Walendeham* DB, *Weleham* 1198 P, 1242 Fees, *Welleham* 13 Fees]. 'HĀM by the *wella* or river' (the Welland). *Walendeham* may contain a form of the river-name Welland.

Welham Nt [*Wellun* DB, *Wellum* 1242 Fees], **W~** YE nr Malton [*Wellun* DB,

Wellum 1173 YCh 1888]. OE *wellum*, dat. plur. of *wella* 'spring'.

OE **welig, wilig** 'willow', cognate with OLG *wilgia*, MDu, MHG *wilge*, Fris *wylch*, *wilig* the same, probably comes from earlier **welg, *wilg*. It is used alone to form the pl. n. WELLOW Wt, and WELWYN, WILLEN, WILLIAN come from the dat. plur. *weligum, wiligum*. It is the first el. of several names, as WEELEY Ess, WELFORD Brk, WILBURY, WILBY, WILDEN (1), WILLEY (1), WILLINGTON, WILLITOFT, WILLOUGHBY, WILLOUGHTON, some WILTONS and others.

OE **well, wiell, wæll (-a, -e)** 'well, spring, stream' is a very common pl. n. el. The two meanings 'spring' and 'stream' are often difficult to keep apart. The latter is certain in river-names, such as BARLE, CHERWELL, IRWELL. The word is used alone in names such as WELL, WELLS, WELHAM (from *wellum* dat. plur.). It is especially common as a second el. The usual ME form is *welle* (OE *wella* &c.). See WEL-, WELL(passim). The WSax form was OE *wiell, will, wyll (-a, -e)*, whence ME *wille, wulle*, as HALWILL, WILCOT, WILTON So, WOOL, WOOLCOMBE, WOOLLEY So. Spellings with *i, u* are common in early forms of names that have now *e*. The WMidl form was *walle* (OE *wælla* &c.), as in COL-, ECCLESWALL He, CHATWALL, WALL Sa, CAVERSWALL St, ETWALL Db, HESWALL Chs, CHILDWALL La. See also WALHAMPTON, WALLOP. Special developments are found in COBHALL, THATTO.

Well K [*Welles* 1242 Ass, *Well* 1314 Ipm], **W~** Li [*Welle* DB, 1234 Ep], **W~** YN [*Welle* DB, 1251 Ass]. OE *wella* 'spring or stream'.

Welland R Np, Le, Ru, Li [*Weolud* 921 ASC, *Uuelod* c 1000 Ethelwerd, *Welund* a 1118 Flor, *Weyland* 1200 Cu, *Weland* 1218 For, 1230 Cl]. A Brit river-name. The first el. may be Celt **vesu-* 'good', Welsh *gwiw* 'dignus', the second being a word for 'river' cognate with OIr *lúaid-* 'to move', OE *flēot* 'stream'.

Welland Wo [*Wenelande* 1189 (1335) Ch, -*land* 1233 Cl]. The first el. may be a stream-name *Wen* from Welsh *gwyn* (fem. *gwen*) 'white' &c. The place is on a stream that joins **Wynd Brook** [*Wenbroc* 963 BCS 1109]. Both arms may have been called *Wen*.

Wellesbourne Hastings & Mountford Wa [(in, æt) *Welesburnan* 840, 862, (æt) *Walesburnam* 872 BCS 430, 503, 535, *Waleborne* DB, *Welesburn* 1177 P, *Wellesbourne Mountford* 1327 AD]. Apparently '*Wealh*'s stream' or 'the stream of the Briton' (cf. WALH).

W~ Hastings was held by Thomas de Hastanges in 1316 (FA). Cf. BURTON HASTINGS.— **W~ Mountford** was held by Petrus de Monte Forti in 1236(Fees) and was given to Thurstan de Mountfort t. Hy I (Dugdale). Mountfort from MONTFORT in Normandy.

Wellingborough Np [*Wendle(s)berie, Wedlingeberie* DB, *Wenlingeburg* 1199 FF, *Wendlingburgh* 1220 Fees]. 'Wendel's BURG' and 'the BURG of Wendel's people'.

Wellingham Nf [*Walnccham* (sic) DB, *Uuelingheham* c 1190, *Welingham* c 1200 Middleton, *Wellingham* 1198 FF, *Welingeham* 1267 Ch], W~ Sx [*Wellingeham* W 2 PNSx, *Willinggehamme* 1307 Ass], **Wellingore** Li [*Wallingoure* 1070–87 RA, *Wel(l)ingoure* DB, *Wellinghoure* 12 DC, *Wellingour* 1202 Ass]. The first el. of all three may be an OE *Wellingas* 'people by a stream or spring', the second being OE HĀM, HAMM and OFER 'edge, slope'.

Wellington He [*Weolintun* a 1038 KCD 755, *Walintone* DB, *-tona* 1150–4, *Welintona* 1155–63 Hereford, *Welintun* 1242 Fees], W~ Sa [*Walitone* DB, *Waletona* 1181, *Welintona* 1220 BM, *Weliton* 1192, *Wolinton* 1196 ff. P, *Weolyntone* 1327 Subs], W~ So [*Weolingtun, Welingtun* 904 BCS 610, *Welingtun* 1065 Wells, *Walintone* DB, *-ton* 1178 Wells, *Wellinton* 1225 Ass]. The three names seem to be identical in origin. If the early spellings are trustworthy, the first el. would seem to be a derivative in *-ingas* of OE *wēo-lēah* 'temple LĒAH'. Cf. WĒOH, WEOLEY, WILLEY Sr. The only alternative to derivation from a pl. n. *Wēolēah* appears to be one from an OE pers. n. *Wēola*, a hypocoristic form of names in *Wēoh-*. Cf. OG *Weila*.

Wellow, East, Ha [(æt) *Welewe* c 880, *æt Welowe* 931 BCS 553, 676, *Weleve* DB, *Welewe* 1212 Fees], **West W~** Ha [*Wilewe* 1242 Fees]. The two Wellows are close together, and the OE forms may include both. The places are on the BLACKWATER, formerly **Wellow** [(on) *Welewe* a 670, 826 &c. BCS 27, 391 &c.]. A Brit river-name derived from Welsh *gwelw* 'pale blue'. The name may refer to the colour of the water. But the original meaning of the adj. was 'that has turned', later applied to milk that has 'turned', i.e. become pale blue. The river-name might mean 'winding'.

Wellow So [*Weleuue* 1084 GeldR, *Welewe* 1225 Ass]. Originally the name of a stream [*Weluue, Welwe* 766 BCS 200]. Identical with prec. name. *Æt Welewestoce* 984 KCD 643 seems to be RADSTOCK.

Wellow Li, old monastery [*Welhow, Wel(le)-hogh* 1314 Ipm]. 'Spur of land with or by a spring.' See HŌH.

Wellow Nt [*Welagh* J (1316) Ch, *Welhag* 1234 FF, *Wellehagh* 1268 Ch]. 'Enclosure by a spring.' See HAGA.

Wellow Wt [(æt) *Welig* c 880 BCS 553, *Welige* DB, *Welega* 1167, *Welewe* 1186 P]. OE *welig* 'willow'.

Wells next the Sea Nf [*Guelle, Guella* DB,

Wellis 1291 Tax], W~ So [(æt) *Willan* c 1050 KCD 837, *Welle* DB, *Welles* 1212 Fees, 1225 Ass]. 'The springs.' Wells monastery is called 'monasterium quod situm est juxta fontem magnum quem vocitant *Wielea*' (for *Wiella*) 766 BCS 200.

Wellsborough Le [*Wethelesberne* 1185 TpR, *Wenlesbergh* 1285, *Whenlesberuwe* 1316 FA, *Whelesbergh* 1300 FA]. 'Hill with a wheel or circle.' First el. OE *hweowol* 'wheel'. The forms *Wenles-, Whenles-* should be read *Weules-, Wheules-*. See BEORG.

Welney Nf [*Wellenhe* n.d. Rams]. OE *Wellan ēa* 'the river Well'. W~ is on Old Croft River, called *aqua de Welle* 1250 Ass, *Oldewelnee* n.d. Rams.

Welton Cu [*Welton* 1354 CWNS xxvii], W~ Li nr Lincoln [*Welletona* 1070–87 RA, *-tone* DB, *Wellatuna* c 1115 LiS], **W~ le Marsh** Li [*Waletune* DB, *Welletuna* c 1115 LiS, *Welleton* 1203 Cur], **W~ le Wold** Li [*Welletune* DB, c 1115 LiS], **W~ le** Np [*Waletone, Welintone* DB, *Welleton* 1199 FF, 1201 Cur], **W~ YE** [*Welletuna, Wealletune* 1080–6 YCh 964, 974, *Welleton* DB]. 'TŪN by a spring (or stream).'

Welton Nb [*Walteden* 1198 (1271), *Waltendun* 1204 Ch, *Welteden* 1242 Fees, 1307 Ch]. First el. probably a stream-name *Welte*, derived from OE *wealt* 'shaky, unsteady', originally 'rolling' or the like. Second el. OE *denu* 'valley'. Cf. WALFORD Do.

Welton So [*Welweton* 1220 Cl, *Welwenton* 1238 Ass]. 'TŪN on R *Welwe*.' See WELLOW So.

Welwick YE [*Welwic* DB, *Wellewyk* c 1195 YCh 852, *-wic* 1219 FF]. 'WĪC by a spring.'

Welwyn (wĕlin) Hrt [(on) *Weligun* 11 E, *Wilge, Welge, Welga* DB, *Welewen* 1220 Fees]. The dat. plur. of OE WELIG 'willow'.

Wem Sa [*Weme* DB, *Wemme* 1228 Cl, 1236, 1242 Fees]. Apparently a derivative of OE *wamm* 'stain'. The meaning would be 'marshy ground'.

Wembdon So [*Wadmendune* DB, *Wemedon* 1227 FF, 1243 Ass, 1257 Bath]. The first el. might be an OE *wǽþemann* 'huntsman', corresponding to ON *veiðimaðr*, MHG *weidemann* 'huntsman'. Cf. *wæðeburna* 972 BCS 1282 (perhaps 'fishing stream'), WEDMORE.

Wembley Mx [æt *Wemba lea* 825 BCS 384, *Wembanlea* ib. endorsement, *Wambeleg* 1249 FF]. '*Wemba's* LĒAH.' Cf. *Wamba*, the name of a Gothic king. The names are derived from *wamb* 'womb'.

Wembury D [*Weybiria* Hy 1 Ol, *Wenbiria* Hy 2 Ol, *-bir* 1238 Ass]. *Wey-* in the first ex. is clearly a mistake for *Wen-*. The first el. may be an OE adj. *wēoh* 'holy', corresponding to Goth *weihs* &c. Cf. WĒOH.

The OE form would be *Wēo-burg*, dat. *Wēon-byrig*.

Wembworthy D [*Mameorde* DB, *Wemeworth* 1207 Cur, *Wemmewrth* 1242 Fees]. '*Wemba*'s worþ.' Cf. WEMBLEY.

Wenden Lofts & Wendens Ambo Ess [*Wendena* DB, *Wandenne* 1207 FF, *Wenden Magna* 1238 Subs, *Great Wenden* 1252 FF, *Wenden Loot* 1255 FF, *Wenden Loutes* 1303 FA]. Second el. OE *denu* 'valley'. The first is no doubt a derivative of OE *windan* 'to wind', e.g. an adj. *wende* 'winding' (cf. WANTSUM, WENSUM) or a stream-name *Wende*.

W~ **Lofts** from an early owner. Robert Louhot had a fee here in 1236 (Fees). The name is spelt *Loholt, -hout, le Hout* 1201 f. Cur. *Lohout* is a Fr pers. n. of OG origin (OG *Hlodowald*). —**Wendens Ambo** 'the two Wendens' were formerly Great and Little Wenden.

Wendlebury O [*Wandesberie* DB, *Wendelbiry* 1219 Ep, *-bir'* 1236 Fees, *Wendlebur'* 1232 Ep, *-beri* 1242 Fees]. '*Wendel*'s or *Wendla's BURG.' Cf. WANDSWORTH and OG *Wandilo*.

Wendling Nf [*Wenlinga* DB, *Uuenlinge* c 1095 Bury, *Wenlingauilla* 1166 P, *Wenlinge*, *Wentlingg* 1254 Val]. '*Wendel*'s people.' Cf. WANDSWORTH.

Wendover Bk [(æt) *Wændofron* c 970 BCS 1174, *Wendovre* DB, *Wandoure* 1195 ff. P]. Originally a Brit name of the clear chalk stream at the place, a name consisting of the words corresponding to Welsh *gwyn* 'white' and *dwfr* 'river'.

Wendron Co [(ecclesia) *Sancte Wendrone* 1291 Tax, 1310 Ep]. 'St. Wendrona', according to Oliver.

Wendy Ca [*Wendeie* c 1080 ICC, 1201 Cur, *Wandei, Wandrie* DB, *Wendeia* 1208 Cur]. The place is in or by a bend of the Cam. The second el. is OE *ēg* 'island'. The first may be an OE *wende or the like, derived from OE *windan* and meaning 'bend'. Cf. WENDEN.

Wenham, Great & Little, Sf [*Wenham* DB, 1199 FF, *Wenham Combusta* 1228 FF, *Parva Wenham* 1254 Val]. First el. very likely OE *wynn in the sense 'pasture, meadow', identical with OHG *wunnia*. Cf. WINN. Second el. OE HĀM or HAMM.

Wenhaston Sf [*Wenadestuna* DB, *Wenhaestun* 1199 (1319) Ch, *Wenhaueston* 1197 FF, 1230 P]. '*Wynhæþ*'s TŪN.' OE *Wynhæþ corresponds to OHG *Wunnihad*.

Wenlock, Much, Sa [*Wen-, Win-, Wynloca* c 1000 Saints, *Wenloch* DB, *Gueneloch* 12 Gir, *Weneloc* 1167 P], **Little** W~ Sa [*Wenloch* DB, *Parva Wenlac* 1232 Cl, *Parva Wenlak* 1291 Tax]. The present name represents a Welsh *gwyn-loc* 'white monastery'. *Wenloca* Saints shows association with OE *loca* 'enclosure'. Little W~ is a good way off on the opposite side of the

Severn. It probably got its name from Much Wenlock. It belonged to Wenlock Priory in 1086. But an earlier name of Wenlock is known. Wenlock church is called *Wimnicensis eclesia* 901 BCS 587, and the name *Wimnicas* is mentioned by Capgrave, Vita Sanctorum, and by Leland. This name seems to be cognate with the Gaul river-name *Vimina*, the source of LA VISMES in France and WÜMME in Hanover. There is no river at Wenlock, but Wenlock Edge might have had a name derived from such a river-name, identical with *Viminiacion* in Gaul &c. Wenlock would be aptly named from the characteristic **Wenlock Edge** [*Egge* 1227 Ch].

Wenning R YW, La [*Wenninga* c 1175, *Wennyng* 1165–77, c 1220 FC, *Wenning* c 1245 CC]. An Engl river-name derived from OE *wann* 'dark'. On the river is **Wennington** La [*Wennigetun, Wininctune* DB, *Wenington* 1212 LaInq, *Old Wenigton* 1227 Ch].

Wennington Ess [*Winintune* 969 Crawf, *Uuinitune* 1065 BM, *Weninton* 1190 P, 1198 FF]. 'The TŪN of *Wynna*'s people.'

Wennington Hu [*Wenintone* c 960 BCS 1061, *Weninton* 1167 P]. 'The TŪN of *Wenna*'s people.' *Wenna* occurs in *Wennan stan* BCS 476.

Wensley Db [*Wodnesleie* DB, *-lega* 1167 P, *Wednesleg* 1212 Fees]. 'LĒAH dedicated to *Wōden*.' Cf. WEDNESBURY.

Wensley YN [*Wendreslaga* DB, *Wandeslee* 1200 Cur, *Wendesle* 1203 Cur]. '*Wendel*'s LĒAH.' Cf. WANDSWORTH. The *l* was lost owing to dissimilation. **Wensleydale** YN is *Wandesleydale* c 1150 Mon v, 568, *Wendesleidal* 1218 FF.

Wensum R Nf [*Wenson* 1096, 1119 ERN, *Wensum* 1250 Ass]. Identical with WANTSUM.

Went R YW [*Weneta* 1160 Selby, *Wenet* c 1200 Gervase, *Went* 13 Pont]. A Brit river-name. From the river were named **Went** vil. [*Wenet* c 1180 YCh 1509, *Went* 13 Pont], and **Wentbridge** [*Pons de Wenet* c 1200 YCh 1642, *Wentbrig* 1360 Ipm].

Wentford Sf [*Wanteford* 1315 Ipm]. First el. ME *went* 'way, passage', found in the East Midland Genesis & Exodus (c 1250).

Wentnor Sa [*Wantenovre* DB, *Wontenor* 1237 FF, *Wontenour* 1255 RH, *Wentenour* 1252 Fees]. The interchange of *e* and *o* (*a*) may be due to the same change of *o* to *e* as in WEDNESBURY &c. If so, the first el. may be an OE pers. n. *Wanta (*Wonta), a weak side-form of *Want* (*Wont*). The second is OFER 'steep slope'.

Wentworth Ca [*Winteworde* DB, *Wyntewrth* 1254 Val, *-worthe* c 1260 Bodl], W~ YW [*Wintreuuorde, Winteworde* DB, *Winterwurða* 1195 P, *Wintewurda* 1194 P].

'Wintra's woрþ' or less probably 'woрþ inhabited in winter'. The first *r* was lost owing to dissimilation. **Wentworth** Sr is a late name, probably derived from the family name Wentworth.

Weobley (-ĕ-) He [*Wibelai* DB, 1187 P, *Webbeley* 1242, *Webbel*' 1250 Fees], **Webton** He [*Webetone* DB, *Webbeton* 1230 P, *-tun* 1242 Fees, *Wibbitone* c 1220 Hereford], **Webtree** hd He [*Webetriehdr*' 1160, *Wibbetrehdr*' 1175 P]. The first el. is a pers. n., e.g. OE *Wibba* (cf. WEBBERY) alternating with *Wiobba, *Weobba.

OE **wēoh**, corresponding to OLG *wih*, ON *vé*, OSw *vi*, *væ* 'holy place, heathen temple', is found in the pl. n. *Cusanweoh* 688 BCS 72 and is the first el. of OE *wēofod* 'altar'. It is the source of WYE K and the first el. of WEEDON, WEEFORD, WEOLEY, WILLEY Sr, WEYHILL, WYVILLE. WYHAM represents the dat. plur. of the word. There is also an OE side-form *wīg*, found in *wiggield* 'idol', *wigbed* 'altar'. OE *wēoh* is a substantivized form of an adj. for 'holy' found in Goth *weihs* &c. An OE adj. *wēoh* may well have existed and be the first el. of WEMBURY and of *Weondune* SD, given as the name of the place where the battle of Brunanburh was fought, also of (æt) *Weonfelda* 946–51 BCS 888.

Weoley Castle Wo in Northfield [*Wileya* 1221 Ass, *Welegh* 1264 Pat]. OE *wēo-lēah* 'LEAH with a heathen temple'. Cf. WĒOH.

OE **(ge)weorc** 'work, fortification'. See WARK, ALDWARK, NEWARK, SOUTHWARK. See also WALKWOOD.

OE **weorf** 'draught cattle'. See WARLEY, WORSLEY Wo.

OE **wer** 'weir, dam' is a fairly common el. in pl. ns. There is also a side-form *wær*, found e.g. in ASC (E) 963 and in *wiredes wær* 891 BCS 565, which explains several names with *a* instead of *e*. There is also an OE *waru* 'weir', found in *mylenwaru* 'mill dam'. See WEARE, WARE, WAREHAM, WARHAM, WAREHORNE, WARFORD, WARGRAVE, WARLEIGH, WARLEY. Second el. in DUNWEAR, EDGWARE. OE *wering*, *wæring perhaps in WARRINGTON La, WARWICK Wa.

Were R. See WARMINSTER.

Wereham Nf [*Wigorham* 11 Wills, *Wigreham* DB, *Wireham* 1203 Ass, *Wirham* 1203 Cur, *Werham* 1251 Ch]. The first el. may be an old name of the Wissey, cognate with Gaul *Vigora* (now VIÈRE); cf. WYRE R.

Wergs, The, St [*Witheges* 1202 FF, *Wytheges* 1306 Ass, *Withegis* 1327 Subs]. OE *wiþigas* 'willows'.

Werneth Chs [*Warnet* DB, *Wernith* 1286 Court], W~ La [*Vernet* 1226–8 Fees, *Wernyth* 1352 FF]. 'Alder swamp, place overgrown with alders.' The name is derived from Brit *verno-*, Welsh *gwern*

'alders', and identical with Gaul *Vernetum* (now VERNET, VERNOIS &c.). Cf. also WARREN BURN.

Werrington D [*Ulvredintone* DB, *Wolverinton* 1284–6 FA]. 'TŪN of *Wulfrēd*'s people.'

Werrington Np [*Witheringtun*, *Wiðringtun* 972 BCS 1280 f., *Widerintone* DB, *Widringeton* 1198 Cur]. 'TŪN of the *Wiþeringas* or people of *Wiþer*.' Cf. WITTERING Np.

Wervin Chs [*Wivrevene*, *Wivevrene* DB, *Weruena* c 1100, *Wiruena* c 1190, *Wiruin* 1157–94 Chester]. 'Fen on R WEAVER.' Wervin is not far from the lower Mersey, which must once have been called Weaver. See further WEAVER.

Wesham (-s-) La [*Westhusum* 1189 LaCh, *Westsum* 1327 Subs]. '(At) the western houses.'

Wessex [*West Seaxna lond* 709, (Readingum on) *West Seaxe* 871 ASC, *Occidentalium Saxonum regnum* c 894 Asser]. Originally a tribal name, 'the West Saxons' [*West Seaxe* 514 ASC, *Wesseaxna kyning* c 690 Laws &c.]. Cf. ESSEX, SUSSEX.

Wessington Db [*Wistanestune* DB, *-ton* 12 Rutland, *Wystantone* 1252 FF]. '*Wigstān*'s TŪN.'

OE **west** adv. 'west' is common in pl. ns. See WESHAM and WEST- (passim). An unrecorded OE *wester* 'west', corresponding to OLG *westar*, OFris *wester*, ON *vestr* &c., is found in WESTERDALE &c. Cf. OE *westerra* 'westerly'. The superlative *westmest* is found in WESTMESTON. OE (be) *westan* 'west of' is probably in some cases the source of WEST-. See e.g. WESTBROOK, WESTPORT.

Westbere K [*Westbere* 1212, 1243 StAug]. 'Western BÆR or swine-pasture.'

Westborough Li [*Westburg* DB, 1199 P, 1202 Ass]. 'Western fort.'

Westbourne Mx [*Westeburne* 1259 FF, *-bourne* 1317 BM], W~ Sx [*Borne*, *Burne* DB, *Westbourne* 1305 Ipm]. The last is 'western stream', while the first seems to be '(place) west of the stream' (OE *be westan burnan*).

Westbriggs Nf [*Westbrigg* 1254 Val]. 'Western bridge.' Second el. OScand *bryggia*.

Westbrook Brk nr Newbury (no early forms found), W~ Brk nr Faringdon [*Westebrok* 1220 Fees], W~ K [*Westbroke* 1241 Ep], W~ Wt [*Westebroc* 1251 AD, *Westbrok* 1287–90 Fees]. The first is '(place) west of the brook'. The others seem to be 'western brook'.

Westbury Bk [*Westberie* DB, *-buri* c 1160 NpCh], W~ **on Severn** Gl [*Wesberie* DB, *Westburia* Hy 2 Glouc], W~ **on Trym** Gl [(æt) *Westbyri(g)* 791–6, *Uuestburg* 793–6 BCS 272 ff., *Hvesberie* DB, *Westbury-upon-*

Trymme 1534 LP], W~ Ha [*Wesberie* DB, *Westberia* 1167 P], W~ Sa [*Wesberie* DB, *Westbur'* 1242 Fees], W~ So [*Westbyrig* 1065 Wells, *-berie* DB], W~ W [*Westberie* DB, *-beria* 1190 P]. 'Western BURG or fort.'
Trym is a river-name, derived from OE *trum* 'strong' (an OE *Trymme*).

Westby La [*Westbi* DB], W~ Li [*Westbi* DB, 1172 DC], W~ YW nr Gisburn [*Westby* 1226 Ep]. 'Western BY.'

Westcot Brk in Sparsholt [*Westcota* 1166 f. P], **Westcote** Gl [*Westcote* c 1220 Berk], W~ Ha nr Alton [*Westcota* 1194 ff. P], W~ Wa [*Westcota* c 1140 BM], **Westcott** Bk [*Westcote* 12 BM], W~ Sr [*Wescote* DB, *Westcote* 1212 Fees]. 'Western COT.'

Westenhanger K [*Ostringehangre* 1212 RBE, *Ostringhangre* 1282 Ep, *Westringhangre* 1316 FA]. See HANGRA. The first el. seems to be a derivative of the *ōster* found in OSTERLEY &c. OE *ō* often becomes *uo* in Kent.

Westerby Le nr Market Harborough [*Westerbi*, *Westrebi* 1206 Cur]. 'Western BY', OScand *vestri bȳr*. W~ was joined with SMEETON to S~ WESTERBY.

Westerdale YN [*Westerdale* 1161-7 YCh 562], **Westerfield** Sf [*Westrefelda* DB, *Westerfeld* 1206 Cur], **Westergate** Sx [*Westgate* 1230 P, *Westregate* 1271 Ass], **Westerham** K [*Westarham* 871-89 BCS 558, *Westerham* 10 BCS 1321 f.], **Westerleigh** Gl [*Westerlega* 1176 P, *-legh* 1228 Ch], **Westerton** Sx [*Westerton* 1242 Cl, *Westreton* 1270 Ch]. 'Western valley, FELD, gate, HĀM, LĒAH, TŪN.' The first el. is OE *wester* 'west' or *westerra* 'westerly'.

Westfield Nf [*Westfeld* c 1050 KCD 907, *-a* DB], W~ Sx [*Westewelle* DB, *-felde* 1107-24 BM]. 'Western FELD.'

Westgate Du [*Westyatshele* 1457 PNNb], W~ on Sea K [*Westgata* 1168 P], W~ Nb at Newcastle [*Westgate* Hy 3 BM]. 'West gate.' In W~ K a gate leading to the sea seems to be meant. Cf. MARGATE.

Westhall Sf [*Westhala* 1169, 1176 P, *-hale* 1212 Fees]. 'Western HALH.'

Westham Sx [*Westham* 1230 Ep, *-hamme* 1252 Ch]. 'Western HAMM.'

Westhampnett. See HAMPNETT.

Westhay Np [*Westhey* 1265 Misc], **Westhead** La [*Westhefd* c 1190 LaCh], **Westhide** He [*Westhyde* 1242 Fees, *-hide* 1252 Ch]. 'Western forest enclosure, hill, hide.' See (GE)HÆG, HĒAFOD, HĪD.

Westhope Sa [*Weshope* DB, *Westhope* 1267 Ipm]. 'Western valley.' Cf. EASTHOPE.

Westhorpe Sf [*Westtorp* DB]. 'West thorp.'

Westhoughton. See HOUGHTON (2).

Westlaby Li in Wickenby [*Westledebi* DB, *Westletebi* c 1115 LiS, *Westladebi* 12 DC], **Westleton** Sf [*Westledestuna* DB, *Westleton* 1202 FF]. '*Vestliði*'s BY and TŪN.' *Vestliði* is an ON pers. n. meaning 'one who has travelled west'.

Westleigh. See LEIGH.

Westley Waterless Ca [(at) *Westle* 1043-5 Wills, *Westlai* c 1050 KCD (907), c 1080 ICC, *Westleye Waterles* 1290 Cl], W~ Sf [*Westlea* DB, *Uuestlea* c 1095 Bury]. 'Western LĒAH.' *Waterless* (*Waterleys* 1483) means 'water meadows', *-less* being the plur. of OE *lēah*; cf. PNCa(S).

Westlinton. See LYNE R Cu.

Westmancote Wo [*Westmonecote* DB, *Westmanecota* 1212 Fees]. 'COT of the Western men', perhaps Welshmen.

Westmeston Sx [*Westmæstun* c 765 BCS 197, *Wesmestun* DB]. 'Westernmost TŪN.' First el. OE *westmest*. Cf. *Westmestecumbe* 1214 FF (Sx).

Westmill Hrt [*Westmulne* 1060 Th, *-mele* DB, *-melne* 1212 Fees]. 'Western mill.'

Westminster Mx [æt *Westmunster* 785, *Westmynster* 972 BCS 245, 1290]. 'Western monastery.' Cf. THORNEY Mx.

Westmorland [*Westmoringaland* 966 ASC (E, F), *Westmeringland* 1175-84, *Westmerieland* 1190 Kendale]. 'Land of the *Westmōringas* or people west of the Yorkshire moors.'

Westnewton. See NEWTON.

Westoe. See WESTOW.

Weston Turville Bk [*Westone* DB, W~ *Turvile* 1303 FA], W~ **Underwood** Bk [*Westone* DB, W~ *Underwode* 1363 Cl], W~ Brk [*Westun* DB], W~ **Colville** Ca [*Westtuniga gemæra* 974 BCS 1305, *Westone* DB, *Weston Colevill* 1332 FF], W~ Chs nr Crewe [*Weston by Bertumleg* 1281 Court], W~ Chs nr Runcorn [*Westone* DB], W~ **Peverel** D [*Westone* DB], W~ **upon Trent** Db [*Westone* DB, *Westona* c 1100 Chester], W~ **Underwood** Db [*Westune* DB, *Weston Underwode* 1301 FF], **Buckhorn** W~ Do [*Westone* DB, *Bokeres Westone* 1285 FA, *Bokereweston* 1288 Ass], **Stalbridge** W~ Do [æt *Westune* 933 BCS 696, *Westone* DB], W~ **upon Avon** Gl [*Westtunniga gemære* 922 BCS 636, *Weston* 1220 Fees, W~ *super Abonam* 1291 Tax], W~ **Birt** Gl [*Westone* DB, *Weston la Bret* 13 BM], **King's & Lawrence** W~ Gl [*Westone* DB, *Weston* 1194 P, W~ *Sancti Laurentii, Kyngesweston* 1285 FA], W~ **Subedge** Gl [*Wæsōæma* (gemære) 1005 KCD 714, *Weston sub-egg'* 1279 Ipm], W~ Ha nr Southampton [*Westtun* c 1000 KCD 713], W~ **Corbett** Ha [*Weston* 1203 Cur, *Westone Corbet* c 1270 Ep], W~ **Patrick** Ha [*Westone* DB, *Weston* 1212 Fees, W~ *Patrik* 1316 FA], W~ **Beggard** He [*Westune* DB,

1242 Fees], **W~ under Penyard** He [*Westune* DB], W~ Hrt [*Westone* DB, 1226 Ep], **Hail W~** Hu [*Weston* c 1150 BM, *Heilweston* 1199 Cur], **Old W~** Hu [*Westune* DB, *Wald Weston* 1227 Ass], W~ Li [*Westune* DB], W~ Nf [*Westuna* DB, *Weston* 1201 Cur], W~ Np nr Towcester [*Weston* 1163 P, *W~ Pynkeny* 1311 Ipm], **Colly** W~ Np [*Weston* DB, *Colynweston* 1309 Pat, *Colliweston* 1329 QW], **W~ Fāvell** Np [*Westone* DB, 1232 Ep], **W~ by Welland** Np [*Westone* DB, *Weston super Wylond* 1377 BM], W~ Hu [*Westone* DB, *Nordweston* 1185 P], **W~ on the Green** O [*Westona* c 1130 Oxf, *Weston* 12 BM], **North W~** O [*Westun* 1209 Fees], **South W~** O [*Westone* DB, *Ueston* 1236 Fees], **Edith W~** Ru [*Westona* 1167 P, *Weston Edith* 1275 RH, *Edyweston* 1315 Misc], **Binweston** Sa [*Westun* 1255 RH, *Binneweston* 1292 Ch, *Bynne Weston* 1327 Subs], **Cold W~** Sa [*Coldeweston* 1291 Ipm, Tax], **W~ Cotton** Sa [*Westune* DB, *Weston and Coton* 1272 Eyton], **W~ Lullingfield** Sa [*Weston Lullingfields* 1324 Eyton], **Priest W~** Sa [*Westune* DB, *Preostes Weston* 1315 Ipm], **W~ under Redcastle** Sa [*Westune* DB, *Weston* 1227 Ch], **W~ Rhyn** Sa [*Westone* DB, 1272 Ipm], **W~** Sf nr Beecles [*Westuna* DB, *Weston* 1212 Fees], **Market W~** Sf [*Westuna* DB, *Weston* 1202 Cur], **W~ So** nr Bath [*æt Westtune* 946, *æt Westune* 956 BCS 814, 1009, *Westone* DB], **W~ Bampfylde** So [*Westone* DB, *Weston juxta Cammel* 1349 Ep], **W~ in Gordano** So [*Westone* DB, *Weston in Gordenlond* 1271 Ch, *W~ in Gordene* 1343 Ep], **W~ super Mare** So [*Weston* 1266, *W~ super Mare* 1349 Ep], **W~ Zoyland** So [*Sowi* DB, *Westsowi* c 1245 Glaston, *Weston* 1263 FF], **W~** Sr [*Westone* DB, 1265 Misc], **W~ Coyney** St [*Westone* DB, *Weston sub Keveremont* 1242 Fees], **W~ Jones** St [*Weston* 1242 Fees, *W~ Johannis* 1236 Fees, *W~ Jhones* 1327 Subs], **W~ under Lizard** St [*Westone* DB, *Weston under Lusyerd* 14 PNSt], **W~ upon Trent** St [*Westone* DB], W~ Wa nr Long Compton [*Weston* 1316 FA], **W~ in Arden** Wa [*Westun* 1002 Wills, *Westone* DB], **W~ under Wetherley** Wa [*Westone* DB, *Weston subtus Wethele* 1428 FA], **W~ YW** [*Westone* DB, *Westona* 1189 (1271) Ch]. All are OE *West-tūn* 'western TŪN, TŪN west of another'.

W~ in Arden Wa, see ARDEN.—**W~ Bampfylde** So was held by Johannes de Baumfeld in 1316 (FA). The name is spelt *Benfield* 1310 Ep. It is English and means 'bean field'.— **W~ Beggard** He from a family name, which may be identical with OFr *begard* used of a religious sect.—**Binweston** Sa was apparently named from a Saxon *Binna*.—**W~ Birt** Gl; cf. BIRTSMORTON (under MORTON). Richard le Bret was tenant in 1242 (Fees).—**Buckhorn W~** Do means 'the Weston of the scribe(s)'. OE *bōcere* means 'scribe'.—**Colly** W~ Np is held to have been named from Nicholas de Segrave (d. 1322). *Colin* is a pet-form of *Nicholas*.—**W~ Colville** Ca was held by William de Colevill in 1203

(Cur). Cf. CARLTON COLVILLE.—**W~ Corbett** Ha was held by Thomas Corbet in 1203 (Cur). Cf. CHADDESLEY CORBETT.—**W~ Cotton** Sa is W~ and Cotton. Cf. COT(T)ON.—**W~ Coyney** St was held by Johannes Koyne in 1242 (Fees). Coyney must be a French family name. For *Keveremont* cf. KIRMOND Li.—**Edith W~** Ru probably from Queen Edith, who had large possessions in Rutland at the time of the Conquest (DB).—**W~ Favell** Np was held by Johannes Fauvel in 1232 (Ep). *Favel* is a nickname identical with OFr *fauvel* 'fallow-coloured', often used as a symbol of hypocrisy.— **W~ in Gordano** So. See EASTON IN G~.—**Hail W~** Hu contains an old name of the R Kym [*Haile* c 1180, *Heile* c 1200 ERN]. It is of Brit origin and derived from the root *sal-* 'dirt-coloured, dirty' in OIr *sal* 'filth', Welsh *halog* 'polluted' &c. It means 'dirty stream'.—**W~ Jones** St from an early owner.—**Lawrence W~** Gl from the dedication of the church.— **W~ under Lizard** St is nr LIZARD HILL Sa.— **W~ Lullingfield** Sa. L~ would seem to mean 'FELD of *Lull*(a)'s people'.—**Market W~** Sf had a market in 1263 (BM).—**Old W~** Hu is Weston in the wood called *Bromswold*; see LEIGHTON BROMSWOLD.—**W~ Patrick** Ha from Patrick de Chaworth, who lived in the 13th cent. (VH).—**W~ under Penyard** He. See PENYARD.—**W~ Peverel** D belonged to the Peverels in 1228 (FF). Cf. BRADFORD PEVERELL. —**Priest W~** Sa must have belonged to some priest or priests.—**W~ under Redcastle** Sa. See REDCASTLE.—**W~ Rhyn** Sa is on a ridge. *Rhyn* is Welsh *rhyn* 'hill, promontory'.—**W~ Subedge** Gl is 'W~ at the foot of the steep ridge'. See ECG.—**W~ Turville** Bk was held by Galfrid de Tureuill before 1174 (P). Cf. ACTON TURVILLE.—**W~ Underwood** Bk, Db is 'W~ in the wood'. See UNDER.—**W~ under Wetherley** Wa from a wood called *Wethele* 1204 Ch (apparently OE *wæþelēah* 'wood for hunting'; cf. WEDMORE, WEMBDON).—**W~ Zoyland** So. See MIDDLEZOY. *Zoy-* from *Sowi* is an old name of a district. Zoyland is 'land belonging to *Sowi*'.

Westoning Bd [*Westone* DB, *Weston Ing* 1377 AD]. Originally WESTON. The manor was held by Chief Justice William Inge in 1303 (Ch).

Westover So [*Westour'* 1225 Ass, *Westovre* 1304 Ch], W~ Wt [*Westovere* 1327 Subs]. 'Western bank' (see ŌFER).

Westow YE [*Wynestowe* 1227 Ch, *Wifestowe* 1237 Ep, *Wiuestowe* 1285 Ipm]. '*Wifa*'s or rather *Wife*'s STŌW or holy place.' Cf. *Wifan stocc* 909 BCS 624 (Ha) and WIVENHOE &c. Identical with **Westoe** Du [*Wywestoue* 1195 (1335) Ch, *Wiuestoue* 13 FPD].

Westport St. Mary W [*Westeporte* Hy 1 Malm, *Westport* 1232 Ch]. OE (be) *westan porte* '(the place) west of the town' (i.e. Malmesbury), as shown in PNW(S).

Westrill Le [no early forms found]. Said to be for *Westerhill* 'western hill'.

Westrip Gl, **Westrop** W [*Westhropp* 1275 RH]. 'West thorp.'

Westward Cu [*Le Westwarde in Allerdale* 1354 PNCu(S), *West Warde* 1569 CWNS xxiii]. 'The western *ward* or division' (of Inglewood Forest).

Westward Ho D. A modern settlement named from Kingsley's famous book.

Westwell K [(æt) *Wyllan* 1004 Wills, *Welle* 11 DM, *Westwell* 1226 Ass], W~ O [*Westwelle* DB, 1192 P]. 'Western spring or stream.'

Westwick Ca [*Westuuiche* DB, *Westwica* 1130 P], W~ Du [*Westewic* 1091 FPD], W~ Nf [*Westuuic* DB, *-wik, -wich* 1254 Val], W~ YW [*Westwic* c 1030 YCh 7, *Westuuic* DB]. 'Western WIC or dairy-farm.'

Westwood K [*Beuuestanuudan* 805 BCS 323, *Westwod* 1206 Cur]. '(Place) west of the wood', i.e. Blean Forest.

Westwood W [*Westwuda* 987 KCD 658, *Westwode* DB], W~ Wa [*Westwode* 1386 AD], W~ **Park** Wo [æt *Westwuda* 972 BCS 1284, *Westwod* 1206 Cur]. 'West wood.'

Wetheral Cu [*Wetherhala* c 1100 WR, *-hal* 1229 Cl, *Wederhala* 1186 P]. 'Haugh where wethers were kept.' See HALH.

Wetherby YW [*Wedrebi* DB, *Werebi* 1190 P, *Wetherby* 1238 FF]. 'Wether farm.'

Wetherden Sf [*Wederdena* DB, *Wetherden* 1197 FF, *-denn* 1201, 1203 Cur]. 'Wether valley.' See DENU.

Wetheringsett Sf [*Weddreringesete* 1023–50 Th, *Wetheringsete* 1043–5 Wills, *Wederingesete* c 1050 KCD 907, *Wederingaseta* DB, *Wetheringeset* 1201 Cur]. Perhaps 'the (GE)SET or fold of *Wedr*'s people'. But the place is not very far from WETHERDEN, and the first el. of the name might be an OE *Weþeringas* 'people of Wetherden'.

Wethersfield Ess [*Westrefelda*, *Witheresfelda* DB, *Wihterefeld* c 1150 BM, *Weðresfeld* 1177 P]. '*Wiohthere*'s (*Wihthere*'s) FELD.'

Wetmoor St [*Wihtmere* 11 PNSt, *Witmere* DB, *Wismera* a 1113, *Withmere* 1114 Burton, *Wichtmere* c 1235 BM]. 'Lake by a bend.' Cf. WIHT. W~ is on the Trent.

Wettenhall Chs [*Watenhale* DB, *Wetenhala* c 1150 Chester, *Wetinhale* 1260 Court]. 'Wet HALH.'

Wetton St [*Wettindun* 1252 Ch, *Wetton* 1327 Subs]. 'Wet DŪN.' Cf. WATTON YE.

Wetwang YE [*Wetwangham* DB, *Wete Wang* 1114, *Wetewanghe* 1145–56 YCh 46, 1238, *Wetewonge* 1190 P]. Probably 'wet field'. OE *wang* means 'plain, field'.

Wetwood St [*Wetewode* 1291 Tax]. 'Wet wood.'

Wexcombe W SE. of Burbage [*Westcumba* 1156 ff., 1173, 1190 P, *Westycumb* 1231 Cl, *Wexcumba* 1168 P, *-cumbe* 1201 Cur, *-cumb* 1231 Cl]. Probably OE *weax-cumb* 'valley where wax was got', later associated with

OE *west* 'west' and sometimes remodelled to *Westcumb*.

Wexham Bk [*Wesham* 1195 Cur, *Wexham* 1219 Ep, 1252 Ch]. 'Homestead where wax was produced.'

Wey R Do [*Waye* 1244, 1288 Ass, *Weye* 1367 Pat]. A British river-name identical with WEY Ha and with WYE. On the river are **Broadway** [*Waia* DB, *Brode Way* 1242 Fees] and **Upway** [*Wai* DB, *Upeweye* 1311 Fine]. Broadway is 'Great Wey', while Upway is 'Upper Wey' or '(place) up R Wey'.

Wey R Ha, Sr [*Waie* a 675, c 890, (on) *Wegan* 956 BCS 34, 563, 955, *Waie* 1190 ff. P, *Wey* 1235 Cl]. Cf. preceding name.

Weybourne Nf [*Wabrunna, Wabrune* DB, *Walbruna* 1158 Fr, *Wabrun* 1177 P, *-brunn* 1228 Cl]. Very likely OE *wearg-burna* 'felon stream' (cf. WEARG) with loss of g between the consonants and of the first r owing to dissimilation.

Weybread Sf [*Weibrada* DB, *-breda* 1187 P, *Weiebred* c 1200, *-brade* c 1205 BM]. 'Strip of acre on the road.' Cf. BREDE. The road seems to be a Roman road.

Weybridge Sr [*Waigebrugge* a 675 BCS 34, *Weybrigga* 1062 KCD 812, *Webruge* DB]. 'Bridge over R WEY.'

Weyhill Ha, also called PENTON GRAFTON [*Leweo* 13 VH, *la Wou* c 1270, *la Woe* 1299 Ep, *Weye* 1412 FA]. OE *wēoh* 'holy place, heathen temple', with late addition of *hill*.

Weymouth Do [*Wai-, Waymouþe* 939 BCS 738, *Waimuda* 1130 P, *Weymuthe* 1258 Pat]. 'The mouth of R WEY.'

Whaddon Bk [*Hwætædun* 966–75 Wills, *Wadone* DB, *Whaddon* 1241 Fees], W~ Ca [*Wadona* c 1080 ICC, c 1150 Fr, *Wadone* DB, *Phwaddune* 1086 IE], W~ Gl [*Wadune* DB, *Whaddon* 1252 Ch], W~ W nr Melksham [*Wadone* DB, *Waddon* 1254 Ipm, *Whaddon* 1428 FA]. OE *hwǣte-dūn* 'wheat hill'.

Whaddon W nr Salisbury [*Watedene* DB, *Hwatedena* 1109–20 Sarum]. OE *hwǣte-denu* 'wheat valley'.

Whale We nr Lowther [*Qwalle* 1278 Ipm, *Qwale* 1345 Cl]. ON *hváll* 'round hill'.

Whaley Chs [*Weyeleye* 1284 Ipm, *Weyleg, Weyelegh* 1285 Court]. 'LĒAH by a road.'

Whaley Db [*Walley* 1230 P, 1255 Ipm, *Whaley* 1540 PNDb(S)], **Whalley** (-aw-) La [(æt) *Hwælleage* 798 ASC (D), *Hweallæge* ib. (E), *Wallei* DB, *Whalegh* 1246 Ass], **Whalton** (wahtn) Nb [*Walton* 1203 P, *Whalton* 1205 Percy, *-tun* 1254 Val, *Hwalton* 1269 Ass]. The first el. is no doubt an old word for 'hill' or the like. It might be an OE *hwæl*, related by Ablaut to ON *hváll* 'round hill'. Or it may be simply OE *hwealf* 'vault, arch; vaulted, hollow'. In forms such as *Hw(e)alf-lēah, -tūn* the *f* would easily disappear. The word *hwealf* would here be used in the sense 'hill'. See LĒAH.

Whaplode (-ŏp-) Li [*Cappelad* 810 BCS 331, *Cope-, Copolade* DB, *Quappelad* 1202 Ass, *-lade* 1212 Fees, *Quapelode* 1254 Val]. 'Eelpout stream or ditch.' The first el. is *quap* 'eelpout', found first in 1598 and held in OED to be a Continental loan-word, but doubtless a native word. The second is OE (GE)LĀD 'water-course'.

Wharfe R YW [*Weorf* 963 BCS 1112, 1352, *Werf* 1158 Selby, *Uuarf* c 1155 BM, *Werue* 1268 Ass, *Hwerf* 1170–89 YCh 1854, *Wherf* c 1155 BM]. A Brit river-name meaning 'winding river' and derived from the root of OE *weorpan* 'to throw', lit. 'to twist', Lat *verbena*. From the river-name is derived *Verbeia*, the name of a deity, found in a Roman inscription at Ilkley. The name was later associated with OScand *hverfr* 'winding' and made into *Hwerf, Wharfe.* **Wharfedale** is *Hwerverdale* 12 SD, *Werverdal* 1204 Ch. The original form is an OScand *Hverfardalr* 'valley of the Wharfe', *-ar* being the OScand gen. ending.

Wharfe YW nr Settle [*Qwarf* 1297 Subs, *Querf* 1406 YInq]. ON *hvarf, hverfi* 'bend', 'corner', very likely also, like Norw *kverv, kverve,* 'a group of homesteads'.

Wharles (wawlez) La [*Quarlous* 1249, *Werlows, Warlawes* 1286 Ipm]. 'Hills with a stone circle.' The elements are OE *hwerfel* 'circle' and *hlāw* 'hill'. The stone circle also gave its name to ROSEACRE close by.

Wharncliffe YW. OE *cweorn-clif* 'millstone hill'. There are still quarries here.

Wharram Percy & le Street YE [*Warran, Warron, Warham* DB, *Warrum* 1126–9, *Warram* c 1180 YCh 1012 f., 1087, *Warham* 1179 ff., *Hwarrum, Hwarhum* 1230 P, *Wharrom Percy* 1291 Tax, *Warrum juxta Strete* 1346 FA]. Apparently OE *hwerhamm* 'enclosure in a kettle-like valley'. This suits W~ Percy. The first el. would be OE *hwer* 'kettle, cauldron, pot, basin', used in a transferred sense also in WHERWELL, though here in a side-form OE *hwær*, analogous to *wær* by the side of *ver* 'weir'. The second el. (*hamm* or rather *homm*) became -(*h*)*um* in the unstressed syllable. Cf. NUNBURNHOLME.

W~ **Percy** was held by William de Perci in 1177 (P). Cf. BOLTON PERCY.—W~ **le Street** is on an ancient road supposed to be Roman.

Wharton Chs [*Wanetune* DB, *Waverton* 1260, 1288 Court]. The place is on R Weaver, and the name may mean 'TŪN on R WEAVER', but the early forms rather suggest identity with WAVERTON Chs. **Wharton** He nr Leominster [*Wavertune* DB, *-tona* 1242 Fees] is certainly a variant of Waverton.

Wharton Li [*Warton* DB, *Wartona* 1139 RA]. Apparently identical with WARTON (I). But the first el. may also be OE *waroþ* 'shore', as in WARWICK Cu.

Wharton We [*Werfton* 1202 FF, *Querton* 1238 P, 1292 QW]. 'TŪN by the shore or embankment.' First el. OE *hwearf, hwerf* 'shore, embankment'. The place is on the Eden.

Whashton YN [*Whasinge-, Wassingetun* c 1170 Marrick, *Wassington* 1208 FF, *Qwassyngton* 1285 FA]. Cf. *Hwessingatun* BCS 1131 (Np). The names seem to mean 'TŪN of the people of *Hwæssa*'. *Hwæssa* may be a short form of an unrecorded **Hwætsige* or a derivative of OE *hwæss* 'sharp'.

Whatborough Le [*Wetberge* DB, *Watebergia* c 1125 LeS, *Whatbergh* 1428 FA]. 'Wheat hill', OE *hwǣte-beorg*.

Whatcombe Do in Winterborne Whitchurch [*Hwetecumb* 943 BCS 781]. 'Wheat valley.'

Whatcote Wa [*Quatercote* DB, *Whatcote* 1240 Bodl], **Whatcroft** Chs [*Wate-, Quate-, Wetecroft* 1287 ff. Court], **Whatfield** Sf [*Watefelda* DB, *Whatefeld* 1205 FF], **Whatley** So nr Frome [*Watelei* DB, *-leg* 1225 Ass], **W~** So in Winsham [*Watelege* DB, *Whetelega* 1176 Wells]. 'COT, croft, FELD, LĒAH or glade where wheat was grown.'

Whatlington Sx [*Watlingetone* DB, *Whatlingetune* c 1195 Penshurst]. Perhaps 'TŪN of *Hwætel*'s people', the first el. being a pers. n. **Hwætel*, a short form of names in *Hwæt-*. But the first el. may be a derivative of a lost pl. n. *Hwǣte-lēah* 'Wheatley'.

Whatstandwell Db [*Watstanwell* 1510 AD]. W~ Bridge was named from *Walter Stonewell*, nr whose house it was built in 1390 (Rutland).

Whatton, Long, Le [*Wact(h)on* c 1125 LeS, *Watton* 1195 P, *Whatton* 1327 Subs, *Whatteton* 1428 FA]. If the earliest forms are trustworthy, the first el. may be as in WAKEFIELD. But perhaps the name is identical with next name.

Whatton Nt [*Watone* DB, *Whaton* 1231 Ep, *Whatton* 1242 Fees]. OE *Hwǣte-tūn* 'wheat farm'.

Whaw YN [*Kiwawe* 1280, *Le Kuawhe* 1285 Ipm]. OE *cū-haga* and *cȳ-haga* 'enclosure for cows'.

Wheatacre Nf [*Hwateaker* DB, *Qwet-, Whetacre* 1254 Val]. 'Wheat field.'

Wheatenhurst Gl [*Witenhert* DB, *-herste* 1195, *Whitenherste* 1197 P, *Hwitehurste* 1220 Fees]. 'White HYRST' or '*Hwita*'s HYRST'.

Wheatfield O [*Witefelle* DB, *Whitefeld* 1197 f. P, *Pwytefeld* 1241 Ep]. 'White FELD.'

Wheathampstead Hrt [*Huuæthamstede*

1065 BM, *Watamestede* DB], **Wheathill** Sa [*Whethull* 1237 FF, 1255 RH], W~ So [*Watehelle* DB, *Whethull* 1331 Ep]. 'Homestead and hill where wheat was grown.'

Wheatley Ess in Rayleigh [*Wateleia* DB, *Watel'* 1181 P], W~ La [*Watelei* DB, *Whetelegh* 1227 FF], W~ **Booth** La [*Whitley* 1502 Whitaker, *Wheyteley* 1516 Ct], W~ Nt [*Wateleie* DB, *Hwetele* 1237 Fees], W~ O [*Watele* 1208 Cur, *Whateleg* c 1265 Bodl], W~ YW nr Doncaster [*Watelage* DB, *Wetelag* 1219 FF, *Whetelagh* 1280 Ch], W~ YW nr Otley [*Hwatele* 1219 Ass, *Weteley* 1314 Ipm]. 'LĒAH where wheat was grown.'

Wheaton Aston. See ASTON.

Wheddon So [*Wheteden* 1243, 1253, *Whetedon* 1253 Ass]. 'Wheat hill or valley.'

Wheelock R Chs [*Qwelok* 1321 AD, *Whelok* 1440 BM]. On the river is **Wheelock** vil. [*Hoiloch* DB, *Wehlok* 1260 Court, *Welok* 13 BM]. Wheelock is a Brit river-name cognate with *Chwilogen* R 1198 Mon v (in Wales) and derived from Welsh *chwel*, *chwyl* 'turn' (from a root *svel-* 'to turn'). Evans gives a Welsh adj. *chwelog* 'having turns'. The Wheelock is a very winding river.

Wheelton La [*Weltona* c 1160 LaCh, *Whelton* c 1200 WhC]. First el. OE *hwēol* 'wheel', e.g. 'water-wheel' or 'stone circle'. In Wheelton was a place called *Whelcroft* 13 WhC.

Wheldale YW [*Qveldale*, *Weldale* DB, *Quelledale* 1242 Fees]. First el. OE *hwēol* 'wheel', which may refer to the windings of the Aire.

Wheldrake YE [*Coldrid* DB, *Coldric* 1176, 1194 P, 1207 Cur, *Coudric* 1218 FF, *Queldric* 1190, 1230 P, *-rike* 1231, *-rick* 1246 FF]. The second el. is no doubt OE **ric* 'stream, ditch'. The first looks like OE *cweld*, *cwield* 'destruction, death'. The name may mean 'felon stream'; cf. WARNBOROUGH.

Whelnetham, Great & Little, Sf [*Hvelfiham* DB, *Weluetham* 1170, *Welfuetham* 1179, *Welueteham* 1198 P, *Welnetham* 1206 f. Cur, *Wheluetham* 1242 Fees, *Welnetham Magna, Parva* 1254 Val]. Cf. THELNETHAM. Probably OE *Elfethamm* 'HAMM frequented by swans', with addition of OE *hwēol* 'wheel' for distinction from Thelnetham. *Whelnet-* for *Whelvet-* is due to misreading of *Wheluet-*.

Whelpington, West, & Kirkwhelpington Nb [*Welpinton* 1176 P, *Whelpinton* 1245 Ipm, 1267 Ch]. 'TŪN of *Hwelp*'s people.' *Hwelp* is a nickname derived from OE *hwelp* 'cub'. Cf. ON *Hvelpr*, OG *Hwelp*, *Welp*, *Welpo* (common).

Whelprigg We [*Whelprigge* 1392 Kendall]. 'Cub ridge' (ON *hvelpr* 'cub' and *hryggr* 'ridge'). Wolf cubs may be meant.

Whenby YN [*Quennebi* DB, *Quenebi* 1202 FF, *-by* 1235 FF]. OScand *Kvenna-bȳr* 'BY of the women'; cf. Sw KVINNEBY.

Whepstead Sf [*Hwipstede* c 1000 BCS 1306, *Hwip-, Hwepstede* 11 EHR 43, *Huepestede* DB, *Hwepstede* 1219 FF]. The first el. is no doubt the OE base of ME *whippe* 'whip', early Mod *whip* 'twig, sprig, slender branches'. The probable meaning of the name is 'place where brushwood grew'. For the change of *i* to *e* cf. SWEFLING.

Whernside YW nr Ingleborough, a hill of 2,414 ft. [*Qwernsyd* 1204, *Querneside* 1251 FC], **Great & Little** W~ YW nr Kettlewell, hills of 2,310 and 1,984 ft. [*Querinsyde* 1307 YInq]. OE *cweorn-side* 'hill-side where millstones were got'.

Wherstead Sf [*Weruesteda* DB, *Warvestede* 1207 Cur, *Weruested* 1207 FF, *Hwerstede* 1275 RH, *Querstede* 1283 BM]. 'Place by the shore or embankment.' Cf. WHARTON. We. The place is on the Orwell.

Wherwell Ha [*Hwerwyl* 955 BCS 912, (to) *Hwerwillon, Hwærwellan* 1048, 1052 ASC (E, D), *Warwella* 1130 P]. 'Cauldron springs', the first el. being OE *hwer* 'kettle, cauldron' &c. Cf. WHARRAM YE. The reference is presumably to some bubbling springs.

Whessoe Du [*Wessou* 12 FPD, *Wessehou* 1304 Pat, *Whessowe* 1307 RPD]. Possibly '*Hwessa*'s HŌH or spur of hill'. Cf. WHASHTON. **Hwessa* would be a derivative of OE *hwæss* 'sharp'. Or the base may be OE *hwæssa hōh* 'steep hill', the change of *æ* to *e* being analogous to that in ESH, HETT. Cf. *Whasseho* 1208 FF (Easby YN). But the hill does not seem to be steep.

Wheston Db [*Whestan* 1231 Cl, *Whetston* 1271 FF], **Whetstone** Le [*Westham* DB, *Whetestan* Hy 2 (1318) Ch, *Hwetstan* 1254 Ipm], W~ Mx [*Whetston* 1466 FF]. Cf. also **Westernhope** Nb [*Whestanhope* 1418 PNNb]. 'The whetstone', OE *hwetstān* 'hone'. The reference may be to a stone used for whetting scythes or the like. Or the word may have a meaning like that of dial. *whetstone*, viz. 'strata of argillaceous and siliceous hazle-stone in the carboniferous limestone formation' (Nb).

Wheyrigg Cu [*Q[ui]rig* c 1220 Holme C (printed *Q[uit]rig*), *Whirig* 1332 Subs]. ON *kvi-hryggr* 'ridge with a fold' (ON *kvi*).

Whicham Cu [*Witingham* DB, *Wintinghaham* c 1125, *Hwithingham* 1187 StB]. 'HĀM of *Hwita*'s people.'

Whichford Wa [*Wicford* DB, *Wicheforda* c 1130 BM, *-ford* 1194 P, *Wuchcheford* 1236 Fees]. OE *Hwicca ford* 'the ford of the *Hwicce*'. The *Hwicce* were a large tribe settled in Gloucestershire, Worcestershire and Warwickshire. Early forms of the tribal name are: *prouincia Huicciorum* c 730 Bede,

Hwicca mægð c 890 OEBede, (of) *Hwiccium* 800 ASC.

Whickham Du [*Quicham* 1196 P, 1201 Ch]. 'HĀM with a quickset hedge.' Cf. OE *Cwichege* c 772 BCS 207 and the lost **Quick** La [*Quike* 1202 FF].

Whilton Np [*Woltone* DB, *Whelton* 12 NS, *Hwelton* 1254 Val]. Identical with WHEELTON La. OE *hwēol* 'wheel' here refers to a round hill.

Whimple D [*Winple* DB, *Wimpoll* 1218 Pat, *Wympol* 1296 FF]. Originally the name of a stream, consisting of the words appearing in Welsh as *gwyn* 'white' and *pwll*, *poll* 'pool, stream'.

Whinburgh Nf [*Wineberga* DB, *Quyneberge* 1254 Val], **Whinfell** Cu [*Wynfell* c 1170 Holme C, *Quinfel* 1308 Ipm], W~ We [*Quynfell*, *Wynfel* c 1180, *Quinnefel* c 1210 Kendale]. 'Hill and fell overgrown with furze.' ME, Mod *whin* 'furze' is common in England. The word may be of Scand origin. It is found also in **Whinnyriggs** We.

Whippingham Wt [*Wipingeham* DB, *Wippingeham* 13 BM, *Wyppingham* 1287–90 Fees]. 'HĀM of *Wippa*'s people.' *Wippa* occurs in *Wippan hoh* BCS 883.

Whipsnade Bd [*Wibsnede* 1202 FF, -*sneda* 1209–19 Ep, *Wibbesnade* 1236 FF]. '*Wibba*'s SNÆD or wood.' Cf. WEOBLEY.

Whisby Li [*Wizebi* DB, *Wiscebi* 1202 Ass, -*by* 1212 Fees, *Whisceby* 1322 Ipm]. OScand *Hvits bȳr* '*Hvit*'s BY'. First el. ON *Hvitr*, OSw, ODan *Hvit* pers. n.

Whissendine Ru [*Wichingedene* DB, *Wissendena* 1176 P, -*den* 1203 Ass], **Whissonsett** Nf [*Witcingkeseta* DB, *Witchingseta* 1191 P, *Wicingesete* 1196 FF]. 'Valley and fold of the *Wicingas*'; see DENU, (GE)SET. *Wicingas* is probably a tribal name derived from pers. ns. in *Wic*-, as *Wicbeorht*. Whissonsett is possibly 'the fold of the people of WITCHINGHAM'. The change of *ch* to *ss* is due to Norman influence.

Whistley Brk [*æt Uuiscelea* 968 BCS 1226, *Wiselei* DB, *Wisselea* 1167 P]. OE WISC 'meadow' and LĒAH 'glade'. Identical with *wisclea* (*geat*) 909 BCS 625 (Ha).

Whiston La [*Quistan* 1190 CC, *Whitstan* 1341 NI], W~ YW [*Witestan* DB, *Wistan* J Ass, *Whiteston* 1291 Tax]. 'White stone.'

Whiston Np [*Hwiccingtune* 974 BCS 1311, *Wichinton* 1060 Th, *Wice*(*n*)*tone* DB, *Hwichentone* 1228 Ep]. OE *Hwiccena tūn* 'TŪN of the *Hwicce*'. Cf. WHICHFORD.

Whiston St [*Witestun* 1002 E, 1004 KCD 710, -*tone* DB, *Whyston* 1251 Ass]. Possibly '*Hwīt*'s TŪN'. OE *Hwit* is not actually evidenced, while *Hwita* is common. Cf. ON *Hvítr*. But the early forms rather suggest a first el. without *H*-. It might be an OE **Witi*, a side-form of *Wita*.

Whitacre, Nether & Over, Wa [*Witacre* DB, -*acra* 1175 P, *Netherwhitacre* 1321 BM, *Over Wytacre* 1276 Ipm]. 'White field.'

Whitbarrow We, a hill [*Whitberg* c 1195 Kendale, *Witeberge* 1196 FF]. 'White hill.'

Whitbeck Cu [*Witebec* 12 StB, *Quitebec* 1240–56 FC], **Whitbourne** He [*Whyteburne* 1241 Hereford, *Whiteburn* 1269 Ipm]. 'White stream.' Cf. BECK, BURNA.

Whitburn Du [*Hwiteberne* c 1190 Godric, *Wytebern* 1303 Ep]. '*Hwita*'s tumulus'; cf. BYRGEN. Less likely '*Hwita*'s barn'.

Whitby Chs [*Witebia* c 1100, -*beria* c 1150, -*bi* c 1190 Chester], W~ YN [*Witebi* DB, c 1150 SD, *Hwitebi* 1104–8 SD, *Quiteby* 1218 FF]. 'White village or town.' Whitby YN is recorded in its OScand form in a verse of the 12th cent. (Heimskringla) as *Hvitabý* (dat.).

Whitchester Nb [*Witcestre* 1221 Pat, *Whitcestr'* 1244 Cl]. 'White Roman fort.' See CEASTER.

Whitchurch Bk [*Wicherce* DB, *Witcherche* 1163 P, *Whitecherch* 1197 P], W~ D [*Wicerce* DB, *Whytecherch* 1242 Fees], W~ **Canonicorum** Do [(æt) *Hwitancyrican* c 880 BCS 553, *Witcerce* DB], **Winterborne** W~ Do [see WINTERBORNE], W~ Ha [*Hwitan cyrice* 909, (to) *Hwitcyrcan* 10 BCS 624, 1161, (æt) *Hwitciricean* 1001 ASC], W~ He [*ecclesia de Albo monasterio* 1291 Tax], W~ O [(æt) *Hwitecyrcan* 990–2 ASCh, *Hwitcyrce* 1012 KCD 1307, *Witecerce* DB], W~ Sa [*Album Monasterium* 1199 Eyton], W~ So [*Hwite circe* 1065 Wells, *Wytchirche* 1230 P], W~ Wa [*Witecerce* DB, *Whitechurch* 1291 Tax]. 'White church', in reality very likely 'stone church'. Bede says that **Whithorn** in Wigtownshire [*Candida Casa* c 730 Bede, (æt) *Hwitan Ærne* c 890 OEBede], 'white house', was so called because it was built of stone.

Whitcliff Gl [*Wyteclyve* Hy 3 Berk]. 'White cliff.'

Whitcombe Do [*Widecome* 939 BCS 738, DB, -*cumbe* 1198 P, -*cumb* 1212 Fees], W~ W [*Widecome* DB, -*cumba* 1242 Fees], W~ Wt [*Witecome* DB, *Wydecoumb* 1287–90 Fees]. Looks like 'wide valley' (see CUMB). But the first el. is sometimes at least OE *wiþig* 'willow'. See WIDCOMBE.

Whitechapel Mx [*la White Chapel* 1344 LoPleas]. 'White chapel.'

Whitecliff Do nr Swanage [*Witecliua* DB]. 'White cliff.'

Whitefield Gl [*Wicfeld* DB, *Whytfeld* 1285, *Wyghtfeld* 1303 FA, 1327 Subs, *Wythfeld* 1291 Tax]. The first el. is OE *wiht* 'bend', which here refers to a recess in a neighbouring hill.

Whitefield La [*Whitefeld* 1292 QW], W~ Wt [*Witesfel* DB, *Witefeld* 1182 P, *Whitefeld* 1287-90 Fees]. 'White FELD.'

Whitegate Chs [*Whitegate* 1545 Ormerod]. 'White gate.'

Whitehaven Cu [*Withofhavene* 12, *Wytofthavene* 13 StB, *Hwithothehavene* 1202 ib.]. The first el. is the old name of the headland at W~ [*Witahoua* c 1125, *Withoue* 12 StB], an ON *Hvíta-hǫfuð* 'white headland'. Near W~ is Swartha Brow [*Suart-houed* R 1 (1308) Ch]. This name means 'black headland'.

Whitehill Du [*Whytehill* 1382 Hatfield]. 'White hill.'

Whitehill O in Tackley [*Wihthull* 1004 KCD 709, *Wistelle* DB, *Wichthulla* c 1130 Oxf]. 'Hill with a curved hollow.' First el. OE *wiht* 'bend'.

Whitehorse Hill, Vale of the White Horse Brk [*Mons Albi Equi* 958 Abingd i. 477, perhaps a later addition, *mons, ubi ad Album Equum scanditur* Hy 1 ib. ii. 125, *le Witehors* 1273 Ipm]. The hill takes its name from a colossal prehistoric figure of a galloping horse on its north-western flank. Near by is Wayland Smith's Cave [*Welandes smiðōe* 955 BCS 908]. *Smiðōe* is 'smithy'.

White Lackington. See LACKINGTON.

Whiteoxmead So nr Wellow [*Witochesmede* DB, *Whittukesmed* 1225 Ass]. '*Hwit(t)uc*'s meadow.' *Hwituc* is found BCS 609. Cf. WHITTINGSLOW.

Whiteparish W [*Wyteparosche* 1319 Ipm, *La Whiteparisshe* 1412 FA]. 'White parish.' The earlier name was *la Whytechyrche* 1278 PNW(S), *Album Monasterium* 1291 Tax, which means 'white church'. The explanation of the later name Whiteparish is obscure.

Whitestone D [(of) *Hwita stane* 1072-1103 E, *Witestan* DB]. 'White stone.'

Whitfield Db [*Witfeld* DB, *Whitefeld* 1226 FF], W~ K [*Whytefeld* 1228, 1286 Ch], W~ Nb [*Witefeld* 1254 Val, 1279 Ass], W~ Np [*Witefelle* DB, *Whitefelde* 1219 Ep]. 'White FELD.'

Whitford D [*Witefort* DB, *-ford* 1168 P, *Whytford* 1228 FF]. 'White ford.'

Whitgift YW [*Witegift* c 1080 YCh 468, *Whitegift* 1232 Ch]. Perhaps '*Hviti*'s or *Hwita*'s gift'. Second el. *gift*, a word of OScand origin. The first is rather OScand *Hviti* than OE *Hwīta*.

Whitgreave St [*Witegraue* 1193 P, *-grave* 1203 Cur, *Whytegrave* 1227 Ass, *Witegreve* 1251 Misc]. 'White grove.' See GRÁF.

Whitkirk YW [*Witechirche* c 1160 YCh 1770, *Whitekirk* 1291 Tax]. 'White church', perhaps a Scandinavianized form of WHIT-CHURCH.

Whitleigh D [*Witelie* DB, *Whytelegh* 1242 Fees], **Whitley** Brk at Reading [*Witelei* DB, *-leia* 1198 P], W~ Chs nr Northwich [*Witelei* DB, *Wytele* 1288 Court], W~ Nb nr Tynemouth [*Wyteleya* Hy 2, *Hwyteleya* 1198 (1271) Ch], W~ W [(at) *Witlege* 1001 KCD 706, *Witelie* DB, *Whitele* 1254 Pat], W~ Wa [*Whitel*' Hy 3 AD iii], W~ YW nr Knottingley [*Witelai* DB, *Wit(t)elay* 1196 P, *Withelai* 1202 FF], **Lower & Upper** W~ YW [*Witelaia, -lei* DB, *Hwitteleia* c 1200 YCh 1701]. 'White LĒAH.'

Whitlingham Nf [*Wislinge-, Wisinlingaham* DB, *Wicthlingham* 1206 Cur, *Withlingham* 1254 Val]. 'The HĀM of *Wihthelm*'s people.'

Whitmore St [*Witemore* DB, *Whytemore* 1227 Ass]. 'White moor.'

Whitnash Wa [*Witenas* DB, *Wihtenassh* 1236 Fees]. '(At) the white ash.'

Whitney He [*Witenie* DB, *Hwytene* E 1 BM, *Whyteneye* 1283 Ch]. '*Hwita*'s island' or 'white island'. See ĒG.

Whitsbury Ha [*Wiccheber* c 1130, *-beria* c 1160 (1332) Ch, *Wicheberia* 1168 P]. 'Wych elm BURG.' See WICE.

Whitstable K [*Witenestaple* (hd) DB, *Whitstapel* 1197 f. P, *-staple* 1258 Ch, *Wystable* 1247 Ch]. '(At) the white staple or post', or '*Hwita*'s staple'. See STAPOL.

Whitstone Co [*Witestan* DB, *Wytestane* 1263 Ep], **Whitstones** Wo in Claines [*Whitstan* 1245 Cl]. 'White stone.'

Whitsun Brook R Wo [*Wixena broc* 972 BCS 1282]. 'The brook of the *Wixan*.' See UXBRIDGE.

Whittingham La [*Witingheham* DB, *Whitingham* 1200 f. P, *Whitingeheim* 1202 FF], W~ (-nj-) Nb [*Hwitincham* c 1050 HSC, *Hwittingaham* 1104-8 SD]. 'HĀM of *Hwīta*'s people.'

Whittingslow Sa nr Church Stretton [*Witecheslawe* DB, *Witokeslawa* 1208 FF, *Whittokeslowe* 1274 Fine]. '*Hwit(t)uc*'s burial mound.' See HLĀW. The same name is *Hwittuces hlæw* 955 BCS 908 (Brk). Cf. WHITEOXMEAD.

Whittington Db [*Witintune* DB, *Whitinton* 1194 ff. P, *Hwytinton* 1231 Ch], W~ Gl [*Witetune* DB, *Wythinton* 1211-13 Fees], W~ La [*Witetune* DB, *Witington* 1212 Fees], **Great & Little** W~ Nb [*Witynton* 1233 P, *Parva Witington* 1242 Fees, *Great Whytington* 1296 Ch], W~ (or Drefwen) Sa [*Wititone* DB, *Quitentona* 1138 Ordericus, *Whitinton* 1237 FF, *Trefwen* 1254 Val], W~ St [(æt) *Hwituntune* 925 BCS 642, *Witinton* 1182 P], W~ Wa [*Wytinton* 1242 Fees, *Whytington, Wetendon* 1316 Ipm], W~ Wo [*Huuitingtun* 816 BCS 357, *Hwitintun* 989 KCD 670]. 'TŪN of *Hwita*'s people', or sometimes '*Hwita*'s TŪN' or 'white TŪN'. Drefwen, the Welsh form of W~ Sa, means 'white TREF'.

Whittle, Welch, La [*Withull* 1221 FF, *Walsewithull* 1242 Fees], W~ le Woods La [*Witul* c 1160 LaCh, *Whithill in bosco* 1327 Subs], W~ Nb nr Felton [*Wythill* 1266 Ipm], W~ Nb in Ovingham [*Wythill* 1242 Fees, *Whitehill* 1316 FA]. 'White hill.'

Welch W~ from the family name Waleys, lit. 'Welsh'. Richard le Waleys bought the manor before 1221.

Whittlebury Np [? (æt) *Witlanbyrig* c 930 Laws, *Wytlebyr'* 12 NS, *Whittlebury* 1316 FA]. If the ex. of c 930 belongs here, which is doubtful, the name means '*Witla*'s BURG'. *Witla is a diminutive of *Witta*. See further WHITTLEWOOD.

Whittlesey Ca [*Witlesig* 973 BCS 1297, *Witesie* DB, *Wittleseia* 1086 IE], W~ Mere Hu [*Witlesmere* 963–84 BCS 1128]. '*Wit-(t)el*'s island and lake.' See ĒG, MERE. *Witil* is the name of a moneyer. *Wittel would be a diminutive of *Witta*. Whittlesey and W~ Mere are close together.

Whittlesford Ca [*Witlesforda* c 1080 ICC, *Witelesforde* DB, *Witlesford* c 1190 Fr]. First el. as in preceding name.

Whittlewood Np [*Whitlewda* Hy 1 Mon iv, *Witlewude* 1196 P, *Whittelwode* 1307 Ch]. W~ is near Whittlebury, and the two names must have the same first el. If the latter goes back to *Witlanburg*, Whittlewood must be '*Witla*'s forest'. But the early forms of Whittlewood often have *Wh*-, and the most natural explanation of the name is 'Whitley Forest' (cf. WHITLEY). If that is right, Whittlebury means 'BURG by Whitley Forest'.

Whitton Du [*Wittune, Witun* 1208–10 Fees], W~ Mx [*Witton* 1300 PNMx, *Whitton* 1353 FF], W~ Nb [*Witton* 1228 Pat, *Wytton* 1233 Ch], W~ Sa nr Tenbury [*Witinton* 1180 Eyton, *Wyttone* 1255 RH, *Witintun, Hwitint.* 13 BM], W~ Sa nr Westbury [*Wibetune* DB, *Whitton* 1316 FA], W~ Sf [*Witton* 1212 Fees, *Wytenton* Hy 3, *Whytenton* 1295 BM]. '*Hwita*'s TŪN', or 'white TŪN'.

Whitton Li [*Witenai* DB, *Witena* c 1115 LiS, *Wihitene* 1179 (1328) Ch, *Whiten* 1276 RH]. 'White island' or '*Hwita*'s island', OE *Hwitan-ēg*. The final syllable was dropped.

Whittonstall Nb [*Quictunstal* 13 Newcastle, 1242 Fees, *Whyttonstall* 1271 Ipm]. '*Tūnstall* or homestead with a quickset hedge.' Cf. WHICKHAM.

Whitwell Db [*Hwitanwylles geat* 942 ASC, (æt) *Hwitewylle* 1002 Wills, *Witeuuelle* DB], W~ Do [*Whitewell* 1197, *Hwytewell* 1230 P], W~ Hrt [*Whitewella* 1209–10 Ep], W~ Nf [*Witewella* DB, *Quitewell* 1205 Cur], W~ Ru [*Witewelle* DB, *Whitewell* 1195 P], W~ We [*Quitewelle* 1246 Kendale], W~ Wt [*Quitewell* 1212 Cur, *Whitewell* 1287–90 Fees], W~ YN nr Catterick [*Witewell* 1201 Cur, *Quitewell* 1219 FF], W~ on the Hill YN [*Witewelle, -uella* DB]. 'White spring or stream.'

Whitwham Nb at Lambley [*Le Whitewhom* 1317 Ipm]. 'White valley or corner.' Cf. HWAMM.

Whitwick (-ĭtĭk) Le [*Witewic* DB, c 1125 LeS, *Whytewyk* 1327 Subs]. '*Hwita*'s WĪC' or 'white WĪC'.

Whitwood YW [*Witewde* DB, *Whitewude* 1197 f. P]. 'White wood.'

Whitworth Du [*Wyteworth* 1291 Tax], W~ La [*Whiteword, -worth* 13 WhC]. '*Hwita*'s WORÞ.'

Whixall Sa [*Witehala* DB, *Whitekeshal* 1241 FF, *Witekeshale* 1242–9 Eyton, *Quixhal* 1327 Subs]. '*Hwit(t)uc*'s HALH.' Cf. WHITTINGSLOW.

Whixley YW [*Cucheslage* DB, *Quyquesle* 1150–4 YCh 185, *Quixeleia* c 1200 ib. 516, *Quixle* 1206 Cur]. '*Cwichelm*'s LĒAH' or '*Cwic*'s LĒAH', *Cwic being a short form of names in *Cwic*-.

Whoberley Wa in Stoneleigh [*Watburleia* c 1144 (1348) Ch, *Whatburgele* 1232 PNWa(S)]. 'The LĒAH of *Hwætburh* (a woman).'

Whorlton Du [*Queorningtun* c 1050 HSC, *Cueorningtun* 1104–8 SD]. 'TŪN by the mill stream.' The old name of Whorlton Beck was no doubt *Cweorning*, a derivative of OE *cweorn* 'mill' analogous to *Kverninga* in Norway.

Whorlton Nb [*Wheruel-, Wherwelton* 1323 Cl], W~ YN [*Wirueltune* DB, *Wheruelton* 1202 Ass, 1202 FF, *Whoruelton* 1202 FF]. 'TŪN by the circle', the reference being to a round hill, that at W~ YN being called Whorl Hill. The first el. is the word *whorl* 'wheel, circle' from OE *hwerfel, hwyrfel* (in pl. ns.).

Wibsey (wĭpsĭ) YW [*Wibetese* DB, *Wybecye* 1283 Ch, *Vilbesaye* c 1220 Bodl]. '*Wigbald*'s island.' Cf. SIBSEY.

Wibtoft Wa [*Wibbetoft* 1002 Wills, *Wibetot* DB]. '*Vibbe*'s TOFT.' First el. OSw *Vibbe* pers. n. But OE *Wibba* is also possible.

OE wīc, an early loan-word from Lat *vicus*, means 'dwelling, dwelling-place; village, hamlet, town; street in a town; farm, esp. a dairy-farm'. The element often appears in the plur. form (see e.g. WICK, WARWICK). It is impossible to distinguish neatly between the various senses. Probably the most common meaning is 'dairy-farm'. The WICKS, WEEKS, WYKES no doubt have it. So also names like CHELVEY, SHAPWICK (with the name of a domestic animal as first el.), HARDWICK, BUTTERWICK, CHES-, CHISWICK. See also BARWICK, BERWICK, and Introd. pp. xvi f. In names of salt-working towns such as DROIT-, MIDDLE-, NANT-,

NORTHWICH *wīc* originally denoted the buildings connected with a salt-pit or even the town that grew up around it. But a special meaning 'salt-works', found already in DB, developed.

The element often appears in the form *wich*, as in DROITWICH &c., GREENWICH, HARWICH, IPSWICH, NORWICH, SANDWICH, SWANAGE. It is particularly common in names of places that became important at an early date. The *w-* was sometimes lost early, as in BARNACK, SWANAGE, WINCH.

Wic alone is the source of the pl. ns. WEEK, WICK, WYKE, WIX, also of WICKEN, ASH-WICKEN, WYCOMB, WYKEN (from *wicum* dat. plur.). See also WICKWAR. As a first el. it is sometimes difficult to distinguish from other elements, e.g. *wice* 'wych elm'. In most cases it is combined with HĀM, STŌW, TŪN and the like, and names such as WICKHAM, WYKEHAM, WISTOW, WEIGHTON, WIGHTON, WITTON, WYTON (from *Wic-tūn*), WYCHBOLD probably mean 'dwelling-place, homestead, manor'. See also WICKMERE, WIGFORD, WIGTON YW, WITCHAMPTON, WY-COMBE.

OE **wice** 'wych elm' is the second el. of HORWICH La and the first el. of some names, as WEEKLEY, WICKLEWOOD, WICHENFORD, WISH-, WITCHFORD, probably WHITSBURY.

Wichenford Wo [*Wiceneford* 11 Heming, *Wicheneford* 1204 Cur, 1208–13 Fees]. 'Ford by some wych elms.' See WICE.

Wichling K [*Winchelesmere* DB, *Wincelesmere* 11 DM, *Wychelesmere* 1236 Fees, *Wichelinge* 1220-4, *-s* 1241, *Winchelinge* 1220-4 StAug]. The original name seems to have been *Wincelesmere*, Wichling, from OE **Wincelingas*, being 'the people of *Wincelesmere*'. *Wincel* may be OE *wincel* 'child' used as a pers. n. or OE *wincel* 'corner'. It is also doubtful if *-mere* is OE *mere* 'lake' or *gemǽre* 'boundary'. If it is *gemǽre*, *Wincel* is more likely a pers. n.

Wichnor St [*Hwiccenofre* 11 PNSt, *Wicenore* DB, *Wichenovere* 1236 Fees]. 'The slope or bank of the *Hwicce*.' See WHICHFORD. The name indicates a settlement of Hwiccians nr Lichfield.

Wick Brk nr Abingdon [*Lawike* 1199 FF], W~ Gl nr Abson [*Wyk* 1253 Ch], W~ Gl nr Berkeley [*Wic*, *Wicha* 12 Berk], W~ (or Wyke) **Champflower** So [*Wike* 1225 Ass, *Wikechampflur* 1280 Ipm], W~ **St. Law-rence** So [*Wike* 1225, *la Wyk* 1243 Ass], W~ **Episcopi** Wo [*Wican* 757–75, *æt Wican* 961–70 BCS 219, 1139, *Bisshopeswick* 1221 Ass, *Wyk Episcopi* 1291 Tax], W~ **by Pershore** Wo [*Wiche* DB, *Wyke* Hy 3 BM]. OE **wīc**, probably in the sense 'dairy-farm'.

For W~ **Champflower** cf. HUISH CHAMP-FLOWER.—W~ **Episcopi** belonged to the Bishop of Worcester.

Wicken Ca [*Wikes* 1203 FF, 1208 Cur, *Wykes* 1232 FF], W~ **Bonhunt** Ess [*Wica*

DB, *Wykes* 1254 Val, *Wykes Bonhunte* 1238 Subs, *Bonhunt in Wykyn* 1412 FA], W~ Np [*Wicha* DB, *Wikes* 1209-19 Ep, *Wiken* Hy 3 BM, *Wyca Mainfein* 12 NS, *Wykadive*, *Wyka hamonis* 1254 Val]. OE *Wīc* plur., dat. *Wicum*, from WĪC 'dairy-farm'. The change of *-um* to *-en* is common in the SE. Midlands and adjoining districts.

On *Bonhunt* see that name.—Wicken Np con-sisted of two manors, one called Wyke Dyve &c., the other Wyke Hamund &c. The former was held by William de Dyva in 1246 (Ep). Dyva is from DIVES-SUR-MER in Normandy. The latter was held by Hamon s. of Mainfelin in 1167 (P), by William s. of Hamon in 1209 (Fees). Cf. STOKE HAMMOND Bk.

Wickenby Li [*Wichingebi*, *Wighingesbi* DB, *Uichinghebi* c 1115 LiS]. 'Viking's BY.' First el. ON *Vīkingr*, ODan *Viking* (*Wich-ing* DB) pers. n.

Wickersley YW [*Wincreslei*, *Wicresleia* DB, *Wicaraslaia* 1148, *Wicareslei* 1157 YCh 179, 186, *Wykerlay* 1218 FF]. One DB form may suggest that the first el. is OE *wicnere* 'steward, bailiff'. ON *Vīkarr*, OSw *Vīkarr*, ODan *Vīkær* might also be thought of. See LĒAH.

Wickford Ess [*Wicford* c 995 BCS (1288), 1196 FF, *-fort*, *Wincfort* DB, *Wichford* 1219 Fees]. 'Ford by a wych elm' (see WICE) or 'ford by a wīc'.

Wickham Brk [*Wicham* 821 BCS 366, 1167 P, *Wikham* 1220 Fees], **West W~** Ca [*Wichamm* 970, 974 BCS 1268, 1305, *Wic-ham* c 1080 ICC, *Wicheham* DB], W~ **Bishops** Ess [*Wicham* DB, *Wycham* 1254 Val], W~ **St. Paul's** Ess [*Wicham* 939 BCS 737, DB, *Wikham Sancti Pauli* 1291 Tax, (of) *Hinawicun* c 1000 ASCh], W~ Ha [*Wichæma mearc* 826 BCS 393, *Wicham* 955-8 Wills, 1167 P, *Wicheham* DB], W~ Hrt in Stortford [*Wicheham* DB], **East W~** K [*Estwycham* 1284 Ch, *Est Wycham* 1292 Ipm], **West W~** K [*Wichæma mearc* 862 BCS 506, *Wichamm* 973 BCS 1295, *Wiche-ham* DB, *Westwycham* 1284 Ch], W~ O [*Wicham* DB, c 1155 RA], W~ **Market** Sf [*Wikham* DB, *Wicham* 1254 Val], W~ **Skeith** (-ēth) Sf [*Wic(c)hamm* DB, *Wicham-Skeyth* 1368 FF]. OE *wic-hām* 'dwelling-place, manor' or sometimes *wic-hamm* 'HAMM with a WĪC'. But spellings with *-hamm* need not point decisively to HAMM as the second el.

W~ **Bishops** belonged to the Bishop of Lon-don, W~ **St. Paul's** to St. Paul's in London.— W~ **Market** must have had a market.—W~ **Skeith** is 'Wickham with a racecourse'. Skeith is ON *skeið* 'racecourse'. *Skeyth* in Wickham is mentioned t. Hy 3 (BM). Cf. HESKET.

Wickham, Childs, Gl [*Childeswicwon*, *Wic-wone* 706, (in) *Uuiguuennan* 972 BCS 117, 1282, *Wiquenna*, *Wicvene* DB, *Wykewanne* 1220 Fees], **Wickhamford** Wo [*Wicwona* 709, *Wigorne* 714 BCS 125, 130, *Wyc-weoniga gemere* c 860 KCD 289, *Wicvene* DB, *Wikewaneford* 1221 Ass]. The two places are near each other, though in different counties. The name is very likely British, the elements being Welsh *gwig* (OW *gwic*) 'lodge, opening in a wood, wood' and *gwaun* (OW *guoun*, Bret *gueun*, *geun*) 'plain, meadow, moor'. The meaning may be 'lodge in a plain or moor' or 'plain in a wood'.

Childs may mean 'of the child' or 'of the young nobleman'.

Wickhambreux (-ōō) K [*æt Wicham*, (to) *Wichám* 948 BCS 869, *Wicheham* DB, *Wykham Breuhuse* 1270 Ass, *Wykham Brewose* 1318 FF]. Originally *Wichám* (see WICKHAM).

The manor was held by William de Brayhuse in 1265 (Misc). The name is from BRIOUZE in Normandy.

Wickhambrook Sf [*Wicham* DB, 13 BM, *Wichambrok* 1254 Val]. Originally *Wichám* (see WICKHAM). The addition is the word *brook*.

Wickhamford. See WICKHAM, CHILDS.

Wickhampton Nf [*Wichamtuna* DB, *-ton* 1206 Cur]. OE WĪC and HĀMTŪN. The meaning is the same as that of WICKHAM.

Wicklewood Nf [*Wikelewuda* DB, *Wiclewuda* 1168 P, *-wode* 1242 Fees, 1254 Val]. *Wickle-* is probably OE *wíc-léah* 'wych elm wood'. See WICE. To this was added OE *wudu*, the name meaning '*Wicléah* forest'.

Wickmere Nf [*Wicmara*, *-mera* DB, *Wikemere* 1166 RBE]. 'Lake by the WĪC or dairy-farm.' There is no lake here now.

Wickwar Gl [*Wichen* DB, *Wykewarre* Hy 3 BM, *Warre Wyke* 1285 Ch]. Originally OE WĪC.

The manor was given to John la Warre by King John and was held by Roger la Warre in 1285 (Ch). Warre is a Norman family name.

Wid R Ess. See WRITTLE, WIDFORD Ess.

OE **wīd** adj. 'wide' is difficult to distinguish from OE *wíþig* 'willow'; see WHITCOMBE, WIDCOMBE, WIDFORD. Several names that seem to contain *wid* probably have *wíþig* as first el. WIDNES certainly contains *wid*.

Widcombe So in Bath [(æt) *Widecume* c 1100 E, *Widecumbe* 1236 FF], **North & South W~** So [*Widecomb* 1303 FA, *Wydecomb* 1321 Ipm], **Widdecombe in the Moor** D [*Widecumba* Hy 1 (1270) Ch, *-cumb* 1199 P, *Wydecomb yn the More* 1461 BM], **Widdicombe** D in Stokenham [*Wythecumb* 1249 Ass]. The last two are certainly OE *wíþig-cumb* 'willow valley'. Widdecombe must have been near the *wiðimor*, *widimor* mentioned BCS 1323. The probability is that Widcombe has the same origin, though 'wide valley' is a possible alternative.

Widdington Ess [*Widi(n)tuna* DB, *Withitone* c 1130 Oxf, *Wyditon* 1238 Subs, *Wydyton* 1254 Val], **W~** YW [*Widetone* DB, *Wyeton* 1175-99 YCh 519]. The first is

OE *Wíþig-tūn* 'TŪN among willows'. The second is more likely a variant of WITTON, WOOTTON, 'TŪN by the wood'.

Widdrington Nb [*Vuderintuna* c 1160 FPD, *Wiðerinton* 1163 P, *Wodringatone* 1166 RBE, *Widerintune* c 1170 FPD]. 'TŪN of *Wuduhere*'s people.' OE **Wudu-*, **Widuhere* has an exact analogy in OHG *Withari*.

Widford Ess [*Witford* 1216 Cl, *Wydiford* 1254 Val, 1291 Tax, *Wydeford* 1280 FF], **W~** Hrt [*Wideford* DB, 1205 Cur, *Widiford* 1212 RBE, *Wydiford* 1291 Tax], **W~** O [*Widiforde* DB, *-ford* 1220 Fees, *Wythiford* 1232 Cl]. OE *wíþig-ford* 'ford by willows'. Cf. WIDCOMBE, WĪPIG. 'Broad ford' (cf. WĪD) seems a suitable name for a ford, but the forms in *Widi-* point to *wíþig*. **Wid R** Ess is a back-formation.

Widley Ha [*Wydelig* 12 VH, 1242 Fees, *Wydelye* c 1270 Ep]. OE *wíþig-léah* 'willow wood'.

Widmerpool Nt [*Wimarspol* DB, *Widmerepol* 1181, *Widmerespol* 1186 P]. The first el. is OE *wida mere* 'wide lake' or rather *wíþig-mere* 'willow lake', to which was added an explanatory OE *pōl* 'pool, tarn'.

Widnes La [*Wydnes* c 1200 WhC, *Wydenes* 1242 LaInq]. 'Wide promontory.' The place is by a headland jutting into the Mersey.

Widney Wa [*Wydenhay* 1342, 1390 AD]. 'Wide forest enclosure', or 'willow enclosure'. Cf. WIPIG, GEHÆG.

Widworthy D [*Wideworde* DB, *Wydeword* 1230 P, *-worthi* 1291 Tax]. Perhaps 'wide WORPIG'.

Wield Ha [*Walde* DB, *Walda* 1167 P, *Welde* 1316 FA]. OE *weald* 'wood'.

Wigan La [*Wigan* 1199 LaCh, *Wygan* 1215 CC, 1237 Ass, *Wigayn* 1245 Ch, *Wygain* 1246 Ass]. Perhaps identical with WIGAN in Anglesey, which appears to be elliptical for *Tref Wigan* 'Wigan's homestead' or the like.

Wiganthorpe YN [*Wichingastorp* DB, *Wykenthorp* 1304 Ch]. 'Viking's thorp.' Cf. WICKENBY.

Wigborough, Great & Little, Ess [*Wicgebergha*, *Wigheberga* DB, *Wiggeberga* 1187 P, *Wygeberwe Magna*, *Wigeberwe Parva* 1254 Val]. '*Wicga*'s hill or barrow.' Cf. *Wicganbeorg* 851 ASC.

Wigford Li in Lincoln [*Wikeford* 1146 RA, 1190 P, 1199 NpCh]. First el. OE WĪC in one of its senses.

Wiggenhall Nf [*Wigrehala* DB, *Wiggehal* 1160 P, *Wiggenhal* 1196 FF, *Wygenhale Sancti Germani, Matris Christi, Magdalene, Sancti Petri* 1254 Val]. '*Wicga*'s HALH.'

Wiggenholt Sx [*Wikeolte* 1195 Fr, *-holt* 1212 FF, *Wygeholt* 1230 FF]. Perhaps OE *wicna-holt* 'wych elm wood'. See WICE.

Wigginton Hrt [*Wigentone* DB, *Wigeton*, *Wigginton* 1201 Cur, *Wykin(g)ton* 1254 Val, *Wynkenton* 1265 Misc], W~ O [*Wigentone* DB, *Wiginton* 12 Berk, *Wigingtone* 1226 Ep, *Wigenton*, *Winginton* 1242 Fees], W~ St [*Wicgintun* 11 PNSt, *Wigetone* DB, *Wyggenton* 1230 P, *Wichintona* 1173, *Wikenton* 1175 P], W~ YN [*Wichis-*, *Wichintun* DB, *Wiginton* 1231 Ass, *Wigington* 1241 FF]. The variation in the early forms is remarkable. Perhaps all the names may be explained as '*Wicga*'s TŪN' or 'TŪN of *Wicga*'s people'.

Wigglesworth YW [*Winchelesuuorde*, *Wiclesforde* DB, *Wicleswrthe* 1202 FF, *Wykelesworth* 1299 FC]. Apparently '*Wincel*'s WORP.' First el. OE *wincel* 'child' used as a pers. n. Cf. WICHLING.

Wiggonby Cu [*Wyganby* 1323 Ipm, 1332 Subs]. '*Wigan*'s BY.' *Wigan* (*Wiganus* 1131 WR, often *Wigayn*) is an OFr pers. n. of Breton origin (OBret *Uuicon*, Welsh *Wigan*).

Wighill YW [*duæ Wicheles* DB, *Wikale* 1219 FF, *Wychall* 1224–30 Fees]. OE *wic-halh* 'haugh with a wīc or dairy-farm'.

Wight, Isle of [*Vectis* Pliny, *Vecta* 4 IA, c 730 Bede, *Wiht* c 890 OEBede, *Wielht* 534 &c. ASC, *Wit* DB; (insula) *Gueith* c 800 HB, Welsh *Ynys Wyth*]. A British name, perhaps identical with Welsh *gwaith* 'turn', a word cognate with Lat *vectis* 'lever' (lit. 'the act of lifting'), OE *wiht* 'weight', derived from the root of Lat *veho* 'to carry'. The meaning might have been 'what has been raised', i.e. 'what rises above the sea', 'island'.

Wighton Nf [*Wistune* DB, *-tona* 1130, *Wihton* 1161, *Wichton* 1165, *Wigton* 1194 P, *Wicton* 1212 Fees]. OE *wic-tūn* 'dwelling-place, manor'.

Wigland Chs [*Wygelond* 1299 Rep, *Wyggelond* 1357 Chamb]. '*Wicga*'s land.'

Wigley Db [*Wiggelay*, *Wikeley* Hy 3, *Wiggelee* E 1 Derby, *Wyggeleg* 1255 Ipm], W~ Ha [*Wigelega* 1188 P, *Wiggeleia* 1198 FF]. '*Wicga*'s LĒAH' or 'LĒAH infested by wigs'. OE *wicga* is used of certain insects. Cf. *earwig*.

Wigmore He [*Wig(h)emore* DB, *Wiggemora* 1165 P, *Uggemore* c 1140, Hy 2 Glouc]. Either '*Wicga*'s moor' or Welsh *gwig mawr* 'big wood'. Welsh *gwig* means 'glade, wood'.

Wigsley Nt [*Wigesleie* DB, *Wiggesle* c 1160 RA, *Wigesle* 1257 Ch]. Apparently '*Wicg*'s LĒAH'. **Wicg* would be a side-form of *Wicga*.

Wigsthorpe Np [*Wykingethorp* 1232 FF, *Wygingestorp* 1278 Cl, *Wygstorp* 1412 FA]. Identical with WIGANTHORPE.

Wigston Magna Le [*Wichingestone* DB, *Wikingeston* 1191 P]. '*Viking*'s TŪN.' Cf. WICKENBY.

Wigston Parva Le [*Wicgestan* 1002 Wills, *Witgestan* 1004 KCD 710, *Wicestan* DB, *Wigestan* 1188 ff. P]. Perhaps '*Wicg*'s stone'. Cf. WIGSLEY. Or the name might mean 'logan-stone', the elements being those suggested for STANWICK Np, though in inverse order.

Wigtoft Li [*Wiketoft* Hy 2 (1316) Ch, 1212 Fees, 1227 Ep, *Wigetoft* 1180 P]. The place is in the fen country, where there may well have been a lake formerly. Hence the first el. of the name may be OScand *vik* 'bay, creek'. See TOFT.

Wigton Cu [*Wiggeton* 1163 P, *Wigeton* 1262 Ch]. '*Wicga*'s TŪN.'

Wigton YW [*Wigdon* c 1140 YCh 1861 f., *Wichdunie* n.d. Kirkst, *Wykedon* 1257 Ch]. OE *wic-dūn* 'hill with a wīc or dairy-farm'.

Wigwell Db [*Wyggewelle*, *-walle* Hy 3 Derby, *Wiggewell* 1251 Ch, *Wiggeswalle* 1287 Misc]. '*Wicga*'s spring.'

OE *wiht* 'bend, curve' is not recorded, but must have existed. It is derived from OE *wican* 'to yield', lit. 'to bend'. It is the base of **Great Whyte** Hu, the name of a street in Ramsey, which covers an old waterway with a bend in its course. See WETMOOR, WHITEFIELD Gl, WHITEHILL O, WITHAM, WITLEY Wo, WYTHAM.

Wike YW nr Birstal [*Wich*, *-e* DB], W~ YW nr Harewood [*Wic(h)* DB, *Wyke* c 1140 YCh 1862]. OE wīc 'dairy-farm'.

Wilbarston Np [*Wiberdestone*, *Wilbertestone* DB, *Wilbertestun* 1156, *Wilberdestun* 1157 f. P]. '*Wilbeorht*'s TŪN.'

Wilberfoss YE [*Wilburcfosa* 1148, *Wilburfoss* c 1170 YCh 179, 836, *Wilburgfosse* 1231 FF], **Great & Little Wilbraham** Ca [*Wilburgeham* c 1000 BCS (1306), c 1080 ICC, 1156 P, *Wilburham Magna*, *Parva* 1254 Val], **Wilburton** Ca [*Wilburhtun* 970 BCS 1268, *Wilbertone* DB]. '*Wilburg*'s FOSS or ditch, HĀM and TŪN.' *Wilburg* is a woman's name.

Wilbrighton St [*Wilbrestone* DB, *Wilbritone* 1166 RBE, *Wilbricton* 1242 Fees]. '*Wilbeorht*'s TŪN.'

Wilburton. See WILBERFOSS.

Wilbury Hill Hrt [(fram) *Wiligbyrig* 1007 Crawf]. 'Willow BURG.'

Wilby Nf [*Wilgeby*, *Willebeih* DB, *Wileby* 1220 Fees, *Willobi* 1254 Val], W~ Np [*Willabyg* c 1067 Wills, *Wilebi* DB, 1167 P, *Wyliby* 1230 P, *Wilweby* 1254 Val], W~ Sf [*Wilebey*, *-bi* DB, *Wyleb(e)ye* 1254 Val, *-beye* 1289 BM]. The first is 'BY among willows' (cf. WILLOUGHBY). The second, if *Willabyg* belongs here, may be rather '*Willa*'s BY' or '*Vili*'s BY'. *Vili* is an OScand pers. n. The third seems to be OE *wiligbēag* 'circle of willows'; second el. OE *bēag* 'ring'.

Wilcot W [(æt) *Wilcotum* 940 BCS 748, *Wilcote* DB]. 'COTS by a spring.' First el. OE *well, wiell, will* 'well, spring, stream'.

Wilcote O [*Widelicote* DB, *Wyuelicote* c 1200 Eynsham, *Wyuli'gcote* 1230 P]. 'COT of *Wifel*'s people.' OE *Wifel* is not evidenced in independent use, but must have been a common name. Cf. OG *Wibil*, ON *Vifill*.

Wilcott Sa in Great Ness [*Vinelecote* a 1210, *Winelecote* 1310 Eyton, *Wyvelecote* 1309 Ipm]. Apparently '*Winela*'s COT'. **Winela* is a side-form of *Winel*. Cf. WILDEN Wo.

Wild or **Wyld** Brk [*La Wile* 1183 P, *la Wile* 1199 FF, *La Wyle* 1242 Fees, 1246 Cl], **Monkton Wyld** Do [*La Wilæ* 1186 P, *la Wyle* 1240 Wells, 1302 Ep]. Apparently a name analogous to CROFT Le (from OE *cræft* 'craft, contrivance'), i.e. late OE *wil* 'trick' used of some mechanical contrivance, as a windmill or a trap. The cognate ON *vél* 'trick' is also used in the sense 'engine'. OE *wil* is no doubt a native word.

OE *wilde* 'wild, waste, uncultivated' is a rare el. in pl. ns. See WILDMORE, WILLAND. **Weald Moors** Sa (see EYTON, PRESTON UPON THE WEALD MOORS) is OE *wilde-mōr* 'waste moor'.

Wilden Bd [*Wildene* DB, *Wilden* 1163, *Willedene* 1167 P]. Probably 'willow valley', OE *wilg-denu*. See WELIG.

Wilden Wo [*Wineladuna, Winelduna* 1182 PNWo, *Wiveldon* 1299 Cl]. '*Winela*'s DŪN.' Cf. WILCOTT Sa.

Wilderley Sa nr Pulverbatch [*Wildredelega* DB, *Wildredeslege* 1201 FF, *Wilderdeleg* 1242 Fees]. '*Wilrēd*'s LĒAH' or '*Wilþrȳþ*'s LĒAH.' *Wilþrȳþ* is a woman's name.

Wildmore Li [*Wildemore* 1198 (1328) Ch, *la W~* 1206 Ass]. 'Waste moor.'

Wildon Grange YN [*Wilema* DB, *Wildon* 1138 Mon v, 1224 FF]. The place is by a round hill. The second el. of the name is OE *dūn* 'hill'. The first might be OE *wilde* 'waste, uncultivated', but several other possibilities may be considered, as OE *wil* (see WILD) or OE *wigel* (or *wigle*) 'divination, heathen practice'.

Wildsworth Li [*Winelesworth* 1199 (1232) Ch, *Wyveleswurth* 1280 Ch, *Wylessworth* 1316 FA]. '*Wifel*'s WORÞ.' Cf. WILCOTE O.

Wiley R. See WYLYE.

Wilford Nt [*Wilesford* DB, *-ford* 1169 P, *Wileford* c 1190 Middleton], **W~ Square** Sf at Woodbridge [*Wileforde* DB, c 1150 Crawf, 1254 Val]. 'Willow ford.' See WELIG.

Wilkesley Chs [*Wivelescle* DB, 1284 (1331) Ch, *Wyvelescle* 1253 Ch]. OE *Wifeles-clēa* or possibly *-clæg* '*Wifel*'s tongue of land or clayey land'. Cf. WILCOTE, HARTLEY We, CLAY. Wilkesley is on a tongue of land between Duckow R and a tributary.

Wilksby Li [*Wilchesbi, Wilgesbi* DB, *Wilghebi* c 1115 LiS, *Wilkesbi* c 1200 NpCh, 1212 Fees]. '*Vigleik*'s BY.' First el. ON *Vigleikr*, OSw *Vighlek*, ODan *Wiglek*.

Willand D [*Willelande* DB, *Wildelanda* 1155–8 (1334) Ch, *-londe* 1315 Ep]. 'Waste land.' See WILDE.

Willaston Chs nr Hooton [*Wilaveston* (hd) DB, *Wilaston* 1305 Chester], **W~** Chs nr Nantwich [*Wilavestune* DB, *Wylavestun* c 1250 Ormerod]. '*Wiglāf*'s TŪN.'

Willen Bk [*Wilinges* 1151–4 Fr, *Wilie* 1208 Cur, *Wylien* 1236 Fees]. The dat. plur. of OE *welig, wilig* 'willow'.

Willenhall St [*Willanhalch* 732 BCS 149 f., *Willenhale* 996 Mon, *Winenhale* DB]. '*Willa*'s HALH.' W~ Wa [*Willenhall* 1167 P, *Wilenhall* 1221 Pp, *Wylnhale* 1257 Ch]. 'Willow HALH.' First el. OE **wilgen* 'of willows'. *Willa* pers. n. is a possible alternative.

Willerby YE NW. of Hull [*Willardby* 1262 Selby, *Wilardeby* 1276 Ch], W~ YE SW. of Scarborough [*Willerdebi* 1125–30, *Wilardeby* 1135–9 YCh 1135, 1144, *Wilardebi* 1205 FF]. '*Wilheard*'s BY.'

Willersey Gl [*Willerseye* 714, *æt Willereseie*, (into) *Wylleresege* 854 BCS 130, 482, *Willersei* DB]. Perhaps '*Wilhere*'s island'. But more likely OE *wylleres īeg* 'the island of the salt-boiler'. OE *wyllere* is found in (æt) *wylleres seaðon* 995 KCD 691 'the salt-boiler's pits'(Cuxham O). Cf. WALLERSCOTE.

Willersley Db [*Wildereslay* 1211 Cur, *Willardesley* 1251 Ch]. '*Wilheard*'s LĒAH.'

Willersley He [*Willaveslege* DB, *Wilageslege* Hy 3 Misc]. '*Wiglāf*'s LĒAH.'

Willesborough K [*Wifelesberg* 863 BCS 507, *-beorg* 993 Hyda, *Wivelesberg* 1270 Ch]. '*Wifel*'s barrow or burial mound.' Cf. BEORG, WILCOTE.

Willesden Mx [*Willesdone, Wellesdune* 939 BCS 737, *Wellesdone* DB, *Wilesdune* c 1185 BM, *Wullesdon* 1248 FF]. 'Hill with a spring' (OE *wiell, well*).

Willesley Le [*Wiuelesleie* DB, *Wivelesleia* 1198 FF, *-le* 1208 FF, 1242 Fees]. '*Wifel*'s LĒAH.' Cf. WILCOTE.

Willett R So [*Willite* 854 BCS 476, *ostium Guellit* 12 Rees]. If the name is English, the elements may be OE *wiell, will* 'spring, stream' and an unrecorded **giete* 'a stream', cognate with OHG *gôz* 'fluid', *giozo* 'running water'. But it is possible the name is British. On the river is **Willett House** [*Willet* DB, *Wellyt* 1285 Ipm]. Cf. also WILLITON.

Willey Chs nr Whitchurch [*Wylileg* 1244 Chester, *Wilileg* 1260 Court], W~ He [*Wylileye* 1276, *Wililege* 1277 Ep], W~ Sa [*Wilit* DB, *Wilileg* 1199 P, *-leia* c 1200 Middleton], W~ Wa [*Welei* DB, *Wilega*

1180, *Wili* 1196 P]. 'Willow wood.' See
WELIG, LĒAH.

Willey D [*Wythelgh'* 1242 Fees, *Wygelege*
1244 FF]. Apparently OE *wiþig-lēah* 'willow wood'.

Willey Sr on the Wey [(to) *Weoleáge* 909
BCS 627, *Weleye* c 1200 Ep]. 'LĒAH with
a heathen temple.' Cf. WEOLEY, WĒOH.
The same name is (on) *Weoleage* KCD 712,
780 (Hinton Ampner Ha).

Williamscot O [*Willemscote* 1232 Ep],
Williamstrip Gl [*Willamesthorp* 1303 FA],
Williamsthorpe Db nr North Wingfield
[*Wilelmestorp* DB, *Willametorpe* 1226 FF].
'*Wilhelm*'s or *William*'s COT and thorp.'

Willian Hrt [*Wilie* 1086 DB, *Wilian* 1212
Fees]. 'The willows.' Identical with WIL-
LEN.

Willicote Gl at Long Marston [*Wilcote* DB,
Wilecote 1176, *Willecota* c 1200 Fr, *Wiligcote* c 1250 PNGl]. 'Willow COT.'

Willingale Doe & Spain Ess [*Willing(h)ehala*, *Ulinghehala* DB, *Willingehale* 1198
FF, *Wylinghehale Spayne* 1269 FF, *Wylliggehaledo* 1271 Ch, *Wilinghale Doo* 1291
Tax]. 'HALH of *Willa*'s people.'
W~ Doe from Hugh de Ou (t. Hy 2). De Ou
is a Norman family name, perhaps from EU in
Seine-Inférieure.—**W~ Spain** was held by
Hervei (de Ispania) in 1086 (DB), by William
de Hyspania in 1236 (FF). The family name
is stated to be derived from ÉPAIGNES in Eure.

Willingdon Sx [*Wille(n)done* DB, *Willindun*
1232 (1320) Ch]. '*Willa*'s hill' or 'willow
hill.' Cf. WILLENHALL.

Willingham Ca NW. of Cambridge [*Uuinlingeham* c 1050 KCD 907, *Wiuelingeham*
c 1080 ICC, *Wevelingeham* 1130 P], **North
W~** Li [*Wifilingham*, *Wiuilingeham* DB,
Wiflingham c 1115 LiS, *Wiuelingham* 1203
Ass], **W~ by Stow** Li [*Wilingeham*, *Welingeham* DB, *Uiflingeheim*, *Wiflingham* c 1115
LiS, *Wiueling(e)ham* 1202 Ass, *Wiuelingeham et Stowe* 1212 RBE]. 'HĀM of *Wifel*'s
people.' Cf. WILCOTE.

Willingham Ca by Carlton [*Wilingeam* 1121
AC, *Welingeham* c 1150 Fr, *Willingham*
1254 Val], **Cherry W~** Li [*Wilingeham*,
Ulingeham DB, *Wllingeheim* c 1115 LiS,
Willingham 1163 RA, *Chyry Wylynham*
1386 AD], **South W~** Li [*Ulingeham* DB,
Wllingeheham c 1115 LiS, *Welingeham* 1121-3,
Wellingeham c 1160 RA], **W~** Sf [*Willingaham*, *Wellingaham* DB, *Willingeham* 1188,
1196 P]. 'HĀM of *Willa*'s people.'
Cherry is probably *cherry* the fruit.

Willington Bd [*Welitone* DB, *Wilitona* c
1150 BM], **W~** Db [*Willetune* DB, *Wilintun*
c 1150 BM, *Wyliton* 1230 P]. 'TŪN among
willows.' Cf. WELIG.

Willington Chs [*Winfletone* DB, *Wynlaton*
12 WhC]. '*Winflǣd*'s or *Wynflǣd*'s TŪN.'
Both are OE women's names.

Willington Du [*Wyvelintun* c 1190 Godric],
W~ Nb [*Wiflintun* c 1085 LVD, *Wiuelington* 1204 FPD]. 'TŪN of *Wifel*'s people.'
See WILCOTE.

Willington Wa [*Ullavintone* DB, *Wullavington* 1287 Cl]. 'TŪN of *Wulflāf*'s or *Wiglāf*'s
people.'

Willisham Sf [*Willauesham* c 1040 Wills,
1176 P, *Wilagesham* 1198 FF]. '*Wiglāf*'s
HĀM.'

Willitoft YE [*Wilgetot* DB, *Wilgetoft* 1190
P]. 'TOFT among willows.' See WELIG.

Williton So [*Willettun* 904 BCS 612, *Wille-*,
Welletone DB]. 'TŪN on R WILLETT.'

Willoughby Waterless Le [*Wilebi*, *Wilechebi* DB, *Wileweby* 1236 Fees, *Wilweby*
1254 Val], **W~ in the Marsh** Li [*Wilgebi*
DB, *Wilegebi* 1191 P], **Scott W~** Li [*Wilgebi*
DB, -*bia* Hy 2 BM, *Scot Wilegeby* 1239 Ep,
Scotwyleweby 1247 Cl], **Silk W~** Li [*Wilgebi* DB, *Wilgebia* 1146 RA, *Wyleghby*,
Sylkby 1303 FA, *Wylughby and Silkeby* 1331
Ch], **West W~** Li [*Wilgebi* DB, *Willegebi*
Hy 2 DC], **W~** Nt nr Newark [*Wilgebi* DB],
W~ by Walesby Nt [*Wilgebi* DB, 1156
BM, *Wilgheby* Hy 2 (1316) Ch], **W~ on the
Wolds** Nt [*Wilgebi* DB, *Wilwebi* 12 DC,
Wilughby super Waldas 1363 BM], **W~** Wa
[*Wiliabyg* 956 BCS 978, *Wilebei* DB, *Wilibi*
1198 Fees, *Wylgheby* 1221 Ass]. 'BY among
willows.' See WELIG, WILBY. It is curious
that this hybrid name is so common. Very
likely W~ is in most cases a Scandinavianized form of OE *Weligtūn*. See WILLOUGHTON, WILTON (1).
Willelmus Scot is witness to a charter relating
to **Scott W~** early Hy 2 (DC).—**Silk W~** is
a contraction of Silkby and Willoughby. Cf.
SILKBY.

Willoughton Li [*Wilchetone* DB, *Wilgatuna*
c 1115 LiS, *Wilgheton* 1220 Ep]. 'TŪN
among willows.'

Willsworthy D [*Wiflevrde* DB, *Wyvelesworth* 1242 Fees]. '*Wifel*'s WORÞ.' Cf.
WILCOTE.

Wilmarston He in Golden Valley [*Wilmestune* DB, *Wulmestun* 1242 Fees, *Wilmeston*
13 BM]. '*Wilhelm*'s or *Wighelm*'s TŪN.'

Wilmcote (wĭng-) Wa [*Wilmundigcotan* (gemæro) 1016 KCD 724, *Wilmecote* DB].
'COTS of *Wilmund*'s people.'

Wilmersham So in Stoke Pero [*Winemeresham* DB, 1212 Fees]. '*Winemær*'s HĀM.'

Wilmingham Wt [*Wilmingeham* DB, 1198
P]. 'HĀM of *Wighelm*'s or *Wilhelm*'s people.'

Wilmington D [*Willelmatona*, *Wilelmitone*
DB]. 'TŪN of *Wilhelm*'s people.'

Wilmington K in Boughton Aluph [? *Wighelmes land* 858 BCS 496, *Wylmingtun* 11
DM], **W~** K nr Dartford [*Wilmintuna* 1089
BM, -*ton* 1241 Ep]. 'TŪN of *Wighelm*'s
people.'

Wilmington Sa [*Wilmitun* 1255 RH, *Wilmynton* 1323 Ipm], W~ Sx [*Wineltone*, *Wilminte* DB, *Wilmetun* 1212 Fees]. 'The TŪN of *Wighelm*'s or *Wilhelm*'s people.'

Wilmington So [*Wynmadun* 931, *Wynlmæddun*, *Wulmæddun* 963, (on) *Wynmeduue* 965 BCS 670, 1099, 1164, *Wimedone* DB, *Wilmedune* c 1087, *Welmendona* 1156 Bath]. The last two elements are OE *mǣd* 'meadow' and *dūn* 'hill'. The first may be OE *Winela* or *Willa* pers. n. or *will* (*wiell*) 'spring'.

Wilmslow (-ĭm-) Chs [*Wilmislowe*, -*lawe* 1260 Court, *Wilmeslowe* 1291 Tax]. '*Wighelm*'s or *Wilhelm*'s burial mound.' See HLĀW.

Wilne, Great & Little, Db [*Wilne* c 1100 Chester, *Wilna* 12 DC, c 1200 Derby, *Wylene* 1295 Ch, *Wylne-juxta-Shardelowe* 1279 FF]. The Wilnes are on opposite sides of the Derwent, in which there is an island just here. This island must have been *Wilne*. The second el. of the name is OE *ēg* 'island'. The first may be OE **wilgen* 'of willows'. Cf. WILLENHALL.

Wilnecote (wĭn-) Wa [*Wilmundecote* DB, -*cot* 1236 Fees, *Wilmecote* 1268 Ipm]. Identical with WILMCOTE.

Wilpshire La [*Wlypschyre* 1246 Ass, *Wlipschire* 1258 Ipm, *Wilpschire* 1311 Ipm]. The first el. may be a nickname formed from OE *wlips*, *wlisp* 'lisping'. The second is OE *scir* used in the same way as in early forms of PINNOCK, i.e. in a sense such as 'manor', 'estate'.

Wilsden YW [*Wilsedene* DB, *Wulsingdene* c 1200, *Wilsinden* c 1210 YCh 1688, 1794, *Wylsiden* 1246 FF]. 'The valley of *Wilsige*'s people.'

Wilsford Li [*Wivelesforde* DB, *Wiuelesford* 1202 Ass], W~ W nr Pewsey [*Wifelesford* 892, 934 BCS 567, 699, *Wivlesford* DB], W~ W nr Salisbury [*Wiflesford* DB, *Wiuelesford* c 1207 BM]. '*Wifel*'s ford.' See WILCOTE.

Wilshamstead (wĭlstĕd) Bd [*Winessamestede* DB, *Wyleshamstede* 1220 Subs]. '*Winel*'s homestead.'

Wilsill YW [*Wifeleshealh* c 1030 YCh 7, *Wifleshale* DB]. '*Wifel*'s HALH.' Cf. WILCOTE.

Wilson Le [*Wifeles Ðorp* c 972 BCS 1283, *Wiuelestunia* Hy 2 DC, *Wyveleston* 1242 Fees]. '*Wifel*'s TŪN.' See WILCOTE. *Wifeles Ðorp* may not actually refer to Wilson, but was at any rate near it and was named from the same *Wifel*.

Wilsthorpe Db [*Wiuelestorp* 1169 P, *Wivelesthorp* 1242 Fees], W~ Li [*Wiuelestorp* DB, 1198 FF, *Wiulestorp* c 1180 NpCh], W~ YE [*Wiflestorp* DB, *Willesthorp* c 1150 YCh 1156], **Wilstrop** YW [*Wiulestorp* DB,

Wiuelestorp 1208 FF]. '*Wifel*'s or *Vīfil*'s thorp.' Cf. WILCOTE.

Wilstone Hrt [*Wivelestorn* 1220 Fees, *Wyvelisthorn* 1279 Cl]. '*Wifel*'s thorn-bush.' Cf. WILCOTE.

Wilstrop. See WILSTHORPE.

Wilton Cu [*Wiltona* c 1210 StB, -*ton* 1211 P, 1294 Cl], W~ He [*Wiltone* DB, -*tun* 1156 P, *Wilton super Waiam* 1190 P, *Wiliton* 1227–57 Ch], W~ Nf [*Wiltuna* DB, -*tona* 1121 AC, -*ton* 1242 Fees], **Bishop** W~ YE [*Wiltone*, *Wilton*, *Widton* DB, *Wiltona* c 1110 YCh 93, *Bysshop Wylton* 1428 FA], W~ YN E. of Middlesbrough [*Wiltune*, *Widtune* DB, *Wilton* 1237 FF], W~ YN in Thornton Dale [*Wiltune* DB]. Most of these are no doubt OE *Wilg-tūn* 'TŪN among willows'. Cf. WELIG. Those with the form *Widton* (-*tune*) in DB possibly have OE *wilde* 'wild, waste' as first el.

Wilton So [*Wilton* 1249 Ass, *St. George de fonte* n.d. Wells 38, *Fons Georgii* 1439 BM], W~ W nr Burbage [*Wulton* 1227 Ch, *Wylton* 1428 FA]. 'TŪN by a well or spring.' First el. OE *wiell* (*will*, *wyll*) 'well, spring'. Wilton So has a remarkable well dedicated to St. George.

Wilton W, town [*Uuiltún* 838, *Wiltun* 854 BCS 421, 469, *Wiltun* 871 ASC, c 894 Asser, *Wiltune* DB]. 'TŪN on R WYLYE', here in a shorter form *Wil*.

Wiltshire [*Wiltunscir* 870, 994 ASC, c 894 Asser, 955 BCS 912, *Wiltescire* DB]. 'Shire dependent on Wilton.' The Wilts people are referred to as *Wilsætan* 800, 878 ASC ('dwellers on the Wylye').

Wimbish Ess [*Wimbisc* 1043, 1043–5 Wills, *Wimbeis* DB, *Wimbiss* 1201, 1204, *Winbiss* 1208 Cur]. The first el. seems to be OE **winn* 'meadow, pasture'. The second can hardly be an OE **bysc* 'bush', for it ought to have given ME *besch* in Ess. It might be a derivative of OE *bēos* 'reeds' (see BEESTON), i.e. an OE **biosic* 'reedy place' or the like. Cf. BESTWOOD.

Wimbledon Sr [(at) *Wunemannedune* c 950, *Wimbedounyngemerke* 967 BCS 1008, 1196, *Wimmeldun* 1212, *Wimbeldon* 1221–30 Fees]. '*Winebeald*'s DŪN.' The OE forms are in very poor texts.

Wimblington Ca [*Wimblingetune* c 975 (12) PNCa(S), *Wilmyngton* 1387 Pat]. Perhaps 'the TŪN of *Winebald*'s people'.

Wimboldsley Chs [*Wibaldelai* DB, *Wumbaldelegh* 1313 BM]. '*Winebald*'s or *Wynbald*'s LĒAH.'

Wimborne, now **Allen,** R Do [*Winburna* 705, 946, 956 BCS 114, 818, 958]. 'Meadow stream.' First el. OE **winn* 'meadow, pasture'. On the river are **Wimborne Minster** [(æt) *Winburnan* 718 ASC, *Winburnan monasterium* c 894 Asser, *Winburnan mynster* c 1000 Saints, *Winborne* DB,

Wymburneminstre 1236 FF], **W~ St. Giles** [*Winburne* DB] and **Monkton Up Wimborne** [*Winburne* DB].

Wimbotsham Nf [*Winebodesham* 1060 Th, c 1140 Bodl, *Winebotesham* DB, *Winebadisham* 1195 Cur]. Possibly '*Winebald*'s HĀM'. Or the first el. is a Continental name (OG *Winebaud*).

Wimpole Ca [*Winepola* c 1080 ICC, -*pole* DB, -*pol* 1183, 1199 P]. '*Wina*'s pool.'

Wimpstone Wa [*Wylmestone* 1313 AD, *Wilmaston* 1417 BM]. '*Wīghelm*'s or *Wilhelm*'s TŪN.'

Winca·nton So [*Wincaletone* DB, *Wynkauelton* 1243 Ass, *Wincaulton* 1291 Tax]. The first el. is the river-name *Wincawel* 956 BCS 923, an earlier name of an arm of the river CALE.

Winceby Li [*Winzebi* DB, *Wincebi* c 1115 LiS, 1167 P, 1212 Fees]. An OScand *Vinds bȳr*. ON *Vindr* (a mythical name) and ODan *Vinder* occur.

OE **wincel** is only found in pl. ns., but may be supposed to have meant 'corner', just as the cognate OHG *winkil*. See ALDWINKLE, WICHLING, WINCHCOMB &c.

Winch, East & West, Nf [*Eastuuininc*, *Estwinic*, -*uuinc* DB, *Estweniz* 1242 Fees, *Estwinch* 1254 Val; *Wesuuenic*, -*uuinic* DB, *Westweniz*, -*winic* 1198 FF, *West Weniz* 1203 Cur, *Westweniz* 1242 Fees, *Westwinch* 1254 Val]. OE *wynn-wic* 'wīC with meadow-land'. Cf. WINN, WĪC. The loss of *w* is partly dissimilatory, partly due to the preceding *n*. Cf. OE *ealneg* from *ealne weg*, *ænetre* from *ānwintre*. The reduction from *Winnic* to *Winc* took place in the compounds *East-, Westwinnic*.

Wincham Chs [*Wimundisham* DB, *Wynincham* 1281 Court]. '*Wigmund*'s HĀM.'

Winchcomb Gl [*Wincelcumba* 811, (æt, ad) *Wincelcumbe* 796–819, 897 BCS 338, 364, 575, *Vallis Winclea* 10 Swithun, *Wincelcumbe* DB]. OE WINCEL 'corner' and CUMB 'valley'. The name may be rendered 'side valley' or 'remote valley'. A counterpart of the name is *Uuincelcumb* 824 BCS 378 (K).

Winchelsea Sx [*Wencles* c 960 coins, *Winceleseia* 1130 P, *Winchelese* 1165 P]. Second el. OE *ēg* 'island'. The first may be OE *wincel, wencel* 'child', used as a pers. n. But as the Brede makes a considerable bend here, OE *wincel* 'corner' seems preferable: 'island in or by the bend'.

Winchendon (wich-) Bk [*Wincandone* 1004 Fridesw, *Wichendone* DB, -*don* 1167 P, 1201 Cur, *Winchendon* 1177 P]. The first el. may be OE *wince* 'reel, roller, pulley' in some particular sense. Cf. WINSUM and other names in Holland [*Winkhem* &c. 10]. Or it may be a bird-name, identical with the second el. of OE *hleapewince*, now *lapwing*.

Winchester Ha [*Ouénta* c 150 Ptol, *Venta Belgarum* 4 IA, (ciuitas) *Uenta, Uintancaestir* c 730 Bede, *Wintanceaster* 744 ASC, c 880 BCS 553, c 890 OEBede, *Wintonia* c 894 Asser]. The old name is identical with *Venta Icenorum* (Caister), *Venta Silurum* (Caerwent in Monmouthshire) and *Gwent*, the name of an old district in Wales. To this was added OE *ceaster* 'Roman fort'. The old name may belong to the root *ven-* 'to enjoy, love' in Welsh *gwen* 'smile', Ir *fine* 'kindred', identical with GWYNEDD, the name of a district in Wales.

Winchfield Ha [*Winchelefeld* 1229 FF, *Wynchefeld* 1291 Tax, *Wynceffeld* 13 (1337) Ch, *Wynchesfelde* 1316 FA]. The name may be identical with *Uuincelfeld* 996 KCD 696 (Hrt); one *l* would easily be lost owing to dissimilation. If so, the first el. is OE *wincel* 'corner'.

Winchmore Hill Mx [*Wynsemerhull* 1319 AD]. Second el. OE *mǣrhyll* 'boundary hill'. The first may be OE *Wynsige* pers. n.

Wincle Chs [*Winchul* 12 BM, *Wynkehull* 1291 Tax]. '*Wineca*'s hill.' Cf. WINKFIELD (1).

Winder Cu [*Wynderge* c 1210, -*ergh* 1271 StB], **W~** We [*Winderge* 1170–84 Kendale]. See ERG. The name means 'shelter against the wind'.

Windermere La, We, lake [*Winendemere*, *Wynandremer* c 1160 LaCh, *Winandermer* 1196 FF]. '*Vinand*'s lake.' The first el. is OSw *Vinnunder, Vinandus* pers. n. (gen. *Vinandar*). The lake gave its name to **Windermere** We [(capella de) *Winandemere* 1203 Cur, *Wynandermer* 1282 Kendale].

Winderton Wa [*Wynterton* 1236, *Wintreton* 1242 Fees, *Wynfreton, Wynterton* 1322 Misc]. If one form of 1322 is trustworthy, '*Winefriþ*'s TŪN'. Otherwise identical with WINTERTON Nf.

Winderwath We [*Vinanderuuat* c 1277 CWNS xx, *Wynanderwath* 1292 QW]. '*Vinand*'s ford.' See VAÐ and WINDERMERE.

Windhill YW [*Windhill* 1208 FF, 1339 BM], **Windle** La [*Windhull* 1201 P, 1202 FF]. 'Windy hill.'

Windlesham (wins-) Sr [*Windesham* 1178 PNSr, *Windlesham* 1227 Ch, *Wyndlesham* 1291 Tax]. Perhaps '*Winel*'s HĀM'. The place is on **Windle Brook** [*Vindeles* 1577 Harrison], whose name would then be a back-formation. Of course, the name of the brook may have been OE *Windol* 'winding brook', the name being a derivative of OE *windan* 'to wind'.

Windlestone Du [*Windlesden* 1196 P, *Wymelesdon* 1304 Cl]. Perhaps '*Winel*'s hill'.

Windley Db [*Winleg* 1251 Ch, *Wynleye* 1297 Ipm]. OE *winn* 'meadow, pasture' and *lēah* 'glade'.

Windridge Hrt in St. Stephens [*Wenrige* DB, 1198 FF, *Winrigge* 1195 P]. 'Pasture ridge.' See prec. name and WINN. In this name the first el. seems to have the form *wynn*.

Windrush R Gl, O [*Uuenrisc* 779, *Wenris*, *Wænric* 949, *Wenric* 969 BCS 230, 882, 1036, *Wenrisc* c 1000 Saints, *Wenrich* 1229 For, *Wenrhis* 1247 Ass]. A Brit river-name, whose second el. is OCelt *reisko-* in Ir *riasg* 'moor, fen', in **Moresk** Co [*Moireis* DB, *Morres* 1205 Fr] and in the stream-name *nant ruisc* c 1150 LL (Wales). The first may be Welsh *gwyn* 'white' or the el. *gwen* in GWENDRAETH (Wales). On the Windrush is **Windrush** vil. [*Wenric* DB, *Wenriz* 1220 Fees].

Windsor (-nz-) Brk [*Windlesóra* 1050–65 BM, *-ora* 1061 ff. ASC (E), *-ore* 1065 BM, *Windesores* DB, *Nova Wyndelesor et Vetus* 1242 Fees], **Broadwindsor** and **Little Windsor** Do [*Windesore, Windresorie* DB, *Winlesore, Parva Windlesor* 1210 FF, *Windleshor* 1219, *Parva Windelessor* 1236 Fees, *Magna Wyndesor* 1249 FF, *Brodewyndesores* 1293 Misc]. See WINSOR D, Ha, which seem to be identical in origin. It is difficult to believe that these contain a pers. n. *Windel*, which is not with certainty evidenced. The second el. is ŌRA 'bank', no doubt in the sense 'place suitable for landing'. Windsor Brk is on the Thames, Windsor Do on a stream, Winsor D nr the mouth of the Yealm, and Winsor Ha in Eling on Southampton Water. A very good etymology would be obtained if we may assume that *windlass* (found from c 1400) is an old word and goes back to OE **windels*, a derivative of OE *windan* 'to wind'. The name would mean 'landing-place with a windlass'.

Winestead YE [*Yiuestode, Yinestede* 1033 YCh 8, *Wifestad, -stede* DB, *Wiuestud* 1238 Ep, *Wynestede* 1256 Ch]. The first ex. is corrupt. The first el. is OE *Wife*, or *Wifa*, as in WESTOW YE, the second being OE STEDE. *Wine-* is due to misreading of *Wiue-*.

Winfarthing Nf [*Wineferthinc* DB, *-ferding* 1165, *-ferðing* 1168 P]. '*Wina's* quarter part.' Second el. OE *feorþung* 'fourth part'.

Winford So [*Wunfrod* c 1000 Wills, *Wenfrod, -fre* DB, *Winforð* 1169, *-frod* 1172, *-fred* 1188 P]. Originally a stream-name, identical with GWENFFRWD in Wales [*Guenfrut* c 1150 LL], which consists of Welsh *gwyn* 'white, holy, happy' and *ffrwd* 'stream, torrent' (= OBret *frut, frot*, Co *fros*).

Winforton He [*Widferdestune* DB, *Wynfreton* 1265 Ipm, 1291 Tax]. '*Winfriþ's* TŪN.'

Winfrith Newburgh Do [*Winfrode* DB, *Wynfrode* 1212 RBE, *Wymfrod* 1244 Ass, *Winford, -frod* 1212 Fees, *Wynfred Neeuburgh* 1288 FF]. Identical with WINFORD. The manor was held by Robert de Novo Burgo in 1212 (Fees). Newburgh from one of the NEUBOURGS in France.

Wing Bk [(æt) *Weowungum* 966–75 Wills, *Witehunge, Withunga* DB, *Wiungua* Hy 2 BM, *Wehenge, Weenge* 1203 Ass]. *Weowungum* is for *Weopungum*. The base is probably OE *Weohthúningas* '*Weohthún's* people'. The loss of *h* may be compared with that in WITTERING Sx. This explanation accounts for the common early form with *u* before *ng*. At Wing is **Wingrave** [*Wit*(*h*)*ungraue* DB, *Wiungraua* 1163 P, *Weengraue* 1203 Ass]. 'The grove belonging to Wing.'

Wing Ru [*Wengeford* 1046 KCD 784, *Wenge* 1202 Ass, 1291 RA, *Weng* 1206, 1208 Cur]. OScand (ON) *vengi*, a derivative of *vang* 'field' and the source of the Sw pl. n. VÄNGE (OSw *Vængia*).

Wingate Du [(æt) *Windegatum* 1071–80 ASCh, *Windegat* c 1150 Finchale, *-e* 1253 Ch, Hy 3 BM], **Wingates** Nb [*Wyndegates* 1208 Percy, *Windegatis* 1236 Fees]. 'Gate for the wind', 'pass where the wind drives through'. Cf. *windgeat* 961 BCS 1066 and **Winnats** Db ('a steep rocky chasm').

Wingerworth Db [*Wingreurde* DB, *Wingerwurth* 1242 Fees]. '*Winegár's* WORÞ.' *Winegár* seems to occur here in the gen. form *Winegára*; cf. OE *Wihtgara burg* 544 ASC. OE *Winegár* is found in *Winagares stapul* 1032 KCD 746.

Wingfield Bd [*Winfeld* c 1200, *Wintfeld* 13 PNBd, *Wynchefeld* 1276 Ass]. See FELD. First el. as in WINCHENDON.

Wingfield, North, Db [*Wynnefeld* 1002 Wills, 1004 KCD 710, *Winnefelt* DB, *Wynnefeld* Hy 3 BM], **South W~** Db [*Winefeld* DB, *Wynnefeld* 1236, *Wynefeld* 1242 Fees, *Sutwynnefelde* 1284–6 FA]. The two are not very far apart and were named from the same FELD or open land. The OE examples may refer to both or only to one of them. The first el. is OE **winn* 'meadow, pasture'. The name means 'grazing-ground'.

Wingfield Sf [*Wingefeld* c 1035 BCS 1020, *Wighefelda* DB, *Wihingefeld* 1185 f. P]. OE *Wigingafeld* or the like, 'FELD of *Wiga's* people'. Or the first el. might be a derivative of OE *wēoh* 'temple'.

Wingham K [*Uuigincggaham* 834 BCS 380, *Wuungham* 946 ib. 811, *Wingheham* DB, *Wingeham* 1165 P]. 'HĀM of *Wiga's* people.' *Wiga* is recorded in the form *Wyga* and corresponds to OG *Wigo*.

Wingrave Bk. See WING.

Winkburn Nt [*Wicheburne* DB, *Winkeburna* c 1150 DC, *Wincheburna* 1167 P]. Perhaps '**Wineca's* stream'. Or the first el. may belong to OE *wincian* 'to wink' in some earlier sense. Cf. OHG *winkan* 'to turn aside, totter'. Perhaps 'winding brook'.

Winkfield Brk [æt *Winecan felda* 942 BCS 778, *Wenesfelle* DB, *Winegefeld* 1167 P, *Wunekefeld* 1242 Fees]. '*Wineca's* FELD.'

OE *Wineca, a diminutive of Wine, corresponds to OG Winicho.

Winkfield W [Wuntfeld 964 BCS 1127, Winefel DB, Winesfeld 1242 Fees]. The form of 964 is in a poor transcript. Perhaps the name means 'Wina's FELD'.

Winkleigh D [Wincheleie DB, Winchalega 1107 (1300) Ch, Winkelea 1182 P, -leg 1219 Fees]. Probably 'Wineca's LĒAH'. Cf. WINKFIELD Brk.

Winksley YW [Wichingeslei, Wincheslaie DB, Winkesle 1198 P, Winkeleie 1231 FF, Winkeresley, Winkesley, Winkerle 1234 Ep]. Perhaps identical with WICKERSLEY. If so, one DB form has been influenced by the Scand pers. n. Vikingr.

Winkton Ha [Weringetone DB, Wyneketon 1236, 1242 Fees]. 'Wineca's TŪN.' Cf. WINKFIELD Brk.

Winlaton Du [Winloctune c 1085 LVD, Winlaketon 12 ib.]. 'Winelāc's TŪN.'

Winma·rleigh La [Wynemerislega 1212 Fees, Winmerleie c 1220 CC]. 'Winemǣr's LĒAH.'

OE *winn 'meadow, pasture', corresponding to Goth vinja, ON vin, OHG winne, MLG winne, must be assumed for some pl. ns. The word is related to OE wynn 'joy', OHG wunnia. OHG wunnia is also used in the same sense as winne, and very likely OE wynn also had the sense 'pasture'. See (East) HEDDON Nb, WIMBISH, WIMBORNE, WINDLEY, WINDRIDGE, WINGFIELD Db, WINNERSH, WINTON We, WOOLLAND, WYNYARD.

Winnall He [Wilehalle DB]. 'HALH with willows.' First el. OE WELIG, wilig or rather *wilgen 'of willows'.

Winnats. See WINGATE.

Winnersh Brk [Wenesse 1190 ff. P, Wenerssh, Wynerssche 1397 AD]. OE WINN, wynn 'pasture' and ERSC 'park'.

Winnington Chs [Wenitone DB, Wynington c 1210, Wynentona c 1215 Chester, Winnington 1278 BM]. 'TŪN of Wine's or Wina's people.'

Winnington St [Wennitone DB, Woninton 1273 Ipm]. 'TŪN of Wynna's people.'

Winsbury So [Wineces burug 963, Winces burch 965 BCS 1099, 1164]. 'Winuc's burg.' Winuc is the name of a moneyer t. Eadmund I, Eadred.

Winscales Cu [Wyndscales 1227 StB, Windscales 1294 Cl]. 'Shelter against the wind.' See SKÁLI.

Winscombe So [Winescome DB, -cumb 1196 P], **Winsford** Chs [Wyneford bridge c 1334, Wynsfurth brygge 1475 Vale Royal], W~ So [Winesford DB, Wynesford 1251 Ass], **Winsham** So [Winesham 1046 KCD 1334, 1065 Wells, DB], **Winshill** St [Wineshylle 1002 Wills, 1004 KCD 710, Wineshulla

a 1113 Burton]. 'Wine's coomb or valley, ford, HĀM, hill.'

Winskill Cu [Wynscales 1292 QW, 1332 Subs]. Identical with WINSCALES.

Winslade Ha [Winesflot DB, -flode 1270–80 Selborne, Wynnesfloud c 1270 Ep]. 'Wine's flōde or stream.'

Winsley Db [Wiuesleia 1197 FF, Wynesleye 1269 FF], W~ W [Winesleg 1242 Fees, Wynesley 1316 FA]. 'Wine's LĒAH.' Windærlæh mæd 987 KCD 658 is now WINTERLEYS in Bradford-on-Avon.

Winslow Bk [Wineshlauu 795 BCS 849, Weneslai DB]. 'Wine's burial-mound.' Cf. HLĀW.

Winson Gl [Winestune DB, Wynston 1220 Fees]. 'Wine's TŪN.'

Winsor D in Yealmpton [Winlesore 1202 FF, Wyndesore 1309 Ch], W~ Ha in Eling [Windlesor 1236 Ass, Windelesore 1286 Ch]. See WINDSOR.

Winstanley La [Unstanesle 1206 P, Winstanesle 1212 Fees]. 'Wynstān's LĒAH.'

Winster Db [Winsterne DB, Winesterna 1121–6, 1155 RA, -tere 12 Derby]. 'Wine's thorn-bush.' Second el. OE ÞYRNE.

Winster R La, We [Winster 1170–84 Kendale]. Identical with Vinstra in Norway, a derivative of ON vinstri 'left' and meaning 'the left one'. Hence **Winster** La [Winstirtwayts 1240–6, Wynster 1377 Kendale].

Winston Du [Winestona 1091 FPD, Wynston 1291 Tax], W~ Sf [Winestuna DB, 1109–31 BM]. 'Wine's TŪN.'

Winston Wt [Winsiston 12 (1313) Ch, Wyneston 1287–90 Fees]. 'Wynsige's TŪN.'

Winstone Gl [Winestan DB, Wenestan 1191 P, Wunnestan 1211–13, Wonestan 1220 Fees]. 'Wynna's stone.'

Winterborne R Do [Winterburne 942 f. BCS 775, 781]. A common river-name, OE winterburna, dial. winterbourne 'an intermittent stream', really 'a stream dry except in winter'. Other examples of the name are Winterburna 775–8 BCS 226 (Eisey W), 855 ib. 467 (K), 930 ib. 667 (Flyford Flavell Wo). Cf. also **Winter Beck** Nt [Winterbek 13] and **Winterbrook** h. Brk. From the Winterborne were named several places, now distinguished as Winterborne Anderson, Clenston &c. [Winterburne 942 BCS 775, Wintreburne DB].

W~ Anderson [Wynterbourne Fiveesse 1284 Cl, Wynterborn Fifhassh, Andreweston 1331 FF]. 'W~ dedicated to St. Andrew.' Now alternatively ANDERSON. Fiveesse means 'five ash trees'. **—W~ Clenston** [Winterborn Clench 1242 Fees, Wynterborn Cleyngestone 1274 Ipm] took its name from a family named Clench. Robert Clench held land here in 1232 (Pat).—**W~ Houghton** (how-) [Hugeton 1176 P, Winterborn Hueton 1247 Ipm] was named from an early owner called Hugh.—**W~ Kingston** [Kinges-

winterburn 1194 P, *Wynterburne Kyngeston* 1306 FF] was held by the King in 1086 (DB).—W~ **Muston** [*Winterborn Musters* 1242 Fees, *Wynterbourn Mustereston* 1354 FF] was named from the family of Musters (*de Mustiers* 1195 P, *de Monasteriis* 1190 P). Musters is from MOUTIERS (< *Monasteriis* 'the monasteries') in France.—W~ **Stickland** [*Winterburn Stikellane* 1203 Fr, *W~ Stikelan* 1244 Cl]. Stickland was a place in the vicinity, referred to as (on ða) *sticelen lane* 1019 KCD 730. The name means 'the steep path'.—W~ **Tomson** [*Winterborn Thom'* 1242 Fees, *Wynterbourn Thomaston* 1280 FF]. Named from an owner called Thomas.—W~ **Whitchurch** [*Winterburn Albi Monasterii* 1202 FF, *Wynterburne Whytchirche*, (town of) *Whitchirche* 1294 Cl]. Cf. WHITCHURCH.— W~ **Zelstone** [*Winterburnia* c 1175 Middleton, *Wynterburne Malreward* 1230 P, *Wynterbourn Selyston* 1350 FF]. Probably named from a family called Zeals (from ZEALS W). Cf. LANGTON MATRAVERS.

Winterborne Abbas, Came &c. Do took their name from a stream which must have been called *Winterburna*. Examples of the name are *Wintreburne* DB, *Winterburne* 1212 Fees.

Winterborne Abbas [*Winceburna* 987 KCD 656, *Wintreburne* 1212 Fees] belonged to the Abbey of Cerne.—W~ **Came** [*Wintreborna* 1190 (1332) Ch, *Wynterborn Cham* 1291 Tax] belonged to the Abbey of Caen in Normandy in 1086(DB).—W~ **Herringstone** [*Winterborn Harang* 1242 Fees] belonged to the Herring family. Cf. CHALDON HERRING.—W~ **Monkton** [*Winterburn Waston* 1213 Cl, *Moneketone* 1285 FA] belonged to the Abbey of St. Vaast in Arras.—W~ **St. Martin** [*Winterburn St. Martin* 1267 Ch] from the dedication of the church.—W~ **Steepleton** [*Stipelwinterburn* 1199 Rot Cur, *Wynterburn Stupilton* 1260 FF] from the church steeple.

Winterbourne Brk [*Winterburninga gemære* 951 BCS 892, *Wintreburne* DB], W~ Gl [*Wintreborne* DB, *Winterburn* 1156, 1190 P], W~ Sx [*Winterburna* 966 BCS 1191, *Wintreburne* DB] were all named from streams once called *Winterburna*.

Winterbourne Bassett & Monkton W [*æt Winterburnan* 964 BCS 1145, *Wintreburne* DB, *Winterburn Basset* 1242 Fees, *Wynterborne Monachorum* 1316 FA] were named from a stream called *Winterburna* 964 BCS 1145.
Winterbourne Bassett was held by Alan Basset in 1220 (FF). Cf. BERWICK BASSETT.— W~ **Monkton** belonged to the Abbey of Bocherville nr Rouen in 1114 (Fr), later to Glastonbury Abbey.

Winterbourne Dauntsey &c. W [*Wintreburne* DB] are on the BOURNE, called *Winterburna* 972 BCS 1286.
Winterbourne Dauntsey [*Winterburn Dauntesye* 1275 RH] was held by Roger Daunteseye in 1242 (Fees). The name is from *de Anesia* (from ANISY in Normandy).—W~ **Earls** [*Winterburn comitis Sar'* 1198 P, *Winterburne Earls* 1250 Misc] was held by the earls of Salisbury.— W~ **Gunner** [*Winterburn Gonnore* 1275 RH] was named from Gunnora de la Mare, who held the manor in 1250 (Fees). *Gunnor* is a Norman woman's name of OScand origin (ON *Gunnvǫr*,

Gunnor).—W~ **Ford** [*Winterbourneford* 1320 PNW].

Winterbourne Stoke W [*Wintreburne-Stoch* DB. *Winterburnestok* c 1180 Fr]. 'STOC on (a stream formerly called) *Winter-burna*.'

Winterburn YW [*Witreburne* DB, *Winterburne* 1155–90 FC]. Identical with prec. names.

Winteringham Li [*Wintringeham* DB, c 1115 LiS]. 'HĀM of *Winter's* or *Wintra's* people.' *Wintra* is well evidenced. *Winter* occurs in *Wintres hlæw* 940 BCS 761. Cf. OG *Wintar*, *Wintrio*, ON *Vetr*.

Wintersett YW [*Wintersete* 1121–7 YCh (1428), 1226 FF]. 'Winter shieling.' Second el. ON SÆTR.

Winterslow W [*Wintreslev* DB, *-lewe* c 1192 Fr, *Winterlawa* 1166 P]. '*Winter's* burial mound.' Cf. WINTERINGHAM and *Wintres hlæw* 940 BCS 761 (Brk), HLĀW.

Winterton Li [*Wintringatun* c 1067 Wills, *Wintrintune* DB, *Wintringtuna* c 1115 LiS, *Wintringeton* 1228 Ep]. 'TŪN of *Winter's* or *Wintra's* people.' W~ is nr WINTERINGHAM.

Winterton Nf [*Wintertun* 1044–7 Holme, *Wintretuna* DB]. 'TŪN used in winter.' The place is near SOMERTON.

Winthorpe Li [*Winetorp* Hy 2 DC, *Wintorp* 1212 Fees, *Winthorp* 1209–35 Ep]. '*Wina's* thorp.'

Winthorpe Nt [*Wimuntorp* DB, *Wimethorp*, *Winetorp* 12 BM]. '*Wigmund's* thorp.' First el. ON *Vigmundr*, OSw *Wighmund* or OE *Wigmund*.

Wintney Ha [*Winteneia* 1139–61 (1337) Ch, *Winteneye* 1251 Cl]. '*Winta's* island.'

Winton La [*Wythynton* 1284 WhC, *Wythinton* 1322 LaInq]. 'TŪN among willows.' Cf. WĪÞIG.

Winton We [*Wyntuna* 1090–7 Kendale, *Wintonia* R 1 (1308) Ch]. OE *Winn-tūn* 'grazing farm'. Cf. WINN.

Winton YN [*Winetune* DB]. '*Wina's* TŪN.'

Wintringham Hu [*Wintringeham* Hy 2 PNBd], W~ YE [*Wentrigeham* DB, *Witer-ingeheim* 1190 P, *Wintringham* 1234 FF]. Identical with WINTERINGHAM.

Winwick (winĭk) Hu [*Wineuuiche* DB, *-wic* 1195 BM], W~(winĭk) Np [*Winewican* 1043 Th, *-wiche* DB, *-wic* Hy 2 BM]. '*Wina's* WĪC.' W~ Np has an alternative form *Winewincle* DB, 1189 (1332) Ch &c. It may be an alternative name in *-wincel* (cf. ALDWINKLE). Or it may contain a diminutive *wicincel* 'little *wic*'.

Winwick (winĭk) La [*Winequic* 1170 ff. P, *Winewich* 1204 P, *Wynequic* 1212 Fees]. '*Wineca's* WĪC.' Cf. WINKFIELD Brk.

OE **wīr** 'bog myrtle'. See WIRRAL, WORRALL, WYRLEY.

Wirksworth Db [*Wyrcesuuyrthe* 835 BCS 414, *Werchesuuorde* DB, *-worda* 1130 P, *Werkewurda* 1182 P]. '*Weorc*'s worþ.' *Weorc* may be found in *Weorces mere* BCS 1282. A pers. n. **Weorc* or the like must be assumed for several pl. ns., as WORKSOP, WORSALL, WORSBOROUGH. OE *Werca* is found. OG has *Werchari*, *Wercrata* and several names in *-werc*. OE *(ge)weorc* is sometimes a possible alternative in these pl. ns., but the regular gen. form points to a pers. n. as first el.

Wirral Chs [(on) *Wirhealum*, (of) *Wirheale* 894 f. ASC, (on) *Wirhalum* 1002 E, *Wirhale* c 1100 Chester]. OE *wir* 'bog myrtle' (found in *wirdenu*, *wirhangra*) and the plur. of OE HALH. The exact sense of *halh* is not clear. It may be 'haugh' or 'river meadow'.

Wirswall Chs [*Wireswelle* DB, *Wyriswall* Hy 2 BM, *Wiriswalle* 1260 Court]. '*Wighere*'s or *Wigric*'s spring.' Cf. WELL.

Wisbech (wĭzbēch) Ca [*Wisebece* 656 ASC (E), *Wisbece* DB, *Wisebec* 1173 P, 1199 Cur, *Wissebeche* 1291 Tax]. 'The valley of R WISSEY.' See BÆCE.

Wisborough Green Sx [*Wisebregh* 1227 Ch, *Wisseberge* 1279 Ass, *Wysberge* 1252 Cl, *Whishbergh* 1307 Ass]. It is suggested in PNSx that the first el. is OE *wisc* 'damp meadow', the second being OE *beorg* 'hill'. This is not quite convincing, but it is not easy to find something better.

OE **wisc** 'damp meadow, marsh', dial. *wish*, is found in some pl. ns., as WHISTLEY, WISLEY, CRANWICH, DULWICH. Cf. WISKE.

Wiseton or **Wyeston** (wĭstn) Nt [*Wisetone* DB, *Wiston* 1212, 1242 Fees]. Possibly identical with WISSINGTON. OE *Wiges tūn* does not quite suit the DB form.

Wishaw Wa [*Witscaga* DB, *Wiðshada* 1166, *Witteshage* 1184, 1190 P, *Witsahe* c 1195 Middleton]. OE *wiþig-scaga* 'willow wood' or possibly *wiht-scaga* 'wood by a bend' (cf. WIHT). A recess in a hill would be referred to.

Wishford W [*Wicheford* DB, *Major Wichef*' (*Wichford*) 1207 f. Cur, *Wichford* 1242 Fees]. 'Ford by the wych elm(s).' Cf. WICE.

Wiske R YN [*Wisca* 1100–15 LVD, *Wisc* c 1180 YCh 946, *Wisk* J Ass]. A river-name cognate with or derived from OE *wisc* 'damp meadow'. The OE form will have been *Wisce* (gen. *Wiscan*). This could mean 'meadow stream' (cf. WIMBORNE), or a meaning 'river' may have developed from that of 'wet place, marsh'.

Wisley (-ĭz-) Sr [*Wiselei* DB, *Wisle*, *Wisseleg* 1204 Cur, *Wysheleye* 1279 QW]. Identical with WHISTLEY.

Wispington Li [*Wipsinton* 1060 Th, *Wispinctune* DB, *Wispingtuna* c 1115 LiS, *Wispinton* 1253 BM]. The name is to be compared with *Wisp* 1218, *Wysp* 1228 For, which occurs in forest rolls dealing with the forest of Rutland. The name is apparently identical with ME *wisp* 'a handful, a bunch of hay' &c. The word may have been used in the sense 'thicket' or the like. Wispington would then mean 'TŪN of the people at the thicket'.

Wissett Sf [*Uuitsede*, *Wisseta*, *Wiseta* DB, *Wicsota* 1162 P, *Witseta* 1165 P, *Wytsett* 1235 Cl]. Second el. OE (GE)SET 'fold' &c. The first might be OE *wiþþe* in the sense 'willow'.

Wissey R Nf [*Wusan* (obl.) 905 ASC, *Wissene* 1257 Ass, *Wissenhe* 1277 ERN, *Wise*, *Wisse* 1314 De Banco]. Cf. *wusan* (obl.) BCS 875, the name of a river in Brk. The OE form was *Wise*, gen. *Wusan*, to which was added OE *ēa* 'river'. The name is related to G *Wiese* 'meadow', OE *wāse* 'mud', ON *veisa* 'a pool', OSw *Visa*, a river-name. The name very likely means 'water, river'. The folk-name *Wissa mægð* in Guthlac is probably derived from the river-name Wissey. Cf. also WISBECH. The Wissey probably once ran past Wisbech.

Wissington (wĭstn) Sf [*Wiswypetun* c 995 BCS 1289, *Wisinton* 1242 Fees]. '*Wigswiþ*'s TŪN.' *Wigswiþ* is an OE woman's name.

Wistanstow Sa [*Wistanestov* DB, *-stowa* 1177 f. P], **Wistanswick** Sa [*Wistaneswick* 1274 &c. Eyton, *Wystaneswyk* 1285 FA], **Wistaston** Chs [*Wistanestune* DB, *Wistanistona* Hy 3 BM], **Wisteston** He [*Wystaneston* 1198 Fees, *Wistaneston* 1241 Ch]. '*Wigstān*'s STŌW or holy place, wīc, and TŪN.'

Wiston (wĭsn) Sx [*Wistanestun* DB, *-ton* 1170, *Winestaneston* 1190 ff. P]. '*Wigstān*'s (or *Winestān*'s) TŪN.'

Wistow Hu [*Kingeston*, i.e. *Wistowe*, *Kyngestune* i.e. *Wicstone* 974 BCS 1310 f., *Wistov* DB], W~ YW [*Wicstow* c 1030 YCh 7, *Wikestowe* c 1160 YCh 36, *Wixtowe* J Ass]. OE *wic-stōw* 'dwelling-place, manor'. W~ Hu was a royal manor.

Wistow Le [*Wistanestov* DB, *Wystanstowe*, *Wistowe* 1254 Val]. '*Wigstān*'s STŌW or holy place.' Cf. WISTANSTOW Sa.

Wiswell La [*Wisewell* 1207 FF, *-wall* 1262 Ass]. The first el. is identical with the river-name WISSEY. The second is OE *wella* 'spring' or 'stream'.

Witcham Ca [(on) *Wichamme* 970 BCS 1268, *Wiceham* DB, *Wicheam*, *Wicham* 1254 Val, *Wytcham* 1282 Ipm]. Probably 'wych elm HAMM'. Cf. WICE.

Witcha·mpton Do [*Wichemetune*, *Wichamatuna* DB, *Wichhampton* 1242 Fees]. OE *Wichæma tūn* 'the TŪN of the dwellers at a wīc'. Perhaps wīc is here 'town' and refers to Wimborne Minster.

Witchford Ca [*Wiceford* DB, *Wyccheford* 1252 Ch]. 'Ford by the wych elm(s).'

Witchingham Nf [*Wicinghaham, Witcingeham* DB, *Wichingheham* 1106–9 Fr, *Wichingeham* 1130 P]. 'HĀM of the *Wicingas*.' Cf. WHISSONSETT, WHISSENDINE.

Witcombe Gl [*Wydecomb* 1220 Fees, 1291 Tax]. Apparently 'wide valley', which suits the topography. Or 'willow valley'. 'Wide valley' is clearly the meaning of *uuidan cumb* 716–43 BCS 164 (in boundaries of Woodchester not very far from Witcombe, but referring to a different valley).

Witcombe So [*Wythicumbe* 1243 Ass, *Wydecumbe* 1285 FA]. 'Willow valley.' Cf. WĪÞIG.

Witham (-t-) Ess [*Witham* 913 ASC (A), *Witanham* ib. (D), *Witham* DB, 1221 Pp], **W~ on the Hill** Li [*Witham* DB, c 1130 BM, 1227 Ch, *Wiham* 1202 Ass], **W~ Friary** So [*Witeham* DB, 1156 P, *Witeham, Wuttheham, Witheham* 1212 Fees, *Witham* 1160 P]. Perhaps '*Wita*'s or *Witta*'s HĀM' or 'the HĀM of the *wita* or councillor'. It is true the early form *Witham*, especially for the Ess name, is curious. OE *Wiht-hām* 'HĀM in a bend' would suit W~ Ess and Li. W~ So is certainly *Witta*'s HĀM.

W~ **Friary** was given by Hy 2 to the Carthusian friars.

Witham (wĭdhm) R Ru, Li [*Wiðma* c 1000 Saints, *Widme, Withma* c 1150 DC, *Widhem* 1243 Cl]. The river is probably referred to by Ptolemy as *Eidoumanios, Idoumaniou* (gen.). This stands for *Widumanios*. *Widu-* may be identical with Welsh *gwydd* 'forest'. The rest of the name may be a derivative suffix or a word for 'river' related to Lat *manare* 'to flow'. On the river are **North & South Witham** Li [*Widme, Wime* DB, *Nordwiéma* 1184 DC, *Suthwyme* c 1250 BM].

Withcall Li [*Widcale* DB, *Uitcala* c 1115 LiS, *Withcala* 12 DC, *Witthcal* 12 BM]. OScand *við-kiǫlr* 'wooded ridge'. See *viðr*. The second el. is identical with KEAL Li.

Withcote Le [*Wicote, Wicoc* DB, *Wittok* c 1125 LeS, *Witecoc* 1167 P, *Withcoc, Wythcoc* 1236 ff. Cl]. The elements are OE *wiþig* or *wiþþe* 'willow' and *cocc* 'heap', here probably in the sense 'a clump' (of trees). Cf. COCK BECK.

Witheridge D [*Wirige* DB, *Wytherigge* 1256, 1263 FF, *Wetherigge* 1249 Ass]. 'Willow ridge.' Cf. WĪÞIG.

Witherley Le [*Witheredel, Witheresdal* 1202 Cur, *Withed'le* 1203 Ass, *Wyrithele* 1247 Ass]. '*Wigþrȳþ*'s LEAH.' Cf. *Wyðreðe cross* BCS 1130. *Wigþrȳþ* is a woman's name.

Withermarsh Green Sf nr Hadleigh [*Hwifer[mer]sc, Wifærmyrsc, Hwifermirsc* c 995 Wills, *Withermers* DB, *Wiuermers* 1190 ff. P, *Wivermers* 1219 Fees]. 'The quaking bog.' The first el. varies between *wiver-* (dial. *wiver* 'to tremble, shake') and *wither-* (dial. *whither* the same). *Wiver-* is better evidenced in early sources.

Withern Li [*Widerne* DB, *Widerna* c 1115 LiS, *Wierne* Hy 2 DC, *Wihernia* J BM]. Perhaps 'house in a wood', the elements being OE WUDU (*widu*) and ÆRN. If so, *th* is due to Scand influence. OE *wiþig* 'willow' might also be the first el.

Withernsea YE [*Widfornessei, Witfornes* DB, *Witfornesel* c 1100 YCh 1300, *Withthornese, Wythorense, Wythornse mere* 1260 Ipm], **Withernwick** YE [*Withfornewinc, Widfornewic* DB, *Wifornewic* 1115 YCh 1304, *Withornwic* 1202 FF]. The two places are a good way apart. The first has as second el. OE *sǣ* or OScand *siōr* 'lake', the other OE WĪC. The first el. can hardly be anything else than a pers. n., which may be an ON *Við-Forni*, i.e. the well-known pers. n. *Forni* with a distinctive addition, e.g. OScand *viðr* 'wood' or *viðr* 'wide'. This man probably lived at Withernsea and had a dairy-farm at Withernwick.

Withersdale Sf [*Wideresdala* 1184 P, *Witheresdale* 1254 Val, *Wetheresdale* 1254 Val, *-dele* 1291 Tax], **Withersfield** Sf [*Wedresfelda* DB, *Wetherisfeud* 1254 Val, *Witheresfeud* 1235 FF, *Wytheresfeld* 1242 Fees]. 'Valley and FELD where wethers were kept.' The OE form of the word *wether* is *weþer*, but cf. OHG *widar*, ODu *wither*. There may well have been an OE form *wiþer*, used in some dialects.

Witherslack We [*Witherslake* c 1190 Kendale, 1267 Ch]. ON *viðar-slakki* 'wooded valley'. First el. ON *viðr*, gen. *viðar*. See SLAKKI.

Withgill YW nr Gt Mitton [*Withikill* 1226 FF, *Wychichil* 1260 Ass]. 'Willow hill', the elements being OE *wiþig* or ON *viðir* 'willow' and ON *kiǫlr* (dat. *kili*) 'ridge'. The place is on a hill.

Withiel Co [*Widie* DB, *Wythiel* 1291 Tax]. Cf. LOSTWITHIEL, which is separated from Withiel by an upland district. Withiel may be a name of this and mean 'wooded upland' or 'wood of the upland', the elements being identical with Welsh *gwydd* 'wood' (= Co *gwydh* 'trees') and *iâl* 'fertile upland region'.

Withiel Florey So [*Wiðiglea* 737, 938 BCS 158, 727, *Wythel* 1237 Cl, *Wythele Flory* 1305 FF]. 'Willow wood.' Cf. LĒAH.

The manor was held by Randulfus de Flury in 1237 (Cl). Cf. COMBE FLOREY.

OE **wīþig** 'willow', especially the osier willow, *Salix viminalis*, is a common first el. in pl. ns., and occurs as a second el. in HOARWITHY. There is also a ME *wiþin*, dial. *withen*, with the same meaning, also 'willow holt, wet land where willows grow'.

There may have been a side-form OE *widig*. At least the el. often appears with a *d* instead of *th*, especially in the south, and *widdy* 'a withy' is common in Scotland and the north. Perhaps we have a change *þ > d* in the word. OE *wiþþe* 'withe, thong' (= OFris *withthe*, ON *viðia*) may also have meant 'willow'. It is found in this sense from c 1340. It would explain names like WITHCOTE, WYTHEMAIL better than *wiþig*. ON *víðir*, OSw *vide* 'willow' occurs occasionally in Engl pl. ns. *Wiþig* is especially common in combination with *-tūn* (WEETON, WEIGHTON, WINTON, WITHINGTON, WIDDINGTON, WYTON), *-cumb* (WHIT-, WID-, WIDDE-, WIDDI-, WITHYCOMBE), *-ford* (WID-, WYTHEFORD), *-lēah* (WEETHLEY, WIDLEY, WITHIEL). See further e.g. WEEDLSADE, WERGS, WISHAW, WITHERIDGE, WITHY- (passim), WYDDIAL, WYTHOP. ON *víðir* is certainly found in WYTHBURN.

Withington Chs [*Widinton* 1186 P, *Withington* 1245-50, *Wythinton* 1267 Chester], W~ He [*Widingtune* DB, *Chircwithinton* 1266 Misc], W~ La [*Wythinton* 1212 Fees, *Witheton* 1219 Ass], W~ Sa [*Wientone* DB, *Widinton* 1192 f., *Widiton* 1194 P, *Withinton* 1267 BM]. 'TŪN among willows.' Cf. WĪÞIG.

Withington Gl [*Wudiandun* 737, *Uuidian-, Wudiandun* 774, *Widiandun* c 800 BCS 156, 217, 299, *Widindune* DB]. '*Widia*'s (*Wudga*'s) DŪN or hill.' This name is found in Widsith and is the first el. of *Widian byrig* 982 KCD 633. It is quite possible that the places were named from the legendary hero *Widia* (*Witege* in German tradition).

Withnell La [*Withinhull* c 1160 LaCh, *Withenhull* 1246 Ass]. 'Willow hill.' Cf. WĪÞIG.

Withybrook Wa [*Widebroc* Hy 2 BM, 1190 P, *Withibroc* 1205 Cur]. 'Willow brook.'

Withycombe Raleigh D [*Widecome* DB, *Wydicumb* 1242 Fees], W~ So [*Hwidigcum* 1065 Wells, *Widicumbe* DB, *Withicumbe* 1225 Ass]. 'Willow valley.' See WĪÞIG.

W~ **Raleigh** was held by Hugo de Ralegh in 1303 (FA). Cf. COLATON RALEIGH.

Withyha·m (-id-) Sx [*Withiham* 1230 FF]. 'Willow HAMM.' See WĪÞIG.

Withyhook Do nr Yetminster [*Widihoc* 1197 FF, 1200 Cur, *La Wytheoc* 1283 Ipm]. 'Willow-grown bend.' Cf. WĪÞIG and HOOKE.

Withypool So [*Widepolle* DB, *Widipol* 1185 P]. 'Willow pool.'

Witley Sr [*Witlei* DB, *Wittelega* 1186 P, *Witle* 1212 Fees, *Wytteley* 1247 Ch]. '*Witta*'s LĒAH.'

Witley, Great, Wo [*Whitele Major* 1275 Ass], **Little** W~ Wo [*Wittlæg* 964, *Witleah* 969 BCS 1135, 1242, *Wihtlega* 11 Heming, *Witlege* DB]. 'LĒAH by the bend.' See

WIHT. The bend is a deep recess in Abberley Hill at Great Witley.

Witnesham Sf [*Witdesham* DB, *Witlesham* 1195 Cur, 1253 Ch, *Witnesham* 1254 Val, *Wytnesham* 1259 Ch]. '*Wittin*'s HĀM.' **Wittin* is a diminutive form of *Witta*.

Witney O [*æt Wyttannige* 969 BCS 1230, *Wittannige* 1044 KCD 775, *Witenie* DB], **Wittenham, Little & Long,** Brk [*Wittanham, Withennam* 862, *Wittanhamm* 892-901 BCS 504 f., 581, *Witeham* DB, *Est-, Westwitteham* 1220 Fees]. '*Witta*'s island and HAMM.' In the boundaries of Witney in BCS 1230 is mentioned *Wittan mor*.

Wittering Np [*Widerigga* [land] 7 BCS 297, (æt) *Wiðeringige*, (on) *Wiðeringaeige*, (wið) *Wyðeringaige* 972-92 BCS 1130, *Witheringham* DB, *Witeringa* 1167 P]. Cf. WERRINGTON Np. An old folk-name, derived probably from OE *Wiðer* pers. n. *Wiðer* is a short form of *Wiðergyld* and similar names. Cf. OG *Widargelt, -olt*.

Wittering Sx [*Wihttringes* 683, *Wystrings* c 770 BCS 64, 211, *Westringes* DB, *Witteringes* 1227 Ch]. '*Wihthere*'s people.'

Wittersham K [(æt) *Wihtriceshamme* 1032 Th, *-ham* 11 DM]. '*Wihtric*'s HAMM.'

Witton Chs [*Witune* DB, *Wittonia* 1200-8 Chester], W~ **Gilbert** (j-) Du [*Wyton* 1195 (1335) Ch, *Wittona* 1248 FPD], W~ **le Wear** Du [*Wudutun* c 1050 HSC, *Wudetun* 1104-8 SD, *Wytton in Werdal* 1300 Misc], W~ La [*Witton* 1246 Ass], **Long** W~ Nb [*Wttun* 1236, *Wotton* 1242 Fees, *Langwotton* 1340 Newminster], **Nether** W~ Nb [*Wittun* 1236, *Witton* 1242 Fees, *Wytton* 1247 Misc], **Witton** Nf nr N. Walsham [*Wittuna, Widituna* DB, *Wittone, Wottone* 1254 Val], W~ Nf nr Norwich [*Witona* DB, *Wuton* 1202 Cur, *Witton* 1242 Fees], W~ Wa [*Witone* DB, *Wichtona* 1169, *-ton* 1170 P, *Witton* 1241 Cl], W~ Wo [*Wittona* 716, *Wictun* 817 BCS 134, 361, *Witone in Wich* DB], **East** W~ YN [*Witun* DB, *Estwiton* 1208 Cur, *Est Wotton* 1316 FA], **West** W~ YN [*Witun* DB, *Widtona* 1166 P, *West Wytton* 1246 FF]. There are at least two sources for the name Witton, OE *Wic-tūn* 'TŪN by a WĪC', certain for W~ Wo, which is at Droitwich, and W~ Wa, and probable for W~ Chs, which is close to Northwich, and OE *Widu-tūn* (*Wudu-tūn*) 'TŪN by a wood'. The latter is certain for W~ le Wear, Long and Nether W~ Nb (which are near each other), the Nf Wittons and the YN Wittons. The rest are doubtful. W~ La may quite well be '*Witta*'s TŪN'.

W~ **Gilbert** Du 'from Gilbert de la Ley (t. Hy 2).—W~ **le Wear** Du is on the Wear; *le* is the French def. art.

Wiveliscombe So [*Wifelescumb* 854 BCS 476, *Wyfelescumbe* 1065 Wells, *Wivelescome* DB], **Wivelsfield** Sx [*Wifelesfeld* c 765

BCS 197]. '*Wifel*'s CUMB or valley and FELD.' See WILCOTE.

Wivenhoe (-ĭ-) Ess [*Wiunhov* DB, *Wiuueho* 12 BM, *Wyvenho* 1246 Ch]. '*Wife*'s or *Wifa*'s HŌH or spur of land.' Cf. WESTOW, WINESTEAD, WIVETON.

Wiverton (wer-) Nt [*Wivretun, Wiuretone* DB, *Wiuertona* Hy 2 DC, *Wyuereton* 1230 P]. '*Wigþriþ*'s TŪN.'

Wiveton Nf [*Wiuentona, Wiuetuna* DB, *Wyveton* 1242 Fees, *Wiventone* 1254 Val]. '*Wife*'s or *Wifa*'s TŪN.' Cf. WIVENHOE.

Wix Ess [*Wica* DB, *Wikes* 1191 FF, *Wiches* 1198 Cur]. The plur. of OE WĪC 'dairy-farm'.

Wixford Wa [*Wihtlachesforde* 962 BCS 1092, *Wichtlakesford* 1154–8 (1340) Ch]. '*Wihtlāc*'s ford.'

Wixhill Sa [*Witekeshill* 1203 Eyton, *Wyne-keshull* 1252 Ch, *Wykeshull* 1327 Subs]. '*Wittuc*'s hill.' Cf. UTTOXETER.

Wixoe Sf [*Wlteskeou* DB, *Widekeshoo* 1205 Cur, *Wydekesho* 1219 FF, 1326 Fees]. '*Widuc*'s HŌH or spur of land.'

Woburn (wŏo-) Bd [*Wóburninga* (gemæru) 969 BCS 1229, *Woburne* DB, *Wauburn* 1200 Cur, *Woborn* 1208 Cur], **Woburn** Sr [*Woburne* a 675 BCS 34, *Woburn* 1198 FF]. Originally the name of the streams at the places. The base, OE *Wŏburna*, means 'crooked, winding stream', from OE *wŏh* 'crooked' and BURNA. Identical are OBORNE and WOOBURN. Cf. also *Wohburna* 901 BCS 596 (Ha), *Woburna* 963 ib. 1114 (Sx), also WAMBROOK, WOMBOURN, WONERSH.

Wokefield Brk [*Hocfelle* DB, *Wokefeld* 1198 FF], **Woking** Sr [*Uuocchingas* 708–15, *Woccingas* c 796 BCS 133, 275, *Wocingas* 777 ASC (E), *Wochinges* DB], **Wokingham** Brk [*Wokingeham* 1227 Ch, 1242 Fees]. '*Wocca*'s FELD', '*Wocc*'s or *Wocca*'s people' and 'HĀM of *Wocc(a)*'s people'. *Wocc* is found in *Wocces geat* 901, 961 BCS 594, 1080 (Ha), OAKSEY, *Wocca* in OCKENDON. Cf. OG *Woco, Wocco* pers. n. Wokefield and Wokingham are not far apart.

Wolborough (-ŏo-) D [*Ulveberie* DB, *Wlue-berue* 1228 FF, *Woluebergh* 1242 Fees]. 'Wolves' hill', OE *wulfa-beorg*.

Woldingham Sr [*Wallingeham* DB, *Wald-ingham* 1232 Cl, *Waldingeham* 1242 Fees]. 'HĀM of the people of the *weald* or wood.' See WALD.

Woldringfold Sx [*Wolfringfolde* 1327 Subs]. 'Fold of *Wulfhere*'s people.'

Wolds, The, Le [*the Wold* 1610 Holland (OED); see WALTHAM ON THE WOLDS], Li [see NEWTON LE WOLD, CUX-, STIXWO(U)LD], Nt [*Wolde* 1252 Ipm, *Waldas* 1363 BM], Y [*Yorkwold* 1325–50 Map, *Yorkeswold* 1472–5 OED; cf. WOLD NEWTON, COX-, EASINGWOLD]. OE *wald* 'wood, wold'.

Wolferlow He [*Ulferlav* DB, *Wulfereslowe* 1242 Fees], **Wolferton** Nf [*Wulferton* 1166 f. P, 1196 FF]. '*Wulfhere*'s HLĀW or burial-mound and TŪN.'

Wolfhamcote Wa [*Ulfelmescote* DB, *Wlf-hamecot* 1236 Fees]. '*Wulfhelm*'s COT.'

Wolfhole Crag La [*Wolfalcrag* c 1350 LaCh]. Wolfhole seems to be OE *wulf-halh* 'wolves' HALH or valley'.

Wolford, Great & Little, Wa [*Ulware, Ulwarda, Wolwarde* DB, *Wlwarda* Hy 2 (1314) Ch, *-ward* 1201 Cur, *Magna, Parva Wulleward* 1242 Fees]. The second el. seems to be OE *weard* 'guard'. If so, the use of the word is unique. It might be used in a concrete sense such as 'arrange-ment for protection, fence'. If the first el. is OE *wulf* 'wolf', the name might mean 'enclosure to protect flocks from wolves'.

Wollaston Np [*Wilavestone* DB, *Wullaues-ton* 1194 P, *Wullaveston* 1220 Fees, *Wil-laueston* 1220 Ep], W~ Sa [*Willavestune* DB, *Wylaveston, Wilaston* 1242 Fees]. '*Wiglāf*'s TŪN.'

Wollaston Wo [*Wollaueston* 1275 Ass], **Wollaton** (-ŏo-) Nt [*Olavestone* DB, *Wul-laueton* 1236 Ep]. '*Wulflāf*'s TŪN.'

Wollerton Sa [*Ulvretone* DB, *Wluruntona* 1130–5 PNSa, *Wulfrinton* 1187 P]. '*Wulf-rūn*'s TŪN.' Cf. WOLVERHAMPTON.

Wollescote Wo [*Wlfrescote* 1275 Subs]. '*Wulfhere*'s COT.'

Wolseley St [*Ulselei* DB, *Wulfsieslega* 1175 ff. P]. '*Wulfsige*'s LĒAH.'

Wolsingham Du [*Wlsingham* c 1150 FPD, *Wulsingeham* 1196 P]. 'HĀM of *Wulfsige*'s people.'

Wolstanton (-ŏo-) St [*Wlstanetone* DB, *Wulstaneston* 1199 P], **Wolstenholme** La [*Wolstonholme* c 1180 Whitaker, *Wlstanes-holme* 1278 FF]. '*Wulfstān*'s TŪN and holm.'

Wolston Wa [*Ulvricetone* DB, *Wlvrich[eston]* c 1180 Fr, *Wuluricheston* 1193 P]. '*Wulfric*'s TŪN.'

Wolsty Cu [*Wolsstibay* 1323 Cl]. 'Wolf path.' See STĪG. W~ is on the sea.

Wolterton Nf [*Ultretune* DB, *Wltreton* 1195 FF, *Wulterton* 1235 Cl]. The first el. is a pers. n., e.g. OE *Wulfþrȳþ*, a woman's name.

Wolvercote (-ŏo-) O [*Ulfgarcote* DB, *Wl-garicota* c 1130 Oxf, *Wolgarcote* 1149 Osney]. 'COT of *Wulfgār*'s people.'

Wolverhampton (-ŏo-) St [*æt Heantune* 985 KCD 650, (into) *Heantune* 10 BCS 1317, *Wolvrenehamptonia* 1074–85 Reg, *Wulfrune-hanton* 1169 P]. Originally OE *Hēatūn*, dat. *Hēantūne*, 'high TŪN'. The manor was given in 985 to a lady called *Wulfrun* (KCD 650), who gave land to Wolverhampton church.

Wolverley Wo [*Wulfferdinleh, Uulfordilea,* (æt) *Wulfweardiglea* 866 BCS 513 f., (æt) *Wulfweardiglea* 11 KCD 766, *Ulwardelei* DB]. 'The LĒAH of *Wulfweard*'s people.' The manor was given in 866 to one *Wulfferd* or more correctly *Wulfweard.*

Wolvershill Wa [*Wulfhereshilla* 12 DC, *Wolfarshull* 1405 AD]. '*Wulfhere*'s hill.'

Wolverton Bk [*Wluerintone* DB, *Wulfrinton* 1195 Cur, *Woluerington* 1227 FF]. '*Wulfrūn*'s TŪN' (cf. WOLLERTON) or 'TŪN of *Wulfhere*'s people'.

Wolverton Ha [*Ulvretune* DB, *Ulfretun* 1156, *Wulfertona* 1168 f. P, *Wulurington* 1230 P]. The first el. is derived from a pers. n. such as *Wulfhere, Wulfrēd* or *Wulfrūn.*

Wolverton Wa [*Ulwarditone* DB, *Wlwarditune* 12 Fr, *Wluuardinton* 1236 Fees]. 'TŪN of *Wulfweard*'s people.'

Wolverton Wo [*Wulfringctun* 977, *Wulfringtun* 984 KCD 612, 645, *Ulfrintun* DB]. 'TŪN of *Wulfhere*'s people.'

Wolvesnewton Monm [*Wolvesneuton* 1311 Ch]. Held by Ralph le Wolf in 1314 (Charles).

Wolveton or **Wolfeton** Do [*Wolueton* 1231 f. FF]. '*Wulfa*'s TŪN.'

Wolvey Wa [*Ulveia* DB, *Wulfeia* Hy 2 BM, *Wulfeie* 1195 P, *Wulfheia* 1221 Ass]. OE *wulf-hege* 'enclosure to protect flocks from wolves or to trap wolves' (cf. OE *wulfhaga*) rather than OE *wulf-ēg* 'island infested by wolves'.

Wolviston (wŏŏstn) Du [*Oluestona* 1091, *Wluestuna* Hy 2 FPD]. '*Wulf*'s TŪN.'

Wombleton YN [*Winbeltun* DB, *Wimbaltuna* 1148 YCh 179, *Wimbelton* 1208 Cur]. '*Winebald*'s or *Wynbald*'s TŪN.'

Wombourn St [*Wamburne* DB, *-burna* 1167 P, *Wanburn* 1175 P, *Womburne* 1236 Fees, *-borne* 1242 ib.]. OE (æt) *wōn-burnan* 'the winding stream'. Cf. WOBURN.

Wombridge Sa [*Wombrugga* 1181 BM, *-rug* 1236 Fees, *Wambrigg* 1207 Cur]. OE *Wambhrycg.* The first el. is OE *wamb* 'womb', here very likely used of a lake; cf. *Vámbsjön* in Sweden. There are several small lakes at the place. Second el. OE *hrycg* 'ridge'.

Wombwell (-ōō-) YW [*Wanbuelle, Wanbella* DB, *Wambewelle* c 1200 YCh 646, 1240 FF]. '*Wamba*'s spring.' Cf. WEMBLEY. Or the first el. may be OE *wamb* 'womb' in some transferred sense.

Womenswold (-ĭ-) K [*Wimlincga wald* 824 BCS 381, *Wimlingweald* 11 DM, *Wymelingwelde* 1291 Tax]. 'The wood of the *Wimlingas*.' *Wimlingas* is a folk-name of doubtful etymology.

Womersley YW [*Wilmereslege* DB, *-ley* 1286 YInq, *Wilmerisley* Hy 3 BM]. '*Wil-*

mǣr's LĒAH.' OE *Wilmǣr* is not with certainty evidenced, but has a counterpart in OG *Willamar.* The first el. might also be an OE *wilg-mere* 'willow lake'.

Wonastow, Welsh **Llanwarw** Monm [*Lanngunguarui* c 1150 LL, *Llanwarwe* c 1566 Peniarth MS 147, *Wonewardstowe* 1284 Charles, *Wonewarestowe* 14 LL]. 'The church of St. Gwnwarwy.' See LANN, STŌW. The Engl name is a translation of the Welsh.

Wonersh (-ŭ-) Sr [*Wigehers* 1198 Cur, *Woghenhers* 1199 Cur, 1287 Cl, *Woghenersh* 1305 Ch, *Woners* c 1270 Ep]. Apparently (æt) *wōgan ersce* 'crooked field'. First el. OE *wōh* (gen. *wōs* or *wōges*) 'crooked'.

Wonford D nr Exeter [*Wynford* 937 BCS 721 f., *Wunford* c 1100 E, 1130 P, *Wenford* DB, *Winfrodhundredum* 1195 f. P]. 'Meadow ford.' Cf. WINN.

Wonford D in Thornbury [*Wenforda* DB, ?*Wainford* 1204 Cur, *Wamford, Wantford* 1230 P, *Wanford* 1230 P, 1244 FF]. Possibly OE *wænford* from *wægnford,* identical with WANGFORD (2).

Wonston Ha [*Wynsigestun* 901 BCS 596, *Wenesistune* DB, *Wonsintone* c 1124, 1243 Ep, *Wensieston* 1205 Pp]. '*Wynsige*'s TŪN.'

Wooburn Bk [*Waburna* 1070–87 RA, *Waborne* DB, *Woburna* c 1145 RA]. Probably identical with WOBURN in spite of the early *a*-forms.

Woodale YN [*Wulvedal* 1223 FF]. 'Wolves' valley.'

Woodbarrow So [*Udeberge* DB, *Wodeberg* 1243 Ass]. 'Wooded hill.' See BEORG.

Woodbastwick. See BASTWICK.

Woodborough Nt [*Ude(s)burg* DB, *Wudeburc* 1169 P, *-burgh* 1231 Cl], W~ W [*Wideberghe* 1208 Cur, 1242 Fees, *Wudeberg* 1241 Cl]. W~ Nt is OE *Wudu-burg* 'fort built of wood' or 'fort in a wood'. W~ W is identical with WOODBARROW.

Woodbridge Sf [*Oddebruge* c 1050 KCD 907, *Wudebrige* DB, *-breg* c 1205 BM]. 'Wooden bridge.'

Woodburn Nb [*Wodeburn* 1265 Sc]. Named from the stream at the place [*Wdeburne* 1225 Ep]. 'Stream in or coming from the wood.'

Woodbury D [*Wudeburge land,* (on) *Wudebirig* 1072–1103 E, *Wodeberie* DB]. See WOODBOROUGH Nt.

Woodchester Gl [*Uuduceastir* 716–43, *Wuduceaster* 896 BCS 164, 574, *Widecestre* DB]. 'Roman fort in a wood or built of wood.'

Woodchurch Chs [*Wodechirche* c 1100, *Odecerce* c 1150 Chester], W~ K [*Wudecirce* 11 DM, *-chirch* 1240 Ass]. 'Wooden church.'

Woodcote Ha nr Bramdean [*La Wodecote* 1212 Fees, *la Wudecot* 1233 Cl], W~ Ω

[*Wdecote* 1109, *Wodekot* c 1250 Eynsham], W~ Sa nr Newport [*Udecote* DB], W~ Sr [*Wodecot* 1202 f. Cur], W~ Wa [*Widecote* DB, *Wudecota* 1190 AC], **Woodcott** Chs [*Wodecot* 1205 BM], W~ Ha nr Whitchurch [*Odecote* DB, *Wodecote* 1242 Fees]. 'COT in a wood.'

Woodcroft Np [*Wudecroft* 1163, 1179 P]. 'Croft in or by a wood.'

Wooden (ōōdn) Nb [*Wolveden* 1265 Misc]. 'Wolves' valley.'

Woodend Np [*Wodende* 1316 FA]. 'End of the wood.' Originally **Little Blakesley** [*little Blacolvesle* 12 NS].

Woodfall Hall La [*Wudefal* a 1230 CC, *Wodefal* 1321 FF]. 'Place where trees have fallen down' or 'place where wood may be felled'.

Woodford Chs nr Handforth [*Wideforde* 1248 Ipm, -*ford* c 1280 Chester], W~ Chs nr Over [*Wodeford* c 1225 Vale Royal], W~ Co [*Wdeford* 1197 FF], W~ Ess [*Wudeford* 1062 Th, *Wdefort* DB], W~ Np nr Thrapston [*Wodeford* DB, 1196 Cur], W~ Halse Np (in manor of Halse) [*Wodeford* DB, 12 NS, *Wudeford* 1224 Ep], W~ W [(to) *þæm ealdan wuduforda* 972 BCS 1286, *Wodeford* 1120 Salisbury]. 'Ford by a wood.'

Woodgarston Ha [*Wodegarstone* 1284 Ep]. 'Grazing-ground (OE *gærstūn*) in a wood.'

Woodhall Park Hrt [*Wudehale* 1198 FF, *Le Wodehalle* 1303 FA], W~ Li [*Wudehalle*, -*halla* 12 DC, *Wudehall* 1212 Fees], W~ YE [*La Wodehalle* 1286 Ch], W~ YW nr Harthill [?*Wdehall* c 1200 YCh 1803, *Wodhall* 1372 AD]. There are several other examples of the name. OE *wudu-hall* 'hall in a wood', at least in some cases no doubt 'hall for a forest court'.

Woodham Bk [*Wodehamme* 1370 BM]. Originally *Hamm* (v. HAMM), later *Wode-hamm* for distinction from Ham or **Field-ham** close by.

Woodham Du [*Wodon* 1091, *Wdum* Hy 2 FPD]. OE *wudum*, dat. plur. of *wudu* 'wood'.

Woodham Ferrers Ess [*Udeham* DB, *Wudeham* 1212 Fees, *Wodeham Ferreres* 1231 FF], W~ **Mortimer** Ess [*Odeham* DB, *Parva Wudeham* 1212 Fees, *Wodeham Mortumer* 1238 Subs], W~ **Walter** Ess [*Wdeham* DB, *Wudeham* 1212 Fees, *Wode-ham Walter* 1238 Subs], W~ Sr [*Wodeham* a 675 BCS 34, *Wudeham* c 890 ib. 563]. 'HĀM in a wood.' *Wuduham* c 995 BCS 1289 probably refers to one of the Ess Woodhams.

W~ **Ferrers** was held by Henricus de Ferreriis in 1086 (DB). Cf. BERE FERRERS.—W~ **Mortimer** was held in 1212 (Fees) by Robert de Mortimer, who (or whose father) had got it from Henry II. Cf. CLEOBURY MORTIMER.— W~ **Walter** was held by Robertus fil. Walteri in 1212 (Fees).

Woodhay (-dī), **West**, Brk [*Widehieie* 1203 Cur, *Wydehaye* 1220 Fees, *Wdehaye* 1228 For, *West Widyhay* 1285 Ipm], **East** W~ Ha [*Widenhai* 1189 BM, *Wodehaye* 1171, *Wyde-haye* 1172 Ep]. OE *wudu-gehæg* 'enclosure in a wood'. For the *i*-forms see WUDU.

Woodhill. See ODELL.

Woodhorn Nb [*Wudehorn* 1178 P, *Wode-horn* 1242 Fees]. 'Wooded point of land.' The place is near a promontory called Beacon Point. See HORN.

Woodhouse Le [*Wodehuses* 1209-35 Ep, -*houses* 1327 Subs], W~ YW in Leeds par. [*Wodehus* 1208 FF, *Wudehus* 1258 YInq], W~ YW nr Saxton [*Wdehus* 1157 YCh 186, *Wodhous* 1402 FA], W~ YW nr Sheffield [*Wdehus* 1200-18 YCh 1279]. 'House(s) in a wood.' Cf. WOODSOME, WOTHERSOME. There are many other examples of Wood-house, especially in combination with an-other name, as HORSLEY WOODHOUSE Db [*Wodhows* 1431 FA].

Woodhurst Hu [*Wdeherst* 1209 FF]. See OLD HURST.

Woodkirk YW [*Wudechircha* 1121-7 YCh 1428, *Wudekyrcæ* c 1150 Crawf]. 'Wooden church', probably a Scandinavianized form of WOODCHURCH.

Woodland D [*Wodelonde* 1328 Ep], **Wood-lands** Do [*La Wodelond* 1303, 1412 FA], W~ K [*Wodlond* 1276 Cl, 1291 Tax], W~ So in Frome [*Wodelond* 1342 Misc]. Self-explanatory. The names are common.

Woodleigh D [(æt) *Wudeleage* 1008-12 Crawf, *Odelie* DB]. 'Glade in a wood.' See LĒAH.

Woodlesford YW [*Wrislesfordia* c 1150 Crawf, *Wridelesford* 1170 P, *Wridlesford* 1201 Cur, *Wriddlisford* 1202 FF]. The first el. is probably an OE *wridels*, cognate with WRĪD and meaning 'thicket' or the like.

Woodmancote Gl nr Dursley [*Wodemone-cote* Hy 2 Berk, -*cota* 1220 Fees], W~ Gl nr Bp Cleeve [*Wodemanecota* 1220 Fees, *Wudemancote* 1221 Ass], W~ Gl nr N. Cerney [*Wodemancote* 1279 Winchc, 1287 QW], W~ Sx [*Odemanscote* DB, *Wode-manecot* 1241 FF], **Woodmancott** Ha [*Woedemancote* 903 BCS 602, *Wdemancote* DB]. OE *wudumanna-cot(u)* 'woodmen's cottage(s)'.

Woodmansey YE [*Wodemanse* 1297 Subs, 1298 Misc]. 'The woodman's lake.' The spelling *Woodmancy* 1573 BM points to voiceless *s* and OE *sǣ* as second el.

Woodmansterne Sr [*Odemerestor* DB, *Wodemerestorn* 1207 Cur, *Wudemeresthorn* 1242 Fees]. 'Thorn-bush by the boundary of the wood', OE *wudu-gemǣres-þorn*.

Woodnesborough (winzbru) K [*Wanes-berge* DB, *Wodnesbeorge* 11 DM, -*berg* 1242 Fees]. 'Wōden's hill', 'hill sacred to Wōden'.

Woodnewton, -perry, -plumpton. See NEWTON, PERRY, PLUMPTON.

Woodrow W [*La Woderowe* 1280 Cl, *Wode-rewe* 1286 Ch], W~ Wo [*Wodrewe* 1505 PNWo]. OE *wudu* 'wood' and *rāw, rǣw* 'row', here perhaps used of a lane in a wood.

Woodsetts YW [*Wodesete* 1324 Ipm]. 'Fold in a wood.' See (GE)SET.

Woodsford Do [*Werdesford* DB, 1194 P, 1212 Fees, *Wyrdesforde* 1291 Tax, *Wirdes-ford* 1318 FF]. Identical with *Wierdes ford* (St) 975 BCS 1312. '*Wigheard*'s ford.' The loss of the first *r* is due to dissimilation.

Woodsome YW [*Wodehus* 1236, 1240 FF]. OE (æt) *wudu-hūsum* 'houses in a wood'.

Woodspring So [*Worsprinc* DB, -*springe* 1243 Ass, -*spring* 1264 Ep]. The first el. is OE *wōr* in *wōrhana, -henn* 'wood-grouse' (= Du *woerhaan*). Cf. WORLE. The second is *spring* 'copse' rather than *spring* 'well'.

Woodstock O [*Wudustoc* c 1000 Laws, *Wudestoke* 1123 ASC (E), *Wdestoc* c 1150 SD]. The name is rendered by Symeon of Durham by 'silvarum locus', i.e. 'place in the woods'.

Woodstone Hu [*Widestun* 973 BCS 1297, *Wudestun* 973 PNHu, *Wodestun* DB]. Possibly 'TŪN in a wood', though the gen. form of the first el. rather suggests some other etymology. The OE forms are in late texts and not authoritative. A better etymology would be obtained if we may assume *Wudes-* to be contracted from an OE *Wud(u)-efes* 'edge of the wood'. MEADS may be compared.

Woodthorpe Db nr Staveley [*Wodesthorp* 1265, *Wodethorpe* 1269 Ipm], W~ Li [*Wde-torp* Hy 2 DC, 1202 Ass, *Uudethorp* 1242 Fees]. 'Thorp in a wood.'

Woodton Nf [*W(o)detuna, Videtun* DB, *Wudeton* 1254 Val]. 'TŪN in a wood.'

Woodyates, East & West, Do [*at Wdegeate* 869 BCS 525, *Besuðan wudigan gæte* c 980 Rit Dun, *Odiete* DB, *Wudiete, Widiate* 1198 FF]. Either OE *wudige geat* 'wooded gate' or '*Wudiga*'s gate'. Cf. WITHINGTON Gl. OE *geat* 'gate' is here used in the sense 'pass'; cf. WINGATE. There is a pass here, through which runs a Roman road.

Woofferton Sa [*Wulfreton* 1221 Ass, 1259 Ipm, *Wulferton* 1221 FF]. '*Wulfhere*'s or *Wulffriþ*'s TŪN.'

Wookey (-ŏŏ-) So [*Woky* 1065 Wells, 1231 Ch, *Wochi* 1178 Wells]. Near Wookey is a famous cavern called **Wookey Hole** [*Wokyhole* 1065 Wells, 1249 FF]. There is an OE *wōcig*, found only in glosses and apparently meaning 'noose, snare'. This seems to be the source of the name, which would then refer to a trap for animals.

Wool Do [(æt) *Wyllon* 1002–12 ASWrits, *Wille, Welle* DB, *Welles* 1212 Fees]. 'The springs', OE *wella, wiella* &c.

Woolacombe D [*Wellecome, Wolnecome* DB, *Wollecumb* 1242 Fees]. If *Wolnecome* stands for *Woluecome*, the name means 'wolves' valley'. But the first el. seems rather to be OE *wella* (*wiella*) 'spring, stream'.

Wollaston Gl [*Odelaveston* DB, *Wullaves-tona* Hy 2 (1307) Ch, *Wollaveston* 1253 Cl]. '*Wulflāf*'s TŪN.'

Woolavington So [*Hunlauintone* DB, *Wil-laveton* 1201 Cur, *Wulavinton* 1248 Ch, *Wollavynton, Wilavinton* 1276 RH]. 'TŪN of *Hūnlāf*'s or *Wiglāf*'s people.'

Woolavington Sx. See LAVINGTON.

Woolbeding Sx [*Welbedlinge* DB, *Wolbed-ing* 1191 P, *Wulfbeding* c 1230 Selborne]. '*Wulfbeald*'s people.' The second *l* was lost owing to dissimilation.

Woolcombe Do in Melbury Bubb [*Welle-come* DB, -*cumbe* 1200 Cur, *Wullecumb* 1236 Cl]. 'Valley of the stream' (OE *wella, wiella*).

Wooldale (ŏŏdl) YW [*Uluedel* DB]. 'Wolves' valley.'

Wooler Nb [*Wulloure* 1187, *Welloure* 1196 P, *Wllovera* 1199 Ch, *Wullouer* 1212 Fees]. Apparently 'bank of the stream', the first el. being OE *wella*. If so, we have to assume that *Well-* became *Wæll-* and *Wull-*. OE *e* often becomes *æ* after *w* in Old Northumbrian. See ŌFER.

Woolfardisworthy (wŏŏlzerī) D nr Credi-ton [*Ulfaldeshodes* DB, *Wolfardesworthi* 1264 Ep], W~ ((w)ŏŏlserī) D nr Hartland [*Olfereordi* DB, *Wulfrideswurd* 1230 P]. Both seem to mean '*Wulfheard*'s WORPIG'.

Woolhampton Brk [*Ollavintone* DB, *Wl-lauintona* a 1162 Oxf, *Wullavinton* 1242 Fees]. 'TŪN of *Wulflāf*'s people.'

Woolhope He [*Hope* DB, *Wulvivehop* 1234 Cl, *Wulveve Hope* 1242 Fees]. Originally HOPE from OE *hop* 'valley'.

The manor was given to Hereford Cathedral by Wulveve and Godheve (see 1242 Fees) and took its surname from the former, a lady called *Wulfgifu*.

Woolland Do [*Wonlond* 939 BCS 738, *Win-lande* DB, *Wunlanda* 1170 f. P, *Wuland* 1212 Fees]. OE *wynn-land* 'meadow land'; cf. WINN.

Woolley Brk nr Brightwalton [*Olvelei* DB, *Wlvelye* 1220 Fees], W~ Brk in White Waltham [*Wolveleye* 1286 Cl], W~ Hu [*Wululeia* 1158, *Wulfelea* 1180 P], W~ YW [*Wiluelai* DB, *Wlflay* c 1125 YCh 1663, *Wuluelegh* 1202 Ass]. 'Wolves' wood.' Cf. LĒAH.

Woolley So [*Wilege, Wllega* DB, *Wllege* 1201 Cur, *Wolleye* 1303 FA]. 'LĒAH by a stream' (OE *wella, wiella*).

Woolmer Forest Ha [*Uulfamere* 970 BCS 1266, *Wolvemere* c 1200 Ep, *Wulvemere* 1274 RH]. 'Wolves' lake.' Originally the name of **Woolmer Pond** [*Wulvemar* 1236 Cl].

Woolpit Sf [*Wlpit* 10 BCS 1013, *Wulpettas* 11 EHR 43, *Wlfpeta* DB, *Uulfpet* c 1095 Bury]. OE *wulfpytt* (cf. *wulfputt* BCS 936) 'pit for trapping wolves'. Identical is **Woolpit** Sr. Cf. *la Wolpette* c 1200 Ep (in bounds of Alice Holt and Woolmer forests in Ha and Sr).

Woolscott Wa in Grandborough [*Wlscote* 13, *Wulscote* 1453 AD]. Perhaps '*Wulfsige*'s COT'.

Woolsington Nb [*Wlsinton* 1204 Ch]. 'TŪN of *Wulfsige*'s people.'

Woolstanwood Chs [*Wolstanwode* 1316 Chamb], **Woolstaston** Sa [*Ulestanestune* DB, *Wlstaneston* 1242 Fees], **Woolsthorpe** Li nr Grantham [*Ulestanestorp* DB, *Uulstanestorp* 1106–23 (1333) Ch, *Wlstorp* 1212 Fees]. '*Wulfstān*'s wood, TŪN and thorp.'

Woolsthorpe Li nr Corby [*Wolestorp* 1185 TpR, *Wlfesthor*[*p*], *Wlfthorp* 1212 Fees]. '*Wulf*'s thorp.'

Woolston D [*Ulsistone* DB, *Wolsingthon* 1242 Fees], W~ La [*Oscitona* 1094, *Ulfitona* 1142 LaCh, *Wolueston* 1246 Ass]. '*Wulfsige*'s TŪN.'

Woolston Ha [*Olvestune* DB, *Wlveston* 1236 Ass], W~ So in N. Cadbury [*Ufetone*, *Ulftona* DB, *Wolston* 1316 FA]. '*Wulf*'s or *Wulfhēah*'s TŪN.'

Woolston So in Bicknoller [*Wolwardeston* 1225 Ass, *Wulvreston* 1236 Dunster, *Wolfarston* 1251 Ass]. '*Wulfweard*'s TŪN.'

Woolstone Bk [*Wlsiestone* DB, *Wulfsiestona* 1188 P], W~ Gl [*Olsendone* DB, *Wolsistun* 1218 Pp, *Wulsiston* Hy 3 Misc]. '*Wulfsige*'s TŪN.'

Woolstone Brk [*Olvricestone* DB, *Wulf-*, *Wulvricheston* 1242 Fees]. '*Wulfric*'s TŪN.'

Woolton, Much & Little, La [*Uluentune* DB, *Wulueton* 1246 Ass, *Minor, parva Wolueton* c 1200 WhC, *Wolueton Magna* 1327 Subs]. '*Wulfa*'s TŪN.' *Wulfa* is found in *Wulfandun* 708 BCS 120.

Woolverstone Sf [*Uluerestuna, Hulferestuna* DB, *Wolferston* 1196 FF]. '*Wulfhere*'s TŪN.'

Woolverton So [*Wulfrinton* 1196 P, *Wolfrington* 1291 Tax]. 'TŪN of *Wulfhere*'s people' or '*Wulfrūn*'s TŪN'. Cf. WOLVERHAMPTON.

Woolwich (wŏŏlij) K [*Uuluuich* 918 BCS 661, *Wulleuic* 964, *Vulwic* 1016 Fr, *Wolewic* 1089 BM, *Hulviz* DB, *Wulewic* 1227 Cl]. OE *Wull-wic* 'farm where wool was produced' or 'town where wool was exported' or the like.

Woonton He [*Wennetune, Wenetone* DB, *Wunetun* c 1222 Brecon, *Woneton* 1316 FA]. '*Wynna*'s TŪN.'

Wooperton (wŏp-) Nb [*Wepredane* 1179, *-den* 1180 P, *Weperden* 1242 Fees, *Weperdon* 1256 Ass]. Second el. OE *denu* 'valley'. The first might be OE *wēoh-beorg* 'temple hill'.

Woore Sa [*Wavre* DB, *Waure* 1256 PNSa, *Wouer* 1327 Subs]. See WAVERLEY &c. The place is on a hill and not on a stream.

Wootton Bd [*Otone* DB, *Wutton* 1197 FF], W~ Brk [*Uudetun, Wudtun* 821 BCS 366, *æt Wuttune* 985 KCD 1283], **Abbott's** W~ & W~ **Fitzpaine** Do [*æt Wudetune* 1044 KCD 772, *Wide-, Wodetone* DB, *Wudeton* 1242 Fees, *Wotton Fitzpayn* 1412 FA], W~ **Glanville** Do [*Widetone* DB, *Wotton Glannvill* 1317 FF], **North** W~ Do [*Wotton* c 1180 Salisbury, *Witton, Wotton* 1228 FF], W~ Ha in Milton [*Odetune* DB, *Wodeton* 1316 FA], W~ **St. Lawrence** Ha [*Wudatuna* 10 BCS 1161, *-tun* 990 KCD 673, *Odetune* DB], W~ K [*Uudetun* 687 BCS 69, *Wudutun* 799 BCS 296], W~ Li [*Udetone* DB, *Wit*(*t*)*una, Wttuna* c 1115 LiS], **North & South** W~ Nf [*Wdetuna* DB, *Nordwitton* 1166, *Sudwutton* 1182 P, *Nort, Suth Wottone* (*Witton*) 1254 Val], W~ Np [*Witone* DB, *Wotton* 12 NS, 1202 Ass], W~ O [*æt Wudutune* 958 BCS 1042, *Wottone* 1219 Ep], W~ Sa nr Oswestry [*Udetone* DB, *Wodeton* 1272 Ipm], W~ **Courtney** So [*Otone* DB, *Wotton* 1274 Ipm, W~ *Courtenay* 1408 Ep], **North** W~ So [*Wodetone* 946 BCS 816, *Utone* DB], W~ St nr Ashbourne [*Wotton* 1274 Ipm, 1322 BM], W~ St nr Eccleshall [*Wodetone* DB, *Wotton* 1253 Ch], W~ **Bassett** W [*Wdetun* 680 BCS 54, *Wodetone* DB, *Wotton Basset* 1271 Ipm], W~ **Rivers** W [*Wdutun* 803–5 BCS 324, *Otone* DB, *Wotton Ryvers* 1332 Pat], **Hill** W~ Wa [*Wutton et Hulle* 1195 P, *Wotton and Hulle* 1259 Ipm], **Leek** W~ Wa [*Witona* Hy 2 (1314) Ch, *Witton* 1200, *Wotton* 1203 Cur, *Lekwottone* 1327 Subs], W~ **Wawen** Wa [*Uuidutuun* 723–37 BCS 157, *Wodton* 1201 Cur, *Waghnes Wotton* 1285 Ch], W~ Wt [*Odetone* DB, *Woditon* 1287–90 Fees, *Wodinton* 1291 Tax]. OE *Wudu-tūn* 'TŪN in or by a wood'.

Abbott's W~ Do belonged to Abbotsbury Abbey.—W~ **Bassett** W was held by Alan Basset in 1230 (Ch). Cf. BERWICK BASSETT.— W~ **Courtney** So was first held by John de Curtenay, who died in 1274 (Ipm). See IWERNE COURTNEY.—W~ **Fitzpaine** Do was held by Robert s. of Robert Fitz Payn (*filius Pagani*) in 1303 (FA). Cf. CARY FITZPAINE.—W~ **Glanville** Do was held in part by Henry de Glaunvyle in 1303 (FA), and the corrupt form *Wottingglayvile* 1291 Tax shows that it belonged to the Glanvilles before 1291. The family name is from GLANVILLE in Normandy.—**Hill** W~ Wa was originally **Hill**. It is in Leek Wootton. —*Leek* in Leek W~ Wa may be *leek* the plant.— W~ **Rivers** W was held by Walter de Rivere in 1211 (Cur). Cf. BUCKLAND RIPERS.—W~ **St. Lawrence** Ha from the dedication of the church.—W~ **Wawen** Wa was named from an early owner of Scandinavian descent. *Wagene de Wotton* is mentioned c 1050 KCD 939. *Wagene* is ODan *Vagn*.

Worcester (wŏoster) Wo [*Uueogorna civitas* 692, (on) *Uueogorna ceastre* 889 BCS 75, 560, *Uuigorna civitas* 716–43, *Uueogerna civitas* 774 BCS 164, 217, *aet Wigorna ceastre* 779, *Wigraceaster* 904 BCS 231, 608, *Wirecestre* DB]. 'The Roman fort of the tribe called *Wigoran* or *Weogoran*.' The tribal name is connected with the name WYRE FOREST [(in) *Weogorena leage* 816 BCS 357], which means 'the forest of the *Wigoran*', and with *Wyre* in WYRE PIDDLE Wo, and may be derived from a river-name identical with *Wyre* La, Gaul *Vigora*. Wyre may be the old name of the Piddle Wo, on which Wyre Piddle is. The present name *Piddle* is English.

Worcestershire [*Wireceastrescir* c 1040 KCD 757, *Wihracestrescir* 1038 ASC (C), *Wirecestrescire* DB].

Worden La [*Werden* a 1250 CC, *Werden, Werthen* 1246 Ass]. 'Valley with a weir.' See WER.

Wordwell Sf [(æt) *Wridewellan* c 1025 BCS 1018, *Wridewella* DB, -*well* 1181 P]. An old name of the Lark, on which is also WORLINGTON Sf. The first el. might be OE *wrid* 'thicket' or the like (see WRID), but the form should have been OE *Wridwella*. It is rather an OE ***wride** 'twist, turn', derived in a normal way from OE *wriþan* 'to twist'. The name would then mean 'winding brook'.

Worf R Sa [*Wrhe* c 1211, *Wurgh* 1227 Ch, *Worgh* 1300 For]. On the river is **Worfield** (wer-) [*Wrfeld* DB, *Werfeld* 1095–8 Fr, *Werrefeld* 1162, *Wurefeld* 1174, *Wirefeld* 1185, *Wurrefeld* 1230 P]. The latter is 'FELD on R Worf'. The river-name seems related to OE *wōrian* 'to wander', OE *wērig* 'tired', ME *wori* 'troubled, turbid', OHG *wuorag* 'drunk', the base being an adj. ***wōrig** 'wandering', i.e. 'winding', or 'turbid'. The vacillation between *o* and *e* in the early forms (see WORFIELD), however, may suggest that OE *wērig* had the same meaning as *wōrig* and that the river-name occurred in two forms, *Wōrig, Wērig* fem. (gen. *Wōrge, Wērge*). The gen. form *Worge* is evidenced in *Worgemynster* (see WARMINSTER), which contains a river-name identical with Worf. The interchange of *o* and *e* may also be analogous to that in WEDNESFIELD and the like.

Worgret Do nr Wareham [*Vergroh* DB, *Wergerod* 1202, *Werghrode* 1227 FF, *Wirgrode* 1285 FA]. OE *wearg-rōd* 'gallows'.

Workington Cu [*Wurcingtun* 946 BCS 815, *Wirchingetona* c 1150, *Wirchintuna* 12 StB, *Wirchintuna* R 1 (1308) Ch]. 'TŪN of *Weorc*'s or *Wirc*'s people.' Cf. WIRKSWORTH.

Worksop Nt [*Werches(s)ope* DB, -*hope* 1187 P, *Wirkesop* Hy 2 DC, 1226 BM]. '*Weorc*'s HOP or valley.' Cf. WIRKSWORTH.

Worlaby Li nr Brigg [*Uluricebi* DB, *Wlfrichesbi* c 1115 LiS, *Wolurikesbi* 1202 Ass], W~ Li nr Louth [*Wlvricesbi* DB, *Wlfrichesbi* c 1115 LiS]. '*Wulfric*'s BY.'

Worldham, East & West, Ha [*Werildeham* DB, 1177 P, *Worildham* 1212, *Werilleham* 1219, *Werldham* 1242 Fees, *Werildham* 1236 Ass, *Estwerldham* 1250 Fees, *Westwerldham* c 1240 Selborne]. The first el. at first sight looks like OE *weorold, worold* 'world', and the name has evidently been associated with that word. But the early forms do not go well with this base. The WSax form of *world* was *worold*, and the numerous *i*-forms offer difficulty. Either the name means '*Wǣrhild*'s HĀM, **Wǣrhild* being a woman's name corresponding to OG *Warehild*. Or the first el. is a compound containing OE *hielde* 'slope'. The first might then be an OE ***wēr**, a word derived from *wōr* 'woodgrouse' or the like (see WOODSPRING) and found in ME *wercock*.

Worle So [*Worle* DB, 1212 Fees, 1225 Ass, *Wurle* 1257 Ass]. OE *wōr-lēah* 'woodgrouse wood'; cf. WOODSPRING. The place is near Woodspring and Worle Hill.

Worleston Chs [*Werelestune* DB, *Weruelestona* c 1100, *Uerulestane* c 1150 Chester, *Werliston* 1260, *Worleston* 1282 Court]. '*Wērwulf*'s TŪN.'

Worlingham Sf [*Werlinga-, Warlingaham* DB, *Werlingeham* 1168 P, *Little Wirlingham* 1249 Bodl]. 'HĀM of *Werel*'s or *Wērwulf*'s people.' OE *Werel* is found in *Werǣles wellæ* BCS 624. The OE base probably had *ē* in the first syllable.

Worlington D [*Ulvredintone, Oluridintona* DB, *Wolvrington* 1242 Fees, *Est, West Wlfrintone* 1261 f. Ep]. 'TŪN of *Wulfrēd*'s or *Wulffriþ*'s people.'

Worlington Sf [*Wirilintona* DB, *Wridelingeton* 1201 Cur, *Wridelincton* 1242 Fees]. 'TŪN of the people at *Wridewella* or Lark.' See WORDWELL. The place is on the Lark not very far from Wordwell.

Worlingworth Sf [(et) *Wilrincgawerþa* c 1035 Wills, *Wirlingaweorð* 11 EHR 43, *Wyrlingwortha* DB]. 'WORÞ of *Wilhere*'s people.'

Worm Brook He [*Guormui* c 1150 LL, *Worme* c 1540 Leland]. A Brit river-name derived from Welsh *gwrm* 'dusky, dun' (= OBret *uurm*- in pers. ns., Ir *gorm* 'blue') and meaning 'dark stream'. On the stream is **Wormbridge** He [*Wermebrig* 1207 Cur, *Wuremebrigge* 1256 Ass]. From it were also named **Wormelow** hd [*Wermelau, Urmelauia* DB] and the lost **Wormton** [*Wermiton, Wirminton* 1208 Cur, *Wurmetun* 1242 Fees].

Wormegay Nf [*Wirmegeie* c 1150 Crawf, *Wermegai* DB, 1162 P, *Wurmegai* 1159, *Wirmingai* 1173 P]. 'The island of *Wyrm*'s people.' Cf. WARMINGHURST.

Wormhill Db [*Wruenele* DB, *Wurmhill* 1185 P, *Wrmenhulle* 1226 FF, *Wurmehill* 1227 Ch]. 'Hill frequented by reptiles' or possibly '*Wyrma*'s hill'. Cf. WYRM, WORMINGTON.

Wormingford Ess [*Widemondefort* DB, *Wiðermundeford* 1186 P, *Wydremundeford* 1198 (1253) Ch]. '*Wiþermund*'s ford.' Cf. OE *Wiþergyld*, OG *Widargelt* &c. and see WITTERING Np.

Worminghall Bk [*Wermelle* DB, *Wurmehal* 1163 P, *Wirmehale* 1229 Ep]. 'HALH frequented by reptiles' (see WYRM) or possibly '*Wyrma*'s HALH'. Cf. next name.

Wormington Gl [*Wermetun* DB, *Wirmiton* 1200 Cur, *Worminton* 1220, *Wurminton* 1236 Fees]. '*Wyrma*'s TŪN.' OE **Wyrma* is a side-form of *Wyrm*. Cf. WARMINGHURST.

Worminster So nr Glastonbury [*Wormester* 946 BCS 816, *Wuormestorr* 1065 Wells, *Weremestorre* 1176 ib.]. '*Wyrm*'s tor or hill.' Cf. WARMINGHURST. In the boundary of BCS 816 is also mentioned *Wormesleighe*. It is just possible that the first el. might be OE *wyrm*, used in the sense 'dragon', as in Beowulf.

Wormleighton Wa [*æt Wilmanleht(t)une* 956 BCS 946, *Wimenes-*, *Wimelestone* DB, *Wilmelayton* 1236 Fees]. In the boundaries of Wormleighton in BCS 946 is mentioned *Wilman ford*, and *Wylman ford* and *Wylman broc* occur 998 Crawf in bounds of places nr W~. The first el. may be a pers. n. *Wilma*, a short form of *Wilmund*. But it may just as well be a river-name *Wilme*, *Wielme*, a derivative of OE *wielm* 'flowing, bursting forth'. The second el. is OE *lēactūn* (see LEIGHTON).

Wormley Hrt [*Wurmeleá* c 1060, *Uurmelea* 1065 BM, *Wermelai* DB]. 'LĒAH frequented by reptiles.' See WYRM.

Wormshill K [*Wodnesell* 1232 Subs, 1242 Fees, *Werneshóll* 1236 Fees, *-hulle* 1282 Ep, *Wornesell* 1275 Ipm]. OE *weorn-gesell* 'shelter for the herd of pigs'. First el. OE *weorn*, *worn* 'band, flock' (used of a flock of birds or swine). See (GE)SELL. Or OE *Wōdnes-hyll* '*Wōden*'s hill'.

Wormsley Bk [*Wodemundesleg* 1219, *Wide-*, *Wudemundele* 1236 Fees]. '*Widmund*'s LĒAH.' *Widmund* is evidenced in *Uuidmundes felth* 695 BCS 87 (Ess).

Wormsley He [*Wermeslai* DB, *Worvesleg* c 1180 Fr, *Wurmeleys* 1242 Fees, *Wrmesl'* 1249 ib.]. '*Wyrm*'s LĒAH' or 'LĒAH of the reptile(s)'. Cf. WORMEGAY, WYRM.

Wormwood Scrubbs Mx [*Wermeholte* 1200, *Wrmeholt* 1290 PNMx]. 'Holt frequented by reptiles.' See WYRM. Scrubbs seems to be *scrub* 'stunted tree, brushwood'.

Worplesdon Sr [*Werpesdune* DB, *Werplesdon* 1215, 1228 Cl, 1242 Fees]. OE *werpels* (cf. WARPSGROVE) and DŪN 'hill'. The meaning may be 'hill with a bridleway'.

Worrall (-ŭ-) YW [*Wihale* DB, *Wirhal* 1218 FF, *Wyrhale* Hy 3 BM]. Identical with WIRRAL.

Worsall, High & Low, YN [*Wercesel*, *Wirceshel*, *Wercheshala* DB, *Wirkeshale* Hy 2 Guisb], **Worsborough** YW [*Wircesburg* DB, *Wircasburc* 1148 YCh 179, *Wirkeburg* c 1175 YCh 1680, *-burc* 1195 P, *Werkesburg* 1219 FF]. '*Weorc*'s or *Wirc*'s haugh and BURG.' See WIRKSWORTH.

Worsham. See WARTLING.

Worsley (wŭslĭ) La [*Werkesleia* 1196 P, *Wyrkedele* 1212 Fees, *Wirkedley* 1219 FF, *Wyrkitheley*, *Wurkythesle* 1246 Ass]. See LĒAH. The first el. might be a pers. n. such as **Weorchæp* or **Weorcgўþ*.

Worsley Wo in Rock [*Worfesleahges gemæra* c 957 BCS 1007, *Werveslega* c 1150 Surv, *Worvesleg* c 1180 Fr]. 'Pasture for cattle.' Cf. WARLEY SALOP.

Worstead (wŏŏstĕd) Nf [*Wrŏestede* 1044–7 KCD 785, *Wrdesteda* DB, *Wrthested* c 1150 Crawf, *Wurdesteða* 1169 P]. 'Site of the WORÞ or enclosure.'

Worsthorne La [*Worthesthorn* 1202 FF, *Wurthesthorn* 1246 Ass], **Worston** La [*Wrtheston* 1212 LaInq, *Wurtheston* 1285 Ass]. '*Wurþ*'s TŪN.' OE *Wurþ* is not evidenced but belongs to OE *weorþ* 'worth' or *wyrþe* 'worthy'. Cf. OHG *Werdo* and WORTHING Sx.

Worston St [*Wyverston* 1227, 1251 Ass, *Wynreston* 1251 Misc]. '*Wigferþ*'s TŪN.'

OE **worþ** (**wyrþ**) corresponds to OLG *wurð* 'soil', MLG *wurt* 'homestead', LG *wort*, *wurt* 'open place in a village'. The original meaning of the word appears to have been 'fence' or 'enclosure', but from this at an early date developed a meaning 'enclosure round a homestead' and 'homestead'. *Worþ* is a common pl. n. el., and is sometimes also used alone as a pl. n. It is a rare first el. (see WORSTEAD, WORTHAM), but common as a second el. Some names in *-worth* are found in very early sources, e.g. ISLEWORTH (695), HILLBOROUGH (710), ASHMANSWORTH (909), and some other names are no doubt pretty old, but *worth* must have long continued to be used in forming pl. ns. Some names of small places, as STANWORTH, WARDLEWORTH La, can hardly be very old. Some names in *-worth*, as TORWORTH Nt, may have a Scand pers. n. as first el. The meaning of the el. probably varied. It seems to have been 'enclosure' or even 'fence' in some names such as SHUTTLEWORTH ('enclosure with bars of a certain construction') or STAWARD, perhaps TWIGWORTH, where the first el. appears to denote the material of which the fence or enclosure was made. Cf. also KEY-, TURNWORTH. 'Homestead' is the probable meaning in the numerous names with a pers. n. as first el. In WARDLEWORTH, whose first el. is the pl. n. WUERDLE, a meaning such as 'dependent farm' or 'cattle-fold' is likely, and this would suit names with a folk-name as first el., as ABINGER, BENGE-, BOBBINGWORTH, and the name WORTH. The first el. is sometimes a common noun, as in BEAU-, BROCK-, RISHWORTH.

As a second member the element usually appears as -worth, but sometimes in other forms. See e.g. BARKWITH, DIBBER-, DUX-, UFFORD, CHEL-, CLAREWOOD, STAWARD, EWART, ABINGER, IMBER, SEACOURT, HILL-BOROUGH.

OE worþig seems to have been on the whole synonymous with WORÞ, and worþign appears to be a variant form of worþig. Meanings such as 'enclosure, yard about a house, open place in a village or town, homestead' seem to be recorded. Worþig is found alone in WORTHY Ha, and worþign in WORTHEN Sa, WORTHING Nf. In early sources worþ and worþig(n) often interchange for the same name (see e.g. TAMWORTH). In modern times -worthy is especially common in the South-West, as in CLAT-, EL-, SELWORTHY Sa, HOLS-, PY-, WOOLFARDISWORTHY D, HAMWORTHY Do, RANGEWORTHY Gl. The form worþign is particularly common in the West Midlands, where the first el. is frequently a word denoting some natural feature, a hill or the like. See e.g. NORTHENDEN Chs, BRED-, LEINT-, LUG-, PEDWARDINE, MARDEN He, SHRA-, STAN-, WROCKWARDINE, INGARDINE Sa, also RUARDEAN Gl.

Worth Chs [*Worth* c 1250 Chester, 1286 Court], W~ **Matravers** Do [*Orde, Wirde* DB, *Wurth* 1236 FF], W~ K [*Wurth* 1226 Pat, 1227 Ch], W~ Sx [*Orde* DB, *Wurðа* 1175 P]. OE WORÞ 'enclosure' &c.
W~ **Matravers** came to John Mautravers after 1335 (Ch). Cf. LANGTON MATRAVERS.

Wortham Sf [*Wrtham* c 950 Wills, *Wortham, Wordham* DB, *Wurtham* c 1200 Bodl]. OE *Worþ-hām* perhaps 'enclosed homestead'.

Worthen Sa [*Wrdine* DB, *Worthyn* 1246 Ch], **Worthing** Nf [*Worthing* 1282 Ipm, *Worthene* 1355 BM]. OE WORÞIGN (see WORÞ).

Worthing (-erdh-) Sx [*Ordinges* DB, *Wurddingg* 1219 FF, *Worthinges* 1288 Ass]. 'Wurþ's people.' Cf. WORSTHORNE.

Worthington La [*Worthinton* 1210 Cur, *Wurthington* 1246 Ass], W~ Le [*Werditone* DB, *Wrdintona* c 1125 LeS, *Wurðinton* 1169 P, *Wirthenton* 1209–35 Ep, *Wurthinton* 1242 Fees]. 'TŪN of the *Wurþingas*' (cf. WORTHING Sx). Or the first el. may be OE WORÞIGN.

Worthy, Headbourne, Kings & Martyr, Ha [*æt Worðige* 825, 868, 901 BCS 389, 520, 596, (æt) *þan twan Worþigum* 955–8 BCS 652, *Ordie* DB, *Ordia* 1156 P, *Hydeburne Wor*[*t*]*hy* c 1270 Ep, *Chinges Ordia* 1157 ff. P, *Wordia le Martre* 1243 Ep]. OE WORÞIG.
Headbourne is the name of a stream, called *hydiburna* 854 BCS 473 (in bounds of *Worðig*). *Hideburninga gemære* is mentioned 909 ib. 620 (in bounds of Chilcomb nr Worthy). The first el. seems to be OE *hid* 'hide, household'.—**Martyr** W~ was held by Henricus la Martre in 1201 (Cur). *Martre* may be OFr *martre* 'martyr' or OFr *martre* 'marten', used as a nickname.

Worting (-er-) Ha [*Wyrtingas* 960 BCS 1055, *Wortinges* DB, *Wrtingge* 1232 Selborne]. An old folk-name of doubtful origin. The base may be a pers. n. related to that which is at the bottom of WARTLING. A pers. n. *Worta* is perhaps found in *Wortan beorg* 958 BCS 1037.

Wortley YW nr Barnsley [*Wirtleie* DB, *Wrtlaye* c 1200 YCh 1803, *Wortelay* 1204 Cur]. OE *wyrt-lēah*, the first el. being OE *wyrt* 'plant'. See LĒAH.

Wortley YW nr Leeds [*Wrchelai* 1166 P, *Wirkeleia* 1189 Kirkst, *-lay* Hy 3 Calverley]. '*Wirca*'s LĒAH.' OE *Werca* is evidenced.

Worton O nr Cassington [*Vurtone* DB, *Wrtona* a 1123 Eynsham, c 1130 Oxf, *Wurtton* 1194 P, *Worthona* 1199 (1320) Ch], W~ W [*Wrton* 1173 Salisbury, *-a* 1175–9 BM], W~ YN [*Werton* DB, *Wirton* 1218 FF]. The last is OE *wyrt-tūn* 'garden' or 'TŪN with a garden'. The same is very likely the etymology of the other two, but OE WORÞ is a possible first el. Cf. WORTHAM.

Worton, Nether & Over, O [*Ortun* 1049–52 KCD 950, *Hortone* DB, *Orton* 1191 ff. P, *Nether-, Overhorton* 1242 Fees]. 'TŪN on a bank or slope.' First el. OE ŌRA.

Wothersome YW [*Wodehuse, -husun* DB, *Wodehus* 1240 FF, *Wodeusum* 1290 Ch]. Identical with WOODSOME.

Wotherton Sa nr Chirbury [*Udevertune* DB, *Wdeverton* 1206 Cur, *Wodforton* 1242 Fees]. 'TŪN by *Wuduford* or ford at a wood.'

Wothorpe (-ŭdh-) Np [*Wridtorp* DB, *Wirthorpe* 12 NS, *Wridtorpe* 1224 Ep]. 'Thorp by a thicket.' See WRĪD.

Wotton Underwood Bk [*Wudotun* 848 BCS 452, *Vittona* 1155 Fr, *Wotton under Bernwode* 1382 AD], W~ (-ōō-) **under Edge** Gl [æt *Wudetune, Wudutun* 940 BCS 764, *Vutune* DB, *Wotton under Egge* 1466 AD], W~ **St. Mary Without** Gl [*Utone* DB, *Wotton* 1220 Fees], W~ (-ōō-) Sr [*Wodetone* DB, *Wudetun* 1157 P]. Identical with WOOTTON, i.e. 'TŪN by a wood'.
W~ **Underwood** means 'W~ in the wood'. *Bernwode* is BERNWOOD FOREST.—W~ **under Edge** is at the foot of a hill. *Grangia del Egge* is mentioned 1291 Tax. See ECG.

Woughton (-ōōf-) **on the Green** Bk [*Ulchetone* DB, *Wocheton* 1163, 1167, *Woketon* 1183, *Wuketon* 1197 P, *Weketun* 1199 FF]. First el. the pers. n. *Wēoca*, found in *Weocan þorn* BCS 1022, a derivative with a *k*-suffix of pers. ns. in *Wēoh*-, as *Wēohstān*.

Wouldham (-ōō-) K [*Uuldaham* 811, *Wuldaham* c 960, *Wuldeham* 10 BCS 339, 1097, 1322, *Oldeham* DB]. '*Wulda*'s HĀM.' *Wulda* pers. n. is not evidenced, but *Wuldric*, *Wuldwine* are the names of moneyers, and the corresponding stem is well evidenced in OG names, as *Wultgar, Ulta*. The name belongs to the base of Goth *wulþus* 'glory', ON *Ullr*, the name of a god, OE *wuldor* 'glory'.

Wrabness Ess [*Wrabenasa* DB, *-nessa* c 1140 Bury, *-nase* 1234 FF, *Wrabbenase* 1274 RH]. The same first el. is found in **Rapton** Sf [*Wrabbatuna* DB, *Wrabetuna* DB, *-tun* 12 BM]. Names in *-ness* often have a pers. n. as first el. Wrabness is probably '*Wrabba*'s NÆSS'. *Wrabba* is easily explained as a nickname cognate with *wrabbed* 'perverse'. The second el. is OE NÆSS 'headland'.

OE **wrǣnna, wǣrna** is found in the sense 'wren'. This may be the first el. of some pl. ns., e.g. WARFIELD. It is probable, however, that OE had the word corresponding to OLG *wrēnio*, OHG *reinno*, ON *reini* 'stallion'. It would be OE **wrǣna*. OE *wrǣne* 'lecherous', a cognate word, occurs. An OE *wrǣna* 'stallion' is very likely the first el. of some names, as WRANTAGE, WRINSTEAD, WARMFIELD, WARNFORD, WARNHAM, WARNINGCAMP. But a pers. n. may be assumed in some of these.

Wrafton D [*Wratheton* 1238, *Wraghton* 1284 Ass, *Wrathton* 1412 Ipm]. First el. possibly OE *wraþu* 'prop, support', but the exact meaning is obscure.

Wragby Li [*Waragebi* DB, *Wrag(h)ebi* c 1115 LiS, *Wraggeby* 1308, *Wragby* 1316 Wakef]. W~ YW [*Wraggeby* 1308, *Wragby* 1316 Wakef]. '*Wraghi*'s BY.' *Wraghi* is an ODan pers. n. Wragby Li is in Wraggoe wap. [*Waragehou* DB, *Wroghehou* c 1115 LiS]. Wraggoe means '*Wraghi*'s burial-mound', the second el. being OScand *haugr*. The same first el. may be found also in **Wragholme** Li [*Wargholm* 13 Gilb, 1276 RH].

Wramplingham Nf [*Wranplincham* DB, *Wramplingham* c 1185 Bodl, 1202 FF]. The first el. is a tribal name, connected probably with words such as *wramp* 'a twist' (17th cent.), *wrimpled* 'wrinkled' (c 1430). It may be derived from a nickname formed from the base of these words.

Wrangbrook YW [*Wrangebroc* a 1194 Kirkst, *-broke* c 1190 YCh 1677]. 'Winding brook.' The first el. is the adj. *wrong* which is generally held to be a Scand loanword. But it is found already in late OE

and seems to be the first el. of (on) *wrangan hylle* 944 BCS 801 (Brk). The word is found also in Continental languages (MLG *wrang* 'sour, bitter', MDu *wrangh* 'bitter, hostile'). An OE *wrang* may well have existed.

Wrangle Li [*Werangle* DB, *Wrengle* c 1200 DC, 1207 Cur, *Wrangl'*, *Wrengl'* 1202 Ass, *Wrangle* 1212 Fees]. A derivative of OE *wrang* or OScand *vrangr* 'bent, crooked'. It may be the name of a stream that has now disappeared. W~ is in old fenland. Cf. the Norwegian stream-name *Rangla*, Norw dial. *vrengjell* 'twisted tree'.

Wrantage So [*Wrentis* 1199 P, 1201 Cur, *Wrentise* 1227, *Wrentisse* 1246, *Wrentyssh* 13 Wells]. The second el. seems to be an OE **etisc*, corresponding to Goth *atisk*, OHG *ezzisch* 'a piece of land' suggested as the second el. of DUNTISH. The first el. may well be OE **wrǣna* 'stallion'. The name would mean 'stallion's pasture'. Cf. WRÆNNA.

Wratting, West, Ca [*æt Wreattinge* 974, *Wrǣttincg* c 1000 BCS 1305 f., *Wrattinga* c 1080 ICC, *Waratinge* DB, *Wrettinges* 1200 Cur], **Great & Little W~** Sf [*Wratinga* DB, *Wrotting* 1206 Cur, *Wretting* 1206 FF, *Wrattinge Magna, parva* 1291 Tax]. A derivative of OE *wrætt* 'crosswort': 'place where crosswort grew'.

Wrawby Li [*Waragebi* DB, *Wragebi* c 1115 LiS, *Wrahebia* 1212 Fees, *Wraweby* 1276 Ipm]. '*Wraghi*'s BY.' Cf. WRAGBY.

Wraxall Do [*Brocheshale* DB, *Wrockeshal* 1196 FF, *Wrokeshal* 1196 Cur], **W~** So [*Werocosale* DB, *Wrokeshall* 1227 FF], **North Wraxhall** W [*Werocheshalle* DB, *Wroxhale* 1316 FA, *Northwroxhall* 1468 Ipm], **South W~** W [*Wrokeshal* 1242 Fees, *Suthwroxhall* 1468 Ipm]. There is one **Wraxhall** nr Castle Cary and one **Wraxall** nr Frome So. Identical are WROXALL Wa, Wt. The same first el. is found in **Roxhill** Bd [*Wrocheshol* 1180 P, *Wrockeshull* 1200 P, 1219 FF] (second el. OE *hyll*), WROXHAM, WROXTON. Cf. also *Wrokcumbe, Wrockumbe* 937 BCS 717, also *wrocena stybb* 944 BCS 801. The first el. cannot well be a pers. n. in view of the common combination with HALH, at least not in all cases. On the other hand the regular gen. form rules out a word with a topographical meaning. An animal's name would suit the case. OE *wroc* may be cognate with and identical in meaning with Sw *vråk* 'buzzard', generally held to be a derivative of *vräka* (= OE *wrecan*) 'to pursue'. Lidén, *Meijerbergs Arkiv*, i. 55 ff., points out, however, that ON *vákr* (Norw *vaak*) corresponds to Sw *vråk*, and takes *vr-* for *v-* in the Sw word to be due to a late change. If that is right, the OE word must be dissociated from Sw *vråk*. OE *wroc* (or rather *wrocc*) can then be referred to Du *wrok*, MLG *wrok, wruk* 'hate, enmity', which have been derived from the root of G *würgen*, OE *wyrgan* 'to strangle'. A word for a bird of prey would naturally be formed from this root, but there is no longer any reason to assume just the meaning 'buzzard'. The common Wraxall (Wroxall) may be compared with ARNOLD, *wrockumbe* with YARNSCOMBE. But the bird-name might naturally come to be used as a pers. n. Cf. cases like *Puttoc* (ME *puttok* a kind of hawk), OE *Hafoc*, ON *Haukr* from the word for *hawk*, OE *Hræfn*, ON *Hrafn* from the word for *raven* &c. Wroxham may quite well be '*Wroc*'s HĀM'.

Wray (rā) **with Botton** La [*Wra* 1227 Ch, 1229 FF], **High W~** (rā) La [*Wraye* c 1535 PNLa], **Wrayton** La [*Wraton* 1247 CC, 1271 Ass], **Wrea** (rā) La [*Wra* 1201 P], **Wreay** Cu [*Wyte Wra* 1235 P]. There are several examples of the name in Cu. One s.c. of Carlisle was formerly *Peterelwra* 1268, *Peytrelwra* 1285 For; named from the river Petteril. Wray, Wrea, Wreay are OScand *vrá* 'corner', here used in the sense 'remote or isolated place'. Cf. VRĀ. Wrayton is TŪN in a remote place or valley'.

Wraysbury. See WYRARDISBURY.

Wrayton, Wrea. See WRAY.

Wreak (-ē-) R Le [*Werc* Hy 2 DC, 1237 AD, *Wrethek* 1276 RH, 1325 BM, *Wrethk* 1320 AD, *Wreyke* 1276 RH]. An OScand **Vreþk*, a derivative of OScand *vreiðr* (ON *reiðr*, OSw *vreþer*) 'wrathful', but originally 'twisted', a sense still to be traced in pl. ns. River-names are often formed with a *k*-suffix in Scand languages. The Wreak is a very winding river, and it runs through a strongly Scandinavian district.

Wreay. See WRAY.

OE wrecca 'outlaw'. See RATCHWOOD, WREKENDIKE, WRETCHWICK.

Wrecclesham Sr [*Wrekelesham* 1282 Ep]. Perhaps '*Wræcwulf*'s HĀM'. **Wræcwulf* may be compared with OG *Wracwulf*.

Wreigh Burn R Nb [*Rye* c 1540 Leland]. Identical with WARNBOROUGH.

Wreighill (rē-) Nb [*Werihill* 13 Newminster, *Werghill* 1293 QW]. OE *wearg-hyll* 'felon hill, hill where felons were executed'. Cf. WARNBOROUGH.

Wrekendike Du, a Roman road [*Vrakendic* c 1150, *Wrakendyk* c 1225 FPD]. Possibly OE *wræccna dic* 'dyke of the fugitives' (first el. OE *wrǽcca*). A possible analogy is **Flendish** hd Ca [*Flamdichdr*. 1158 P], whose first el. may be OE *fliema* 'fugitive'.

Wrekin (rēkĭn), **The**, Sá, a hill [(on) *Wrocene* 975 BCS 1315, *La Wrekene* 1278 Ipm, *La Wrokene* 1284 Ipm]. The name also occurs in combination with -*sǽtan* to denote the people dwelling near the Wrekin [*Wocensætna* (land) 7 BCS 297, (in) *Wreocensetun* 855, (provincia) *Wrocensetna* 963 BCS 487,

1119]. Wrekin is really the Brit name of WROXETER, which was transferred to the hill. The interchange of forms with *e* and *o* may be due to OE *eo* (as in *Wreocensetun* 855), which could give ME *e* and *o*. OE *Wreocen* would be the correct descendant of OBrit *Viriconio-*.

Wrelton YN [*Wereltun* DB, *Werl-*, *Wrelton* 1246 FF]. The first el. may be identical with WREIGHILL.

Wrenbury Chs [*Wareneberie* DB, *Wrenne-bury* 1230 (1331) Ch]. 'Old fort inhabited by wrens' (cf. WRÆNNA) is possible. A pers. n. *Wrenna* or OE **wrǽna* 'stallion' would be preferable. Cf. next name.

Wreningham Nf [*Wreningham* c 1060 Wills, *Urnincham* DB, *Wreningeham* 1197 P]. Apparently 'HĀM of *Wrenna*'s people'. **Wrenna* may be a nickname formed from *wrenna, wrænna* 'wren'.

Wrentham Sf [*Wretham, Uuereteham* DB, *Wrentham* 1228 FF, 1254 Val, *Wrantham* 1272 Ch]. Perhaps '*Wrenta*'s HĀM'. **Wrenta*, which may be the first el. also of **Wrentnall** Sa [*Werentenehale* DB], might be a nickname cognate with MLG *wranten* 'to sulk', MDu *wrant* 'sulky person'.

Wrenthorpe YW [*Wyverunthorp* 1284 AD, *Wyrnthorpe* 1350 BM]. '*Wifrūn*'s thorp.' **Wifrūn* (*Wiuerona* 1130 P) is a woman's name; cf. OG *Wifhildis, Wiblind* &c.

Wrentnall. See WRENTHAM.

Wressell YE [*Weresa* DB, *Wresel* 1183 P, 1226 FF]. A derivative of OE *wrāse* 'knot, lump'. OE **wrǽsel* or the like in a sense such as 'thicket' may be assumed. A wood in *Wresel* is mentioned in 1226 (FF).

Wrest Park Bd [*Wresta* 1185 f. P, *Wrast* 1276 RH]. The name presupposes an OE **wrǽst* 'knot, thicket' or the like, related to OE *wriþan* 'to twist', *wrāse* 'knot' &c. The place is situated by a spur of land.

Wrestlingworth Bd [*Wrastlingewrd* c 1150 BM, -*worde* Hy 2 ib., 1198 FF, *W[r]estlinge-wurda* 1194 P, *Wrastlingeworth* 1227 Ch]. *Wrǽstles hyll* BCS 789 (Brk) seems to contain a pers. n. **Wrǽstel*, related to the word *wrǽst* postulated for WREST PARK and OE *wrǽste* 'delicate'. But it is possible that the first el. of the name is derived from a pl. n. cognate with Wrest Park.

Wretchwick O [*Wrechewic* 1211 Cur, 1272 Ipm, -*wik* 1291 Tax]. 'WĪC of the *wrecca* or outlaw.'

Wretham (rĕt-), **East & West**, Nf [*Wret-ham, Weretham* DB, *Wretham* 1177 P, *Est-*, *Westwretham* 1212 Fees, *Wrottham* 1199 FF, *Wrotham* 1244 Cl], **Wretton** Nf [*Wret-ton* 1198 FF, 1249 Ipm, 1254 Val, *Wrottun* 1251 Ch]. 'HĀM and TŪN where crosswort grew.' Cf. WRATTING. Crosswort was a medicinal plant.

Wribbenhall Wo [*Gurberhale* DB, *Wrubben-hale* c 1160 PNWo, *Wurbiehal* 1221 Ass]. Apparently '*Wrybba*'s HALH'. **Wrybba* would seem to be an umlauted side-form of *Wrobba* in RABLEY &c.

OE **wrīd, wrīþ, gewrid**, also in *hæslwrid, -wriδ*, is the source of dial. *ride, wride* 'a shoot, stalk, or stem, a group or bush of stalks, &c. growing from one root'. The words belong to OE *wriþan, widan* 'to put forth shoots'. In pl. ns. the word may mean 'bush' or 'thicket'. See WO-THORPE, WRITHLINGTON, also WORDWELL, WORLINGTON Sf., which latter seem, however, to contain a different word. **Wryde** Ca goes back to OE *wrid*. The name is fully exemplified in PNCa(S); it is (le) *Wride* (*Wryde*) 13 &c. It is taken to be a stream-name derived from an OE *wride* 'twist, turn', as in WORDWELL. But this does not account for the long *i* apparently pre-supposed by the spelling *Wryde*. A related word is found in WOODLESFORD.

Wrightington La [*Wrichtington* 1202 FF, *Wrictinton* 1212 Fees]. OE *wyrhtena tūn* 'TŪN of the wrights'.

Wrinehill St [*Wryme* 1299 Ipm, 1332 AD, *Wrinehull* 1225 Cl, *le Wrimehull* 1486 AD]. The first el. is found in *Wriman ford* 975 BCS 1312, *Le Wrineford* 1322 Ipm. It may be an old name of Checkley Brook. But it is equally possible that it is an old name of the hill from which Wrinehill was named. The name may be a derivative with a suffix *-man* from OE *wrigian* 'tend, go forward, bend', a word related to Mod *wry*.

Wrington So [*at Wring*' 904 BCS 606, *Weritone* DB, *Wringeton* 1243 Ass]. 'TŪN on R **Wring**' (now YEO). The river-name is *Wring, Wryng* 904 BCS 606, *Wrynge* 1276 RH. It may be from earlier **wrīoing*, the base being a word cognate with OE *wrīgian* (see WRINEHILL), MLG *wrîch* 'perverse', Engl *wry*. What suggests such a possibility is the fact that the charter where Wring occurs mentions a locality *wryoheme. Wryo-heme* might be corrupt for *Wrēohǣma* 'of the dwellers at *Wrēo*'.

Wrinstead K [*Wrensted* 1111 StAug, *Wran-sted* 1236, *Wernnestede, Wrennested* 1242 Fees]. Probably 'place where stallions were kept'. Cf. WRÆNNA and HORSTEAD.

Writhlington So [*Writelinctone* DB,*Writhel-ington* 1225, *Writhlington* 1243 Ass]. The first el. is perhaps an OE *wriþ-hlinc* 'hill with a thicket'. Cf. WRĪD.

Writtle Ess [*Writelam* DB, *Writela* c 1136 BM, *Writelea* 1173, 1190 P]. Originally the name of the **Wid** [*Writolaburna* 692 BCS 81]. The name presupposes an OE **writol* 'babbling', cognate with OE *writian* 'to chirp, chatter', OLG *writolôn* 'to chatter'.

Wrockwardine Sa [*Recordine* DB, *Wroch*

Wurδin 1169, *Wrocwurδin* 1196 P]. 'WOR-PIGN by the WREKIN.'

Wroot Li [(insula de) *Wroth* 1157 YCh 354, *Wrot* 1193 f. P, 1212 Fees]. OE *wrōt* 'snout, trunk'. The spur of land or earlier island must have been thought to resemble a pig's snout.

Wrotham (rōotm) K [*Uurotaham* 788, *Wrot-, Wroteham* 10 BCS 253, 1321 f., *Broteham* DB]. '*Wrōta*'s HĀM'. **Wrōta* is a nickname derived from OE *wrōt* 'snout'.

Wrottesley (rŏtslī) St [(æt) *Wrotteslea* 10 BCS 1317, *Wrotolei* DB, *Wrotteslega* 1167 P]. See LĒAH. The first el. looks like a pers. n. *Wrott*.

Wroughton (-aw-) W [*Wervetone* DB, *Worfton* 1195 Cur, *Wrftona* 1242 Fees]. The first el. is an old name of the river RAY [*Worfe* 796, *Wurf* 943, 956, *Worf* 962 BCS 279, 788, 983, 1093, *Werfe* 1228 Cl]. The name is identical with WHARFE.

Wroxall Wa [*Wrocheshal* 1162 P, *Wrokeshal* 1204 Cur, 1255 Ch], W~ Wt [(æt) *Wrocces-heale* 1038–44 KCD 768, *Warochesselle* DB, *Wrocheshala* 1186 P]. See WRAXALL.

Wroxeter Sa [*Ouirokónion* c 150 Ptol, *Uri-, Viroconium* 4 IA, *Rochecestre* DB, *Wrox-cestre* 1155, *Wroccestre* c 1175 Eyton]. 'The Roman fort of *Viriconion* or *Viroconion*.' The etymology of the Brit name is obscure. It resembles the old name found in ARCHEN-FIELD.

Wroxham Nf [*Vrochesham* DB, *Wrokesham* c 1220 Bodl]. '*Wroc*'s HĀM or HAMM' or 'HAMM frequented by birds of prey'. See WRAXALL.

Wroxton O [*Werochestan* DB, *Wrucestan* 1204 Cur, *Wroxstan* 1242 Fees, *Wrocstan* Hy 3 BM]. See WRAXALL. First el. the bird-name or pers. n. suggested there. Second el. OE STĀN 'stone'.

Wryde Ca. See WRĪD.

Wrynose Cu, We, La [*Wreineshals* 1157–63 LaCh]. 'The stallion's col or neck of land.' The elements are ON (*v*)*reini* 'stallion' and *hals* 'hause'.

OE **wudu** 'wood' occurs in pl. ns. both in the sense 'forest' and in that of 'timber'. The latter is found in names such as WOOD-BRIDGE, WOODCHURCH, and is possible in WOODBURY, WOODCHESTER and the like. The most common meaning is no doubt 'forest'; it is, of course, always that when the word is used as a second el. OE *wudu* comes from *widu*; cf. ON *viδr*, OHG *witu* &c. The form *widu* occasionally occurs in OE (cf. e.g. WOOTTON WAWEN Wa), and the gen. and dat. form *wida, wyda* is not rare (e.g. *to wida* 854 BCS 468, *gauolwyda* 901 ib. 594; see also SELWOOD). In pl. ns. the word often appears with *i*, as in several WITTONS, and the name WOOTTON often has *i* in early forms. See further WOOD-

(passim), WOOTTON, WOTTON, ODIHAM, UDI-
MORE, WOTHERSOME, WOTHERTON, COQUET
and others.

Wuerdle (wōōdl) La [*Werdull* c 1180, *Worde-hull* 13 WhC, *Wordhull* 1299 FF]. The first el. may be OE *weorod* 'troop, host'. The place is near WARDLE. Wuerdle may be 'hill where the host was stationed'.

OE **wulf** 'wolf' is a common first el. of pl. ns., and is combined especially with words such as *dale* 'valley' (WOODALE, WOOLDALE), LĒAH 'wood' (WOOLLEY). See further WOLF- (passim), WOLBOROUGH, WOLVEY, WOOLMER, WOOLPIT, and cf. ULFR.

Wyaston Db [*Widerdestune* DB, *Wyardestone* 1244 FF, *-ton* 1288 Cl]. 'Wigheard's TŪN.'

Wyberton (-ĭ-) Li [*Wibertune* DB, *-ton* 1212 Fees, *Wibretona* 12 DC]. The first el. is a pers. n. such as *Wigbeorht* or the woman's name *Wigburg*.

Wyboston (-ĭ-) Bd [*Wiboldestone* DB]. '*Wigbeald*'s TŪN.'

Wybunbury (wĭm-) Chs [*Wimeberie* DB, *Wybbenburi* 1227 Ch, *Wybonbur'* 1243 Cl, *Wybbunburi* 1276 BM, *Wybirisbiry* 1290 Court]. '*Wigbeorn*'s BURG.'

Wychbold Wo [*Uuicbold* 692, *Wicbold* 831 BCS 77, 400, *Wicelbold* DB, *Wichebald* 1160 P]. 'Manor, dwelling-house.' *Wicbold* is called *villa regalis* in BCS 400. But as Wychbold is near DROITWICH the name may mean 'dwelling by the WĪC'. See WĪC, BŌPL.

Wychwood O [*Hwicca wudu* 872 BCS 535, *Huchewuode* DB, *Wicchewude* 1204 Cur]. 'Forest of the *Hwicce*'; see WHICHFORD. Wychwood Forest was formerly a large forest district.

Wycliffe (wĭ-) YN [*Wigeclif* c 1050 HSC, c 1130 SD, *Witcliue* DB, *Wycheclif* 1260 Ass, *Wittecliff* 1275 PNNR]. Hardly 'white cliff'. The place is on the Tees, which makes a bend here. The name might possibly be OE *wiht-clif* 'cliff by the bend'. See WIHT.

Wycomb Le [*Wiche* DB, *Wicham* 1316 FA]. OE *wicum*, dat. plur. of WĪC 'dairy-farm'.

Wycombe (wikum), **Chipping**, **High**, & **West**, Bk [?*Wichama* 799–802 BCS 201, (æt) *Wicumun* c 970 ib. 1174, *West-wicam* 944–6 ib. 812, *Wicumbe* DB, *Wycombe Marchaunt* 1340 Cl, *Chepingwycomb* 1478 Ipm, *West Wicumbe* 1195 Cur]. If the examples of 799–802 and 944–6 belong here, as seems probable, the elements are OE WĪC and the plur. of OE HĀM or HAMM, i.e. OE (æt) *Wic-hāmum* or *Wic-hammum*. The meaning is 'dwellings' or 'HAMMS with or by a WĪC'. The river-name Wye is a late back-formation. The name was at an early date associated with OE *cumb* 'valley'. *Chipping* means 'market'.

Wyddial Hrt [*Widihale* DB, *-hal* 1208 Cur]. 'Willow HALH or nook.' See WĪÞIG.

Wye K [*an Uuiæ* 839 BCS 426, (to, on) *Wii* 858 ib. 496, 1043 Th, *Wi* 1082–7 BM, DB]. Cf. *Wistræt* 868 BCS 519, *Wiwarawic* 858 ib. 496 (both in Canterbury), *Weowera-weald* 724 BCS 141 ('the weald of the Wye people'). OE *wēoh*, *wig* 'holy place, heathen temple'. The original name was probably *Wioh*, of which *Wi* is a locative form (from *Wihi*).

Wye R Bk. See WYCOMBE.

Wye R Db [*Wey* 1235 Ch, *Weye* 1286 Rutland], W~ R Wales, He, Welsh Gwy [*Guoy* c 800 HB, *Gui*, *Guy*, *Guai* c 1150 LL; (on) *Wæge* 956 BCS (928), c 1000 Saints, *Waia*, *Waie* DB, *Waye* 1227 Ch, *Weye* 1200 Ch, *Weȝe* 1205 Layamon]. A Brit river-name identical with WEY.

Wyegate Gl in Forest of Dean [*Uuiggangeat* 972 BCS 1282, *Wigheiete* DB]. '*Wicga*'s gate or pass.' The name was later associated with the name WYE. The place is near the Wye.

Wyfordby Le [*Wivordebie* DB, *Wyfordebia* c 1125 LeS, *Wyvordeby* 1254 Val]. The place is on the river Eye and the name may well contain an OE name in *ford*, to which was added OScand BY. The base may be an OE *Wigford* 'ford by a *wig* (*wēoh*) or temple' or 'battle ford'.

Wyham Li [*Widun* DB, *Wihum* c 1115 LiS, 1229 BM, *Wium* Hy 2 DC, 1202 Ass]. The dat. plur. of OE *wēoh*, *wih* 'heathen temple'.

Wyke Regis Do [(to) *Wike* c 988 KCD 1284, *Wik* 1212 Fees, *Kingeswik* 1242 Cl], **Wyke** Sr [*Wucha* DB, *Wicha* 1170 P, *Wike* 1198 FF, *Wykes* 1242 Fees]. OE WĪC 'dairy-farm'.

Wykeham Li in Nettleton [*Wiham* DB, *Uicheim* c 1115 LiS], **East W~** Li [*Wicham* DB, *Wicheim*, *Parva Wicheham* c 1115 LiS, *Estwicham* 1228 Ep], **West W~** Li [*Wicham* DB, *Wic(he)heim* c 1115 LiS, *West Wicham* 1228 Ep], **W~** YN [*Wicham* DB, 1160–76 YCh 383]. OE *wic-hām* 'dwelling-place, manor'. Cf. WĪC.

Wyken Wa [*la Wicha* 1175 ff. P, *la Wyke* 1236 Fees, *Wykene* 1306 AD], **Wykin** Le [*Wich* 12, *Wiken* c 1200 DC, *Wychen* 1209–35 Ep]. OE *wicum*, dat. plur. of WĪC 'dairy-farm'.

Wylam Nb [*Wylum* Hy 2 (1271) Ch, 1254 Val, *Wilum* 1198 (1271) Ch, 1201 FF, *Wilham* 1204 Ch]. Apparently the dat. plur. of OE *wil* 'trick', here used of some mechanical contrivance, e.g. a water-mill or a trap. Cf. WILD. But *Wilhamm* is a possible base. The first el. is then OE *wil*.

Wyld. See WILD.

Wylye or **Wiley** (wī-) R W [*Wileo* 688 BCS 70, *Guilou* c 894 Asser, *Wilig* 901, 943 BCS

595, 783, *Wyly* 1268 Ass]. A Brit river-name identical with GWILI in Wales. The name is probably derived from the obsolete Welsh *gwil* that occurs in several compounds and seems to mean 'trick' or the like. The word is cognate with OE *wil*, ON *vél* 'trick'. The name means 'tricky river', i.e. one liable to floods or the like. A shorter form of the river-name is found in WILTON, OE *Wilsætan*. On the Wiley is **Wylye** vil. [*Biwilig*, (æt) *Wilig* 901 BCS 595, æt *Wilig* 977 KCD 611, *Wili* DB].

Wymering (-ĭ-) Ha [*Wimeringes* DB, 1185 P, *Wymeringes* 1242 Fees, *Wymeringges* c 1270 Ep]. '*Wigmǣr*'s people.'

Wymeswold Le [*Wimundewal(l)e, Wimundeswald* DB, *Wimundewald* 1166 P]. '*Wigmund*'s WALD or wood.'

Wymington (-ĭ-) Bd [*Wimentone* DB, *-ton* 1165 P, *Wimunton* 1169 P, *Widminton* 1195 Cur]. '*Widmund*'s or *Wigmund*'s TŪN.'

Wymondham Le [*Wimundesham* DB, *Wimundeham* c 1125 LeS, 1195 P], W~ (windam) Nf [*Wimundham* DB, *Wimundehamia* c 1150 Crawf, *-ham* 1168 P]. '*Wigmund*'s HĀM.'

Wymondley, Great & Little, Hrt [(æt) *Wilmundeslea* 11 E, *Wimundeslai* DB, *Wilemundeslee* 1197 FF, *-lea* 1212 Fees, *Wilmundele Magna* 1199 Cur, *Parva Wymundele* 1219 Ep]. '*Wilmund*'s LĒAH.'

Wynd Brook. See WELLAND Wo.

Wynford Eagle Do [*Wenfrot* DB, *Winfrod* 1227 Ch, *Wynfrod* 1232 Cl, *Wynford Aquile* 1275 RH, *Wymfrodegle* 1291 Tax]. Identical with WINFRITH.

The manor was held by Gilbert de Aquila in 1227 (Ch). The Norman house of Laigle took its name from LAIGLE in Normandy.

Wynyard Du [*Winyard* 1208–10 Fees, *Wyneiard* 1237 Pat]. WINN 'meadow' and GEARD 'enclosure'.

Wyrardisbury or **Wraysbury** (rāz-) Bk [*Wirecesberie* DB, *Wiredesbur'* 1195 Cur, *Wyredebiria* 1209–19 Ep, *Wyrardebury* 1274 Fine]. '*Wigric*'s BURG' originally. Later apparently changed to '*Wigrēd*'s BURG'.

Wyre R La [*Wir* a 1184 CC, c 1195 LaCh, *Wyr* c 1200 CC, *Wira* 1205 P]. A Brit river-name, very likely identical with Gaul *Vigora* (now VIÈRE and VOIRE in France) and derived from the root *vig-* in Sanskrit *vijāté* 'recoils', OE *wican* 'to yield' &c. The name would mean 'winding river'. Cf. WEREHAM. On the Wyre are **Over Wyresdale** La [*Wyresdale* 1246 Ass, c 1250 CC] and **Nether Wyresdale** La [*Wiresdale* 1190 CC]. 'The valley of the WYRE.'

Wyre Forest Wo, Sa [(in) *Weogorena leage* 816 BCS 357, (forest of) *Wyre* c 1080 Fr, *foresta de Wira* 1177 P, *Werewud* 1239 Cl]. The OE name means 'forest of the *Weogoran*', i.e. of the tribe that gave its name to WORCESTER. Later the name was changed

to *Wire-wude* and finally to *Wyre Wood* or *Wyre Forest*. Wyre Forest was in the early 9th cent. a large district on the western bank of the Severn west and north-west of Worcester, though it has later dwindled into the small Wyre Forest on both sides of Dowles Brook. See further WORCESTER.

Wyre Piddle. See PIDDLE Wo.

OE **wyrhta** 'wright'. See RIGBOLT, WRIGHTINGTON.

Wyrley, Great & Little, St [*Wereleia* DB, *Wirlega* 1170, 1176 P, *Great Wyrleye* 1300 For, *Little Wyrle* 1293 Ass]. OE *wir-lēah* 'bog myrtle glade'.

OE **wyrm** 'reptile; serpent; worm'. See WORM- (passim), WARMLEY.

OE **wyrt** 'plant, vegetable'. See WORTLEY (1), WORTON (1).

Wysall (-īs-) Nt [*Wisoc* DB, *Wisho* 1199 P, *Wisou* 1236, *Wisow* 1242 Fees]. Second el. OE *hōh* 'spur of hill'. The first may be OE *wēoh (wih, wig)* 'heathen temple' in the genitive form.

Wytham (wītam) Brk [*Wihtham* 957 BCS 1002, *Uuihtham* 968 Abingd, *Wyhtham* 1291 Tax]. 'HĀM by the bend.' See WIHT. W~ is in a sharp bend of the Thames.

Wythburn (-ĭdh-) Cu [*Withebotine* c 1280 FC, *Wythebocten* 1303 Ipm]. 'Willow valley.' The elements are ON *víðir* 'willow' and *botn* 'the innermost part of a valley'.

Wytheford, Great, Sa [*Wicford* DB, *Widiford* 1195 Cur, *Magna Wythiford* 1285 FA], **Little W~** Sa [*Wideford* DB, *parva Withiford* 1242 Fees]. 'Willow ford.' See WĪÞIG. The places are opposite to each other on the Tern.

Wythemail Np in Orlingbury [*Widmale* DB, *Wymale* 12 NS, *Wismalua* 1130 P, *Wizmalua* 1156 P]. The second el. is the OE **malu* (dat. *malwe*) 'gravel ridge' or the like that is the first el. of MAWSLEY Np, MANSELL. The first el. may be OE *wiþþe* 'willow'.

Wythop (-ĭdh-) Cu [*Wizope* 1195 FF, *Wythope* 1308 Ipm]. 'Willow valley.' Cf. WĪÞIG, HOP.

Wyton (-ĭ-) Hu [*Witune* DB, *Wictun* 1253 BM]. OE *wic-tūn* 'dwelling-place, manor'. Cf. WĪC and WITTON.

Wyton Hall YE [*Widetun* DB, *Wyuetona* 1297 Subs, *Wyveton* 1344 FF]. Either OE *Wiþigtūn* 'TŪN among willows' or '*Wifa*'s (*Wife*'s) TŪN'. Cf. WESTOW, WINESTEAD. Wyton is not very far from Winestead.

Wyverstone Sf [*Wiuerthestune* DB, *Wiverdeston* 1203 Cur, *Wyverdeston* 1231 Ch]. '*Wigferþ*'s TŪN.'

Wyville Li [*Huuelle* DB, *Uuiuuella* 1106–23 (1333) Ch, *Wiwel* 1212 Fees]. OE *wēoh (wih)* 'holy place, heathen temple' and *wella* 'spring or stream'.

Y

Yaddlethorpe Li [*Iadulf(es)torp* DB, *Edoluestorp* c 1115 LiS]. '*Ēadwulf*'s thorp.'

Yafforth YN [*Eiford, Iaforde* DB, *Iaford* 1198 FF]. OE *ēa-ford* 'ford over the river (Wiske)'.

Yagdon Sa nr Ross Hall [*Iagedone* DB, *Jagedon* 1255 RH, *Yakedon* 1311 Ipm]. Perhaps OE *gēaca-dūn* 'hill of the cuckoos'.

Yaldham K in Wrotham [*Aldeham* 1212 RBE, *Eldham* 1215 Cl, *Ealdeham* 1346 FA]. 'Old HĀM' or '*Ealda*'s HĀM'.

Yalding K [*Hallinges* DB, 11 DM, *Ealding*' 1207 Ep, *Elding* 1263 Ipm]. '*Ealda*'s people', an OE *Ealdingas*.

Yantlet Creek K [*Iaenlad* 779, *Iaegnlaad* 789, *Genlad* 808 BCS 228, 257, 326, *la Yenlade* 1277 RH], **Yenlet** K, an alternative name of the northern arm of the Wantsum [*Genladae* (gen.) c 730 Bede]. An OE *gegnlād* 'backwater' or the like, very likely the source of the word *inlet* 'arm of the sea, creek'. The elements are OE *gegn, gægn* 'again' and *lād* 'water-course'. Yantlet may have been an old name of the lower Medway. At present the name Yantlet Creek is applied to the arm that separates the Isle of Grain from the mainland, while South Yantlet Creek is another arm of the river.

Yanwath We [*Euenewit* 1150–62 YCh 1241, *Yafnewid* c 1244 Kendale]. A Scandinavianized form of EVENWOOD, the meaning being 'flat wood'. OScand *iafn* and *viðr* have replaced OE *efn* and *wudu*.

Yanworth Gl [*Teneurde* DB, *Ianeorþe* c 1162, -*wurd* c 1201, -*worth* 1202 Winchc, *Yanewrth* 1251 Ch]. '*Geana*'s WORÞ.' The first el. is an OE pers. n. *Geana (*Gæna), a short form of names like *Gænbald, Geanburh, Iaenbeorht*.

Yapham YE [*Iapun* DB, *Yapum* c 1155 YCh 442, 1230 Ep, *Yapom* 1299 BM]. Apparently the dat. plur. of some OE noun. It may be a substantivized form of OE *gēap* 'steep, lofty', used in the sense 'eminence'. The place is on a hill of irregular shape. Alternatively it might be OE *hēap* 'heap'. Cf. SHAP.

Yapton Sx [*Abinton* 1197 P, *Ebinton* 1235 FF, *Yapeton* 1295 Ch]. 'TŪN of *Eabba*'s people.' *Eabba is presupposed by *Eabbincg-wyll* 854 BCS 480 and is a short form of names like *Ēadbeald*. *Eabbe* fem. is on record.

Yar R Wt. No early forms have been found. On the stream is **Yarbridge**. Perhaps the name is a back-formation from the name of a place near St. Helens at the mouth of the Yar, referred to as *Yarneforde* 1324 Misc. This seems to be OE *earnaford* 'eagles' ford'.

Yarborough Camp Li. The place gave its name to **Yarborough** wap. [*Gereburg* DB, *Ierburc* c 1115 LiS, *Ierdeburg* 1254 Val]. The source is OE *eorþburg* 'earth fortification'; cf. ARBURY. Identical in origin with **Yarburgh** Li, which is not in Yarborough wap. [*Gereburg* DB, *Ierburc* c 1115 LiS, *Jerdeburc* 12 DC, *Jerdeburch* Hy 2 BM, *Jerdburg* 1212 Fees, *Yardbury* 1209–19 Ep].

Yarcombe D [*Erticoma* DB, -*cumba* 1156 Fr, *Yarte(s)cumbe* 1281 QW, *Yeartecomb* 1332 Misc]. 'Valley of R YARTY.'

Yardley Ess in Thaxted [*Gerdelai* DB, -*leg* 1236 FF], **Y~ Gobion** Np [*Gerdeslai, Jerdelai* 1167 P, *Yerdele Gobioun* 1353 Ipm], **Y~ Hastings** Np [*Gerdelai* DB, -*le* 12 NS, *Ierdele* 1265 Misc, *Yerdele Hastinges* 1316 FA], **Y~ Wo** [*Gyrdleah* 972 BCS 1282, *Gerlei* DB], **Yarley** So nr Wells [*Gyrdleg* 1065 Wells]. OE *gyrd-lēah* 'wood where yards, i.e. spars &c., were got'. First el. OE *gierd, gyrd* 'yard, pole' &c.

Y~ Gobion was held by Henry Gubyun in 1228 (Cl). Cf. HIGHAM GOBION.—**Y~ Hastings** is mentioned as having belonged to Henry de Hasting' in 1250 (Cl). Cf. BURTON HASTINGS.

Yare R Nf [*Gariénnos* c 150 Ptol, *Gerne* c 1150 ERN; cf. YARMOUTH]. A Brit river-name derived from the root *ger* in Welsh *gair*, Bret *ger* 'word', MIr *gairm* 'shout', Welsh *garan* 'crane'. The meaning may be 'babbling river'. From the Yare was derived the Brit name of **Burgh Castle** [*Garianno* c 425 ND; cf. (equitum) *Gariannonensium* ib.].

Yare R Wt. See YARMOUTH Wt.

Yarkhill He [(æt) *Geardcylle* 811 BCS 332, *Archel* DB, *Archil* 1190 P, *Iarculn, Yarchulle* 1242 Fees]. OE *geard* 'yard, enclosure' and *cylen* (dat. *cylne*) 'kiln'. The assimilation of *ln* to *ll* is somewhat early, but OE *mylen* 'mill' is *myll* c 1020 (OED), and *cyllriðe* 1032 KCD 746 is doubtless from *cylnriðe*.

Yarlet St [*Erlide* DB, *Erlida* 1167 f. P]. OE *ēar-hlid* 'gravel slope' or *earn-hlid* 'eagle slope'. Cf. ERITH, HLIP. The place is at **Yarlet Hill**.

Yarley. See YARDLEY.

Yarlington So [*Gerlincgetuna* DB, *Gerlingatune* Hy 1 Montacute, *Gerlingeton* 1212 Fees]. A possible etymology is 'TŪN of the people of YARNLEY'. Yarnley would be OE *earn-lēah* 'eagle wood'; cf. ARELEY, EARLEY. But no Yarnley is known in the vicinity. More likely the first el. is an OE folk-name *Gerlingas*, derived from an OE *Gerla, a hypocoristic form of names in OE *Gearo-* and cognate with *Gerling* DB (*Lerlincus* ExonDB), the name of a pre-

Conquest tenant in Do. Y~ is near the Dorset border.

Yarm YN [*Iarun* DB, 1155–65 YCh 654, *Gerou* DB, *Iarum* 1208–10 Fees, *Garum* 1218 FF]. The dat. plur. of OE *gear* 'dam, enclosure for catching fish'.

Yarmouth, Great, Nf [*Gernemwa* DB, *-muda* 1130 P, *Gernemuta Magna* 1254 Val, (*við*) *Járnamóðu* 13 Hákonarsaga (Chr. & Mem. 88)], **Little Y~** Sf [*Gernemutha* DB, *Parva Gernamuta* 1219 Fees]. 'The mouth of R Yare.'

Yarmouth Wt [*Ermud* DB, *Eremue* 1196 P, *-mua* 1206 BM, *Errenemuth* 1235, *Ernemuth* 1243 Cl]. The place is at the mouth of the R Yare and the name would seem to mean 'the mouth of R Yare'. But no early forms of the river-name have been met with, and it may be a back-formation. Yarmouth may mean simply 'gravel harbour'. Cf. ERITH. The first el. would then be OE *ēar* 'gravel'. OE *mūþa* occurs in the sense 'harbour'. The occasional forms with *n* (*Errenemuth* &c.) may be due to influence from Yarmouth Nf. An OE *ēaren* 'gravelly' is possible.

Yarnfield So, now W [*Gernefelle* DB, *-feld* 1225 Ass, (boscus in) *Gernefelde* 1274 RH], **Y~** St nr Stone [*Ernefeld* 1266 Ass, *Ernefen* 1327 Subs]. Probably OE *earna-feld* and *earna-fen* 'eagles' FELD and FEN'.

Yarnscombe D [*Hernescome* DB, *-cumbe* 1228 FF, *Ernescumbe* 1270 FF, *Jernescom* 1275 RH]. OE *earnes-cumb* 'eagle valley'.

Yarnton O [*æt Erdintune* 1005 KCD 714, *Hardintone* DB, *Erdentuna* 1090, *Hardingtona* 1146 RA, *Erdington* 1236 Ep]. Either OE *earding-tūn* 'dwelling-place, manor' (cf. OE *eardinghūs, eardingstōw* 'dwelling' from *eardian* 'dwell') or 'TŪN of *Ēanrēd*'s or *Earda*'s people'; cf. ARDELEY.

Yarpole He [*Iarpol* DB, *Garepolla* c 1145 Oxf, *Yarepol* 1212 RBE, 1278 Ep]. 'Pool formed by a dam' or 'pool with an enclosure for catching fish'. See GEAR.

Yarrow R La [*Yarwe* c 1190 CC, 1292 Ass, *Earwe* 1203 FF, *Yarewe* 1246 Ass]. Either identical with ARROW Wo or identical with GARW in Glamorgan [*Garewe* 1207 Ch] and cognate with YARROW in Scotland [*Gierua* c 1120]. In the latter case it is derived from Welsh *garw* (= Ir *garbh*) 'rough'.

Yarty R D [*Jerti* 1238 Ass, *Yearte* 1467 Ct, *Yartey* c 1540 Leland]. Possibly a Brit river-name derived from Celtic *arto-* 'a bear' (Welsh *arth*). Cf. AFON ARTH in Wales and ARCE in France [*Artia* 1263]. Holthausen derives the name from OE *earte* 'wagtail'. This is a better explanation.

Yarwell Np [*Yarewell* 12 NS, *Jarewelle* 1166 RBE, 1194 BM, *-well* 1220 Fees]. 'Stream with an enclosure for catching fish' (see GEAR). An arm of the Nene may be referred to.

Yate Gl [*aet Gete* 779 BCS 231, *Giete* DB, *Iete* 1196 P]. OE *geat* 'gate'.

Yately Ha [*Yatele* 1248, *-leghe* 1281 Crondal]. 'LĒAH by or with a gate' (OE GEAT).

Yatesbury W [*Etesberie* DB, *Hyatebir'* 1199 FF, *Jetesbiri* 1205 ib., *Getesbir* 1226, *Gytesbyre* c 1230 Sarum, *Yatesbur', Yitesbir'* 1242 Fees]. The first el. might be OE *geat*, here 'pass', but some forms tell in favour of a word with *ēa*. If so, the probable explanation is '*Ēat*(*a*)'s BURG'.

Yattendon Brk [*Etingedene* DB, *Etyng*(*e*)*den* 1258 Ch, *Yetingeden* 1220, 1242 Fees, *Yatingedene* 1242 Fees, *Yatingden* 1252 Ch]. Second el. OE *denu* 'valley'. The first is a name in *-ingas*, derived from OE GEAT in the sense 'pass' or from *Ēata* or *Gēat* pers. ns.

Yatton He [*Getune* DB, *Yatton* 1307 Misc], **Y~ Keynell** W [*Etone, Getone* DB, *Iatton* 1258 Ipm, *Iattone Kayngnel* 1317 Ipm]. OE *Geat-tūn* 'TŪN in a pass'. Cf. WINGATE.

Y~ Keynell was held by Henricus Caynel in 1242 (Fees). Caynel is 'man from Cahagnes' in Normandy; cf. OFr *espaigneul* 'Spanish, Spaniard' from *Espagne*. Cf. ASHTON KEYNES.

Yatton So [*Iatune* DB, *Jatton* 1178 Wells, 1225, 1243 Ass, *Jactun* 1227 FF, *Jacton* 1256 FF, 1276 RH]. Perhaps identical with the other Yattons; *ct* may be an incorrect spelling for *tt*. If the real form was Yacton, the first el. appears to be OE *gēac* 'cuckoo'.

Yaverland Wt [*Evreland, Everelant* DB, *Awerlond* 1287–90 Fees, *Yoverlond* 1311 Ipm, *Yaverlond* 1412 FA]. 'Land where boars were kept.' OE *eofor* means 'boar'.

Yawthorpe Li [*Iole*(*s*)*torp* DB, *Iolthorp* c 1115 LiS, *Hioltorp, Yoltorp* 1212 Fees]. '*Iōli*'s thorp.' First el. ODan *Ioli* (coins); cf. OE *Iola* (coins), ODan *Juli*, OSw *Iule*.

Yaxham Nf [*Jachesham, Jakesham* DB, *Iakesham* 1254 Val]. '*Gēac*'s HĀM.' OE *Gēac* pers. n. is not evidenced, but cf. ON *Gaukr*. OE *gēaces-hamm* 'cuckoo meadow' is a possible alternative.

Yaxley Hu [(æt) *Geaceslea* 963–84, *Geakeslea* 973 BCS 1128, 1297, *Iacheslei* DB], **Y~** Sf [*Jacheslea, Iachelea* DB, *Iakeslea* 1170 P, *Iachesle* 1198 FF]. 'Cuckoo LĒAH or wood.'

Yazor He [*Iavesovre* DB, *Jagosoure* c 1170 Hereford, *Eausore, Iagesoure* 1242 Fees, *Yakesour* 1265 Ch, *Yavesore* 1303 FA]. Apparently '*Iago*'s *ofer* or hill slope'. First el. Welsh *Iago*, OW *Iaco* from *Iacob*.

Yeading (-ĕ-) Mx [*Geddinges* c 757 BCS 182, *Geddingas* 793, (æt) *Geddincggum* 825 BCS 265, 384, *Geddinges* 1212 RBE]. '*Geddi*'s people.'

Yeadon YW [*Iadun* DB, *Iadon* DB, 1167 P, *Jeaddun* 12, *Jhadun* c 1190 Calverley, *Jedona* c 1180 YCh 1873, *Yaddon* 1195 P]. The place is in a high situation on one of the spurs of the Chevin. The name may be

OE *hēa-dūn* 'high hill'. OE *geat-dūn* 'hill with a pass' is a possible alternative.

Yealand (-ĕ-) Conyers & Redmayne La [*Jalant* DB, *Yeland* 1190 CC, 1208 FF, *Hielande* 1202 FF, *Yeland Coygners* 1301 FF, *Yeland Redman* 1341 NI]. OE *hēa-land* 'high land'.

Y~ **Conyers** was held by Robert de Conyers in 1242 (LaInq). Cf. HUTTON CONYERS.— Y~ **Redmayne** came to the Redman or Redmayne family in the 12th cent. The family may have come from REDMAIN Cu.

Yealm (yăm) R D [*Yhalam* 1309 Ipm, *Yalme* 1414 Ep; cf. *Yalmmue* 1297 Cl]. On the river is **Yealmpton** (-ămp-) [*Elintona* DB, *Almenton* 1249 Ass, *Yalmeton* 1238 Ass, *Yealmintton* 1224–44 Ep, *Ealmintone* 1270 Ep]. If Yealm goes back to OE *Gealme* or the like, it is a derivative of OE *giellan* 'to sound' and related to OHG, OLG *galm* 'sound'. But forms without a Y- are very common in early records, and the base is perhaps rather OE *Ealme*. Such a name would probably be Celtic, perhaps identical with *Alma* in Italy and cognate with Lat *almus* 'kind'.

Yearby (-ēr-) YN [*Uverby* c 1275, *Overby* 1270 Guisb]. Apparently OScand *Efri-bȳr* 'upper BY', in the early forms anglicized to *Uverby* &c.

Yeardsley Chs [*Urdisl'*, *Hurthesle* 1285–8 Court, *Urdesley* 1502 Ormerod]. '*Eorēd*'s LĒAH.'

Yearsley YN [*Eureslage* DB, *Euereslai* 1176 P]. Identical with EVERSLEY.

Yeaton Sa [*Eton* 1327 Subs]. OE *Ēa-tūn* 'TŪN on the river' (the Perry).

Yeaveley Db [*Gheveli* DB, *Yeueleye* 1277 BM]. '*Geofa*'s LĒAH.' *Geofa* is found in *Geofanstig* 961 BCS 1074, *Geofandene* 706 BCS 116.

Yeavering (-ĭ-) Nb [*Adgefrin, Adgebrin* c 730 Bede, *Ætgefrin* c 890 OEBede, *Yever* 1242 Fees, *Yeure* 1329 Misc]. *Gefrin* is evidently the old name of **Yeavering Bell**, a prominent hill at the place. The name is derived from Welsh *gafr* 'goat' or a compound containing the word, e.g. a name with Welsh *bryn* (mutated *fryn*) as second el.

Yedingham YE [*Edingham* 1170–5, 1185–95 YCh 390, 395, 1218 FF, *Yedingham* 1185–95 YCh 392, 1219 FF]. 'HĀM of *Eada*'s people.'

Yĕlden Bd [*Giveldene* DB, 1272 Ipm]. 'The DENN or swine-pasture or the DENU or valley of the *Gifle* or Ivel people.' See IVEL.

Yeldersley Db [*Geldeslei* DB, *Yldreslee* 12, *Yhildrisleye* Hy 3 BM, *Gildreleg'* 1212 Derby, *Yhildirleg* 1242 Fees]. Apparently '*Geldhere*'s LĒAH.' OE **Geldhere* has a counterpart in OHG *Gelther*. Cf. OE *Geldwine* and see YELVERTOFT.

Yeldham, Great & Little, Ess [*Geldeham* DB, *Geldham* DB, 1194 P, *Great, Little Gelham* 1265 Misc]. First el. OE *gield* 'payment, tribute, tax'. The name would refer to a homestead or village which had to pay a certain tax.

Yelford O [*Aieleforde* DB, *Eleforde* 1221 Ep, *Eleford* 1242 Fees, *Eillesford* 1245 Ch]. The first el. is identical with that of *Æglesuullan broc* 958 BCS 1036, which denoted a brook close to Yelford. The first el. seems to be OE *Ægel* pers. n. Cf. AYLESBURY &c.

Yelling Hu [*Gilling', Gillinge* 974 BCS 1310 f., *G(h)ellinge, Gelinge* DB, *Gellinches* c 1150 BM, *Gillynges* 1179 RA]. The name is ultimately a derivative of OE *giellan* 'to scream', the immediate base being very likely a pers. n. cognate with ON *Gellir*. Cf. OHG *Gellingin* pl. n.

Yelvertoft Np [*Celvrecot* DB, *Chelvertoft* 1206 Cur, *Gelvrecote* DB, *Gelvertoft* 12 NS, 1224 Ep, *Ielvertoft* 1254 Val]. Some forms point to OE *Cēolferþ* pers. n. as the first el., but the sound-development would be unparalleled. Possibly we may postulate an OE *Geldfriþ*; cf. YELDERSLEY. See TOFT.

Yelverton Nf [*Ailuertuna* DB, *Ielverton* 1198 FF, *Gelvertone* 1254 Val]. Possibly '*Geldfriþ*'s TŪN'. Cf. prec. name.

Yen Hall Ca nr West Wickham [(to) *Eanheale* 974 BCS 1305, *Enhale* 1242 Fees, 1291 Ch]. 'Lambs' HALH or valley.' Cf. ENHAM.

Yenlet. See YANTLET.

Yeo R D, a trib. of the Creedy [(on) *Eowan* 739 Crawf, *Jouwe* 1238, *Jou* 1244 Ass]. A derivative of OE *ēow, iow* 'yew', the name meaning 'river on whose banks yews grew'. On the Yeo is **Yeoford** [*Ioweford* 1242 Fees].

Yeo R Do, So [*Yevel* 878, (oth) *Gifle* 933, (on) *Gifle* 946–51 BCS 546, 695, 894, *Givell* 1243 Ass]. Identical with IVEL. See YEOVIL, -TON.

Yeo is a common name of streams in D, So. It is OE *ēa* 'river', which in the dialects of these counties has given *yeo*. Yeo has often displaced earlier names. See e.g. ASHBURTON, NYMET, WRINGTON.

Yeolmbridge (yōm-) D [*Yambrigge* Hy 3 PND, *Yombrigge* 1308 Ep]. The first el. may be an OE *ēa-hamm* 'river meadow'.

Yeoveney (-ē-) Mx in Staines [*Giueneya* 1042–66 PNMx(S), *Yvenay* 1204 Cur, *Yveneye* 1219 FF, *Jeveneye* 1277 Misc, *Yeveneye* 1382 FF]. '*Geofa*'s island.' Cf. YEAVELEY.

Yeovil (yō-) So [(æt) *Gifle* c 880 BCS 553, (to) *Gyfle, Gifle* c 950 Wills, *Givele* DB]. Derived from the river-name YEO, earlier *Gifl*.

Yeovilton So [*Geveltone* DB, *Giueltona* DB, 1179 P]. 'TŪN on R YEO' (earlier *Gifl*).

Yetlington Nb [*Yettlinton* 1187, *Yatlinton* 1196 P, *Yetlingtun* 1236 Fees]. Possibly 'TŪN of *Gēatela*'s people'. '*Gēatela* would be a diminutive of *Gēat*. But the first el. may be a derivative of a lost pl. n. identical with YATELY.

Yetminster Do [*Etiminstre* DB, *Eteministre* 1212 Fees, *Ettemunstr'* 1230 P, *Gateministre* 1243 BM, *Yeteministr'* 1252 Cl]. '*Ēata*'s mynster or church.'

Yettington D [*Yethemeton, Yetematon* 1242 Fees]. OE *Geathǣma-tūn* 'TŪN of the dwellers in the pass'. See GEAT.

Yewdale La [*Ywedalebec* 1196 FF]. 'Yew valley.'

OE **yfer** fem. is only found in charters and pl. ns. It is formally identical with Goth *ubizwa* 'hall' and related to OE *efes* 'eaves'. It seems to mean 'edge, brow of a hill, escarpment'. Examples in charters are *beneapan yfre* 940 BCS 756 (W), *be yfre* 944 ib. 802 (Brk). See HEVER, IVER, RIVER Sx, UNDERRIVER.

Yiewsley (ūz-) Mx [*Wiuesleg'* 1235, *Wewesley* 1593 PNMx(S), *Wyneslee* 1382 FF]. The first el. is supposed in PNMx(S) to be perhaps an OE *Wife* pers. n. *Wifel*, either *wifel* 'beetle' or *Wifel* pers. n., would be better; the first *l* would disappear early in *Wifeleslēah*. *Wives-* became *Wiwes-*, and when *iw* passed into *ū* (yōō) initial *W-* disappeared. A *w* is never found before *ū* (yōō) in English.

Yockenthwaite YW [*Yoghannesthweit* 1241 Percy, *Yokenthwaite* 1499 Whitaker]. '*Eogan*'s thwaite.' *Eogan* is an OIr pers. n.

Yockleton Sa [*Ioclehvile* DB, *Lokelthulla* c 1100 Mon, *Yokethil* 1246 Ch, *Yokelthul* 1274 Ipm, *Yokolton* 1327 Subs]. The elements are OE *geocled, iocled, -let* 'a small manor' (chiefly found in Kent) and *hyll* 'hill'. The later form Yockleton seems to be a modification of *Yokelthull*.

Yokefleet YE [*Iucu-, Iugufled* DB, *Jukeflet* 1165–85 YCh 988, *Yokeflet* 1189–95 ib. 987, *Yoclesfliet* 1199 P]. Second el. OE *flēot* 'stream, creek'. The first seems to be OScand *Iōkell*. This name often appears with *u* in the first syllable, as *Iukil* KCD 519, *Iuchil* SD. But the early spellings *Iucu-, Iugu-* offer difficulty, and possibly the first el. is rather ON *Iǫkull*, a common name, or original *Iōkell* may have been mixed up with *Iǫkull*. The loss of the first *l* is due to dissimilation.

Yordale YN is 'the valley of the URE'. It preserves the old form of the river-name.

York Y [*Ebórakon* c 150 Ptol, *Eboracum* Cassiodor, c 730 Bede, *Eburacum* 4 IA, c 730 Bede, *Eoforwicceaster* 644 ASC, c 890 OEBede, *Eferwic* 10 Ælfric, 1070 ASC, *Eforwicceaster* c 893 Alfred Or, *Euruic* DB,

Euerwik 1297 Rob Gl, *Eoverwik, ʒeorc(ʒorc)* 1205 Lay, *ʒork* 1338 Rob Br; *Cair Ebrauc* c 800 HB, Welsh *Caerefrog*). The Brit name is held to be derived from a pers. n. *Eburos* (Gaul *Eburos*, Welsh *Efwr*). But this name is supposed to be a derivative of Gaul *eburos* (Ir *iubhar*) 'yew', and *Eburacon* might then well be derived directly from the tree-name. Owing to popular etymology the Brit name was changed into OE *Eoforwic*, which may have been supposed to contain OE *eofor* 'boar'. Scandinavians at an early date came to know the name, and in their speech it became *Iorvík*, found in Egill's Arinbjarnardrápa of 962. A later development of this is *Iork*, found in later ON sources, as in Fagrskinna. In this form the name was re-adopted by the English.

Yorkshire is *Eoferwicscir* c 1050 KCD 1343, 1065 ASC (C), *Evricscire* DB.

York Town Sr was named from Frederick, Duke of York, who founded Sandhurst College in 1812.

Yorton Sa [*Iartune* DB, *Iyartun* 1255 RH, *Yorton* 1327 Subs]. OE *Geard-tūn* 'TŪN with a yard or enclosure'. OE *geard* sometimes appears in ME as *yord*.

Youlgreave Db [*Giolgrave* DB, *Hyolegrave* 1208 FF, *Yolegrave* 1259 BM, *Yolgreue* 1285 FF]. 'Yellow grove' (OE *geolu* 'yellow' and GRĀF, GRǢFE).

Youlston D [*Yoldeston* 1451 Ipm]. OE *ealda stān* 'old stone'.

Youlthorpe YE [*Aiul(f)torp* DB, *Joel-, Joilthorp* c 1155 YCh 828, *Yoltorp* 1194 P, *Joltorp* 1204 FF]. The original name seems to have been '*Eyiolf*'s thorp', the first el. being ON *Eyiolfr* pers. n. The later forms may not be a continuation of the old name, but represent a new name containing the name of some early Norman or Scandinavian owner. Cf. JOLBY, YAWTHORPE.

Youlton YN [*Ioletun* c 972 BCS 1279, *-e* DB, *Yolton* J Ass]. '*Iōli*'s TŪN.' Cf. YAWTHORPE.

Yoxall St [*Iocheshale* DB, *Yoxhal* 1222 Ass, *Iokeshale* 1242 Fees], **Yoxford** Sf [*Gokesford, Iokesfort* DB, *Jokeford* 1203 Cur, *Iokesford* 1254 Val]. The first el. is OE *geoc* 'yoke, yoke of oxen, a measure of land'. The second el. is HALH and FORD. The probable meaning of YOXFORD is 'ford that could be passed by a yoke of oxen', while that of YOXALL is not apparent.

OE **yppe**, a derivative of *upp* 'up', seems to have meant 'a raised place, look-out place', but may also have denoted 'a hill'. It is found in some pl. ns., but is sometimes difficult to distinguish from *Ippa* pers. n. See IPLEY, IPSDEN &c., TIPALT, EPPING, UPPINGHAM.

Z

Zeal Monachorum D [*at Seale* 956 BCS 968, *Sele* 1228 FF, *Sele Monacor'* 1275 RH]. OE *sealh*, dat. *seale* 'sallow'.

The manor belonged to Buckfast Abbey. In the OE charter *at Seale* is mentioned in connexion with *at Dunnynghefd*, which is **Dunheved** at Launceston Co [*Dunhevet* DB, *Dunehavede* 1140–75 Ep]. Cf. DONHEAD. According to PNW(S) *at Seale* belongs to ZEALS, *Dunnynghefd* being DONHEAD W. No certainty is possible.

Zeal, South, D [*La Sele* 1168 P, *Zele Tony* 1299 Ch]. OE *sele* 'hall', or more likely OE *seale* from *sealh* 'sallow'.

Zeals W [*Sela, Sele* DB, *Selis* 1167 P, *Seles* 1242 Fees]. Identical in the main with ZEAL MONACHORUM, though rather the plur. *sealas* of *sealh* 'sallow'. Zeals is near SELWOOD FOREST.

Zennor Co [(ecclesia) *Sancte Senare* 1291 Tax, c 1300 Ep]. *Senara* is stated to be the name of a woman saint.

Zoy. See MIDDLEZOY, WESTON ZOYLAND.